UM Fundamental Questions and Where to Find the Answers

Do modern scientists indirectly acknowledge a Universal Flood? (p478-479)

How do we know that oceans covered the entire Earth at one time? (p479, 659)

What caused the mass extinction of the dinosaurs? (p480-482, 654)

What is the origin of the Universal Flood waters? (p282-286, 492-499)

What mechanism caused the Universal Flood? (p493-499)

What evidence demonstrates the Universal Flood? (p528, 546, 574, 595, 686)

What are hydrofountains and what evidence shows they exist today? (p506-528)

What evidence shows the Flood's recent timing? (p546-560, 648-649, 665)

What evidence shows most of the Earth's sediment is Flood based? (p528-564)

Can petroleum and coal form in a matter of hours? (p610-621)

Are pyrite, ore, geodes, and diamonds, evidence of the Flood? (p622-686)

Can we understand weather without knowing its true origin? (p700-715)

Why do cumulus clouds form in minutes over dry land? (p716-722)

In the last 50 years, have melting glaciers caused sea levels to rise? (p780-782)

How can Earth's magnetic field exist if heat destroys magnetism? (p114-117, 739)

What is the true origin of the Earth's energy (magnetic) field? (p740-744)

Michael, 4.6.17

So nice to meet you! May you enjoy the new scientific discoveries found in this in! Your brother in truth,

Dean

UM

Universal Model
A New Millennial Science

Volume I
The Earth System

Dean W. Sessions

Edited By

Russell H. Barlow

Copyright © 2017 by Dean W. Sessions

Published by Millennial Science, a division of AnQuest, LLC.

First Print Version, January 2017

All rights reserved. Research, writing, and publishing finalized in the United States of America. No part of this publication may be reproduced, stored in a retrieval system, or transmitted, in any form or by any means, electronic, mechanical, photocopying, recording, or otherwise, without the prior written permission of the publisher.

The scanning, uploading, or distribution of this work via any electronic means without the written permission of the publisher is illegal and punishable by law. Please purchase only authorized electronic editions which can be found at UniversalModel.com. Thank you for your support. This will help in maintaining and promoting Millennial Science.

Although the author and publisher have made every effort to ensure that the information in this book was correct at press time, the author and publisher do not assume and hereby disclaim any liability to any party for any loss, damage, or disruption caused by errors or omissions, whether such errors or omissions result from negligence, accident, or any other cause.

All references and quoted material found herein were used in compliance with fair use laws and are considered correct with respect to their content. Neither the publisher nor author assumes any responsibility for inaccuracy, errors or changes in the material before or after publication.

Layout, cover, and text design by Dean W. Sessions

Consulting by Legends Library

UM logo is a registered trademark in USA and is pending.

ISBN: 978-1-944200-91-6   Library Of Congress Control Number: 2016955721
1. Science 2. Earth Science 3. Life Science 4. Space Science 5. Natural Science

Printed in Hong Kong

For permissions email:
support@universalmodel.com

For information or to order more copies please go to:

UniversalModel.com

To My Sweetheart — Danette

Preface

Welcome to the Universal Model

In all educated societies where science is taught, we learn about the atom and the Universe. We learn about what is inside the Earth and what is on its surface. We learn about things that are living and things that have died. After reading the Universal Model, (UM)—many of these things will have new meaning. The very definition of some of the most fundamental aspects of our Universe will change due to the hundreds of new scientific discoveries found in the Universal Model.

It may seem an impossible task—to change the entire fabric of modern science by replacing the very foundation that it has been built upon. Yet, this is exactly what the UM is destined to do. The exciting part is that you are not asked to simply believe or trust the extraordinary claims in this book. You can discover for yourself whether they are true or not. You will be able to evaluate the many evidences contained in this book and come to *know*, through first-hand observations, experiments, and evaluations what is true and what is false.

The Universal Model is not only an iconoclastic scientific work—it contains a collection of one-of-a-kind images and diagrams that will further enhance your learning experience. The UM is literally an effort to pioneer a new Millennial Science—a science that will carry humanity through the current millennium to heights never before imagined. It may be the scientifically minded that have the most to gain or lose from the new discoveries in the UM, however, this book was not written to the scientists directly—it was written to you, and for you; the individual who simply wants to know how Nature works and why we are such an important part of all the beauty that surrounds us.

UM
Brief Contents

	Chapter		Page
Volume I		**Introduction to the UM**	
	1	Introduction to the Universal Model	5
	2	Methods Models and Pseudotheory	23
	3	The Dark Age of Science	39
	4	Scientific Revolutions	61
		The Earth System	
	5	The Magma Pseudotheory	69
	6	The Rock Cycle Pseudotheory	131
	7	The Hydroplanet Model	229
	8	The Universal Flood Model	475
	9	The Weather Model	699
Volume II		**The Living System**	
	10	The Age Model	791
	11	The Fossil Model	915
	12	The Evolution Pseudotheory	1009
	13	The Living Model	1081
	14	The World History Model	1129
	15	The Clovis Model	1216
	16	The Human Model	1284
Volume III		**The Universe System**	
	17	Essential Measurement	1379
	18	The Mass-Weight Model	1389
	19	The Length Model	1427
	20	The Time Model	1465
	21	The Relativity Pseudotheory	1491
	22	The Energy-Matter Model	1554
	23	The Air-Water Model	1629
	24	The Chemical Model	1654
	25	The Universe Model	1699
	26	The Meteor Model	1759
		Futurity	
	27	UM Extraordinaries	1791
	28	The Reaction	1806
	29	Millennial Science	1821
	30	Millennial Education	1856
	31	Advancement of Science	1891
	32	Ending with the Beginning	1906
		Acknowledgments	1926
		Notes	1930
		Bibliography	1990
		Image Credits	1998
		Index	2004

UM Detailed Contents

Preface
Contents

Introduction

Chapter 1 Introduction to Universal Model
    Page
1.1 Answers to Questions   6
1.2 The Question Principle   9
1.3 The Learning Process   12
1.4 The Importance of Definitions   15
1.5 Truth Defined   15
1.6 Scientific and Absolute Truth   19
1.7 The Human Touch   19
1.8 The UM Book Format   20

Chapter 2 Methods Models and Pseudotheories
2.1 The Line Between Theory and Natural Law   23
2.2 Against Method   24
2.3 The Modern Science Method of Confusion   25
2.4 Controversial Science   27
2.5 The Universal Scientific Method (USM)   28
2.6 Scientific Models   34
2.7 Pseudotheory and UM Models   35

Chapter 3 The Dark Age of Science
3.1 The History of Science   39
3.2 The Dark Age of Science   43
3.3 Modern Science Today   48
3.4 Modern Science and Truth   50
3.5 The Big Picture of Modern Science   51
3.6 How Popular is Modern Science?   54
3.7 Science, Technology and Mathematics   57
3.8 Final Review of Modern Science   58

Chapter 4 Scientific Revolutions
4.1 What is a Scientific Revolution?   61
4.2 The Steps for a Scientific Revolution   63
4.3 Why Will a Scientific Revolution Take Place?   65
4.4 Where Will the Next Revolution Come From?   66

Earth System

Chapter 5 The Magma Pseudotheory

5.1	Magma Defined	70
5.2	The Magmaplanet Belief	73
5.3	The Lava-Friction Model	77
5.4	Magma Defies Heat Flow Physics	90
5.5	The Accretion Theory	96
5.6	The Radioactive Myth	97
5.7	Quartz is Not Glass	101
5.8	The Piezoelectric Evidence	105
5.9	The Non-Iron Core Evidence	106
5.10	Deep Earthquake Evidence	109
5.11	The Drilling Evidence	109
5.12	Earth's Magnetic Field Pseudotheory	114
5.13	The Continental Uplift Pseudotheory	117
5.14	Other Magma Pseudotheories	122
5.15	The 'Smoking Gun'	126
5.16	The Magma Freeze	130

Chapter 6 The Rock Cycle Pseudotheory

6.1	The Most Abundant Mineral Mystery	132
6.2	The Real History of Geology	145
6.3	The Sand Mystery	148
6.4	The Quartz Mystery	161
6.5	The Basalt Mystery	170
6.6	The Obsidian Mystery	175
6.7	The Iron Mystery	177
6.8	The Ore Mystery	180
6.9	The Carbonate Mystery	183
6.10	The Loess Mystery	198
6.11	The Erosion Mystery	205
6.12	The Earth Crust Mystery	217
6.13	Geotheoretical to Geological	226

Chapter 7 The Hydroplanet Model

7.1	Magmaplanet to Hydroplanet	230
7.2	Celestial Water	234
7.3	Hydrospheres	243
7.4	The Crystallization Process	253
7.5	A New Geology	273
7.6	The Hydroplanet Earth	279
7.7	Hydrology Redefined	287
7.8	The Hydrocrater Model	305
7.9	The Crater Debate	318
7.10	The Meteorite Model	358
7.11	The Arizona Hydrocrater Evidence	382
7.12	The Impact to Hydrocrater Evidence	414
7.13	The Hydromoon Evidence	437
7.14	The Hydrocomet Evidence	453
7.15	The Hydroid Evidence	458
7.16	More Hydroplanet Evidence	463
7.17	The Hydroplanet Frontier	471

Chapter 8 The Universal Flood Model

8.1	The Universal Flood History	475
8.2	The Acknowledged Flood	478
8.3	The Universal Flood Mechanisms	492
8.4	The Hydrofountain Mark	506
8.5	The Sand Mark	528
8.6	The Erosion Mark	546
8.7	The Depth Mark	560
8.8	The Carbonate Mark	574
8.9	The Salt Mark	595
8.10	The Oil and Gas Mark	610
8.11	The Coal Mark	615
8.12	The Pyrite Mark	622
8.13	The Ore Mark	635
8.14	The Surface Mark	645
8.15	The Diamond Mark	682
8.16	The Inclusion Mark	686
8.17	The UF Summary	692

Chapter 9 The Weather Model

9.1	The Mystery of Weather	700
9.2	The Origin of Weather	703
9.3	The Earthquake Cloud Evidence	716
9.4	The Global-Weather System Evidence	728
9.5	The Geofield Model	737
9.6	The Geofield Evidence	744
9.7	The Aurora Evidence	762
9.8	The Ozone-CFC Pseudotheory	769
9.9	The Global Warming Pseudotheory	772
9.10	Weather Geofield Prediction	784

Acknowledgements	791
Editor's Note	793
Notes	795
Bibliography	815
Image Credits	823
Index	825

The **UM Brief Contents** listed all 32 UM chapter titles; the **UM Detailed Contents** includes chapter headings with subchapters pertaining only to the first nine chapters included in the first volume. At UniversalModel.com you can find a list of the remaining chapters and subchapter details, although they are subject to change because the work is still in progress. See UniversalModel.com for more information about upcoming chapter releases, available initially in digital form with a printed version arriving later.

1

Introduction to the Universal Model

UNIVERSE SYSTEM

1.1 Answers to Questions
1.2 The Question Principle
1.3 The Learning Process
1.4 The Importance of Definitions
1.5 Truth Defined
1.6 Scientific and Absolute Truth
1.7 The Human Touch
1.8 The UM Book Format

There is an order in Nature, which at first may not be apparent. The stars and galaxies seem carelessly strewn throughout the heavens, and rivers look as if they meander aimlessly across the countryside. Sand and clay deposits appear randomly placed, isolated, and different from the mountains surrounding them; plants and animals appear as though scattered without purpose throughout the Earth. However, as we shall learn in this book, *there is order*, direction, and purpose throughout the Universe; a state of order that has remained largely unrecognized, until now.

We attempt to observe and understand Nature's orderliness by arranging and categorizing what we see and what we think we know into manageable parts. We study, research, and experiment on many individual facets of Nature, but we must never forget there is synergy among complementary processes and symbiotic relationships that abound in the natural world. To appreciate Her secrets, we need to study and consider all things together as one whole.

This book, the **Universal Model** recognizes that many aspects of science and discoveries made over the past several centuries are true. Knowledge about the temperature and pressure at which water freezes, the existence of cells, the discovery of DNA in all living organisms and many other veritable facts exist because of experimental work and empirical evidence; these are not among the subjects addressed here in the UM. However, there are some modern science theories that have no realistic, first hand proof, although science leaders and researchers teach them as fact, lending those theories an air of credibility because of their own prestige and standing. These unproven theories form the foundation of modern science.

Any structure built on the sandy foundation of falsehood must eventually fail when subjected to the stresses of time and truth. The UM shines a bright light on many of these centuries-old false theories, focusing on the discovery of natural law, and establishing a foundation for the advancement of a new **Millennial Science**. Built on the Universal Model, Millennial Science will encourage the development of science with a new view of Nature that correctly describes the Universe in which we live. Over the last century, science has largely abandoned the whole-science approach in favor of field specialization. The UM returns to the intact-science approach, answering many questions people have asked for millennia and asking fundamental questions never before imagined.

There are many who oppose modern science's theory-based ideology today, especially those affiliated with creationist science. Admittedly, the UM may seem to some as just another creationist's work, but it is far more revolutionary in scope. The Universal Model challenges the false theories in all areas of modern science by presenting new natural laws that have come from unprecedented scientific discovery.

The Universal Model examines Nature in three primary segments, each focusing on a 'system' within the whole. The **Earth System**, the **Living System**, and the **Universe System**, encompass different parts of Nature, each one filled with many thought provoking questions and answers, and with experiments, observations, and empirical evidence.

We learn that asking the right questions, within the proper mindset, is of the utmost importance. You may at first find the ideas presented here unbelievable, the scope, and magnitude too incredible. However, if you persevere, keeping an open mind,

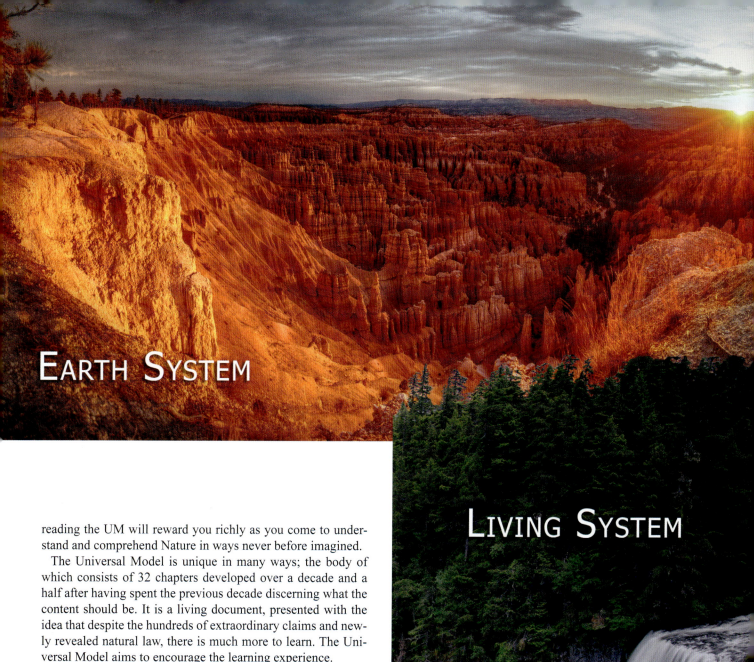

EARTH SYSTEM

LIVING SYSTEM

reading the UM will reward you richly as you come to understand and comprehend Nature in ways never before imagined.

The Universal Model is unique in many ways; the body of which consists of 32 chapters developed over a decade and a half after having spent the previous decade discerning what the content should be. It is a living document, presented with the idea that despite the hundreds of extraordinary claims and newly revealed natural law, there is much more to learn. The Universal Model aims to encourage the learning experience.

1.1 Answers to Questions

It is a known fact that water exists on many of the planets and their moons in our Solar System and science has detected water in the stars and even on our own Sun. Just how abundant is water in the Universe and how important is it? Geologists widely accept that the salt in the oceans came from the erosion of mountains, but is this true—is that process happening today? We find petrified wood—trees turned into solid rock—throughout the world on every continent. How did once-living trees become silicified crystalline quartz rock before they decayed? If the Universe was the result of a big bang explosion, why are the cosmos not filled with randomly scattered matter? Why do stars, planets and galaxies all rotate in an organized pattern?

The correct answers to questions like these mark the beginning of a new understanding in science that will have a far-reaching effect on our everyday lives. We answer questions like these and hundreds of others like them in this book. Questions form the basis—the foundation of a proper science and it takes effort to ask them properly.

How important are questions? Questions stimulate the human mind. They initiate inward reflection and outward expression. Most importantly, answers come by asking questions. Questions are the foundation of learning and there is much to say of this principle. We introduce many correct principles in the Universal Model, and this is one of the most important; **The Answer Principle**:

Answers come from questions.

A quest is a journey toward a goal just as a question is a journey toward knowledge and wisdom. Questions come in many forms and often have very different objectives. A question framed improperly may produce an erroneous answer or maybe no answer at all. Oftentimes the question requires thinking 'outside the box' to reach the right answer. In nearly all cases, the key to the answer requires a correct *understanding of the question*.

The Key to a Puzzle

The solution to a puzzle is its key. If you know the key to a frustrating, brainteaser type puzzle, solving it is a simple matter. One popular puzzle—the Nine-Dot Puzzle—will serve to illustrate this point. Arrange nine dots in a 3 x 3 grid pattern (see Fig 1.1.1), then without removing the pencil from the paper, draw through all nine dots with four straight, connected lines. The solution must include these parameters; the lines must be straight and connected contiguously, and there can be only four lines. You are encouraged to try this!

It may seem impossible if you are unfamiliar with this puzzle, and your attempts may include drawing lines around the perimeter, drawing a triangle, or maybe a cross pattern, each time leaving one unconnected dot. The solution requires looking beyond the dots; the key to the puzzle is to *understand the question and have an **open mind***.

This puzzle requires thinking outside the confines of the grid pattern itself. The arrangement of the dots suggests a barrier to the mind in which it assumes it must stay. However, allowing the pencil to venture into empty space beyond the grid opens up many new possibilities, including the solution possible—we *can* connect all nine dots with only four straight lines.

We can compare the key to this puzzle with the key to Natures puzzle. Fancier equipment or a longer equation does not necessarily lead to the key that unlocks the answers to many of Nature's questions. Answers come by thinking with an open mind and by asking the right questions, in the right way.

Seeing With an Open Mind

The most difficult part about comprehending the Universal Model will not be the math, the language, or the experiments, but the struggle of overcoming a **closed mind**. It is natural to resist new information that appears to contradict our experiences and our previous learning, but sometimes, what we think are our own personal experiences may not actually be *our own firsthand* experiences. The sum of our ideas and thoughts that came from seeing or hearing movies, television programs, textbooks, teachers, and a myriad of other information sources forms our conscious experience, so it is critical that both the layperson and the scientist seek personal *firsthand* experience with the experiments and observations contained in the UM. In this way, your mind will be open to understanding Universal Model concepts and ideas about Nature.

The well-known science figure Carl Sagan affords us an example of how a closed mind can shut out important new ideas. In the 1960s, amateur astronomer Charles Boyer took photos that he said demonstrated a four-day rotation in the upper atmosphere of the planet Venus. Sagan, the professional astronomer had this to say about Boyer's claim:

"The four-day rotation is **theoretically impossible, and shows how foolish the work of the inexperienced amateur can be**." Note 1.1a

Just a decade later in 1974, the spacecraft, Mariner 10 performed a fly-by mission collecting data and making observations of Venus. The mission revealed physical evidence of Boyer's earlier assertion:

"The Mariner 10 spacecraft, encountering Venus in February of that year, imaged the planet in the ultraviolet during its approach. When these images were combined into a movie sequence, **the four-day retrograde rotation of the upper atmosphere was dramatically confirmed**." Note 1.1a

Although the Mariner 10 data eventually vindicated Boyer, the lack of an open mind on the part of the 'professionals' consequently dismissed truth and stifled scientific advancement. Notably, Sagan boasted of modern sciences' 'theoretical' limitations while condemning the foolish amateur's empirical observations. When a new idea comes backed with evidentiary support, even if such an idea runs up against accepted theory, the professional's best ally should be an *open mind*.

This, unfortunately, is too often not the case and theoretical presumptions regularly supplant actual evidence because it contradicts the latest theory. The result of such confusion is that science dismisses or misplaces many pieces of Nature's puzzle, some altogether lost.

Modern Science and Nature's Puzzles

Two views of Nature's puzzle in Fig 1.1.2 illustrate the contrast between theoretical modern science ideas and the actualities in Nature. Modern science possesses some truths, and there are instances of properly placed pieces, even whole assemblages, but there are many misplaced pieces and some are missing entirely, leaving the viewer confused and the picture unclear. However, by properly interpreting the pieces of the puzzle, and by connecting them in their right order, we see an understandable and comprehensible picture of Nature, ready for the addition of new pieces, each one fitting in a cohesive, ever-expanding model.

Throughout the years, many valid discoveries did not seem to fit or connect to the modern science theories. Sometimes, discoveries fit within a group of properly assembled pieces, but the overall assemblage seems out of place within the big picture. The Universal Model takes many

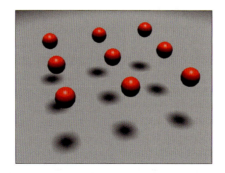

Fig 1.1.1 – Is it possible to connect all nine dots by drawing four straight, connected lines, without removing the pencil from the paper? The key is to understand the question and to have an open mind. The answer is on a later page.

Modern Science Puzzle   # Nature's Puzzle

Fig 1.1.2 – Two identical puzzles assembled from two different points of view. The pieces of the puzzle represent Nature's processes, laws, and observations. We see a disconnected and confusing modern science puzzle because improperly placed pieces, force fit where they do not belong. Nature's puzzle renders an understandable and cohesive image when answers are not forced to fit false theories.

of these anomalous scientific discoveries and by reevaluating them, places them properly into a coherent view of Nature so that others can see, test, and understand the new ideas.

True science requires that we ask questions, test the assumptions, and reproduce the experiments. If we ask students to memorize the discordant arrangement of modern science's version of Nature's puzzle, how likely are they to succeed in reproducing it in the exact same order? Certainly only a few could accomplish the task, and then only if they possessed superior memory skills. Yet this is how we measure success in the classroom today. Both teachers and students must rely heavily on memorization because of untestable assumptions and because of the confounding effects of misplaced puzzle pieces.

Conversely, show the same group of students the complete puzzle representing a correct view of Nature (Fig 1.1.2, right), then ask them to duplicate it. Without doubt, every student would find greater success in reproducing the correct version of the puzzle because the pieces fit together in a cohesive and orderly manner; everyone can understand what the image portrays.

The Simple Truth Principle

The *real* picture of Nature exhibits astounding variety, complex but not necessarily complicated; that which at first seems disorderly embodies beautiful order after further examination. The simplicity and elegance of the double helix discovered in 1953 exemplifies the simplicity in Nature. Known today as DNA, four little nucleotides arranged along two strands store an enormous quantity of information and instruction, which they can replicate and execute in relatively simple processes.

The niversal Model's goal is to discover Nature's simple truths and to assemble those pieces of truth in their proper order. A foundational UM principle is that truth is not complicated.

The Simple Truth Principle:

In Nature, the simple truth is that truth is simple.

Even a single step into a forest or a park and Nature's order becomes apparent to all who open their eyes. The Sun rises every day, warming the Earth and the air around us. Buds appear on dormant plants each spring and birds sing their beautiful melodies to all who will listen. Microbes in the soil decompose dead foliage in preparation for new plant growth. Spring and fall precipitation waters the microbes and plants upon which the animals feed. Then comes winter, a time for rest and renewal, where the temperature drops and the cycle of life changes for a season. How easy it is to take for granted the summer that comes each year, melting the snow, watering and warming the ground, giving us our year's longest days.

Even seemingly chaotic events have an order about them. Nature is ordered; it must therefore follow correct principles and natural laws, which are simple once we understand them. Imaginary theory is almost *never* as simple as the truth.

The Importance of Principles

We introduce UM principles throughout the Universal Model, especially in the introductory sections of the book. We have already seen two of them, but there are more to come.

Correct principles assist our learning through the application of *fundamental truths*. Although scientific laws identify physical processes in Nature, comprehending the laws becomes easier if we understand the intangible principles that bind them together. Correct principles are what bind the pieces of Nature's puzzle together, allowing us to apply the powerful tools of reason and logic in our discovery of truth.

Some familiar principles often have deeper meaning. One reason many of Nature's laws remain hidden is our misunderstanding or disregard of the correct principles that underlie the laws. Building The Universal Model on correct principles allowed the discovery of new natural laws, so we must first share those principles before advancing into the science.

The key to any puzzle, including Nature's Puzzle, is an open mind.

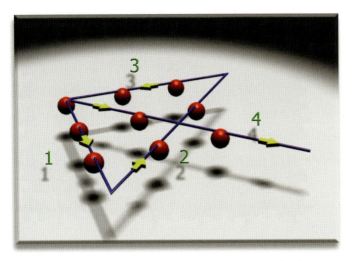

Fig 1.1.3 – We discover the key to the Nine–Dot puzzle only after overcoming natural mental barriers. By drawing lines outside the perceived boundary set by the grid, we see clearly the solution (follow the numbers and arrows). New scientific discoveries await those who look at Nature with an open mind.

1.2 The Question Principle

Of all the principles in the Universal Model, *the single most important principle* is the Question Principle:

Question everything with an open mind.

Without this principle the UM would not exist. An example of someone who applied this principle was Sir Isaac Newton. The following account describes how Newton began to question everything.

"Like many of the more conscientious students, he had been following the latest philosophical developments and was reading 'fashionable' philosophers, such as Descartes and Galileo, whose works were gradually becoming available in England. As a result, sometime in early 1633, Newton underwent a radical change of approach. During a lecture, while making meticulous notes on Aristotle's teachings, mid-page he stopped abruptly. Then, after leaving dozens of pages blank, he wrote at the top of a fresh page, 'Some Problems in Philosophy.' Beneath this he wrote… **'I am a friend of Plato, I am a friend of Aristotle, but truth is my greater friend.'**

"This collection of 'Quaestiones' – or the Philosophical Notebook, as it is sometimes called – **marks the point at which Newton stepped away from tradition and began to question what he was taught**. He began by creating forty-five headings in the notebook – topics concerning the nature of the universe which he would attempt to investigate and answer. These included 'Of Water and Salt', 'Attraction Magnetical', 'Of the Sun Stars & Planets & Comets' and 'Of Gravity & Levity'." Bib 19 p53

If we learn to question everything with an open mind like Newton did, it will aid us in judging between modern science and the Universal Model. Another great example of the importance of re-examining the foundations of the "whole structure of knowledge" comes from the science philosopher, Hans Reichenbach (1891-1953):

"It seems that **progress in the knowledge of nature** can be made only through conflict between two successive generations. What is considered at one time as a revolution of all thinking, a tempest in the brain, is for the next age a matter of fact, a school knowledge acquired under the influence of one's environment and believed and proclaimed with the certainty of everyday experience. Thus, possible criticism to which even the greatest discoveries should be continuously submitted, is **forgotten**; thus we lose sight of the limitations holding for the deepest insights; and thus **man forgets in his absorbing concern with the particulars to re-examine the foundations of the whole structure of knowledge**. We shall always have to depend on men like Copernicus **who question obvious matters and whose critical judgment penetrates deep into the foundations of truth**." Bib 11 p24

We discuss the significance of the Copernican Revolution in a future chapter but recognize here that a revolution came because Copernicus chose to "question obvious matters." Copernicus applied the Question Principle. Today, we herald him as a great forbearer of science because he challenged the prevailing notions of his day.

Why should it be different now? The Copernican and other scientific revolutions came by questioning the *foundations of knowledge* during their time. Future revolutions will not come without doing the same; the discovery of truth demands it. Once we find a new truth, we should never stop questioning our understanding of it.

Truth is the same every time we question it—or it is not truth.

Hard to Question

Today's scientists find it extremely difficult, even dangerous to venture outside the current scientific paradigm. If one oversteps too far the bounds of mainstream scientific theory and practice, he or she does so at the peril of having derailed a career. Even after convincing a number of one's peers about the correctness of a new idea or theory, there is no assurance that the daring scientist will escape ridicule or the barring from future opportunities. Ben Stein's 2008 movie, *Expelled - No Intelligence Allowed*, Note 1.2a documented the ostracization of scientists expelled by their peers and employers for simply questioning the validity of parts of evolutionary theory.

Many are unwilling to question the obvious because of the difficulty one encounters when challenging the long held and 'established' scientific dogma. However, if we want to learn the truth, we must begin with the Answer Principle—answers come from questions, and its companion, question everything with an open mind. Our questioning of the obvious must continue until we gain the personal experience that an answer is true.

Old ideas pose stubborn obstacles to progress. In today's highly specialized modern science environment, isolation insulates the researcher from critical information observed in other fields of study. The

Correct principles are like icebergs, their meaning lies far below the surface.

CHAPTER 1 INTRODUCTION TO THE UNIVERSAL MODEL

answer to a particular geology question might not come from geology, but from astronomy; we may find the answer to an astronomy question in physics, and from chemistry may come the answer to a physics question, from biology, an answer to chemists' question, and so on through all of the sciences.

All of science is interrelated and universal; the natural Universe is not neatly divided into a myriad of 'ologies' as found in today's academia. Biology, geology, and archeology represent man's attempt at separating and cataloging areas of study, sometimes arbitrarily. Nevertheless, all fields of science connect to all other fields; the 'pillar' in one field is the foundation of another. Indeed, what would happen to the world of chemistry or biology if one of geology's fundamental 'pillars' proved untrue?

Experimentation and observation should be the hallmark of good science instruction, but many teachers have the attitude that there is not enough time to conduct the 'old experiments' so one should just accept the statements in the textbooks as fact. However, by testing many of the accepted 'facts' taught in various science fields, we discover that many of them have never actually been proven. How is this possible, we might ask?

The problem is one of repetitive dogma. The student who never reproduced the original experiment becomes the teacher of the next generation, teaching the same old dogma. Seldom questioned, long-tenured professors direct their respective departments and junior teachers, so that after several student-to-teacher generations, we have theories taught and accepted as fact, even if no one ever proved them true in the first place.

To overcome this recalcitrant attitude, we must return to the basics by performing simple experiments and by questioning even the 'foundations' of science with an open mind.

The Textbook Question

To show the importance of questioning everything with an open mind, we turn our attention to a common example from chemistry. Most chemistry textbooks contain a formula that states $2Na(s) + Cl_2(g) = 2NaCl(s)$ (two solid sodium atoms and two chorine gas atoms equals two solid sodium chloride or salt molecules). This formula, the books explain, illustrates salt's origin. Fig 1.2.1 shows the components of the formula in an illustration similar to what many textbooks show. From one high school textbook, *Heath Chemistry*, we read the description of the sodium and chlorine elements and the reaction that happens when combining them:

"Left: The bottle contains chlorine, a pale yellow-green, poisonous gas. Middle: Sodium metal is kept under oil to prevent its reacting with substances in the air. Right: A vigorous reaction occurs when sodium and chlorine are mixed. **White, solid sodium chloride, NaCl, is formed**." Bib 45 p99

Look closely at all three photos in Fig 1.2.1. We see a chunk of sodium, a beaker of chlorine gas, but the third image is not "white, solid sodium chloride." Why did the authors not wait for the sodium and chlorine to actually form salt, and then photograph it?

Is it possible that the textbook authors portrayed this experiment incorrectly? After all, the experiment as explained has never produced the NaCl salt crystals we are accustomed to seeing on our kitchen table.

The textbook fails to mention the fact that in the third photo, the added component of water caused the reaction. There are many parts to salt's story, which we visit throughout the UM, but Nature does not employ the simple formula of Na + Cl to make salt; sodium (Na) and chlorine (Cl) do not exist by themselves in Nature.

Salt represents only one of dozens of UM subjects of which we demonstrate its origin. This new origin will affect the fields of geology and biology in ways never before imagined, but to do so, we must first be willing to question the science textbooks we studied in school. When new discoveries challenge long-held important theories, we can find it difficult to change old thinking—even though this is the only way to progress and gain intelligence.

It is not easy to confront the experts or the authority figures, as Halton Arp, a well-known modern astronomer discovered. He recalled a personal experience where he challenged his teacher after discovering an error in his science textbook:

"I would suggest that this (**training rather than learning**) starts in grade school and accelerates as the degrees become more advanced. I had only one year of formal schooling up to the seventh grade when I discovered a wrong answer in the back of the book. I was amazed at the reaction of the teacher and the class who could not believe that the answer in the book was not correct. Right from the beginning in science, **authority tends to override independent judgment**." Bib 65 p258

The Wisdom of Discernment

Theodore Schick Jr., professor of philosophy at Muhlenberg College, Pennsylvania, USA once made the following insightful statement regarding the importance of questioning the answers we think we already know:

"Scientists and educators alike need to realize that **the edu-**

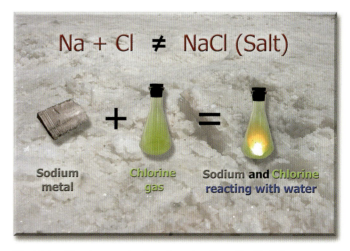

Fig 1.2.1 – Chemistry textbooks tell us that Na + Cl = NaCl (common table salt). Photos from school textbooks even attempt to demonstrate this process by showing that sodium metal and chlorine gas ignite when combined. This however, is not true. Sodium and chlorine react only if we add *water* to the flask; water and sodium react violently in an exothermic reaction. Salt is not made naturally by combining Na with Cl.

cated person is not the person who can answer the questions, but the person who can **question the answers**. In our age of rapidly changing information, knowing **how** to **distinguish truth from falsity is more important than knowing what was once considered true and false**." Note 1.2b

Does having an open mind mean being open to every possible thought or idea that comes along? If we let *everything*, including errors into our minds, it must eventually become corrupt. Thus, we must answer this Fundamental Question:

How do we distinguish truth from error?

The answer to this question is the **Wisdom of Discernment**. Questioning whether something is true or false is simply a matter of discovering whether it passes the *Test of Time*. Throughout this book, and throughout our entire lives, distinguishing truth from error will prove one of the greatest challenges we face.

How does one distinguish truth from error
if we do not know what truth is?

We discuss the definition of truth and the reason why its definition is so critical in Chapter 1.5, Truth Defined. Without the correct definition of truth, we cannot distinguish between truth and error. By knowing what truth is, we can employ the Wisdom of Discernment.

To discern with wisdom takes research, forming good questions, and conducting our own experiments. Leaning on another's unproven research, which may be false can prove detrimental. We should prove science's fundamental answers by applying the Test of Time through experimentation and observation.

How to Question

When we ponder the idea of 'how to question,' we begin to walk the road of wisdom. Here we connect the concepts of open-mindedness and questioning. When we pose a question, we contemplate what the answer might be, but if we *think* we already know the answer, we may close our minds to new answers; no doubt, we have all experienced this at one time or another!

Indeed, in our search for truth, what we discover is going to depend on *how* we question. Our questioning *attitude* will determine our degree of open-mindedness.

These four words describe four different *attitudes* we ascribe to when asking questions:

1. Critical
2. Skeptical
3. Partial
4. Objective

We used *Webster's Dictionary* Bib 7 to define the words that identify the questioning attitudes. If we question **critically**, we are inclined to *find fault*, or in other words, we are critical in our approach. If we question **skeptically**, we employ a *doubting attitude*. If we question with **partiality**, we are *prejudiced* in our approach, leading us to *partial blindness* toward the answer. However, if we question **objectively**, we *lessen the influence* of our personal feelings and prejudices; we remove as best we can our personal *bias*.

Imagine you have a new theory you wish to share with your esteemed colleagues; how would you hope they question *your* new theory?

Although popular in many professional circles, the attitude of reviewing as a 'critic' or as a 'skeptic' creates a sense of negativity or cynicism, which fosters doubt and distrust. A biased approach, which only allows for partial consideration stifles objectivity. These three attitudes carry a negative connotation, and as a result, are unlikely to encourage the discovery or acceptance of truth. The objective attitude carries with it a sense of positivity, so that one can ask:

Which of the answers presented is best?

Of course, the objective method requires more work as we evaluate alternatives; nevertheless, objectivity allows us to consider new possibilities, and it helps develop the essential ability with which to judge truth. With this mindset, when a new answer better fits the observation than any other answer, we can let go of old ideas as we embrace new information, advancing our understanding.

In an effort to counter the longstanding mentality that one must be a critic or a skeptic to review an idea rigorously, the Universal Model refers to the person who strives for objectivity as an **objectic**.

We best accomplish an open mind if we have an agenda of truth. Other agendas encourage bias and decrease objectivity. The greater the bias the more closed the mind.

We should govern all of our questioning with **The Objective Principle**:

We best achieve discovery with objective questioning.

We might find putting the Objective Principle into practice difficult, especially if we are accustomed to using the other questioning attitudes. However, the objective attitude most easily inspires scientific discovery and it remains essential in the continuing quest for truth. Science has many incorrect theories, but it has also discovered many truths. The objectic can evaluate and determine the truthfulness of those theories and can properly apply them in context with greater surety. Therefore, the critic, the skeptic, and the biased inquisitor must give way to the objectic when contemplating the questions from both modern science and the Universal Model.

What Do We Question First?

Questioning everything can lead to chaos if there is not an order to the questions we ask. We cannot question everything at once and so we should establish a priority of what to question first.

In scientific inquiry, several steps help scientists in their discovery process. These steps form the scientific method that we will discuss in the following chapter. One of the final steps includes the formulation of *natural law*, which should rightly endure the most rigorous testing, and which once proven, is the least likely for us to question; however, we constantly verify the law by questioning the results of the law's statements about Nature.

Other steps in scientific inquiry include *prediction*, *testing*, *observation*, and subsequent *evaluation*. To question these steps is a matter of employing one's own senses, by observing and evaluating first hand. If many people have previously observed the phenomenon, there may not be any new questions. However, if only a few—or maybe none at all—have actually made observations or evaluations, there may be a goldmine of questions to answer.

Before any type of scientific inquiry occurs, the inquisitor must pose an idea or a *theory*. Thus, the hypotheses or theory represents the most questioned step in the scientific process. Many people have thought that Nature's fundamental questions were too difficult to understand so they left those questions to the 'experts.' Nevertheless, there is an answer to every question, and it becomes a matter of changing ones attitude to find that answer. One of the UM's simplest principles is the **Hard Question Principle**:

> Hard questions are easy, when you know the answer.

Learning how to question and what to question are the first steps in the **Learning Process**, a critical discovery during the developmental period of the UM.

1.3 The Learning Process

The Importance of Questions

We place a great deal of emphasis on the method, attitude, and principles of asking questions. This is a good time to recall two important, previously stated principles:

> Answers come from questions.
> Question everything with an open mind.

If we do not ask questions, answers cannot come. It is through questions that we learn to think. The proper question will open the mind and can lead to new scientific discovery. It is therefore necessary to think carefully about the questions we ask, and learn to ask the right questions. Although questions are found in many forms, they can be simplified into one of six different types—**what, where when, who, how,** and **why**. Recognizing the *type* of question we are asking is very important.

Most questions in science are those from which we derive the basic concepts that shape our ideas of the world around us. If we have not answered the essential questions correctly, then we are the bearers of a flawed understanding.

A question about a basic concept in Nature is a **Fundamental Question (FQ)**. An answer to a question that provides a fundamental understanding of Nature is a **Fundamental Answer (FA)**. Throughout the UM, we employ the use of FQ's to aid in the discovery of Nature's secrets. There are some completely new questions and some you may have already asked.

The Learning Process

During the development of the Universal Model, we discovered a new learning process (Fig 1.3.1). If asked the difference between *knowledge* and *wisdom*, nearly everyone has diverse ideas and there is no consensus from one dictionary to the next about their meaning. Apparently, no one has ever made clear the definitions of these two words. We wondered why. Knowledge and wisdom form the basis of human intellect, and yet it seems there is no satisfactory definition, nor are the terms self-evident. The realization of how the words knowledge and wisdom differ from each other and their true definition was an enlightening discovery that led to the development of *The Learning Process*.

In the Universal Model, we define **knowledge** and **wisdom** as follows:

> Knowledge comes from answers to the questions:
> "what, where, when, and who"
>
> Wisdom comes from answers to the questions:
> "how and why"

We can divide answers to the six primary types of questions into two groups—those that provide knowledge and those that provide wisdom. We begin learning from answers to basic questions best described in Fig 1.3.1, The Learning Process diagram. Answers to the physical questions of what, where, when, and who provide *knowledge*. They *describe* tangible things. We gain *wisdom* by answering the *intangible* questions of how and why, which *explain* things.

Knowledge deals with the characteristics of physical or material things. The questions, 'what,' 'where,' and 'who,' *describe* a tangible object, a place, or a person. The question, 'when' classifies the physical attribute of time. Knowledge then provides *understanding* by describing material or physical objects.

Wisdom is of a less materialistic Nature. Answers to the wisdom questions of 'how' and 'why' *explain* rather than describe an event or an object. We *comprehend* answers to wisdom questions rather than understanding them.

To comprehend, we must *first understand*. We gain understanding, which is knowledge, by answering the tangible questions, and then we add comprehension by answering the intangible wisdom questions.

> Increase **Understanding** by answering knowledge questions; increase **Comprehension** by answering wisdom questions.

An increase in our understanding and the expansion of our ability to comprehend allows us to *describe* and *explain* Nature more completely. This is exactly what science is about—to *describe* and *explain* Nature!

The Three Laws of Learning

With the Learning Process explained, we can now discover the Three Laws of Learning. Although we discuss the particulars of natural law in the next chapter, we introduce the first three UM laws here as part of the Learning Process in order present a congruent and harmonious path to their understanding. We view natural law as a cause and effect process, that there is a predicable consequence from a stated action.

Although most of the natural laws presented in the UM consist of physical laws that describe and explain Nature, these three learning laws involve the intangible human experience of discovery, and we use them extensively throughout the UM. Although modern science supposes that human beings rank as just another member of the animal kingdom, the UM understands that being human is very special, and that we can increase in intelligence beyond what some have thought possible. These are the Three Laws of Learning:

Law of Knowledge

Knowledge and therefore understanding increases when we correctly answer the questions what, where, when, and who.

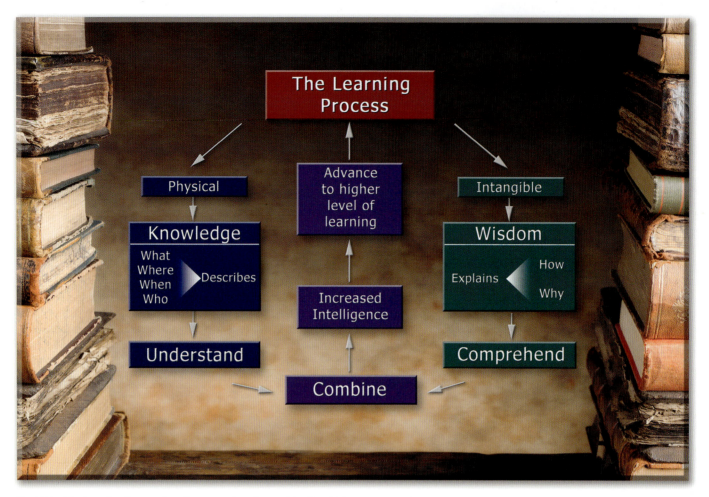

Fig 1.3.1 – The Learning Process describes a new learning model with which every human being can benefit. Answers to knowledge questions give us tangible information from which we derive an understanding. Answers to wisdom questions give us comprehension, and when combined with knowledge and understanding, increase our intelligence, leading us to a higher level of learning. Learning to ask the right questions is an important step in the learning process because we cannot comprehend without first seeking to understand.

Law of Wisdom
Wisdom and therefore comprehension increases when we correctly answer the questions how and why.

Law of Intelligence
When descriptions and explanations are correct, a true understanding of knowledge and comprehension of wisdom increases intelligence.

The Learning Process Diagram in Fig 1.3.1 accompanies these laws, explaining how knowledge, wisdom, and intelligence increase in a systematic and orderly fashion.

These laws emphasize the Answer Principle and Simple Truth Principles in that answers come from questions and that we merely need ask the correct questions first to discover the answers to Nature's most simple and basic questions. The simple questions—the Fundamental Questions (FQ's)—are the most profound.

How can we know that the Three Laws of Learning will hold up under scrutiny? Natural law is simply a statement of truth, so while some truthseekers may recognize the laws as true immediately, for most it will take time and experience. One reason for writing the UM was to demonstrate how we can increase in intelligence.

Lee Smolin, a physicist and well-known author once wrote:

"These **fundamental questions are the hardest to answer** and progress is seldom direct. Only a handful of scientists have the patience for this work. It is the riskiest kind of work, but the most rewarding: **When someone answers a question about the foundations of a subject, it can change everything we know**." Bib 178 pvii

The Three Laws of Learning are the first three of some 50 new natural laws, each testable and available for analysis by anyone, including the non-scientist. As we learn new laws that answer fundamental questions, many for the first time in their respective fields of scientific inquiry, they will as Smolin suggested, "…change everything we know."

In today's age of information, we can find answers and share them quickly and easily, no longer restrained by the expense of books and journals or by the inaccessibility of knowledge. Now, anyone can access information once hidden away at a few prestigious universities, in governmental archives, or private institutions, out of the reach of the average person. The internet dramatically changed the access to knowledge; from the proverbial little dirt road to a ten-lane superhighway, the internet has become the medium of exchange for an enormous amount of

information and the dissemination of knowledge, easily accessible by the general population. Access to such an abundance of knowledge is one reason *why* The Universal Model came about during this new millennium.

We have written The Universal Model to help increase our knowledge and to learn wisdom about Nature and Her secrets. The Learning Process is an important step in gaining an understanding and a comprehension of science, but The Learning Process is not limited to science alone; one can apply it to all aspects of life. Toward the end of this book, in the Millennial Education chapter in volume 3, we revisit the Learning Process and its application in other areas of education.

1.4 The Importance of Definitions

To communicate an idea effectively, it is important to have a common definition of words. For example, if someone said the words "light is *c*" would we know what they meant? Did the person mean 'sea' as in the ocean, or sí, the Spanish word for yes, or '*c*' as in the letter. In this example, the lowercase letter *c* denotes the speed of light in a vacuum, but we could not know this if we did not have the correct definition of '*c*.'

In this book, we define important words, less common scientific words, and words that have unclear or contradictive meanings. **The Definition Principle** is an essential principle.

*If we don't say what we mean,
we don't mean what we say.*

By using common definitions, everyone can better understand what the author is trying to convey with a particular word or idea. One of the best ways to communicate with others is to share similar experiences. Learning can take place through such sharing and it is essential that the definitions of the words are the same for all parties.

There is no better way to learn than direct, hands-on experience. You are encouraged to *experience* as many of the experiments and places described in the UM as you can. Most of the experiments and the new observations about Nature are simple enough to perform or easy enough to visit, which will contribute to a deeper understanding of the concepts presented herein.

Incorrect Definitions

With *incorrect definitions,* confusion and disorder ensue. We can reach no agreement if definitions are not correct. Without the most basic definitions of knowledge and wisdom correctly defined, how can we advance in our scientific learning?

As we progress through the Universal Model, we will discover that *incorrect definitions of scientific words* have inhibited humanity from understanding Nature's truths. With incorrect definitions of natural phenomenon, we cannot understand or predict its occurrence. Many incorrect definitions have come from science's acceptance of theories as fact, with no experimentation or observation to back them up.

When we teach incorrect scientific definitions to students who later become the teachers and professors, we allow the building of new theory upon old theory, which incorporates incorrect definitions. What is the consequence of relying on incorrect definitions and theories for more than a century? We are about to find out. But first, we must establish a solid foundation for the discovery of new scientific truths. Before anyone can understand *true* science, we must know what *truth* is.

1.5 Truth Defined

We must clearly define **Truth** just as we defined knowledge and wisdom so that real progress can happen. Furthermore, it is essential for the definitions of all words based on truth that Truth have a true and correct definition. In essence, Truth is the foundation upon which we should build all other things, and it well deserves our attention in defining it.

Truth is knowledge of what is, was, and will be.

How Important is Truth?

Before we define truth, we need to establish just how essential truth is in our daily lives. If we can gain a sense of that, we can understand the need to define truth correctly.

At one time or another everyone will set a goal. That goal may be to finish school, it may be a goal to find a job or buy a car, or it may be as simple as getting a drink of water. The goal stimulates questions such as; what classes should I take, what hours will I work, or where is a cup? By asking these questions, we in essence seek for truth. *Who wants to know the wrong answer*? Who pays the consequence if we do not obtain the truth? Life is a quest for truth.

Realizing the role of truth in our lives gives us a basis for comprehending the importance of it. Such wisdom persuades us to ask this Fundamental Question:

How do we find truth?

To find truth, we must start by understanding what truth is.

Dictionary Truth

Doesn't the dictionary define truth already? If you consult a recent dictionary, the entry for 'truth' will include descriptive words such as fact, real, or actual. These words *do not define truth* and they do not actually convey what truth ***is***.

Look in the dictionary under 'fact' and the definition reads truth, or reality. 'Real' is defined as; true, actual. And finally, 'actual' is defined as; real, fact. [Bib 7 p9, 286, 656] So we see that turning to the dictionary for the definition of truth is merely a tautological thesaural list of words. This will not do.

Defining truth is more than a general description of the word. To define truth we must use a definition that is specific to *no other word* otherwise confusion is certain.

Furthermore, the words fact, real, and actual do not properly define truth because these words deal only with the present or the past. They do not account for the future. Truth involves not only the past and present—it implies involvement with the future.

The Test of Truth

From what foundation can we then build upon to define truth? There is a single foundation, firm and immovable, from which we can perform the *Test of Truth*.

Five hundred years ago, some people 'knew' the world was flat. The mathematicians of the day 'knew' that the product of 2 x 3 was equal to 6. Today, we know that the Earth is *not* flat and that 2 x 3 still equals 6. What caused one idea to prove false and the other true? *Time*.

Time is the test of truth.

Truth is the same yesterday, today, and in the future. This is why truth is so important to humanity. Truth will never cease to enlighten humankind. *Truth is simply knowledge that does not change*. It consists of the physical and tangible, or *knowable* aspects of Nature. With this understanding and with an appreciation for the Test of Time in discovering truth, we can now properly define **Truth**:

Truth is knowledge of what is, was, and will be.

Truth is knowledge of facts, but the facts must stand the test of time. Truth is reality, even if we do not yet know that reality. To possess truth is to possess knowledge.

The reality that craters exist on the Moon was a reality before Galileo knew about them. Science did not possess the truth about Moon's craters until Galileo discovered them, although it was fact that the craters existed long before Galileo's time.

We learn in Chapter 7, the Hydroplanet Model that science still does not correctly understand how the majority of the Moon's craters formed. Once we reveal the true origin of the

 lunar craters, our understanding and comprehension of the Earth's craters and the craters on other solar system bodies can increase, as will our ability to understand the correct formation of planets and moons.

In science, when a statement of *cause* gives a predicted *effect* over a continuous period, we can say the statement is true. Thus, with the fundamental measurement of time, we can define truth.

To find a dictionary that included a reference to time in its definition, we had to look through many current and old dictionaries. Finally, in a pre-1960s unabridged Webster's dictionary we found a definition of truth with an important reference to time.

One wonders why there are so many different definitions of truth. Apparently, because we have no clear definition of truth, especially in science, and because what we have today has no connection with time, the world overlooks the importance of defining truth.

Truth is Knowledge

In science, perhaps truth is the ultimate word to define. Because truth tells us what is, what was, and what will be; having truth we are able make predictions about the future. Predictions that correctly occur such as when the Moon will be at full phase, or what time of the year Spring will transpire, which reaffirms our knowledge of the truth. Throughout humanity's history, the discovery of truth has led to knowledge of the future. America was founded on the discovery of truth. The Universal Model chooses to make the same declaration made by the Founding Fathers of the United States of America when they wrote the Declaration of Independence:

"We hold these **Truths** to be self-evident, …"

The Founding Fathers declared that Truths do exist and that some are self-evident, and they listed them. We can see that the truths chosen by the Founding Fathers have passed the 'test of time.' After just two years as a nation, George Washington, the first President of the United States, said:

"[T]he United States enjoy a scene of prosperity and tranquillity under the new government, that could hardly have been hoped for…" Note 1.5a

Over 200 years later, the US Constitution remains the shortest and the oldest written national constitution still in use.

The right to "Life, Liberty, and the Pursuit of Happiness" is a right to which every human being is entitled. It was one of the truths declared 'self-evident' in the Declaration of Independence, and sometimes it takes time to assimilate new truths. In the United States, it took many decades for all Americans to receive the rights of life, liberty, and the pursuit of happiness, but this ideal is what has made America the country it is, unique in the world with a Statue of Liberty reminding us all of what made America; a place where dreams can come true.

Examples of Truth

There are truths in science that are also self-evident. For example:

1. A living human body contains water.
2. A solid cube has six sides.
3. The order of color in a primary natural rainbow is always the same.

Each of these statements is true because they state what is right now, what was, and what will be. Scientific truths like these are examples of The Self-Evident Principle:

It is, because it could be no other way.

The Science Teacher That Did Not Exist

Perhaps you might be asking why place such an emphasis on truth? After all, if something were self-evident, wouldn't it just be accepted? Who doubts that a solid cube has not always had six sides, or that it will ever have more than six sides? A personal example may better illustrate the point.

Some time ago, having graduated from college years earlier, I took an anthropology course called *Human Origins* with the intent to learn what scientific evidence existed about the origin of human beings. The professor appreciated my inquisitiveness and I thought the class was very interesting. In one of our assignments however, I took issue with a statement made in our class textbook. The author had said that there were 'two truths,' absolute truth and relative truth. I wanted to know how something could be both true and false at the same time. I wondered what the professor thought about 'relative truth.' I was in for quite a surprise and I did not know at the time that her answer would be a catalyst toward the research culminating in the Universal Model project.

This was her written response to my query:

"A major topic of discussion in science, including archaeology is whether or not the world can be 'known' (understood)."

I was excited to find out more about this "major topic of discussion" when I received an invitation to the professor's office to discuss it. It was an illuminating experience. Put very simply, the professor questioned, "whether or not the world can be known," which included her own existence! This articulate, educated individual sitting directly in front of me was questioning whether she existed or not. I was baffled, as I had never been exposed to someone who doubted his or her own existence. (For more details on truth in science being rationalized, see note 1.5b.)

Truth Denied

As it turns out, this particular 'philosophy of science' has been around for a long time and has gained great momentum in recent years. The Nobel Laureate H.J. Miller once said the following:

"The honest scientist, like the philosopher, will tell you that **nothing whatever** can be or has been proved with fully 100% certainty, **not even that you or I exist**, nor anyone except himself, since he might be dreaming the whole thing." Note 1.5c

Because of ideas like these, some scientists and professors have denied the existence of truth. How does this 'scientific philosophy' affect the minds of our students?

One student's experience in a humanities class at Arizona State University about a decade ago illustrates the point. The teacher had asked the class of 185 students how many believed in absolute truth. One hundred seventy five students said they did *not* believe in 'absolute' truth. **Only** 10 stated that they could believe in absolute truth. Why?

If we do not believe in absolute truth, there must be another answer to 2 x 3 besides 6 in modern mathematics. How

would the 175 students who did not believe in truth explain that 2 x 3 had another answer besides 6? Next, we explore why truth in science remains misunderstood.

1.6 Scientific and Absolute Truth

The First Six Senses

How do we come to 'know' and learn? In life as in science, we gain knowledge through our senses. A broad definition of this process is:

> Senses are the means by which we increase our wisdom and knowledge.

There are **five physical senses**: *sight, hearing, smell, taste, and touch*. Additionally we increase wisdom and knowledge by using the **Sixth sense—logic and reason**. The employment of this sense gives us a basis for understanding and comprehending mathematics.

If 2 x A = 6, it is *logical* that A = 3. Science has the great advantage of being logical and relies on logic to convince not just other scientists, but also all of us that through reasoning, one's ideas are sound. True science uses reason and logic, in conjunction with prediction, testing, observation, and evaluation to establish knowledge. We use the first six senses regularly in science.

The Seventh Sense—Intuition

We know the **Seventh Sense** by many names, but the word most often used to describe it is **intuition**. John Horgan, a popular science writer who has interviewed numerous scientists, said:

"It has become a truism by now that scientists are not mere knowledge-acquisition machines; they are guided by emotion and **intuition** as well as by cold reason and calculation. … The greatest scientists want, above all, to discover truths about nature … they want to *know*." Bib 20 p5

Another author-scientist explains how we apply intuition in science:

"It is **intuition** that propels science beyond the edge of its frontier, like a long forward pass in football; but the ball has to be caught—**the intuition confirmed**—to be counted as a touchdown." Bib 18 p61

Scientific Truth and Absolute Truth

Although many scientists acknowledge the existence of the Seventh Sense, and certainly many have been the benefactor of it, intuition is, by itself, non-transferable. One cannot give intuition to another. Individuals must discover it by themselves. However, outcomes derived from the first six senses are transferable. Show someone a glass of lemonade and they can feel it, taste it, smell it, hear it as it is poured and they reason that it must have come from lemons. *This is science.*

> We demonstrate science using **only** the five physical senses and the sixth sense of logic and reason.

Individuals may develop and employ the Seventh Sense in their work, but for others to accept their discoveries they must employ a process by which they can all know. This is why it is with **only** with the first six senses that we can verify scientific discovery.

Because of this limitation, science cannot prove via the first six senses that something will *always be*; we cannot directly experience the future. Since science cannot show that something will *always be* in the future, we must define **Scientific Truth** by including the words, "so far as is known."

> Scientific Truth is knowledge of what is, was, and, **so far as is known**, what will be.

To gain absolute truth, one must look beyond the first six senses. The use of the Seventh Sense of intuition allows us to understand and define **Absolute Truth**:

> Absolute Truth is knowledge of what is, was, and what **always** will be.

We acknowledge the existence of the 7th Sense, and recognize that without it, there would have been no Universal Model. Many readers will undoubtedly use intuition to help them understand and comprehend the Universal Model. However, the Universal Model does *not* nor can it use the 7th Sense to prove new scientific truths. In the UM, we use only the Universal Scientific Method discussed in Chapter 2.6 to demonstrate new scientific truths.

> Scientific information is **only** demonstrated by the first six senses.

Truth Restored

When Columbus and others discovered lands unknown to the Old World, what they actually did was discover truth. The discovery of this 'truth' had a universal impact on the entire world. The Americas became known as the 'New World' because of one person's intuition, insight, and perseverance. With a literal new world opened to civilization, Columbus' intuition and the resulting discovery of truth forever changed humanity. Could a discovery of such magnitude happen today?

If we restore truth to many aspects of science in our day, it must surely come with the discovery of new scientific truths. The Universal Model demonstrates the existence of truth by the discovery of truth. This is the primary purpose of the Universal Model:

> **Purpose of the UM**: To restore truth and order in science by identifying new natural laws.

Could the definition of this one word—*Truth*— have a universal impact on humanity? We will see.

1.7 The Human Touch

A Clumsy Beginning

In the UM, 'the human touch' is our way of describing the fallibility of the human race. We are all ignorant, just in different areas. Try as we may, we cannot produce perfect books and The Universal Model is no exception.

Sir Isaac Newton gained notoriety with the publication of his first book, the *Principia*, which described the laws of motion. Even a renowned book such as the Principia fell victim to 'the human touch'.

"The *Principia*, for example, did not always prove an easy work to apply, partly **because it retained some of the clumsiness inevitable in a first venture** and partly because so

much of its meaning was only implicit in its applications." Bib 33 p33

The Universal Model, the first book to venture into this new territory, will no doubt include similar clumsiness. Regardless of this inherent weakness, with time and effort the new scientific truths outlined in the UM will develop into a Millennial Science on which science can build for centuries.

Moreover, individuals with no academic affiliation and limited funds developed the Universal Model. We acknowledge our weaknesses and claim the errors that exist herein as our own, and we hope that you will temper your judgment of these infirmities until you have completed the reading of the book. After reading the book and performing some of the experiments and by making your own observations, you will have a more accurate perspective with which to judge the work.

Nature's Puzzle is Simple

We began this chapter with a reference to a puzzle used to illustrate the difference between the Modern Science Puzzle and Nature's Puzzle. The complicated and chaotic Modern Science Puzzle has endless combinations because of the confusion that exists. This is not so with Nature's puzzle.

There is only *one* correct way the pieces of the puzzle fit together; only after we assemble them correctly can we see the true picture. *This is the nature of truth.*

It is one of the UM goals, to assemble the core pieces of Nature's Puzzle. With the placement of each new piece, we move on to the next. Unfortunately, we did not have the luxury of time or resources to confirm the placement of every piece multiple times. Then again, any project with a scope the size of the UM, or which replaces any number of well-entrenched theories in multiple fields of science would have *many limitations.* The time and resources needed to repeat so large a number of experiments by other scientists was not available. Neither was the time available to research *every* article that exists on every subject in this book. What the Universal Model has accomplished:

The UM has framed Nature's Puzzle so that for the first time, we see a clear and distinct image.

We believe that more importantly than multiple confirmations provided by pieces of the puzzle, the interconnectivity of the whole, with each piece in its proper order, confirms powerfully the truth of this work.

The Human Touch of the UM

With respect to the limitations expressed herein, the author takes full responsibility for Universal Model's 'human touch.' It is hoped that the UM will plant the seed of discovery in every individual, so that once watered, it will grow, bear fruit, and carry on through others who will engage in scientific discovery that will contribute to the growing Millennial Science.

1.8 The UM Book Format

The Big Picture of the Universal Model

The diagram in Fig 1.8.1 represents the Big Picture of the Universal Model, which contrasts with the Big Picture of Modern Science in chapter 3.5. The book is divided into six main sections:

1. Introduction
2. The Earth System
3. The Living System
4. The Universe System
5. Futurity
6. Notes

The Introductory chapters of this book (1-4) build on each other, providing extremely important prerequisites for later chapters. One will not be able to absorb and understand the material in the main body of the book without reading the introductory chapters.

The **Introduction** section that you are now reading, which encompasses the next three chapters, includes an introduction to the UM, the universal scientific method, a short history of science, and a word on scientific revolutions. Although many people express an interest in Nature and science, only a small percentage of people pursue some form of science study on a regular basis. One of the Universal Model's goals is to bring back simplicity and understanding to science so that a larger percentage of people will want to pursue and learn about Nature. Keeping this in mind, we will define and explain many unfamiliar scientific terms as we introduce them.

You will find most of the science in the UM in the main body of the book under one of the three primary sections, the **Earth System,** the **Living System**, and the **Universe System**. Each system rests on a firm base of new, natural models derived from empirical evidence that almost anyone can observe firsthand. Of all the scientific questions we asked, what is more important than those about our origins? Understanding our origins requires a correct knowledge about the Earth, and Nature's Living and Universe Systems.

When we have the true Universal Model picture of the Universe—each new observation grants to us new understanding and every new explanation expands our comprehension. This is how we as human beings can truly progress and gain intelligence.

The section of the book entitled **Futurity** takes a forward look into the future of science based on the new UM discoveries. This section discusses the impact the Universal Model will have on Millennial Science and Millennial Education. Toward the end of the book, a **Notes** section lists the thousands of references for the blue quotes contained in the body of the text and will provide additional details or other information on selected topics. Notes come generally from scientific journal articles or websites; bibliographies comprise scientific books from which we drew a number of quotes.

We divide the main sections into numbered chapters and subchapters, referring to the black bolded subheadings as **subsubs**, all of which we frequently connect.

Book Format—Nuts & Bolts

The UM contains hundreds of extraordinary claims and discoveries, and it identifies many new facts. We do ***not*** expect the reader to take any of these at face value. Although we support and verify them with extraordinary evidences, if they fail future tests, we will discard them.

To demonstrate modern science beliefs and ideas, or to present discoveries and direct observations, we quote thousands of statements and citations from scholars and professionals from a variety of fields in blue for easy recognition. The portions of the blue quotes that are **bolded** by the author intend to illustrate and emphasize particular points of the cited statement.

We draw many scientific statements from well-known and prestigious scientific journals and magazines such as *Nature*, the premier European journal, and *Science*, its American counterpart. Some of the popular scientific magazines cited include *Scientific American*, and the *New Scientist*, published in Europe. There are over 200 books included in the bibliography, from which we pull multiple citations. Many other books, scientific journals, and literally thousands of references comprise the body of research, of which not all made their way into the final edit. When possible, references include page numbers and the full reference to encourage verification and further research on the subjects.

It is worth noting that we have taken great lengths to represent scientist's statements and observations (as 'blue quotes') in a manner not intended to mislead, although in many cases we draw contrasting conclusions from their statements. Many of the quotes cited in the UM come from those who made the actual observations, which we deem highly important to the new discoveries and claims made herein. We also recognize that the scientists who made the statements *may not agree* with the way we use their words to confirm the Universal Model's extraordinary claims. Nevertheless, we consider this a necessary part of scientific inquiry. We have attempted to follow all fair use laws on all quoted material.

Bibliographies are referenced accordingly—Bib 20, p5 means page 5 of the 20th bibliography reference section.

Figures in the book are labeled accordingly—Fig - 12.6.4 is chapter 12, subchapter 6, Figure 4.

Black bolded words used within a paragraph comprise newly introduced or defined words. We use *italicized* words for em-

Fig 1.8.1 – This diagram representing the **Big Picture of the Universal Model** illustrates a number of the scientific research areas and discoveries included in various UM chapters. We found it difficult to describe all new UM science within a single diagram, but the collection shown here gives some idea of the scope and breadth of the project, and an indication of where Millennial Science will lead us.

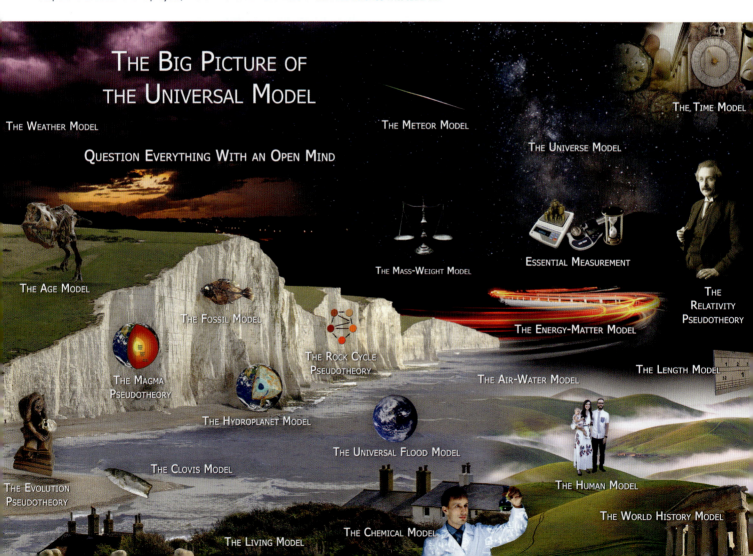

phasis—***bold italicized*** words provide further emphasis, and are used throughout the book.

Single quote marks ' ' around a word or phrase are used to denote special meaning to words used in the context of the sentence.

Green words identify principles, important statements, fundamental questions, answers, extraordinary claims, and discoveries throughout the book. Gold words highlight special quotes or phrases and new UM natural laws.

Extraordinary Discoveries and Claims

A common statement in science and one used frequently in the UM is:

Extraordinary claims require extraordinary evidence.

We define an extraordinary discovery as:

Extraordinary Discovery: a new scientific discovery that affects the world in a significant way.

What is the standard by which we can judge whether a new scientific discovery, evidence, or claim is 'extraordinary' or not? For instance, a dissertation on a newly discovered species of grass would not qualify as extraordinary, but the discovery of a new planet around our Sun would. The difference is that we would teach the discovery of a new planet in *all* future science textbooks, whereas we would not include the discovery of the new species of grass.

Significant scientific discoveries are those that influence whole segments of science that becomes part of the science curriculum taught in the science classroom. Of course, anyone can make extraordinary claims (and many have). To validate extraordinary discoveries or claims, **extraordinary evidence** must accompany it, which we defined as:

Extraordinary Evidence: a significant demonstration of a scientific claim that includes the observation and evaluation of physical phenomena.

In the past, many extraordinary claims have been taught to students *without* evidence to back them and these 'claims' have become the important *theories* in modern science today. It is critical that we substantiate extraordinary claims with extraordinary evidence before we teach them as fact. When we embed false extraordinary claims into the textbooks, like barbed hooks, they are difficult to remove.

The Universal Model is unique because of the number of extraordinary scientific claims *and* evidences it contains. An extraordinary discovery from one field often affects the whole of science. This initiated an examination early on of many science fields, often leading to significant discoveries in those other fields as well. Researchers today rarely study across multiple fields, which has caused many major discoveries to go unnoticed. Some think there are no more big discoveries left to make, but this has proven far from true. Some of the greatest discoveries are yet to come, once a true foundation is established.

Verifying Extraordinary Claims and Discoveries

The UM includes *hundreds* of extraordinary claims and discoveries, of which we provide a list in Chapter 27, UM Extraordinaries. We encourage the reader to test these against the extraordinary *evidences* and the extraordinary discoveries to verify the claims; there are two ways to do so.

The first comes from the statements, observations, and experiments of many scientists, journals, books, and other sources identified throughout. If we find that a previous discovery supports the UM, we make every effort to credit and reference that discovery and its original bearer.

The second comes from our own observations of Nature or from experiments performed by those involved with the UM. We lay out these new scientific claims in a simple format *so that you* can test the claims yourself by repeating the experiments or by experiencing the natural phenomena. *We cannot overemphasize the importance of this step!* What a wonderful opportunity it is to discover what you read is true, without having to trust someone else. By choosing to do the experiments, or by observing firsthand the phenomena, you can better judge the validity of the UM's extraordinary claims and discoveries.

Q & A about the UM

Q: What is The Universal Model?
A: The Universal Model is a study of demonstrated truths and natural laws that describe and explain Nature.

Q: What is Millennial Science?
A: Millennial Science is a new enduring science based on the new natural laws and observations in the Universal Model.

Q: What is the purpose of the UM?
A: To restore truth and order in science by identifying natural law that helps describe and explain Nature so that all can more clearly understand and comprehend it.

Q: Who developed the UM?
A: The Universal Model is a collaborative research project developed by private individuals and written by the author independent of any educational or governmental institution.

Q: Who was the UM written for?
A: The Universal Model is written for everyone. Although written to an entry-level college audience, the concepts are within the grasp of most grade-school students.

Q: Why is the Universal Model not written or published from within the scientific establishment?
A: Today, most people submit their new scientific theories and papers to the scientific establishment to conduct a rigorous review process wherein establishment-trained peers decide whether the content is worthy for publishing, and that it falls within the confined views of its respective field. In the past, large-scale changes in science come from outside the establishment from individuals with an outside perspective, unfettered by widely accepted dogma. We noticed that new theories and papers within the establishment, which build on old theories written to well-schooled peer groups who hail to the old theories, come in the form of complicated explanations and techno-speak, that lies beyond the grasp of the average person. The Universal Model is *not* about complicated theories; it is about simple models that demonstrate Nature's laws that all can observe and experience. This is how scientists inquired and how they conducted their study centuries ago.

2

Methods Models and Pseudotheories

2.1 - The Line Between Theory and Natural Law
2.2 - Against Method
2.3 - The Modern Science Method of Confusion
2.4 - Controversial Science
2.5 - The Universal Scientific Method (USM)
2.6 - Scientific Models
2.7 - Pseudotheory and UM Models

This chapter is about what scientists do. You may not think you do science yourself, but everyone does science. Checking the weather by looking outside or baking a cake is a form of doing science. The methods used to investigate and research the natural world and the models resulting from the experiments and observations are immensely important. To conduct scientific research properly requires following a Scientific Method, identifying and building on Natural Laws and Correct Principles, and recording the outcomes and evaluating the results. We developed the Universal Model in part because of knowledge contained in this chapter. As you read, we think you will come to appreciate the importance of this information.

The primary thrust of this chapter is to identify a *universal* scientific method and to introduce Scientific Models. We will demonstrate that there is no universally accepted method of conducting science in the modern world and we discuss why it happened and why it is important. Identifying a Universal Scientific Method (USM) proved essential in the development of the UM, and continues to reinforce the discovery process. We begin by examining the difference between theory and natural law.

2.1 The Line Between Theory and Natural Law

Theory and Natural Law Defined

Before we discuss the Scientific Method, there are two important terms often used in relation to science that we must understand and define correctly, or we run afoul of the **Definition Principle** from chapter one; if we don't say what we mean, we don't mean what we say.

Over the last several decades, modern science has seemingly attempted to 'redefine' the words 'theory' and 'law' by making them appear evermore synonymous. Notice in the following quote from the book, *Truth in Science*, how these two words appear together:

"…the line that separates the **laws and theories** of science from the factual evidence **is not sharp**." [Bib 18 p101]

There is a tendency in modern science to associate *theories* and *laws* as if they are equivalent, or at least nearly so. Why try to lump these two very different words and meanings together? When we examine the History of Science in chapter three, we learn there have been no significant natural laws added over the past 100 years (see figure 3.2.1 History of Science Table). However, during that same time, there was an abundance of new 'theories,' some of which science touts as *fact* today. Perhaps because 'theory' has been associated with 'law' long enough we presume they are the same.

In *Webster's Universal College Dictionary*, [Bib 7 p815] there are several definitions for the word **theory**. Among the more common are:

1. A proposed explanation whose status is still conjectural
2. A guess or conjecture
3. Contemplation or speculation

The dictionary also includes the following definition for 'theory' when used in a scientific sense:

4. A coherent group of general propositions used as principles of explanation for a class of phenomena: *Darwin's theory of evolution*.

We read that this dictionary assigns to the term, 'theory' a guess, or speculation, but as it relates to science—Darwin's theory specifically—it proposes we substitute for principles a group of propositions. Nevertheless, even the definition including scientific theory falls short of ranking as a 'law.'

In Webster's Seventh New Collegiate Dictionary, Bib 8 p478 the definition of the word **law** with respect to Nature and science asserts:

1. A statement of an order or relation of phenomena that, **so far as is known**, is invariable under the given conditions.
2. A relation proved or assumed to hold between mathematical or logical expressions.
3. The observed regularity of Nature.

Note the words, *order, invariable, mathematical,* and *regularity*. What do these words have in common? If we describe a thing in Nature as orderly, invariable, or exhibits regularity, we do so because it is the same yesterday, today, and in the future. When something like this remains unchanged over time, it denotes truth.

A natural law is a statement of truth.

It is no coincidence that many modern scientists deny the existence of truth (see subchapter 1.5) while at the same time, science has discovered no new laws. We witness this reoccurring theme throughout the UM, affirmed by the science leaders' own words. Moreover, we find it interesting that the definition of natural law that we read in Webster's and other dictionaries is absent in modern science textbooks. How can science hope to fulfill its purpose by helping us understand and comprehend Nature without knowing what a natural law is? Future generations of scientists can never hope to discover natural law if they do not know what a law is, or how to recognize one. We provide the UM definition of natural law later in this chapter; but first, a little more backstory.

Blurring the Line between Theory and Natural Law

By degrees for several generations, scientists abandoned the idea of natural law, replacing it with theory. By ignoring the definition principle, the incorrect usage of the terms theory and law brought confusion, and in many textbooks, the word theory stands in lieu of the word 'law,' blurring the line between *speculation* and *fact*.

A Natural Law is "a statement of an order or relation of phenomena that, **so far is known**, is invariable under the given conditions."

Scholars point out that science is 'progressive' and is an ongoing quest, but a quest for what—more theory? Modern science would have us think so, but without natural law, we are ever learning but never able to come to the knowledge of truth.

Another reason for the abandonment of the use of the term 'natural law' is the possibility that a law may change. We expect theories to change and so there is a measure of safety by calling everything a 'theory.' On the other hand, laws by definition, including scientific law, endure without change. At least if they are true—they stand the test of time. As our knowledge and understanding changes because of new scientific discovery, we may modify or even replace a scientific law because, as we learned in chapter 1.6, **Scientific Truth** is knowledge of what is, was, and, *so far as is known*, what will be. Natural Laws far outlast theories; they are the pinnacle of scientific discovery, and as we make more discoveries, we derive a clearer perception of those laws. The UM reconditions several existing natural laws, adding updated observations and definitions that better explain the phenomena at hand.

There is a very important distinction to know about the difference between theory and law. Whereas theories *always* change, true natural law does not. Most people are realists—they believe in truth and reality and that Nature's true laws describe and explain Nature in Her truest sense.

Ask yourself whether the following statements are true:

- 2 x 3 = 6
- A living human body contains water.
- A solid cube has six sides.
- The order of colors in the primary bow of a natural rainbow on Earth is always the same.

If you believe these statements *are not* subject to change, then you are probably a **realist** and you find the line between theory and law distinct. If you believe these statements *are* subject to change, you most likely are a **theorist**; the line for you is not clearly drawn and things are blurred. In the next discussion, we define the line between theory and law more clearly.

2.2 Against Method

A False Philosophy

For three decades prior to 1989, Paul Feyerabend was a professor of philosophy at the University of California, Berkeley. He, along with Thomas Kuhn and Karl Popper were the prominent scientific philosophers of the twentieth century, and his most influential work, *Against Method*, remains required reading in many college philosophy courses today. The online encyclopedia, Wikipedia says this about him:

"Feyerabend became famous for his purportedly anarchistic view of science and his **rejection of the existence of universal methodological rules**. He is an influential figure in the philosophy of science, and also in the sociology of scientific knowledge." Note 2.1a

Amazingly, Feyerabend became famous for his "rejection" of a universal scientific method! Although not all scientists agree with Feyerabend's philosophy, his influence and his ideas have crept into *many* of the most influential scientific writings of the twentieth century. Because of this, it is imperative that we become familiar with his ideology. The book, *Against Method,* opens with the following sentence:

"This book proposes a thesis and draws consequences from it. The thesis is: *the events, procedures and results that constitute the sciences have no common structure*; there are no elements that occur in every scientific investigation but are missing elsewhere." Bib 185 Introduction

Feyerabend was correct in his assessment of 'modern' science when he stated that there is "no common structure" in scientific procedure today. Feyerabend cherished this idea and proclaimed that this is how science should be—without "reason" or "rationality":

"A theory of science that devises standards and structural elements for *all* scientific activities and authorizes them by reference to '**Reason**' or '**Rationality**' may impress outsiders—**but it is much too crude an instrument** for the people on the spot, that is, **for scientists facing some concrete research problem.**" Bib 185 Introduction

Is Feyerabend suggesting there is no reasoning or rationality in science? Indeed, he is. Feyerabend pens his thoughts in another collection of papers, *Farewell to Reason,* where he continues to develop his philosophy. Here is another example of his premise:

"Scientists are like architects who build buildings of different sizes and different shapes and who can be judged only *after* the event, i.e. only after they have finished their structure. It may stand up, it may fall down—**nobody knows**." Bib 185 Introduction

Ultimately, this is where Feyerabend's thinking finds its end—*nobody knows*. Of course, everyone knows if a building stands or falls, but how do we test a philosophy like Feyerabend's idea, which essentially is, there is no truth?

Does it stand the test of time?

In subchapter 1.5, we explained that *time* is the test of truth. This is the test science must use, along with reasoning to discover how Nature works. Some philosophers, like Feyerabend seem averse to the certainty that real science can provide. He would have us believe that there are no rules (natural laws) in Nature that "remain valid" and he supposes that without chaos there is "no knowledge." Feyerabend's last words in *Without Method* clearly show a lack of reasoning or logic in his philosophy:

"Without 'chaos,' no knowledge. Without a frequent dismissal of reason, no progress. Ideas which today form the very basis of science exist only because there were such things as prejudice, conceit, passion; because these things *opposed reason*; and because they *were permitted to have their way*. We have to conclude, then that *even within* science reason cannot and should not be allowed to be comprehensive and that it must often be overruled, or eliminated, in favor of other agencies. **There is not a single rule that remains valid under all circumstances** and not a single agency to which appeal can always be made." Bib 185 p158

For Feyerabend, the idea of no rules in science implies that there are no laws in Nature. One could make a case that the philosophies of Feyerabend are the antithesis of the Universal Model. Although riddled with error, we see the tendrils of his philosophy deeply entwined in modern science. How do you place Nature's secrets out of reach? Take away its simplicity. It seems this is the purpose of *Against Method*, as stated in its introduction:

"One consequence of the thesis is that *scientific success cannot be explained in a simple way*." Bib 185 Introduction

The Universal Model refutes this notion flatly. Scientific truth and reason *can* explain the Universe in a simple way and there is a method—a single method—we can employ to do so.

Multiple Methods Create Chaos and Confusion

Stated simply, a **scientific method** is a method of conducting scientific research. It is a basic concept, but there is more to the story. First, we ask; is there one universal method that everyone can use, and if not, how many are there? To establish order we must identify and determine what method is best. In the profession of law, for example, an attorney may employ his own style of questioning but he must follow protocol as to when he can ask the questions and what questions are allowed; he must also adhere to standardized procedures. There would be no order without procedures and protocol; chaos and confusion would reign.

Another example is the accountant. An accountant has the latitude to establish any number of categories and procedures, but they must align with Generally Accepted Accounting Principles to withstand the scrutiny of an audit. The format of financial reports, regardless of the size or type of company, follow prescribed patterns. This assures uniformity and reliability.

Attorneys and accountants must observe their respective prescribed methods to establish trust and reliability; we should expect the same of the science profession. However, modern science does ***not*** have an established 'universal method' to which its professionals can adhere. The absence of a Universal Scientific Method leads to a deficiency in trust among peers and among the public.

Science must ascribe to a Universal Scientific Method to establish trust and credibility.

2.3 The Modern Science Method of Confusion

Textbook Confusion in the Scientific Method

How do we define the Scientific Method today? The short answer—in **countless ways**.

There is no **universal** scientific method.

In the following example, taken from the San Diego public school's *Science Content and Performance Standards* we read that the scientific method has "no single set of steps":

"**There is no single set of steps that constitutes 'the scientific method'**; rather, there are different traditions in different scientific fields about what is investigated and how." Note 2.3a

How does a school teach a standardized method if scientists do not have one? We reviewed the "Scientific Methods" from a number of science textbooks and included eight of them here, each one illustrating the *absence of a* **standard methodology** *in modern science*. We begin with *Psychology in Action*. Here the author refers to the scientific method as "research methodology," beginning the process with theory. From there we see various methods in geology, general science, and chemistry. Even among fields, there is no consensus. There are common elements among the books such as Theory, Laws, Hypothesis, and Observation, but there is no uniformity in how they are applied. Compare them all, paying close attention to where Law and Theory falls in the steps they list:

A) **Textbook: Psychology in Action**
"… psychologists, like scientists in biology, chemistry, or any other scientific field, need to conduct investigations … psychologists must follow standardized scientific procedures, in conducting their studies. These procedures are collectively known **as research methodology**." Bib 49 p13

1. The Theory
2. The Hypothesis
3. Independent and Dependent Variables
4. Experimental Controls

5. Assigning Participants to Groups
6. Bias in Research
7. Naturalistic Observation
8. Surveys
9. Case Studies

B) **Textbook:** *Physical Geology Exploring the Earth*
"Theories are formulated through the process known as the **scientific method**. This method is an orderly, logical approach that involves gathering and analyzing the facts or data about the problems under consideration. Tentative explanations of **hypotheses** are then formulated to explain the observed phenomena. Next, the hypotheses are tested to see if what they predicted actually occurs in a given situation. Finally, if one of the hypothesis is found, after repeated tests, to explain the phenomena, then that hypothesis is proposed as a **theory**." Bib 54 p13
1. Gathering and Analyzing the Facts
2. Hypotheses formulated
3. Hypotheses Tested to see if Predict
4. Theory

C) **Textbook:** *General Science*
"It is important to remember that a scientific **theory** is not an opinion, or an untested guess. It is based on the results of many experiments by different scientists throughout the world. A scientific theory is the best scientific explanation presently available to explain some natural event." Bib 53 p11
1. Identifying a Problem
2. Gathering Data
3. Forming a Hypothesis
4. Performing Experiments
5. Scientific Theories

D) **Textbook:** *Heath Chemistry*
"Science says something has been explained when over time the explanation or hypothesis accounts for past events and accurately predicts future ones. When the hypothesis or explanation can do this, it is then called a **theory** in science." Bib 45 p7
1. Hypothesis
2. Theory

E) **Textbook:** *Investigating the Earth*
"When a hypothesis has been tested in every way possible but has not been proven, it can safely be called a **theory** but not a **fact**." Bib 52 p6
1. Observing
2. Investigating
3. Hypothesis
4. Theory

F) **Textbook:** *Chemistry - Zumdahl*
"A **theory** is an *interpretation*—a speculation—as to why Nature behaves in a particular way." Bib 48 p3
1. Observations
2. Law
3. Theory
4. Theory tested for further experiments

G) **Textbook:** *Chemistry - Birk*
"Although the scientific method is often outlined as consisting of sets of steps and procedures, it is as **much a frame of mind** as it is anything else… There is **no one universal definition of the scientific method**…" Bib 55 p10
1. Observation
2. Hypotheses
3. Laws
4. Theories

H) **Textbook:** *Earth Science*
"The study of earth science, like that of all the other sciences, is conducted according to the **scientific method** – an awesome-sounding concept, but one which has a beautiful simplicity… If the hypothesis later proves to be universally applicable, it becomes known as a **law**." Bib 51 p6
1. Observations
2. Predictions
3. Hypothesis
4. Law

The textbooks we included represent only a small sample of the books we reviewed, but one thing was certain: our research showed that no two textbooks agree on the scientific method. This inconsistency is not limited to published textbooks. We found it across the Internet; in the worldwide classroom where all scientists can post their diverse views and methodologies. A search of the words "scientific method" renders literally thousands of opinions. Here are two examples that illustrate the confusion:

The Scientific Method: A Model for Conducting Scientific Research
Note 2.3b
1. Defining the Question
2. Locating Resources/Gathering Information
3. Forming a Hypothesis/Hypotheses
4. Planning Research Collection Methods
5. Collecting Data
6. Organizing & Analyzing the Data
7. Interpreting Data & Drawing Conclusions
8. Communicating the Results

"There is no single set of steps that constitutes 'the scientific method', rather, there are different traditions in different scientific fields about what is investigated and how."

"The" Scientific Method [Note 2.3c]
1. Gathering data
2. Develop hypotheses
3. Test hypotheses with specific data
4. Negate hypotheses, then develop new hypotheses
5. Elevate to theory
6. Elevate to law

Evidently, there is no single set of steps that constitute the scientific method. Each method came from a variety of traditions, habits, opinions, and a host of other contributing factors, all of which led to confusion, chaos, and inconsistency in science today. We would scoff at the builder that began building a house by constructing the roof first and ending with the foundation, yet this is what we see in some examples of the methods.

> Imagine a homebuilder constructing the roof first and ending with the foundation.

It is apparent that there is no established, standardized method in modern science. As one textbook says:

"It is as much a frame of mind as anything else." [Note 2.3d]

When scientists use different methods to conduct their research, it is like having everyone pull a rope in a different direction.

> How can modern scientists follow a **universal scientific method** if they do not know what it is?

Of course, true science does exist and there must be a universal scientific method, which all scientists can follow. The question is—where is it?

One final thought about how modern science has defined the scientific method comes from a well-known science contributor, Norman W. Edmund, founder of the Edmund Scientific Company, a decades-old purveyor of all things scientific for amateurs, teachers, and professional scientists alike. He wrote a detailed report on the scientific method from which we take the following:

"*It has been stated that 'the greatest discovery in science was the discovery of the scientific method of discovery.*' This method is most often called **The Scientific Method, a Scientific Method, and Scientific Method**. Actually, it was a method that evolved slowly over the centuries. Numerous formulas were proposed. There were debates, discussions, meetings, papers presented, philosophical theories expounded, and much literature published. Even today, our leading scholars are still arguing as to what it is. **Some are going so far as to claim that it does not exist**—" [Bib 56 p34]

2.4 Controversial Science

What Makes Real Science

Sir Isaac Newton gave us a clear answer of what to expect when science uses only 'hypotheses' or theories and not a properly unified scientific method.

"If any one offers conjectures about **the truth of things from the mere possibility of hypotheses**, I do not see how anything certain can be determined in any science**; for it is always possible to contrive hypotheses, one after another, which are found rich in new tribulations.**" [Bib 2 pxxiii]

Newton and others struggled with the same problem during their lifetimes that we struggle with some 300 years later—scientists trying to determine truth with only a hypothesis or a theory.

A former professor of zoology and an author of a number of influential 20th century volumes on evolution had this to say about the role of experiments and laws in biology when he received the Crafoord Prize from the Royal Swedish Academy of Science on September 23, 1999:

"Another aspect of the **new philosophy of biology** concerns the role of laws. **Laws give way to concepts in Darwinism**. In the physical sciences, as a rule, theories are based on laws; for example, the laws of motion led to the theory of gravitation. In evolutionary biology, however, theories are largely based on concepts such as competition, female choice, selection succession and dominance. These biological concepts, and the theories based on them, **cannot be reduced to laws and theories of the physical sciences**… Observation, comparison and classification, as well as the testing of competing historical narratives, became the methods of evolutionary biology, **outweighing experimentation**." [Note 2.4a]

This citation illustrates that some practitioners of modern science have even gone so far as to say their "new philosophy" of science "outweighs experimentation," and that "laws give way to concepts."

If we examine the words of these two scientists, Newton who spoke of and discovered "the truth of things" and the author of the above article, who speaks of "concepts" that replace laws, the difference is quite apparent. Just as Newton said, "If any one offers conjectures about the truth of things from the mere possibility of hypothesis, I do not see how anything certain can be determined in any science."

What kind of foundation has science built its house upon if

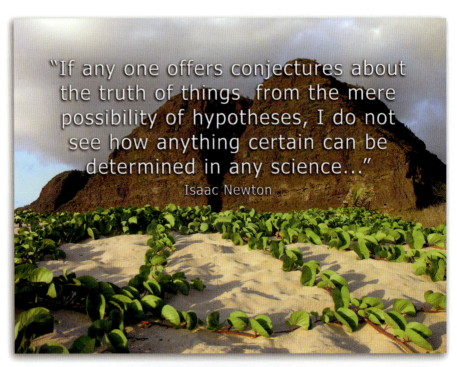

"If any one offers conjectures about the truth of things from the mere possibility of hypotheses, I do not see how anything certain can be determined in any science…"
— Isaac Newton

CHAPTER 2 METHODS MODELS AND PSEUDOTHEORIES

some of its most prominent member's methods are to abandon the use of experiments? Two social scientists, Collins and Finch, performed a detailed survey of experiments conducted by other scientists to review their controversial interpretations. They made this observation:

"If we stopped doing experiments we would no longer be doing science." Bib 76 p175

This seems self-evident; however, the social scientists assert the following after noting how many scientists actually **perform** experiments:

"In the end we find that those who have the kind of direct experience that, according to the ideals of science, should be the only thing that leads to truth **are very few in number**..." Bib 76 p174

The complexities of modern science and the proliferation of data have led to an increase in the number of scientists founded on 'theoretical' ideas and to fewer 'experimental' scientists, producing a generation of professors whose careers are limited to teaching the theories of others. It has become commonplace for science to build on *ideas* but not on empirical evidence.

Controversial Science

Without experiments that produce empirical evidence, science becomes 'controversial.' During their survey, Collins and Finch came up with a definition of scientific truth to explain the reality as it exists for some:

"For the sociologist, as sociologist, scientific truth is that which is **affirmed** by scientists." Bib 76 p172

Here we see that truth is not to *know* what really is, was, and always will be, but it is merely "what is affirmed by scientists." If the scientists *affirm* that 2 x 3 = 7.148 or some other number, then accordingly, that then signified truth. In their study, the sociologists comment that truth is *not:*

"... forced on us by the inexorable logic of a set of crucial experiments." Truth is "... brought about **by agreement to agree about new things**". Bib 76 p54

Without the confirmation of 'the test of time' that comes from repeatable experiments, do we find that scientific truth is simply that which scientists decide to agree upon? It seems so; however, scientists often do not agree. Experiments that contribute to the understanding and establishment of natural law have taken a back seat to the science of theory.

Of course, there are many experimental scientists performing a host of valuable and informative experiments, but to what end? They are not for the understanding or identification of natural law today because the science leaders have declared that theories are the 'end points' of science:

"In science, theories do not turn into facts through the accumulation of evidence. Rather, **theories are the end points of science**." Note 2.4b

If we perform experiments and then interpret the results within the confines of incorrect theory, it must lead scientists down a road of continual failure.

Experiments in the UM

It therefore becomes vitally important to experiment and to verify the claims of others by reproducing their experiments, even, and perhaps especially, for those first conducted hundreds of years ago, upon which much of modern science rests. Today, unfortunately, students are *not* conducting those fundamental experiments and we just teach them to accept the outcomes arrived at long ago without question. What if there are flaws in some of those crucial experiments from the past, or what if we cannot duplicate them? What does this say about a science built on the false conclusions of flawed experiments?

We routinely hear scientists make statements of fact even if such claims have never been proven. Take for example sugar, one of the most basic organic substances known to man. Medical and nutritional sciences assume that sugar is made from water and carbon, two of the commonest and simplest components in Nature. Are there any experiments conducted in the classroom that support this notion? Has *the instructor* ever made sugar from just water and carbon? Bib 76 p175 Even though we can dissect the sugar molecule into its elemental components of Hydrogen, Oxygen, and Carbon, if we cannot actually make sugar *from those elements* we must ask, do we really know how sugar is made?

If we are not conducting basic experiments in the classroom, and if we base some of the accepted "facts" on statements attributed to experiments that have never actually proven successful, what scientific *method* are we using? Taking that thought a little further, if the standard in the classroom is not experimental proof, what kind of *science* are we teaching?

2.5 The Universal Scientific Method (USM)

Therefore, we see, a 'universal' standard method of doing science does not exist in modern science. The UM establishes a Universal Scientific Method or USM by making significant discoveries in all the main fields of science with this new method. Here we now discuss the USM and its steps that lead to the ultimate goal of science—the discovery of natural law.

The Lack of a Standard Method in Science

One difficulty in establishing a Universal Scientific Method is its inherent lack of flexibility. No scientist wants someone or some method to confine him or her to a rigid set of rules that stifle creativity. Although their concern is valid, there must be a set of guidelines leading to truth. Just as the attorney and accountant must adhere to a code of standards and conduct, following prescribed methods accepted in their respective fields, so to must science adapt and follow a standardized, universal method to ensure credibility, accuracy, and accountability.

During the 1600s, Isaac Newton's era, scientists enjoyed inventing theories without proper experiment-based work, sometimes basing their ideas on casual observation:

"An indication of the revolutionary nature of Newton's 'Theory of Light and Colours' is that, before its publication in the February 1672 edition of the *Transactions*, **not one of seventy-nine issues of the journal previously published had contained an experiment-based radical revision of an accepted scientific theory**. The *Transactions* was full of articles **based entirely upon casual observation or speculation**. These ranged from medical reports such as 'An Observation Made upon the Motion of the Hearts of Two Animals after Their Being Cut Out', to 'New Observations of Spots in the Sun'… **but nothing that produced a verifiable challenge to orthodoxy**."
Bib 19 p173 Author quoted Per Bak

How does today's science arena differ from Newton's day? Although researchers often invoke the term 'rigors of scientific research,' we do not hold modern science to a universal

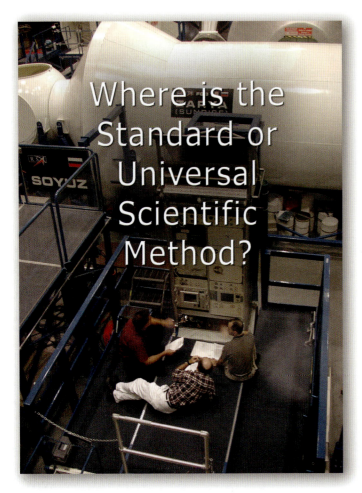

Where is the Standard or Universal Scientific Method?

standard of conduct. Our understanding of Nature over the past 100 years has come down to us based more on theory than on natural law. Despite many new discoveries during the past 100 years, they have not contributed to the establishment of new natural law. (See The History of Science Table in chapter 3.2). What has science become if it lacks a universal scientific method based on experimentation intent on discovering natural law? It has digressed into the casual 'philosophy' that existed before Newton's day.

The Unaware Public

Perhaps another reason there is no standard method in modern science is that the public has not demanded it. Until the public recognizes the need of a universal standard and demands that science follow those standards, millions of taxpayer dollars will continue to be spent on fruitless projects. An insightful opinion about 'what science does' comes from a distinguished physicist from Brookhaven National Laboratory.

"'In science **there is no meaning to anything**. It just observes and describes… The whole idea is **we cannot predict**.'"

Bib 20 p205 Author quoted Per Bak (Danish physicist)

In reality, we may not be far from reaching this prognostication. In an interview with the host of the popular television show, Star Gazer, the interviewer asked, "What else have you learned from the public?"

"There are millions and millions of people dying to know about the universe, dying to know about themselves. They're looking for answers. One of the places you look to find the ultimate answer is in the stars, but if you contemplate the stars you will realize there is no answer, there isn't even a question. That's what you learn. **There are no questions**. **There are no answers**. There's only self-discovery through the discovery of the cosmos." Note 2.5a

In the modern science era where everything came from nothing, if there are no questions and no answers, what are we to expect of the scientific 'method'?

The Answer to a Universal Scientific Method

The third and probably the most important reason there is no standard scientific method is that no one has been able to propose and gain consensus for one that satisfies the leaders of every major science field. There is such a wide range of opinions and traditions of how to conduct science (refer back to the textbook examples in subchapter 2.3), that it seems almost impossible to agree on just one.

That became one of the important initial purposes of the UM, to develop a Universal Scientific Method. To accomplish this meant identifying and developing a standard that would apply across all scientific fields, and it meant doing something that no one had ever done before—develop a single, universal scientific method that all the sciences can use.

The Six Steps of the Universal Scientific Method

The Universal Model is a byproduct of the Universal Scientific Method (USM). Although science has had some moments of well-organized scientific research, scientists have often simply stumbled upon important discoveries.

Over the last several years, utilizing the organized approach of the Universal Scientific Method, the UM has made many discoveries. However, the method's six steps are not new; we can see them in use when long-ago scientific discoveries led to the identification of natural laws.

The Universal Scientific Method embodies the self-evident principle from subchapter 1.5—it *is*, because it could be no other way. It also follows the simplicity principle—the simplest correct explanation is the best. Importantly, the scientific method *must* contain the following six steps if we expect the method to lead us to the discovery of Nature's laws.

Although we may use additional steps, we include here the necessary basic steps intended to preserve the integrity of the process. The order of the steps ranks highly important, because theory represents the beginning, not the end of true science.

The six steps of **The Universal Scientific Method (USM)**:

1. Theory
2. Prediction
3. Testing
4. Observation
5. Evaluation
6. Natural Law

We should look at the USM as a **method of public accountability** for use in establishing the credibility of day-to-day research and for verifying the truth and accuracy of scientific discovery. Easily understood, each of the steps within the method allow for the reproduction and evaluation by both scientists and laypersons alike.

We must realize that a person can only develop a theory *based on what they already know*. This idea may sound simple, but it

CHAPTER 2 METHODS MODELS AND PSEUDOTHEORIES

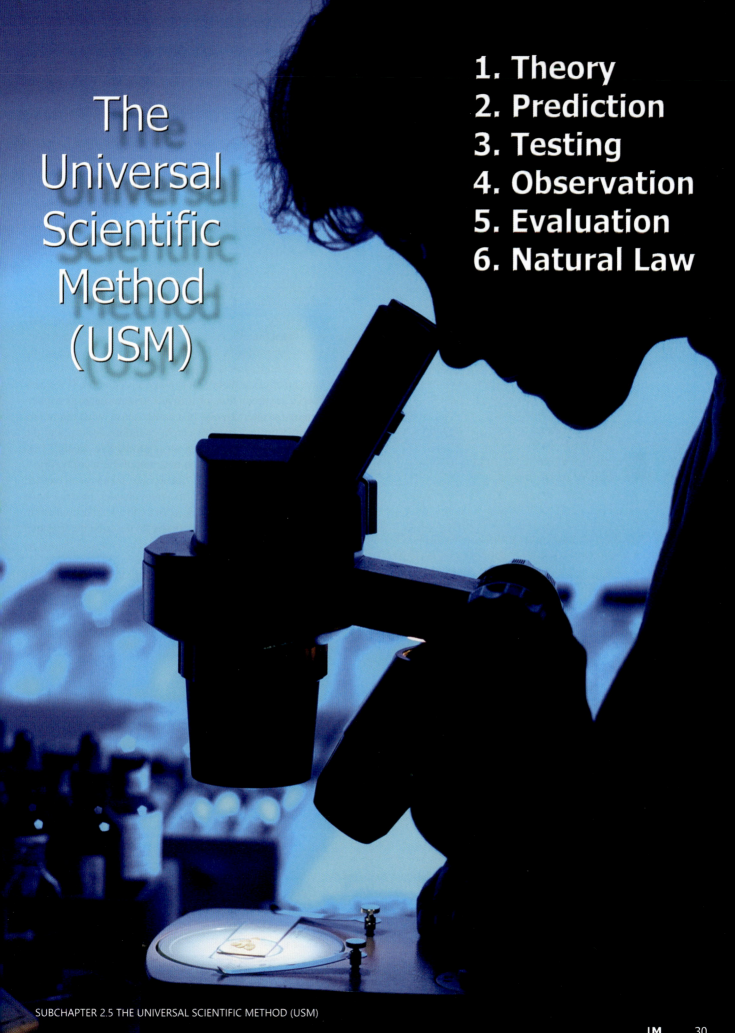

The Universal Scientific Method (USM)

1. Theory
2. Prediction
3. Testing
4. Observation
5. Evaluation
6. Natural Law

is the 'key' to discovery. Our life's experiences give us the ability and the framework within which to formulate **correct** theories. If we have had no experiences that afford us actual observational opportunities, we find our ability to construct a theory severely hampered. To devise a theory about an observation or to make a prediction, to test, or to observe and to evaluate that prediction requires creativity and open-mindedness. To a large degree, the *discovery processes* is very individual and fosters the development of a **personal method**.

Each person's individual life experiences shape the personal method of discovery, and this part of the process is not something to ignore. The personal method has at its core, the 'intuition' discussed in subchapter 1.5. Personal, nontransferable, and unique, intuition comes from the journey of one's own life, making the personal method very important in the discovery process, but it remains beyond the scope of the Universal Model to define this inimitable human characteristic.

The Scientific Learning Process

Applying the Learning Process presented in Chapter 1 with the Universal Scientific Method and with one's own personal method, a new level of understanding and comprehension of science can happen. We outline the Scientific Learning Process in Fig 2.5.1, which illustrates where the six essential steps from the USM fit within the learning process and how they lead to scientific truth.

Understanding each of the six USM steps is necessary to engage in the Scientific Learning Process.

1. Theory

How does an artist begin a painting without an 'idea' of what to paint? Does a farmer plant seeds without knowing something about the crop he hopes to produce? We all formulate theories, based in part on the sum of our life's experience. A 'theory' is simply a proposed explanation of an occurrence or phenomenon. With the Simplicity Principle as our guide, we use the following definition of a **scientific theory** in the Universal Model:

Scientific Theory: a proposed explanation for natural phenomena that is conjectural and subject to verification.

The words "conjectural" and "subject to verification" are significant parts of this definition.

A scientific theory or 'theory' for short is not scientific if we cannot evaluate and verify or falsify it. Many modern science theories fall short because there is no way to verify or falsify them, yet some we ***still teach*** in our classrooms as though they were fact.

Without the ability to verify or test the theory, we must apply the weaker definition of theory, relegating the theory to nothing more than a mere guess or speculation; the end of real science *should never* be guesses and speculation.

In the USM, the scientific theory is the first step, and probably the easiest to take. Formulating an idea in the mind is easy. Uncovering empirical evidence, sound enough to convince fellow scientists and the public to recognize a new law is quite another story. However, the truer our starting point the more correct our theory *can* be. For this reason, we are beholden to accept truth no matter its source, and to apply that truth to the theory-making process.

Until a theory advances through all of the steps of the USM, we must accept that it is *only* a theory, regardless of time, effort, and money put toward it. Until and unless we prove a scientific theory demonstrably, it is nothing more than an idea.

To have an effect, a theory must make a prediction. If we cannot observe the prediction, or if it fails to materialize as expected, what are the alternatives? In a darker sense, what might one do after investing years of research, countless dollars, and the prestige of fellow scientists on an unproductive theory? Perhaps one might hope to alter the meaning of the word theory.

In modern science, the word 'theory' has taken on an exalted meaning, which, as we explain in a later chapter, *theories* have become 'the ***end points*** of science.' To try to alleviate confusion, science texts use the word 'hypothesis' to describe the first step in many of their scientific methods, but they *end with theory*, as we have seen in some previous examples.

In the UM, the end points of Millennial Science are natural laws—not theory. In this book, we generally refrain from the

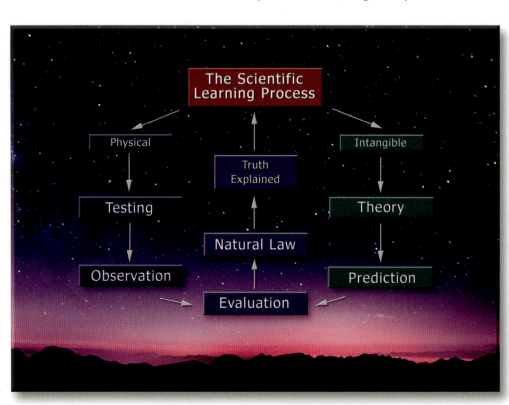

Fig 2.5.1 – The Scientific Learning Process incorporates the six steps from the Universal Scientific Method (USM). Those steps include Theory, Prediction, Testing, Observation, Evaluation, and Natural Law. Scientific progress and learning requires the use of these six steps, without which we fall victim to same thing that happened during the Scientific Dark Age.

CHAPTER 2 METHODS MODELS AND PSEUDOTHEORIES

use of the word 'hypothesis' to avoid confusion because it means essentially the same thing as 'theory.' The Universal Model restores the original meaning of the word 'theory' as part of the definition of a scientific theory, which is a proposed explanation for natural phenomena that is conjectural and subject to verification.

2. Prediction

Following theory, the next step in the USM and in the scientific learning process is prediction.

"**The issue of *prediction* now enters as a powerful additional requirement for a *scientific* explanation**, which is why physical scientists put so much emphasis upon it. When we think we understand a process scientifically, we ought to be able to use it in order to make a precise and unambiguous prediction; no other mode of understanding has this feature. The precision of a prediction will depend on the manner and degree of comprehension embodied in the explanation, but **without prediction a scientist's understanding is regarded as deficient**." Bib 18 p51

This is a powerful statement! To understand, we *must* be able to predict. We define a **scientific prediction** in the UM as follows:

> **Scientific Prediction**: the ability to predict a consequence from a stated action.

3. Testing

The third step of the USM is testing. This step may seem obvious, but many of today's science fields conduct no testing, yet tout many of their theories as fact. Some of the biggest names in science, like Einstein, never performed experiments or conducted any tests on their theories. Surprisingly, this is typical throughout science. Some scientists feel they are above the essential step of testing, leaving the mundane task of experimentation to technicians and engineers.

This attitude has bred several fields of 'theoretical' sciences where the so-called scientists perform no experimentation at all. Thanks to men like Einstein and his "imagination experiments," many investigators have fallen into this trap. Not experiencing Nature through experiment and observation, they remain unaware of the way things really are. This has led to the building of false theories upon false theories, carrying us finally to a coming *crisis* in science:

"The **current crisis** in particle physics springs from the **fact** that the theories that have gone beyond the standard model in the last thirty years fall into two categories. Some were falsifiable, and they were falsified. The rest are **untested**—either because they make no clean predictions or because the predictions they do make **are not testable** with current technology." Bib 178 pxiii

This statement, from a 2006 book, *The Trouble with Physics*, documents the crisis this particular field experienced over the last three decades. However, the crisis extends to all fields of science, a concept developed more fully in Chapter 4, Scientific Revolutions.

The test of a prediction must be falsifiable, or in other words—vulnerable to failure. This is necessary to discover the truth of the prediction and ultimately, the theory. The UM defines **scientific testing** as follows:

> **Scientific Testing**: repeatable and observable experimentation designed to verify or falsify a prediction based on scientific theory.

This critical USM step has not happened with well-known theories such as evolution, despite its immersion in continual controversy since its introduction to the world.

4. Observation

"Science is a way of describing reality; it is therefore limited by the limits of **observation**... Anything else is not science; it is scholastics." Bib 30 p 72

Without observation, there is no science. This self-evident statement forms the basis of **scientific observation**:

> **Scientific observation**: A description of Nature or phenomena based on the five physical senses and logic.

There were many observations while performing the experiments in this book, which usually included the five following contributors—**materials, procedures, results, discussion and variable checks**. Although we find these in nearly every experiment, there may be many others. Materials are the objects related to the experiment. Procedures explain how to perform the experiment. Results summarize the data obtained. Discussion represents the interpretation of the data, which may include new questions and conclusions. Variable checks are those factors that the experimenter must account for while performing the experiment to ensure that the outcome or results come from the intended cause.

Observations come in two forms—direct and indirect. Webster's defines **direct** as "absolute or exact" whereas **indirect** is defined as being "not straightforward." Bib 7 p227, 416 While indirect observation may *help* establish theories, ***only direct observation*** can ever ***validate*** Natural Law.

During the Dark Age of Science, and because of the lack of focus on natural law throughout the last century, science frequently makes use of the word '**infer**.' As defined in Webster's dictionary, to infer is "to derive by reasoning or to guess, speculate or surmise." Bib7 p418 Inference then indicates only indirect

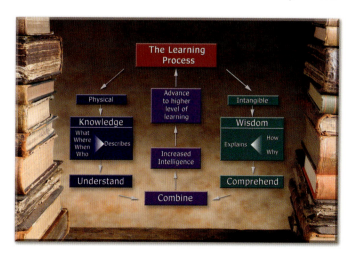

Fig 2.5.2 – The Learning Process from subchapter 1.3 forms the basis of the Scientific Learning Process in Fig 2.5.1. 'Testing' and 'observation' ***describe*** the physical side of Nature while 'theory' and 'prediction' ***explain*** intangible things. 'Evaluation' precedes 'natural law' and leads to working models that explain scientific truth.

observation, which only applies when formulating theories but not natural law.

5) Evaluation

Evaluation must accompany observation, which must include the question; "is the observation consistent with what was predicted?" If yes, we may consider the theory validated; if not, the theory is invalidated and must be modified, re-tested or eliminated. We define **scientific evaluation** here as follows:

> **Scientific Evaluation**: the acceptance or rejection of a theory, based on observable results and outcomes, as predicted by the scientific theory.

The correct evaluation of a theory distinguishes scientists—those who question everything with an open mind and accept theories consistent with observable facts—from pseudo-scientists, who accept unverified scientific theory and who remain closed-minded due to educational bias. True science can *only* progress when pseudo-science theories no longer dominate the sciences; the UM exposes dozens of pseudotheories, many of which most people are unaware of.

6) Natural Law

In 1964, Time/Life published the *Life Science Library*. The volume *The Scientist* included a statement that clearly establishes the importance of natural law:

"As he formulates his final **theory**, the scientist subjects it to intensive criticism. Seeking to make it as useful as possible, he asks himself: **Is this proposed law universal throughout the extent of space and the passage of time**? Does it lead anywhere? Does it **predict one state of affairs as arising out of another**? …

"Now comes the moment of verification and **truth**: testing the theory back against protocol experience to establish its validity. If it is not a trivial theory, it suggests the existence of unknown facts, which can be verified by further experiment… If the theoretical predictions do not jibe with **observable** facts, then the theorist has to forget his disappointment and start all over again. This is the stern discipline, which keeps science sound and rigorously honest.

"If a theory survives all tests and is accepted into the canon of **scientific law, it becomes a fact in its own right and a foundation for higher spires of thought**." Bib 14 p57-8

It is noteworthy that the publisher included this statement over half a century ago, especially since most modern scientists today have never witnessed the discovery or addition of a significant new natural law. They no longer write this way because no one really expects to find a new natural law today.

Natural laws are simple. They are simple because once discovered, we now know the truth. As noted previously:

> A natural law is simply a statement of truth.

Why are most of the scientific leaders today unable to discover new natural law? Because as we will find out later in the UM, the typical modern science leader does not believe in truth—and because natural law is a statement of truth, they are unable to even look for truth or natural law. A new natural law is the pinnacle of scientific discovery and becomes a "fact in its own right and a foundation for higher spires of thought."

However, the proposed law *is only universal* in the sphere or domain in which it pertains. Contrary to what *The Scientist* statement claimed, a natural law is not necessarily universal in nature. We may use a natural law to describe the way water behaves; however, in a colder climate water may become ice and not behave this way. Therefore, we adjust natural laws as we make new observations, although some laws may never need adjustment. Likewise, laws that describe and explain Nature in one domain or part of our Universe may not do so in another location. The 'universal' nature that modern science assigns to all laws is inappropriate since we have not been to all places in the Universe.

Human beings learn day-by-day, principle upon principle, and we cannot expect to think we know all things; why then should we expect one of our scientific laws to apply to all things in all situations since we have not experienced all things? For this important reason, scientific truth includes the caveat, "so far as is known."

We emphasize the importance of a natural law's domain in subchapter 25.6, the Revolutionary Universe Model, where we show that physical laws change in various parts of the Universe.

To verify its scientific truth, a proposed law must "predict one state of affairs as arising out of another," thus; *cause* and *effect* are the expected results of natural law. We can then test the effect "against protocol experience to establish its validity."

Without knowledge of natural law, it would not be possible to demonstrate and explain how Nature works. The UM defines **Natural Law** as:

> **Natural law**: a predictable effect from a stated cause.

We can also state that a natural law is a predictable consequence from a stated action and it follows the test of time but only so far as is known. The pinnacle of science is to discover natural laws that predict the behavior of Nature.

USM Summary

"The primary tools of the physical sciences, whose **central aim** is to understand and explain natural phenomena, are **theories** and **laws** leading to successful **predictions** subject to **observational** or experimental **test**." Bib 18 p200

Coined by Roger Newton, a scientist, these words succinctly state the objective of the UM. The "central aim" of science is well stated, and five of the steps making up the Universal Scientific Method are included.

Every scientist should be aware that the six steps of the USM are a necessary part of the scientific method. However, some scientists do not *believe* these steps will work. The Dark Age of Science came about in part because science abandoned belief in the six simple steps of the USM, which resulted in 100-year dearth of natural law discovery. We discuss this concept thoroughly in the next chapter. At the beginning of this chapter, we identified that natural law is a statement of truth. Is it possible that, although science knows laws are essential, they avoid the identification of new laws because the modern scientist does not believe in the existence of truth? Consider the telling words of the scientist-author

> The pinnacle of science is to discover natural laws that predict the behavior of Nature.

Cause and effect are the natural results of natural law.

from the previous quote:

"**In the end, the relentless search for truth, never found**, with the beacon of objectivity as the ideal, **never quite achieved**, engenders an *attitude* that **pervades the lives of the men and women of science**." Bib 18 p222

The scientist knows what the tools of the trade are; the steps of the Universal Scientific Method are not a secret, yet these are often unused in the effort to find truth. Why—because these tools help uncover truth. If it is an "*attitude* that pervades the lives of the men and women of science" then this belies their own internal notion that they can *never* find truth. This should shock the layperson, but to today's scientist, it seems as though it is a matter of fact.

With an attitude that they can never find truth, there can be no certainty or prediction.

Without prediction, no effect can have a cause.

There is only one way in which a Universal Scientific Method can exist. It must work, and it must work in all fields of science. The USM aids in the discovery of truth—of what was, is, and what will be. The purpose of the Universal Model, through the establishment and use of the USM is to restore truth and order in science by establishing new natural law that helps describe and explain Nature so that we can understand and comprehend it clearly.

2.6 Scientific Models

Defining a Scientific Model

Over the course of time, the word 'model' has had many meanings. Currently, many scientists look at the word 'model' in a completely different way than defined here in the Universal Model. Below is a good quote explaining a modern scientist's view of a model:

"But whenever physicists want to emphasize their lack of commitment to the reality of what is described by a **theory**, or to express their consciousness of its limitations, **they simply call it a model**." Bib 18 p65

Fig 2.6.1 – A model is like a map. Both are an organized representation of facts that can demonstrate and explain the truth about the world around us.

Fig 2.6.2 – Models include cars, trains and any other things that represent the world around us in such a way that they give us a perspective and framework that allows us to observe the subject in new ways.

Note how the experts give theory more weight than a model. The weak definition of the word 'model' may have come from the frequently referred to 'computer generated models' that simulate possibilities but that are not directly related to reality. In a more practical sense, most people can relate to model airplanes, cars, or trains, or to the models used in architecture for renderings, or perhaps a model of the solar system. These types of models allow us to walk around and inspect a three-dimensional structure in a way not normally possible. It is this application of the word 'model' that most fits its use in the Universal Model.

There are 18 models, developed as chapters in this book, including the Universal Model itself, which is essentially a collection of many models. A **model**, as used in the UM, is:

Model: an organized representation of facts in a framework that we can demonstrate and explain.

One of the best ways to understand the various models contained in this book is to view them as maps. A map is an organized representation of facts that we can use to describe our present location as well as our desired destination.

If the roads and features identified on a street map are in fact where they say they are, we can easily locate them by keeping track of distance and direction traveled.

As we trace our course on the map, we can look back and say, "been there, done that!" We can apply this to the Universal Model as each model builds on previous ones. Through the journey, we will come to understand the Universal Model as a whole.

Simplicity is another important aspect of a successful model. Many modern science models and theories fall short because they are too complicated, and many scientists agree with this assessment:

"In cosmology, as in particle physics, scientists have come to believe that a simple and elegant theory is a better explanation—closer to the truth—than a complex and arbitrary one. **Simple models are credible; complicated models are more incredible the more complicated they become.**" Bib 21 p131

Of course, this is a manifestation of the Simplicity Principle—the simplest correct explanation is the best.

Testing the Road Map

Every time we use a road map, we put it to a truth-test. Testing maps or models is not difficult, especially when they are simple. What do we do with a map that does not represent the roads and landmarks we encounter? We discard it and we no longer use the defective map. This is what should happen to incorrect science models and theories; if we cannot confirm their predictions, we should discard them. That is not what happened over the last century, and we document that in the next chapter.

We also find that modern science is actually not very simple nor is it straightforward and most of their models are complex and convoluted, at best, a representation of facts. To confirm this, just look up the latest model of the atom or the Universe and see for yourself. Here is an example from a college chemistry textbook where the text states the model is "doomed to fail":

"It is important to understand that **a model can never be proved absolutely true**. In fact, *any model is an approximation by its very nature and i*s **doomed to fail** at some point." Bib 48 p175

Imagine being a college chemistry student and reading this in the textbook you are to study. If the models you are researching are "doomed to fail," why study them at all?

Many people enjoy science because of the factual nature of the sciences. There are natural science laws that, as far as we know, have not changed. This foundation has attracted scholars throughout the ages.

However, over the past 100 years we have seen a complete shift in the models presented. Because we no longer expect scientific theories and models to last or to state 'truth,' there is no foundation on which to build a structure of demonstrable facts. Many young people clamor for truth in science. They want something that is real, something secure—something that was, is, and will be. At least they want it until they succumb to the relentless message that there is no truth, that they cannot find it in modern science today. 'Modern' science has its roots in modernism, which means to be without certainty, and in our next chapter, we examine how **modernism** fails.

The interesting point here is that even though many scientists believe that there is no truth and that a model can "never be proved absolutely true," intuitively they know differently. The same author of the quote about models never proving true also wrote:

"The **true test of a model** is how well its predictions fit the experimental observations." Bib 48 p175

How does one believe the "true test of a model…" and then believe that one can never prove a model "absolutely true"?

The Solar System model predicted that the Earth and other planets orbit the Sun. How "well" did this model's prediction "fit the experimental observations?" Do the planets still orbit the Sun? Even if the planets were to stop orbiting the Sun in some future time, they orbited the Sun when the solar system model was first recognized. The predictions fit the innumerable "experimental observations" regarding the planets' orbits, verifying that the planets did indeed orbit the Sun.

How is this model "doomed to fail?" It already has passed the test of time for hundreds of years, and has been true since we have been observing planets, and it is true now. Just because we cannot see all of the future, does not mean that Scientific Truth (knowledge of what is, was, and, as far as known, will be) does not exist.

If we look closely at the scientific literature, we find this contradiction of "no truth" and "the true test is…" interwoven throughout science. Why is this?

> Science today is ever looking—but never expecting to come to the knowledge of the truth.

Without truth, there is no framework or structure to build upon, making the 'modern' science models more fiction than reality. True maps and models have, and always will, take us to our intended destination.

2.7 Pseudotheory and UM Models

Throughout this book, we compare pseudotheories to UM Models. A UM Model seeks firsthand experiences and observations whereas Pseudotheories rely on inferences. A short study of reports and inferences can help in our evaluation of UM Models and Pseudotheories.

Defining a Universal Model

Having defined the term 'model,' let us now look at the word 'universal.' The Universe includes everything here on the Earth and in space. Therefore, in the broad sense, the word **universal** is inclusive of *all things* and we define it thus:

> **Universal**: applicable to all people and all things.

Looked on by many as the 'father of modern empirical science,' Isaac Newton left us a model or map for future scientists to follow:

"For many historians of science, Newton's ideas about how matter behaves and how energies and forces operate can be seen as a watershed in the development of physics. Indeed, some perceive his work as making possible the Industrial Revolution. Newton provided a focus: he was an individual scientist who drew together the many threads that led from ancient times to his **fathering of modern empirical science** (a study based upon mathematical analysis as well as experimental evidence). Behind Newton lay some 2,000 years of changing ideas about the nature of the universe; **his great achievement** was to clarify and to bring together the individual breakthroughs of men like Galileo, Descartes and Kepler and **to produce a general overview - a set of laws and rules that has given modern physics a definite structure**." Bib 19 p29

If any model hopes to provide a "general overview" of science, it would do well to follow Newton's example. Newton

Fig 2.6.3 – The more correctly a scientific model describes Nature, the more simply we can understand and explain the portion of Nature the model describes.

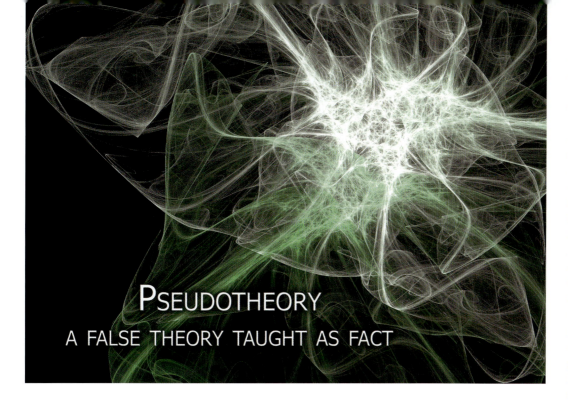

Pseudotheory
A false theory taught as fact

gave us for the first time "a definite structure" for science to follow—not a science based on inference. How was he able to do this? Newton turned science into an empirical pursuit. Mathematical analysis as well as experimental evidence became the method for science, not just theoretical concepts. Additionally, he left a set of laws and rules to guide that pursuit. The UM strives to restore Newton's empirical concepts to science through a Universal Model that reveals new natural law by using the Universal Scientific Method.

Pseudotheory Defined

"Pseudo" is a Greek prefix attached to words to indicate 'false' or 'unreal.' Pseudoscience, commonly used to reference astrology and Ufology, lacks the repeatable experiments or predictions that others can test and observe, and so they earn the title of **pseudoscience**; they claim the authority of real science, but are unable to make or repeat actual predictions or experiments. Thus, they remain a false science until proven otherwise.

A *pseudotheory* is similar to pseudoscience except that it is a false *theory* that claims the authority of fact while it is not so. We define a **pseudotheory** as follows:

> **Pseudotheory**: a false theory taught as fact.

There are a number of pseudotheories presented in the UM. We present each pseudotheory briefly, as modern science teaches it today. After the brief introduction of the false or unfounded theory, we examine the claims of the pseudotheory by asking FQ's (fundamental questions). In most cases, we answer these questions with FA's (fundamental answers) in the nearby text. A pseudotheory is a false idea that when replaced with a true model that demonstrates reality as it "is, was, and will be," helps us understand and comprehend Nature in its true light. Following the pseudotheory chapter discussions, we introduce the complementary UM models that explain the observation in new light.

We do not expect that the models presented herein paint a perfect picture of Nature; all scientific explanations have shortcomings, but the new model accounts for and explains phenomena with surprising cohesiveness.

If we open our minds to possibilities coming from new discovery instead of just believing established pseudotheory, a completely new universe of ideas awaits, opened by reading the Universal Model. Those who think they already 'know' and choose to remain closed-minded will likely choose to see the new UM models as pseudotheories from their standpoint, but if they do, they deny themselves the opportunity of understanding and comprehending natural processes never before known.

The Pseudotheory Mingle Process

Observation and verification establishes truth, and many scientists and researchers have contributed significant observations and identified many correct findings, which we talk about throughout the UM. However, when a new, correct observation does not match entrenched popular pseudotheory—scientists too often ignore the observation, or they rationalize it by mingling truth with pseudotheory. The UM pseudotheory chapters demonstrate this mingling of truth and error; errors we can confirm by asking fundamental questions about the theory. In many cases, the pseudotheory cannot easily answer such questions, and in some instances, the actual scientific evidence confirms that the theory is false. Throughout the UM we read about scientists who ignore many fundamental questions because of the mingling of truth with error, and this can cloud our sense of reason and suppresses our desire to ask new questions.

Reports and Inferences

In 1949, Professor Samuel Hayakawa wrote the book *Language in Thought and Action* which quickly became a best seller and continues to influence the semantic language and thought of millions in academic circles today. Hayakawa discusses the importance of 'reports' and 'inferences' as they pertain to science. Reporting with things we can all agree on is an essential ingredient of science, Hayakawa says:

"At its highest development, the **language of reports is known as science**. By 'highest development' we mean greatest general usefulness. Presbyterian and Catholic, worker and capitalist, German and Englishmen, **agree on the meanings** of such symbols as 2 x 2 = 4, 100°..." Bib 170 p24

If we build real science on the language of 'reports,' those reports are only useful as we verify them. Verified reports, as Hayakawa describes them, are the **Models** in the UM, and since the reader can verify those models, there is a greater chance for agreement with the general content of each Model. Certainly, scientists should never agree on a scientific report or model without verifying its authenticity with observational data and empirical evidence.

A statement made about an unknown is an *inference*. According to Hayakawa, there is a clear and significant difference between "reports" and "inference":

"The **reports should be about firsthand experience**—scenes the reader has witnessed, meetings and social events he has taken part in, people he knows well. The reports should be of such a nature that they can be **verified and agreed upon**. For the purposes of the exercise, **inferences are to be excluded**.

SOLAR SYSTEM PLANETS SIZE COMPARISON

"Not that inferences are not important. In everyday life and in science, we rely as much on inferences as on reports. **Nevertheless, it is important to be able to distinguish between them**.

"An **inference**, as we shall use the term, is *a statement about the unknown based on the known*." Bib 170 p24

In scientific research across all fields during the past two decades, we discovered one common denominator in every false report—*inference*. Hayakawa recognized the difference between reports (models) and inferences (theories), and he seemed to have been aware of just how pervasive *inferences* had become in modern science and how much the experts rely on them:

"In some areas of thought, such as geology, paleontology and nuclear physics, **reports are the foundations, but inferences—and inferences upon inferences—form the main body of the science**." Bib 170 p24-25

The inferences that form the "main body" of existing science today are actually just the theories that form the foundation of the modern scientific establishment. The UM makes it extremely clear—and will demonstrate with repeatable empirical evidence—that over the last century, modern science's primary theories are false, and as they gained wider acceptance, they displaced natural law and stifled its discovery.

In the UM, we show how science disciplines are immersed completely in Pseudotheories—false theories taught as fact; conversely, new **UM Models** verify the facts and teach the truth:

UM Model: a verifiable model that teaches fact.

Unified Field Theory Compared to the Universal Model

Modern physics' highly sought after Unified Field Theory and the Universal Model seem to have similar objectives intended to bring together separate fields of science to foster a greater understanding about Nature, but that is where the similarity ends.

Presumably, Einstein worked on the Unified Field Theory for the last 30 years of his life, but never found it. For decades, physicists pursued the theory with the hope that it would encompass both the atom (quantum mechanics theory) and the Universe (big bang theory) in a grand 'unifying' theory of everything. However, try as they might, they seem further from that goal, and even farther from reality. Complex and radical by design, unified theorists have strayed far from the "compelling naturalism" well-known physicist, Roger Penrose insisted upon during an interview with science writer John Horgan:

"'If there is going to be such a kind of total theory of physics, in some sense it couldn't conceivably be the character of any theory I've seen,' he said. Such a theory would need a '**kind of compelling naturalism**.' **In other words, the theory would have to make sense**." Bib 20 p175

Anyone familiar with physics today knows that the field depends heavily on mathematical and highly theoretical constructs, relying on multi-million dollar computers to crunch numbers and spit out theoretical models as well as billion-dollar atom-smashing super-colliders and billion dollar space telescopes in hopes of reaching farther into space than ever before. Despite having spent billions of dollars to build ever-bigger and better equipment, no one has identified any 'unifying' theory and the future is not promising. For some scientists like Penrose, they see no future in the unifying of *any* existing theories.

Technology keeps pushing forward at mind-numbing speeds, but compared to the technological advances, our understanding of Nature is at a virtual standstill. Are the theories upon which modern science builds so wrong as to stifle progression? Is the lack of success in the unification of theories due to the incorrectness of the theories themselves? If the foundational theories are not true, they will never be able to combine in demonstrable ways to show how Nature really works.

Rather than attempting to unify old established pseudotheories, the Universal Model takes the approach of challenging foundational theories, re-evaluating the data, performing new experiments, and identifying new models with the intent of recognizing new natural law. In so doing, it develops a new Millennial Science based on natural law expected to last well into the current millennium.

Science today is ever looking— but never expecting to come to the knowledge of the truth

3

The Dark Age of Science

3.1 The History of Science
3.2 The Dark Age of Science
3.3 Modern Science Today
3.4 Modern Science and Truth
3.5 The Big Picture of Modern Science
3.6 How Popular is Modern Science?
3.7 Science, Technology and Mathematics
3.8 A Final Review of Modern Science

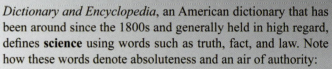

In this chapter, we shine as it were, a light in darkness as we illuminate science in a new way. We discuss briefly, the history of science and the individuals who influenced it. We examine the controversial Dark Age of Science; many readers may be surprised to discover what has become of today's modern science.

The Big Picture of Modern Science describes the foundation of science in our day. We learn how that view contributed to the reign of a Dark Age of Science during which an enigmatic comingling of science and technology occurred, blurring the line between them and obscuring their critical differences. The relationship between math and science and the real agenda behind modern science comes keenly in view as we reveal its future.

3.1 The History of Science

Science Defined

During the early 1800s, the term 'scientist' did not yet exist. Men of letters who studied the natural world were, known as philosophers until Cambridge historian and philosopher, William Whewell changed that in 1840:

"'We need very much a name to describe a cultivator of science in general. I should incline to call him a **scientist**.'" Bib 14 p29

Though 'scientists' did not earn their new moniker until the 1800s, science's beginnings reach back at least to the 1400s, when men like Copernicus and Galileo expressed their views about Nature. Today, 'scientist' is a common household word generally applied to anyone who works in or is associated with any field of science.

The word science comes from the Latin word *sciens*, which means 'knowing.' In the 1960s edition of *Webster's Unified Dictionary and Encyclopedia*, an American dictionary that has been around since the 1800s and generally held in high regard, defines **science** using words such as truth, fact, and law. Note how these words denote absoluteness and an air of authority:

"A branch of study which is concerned either with a connected body of **demonstrated truths** or with **observed facts** systematically classified by being brought under **general laws**, and which include trustworthy methods for the **discovery of new truths** within its own domain." Bib 98 p3783

Compare this old definition with *Webster's 1997 Universal College Dictionary* definition of **science** and we see little change:

"1. A branch of knowledge or study dealing with a body of **facts or truths** systematically arranged and showing the operation of **general laws**. 2. Systematic knowledge of the physical or material world gained through observation and experimentation." Bib 7 p703

However, a look at the *McGraw-Hill Concise Encyclopedia of Science & Technology* from 1998 reveals a clear *lack of absoluteness* in the words they used to define **science**:

"In common usage the word science is applied to a variety of disciplines or intellectual activities which have certain features in common. **Usually** a science is characterized by the **possibility** of making precise statements which are susceptible of some **sort of check or proof**. This **often** implies that the situation with which the special science is concerned can be made to recur in order to submit themselves to check, **although this is by no means always the case**." Bib 12 p1719

We read a very unique definition of science, substantially different from Webster's earlier definitions as they eliminated the use of the words truth, fact, and law. These words are also

missing in the 2006 online *Wikipedia Encyclopedia* definition of science:

"**Science** (from Latin *scientia* - knowledge) refers to the system of acquiring knowledge – based on empiricism, experimentation, and methodological naturalism. The term *science* also refers to the organized body of knowledge humans have gained by such research." Note 3.1.a

In the 1995 *New York Public Library Science Desk Reference*, we find the following comment about science in its opening paragraph:

"**Science, at its very heart, is never constant**. The study of any science entails looking at things, measuring and defining them, analyzing their properties, or figuring out how they work. Along with the observing and measuring are the ever-changing new technologies to accomplish these goals. Add all of these factors together, and it is easy to see why science is constantly changing and growing." Bib 145 pxix

Of course, knowledge of all things scientific should always grow, but what happened to the idea of discovering facts and truth? If science is as they say, "never constant," how can it claim anything as fact? Perhaps this is one reason why so many people, including scientists, do not understand *what* science has become today.

It is apparent that we need a solid definition of science that is in harmony with the maxims of the Universal Model. Thus, we define **science**:

> **Science**: A study of demonstrated truths and natural laws that describe and explain Nature.

After reading this book in its entirety, the importance of this definition will be especially apparent. While we are in the mode of clarifying words, we use the following broad definition of **Nature** to represent all of the natural Universe and its workings:

> **Nature**: The Universe and its workings.

Philosopher Versus Scientist

Before the 1800s and the advent of the *scientist*, philosophers worked on unlocking Nature's secrets; a philosopher then was very different from the philosopher of today. Many of the ancient philosophers performed *experiments* and were in fact scientists by another name. Conversely, modern philosophers are merely theorists. *Webster's Universal College Dictionary* defines a **philosopher** as:

"A person who offers views or **theories** on profound questions…" Bib 7 p594

How do science and philosophy differ? The distinction between the two is very important *and* very simple. *Science* deals with *truth, fact,* and *natural law* whereas *philosophy* deals only with *theories*. The trouble is that modern science has blurred these two terms and their definitions, altering the foundation of real science; that is one of the subjects of this subchapter.

In the Universal Model, a **scientist** is one who describes and explains Nature using demonstrated truths and natural laws:

> **Scientist**: One who describes and explains Nature using demonstrated truths and natural laws.

How does a scientist demonstrate truths and natural laws? By observation and experiment.

There is no way to complete the scientific method except through observation and experimentation. Unfortunately many modern, so-called *theoretical* 'scientists' do not perform experiments; some of them *never* performed experiments. How can one be a scientist without performing experiments?

Why do we find it necessary to define scientists? It would seem that everyone already knows what a scientist is, but because science has changed so dramatically throughout the last century, the foundation of science has shifted from the discovery of new natural law to the pursuit of theory—scientists have become *philosophers* according to Webster's definition.

What is the Purpose of Science?

A student once answered this question by saying, "To me the purpose of science is to help us understand the world around us." Most people would probably agree with this answer, including professional scientists. Here is one example:

"For the vast majority of basic scientists, …the ultimate purpose of their calling is to **understand** the world around them and to **explain** its workings." Bib 18 p45

As we learned in Chapter 1.2, *understanding* comes with an increase of knowledge. Knowledge comes as we gain answers to the tangible questions of what, where, when, and who. These answers *describe* Nature. Answers to the questions, how and why, affect our *comprehension* of Nature by *explaining* its workings. This describes the **Purpose of Science**:

> **The Purpose of Science**: To describe and explain Nature so that we can understand and comprehend it.

If we can agree on the purpose of science, we can move on to other Fundamental Questions (FQs). How do we all agree on which description or explanation of Nature is the correct one?

Unlike democratic politics where we come to decisions by the voice of the majority, science supposedly rests on the maxims of 'truth' and 'law,' which is the only foundation in which we can reach agreement.

One method used by science to describe and explain Nature is the scientific method, but as we learned in the previous chapter, there is no consensus for what this is in modern science. We cannot expect to agree on the workings of Nature unless the method science uses relies on the five physical senses and the sixth sense of reason. Only through these **common senses** can we reach a **common consensus** concerning a proper scientific method from which we can learn all about Nature.

Who is Science For?

Some think science is only for the educated. Some scientists have taken it upon themselves to decide that they alone are able to understand and comprehend Nature. Others believe that there are few who can learn the mysteries of the Universe.

Once, the icon of science, Einstein, expressed the opinion that scientists themselves believe very few people can really understand science, and there are many instances of this elitist attitude in science today.

However, science is for everyone. In the not-too-distant past, people spent most of their time working to provide for the necessities of life. Today, opportunities and available resources have allowed a greater part of society to participate in scientific learning and discovery. More than ever before, science is for the common person and the whole of society. As one educator noted:

"Science is the activity of learning by **a whole society**… the laws of science are those principles of prediction and adaptation to the future which apply to the whole society, and can be learnt by all its members in explicit form." Bib 30 p115

The history of natural science is really *a history of people making discoveries and performing experiments*. In fact, many historically significant discoveries were not made by professional scientists as we might think of them today, but rather by individuals who simply had a great desire to learn—people like Benjamin Franklin, who in addition to his statesmanship duties, expressed a deep interest in Nature, conducting serious scientific research in his spare time.

In the next subchapter, we briefly examine a study of 100 scientists who shaped world history, five of which we find worthy of special note; they were the leaders in the science-by-experiment movement.

Nicholas Copernicus (1473-1543)

One of science's first reformers, Copernicus, developed the innovative helio-centric (sun-centered) solar system, countering the prevailing thought at the time that the Sun orbited the Earth because observation included a Sun rising in the east and setting in the west. Copernicus explained effectively why this was not the case using mathematics and planetary

Nicholas Copernicus

observations as he introduced his idea that the sun and not the Earth lay at the center of the Solar System.

A great example of independent thinking, Copernicus challenged the accepted philosophy of his day, even though he did so at the peril of his own life. Despite risks, he published his ideas, and encouraged others to question unproven concepts taught by the political and religious leaders of the day.

Galileo Galilei (1564-1642)

Probably one of the all-time greatest contributors to science, Galileo and his scientific method included experimentation and observation. Galileo believed in natural law and that observation as well as prediction was essential in proving a theory. Galileo developed a telescope for looking into the heavens and was instrumental in the development of a 'new astronomy' that sprung from this technology. The first to view craters and mountains on the moon, the phases of the

Galileo Galilei

planet Venus, and the four largest satellite moons of Jupiter, Galileo helped pave the way for Kepler and Newton to develop the laws of motion. Galileo demonstrated that the weight of an object does not determine the speed of its descent, again challenging the current belief of the day. He was not without his critics and many close-minded 'learned' men found his new ideas unfavorable because it contradicted the accepted dogma, even though his ideas could be proven. Ultimately, an inquest convicted Galileo of heresy, condemning him to live out his life in confinement.

Johannes Kepler (1571-1630)

Kepler, a contemporary of Galileo shared with him the sun-centered belief that Copernicus had introduced them to. Years spent observing the movement of the planets led Kepler to discover the three laws of planetary motion. Based on the understanding of these laws, science could predict the future location of the planets.

Many recognize that Kepler's discoveries came because the world worked according to natural *law* and he believed he could discover them:

Johannes Kepler

"While he attained immortal fame in astronomy because of his **three planetary laws**, Kepler also made fundamental contributions in the fields of optics and mathematics." Note 3.1b

Kepler achieved prominence in part because he believed that logical principles led to discovery, leaving one to wonder if he was alone in this belief outside of his close circle of friends like Galileo. Are we seeing a repeat of history where logical principles and experimentation have given way to theory and establishment dogma?

For Kepler to have had such a great influence upon humanity, he had to do something that would outlast time. He had to discover *truth* in Nature's laws, and that is exactly what he did.

Robert Boyle (1627-1691)

A champion of the experimental method, Boyle promoted the idea that ancient philosophical doctrine was not correct until proven. He rejected the notion that mere authority should dictate scientific thought. He led the effort to establish the Royal Society of London, a remarkably influential body of scientists that led the international study of science. Boyle helped develop the air pump and worked actively in the development of the fields of chemistry and physics, discovering the relationship between pressure and gas volume, now known as Boyle's Law. Boyle published works on air and chemicals, and his work led him to define the 'element,' a significant contribution to today's chemistry.

Robert Bolyle

Sir Isaac Newton (1642-1727)

Acknowledged by many scientists as probably the world's greatest contributor to scientific discovery, Newton did not gain such notoriety by his 'belief' in something, but by what he *demonstrated so others could believe*. The following quotes from three different sources give us some insight into the importance of the man:

"No man of science, no man of thought has ever equaled the reputation of Isaac Newton." Bib 30 p15

"No other work known to the history of science has simultaneously permitted so large an increase in both the scope and precision of research." Bib 33 p30

"His law of gravitation, regarded then and now as the greatest of all scientific discoveries, was held to be ultimate and unassailable, the typical law according to which all other laws must be fashioned…" Bib 2 plxi

Sir Isaac Newton

There is no doubt that the history of science will always remember Newton for his great insight into Nature and for the experimentation that he brought to science. The application of Newtonian physics ultimately proved true many of Newton's insights.

The Forefathers of Science

These five individuals and others gifted to us the traditional roots of true science. Responsible for inspiring many to *question everything with an open mind*, these men challenged established philosophical-scientific authority and theory to establish newly identified natural laws and to prove them using observation and experimentation. From their efforts has come new technology and a greater understanding and comprehension about Nature. For the student to truly grasp science, they must know the history of science.

The America-Science Connection

Not many people realize that the experimental, observational attitude in science carried into the lives of American founding fathers. Many know of Benjamin Franklin's electricity experiments, but few know that we sometimes refer to Thomas Jefferson as the 'father of archaeology.' Jefferson was the first to systematically excavate burial mounds near his home, making a meticulous record of his observations. Today, archeology still uses stratigraphic excavation processes used by Jefferson. Bib 67 p29

Some of the founding fathers were scientists; most had an attitude favorable to scientific experimentation. The birth of America and the subsequent constitutional movement, dubbed 'The American Experiment,' came from the mindset implanted in the character of the American founding fathers and other highly influential individuals that helped shaped America's beginnings.

Noah Webster (1758-1843)

Perhaps one of the most influential early Americans, Noah Webster, authored the *Webster Dictionary*, an authoritative and reliable book in the early 1800s. Unfortunately, most people today know little about the man behind the name. Webster owned and ran a daily newspaper in New York City that published an influential pamphlet on the Federal Constitution. At the time, America was still young and it did not have its own language. Most books came from England, including spelling books and dictionaries, but Webster saw the need for an American version, so he worked for decades, drawing from twenty languages to develop the *American Spelling Book* and the *American Dictionary of the English Language*. Amazingly, the *American Spelling Book* is still one of the most widely printed books in history.

Noah Webster

How did Webster author a dictionary defining new American words, some of which had never been defined before, unless the author could relate personally through the direct experiences he had acquired throughout his life? This is where Webster stood out. Webster had extensive life-experience and had written on many subjects, including economics, politics, epidemics, insurance, the French Revolution, philosophy, orchardry, law, banking, and many others. He practiced before the United States Supreme Court, was a member of the General Assembly of

Connecticut, and was on the General Court of Massachusetts. He was a councilman and an alderman in New Haven and he was a county judge. He was the vice-president of the Hampshire and Hampden Agricultural Society, helped establish Amherst College, and had many other accomplishments. Webster had a direct connection to science by founding the *Connecticut Academy of Arts and Sciences*. Amongst everything else in his busy life, he was an ***experimental scientist***.

The 1957 the unabridged Webster's dictionary, seen in figure 3.1.1, is full of scientific words and formulas. Webster placed a great deal of importance on science. America became a leader in the scientific world, thanks to people like Webster who saw the importance of what we defined in the first chapter as the Definition Principle: If we don't say what we mean, we don't mean what we say.

Webster helped move American science forward by exposing the common people to scientific language contained in Webster's Dictionary.

Traditional, Modern, and Millennial Science

There are primarily three scientific eras identified in the UM. These do not cover a definite period but they cover an era in which a particular attitude toward science prevailed.

1. Traditional Science – 1500 to 1850
2. Modern Science – 1850 to Universal Model
3. Millennial Science – The new science of the UM

Traditional Science is the period of science that includes the likes of Copernicus, Galileo, Kepler, Newton and others up to about the mid-nineteenth century. During this approximately 350-year period, the founding scientists came to emphasize experimentation and empirical evidence as opposed to just 'rational or logical' thinking. They discovered natural laws that they used to describe and explain Nature in simple ways without the use of mythological figures and ideas. Discovering natural law was the driving force behind scientific endeavors of the day.

Although Traditional Science *began* as an experimental, empirical-based science, it underwent changes during the mid-1800s. The History of Science Table on the following page illustrates how Traditional Science evolved into something very different from what the founding scientists started—it had become the ***modern*** science of the 1800's. Traditional Science founded on *natural law* had given way to Modern Science, built primarily on *theory*.

Modern Science originated during the early nineteenth century, based on the philosophies of key figures like Charles Darwin, Karl Marx, and Charles Lyell. Darwin's theory of evolution brought human beings down to the level of the animals and Marx's political science would eventually bring about the rise of socialism and communism. Modern science would become the largest instigator in the destruction of religion and would undermine humanities great economic and social achieve-

Fig 3.1.1 – This 1955 Webster's Dictionary contains many of the newest scientific terms and words that shaped the world in the twentieth century.

ment—free market capitalism.

Sigmund Freud and Albert Einstein would later become key players on the stage of this new 'modern' science with Freud building on Darwin's 'human animal's' basic drives and instincts and Einstein's theory questioning the very nature of reality and ultimately our own existence.

> Modernism rejects Traditionalism.

From its foundation, modernism rejects traditionalism and certainty (see modernism at Wikipedia.org). However, certainty is merely another word for truth, or things that do not change. Thus, modernism by its own definition rejects truth and it rejects the knowledge of things that are, were, and will be. The *rejection* of certainty and truth is the hallmark of modernism and explains why modernism and modern science are incompatible with religion or a Deity that knows all things.

We use the term '**modern science**' throughout the UM to refer to the science that began during the early mid-1800s through the beginning of the 21st century, or to the advent of the Universal Model. For about 150 years, modern science leaders deceived the scientific world into thinking science could discover what is true without believing in truth.

Millennial Science marks its beginning with the publication of the Universal Model in 2016. Newly discovered scientific truths contained in the UM set the stage for the next millennia of scientific discovery. As scientists and others investigate and identify additional evidences for the extraordinary discoveries listed in the UM, our understanding about Nature's puzzle can continue to grow.

One final thought before we leave the subject at hand. Technology and science are not the same. Significant technological advances have been made throughout the last 150 years but there are different forces driving that success. We'll talk more about that in subchapter 3.7

3.2 The Dark Age of Science

What is the Dark Age of Science? After all, hasn't modern science *progressed* since the mid-1800s? In the History of Science Table (Fig 3.2.1), we arrange the top 100 scientists and their observations, technologies, theories, and natural laws for which they are credited in chronological order. By doing so, we see the development of an interesting pattern, a relationship that identifies a sharp decline in the discovery of natural law—a **Dark Age of Science.** This period of scientific regression is not mere conjecture, and this historical fact seems to have escaped notice in modern times.

The History of Science Table

We adapted the History of Science Table in Fig 3.2.1 from the book *100 Scientists Who Shaped World History* [Bib 22] (Fig 3.2.2) published by Bluewood Books as part of their *100 Series* collection. The books each contain concise summaries of the

HISTORY OF SCIENCE TABLE

Year	New Observations	New Technologies	New Theories	New Laws
2000 1900's	James D. Watson – DNA double helix Har Gobind Khorana – DNA protein codes Dorothy C. Hodgkin – structure of penicillin, insulin, vitamin B12 Barbara McClintock – gene transposons Margaret Mead – cultural anthropology Enrico Fermi – nuclear reactions Irene Joliot-Curie – radioactive elements Arthur H. Compton – Compton Effect Edwin P. Hubble – red shift of light Selman A. Waksman – microbiology Alexander Fleming – penicillin Nettie M. Stevens - heredity	Rosalyn S. Yalow – radio immunoassay Rosalind E. Franklin – x-ray diffraction, photography of DNA Frederick Sanger – structure of proteins Charles H. Townes – maser, laser Jacques Y. Cousteau – aqualung, submersibles Shockley, Bardeen, Brattain – transistor Grace B.M. Hopper – supercomputer, COBOL language Enrico Fermi – nuclear chain reaction Robert Watson-Watt – radar Selman A. Waksman - antibiotics	Steven Hawking – black holes Tsung-Dao Lee – overthrew parity theory Luis W. Alvarez – bubble chamber theory Richard P. Feynman – nuclear shell theory Marie Goeppert-Mayer – nuclear shell theory Werner Heisenberg – quantum theory Linus C. Pauling – resonance, electro negativity theory Edwin P. Hubble – red shift Niels Bohr – quantum theory Albert Einstein – theory of relativity Lise Meitner – nuclear theory	**The Scientific Dark Age**
1800's	Ernest Rutherford – radioactive half-life, atom's nucleus and proton Henrietta S. Leavit – Cepheid stars, period-luminosity Marie S. Curie – polonium and radium George Washington Carver – soil nutrients Joseph J. Thomson – sub-atomic particles Albert A. Michelson – measured speed of light Antoine H. Becquerel – spontaneous radioactivity William Ramsay – discovered argon, krypton, neon and xenon gases Ivan P. Pavlov – reflex action conditioned Luther Burbank – plant hybridization and grafting Wilhelm K. Roentgen – x-rays Dmitri I. Mendeleyev – periodic table Joseph Lister – antiseptic surgery Louis Pasteur – pasteurization Jean Bernard Leon Foucault – speed of light in water, pendulum Matthew F. Maury – oceanography Michael Faraday – electromagnetic induction Jons Jakob Berzelius – cerium, selenium, silicone, thorium Humphry Davy – potassium, sodium, barium, trontium, calcium and magnesium discovered through use of electricity Alexander von Humboldt – explorer, scientific encyclopedia Georges Cuvier - paleontology	John A. Fleming – diode Thomas A. Edison – phonograph, lightbulb, movie projector William H. Perkin – synthetics James C. Maxwell – color photograph Jean Bernard Leon Foucault – gyroscope, silver glass mirror Augusta Ada Byron – binary notation Joseph Henry – electric doorbell, assisted with telegraph Charles Babbage – mechanical calculator, speedometer, ophthalmoscope Michael Faraday – electric generator, motor, transformer Karl F. Gauss – heliotrope, first telegraph	Sigmund Freud – theory of psychoanalysis James C. Maxwell – kinetic theory of gases Friedrich A. Kekule – molecule theory, organic compounds William Thomson – Kelvin temperature Charles R. Darwin – theory of evolution Louis Agassiz – ice age theory *During These Time Periods, Theories Were Tested*	William Thomson – second law of thermodynamics James C. Maxwell – electromagnetism Gregor Mendel – heredity laws James P. Joule – Joule-Thomson effect, Joule's Law, Law of conservation of energy Jons Jakob Berzelius – law of definite proportions Joseph L. Gay-Lussac – Charles law of gases John Dalton – law of partial pressures, law of definite proportions
1700's	Edward Jenner – vaccinations William Herschel – astronomy Joseph Priestley – carbon dioxide and oxygen Henry Cavendish – hydrogen Carolus Linnaeus – binomial nomenclature Leonard Euler – mathematical nomenclature Benjamin Franklin – lightning is a static discharge Daniel Bernoulli – Bernoulli's effect	Alessandro Volta – electric battery Benjamin Franklin – cast iron stove, lightening rods, bifocal glasses	*and Either Became*	Antoine L. Lavoisier – law of conservation of mass Benjamin Franklin – static electricity laws
1600's	Edmund Halley – star positions, Halley's comet Isaac Newton – white light contained all colors of the spectrum Robert Hooke – discovered cells Anton van Leeuwenhoek – micro-observation Christian Huygens – astronomy, Huygens' principle Robert Boyle – vacuum pump, chemistry Blaise Pascal – Pascal's principle, mathematics Rene Descartes – mathematics William Harvey – the heart pump's blood	Robert Hooke – weather instruments Anton van Leeuwenhoek – microscope Christian Huygens – pendulum clock Robert Boyle – first match from phosphorus	*Natural Laws*	Isaac Newton – law of gravity, three laws of motion Robert Hooke – law of elasticity Robert Boyle – Boyle's gas law Rene Descartes – law of reflection Johannes Kepler – Kepler's three laws of planetary motion
1500's	Galileo Galilei – astronomy, gravity Andreas Vesalius – human anatomy Nicolaus Copernicus – astronomy	Galileo Galilei – telescope used in astronomy	*or Were Discarded.*	Galileo Galilei – law of inertia
1000 500 B.C	Hakim ibn-e-Sina (Avicenna) – medicines Galen – human anatomy Eratosthenes – scientific history Euclid – geometry Aristotle – treatises on logic Hippocrates – diseases come by nature, medicine Pythagoras – mathematics, Pythagorean theorem	Archimedes – Archimedes' screw for raising the level of water, catapult		Archimedes – law of buoyancy, law of simple machines Euclid – geometric laws

Fig 3.2.1

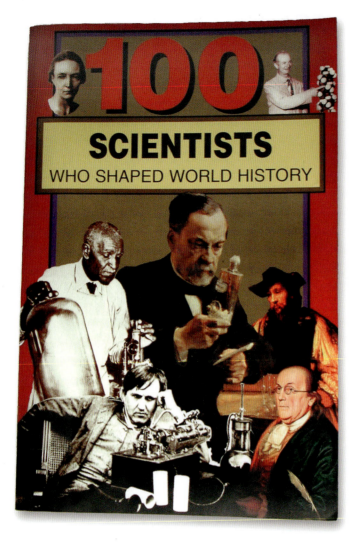

Fig 3.2.2 - This book, 100 Scientists Who Shaped World History, is a wonderful example of the observations, technologies, theories and natural laws that scientists and technologists have contributed to society throughout time. From these contributions we organized the History of Science table in Fig 3.2.1.

men and women most influential in shaping world history. The 100 scientists chosen by the author, John Tiner, provides for us a solid summary of scientific discovery during the past 2,500 years.

By arranging these notable characters both chronologically and categorically, according to their contributions to science, we gain a new perspective of the direction modern science is moving. The table is not inclusive of all scientists or all advancements in science, nor is it a UM production; we list only those who appear in the afore-mentioned book. Some individuals appear more than once in the table as each contribution is assigned a category—observation, technology, theory, or natural law.

One of the first patterns to emerge is the consistency of entries in the **New Observations** and **New Technologies** columns. In these columns, especially the New Observations column, contributions show continuity *throughout* history.

This reveals an underlying fact that for these categories, **Observation** and **Technology,** discovery endures, and new knowledge abounds. Cleary, they passed the **Test of Truth,** which is *time*, as described in Chapter 1.5. Hundreds of years ago, Galileo observed for the first time that the Moon had craters. We can still observe them today because they are real—they are *true*. Thomas Edison invented the light bulb. It worked then, it works now, and we have no reason to expect that it will not work tomorrow.

In the third column listing **New Theories**, we see an increasing number of them from the mid-1800s forward. Where are the theories prior to this time? Before the mid-1800s, during the traditional science period, scientists tested and proved their theories and established natural law or they discarded them.

Compare the fourth column, **New Laws**, with the column listing **New Theories** to see a very clear relationship; as new theory proliferated, the addition of new laws subsided, even to a point of notable absence.

Why is the twentieth century
void of new natural laws?

This question becomes even more bewildering as we follow it with this line of questions:

1. Are there more scientists today than 100 years ago?
2. Are we spending more research money than we did 100 years ago?
3. Do we have easier access to advanced technology than we did 100 years ago?

Since the obvious answer to all of these questions is yes, then:

Why is **less natural law**
discovered today than 100 years ago?

Even with countless more scientists, the annual spending of billions of dollars, and bewildering technological advances we see a veritable dearth of natural law discovery. Does anyone wonder why this happened? The picture as to why began to take shape as we compiled an answer for this phenomenon.

Because technology makes it possible to pick Nature apart like never before, and because of the remarkable innovations we use every day, we think science has made significant progress in their understanding of the natural world, but the UM discoveries suggest otherwise.

The Dark Age of Science

The shaded area in Fig 3.2.1 represents a period we recognize as the **Dark Age of Science,** or the **Scientific Dark Age**, because the discovery of natural law suddenly ceased. Throughout the twentieth century, we witnessed a dearth of natural law, yet few scientists seem aware of this Scientific Dark Age, and even fewer know why it exists. In fact, because so much time has passed since the discovery of any significant natural law, few scientists even look for them! Instead, as we saw in the last chapter, theory became the end point of science.

Of the hundreds of extraordinary claims made in the Universal Model, one of the boldest is the claim that modern science flounders in a Dark Age today, and it has for more than a century. One of the principal purposes of the UM—which demonstrates the reality that modern science remains stuck in a Scientific Dark Age—is to restore order in science.

We are not entirely alone in our assessment as keen observer and well-known physicist Carver Mead says:

"It is my firm belief that the last seven decades of the twentieth century will be characterized in history as **the dark ages** of theoretical physics... **it doesn't make sense**, and it isn't the way science works **in the long run**. **It may forestall people from doing sensible work for a long time**, **which is what happened.** They ended up derailing conceptual physics for the next **70 years**." Note 3.2a

Although Mead saw a dark period of only about 70 years back in 2000, he wasn't very far from the century and a half recognized by the UM. Mead's years of work astride the fields of both technology and science led him to the realization of just how far theoretical physics had diverged from reality. What Mead and other scientists are not so aware of is how far *the whole* of modern science in general has diverged from reality.

Look back at the History of Science Table's third column (New Theories). The column appears incomplete because there are no theories listed prior to the year 1800. Does this actually mean there were no theories prior to this time? Of course not! There were many theories, but a theory is only an idea or a conjecture in need of testing. Over *time, researchers tested theories until they reached the level of natural law or died*; this was the fate of pre-1800 theories.

What will happen with the theories currently showing in the New Theories column? Eventually, they must prove true or we must discard them, but modern science seems extraordinarily reluctant to abandon their highly favored notions.

Can we ever move theories into the New Laws column if we cannot prove them?

A scientific theory can follow only one of two courses if retained; *First*, scientists must prove the theory true over time so that it becomes natural law through the application of the scientific method as they test, observe, and evaluate all of their predictions. Though rare, this does happen. Long ago, this step happened relatively consistently, at least until the last century. The *second* possibility is that the theory stays a theory. We might not discard it because of its popularity, because of its deep entrenchment into the stream of education where we continue to teach it as fact, even though it *never* proved true. We sometimes see its replacement by new or improved versions of the same theory, but they are still unproven. As a result, the relationship between New Law and New Theory remains glaringly apparent in the History of Science Table.

The Dark Age of Scientific Truth

Another reason modern science fell into a dark age is that a number of modern scientists did not *want* to discern the truth from error. Over the past two decades while developing the Universal Model, it astonished us that so many 'establishment' scientists either lost the desire to find truth or never cared about it in the first place. The professor who possesses this attitude places the student in the uncomfortable position of having to study what the professor thinks, not what he or she can demonstrate.

A professor may devote years of research to a particular field, eventually gaining high esteem, well-known among peers in their field, so we can imagine how this might affect the outcome if a new discovery completely wiped out years of work, erasing the long-earned stature.

We once asked a professional geologist what he thought about some of the concepts presented in the UM. He said simply, "If true, a lot of scientists will be *very* disappointed." Why would they be disappointed? *Is there an opposition to truth?* In the utopian idea of science, scientists that spend years developing their theories only to discover they are wrong are supposed to step aside and let truth prevail, but this rarely proves true in the real world. After dedicating considerable effort toward a project, many find themselves unwilling to give it all up—even after proving the theory false. Most scientists oppose anything that challenges their work, especially if their works span many years. Perhaps they reason, too much time and effort has been invested; it can't be wrong!

This leads us to the chief deterrent of truth—*pride*. When new scientific discoveries challenge current scientific theory, we need to ask new questions. The mode, or questioning attitude, whether it be critical, skeptical, partial or objective will determine the outcome—the answer to those questions. As we learned in the first chapter, the critic and the skeptic approach the question with negative sentiment, leading to a closed mind. This pride-based closed mindedness has had grave consequences for humanity.

There are several historical examples that illustrate how pride and a lack of objectivity dissuaded medical science from discovering scientific truth that was right in front of their eyes. Unfortunately, this opposition to truth has contributed to a tremendous loss of human life.

Andreas Vesalius

About 2,000 years ago, a Greek doctor named *Galen* ranked as the undisputed authority in medical science. He wrote over 50 books including an encyclopedia, *On Anatomical Procedure*, a book that earned its place as the seminal text used by doctors for almost 1,300 years. Then during the mid-1500s, *Vesalius*, a knowledge thirsty 17-year-old, began his medical studies in Paris. Vesalius' professors taught from Galen's ancient works but Vesalius decided to find out for himself what the human body was really like and secretly secured a corpse for his own personal study. To his astonishment, he discovered that some of Galen's descriptions of the human body had been based on monkey skeletons. Vesalius cataloged over two hundred mistakes in the Galen-based medical textbooks of the day, publishing his own work in 1543. His book, *On the Workings of the Human Body,* described the errors in detail with accompanying illustrations, revealing many previously unknown facts.

Although Vesalius's book contained accurate depictions of the human body, the doctors flatly rejected it because their reputations were at stake and nobody wanted to refute the undisputed authori-

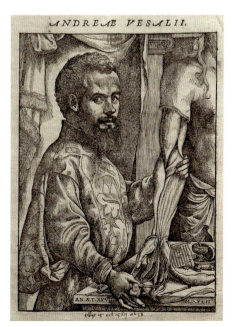

Andreus Vesalius

ty; however, they all *knew* Galen was correct. Scorned to shame and threatened, Vesalius had to flee Paris. He died at the age of fifty, never seeing his work fully accepted. However, the test of time has confirmed the truth of Vesalius's discoveries, and today many consider him the father of anatomy. Note 6.3a

Interestingly, in the same year Vesalius's book challenged the foundations of medical science, Copernicus published his book, *On the Revolutions of Heavenly Spheres,* which challenged astronomies philosophies.

William Harvey

About twenty-five years after Vesalius died, *William Harvey* attended the same prestigious European medical school where Vesalius had begun his studies. The professors were still teaching Galen's ideas, including the idea that the blood flowing through the heart caused the heartbeat, and that there was no connection between arteries and veins. Harvey established a successful medical practice, counting King James I of England among his patients. Harvey's experiments led to the discovery that the heart circulates blood through veins and arteries. He speculated about the role of capillaries, but was unable to see them because of the limitations of the tools at his disposal. He received sharp criticism for his book, *An Anatomical treatise on the Motion of the Heart and Blood in Animals* (1628), and doctors refused to repeat his experiments. It was not until several years after his death that researchers discovered capillaries and confirmed the truth of his work.

William Harvey

Ambroise Pare

16th-century medical doctors usually left the messy work of operating to the knife wielding experts—the barbers. Born in humble circumstances in 1510, *Ambroise Pare* never received formal schooling. Without knowledge of Latin or Greek, he found it impossible to study medical books in his home country of France. However, Pare found work as a barber-surgeon in the first hospital in the world, the Hotel-Dieu, which treated the poor. Pare's personal experiences led him to believe he was meant to ease suffering—not to follow the orthodoxy of the physicians. He spent over 30 years in the French army treating gunshot wounds when he attempted to publish his book, *Treatment of Gunshot Wounds*. The doctors of the day opposed the uneducated barber-surgeon's practices, ridiculing him for his ignorance. Many gunshot victims suffered wounds so sever as to require amputation, and the treatment at the time was to use boiling oil and a hot iron to staunch the bleeding. This was so painful that many soldiers expired before doctors could complete their operation.

Ambroise Paré

Pare set out to find ways to ease suffering, so instead of boiling oil, Pare prepared an ointment, which he applied to the wound. Eventually his method became the preferred treatment. Pare also started the practice of using silk thread to tie the torn blood vessels closed. He made many new kinds of artificial limbs for those who had lost theirs in battle, and he finally got his book published in French. It became so successful that it was reprinted in many languages. Pare became the practicing surgeon to four kings of France and a hero for the thousands of soldiers' lives he saved. Although rejected at first, Pare's medical practices passed the test of time, and he contributed significantly to the field of amputational surgery and prosthetics.

Dirty Handed Doctors

In 1844, after *Ignaz Philipp Semmelweis* graduated from medical school, he became interested in the terrible killer of young mothers, childbed fever. He took a position at the Vienna Maternity Hospital under the direction of Johann Klein. For the 40 years preceding Klein's administration, the mortality rate for expectant mothers at the facility was about 1-2%. However, while Semmelweis worked at the hospital under Klein, he saw the mortality rate soar to over 15%. Semmelweis also noticed that the "street births" mothers, women who gave birth to their babies in dirty city streets, had a lower mortality rate than the mothers treated at Klein's hospital. The mortality rate for the ward at the hospital where student midwives attended the mothers but did not work on human cadavers was four times lower than the ward where the 'educated' doctors worked. Semmelweis observed that Klein allowed his student-doctors to work on cadavers between times attending to living patients, and they did not wash their hands adequately.

Ignaz Phillipp Semmelweis

Realizing the possible connection, Semmelweis suggested that washing hands in a chlorinated lime solution would remove "cadaverous particles," a practice that had remarkable success. However, the world will perhaps ever know Johann Klein for his opposition to truth. His pride stood in direct opposition to saving thousands of lives. Klein would not accept the evidence of 'clean hands,' rejecting everything Semmelweis tried to do to get doctors to wash their hands.

Some 10 years later, a surgeon in Scotland, *Joseph Lister* observed a similar problem; patients were dying after routine operations. After reading Semmelweis's paper and the works of Louis Pasteur, he realized that **germs** might be the cause of infections and he began the practice of washing. He washed everything in the operating room including the patient and the

CHAPTER 3 THE DARK AGE OF SCIENCE

results were astonishing. Previously risky operations became safe.

Yet medical doctors still opposed the truth! *They* could not imagine the possibility that they might be causing their patients' death. This was just too much for a learned doctor to accept, some actually believed their gentlemanly hands could not possibly be unclean.

Lister moved to London but the doctors there resented his success. However, Lister had the test of time on his side and doctors and patients alike soon discovered that Lister's clean room reduced mortality following operating procedures. Eventually hospitals became sparkling clean. Lister was elected to the House of Lords in government and to the president of the Royal Society for scientists. He became the personal physician of Queen Victoria.

Sadly, the success of Lister did not cross the ocean for some time:

"**Initially, most American physicians were loath to buy into germ theory,** seeing it as a European phenomenon incompatible with the 'truth' of spontaneous generation and as a threat to the general practitioner from the growing cadre of scientifically trained laboratory microbiologists and specialist physicians." Note 3.2b

In the end, it was not so much the doctors' dirty hands, but the pride of men and women that cost people their lives. Fortunately, for humanity, time is on the side of truth. The test of time will eventually overcome the opposition to truth and the reigning darkness will be replaced with light.

3.3 Modern Science Today

Modern Science is a topic that could easily fill several volumes as large as the one you are now reading, but we will touch on a just few items, as follows:

1. The Fields of Modern Science
2. Modern Science is Not Simple Science
3. A Modern Science Attitude
4. Lack of Order
5. Science in a Nutshell

The Fields of Modern Science

Today there are many highly specialized science fields and some of those have many layers of specialty within them. For the layperson, the complexity can seem bewildering. For simplicity, we group them into broad categories in the UM:

Physics - The study of matter and energy.
Chemistry - The study of the chemical nature of matter.
Astronomy - The study of outer space.
Geology - The study of the Earth.
Meteorology - The study of the weather.
Biology - The study of living things.
Paleontology - The study of fossils.
Archeology - The study of human artifacts.
Anthropology - The study of Mankind .
Psychology - The study of human behavior.
Sociology - The study of societal behavior.

Field specialization has the unintended consequence of isolation and disunification, which contributes to reliance on unproven or false foundational pillars in other fields. The UM's unique position comes because of its science-wide review of some of science's foundational pillars. The research turned up a number of inaccuracies and unproven assumptions, and it uncovered new 'pillars' upon which to base a new foundation. We summarized the results of that research in the book series you are now reading.

The undertaking of this enormous endeavor, the voluminous information available from each field and the time it took to review such a vast store of literature led to one incontrovertible conclusion; we must establish the truth about the foundational pillars in science based on actual empirical evidence. And we must make that evidence available to anyone who desires to see the proof.

Modern Science is Not Simple Science

Most of us would rather carry a single paper U.S. dollar in lieu of 100 copper-coated pennies. Who would depart on an hour-long detour from a hurried 15-minute commute home to dinner just to make things more complicated? There is power in simplicity. Keeping things simple in science, sometimes expressed using 'Occam's Razor' resists the temptation to pile up assumptions:

"The most influential aesthetic principle in science was set forth by the fourteenth-century British philosopher William of Occam. He argued that **the best explanation** of a given phenomenon is generally the simplest, the one with the fewest assumptions." Bib 20 p70

We call this **The Simplicity Principle** in the UM, which we define:

The Simplicity Principle:
The simplest, correct explanation is the best.

Modern science understands the Simplicity Principle and though it heralds 'Occam's Razor,' in practice it seldom follows the principle. In modern scientific journals—where scientists report cutting-edge research—we read articles sometimes so subjective that one can only guess at its meaning. After attending a symposium on quantum mechanics (the study of atomic particles), one science author has written about how the scientists at the gathering understood their own science:

"… each speaker seemed to have arrived at a private understanding of quantum mechanics, couched in idiosyncratic

language; no one seemed to understand, let alone agree with, anyone else. The bickering brought to mind what Bohr once said of quantum mechanics: '**If you think you understand it**, **that only shows you don't know the first thing about it.**'" Bib 20 p91

Another physicist who spent his career among professional physicists reflected on the insensible disharmony within his field of study:

"You should realize that in general only about 90% of professional physicists are able to make sense of less than 10% of what other physicists say." Bib 20 p91

These examples illustrate the difficulty of understanding modern science. The layperson is not alone in his or her angst; even the experts have difficulty with understanding! Why do we find modern science so difficult and complex? If the purpose of science is to understand Nature, and if Nature is simple, why does science have such a hard time explaining it? An examination of science's foundations helps us to determine why.

A Modern Science Attitude

Perhaps the phrase most well known in all of modern science literature is, 'we now know.' However, many instances prove a great chasm exists between what modern science '*knows*' and what reality presents. For example, NASA listed the *Top Ten Scientific Discoveries Made During The Apollo Exploration of the Moon*. The article included:

"Before Apollo, the state of the Moon was a subject of almost unlimited speculation. **We now know** that the Moon is made of rocky material that has been variously melted, erupted through volcanoes, and crushed by meteorite impacts." Note 3.3a

The top scientific discoveries from NASA's lunar program conflict with the empirical evidence in Chapter 7, The Hydroplanet Model, which shows that the Moon did not come from a melt nor was most of it pulverized by meteorite impacts.

"We now know" statements are a phenomenon of modern science culture and an attitude not necessarily infused with truth. We discern the modern science *attitude* in this excerpt from one science writer:

"Scientists must endure not merely Shakespeare's King Lear, but Newton's **laws** of motion, Darwin's **theory** of natural selection and Einstein's **theory** of general relativity. **These theories are not merely beautiful; they are also true, empirically true**, in a way that no work of art can be." Bib 20 p6

We make two points about modern science's *attitude* today; first, they use 'theory' and 'law' interchangeably, giving us the impression the two mean the same. Demoting natural law to theory or placing theory on equal status with natural law, violates the maxims of truth.

One real problem that occurs when we teach theories as laws is that people come to believe that those theories are 'empirically' true. **Empirically**, meaning verifiable by experience or experiment. If something is empirically true, we must observe and repeat it. Many modern science theories are neither observable nor repeatable.

A second problem relates to modern science's *general* attitude. The scientific establishment's invocation of the all-too-familiar 'we now know' cliché, supposes that the theories they espouse are absolute fact, and in so doing, they actually impede the discovery of truth.

Lack of Order

We look next at three quotes typical of modern science.

1. "In general, **science deals with repeatable events** and with answering "**how**" rather than "**why**" questions." Bib 46 p6
2. "In science there is no meaning to anything …The whole idea is **we cannot predict**." Bib 20 p205
3. "However, although laws summarize observed behavior, they do not tell us *why* nature behaves in the observed fashion. **This is the central question for scientists**" Bib 48 p175

Couple these contradictory statements that are indicative of the confusion that exists with this quote from earlier in the chapter:

"For the vast majority of basic scientists,… the ultimate purpose of their calling is to *understand* the world around them and **to explain its workings**." Bib 18 p45

Or this idea from a typical graduate student:

"Unfortunately, **it is not the task of modern physics to *explain* nature**, but just to *describe* it in the most simple terms." Note 3.3b

These statements reveal the confusion that exists at the *very foundation* of science—*its purpose*. A lack of order persists in every science field, but maybe, most scientists are ignorant of the disorder; perhaps because they have become so accustomed to it.

If the scientific community cannot agree on the purpose of science, there is little chance for order within their ranks. The lack of order in modern science contributed to the Dark Age of Science, and it helped modern scientists envision and create the illogical big picture of science.

Modern Science in a Nutshell

Modern science cannot define itself using natural law because they have taken natural law out of the picture. Mainstream science practitioners of the day define science with theories. In a 'nutshell,' modern science is not a simple science; it is a geo-politically influenced body, out of order and with an atti-

CHAPTER 3 THE DARK AGE OF SCIENCE

tude biased toward theory. It boasts of inconsistent definitions and paradoxical objectives. One might ask, into what has modern science evolved?

Modern science has evolved from
science to **philosophy**.

The dictionary defines science as dealing with facts, truth, and law. Modern science literature demonstrates that modern scientists today are more interested in 'offering theories on profound questions' and few see anything wrong with this assessment. As noted previously, 'offering theories on profound questions' is the dictionary's definition of a *philosopher*. Indeed, the title 'PhD' literally stands for Doctor of Philosophy—not science.

From the book titled *The Common Sense of Science* we read:
"Science is a great many things… but in the end they all return to this: **science is the acceptance of what works and the rejection of what does not**. That needs more **courage** than we might think." Bib 30 p146

It certainly will take courage to accept empirical evidence that challenges long held scientific theories. If the evidence is real, one cannot simply walk away from it. The beauty of scientific truth is that eventually, the evidence will come out that supports true models and natural law. We may have to adjust some of the details, but if we can apply the first principle of the UM, to question everything with an open mind, truth will prevail within our own constructs.

3.4 Modern Science and Truth

What is the relationship between modern science and truth? The dictionary defines science as a "body of demonstrated truths" yet some modern science *leaders* think, "truth is always unknowable." Thus, modern science and truth are at odds with each other when they should be inseparably connected.

Modern Science and Truth

Distinguished professor emeritus Roger Newton, at Indiana University department of physics wrote the book *The Truth of Science*. He favored another well-known physicist's assessment of truth:

"Truth is always **unknowable**, the only certain knowledge is that of **error**." Bib 18 p114

One cannot know what is true in science if truth is unknowable. This 'philosophy' of modern science is at odds with the great scientific discoveries of the 1500s and 1600s that came from Newton, Copernicus, Boyle, Galileo, Kepler, and the like. These early scientists believed in scientific truth and they believed in absolute truth.

Today, the majority of the leaders in the scientific community and those of the educated elite, deny the existence of truth. This denial has led to a Scientific Dark Age and an absence of newly discovered natural law.

Perhaps you might ask, "Why would science deny the study of truth, which is by definition what science is?"

The first step is the nullification of truth's definition (we discussed this in Chapter 1.5). Next comes the practice of combining truth with error.

An example of a 'philosophical truth-trap' comes from 1976 Nobel Prize winner, Baruch Blumberg. His statement made in the year 2000 sums up common dogma held by many in science today:

"**By definition** science welcomes new evidence, new ways of thinking. **It has no final truths**. It is a continuous quest and exploration." Note 3.4a

His declaration deftly sandwiches a falsity between two true statements. The comingling of false and true leads the reader to assume all three are true. The first and last parts are self-evident. If science is to be "a continuous quest and exploration," it must "welcome new evidence, new ways of thinking." However, his five little words, "It has no final truths" is *key* to understanding what has become of science.

If science has no final truths, how can it have facts?

Scientists and intellectuals alike have fallen into similar philosophical truth-traps, wherein modern science claims there are no final truths.

The **Dark Age of Science** began
with the false philosophy that
there is **no truth** in science.

This modernistic false philosophy of no final truth has led to grave consequences. Without truth in science, there can be no natural law, and without natural law, there can be no science. The trend has continued through several generations of university professors schooled in this dogma when they were students. The following excerpt from the 1964 *Life Science Library* book *The Scientist* expresses the decades-old modern science philosophy of ignoring reality and not expecting to see truth:

> Without truth in science, there is no natural law.
>
> Without natural law, there is no science.

> "Truth is always unknowable, the only certain knowledge is that of error."

"In a manner of speaking, the scientist stands in a house of mirrors—mirrors which he himself has built, and which he himself keeps improving so that they give increasingly detailed and accurate reflections. **Reality continues to elude him**, but he can see it darkly in one or another of the flawed glass surfaces before him. Some of the mirrors give one distortion, some another; all offer only reflections. By piecing together these reflections, the modern scientist seeks ever better approximations of truth, but **he no longer expects to see truth naked**." Bib 14 p62

The famous physicist, Niels Bohr (1885-1962), instrumental in developing atomic 'theory,' expressed this insightful opinion on the task of modern physics:

"It is **wrong** to think that the task of physics is to find out **how Nature is**. Physics concerns what we can **say about Nature**." Bib 18 p176

By degrees, the most influential minds in science have changed the very definition of science and its purpose. Suppose you go to your auto mechanic to have your car repaired. The mechanic "says" things about your car, but never identifies what the problem "is." Would the car ever get fixed? Maybe—or maybe not. How can science discover anything in Nature if its task is to *just* 'say' what they 'think' about Nature?

While **not all** *scientists believe this* way, almost all of the educated elite, including influential leaders, professors, heads of universities, professional science organizations and government institutions across the world, teach modernism, denying certainty with the claim that "truth is unknowable." Examples of how science denies truth will appear throughout the Universal Model; the real evidence of it comes from the textbooks, magazines, and literature of the professional science organizations. Most of these have taken a stand against truth by denying its existence, promoting their own agendas instead.

Newton the Truthseeker

Newton once said he was a friend of two of the greatest thinkers of all time—Plato and Aristotle. He gave us the key to his success when he said, "...but *truth* is my greater friend."

Isaac Newton's biography makes it apparent that truth may have quite literally been his "greatest" friend. Newton led a rather reclusive life. He had no father figure, had few close friends, and did not have a family of his own. What he did have was an unquenchable thirst for truth. Newton was a **truthseeker**.

Sir Isaac Newton

Newton lives on today in the world of science because of the laws of Nature he discovered while searching for truth. Newton realized he could not find truth solely in the realm of science and invested significant time searching for truth in areas such as alchemy and theology; it was one of Newton's powerful yet oft overlooked traits. He was human and he made mistakes as we all do, but he believed he could find truths in science, and that attitude led him to many great discoveries.

How do we know Newton discovered a scientific truth? We can apply the *test of time*. It has been over 300 years and we still employ most of Newton's laws (we will discuss this further in Chapter 18, the Mass-Weight Model). They passed the test of time because they work. Today we hold the laws he developed as self-evident.

Truthdeniers

Some people are not truthseekers. We call one who chooses to deny the existence of truth a **truthdenier**—one who denies truth. Because all things have their opposite, truth's is error. Therefore, a truthdenier seeks out error, and the error seeker like the truthseeker, will find exactly what they are looking for.

Many of the influential doctors who opposed Vesalius, Harvey, Pare, Semmelweis, and Lister could have validated the new discoveries, but because of their education and their pride, these truthdeniers could not recognize the truth. On the other hand, many of the men and women who made remarkable discoveries were *neither* doctors nor scientists, but they were truthseekers. The same pattern appears among the people involved with the Universal Model project—truthseekers not necessarily educated by the scientific establishment, applied their objectivity and a fresh outlook into science with surprising results.

Truthseekers will better judge the merits of the UM objectively, whereas the establishment-trained scientist may find it difficult to accept the challenging new UM paradigm. Students and those who seek to understand the truth about Nature and who are willing to perform new experiments, make new observations about Nature and review newly discovered phenomena with an attitude unfettered by the dogma of modern science will become leaders in the new Millennial Science movement. Those Millennial Science students will probably have the opportunity to *teach their teachers* of things their teachers never knew.

3.5 The Big Picture of Modern Science

The Overall Vision of Modern Science

Most people are familiar with the words 'evolution,' 'relativity,' and 'big bang.' They understand that they are the leading scientific theories but are unsure just what they mean or how they fit together. Professional scientists talk about their *part* of the 'Big Picture' but rarely about how their part connects to the other parts of the picture. What is the overall message modern science tries to get across?

To be sure, most scientists specialize in a particular field of study and it is common for them to speak voluminously about theories in their field. When asked about the discoveries and theories from other disciplines, they may not be so sure, because it is not within their field or specialty.

The Big Picture of Modern Science is not clear to the professional scientist, nor is it clear to the general population. Is there a hidden agenda? Perhaps science can best accomplish the purpose of the **Big Picture of Modern Science** (Fig 3.5.1) when no one sees it all together, by not viewing it as a *whole picture*. It is controversial and its foundation defies logic.

Presenting the Big Picture of Modern Science astounds most people when first revealing this doctrine to them. Most professional scientists and modern science adherents believe humans evolved from an ape-like common ancestor that evolved from single-celled bacterial microbes born of a chemical reaction, which began with a Big Bang; which came from nothing billions of years ago. Therefore, we humans and everything else in our Universe came from nothing!

Hearing these words for the first time evokes a look of bewilderment for persons unschooled in the views of modern science. The answer to their question, "Do they *really* believe that?"—which is, of course yes—leaves them perplexed. Yet when we open the science textbooks in today's classrooms, we see countless examples referencing portions of the "Big Picture of Modern Science." Researchers examine pieces of the big picture and then report in science magazines, journals, and in the media, but never the *entire* 'Big Picture' presented here because it seems just too ridiculous. However absurd it may sound, scientists continually try to describe and validate the individual parts of the whole picture shown in Fig 3.5.1.

Science explains the basic premise of evolution theory as microbial life forms evolving into ape-like creatures and ultimately into humans, with every other lifeform in the mix too. However, few understand that Einstein's theory of relativity forms the basis of modern cosmological theory. In the theory of modern cosmology (origin of the Universe), chemical elements (including those in the stars and planets) all came from a Big Bang billions of years ago, and that sudden emergence of all matter—a 'singularity'—came from nothing. When presenting the Big Picture of Modern Science as a whole, inevitably there are those who question how logically minded scientists could

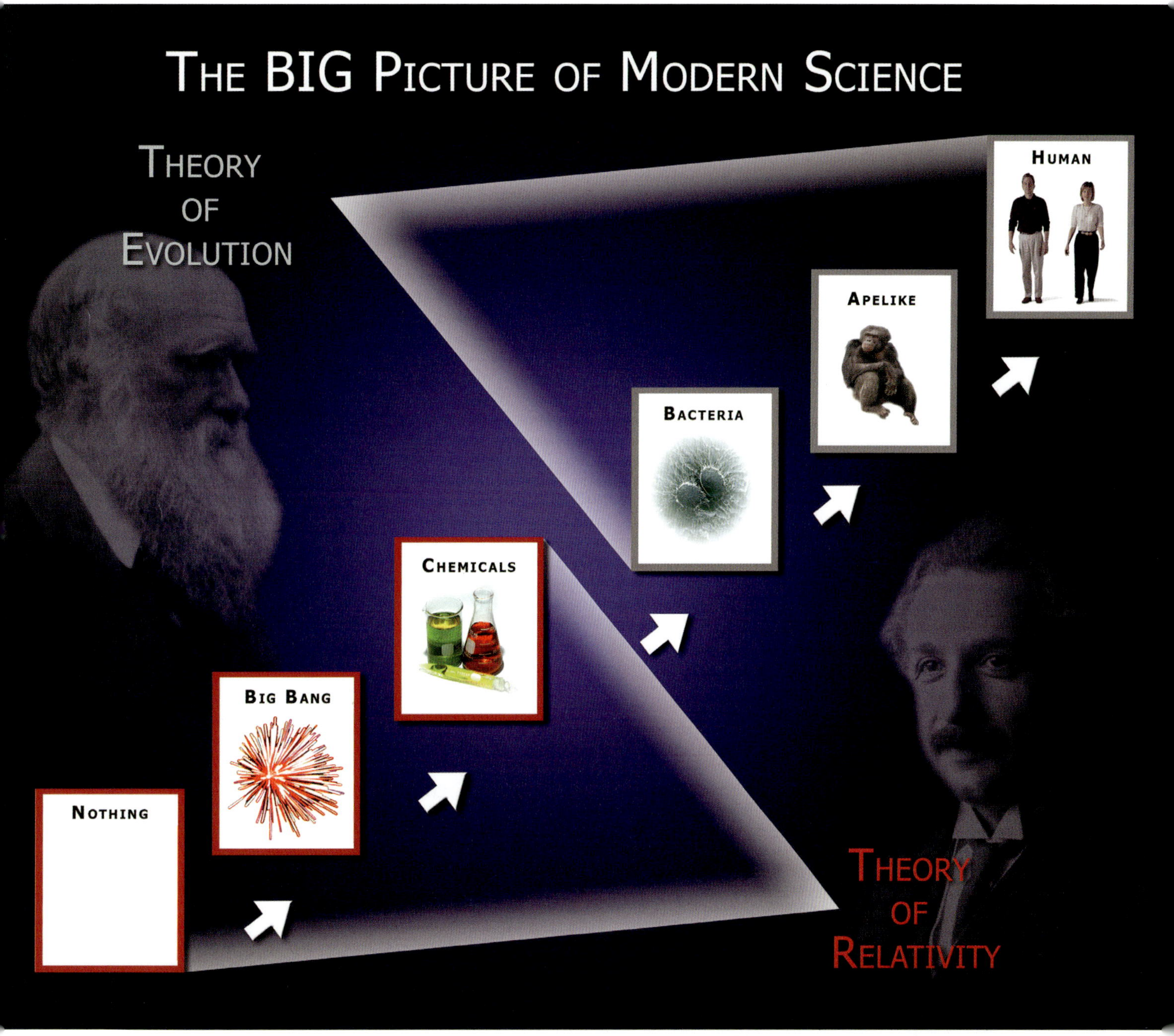

THE BIG PICTURE OF MODERN SCIENCE

Fig 3.5.1 – This diagram illustrates the Big Picture or overall view of the modern science origin of humanity. Through the theory of evolution and the theory of relativity, modern science teaches that human beings came from an ape-like creature, which came from bacterial microbes, which came from chemicals, which came from the Big Bang, which came from a singularity, or simply from nothing. This picture shows the logic of the scientific establishment, but the majority of the people we showed it to knew intuitively that it was not true. The Universal Model demonstrates, with empirical evidence that the Big Picture of Modern Science is false.

SUBCHAPTER 3.5 THE BIG PICTURE OF MODERN SCIENCE

actually *believe* the Big Bang and its origin from *nothing*.

We include statements from scientists on each of the five steps illustrated in the Big Picture of Modern Science to provide a sense of where modern science stands. Anyone can find countless additional quotes throughout the scientific literature, as this represents common thinking today.

Humans from Apelike Creature

"In a series of forms graduating insensibly from some **apelike creature** to man as he now exists, it would be impossible to fix on any definite point when the term 'man' ought to be used... We thus learn that **man is descended from a hairy**, **tailed**, **quadruped**, probably arboreal in its habits, and an inhabitant of the Old World." (Charles Darwin, The Decent of Man) Bib 72 p188, 632

Ape from Bacteria

"'**Everything that's alive today traces its ancestry back to the common bacteria**,' she tells me, pausing so I can consider my lowly origins. 'We always think we came from the apes. But our cells really come from the **bacterial** world.'" Note 3.5a

Bacteria from Chemicals

"The most common scientific theory of how life began on Earth is that sometime after the planet formed 4 billion years ago, the right combination of chemicals and conditions formed a rich soup that eventually led to the emergence of life." Note 3.5b

"...the first 2 billion years of our planet's history saw the **first stirrings of life, when systems of molecules began reproducing themselves and deriving energy from chemicals** and from sunlight." Note 3.5c

"Nature worked hard for billions of years to build complex organisms like the one who wrote these lines and those who will read them, **beginning with simple chemicals** formed on the Earth when it was quite young." Bib 71 p14

Chemicals from Bang

"Matter in the universe was born in violence. Hydrogen and helium emerged from the intense heat of the **big bang** some 15 billion years ago... Once formed, violent explosions returned the elements to the space between the stars. There gravitation molded them into new stars and planets, and electromagnetism cast them into the **chemicals of life**." Note 3.5d

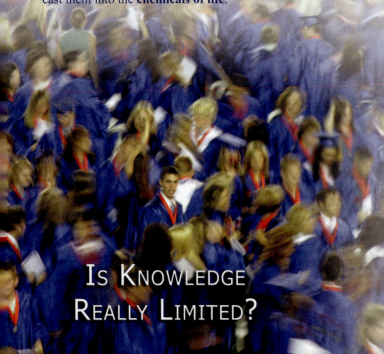

IS KNOWLEDGE REALLY LIMITED?

Bang from Nothing Quotes

"Physicists have the outline of a plausible mathematical theory explaining how a universe can originate from **nothing**." Paul Davis, Bib 31 p52

"About 10 to 20 billion years ago all of the matter and energy in the Universe was concentrated into an area the size of an atom. **At this instant, matter, energy, space and time did not exist**. Then suddenly, the Universe began to expand at an incredible rate and matter, energy, space and time came into being (the Big Bang). As the Universe expanded, matter began to coalesce into gas clouds, and then stars and planets." Note 3.5e

Of course, not all scientists believe that everything came from nothing, but many of the leaders do, especially those in physics and cosmology. Some say the beginning of the Universe was a "singularity," which is described as being smaller than an atom but containing everything in the Universe. Whether a "singularity" or "nothing," both are the same—and *neither* are science!

Imagination vs. Understanding

Who can *understand* the Big Picture of Modern Science? To have understanding, some sense of observation concerning that which we are to understand must exist. Regrettably, for the modern scientist and as the Universal Model will demonstrate, no one has ever observed any of the Big Picture's five steps in figure 3.5.1. This certainly does not follow the Simple Truth Principle—In Nature, Truth is Simple—but it does require a lot of *imagination*.

Try to imagine how an explosion came from nothing. Then imagine how all matter, stars, galaxies, planets and the chemicals found in Nature, came from this and subsequent explosions. Next, try to imagine how living organisms came from non-living chemicals. The imagination leads from microorganisms that randomly mutate into all kinds of animals and plants, with both male and female mutations happening simultaneously, eventually leading to an apelike creature that one must imagine is a great grandparent of the human race. There is only one thing 'clear' about the Big Picture of Modern Science; it requires *a lot of imagination*. No one said it clearer than Sir Isaac Newton:

"'A man may **imagine** things that are **false**, but he can **only understand** things that are **true**.'" Bib 19 p5

What is clear from Newton's statement is that *modern science is not what science once was*. This statement identifies a core principle of the UM and one that we will discuss throughout the book.

Of the main factors driving the Big Picture of Modern Science, the acceptance of Einstein's Theory of Relativity rates highest. In conflict with Newton's wisdom, Einstein popularly effused:

"**Imagination** is **more** important than **knowledge. Knowledge is limited**." Note 3.5f

We know Einstein for his imagination, but as Newton rightly expressed, we can imagine things that are *false* and as the UM will demonstrate, that is exactly what Einstein did. Moreover, knowledge has never been *limited*. Every day, we make new discoveries; we learn new things. It is truly surprising that such statements ever became popular. According to Webster's dictionary, **imagination** is:

"The action or faculty of forming mental images or concepts

of **what is not actually present to the senses**." Bib 7 p407

In subchapter 1.3, we established that the senses represent the means by which we gain wisdom and knowledge. Thus, how can something "*not* actually present" increase comprehension and understanding? In real science, knowledge—the source of understanding—is vitally more important than imagination.

We often confuse the word 'imagination' with intuition. Indeed, intuition, as discussed in chapter 1.6 "propels science beyond the edge of its frontier." Webster's dictionary defines **intuition** as a:

"Direct perception of **truth**." Bib 7 p431

Perhaps this is what Newton meant when he said that we can *only* **understand** what is *true*, but we *may* **imagine** things that are *false*. After all, who hasn't had the experience of imagining unreal situations and otherworldly events? The imagination does not always employ the senses, does not deal with reality, and therefore, may not be true. **Intuition**, on the other hand, is a direct perception of truth. Obtaining new scientific truth is what real science should be about. Imagination is great for science fiction. Unfortunately, for modern science, it has helped create the theoretical-imaginative mess modern science is in.

Intuition: the direct perception of truth.

There is a logical sequence to where the ideology of 'everything from nothing' came from and imagination is at the base of this sequence. The well-known science philosopher Paul Feyerabend said it well:

"It can't just be that the universe — Boom! — you know, and develops. **It just doesn't make sense.**" Bib 20 p55

To "understand" science, it must "make sense," if it is real science. John Wheeler, a distinguished American physicist, got it right when he said this:

"I don't know whether it will be one year or a decade, but I think **we can and will understand**. That's the central thing I would like to stand for. **We can and will understand**." Bib 30 p83

It is illogical to think "we can and will understand" how everything came from nothing or how the Universe could have been packed into a singularity.

3.6 How Popular is Modern Science?

The Bookstore Experience

A number of years ago my wife and I were strolling through a bookstore in the local shopping mall. While she browsed books on gardening, I perused the science section. To my surprise, there were three aisles of science *fiction*, but no sections on the hard sciences, physics, chemistry, astronomy, geology, archeology, or biology. The clerk at the front counter informed me they didn't have a science section. "…I wonder why" she said, "it's sad that so few people are interested in science."

I continued looking and eventually found a few books on minerals and astronomy tucked away on a back shelf, but not much else. The bookstore boasted an entire section on witchcraft/sorcery, besides the three aisles of science fiction! Another bookstore in this relatively large shopping mall had books on many subjects, but it too did not have a 'science section.'

Years of browsing bookstores throughout the United States revealed to me a general lack of interest for books on science as compared with other types of books. Larger bookstores usually had a science section, but it seemed *small* in comparison to other topics. Was I witnessing indifference toward science?

Why is Science so GRIM?

Jere H. Lipps, a paleontologist at the University of California wrote an article titled *The Decline of Reason*? In it, he reflected on the poor rating given Americans when the NSB (National Science Board) conducted a survey on science literacy in 1996:

"Most people, the NSB survey revealed, believe that science is a good thing. But this is based more on their perception that technology and medicine have benefits rather than a clear understanding of how science works or even what it is. **People in general find science grim and seem to fear it**. It works in esoteric ways. It is too difficult and too complicated for an average person to understand. Einstein, the name most frequently associated with science, was a genius. Noble prizewinners, even if we cannot recall their names, are very superior people. Movie scientists are either mad or unintelligible — strange folk at the very least. **Most people think that TVs, VCRs, and computers are science**. Even Time Magazine, in listing the 10 most significant science developments of the year, included **more technological developments than science. Its editors are terribly confused as well. No wonder the average person fails to understand science**. No wonder they don't even want to try! No wonder reason among the common folk is in decline. **Why is it so GRIM**?" Note 3.6a

Many scientists and the highly educated understand that things are "grim" for science as a whole, but they don't know why, except to suppose that "common folk" just don't understand. Is modern science *unpopular* simply because of the theories and concepts it teaches?

In Fig 3.5.1, The Big Picture of Modern Science illustrates two scientists and the theories they directly influenced—evolution and the big bang theory. The basic theory of evolution, popularized from Charles Darwin's book, *The Origin of Species* was published in 1859. The big bang theory, so named by Fred Hoyle in the late 1940s, was an idea developed in the late 1920s and early 1930s, based heavily on Einstein's theory of general relativity. These twin theories added together average more than a century and a half of thought which modern science has tried for generations to convince the populous that they are true. Today, one cannot view a Nature program, or read a Nature magazine without encountering the 'assumption' that these theories are fact. After decades of exposure to both theories, how

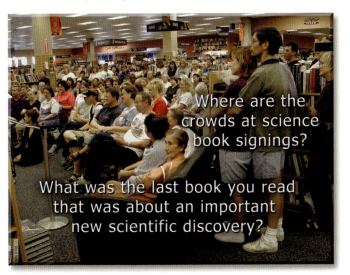

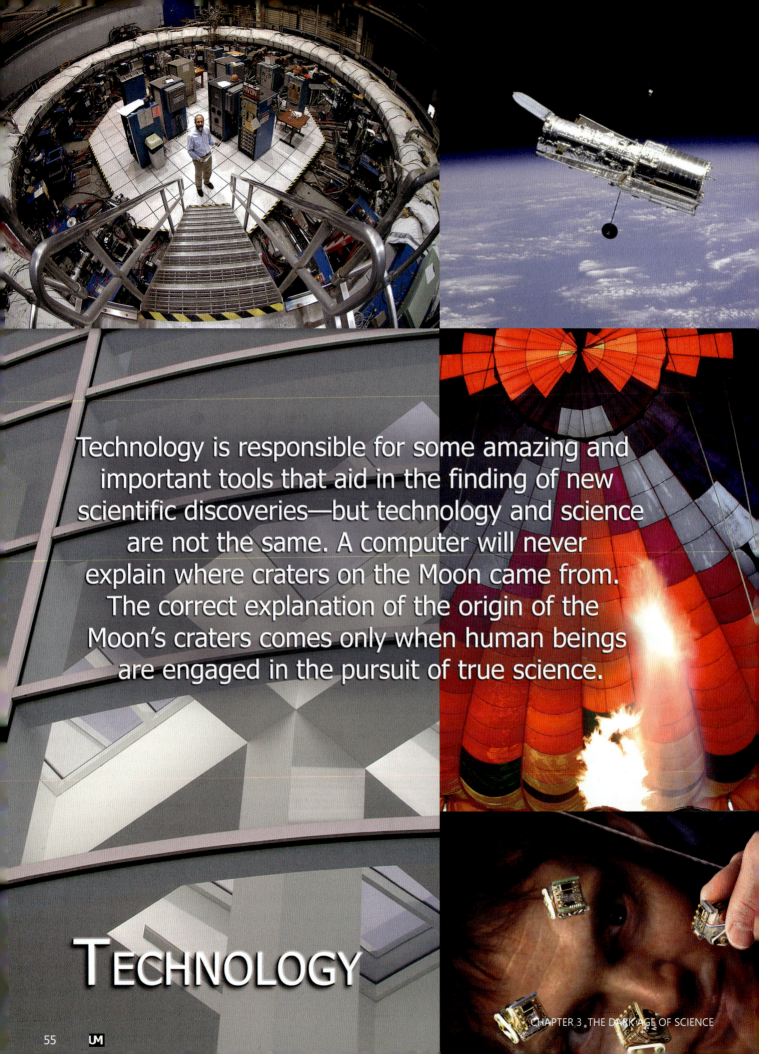

Technology is responsible for some amazing and important tools that aid in the finding of new scientific discoveries—but technology and science are not the same. A computer will never explain where craters on the Moon came from. The correct explanation of the origin of the Moon's craters comes only when human beings are engaged in the pursuit of true science.

TECHNOLOGY

SCIENCE

True science describes and explains Nature using natural laws. In contrast, modern science has fallen into a dark age in the last century by relying on scientific theories that have been unable to correctly describe and explain Nature. This has led to the mingling of technological success with scientific theory, lending an air of credibility as compensation for failures in science.

SUBCHAPTER 3.6 HOW POPULAR IS MODERN SCIENCE?

many Americans believe them? Modern science itself provides the answer.

Every other year since 1972, the National Science Foundation (NSF) has conducted a survey of America's attitude towards science.[2.5b] Approximately 2,000 randomly selected Americans were asked to answer whether a statement was true or false. The survey produced these results on two of the 2006 survey questions:

Statement: The Universe began with a huge explosion.
Belief: 67% of Americans said this was FALSE

Statement: Human beings, as we know them today, developed from earlier species of animals.
Belief: 57% of Americans said this was FALSE

From this study, it is clear that the majority of Americans do not support the theories of evolution and the big bang, which is the Big Picture of Modern Science. These theories represent the 'heart' of modern science and if they do not have the support of the people after decades of almost forced teaching, do we really wonder why modern science remains unpopular.

3.7 Science, Technology and Mathematics

From the citation in the last section, Lipp's article identified that most people did not correctly distinguish between science and technology. What is the difference between science and technology and how are they related? Have both experienced equal success? These are important questions and the answers will help clarify why there is a Dark Age of Science.

How important is math to science? How much math is required to understand science? Can we hinder scientific advancement by using too much mathematical modeling? Science intimidates many folks because modern science connects complicated math with science. Of course, it is necessary to have a fundamental knowledge about basic mathematical operations and it is essential to have at least a rudimentary knowledge of geometry because both provide the means to explain data and observations. However, the answer as to how much math is too much may come as a surprise.

Science and Technology are NOT the Same

We previously established the fact that some key, modern science theories are not held in high regard by the majority of the people. Leaders recognize the difficulty in getting the public excited over natural science but they also understand the importance of gaining their support. Popularity drives the funding for new research, a multi-billion dollar proposition in America today. How do you convince the public to fund unpopular theoretical research? Over the past several decades, there has been a growing trend to group science and technology into the generic category of 'science.' This largely successful effort has effectively blurred the line between them and few people recognize their important *differences*. One difference remains significant—*science has not enjoyed the success that technology has.*

We recognize one example in the bestselling American science magazine *Scientific American*. During the last two decades, it has become increasingly difficult for *science* journals and magazines to fill their pages with purely scientific discoveries and stories to maintain readership. On the other hand, *technological* journals and magazines have grown exponentially, graced with abundant technological discoveries and stories. *Scientific American* recognized this, and, several years ago, the magazine announced to its readers it was changing its format to include new sections dedicated to technology.

In one instance, they presented twenty *World Changing Ideas* in the December 2009 issue, to "build a cleaner, healthier, smarter world" (page 50). Of the twenty ideas, only three were *science-based*. The remaining seventeen were *technology-based*. This ratio of 3/20 science-based ideas is probably typical of the science-verses-technology circumstance that exists today. There are many more successful technological inventions than there are new scientific discoveries. Most new scientific discoveries (such as those in astronomy and biology) are often a direct result of technological advances in telescopes, DNA testing, and other equipment and procedures. Often times, these new techno-related discoveries contradict existing theories, a topic we approach in later chapters.

In another text, the "*power* of science" is mingled with "technologies":

"The **power of science** cannot be denied: it has given us **computers** and **jets**, vaccines and thermonuclear **bombs**, **technologies** that, for better or worse, have altered the course of history."
Bib 20 p4

Computers, jets, and bombs are *not* science, but are technological inventions. We can find examples like this, where science and technology incorrectly mingle outcomes in almost every science related book, magazine, or newspaper.

In order to separate these two more clearly, the UM recognizes the **Science-Technology Principle**:

We base Natural laws on **science**.
We base inventions on **technology**.

Another reason science and technology often get lumped together is that the experts define **technology** as *applied science*. However, many new technologies *cannot* explain why or how the invention works—it just does. The following quote from a water conditioning company describes the "substantial research" behind the "science" of their product:

"There has been **substantial research** done to determine the **science** behind the process of electronic water conditioning."
Note 3.7a

Despite all the "research" that they claim happened, no one knows the correct science, nor has science driven the development of the system. Admitting that they did not know what happened behind the scene, they said:

"…may **not** be fully understood at this time. **The key thing is that the process works!**" Note 3.7a

Science does not understand why a little magnetic invention reduces 'hard' water because they do not understand the origin of hard or calcified water itself. We explain this in detail in the Carbonate Mystery (subchapter 6.9) and answer questions about it in the Carbonate Mark, (subchapter 8.8). We explain the calcified water has a microbial origin, one that is not strictly chemical or geological.

For many people modern science remains grim because it cannot correctly explain Nature, while the popularity of technology persists because stuff works! People will not buy a product that does not work and for this reason, technology experiences greater success than does science. Technology is a capitalistic driven proposition; if a product does not produce the predicted results, it will not sell and no one will invest any further. When we press the power button on the TV remote, we expect the TV to turn on. If an automobile is unreliable and works only part of the time, we would abandon it in favor of a better one; that is the nature of technology.

Although science and technology are not the same, we should hold both to the same standard to measure their success—both must work! Today, nothing requires success in science; funding comes even when science fails to explain Nature. Of course, there are many opportunities to misuse or misapply scientific funds.

Another reason for combining science and technology is the title bestowed on the people working within both sectors. We know these men and women universally as scientists, but it is very important to distinguish between the two. We'll refer to those people who work with technology as **technologists**. Computer 'science' and rocket 'science' are misnomers and should instead be dubbed *computer technology* and *rocket technology*.

A technologist is a person who works with technology.

Understanding that science and technology are not the same is *vital* if we are to increase knowledge and wisdom about Nature. Finding the answers to Nature's fundamental questions is not the same as inventing a faster airplane. Technology plays an important role in *helping* to discover the answers to Nature's questions and science may assist the technologist, but technology is ***not*** science.

Technology is a powerful tool but we should never confuse technological advances with discovering how and why Nature works. It is important to remember that natural laws come from science—inventions come from technology.

Is Mathematical Rigor Essential in the Search For Truth?

David Lindley wrote a book he called *The End of Physics—The Myth of a Unified Theory*. In his book, he made this excellent statement concerning mathematics and science. He added, what we must do, if science is to "work properly":

"The lure of mathematics is hard to resist. When, by dint of great effort and ingenuity a previously vague ill-formed idea is encapsulated in a neat mathematical **formulation it is impossible to suppress the feeling that some profound truth has been discovered**. Perhaps it has, but **if science is to work properly** the idea must be tested and thrown away if it fails.

"Not all of science is mathematical. Not everything mathematical is science. Yet there persists in theoretical physics **a powerful belief that a sense of mathematical rigor and elegance is an essential, perhaps overriding element in the search for the truth**." Bib 21 p13

Indeed, mathematical rigor plays an essential part in the search for truth. Without mathematics, we would find it difficult, even nearly impossible, to obtain measurements and describe physical attributes such as mass, length, and time of an object or its relationship to other objects. Scientific theories that lack a mathematical component may not be testable and risk not having the sound reasoning mathematics can contribute.

In the book *The Truth of Science* the author commented how theories register as often controversial because of their inapplicability to the real world and because those theories are not "based on reliable mathematical reasoning":

"For other sciences, model building is also very common; it is particularly **the theory of evolution** that invites such efforts in biology. These models **are often controversial**, not only because **their applicability to the real world is not well established**, but because the conclusions they lead to may **not be based on reliable mathematical reasoning**." Bib 18 p69

In some cases, a lack of mathematical rigor and reasoning leaves many decades-old theories mired in ambiguity.

Too Much Math in Science Can Be Harmful

As in almost everything, too much of a good thing can make it bad—the fire cooking our food can burn our house down. In today's scientific world, physicists have formulated complex mathematical equations to build theoretical models only the 'elite' can understand. The theories of relativity and quantum mechanics are two examples where mathematics is the primary method of 'proof' instead of physical observation. By itself, mathematics will *never* prove the truth of a scientific theory. Regardless of how 'elegant' the numbers may appear, proof must come from observation and empirical evidence. And others must be able to evaluate and substantiate those observations and evidences.

Science regularly uses complicated math to describe things we do not understand.

Applying the Simple Truth Principle, we do not need tricky math to understand Nature. An understanding of calculus is not required to read the Universal Model nor should it be necessary to comprehend how and why things are the way they are in Nature. Nature is beautifully simple and simply beautiful *when we understand* and *comprehend it*. The UM will demonstrate that the principles and laws of Nature are simple, once they are known.

Natural science is not 'rocket science'.

In the end, mathematics plays an important role as scientists work to explain and quantify observations, make analysis, draw conclusions, and share the results. We should subject theories to mathematical rigor, but not go so far as to build theories or models solely on a foundation of math.

3.8 A Final Review of Modern Science

Exposing Modern Science For What it Is

From the inception of the Universal Model project, some reviewers felt that the UM did not make a strong enough emphasis about the false theories that are taught as fact in science today. Others expressed frequently that the UM seemed too negative towards modern science in general. The majority of the reviewers held this opinion as they began their evaluation, but as they continued their reading, their views underwent a marked change as it became obvious to them that there was indeed, a *scientific agenda*.

David Berlinski a well-known writer and professor of mathematics and philosophy sums up well the agenda of modern science:

"A great many scientists are satisfied that **at last someone has**

said out loud what so many of them have said among themselves: *Scientific and religious belief are in conflict.*

"*They cannot both be right. Let us get rid of the one that is wrong.*" Bib 174 p3-4

Modern science views religion as a competitor in the realm of intellectual advancement and has itself drawn a hard line.

Rather than focus on the discovery of scientific truth and natural law, the "end points" of modern science are theories that attempt to eradicate all belief in deity. The views of the scientific elite and modern science's agenda are very often diametrically opposed to the views of the general populace. To learn more of the details about this scientific agenda and to understand why we find it necessary to hold such a hard line against modern science, review some of the science leaders' statements will be included in subchapter 12.10, the Agenda.

The most difficult challenge the UM project faced had nothing to do with making and understanding new scientific discoveries. It was the challenge of questioning the status quo and sharing new ideas with other scientists who presumably were interested in finding truth. Being a truthseeker made it difficult to interact with intellectuals uninterested in knowing the truth or with those deluded souls who think there is no truth. After studying these kinds of people and their research for more than two decades, it became clear that many false scientific theories had inflicted serious consequences on society.

The vast majority of scientists are hardworking, honest, and intelligent people, but few will be able to lay aside years of indoctrination to deliberate on what they think are ridiculous extraordinary claims in the Universal Model. We already have found this in most situations.

We wrote The Universal Model for the layperson, to dispel the falsehoods that led to science's Dark Age. It is poised to encourage a great debate as to the future purposes of science and to foster the discovery of new truth. The public will have the last word in this debate.

Four Categories of People and Science

Throughout this text, reference is made to the **public** and to **scientists**. The term 'scientist' is a general term referring to pure research scientists and to the university professors. This group represents less than 1% of the population yet wields one of the most influential positions in society—the teaching of future generations.

The teaching of false theories as if they were fact and the proliferation of false philosophies, also taught as fact, has been a source of great confusion, and has allowed many incorrect ideas to take root, especially among the biological and social sciences. No educated society in the world is immune to these false teachings. When society allows the teaching of false scientific ideas and philosophies to the rising generation, it is only a matter of time before those and other false ideas creep throughout society and corruption becomes commonplace. The degeneration of many of the world's societies is evident today.

To comprehend further how the public and scientists have interacted in the past, we separated each of these groups into groups or categories so that the reader can see clearly how and why the ongoing drama of science has happened historically. This wisdom will help prepare us for what will take place in the future.

1. **Followers** – This group represents the largest segment of the population, and it includes the majority of the educated. Followers received a basic science education and generally have no reason to question what they were taught. They accept (or adopt) the teachers' "trust me" attitude because they are the authority figure. As a student, they have no incentive to question the foundations of what they learn and have little interest in pursuing truth as based on their own empirical observations. With the new discoveries in the UM, Followers will have to decide whether to follow the false theories of the past, or to accept new natural laws.

2. **Truthseekers** – This is the second largest group made up primarily of the general populous. These are they who seek truth regardless of past training. Truthseekers question what they are taught and are willing to lay aside scientific theories that lack evidence, even highly favored ones. Truthseekers tend to have a deeper background in science. Many scientists are Truthseekers and hopefully, many more will join their ranks. This group will take up the task of transforming the Universal Model into a new Millennial Science.

3. **Sentinels** – This group is comprised of scientific leaders, professors, well-known research scientists, and others who have reputations to uphold and who have a financial stake in the acceptance or rejection of scientific theories. Long tenured and highly respected, the Sentinels of science passionately defend their fields of science and the foundation upon which it is built. For years, they have accepted status quo and have integrated with other fields, building a net of false security. Sentinels who choose to ignore new, verifiable scientific truth become truth-deniers. No longer true scientists, the Sentinels become philosophers, perhaps unaware they have been led astray by the Deceivers.

4. **Deceivers** – The Deceivers are those few scientists that are aware of actual research and experiments and *know* there is no evidence to support the false theories, but still teach them as though they were fact. In some instances, Deceivers have committed fraud. This *small* group of individuals encompasses the authors of the modern science agenda and are the ushers of the Dark Age of Science. Some have changed the very purpose of scientific discovery while others have been instrumental in persuading the public into believing well-known theories are fact. Although we can find Deceivers throughout the various fields of science; the Age Model, Evolution Pseudotheory, World History Model, and the Relativity Pseudotheory chapters expose a number of these individuals.

It is the goal of the UM to expose the false theories and traditions of modern science and to herald an era of freedom in the search for scientific truth. We must bring the era of the Deceivers to an end. The level of deception and the perpetration of lies demands that we take a hard stand against modern science, and this book, with new empirical evidence and observations, will do just that. The UM has been written to be measured on its own merits and is here for all to review.

Will Modern Science End?

Interestingly enough, some science writers have written books addressing just such a question. When asked, "Where is science going?" some expressed a belief that modern science has all the major discoveries about Nature in their possession and all that is left is to mop up the details.

In 1996, John Horgan, a popular science writer, wrote the

book *The End of Science*. Among his interviews with some of the leading scientists of the day, we can gain some insight into the mind of the modern scientist. In his assessment, Horgan expresses his opinion about modern science's knowledge:

"My guess is that this narrative that scientists have woven from their knowledge… will be as viable 100 or even 1,000 years from now as it is today. **Why**? **Because it is true**. Moreover, given how far science has already come, and given the physical, social, and cognitive limits constraining further research, science is unlikely to make any significant additions to the knowledge it has already generated. **There will be no great revelations in the future comparable to those bestowed upon us by Darwin, or Einstein or Watson and Crick**." Bib 20 p16

Horgan echoed the feeling of many of the scientists he interviewed—because they believe that modern science in general "is true," and that "there will be no great revelations in the future…"

The UM stands as a testament against the belief that the future holds no new great revelations.

Roger Newton, former physics professor at Indiana University expressed the confidence also shared by many scientists as he reflected on the stability of science:

"While it is a great oversimplification to say that theories and laws are finally based on independently verifiable facts, it is nevertheless the case—and this needs emphasizing—that in physics and in most other areas of science the combination of laws and facts, **theory-laden though many of the latter may be**, **has an enormous amount of stability**. Just as the stability of a large building depends not simply on the firmness of its corner posts but is due in large measure to the ubiquitous cross-bracing in its walls and floors, so the structure of science is secured by the dense network of interdependencies that are established among its various parts. Pointing out, when necessary, that **the evidence for certain specific parts of science may be weak** is an important function of scientists and knowledgeable commentators, but has little bearing on the **safety of the edifice as a whole**." Bib 18 p100

Is this true? Does modern science with all its theories have an "enormous amount of stability?"

Unfortunately, Roger Newton like many researchers has a false sense of security. Even though they may acknowledge areas of weakness, the whole of modern science could only be stable and secure *if* the Big Picture of Modern Science were correct—that everything came from nothing. Nevertheless, nothing comes from nothing, not everything. This is the *philosophical* argument that we find modern science making with Darwinism and Relativity which is false, and the scientific evidence presented here in the UM shows it.

Could there be a different end to modern science? One way to answer this question is to ask whether modern science is fulfilling its purpose. If it is not, *it is only a matter of time* before it must end as we now know it.

Is modern science fulfilling its purpose of describing and explaining Nature to enable us to understand and comprehend it?

The majority of the public will find that 'modern' science is not fulfilling the true purpose of *science* and is indeed ripe for a revolution.

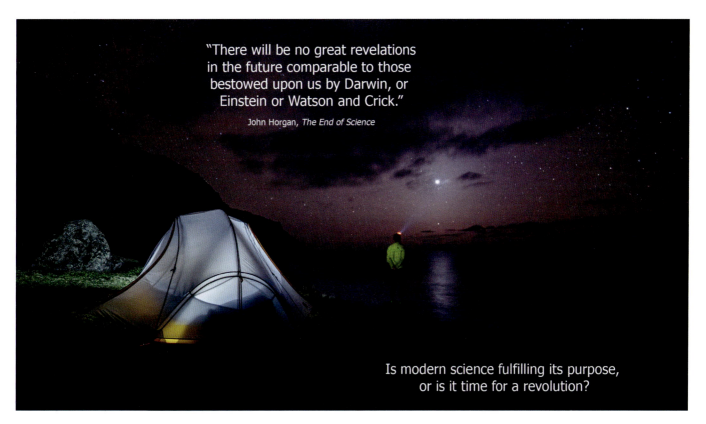

"There will be no great revelations in the future comparable to those bestowed upon us by Darwin, or Einstein or Watson and Crick."
John Horgan, *The End of Science*

Is modern science fulfilling its purpose, or is it time for a revolution?

4 Scientific Revolutions

> "Led by a new paradigm, scientists adopt new instruments and look in new places. Even more important, during revolutions scientists see new and different things when looking with familiar instruments in places they have looked before…"
>
> Thomas S. Kuhn

4.1 What is a Scientific Revolution?
4.2 The Steps for a Scientific Revolution
4.3 Why Will a Scientific Revolution Take Place?
4.4 From Where Will the Next Revolution Come?

Scientific Revolutions shape the world in which we live, they represent an integral part of the history of science, and in this chapter, we talk about how important they are. The last Scientific Revolution happened a long time ago; it's very likely no one knows what a revolution in natural science even looks like. Revolutions in *technology* happen frequently, and just about everyone can relate at some level. The computer, the internet, smart phones, and drones each revolutionized the world, imposing permanent change as they disrupted societal norms. When will we next experience a Scientific Revolution?

This chapter answers this and other Scientific Revolution questions, because a revolution is coming—a Universal Model revolution!

4.1 What is a Scientific Revolution?

Thomas S. Kuhn & Scientific Revolutions

Well-known scientific historian and philosopher, Thomas Kuhn wrote *The Structure of Scientific Revolutions* in 1962, explaining the process of how science advances, or why it does not. Gaining surprising notoriety, Kuhn's book became a standard when discussing scientific progress. Now available in 16 different languages and at over a million copies sold, it remains required reading for many science history courses. In his book, Kuhn coined the now common word '**paradigm**' to define a collection of beliefs that form a model or a perspective of any subject. It represents consensus thinking in a general sense.

Figures 4.1.1 and 4.1.2 illustrate two ways in which science advances, 'Linear Science' or 'Scientific Revolutions.' Both include the discovery of scientific knowledge and an increase in wisdom, including new natural law. Linear scientific advancement implies a constant, steady increase of knowledge based on discovery. Today, most textbooks portray modern science advancing in knowledge based on a steady linear progression. The Science Revolution process illustrates discovery at significant points along the way, which brings rapid and significant change only occasionally, each time raising the aggregate of human knowledge through a **paradigm shift**. Kuhn reveals that the development of natural science coincides with these paradigm shifts, which he calls ***revolutions***:

"By disguising such changes, the textbook tendency to make the development of science **linear** hides a process that lies at the heart of the most significant episodes of scientific development." Bib 33 p140

CHAPTER 4 SCIENTIFIC REVOLUTIONS

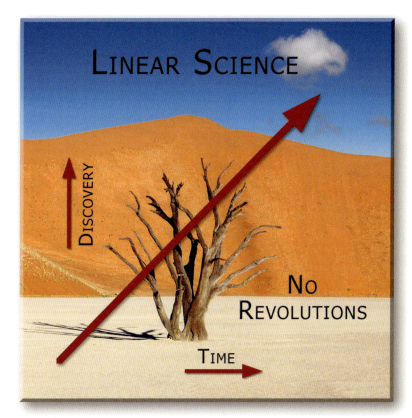

Figure 4.1.1 – Many believe scientific advancement happens incrementally, a little at a time, a method called Linear Science. However, evidence suggests that scientific learning happens primarily in sudden paradigmatic shifts (see Fig 4.1.2) when new learning or new discoveries revolutionize current thinking.

Kuhn mentioned some of the major scientific revolutions in his book:

1. Copernican Revolution

Nicolaus Copernicus gathered evidence to show that the Earth travels around the Sun in a planetary system. This heliocentric cosmology displaced the idea that the Sun occupied the center of the Universe.

2. Newtonian Revolution

Isaac Newton demonstrated laws about motion and gravity; he also developed a model of light, designed the reflector telescope, and created calculus. Many people consider Newton one of the most influential people in human history. We know him for bringing scientific principles into 'universal' acceptance through experimental demonstration.

3. Chemical Revolution

Three individuals responsible for our understanding of air and water, Cavendish, Priestley, and Lavoisier, helped discover many now-common elements.

When discussing scientific revolution in his book, Kuhn related how a change in paradigm must happen for a revolution to occur:

"Examining the record of past research from the vantage of contemporary historiography, the historian of science may be tempted to exclaim that **when paradigms change, the world itself changes with them. Led by a new paradigm, scientists adopt new instruments and look in new places. Even more important, during revolutions scientists see new and different things when looking with familiar instruments in places they have looked before**. It is rather as if the professional community had suddenly transported to another planet where **familiar objects are seen in a different light and are joined by unfamiliar ones as well**." Bib 33 p111

What happens after a paradigm shift? Kuhn explained the development of science between paradigms:

"In the development of any science, the first received paradigm is usually felt to account quite successfully for most of the observations and experiments easily accessible to that science's practitioners. **Further development, therefore, ordinarily calls for the construction of elaborate equipment, the development of an esoteric vocabulary and skills, and a refinement of concepts that increasingly lessens their resemblance to their usual common-sense prototypes. That professionalization leads, on the one hand, to an immense restriction of the scientist's vision and to a considerable resistance to paradigm change. The science has become increasingly rigid**." Bib 33 p64

Kuhn recognized that between paradigm shifts, science becomes rigid and resistant to change, which is precisely what happened to modern science during the last century, during the Dark Age of Science.

When the next new scientific revolution happens how will important new discoveries in astronomy or geology affect the average person? There are many incremental discoveries today, and sometimes they are of some significance, but these do not constitute a revolution.

However, hundreds of new discoveries and extraordinary claims spanning the scientific spectrum would definitely represent a **universal revolution**.

English mathematician J. Bronowski noted that there was a time in history where the world experienced a "universal revolution" that "reached into all forms of culture":

"About 1660, therefore, Europe was in the course of a great revolution in thought. This was the **Scientific Revolution**, and **it reached into all forms of culture**... The Scientific Revolution in the seventeenth century was a **universal revolution**." Bib 30 p21

One can easily see how the automobile, the computer, and the smart phone changed people's lives in the developed world. However, the revolutions spawned by or related to these are *technological revolutions*—the industrial and information revolutions and the social media revolution—they are *not* natural science revolutions. *The world has been so long without a major natural science revolution that neither the scientist nor the layperson expects to experience one.* When it does happen, new understanding about Nature will catalyze new technological advances, but even more importantly, our perception of the world and the Universe will change.

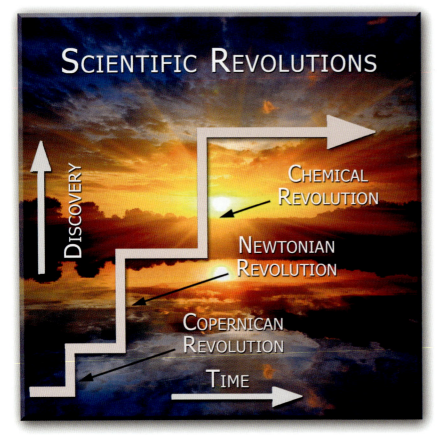

Figure 4.1.2 – Paradigm shifts in science encompass the revolutions of the past and future, destined to change the entire scientific landscape. These Scientific Revolutions or paradigm shifts bring new explanations and change our understandings of the Universe.

4.2 The Steps for a Scientific Revolution

In Kuhn's book on scientific revolutions, he lists four steps that happen on that path:

1. Crisis
2. Claims
3. Demonstration
4. Revolution

1. Crisis

A Crisis represents the first step in the scientific revolution process. While most scientists may experience a level of confidence with current modern science theories, there are signs of an impending crisis. In the following examples from various fields, we read "the most basic assumptions" are "open to question." This example comes from the article *Crisis in the Cosmos*, published in *Discover* magazine:

"Too Little mass, too little time – either problem alone would be disturbing. Taken together they raise the specter of a **scientific revolution, a shift in the cosmological world view** in which some fundamental assumptions in cosmological theory – perhaps even the Big Bang itself – will have to give…

"And no one, it seems clear, has good explanations right now, not for any of the problems that bedevil cosmology, and certainly not for all of them at once. The field is in a troubled state, a disconcerting or an exciting one, depending on your personality – **a state in which even the most basic assumptions seem open to question**. Maybe the microwave background has nothing to do with the Big Bang after all but is due to some entirely different phenomenon. Maybe redshifts aren't really due to recessional velocity, and the universe isn't expanding. Maybe we don't really understand gravity, which would throw all cosmological theories into the trash – even the Big Bang itself. Few cosmologists think that's at all likely. But no reasonable cosmologist would claim that the Big Bang is the ultimate theory. With enough contradictions and inconsistencies, **it could eventually be overthrown**…" Note 4.2a

In future chapters, we actually do discover that the microwave background has nothing to do with the Big Bang after all, redshifts are not actually due to recessional velocity, the Universe isn't expanding, and modern science doesn't really understand gravity. All of these actually *does* "throw all cosmological theories into the trash—even the Big Bang itself."

The field of cosmology is not the only field of science in a "troubled state." The following account in *Scientific American* reflects information obtained from the Hipparcos satellite, decrying the state of astronomy:

"Some observers, on the other hand, aren't so quick to pronounce the age paradox as solved. Rather they are suggesting that **the Hipparcos's most profound result is to show that scientists don't understand stars very well at all**.

"One of the more cautious is the European Space Agency's Michael Perryman, project scientist for Hipparcos. 'I would be reluctant to accept any of these results as the final word, there's too much massaging to get things to fit.' None of the models, he says, accurately explain all the observed properties of stars and thus do not inspire great confidence. Worse, Hipparcos data have shown that some stellar models—including those that seemingly encompass the sun—**are spectacularly wrong**." Note 4.2b

These remarks leave little doubt that much uncertainty exists in the scientific community. Although the scientist might be the last to admit of a crisis in science, especially in his or her field, it is not difficult for the science writers to note the uncertainty of the scientists they interview. Thomas Kuhn points this out in his discussion about the history of science and its development. He says the "awareness of anomaly"—the observance of the abnormal or unusual—in Nature precedes crisis, and he claims there is "a period of pronounced professional insecurity" amongst the scientists:

"If awareness of anomaly plays a role in the emergence of new sorts of phenomena, it should surprise no one that a similar but more profound awareness is prerequisite to all acceptable changes of theory. On this point historical evidence is, I think, entirely unequivocal. The state of Ptolemaic astronomy was a **scandal** before Copernicus' announcement. Galileo's contributions to the study of motion depended closely upon difficulties discovered in Aristotle's theory by scholastic critics. Newton's new theory of light and color originated in the discovery that none of the existing pre-paradigm theories would account for the length of the spectrum… Furthermore, in all these cases except that of Newton **the awareness of anomaly had lasted so long and penetrated so deep** that one can appropriately describe the fields affected by it as in a **state of growing crisis**. Because it demands large-scale paradigm destruction and major shifts in the problems and techniques of formal science, the emergence of new theories is generally preceded by **a period of pronounced professional insecurity**. As one might expect, that insecurity is generated by the persistent failure of the puzzles of normal science to come out as they should. Failure of existing rules is the prelude to a search for new ones." Bib 33 p67

This book contains literally hundreds of examples like this, identifying and discussing the "anomalies" in every major science field.

Current theory does not match current observation.

Because current theory does not match current observation, the undercurrent of crisis in the scientific community continues to flow even though most researchers think all is well in their individual fields. When science goes for decades without the discovery of new natural law, it reflects an obvious problem.

We already read of direct evidence of the impending crisis in subchapter 3.5 with the Big Picture of Modern Science, revisited in Fig 4.2.1. As we dig into the festering underbelly of the enormous black money hole through which science has poured billions of taxpayer's dollars in an attempt to discover what happened in that very first moment when everything—including you—came from nothing, we realize that the looming crisis is a whole lot bigger than anyone ever imagined. Whether or not many see it, the impending crisis will change everything in modern science today.

A theory is only an idea, unless it can be proven.

2. Claims

Every science field has its theories, but a theory is only an *idea* unless the claimant can prove it. Of the infinite number of ideas one can invent, we can divide all of them into two groups—testable and untestable claims. Many modern science theories fall into the untestable category. Kuhn explained that for a claim to be effective, it must be testable and it must *solve* the problems that led to the crisis.

"Probably the single most prevalent **claim** advanced by the proponents of a new paradigm is that they can **solve the problems** that have led the old one to a crisis. When it can legitimately be made, **this claim is often the most effective one possible**. In the area for which it is advanced the paradigm is known to be in trouble. That trouble has repeatedly been explored, and attempts to remove it have again and again proved vain. '**Crucial experiments**'—**those able to discriminate particularly sharply between the two paradigms**—have been recognized and attested before the new paradigm was even invented…

"Claims of this sort **are particularly likely to succeed** if the new paradigm displays a quantitative precision strikingly better than its older competitor." Bib 33 p153

Kuhn further explained *how* "crucial experiments" affect a paradigm shift. Paradoxically, we might discover a shift unnecessary had we always experimented and approached science in this way. In fact, proper experimentation and evaluation may have prevented the onset of science's Dark Age.

One unique and overarching Universal Model claim is that modern science continues in a state of crisis as it has for some time, and that *all* fields are ready and poised for revolution. This will be the first time in history that a *universal* scientific *model*, including new crucial experiments and observations that are attainable by both scientists and laypersons, will revolutionize all of science.

3. Demonstration

The next step in a scientific revolution is the *demonstration*. In the UM, the Universal Scientific Method (USM) embodies the demonstration step. The first step in the USM described in chapter 3 involves presenting a *theory*. That new theory must *predict* phenomena, says Kuhn:

"…particularly persuasive arguments can be developed if the new paradigm permits the **prediction of phenomena** that had been entirely unsuspected while the old one prevailed." Bib 33 p154

The steps following prediction include *testing* and *observing* the results, followed by *evaluation* of the data. Where applicable, new natural *laws* appear, formulated by repetitive testing and observation. A new natural law demonstrates the order in Nature.

One can judge the new science in the UM by evaluating the data and demonstrations it presents. Individuals can observe the natural laws and compare both the new and the old paradigm:

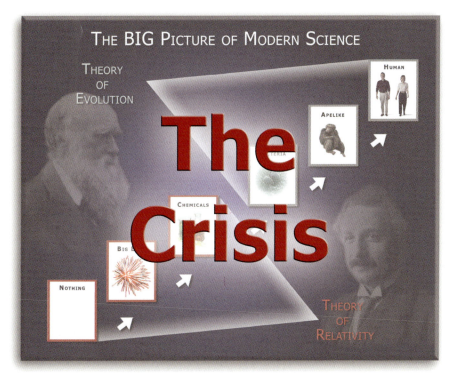

Fig 4.2.1 – A Crisis is the first step necessary for a scientific revolution to occur. Today's "Crisis" is evident by the acceptance as fact, the two theories of *The Big Picture of Modern Science*. Individual chapters in the UM will explain these theories and why they were accepted as fact, even though neither of them has ever been demonstrated to be true, using empirical evidence.

"... the **act of judgment** that leads scientists to reject a previously accepted theory is always based upon **more** than a comparison of that theory with the world. The decision to reject one paradigm is always simultaneously the decision to accept another, and the judgment leading to that decision involves the **comparison of both paradigms with nature *and* with each other**."
Bib 33 p77

4. Revolution

This last step in the scientific revolution is the revolution itself. The general public often assimilates newly demonstrated scientific truth, but the professional scientist must feel great apprehension, especially if the scientific revolution challenges his or her own field. The physicist, Roger Newton, commented about the scientists' response to such a paradigm change:

"It is possible, of course, that a startling, discordant fact, long ignored by the '**establishment**,' may someday be discovered, producing a new paradigm with its own coherence. The body of scientific knowledge, however, is by now so large that this scenario is **extremely unlikely**, and examining seriously and carefully every assertion that falls far outside the coherent web of scientific learning would hardly be productive. **An entirely new effect would require an enormously convincing demonstration to persuade scientists to pay attention, but if the effect is real, it will eventually be embraced. Overcoming the natural conservatism of scientists is difficult but not impossible**." Bib 18 p208

The well-educated scientist would consider it "extremely unlikely" that the "establishment" could have missed any discordant fact—let alone hundreds of them. It will, no doubt require an "enormously convincing demonstration" for the scientist to pay attention.

However, *if* the scientists can see the *crisis* in science, and *if* he or she considers the extraordinary *claims* that are coupled with extraordinary evidence, then the scientists' natural conservatism might be overcome and they will witness a *scientific revolution*.

Kuhn explains that mainstream science and education must assimilate the new evidence, discoveries, and observations; they must displace the old theory with the new scientific model. Moreover, the new model must account for and explain natural phenomena better than previous theories or models:

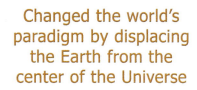

Copernican Revolution

Changed the world's paradigm by displacing the Earth from the center of the Universe

Nicolaus Copernicus

"After the discovery had been assimilated, scientists **were able to account for a wider range of natural phenomena or to account with greater precision for some of those previously known**. But that gain was achieved **only by discarding some previously standard beliefs or procedures and, simultaneously, by replacing those components of the previous paradigm with others**." Bib 33 p66

As we understand that scientific revolutions include both the assimilation of new discoveries and the ability to account for and explain, with greater precision, a wider range of natural phenomena, our ability to ask new questions increases.

4.3 Why Will a Scientific Revolution Take Place?

Lack of Simple Unanswered Questions

Developing the UM meant asking countless questions to a variety of scientists; while some answers proved helpful, many came back clearly indicative that the scientists could not adequately explain or clearly articulate accurate answers. Instead of simply saying they did not know, the scientists' often provided explanations with the color of fact when in reality, the explanation was only an opinion.

We found it amazing during the development of the UM that many nonscientists recognized the same behavior. In fact, many people with whom we spoke shared with us a sense that modern science's theories lacked integrity, which in turn affected their belief about those theories.

If modern science focused on discovering and teaching demonstrable natural laws instead of untestable scientific philosophy, science might have continued on its path uninhibited. However, by expressing the dictum that "Truth cannot be known" along with the philosophy that humans descended from an unknown ape-like creature, which sprung from simple bacteria, which arose from an unknowable chemical soup, which materialized in an explosion of essentially nothing, modern science's bizarre path must end.

Time For a Scientific Change

Over fifty years ago, in 1960, the experts developed a 'philosophy of science,' which had the misfortune of becoming mainstream modern science

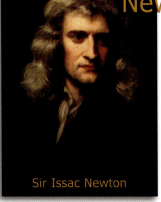

Newtonian Revolution

Revealed new laws of motion, gravity and changed our paradigm with respect to light and the solar system

Sir Issac Newton

CHAPTER 4 SCIENTIFIC REVOLUTIONS

thought. Karl Popper, who helped establish this philosophy later wrote:

"For us, therefore, **science has nothing to do with the quest for certainty or probability or reliability**. We are **not** interested in establishing scientific theories as **secure, or certain, or probable**. Conscious of our fallibility **we are interested only in criticizing them** and testing them…"
Bib 40 p189

If mainstream science proposes *to make theories and criticize them* while disbelieving truth, what options might they have? Proffering predictions with the intent to discover natural law means knowing, or attempting to know the future. Knowing the future is to possess truth. Therefore, by painting itself into a corner without truth, "science has nothing to do with the quest for certainty," but instead, relies on making theories and "criticizing" them; and that is what modern science is all about.

In 1690, thought provoking English philosopher John Locke (1632-1704) said:

"It is one thing to show a man that he is in error, and another to put him **in possession of truth**."
Note 4.3a

The inexorable march of time has brought us to the point of an inevitable revolution. The internet opened the floodgates of knowledge with the dissemination of information and the results of scientific inquiry. Suddenly researchers could share observations instantaneously, anywhere in the world. The flux of knowledge means that all of humanity is ripe for revolution.

Figure 4.4.1 – From where will the next scientific revolution come? If history repeats itself, the next revolution will come from individuals outside the traditional scientific establishment.

4.4 Where Will the Next Revolution come from?

Where to expect "Surprising Discoveries"

One of the great scientists and technologists of our day, Charles Townes, developer of the laser, provides a wonderful example of real scientific discovery. From his book, *How the Laser Happened*, Townes shared where he thought the next "surprising discoveries" would come:

"Individuals sometimes ask me whether a story such as that of the laser can happen in science today. They raise the question partly because there's an impression that at least the so-called physical sciences are almost finished—'we know all the basic science'—and partly because much of science appears to come nowadays from large teams of people—with large and expensive equipment. Is there still room for individual discovery of importance? My answer is that every scientific discovery is different in detail, and essentially unpredictable, but that **there will be many more**. There is much that we didn't understand; in many cases **we don't understand that we don't. And the really surprising discoveries will probably depend primarily on individuals**, **not teams or committees** even though the individual may be part of a team." Bib 26 p190

Kuhn reveals additional insight into where the next scientific revolution will begin; he says it will come from *individuals*, but *not* necessarily those working on "normal research problems":

"The scientific enterprise as a whole does from time to time prove useful, open up new territory, display order, and test long-accepted belief. Nevertheless, *the individual* engaged on a **normal research problem** *is almost never doing any one of these things*." Bib 33 p38

Kuhn shares a deeper understanding as to *why* "normal-scientific work" will not produce the next scientific revolution:

"But for normal-scientific work, for puzzle-solving within the tradition that the textbooks define, the scientist is almost perfectly equipped. Furthermore, he is well equipped for another task as well—the generation through normal science of significant crises. When they arise, the scientist is not of course equally well prepared. Even though prolonged crises are probably reflected in less rigid educational practice, **scientific training is not well designed to produce the man who will easily discover a fresh approach**. But so long as somebody appears with a candidate for paradigm—usually a young man or one new to the field—the loss due to rigidity accrues only to the individual." Bib 33 p166

Finally, Kuhn points out that the next scientific revolution will *not come from those well trained in the current, rigid educational system*, that it will likely come from an independent, unbiased source. Only with true independence, free of agenda-driven universities and power-hungry politicians, propelled by an unbiased approach can one hope to accomplish the unenviable task of dislodging the modern science behemoth.

What Kind of an Individual Needs to Practice Science

Some think that to practice science, one must be highly trained and skilled in the field. The authors of the book, *Crystals and Crystal Growing* understood this a little differently:

"It is a pity that most people think a scientist is a specialized person in a special situation, like a lawyer or a diplomat. To practice law, you must be admitted to the bar. To practice diplomacy, you must be admitted to the Department of State. **To practice science, you need only curiosity, patience, thoughtfulness, and time**." Bib 44 p11

It is unnatural to think that someone will discover something of importance within a specialized scientific field without years of training, and history proves that anyone who claims otherwise feels the sting of ruthless establishment censure. A personal story sheds light on how the 'independent research' that led to the UM began, and how anyone can contribute to science.

One of the first scientific fields to involve UM science was archeology. After several years of following archeology as an avocation, it became apparent that what was in the ground did not always match what the textbooks taught. It looked as though anyone could make new discoveries in this field and this piqued my interest. After talking with the professors about enrolling in an advanced archeology program at a nearby university, I shelved the paperwork and pursued other interests; it just did not feel right at the time.

Later, a family friend and physics professor taught me something of great importance when I asked him what a person should do if they discovered something of importance in a particular scientific field, but did not have the advanced degree in that field to garner credibility. Sure that he would advise me to obtain a degree before publishing, he surprised me with his answer:

"If what you've found is *important and true*, it doesn't matter where it comes from."

His words changed my entire paradigm. I would discover later that this would have a significant impact on my research. Had I continued my original course, with a focus limited to the confines of a single, narrow field of 'normal science,' the vast majority of the discoveries in this book might never have seen the light of day.

I knew the next scientific revolution would not come from the current rigid educational system. Requiring a fresh approach, the new paradigm must come from outside of modern science, and this was how the UM began.

The following quote from a distinguished physicist describes the relationship between true science and its origin. Read the following by substituting the words, "idea" and "theory" with 'model' and 'law' to render a truer statement:

"The validation of an idea or a theory…**is entirely independent of its origin**." Bib 18 p105

In other words, truth is truth, regardless of its origin.

Replacing Only the Foundation of Modern Science

Throughout the first four introductory chapters, some reviewers have felt that we treat modern science, as a whole somewhat harshly. However, we think it necessary that the public become aware of what kind of science we teach in our schools today, and to *contrast* the differences between modern science and the revolutionary ideas in the UM. Furthermore, understand that the UM generally intends to displace modern science's foundations that are based on false theory; but it also recognizes that not all science is based on false theory. Much of today's science has some basis in correct principles and natural law, and so the UM does not tackle these issues. The enduring maxim that what is true will continue to stand the tests of time remains a viable part of scientific knowledge.

Because some of the foundational pillars in modern science form the edifice upon which modern science rests, the science observer will witness a substantial shift in the scientific paradigm. By replacing false theory with scientific truth, we establish the foundation of Millennial Science, a new science destined to rise with the Millennials, and to last for a thousand years. They will continue to develop alongside anyone from around the world who is interested in *studying demonstrated truths and natural law that correctly describe and explain Nature*—the UM definition of **science**.

TRUTH IS INDEPENDENT OF ITS ORIGIN

CHAPTER 4 SCIENTIFIC REVOLUTIONS

THE UNIQUE UM CLAIM IS THAT MODERN SCIENCE LANGUISHES IN CRISIS WITH ALL FIELDS POISED FOR A UNIVERSAL REVOLUTION

5

The Magma Pseudotheory

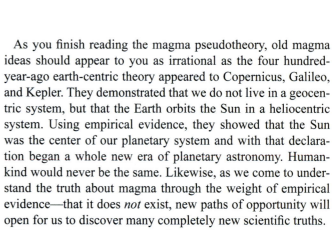

5.1 Magma Defined
5.2 The Magmaplanet Belief
5.3 The Lava-Friction Model
5.4 Magma Defies Heat Flow Physics
5.5 The Accretion Theory
5.6 The Radioactive Myth
5.7 Quartz is **Not** Glass
5.8 The Piezoelectric Evidence
5.9 The Non-Iron Core Evidence
5.10 Deep Earthquake Evidence
5.11 The Drilling Evidence
5.12 Earth's Magnetic Field Pseudotheory
5.13 The Continental Uplift Pseudotheory
5.14 Other Magma Pseudotheories
5.15 The 'Smoking Gun'
5.16 The Magma Freeze

This chapter begins our study of the Earth System. A true and correct understanding of the Earth's geology will help us answer questions of what, where, and when: questions such as what is quartz, where does the Earth's heat originate, and when do volcanoes erupt? The answers to questions like these help us *understand* the Earth's complex processes and can allow us to *explain* how and why they happen.

As we apply the Question Principle—to question everything with an open mind—to this chapter, the Magma Pseudotheory, and the upcoming Rock Cycle Pseudotheory, it will allow us to present a number of extraordinary claims. Throughout these chapters, we will establish that many of today's theories, which modern science teaches as fact, are actually false. In this chapter, we ask many new fundamental questions, consider current scientific data with a new paradigm, and present new scientific evidence to show the errors in these pseudotheories. Following these two chapters, in the Hydroplanet and Universal Flood Models, we evaluate the empirical evidence and observations that answer many of the questions about our planet that we raise in this chapter.

As you finish reading the magma pseudotheory, old magma ideas should appear to you as irrational as the four hundred-year-ago earth-centric theory appeared to Copernicus, Galileo, and Kepler. They demonstrated that we do not live in a geocentric system, but that the Earth orbits the Sun in a heliocentric system. Using empirical evidence, they showed that the Sun was the center of our planetary system and with that declaration began a whole new era of planetary astronomy. Humankind would never be the same. Likewise, as we come to understand the truth about magma through the weight of empirical evidence—that it does *not* exist, new paths of opportunity will open for us to discover many completely new scientific truths.

The next time you go out and enjoy the great outdoors, take the time to observe and experience Nature and the Earth's geology with the new concepts you will learn in the Universal Model. Because the new geology model, presented in the next few chapters, builds upon observable evidence and not the imaginary magma pseudotheory, it is easier to grasp, and easier to gain a more complete understanding of the great many geological mysteries that surround us.

5.1 Magma Defined

The latest films about Nature depict world-changing super-volcanoes with massive magma chambers that blast explosively skyward while fiery hot lava pours from the caldera and spreads across the landscape. Are magma and lava the same thing? Scientists often use the terms interchangeably, but there is a significant difference. Lava is the molten rock on the surface, whereas magma is the theoretical molten rock inside the Earth; but even then, we need to define the two more clearly. If magma is molten rock beneath the surface, that could include molten rock ten feet, ten thousand feet, or even a thousand miles below the surface. This subchapter draws a clear distinction between the terms lava and magma and defines both. We also learn where the ideas about magma originated.

Magma and Lava Defined

"The term *magma*, and concept of a single dominant *primary magma* was introduced by G. Poulett Scrope in 1825." Bib 87 p3

Magma was introduced during the early 1800s. Today we first learn about magma in elementary school, and we continue to learn about it in detail throughout our schooling years. Most people today have seen the textbook diagram of a deep, red-hot molten interior Earth. Magma and its heat of thousands of degrees is geology's icon of the deep Earth. Students learn that lava comes from magma, and most of us remember photographs or movies of volcanoes erupting and spewing molten red-hot lava. The interchangeable use of the two terms, lava and magma, often creates confusion.

Are lava and magma the same? We first turn to Webster's Universal College Dictionary for help:

"**Lava** is defined as the molten, fluid rock that issues from a volcano or volcanic vent." Bib 7 p459

"**Magma** is defined as a molten material beneath or within the earth's crust, from which igneous rock is formed." Bib 7 p488

According to a 2003, Essentials of Geology 8th ed. textbook:

"**Magma** is molten material that forms in certain environments in Earth's interior where temperatures and pressures are such that rock melts. Once formed, magma migrates upward into Earth's outer layer or crust." Bib 172 p12

Now, there is no question that *lava*, the molten material that issues from a volcano and flows onto the surface is real—we see it and can experience it. Obviously, the molten material must also have existed below the surface, and no doubt, it is there where melting occurs. However, the term magma has a broad definition in today's geology, and it includes a liquid molten core, a molten or partially molten mantle, and the molten rock in the crust of the Earth. Additionally, geologists believe that most of the rocks and minerals that make up the crust of the Earth come from a magma melt. The mingling of observable truth with theory while defining magma and lava leaves us with a vagueness that we must change if we are to understand the processes involved.

For the purposes of the UM, we define magma and lava as follows:

Magma is **only** a **theoretical** molten rock material generated deep within the Earth in the outer core or mantle.

Extrusive lava is molten or fluid rock that is manifest on the surface of the Earth.

Intrusive lava is the molten sub-surface rock that occurs within the crust of the Earth.

If magma, as defined by modern science, included only the molten material within the Earth's crust, magma would not be a pseudotheory (a false theory taught as fact) because this part is true, but it does not; the magma theory includes the entire molten-Earth theory, the idea that our current Earth came from an ancient molten magma planet. This idea is deeply entrenched in modern science today, receiving almost universal acceptance by scientists and the general public. Actual lava is easily photographed and sampled; intrusive lava is detectable and evidenced because of modern technology, but *magma*, and the current belief that it resides deep in the Earth is mere speculation. As we will see, the Earth could not have come from a magma planet. Beyond the artistic conception of magma, has there ever been any real evidence of magma?

Magma is Only a Theory

We all learned in school that the Earth is a sphere because it was once a molten ball of magma and that the interior of the Earth still is magma. Our so-called magnetic field supposedly also comes from magma. However, geologists have long known that there are major problems with this theory.

In the scientific article *What Lies Beneath* we read a summary of what science *really* understands about what lies under our feet:

"Because so much is known about planets that are light-years away, one might assume that science has unearthed everything worth knowing about the one beneath our feet. **To the contrary, Earth's innermost reaches remain, in many ways, as mysterious as the cosmos at large**.

"Why, precisely, is Earth's interior hot? What, exactly, is the source of the magnetic field, and why does it sometimes reverse? What, specifically, is the core made of? The obvious way to answer such questions would seem to be: Dig a hole and take a look. But the world's deepest excavation, a research project on Russia's Kola Peninsula, has drilled only 7.5 miles into the crust, or less than 0.2 percent of the 3,959 miles to Earth's center. Learning about the deep interior from this hole is roughly

Fig 5.1.1 – This is an *actual*, observable extrusive lava flow on the surface of the Earth. We can see it, and test it. Magma in the Earth's core and mantle is only a theory; we cannot see it or test it.

Fig 5.1.2 – This is an actual lava flow in Hawaii. How does science really know that this lava came from magma deep inside the Earth? Is there another source for lava?

comparable to learning about Alaska by driving from St. Petersburg, Florida, to Tampa. **So geophysicists infer**. Seismology, gravitational variation, high-pressure mineral experiments, and chemical analysis of rock samples and meteorites all provide clues. **As with all circumstantial evidence, however, interpretations vary**." Note 5.1a

Note that in this article the "geophysicists infer" because they do not know. Continuing with the same article:

"Geophysicists contacted for this story conceded that alternatives to the mainstream scenario [magma Earth] are possible: **The cooling iron-nickel-ball inner core, they admit, is still just a theory**, if a dominant one." Note 5.1a p41

It is *rare* to read such an admission in print. In fact, today's modern Earth Science textbooks stop far short of the declaration that magma "is just a theory."

Sometimes, older geology textbooks, like one written over three quarters of a century ago, included statements less dogmatic than modern textbooks:

"The problems connected with the Earth's interior are fascinating but difficult... It is well to keep in mind that some of the present views and conclusions are tentative and may be changed by continued investigation. **Speculations in this field are numerous, and these should be kept distinct from legitimate inference and proved fact**." Bib 125 p403

Although geology is the study of rocks, observed first person as we hold them in our hands, we don't see the words "proven fact" often in modern geology today. Instead, we find throughout the 'modern' geology literature, "inferences" and supposition.

A Theory Taught as Fact

The following two quotes from *Magmas and Magmatic Rocks,* a mainstream igneous petrology textbook, give us a glimpse at the contradiction in modern geology. Appearing only two pages apart, these statements provide an excellent example of how geology teaches magma *theory* as *fact*.

"**Magmas properly belong to the realm of theoretical petrology**. They are the most important concept in igneous petrology, yet they **cannot be examined in the field, collected, studied or directly experimented with**." Bib 87 p8

Two pages later:

"There is, however, **irrefutable direct evidence that materials with the physical properties of magma exist within the Earth**." Bib 87 p10

In the first statement, magma ***must be defined*** as "***theoretical***" because it ***cannot "be examined,"*** however the researcher states that there is "irrefutable direct evidence" that materials with the physical properties of magma exist! How can this be? How was this direct, irrefutable evidence obtained?

"Theoretical petrology" is the study of rocks we cannot see. Geology must relegate magma to this realm. Why does modern science believe so strongly that magma exists, even though they admit it cannot be examined, collected, studied, or established by experiment? The answer is simple—the idea of magma, interwoven into the fabric of so many scientific fields for so long, exists such that any other explanation, however true, would disrupt the entire scientific community.

If we have never observed it, can we concede that magma may not be real? One way to falsify this question is to **prove** magma exists—to identify reproducible physical evidence of magma; scientists have yet to do this. Volcanologists have studied lava and searched for clues about the illusive mystical liquid, magma, for their entire careers, searching for the origin of magma and the forces that generate it. Here is what they have to say:

> "**Magmas properly belong to the realm of theoretical petrology.**"
>
> Eric K. Middlemost (petrologist)

"**The question of where the magma comes from and how it is generated are the most speculative in all of volcanology**. We cannot see to any appreciable depth below the surface of the earth and have few direct measurements of the nature of the materials in the earth's interior." Note 5.1b

Think for just a moment that nearly all of the natural sciences are based on geology, and that all of geology is based on magma. It is astounding to read the truth from the geologists themselves—that the questions of where magma comes from and how it is generated are "the most speculative in all of volcanology".

Two facts are clear when we consider these things:
1. Lava can be observed and is real.
2. Magma has never been observed and is speculative in Nature.

This ought to lead scientists to ask a very fundamental question:

What if there is **no** Magma?

James Hutton—the Father of Geology

James Hutton (1726-1797) became known as the 'father' of modern geology, but was a philosopher more than he was a scientist. Hutton, who despised experimentation, left us *ideas,* not experimental facts that proved anything:

"...Hutton's **open dislike of experimentation**. Hutton, **a great skeptic of experimental methods in general**..." Bib 136 p152

He proposed the idea that granite, the basement rock underlying the continents came from a "hot molten material":

"...Hutton proposed that **granite had formed from a hot**

molten material that solidified deep in the Earth. The **evidence** was conclusive, as no other **explanation** could accommodate all the facts." Bib 59 p78

However, Hutton's "evidence" was his own "explanation"; he had no direct proof of his assumption.

Hutton, the father of geology was among those influential individuals who impeded real scientific discovery because of his supposed superior 'ideas.' His new ideas lacked the crucial experiments and observations that might have disproven or changed them, and as a result, he left modern science, at least modern geology, poised for a leap into a Scientific Dark Age.

Besides the idea of "a hot molten material...deep in the Earth" (see Fig 5.1.4), Hutton espoused two other significant theories that have undeniably affected all of modern science. The first is his theory about **Geological Time**. His geological time-theory gave rise to an Earth age stretching first to millions and then to billions of years, the age which we hear so often in science today. We will examine this idea and the actual age of the Earth in detail, later in the Age Model chapter. The second of Hutton's significant theories, his so-called **Uniformity Principle**, reinforced his geological time scale, opening the door for Darwinian evolutionary theory's rise.

The Uniformity Myth

Probably the most significant of Hutton's philosophies, his Uniformity 'Principle' or Uniformitarianism, referred to at the beginning of almost every geology textbook printed in the last 100 years, is the guiding force behind many other theoretical ideas such as geological time and evolution. Hutton's **Uniformity 'Principle'** in its simplest form today says that:

"The present is the key to the past." Bib 172 p3

According to Hutton, one need only observe Nature as it exists today to understand all past events. He abhorred the idea of catastrophism. Originally, scientists accepted uniformitarianism as a way to defeat the idea that cataclysms had a part in shaping the Earth, as one scientist noted:

"It was the emergence of the **principle of *uniformitarianism***—one of the basic principles of modern geology, **which insists that the rocks and landscapes we see around us have been formed not by special cataclysms of the past but by the ordinary familiar processes that we see going on around us today**. The principle was eventually **firmly established by James Hutton** and succinctly set forth by the brilliant English geologist, Charles Lyell, in the words, '**The present is the key to the past**.' Note 5.1b p31

The obvious error is the implication that the present *is always the key to the past*; that the geological processes such

Fig 5.1.3 - James Hutton, father of geology, uniformity, and magmaplanet theory.

as sedimentation and erosion as we currently experience them can explain and gauge ***all*** geologic history. It is a simple fact that some past events, such as the formation of the Earth, *never* happened in historical times, which means that at best, Hutton's claim should state that the present is the key to ***only a part of*** the past.

Here we see a mingling of truth and error, so that when modern geology says, "The present is the key to the past" it *implies* that it was *always* so, that present processes always operated as they do today. In the next several chapters, we show how obvious this error is.

We cannot give Hutton all of the credit for modern science's adoption of his assumptions. Hutton was not all that well known, and though he published, his ideas did not take full hold until another geologist, British born Charles Lyell (1797-1875), famous and well liked in his own day, picked up Hutton's torch, publishing his theory in the first volume of the wildly successful, *Principles of Geology* in 1830:

"Hutton's uniformitarianism (see 1785) was now nearly half a century old but had not made much headway. For one thing, Hutton was not an inspiring writer and his chief opponent, Cuvier, the supporter of catastrophism (see 1812), was.

"In 1830, however, the first volume of *The Principles of Geology* appeared. Written by a British geologist, Charles Lyell (1797-1875), it was to run to three volumes. All were attractively written and explained the uniformitarian theory so well that it finally achieved popularity. The extreme catastrophism of Cuvier and his followers thus went down to defeat. This is not to say that the pendulum doesn't swing and that certain catastrophes in the course of Earth's history did not take place. Earth's development is now viewed mainly, but not entirely, uniformitarian." Note 5.1c

The test of any principle is time, and if Hutton's uniformity represents a *true* principle, it will pass that test, but if it does not, we must modify it or throw it out. As the geologic evidence mounts, with events like the impact of comet Shoemaker-Levy 9 into Jupiter in 1994, geologists continue to back away from the crumbling uniformity pillar. What we need now is a com-

Fig 5.1.4 - James Hutton's drawing of a *theoretical* magma chamber below a mountain.

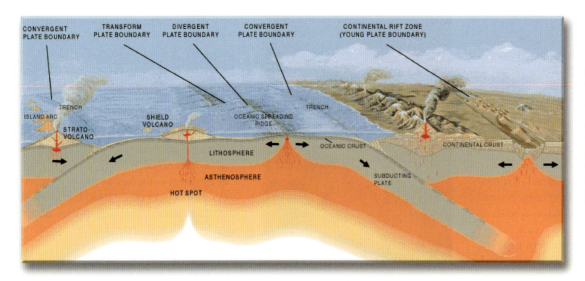

Fig 5.1.5 – This diagram of the Earth's crust is typical of the modern geology viewpoint. There is no suggestion that this is one theory among many; it is shown as if it is known fact. In reality, this stylized depiction is the ***theoretical*** 'magmaplanet' concept of the Earth. Courtesy of NOAA.

plete replacement of uniformity theory, one with a new foundation based on true and empirical geologic evidence.

The Catastrophic Principle

The modern science uniformity principle precluded any major catastrophes, but that notion has since given way as geologists are forced to consider new catastrophism-based theories. The Catastrophic Principle includes major catastrophic events that initiated processes that are not happening today,

> **The Catastrophic Principle**: Major geologic events occurred in the past, including processes that are not happening today.

If we are to discover geologic truth, we must replace uniformity as we seek to understand the many processes that occurred in the past but that are not happening today. These catastrophic events affected and shaped the Earth in ways never observed by modern science.

Why is Magma so Important to Modern Science?

The answer to this question will become clear throughout the next few chapters. One common interest often shared among people is to understand something of our origins. Modern science claims that man is the result of eons of evolution. Biologists hang evolutionary theory on fossils and the geological time scale; both fossils and the long spans of geologic time find their origins in the theory of how rocks form. Geologists assume all rocks came originally from a melt, on a molten magmaplanet. Therefore, the theory of magma is the foundation for much of modern sciences' understanding of the Universe, including our own origins.

No doubt, some think magma is not so important as to link together the disciplines of Earth science, astronomy, physics, chemistry, and even biology. However, if all rocks supposedly originated from a hot, molten, magmaplanet Earth, and because the experts in those fields reckon geologic time according to those rocks, then any branch of science that relies in any way on rocks or the geologic time scale must admit a *direct connection* to magma.

Researchers identified the framework linking the subdisciplines of Earth science in this journal article:

"**Mantle convection** [circulating magma inside the Earth] and plate tectonics **provide the central framework linking the subdisciplines of solid Earth science**, including geochemistry, seismology, mineral physics, geodesy, tectonics, and geology. **A successful model must thus satisfy constraints from all of these fields**." Note 5.1d

Accordingly, a successful model of our Earth must satisfy many constraints from all the various Earth science fields, which is why we integrate each of the Earth System chapters in this book; they have to be, and as such, they cross-confirm each field's claims and discoveries, supporting new points of view. Magma is the foundation of so many modern scientific fields, and as our confidence grows in the new models presented, we will begin to see connections that will enable us to shift our foundation to the new universal models.

5.2 The Magmaplanet Belief

A *key* to understanding modern geology is to understand that *magma is a **belief**, not a fact*. Once we accept this truth, we can consider the evidence and the observations with an open mind and we can pursue other possibilities and explanations for the anomalies, mysteries, and phenomena of our Earth. This will enable us to ask new questions about the true origin of the Earth and the events that shape it, both now and in the past.

Why Do Geologists Today Believe in a Magmaplanet?

The primary reason why today's geologists believe in a magmaplanet is that their teachers taught them that the Earth is a magmaplanet. They, in turn, learned from their teachers and professors who learned that Earth is a magmaplanet from their teachers and professors. Although we have already established by the geologists' own words that magma is only a *theory*, it is today still taught as if it is a fact. The magmaplanet idea has been around a long time. Traditions like this carry a great deal of weight and change is not easily accepted.

Another reason geologists believe in the magmaplanet theory is *lava*. Lava is said to be dramatic proof that magma exists. When lava erupts from volcanoes, geologists assume it came from deep inside the Earth, and although science has never demonstrated the existence of deep-earth magma, they have no other good explanation of lava's origin. They deduced that it must come from deep inside the Earth. The idea is not new:

"Strabo, the Greek geographer and historian (63 BC-AD 20), described volcanic activity associated with Etna, Somma-Vesuvo and the Lippari Islands. He also suggested that **volcanoes were natural safety valves through which fluids escaped**." Bib 87 p2

The third and perhaps most important reason geologists believe in a magmaplanet is that the so-called 'evidence' for this pseudotheory is so convincing that scientists think not to question it. There is no other explanation supported with scientific proof that might persuade scientists to believe otherwise.

Why does Science Believe the Interior of the Earth is Hot?

For Earth scientists, the question of why the interior of the Earth is hot is not an easy one to answer, and so far seems impossible for geologists to prove. The *Understanding Earth* geology textbook states:

"Earth's interior is hot for several reasons... the violent origin of Earth by the **infalling of chunks of matter** made the interior hot, and disintegration of the **radioactive elements** uranium, thorium, and potassium also produced a significant amount of heat." Bib 59 p495

Based on this textbook, magma's heat comes from a combination of material infall as the young orbiting Earth accreted matter from a protoplanetary disk during the early days of the solar system and from the decay of radioactive elements. However, both of these scientific alibis (sci-bis) are purely theoretical; no one has ever seen it or demonstrated it experimentally. Perhaps the more common reason scientists believe that the center of the Earth is hot is the visibly incandescent extrusive lava erupting from volcanoes. The reasoning is that lava comes from a sea or plume of magma that rises from deep within the Earth before it erupts onto the surface. One geophysicist lists evidence confirming this mindset, explaining that the interior of the Earth is "clearly hotter than its surface":

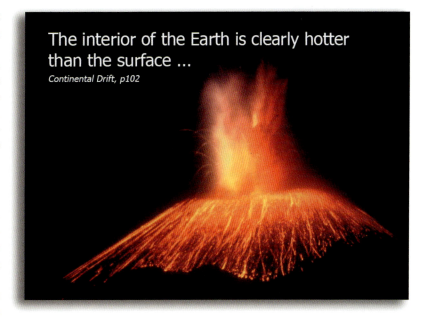

The interior of the Earth is clearly hotter than the surface ...
Continental Drift, p102

"The interior of the Earth is clearly hotter than its surface, as shown by **volcanoes** and the temperatures within **mine shafts** and **drill holes**." Bib 103 p102

For the modern scientists, the "evidence" of internal heat is "everywhere" and bolsters their belief in magma:

"The evidence of Earth's internal heat is **everywhere**: volcanoes, hot springs, and the elevated temperatures in mines and boreholes. Even **global plate motions, earthquakes, and the uplift of mountains** are driven by this internal heat." Bib 59 p495

Both horizontal plate motions and earthquakes are real; we can observe them, but we cannot definitively observe the internal heat that science says drives these processes. Later in this chapter, we will discover why magma does not drive these processes and how science extrapolated the "uplift of mountains" into an uplift of continents based on theoretical concepts founded on the magmaplanet model.

Another proposed source of magma is the intense pressure deep inside the Earth:

"Melting, which is accompanied by an increase in volume, *occurs at higher temperatures at depth* because of greater confining pressure. Consequently, an increase in confining pressure causes an increase in the rock's melting temperature. Conversely, reducing confining pressure lowers a rock's melting temperature. When confining pressure drops enough, **decompression melting** is triggered." Bib 172 p58

We need to clarify a couple of points about this concept of melted rock under pressure so that we can better understand the forces at play. Although rocks will melt at lower temperatures near the surface than they will at depth because of pressure, the fact that there is pressure does not account for the heat:

1. Pressure by itself does not create heat.
2. Pressure decreases as melted rock moves upward or by the removal of material above the rock.

Applying friction and pressure creates heat

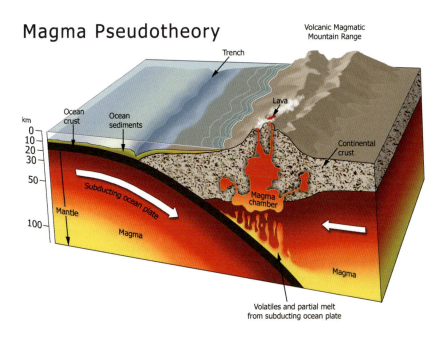

Fig 5.2.1 – This diagram depicts the modern science Magma Pseudotheory. Scientists believe observable surface lava comes from magma far below.

SUBCHAPTER 5.2 THE MAGMAPLANET BELIEF

and the increase of pressure also increases the required temperature at which rocks melt; reducing pressure reduces the required melting temperature.

Geologists turn to other places for evidence about the Earth's interior, as explained in the text, *Magmas and Magmatic Rocks*:

"In the past, **geophysical data**, information from **experimental studies at high pressures**, and chemical data obtained from the study of **meteorites**, have all been used to construct models of the chemical composition of the core [of the Earth]." Bib 87 p176-7

Researchers use the data from experiments on meteorites to construct theoretical models, all on indirect evidence. The geophysical data mentioned consists of collected rocks lying on the crust of the Earth that the geologists 'think' came from an ancient magmaplanet core; one that had an environment similar to the Earth's core. Nevertheless, since no one has ever directly observed magma, we should remain open to the idea that magma is not credible. The "experimental studies at high pressures" actually supports the scientific evidence refuting the existence of magma. In Chapter 7.10, the Meteorite Model subchapter, we discuss the origin of meteorites and the extrapolation that science employs as it relates them to the core of the Earth.

On a much simpler note, a question appeared on the public website, *Ask an Expert*; "Why is the Earth's core so hot?":

"**How do we know the temperature? The answer is that we really don't**—at least not with great certainty or precision. The center of the earth lies 6,400 kilometers (4,000 miles) beneath our feet, but the deepest that it has ever been possible to drill to make direct measurements of temperature (or other physical quantities) is just about 10 kilometers (six miles).

"As a result, **scientists must infer the temperature in the earth's deep interior indirectly**." Note 5.2a

Here again we see that despite not knowing the Earth's temperature, even though elementary, junior, and high schools from around the world teach children the so-called 'fact' that the

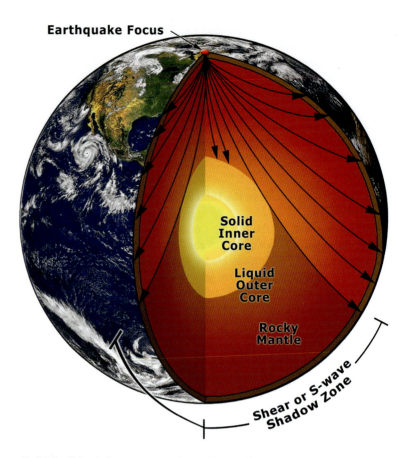

Fig 5.2.2 – Seismic S-waves create observable sound patterns as they travel through the Earth. Because these wave patterns travel differently through solids than they do through liquids, it is known that there is a liquid outer core and a solid inner core. However, the type of liquid is unknown, but science *assumes* that liquid is magma because of the theoretical high temperatures at the core.

Earth's core is molten, the *real* answer to the question, "How do we know the temperature?" We really don't know.

Seismic Waves and Magma

Earthquakes produce seismic waves that travel through the planet's interior and along her outer crustal layers. Some seismic waves penetrate the Earth's body, passing through its center, recorded by seismometers on the other side of the world. Because we know the start time of the earthquake and the diameter of the Earth, we can reasonably calculate the speed of the waves as they travel through the Earth. There are two primary types of waves; P waves, or primary waves, are compressional waves that travel faster and can pass through solids or liquids, whereas S waves, known as secondary, or shear waves travel much slower traveling only through solid matter.

Fig 5.2.2 shows the curved paths of S waves as they travel from an earthquake's focus across and through the Earth to distant points where seismometers record their data, as seen in Fig 5.2.3. A *shadow zone* created by the S waves as they bend around the liquid core of the Earth provides an important clue:

"Discovery of the shadow zone led geologists to surmise that the Earth has a core made of a material different from that of the overlying mantle. **They could even conclude that the core is liquid, because the waves are bent downward rather than upward when they enter the core, much as a light beam bends downward after it enters water**. This means that the waves travel more slowly in the core than in the mantle. P waves are known to move much more slowly in liquids than in

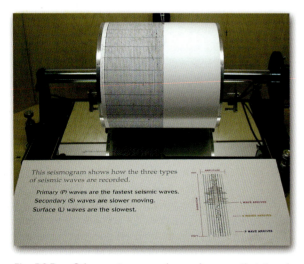

Fig 5.2.3 – Seismometers record sound waves that travel through the Earth when earthquakes happen around the globe. These sound waves help determine the structure of the Earth's interior.

CHAPTER 5 THE MAGMA PSEUDOTHEORY

solids, **so it was reasonable to guess that the existence of the shadow zone implies a molten core**… Only **fluids** are known **not** to transmit shear waves…" Bib 59 p487

Seismic waves *do establish* that a large portion of the interior of the Earth is liquid but it does not establish what that liquid is. A simple question one could ask; is magma *the only liquid* found in Nature? The answer—no.

The geologists themselves state in the foregoing statement that they have only been able to "**guess**" what liquid occupies the Earth's underworld. Their research "***implies a molten core***" but they do not know this. They do know that there is a shadow zone caused by the liquid in the outer core of the Earth as illustrated in Fig 5.2.2. The shadow zone appears repeatedly, when earthquakes occur.

From the different magnitudes and arrival times of the different waves, researchers in the early twentieth century were able to develop a rough picture of the interior of the Earth. As technology improves, the picture is ever clearer, and one of the most convincing evidences that magma does not exist comes from an understanding about these seismic waves.

The "Old Idea of a Universal Magma"

Ideas change. Magma was once a new idea, and as it developed, geologists imagined a great ocean of magma deep inside the Earth: an all-encompassing body that supplied the heat and lava to all volcanoes, but that idea fell out of favor during the early 1900s. Quoting from a 1911 encyclopedia:

"The **old idea** of a **universal magma**, or continuous pyrosphere, **has been generally abandoned**." Note 5.2b

It fell out favor because the empirical evidence did not support it. The lava ejected from different volcanoes, even nearby volcanoes, was different. Again, from the same 1911 encyclopedia we read:

"Whatever may have been the case in a primitive condition of the interior of the earth, it seems necessary to admit that the magma must now exist in separate reservoirs. The independent activity of neighboring volcanoes strikingly illustrated in Kilauea and Mauna Loa in Hawaii, only 20 m. apart, suggests a want of communication between the conduits; and though the lavas are very similar at these two centres, it would seem that **they can hardly be drawn from a common source**." Note 5.2b

In addition to the problem of dissimilar melt composition, other problems, such as the origin of heat and the force driving the rising magma led to the abandonment of this common source idea:

"In discussing the cause of vulcanicity **two problems demand attention**: first the **origin of the heat** necessary for the manifestation of volcanic phenomena, and **secondly the nature of the force by which the heated matter is raised to the surface and ejected**. According to the old view, which assumed that the earth was a spheroid of molten matter invested by a comparatively thin crust of solid rock…

"Neighboring volcanoes **seem** in some cases to draw their supply of lava from independent sources, favoring the idea of local cisterns or intercrustal reservoirs. It is probable, however, that subterranean reservoirs of magma, **if they exist**, do not represent relics of an original fluid condition of the earth, **but the molten material may be merely rock which has become fused locally by a temporary development of heat or more likely by a relief of pressure**." Note 5.2b

It is interesting to note that although the idea of a universal magma-ocean was discarded, the scientist's musings that "molten material may be rock which has merely become fused locally by a temporary development of heat" might have, if the observations had been pursued further, changed the whole idea of magma itself, even to the point of discarding it. No doubt, it would not exhibit the popularity it has today.

Magma Summarized

For almost 200 years, magma has been a pillar that underpins nearly every branch of science. Despite scientist's inference from observable volcanic activities and lava flows that the Earth has a hot melted interior, the theory remains unproven, without observation, and untested, yet science professors still teach it as if it were a fact. Because of the magmaplanet paradigm, scientists presumed that processes observed today act as they have for eons (uniformitarianism), and that global, catastrophic, Universal Flood-like events did not happen. We have seen the mingling of intrusive lava truth with the speculation of deep-earth magma. As we summarize the Magmaplanet Belief, let us look at the summary of magma generation according to the textbook, *Essentials of Geology, 8th Ed*:

"In summary, **magma can be generated under three sets of conditions**: (1) *heat* may be added; for example, a magma body from a deeper source intrudes and melts crustal rocks; (2) *a decrease in pressure* (without the addition of heat) can result in *decompression melting*; and (3) the *introduction of volatiles* (principally water) can lower the melting temperature of mantle rock sufficiently to generate magma." Bib 172 p59

The summary fails to identify the **source of heat** that *generates magma*; the first condition mentions a magma body (mag-

Fig 5.2.4 - The origin of the heat for lava flows like this one in Hawaii remains a mystery. The idea of a universal magma was discarded in the early 1900s, but a revised version of it is still central in nearly all Earth science texts.

ma already exists) coming from a deeper source. The second condition does not add any heat, but just reduces pressure, and the third condition simply adds water (or other volatiles) again, without revealing any source of heat. There is no identified heat source within this summary; it relies on circular reasoning to explain magma generation.

The bottom line is that the three conditions *do not actually show* how magma is generated! Although many science textbooks point the reader to the theoretical heat producing conditions of radioactivity, accretion, or impact by external bodies, these are not included in the magma-generating conditions shown above. The truly theoretical modern science concept of magma fails to identify correctly a source of heat and cannot stand under experimental scrutiny.

5.3 The Lava-Friction Model

Having recognized the theoretical nature of deep-earth magma and having made the extraordinary claim that it does not exist, we must then answer the question:

Where does lava come from?

In this subchapter, we introduce and outline the Lava-Friction Model, which consists of three parts: the origin of heat (Frictional-Heat Law), the origin of the movement that creates friction (Gravitational-Friction Law), and the effects of these two laws. In support of this model, we evaluate the significance of earthquakes and Earth movements, and their correlation and connection to the origin of lava.

Skin-Friction Analogy

A simple friction-heating experiment visually demonstrates the mechanism behind the origin of both intrusive and extrusive lava.

With your hands pressed firmly together in front of you, begin rubbing your hands briskly, in a back-and-forth motion for a few seconds. Most people notice an immediate rise in temperature, and some observe bits of balled-up dirt and dead skin (skin lava) on their palms, if they rubbed hard enough. (Figure 5.3.1)

Using the skin friction analogy as an illustration of the mechanism that generates real lava, we begin to understand in a very small way, just how much heat the Earth's crust is capable of generating. Riddled with tens-of-thousands of faults, the Earth's plates experience enormous pressure, and they need only rub against each other a very small amount to produce surprisingly high temperatures capable of subsequently melting the surrounding rock. The Earthquake Friction Lava Diagram in Fig 5.3.2 shows how plate movement forms intrusive-lava at or near the earthquake focus (the initial point of rupture). The newly melted intrusive lava, although under pressure, can be buoyant compared with surrounding rock and it pushes fluidly toward the area of least resistance, which is generally toward the surface.

Correlating the relationship earthquakes and lava share confirms lava's origin on display with this simplified skin friction illustration. Establishing a connection between earthquakes and lava flows by recording seismic activity incident to lava flows or a volcanic eruption event can do this.

The Lava-Friction Model

In this subchapter, we discuss how the movement of Earth's crust can cause melting through frictional heating. To understand lava's origin, we must establish a foundation from which to view the evidence as we begin to structure a new understanding of the processes we observe. Based in part on the fact that friction generates heat, the new Lava-Friction Model is the basis for explaining the origin of lava, both its intrusive and extrusive forms. The **Lava-Friction Model** consists of three principles:

1. Lava originates from frictional heat (The Frictional-Heat Law) generated by movement within the crust.

2. Crustal movement is attributable to the solar and lunar cycle's diurnal effects. (The Gravitational-Friction law)

3. The resulting melted rock moves along paths of least resistance, including faults, subjecting the rising melted rock to further decompressional melting.

Simply stated, the concepts include; the origin of heat in the Earth's crust, the origin of the movement that creates the heat and the combined results of both of these concepts. The geosciences community has yet to comprehend just how much frictional heat transfers through the faults and has yet to consider frictional heat as a source or origin of lava-making heat for making lava. To understand lava, we must then first understand earthquakes, and the origin of the movement that causes earthquakes, including how astronomical cycles affect this movement. With this understanding, we will see that there is a definite connection between earthquakes and lava.

Do Earthquakes Cause Volcanic Activity?

According to most scientists, the "internal heat" associated with magma is the cause of earthquakes. From a USGS (United States Geological Survey) government web site, we read the answer to the following question:

"Q: Do earthquakes cause volcanoes?"

"A: **No**, there are different earth processes responsible for volcanoes. Earthquakes may occur in an area before, during,

Fig 5.3.1 – Rubbing hands together briskly produces frictional heat, and sometimes "skin lava" of dead skin or dirt, seen in the inset.

and after a volcanic eruption, but they are the result of the active forces connected with the eruption, and **not the cause of the volcanic activity**." Note 5.3a

Commonly associated with volcanoes, volcanologists consider earthquakes a signal of impending volcanic activity on both active and dormant volcanoes. Although nearly all volcanoes and lava flows have a great deal of associated seismic activity when they are active, the sudden quaking of long-dormant volcanoes gets the scientists' attention and causes alarm in surrounding communities. This is because seismic activity is the harbinger of imminent volcanic activity. However, scientists in general have not believed that earthquakes *cause* volcanic activity.

What Do Scientists Think Causes Earthquakes?

On the USGS web site, scientists answer the question as follows:

"Earth scientists **believe** that most earthquakes are **caused by slow movements inside the Earth** that push against the Earth's brittle, relatively thin outer layer, causing the rocks to break suddenly." Note 5.3b

Scientists "believe" the slow movements inside the Earth are caused by the movements of crustal plates:

"The **plate tectonics theory** is a starting point for understanding the forces within the Earth that **cause earthquakes**." Note 5.3b

The plate tectonics theory proposes crustal movement based on convective magma, one facet of the magmaplanet model. The subject of plate movement is an important one and we will get into the subject in some detail further in the chapter, but first, we continue to explore how earthquakes and lava are connected. The following few examples provide a starting point different from the one that is widely accepted.

The Earthquake-Lava Connection

First, let us answer the question, "What do we find accompanying lava eruptions all around the world?"

"Last October [2002] about 1,000 Italians fled their homes after Mount Etna, the famous volcano on the island of Sicily,

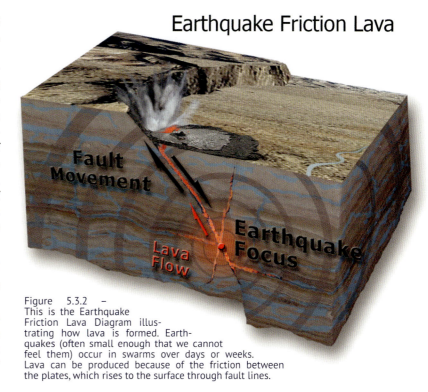

Figure 5.3.2 – This is the Earthquake Friction Lava Diagram illustrating how lava is formed. Earthquakes (often small enough that we cannot feel them) occur in swarms over days or weeks. Lava can be produced because of the friction between the plates, which rises to the surface through fault lines.

rumbled to life. Shooting molten rock more than 500 meters into the air, Etna sent streams of lava rushing down its northeastern and southern flanks. **The eruption was accompanied by hundreds of earthquakes measuring up to 4.3 on the Richter scale.**" Note 5.3c

Where there are eruptions, there are earthquakes, and despite the USGS statement that scientists believe that most earthquakes are caused by "slow movements [of magma] inside the Earth," can researchers demonstrate that 'theoretical magma' is the cause of the earthquakes? When an earthquake occurs, we often know only because seismic activity recorded on seismometers provides a visual record. What is actually being recorded are vibrations, or 'sound,' but researchers noted:

"**Seismic investigations have shown** that the rising magma **produces little noise and appears to move rather smoothly**, without encountering major obstacles." Note 5.3c p63

On the one hand, slow movements of magma inside Earth are said to cause earthquakes, but on the other, rising magma produces little noise and moves smoothly enough not to register on seismic instruments, or if so, very little. This is a significant, yet contradictory statement, especially as we examine, which came *first* as recorded in the following description of Mauna Loa, a Hawaiian eruption:

"Following the flank eruption of 1935, Mauna Loa was in repose for

Fig 5.3.3 – Lava flow along a fault line in Hawaii. Courtesy of USGS

Fig 5.3.4 – Fault in a road-cut near Kingman, Arizona, USA.

SUBCHAPTER 5.3 THE LAVA-FRICTION MODEL

more than 4 years. During 1939 and early 1940, however, **an increasing number of earthquakes indicated that the quiet would soon come to an end.** At 11 P.M. on April 7, 1940, volcanic tremor began recording on the seismographs at the Volcano Observatory, and at 11:30 people in Kona saw the orange glow of eruption at the summit of the mountains." Note 5.3d

The Mauna Loa earthquakes provide a clear example where seismic activity preceded eruptions. However, one example does not solve the Earthquake-Lava question:

Does magma cause earthquakes
or do earthquakes cause lava?

The Earthquake Swarm-Lava Evidence

We examined a large study of earthquakes and volcanic eruptions to answer this question. Of course, not all earthquakes produce eruptions because they are usually of short duration. Earthquake 'swarms' typically last for days at a time and usually accompany eruptions. Because of new technology, John Benoit and Steve McNutt of the University of Alaska, Fairbanks Geophysical Institute, were able to examine and analyze hundreds of earthquake swarms and correlate them with volcanic eruptions over a 10 year period. We found this to be the most extensive study of its kind anywhere, to date. The study included over 600 earthquake swarms recorded worldwide, which took place from 1979 to1989. In Fig 5.3.5, the Earthquake Swarm-Volcanic Eruption Diagram shows the results of their study. They separated activities into three types of swarms. Type 1 swarms (the most common) are those that *preceded* eruptive activity; Type 2 swarms *accompany* the eruptive activity; and Type 3 swarms were *not associated with eruptive activity*. Type 3 swarms accounted for 39% of the record but apparently, the seismic activity did not have enough time to produce an amount of lava needed for an eruptive event, or it occurred too deep for the lava to work its way to the surface. On average, Type 3 swarms lasted for only 3.5 days, while Type 1 swarms lasted up to 9 days, almost three times as long.

The Earthquake Swarm-Volcanic Eruption Diagram in Fig 5.3.5 demonstrates the specific relationship between swarms and eruptions. Notice the timing of type 1 & type 2 swarms relative to the time of the eruptions. Note 5.3e

Every swarm that accompanied a volcanic eruption preceded the eruption, or occurred during the eruption. **No earthquake swarms started immediately after volcanic eruptions.**

Another significant relationship can be seen in Type 1b swarms. The greatest number of swarms was of this type, and it is the only type where earthquakes occur **right before** the eruption. This study puts us one step closer to verifying that earthquakes are the cause of lava eruptions. Note 5.3f

As time passes and with the recording of a greater number of volcanic eruptions and with ever-greater seismic data, the information continues to support the Earthquake-Lava Connection. In addition, further studies of temperature conditions and temperature changes within faults and fault systems produce data supporting the frictional heating-earthquake relationship. If magma caused the earthquakes, heat from deformation and intruding magma would appear *before* the earthquake, not after. At present, it seems there is little firsthand knowledge about heat and heat production in faults.

How Much Does Science Know About Frictional Heat Generated by Faults?

If modern geology recognized the possibility that earthquakes were causing, or at least contributing to volcanic eruptions of molten extrusive lava, one would think there should be extensive studies on the matter, and with such studies would come the knowledge of just how much frictional heat actively moving faults generate. Additionally, the more the geologists' know about frictional heat from seismic activity, the less inclined they would be to dismiss it. Surprisingly little detail exists when researching those who published journal articles discussing frictional heating via faulting. It seemed almost as though there was a 'don't go there' attitude;—as if they were saying we 'already know the heat comes from magma' so why look elsewhere? However, our search did uncover this 1998 article in the journal of *Science*:

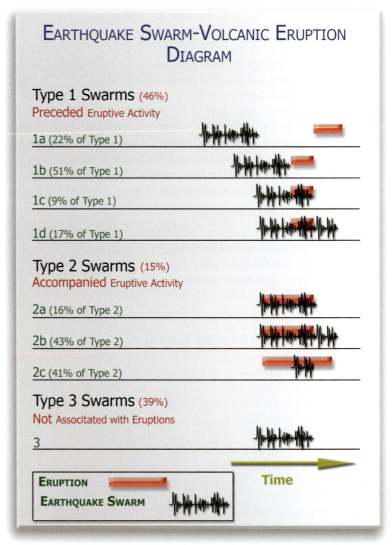

Fig 5.3.5 – Earthquake Swarm-Volcanic Eruption Diagram showing over 600 earthquake swarms occurring during or to volcanic eruptions. Adapted from Benoit and McNutt.

"The **possibility** of frictional melting during faulting has been **suggested** by several investigators." Note 5.3g

Perhaps at that time, it stretched the imagination too far to suggest the possibility that friction might be the cause of melting because few actual measurements of heat generated in faults had been taken. As other experts in the field note in the following quote, they simply do not know:

"**The problem of heat generation on fault surfaces has yet to be satisfactorily resolved**. It appears likely from the above discussion that different faults may exhibit different behavior in this respect, perhaps because of different degrees of lubrication related to pore-fluid pressure. As numerical modeling techniques improve, **and more heat flow data are collected from the vicinity of large faults**, the question may be answered. **However, for now there is no simple solution as to how much frictional heat is generated by faults**." Note 5.3h

Despite the geologists' admission that they ***do not know*** how much frictional heat faults generate, some claim that faults do not produce heat sufficient to create lava:

"The presence of faults, however, accounts **only** for the ability of magma to reach the surface; **it does not explain why the magma [intrusive lava] is produced in the first place**." Note 5.3i

From the *Science* article previously cited, researchers recognized significant heat generation during a seismic event in Bolivia, in 1994:

"The amount of nonradiated energy produced during the Bolivian rupture was comparable to, **or larger than, the thermal energy of the 1980 Mount St. Helens eruption** and was sufficient to have melted a layer as thick as 31 centimeters." Note 5.3j

The enormous 1980 Mount St. Helens eruption, compared by some to an atomic blast, generated an immense quantity of heat energy, so why is it that questions remain unasked about how heat impacts melting during earthquake events? Is this an important factor or not? Here is the response from the same journal article:

"These studies indicate that frictional melting **can occur** if the stresses involved in faulting are sufficiently high. **Despite these studies, frictional melting is not generally regarded as an important process during earthquake faulting because of uncertainties in the stress levels…**" Note 5.3j

Amazingly, these scientists observed an astonishing amount of heat generated in the fault area of the Bolivian quake where the melted thickness was only 3.7 mm:

"If the thermal penetration depth, Delta d = 3.7 mm, is used, the local temperature rise is of the order of **52,000 Celsius**." Note 5.3j p840

It only requires 1,700°C (3,100°F) to melt silicate rocks, including quartz, and the researchers determined that this earth-

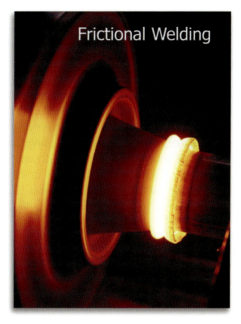

Fig 5.3.6 – Modern 'frictional welding' welds dissimilar metals with frictional heating in a fraction of a second. Courtesy of American Frictional Welding, Inc.

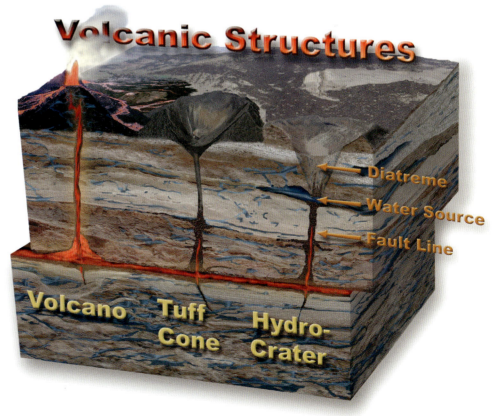

Fig 5.3.7 – This diagram illustrates different types of Volcanic Structures resulting from frictional heating. The most recognizable structure, volcanoes are not the only example. Tuff cones and hydrocraters remain less known due to a lack of viewable eruptions in modern times. Large earthquakes in the past caused massive steam explosions, which formed the various craters, and mountains.

SUBCHAPTER 5.3 THE LAVA-FRICTION MODEL

quake had generated more than 30 times that amount of heat!

In another research article, published in 2005, one investigator explained:

"Melt lubrication can explain the scaled energy of large earthquakes being 10 to 100 times more than that of small earthquakes." Note 5.3k

He noted that by rotating granite under pressure, kinetic friction increased in direct proportion to temperature until it reached a critical value corresponding with the melt temperature of feldspar (~1150°C), an important constituent of granite. If this critical temperature:

"…is exceeded in natural slip systems, which will depend on rock type, **this can result in the generation of friction melt.**" Note 5.3k

Geologists generally regard faults as conduits through which magma reaches the surface, but they are reluctant to think of them as the source of heat able to produce lava. When lava does flow from faults, they recite the sci-bi that the fault simply provided the avenue for 'magma' to escape. Earthquakes occur frequently without an occurrence of lava, leading some investigators to make the assertion that there is no melting from fault movement. However, volcanic eruptions are rarely, if at all, the result of large-scaled, single earthquakes, but rather are the result of *earthquake swarms, often lasting a week or more at a time*. Quick releases of energy under pressure produce bursts of high temperatures and some amount of residual heat, but it is the prolonged and relentless movement of the crust that stores energy and produces the frictional heat sufficient to melt rock.

The Frictional-Heat Law

Having considered the evidence for the origin of the Earth's internal heat, we can now recognize a new natural law identified with the Lava-Friction Model:

The Frictional-Heat Law
Frictional heating produces lava from pressure and movement in fault planes.

There are many variables contributing to the production of lava and nearly all of them connect either directly or indirectly to heat generated from friction.

Earthquakes are often an indicator of Earth movement, but not always. Over the past several years, scientists discovered a *new kind of earthquake* heretofore unknown—**silent earthquakes.** Slow-moving shifts in the Earth's crust remained undetected until only recently:

"In early November 2000 the Big Island of Hawaii experienced its largest earthquake in more than a decade. Some 2,000 cubic kilometers of the southern slope of Kilauea volcano lurched toward the ocean, releasing the energy of a magnitude 5.7 shock. Part of that motion took place under an area where thousands of people stop every day to catch a glimpse of one of the island's most spectacular lava flows. Yet when the earthquake struck, no one noticed—not even seismologists.

"How could such a notable event be overlooked? **As it turns out, quaking is not an intrinsic part of all earthquakes.** The event on Kilauea was one of the first unambiguous records of a so-called **silent earthquake, a type of massive earth movement unknown to science until just a few years ago.**" Note 5.3l

Investigators observed that over a period of 36 hours, the ground shifted 10 centimeters, detected because of newer, more sensitive instruments. The discovery of these massive, silent Earth movements ensures that scientists must re-examine "long-held doctrines about all earthquakes:"

"**If future study reveals silent earthquakes to be a common feature of most large faults, then scientists will be forced to revisit long-held doctrines about all earthquakes**. The observation of many different speeds of fault slip poses a real challenge to theorists trying to explain the faulting process with fundamental physical laws, for example. **It is now believed that the number and sizes of observed earthquakes can be explained with a fairly simple friction law.**" Note 5.3l p91

The connection between earthquakes, friction, and lava is clearer because slow, silent earthquakes, which result in massive, albeit quiet Earth movements, generate incredible heat by friction. That quiet movement is the result of the gravitational influence exerted by other astronomical bodies.

Earthtide—An Origin of Earthquakes and Lava

At the beginning of this subchapter, we used the analogy of rubbing our hands together to illustrate the effects of friction. Imagine that experience extrapolated for 24 hours a day, 7 days a week. The Earth experiences a phenomenon very much like that, a scenario called **Earthtide**:

Earthtide: The daily tidal movement of the Earth's crust.

Almost everyone understands something of the constant cycle of tidal motion that affects the oceans, resulting in high and low tides twice daily, and it is well established that the Moon drives these ocean-tidal events. A much lesser known fact though, is that there are tidal movements in solid ground too. From the *Glossary of Geology* (2005):

"…Earth tides have a fluctuation of between **seven and fifteen centimeters**" Bib 173 p200

This means the ground under your feet travels up and down from three to six inches every day! The same force that causes the ebb and flow of ocean's tides triggers *Earthtide*, but we cannot feel this movement because it happens so slowly. It is detectable using advanced scientific instruments and satellite technology. Three to six inches may not sound like a lot, but when we consider that there are thousands of feet of solid rock beneath our feet, and that it moves up and down every day, we can begin to appreciate the magnitude of this geological fact. Few people, including many scientists, know about Earthtide. Science organizations, like NASA recognize Earthtide, but seem to minimize its effects and its contribution to the production of lava and earthquakes:

"Earth has solid ground tides too, but they amount to less than 20 centimeters (about 8 inches)." Note 5.3m

Over time, stress builds in the crust of the Earth because

> "It is now believed that the number and sizes of observed earthquakes can be explained with a fairly simple **friction law**."
>
> Peter Cervelli

of the constant, daily up-and-down movement caused by the gravitational influence of the moon, accumulating until released in an earthquake. Frictional heating also happens, but the amount of heat generated depends on the magnitude of the seismic activity and the pressure present at the movement location; increased pressure equals increased heat. Therefore, we can expect higher temperatures as the depth of the seismic event increases. Later in this chapter, we will see physical evidence for this phenomenon.

Lunar Earthquake-Eruption Cycle Evidence

If the Moon's orbit affects the movement of the Earth's crust, then evidence of the connection between the Earth's diurnal (daily) rotation, the lunar cycle, and earthquakes should exist. We see evidence of this association at the most active volcano in the world, which is in Hawaii. In 1988, scientists announced in the *Journal of Geophysical Research*:

"Between 1967 and 1983, four earthquake **swarms** occurred on Kilauea Volcano, Hawaii, with durations ranging from 68 to 156 hours. Plots of the number of events per hour **show a remarkable modulation having diurnal and semidiurnal periodicities**…tidal influences appear to be **the best explanation for the modulation of the activity**." Note 5.3n

The April 2003 issue of *Scientific American* included an article on another of the most active and impressive volcanoes in the world, Mt. Etna. This volcano, an active participant in world history by destroying human settlements over many centuries, is located on the island of Sicily, in Italy. Here is what researchers had to say about the discharge of steam from Etna:

"Besides lava flows, Etna produces an almost constant, **rhythmic discharge of steam**, ash and molten rock." Note 5.3o

Could there be a cycle hidden in Etna's volcanic activity connected to the Earth's rotation period or to the orbit of the Moon?

"In February 2000, violent eruptions at Etna's southeast crater were occurring at **12- or 24-hour intervals**." Note 5.3o

This is strong evidence of a cyclical gravitational component affecting directly lava flows and earthquakes. The cycles of earthquakes and lava eruptions baffle researchers because they think within the magmaplanet paradigm. Etna's eruptive cycles are not mere 'coincidence' because similar cycles show up elsewhere, such as at Merapi, a volcano in Java. From the book, *Volcanic Seismology*, the author records the following observation respecting the October 1986 eruption of Merapi:

"Before the building of the lava dome, **a significant periodicity at 24 and 12 h in the occurrence of the shocks was noted**. This phenomenon was greatly reduced with the accumulation of outcoming magma [lava]." Note 5.3p

The Magmaplanet theory does not account for 'cycles' and indeed, it cannot because no mechanism exists to account for the cyclical movement of matter deep in the Earth, especially when the movement presumably happens very slowly—only centimeters *per year*. This is why earthquake cycles remain a mysterious puzzle for many researchers. In 2002, a husband-and-wife research team set out to find the connection between the Moon and volcanic activity:

"If predicting eruptions is a confusing puzzle, volcano hunters Steve and Donna O'Meara believe that they may have iden-

> "Earth has solid ground tides too, but they amount to less than 20 centimeters (about 8 inches)."
>
> NASA

tified a key piece. The husband-and-wife team are investigating **a connection that some volcano watchers have noted since early times**, but **none has adequately studied—the role of the moon in affecting volcanic activity**…

"The team's task was to determine when the greatest peaks in eruption activity occurred, and what connection the increased activity might have with the moon's gravitational pull. Following the patterns they had seen in the past, the O'Mearas predicted that during the volcano's ongoing eruptions, there would be peaks in volcanic activity at perigee and at full moon. In this case, events bore out that hypothesis and **in fact the greatest spike in volcanic activity occurred at a point in time just between full moon and perigee**." Note 5.3q

The team discovered a ***factual***, *direct link* between the position of the Earth and the position of the Moon, related to volcanic activity. One would think that with this earthquake-eruption cycle evidence, science should take a hard look at the so-called 'cause' of volcanic eruptions. The evidence of predictable periodicity continues to mount, especially with the discovery of "silent earthquakes":

"…the discovery of **silent earthquakes** is forcing scientists to reconsider various aspects of fault motion…One **curious feature of these silent earthquakes is that they happen at regular intervals**—so regular, in fact, that scientists are now predicting their occurrence successfully." Note 5.3r

Recent discoveries of Earthtide and silent earthquakes mark just the beginning of new discoveries that help us better understand the gravitational influence of celestial bodies and their effects on Earth movements.

The Earth Breathing Evidence

Is the Earth breathing? In a geological sense—yes:

"Now, some suspect that **Earth is also 'breathing,'** compressing its crust and extending it once each year. This **cycle** is most evident in Japan, geophysicists told the meeting, where it may be responsible for that country's **'earthquake season.'** Elsewhere, it may lead some volcanoes to erupt **almost solely between September and December**." Note 5.3s

Fig 5.3.8 – This Mt. Etna volcanic eruption plume over the island of Sicily, Italy, comes to us courtesy of a NASA satellite. Contrary to what modern geology teaches, there are cycles in volcanic activity, like 12 and 24-hour event intervals present on Mt. Etna.

This article, *Earth's Breathing Lessons*, from the journal, *Science*, demonstrates just how strong the periodic correlation can be. In the case of Pavlof, an Alaskan volcano, there is a **99% correlation**:

"GPS and strainmeters aren't the only things that seem to pick up planetary breathing. McNutt reported identifying four volcanoes—Pavlof in Alaska, Oshima and Miyake-jima in Japan, and Villarica in Chile—**that have a strong preference for erupting between September and December a preference significant at better than the 99% level in the case of Pavlof.**" Note 5.3s

Because geophysicists, stuck in a magma paradigm, think earthquakes come from magma, they seem reluctant to consider the earthquake-gravitational connection, making it difficult for them to be receptive to the significance of cyclical occurrences:

"**Geophysicists have traditionally shied away from making such connections**. 'In the past, we've tended to dismiss things without an obvious physical mechanism,' says volcano seismologist David Hill of the U.S. Geological Survey in Menlo Park, California. **But** after California's 1992 Landers earthquake reached out hundreds of kilometers to trigger quakes by some still mysterious means, Hill, for one, **became more receptive**. At the meeting, he says, '**I was struck that there may be something' to Earth breathing and eruptions or earthquakes.**" Note 5.3s

The Earth is not the only celestial body on which we know quakes occur. Apollo astronauts left four seismometers on the Moon, which registered around 12,500 'moonquakes' or seismic events during the 1970's. What evidence is there of a gravitational connection with the Moon?

Moonquake Tidal Evidence

Geologists rely on the magma theory to explain the origin of quakes on Earth, supposing that slow moving magma inside the Earth causes those earthquakes. The force behind the movement is the supposed upwelling magmatic currents. The Moon also has quakes, but, unlike the Earth, geologists think that the Moon has little or no remaining internal heat:

"The Moon, a body much smaller than the Earth, **lost its internal heat relatively early in its history**. As a result, **it ceased to be an internally active planet about a billion years or more ago.**" Bib 133 p193

There are no volcanoes or active lava flows on the Moon—but there are moonquakes, therefore, if the Moon has no internal heated magma to cause quaking, why do they exist? From the book, *Melting the Earth,* the author states that the Moon is "dead" inside, and that "tidal forces exerted by the Earth" cause cycles of moonquakes:

"When the Apollo 12 seismometers detected the first moonquakes in November 1969, scientists got a direct confirmation that **the Moon is 'dead' inside, harboring no volcanic energy**. Moonquakes, it was found, originate about 600 to 800 km (375 to 500 mi) below the surface, are highly localized, and **occur at intervals of about fourteen days. Apparently they are triggered by the tidal forces exerted by the Earth.**" Bib 136

These researchers said that tidal forces "apparently" cause deep moonquakes. Thousands of moonquakes recorded over several years, prove without a doubt that deep moonquakes happen, and they must be the result of tidal forces because they happen on *predictable cycles*. From the Fourth Lunar Science

What causes lava and the heat in the crust?
Answer: the daily Earthtide.

Conference:

"Comparison has revealed that many of the long-period lunar seismic signals match each other in nearly every detail throughout the entire wavetrain. Forty-one sets of matching events have been identified thus far. Matching signals of each set are generated by **repetitive moonquakes which occur at monthly intervals at one of forty-one moonquake hypocenters**…

"Each hypocenter is active for only a few days per month at a characteristic phase of **the tidal cycle**. The number of moonquakes observed in an active period ranges from none to four. Approximately equal numbers of hypocenters are active at nearly opposite phases of the **monthly tidal cycle**, thus accounting for **the observed semi-monthly peaks in moonquake activity.**" Note 5.3t

The researchers described other tidal cycles connected with the lunar orbit, stating that the "dominant source of energy released as moonquakes" consists of *"tidal energy":*

"This long-term decrease in activity appears to correspond to the 6-year variation in tidal stress which results from variations in the relative phase relationships among several of the **lunar orbital parameters. The strong correlation between moonquake occurrence times and energy release and lunar tidal amplitudes and periodicities suggests that tidal energy is an important, if not dominant source of the energy released as moonquakes.**" Note 5.3t p2522

Therefore, the Moon's quakes, which are similar to the quakes on our Earth, are not caused by magma inside the Moon but by "tidal energy" caused by the celestial dance of the Moon and Earth.

Perhaps you may be wondering why moonquakes do not

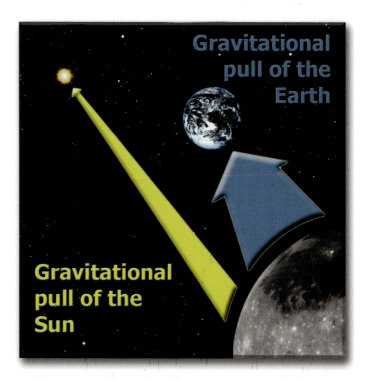

Fig 5.3.9 - The gravitational tug of Earth and Sun, not magma, causes 'moonquakes' on the Moon.

cause lunar volcanoes. Compared with the Earth, moonquakes are surprisingly infrequent based on seismic data gathered from seismometers placed on the Moon by Apollo astronauts between 1969 and 1972. Of four types of moonquakes, only the shallow quakes, within 20-30 kilometers of the surface appeared significant. The seismometers counted 28 of them over a five-year period, from 1972 to 1977, with the largest registering 5.5 on the Richter scale, according to NASA. While incomplete, the count, limited to the areas where Apollo missions allowed instrument placement, demonstrated the significantly less frequent number of moonquakes as compared with Earthquakes, which occur at the rate of **several million each year**, according to the USGS. Such reduced seismic energy apparently does not elicit visible volcanology on the Moon today. One reason for this is that the Moon's more solid geological interior is different from the Earth's more liquid interior. In addition, the Earth rotates once every day on its axis whereas the same side of the Moon always faces the Earth. These and more lunar topics will be explained in subchapter 7.13, the Hydromoon Evidence.

The Solar-Cycle Evidence

Although we have repeated this many times, recall the modern geology belief of earthquake origins:

"Earth scientists **believe** that most earthquakes are **caused by slow movements inside the Earth** that push against the Earth's brittle, relatively thin outer layer, causing the rocks to break suddenly." Note 5.3u

What causes the slow movements inside the Earth? Presumably, magma, a substance investigators have never actually seen. From a scientific journal article in *Nature*, we read of earthquake researchers' difficulties understanding what really triggers earthquakes:

"The mechanism responsible for the triggering of earthquakes **remains one of the least-understood aspects of the earthquake process**." Note 5.3v

There is a discrepancy between what the geological community *believes* generally (that magma causes earthquakes) and what hands-on earthquake research shows. A major difference is due to the cycles that are associated with earthquakes. Not only do lunar cycles exist, annual solar cycles exist too. The same scientists that authored this article found an *annual cycle* in the earthquakes they studied:

"The magnitude-7.3 Landers, California earthquake of 28 June 1992 was followed for several weeks by triggered seismic activity over a large area, encompassing much of the western United States. Here we show that this triggered seismicity marked **the beginning of a five-year trend**, consisting of an elevated microearthquake rate that was modulated by an **annual cycle**, decaying with time. **The annual cycle is mainly associated with several hydrothermal or volcanic regions** where short-term triggering was also observed." Note 5.3v

Monthly and yearly cycles that affect earthquakes make no sense within the magmaplanet paradigm; the magma model cannot account for them. Deep earthquakes and other evidences demonstrate that there is no molten liquid moving in daily cycles. Knowledge of the Gravitational-Friction Law, the association of earthquakes with the daily, monthly (lunar), and yearly (solar) cycles makes it possible to better understand the whole picture.

Other astronomically related factors affect earthquake movement. One of those factors is the annual snowmelt that occurs in the spring and summer:

"'Its been known for 20 years that we have **more inland earthquakes in spring and summer**,' says geologist Kosuke Keki of the National Astronomical Observatory in Iwate, Japan." Note 5.3w

Melted snow water lubricates fault systems, which makes them susceptible to slippage. The weight loss that occurs with snowmelt also contributes to movement. The *Nature* article continues:

"'It's just like taking a weight off a spring,' says Heki.

"In snowy areas, powerful earthquakes were **three times more likely to occur in spring and summer** than in autumn or winter, he found." Note 5.3w

Here again we have evidence of cyclical events that affect earthquake activity.

Rock-burst Tidal Evidence

Evidence connecting quakes on the Moon and volcanic eruptions on the Earth with the Earth-Moon orbital system are now more evident with acceptance gaining some ground. The cyclical, violent eruptions of the Merapi volcano in Java, Indonesia, one of the most active volcanoes in the world, provides us with new evidence about cyclical, silent earthquakes that are undeniably tied to the tidal forces of the Earth-Moon dance. And the evidence does not stop with these 'coincidences.' If lava actually comes from frictional heating, which originates from earthquakes caused by Earthtide, which lunar and solar cycles affect, we should see evidence everywhere we look.

One place this evidence appears is in subterraneous mines. The constant threat of injury or death to a miner comes in many forms: bad air, ceiling collapse, explosion, and fire to name a few. Another threat of injury happens when solid rock suddenly bursts, hurtling missiles of rock shrapnel, caused by tidal flexure:

"There is a problem of predicting the impulsive, rapidly changing stressed state of the Earth crust. In some cases **these changes are accompanied by earthquakes**. **Numerous observations have shown that rock bursts** (for instance, in the mines at the Khibiny Massif) **correlate with lunar-solar tides**. Rock bursts often bring considerable threat to life and property and sometimes cause fatal accidents." Note 5.3x

The Russian geologists who studied these events, correlated the *earthquakes and rock bursts with lunar-solar tides*.

Juan de Fuca Ridge Tidal Evidence

Off the northwest coast of the United States lies the Juan de Fuca Ridge (Fig 5.3.10), an intensely studied area of seismic activity. An article published in the October 15, 2001 *Geophysical Research Letters* included evidence of daily low tidal triggering, related ocean tides, and micro-quakes with seasonal periodicity:

"**Earthquakes occur more frequently** near low tides, especially the lowest spring tides, when the extensional stresses are at maximum in all directions." Note 5.3y

This evidence adds further proof correlating earthquakes and tides with solar and lunar cycles. Apparently the evidence had been overlooked in the past but is now clear because of improved technology and measurements.

Large Earthquake Tidal Evidence

Having identified direct evidence confirming the association between micro-earthquakes and tides, we ask, "Are large earthquakes also affected by tides?" A 1995 study involving 988 globally distributed earthquakes with a magnitude of 6.0 or greater as reported in the *Geophysical Journal International*, appeared to be the first in which science incorporated the effects of ocean tide loading at the depth of the hypocenter into the research. In this comprehensive report, the researchers noted:

"The highest population of normal-fault-type earthquakes appears at the time of maximum extensional stress, implying that a decrease in the confining pressure **due to the earth tide is responsible for triggering earthquake occurrence**." Note 5.3z

The researchers also noted in the same study that:

"Indeed, there are **many studies that have reported a positive correlation between the earth tide and earthquake occurrence as well as correlation with volcanic eruptions**."
Note 5.3z

Next, we examine the correlation between earthquakes, Earthtide, and geysers.

Geyser Tidal Evidence

Fig 5.3.11 depicts famous Old Faithful Geyser at Yellowstone National Park in Wyoming, USA. Truly an amazing place, Yellowstone is the most geologically active location in the continental US with more than double the geysers anywhere else in the world, including twice that of a geyser park in Russia, and more than ten times the geysers at the number three spot, Geyser Park Rotorua, New Zealand.

For decades, scientists observed the mechanics of Yellowstone's erupting geysers. Most geysers fill empty cavities before an eruption can take place, but often forgotten (and certainly not well understood) is the heating of the water. In the book, *The Geysers of Yellowstone* we find:

"Once the plumbing system is full, the geyser is nearly ready for an eruption. **Often forgotten but extremely important is the heating that must occur along with the filling**. Only if there is an adequate store of heat within the rocks lining the plumbing system can an eruption last for more than a few seconds. (If you want to keep a pot of water boiling on your stove, you have to keep the fire turned on. The hot rocks of the plumbing system serve the same purpose.)"
Bib 134 p5-6

If magma causes the heating then it should behave somewhat like a stove, slow and constant. However, this is not how it seems to work.

At 11:37 P.M. on the night of August 17, 1959, a large 7.5 earthquake rocked Yellowstone. The earthquake and the tremors following it caused hundreds of geysers to erupt:

"One of the greatest and longest-lasting **reminders of the quake** was its effect on the geysers and hot springs. On the night of the tremors and within the next few days, **hundreds of geysers erupted**, including many hot springs that had not previously been known as geysers." Bib 134 p14

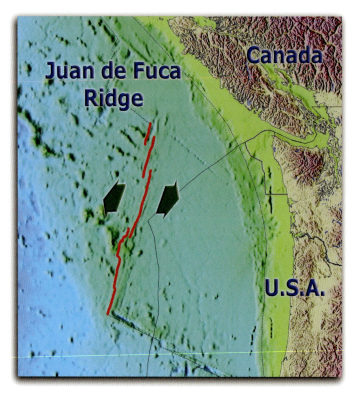

Fig 5.3.10 – This is the Juan de Fuca Ridge off the northwest coast of the US, where spring tidal movements trigger earthquakes which occur more frequently at low tide.

This was not the only year when large earthquakes caused or changed geyser eruption patterns in the area. In the years 1975, 1983 and 1994, earthquakes released pressure and tension in the area's faults affecting geyser activity. No doubt, the seismic activity generated heat in the faults. Moreover, the frictional heat from Yellowstone's earthquakes affects the area's hydrothermal waters. Surely, scientists should study this heat source more thoroughly. When the 1959 earthquake occurred in the Yellowstone area, causing hundreds of geysers to erupt, what caused the eruptions?

"Exactly what caused these eruptions is **difficult to say**..."
Bib 134 p15

Small earthquakes are not rare at Yellowstone; they are, in fact, quite common:

"**Every year, in the Yellowstone Plateau, up to 2,000 tremors are recorded** by seismographs. Mostly far too small to be felt by people, these quakes are **normal events**."
Bib 134 p17

Why is it so "difficult to say" what causes eruptions?—Modern geology is simply not aware of the Gravitational-Friction Law.

With the Lava-Friction Model, geologists can now begin to unravel the "mysteries" that previously had no identifiable "cause." Consider the eruption behavior of Geyser Hill in Yellowstone:

"Most, if not all, of the hot springs on Geyser Hill are connected with the other members of the group. Exchange of function is extremely common. There are also **two mysteries** that have only recently been dis-

Fig 5.3.11 – Old Faithful Geyser in Yellowstone National Park, USA.

CHAPTER 5 THE MAGMA PSEUDOTHEORY

covered, **neither of which has a clearly identified cause. One is a tendency for several of the geysers to exhibit diurnal behavior** – in general, short intervals by day and substantially longer intervals at night. Superimposed upon this is a weekly cycle known informally as the 'Geyser Hill Wave,' which causes **fairly regular weekly increases and decreases in the intervals of many geysers in the group**. It takes a great deal of observational experience to become familiar with these changes, but understanding them makes the activity on Geyser Hill anticipatable if not **outright predictable**." Bib 134 p30

Here again is evidence that links geysers to earthquakes, that geysers have daily and weekly cycles that make their activity sometimes "outright predictable." Curiously, the 'predictability' makes geysers more mysterious to the experts. Adding to the mystery, scientists do not think geyser water heating is on a cycle, yet there it is, in a predictable pattern.

It appears that the geyser's tidal cycles, which directly link to earthquake activity, remain almost unnoticed in the geological community. The data do not fit within the magma model, where scientists suppose that a large pocket of molten magma resides beneath the Yellowstone crust. Perhaps investigators are not aware that the gravitational forces of the Sun and Moon trigger earthquakes, and that when these gravitational forces are at their peak, or at their minimum, stress changes within the crust become even more evident, which demonstrates how observable cycles connect the earthquakes:

"It is well known that strains associated with solid body tides influence geophysical phenomena (Melchior 1966), such as moonquakes (Latham et al. 1971), volcanic activity (Wood 1917), tilt (Rinehart and Stepp 1973), and geyser activity (Rinehart 1972a). While most tidal effect studies have concentrated on the semidiurnal and diurnal components, the effects of the longer components, **fortnightly, semiannual, and 4.4 and 18.6 years, appear to be major agents in bringing about change**.

"Observations indicate that individual geysers correspond differently to the various tidal components. The performance of Riverside Geyser in Yellowstone National Park, U.S.A. is influenced by the fortnightly and semiannual components, an increase in variation in gravity causing the geyser to erupt more frequently. The high correlation between eruption interval and the variation in earth tidal forces during May and early June 1967, and again from late July through August occurs when the fortnightly component associated with the new moon and full moon is especially evident…

"The strong response of the Calistoga, California, geyser to the **4.4 year tidal component…** high tidal forces again shortening the interval.

"The **18.6-year tidal component can influence dramatically the action of a geyser**. The interval between eruptions of the large and spectacular Grand Geyser, Yellowstone National Park, has varied seasonally over some 4 decades of observations. During times of high tidal force, around 1930, 1955, and 1970, it erupted 2 to 3 times daily, only to become almost dormant during the years 1943 and 1960 when the tidal force was the lowest. Two other large geysers in Yellowstone, Beehive and Giant seem to respond similarly to Grand." Note 5.3aa

The 4.4 and an 18.6-year lunar nodal cycle appears in other natural occurring phenomena that also correlates with Sun-Moon-planet orbits. Rinehart's 1976 paper includes excellent graphs that illustrate the 4.4 and 18.6-year cycle of both geyser eruption and the Moon. The researchers then list 19 such scientific studies with positive correlations between Earthtide, earthquakes, moonquakes, volcanic activity, and geyser activity.

The Gravitational-Friction Law

The Frictional-Heat Law identified the *origin of heat*, explaining how lava comes from heat generated by friction. The Gravitational-Friction Law identifies the *origin of movement* that creates friction.

> ### The Gravitational-Friction Law
> Frictional heating in the crust of celestial bodies is caused by the gravitational pull and release of the crust by other celestial bodies.

The constant daily movement of the crust is a result of Earthtide. This has huge implications. The Gravitational-Friction Law illustrated in Fig 5.3.13 shows how friction between plates generates the lava for volcanoes. The daily upward and downward movement of miles of crust causes the buildup of tremendous stress. With the alignment of the Sun and Moon during the full or new Moon phases, the movement is greatest, and when the Sun or Moon is at either its closest or furthest point from the Earth, the movement can be even greater, possibly triggering a release of accumulated stress energy in the form of a large earthquake. Earthtide provides evidence of how increased pressure and frictional grinding can produce heat in the crust.

Is it coincidence that at the end of 2004, Mt. St. Helens, the Hawaiian Volcanoes, Mt. Etna, and others began erupting more than usual, and then on December 26, 2004, a 9.1 earthquake, the largest in decades, struck the west coast of Sumatra triggering a series of devastating tsunamis? There are cycles throughout Nature, and they connect earthquakes and volcanic activity with astronomical cycles. We will discuss additional evidence of astronomical cycles in the Weather Model chapter, where we also discuss the worldwide eruptions and earthquakes that occurred toward the end of 2004, as well as the subsequent increase in hurricane activity the following year, in 2005. Were these events also coincidental? Once we understand the connection between the astronomical cycles, earthquakes, and frictional heating, we can begin to predict, at least generally when and where these events might take place. If we know the origin of movement, then we can understand why

Fig 5.3.12 – Aurum Geyser in Yellowstone National Park, USA. Courtesy Jim Peaco

movement occurs and we can begin to predict the results of that movement also.

Earth Movement by Water

Is there a liquid other than magma that could cause earthquakes and the ground movement beneath our feet? From an article in *Nature*, we read:

"Geologists knew that the use of rocks to store water (aquifers) could cause ground movement, but not to this extent. '**It's the magnitude of the deformation that's so overwhelming**,' says John Shaw, who studies the geology of the Los Angeles basin at Harvard University in Cambridge, Massachusetts…

"The largest area of moving ground – measuring some 800 square kilometers of greater Los Angeles – is **rising and falling by as much as 11 centimeters each year**, Bawden's group reports…

"Bawden began to suspect that the movement was due to thirsty people **rather than tectonic plates** when he noticed certain GPS sensors in the array moving **far more than a fault can in a year**." Note 5.3ab

The cycle noted by Bawden shows up in many other areas where GPS systems have allowed scientists to record similar observations. With this, we can see that water has a known effect on crustal movement because of cyclical wet and dry conditions.

Another example of how water can affect movement within the crust of the Earth comes from research data at the world's deep boreholes where researchers produced earthquakes by pumping water into them. In 1997 at the German KTB borehole, one of the deepest holes ever drilled, there were almost 400 micro-earthquakes induced at an average depth of 8.8 km when scientists injected water into the borehole. Note 5.3ac

Not only have relatively small amounts of water been shown to induce seismicity in boreholes, scientists frequently observe Earth movements that are associated with water. Naturally, the next question to ask would be; "are there events that happen in connection with the *displacement of huge amounts of water*?" The mechanism that facilitates the movement of large amounts of water, both on the surface and within the crust certainly exists. What is this mechanism? On the Earth's surface water, it is the ocean's *tides*, but what of the subsurface. Shouldn't we expect the movement of so much water to affect the entire area in which it moves?

The Ring of Fire Evidence

The 'Ring of Fire' makes up a semi-circular region of plate boundary bordering the Pacific Ocean; it is the most seismically active region in the world. It is extraordinarily active volcanically and is the location where almost half of the world's volcanoes sit. Fig 5.3.14 illustrates the intensity of the area. Red triangles represent volcanoes and black dots represent earthquakes. It is easy to see the geographic relationship between them. It is also notable that a great deal of volcanic and seismic activity occurs along the northern and eastern edge of the Pacific Plate where considerable friction occurs.

"Only about 10 percent of the world's earthquakes occur along the oceanic-ridge system, and they contribute only about 5 percent of the total seismic energy of earthquakes around the world. In contrast, **earthquakes occurring where plate boundaries converge, such as at the trenches, contribute more than 90 percent** of the world's release of seismic energy from shallow earthquakes, as well as most of the energy from intermediate and deep-focus earthquakes." Note 5.3ad

Where plate boundaries converge, a great amount of friction is present. It is here that frequent and massive earthquakes occur and as a result, significant lava production by frictional heating manifests. In Fig 5.3.14, the black lines that appear to delineate the plate boundary are in fact closely packed black dots, markers of measurable seismic events. Notice that earthquakes and volcanoes correlate well with plate boundaries, obvious evidence that frictional heating occurs there. But what of the mechanism that moves the plates? Could it be associated with the same mechanism that causes the tides?

Intraplate Earthquake Volcanic Evidence

On March 25, 1998, one of the world's largest earthquakes shook the ocean bottom between Antarctica and Australia. The quake 'befuddled' researchers because it violated the "usual rules" (theories) of modern geology:

"'It's really kind of a befuddling earthquake **because it seems to violate a lot of the usual rules**', says Douglas A. Wins, a seismologist at Washington University in St. Louis. Most giant earthquakes occur in distinct seismic zones, where two of Earth's surface plates scrape against each

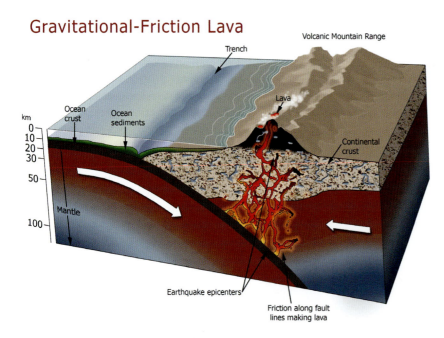

Gravitational-Friction Lava

Fig 5.3.13 – This diagram illustrates lava formation resulting from friction between continental plates, not from magma deep inside the Earth. Although magma has long been theorized as the source of lava, researchers are beginning to recognize that Earth's tidal forces cause frictional heating, and that heating is producing lava beneath volcanic centers.

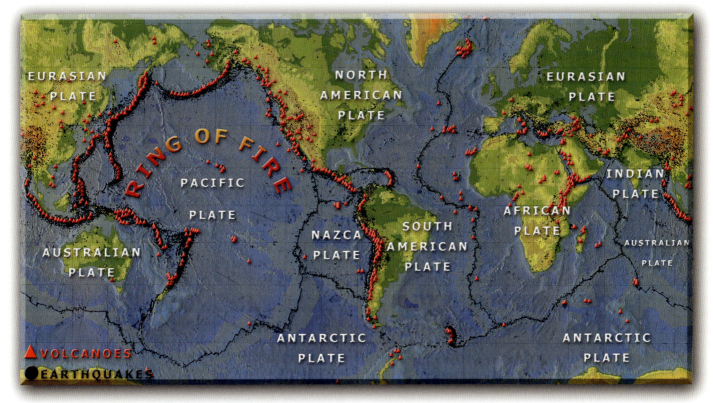

Fig 5.3.14 - The 'Ring of Fire' is a ring of earthquakes and volcanoes along Pacific Ocean plate boundaries. The lines are actually thickly clustered black dots, which are major earthquakes, the red triangles are volcanoes, both occurring where most of the friction takes place—along the plate boundaries, marked out by the lines of seismicity. It is there where the greatest daily rubbing of the daily Earthtide takes place.

other. The March quake, however, struck within the Antarctic plate nearly 350 kilometers from the nearest border with another plate, says Wiens. Seismologists call these sorts of events intraplate earthquakes.

"The researchers raise a number of possible theories to explain the quake, but **'none of the ideas are really that attractive,'** says Wiens." Note 5.3ae

We can see in Fig 5.3.14 that most earthquakes occur along fault lines where plate boundaries meet. The usual assumption is that magma oozes through cracks at the point where these zones or plates meet. However, there is no satisfactory explanation for ***intra-plate*** earthquakes, nor is there an explanation for mid-plate *volcanic* activity. The following is the sci-bi that attempts to explain how magma normally reaches the surface at plate boundaries:

"Because hot magma is lighter than the cooler, overlying rocks, it rises through buoyancy, welling up through great cracks or rifts that are produced in the ridges by the unremitting forces that pull apart the plates." Note 5.3af

What if there were no 'great cracks' or rifts like those found along plate ridges? How then would magma rise to the surface? One such example where mid-plate magma supposedly exists is the Yellowstone National Park caldera in the North-Western United States. Located high in the mountains and hundreds of miles from any plate boundary, the Yellowstone continental crust is relatively thick, roughly six times thicker than oceanic crust and there is no known 'great crack or rift' providing a path for the theoretical rise of magma to the surface. Other than the presence of heat, what evidence does science have that magma is burning its way through to the surface, or that a giant magma plume exists? They have none.

One has to stretch the imagination to see how or why magma would rise through miles of continental crust. With the Lava-Friction Model and the Frictional-Heat Law, the answer is quite simple. There is an abundance of intersecting and converging faults at Yellowstone that experience continual grumbling, even though it is far from plate boundaries. Of course, there are faults everywhere around the world, but what makes those at Yellowstone unique is the high degree of seismicity on *converging* and intersecting faults. Frequent *stress* build up results in numerous, small, continuously occurring earthquakes, which generate enormous frictional heat in the surrounding area, which in turn heats underground water fueling the geysers and other geothermal features. Constant tidal movement within the crust builds up stress along faults, and sometimes releases that stress in a large earthquake, creating even more heat through friction, which results in a rise in both magnitude and multitude of the spectacular displays around Yellowstone Park.

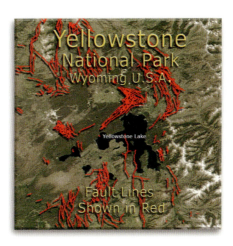

Fig 5.3.15 - Yellowstone National Park, USA boasts of intersecting faults that produced heat by earthquakes in the middle of the North America Plate.

Frictional Heat Realization in Taiwan

On September 21, 1999, 2,400 people died in the 7.6 Chi-Chi earthquake, the largest in over a century in Taiwan. This quake is significant for a number of reasons that we will refer to later on. Taiwanese researchers who studied the quake reported an interesting observation in the journal, *Geophysical Research Letters*. Instead of crediting magma as the heat source associated with this massive earthquake, they described frictional heating:

> "Even though the trapped water in the interconnected fractures was small, **once vaporized by frictional heat during the earthquake** it could cause a very large amount of pressure. When the pressure exceeded the lithostatic pressure or strength of rock, the huge rock eruption occurred." Note 5.3ag

The reoccurring theme is that friction can and does generate significant heat. Importantly, frictional heat occurs to some degree during *every* seismic event, and there are hundreds of seismic events every day, around the Earth. Hopefully, we are beginning to understand where intrusive and extrusive lava originate without looking to magma.

Denying the Earthquake Origin Evidence

Thus far, we have presented the following thirteen independent scientific evidences of the **Lava-Friction Model** and the **Gravitational-Friction Law**:

1. **Earthtide**
2. **Earthquake-Eruption Cycle Evidence**
3. **Earth Breathing Evidence**
4. **Moonquake Tidal Evidence**
5. **Solar Cycle Evidence**
6. **Rock-burst Tidal Evidence**
7. **Juan De Fuca Ridge Tidal Evidence**
8. **Large Earthquake Tidal Evidence**
9. **Geyser Tidal Evidence**
10. **Earth Movement by Water**
11. **The Ring of Fire Evidence**
12. **Intraplate Earthquake Volcanic Evidence**
13. **Frictional Heat Realization in Taiwan**

Each of these evidences demonstrates that astronomical cycles affect and are associated with earthquake behavior. This means that we might be able to calculate where or when earthquake events are likely to occur using lunar or solar orbital calculations. When the Sun and Moon align and the Earth is at perigee (closest point to the Sun in its orbit), substantially higher gravitational stresses affect the Earth's water tide and Earthtide, more than is present when the Moon is not aligned with the Earth and Sun, or when Earth is at apogee (furthest from the Sun). We can use these actual evidences to apply a prediction-observation learn-

Fig 5.3.16 – Actual lava flows on the surface of Io, one of Jupiter's four largest moons. The lava comes not from magma, but from the Gravitational Earthquake Friction Mechanism. Courtesy of NASA

ing process and thus utilize the Universal Scientific Method to validate the Lava-Friction Model.

No one should think the evidences presented here are all-inclusive nor are they representative of all the facts that support the Lava-Friction Model. There are many more examples available and many yet to find for those willing to look.

What is so significant about these facts? By examining this evidence, it *should be apparent* that both earthquakes and moonquakes connect directly the tidal effects of the Earth and Moon, and that there is evidence correlating seismic and geothermal activity with orbital periodicities and other cycles. On the other hand, magma has no cycles and as such, at least from the geologist's tectonic-magma point of view, ***there should be no cycles in earthquake activity***.

> *There is no such thing as coincidence—we just don't comprehend the purpose yet.*

From the USGS' frequently asked questions website, we learn what geologists think about earthquakes and how they associate them with cycles:

> "Q: Are there more earthquakes in the morning/in the evening/at a certain time of the month?
>
> A: Earthquakes are **equally as likely to occur at any time of the day or month or year**. The factors that vary between the time of **day, month, or year do not affect the forces in the earth that cause earthquakes**." Note 5.3ah

> "Q: Can the position of the moon or the planets affect seismicity?
>
> A: **No significant correlations have been identified** between the **rate of earthquake occurrence and the semi-diurnal tides** when using large earthquake catalogs." Note 5.3ah

Although empirical evidence is available to everyone, if it does not fit the current magma model paradigm, modern science seems to reject it. Why is this so?

The Unequivocal Io Evidence

We have considered evidence of lava's origins by examining 13 events that happen on the Moon and the Earth. If it happens in these places, it should occur in other places as well, and we should be able to see evidence of it happening. An additional 14th evidence now discussed here seperately, is the Unequivocal Io Evidence of earthquake origin. We find this lava evidence as we look across the solar system to see one of the finest examples of the Lava-Friction Model at work. This incredible place, *the most active volcanic celestial body in our solar system*—is Jupiter's moon, Io (see Figs 5.3.16-18). Io is the innermost moon of Jupiter's four Galilean moons and has the greatest and most extensive volcanic activity known in the solar system. What is the secret as to why this volcanic activity is so remarkable on Io?

100-meter tidal bulges!

As NASA states it:

> "Solid tidal bulges on Io are about **100m high**, taller than a 40-story building!" Note 5.3ai

The surface of Io is rising and falling 100 meters—the equivalent of an entire football field—every day, which is about 42 hours on Io. Fig 5.3.17 illustrates the forces at work on Io, which can come from many directions depending on the positions of Jupiter and the other Galilean moons. Scientists know that tidal frictional-heating occurs every day on this Jovian moon because they can see it happening. From where did the force come that causes such surface movement?

The following quote about Io comes from one of NASA's educational web sites. Note the *cause* of Io's 100-meter tidal bulge:

"Here the **gravity of Jupiter** and large moon Ganymede (with help from moons Europa and Callisto) play tug-o'-war, with Io playing the part of the rope! **Io bulges on two sides like a football**." Note 5.3aj

The events that play out on Io provide an excellent example of the **Gravitational-Friction Law** at work. Lava production there is not from theoretical magma; it is a direct result of frictional heating. Tidal forces exerted by other celestial bodies cause that heating. Researchers note the effects of gravitational tidal bending on Io (see Fig 5.3.18):

"At this time, **Jupiter and all three of the other large moons pull on the same side of Io**. Its orbit bends to pull it closer to Jupiter. Io is again squished like a football." Note 5.3aj

What effect does 'squishing' have?

"**All this bending causes heat to build up inside Io**. Io gets so hot inside that some of the material inside melts and boils and tries to escape any way it can. So it blows holes in the surface! **That's what volcanoes are**. Some on Io have shot their hot gas plume 300 kilometers (about 200 miles) into space!" Note 5.3aj

How much heat does Io actually release? It may be the most active volcanic body in the Solar System, but how much melt-

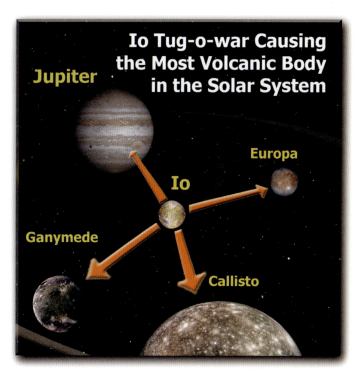

Fig 5.3.17 – Gravitational tidal forces act on Jupiter's moon Io, pulling it like a football, causing the greatest amount of volcanism in the Solar system.

ing is actually happening on Io's surface? Researchers studying data from the Galileo spacecraft reported in 2004:

"Io may be giving off **so much total heat**, the best explanation would be that **virtually the whole sphere is covered with lava spewed so recently it is still cooling**, new calculations suggest." Note 5.3ak

In summary, the **Frictional Heat Law** provides us with the *origin of heat*, and the **Gravitational-Friction Law** provides us with the mechanism, or the *origin of movement* within the Earth's crust that drives the frictional heat engine. We also considered the *results* of these two Laws, including 13 verifiable evidences. These three concepts make up the **Lava-Friction Model**, a model that explains where lava really comes from on Earth, and as a further 14th witness, Jupiter's moon, Io provides unequivocal evidence of the Lava-Friction model.

5.4 Magma Theory Defies Heat Flow Physics

In the previous subchapter, we demonstrated lava's origin through frictional heating within the crust as defined by the Frictional-Heat Law. In this subchapter, we turn our attention to the scientific evidence regarding the flow of heat through Earth's crust. An examination of the heat gradient from the surface toward the center of the Earth casts further doubt on the existence of magma. As we will show, this evidence, which was originally purported to prove the magma theory, actually refutes it. The observed facts show that the heat-flow of a *theoretical magmaplanet* does not follow the simple laws of heat flow physics.

Thermal History of Earth – A Problem of 'Enormous Difficulty'

Have you ever wondered where the thousands of degrees of heat comes from that keeps rocks melted? O. M. Phillips, a former professor of geophysical mechanics at Johns Hopkins Uni-

Fig 5.3.18 – Jupiter's moon Io experiences a 100-meter tidal bulge (vertical crustal movement) each day during its daily rotation and orbit around Jupiter. This is direct, empirical evidence of how the Lava-Friction Model works.

versity, wrote in the *Origin of the Earth*:

"It has already become apparent that **the thermal history of the earth**, its constitution and the distribution of various materials with depth, are reflections of the way in which the earth was formed. The contemplation of these matters has occupied many of the great geophysicists and mathematicians of the last hundred years, among them Poincaré, Jeans, Eddington, Jefferys, Lyttleton and Hoyle, **but no finality has been reached – cogent objections have been raised for every model proposed. The problem is one of enormous difficulty.**" Bib 63 p.151

The thermal history of the Earth and how the Earth was formed is a problem of enormous difficulty and has been ever since the magma pseudotheory was adopted many years ago. How could something *so basic* to the science of geology be of such *enormous* difficulty unless geologists simply do not know?

Physics is the study of matter and energy, including laws that describe and predict how heat energy flows. A magmaplanet, or any planet for that matter, should be subject to these simple physical laws.

Heat Flow Physics Defined

Heat flow is something we experience every day. Whether being warmed by a fire, feeling the touch of a loved one's hand or suffering the blast of a hot afternoon wind, we are witness to the constant flow of heat on Earth. Heat transfers by radiation, by conduction, by convection, or by some combination of these. Figure 5.4.2 provides an example of all three. Heat reaches the handle through conduction as heat transfers up through the glass. The water, heated first by radiation, is then subject to convection or circulation as water in the bottom heats and rises to the top. These methods of heat transfer have seen extensive study and they follow well-known heat transfer laws. In the Magma Pseudotheory, heat coming from deep inside the Earth transfers to the surface through convection currents from the inner core toward the mantle and ultimately to the crust. It is then conducted to the surface, through the crust. The flow of heat should be easily predicted and follow known patterns of heat transfer if the Earth's heat is actually coming from magma. However, if the Earth's heat is coming from some other source, such as through frictional heating in the crust, we shouldn't be surprised at the general confusion in modern geology because data was interpreted according to an incorrect theory.

How Magma Defies Heat Flow Physics

To understand how magma defies heat flow physics, there are two concepts to consider:

1. **Global** heat flow—theory (assumption) versus reality.
2. **Crustal** heat flow—theory (assumption) versus reality.

Thermal history of the Earth—
a problem of "enormous difficulty"
O. M. Phillips

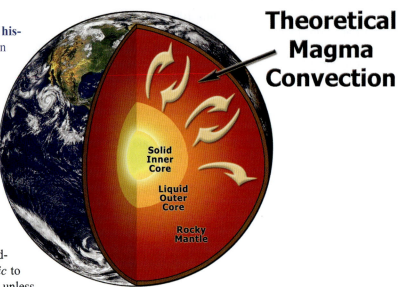

Fig 5.4.1 – Theoretical convection of heat from Earth's core to the surface as held in the Magma Pseudotheory has always been a problem of enormous difficulty and there is no empirical evidence to support the theory. As we read through this chapter, we find the physical evidence actually supports cooler temperatures as we move towards the center of the Earth— just the opposite of what the magma theory predicts.

Global Heat Flow Assumption Verses Reality

Fig 5.4.3 shows a map based on the Earth's crustal thickness. The thickest continental crust, shown as brown and yellow, contrast the thinner parts beneath the ocean showing as mid-blue and dark blue. The in-between areas of green represent an approximate median thickness.

How much heat should flow through the thinnest areas of crust as compared to the thick areas? With magma, heat should flow via conduction through the thinnest parts of the crust *more easily,* and those areas should register the highest amount of outward heat flow.

In Fig 5.4.4, the thinner oceanic crust appears red, based on the crustal thickness seen in Fig 5.4.3. These areas are expected to represent the area of highest heat flow. The thickest areas of Earth's crust are the continents, shown blue in this illustration, depicting what the heat flow of the Earth *should* look like *if* magma was the heat source. This is the **Magma Pseudotheory**

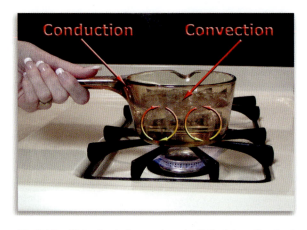

Fig 5.4.2 – Water heated on a stove (radiation) transfers heat through conduction (through the handle) and through convection (circular motion of water).

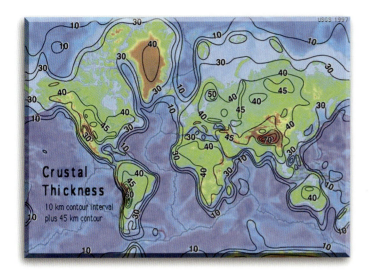

Fig 5.4.3 – The world's crustal thicknesses, indicated in kilometers and shown here as green and brown on the continents, average about 36km whereas the oceans, shown in blue, average only 6km. We used this map as a basis for the heat flow map in Fig 5.4.4. *Courtesy of USGS*

Heat Flow map.

For many years, geophysicists have taken actual detailed heat flow readings from around the world. From that data, a map was drawn showing the **Actual Heat Flow**, which appears in Fig 5.4.5. Compare this map with the Magma Pseudotheory Heat Flow map. Do you notice the surprising contrast between the two? The magma-model map predicts heat flow based on crustal thickness, as it should, because more heat should be emanating from the thinner crust, nearer the heat source. The actual heat flow map renders actual collected data and represents what really exists. The importance of this is clear when we notice *where* the hottest areas on the surface of the Earth occur.

The hottest areas on the Actual Heat Flow map correspond to plate boundaries—right where the greatest amount of gravitational friction occurs. The actual flow of heat from the Earth does not match the theoretical flow of convective magmatic heat.

The Actual Heat Flow data have been around for more than a decade, but apparently, no one before realized the significance of this comparison. Within the magmaplanet paradigm, the actual flow of heat makes no sense, but with the application of the Frictional Heat Law and the Gravitational Friction mechanism, the patterns of Actual Heat Flow make perfect sense.

The Crustal Heat Flow Mystery

Figures 5.4.6 and 5.4.7 both illustrate another crustal heat flow mystery. Geophysicists have not been able to explain why heat flow through the thin oceanic crust is less than the heat flow through the thick continental crust. The thicker and more insulated continental crust areas should have a *significantly lower amount* of heat flow whereas the thicker continental crust should theoretically be cooler than oceanic crust because of the distance from the heat source as predicted within magma theory.

The physics of heat flow tell us that heat travels or flows across a gradient from hot to cold and that the flow will be greater when the gradient is steeper. In other words, more heat flows from hot toward cold than flows from hot toward warm. The bottom of the ocean, at around 2°C (35°F) is much cooler than the surface of the continents, which averages approximately 14°C (57° F). Because of this, we should expect heat flow *to be greater* through the oceanic crust than through continental crust. Note that we are not referring to why the *surface* of the Earth is warmer than the bottom of the ocean—that is primarily due to solar heating. We are considering the flow of heat *through* the crust. Let us look at the known empirical facts.

The oceanic crust is a thin layer of dark basaltic rock between the bottom of the ocean and the mantle as illustrated in the heat

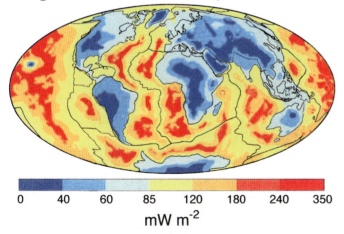

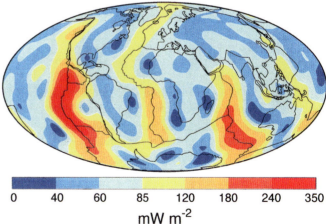

Fig 5.4.4 – This map illustrates the **assumed** heat flow through Earth's crust, based on the magmaplanet theory and the USGS map of crustal thickness (Fig 5.4.3). This map was produced using a color gradient derived from crustal thicknesses of the USGS map. Thinner crustal areas, those that should show the highest amount of heat flow are red while the thicker crustal areas, (brown and green areas on the USGS map) are colored blue, indicative of a lower heat flow. The theoretical Magma Pseudotheory Heat Flow map makes it possible to compare theory with observed data, shown on the Actual Heat Flow Map, Fig 5.4.5.

Fig 5.4.5 – This map illustrates **the actual** *measured heat flow* through the crust of the Earth. The greatest concentration of heat, shown in red and yellow, land on plate boundaries where gravitational frictional heating is highest. Compare the Actual Heat Flow map with the Magma Pseudotheory Heat Flow map in Fig 5.4.4. This demonstrates unequivocally that the Earth's heat flow through the crust cannot originate from a theoretical magma heat source beneath the crust, confirming the Frictional Heat Law and the Gravitational Friction Law.

Courtesy of H. N. Pollack, S. J. Hurter, and J. R. Johnson. – *Heat Flow from the Earth's Interior: Analysis of the Global Data Set, Reviews of Geophysics* 31(3), p267-280, 1993

flow diagrams. The average thickness of the oceanic crust, which is typically composed of three layers, is only around 6.4 kilometers (4.0 miles) thick:

"Seismic studies have shown that the **oceanic crust** is layered, and in most oceanic basins there are three main layers. Layer one is approximately 0.3 km thick, and it is essentially composed of sedimentary materials. Layers two and three have mean thicknesses of 1.4 and 4.7 km respectively…"
Bib 87 p92

Continental crust is around *six or more times thicker* than the oceanic crust:

"Average thickness of continental crust: **36 km**." Bib 140 p62

If you have ever slept under a blanket, you know that a thicker blanket means less heat loss and a warmer bed, because of the increased insulation. Fig 5.4.6 shows the six-times-thicker continental crust as compared with the thinner oceanic crust and the resulting hypothetical heat flow. Accordingly, the heat flow should be six times greater through the oceanic crust.

How does the **Hypothetical Heat Flow** of the magmaplanet model compare to the **Actual Heat Flow** of the Earth? Looking at the second illustration, Fig 5.4.7, one can clearly see the results are completely opposite each other. Researchers discovered that *more* heat flows from the *thick* continental crust than from the thin oceanic crust:

"For **oceanic crust**, where the temperature rises **about 15 degrees C per kilometer** of depth… For **continental crust**, where the temperature is often near 20 degrees C at the surface and typically increases by about **25 degrees per kilometer**…"
Note 5.4a

Another source of data where researchers measured continental crust heat flow is the German KTB borehole, a hole that is 9.1 km (5.7 miles) deep. The borehole exhibits a 27°C/km temperature gradient, which means the temperature rises 27°C for each kilometer of depth. Note 5.4b

How can this be? For decades, researchers have been trying to figure out how convective magma can generate *more heat* under a thick continental crust than it does through the thin oceanic crust. The following statement from the *Textbook of Geology*, published in 1939 describes how heat *should* flow out of the continental crust, very slowly as is depicted in Hypothetical Heat Flow diagram above:

"Rocks such as granite are extremely poor conductors of heat. Therefore if the temperature at a depth of several miles **should be** high, say 1000° C., **heat would flow out very slowly** and the change in temperature for each 100 feet **would be considerable**." Bib 125 p398

Of course, this isn't what actually happens, but why not?

Heat is generated *in the crust* because of frictional heating, which is driven by the gravitational friction mechanism. Daily Earthtide movement generates heat in the crust. More heat is generated in the thick continental crust than is generated in the thinner oceanic crust. Because knowledge of this process remains unknown, researchers refer to this excess heat in mountain ranges as the "orogeny paradox." Referring to the mountains of Tibet, geoscientists state:

"Like the crust, the upper mantle portion of the lithosphere beneath the plateau *should* thicken as the continental plates collide, which *should* **make the lithospheric mantle colder** and stronger. Yet, the upper mantle in this region, in contrast,

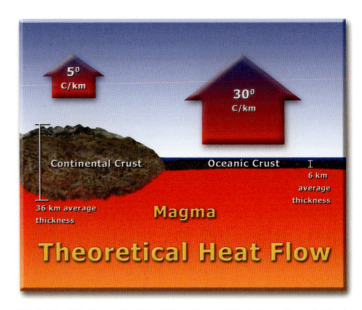

Fig 5.4.6 – This theoretical Heat Flow diagram illustrates a **Hypothetical** flow six times higher through the thin oceanic crust verses continental crust. This is what should take place if the heat source below the crust was magma. The thicker continental crust should act as an insulator as compared with the thinner oceanic crust, which should shed heat six times faster than the thicker continental crust.

appears not only to be weak, **but is also relatively hot**, as evidenced by the uppermost Tibetan mantle and the presence of active volcanism throughout much of the plateau. **That is hardly what we would expect from thickened, cold lithosphere**.

"This unexpected heat, common to many mountain-building regions, has been termed the orogeny paradox." Note 5.4c

Objectively questioning the magma theory, we consider that miles-thick rock in the crust acts as an insulator from the heat of magma, which could cause anomalous temperature readings. However, this does not hold true because there is a *consistently* greater flow of heat from continental crust than from oceanic crust.

The theoretical heat flow idea is further invalidated because

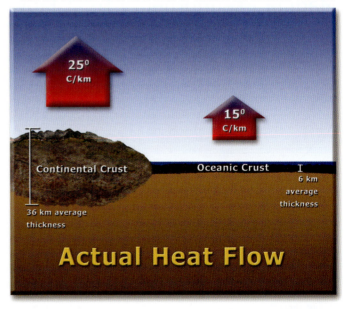

Fig 5.4.7 – The **Actual** Heat Flow Diagram shows how heat flows through the crust; oceanic heat flow is less than continental heat flow, contrary to magma theory. Thicker continental crust allows for increased *gravitational frictional heating*, which is confirmed by measured actual heat flow numbers.

of the presence of water. The crust is not made up of only rock. There is a great deal of water present. Vast worldwide aquifers run through the crust and, as we will detail in the Hydroplanet Model chapter, there are aquifers deep in the mantle. What does water do when it encounters heated rocks? It has an affinity for heat and loves to aid in its transfer. There is actual direct evidence of heat transfer from deep within the crust to the ocean floor. Once again, it is not heat from magma. It is the result of earthquakes and frictional heating:

"A swarm of microearthquakes on the East Pacific Rise **increased fluid temperature by 7° C at a hydrothermal vent located directly above the swarm—but only after a delay of 4 days**. Vertical cracks propagated into a previously isolated high temperature region of lower crust, and it took days to transfer this heat upward to the seafloor." Note 5.4d

This example demonstrates what happened to the temperature of the ocean water when a swarm of microearthquakes heated the oceanic crust. *If* hot melted rock existed everywhere under the oceanic crust, ocean water on the seabed should not be as cold as it is and there should be evidence of ocean-floor heating. Ascending toward the surface, the water temperature *increases* due to solar heating and because the atmosphere is warmer than the ocean's seabed, *except* when there are oceanic earthquakes.

The Geotherm Belief

The **geotherm** is a temperature-depth graph that describes how temperatures in the Earth change with depth.

> If there is a heat source in the center of the Earth, and if liquid around that heat source is moving heat toward the crust through convective currents, Earth's geotherm **must follow** convection heat laws.

Geologists admit they have very limited information about the temperature of Earth's interior:

"At present, **all geologists can do is draw certain conclusions from the limited information they have about temperature**." Bib 59 p497

Simply said, *geologists do not know the temperature of the mantle or of the core of the Earth.* Because of this, researchers can only infer what the temperatures are. Here is an example from a college geology textbook:

"They combined the temperature of lava that originates in the mantle and emerges from volcanoes, laboratory data on the temperatures at which rocks and iron begin to melt, and information from seismology to **infer** the geotherm from the surface to the very center of the Earth, where they **believe** the temperature rises to between 4000° and 5000° C." Bib 59 p498

The astonishing fact is that:

All of geology has been built on the Hot-Earth belief!

Why should science build on a basis that can only be inferred? Repeatedly, modern science has built complex theories and models on the shaky foundation of inference, the result being stifled growth and relatively little increase in the broad knowledge of how Nature works.

Deep Borehole Drilling and the Geotherm

Most scientists agree there is no substitute for direct observation. Seeking to learn more about the Earth, Russian scientists spent over a billion dollars (in US$) to drill the world's deepest borehole on the Kola Peninsula. Ultimately, they reached a depth of 12,261 meters (7.6 miles). What did the deepest hole drilled in the world show about modern magma planet concepts and ideas?

"Direct measurement of temperatures in the well **compels revision of ideas about the distribution and flow of heat in the earth's interior**." Note 5.4e

Why did researchers feel *compelled* to revise their "ideas about the distribution and flow of heat in the Earth's interior?" Data from the deepest hole drilled in the Earth contradicted heat flow predictions deduced or inferred from the standard magma-theory.

Too Hot too Fast

One factor that initially seemed to support the magma pseudo-theory was that boreholes showed an increase in temperature as depth increased. However, it turns out the evidence *does not support* the magma theory at all. Why is this? Remember from where the heat in the crust actually originates. It comes from the gravitational friction mechanism, which is the pushing and pulling of the Moon on Earth's crust, creating frictional heat—heat in the crust, **not** deep inside the Earth.

A heat source generated in the crust because of frictional heating would increase the thermal gradient (a more rapid increase) as we approach the heat-source, which is the place of frictional heat, or an earthquake epicenter. Whereas the thermal gradient from a magmatic heat source should be smaller, steady because the heat source is supposedly thousands of kilometers below the surface.

To understand this concept, consider this example; Turn on a stove to 400°F. Move one foot away from the burner, the temperature drops to around 100°F. We understand this because practical experience tells us the heat will decrease as we move away from the heat source. The heat gradient within the Earth acts in a similar way. Heat generated in the crust through frictional heating dissipates away from the heat source. In other words, excluding the variable of pressure, we would encounter the greatest amount of heat at the source of the heat, or the place of frictional heating. Based on our heated stovetop, how hot must the burner be to generate a temperature of 100°F out the door and down the street, a full block away? Obviously, the burner would have to be much hotter, perhaps thousands of times hotter than the 400°F we started with. This is the geophysicists' problem; the Earth's actual temperature gradient, as observed in the borehole gets too hot, too fast!

And this is not a new problem either. In the 1930s, one geophysicist put it this way:

"The average change in temperature in the Earth for a given unit distance is known as the *geothermal gradient*. If the gradient determined in mines and bore holes should continue downward unchanged, **the temperature at the center would exceed 350,000° F.**; but for several reasons the average rate of change in the shallow zone cannot be used with confidence for great depths." Bib 125 p398

No one thinks the center of Earth is thousands of times hotter than the surface of the Sun. How can heat be so high in the crust and still be thousands of kilometers from the source of heat, residing at or near the core? This too-hot-too-fast problem just does not follow the physics of heat flow. We would likely boil away if the Earth's geotherm followed the assumed gradient of

Magma Defies Heat Flow Physics

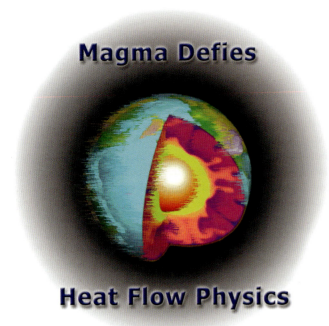

the magma model pseudotheory. A temperature of 350,000°F at the core just does not work for any theory!

The magma model has not changed much over the last seventy years. One would suppose the scientists would have gotten closer to figuring out *why* it gets too hot too fast. In 1989, the Germans started work drilling the second deepest borehole in the world, the KTB superdeep borehole. They found out immediately that what they thought they knew about the temperature gradient was incorrect:

"But a recently completed pilot hole down to 4 km **showed the scientists how little they actually know about the rock underground**. The bottom temperature in the 4 km hole was 118 [degrees] C rather than the 80 [degrees] C. 'That is really a big difference,' admits Rolf Emmermann of the University of Giessen. '**What we got was a very wrong result for a depth of only 4 km**.'" Note 5.4f

Why were the borehole results wrong? It simply did not fit the theoretical magma paradigm. For decades, the geothermal gradient problem continued to elude researchers, and then in 1997, researchers at the deepest borehole in Russia admitted:

"The prediction of deep temperatures for the KSDB-3 [Russian borehole] based on measured heat flow data in shallow holes **failed**, as did the analogous attempts for German KTB-hole. The unexpected increase of the temperature gradient with depth in both of the holes regardless of a number of contingent possibilities, **has not been satisfactorily explained yet**." Note 5.4g

Something is wrong with the geothermal gradient model and the scientists have known about it for a long time, but they had no means to explain what they had encountered. Now through a completely new paradigm, as outlined here in the Universal Model, and by employing the Universal Scientific Method, there is an explanation of Earth's geotherm and there is a way to predict what new drilling experiments will show.

Fourier's Heat Flow Law Violated

Fourier's Heat Law states that heat will flow from hot to cold and that the change in heat is proportional to the cross sectional area, the change in time, and is derivative of temperature. This is a well-established process. If the Earth's heat actually came from a deeply occurring magmatic melt, then as we drill toward the center of the Earth, natural laws, such as Fourier's Heat Law *should be predictive*. From a recent study conducted at the Kola borehole, researchers employed the traditional assumption that the temperature gradient direction was vertical. This meant they assumed heat would come from directly below, from the center of the Earth:

"We used the **traditional assumption** that the direction of the temperature gradient vector in the massif [rock sequence] is vertical." Note 5.4h

If magma existed, using the traditional assumption that heat comes from below, this would pose no problems and the thermal gradient would be consistent with Fourier's Law. All would be well, except:

"…the thermal conductivity and thermal gradient variations **are not consistent with the Fourier law**…" Note 5.4h p358

The scientist's observations of heat flow did not match their assumptions. What they thought was that heat flow in the borehole:

"…**should be observed** in the case of stationary conductive heat transfer…" Note 5.4h p358

However, it was not!

The heat in the crust of the Earth did not flow from a stationary conductive heat source—magma; once again, their predictions had "failed":

"The prediction of deep temperatures for the KSDB-3 [Russian hole] based on measured heat flow data in shallow holes **failed**, as did the analogous attempts for German KTB-hole." Note 5.4i

Geologic time is often used to explain away anomalies that defy answers. In the case of the unexplainable anomalies encountered in the boreholes, time is of no consequence. Researchers recorded how long it took for the heat generated by drilling activity to reach equilibrium:

"Our results from SG-3 and the Vorotilovo borehole show that the temperature gradient may attain equilibrium in a very short time span after drilling. Occasionally, **only a few days are required to reach equilibrium**." Note 5.4j

As we can see, the temperature gradient reached equilibrium within just a few days. Another factor researchers associated with the anomalous heat flow was a "fluid motion," which they said could come from rain or meteoritic water:

"It may be **inferred** that the second factor, i.e. **fluid motion** in the rock massif, affects the temperature field to a greater extent." Note 5.4j. p361

However, the researcher's calculations showed that only 10 mm of meteoritic water each year percolated from the ground surface to the permeable part of the rock sequence. This was not enough water to change the temperature of a few dozen meters of rock, and certainly not enough to affect 9000 meters of rock along the borehole. It took only a few days for the rock to come to temperature equilibrium after an increase in temperature due to drilling, with no noted continuous fluid flow along the borehole that would prevent the temperature gradient from following Fourier's Heat Flow Law. Furthermore, the researchers had no data to explain fluid motion:

"…the data for quantitative modeling of transient **fluid motion in the rock massif are not available**." Note 5.4j p363

The following is the full quote of the researchers' notes recounting the difficulty they had trying to create a detailed predictive model without relevant data:

"It is likely that both types of fluid motion affect the temperature field, but presently it is difficult to create a more detailed model because **the data** for quantitative modeling of transient fluid motion in the rock massif **are not available**." Note 5.4j p363

In other words, they simply could not explain why the measured heat in the deep boreholes did not match their predictions. This is to be expected if the theories upon which the predictions are based are incorrect. However, should we make a prediction using the new UM model, the Frictional Heat Law would suggest that the higher temperature variations in the temperature gradient are possible and would likely occur where fractured zones or faults are encountered in the borehole:

"The intensity of temporal variations of T [temperature gradient] values in different depth intervals is related to specific hydrogeological features of the rock…**Higher than average temporal variation of T are observed in zones of enhanced fracturing with water occurrences**." Note 5.4j p364

It was no surprise that the most prominent, local heat flow 'anomaly' in the Kola borehole occurred where researchers encountered a large fault structure—the Luchlompol fault.

"**It is easy to show that the most prominent local anomaly of heat flow, the increase in q [heat flow] near the Luchlompol fault** (depth interval of 4.7-5.0 km), cannot be attributed to the refraction of the heat flow vector." Note 5.4j p363

In summary, the magma theory leaves many unanswered questions and anomalies, and seems to defy basic heat flow physics. It provides no answers as to why the Earth gets so hot so quickly, and cannot explain why heat flow is greater through the thicker, insulated continental crust, and it has no method to account for the erroneous thermal gradient. The Lava-Frictional model predicts that the highest amount of heat and hence the greatest amount of heat flow will occur in the crust at fault zones or zones where friction is present—and it does. It also predicts that the thermal gradient is not necessarily vertical, because the heat source is not always from below, it may occur laterally in relation to the point of measurement—this is also true. The Lava-Friction Model also predicts that thicker continental crust will exhibit greater heat flow than thinner oceanic crust because the generation of heat in the crust comes from frictional heating—which it does. The Frictional-Heat law does not violate heat flow laws and because it predicts no distant, deep-earth heat source, the observed thermal gradient is not anomalous.

5.5 The Accretion Theory

Scientists believe that four-and-a-half to five billion years ago, leftover debris from the formation of the sun formed an enormous disk of matter called the solar disk. The material coalesced and collided to form the Earth and other planets through a process called accretion. The constant impact of infalling material heated and melted the Earth, giving rise to the idea of a completely melted Earth. However, it would require an extraordinary number of hyper-velocity impacts to create enough heat to melt so much material and no mechanism exists to explain how this would work or how the rocks would stick together in the first place. Additionally, a melted planet would have cooled long ago in the dead cold of outer space.

Accretion Defined

Oxford's Dictionary of Astronomy defines **Accretion** as:

"The process by which the mass of a body increases by the accumulation of matter, in the form of either gas or small solid bodies which collide with and **adhere** to the body. The bodies in the Solar System **are thought to have** grown by accretion…" Bib 108 p3

The Accretion Theory has not earned the pseudotheory designation because it has not been taught as fact in many scientific circles. However, with each passing year, more scientists and schools accept accretion as fact over theory. The notion that our solar system grew by accretion from a solar disk is important to modern science as this is where the Magma Pseudotheory story begins. From *Sky and Telescope*:

"This hypothetical disk, the *solar nebula*, is where any discussion of the origin of our solar system **must begin**." Note 5.5a

Magma was born in the minds of early modern scientists as they imagined objects in great abundance colliding in a heavy bombardment of materials some 4.5 billion years ago. The idea that a period of heavy bombardment formed the Moon's craters further bolstered the assumption. *In Scientific American* we find:

"Studies of moon craters revealed that these gouges were caused by the **impact of objects** that were in great abundance about 4.5 billion years ago. Thereafter, the number of impacts appeared to have quickly decreased. This observation rejuvenated the **theory of accretion** postulated by Otto Schmidt. The Russian geophysicist had suggested in 1944 that planets grew in size gradually, step by step." Note 5.5b

The 'magma ocean' was then born as very large objects smashed into the planet:

"Large bodies slamming into the planet produced immense heat in the interior, melting the cosmic dust found there. The resulting furnace—situated some 200 to 400 kilometers underground and called **a magma ocean**—was active for millions of years, giving rise to volcanic eruptions." Note 5.5b

To produce such immense heat, the impactors must be traveling at hypervelocity, thus the Accretion Theory demands high-speed collisions to be valid. Despite the accretion model's conflict with the laws of physics, some astronomers continue to describe the origin of celestial bodies in terms of "cosmic billiards":

"Like **cosmic billiards** on a warped pool table—the warp being gravity—these spheres can hit one another, rebound and slow down because of friction and other forms of energy dissipation." Note 5.5c

Billiard Balls Do Not Stick Together

As Fig 5.5.1 illustrates, in the game of billiards, the balls do not clump together when they hit one another; they bounce off each other and remain on the table only because of the contain-

> The predicted magma "…thermal conductivity and thermal gradient variations are not consistent with the Fourier law…"

ment afforded by the side rails. In space where there is little friction and no containment rails, there is nothing to stop the billiard balls from flying infinitely apart.

Large rocks (asteroids) and planets supposedly accreted first from small particles until they formed an aggregate assemblage of rocks and debris—a pile of rubble. Small particles lack sufficient gravitational attraction to have a significant effect on other particles to get the process of accretion started, and collisions that occur at almost any speed result in rebound.

Since accretion is the favored theory at the moment in modern science, can it explain how small rocks 'accrete' or join in space? Where do we see such clumps of rocks in space?

The 21st century dawned with astronomers observing asteroids up close with new satellite technology. What they saw surprised them. In fact, the whole rubble-pile hypothesis was "troublesome":

"Yet **the rubble-pile hypothesis is conceptually troublesome**. The material strength of an asteroid is nearly zero, and gravity is so low you are tempted to neglect that, too. What's left?" Note 5.5c p53

The asteroids did not appear like rubble-piles at all and studies showed that even the gentlest of collisions failed to support the accretion theory:

"It turns out to be surprisingly difficult for planetesimals to accrete mass during **even the most gentle collisions**." Note 5.5c p54

Researchers from this *Scientific American* article concede that not only is the topography of impact craters on asteroids ambiguous, but that the underlying science of the impacts and the impact craters is also "uncertain":

"**The ambiguity is a sign that the underlying science is uncertain**." Note 5.5c p54

Asteroids, comets and other matter travel through space at speeds of kilometers-per-second, *not per hour*! If those projectiles collided they would not stick together, according to one *Nature* article discussing impact physics:

"From a physics perspective, the simple **billiard-ball analogy** of collisions between rocks knocking each other around the inner Solar System **does not stand up to elementary scrutiny**. It is no easier to 'bump' icy, rocky or even metallic objects, with finite material strengths, from the asteroid belt into Earth-crossing orbits **than it is to hit eggs around the fairways with a golf club**." Note 5.5d

Furthermore, according to the accretion theory, everything in the solar system came first from atoms and molecules that combined to form gas and dust. This gas and dust then coalesced into larger and larger bodies, presumably through impact. Physics does not agree that this works and scientists have yet to identify the 'glue' that would hold the accumulated bodies of rock and rubble together. Finally, the high velocity impact of large bodies and the enormous heat generated by the impact should be evident in the impact body.

The Melting Pseudotheory

According to the popular accretion theory, in the beginning, small particles of gas and dust stuck together eventually forming small rocks. These small rocks then hit each other at incredible speeds, and still stuck together, getting bigger; many small rocks accumulated to make big rocks, and bigger rocks

Fig 5.5.1 – Billiard balls obviously do not clump together. What if they were traveling hundreds of times faster in space? There would still be no clumping; just pieces of billiard balls would result.

began to slam into one another too. Instead of flying apart (keeping in mind that these rocks were speeding along at many kilometers-per-*second*), the big rocks stuck together because they **melted**. However, as subchapter 5.7 demonstrates, the vast majority of continental rocks are crystalline quartz based, not glass, which is what comes from a melt. Chapter 7, the Hydroplanet Model that most minerals in Nature, including quartz, form from water not from a melt.

According to the accretion theory, the hypothetically melted rocks became *spherical* planets because spheres form naturally in space when matter is in liquid form, and not heavily influenced by another body. Presumably, they say, the Earth's round shape is the result of many large rocks crashing together, melting, and forming a magma sphere. This Melting Pseudotheory has no physical evidence; no one has every observed this happening and there are no glass meteorites coming from space showing that they were once melted accretion bodies. The Meteorite Model in subchapter 7.10 introduces a new model with physical evidence for the origin of meteorites.

Space is cold, and a collision-heated Earth would have cooled long ago, say some scientists. Since geologists believe the center of the Earth is still molten, there had to be another theory as to why the interior of the Earth is still hot. One idea is radioactivity, or the radioactive theory, which infers that the heat in the Earth's center comes from the decay of radioactive matter deep inside the Earth.

5.6 The Radioactive Myth

Despite its widespread acceptance, no known mechanism exists to supply enough heat to keep the Earth's interior hot. Since modern science does not recognize frictional heat in the crust as the source of lava or interior heat, some other source of energy had to be proposed. Thus was born the idea of heat from radioactivity. However, naturally occurring radioactive rocks are weak and generate very little heat. The most abundant, naturally occurring radioactive rock is uranium, which is found only near the surface of the Earth. Moreover, there are no known radioactive lava flows. Despite all the evidence to the contrary, modern science insists on perpetuating the myth of a radioactively heated Earth. In this subchapter, we explore some of the evidence for this myth.

Why is the Interior of the Earth Hot?

The answer to this piece of the magma puzzle has not been easy for geologists to solve, and as we shall see, it is impossible for them to prove. The *Understanding Earth* geology textbook states:

"Earth's interior is hot for several reasons… the violent origin of Earth by the **infalling of chunks of matter** made the interior hot, and disintegration of the **radioactive elements** uranium, thorium, and potassium **also produced a significant amount of heat**." Bib 59 p495

Geologists have known that the Earth would have already cooled *if* the molten core had been originally heated from infalling "chunks of matter" that melted everything. To keep the center of the Earth hot, the experts proposed a heat-from-radioactivity hypothesis and thus came the **Radioactive Myth**.

Do Radioactive Rocks Come From Magma?

Geologists list uranium as one of the radioactive elements that produce a significant amount of heat in the interior. So let's ask another Fundamental Question:

Where do we find 'hot' uranium, in Nature?

It turns out that natural uranium ore is no hotter than any other typical rock, sampled in the same area. Radioactive rocks and minerals come from ores found *near the surface* of the Earth. Natural uranium rocks set off a Geiger counter due to their internal radioactivity, which is the release of unstable alpha, beta or gamma particles. If you hold a uranium rock in your hand (see Fig 5.6.1), you will feel no difference in temperature. Fig 5.6.2 shows a once highly productive uranium mine out of which miners drew uranium ore, but the temperature inside the mine is not "warm" at all. *Pure* uranium is silvery in color,

Fig 5.6.1 – 'Hot' radioactive uranium ore sets a Geiger counter buzzing, but is no hotter (temperature wise) than any other rock.

looks like aluminum, is slightly warm to the touch, and most importantly is *not* found in Nature. It takes the extraction and processing of thousands of tons of uranium ore to produce a very small amount of pure uranium, about the amount that fits on the tip of your finger. Even this highly concentrated artificial uranium element is *not hot!*

Radioactive minerals found in Nature are actually rather cold, about the same as the rocks among which they lay. None produces temperatures anywhere near the thousands of degrees magma theorists say exists in the center of the Earth. It is only in man-made nuclear reactors, with *artificially enriched* uranium and other elements, (also not found in Nature) where 'hot' temperatures exist because of radioactivity. Geologists know of no place in Nature where events are happening mirroring those inside nuclear reactors.

Surface rocks, uplifted from a hot radioactive core after they formed, must naturally retain their *radioactivity*. According to modern geology, all metamorphic rocks are altered by high temperature and pressure at depth. Therefore, they *should* also exhibit some radioactivity, but observations show they do not. When examining core samples from deep boreholes, findings show those samples taken at depth are no more radioactive than rock samples taken near the surface.

The idea that radioactive substances in Nature melt rock is total speculation and all available observational data testify that this theory is incorrect.

Radioactive Dogma

A relatively recent college geology textbook, *Understanding Earth*, tells students where the Earth gets its "internal heat":

"Earth's internal heat engine is powered by the heat generated by **radioactivity**." Bib 59 p11

This statement and many others like the following one found in *Magmas and Magmatic Rocks*, a common source book for geologists, assert the fields' current ideology:

"At present, **radioactivity** is the major source of internal energy within the Earth." Bib 87 p13

Fig 5.6.2 – Radioactive uranium ore near Moab, Utah, USA. Ores are found near the surface in veins and pipes, usually in sedimentary material that has not been melted.

A university sponsored public web site; *Volcano World* gave this answer to the question: How is lava made inside the world?

"So, you want to know how magma is made? There is a lot of heat within the earth, and **this heat is produced by radioactive decay of naturally-occurring radioactive elements within the earth**. It is the same process that allows a nuclear reactor to generate heat, but in the earth, the radioactive materials are much less concentrated. However, because the earth is so much bigger than a nuclear power plant it can produce a lot of heat." Note 5.6a

From these quotes, it is apparent that the idea of radioactive heat is deeply entrenched in the field of geology. Why is this belief so deeply rooted today when there isn't any real evidence of it?

As we have noted, naturally occurring radioactive uranium is not concentrated enough to feel warm. It certainly is not hot enough to boil water, and no matter how big the pile of uranium is, unless it is highly *concentrated*, through the manmade refining process, it *generates no heat*. *If* natural radioactive elements were concentrated enough to generate the heat necessary to melt rock, then the resulting molten rock (lava) *must also be* **highly** radioactive.

As indicated earlier, highly concentrated, refined and enriched uranium does generate heat inside modern nuclear power plants, and that heat is used to generate steam, which turns the turbines that create electricity. The presence of this heat has been cited as proof for radioactive heat deep within the Earth. However, the heat required to produce steam—100°C at sea level, is significantly less than the heat required to melt rock (around 1700°C) at sea level, and heat required to melt rock increases dramatically as depth increases, because of pressure.

If heated, highly radioactive uranium does not occur naturally and if there are no other known naturally occurring highly radioactively heated minerals, how certain can scientists be of the radioactive heat in the core? In a 2008 *Science* article, the author admits:

"Current **uncertainties** prevent a definitive assessment of the radioactive heat sources in the core." Note 5.6b

Now, think about our next, very compelling Fundamental Question:

How can lava from radioactively melted rock, **not** be radioactive?

We posed this question to several geologists and geophysicists but have yet to receive a clear and simple answer. The following is one geophysicist's personal answer to our question of *why lava is not radioactive*:

"With respect to the magma issue, I think that the matter is more **complicated** than simply radioactive melting magma. There is heat coming out of the earth. South African gold miners have to deal with that heat, more and more, the deeper they dig. But generally, that heat is not sufficient to melt rocks. Volcanic activity seems to occur near plate boundaries and, dif-

Fig 5.6.3 – These nuclear power plant images conjure ideas of the heat produced in a hot radioactive planet core. Although high heat comes from artificially enriched man-made radioactive fuel cells, we do not find this kind of reaction in nature. Hot radioactive minerals capable of melting other rocks have never been shown to exist.

ferently, in certain hotspots which are often ocean islands like Hawaii. **I don't think the final chapter has been written on magma formation**." Note 5.6c

Where is the Radioactive Lava?

Children, naturally inquisitive, often ask fundamental questions about Nature that adults overlook. When one of our daughters was in junior high, she asked her science teacher this question: *Why aren't the lava flows we walk on radioactive?* She reasoned that if magma inside the Earth is supposed to be hot because of radioactivity, then the lava we walk on should also be radioactive. The science teacher did not know but agreed to find out, and so she asked a geologist friend from Arizona State University to explain. He responded:

"Lava flows have an amount of radioactive elements that are typical for the depths from which they are derived. There are amounts of these elements in the mantle and more in the crust (since elements like U [Uranium], Th [Thorium], and K [Potassium] tend to be concentrated in the crust). **It's not like the rocks at depth are highly radioactive**, but they contain enough of these elements to generate heat **over time**."

The geologist's answer that; "Lava flows have an amount of radioactive elements that are typical for the depths from which they are derived," left the teacher confused. After all, surface lava flows; both new and ancient have almost no radioactive elements. The geologist went on to say that, radioactive elements are concentrated in the crust. Why then, isn't the crust melting if radioactive elements are concentrated there?

The geologist said the rocks at depth were not highly radioactive but they were radioactive enough to generate heat "*over*

CHAPTER 5 THE MAGMA PSEUDOTHEORY

time." Thus, the geologist ended with the most common sci-bi in science today—events occur 'over time.' The 'over-time' sci-bi simply does not work! If a stove is left on low all day, does it continue to get hotter? Trying to say that rocks at depth are melted because the radioactive elements could "generate heat over time" is no different. More time does not make cold rocks any hotter.

Obviously, any rock that encountered enough heat to melt from radioactivity must be highly radioactive itself, but no one has ever seen these highly radioactive rocks from inner-Earth because they do not exist.

The lack of radioactive lava is direct empirical evidence that magma, *as theorized*, does not exist. If lava really formed because of radioactivity, there would be '*Caution, Radioactive Lava Flow*' signs over Hawaii and every other lava flow in the world—and no one would want to live in those places.

Missing Radioactive Evidence from Deep Boreholes

Let us look again at the empirical evidence from the deepest borehole in the world—the Kola Superdeep Borehole. The Kola project, a scientific drilling expedition designed to drill as deep as possible into the Earth's crust, reached its deepest point of 12,262 meters (40,230 ft) in 1989. This is the deepest hole ever made by humans and the closest science has been to so-called 'radioactive magma.' Significant increases in the radioactivity of the rocks as compared to surface samples should have been apparent, but were not:

"Since the **radioactivity of the rocks** traversed by the well can **make only insignificant contribution to this heat flow**, it [the heat flow] must plainly come from the mantle below." Note 5.6d

Scientists, were only able to say that radioactivity made an "*insignificant contribution*" to the heat flow in the rocks along the borehole after examining core samples taken from the deep borehole and recognizing that they were similar to rocks found at the surface. Without hot radioactive rocks, the researchers concluded that heat came from "the mantle below" but had no direct evidence upon which to base their supposition.

Conversely, the data refuted another long held idea that heat-producing radioactive elements were concentrated in the crust but decrease exponentially with depth. The scientists had this to say about this "widely held *assumption*"

Fig 5.6.4 – Signs like this one would be seen at lava flows all over the world, if lava really came from radioactive magma. Radioactivity from heated lava does not just disappear—it never existed.

in a journal article from the highly studied German Continental Deep Drilling program, the KTB borehole:

"The large amount of data from logging and from laboratory measurements **clearly refuted the widely held assumption that heat-producing elements are strongly concentrated in the upper crust and then decrease exponentially with depth**." Note 5.6e

If the core samples taken from deep within the Earth had been heated by radioactive elements, they would remain hot for a long period and would require special handling.

However, they did not remain hot when brought to the surface, and *they were **not** radioactive*. The rock cores brought from deep underground *were no more radioactive than any other average surface rock!*

Thanks to the data from the KTB borehole, informed geophysicists can lean no longer on radioactivity to explain how the crust gets its heat. However, no one yet has explained with empirical evidence how rocks 9 km below the surface can get so hot, as was present deep in the borehole where rock temperatures reached 260° C (500°F).

Heavy Elements did not Sink to the Center

According to the accretion theory, heavier elements like Uranium and Thorium, both radioactive, sunk to the center of the melted, spherical magmaplanet Earth, but according to the geology book, *The Heart of the Earth*, this is not reality:

Fig 5.6.5 – Lava flows around the world have no, "Danger Radioactivity" signs.

"Again, **why *should* the radioactive materials be concentrated in the surface layer?** The elements involved are very dense; if the earth cooled from a liquid mass, one would expect them to settle to the center. **But no: they are apparently found almost entirely at the surface—why?**" Bib 63 p151

One important point to consider here is that the heavy elements ***should not*** even be on the surface ***if the Earth was once completely melted***. Because they are the heaviest, these elements should have sunk to the center of the Earth, but apparently, they did not. Later, in the Ore Mark subchapter, 18.13, we discuss why we find the heaviest elements near the Earth's surface.

No Radioactivity—No Magma

By the early 1900s, science understood that the theoretical magma-planet could not stay hot after accretion waned. Heated objects cannot stay melted in the frozen depths of space and the radioactivity myth that came along, to rescue the magma-planet theory by providing a mechanism to supply heat to keep the planet melted, failed. There is no substantiated scientific evidence that radioactivity generated magma at all, even though many science textbooks still teach this untruth. When we dig into the scientific journals, we discover scientists acknowledging that the supposed massive amounts of radioactive heat producing elements presumed to be present in the lower mantle do not exist:

"Mass balance, heat flow and Pb-isotope data suggest **that most of the lower mantle is depleted in the heat producing elements**." Note 5.6f

Further, if the lower mantle is depleted of these heat producing radioactive elements, then there is no mechanism left to produce magma from radioactivity anyway. Finally, if magma does not exist in the first place, then there is no need to find or account for an internal source of heat, and there is no need to search for missing radioactive materials.

5.7 Glass is **Not** Quartz

In this and the next two chapters, we will discuss quartz and examine it in detail. No other mineral holds more clues and is more central to understanding the formative processes of the Earth than quartz. Although glass and quartz form from the same molecule, SiO_2, glass comes from a melt, Quartz does not. The environment in which quartz

Fig 5.7.1 – This man-made synthetic quartz crystal has the same properties as natural quartz; neither (natural nor man-made) is grown from a melt.

Fig 5.7.2 – This natural quartz crystal cluster with amethyst tips was grown the same way all quartz minerals are grown—from an aqueous solution, not a melt.

forms includes high pressure, heat, and water and we must thoroughly explore each of these components and their environment. When we comprehend how to make quartz and in what conditions it grows, then we can learn where and when those conditions existed on the Earth, and why the Earth could not have formed from a melt.

The Importance of Quartz

Geology courses today teach that the mineral SiO_2, the constituent molecule of quartz, is the most common on Earth but they say little of its incredible properties. Nor do they explain how to make quartz. Most references state that among the Earth's continental rocks, no mineral is as common as quartz:

"Quartz is the **most common mineral on the face of the Earth**. It is found in nearly every geological environment and is at least a component of almost every rock type. It frequently is the primary mineral, >98%. It is also the most varied in terms of varieties, colors and forms. This variety comes about because of the abundance and widespread distribution of quartz." Note 5.7a

As we stated earlier, no other mineral holds more clues about the Earth, and if we do not comprehend how to make quartz or the environment in which it forms, how can we explain the formation of the Earth?

Amazingly, the typical geologist has had no practical experience making quartz. Why would geologists and geophysicists who study minerals, how they form, and how those minerals interact, not have any real hands-on experience making quartz? The way in which quartz actually forms is not taught in school. In fact, as we will see in the next chapter, the Rock Cycle Pseudotheory, the field of geology rests entirely on false assumptions about the origin of rocks. We must completely overhaul this system of beliefs if we are to understand the true process behind how rocks form, and if we are able to understand the process that forms most of the minerals found on Earth. This is ***why*** understanding quartz is so important and cannot be overemphasized. By comprehending how quartz grows in Nature, we will comprehend how Nature makes almost all minerals, and we will learn how they ***are not*** made.

Why Glass is not Quartz

To understand Nature more clearly and to understand more about how our planet

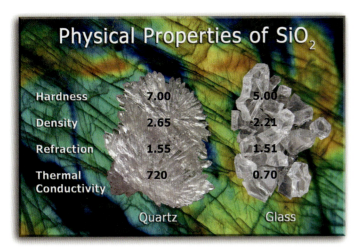

Fig 5.7.3 – These are the Physical Properties of two SiO_2 minerals, Quartz and Glass and their characteristic differences. Note 5.7b

If we do not comprehend how quartz, the most common mineral in the crust of the Earth forms, how can we explain the formation of the Earth?

In Chapter 7, the Hydroplanet Model, we discuss further the process of how natural quartz forms in a water solution, but first we must understand how quartz is *not* formed—quartz does not form from a melt.

Silica Phase Diagram

The Silica Phase Diagram in Fig 5.7.6 illustrates various forms of silica-based quartz minerals and the specific pressures and temperatures at which they form. Although we refer to the diagram as a 'phase' diagram, it does not depict changes from one mineral to another mineral *after* they form, but rather, it reveals the relationship between pressure and temperature *when* they formed. Unlike an actual phase-change diagram where a substance such as water undergoes changes from a solid (ice) phase to a liquid phase, and then to a gas (steam) phase, this phase diagram only illustrates the *differences* that exist in the solid phase of silica minerals. Natural quartz, sometimes called alpha or α-quartz, exists in great abundance worldwide. A small amount of glass occurs around volcanic areas, whereas coesite and the other phases of SiO_2 are extremely rare.

Silica or SiO_2 is the element that forms quartz and glass, but the specific variety of SiO_2 mineral depends greatly on many environmental factors and forces that were present during the formation stage of the siliceous mineral.

The lower left-hand corner of the diagram features natural, alpha state quartz; it depicts a type of quartz formed under very strict physical parameters. During formation, if the pressure or temperature changes only slightly, a different silica mineral results.

Many laboratory experiments helped to establish the Silica Phase Diagram, which demonstrates one very important geological fact:

Natural quartz does **not** form within 'magma conditions' (high temperature-high pressure).

Natural Quartz Mystery

If the Earth formed originally as a magmaplanet and if natural quartz cannot form under 'magma conditions,' from where did all the natural quartz come? Natural quartz rocks and minerals are found everywhere around the world on every continent, their presence testifying that the Earth could not have been melted. Let us explore why.

High Temperature-Pressure Mineral Mystery

We know the physical properties of coesite and other high pressure, high temperature, silica-based minerals depicted in the Silica Phase Diagram, because of laboratory experiments conducted by scientists who were able to produce these minerals. After mineral formation, temperature and pressure return to normalized conditions and researchers observe and measure the physical properties of the minerals, such as density and crystal structure. Once heated, the minerals *do not revert to **natural** quartz* after they cool and/or with pressure reduction; the prop-

formed, we must know the difference between glass and quartz. Although pure glass and pure quartz are both clear and chemically defined by the same formula, SiO_2 (silicate or silicone dioxide), their physical makeup and origin are completely different. Each has very different characteristics because of the way each forms in Nature.

The chart in Fig 5.7.3 compares the physical properties of these two minerals. Notice that quartz is harder, denser, has a higher index of refraction, and is over 1000 times more thermally conductive than glass. The reason for this has to do with the crystalline structure of quartz as compared to glass. Natural quartz has an orderly geometric structure that grows in a pressurized liquid (not a melt) environment and at temperatures much lower than that of glass. Glass forms when a hot, liquid SiO_2 *melt* cools. Glass remains amorphous and does not form an orderly geometric crystalline structure. For clarity, we define the term **melt** as a mineral or minerals liquefied (melted) by heat only, without any significant water. In the natural quartz-growing environment, quartz crystallizes or precipitates out of a highly siliceous aqueous solution much like salt or sugar does when dissolved in water and the water evaporates.

The high thermal conductivity of quartz is a result of the orderly growth of micro-crystals under pressure as opposed to just the cooling of a hot liquid *melt*. Diamonds are another example of this crystal growth process. They have the highest thermal conductivity of any natural material known to man. This is because of the very high pressure and high temperature *aqueous* environment in which they formed. We will discuss the environment in more detail later.

The comprehension of the process, which formed many of Earth's rocks (like granite) becomes clearer after we understand how the Earth's most common continental mineral —quartz—forms.

Fig 5.7.4 – Manufactured glass items formed from melted sand; even though they are made of silica, they are not quartz.

SUBCHAPTER 5.7 GLASS IS **NOT** QUARTZ

erties and crystalline structure of the minerals are preserved, remaining as they were when formed.

Diamonds form in a process similar to quartz but with carbon, which also does not revert to graphite or elemental carbon when exposed to surface pressure and temperatures. The same appears with all high pressure, high temperature, quartz-based minerals including stishovite, coesite, β-quartz (beta quartz) and other rare, silica-based minerals. Where do we find high-temperature, high-pressure minerals in Nature? Ideally, if rocks formed according to modern rock cycle theory and all rocks formed in magmatic heat and pressure, the silicate varieties of stishovite, coesite, β-quartz (beta quartz), tridymite, and cristobalite rocks should be common on the Earth.

Coesite occupies the largest area of the Silica Phase Diagram with the greatest range of pressure and temperature, suggesting that it should be among the most abundant of all the silicates, based on the modern assumptions of planet formation. However, this is not what researchers find, as reported in the *Journal of Geophysical Research*:

> "**Coesite has not been found preserved** as a polycrystalline aggregate in **natural rocks**…" Note 5.7c

No one knows if large single crystals of coesite or common rocks made of coesite even exist in Nature because we have only seen tiny crystals, around a millimeter or two, or less, and even they are rare. In addition, the other types of silica minerals (Beta quartz, Stishovite, Tridymite, and Cristobalite) on the Silica Phase Diagram are all *extremely rare* natural minerals!

Where are all the stishovite, coesite, beta quartz, and tridymite and cristobalite rocks? What does this all mean? It means that high pressure-temperature silica minerals formed in the laboratory testify that magma could not have formed all the rocks on the surface of the Earth—because high pressure-high temperature silica rocks are so rare.

The Rock Melting Experiment

According to modern science, rocks and debris crashed together in space and melted until the primordial hot Earth, a boiling molten caldera began to cool. What did the rocks formed from the cooling mass on the surface of the Earth look like, having *been under very low surface pressure* as they cooled. What properties would those surface rocks, many of which would be around us right now, exhibit? The intuitive answer is easy, and it reveals one of the accretion theory's major flaws—that melted rocks become *glass* not quartz! The vast majority of rocks, including rocks and sand on the surface of the Earth are quartz. If the rocks had been melted as the accretion theory supposes, the world's beaches and sand dunes would be glass.

The **Rock Melting Experiment** reveals what melted rocks look like. With adequate heat, such as an oxy-acetylene torch, and a few rocks or small pile of sand shows the effects of melting. The results of our experiment in the photos in Fig 5.7.9 produced rocks not seen in Nature, and ap-

Fig 5.7.5 – Natural glass tubes known as fulgurites form when lightning strikes and melts sand. Neither natural nor synthetic glass can grow into crystals like natural quartz.

parently not seen by many researchers either. Our search of the internet and hundreds of geology books and science journals produced no references to the melting of common rocks.

Ample references by geologists referring to Bowen's reaction series, which identifies the melt temperature of different minerals, implies that rocks form once temperatures cool to below specific melt points. However, the series, based primarily on melting rock powder followed by slow cooling, does not produce crystalline rock like their counterparts in Nature. Re-

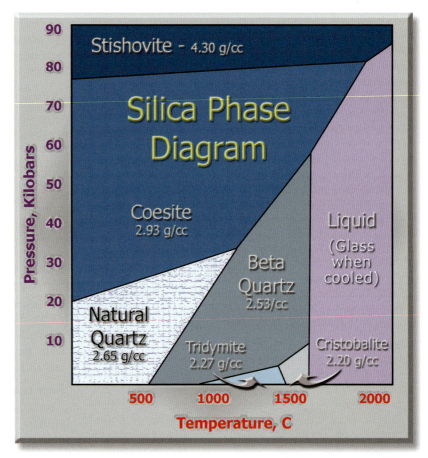

Fig 5.7.6 – Several researchers contributed to the development of this silica phase diagram, which illustrates how different silica minerals form in various temperature/pressure combinations. Natural quartz forms only at the low end of the combined temperature/pressure environment, yet it exists in **abundance** all over the Earth. With the application of high temperatures to silica, liquid glass forms (on the right). Temperature and pressure extremes produce coesite and stishovite, both silica minerals, similar but different from quartz. It is important to note that the high temperature/pressure minerals, coesite and stishovite, which should be plentiful on a magma-formed planet, occur only **rarely** in nature.

searchers erroneously deduced that natural rocks crystallize in a certain order from *melted* rock, but could not make it happen experimentally.

One well-known gemologist, Kurt Nassau, melted rubies in what he called a 'reconstruction attempt' in an attempt to produce larger rubies by melting smaller rubies. He heated the rubies to 1800°C (3300°F) with an oxygen-hydrogen blowtorch, but in the end, he demonstrated that smaller, melted natural rubies could not make larger rubies:

"These experiments conclusively demonstrated that the most popular explanation [melt technique] for the production technique of 'reconstructed' rubies **cannot be correct.**" Bib104 p48

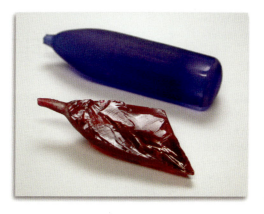

Fig 5.7.7 – Synthetic quartz boules grown from a melt, which are different than natural quartz. These boules do not have the natural crystalline structure or shape as quartz found in nature.

Our own Rock Melting Experiment, simple enough that almost anyone could reproduce—although with extreme caution because it requires substantial heat—demonstrates that the Earth's crust did ***not*** *form from melted rock.* The accretion theory is wrong; if not, where are the glass beaches? Where are the glass-covered deserts, or the glass mountains? All of these should be here if the planet was originally molten.

Two Different Camps

During our magma research, it became quickly apparent that there were two camps when it came to investigating and explaining mineral growth. The first camp—the theoretical group—consisted of the 'magma theorists,' the geologists. The other camp, the 'capitalists' included the mineralogists and engineers driving the discovery of new marketable technology worthy of exploitation. The two camps not only had different goals, they worked on very different experiments in markedly different social environments under completely different paradigms, for completely different reasons.

The magma theorists/geologists labored at the universities, while the mineralogists and technologists occupied technology and research centers such as the well-known Bell Laboratory. Both published technical papers, the magma theorists' latest *theories* attempted to explain how natural minerals *should* form from cooling magma while the technologists actually produced the minerals, and got better at it. The lab-grown minerals they made were nearly identical to natural minerals and they kept getting better.

One primary difference between the two camps is how they received their funding. Theorists develop new theories, and solicited grant money through government directed organizations to obtain funding. The technologists, on the other hand, worked in privately funded labs seeking ways to produce higher quality minerals faster and at lower prices. It is not difficult to guess which group provided the best information about how minerals actually form. Our next section investigates the findings of the mineralogists who worked with quartz.

Glass—*Not* Quartz—Comes From a Melt

The techniques describing the reproduction of natural minerals fall generally into one of two groups.

1. Melt – no water required in process.
2. Solution – water required in process.

Both the Magma chapter and the next-up Rock Cycle chapter are about *pseudotheories,* false theories taught as fact because they build on the unproven and incorrect basis that the Earth and rocks formed from *melt* processes. In the chapters following these two, we will discuss and demonstrate processes involving an ***aqueous solution,*** processes responsible for the formation of many, if not most natural minerals.

When minerals melt on the surface of the Earth under normal atmospheric pressure, any water in the melted rock would quickly turn to gas and escape. This is due to the low vaporization temperature of water and the high melting point of the minerals. One forerunner in modern mineral synthesis, Kurt Nassau, wrote the book, *Gems Made by Man,* which quickly became a standard in the mineral growing industry. Through trial and error, *technologists* like Nassau, discovered long ago that the melting technique does not produce quartz:

"Although the melt growth techniques provide rapid growth and are basically simpler and easier to control than growth from solution, **there are certain materials for which melt techniques cannot be used**. This is the case when the melt is so viscous that a **glass would form**, as happens with quartz…" Bib 104 p6

Elsewhere, in the *Chemical and Engineering News*

Fig 5.7.8 – Each type of silica in the Silica Phase Diagram is unique in its geometric shape. This Silica Crystal Forms diagram shows the various shapes of each crystal, which are dependent on the temperature and pressure that existed when the crystals grew. In nature, *only* natural quartz is found in any significant quantity—confirming the UM extraordinary claim that the natural quartz-based rocks found all around us did not come from magma.

SUBCHAPTER 5.7 GLASS IS **NOT** QUARTZ

journal, a special report authored by mineralogists explained what most geologists have yet to understand:

"**Quartz cannot be grown from a melt** ... because silicon dioxide [quartz] melts are so viscous that **they form glasses rather than crystals** when they are cooled." Note 5.7d

The fact that "quartz cannot be grown from a melt" is *one of the most important geological facts* that modern geology seems to have completely overlooked. The technologists understand that quartz cannot grow from a melt, and they accept it because of experimental successes and failures in the laboratory. Their paradigm, based on technology, not geology, becomes necessary because their discoveries must work in order to sell their wares, and so empirical facts and not theories guide their work. This is not the Paradigm guiding the academic geologist in his world filled with imaginative ideas and theories yet unproved, leaving little room for natural law.

Even mineralogists do not necessarily recognize that they know of the *only* process capable of natural quartz growth. Steeped in the magma tradition because of their book learning, they wonder why their geologist colleagues struggle to create quartz using high pressure and very high temperatures.

Glass Planets Do Not Exist

Another flaw in the accretion theory is the problem that planet formation from melted rock must include glass surface rock. The Moon and other planets, Mars and Mercury, must show the telltale signs of a glass-like crust if they originated from a magmatic melt. The Moon has no tectonic plate movement, and it does not appear that Mars does either, and apparently, neither of their surfaces experiences much erosion or modification. If they formed from molten rock, both must have an abundance of glass or glass-like rock.

If the Moon's crust formed from a melt, the rocks the Astronauts brought back should all be lava-like, and if meteorites impacted the Moon's surface, the resulting debris must include the broken fragments of previously melted glassy material.

If the Moon cooled from molten magma, its crust must be glass, but it is not.

The lunar rocks collected during the Apollo missions are not glass and by all known research, neither is the surface of Mars glass. There are no known glass planets!

"Quartz cannot be grown from a melt..."

5.8 The Piezoelectric Evidence

The piezo (**pee**-ay-zo) electrical property of quartz-based (silicates) rocks is one of the most important components in the Earth's energy field. This property, the piezoelectric effect, di-

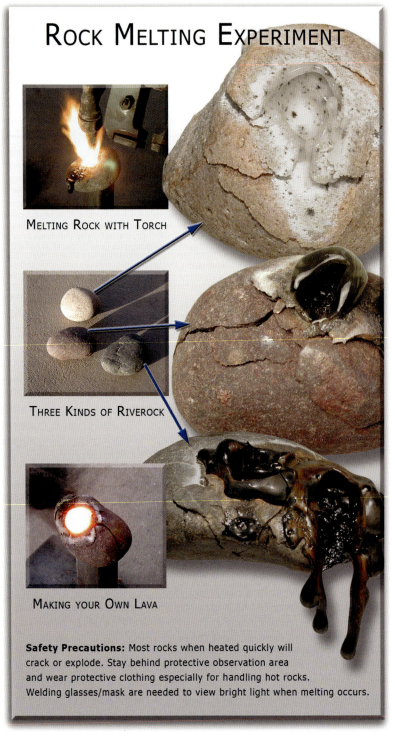

Fig 5.7.9 – The Rock Melting Experiment melts average quartz rocks, vividly demonstrating that melted rock makes glass, not quartz.

minishes almost entirely by heating quartz rocks above 570ºC. This is direct evidence that the quartz-based rocks, ubiquitously abundant on the continents, could *not* have formed from melted rock because that requires temperatures exceeding 1200ºC.

Piezoelectric Rocks

One very interesting property of natural quartz is its **piezoelectricity,** a fascinating electrical property found in many rocks. To illustrate this effect, refer to Fig 5.8.1. Two quartz river rocks, seen in the foreground, smoothed by the tumbling effects of water transport, produce tiny bursts of light when struck or rubbed firmly together! With the right rocks, anyone can perform this simple experiment. All it requires is two rocks with a high percentage of quartz, such as the two white quartzite stones seen in Fig 5.8.1. The effect of rubbing firmly or striking the two rocks creates a mechanical action where quartz crystals are squeezed rapidly and then released, producing a tiny electrical current, seen briefly as a burst of light. Additionally, electricity directed through quartz causes the quartz crystal to move or vibrate at a predictable rate, making it a very dependable timekeeping component. In subchapter 9.5, we show how this electrical phenomenon is the origin of the Earth's energy.

Industry utilizes the piezoelectric properties of quartz in technology in many ways, including one of the most common applications, the modern wristwatch, worn universally since its inception during the 1970s. They keep exceptionally accurate time because of the consistent 32,768 oscillations per second of a tiny quartz crystal, such as the one shown in Fig 5.8.2. Because quartz behaves according to natural laws and provides consistent and predictable results, technologists are able to produce cheap electronic watches that are more accurate than expensive, sophisticated mechanical watches that depend on intricate gears to keep time.

The Curie Point Evidence

The importance of piezoelectricity will be outlined in future chapters; its importance here in the Magma Pseudotheory chapter simply relates the existence of the piezoelectric property of quartz-bearing minerals so as to rule out the possibility that natural quartz could have grown from a melt:

"**Quartz cannot be grown from a melt** because the **piezoelectric phase** necessary for formation of electronic crystals is **not stable at the melting point**…" Note 5.8a

Melt quartz or even heat it to a high temperature and it will lose its piezoelectric effect. We do not find these answers in geology; instead, we find that *technology* is responsible for identifying the piezoelectric properties of quartz including its loss through heating.

Although technologists discovered that a small, nearly perfect quartz crystal would run a quartz watch much more accurately than a mechanical watch, they knew perfect crystals are not always available. They needed a reliable source of perfect quartz crystals for use in a variety of electronic devices. Demand increased rapidly during World War II, and Brazil supplied the insatiable American and Russian demand for high quality quartz. And then they stopped exporting and a frantic rush to supply quartz resulted in the discovery of the process for growing quartz crystals.

The technologists realized there was a specific temperature at which quartz loses its piezoelectric properties, a temperature at which quartz transitions from its alpha state to a beta state, now known as the **Curie Point.** That point is 570°C and engineers knew that the quartz growth process must be below that temperature to enable natural quartz (alpha phase) to retain its piezoelectricity:

"**Any growth process for quartz must be effective below 570°C**, the α→β quartz transition, if it is to produce the piezoelectrically useful alpha phase. The melting point of silicon dioxide is above 1700°C, **ruling out melt growth**." Note 5.8a p33

This means also that common quartz as found in Nature did

Fig 5.7.10 – The Moon is not glass. Modern Moon origins suppose that it formed from magma along with the Earth, but it clearly could not have come from a melt.

not form in temperatures exceeding 570°C because it retains its useful piezoelectric properties. Most of the Earth's rocks are primarily natural quartz or similar derivatives, and almost all natural quartz minerals have piezoelectric properties, therefore, such rocks simply did not form in a process that had temperatures exceeding 570°C! This means that magma was not the environment in which natural quartz formed.

Heating natural, alpha-state quartz above 573° C begins a transition to beta-state quartz where it loses most of its piezoelectric properties. As such, technologists realized that they could not use beta quartz crystals in the manufacture of electronic units, which are almost non-existent in Nature anyway:

"At a temperature of approximately 573° C, quartz transforms from **Alpha to Beta quartz**. During the transformation, most of the piezoelectric characteristics are lost, **rendering Beta quartz unsuitable** for the manufacture of crystal units." Note 5.8b

Today most of the quartz used in industry is laboratory-grown quartz, grown in an aqueous environment at temperatures considerably lower than the melting point of quartz. This makes the piezoelectric property of quartz another physical property that helps us understand the environment in which natural quartz minerals grew. The Curie Point establishes the point where quartz loses its piezoelectricity, showing that growing alpha-state quartz from a melt in a magma environment is not possible.

5.9 The Non-Iron Core Evidence

Density calculations, based in part on gravitational measurements, inferred that the Earth's average density is 5.52 g/cm³. Because observed crustal density is 2.7 g/cm³, the apparent need for a significantly greater density convinced scientists that the Earth has an iron or nickel-iron core. Inconsistencies with the gravitational constant reported in Chapter 18, the Mass-Weight Model, leave scientists with few options, whereas researchers try to explain how pure iron exists at the core, even while knowing pure iron does not exist in Nature and is unstable at the pressures expected at the center of the Earth. In reality, the material that actually exists at the core is as extraordinary as it is simple.

The Earth's Core Has Not Been Directly Observed

Turning our attention to the Earth's core, the first fact to establish is that the popular belief about the Earth's iron core is not a proven fact at all. As seen in a popular college geology textbook:

"Geologists **still know nothing** about the composition of the **core from direct observation**." Bib 59 p492

Without direct observation, we have only indirect evidence to determine what is at the core of the Earth. One of the indirect evidences used in determining the Earth's core composition is density. From where did the inferred average density of 5.52 g/cm³ come? The answer comes from one experiment described in subchapter 18.4, the Cavendish Experiment. In 1798, Henry Cavendish constructed an apparatus similar to a pendulum but designed to measure the faint gravitational attraction between two large lead balls and two small lead balls. The two sets of balls suspended independently allowed Cavendish to obtain accurate measurements of the twisting suspension wire as the balls oscillated back and forth past each other. The whole process of this experiment, fascinating as it is, gets duplicated and retested by others in physics labs today. However, there is one major flaw in the experiment leading to the Cavendish Error. Unlike the Earth, the lead balls are not in outer space, and thus, the balls, *restricted* by the air and influenced by the Earth's gravity rendered incorrect data. Their attraction should have been measured in a vacuum, in low gravity. Air, a denser medium than the vacuum of space, along with the attractive gravitational force of the Earth, slowed the balls' oscillation rate. Cavendish neglected to account for the reduced oscillation in the original experiment, leading to an incorrect gravitational constant and errors in the Earth's density estimates.

As we will learn in subchapter 18.4, the New Mass of the Earth, the Earth's density, recalculated to approximately **2.3 g/cm³** using the physics of gravitational attraction and the new geological discoveries outlined in this and other chapters, renders a truer density of the Earth that aligns with empirical observations. We next examine the geological nature of the Earth's density.

The Earth's Density Pseudotheory

In Chapter 17, the chapter on Essential Measurement, we will explain that without true and accurate measurements, there can be no science. Measurements must be accurate or they are meaningless. The fact that the gravitational constant is not at all constant is one of the discrepancies within science identified here in the UM. A change in the gravitational constant means a change in how density calculations of the Earth are made. Science produces many "precise" gravity measurements, but they are not necessarily accurate, and that difference led to the assumption that the Earth's density is 5.52 g/cm³:

"By **precise** physical experiments we determine the *constant of gravitation*; the method involves essentially the measuring of the force with which the Earth attracts a body of known mass. The result gives a basis for calculating the total weight of the Earth; and since its size also is known, it is a simple matter to compute the average density. This value, arrived at by many experimenters, is **5.52**; that is, an average sample of the Earth weighs about five and a half times as much as an equal volume of water." Bib125 p390

Geophysicists do not explain what type of naturally occurring minerals actually achieve the density of 5.52 g/cm³ after accounting for the crustal density of only 2.7 g/cm³:

"Direct determinations of density, using rocks of all kinds known at the surface, give an average value of **2.7**. As this is **less than half the density of the whole Earth**, the interior **must** consist of much heavier material than the outer part." Bib 125 p390

If surface rocks have an average density of just 2.7 g/cm³, and if the Earth's overall density is 5.52 g/cm³, there must be a relatively **common** mineral assemblage (not a single element because those are not common in Nature) in the Earth with a density *significantly* greater than 5.52 g/cm³. But, therein lies the problem—it does **not** exist.

Geology textbooks claim that the Earth has an 'iron core.' Iron has a known density of 7.87 g/cm³, so science reasons that this solves the average density dilemma. There are problems with this idea, the first being the assumption that pure iron ex-

Fig 5.8.2 – The blue arrow is pointing at the quartz oscillator in a quartz watch, which uses the piezoelectric effect to keep time.

Fig 5.8.1 – Two quartzite rocks as seen in the dark when struck against each other produce light.

CHAPTER 5 THE MAGMA PSEUDOTHEORY

ists at the core, an idea based in part on the Element Pseudotheory described in subchapter 24.2. The problem; *pure iron is **not** found in Nature*. Thus, the 7.87 density figure accepted for pure iron is worthless; science *cannot* use it to calculate or explain the density of the Earth.

Science also explains that the Earth's density is the result of heavy radioactive minerals near the core. However, as we read earlier, in the Radioactive Myth sub-chapter, there is no evidence that radioactive minerals exist in the core, but only in the crust. In fact, later on in the book, we discuss the biologic origins of all ores, including radioactive minerals. They did not originate in a heated magma core.

The two heaviest commonly occurring iron ores, magnetite and hematite, both share a density of about 5.2 g/cm^3, falling far short of the Earth's theoretical 5.5 g/cm^3 density. Pentlandite, the most commonly occurring nickel ore, usually found with magnetite, has a density of only 4.8 g/cm^3, and so it does not work either. The only other common metal-minerals are aluminum minerals, and their densities are half that of the iron minerals, so they also, do not work.

What does work? To find the answer will require a trip back to the drawing board, taking a completely new view of the real density of the Earth in general; a view without the magma-planet paradigm at its core.

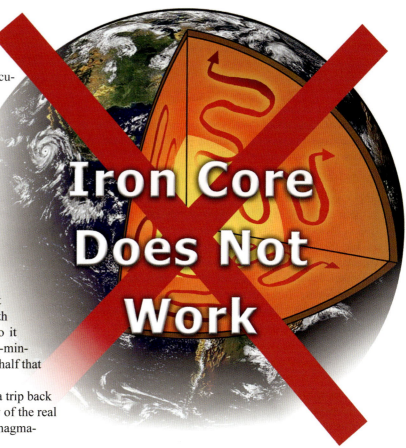

Inner Iron Core Does Not Work

Extraordinary claims require extraordinary evidence, and new scientific discoveries rarely ever stand-alone. If pure iron does not exist and so cannot solve the Earth-density problem, is there other significant evidence supporting a non-iron core?

Investigators determined that if iron was at the Earth's core, the iron must be in a solid, stable crystalline structure at high pressure and temperature because seismic waves established that the inner core is solid. In recent years, experiments to find a stable iron structure resulted in the development of an iron phase diagram. Just as liquid water becomes solid ice or gaseous steam at certain temperatures and pressure, so does iron. What did researchers from the Mineralogical Society of America find?

"To the extent that the inner core consists of pure, or nearly pure iron, its crystalline structure is determined by the iron phase diagram. While there has been considerable progress in experimental determination of the phase diagram at pressure approaching the inner core, **the stable phase of iron at inner core conditions cannot yet be uniquely identified on the basis of phase equilibrium measurements**." Note 5.9a

Another report states:

"The stable phase of iron at inner core conditions is **unknown**." Note 5.9b

It is "unknown" because as far as anyone knows, it does not exist! Although stable iron phases exist at *low* pressure, empirical evidence from actual experiments proved that at high temperature and pressure, iron is not stable:

"[We]…find **no evidence for phases other than those long known from low pressure work**…" Note 5.9b p273

No doubt, the researchers are not standing idly by, scratching their heads, and pushing this important issue under the rug. They recognize that it is central to their understanding of the magma theory and of the Earth. The same researchers continue in the journal of *Mineralogical Society of America, Reviews in Mineralogy*:

"**This is an important issue** from the geophysical and geochemical point of view for a number of reasons. **First, it is central to our understanding**…" Note 5.9b p273

Other evidence found in this subchapter supports the new concept that there is no iron core, and we will continue to expose many related errors throughout the UM, but the details of the true nature of the Earth's interior we saved for later examination, in subchapter 7.6, the Hydroplanet Earth. Nonetheless, the truth about the Earth's core is so surprisingly simple, that by now, it probably is no surprise that the extraordinary answer to this perplexing problem is—*Ice*!

We know through seismic waves that the Earth's core has both liquid and solid components. We also know that **H_2O ice** proved in recent experiments that ice forms a stable crystalline structure under high pressure, says *Reviews in Mineralogy*:

"The prototype system is **H_2O ice**, where **recent experiments have shown** that a symmetric hydrogen-bond state of **ice forms at 60 Gpa, and persists to at least 210 Gpa**." Note 5.9b p628

At these high pressures, ice did not melt even up to 50° C (122°F). We will soon explore other exciting details about water and the role it plays in planet making in the Hydroplanet Model, Chapter 7, but first, we must understand that the researchers themselves recognize the "considerable uncertainty" that shrouds their current understanding of Earth's core:

"**Considerable uncertainty** still shrouds the outer core and

SUBCHAPTER 5.9 THE NON-IRON CORE EVIDENCE

inner core from our complete **understanding**. Much of this uncertainty is associated with our **ignorance** of the physics and chemistry of iron and iron alloys at the extreme conditions that are relevant." Note 5.9b p273

Such ignorance is due, not only to a lack of understanding about the physics or chemistry of iron; it is due to the flawed ideas of the magma pseudotheory and the errors relating to the density of the Earth.

No Iron Core—No Iron Meteorite Origin

Another corollary associated with Earth's not having an iron core is the mistaken idea of the origin of iron meteorites. We will discuss Meteorites and craters in detail in the Hydroplanet chapter, but for the moment, it is important to recognize that the evidences for the Earth's iron core and the presumed iron cores of other celestial bodies, such as asteroids, are merely theoretical in Nature. It is the theoretical basis, upon which the magma and rock cycle theories are built that causes so much confusion regarding the iron core, iron meteorites, craters, and their origins.

5.10 Deep Earthquake Evidence

Deep earthquakes occur at depths from 300 km to 700 km. First discovered in the 1920s, deep earthquakes remain a subject of contention today in part because they are not supposed to happen, according to Magma Pseudotheory understanding, yet they account for more than 20 percent of all earthquakes. They should *not* take place because the *inferred* magma-model temperatures at those depths *would* make the rock ductile, and not brittle enough to allow for quaking.

The Deep Earthquake Controversy

"In 1922 H. H. Turner, who directed the clearinghouse of seismological data that later became the ISC [International Seismological Centre], applied this method in a stimulating and controversial paper. Based on an analysis of data from stations around the world, Turner proposed that earthquakes occur in three depth ranges. 'High focus' events have sources near the surface, but normal earthquakes, the most plentiful kind, take place roughly 150 kilometers down. **'Deep focus' events have focuses at depths of as much as 650 kilometers.**" Note 5.10a

Using the now common procedure of recording and interpreting primary and secondary seismic waves to determine earthquake depths, Turner's proposition about "deep focus events" did not sit well with those steeped in the hot magmaplanet theory. With the adoption of the magma pseudotheory as accepted fact, alternative ideas such as Mr. Turner's deep earthquakes, were disallowed. Harold Jeffrey, a leading theoretician of Seismology concluded that earthquakes "could not take place at such depths":

"Harold Jeffreys of the University of Cambridge put forward a more fundamental objection: he argued that earthquakes simply could not take place at such depths.

"Below a depth of about 50 kilometers, Jeffrey contended, heat and pressure change the mantle rock from a brittle material, capable of fracturing, **to a ductile one**...He also cited laboratory work confirming that at high temperatures and pressures rock deforms gradually in response to stress instead of

> "Considerable uncertainty still shrouds the outer core and inner core from our complete understanding."

fracturing suddenly." Note 5.10a

Although Jeffrey employed reasonable logic, he missed an important point because he based his reasoning on the magma planet theory, tainting his interpretation of the observed facts. When we make predictions based on false theory, eventually those predictions will fail, which is precisely what happened to Jeffrey's argument that "earthquakes simply could not take place at such depths."

Deep Earthquake Evidence is Real

If it was as hot at depth as the geophysicists' suppose, the ductility of the rock would prevent or at least retard the fracturing manifest in brittle rock. Other independent investigators have verified this. Because of Japan's earthquake history, that country established one of the best seismograph station networks in the world. In 1935, data from the Japanese seismograph network led to an interesting truth:

"Kilyoo Wadati, a 25-year-old employee of the Japan Meteorological Agency, did not refute Jeffrey's argument; **he simply presented convincing evidence that some earthquakes are very deep**...

"Other workers applied Wadati's techniques to data from earthquakes in other geographic areas and **confirmed his results**: 'normal' earthquakes had a focal depth of 50 kilometers or less, and **yet a few events had much deeper origins—as deep as 600 kilometers or more**. Turner had been wrong about the depth of normal earthquakes [50 instead of 150 km], but **deep events did exist**." Note 5.10a p48-9

Today, with the observation and recording of many deep earthquakes, no one questions their existence. However, within the magma paradigm, the understanding of how and why such deep earthquakes occur remains unanswered.

Magma Theory Leads Only to More Speculation

The beauty of science should be in its progress; however, an Age of Scientific Darkness clouds our views about the Earth, regrettably holding back much advancement, especially during the last century. Real progress happens when old, false theories die as new knowledge and understanding replace the myths with new scientific truths. Actual seismic observations *refuted Jeffrey's* **theoretical** *argument that high temperatures would not allow for deep earthquakes,* yet today, because magma is still so entrenched in the minds of scientists and researchers, they know little about the *mechanism* driving deep earthquakes:

"At present the **mechanism** for these very deep focus earthquakes remains **speculative**..." Note 5.10b

In the end, according to modern geology, what actually drives those seismic events recorded at depths around 650 kilometers, or why they happen at all in the very heart of the subterranean magma realm, remains speculative.

5.11 The Drilling Evidence

One definite way to 'know' what the inside of the Earth looks like is to go there, but that's not easy, and as of now, *drilling* is the closest we have come to getting there. A variety of reasons compels us to drill into the Earth. Whether for water, oil, or natural gas, or exploratory mining and scientific discovery, there are drilling projects everywhere, on every continent. Data from bore sites and core samples taken from the boreholes provide information about temperature, pressure, and other data which comprises the only direct empirical evidence we have of the interior of our planet. In this subchapter, we examine the data from two of the world's deepest holes, the KTB borehole in Germany and the Kola borehole in Russia. In addition, another drilling project was attempted to reach a magma chamber believed to exist below Long Valley, California (see Fig 5.11.1). Although they never found the magma chamber, what researchers observed was in conflict with their expectations.

Drilling to Know

In science classes around the world, students hear the words, 'we now know' preceding newly presented data and redefined theories, whether true or not. Sometimes the new information is mere conjecture, answers drawn from inference and speculation. Other times it comes from hard evidence, from data gathered on projects such as borehole drilling deep into the Earth's crust. Surprisingly, not all researchers see such projects as worthy ventures:

"I never was a big enthusiast for drilling...But the more we drill, the more we find out **how little we know**." Note 5.11a

The direct evidence from the interior of the Earth shows how little geologists know; the findings refute their theories, and it is time to know why.

The Magma Energy Source

An important topic in today's world, energy resources and the vast stores of heat in the interior of the magmaplanet Earth tantalize energy researchers. This type of heat, known as geothermal heat has the potential to efficiently heat and cool buildings as well as producing electrical energy through steam driven generators. When liquid water contacts sufficiently hot rock, it expands rapidly as it changes to steam, forcing its way through fissures and permeable rock under high pressure. Captured and directed through turbine generators, it produces electricity. With ever-increasing energy demands, one would think modern geology plays a vital role in the development of this natural resource: apparently, not all agree:

"Geothermal energy probably will **not** make large-scale contributions to the world energy budget until well into the next century, **if ever**." Bib 59 p578

Why would scientists *not 'ever' consider* this energy source, probably the Earth's largest based on magma theory, a serious contributor to the world's energy bank? If magma actually existed, geothermal energy should play a major role in the world's energy plan. Perhaps the results from drilling efforts over a supposed magma chamber had something to do with the dismal view of this resource.

The Long Valley Magma Myth

Nestled along an area east of the Sierra Nevada Mountains in California, Long Valley boasts of several geothermal springs, many tapped and used to generate electricity by pumping the steam through generators. The 1980's energy crisis drove scientists to convince California and federal authorities that they could produce more energy if they were to drill into the magma

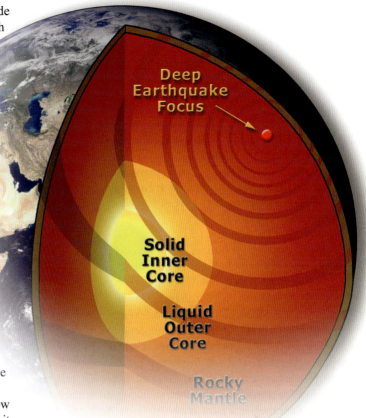

source lying beneath the Long Valley caldera, inject water into the well and recover the naturally heated steam. From *the Sandia National Laboratories Fact Sheet,* the objective reads:

"The intense heat contained in magmas, pockets of molten rock beneath the earth's surface in tectonically active areas, is a potential source of geothermal energy. The United States Geological Survey (USGS) estimates that magma less than six miles beneath the continental United States contains 50,000-500,000 quads of energy, compared to the total U.S. annual energy consumption of less than 100 quads. **Theoretically**, wells completed to depths near a magma body could serve as heat exchangers to bring that energy to the surface.

"The caldera at Long Valley was selected from among 22 potential sites **because extensive geophysical evidence indicated the existence of a magma body** at a depth of 6-7 km. Seismic data comprised much of the evidence, but the site had also risen almost a meter in fifteen years, suggesting the injection of fresh magma into the underlying chamber." Note 5.11b

Excitedly, scientists selected the Long Valley drill site from among 22 other sites "because extensive geophysical evidence indicated the existence of a magma body." They would invest millions of dollars to find magma and this seemed the most likely place to find it. The diagram in Fig 15.11.2 illustrates the drilling platform located atop the magma chamber which scientists claimed, based on "extensive evidence," existed just a few kilometers below the surface.

Magma Drilling Comes Up Short

The September 25, 1998 issue of *Science* reported the end of the drilling activity at the Long Valley magma-well. The article *Magma Drilling Comes Up Short* said in part:

"Prospects have cooled for tapping a sleeping volcano as an

energy source in California." Note 5.11c

After investing several million tax dollars searching for the elusive magma-powered energy, the prospects had cooled, as co-project manager John Finger noted:

"We haven't produced any results that point to a big energy resource payoff, and that's disappointing." Note 5.11c

The results of the magma-well drill project did provide some valuable information. As we dug deeper into the reports of the drilling operation, we discovered empirical evidence that contradicted the magma theory:

"The Phase 3 science studies to date **provide no evidence for a hydrothermal system or magma from which heat can be exploited within the central part of the resurgent dome of the Long Valley caldera.**" Note 5.11d

The researchers continue:

"The **observed temperatures** favor a model in which **there is no massive magma chamber in the upper 10 km**…" Note 5.11d

Extensive evidence and millions of dollars spent drilling and, "…*there is no massive magma chamber*…" How did researchers come to this conclusion? The first 6,500 feet rendered temperatures of only 100° C, well below expectations, but the next 3,300 feet really confused the geologists when they observed that the temperature *did not increase at all!* We wrote to one USGS researcher and asked, "If the well is getting progressively closer to the magma chamber, shouldn't the temperature continue to rise even in water? Why did it not? His response:

"We do not fully understand why the temperature in the well stopped increasing below about 6500'." Note 5.11e

Their confusion is understandable, as the geologists had intently studied the nearby Casa Diablo Hot Springs, which produced water temperatures of 170° C near the surface. At the drilling site, investigators wondered why the deep borehole did not exhibit similar heat as it approached the magma source. The answer lies *beneath* the hot springs:

"Hot Creek is the lower reach of Mammoth Creek where several vigorous hot springs discharge hot water to the surface. The boiling pools along the creek (93° C at this elevation), **commonly change in vigor and location in response to local earthquakes** and the seasonal rise and fall of the water level in the creek. **The largest and hottest springs are located at the intersection of Hot Creek and two faults that are about 1 km apart.**" Note 5.11f

The secret beneath the hot springs consists of a series of continuously active faults, linking hot springs' discharge with local earthquakes, with the hottest of them located on the most active faults. Examples of these fault-driven geothermal springs suggest further investigation within the paradigm of the Frictional-Heat Law.

It clearly was no coincidence at the Long Valley drill site that the temperature stopped rising below 6,500 feet because there were no deep earthquakes below the dome.

"Earthquakes under the resurgent dome occur no deeper than 3 miles…" Note 5.11g

Two strikingly different papers published in 1999 interpreted the findings from the drilling project, but reached very different conclusions. The first group, J. T. Finger and R. D. Jacobson, wrote the *Phase III Drilling Operations at the Long Valley Exploratory Well* LVF51-20 in **June 1999.** Their straightforward report made it very clear that the temperature stopped increasing in the last 3,300 feet and that no exploitable heat-energy existed:

"… the local temperature gradient in this hole is ambiguous with respect to inferring the depth to a **postulated magma chamber**…

"The observed temperatures favor a model in which there is **no massive magma chamber in the upper 10 km**…

"The Phase 3 science studies to date provide **no evidence** for a hydrothermal system or magma from which heat can be exploited within the central part of the resurgent dome of the Long Valley caldera." Note 5.11d p14

It seems clear from the directly observed data included in the June '99 report on Phase 3 that the "postulated magma chamber" did not exist. However, after spending millions of dollars looking for the non-existent magma heat source, and perhaps in an attempt to defer ridicule, the second team published an article expressing an entirely different outcome. In the journal, *Science*, 24 September 1999 p2119-2122, M. Battaglia, C. Roberts and P. Segall published the article, *Magma Intrusion beneath Long Valley Caldera Confirmed by Temporal Changes in Gravity*. In this highly technical and ambiguous article, the researchers contend that the magma chamber still exists:

"Precise relative gravity measurements conducted in Long Valley (California) in 1982 and 1998 reveal a decrease in gravity of as much as -107 +/- 6 microgals centered on the uplifting resurgent dome… **Assuming** a point source of intrusion, the density of the intruding material is 2.7×10^3 to 4.1×10^3 kilograms per cubic meter at 95 percent confidence. The gravity results require intrusion of silicate magma and exclude in situ thermal expansion or pressurization of the hydrothermal system as the cause of uplift and seismicity." Note 5.11h

The paper proposed "a depth of 10.6 km and a mass of $7.4 \times 10(11)$ kg" for the postulated magma. Essentially, Battaglia and his colleagues suggested there was a magma body underneath the caldera (from micro-gravity measurements) just three months after Finger and Jacobson reported from Sandia that there was no evidence for magma underneath the caldera "in the upper 10 km." Finger and Jacobson depended on the borehole data, which constituted **direct evidence**, whereas Battaglia and his group used **indirect gravity measurements.** The indirect measurements required considerable speculation and interpolation to

> "The more we drill, the more we find out how little we know."
>
> Alfred Duba

Fig 5.11.1 The Long Valley California, USA, borehole. An attempt to reach a magma body and exploit the heat of magma for energy production failed. No magma was found.

make the data 'fit.'

This example of direct contradictions come about because researchers use very different methods of analysis; one empirical, one theoretical. Because they found no magma while drilling, the direct borehole evidence should outweigh inferred gravity evidence, but what can researchers do when other researchers publish conflicting reports? We wrote requesting an update, but were unable to get a response.

This research is an example of wasted taxpayers' dollars because we allow long-held, unconfirmed theories to stand even after the evidence shows that we should change them. No doubt, with millions of grant-money dollars at stake, intellectual dishonesty and the temptation to hide the truth influences some of the parties involved.

Bore Hole Fact—No Magma Chambers

One simple fact remains; no borehole ever penetrated into a magma chamber. Despite this fact, geology textbooks claim:

"Geochemical, geophysical and petrologic **evidence** all support the **concept that reservoirs of magma** lie beneath many volcanoes." Bib 87 p19

In this quote, concept is just another term for theory, and the 'evidence' claimed as support remains tenuous and inferred. If only the experts could answer one of these questions: Where are the 'reservoirs of magma' and, why have we not seen them? The experts selected the Long Valley, California drill site as *the best* of 22 possible 'magma chamber' sites in the United States, yet the Long Valley project failed to locate the magma chamber, and the temperature stopped rising as they continued to drill deeper. Magma is simply a theoretical concept. Researchers can only infer the size and existence of magma chambers by researchers using computers to 'model' the imaginary chambers:

"**In actuality**, geological and physical **models of magma chambers** cited in literature **visualize them** in general to be small volumes (2-3 km dimensions) of total melt. On the other hand, geophysically **inferred magma chambers** seem to have volumes of tens to hundreds of km^3..." Note 5.11i

While the evidence shows no magma chambers under any of the Earth's volcanic regions, there are areas of intrusive lava. One example is the so-called magma body geologists study in the Socorro region of the Rio Grande rift. From the journal *Volcanic Seismology* we read:

"**The best evidence for such a chamber** comes from the Socorro region of the Rio Grande rift..." Note 5.11h p314-16

At the Rio Grande rift, researchers identified what they termed a thin and sill-like layer of magma:

"In a later study using a **micro-earthquake swarm** in the Socorro area, Aki and Sanford (1988) identified P to P reflections from the magma body, and by synthetic seismogram modeling of these and the S to S reflections, showed the magma **body to be thin and sill-like, with a 70-m layer** of magma (low velocity) underlain by about 60-m layer of slightly higher velocity interpreted to be crystalline mush." Note 5.11h p316-18

This "sill-like" layer, a horizontal tabular (table like) structure, describes exactly the type of molten rock we identified earlier as intrusive lava. Magma is the stuff science supposes comes quietly from deep in the Earth, Intrusive lava is the result of frictional heating. At the Socorro site, there was no vertical magma chamber, but there was a horizontal manifestation of molten rock accompanied by an earthquake swarm. The melted rock at Socorro was not magma; it was frictionally heated intrusive lava.

Borehole Evidence

One of Earthtide's aftereffects is the generation of heat in the crust, precisely what we expect to see in borehole temperature logs. The temperatures observed in the boreholes remain dependent on the type of minerals present in the crustal layers, the amount of micro-movement, pressure directly related to the depth of the borehole, and the amount of water present in the rock. With the new 'Earthtide' criterion, previously unexplained 'temperature anomalies' can now be explained. Instead of assuming that aquifers cooled the rock below the anomalous heat spike, researchers can identify the seismicity relationship with the source of the heat anomalies. Water is an excellent conductor of heat and we expect deep water near fault lines to be very hot, increasing with depth or frictional pressure.

Caves provide another real time experience of subterranean temperature. Those who have visited deep cave systems know that a jacket or coat is requisite gear. Caves are often cold and usually maintain constant, year-round temperatures, with only minor seasonal variation. There are instances of ice caves even in very hot, relatively arid environments. Why do the vast majority of caves *not* get hotter as they get deeper? One reason is that natural caves often exist in environments where there is little or no seismic activity, and we typically do not find them in solid crystalline rock. In fact, the rock most commonly associated with deep caves is limestone, which is relatively stable seismically. By contrast, most mines are excavated in brittle, crystalline rock. Friction from movement, along the faults in crystalline rock creates measurable heat.

Friction from the Earth's shifting crust, a result of tidal movement causes a buildup of *stress* in fault areas, which causes an increase in the temperature of the rock surrounding the faults. When stress from tidal movement, or any movement in the surrounding rock reaches its maximum, a sudden shift along planes of weakness (faults) generates an earthquake, which releases stress. Although all earthquakes generate frictional heat, most do not melt rock because they are of too short a duration or of too little magnitude. When there is enough stored stress, multiple earthquakes or earthquake swarms can produce

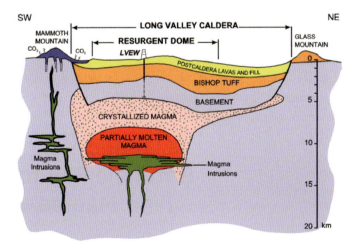

Fig 5.11.2 – Long Valley California, USA, drill rig located above the theoretical magma chamber that drillers never found.

enough heat to melt rock. Lava (extrusive lava) may then travel to the surface in the form of an eruption. This sometimes happens when there are earthquake 'swarms' or many small quakes in the same location for days or weeks at a time (see Fig 5.3.5).

If there is a cause and effect relationship between earthquakes and heat, we should also find a cause and effect relationship between earthquakes and heat in boreholes themselves, and it just so happens, there is.

The 'Christmas-Present' Quake Evidence

Because the magmaplanet paradigm is so ingrained in modern geology, geologists apparently put little effort into researching the connection between earthquakes and heat production in the crust, although some research discusses the heat-earthquake connection.

One example comes from Russia, from the deepest borehole in the world. While logging data from the Novo-Yelhovskaya 20009 superdeep borehole, researchers noted 'temperature anomalies' within the strata. A temperature anomaly is an abnormal temperature rise or fall that occurs outside the temperature gradient within the borehole. If magma was the source of heat that caused the increase in temperature as depth increases, the data should reveal a relatively constant rate of change. In the case of the temperature anomaly in this particular borehole, there *was an **increase** in temperature* followed by a return to the gradient, which occurred at "...a depth range of 5284 to 5322 m." Note 5.11j

What is the significance of this and other similar temperature anomalies? It establishes a source of heat other than just heat from pressure, and that the source of heat did **not originate below** the anomaly. Rocks at this depth exhibited a unique characteristic, reported the research team; "The drilling in the hole revealed **heavily shattered rocks** (fault breccia) and tectonic high-dipping **fault**." Note 5.11j Here we see that the increase in temperature (the anomaly) occurred contemporaneously with a fault zone of shattered rocks—thus establishing the strong possibility of a connection.

Then with seismic instruments in place, the research team received a 'Christmas' present:

"In 1998, 25 December the local shallow earthquake occurs near the 20009 borehole. Geophysical measurements revealed that the **earth-quake focus is at a depth of 5300 m**." Note 5.11j

This earthquake had occurred near the fault associated with the previously known temperature anomaly. One month later, a reinvestigation confirmed the location of the earthquake:

"Our temperature measurements one month later, in 1999, 22 January showed that there was caving of disintegrated rocks in the zone of tectonic fault at a depth of about 5320 m that was apparently responsible for the observed earthquake." Note 5.11j

Six months after the event, borehole temperatures revealed that the temperature anomaly had *disappeared*:

"...repeated measurements 0.5 years later (1999, July) argue **that temperature anomalies are not present in this depth range**. The important change in the behavior of the temperature from the large anomaly to the normal distribution is evident." Note 5.11j

What caused the temperature anomaly to go away? It certainly was not the disappearance of the 'magma chamber.' The release of *stress* that had built up in the fault zone caused an increase in temperature due to frictional heating during the seismic release event—the earthquake. This confirms the connection between earthquakes and heat, and it identified the source of short-term heat.

The KTB and Kola Borehole Evidence

The deepest humankind has ever penetrated the crust of the Earth is the Kola superdeep borehole, located on the Russian Kola peninsula near the Norwegian border. Drillers reached the deepest point at just over 12 km, or about 7.5 miles (40,230 ft). Despite the incredible engineering feat, figure 5.11.3 illustrates just how small this distance is in comparison with the entirety of the Earth. Thus, we see the magmaplanet we learned about in school consists of theoretical melted rock that lies far below anything humankind has ever been able to drill—and we have never observed that magma.

In Chapter 5.4, Magma Defies Heat Flow Physics, we noted researchers' statement that, "Direct measurement of temperatures in the well compels revision of ideas" concerning the outward flow of heat from the Earth's interior. The data showed the crust getting 'too hot too fast' to attribute the heat source to magma, but it matched precisely with the Gravitational-Friction Law outlined in Chapter 5.3. The second deepest and most studied hole in the world is the KTB borehole, near Nürnberg, Germany. Again, the deep boreholes should be one of the first

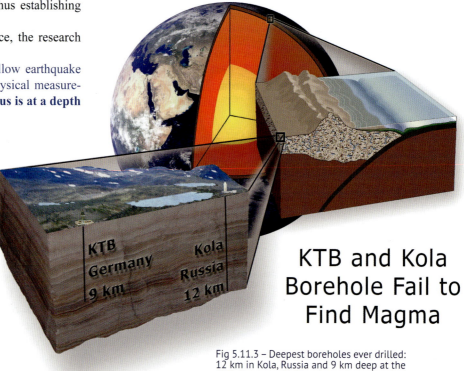

KTB and Kola Borehole Fail to Find Magma

Fig 5.11.3 – Deepest boreholes ever drilled: 12 km in Kola, Russia and 9 km deep at the KTB, Germany. Neither has ever supported any evidence for the existence of magma.

places to find the evidence of underground magma, if it really existed. At KTB, researchers hoped to establish what they called "ground truth" with respect to what lies below our feet, reported the journal, Science:

"One object of the hole was to provide a kind of **ground truth** for seismic reflection profiling, the radar-like technique that creates images of the subsurface structures from the manmade seismic waves that they reflect. 'You see all kinds of [seismic] reflectors around the world,' says KTB operations leader Peter Kehrer, 'but we don't know what they are.' Distinguishing among faults, changes in rock type, fluid-filled cracks, or other **possibilities has been largely guesswork—and the KTB hole suggests an extra measure of caution.**" Note 5.11k

In reality, the "ground truth" concerning many past claims about rock types and fluid-filled (presumably by magma) cracks routinely made by geologists amounts to, "***largely guesswork.***" Interestingly we see the words 'truth' and 'guesswork' in the same paragraph in this modern science journal, a problem that belies much of modern geology's understanding; merely guesswork.

Referring to the Kola project, lead researcher at the site, Peter Kehrer confesses:

"The science coming from the hole means 'the geology textbooks will have to be rewritten'." Note 5.11l

We need a geology textbook rewrite not just because the data from the borehole refutes most of what modern geologists think about magma, but also because the whole foundation of modern geology sits on theoretical magma. The drilling evidence sited here amounts to a mere drop in the ocean of data accumulated from the worlds deep boreholes, and with newer technology coming available every day, ever more evidence emerges to refute the current geologic paradigm, setting the stage for a paradigm shift in geology.

5.12 Earth's Magnetic Field Pseudotheory

Essential to all life, the invisible field of electrical energy surrounding Earth protects and shields her from harmful cosmic radiation, primarily from the sun, but also from other cosmic sources. This energy field allowed both ancient and modern man to navigate around the world using a simple, age-old tool—the compass. This energy field, although designated a magnetic field in modern science, consists of complexities not at first obvious. The field, oriented generally in a north-south direction, does not remain fixed; it fluctuates, oscillating daily and migrating, ever wandering across the landscape. We know Magnetic North always moves, but science does not know why. One of the more popular explanations is that a spinning molten nickel-iron core, a dynamo, creates the energy field, but no experiment or even a simple explanation satisfactorily demonstrates this highly complex and mystical electric/magnetic field phenomenon. Could it be from something else?

Fig 5.11.4 – This is the site of the German KTB 9 km deep borehole Scientific drilling project carried out 1987-1995.

Magma Pseudotheory Relic—Earth's Magnetic Field

Most of us have had some experience with magnets and magnetic fields. In Fig 5.12.1, iron filings organize in a twin radial pattern on a piece of paper because of a magnetic field produced by a magnet held beneath the paper. The iron filings appear to bend toward the poles, of which every magnet has two, referred to as north and south poles. The magnet generates an organized energy field. Certain substances like iron, strongly affected by the energy field of the magnet, try to align with the invisible field.

Note the magnets in Fig 5.12.2. The four compasses in the lower right image each point toward the magnet. The compass needles respond to the bar magnet because they are also magnetized. Opposite poles attract (north points to south) and in this way, the compass responds to changes in the Earth's energy field (compasses in upper left image all point to the Earth's magnetic North). Watch a sensitive compass and you will notice that over a period of time, the needle will fluctuate. The fluctuation appears random, yet continuous every few minutes. Why does it do this?

Usually referred to as the Earth's 'magnetic field,' in most explanations, the Earth's energy field is compared to the field surrounding a bar magnet. The illustration in Fig 5.12.3 represents Earth's magnetic field according to most science textbooks. How do scientists explain the origin of this field?

"The **magnetic field is created by** the flow of **molten iron** inside the Earth's core."
Note 5.12a

In one form or another, modern science consensus says Earth's "magnetic field" results from the motion of molten iron at the core of the Earth—a dynamo. There is one glaring problem with the field illustrated in Fig 5.12.3—it is incorrect.

The Earth's energy field does not form a smooth, circular field surrounding our globe. The actual field looks like the image in Fig 5.12.4, writhing and twisting, turning in and out, it consists of a mix of high and low energy levels within the field. The field is anything but smooth and circular as illustrated by the permanent magnet illustration.

The Earth's Magnetic Field Mystery

Back in 1269 A.D., Peter Adsiger noticed a continually fluctuating energy field around the Earth. Since then, science still wonders what causes the magnetic needle in a compass to fluctuate as it moves around the globe, or why the needle changes where it points over time, even when the compass remains stationary. An 1820's science encyclopedia included the following statement about the intriguing energy field surrounding us:

"Since that time [1640] it has been found that the magnetic needle not only varies after a considerable period, but that it is **continually fluctuating**, so that the variation of it may generally be observed within the period of an hour or two, and often in a much shorter time.

"The declination is not only subject to

a continual variation in the same place, **but it is different in different parts of the world**. It also varies differently in each particular place; so much so, that notwithstanding the exertions of the greatest philosophers and mathematicians, **no theory nor rule** has been discovered which might furnish the means of foretelling with accuracy the declination of the magnetic needle, for any future period, at any particular place." Bib 123 under "Declination"

That was nearly two centuries ago; have we solved the mystery of the Earth's magnetic field in modern times? Just how important is this mystery to physics? A modern textbook on Earth's magnetic field gives this answer:

"In more modern times Einstein, shortly after writing his special relativity paper in 1905, described **the problem of the origin of the Earth's magnetic field as being one of the most important unsolved problems in physics**." Bib 36 p17

Einstein made his comments over a century ago, yet here we are today, with over a hundred years of research involving many academics and scientists, investing millions of dollars and employing the latest technology. Should we not expect a working geophysical model?

In one attempt to describe the Earth's complex magnetic field, the International Association of Geomagnetism and Aeronomy, working since 1968 on a model dubbed the International Geomagnetic Reference Field (IGRF), derived a model based on the belief that the magnetic field originates from a fluid core and magnetized rocks within the Earth's crust. The following quote comes from one of the contributors to the current model.

"The IGRF is **inevitably an imperfect model**. Firstly, the numerical coefficients provided will not be correct: the model field produced will differ from the actual field we are trying to model - … If you measure the magnetic field at a point on the Earth's surface, **do not expect to get the value predicted by the IGRF!**" Note 5.12b

Here we see the researchers acknowledge that their magnetic field model does not work. If we take that thought further, we should expect that *any* model of Earth's magnetic field based on the magma theory must fail, because magma does not exist.

In the April 2005 issue of *Scientific American*, an article titled *Probing the Geodynamo* has one geophysicist admitting:

"At last count, more than a dozen groups worldwide were using them [computer dynamo models] to help understand magnetic fields that occur in objects throughout the solar system and beyond. **But how well do the geodynamo models capture the dynamo as it actually exists in the earth? The truth is that no one knows for certain**." Note 5.12c

An interesting answer considering the widespread acceptance of this theory!

Any theory or model proposed to explain the geology of inner Earth must include an explanation of Earth's energy field. Incorrect theories are difficult and ultimately impossible to combine with each other because of the disconnect that exists between the ideas. However, truth is easily connected. This is the beauty and universality of truth.

> "The science coming from the hole means the geology textbooks will have to be rewritten."
>
> Peter Kehrer

There is an explanation here, in the Universal Model in subchapter 9.5, using empirical evidence to show the source of Earth's energy field. It is surprisingly easy to understand once we abandon the magma paradigm.

Magnetic Fields and Magma Don't Mix

The existence of the Earth's magnetic field provides us simple, yet powerful evidence that the magma model is incorrect. Why? ***Heat destroys magnetism.*** There should not be a magnetic energy field emanating from inside the Earth under the current hot-interior model. Geologists recognize this and describe it as a "fatal defect":

"Unfortunately, although a good description of the magnetic field can be given if we assume a permanent magnet at the center of the Earth, this model has **a fatal defect**. Laboratory experiments show that **heat destroys magnetism**, and materials lose their permanent magnetism when temperatures exceed about 500° C. Material below depths of about 20 or 30 km in the Earth, therefore, **cannot be magnetized because the temperatures are too high**." Bib 59 p498

With such knowledge, one wonders 'why do scientists keep describing Earth's energy field as though it were a magnet?'

> **A heated iron core does not make a magnetic field. Heat annihilates a magnetic field, it does not facilitate one.**

The Dynamo Theory That Does Not Work

The theory that attempts to explain the Earth's magnetic field is the dynamo theory. The idea explained states that, "Earth's main magnetic field is sustained by a self-exciting dynamo action in the fluid core." Bib 173 p198

In 1996, three accomplished geophysicists wrote *The Magnetic Field of the Earth* which was published as part of the International Geophysics Series. In the book, they explain the tests of the geodynamo theory:

"Unfortunately, proposed tests of geodynamo theory **are few and the results of those tests ambiguous**." Bib 36 p423

The authors acknowledge the geodynamo idea as ***only*** a theory and that the test results were too "ambiguous" to prove anything and there may not even be a way to test the dynamo theory:

"… a convincing test will probably **not** be possible in the near future." Moreover, "This does **not** appear to be able to provide a test of dynamo theory." Bib 36 p423-4

They also concluded

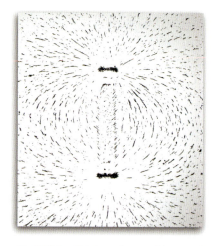

Fig 5.12.1 – This is an image of iron sand on paper with a magnet below the paper showing a magnetic field with two magnetic poles.

that the evidence of magnetic field reversal was not proof of a dynamo:

"The reader might think that there must be far better tests of geodynamo theory than those given above, such as the existence of magnetic field reversals. However, the prediction of the existence of magnetic field reversals was made before there was any dynamo theory and, perhaps surprising to many, **their existence is not in fact a test of dynamo theory**." Bib 36 p424

Scientifically objective thinkers must raise a red flag when it comes to the theory's inability to explain the Earth's energy

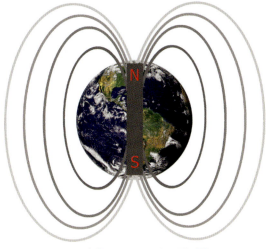

Earth's Magnetic Field

Fig 5.12.3 – The typical diagram of Earth's Magnetic Field in most modern science textbooks represents the energy field as a permanent bar magnet centered inside the Earth. This representation and associated theory is incorrect.

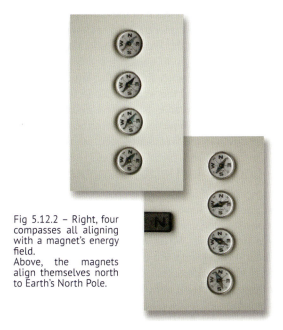

Fig 5.12.2 – Right, four compasses all aligning with a magnet's energy field.
Above, the magnets align themselves north to Earth's North Pole.

field. The obvious flaws of the dynamo theory are one reason why scholars have had to admit that the thermal history of the Earth is a problem of "enormous difficulty." A heated iron core does not allow for a magnetic field. Heat annihilates a magnetic field it does not facilitate one.

Researchers frequently make claims regarding the dynamo theory without any substantiative proof. For example, in *Nature*, geophysicists state at the beginning of their article:

"In the Earth's fluid outer core, a **dynamo process** converts thermal and gravitational energy into magnetic energy." Note 5.12d

Nevertheless, they cited no observations or experiments to support their claim. In *Surveys in Geophysics*, researchers begin their article, *The Riga Dynamo Experiment* with this statement: "Cosmic magnetic fields, including **the magnetic field of the Earth, are produced by the homogeneous dynamo effect** in moving electrically conducting fluids." Note 5.12e

In their experiment, the temperature does not exceed 300° C, which is more than five times lower than the melting point of even quartz-based minerals, which is still far below the melting point of iron or magnesium based rocks. Put simply, this experiment was nowhere near the presumed natural dynamo conditions as theorized in magmaplanet theory. Another paper, *Laboratory experiments on hydromagnetic dynamos* Note 5.12f gave a good summary of dynamo experiments and provided an overview of past results. Even though the researchers from this article claim some success with the Riga and Karlsruhe experiments, the experiments included temperatures considerably less than the hot temperatures implied in the magmaplanet model.

Low temperature sodium melts cannot take the place of a core that supposedly consists of molten nickel and iron. Finally, even *if* molten iron could produce a 'magnetic field', it does not prove that Earth's energy field comes from a molten core.

Magnetic Field Movement

Geology students learn from textbooks that say something like:

"Earth's magnetic field behaves as if a small but powerful **permanent** bar magnet were located near the center of the Earth…" Bib 59 p498

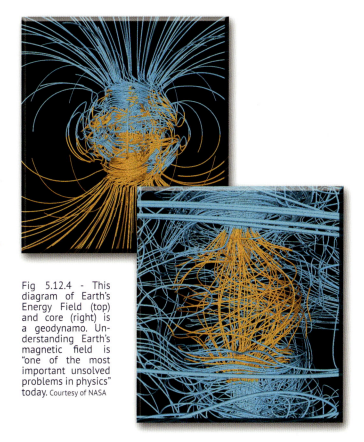

Fig 5.12.4 – This diagram of Earth's Energy Field (top) and core (right) is a geodynamo. Understanding Earth's magnetic field is "one of the most important unsolved problems in physics" today. Courtesy of NASA

However, a 'permanent' magnet's energy field does not change or oscillate. Iron filings held in place on paper by a permanent magnet do not simply 'drop off' or become more strongly attracted. We noted earlier, "…the magnetic needle not only varies after a considerable period, but that it is **continually fluctuating."** Bib 123 under "Declination"

With the contradictory terms, "continually fluctuating" and "permanent," describing Earth's energy field as a 'magnet' is both wrong and misleading. The foundation of the dynamo theory presumes permanence and scientists describe the energy field of the Earth as permanent, basing the idea on workings of a magma core. There is no known large-scale fluctuating or cyclical movement of magma as measured in terms of days, hours, or seconds. However, Earth's energy field experiences many such fluctuations and movements on these time scales.

Just as many other things observed in Nature, the Earth's fluctuating energy field is cyclical as is all matter, which we note in the Time Model in Chapter 20. There are daily oscillations, and annual movement of the Earth's magnetic field, which any true model of the Earth's energy field must explain.

Finding the True Source of the Earth's Energy Field

As noted earlier: "…the **problem of the origin of the Earth's magnetic field** as being one of the most important unsolved problems in physics." Bib 36 p17

In addition to physics, the problem is of equal importance to geology, astronomy and biology. Learning the true source of the Earth's energy field and that of other celestial bodies is critical in comprehending how celestial bodies form. Moreover, the Earth's energy field affects all living things that exist within it. A model that can explain and predict the Earth's energy field will certainly affect all of us and must explain the following:

1. What causes the magnetic field and why does it exist?
2. How does the field fluctuate instantaneously?
3. What causes the magnetic poles to experience daily cycles?
4. Why do the magnetic poles exhibit annual migration?

No modern science theory that we know of, including the dynamo theory, satisfactorily explains any of these four critical energy field questions. Currently, the magmaplanet model forms the basis for all of the Earth's magnetic field theories.

The UM provides answers for each of these questions, in detail, in the Weather Model (Chapter 9). Why include the origin of the Earth's energy field in the Weather Model chapter? Because, as we will see, the origin of the Earth's weather, *as well as* the origin of the Earth's energy field, is the same! The true source of the energy field *and* the weather, along with several new natural laws are discoverable using the Universal Scientific Method to establish the evidence and describe the mechanism for their operation in the Weather Model.

5.13 The Continental Uplift Pseudotheory

Some continents exhibit higher topography relative to other continents and geologists have long wondered why. The crust of the Earth consists of giant continental and oceanic plates, and those plates are moving. We observe and measure the horizontal movement of the plates, but there is no recorded vertical movement of large landmasses over geological time. This is not to be confused with local uplift or subsidence or faulting and mountain building episodes. Continental uplift as inferred among geologists includes the Continental Uplift Pseudotheory. It causes more confusion and error than probably any other geologic theory, other than the Magma Pseudotheory itself.

Uplift Defined

The uplift pseudotheory extensively mingles uplift fact with uplift fiction so that almost everyone on the planet who studies the Earth believes that ongoing uplift of large landmasses and continents is real. *Uplift facts* consist of observed phenomena and demonstrable processes, which we define here in the UM as **micro-uplift**. *Uplift fiction* is the *inference* of ongoing gradual uplifting of large landmasses or continents over long periods of geological time. We define this as **macro-uplift**.

Micro-uplift observed – Scientists observe uplift on the local level of small landmasses such as volcanoes, mountains, domes, scarps, etc. These usually coincide with earthquakes. We measure and observe current and past micro-uplift events, and we know such events are part of the Earth's ongoing processes. It is real, reproducible, demonstrable, and with the right model, perhaps even predictable.

Macro-uplift inferred – Scientists have not observed uplift on a continental or global scale. They have no first hand data showing ongoing continental uplift. Every modern geology textbook includes diagrams that illustrate and explain the *horizontal* movement of tectonic plates (see Fig 5.13.1). Actual numbers accompany the depictions of horizontal plate movement because the data exists. Scientists record actual year-by-year movement from data gathered from monitoring stations around the world.

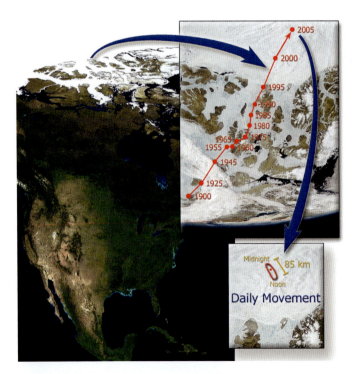

Fig 5.12.5 – This diagram shows both the daily oscillation and the yearly migration of the Earth's Magnetic North Pole. Every day, magnetic north oscillates as much as 85 km, and the central point, or the magnetic pole migrates dozens of kilometers each year. The rate of migration shows an increase during past few years. Such movement is in conflict with the molten iron dynamo theory.

Micro-uplift: the **actual** rising or lifting of hills or mountains above the surrounding landscape over a **short** period of time.

Macro-uplift: the **theoretical** lifting of large landmasses or continents above the surrounding landscape over **long** periods of geologic time.

Vertical Movement Map Myth

Why do modern geology textbooks omit maps or diagrams of vertical tectonic plate movement and accompanying data, as suggested in Fig 5.13.2? It seems horizontal plate movement gets all the glory while vertical movement gets little, if any mention!

According to modern geology, continental uplift, or the vertical movement of entire landmasses and continents, comprises an ongoing process, and one that happened to every continent in the past. Marine deposits at the top of nearly all mountain ranges and continents bolster the geologists' assertions that continental sized landmasses once resided below sea level before they were 'uplifted' over long periods of geologic time. Scientists suggest that the Grand Canyon and surrounding area experienced uplift and subsidence as many as five times! Where is the evidence of ongoing vertical movement of tectonic plates today?

Knowing that a pseudotheory is a false theory that is taught as fact; in this subchapter, the Continental Uplift *Pseudotheory* includes geology's claims of uplift of continental landmasses over geologic time yet, there has been no physical evidence of the process. The geologists we spoke to have no basis for their "knowledge" other than to refer back to the 'experts' or to the textbooks written by the 'experts.'

To illustrate how geology textbooks mislead students with respect to uplift, we refer to one popular geology textbook in the United States—*Understanding Earth*. This textbook actually included a map showing vertical uplift and it claimed the uplift was the result of actual observations, even citing the NOAA as the source for the claim.

Since this was the first and only evidence we encountered, we excitedly contacted the author to ask for the actual reference for the uplift observations since we were unable to find the data directly at NOAA or for that matter, anywhere on the Internet. Sadly, he informed us that:

"Unfortunately the notes and background materials for that edition have since been destroyed." Note 5.13a

The author to whom we had written did find a "brief reference" to the original researcher, S. P. Hand, on the internet. He told us that the map used in the textbook had come from Hand's 1974 book, *Earth*. The web reference to which the author referred stated the following about the 'uplift map':

"The first compilation of its kind for the United States, the map reveals **probable** annual rates of crustal movement over large regions… much of it is based on **interpolation**…The map is based on measurements made over the past 100 years…"
Note 5.13b

The reference cited observable phenomenon as related to subsidence along the Gulf coastal plain and Central Valley California. However, it explained that the movement was due to the removal of underground resources, such as water drawn from thousands of wells, resulting in subsidence—*local* subsidence. This was the extent of the evidence used to justify vertical uplift!

Elevation measurements made since the 1800's have gone through major modifications with the advent of new technology. What was surprising from the web article, *Crustal Movement Map of USA*, was the use of the word "subsidence." *Seven times* it described actual measurements of areas in the US that '*subsided*' or '*lowered*,' but did not mention *Uplift* even once.

We searched in vain for a map or diagram showing Vertical Plate Movement based on real data or actual measurements of uplift of any large landmasses. The reason proving or disproving vertical movement of large landmasses is important is the fact that modern geology bases nearly its entire ideology on the uplift pseudotheory, including the entire concept of 'metamorphism,' which requires the upward and downward movement of the continents over 'geologic time.' For more than a century, geologists turn to metamorphism to explain the origin of many rocks and rock formations, but as we will soon see, their explanation is in error, and that is just the tip of the iceberg.

Another aspect of the continental uplift pseudotheory is the subduction and exhumation pseudotheory. This theory states that the crust of the Earth raises or sinks along crustal plate boundaries. There are no real vertical plate movements on which to base the theory and no high-pressure metamorphic rocks actually substantiate it. The following quote comes from the *Journal of Geophysical Research*:

"However, the large-scale flow patterns involved in **subduction and exhumation** of continental crust, with preservation of the UHPM [ultrahigh pressure metamorphic] record, **remain poorly understood**." Note 5.13c

It is always difficult to see outside of the established paradigm because of long-held theories that science has accepted as true for so many decades. Scientific advancement cannot happen if the paradigm upon which it bases its models is not true.

Continental Uplift & Magma Pseudotheory Connection

The source of energy supposedly driving the uplift and subduction of the continental crust connects the two grand theories of uplift and magma. That energy source is the internal heat of the hot magmaplanet Earth:

"The **internal heat** melts rocks, forges volcanoes, and **supplies the energy to build and move continents** and to thrust mountains upward." Bib 59 p11

Without the heat of magma to drive it, the uplift theory has no credibility. Furthermore, uplift is the support behind the uniformity myth—that small continual uplift and subduction of only centimeters-per-year over *millions of years* forms areas such as the western United States' Colorado Plateau. Actual observational data supporting this claim does not exist. The continental uplift pseudotheory is another failed example of the uniformity myth.

Of course, uplift does occur in micro-scale. When it does, it happens rather quickly, not slowly over 'geological time.' We can understand this better by considering processes related to the Catastrophic Principle. A recent dramatic example occurred on December 26, 2004 off the west coast of Sumatra, Indonesia. An earthquake with a magnitude greater than 9.0 triggered massive tsunamis throughout the coasts of the Indian Ocean and it triggered other earthquakes halfway around the world in Alaska. This earthquake, among the largest ever recorded since accurate tracking began in the year 1900, produced documented uplift of as much as 2.5 meters (8 feet) in the exposed coral reefs along the affected area. This event occurred on a catastrophic time scale, not on a 'geologic' time scale. Note 5.13d

The Continental Uplift Pseudotheory—The "Unresolved Controversy"

One of the seven natural wonders of the world, the iconic Grand Canyon remains the center of an unresolved controversy. From the book *Geology of the Grand Canyon*:

"**Without uplift and erosion, there would be no Grand Canyon**. Up until the close of the Cretaceous Period 60 million years ago, the area that is now northern Arizona was for most of its existence a low flat-lying plain. **Sometimes it was slightly above sea level receiving deposits from rivers and wind-blown sand; at other times the area was below sea level**. It was not until this whole area was **uplifted over 10,000 feet**, then eroded and sculptured to its present form, that the Grand Canyon as we know it today, came to be." Note 5.13e

The authors state "…this whole area was uplifted over 10,000 feet" as if it were a *fact*. The only *real fact* to come from their proposed 'uplift theory' is that it leads only to more "unresolved controversy." The same Grand Canyon geologists admit this in their book:

"…**unresolved controversy that still continues in regard to exactly how the Grand Canyon was formed**. One thing is certain: the history of the formation of Grand Canyon is **not a simple matter of the land rising** with the river cutting down through it like a knife through a layer cake!" Note 5.13e

Why isn't the formation of the Grand Canyon a simple matter of the land rising and a river cutting down through it? If the canyon actually formed this way, *it should be a simple matter*, but no one has ever observed any of the land rising! We comprehend the Canyon's formation or alteration *by erosion* because we see this happening—we have the data. *How* the erosion happened remains an unanswered question, but we at least know by direct observation that erosion fits in somehow. When we discover how the Grand Canyon actually formed (subchapter 8.3), it *will be* a simple matter, but before we can discover this wisdom, we have several things to consider, including some very important new geological ideas. These ideas set the stage for a new and incredible understanding about how our Earth formed.

Mount Everest Evidence

Real science deals with real observation, or at least it should. What do observations of the highest mountain in the world tell us about vertical movement? If geology had a 'poster-child' for the continental uplift pseudotheory, it would probably be Mt Everest because the experts herald it as the 'example' for large vertical movement or uplift. The U.S. National Imagery and Mapping Agency and China's National Bureau of Surveying and Mapping recently gave "enthusiastic approval" of the latest data received on the height of the highest peak in the world. What **change in height** did the surveyors record when they measured Mt Everest?

"Washburn said the reading of 29,035 feet (8,850 meters) showed **no measurable change in the height of Everest** calculated since GPS observations began at the South Col four years ago. **But** from GPS reading from the South Col over the past four years, with the receiver being attached to the same steel bolt fixed permanently into the rock face, it appears that **the horizontal position of Everest seems to be moving steadily and slightly northeastward—between 6 centimeters (2.4 inches) a year**." Note 5.13f

Here we see again that while geology texts include maps depicting observable *horizontal* movement, there are no maps of *vertical* movement. Clearly, one type of movement is fact while the other is fiction!

Another piece of the uplift theory puzzle is the missing continent-wide faults and ridges. If only *part of the continent* is uplifted or buried, there should be visible faults and ridges, and these thousand-foot landforms would stretch across entire continents, at the uplifted margins. There is no reasonable way to explain how a large section of continental plate moves vertically against the remaining section without leaving evidence of such colossal faulting.

Seismic activity on the rigid continental plate makes it easy to identify the faults, and large, continent-sized faults would be exceptionally easy to spot. The question is where are these thousand-foot-high *faults* and ridges? Since they do not exist, uplift theorists claim **entire continents** move up and down, which is what we term Macro-Uplift, which defined means the **theoretical** lifting of **large landmasses** or **continents** above the surrounding landscape.

Other Geological Evidence Opposing Uplift

Students learn the continental uplift pseudotheory as if it was fact, and scientists in general just accept the geologists'

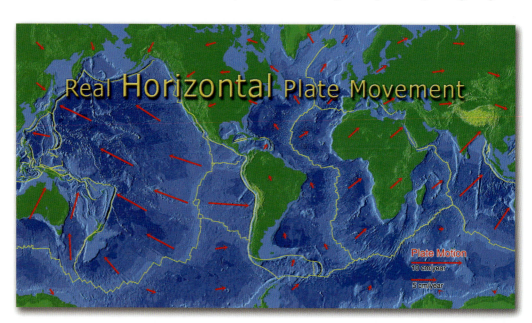

Fig 5.13.1 - World map of *actual* **horizontal** plate movement in centimeters per year. Modern technology makes it easy to measure the small horizontal movements of the plates. Courtesy of USGS.

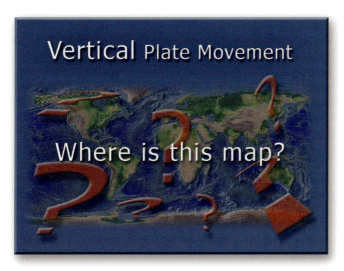

Fig 5.13.2 – Where is the world map of **vertical** plate movement? A real map showing continental uplift does **not** exist because modern geology has no data showing uplift exists. This is important because the Uplift Pseudotheory is continually taught in the classroom and found in scientific literature.

tioned, do not realize just how unsupported the uplift theory is, although they did include this disclaimer:

"Although this map is not well constrained, it illustrates that epeirogeny as a dynamic lithospheric process is **still so poorly known that it remains a poorly defined element in introductory geologic instruction**." Note 5.13j

In reality the continental uplift pseudotheory that attempts to explain the theoretical uplifting and lowering of continents remains not only "poorly defined," it has no basis in fact. Yet, from professor to student, the uplift myth passes down the educational hierarchy as unquestionable fact; teachers teaching teachers all the way back to James Hutton, the father of pseudo-geology. Like magma, modern geologists cannot produce the proof that continents uplifted or lowered over geologic time allowing the oceans to cover the land.

In light of today's global warming controversy, some geologists even express the idea that glaciers account for sufficient water to inundate whole mountain ranges, allowing the deposition of limestone on high mountaintops. In discussing the lack of evidence for continental uplift, one geology graduate student suggested a process requiring no uplift if all the glaciers melted and the sea levels rose. However, the evidence refutes this:

"In addition, approximately **75 m of potential sea-level rise** is currently sequestered from the global hydrologic cycle **by glacier ice 'stored' on land**." Note 5.13k

A 75-meter potential rise in sea level falls far short of covering the world's limestone-layered mountains. No empirical evidence of a rising ocean due to retreating glaciers exists. If evidence did exist for the complete disappearance of all glaciers and snow-pack, the subsequent rise in sea level does not come close to accounting for the formation of many large landforms like the Grand Canyon.

consensus in this area. What do the experts who are involved in measuring uplift have to say about the evidence for vertical motion? One geologist notes the following about the evidence of vertical change:

"For much of the western U.S. and Alaska, surface elevation change at the decadal timescale may be measured by continuous GPS of the PBO and InSAR, depending on the magnitude of the motion and the timeframe studied... In almost all cases, we would expect such change to be unmeasurable, given the lack of geological evidence for sustained vertical motion."
Note 5.13g

> "...the **horizontal** position of Everest seems to be moving steadily and slightly northeastward —between 6 centimeters (2.4 inches) a year," but "**no measurable change in the height** of Everest" has been observed!
>
> Bradford Washburn – Head Researcher

The popular college geology textbook, Understanding Earth Note 5.13h includes a map of a process dubbed **epeirogeny,** which describes *the upward or downward movement of large areas of land or ocean bottom*. This is the article referenced earlier in this subchapter for which no reference or origin of the map could be found. The author of the textbook referred us to a website that cited a reference leading to the journal, *California Geology*, March 1974, Vol. 27, No.3. Note 5.13i

> Why use a 1974 study based on manual surveys over a 100-year period when we have the capability to collect measurements via GPS today?

Considering the fact that we have advanced GPS technology, it is astonishing that they used the measurements and data from a four-decade old report, especially since it included no figures relating to uplift in the survey. In other words, the illustration depicting uplift in the textbook was completely imaginary! All of the specific examples of crustal movement in that 1974 survey indicated **subsidence or the lowering** of regional areas only, ascribing that primarily to the removal of underground resources, such as water for irrigation.

Sadly, geologists like author J. A. Spotila, though well inten-

The Isostatic Adjustment Myth

Even though uplift is an integral part of modern geology, finding quantitative or numerical data to support uplift is another matter. Without the physical evidence, researchers remain unwilling to state how much uplift there is. One geology textbook expressed the popular notion of balance between uplift and erosion. The textbook included a diagram explaining how *two meters of erosion over a million years* conveniently balanced two *meters of uplift over those same million years*:

"Tectonic uplift causes an increase in erosion rate, which in turn lowers the surface elevation. **The elevation is thus a balance between tectonic uplift and erosion rate**." Bib 59 p410

The concept of **Isostatic Adjustment** states that weight added to the crust causes it to sink, and the removal of weight causes rebound. Hence, uplift supposedly occurs as erosion happens so that mountains stay in balance. From *Essentials in Geology*:

"...**isostatic adjustment** can account for considerable vertical crustal movement. Thus, we can understand why, as **erosion slowly lowers** the summits of mountains, the **crust slowly rises** in response to the reduced load." Note 5.13l

The text goes on to explain that eventually, mountains continue to erode until they reach normal crustal thickness, near sea level, exposing the bottoms of the mountains. The material from the eroded mountain, deposited on the continental margin (already below sea level), causes the margin to sink further.

The idea of isostatic adjustment spawns many questions: If the surface elevation of a mountain experiences continual balance by uplift and erosion, meaning erosion offsets uplift, how does the mountain *ever* get uplifted in the first place? Shouldn't a mountain built during a "mountain building phase" just sink from the added weight? Do erosional materials from mountains cause the already below sea level continental margin to sink even further? Questions like these lead one to think that the concept of mountain balancing by offsetting uplift with erosion is imaginary thinking, yet science has persisted in the acceptance of this unsupported sci-bi for more than 75 years, as evidenced in this seven-decade old example:

"Since marine sedimentary rocks underlie large areas of the present continents, even in the highest mountains and plateaus, manifestly there are powerful forces within the Earth that cause widespread **uplift and thus tend to offset the wearing away** of the lands by destructive surface agents. **These internal forces, whose origin and mode of operation are still in large part mysterious**…" Bib 125 p12

Therefore, the mystery stands unchanged, as it must as long as researchers continue thinking large-scale uplift actually happens.

Other geological evidence refutes the uplift theory. Suppose a mountain range is around 200 to 400 million years old. At two meters per million-year rate of erosion, *the top 400 to 800 meters* of the mountain range would be *eroded away* since it last rose above sea level, yet we ***still*** find pillow lava and limestone fossils at the *tops* of these mountains, which formed *at or near the floor of a* **marine** *environment*. In other words, even the top

"One thing is certain: the history of the formation of Grand Canyon is **not** a simple matter of the land rising with the river cutting down through it like a knife through a layer cake!"

few meters of soil and rock have *not* eroded away since they first formed in a marine environment. The Erosion Mystery in subchapter 6.11 and the Erosion Mark in subchapter 8.6 show that what the erosion scientists *assume* occurred over hundreds of millions of years has no basis in fact.

The Micro-Fracture Evidence

The following quote is a great example of the evidence that confirms the extraordinary claims made in the Universal Model. Researchers at the Russian Kola borehole brought rocks up from a depth of 12 km that had been under great pressure. They noticed that as they brought the samples to the surface, *micro-fractures* began occurring in the rock matrix because of decreased pressure. When sound waves passed through the fractured rocks that drillers brought to the surface from such great depths, the rocks exhibited slower P-wave velocities—as slow as **.6 km per second** whereas, surface rocks exhibited velocities of a whopping **6.2 to 6.75 km per second**:

"Longitudinal wave velocity determinations in core samples indicated that the P-wave velocities are even more depth-dependent, and this especially pertains to the leucocratic rocks. While **near the surface the P-wave velocities vary from 6.2 to 6.75 km/s, the rocks extracted from the depth of 12 km range in P-wave velocities from 0.6 to 6.75 km/s**. There are **no known analogues** of metamorphic rocks having Vp [P-wave velocities] of about **0.6 km/s on the surface**… The lower Vp values in rocks extracted from great depth are a **result of disintegration, which is a process of formation of numerous micro-fractures** at the boundaries of anisotropic mineral grains. This phenomenon takes place when a core sample is lifted to the surface and **the mineral grains are released from the compression stress**." Note 5.13m

So now, we can ask the FQ:

If rocks now on the surface of the Earth were uplifted from great depths as geology claims, why do none of the micro-fractured rocks that exhibit slow wave velocities exist on the surface?

On what basis can we suppose that continental uplift happens if rocks deep in the crust begin a process of "disintegration" and when brought to the surface, form micro-fractures that are not comparable to other rocks already on the surface? Unanswered questions like these are the reason modern science should question the Macro-uplift theory.

The Missing Continents

Perhaps one of the most striking evidences that there is no continental uplift is the evidence we don't see. The uplifting of the continents is only half of the story. According to modern geologists, continents are sometimes rising and sometimes subsiding so that seawater can inundate the land before rising again. Almost all geology books

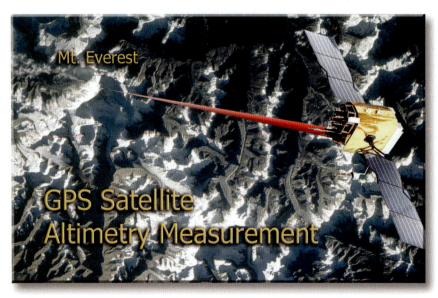

Fig 5.13.3 – Modern GPS measurements accurate to within a millimeter confirm that over the last several years Mt. Everest is not moving up or down. This direct evidence refutes the Uplift Pseudotheory but remains suppressed or ignored with no alternative in Modern Science. For the first time in history, models in the UM are able to clearly and simply explain the workings of geology where the theories of modern geology have failed.

Fig 5.13.4 – According to the continental uplift theory, continents experience both uplift **and subsidence.** If so, where are the submerged continents? This stylized map is an artistic depiction of randomly inserted submerged continents that, according to modern science, should exist. None do-. The scientific literature emphasizes the "uplift side" of the pseudotheory, but ignores the fact that submerged continents must also exist in the uplift/subsidence cycle. This fact alone should cast doubt enough to require a complete reassessment of the current continental plate theory.

written since the days of Charles Lyell in the late 1800s talk about the **uplifting** of continents and speak as though uplift is fact. However, it is difficult to find any details about the **lowering** of continents. We are not referring to subduction, a geologic term that suggests the movement or sliding of oceanic plates underneath continental plates.

In geology today, the experts infer that the formation of many geologic structures is the result of multiple subsidence and lifting events. Although no real explanation exists to explain how it happens, during periods of subsidence, seas move in and cover large parts of the continent, which subsequently rise again, sometimes thousands of feet above sea level. It is reasonable then to suppose that in addition to the Earth's continents visible today, some landmasses in the throes of subsidence and inundated with seawater right now should rise out of the ocean,

Technology has allowed us to map the entire planet in high definition. Once shrouded by abyssal darkness, the mid-ocean ridges, mountains, valleys, trenches and a host of landforms no longer lay hidden beneath the waves.

> Where are the submerged continents currently under the ocean's surface?

This is the one thing noticeably absent—***there are no large continental landmasses hiding below the ocean's surface***.

5.14 Other Magma Pseudotheories

There are many theories built on the magmaplanet belief system. In addition to those covered already, we consider these next four important because of the status they hold in modern geology. Designated pseudotheories because geology teaches them as fact, they remain unproven, such as is the case of Bowen's reaction series. His ideas about how rocks and minerals crystallize from melted crust form an important primer of rock formation in most high school and college geology classes today. His teachings not only form the basis for explaining how all rocks form, they underlie modern geology's ideas about how all continents form. Other pseudotheories in this subchapter include the movement of magma, or rather, the movement caused by magma, as in the case of plate tectonic movement. Continental and oceanic plates do indeed move, but they do not move from magma-based heat currents because those theoretical currents do not exist. In all of these ideas, long-held beliefs about magma continue to prevent an open-minded evaluation of the observed data. This results in a mingling of truth (observations) and error through incorrect interpretation.

Other Magma Pseudotheories Identified

The long-time acceptance of theoretical magma has spawned many other theories that owe their existence to the magmaplanet paradigm. These theories also enjoy long tenure among many fields of science. If the existence of magma (correctly identified as sub-crustal melted rock) is an incorrect idea, we should expect many of the theories based on it to change because they are also incorrect. So important are some theories in the field of geology that they rank among the "...all-encompassing synthesis of geological knowledge." [Bib 59 pxi]

Thus far, we examined three major pseudotheories, magma, the Earth's magnetic field (as based on the geo-dynamo) and continental uplift. Four other significant aspects of the magma theory keep scientists from discovering solid truth about geology. They are:

1. **Bowen's Reaction Series Pseudotheory**
2. **Magma Convection Pseudotheory**
3. **Mantle Plume Pseudotheory**
4. **Magma-based Tectonic Plate Pseudotheory**

1. Bowen's Reaction Series Pseudotheory

Earlier in this chapter, we noted that for the Earth to take the shape of a sphere, it had to be liquid at some time. Today's geologists suppose that liquid was magma, and that the Earth was melted. Thus, igneous rocks born of the primordial melted magmaplanet crystallized from the melt, and from this idea, hundreds of scientists have spent millions of dollars and many years, some their entire lives, trying to understand how a magmatic melt transforms into crystals, forming rocks such as granite and basalt. With so much invested, they should have had it figured out by now, but it all remains shrouded in mystery. Ask any geologist if they know of anyone who has made granite and they cannot say. Science can reproduce the temperature and pressure required to make granite according to the modern magma theory, yet scientists cannot synthesize it.

In 1928, petrologist, Norman L Bowen devised a theory about how magma crystallized into rocks and minerals, such as granite. The still widely accepted process, referred to as Bowen's Reaction Series of magmatic differentiation, provides an explanation in geology textbooks describing how rocks form, or in other words, how crystals and crystalline rock structures form in cooling magma bodies. From Wikipedia, the online encyclopedia:

"He experimented in the early 1900s with powdered rock material that was heated until it melted and then allowed to cool to a target temperature whereupon he observed the types of minerals that formed in the rocks produced. He repeated this process with progressively cooler temperatures and the results he obtained led him to formulate his reaction series which is **still accepted today as the idealized progression of minerals produced by cooling magma**." Note 5.14a

Although nearly all geology texts and teachers today talk of Bowen's reaction series as accepted fact, some are not so sure:

"At first Bowen's theory of magmatic differentiation **seemed** to be a great success." Bib59 p93

After conducting further experiments, some of the most simple and abundant rocks, like granite, seemed *not* to originate as Bowen envisioned:

"The biggest problem, however, was the source of granite. The first sticking point is that the great volume of granite found on Earth **could not have been formed as Bowen's reaction series suggests**."

"Bowen's original theory of magmatic differentiation **has been supplanted** since he proposed it many decades ago." Bib 59 p94

In this geology text, "supplanted" seems to imply a new process or idea replaced or even eliminated Bowen's ideas, but not so. Geology saw no other means by which crystallized rocks and minerals formed from liquid magma so they built on the foundation of an already failing idea:

"Nevertheless, as we noted earlier, **most later work** on the differentiation of igneous rocks **was built on the foundation of Bowen's ideas**." Bib 59 p94

As with all theories, the 'foundation' upon which scientists build determines whether it succeeds or fails. Bowen's ideas never actually panned out in the laboratory; although some journal articles imply success, the fact remains that granite did not come from a melt. The result of the geological work based on Bowen's false theory adds up to a lot of wasted time, money, and effort expended looking for answers shrouded in mystery because the work was built on an erroneous foundation.

Bowen based his theories and experiments on James Hall (1761-1832), who based his theories and experiments around the ideas proposed by James Hutton (1726-1797). Hutton placed greater emphasis on theoretical reasoning than he did facts or observation to bring others to his side of thinking, and men like Hall spent many years trying to prove Hutton's theories. Unfortunately, Hall did a relatively poor job demonstrating how heated minerals became other minerals. The details about his limestone experiments remain sketchy, he does not discuss water, which is an important part of all natural limestone, and the heat he used far exceeded temperatures found in the Earth's crust, limiting the significance of his ideas about how limestone changed to marble. Note 5.14b

Hall's physical descriptions derived from his experiments involving the slow cooling of melted rock say nothing other than "stony" when describing his cooled, melted rock. He describes a rock type he called whinstone (a type of basalt) as merely "resembling" the real thing but only visually. Bowen's ideas about crystal differentiation amount to how different rocks and minerals melt, and very little about how they form as natural minerals. The heart of the natural mineral formation process involves water and a specific recipe identified in subchapter 7.4, the Crystallization Process, and in 11.4, the Fossil Experiments. Note 5.14c

It is unknown if his experiment has been reproduced and studied in detail in modern geology. Perhaps one reason why so little interest has been shown is that today, the environment for making sandstone on the continents does not exist. The myth of uniformity tells us that the present is the key to the past, and at present, this is not happening so there is little point in understanding and conducting Hall's experiments in modern geology. With the new evidence presented here in the UM, perhaps we should become interested!

2. The Magma Convection Pseudotheory

To explain the means behind the uplift and lowering of large continental landmasses, geologists turn to the Magma Convection Pseudotheory. For geologists, this theory also explains metamorphism or the cycling of rocks within the mantle, which is a well-entrenched theory and the basis of the next chapter—The Rock Cycle Pseudotheory. The idea of internal convection is not new, as we see from the 1939 *Textbook of Geology*:

"One hypothesis that has received considerable attention attributes crustal deformation to slow-moving convection currents within a thick shell of the Earth. At first thought, it would appear that such currents are impossible, in view of abundant evidence proving rigidity in the Earth. It is urged, however, that a thick zone below the crust may be nearly devoid of strength, **because of high temperature**. Some mathematical physicists agree that slow convection may operate in such a zone, **provided an adequate source of heat exists**. Advocates of the hypothesis **assume** that minute quantities of **radioactive elements** are distributed to depths of several hundred miles in the Earth, and that disintegration of these elements **provides sufficient heat to set up convection**." Bib 125 p429-30

Initially, the 'convection' theory appears to provide a solution to major geological problems, but as we read on:

"This **hypothesis appears** to offer help in the solution of **major geologic problems; but the concept is highly speculative, and is open to serious objections**. However, it must be admitted that the cause of diastrophism [all movements of crust] is one of **the great mysteries of science and can be discussed only in a speculative way**. The lack of definite knowledge on the subject is emphasized by the great diversity and contradictory character of attempted explanations" Bib 125 p430

For more than 65 years, the explanations have not improved. Like many other geology pseudotheories, modern geology textbooks don't tell the whole 'story' respecting magma convection, although they present it as presumably simple fact. Nevertheless, from an article in the journal, *Science*, June 2000 we find real "observational and dynamical constraints" that cannot be "reconciled" with whole-mantle convection:

"The second major question is how the differing chemical compositions of erupted magmas, which **require several chemically distinct reservoirs** within the mantle, **can be reconciled** with other observational and dynamical constraints favoring whole-mantle convection, **which should mix** chemical heterogeneities." Note 5.14d

In other words, how can erupted magmas (lava) be of such different compositions if the magma source is a homogeneous mix because of convection? Some geologists recognize magma convection as highly speculative, and open to serious debate. Others hold tight to the theory, supposing and referring to it as fact. Answers about melted rock heterogeneity could be simpler, except that under the current magmaplanet paradigm *no other explanation* accounts for volcanoes or uplifting mountains. Without a paradigm shift, we can only discuss these great mysteries in a speculative way.

"Fixed Hotspots Gone With the Wind"

It is easy to understand volcanism along plate boundaries where friction from plate movement is evident, but mid-plate volcanoes and volcanic island chains remain problematic. According to today's geology textbooks, volcanic islands such as the Hawaiian chain form over 'hotspots' consisting of mantle plumes, or upwelling blobs of hot magma from deep within the Earth. Plates move over these plumes on a slow, 'geologic time' frame allowing the out flowing magma to create each successive island. Hawaii, the big island, is a recent growing island. This facet of the Magma Pseudotheory represents an important aspect of the plate tectonic model:

"Such mantle plumes are deeply **entrenched** in the geological 'standard model', going hand-in-hand with **plate tectonics**." Note 5.14e

Technological advances often shine new light on old problems, revealing the truth of what is actually happening. That happened in 1998 when new light illuminated a portion of the mantle plume theory. In the journal, *Nature,* the article *Fixed Hotspots Gone With the Wind* explained data gathered using new seismic tomography technology, which uses earthquake-generated waves to image the Earth's internal structure. That data verified that there are no hot plumes emanating from the mantle feeding the islands, causing researchers to conclude::

"It seems that **we must abandon** the convenient concept of **fixed hotspots** as reference points for past plate motions." Note15.14f

In another science journal, *Tectonophysics*, an article had this to say about mantle plumes in 1999:

"Hypothesized mantle plumes do not appear responsible for most large igneous provinces; instead, their very existence is questionable. **No geological evidence of any kind – geochemical, petrological, thermal, topographic – requires mantle plumes**." Note 5.14g

In an uncommon challenge to an important aspect of the magma theory, this particular article emphasized a position of outright abandonment of the theory:

"All the evidence that has been used so far to support the plume model – geochemical, petrological, thermal, topographic – is equivocal at best, if indeed not contrary. **The plume idea is ad hoc, artificial, unnecessary, inadequate, and in some cases even self-defeating, and should be abandoned**." Note 5.14g p23

Abandoning this theory would have a profound effect on radiometric dating of rocks, a subject covered in the Age Model chapter. Radiometric dating experts calibrate *based* on the Hawaii-Emperor Seamount Chain's formation by a *mantle plume*. If a mantle plume did not form the island chain, then the calibration method based on that theory is of no value.

3. Mantle Plume Pseudotheory

By 2003, according to a *New Scientist* article, new observations of the Earth's interior revealed startling new facts making the case against both the mantle plume theory and the existence of magma itself stronger:

"Although there is a chain of progressively older volcanoes marching from Yellowstone toward Idaho, a variety of seismic studies from a number of plume-hunters have **shown no signs of hot magma below a depth of 200 kilometers and no disturbances in the mantle below that**. 'Yellowstone was supposed to be the granddaddy of all plumes. It's a huge volcanic centre **but its status as a plume has evaporated**,' says Dean Presnall, a petrologist with the University of Texas at Dallas." Note 5.14h

Having shown in this chapter that the physical evidence overwhelmingly demonstrates that there are "no signs of hot magma" or magma plumes, how do we account for the elevated temperatures in the first 200 km of crust in Yellowstone? The area below 200 km is cooler, simply because there is no magma—and no magma-driven mantle 'plumes.' The researchers themselves tender strong evidence against mantle plumes as noted in the article:

"One of the more **damning** pieces of evidence **against man-

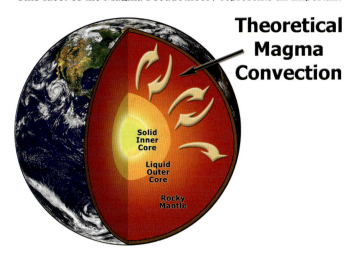

Fig 5.14.1 – The decades-old Magma Convection Pseudotheory has no basis in observable fact. Several researchers suggest that the theory has "major geologic problems" and that it is "highly speculative"– yet it remains widely accepted and taught as fact.

tle plume theory** is that regions of the crust above suspected mantle plumes **don't actually appear to be hot**—despite the fact that huge fountains of magma from the hot **core should be rising directly beneath**... In Hawaii, however, they found the temperature below the sea floor to be much the same as everywhere else—**there is no anomalous heat flow**." Note 5.14h

To make the false plume theory work, scientists concede:

"'You just have to keep **making up excuses and modifications to make plume theory work**,' says Foulger." Note 5.14h

Perhaps geologists finally see that what they thought they knew may be wrong after all:

"And there are no samples of the inner Earth being spat out of the Hawaiian volcanoes as **we once thought. Everything is up for grabs**.

"'**We'll have to acknowledge we know far less about the centre of the Earth than we thought we did**,' says Foulger." Note 5.14h

Although the interpretation of everything may be up for grabs, the old and firmly entrenched Magma Pseudotheory leaves scientists unable to imagine any other model. For modern geologists, the magmaplanet Earth is the only possibility that describes the interior of the Earth and for more than a century, this is all that the scientists know; it is all they were taught. Dumping the mantle plume idea did not come easily and it still enjoys prominence in many circles; it will be even more difficult to discard the entire concept of magma. Nevertheless, some encouraging new observations made possible by technological advances push the geological community in part, to consider revolutionary ideas:

"Most aren't convinced that mantle plumes should be dumped entirely. But they're willing to admit that the geological community is **standing on the brink of a radical shift in thinking that could completely change our ideas about the inner workings of the Earth**." Note 5.14h p34

4. Magma-based Tectonic Plate Pseudotheory

A few decades ago, seismographic technology enabled scientists to collect images of seafloor plates, giving us a greater understanding of all the Earth's plates. Now we can actually view and measure the movement of those plates on a global scale. The outcome of this technological advancement included the development of the **Plate Tectonic Theory**, a modernized version of Alfred Wegener's 1912 continental drift hypothesis. This modern magma-based plate tectonic theory attempts to explain where these plates originated and by what means the plates move. However, the new proposed origin of the plates and the mechanism driving their movement mingles observed fact with the false magma theory, causing nothing short of total confusion.

Consider the following statement from a high school teacher's web post:

"Do continents drift? We now know that Wegener was correct, the continents have moved through time. As a result **continental drift is an observational fact**. Plate tectonics is a more **useful theory** because it includes an explanation of the driving mechanisms of continental drift: mantle convection. What makes plate tectonics a particularly interesting theory is that it explains virtually all the major questions posed by early geoscientists, such as the distribution and timing of earthquakes, volcanoes and mountain-building events." Note 5.14i

The teacher correctly assessed continental drift as an observable "fact" but then assumes that plate tectonic *theory* is more "useful" because it explains earthquakes, volcanoes, and mountain-building events. However, because plate tectonic theory builds on magma theory, how useful can it be?

When researchers began to drill the world's second deepest hole, the KTB super deep borehole in Germany, one of the main attractions was the opportunity to drill right through two supposed tectonic plates. However, the second plate did not show up:

"The other attraction seemed to be an opportunity to drill through the buried boundary between two tectonic plates that collided 320 million years ago to help form the present Eurasian plate. But the suture, first predicted to slant under the KTB site at a depth of about 3 kilometers on the basis of surface geology, **failed to show up at 3 kilometers, or at 5 kilometers as later hoped**. And at 7.5 kilometers, researchers still 'haven't seen any sign of a dramatic change' that would mark the boundary between the two plates, according to Jörg Lauterjung of the KTB project." Note 5.14j

Why did the geologists' predictions about the plate's existence fail? Could it simply be that the magmatic origin of plates is wrong? The concept of Plate Tectonics is very important today, and we read why in the introduction of this college geology textbook:

"For the first time in the history of the discipline, an all-encompassing synthesis of geological knowledge was being advanced. **Plate-tectonic theory gave us a framework for the learning about the immense forces turning cyclically in the core**, in the surrounding mantle, in the crust, and in the air, oceans, and biosphere, keeping our planet in a constant state of change. This new picture of Earth as a dynamic, coherent **system was central to our writing of that book and its successor, *Understanding Earth*.**" Bib 59 pxi

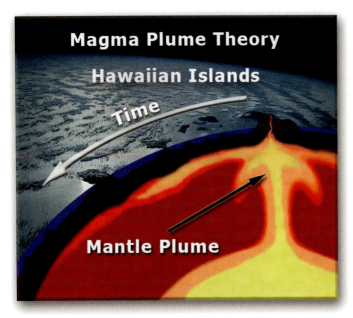

Fig 5.14.2 – For many decades, geologists held to the idea that the Hawaiian Islands formed over a Mantle Plume. Now the mechanism that supposedly brought "magma" to the surface is "gone with the wind" say researchers. They acknowledge that geology was "making up excuses and modifications to make plume theory work".

The authors of this widely used college textbook admittedly rely on plate-tectonic theory, and nearly every modern geology textbook expresses the same sentiment. Of course, it would be just fine if this "all-encompassing synthesis of geological knowledge... about the immense forces turning cyclically in the core" was true, but it is not. How do we know? From first-order unresolved questions asked by geologists.

"Unresolved" Questions of Plate Tectonic Pseudotheory

If the magmaplanet-based theory of plate tectonics was true, then, based on the Universal Scientific Method, scientists can make predictions, devise tests, observe and verify their predictions, and thus substantiate new natural law about it. Instead, because **magma theory** forms the foundation of plate tectonics, many unresolved questions remain unsolved. While the school textbooks paint a pretty "we now know" picture for the student, there are scientific journals that tell a different story, such as the following in the journal, *Science*:

"Despite three decades of research, some **first-order questions** regarding the plate-mantle system **remain largely unresolved**." Note 5.14k

By "first-order questions," the researchers mean *fundamental questions* that lie at the very base of the plate-mantle system, and even the entire magma pseudotheory. The unresolved first-order questions include the following:

"The first of these is **why** Earth developed plate tectonics at all, given that other terrestrial planets such as Mars and Venus do not currently exhibit this behavior." Note 5.14k

The answer to the question as to why Mars and Venus do not exhibit plate tectonics has much to do with the gravitational force of a large moon, which neither of them has. We can better understand the workings of the Earth's plates by examining Jupiter's moon, Io, the most volcanically active planetoid in the entire solar system with over 400 active volcanoes. Io is a violent example of the **Gravitational Friction Law** at work.

Io, the closest of Jupiter's Galilean moons, endures tidal forces that cause its surface to rise and fall by 100 meters every Ioan day as it turns on its axis. Likewise, the Earth's relatively large moon generates a lot of tidal activity, and there are dozens of earthquakes occurring somewhere on the Earth at all times. Mars has two small moons and Venus has no moon, so neither planet feels the gravitational pull-and-tug that the Earth and Io experience. This is the primary reason why the Earth's two neighboring planets exhibit no tectonic activity.

Before concluding this subchapter, let us consider one more fundamental question pertaining to the plate tectonic theory. According to prevailing theory, the area between the crust and the outer core of the Earth, also referred to as the 'mantle,' consists of 'convecting' magmatic fluid, which implies a degree of liquescence. However, Earthquake produced S-waves are present in the mantle but *cannot occur in liquids!* So which is it? Is the mantle liquid or solid? Modern geology does not have a clear answer.

5.15 The 'Smoking Gun' - Tomography

While it remains impossible to visit physically the Earth's interior, we can see into the interior through pictures. How do we get pictures of Earth's interior? An expectant mother has the same challenge, wanting to see her unborn baby, and today she can 'see' her unborn child with ultrasonic waves, which are sound waves converted into a digital image viewable on screen.

In much the same way, we peer into the interior of the Earth using seismic sound waves emitted every time an earthquake happens. During an earthquake, much of the damage comes from shockwaves moving through the ground. These waves register on seismometers placed around the world. Thousands of seismometers recording thousands of seismic events have enabled scientists to begin to see a picture of the Earth's interior. New and sophisticated technology reveals that sound moves differently through solid or liquid matter and that the temperature of the material also greatly affects the speed of the sound waves. The latest seismic tomographic images of the inner Earth show a picture far different from the one we are used to seeing in today's geology textbooks.

Missing 'the Boat' on Critical Observations

Over the years, scientific investigators missed the boat when it comes to identifying the real cause of lava, probably because they ignore critical observations. For example, the textbook, *Understanding Earth* notes what researchers say they have learned through experience:

"Long ago, miners and geologists **learned through experience** that the deeper we go into the Earth, the hotter it gets..." Bib 59 p197

The cover of that textbook depicts a three dimensional Earth with a slice removed to illustrate the red-hot 'magma' center. For over a hundred years, this is the 'picture' painted in our imagination of what the inside of our Earth looks like. Do miners and geologists *really* learn from experience that every time they go deeper into Earth, it gets hotter?

Have you ever visited a deep cave yourself? Most of us have, and we probably noticed that the temperature was quite cool. In fact, the vast majority of caves and most mines do *not get appreciably hotter* as they get deeper. Excluding mines that experience seismic activities and caves with geothermal features, underground places are usually quite cool and remain at a surprisingly constant temperature (see 5.15.1). Why does it seem that no one noticed this critical observation?

The warm mines and caves, exceptions to the rule, have the Gravitational Friction mechanism to explain why they are warm. Frictional-heat generation continues to confuse geologists who think magma

> "Everything is up for grabs. 'We'll have to acknowledge we know far less about the centre of the Earth than we thought we did...'"
>
> Gillian Foulger

> "...the geological community is standing on the brink of a radical shift in thinking that could completely change our ideas about the inner workings of the Earth."
>
> Nicola Jones

below ground heats and warms caves, mines, and powers geothermal features. There is a whole lot more to the picture.

The 'Smoking Tomography Gun'

Another recent technologic advance, *seismic tomography* provides critical observations into the inner Earth, yet few people and not many scientists are aware of this technology. Seismic tomography comprises actual, empirical evidence, and it is the 'smoking gun' that demonstrates the fallaciousness of the magma theory.

With advanced technology, researchers continue to collect increasingly accurate temperature data from the interior of the Earth. Super sensitive seismic monitors collect and record data from many worldwide earthquake events of all magnitudes, every day. Each year, new countries join the process and new locations are added, dramatically increasing the number of seismic monitoring stations. The data, downloaded into the latest computers, produce three-dimensional CAT images of the Earth, similar to medical CAT scan images. The data show that slower P waves indicate warmer temperatures while faster P waves signify cooler regions of the Earth's interior. Researchers assemble actual diagrams illustrating the hot, warm, and cold areas of the Earth's interior. In 1993, Bruce A. Bolt, a researcher involved in early tomography exclaimed:

"**It is not easy to overdramatize the radical improvements** that the x-raying of the Earth's interior **using three-dimensional tomographic methods** has brought to the process of **geological discovery**. The heavy constraints imposed by using models of the Earth that allowed only radially symmetric changes in physical properties **have now been broken**." Note 5.15a

Radical improvements using tomographic methods are for the first time providing new insights and geological discovery. Bolt went on to comment about the future of this new technology and what we should expect to know about the inside of the Earth.

"**The future of this type of research appears bright**. It is well known that the additional tomographic resolution needed can be obtained by installing more broad-band digital seismographs around the globe, particularly in the spaces in the deep oceans from which there are now no recordings. **We can expect that within the next decade the main patterns of flow inside the Earth will be well understood**." Note 5.15a

In the more than two decades since Bolt made his prediction in 1993, the first half has come true. Tomographic research has proven to be a bright light exposing hidden truths inside the Earth—every bit as informative as he had hoped. However, the second half of the prediction, understanding the patterns of flow inside the Earth, has not happened.

Seismic Tomography Evidence

Before we examine the Earth's *actual* geotherm, look at Fig 5.15.2 which illustrates the widely accepted **theoretical *magmaplanet* geotherm**. This simplified diagram depicts primarily what geophysicists believe the inside of the Earth looks like—a partially molten mantle, hotter near the core and cooler toward the crust.

The empirical evidence in Fig 5.15.3 gives us a view of what the inside of the Earth actually looks like. It takes little imagination to see that the *Earth's* **actual** *geotherm* refutes the magmaplanet geotherm. Where are the hottest areas, shown in

Fig 5.15.1 – As researchers descend inside the Earth through caves, have they really "learned through experience" that the deeper we go the hotter it gets? Most people wear jackets when descending into caves because they are colder than the outside temperature! Very few caverns are warm and if they are, researchers are bound to find continuous small earthquake activity near the cavern, which would account for the frictional-earthquake heating. Carlsbad Cavern photos courtesy of NPS and Peter Jones.

Theoretical Magmaplanet Geotherm

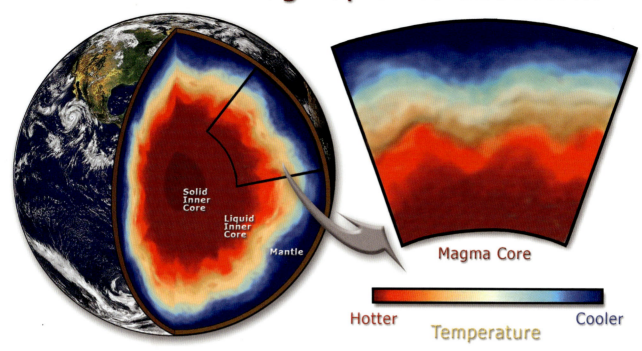

Fig 5.15.2 – This *theoretical* Magmaplanet Geotherm shows a cross-section of Earth with a hotter interior (red) and cooler (blue) areas in the outer mantle of Earth. The inset section on the right shows colder temperatures with a gradient towards a hotter interior. If the Earth's heat comes from the molten core, this is what the tomographic evidence should reveal. See Fig 5.15.3 for the **actual** geotherm of the Earth as derived from the seismic tomography.

red located? They are not near the so-called magma core; they lay closer to the surface of the Earth, near the plate boundaries where earthquakes occur frequently, just as the Frictional-Heat Law predicts.

The cross section running through Central America in the upper right hand corner of Fig 5.15.3 shows the hottest area just below the surface toward the left side of the cross section. The red-arced line with control markers across the top of the tomographic section corresponds to the red-arced arrow drawn through Central America and Florida. On the left of the arced arrow, black lines represent the plate boundaries where the Cocos plate meets the Caribbean plate, the location of the largest earthquakes in this area, and the place where heat is highest.

All five cross sections in the illustration show a similar pattern of hot and cold areas in the mantle. The important point to note is that the geotherm cross sections depicted here are ***not*** theoretical or imagined like the magma geotherm illustration in Fig 5.15.2. Investigators derived this diagram from actual measurements collected by a number of researchers, including Rob Van der Hilst of the Massachusetts Institute of Technology, a leader in the field of seismic tomography. This budding new field, operational for only about the last two decades, scans the mantle with sound waves and calculates hotter and colder areas based on the speeds of the waves, which travel at different speeds based on temperature.

It surprised us to see seismic tomography diagrams in geology textbooks because the depicting of cooler and hotter areas represented in the diagrams refute the established magma paradigm. The textbook we reviewed said almost nothing about the incredible discrepancy:

"The future holds even greater promise for exploration of Earth's interior as new tools are developed. The newest tool is an adaptation of the computerized axial tomography (CAT) scanners used in medicine to reconstruct images of organs..." Bib 59 p495

While admitting that the technology allows for better imaging of the Earth's interior, the lack of discussion about the discrepancy reveals a problem for both researchers and teachers. What could they say? In today's modern geology, scientists might earn the brand of heretic by suggesting mantle areas near the surface are the hottest and that temperatures nearer the core are cooler, even though the empirical data suggests *exactly* such an environment. Why this is has much to do with edifice science built on several key pseudotheories, including magma, one of the most important. Without magma, the Big Picture of Modern Science falls apart. The comparison between the two theories as depicted in the diagrams—the actual observed geotherm and theoretical magmaplanet geotherm—we see why seismic tomography is the Magma Pseudotheory's 'Smoking Gun.'

With so much evidence refuting magma, how can we continue to teach this pseudotheory?

Where the Earth's Heat Really Comes From

As far back as 1901, scientists recognized a problem with the origin of the Earth's heat, recorded in the 1901 book, *Lessons in Physical Geography*:

"The **fact** that while the temperature of the earth-crust increases downward, the temperature of the sea **decreases** in the same direction, constitutes **one of the most interesting problems of oceanic geography**." Bib 142 p252

SUBCHAPTER 5.15 THE 'SMOKING GUN' - TOMOGRAPHY

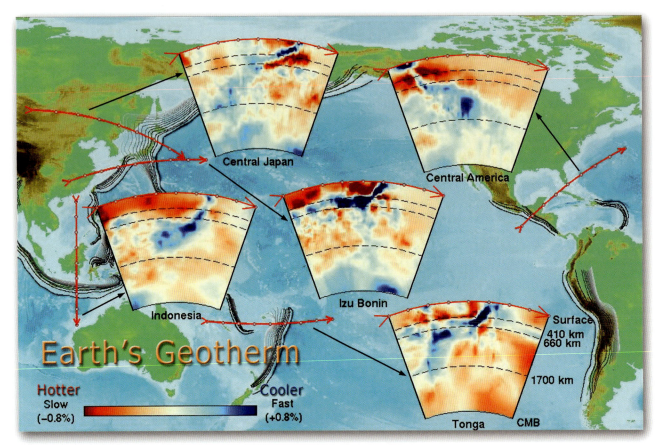

Fig 5.15.3 – This diagram portrays several cross-sections corresponding with the curved red arrows. Compare this chart to the Theoretical Magmaplanet Geotherm in Fig 5.15.2. These **actual** geotherm profiles are *complexly opposite* the predicted profile. Instead of the hotter areas being at the base of the cross-sections as displayed in the theoretical Magma Geotherm, the red areas are at the **top** of these profiles where frictional heating occurs, clearly near plate boundaries. Look, for example, at the Central American profile. The area showing the hottest is at the top left of the cross-section, exactly where the corresponding arrow crosses the Central American continent and plate boundaries, and right where earthquakes and frictional heating happen. Courtesy of and adapted from the work of Rob Van der Hilst at the MIT.

Science experts still have no rational explanation as to why the oceans get colder as depth increases; clearly, the heat cannot originate beneath the ocean floor where magma supposedly resides. The heat flow map in Fig 5.15.4 identifies the location of heat generation, along continental plate boundaries where friction is greatest.

After a century of studying the crust, magma models still contain uncertainties because they are founded on incorrect hypotheses. A 1997 article in *Geophysical Research Letters* reveals:

> "Despite the availability of data for the upper third of the crust, **models to predict temperatures for greater depths still contain uncertainties**." Note 5.15b

Why can the geophysics models not predict correct temperatures at great depths? The correct model should be able to predict these things but the magmaplanet model cannot because it assumes heat comes from the center of the Earth, not from the crust. The Frictional Heat Law and the tomographic evidence demonstrate that the Earth's heat resides at the surface, *in the crust*, generated by frictional heating along faults.

All the way back in 1968, geophysicists knew the magma model could not explain what researchers measured below their feet. After relating all the problems with magma, one researcher O. M. Phillips, wrote in the book, *The Heart of the Earth*:

> "The model that emerges is one in which the internal temperature of the earth is governed largely by **the generation of heat in the earth's crust and possibly a little below it**." Bib 63 p150

Phillips aptly identified the truth:

> "...the 'fire' is **not** concentrated **deep in the heart of the earth** but is an encircling **sheet near the surface itself!**" Bib 63 p149

In a final summation, the problems and uncertainties will never go away until there is a shift away from the magma paradigm. Such a shift will affect *every* branch of modern science,

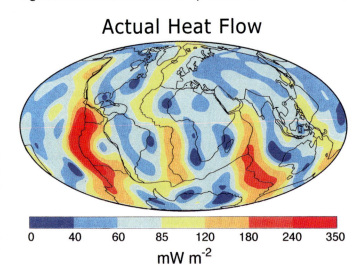

Fig 5.15.4 – For the first time, the Actual Heat Flow map of the Earth as seen here makes sense (first shown in Fig 5.4.5). The oceans are not evenly heated by magma because magma does not exist. Oceanic heat comes from the vertical plate boundaries that experience continual friction because of the Moon's gravitational force and Earthtide. The areas that are the most active show as red and orange on the map.

CHAPTER 5 THE MAGMA PSEUDOTHEORY

and so we cannot take it lightly, that is why the observation and empirical evidence must lead the research.

5.16 The Magma Freeze

This chapter provided a brief overview of a few of the many physical evidences that refute the magmaplanet theory. From the evidence, it is clear that magma is a pseudotheory—a false theory taught as fact in science today. It is not possible to contain all of the available information and observations that disprove magma in this book. Space constraints allow us to treat topics only briefly. Think of the subchapters herein as a piece of a larger puzzle, each one addressing a different segment of magma theory. Questions, discussions, and the scientific journal articles we included show that magma is a fictitious idea, and we laid the groundwork for new ideas. If researchers can shed old dogma and take a different view of the Earth and the data they have collected, new and refreshing discoveries await them!

As modern science employs new technology and makes better observations from within the correct paradigm, we will know the truth. After reading the UM and analyzing the observed data from this new vantage point, one might wonder why magma remains so strongly entrenched in the scientific community. But now the time has come to 'freeze' the magma planet concept and look for the truth; the truth as it existed in the past and as it exists today—because truth does not change.

Discarding the "Tentative Hypothesis"

In the first subchapter of this chapter, we defined magma and we explained that lava, which is melted rock on the surface and below the surface in the crust, is real, it is observable, and it is different from magma. We also showed that magma, the supposed melted subsurface rock in the mantle or outer core of the Earth is merely theoretical. No doubt, the magma theory will continue, hotly debated in the coming years because of the many false ideas science constructed on this all-encompassing incorrect theory. This chapter is full of evidence and compelling testimony that the Earth is not a magmaplanet, suggesting that we must question the origin of magma and even its very existence. One geophysicist gives us good advice for what to do with the old magma model with our new knowledge:

"So let us make the **tentative hypothesis** that the core of the earth is made mostly of iron and perhaps some other lesser ingredients. This fits the few facts that we know at this stage and seems a sensible choice. **Remember, however, that it is still tentative, and should we discover facts with which this model cannot be reconciled, it will have to be discarded**." Bib 63 p69

New and improved technology heralding new discoveries will establish the truth. Perhaps there is already a shift from magma planet thinking amongst some researchers, as we read in an online geology blog for a course taught at Yale University:

"…in an average slice through the Earth, **there is no molten rock between the core-mantle boundary and the surface**, save perhaps in the asthenosphere at the base of the tectonic plates (100-200 km), where a partial melt of **a few percent may exist**. As a result, **unusual conditions must prevail in regions that produce magma for surface volcanism**." Note 5.16a

If there is no molten rock between the surface and the core

boundary, except for the few percent they say may exist because of tectonic activity, then **no magma exists** at all throughout the *vast interior* of the Earth! The quote also says that it takes unusual conditions for magma to produce surface volcanism, referring to what we earlier established as extrusive surface lava.

Once the truth about magma is established, geologists will no longer think lava comes from magma deep inside the Earth. Seismic tomography confirms that deeper parts of the Earth are not as hot as the crust. There is simply no scientific evidence for the existence of magma. Furthermore, since magma is the supposed *source of all rocks*, and since it does not exist, then the scientific theory about the way rocks form and the cycle through which they pass must also prove faulty. This means that the modern geology rock cycle theory, which states that all rocks are of igneous (melted) origin before transitioning through metamorphism and sedimentation—is false.

Just as we learned in the Magma Pseudotheory Chapter, we will now review the false Rock Cycle, another pseudotheory.

> "So let us make the tentative hypothesis that the core of the earth is made mostly of iron… Remember, however, that it is still tentative, and should we discover facts with which this model cannot be reconciled, it will have to be discarded."
>
> O. M. Phillips

6

The Rock Cycle Pseudotheory

The Rock Cycle

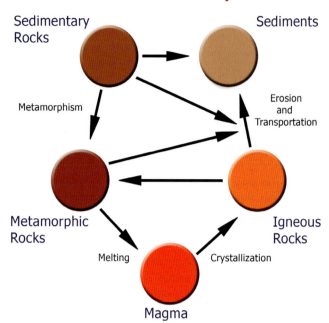

Fig 6.1.1 From modern science's father of geology, James Hutton, sprang the Magma Pseudotheory and the basis for what would become known as the Rock Cycle theory. This theory is taught in elementary schools, high schools and colleges worldwide. Notice how magma is at the base of the Rock Cycle. How could this theory be correct if its foundation—magma, does not exist? The founding pillars of modern geology are the existence of magma and the Rock Cycle.

6.1 The Most Abundant Mineral Mystery
6.2 The Real History of Geology
6.3 The Sand Mystery
6.4 The Quartz Mystery
6.5 The Basalt Mystery
6.6 The Obsidian Mystery
6.7 The Iron Mystery
6.8 The Ore Mystery
6.9 The Carbonate Mystery
6.10 The Loess Mystery
6.11 The Erosion Mystery
6.12 The Earth Crust Mystery
6.13 Geotheoretical to Geological

James Hutton, called by some the father of geology, came to the idea that the Earth was shaped by a long succession of re-occurring events, driven by heat from the inner Earth. Charles Lyell further advanced Hutton's theory, which he published in his three-volume work, *Principles of Geology* in the early 1830s. The ideas and tenets of uniformitarianism and the endless cycle of once melted rocks being recycled, eroded, crushed and melted again, sent geology headlong into the future, with a belief that Earth was formed, and continues to be shaped by excruciatingly slow processes. The influence exerted by these ideas was not limited to modern geology. Darwin read of the uniformity theory in Lyell's first volume while on the HMS Beagle in 1831. It profoundly influenced his ideas, satisfying in his mind that the Earth was immeasurably ancient, allowing time for life to evolve.

While exploring St. Jago, off the coast of Africa, Darwin observed a seemingly undisturbed layer of rock formed of shells and coral, 30 feet above the level of the sea. To him, this proved what Hutton and Lyell had said; that slow and gentle uplift, combined with erosion, was responsible for the stratum he observed.

In the Magma Pseudotheory (Chapter 5) we established that continental uplift is not occurring. In this chapter, we will investigate how sedimentary rocks and other rocks are said to have been formed, and we will show that much of the sedimentary rock does not come from erosional processes observed today. Because of this, the modern-science rock cycle must be reevaluated. Magma, the foundation of geology, is not a sound foundation. This makes the rock cycle also unsound. Of course erosion does occur. Rocks do break down by the abrasive action of wind and water, by the effects of heating and freezing, and by chemical and biological means. However, there is much more to the story than what is proposed in modern geology!

This chapter is similar to The Magma Pseudotheory chapter in that it continues to lay a foundation for a new paradigm. We will find that some of the most fundamental questions that should be asked in modern geology remain unanswered by current theory. These FQs will identify errors in modern science's current theoretical rock cycle. Although it is somewhat difficult to present these fundamental questions, one after another without answers, this course will allow us to understand that there are not just one or two errors in modern geology, but that the whole branch of science is in need of a revolution. The answers will be better understood when presented as part of the whole model. This will be accomplished in the following chapters.

To begin, we look at the Rock Cycle diagram at the top of the page. What are we to find at the base of the modern science concept of the rock cycle? In the previous chapter, we learned that magma is an unproven theory, and now we see the Rock Cycle theory being built upon that theory. Theories built upon theories, make the field of geology, geo-theoretical.

Geology has become geo-theoretical instead of geo-logical.

This chapter is made up of more than a dozen geological 'mysteries.' Each of these mysteries in geology will comprise a subchapter and will exhibit the foundations of geological thought today. These mysteries exemplify the geo-theoretical point of view in modern geology and lead us to ask simple, fundamental questions that modern geology has left unanswered or has never even asked. You the reader will be the judge as to why such simple questions, which lie at the very root of geology, have not previously been asked or answered.

6.1 The Most Abundant Mineral Mystery

In this subchapter, we will discuss one of the Earth's most abundant minerals. Wars have been fought over this mineral and at times, this mineral has been valued more than gold. It is the only *rock* we eat and is in fact necessary for us to live, yet too much of it and we would die. There is a great mystery surrounding this abundant mineral, the depth and breadth of which is almost unfathomable, as you will soon see. Neither geologists nor modern science have found acceptable answers for its origin. Although they claim that all modern deposits of this mineral are evaporites that came from desiccated shallow seas, they do not address where the actual mineral itself comes from. It forms the largest single rock mineral formation in the world, yet it remains a puzzling mystery in science today. Hopefully, we will soon see this all-important, life-giving mineral in a whole new light, and just as importantly, we will recognize how the modern science theory of the origin of all rocks has no answers for the origin of the mineral, **Salt**.

A Most Fundamental Geological Question

The following question is perhaps one of the most important scientific questions of our day.

What is the most abundant mineral on the **entire surface** of the Earth?

Such a simple question would seem to have a simple answer, yet there is not one to be found in today's geology textbooks. The key is to understand the question. Most often, silicates or feldspars are cited as being the most common mineral in the *crust*. However, that answer does not address the specific question that was asked! Our query references *"the entire surface of the Earth"*, which includes the oceans. This is more than just a simple question—and *its answer lies at the very heart of our understanding of geology*.

Over the last several years, we asked many geologists and other scientists this very question. Not one could answer the question correctly. What is the mystery mineral?—**Salt!**

When we told them the answer and explained the facts, every scientist agreed with the answer we gave them. Why would such a simple answer to such a fundamentally important scientific question remain so wholly unknown, even by the experts? Could it be that the bedrock of modern science—geology, has been built literally on a faulty foundation?

Without the wisdom of why the oceans are salty or the knowledge of where the salt comes from, science can never really

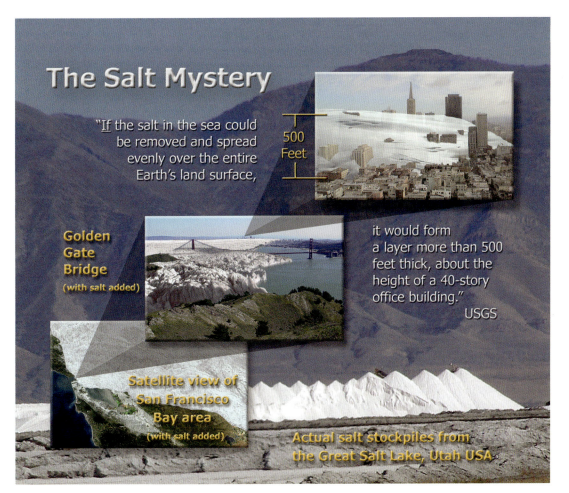

Fig 6.1.2 – The Salt Mystery diagram illustrates one of the mysteries in the origin of salt. Geologists say, "Salt in the ocean comes from rocks on land." However, if the average thickness of land surface is only 410 feet (125 m), how can 500 feet (152 m) of salt come from these 410 feet? Understanding the mystery about the origin of salt is an important key to understanding the Earth.

It is a myth that sodium (Na) + chloride (Cl) equals salt. This false assumption has lead geologists to misinterpret the origin of the most abundant and perhaps even the most important mineral on the entire surface of the Earth.

What is true is that the individual elements, sodium and chloride often come from salt. However, the individual elements are not found by *themselves* in nature and therefore, cannot combine *by themselves* to form salt.

understand the crust of the Earth, its landforms, or their origins. Knowing *why* the ocean is salty *is* that important. Probably because of the magma paradigm, geologists have ignored this material fact. Moreover, the Rock Cycle Pseudotheory has blinded the eyes of researchers to the point they completely miss important, critical questions like this question about salt. If we are to understand the origin of the Earth, questions like this one and many more must be asked and answered.

The Modern Science Salt Answer

We can establish what modern science says about salt in the oceans from the popular website "Ask A Scientist" by asking: How did the oceans get salty? This answer, although sounding a bit oversimplified, is the predominant view of science:

"**Salt in the ocean comes from rocks on land**." Note 6.1a

The answer sounds reasonable. After all, where else would all that salt come from? There must have been a lot of salt in the ground if the salt in the oceans came from land. There does not seem to be a great deal of salt lying around today. When water wells are drilled, the water is rarely found to be salty. How much salt do you see on a hike through the mountains, or when exploring caves or examining a road cut into the landscape? How many salt deposits are in the mountains, forests, deserts, or valleys around where you live? Just how much salt is in the oceans? Although there are some places like Death Valley, CA or The Great Salt Lake, UT that have seemingly vast amounts of salt, they are small in comparison to the amount of salt in the oceans. To comprehend the scope of the question, "how much salt is in the oceans," we need to do a little visualization. Imagine taking all of the salt out of the oceans and placing it on the land surface of the Earth. How deep do you think the salt would be? Having asked dozens of people this question, the usual response was several inches, though some thought it might be a foot or two. Keep this in mind as we explore the real answer, it usually comes as quite a surprise to everyone.

Real Numbers Give a Completely New Answer

Thanks to modern technology, an accurate map of the surface area and depth of the ocean is available, from which we can easily estimate the volume of the ocean. Seawater contains a surprisingly consistent amount of salt. The answer as to how much salt is in the sea can be reasonably calculated. From a US Geological Publication, we find this remarkable statement:

"If the salt in the sea could be removed and spread evenly over the Earth's land surface, **it would form a layer more than 500 feet thick, about the height of a 40-story office building**." Note 6.1b

I have to admit, the first time I read this and visualized how much salt was *really* in the sea, I was astonished! To comprehend just how much salt this equates to, examine the inset images in Fig 6.1.2. Imagine every city, desert, forest, plain, mountain, and valley covered to the height of a 40-story building. *500 feet of salt* is almost as high as two football fields, stacked on end. Incredible as this is, it was only the beginning of many discoveries about salt! The 500 feet of salt was *only half of the* equation. The other half comes from knowing how high the *average* continental crust rises above sea level. In other words, if we were to level all of the mountains and fill all the valleys, what would the average surface elevation of Earth's continental crust be, above sea level? Two leading researchers in the field of crustal evolution as published in *Scientific American* said this:

"**Continental crust rises on average 125 meters [410 feet] above sea level**..." Bib 37 p15

If the continental crust is on average only *410 feet* above sea level, what is the answer to this very important question:

How could 500 feet of salt come from 410 feet of crust?

This math just seems too *simple*. Obviously, 500 feet of salt cannot come from 410 feet of land. This would not be possible even if all of the land were pure salt, which we know it is not. In fact, the land has relatively little salt spread throughout the crust.

Water that flows across the land in rivers like the Mississippi, the Amazon, the Zambezi or the Columbia are each emptying tens of billions of gallons of ***fresh water*** into the Gulf of Mexico, the Atlantic Ocean, the Indian Ocean and the Pacific Ocean, every day. The small amount of salts in these fresh water rivers has not been able to account for all the salt in the oceans.

Some geologists have reasoned that continents were once much higher and have since eroded away, uplifted, and eroded again. However, remember from our discussions in the Magma Pseudotheory chapter that uplift is not really happening, therefore, the uplift sci-bi cannot be used to explain how in the distant past, there may have been more land above sea level.

For many geologists and students of geology alike, the answer to the question of where the salt in the ocean came from *seems* so simple. Nevertheless, there is much more to the story than "Salt in the ocean comes from rocks on land" and it leads us to ask another fundamental question about the field of geology itself:

Fig 6.1.3 – This table shows the percentage of the elements contained in igneous rocks, according to Bowen. Sodium chloride (NaCl) is the most abundant mineral on the entire surface of the Earth (including oceans). If all elements, including salt, came from igneous rocks, why is there **no chlorine in the igneous rocks?**

*How much can geologists **really** know about the 410 feet of Earth's crust, if they do not understand where the equivalent of 500 feet of salt in the oceans came from?*

This is an intriguing thought and a critical question that *must be answered, if we are to come to an understanding of the origin of Earth*. Why is this so critical? Until we learn how salt is actually made in Nature, we will not know the true history of this planet. Imagine for a moment trying to comprehend how an automobile works without knowing anything about the engine or the fuel it needs.

Salt, which is the most abundant mineral on the *surface* of the Earth—*is the engine*. In the bigger picture, it is not just *a mineral*; it is so essential that without it—there would literally be no life.

Chlorine—the Missing Element

As mentioned earlier, the origin of salt is one of, if not, *the* most important geological questions needing an answer in today's modern science. A major problem is that geologists don't understand where this abundant mineral comes from. In 1935, geologist Norman L. Bowen, creator of Bowen's Reaction Series, (discussed in subchapter 5.14) notes this:

"Now there seems no escape from the conclusion that **all substances in or upon the earth**, including even the atmosphere and the material parts of living organisms, **must have their ultimate source in the igneous body of the earth. We should expect the igneous rocks to contain all these elements**... Only eight elements occur in rocks in an amount exceeding 1 per cent, and here it must be remembered that we speak of rocks on the average and not of what may be found in an individual specimen. In Table 1 the ten most abundant elements in the rocks are listed in the order of their abundance. They make up 99.4 per cent, so that the other eighty–two total only 0.6 per cent." Note 6.1c

Salt consists of sodium and chlorine; notice in Bowen's Table 1 (Fig 6.1.3) how much sodium is present in the composition of igneous rocks. He lists only 2.85% sodium, but more importantly, chlorine is entirely absent. That's right, *no chlorine!*

If igneous rocks have no chlorine, how can the most abundant mineral on the entire surface of the Earth, sodium chloride, come from igneous rocks?

Ponder for a moment; how is it that the most abundant mineral on the surface of the Earth—salt, has no real origin in geology today? The only reason this could be is if the foundational theories of geology—Magma and the Rock Cycle, are faulty.

This chapter is full of Fundamental Questions (FQs) we must ask, some for the very first time. It is difficult to consider so many questions without providing the answers; rest assured we will answer them in the next few chapters; but before we do, it is necessary to see the bigger picture geology has painted and the fundamental errors that exist in it. We will continue by asking this simple FQ;

What is the largest single-mineral rock formation in the world?

The Largest Single-Mineral Rock Formation in the World

It would seem the answer to our last question would elicit a quick and simple answer. During the years spent researching the UM, we often asked geologists to name the largest rock, defining that to mean the largest rock as a single mineral. How many were able to give us the correct answer? *None.*

Why is this?

Ask an architect where the tallest building in the world is and he will likely tell you it is the Burj Dubai (topped-out on 17 January 2009). A biologist will surely identify the blue whale as being the largest living animal in the world today. Why do geologists not know the answer to the question, what is the largest single-mineral rock formation?

Many of the geologists we asked would not guess an answer. Some thought it might be a continental lava flow. When we added the caveat that the mineral body is *taller* than Mt. Everest, we received a great many puzzled looks. After all, the summit of Mt. Everest is over 29,000 feet, or about five miles above sea level. The largest single-mineral structure in the world has a

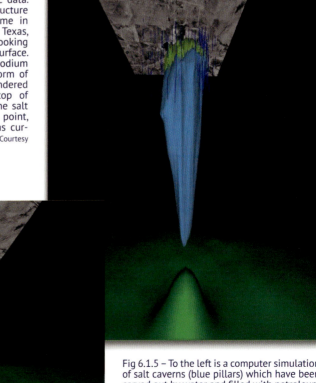

Fig 6.1.4 – To the right is a salt dome as imaged by computer modeling using drilling and seismic data. This salt dome structure is the Big Hill dome in Jefferson County, Texas, USA. This view is looking up towards the surface. The blue area is sodium chloride. Another form of salt, anhydrite (rendered green), caps the top of the dome. Notice the salt column comes to a point, not a bed of salt as current theory states. Courtesy of Continuum Resources.

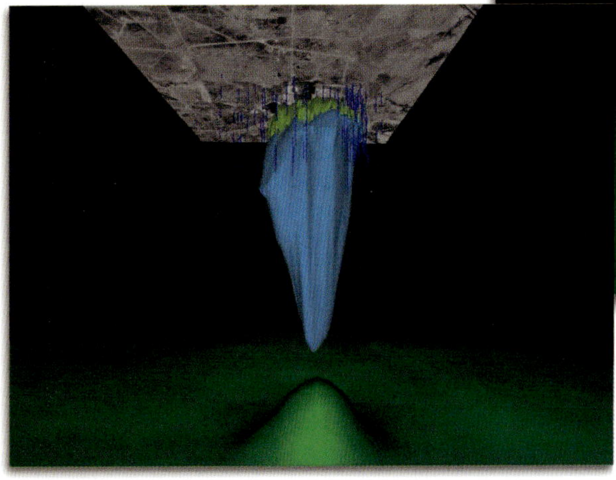

Fig 6.1.5 – To the left is a computer simulation of salt caverns (blue pillars) which have been carved out by water and filled with petroleum products. These caverns store petroleum for the Strategic Petroleum Reserve. Courtesy of Continuum Resources.

height of more than 7 miles! ***Where*** is this single-mineral rock formation that is taller than the tallest mountain on the Earth? The answer—it is *underground* and it is called a **salt dome.**

One reason many geologists are not aware of the significance of salt domes may be the lack of discussion in modern geology textbooks. One popular college geology textbook we reviewed did not even have 'salt dome' listed in the index (Bib 59 p678). All it had to say about domes and basins was this:

"It is not entirely clear **why** domes and basins form." Bib 59 p253

This is an example of how a lack of wisdom about key facts has kept geology in a dark age. Knowing the true nature of domes and basins will enlighten us and will help explain the real geological history of the Earth.

The Salt Dome Mystery

An island once named Petite Anse rises 163 feet above sea level in the dark bayous 140 miles west of New Orleans. Salt works were established on the island in the late 1700s, when a local hunter discovered a brine pool. Salt had been made here off and on for hundreds of years. Salt production was resumed during the war of 1812, and continued well into the 1800s. Salt had been made from brine, collected from wells dug in the surrounding area.

On May 4, 1862, a slave was attempting to clean and deepen one of the brine-wells. He was at the bottom of the sixteen-foot deep hole when he hit a log he could not remove. The 'log' turned out to be very pure, very dry solid rock salt. Unknown at the time, this was the top of a massive salt dome. Initial estimates were that this valuable find was at least forty feet deep.

Many years would pass before the enormity of the find was fully realized. Today, the island is known as Avery Island (Fig 6.1.7) and is world-famous for the pepper sauce produced there–Tabasco®. Only salt from the mine located on the island is used in the McIlhenny family pepper-sauce recipe. There is still much to be understood about the dome-shaped salt structure sitting below Avery Island.

A **salt dome** is a deep, vertical, pipe-structure of mostly pure sodium chloride—common table salt. The pipe formation is often capped by anhydrite, another type of salt. Sometimes, limestone or other sedimentary rock lies atop the salt dome

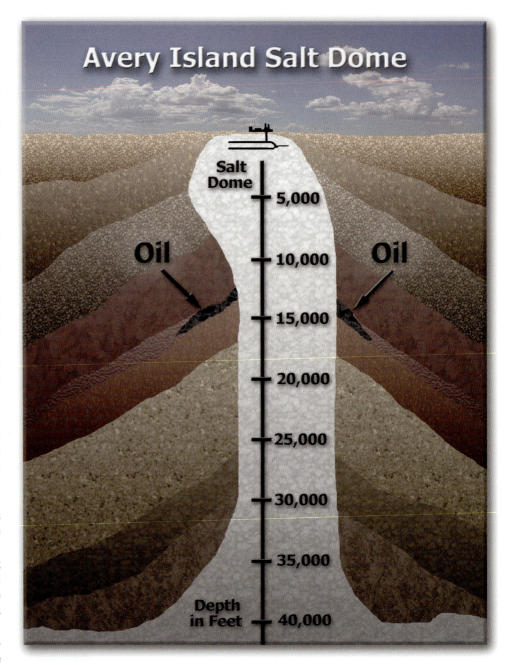

Fig 6.1.6 – This diagram shows the vertical pillar of salt and some of the oil deposits found at the Avery Island salt dome. This illustration is typical of how many salt domes appear. Modern salt dome 'theory' states that the salt originated as an evaporite bed of salt. Eventually it was pushed upward because it is less dense than the surrounding rock. However, no drill has ever penetrated the bottom of the deepest salt pillars and the domes show no evidence that they are rising. The flat 'salt bed' (seen at the bottom of this diagram) has not actually been seen. Its existence is *only* a theory.

and can seem as if it has pushed up through the surrounding landscape. On the surface, this may appear as a circular-shaped dome structure. Although salt domes are found in many parts of the world, some areas like China and Korea have no known salt domes. Sometimes salt domes occur in great abundance, in places like the Gulf Coast Basin in south-central United Sates and around the northern coasts of Germany. In areas where dome structures exist, there may be several dozen buried pillars of salt in a single salt-dome field.

In the United States, salt domes play an important part of the U.S. Department of Energy's critical emergency plan by providing storage space for crude oil. The Strategic Petroleum Reserve, the world's largest emergency fuel stockpile, uses the

Big Hill salt dome (Fig 6.1.4) and four other salt domes for the storage of 727 million barrels of crude oil. This space is created by pumping water into the salt formations and removing the resulting brine. See Fig 6.1.5. Oil and other petroleum products can be safely stored in the enormous salt caverns because they are impermeable and relatively leak-proof.

How do salt domes form? This is still a mystery. From the book *Salt Domes, Gulf Region, United States & Mexico*, written by the leading authority on the subject, Michel T. Halbouty, we read the 'modern' theory of salt dome origin:

"The modern **theory** of salt dome origin **postulates** that salt has flowed into structures by means of plastic deformation in responses to density differences between the salt and surrounding sediments. This concept, **now generally accepted by geologists the world over, has evolved from 100 years of speculation and interesting controversy**." Bib 144 p27

Speculation and controversy do not make sound science. Nothing in science should be accepted and taught as fact, unless it passes the scrutiny of the Universal Scientific Method. Since modern geology accepts and teaches the 'plastic deformation' theory of salt dome formation as though it were fact, we must identify it as a pseudotheory. As we continued our research, we uncovered many more examples where unproven theory is being taught as fact.

Returning to our story, note how deep salt domes are estimated to be:

"All stages of piercement up to a maximum of approximately **50,000 feet** occur in the gulf region. Sedimentary strata which have been ruptured or pierced **by a rising salt plug** may be gently to steeply upturned or completely overturned." Bib 144 p65

Although the importance of the Gulf region salt domes is recognized, we see that they are not understood:

"For example, **it is recognized that the most important geological structures in the Gulf region are the salt domes**. Yet these features reflect subsurface geology in its most difficult form: **highly complicated and not wholly understood**." Bib 144 pxiv

The foremost expert on salt domes had this to say:

"Perhaps **if there is one phase of geology on which we have accumulated more data, yet have more to learn about, it is salt dome geology... Ideas on origin** and mode of development have **not progressed significantly from the stage of theory, despite widespread acceptance today** of a few basic concepts." Bib 144 p1

This is why the origin of salt domes is misunderstood—there has been widespread acceptance of salt dome *theory*, without backing it with empirical evidence.

The *McGraw-Hill Concise Encyclopedia of Science & Technology* gives this definition of a salt dome:

"**Salt dome**—An **intrusive body of rock salt which has penetrated large thicknesses of overlying sedimentary rock**." Bib 12 p1704

After defining what a salt dome is, we can begin to explain why it is not wholly understood. The salt dome theory includes a process called 'creep.' This is the slow movement of a salt body, sometimes observed in the caverns within the salt dome formation. In older salt mines, it was observed that the surrounding rock salt would 'creep' slowly over time, sometimes sealing off old mine entrances. This observed movement *within* the salt body became the basis for the belief that salt domes formed when pressure from overlying rock and sediments pressed down on vast expanses of ancient, evaporated sea beds, forcing less-dense salt upward. The theory further states that the compressed salt begins to move, creeping up through miles of sediment.

Several problems are identified with the assumption that creep, as is observed in the caverns within the salt mines, is the mechanism responsible for salt dome formation. First, creep has been observed only *within* the salt body, in the **open air spaces** within the mine. There is little resistance from the air in the cavities in the mine. In these spaces, the salt seems to flow plastically, filling the void. However,

Fig 6.1.7 – Avery Island, Louisiana is home to the McIlhenny Company, makers of the world-famous Tabasco sauce. The island, once named Petite Anse, is a salt dome rising 163 feet above sea level. The salt used in Tabasco sauce comes from a mine located on the island. The large salt crystal in the picture came from the mine. Oil deposits are also found to be associated with salt domes but their origin has yet to be clearly explained. The true connection between oil and *salt*—both important commodities in society, is not wholly understood. This must change for it to be correctly taught. The unproven theories of salt formation must be abandoned if we expect to discover the true origin of salt.

SUBCHAPTER 6.1 THE MOST ABUNDANT MINERAL MYSTERY

this does not occur to the salt dome as a whole, as has been suggested. There are no voids or hollow spaces in the overlying sediment into which the salt body can flow.

Humidity in the cavities within the salt dome is another factor related to observed salt creep. Salt readily absorbs moisture from the air, and when combined with the pressure of overlying and surrounding sediment, the salt becomes more pliable. When this happens, it slowly moves along or 'creeps,' filling the cavity. While we may not know how much air and humidity is trapped deep underground, near the base of the salt formations, we know by observation that creep occurs only when the salt body is malleable and has a space to fill.

Finally, we know of no data supporting the view that salt domes are rising. GPS technology should easily detect any rising salt domes, but there are none known at this time.

The Cap Rock Mystery

The Avery Island salt dome does not have a protective cap of clay or anhydrite, and as a result sinkholes abound where ground water dissolves salt, creating brackish pools. However, in most salt domes, ground water is sealed out of the salt structure by a cap rock of anhydrous and other salts as well as other sediments.

The cap rocks over salt domes are another mystery in geology. Fig 6.1.8 illustrates the typical geology makeup of a salt dome cap rock. Starting from the bottom, a generally thick layer of anhydrite (anhydrous calcium sulfate) is overlain with a layer of gypsum (calcium sulfate) and then a carbonate layer such as limestone (calcium carbonate). These are each a type of salt that form differently and in distinct environments that we will discuss in more detail shortly. Each of these minerals are a unique part of the puzzle and holds specific clues to the origin of the salt domes.

The Cap Rock structure is usually covered by loose unconsolidated sediments. Shallow domes often have a small amount of sediment covering the cap rock whereas deep domes may reside beneath thick layers of sediment. Note where cap rocks are commonly found:

"A cap rock, composed pri-

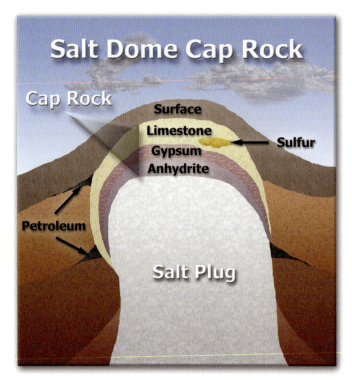

Fig 6.1.8 – Salt domes typically have Cap Rocks, which are comprised of salts other than sodium chloride (NaCl). The existence of this Cap Rock complicates the theory that an intrusive body of rock salt penetrated harder sediments.

Fig 6.1.9 – Rock-salt from the underground Redmond mine in Utah, USA is on the left. On the right is an example of rock salt from a salt dome in Russia. A U.S. quarter is used for scale.

marily of anhydrite, gypsum, and limestone is found on top of many salt domes. **It is commonly found on shallow domes, and is thin or nonexistent on deep domes**. Although it is normally 300-400 ft thick, the cap rock may exceed 1000 ft in thickness."
Bib 143 p3

Why is the cap rock found on shallow domes and not on deep domes? According to current theory, salt plugs (diapirs) move upward through miles of sediment, yet the cap rock somehow remains atop the structure rather than being fractured and pushed off. This fact directly contradicts the explanation that the intrusive salt body has penetrated great thicknesses of overlying sedimentary rock. Consider the relative hardness of the different cap rock layers; Gypsum is very soft, with a Mohs hardness of 1.5-2; Anhydrite lies below the gypsum, yet is harder at 3.5; Limestone is approximately 3.0 and the Salt Body is about 2.5. (The Mohs hardness scale is a way of defining a mineral's hardness on a scale of 1 through 10. Talc, with a Mohs of 1, is an example of the softest of minerals while diamond, with a Mohs of 10 is the hardest).

> How does a cap rock of comparatively soft material stay on top of a rising salt body, and not be pushed aside along with other sediments?

Researchers involved in mining and resource development around salt domes recognize that the origin of the cap rock is *not fully understood*:

"Of totally different origin are those calcium sulfate [anhydrite] deposits which constitute part of the cap rock of salt domes in the Gulf Coast basin. **The exact origin of these domes is not fully understood** but most prevailing views involve flowage and upward movement of plastically deformed salt in response to overburden pressure from the overlying sediments...

The most widely accepted **hypothesis** for the **origin of the cap rock** is that it represents a residue of anhydrite and other relatively **insoluble minerals, which accumulate at the upper surface of the salt dome as it rises through water-bearing sediments** and the salt is dissolved. The anhydrite crystals, which make up about 99% of the water insoluble residue, are then recrystallized and compacted to form a massive rock which may be locally hydrated to form gypsum." Note 6.1d

Under this hypothesis, the anhydrite, which is insoluble, collects above an uprising salt body. Presumably, the salt dissolves away as the body continues its upward movement. However, there were no cited observations or explanations of how a salt structure could continue to rise through water-bearing sediment without being totally dissolved. Neither is there an explanation for the dewatering, or removal of the brackish water. Salt domes typically exhibit a distinct contact interface with surrounding sediment, not what one would expect if the salt had been dissolved into nearby strata.

Another important physical feature that any theory on salt dome origin must account for is the existence of the 'false cap rock':

"Associated with many domes is a zone of hard, **secondarily cemented sediments that immediately overlies the cap rock or salt where cap rock is absent**. The hard strata, commonly called '**false cap rock**,' may grade into the cap rock itself. The sands of the false cap rock are usually **cemented by calcite or other water insoluble minerals such as pyrite**." Bib 144 p45

As we have already discussed, the cap rock should have been pushed off the dome if the salt diapir had risen many miles to the surface. The false cap structure should also have been pushed aside. Both cap rock and false cap rock are distinctively different from the materials and sediment surrounding the salt column, suggesting *their formation is related* to the formation of the salt dome. No theory has explained how the sedimentary rock found directly above the salt body was formed or where it came from. No mechanism is offered for the cementation of sands and for the forming of pyrite in false cap rock, miles below the surface. Many questions are left unanswered in the modern salt dome theory.

There is even a bigger mystery within the modern salt dome theory—the oil mystery.

Fig 6.1.10 – This Jar contains evaporates from one liter of seawater. The evaporite or salt does not consist of large crystals like most natural salt deposits. Evaporated salts are thin and crusty and take a lot of time to form, allowing silt, sand, and other impurities to be blown or washed in. These impurities are typically not found embedded in salt dome deposits.

Fig 6.1.11 – Salton Sea salt from Southern California, USA. The pink color is common in salt-water evaporites. It comes from microorganisms that live in the brine. Large quantities of this type of pink salt are uncommon in most geological salt deposits.

Refer again to the illustration in Fig 6.1.8. Notice that pockets of petroleum (oil) are found on the perimeter of salt diapirs. Often, these are relatively near the surface. If salt plugs or pipes move plastically up through the crust, through denser overlying rock, why would oil accumulate there? Many of the oil fields found in the US exist near salt domes. What part does salt play in the existence of oil? How do such vast quantities of oil and such large bodies of salt end up together? The mystery of how oil and salt are related must be solved before the true origin of salt domes and oil can be understood.

The Salt Evaporation Mystery

All of the salt deposits found in the Earth's crust today are said to have come from the evaporation of ancient seas or saline lakes. If this is true, salt samples taken from mines or salt domes should compare to samples of evaporated salt from modern-day saltwater bodies. Fig 6.1.9 shows representative samples of salt deposits. One sample is from a salt mine in Redmond, Utah. The other sample is from a salt dome in Russia. These samples are quite small in comparison to the size of the deposits, which can consist of hundreds, thousands or even millions of cubic feet of salt.

At many salt domes, the salt is 99% pure sodium chloride, or halite. The Avery Island salt dome for example is 99.25% NaCl. The Dundee and Sylvania salt body in Michigan is 99.89% NaCl.Bib 143, p13, 50 Salt from the Redmond Salt Mine in Utah is 97.97% sodium chloride.

The jar in Fig 6.1.10 is an example of the evaporite from a liter of seawater. The evaporite is thin and crusty and the salt crystals are very small. Evaporation is a slow process. In the time it takes water to evaporate and crystals to form, silt, sand, or other detritus can be blown or washed over the deposits. If the evaporite theory were true, thousands of layers of impurities should be evident, as they would have accumulated in these natural deposits. However, these layers are generally non-existent in salt domes. Furthermore, when the seawater evaporated from this jar, more than five salts were present.

Look at Fig 6.1.11 to see an example of salt minerals formed from natural evaporation. This example is from the Salton Sea, an inland saline lake in Southern California, USA. Like many evaporative basins around the world, this salt has a distinctive pink color, which it gets from bacteria living in the brine. The bacteria and the pink color it causes are absent in salt domes and most large salt bodies, suggesting they were formed in an environment different from an evaporative environment.

Crystallization from evaporation forms a thin crust on the shore as water recedes. If the water level rises and covers the newly formed crust, the salt crystals dissolve. The accumulated salt crusts on the banks of saline bodies of water are relatively thin, usually only a couple of inches or less.

Zuni Lake in New Mexico, USA, has some of the thickest deposits of *observed* evaporite salt found in America. Fig 6.1.12 is a photo of the shore around the salty Zuni Lake. Although these are some of the thickest observed evaporite salt deposits, they are still only a few inches thick. Where are the evaporite basins with miles-thick deposits of salt being made today?

One geologist who was working with salt evaporites noted that the "rules of the game" were set down over a century ago:

"It is almost a century since Ochsenius (1877) **first set down the 'rules of the game,'** when he wrote: 'The origin of marine evaporites **must be explained through a comparison with modern occurrences**, despite their difference in scale'." Note 6.1e

Marine evaporites **must be explained** by modern occurrences of evaporation—there is no other way around it, and as the geologist points out, it **should not** constitute any particular problem:

"**The origin of evaporites should constitute no particular problem**: evaporites will form where evaporation exceeds influx, and every little pond, lagoon, or man-made depression in an arid environment is an actual or potential salt pan. **The puzzle presented itself when we discovered evaporite deposits covering** hundreds of thousands or **millions of square kilometers. Where do we find such large salt pans?**" Note 6.1e p371

The researcher poses an intuitive fundamental question: Where do we find such large saltpans? If worldwide salt deposits were all formed by evaporation of marine waters, there would be no "puzzle." The mystery remains because there are no large saltpans where immense evaporite deposits are being formed.

In some parts of the world there are salt bodies measuring thousands of meters thick. Some have been identified as underlying areas the size of whole countries. From the *Handbook of World Salt Resources*:

"The principal evaporite sequence in Germany is the Permian Zechstein, well known throughout the world not only for the voluminous production of potassium salts at Stassfurt, but also for the repeated sequences of evaporites which include the Stassfurt series. **This is the standard the world over**. The Zechstein basin occupies an extended area of sedimentation encompassing northwest Germany, the Netherlands, and the large part of Denmark, the sediments reach a total thickness of **6000-8000 m [5 miles]**." Bib 143 p192

"Where do we find such large salt pans?"

K. J. Hsü

Real Evaporation in Peru

To put these vast salt deposits in perspective, we will look at a well-studied case of evaporation taking place today. One of the best studies on natural evaporation was conducted in Bocana De Virrila, Peru, where researchers recognized that *very little* deposition of salt through evaporation is taking place. This is in contrast to *thick and extensive* salt bodies found in the geological record:

"**In contrast to the thick and extensive marine evaporate deposits common in the geologic record** of the Phanerozoic, **very little evaporate deposition is occurring at the present time**. In order to explain thick salt deposits, which in many cases correspond to the evaporation of **hundreds or thousands of meters of seawater**, various theories of evaporate deposition have been proposed." Note 6.1f

The salt-by-evaporation theory assumes that deposition of salts occurred over long periods of time. This poses a problem of salt purity. During these indefinite periods, environmental conditions would surely have had a negative effect on the purity of the evaporite deposits. Within the geologic salt deposits, sand, clay or other organic impurities **should be** present, interbedded in layers suggestive of the sequence of deposition. **No interbedded layers** of sand, gypsum, or organic matter are found in salt domes. They are however, found in modern marine evaporite environments, as was reported by the Peruvian research team at Bocana De Virrila:

"The gypsum prisms radiate out from the axial planes of the folds, which are usually filled with **a mixture of quartz sand, gypsum, and organic matter**. The scale of relief over the estuary floor [salt thickness] is on the order of **tens of centimeters**, and the **average wa-**

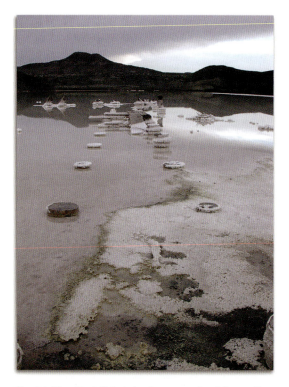

Fig 6.1.12 – Zuni Salt Lake lies near the Arizona, New Mexico border, USA. This lake lies within the depression of a steam explosion crater. The lake is continuously fed by underground springs. Surface salt deposits like this one are only inches deep. However, below the surface, very large deposits of salt exist. Where are the saltpans that are forming miles-thick salt deposits today?

CHAPTER 6 THE ROCK CYCLE PSEUDOTHEORY

ter depth ranges from **35 to 70 cm**. The majority of the crystals are **pink in color**, apparently as a result of red halophilic bacteria." Note 6.1f p451

The geochemists refer to a core sample taken by previous researchers, which confirms the stratification of salts, sand and other impurities:

"A 1-m core taken in this region by Morris and Dickey (1957) consisted of interstratified beds of **halite, gypsum, and unconsolidated sand**." Note 6.1f p451

Another observation from the Peruvian research identified a lack of bittern salts during historical precipitation of seawater:

"The precipitation of bittern salt would be expected in the Bocana if dense K- and Mg- rich brines build up as the inland waters continue to concentrate along the seawater evaporation pathway. Even if the Bocana precipitates are periodically flushed by El Niño events, K- and Mg- salts would accumulate in the intervening years. In the halite-precipitating regions of the estuary, **we found no evidence of bittern salts**." Note 6.1f p459

Bittern salts (KCl, MgCl, $MgSO_4$) are found both in seawater and in geological salt deposits (See fig 6.1.15), yet they are not forming in marine evaporite environments today. This observational fact alone should raise serious questions about the Salt Evaporation Pseudotheory.

One final but crucial observation deals with anhydrite. Anhydrite is chemically similar to gypsum but does not contain the water molecule, H_2O. Anhydrite is found abundantly in the crust—even to a depth of several thousand feet. However, actual observations from the Peruvian experiment found:

"**No anhydrite** was identified in the estuary." Note 6.1f p454

How could this be if evaporation was the mechanism for all salt deposits?

Another study in the Bahamas showed the following precipitated salts:

"Brines and salt were sampled at the Morton Bahamas solar salt production facility on Great Inagua Island in the Bahamas. The brines were analyzed by ion chromatography to define more precisely than heretofore the evaporation path of seawater to the end of the halite facies. At Inagua, **calcium carbonate** begins to precipitate at a brine concentration of about **1.8** times that of seawater. **Gypsum** begins to precipitate at a brine concentration of **3.8** times seawater, and **halite** at a concentration factor of **10.6**. Three of the most concentrated brines from Inagua (40 times seawater) were evaporated further in the laboratory. **Magnesium sulfate** first precipitated at brine concentrations about **70** times those of seawater, and **potassium-bearing** phases began to precipitate from these brines at concentrations greater than **90** times those of seawater." Note 6.1g

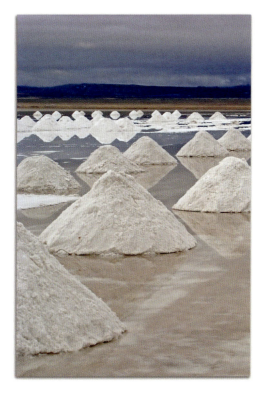

Fig 6.1.13 – These are commercial evaporate salt pans in Bolivia. Salt has no origin in modern science.

This study identified a specific order of salt precipitation. Calcite was first, followed by gypsum, halite, epsomite and potash. However, no anhydrite was observed. This order is not found in ancient geological deposits in which anhydrite is included. These observations, and others, testify that the ancient geologic salt deposits did not come from evaporation, but by another mechanism altogether.

"**No anhydrite** was identified in the estuary."

Salt With No Known Origin

The mysteries and anomalies surrounding salt are not widely known and answers about them have seemed to elude modern science. As we will learn in the Chemical Model chapter, sodium (Na) does not combine naturally with chlorine (Cl) to make salt (NaCl). Given the current knowledge about these mysteries, they will likely remain unknown because teachers cannot teach what they do not know. It may be that the origin of salt remains elusive because the true abundance of salt has generally not been realized. Without a knowledge of where salt comes from, science cannot explain why the oceans are so salty, nor can it explain where the enormous formations of salt such as salt domes have come from. The truth is, there never has been an explanation for the *origin* of salt itself. Modern science recognizes this.

Here is what one USGS publication has to say about what modern science really knows about the seas:

"What arouses the scientist's curiosity is not so much why the ocean is salty, but why it isn't fresh like the rivers and streams that empty into it. Further, what is the origin of the sea and of its 'salts'? And how does one explain ocean water's remarkably uniform chemical composition? **To these and related questions, scientist seek answers with full awareness that little about the oceans is understood**.

"There are several theories of the origin of the seas, **but no single theory explains all aspects of this puzzle**." Note 6.1h

Perhaps we can begin to see how the Rock Cycle Pseudotheory is woefully inadequate in explaining one of the most basic geological questions about salt's origin. Answers about these salt mysteries are contained in the following chapters, but first we must continue to discuss *why* so little about the oceans is understood. The first fact to recognize is that the so-called evaporite deposits are *not* evaporites.

The vast geological 'evaporite' deposits are **not** evaporites.

Geological Salt Deposits Are Not Evaporites

According to the studies conducted in Peru and the Bahamas, *precipitation of modern seawater* into salts was shown to have the following characteristics:

1. Precipitation occurs in two meters of water or less, because 90% of the seawater must evaporate for crystals to form.
2. The evaporite salts are less than a meter thick.
3. Sand and clay from erosion are mixed within the meter-thick salt layers.
4. Bittern salts (Mg and K salts) were not found to precipitate readily in modern natural evaporative settings even though they are more abundant in ocean water than are gypsum and calcite.
5. No anhydrite was found in the modern evaporite environment.

Natural deposition of salts by the evaporation of seawater in thousand-square mile saltpans is not taking place anywhere today. If we compare these five characteristics of observed evaporite deposition of salts with geologic salt deposits, many areas of conflict can be identified. Evaporite salts are less than a meter thick whereas geologic salt deposits are thousands and even tens-of-thousands of feet thick. Erosional materials are mixed throughout the thin evaporite salt layers but they are not in the thick geologic deposits. These facts directly contradict the evaporation pseudotheory as being an origin for geological salt deposits.

To generate massive amounts of salt through an evaporative process, some have proposed the existence of deep evaporite basins. Are these basins found on the Earth today?

> "Thick deposits are **really** hard to explain!"

No Deep Evaporite Basins

Researchers have known for more than four decades that no active, deep-evaporite basins exist in the world today. If the world's extensive salt deposits are evaporites, deep basins were hypothesized to have been a necessary part of the process. Salt depths of hundreds or thousands of feet would require the evaporation of tens-of-thousands of feet of seawater. From that realization came the idea of deep basin accumulation. In the scientific article *Origin of Saline Giants* one geologist notes:

"**'There are no active deep evaporite basins today'**, quoting R.F. Schmalz (1969, p822), who nevertheless proceeded to formulate a most elaborate hypothesis for deep-water evaporite deposition. **What motivated him and others before and after him to choose such a radical avenue?** An analysis of their writings revealed that their hypothesis is based upon considerable evidence for the existence of deep evaporite-depositing *basins*, and a tacit **assumption** that the water level within such topographical depressions **was only slightly lower than the world-wide sea level**." Note 6.1i

This basic assumption was never proven. The myth of uniformity convinced early researchers and many modern scientists that observed evaporation, extrapolated over time, and in deep basins, would form the miles-thick salt deposits. No deep *evaporite* basins exist today.

However, deep **non-active** basins of salt do *exist*. Researchers reported in 1972 that massive, deep basin salt deposits were found below the Mediterranean Sea:

"The existence of salts, not only halite, but also gypsum, anhydrite, and dolomite, under the Mediterranean was **proven by drilling**." Note 6.1i p385

How and when did these thick salt deposits form and by what process? These questions are still a mystery today.

No solid answers have come from the deep evaporite basin idea. Other researchers have developed a theory where extensive *shallow basins* evaporated and formed salt deposits. This is sometimes referred to as the Sabkha Hypothesis. This idea has also left many questions unanswered as researchers report:

"However, **evidence is completely lacking** to relate the origin of bedded halite or potash salts to the Sabkha diagenesis." Note 6.1i p385

In the sabkha model, lagoons or playas form shallow basins of evaporative salt. The massive kilometer-size worldwide salt deposits exceed the scope of this theory:

"A salt basin covering millions of square kilometers would neither be a lagoon nor a playa, but a desiccated inland sea!" Note 6.1i p385

Inland seas like the Mediterranean cover millions of square kilometers and are not shallow. Therefore, researchers are stymied back into deep basins thinking. It is easy to see why the Earth's salt deposits are such a mystery in geology.

Fig 6.1.14 – The Deep Basin Salt Mystery. Massive salt layers are not being formed in deep ocean basins anywhere today. Geology textbooks routinely state that massive salt deposits are formed this way. The true origin of these salt deposits has eluded modern science and remains hidden today.

Ocean Depth Mystery

Another troubling mystery related to the formation of salts is what we will refer to as the Ocean Depth Mystery. How much seawater would be required to account for the salt deposits found around the globe? Researchers realize that literally thousands of meters of seawater would have had to evaporate to explain thick salt deposits:

"In order to explain thick salt deposits, which in many cases correspond to the evaporation of **hundreds or thousands of meters of seawater**, various **theories** of evaporate deposition have been proposed." Note 6.1j

It is a commonly held idea among geologists that "shallow seas" covered all the continents at one time or another. What about the hundreds or thousands of feet of seawater needed to account for such thick salt deposits? One geology professor, in his syllabus declares to his students:

"Thick deposits are *really* hard to explain!" Note 6.1k

The professor identifies why they are hard to explain:

"1000 feet of water will produce .3' gypsum; 11.5' halite; 3' other salts." Note 6.1k, For a similar reference, see Bib 144 p7

He then asks the question: "How do we get 1000' thick gypsum deposits?" before asserting "A full understanding of evaporites will take more work!" Note 6.1k

Let us explore how big of an understatement this is. If it takes 1000 feet of seawater to produce one third foot of gypsum, it would take 3,333,333 feet, or 631 miles of seawater to make 1,000 feet of gypsum. Contrast this with the fact that the deepest parts of the ocean are less than 7 miles deep. This however, is only the beginning of the mystery.

In 1972, the Humble Oil & Refining Company drilled a 10,179-foot hole east of Eloy, Arizona, USA. They found an astonishing 6,000 feet of anhydrite:

"According to Pierce (1973), the **6000 ft of anhydrite** in the Humble hole might be the thickest sequence of anhydrite penetrated in the world." Note 6.1l

Returning to our example, three million-plus feet or 631 miles of seawater per 1000 ft of anhydrite multiplied by six to equal the anhydrite found in the Humble hole equals **3,786 miles of seawater.** That is roughly the distance to the center of the Earth. The professor was barely touching the tip of the iceberg when he said, "Thick deposits are *really* hard to explain!" The Ocean Depth mystery is of monumental size when trying to understand the problem of evaporites.

The Sea Salts Mystery

Ocean water tastes salty. Although the most common salt found in seawater is sodium chloride (NaCl), there are five other salts, listed in Table 1 in Fig 6.1.15, which come from seawater.

As was mentioned previously, if geologic salt deposits came from evaporated seawater; all six salts should be present in the geologic record at roughly the same ratio as they occur in the seawater. A quick look at the table shows this is not the case.

Bischofite, the first salt on the table is 74 times more abundant in the ocean than it is in ancient geological deposits. Conversely, calcite is 17 times more prevalent in rocks than it is in seawater. Both are said to come from seawater, through evaporation.

Each of the six salts listed in table 1 has a different density. When seawater evaporates, each salt in the solution should precipitate out in a predictable order. Geologist have been unable to evaporate seawater where salts precipitate in the order they are found in naturally occurring geologic salt deposits. This problem was recognized by the author of this article:

"Relevant to our discussion is, however, the **fact** that a desiccation [dehydrate] of a marine lagoon **should produce not only gypsum and halite, but also other salts**.

"The volume of salts that **should be** precipitated from the isochemical evaporation of seawater is shown by Table 1. In comparison, the ancient evaporites, including the famous Zechstein, are richer in calcium sulphates and poorer in magnesium and potash salts than such an isochemical precipitate... When Branson (1915) noted the absence of halite above gypsum (or anhydrite) deposits tens or even hundreds of meters thick, **the problem could no longer be evaded**." Note 6.1m

The author states the problem could no longer be evaded, but it certainly seems to have been avoided. There is simply no clear answer for the origin of salt and salt formations in the modern Rock Cycle paradigm.

It is beyond the scope of this book to discuss each of the salts in the ocean. The intent is to identify that the origin of salt is a *mystery* in geology today. What we find amazing about this glaring discrepancy in the sea-salts mystery is how little discussion is found in geological literature. It is simply illogical

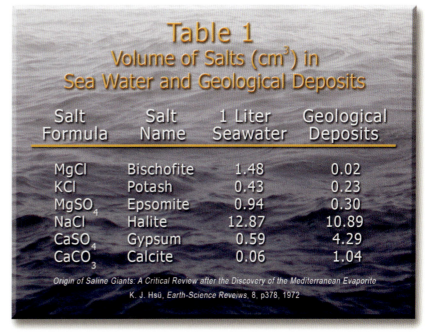

Table 1
Volume of Salts (cm³) in Sea Water and Geological Deposits

Salt Formula	Salt Name	1 Liter Seawater	Geological Deposits
MgCl	Bischofite	1.48	0.02
KCl	Potash	0.43	0.23
MgSO₄	Epsomite	0.94	0.30
NaCl	Halite	12.87	10.89
CaSO₄	Gypsum	0.59	4.29
CaCO₃	Calcite	0.06	1.04

Origin of Saline Giants: A Critical Review after the Discovery of the Mediterranean Evaporite
K. J. Hsü, *Earth-Science Reveiws*, 8, p378, 1972

Fig 6.1.15 – This table lists the volume of six salts found in a liter of seawater and compares that to the average ratio found in geological (crustal) deposits. If the geological salt deposits came from evaporated seawater, the ratios should be similar. However, 17 times more calcite (CaCO3) is found in geological deposits than is found in seawater, and 74 times more bischofite (MgCl) is found in seawater than is found in geological deposits. This glaring discrepancy means the origin of the geological deposits should be questioned. Contrary to what modern geology says, they *do not come from evaporated seawater*.

Fig 6.1.16 – White Sands National Monument in New Mexico, USA is a unique deposit of gypsum sand whose existence is a mystery in modern geology. Notice the white clouds (look for shadows beneath them) on the left side of the inset photo, top left. The large white area is the natural color of the monument. The white sand in the valley could not have come from the dark mountains to the left. Viewed close up, the sand contains very few other minerals that should be present should the sand have come from the nearby mountains via erosional processes. From where did the white gypsum sands come?

to think that all of the worldwide salt deposits could have come from evaporated seawater. Modern science seems to have ignored this fact and offers no alternative to the evaporite model.

The Pure Salt Evaporation Mystery

There is another reason why thick geological salt deposits could not have formed from evaporation. Consider the maximum known evaporation rate per year:

"At the present time **the maximum evaporation from large water surfaces in very arid regions is 2 m [meters]**… Furthermore, in the absence of wind, halite precipitates at the water surface, reducing evaporation still further, so that the present-day evaporation from halite-saturated solutions may be taken as about **1 ½ m per year**." Note 6.1n

Using the figure of 1½ meters per year of seawater evaporation, we can approximate how much deposition might occur during each yearly cycle. Each year, 1½ meters of seawater will evaporate which will produce about ½ millimeter of gypsum. After a decade, five millimeters (less than one quarter of an inch) of gypsum will have accumulated. 100 years would net 50mm (about two inches) and 1000 years would see approximately 500mm, around 20 inches of pure pristine gypsum, or anhydrite or sodium chloride. However, is it possible to have *pure* salt deposited over a thousand or even a hundred years? What seashore exists where storms or winds never visit, where nothing organic serves to sully pristine salt deposits?

The fact is, even ten years of accumulation would see sand and clay particles interbedded in the salts. Organic material would be present, and even some organisms may be living and dying within the accumulation. The remains of marine organisms would likely become fossils, found deposited among the layers of gypsum and other salts. It certainly would not be pure.

How is it possible that only white gypsum sand and no other minerals are 'transported by wind' from the surrounding mountains?

Interestingly, while marine fossils are almost non-existent in geologic salt deposits, some terrestrial or **land** fossils *have been found* in salt deposits, further muddling the seawater evaporation theory.

There are simply too many *pure salt deposits, hundreds, or thousands of feet thick* for the evaporation theory to be valid.

The Gypsum/Anhydrite Mystery

Gypsum is a widely used mineral in the world today. Among its many uses, gypsum is found on the walls of most homes and in toothpaste. It is used in gardening and in the manufacture of sterling silverware, and it is commonly used in dentistry.

Gypsum is also used to make the commonly known Plaster of Paris, which is used by many industries to make molds. It is a favorite of artisans and hobbyists because of its unique properties.

The difference between gypsum and anhydrite is that gypsum has water within its crystal structure, while anhydrite is waterless or *void of water*. How is it possible for both forms of this mineral to come from the evaporation of seawater as geology claims?

"Most deposits of massive **gypsum and anhydrite were formed by evaporation of sea water** in basins that had one or more restricted openings to the sea. The basins range in diameter from a few miles to many hundreds of miles." Note 6.1o

It is quite a stretch for geologists to say both anhydrite and gypsum are formed by evaporation. We have yet to find any reports where seawater was evaporated producing anhydrite. Yet this is exactly what has been propagated throughout the geological community. In fact, we have been unable to find anything or anyone with evidence to dispute this specific UM claim:

Anhydrite does not come from evaporated seawater.

Recall from the detailed Peruvian salt study that: "**No anhydrite** was identified in the estuary." Note 6.1p

The UM claim that anhydrite does not come from evaporation will be supported in the Hydroplanet and Universal Flood Model chapters. There, we will show exactly how anhydrite is made today and how it was made in the ancient past.

White Sands Gypsum Salt Mystery

There is a very interesting gypsum deposit located in New Mexico, USA. (Fig 6.1.6) It is another excellent example of how the Rock Cycle Pseudotheory fails to explain even the simplest salt deposits. The white sands of the National Monument are said to have come from nearby mountains. However, as one geologist noted, the dunes are of "remarkable purity":

"Gypsum sands are made up of **crystals that have been transported by wind and deposited in dunes of remarkable purity**." Note 6.1q

In fact, the white sands are **too pure**. That is, too pure to be eroded from the mountains. How is it possible that only the white sand and no other minerals were transported by wind from the surrounding mountains? We were given no answer from the staff at White Sands National monument for this significant FQ. Neither were there any research articles located which addressed this question. The visitor center at the White Sands National Monument tells of water and wind erosion, but clearly, no one has thought this process through.

Where are the rivers that carried gypsum from the mountains?

Fig 6.1.17 – Goshen Utah, USA salt flat showing precipitated salts from evaporation. Note how a thin crust is formed, nothing like the White Sands Gypsum sand deposit, which could not have come from evaporation.

The sand is reported to consist of similar sized "crystals," not broken pieces and fragments from wind erosion. Furthermore, if wind borne salt (gypsum) crystals are the source of the White Sands deposit, a trail of differing sized particles should be seen emanating from the mountain source. This trail of larger sized particles should be evident because the larger particles are heavier and would fall out first.

Fig 6.1.17 is a salt deposit located in Goshen, Utah. This deposit *formed from evaporation and erosion*. This is known because it is still taking place today. Notice how the dissolved salt in the water left a fine crust on the ground, mixed with clay and mud. It does not form crystal sand grains.

Fig 6.1.18 shows the results of a dish of evaporated saltwater

Fig 6.1.18 – Evaporated NaCl salt from the Redmond salt mine in Utah, USA. Even if enough gypsum could dissolve into surface or rainwater to precipitate crystals similar to those seen in this dish, they would not all grow to the same size. The grains of White Sands are essentially all the same size. Why is this?

SUBCHAPTER 6.1 THE MOST ABUNDANT MINERAL MYSTERY

from the Redmond Salt Mine in Utah, USA. Notice how the salt crystals are of various sizes. At White Sands, NM, we find only sand sized particles in the dunes deposit. Similarly, if gypsum had precipitated from mountain streams, evidence of evaporation would be found on the ground, along the streambeds. Evaporite crystals would be found in various sizes.

The white gypsum sand is a very rare deposit. It did not come from evaporation; otherwise, we would find examples around the world where seawater is being evaporated and forming this type of 'salt' deposit.

> The White Sands National Monument is a one-of-a-kind geological mystery in science today. Nowhere else in the world is a similar deposit found. Modern geology has no clear explanation for where the white sands came from.

The Evaporite Crisis

Scholars have been researching the world's salt deposits for a long time. Some realize there is a crisis in geology because of the disparity between theoretical ideas and the actual observations of the salt deposits:

"Unfortunately, in this respect the study of salt deposits is very backward when compared with the petrography of igneous rocks, and **there exists a shocking disparity between theoretical knowledge and the quantitative compositional survey of the salt deposits themselves... Today we can and must demand of geological investigations**, insofar as they have pretensions to pose and answer questions concerning the genesis of salt deposits, adequate documentation of their qualitative and quantitative compositions and, over and above this, an understanding of the underlying physico-chemical principles." Note 6.1r

An understanding of salt cannot happen as long as modern science continues to accept evaporation as the only origin for salt. There must be a shift away from the current geological paradigm. We can begin by asking this simple question: If the geological salt deposits did *not* come from evaporation, where did they come from? The extraordinary truth is:

> Salt is formed by **methods other** than evaporation.

The knowledge that salt deposits have an origin other than evaporation leads us to ask this fundamental question:

> How were ancient salt deposits formed?

A review of different evaporite theories has identified a great controversy:

"Looking back, we have seen that the development of theories on the genesis of evaporites has been evolutionary during the last century…

"As competing and mutually exclusive theories began to take shape, **controversies began to rage, leading to a crisis**." Note 6.1s

Before we can consider the answer to the question of alternative salt deposit formation, we must continue to set the stage for discovery by examining other mysteries and inconsistencies in the Rock Cycle Pseudotheory. We will then be ready for the essential truth.

In subchapter 4.2, we identified the elements leading to a scientific revolution. The first step in the process is to recognize that there is a crisis. Our discussions about salt indeed show that there is a *crisis* in modern geology today. Geologists cannot identify the most abundant mineral on the surface of the Earth, nor do they recognize that the largest occurrence of a single rock mineral is a salt dome. The field of geology talks of salt deposits being the accumulation of evaporites, but has **no actual origin for the salt itself**. This fact among others is why the whole of geology is ready and poised for a scientific revolution.

6.2 The Real History of Geology

In the Magma Pseudotheory chapter, we established that the Earth's crust and the crust of other planets could not have come from melted magma and there is no physical evidence that the center of the Earth is magma.

James Hutton, who is called by some the father of geology, came to the idea that the Earth was shaped by a long succession of reoccurring events, driven by heat from the inner Earth. His ideas profoundly influenced modern geology. The tenets of uniformitarianism and the endless cycle of once melted rocks being recycled, eroded, crushed and melted again, sent geology headlong into the future, but on the wrong path.

Here we will revisit some of the mysteries resulting from the now well-accepted theories and evaluate what defines modern geology. Does modern geology have the correct answers for the origin of rock? Do they know the origin of the moon?

The Rock Shop Professor

One summer vacation, we were touring the American Southwest and we stopped at a rock shop. There had been many stops at rock shops on this trip, but this one was different. The owner of this particular shop had been a professor of geology at a local university earlier in his life. The rock-shop professor,

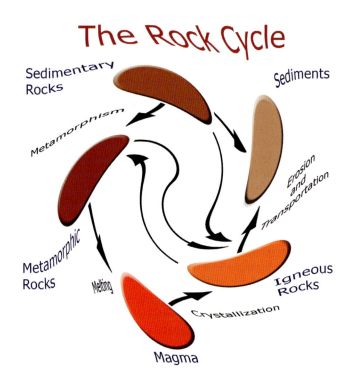

Fig 6.2.1 – Because the Rock Cycle Pseudotheory Diagram is based on magma, which does not exist, the definitions of both metamorphic and igneous rocks are incorrect and must be replaced. So too must the origins of the sedimentary rocks be more closely examined where it will become known that they did not all come from erosional processes as we know them today.

who was very informative, commented on a number of rocks on display, and discussed the details about how they were formed. As we chatted about various specimens, I asked him why he left school. He declared in rather colorful language that geologists do not know anything about rocks and what they think they know is "all just theoretical."

His comments were surprising, but over the years, I began to realize his claim about modern geology was astonishingly accurate.

Basic Modern Geology

Every basic geology textbook includes a discussion on the origin of rocks and the processes they go through. In all cases, rocks are said to transform over time from one type of rock to another. The modern science concept of the rock cycle was developed from the ideas of James Hutton and others in the late 1700s. Fig 6.2.1 is representative of the Rock Cycle Pseudotheory as it is found in all introductory geology textbooks today. The author of the article from which this particular diagram was adapted begins his rock cycle discussion by saying:

"Mineral creation **is now fairly well understood**." Note 6.2a

A few sentences later:

"**Not that mineral creation is simple and straightforward**. They are continuously being destroyed and recreated as in the 'Rock Cycle' chart..." Note 6.2a

Geologists say they understand mineral creation, yet in reality the mineralization theory is complicated and is not supported through observation. This has been a challenge for researchers since Hutton and Lyell first espoused their ideas about geology. Although rock and mineral formation is presumed to be "fairly well understood" in modern geology, we will demonstrate how misunderstood the process really is.

Take away the very foundation of the Rock Cycle—magma, and both metamorphic and igneous rock definitions come under question. In this chapter, we question the very definition of these rock-forming processes, and show the lack of empirical evidence that exists to support them.

To understand truth in any field of science, carefully *examine the primary definitions of words that define that field*. We will do just that. To begin, we look at the definitions of the rock classifications. From the geology textbook *Understanding Earth*, there are three types of rocks defined in the Rock Cycle:

"**Igneous Rock** – a rock formed by the solidification of a magma." Bib 59 p655

"**Metamorphic Rock** – a rock whose original mineralogy, texture, or composition has been changed by the effects of pressure, temperature, or the gain or loss of chemical components." Bib 59 p657

"**Sedimentary Rock** – a rock formed by the accumulation and cementation of mineral grains by wind, water, or ice transportation to the site of deposition or by chemical precipitation at the site." Bib 59 p661

The definition of the rock cycle theory from the same text:

"The **Rock Cycle** is a set of geological processes by which each of the three great groups of rocks is formed from the other two. The Scotsman **James Hutton** described this cycle in an oral presentation in 1785 before the Royal Society of Edinburgh; ten years later he presented it in more detail in his book *Theory of the Earth with Proof and Illustrations*." Bib 59 p68

At first glance, Hutton's rock cycle theory seems realistic and plausible. However, when we look closer, Hutton made three *false assumptions* that doom this theory to failure. The first assumption is that the cycle occurs indefinitely and in a uniform way—this is referred to as uniformitarianism. The second assumption is that the uplift of large, continent-sized landmasses allows erosion and transport of all the sedimentary material seen. The third assumption is that all rocks originally came from magma, or that they were melted at first. Although we discussed each of these false assumptions in the previous chapter, we will add many more details and examples of why they are false and why these theories will continue to fail.

The Rock Tree Mystery

Geology has a 'rock tree' within the rock cycle theory that attempts to identify the processes and environments in which minerals form. Bowen's reaction series (See subchapter 5.14) is a part of the geologic 'tree' and is somewhat analogous to the Tree of Life found in biology. Both are attempts to classify and order their respective elements: rocks and minerals in geology and living organisms in biology. Both portray branches within a tree that originate from a single trunk. A shared commonality between biology and geology is that both fields are *still looking* for the single trunk and many of the missing branches.

The author of the textbook, *Magmas and Magmatic Rocks* wrote about the difficulty geologists have defining "real rocks":

"Igneous rocks are **usually defined** as those rocks that so-

Fig 6.2.2 – These minerals are on display at one of the largest rock and gem shows in the world, held each February in Tucson, Arizona, USA. Dealers from around the world gather to display and sell some of the most beautiful minerals that nature has to offer. How did nature make these minerals? Knowing the truth about the processes that form minerals is one of the central themes of this and the next two chapters.

SUBCHAPTER 6.2 THE REAL HISTORY OF GEOLOGY

lidified from a hot (usually greater than 600 °C), molten, or partly molten condition. When one attempts to apply **this definition to real rocks**, one discovers that in the plutonic [below the surface] **realm it is often difficult to establish** criteria for deciding whether particular crystalline rocks are either **igneous or metamorphic**; whereas in the surficial [at the surface] realm it may be **difficult to decide** whether pyroclastic materials are of **igneous or sedimentary** origin." Bib 87 p71-2

When geologists try to apply the definitions of the Rock Cycle Pseudotheory to rocks, they often find it 'difficult to decide' the origin of the rocks. Many people can identify sandstone, limestone, and granite. However, with other less common rocks, geologists often disagree whether a rock is igneous or metamorphic.

The *Handbook of Rocks, Minerals & Gemstones* by Walter Schumann is a standard for people interested in the rocks and minerals of the Earth's crust. It has over 600 beautiful photographs of rocks and minerals. The author states:

"The text, based on the latest scientific insights, is composed in such a way that it can appeal both to the layman and to those who already have expert knowledge, namely, anyone who deals with rocks and gemstones professionally, as well as to students." Bib 15 Forward

What is interesting about the book, which was written to both nonprofessional and professional alike, is how the author explains that geology has been unable to come to an agreement on classification:

"In no rock group is it **so difficult to gain an overview** as it is with **metamorphic rocks**. There is **no generally valid classification** and in consequence a unified terminology **is lacking**." Bib 15 p305

If the true origin of rocks is known, we should be able to classify each kind of rock according to how it is made. The author even suggests further:

"The scientific grouping of metamorphic rocks **is not acceptable for anyone doing field work or working practically with rocks or for the broad field of the sciences allied to geology** since it presupposes a great deal of expert knowledge **which is not in fact available**." Bib 15 p306

This is an interesting statement considering the *Handbook of Rocks, Minerals & Gemstones* is a guidebook to aid in the classification and grouping of rocks. Although the handbook is intended to be taken into the field to help classify rocks, it "is not acceptable for anyone doing field work or working practically with rocks" to classify or group them. We find there is only one reason why geologists have not been able to classify or group metamorphic rocks according to their origin. *Their origin is unknown*.

The Origin of Rocks is Unknown

Sedimentary rocks on Mars are a remarkable example of how the origin of rocks is misunderstood, due to the Rock Cycle paradigm. Even though the evidence of past oceans on Mars is overwhelming, the water-related sedimentary minerals found on Earth such as carbonates (limestone), clays and salts, are almost totally absent from Mars. (See Fig 6.2.3). This has left scientists "*more baffled than ever*":

"The Mariner and Viking missions of the 1960s and 1970s revealed a cold, dry and lifeless world, but one etched with remnants of past vigor: delicate valley networks from the distant past and vast flood channels from the intermediate past. **Researchers expected that when new space probes assayed the planet, they would find water-related minerals: carbonates, clays, salts**.

"Over the past six and a half years, the Mars Global Surveyor and Mars Odyssey orbiters—bearing duplicates of the instruments that the ill-fated Mars Observer carried—**have looked for and detected essentially none of those minerals**. They have found layers of olivine, a mineral that liquid water should have degraded. And yet the orbiters have also seen fresh gullies, old lake beds and shorelines, and an iron oxide mineral, gray hematite (as opposed to red hematite, otherwise known as rust), that typically form in liquid water. The planet holds extensive reservoirs of ice and bears the marks of recent geologic and glacial activity. **Scientists are more baffled than ever**." Note 6.2b

More baffled because the carbonates, clays, and salts on Earth are not formed the way they have come to accept. Mars does not have carbonates, clays and salts like Earth because Mars and Earth do not have the same environment. Mars has never had an environment like that of Earth. Most carbonates and salts have a biological origin, which we will expound upon, in the Universal Flood chapter along with the origin of clays and most other sediments.

The Perfect Geological Model—the Moon

The Moon is the closest celestial body to the Earth—near enough that we can observe it almost every day. The primary difference between the Earth and the Moon is the lack of an atmosphere. Without an atmosphere, there is no weather to alter its surface and it remains much as it was when it was created. The Moon is a perfect geological model of planetary formation. Any planet-forming model including a model of the Earth's formation *must logically include the origin of the Moon*. Because there is no atmosphere and no erosion, the Moon is a unique laboratory and has been preserved for us to observe and to advance our understanding. If the Moon is a Perfect Geological Model, consider this Fundamental Question:

Does modern science understand
the Moon and its origin?

A Moon Without an Origin

Geology is not just about rocks and minerals on the Earth. It is about the materials and formations that exist on the planets, moons and any other non-stellar body in our solar system, and

Fig 6.2.3 – The Mars landscape as seen by the Spirit Rover. There is overwhelming evidence that Mars was once covered by great seas, yet scientists are "baffled" as to why Mars does not have sedimentary minerals such as limestone, clays, or salts like those found on the Earth.

Fig 6.2.4 – This is not the Moon we are accustomed to seeing. Because we see the same side of the Moon all the time, this photo, which includes part of the far side looks strange. The origin of the Moon is even stranger to modern science.

maybe beyond. The planets and moons in our solar system certainly have many geologic forms, some of which may not be seen or exist on Earth. However, there may be similarities from which we can learn much. That inspiration led mankind to visit the moon in the 1960s and early 1970s. Man went to the Moon on a mission of discovery, hoping to find clues to its origin and to unlock clues regarding Earth's origin too. Hundreds of pounds of rocks, minerals, and lunar dust were collected from many areas on the Moon. Many creative and interesting experiments were carried out, all in the hopes of understanding more about the origin of the moon and the Earth. How much do we know after all these years? We read from an article in the journal *Nature* the following:

"In astronomical terms, therefore, **the Moon** must be classed as a well-known object, **but astronomers still have to admit shamefacedly that they have little idea as to where it came from.** This is particularly embarrassing, because the solution of the mystery was billed as one of the main goals of the US lunar exploration programme." Note 6.2d

Astronomers admit the origin of the Moon has eluded them. Geologists are no different and acknowledge they too, do not know where the moon originated. The book, *The Age of The Earth,* discusses the following hypotheses of the Moon: the fission hypothesis, the capture hypothesis, the double-planet hypothesis, the collision hypothesis, and finally states:

"As should be obvious from this brief summary, **we do not know how the Moon originated** or its exact genetic relationship to the Earth." Bib 133 p199

While it may not be apparent from the pictures shown on the Discovery channel, the truth is, the answer to most geological questions today is "we do not know." Because the foundation of modern geology is the Magma and Rock Cycle Pseudotheories, we often see a flip-flop in current thinking on the origin of the Moon according to modern science. In a 1970s Life Science Library Book we read:

"A more dramatic but now discredited view held that the Pacific basin is the hole left behind when the moon was torn free from the earth. **That idea was given up** when scientists got a chance to examine the rocks the astronauts brought back from their lunar exploration. The moon rocks proved to be **totally different from—and much older than—any found on earth. Clearly the moon could never have been part of the earth**." Bib 13 p36

This statement was made after the Apollo astronauts returned from the Moon. As the article states, the idea that the two bodies, the Earth and the moon, could have ever come from the same source was given up. Then in 1991, a respected authority on the age of the Earth, wrote:

"Unlike the other hypotheses for lunar origin, **there is as yet no substantial negative evidence against the collision hypothesis**, and so it is currently, if tentatively, **favored**." Bib 133 p199

There are many variations of the collision hypothesis but all share one similarity; any collision involving the Earth and the resultant Moon, included substantial mixing of native rock, melted or otherwise. Yet in the 1970s, it was said the rocks from the Moon "proved to be *totally different*" from any found on the Earth. The latter author further noted there was "no substantial negative evidence against the collision hypothesis," seeming to imply that the rocks of both the Earth and the Moon must be similar in age. Of course, the rocks should be similar in age based on the dating method geologists' use, which is based on the time the rocks originally melted. In any case, if there had been an impact, both Earth rocks and Moon rocks would have been melted at the same time and would be of the same age. Again, from the 1970 quote: "The moon rocks proved to be... *much older than—any found on earth*". Major discrepancies like this occur throughout modern geology and must be reconciled by whatever model that seeks to explain the origin. So far, this has not happened.

> "The Man has not yet lived who can adequately describe a grain of sand."
>
> Charles R. Van Hise

6.3 The Sand Mystery

Mark Twain once said, "Our best built certainties are but sand-houses and subject to damage from any wind that blows." Next to water, there probably exists no natural material for humankind to ponder over more than sand. Sand can be soft and warm on the beach, or hot and unforgiving in the desert. Language is replete with references to sand that conjure poignant feelings. Sayings such as, the sands of time; shifting sands; built upon a sandy foundation, and many more such sayings, each conveying volumes of thought in the utterance of just two or three words. Sand is like that. The true origin of sand is as powerful and meaningful as the sayings it is in, but that

6.3.1 – These well rounded small rocks, pebbles, and sand grains were collected along the shore of Lake McDonald in Glacier National Park, USA. They exhibit a heterogeneous mixture of various types of minerals. The pile on the right is typical of what the vast majority of sand deposits *around the world* **should** look like – a heterogeneous (dissimilar minerals) mixture of rock fragments from nearby parent mountains. Why don't they?

origin remains unknown in modern science today. It is unknown because the magmaplanet foundation upon which the rock cycle is built is even more unstable than *sand*.

Sand Defined

What could be so mysterious about sand? It is a common assertion among geologists that sand is simply small fragments of rock, which were eroded from larger rocks. This idea has been taught for generations. Now however, new Fundamental Questions about sand will likely cause a permanent shift in the sand paradigm.

In the *Journal of Geology,* sand is defined:

"…**sand is qualitatively defined as** any particle that is light enough to be moved by the wind but too heavy to be held in suspension in the air. Very fine particles that can be held in suspension are therefore classified as silt or dust, while heavier particles unaffected by wind are classified as pebbles or gravel.

"While sand can be composed of various minerals, **quartz makes up the bulk of the world's sand grains**." Note 6.3a

Notice that sand is a particle that is able to be moved by wind but is too heavy to stay suspended in the air. We will return to this later, but first a Fundamental Question:

What is the true Origin of Sand?

The Sand Mystery—Its Origin

Where does sand come from, or what is the origin of sand? This seems like it should be an easy question to answer:

"Whenever the sand arrived, in different parts of the Great Plains, it has clearly come from different places, **although it all originally was mountain rock ground into sand by glaciers, weather and plunging streams**. Note 6.3b

Geologists believe sand is simply "mountain rock" ground down through erosional processes. Fig 6.3.1 depicts rocks of successively smaller size as they might occur through erosion. However, if sand is made simply by the erosion of mountain rock, why does the researcher in the last quote say this in his next paragraph?

Fig 6.3.2 – These eight slides are samples from around North and Central America. Notice the different colors and textures of the sand. Some sand grains appear to be homogeneous, of the same material, while others appear to be of a heterogeneous makeup. How do billions of sand grains erode from many different mountains which consist of many different minerals and yet all look the same? How can whole beaches be made of clear sand that looks like glass beads eroded from faraway mountains?

"Muhs and other **geologists don't know exactly how all that sand wound up in the Nebraska Sand Hills**." Note 6.3b p77

It may be mysterious how sand was transported to its final destination, but there is an even *deeper mystery* that lies within the sand grains themselves.

In 1960, a respected researcher of sand, P.H. Kuenen, noted:

"The little grains of sand have been somewhat **neglected by geology** until recent years, although they have played a mighty role in the history of the continents.

"The very abundance of sand has made it so familiar that **even geologists did not stop to ask how it came to occupy its place in the landscape and why it tends to accumulate in masses of uniform grain size**." Note 6.3c

Kuenen was one of only a few researchers that came close to realizing how sand literally holds "the key to many geological questions still unanswered":

"Now that the advance into the unknown has broadened, **geologists are realizing that sandstones and sands hold the key to many questions still unanswered**." Note 6.3c

After half a century, something as simple as the origin of sand and sandstone is still being debated. One investigator who witnessed the development of the science of sand over the last five decades notes in the *Journal of Geology* in 2003:

Fig 6.3.3 – The Microscopic Sand Mystery (next page), illustrates several different types of sands from North and Central America. All shown at 40X their actual size. Nature has made beautiful grains of sand like these that exist around the world. Many people have never seen the elegance of sand up close, and have therefore not been inclined to ask where sand came from. Modern geology has only one answer for where sand comes from—erosion. Nevertheless, the UM will demonstrate that the vast majority of sand and clay in the world *could not* have come from erosion. No mountains exist from which sands like those at Daytona Beach, or Cancun Mexico Spheres could have eroded from. They are not known to exist anywhere.

The Microscope Sand Mystery

Fig 6.3.3

"When I was a student half a century ago, **the origin** of pure quartz sheet sandstones, then called orthoquartzites, was considered **a major puzzle**. Together with the origin of dolomite, red beds, black shale, and banded iron formation, they made up a group of **seemingly intractable geological problems. Even now, 50-odd years later, their origins are still being debated**." Note 6.3d

It is only because of the obscurity brought about by the dark age of science that something as familiar and simple as a grain of sand could be so completely misunderstood. To comprehend why the debate is still raging in modern geology, we have to look *more closely* at the grains of sand themselves.

The Mystery of Microscopic Sand

To understand the mystery of Microscopic Sand, look at Fig 6.3.2. Eight different samples of sand are shown. See these samples magnified on the following page, in Fig 6.3.3. If you have never peered through the lens of a microscope or seen magnified objects on a screen, a whole new world of splendor awaits your eyes. As we take in the scene, new fundamental questions begin to form. These questions uncover a completely new dimension about sedimentary material and the erosional pseudotheories that exist in modern geology.

The samples in figure 6.3.3 are seen 40 times their original size. The first two slides are sands from the Rocky Mountains at Young Creek and the Colorado River in Arizona, USA. These sands are examples of actual eroded sediment caused by running surface water. They are referred to as **heterogeneous sand**, which is defined as a random mixture of eroded grains and rock fragments from the nearby mountains. It is easy to see these sands are not made up of the same mineral but are a mixture of many different minerals.

The other examples of sand reveal sands that are *not* heterogeneous! Notice the similarity of the sand grains from Coral Pink Dunes and Daytona Beach. It may come as a surprise to know that the *majority* of the world's sands are of this type. These types of sand are defined as **homogeneous sand** and exhibit similar grain size and mineral composition. These and other homogeneous sands will be discussed in detail in the next chapter. For now, we have the knowledge from our observations to ask two Fundamental Questions:

*If all sand is the product of erosion, why are many sand deposits **homogeneous** and do not contain a mixture of all of the minerals from the supposed parent mountain sources?*

*Why have geologists not recognized that most of the sand in the world is **homogeneous** sand?*

These two questions go hand in hand. Understanding that much of the world's sand is of a homogeneous nature is necessary if we are to ask the right questions about the sand's origin. If it is assumed that all sand comes from erosional processes at work today or similar but accelerated erosional processes in the past, the right questions about sand's origin might never be asked. The homogeneous nature of sand is Nature's way of encouraging us to ask those questions.

Fig 6.3.4 – These are the Imperial Dunes in Southern California, USA. The close up photo shows many clear quartz grains and some iron oxide grains. Where are all of the other minerals that should have eroded from the mountains in the background?

The Imperial Dunes Mystery

Look at the sand in the left column, second row down in Fig 6.3.2 and 6.3.3—the Imperial Sand Dunes of Southern California. These dunes are some of the largest dunes in North America. Notice how most of the grains in the photo are clear. This is because many of the grains are pure quartz. The black grains are mostly a mixture of hematite and magnetite, (iron oxides). Quartz sand has as density of only 2.6 g/cm^3 while the iron grains have a density of 5.2 g/cm^3, nearly twice the density of the quartz grains. This leads us to ask another FQ:

Where are all the grains of sand with densities between the quartz (2.6 g/cm^3) and iron (5.2 g/cm^3)?

This question takes on additional meaning when we compare the sand grains taken from the Imperial Dunes with sand grains taken from the nearby mountains, located toward the northeast of the dunes. The mountains can be seen in the background of Fig 6.3.4. At the base of these mountains, adjacent to the dunes, we collected sand from a dry creek bed. Both types of sand grains can be seen in Fig 6.3.3. (See the photos labeled Imperial Dunes and Imperial Dunes Mtns).

The mountains contain many minerals that are not found in the Imperial Sand Dunes. The most intriguing difference is the lack of magnetite in the mountain talus sand. Using a magnet, *no* iron minerals were accounted for in the rock fragments or talus sand collected from the nearby mountains, yet an enormous amount of magnetite (iron-ore sand) was collected from the Imperial Dune sand. How could this be if the dune sand came from the nearby mountains?

The granitic mountains far to the northwest contain large amounts of feldspar, yet the mineral feldspar is almost non-existent in the dunes. A trained geologist understands that there must be a source for the sand. What seems to have been overlooked in this case is the purity of the sand in the dunes. Where does a source rock of such purity exist? By all appearances, the sand that makes up the Imperial Dunes could not have come from the surrounding mountains, which contain a variety of minerals in addition to quartz and iron. The geologist has to ask, where are the mountains consisting *solely of quartz and iron minerals?*

We described the Imperial Dune sand as seen under a microscope to researchers at the University of Arizona, and asked them questions about the sand's origin. They could not give us an answer. We asked an Arizona Geological Survey geologist similar questions but he could not answer them either.

Why is it that modern geology does not know the origin of the largest dunes in the Western United States? It is easy to understand why they do not have answers because they are seeking them from within the paradigm of the Rock Cycle Pseudotheory.

The Crystal Clear Mountain Mystery

Refer again to Fig 6.3.3 and look at the two images of almost pure clear quartz sand. The first is from Daytona Beach, Florida and the other is from an area near Clearwater, Florida. These types of nearly pure, clear sands are quite common and can be found worldwide. Beaches of this sand have been admired for their beautiful white color, but they have much more to offer science than just beauty. Their origin holds keys to the understanding of our planet's geological past. As we will see in the next few chapters, sand is one evidence that is found everywhere on Earth that can help explain both the history of Earth and even human history.

The Daytona Beach and Cancun, Mexico, sands are found on beaches near where vast amounts of limestone are found. In fact, much of Florida and almost all of the Yucatan Peninsula in Mexico sits atop massive limestone deposits. The photograph of Cancun Limestone sand on the bottom left of Fig 6.3.3 shows how finely polished the limestone sand can be. Since limestone is the primary mineral in the area and occurs in great abundance, why are clear grains of quartz sand deposited in such huge quantities along these beaches?

Where are the crystal-**clear** mountains from which clear sands erode?

This is a simplistic yet profound Fundamental Question. Certainly, the beautiful clear sands found on beaches around the world have an origin. Just what is that origin? Geologists try to explain that the origin is through a complicated erosional process, but the process cannot be demonstrated. Clear, quartz mountains, like the artificial one shown in Fig 6.3.6, simply do not exist. If clear quartz grains eroded from mountains, there should be a trail of pure quartz sand of ever-progressing siz-

Fig 6.3.5 – Most sand dunes around the world do not have an origin that is clearly explained in modern geology. Erosion cannot explain the missing minerals that should be present.

es from the beach to the source. No evidence of this has been found and it does not seem to be taking place anywhere.

What is observed is that nearly pure, clear grains of quartz sand from many Florida beaches have no identifiable mountainous sources at all! There is no trail of clear sand leading anywhere. The surrounding landscape is flat and consists of limestone and clay, not clear quartz.

The "Strange Mounds" of Sand and Clay

Sometimes, scientists make observations and recognize their importance, even though they lack understanding. Thomas Gold made one such observation, which dealt with clay and sand mounds that did not have a clear origin. These 'strange' mounds, sometimes referred to as badlands, are found all around the world. Because modern geology has assumed there is only one source for clay and sand—erosion, few investigators have considered these mysterious formations. Wind cannot account for them; rivers or ancient seas cannot account for them; rain cannot account for them. Their origin remains a mystery:

"This is the presence of **clusters of earth mounds** that stand abruptly out of the alluvial plain. From a few feet to 40 feet in height and up to 200 feet or so in the horizontal dimensions,

Fig 6.3.6 – While this is not a real picture, it does make one wonder where all the clear sand crystals came from. Where are the clear crystal mountains from which clear sand grains eroded? Clear, crystal mountains do not exist so there must be another explanation for the origin of clear sand grains.

"The linear (or seif) dune is the most common of all inland dunes and the **most controversial in origin**.

"The great diversity of location, surface texture and wind regimes in areas of linear dune development suggests that **more than one mechanism may be responsible for the formation and growth of these dunes, and/or that a combination of two or more mechanisms may be operating at the same time.**"
Note 6.3f

This statement from the *Journal of Geology* represents early researchers' observations, which remain largely unnoticed today. The mechanism that formed many of the sand dunes around the world is not clearly understood. Textbooks suggest wind and water as the mechanisms for dune formation, but close observation of dune geometry and location do not always support this view. The shape and pattern of some sand dune areas do not always coincide with local wind and water patterns. Couple this with the absence or presence of unaccountable minerals and many dunes remain a mystery.

Fig 6.3.7 and 6.3.8 show examples of so-called "strange mounds" of sand and clay that exist on continents around the world. Fig 6.3.8 looks as if it should be in a science fiction novel, a landscape from another world. This is Goblin Valley, Utah, USA. The strange looking, almost eerie hoodoos are even stranger when we hear the explanation for their origin according to modern geology. The red and green clays in Goblin Valley could not have been laid down by a river, lake, or ocean. Different colored mounds are found next to one another and consist of different minerals. The mounds are homogeneous and could not have been formed by erosion. Erosion produces a heterogeneous mixture of sediment from the many types of minerals found in the surrounding mountains.

Erosion is a simple concept, but it cannot explain why red clay is found beside gray granite, as seen in the landscape shown in figure 6.3.7. This is another example of an 'unaccounted for' strange mound of red clay that exists at the foot of a granite

they are **composed internally just of the clay and sand** of the local alluvium, and **no good reason has been offered to account for their origin**.

"The association in both areas of **these strange mounds with locally concentrated seismic activity cannot reasonably be ascribed to chance**. While such mounds do occur elsewhere, dense clusters of them are extremely rare, and **an explanation for them is required**." Note 6.3e

An explanation for them is required, but it cannot be found in the modern science Rock Cycle Pseudotheory because they are thought to come from erosion. A paradigm shift is required before the mounds of clay and sand can be explained. One key to the origin of these mounds is noted in the previous quote. It has to do with seismic activity. There is a direct connection between earthquakes and sediments found on the surface of the Earth and we will explore this connection in detail in the next two chapters.

Many *sand dunes* also share a controversial origin:

Fig 6.3.7 – "Strange Mounds" and even mountains of clay exist all around the world, which cannot be explained by the Rock Cycle Pseudotheory. Left is a beautiful view of Yosemite in the Sierra Nevada Mountains, which are made up of primarily grey granite. However, descending from the park driving west into California, red clay can be seen. (Below) Where did the red clay come from?

CHAPTER 6 THE ROCK CYCLE PSEUDOTHEORY

mountain. How did red clay erode from the gray granite? Where did the red clay come from? Questions like these can be asked about every desert and mountain range in the world. Erosion cannot answer these questions.

The Sandstone Origin Mystery

What exactly is sandstone? Sandstone is simply sand grains that are cemented together to form solid rock. It is very porous and is found worldwide. Sandstone is one of the most easily recognized rocks and can include a variety of minerals. However, most sandstone consists primarily of clear grains of quartz that have been stained or colored by other minerals.

The definition of sandstone from Wikipedia, the online encyclopedia has this to say about the origin of sandstone:

"The formation of sandstone involves two principal stages. First, a layer or layers of sand **accumulates** as a result of sedimentation, either from water (as in a stream, lake, or sea) or from air (as in a desert).

"Finally, once it has accumulated, the sand becomes sandstone when it is compacted by pressure of overlying deposits and cemented by the **precipitation** of minerals within the pore spaces between sand grains." Note 6.3i

A student reporting on sandstone would likely understand this explanation to be fact. After all, this definition is the standard and widely accepted 'fact' in modern geology. However, our discussion about the origin of sand revealed that science has not been able to explain how sand of nearly pure quartz can accumulate in sand beds.

The next step in the formation of sandstone involves the precipitation of minerals within the pore spaces of the sand. This is as much a mystery as the accumulation of sand. Modern science has not identified how quartz or other minerals can precipitate out of natural rivers, lakes or oceans to form sandstone.

If the encyclopedia is correct and the formation of sandstone is settled, then there should be no debate. This is not the case. At a conference of the International Association of Sedimentologists in 2000, delegates from 12 countries met and heard this statement during the opening presentation:

"Despite quartz cement being the most important pore-occluding mineral in deeply buried (>2500m) sandstones, **its origin and the controls on its distribution are still subject to disagreement and debate.**" Note 6.3h

How long has the debate concerning the origin of sandstone been going on? A report by three sedimentologists at the world conference in 2000 said this:

"The origin of this [quartz sandstone] cement has been studied **for over a century** (e.g. Sorby, 1880) yet there still exists **considerable debate** as to which diagenetic processes are responsible for its **distribution and abundance**." Note 6.3h p129

Why has there been such a debate over the last 100 years? The debate is about the diagenetic process, which is the chemical, physical, or biological change of the sandstone. Sandstone deposits are extraordinarily thick in many areas of the world, but deep or thick sand alone, does not form sandstone. Sand does not naturally stick together. It is not melted together or we would see the evidence of heat such as glass connecting the grains. Another question is then about the origin of the cement and the cementation process. The cement is commonly iron oxide, calcium carbonate, or silica–quartz. What is the *origin* of the cementing agents?

Red sandstone is probably the most recognized and ubiquitous

6.3.8 – These mysterious clay and sand hoodoos are located in Goblin Valley, Utah, USA. Here we see unparalleled beauty and the hand of nature. This is an opportunity to sit back and wonder how erosion from wind, rain, rivers or shallow seas created this magnificent landscape. There is no evidence of rivers in the area. The red clay in the foreground is very soft and is not mixed with the green clay deposits. Truly, a completely new explanation for how this landscape formed will be required before we can ever begin to comprehend its origin.

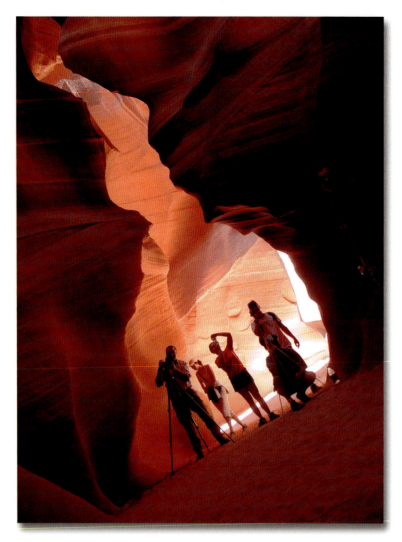

Fig 6.3.9 – Antelope Canyon in northern Arizona, USA. Photographers and researchers are mystified by these beautiful formations. What is their real origin?

sandstone known. It consists of clear quartz grains, cemented and stained by iron oxide and is found in layers thousands of feet thick. Ask yourself this simple question:

Where are the original red mountains from which red sandstone came from?

Another, more important Fundamental Question is this:

How do 100-meter thick layers of homogeneous uncontaminated sandstone form?

There are no soil layers in the massive, thick sandstone deposits, found on the Colorado plateau. Most deposits do not have pebbles, cobbles, boulders, or petrified trees. Weather driven erosional processes bring organic material into streams, rivers, lakes, and oceans. The sandstone layers of the Colorado Plateau are not contaminated with organic material. Another FQ we can ask is:

Where can we find sand dunes that are 2500 meters (1.5 miles) high?

Sand dunes of this magnitude are not found on the Earth today. One commonly held idea, as sometimes told in geology textbooks, is that sandstone was formed in a shallow sea. How does sand accumulate in a shallow sea and eventually become hard? The first problem that can be identified is one of contamination. A deposit in a shallow sea would be contaminated with organic matter, salts, and other contaminants. What is the mechanism that would allow a shallow sea to transform sand into sandstone? No known oceanic basins exist where sandstone formation is being observed today.

The mechanism for creating sandstone on a massive scale has n*ot been seen or demonstrated*. The uniformity theory would have us believe that massive sand formations accumulated slowly over time. The uniformitarian adherent should answer this Fundamental Question:

Where are massive sandstone deposits forming today?

The truth is there are no new sandstone formations being created today.

According to geologists, the processes where sediment or other material is transformed into rock is called **lithification**. The lithification process is explained as being a destruction of porosity (pores in the rocks) through compaction and cementation. Understanding this process has been a difficult challenge for geophysical scientists to explain because all current ideology is based on the presumed cycle through which all rocks pass, which cycle is ultimately based on magma.

The Rock Cycle Pseudotheory cannot be taken too seriously if the process of lithification cannot be demonstrated. In spite of many attempts over the past century, this has yet to be shown–until now. In the next chapter, the Hydroplanet Model, we will demonstrate how sandstone can be made. But first we must continue to identify other mysteries that have defied explanation in modern geology.

Fig 6.3.10 – This is a close up of a unique piece of multi-colored sandstone surrounding a quartz crystal. How does loose sand become cemented together with silica to form sandstone? Pressure alone cannot do it. Water does not do it. This is referred to as the Lithification Anomaly, a mystery in modern geology.

The Colorado Plateau Sand Mystery

The formation of the Colorado Plateau is one of the biggest mysteries in modern geology. The Colorado Plateau covers an area in the Southwestern USA of over 130,000 square miles (337,000 km^2). Elevations range from 1,200 ft (370 m) within the Grand Canyon to 12,700 ft (3,870 m) in the La Sal Mountains. Among the many natural resources and features found on the Colorado Plateau, the most recognized and most abundant are the sandstone formations. World renowned for their beauty, the sandstone arches in Arches National Park, the sandstone monoliths of Zion National Park, the deep red sandstone plateaus of Moab, Utah, and the thousand-foot sandstone cliffs of the Grand Canyon, are a few of the treasures found there. The big question geologists ask when studying these formations–what is the origin for all of the sand?

"Traditionally, geologists have looked at a sandstone's grain types to discern its rocky parentage. Other clues, such as which way the wind or water that deposited the grains was flowing, pointed them in the right direction. **But for some rocks, such as the ancient dunes of the western United States, these methods cannot narrow the possibilities much**." Note 6.3i

Looking for parental source rocks, geologists have recently established a new *theory* to account for the sandstone that covers a huge area of the western United States, which until now has been a mystery:

"**Until now, the origin of the sands that covered approximately 350,000 square kilometers of the western United States and solidified into sandstone between 150 million and 300 million years ago has been a mystery**." Note 6.3i

Fig 6.3.12 – How did nature make this interesting non-layered sandstone landscape in the Valley of Fire, Nevada, USA? No clear answer exists in geology today.

What was this new theory?

"Like many of their tourist visitors, **some of the rocks that make up the Grand Canyon came across North America from the East Coast**, a new study reveals." Note 6.3i

Because researchers have been hunting for 'parentage' rock and for answers to the Colorado Plateau sands, they have had to look far beyond the Plateau to find anything close to an answer. Their hypothesis required great imagination to explain how uniformly sized sand grains from the East Coast of the US could be carried all the way over the southern Rocky Mountains of New Mexico and Colorado to be deposited on the Colorado Plateau:

"Now Dickinson and his Arizona colleague George Gehrels have discovered that around half of the sand was once part of the Appalachian Mountains, thousands of kilometers away. They **propose** that **huge rivers carried the sand westwards,**

> Where are the solid red rock mountains that eroded, and that are eroding, to make up all the beautiful red **sandstone** formations?

Fig 6.3.11 – The Colorado Plateau in Northern Arizona. Many true and fictional stories about Cowboys and Indians are told in the setting of the Wild West. Where are the true stories of how this red sandstone country was made? The Uniformity Principle cannot answer this because no red sandstone mountains are being formed today.

depositing it on an ancient shoreline in Wyoming, from whence winds blew it south into the dune fields. Note 6.3i

One would think that a theory such as this would be controversial, but it seems it is not:

"Geologists are **enthusiastic** about the results. 'I'm very pleased,' says Bob Dott, emeritus professor at the University of Wisconsin-Madison who has studied the geological history of the western US for decades. '**The big question was where did all the sand come from**, and this paper has been the first to document it with **hard data**.'" Note 6.3i

It seems the origin of the Colorado Plateau sand had been a mystery for so long, any theory, no matter how incredible, is enthusiastically received. What is the hard data cited by the researchers? It was not the discovery of an ancient river. No new discoveries of sand deposits in Wyoming have been made. The so-called hard data comes from the mineral zircon, which is supposed to crystallize from molten magma:

"...zircon, a uranium-based mineral, in the sandstones. As soon as zircon '**crystallizes from *molten magma***' gave *dates* for "around 1.2 billion years ago or around 500 million years ago." Note 6.3i

The dating of sandstone minerals is based on its origin being from theoretical magma. The dates are then compared to other theoretical dates from surrounding mountains. The closest mountains to which the researchers could link their zircon dates were the Appalachian Mountains in Eastern USA. One problem with this theory is the assumption that the sand *came from magma or that it was eroded from mountains*.

The Classic Geologic Anomaly

Richard Allen, author, geologist, and tour guide in Phoenix, Arizona has an interest in geology and rock formations and is well versed in the magma and rock cycle theories. However, he is puzzled about a formation commonly found in the local red sandstone and says its:

"...formation stretches my imagination, as well as that of other geologists." Note 6.3j

Dubbed a "classic geologic anomaly" by Allen, he asks an excellent question about the boulders found in the red sandstone in Fig 6.3.15:

"Here was a **classic geologic anomaly**. How is it, that this big rock could have been deposited right down into the depth of the fine sand? **Think about it**. Sand like this is usually laid down by relatively slow moving water, or maybe even wind, as in sand dunes. **How did the heavy boulder get carried into this setting and just dropped off, before being buried by even more sand**?" Note 6.3j

The mountain in the background of Fig 6.3.15 is Red Mountain, which is located in Mesa, Arizona. Besides being at a loss to explain the boulders imbedded in the sandstone, geologists have no solid explanation for the origin of the red mountains in Mesa and the surrounding Phoenix, Arizona area.

Red sandstone hills and mountains follow a narrow corridor running through the Phoenix area. Red Mountain and other popular hills such as Camelback Mountain and Papago Park Mountains are included in this group. How did these mountains form? The surrounding mountains are rich in many minerals, but not red sand.

Where did Red Mountain come from?

The more we think about anomalous geology like this, when viewing the world from within the Rock Cycle Pseudotheory paradigm, the more confusing it seems to be. Modern geology has only a weak explanation for ordinary sandstone. Sandstone formations with inclusions like those in figure 6.3.15 are found in many places worldwide and truly stretch the imagination of the geologist. They will continue to do so until we discover their true origin.

The Sand Crystal Mystery

How does ordinary sand become a large and beautiful sand crystal? Fig 6.3.16 is an example of these mysterious novelties of Nature, sometimes referred to as desert rose. They are mysterious because it is not known how they form.

The delicate looking crystals are actually quite hard and are made of gypsum or barite salts. Researchers do not know how gypsum is made. There are no large deposits occurring anywhere in the world today. It would be nice if scholars could point to and say; 'this is how this or that salt deposit forms, and we can see it in process right here.' Because this is not the case

Fig 6.3.13 –This layered sandstone landscape is in the Valley of Fire State park in Nevada, USA. How could an erosional process deposit both red and yellow sand layers next to each other without a mixing of the layers?

and because there is a lack of real observational data, geologists can only infer or theorize. When those theories are based on the wrong premise, confusion and mystification are the result.

The Mudstone Mystery

Fig 6.3.17 depicts two examples of mudstone combined with sandstone. Nearly pure white grains are embedded within various colors of red clay and sand. It defies the imagination how white and red sand particles where brought together by aeolian or fluvial processes and did not intermix.

Perhaps these deposits are the work of complicated and imaginative depositional environments. Whatever the case of their origin, it is a mystery to modern geology. Look at Fig 6.3.18. These rock samples resemble potsherds from an ancient civilization. They are not painted by human hands. They are the handiwork of unknown natural processes from Western Australia. They are called **Zebra Rock**.

The Zebra Rock Mystery

Zebra Rocks are intriguing rocks. The striking detail and contrast between the different colors of the stone amazes everyone who sees them. Zebra Rock (Fig 6.3.18) consists of patterned siltstone (siliceous argillite) in an undulating-like pattern. The finely grained stone is colored by ferric (iron) oxide. Like mudstone, geologists have no valid explanation for the origin of Zebra Rock. A distributor of these unique rocks describes them:

"The origin of this fined grained silt or clay stone, called Zebra Stone with its characteristic red banded pattern **has baffled geologists since its discovery**. The only deposits in the world are in the East Kimberlys, Western Australia. **It is not known how the regular patterns were formed** but the red stripes are colored by iron oxide." Note 6.3k

A process as simple as mud deposition should be better understood. The only logical conclusion is that sandstone, siltstone and mudstones came from depositional processes that are not currently understood and completely different from the ideas held today in modern geology.

The Wonderstone Mystery

Sandstone is a 'sedimentary' rock and scientists describe it as forming in layers. From the book *Rocks, Minerals & Gemstones* we read:

Fig 6.3.14 – The beauty and extent of the Colorado Plateau can best be seen from the air. How did thousand-foot-tall cliffs form in canyons where water does not consistently run through them? If a shallow sea left these sands, how were pure, homogeneous white and red sand grains deposited and why didn't they mix? How were these sands lithified? These basic FQs should have been answered long ago by sedimentologists who claim to understand the Colorado Plateau.

Fig 6.3.15 – Large boulders included in red sandstone (inset) are from a road-cut near Red Mountain, Mesa, Arizona (seen above). One geologist calls this the "classic geologic anomaly" asking, "How did the heavy granite boulder get carried into this setting". Red Mountain is without an origin in modern geology.

Fig 6.3.16 –These beautiful sand crystals include ordinary sand in the naturally grown gypsum crystals. They are found worldwide in arid sandy locations. They are not known to be forming today. How were they formed?

"Sandstones are **always layered**." Bib 15 p272

In Fig 6.3.19, several examples of a type of sandstone called **wonderstone** are shown. This sandstone comes from near Kanab, Utah, USA. It looks almost surrealistic when first viewed. Enormous formations containing seemingly every color in the rainbow exist. How did they form?

The lower right image in Fig 6.3.19 is a close-up of the sand grains that make up the wonderstone. Veins of brown hematite are seen running through and defining areas of the sandstone. Understanding the origin of the hematite is vital to understanding why the sandstone exhibits so many colors

There has never been an explanation for how sandstone can form in vertical formations. Geology says sandstone could have been produced when natural waters percolated down from the surface. In fact, the opposite is true—horizontal and angled layers in many cases can be seen in natural formations around the world. Quartz cementation of sandstone and mudstone has not been observed in Nature and has nothing to do with time. It is simply a matter of environment. As we will see, with the proper environmental factors in place, silica cementation does not take days to occur—it only takes hours to make beautiful formations like these.

The Cross-Bedded Sandstone Mystery

In Fig 6.3.20, an example of cross-bedded sandstone shows the intriguing crisscross pattern that led many geologists to speculate that it was of sand dune origin. This mystery has been around for a long time. From a 1931 publication, Grand Canyon naturalist Edwin D. McKee says this about the steep-angled, hardened sand dunes:

"The light-colored formation which appears as a conspicuous ribbon-like band around the upper part of the Grand Canyon **has long presented a puzzle concerning its origin**. The grains

Fig 6.3.18 – This unique rock is called Zebra Rock and is from Australia. Circular columns of fine red claystone are evenly distributed throughout a white claystone. Zebra Rock has "baffled geologists" since its discovery.

of white sand of which it is composed apparently were deposited at steep angles, for the many and varied slopes which were formed may be readily seen today on the surface of the rock. These slopes were **probably once the lee sides of sand dunes** deposited by winds in an area bordering the sea." Note 6.31

One reason the origin of cross-bedded sandstone has been such a puzzle is that no one has seen anything like it being formed today. No process has been observed which can explain its formation. The cross-bedded sandstone appears to be similar to sand dunes and so the idea was put forth that they must have formed slowly over time.

Refer again to Fig 6.3.20. Notice the cross bedding and the wavy formation toward the top of the image. The inset image is that of a present-day sand dune. How did both, the cross-bedded layers and the bent layers form in the example of the sandstone? They do not exist in the sand dune. It is a mistake to think that

Fig 6.3.17 – These mudstone-sandstone deposits consist of various colors of clay and sand. They exhibit no sedimentary layering. They are found worldwide and have been a source of wonder for both traveler and scientists alike. These examples are from Kane Canyon, Utah, USA.

Fig 6.3.19 – These may look like fine art paintings but they are not. They are slabs of natural sandstone called 'Wonderstone'. These come from southern Utah, USA. The lower right image shows a close up view of the naturally colored sand grains in a magnified section of one the Wonderstone specimens. The dark brown vein is the mineral hematite (iron oxide).

processes that are taking place today could have created the interesting fluid-looking formations in the sandstone. Hutton's uniformity myth has kept science in a dark age when it comes to understanding formations like these. It is apparent that the uniformity principle has failed to answer many of the pressing questions we have asked.

The mystery of cross-bedded sandstone can only be understood by reconsidering how they must have formed, and that they formed in an environment and by processes completely different from those observed today. This same idea holds true for all of the sand mysteries we have discussed thus far.

The Sand Mystery Summary

Only the tip of the 'Sand Mystery' iceberg has been presented here in the UM; we simply cannot fit it all into one book. Enough examples have been presented for the reader to experience the tremendous number of unanswered, sand related questions in modern geology. Sand and clay are at the heart of the sedimentary processes of the Rock Cycle Pseudotheory. The origin of the natural formations of sand and clay *must be clearly understood* if we are to comprehend Nature's **real Rock Cycle**. The trend of unanswered questions continues as we move through the chapter. About 70% of the Earth's sand is quartz and so we will continue to follow this ubiquitous mineral quartz in the next section and in many sections throughout the book. In Chapter 5, we learned that quartz is not glass, but that was only the beginning of the mystery. There is much more of that story to be told.

Fig 6.3.20 – Cross-bedded sandstone (background) and present-day sand dunes (inset) from Southern Utah, USA. Note the wavy, folded formations near the center of the sandstone. Geologists have never demonstrated how these or the cross-bedded layers are formed. If they were sand dunes, how were they buried? Once buried, how was the sand dune lithified (turned into stone)?

SUBCHAPTER 6.3 THE SAND MYSTERY

6.4 The Quartz Mystery

The Newton Principle states, "We may imagine things that are false, but we can only understand things that are true." Science has imagined that all sand comes from the erosion of rocks by the forces of wind and water. Whether the sand is loose on the beach, piled in a dune or entombed in thousand-foot-thick monoliths, it consists mostly of clear quartz, stained or colored by contemporaneous minerals. From where did the quartz come? To understand the truth about quartz, we will examine many of its forms and properties. Modern science has supposed, because of the Rock Cycle Pseudotheory, that quartz comes from a 'melt' and that melt cooled slowly enough to crystallize.

Petrified wood, fossil bones, geodes, and chalcedony are a few of the many quartz rocks whose origin defies understanding in modern geology today. If we discover the true origin of quartz, a host of unanswered questions in geology will suddenly become clear. In this subchapter, we will explore some of the mysteries surrounding quartz.

Quartz With No Known Origin

In the Magma Pseudotheory chapter, we presented evidence showing why magma is a myth and why natural quartz did not come from a melt. It is important to realize that the science of geology has been unsuccessful, from the magma perspective, in producing most of Nature's minerals. Granite is considered a plutonic rock (cooled magma-sourced rock made deep in the crust). It often exhibits large crystals of quartz and feldspar. For over a century, geologists have tried unsuccessfully to make granite from a melt:

"Plutonic textures have **not** been duplicated in the laboratory, however. The complete crystallization of the interstitial liquid as large crystals **has not been achieved in granitic melts.**"
Note 6.4a

Any rock that includes quartz as part of its makeup could not have come from a melt because natural quartz does not come from material in a molten state. Because modern geology has snubbed other explanations of how quartz is grown naturally, the modern science Rock Cycle Pseudotheory is left with many unanswered questions and mysteries. In this subchapter we consider some of the quartz formations and how their origins remain unknown today.

From the many anomalous rock formations consisting primarily of quartz, we have selected seven for consideration in this chapter. Modern geology has attempted to explain that all minerals are essentially formed from magma, through processes explained in the pseudotheory of the rock cycle. These geological formations remain mysterious and their origin misunderstood. How these seven quartz-based rock types formed will become more clearly understood in the next two chapters, where we will begin to learn of the critical role that water plays in all mineral formation.

1. Agates and Geodes
2. Surface Chalcedony
3. Limestone Quartz Chert
4. Dike Intrusions
5. Pegmatites
6. Granites
7. Double Terminated Crystals

Fig 6.4.1 – The foreground rock is a septarian nodule with a bone inclusion. A large dinosaur bone is in the background. The bone is filled with agate. Fossil specimens like these have forced researchers to accept the fact that geodes and nodules with large calcite and quartz formations around the world are surface phenomena. They were formed on or near the surface in a manner that is not known to geology today.

1. **The Agate and Geode Mystery**

Agates and geodes are exquisitely beautiful rock formations, commonly displayed for sale in rock shops around the world. Both are made primarily of quartz. **Agates** are generally solid, microcrystalline quartz, often exhibiting banding or layering. Various mineral impurities present during the formation of the specimens contribute to a wide range of colors. **Geodes** are similar to agates but have a hollow center and include fantastic crystals of many colors. They are tremendously popular and have been used for centuries as adornments by many people. Each geode is unique, and when opened, reveals a complex and delicate creation of Nature. Fig 6.4.2 illustrates several examples of geodes from Brazil.

Fig 6.4.2 – These colorful geodes are from Brazil. Geodes like these are found worldwide. Like so many of the quartz-based rocks, the origin of geodes and agates remains a mystery in geology today. Why is the process that forms agates and geodes "enormously complicated," as the expert's state? The modern science rock cycle does not have the answer.

With agates and geodes found worldwide, and being so popular, one would think that their origin would have been discovered long ago. However, as one expert we talked with states:

"'Agates are **enormously complicated**... Even in mineralogical terms, **this is complicated stuff—or it would have been solved a long time ago**.'" Note 6.4b

Is the origin of geodes really that complicated, or have geologists overlooked clues that might provide real answers?

"Geologists have **proposed** several theories to explain the conditions and processes **that form geodes, but none seems to be entirely adequate to explain all geode features**...

"As limey sediments accumulated in shallow midcontinental seas, rounded cavities that are characteristic of geodes could not have existed at the interface or contact of water and sediments. Nor could they have existed during the earliest stages of sediment compaction and cementation. Therefore, **some feature** of a different texture than the host limestone had to be present. This feature either caused geodes to form or was transformed into a geode." Note 6.4c

A *very different environment than is currently envisioned by researchers* must have been present for the complicated process of making agates and geodes. If not, the process would have been solved long ago. The observation that agates form 'close to the surface of the Earth' has led to confusion:

"'**We know agates form close to the surface of the earth**, at low pressures and temperatures,' Heaney says, 'and not only in volcanic rock, **but in dinosaur bones**.'" Note 6.4d

The confusion comes because scientists have thought that low pressures and low temperatures were the conditions in which agates form. They reasoned that because fossils, such as dinosaur bones (see Fig 6.4.1) or petrified wood are made of agate, and because they must have formed near the surface, the temperature and pressure must have been low. However, it must be remembered that the mineral we are talking about is *quartz*:

"Studying agates with transmission electron microscopy and by x-ray diffraction, Heaney found that **90 percent of an agate is quartz**." Note 6.4d

To understand how agates and geodes form, we must understand how quartz is formed. In the next chapter (The Hydroplanet Model), we will examine the environment in

Fig 6.4.3 – Geodes are found only on or near the surface of the Earth. If the geological time scale was true, geodes would have formed during many different times and would be found in many different layers of the crust. Since they are not, the environment in which they formed must be very different from the one alleged by modern science.

SUBCHAPTER 6.4 THE QUARTZ MYSTERY

Fig 6.4.4 - This is a typical specimen of surface chalcedony found in place where it likely formed (in situ), in the Arizona, USA desert. Throughout the world, rocks like these grew directly on the surface. We know the rocks formed on this desert floor not long ago because of the **lack of erosional damage** to the specimens.

which natural quartz grows. Although quartz does not come from a melt, it *does require* a high pressure and a high temperature environment. With this new understanding, we can begin to comprehend how agates and geodes were formed, and are found on the surface of the Earth. Lacking the wisdom of how quartz is grown, one could not make an agate and so far, no one has:

"'You'd have to make an agate, and **no one has ever made an agate**...'" Note 6.4d

2. The Surface Chalcedony Mystery

Chalcedony is a form of agate that usually has a milky or wax-like luster. Like most agates and geodes, it is also quartz based. Chalcedony is a cryptocrystalline form of silica, where

Fig 6.4.5 – Surface Chalcedony as it grew in place on the surface of a rock. This whitish, hard quartz-based mineral is found worldwide, directly on the surface of the ground. It is commonly found where there has been little or no erosion over the past several thousand years, typically in dry desert climates.

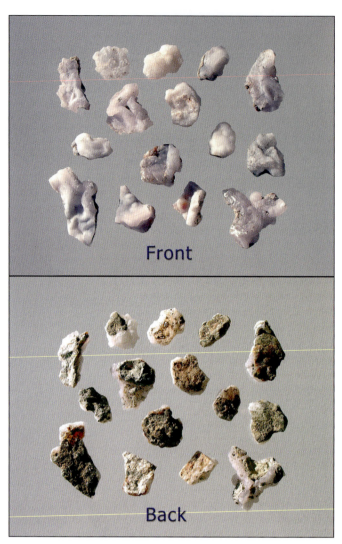

Fig 6.4.6 – The Surface Chalcedony Mystery consists of these swirly quartz rocks formed on the surface of the Earth with no origin in modern geology. They are not forming anywhere in the world today—how did they form? The specimens shown have been completely washed. Notice how the backside (bottom) of each piece is rough where it was embedded in the soil or on the rock upon which it was formed. The embedded rock or soil matrix does not usually wash off. In some cases it does, leaving an imprint on the back of the chalcedony specimen.

the crystalline structure is so minute that it is difficult to discern even microscopically. It is found in many geological settings. As far as we know, this important geological mystery has previously gone largely unnoticed. The outer walls of the geode in Fig 6.4.3 are made of chalcedony. **Surface Chalcedony** is a form of chalcedony found on the surface of the Earth, often on unconsolidated materials.

What makes this mystery unique is how simple it is to find and observe this rock in its natural setting. This rock can be found worldwide, and anyone that goes looking should find evidence of Surface Chalcedony. In dry desert climates where precipitation is low and erosion is less common, there is ample opportunity to find this interesting mineral. Fig 6.4.4 is an example of Surface Chalcedony found in eastern Arizona, USA. Although chalcedony is usually milky-white, colors can vary widely with the addition of other minerals that were present during its formation. Many rock collectors prize these chalcedony specimens because of their beauty and swirl-like nature.

The simple nature of this beautiful mineral and the environment where it must have formed has gone largely unrecognized by the modern geological community. One professor of geosciences, Peter Heaney, has been dubbed by his peers as one of the world's foremost authorities on agate geochemistry. When asked if he knew about Surface Chalcedony, he said he had seen specimens, but had personally never seen any in the field. This seemed quite amazing considering he has published more on agates and geodes than anyone else. Although thousands of these specimens have been collected, their significance has not been recognized.

Fig 6.4.7 - Limestone-quartz nodules are found worldwide, near the surface of the Earth. How do they form? Quartz-containing limestone nodules like these are not being formed anywhere in the world today.

Fig 6.4.5 is a fine example of a complete white chalcedony swirl (most specimens are only part of a complete swirl) formed directly on a rock. This helped keep the chalcedony specimen intact. Similar deposits can be found inside cracks and voids in the rocks or any other open space where the silica responsible for the formation of this unique crystal collected and grew.

The scientific importance of Surface Chalcedony can be appreciated by answering the following FQ:

> How can chalcedony be formed on the **unconsolidated surface** of the Earth and still be resting in its original position, if it formed millions of years ago?

We remember from the Answer Principle that answers come by asking questions. Because the modern geologist would not likely ask such a question, there is no reason to expect them to have an answer. However, now that we have asked the question, let's consider the science behind it.

Fig 6.4.6 is a fine assortment of Surface Chalcedony collected in Arizona, USA. The front or topside of each specimen is very different from the backside. Darker rocks and minerals are solidly attached to the crystalline silica mineral. The white upper or top side of the specimens grew in an open environment, allowing the unique open banded appearance. The bottom of Surface Chalcedony was attached to the matrix, often loose and unconsolidated; it could have only formed directly on it.

When cataloging the observed specimens, an interesting pattern begins to emerge. In rural, dry desert areas where the ground is little disturbed by animals and weather, *the white side of the chalcedony specimens is found facing upward*. This is how they laid *when they were formed*.

If the chalcedony specimens had been formed somewhere else and were then transported through some erosional processes to the place where they now lay, there would only be a 50/50 chance of finding the white side up. Erosion would also have rounded the corners of the specimens or broken them altogether. *This is not what is observed* when specimens are collected at undisturbed areas.

There is other direct evidence that this type of chalcedony is a surface phenomenon. Like geodes, surface chalcedony is not found throughout the geologic column of the Grand Canyon. Surface chalcedony is only a surface phenomenon and understanding how they formed is an important clue to understanding the processes that shaped the planet.

Incidentally, if you find yourself in an area where Surface Chalcedony exists in an upright (undisturbed) position, you may be walking on a surface relatively unchanged for thousands of years. A surface unchanged from the time when the conditions existed in which rocks like Surface Chalcedony were formed. Those conditions will be discussed in Chapter 8, the Universal Flood Model.

3. Limestone-Quartz Chert Mystery

Limestone is a carbonate rock ($CaCO_3$) whereas quartz is a silicate (SiO_2). Another of the great mysteries in geology is how the two of them form in the same proximity. It is even more puzzling when the two combine to form rocks like those seen in Fig 6.4.7. In the image, the colored, quartz-based core of the nodules can be clearly distinguished from the carbonate outer shell. This type of quartz is a densely packed, microcrystalline form of quartz, commonly called chert. Nodules of chert are found within limestone, relatively near the surface of the Earth.

Flint is quite similar to chert and is found in chalky or limestone layers. Flint, like chert is microcrystalline quartz. It has been used by man for thousands of years to make spear points, knives and other tools because of its unique ability to break conchoidally, leaving extremely sharp edges. Under a skilled hand, it can be fashioned into beautifully shaped arrow points. Flint was widely used in the manufacture of early firearms because of its ability to create a spark when struck against steel.

How rocks like Flint and Chert formed is a mystery—and the Rock Cycle Pseudotheory has no sound explanation for the origin of the chert and flint nodules. They are not forming on or near the surface of the Earth today, nor have they been found forming within oceanic sediments. Where does modern geology say they come from? The widely accepted sci-bi says chert

Fig 6.4.8 – This is a dike in the Sierra Nevada Mountains at Yosemite National Forest. How did this dike form from an igneous molten source and not melt the outside walls next to it?

they were factual is an abuse of authority.

Some publications give us a more accurate portrayal of what is known, and what remains unexplained, with respect to the origin of chert. The *Handbook of Rocks Minerals and Gemstones* says this:

"**Opinions differ** as regards mode of origin… The **origin of chert** nodules **has not been completely explained**." Bib 15 p292 and 294

4. The Dike Intrusion Puzzle

Modern geology defines a **dike** as "a long, narrow, cross-cutting mass of igneous rock intruded into a fissure in older rock." Bib 7 p226

In modern geology, the definition has been expanded to identify different types of intrusions. Those that parallel the rock layers are referred to as sills and are said to be concordant. Those that cross the rock layers are said to be discordant. In either case, they are believed to be igneous intrusions, which by definition means they originated from molten magma. However, you will remember from the Magma Pseudotheory chapter, that *magma is not lava* and that magma does not exist. The puzzle lies in understanding the true environment in which dikes like those in Fig 6.4.8 formed.

Fig 6.4.9 shows melted lava typically associated with volcanoes like those found in Hawaii. When the lava cools, its appearance is not at all like the rock seen in most dikes. Dikes are not composed of melted glass-like minerals. How does modern geology explain this?

The Uplift and Erosion Sci-bis

There are conflicting ideas in modern geology about how a dike can form. All of the ideas have one commonality in that they begin with a mag-

comes from organisms buried in the ocean floor, which undergo diagenesis, eventually becoming chert. We read this authoritative sounding statement in a geology textbook under the subheading, "Source for Chert":

"Like calcium carbonate, much silica sediment is precipitated biochemically, secreted as shells by ocean-dwelling organisms. When these organisms die, they sink to the deep ocean floor, where their shells accumulate as layers of silica sediment. After these silica sediments are buried by later sediments, they are **diagenetically cemented into chert**. Chert may also form as diagenetic nodules and irregular masses replacing carbonate in limestones and dolomites." Bib 59 p190

This explanation is purely theoretical. No observations have been made of chert in the process of forming diagenetically in Nature. No known lab-formed chert has been made, diagenetically or otherwise. Without empirical evidence or experimental data to back up a claim, stating theories like these as though

Fig 6.4.9 – Melted rock at the surface like this lava in Hawaii does not look like many dikes which are said to be melted igneous rock.

Fig 6.4.10 – Shiprock Mountain in New Mexico, USA and dikes. Notice the mountain and dikes are intrusions that have pushed up the sediment on each side of the dike in the recent past. Only a small amount of erosion has taken place. This can be seen by the lighter colored roads along the landscape that only recently have been made. Yet, when has geology ever observed dikes like these forming or a mountain like Shiprock rising above the surface? Hutton's Uniformity Principle fails again.

matically driven intrusive body. That is, all intrusions started out as heated, melted rock, deep in the Earth. As it rose, it found its way through cracks, veins and other weak areas in the crust, where it eventually cooled. Many questions remain unanswered in all of the ideas put forth by modern geologists. Questions like, where did the space come from for such large intrusions. How is the boundary between the intrusion and the host rock so defined? These and many other questions lead to confusion over the origin of the dike intrusions.

Most dikes are *believed* to have been formed deep in the crust. It is *believed* that the only way we can view them today is through uplift and subsequent erosion, which left the dikes exposed. See (Fig 6.4.8).

Other dikes, such as those radiating from Shiprock, a so-called volcanic neck, (Fig 6.4.10) are *believed* to have formed underground and then exposed, after long periods of massive erosion. Geologists report the Shiprock formation as being 27 million years old and that over 1,583 feet (482 meters) of sediment has eroded away over the millions of years. Note 6.4e

As we discussed in Chapter 5, there is a *lack* of scientific evidence to support the uplift pseudotheory, where continent-sized landmasses move vertically up or down, or where mountain ranges are formed over longs periods of geologic time. Without the support of the uplift pseudotheory, the associated theories of dike formation and exposition must be abandoned.

Further, the idea that 1,583 feet of surrounding sediment eroded away cannot be substantiated. There is no evidence of significant erosion around the peak itself. There are no deep valleys or riverbeds supporting the idea of massive fluvial transport. Simple precipitation events like rain or snow do not produce sufficient water to erode the hundreds of feet of overlying sediment. There is evidence of erosion, but only small-scale events as can be seen in Fig 6.4.10. Look for the black rock along the dike and the gentle wash located in the lower right of the image. The observed erosion suggests only a few thousand years of erosion activity.

Finally, the vast flat plains surrounding the Shiprock monolith are indicative of forces and events far larger than modern geology ever imagined.

Dike Walls Not Melted

Another puzzling factor is that dikes do not exhibit melting along dike walls where it is bounded by the host rock. In one modern geological theory, the molten material melted the host rock to create room for the intruding dike. This claim is unfounded because partial melting is not seen at the dike boundary. In some cases, there are clastic inclusions within the dike that are connected with the bounding host rock, but they show no signs of melting either. If magma was the source for the intruding rock, there should be signs of melting evident in the surrounding rock. This is rare, if existent at all. Referring to figure 6.4.8, the lighter colored dike cuts directly across the darker host rock, which shows no signs of melting or mixing with the intruding dike minerals. This would not be expected if the dike intrusion were molten.

Unanswered Dike Questions

Dikes that consist of many types of minerals are found in a wide variety of geological formations in every corner of the world. Yet there is no evidence they are forming today. What modern science must consider is that there is some other origin for the dikes. The Rock Cycle does not have answers about the formation of the dikes because it does not have the correct answers about the formation of quartz, the most common mineral found in many dikes. Most dikes are referred to as being either felsic, which are silica rich and lighter in color, or mafic, a darker silicate, rich in iron and magnesium. Silicate minerals, like those that make up the intrusive dikes, are formed in an environment similar to other silicate or quartz-based rock. Modern geology has tried to explain that quartz comes from magma but has provided no evidence that it does. To understand how dikes form is to understand how quartz forms. With this understanding, it will be possible to know how other minerals form like those found in formations similar to dikes—pegmatites.

5. Mysterious Pegmatites

Geologists have recognized that mineral growth often involves a hydrothermal process. They also have thought that most of the Earth's rocks involved magma. However, direct evidence has never been obtained for this inferred magmatic process. Scientists were hopeful that a type of deposit called a *pegmatite* would be the 'missing link' between theoretical magmatic deposits and observable hydrothermal deposits.

Pegmatites are often massive intrusions that propagate

throughout the crust. They are found in both concordant (sills) and discordant (dikes) formations. Fig 6.4.11 shows the large and beautiful beryl crystals (dark green beryl is known as emerald) that came from a pegmatite deposit in Buckfield Maine, USA. Although researchers have long known that if they could discover the origin of pegmatites, they could understand the origin of many of the Earth's minerals. So far, the mystery has not been solved:

"**Despite the extensive study, pegmatites have yielded little information** about the transition from magmatic to hydrothermal fluids and in recent years interest in these deposits has waned."
Note 6.4f

Although interest in the pegmatite deposits may have "waned" in modern geology, they have proven to be of great interest in the Universal Model.

Pegmatite Evidence for UM

Why did the pegmatite deposits not provide the answers to mineral formation that modern geology had hoped? The answer was not there because pegmatites contain abundant quartz, and like all quartz-based rocks, their origin is not from magma.

There are *many* natural minerals, which mineralogists would love to discover the origin of:

"... muscovite, biotite, apatite, lepidolite, spodumene, beryl, tantalite, columbite, cassiterite, petalite, tourmaline, pollucite, zinnwaldite, uraninite and other minor minerals too numerous to mention."
Note 6.4f

The fact that all of these minerals are in pegmatites is very important. Geologists recognize that the pegmatites were formed as "envelopes." Thus, the *same geological process* involved in making the quartz was the same process that *made all the other minerals in the pegmatite, and the processes happened at essentially the same time.*

Because we recognize that quartz does not come from a melt, and because all of the other minerals found in the pegmatites were made by the same process as quartz, these other minerals also *did not come from a melt. In other words, they did not come from magma.* There is a different process in which all the minerals were made. This process is one of the topics in our next chapter, the Hydroplanet Model.

Fig 6.4.11 – This is the Orchard Quarry pegmatite deposit in Buckfield, Maine, USA. The map shows a cross-section pipe deposit with green beryl locations seen throughout the vertical pipe. Beautiful beryl crystals can be seen in the inset photos. Pegmatite formations are important because they reveal how the vast majority of the Earth's minerals are formed. Instead of originating from a 'melt' or magma, they were formed in an environment involving heat, pressure, and *water*.

6. The Granite Mystery

One of the most common rocks in the Earth's crust is granite. It is the most abundant basement rock, and underlies much of the continental crust. According to the Rock Cycle theory, granite is an igneous rock formed from cooled plutonic magma. It seems the idea of how granite is formed is perfectly satisfied in modern geology, or is it:

"These are matters I must continually return to, but clearly **the full understanding** of this gassy, multicomponent system representing **real granites lies ahead in difficult experimental terrain,** bedevilled by the problem of representing the results in any easily comprehensible form." Note 6.4g

As is often the case, trying to understand and devise experi-

Fig 6.4.12 – This lighter specimen of granite is sitting on a polished black granite slab. Both granites contain quartz crystals, feldspar, biotite mica and other minerals.

ments to explain the origin of rocks through incorrect processes can be incredibly frustrating:

"The generation and evolution of **granitic magmas** are so much matters of physical chemistry that **we might hope** that the processes involved will **eventually be fully understood.**" Note 6.4g p32

One of the biggest, perhaps unknown, problems facing researchers is that the primary constituent of all granite is *quartz*. The Rock Cycle theory says granite is formed from magma, therefore the quartz in the granite must have also come from magma. We remember from earlier discussions that quartz cannot come from a melt. Research into the history of geology has revealed that scientists knew quartz was formed differently than is believed by modern geology. Early researchers discovered that a molten, dry melt (magma) process does not produce quartz, but that there was another way that did. A 'wet' process—precipitation from heated water does indeed produce natural quartz. Davis A. Young, a well respected geologist, in his 2003 book *Mind Over Magma, The Story of Igneous Petrology* shared an excellent history of the origin of granite:

"Several arguments were leveled against the idea of dry fusion **origin of granite**. These included the claims that the order of mineral crystallization in granite excludes an origin by fusion; that quartz-bearing volcanic equivalents of granite do not exist; that **experimental evidence confirmed the production of quartz only in the 'wet way'**..." Note 6.4h p82

Young recognized that geologists have observed that slowly cooled lava does *not* make quartz:

"Scheerer (1847), for example, **observed that not even very slowly cooled lava flows erupted from the volcano Jurullo had produced quartz**. Noting that although volcanic products like obsidian and pumice have granite-like chemical compositions, he pointed out that **they do not contain free quartz**." Note 6.4h p85

Young further reflects upon the history of granite research and touches on another important point in the history of quartz research—one that opponents of the dry fusion quartz origin stressed; no one had created quartz from a melt. However, several investigators had success forming quartz using heated water:

"A third argument **against the production of granite by dry fusion stemmed from experimental evidence**. Opponents of dry fusion insisted that no one had ever produced granite by fusion in a furnace despite the assertion of James Hall that he had recreated substances with stony texture from slowly cooled melts. **Nor, opponents alleged, had anyone crystallized quartz from silicate melt. In contrast, several investigators formed quartz from hot water**, corroborating the idea that granite was produced in the 'wet' way. In 1845 Schafhäutl showed that water vapor heated above 100°C dissolves silica. Upon cooling the vapor, Schafhäutl precipitated hexagonal dipyramidal crystals of quartz." Note 6.4h p85

Although observations and data from experiments were had by the researchers, logic and reason did not prevail, as is often the case. The idea of magmatic granite continued to gain 'widespread acceptance':

"The nineteenth-century **debate over the origin of granite** largely fizzled out with the dispute between Hunt and Forbes in the 1860's. In the 1870's, the growing acceptance of both the results of microscopic petrography and the solution theory of magma also led to **widespread acceptance** of igneous, eruptive, **magmatic granite.**" Note 6.4h p103

The idea that granite is of magmatic origin still holds sway in modern geology even though the debate was never really settled. No observational data has come to light to support the now widely accepted view and:

"**Into the 1860's, no consensus had been reached**, and the views of geologists about granite played a major role in the manner in which they **classified the entire spectrum of rocks.**" Note 6.4h p103

We will discuss more of this debate in the Hydroplanet Model chapter. Was there an agenda among the scientists who did not want to accept the experimental data of a 'wet' origin of granite and other similar rocks? Whatever the reason, because modern geology has chosen to follow the magma road, the true origin of granite has not yet been discovered.

7. Double Terminated Crystal Mystery

Quartz crystals generally grow from the surface of surrounding host rock. See Fig 6.4.13 for an example of quartz grown on surface matrix. Quartz grows similarly inside geodes and in veins and pockets in the host rock. Crystals require open space if they are to grow as individually defined crystals, exhibiting their unique geometric habit. When crystals break off the base from which they were attached, they look like those in Fig 6.4.14. At least one end of the crystal is broken or fractured where the crystal was separated from the growth surface.

Some crystals have been found where the crystal growth occurred at *both ends of the crystal*. These crystals show no evidence of being attached during growth. These types of crystals are referred to as **Double Terminated** crystals. See an example of this type of this special type of quartz crystal in Fig 6.4.15. The faces on each end of the crystal are where the crystal lattice was added upon while it was growing in a free-floating environment. Notice there are no broken ends or rough 'attachment' scars on the crystals. Consider this significant FQ:

How do double terminated crystals form in Nature?

What was the physical condition in which these double terminated quartz crystals grew? This is a mystery in geology to-

Fig 6.4.13 (left) – Quartz crystals generally grow from a surface rock. When crystals are broken off the base, the broken end is rough and fractured and does not exhibit the clean geometric habit of the unattached end.

Fig 6.4.14 (right) – These quartz crystals were found near Quartzite Arizona. Like most crystals found in nature, these were broken from off the base they grew from.

Fig 6.4.15 (left) – These quartz crystals from China are referred to as **Double Terminated** quartz crystals. It can be clearly seen that these specimens have no contact point where they were attached to a base rock. They exhibit the geometric growth pattern at each end of the crystal. This establishes that these crystals did not grow while attached to another rock. How were they formed? Modern geology does not have an answer.

Fig 6.4.16 (right) – **Double Terminated** crystals can form from a variety of minerals and in a variety of orientations. These **multi-terminated** or twinned quartz crystals remain a mystery in modern geology. There is not a clear explanation for their origin in current scientific literature.

day. After much research and perusal of scientific literature, and after personal discussions with many geologists, we have no satisfactory explanation from modern geology to account for the origin of Double Terminated quartz crystals.

If you know where to look, pockets of double terminated crystals can be found in many places worldwide. Where do we find these beautiful crystals of Nature growing today? Nowhere. This is another example of how the Uniformity Principle fails again and how the Rock Cycle Pseudotheory cannot explain the origin of another of Nature's minerals.

6.5 The Basalt Mystery

What could possibly be so mysterious about a common rock such as basalt that we would dedicate an entire subchapter to it? Basalt lies at the center of both the Magma and Rock Cycle Pseudotheories. Within these two false theories-taught-as-fact, the very definition of basalt should be questioned. Basalt is a common, planet-wide, so-called volcanic rock. Volcanic rocks come from volcanoes, or do they? (See Fig 6.5.1) In this subchapter, we will ask new questions about the origin of basalt and other similar rocks and discuss why they did not come from volcanoes on the Earth's surface.

Basalt Defined

For many millennia, basalt has been used in the construction of buildings and roads because of its resistance to weathering. **Basalt** is a common *extrusive lava* rock, usually fine-grained and dark in color. Basalt usually contains a fair amount of silica but is rich in magnesium and iron. These minerals contribute to its characteristic darkness and density. Basalt often contains many minerals such as sodium, calcium, aluminum, tin, and others.

The definition of basalt can be found on wikipedia.org:

"**Basalt** is a common gray to black **volcanic rock**. It is usually fine-grained **due to rapid cooling of lava** on the Earth's surface." Note 6.5a

In Fig 6.5.2, we see basalt that was deposited on a red clay/sandstone base where it exists in its original, tabular state. Fig 6.5.3 is of basalt and shows both fine-grained and vesicular examples of this dark and common rock. Vesicles are the voids or holes left behind because of expanding gas bubbles during the rock's formative period.

Basalt is a surface volcanic rock that can be seen on every continent. Its abundance is noted in the *Handbook of Rocks, Minerals and Gemstones*:

"Basalt is the **most widely distributed of all volcanic rocks**."
Bib 15 p246

The Basalt Mystery

Notice from the previous definitions of basalt that basalt is a "volcanic rock" and that its texture is due to rapid cooling from lava. However, if we look a little closer, we will see a number of problematic questions that remain unanswered. Although there have been many volcanic eruptions observed and recorded in historic times, no one has ever observed fine-grained, non-vesicular basalt resulting from any *continental* volcanic activity. To make this clear, let us state this extraordinary claim once again:

> **No one** has ever observed fine-grained basalt resulting from any continental volcanic activity.

Historical Lava Flows

Where is the Basalt?

Fig 6.5.1

This certainly presents a 'problem' when it is said that basalt is the most widely distributed volcanic rock, and yet no observations of newly forming fine-grained non-vesicular basalt have been made.

Fig 6.5.1 shows examples of what various historical lava flows look like. The upper right and lower left are examples of pahoehoe lava flows. These types of lava are smooth and flowing. Middle right and middle left are examples of 'aa lava,' noted for its sharp, jagged edges. In the background, cinder cones can be seen. The Fundamental Question to ask is:

Where is the Basalt from historical volcanic activity?

When we think of lava, most of us recall images we have seen from televised documentaries—red-hot molten rock spewing or flowing from a volcano. When this flowing lava cools, it looks like the flows depicted in Fig 6.5.1. It does not look like the basalt formations as seen in Fig 6.5.2, 3 & 4. Although lava is dark in color, similar to basalt, when lava cools, it does so in flowing swirls or irregular, jagged shapes. Basalt on the other hand, forms solid, distinct masses or columns. Up close, the cooled lava has a glassy appearance. The erratically shaped lava fractures and cracks irregularly. Basalt commonly exhibits tabular or column-like patterns when it cracks. Basalt may *look* like it came from a volcanic source on the surface of the Earth, but unlike lava, some basalts contain quartz phenocrysts (quartz crystals). Quartz crystals are not formed in the cooling surface lava; they are only formed in an environment that will be explained in the following chapter.

Almost 70 years ago geologists recognized what many today do not—that there are many "fundamental problems" with the source and origin of magma and igneous rocks:

"The **igneous rocks**, and their mode of occurrence in the crust as well as on the crust, present **many profound problems**. **The source and origin of the various kinds of magmas and the cause of their rise from the depths are fundamental problems**." Bib 125 p267

This statement is taken from the 1939 *Textbook of Geology*, a primary resource book used in the education of the last generation of geology teachers. We can learn what has changed since that time by answering this question:

What new discoveries have solved the 'profound' and 'fundamental problems' identified in 1939?

The answer is—there have been no new discoveries, that we know of, that solved these problems.

Since basalt is considered an 'igneous' rock, and is recognized as being the most widely distributed of all volcanic rocks, surely a volcanic eruption of fine-grained basaltic lava in historic times should have been documented. Even our research team was puzzled by how something so seemingly apparent proved so difficult to substantiate. In the field of geology, everyone seems to have concluded that basalt is lava and they have not thought to question the validity of that belief. This is where application of the *Question Principle* is so important. We

Fig 16.5.2 — This basalt rock deposit formed above red clay and sandstone near St. George, Utah, USA. Note the fine grained texture and sharp, distinct fracturing typical of basalt.

must question everything with an open mind.

We asked several geologists if they knew where fine-grained basalt, known to have come from historic lava, could be found. None knew of any such place. The terms lava and basalt are often used interchangeably, causing a great deal of confusion. For example, the Hawaii National Park web site, *Frequently Asked Questions* #8 says that in Hawaii; "The lava is basalt." However, there is a significant difference between lava (Fig 6.5.1) and basalt (Fig 6.5.2 & 4). The volcanoes in Hawaii National Park and other lava producing volcanoes around the world produce *lava*, not basalt. Recognizing this difference is very important!

These and other inconsistencies confirm that basalt is not from a continental volcanic source. At the very least, the source is unlike anything seen in historic times on the surface of the Earth, where we can walk to and observe lava flows today.

The Basalt Column Mystery

Another interesting characteristic of basalt is its tendency to form in columnar blocks. Fig 6.5.4 depicts columnar basalt atop well-rounded cobbles and boulders near Bozeman, Montana, USA. Basalt column formations are found worldwide, and on every continent. During the 1800s, they were part of a huge debate in geology. The debate centered on whether basalt columns came from volcanoes or from an aqueous source.

One modern-day geologist, Haraldur Sigurdsson, wrote one of the most comprehensive histories of magma in 1999, in the book *Melting the Earth*. Sigurdsson made the claim that:

"The geologist who first **demonstrated** the volcanic origin of basalt was Nicholas Desmarest (1725-1815), opposing Guettard's claims that basalt columns in the Auvergne were of aqueous origin.

"While climbing the plateau of Prudelle near Mont d'Or, Desmarest noted some loose prismatic columns of dark basal-

Fig 6.5.3 – Basalt rock from the Columbia River Gorge, Oregon, USA. Although some basalts have vesicles (voids from expanding gas), most basalt is dark, fine-grained, and exhibits no vesicles. The origin of Basalt continues to remain a mystery in modern geology.

Fig 6.5.4 – Basalt columns are found worldwide. These columns of rock have been formed over sedimentary material near Bozeman, Montana, USA. Basalt columns can reach hundreds of feet high. When was the last time anyone saw lava cool and form these types of columns?

tic rock that had fallen from the cliff above. Tracing the rocks to their source, he found similar columns standing vertically in that cliff, grading upward into the scoriaceous top of the lava flow. **His simple observation of the association of columnar basalt as a component of a lava flow stands as a major advance in science.**" Bib 136 p134

Sigurdsson then cites Desmarest's *actual observational notes*:

"I traced out the limits of the lava, and found again everywhere in its thickness the faces and angles of the columns, and on the top their cross-section, quite distinct from each other. I was thus **led to believe** that prismatic basalt belonged to the class of volcanic products, and that its constant and regular form **was the result of its ancient state of fusion**." Bib 136 p134

As we can easily see, Desmarest did not really "demonstrate" that basalt columns came from volcanic lava. What Desmarest observed was ancient basalt in its columnar habit below a "scoriaceaous lava flow." Through his own reasoning, he was "led to believe" that the basalt was connected to the volcanic products laying on it. He *did not directly observe the basalt as being connected with the lava flow.* Desmarest's observation **should not stand** "as a major advance in science" as Sigurdsson suggests.

The volcanic rocks at Prudelle are of "ancient" date and geologists must recognize that different environments existed at different times at this site. Therefore, multiple eruptive events, each in a different environment, can account for the existence of both basalt columns and lava at the same site.

Sigurdsson then quotes Jean-Etienne Guettard (1715-86):

"'If a columnar basalt,' he wrote, 'can be produced by a volcano, why do we not find it among the recent eruptions of Vesuvius and other active volcanoes?' This was a fair argument, given the state of observations of lavas from active volcanoes at the time." Bib 136 p134

This was not just a "fair argument" in the 1700s. Today it remains a fair argument, and in fact it is critical to our understanding of basalt. Ask this FQ:

> If basalt is produced by volcanoes, why do we not find columnar basalt linked to historical eruptions, or from the eruptive events of active volcanoes?

What we find astonishing is the total lack of empirical evidence supporting the theory that prismatic basalt comes from volcanoes as Desmarest claimed. Today, geologists routinely assert that basalt is formed from volcanic land sources. If we had not questioned this premise, no doubt we would have continued along, like everyone else, believing someone had already figured it out.

The Devils Tower Mystery

Basalt is not the only mineral to form columns. One example is the rock making up Devils Tower, which is classified as phonolite porphyry. This rock is rich in soda feldspar and like basalt, is considered an *igneous* rock. The impressive rock formation beckons the question—how did it form?

In the hundred years since this National Monument was first set aside, many ideas have been suggested about how the Tower was formed. The debate continues because of the Uniformity Myth, where it is assumed that erosion is the chief cause of the monuments exposure. The National Park Service's website for Devils Tower National Monument notes:

"**The Tower is Formed: An Ongoing Debate**

"Geologists agree that Devils Tower was formed by the intrusion (the forcible entry of magma into or between other rock formations) of igneous material. What they cannot agree upon is how that process took place and whether or not the magma reached the land surface...

"Other ideas have suggested that Devils Tower is a volcanic plug or that it is the neck of an extinct volcano. Although **there is no evidence of volcanic activity** - volcanic ash, lava flows, or volcanic debris - anywhere in the surrounding countryside, it is possible that this material may simply have eroded away.

"The simplest explanation is that Devils Tower is a stock—a small intrusive body formed by magma which cooled underground and was later exposed by erosion." Note 6.5b

The Tower rises above its surroundings, like a lone sentinel, invoking awe and challenging all who view it to wonder at its origin. Because geologists have tried to understand how Devils Tower formed within the paradigm of modern geology's Rock Cycle, they only agree on the obvious:

"Geologists agree that the Devils Tower was **formed by the intrusion**... **What they cannot agree upon is how that process took place**..." Note 6.5b

The wisdom of how Devils Tower formed still eludes researchers today. No historical volcano has ever produced or formed columns such as the ones at Devils Tower. There are other columnar formations around the world whose origins still stymie researchers. The principle of Uniformity fails again, be-

Fig 6.5.5 – Devils Tower in Wyoming, USA is America's first national monument, and has dazzled travelers for centuries. Today, climbers scale the thousand-foot columns. After more than a century of research under the Rock Cycle Pseudotheory paradigm, no one yet has a definite answer as to how Devils Tower was formed.

Fig 6.5.6 - These columns are part of a formation called Ha-Minsara in the center of the Ramon crater in Israel. They appear similar to basalt columns but are made of quartzite. Could a similar process have formed the prismatic columns here as well as those at Devils Tower and other similar formations around the world? We do not see anything like them being formed today and so must conclude that the conditions in which they formed were far different from those that exist today.

cause we do not see columns like these forming anywhere in the world today. Our minds should be open to the idea that events and processes that took place in the past are unlike anything ever seen by modern science.

Lunar Basalt Mystery

By 1976, astronauts had been to the Moon and scientists had come to accept the belief that no water existed there. Today, that idea is being challenged by some scientists, and in the next chapter, the Hydroplanet Model, we will conclusively show how water has indeed, affected the Moon. In the meantime, we read from Patrick Moore's book, *Guide to the Moon*, the widely held belief that the great dark areas on the moon, the maria, are basalt, which originated as lakes of "liquid lava":

"**We know, of course**, that there is no water on the Moon now, and the so-called seas are dry plains without a trace of moisture in them. Analyses of the rocks brought home by the astronauts **seem** to show that there has never been much water on the Moon; the maria were once lakes of **liquid lava**, and this has solidified to make the volcanic rock known as **basalt**."
Bib 17 p99

Because basalt does not form from surface volcanic lava flows on Earth, can researchers conclude that surface lunar basalt formed from volcanic activity on the moon? No, they cannot. The formation of basalt on the moon remains as mysterious as its formation here on the Earth.

Vesicular Basalt Nodule Mystery

The Basalt Mystery is not just about the origin of fine-grained basalt. It includes the origin of vesicular basalt nodules found in many arid deserts, far from any known volcanic source. Rounded, vesicular basalt rocks as seen in figure 6.5.7, are a common surface phenomenon. They are found in many places worldwide. Their existence defies any modern geological explanation. The source of the supposed volcanism is absent and the mechanism for erosional transport is lacking. In Arizona, these ubiquitous rocks can be found all over the state. They are found in the deserts, on the plateaus, at the tops of the mountains, and almost everywhere one looks. Although Arizona has many volcanoes, they are not common enough to be a satisfactory source of these rocks. Nor is there sufficient evidence to support the idea of fluvial (river) transport of these large basalt rocks that would leave them so evenly spread across the state.

When the Grand Canyon stratigraphy is viewed, there is no evidence of these vesicular basaltic rocks occurring in the various layers, confirming they are a more recent surface phenomenon. Since they seem not to be a normal artifact of volcanic activity, we can ask this FQ:

> What is the origin of vesicular basalt nodules, found on the surface of the Earth?

Much of modern geology's idea of the origin of basalt is based on the experiments conducted by James Hall 200 years ago. Hall's results are sketchy at best, and we could find no reports of his experiments being reproduced today. There are no examples of erupting volcanoes forming basalt, granite, or obsidian. Hall is cited in the book, *Melting the Earth*:

"**Demonstrating that melts of basaltic rocks precipitate silicate crystals on cooling** was, of course, of fundamental importance in establishing the **volcanic theory on the origin of basalt**—and thus a death blow to the Neptunist theory of aqueous origin **for basalt, granite, and other rocks of igneous origin**. His [James Hall] experiments thus reproduced in miniature the processes that Hutton held responsible for the formation of rocks when he **demonstrated that mixtures of melted minerals may be solidified either in glassy, slaggy, or wholly crystalline form by varying the rate of cooling**." Bib 136 p153

Fig 6.5.7 – Vesicular Basalt rocks like these are found lying across hundreds of square miles of desert near Yuma, Arizona, USA, however there are no volcanoes in the area. Where did they come from? Notice that they are somewhat rounded. This indicates that there must have been some erosional activity at work. Why do these rocks appear so similar in size? Basalt Mysteries like this exist in almost every desert we have visited.

Even with the great modern technological advancements, geologists have yet to repeat and confirm Hall's results. True basalts do not contain glass and are not glassy, and therefore were never a surface melt.

Many types of basalt contain crystalline quartz, which suggests basalt must share a similar origin with quartz. At the very least, the environments where the two formed must have been comparable. It is in the Hydroplanet and Universal Flood Models chapters where we will detail the origin of the Earth's basalt.

6.6 The Obsidian Mystery

In times past, Obsidian was considered a sacred stone to many of the world's people because of its many uses. Carefully shaped blocks of obsidian were the trader's currency. Obsidian is a witness of the intense heat generated in Earth's crust and scientists have assumed that this once-melted silicate rock exists only because rapid cooling did not allow the gasses and water trapped within it to escape. Heat alone cannot account for this form of Nature's glass, for if heated to 1000° C, it inflates to become pumice. The formation of Obsidian has not been observed and its origin remains a mystery in modern geology.

Obsidian Defined

Many people have seen the naturally occurring glass rock called **obsidian**. It is usually dark in appearance, being black, brown, or even green with occasional specimens being colorless. Obsidian does not have a crystalline structure and the rock is therefore described as being amorphous, and glassy. Because of this, it is not considered a true mineral. Sometimes it is referred to as a mineraloid. Obsidian consists mainly of silicone dioxide (SiO_2) with impurities of iron, magnesium and other minerals that affect its color.

When struck, obsidian fractures conchoidally, capable of leaving an extremely sharp edge. Historically, obsidian has been used to make arrow points (Fig 6.1.1) and tools. Even in modern times, obsidian is used to make surgical blades because it can create an edge many times sharper than surgical steel.

According to modern science, obsidian forms when highly viscous lava cools rapidly. Fig 6.6.2 is a photograph of obsidian on display in a museum with the caption stating; "Obsidian forms when lava cools very quickly…"

Probably *all* geologists think obsidian comes from lava flows because that is the commonly held notion. For an example, we read from *Geysers of Yellowstone*:

"Obsidian is a natural glass. It is formed from **volcanic lava** that cooled too quickly for significant crystallization to occur." [Bib 134 p136]

Fig 6.6.2 – This natural obsidian specimen, on display in a mineral museum with the caption explaining the widely accepted definition of obsidian's origin, that it was lava cooling very quickly.

In Walter Schuman's book *Rocks, Minerals and Gemstones*, we learn that Obsidian is a natural glass. It is formed from **volcanic lava** that cooled too quickly for significant crystallization to occur. Obsidian is not uncommon and can be found worldwide. Locations where it has been identified include such diverse places as:

"Lipari Islands/Italy, Anatolia/Turkey, Iceland, Hungary, New Mexico and Wyoming/USA, Java, Japan." [Bib 15 p238]

Why is there a mystery about obsidian? There are actually many similarities between obsidian and basalt. Just as we learned in the Basalt Mystery, there are perplexing problems relating to the origin of obsidian also.

The Obsidian Mystery

Like basalt, the origin of obsidian is a mystery because its formation has never been observed. The principle of uniformitarianism says that the present is the key to the past and therefore, we should find events where we can view Nature's processes, like the formation of obsidian, in action. Earth remains volcanically active, and there must have been a historic event sometime over the last few hundred years where obsidian is known to have occurred. However, from our research we have been unable to find any record of an observed obsidian flow.

No one has ever observed a historical obsidian flow.

Obsidian is found worldwide on the continents. There are

Fig 6.6.1 – Artifacts like these bird point arrowheads were made from obsidian originating from the Columbia River Gorge, Oregon, USA. Obsidian was a favored material of ancient peoples because it is easy to shape and can be extremely sharp. Ceremonial specimens often exhibit meticulous detail. Obsidian was widely traded amongst different cultures, sometimes thousands of miles from its source. Examples like these are found worldwide.

CHAPTER 6 THE ROCK CYCLE PSEUDOTHEORY

Fig 6.6.3 – This is Glass Mountain Obsidian Flow in California, USA. Notice both the glassy nature of the deposit and the congruent sized blocks of obsidian that are very similar to deposits of basalt previously discussed.

Fig 6.6.5 – This brown rhyolite and black obsidian on the bottom left of this photo is from Obsidian Dome in California, USA. Beautiful lichens make their home on these types of formations and come in all colors of the rainbow. Why does obsidian mainly form only at the edges of these dome deposits?

whole mountains of obsidian, like Glass Mountain in California shown in Fig 6.6.3. Lava and lava flows have been studied extensively by volcanologists at universities around the world. Lava flows are well known and well studied and there are many modern flows occurring somewhere on the Earth continually. Volcanic eruptions are not uncommon and many have been recorded and observed.

The Obsidian Mystery is this – why is obsidian not being formed in any modern lava flow?

Over the years, there have been thousands of scientists and researchers studying and observing lava flows. Billions of people have been living on the planet for many, many years. Surely, someone should have observed the formation of obsidian, if it had formed in the manner modern science says it does. Perhaps we should answer this Fundamental Question:

> Does modern geology **really know** how obsidian is formed?

The Rarity of Rhyolite

We found no modern day researchers reporting having actually seen obsidian form, however there was an observation of rhyolite (not obsidian but a natural, glass-like rock, sometimes found with obsidian) from a lava flow, as recorded in *Volcanoes, A Planetary Perspective*:

"**Only one true rhyolite flow has been observed to erupt in modern times**. This was during the 1953-7 activity which formed the Tuluman Islands, two new islands in the St Andrew Strait off the north coast of Papua, New Guinea." Note 6.6a

The rarity of this event is noted as being the "Only one true rhyolite flow…" that has been observed in modern times. We will discuss this particular flow in the Universal Flood chapter, where it will be seen that this eruption did not actually occur from a lava flow on dry land, but rather, in water.

There is another unique and interesting form of obsidian that should be mentioned—Apache Tears. They too have many unanswered questions pertaining to their origin.

The Apache Tears Mystery

Apache Tears are a nodular form of obsidian ranging in size from about .5 to 5 cm (1/4 to 2 inches). (Fig 6.6.4) Some geologists believe these obsidian nodules are shaped by fluvial (river) transport and further polished by the abrasive action of sand:

"The rounding is due to transport by rivers and polishing by the sand." Bib 15 p238

However, the nodules rarely exhibit the percussion marks and fractures that would be evident if they were rounded by erosional processes. Moreover, they are often found in perlite deposits, not in streambeds of gravel. Finding obsidian in perlite is interesting too, because perlite has a relatively high water content, and when it is heated to a temperature of around 875°C (1600°F) at surface pressures, the perlite will expand 10 to 15 times its original volume. How well rounded obsidian nodules came to form in a perlite matrix is indeed a mystery.

In summary, modern geology has concluded that the origin of obsidian can be traced to an event no one has actually observed, and for which no certain evidence exists. This false conclusion has come from the uniformity principle, the foundation upon

Fig 6.6.4 – These obsidian nodules are sometimes called Apache tears. Some geologists have reasoned these nodules were rounded during transport but the material that they are deposited in does not support this. Rather, they were formed in this rounded, nodule shape. Is flowing lava really involved in the formation of these little specimens?

conclusion has come from the uniformity principle, the foundation upon which geology built the Rock Cycle Pseudotheory. Obsidian has not been observed in any modern lava flows and the formation of obsidian nodules by erosional processes is unfounded. Modern geology can only *guess* how obsidian is formed. Chapter 8, The Universal Flood, will shed new light on the mystery of obsidian formation, and we will uncover and discuss the actual processes of rock formation even though they are not occurring naturally in modern times.

6.7 The Iron Mystery

Iron is essential to life and an extremely important mineral to humanity. Technologically speaking, iron is what humans have used to built much of the modern world. Iron accounts for approximately 95% of worldwide metal production. Why it associates with some minerals and how and where it forms is a mystery. Iron is significantly heavier than most other minerals, and scientists consistently assert that, when the Earth was molten, the heavier iron sank to the center of the Earth. Yet surface incidences of iron are not rare and we do not find iron becoming more abundant as we descend into the crust. Where did this abundant mineral come from? That is the Iron Mystery.

The Pyrite Mystery

The name pyrite is of Greek origin meaning 'of fire' and is known for the sparks created when struck against steel or flint. Throughout the ages, many a gold prospector has been fooled by the teasing wink of pyrite. While it is known as fool's gold because of its similar appearance, pyrite is not at all similar to the valuable precious metal, gold. **Pyrite** is iron sulfide (FeS_2) and has a density about one fourth that of gold. Because it is heavier than sand, it is commonly found among the sediments in streambeds. The light colored gold flecks of this mineral can be seen in Fig 6.7.1.

Pyrite is an important mineral in Nature. It is an important mineral in the Universal Model because of its association with many other minerals, and because it is commonly found among fossils. Pyrite occurs in many shapes and exhibits many unique properties. Fig 6.7.2 is an example of the cubic form of pyrite, where the pyrite cubes are still a part of the matrix in which they were formed. These cubes have not been cut from the native rock nor are they man-made. The individual, cubic-shaped crystals grew naturally into these almost perfect cubes.

Pyrite is a replacement mineral in fossils. In some cases, the fossil may become completely 'pyritized.' Of special significance, pyrite is also associated with quartz.

The pyrite mystery, like the mysteries in the previous sub-chapters, is a mystery of the mineral's origin. How did these beautiful crystals grow? From where did the iron come to grow them? Why are these iron sulfide mineral deposits found only near the surface, and not deep underground.

Fig 6.7.1 – The sparkling gold-looking metal in this stream is the common 'fools gold' known as pyrite. It is not as heavy as real gold and is a square shaped iron sulfur sand that is always associated with other black iron sand as seen in the photo. Where does all this iron sand come from? The origin of pyrite and black iron sand is a continuation of the Sand Mystery discussed earlier.

There is no evidence that pyrite forms from a 'melt,' and there are no known places where crystals like those seen in Fig 6.7.2 are forming at the present time.

The Pyrite Sun Mystery

How did the Pyrite 'Sun' rock in Fig 6.7.3 form? Geologists simply do not know. Pyrite Suns can sometimes be several inches in diameter. They are usually found in coal bed laminations, near the surface of the Earth. The specimen shown in Figure 6.7.3 is from a coal mine in Sparta, Illinois. Pyrite suns are not known to be forming today. No clear explanation has been offered for how they were formed in a peat bog or in a coal formation. To understand how pyrite suns were formed, we must first know the origin of the iron and sulfurous minerals they are made of.

There are many critical questions that have not been adequately answered. If these questions seem difficult for geologists to answer, perhaps it is because they are looking in the wrong places. Hard questions are easy, when we know the answer. To learn the answers, we have to look in the right areas and ask the right questions.

The Iron Soil Mystery

Running a magnet through typical

Fig 6.7.2 – These cubes are not man-made but actual crystals made out of pyrite, an iron mineral. The largest cube is about ½ inch or 1.5 cm. They are found in this material that they are connected to and surrounded by it. How did large deposits of these minerals form and why do we only find them near the surface, not deep underground?

Fig 6.7.3 – This is a Pyrite Sun on a piece of slate found in a coal mine of Sparta, Illinois, USA, 300 feet below the surface. Miners bring these out by hand in their lunch buckets when they come across them, otherwise they would not be preserved. No rational explanation for their formation has ever been provided. Note that the environment where they were formed is next to coal, which does not provide an answer to how these pyrite crystal deposits grew.

sands or soils will yield an appreciable amount of black iron-like material, clinging to the magnet. Children are fascinated when this iron soil, called **magnetite**, is made to dance and move about by the influence of a magnet. Magnetite is found in most unconsolidated sediment.

Iron-bearing sand has a special place in Nature and here in the UM. Besides being magnetic, the presence of iron in so many sedimentary materials establishes one of the most intriguing but unanswered fundamental questions in geology today:

Where do iron minerals come from?

One of the first facts to note is that iron minerals are twice as heavy as the quartz minerals with which they are commonly found. If the iron minerals were present during the early stages of the formation of the Earth, assuming for the moment that the Earth was a magma-planet, this loosely occurring iron should have sunk to the center of the magmaplanet when it was molten. Neither the magma pseudotheory nor the uplift pseudotheory of modern geology have explained the mechanism that would bring heavier minerals to the surface. We learned in subchapter 5.9, the Non-Iron Core Evidence, that there is a lack of evidence supporting the theory that an iron core even exists. The modern science rock cycle has no explanation for where iron minerals came from originally, or for how they were supposed to have reached the center of the Earth, or how they found their way back to the surface. When answers to questions of the origin of minerals like magnetite are not known, scientists often resort to the ever-common sci-bi, erosion. Let us explore that possibility.

Iron sand or magnetite is said to have come from rock outcroppings, mountains and larger iron-bearing rocks. This should be a relatively easy theory to confirm and one that should identify the source rock from which magnetite came. To begin, we refer to this quote from the *Handbook of Rocks, Minerals & Gemstones* concerning sedimentary deposits:

"Sedimentary deposits originate from the weathering of rocks, through the agency of water or **through chemical processes under certain climatic conditions**." Bib 15 p97

Everyone probably understands a little about the weathering of rocks. Wind blows relentlessly, whisking away fragments of rock. Ever-present water freezes, cracks, and washes away sediment. What of the 'chemical processes' that happen 'under certain climatic conditions?' Some chemical processes are easy to understand. For example, hydrochloric acid dissolves carbonate rocks. The phrase 'under certain climatic conditions' gives us a hint as to the depth of the mystery that exists. There are many such "conditions" that remain unknown in science today, and so it remains a mystery as to how or when those conditions occurred.

For now, we are interested in the processes that explain how iron ores, like magnetite, come to be in sand deposits simply from erosion. This process is not clearly explained in any of the geological literature we researched and we found no one conducting fieldwork or experiments that would explain how iron sand, like the type shown in Fig 6.7.4, accumulated.

Since we could find no hands-on experiment to explain this, we decided to conduct one. This is an easy way to demonstrate what materials will come from erosion of a rock or an outcrop or even a mountain. All we need to do is simply speed up the erosion process! What we learned from this simple experiment is that the massive amounts of iron present in worldwide sediments *did not come from erosion*.

The Iron Sediment Experiment

A large portion of the sedimentary deposits in Nature, including the iron filings in the soil did not come from erosion. The Iron Sediment Experiment (Fig 6.7.5) is a simple and fun exercise that will demonstrate this. It is simple enough that a child can do it, under adult supervision of course. Be sure to wear eye protection and do the crushing under a towel. The goal is to acquire sediment from a *streambed* where the soil and rocks are thought to have eroded from nearby mountains. For this experiment, the sediment must have iron filings so use a strong

Fig 6.7.4 – This is a desert streambed in Wickenburg, Arizona. We can see a variety of rocks and a large amount of black sand. How many of the rocks look black or appear to contain black iron minerals from which this iron sand could have eroded? The origin of this black iron sand is a mystery. By performing the Iron Sediment Experiment, we can demonstrate that sand like this does not come from erosion.

Fig 6.7.5

magnet to test the collected sediment for iron first.

Once a suitable location is found, collect a quantity of the sediment, about 2 to 4 liters (1/2 to 1 gallon). Then, gather larger rocks from the sediment itself or directly from the mountain source or outcropping that looks most like the sediment in the streambed. Wash the larger rocks to remove surface dirt and soil. Collect enough rock so that, when crushed, you will have approximately the same amount of crushed rock as you have soil. The crushing of the rocks can be completed in the field or later if field conditions are not suitable.

Using a large magnet, drag the magnet through the collected sediment. Pull a good quantity from the soil sample and place it in a vial as shown in Fig 6.7.5. Next, (complete the crushing if you have not already done so) drag the thoroughly cleaned magnet through the newly crushed rocks. In most of the locations that this experiment was performed, there was almost no iron attached to the magnet from the crushed rocks. In many cases, there was not any iron at all. This is simply because most mountains and rocks do not contain *magnetite*, the iron mineral in the soil that is attracted to the magnets.

> How could the large amount of iron in the soil be eroded from larger rocks, if the rocks contain almost no iron?

The Iron Sediment Experiment is a powerful tool, which confirms there is a mystery about the origin of the iron in the sediment.

6.8 The Ore Mystery

Gold! Silver! Uranium, Copper, Magnesium, Lead, Iron, Platinum, and many other precious, semiprecious, and ordinary metals all begin as mineral bearing rock—ore! Prospectors once read the lay of the land to find what they hoped would be the next rich pocket of ore. They learned quickly that ore bodies occurred in proximity with other ordinary rock. Geologists later tried to understand why these rocks and ore bodies formed together but were unable to grasp the concept because of the blinding influence of the Rock Cycle Pseudotheory and because they assumed all ores precipitated from melted rock. The relationships of heat, pressure and water in the formation of ore is still a mystery today. More especially, the role of microbes in the origin of ore is all but unknown. In this subchapter, we discover that ores do not come from magma.

Ore Defined

What comes to mind when you hear the word 'ore'? Do you imagine giant trucks and colossal earth-moving machinery shuffling massive amounts of rocks from deep mines as seen in Fig 6.8.2?

Ore is a mineral-bearing rock. The rock may contain one or many types of minerals or gemstones. In most cases, there are many minerals found together in an ore body. The type and concentration of the mineral or minerals is determined by assaying or analyzing the ore. The grade and concentration of the mineral determines the value of the ore, and that determines whether the ore can be extracted profitably or not. Most ore deposits re-

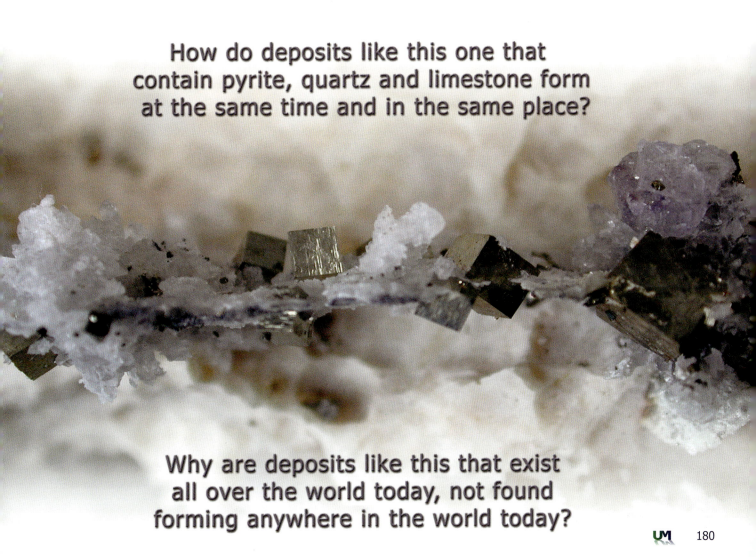

How do deposits like this one that contain pyrite, quartz and limestone form at the same time and in the same place?

Why are deposits like this that exist all over the world today, not found forming anywhere in the world today?

quire large quantities of rock burden to be transported, crushed, processed, and smelted to get a relatively small amount of the end product we use.

"**Ore deposits** are naturally occurring geologic bodies that may be worked for one or more metals." Bib 12 p1364

There are many types of ore. Manganese ore often occurs with iron, whereas lead is usually found with zinc, silver, and especially copper. Gold ore is also found with silver, copper, lead, and a host of other metals.

Because the Rock Cycle Pseudotheory is the basis upon which all geologists are trained, this paradigm, or thought process, has restricted the geologists' thinking in a field where more money has been spent than any other—the mineral-ore industry.

The world's largest mining and mineral exploration companies hire highly trained geologists to determine where ore bodies are located. They are also hired to determine the quality of the ore and the extent of the ore deposits. In *1971*, a Senior Research Officer from the Department of Geophysics and Planetary Physics at the University of Newcastle, who has had experience with mineral resources on many continents, made this statement with respect to the world's copper resources:

"With the mineral resources of the continents rapidly diminishing (**on current estimates the world's present copper mines will be exhausted within the next twenty years**) it will become increasingly necessary to tap the mineral resources of the ocean floors." Bib 103 p10

Having the benefit of hindsight, we see twenty years later in *1991*; the copper mines were far from exhausted. It has been a practice for decades, whether intentional or not, to bemoan a lack of resources. The perpetuation of myths like these can affect the economics of the mineral resource. Of course, we should conserve our natural resources, which are vital to society now and in the future. How do geophysicists make predictions like this and be so far off? This is not an isolated case. Estimates from many 'experts,' who have been trained with modern science's Rock Cycle Pseudotheory paradigm, have been completely wrong in a wide variety of disciplines.

Ores make up a large part of the Earth's crust. How could geology be so far off base about the ores in the crust and still understand the basics of crust formation? Could there be *a fundamental problem* with the entire current geological paradigm?

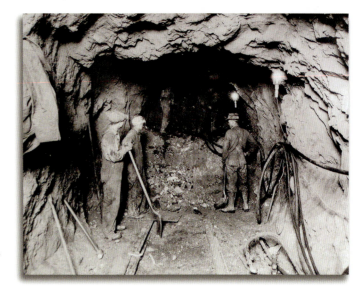

Fig 6.8.1 – Ores are naturally occurring geological bodies that contain metals, which are separated by smelting. Mankind has known about the smelting process for thousands of years.

Is Magma the Ore-Bringer?

According to modern geology, the origin of ore has always been directly from or associated with magma. This presents quite an obvious problem if there is no magma. Since we now understand that there is no magma, we must be prepared to rethink the entire premise of the origin of ore. Add to the confusion the Rock Cycle term igneous. Because 'igneous' rocks are said to have cooled from magma, any theory or association geology has made with igneous rocks must be discarded, including the origin of ore. These quotes from an old geology textbook are still held to be true today:

"The association of mining districts with **igneous** rocks early [on] led to the **idea that there is a fundamental relation between the igneous rocks and the origin of ore deposits**." Bib 125 p474

The idea that they were igneous rocks at all is an artifact of the Rock Cycle mindset. Because the rocks that are associated with ore were *assumed* to have been associated with magma, geologists have said this:

"The magma is the **ore-bringer**." Bib 125 p474

Many investigators of ores have recognized a *part* of the true origin of ore, and it has nothing to do with magma. It has everything to do with water:

"The solutions by means of which metals leave the magma are at high temperatures, so high that they are gaseous. As these solutions move upward through fissures toward the Earth's surface, they are steadily losing temperature, they are reacting with the wall rocks of the fissures, they are **becoming admixed with more or less ground water**, and the pressure on them is diminishing. As a result the composition of the solutions changes and certain constituents become insoluble and consequently are deposited in the fissures. Loss of temperature is probably the most potent factor in thus causing precipitation. As the solutions leave the magma they are gaseous, consisting mainly of steam, but during their journey upward they condense to water.

Fig 6.8.2 – This is the largest open pit mine deposit in the world—the Bingham Canyon Kennecott Copper Mine, located in Utah, USA. The trucks are so large that a man, standing next to its tire barely reaches the wheel. Ore deposits are thought by modern science to have come from magma. They have been called "freaks of nature" by some geologists.

CHAPTER 6 THE ROCK CYCLE PSEUDOTHEORY

Such hot-water solutions are the agents by which most ore bodies have been formed, and **the ores thus deposited are said to be of *hydrothermal* origin.**" Bib 125
p474-5

This idea that ores are of hydrothermal origin edges closer to the real origin, but it still falls short. In modern geology, hydrothermal fluids are a part of the magma-planet paradigm. Essentially, geologists believe that the ore still originates from magma but is being redistributed and concentrated by hydrothermal action, and that the hydrothermal fluids are heated by magma. As we have pointed out before, if magma is the source of the mineralization of ore bodies, we should expect to see ores forming somewhere on the continents today. The answer to this FQ is of great importance:

Fig 6.8.3 – This is the Kennecott Copper Mine as seen from the air. Quite literally, an entire mountain has been removed. The insets are cross-cut diagrams illustrating a vertical view of the ore layers—before mining on the left, today on the right. Notice the ore bodies are in vertical or near-vertical orientation, showing they had to have formed from rising material. If this material was not magma, what was it? How did ore bodies like these really form?

Where on the **continents** are
ore bodies forming today?

There are no new bodies of ore being formed *on the continents* today. Ores and ore-like minerals make up much of the ground we walk on. Ore deposits are found worldwide, on every continent and in every country. Ore bodies may be miles wide and thousands of feet thick. If they were formed hydrothermally, according to the theory of uniformity, we should see mineralized hot water coming from hydrothermal fountains, proportionate to the ore bodies that exist. Of course, this is not the case.

The uniformity myth, coupled with a magmaplanet belief, has done much to keep geology trapped within a Dark Age of Science. Ore is not being formed on the continents today because the environment in which it forms no longer exists. Although modern geology agrees with this assessment, geologists cannot agree on what the environment was like.

Ore Deposits, Freaks of Nature?

Ore bodies contain various metals that make the ore bearing rock heavier than most other rocks. Generally, ores have a density of at least 4 g/cm^3 whereas average rocks have a density of less than 3 g/cm^3. In other words, ore-bearing rock is significantly heavier than silica or quartz rocks.

Uranium is one of the heaviest elements found in Nature. If the Earth was once molten, the heavier elements would have been allowed to sink. Uranium would then be found deep in the Earth. However, uranium is not mined from deep in the Earth, it is found in the upper crust. It may be a surprise that half of the Earth's uranium is *believed* to be found in the crust:

"The crust contains more than 50% of the terrestrial inventory of U[uranium]..." Note 6.8a

As we discover the true nature of how uranium forms, it will be evident that *far more* than 50% of the Earth's uranium is likely to be found in the crust. Before we investigate this further, let us ask this Fundamental Question:

Fig 6.8.4 – This Arizona Copper ore is in its beautiful crystalline form. Among these specimens are some nearly pure, rare copper crystals. Copper ore is still plentiful. It does not come from volcanoes nor does it form in shallow, cool seas. Its origin in geology is a mystery. Where did the massive amounts of copper ore, found on the continents worldwide, come from?

SUBCHAPTER 6.8 THE ORE MYSTERY

If the Earth was once molten, why didn't all the heavy metals sink deep into the Earth?

If magma is real and the Earth was once liquid, the dense heavy minerals should have sunk to the center of the Earth, near the so-called iron core. This does not match the evidence. Where are the ore bodies and ore bearing rock found? *On or near the surface of the Earth!* In most cases, the rocks that are found beneath the ore are lighter and less dense than those in the ore body.

The book, *Ore Deposits of the Western States*, published in 1933, was *the* authoritative work on the subject of ore deposits at that time. What was said then, with respect to the knowledge of the origin of ore, is still true today:

"Notwithstanding all of the studies of ore deposits in the western United States during the past quarter of a century, **it must be admitted by any impartial judge that the general problem of the origin of ore bodies is still far from [a] complete solution.**" Bib 122 p6

Here we see the foremost authorities admitting in 1933 that they really have no idea where the ore bodies came from. A geophysicist in 1995 wrote this about what ore bodies really are:

"Minable ore deposits are **really freaks of nature** created by the combination of exceptional factors." Bib 120 p146

Could the exceptional factors that make ore deposits freaks of Nature simply be the result of theories based on the incorrect premises of the Rock Cycle Pseudotheory? We have only two choices; pursue complicated theories based upon the modern geologic rock cycle or consider an entirely new concept. In the next chapter, the nature and origin of ore will no longer appear as a freak of nature. We will discover its true origin.

The Mystery of Paragenetic Ore

Geologists commonly refer to ores by their elemental mineral name or by a chemical formula name, but this can be misleading. Mineralogists have learned that in Nature, minerals are never found *by themselves,* in ore:

"…actual ores, which are **always composed of several minerals.**" Bib 15 p95

This is important because, as will be explained in the Chemical Model chapter, if we want to comprehend how Nature works, we need to think naturally—not technologically. For important reasons, ores are made from specific elements, found in particular associations with other elements. The order in which these associated elements formed is called **paragenesis**. Random ore minerals could not have come from a magmatic mix to be deposited with whatever other minerals happened to be in a 'fissure.' If ores truly came from magma, we would expect ores to be associated by their density. This is not how we find mineral ores in Nature. Any natural law that explains the origin of ore must have an answer to the following FQ:

Why is it that certain specific ores are always found associated with other ores?

In the next chapter, the Hydroplanet Model, we will reveal evidence of biological transformation occurring in ore body formation. This will help us understand the mystery of gold's origin as well.

The Gold Mystery

Without a doubt, the most famous and infamous metal throughout history is gold. More resources have gone into prospecting, mining, recovering, and discovering the origin of this shiny heavy mineral than any other. Gold has built and toppled many empires and has been the source of countless schemes. For thousands of years, ancient alchemists attempted to transform common minerals into gold. Today, scientists are still puzzled about the origin of gold. From the journal *Science*, September 2002 issue, we read:

"Despite its enormous economic significance and hundreds of research papers over the past decades, **no consensus has been reached on the origin of the gold**." Note 6.8b

Why is there no consensus among the experts on the origin of this important mineral? The rock cycle theory, from which the modern science origin of all ores comes from, fails to provide answers to the mysteries we have identified. We are left to ask this final question about the ore mystery:

What are the true geneses of paragenesis ores?

6.9 The Carbonate Mystery

Carbonates are to the Earth as wheat is to bread. Carbonates, like wheat, are found worldwide and the carbonates are an ingredient in countless rock recipes. Like wheat, the temperature, pressure and other ingredients, including the effects of biological activity, influence the density, texture, look, and many other physical properties of the final artifact.

The mystery of the ubiquitous carbonates lies in their origin. Carbonates are known to be associated with biologic activity. However, in modern geology, much of the carbonate rocks are thought to come from inorganic processes. In this subchapter, we discuss many such mysteries and inconsistencies that exist within the geologic mindset on the origin of the world's carbonate deposits. As we have come to understand from previous subchapters, trying to understand rock-forming origins and processes by looking through the modern geology Rock Cycle Pseudotheory lens leads to vague and misinterpreted observations.

Carbonate Defined

If you fill a glass with unfiltered water from your sink, a lake or a stream, and then let the water evaporate, there will often be a white deposit left in the bottom of the glass. The same deposits, often called 'hard water' deposits, can be seen on shower-

Fig 6.8.5 – Finding a gold nugget like those shown above was every prospector's dream. They are usually found in streambeds. These nuggets eroded from a fissure or vein of gold where all pure gold deposits are found. Why do we find pure gold in veins between other rocks? What is the true origin of gold?

heads, glass enclosures, and dripping faucets. What is the hard, scaly white stuff? It is a form of **carbonate** (CO_3).

Most likely, the white, hard material is a form of **calcium carbonate**, which is made of calcium, carbon, and oxygen atoms. There are many types and uses of carbonates, such as magnesium carbonate ($MgCO_3$), used in antacids, zinc carbonate ($ZnCO_3$), used in making shampoos, and Sodium Bicarbonate (Na_2CO_3), an important ingredient in making glass. However, the most common type of carbonate is calcium carbonate ($CaCO_3$), the source of most of the hard water scale buildup we are probably familiar with. Calcium carbonate is known as calcite when it is in a crystalline form. Limestone, chalk, and marble are made of calcium carbonate. There are many uses of calcium carbonate, especially in the construction industry. It is a primary ingredient in concrete; it is used to purify iron, adds rigidity to plastics, and is an ingredient in a host of other interesting products. It would be hard to imagine life without calcium carbonate!

Just where does all the calcium carbonate come from? Because a large portion of Earth's crust is made of carbonate minerals, knowledge of the origin of the carbonates is essential to understand how the surface of the Earth, as we see it today, was formed.

The Carbonate Mystery subchapter touches upon the mysteries and incongruent ideas held in modern science concerning the thin white deposit at the bottom of the glass, and the miles-thick layers of limestone found worldwide. First, we will discuss something many people, especially those from the southwestern United States, are very familiar with; those stubborn white deposits left in the bottom of the glass—hard water.

Hard Water Pseudotheory

Hard water is water that has a high concentration of minerals in it. Hard water minerals can include magnesium, calcium, sulfates, and other minerals, but the most common type of hard-water mineral is calcium carbonate. This mineral, sometimes referred to as lime, is said to come from limestone. The typical modern science view is that rainwater (H_2O) becomes acidic by combining with carbon dioxide (CO_2) in the air. The result is carbonic acid (H_2CO_3), a weak acid that passes through limestone and other rocks, dissolving the calcium carbonate and carrying the molecules away in a suspension. From www.hardwater.org we find:

"Water is a good solvent and picks up impurities easily. Pure water -- tasteless, colorless, and odorless -- is often called the universal solvent. **When water is combined with carbon dioxide to form very weak carbonic acid, an even better solvent results**.

"As water moves through soil and rock, it dissolves very small amounts of minerals and holds them in solution. Calcium and magnesium dissolved in water are the two most common minerals that make water 'hard.' The degree of hardness becomes greater as the calcium and magnesium content increases." Note 6.9a

According to modern science, there are three variable factors involved in making water 'hard':

1. The amount of carbon dioxide (CO_2) in the air.
2. The amount of water (H_2O) falling through the air as rain.
3. The availability of limestone ($CaCO_3$) source rock which 'carbonic acid' rainwater can dissolve.

From these three variables, can we determine where the hardest water in the United States *should* be? For our prediction, we will eliminate the first of the three factors because the amount of carbon dioxide in the air is relatively constant across the United States. The second variable factor is rainfall. Across the United States, we generally find the northwest coast and the east coast as having the greatest amount of precipitation (see Fig 6.9.3). The third variable is limestone. Where are the largest concentrations of limestone in the United States? They predominate in the Eastern and Mid-Eastern United States, especially in Florida, the Appalachian Mountain region and in parts of Texas.

From these three simplified variables, where *should we expect* to have a greater degree of hard water in the United States? Where there is a *lot of rainfall and a high concentration of limestone—the Southeast*.

Where do we *actually* find the highest concentration of hard water?

Fig 6.9.2 is a USGS map of water hardness. On this map, the location where the hardest water is found is shown in red and the area where the least amount of hardness occurs is shown as dark blue. You can see clearly that the mid-Southwest and western plains are the areas with the hardest water. On the other hand, the dark blue area in the Southeast, which has abundant rainfall and much limestone, is among the areas with the *lowest* water hardness.

As we can see, attempting to predict water hardness based on the factors attributed by modern science does not work too well. In fact, the scientific evidence seems to directly contradict this theory. If water hardness cannot be predicted based on these criteria, perhaps there is another source of the carbonates and

Fig 6.9.1 – A residential tap water distiller used to purify water shows 'hard water' minerals forming from a leak on the outside of the distiller. Calcium carbonate is the primary mineral that makes water hard. Where does this white mineral come from?

Water Hardness
(Based on Concentration of Calcium Carbonate)

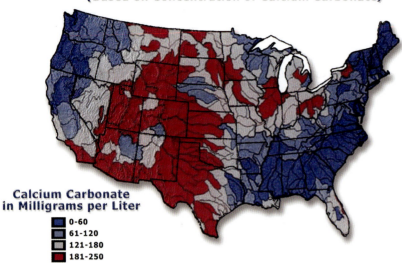

Fig 6.9.2 – Water Hardness as measured at NASQAN (Stream Quality) stations during the 1975 water year. Red areas identify the hardest water and dark blue areas the least hard. If Water Hardness was really a result of limestone being dissolved by carbonic acid, the Southeast should have the hardest water. In fact, it is the Mid-Southwest that has the hardest water in the United States. Why?

another way to predict what makes the water hard. To answer this question, we need to look deeper into this mystery.

The Hard Water Mystery

Modern science implies that hard water is simply an inorganic dissolution process. This brings us to ask another FQ:

Is there a biological process connected to the origin of Carbonates?

To determine if there was a direct link between a biologic process and water hardness in the culinary water supply of Mesa, Arizona, we monitored the water delivered to our house through the city's distribution system. For a period of over one year, daily observations of temperature and hardness were made and recorded each morning. During the test period, we used a Hanna Conductivity model 9635 to measure the dissolved solids (water hardness) in the water.

We discovered that there was a direct correlation between the outside ambient temperature and the hardness of the water. The recorded data showed that the hardness of the water increased during the hot months and subsequently decreased during the cool months.

We contacted a water-monitoring expert from the city's water department to discuss our findings. The specialist we spoke with had a background in chemical and substance analysis. He was very familiar with the hardness of the Mesa water supply. We explained the observations and noted the direct correlation between water hardness and temperature. We discussed how increased temperatures typically encourage increased organic activity in water. The specialist was surprised. He expressed having never considered any reason

water would be 'hard,' other than from dissolved minerals in the soil.

Mesa City's drinking water brochure is distributed annually to every water customer. It states this about water hardness:

"Two common minerals in the Arizona soil, calcium and magnesium, dissolve in the water to create 'hard water.' Note 6.9b

The water department specialist could recall no specific scientific evidence to support the statement made in the City's report. He began to question the validity of the statement himself as we explained that even if the natural water supply from which the City of Mesa draws its water was acidic (it is not), the magnesium, purported to come from dolomite, would not have dissolved. It is quite well known that dolomite does not easily dissolve, even in *highly acidic* water. The city water brochure goes on to state:

"The City of Mesa can experience a **seasonal** taste and odor problem associated with the drinking water. The safety of the water is not at risk; however, there is a perception that the water is questionable because it has an unpleasant smell or taste. Geosmin and Methylisoborneol (MIB), **non-harmful naturally occurring compounds associated with algae growth in lakes and canals, are the primary causes**." Note 6.9b

This 'seasonal' change in taste that comes from 'naturally occurring compounds' is caused simply because of an increase of natural microbes, which occurs as the temperature gets hotter. This is *why* our water-monitoring experiment showed an increase in water hardness during the summer months. This also explains why the majority of 'hard water' is found in the mid-Southwestern part of the United States. Increased temperatures allow for increased microbial growth, which increases water hardness.

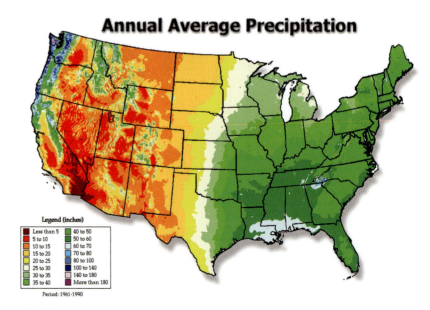

Fig 6.9.3 – Average annual precipitation in inches per year from 1961 to 1990. Red areas have the least precipitation, darker blue the highest. Compare this map with the water hardness (16.9.2) and limestone deposit (16.9.11) maps. They do not corroborate the modern science explanation for the origin of 'hard water.'

CHAPTER 6 THE ROCK CYCLE PSEUDOTHEORY

We will go into more detail concerning this microbial growth and its connection to the carbonates in the Universal Flood Model chapter, but for now, it is important to be aware that the microbes in natural waters play a much bigger role than has been generally realized.

Fossil Creek Formation Mystery

Nearly a century ago, when Phoenix, Arizona was a much smaller community, a dam was built on Fossil Creek and a flume to transport water to a small power plant was constructed. The abandoned flume, shown in Fig 6.9.4, located in Strawberry, Arizona, has been dismantled. Fortunately, we were able to make some interesting discoveries before this happened!

Along the flume, leaks from the crudely fashioned metal aqueduct resulted in many small feeder streams flowing down the canyon walls. A composite of several photos, Fig 6.9.5 depicts one of these Fossil Creek Canyon streams and a calcium carbonate formation called the 'Yellow Brick Road.' This formation is a *rare, modern-day limestone deposit*. After visiting and investigating dozens of rivers and streams across Arizona and the Western United States, this was the only modern-day limestone deposit we came across. There certainly must be others, but they are indeed, very rare. In the bottom right hand corner of Fig 6.9.5, the thickness of the deposit, which was well over a foot, can be seen. One of the unique features of this deposit is the fragments of the original wood beams and other construction material, like the wire shown in Fig 6.9.5, encapsulated almost completely by the carbonate deposits.

The Fossil Creek limestone deposits are unique. Along the creek and accompanying flume, there were other leaks but not all of the leaks produced limestone deposits. Why did some leaks result in massive limestone deposits while others had none? We decided to look a little further. The water source for Fossil creek is Fossil Springs, further up in the canyon. Limestone is found deposited sporadically along the creek path. Why would the limestone deposits be so seemingly random if they were formed by the precipitation of calcium carbonate out of the water? This is in part why the limestone formed at Fossil Creek is a mystery today:

> Why are there so few natural streams where limestone deposits are presently being formed?

If the lime deposits were simply dissolved calcium carbonate that precipitated out over time, we should expect to see thin, rel-

Fig 6.9.4 – This flume was built nearly 100 years ago to carry water to a small power plant in Fossil Creek Canyon, Arizona, USA. Leaks in the flume created many small streams that flowed down the face of the canyon walls. Some have created a modern day limestone formation, referred to as the 'Yellow Brick Road.' It can be seen in figure 6.9.5. The flume has been removed, but when it was in operation, water was transported down the flume and was exposed to the hot Arizona sun. The Sun stimulated increased algae growth, which in turn affected the growth of the limestone.

atively consistent deposits on the rocks and shoreline along the stream where water evaporates. What we found instead were rather thick deposits in some locations and sparse or nonexistent deposits in other locations in the same vicinity. We found that where more dense deposits occurred, carbonates accumulated on the rocks and anything else that was in the path of the water, including small twigs and even blades of grass.

This seems to have happened only when precipitation was rapid. Oral reports from people living in the area noted rapid deposition of lime after periods of heavy rain or floods.

The *mystery* of hard water can only be solved if we understand the *origin* of hard water. The limestone at Fossil Creek was **not** from local limestone dissolved by carbonic acid. Fossil Creek water, like most natural water, has a basic pH, not an acidic pH and therefore cannot dissolve the limestone. The limestone deposit at Fossil Creek is from *a biological process, not a chemical process*. This is why some areas along the creek have more substantial deposits than other areas and it explains why some deposits can happen rapidly and at different times.

There are several species of microorganisms responsible for limestone deposits, and biological growth occurs in cycles. These are dependent upon several variable factors such as water temperature and specific mineral catalysts that must be present for the microbes to reproduce. Because these factors are often seasonal and can change quite rapidly, these variables can affect the timing and extent of the deposit. Furthermore, the locations of deposits are directly related to the existence of dormant microbes. When dormant microbes exist in sufficient number, and then if the right environment is present, there is an opportunity for a microbial bloom and exponential growth rates, which can produce limestone deposits like those we see at Fossil Creek. We will explain much more about this concept in the Universal Flood Model chapter.

Limestone Origin Mystery

In The Carbon Cycle Pseudotheory, found in the Chemical Model, we determine that the *origin* of carbon in the carbon cycle is not actually known. Though modern science holds that carbon originally came from the atmosphere and from the oceans, the amount of carbon in both the atmosphere and the oceans is thousands of times less than the carbon found in the carbonate deposits. Indeed, there must have been a source, other than just the atmosphere, for such massive carbon-compound deposits.

For many years, scientists have essentially claimed that carbonate rocks have come from 'themselves' by using circular reasoning. This reasoning includes the so-called 'carbon cycle'

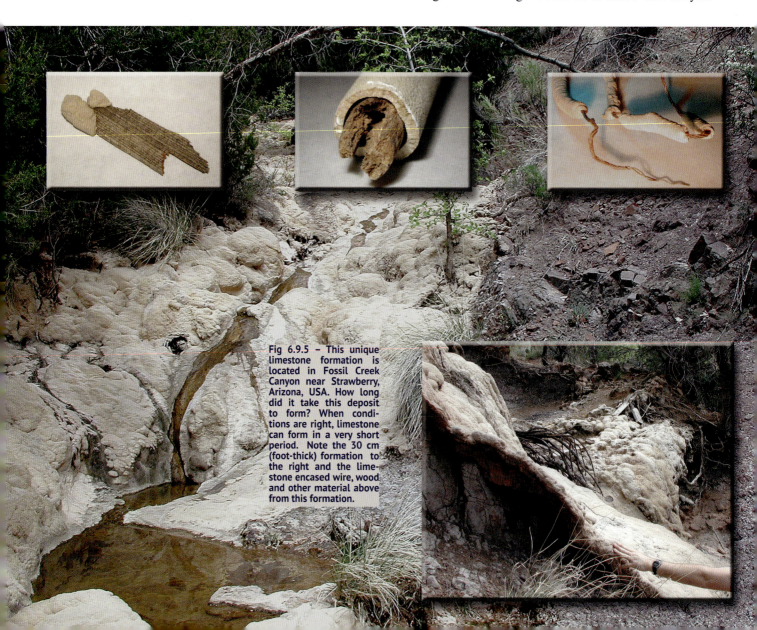

Fig 6.9.5 – This unique limestone formation is located in Fossil Creek Canyon near Strawberry, Arizona, USA. How long did it take this deposit to form? When conditions are right, limestone can form in a very short period. Note the 30 cm (foot-thick) formation to the right and the limestone encased wire, wood and other material above from this formation.

and the idea that the carbonate rocks are simply the result of a chemical process.

Certainly, the largest of the carbonate deposits is limestone, which consists largely of calcium carbonate ($CaCO_3$), also commonly called calcite. Much of calcite comes from living organisms. Just how do the organisms obtain the necessary materials from which many fashion elaborate outer coverings like those in Fig 6.9.6? This is the basis of the Limestone Origin Mystery. Typically, students are taught that these animals make their shells from the calcium carbonate that is dissolved in seawater. However, that isn't so easily done. Seawater has a slightly basic pH in the range of 8.0 to 8.4, which would make it very difficult to dissolve calcite. To dissolve calcium carbonate, a much lower pH is required.

We can perform a simple experiment to demonstrate how difficult it is for limestone to dissolve in water. Take two pieces of limestone and rub them together to create a powder. Add the powder to a glass of ordinary tap water observe how the finely ground limestone floats on the top of water. Tap water usually has a slightly lower pH (slightly more acidic) than seawater, which should actually help it dissolve a little easier. As you will see, it is very difficult to get the limestone to dissolve or to *go into solution*.

Later in this subchapter, we will discover that the vast majority of rivers and all of the oceans are not acidic at all but are basic and do not chemically dissolve the carbonate rocks like limestone to the degree maintained in geological textbooks.

Fig 6.9.7 is a section of limestone with a crystalline calcite inclusion. This particular specimen originated from Israel. Geologists believe that limestone forms in a shallow sea, but how do they explain the large calcite crystals that are common in limestone? The only way limestone and calcite crystals can coexist is if they formed in the same environment, *together*. This idea has confused many geologists because large calcite crystals require a special environment in which to grow. Such an environment includes a significant amount of pressure and heat, and as we will soon learn, water.

Fig 6.9.8 is a beautifully colored calcite specimen from a carbonate deposit in Mexico. Crystalline calcite like this is direct evidence that carbonate deposits could not have formed in shallow seas, because there would not have been enough pressure or heat to allow them to form. This fact should have been realized long ago, but as we will see in the next chapter, a strong bias against any explanations involving unconventional water depths in geological formations has kept geology from the truth.

The Mystery of the Vast Limestone Deposits

The origin of limestone is not the only perplexing mys-

Fig 6.9.6 – Seashells and other marine organisms are made of calcium carbonate. The currently held belief is that these animals obtain the materials for their outer coverings from dissolved solids in seawater. The real mystery is; where does the calcium carbonate material really come from?

Fig 6.9.7 – This is a chunk of limestone from a large Israeli deposit with calcite crystal inclusions. Limestone deposits with this type of a calcite geode embedded into them provide evidence that the formation did not occur in a 'shallow sea' as modern geology suggests.

tery. The idea that limestone forms in layers thousands of feet thick and is found worldwide is an amazing, yet unanswered puzzle. We learn just how thick limestone deposits can be from the Encyclopedia of Caves and Karst Science:

"Bedded limestones can range from single beds within sequences of clastic rocks such as sandstones or shales, **to continuous carbonate deposits up to five kilometers in thickness**."
Note 6.9c

Five kilometers is a little over three miles. How can a miles-thick, continuous deposit of limestone form from ocean water without clay, sands, or organic debris being intermixed in the layers? (See Fig 6.9.10) This same challenge was identified in the Sandstone Mystery, where thousand-foot-thick beds of pure sandstone were deposited with no erosional debris exhibited within the facies. In the cases of both sandstone and limestone, the thick deposits are sometimes *single layers*!

Fig 6.9.8 This is a beautiful calcite specimen from a limestone formation in Mexico. Large calcite crystals like this do not grow in 'shallow seas'. So how did they form?

Any such layer, *no matter how thick*, must have been created in a relatively **short time-period**. There is *no known reasonable method* in modern science where thick, continuous layers of sandstone or limestone could have formed over millions of years, without foreign matter being interspersed throughout the formations from erosional activities.

Large limestone deposits are ubiquitous and are found worldwide, on every continent. If the Rock Cycle theory is credible, and if the principle of Uniformity is true, we should be able to see large limestone deposits in some stage of formation today. These should be especially evident on the coastlines of the world in the conditions where geology says limestone formed. However, on what beaches along the coasts of America, Europe, Africa, or Australia can we see limestone forming?

Although the formation of limestone is not being observed, there are instances where carbonate "mud" is being formed. Even there, the cause or origin of the fine white limey material remains "uncertain":

"Calcareous sediment in the form of a fine mud is now being precipitated in **parts** of the sea, notably on the Bahama Banks, south of the Florida Straits. This great shoal exceeds 7000 square miles in area, and the water over it averages less than 20 feet deep. The shoal is formed of limestone, and much of its surface is mantled by **fine, white, limy mud. The cause or causes of deposition of this limy material are still uncertain.** Coral reefs are limited to the margins of the shoal and do not appear to contribute much to the deposit at the present time. It is thought that **bacteria**, chiefly those of the group that produce ammonia, cause the precipitation of the calcium carbonate; but purely inorganic processes, namely evaporation and increased temperature during the summer months, may have co-operated. **The relative importance of inorganic and biological agents in the deposition of the carbonate sediments is therefore still unknown.**" Bib 125 p233

The Bahama Banks, seen in Fig 6.9.12 are said to be submerged carbonate platforms. Here we can observe some parts of the process of how limestone is made. It is only a part because we can only observe how the lime Note 6.9d in limestone is made, not how it actually becomes lithified, or turned into rock. Remember from chapter 5.13, The Continental Uplift Pseudotheory, (Fig 5.13.4) there are no *submerged continents*

Fig 6.9.10 – Eight to ten million tons of stone per year are extracted from Reed Quarry, U.S. Enormous miles-deep and miles-long carbonate deposits occur worldwide. Where are deposits like these being formed today?

Fig 6.9.9 - Reed Quarry in Kentucky, USA, is the largest limestone quarry in the U.S. It is an example of the massive limestone deposits that can be found worldwide.

anywhere in the world.

Continental limestone is not being eroded or dissolved and then re-deposited along Florida's seashores. Natural fresh-waters have a typical pH of 7.3-9.0 Note 6.9e and are ***not*** acidic and so do ***not*** dissolve land-based limestone. Seawater has a very stable average pH of about 8.1, which limits its ability to dissolve limestone.

What is observed in select areas like the Bahama Banks is an environment in which certain bacteria can thrive. When the bacteria die, their microscopic carcasses fall out of suspension and form a white, chalky lime-like mud. What is not observed are the progressive stages of hardening, or lithification, of the mud into limestone. Can the Bahama Banks carbonate sediment harden into limestone in cool water less than 20 feet deep? It cannot. Both pressure and temperature are far below what is necessary to form limestone. The Bahama Banks carbonate platforms may be a good example of how carbonate *sediments* are being formed today but they are not an example of how limestone is being formed.

Geologists have been uncertain about limestone and dolomite formation for many years. One reason for this uncertainty may be their acceptance of the Carbon Cycle Pseudotheory. There has not been a substantiated origin for the carbonates or even carbon itself. It is certainly not logical to say that carbon comes from itself. Diagenesis is modern science's way of trying to explain some of this uncertainty.

The Diagenesis Pseudotheory

Many textbooks today state that carbonate sediments come from both organic and inorganic sources:

"Carbonate sediments and sedimentary rocks are formed from the accumulation of carbonate minerals precipitated **organically or inorganically**. The precipitation may occur during **sedimentation or diagenesis.**" Bib 59 p185

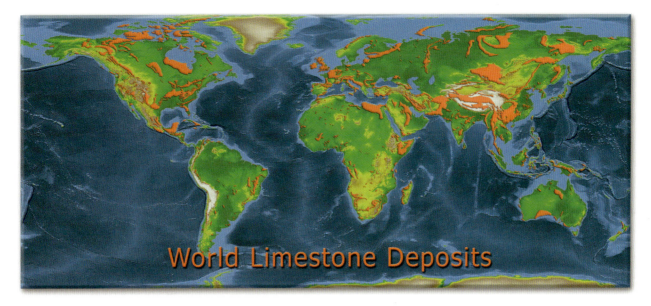

Fig 6.9.11 – This map of World Limestone Deposits (represented in orange) shows how extensive these geological carbonates are. However, without a clear explanation for the origin of carbon, these deposits and their origin remain a mystery in geology. How did they form? None are forming today. How did soft, limy mud turn to stone and from where did the large calcite crystals found in many of these deposits come?

Although we can observe the *organic* precipitation of calcium carbonate, the evidence for the *inorganic* process that is described as diagenesis has not been substantiated by modern geology.

Diagenesis is defined in the geology textbook *Understanding Earth* published in 1998:

"DIAGENESIS: The chemical and physical changes undergone by buried sediments during **lithification** and compaction into sedimentary rock." Bib 59 p652

Lithification as defined in the same textbook:

"LITHIFICATION: The chemical and physical **diagenetic processes** that bind and harden a sediment into a sedimentary rock." Bib 59 p656

In this example, each definition relies on the other to define itself. Circular reasoning like this often obscures the truth and leads to incorrect thinking.

Wikipedia gives us a somewhat clearer idea of this modern science tenet:

"It [**diagenesis**] is any chemical, physical, or biological change undergone by a sediment after its initial deposition, after its lithification. This process excludes surface alteration (weathering) and metamorphism." Note 6.9f

Does diagenesis really explain how changes really happen? Over one hundred years ago, sedimentologists came up with an idea, a sedimentation sci-bi, to explain how shale, sandstone, and limestone form. From the 1901 textbook, *Physical Geography* we read:

"By **pressure** and the cementing action of **sea water**, probably assisted in the deeply buried layers by the **internal heat of the earth**, the beds of mud and sand near the shore are gradually consolidated into strata of shale and sandstone, while farther from the shore, beyond the reach of the coarser sediment from the land, limestones are formed by the rain of shells upon the bottom." Bib 142, p177

Textbooks today say much the

Fig 6.9.12 – This is a satellite photo of the Bahama Banks carbonate platform. The light blue and green areas are where the ocean is shallow and abundant carbonates are formed from microbes that shelled animals use to build their elaborate carbonate shells. No Limestone is being formed here.

SUBCHAPTER 6.9 THE CARBONATE MYSTERY

same thing. Essentially they say sediments are deposited and compacted by succesive layers of sediment and then cemented by minerals precipitating from solution. However, the origin of the 'cement' or minerals in the solution is still a mystery.

Many limestones contain *terrestrial* fossils (animals and plants that lived on land) throughout the limestone body. Animals such as lizards and wasps, plants like ferns and pinecones are preserved in every detail, buried in limestone deposits found on land. How does this happen, if limestone is formed only in a marine environment?

Flint is a cryptocrystalline form of quartz, made of extremely fine silicone dioxide. It typically forms as a nodule, and is found in limestones and chalk. How did quartz-based flint nodules come to be in carbonate based limestone and chalk? Wikipedia explains:

"The exact mode of formation of flint **is not yet clear** but it is thought that it occurs as a result of chemical changes in compressed sedimentary rock formations, during the process of diagenesis." Note 6.9g

Again, we see how the formation is said to occur during diagenesis. However, geologists have yet to show where or how diagenesis is actually happening. The reason diagenesis is a pseudotheory is simple:

> Without observation, a theory is only philosophy—not science.

Diagenesis is said to occur at temperatures ≈ 200° C and at pressures ≈ 1Kbar (about 14,000 psi), yet there is no agreement in the field of geology for the temperature and pressure of diagenesis. When pressures and temperatures are higher than these, science calls it metamorphism. Simply burying sediments under great pressure at high temperature does not form the minerals modern geology claims.

The Marble Mystery

The textbook answer to the origin of marble is simply that it is metamorphic limestone. In other words, limestone, created near the surface of the Earth, was then buried by subduction deep inside the Earth where it was heated and pressurized and then, without any apparent geological mechanism, was uplifted back to the surface:

"Marbles are the metamorphic products of **heat and pressure** acting on limestones and dolomites. They may result from either contact or regional **metamorphism**. Some white, pure marbles, such as the famous Italian Carrara marbles prized by sculptors, show an even, smooth texture of intergrown calcite crystals of uniform size. Other marbles show irregular banding or mottling from **silicate** and other mineral impurities in the original limestone." Bib 59 p204

Calcite crystals, flint nodules, silicate inclusions—these all tell us that a lot more was happening than just heat and pressure. Researchers state the temperatures and pressure involved in the metamorphic process that changes limestone to marble are generally *assumed* to be:

"...at temperature and pressure ranges varying from **500°C to 800°C and 5 Kbar to 10 Kbar respectively.**" Note 6.9h

We learned in Chapter 5.8 that the piezoelectric effect of alpha-state quartz is eliminated when temperatures surpass 570°C. From this, we can see that the temperature limit of piezoelectric minerals was less than the heat at which marble supposedly is made. Because carbonate rocks like marble are often in proximity with quartz, marble could not have been in an environment as hot as geologists have stated.

We have not known of anyone actually being able to produce metamorphosed marble from limestone in the laboratory. Nor have we been able to find record of anyone observing the process in Nature.

There are other problems associated with marble that are not clearly addressed or answered in the fields of geology or paleontology. One of which is the massive accumulation of fossils found in many marbles. Fig 6.9.13, lower left is an example of the many limestones and marbles that contain great quantities of fossils. In some cases, the soft bodies as well as the hard shells of the organisms are preserved. Today, there are no clear answers how this happens. We will go into much more detail on fossil formation later in the Fossil Model chapter, but for now it is important to recognize that there are many holes in the metamorphic process as explained by geologists.

Another of the problems associated with this process is identifying the difference between limestone and marble:

"Distinction between the two [limestone and marble] is

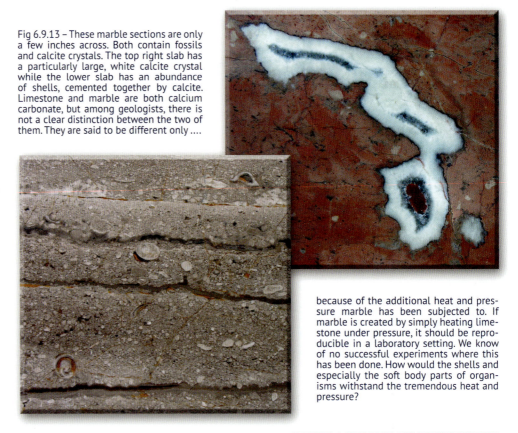

Fig 6.9.13 – These marble sections are only a few inches across. Both contain fossils and calcite crystals. The top right slab has a particularly large, white calcite crystal while the lower slab has an abundance of shells, cemented together by calcite. Limestone and marble are both calcium carbonate, but among geologists, there is not a clear distinction between the two of them. They are said to be different only ....

because of the additional heat and pressure marble has been subjected to. If marble is created by simply heating limestone under pressure, it should be reproducible in a laboratory setting. We know of no successful experiments where this has been done. How would the shells and especially the soft body parts of organisms withstand the tremendous heat and pressure?

sometimes difficult, **for the transition from limestone to true marble is not well defined.**" Bib 15 p284

The transition from limestone to marble is not well defined, probably because the difference between them is not clearly understood. There is a real lack of observational data showing *how the vast quantities of marble at the surface of the Earth were made*. Otherwise, the transition from limestone to marble would be more easily recognized and the differences more clearly defined.

The Carbonic Acid Cave Formation Pseudotheory

At the beginning of this sub-chapter, we discussed how water hardness is believed by modern science to be the result of the dissolution of carbonates through the action of carbonic acid. We showed how that theory failed to predict where the areas of hard water should be and we explained how natural fresh water does not dissolve carbonate rocks. There is a great deal more to discuss about this supposed process and how it relates to caves and the belief of how they formed.

There are thousands of caves worldwide. Caves can be found in many types of material, such as lava, salt or ice, but the largest percentage, by far, are caves that occur in karst (limestone) formations.

How were massive limestone cave systems around the world created? At first, geologists seem to have an easy answer—they formed by the action of acidic waters. That is, a source of natural water with a pH of less than 7, dissolved rocks leaving the caverns behind. The dissolved carbonates were transported away or re-deposited as various speleothems over long periods of geologic time. We call this the Carbonic Acid Cave Pseudotheory, because the scientific establishment universally teaches this process as though it were a proven fact. Modern science says that carbon dioxide in the air combines with rain to make carbonic acid, which subsequently erodes the limestone, forming the caves. From wikipedia.org:

"Limestone dissolves under the action of rainwater and groundwater charged with H_2CO_3 (carbonic acid) and naturally occurring organic acids." Note 6.9i

The primary problem with this pseudotheory:

Almost all natural waters in and out of caves are **alkaline, not acidic!**

A pH of 7 is considered neutral and is neither acidic nor basic. The vast majority of all fresh water does not dissolve limestone or dolomite because it generally has a pH of 7 or higher, which is not acidic. The ocean, with a pH level higher than 8 is also *alkaline, and **not** acidic*. The following citations provided by various researchers, or published in scientific journals, give us an idea of the pH levels found in typical caves from around the world.

<u>1. Tamaulipas, Mexico Caves</u>
"Ranged from **pH 7.1 to 8.5** in March/April 1980." Note 6.9j

<u>2. Carlsbad Cavern, New Mexico, USA</u>
"pH ranges from **pH 7.23 to 9.14**…most values being close to **8.5**". Note 6.9k

<u>3. Timpanogos Cave, Utah, USA</u>
Mean pH – 8.07 Note 6.9l

<u>4. Representative Cave Waters Found Throughout the World</u>
"For purposes of illustration, we used analyses of three water samples taken from **Scott Hollow Cave**, West Virginia. **These waters are representative of those found in temperate karst regions throughout the world.**" Note 6.9m

Most of these statements were taken from the comprehensive scientific journal on cave analysis—the *Journal of Cave and Karst Studies*. In the journal, three measurements of the pH in the Scott Hollow Cave were cited as being 7.78, 7.41, and 7.8.

None of the pH readings were below 7.0, clearly showing the cave waters are not even weakly acidic. Moreover, alkaline or basic waters were not observed to exist only in the cave environment. The researchers go on to make this *significant* statement about the pH of "most karst waters:"

"Most karst waters are in the pH **7 to 8** range…" Note 6.9n

It seems particularly amazing that, although the researchers *know* that the waters in the caves are not acidic, they continue to assert that acidic waters were responsible for cave formation.

Digging deeper into the scientific journal articles, we found cave scholars getting just a little closer to the truth when they hint at the role microbes play in cave formation. In their statement, they recognize a reduced influence of carbonic acid when they say it "***may*** contribute to the dissolution of the bedrock carbonate." Regarding the Spider caves, near the Carlsbad Caverns area in New Mexico, USA, researchers said:

"Both chemical and microbial processes **probably** influence the formation of these cave deposits. Chemical weathering from the condensation of weak **carbonic acid** in the cave atmosphere **may contribute to the dissolution of the bedrock carbonate**, **but** microbial breakdown of bedrock **probably** plays a more significant role, as the microorganisms themselves may release organic acids." Note 6.9n

Researchers cannot say with certainty that carbonic acid was the agent that formed the Spider caves, or any other cave. It has not been scientifically *demonstrated* that caves are created this way.

Sulfuric Acid Cave Formation Pseudotheory

In some areas, the Carbonic Acid Cave Formation Pseudotheory has taken a back seat to another dissolved-by-chemical-acid theory. This alternative theory involves an elaborate 5-step process leading to sulfuric acid as the agent to explain the dissolving away of the limestone, and the leaving behind the caverns, found in some areas of the world:

"The following sequence of steps is **well accepted** for the origin of Guadalupe caves: (1) reduction of sulfates (gypsum and anhydrite) at depth within the Delaware Basin to produce hydrogen sulfide (H_2S); (2) ascent of H_2S, either in solution or as a gas, into the carbonate rocks of the Guadalupe Mountains; (3) oxidation of H_2S to sulfuric acid, either where oxygen-rich groundwater mixes with the H_2S or at the water table; (4) dissolution of carbonate rock by sulfuric acid; and (5) removal of the dissolved solids to springs by groundwater flow." Note 6.9o

This "well accepted" theory makes some big assumptions. It is true there are gypsum deposits inside many of the Guadalupe caves, and there are curiously abundant concentrations of gypsum in the surrounding area. There are also large quantities of hydrogen sulfide nearby in the oil reservoirs of the Permian Basin. Oil from this area is known as 'sour crude oil' because it contains significant amounts of sulfur, and hydrogen sulfide. What is *not clear*, is how oxygen-rich water came to be available and how complicated chemical reactions occurred to

facilitate this process. The production of hydrogen sulfide by the reduction of sulfates is accomplished by anaerobic (without oxygen) bacterium. However, the authors of the journal article know this "is not a simple matter" because it is not observed happening today:

"Sulfuric acid cave origin **is not a simple matter** of dissolution of carbonate rock." Note 6.9o p48

The dissolution, transport, and removal of the dissolved solids by ground water flow leave the question of where it was transported unanswered. In other cases, where the sulfuric acid theory has replaced the carbonic acid theory, the source of sulfuric acid is even more 'imaginative.' Geologists for the National Park Service explain the development process that formed Wind Cave in South Dakota:

"Since acid-rich water dissolves limestone, a chemical change in the groundwater **had to occur for the cave to form**. The oceans receded allowing fresh water into the region. As gypsum was converted to calcite, sulfur was chemically freed to form either **sulfuric or sulfurous acid**." Note 6.9p

Because modern science does not have a clear origin for carbonate caves, theories of chemically derived acid have been invented and are believed to have occurred over millions of years, forming the caves we see today. It is a concern that these acid theories are considered "well accepted" and offered as fact to the uninformed public. The simple reason the sulfuric acid cave formation theory fails, just as the carbonic acid dissolution the-

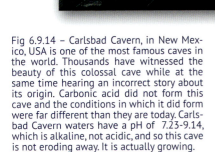

Fig 6.9.14 – Carlsbad Cavern, in New Mexico, USA is one of the most famous caves in the world. Thousands have witnessed the beauty of this colossal cave while at the same time hearing an incorrect story about its origin. Carbonic acid did not form this cave and the conditions in which it did form were far different than they are today. Carlsbad Cavern waters have a pH of 7.23-9.14, which is alkaline, not acidic, and so this cave is not eroding away. It is actually growing.

ory fails, is that these formative processes are not taking place today, as noted in the *Journal of Cave and Karst Studies*:

"Most caves in the Guadalupe Mountains have ramifying patterns consisting of large rooms with narrow rifts extending downward, and with successive outlet passages arranged in crude levels. **They were formed by sulfuric acid** from the oxidation of hydrogen sulfide, **a process that is now dormant**." Note 6.9q

Why such processes would suddenly be dormant is a mystery in modern geology. The gypsum is still present; the hydrogen sulfide rich oil deposits are still present, and ground water is present. What are not present are the residual hydrogen sulfide gases or residual evidence of sulfuric acid. Consequently, while they correctly conclude that the process that formed the caves is now dormant, evidence of cave formation by acid dissolution *over long periods of time—does not exist.*

Kartchner Caverns Mystery

Kartchner Caverns, located in Southern Arizona, is a beautiful example of Nature at its finest, exhibiting many colors and types of speleothem formations. How this wonderful cave system formed is a mystery, and what is surprising is how this simple mystery has eluded researchers.

The Kartchner Caverns State Park brochure tells us the typical cave-formation sci-bi:

"It all began with a drop of water…A **shallow inland sea** covered this area 330 million years ago, depositing layers of sediment that eventually hardened into limestone. Millions of years later this Escabrosa limestone along with other rock layers **uplifted** to form the Whetstone Mountains. The Escabrosa limestone, due to a type of tremor or fault, **down-dropped** thousands of feet relative to the mountains above." Note 6.9r

Once again, we are told a shallow inland sea was responsible for the limestone formation and was supposedly exposed by the processes defined in the Uplift Pseudotheory. Moreover, the Carbonic Acid Cave Formation Pseudotheory is cited as being the cause for the formation of this cave, as is noted in the brochure distributed to each visitor upon entering the cave. The Kartchner Caverns website says this:

"Rainwater, made slightly **acidic by absorbing carbon dioxide** from the air and soil, penetrated cracks in the down-dropped limestone block and slowly dissolved passages into it. Later, lowering groundwater levels left behind vast, air-filled rooms." Note 6.9r

Remember that one of the most important principles in the Universal Model is the Question Principle, which is to *question everything with an open mind.* So it was, that after our tour of the cave, we went to the museum office seeking the chief geologist. We asked him a simple question regarding the pH level of the cavern waters. Based on the explanation of the cave's formation we read in the Park brochure, we expected to hear that the water would be slightly acidic. Amazingly, this was not the case. The geologist reported to us that the pH of the water in the caverns was neutral to basic, and not acidic at all! Of course

Fig 6.9.15 – Kartchner Caverns, located in Southern Arizona, USA, are unique in many ways. Discovered in 1974, the caves were kept secret until 1988 when the state of Arizona converted the area into a State Park. The caves are essentially macro-scale geodes, formed at the Earth's surface within a limestone deposit. They were not formed by acidic waters, as is commonly taught in the scientific community. When you visit the cave, you will notice the entrance leads straight into the hillside and as the cut-away model shows, only a thin layer of limestone rock forms the upper shell of the cave.

Kartchner Caves

Cut-away Model

Arizona USA

Cave Entrance

SUBCHAPTER 6.9 THE CARBONATE MYSTERY

we were puzzled and found ourselves asking; "Why do they tell everyone the opposite?"

There are two additional reasons Kartchner Caverns are a mystery. The first is the recognition that the cave is still growing. There is no acidic water in or around the cave, and the cave is not dissolving:

"Kartchner Caverns is a 'living' cave; the formations are still growing!" Note 6.9r

The second reason this cave system is a mystery is because it was formed 'above ground.' Kartchner Caverns are unique in that they do not exist below the average surface elevation of the area! The Cave Entrance, seen in Fig 6.9.15 leads upward to the above surface-level hollow mountain. Once inside, one looks upward at all the wonderful formations. The cutaway model in Fig 6.9.15 shows how relatively thin the rock above the caverns is. The mean thickness of the cave ceiling is reported to be only about four to five feet (a meter and a half) and shows no signs of erosion. Upon further investigation of the surrounding area, we found no evidence of 'uplifting' or 'subduction' and there seemed to be an absence of faults in the area. It is highly likely the fragile formations in the caves would have fallen or broken if any significant earthquakes had occurred.

How can such a thin roof structure remain intact after hundreds of thousands of years of uplift, subduction and erosion?

Just as we explained in the Sand Mystery (subchapter 6.3), the vast majority of sediment was not formed from erosion, so it is that the vast majority of cave systems are not formed from erosion. These caves can more correctly be understood as macro-scale carbonate geodes, formed at the Earth's surface! The evident lack of erosion of the thin outer layer of rock, above the cave, verifies that these caves were not uplifted under layers of now-gone sediment that was eroded away. These fragile caves would have collapsed under the pressure induced during an uplift event. Researchers do not understand this because they do not comprehend how geodes are formed, as we saw in the Quartz Mystery (subchapter 6.4). The Kartchner Caverns cave system grew to its present condition in much the same way geodes grew. Although the cavern formations are still growing today, they are growing at a much slower rate than when they were first formed. The reason for this will be presented in the Universal Flood Chapter.

As you have probably begun to notice, the modern science pseudotheories have had a great influence on the scientific truth that is waiting to be discovered. Because the water in the cave is not acidic, the limestone in the caverns is **not** being dissolved or worn away by carbonic acid water, as every visitor to the cave is told. However, relinquishing the Carbonic Acid Pseudotheory is not so easy to do because it leaves geology without any explanation of how caves are formed.

There are several reasons why geologists should not say that carbonic acid made the caves a long time ago. Most caves are discovered in a pristine condition, showing little evidence of being 'caved in.' Stalactites and stalagmites are seldom observed to have been broken off, existing just as they did when they were formed. If they had been formed over hundreds of thousands of years, they would have been subjected to countless earthquakes and millennia of uplift and subduction. Old theories die hard.

Why was a **hollow** hill left behind instead of a valley after eons of supposed erosion?

From the cave formation theories within the Rock Cycle Pseudotheory, there can be no easy answers to questions like these. Caves did not form by simply eroding away. If the cave formation did not happen by the action of acidic waters within the cave, what was the mechanism that formed them? The wisdom of how geode-like caves formed will be explained in the Universal Flood Model, Chapter 8.

The Helictite Mystery

Most likely, you have seen stalactites and stalagmites, the common formations that hang from the ceilings and grow from the floors of most caves. Have you ever seen a helictite? See Fig 6.9.16 for an example of a helictite from Timpanogos Cave in Utah, USA. **Helictites** are remarkable speleothems, twisting and curving almost grotesquely from the walls and ceilings of some caves. The origin of stalactites and stalagmites seems easy to explain, especially when water can be seen *dripping from stalactites, or onto stalagmites*, but this is not the case

Fig 6.9.16 – Inside Timpanogos Cave in Utah County, Utah, USA, are unique formations called **helictites**. Because they appear to grow without dripping water and do not share the vertical habit of stalagmites and stalactites, these twisted, oddly formed and branched formations are 'curious' to researchers. Often a needle-like form of calcite called aragonite; helictites are some of the most delicate speleothems. They grow in all directions, seeming to defy gravity. Why do they form this way? The answer is to be found by understanding the origin of all limestone – microbes.

with helictites. The formation of these exquisite specimens remains a complete mystery:

"Perhaps the oddest and most interesting of all cavern formations are called helictites. These are curious twisted and branched formations that occur on the side walls of caverns or even on the sides of stalactites. **The mechanics of helictite formation are not clearly understood**. The delicate branching is perhaps a combination of released carbon dioxide in the water and shifting air currents. The sheer beauty of helictite formations is difficult to describe, particularly if the waters contained mineral impurities and tinged them faintly in blue, green, yellow, or red hues.

"Some wall surfaces in the Queen's Chamber of Carlsbad are so heavily encrusted with helictite formations that they resemble thorny thickets. Similarly, **some of the world's most beautiful mineral specimens are the fabulous helictite-encrusted stalactites on display in Timpanogos Cave**." Note 6.9s

To understand the origin of helictites, we must first learn the real origin of the carbonates that grow in caves. Carbonate deposits are primarily microbial in origin. They are not just a chemical precipitate of acid-dissolved carbonates. As in other caves previously mentioned, this also includes helictite cave formations.

Fig 6.9.17 – This calcium carbonate Lava Flow from the Ol Doinyo L'Engai African volcano contains a large amount of calcite. As we read in the magma pseudotheory chapter, we recognized that lava does not come from magma. Mysteries like these are far easier to solve when we learn the true origin of the carbonates, combined with the correct understanding of the origin of lava.

The mystery of helictites can be solved when we understand that the caves they grow in are very high in humidity, nearing 100%. The next fact is also a key to the Helictite Mystery—all natural waters, including humid air, contain microbes. The air in the cave contains microbes and nutrients along with a water medium, allowing for the multi-directional growth of the helictites. Air currents change, as do the quantities of microbes and dust in the air, thus changing the growth environment. This in turn, can change the direction of growth of the small fingers on the helictite.

Instead of an acid-chemical process dissolving limestone and depositing it, the *biological origin* of helictites is one key to comprehending all carbonate deposits.

The Calcite-Lava Mystery

In 1954, the Ol Doinyo L'Engai volcano in East Africa (Fig 6.9.17) erupted. A very interesting observation was made:

"A specimen of soda crust collected shortly after **an eruption** was found to contain about 50 per cent of material insoluble in water of which 70 per cent was identified as **calcium carbonate**…" Note 6.9t

Magma is supposed to produce silica-rich lava, presumably melted rock *from deep inside the Earth*. Instead, what was observed here was silica-poor lava, rich in calcite. This sodium and calcium rich lava crumbles to a damp white powder after a few weeks:

"The fresh rock is soft, dark grey, holocrystalline medium- to fine-grained and granular in texture. **On exposure to the air the surface becomes**

Fig 6.9.18 – An excavated swimming pool showing clay soil above a thick caliche bed. Caliche is Nature's concrete and consists of fine grain sized rock, pebbles, gravel, and sometimes large cobbles, all cemented together with calcium carbonate. It is extremely difficult to dig through and is generally impermeable, which means water cannot flow through it. The mystery is illustrated here, where the layer of soil above the caliche is clay and does not contain the carbonate minerals that cement the caliche bed together. How does caliche form from minerals that leach from the upper soil layer if the upper soil layer does not contain those minerals? To understand how caliche is formed, we must understand the true origin of the carbonates.

white within two days. Left for a few weeks it crumbles to a damp white powder which X-ray analysis shows consists of a mixture of calcite and trona." Note 6.9t

"Dolomite at present time, does **not** form on the surface of the Earth; yet massive layers of dolomite can be found in ancient rocks."

This is certainly not a typical lava flow. Researcher Frederick A. Belton has studied the Ol Doinyo L'Engai African volcano extensively, and made this comment:

"No one knows what causes the lava to flow out at any particular time or how the various vents in the crater are interconnected. Mineralogists would like to understand how the lava evolves under the surface and **why it has its unusual chemical composition, but that also is unknown**..." Note 6.9u

Where do these lava flows, which contain a large amount of calcium carbonate and very little silica, originate? Calcite lava flows become easy to explain if we recognize the true source of lava, which we learned in the Magma Pseudotheory chapter and when we comprehend how calcite is formed.

The Caliche Mystery

What is caliche? It is the bane of construction and excavation companies and requires the use of expensive equipment and is often time consuming to remove. Caliche is essentially Nature's concrete:

"**Caliche** is a hardened deposit of calcium carbonate. This calcium carbonate cements together other materials, including gravel, sand, clay, and silt." Note 6.9v

Fig 6.9.18 is an example of caliche that exists right at the surface under just a few inches of soil. Many landowners in the Southwestern United States quickly become aware of the 'caliche problem' when they are digging building foundations or excavating for a swimming pool, like this residence in Mesa, Arizona, USA. How do geologists think this natural concrete formed? From wikipedia.org we read:

"Caliche generally forms when minerals are **leached from the upper layer of the soil (the A horizon) and accumulate in the next layer (the B horizon)**, at depths of approximately 3 to 10 feet under the surface. Caliche generally consists of carbonates in semiarid regions, while in arid regions, less soluble minerals will form caliche layers after all the carbonates have been leached from the soil. The calcium carbonate that is deposited accumulates, first forming grains, then small clumps, then a discernable layer, and finally a thicker, solid bed." Note 6.9v

If we examine the caliche-formation theory a little closer, we will clearly see it *does not work*. Notice the brown layer of clay in Fig 6.9.18. This top layer would be considered the A horizon in a soil profile. It does not contain the minerals nor does it appear to have ever contained the minerals responsible for the cementation of the caliche formation. Although this is a residential area, the soil profile is very typical of undisturbed land in the area. The upper layer of soil has likely covered the caliche for *hundreds of years*. No natural process has been observed where carbonates are 'leached' into the layers below the soil to form caliche.

Caliche often contains large cobble sized rocks and sometimes even large boulders. Geologists cannot account for how large rocks and boulders become cemented in thick beds of caliche, sometimes covering hundreds of square miles. It would require a considerable amount of water to move such large rocks. Water of that magnitude would create great valleys, but the caliche deposits show no evidence of rivers or ravines. In order for minerals to leach from the A soil horizon to the B soil horizon, those minerals must first be present in the soil. The minerals must be soluble, or rather they would need to be dissolved in order to 'leach' into the lower layer. However, the mineral source rocks, like limestone, are often absent. Of course geologists like to think these processes, now obviously dormant, took place millions of years ago. This is one more proof that the foundation of modern geology—the principle of uniformity—is only a myth.

The caliche mystery is similar to the "classical geological problem" discussed previously in the Sand Mystery subchapter. In that section, we saw granite boulders (Fig 6.3.15) in red sandstone. The red sandstone formation is very near great caliche deposits, and it is not far from the residence in Fig 6.9.18. If caliche was really formed by leached carbonate minerals, and if the geological time concept was real, then we should find layers of caliche throughout the Grand Canyon series. In the Grand Canyon, there are no less than five layers of limestone

Fig 6.9.19 – The beautiful Dolomite Alps in Italy are over 4,000 feet thick. These carbonate massifs are a part of the "Dolomite Problem." Dolomite, which is resistant to acid, is often formed together with limestone but is not forming anywhere today. Thus, the origin of dolomite deposits remains a "problem."

and the canyon is said to be millions of years old. If the classical theory of caliche formation, as it is expressed in the modern geological Rock Cycle is correct, other canyons throughout the world should exhibit layers of caliche. Few of them do, and there is no caliche in the Grand Canyon.

Like all of the carbonate mysteries, to understand the origin of caliche, we must understand the origin of the carbonates, and the importance microbes play in the process.

The "Dolomite Problem"

Could there be even stronger evidence that acid did not form the caves in limestone and the formations within them? The answer is yes—and can be found by examining a close cousin of limestone—dolomite. **Dolomite** has nearly the same chemical formula as limestone with the added element of magnesium. This extra ingredient changes the carbonate in such a way that it becomes harder, denser, and *resistant to acids*. Calcium magnesium carbonate, $(CaMg(CO_3)_2)$ is a common mineral, forming great massifs such as the Dolomite Mountains of Italy, seen in Fig 6.9.19.

Why is there a Dolomite Problem?

"Dolomite, … is a common sedimentary rock-forming mineral that can be found in massive beds several hundred feet thick. They are found all over the world and are quite common in sedimentary rock sequences. These rocks are called appropriately enough dolomite or dolomitic limestone. **Disputes have arisen as to how these dolomite beds formed and the debate has been called the 'Dolomite Problem'. Dolomite at present time, does not form on the surface of the earth; yet massive layers of dolomite can be found in ancient rocks. That is quite a problem for sedimentologists… Why no dolomite?**"
Note 6.9w

Here we see, once again, how the foundation of modern geology—the uniformity principle—fails. Although there are massive layers of dolomite in many so-called ancient rocks, none are forming at the present time. Moreover, the realization that most dolomite formations include a *combination* of both limestone and dolomite, serves to seriously discredit the whole rock cycle pseudotheory on it's own. What we must understand is that **both,** the limestone and dolomite carbonate formations found worldwide, share a similar geological origin. That origin, and the origin of the caves within them, must have been by some process other than chemically acidic waters. This is part of the reason why dolomite deposits are such an enigma in modern geology.

Throughout this subchapter, we have alluded to the answer, which will be more fully explored in Chapter 8, the Universal Flood Model. Did you happen to notice the common link connecting all of the carbonate mysteries? The link that is the key to understanding the formation and origin of all of the carbonates is microbes. But, we're getting a little ahead of ourselves!

6.10 The Loess Mystery

Loess is one of the most abundant surface sediments in the United States. It covers a large part of the great plains of the Midwest and is abundant in Alaska. Worldwide deposits cover thousands of square kilometers and yet its true origin remains unknown. Modern science claims this agriculturally important sediment comes from aeolian (wind) erosion processes, but cannot duplicate it in their own wind-erosion experiments. By geologists' own account, the particle sizes that make up loess sediments are missing from observed erosional processes. The method by which loess is formed is not known in science today. Its origin will be revealed in the Universal Flood chapter.

Loess Defined

If we were to ask a group of geology students what natural sediment covers approximately one third of the United States and is not limestone, sandstone, clay or sand; there would be a lot of confused looks. The answer is a material most people are completely unfamiliar with—*loess* (this German word is pronounced several ways, but *luss* is preferred). **Loess** is a fine, silty deposit, unique in composition and origin to other sedimentary material. Fig 6.10.2 shows what a typical loess deposit looks like. The geology textbook *Understanding Earth*, has the following definition of loess:

"**Loess:** An **unstratified, wind-deposited**, dusty sediment rich in clay minerals." Bib 59 p656

This definition of loess says that loess is unstratified, meaning there are no discernable horizontal layers, and that it is allegedly wind deposited. These definitions do not identify anything special about Loess. Nothing could be further from the truth!

As we begin our account of loess, we ask this question; Do the geological formations of Loess fit the aeolian (wind) deposited explanation? Reading further in the textbook *Understanding Earth*, we are informed:

"As the velocity of dust-laden **wind** decreases, the dust settles to form **loess**, a blanket of sediment composed of fine-grained particles. Beds of loess **lack internal stratification**; in compacted deposits more than a meter thick, loess tends to **form vertical cracks and to break off along sheer walls** during erosion. The vertical cracking may be caused by a combination of

Fig 6.10.1 – These are mountains of loess in China. Local residents carved caves in the sides of many of these mountains for their homes. The origin of loess has been a mystery in geology ever since it was discovered to be a unique sedimentary deposit. Surprisingly, other than for agricultural purposes, little research into the origin of loess has been conducted.

Fig 6.10.2 – This loess deposit in Eastern Arizona illustrates how loess forms vertical columns, and breaks along sheer walls. The deposit is far from supposed glacial related activity that geologists have inferred is the origin of loess. Vertical columns from wind-blown silt deposits seem to defy explanation! At the top of this photo, we see an example of the Surface Gravel Mystery, which will be detailed in the Universal Flood chapter. There are no rivers in the region to explain how this gravel, which covers a very large area, was deposited. Moreover, the gravel did not come from erosion of the topsoil. As can be seen in this image, the soil directly below the gravel is not mixed with any gravel. Where did all the surface gravel come from and what is the origin of loess?

root penetration and uniform downward percolation of groundwater, **but the exact mechanisms are still unknown**." Bib 59 p358

Notice that the horizontal stratification, that should prevail in sediment laid down by wind, *does not exist*. How is it possible that wind-blown silt tends to form vertical cracks and sheer walls? The author speculates but declares there is no known mechanism.

After more than 60 years, researchers still have not been able to identify the true origin of loess. One reason is that no one has documented any modern deposits where Loess is currently being accumulated. Another is that geologists have been unable to explain why Loess, an aeolian sedimentary formation, does not exhibit horizontal stratification as ordinary sedimentary formations do:

"Typically it has **no horizontal stratification**, **like that in ordinary sedimentary formations**, but occurs in **a single massive layer**, 20, 50, or even more than 100 feet thick." Bib 125 p187

One particular type of loess, found in Siberia, is known as **Yedoma**. This deposit has been found in some areas to be tens of meters thick and covering an area of over one million square kilometers. Yedoma is a permafrost loess with a water-ice content of 50-90%.

Geologists long thought that Loess came from glacier activity. It is commonly held that Loess was formed in colder regions of the Earth by the grinding of glaciers during the last glacial maximum. Wind later carried the finely ground rock powder and piled it up in vast deposits. However, loess is found all over the Earth, including many deserts like the Sahara Desert in Africa. The Loess geological mystery has generally been overlooked and not much is said about it in the classroom today.

The Loess Map

When we tried to find a map identifying major worldwide loess deposits, we found out just how overlooked loess is within the geological community. We were unable to locate a definitive map. We did locate various local maps showing known deposits for some parts of the world. Nobody had a single, consolidated map of loess deposits for all areas of the world. Our research identified several international conferences discussing the agricultural aspect of loess and its impact on crop growth, but nothing concerning the origin of what is the third most common sediment in the world. The Loess Mystery is indeed very real. Note 6.10.a contains the six source maps used to compile Fig 6.10.3 – *Worldwide Loess Deposits*.

It is Important to know that the Loess deposit map is not complete. There are large areas on the map that show no loess deposits. This does not mean that they do not exist in these areas; it simply means nobody has documented them yet.

Loess Lacks a Source or Origin

One of the mysteries surrounding loess is that *it has no reasonable, verifiable origin* in modern science. In Alaska, loess is *the largest surface deposit* in the state and yet little attention has been paid to its origin. It was in the 1950s that geologists settled on the idea that loess had a wind-deposited origin. From a USGS research paper on a project in Alaska, the idea that loess is of an eolian origin is conveyed:

"**Loess** (eolian silt) **is geographically the most extensive surficial deposit in Alaska**. In central Alaska, near Fairbanks, loess is several tens of meters thick…Nevertheless, there have been few studies of loess origins in Alaska **and it was not even until the mid-1950's that an eolian origin for these deposits was finally accepted**, based on Pewe's (1955) pioneering study." Note 6.10b

Although this origin has been accepted, it has not necessarily been proven. Still today, the source areas of the loess are not apparent as the researchers identify:

"**The source areas of loess in central Alaska** (near Fairbanks and Nenana) **are not immediately apparent**…The high mica content suggests that local, schist-bedrock-derived particles, carried by first-order streams, is a possible source. However, Pewe (1955) **showed that there are many accessory minerals in the loess that do not occur in the local bedrock**." Note 6.10b

If the loess deposits are eolian as has been inferred, where did the accessory minerals, noted in the article, come from? Reading further, we are informed:

"Local schist bedrock has a highly variable composition, typical of schists in many other regions. **Because the loess has a**

Fig 6.10.3 – Worldwide identified loess deposits (orange areas) as compiled from various maps. (See note 6.10.a for references) Although this is the most complete map of worldwide loess deposits we were able to compile, it is incomplete due to the lack of field research in many countries. Many of the areas showing no loess deposits simply had no data from which we could draw. Loess is far more abundant than what is shown here and surprisingly little research is available on this important surface material.

much more uniform composition, schist is unlikely to be the sole source of the loess, which supports Pewe's (1955) earlier mineralogical work." Note 6.10b

We ask the question—How is it that loess sediment has a "uniform composition" throughout Alaska, while the surrounding mountains are obviously of a different composition?

Fig 6.10.4 identifies a number of loess deposits found in Alaska, but shows none throughout Canada, the vast region to the right of Alaska. This is not because there is an absence of loess or a significant change in geology. Sufficient research into the abundance of loess has yet to be conducted in many areas of the world. However, in many areas like Alaska, where loess deposits have been more thoroughly investigated, loess is recognized as being the **largest surface deposit**!

We learn how abundant loess is from a statement in a relatively current geology textbook:

"Loess covers a huge amount of the Earth's land surface, **perhaps as much as 10 percent of it**. Of this, more than a million square kilometers are in China. The great loess deposits of China spread over wide areas in the northwest; most are 30 to 100 m thick, although some exceed 300 m." Bib 59 p358

After viewing the maps identifying loess deposits, from areas that have had extensive geological mapping one, it would seem that the figure of 10% is very low. Nearly a third of the United States is overlain with loess deposits. These can be seen in the Worldwide Loess Deposit map, Fig 6.10.3.

Fig 6.10.4 – Compare Alaskan Loess Deposits (in red) with those in Canada. Where are the Canadian loess deposits? The geology is not necessarily different. Research and mapping of Canadian Loess is incomplete or nonexistent, as it is in many parts of the world.

The Untold Story of Sedimentology

An important, yet largely untold, story about sedimentology will be unfolded in the next couple of pages. We will learn why sedimentologists in general, have never understood the erosion of sedimentary material on the surface of the Earth. We will discover the real effects water and wind have on sand, silt, and clay, and we will learn how all of this correlates with the Gravel, Sand and Silt Mystery.

Our story begins during the year 1960, with Dutch geologist, Philip Henry Kuenen. Although Kuenen was well known for his work in marine geology, his not-so-well known experiments involving the rounding and abrasion of sediment particles is unparalleled. Kuenen asked tough questions and challenged the prevailing theories of erosion and of loess' origin.

Sand Abrasion by Water

Many scientists have supposed that rivers and streams were responsible for the production and rounding of small grains of sand. Kuenen however, realized this was not the case:

"Since the gravel in a stream bed shows progressive rounding downstream from the source, it has been thought that sand grains also are rounded by transport in running water. Surveys along river courses that showed the roundness of sand grains increasing slightly downstream **seemed** to support this deduction. **Curiously enough, however, a decrease in roundness downstream was detected in some rivers**." Note 6.10c

SUBCHAPTER 6.10 THE LOESS MYSTERY

Kuenen did extensive experiments, showing just how little effect the river currents had in rounding quartz sand particles, and that earlier reports attributing the rounding of sediments to river transport were "exaggerated":

"An extensive series of runs showed that larger grains in a fast current lose about .2 percent of their mass per 100 miles of travel and that medium-sized grains lose only .01 per cent. **This means that to round a .5-millimeter cube to a sphere, the particle would have to be rolled 50 times around the Equator**. The first dulling of sharp angularity requires a 1- or 2-per-cent loss and can happen in the first few hundred miles. But thereafter mechanical abrasion of this kind has little effect on quartz sand with medium-sized grains…

"By using a current in a circular moat, it was demonstrated that the mechanical abrasion of sand-sized particles by rivers is so slight as to have little significance **even for coarse sand and even after the cumulative effect of a succession of sedimentary cycles**…

"Earlier experimental investigations have given an **exaggerated impression of the mechanical action of river transport**." Note 6.10c

Kuenen further noted that the age of the sand had no bearing with respect to the abrasion caused in running water:

"**Abrasion in running water, however, could not in itself account for the roundness of even the most ancient grain of sand**." Note 6.10b p102

This should have been *big news* to the scientific world, especially to the geologists. Yet modern geologists we talked to were unaware of the facts Kuenen uncovered. Abrasion by water cannot account for sand and smaller particles! What other mechanisms are there that can abrade rocks and make sand, loess, and clay sized particles?

Sediment Abrasion by Wind

In the mid 1900s, the 'Great Wind-or-Water Debate' was raging as to *how* loess was created. Investigators believed the source of the loess sediments was either erosion by wind or erosion by water. Actually, no clear physical evidence for either of the erosional processes was evident. Geologist R.J. Russell observed that loess deposits "couldn't possibly be explained by a hypothetical wind":

"Of course, not everyone was transfixed by the powerful image of a continent besieged by dusty wind. In 1944, two years after Smith's publication, respected geologist R.J. Russell declared that loess was ordinarily as well developed on one side of a ridge as on the other, **so it couldn't possibly be explained by a hypothetical wind blowing in a prevailing direction**. He further claimed that **the sorting of loess particles was too uniform to be explained by either the deposition of wind or water**, and that, based on his own field observations, especially with regard to stratigraphic relationships, **wind just didn't seem to have anything to do with the whole matter**." Note 6.10d

So it is that the sorting of loess cannot be explained by "either the deposition of wind or water." Still, no discovery has settled the issue:

"Years passed, but no single great discovery ever settled the issue. Not even to this day." Note 6.10d

Returning to Kuenen's story, he performed experiments that *should have* settled the issue. He dismissed water as the agent responsible for abrasion and turned to the next logical means of change—*wind*:

"The cause of roundness in sand grains **must therefore be sought in some other mechanism. Perhaps the wind is the agent**. Well-rounded grains are abundant in deserts and dunes, and geologists have long suspected that transport by wind powerfully abrades the grains." Note 6.10e

Kuenen's extensive experiments on the effects of wind on sand grains and abrasion produced "unexpected" observations:

"By this means we have made the **unexpected finding that transport by wind causes quartz grains to lose 100 to 1,000 times more mass than water transport cause them to lose over the same distance**. Moreover, wind abrasion reduces quartz almost as rapidly as it does feldspar or limestone. Apparently the brittleness of quartz causes it to flake off in the impact of a bouncing grain against a stationary one, on the other hand, we found that well-rounded and polished quartz-grains remain perfectly intact even after prolonged, violent wind action. It must be that they rebound elastically, as billiard balls do." Note 6.10e

While reading this, one can see how scientific investigators assume wind is the agent responsible for the erosion of sediment, including fine sediments like loess. However,

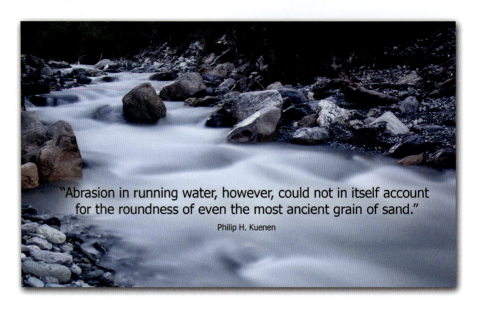

"Abrasion in running water, however, could not in itself account for the roundness of even the most ancient grain of sand."
Philip H. Kuenen

"This means that to round a .5 millimeter cube to a sphere, the particle would have to be rolled 50 times around the Equator."
Philip H. Kuenen

"Particles with a diameter 0.1 mm or less show no abrasion at all."
Philip H. Kuenen

as we read on:

"In tests of samples containing a wide assortment of grain sizes we found that **the smaller the particle the less abrasion it suffers. Particles with a diameter 0.1 mm or less show no abrasion at all**. Apparently these smaller particles cannot hit one another with sufficient momentum to cause a crack in a grain." Note 6.10d p105

Kuenen's scientific observations were especially significant when we understand that particles of 0.1 mm and smaller include these two groups of material:

1. Very fine sand.
2. Silt, which includes *loess*.

Naturally, we are led to ask this Fundamental Question:

> If neither water nor wind rounded the fine particles of which loess is comprised, how did loess grains become rounded?

Remarkable Results With Respect to Loess

An observation Kuenen made during his aeolian experiments was that tiny sand chips that are residual from wind abraded quartz sand, are not found in loess. Loess is primarily quartz, but the particles in loess are smaller than those produced by wind abrasion. Kuenen actually showed that loess must have been formed by some method other than by wind:

"**Curiously** one does **not** find the tiny sand chips in the loess laid down by ancient dust storms. The bulk of loess is quartz, but **its particles fall outside the size range produced by wind action on sand. This shows that the loess must have been formed in some other way**…" Note 6.10e p106, Note 6.10f

Kuenen observed that loess had to have been formed in some way other than wind erosion because the larger sediment particles, when blown and abraded by the wind, produced particles either *smaller than 10 microns or larger than 50 microns*. Particles *between 10 and 50 microns* just happen to be the size of the particles that make up the largest percentage of loess:

"These **results are remarkable** in connection with the **origin of loess**. The bulk of these eolian [wind] deposits is quartz of **grain sizes between 50 and 10 microns**. **This is just the size that is absent in the 'experimental quartz loess**.'" Note 6.10g

Experiments designed to produce loess-sized particles of between 10 and 50 microns failed. The most common size range of particles found in loess was noticeably absent in Kuenen's eolian experiments. What Kuenen did notice was that only the larger particles (larger than 50 microns) and the smaller particles (smaller than 10 microns) came from wind abrasion and he realized that the grain sizes between 10 and 50 microns "must have been available from the start":

"The following deduction can be made. Part of the major-size fraction of loess is formed by eolian attack of feldspar, but **the abrasion products of quartz are only represented in the finer and coarser elements of loess**. All the quartz and a significant part of other minerals in the major-size fractions **must have been available from the start**." Note 6.10g

If the individual particles were available "from the start" as Kuenen concludes, we can further reason that those individual loess particles must have *formed as they accumulated* in the massive deposits around the world. In other words, the initial "start" was not from rocks or mountains eroded by wind. Moreover, it could not have been a river that laid down such massive loess deposits. Some areas, such as the loess plateau in China are known to be hundreds of meters thick and cover thousands of square kilometers. Transport of massive amounts of loess sediment must have been a very unique process as the particle sizes are very similar, even across great distances, showing "relatively little variation in the median diameter":

"They [loess grains] have been concentrated by selection during transportation. This conclusion finds strong support in **the fact that there is relatively little variation in the median diameter of loess from different localities**. This holds even for the Chinese loess, although it is derived from a desert, in contrast with the periglacial loess of Europe and North America." Note 6.10g

We observe that the median size of loess particles is relatively uniform worldwide. This holds true whether the loess is found near where glacial activity is thought to have occurred or whether it is found in the desert where no glacial activity has been identified. These four geological facts have been overlooked:

1. Loess exhibits a vertical columnar formation and not one of horizontal layering.
2. The unique combination of minerals is not linked to the minerals of neighboring mountains.
3. The particle sizes of which loess is formed must have come about by some means other than erosion.
4. There is a relatively minor variation in the size of loess particles worldwide.

With these observational constraints, a *feasible mechanism*, capable of forming the world's loess deposits must be identified. Since loess deposits are found only on the surface, this now-dormant deposition event must have been relatively recent and must have occurred worldwide. It doesn't take much imagination to see where this and a host of other geological facts surrounding sediments are leading us, the details of which will be outlined in the Universal Flood chapter.

> "This shows that the **loess must have been formed in some other way**…"
>
> Philip H. Kuenen

Experimental Confirmation from 2004

Since the 1960s, when Kuenen conducted his detailed investigations, there has been little research successfully contradicting his experimental results. One particular account was published in the journal, *Nature*, in 1982. In the article, three researchers were attempting to explain how Kuenen's results were incorrect by indicating that the loess sized particles (10-50 microns) *were* being created in wind abrasion. Note 6.10g

However, the experiment and the resulting misguided paper were unscientific and lacked quantitative observations whereby a real assessment of the particle sizes from their experiment was shown. This is apparent in their conclusion, which states no convincing evidence whatsoever:

"Finally, it **seems** that an aeolian abrasional origin for **some** loessic material quite **possible**. It is **unlikely** that much European **loess is produced in this way** even though long distance travel from the Sahara is **possible**." Note 6.10h

Without a Universal Scientific Method, science has no credibility. The authors of this article disagreed with the facts Kuenen had discovered, even though Kuenen used quantitative experiments that were repeatable and predictable. They made claims discrediting Kuenen's experiments using only their own statements, without including specific data to support their claims and they showed no repeatable experiments.

Not much research seemed to happen over the next 22 years. Then in 2004, a team of UK and Australian researchers conducted a well-documented experiment with qualitative data. What was their conclusion? Kuenen's test results were corroborated. The essence of this recent experiment on loess-sized particles generated by aeolian abrasion showed the resultant material to be of the same size range Kuenen identified; smaller than 10 μm (microns) and larger than 50 μm:

"Samples with a mix of particle sizes, including those <250 μm, **produce fine [grain] particles in one of two main size classes, < 10 μm and > 50 μm**." Note 6.10i

At the microscopic level, both air and water mediums *are too thick for abrasion to take place* between certain sizes of particles. This is why particles under 50 microns are not easily abraded. When they are larger than 50 microns, pieces are broken off until the particle is approximately 50 microns, at which time it no longer sustains abrasion. These *broken pieces* account for the particles that are less than 10 microns.

In 2004, better technology was available to the researchers and they were able to identify even more clearly that when larger sized particles are blown against each other, they do not produce sufficient amounts of loess-sized particles to account for the abundance of loess in Nature. Thus, Kuenen's results still stand, and the processes that formed loess must have been formed by a yet unidentified method.

The Missing Sediment Mystery

Because modern geology has no known mechanism from which to derive the origin of loess, it has been routinely asserted that erosion, more especially aeolian, or wind erosion is the source of loess sediments. However, we have shown how aeolian studies and lab experiments have failed to produce evidence that loess-sized sediments (10 μm (0.01 mm) to 50 μm (0.05 mm)) are created by erosional forces. When we look for corroboration in Nature, we are confronted with another perplexing mystery of missing sediment sizes. The **Sediment Size Table** in Fig 6.10.5 identifies two groups of sediment sizes that are scarce or missing in the geologic record, in areas of *observed aeolian or fluvial erosion*. In other words, the two groups of sediment are

> "So far as **quartz sand is concerned... abrasion plays no effective part in its formation.**"
>
> I. J. Smalley

noticeably absent in sediment observed to have been created by erosion. This is the mystery of the **Missing Pebbles** and the **Missing Sand and Silt** segments.

Recognizing that these particle sizes are missing in the sedimentary record of the Earth, geologist I.J. Smalley said:

"There seems to be some **confusion over the role of abrasion in the formation of sediments**." Note 6.10j

Smalley, after observing the deficiency of certain sized sediments, continues with more detail about these two missing sediment particles:

"**It is observed that there is a scarcity of certain particle sizes in sediments**. Tanner quotes **0.03-0.12 mm and 1.0-8.0 mm as the particular values. These, effectively divide sediments into different types: gravel, sand and silt**. The relevant units may be called blocks, grains and flakes. The experiments of Rogers *et al.* were designed to explain the scarcity of 0.03-0.12 particles by proving that there are two distinct modes of abrasion, one giving sand (grains) and the other giving silt (flakes). **So far as quartz sand is concerned, it seems more reasonable to suggest that abrasion plays no effective part in its formation**." Note 6.10j

Smalley reasoned that abrasion played no part in the formation of quartz sand! This is an interesting comment about the most common sediment on the face of the Earth.

Boulders, pebbles, sand, and clay are typical of many rivers. The river in Fig 6.10.6 is strewn with various sediment materials, ranging in size from large boulders and cobbles to sand and fine clays. Unless the individual particles sizes were cataloged, it might never be observed that there are two groups of sediment that are missing—and this is true of rivers around the world.

If all of the sedimentary materials on the Earth have come from the slow, millions-of-years processes of erosion and abrasion, this record should be easily verified in the geological strata. However, *it is not*.

Geologists have assumed all sedimentary materials are essentially formed when boulders break into smaller sized cobbles and successively smaller gravel, pebbles, and then sand. However, because of the missing particle sizes identified in the Sediment Size Table, considerable doubt is cast on the erosion sci-bi of geology, and the entire geological time scale. This will be discussed in more detail in the Age Model chapter.

Because we employed the fundamental UM principle of questioning everything with an open mind, we became aware of the rarity of pebbles in typical fluvial sediments, and began a search to see what other researchers had observed. Other researchers had noticed this anomaly in the geological record, but to our surprise; the geology textbooks remain silent on this fact. These books still advance the sci-bi of erosion as the origin of all of the sand, silt, and clay, and students are taught the theories of erosion and abrasion without the support of facts or field observations.

The peculiarity of the missing sizes has been known for some time. In an article from 1963, in the journal *Science*, investigators acknowledge the small particles are not and have not been formed by "normal stream abrasion." They further recognized that a crushing process, typical of what one would expect from weathering, does not create these small particles either:

"The general ineffectiveness of normal stream abrasion on particles smaller than 1 to 2 mm **precludes the formation of silt and clay by continued size reduction of coarse materials**. Furthermore, lognormal distributions are **not** formed by single-stage **crushing processes**." Note 6.10k

It seems amazing that the knowledge of the missing sediment sizes of pebbles, sand and silt has been a known 'fact' amongst geologists for many decades:

"The **fact** that detrital sediments in general appear to be deficient in certain sizes **has been known for many years**. Pettijohn (1949, p. 41-45) summarized previous work on this problem. His summary, covering **thousands of samples ranging from silt to gravel, made clear the deficiency of materials in the 0.03-.12 and 1.0-8.0 mm sizes**." Note 6.10l

Francis J. Pettijohn, in his book, *Sedimentary Rocks*, published in 1957, cites Udden's (Notes: Udden's 1914 work page 741) work from 1914 where he noted the scarcity of certain sized particles in stream deposits and aeolian sediments. Unfortunately, like so many scientists who have spent their entire careers in fields like this, they have been blinded by the Rock Cycle and Erosion Pseudotheories. As a result, they find:

"**It is difficult to see, under any of the hypotheses** accounting for the apparent deficiencies of certain sizes, if these sizes were produced by weathering or abrasion, **what has become of them**." Note 6.10m

Few scholars in sedimentology have seen what Kuenen recognized; that because erosion was not forming all of the sediment on the Earth, the loess sediments "*must have been formed in some other way.*"

The Loess Mystery Summarized

In this subchapter, we have shown:

1. No clearly defined origin for loess can be found in modern geology.
2. No verifiable mechanism of how loess was deposited.
3. No comprehensive study of loess has been performed.
4. A complete worldwide map of loess deposits has not been made.
5. Loess is a part of the Untold Story of Sedimentology.
6. Loess is a part of the Missing Sediment Mystery.

Before we leave the subject of loess, there are a few other important questions to consider about this widespread mineral. Although we do not go into too much detail here, answers to these questions will be discussed in the next two chapters.

1. Where are the 'in process' loess deposits occurring to-

Fig 6.10.5 – This Sediment Size Table has been adapted from the Wentworth Scale and includes Universal Model related comments. The table illustrates the different size categories of sediments and identifies two missing segments that are not accounted for by modern geology. We identify these segments as Missing Pebbles and Missing Sand. They are missing from observed river and aeolian sediment deposits. This mystery must be accounted for by any geological model that is to be held as scientific truth.

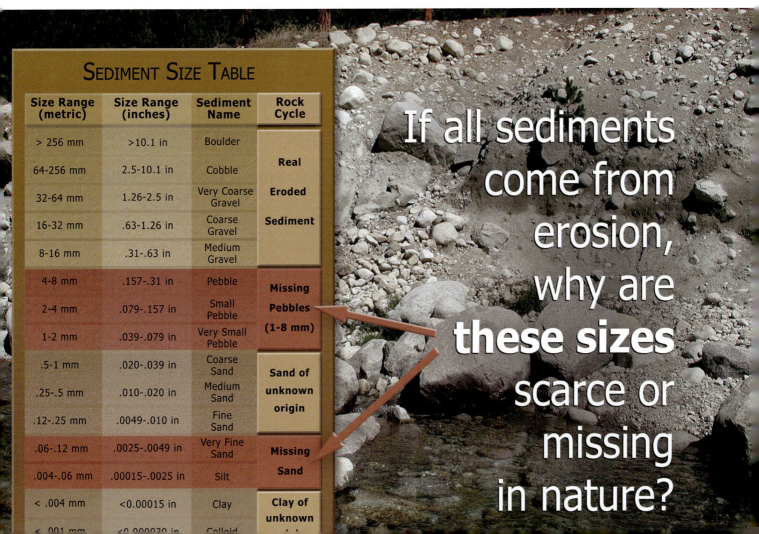

Figure 6.10.6 – This riverbed exhibits various sizes of sediment. Although it is not apparent at this scale, if we were to observe and compare the sediment particles and divide them into the size groups of the Sediment Size Table, two sizes of pebbles and sand would be noticeably missing. If geologists are aware of this fact, they seem to have swept it under a 'geological rug!'

day? If the deposition-by-wind theory of loess is correct, there should be evidence of loess being deposited worldwide today. Where is this evidence?

2. If loess is derived from wind, why does the loess in Siberia (Yedoma) consist of up to 90% frozen water?

3. How do loess columns form by wind, over time? Loess columns break apart in a vertical, 'crystalline' form. If sand, silt, or dust were piled up by wind, there would be a pyramidal form of various particle sizes—not columns of a comparatively crystalline form. When natural water runs through loose loess grains, it does not cement the loess or form the types of structures seen in SEM (Scanning Electron Microscopy) micrographs of loess. Pressure alone does not create stable loess deposits, yet existing loess is stable enough that homes have been carved into the loess mountains on the loess plateau in China.

4. If loess is deposited by wind, why is the worldwide mineral composition of loess so similar? This similarity has been known for many years:

"…**peculiar** yellowish, fine-grained sediment that covers vast areas in Asia, Europe, and North and South America… This sediment, **so similar in widely separated continents**, is known by the German names *loess* (pronounced *lûs*)." Bib 125 p187

Loess can be identified by both its physical and chemical properties. The landscapes where deposits occur are obviously not the same. No two of the worldwide locations where loess is found have the same geology. Thus, it is physically impossible for loess to have been eroded from these very different geological landscapes, yet retain the same unique physical and chemical characteristics. This fundamental question cannot be answered scientifically by the wind erosion sci-bi.

5. Why is loess unstratified? Where are the layers within the loess deposits? If loess was deposited by wind, it would have taken thousands of years to accumulate the amounts of sediment found today. Are we to assume that loess was deposited over thousands of years, and not a single flood event or adverse weather condition caused a discontinuity in the loess deposits, where new sediment would have be deposited? According to modern geology, loess *should* be deposited, eroded, and re-deposited in an endless cycle of erosion and accumulation, and this should be evident by viewing stratified layers. Where are those layers?

6. Where are the multiple layers of loess, deep within the Earth's crust? Did wind blow only in the last several thousand years of the supposed billions of years of Earth's existence? Why are loess deposits found **only** on the surface of the Earth today? Why are there no layers of loess in the Grand Canyon or any other deep sedimentary layers?

7. Why is worldwide loess found most often in lower elevations and in basins? If loess is silt formed from glaciers in the mountains, where is the trail of loess from the high glacier riverbanks in the Rocky and Canadian Mountains, leading to the Mississippi Valley?

6.11 The Erosion Mystery

In the 1850s, gold miners in California employed the use of water to speed the collection of gold in the gravel foothills of the western Sierra Nevada Mountains. One hundred and eighty-five thousand cubic feet of water per hour was piped from upstream reservoirs and forced through nozzles with spectacular force, stripping and laying waste everything in its path. This

Fig 6.11.1 – This eroded valley is typical of what geologists might easily say took thousands or even millions of years to form. This valley and its accompanying landforms were shaped in a matter of weeks. The eruption of Mt. St. Helens occurred here in 1980 bringing many erosional forces to bear over a short period. We will look at specific examples of erosion in this subchapter and will find that the long geological time scales and erosional sci-bis of modern geology are not valid explanations for many of the landforms found in nature.

CHAPTER 6 THE ROCK CYCLE PSEUDOTHEORY

Joshua Tree Granite Mystery

Where are the piles of granite under these rounded boulders?

Fig 6.11.2 – Well-rounded granite formations can be seen in Joshua Tree National Park, California, USA. When the granite rocks do erode, pebble-sized chunks of the granite are left behind. Examples of these can be seen on the hiking trail (inset top). Where are the piles of granite that should exist beneath every well-rounded rock? The eroded rocks are far too large for wind to move them and there are no rivers in the area that would have carried the material away.

---

incidence of man-made erosion was an ecological disaster and was finally banned in 1884.

Erosion by wind and water is an ever-present and active landscape shaping force. Observed erosion has been extrapolated by scientists and mingled with theory as they attempt to explain Earth's landforms. The truth is that erosional processes observed or recorded in historic times cannot account for many of those landforms. We will consider seven of them in this section.

Erosion Defined

Before we consider the processes of erosion, some clarification about what erosion means is in order. According to modern geology, erosion is defined as:

"The general process or the group of processes whereby the materials of the Earth's crust are loosened, dissolved or worn away, and simultaneously moved from one place to another by natural agencies, which include weathering, solution, corrasion, and transportation, but usually exclude mass wasting." [Bib 173 p217]

Webster's Dictionary defines **erosion**:

"…the process by which the surface of the earth is worn away by the action of water, glaciers, winds, waves, etc." [Bib 7 p272]

The term **denudation** is broader in scope than erosion, but is sometimes used synonymously with erosion in geology. These definitions seem straightforward and simple and they define the processes we see taking place around us today. For example, when heavy rainfall fills a river to overflowing; we see erosion on its banks.

Why is erosion a mystery in geology today? Both the Sand Mystery (subchapter 6.3) and the Loess Mystery (subchapter 6.10) are a part of the Erosion Mystery. The Missing Sediment Mystery is a dramatic example of how erosion cannot account for all of the sediment found on the surface of the Earth. Erosion is real and erosional processes can be observed easily. However, the fundamental question left to answer is whether the erosion we observe can explain all of the Earth's landscape or not. We already know from the previous subchapter on loess that there are some sizes of sediment missing in active erosional environments. The Sediment Size Table indexes sedimentary particles from the smallest colloids to the largest boulders. Currently, there is no erosion model complete enough to explain all of the sediment sizes that exist, and that can account for the sizes that are missing.

The Erosion Mingle

The simple fact of erosion has been mingled with the false idea that erosion is the cause of all of the Earth's sediment. Erosion is an actual physical process that is taking place all around us. Wherever we live, the soils and rock landscapes are in a constant state of change. Rocks do break down over time, wind etches and abrades particles, both large and small, and rivers of water continue to shape the valleys where they flow. However, when these facts are mingled with theories to explain how ***all*** geological landforms are created, the theories fail miserably.

In Fig 6.11.1 we see an example of a heavily eroded valley. It is apparent that water was the active agent in the carving of this valley. How long did it take this valley to erode into its present condition? It did not take millions or even thousands of years, but *only a few weeks*. The destructive forces unleashed by Mt. Saint Helens in 1980 moved millions of tons of dirt and rock across whole valleys. Mudflows carried enormous quantities of thick, sediment-laden water through the valley, carrying material some 60 miles to the Columbia River, where it blocked transportation. Today, the Tootle River continues to carry sediments from the site of the eruption through the valley to the Columbia River.

Since the eruption of Mt. Saint Helens over two decades ago, some geologists have begun to recognize that the principle of uniformity might not hold up as well as they had maintained. Long held pseudotheories die hard, and so it is with uniformity.

There are numerous of examples of how geology has assumed that the erosion processes active today shaped all of the Earth's landforms. To assume this requires the geologist to simultaneously ignore some of the facts. Although there are many mysterious erosion related events, we have chosen seven to represent the Erosion Mystery. Keeping an open mind while questioning the current beliefs about these mysteries will open a door of understanding and further knowledge.

Fig 6.11.3 – Beautiful and massive granite boulders adorn the area around Prescott Arizona, USA. Where is the loose, decomposed granite between and below the well-rounded boulders? The inset right image shows loose granite at a road cut in the mountains near San Diego. Several feet of loose granite has accumulated at the bottom of the road cut.

no rivers running through this desert. Many rounded granite formations reside at the tops of the hills and are only subject to occasional rain. Water to erode these granite boulders is certainly scarce. Wind cannot transport the material away, due to the large size of the granite detritus that occasionally breaks off the blocks. If water or wind were able to move large pieces of granite from beneath the boulders, where did the material go?

Many well-rounded granite formations exhibit a hardened outer coating, leaving them more resistant to erosion. They probably have not changed very much since they were originally formed. There is no evidence of larger granite cobbles weathering into smaller pebbles on a significant scale today.

Some of the granite boulders in Joshua Tree National Park are composed of a particularly 'soft granite,' that can be rubbed off with your hands. In several areas, small amounts of eroded granite are on the ground beneath the boulders, but these typically amount to less than a few centimeters in depth. This material often lies atop a clay deposit. The small amount of eroded granite makes up less than 1% of the granite material that presumably came off the original unrounded boulders.

There is no evidence of a process that would move the hundreds of feet of decomposed granite away from the parent formation, and there is no 'trail' of granite debris emanating

Seven Examples of the Erosion Mystery
1. Granite Boulders
2. Arch Formation
3. Soil formation
4. Skipperocks
5. Planation
6. Pedestal Formations
7. Alluvial Fans

These seven were chosen because examples of them can be found almost everywhere. You are encouraged to observe these mysteries first hand, and you can easily do so at sites similar to those mentioned here.

1. The Granite Boulder Mystery

Granite is said to be subject to spheroidal weathering. This type of erosion is a form of chemical weathering where the surface of the granite is loosened by water and separated from the block. If this was really the method that formed the rounded boulders, where are the 'piles' of weathered granite beneath the boulders?

Joshua Tree National Park (Fig 6.11.2) is situated in the Mojave Desert, in Southern California, USA. There are

Fig 6.11.4 - Sandstone arches in the Arches National Park, Utah. Still forming modern arches have large rocks and debris below them. Where did the fallen sandstone blocks from the ancient arch formations go? This is a mystery in geology.

CHAPTER 6 THE ROCK CYCLE PSEUDOTHEORY

from the hundred-foot high granite outcrops. This suggests they were formed recently and quickly, not over millions of years as has been supposed. If the rocks had laid here for any amount of time, an accumulation of eroded material would certainly be present. The amount of eroded material that does exist around the boulders at Joshua Tree National Park would have taken only a few hundred or maybe a few thousand years to accumulate.

The erosion mystery in Joshua Tree National Park is not an isolated example. Many instances of rounded granite outcrops and boulders, like those found in the Park, can be identified all over the world. Another example of rounded granite boulders occurs near Prescott, Arizona, USA (see Fig 6.11.3).

The lower right inset of Fig 6.11.3 is of eroded granite, exposed when a portion of the mountain was removed to construct a road in the mountains near San Diego, California, USA. The less than 100-year-old road cut shows several feet of loose material built up from the exposed granite and is still lying below the original deposit. Bear in mind that this material is on a sloped hill, whereas the granite boulders in Joshua Tree Park exist on relatively flat plains.

2. The Arch Formation Mystery

Arches National Park in Utah, USA is said by many to be one of the most inspiring places on the planet. Hundreds of unusual and unique geological formations exist there. The highest density of natural stone arches in the world is found at Arches National Park. When one first encounters the majestic sweeping arch of Double Arch (Fig 6.11.4 Background), no words can be found, and a moment of awe-inspired silence takes hold.

Naturally, the first question that enters the mind is the same one so many before have asked:

How were the majestic arches formed?

Simple questions deserve simple answers. However, this is not what is found in the Arches National Park brochure or on the official web site. The answer to how the arches formed, according to the National Park Service web site:

"Salt under pressure is unstable, and the salt bed below Arches began to flow under the weight of the overlying sandstones. This movement caused the surface rock to buckle and shift, thrusting some sections upward into domes, dropping others into surrounding cavities, and causing vertical cracks which would later contribute to the development of arches.

"As the subsurface movement of salt shaped the surface, erosion stripped away the younger rock layers. Water seeped into cracks and joints, washing away loose debris and eroding the 'cement' that held the sandstone together, leaving a series of free-standing fins.

"Ice formed during cold periods and its expansion placed pressure on the rock, breaking off bits and pieces, and sometimes creating openings. **Many damaged fins collapsed**. Others, with the right degree of hardness and balance, have survived as the arches we see today." Note 6.11a

To make a point, the Park Service's answer to arch formation is analogous to explaining that $2 \times 3 = 7.418$.

In Nature, the simple truth is, that *truth is simple*.

The 'official' answer is absurd and requires so much imagination that we can rest assured the 'experts' do not really know. The truth is just not that complicated.

Fig 6.11.5 – The Soil Mystery is well represented in these photos. In each image, many layers can be seen. Some contain thick beds while others contain many thin laminations. Each is thought by modern science to have been laid down over long periods of geologic time as seas came and left and environments changed. One commonality among formations like these is that there are no organic layers within them. This is "not what one would expect" as one geologist put it. Almost every type of soil can be recognized as a mystery in today's geology—if we ask the right questions.

From what we have been able to find, geologists have no empirical evidence that underground salt has had anything to do with arch formation. Geologists and park rangers have observed that most arches occur in the fins where the sandstone walls are thinner, but this is not always the case. Double Arch is seen in Fig 6.11.4. These arches formed at nearly ninety-degree angles to each other, demonstrating that arches do not form exclusively along fins. The way in which natural stone arches form can be seen in the leftmost image in Fig 6.11.4. Large blocks of sandstone lie at the base of the developing arch. The sandstone blocks are evidence of real erosional forces at work today.

> Where is the debris that fell from Double Arch or Delicate Arch in Arches National Park?

If we look beneath most of the natural arches, **little or no debris can be seen**. Why is this? Geologists would like to know too. Speculations abound about how this debris disappeared or where it went. Before real answers about these landforms can be identified, we must have a correct knowledge of how sandstone is made.

3. Soil Mystery

Pedology is the scientific study and classification of soils and their origin. There has been a great deal of research devoted to the study of soils and their characteristics, but like all fields of modern science, the ideas of soil *origin* are based on the Rock Cycle Pseudotheory.

According to the Rock Cycle Pseudotheory, a hard crust formed over the Earth as magma cooled. Sometime during a multi-billion year life, the oceans appeared from an unknown source. This started the weather processes that began to erode the crust of the Earth. The now-buried bedrock would have been higher in elevation where it would have broken apart to move to lower areas. Water might transport some of this material to lower basins. As erosion continued, small-sized sediment and material from great distances would be transported by storms, leaving behind thousands of layers of heterogeneous material, over millions of years.

The Earth is a dynamic planet, and with few exceptions vegetation can be found on nearly every square foot of the continental crust. Dead vegetation accumulates to form an organic layer of soil. In some highly active areas like the flood channels of the great rivers, the organic layers are constantly being moved about. In other areas of the world, where vast plains, deserts, and mountains are found, the soil is left undisturbed for long periods. It is in these latter areas that we will look to identify a puzzling mystery in modern geology.

The mystery to which we refer is the Soil Mystery. Where are the layers of organic soil in Fig 6.11.5? Landforms like these in Southern Utah can be seen worldwide showing similar layering. Thousands of thin laminations of sediment have been laid down—with no organic layers between them. This is not what we should expect to find if the layers had been laid down over millions of years.

> Where are the multiple **missing** layers of organic soil? These would **have to exist** between the layers of eroded sediment, subsequent to each of the numerous environments that supposedly happened over geological time.

In the book *Soils of the Past* we read the unexpected observation of what really exists. The researcher explains that ancient soils (Precambrian paleosols) are not "what one would expect" because they appear to be "stable," and not subject to the normally expected erosion and weather:

"The Jerico Dam paleosol below the 3000-million-year-old Pongola group in South Africa, for example, now has about 50 vol. % clay to a depth of **at least 6 m** [meters]. The 2300-million-year-old Denison and Pronto paleosols of Ontario, Canada, are also impressively well developed. **These represent landscapes that were stable for hundreds of thousands, if not millions, of years. This is not what one would expect on an**

Fig 6.11.6 – Organic soil layers can be easily seen at excavation sites or at road cuts like this one in Deland, Florida, USA. Wetter environments usually have deeper and darker organic layers. In this photo, the organic layer is a thin, dark layer just at the surface. Notice there are no other organic layers in the profile. Where are they? If the environment was constantly changing as explained in modern geology, there should be many layers of tens-of-thousands of years old organic soil. They do not exist.

> "These represent landscapes that were **stable** for hundreds of thousands, if not millions, of years. **This is not what one would expect** on an abiotic [absence of life] landscape."
>
> G. J. Retallack

abiotic [absence of life] landscape. Under abiological conditions each mineral grain, as it was loosened from bedrock by chemical and physical weathering, should quickly have been removed by surface erosion if it were appreciably above the water table. As a consequence, **Precambrian paleosols on bedrock should be sandy and thin rather than clayey and thick** as some of them were. Under abiotic well drained conditions also, soils formed on clay deposits **should have been thin and shallowly cracked** like those forming in desert badlands of the western United States. Yet Precambrian paleosols on shales such as the Waterval Onder paleosol **are not of this kind**." Bib 118 p 369

The author goes on to say this is not a rare occurrence, but states "many" of the ancient soils on bedrock are thick and clayey:

"At the simplest level **a soil can be regarded as** a natural body formed by the **accumulation of organic and inorganic materials at the Earth's surface which differs from the underlying material** in its morphology and its physical, chemical and biological properties." Note 6.11b

This definition of soil is from a book written about ancient buried soils and the environments in which they exist. Fossil soils are referred to as **paleosols**, and they consist of organic and inorganic materials. We are interested in the organic layer of soil. Usually, the organic part of the soil can be easily distinguished from inorganic sedimentary material. When organic vegetation dies, it leaves behind decaying debris that accumulates into an organic layer. This layer does not just disappear.

Road cuts are an easy way for everyone to observe soil layers. Soil profiles can also be seen at excavation sites where basements or building foundations are being dug. Look for a dark layer of organic material at the surface. It will usually be above lighter layers of dirt.

At a road cut in Deland, Florida, USA, (Fig 6.11.6) we see a vertical soil profile of approximately 20 feet. On the surface, a dark, distinct layer of organic soil can be seen. Where are the paleosols, or ancient organic soil layers below the surface?

Geology teaches that continents were uplifted and lowered over millions of years. During that time, seas are supposed to have come and gone and a variety of environments should have existed, where all sorts of vegetation would have been allowed to grow and die. In these varied and multiple environments, the vegetation should have produced *many layers of organic soil material,* in various stages of preservation. Is this what we observe in Nature?

The Grand Canyon in Arizona exhibits many different layers of sandstone, limestone, shale, and other rock. Many of these layers are themselves layered with hundreds or even thousands of thin laminations, each with a story about how the depositional environment changed. Modern science has looked to the Grand Canyon as the seminal record of tens-of-thousands, even millions of years of geological history and as a record of environmental change.

Where are the layers of organic sediment in the Grand Canyon series that must have been present at times in the modern science version of the canyon's history?

Every geologist who has spent time in the field, studying the Grand Canyon or a myriad of other similar landforms, should

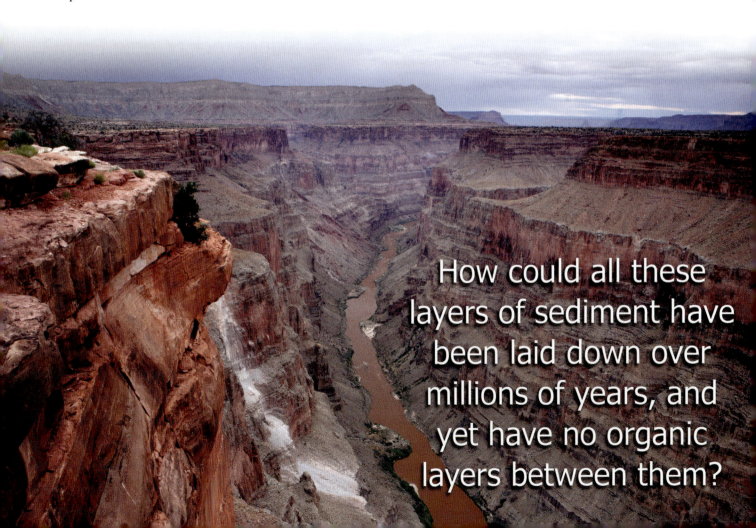

How could all these layers of sediment have been laid down over millions of years, and yet have no organic layers between them?

The Skipperock Mystery

Fig 6.11.7 – 'Skipperocks' are oval-shaped, relatively flat and smooth. These rocks are found in a band of sandy sediment above the shoreline at Carlsbad Beach, California, USA. They are a mystery because of the fact that no riverbed anywhere contains just Skipperocks. So where did they come from and how were they formed? Although the answer is not found in modern geology, it is given in the next two chapters—the Hydroplanet and Universal Flood Models.

have misgivings about the story of the canyon's formation because of the Soil Mystery.

4. Skipperock Mystery

Perhaps you recall skipping rocks across a pond as a youth. The search was on to find the smoothest, flattest rock to achieve the most 'skips.' The perfect rock was often hard to find. However, in some areas of the world there are millions of 'Skipperocks'—all found in the *same* deposit! Fig 6.11.7 shows photos of Skipperocks at Carlsbad Beach in California. These images dramatically illustrate one of the most interesting mysteries in the Rock Cycle Pseudotheory.

How did Nature deposit all the Skipperocks together? To recognize how unique this deposit is, compare the image of a typical streambed in Fig 6.11.8 to the photos of the Skipperocks. Although a few of the rocks are rounded, notice the vast array of shapes, sizes, and textures of the rocks. There is a significant difference between the rocks in this riverbed and the skipperocks.

In addition to the unique shape, size and texture of skipperocks, there is another huge mystery right in front of our eyes! In

Fig 6.11.8 – Rounded rocks found in a typical streambed. Notice how the size, shape, and texture differ from the Skipperocks. The Rock Cycle cannot explain where Skipperocks came from.

Fig 6.11.9 – These California, USA, foothills west of Carlsbad, from which the red and grey sandstone and Skipperocks in Fig 16.11.6 should have eroded, are actually yellow granite. This is an example of a mystery that remains unexplained in the Rock Cycle Pseudotheory.

what sediment are the skipperocks deposited? They are deposited only in sand.

These observations lead us towards two critical FQs:

> Do we find smooth, flat, oval shaped rocks forming unaccompanied in any riverbed, anywhere?

> How could Skipperocks have eroded from a mountain, and then be deposited with only sand sized sediments?

What happened to the rest of the rocks and sediment with sizes between the sand and the skipperocks? If you look closely at the photos in Fig 6.11.7, you will see they are not there! How can this be? How does water deposit 3-6 inch sized rocks in *sand* with no other sizes present?

The Skipperock photos illustrate another peculiarity unanswered by the modern science erosion story. There are two *completely different* layers of sandstone in the top-right photo, Fig 6.11.7. How did the layer of red sand come to lay on top of the grey sand? This question is even more troubling for geologists when we recognize that the California foothills to the east of the beach are *yellow granite!* It is from these granite mountains (see Fig 6.11.9) that presumably the red and grey sandstone and the skipperocks came. The sandstone minerals are quite different from the granite, yet the processes of erosion say the sand must have come from some place nearby.

Instances like this abound, all over the globe. Many geologists acknowledge that there are unanswered problems in their individual fields of study, but they often have no idea how serious and widespread the issues really are. Answers to many of these problems would have a significant impact on the widely accepted theories of the day.

Near Carlsbad Beach, there are a number of respected universities with well-developed geology departments, but we were unsuccessful in identifying anyone there who had observed the Skipperock anomaly. Personal interviews with Carlsbad Park Service geologists and exhaustive scientific journal searches revealed no related articles.

5. Planation Mystery

In Nature's landscape, some of the most dominant features are the great, flat plains. These vast landforms are found on every continent. In the geological sense, this is called **planation**.

Fig 6.11.10 – The Grand Canyon series consists of many horizontal layers that were themselves once vast, flat plains. Planation or the process of how these plains were formed remains a mystery to geology today because of the Rock Cycle paradigm.

"At present, the cause of the observed high rate of planation remains a mystery."

Fig 6.11.11 – This view of Ayers Rock in Australia gives an aerial perspective of the Sandstone and Planation Mystery. Most scientists have never considered how these vast plains came to be. However, open-minded investigators are now asking the right questions, and the answer is a piece of Nature's puzzle that does not fit the Rock Cycle paradigm.

According to Wikipedia, "planation is a geomorphic process which creates nearly flat surfaces by fluvial (river), alluvial (wind) or marine processes." Note 6.11c

Historically, it has been said that many of the vast plains were formed over millions of years as seas moved in and then later retreated. However, the abyssal plain at the bottom of the ocean today is not necessarily comprised of large, flat plains like those found on the continents.

How are large flat plains such as those seen in Fig's 6.11.10, 11, 12 and 13 formed? If the crust of the Earth was subjected to uplift and subsidence several times over millions or billions of years, the desiccated sea beds would have been deeply eroded. This is not what we see.

When we look at the Grand Canyon for example, we see many horizontal layers exhibiting a great degree of flatness over a large area. In fact, we see this throughout nearly the entire Colorado Plateau (Fig 6.11.10). This is also seen across thousands of square miles of the Australian outback, surrounding Ayers rock (Fig 6.11.11), and the valleys of Phoenix,

In modern geology's theoretical process of uplift and subduction, large, level plains would never be created.

Arizona (Fig 6.11.12) and Death Valley, California (Fig 6.11.13).

In modern geology's theoretical process of uplifting and subsiding, no large horizontal horizons or plains could be created because there is no mechanism to do so. A completely different, yet unidentified process must be responsible for creating the vast horizontal plains of the world.

Some astute researchers reached a remarkable conclusion regarding this long-standing mystery of the origin of the world's plains. In the book, *The Origins of Mountains*, these geologists, who have perhaps spent a lifetime observing plains and mountains, said this:

"**The remarkable thing is that plains of great perfection are ever made**, despite all the obvious possibilities of complication. But they are real…" Bib 141 p302

These geologists realized what so many others have missed! The enormous flat plains that exist around the world are a ***real*** "mystery" in the Rock Cycle Pseudotheory paradigm:

"At present, **the cause of the observed high rate of planation remains a mystery**.

"It is even more difficult to make a planation surface **if the land is rising tectonically**, yet the planation surfaces are there. This suggests tectonic quiet in many different places. **It is virtually a global tectonic quiet period. Why should this be? And why should a period of tectonic quiet be followed so rapidly by a period of great uplift?** Furthermore, in many regions the planation surfaces cut structures that indicate **high**

Fig 6.11.12 – This aerial photo is from North Phoenix, Arizona, USA and shows how the surrounding area is extremely flat, except for the darker volcanic hills. These flat plains are typically *two miles* thick and are without any clear origin. The volcanic hills shown here are also *not* made up of volcanic lava glass and are without a clear explanation in geology.

Fig 6.11.13 – This road heads north into Death Valley, California, USA, through an example of the great plains that are found worldwide. In some places, no mountains or valleys exist for hundreds of miles. How can this be? The sedimentary materials that make up the floor in the foreground of this photo did not come from the mountains in the background. Where did this material come from and what was the source of water that brought them? Most of the sediment is far too big for wind transport, and there are no major rivers or beds of ancient rivers in Death Valley. There is a lack of marine sediment and fossils in the sediment, and last but not least, there is no geological evidence for uplift or subduction that supports the shallow sea sci-bi. Perhaps this is why some researchers have begun to recognize the truth, that no current geological process can account for the creation of these vast plains. Planation is a complete mystery within the Rock Cycle paradigm.

In fact, **tectonic activity just before the planation**." Bib 141 p302

These perceptive researchers had asked an excellent FQ. How could there be a period of "global tectonic quiet" (no tectonic or plate movement), which would have been required for the creation of the plains?

The plains would have had to be uplifted rapidly, and there is no mechanism for this. Most geology textbooks have ignored the Planation Erosion Mystery, even though the wisdom of how plains are formed is critical. Simple answers to these questions are found within the Universal Flood Model chapter.

The Grand Canyon has five layers of limestone. To account for these layers, the canyon is said to have been covered five different times by 'shallow seas.' However, we do not find other large areas between the Grand Canyon and the ocean that have five layers of limestone.

when researchers look into the basins between the Colorado Plateau and the Pacific Ocean, all they find is more mystery. The Safford Basin in Arizona illustrates the mystery geologists find when they investigate basins and valleys. Shallow seas or lakes would leave salts and clay sediments behind as they dried up. To investigate this, core samples were taken from a borehole, twelve miles south of Safford. Findings from the 8,509 foot Phillips hole were reported by the Arizona Geological Survey:

"The Phillips hole, drilled to a depth of **8509 ft, penetrated sand and conglomerate from the surface to total depth. No gypsum, salt or clay was reported**." Note 6.11d

This borehole is *a mile and a half deep,* and the drill did not hit bedrock. Where are the clay and other organic materials? If oceans covered the Colorado Plateau, they would have left salt

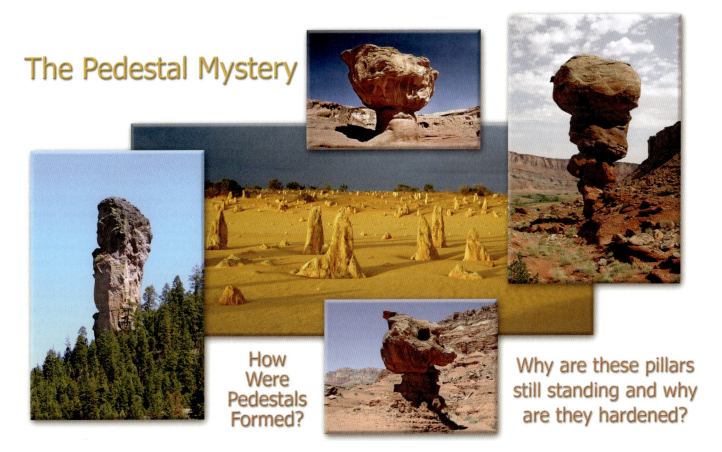

The Pedestal Mystery

How Were Pedestals Formed?

Why are these pillars still standing and why are they hardened?

Fig 6.11.14 – Pedestals and pillars are found worldwide in various shapes and sizes. The yellow pillars in the background-center are in Nambung Australia, to the left is Steins Pillar in Oregon, the pedestal on the right is in Kane Canyon Utah, USA, the pedestal at top-center is in Israel and the bottom-center pedestal is near the Grand Canyon, in Arizona. These structures truly are a symbol of mystery. Where do we see them being formed today? Why are the columns of hardened sediment still standing while the surrounding sediment is long gone? Until these and many other FQs are answered, their origin will stay a mystery. Clearly, it is time for a new geological model that will answer questions like these simply and correctly.

SUBCHAPTER 6.11 THE EROSION MYSTERY

Fig 6.11.15 – These are examples of **Alluvial Fans** (upper left, center inset, and lower right) in Death Valley, California, USA. The alluvium sediments (background and upper right) include gravel, rocks and boulders that were transported by water. Look closely (lower left) and see how the dome-shaped fans rise at steep angles relative to the valley floor. How did these boulder-laden, domed-shaped fans form? When modern rivers are observed carrying boulders downstream, they tend to carve out deep riverbeds and valleys, not built up alluvial fans. Fans like these in Death Valley *do not originate from large river channels that would have to exist if they formed from the action of a robust river*. No one has ever observed the formation of Alluvial Fans like those in Death Valley. The Alluvial Fan Mystery is another of the Erosion Mysteries that has been largely unnoticed in the geological community.

behind when they receded. Why is there no salt found in the Grand Canyon?

Fig 6.11.12 is an aerial photo of North Phoenix, Arizona. Small volcanic hills, made of basalt, dot the landscape. Basalt does not come from the type of lava flows seen in modern times like those further north in Flagstaff, Arizona. How did the surrounding plains become so flat? The large plains in Arizona are similar to plains found in other parts of the world. The plains are extraordinarily flat, and there is no explanation for their origin in geology today.

We conclude the Planation Mystery by quoting once again, the geologists who recognized this important idea:

"The remarkable thing is that plains of great perfection are ever made, despite all the obvious possibilities of complication. **But they are real**…" Bib 141 p302

These plains seem "remarkable" because they do not fit within uniformitarianism. The formation of plains like these has not been seen or recorded. What is remarkable is that geologists say they actually believe that the present is the key to the past.

6. The Pedestal Mystery

Rock pedestals and pillars have fascinated human beings for thousands of years. Early travelers wrote of their mysterious appearance and natural beauty. In Fig 6.11.14 we see pedestals and pillars, represented from around the globe, in their splendor. Many of these geologic anomalies are huge. Steins Pillar can be seen on the left in Fig 6.11.14, and is over 350 feet tall. Certainly, anyone who looks upon these magnificent symbols of Nature must wonder how they were formed.

Not surprisingly, geology has spoken very little about them. The story that is generally told is that erosion over geological time made them. By now, we should be objective enough to ask fundamental questions that will illustrate why these structures are a mystery in geology today. Geology says that the boulders resting atop the pillars protected the pedestal columns below them from erosion by rain, snow and wind. But not all of the pillars have boulders to protect them.

If boulders are needed to protect the pillars from weather, why do many of them not have boulders?

Another question to ask is this:

How did the pillars become hardened in the first place and why was the sediment surrounding them not hardened?

Until simple questions like these can be answered, the validity of the modern geological model and the Rock Cycle must be questioned. These failed theories should be replaced with a model that has simple answers and exhibits real empirical evidence.

7. Alluvial Fan Mystery

The first step in understanding the Alluvial Fan Mystery is to define what alluvial is. **Alluvium** is a general term describing unconsolidated sediments including silt, sand, gravel, rocks and boulders that are deposited by running water such as in a time of flood. The mystery we are to consider does not include the alluvial fans of the finer sediments of sand or clay. Here in the UM, the Alluvial Fan Mystery concerns only those fans that have *large rocks and boulders*, like those as seen in Death Valley, California (Fig 6.11.15).

As the Mississippi River empties into the Gulf of Mexico, the sediment-laden water slows, allowing the sediment to settle, forming a delta, similar to an alluvial fan. River deltas are found at the mouth of many of the world's rivers where they form similar fans of sediment that can cover large areas. The sediments that make up most of these delta fans are fine silts, clays and sand with some smaller pebbles, material easily carried by a river. Alluvial Fans on the other hand often contain large rocks and boulders that were deposited in a characteristic dome shaped fan, emanating from a valley opening in a mountain range.

As we consider the difference between river deltas seen forming at the mouths of many rivers and the Alluvial Fans in Fig 6.11.15, we ask these two Fundamental Questions:

Have there been any observations of Alluvial Fan formation in the past?

Are there any Alluvial Fans being formed anywhere today?

Alluvial Fans should be forming today—but they are not. Geologists often turn to river deltas and fine sediment fans as examples of Alluvial Fan formation. Although they appear to share some similarities, the size of the deposited material and the setting where they are found is vastly different. When flash floods with enough energy to move boulders occur, the rocks and boulders are not deposited in alluvial fans. Instead, deep channels are carved and the boulders are pushed out, along the sides of the channel.

Mudflows are another form of mass wasting. Mudflows usually exhibit rapid downhill movement of loose sediments such as dirt and rock mixed with water. Are the boulder-strewn alluvial fans the result of large mudflows during periods of increased precipitation? Looking at the events of the May 1980 eruption of Mt Saint Helens in Washington State, we can examine the results of the largest debris flow or mudflow in recorded history. How many Alluvial fans were formed during this epic event? None.

Moreover, this mystery is even more perplexing when we examine the differences between alluvial fans. Look closely at the center panoramic photo in Fig 6.11.15. The larger rocks in this photo are *sharp and angular*, indicating little erosion occurred as the rocks moved from their origin. Compare the rocks in the center image with those in the top right image. Notice here that the rocks are somewhat rounded, indicating they traveled further or within a higher energy environment, allowing them to be more eroded. These are two different alluvial fans in Death

Fig 6.11.16 – This Colorado Riverbed is a good example of how a strong, high-energy river carves out channels—it does not form fans. The formation of channeled riverbeds and valleys can be observed as being directly associated with water erosion processes. Why have there been no observations of Alluvial Fans being formed, especially those that have large rocks and boulders?

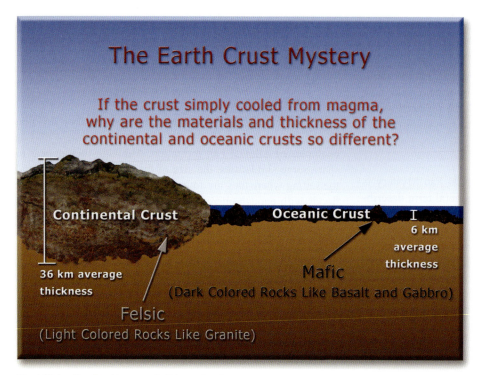

Fig 6.12.1 – The Earth's outer shell is made of two distinctly different types of crust. The continental crust is significantly thicker, lighter colored and consists of primarily felsic rocks, whereas the oceanic crust is much thinner and made up of darker mafic rocks. Within the Rock Cycle theory there is no clear explanation for why this is. What is the true origin of these two strikingly different crusts?

Valley, each of very different composition.

For a model to define correctly the genesis of Alluvial Fans, it must be able to account for the origin of *different fans.* Fans with angular verses smooth large rocks; the methods of transportation of large rocks must be accounted for and the difference in material composition and origin must be answered. In the alluvial fans we observe, the fan materials exhibit different hardness not necessarily coincidental to their roundness. The standard erosion sci-bi—that the rocks were simply washed down slope, does not account for all the evidence.

The Erosion Mystery Summary

Unfortunately, we can only present a small sample of the thousands of Erosion Mysteries that exist worldwide. However, after examining the seven Erosion Mysteries we have chosen to discuss, it should be evident that the true origin of many of the sediments on Earth's crust remains unknown. The concept of erosion over so-called geological time is a sci-bi taught in every modern science geological textbook and it fails to explain the erosion mysteries we have discussed.

6.12 The Earth Crust Mystery

In the scheme of things, the Earth's crust is relatively thin. If we imagine the eight-thousand mile (12.8 km) diameter Earth as a basketball, the crust would be the thickness of a single sheet of paper. Yet within this thin shell lies a great many unanswered mysteries. Why are there differences between the crust of the continents and the crust of the oceans? Geologists have assumed that slow processes of uplift, erosion, and the inundation of seas deposited thousands of miles of sediment, which were then carved into canyons like the Grand Canyon. But there are problems with their theories. The layers of salt from the receded seas as well as many other layers are missing from many canyons' records. Coal and oil provide additional mysteries because of their great worldwide abundance, yet 'new' coal or oil is not being formed today. In this subchapter, we explore a few of Earth's crust mysteries.

The Crust Material Mystery

For many years, geologists have used seismic waves and other technology to discern the Earth's crust. However, the most basic questions regarding the crust remain unanswered. In Fig 6.12.1, we see this mystery clearly identified. When the crust materials are compared, the continental crust and the oceanic crust are completely different from each other. Continents are essentially **felsic** type rocks, and are lighter in color. They consist primarily of granitic rocks and include an abundance of feldspar and quartz. Oceanic crust consists typically of dark colored, **mafic** rocks like basalt and gabbro.

Though drilling and seismic studies have provided detailed knowledge about what these crusts are made of, the wisdom of how they were formed or why they are different has continued to elude geologists.

If the crust of the Earth simply cooled from magma, why are the materials of the continental crust markedly different from those of the oceanic crust?

The Crust Thickness Mystery

Different parts of Earth's crusts are made of different minerals, and they are of different thicknesses. The continental crust averages six times the thickness of the oceanic crust.

If the crust of the Earth simply cooled from magma, why is the continental crust on average six times thicker than the oceanic crust?

Clearly, the Earth Crust Mystery exemplifies another major flaw of the Rock Cycle Pseudotheory. This mystery shows how even the basic questions remain unanswered. Scientific literature revealed no consensus among scientists and no simple explanation was found to account for how these two different crusts formed. Surely, the truth of it is simple, but must be found within a wholly new paradigm.

The Pangaea Mystery

In 1912, German scientist Alfred Wegener advanced his theory of 'continental drift,' stating that all modern continents drifted apart from an ancient supercontinent, Pangaea. Few theories have had such impact on geology, and over the last century, even fewer have been scientifically demonstrated to be true. Wegener's publication was at first rejected because he had not identified a mechanism for continental drift to work. It took more than 50 years for his ideas to be accepted:

Fig 6.12.2 – The supercontinent Pangaea broke apart into the many continents we see today. Alfred Wegener proposed the existence of this supercontinent when he drafted his continental drift theory. Science was slow to accept this theory. Eventually, evidence was found that fossils and minerals from different continents matched up. Later, seafloor spreading was observed and the basis of the theory became more generally accepted. However, no one has identified the mechanism for the dividing of all these continents. How and why did they become separate bodies of land?

"The one American edition, published in 1924, provoked such hostility that it was not revised. Many geologists focused on a lack of a demonstrable mechanism and **rejected and ridiculed Wegener for his ideas**, noting that he could not explain how continents were able to move. The theory received support through the controversial years from South African geologist Alexander Du Toit as well as from Arthur Holmes. Only after the mid-20th century discovery of seafloor spreading did Wegener receive credit, as an early developer of the theory of plate tectonics. It took **more than 50 years** before adequate evidence was acquired and presented to convince mainstream geologists to acknowledge that the continents were actually in motion; and the fit between the coasts of Africa and South America was more than just illusionary." Note 6.12a

In Patrick Hughes' article, *The Meteorologist Who Started a Revolution*, we get a glimpse of what bold theories must endure when they go against the proverbial tide. No one wants to start over again, especially when we think we have already 'learned' so much. Nevertheless, when truth is revealed, the tide must surely change and paradigms must shift:

"'Utter, damned rot!' said the president of the prestigious American Philosophical Society.

"'If we are to believe [this] hypothesis, **we must forget everything we have learned in the last 70 years and start all over again**,' said another American scientist.

"Anyone who 'valued his reputation for scientific sanity' would never dare support such a theory, said a British geologist.

"Thus did most in the scientific community ridicule the concept that would **revolutionize the earth sciences and revile the man who dared to propose it**, German meteorological pioneer and polar explorer **Alfred Wegener**. Science historians compare his story with the tribulations of Galileo." Note 6.12b

Like many readers, we may see several parallels between Wegener's Continental Drift and the ideas here, in the Universal Model. Although Wegener was not the first to suggest that the Earth's continents were once connected, he was the first to present evidence from *several fields of science*. This has been the pattern the UM has followed in presenting evidence for many of the new models contained herein. As Alfred Wegener once noted, "It is only by combining the information furnished by all the Earth sciences that we can hope to determine 'truth' here…":

"'**Scientists still do not appear to understand sufficiently that all earth sciences must contribute evidence** toward unveiling the state of our planet in earlier times, and that **the truth of the matter can only be reached by combing all this evidence…It is only by combining the information furnished by all the earth sciences that we can hope to determine "truth" here**, that is to say, to find the picture that sets out all the known facts in the best arrangement and that therefore has the highest degree of probability. Further, we have to be prepared always for the possibility that **each new discovery, no matter what science furnishes it, may modify the conclusions we draw**.'" Note 6.12c

Wegener found truth because he was looking with an open mind. This is what stands between modern science and millennial science. Investigators will once again have to look for predictable outcome from stated action, if they are to discover new natural laws.

Most continental plates are known (as tracked by GPS) to be moving horizontally at a rate between 5cm and 10cm per year. How are the plates moving and what is the mechanism that moves them? This is the Pangaea Mystery. Wegener did not know in his day and modern geologists, looking from within the magma-planet paradigm, cannot explain what causes the plates to move either. Remove magma from the equation and geologists are in no position in which to make a predictable outcome. To fully comprehend why these plates are moving, we will need to read the Universal Flood Mechanism in Chapter 8.

The Grand Canyon Crustal Layers Mystery

The Grand Canyon is probably the most visited canyon in the world. It is likely the most scientifically studied canyon in the world as well. Because the Grand Canyon is so deep and covers *several hundred square miles*, we have been blessed with both a beautiful panorama of Nature and a *perfect laboratory to study the Earth's crust*. Nowhere else on Earth do we find a better example of Mother Nature exposing the layers of her crust from which we can derive scientific insight and inspiration. If the theories and processes in the modern geologic Rock Cycle were true, there would be no mystery. However, this is not the case. Great mystery still shrouds the canyon:

"Yet, in one of the crowning ironies of our day, **today's scientists concede that most of the basic facts about the Canyon's origins and history are an unsolved mystery**. Despite decades of research, and a flood of theories, **a factual abyss exists as deep as the celebrated Canyon itself**. Nary an expert has yet to prove beyond a reasonable doubt **how the Canyon was formed**, the identities of its first inhabitants, and when they arrived.

"Anyone reviewing the research will quickly encounter this **vexing enigma**: Many features in the vast wilderness of rocks, deep gorges, cliffs, buttes and pinnacles in pastel shades of orange, purple, green and pink **simply do not make sense**." Note 6.12d

The insightful comment—"a factual abyss"— is an accurate description of where the whole of geology is today. The writers further remarked "Anyone reviewing the research will quickly encounter this vexing enigma"—that geological features within the purview of the Rock Cycle Pseudotheory, "simply do not make sense."

Geologists from the Arizona Geological Survey at first acknowledge they know little about *why* the Canyon exists:

"While the rocks in the canyon record much of the geological history of northern Arizona over the past 1.8 billion years, **they reveal little about why the canyon exists**." Note 6.12e

Then immediately in the next sentence, make this proclamation:

"Of course, the Grand Canyon **was cut by the Colorado River**." Note 6.12e

There are examples throughout the scientific community, where closed-mindedness forces a reality check. Honest introspection reveals that in all the fields of science, many are perplexed and are groping with theories. The investigators are blinded because they have accepted theory as fact, *which has caused* them to admit they do not understand the observations they find.

The reason the rocks "reveal little about why the canyon exists" is that geologists *believe* "the Grand Canyon was cut by the Colorado River." But what if the river did not carve the Canyon? This would help explain why geologists do not know "why the canyon exists." Empirical evidence of an alternative origin of the Grand Canyon will be presented in subchapter 8.3.

One way to obtain an overview of a scientific subject is to interview an investigator who has been around that subject for much of his life. Dr. Robert Euler was just such a person. As he pondered his life's work, he reflected upon the collective knowledge of the Grand Canyon:

"The late Dr. Robert Euler, Grand Canyon National Park's longtime chief anthropologist, made a telling admission in an interview in Prescott several years ago: '**It's disgraceful, but we really know so very little about the Canyon's prehistory —no more than 1 percent of what we could and should know.**'" Note 6.12f

Euler was the Park's anthropologist and was well versed in the geology of the Park and the surrounding Colorado Plateau. He spent a great deal of time with scientists studying the Grand Canyon and was in the unique position of seeing firsthand what science really understood about this deeply mysterious canyon. Euler's statement—"we really know so very little about the Canyon's prehistory," was right to the point in identifying this woeful condition.

Of course, there is a valid reason we know so little. It is hard to see the picture of a puzzle when there are so many missing pieces. The geologic story of the Grand Canyon has many missing pieces, and we are about to find out just what some of them are. Although we will identify some of these new pieces, we will not fit them together and solve the puzzle just yet—that will come in the next two chapters. Nevertheless, identifying

"...a factual abyss exists as deep as the celebrated Canyon itself."

"It's disgraceful, but we really know so very little about the Canyon's prehistory— no more than 1 percent of what we could and should know."

Grand Canyon - Missing Layers

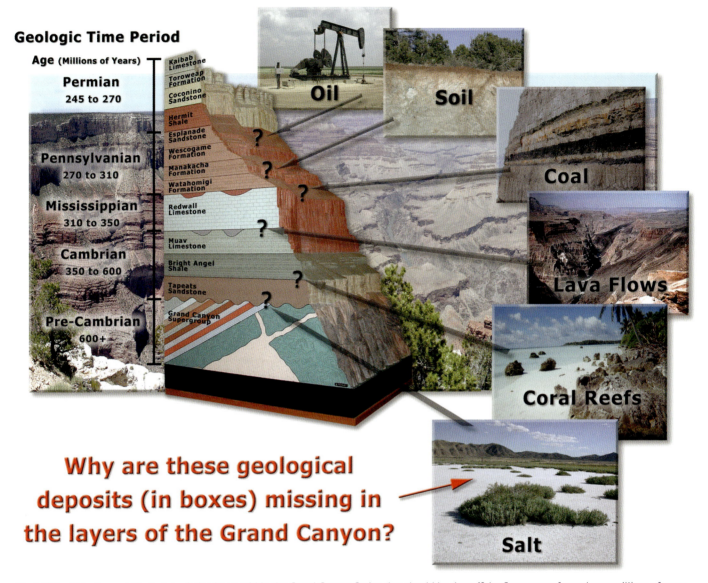

Why are these geological deposits (in boxes) missing in the layers of the Grand Canyon?

Fig 6.12.3 – This diagram illustrates *missing layers* within the Grand Canyon Series that should be there *if* the Canyon was formed over millions of years, as geology has claimed. The missing layers of the Grand Canyon are another mystery of the Rock Cycle Pseudotheory that has either been overlooked or not recognized by modern geology. Most geologists have never even asked why these missing layers are not present. The implications are profound. These missing layers should be present in the Canyon, unless of course, the Canyon was not formed as science has supposed.

these new pieces will open the eyes of many as to why we "really know so very little about the Canyon's prehistory."

Fig 6.12.3 is a cross section of the Grand Canyon identifying many of the *actual* layers found in the canyon's geologic column. The geological time span depicted on the left shows the layers of the Canyon purportedly being *several hundred million years old*. Surrounding the cross section are photos of layers that *should exist,* but do not. These missing layers should be there, according to the Rock Cycle Pseudotheory. The fact that these layers are missing should raise serious questions about the validity of the age and origin of the canyon. Fundamental Questions, like:

Do the layers of the Canyon display and support geological time, as modern science holds it to be?

Over such a span of a time, many periods of vegetation growth would have occurred, leaving the organic remains to be buried. Forests, swamps, deserts, and other environments should have come and gone. In addition, geologists believe that over millions of years, seas covered the Canyon area not once, but *five times*. Coral reefs and salt deposits must surely have been left behind, and in several such layers. If this was true:

Where are the salt deposits? They do not exist.
Where are the coral reefs? They are also missing.
And the coal deposits? Not found.
Oil? Doesn't exist.
Multiple organic soil layers? Not present.

These missing layers should exist in the geologic column of the Canyon, but they are nowhere to be found. Lava flows are found in the canyon (see center right inset, Fig 6.12.3) but are of a recent date, having flowed from the Canyon's surrounding plains and are *surface* flows. These flows surged over the rim and down into the canyon. There are lava flows within the *bottom* layers of the Canyon as well. These provide an important clue as to how the Grand Canyon was made. The Grand Canyon lies along several fault lines and is a part of the most active volcanic region in Arizona. Thus we see, lava flowed *before* the Canyon's horizontal layers were deposited *and again after*. What about the millions-of-years time span throughout the rest of the layers? How many occurrences of lava do we find in the middle layers of the Grand Canyon Series that spanned this immense period of geological time? Should it be a surprise to find that there are none?

If the Grand Canyon layers were laid down over millions of years, why are there no layers of: salt, coral reefs, coal, oil, organic soils, and lava?

The physical geologic evidence observable in the Grand Canyon contradicts the notion that the Canyon's layers were laid down over millions of years. What it testifies of is an origin involving processes not related to typical erosion.

We have barely touched the surface of many crustal mysteries in the Grand Canyon. In the Canyon, there are massive layers of homogeneous (same kind of mineral) sandstone and limestone and many missing layers of sediment. The Loess and Erosion Mystery discussions earlier in this chapter, demonstrate that the Earth's crust does not represent millions of years of erosional processes. The missing segments identified in the Sediment Size Table also indicate that some mechanism other than the erosion-over-time pseudotheory was responsible for forming many of the world's crustal landforms. Fossils are ***not*** found *in many layers of the Grand Canyon* in the proportion they are expected, or not at all. These and many more examples are why "we really know so very little about the Canyon's prehistory." This cannot change under the current rock-cycle paradigm. The Universal Model and a new Millennial Science will replace theory with natural law.

The Coal Mystery

Coal is identified as one of the 'missing' layers not found in the Grand Canyon, but the mystery does not end there! Coal is itself another mystery within the Earth's crust. It seems amazing that a substance as common as coal, that has been found worldwide and used extensively by humankind for millennia, is still such a mystery in geology today. Although it does not seem to be a mystery when reading the latest textbook, or watching a televised nature documentary, let's look a little deeper into the facts.

For generations, the story has been that coal beds were ancient swamps, and that millions of years of accumulation of organic debris lead to eventual coalification. Here is how one typical geology textbook portrays the formation of coal:

"Organic sedimentation is yet another type of biochemical precipitation. **Vegetation may be preserved from decay in swamps** and accumulate as a rich organic material, peat, which contains more than 50 percent carbon. **Peat ultimately is buried and transformed by diagenesis to coal**." Bib 59 p171

We learned previously in this chapter that 'diagenesis' is a pseudotheory, a process not observed by science. Thus, we must group the coal-from-swamp sci-bi into this same category—a false theory taught as fact. How can we know this theory of coal's origin is false? Ask this Fundamental Question:

Where are the swamps with accumulated vegetal matter or peat that are in the process of becoming coal today?

If the commonly accepted origin of coal is valid, then there should be observable examples of the 'diagenesis to coal' process. We find however, there are no swamps anywhere in the world that are beginning the process of becoming coal or that are partially through it, simply because this is not how coal is formed.

The Four-Stage Coal Formation Pseudotheory

Coal *is* formed from vegetation and it is found worldwide in different states of hardness and density. Coal is usually graded according to density and quality of heat when it is burned. The least dense variety of coal is referred to as **lignite**. Lignite burns at a lower temperature and creates excessive smoke. **Bituminous** coal is the most common form of coal and is denser and cleaner burning than lignite. The densest form of coal is **anthracite**, a heavy, dense form of coal, which produces the highest temperature and therefore the most energy when it burns. **Peat** is similar to coal in that it is vegetal matter and has been used as a fuel source for centuries.

The Four-Stage Coal Formation Pseudotheory, also known as the peat-to-anthracite theory, states that coal begins as dead vegetation accumulates and becomes:

1) **Peat**, which changes to become lignite.

2) **Lignite** undergoes further change into soft coal, called bituminous coal.

3) **Bituminous** coal eventually becomes hard coal, called anthracite.

4) **Anthracite** is the final stage in the coal formation theory as taught by geologists.

We read from a recent geology textbook, *Understanding Earth*:

"The process by which coal beds form begins with the deposition of vegetation. Protected from complete decay and oxidation in a wetland environment, the deposit is buried and compressed into **peat**. Subjected to further burial, peat undergoes mild metamorphism, which transforms it successively into **lig-**

Fig 6.12.4 – Coal is a major source of energy in the world. Although many people living in the USA never see coal today, over one billion tons of coal a year is mined and used to produce approximately half of the electricity in the United States. Less than a century ago, people used coal every day for heating and cooking. The USA has more coal than any other country in the world. According the U.S. Department of Energy, "The United States has more coal that can be mined than the rest of the world has oil that can be pumped from the ground."

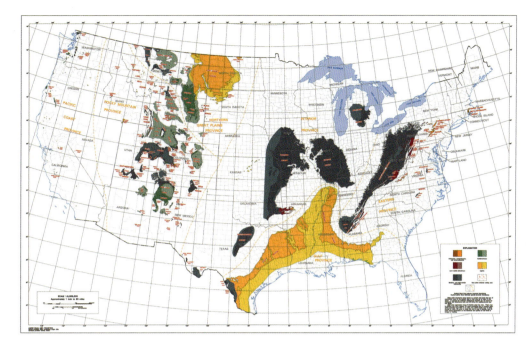

Fig 6.12.5 – Coal deposits in the USA per USGS map. Deposits are identified as either lignite (yellows), bituminous (reds, grays and greens) or anthracite (oranges). Notice that there is a correlation between the least dense, lignite and the most dense, anthracite with no mid-density coal deposits associated. There are no areas where two or three of the different types of coal are found. This simple fact directly contradicts the method which geologists say forms coal.

nite, subbituminous and **bituminous** (soft) coal, and **anthracite** (hard coal), as **the deposit becomes more deeply buried, temperature rises, and structural deformation progresses**."
Bib 59 p570

The story is that the Four-Stage Coalification process simply depends on the depth of the deposit and the pressure and temperature that exists. The higher grades of coal are formed at higher pressure and temperature. It would seem all we need to do is bury vegetation in a swamp and after a few million years—presto, we will have coal! Is it that simple?

The first strike against the Four-Stage Coal Formation Pseudotheory is that nowhere can coal be seen in the formative process today. This fact seems to have been missed by the scientific investigators involved in the research. If this is a natural process, according to the uniformitarian, it should be seen in many places *all over the Earth today*—it is not.

The second strike may come as a bit of a curve to many geologists. We sent email to researchers working in the coal industry and various coal mines and asked where they had seen the different types, or grades, of coal layered within the individual deposits. That is, coal beds that exhibit the stratification from the softest coal, lignite, through hard coal, anthracite. We asked this question:

> Do you know of any locations where coal beds grade vertically, from lignite to bituminous or from bituminous to anthracite?

The responses were not really a surprise. Stratified beds of different grades of coal do not exist that we know of. This typical response came from one research company we questioned:

"I'm sorry but I have **not been able to find** info on any seams/mines that fit this." Note 6.12g

If coal was really formed over 'geological time' as geology has taught, we would find layers of different grades of coal, one atop another in at least some of the coal mines around the world—but we do not.

The third strike against the Four-Stage Coal Formation Pseudotheory is the peat itself. Just as the origin of coal is a mystery—so too, is peat.

The Peat Mystery

Although peat is said to be an accumulation of partially decayed vegetation, it has a very distinct makeup and is found in large quantities at specific locations around the world:

"Peat deposits are found in many places around the world, notably in Russia, Ireland, Scotland, northern Germany and Scandinavia, and in North America, principally in Canada, Michigan and the Florida Everglades. In Beauchene Island, it forms at around ten times the rate of anywhere else in the world. **The process by which it forms is not known.**" Note 6.12h

The origin of the very material coal is claimed to have evolved from is a piece of Nature's puzzle. Consider the mechanics of the four step process. To begin, a vast amount of peat would be required to produce a small amount of lignite, which would then be compressed further into a lesser amount of bituminous coal, which would then be pressed into an even lesser amount of anthracite. Presumably, this all begins with an incredibly vast amount of peat.

If we look to see where peat is found in the world, we will notice that about 80% of the world's peat is found at higher latitudes, where it is colder and has less vegetation as compared to the equator. (See Fig 6.12.7) Why is it that peat is not forming where there is abundant vegetation and ample water? We ask

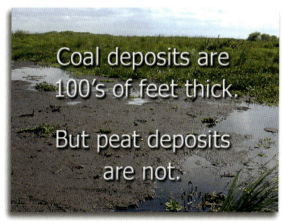

Fig 6.12.6 – This is a native peat deposit in Florida. If coal really came from peat, why are peat deposits so much thinner than coal deposits? Presumably, a significant amount of peat must be compressed to form coal.

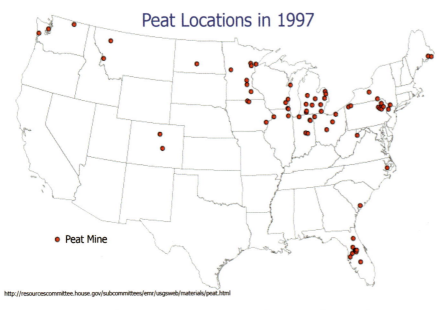

Fig 6.12.7 – These are US peat mines across the country. Compare the location of peat deposits with the coal deposits identified in Fig 16.12.5. Notice how peat deposits are located where coal is not. Why is there no correlation between the peat deposits and the coal beds if coal comes from peat? For more than a century, the peat deposits and coal deposits have been a mystery in geology. This will never change unless there is a paradigm shift toward recognizing the true origin of coal.

this because it takes a lot of peat, according to modern science, to make coal.

There are tens-of-thousands of square miles of coal deposits in the US holding nearly 1000 billion tons of coal. Such a vast amount of coal would require enormous peat deposits, if the coal is, as they say, derived from peat. Let's find out just how *deep* the Peat Mystery is. From the US Department of Energy's website on coal, we are informed how much peat (decayed plant debris) is necessary for the making of coal:

"**Ten feet** of prehistoric plant debris was needed to make **one foot of coal**." Note 6.12i

To comprehend just how much vegetation this is, we have to do a little math. The example we will use is the famous coal mine in the Latrobe Valley, Australia. Here, there is a coal bed 65 km long, up to 30 km wide and between 60 and 150 meters thick. If we multiply the thickness of the coal times a conservative 10 meters (using the 10/1 ratio from the above quote), we arrive at 1500 meters. This means a peat bed of *nearly one mile thick* would have been required to produce the Latrobe Valley coal bed, according to *modern geology's coal sci-bi.*

Where do we find, anywhere in the world today, a mile-thick peat bed?

Sometimes FQs are powerful because they are so simple. The coal formation pseudotheory, which states that coal comes from peat, cannot account for how coal can be found in layers up to 150 meters thick because there is no evidence for the existence of peat beds hundreds of meters thick. Neither can the theory account for how the peat was buried without any interbedding of sand or other sediments. Researchers recognize that there is a bit of a problem due to the lack of sediments in coal. They assume that the environment must have remained "steady for prolonged periods of time" (i.e. millions of years) in order for the swamps to "remain essentially free of sediment":

"For masses of undecayed organic matter to be preserved and to form economically valuable coal the **environment must remain steady for prolonged periods of time**, and the waters feeding these peat swamps **must remain essentially free of sediment. This requires minimal erosion in the uplands** of the rivers which feed the coal swamps, **and efficient trapping of the sediments**." Note 6.12j

Here again we see that geology is divided against itself. Weather and the environment do not remain static for even dozens of years. How is it possible for these dynamic forces to remain steady for millions of years? The geological evidence strongly contradicts both the Uniformity Myth and the Uplift Pseudotheory, both of which are essential for the modern science version of coalification to occur.

In modern geology's view, we envision an environment where hundreds of meters of dead vegetation accumulate over hundreds of thousands of years without the impact of sediment laden rivers. The swamp's feeder rivers are assumed to be sediment free because of minimal erosion up stream. Then suddenly, thousands of square miles of swamps are covered, seemingly in an instant, with such great quantities of sediment that the decayed vegetal matter, or peat is compressed into coal. Later, mountains of sediment are eroded away, leaving a just a few meters of overburden (rock) above the coal.

> "Again, nowhere in the world, at present, can accumulations of vegetable matter be found which are quantitatively commensurate with any of the major coal deposits of past geological time."
>
> W. G. Woolnough

To summarize the three strikes against the Four-Stage Coal Formation Pseudotheory, they are:
1. Coal formation is not occurring anywhere today.
2. The different densities of coal are not found anywhere to be formed successively, one atop another.
3. Peat is also a geological mystery and does not occur in sufficient quantity to justify the explanation that coal comes from peat. No observations have been made that empirically support the idea that peat makes coal.

This particular coal formation theory has not been proven and lacks factual observation of its claims; therefore we must conclude that this long held belief is just a pseudotheory.

The "Knife-Sharp" Sediment Contact Mystery

Look at the image in Fig 6.12.8 to see what typical coal seams look like. Each dark layer of coal has a clearly defined contact point with the overlying and underlying sediment. How this happens is not explained in the traditional geological theory:

"Eventually, and usually due to the initial onset of orogeny [mountain building] or other tectonic events, the coal forming

Fig 6.12.8 – The coal deposit seams with the arrow sign are near Price Utah, USA, and an open coal mine with the loader is in Wyoming, USA. Notice the "knife-sharp" sediment contact above and below the coal seams. How does sediment from a river "quickly" make such "knife-sharp" level layers of sediment over thousands of square miles? Why can geologists not show us where this has ever been observed to happen?

ronments. Most pine forests are found in mountainous terrain. The accumulated vegetal debris from these forests would have been left where the forests grew—among hills and mountains, *not in flat plains*.

Here a geology textbook outlines *the fact* that "coal is a *special* kind of sedimentary rock," not a typical sedimentary layer:

"Coal is a **special kind of sedimentary rock** formed from accumulated vegetable matter, and petroleum is generated from other organic substances — both plant and animal — **that are trapped in marine sedimentary deposits**." Bib 125 p12

Marine fossils are common in many coal deposits around the world. Fish, brachiopods, mollusks, and other fossils have been found. In what environment do we see significant numbers of fish or other animals dying and being preserved en masse before decomposition? None are known and none have ever been recorded.

Are any coal beds being formed anywhere, under any ocean, where marine fossils are included?

environment ceases. **In the majority of cases this is abrupt, with the majority of coal seams having a knife-sharp upper contact with the overlying sediments**. This suggests that the onset of further sedimentation **quickly destroys the peat swamp** ecosystem and replaces it with meandering stream and river environments during ongoing subsidence." Note 6.12j

Common sense says that if a peat swamp environment were to change 'abruptly,' substantial deposition of sediment must have occurred, indicating the presence of a large amount of sediment-laden water. It would have been an environment something akin to a high-energy river, not a quiet 'meandering stream.' Keep this in mind as you imagine thousands of square miles of horizontal coal seams, all with 'knife-sharp' boundaries and no sign of stream-cut channels in the layers of coal.

How does a large river deposit sediment over thousands of square miles of swamp, leaving a "knife-sharp" boundary, without depositing sediment 'in' the swamp?

It is known that *coal is **not** being formed today from accumulated vegetable matter*. There are surprisingly few geologists that agree with what respected geologist, the late W.G. Woolnough stated back in 1937:

"Again, **nowhere in the world**, at present, **can accumulations of vegetable matter be found which are quantitatively commensurate with any of the major coal deposits of past geological time**." Note 6.12k

More Coal Mysteries

It would take a very large book to contain the myriad of coal mysteries, yet few people have heard of them. We touch on only a few here. One important observation is that many coal deposits are made of primarily coniferous trees (evergreen pines). Conifers are not typically found in swampy low-lying envi-

Again, if continents were really subject to uplift and subsidence over geological time, as the uniformitarian would have us believe, there should be partially submerged continents with coal reserves on them. Of course, these do not exist.

Another mystery in the world of coal, is **coal balls**. They are spherical masses of vegetal matter and various animals including marine creatures that have been fossilized. Coal balls range in sizes from two inches (5cm) to two feet (61cm) in diameter. Fig 6.12.9 is a small coal ball from Illinois, USA. How did these balls form? They are not seen forming today.

Another coal mystery few geologists are aware of; **boulders in coal**. Rock boulders were first discovered in the coal mines of Europe and then in America in the 1870s. The best research we could find on these mysterious boulders was a journal article from 1932 titled *Erratic Boulders in Sewell Coal of West Virginia*, by Paul H. Price. Price commented on the size of the boulders:

Fig 6.12.9 – This is a **coal ball** from the Carbondale Formation in Illinois, USA. It is a concretion and like other concretions, is a mystery in geology because the mechanism that formed it is not understood.

SUBCHAPTER 6.12 THE EARTH CRUST MYSTERY

"...that they vary in weight from **161 ½ pounds to ½ ounce, the average being approximately 12 pounds**." Note 6.121

The mystery of these boulders lies in their origin. How did they come to be lodged in the coal? Price continues:

"The character of the boulders, as shown by the preceding descriptions, marks them, with one or two exceptions, as being **unlike rocks known to occur in West Virginia**. The nearest source of similar rocks is the crystalline area of Blue Ridge, Virginia, a distance of some **60 miles to the east**." Note 6.121 p70

Sixty miles is a long way for boulders to have traveled in a 'swamp.' Obviously, rivers carry more than just boulders and Price was keenly aware of this. Where was the other sediment that should have accompanied and been more abundant than the boulders? The only explanation geologists deduced, was that the boulders must have been 'rafted' to their final location. However, Price explains why neither of the two proposed rafting processes are plausible. The first method he discusses involves the roots of the trees in the swamp:

"To account for the presence of boulders in coal, the view previously expressed by most geologists is that they were held in the roots of trees and rafted to their present position. To assume that all of these boulders, especially the larger ones could have been carried to their present location without considerable quantities of other foreign material, **calls for a stretch of the imagination**. That some may have been so transported is not doubted. It does not seem logical, however, that a stream sufficiently large to raft trees that could carry some of the boulders here noted would be found in a coal-forming environment. **A stream of such size would certainly 'wash out' the peat bog itself**. Furthermore, the presence of boulders **at various horizons in the coal would necessitate the presence of the stream throughout the entire time of the accumulation of the coal-making material**." Note 6.121 p72

Next Price discusses the theory of transportation by ice, which also fails scrutiny:

"The present boulders, however, **do not show the characteristics common to those transported by ice, such as faceted faces or striations**. It does not follow, however, that river or shore ice may not have carried these boulders from beaches or along the banks of streams into the Pottsville basin. This, however, would be expected **prior to or following the coal accumulation, and could account for the boulders only in the underclay or the overlying sediments**. It has already been pointed out that the boulders **do not occur at any one particular horizon in the seam, but may be found at any level from the underclay to the roof shales**. It should be stated, however, that the majority are found in the lower part of the seam." Note 6.121 p72

You may recognize some similarities between this coal mystery and one we discussed earlier in this chapter, under the Sand Mystery Sub Chapter. In that section, we introduced the "classic geologic anomaly," wherein light grey granite boulders were found in the red sandstone formations near Mesa, Arizona. Are these similarities a coincidence? We will see just how connected they are in Chapter 8, the Universal Flood Model.

The Oil Mystery

Oil is found both on and off shore in great abundance, all around the world. However, the origin of oil and petroleum has long been a mystery in geology. The origin of oil is one of the few mysteries we address in this chapter where many geologists recognize and agree that strong debate still exists among researchers. This is in stark contrast to most of the concepts

Fig 6.12.10 – These photos are from an oil field on the coast of the Gulf of Mexico, USA. Oil rigs such as these can be set up and taken down in just a few days and can drill to depths of several thousand feet. At the time these photos were taken, the operator of this rig stated that they were drilling through hundreds of feet of salt, a common layer in the region. Salt is often encountered before the oil is reached. After the oil piping has been placed, oil pumps are installed above the well. The pumps will pump the oil continuously to downstream storage facilities. Salt-water slurry from drilling operations is being discharged from a pipe in the inset image. If oil and salt were simply left behind as seas dried up, how would the deposits be so pure, often without contamination that would have been present from such a deposition? We will have to do more than just drill if we want to find the truth—we will have to start from a completely new paradigm.

taught in geology today. From a recent geology textbook we read:

"In both **lake and ocean waters**, the remains of algae, bacteria, and other microscopic organisms **may accumulate in sediments as organic matter that is transformed to oil and gas**." Bib 59 p171

In what lakes or oceans have researchers observed the transformation of organic matter into oil? In some circles, it is even still questioned whether oil is biogenic or abiogenic (meaning not from organisms) in origin. Although geology textbooks keep telling the tale about the origin of oil, open-minded researchers know the truth, which is:

"The origin of petroleum still, **despite the immense amount of research devoted to it, has more uncertainties concerning it than any other common natural substance**." Note 6.12m

One would be hard pressed to say that there isn't quite a bit of uncertainty around the origin of oil. One of the mysteries surrounding oil is its close association with *salt* (see Fig 6.12.10). Why is oil so *intimately* associated with salt deposits? At the beginning of this chapter, salt was identified as being the most abundant mineral on the surface of the Earth, and yet no real explanation for the origin of salt exists in modern geology. Because oil and salt are so intimately connected, when we identify the origin of salt, we will also be able to identify the true origin of oil.

Earth's Crust Mysteries Abound

We should not be surprised to find more mysteries about the Earth's Crust each time we look. Once one sheds the shackles of the modern geological paradigm and begins to examine the world with an open mind, many geological anomalies will be seen. We have discussed only a few of them, enough to substantiate that there is a crises in modern geology today and a need for a paradigm shift.

The next two chapters are the longest chapters in this book and completely change the foundation of what we call 'geology.' They are extensive and along with details about the crust of the Earth, significant new evidence about how the crust was formed will be presented. Because so much of modern science is based on modern geological concepts and theories, if they were to change fundamentally—every field of scientific study would be affected. This is one of the primary reasons the Universal Model was written. These chapters will be an interesting foray into the new science of the Universal Model.

6.13 Geotheoretical to Geological

The Doctrine of Uniformity—A Geotheory

"**One of the most fundamental conceptions in geology is that great generalization**, largely formulated by Sir Charles Lyell, to which is applied the name of the 'DOCTRINE OF UNIFORMITY.' **Without this, the development of geology as a science could not have taken place**." Note 6.13a

Woolnough, a distinguished geologist made this statement back in 1937. Uniformity is *so important*, that geology would not exist today without it—wow. This is quite a statement. Could this statement be true? Consider this; if the generalized Doctrine of Uniformity is not true, modern geology, as it is now known, would be gone—Poof! This is not a comforting thought to the well-versed uniformitarian-trained geologist.

Troubling inconsistencies to the uniformity story abound. In 1980 and again in 1982, Mt Saint Helens erupted, spewing massive amounts of material—but no lava. What *was* ejected were enormous quantities of steam and ash, debris and mud. For the first time, and on an unprecedented and colossal scale, geologists witnessed almost instant changes to the landscape. Those changes involved sediment and the power of water. Nothing quite like this had ever been observed or recorded. Ideas changed and textbooks had to be rewritten. The Doctrine of Uniformity had failed.

Again, and again in this chapter, we have shown how the fundamental principle of modern geology—Uniformitarianism—fails. Despite the evidence, modern colleges of geology still hold this tenet as a guiding principle and teach from textbooks based on it. The foundation, upon which the Rock Cycle Theory is built, is the principle of uniformity. From the textbook *Understanding Earth,* we see the emphasis uniformitarianism carries as being the "great guiding generalization of modern geology" and that sedimentary rocks are essential to understand this:

"In Chapter 1, we introduced the **principle of uniformitarianism**—the idea that geological processes have operated in the past much as they do today—as one of **the great guiding generalizations of geology. The principle applies even though the rates of some processes may have changed and catastrophes such as cometary impacts have occurred in the past**. Since the beginning of the nineteenth century, **sediments and sedimentary rocks have been central to explaining uniformitarianism**." Bib 59 p164

Contrary to this statement, our discussions in this chapter on the Sand Mystery, the Iron Mystery, the Carbonate Mystery, the Loess Mystery and the Erosion Mystery all serve to show that

Fig 6.13.1 – The Uniformity Myth, which states that, "The present is the key to the past" has failed. Several dozen geological examples have been given in the last two chapters that demonstrate unequivocally that most of the geology we see around us today was **not** formed by presently operating geological processes.

sediments do not prove uniformitarianism. In fact, the very reason each of these sediments remain *a total mystery in geology today,* as we have shown in many cases to be in the geologist's own words, is that they are **not uniformly deposited**!

Admitting the Undeniable

Have *all* geologists missed the empirical evidence presented in this chapter? Have they all fallen victim to the false uniformity doctrine? Certainly not all have—but probably every geologist today is at least influenced by it.

We find some scientists and researchers who are recognizing the flaws in the Uniformity Doctrine, as we see from statements made by the late W. G. Woolnough, one of the first honorary members elected by the Geological Society of Australia. Woolnough was a professor who spent much time in the field conducting research in oil and salt. He wrote papers on a variety of geological subjects and was influential in oil exploration circles. In one such paper, *Sedimentation in Barred Basins, and Source Rocks of Oil*, published in the American Association of Petroleum Geologists journal, he begins:

"While the DOCTRINE OF UNIFORMITY must always form the basis of geological theory, it is necessary to **admit** that **phenomena have occurred during geological time which have not been observed at all,** or not in their entirety, during the very limited period of modern scientific investigation.

"Reasons are given for the belief that **major coal measures, major primary salt deposits, major fresh-water series of sediments, and sources of oil deposits are not now being deposited anywhere on the earth**."
Note 6.13a p1101

Woolnough recognized that major coal, salt, sediments, and oil deposits, which abound in the Earth's surface "*are not now being deposited anywhere on the earth*"! In making these assertions, he must have known he was bucking tradition. Others have recognized the fallacy of uniformity. Hopefully, every geologist that reads this book will question what he or she has been taught and will recognize how the field of geology has immersed itself in theories, becoming geo-theoretical. Note 6.13b

From Uncertainties to Certainties

Claude J. Allegre, a geophysicist who wrote the foreword to *The Mid-Oceanic Ridges*, explains what modern science is *really* built upon:

"Science is **not built on certainties but relies on uncertain hypotheses** assembled into a coherent framework." Note 6.13c

This statement epitomizes the Dark Age of Science, where science is not built upon natural laws or certainties, but on uncertainties and uncertain hypotheses. This is essentially, what the last two chapters have been about—scientific *uncertainties*. We will find the seeds of uncertainty were planted in the Heisenberg's Uncertainty Pseudotheory in the Universe Model Chapter. He convinced his followers in the 1920s that because technology could not predict where and when an individual electron would be, modern science could never really 'know' anything.

Although the 'uncertainty mentality' began in the field of physics, it has, unfortunately, spread throughout all of the sciences, including geology. Returning to the foreword mentioned above, Allegre, after stating how uncertain modern science really is, concludes his foreword by encouraging aspiring geologists:

"May this book **inspire** many young people to practice and enjoy the profession of geology." Note 6.13c

What student is going to be 'inspired' when they are told to memorize "uncertain hypotheses," which by their very definition are going to change? Is it any wonder why so many students struggle learning modern science concepts?

What is amazing is how well modern science has connected the host of uncertainties into a broad, 'coherent framework' that on the surface seems so tidy. Because of this, it is extremely difficult for any scientific investigator, including those working within the Universal Model, to penetrate this framework that exists armor-like in every piece of scientific literature we read. It has taken many years wading through the muddy waters of data and theory to begin to disassemble the framework constructed by modern science, enabling the discovery of truth.

Every once in a while, a diamond of truth is found buried among the papers and journals of the scientific establishment, but too often it is little recognized and if it challenges the current dogma, it is tucked away into oblivion. Nevertheless, using the Universal Scientific Method as our guide, we can navigate through scientific literature and observe in Nature, first-hand, many mysteries like those presented in this chapter, even though these important scientific discoveries were overlooked by the scientific establishment.

It is time to contrast the two geo-theoretical chapters of modern science, the Magma Pseudotheory and the Rock Cycle Pseudotheory, with a new *geo-logical* model. Certainly, every field of science must be built on a solid foundation. If that foundation is not solid, basic ideas and concepts, as we have discovered in the last two chapters, will be replete with mysteries, anomalies, and uncertainties. That being the case, one could never hope to fully understand or comprehend Nature.

We must leave the realm of scientific uncertainty and move on to discover Nature's *geo-logical laws* and scientific **certainties**. The new model, the Hydroplanet Model, presented in the next chapter, is built on a completely new foundation. New technology and new experiments will be required to fully understand and comprehend the formation of the Earth and other celestial bodies. It is hoped that a new generation of students will be truly inspired to learn and develop this new model of natural laws and certainties.

> "Reasons are given for the belief that **major coal measures, major primary salt deposits, major fresh-water series of sediments, and sources of oil deposits are not now being deposited anywhere on the earth**."
>
> W. G. Woolnough

> "Science is **not** built on certainties but relies on **uncertain hypotheses** assembled into a coherent framework."
>
> Claude J. Allegre,

7

The Hydroplanet Model

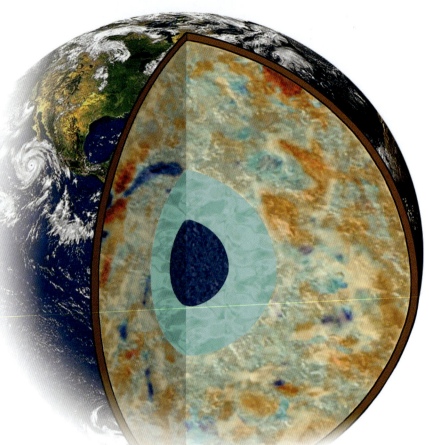

7.1 Magmaplanet to Hydroplanet
7.2 Celestial Water
7.3 Hydrospheres
7.4 The Crystallization Process
7.5 A New Geology
7.6 The Hydroplanet Earth
7.7 Hydrology Redefined
7.8 The Hydrocrater Model
7.9 The Crater Debate
7.10 The Meteorite Model
7.11 The Arizona Hydrocrater Evidence
7.12 The Impact to Hydrocrater Evidence
7.13 The Hydromoon Evidence
7.14 The Hydrocomet Evidence
7.15 The Hydroid Evidence
7.16 More Hydroplanet Evidence
7.17 The Hydroplanet Frontier

During the 1700s, science notables; Newton, Hooke, Halley, and others were rising stars. New discoveries abounded, but some ideas, like the *knowledge* that there were *only* five extraterrestrial planets orbiting the Sun, were solidly established and no one expected any revision to that well-known *fact*. Then, on March 13, 1781, Sir William Herschel made the spectacular discovery of the planet Uranus. He had not been the first to see the planet, but because of accepted tradition, it had escaped recognition until that fateful day.

We are not so different today, self-satisfied in our belief that 'we now know' so much about nature that we sometimes miss the obvious. Science writer Bob Berman concluded:

"…to duplicate the shock of Uranus's discovery, we'd have to stumble upon something that would shatter cherished, long-held beliefs, instantly revealing that we had been dead wrong

"The history of how Earth's interior evolved, and how it accounts for many aspects of our planet's behavior, remains largely unwritten. Taking **water** into account could well help to explain a great deal more."

David Stevenson

about some universally accepted tenet of reality. But as science and technology advance with **exponential rapidity**, our capacity for astonishment shrinks to **near zero**." Note 7.0a

We have a false sense of security because of the mingling of science and technology. If modern science is moving in "exponential rapidity" or with great speed in the **wrong** direction, there is no reason that our capacity for astonishment could ever *shrink* to "near zero." Clearly, modern geology has missed something even more momentous than the discovery of a new planet, the realization that the Earth is *not* a magmaplanet.

The idea of a hot, molten Earth is not a terribly old idea; it was first seriously considered just over two hundred years ago by James Hutton. Since then, it has become one of the principle pillars that all of modern science is built upon. Today, this dogma is so deeply entrenched and so completely accepted that *no modern scientist* questions the revered magmaplanet tenet, and nobody is looking at the mountain of evidence that demonstrates otherwise.

Is it possible that such a well-known theory is wrong? Mary Hill, senior geologist with the California Division of Mines and Geology, said the following in her book *Geology of the Sierra Nevada*:

"Some of the theories we have today may seem as ridiculous tomorrow as the idea that earthquakes are caused by a great turtle shaking the Earth on its back. Such an idea of the Earth was once held by millions of people. We call it primitive; **yet our current theories may be just as far from the truth**." Note 7.0b

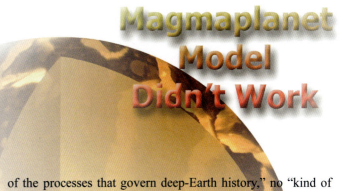

Magmaplanet Model Didn't Work

In this chapter, new evidence and a reevaluation of old data will shed new light on the important roll water plays on rock forming processes, landforms and surface morphology. Water is more than an erosional process; it is the central ingredient of planet and moon formation, the primary cause of cratering, and the most essential part of rock and mineral crystallization. Truly, there is water, water everywhere and we are about to discover that there is more water in the universe and right here on our planet than modern science has ever imagined.

The Crater Debate is one of the greatest examples in the history of science, where modern science strayed so far from the truth; it metaphorically 'concluded' that 2 x 3 was now equal to 7.159 rather than 6. The world's most famous crater, the Arizona Crater holds clues that will forever change the way science sees craters. Meteorites hold astonishing secrets that will also be revealed here.

This is the longest chapter in the book, and for good reason—the Hydroplanet Model is the new, rock-solid foundation upon which Millennial Science will be built.

7.1 Magmaplanet to Hydroplanet

Everywhere we look on our Earth, we see the effects of water. We see similar effects on other planets and moons in our solar system and we expend great quantities of intellectual and physical resources looking for water in the universe. However, the primary reason science has been looking for water was to find possibilities of life—not to learn how rocks or planets were made. The empirical evidence in the last two chapters showed that the magmaplanet idea is a pseudotheory, and that it was time for a paradigm shift, a complete reversal of thinking where water is and how rocks are made—a move toward a hydroplanet model.

A New Water Model Needed

In February of 2004, David Stevenson attended a National Science Foundation program workshop in La Jolla, California dubbed the *Cooperative Study of Earth's Deep Interior*, where scientists met to discuss questions about the Earth's interior. Stevenson, a leading scientist reported in a *Nature* journal article, April 2004:

"The history of how Earth's interior evolved, and how it accounts for many aspects of our planet's behaviour, remains largely unwritten. **Taking water into account** could well help to explain a great deal more." Note 7.1a

Stevenson drew this conclusion:

"From the talks and ensuing discussions among deep-Earth devotees, it is evident that **we need a better knowledge** of the processes that govern deep-Earth history, and the material parameters that control those processes, **before any kind of 'standard model' can be constructed**..." Note 7.1a

He went on to say:

"**We do not know the total amount of water in the mantle** (although it is at least comparable to that in Earth's oceans), or the rate at which this reservoir is tapped and replenished. Nonetheless, the emergence of provocative ideas on the topic illustrate a growing willingness to tackle the central questions of **Earth's interior ocean of water**." Note 7.1a

Perhaps without knowing it, Stevenson hit the nail squarely on the head by recognizing that without "a better knowledge of the processes that govern deep-Earth history," no "kind of 'standard model' can be constructed."

Throughout the last two chapters, we examined the theories and processes that are ***not*** at work in the interior of the Earth. By dissecting the modern science Magma Pseudotheory and The Rock Cycle Pseudotheory, we are able to understand *why the Earth could not have* formed from magma and *why there is no* magma inside the Earth today. These two chapters established, using logic, observation and experiment, why magma has never been anything more than a 'hot' theoretical idea. Ideas by themselves, even if accepted by a consensus of the masses, do not make the ideas true.

Eventually, every magma borehole driller reaches the conclusion that he has run a dry hole and has come up empty. Magma is a dead-end road and there are signs that some are questioning its veracity. In a 2002 article *A cool early Earth* from the journal, *Geology,* researchers questioned the very existence of magma oceans:

"The nature and timing or **even existence of magma oceans is uncertain**." Note 7.1b

We can expect to see more and more sentiments like this as the true nature of the Earth is revealed. Shifting from the old magmaplanet paradigm to a water-based, Hydroplanet model will entail abandoning old dogma for new truth. The dogma of magma, or **dogmagma** as we shall refer to it, is the long-held doctrines and ideologies of a magma-based institution of geology. This dogmagma has been responsible for the inhibition of geological discovery for over a century.

Dogmagma: the dogma of magma.

Stevenson partially identified the *type of knowledge* needed for constructing a new standard model of the Earth when he noted, "We do not know the total amount of water in the mantle…" Stevenson was not alone in his recognition for the need for better knowledge. Just prior to the Deep Interior workshop in 2004, participants from 12 nations gathered in Hveragerdi, Iceland during the summer of 2003 to attend the Penrose Conference, to discuss why the magma models are not working:

"Over 60 geologists, geophysicists, and geochemists from 12 nations gathered to brainstorm the fundamental evidence that constrains the origins of volcanism…**This meeting was the most significant gathering ever held of scientists working on alternative mechanisms for anomalous volcanic regions**." Note 7.1c

This was purported to be the first time such a large gathering of Earth scientists came together in an attempt to understand the origins of volcanic regions. They recognized problems with global hot spots and the assumption that they derived their heat from the mantle below. Speakers at the conference noted, "heat flow measurements provide no evidence" for this and actually prove it to be a wrong assumption:

"It is generally **assumed** that 'hot spots' are hot. They are hot in the sense that volcanic activity **occurs at the surface**, but the important point is the **relative potential temperature of the mantle beneath**. Three sessions focused [on] this **critical issue**. Marine heat flow measurements provide little evidence for enhanced heat flow around 'hot spots'. **Two spectacular examples of this are Hawaii and Iceland where heat flow measurements provide no evidence for elevated sub-lithospheric temperatures**. Such findings are typical of 'hot spots'." Note 7.1c

From this significant gathering of scientists, each working on alternative ideas to support the magma model, came the revelation of an important new insight:

"Various aspects of volcanism in the Pacific appear to **require new models**." Note 7.1c

This might have been a moment of serendipity, where old paradigms are challenged, discarded and new insight brought to bear on an old problem. Scientists working together to understand the nature and origin of anomalous volcanic regions:

"This Penrose conference brought together **for the first time scientists who still seek to understand the fundamental origins of volcanic regions**. The full range of ambient ideas in this embryonic field was laid out, brainstormed, criticised and challenged. The problems, needs and tasks ahead were brought into focus. **We are at the start of a long and exciting journey**." Note 7.1c

The **Hydroplanet Model** *is the new model*, but the scientists who sought to understand the fundamental origins of the Earth's volcanic regions were far too steeped in dogmagma to make this leap. For those who dare to look beyond the dogmagma for the truth of the Earth's interior and its origin, this chapter is *truly* the "start of a long and exciting journey!"

The Floating Continents

Our first step on the Hydroplanet journey is to explore the fluidization beneath the continents. It is important to note that modern geology already has *empirical evidence* establishing that the Earth's continents are *floating*. In fact, children are taught in grade school about floating plates, along with other not-so-proven concepts.

Here is the website geography4kids.com, which explains how the Earth's continental plates float:

"THEY REALLY FLOAT?

"These plates make up the top layer of the Earth called the **lithosphere**. Directly under that layer is the **asthenosphere**. It's a flowing area of molten rock. There is constant heat and radiation given off from the center of the Earth. That energy is what constantly heats the rocks and melts them. The tectonic plates are **floating** on top of the molten rock and moving around the planet." Note 7.1d

We previously discussed Earth's continental plates and their observable movement of several centimeters per year (see Fig 15.13.1), but we have taken the position that there is no magma, and therefore, no molten rock upon which the plates ride, so we naturally have to ask; what *are* the plates floating on? This is a truly fundamental question.

FQ: What are the Earth's plates floating on?

The Sea Mystery

In the Universe Model, we will discuss the Hydrogen Abundance Pseudotheory which reveals that the most abundant substance in our universe is water, not hydrogen. Moreover, the Air-Water Model will lay out completely new ideas with respect to the nature of water. In this chapter, we continue to explore the origin of water, especially the water here on Earth.

First, let's consider a question you've probably wondered about: where did the Earth's seas come from? If today's theories of planetary science were *true*, the ever-accumulating research of the Earth's oceans should be drawing an ever-clearer picture of the ocean's origins, but in the March 2002 issue of *Science News,* we find quite the opposite:

"**How did Earth come to possess its seas**?

"Over the years, planetary scientists have proposed several possible answers to that question, but until recently they've had little data for testing their hypotheses. As research in the field progresses, however, **the picture is getting more complicated— not less**." Note 7.1e

Working with an incorrect theory about the origin of the seas, researchers have made those origins more complicated than they need be and are "struggling" to reconcile the real data with their

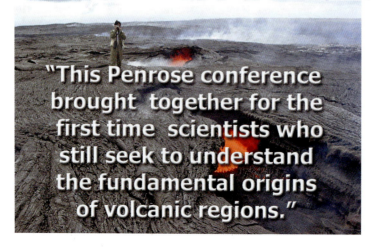

"This Penrose conference brought together for the first time scientists who still seek to understand the fundamental origins of volcanic regions."

CHAPTER 7 THE HYDROPLANET MODEL

hypotheses. Trying to make sense of the data under the magmaplanet paradigm just will not work. Many geochemists and astronomers have come to a consensus that the "celestial pantry is now *empty of a key ingredient*" in the recipe of the Earth—***water***. The *Science News* article continues:

"Analyses of the geochemical properties of various bodies in the solar system and computer modeling of the dynamics of ancient planetary interactions have undermined a formerly popular theory, **which attributes Earth's water to a bombardment by comets late in the planet's formation**.

"New hypotheses are emerging as **that theory's plausibility fades**, and planetary scientists are **struggling** to reconcile data with these alternative scenarios. There's one thing on which most geochemists and astronomers agree: **the celestial pantry is now empty of a key ingredient in the recipe for Earth**." Note 7.1e

Is the celestial pantry *really* empty of this key ingredient?

The Liquid Sphere

Pondering a drop of water can evoke a poet's pen or an artist's brush; it is a source of endless curiosity. What does a drop of water look like? The teardrop shape is a compelling icon used by many water companies to signify their association with water, and we will examine why this shape seems to remind us of water.

At the surface of the Earth, the attractive force we have come to know as gravity pulls liquid water downward, forming the water droplet into its characteristic teardrop shape. (Fig 7.1.1) Leave the Earth's surface and travel well beyond its atmosphere and that force becomes very weak. The water droplet is no longer shaped by gravity and becomes subject to its own internal attractive force. This internal force exerts an equal force in all directions, forming the water drop into natures most perfect and balanced shape—a sphere.

Pour water onto the ground at the surface of the Earth and it will puddle. This is not so in space. Water poured onto the floor of the orbiting space shuttle does not puddle—it forms into countless spheres of water.

Because of our every day experience with the properties of water as it behaves on the surface of the Earth, we tend to overlook the most natural shape water seeks to exist in. However, with the advent of space travel and the ability to remove the external forces acting upon water, we began to learn much about waters' natural habit. In the inset images of Fig 7.1.3, we see what water in space looks like. NASA says this about water:

"In microgravity, water droplets form free floating **spheres**." Note 7.1f

When water drops approach one another on a plate of glass, they combine to form larger drops. In space, microspheres

Fig 7.1.1 – The teardrop shape of water at the surface of the Earth is the result of the downward force of gravity exerted on the water drop.

Fig 7.1.2 – This sphere is a large air bubble formed on a hotplate, underwater in a microgravity environment, in orbit above the Earth. It illustrates how spherical shapes form when the forces of gravity are neutralized. This applies to both liquids and gases. Courtesy of NASA.

of liquid water collect to form larger spheres, as they encounter other spheres. Over time and in the right environment, (we will discuss the effects of temperature shortly), water will continue combining, forming ever-larger spheres.

Within the magmaplanet paradigm, researchers had only the accretion theory to explain the formation of planets in space. Nonetheless, this theory never fully satisfied many open-minded astronomers we talked to, for reasons explained in the Magma Pseudotheory. Even if one could imagine dust and rocks combining and melting, forming a sphere of molten rock in space, the resulting sphere would be glass. Such a sphere has never been discovered. On the other hand, one can easily see how small spheres of water can join to form larger spheres, a process one can observe. Conveniently, it happens right here on earth.

From Magma to Water

Now we can ask two very important questions. They are fundamental questions and lay at the foot of all modern science, especially geology. Answer these questions and a new foundation will be established. We already know that the continental plates are floating and that seismic evidence has established that there is abundant liquid inside the Earth.

FQ: What is the liquid inside the Earth?

From this line of questioning, we are naturally led to ask the next question about the formation of the Earth. Recognizing that the Earth is spherically shaped and therefore formed from a liquid, we need to answer this FQ:

In what liquid was the Earth formed?

At first glance, the idea of a "hydroplanet" or that *a planet formed out of water* may seem absurd. However, it will only take a little pondering to see that there is no other alternative, no matter how absurd it may *first appear*. Copernicus's idea and Galileo's support that the Earth revolved around the Sun was once also considered absurd, even heretical when it was first encountered. Certainly, the spherical Earth had to have come from a liquid state, and the interior of the Earth contains abundant liquid. We have determined that there is no scientific evidence for magma's existence. This leaves little else from which to choose. By the end of this chapter, after examining the empirical evidence, the reality of the Hydroplanet Model will become self-evident.

From Neptunism to Plutonism and Back

The idea of a water-born planet is hardly groundbreaking. This was a hotly debated topic in the scientific circles of the mid 1700s. Those supporting the existence of a primeval ocean were the Neptunists

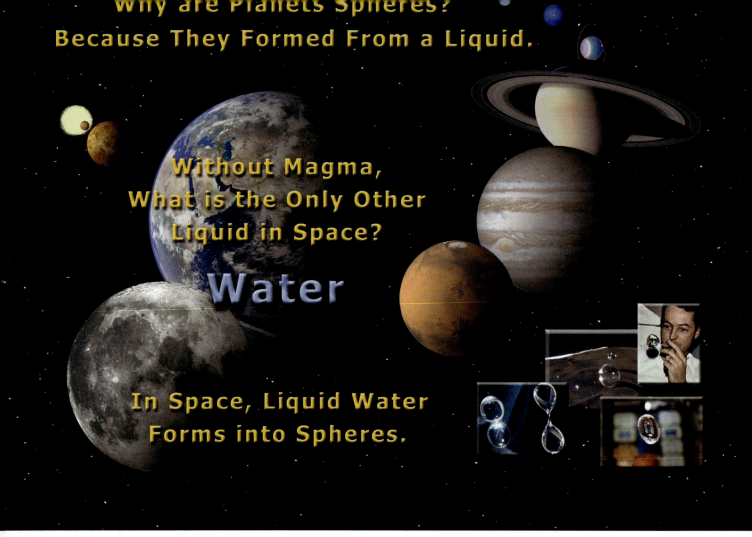

Fig 7.1.3 – This collage of celestial bodies illustrates the dominant shape of planets in our solar system. These Celestial Spheres had to have been formed as liquid. No magma-like spheres have ever been observed in space, but spheres of water and ice have been. The inset photos, bottom right, show water as photographed in space by astronauts. The spherical shape of liquids in space is made possible outside where gravitational forces are near neutral allowing the water's own cohesive force to apply equally in all directions.

while those who proposed a volcanic origin of the Earth were Plutonists, or Vulcanists:

"During the eighteenth century there were many passionate debates about the nature of origin of rocks. One of the major debates was between the *Neptunists* (e.g. Abraham Gottlob Werner, 1749-1817) on one side, and the *Vulcanists* (e.g. Jean Etienne Guttard, 1715-86 and Nicolas Desmarest, 1725-1815) and later the *Plutonists* (e.g. James Hutton, 1726-97) on the other. The **Neptunists** supported the ideas of Tobern Bergman (1735-84) that the rocks of the crust had all been deposited sequentially from, or precipitated out of, **a primeval ocean**. Both Basalt and granite were regarded as being rocks that had precipitated out the universal ocean." Bib 87 p2

A significant obstacle to the Neptunists' idea was that they could not adequately explain where lava came from or why volcanoes erupted. The Plutonists seemed to have the best answer. What they did not know then, in fact what modern science still does not know, until now, are the Lava-Friction and Gravitational-Friction Laws presented previously in the Magma Pseudotheory chapter of this book. Werner, the lead contender of Neptunism, made a critical error at the time, thinking coal was the source of volcanic activity:

"Werner considered volcanoes to be curiosities caused by the subterranean combustion of coal." Bib 87 p2

In Werner's day, sensitive earthquake measuring devices had yet to be developed that would later show volcanic activity and earthquakes are associated. Werner was not the last to err on the internal source of Earth's heat. It is actually far *more* shocking today, with the myriad of modern technological advances and thousands of researchers working in many fields of study, that science has overlooked the frictional source of heat responsible for the production of lava. Instead, they continue to look to radioactive magma as the source of heat in the Earth. At least Werner and his associates had *observed* coal burning underground. Radioactive magma has *never* been observed and neither has radioactive lava.

Nevertheless, Werner was not completely wrong in his coal-combustion theory. Subsurface fires do exist, but they are poorly understood and are relatively rare:

"Subsurface fires are well known as a phenomenon, but are poorly documented in the literature." Note 7.1g

In the July 2003 journal of *Geology*, researchers reported

CHAPTER 7 THE HYDROPLANET MODEL

finding subsurface organic fires in western Africa that were originally thought to be volcanic in previous studies:

"A 2.5-m-deep trench was dug into the heat front. We located a combusting organic-rich layer at 60 cm depth where temperatures reached 830° C. Visible flames emerged from the organic-rich layer." Note 7.1g p582-3

Moreover, by measuring the heat in a layer beneath the combusting material, the heat source was confirmed to be the combusting organic-rich layer rather than being volcanic.

"The temperature was as low as 40° C in a sand layer 0.75 m below the combusting layer." Note 7.1g p583

From this, we can determine that coal-like material can burn underground, as the Neptunists' thought. However, events like these are rare and they do not account for the heating of the Earth's crust, nor do they explain volcanoes. The Neptunists clearly missed the mark when attempting to explain the source of Earth's heat, but so too did the Vulcanists and Plutonists. Interestingly, the Neptunists' theory of rocks precipitating from a universal ocean was actually quite close to reality.

From the Magma Pseudotheory, we learned that a magmatic melt does not produce quartz, basalt or granite. Of course, if it did researchers would have proven this long ago. Melted rocks produce *glass*, not crystalline rock. The following historical account from the book, *Melting the Earth,* tells of how the Neptunists' ideas of crystal formation were supplanted by the Plutonists' ideas. This is a clear example of how false ideas, in this case that "silica melt" crystallizes into quartz, can drive an entire field of science down a dead-end road:

"Because Werner and his followers had recognized, **quite correctly, that basalt was a crystalline rock**, it was perhaps not unnatural that they also deduced its origin as a precipitate from an aqueous solution. Not until after the middle of the eighteenth century did **the idea slowly begin to spread that crystallization could also occur as a result of the removal of heat from a silicate melt as well as from an aqueous solvent**. In 1761 Pierre Clement Grignon complained that chemists overemphasized aqueous systems, pointing out, after he managed to extract crystals from glass slags, that the products of glass furnaces compared to the products

Neptunists could not explain where lava comes from.

of volcanoes. **The idea that magma was a solution that could produce crystals was beginning to emerge**, and a group of geologists embraced the view that basalt owes its origin to solidification of magma or molten rock, either bought up to the surface during volcanic eruption or intruded into the Earth's crust. The Plutonists, as the adherents of this theory became known, found their intellectual leader in **James Hutton** in Scotland. **A profound difference between the Neptunist and Plutonist theories related to the quantity and role of heat in the planet's interior**. The Neptunists saw a negligible role for heat as a geological agent and considered volcanoes minor phenomena related to shallow-level processes—**not as an indication of deep-seated heat. The Plutonists, on the other hand, regarded heat as the fundamental driving force for mountain uplift, folding, and deformation, pointing out the abundance of volcanoes, basalts, and granite as evidence of the melting of rocks at high temperature within the Earth**." Bib 136 p118-9

From the Plutonists' intellectual leader, James Hutton, the primary building blocks of modern geology and the false ideas of the Magma Pseudotheory came forth. From this came the incorrect assumptions of continental uplift, and that basalt and granite came from "the melting of rocks at high temperature within the Earth."

Two and a half centuries later, we return to the ideas of the Neptunists and will soon see the important and often overlooked role that water has played throughout the celestial universe.

7.2 Celestial Water

To begin, we will discover how water behaves in space and beyond our solar system, in the galactic void. Following this, we will give a quick overview of water in the neighborhood of our Sun, with more watery details about the Moon, the planets, comets, and asteroids.

Paradigm shifts of this magnitude are rare and would certainly be difficult to accept. To accomplish a feat such as identifying and explaining a model like the Hydroplanet Model so that it can be understood and accepted, based on supportive and empirical evidence, is no small task. In order to dislodge the pillars of scientific theory, significant and extraordinary, perhaps even bizarre evidence must be brought forth and presented; evidence such as water in space, in the stars and even on the Sun.

The Celestial Water Universe

The Earth as a hydroplanet came from water, as did the planets, moons and other celestial bodies. Even the stars themselves came from water. Because water is so abundant, we should expect to find it nearly everywhere we look, from the other planets in the solar system to intra-solar space and far out into interstellar space. Until now, the Periodic Element Pseudotheory has been the source of the assumption that hydrogen is the most abundant element in the universe. It has been touted as *the element* from which all other matter has come. Thanks in part to new technology, developed over the past several years, we are about to uncover what is really the most abundant molecule in space, and the true origin of all matter.

Does Liquid Water Exist in Space?

One of the first conceptual hurdles to overcome when envisioning liquid water in space is the extreme cold. Space is very cold—close to absolute zero, which is colder than minus 450°F (-267°C). This is certainly too cold for water to be a liquid on Earth. But water behaves differently when it is under different *pressures*. New research shows that in space, even though it is only a few degrees above absolute zero, heat radiating from distant stars is enough to cause water to *flow* in a vacuum environment:

"Perhaps the most exciting property of interstellar amorphous ice is that **when exposed to radiation such as that found in deep space, it too can flow—even though its temperature is a scant few degrees above absolute zero** (which is equivalent to –273 degrees Celsius)…

"The discovery that amorphous ice is **more like liquid water** than it is like crystalline ice came as **a huge surprise**." Note 7.2a

We have discovered that today, most researchers are still unaware that liquid water exists in space. One key factor many investigators missed, a fact in most of the previous experiments, is that water, observed in low atmospheric pressures (such as found in outer space), was observed at *higher* ambient temperatures. Very little research had been done dealing with water at very low pressure *and* at temperatures approaching absolute zero.

To illustrate how little-known this discovery remains, we read from a 2004 *National Geographic* article, published three years after the discovery of liquid water in space. The writer errs, as most scientists have, by assuming that there is no liquid water in space:

"In the near vacuum of space, **water at any temperature boils away because there isn't enough pressure to keep it liquid.**" Note 7.2b

Contrary to both past and current beliefs, the researchers in *Scientific American* discovered that liquid water could indeed exist in the cold temperature, low-pressure environment that exists in space:

"**Water**, it seems, was present **at every step in the creation and processing of molecules necessary for life.**" Note 7.2c

In summary, liquid water does exist in space. Moreover, if water was present at *every* step in the processing of molecules necessary for life, would not this include the *first step, the making of minerals for that life?*

The discovery of liquid water in space is extremely important, and as we will shortly find out, it supports the idea that the most abundant substance in the universe is indeed, *water*.

How Much Water is in Outer Space?

Charles Townes was the co-inventor of the maser, a device used to produce or collect and amplify microwave radiation. The laser is a familiar example of this device that operates in the optical range. Townes was also an originator of spectroscopy using microwaves and was a pioneer in the study of gas clouds in galaxies and around stars. His research led to the discovery of astrophysical masers, natural sources of microwave energy in space.

In space, the area between stars is extraordinarily vast and often touted as empty. Nevertheless, space is not as empty as was once thought. Materials from which stars are formed are found in various states of dispersion in the interstellar medium. The question is; just what are these materials? If the Big Bang theorists would have their way, the most abundant material would be hydrogen.

Joseph von Fraunhofer first mapped the Sun's spectrum in the early 1800s. He observed intermittent dark lines in the broad,

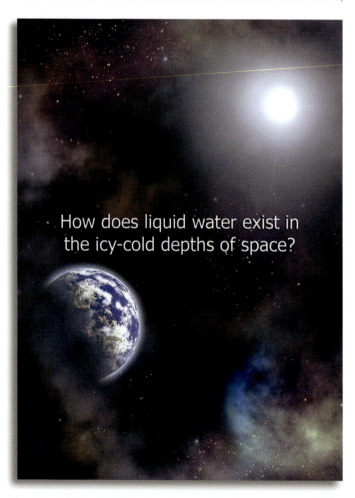

Fig 7.2.1 – Everyday experience tells us that water changes from a liquid to a solid in very cold conditions. Based in these observations, scientists believed for a long time that water in deep space could only be solid. Recently, new observations have shown that under the low pressure and low temperature conditions that exist in space, the faint radiant heat of stars is sufficient to liquefy water, "a huge surprise" to researchers.

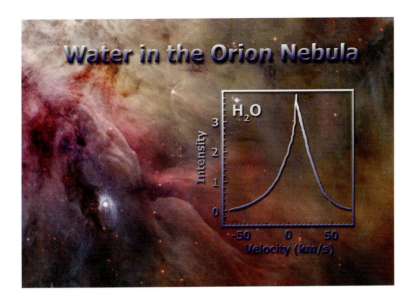

Fig 7.2.2 – When researchers looked for water inside the Orion Nebula, one of them declared, "It must be raining in Orion." This was due to the strong water line found with the maser. This water signal was stronger than elemental hydrogen, the 'supposed' most abundant substance in the universe.
Image and graph courtesy of NASA HST and SWAS.

colored spectrum and assigned letters to the most prominent ones. These dark lines, called absorption lines appeared where light of certain wavelengths was blocked. Among other substances, hydrogen was easily identified and appeared rather dominant. This observation led scientists to believe that hydrogen was the most abundant element in the universe. Today, this pseudotheory is presented as *a fact* in just about every textbook that discusses it.

Townes' discovery of natural masers (microwave energy that could be magnified) allowed him to search for and identify substances in space, which he did with enthusiasm. If hydrogen was the most abundant element, then the largest maser source would be hydrogen. Let us read what he found:

"What molecules to look for next? There were no obvious, easy targets. Carbon monoxide and hydrogen cyanide produced wavelengths of only a few millimeters or shorter; they would take some special effort because we had no equipment already suited to work with such wavelengths. Water had a spectral line very near those of the ammonia we had found, but the excitation of water energy levels that produced this wavelength would take much higher **densities** and **higher temperatures** than seemed likely in interstellar clouds, even after the new ammonia results. However, our group—Jack Welch, Dave Rank, Al Cheung and I—thought **we might as well look for water**. After all, further surprises could happen. **And happen they did—the water line also showed up**. We found our first water in Sagittarius B2, the same place where we had first found ammonia..."

"After water had turned up in Sagittarius B2, we of course wanted to search other sources to see **if it was more widespread**." Bib 26 p174-5

Townes describes his associate, Al Cheung's reaction to the discovery made at the location the team searched next for a water absorption line:

"'**It must be raining in Orion! It has a very strong water line**.' **He had found water, lots of it**. The Orion water line was 20 times stronger than the previous one, much more than we would have expected." Bib 26 p175

The researchers were so shocked that they quickly arranged to repeat the observations on the larger, 85-foot Navy antenna to verify their data was correct. They found even more water:

"There were a number of other sources as well. The nebula W49, a cloudy region in which **new stars were forming**, had water radiation about **20 times still brighter than Orion**." Bib 26 p176

Within the traditional Big Bang Pseudotheory, elemental hydrogen, helium and oxygen should have been considerably more abundant than water, and thus should be producing the most powerful masers in space. Researchers found just the opposite to be the case:

"**No space masers have been found more powerful than those based on water molecules**. They might be considered the most powerful radio stations known. A single astrophysical water maser can put out much more power than the total radiation from the sun, all at a single spectral line or frequency. Some are so intense they are called megamasers—enormous in power, though perhaps not much larger in size than the solar system." Bib 26 p176

What substance is the most abundant in space, based on the power of its cosmic radiation?—**Water**!

In the laboratory, nature's molecules and elements have been tested to examine the absorption and emission spectra of the substances. Testing and observing helps confirm the abundance of a substance by comparing its strength of emission or absorption. It is because of this that water has been found to be the *most abundant substance* in space because it has created the most powerful cosmic radio signals received to date. This discovery is not commonly known among the scientific community, and discussions about it will surely lead to much conversation and debate—all necessary to bring about further scientific truth.

> "No space masers have been found **more powerful than those based on water molecules**."
> Charles Townes, inventor of laser and maser

Water in the Stars

An online encyclopedia indicates that water molecules create the "brightest spectral line in the radio universe":

"Maser-like stimulated emission also occurs in nature in interstellar space. **Water molecules in star-forming regions** can undergo a population inversion and emit radiation at 22 GHz, **creating the brightest spectral line in the radio universe**." Note 7.2d

Although this information is not new, it is *not* widely known. More importantly though, modern science has not put two and two together to figure out *why* water has such a bright spectral line and *why* there is so little elemental hydrogen and oxygen in interstellar space in comparison to molecular water, especially where star-forming regions exist. As we will discover in the Air-Water Model chapter, the individual elements of hydrogen and oxygen in and of themselves, cannot be merely forced together to make water, even at high temperature and pressures. This is a reoccurring observation throughout the UM, such that abundant molecules, such as salt (NaCl) or water (H_2O) can have individual elements liberated from their parent molecule by technological processes, but this does not usually happen in nature.

Water came first, not hydrogen, not helium, nitrogen, or oxygen. This is the wisdom of why water is found almost everywhere we look, even in the spectra of stars:

"**The hot water molecule is the most important source** of infrared opacity in the spectra of oxygen-rich late-type **stars**. **Water is particularly prominent** in the spectra of **variable red giant stars** (Mira variables) and in other cool **M-type stars**." Note 7.2e

Detecting water in the spectra of stars was an astonishing discovery. Even more amazing was the role water plays in star formation. The article *Water's Role in Making Stars* in the journal, *Science* (November 2000) begins:

"The **water molecule plays a fundamental role** during the first stages of **star formation**." Note 7.2f

Detection of water in space had been hampered by water vapor in the Earth's atmosphere, even at high altitudes accessible only by aircraft and balloons. It was not until the Infrared Space Observatory (ISO), launched in 1995, and the Submillimeter Wave Astronomy Satellite (SWAS), launched in 1998, that the *true abundance of water in space* became known:

"The discovery of **huge amounts of water** streaming away from an aging, swollen **red giant star** in the constellation Leo shows that our own planetary system is not alone in harboring a key ingredient of life as we know it…

"**SWAS detected 10,000 times more water than the star [CW Leonis] could have been giving off**." Note 7.2g

The discovery of water was not limited to cooler, older red giants. It was detected on young hot stars as well. Astronomers have actually found stars similar to our Sun with water in their light spectra:

"ISO **detected water lines** not only in **high-mass, luminous young stellar objects but also in low-luminosity protostars**, which are precursors of stars **like our sun**." Note 7.2h

Perhaps this may come as a surprise to many. There is water on the Sun!

Water On the Sun

What, water *on the Sun*? How could this be?

"High-resolution infrared spectra of sunspot umbrae have been recorded with the 1-meter Fourier transform spectrometer on Kitt Peak. **The spectra contain a very large number of water absorption features originating on the sun**. These lines have been assigned to the pure rotation and the vibration-rotation transitions of hot water by comparison with high-temperature laboratory emission spectra." Note 7.2i

By comparing spectra of high-temperature water in the lab-

Fig 7.2.3 – The last place one would expect to find water would be on the Sun—but there it was. Researchers confirmed this discovery by comparing water emission spectra from hot water in the lab to those observed on the Sun. Graph is courtesy of Peter Bernath.

oratory with spectra from the Sun, scientists had indeed confirmed that water exists on the Sun. The researchers then summarized their findings by stating just how much water showed up in the spectra of the Sun:

"We conclude that **the remarkably dense lines in the sunspot spectrum are due mainly to water**. Moreover, **most of the weaker unmarked lines in this spectrum are probably also due to water**." Note 7.2i p1157

An article in the July 1997 issue of *Science* was actually titled—*Water on the Sun: Molecules Everywhere*. Here are some of the highlights:

"Ordinary polyatomic molecules like **H_2O are not supposed to exist on the surface of the sun**; at the high temperature of ~5800K, a water molecule should dissociate into a free-radical OH and an H atom or further into an O atom and two H atoms. I therefore found the title of the 1995 report by Wallace et al. 'Water on the Sun' quite incredible, although completely believable after seeing the data…the laboratory infrared spectrum of hot water taken by Bernath and his co-workers **matched exactly with the observed solar spectrum**.

"The two spectra may look different at a glance, but closer examination shows exact coincidence in the frequencies of the spectral lines. **Matching frequencies between astronomical and laboratory spectra has been the cornerstone in discoveries of molecules in space…The matching of the frequencies of tens of spectral lines, as in the figure, is foolproof evidence of the H_2O detection**." Note 7.2j

> "The **water molecule plays a fundamental role** during the first stages of **star formation**."
>
> *Science*, Vol 290, 24 November 2000 p1513

According to theory, water was "not supposed to exist on the surface of the sun," but the evidence showed otherwise. Water was there and must now be accounted for in all stellar models. In fact, all astronomical theories must consider water as a fundamental element within the theory. Identification and detection of "huge amounts of water" associated with stars no longer constitutes new news, yet this has *not* stimulated the exploration or development of a new 'water model' of stellar or planetary formation.

Water on the Planets

Abundant water has been detected in space, in interstellar gas clouds, on distant stars and even on the nearby Sun. Is there detectable water on the planets?

Most of us learned while growing up that there are four inner rocky planets, which include the Earth, and four outer gaseous planets as well as a distant rogue planet, Pluto, also rocky. Never did we hear a word about 'water' being on **all of the planets!** Nevertheless, this is exactly what astronomers have found.

Probe data from the Mariner missions in the mid 1970s indicated that there might be water ice on **Mercury**:

"Radar data suggest that **Mercury**, like the Moon, **has deposits of water ice** in permanently shadowed areas at the poles." Note 7.2k

Then in 2012, NASA's Messenger spacecraft found:

"Indeed, **Mercury**, the closest planet to the Sun, **possesses a lot of ice — 100 billion to one trillion tons** — scientists working with NASA's Messenger spacecraft reported on Thursday. Sean C. Solomon, the principal investigator for Messenger, said there was enough ice there to encase Washington, D.C., in a frozen block two and a half miles deep. That is a counterintuitive discovery for a place that also ranks among the hottest in the solar system. At noon at the equator on Mercury, the temperature can hit 800 degrees Fahrenheit. But near Mercury's poles, deep within craters where the Sun never shines, temperatures dip to as cold as minus 370." Note 7.2l

Evidence for abundant water has been found, perhaps even an ocean on **Venus** at some time in the past:

"*Venus*. The very high ratio of heavy hydrogen (deuterium) to the common sort in what little water vapor remains says that **there was once 100 times or so more**, **enough to make a wading pool over the surface**, if not an ocean." Note 7.2m

There are so many observations and reports of water on **Mars** that we will discuss this in a section all its own, following this brief enumeration of the remaining planets of our solar system.

In 1995, the Galileo spacecraft visited **Jupiter** and was able to detect water vapor in Jupiter's atmosphere at up to eight times the pressure found on Earth:

"The Near Infrared Mapping Spectrometer (NIMS) on the Galileo spacecraft continues to provide jovian atmosphere spectral images, covering wavelengths between 0.7 and 5.2 micron with a spectral resolving power of 200 at 5 micron. The spatial resolution is on the order of several hundreds of kilometers. Here we analyze NIMS data cubes that contain spectra in Jupiter's 5-micron window. **From the spectra we obtain water vapor relative humidity between 4 and 8 bar and the cloud opacity for pressures lower than about 2 bar**." Note 7.2n

The thick atmosphere of Jupiter impedes our ability to investigate the existence of water on its surface even by dispatching the Galileo probe through the atmosphere. The most enlightening news about Jupiter's association with water involves its moons. There are many details to discuss about Jupiter's moons, but we will save that discussion for the following subchapter.

The first time anyone looks at **Saturn** through a telescope, it is a sight not to be forgotten. The beautiful rings of Saturn are so unique and inspire such awe that it leaves the viewer spellbound. Few things in the entire universe have the impact Saturn does on the first-time viewer.

In 2004 the Cassini-Huygens' spacecraft took the spectacular photo seen in Fig 7.2.4. As it turns out, Saturn's rings are made of water ice:

"**Saturn's rings are made primarily of water ice**. Since pure water ice is white, it is believed that different colors in the rings reflect different amounts of contamination by other materials such as rock or carbon compounds." Note 7.2o

Saturn's moons are no less spectacular than Jupiter's moons and they too have much to share about their association with water. These will also be discussed in the next subchapter. The wisdom or comprehension of how Saturn's inspiring rings were formed will only come after understanding how Saturn's moons were formed.

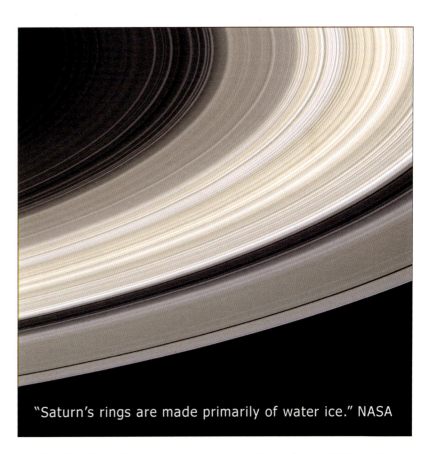

"Saturn's rings are made primarily of water ice." NASA

Fig 7.2.4 – The rings of Saturn as seen in this image taken by the Cassini-Huygens' spacecraft are truly magnificent. They are made primarily of water ice from hydrofountains found on Saturn's moons, which will be covered more in depth later in this chapter.

A common aspect of all four 'gas giants' is water. Similar to the other gas planets is the planet **Uranus.** Although we know little about this planet, it appears that it too, has lots of water—perhaps even oceans:

"**Uranus** is a giant ball of gas and liquid. Its diameter at the equator is 31,763 miles (51,118 kilometers), over four times that of Earth. The surface of Uranus consists of blue-green clouds made up of tiny crystals of methane. The crystals have frozen out of the planet's atmosphere. Far below the visible clouds are probably thicker cloud layers **made up of liquid water** and crystals of ammonia ice. Deeper still -- about 4,700 miles (7,500 kilometers) below the visible cloud tops -- **may be an ocean of liquid water** containing dissolved ammonia. At the very center of the planet may be a rocky core about the size of Earth." Note 7.2p

Neptune is another blue water-planet. Although we do not know the full extent of how much water Neptune has, we see that water is found even in the furthest reaches of the solar system:

"Scientists believe that Neptune is made up chiefly of hydrogen, helium, **water**, and silicates. Silicates are the minerals that make up most of Earth's rocky crust, though Neptune does not have a solid surface like Earth. Thick clouds cover Neptune's surface. Its interior begins with a region of heavily compressed gases. Deep in the interior, these gases blend into a **liquid layer** that surrounds the planet's central core of rock and **ice**." Note 7.2q

The Most Important Discovery in the last 25 Years

Toward the end of 2005, a think-tank of senior scientists was asked to pinpoint the *most critical developments in space over the previous 25 years*. Numerous advancements in spacecraft development and space exploration have allowed us to visit other celestial bodies. Advances in telescope technology, both in space and on the ground, using sophisticated optics, allow us to peer into the deep reaches of the universe. Discoveries of phenomena and events never before seen or imagined were made. The conclusion reached by several researchers—the confirmation of large quantities of water ice on Mars, was deemed the most important:

"**The most important discovery in space exploration in the last 25 years was the confirmation of large quantities of water ice on Mars**. Geomorphological evidence suggests that almost everywhere on the planet there is a water table—or, more precisely, an ice table—and if you dig down deep enough, you can get to it." Note 7.2r

The assessment that water on Mars was the most important discovery over the previous 25 years should excite those seeking to understand the Hydroplanet Model. The reason this discovery trumped other fantastic discoveries over those two and a half decades is that scientists accepted the significant role water played on Mars.

Recognizing water as the precursor to life, scientists have been driven in their search for water with a biological mindset. Far from being simply a place to find life, or an erosional force on the planet, water plays host to more than modern science has imagined. Water is *the basis* for the geological formations on planets like Mars, just as it is for the geological formations on Earth. Recognizing the discovery of water throughout the solar system is an important step in establishing the New Geology discussed later in subchapter 7.5.

Over the past several years, hundreds, if not thousands, of articles have been written about water on Mars. It is beyond the scope of this book to detail all of the observations and accumulated evidence that attest to water on Mars, but in order to capture the essence, we have included some. These several statements reveal the magnitude of this important discovery—there are large quantities of water on mars.

Fig 7.2.5 graphically illustrates just how much water exists in the topmost meter of Martian soil. January 2004, *National Geographic* included an article on the new missions to Mars and discussed with researchers what they had learned. By this time, water was known to exist at the poles, but the scientists were amazed at the extent of the water:

"It was no surprise to see proof of plenty of water around the north pole, and researchers had theorized that frozen water was locked beneath the CO_2 ice around the south pole. But the seeming abundance of water ice extending to the midlatitudes was news to scientists. And **what amazed them was the amount of water-equivalent hydrogen at low latitudes and even the equator**." Note 7.2s

Although researchers were amazed at how much water existed just under the surface of Mars, they could not begin estimating how much water was below the topmost meter! In truth, we have only just begun to learn how much water Mars really has. Where did this water come from?

"Mars seems to have a frozen lake on its surface, according to images obtained by the European Space Agency's Mars Express satellite. One hypothesis has it that **Mars sporadically belches up volcanic gases and floodwater, leaving behind huge seas**, which then evaporate. Researchers have identified the apparent remnants of such an outburst: a frozen

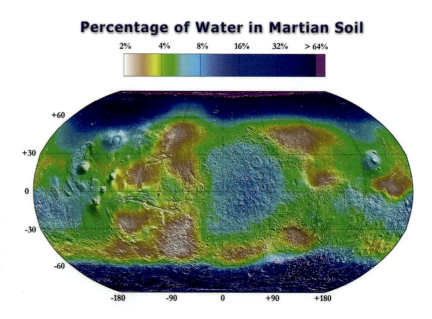

Fig 7.2.5 – This is a map of the surface of Mars illustrating the abundance of water in the topmost meter of Martian soil. The key represents the percentage of water in the soil by weight. Data for this map came from the neutron spectrometer onboard the Mars Odyssey spacecraft in 2003. The blue areas have enormous amounts of water. Courtesy of NASA/JPL.

body of water the size of Earth's North Sea." Note 7.2t

Scientists are just beginning to recognize the direct evidence of enormous floods that *came from **within** the planet.* These floods are indirect evidence that similar events happened here on Earth, a topic that will be examined in detail in the next chapter.

Water on Exoplanets

An **exoplanet** or extrasolar planet is a planet that does not orbit the Earth's Sun, but instead orbits another star in the Universe. In the 1990s, exoplanets were confirmed to exist and raised the high possibility that life, as we know it could exist in other star systems. Water is a precursor to life and therefore the discovery of water on one of these exoplanets would be important. On the 24th of September, 2014, researchers found a planet four times the diameter of the Earth orbiting a star in the Cygnus constellation 124 light-years away with "significant quantities of water vapour".

"A cloud-free atmosphere has allowed scientists to pick out signs of **water vapour on a distant planet the size of Neptune**: the smallest '**exoplanet**' ever to reveal its chemical composition." Note 7.2u

On the following day, another article published had the title; *Water on Earth is OLDER than the sun,* and explained that scientists studying ancient ices preserved in comets and asteroids found that the water in these celestial objects "pre-date the sun itself" which implies that "water is a common ingredient during the formation of all planetary systems":

"Ilsedore Cleeves, a PhD student at the University of Michigan and lead author on the paper, explained:...'The implication of these findings is that **some of the solar system's water must have been inherited from the sun's birth environment, and thus pre-date the sun itself**. If our solar system's formation was typical, this implies that **water is a common ingredient during the formation of all planetary systems**.'" Note 7.2v

Thus, the recent discovery of water on exoplanets is another evidence of celestial water being everywhere we look in the Universe.

Water on the Moon

Prior to the Apollo Lunar program, astronomers and geologists had made up their minds that there was no water on the Moon, despite the large, smooth plains or maria that exist there. We take this quote from a 1946 encyclopedia:

"When we look at the moon through a telescope and thus magnify its surface features, we see the areas which are dark to the naked eye as **smooth plains**. Early astronomers thought they were seas and so called them *mare*, a name which persists in spite of **the fact that there is no water on the moon**." Note 7.2w

Investigate the vast, smooth plains of the Earth and you will find geologists explaining that *water* was the primary force involved in the smoothing of the plain. On the Moon however, geologists explained that lava flows rather than water were the source of the lunar plains. This sci-bi is still in use today. Lava flows large enough to have produced the enormous, circular-shaped maria found on the Moon have never been observed.

Even after the Apollo program in the 1960s and 70s, geologists still held that no water existed on the Moon. This idea must certainly have been due in part to the deeply entrenched magma pseudotheory and geological dogma that existed because of it. During the 1990s, with the advent of new technology, the truth about water on the Moon was finally recognized. January of 1998, NASA launched the Lunar Prospector satellite. Among its mission objectives, it was to 'prospect' for potential resources, including water ice. Results were reported in the journal, *Science*:

"After only a month in orbit, the $63 million, five-instrument spacecraft gathered 'the first **unquestionable result that there is water on the moon**'... researchers '**were certain water is there**'... '**The uncertainty is in how much**.' About 0.5% to 1% of the lunar soil near the poles appears to be fine particles of ice. That means a cubic meter of soil would contain about 5 to 20 liters of water, which might add up to a total of **10 to 300 million metric tons of water**..." Note 7.2x

Water in Comets

People have observed comets for millennia and though they are not a common phenomenon, most people are familiar with what a comet looks like. Aristotle once thought comets were an artifact of the Earth's upper atmosphere, an idea that held sway for almost two thousand years among some people. Comets have their place rightly established in the heavens as a part of our solar system. However, do we really know what a comet is?

The characteristic tail originating from the comet body is made of mostly water, streaming away from inside the comet as it approaches the Sun. Heat is produced either by solar radiation or frictional heating due to gravitational interaction with the Sun and nearby planets, which forces water out through steam venting explosions.

We will discuss more about comets later in this chapter. For now, our discussion is limited to the abundance of water in comets. An article in Sky & Telescope magazine reported on water in comets:

"***Comets***. The rate they shed gas as they approach the Sun says that **water ice is the most abundant substance in nearly all of them**." Note 7.2y

Water above the Earth-star

In a collaborative effort between NASA and ESA, the Solar & Heliospheric Observatory, SOHO, was launched in 1995, to study the Sun. The $1 billion solar satellite telescopes spun out of control in 1998, and engineers worried that sensitive instruments would be harmed by the extreme heat and cold of space. All three gyros failed but through the efforts of the SOHO recovery team, the spacecraft was eventually returned to full scientific operation. The fears of equipment damage proved to be unfounded, and in fact:

"...one sophisticated telescope is working better than before the accident..." Note 7.2z

Why did this equipment's performance improve? As it turns out, the Extreme Ultraviolet Imaging Telescope, used to study the Sun's corona and other features, had been plagued by blurring contaminants such as *water vapor*. Where did this frozen water vapor come from?

SOHO is locked in an Earth-Sun orbit about 1 million miles

from Earth, which is around four times farther than the Moon. It is in this seemingly desolate area of space that SOHO found itself amidst so much water, that it had become frozen on the telescope's optics. When the satellite spun out of control in 1998, previously shadowed areas became bathed in sunlight, melting the frozen water. The performance of the scope was significantly improved:

"…improved the scope's sensitivity by 60% over its preaccident performance." Note 7.2z

Thus, we see that space is not as empty and dry a frontier as many have thought. Water exists in space, and all satellites must be prepared for moisture contamination that will affect the performance of the spacecraft.

Water, Water Everywhere

Water, Water Everywhere is the title of an article in the February 1999 edition of *Sky & Telescope* magazine. It is a story about astronomers finding water, well, as it states—*everywhere*. As we have seen in this subchapter, it is hard for astronomers to look anywhere without finding water. What makes this more astonishing is that we were not supposed to find such an abundance of water, according to the modern science Big Bang Hydrogen Abundance Pseudotheory. Hydrogen, not water is purported to be the 'most abundant' substance. However, the reality of the subject is before us:

"The inviting blue mantle of water that nearly covers the Earth has always seemed our planet's most distinguishing characteristic. But as astronomers look closer at the rest of the solar system and beyond, they are arriving at the conclusion that Earth is really not that special after all. **Water, one of the most abundant chemical compounds in the universe, turns up almost everywhere we've searched: on the moon, Mars, Jupiter, Saturn, Uranus, Neptune, and Pluto, on many moons of those planets, in comets and meteors, in the sun and perhaps on Mercury, all over interstellar space, and, most likely, on planets circling other stars**." Note 7.2aa

It seemed like water was being found everywhere. This article was published in *Air & Space* in 1997. Since then, astrophysicists have been finding more of this ubiquitous substance with every new satellite sent into space. More water, not hydrogen. In another article, five years later in *Sky & Telescope*, we read almost the same words:

"**Water, Water, Everywhere**

"To date SWAS has inspected comets; the atmospheres of Mars, Jupiter, and Saturn; and about 120 molecular clouds selected largely on the basis of their interesting radio characteristics. Like ISO, SWAS found '**water, water everywhere**,' says

Fig 7.2.6 The SOHO satellite, designed to take images of the Sun, is in orbit about 1 million miles from the Earth. The telescope was severely impaired because of water accumulation on its optics. Performance improved after temporary loss of control of the spacecraft turned the telescope in such a way that the frozen water was melted away. This serendipitous event proved there was water in space.

Melnick, the mission's principle investigator." Note 7.2ab

Considering the abundance of water in space and that it has been found in so many places, the average person would think that water would have become a central topic of celestial chemistry and star/planet formation. Unfortunately, this is not the case.

Space Chemistry Theories Fall Short

In the early 1800s, an English chemist and a German physicist independently observed dark lines in the spectra of the Sun. These lines, called Fraunhofer lines after Joseph von Fraunhofer who studied them in detail, are absorption lines, later correlated with certain elements such as hydrogen and helium. From these and other observations, it was deduced that the four elements of hydrogen, helium, oxygen and carbon make up 99.88% of the Sun's matter, and from that, it is assumed that the "universe as a whole" also contained these same elements in similar proportion:

"The Sun is clearly mostly hydrogen and helium, with only a trace of heavier elements. **This is also true of the Universe as a whole**: **most of the Universe is hydrogen, with some helium, and the remainder of the elements occur only in trace concentrations**." Note 7.2ac

However, as we will learn from the Solar Model in subchapter 25.9, astronomers have never been able to peek below the surface of the Sun—a little known fact. This means that the lighter gases seen in the Sun's spectra, responsible for many of the dark absorption lines may be merely the outermost gases of the sun, and not the inner gases, liquids or even solids of the Sun itself.

Moreover, as outlined in the Hydrogen Abundance Pseudotheory, in subchapter 25.3, there has never been any empirical evidence that hydrogen, helium or carbon is as *abundant in the universe* as is water. The more astronomers look, the more matter they find between the stars. This has challenged the idea that the majority of the mass in the universe is comprised of the stars themselves. The observation of hydrogen absorption lines on the Sun was extrapolated and applied to the universe, without scientifically verifying that this was so.

Turning to the subject of oxygen, measurements taken by the SWAS spacecraft showed no sign of oxygen in the molecular clouds it was pointed toward. The lack of oxygen and the lack of water in a gaseous state raise some interesting questions and have left scientists wondering:

"The ostensible shortage of water and oxygen, says Goldsmith, 'might be a hint **that something is wrong with our picture of the structure of these giant clouds**.' He was ini-

CHAPTER 7 THE HYDROPLANET MODEL

Water, Water Everywhere

tially disappointed by the results, as they suggested **a big gap between existing theories and the first observations of this kind ever made**. But upon reflection, he says, 'In some sense it's much more interesting when your observations contradict prevailing wisdom.'

"While the standard theory appears to be off in several crucial details, the general picture is not entirely shaken. 'If our interpretations are correct,' Melnick observes, '**water is not a rare species**; **apparently**, **it's just rare in the gaseous phase**.'"
Note 7.2ad

Importantly, these observations were made anticipating the detection of water in a gaseous phase. Liquid water in interstellar space had not been known to exist until recently, as we discussed at the beginning of this subchapter. Future research on the molecular clouds should focus on liquid water, and in doing so, many new discoveries are likely to occur.

Water has been detected on the Sun, in the stars, in the molecular clouds where stars are born, and in the void of space. If water was simply the combination of hydrogen and oxygen, then there should be an appreciable amount of oxygen in the molecular clouds and in other stars. What we have learned is that star and planet formation based on the revered Big Bang hydrogen abundance theories do not work. One example is the expected lack of water on carbon stars because of carbon's affinity for capturing atomic oxygen. The SWAS research team reported on their findings in *Sky & Telescope*:

"'In this kind of environment, **you wouldn't expect to find water**, because carbon will **snatch up oxygen** leaving little available to form H_2O,' Melnick explains.

"But SWAS found the star's water to be 10,000 times more abundant than predicted." Note 7.2ag p37

As we will find in the Air-Water Model, Chapter 23, hydrogen does not just 'snatch up oxygen' to make water and neither does carbon. Any model of the Universe will fail when scrutinized with actual observations—unless water is at the *center* of the model.

The Models "Surely Have To Be Revised"

The European Space Agency's ISO satellite gave us unprecedented new information on deep-space chemistry and aided in the discovery of a chemical substance so mind-boggling, researchers hardly knew how to write about it. The surprise was the *amount of water* that existed:

"**Water throughout the universe**!

"Because of the Earth's atmosphere acting as an opaque screen, only with ISO have astronomers been able to study the chemistry of the Universe in detail. **One of the most mind-boggling discoveries they've made is that there's a lot of water vapour in the universe**.

"Indeed, ISO found the telltale signature of water in starforming regions; in the vicinity of stars at the end of their lives; in sources very close to the galactic centre; in the atmospheres of planets in the Solar System. **On the whole, scientists estimate the total amount of water in our own galaxy to be millions of times the mass of the Sun**." Note 7.2ae

The existence of water in this quantity directly contradicts all

7.2 CELESTIAL WATER

existing Big Bang chemical and star forming theories. In fact, researchers admit that the models explaining water in the solar system, on Earth and in star-forming clouds, "will surely have to be revised":

"Scientists knew water vapour was present in clouds close to starforming regions -like the Orion Nebula- **but they have no explanation yet as to why there should be such an enormous concentration of water vapour there**. Scientists believe the 'huge chemical factory' observed in Orion generates '**enough water molecules in a single day to fill the Earth's oceans sixty times over**.'

"This discovery may have also implications for the origin of water in the Solar System and on the Earth itself, and models explaining the evolution of molecular clouds where stars are born **will surely have to be revised**." Note 7.2ae

With so much evidence of celestial water, the Hydroplanet Model is a model whose time has come. It has been written to replace the existing, failed models of the origin of the solar system and of star formation that scientists acknowledge must "surely have to be revised".

The Universal Concept of Water

More water than anyone ever expected to find has been found in every corner of the universe. Water exists on the planets and their moons. It has been found on our Moon, in the Sun, and in the stars. It is in the vast interstellar reaches of our galaxy, and literally everywhere we look in the Universe.

There is no substance more universal, than water. The Universal Concept of water is discussed in Chapter 23, the Air-Water Model, but bears review here, since it is so important, and so central to the comprehending of the true nature of the universe; it must be internalized by the reader. Water is not simply a combination of hydrogen and oxygen atoms. It is true that we commonly refer to water as H_2O, and it is possible to isolate both the hydrogen and oxygen elements from a water molecule. However, in nature, evidence has yet to be found where the *individual elements* of hydrogen and oxygen combine to form water. Essentially, we have been taught about water, backwards—that hydrogen and oxygen combine to form water, but no one has demonstrated that this takes place in Nature.

We cannot overemphasize the importance of water and its effect on our lives and the processes of creation.

> Water is a universal concept
> and is applicable to all things.

The four **Universal Laws of Water** as described in the Air-Water Model upon which the Hydroplanet Model is based are:

1. The Law of Primordial Matter:
   Water is the primordial matter in the Universe.

2. The Law of Hydrogenesis:
   All other matter originated from water.

3. The Law of Hydroformation:
   All natural crystalline minerals formed in water.

4. The Law of Hydrobiogenesis:
   All organisms are born of water.

From the evidence in this subchapter, we see that water is abundant in the Universe and that it seems to be everywhere, especially on our planet, perhaps even *in and through all natural things on Earth;* just what one might expect since water is the primordial matter. To show that all other *observed matter* comes from water, we will discuss how rocks actually form in nature and we will look at the rocks formed here, on our own planet. First, let's examine the effect of water on planetary formation, the topic of the next subchapter, Hydrospheres. Following that, we'll move to the subject of how natural matter is organized—a water-crystallization process, which includes the creation of the Earth.

7.3 Hydrospheres

Hydrosphere Defined

We have seen scientific evidence of water found throughout the universe. Now we turn our attention to **Hydrospheres**—spherically shaped celestial bodies with *large amounts of water and ice*. Then we will look at other evidences of water in the solar system; water left behind after these celestial bodies were formed.

Celestial Body Formation

Why are some celestial bodies irregularly shaped while others are spherical? What determines whether a planet or moon contains water or ice on its surface or subsurface?

To answer these questions we must first remember that we are approaching the subject from within a completely new paradigm. The Magma and accretion pseudotheories failed to answer these questions, so we must look in a new direction for the answers. From the last subchapter, we found that water was the most abundant substance in our universe.

We also learned that when water is beyond the reach of nearby gravitational fields, unaffected by other bodies of significant mass, it retains its *spherical shape.* However, when small amounts of water are gravitationally affected by a larger mass, an irregular shape can occur. Smaller celestial bodies such as asteroids and comets are *irregularly shaped.* They are shaped this way partly because of the external influence of a larger gravitational force, and because they lack the internal force to form a sphere. Moreover, had they formed simply by impact, they should exhibit the sharp, angular forms of broken rock—but they do not. Later in this chapter, we will discuss the details of why these are not formed by impact or accretion and we will examine a number of photos of these unusual celestial bodies.

When a sufficient quantity of water is present, the internal force of a large sphere of water is capable of holding its shape even while other significant forces are acting upon it. However, in some cases it is altered, such as the Earth as it is pulled into an oblate sphere by the tug of the Moon.

We must also bear in mind that although we do not know the exact position each planet held during the formation of the solar system, we can conclude that they were once in different positions than they are now. It is likely that the hydrospheres were closer together at one time, possibly heated by the Sun, during which time the cumulative forces of gravity, heat and pressure contributed to the crystallization of minerals as they became the planets, moons, asteroids and comets we see today.

The second question as to why some of these celestial bodies

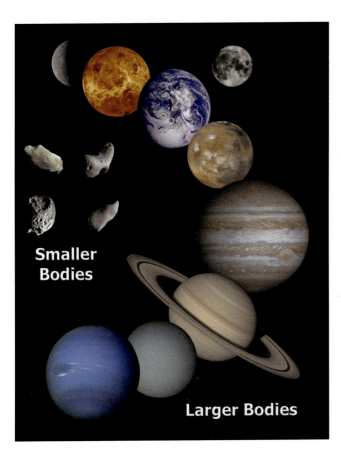

Fig 7.3.1 – In our Solar System the shape of celestial bodies can be determined from their size. Larger bodies are spherical whereas smaller bodies have irregular shapes. Why is this? The smaller bodies are not simply pieces of broken planets from ancient impacts, as astronomers have imagined. They were formed in these irregular shapes due to their smaller size, and because other, larger masses were able to pull them into irregular shapes. Larger bodies have a higher internal force of gravity that pulls the water and minerals of which they are formed equally in all directions, towards the center of mass, resulting in a spherical shape.

have icy oceans, water oceans, or no oceans at all can also be answered by recognizing that there are two important factors that contribute to how much water is present on the bodies in our solar system—the size of the body and its distance from the Sun.

When hydrospheres are of a size ranging from approximately the size of our moon to about the size of Mars, they have enough gravity to retain large, frozen water molecules, but not enough gravitational force to hold water vapor. Thus, when water becomes vaporized on the surface of bodies of this size, gravity looses its hold and it escapes into space. It is for this reason that the Moon and Mars do not have much atmosphere, if any. Larger planets like Earth and Venus have the gravitational force to hold water vapor and as a result, both retain thick atmospheres.

The second factor affecting how much water can be found on the solar system's bodies is their distance from the Sun. Proximity to this giant heat source affects the temperature of the body, therefore, the closer the body is to the Sun, the higher the temperatures it will experience. As it happens, the Earth is not only the right size to retain an atmosphere, it orbits the proper distance from the Sun, which allows temperatures to remain moderate, an aspect of Earth life to which we have become greatly accustomed.

Mercury is the closest planet to the Sun, experiences the highest temperatures, and is far too small to be able to retain water vapor because of its weak gravity. Moving out toward the gas giants, we encounter spheres large enough to capture and hold many types of gases. Temperatures may be high but gravity is sufficiently strong to retain the gaseous atmosphere. Beyond the atmosphere of the gas-giants, a number of small satellites are locked in orbit around them. Many of these moons have water, but because of the deep cold, it is preserved in the form of ice on their surfaces.

Examining Saturn and Jupiter's hydrosphere moons, their icy surfaces reveal that they have been broken and refrozen many times. Tidal forces of their parent planet and of other moons that share nearby space pull and tug at the crusts of these bodies. Gravitational-Friction heat is generated beneath the surface of the moons, melting, deforming and reshaping their surfaces.

Let's look more closely at the melted and deformed surface structures of these hydrospheres. They have a story to tell about their icy, water formation.

Saturn's Hydrospheres Evidence

Tethys is the fifth largest satellite of Saturn—about a third the diameter of the Earth's Moon. It is a hydrosphere. Amazingly, astronomers already recognize that this moon is composed "mainly of water ice":

"Having now passed closer to **Tethys** than the Voyager 2 spacecraft, Cassini has returned the best-ever natural color view of this **icy Saturnian moon**.

"This moon is known to have a **density very close to that of water**, indicating it is likely composed mainly of water ice." Note 7.3a

In Fig 7.3.2, Tethys' gray form is dwarfed by Saturn. These photos have only recently been made available to us although we have known about the water-ice composition of this hydrosphere since the Voyager spacecraft visited Saturn several decades ago. Because the density of Tethys was so close to that of water, it was easy to determine its makeup. Tethys' water-ice constitution did not fit within the accretion pseudotheory and as a result, its origin has stumped scientists. Where was all the planetary dust, meteorite ejecta and other debris?

The bottom line is that hydrospheres do not fit into the prevailing solar system dust-formation theories. Tethys is a beautiful example of a *water sphere formed in space*, one of many we can explore.

Icy Dione

The Dione Hydrosphere, Saturn's fourth largest satellite, is slightly larger than Tethys. It has a density of 1.4 g/cm^3, just above that of water. One of the photos in Fig 7.3.3 was captioned:

"Speeding toward pale, **icy Dione**, Cassini's view is enriched by the tranquil gold and blue hues of Saturn in the distance." Note 7.3b

In the bottom left image of Fig 7.3.3, large, wispy streaks cut across the surface. These features stunned the first humans who observed them. Scientists recorded how intriguing they are:

"Scientists continue to be **intrigued** by the strikingly linear features seen crisscrossing the southern latitudes. The fine latitudinal **streaks appear to crosscut everything**, and appear to be the youngest feature type in this region of Dione." Note 7.3c

As the Cassini spacecraft neared the moon, high-resolution

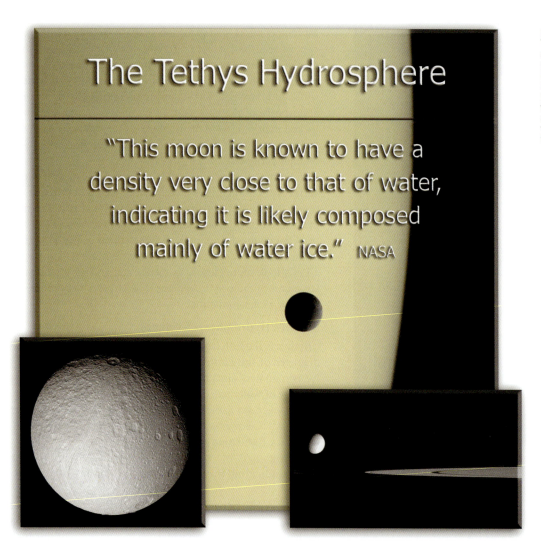

Fig 7.3.2 – The Tethys Hydrosphere. Tethys is one of Saturn's moons acknowledged by researchers to consist primarily of water ice. Perhaps the only thing to surpass the beauty of Saturn and her moons is the understanding that celestial bodies formed from the most abundant substance in the Universe—water. They did not form from melted rock.

photos of these features made it apparent that the surface of this moon was far from typical. The wispy lines were not simply deposits of ice; they were huge ice cliffs, formed by *tectonic* forces:

"Five narrow angle frames comprise this view of the 'wispy terrain' on the anti-Saturn side of **Dione**. To the surprise of Cassini imaging scientists, the wispy terrain does not consist of thick ice deposits, **but rather the bright ice cliffs created by tectonic fractures**." Note 7.3d

Tectonic fractures on Dione are the result of 'dionequakes,' which are equivalent to the earthquakes on Earth. However, modern geology textbooks say earthquakes come from magma. What is the cause of the quaking on ice moons like Dione?

Is there evidence of magma on hydrospheres of ice?

No, while there may be evidence of *surface* lava on some moons, like Jupiter's Io, there is no evidence of magma on any hydrosphere of ice.

Icy Volcanoes on Titan

The discovery of water ice on Saturn's largest moon Titan shocked the scientific community. Why did this newfound composition cause such a disturbance?

"The composition of **Titan's** solid ground has **thrown up a shock**: it's made of **water ice rather than rock**." Note 7.3e

We have become accustomed to seeing the rocky surfaces of the nearby planets of Mars, Venus and Mercury and incredible images of the Moon. We have visited the surfaces of these bodies with robotic probes and in the case of the moon, set more than one human foot on its soil. Water was not at first highly visible on these spheres as most of their surface water had boiled away, due in part to heating by the Sun. The obvious connection to the hydro-formation of these celestial bodies had been hidden from science. It was not until we ventured further out into the colder regions of the solar system that we were able to observe hydroplanets with water and ice still present. Saturn's Titan was one of those bodies.

There was a great deal of water on Titan. In Fig 7.3.4, NASA compares Titan's largest lake with Lake Superior and comments:

"This side-by-side image shows a Cassini radar image (on the left) of what is the **largest body of liquid ever found on Titan's north pole**, compared to Lake Superior (on the right). This close-up is part of a larger image (see PIA09182) and offers strong evidence for **seas on Titan**." Note 7.3f

The evidence and amount of liquid found on Titan surprised researchers. Other water related features observed included what appeared to be dry lakebeds with rocks that seem to have been in a stream channel. Eroded looking stones were presumed to be water ice, but of this, they were less certain:

"'Stones' in the foreground are 4 to 6 inches (10 to 15 centimeters) in size, presumably **made of water ice**, and these lie

Fig 7.3.3 – The Dione Hydrosphere. The bottom left photo of Dione shows a light colored wispy terrain. Scientists now know that they are huge ice cliffs and fractures. They are created by tectonic forces or 'Dionequakes,' similar to earthquakes on Earth. However, Dione has no magma. What causes the movement? Dione-quakes occur when its floating crust is fractured by gravitational flexing.

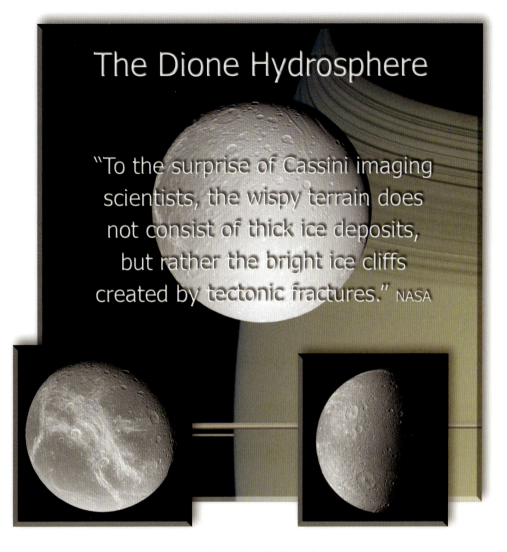

on a darker, finer-grained substrate. A region with a relatively low number of rocks lies between clusters of rocks in the foreground and the background and matches the general orientation of channel-like features in the panorama of PIA06439. The scene evokes the possibility of **a dry lakebed**." Note 7.3g

The surprises on Titan were not limited to large bodies of water, dry lakebeds and stream channels. Fig 7.3.4 shows what scientists observed as being an icy volcano:

"This false-color mosaic of Saturn's largest moon **Titan**, obtained by Cassini's visual and infrared mapping spectrometer, shows what scientists interpret as an **icy volcano** (see inset). The mosaic was constructed using six medium-resolution infrared images, obtained during Cassini's flyby of the hazy moon on Oct. 26, 2004." Note 7.3h

The term volcano evokes images of hot molten rock spewing from a smoking vent. In reality, there is no such thing as an "icy volcano." Researchers simply applied a common name to explain this newly observed phenomenon. Nevertheless, water eruptions like these are far more common than has been recognized. Understanding them is important and we will refer to them frequently in the Universal Model as hydrofountains. **Hydrofountains** are large water geysers that may discharge liquid water, solid and semi-solid water ice and other related materials. Hydrofountains will receive special attention both in this chapter and in the following chapter.

Other Saturnian Hydrospheres

Fig 7.3.5 contains images of four of Saturn's moons. The mosaic was assembled by NASA and captioned in part by the following:

"This montage shows **four major icy moons of Saturn** that the Cassini spacecraft visited while surveying the Saturnian system during 2005. Even though all of **these bodies are made largely of ice**, they exhibit remarkably different geological histories and varied surface features." Note 7.3i

Modern science now acknowledges that many celestial bodies in our solar system are "made largely of ice." How do they explain their spherical shape, and from where do they claim the origin of their water makeup? Is there a relationship between these spherically shaped hydrospheres and the spherically shaped inner planets of rocky composition such as the Earth and the Moon? How would the rocky silica surface of Titan differ from the Moon or Earth if the outer shell of water vapor were removed? Questions like these are often left unasked in modern science because everyone is still thinking the moons and planets were formed by accretion. The concept that these celestial bodies are Hydrospheres is foreign to modern astronomers, and this must change if we are to come to a correct understanding of their origin.

The Amalthea Hydromoon Evidence

Ten years ago, if someone were to have suggested that a ce-

The Titan Hydrosphere

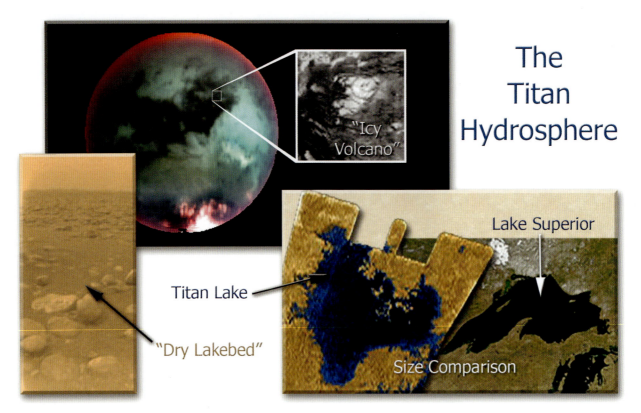

Fig 7.3.4 – Titan is the second largest natural satellite in the Solar System, second only to Jupiter's moon, Ganymede. Titan is larger than the planet Mercury. Researchers were "shocked" to see water ice and huge lakes, as big as Lake Superior on Earth. Many other evidences of water were observed, including "icy volcanoes." These were seen ejecting water and ice and since volcanoes, by definition, spew molten rock, these features on Titan are more properly termed hydrofountains. Below Titans icy surface, liquid water oceans are said to exist and below that, a silicate (quartz based) core.

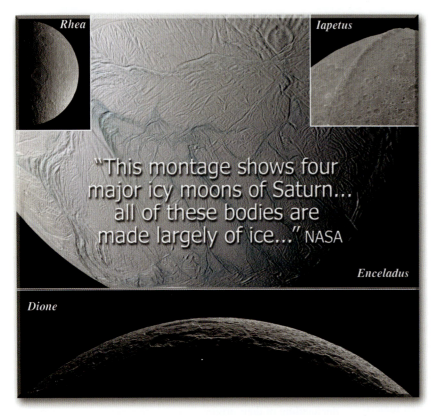

Fig 7.3.5 – In this image are four celestial bodies NASA describes as being "four major **icy moons** of Saturn." Icy bodies like these are not the exception in the solar system. As we look further from the Sun, there is an abundance of bodies "made largely of ice." As we extend our reach into depths of space, one substance consistently shows up, everywhere we look, and that is water.

lestial body could have a density *less than* that of water, they might surely have been laughed at. Today, this is an accepted fact. An article appeared in the May 2005 issue of *Science*, entitled: *Amalthea's Density is Less Than That of Water*. (See Fig 7.3.6) Discovered in 1892, Amalthea, also known as Jupiter V, was the fifth of Jupiter's moons to be discovered. After a relatively recent flyby, scientists determined the composition of this celestial body to be *porous*, consisting of water and ice:

"The satellite thus has a density of 857 ± 99 kilograms per cubic meter. **We suggest that Amalthea is porous and composed of water ice, as well as rocky material, and thus formed in a cold region of the solar system, possibly not at its present location near Jupiter.**" Note 7.3j

Scientists suggest Amalthea is porous and in light of these new observations, have been forced to reconsider the origin of celestial bodies like this, and develop a completely different model of their formation. They conclude their formation must have occurred far from the supposed heat of Jupiter's origin—in some colder region of space.

Hydromoons formed in a **water** environment.

However, like their larger hydrosphere coun-

CHAPTER 7 THE HYDROPLANET MODEL

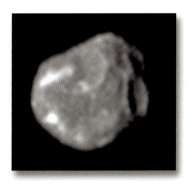

Fig 7.3.6 – Amalthea, the fifth of Jupiter's satellites to be discovered, is seen here in this inferior, yet important, image from a flyby in 1999. Amalthea's density has been calculated to be *less* than that of water indicating a hollow core geode-like structure. Courtesy of NASA.

terparts, these small satellites are intimately connected to water, and so should more correctly be named **Hydromoons**. Large Hydromoons that exist far from the Sun may be made of water and ice, much like the icy moons orbiting Saturn, or they may have significant rocky material combined with the ice and water, like the Hydromoons we see around Jupiter. Closer to the Sun, Hydromoons such as Earth's Moon have much less water due to vaporization. The Moon's weak gravitational pull allows the water vapor to escape. The commonality among all of the Hydromoons is that they were created in a water environment.

The discovery of many new, icy Hydromoons caused quite a stir in the scientific community. As stated in the 2005 journal article in *Science*, current models are "inconsistent with their formation from water ice":

"**Current Jupiter subnebula models imply that temperatures were high** at Amalthea's current position when the Galilean satellites formed, **inconsistent with their formation from water ice**. We suggest that Amalthea **was formed in a colder region**. One possibility is that it formed later than the major satellites; another is that it formed farther from Jupiter, either beyond the orbit of Europa or in the solar nebula at or beyond Jupiter's position, and was then dynamically transported to or captured in its current inner orbit. **Either alternative poses challenges for models of giant planet satellite formation**." Note 7.3j p1293

Challenges like these exist for scientists because of the magmaplanet paradigm. On the other hand, these challenges can be more easily answered with a Hydroplanet Model.

Jupiter's Three Large Hydromoons

Although the Hydroplanet Model contains many 'smoking gun' evidences, clearly, the strongest are the evidences drawn from Jupiter's Hydromoons. Quite surprisingly, the illustration in Fig 7.3.7, showing three of the Galilean moons orbiting Jupiter, was *not* drawn by the Universal Model, but was drawn by NASA. All three moons are depicted as having suboceans!

> A **subocean** is a large body of water beneath the surface of a celestial body.

The following is an excerpt from the caption released with the image:

"…the rock layers of Ganymede and Europa (drawn to correct relative scale) are in turn surrounded by **shells of water in ice or liquid form** (shown in blue and white and drawn to the correct relative scale). Callisto is shown as a relatively uniform mixture of comparable amounts of **ice and rock**." Note 7.3k

In recent years, planetary scientists have begun to realize the important role that water has played in planetary formation. Every time a planetary probe ventures into new and unseen territory, more evidence for a hydroplanet model is discovered. The icy crusts, the icy-rock interiors and the existence of suboceans sparked the development of new models among planetary scientists. These models were affected by these two observations. First, the densities of Ganymede and Callisto have been calculated at only 1.9 g/cm^3, which are only twice that of water, much less than most rock. Secondly, as we will see in the following pages, these moons are covered with water ice.

The Ganymede Hydromoon

Ganymede is the largest of the Jovian moons and the largest moon in the solar system. It has an enormous volume of water. When viewed in the infrared spectrum, as seen in the right two images of Fig 7.3.8, the abundance of water is clear-

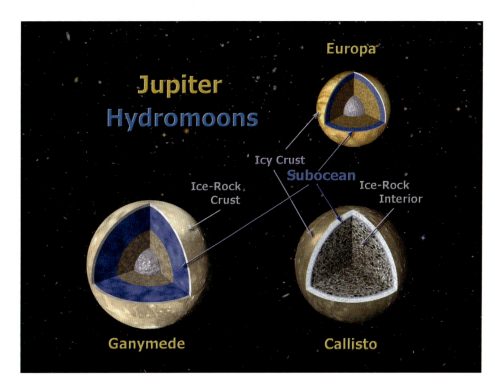

Fig 7.3.7 – Three hydrospheres, also known as Hydromoons, orbit Jupiter. These Hydromoons each contain large amounts of water. From drawings adapted from NASA, each sphere is shown representative of how much water each hydrosphere is thought to contain. The amount of water researchers discovered in the moons is massive. The percentage of water they hold is even higher than is proposed for the Hydroplanet Earth Model in this chapter. Image adapted from NASA/JPL

Fig 7.3.8 – The abundance of water on the Ganymede Hydrosphere is shown in this diagram. Imaged in the near infrared, these false color illustrations show significant water, which are the areas of both green and blue.
Courtesy of NASA (PIA47903).

ly visible. Areas of high water content are the green and blue regions, whereas red signifies the location of minerals. Clearly, the existence of large amounts of water on Ganymede is now well-attested.

Although *dogmagma* has had a firm grip on the minds of most planetary researchers, seeing these hydromoons in this new light, and for the first time observing the abundance of water, has changed opinions on the variety of materials that are spewed from volcano-like landforms. Eruptions of water or ice are now being considered:

"In its warmer past **Ganymede's surface may have surged with volcanoes not of rocky lava but of ice or water**. Such eruptions could have sculpted channels into the terrain." Note 7.3l

Ganymede is a clear example of a hydrosphere, a celestial body made mostly of water and or ice. The tectonic forces that shaped and sculpted the surface of this moon did not come from magma—and the scientists today agree. One has to wonder why they continue to think tectonic forces on Earth are resultant from magma. Tradition and old theories are hard to let go.

The Callisto Hydromoon Evidence

Callisto is the second largest satellite of Jupiter and has a surface unlike any of Jupiter's other moons. In Figure 7.3.9, the Callisto Hydromoon is represented in several plates. Inset plate number one is described by NASA:

"Callisto's cratered surface lies at the top of an **ice layer**, (depicted here as a whitish band), which is estimated to be about 200 kilometers (124 miles) thick. Immediately beneath the ice, the thinner blue band represents the possible **ocean**, whose depth must exceed 10 kilometers (6 miles), according to scientists studying data from Galileo's magnetometer. The mottled interior is composed of rock and **ice**." Note 7.3m

On this hydromoon, we see a celestial body with significant amounts of ice both on its surface and in the interior. Plate number two in Fig 7.3.9 shows a comparison of the surfaces of the three largest moons of Jupiter. From left to right are Europa, Ganymede and Callisto:

"These images show a comparison of the surfaces of the three icy Galilean satellites, Europa, Ganymede, and Callisto, scaled to a common resolution of 150 meters per picture element (pixel). Despite the similar distance of 0.8 billion kilometers to the sun, their surfaces show dramatic differences. Callisto (with a diameter of 4817 kilometers) is 'peppered' by impact craters, but is also covered by a **dark material layer of so far unknown origin**, as seen here in the region of the Asgard multi-ring basin." Note 7.3n

Note here that the "dark material" is not thought to be from impact ejecta. There is *no indication where an impact may have occurred and there is no evidence of ejecta pointing toward an impact crater*. This dark material appears to be similar to the material found on Earth's Moon, and—as we will see shortly—to the material covering some of the cracks on Europa. If this dark material laying on the surface of Callisto did not come from space, (many other moons lack a similar covering of material) where did it come from? The most likely origin is that it came from below the surface of Callisto. The meteor-impact theory cannot explain the origin of sediment-like materials blanketing such large areas. The surface is also void of volcanic cones or circular landforms suggestive of volcanic activity.

How did subsurface material become spread so evenly across the surface of these celestial bodies? Water—oceans of water with underwater eruptions of subsurface sediment, carried by the currents and spread evenly around a globe.

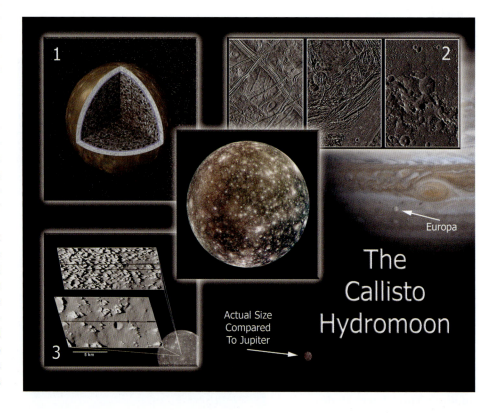

Fig 7.3.9 – In this diagram are various images of the hydromoon, Callisto, the second largest of Jupiter's moons. Plate 1 is a NASA illustration showing an icy crust, subocean and an ice-rock interior. Plate 2 compares the surfaces of Europa, Ganymede ad Callisto, showing a "dark material layer" that is easily accounted for in the Hydroplanet Model. Plate 3 shows icy spires, landforms difficult to explain without a water origin.

CHAPTER 7 THE HYDROPLANET MODEL

Plate three in Fig 7.3.9 is of large icy spires, similar to steep hills or mountains. NASA describes these landforms in the original image as follows:

"The knobby terrain seen throughout the top inset is unlike any seen before on Jupiter's moons. The **spires are very icy** but also contain some **darker dust**. As the ice erodes, the dark material **apparently slides down and collects in low-lying areas**." Note 7.3o

If the ice simply melted allowing the dark material to slide down and collect in low-lying areas, why are there no channels emanating from the spires and connecting to the low-lying areas? On Earth, ice spires do not usually form by melting (they would be smooth and rounded). Sharp spires are fashioned when ice is pushed up through subsurface heating and then refreezing. The dark material would slide down the spires as the ice pushed through the surface.

The Europa Hydrosphere

Amazingly, although Europa is smaller than the Earth's Moon, researchers acknowledge that:

"'To understand scientists' fascination with Europa, you need to know just one fact: **Europa holds more water—possibly liquid water—than all the oceans on Earth**." Note 7.3p

Moreover, space probe data has shown this hydrosphere to have a watery crust 150 km thick:

"Ground-based spectroscopy of Jupiter's moon Europa, combined with gravity data, suggests that the satellite has an **icy crust roughly 150 km [93 miles] thick** and a rocky interior."
Note 7.3q

Europa has a diameter of less than one fourth that of Earth, and yet, Europa's icy crust, possibly its oceans, are over 37 times *deeper* than the Earth's oceans, which average about 4 km deep. Even then, Europa's subocean is less than 10% of the volume of water on Ganymede, Jupiter's largest moon. The sheer amount of water and the depth of the suboceans **ought to change our entire paradigm** as it relates to water on planetary bodies in our solar system, including the Earth. The philosophers that came up with the magmaplanet ideas in the late 1700s didn't imagine that such a vast amount of water could exist.

In Fig 7.3.10, enormous blocks of Europa's frozen Hydrosphere are evident. As the blocks of ice crack and shift, sediment is ejected onto the surface through geologic activity:

"In this false color image, **reddish-brown areas represent non-ice material resulting from geologic activity**." Note 7.3r

Europa is only one example where sedimentary materials, originating from below the crust, find their way to the surface through thermal waters. These sediments are not cycled by heated magma, but are an example of the Gravitational-Friction Law at work. The material is ejected by hydrofountains, a subject of great importance and one that we will discuss in detail throughout this chapter. Hydrofountains are a key component of the Hydroplanet Model.

Galileo Spacecraft Photos—"blowing us away"

After a journey of more than six years, the Galileo spacecraft, launched October 1989, reached its intended destination and ultimately completed 35 orbits around Jupiter. There is no question that the resulting detailed photos taken by the Galileo spacecraft stunned the world and the scientific community. What made them so stunning? It was ultimately not the images, although they were fantastic, it was the facts those images portrayed! It seemed there was water, water everywhere. Note the investigators reaction to the images of Europa:

"The case got stronger after the Galileo spacecraft began transmitting high-resolution images of Europa late last year. The pictures were **dramatic**. They showed close-ups of huge rafts of ice that appear to have broken up and sloshed around like ice cubes in a glass of water. Perhaps the most striking thing about the photographs is that they seem so common, so familiar, that someone without the slightest bit of astronomical expertise can take one look at them and quickly conclude: **Icebergs. Ocean.**

"'They're just blowing us away,' says Paul Geissler, a planetary scientist at the University of Arizona's Lunar and Planetary Laboratory and a member of the Galileo imaging team charged with interpreting the spacecraft's pictures. Perhaps the only way Galileo could **document a Europan ocean beyond a reasonable doubt** would be to photograph ice floes that have actually moved since Galileo's last visit, **or geysers of water jetting from the surface**." Note 7.3s

Although the researchers were astonished at the discoveries they were making, they still held reservations that what they were indeed seeing was water. It was surmised that to remove that doubt, live evidence would be needed. An event such as a

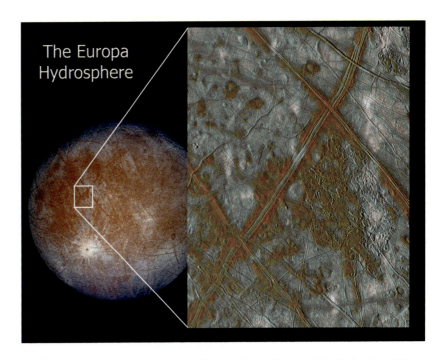

Fig 7.3.10 – The Europa Hydrosphere offers amazing evidence of the Hydroplanet Model. Planetary scientists calculate that Europa—smaller than Earth's Moon—holds an ocean 150 km (93 miles) deep! (The Earth's oceans average only 4 km (2.5 miles) deep). The enlarged section of Europa's crust shows an icy surface that has been broken and fractured by the tidal action of the Moon's nearby parent, Jupiter. Brown areas are sediments blown onto the surface by steam and water, carried from below the surface. They are an important part of the 'hydrofountain' concept introduced in this chapter.

Enceladus Hydrosphere

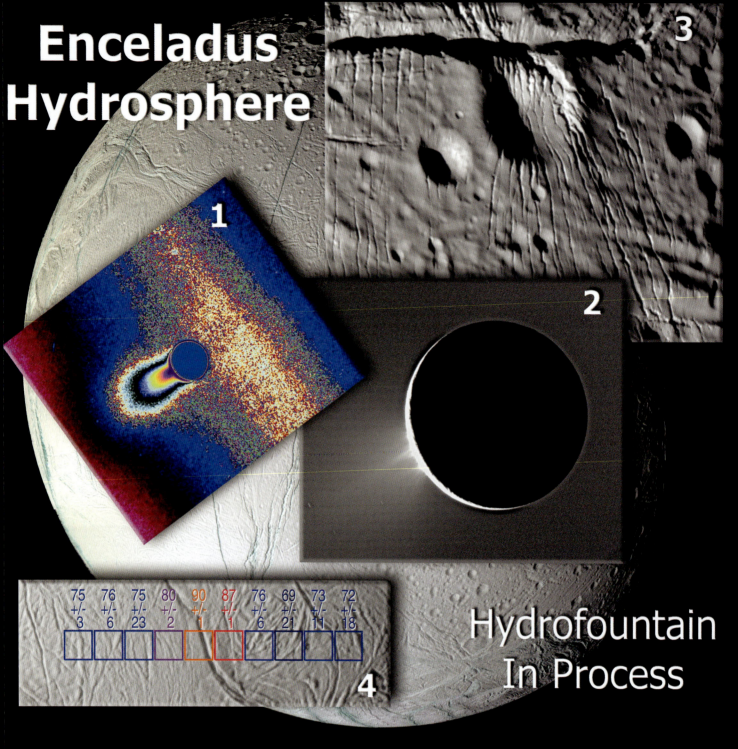

Hydrofountain In Process

Fig 7.3.11 – These images are of a gigantic water jet, a 'Hydrofountain' blasting from the surface of Enceladus, Saturn's sixth largest moon. Researchers were stunned to have observed such an event. This real-time eruption unquestionably demonstrates the existence of active hydrospheres in the Solar System. At the time the Cassini spacecraft was able to image this heretofore-unimagined event, Enceladus' hydrofountain was spewing water into space at a distance greater than the entire diameter of Enceladus (see inset plate 1 for a spectrographic image of the hydrofountain). Plate 2 is a visual light photo of this event in its beginning stages. Plate 3 is a close up of Enceladus' surface, showing large fissures and canyons from previous ruptures. These would also have released enormous amounts of water and ice into space, ultimately becoming a part of the Saturnian ring system. Scientists have discerned that Saturn's rings are mostly water ice and Enceladus is known to have been a major contributor to them.

Inset 4 is a close-up of Enceladus' surface. The squares represent temperatures on the surface. Red squares are hotter, blue are cooler. Notice the relationship of heat with the surface cracks (faults). This is an example of the Gravitational-Friction Law, first described in the Magma Pseudotheory chapter. Friction along faults in the crust heats subsurface water, which rises toward the surface. Sometimes, as was the case during the Cassini flyby, eruptions occur with such force that huge jets of water and ice are observed, spewing matter far into space. The Enceladus hydrosphere is reported to have a density just over one and a half times that of water and is likely a combination of water, rock and other minerals.

If you ever have the chance to look at Saturn through a telescope—everyone should do this—expect to be inspired, not only by the beauty of Saturn's rings—but also by the fact that they are made mostly of water!

CHAPTER 7 THE HYDROPLANET MODEL

geyser, a planetary water jet, observed from space would certainly prove the existence of a sub-ocean. Amazingly, in 2004, another spacecraft with a different mission to explore Saturn—did just that. The Cassini-Huygens spacecraft photographed a gigantic geyser of water, jetting from the surface of Saturn's moon, Enceladus. Just what scientists said they needed to prove the existence of suboceans beyond a reasonable doubt!

The Enceladus Hydrofountain

Saturn, the sixth planet from the Sun is one of the most striking sights in the sky, even when viewed through a small telescope. However, Saturn's intrigue is not limited to its rings. Saturn has numerous diverse moons and moonlets (61 at the time this book was written). Of its moons, Enceladus is Saturn's sixth largest, and it yielded a startling discovery. On the following page, Fig 7.3.11 are a number of images of this amazing hydromoon. In the background, the sphere of Enceladus can be seen. Its surface does not consist entirely of rock. Like other hydrospheres, it contains water-ice:

"This color Voyager 2 image mosaic shows the **water-ice-covered surface of Enceladus**, one of Saturn's icy moons. Enceladus' diameter of just 500 km would fit across the state of Arizona, yet despite its small size Enceladus exhibits one of the most interesting surfaces of all the icy satellites. Enceladus reflects about 90% of the incident sunlight (**about like fresh-fallen snow**), placing it among the most reflective objects in the Solar System." Note 7.3t

This statement was made in 1998 when researchers were still trying to figure out just when Enceladus had been geologically active and how recently its icy surface had been melted. Though researchers knew that Enceladus had a "heat engine," they did not attribute it to a molten magmatic core, nor did they consider it radioactive:

"The implication carried by Enceladus' surface is that this **tiny ice ball** has been **geologically active** and perhaps practically **liquid in its interior for much of its history**. The **heat engine** that powers geologic activity here is thought to be **elastic deformation caused by tides induced by Enceladus' orbital motion** around Saturn and the motion of another moon, Dione." Note 7.3t

Although researchers did not realize it at the time, Enceladus' heat engine was evidence of the Gravitational-Friction Law, first introduced in the Magma Pseudotheory chapter of this book. The old magma-planet theories still hold fast for Earth's heat engine, but the moons in the solar system are too small to contain a magma core, and there is no evidence of radioactive heat. That leaves only one other explanation to account for the 'heat engine' in these small bodies—gravitational friction.

Cracks in the surface of Enceladus can be seen in Fig 7.3.11, in plates 3 and 4. Located near these cracks, gravitationally generated friction produces heat. Plate 4 shows a series of overlaid squares revealing the location of surface heat. Red signifies the hotter areas whereas blue and black signify cooler areas. It is easy to identify that increased heat correlates with the surface cracks.

A fascinating aspect of the relationship between the temperature measurements of the moon's surface and the cracks and fissures they relate to is they confirm the existence of hydrofountains, an important Hydroplanet feature. In November of 2005, NASA reported seeing a "spray above Enceladus." The researchers used spectrometers to discover that ice particles had indeed been sprayed into space:

"Cassini's visual and infrared mapping spectrometer measures the spectrum of the **plumes** originating from the south pole of the icy moon, capturing a very clear signature of **small ice particles**." Note 7.3u

This observation was a crucial key in our understanding that these moons are currently *active* and the surface structures are not billions or millions of years old. Recent observations like these have demanded that scientists reevaluate many of the surface features on Enceladus and other celestial bodies that were previously thought to be 'geologically old.' If we are able to observe colossal events on the active surface of Enceladus in real-time, the idea that million-year-old impact craters remain evident, hardly seems plausible. Moreover, if eruptions like this took place very often, after a billion or two years passes by, there would be no water left! Obviously, there is a lot of water left and only a young solar system, measured in terms of thousands-of-years, not billions-of-years can explain this. We will discuss more about this later in this chapter and in some detail in the subchapter 20.9, Astronomical Dating Evidence.

No one imagined the spectacular event seen in Fig 7.3.11, inset plate 1. This photo is an enhanced and colorized spectrographic image of a plume of water and ice *extending farther into space than the diameter of Enceladus itself*! This astonishing feature received surprisingly little attention in the scientific community. In 2006, researchers had to acknowledge that it was liquid water erupting from jets on Enceladus:

"Imaging scientists, as reported in the journal Science on March 10, 2006, believe that the **jets are geysers erupting from pressurized subsurface reservoirs of liquid water** above 273 degrees Kelvin (0 degrees Celsius)." Note 7.3v

This inconceivably immense water geyser does not fit well within any of the magma/accretion/geological time theories in science today. However, the UM will demonstrate that hydrofountains play a crucial role in the shaping of the surface of planets and moons millions of miles away, and that hydrofountains have been a major factor in the formation of the surfaces of both the Earth and its Moon.

More Hydrospheres

With the advent of digital photography, driven and processed by ever-faster and smaller computers, technology has afforded us views of the solar system like never before. During the 1990s, advances in spacecraft technology helped discover and identify worlds no one knew existed. Most of these objects were located beyond the orbit of Neptune in an area referred to as the Kuiper belt. This area can be thought of as being somewhat similar to the asteroid belt that lies between Mars and Jupiter. It is interesting how these new worlds are described:

"Most of us grew up with the conventional definition of a planet as a body that orbits a star, shines by reflecting the star's light and is larger than an asteroid. Although the definition may not have been very precise, it clearly categorized the bodies we knew at the time. In the 1990's, however, a remarkable series of discoveries made it untenable. **Beyond the orbit of Neptune, astronomers found hundreds of icy worlds**, some quite large, occupying a doughnut shaped region called the **Kuiper belt**." Note 7.3w

More than a thousand Kuiper belt objects (KBOs) have been

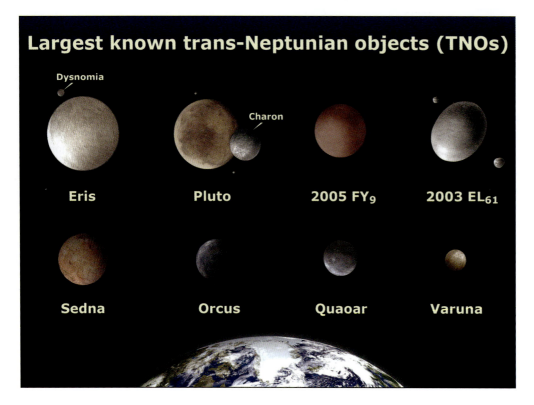

Fig 7.3.12 – This NASA diagram illustrates comparative sizes of celestial bodies outside Neptune's orbit. These objects are known trans-Neptunian objects (TNOs) or Kuiper belt objects (KBOs). To date, more than 800 of these objects have been discovered using digital technology and computers available in the 1990s. Note that Pluto, which is no longer designated as a planet, is not the largest object among the group of KBOs. Although accurate in its portrayal of size, the colors and textures are only artist's conceptions. Located far from the Sun, these icy worlds retain water from their formative period. Their frozen and reflective surface aids in their detection.

discovered, most commonly described as *"icy worlds."* Once carrying the distinction of being the tenth planet, Pluto is now considered a KBO, or sometimes a TNO (trans-Neptunian object) and is described as having a water-ice surface. As we venture further into space, many new observations of yet-unknown worlds will be made and the photos and images taken will only be understood by using correct models.

How do the observations and data we have discussed, fit within the Hydroplanet Model so far? The discovery of abundant water everywhere in the solar system and the continued discovery of icy worlds and water-ice phenomena on planets and moons continue to validate and support the Hydroplanet Model. As we turn our focus closer to home, a new dimension in our understanding of the role water plays in the formation of rocks and minerals that make up the crust of the Earth will become apparent. This understanding will add to our ability to interpret correctly the origins of material on other planets and moons.

7.4 The Crystallization Process

Having established that water is found throughout the universe, in galaxies, stars, the Sun, planets, and moons, we turn now toward the Earth. Just how much water is here, inside our own blue planet? About 70 percent of the planet is covered in oceans and a much smaller amount is locked away in the glaciers and polar ice caps. But there is more, far more to the story of water. The Earth, following the Universal Law of Water, was formed from primordial water. Then, a Crystallization Process formed the rocks and minerals we see all around us as mountains and continents. This subchapter, for the first time in modern science, sets forth a comprehensive model of water-based rock and mineral formation.

The Dark Age of Science prevented a true understanding of the Universal Concept and Laws of Water and the real process of the crystallization of natural minerals. But the rocks themselves have a story to tell, and nearly all testify of their water-born existence.

Crystallization—Making Rocks

There are three basic forms of matter: gas, liquid and solid. Solids can crystallize from a gas or liquid, but in nature, the vast majority of minerals form as the result of a liquid process. Further, in nature, there are only two inorganic-liquid processes from which minerals form. These are:

1. **Water processes**
2. **Melt processes**

Nearly all of the Earth's minerals were formed in processes involving water. We know this because nearly all natural minerals are crystals, the exception being a small percentage of rocks that were melted, which formed *glassy* or *glass-like* rocks such as pumice, obsidian or other lava rock. Even so, water played a pivotal role in the formation of these glass-like rocks as well. Here is how crystals differ from glass:

"In chemistry and mineralogy, a **crystal** is a solid in which the constituent atoms, molecules, or ions are packed in a **regularly ordered, repeating pattern** extending in all three spatial dimensions." Note 7.4a

The "regularly ordered, repeating pattern" of crystals is a simple key to understanding the origin of nearly all of Earth's natural minerals. In the Magma Pseudotheory, we discussed how quartz *glass* and quartz *crystals* have very different physical properties. The most important distinction is that *glass is amorphous and does not have a regular crystalline lattice or structure.* Quartz *crystals* do have a very orderly structure.

Even though pure quartz (SiO_2) and glass (Also SiO_2) can be chemically the same, quartz is over 1000 times more thermally conductive. This is because heat can travel quickly through the *ordered* quartz crystal lattice, whereas the *disordered* structure of glass is more like a maze.

Minerals, like quartz are good examples of the Universal Law of Order and the Law of Crystallization which will be presented in Chapter 22, The Energy-Matter Model. Quartz is an example of the third law of the **Universal Concept of Water,** the Law of Hydroformation:

All natural crystalline minerals formed in water

Water is the universal substance that provides a medium in which natural crystals will form and in this subchapter, we examine some of the evidences for this. The Second Universal Law of Water explains *why* geologic research based on the magma paradigm has been unsuccessful at reproducing natural minerals from melted rocks without water. Researchers have been unsuccessful because nature does not work this way. For this reason, the vast majority of minerals are formed from water—because it is *only within a water medium* that the crystal structure of a mineral will form.

The Crystallization Process is not limited to just a small percent of Earth's minerals. This can be better understood as we examine the following quote from a popular book, *Crystals and Crystal Growing*. This statement was taken from that book under the subsection *The Genesis of Minerals*:

"If someone suddenly said to you, "Find a crystal, and be quick about it," you would probably forget the sugar in the bowl, the salt in the shaker, and leap out of doors to hunt for a glittering rock. The minerals of which rocks are made furnish the most familiar examples of crystals; everyone recognizes quartz, gems, and most semiprecious stones as crystalline. **But it is less familiar that the entire solid crust of the earth is crystalline, with little exception**. Indeed, most of the crust will show this to a sharp eye, aided here and there by a small magnifying glass." Bib 44 p46

Recognizing this is the first step in gaining wisdom of ***how*** the Earth was formed—that is, with little exception the Earth is a *crystal*. Modern geology has failed to comprehend how the Earth formed because it has not recognized the fact that natural crystals come from water. The authors of *Crystals and Crystal Growing* go on to make this insightful declaration:

"**Just how the earth arrived at the form in which we find it is a question still far from settled.**" Bib 44 p46

Why is this question far from settled? We identified many of the reasons in the Rock Cycle Pseudotheory chapter—geology is literally full of mysteries that will never be answered under the magma paradigm. It is time for science to candidly accept this fact and look toward a new model, a new paradigm that is true and can answer these mysteries.

> "Just how the earth arrived at the form in which we find it is a question still far from settled."
>
> *Crystals and Crystal Growing*
> Alan Holden and Phylis Morrison

With the understanding that the vast majority of the terrestrial rocks are crystalline based, and that natural crystals are formed only from water, we can begin to discover *how* rocks and minerals come out of water in a crystalline state. To comprehend this we must revisit and clarify the meaning of the word **precipitation**.

Precipitation Redefined

Most people have heard of 'precipitation' as it refers to the weather. Snow and rain are 'precipitates.' Nevertheless, to the chemist or the physicist, this word takes on a different meaning. In Webster's Dictionary, we find the general definition for **precipitate**:

"Precipitate – to separate (a substance) in solid form from a solution." Bib 7 p622

However, this definition is far *too* general in its scope. For example, a kitchen 'strainer' also separates a substance in solid form from a solution. Thus, we need a more refined definition for our purposes. We found this one in *The Facts on File Dictionary of Chemistry,* where **Precipitate** is defined as:

"A suspension of small particles of a solid in a liquid formed by a **chemical reaction**." Bib 83 p199

In a chemistry lab class, a precipitate is often demonstrated by mixing two clear liquids together and watching a solid form out of the clear solution at the bottom of the test tube. Is this the *only* way solid particles can come out of a solution—by a chemical reaction? To answer this question, we turned to the following definition of precipitation from Wikipedia:

"**Precipitation** is the formation of a solid in a solution during a **chemical reaction**. When the **chemical reaction** occurs the solid formed is called the precipitate. This can occur when an insoluble substance, the precipitate, is formed in the solution due to a reaction or when the solution has been *supersaturated* by a compound. **The formation of a precipitate is a sign of a chemical change**. In most situations, the solid forms ("falls") out of the solute phase, and sinks to the bottom of the solution (though it will float if it is less dense than the solvent, or form a suspension)." Note 7.4b

Note the bolded words, "chemical reaction" and "chemical change" in this definition. After looking at several resources and talking to a number of chemistry professors, we concluded that this definition of precipitation was not entirely correct. The reason was that crystallization could take place *without* a chemical change, but simply by a **physical change** to the solution.

Back in 1958, researchers discovered that the most abundant mineral on Earth's continents, silica (quartz), could crystallize out of solution by altering the physical properties of *temperature or pressure* of the solution. From an article in *The Geological Society of America*:

"Crystallization in silicate systems is so intimately tied to temperature that the possibility of isothermal [constant temperature] crystallization is **rarely** considered; **yet complete crystallization, starting with a liquid containing no crystals, can take place with no drop in temperature**. This is possible because of the manner in which the water affects the liquids and because the amount of water held in the silicate melts is a function of **pressure**…

"Crystallization in these hydrous systems can be promoted by **temperature lowering, pressure lowering, or pressure increase**." Note 7.4c

The importance of clarifying the definition of 'precipitation' cannot be overstated. The significance of this change allows for a mechanism for natural mineral growth, one that has heretofore gone unnoticed. Although researchers realized many years ago that quartz crystallization can take place by altering temperature or by raising or lowering pressure, they have been unable to see the connection between this important physical fact and the fact that all minerals grow in a water solution.

Let us summarize the redefined process of precipitation, with the added mechanisms known to be responsible for the formation of crystal solids:

> **Precipitation** is the formation of crystalline solids out of solution either chemically or by changes to **temperature and/or pressure**.

Evaporate Rock Pseudotheory

What effect has the lack of a true definition of **precipitation** had on science? As noted, the true and complete definition of precipitation is not being taught in the physical sciences because students are taught that precipitates come only from chemical reactions. They also come from temperature and pressure changes. This clarification opens a completely new opportunity for discovery. When we are in possession of this truth, new technology and natural processes can be evaluated and better understood.

One natural process of precipitation with which we are all familiar is **evaporation**. When a solution of salt and water evaporates, the salt minerals precipitate out of the solution, leaving behind a crystallized mass called an **evaporite**. This is the process alleged to be the origin of many mineral deposits, such as salt, encountered in nature. We read of this process in Wikipedia:

> "Although **all** water bodies on the surface and in aquifers contain dissolved salts, **in order to form minerals from these salts, the water must evaporate into the atmosphere in order to precipitate the minerals**." Note 7.4d

Is this statement true? Is evaporation the *only* way for salts to precipitate into minerals? Not according to The Geological Society of America, (see the earlier quote). In the Rock Cycle Pseudotheory chapter, statements from salt researchers were cited, showing the real origin of salt is absent in today's geology. Moreover, many scientific evidences were given in the Rock Cycle chapter, demonstrating that "the vast geological 'evaporite' deposits are not evaporites" at all. What then is the process?

Precipitate Salt Deposit Model

A tasty treat called Rock Candy, seen in Fig 7.4.2, is made of sugar crystals grown out of a supersaturated solution of sugar and water. Evaluating this process identifies one of the key factors in the precipitation process.

To begin, sugar is added to water and stirred until it dissolves. Applying heat to the water will allow more sugar to dissolve into the solution. Eventually, no more sugar will dissolve because the solution will have become supersaturated. At that point, the solution is removed from the heat source and allowed to cool. Pour the cooled solution into wide-mouthed bottles. Place a stick in the solution, cover with an airtight lid to prevent evaporation from taking place, and wait. Eventually, small sugar crystals will begin forming on the stick. Colored dye or flavors can be added to the solution to alter the appearance and taste of the rock candy crystals.

Sugar crystals formed as the supersaturated sugar-water **cooled**. *Temperature reduction is a method of precipitating minerals*, a process quite different from that of evaporation. Lowering the temperature of a solution is a common occurrence in physics and chemistry labs, but the same cannot be said of geology. Today, there are no large bodies of cooling, supersaturated salt water allowing the precipitation of massive, kilometer-thick salt formations. However, there *was a time* when huge, supersaturated hot oceans cooled producing enormous salt deposits, just like the ones we see in many parts of the world today.

Fig 7.4.1 – Natural Fluorite crystals are not formed from a melt—they are formed in water.

The most common and the most important salt in both geology and biology is NaCl—common table salt. When water temperature is increased from 0° to 100° C, the solubility of NaCl increases from 35g/100 mL to nearly 40g/100mL. Like our sugar crystal example, supersaturated NaCl solutions will allow salt crystals to precipitate from the solution when the temperature drops (although not as dramat-

Fig 7.4.2 – These are sugar crystals formed on strings suspended in supersaturated sugar water. As water is heated, sugar will dissolve more readily into solution until it becomes 'supersaturated.' As the high-temperature, saturated-sugar solution is cooled, sugar crystals precipitate out of the water onto the strings. Blue dye provides added color. This is the process for making this tasty 'rock candy' treat. It is essentially by the same process that massive, natural salt formations are formed.

ically as the sugar solution does). In general, geology has not considered the real possibility that large hot oceans covered the Earth and as a result have not considered that massive salt deposits could have grown from a supersaturated solution because of temperature reduction. Each of the six salts in the ocean, identified in the Rock Cycle Pseudotheory, will crystallize out of solution, *at different temperatures and pressures*. This is *how* massive, pure salt deposits were formed!

Similar to the formation of rock-candy sugar crystals, mysterious salt deposits, like those discussed in the Rock Cycle Pseudotheory chapter, formed as salt precipitated *under* water. This is *why* vast geological salt deposits are not 'evaporites.' This is also the reason why researchers have never been able to reproduce or adequately explain large-scale salt deposits formed by evaporating seawater.

> Precipitate Salt Deposit Model
> Thick salt deposits were formed when salt precipitated out of an aqueous solution because of changes in temperature and pressure.

Salt Origin Without Evaporation Confirmed

Is there laboratory or field evidence of the Precipitate Salt Deposit Model? The idea that massive salt formations were formed by precipitation of salt resulting from temperature and pressure changes came about during the year 2000. Since then, there have been few if any experiments or observations in the geological community, investigating whether salt deposits came from anything other than evaporation. That changed in July 2006 when a group of Norwegian scientists published a groundbreaking research article in the journal of *Marine and Petroleum Geology*. This research was also reported in the *Oil & Gas Journal*:

"A group of authors led by a Statoil ASA specialist in marine geology has proposed an unconventional theory for **the origin of salt** that could have **far reaching implications** for oil and gas exploration.

"**Masses of solid salt may form and accumulate underground, independently of solar evaporation of sea water**, Martin Hovland of Statoil and four other authors have suggested." Note 7.4e

Here we have researchers proposing an "unconventional theory" having "far reaching implications." However, the implications reach beyond the field of oil and gas exploration and affect *all of geology*. The article further stated that the research team "demonstrated how solid salt forms." They also said it was the physical properties of water that caused the precipitation:

"The Norwegian research team **demonstrated how solid salt forms in high temperature/high pressure (HTHP) conditions when seawater circulates in hydrothermal systems in the crust or under piles of sediment**.

"It is the **physical properties of supercritical water that stimulate the precipitation**." Note 7.4f

The Norwegian team demonstrated how large salt deposits form through laboratory experimentation and actual observa-

> "Crystallization in these hydrous systems can be promoted by **temperature lowering, pressure lowering, or pressure increase**."
>
> *Origin of Granite in the Light of Experimental Studies in the System*, O. F. Tuttle and N. L. Bowen, *The Geological Society of America*, 1958, p67-69

tions in the field. Details from their observations will be further outlined in the Salt Mark subchapter of Chapter 8. Returning to the *Oil & Gas Journal* report, the authors charge geologists with overlooking this important process:

"Geologists, whose current model for salt deposition and accumulation relies **only on solar evaporation of seawater, have overlooked this novel hydrothermal outsalting mechanism**." Note 7.4g

A hydrothermal mechanism, utilizing high pressure and temperature is the foundation for the processes that formed large salt deposits, including the salt domes discussed in the Rock Cycle Pseudotheory chapter. There are large salt deposits on all the continents, yet there are no signs of gradual continental uplift. To determine how these salt deposits came to be in their current locations, we must identify and examine the physical evidence of increased ocean temperatures and ocean depths of former times. The main objective of the next chapter, the Universal Flood, is to do just that. In that chapter, we will discover details about the *origin* of massive solid salt deposits and the origin of sodium and chlorine (these do not exist in elemental form in nature) from which the salt deposits and the salt of the seas are made.

The Prethermation Process

Here for the first time, the scientifically demonstrated precipitation process of "hydrothermal outsalting" is recognized as being able to explain the formation of the largest single-mineral rock formations in the world—salt domes. How do these ideas apply to other rocks and minerals?

It is commonly known that salt and sugar will *dissolve* in a glass of water—but when was the last time anyone saw a rock like granite *dissolve in water?* Most will agree that rocks, especially quartz rocks like granite do not typically dissolve in water. However, this is the first step in understanding how quartz rocks *grow out of water*. The *fact* that common rocks can, and do dissolve in water has been overlooked in the geosciences community. This is a central theme of the this chapter.

Obviously, common minerals such as quartz sand are not seen dissolving and precipitating out of natural waters by evaporation today. However, the present is *not always* the key to the past. Because we do not experience or observe certain processes today does not mean it did not happen in the past. Many such processes can be duplicated to some degree in the laboratory.

At a pressure typical of sea level (about 14-psi) and at a temperature of 1,700° C, quartz will *melt*. However, at 375° C and at the high pressure of 14,000-psi, *quartz will dissolve in mineralized water*. This is not a new observation but it is not well known.

The process in which salts dissolve and precipitate through evaporation is so commonly seen and understood, naturalists have for centuries, used it to explain all salt deposits, even though it doesn't satisfactorily account for most of the salt deposits. By understanding that *rocks* can be dissolved in water, we can also begin to understand that crystal or mineral growth out of water can happen by methods *other than by evaporation*. The crystallization of minerals such as quartz takes place in one

of three ways:
1. Temperature lowering
2. Pressure lowering
3. Pressure increasing

If we discard the Uniformity Principle, established as being incorrect in previous chapters, and open our minds to the possibility that the Earth, in former times, experienced elevated temperatures and high pressures from deep water, we will be rewarded with the true knowledge of the processes of rock and mineral crystallization.

High pressure can come from being deep under water. As depth increases, pressure increases exponentially. It has only been within the last two decades that researchers have had the technology to withstand the crushing pressures found at the bottom of the ocean. Elevated temperatures come from sources such as frictional heating, first discussed in subchapter 5.3.

This newly recognized environment is one in which minerals are *known to form* by processes yet unnamed. We will define some new words to describe these processes. The first, **Prethermation** [prĕh· thûrm·ay·shun], is the process of obtaining a *precipitation* of solids from a solution or gas by changes of pressure or temperature.

Just as we use the term 'evaporation' to describe how some minerals crystallize following the evaporation of water, prethermation describes how minerals crystallize because of a change in pressure, temperature, or both. The word prethermate is formulated from the words '**pre**ssure' and '**therm**al' and is used to describe the solids left behind from the **Prethermation Process**. These are called **prethermites** [prĕh·thûrm·īte]. A prethermite is similar to an evaporite. An evaporite is described in geology texts as being salts left behind after evaporation whereas a prethermite mineral formed because of pressure, and/or temperature changes.

Prethermite is the **precipitate** from a solution/gas when the pressure or temperature changed.

Evaporation and *prethermation* are related processes. Both are crystallization processes involving solids that have been dissolved in an aqueous solution. Water is said to evaporate when it changes from a liquid to a gas and escapes, leaving behind previously dissolved solids. When the solution is exposed to temperature and/or pressure changes, prethermation can occur. In nature, there are no 'pure' waters and all evaporated water will leave behind a residue of solid material. When crystals form because of a change in temperature (like rock candy) or a change in pressure, a prethermite is formed.

Prethermation is the process of obtaining a **precipitation** of solids from a solution/gas when the pressure or temperature changes.

Evaporites and prethermites are both precipitates that crystallize from solids dissolved in a solution. A solution exposed to evaporation, like seawater, will leave behind evaporites made from minerals that were in the solution, but only shal-

The Enhydro Evidence

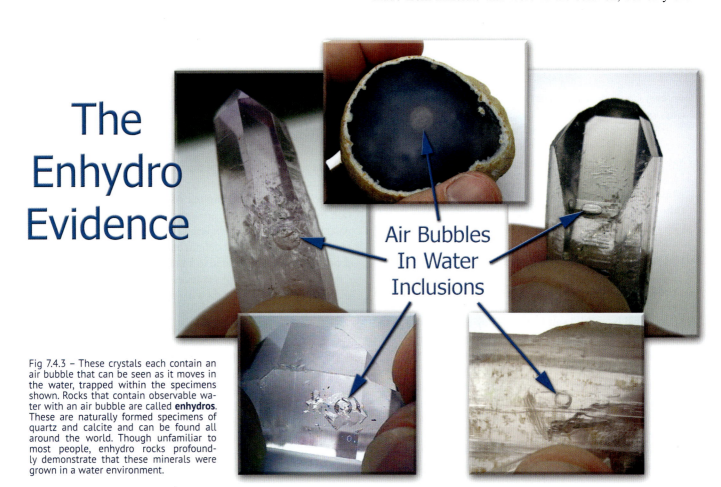

Air Bubbles In Water Inclusions

Fig 7.4.3 – These crystals each contain an air bubble that can be seen as it moves in the water, trapped within the specimens shown. Rocks that contain observable water with an air bubble are called **enhydros**. These are naturally formed specimens of quartz and calcite and can be found all around the world. Though unfamiliar to most people, enhydro rocks profoundly demonstrate that these minerals were grown in a water environment.

CHAPTER 7 THE HYDROPLANET MODEL

ENHYDRO ROTATED

Fig 7.4.4 – These four photos were taken as this Brazilian quartz crystal was rotated. The photos show how the bubble in this specimen moves and changes shape as the rock is rotated.

low seawater (around 1 meter or less) has ever been observed to evaporate and leave behind solids. When this happens, all six salts within the seawater will be present, often deposited in layers. There is no possibility of large homogeneous deposits of a single type of salt, forming from evaporation. It is because of this the Salt Mystery exists. Now, with the understanding of the prethermation process, we can describe the origin of the massive salt deposits.

Most of us understand that salts dissolve in water, and then recrystallize when the water evaporates. We also know they will dissolve again, if water is added back in. Other minerals like quartz do not dissolve in natural waters. Because we are unfamiliar with the idea that quartz crystals grow out of water, it is quite amazing to actually 'see' water in rocks.

The Enhydro Evidence

It is not surprising that most people have never heard the word 'enhydro.' Also surprising is the fact that many graduate students and professors of geology do not know what an enhydro is either. Although the word is not listed in the typical dictionary, it is defined in the *Glossary of Geology* and they are known by rock enthusiasts. The knowledge of these unique gems, and their importance is about to change.

A number of enhydros are shown in Fig 7.4.3. An **enhydro** is a rock that contains observable water, (sometimes, large amounts of it) and a gas bubble. Many specimens contain multiple water pockets, each with their own bubble. Often, the gas bubbles move freely about as the rock is tipped and turned. Fig 7.4.4 shows how a bubble trapped in a quartz crystal moves and changes shape as the rock is rotated.

During the formative years of the UM, interested parties were shown enhydros and nearly all reacted the same. As they handled the water-included rock with an observable moving bubble, their first reaction was one of bewilderment. Having never seen a rock like this before, it was intriguing and necessitated explanation. As the realization set in that water was trapped in the rock, astonishment turned to confusion.

Because incorrect theories of rock formation were taught in school and because pop-culture shows the hot, molten Earth, the last thing one would expect to find inside a rock, is water. It is a real paradigm shift when we examine one of these rocks for the first time. This naturally leads to the critical question:

How did water get inside enhydros?

The answer is quite simple. When crystal growth is rapid, growth protrusions form and trap some of the liquid-gas and solvent in which *the mineral was growing*. What is the liquid-gas in which the crystals grew? It was water. You can actually grow your own Ice Cube Enhydro in a freezer, if the water can be frozen fast enough. Fig 7.4.5 is an example of what an Ice Cube Enhydro looks like. Of course, there should be no confusion as to how water became trapped in the frozen ice cube. The bubble is evident because the water was not completely frozen. The existence of water within these crystals clearly establishes that a water environment was present when the ice cube was frozen. In the same way, mineral enhydros inform us of the mineralized solution in which the enhydro grew.

To understand how crystals can grow in water saturated with minerals that do not readily dissolve in natural surface water is

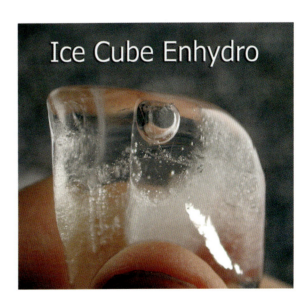

Fig 7.4.5 – We can understand how enhydros are made by observing Ice Cube Enhydros. When water freezes fast enough, air is trapped and is encapsulated within the ice cube. The gas (air bubble) and liquid (unfrozen water) trapped within the ice cube ***clearly*** came from the solution that made the ice cube. In the same way, mineral enhydros entrap liquid and gas of the same material in which they were formed. They are a testament to the water environment in which they crystallized.

a matter of understanding how pressure and temperature alter the solution's ability to become saturated.

The Enhydro Sci-bi

Some researchers have proposed that enhydros formed when ground water seeped into the open cavities of rocks. For example:

"The term *enhydros* refers to a **water-filled geode. Since geodes are essentially produced by mineral-laden waters percolating into a cavity**, it is not uncommon to find the growth process still taking place." Note 7.4h

Recall that in the Quartz Mystery (subchapter 6.4 in the Rock Cycle Pseudotheory) statements by geologists studying geodes clearly established that although several theories had been proposed, "none seems to be entirely adequate to explain all geode features." They have not adequately explained geode formation and they certainly have not been able to explain how water entered the geode. The theory, or more correctly, the sci-bi that mineral-laden water percolated into a cavity to create a geode has not been observed, nor will it. Why? Quartz crystals larger than a couple millimeters do not form in low pressure, room temperature, mineral-laden water.

Enhydros can lose their water once they are removed from their natural environment and exposed to freezing temperatures, extreme heat, or damage. However, many do not lose their water. If specimens have thick mineral walls, they can withstand some changes in pressure and temperature. Many have been in the possession of collectors for decades without losing their water. They attest to how well sealed enhydros can be. Moreover, solvents and other substances that have been preserved inside the enhydros can tell us a great deal about the water environment where the crystal geodes grew. This evidence shows unequivocally that these rock crystals grew in water and did not come from a melt.

Learning From Enhydros

We can learn much from the enhydros that *lose their water* once they have been removed from their native environment in Earth's crust. When rocks with water inside them are exposed to freezing conditions, the water inside will expand causing the chambers to burst. In Fig 7.4.6, a large cavity can be seen on the side of the quartz crystal. There are other chambers in this specimen that are surrounded by thicker walls that still contain water and can be seen moving about inside the crystal. One important detail that intact enhydros reveal is that the environment in which they formed could not have changed dramatically since the crystal was originally formed. This fact becomes more significant when we recognize that most water-containing enhydros are found on or near the surface. Had they been subjected to the freezing conditions of the so-called ice ages, far more of them would have burst, leaving empty, broken fragments.

Why do we find enhydros near the surface if crystal rocks and minerals came from deep with-

Bursting Enhydros

Fig 7.4.6 – This specimen is a large, naturally formed quartz crystal enhydro. It has a large amount of freely moving water and a large open and exposed cavity. The exposed empty cavity, indicated in the diagram, shows where a large water inclusion once existed before it escaped the crystal, probably from the pressure exerted by the freezing or heating of the water. Enhydros can also indicate specific environmental forces that existed at the surface locale of the Earth where the specimen was originally found.

in the Earth, as modern geology teaches—why don't rocks at the bottom of the Grand Canyon contain enhydros? In the next chapter, we will discuss why these enhydro rocks are found near the surface—because they *formed on or near the surface*. This suggests that wherever they are found, enhydros provide direct physical evidence that the surface of the Earth must *not* have changed dramatically since the formation of these water-filled rocks.

To the mineralogist, enhydros are an oddity and it was not surprising to find relatively little research on them. In fact, only a handful of researchers discuss these types of rocks in the geology journals. We did find some mention of them in engineering research papers that addressed the subject of growing quartz. Later in this chapter, we will discuss why engineers would be interested in water-bearing quartz rocks. Geodes are water-bearing rocks and can contain quite a bit of water. To illustrate how much water can be entombed inside them, we refer to the book *Oddities of the Mineral World*:

"Perhaps **the most prolific enhydros sources** in the United States are the various geode beds already mentioned in the states of Illinois, Missouri, and Iowa. Several of the localities in this area yield geodes that spill water when cracked open.

What would you say if someone told you **all natural rocks** *contained water?*

The most notable is the St. Francisville, Missouri bed, where **very large geodes were found containing over a quart of water**." Note 7.4h p51

You could probably have a nice drink from an enhydros—but the water may not be all that pure when you taste it! It is extremely common to read that enhydros have water in them that is 'millions' of years old, but there is zero empirical evidence to support that claim. We shall soon discover that the millions-of-years old dating method stems from dating melted rocks—and natural quartz crystals were never melted.

As we noted previously, water-filled geodes subjected to the freezing temperatures of an ice age would have caused the geodes to burst. How is it that the geode fields in Illinois, Missouri and Iowa contain such an abundance of unburst geodes? Today, the enhydro specimens taken from the Eastern United States must be kept from freezing temperatures once removed from the ground because they will break open if frozen. This repudiates the idea of the ice ages, whether they occurred 10,000 years or millions of years ago. There are so many questions left unanswered.

The Unseen Water in Rocks

What would you think if you heard that *all rocks contained water*? Although generally unknown to the public, researchers have known for decades that rocks and minerals contain water. This unsolved piece of Nature's puzzle remains safely tucked away. This might sound bizarre at first, but this little-known fact that rocks have water has far-reaching implications.

Enhydros are unique because they often have large, water-filled chambers inside the rock or crystal, filled with an easily observable liquid/gas solution. Other rocks and minerals also have water. Trapped inside the rock beyond the limits of visible observation, deep within the molecular or crystal lattice structure, water exists.

Evidence that rocks came from water and not from a melt can be demonstrated easily by simply heating rocks. Because rocks contain water in their microstructure, when they are heated, the water will expand, vaporize and escape. This can be verified by comparing the weight of a rock prior to heating, to the weight of the rock after heating.

In one experiment detailed in the *Journal of Geology*, jasper, flint, and other common quartz rocks lost weight after heating. Reported here, jasper and flint sustained losses of .5% and 1.5% respectively, when heated:

"The density of the jasper is 2.68 gm/cc, and the **weight loss** after vacuum heating at 800° C for 12 hours is **0.5% wt %**."

"The density of the flint is 2.59 gm/cc, and the **weight loss** after vacuum heating at 800° C for 12 hours is **1.5% wt %**." Note 7.4i

From the *Handbook of Rocks Minerals & Gemstones,* obsidian is reported as having "up to **3% water**" Bib 15 p238. Three percent might not seem significant until we understand that it is 3% by weight. Fig 7.4.7 is a good way to visualize just how much 3% by weight represents and how one's perspective can change! The water in the beakers represent approximately (3% by weight) how much water is contained in a piece of obsidian of the size shown. This would make a nice enhydro if all the water had been concentrated in one cavity instead of being dispersed throughout the microstructure of the rock!

A water content of 3% by weight may seem to be a surprisingly large amount of water, but there is a class of minerals that has as much as three times the water contained in obsidian. These water-laden rocks are some of the most unlikely rocks thought to contain water—meteorites:

"**The water content (by weight) of the meteorites is about 11 percent** for type 1 chondrites, about **9 percent** for type 2, and **2 percent** or less for type 3." Note 7.4j

Eleven percent is *an enormous quantity of water* for rocks that presumably came from asteroids that were supposedly at one time melted and should have been void of water. However, if the meteorites came from asteroids or other fragmentary sources that originated in water, then water content of 11% makes sense.

There is another mineral that can contain even *more* water—opal. Opal can have an astonishing 30% of water by weight. Like many of

Fig 7.4.7 – This glassy looking rock is obsidian and the amount of water shown in the two beakers (18g) is the amount of water contained in the obsidian rock shown (617g). Yes, this rock actually has up to this much (3% by weight) water in it! Why do we not see the water? For the same reason we do not see germs. The water is in the microstructure of the minerals in portions too small for the naked eye to see. However, we can heat rocks slowly then weigh them after they have cooled to see how much weight, (in water) was lost. Why were we not taught this in school? For the simple reason that the unseen water in rocks has always been a mystery to geology in general and did not fit in well with the magma Earth theory.

Fig 7.4.8 – This rainbow colored rock specimen is natural opal. Opal is one of the wettest rocks on Earth, holding formative water of up to 30%. Most high quality opal comes from mines located in Australia, but it can be grown synthetically. In nature and in the laboratory, water is essential in opal formation.

nature's crystals, opal has been reproduced synthetically—in a water environment of course. Fig 7.4.8 is an example of a beautiful natural opal specimen in a sandstone matrix, showing off its rainbow of colors.

Even if the origin of unseen water in rocks remains a mystery in modern geology, the fact that all rocks have water in them is not disputed. On the Smithsonian *National Museum of Natural History's* web site, http://geogallery.si.edu/index.php/10026437/water-in-rocks, a new section entitled "All Rocks Contain Water" illustrates this important fact by showing various minerals and the amount of water they contain.

Examining the chemical formulas of a few minerals listed in the book *Rocks, Minerals & Gemstones* there is a common molecule included in many of the formulas—it is H_2O. For example, Gypsum has the chemical formula $CaSO_4 + H_2O$. Analcite is $Na[AlSi_2O_6] \cdot H_2O$, Natrolite is $Na[Al_2Si_2O_{10}] \cdot 2H_2O$, and Autunite is $Ca[UO_2|PO_4]_2 \cdot 8\text{-}12 H_2O$.

The H_2O designation at the end of a mineral chemical formula means that water is known to be part of the mineral *structure*. Without the water molecule, the mineral would not exist in that form. Remove the H_2O from the $CaSO_4 \cdot H_2O$ formula for gypsum and it is no longer gypsum. It becomes anhydrite, or $CaSO_4$ - without water. Of course, not all mineral formulas listed by mineralogists contain water, but that does not mean that they do not have water in their sub-microscopic makeup.

Perhaps the authors of some mineralogy texts did not realize this, or simply chose to exclude the fact that all rocks have water. Water contained in all rocks can be likened to the germs on a doctor's hands. Until medical doctors came to understand the role germs played in the health of the human body, they could not advance in medical wisdom. The same holds true in geology. Until we come to a realization of the role water plays in the origin of rocks, we cannot advance in geological wisdom.

Volcanic Rocks Contain Water

A crystalline rock or mineral exhibits an order in its makeup whereas an **amorphous** mineral does not. Glass is an **amorphous** substance because it exhibits no apparent crystalline form. Volcanic rocks are commonly amorphous and glassy. Water plays an important role in volcanic eruptions and in the formation of rock, commonly called 'volcanic rocks.' Geology has long known that these types of rocks "contain some water bound up in the minerals of the rock," as we read from the book *Melting the Earth*:

"The importance of water was also supported by chemical analyses; by 1824 Knox had established **by experiment that all volcanic rocks contain some water bound up in the minerals or the rock**." Bib 136 p220

Thus, both volcanic rocks (from melt) and crystalline rocks (from precipitate) contain water. This fact, that all natural rocks on Earth contain water, is confusing to one trained in the magma and rock cycle paradigm. Where did the water come from and why was it not released into the atmosphere or into space during the Earth's supposed hot, melted formative years? These questions have no easy answer in geology today. However, if volcanic rocks originated from frictional heating in the Earth's crust, which contains water, the origin of water in volcanic rocks is easily understood.

Fig 7.4.9 is an image of several pieces of scoria, pitted lava rock. The pits, or holes, are called **vesicles,** and were formed

Fig 7.4.9 – These rocks are typical of volcanic rocks. They are amorphous (glass-like) and exhibit characteristic vesicles or 'holes' caused by escaping steam. Researchers have long known that "all volcanic rocks contain some water bound up in the minerals or the rock". This can be easily demonstrated by weighing the rock, slowly heating it and letting the rock cool, then weighing the rock again. The heat causes the water to expand and escape through micro fractures in the rocks.

when volatiles such as superheated water escaped from the rock during and after its ejection. Certainly, water played a crucial role in the formation of all rocks exhibiting such vesicles. As we become aware of this, we see substantial evidence of the presence of water in former times. Rocks exhibiting vesicles were not formed from erosion, and the presence of vesicles is not limited to volcanic rock. Many instances of vesiculated rocks, such as sandstone, basalt, and others are evident in nature.

Mind Over Magma—The Origin of Granite

In 2003, Davis A. Young published the book; *Mind Over Magma.* In it, Young tells the story of igneous petrology and the effort to understand, from the perspective of a magmaplanet paradigm, the mysteries of rocks and minerals believed to have come from magma. The ideas from one of the chapters in his book, *Wet or Dry?—The Origin of Granite,* was previously touched upon the in Rock Cycle chapter. Young concludes the chapter by stating that the debate of whether granite comes from a wet or dry environment had not been settled and that "no consensus had been reached." Amazingly, the entire spectrum of rocks in geology today has been classified in large part by granite research from the 1800s.

Here again, we see that modern science made a serious mistake by building modern geology on 'theories' of granite formation without any hard supporting evidence. As 'natural laws' have fallen out of favor over the last 100 years, the theories of people—albeit highly intelligent people—have taken precedent.

In his book Young recounted the story of the French experimentalist Gabriel-Auguste Daubrée (1814-1896), a mining geologist and professor at the University of Strasbourg, who demonstrated how minerals grew the *wet way*. Daubrée was probably the first person to grow a quartz geode. Unfortunately, the significance of this event was not recognized:

"Although Daubrée's primary contributions focused on metamorphism and meteorites, his studies of **silicates in the** presence of superheated water demonstrated that many of them crystallized from water temperatures far below their fusion points. He grew quartz, feldspar, and pyroxene in the wet way.

"In experiments on metamorphism, Daubrée (1857) examined the behavior of glass tubes filled with a small quantity of water heated to 400°C for at least a week. The water became charged with alkali silicate. The glass was transformed into an opaque white mass composed of various crystalline substances, one of which was **quartz**, which lined the tube walls **much like a geode**. After a month of heating, he produced quartz crystals as much as **two millimeters long**." Note 7.4k

Daubrée also discovered that the second most important mineral in granite, feldspar, also grew in the "wet way":

"Daubrée knew that feldspar had previously been observed in the upper parts of copper smelting furnaces. Rather than conclude that the feldspar crystallized during cooling of melted slag, however, he suggested that the feldspar was deposited on the furnace walls by **vapor**. After all, he noted, the most skillful chemists had been **unable to produce feldspar synthetically by dry fusion**, but his experiments on obsidian demonstrated convincingly that feldspar readily formed in the wet way." Note 7.4k p86

Despite Daubrée's experiments and the empirical evidence he provided, the intellectual tide of the time opposed the idea of a Hydroplanet Earth, a concept still opposed in modern science. Other professors of that era, Heinrich Rose of Berlin, and American geochemist, T. S. Hunt, argued in vain to convince other scientists that quartz became an amorphous glass when fused or melted, and that natural *quartz would not grow without water*:

"Heinrich Rose (1795-1864), a Professor of Chemistry at the University of Berlin and older brother of mineralogist Gustav Rose, **showed that after fusion [melt], quartz is converted into amorphous silica [glass] accompanied by a decrease in specific gravity from 2.6 to 2.2**. Rose (1859) argued that the **quartz in granitic rocks could not have separated from a dry fused mass and could never have experienced elevated temperature because it always has a specific gravity of 2.6**. The American geochemist, **T. S. Hunt** constantly appealed to the experimental work of Rose and others to support his claim that **quartz was never known to be formed in any other way than in the presence of water at temperatures far below those of its fusion temperature**." Note 7.4k p86-87

These quotes came from the text *Mind Over Magma*, published in 2003, and are cited as being the first-ever comprehensive history of the study of such igneous rocks. Rose and Hunt demonstrated that "quartz in granite rocks" could not have come from magma, and "was never known to be formed in any other way than in the presence of water" at temperatures far below melting—*and no one has ever proved them wrong!*

True science is supposed to progress based on observed facts, not prejudiced theories. However, the authoritative control of most scientific publications of the 1800s rested in the hands of a few influential scientific leaders. The pendulum of scientific thought about the Earth's creation swung toward the dry side. Those dry-side scientists envisioned a hot molten Earth and rejected anything related to the 'wet world' of Neptunism, regardless of what the empirical evidence showed. Thus, empirical evidence supporting the "wet way" was laid aside as researchers adopted the "dry way." The 'dry world' of Plutonists prevailed without solid experimentation to support it.

All natural rocks on Earth contain water.

After a century and a half of experimentation and study, geologists have yet to discover how to make natural granite from a dry melt.

Without Water—No Continents

If you were to embark on a modern geology course of study today, you would be taught that the Earth has a molten magma interior and that granite is the primary basement rock of the continents. It is highly unlikely that you would hear that "***Water is essential** for the formation of granite.*" However, according to several maverick researchers at the Research School of Earth Sciences at Australian National University, who published an article in 1983, water is absolutely essential. Their article, *NO WATER, NO GRANITES, — NO OCEANS, NO CONTINENTS* begins:

"**Water is essential for the formation of granite** and granite, in turn, **is essential for the formation of continents**." Note 7.4l

In 1983, the importance of water in the crystallization of minerals was becoming known, yet as of the printing of this book, it is still heavily resisted. It is amazing how long it has been known that water is an essential ingredient in the crystallization of rocks and the formation of our continents, and it is just as amazing how magma thinking allowed this knowledge to be laid aside. For example, in 1958, some researchers knew that

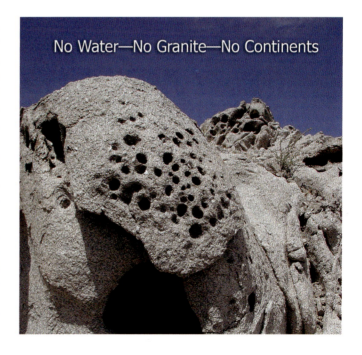

Fig 7.4.10 – This unique granite outcrop is located in Sonora Mexico near the Gulf of California. Most granite deposits do not exhibit holes like these. Researchers have attempted to form granite through experimentation of many pressure/temperature environments, all without water. They had no success. Eventually, they discovered that "the water content" was the "most critical factor" to simulate nature in growing granite, and without granite, there would be no continents.

without water, no quartz or feldspar crystallization would take place in granite:

"...quartz and feldspar nucleate only with the greatest difficulty. **Dry granite** liquids have been held at temperatures below their liquidus for periods of **ten years and longer with no hint of crystallization** (Tuttle and Bowen, 1958)." Note 7.4m

Dry granite liquids are a *melt*, and produce *no crystallization*. Of course, all natural granite is crystallized. How do researchers account for this? The idea that water is essential for granite formation is not a new idea but it is not widely known among the ranks of researchers. In some cases, the dry magma mindset has induced investigators to discount their findings. For example, in a book published in 1997, *The Nature and Origin of Granite*, the author notes how one researcher, Piwinskii, wrote that previous experiments involved too much "excess water" to be related to any "processes occurring within the earth":

"Luth's review of 1976 showed how these early investigations engendered a plethora of similar work from which both experimentalists and petrologists made extravagant claims as to their relevance to nature. Although the use of synthetic starting materials was appropriate enough in the investigation of a specific phase relationship, and the carrying out of the experiments under **water-saturated conditions** was necessarily dictated by the existing technology, there was a growing realization that such experiments might bear **little relationship to natural processes**. Piwinskii was compelled to write: '**Because the experimental investigation was undertaken in the presence of excess water, no direct link can be established with physico-chemical processes occurring within the earth**.'" Note 7.4n

However, the author of *The Nature and Origin of Granite*, Wallace S. Pitcher, exclaims that the "water content" proves to be the "most critical factor" in the experiments involving natural granite:

"It is the **water content** that proves to be **the most critical factor** in all these experimental attempts **to simulate nature**."
Note 7.4n p34

Water content and being formed *out of water* represent two different things. Remember that Pitcher is forming his ideas as one who views the science from within the magma paradigm. In that mindset, granites come from "dry metamorphic rocks":

"It was quickly realized that **if** granites were generated by the melting of **dry** metamorphic rocks deep in the crust, their melts were likely to be relatively **water-deficient and vapour-absent**." Note 7.4n p34

Thus, researchers have searched in vain for a process in which to create "fully crystallized granite" from a relatively water-deficient source:

"This view of the likely **water-undersaturation of natural granitic magmas was amply confirmed** by the new approach initiated by Piwinskii and Wyllie in 1968 and adopted by Wyllie and his co-workers in a long series of experiments (reviewed in 1983). In essence the method involved the **complete melting of natural granite under various pressures, temperatures and concentrations of water**, with the conditions adjusted so as to reproduce on cooling the original mineral assemblage in its natural order of crystallization. Despite all the possible reservations, coupled with the realization that such experiments **have never yet been carried out to completion in reproducing a fully crystallized granite**, the results are **highly significant** and, I am tempted to add, geologically realistic."
Note 7.4n p34

> "**Water is essential for the formation of granite** and granite, in turn, **is essential for the formation of continents**."
>
> *Geophysical Research Letters*, Vol. 10, No. 11, November 1983,
> I. H. Campbell & S. R. Taylor, p1061

The only thing "geologically realistic" about these water-*under*saturated experiments and observations is that they were *not* successful. When one comes to realize that each of the individual minerals that make up granite has its origin in water, it is easy to grasp the concept that granite too must crystallize from water.

Confirming the Law of Hydroformation

Continuing in this chapter, we will uncover considerable evidence to support the new Third Universal Law of Water, a law that should be the basis of all geology. This new law is the natural result of the process from which the magmaplanet model is taken from its pseudo-theoretical position and replaced with the Hydroplanet Model and where the truth about the liquid inside the Earth can be known. The outer core of the Earth is liquid, and since this liquid is not magma, there is *only* one other natural liquid that it could be—*water*.

That there is liquid near the center of the Earth is not in question because there have been many seismic observations and data has been collected from all around the world for more than a century. Sound waves generated by earthquakes and the knowledge that such waves do not travel through liquid and solid materials in the same manner, has allowed the mapping of the boundary between the solid and fluid areas of the outer core of our planet. The Earth is essentially an enhydro—a large geode with a watery center. This will become far more self-evident as we understand how much water is inside the Earth and as we learn how geodes are made.

We can apply the Self-Evident Principle: Water must be at the Earth's core because it could be no other way. If this statement is true, we should be able to confirm it with observations and empirical, scientific evidences.

One of the evidences is that the Earth is not glass. Had it formed from a dry magmatic melt, the crustal rocks would have been amorphous (without crystal structure) and glassy. Instead, it is made of minerals that form only in an aqueous solution, which supports the Law of Hydroformation—all natural crystalline minerals formed in water.

Having already established that water exists in all minerals, we will investigate the hydro-forming processes involved in the making of all natural minerals, and in doing so, confirm the Law of Hydroformation.

The Hydrothermal Process

In *Webster's Universal College Dictionary*, **hydrothermal** is defined as:

"Of or pertaining to the action of **hot aqueous solutions** or gases within or on the surface of the earth." Bib 7 p400

This is how the term *hydrothermal* is typically used today, the combination of water and high-temperature. Hydrothermal activity is further defined in the geology textbook *Understanding Earth*:

"Any process involving **high-temperature** *groundwater*, especially the **alteration and emplacement of** *minerals* and the formation of **hot springs and geysers**." Bib 59 p655

Notice that both modern definitions refer only to high-temperature water. The "alteration and emplacement of minerals" mentioned in the definition cannot be referring to the rocks and minerals we walk on every day because common rocks are not formed this way. Hot spring minerals formed around geothermal vents in places like Yellowstone include the mineral **geyserite**. Geyserite is a soft, silica-based mineral that can be seen around the edges of geysers and hot springs (see Fig 7.4.11). Although geyserite is a silicate based mineral, it is *not* quartz. It is not quartz because it was not under high pressure when it precipitated out of the water. Without high pressure, the silica, dissolved in subterranean hydrothermal waters from surrounding quartz-based rocks, forms the softer mineral geyserite. This story is told in the book, *Geysers of Yellowstone*:

"At depth, the water is heated by contact with the enclosing volcanic rocks. Once heated, it dissolves some of the quartz from the rocks. All of this takes place at very high temperatures—over 400°F (205°C) in many cases; and 460°F (237°C) was reached in one shallow research drill hole. This silica will not be deposited by the water until it has approached the surface and cooled to a considerable extent.

Now an interesting and important phenomenon occurs. Although it was the mineral quartz that was dissolved out of the rocks, the deposit of geyserite is a non-gem form of opal." Bib 134 p3-5

The author notes that the process that involves the dissolving of silica from quartz-based rocks and then redepositing it as geyserite is an "important phenomenon" but it is perplexing in modern geology because the silica from dissolved quartz rocks does not become a quartz rock again. Geologists have largely missed the all-important component of *pressure* in mineral formation. In fact, the very word *hydrothermal*, previously defined using modern texts, was changed some years ago. The original definition for the term hydrothermal, as it referred to the formation of rocks and minerals included the very important factor—*pressure*:

"The **term hydrothermal** is purely of geological origin. It **was first used by the British Geologist, Sir Roderick Murchison (1792-1871), to describe the action of water at elevated temperature and pressure** in bringing about changes in the earth's crust leading to the formation of various rocks and minerals." Note 7.4o

The omission of pressure from the original definition is due to the *dogmagma* that exists in geology today. It is important to recognize that geology has taken a back seat when it comes to understanding natural crystal growth in "water at elevated temperature and pressure." On the other hand, technologists have embraced the idea of crystal growth using a pressurized hydrothermal process. This has resulted in the development of many new synthetic minerals that are very similar to naturally grown rocks and crystals.

We found no term to describe the specific environment that includes the components of water, pressure, and heat (thermal) in the mineral growing process. Because the *hydrothermal* process has come to mean **without** pressure, it is inadequate for describing the pressurized environment where most rocks are grown.

The Hypretherm

In order to distinguish the environment of water-pressure-heat from the water-heat only environment, we will use the term **hyprethermal**. It is to be differentiated from hydrothermal as follows:

1. Hydrothermal – minerals formed in a thermal water environment without pressure.

Hydrothermal is Without Pressure

Geysers and hot springs do not produce quartz rocks and minerals because they are not under pressure.

Fig 7.4.11 – Everyday rocks we walk on did not come from geysers or hot springs because there is negligible pressure in these geothermal springs. Geyserite is a form of opal and is a mineral formed in or near hydrothermal springs.

2. **Hypretherm** – minerals formed in a *pressurized* thermal water environment.

To understand the origin of rocks and minerals on the Earth and on other celestial bodies requires a clear understanding of the hyprethermal environment. Another new term, **hypretherm,** will be used frequently throughout the remainder of the UM to describe this extraordinary environment.

> A **Hypretherm** is the physical environment in which mineral crystals grow. It includes the components of water, heat, pressure, a mineralizer and a gas.

Note that the *hyprethermal environment* requires all five of the following ingredients:

1. Water
2. Heat
3. Pressure
4. Mineralizer
5. Gas

As with any recipe, if the right ingredients are not present, the results will not be produced as expected. If we learn the correct recipe and baking instructions, we can easily reproduce our favorite cake, but if the recipe is incorrect, the outcome will fail to produce the favored dessert. It is the same with rocks and crystals.

The hyprethermal process has been made simple with modern technology. Inexpensive autoclaves can be purchased or built and reused to make the many attempts that are needed to find just the right recipe of temperature, pressure, mineralizer and air for the type of crystal to be grown.

Fig 7.4.13 is a collection of images taken to illustrate the process of how to grow a quartz crystal to double its size—in one day—using a hyprethermal process. The device used in this example was a high-pressure reactor, also called an autoclave, which was placed vertically in the oven. The oven was heated to 400°C at the bottom and about 50°C cooler near the top. This induced a natural convection or circulation of the liquid/gas mixture inside the pressurized vessel (the autoclave). The oven was heated using electrical heating elements controlled by a programmable thermostat that increased the temperature gradually. It is important not to increase the heat too quickly. Approximately two inches of pulverized quartz was placed in the bottom of the autoclave. The quartz will dissolve when the temperature and pressure reach a specific level. At the top of the autoclave, a hanger was added to hold small quartz crystals. These crystals will act as the seeds for the dissolved quartz to grow on. The quartz seeds can be as small as microscopic crystals, sand grains or they can begin as very large crystals, depending on the size of the available autoclave. If the hyprethermal environment was large enough, crystals the size of mountains could be grown. We started with small seeds, about 1-2cm since we did not have access to an Earth-sized autoclave.

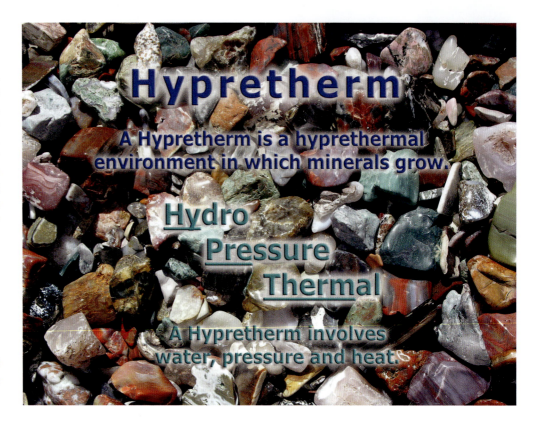

Fig 7.4.12 – These colorful natural rocks are mostly quartz based and were all grown in a Hypretherm. This is a new word developed with its definition in mind. Minerals in nature are crystalline and require a water (hydro) solution to grow. The rocks also require pressure because most of the rocks seen here are quartz based. The harder the rock the higher the pressure required to grow the crystal. Diamonds are one of the hardest minerals and require the highest pressure to be grown. Salts are considerably softer and dissolve quite readily in water. Unlike most of the rocks seen here, the salts did not grow under pressure. Finally, these rocks require around 350°C – 500°C temperatures (thermal) for the silica to dissolve in the water solution to enable them to crystallize. Putting the words together gives us hy-pre-therm, the environment in which these minerals grow. The hyprethermal environment emphasizes a higher pressure to grow the harder minerals that most of the Earth's crust is made of. Additionally, a mineralizer and a gas are generally involved for the crystal growing process to take place.

Once the crushed quartz and quartz seeds were placed in the autoclave, it was filled with a sodium hydroxide (NaOH) water solution that acted as the natural mineralizer in which the quartz will dissolve once the temperature reaches 350-400°C and the pressure reaches 12-17,000 psi. A specific amount of air was left at the top of the vessel, which helps control the pressure. Once the variables were in balance, (it took a number of runs and many years to accomplish this) our patience was rewarded with a quartz crystal, double the size of the original seed crystal and in only one day. This experiment can be repeated quite easily, once the processes and ingredients are known, and today, it is carried out commercially on a large scale, although in only a few places around the world.

Surprisingly, we found no evidence of geological research being conducted using this type of autoclave. How

> Geoscientists will not obtain the wisdom of how quartz-based rocks are formed— **unless** they grow quartz rocks.

do geoscientists ever expect to gain the wisdom of how quartz-based rocks are made if they do not grow quartz rocks?

Commercial Quartz Growth

During the first half of the 20th century, the majority of crystal quartz used in industry came from Brazil. As demand increased, concern about the reoccurrence of the quartz shortage crisis during World War II led to the development of synthetically produced quartz. Commercial quartz production began in 1958 and by 1971, cultured quartz consumption surpassed natural quartz usage.

Today, commercial autoclaves range in size from 10" diameter by 15' long to those with diameters of over 1 meter (3+ Ft) and over 12 meters (32+ Ft) in height. Modern quartz producers may have several hundred vessels producing quartz for use in industry. The synthetic quartz produced in these autoclaves is produced in a hyprethermal environment.

Fig 7.4.14 is an example of commercially produced quartz. The clear stripe seen in the center of the lower left image is the 'seed' that was hung in the autoclave. The wire hanger can also be seen. This specimen is blue because other minerals were added to the crushed quartz. Varying the environment and changing the ingredients can alter the physical appearance and the properties of the final crystal.

One would think geoscientists would be very interested in this method of mineral growth, but it seems it is not so. It has been the engineers and the technologists that have taken up the slack and reported the similarities between the lab and nature:

"Brazilian quartz is vein quartz. **It is deposited in cavities from supercritical hydrothermal fluid in much the same way that quartz crystals are grown in the lab.**" Note 7.4p

Fig 7.4.15 is natural sandstone matrix with natural white quartz and a layer of synthetic amethyst (purple crystals) quartz

Fig 7.4.13 – This diagram illustrates the hyprethermal quartz-growth process. The word **hyprethermal** is a merger of the words "hydro" (water), "pre" (pressure), and "thermal" (heat). A combination of these three physical properties creates a pressurized thermal environment in which quartz crystals can grow. In this diagram, a hanger holding quartz crystal seeds is placed in the high-pressure reactor. A water solution is added to the reactor and it is placed in an oven and heated until the solution reaches 350-400° C. Compare the images of the crystal before growing and after growing. The crystals experienced a rapid growth rate approximately doubling in size in **one day**, not over millions or even thousands of years.

Fig 7.4.14 – This is a man-made quartz crystal grown for technological purposes. The clear strip seen in the bottom photo is the quartz 'seed' while the blue material is the grown quartz. The addition of the element chromium is responsible for the blue color.

on top. With the addition of other natural minerals to the water solution in the autoclave, various colors can be achieved, like this stunning display of amethyst atop natural quartz. Comparatively little research in this segment of synthetic quartz has taken place except in Russia, where for many years much of that work was kept secret under the control of the Soviet Union.

Why Such a Focus on Quartz?

Why focus on quartz in this and the last two chapters? There are so many minerals and variations of those minerals that it may at first seem limiting to direct such attention to quartz. Concentrating on quartz keeps us from becoming too entangled in the specific mineralogical technicalities and allows us to focus on the one change that must take place in modern geology—that is the paradigm change from a dry-melt origin to an origin-with-water.

Another reason to turn our attention to *quartz and other silica-based rocks* is that there are more silicates on the Earth than all other minerals put together:

"Silicates are metals combined with a silicate group (silicon and oxygen) and are the most common of all minerals. **There are more silicates than all other minerals put together**, both in mass and number. Almost a third of all minerals are silicates, and **they make up 90 per cent of the Earth's crust**. Quartz and feldspar alone make up a huge proportion of most rocks." Bib 151 p59

That said, it is interesting to understand the importance of water in other minerals and that water affects far more than just quartz rocks. One instance where water is apparent in crystal formation is the **natural** emerald:

"**Water is present within the structure in natural and hydrothermal but not in flux-grown [melt] emeralds**... The water can be **incorporated into emerald only during growth** and cannot be added or removed at a later stage without destruction of the emerald. This can be demonstrated by a test (which obviously cannot be recommended!): the emerald is slowly heated to red heat. Flux-grown [melt] synthetic emeralds are **not** affected **but natural and hydrothermal emeralds shatter** or turn cloudy **owing to the release of their water content**." Bib 104 p128

It is evident why **natural rocks** and gems could not have come from magma or a melt. They all have water "incorporated into" their structure—through a hyprethermal process. This can be quite dramatically displayed by the sudden fracturing of rocks when they are heated, due to the rapid expansion of the contained water.

"Indistinguishable" From Natural Quartz

Questioning everything in science with an open mind led us to ask the question "just how similar is hyprethermally formed synthetic quartz to natural quartz?" The answer to this question is important as it will either support or refute the Identity Principle that states:

> Identical results come from duplicating processes found in nature.

When the duplication of a natural process can produce synthetic quartz indistinguishable from natural quartz, it is at that moment we become the possessor of a pearl of nature's wisdom. We will know we have discovered a *scientific truth*. Once we determine exactly how quartz *is* made, we can know with a certainty how it *was* made in the past:

"No consistent identifying features are known at present for the reliable differentiation of **synthetic from natural quartz and the two types are so far indistinguishable**." Bib 104 p99

This important statement has profound consequences. A gemologist, not a geologist identified the two types of quartz as being "indistinguishable." As is often the case, the researchers

Fig 7.4.15 – This is a cross section showing natural quartz (white) on a sandstone base with synthetic amethyst quartz (purple) grown on top of the natural quartz. This specimen was made in Russia where much research involving the use of autoclaves was conducted, prior to the collapse of the Soviet Union. This specimen helps illustrate how natural quartz grows in a hypretherm.

working in the side of technology are responsible for this advancement. Technology succeeds where science fails because technologists are pragmatic and practical whereas pure scientists often operate in the theoretical realm.

This discovery validates the Identity Principle and it should have happened a long time ago. Almost two centuries ago, scientists were on the right track:

"In 1822 Humphrey Davy examined natural quartz crystals containing **liquid inclusions** and found that these inclusions contained **mostly water with some salts** such as alkali sulfates. This observation **correctly established the direction for almost all subsequent quartz growth attempts**." Bib 104 p100

Davy's discoveries and correct attempts at making quartz centered on the observation of enhydros—rocks with liquid water inclusions. Unfortunately, quartz research was derailed because of the dogmagma of the melted Earth mindset. Successful quartz growth happened only when investigators returned to the correct paradigm—growing minerals from water.

Natural Hypretherm Growing Conditions Known

It is critical to understand that the hyprethermal process is the ***only proven process*** for making quartz and that quartz-based minerals make up the majority of the *crust* of the Earth. Moreover, the piezoelectric properties of many of the silicates provide another important clue as to the approximate conditions of the hypretherm in which these quartz-based rocks grew. We discussed the piezoelectric effect in subchapter 5.8 and go into much more detail in subchapter 9.5, the Geofield Model. Applying and releasing pressure on α-quartz (alpha-quartz) rocks generates electricity. However, this piezoelectric property begins to diminish as the mineral reaches 570°C. Because nearly all natural quartz-based rocks are piezoelectric, we *know* they did **not** grow at temperatures above 570°C. There is no evidence of natural α-quartz being produced above 570°C. This is the limiting temperature of α-quartz. The limiting temperature is also an indicator of the pressure at which the minerals were grown because temperature and pressure are directly proportional.

Fig 7.4.17 shows various forms of crystalline quartz and the Silica Phase Diagram. These were introduced in the Magma Pseudotheory chapter. They illustrate the differ-

Fig 7.4.16 – Flux-grown (melt) synthetic emeralds have no water in their crystal matrix, whereas all natural emeralds do. Natural emeralds grow in the same manner as quartz but higher pressures and temperatures.

ent states of quartz minerals and the variations of temperature and pressure to which they were subjected. The quartz minerals from the highest temperatures and pressures are very rare in nature. Natural α-quartz makes up most of the quartz found in nature and it is grown at *lower* temperatures, not the theoretical temperatures assumed to exist deep in a magmaplanet Earth.

The Hydrothermal History

Although the influence of James Hutton and Charles Lyell's magmaplanet theories ruled the 19th and 20th centuries, there were experimentalists having success growing minerals in a hyprethermal environment. Over 80 minerals were reported to have been grown:

"According to Morey and Niggli (1913), **over 80 mineral species** are supposed to have been **synthesized during 19th century**. The list includes quartz, feldspars, mica, leucite, nephelite, epidote, hornblende, pyroxene of minerals from the silicate group and several non-siliceous minerals like corundum (Al_2O_3), diaspore ($Al_2O_3 \cdot H_2O$), and brucite ($Mg(OH)_2$)." Bib 156 p58

Even with successes like these, the resources spent understanding mineral growth was negligible and eventually, false theories and theoretical research edged out previous experimental work, ushering in a Dark Age of Science. Today, the best and brightest of the new crop of scientists seek to be theoreticians, especially those interested in the fields of physics and cosmology. In the geosciences, theoreticians hold great sway and have literally directed research

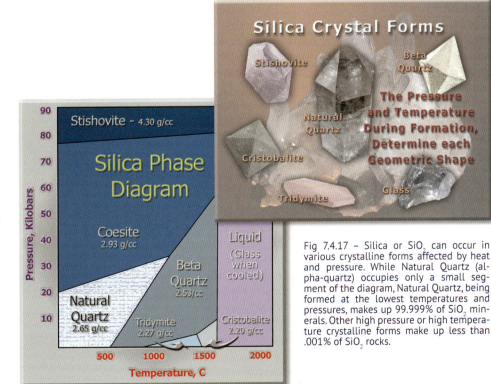

Fig 7.4.17 – Silica or SiO_2 can occur in various crystalline forms affected by heat and pressure. While Natural Quartz (alpha-quartz) occupies only a small segment of the diagram, Natural Quartz, being formed at the lowest temperatures and pressures, makes up 99.999% of SiO_2 minerals. Other high pressure or high temperature crystalline forms make up less than .001% of SiO_2 rocks.

and government grant money toward many fruitless, endeavors resulting in little or no new comprehension about the world around us.

In the first half of the 20th century, Einstein and Darwin's pseudotheories, coupled with the dogmagma of the Rock Cycle and Magma pseudotheories, became the driving force behind modern science. In 1928, Norman L. Bowen published *The Evolution of the Igneous Rocks*; it became **the** petrology handbook. In his book, the author propounded Bowen's Reaction Principle, which tried to explain a series of reactions based on the theory that all rocks come from a melt. It is still used in textbooks today, even though *Bowen never showed* his Reaction Principle to be true. Of course, one reason Bowen's experiments failed and ultimately the reason his theory is off the mark was that the temperatures he used were too high and he neglected to recognize the crucial role of *water* in the crystallization process.

By the middle of the 20th century, other geologists followed Bowen's lead in neglecting water, and abandoned the hypretherm environment in favor of the prethermal environment—high pressure and temperature. These experiments yielded little with respect to how nature works. By the 1960s and 1970s, attention was again being focused on the role of water:

"During the 1960s, an intensive study of the hydrothermal process of synthesis and growth of single crystals which did not have the analogues in nature began. During the 1970s, there was a quest for the search and growth of hitherto unknown compounds of photo-semiconductors, ferromagnets, lasers, piezo- and ferroelectrics and, **in this regard hydrothermal technology attracted a great attention**." Bib 156 p74

Again, the credit for understanding and developing the processes for growing large quartz crystals goes not to the geologists, but to the technologists, driven by the demand of the electronics industry. They pursued the development of the hyprethermal process with vigor. By the 1930s and '40s, technologists had made new discoveries using autoclaves and had gained critical knowledge about the roles of the mineralizer, temperature, pressure and gas levels and other variables necessary for optimal growth. The United States, the Soviet Union, and the UK became leaders in the production of synthetic minerals. Eventually, Japan built the largest autoclaves in the world and became the largest producer of commercial, synthetic quartz.

Vein and Geode Crystals

There are excellent opportunities to view nature's own hypretherm by examining geodes and quartz veins. Amazingly, this is an understudied topic in petrology, which is unfortunate, as vein and geode crystals provide some of the best evidence about the origin of *all* minerals. We found only one investigator who had spent time researching geodes, and even he spent little time on the subject. He knew of no other scholars working specifically with geodes. Finding the geosciences unable to offer information about quartz veins and geodes, we drew upon other fields of science and searched their journals for research on these topics. The only reports we found, mainly from the 1980s, offered little new insight.

These reports explained what seems to have become the consensus in geology, that is quartz veins and geodes are presumed to form *hydrothermally*, in an environment of heat and water:

"**Quartz crystals form in nature when a hydrothermal solution saturated with silicon dioxide fills** (or, perhaps, creates) a cavity in a mineral matrix. Crystallization within these cavities forms **veins** of quartz when the cavity is completely full of crystals or forms **geodes** when it is only partially filled. Quartz often is found in cavities in other minerals. When opened, geodes provide remarkably beautiful mineral specimens." Note 7.4p p40

Missing in the formative process explained in the *Chemical & Engineering News* article, is the all-important factor of **pressure**, which is required for crystal growth. The engineer's utilized pressure in their quartz growth experiments, but the connection that pressure is necessary for natural quartz formation was not made.

How were millions upon millions of veins and geodes formed in the surface layers of Earth's crust? They are found in hard rock and in soft clay-like materials. Geodes are found relatively near the surface and are ***not*** found deep down in sedimentary materials. Many of the veins and cavities found on the Earth's surface are filled with quartz crystals. As evidence that they were not formed deep inside Earth, many of the crystal-filled areas remain undisturbed in unaltered horizontal sedimentary material. Because quartz veins are so common and plentiful, we should ask this FQ:

Are quartz veins and cavities being created today?

Quartz veins are found in the highest mountains, and in many rock types. Some of these were quite obviously cut in after the mountain or rock formation was formed. Geodes and quartz vein growth is not known to be happening anywhere on Earth today, yet it did occur *all over the surface of the Earth a rela-*

Fig 7.4.18 – We find beautiful crystals like these in veins and geodes around the world. They can only be reproduced by man in a lower temperature/high pressure water environment verses a melt. However, these crystals are no different from others found in so-called igneous and metamorphic rocks that were supposed to come from much higher pressures and temperatures and without much or any water. The crystals we can hold in our hands actually testify to how the Earth's rocks were originally formed—in a hypretherm.

tively short time ago. We know how quartz is made, and therefore can recognize the missing component of *pressure* in the previous quoted article.

There had to be a time when significant pressure existed on the surface of the entire Earth. Pressure, combined with other components present during crystal growth was all a part of the environment we will discuss in the next chapter.

Pegmatite Mystery Explained

In the Rock Cycle Pseudotheory chapter, we introduced the Pegmatite Mystery (Mystery #5, subchapter 6.4) with the following quote:

"Despite the extensive study, pegmatites have yielded little information about the **transition from magmatic to hydrothermal fluids** and in recent years interest in these deposits has waned." Note 7.4q

Previously, we discussed how scientists had hoped that the pegmatite deposits would be the missing link between magmatic or molten rock to the hydrothermal environment believed to have been responsible for the formation of the pegmatites. Interest in the pegmatite deposits "waned" because the physical evidence from studies of them did not provide the hoped-for link. Although the evidence does not support the Rock Cycle Pseudotheory, it is convincingly strong evidence for the Hydroplanet Model.

We will explore a small amount of the vast collection of data showing that all inorganic rocks, if not melted by frictional heating, are formed in a *hyprethermal* process. Because the basement rocks of the continental crust are made primarily of granitic rocks, an understanding of the origin of this mineral and its connection to pegmatite formation is critical, if we ever expect to comprehend the original formation of the Earth. From *Origins of Igneous Rocks*, the author says this of the connection between granites and pegmatites:

"The genetic connection between granites and pegmatites **is well established**.

"**An aqueous-rich fluid** is regarded as the **critical element in the genesis of pegmatites**... Although large portions of the pegmatite approximate a granite mineralogy, the quartz cores and veins are most certainly **formed from precipitation from an aqueous fluid that is highly siliceous**." Note 7.4r

Researchers thought that if they could figure out the origin of pegmatites, the origin of the Earth's minerals, including granite would be solved. Remembering that quartz is *a primary constituent of all granite*, we realize that quartz, pegmatites and the granites would have "formed from precipitation from an aqueous fluid." This is precisely what the Hydroplanet Model is.

As you might imagine, researchers steeped in dogmagma would be highly reluctant to admit that the origin of granite and pegmatites are directly connected. In fact, the author from the previous quote went on to say the following regarding the relationship between granites and pegmatites:

"This discussion must remain on a **sketchy** and highly qualitative level." Note 7.4r p244

This statement is only true if you have a paradigm that Earth is a magmaplanet. Consider the data from the Hydroplanet paradigm and the *obvious* answer is "an aqueous-rich" solution.

Many times, researchers come close to bumping into the Hydroplanet Model. A 2002 paper, *Contributions to Mineralolo-*

Fig 7.4.19 – Pegmatites often contain large crystals like this beryl crystal. They are important because they hold clues to how all rocks were made. All of the crystals in a particular pegmatite were made from the same materials at essentially the same time and in the same way. Researchers have finally begun to recognize that "An **aqueous**-rich fluid is regarded as the **critical element** in the genesis of pegmatites..."

gy and Petrology, several researchers described experiments with "synthetic granite pegmatite" at hyprethermal pressures of 14,000 & 28,000 psi (0.1 & 0.2 Gpa). Note 7.4s

They considered their experiment to include an "extreme enrichment in H_2O" because water accounted for about ¼ of the total weight of the minerals and solution in the pressure vessels. Six natural minerals were grown simultaneously over a period of less than three weeks. These included berlinite, muscovite, quartz, amblygonite, lacroixite and Cs-bearing alumionosilicate. The best run in their series of experiments occurred at lower temperatures of 450 °C and with a pressure of 14,000 psi. The pressure, temperature and water levels are within the range of the hyprethermal experiments we performed, as outlined a few pages ago.

These experiments, as performed by Veksler and Thomas (2002) are important because they show that the dissolved minerals formed new fluids and separated into different layers inside the pressure vessel, much as oil and water do. This is called **immiscibility**, and it allows *individual crystal growth* of different minerals *out of the same hypretherm solution*:

"Experimental evidence of the liquid **immiscibility** and mineral reactions documented in our study **offers new explanations of many enigmatic features of natural pegmatites**." Note 7.4s p675

The pegmatite mystery could have never been solved using a 'dry' magma-melt process because all pegmatite minerals are formed in a hyprethermal environment.

Dolomite Hypretherm Evidence

Silicates like quartz are far from being the only rocks to be made in a hypretherm. Dolomite, a very common carbonate based mineral is also not surprisingly formed in hyprethermal conditions:

"Dolomite presents a different challenge. Its nucleation and crystal growth are strongly inhibited at room temperature and, so far, **successful precipitation of dolomite from a fluid or by replacement of a $CaCO_3$ precursor has only been produced experimentally near or at hydrothermal conditions**." Note 7.4t

Here again we recognize that conditions are represented by the researchers as being hydrothermal and as before, they have neglected the pressure component. The dolomite "problem" was mentioned toward the end of the Rock Cycle Pseudotheory (subchapter 6.9). Massive beds of dolomite exist all over the world, in areas that currently are not hydrothermal. Moreover, calcium-magnesium rich hydrothermal waters are not producing modern instances of dolomite. Dolomite does not dissolve readily in meteoric waters that flow naturally through the dolomitic beds and it is highly resistive even to acidic water, whereas limestone is not.

In the journal of *Sedimentology,* we found a paper discussing experiments wherein small dolomite crystals were formed and upon inspection, were found to be similar to natural dolomite:

"Comparison of synthetic dolomites with natural dolomites demonstrates (1) **similar** nanotopography on natural and synthetic dolomites and (2) both natural planar and non-planar dolomite may have island nanotopography." Note 7.4u

How did the research team produce these near-natural synthetic dolomites? An aqueous (water) solution infused with natural calcite ($CaCO_3$), magnesium chloride ($MgCl_2$), and calcium chloride ($CaCl_2$) was placed in a high-pressure "bomb" or autoclave and heated to 200 °C. In the hyprethermal conditions, the experimenters did not need millions of years to grow dolomite—they only needed about 40 hours. Had slightly higher temperatures been used, the growth process would have likely been even further accelerated.

Geologists have known for many decades that calcite-dolomite ($CaMg(CO_3)_2$)—true dolomite, and magnesite ($MgCO_3$)—dolomite without calcium are almost indistinguishable from true dolomite. Both are found in large quantities and can be transformed from one to the other. In a 1964 *Science* report, researchers were able to dissolve these minerals and grow new crystals. Note 7.4v

The environment researchers used to conduct this experiment was one of low temperature but *not* low pressure. Once again, they achieved success and minerals were grown using hyprethermal conditions with temperatures of 275-420 °C, *high-pressure*, water and CO_2 gas.

The Dolomite and Calcite Hypretherm Evidence (Carbonate Mark in the Universal Flood chapter) both come from the physical and chemical requirements that are used to grow these minerals. Once again, *only in water at slightly elevated temperatures and under high pressure* do large dolomite crystals form.

They are not formed by dissolving other rocks or minerals in water at room temperature as modern geology suggests. Dolomite forms in a hyprethermal process that involves limestone, other solutions, and organics, all of which will be detailed in the following chapter.

Calcite Hypretherm Evidence

Calcite crystals like the one seen in Fig 7.4.20 require hyprethermal conditions for growth. Many researchers have tried to grow calcite crystals in an ordinary surface-water environment with neither elevated temperature nor pressure—those attempts all failed. This has been a source of great frustration because this environment (average atmospheric pressures and temperature), envisioned by the believers of uniformity fail to reproduce calcite crystals.

Another sci-bi that attempts to explain the origin of calcite crystals is diagenesis. Diagenesis is a chemical or physical change that sediment undergoes after being initially deposited, during its lithification (becoming rock). Presumably, this occurs at low temperature and low pressure and is modern geology's proposed transition stage prior to metamorphism. These two terms, diagenesis and metamorphism, have been used to explain changes sedimentary rocks undergo, but a distinction between them is not clear because rock formation as a result of these processes has not been observed:

"There is not a clear, accepted distinction between diagenesis and metamorphism, although metamorphism occurs at pressures and temperatures higher than those of the outer crust, where diagenesis occurs." Note 7.4w

Without observation, there can be no true science. The growth of large calcite crystals *has been observed* in the lab, but it was not from the field of geology that we found it. In the field of optoelectronics, a paper entitled *Hydrothermal Synthesis and optical Properties of Calcite Single Crystals*, published in 2003 contained details on the growth of optical grade calcite crystals. The author, I.V. Nefyodova and his colleagues reportedly grew optical grade calcite crystals in hyprethermal conditions. They reported temperatures ranging from 250-300 °C, and pressures of 50-100 Mpa (7,250-14,500 psi). Note 7.4x

The results obtained from the experiments conducted by Nefyodova and others were calcite crystal growth of approximately 10mm (3/8") in 30-85 days. The results of these experiments can also be found in the *Journal of Crystal Growth*. Although they were conducted for the purposes

Fig 7.4.20 – This is a specimen of **calcite**, the second most common mineral type on the continents next to quartz-based minerals. This colorful piece came from Mexico, with the different colors representing various mineral substances that were in the calcite solution when this crystal grew. Large calcite crystals such as this have only been found to grow in a hypretherm.

of making optically pure crystals for use in the optical and electronics industry, why would the same processes not have application in discovering the true origin of *natural* calcite crystals? Indeed, they are very applicable, and geologists should apply these discoveries to understand the natural world and the true origin of calcite crystals.

Olivine Hypretherm Evidence

The mantle of the Earth is believed to be primarily composed of the silicate mineral **olivine,** which is said to be derived from magma. However, when we investigate the findings of geological researchers, they tell a different story. From the *American Mineralogist* journal, laboratory experiments demonstrated that olivine crystallizes *out of water*. Researchers found that using "air-saturated water" inside an autoclave heated and under pressure that, "...at 300° C and 300 bars for 1368 hours...clear signs of dissolution, and growth" of olivine minerals occurred in the laboratory experiments." Note 7.4y

This is hyprethermal growth similar to the quartz growth process already discussed. Moreover, at higher temperatures (yet far less than melt temperatures) and greater pressures, much less time is required for crystal growth. No experiments were found documenting any evidence that crystalline olivine minerals came from a melt. Later in subchapter 7.10, we will discover what actually happens to olivine when it is melted. Hyprethermal experiments such as the one noted in *American Mineralogist* in 2002 clearly establish that olivine, one of the primary ingredients of the mantle, precipitated from a watery solution.

Because olivine is such an abundant crustal mineral, there are other evidences connecting olivine to the watery nature of the Hydroplanet Model. A study reported in *Science* in 2001 documented "puzzling observations of seismic anisotropy" and attributed it to the presence of abundant water:

"The interpretation of seismic anisotropy in Earth's upper mantle **has traditionally been based on** the fabrics (lattice-preferred orientation) of relatively **water-poor olivine**. Here we show that **when a large amount of water is added to olivine**, the relation between flow geometry and seismic anisotropy undergoes marked changes. Some of the **puzzling observations** of seismic anisotropy in the upper mantle, including the anomalous anisotropy in the central Pacific and the complicated anisotropy in subduction zones, **can be attributed to the enrichment of water in these regions.**" Note 7.4z

Seismic anisotropy describes the movement of directionally dependent sound waves through the crust. In the past, geophysicists found this difficult to explain with "water-poor olivine" but "when a large amount of water is added to olivine," it was easy to understand. As we continue further through the Hydroplanet Model, olivine will afford much more evidence about the watery interior of Earth.

Hyprethermal Solution is the *Only* Solution

Just as every fingerprint is different, so to, is every mineral specimen. No two are the same. However, all crystalline minerals require the watery environment of the hypretherm to grow. Some technologists and mineralogists have long recognized this, but it has had seemingly little effect on the theories of the geosciences.

In the book *Gems Made by Man*, author Kurt Nassau explains "the processes by which quartz is grown by man" and the processes in nature "are essentially the same":

"**Interesting enough**, the processes by which **most quartz crystals grew in nature and the processes by which quartz is grown by man are essentially the same**. Although quite insoluble in water under ordinary conditions, quartz becomes soluble if the temperature is high enough. When water is heated under pressure to well above the boiling point, quartz dissolves in one region and deposits in another. This is called **hydrothermal growth**." Bib 104 p100

Why did the author find this fact "interesting enough"? Perhaps because it is not widely known—or perhaps it is because it just makes good sense. To the modern geologist it does not make sense because *entire mountains of crystalline ore exist* and they cannot conceive of the idea that an Earth-sized hypretherm could have existed. The mountains of ore consist of far more than single mineral crystals; they are a complex assemblages of many minerals and crystals that all grew in a hypretherm. Researchers come near to recognizing this as they identified many minerals as being "hydrothermal":

"Among the more important economic minerals found as **hydrothermal deposits are tin (cassiterite); tungsten (scheelite); molybdenite; sulfides of iron, lead, zinc, and copper; and silver and gold ores; as well as quartz, mica, tourmaline, and topaz.**" Note 7.4aa

How did ore deposits form in the crust and on the surface of the Earth? The *only* solution is in a Hyprethermal Solution. A sci-bi commonly cited throughout modern science is that shallow seas contributed to the ore deposits, but quite the opposite is true. For a hypretherm requires *deep* oceans and *hot* water at depth. Only then are all the necessary components of heat, pressure and water brought together to set the stage for crystallization. Even today, some mineral production is occurring at hot spots deep in the ocean.

What about the metals and minerals that form the ore and ore-bearing rock, from where did they come? They did not come from the basement granite and basalt rocks. There is a general idea among geologists that ore is the result of magma convection but there is no *observational* data supporting the theory of convecting magma. Magma has never been observed and seismic studies actually disprove the existence of magma plumes. The elusive answer is that ores owe their origins to—organics.

Role of Organics Not Understood by Geoscience

With technological advances and as "more and more new findings" are made, the magma and rock cycle crisis will continue to escalate. As each new problem arises, the "geological thinking" must be twisted further and further until it eventually breaks:

"**More and more new findings** and applications in this field [hydrothermal technology] are not only contributing to the scientific knowledge of the hydrothermal technique, but are **also posing new problems**. The role of organics in hydrothermal systems for example, has to be studied more seriously, which would **definitely twist the geological thinking to a greater extent**." Bib 156 p42

This statement was made by scholars in the field of hydrothermal technology because they have seen firsthand what impact new technologies have had on the geosciences. Many new problems are identified because of observations in hyprethermal environments, which are rooted in the fact that the geologists are trying to understand the observations while being influenced by an incorrect paradigm. Technologists are pushing the crisis to its breaking point by discovering ways of making near-perfect synthetic minerals using hyprethermal processes and **organics**. Organic material that is not supposed to be in rocks from the center of the Earth. Continuing from the *Handbook of Hydrothermal Technology*:

"It is well known that the **organics** in hydrothermal systems not only lead to the formation of new phases or new structures, and stability of metastable phases, but **can also bring down the pressure and temperature conditions of crystallization**. In nature, we can expect a very wide range of chemical components including hydrocarbons, which greatly contribute to the crystallization of rocks, minerals, and ore deposits. **The role of these organics in the earth's crust has not been understood properly by geoscientists in the context of thermodynamics, and kinetics of crystallization**. It is expected that such studies will definitely propose **much lower temperature and pressure conditions of crystallization of various rock bodies**, and add many more new questions to experimental petrology." Bib 156 p42

Age-old theories of igneous and metamorphic rocks being formed at high temperatures and enormous pressures are facing the scrutiny of actual observations. Lower temperatures and pressures are being successfully employed along with water to synthesize an ever-expanding assortment of minerals and crystals. Add to this the knowledge that organics play a vital role in the creation of ore and ore-bearing minerals—*organics* that would not have survived a trip from a molten magma core of the Earth—and the inescapable conclusion remains, the whole field of geology is in dire need of review.

The true significance of organics in the origin of rocks cannot be overemphasized and their role will become increasingly clear in future sections of the UM, especially in the Universal Flood chapter.

7.5 A New Geology

This subchapter discusses the Old Geology and why it is outdated. A New Geology is presented with a new set of mineral classifications based on actual mineral origins rather than theory. The Law of Paragenesis and a New Geology Time Scale is also introduced.

The Old Classification of Rocks

Today, everyone from grade school children to post-doctorate college students are taught that rocks can be divided into three traditional classifications or groups that focus on their origin according to modern geology. The three types of rocks are:

1. **Igneous** – rocks that form by the solidifying of molten rock, or melt.
2. **Metamorphic** – preexisting rocks that are changed into new rocks when subjected to higher pressures and/or temperatures.
3. **Sedimentary** – minerals that eroded from igneous or metamorphic rocks.

> "The role of these organics in the earth's crust has not been understood properly by geoscientists in the context of thermodynamics, and kinetics of crystallization."

These three classifications are based on the *origin* of minerals as perceived by modern geology. These classifications can only be valid if the origins they are based upon are correct. James Hutton, the ostensible father of modern geology, was the main character in the contrivance of this old classification of minerals. Unfortunately for geology and for the rest of the world, Hutton made a *critical* error while conducting his observations of rocks and deducing their origins. During Hutton's time, most scientists thought that rocks precipitated from the universal ocean of water. Hutton rejected this idea because he thought that if rocks came from water they must be able to be dissolved in water:

"Most mineralogists believed that all visible rocks—granite, basalt, stratified rocks, marbles, and so forth—were precipitates (mineral remnants) from the universal ocean. Hutton found this difficult to believe because he had observed through his experiments that every possible substance appeared in rocks, even substances that could not be dissolved by water. If every rock on the surface of the earth had precipitated from water, then **water must be able to dissolve every substance found in rocks**." Bib 154 p133

Some minerals, salts for example, dissolve readily in water, but most other rocks and minerals do not. Hutton rejected the water-origin for rocks because he could not observe the dissolution of all of the minerals in water, and he was not acquainted with the concept of a hypretherm. Today, researchers have observed just such an environment at the bottom of the ocean. Now after 200 years, technologists regularly employ the use of heat, pressure and water to duplicate nature's hypretherm and produce a host of synthetic minerals.

Over the past two centuries, two critical errors have lead to the Dark Age in geology. First, Hutton's uniformitarian mindset—if I cannot see rocks dissolving in water today, they did not dissolve in water in the past. Many geologists have fallen into this same trap. Ultimately, nearly all rocks will dissolve if they are subjected to hyprethermal conditions. It requires a very specific amount of water, pressure and temperature to grow minerals and it takes a similar environment to dissolve them.

The second critical error stems from the magma/uplift dogma that has held sway for many decades. Because hypretherm conditions exist at the bottom of the ocean today, some researchers have been led to believe that continental rocks grew at the bottom of the ocean and were then uplifted over millions of years. As we have already learned, there is no empirical evidence that slow and steady uplift occurs.

Because these three rock classifications have been around for so long, they have become almost Aristotleon, that is to say, they have become accepted as unquestionable fact based purely on their long stance in scientific history. However, just as Aristotle's ideas of matter being earth, air, water, and fire gave way and failed the test of time, so too shall the 'modern' geological processes and classifications fail when confronted by scientific truth.

The Old Geology Does Not Work

Obviously, if the rock cycle theory is not correct, the three

Since the Old Geology Does Not Work

A New Geology is Needed

classifications of rocks based on rock-cycle theory must be suspect. Why might these old classifications not provide a correct origin of rocks? From the *Handbook of Rocks, Minerals and Gemstones,* in the section "The Origins of Minerals" the expert speaks of igneous, metamorphic and sedimentary rocks and lists the three origins minerals are believed to have come from. The origin of **igneous** minerals is described as follows:

"Many minerals are formed **directly from the magma**. Feldspar, mica and **quartz**, for example, **form as the magma cools down**, deep in the Earth's crust, at temperatures from **1100° C to 550° C**." Bib 15 p12

In The Magma Pseudotheory, we learned why this statement from the *Handbook* is false. Natural quartz cannot come from a cooling magma and this is evident for many reasons including:

1. **Quartz is not highly radioactive (the predominant theory of heat in the Earth is radioactivity).**
2. **Quartz is not a glass (quartz has an ordered crystalline structure whereas glass does not).**
3. **The Quartz would not be piezoelectric (natural quartz looses this property when heated above 570°C).**

The most widely accepted source of heat in the inner Earth from which magma is generated is radioactivity. This would naturally suggest that all rocks originating from magma should be radioactive, but they are not. Another obvious shortfall is that when quartz is melted, it becomes an amorphous glass and cannot become quartz again by simply cooling. Additionally, common, natural quartz *is* piezoelectric. Since heating quartz beyond 570°C causes the mineral to lose its piezoelectric effect, quartz, like all the other hyprethermal minerals, *cannot* have their origins traced back to an *igneous* magmatic source. Therefore, the **igneous** classification of minerals of igneous origin is flawed and must be replaced.

The next classification, **metamorphic** is said to be rocks formed in a process involving high temperatures and high pressures, generally without water that results in the formation of new minerals through the reconstruction of existing minerals. This process has never been observed in nature. We might be able to adapt and use this term if petrologists were able to reproduce metamorphic rocks from igneous or other rocks in the laboratory by applying just high temperature and/or pressures. However, researchers—even after centuries of trying—have been unable to do this. The main ingredient—water, has been left out of their equations and thus, the metamorphic classification must be replaced.

The **sedimentary** rock mystery was discussed throughout the Rock Cycle Pseudotheory chapter. The true origin for the vast majority of sedimentary material is *not from erosion*. Although geologists have held firmly to this view, the evidence against erosion is undeniable. Geologists cannot adequately explain the origin of the majority of the Earth's sediments and though it will not be fully explained until the next chapter, the origin of sedimentary rock is one of the most significant changes to be introduced in the New Geology.

Over 50 years ago, a premier scholar of sedimentary rocks wrote a book on the subject called *Sedimentary Rocks*. The author, F. J. Pettijohn noted:

"The origin and accumulation of sedimentary rocks might, at first thought, seem relatively simple." Bib 159 p2

Pettijohn lays out the truth for us to see:

"Unfortunately the matter is **not** so simple. Not all of the formative processes **can be seen**." Bib 159 p2

Geoscientists have neglected this fact, and for the past five decades have continued building theories about something no one has ever seen. This led to the development of theories advocating "long-time trends," of which Pettijohn makes very clear--are "not even known" to exist:

"Not only is it of interest to determine what happened to a particular bed or layer, but the *geologist* is interested also in the **long-time trends** which **may** be a clue to the chemical and physical evolution of the earth's crust. **About these secular trends little is known—it is not even known certainly whether or not there are any such trends**." Bib 159 p2

This may come as a shock to many who thought that if geology had gotten anything right, it would be that sedimentary materials came from erosion. Nevertheless, as the Loess and Erosion Mysteries in the Rock Cycle Pseudotheory chapter demonstrated, they haven't gotten it right. Moreover, in subchapter 6.2, The Real History of Geology, we learned that it was "difficult to decide" which of the three classifications a rock should be assigned:

"Igneous rocks are **usually defined** as those rocks that solidified from a hot (usually greater than 600 °C), molten, or partly molten condition. When one attempts to apply **this definition to real rocks**, one discovers that in the plutonic [below the surface] **realm it is often difficult to establish** criteria for deciding whether particular crystalline rocks are either **igneous or metamorphic**; whereas in the surficial [at the surface] realm it may be **difficult to decide** whether pyroclastic materials are of **igneous or sedimentary** origin." Bib 87 p71-2

Why should it be "difficult to decide" which class "real rocks" belong too? It is because rocks have traditionally been placed into one of the three classes, based on origin theory and not on observation:

"Classifying rocks is **not** simply a matter of identifying rocks and sorting them. Each classification system **depends on a theory of how rocks are made**." Bib 151 p56

What if the theories on "how the rocks are made" were incorrect? If this were the case, it would certainly describe the current situation in geology—geologists having a "difficult" time applying theoretical classifications to "real rocks."

A New Geology—A New Mineral Classification

It is time to replace the Old Geology with a New Geology—the magmaplanet theory has been replaced by the Hydroplanet Model. Attempting to present the Hydroplanet Model before presenting the Universal Flood Model in the next chapter is a little like the old adage of which came first, the chicken or the egg. It is not entirely possible to understand how the vast majority of the rocks have a watery origin without the knowledge of the Universal Flood. On the other hand, to begin to understand how a worldwide flood could have occurred requires the knowledge that the Earth is a Hydroplanet.

In this short subchapter, we can only begin to touch on the general mineral classifications. It will be the job of Millennial Scientists to define and further clarify these new mineral classifications so that every mineral's specific origin can be identified. One of the chief highlights of the UM is the opportunity of actually presenting the origin of where the solid matter of the Earth originated. Explaining the *origin* of minerals is no different than explaining the origin of earthquakes, or weather, or humankind. Understanding each origin is the very foundation and purpose of science—to describe and explain nature so that man can understand and comprehend it. The simple reason why so *many* mysteries exist in geology today is that the origin of minerals is not understood.

Reclassifying the Origin of Minerals

Technically speaking, there is a difference between a rock and a mineral, but that difference is not always clear. Generally, a rock is a naturally occurring assemblage of minerals, but we use the terms interchangeably throughout the UM.

A fundamental step in defining the specific origins of rocks and minerals will be to replace the old igneous, metamorphic and sedimentary classes of minerals with new classes, based on the Law of Hydroformation, which is; **All natural crystalline minerals formed originally in water.**

The first four new mineral classifications begin with the prefix "hydro" or "hy" representing that these types of minerals were formed or changed in a hydrous, or water solution. The next two new classifications are those minerals that are formed or changed with minimal water. It is notable that the first six classes involve heat, hence the suffix "thermal" or "thermic" in their word form. Pressure is a component of three of the first six classes. Classes 7 and 8 represent rocks of sedimentary origin and class 9 comprises minerals of organic origin.

Except for the so-called basement rocks like granite, most minerals on Earth's crust have either a direct or an indirect connection to biogenic activity. That is to say, the crustal rocks were made in an ocean rich with organic material and this organic material played a vital role in their formation. Organics were left out of the old mineral classification because, as the *Handbook of Hydrothermal Technology* notes, "The role of these organics in the earth's crust **has not been understood properly** by geoscientists…" Bib 156 p42

This is quite amazing when one considers that some of the largest landforms on the Earth's crust (limestone, dolomite and salt) are *from organic sources*. The origin of these biogenic minerals will be discussed in detail in the next chapter but for now, it is important to recognize that organics play an undeniable role in mineral formation, a role that was previously unknown.

Now, we will outline the Earth's minerals according to their *origin* with empirical evidence in support of each classification.

The Nine Classifications of Minerals

1. Hydrothermal – minerals formed in a *thermal water* environment.

2. Hydrothermic – minerals *changed* in a *thermal water* environment

3. Hyprethermal – minerals formed in a *pressurized thermal water* environment.

4. Hyprethermic – minerals *changed* in a *pressurized thermal water* environment.

5. Igneothermic – minerals *formed* or *changed* to glass from *heat* with minimal water.

6. Endoprethermic – minerals *changed* through *pressure and heat* with minimal water.

7. Hydrosediment - sediment formed in a *water* environment.

8. Erosionary Sediment – sediment formed from *erosion*.

9. Biogenic – minerals of *organic* origin.

1. Hydrothermal Minerals – These minerals can be seen forming around thermal geysers. The high temperature of subterranean water dissolved some minerals into solution. When the water is ejected through the geyser, it cools allowing the dissolved minerals to precipitate forming prethermite minerals. Mineral formation around hydrothermal vents may contain micro crystals. They are very small because of the short time period (usually hours or less) they had to develop and because of the *lack of pressure*. Another common Hydrothermal Mineral is salt, which prethermates from a cooling solution. Very few examples of hydrothermal salt are seen actively forming in nature today, so investigators have missed, or ignored this important Hydrothermal Mineral.

2. Hydrothermic Minerals – These minerals originally formed during a hydrothermal or a hyprethermal crystallization process but were later altered or changed in form in a low-pressure thermal water environment. Once crystals are formed hydrothermally, they can typically be dissolved and recrystallized again through the same hydrothermal process. Complementary crystals may form on the original minerals. Any change to the original mineral or crystal via heat and water is called a *hydrothermic change*. This change can occur anytime the temperature and water components of hydrothermal crystallization are in place. In nature, hydrothermic minerals such as geyserite will be altered as the temperature rises and falls and new layers form and old ones are dissolved.

3. Hyprethermal Minerals [hī·prĕh·**thûrm**·l] – In the previous subchapter, the hyprethermal crystallization process was outlined and was recognized as being the origin of *most* minerals. Both hydrothermal and hyprethermal minerals are grown in water, but because hyprethermal minerals are also grown under *pressure*, they exhibit larger, denser, and harder crystals than hydrothermal minerals. Predictably, even higher pressures produce harder and denser minerals, like diamonds. Hyprethermal Minerals may or may not include biogenic minerals. Many of the sedimentary minerals in nature are of hyprethermal and/or biogenic origin. A hyprethermal mineral is a general classification that includes minerals that have not been altered since their formation. Examples include quartz and other quartz-based minerals, granite, basalt, salt domes, most marble, ores and many of the minerals traditionally dubbed 'igneous' in the old classification system.

4. Hyprethermic Minerals [hī·prĕh·**thûrm**·ĭc] – These minerals are similar to hyprethermal minerals. This type of mineral is created when existing minerals are subjected to a hypretherm where pressure and temperature cause some dissolution of minerals into water or affect a change in the rock structure. It may be that the original mineral becomes deformed and is no longer the same type of mineral. Some of the 'metamorphic' rocks of the old classification fall into this category. Hyprethermic changes occur deep within the Earth's crust where hyprethermal conditions are present. Hyprethermic rocks that were formed on or near the surface of the Earth required a hypretherm to crystallize, an event not common in Earth's history. Dikes (See Fig 6.4.8) are commonly of Hyprethermic origin. Recognizing that dikes formed in a hypretherm answers the Dike Intrusion Puzzle introduced in subchapter 6.4.

5. Igneothermic Minerals [ĭg·nee·ōh·**thûrm**·ĭc] – When minerals are melted, they are unable to assume a crystalline form, especially if they cool quickly. With little or no water present, an orderly crystal growth process cannot take place and an amorphous (without form) *glass* forms as the melt cools. Igneothermic Minerals are natural forms of glass and have irregular structures. Examples of igneothermic minerals include common volcanic rocks that cooled from extrusive lava without pressure, such as scoria, pumice, and tuff. When melted lava cools at the surface, the absence of pressure allows any trapped water to expand and escape as a vapor, leaving behind small holes or vesicles in the rock. If there was abundant water in the melted rock, the vaporized water escapes leaving behind a lightweight frothy rock like pumice. Obsidian is an Igneothermic Mineral that cooled rapidly from melted silica but under high water pressure (discussed in subchapter 8.7). All obsidian contains some water and has a characteristic smooth and glass-like appearance. Thus, water may be present in igneothermic rocks but it is *not* essential in its formation.

6. Endoprethermic Minerals [en·dō·prĕh·**thûrm**·ĭc] – These rarely seen minerals generally form deep in the Earth's crust where *pressure* and temperature are present but where water is scarce. They will be concentrated in faults where heat is generated by movement in the crust (See Friction-Heat Law, subchapter 5.3) causing new minerals to form with heat and pressure but with minimal water. Endoprethermic Minerals are rarely seen, requiring significant movement of the crust to

Fig 7.5.1 – The old classifications of minerals—igneous, metamorphic, sedimentary—are not valid because they are based on flawed theory. The New Geology of the Hydroplanet Model includes nine new rock/mineral classes based on the nine ways rocks are formed. Five of the classes, listed in blue, require water in their formative processes. They are of hydro-origin. Two classes, listed in red, are of a thermal only origin and include little or no water. The last two items classify rocks and minerals that have their origins based on erosion or organic activity.

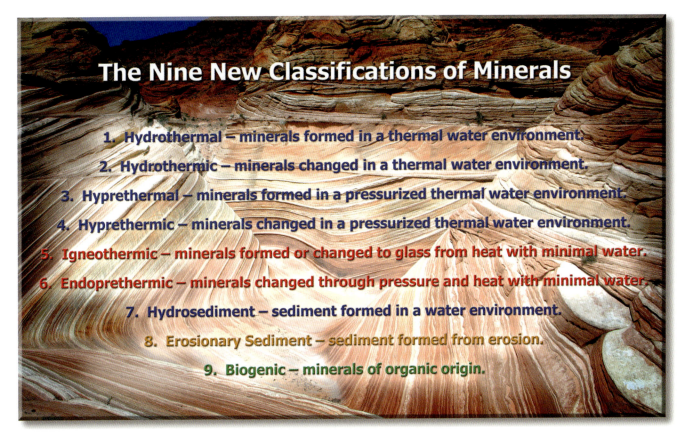

The Nine New Classifications of Minerals

1. **Hydrothermal** – minerals formed in a thermal water environment.
2. **Hydrothermic** – minerals changed in a thermal water environment.
3. **Hyprethermal** – minerals formed in a pressurized thermal water environment.
4. **Hyprethermic** – minerals changed in a pressurized thermal water environment.
5. **Igneothermic** – minerals formed or changed to glass from heat with minimal water.
6. **Endoprethermic** – minerals changed through pressure and heat with minimal water.
7. **Hydrosediment** – sediment formed in a water environment.
8. **Erosionary Sediment** – sediment formed from erosion.
9. **Biogenic** – minerals of organic origin.

expose them. Since subterranean water is abundant and substantial uplift is an uncommon occurrence, these minerals are not frequently observed.

7. **Hydrosediment** – This is the first of two different sedimentary mineral types and is sediment from a hyprethermal or hydrothermal process. An incorrect application of the word sediment in modern geology occurs when sediment is defined as being "solid fragmental material that originates from weathering of rocks..." - This is incorrect because it is assumed that *all* "fragmental material" is from erosion processes. However, the majority of the sediment on the Earth is not from the "weathering of rocks" but from heretofore unknown hydrous processes and is more correctly identified as Hydrosediment (further explained in subchapter 8.5). Hydrosedimentary Minerals formed chiefly in a hypretherm environment, are crystalline in form and can be recognized as loose clay particles, sand or solid claystone, and sandstone. They make up the *majority* of the sediment on the surface of the Earth, which will be explained in the following chapter. Hydrosedimentary rock is formed when sediment is subjected to a hyprethermal environment.

8. **Erosionary Sediment** – Weathering of rocks by wind, water and other chemical means does occur and is the source of some of the Earth's sediment. This type of sediment, properly classified as Erosionary Sediment, includes the materials fitting the definition of the old sedimentary term—"solid fragmental material that originates from weathering of rocks..." In the geology of the 1900's and early 2000's, geologists thought nearly all of the sediment on the Earth originated from erosional processes. This error in their thinking led them to believe that nearly all sedimentary rocks owed their origins to worn down old mountain ranges. Actually only a very small percentage, perhaps even less than 1% of the Earth's sediment came from weathered rocks. The majority of the Earth's sediment actually comes from erosional processes taking place *below* the surface, in the Earth's crust (endoerosion - erosion in underground aquifers). The material is then brought to the surface by hydrofountains, a topic we will cover in detail later in this chapter. Erosionary Sediment is *usually a loose material* found in places such as streams, rivers and at the foot of glaciers.

9. **Biogenic Minerals** – Most organic mineral formations, such as mountain-sized limestone and dolomite deposits, originated from Biogenic sources. These organic sources, which will be described in detail in the next chapter, contributed impurities in a host of other rocks and minerals found near the surface of the Earth. Many hyprethermal rocks, formed in a hyprethermal crystallization process are partly Biogenetic, due to their atomic inclusion of organic minerals. The world's ore deposits are often formed hyprethermally and biogenically. Examples of Biogenic Minerals include limestone, calcite, salts, sulfur, metal ores, and pyrite.

In subchapter 6.10, the mystery of the missing sizes of pebble sediment and sand/silt sediment was introduced, where certain sizes of sediment were noticeably absent at places where *ob-*

Fig 7.5.2 – Hydrothermal minerals like this cone and surrounding area are formed when hot thermal waters become cool and form prethermite. This is the Beehive Geyser in Yellowstone National Park, USA.

servable erosion is happening. In addition, the origin of loess remains shrouded in mystery in modern geology as does the origin of almost all sediment. Recognizing and understanding Hydrosediment and Hyprethermal conditions answer these and many more mysteries. Using these nine new classes of minerals, we have begun to identify and define the New Geology, centered on the Hydroplanet Model.

The Paragenesis Fundamental Answer

In subchapter 6.8, we touched on the Mystery of Paragenetic ore and asked why specific types of ores are always associated. Paragenesis is a geological process of utmost importance. Here is the definition of paragenesis from the field handbook *Rocks, Minerals and Gemstones*:

"Numerous minerals **appear in regular associations because of the same or similar process of formation**; this is known as **paragenesis**. Conversely, other minerals never occur together naturally as their formation processes are so different. **Knowledge of paragenesis is an important aid both in searching for and in identifying** minerals, and especially in the science of mineral deposits. Thus, for example, barite, fluorite and galena **always occur together in certain rocks**. On the other hand, feldspar and halite **never appear in the same crystal specimen**." Bib 15 p12

The word "paragenesis" comes from the Greek *para* ("beside"), and the Greek word *genesis* ("origin"), meaning many minerals form side-by-side in the same environment. Understanding this process is fundamental as we learn about the environments in which rocks and minerals form. Surprisingly, this process is not outlined in a typical geology textbook. In fact, the word paragenesis is usually *not even in* standard geology textbooks, and that is unfortunate. Geoscientists have long known that certain minerals are often deposited in an assemblage with other minerals that formed at the same time. Mineralogists have seen a paragenetic relationship among many minerals. However, because geology has taught the magmaplanet model for

so long, the importance the hyprethermal environment has on minerals has been totally missed.

In the pages to come, we will discuss how geoscientists have come to recognize that ores are formed in an environment consisting of heat, pressure and water and that many types of ores form paragenetically, which is to say different ores form in association with one another. Minerals are *not* found to be associated based on the mineral density, which would seem to be the case if "magma chambers" were the source of the mineral assemblage. In that case, the heaviest minerals would sink to the bottom and at least some of the assemblage would be glass or glass-like—but this is not what is found.

We find instead, a host of mineral assemblages, void of any glass with a variety of minerals and crystals, all of various densities, interspersed throughout the assemblage. *Pegmatites* (Fig 6.4.11) are a wonderful example of paragenesis. Pegmatites contain many of the minerals found throughout the Earth's crust. These minerals were formed in a hypretherm in an assemblage small enough to study the entire group. *Dikes* (Fig 6.4.8) are another example of paragenesis. Dikes formed in cracks and in seams of preexisting rocks. These cracks varied from a centimeter or two to tens-of-meters across. The mineral formed without melting the walls of the surrounding rock. Just like in an autoclave, hyprethermal conditions affected mineral growth in place and without melting the host rock.

In some cases, the mineral assemblage is literally the size of a whole mountain. It may seem impossible to visualize how an entire mountain could have been subjected to a hypretherm, but many were, deep in the ocean where hot, pressurized water was present. Many mountains are simply large crystalline assemblages of a countless variety of minerals, and have exhibited relatively little erosion since their formation. Once it is recognized that there was a colossal worldwide hypretherm, the whole concept of mountain building will change.

The Law of Paragenesis

Every scientific law relies on a predictable and observable sequence of events that validate that law. A repeatable sequence of events lies at the heart of the Universal Scientific Method. Understanding and duplicating the sequences that occur in nature lead us to a true and correct awareness of nature itself. In the field of geology, recreating the sequence of events that lead to rock formation is paramount to understanding the rocks themselves. In the obsolete magmaplanet paradigm, experiments with melts failed to reproduce the minerals or mineral assemblages that exist in nature. However, many of those minerals have been reproduced in a hypretherm, and the results have been predictable and repeatable. Paragenesis is the simultaneous growth of multiple types of minerals in a predictable order. The **Law of Paragenesis** is predictable, repeatable and it is observable.

<u>Law of Paragenesis</u>
Hyprethermal minerals prethermate from mineralized water at the same general time and in chronological order.

Already in this chapter, we have shown that rocks and minerals do not take a long time to grow. All that is needed is a sufficiently large autoclave or hypretherm in which to grow them. The vast majority of minerals, including the salts, grew according to the Law of Paragenesis, which is, they grew in a hypretherm and prethermated or crystallized out of solution in a chronological order. This is why large salt formations, like the Avery Island salt dome, (Fig 6.1.6) do not support the evaporite theory. The salt in salt domes is a prethermite, and it did not come from evaporation but by prethermation from hot, super-saline water. It is the product of crystallization resulting from pressure and temperature changes.

We live on a Paragenetic Earth. The Law of Paragenesis follows the Natural Law of Order set forth in Chapter 22, the Energy-Matter Model.

All things in the natural world exhibit order, and that includes minerals. Minerals will continue to form and crystallize in the same manner as long as the water solution they are growing in contains the particular chemicals for that mineral, and the physical attributes of temperature and pressure are present. Every mineral will in turn crystallize according to its physical properties and in its proper order, in a predictable manner.

A New Geologic Time Scale

With the introduction of new mineral classifications and the Law of Paragenesis, a discussion of geologic time scale is necessary. An entire chapter (Chapter 10, the Age Model) is devoted to the age of rocks and fossils and draws evidence from many scientific disciplines to establish a more correct age of the Earth. However, one 'geo-legend' needs to be addressed now—that of the "long periods of slow cooling and crystallization" presumably

These layers of sediment now have for the first time an origin for cementation - a hypretherm. They are called Hydrosediment Minerals and formed in a hyprethermal environment.

'*required*' for the growth and formation of the igneous rocks.

Although textbooks teach that long time periods are required for igneous rock growth, researchers learned decades ago that this is not the case:

"This study is concerned with the concept of **time of crystallization and texture in igneous rocks**. Conceptually, plutonic rocks are coarse-grained because of **long periods of slow cooling and crystallization**, whereas volcanic rocks are fine grained because of **very rapid cooling and crystallization**. Nucleation density and growth-rate plots... offer an **alternative explanation** for the development of plutonic and volcanic textures. Crystal-growth rate within these systems reach maximum values on the order of **several millimeters per day. Obviously, at this rate long time periods are not needed to produce the type of crystals characteristic of plutonic rocks.**" Note 7.5a

The conclusion the authors of this 1977 *American Mineralogist* article reached was that "*Obviously*... long time periods are **not** needed" to grow rocks! Why has this important piece of nature's puzzle been swept under the rug for so long? Unless it simply does not support the Geological Time Pseudotheory, which of course is 'complicated' because of its many 'missing' pieces.

The New Geology Time Scale does not require millions of years for rock growth or the laying down of the sedimentary layers of the crust. Moving forward in this chapter and into the next chapter, we will introduce additional empirical evidence establishing a much shorter formative period for the Paragenetic Earth. Let us now examine the evidence for a Hydroplanet Earth on a large scale—the Earth itself.

7.6　The Hydroplanet Earth

The concept that the Earth is a Hydroplanet instead of a magmaplanet is one of the key components of the UM. This subchapter will discuss the "last thing you would expect to find" throughout the Earth—water. We will show that researchers have known that the "textbook view…could be wrong and that there is no known origin for the Earth's oceans with modern geology. Abundant evidence of the Hydroplanet Earth comes from the drilling of deep boreholes and other research projects. The equatorial bulge and the distribution of water throughout the Earth, along with a host of other evidence will confirm that geology textbooks need to be rewritten, based on the new scientific discovery of the Hydroplanet Earth.

Underground Water, the Textbook Answer

Fig 7.6.1 illustrates a typical geology textbook's answer to how much water is underground. In this example, abundant water is shown near the surface but as depth increases, permeability decreases and pressure squeezes the water out of the pore

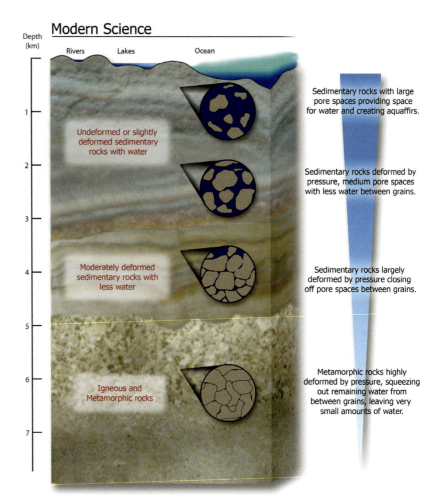

Fig 7.6.1 – This diagram represents a generalized cross section of the Earth's crust, based on modern geology's magma pseudotheory. Accordingly, pressure increases with depth, squeezing water from the pore spaces, making the deepest rocks impermeable. Surface water would be unable to penetrate the deepest layers, leading to the long-held idea that subterranean water existed only in the upper crust.

Recently, deep boreholes, seismic data, and the discovery of water on other planets and moons, have lead to a complete reevaluation of the role water played in the formation of the Earth and in how much water exists inside her. Actual observations are challenging old notions; there is a lot of water deep in the Earth, rendering the above diagram obsolete. Diagram adapted from Bib 59 p311.

spaces, rendering the rock impermeable. The resultant idea is that the deeper we look, the less water we can expect to find.

This dogmagma relic assumes there is no room for water below Earth's thin upper crust. What if there was *water inside the Earth*? What if the Earth's origin was from water and the Earth is essentially an enormous enhydro?

Hydroplanet Model of the Earth

The magmaplanet Earth has an oceanic crust averaging approximately six miles thick and a continental crust of thirty or more miles, as we discussed in the Magma Pseudotheory. Below that, the area between the crust and the outer core is a mystery to geophysicists. Referred to as the mantle, it is said to support "convection" currents, so it *cannot be solid*, yet seismic waves travel through the mantle, which *cannot happen if it is liquid*. Is the mantle liquid, or solid?

Is the mantle liquid or solid?

This longstanding geological mystery can now be solved, as we review the Hydroplanet evidence. The mantle *is both liquid and solid*. Think of it like Swiss cheese, where water occupies the cavities and voids between various solid minerals.

Refer to Fig 7.6.2 for a simplified illustration of the Earth's Hydroplanet interior, based on observations and data from deep boreholes, seismic studies and seismic thermographic studies. The outer core of the Earth is liquid, as S-waves do not pass through it. There is only one liquid in nature of sufficient quantity to occupy the outer core and it is not theoretical magma, it is *water*.

The inner core has also been an enigma for geologists. According to the magmaplanet paradigm, the inner core of the Earth is hot and radioactive and should be liquid, because of its presumed heat. How the core can be both a solid and a liquid—Researchers are in "disagreement." Note 7.6a The closer to the heat source an object is, the higher the temperature and the more liquid it will be. Yet seismic observations have established just the opposite—*the inner core is solid and the outer core is liquid!*

The inner core is solid, which we will confirm shortly, with empirical evidence, but let us first look at some specific physical evidences for a Hydroplanet Earth Model. Like a puzzle, no one piece of evidence will show us the whole picture, but with **many** pieces, we can begin to assemble a new conceptual view of what the interior of the Earth is really like.

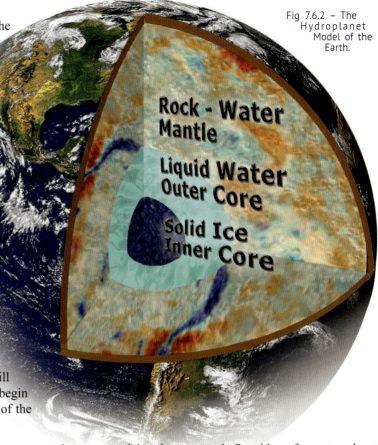

Fig 7.6.2 – The Hydroplanet Model of the Earth.

The Rotating Core Evidence

Because certain seismic waves are not able to penetrate the outer core, science has speculated that the outer core of the Earth is ***liquid***, but additional evidence for a liquid outer core has been recently discovered:

"Moreover, **there is remarkable evidence** that the Earth's **inner core rotates at a different rate** than the rest of the Earth." Note 7.6b

Certainly, the inner core could not rotate at a different rate than the outer core if the entire core was solid, and it is reasonably obvious that the outer core must be liquid and the inner core solid for the inner core to move as it does. Further, if the Earth's outer core was liquid magma; the melted rock would likely be too dense for the inner core's observed rotational rate. One also wonders how the hot inner core does not melt in such conditions, unless there is another possibility.

What are the implications if the core of the Earth consisted of simply water and ice? There would be no problem with the inner core being a solid—solid water is ice—and the geologists' enigmatic rotation 'problem' goes away. Water works—magma does not.

"The Last Thing You Would Expect To Find"

Physical evidence will continue to be revealed and truth will be established if we ask the right questions. We need correct models to help shape questions so that new observations can validate the truth of what has always been there. The Hydroplanet Model of the Earth in Fig 7.6.2 is one such model, but for most scientists, that idea is just too fantastic.

This is not the first water-based planetary model to be proposed. One idea of a water-planet model was published in the online journal *NatureNews* in 2003. Titled; *Making other Earths: Dynamical simulations of terrestrial planet formation and water delivery,* Raymond, Quinn, & Lunine, made this interesting observation:

"If there are planets like Earth around other stars, **they'll probably be water worlds, with awesomely deep, pole-to-pole oceans and no land in sight**. So say Sean Raymond, of the University of Washington in Seattle, and colleagues. They have computer-modeled the late stages of planet formation, when Earth probably acquired its oceans." Note 7.6c

Using a computer program, the researchers simulated planet formation by scattering planetary embryos inside the orbit of large Jupiter-sized planet in a mock solar system. Instead of the traditional rock-impact-accretion simulation, small pieces of rock and ice were subjected to the pull of the large planet to see what kind of planetary bodies would form. To their surprise, hydroplanet formation dominated the outcome:

"These simulated planets acquired water masses ranging from almost zero to 300 times the amount on the Earth's surface. **The formation of wet planets like ours 'seems easy', say the researchers. More surprising still is the fact that more than half the worlds were much more water-rich. Land dwellers like us would be impossible on planets entirely covered by ocean**.

"In fact, more than 400 kilometers inside the Earth there may be enough water to replace the surface oceans more than ten times."

"The existence of planets like this, for which there is no analogue in our Solar System, **must be seriously considered**, the team concludes." Note 7.6c

Their model predicted planets with more water than the Earth, but curiously, they do not see the real-time evidence, right here, in our own solar system. Some of Jupiter's hydromoons and other celestial bodies have pole-to-pole oceans, lending credibility to their water model.

In 1997, *Deep Waters,* an article published in the *New Scientist* clearly identifies new "strands of evidence" that point to "hidden water":

"Deep inside the Earth, the pressure is excruciating. Squeezed into strange shapes and forms, the rocks are so hot that they crawl like super-thick treacle. It is an inferno worthy of Dante, but it also contains something surprising. **What's the last thing you would expect to find in this hellish environment? Water. Vast amounts of the stuff. In fact, more than 400 kilometers inside the Earth there may be enough water to replace the surface oceans more than ten times**.

"How much there is down there is still fiercely debated. But these inner 'oceans' **could help to explain long-standing puzzles about Earth's formation**, the causes of deep earthquakes hundreds of kilometers inside the Earth, and why massive volcanic outbursts suddenly flood hundreds of thousands of square kilometers with lava. They may even give a glimpse of what the future holds for the Earth's climate—and if we might ever be drowned from below.

"Many different strands of evidence point to the existence of this hidden water." Note 7.6e

Geophysicists who view the center of the Earth as a hot "hellish environment" are not the only ones who will have to rethink the magmaplanet mindset. Theologians and philosophers may have to reevaluate their thinking as technology allows us to dig deeper into our planet's interior. The *New Scientist* article continues:

"Frost says that solidified lava that has erupted at mid-ocean ridges contains **glass** that can be analyzed **for water content**. His research team has calculated **how much water** the lava's parent material **in the mantle must have contained**. 'It ends up being between 100 and 500 parts per million,' he says. And if the whole mantle contained 500 parts per million of water, Frost calculates that would be the equivalent of **30 oceans of water."** Note 7.6e

Imagine the equivalent of up to "30 oceans of water" beneath Earth's surface. The more the researchers look, it seems, the more water they find. Five years later, in 2002, an article appeared in *Earth and Planetary Science Letters* titled: *Is the transition zone an empty water reservoir?* In it, the author talks about interior water:

"…addresses the question of stability of **free water in the lower mantle** and its mobility by percolation process, which could be a very efficient transport process, **previously unconsidered** in this field of research." Note 7.6f

Just how much "free water" do the researchers think exists in the inner Earth? From their research, they estimate as much as "100 times the water contained in the hydrosphere":

"The total water content of the inner Earth could be less than 10% or **as high as 100 times the water contained in the hydrosphere**. In this case, **water** will be essentially **present in the core…"** Note 7.6f p38

Their estimates include the water of a 100 oceans inside the Earth, even imagining water being "present in the core!" Remember that from their paradigm, the core is supposedly magma and they suggest that water "will behave like magma." Perhaps these scholars were close to recognizing Earth as a Hydroplanet:

What if the Earth's core was simply water and ice?

"Further work is needed to establish the distribution induced by free water percolation, but we can guess that **water in a porous medium will behave like magma** described by Stevenson, in 1989. **Water or magma would migrate** along the direction parallel to the axis of minimum compressive stress and accumulate in veins. Then buoyancy would drive the water back up to the transition zone." Note 7.6f p50

Empirical seismic evidence cannot be forever ignored and better images will show that *only water* can explain what these researchers are seeing:

"The presence of a **significant amount of water at the core-mantle boundary** could explain some seismic characteristics of the D" layer." Note 7.6f p49

After reading reports that reveal evidence of massive amounts of water inside the Earth, one might assume that the Hydroplanet Model would soon be accepted, but this is not the case. Magma planet heat still strongly influences interpretations of present day observations.

Ongoing research into ice has revealed that it occurs in a variety of crystalline forms, each based on changes to temperature, pressure or other variables. See the discussion in the Mass-Weight Model subchapter 18.4 about how the Earth's water/ice core and underground oceans changes the Earth's density to be around *one third* of the previous magma estimate. A recently announced discovery of ice-XV, created in the University of Oxford's lab June of 2009 displays the bias researchers have about the Earth's interior. Christoph Salzmann, coauthor of a *Physical Review Letters* paper says this:

"The only places on Earth with high enough pressure to sustain ice XV are also extremely hot, so ice XV can't form there, he says." Note 7.6g

Possibilities have been clouded by hot Earth theories that otherwise would provide answers to the evidence that the Earth does not have an iron core (see subchapter 5.9), but rather, it has a liquid water and/or ice core.

"The Textbook View… Could Be Wrong"

Today, students are taught from textbooks with magmaplanet Earth theories and diagrams between their covers. Gradually, researchers have made observations bearing empirical testimony of a hydroplanet model. Some leading professors have even begun to accept the notion that instead of the early Earth being a roiling ocean of magma, it had "oceans and continents":

"Analysis of the object [zircon crystal] in 2001 by **John Valley**, a UW-Madison professor of geology and geophysics, **startled researchers around the world** by concluding that **the early Earth, instead of being a roiling ocean of magma, was cool enough to have oceans and continents – key conditions for life**." Note 7.6h

Although the authors of this article are in the magma paradigm, they refer elsewhere in the article to the early Earth as being a "hydrosphere," based in part on the evidence that zircon crystals were formed in unique conditions on the early Earth.

As technology progresses allowing more experimental research, the error of magma theory will become more evident.

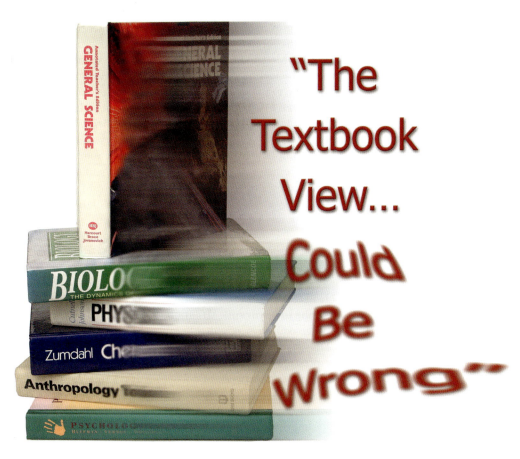

"The Textbook View... Could Be Wrong"

Forward-looking researchers such as John W. Valley have recognized that crystals used in dating the early Earth did not come from a melt, but from hydrothermal conditions. Valley recognizes the possibility that water may have been a much bigger part of the early Earth and that the textbooks may be wrong:

"**The textbook view** that the earth spent its first half a billion years drenched in magma **could be wrong**. The surface may have cooled quickly—**with oceans**, nascent continents and the opportunity for life to form much earlier." Note 7.6i

The UM may surely attract many skeptics, yet open-minded researchers have come near to the realization of the Hydroplanet model already, and to the realization that the modern textbook view could actually be wrong. The UM takes the position based on the scientific evidence, some of which is presented here, that the textbooks *are indeed* wrong.

Oceans Are Not from Comets or Meteorites

What do we know about the origin of the Earth's oceans?

This question was posed on the *Scientific American Ask The Experts* web site. Tobias C. Owen of the *Institute for Astronomy* in Honolulu, Hawaii first answers the question by saying:

"This is a very good question, because **we do not yet have an answer that everyone accepts**." Note 7.6k

There are several reasons why science does not have a clear origin for the oceans. One idea lists comets as a source for the Earth's oceans, but comets don't work, as Owen says:

"The composition of the ocean offers some clues as to its origin. If all the comets contain the same kind of water ice that we have examined in Comets Halley and Hyakutake—the only ones whose water molecules we've been able to study in detail—**then comets cannot have delivered all the water in the earth's oceans**. We know this because the ice in the comets contains twice as many atoms of deuterium (a heavy isotope of hydrogen) to each atom of ordinary hydrogen as we find in seawater." Note 7.6j

Another theory is that the oceans came from meteorites, but Owens dismisses this idea as well:

"At the same time, **we know that the meteorites could not have delivered all of the water**, because then the earth's atmosphere would contain nearly 10 times as much xenon (an inert gas) as it actually does. Meteorites all carry this excess xenon." Note 7.6j

Theories like these are the reason the origin of the oceans is still a mystery—researchers keep looking somewhere else for the water's origin, hoping to identify the object that *brought* water to Earth. Without the knowledge of Celestial Water or the Crystallization Process,—the process where all minerals come originally from water—one cannot see that the Earth's water was here from the beginning!

The Long Valley California Water Evidence

In the Magma Pseudotheory, subchapter 5.11, the Long Valley California drill site illustrated that science's best prospect for drilling into magma came up short. During our discussion, we never fully explained *why* the temperature stopped increasing, even as they drilled deeper. A representative of the U. S. Geological Survey said this:

"We do not fully understand why the temperature in the well stopped increasing below about 6500'. It is unusual for there to be no positive temperature gradient over large depth intervals in the Earth. An isothermal depth profile like the one found in LVEW likely requires a combination of **lateral flow and upflow of water**." Note 7.6k

An *upflow* of water was suggested by this researcher. From where did he imagine the water originating, one wonders. The 'fact sheet' for the Long Valley drill site notes:

"The highly fractured rock drilled for most of this phase has **many open veins** containing crystals **that show evidence of present or past water flow**." Note 7.6l

From the government 'fact sheet,' we learn that the Long Valley California borehole never reached its intended purpose, magma, but it did produce evidence of water.

There may not be a better scientific example to illustrate that magma is a pseudotheory and that the Hydroplanet Model is a reality than the example taken from the Long Valley California project. This drill sight was selected as the best opportunity to penetrate a magma chamber. What they found instead, was water, and that the temperature stopped increasing below 6500 feet, the last thing researchers expected to find.

KTB Evidence for Water Boundary Layers

We are in a unique position to see real scientific data on the Earth's interior at what is probably the most researched deep borehole in the world, the German KTB borehole:

"It was predicted that the drill bit would meet the most prominent, steeply dipping, crustal reflector at a depth of about 6500-7000 m, and indeed, the borehole penetrated a major fault zone in the depth interval between 6850 and 7300m. **This reflector offered the rare opportunity** to relate logging results, reflective properties, and **geology to observed and modeled data**."
Note 7.6m

This did provide a "rare opportunity" to compare actual "geology to observed and modeled data," or in other words to determine if the geologic predictions matched the actual observations. Understanding the reflective signals is *very important* because there are *similar*, strong signals coming from the crust-mantle boundaries. Thus, the material reflecting seismic waves at the crust-mantle boundaries must be similar to material found in the KTB borehole. From where did the "strongest reflect signals" originate? The KTB researchers noted:

"**The strongest, reflected signals** originated from **fluid-filled fractures** and cataclastic fracture zones rather than from lithological boundaries (i.e., first-order discontinuities between different rock types)..." Note 7.6m

The fluid referred to was *water*. Researchers often use "fluid" rather than water, perhaps for many reasons. Dogmagma has influenced scientific thinking for so long that evidence of a water-filled Earth has been 'off-limits' for more than a century. To earn a PhD in geology, one must not venture too far astray from the magma mindset. Researching Earth as a hydroplanet would likely earn one the title of heretic, and so the bias against 'water' continues. But slowly, things will change.

The researchers at the KTB site found that inside the Earth, between different layers of material, the greatest seismic reflections (which tell us how far down these boundaries are) came *not* from different rock types (lithological boundaries), but from "fluid-filled fractures." The evidence from the KTB borehole is that *fluid-filled fractured areas and zones* encountered there appear similar to those detected at the boundaries seen between the crust, mantle and core. Other *direct evidence* of water zones like these will be presented in the next subchapter, as we redefine hydrology.

Significant data were gathered from Germany's KTB borehole, but in some ways, the project was a disappointment for many researchers whose theories were not proven by observations at the drill site. Nevertheless, the KTB has been a wealth of confirming evidence for the Hydroplanet Model. Entire books can now be written showing how real scientific observations match a water model of the Earth. One example comes from a 2000 article in the *Geophysical Journal International*, where geophysicists acknowledged:

"**A fundamental understanding of the origin, geometry and extension of the fluid pathways in crystalline rock is still incomplete**. We analyzed a broad spectrum of stochastic fracture networks with varying fracture apertures, fracture lengths and distribution function in order to estimate fracture permeability at the KTB site. **The problem is highly underdetermined**, and therefore additional assumptions were used referring to the state of stress and the percolation threshold. Calibrations using measured permeability values from hydraulic tests were also included. Possible stochastic fracture networks and estimations of fracture apertures and fracture lengths were obtained. **Under the assumptions made, these networks can describe the hydraulic situation at the KTB site**." Note 7.6o

Water pathways posed a "problem" for researchers at the KTB site, but the "hydraulic" nature—water movement under pressure—of the pathways meant that water was moving through a vast network of suboceans inside the Earth. Astronomical cycles cause gravitational flexing and heating of subcrustal waters, leading to increases in pressure. Knowledge of water movement beneath the crust provides an opportunity to reexamine other long held notions. Effects on mineral growth, erosion, and even the weather can be recognized from processes never before known. We will continue to explore many of these new ideas.

The Underground Slabs Evidence

In the 1998 American Geophysical Union's (AGU) annual meeting, something totally unexpected, or as one seismologist described it, "something funny" was going on inside the Earth. New seismic images showed huge slabs of Earth "accumulated in broad piles on the mantle floor." This is troubling because

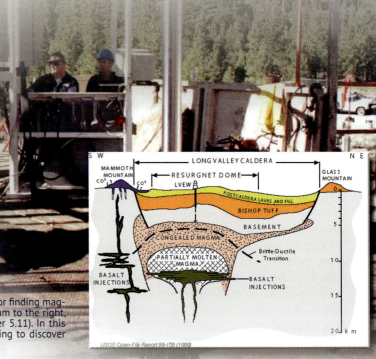

Fig 7.6.3 – The Long Valley California borehole, geologists' best prospect for finding magma, ended up finding evidence for *water flow* but no magma. In the diagram to the right, the "Partially Molten Magma" chamber was once red (seen in Subchapter 5.11). In this revised diagram, USGS researchers changed the color to white after failing to discover magma.

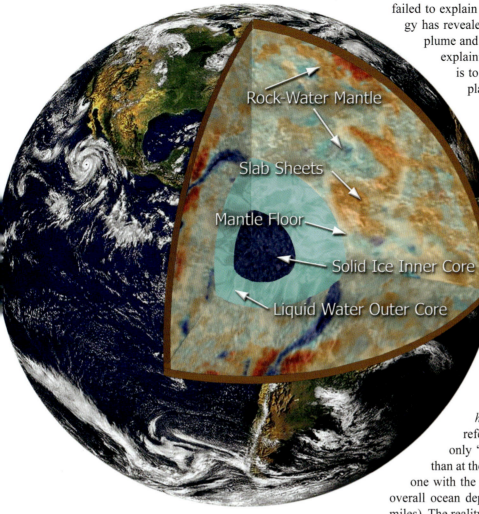

Fig 7.6.4 – With a magmaplanet paradigm, researchers find that "something *funny* does happen about 2000 kilometers down…" New observations, like giant slab sheets under the continents have only continued to support the Hydroplanet Earth Model. The arrows identify boundaries between different water and ice-filled areas inside the Earth.

they were not melted, even being near the 'hot' core:

"The sharper seismic view confirms that **two great slab sheets**, one hanging **beneath the Americas** and the other **beneath southern Eurasia**, plunge deep into the mantle, reaching depths of at least 1600 kilometers. And as in the earlier images, the bottom 300 kilometers of the mantle seems to hold **slab material that has accumulated in broad piles on the mantle floor**. In both the old and new images, the slabs seem to vanish between these two layers. Some researchers blamed the disappearance on poor resolution in the earlier images. But the new images suggest that the massive slabs do disrupt about 1800 to 2000 kilometers, melting away into smaller scale features, **only to reappear near the mantle floor**. 'Something funny does happen' about 2000 kilometers down, says seismologist Kenneth Creager of the University of Washington, Seattle. **'It's suggesting some new phenomenon.**'" Note 7.6p

"Something funny" is going on 2000 kilometers down and modern science's magma theories are unable to explain it. The tectonic plate theory has failed to explain many of the latest discoveries new technology has revealed about the interior of the Earth. The mantle plume and uplift theories have also been unsuccessful at explaining observations like these. What is needed is to reexamine the evidence based on the Hydroplanet Model.

The Equatorial Bulge Evidence

Up to this point, evidence supporting a Hydroplanet Earth has been based on the presence of substantial intercrustal and sub-crustal water as well as a water-based core. The misshaped Earth provides us additional evidence with a phenomenon called the Equatorial Bulge. One geophysicist explained:

"The assumption of **a liquid earth** is, even on the face of it, **far from silly**, since a good fraction **of the earth's surface** is covered by water. This **bulge** would have the effect that a body at sea level at the equator is a **little farther from the center of the earth than one at sea level at the poles**…" Bib 63 p58

What this geophysicist underplays is just *how big the bulge is*. Notice the geophysicist referred to the sea level at the equator as being only "a little farther from the center of the earth" than at the poles. This underwhelming statement leaves one with the idea that the bulge is a *small fraction* of the overall ocean depth, which averages about 4 kilometers (2.5 miles). The reality is that *the Earth's bulge is **43 kilometers** (27 miles)!*

"**This buldge results from the roation of Earth, and causes the diameter at the equator to be 43 kilometers (27mi) larger than the pole-to-pole diameter.**" Note 7.6q

How does a 43-kilometer bulge near the equator exist if the bulge only came from a 4-kilometer deep ocean? You will recall the tallest mountain on Earth (Mt. Everest) is 9 kilometers (5 miles). The 43-kilometer bulge is obviously not just

Spherical Earth

Equatorial Bulge

Fig 7.6.5 – The Earth is not a perfect sphere. The image on the right depicts the 27 mile (43 km) Equatorial Bulge of the Earth. The bulge has been exaggerated to make it easier to see. The oblate spheroidal shape of the Earth is evidence of liquid in its interior. The question is of course—what liquid? The Magma Pseudotheory chapter challenges the magma paradigm. Newly discovered, large-scale mass redistributions testify that the Earth's liquid interior is water.

surface-ocean related. Moreover, the bulge would not be present if the Earth was solid. An Earth-sized solid object spun at the relative speed of the Earth would not exhibit bulges. The Earth's Equatorial Bulge *must* originate from liquids inside the Earth.

> "The Kola well and our own have shown that **a deep crust of dense, hot rock is definitely not the case**... **There are large amounts of highly saline brine in the crust that migrate**, carrying metals around and depositing them as minerals."
>
> *Science*, Vol. 261, 16 July 1993, p296

Scientists have not recognized the true nature of the liquid and have tried to solve Nature's Puzzle using molten magma. One such unanswered question about the Equatorial Bulge is troubling from a magmaplanet point of view because the Bulge does not stay constant. We will learn later in the UM that the Earth experiences "large-scale mass redistributions," and that Earth's gravitational field is constantly changing as large masses move, redistributing weight. Researchers are confused with new observations from sensitive satellites because the new data does not harmonize with the magmaplanet theory.

The Earth is an oblate spheroid, or 'pear shaped,' with the equatorial bulge lying just south of the equator. Why is this? The answer is easily understood as we take into account the composition and location of the continents with respect to the Earth as a whole. Most of the Earth's landmasses lie in the northern latitudes. To compensate for the additional mass north of the equator, more water must move below the equator, pulled by centrifugal force to an equatorial mass balance. The laws of physics dictate that a free moving material, such as water, both surface and subterranean, will shift to balance the mass of the entire Earth.

Does the balance ever change? It would not if the external gravitational forces on the Earth were constant, but they are not. The Sun's rotation, the Earth's elliptical orbit around the Sun, the Moon's elliptical orbit around the Earth and other celestial bodies cause Earth's gravitational field to be in a constant state of change. Visual evidence of this ever-changing gravitational environment is well known to Earth's oceanside dwellers and mariners. It is long known that the ocean's tides are caused by the tug of the Moon, and recently discovered is Earthtide, the daily 8-12 inch (~20 cm+) swelling and subsiding of the crust. Another overlooked and vastly important **subocean tide** occurs deep within the Earth. The subocean tides' impact the Earth's weather system in ways meteorologists have never imagined. The **subocean tide** will be discussed in more detail in the Weather Model chapter, later in the book.

The Deeper We Go—The Wetter It Gets

Fig 7.6.1, located at the beginning of this subchapter, is a representative diagram typical in geology textbooks illustrating how underground water decreases with increased depth. The supposition is that depth increases pressure and closes pore spaces, squeezing out water and preventing the influx of new water. The following dramatic example proving that the 'dry Earth' theory is totally incorrect comes from the journal of *Science*:

"**Water from the stone.** The **brines** are another **surprise that is opening researchers' eyes** to the merit of deep drilling. 'When I started 25 years ago, **the idea was that the deeper you go into the crust, the drier it gets**,' says Kehrer. Conventional wisdom had it that kilometers of overlying rock squeeze shut any cracks, cutting off the fluid flows that deposit ores and chemically alter the rock at shallower depths. But after the drill bit had penetrated more than 3 kilometers of dry rock, it broke into **water aplenty**. Core samples retrieved from 3.4 kilometers were veined with open cracks more than a centimeter wide that had presumably carried fluids. That was only a hint of what was to come at 4 kilometers, where more than half a million liters of a gas-rich, **calcium-sodium-chloride brine** twice as

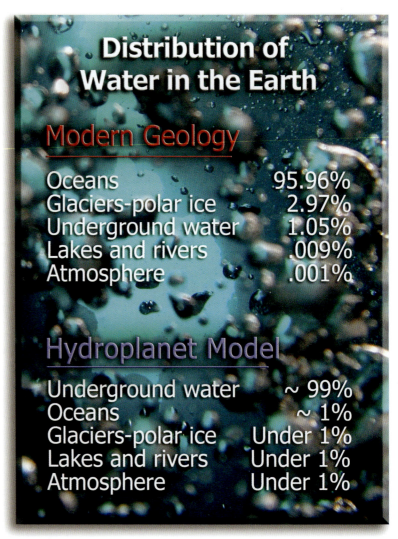

Fig 7.6.6 – What is the distribution of the water in the Earth? Modern Geology textbooks (Bib 59 p288) show that oceans make up almost 96% of Earth's water with glaciers, underground water, lakes and atmosphere making up the rest. However, the latest scientific data support a Hydroplanet Model where the vast majority of Earth's water is underground water (over 99%). With a water/ice core and water or ice disbursed throughout the mantle and crust, significant amounts of water can be accounted for.

concentrated as seawater poured into the well. Abundant fluids gushed from depths as great as 6 kilometers. 'This has been a real sensation,' says Kehrer. 'The surprise is that there are fluids of *that* amount.'

"Geophysicists had had some earlier hints from the Kola Hole [Russian hole], where drilling 'mud' pumped into the hole to lubricate and cool the bit returned to the surface with its chemistry subtly altered. The Soviets took the changes as evidence that brines were flowing into the hole from surrounding rock, but western scientists remained unconvinced. Now the obvious inflows in the KTB hole give more weight to the Kola claims. **'The Kola well and our own have shown that a deep crust of dense, hot rock is definitely not the case,'** says Kehrer. **'There are large amounts of highly saline brine in the crust that migrate, carrying metals around and depositing them as minerals.'"** Note 7.6r

Claims of "large amounts of highly saline brine in the crust" or that a "deep crust of dense, hot rock is definitely not the case" are certainly not what one would expect to hear on the Discovery Channel! One researcher affiliated with the KTB borehole noted:

"'You normally think the overburden [of rock] squeezes the cracks closed,' says Karl Fuchs of the University of Karlsruhe, **'and the hole would become drier [with depth], but it was just the opposite.'** Because such fluids carry metal-bearing minerals, the finding is altering many geophysicists' picture of ore formation." Note 7.6s

How much more evidence is needed to lay to rest dogmagma—the dogma of magma—for the *reality* that the deep Earth is "just the opposite," a Hydroplanet. Because of data and observations acquired from deep boreholes, researchers recognize that reality and current ideology expressed in geology textbooks are not the same:

"The science coming from the hole means **'the geology textbooks will have to be rewritten'** – Peter Kehrer" Note 7.6s

Rewriting the Geology Textbooks

How exhaustive will the textbook rewrite need to be? Most of modern geology has been based on pseudotheories, so large portions of the geology textbooks are wrong. Take for example, the distribution of water in the Earth. Fig 7.6.6 illustrates how substantial the difference in water distribution is between the Hydroplanet Model and modern geology. The amount of deep-Earth water dramatically changes the percentages and while the exact nature of the distribution is not yet known, it is significant enough to affect every field of natural science that deals with our planet.

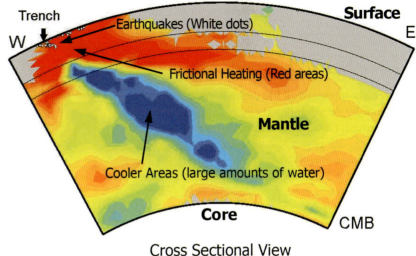

Tomography of Central America

Fig 7.6.7 – This is a crosscut view of the Earth's Mantle. In the Magma Pseudotheory, tomography evidence proved to be a 'smoking gun' showing that magma could not exist because temperatures inside the Earth do not rise from the core to the surface. Temperatures are seen to rise at plate boundaries, where friction is taking place (red areas). This is evident in the above diagram. White dots identify earthquake activity and occur where continental plates meet. Seismic activity and Earthtide heats the area through gravitational friction. This adapted diagram comes courtesy of the detailed work of Rob Van der Hilst.

The Tomography Evidence

The Smoking Gun of the Magma Pseudotheory was the evidence from Seismic Tomography. Sound waves help scientists recognize characteristics and relationships between liquid and solid material in the Earth's mantle. The Tomographic Evidence does not substantiate a magma mantle. One reason is that magma—molten liquid rock—presumably moves via convective currents and magma plumes. However, seismic evidence shows the bulk of the mantle is *not* a liquid:

"Seismic and petrological evidence indicates that **the bulk of the mantle is a crystalline solid.**" Note 7.6t

This apparent conflict between observation and pseudotheory has given way to many anomalies that remain a mystery in modern geology, but can be explained through the Hydroplanet Model. There are large areas of the mantle that are not solid and are simply water.

Many years ago, seismic waves originating from earthquakes demonstrated that some parts of *the center of the Earth* were *liquid*. Some seismic waves—P waves—will travel through both liquid and solid while others—S waves—will not travel through liquid. When researchers observed that certain seismic waves did not pass through the outer core, they knew it was liquid—they just did not know what kind of liquid. Wave velocity and other attributes have made it possible to discern more about the nature of the liquid deep in the Earth. Questions such as the viscosity of the liquid could be answered. Scientists reporting in the *Journal of Science* drew this compar-

> "The viscosity of the **liquid outer core** is comparable to that of **water**..."
>
> *Science*, Vol 288, 16 June 2000, p2007

ison about the "liquid outer core" of the Earth:

"The viscosity of the **liquid outer core** is comparable to that of **water**..." Note 7.6u

The answer seems so easy, as we think about Earth's several thousand cubic miles of naturally occurring and observable liquid.

7.7 Earth's Hydrology Redefined

Water is a substance all life depends upon. Too little and death by dehydration sets in, too much and lives are destroyed. Humans have learned to appreciate waters' unforgiving power and have learned to live with its sometimes-unpredictable whims. Hydrology is the study of water's distribution, movement and impact on humans, but in modern science, it only addresses the water in the atmosphere, on the surface of the Earth, and water just below the surface. Having been unaware of the existence of water in the mantle and Earth's core, Earth's Hydrology has been only partially defined.

Water's journey runs longer and deeper than previously known and this chapter continues by redefining hydrology to include those heretofore-unknown elements of the hydrologic cycle.

Hydrology—a New Definition

From an encyclopedia, modern hydrology is defined as follows:

"**Hydrology** is the scientific study of the movement, distribution, and quality of water on Earth and other planets, including the hydrologic cycle, water resources and environmental watershed sustainability." Note 7.7a

This definition defines the water cycle of a planet with a supposed liquid center of molten rock, but without magma, it must be reexamined. Previous understanding about the hydrologic cycle has been based on the assumption that water resources are limited to water in the atmosphere, surface water held in oceans, lakes, rivers, and ice, and groundwater. No provision has been made for water in the mantle or core, and that must change. This subchapter expands the original definition of hydrology to include this water source.

> **Hydrology:** The study of Earth's water, its movement and distribution, including atmospheric water, surface water, groundwater, and sub-crustal water in the mantle and core, and the study of the changes and effects of water, both past and present.

In addition to the commonly known processes of **Hydrology,** such as precipitation, evaporation, and runoff, the study must include hyprethermal environments and redistributions of deep-mantle water.

It is Difficult to Fracture Honey

The first step in understanding Hydrology is to recognize the origin of earthquakes. Throughout the crust, many water channels or aquifers were created or changed by earthquakes, some of which extend deep into the Earth. In the Magma Pseudotheory, subchapter 5.10, one of the evidences that disproved the existence of magma was the occurrence of deep earthquakes. The mechanism for deep earthquakes "remains speculative"

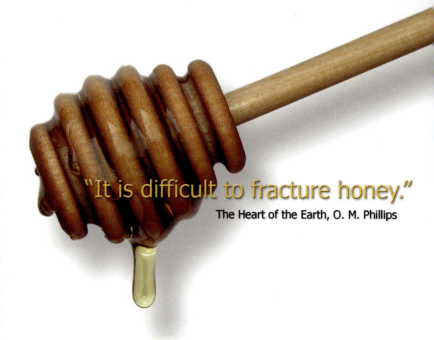

"It is difficult to fracture honey."
The Heart of the Earth, O. M. Phillips

due to the assumption that high pressures and temperatures existent in the Magma Pseudotheory *infer* a 'honey-type' of thick, molten rock at depth. Geophysicists, long knowing that molten rock does not crack have been perplexed how soft magma could break and fracture. In 1968 one geophysicist, O.M. Phillips wrote:

"If, as we have surmised earthquakes are the result of a sudden local fracture, a release of accumulated stress, then it follows that **considerable rigidity is required** of the material before an earthquake can be generated. **It is difficult to fracture honey. The fact that earthquake centers are sometimes found to depths as great as 700 km indicates that the material at these depths must still be rigid enough to support a gradually accumulated stress without flowing, until finally the stress concentration becomes so great that it breaks.**" Bib 63 p111-2

Not unlike honey in its viscosity, magma would not break or fracture. The geophysicist didn't explain how Earth's deep-centered earthquakes occur in magma. Other researchers have come up with various sci-bis in an attempt to explain the unexplainable. In each case, deep earthquakes are not earthquakes "as we know them." One researcher reported in the journal *Science*:

"Faults form by brittle failure near the surface, but increasing pressure (hence friction) with **depth limits faults to the upper ~15 km of the plates**, which are some tens of kilometers thick." Note 7.7b

Through personal email (October 7, 2003), the researcher was asked to confirm his statement that "faults and thus earthquakes only happen in the upper ~ 15 km of the crust," to which the geophysicists responded:

"**This is a complicated issue**; earthquakes do occur deeper, but probably not on faults 'as we know them'.

"How they can occur so deep **is mysterious**..." Note 7.7c

Deep earthquakes remain "complicated" and "mysterious" because it is impossible to imagine how to fracture magma. One incident that continues to confuse researchers is the observation that deep earthquakes exhibit many of the same characteristics present in shallow earthquakes, namely slippage:

"Actually the first motions of a deep earthquake are downward in some areas and upward in others, **just as they are in** shallow earthquakes. The upward and downward motions are segregated, **as if part of the earth had moved in one direction along a slip plane and the other part had moved in the opposite directions**; it is the same pattern observed in seismograms of shallow events. Furthermore, in deep earthquakes as in shallow ones the S waves are much stronger than the P waves, **which points to slip** rather than an implosion as being the source." Note 7.7d

Of course, liquids do not "slip," and so for decades, deep earthquakes are unexplained and remain a mystery. During the last decade or so, new laboratory technology and field observations have allowed researchers new insight into the cause of deep fractures from an unlikely source—*Icequakes*.

Icequakes

Without magma, what might the temperature be deep inside the Earth? Humans are accustomed to the cozy 'room temperature' of Earth's surface, warmed primarily by the Sun's rays and insulating atmosphere. Without the atmosphere or the Sun's life-giving rays, the temperature would quickly drop several hundred degrees. When the space station drops into the Earth's shadow zone, temperatures plummet to minus two hundred degrees, as the Sun is taken out of the picture.

Without the Sun, the temperature of Earth's interior would drop to several tens of degrees below zero, except for places where gravitational-friction heating is taking place. There is no evidence of radioactivity sufficient to heat the interior, (see subchapter 5.6) and pressure alone cannot justify a hot interior. Pressure is an important factor to account for inside the Earth and future research will likely show a *reduction* in pressure near the center of the Earth as the gravitational forces come to near equilibrium. This change in gravity and pressure may well be responsible for the phase change of the water from solid ice to liquid, at the core.

By discarding the magmaplanet philosophy, cold temperatures inside the Earth should be self-evident, although the thought may never have occurred before. Magma theory cannot explain earthquakes at depths of 680 km, but some research and experiments have demonstrated how deep-Earth fractures *can* occur. One such process was reported in the *Journal of Geophysical Research*, in the article ... *Physical Mechanism of Deep Earthquakes*. The research explains how deep earthquake fractures can occur *in ice*:

"*Durham, Heard and Kirby* [1983] discovered a very unusual form of high-pressure faulting in constant-strain-rate triaxial compression tests on polycrystalline **ice I_h deformed at low temperatures and comparatively high pressures. At low confining pressure P_c, ice I_h fails by ordinary fracture and the failure strength increases steeply with increasing confining pressure**. Above a critical pressure, which depends on temperature and strain rate, **ice I_h fails by faulting having very unusual characteristics...**" Note 7.7e

Not surprisingly, the behavior of ice under pressure and low temperature, explains deep earthquakes. In addition, the phase changes of ice under pressure can also explain the "maximum depth of earthquakes at about 680 km." The *Geophysical Research* article continues:

"Finally, if this is the earthquake source mechanism in deeply subducted lithosphere, then earthquakes should shut off when all the polymorphic phase changes have occurred. **This may explain the maximum depth of earthquakes at about 680 km**." Note 7.7e

It seemed that for the first time, geology had a real mechanism for explaining deep earthquakes. Ice under pressure has unique properties that allow for fracturing even with increasing pressure. Solid water-ice at pressures found deep inside the Earth, explained deep earthquakes. Moreover, when phase changes occurred or during recrystallizing events, researchers noticed "cracking or snapping noise—laboratory analogues of earthquakes":

"Stephen H. Kirby of the U.S. Geological Survey has proposed a deep-earthquake mechanism that depends **on phase transitions** but, in contrast to earlier proposals, results in **slip** rather than implosions. As a surrogate for actual mantle rock Kirby and his colleagues studied **ice**...

"When the workers compressed each material to a pressure slightly below that of the normal phase transition and subjected it to shear stress, they found that the phase transition was

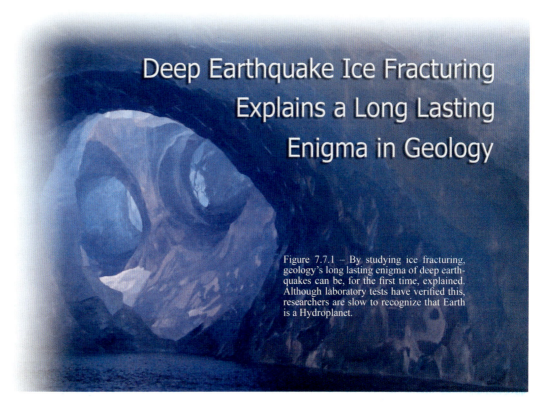

Figure 7.7.1 – By studying ice fracturing, geology's long lasting enigma of deep earthquakes can be, for the first time, explained. Although laboratory tests have verified this, researchers are slow to recognize that Earth is a Hydroplanet.

triggered along a thin layer parallel to the stress, the sudden rearrangement of crystal structure along the layer apparently weakened the material, allowing it to slip. Kirby and his colleagues noticed that in the process their samples emitted **cracking or snapping noises—laboratory analogues of earthquakes."** Note 7.7f

With such solid results, it would again seem the Hydroplanet model would become reality, but it was not to be. The ice was studied to understand its mechanical properties as it exists elsewhere in the solar system but it served only as a substitute for mantle rock on the Earth:

"**No one has yet shown** that shear stress has the same effect on phase transitions in **actual mantle rock** that it has on transition in **ice**..." Note 7.7f

Ice is the most promising substance to explain events occurring deep in the Earth. When compared with solid mantle rock, *ice* is the *only* substance that exhibits phase transitions and evidences the "shear stress" seen in deep earthquakes. Other rock minerals, melted or not, do not have the phase transitions ice does under high pressure and low temperature. The Deep Earthquake Evidence in subchapter 5.10 refuted the magma theory, but that same evidence supports the presence of ice in the Earth's interior.

Breaking the Ice Barrier

With geophysicists so fully entrenched in dogmagma, it was curious to see a discussion centered on ice in Earth's interior in a major science journal the likes of *Nature*. Yet it was:

"**Large amounts of pure ice** VII might accumulate **during subduction and, as a sinking slab warms**, eventual melting of the **ice** would **release large amounts of water in a small region over a short period of time**, with a significant positive volume change." Note 7.7g

Perhaps this was the 'Breaking of the Ice Barrier' that long impeded advances in the discovery of the makeup of Earth's interior. This article was published in 2000 and it will take time for others to see the world as a Hydroplanet—but it will happen. Discussing ice is a good first step. Once the 'ice is broken,' it will be much easier to make new discoveries. Before long, the Earth's Hydroplumbing System will be known to contain massive amounts of both water and ice. Just as the researchers noted, when areas melt because of frictional heating, it is only "a short period of time" before water pressure increases, affecting other areas of the crust. This in turn affects a change in the whole water cycle and in the weather we experience. Details of this process will be discussed in the Weather Model.

7.7.2 – The earth has a hydroplumbing system that is influenced by earthquakes that occur throughout the crust. Yellowstone Park in Wyoming is a good example of how frequent earthquakes heat subterranean and produce geysers. There are many examples like this, worldwide.

The Earth's Hydroplumbing System Confirmed

Since the Earth is a hydroplanet as opposed to being a magma-planet, there should be evidence with respect to how the Earth's inter-crustal thermal water is heated through earthquake activity instead of magma. If magma was the cause of the heating of subaerial geothermal water sites, how does it cause an *immediate* region-wide increase in geothermal water temperature? If earthquakes heat underground waterways through Gravitational-Friction, rapid and widespread effects will be apparent and could affect *large* areas of geothermal water over a very short period of time. *Science* reported such an event in May of 2002:

"**Within a few minutes** of the 1992 magnitude 7.3 earthquake in Landers, California, hydrological changes occurred at subaerial **geothermal sites throughout the western United States**, including changes in reservoir level, fluid temperature, and discharge rate." Note 7.7h

The hydroplanet model, earthquakes, and gravitational heating can explain geophysical observations happening just below our feet, but how extensive is the Earth's Hydroplumbing System? The system stretches over such large distances it encompasses whole continents. In November 2002, a huge 7.9 earthquake struck Alaska's Denali fault causing water pressure to increase near the fault in Alaska, and all the way to California and Wyoming. The story was told in *Scientific American*:

"The Denali temblor is the third major earthquake in the West in the past 10 years known to have caused smaller quakes. The other two were in southern California: the Landers earthquake in 1992 and the Hector Mine quake in 1999. **All three quakes affected the same geothermal volcanic fields in Wyoming's Yellowstone National Park, Mount Rainier in Washington State, and several sites in California**. These fields, which are hot springs **fueled by magma** roiling deep underground, normally rumble at low levels. But the secondary quakes that were triggered far exceeded the background seismicity, and researchers aren't quite sure why." Note 7.7i

Researchers have long thought the hot springs in these areas were "fueled by magma" but there is no magma. What is evident is that there is a worldwide network of aquifers. Changes in fluid pressure come from seismically heated water, not magma chambers. The article in *Scientific American* continued:

"In Greece, Brodsky has found that **hot springs are fueled not by a magma chamber but by changes in the pressure of fluids coursing through underlying crystalline rocks**." Note 7.7i

In 2005, researchers from the University of New Mexico and Scripps Institution of Oceanography wrote an article titled *Continental-scale links between the mantle and groundwater sys-*

tems of the western United States. Here they described a mantle-connected Hydroplumbing system that extends throughout the western United States:

"Thus, these springs are the surface expression of groundwaters altered by **complicated mixing pathways** as well as the introduction of **mantle-derived fluids**, and we argue that **these processes are occurring within groundwaters throughout the western United States**." Note 7.7j

The new Hydroplanet Geology recognizes that geothermal systems are heated by earthquake activity, instead of magma. The Earth has a well-connected, Hydroplumbing System, and seismic events in one area can affect whole continents. One dramatic example occurred after the famous Sumatra earthquake of 2004. Researchers observed how this large earthquake triggered other earthquakes *across the world*:

"As surface waves from the **26 December 2004 earthquake in Sumatra swept across Alaska**, they triggered an 11-minute swarm of 14 local earthquakes near Mount Wrangell, almost **11,000 kilometers away**. Earthquakes occurred at intervals of 20 to 30 seconds, in phase with the largest positive vertical ground displacements during the Rayleigh surface waves." Note 7.7k

Silent earthquakes, Earthtide, and earthquake swarms constantly heat the Earth's geothermal systems, but now we can see that large earthquakes can affect geothermal and volcanic systems around the world:

"The recent Wrangell episode demonstrates that **great earthquakes can perturb geothermal and volcanic systems around the world**." Note 7.7k

Nearly everywhere we turn, geophysical studies supply evidence of Earth's Hydroplumbing System. One could easily write an entire book on the subject, based solely on the evidence that already exists.

The Tibetan Hydroplumbing System Evidence

The Tibetan plateau covers an area four times the size of Texas, 2.5 million square kilometers:

"The **Tibetan plateau** is the largest area of thickened and elevated continental crust on Earth..." Note 7.7l

With an average height of almost 15,000 feet (4500 meters), we should expect to find widespread evidence of water here, if the Hydroplanet Model is true. It takes a *liquid content* of *4 to 30%* in various rock types to stop seismically generated shear waves from traveling through the Earth's interior. Interestingly, a 1996 article in the journal *Science* gives serious consideration to liquid water:

"The **minimum fluid-filled porosity needed** to reduce the shear modulus of a rock to zero is about **4%**, in the case that the fluid is distributed in thin grain-boundary films or high aspect-ratio cracks. Other pore geometries require a substantially greater minimum fluid fraction (**10 to 30%**) to drive the effective shear modulus of the material to zero. This condition implies that the material producing the strong P_xS reflections beneath the Yadong-Gulu rift is either partially molten [that is, contains magma] or **has significant water content**." Note 7.7m

Geophysicists acknowledge that there is either magma or a significant amount of water in the Tibetan crust, by way of seismic evidence that cannot be ignored. A water quantity of this magnitude is thousands of times more abundant than has ever been stated before. These researchers have certainly not read the Magma Pseudotheory chapter, but they have begun to see the writing on the wall that water plays a role magma cannot fill. The researchers continue:

"If the fluid is **meteoric water**, then the **hydrothermal system**, active at the surface, **has to extend to depths of ~15 km over a wide area of the Yadong-Gulu rift**..." Note 7.7m p1691

Researchers cite no mechanism for surface water to reach depths of 15 km (9 miles). In actuality, there is no need to, as water would already be there. Geophysicists have missed this because of dogmagma. Explanations of surface water seeping into a near impenetrable Earth are unnecessary. There is ample evidence of deep water coming to the surface.

Additional proof of the Tibetan Hydroplumbing System comes from a group of researchers reporting in the journal of *Science*, in an article titled *Detection of Widespread Fluids in the Tibetan Crust by Magnetotelluric Studies*. Fifteen researchers from universities around the world used the latest technological instruments to measure electrical conductivity in the crust. Their findings conclude that the thickest crust in the world contains *widespread fluids*:

"Such pervasively high conductance suggests that partial melt and/or **aqueous fluids are widespread within the Tibetan crust**." Note 7.7n

The author suggests the fluids within the Tibetan crust are "widespread" and consist of either "partial melt" or "aqueous fluids." His thoughts are to be expected, given the magma paradigm. However, the article continues by acknowledging that aqueous (water) fluids are more than ten times as conductive as any partially melted rock would be. They explain that while

Fig 7.7.3 – Fluid flowing into faults is another factor that can trigger earthquakes. The injection of water into boreholes has been shown to cause shallow earthquakes. This photo captures damaged buildings around the Balaju Area during the Gorkha Earthquake in Nepal in April 25, 2015.

1.6 km (~1 mile) of rock containing 10% water would generate the observed conductivity of 10,000-S, more than *16 km of rock containing a 10% melt* would be required to achieve the same conductivity. Such a large quantity of melted rock has never been observed before.

Another reason the melt alternative fails; the heat flow readings do not match the conductivity readings. Areas of greater melt would exhibit greater heat flow. In the Magma Pseudo-theory chapter, this was confirmed not be the case. Moreover, magma-theory plumes thought to originate from near the core of the Earth have been disproven, leaving no real alternative for the 'melt' option.

Left with the most probable choice—widespread aqueous fluids—researchers confirm, with mounting evidence, the Earth's Hydroplumbing System.

Hydroquakes

Researchers are finding evidence of trapped fluids, i.e. 'water' in the crust. This has lead to research into how water is involved with earthquakes. At shallower depths, water acts as a lubricant in fissures, allowing for fault slippage:

"Laboratory work has shown that at pressures equivalent to **shallower depths, fluids trapped in rock pores can counteract the forces binding a potential fracture, allowing it to fail at a lower shear stress than before**. In at least one case, at the Rocky Mountain Arsenal near Denver, a sequence of **shallow earthquakes occurred after fluid wastes were injected into the earth**, apparently lowering confining stresses enough for rock layers to **slip**." Note 7.7o

Water is a contributor to earthquakes and earthquakes change the water below the ground and above the ground. In Chapter 9, in the Weather Model, we will discuss related scientific phenomena—Earthquake Clouds.

At present, we do not know how much rock, water, and ice is in the Earth's interior, but as technology improves, new discoveries will give us a much more detailed picture of the Earth as a hydroplanet. One fact is certain, water played a pivotal role in the formation of the Earth's minerals and it continues to play a critical role in the Earth's seismicity.

Kobe Earthquake Evidence—Fluids at Hypocenter

In subchapter 5.4, we reported that geophysicists speculate that earthquakes are *caused by magma*. Since magma does not exist, it cannot cause earthquakes. Instead, we should find evidence of water being associated with earthquakes:

"Near Kobe there is **no active volcano**, and heat flow studies revealed **no significant lateral changes in temperature before the earthquake**. Therefore we suggest that the anomaly at the Kobe hypocenter is **not related to a magma reservoir, but rather to the presence of fluids in the crust**." Note 7.7p

The 1995 Kobe, Japan Earthquake has been one of the most-studied quakes because it took place in a populated area that was monitored with the latest in seismic technology. Researchers observed and recorded data from thousands of aftershocks and microquakes. They created a 3-Dimensional model to understand what was taking place below the ground of this devastating quake. What they found was not magma—but "fluids":

"Potential **sources of the fluids** may be dehydration of minerals, **fluids trapped in pore spaces, and meteoric water**."
Note 7.7p p1893

They speculated about the origin of the water but unfortunately did not consider the possibility of fluid transport through hydroplumbing sources. Each year yields additional evidence establishing the Earth's hydroplumbing system. The fact that water, not magma was recognized as being the chief mechanism in the cause of the earthquake is an amazing turnaround in the geosciences.

The Japanese are not new to water related earthquakes. Because of population density in this seismically active earthquake zone, Japanese earthquakes are highly researched. It is known that earthquakes follow a seasonal cycle that corresponds to snowmelt and the lifting of the crust from the reduced weight. Perhaps Japanese seismologists will recognize the effect of water beneath the crust.

Hydrothermal Precursors to Earthquakes

One question that researchers diligently seek to answer is whether earthquakes have an observable precursor or some geological sign before they occur. In an article in *Science* titled, *Detection of Hydrothermal Precursors to Large Northern California Earthquakes,* investigators recognized just such a precursor:

"During the period 1973 to 1991 the interval between eruptions from **a periodic geyser** in Northern California **exhibited precursory variations 1 to 3 days before the three largest earthquakes within a 250-kilometer radius of the geyser**. These include the magnitude 7.1 Loma Prieta earthquake of 18 October 1989 for which a similar preseismic signal was recorded by a strainmeter located halfway between the geyser and the earthquake. These data show that at least some earthquakes possess observable precursors, one of the prerequisites of successful earthquake prediction." Note 7.7q

If we can begin to understand that instead of a non-cyclical magma interior, the Earth's hydroplumbing system responds to the diurnal activities of the Moon, Sun and other periodicities, and that earthquakes are directly linked to this hydro-system, we will see that periodic geysers and earthquakes are related.

"Near Kobe there is no active volcano, and heat flow studies revealed no significant lateral changes in temperature before the earthquake. Therefore we suggest that the anomaly at the Kobe hypocenter is **not** related to a magma reservoir, but rather to the presence of fluids in the crust."
Science, Vol. 374, 13 December 1996, p1892-3

From Magma to Water Boreholes

The bold attempt to reach magma in California's Long Valley—introduced in subchapter 5.11—fell short of its goal, producing a 'water borehole' instead. The published article, summarizing phase III of the project spoke of losing drilling fluid circulation at a depth of 8540 feet. The article spoke also of rocks with 'open veins containing crystals that show evidence of present or past water flow.' When asked if the drilling fluids flowed into some kind of underground water system, a researcher affiliated with the project responded:

"The well encountered **a large, open fracture system at this depth**. We do not know the dimensions or extent of the system. The water in the fracture is relatively cool—102 C. **Open fractures are not uncommon**, even at such a great depth, as we have learned from the German deep drilling experience." Note 7.7r

According to good old-fashioned magma thinking, high temperatures and pressure would close off all such open spaces deep within the crust. However, open fractures are not only common; they are frequently filled with water.

As the drilling drew progressively closer to the supposed magma chamber, the temperature should have continued to rise, even with the presence of water. It did not and we wondered why. The previously consulted researcher answered our query:

"**We do not fully understand why** the temperature in the well stopped increasing below about 6500'. It is unusual for there to be no positive temperature gradient over large depth intervals in the Earth. An isothermal depth profile like the one found in LVEW likely **requires a combination of lateral flow and upflow of water**." Note 7.7r

Why do the researchers "not fully understand" what they are seeing? Their conclusion regarding the temperatures in the borehole suggested a lateral flow and an "*upflow of water,*" concepts not compatible with a magma chamber mindset. The Earth's relatively cool water hydroplumbing system is not understood—yet.

In subchapter 5.5, an extraordinary claim that magma would have to defy heat flow physics (see Fig 7.7.4) and that established laws of thermodynamics refutes the existence of magma but support a hydroplanet Earth. Heat from magma should be coming more readily through the thin oceanic crust in comparison to the thick continental crust, but just the opposite is true. Geologists have no clear answer why. In a *Geophysical Research Letter*, researchers acknowledge:

"...**observed seafloor heat flow anomaly**, the '**missing' conductive heat flow** through oceanic lithosphere..." Note 7.7s

The only reason a "seafloor heat flow anomaly" exists is *dogmagma*. If the thin sub-oceanic crust resided atop a magmatic mantle, conductive heat flow should be widely evident, but it does **not** exist, and researchers know it is "missing." If crustal heat were derived from *earthquake-induced friction* instead of magma, thicker continental crust would produce greater heat than oceanic crust. Interestingly, the same researchers made other confusing observations with respect to heat flow when they drilled a borehole 2500 meters below sea level off the coast of California. After drilling through 250 meters of sediment, hard basalt basement rock was encountered. Ten meters further found temperatures consistently *decreasing* instead of rising:

"The temperature data indicate that the source for this warm fluid (the basement 'aquifer') was the upper 10 m of basement. **Borehole temperatures immediately below this interval are consistently lower**, indicating that warm fluid did **not** flow into the borehole **from greater depths**." Note 7.7s p1312

Just as they did in the Long Valley California borehole, researchers accumulated empirical data revealing Earth's enormous hydroplumbing system. The question again is—how big is the Earth's hydroplumbing system?

Oceanic Crust Hydroplumbing System

Could there be hydrothermal fluid beneath the oceanic crust? Geology holds that deeper in the crust; pore space is reduced or nonexistent due to increased pressure, severely hampering the water carrying capacity of the rock. This theory remains plausible only if the origin of crustal water is *above* the crust, not below it. Considerable evidence has been amassed that water exists deep within the crust.

Other evidence of a large underground hydrothermal water system comes from water beneath the ocean floor, frictionally heated by seismic activity. The extent of oceanic sub crustal aquifers is astonishing:

"Earthquakes also appear to influence this normally quies-

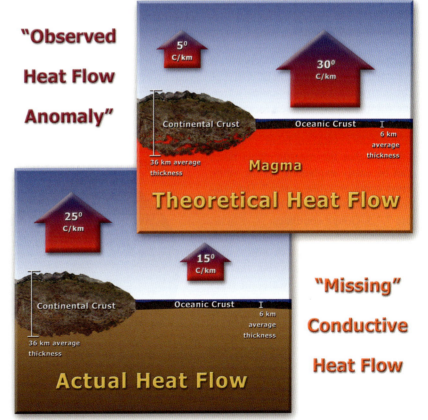

Fig 7.7.4 – Researchers have acknowledged that the observed temperature flow per kilometer through the oceanic crust is an "anomaly" and that the expected heat flow is "missing." This was shown to be the case in the Magma Pseudotheory chapter. Actual observations match Hydroplanet Model predictions.

cent environment. In the eastern Pacific Ocean, **earthquakes directly below and tens to hundreds of kilometers distant from active marine hydrothermal systems can change the temperature and flow rate of hydrothermal circulation, altering the biological and geochemical environment of the ocean crust**. …hydrothermal fluid circulation in the porous **ocean floor** is a thermally driven process within an aquifer that is more uniform and can be continuous **over thousands of kilometers**." Note 7.7t

This amazing report from the journal, *Science* attributes changes in both "temperature and flow rate" across thousands of kilometers of deep-ocean crust to earthquakes. This is only a part of Earth's worldwide Hydroplumbing system.

Continental Crust Water Evidence

The Earth's oceanic and continental crusts are linked through the Hydroplumbing system. Figure 7.7.5 represents an adaptation of a diagram that appeared in the journal, *Nature* in an article entitled *Seismic reflectors, conductivity, water and stress in the continental crust*. Researchers identified "two remarkable features of the continental crust" during research in Western Europe and North America. Deep seismic profiles show that there are many **reflectors** (short blue lines in Figure 7.7.5) below the 10-15 km depth in contrast to the upper crust, which exhibits fewer reflectors. Research at the KTB borehole identified similar patterns that proved to be *water-filled cavities*.

D. Ian Gough summarizes research reported in *Nature* in 1986 with this opening statement:

"The uppermost 10-15 km of the Earth's continental crust **differs in several geophysical properties from the lower crust**. The upper crust is electrically resistive, seismically transparent, contains nearly all intracontinental earthquake hypocenters and responds to stress elastically, with brittle fracture. The **lower crust** is **electrically conductive, contains many seismic reflectors**, is aseismic and shows ductile response to stress. I show here that the characteristics of both regions **can be explained if the entire crust contains saline water**, in separated cavities in the compressively stressed rocks of the upper crust, but in the lower crust forming an **interconnected film on the crystal surfaces**." Note 7.7u

Notice that the measurements taken can be easily explained, "if the entire crust contains saline water." The relative amount of water can be seen in Figure 7.7.5. Gough continues:

"The low resistivities at **mid-crustal depth** could be associated with carbon or with metal sulphides, but there is good reason to expect **water in fractures and pores throughout the upper crust, so that electrolytic aqueous solutions in interconnected cavities are the most likely widespread good conductors in the continental crust**. The fluid will be called **brine**, for brevity." Note 7.7u

Here is clear evidence that "aqueous solutions in interconnected cavities" exist throughout the crust and are widespread. Gough continues by explaining, "much of the lower continental crust may be a saturated environment":

"Geophysical evidence for the subduction of wet sediments has recently been published. It is thus reasonable to suppose that **much of the lower continental crust may be a saturated environment**, with excess water in equilibrium with a hydrated mineralogy unfamiliar at the surface. The suggestion that wet granitic rocks in the lower crust could account for high electrical conductivity there is not new." Note 7.7u p144

As Gough mentions, the recognition that water accounts for high conductivity readings in the crust is not new, but it is not widely accepted—yet. Dogmagma has been around for a long time and has influenced the interpretation of seemingly obvious phenomena. Much of geology is built on the foundation of magma and it will be a hard transition to move away from it. Any of the sciences—geology, paleontology, biology, cosmology, particle physics, astronomy—built on the faulty foundation of pseudotheories must eventually give way to new natural laws that can predict and describe nature better than the old theories.

The Empty Cavity Evidence

Most people have explored the dark depths of a cave or have seen pictures of them. The discovery of underground caverns is an ongoing and worldwide experience. The underground world of open-air (and other gases) cavities also can be home to underground lakes and oceans. The origin of underground water reservoirs will be discussed in the next chapter, but to visualize in a small degree the size of such cavities, we will consider the mystery of disappearing lakes. Worldwide, there are accounts of whole bodies of water suddenly vanishing. Peaceful and serene one day and gone the next. One example was White Lake in Bolotnikovo, Russia.

One day in 2005, in the early morning hours, 74 year old Fyodor Dobryakov went to fish in nearby White Lake. As he reached the lake, he could not believe what his eyes were seeing, and others he told would not believe until they too saw the vacant, muddy lakebed. He explains:

"'The ice was just hanging over an empty lake. I heard a

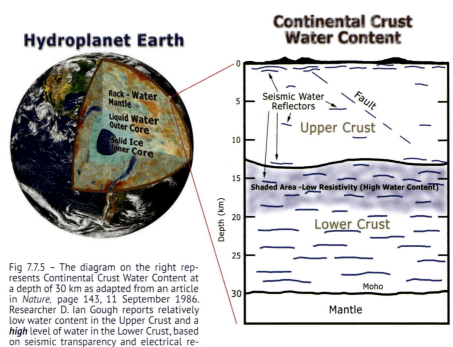

Fig 7.7.5 – The diagram on the right represents Continental Crust Water Content at a depth of 30 km as adapted from an article in *Nature*, page 143, 11 September 1986. Researcher D. Ian Gough reports relatively low water content in the Upper Crust and a **high** level of water in the Lower Crust, based on seismic transparency and electrical resistance data.

Miyake-jima Volcano Hydrothermal System

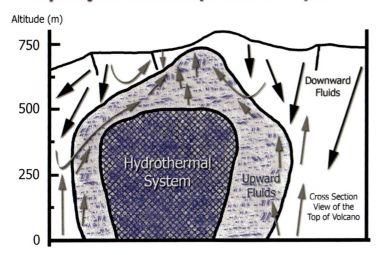

Fig 7.7.6 – This diagram is a cross section of the top of the Miyake-jima, Japan volcano prior to its eruption on July 8, 2000. Using electromagnetic instruments, researchers determined that the volcano did not have a 'magma chamber' but had a water chamber. The water chamber included a hydrothermal component evident at the top of the volcano, fed by water coming up from below. This diagram was adapted from the *Earth and Planetary Science Letters*, 205 (2003), p150.

noise, and when I looked right, I saw there was an abyss, and the water was rushing into the abyss like mad. The trees were falling into the lake and getting sucked in too," Dobryakov said at his home in this small village of pensioners, about 240 miles east of Moscow near the Oka River.

"Within minutes, all that was left of the 48-foot-deep lake was a silent expanse of mud a quarter of a mile wide. Small fish were flapping their death throes near the small, deep pool into which the water had disappeared." Note 7.7v

For so much water to drain so quickly, as it did down one hole, there had to have been a large channel leading to an even larger one. There are many deep cavities in the Earth and some are hard to find, even with seismic or sonar waves, but their existence is evident by the movement of vast quantities of water.

Many of the world's lakes are actually water-filled hydrocraters, remnants of once active hydrofountains and lying astride faults connecting a massive hydroplumbing system. Earth's constantly changing hydroplumbing system of underground open spaces can create vacuums and voids into which surface lakes sometimes drain.

The Miyake-Jima Hydrovolcano Evidence

On July 8, 2000, Japan's Miyake-jima volcano erupted, forcing the evacuation of several thousand residents from an island east of the Japanese mainland. Being in the midst of a highly active volcanic and seismic region, the Japanese have embarked on an extensive study of earthquakes and volcanoes. This has afforded the scientific community valuable data about **Hydrovolcanoes**.

Detailed monitoring of the island revealed to geophysicists that it was not a 'magma chamber,' but a **water chamber** that sat on top of the volcano's caldera:

"In 1995 a large-scale SP mapping of Miyake-jima has **revealed a powerful hydrothermal system centered on the Hatch-Tairo caldera**." Note 7.7w

Fig 7.7.6 details the identified hydrothermal system.

Investigators were beginning to see the importance water played in volcanic eruptions. Just four days before the 2000 Miyake-jima eruption, a "new hydrothermal source" appeared, setting the stage for the eruption:

"Four days before the July 8, 2000 eruption, SP surveys made along some of the 1995 ones **clearly show that a new hydrothermal source**, superimposed to the stable and global one observed in 1995, has taken place to the south of the O-Ana crater." Note 7.7w p152

This newly identified *water inside the volcano* did not appear suddenly but began increasing in the months prior to the eruption with "drastic changes" occurring during the 2 weeks immediately preceding the summit's collapse:

"On a larger scale, the comparison between 1995 and 2000 surveys **has shown a global increase of the hydrothermal activity beneath the volcano**. Its source could have been 250 m to the south of the crater. These observations suggest that the **hydrothermal system** was slowly disturbed in the months preceding the eruption while **drastic changes have occurred during the 2 weeks before the summit collapse when tectonic and volcanic swarms have appeared**." Note 7.7w p139

It was not coincidental that the actual volcanic eruption took place following an earthquake of magnitude 6.4, which released water, allowing it to flow down toward the epicenter, heated by *earthquake swarm friction* that appeared weeks before the eruption:

"The 550 Ω m Resistivity measured on the 1400m long line can be interpreted by the **opening of cracks and fissures in the**

Fig 7.7.7 – Miyake-jima is a Hydrovolcanic island located southeast of Tokyo, Japan. It has been studied extensively. Before the eruption on July 8, 2000, a large water chamber was discovered inside the volcano, above sea level. This water migrated down toward hot rocks and lava, heated by an earthquake swarm. This caused the steam explosion and plume seen in this NASA photo from the Miyake-jima island.

volcano, and the formation of a void above the sea level, between June 26 and July 3, when strong tectonic earthquakes of magnitude up to 6.4 occurred. **The ground water trapped in the hydrothermal system was progressively flowing downwards through the fissures and the cracks till a few kilometres depth where earthquakes were located**." Note 7.7w p153

Volcanologists have long wondered where the steam coming from volcanoes originates. The answer can be found as we analyze the observations of this and other eruptions. The Miyake-jima volcano erupted when water flowed toward the area "where earthquakes were located." When we understand how much pressurized water exists underground, and what happens when earthquake swarms occur, it is easy to comprehend what happens when the two are mixed.

The Miyake-jima Hydrothermal System in Fig 7.7.6 is an extraordinarily intuitive example of what takes place in most Hydrovolcanoes, an observation largely unnoticed in the geological community.

The Mt. Pinatubo Hydrovolcano Evidence

The largest explosion witnessed during the last 75 years was not a nuclear blast; it was the 1991 volcanic eruption of Mount Pinatubo in the Philippines! It has been only recently, since the western American eruption of Mount Saint Helens in 1980, that many scientific investigators and volcanologists have personally witnessed the largest eruptions known to mankind—Hydrovolcanoes.

Worldwide, volcanic activity is relatively commonplace. The words "volcanic eruption" conjures images of red-hot lava flowing down the side of a conically shaped mountain or splashes of molten rock. However, melted rock or lava is only a small part of the ejecta of the largest volcanic eruptions.

Hydrovolcano eruptions are violent because of the interaction of water and heat. When groundwater—subcrustal water at any depth—encounters significant heat, a phreatic explosion occurs. When water is heated sufficiently it expands exponentially, some 1,700 times its original volume.

Modern science has a term—phreato-magmatic—which is that rising magma contacts and heats ground or surface water, causing the characteristic phreatic eruption. Since magma does not exist, the superheating of phreatic water must be by some other source, frictional heating, a process introduced in subchapter 5.3, the Lava-Friction Model.

The enormous eruption of Mount Pinatubo showed evidence of the most important element of the Frictional Heat Law with the occurrence of "thousands of small earthquakes" just prior to the June 15, 1991 eruptive event:

"**Thousands of small earthquakes occurred beneath Pinatubo through April, May, and early June, and many thousand tons of noxious sulfur dioxide gas were also emitted by the volcano.**" Note 7.7x

In addition to the earthquake activity, scientists point out that sulfur dioxide gas was released. The discussion on the USGS web site made no mention of how much water was expelled as steam. In this quote from the same website, notice the ***type*** of ejected material that was emphasized:

"The second-largest volcanic eruption of this century, and by far the largest eruption to affect a densely populated area, occurred at **Mount Pinatubo** in the Philippines on June 15, 1991. The eruption produced high-speed avalanches of hot ash and gas, giant mudflows, and a cloud of volcanic ash hundreds of miles across." Note 7.7x

Again, there was no mention of water or steam. The only references to "steam explosions" are those that occurred several months before the primary eruption in June:

"… powerful **steam** explosions that blasted three craters on the north flank of the volcano." Note 7.7x

In the past, public references to volcanoes have focused on the 'red stuff;' attention grabbing and exciting, it makes for great Hollywood movies, but molten, flowing lava is actually rare.

Not so rare is the fact that many active volcanoes regularly exhibit steam plumes. Clues like these contribute to our comprehension of how volcanoes function and how heated water in the crust beneath our feet may be connected to the world's largest volcanoes.

Given the abundance of water in the Earth's crust, daily Earthtide, and frictional heating, what else is happening beneath our feet? The heating of subcrustal water can increase the humidity in the air and knowledge of this will have a tremendous impact on our ability to understand weather, in a way never before known—but that's another story, one we will cover in the Weather Model.

Volcanic Water Emission Rates Unknown

Melted rock will flow but it does not explode, unless there is gas under pressure within the lava flow. What gas would you guess to be the most common? If you guessed steam, you would be right! Water expands 1700 times when it becomes steam, as it is released from a volcano. This may seem so elementary, yet its implication is far from

Fig 7.7.8 – The 1991 Mount Pinatubo eruption was the largest explosion mankind has witnessed in the last 75 years, including nuclear explosions. Unlike the dust from dust storms, volcanic ash can stay suspended in the atmosphere for days because of **steam**. Until quite recently, the amount of water in volcanic emissions has been unknown. Scientists have yet to identify the source of the water emitted from hydrovolcanoes and in most cases, have not taken measurements of the water quantity. Hydrovolcanoes are another evidence of the vast amount of water lying within our planet.

understood.

Detecting the amount of water being emitted from volcanoes is not easy and most estimates are approximations. It was not until the year 2000 that new techniques were being successfully employed to more accurately estimate the amount of water in volcanic emissions. The October 2000 edition of the journal *Geology* gave this report:

"However, while **H_2O and CO_2 are by far the most abundant volcanic gases**, degassing rates are **difficult to measure in practice**. Global volcanic emission rates for CO_2 have only been approximated; **those for H_2O are unknown**." Note 7.7y

The researchers explained the difficulty in obtaining H_2O and CO_2 data was due to their high atmospheric background levels and the inability to get near enough to the plume. The four researchers who authored this article, Burton, Oppenheimer, Horrocks and Francis, developed a new technique to quantify the emission gasses and took measurements from several volcanoes.

Before this study, the percentage of water in volcanic gases was relatively **unknown**. This is one reason the USGS description of Mount Pinatubo's water emissions was so meager and one reason why volcanologists in general have not been able to comprehend the mysterious and violent workings of Hydrovolcanoes. The Hydroplanet Model affords an opportunity to place observations like this in context with other observations, offering a look into how water is the driving force behind the fury of super-volcanoes.

When these researchers recorded actual measurements of released gases, they revealed a surprising amount of water. Note 7.7y p917

Percentage of Water in the Gases of Three Volcanoes

Masaya 1999	Poas 1981	Momotombo 1980
94.26%	96.16%	97.11%

As can be easily seen, the vast majority of the gas emitted from these volcanoes was steam. This new observation led investigators to state that:

"**The ultimate source of the H_2O and CO_2 in the oceans and atmosphere is in the mantle**." Note 7.7z

This is a departure from the traditional idea that the mantle is composed of magma or melted rock. Water is not the first thing volcanologists expect to discover, but things are changing and knowledge of Earth's true hydrology is more important than ever.

Mt. Saint Helens Hydrovolcano Evidence

The catastrophic eruption of Mt. Saint Helens on May 18, 1980 changed the landscape of the state of Washington, and forever changed our understanding of volcanoes. Each year, thousands of visitors travel to Johnston Ridge Observatory to learn more about Mount Saint Helens. One of the displays inside the building is titled: *Where is the Lava?* No doubt this a question asked every day by many who visit the volcano. The answer given—*Characteristics of Volcanic Rocks*—goes on to describe

> "Global volcanic emission rates ... for H_2O are **unknown**."
>
> *Geology*, October 2000, p915

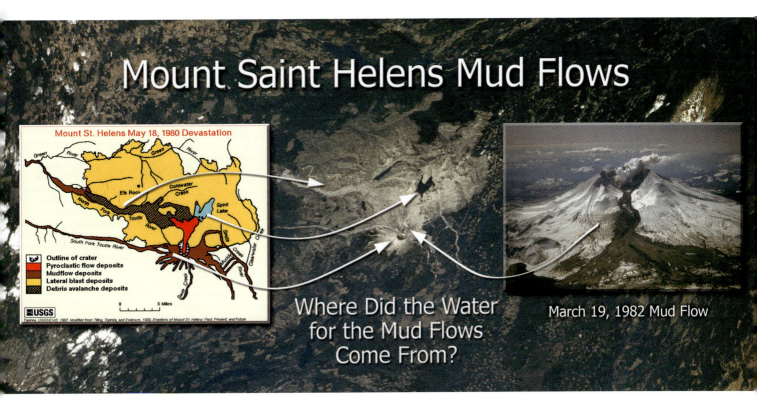

Fig 7.7.9 – A NASA satellite image of the Mt. Saint Helens area shows the scope of the devastation from the May 1980 eruption. Inset diagram on the left identifies different parts of the post-eruption landscape, including the mudflow of 1980. The photo is of the mudflows that took place in 1982. Mudflows have a consistency similar to concrete and require significant water. Millions of cubic yards of glacial ice and snow was lost during the initial blast of 1980, but the 1982 mudflow was less violent, leaving most of the snow and ice intact. Where did the water originate for either flow? The answer can be found in the Hydroplanet Model—it originated from inside the Earth.

Visitors often leave with more questions than when they came.

and name various types of volcanic rocks. In another display, Mt. Saint Helens Lava, photos of steam explosion craters, long since buried are shown, but no lava.

The association of molten lava with volcanoes has been so overemphasized, and the impact of water so overlooked that the typical visitor leaves with more questions than they came with. In fact, they often leave not knowing that mud flowed down Mount Saint Helens, but lava did not. To get a sense of how much mud flowed from the volcano, see Fig 7.7.10.

Mud needs water, and where did the water come from? The easy answer seemed to be the ice and snow accumulated on top of the mountain, and there certainly was a lot there:

"USGS glaciologist Melinda Brugman reports that about 70 percent, or 170 million cubic yards, of Mount St. Helens' **glacial ice mass** was lost during the May 18 eruption." Note 7.7aa

170 million cubic yards is equivalent to tens-of-trillions of gallons of water, most which was certainly a casualty of the eruptive force and was blown away. Some of that ice, according to one article, melted and became the source of water in the mudflows:

"The blast quickly melted the ice on top of the mountain, sending an estimated **46 billion gallons** (174 billion liters) of water down its side in a combination of mudflow and flooding."
Note 7.7ab

With the answer to the water origin seemingly 'in the bag,' researchers worked the numbers, perhaps backward or forward, to arrive at their conclusions. Fig 7.7.9 is a satellite view of Mt Saint Helens with a USGS diagram (inset left) identifying the scope of the devastation from the initial eruption. The mudflows actually extended beyond that map, all the way to the Columbia River some 65 miles distant, where it filled shipping channels and disrupted navigation for days. Material several feet thick was deposited over an area of approximately 100 square miles, by multiple rivers of mud, or **lahars,** as they are known:

"**Lahars have the consistency, viscosity and approximate density of wet concrete**: fluid when moving, solid at rest."
Note 7.7ac

Mudflows like those at Mt St Helens are commonly compared with concrete, which contains 15-20% water. Water totals can be estimated from the quantity of mud and pyroclastic flows, allowing an extrapolation of the amount of water needed, and hence, the estimate of 46 billion gallons of melt-water.

What we do know is this; the overall amount of water in volcanic eruptions was not well known for at least the twenty years after the May 1980 eruption. We also know that even if there were trillions of cubic feet of glacier ice prior to the eruption in 1980, there was relatively little accumulated during the two years following the initial eruption. The average annual snow

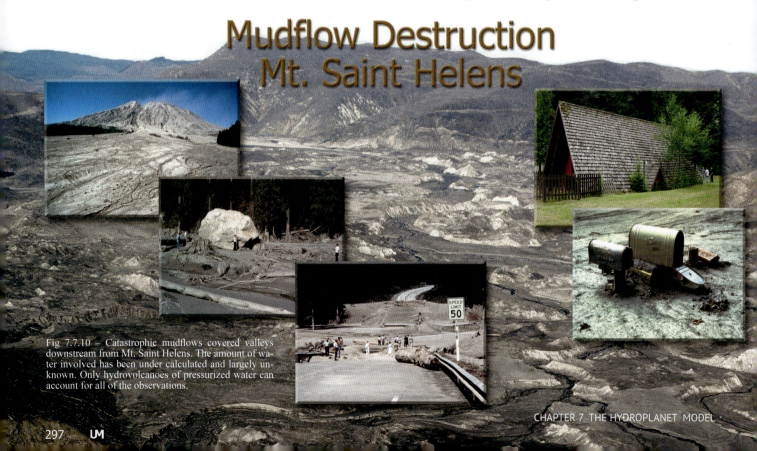

Mudflow Destruction
Mt. Saint Helens

Fig. 7.7.10 – Catastrophic mudflows covered valleys downstream from Mt. Saint Helens. The amount of water involved has been under calculated and largely unknown. Only hydrovolcanoes of pressurized water can account for all of the observations.

CHAPTER 7 THE HYDROPLANET MODEL

accumulation is about 140" above 3000 feet, certainly not enough to answer for the amount of water present in the voluminous mudflows of March, 1982 (See Fig 7.7.9, right inset—note that snow is still present around the mudflow).

FQ: From where did the mudflow water come?

The mudflows of 1982 (the two inset right photos in Fig 7.7.10 are of the 1982 flow) were associated with an eruptive event, but as is the case for most lahars, pyroclastic flows, and mudflows, scientists seemed only to look at the surficial water of snow and ice as being the source of volcanic water and did not recognize that there was substantial subterranean water within the volcano.

Superheated water under pressure would undergo significant decompressional expansion as it rises through preexisting hydroplumbing conduits inside the volcano. We cannot make specific claims as to the exact amount or nature of the expansion rate as that is neither known nor demonstrable at this time. Given the scope of this book, we consider the explosive nature of rising steam only in general terms.

To be sure, water played a substantial role in ***all of the initial components*** of the 1980 Mt St Helens event, as well as those that followed. The swelling of the north flank prior to the eruption was presumed to be from rising magma, but no lava was ever manifested, at least not in the form of a visible flow. Superheated water, heated by extensive earthquake activity and trapped beneath the crust would exhibit the characteristics observed and reported by scientists. The steam and ash in the initial 'ash plume' held considerable water, evidenced by microvesicular pumice and the electrically conductive environment that existed, as well as the subsequent research on water content cited previously. The liquefaction and displacement of the lateral blast deposits had a water component and finally, there were the lahars, and mudflows that drew water from a source sufficient enough to facilitate the transport of trillions of cubic feet of liquescent mud as far as the Columbia River, many miles distant.

Add to this story, the fact that the lahars and mudflows two years after the initial eruption were again powered by water, as if drawn from deep within the Earth. Plumes of steam continue to erupt from the volcano and are an ever-present reminder that ***water*** is the obvious cause and culprit of explosive hydrovolcanoes like Mt. St Helens, ***not magma***.

The thrust of this section is that water played a much bigger role in the Mt. St Helens eruption and in the formation and shaping of the Earth than ever before realized. Mt. Saint Helens

Fig 7.7.11 – Why does geology describe volcanic plumes as being "ash"? What about all the steam? While most large eruptions contain ash and small sediment, many small eruptions are mainly ***water vapor***. What was the origin of the water that produced numerous steam explosions before, during and after the primary Mt. Saint Helens eruption? These represent a small fraction of the many steam explosions that occur regularly on a worldwide basis. Amazingly, geologists have done little to quantify the water content of the plumes.

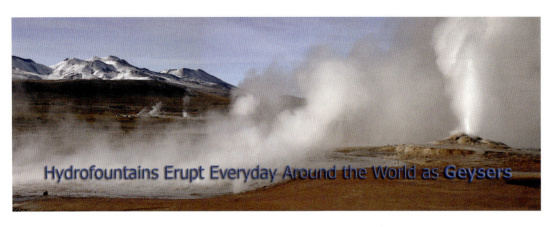

Hydrofountains Erupt Everyday Around the World as Geysers

also proved that large areas of the Earth can be affected suddenly, forming and shaping the surface of the Earth in days. We will be back to visit Mt. St Helens several times in later chapters as she holds many clues that confirm other extraordinary UM claims.

Hydrofountains Defined

We have already mentioned hydrofountains several times throughout this chapter. The important role that hydrofountains have had in shaping planetary bodies has gone unnoticed by modern science due to dogmagma and other anti-water orthodoxy. The essential role of hydrofountains is part of the core context of the Universal Model and their importance will be established.

A **Hydrofountain** is a phreatic eruption of superheated water ranging in size from small to extremely large. Steam, such as that seen in water geysers is the common component of Hydrofountains. The Frictional-Heat Law discussed in subchapter 5.3 is the most common source of heat, but other factors can contribute. Heat from the Sun heats the surface of otherworldly visitors—comets—causing hydrofountains to erupt, ejecting steam and other materials into space. The hydrofountain ejecta streams behind the comet becoming a wispy tail, thousands of miles long and visible to viewers on the Earth.

Geysers are common hydrofountains, erupting daily at many places worldwide. Groundwater is heated, causing it to expand, rise and burst into the atmosphere. As the superheated water rises, pressure is reduced allowing rapid expansion and creating a powerful frenetic explosion.

When superheated water erupts from extremely high pressures deep in the crust, it can bring with it much more than water. The amount of sediment released in a hydrofountain can be linked to the quantity of water and steam forced to the surface, an association not well understood and largely underestimated. We will be investigating examples of this material, from clay-sized sediments to huge boulders, all brought to the surface by worldwide hydrofountain activity.

In areas where there is ample water and heat, hydrofountains have been observed for a long time. One such event occurred February 2, 1884 at the base of the Hawaiian volcano Kilauea:

"A column of water, like a dome, shot several hundred feet up into the air, accompanied with clouds of smoke and steam." Note 7.7ad

A steam column at the base of a volcano comes as no surprise, but the interaction of water and heat at the center of a continent has always been problematic for modern geology because there is no proven mechanism allowing magma to rise through the mid-continent crust.

One well-known mid-continent hydrothermally active area is Yellowstone Park, in Wyoming, USA. There, *geysers* are a common sight. Geysers are hydrofountains with unique and often predictable eruptive cycles. Empty cavities deep in the crust fill with water. The water becomes superheated as it absorbs frictional heat from seismic activity in the surrounding rocks. Pressure increases until the containment point is breached. It is relieved as steam and water rush to the surface in a sometimes-spectacular phreatic eruption (see Fig's 5.3.11 and 5.3.12).

In contrast to the somewhat predictable behavior of gey-

Fig 7.7.12 – Earth's Hydrofountains—water geysers from across the planet are phreatic eruptions associated the gravitational friction of earthquakes. Other planetary bodies, such as the moons of Jupiter and Saturn experience similar gravitational friction resulting in hydrofountains. Comets are another example of hydrofountain activity, showing increased activity as heat from friction and the Sun warm the surface, leaving behind a wispy tail of debris.

sers, many hydrofountains were one-time events. The geysers of today are limited in their size, being subjected to frictional heating of a relatively quiescent Earth. It was not always this way and over the last few decades, geologists have begun to recognize that much larger events took place in the past, far greater than anything seen historically.

Hydrovolcanoes such as Mt. Pinatubo and Mt Saint Helens are themselves, mammoth hydrofountains, spewing enormous quantities of water and other gasses into the atmosphere. In fact, hydrovolcanoes can spew more water than anything else into the environment; a fact once overlooked but now recognized, thanks to ongoing research.

Enceladus 300-Mile High Hydrofountains

Hydrofountains exist in the nethermost reaches of the solar system too. In 2005, NASA's Cassini spacecraft, captured images of one in the process of erupting on Enceladus, one of Saturn's moons. Enceladus' story was told, in part, earlier in this chapter. One important aspect not discussed was the magnitude of the Enceladus Hydrofountain, seen in the upper right inset of Fig 7.7.13.

New celestial hydrobodies are discovered each year and researchers were at first surprised to see evidence of volcanic activity on small celestial bodies made of mostly ice and water. It seemed obvious that there was no magma to melt the ice, and the moons were too far from the Sun to realize any effect from its rays. With surface temperatures far below freezing, where was the heat coming from? NASA reports:

"'**Enceladus** is the smallest body so far found that seems to have active volcanism,' said Dr. Torrence Johnson, Cassini imaging-team member at NASA's Jet Propulsion Laboratory, Pasadena, Calif. 'Enceladus' localized water vapor atmosphere is reminiscent of comets. **"Warm spots" in its icy and cracked surface are probably the result of heat from tidal energy like the volcanoes on Jupiter's moon Io**. And its **geologically young surface of water ice**, softened by heat from below, resembles areas on Jupiter's moons, Europa and Ganymede.'" Note 7.7ae

Tidal energy is the Gravitational-Friction Law in action. The Enceladus hydrofountains are spectacular examples of this process and it is truly amazing how big they are, literally larger than the diameter of the moon itself:

"The fainter, **extended plume stretches at least 300 miles above the surface of Enceladus, which is only 300 miles wide**. Cassini flew through the plume in July, when it passed a few hundred kilometers above the moon. During that flyby, Cassini's instruments measured the plume's constituent **water vapor and icy particles**." Note 7.7ae

Researchers were surprised to see such warm temperatures and "jets of water vapor":

"Imaging team member Dr. Andrew Ingersoll from the California Institute of Technology in Pasadena, said, '**I think what we're seeing are ice particles in jets of water vapor that emanate from pressurized vents**. To form the particles and carry them aloft, the vapor must have a certain density, and that implies **surprisingly warm temperatures for a cold body like Enceladus**.'" Note 7.7ae

If you have ever wondered how Saturn got its rings, gravitational friction, hydromoons, and hydrofountains are the answer:

"Cassini has also confirmed **Enceladus is the major source**

Fig 7.7.13 – Saturn, its beautiful ice rings and its Hydromoon Enceladus. Saturn's rings have been shrouded in mystery for ages, their icy origins unknown, until now. Firsthand observation of Enceladus and its massive hydrofountain have identified it as being a major source of Saturn's largest ring. Enceladus' enormous, 300-mile high Hydrofountain erupted because of the effects of the Gravitational-Friction Law and there is no doubt about the effects of such supersized events, looking at the giant canyons that crisscross the moon. Could Earth's own giant canyons have been formed by hydrofountains? The craters on Enceladus' surface show no signs of impact ejecta material, neither do the craters appear to be eroded. These are likely hydrocraters, formed by the violent discharge of subsurface water, which completely vaporized, left little trace of eruptive materials on the surface.

of Saturn's largest ring, the E-ring". Note 7.7ae

Saturn's rings have been a mystery for ages. Modern science has known for some time that they consist of mostly water ice and dust, but their mysterious origin can now be determined—*hydrofountains*. Researchers compared ice and minerals from Enceladus and found a connection to Saturn's E-ring. There is no doubt that hydrofountains exist on other celestial bodies. They are obviously associated with catastrophic sized events, the likes of which we have not seen on Earth during the modern science age.

Fig 7.7.13 includes an inset photo of the surface of Enceladus showing several craters that do not appear to have any impact ejecta. The moon's surface is young and the craters do not appear to be deformed or marred by surface cracks, therefore, we can deduce that post-cratering erosional processes are not hiding ejecta. Craters with these attributes are more likely to be Hydrocraters. We will revisit Enceladus again in subchapter 7.9, where we will investigate the craters more fully.

Hydrosand Fountain Evidence

Hydrosand Fountains (Fig 7.7.14)—Sand Blows in today's geology—are not a common topic in scientific circles. This type of hydrofountain leaves surface sediment deposits discharged by relatively benign hydrofountains. Sand Blows usually occur after an earthquake, in which the shaken and heated ground forced the sediment to the surface. Recent discoveries made at the site of the New Madrid earthquakes, located in the Mississippi Valley, revealed:

"During the past 12 years, geologists found a record of New Madrid events in the form of earthquake-related features, known as **sand blows**. The sand blows formed as a result of liquefaction, a process by which water-saturated sandy sediment below the surface is liquefied and **vented on the ground in response to strong earthquake shaking**." Note 7.7af

Geology has made no apparent connection between Hydrosand Fountains and hydrovolcanoes, although they share the common denominators of earthquakes and water. This type of volcanic activity is actually more common than the lava-spewing type. While frictionally melted rock can be forced to the surface, superheated water in the form of steam can bring clay-sized particles, sand, even boulders to the surface.

Having established the reality of Hydrosand Fountains, ponder the question:

What would it be like if a large-scale hydrofountain erupted?

This question should open the mind to a host of *new possibilities in geology*. Sediment made or eroded deep in the Earth and brought to the surface through hydrofountains would change the whole of modern geology. Instead of traditional surficial fluvial erosion of rivers and streams, erosion takes place *underground*. The Rock Cycle Pseudotheory established that many types of sediment **do not fit** standard erosion explanations. The Loess and Erosion Mysteries of that chapter clearly illustrated that there are sizes of sediment missing in the sediment record of the Earth's crust. In subchapter 6.11, seven examples of the Erosion Mystery were presented, showing how the sci-bi of weather erosion was not adequate to explain these mysteries.

A large-scale Hydrosand Fountain would require a powerful earthquake, and there is evidence of such events.

Hydrorock Fountain Evidence

Modern geology textbooks contain no mention of hydrofountains as being a major source of sediment. Large or small, hydrofountains change the landscape on a time-scale measured in hours, not billions or millions of years and as such, they do not fit the magma, geological rock-column, or geological timescale paradigm.

Fig 7.7.14 – Sand Blows exhibit sediment brought to the surface with water; they are a form of hydrofountain. Sand Blows can occur after earthquakes, the result of ground vibration and heat. These examples come from California and Japan; others like them and the larger, previously discussed examples have deposited massive amounts of all types of sediment on the Earth's surface. They are a worldwide geological phenomenon overlooked and underestimated by traditional geology. Ancient hydrofountains of unimaginable size are associated with the majority the Earth's sediment.

Fig 7.7.15 – The 7.6R Chi-Chi, Taiwan earthquake of 1999 produced Hydrorock Fountains strong enough to hurl "huge boulders" into the air. Investigators reported that, "When the dust settled, **deep holes pitted the ground, as though columns of rock had been blasted out**." Deep holes like these are remnants of Hydrorock Fountains and though rare, can be found in the landscape if one looks with the paradigm of the Hydroplanet Model. Erosion and time has erased most of them, but some have been preserved for us to see. The following chapter will share some examples.

Some hydrofountains engender material large enough to warrant special recognition—**Hydrorock Fountains**. Hydrofountains of this size bring more to the surface than clay, sand or other small sediment; they produce boulder-sized material. This material is not necessarily of volcanic origin, as we read in the account of eruptions in the islands of Martinique and St. Vincent in the West Indies in 1902:

"Those volcanic products which are solid when ejected, or which solidify after extrusion, tend to form by their accumulation around the eruptive vent a hill, which, though generally more or less conical, is subject to much variation in shape. It occasionally happens that **the hill is composed wholly of ejected blocks, not themselves of volcanic origin**. In this case an explosion has rent the ground, and the effluent vapours have hurled forth fragments of the shattered rock through which the vent was opened **but no ash or other fragmentary volcanic material has been ejected, nor has any lava been poured forth**. This exceptional type is represented in the Eifel by certain monticules which consist mainly of fragments of Devonian slate, more or less altered. In some cases the area within a ring of such rocky materials is occupied by a sheet of water, forming a crater-lake, known in the Eifel as a *maar*." Note 7.7ag

This is extraordinary evidence for one of the most extraordinary UM geological claims; volcanic-like explosions rending the ground and spewing all types of non-volcanic sediment.

As we look across the landscape, there is ample Hydrorock Fountain evidence. It is hard to imagine the power of an active Hydrorock Fountain and they seldom occur. One did occur in 1999, associated with Taiwan's largest earthquake in a century, which took the lives of twenty three hundred people. An article in the journal *Nature* reported that the magnitude 7.6 quake flung boulders into the air and left "deep holes pitting the ground":

"It was no ordinary earthquake that struck near the town of Chi-Chi in central Taiwan on 21 September 1999. **The ground seemed to explode as huge boulders were flung into the air**, and flashes of light lit the night sky.

"Taiwanese geologists have now explained this unusual and terrifying geological event. The boulders were powered by high-pressure steam, they say, as sliding rocks heated ground water to boiling point.

"When the dust settled, **deep holes pitted the ground, as though columns of rock had been blasted out**.

"The friction generated by the rock faces slipping over each other in **the quake made the ground so hot that water turned almost instantly into steam**..." Note 7.7ah

Although it was Taiwan's largest, this event is not listed among the ten largest earthquakes of the past century, all of which exceed 8.0 on the Richter scale. Larger quakes can produce larger amounts and sizes of sediment, non-volcanic and diverse in appearance.

It is common to see fluids being discharged from underground. Bubbling springs, artesian water wells, oil seeps, and sometimes lava are examples of liquids seen emerging from beneath the surface of the Earth. Other than from volcanoes, can you recall ever hearing of rocks coming out of the ground? If not, you are not alone, because Hydrorock Fountains are so rare today. However, there was a time when they were not.

Fossil Hydrofountains

Fossils are the lithified (turned to rock) remains of once living things; petrified trees and dinosaur bones come to mind. In this case, we find the hardened remains of once-active hydrofountains. These peculiar formations are found in many places around the world and tell of a time in the geologic past when many hydrofountains were energetically changing the landscape. They are called **Fossil Hydrofountains**, pipes of hardened rock that formed at the quiet end of their dynamic life as heated water and sediment came to rest in a *hyprethermal environment*. They tell of a past more active with earthquakes and steam eruptions than is present today. Most of these pipes, or diatremes as they are sometimes called, were formed during the same time and by the same process that forms the fossils

Fig 7.7.16 – This Fossil Hydrofountain, located in the Kodachrome Basin, Utah, USA, is a vertically orientated 'pipe' of hardened rock that was formed in a hypretherm. High pressure and hot water along with interstitial mineral growth bind sand grain sediment together, making the pipes harder and resistant to erosion. Light colored layers of rock, seen in the upper left hand corner, are sandstone made of materials ejected from hydrofountains, not carried here from some distant eroded source. Hydrofountains account for the origin of many unique sandstone deposits, long held mysteries in modern geology. In the distant background, to the right of the main pipe, another Fossil Hydrofountain can be seen. There are dozens of these in this state park and surrounding countryside.

of once living things, a process detailed in the Fossil Model chapter.

A well-defined example of a Fossil Hydrofountain can be seen in Fig 7.7.16, from Kodachrome Basin State Park in Utah, USA. The pipe is made of lighter colored sandstone sediment that is harder than the surrounding red sediment, making it resistant to erosion. A second fossil hydrofountain can be seen in the background. These pipes once carried the white material making up the sandstone layers seen at the top left corner of the photo.

These vertically oriented pipes hardened in a hypretherm, an environment of high pressure and hot water, which allowed additional mineral growth to occur between the grains of sand, binding the sediment together.

Another example of a Fossil Hydrofountain comes from the Bushveld Complex in South Africa, noted in the *Mineralium Deposita* from 2003:

"**Blocks** with high aspect ratios are generally **oriented vertically, with their long axis perpendicular to the general stratigraphy of the complex and parallel to the long axis of the pipe, indicating vertical fluid movement.** Seeing that the blocks maintain this orientation, one can infer that there was **a significant solid fraction to the material which moved up the pipe during emplacement**, such that the blocks did not significantly shift during cessation of fluid flow." Note 7.7ai

This Fossil Hydrofountain tells us much because it includes more than just one type of sediment. The large, elongated, and vertically oriented blocks embedded within surrounding sediment tells us many things about the movement of the water and sediment in the pipe. The flow rate was very high, the researchers propose, because the blocks were carried a considerable vertical distance. Moreover, to maintain the vertical orientation of the blocks, the flow would have to be relatively constant for an extended time. Rapid fluctuations in fluid velocity would allow the disorientation of the blocks from their vertical arrangement.

A road-cut near Snowflake Arizona, USA affords us a look into another Hydrofountain. Fig 7.7.17 shows the contents of a hill that contains a variety of **Hydrofountain Sediment**, unrecognized as such in the geological community today. Hydrofountain Sediments remain a mystery because of the magmaplanet-erosional processes that influence the minds of those who try to explain them. Similar examples of this type of Hydrofountain Sediment can be found in many locales around the world.

The entire hill in the photos of Fig 7.7.17 shows several stages of hydrofountain activity. At the surface, outcroppings of fractured rock, stained dark red because of surface varnish, are of the same type of rock buried within the fine-grained sediment. Unlike the well-rounded boulders, they show no sign of heavy erosion.

Fluvial erosion, not by surface rivers long since gone, is apparent on the boulders. Abraded, ground, and tumbled, the large rocks moved through underground waterways to their final resting place. Heated by earthquake friction, pressure increased in subterranean fluid-filled aquifers, forcing sediment-laden water upward. This particular formation shows horizontal bedding of sediment and rock, later pierced by vertically moving material—small hydrofountains.

The physical features of this formation, along with observations of other formations sharing similar characteristics led us to ask this important question:

What effect would the eruption of thousands of Hydrofountains have on the crust?

Certainly, our present-day understanding of the geological column (rock layers) would change as we come to understand the heightened effect of water on crustal formation.

There is no disagreement among scientists that water is one of the most erosive and corrosive forces in nature; fracturing the landscape through freeze/thaw cycles, the dissolution and movement of billions of tons of sediment and a host of other processes are the well known hallmarks of water's power. The accepted hydrological cycle of modern science portrays a generalized sequence of water's movement beginning with evaporation—over oceans, freshwater lakes, and agrarian landscapes—precipitation over land (or oceans), absorption into subterranean aquifers or surface rivers, both of which eventually return the water to the oceans, where the cycle begins again. Certainly, all of these processes are at work today, but there is a significant part of the hydrological cycle missing in their story.

Hydrofountain Sediment

Heavily eroded rocks within the Hydrofountain Sediment are the same type of rock seen at the surface, fractured but showing no erosion.

Well-rounded Boulders in Matrix of Sand/clay Sized Sediments

Evidence of Verticle Movement of Hydrofountain Fluid and Sediment

Enlarged Image

Horizontal bedding of Hydrosediment

Sand and Clay Hydrofountain Sediment

Fig 7.7.17 – This road cut near Snowflake, Arizona, USA illustrates a type of Hydrofountain Sediment that remains unrecognized by the geological community. There are similar instances found worldwide and each is a mystery as researchers employ magma and erosion models to explain them. Notice that the rocks on the surface are made of the same material as the rounded boulders that lie within the complex. The surface rocks are varnished, fractured, and angular whereas the buried rocks are well rounded, a result of fluvial transport. The varnish coating is not forming on the boulders that are now exposed in the road-cut.

This entire hill appears to be a Fossil Hydrofountain that experienced many processes in its formation. The core, probably deep below the road, included the harder darker minerals that were forced to the surface, now crystalline rock. The rounded rocks and boulders appear to have been subjected to erosion from underground rivers and mixed with smaller sediment in a fluid mixture, which was heated by earthquake friction and subsequently thrust upward. This particular deposit also exhibits small hydrofountains, both vertical and horizontal that would have formed after the emplacement of the sediments. In the past, horizontal layers of rock were the presumed consequence of river transport; however, the vertical pipes of rock cannot be explained this way. Sometimes dubbed anomalies, these vertical pipes of rock are not simply 'quirks of nature,' but are important evidences of how Hydrorock Fountains influenced the formation of the landscape.

Earth's hydrological cycle must be redefined to include the water that exists throughout the Earth—Earth's inner 'oceans' and its hydroplumbing system. It must include the study of processes that transport Earth's deep waters; processes introduced in this subchapter.

Many of the Earth's volcanoes like Mt. Saint Helens are indeed Hydrovolcanoes, powered by water from deep within the Earth. Hydrovolcanoes are common occurrences on the Earth and on other worldly bodies, like Saturn's moon Enceladus. Exhibitions of Earth's well-connected hydroplumbing systems abound, some of which tell a story of a time when the Earth witnessed hydrological processes on a cataclysmic scale.

7.8 Hydrocrater Model

Craters are ubiquitous landforms on the moon, the Earth, and other celestial bodies in our solar system. How they got there has been the subject of some debate, but with no other plausible model, most modern scientists have settled with the impact origin theory. This subchapter answers the unsolved puzzle; if not by impact, how were the craters formed. Of course, steam eruption craters are not new, and there are dozens, if not hundreds of them already identified as such. But they are far more common and much more important than was heretofore known.

After clarifying what a hydrocrater is and how to recognize one, we embark on a survey of a number of craters in the southwestern United States and other areas, ending with a brief look at the Earth's ocean floor and the clues it holds.

Hydrofountains Create Hydrocraters

With the dismissal of the magma, and the rock cycle pseudotheories, and with a new understanding of the hydroplanet process, we are prepared to learn how the vast majority of craters were formed. The Hydroplanet Earth and Earth's Hydrology redefined subchapters contained information critical to the understanding of what a Hydrocrater is. **Hydrocraters** are bowl shaped depressions in the Earth, the scars of past phreatic eruptions—hydrofountains.

Without knowledge of the vast amount of water deep in the Earth and the processes of gravitational frictional heating, it would be unrealistic to comprehend how hydrofountains work, and thus, unrealistic to comprehend how hydrocraters formed.

In Fig 7.8.1, we see the components of a steam explosion and formation of a hydrocrater. Melted rock—lava—produced by frictional heating, moves along faults and encounters underground aquifers, heating them. The superheated water decompresses as it rises, exploding through surface layers of the crust, carrying sediment and debris out onto the surface. The re-

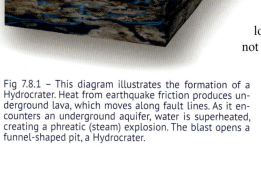

Fig 7.8.1 – This diagram illustrates the formation of a Hydrocrater. Heat from earthquake friction produces underground lava, which moves along fault lines. As it encounters an underground aquifer, water is superheated, creating a phreatic (steam) explosion. The blast opens a funnel-shaped pit, a Hydrocrater.

sulting conical-shaped depression and surrounding ejecta, all part of the Hydrofountain, is left behind for later discovery.

Our foray into the world of craters begins with a definition.

A Crater Without a Definition

The word **crater** is of Greek origin and means cup or bowl. *Webster's Dictionary* contains only *two* definitions of a natural **crater**:

"a bowl-shaped depression with a raised rim, formed by the **impact** of a meteoroid." Bib 7 p190

This definition describes crater formed by *impact*, not what we are looking for, so we read on:

"the cup-shaped depression or cavity on the surface of the earth or other heavenly body marking the orifice of a **volcano**." Bib 7 p190

The second definition of a natural crater describes a crater with a *volcanic* origin. Obviously, hydrocraters are not impact craters, so we accept the second definition, which marks the "orifice of a volcano." Looking further in *Webster's Dictionary*, we read that a volcano is defined as:

"1. a **vent** in the earth's crust through which **lava, steam, ashes, etc., are expelled**, either continuously or at irregular intervals. 2. a **mountain or hill**, usually having a **cuplike crater at the summit**, formed around such a vent from the **ash and lava expelled through it**." Bib 7 p879

It was probably the definition of volcanic craters that caused confusion. Everyone seems to recognize the "mountain or hill" with a crater at its summit was a volcano, but many Hydrocraters are not hills—they are holes. Moreover, not all Hydrocraters have lava or ash associated with their eruptions. Most curious of all, many Hydrocraters appeared to be one-time events, not prone to eruptive events that occurred "continuously or at irregular intervals."

Just as we learned that many hydrofountains eject *sediment* instead of melted volcanic material, Hydrocraters are the byproduct of hydrofountains that, in many cases, exhibit non-volcanic ejecta. Rising steam, decompressed and explosive, can propel sediment from deep below the surface. This then is a form of crater not yet defined:

Hydrocrater: A circular, surface depression formed by phreatic explosion of underground water through a diatreme.

This new classification of crater helps explain the New Geology discussed earlier. Hydrocraters exist where there are no volcanoes, and they can expel great quantities of sediment and water, creating channels and landforms previously misunderstood in geology.

Significance of Planetary Craters

Up until the 1980s, the debate over the origin of craters was one of the most hotly debated topics in geology. Crater origins affect many fields of study and research on them has influenced theories of how planets and moons formed. Since many craters seemed to have no evidence of volcanism, geologists were left with only one option; the craters must have been of impact origin.

One reason the debate lasted for so long was the misconception that the Earth was a magmaplanet. This misconception, along with faulty logic and a lack of physical evidence was the cause of the debate, leaving fundamental questions about craters unanswered. Neither of the traditional crater types—volcanic nor impact—fit the physical data.

With a shift in thinking toward a hydroplanet model, it is clear that the vast majority of natural craters are not formed by traditional volcanic activity or by meteorite impact. We can apply the Universal Scientific Method by postulating whether or not a crater is a hydrocrater and by examining the evidence.

Predicting Hydrocraters

One sure way of knowing without a doubt whether a crater was produced by phreatic eruption or by a meteorite impact is to study the geology under the crater. All Hydrocraters have a plumbing system of faults and aquifers through which steam, sediment, rocks and even liquid water is carried to the surface.

There are actually three significant requirements for the formation of a Hydrocrater.

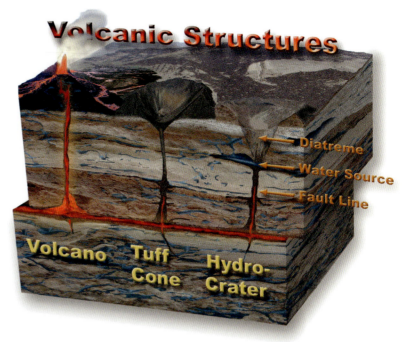

Fig 7.8.2 – This cross section shows three types of volcanic structures. Volcanoes include both cone shaped and shield or flat shaped structures, formed from melted rock, cooling on the surface. Tuff cones are formed of ejected melt-rock, often with steep walls of loose material and a small diatreme. The third type of volcanic structure is a Hydrocrater. This crater is the result of a steam explosion that ejected a large amount of sediment and other material onto the surface. Sediment forming the rim and surrounding ejecta blanket may or may not have been melted. The three requirements that identify a Hydrocrater are faults, water source and a diatreme.

Three Requirements of Hydrocrater Formation

1. Water
2. Fault Line
3. A Diatreme

Fig 7.8.2 illustrates the three requirements of Hydrocrater formation. The water, an obvious necessity to create the steam necessary to drive the explosion, is heated by frictional heating, perhaps even sufficient to melt rock. Lines of faulting provide avenues for the release of pressure and the movement of fluid toward the surface.

Many Hydrocraters exhibit bisecting faults beneath the crater. There are real-time examples of bisecting faults where frictional heat is regularly generated. Yellowstone National Park in Wyoming, USA is one such place, where active faults intersect and cross, producing ample hydrothermally heated water.

A **diatreme** is a conically shaped area directly beneath the crater, usually sediment filled and hidden from view by debris left over from previous eruptions or subsequent erosion.

Central Peak Hydrocraters contain a centrally located peak of either volcanic rock or sediment. There may be multiple peaks and in large hydrocraters, those peaks may be mountain-sized, each from different pipes

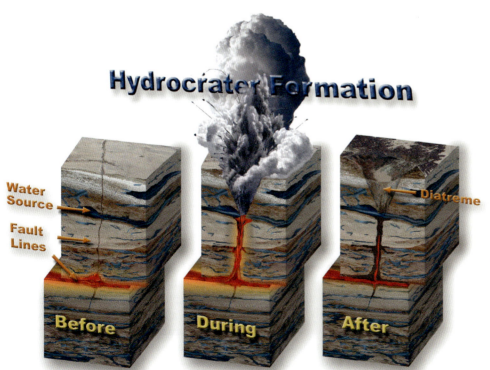

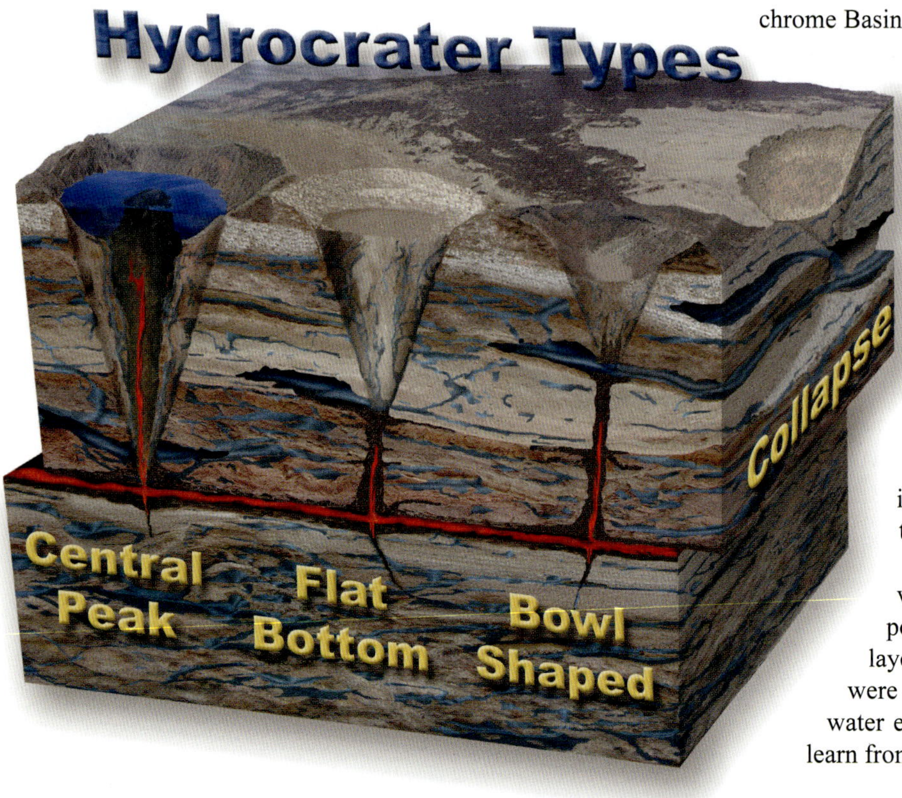

Hydrocrater Types

Central Peak   **Flat Bottom**   **Bowl Shaped**   **Collapse**

Fig 7.8.3 – This diagram depicts the cross sections of four types of **Hydrocraters** and associated diatremes. Central Peak Hydrocraters have a body of melted rock or sediment that was forced upward after the initial eruption that formed the crater; Crater Lake in Oregon, is one example. Flat-bottomed and bowel-shaped craters are created when sediment fills the diatreme, subsequent to the primary eruption. This material may be soft or hard. The forces that shaped the flat floor of a Hydrocrater are easily comprehended, whereas flat-floored impact craters are not. Collapse craters are also a form of Hydrocrater; eroded sub-surface sediment cause the surface to collapse.

within the diatreme. There are several examples of this on the Earth and on the Moon.

The size of the diatreme can affect its final shape, as represented in Fig 7.8.3, where four different types of Hydrocraters are shown. **Flat** and **Bowl** shaped **Hydrocraters** can be formed with different sediments; softer materials tend to flatten out, forming a floor at the bottom of the crater, while harder materials in the diatreme tend to leave the crater bowl-shaped.

The craters shown in Fig 7.8.3 are representative of those formed on a dry surface environment. Hydrocraters formed under water take on slightly different shapes, usually displaying a more rounded rim and smoother surface details due to the effects water has on the velocity of the explosion.

A **Collapse Hydrocrater** is formed when the surface collapses into an open void left by erosion or other means. Common examples of these are the cenotés of Central America and other areas underlain with limestone. Collapse craters are commonly found along channels or rilles (these can be seen on the Moon and other planetary bodies.)

Open Fossil Hydrofountain Evidence

As you might imagine, seeing the profile of an actual hydrofountain is next to impossible so diagrams have had to suffice. In spite of this difficulty, there are opportunities to see part of the hydrofountain profile through ground imaging and other technological tools and processes. Even more amazing is the opportunity to see a hydrofountain profile first hand.

Traveling west on Cottonwood Road just outside Kodachrome Basin State Park in Utah, the flat-topped vermillion hills provide a panorama of ancient hydrofountain activity. This Open Fossil Hydrofountain record is seen in Fig 7.8.4. From a distance, one can easily discern the conical shape of the diatreme. Within the diatreme formation are two *open* vertical pipes. Other vertical pipes can be seen elsewhere in the sandstone cliffs. The uppermost layer of the plateau consists of a cemented conglomerate of well-rounded rocks (inset right). This same material once filled the open pipes. It has since succumbed to the forces of erosion and gravity and now lies in the debris at the foot of the mountain.

This fossil diatreme allows us to see the various sediments capable of being transported in a hydrofountain event. Sediment layers and the shape of the pipe show that there were changes in the flow rate and velocity of the water exiting the diatreme. There is still much to learn from this unique formation.

As was evident in the Kodachrome Basin hydrofountain, many hydrocraters have several pipes and contain evidence that there were multiple eruptions. Several types of non-volcanic material are present in the surrounding layers, on the surface, and at different locations around the rim of the crater. Eruptions can be of different intensities, and be powered from water heated at different levels underground, affecting the type and size of sediment expelled.

Another element of the Kodachrome Basin Open Fossil Hydrofountain is the sediment type making up the **cap rock**; coarse pebbles and cobblestones cemented together at the top of the hill were the last material to be ejected from the now-dormant pipes. This is an example of an important feature, or Surface Mark that will be discussed in detail in the next chapter.

The importance of understanding hydrofountains and hydrofountain sediment as *the source* of much of the Earth's sediment cannot be overstated. Fossil Hydrofountains also provide evidence and answers to the Sand, Loess, and Erosion Mysteries from the Rock Cycle chapter.

The Hydrocrater Survey

To determine how common large hydrocraters are, we conducted a survey of craters in the Southwestern United States and parts of northern Mexico. Several of the hydrocraters and their locations are shown in Fig 7.8.5. Most people, including many geologists, are unaware of these craters, except for one—the so-called 'Arizona Meteor Crater.'

This deserves a little explanation and a personal story will help illustrate. One day, after photographing and documenting the Inyo Hydrocraters in California, I was greeted by an enthusiastic tourist who, awestruck by the sight before him, exclaimed; "Wow, that must have been a **huge** meteorite that made this crater!"

CHAPTER 7 THE HYDROPLANET MODEL

Astonished, I said nothing as the realization set in that *meteorite craters* had become so deeply ingrained into our 'cultural understanding' of nature. The romantic notion of dinosaur-killing super-meteors and the influence of Hollywood have served to solidify this erroneous belief by portraying impact craters as common, although they are not. The majority of the known craters on the Earth are acknowledged by modern geology to be of volcanic or subsurface origin, including the one before me and my new friend. His ignorance was understandable as he looked to the sky, rather than below his feet for the source of this giant hole in the Earth. Meteorites are certainly more fanciful than an austere hydrocrater.

Why science continues to endorse the impact crater notion when the majority of them cannot be substantiated as such is a mystery all its own, and we will return to it later. For now, we have a survey to review.

The **Panum Hydrocrater** (upper left in Fig 7.8.5) is located on the south side of Mono Lake in California. Mono Lake is a water-filled hydrocrater with a central peak. Many deep North American lakes are hydrocraters. The raised peak or dome of Panum Hydrocrater has both volcanic and unmelted sediment. It lies along a fault that includes many other earthquake related geological features. In addition to the Panum and Mono Lake hydrocraters, other craters occur along this fault line, such as Crater Mountain, the Devils Punchbowl, and the Inyo Hydrocraters (below Panum in Fig 7.8.5). Other related features include Obsidian Dome, Earthquake Fault (an open fault), and Mammoth Mountain. All these geological features share a common connection; volcanic activity and fault lines.

One of the three features identifying a hydrofountain is the presence of a fault, the occurrence of which indicates a higher probability that a phreatic explosion caused the crater, not an impact. Further research will undoubtedly confirm this.

Upheaval Dome (top right inset Fig 7.8.5) in Southeastern Utah is a crater that was declared by some researchers around 1995 to be of impact in origin. This is surprising given the fact that the crater has a large outflow channel obviously *not* of impact origin, but common in many hydrocraters.

The **Chevelon Collapse Hydrocraters** are thought to have formed when underground aqueducts eroded salt deposits deep beneath the surface. Indeed, there are deep salt deposits across Arizona and the Southwest with no clear explanation as to how they formed. Note the absence of a raised rim around these craters, suggesting an explosive eruption was not involved.

The **Colton Hydrocrater,** (second from bottom, left in Fig 7.8.5) located just north of Flagstaff, Arizona, is a typical bowl hydrocrater. The lack of sufficient fine-grained sediment and minimal water following the eruption probably contributed to the bowl shape of the crater. Tuff, a light volcanic material is common at the Colton hydrocrater site. There are several other hydrocraters and volcanic cones in the vicinity.

The **Zuni Hydrocrater** lies inside New Mexico just east of the Arizona border and is unique in that it has a salty lake in the crater, continuously fed by underground springs. The Zuni Hydrocrater has several off-center volcanic peaks, evidence of secondary eruptions. Although scientists acknowledge Zuni crater as being formed by phreatic explosion, the hydrocrater is still considered "anomalous in character":

"Zuni Salt Lake crater was visited in 1873 by E. E. Howell, a geologist with the wheeler expedition, who described its basin as anomalous in character and felt that with data then available its origin was inexplicable. Other geologists were similarly puzzled." Bib 119 p21

Without a Hydroplanet Model, hydrofountains, hydrocraters, and related volcanic action are of "little-known kind":

"It was eventually agreed that if volcanic action had indeed been involved, it was of a **little-known kind**." Bib 119 p22

This remains true even today, and explains why the public is relatively ignorant about hydrocraters and why the Moon's craters are still misunderstood. Kathleen Mark the author of the well-researched book, *Meteorite Craters*, relates this story about Zuni Crater, further evidence of the cultural bias of crater

Fig 7.8.4 – This Open Fossil Hydrofountain is near Kodachrome Basin in Utah, USA. It is a unique naturally occurring cross-section of a hydrofountain and an in-filled hydrocrater. At least two open pipes (diatremes) are visible. These types of formations exist in large numbers but are usually underground and not easily seen. The study of this hydrofountain helps to understand a part of the Sand Origin Mystery; it illustrates how most of the Earth's sediment was deposited on the surface. Open pipes like these brought water and sediment to the surface, where it was ejected in a hypretherm and became cemented. The diatreme and the surface cap-rock are easily seen, as is the ejected, cemented Sediment.

The Open Fossil Hydrofountain

Fig 7.8.5 – A survey of Southwestern USA Hydrocraters was conducted to understand the frequency of Hydrocraters. Various types of are represented here, but this is only a fraction of the dozens that exist in the area. Many geoscientists are not even aware of these craters, yet they are the most numerous type of crater in the solar system. Popular science culture and Hollywood have sensationalized the impact crater, claiming them as being the most common type of planetary crater, and that without such impact, the planets and moons of our solar system would not exist. Here on Earth we can study the geology of these craters directly and scientific investigators have acknowledged that perhaps 99% of Earth's craters are steam-explosion craters. The locals and many scientists are only familiar with the famous Arizona 'Meteor' Crater made popular because it is said to have been caused by a meteorite. But what if this crater, like others in the survey, is actually a Hydrocrater? We will explore that possibility later in this chapter.

formation:

> "However, as often happens in such cases, information does not spread very fast. In the early 1970s, some men working at the salt pans in the Zuni Salt Lake crater were asked if they knew how the strange depression had formed. 'Sure,' one of them replied without hesitation. '**It's a meteor crater. Like the one in Arizona**.'" Bib 119 p24

The **Montezuma Hydrocrater,** better known as Montezuma Well, is a *spring fed* circular depression in Central Arizona. The pool is maintained by the constant flow of warm spring water in the bottom of the pool. The water is highly carbonated and flows out of the crater into Beaver Creek at a rate of 1100 gallons a minute.

The Ubehebe Hydrocraters

The Ubehebe Hydrocraters are located in one of the hottest spots in North America—Death Valley, California. These craters help illustrate the concept that a volcanic eruption can occur with little to no volcanic rocks. How do geologists interpret the fact that there is only a modest amount of volcanic debris? In the book *Geology Underfoot in Death Valley,* the authors call the Ubehebe Craters "unusual":

> "At the north tip of the Cottonwood Mountains, west of the northern Death Valley trough, lies a small but **unusual** volcanic field. **Unusual**, in that it consists largely of an assemblage of **explosion craters with only a modest amount of fragmental volcanic debris**." Bib 128 p153

The right inset photo in Fig 7.8.6 is of surface gravel lying on the rim, sorted as if it had been run through a sorting machine. The authors make special note of the non-volcanic "pebbles and cobbles" around the craters:

> "The rocks that make up the crater wall include beds of **stream-deposited conglomerate containing smooth, well-rounded pebbles and cobbles, up to 8 inches in diameter**, of quartzite, limestone, fine-grained lard mudstone, pre-Ubehebe lavas, and fine- to coarse-grained crystalline rocks." Bib 128 p 156

The similarities in size and composition of the rocks raise the question of how this well-sorted, well-rounded conglomeration of rocks came to be located on the rim of this crater. Silt and sand are easily moved and sorted by water, but larger rocks, in such *large quantities* are not sorted like this in surface rivers. However, mass quantities can accumulate in underground water channels, including vertical pipes, which sort materials according to size and density, based on pressure in the pipe or channel.

As we explore other examples of hydrofountains that brought fluvially eroded sediment from underground, we can begin to comprehend how sediments of certain sizes are abundant while others are not, a puzzle introduced in the Missing Sediment Mystery, in subchapter 6.10. One can begin to appreciate the enormity of this concept as we realize how big and how numerous hydrofountains once were. Hydrosedimentary material ejected from hydrofountains exists everywhere around the world and in such abundance that it is essential to identify it separately from conventionally eroded material. In the New Geology section (subchapter 7.5), sedimentary material was divided into two classes; hydrosediment—sediment formed in a water environment, especially subterranean water, and erosionary sediment—material from traditional erosion processes. The UM will document that *most* of the Earth's sediment is hydrosediment that came from events deep underground, in hydrothermal and hyprethermal conditions.

Ubehebe Hydrocrater has an exposed fault, a great example of one of the elements necessary for the formation of a hydrocrater. The "near-vertical fault" can be clearly seen in Fig 7.8.6 between the yellow and red sediments. The authors of the *Geology Underfoot* text include this comment:

> "A **near-vertical fault** that trends a little west of north and has a probable cumulative vertical displacement of **at least 400 feet** cuts through the sedimentary rocks at the crater's northeast corner." Bib 128 p157

The authors mention but make no further comment about the scope of the apparent displacement of four hundred feet! Imagine the effects of an earthquake with a displacement of 400 feet by comparing it to the Sumatra earthquake of 2004. This was the second largest earthquake ever recorded and had a magnitude in excess of 9.1, producing a shift in the crust of approximately 65 feet (20 meters).

One last geological feature of the Ubehebe crater, probably the most important, is the **Fossilized Silica Hydrofountain** seen in the left inset of Fig 7.8.6. Ignorant to Hydroplanet Model concepts, investigators completely miss the significance of this silica dike on the west side of the crater. The Sand Mark in subchapter 8.5 will outline how the Earth's sand, including the massive formations of homogeneous sand came from a hyprethermal process, not from the erosion of mountains. In underground hyprethermal processes, sand, such as the white silica in the dike, was formed and brought to the surface through hydrofountains, like Ubehebe.

The relationship between silica deposits and Hydrocraters

Fig 7.8.6 – The Ubehebe Hydrocrater is one of a dozen hydrocraters located in Death Valley, California, USA. Ubehebe provides an excellent illustration of faulting that is associated with all hydrocraters. The crater also offers a fine example of the ***non***-volcanic debris expelled from hydrofountains, the material being "unusual" to the traditional geologist, who is unaware of the Hydroplanet Model. The most notable feature of Ubehebe is its Fossil Silica Hydrofountain remains, seen in the left inset. This "white silt" dike, as investigators call it, demonstrates that this crater has other faults or diatreme-pipes that carried the siliceous silt from underground. The Silica Hydrofountain—a fossil because of how it was preserved—shares commonality with another southwestern US crater, the Arizona 'Meteor' Crater. The silica deposit demonstrates the similar origin of both craters.

Buell Hydrocrater

Fig 7.8.7

Kimberlite Knob (Fossil Hydrofountain)

Ring Dike (Minette)

Semi-Precious Gemstones

Kimberlite Crater (Small Hydrocrater)

Satellite View
3 miles

Aerial View

"Buell Park is the largest kimberlite diatreme in the field, and probably the world."
Michael F. Roden

Surface Chalcedony

Kimberlite Knob

Buell Mountain

CHAPTER 7 THE HYDROPLANET MODEL

will be highlighted further when we discuss the Arizona 'Meteor' Crater. It too, contains a large silica deposit that has been almost completely ignored by impactologists.

The Buell Hydrocrater

Michael F. Roden wrote his thesis on the **Buell Hydrocrater** (known as Buell Park) and said this:

"Buell Park is the **largest** kimberlite diatreme in the field, and probably the world." Note 7.8a

It would seem that the world's "largest" kimberlite diatreme would be more fully researched, given its relative ease of accessibility, but it has not. The Buell structure is a three-mile wide circular hydrocrater located on the border between Arizona and New Mexico, USA (see Fig 7.8.7). It is on the Navajo Reservation and has been well preserved for geological study. Roden and others trained in the magmaplanet paradigm investigated the Buell Park geologic features, but focused on the small amount of volcanic material, generally ignoring the non-volcanic, non-melted sediment. Perhaps ignorant of the fact that Buell is a Hydrocrater, the remnant of a massive phreatic explosion; researchers dismissed the sediment as typical of erosion. However, the Buell sediment is an excellent example of **hydrosedimentary** material *formed* in a water environment.

Buell lies in the Navajo volcanic field 80 km (50 miles) southwest of the Hopi Buttes, a collection of mostly volcanic pipes. Contrary to popular thinking, there has been only minimal erosion of both the pipes and the area surrounding the Buell Hydrocrater. The entire area of northern Arizona however, experienced significant earthquake and volcanic activity only several thousand years ago.

Fig 7.8.8 is a photo of Green Knobs, a fine-grained sediment deposit located a few miles northwest of Buell Hydrocrater. In the photo, two completely different types of sediment can be seen, red sandstone and green clay, right next to each other. Neither the green nor the red sediment could have been formed by traditional erosional processes because they are both homogeneous. Sediments of this type are typical of hydrosedimentary material from hyprethermal processes deep in the Earth. Moreover, very little of the two types of sediment are mixed together, verifying the young age of the formations.

The flat floor of the Buell Hydrocrater, as seen in the Fig 7.8.7 photos, clearly indicates its young age of only several thousand years. One small dry creek bed, perhaps a few feet wide, runs through the crater, typical of what might be expected from surface runoff.

Referring back to Roden's research, consider this example of the researcher's actual observation at the Buell Hydrocrater—and show how the observed facts confirm the Hydroplanet Model:

"Xenoliths are **dominantly sedimentary**; especially conspicuous is a **white quartzite. Crystalline samples from the crust and upper mantle are also present**. The xenoliths are up to 75 cm [29 inches] in diameter, but generally less than 25 cm [10 inches] in diameter, and the **crystalline rocks are generally rounded**." Note 7.8a p1

Buell Crater's *xenoliths* are rocks brought to the surface during a steam explosion and embedded in the crater's rim. The researcher explained that the crystalline rocks came from the crust or upper mantle—perhaps 20 or 30 miles deep, and are "quartzite" and "rounded."

FQ: How did rounded quartzite boulders come from the mantle?

Again, we see how the magma pseudotheory has influenced researchers and kept them from discovering important truths. The definition of a xenolith:

"In geology, the term *xenolith* is almost exclusively used to describe **inclusions in igneous rock during magma emplacement** and eruption." Note 7.8b

We established that glass, not quartz, comes from magma, so "inclusions in rock during magma emplacement" is an erroneous idea. The Earth has large subterranean aquifers capable of rounding countless millions of boulders. When conditions are right and water in the aquifer is heated by earthquake friction, increased pressure forces superheated water upward where it is released from its containment in a furious burst of energy, carrying a massive sediment load, including the previously rounded boulders, onto the surface. The Hydroplanet process can actually be seen taking place in nature, and in examples shown in this chapter. On the other hand, a magmaplanet explanation requires a great deal of imagination with no laboratory or real-life examples to support it.

The Buell Hydrocrater is important because it provides us with an opportunity to understand that green and red hydrosediment was brought to the surface through a diatreme. With this understanding, we can begin to perceive the origin of many similar deposits on the Colorado Plateau and in other places around the world; hydrosedimentary material ejected onto the surface via hydrofountains.

Fig 7.8.7 includes a photo of a kimberlite knob and crater, off center of the main crater. Kimberlite is a unique mineral identified in many diatremes. It has economic significance because some kimberlite pipes contain diamonds. The connection between diamonds and diatremes is an important part of the

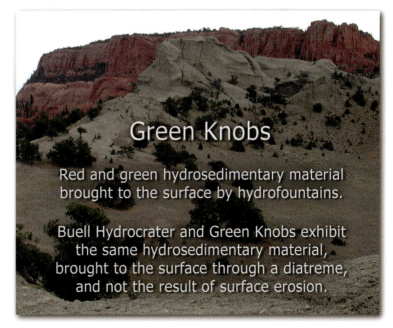

Fig 7.8.8 – The Green Knob deposit is located several miles northwest of Buell Hydrocrater in northeast Arizona. The origin of this sedimentary material is unknown in modern geology. Only with the Hydroplanet Model can it be understood.

UM and will come up again. Many semi-precious gemstones have been found at Buell Park, as well as surface chalcedony. You may remember surface chalcedony from the Rock Cycle Pseudotheory chapter. The surface chalcedony at Buell consists of a white silica deposit on the sandstone matrix around the crater rim. Its existence establishes that the entire area experienced hyprethermal conditions, which were necessary for these mineral deposits to form on the surface.

Many types of **breccia**, a rock made of angular fragments of broken rocks and cemented together, formed in underground hyprethermal environs. Some of these deposits have biogenic rocks and minerals containing organic material present in the underground aquifers. As the underground biospheres became subjected to hyprethermal conditions and the rocks were subjected to tremendous heat and pressure, semi-precious minerals, like those found at Buell form and take on colors resultant from the various impurities present in the water. Another famous gem, found in diatremes worldwide is formed in similar manner—the diamond.

Pinacate Hydrocraters

The Pinacate volcanic field, located south of the Arizona Mexico border, contains examples of both volcanic and sedimentary hydrocraters. The Elegante Hydrocrater, seen in Fig 7.8.9, is covered with tuff, as are other craters in the dark volcanic field in the background of this satellite photo. However, the Colorado Hydrocrater, located on the northwest border of the volcanic field has very little tuff and ash in comparison to the other craters. It contains primarily sediment and non-volcanic debris similar to the sediment in the Ubehebe Hydrocraters.

So alluring was the notion of meteorite impact that early investigators proposed these craters were of impact origin:

"Surprisingly, however, one of the first modern papers to speculate about their origin **proposed that Elegante was caused by a meteorite impact!** Amateur naturalist Allan O. Kelly, inspired by a 1951 popular article on Elegante by Wilson McKenney, visited the crater late that year. **He pointed out the similarities to the shape of Meteor Crater, about 260 miles to the northeast, and noted that a magnet would pick up magnetic material on both craters' rims**. Unfortunately for this idea, the shape similarity is coincidental, and Ives pointed out in 1964 that magnetic rock fragments are 'a normal desert phenomenon.' Interestingly, my reprint of his article, which I believe arrived in the 1960s, contains a handwritten note signed by Kelly: 'I have long since given up hope that Crater Elegante is... meteoritic...'" Note 7.8c

Although scientists today have learned much concerning these types of volcanic features, there is still more to be learned with respect to the hydroplanet Earth and the role hydrofountains play in planet formation. Many hydrofountains once had craters that have since eroded away or have been filled in by recent activity.

Crater Lake Evidence

Crater Lake in Oregon, USA; one of the most beautiful lakes in the world, has been deemed a natural treasure and an "outstanding laboratory." For more than a century, it has been preserved as a National Park and has been the subject of much study. According to the National Park Service via their public web site, we are told that this crater formed because of a *collapsed* volcano, and that lake water levels remain relatively constant due to a balance of *evaporation and seepage*:

Fig 7.8.9 – The Pinacate Hydrocraters are located south of Arizona in Mexico. They are part of a group that included several-dozen steam eruption craters. The background photo is a satellite view of the top half of the crater field. Elegante Crater is covered in a volcanic rock called tuff and is located near the center of the volcanic field. Colorado Crater lies on the outskirts of the volcanic field and is considered an "unusual" type of crater because it has very little tuff in its ejecta material, except for a small amount of volcanically melted rock from other, nearby steam explosions. The non-volcanic debris found here is similar to the material at Ubehebe, except that it is more angular. Colorado Crater has almost exclusively non-volcanic sediment, and is a good example of a hydrocrater within a volcanic field. Because the craters are similar to the Arizona 'Meteor' Crater, at least one investigator thought they could be impact craters too.

"Crater Lake has inspired people for hundreds of years. No place else on earth combines a deep, pure lake, so blue in color; sheer surrounding cliffs, almost two thousand feet high; two picturesque islands; and a violent volcanic past. It is a place of immeasurable beauty, and an **outstanding outdoor laboratory and classroom**.

"Crater Lake is located in Southern Oregon on the crest of the Cascade Mountain range, 100 miles (160 km) east of the Pacific Ocean. It lies inside a caldera, or volcanic basin, **created when the 12,000 foot (3,660 meter) high Mount Mazama collapsed 7,700 years ago following a large eruption**.

"Generous amounts of winter snow, averaging 533 inches (1,354 cm) per year, supply the lake with water. There are no inlets or outlets to the lake. Crater Lake, at 1,943 feet (592 meters) deep, is the seventh deepest lake in the world and the deepest in the United States. **Evaporation and seepage prevent the lake from becoming any deeper**." Note 7.8d

The reason scientists think this crater is a collapse 'caldera' is expressed in a 1939 geology textbook. Geologists believed that the 5-mile wide crater could not have been formed by explosion because of the "absence of the debris":

"The reason for believing that the caldera was **formed by the collapse** and engulfment of the top of the former cone **rather than by explosion, lies in the absence of the debris—the 15 cubic miles of material**—that so gigantic an explosion would have spread over the immediately adjacent outer slopes." Bib 125 p285

Perhaps the beauty of the lake distracted the early scholars or they did not understand the topography. The images of Crater Lake in Fig 7.8.10 clearly show a *raised rim and ejecta encompassing the crater*. The top left inset image is an adapted USGS topographical map illustrating elevated areas, shaded brown and yellow. These rim walls rise as much as 2,000 feet above the level of the water in the crater and over 4,000 feet above the farmlands to the south. Other examples of collapse craters, such as the Chevelon Craters in Arizona (See Fig 7.8.5) do not have elevated rims like those around Crater Lake. Additionally, there appears to be a number of rather obvious drainage channels emanating from the crater rim. The drainage channels occupy wide valleys, formed when water and mud overflowed the crater after the initial eruption. In the center of some of the drainage valleys, particularly on the south of the Crater, small narrow channels are evidence of recent erosional processes, sometime over the last several thousand years. Observation of the drainage channels is important as they are found all over the globe and they give us real clues as to the age of the landscape and the type of geological events that transpired.

Other evidence that the Crater Lake caldera is a hydrocrater is the hydrothermal springs that lie on the floor of the lake. Discovered in 1983, researchers identified both hydrothermal springs and faults on the crater floor (see Fig 7.8.10):

"We located **two thermal springs** on the lake floor. These springs discharge warm dense water through **faults or fractures** that bring the water up from a sub-lake floor **hydrothermal reservoir**." Note 7.8e

Crater Lake's constant water level is not wholly attributable to snow, evaporation and seepage as the Crater Lake NPS website

Fig 7.8.10 - Crater Lake in Oregon is the deepest lake in the USA and one of nature's treasures. The crater has a rim rising more than 4,000 feet from the valley floor surrounding the crater. Scholars continue to teach that this crater formed when subterranean material was evacuated and the ground collapsed. However, collapse craters do not normally have such high rims, and Crater Lake has large drainage channels, seen in the aerial photo and topographic map on the left. These channels, carved by overflow water, carried sediment from the crater after the eruption. In addition, volcanic Wizard Island came from a diatreme extending below the crater as noted in the text. This crater also lies on an active fault that produces hydrothermal water, discharged through rising springs sufficient to heat the water at bottom of the 5.6-mile wide lake 6°C (11°F) warmer than the surrounding water. All of this evidence establishes that the deepest lake in North America—Crater Lake, is a hydrocrater that was formed by a steam explosion.

states. Deep, thermal springs keep this lake filled with exquisite blue waters and a unique set of microorganisms that contribute to the lake's beauty. The thermal springs have an effect on other aspects of the lake, as researchers report:

"Thermal springs discharge into the deep water and **have a dramatic effect** on the temperature stratification, circulation, and chemistry of these waters." Note 7.8e p1099

The spring waters discharged into the lake are 6 °C (11°F) warmer than water at the bottom of the lake:

"This water discharges at 9.5 °C or about **6 °C warmer than the bottom waters of Crater Lake**." Note 7.8e p1097

One final comment on the Crater Lake Hydrocrater is about Wizard Island, the volcanic island that lies in the lake, seen in Fig 7.8.10. This island confirms the existence of a diatreme beneath the lake (see left diagram in Fig 7.8.3). Born of an eruption more recent than the crater itself, Wizard Island exemplifies that reoccurring eruptions are common in hydrocraters, an important fact we will revisit later on.

All of the observations at Crater Lake can be easily explained with the Hydroplanet Model. Although this caldera is not of impact origin, many details about this "outstanding outdoor laboratory and classroom," including the documented hydrothermal spring activity, have surprisingly been left out of the textbooks. Meanwhile, the unsubstantiated *collapse theory* remains the dominant story of Crater Lake's formation.

Hydrocraters in Process

Several examples of Hydrocraters were examined in the Hydrocrater Survey, rendering a view that hydrofountains and hydrocraters are common in nature and an integral part of Earth's geological processes. Many Millennial Science books may be written on this subject alone as we begin to understand that the vast majority of the Earth's sediment does not come from surface erosion, but through erosional processes deep underground, in heretofore-unknown environments, including the hypretherm.

It is quite possible that the reason modern geologists have overlooked the significance of hydrofountains and hydrocraters is because the formation of hydrocraters the size of those in our survey has never been witnessed. Based on Hutton's uniformity principle—a guiding principle taught to all aspiring geologists—we should see their formation occurring somewhere, but we do not. The present has not been the key to the past, as Hutton imagined.

There are a few evidences of the ongoing processes of hydrocrater creation, some of which have been actually observed. A maar is a type of hydrocrater, defined as:

"A low-relief broad volcanic crater formed by multiple shallow explosive eruptions. It is surrounded by a *crater ring*, and may be filled with water." Bib 173 p386

According to the USGS, scientists interpret maars as being formed above diatremes by the expansion of steam. One particular account occurred in Chile in 1955:

"Theories of the volcanic action of maars were vividly confirmed in 1955, when the creation of a maar **was actually witnessed** in a volcanic region in the southern Andes in Chile. Beginning in July of that year, **violent gaseous discharges occurred**, accompanied by smoke and ash. They lasted for twenty or thirty minutes and were separated by similar periods of quiet. Both the explosive periods and the intervening pauses grew gradually longer. By November 1955, all activity had ceased. A few months later, water in the new maar (now known as Nilahue Maar) reached its approximately constant level." Bib 119 p23-24

The report of multiple eruptions and differing sediment load, including both volcanic and non-volcanic material is significant. The Nilahue Hydrocrater in Hawaii was reported in 1957 to be 1,100 meters in diameter and 350 meters deep—curiously *similar* in size and shape as the Arizona 'Meteor' Crater. The magnitude of the explosion was report by eyewitnesses:

"At 10 p.m. the same day a strong earthquake from 2nd to 4th grade shook Valdivia, Osorno and other towns within a radius of several hundred kilometers, and the approximately **7,000 meters high ash cloud of the eruption was well visible at distances of up to 300 km**. Subsequently the seismic activity rapidly diminished." Note 7.8f

Another example is the Alaskan Ukinrek Hydrocrater, seen in Fig 7.8.11. The inset USGS photo of the steam explosion was taken in 1977. The resultant crater is seen in the background. Ukinrek Crater was formed in a region of loose volcanic material from nearby volcanic activity. Because of the unconsolidated material, the eruption did not produce much of a rim around the crater, as it would have if the ground had been more solid.

Phreatic, or steam explosions have not enjoyed the notoriety they deserve; they are not as alluring as a meteorite from space. There are however, popular destinations where visitors can see remnant hydrofountain activities, the result of the ongoing interaction between gravitational friction, earthquakes, and water:

"Geologic study has shown that the surface opening of **every**

Fig 7.8.11 – Ukinrek Hydrocrater is located in Alaska. It is one of the few large, active hydrocraters to have been observed by scientists. There are tens-of-thousands of inactive hydrofountains and hydrocraters worldwide, but the uniformitarian notion that the "present is the key to the past" has stifled modern geology's ability to see the importance of this geological process and its contribution to the formation of the surface of the planets and moons of the Solar System.

hot spring in Yellowstone, be it quiet pool or geyser, probably **formed as a result of steam explosions**, **which are commonly associated with earthquakes**." Bib 134 p15

The hydrothermal attractions at Yellowstone are comparatively quiet today. The massive explosions where vast quantities of sediment were carried to the surface remain a part of a bygone era, unknown to modern man. Evidence of hydrocraters is sparse or nonexistent in the geological column, suggesting most of these events happened during a single, violent period, only a few thousand years ago.

Mt. Saint Helen Hydrocraters

In May of 1980, Mt. Saint Helens experienced the largest steam explosion in American history. That event forever changed the way scientists look at volcanoes. Yet few people, including most who visit the Mt Saint Helens Visitor Center, know that this eruption also produced hydrocraters, seen in Fig 7.8.12. These steam explosion craters testify how hydrocraters can form anywhere, given sufficient water and heat.

All volcanic eruptions are tied to frictional heating from seismic activity. Frictional heating heats rock and water alike. In the case of Mt. Saint Helens, frozen surface water as well as rising subterranean water—heated from friction or otherwise—reached a critical point and exploded, leaving behind a hydrocrater.

The Mt. Saint Helen Hydrocraters are, as of this writing, the largest known hydrocraters formed in the USA while under scientific observation. Today, there is little evidence of them, they have since been buried by mud and debris flows. They remain relatively unstudied; pushed aside as insignificant in geology today.

In addition to the *continental* hydrocraters we reviewed, there is a lot to learn by looking under the ocean.

The Pockmark Evidence

During the last decade or so, new technology has allowed us to peer into the depths of the ocean like never before. As is often the case, new observations challenge old ideas.

It was *after* we had gone to the Moon and the romantic notion of *impact* craters had a firm grip on the sciences, that a large number of hydrocraters, some still being formed, were discovered off the continental shelf of Nova Scotia, Canada.

These 'morphological' ocean-floor features, dubbed "pockmarks," were first observed in 1970. They drew little attention until recent technological advances allowed researchers to view them in deep water. In a 2003 *Marine Geology* journal article, "pockmarks" were explained:

"Nowadays it is widely accepted that **pockmarks** originated by expulsion of gas from over pressured shallow gas pockets, dispersing the sediment into the water column or by **intensive continuous fluid discharge** hindering sediment deposition in and around the seep." Note 7.8g

Pockmarks (Fig 7.8.13) closely resemble hydrocraters, except that they occur under water. Although a steam eruption can occur—influenced by pressure from overlying water—most eruptions are of a water-discharge nature. Gases, such as methane from biological waste also play a role in underwater hydrocrater formation.

In addition to "shallow gas pockets," deep vents, in excess of several kilometers can carry sedimentary material to the ocean floor surface; *Marine Geology* continues:

"Pockmarks [hydrocraters] most frequently occur **near mud**

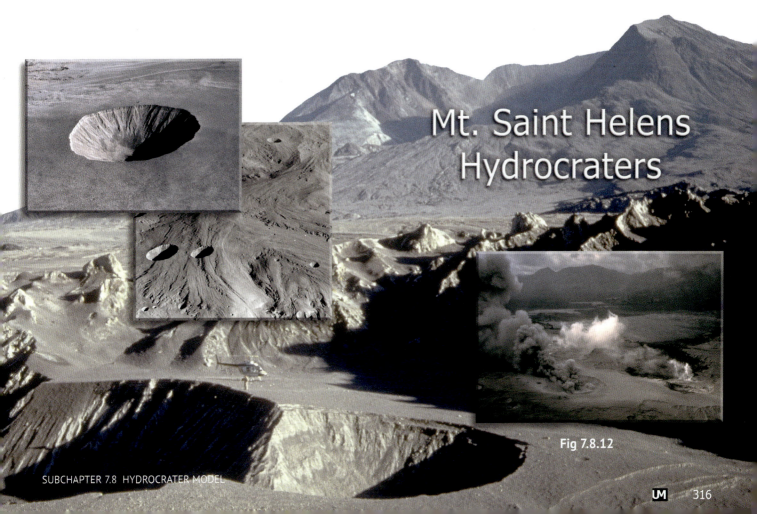

Fig 7.8.12

SUBCHAPTER 7.8 HYDROCRATER MODEL

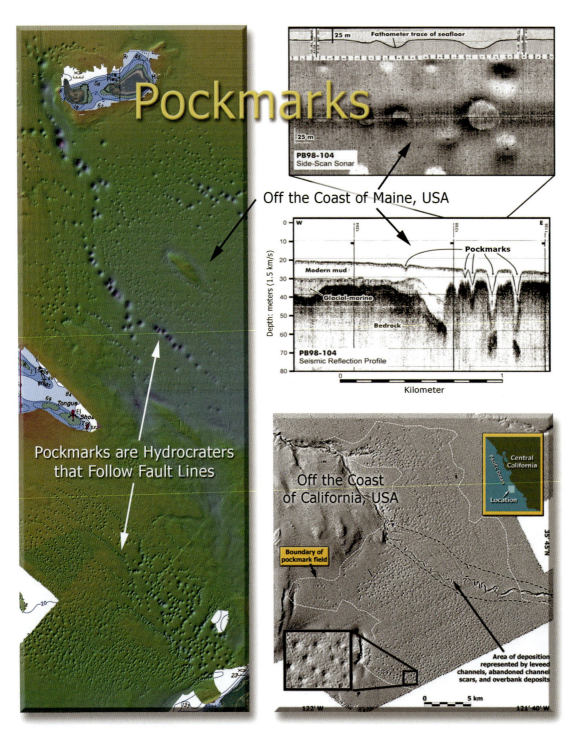

Fig 7.8.13 – Pockmarks are a worldwide phenomenon under oceans and lakes; only recently discovered in the past two decades, deep-diving rovers regularly study them. In the Seismic Reflection Profile to the left, the diatremes below the pockmarks can be seen. Still, scientists have yet to make the connection between these underwater hydrocraters and the hydrocraters formed on the Moon and other planets. Once the realization is had that planets are hydroplanets, water-formed, the hydrocraters beneath Earth's oceans and on the surface of dry celestial orbs will take on new meaning, and be entirely comprehensible.

volcanoes and along fault lines in areas of active fluid discharge." Note 7.8g p274

The ability of hydrocraters to eject non-melted rock on continental land surfaces and in the deep ocean has been a *missing key* in the understanding of how planetary surfaces are shaped. With this key, craters on the Moon, Mars and other planets can be seen in a new light, connecting more of nature's puzzle pieces.

The similarities between the pockmarks at the bottom of the ocean in Fig 7.8.14 and the craters on the surface of the Moon are undeniable, an observation not lost to researchers:

"Pockmarks [hydrocraters] appear on the sonographs as **so-called moon (lunar) patches**..." Note 7.8g p264

Of course, they refer to the similarity as being "so-called" because in their mind, there is no possibility that the Moon could have ever been covered with water. Fortunately, truth and time are on the same side. A shift toward the hydroplanet model and improvements in technology will render old theory incompatible with new observations. Scientists once wondered if Mars was ever covered with an ocean. Then, several years ago, the robotic Mars Exploration Rover mission collected abundant data in support of Mars being once covered by seas.

The origin of Mars and the Moon is not yet connected to water but future exploratory missions will continue to provide supporting evidence of their Hydroplanet Model origin.

Modern Day Concerns With Hydrocraters

Thus far, the primary emphasis has been the formation of hydrocraters and the role they play in shaping the Earth's surface and other planetary bodies, and how their importance has been

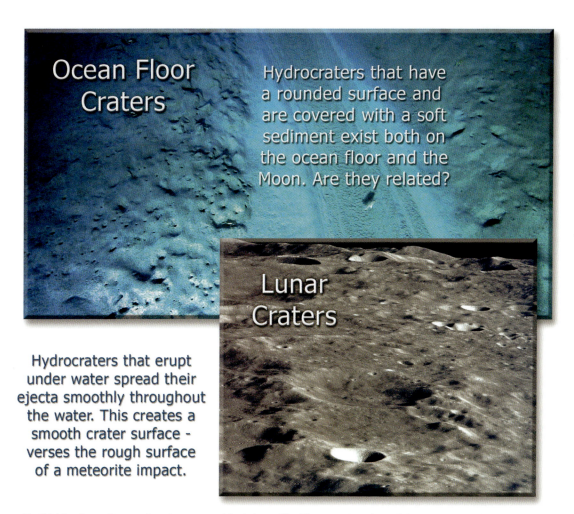

Fig 7.8.14 – Ocean floor pockmarks are surprisingly lunar-like. The smooth surface of the craters took shape as ejected sediment slowly settled through the water, spreading evenly across the surface. Could both lunar craters and deep-ocean pockmarks share a similar origin? Hydrocraters are a newly recognized geological phenomenon, becoming more earnestly studied as our ability to reach the ocean's abyss improves. With only a fraction of the ocean surveyed, investigators estimate that at least 100,000 exist.

overlooked in modern geology. Researchers have misunderstood hydrocraters and comparatively little research has been focused on them.

Discovering hydrocraters in the world's oceans, learning more about the significant role water plays in continental hydrocrater formation, and a host of other new discoveries is slowly changing the scene. New concerns are surfacing about the connection of biological activity in hydrofountains and hydrocraters. Underground gas production via biological agents is closely linked to hydrofountain activity, a topic developed in the next chapter. One recent example involved the release of excessive CO_2 gas from hydrocrater Lake Nyos, Cameroon, Africa in 1986. Seventeen hundred people living in a village near the crater lake died:

"Although carbon dioxide is usually harmless, a large, rapid release of the gas is worrisome because high concentrations can kill. Planners are well aware of the terrible natural disaster that occurred in 1986 at Lake Nyos in Cameroon: carbon dioxide of volcanic origin slowly seeped into the bottom of the lake, **which sits in a crater**. One night an abrupt overturning of the lake bed let loose between 100,000 and 300,000 tons of CO_2 in a few hours. The gas, which is heavier than air, flowed down through two valleys, **asphyxiating 1700 nearby villagers and thousands of cattle**." Note 7.8h

Hydrocrater science is in its infancy with a bright future ahead of it. With this introduction, we are better prepared to understand the centuries-old crater debate among a relatively small group of geophysicists. Understanding this debate is an important step in understanding the whole hydroplanet model. Approaching the crater debate with the luxury of the hydrocrater model knowledge will add a completely new dimension to the picture, and open new doors of discovery.

7.9 The Crater Debate

Venice, summer of 1609, Galileo Galilei was intrigued by a newly created apparatus called a perspicillium; two lenses in a tube that made a faraway steeple appear quite close. He immediately made one of his own, which was nearly ten times stronger than the original perspicillium, and called it a telescope. Aiming his new device toward the moon, he saw craters, mountains, and seas, objects that profoundly affected him and the astronomers and scientists who followed after. Considerable debate ensued throughout the coming centuries on the matter of the craters' origins.

During the 1960s, America was in a heated world race to send a man to the moon and learning as much as could be known about the lunar surface became paramount. The nations rallied behind their astronauts—cosmonauts if you were Russian—and scientists involved in their respective programs, bestowing celebrity status upon many of them.

Outer-space mania gripped America as the first episodes of the hugely successful and long-running Star Trek television series hit the airwaves, and the thrill of seeing an object made by an unearthly traveler, such as a meteorite crater was fantastic.

Since then, science has romanticized the notion of impact craters and 'discovered' countless ancient impact crater sites, some famously blamed for Earth-wide extinction events, such as the dinosaur-ending Chicxulub event off the Yucatan peninsula in Mexico.

The Crater Debate is said to have been settled, but the evidence in the last subchapter has demonstrated otherwise, showing that modern science has never really understood hydrocraters. Can the Crater Debate be settled if no one has a true sense of *how* the majority of the craters were formed? It is far from settled, and there is much to learn about the Crater Debate in this chapter, where we will visit and evaluate evidence that seems to have been left undiscovered.

The Origin of Craters

The last two subchapters introduced the concept of hydrofountains and hydrocraters and the extraordinary claim that these processes have had substantial influence in changing the face of the Earth, perhaps more than any other geological process, *including surface erosion*. In our ongoing quest, the next four subchapters address the topics of impact craters, meteorites and their relationship to hydrocraters.

The Earth's moon, moons of other planets, and even some of the smaller planets have little or no atmosphere and provide a wonderful opportunity to study the effects of celestial body formation. Because they lack an atmosphere, these bodies have no weather to deface their surfaces, affording us an un-eroded look into their past, an invaluable help as we attempt to comprehend how our own planet was created.

Without question, the crater is a dominating feature on viewable planetary surfaces, where they occur with more frequency than any other landform. Until we can correctly discern how these craters formed, their origin, as well as the origin of the planets and moons upon which they reside *must remain* a mystery.

There are two possibilities; the craters were formed either by a downward force from above the surface, or by an upward force from beneath the surface, a fact not lost on the earliest scientists. A contemporary of Isaac Newton, Robert Hooke made

"...the history of lunar science is largely the history of the debate on the origin of craters."
Paul D. Spudis - The Once and Future Moon

these observations over 300 years ago:

"In 1667 Robert Hooke had dropped bullets into a stiff clay, creating little **impact craters**. However, he had also boiled a mixture of powered alabaster and water, and the **bursting bubbles had also formed craters**." Note 7.9a

Two Competing Theories

It is curious that given the importance of crater formation, the debate continues today with the two strongest contenders being:

Volcanic Theory and Impact Theory

The volcanic theory is of course based on an explosive force emanating from beneath the surface while the impact theory assumes an explosive force from above the surface. After analyzing these two theories, we can begin to understand why there has been such a long-standing debate about crater formation.

Volcanism is traditionally thought of as magma rising through the crust where it flows out and onto the surface from the caldera of a volcano. Conically shaped volcanic structures are not commonly discernible on the Moon. Without an understanding of hydrofountains and hydrocraters, volcanic activity on the Moon does not seem to make much sense and most scientists have abandoned this idea.

The impact theory found one of its first devotees in the early 1940s with Ralph Baldwin, an amateur astronomer. At first rejected by three astronomical journals, he continued his lunar studies and found observational support for his ideas by examining bomb craters from the aftermath of Allied bombing in Germany. He graphed his observations and determined the relationship "too positive, to be fortuitous…" Although he came close to a truly "fortuitous" discovery, the scientific knowledge of the day—absent the hydroplanet model—left him with little else to choose:

"The only reasonable interpretation of this curve is that the craters of the moon, vast and small, form a continuous sequence of explosion pits, each having been dug by a single blast. **No available source of sufficient energy is known other than that carried by meteorites.**" Note 7.9b

The impact theory continued to gain momentum into the 1970s with the ongoing space exploration effort. Knowledge about high-speed impacts would come later, in the mid 1990s, until then, through the 1970s, 80s and into the 90s, there were few direct observations of high-speed impact events in the laboratory or in nature. Terminal velocity experiments, bombs, even bullets shot from a rifle produced nothing near the velocity required to understand high-speed or hypervelocity impacts. At the low end, a hypervelocity simulation event requires a speed of 10,000 meters per second (22,700 miles per hour) whereas the speediest bullet moves at the sluggish rate of only 1,500 meters/second. Observed high velocity impacts create an entirely different explosion and result in crater properties not previously expected.

Even though most researchers think the erstwhile Crater Debate is settled, it never really was. The UM will for the first time, present scientific evidence gathered from many sources to demonstrate that neither the volcanic magma theory nor the impact theory correctly accounts for the vast majority of the craters on the Moon or other celestial bodies.

The Great Crater Debate

Paul D. Spudis, a geologist with the Lunar and Planetary institute in Houston, said it best in his book, *The Once and Future Moon*:

"...the history of lunar science is **largely the history of the debate on the origin of craters**." Bib 129 p14

The Moon is a planetary 'geological model' due to its lack of weather-induced erosion and the general belief that the Earth and Moon were formed together. Much of modern geology is based on our understanding of lunar science, and lunar science in turn, is based on the presumed origin of the *craters*.

In the Rock Cycle Pseudotheory, subchapter 6.2, we learned from a 1987 article in *Nature*:

"In astronomical terms, therefore, **the Moon** must be classed as a well-known object, **but astronomers still have to admit shamefacedly that they have little idea as to where it came from**. This is particularly embarrassing, because the solution of the mystery was billed as one of the main goals of the US lunar exploration programme." Note 7.9c

Since the origin of the Moon is directly connected to the origin of the craters on the Moon, and since we have **little idea** as *to where the Moon came from*, how can we be sure we know where the craters came from? Most lunar scientists believe as Spudis does, who wrote this in his book, *The Once and Future Moon*:

"After an extended debate lasting over 200 years, **we now know that the vast majority of the craters on the Moon are formed by the impact of solid bodies with the lunar surface**." Bib 129 p24

How do scientists "now know" where craters came from? Looking further in the scientific literature, we find the debate not necessarily settled:

"When NASA first considered landing man on the moon, a **great debate about the nature of lunar craters fired up**. As understanding of the differences between impact and volcanic craters developed, **one class of volcanic craters (maars and tuff rings) were found to have similar profiles as impact craters**.

"**This debate still continues to this day...**" Note 7.9d

It is rare to read a statement like this one, where the origin of Moon craters is *still* considered a debate. Almost every science textbook and commentary states the Moon's craters are of 'impact' origin. This almost universal acceptance of impact origin has grown over the last two decades; nevertheless, a small minority of researchers say the debate still continues to this day; why? Because the unanswered questions from both theories (impact and volcanism) keep stacking up as *new* technology and *new* observations add more to the picture.

From the 1998 McGraw-Hill encyclopedia of science and technology, we see the old paradigm of these two theories has become more and more "complicated":

"Basins on the Moon's near side, namely, Imbrium, Serenitatis, and Crisium, appear fully flooded. These were maria created by giant impacts, followed by subsidence of the ejecta and (probably much later) upwelling of lava from inside the Moon. Examination of small variations in Lunar Orbiter motions has revealed that each of the great circular maria is the site of a positive gravity anomaly (excess mass). **The old argument about impact versus volcanism as the primary agent in forming the lunar relief appears to be entering a new, more complicated phase** with the confirmation of extensive flooding of impact crater by lava on the Moon's near side, while on the far side, where the crust is thicker, the great basins remain mostly empty." Bib 12 p1247

No matter how hard similar pieces of the puzzle are forced together, they will not contribute to the overall picture of the puzzle if they do not belong there. Force-fitting incorrect pieces only serves to make the picture more "complicated," which is what the new lunar gravity measurements have done to the crater debate. Truthful geologists, like John W. Valley acknowledge:

"...although the **impact model** for the forming the Moon is widely accepted, **significant questions persist**." Note 7.9e

Others claim there is no debate, and even go so far as to say there is no doubt:

"Another conclusive observation is that nearly every rocky or icy planet and moon visited by spacecraft shows the same kind of cratering that we see on our Moon. This multiplanet evidence demonstrates overwhelmingly that nearly all craters in the solar system (including the Moon's) were formed by high-speed collisions caused by impacting asteroids and comets. Volcanism did occur on the Moon as vast lava flows that formed the maria, **but almost all lunar craters are**, **without a shred of doubt**, **impact scars**." Note 7.9f

The UM revisits the crater debate with new, never before considered evidence that proves science does not know how craters were formed, on the Moon or on the Earth! Historically, scientists have claimed "we now know" that man could not fly, that infection did not come from gentleman-doctor's hands, and that the Moon did not have craters, yet unarguably, physical evidence proved otherwise. This has proved to be both a blessing and a curse in science. No three words have gotten scientists into more trouble than "we now know." As a result, and for this very reason, the truth—to know what is, was, and will be—is covertly avoided in modern science. The key to discovery is not just knowing *what is*—it must include knowing what was, and what will be too. If the test of time is not adhered to, every scientist that "now knows" will eat his or her words for lunch.

The Shoemaker Impact

Eugene M. Shoemaker was one of the strongest proponents of the impact theory. Shoemaker, a renowned 'expert' on craters appeared on many television series, and was well-known as NASA's official geologist who prepared the astronauts for their visit to the Moon, made this insightful, yet dogmagmatic claim in 1976:

"... 'the impact of solid bodies **is the most fundamental of all processes that have taken place** on the terrestrial planets... **Without impact, Earth, Mars, Venus, and Mercury wouldn't exist**. Collision of smaller objects is the process by which the terrestrial planets were born.'" Bib 119 p235

Unfortunately, what Shoemaker did not realize when he made this statement is that the impact theory he referred to was based on the incorrect assumption that the planet-forming accretion theory had been proven. The reader will recall that In subchapter 5.5, several statements from leading astrogeologists showed that the accretion theory "does not stand up to elementary scrutiny."Note 15.5d Shoemaker made it very clear that he believed craters were not just a fundamental process, but they were *the most fundamental process* to have taken place on the terrestrial planets. In other words, if we cannot clearly understand craters, we can have no hope of understanding the origin of the planets themselves. Shoemaker was actually right about that!

Is Shoemaker's claim that—"without impact," the terrestrial planets "wouldn't exist"—*true?* If the majority of the craters on the Moon and on the planets were formed by a *process* other than impact, how might that affect the entire modern science paradigm of Earth's creation as well as the formation of Mars, Venus, Mercury, and the host of satellite moons? It certainly would be profound.

As the UM has shown in almost every chapter, the reason science cannot answer many fundamental questions is that many of its fundamental theories are wrong. We will see the application of the Hard Question Principle many times in this subchapter. Hard questions truly are easy once we know the answer, but when scientists don't know the answer, they often resort to scibabble— incoherent musings.

One such musing is Shoemaker's claim that without impact,

Fig7.9.1 – Eugene M. Shoemaker, seen here in the 1960s, greatly influenced modern science with his impact theory. To his credit, he brought together geology and astronomy like no other scientist. He became famous during the Apollo space program as NASA's lead geologist. The co-discovery of comet Shoemaker-Levy and its eventual plunge into Jupiter in 1994 returned Shoemaker and planetary science to the public eye. Unfortunately, Shoemaker's ideas that *impact* was responsible for the creation of the planets and moons and for the Arizona 'Meteor' Crater are unproven; the empirical evidence revealed in the UM confirms that he was wrong.

Shoemaker Impact Statement

"Without impact, Earth, Mars, Venus, and Mercury wouldn't exist."

the planets would not exist. We return to **The Shoemaker Impact** later in the discussion.

First, we will consider the empirical evidence that does exist in support of the Impact Pseudotheory. (Recall that a pseudotheory is a false theory taught as fact, clearly the case with the impact theory as it is in every modern science textbook.)

The Impact Pseudotheory

Previously we noted how the impact theory is unsubstantiated because of the lack of observational data. The observation to which we refer is simply that:

> **No one** has actually seen a large hypervelocity impact event **and performed a close up observation of the impact crater**.

The consequence of this fact is more far-reaching than one might think. An entire new field of science was invented over the last several decades because researchers believed the Earth had hundreds, perhaps thousands of large impact craters and that the Moon's craters are impact craters. *No one* has studied an actual impact crater from a large, hypervelocity impact event *known to be such, by eyewitness account*. This means investigators have **assumed** to know what **large** hypervelocity impact craters look like. There are at least three known impact events available for our review. Unfortunately, they are not close enough for detailed or first-hand investigation. They are:

1. Deep Impact (2005)
2. Shoemaker-Levy comet impact into Jupiter (1994)
3. Mars Impact Events (2006)

Deep Impact (2005)

One of the latest research projects involving hypervelocity impact was the 23,000 mile per hour (10 km/sec) impact of Comet Tempel 1 by a copper satellite impactor dubbed Deep Impact. Fig 7.9.2 is a photo of the actual impact event, an amazing engineering accomplishment. The mission was able to provide much insight into the makeup of hydrocomets; *no images of the actual impact crater were taken*. Researchers were unsure of the density of the surface that was impacted. Even if images of the impact crater could be obtained, the surface of Comet Tempel 1 is not likely to be similar to the crust of the Moon or the Earth and therefore, not a suitable example of what a high-speed impact crater would look like on either of their surfaces. Perhaps a future flyby will allow close-up images of the impact crater for further evaluation. At the time of the impact, the resulting dust cloud was too large and lasted for too long to obtain any meaningful surface images of the new crater.

The 1994 Jupiter Impact

Another astonishing high-speed impact event was the 1994 impact of Comet Shoemaker-Levy into the planet Jupiter. Images of the impact are captured in Fig 7.9.3, taken by the Hubble Space Telescope. The infrared photo came from the University of Hawaii's 2.2-meter telescope. In an article chronicling *Hubble's Top 10* best discoveries, Jupiter's impact by the comets ranked number one:

"From a cosmic perspective, the impact of Comet Shoemaker-Levy 9 into Jupiter was unremarkable: the cratered surfaces of **rocky planetary bodies** and satellites **already bear testimony** that the solar system is a **shooting gallery**. From a human perspective, however, the collision was a once-in-a-lifetime event: it is **thought** that on average a comet plows into a planet only **once every 1,000 years**." Note 7.9g

It may be that some find comfort that the cratered surfaces of rocky planets "bear testimony" that we live in a shooting gallery, but the researchers studying the spectacular Jovian Impact have largely forgotten that craters are formed in more than one way. On Earth, for example, where one can easily observe craters first-hand, researchers have to acknowledge that hydrocraters outnumber the assumed impact craters by more than 100 to 1—yet on celestial bodies such as the Moon, they have been lulled into thinking just the opposite—that almost all the craters are from impact.

Additionally, we see it is common for writers to insert a little theory into their comments, ideas that have no basis in fact. Since this was the only time humankind has seen and recorded a heavenly body colliding with another heavenly body, we have no clue how often this can actually happen. There is not enough data to make a statistical guess on how often impacts occur and the comment of "once every 1,000 years" is simply pure science writer embellishment.

Two important facts about the Jupiter Impact can be stated. First, no craters were observed in the event and second, there was an enormous amount of heat produced, which lasted for days, clearly establishing that an impact of similar magnitude on the Earth or the Moon would produce tremen-

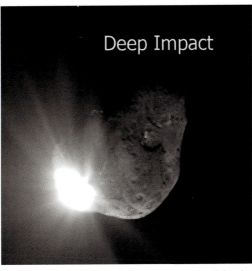

Fig 7.9.2 – In 2005, Comet Tempel 1 was struck by the Deep Impact space probe, a satellite designed for high-speed impact studies. Although the project revealed some of the comet's secrets, no photos of the impact crater were obtained. Thus, the mission provided no evidence of what a high-speed impact crater would look like.

dous heat, likely resulting in a significant amount of melted material. The importance of this will become more apparent shortly.

Mars Impact Events (2006)

In a 2003 paper on hypervelocity impacts, astrophysicists stated:

"...**no large-scale impact crater forming event has ever been observed** on a solid target planetary body..." Note 7.9h

This confirms what has been previously stated; however, in 2006, images from Mars were obtained that showed a dozen or more impact events. These were discovered when images of previous Martian flyovers, just a year or two earlier, were compared to more recent ones. The Martian surface had been impacted by an otherworldly traveler. Fig 7.9.4 illustrates the before and after images of one of the newly discovered impact events. Unfortunately, the crater is too small to see much detail, but a large ejecta blanket, approximately 20 times the diameter of the crater can be seen. Moreover, ejecta rays 50 times the diameter of the crater

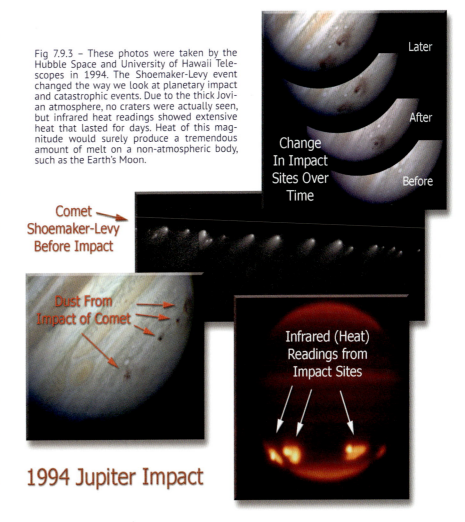

Fig 7.9.3 – These photos were taken by the Hubble Space and University of Hawaii Telescopes in 1994. The Shoemaker-Levy event changed the way we look at planetary impact and catastrophic events. Due to the thick Jovian atmosphere, no craters were actually seen, but infrared heat readings showed extensive heat that lasted for days. Heat of this magnitude would surely produce a tremendous amount of melt on a non-atmospheric body, such as the Earth's Moon.

7.9 THE CRATER DEBATE

Fig 7.9.5 – Left are two Slow-Speed impact events, the Mars Opportunity heat shield and the Genesis satellite. The Martian impact produced a very shallow crater that clearly shows the direction of impactor. Travelling at terminal velocity, the Genesis impactor left no crater, even in the softer soil.

are clearly visible.

In contrast, Fig 7.9.5 shows what slow-speed impacts look like. The Genesis satellite landed on Earth at terminal velocity, a speed generally under 200 miles/hour. Earth's atmosphere contributes to the slowing down of falling objects. Skydivers experience this when jumping out of an airplane, and similarly, the effect of air on speed is evident when a cotton ball is thrown simultaneously with a baseball. As objects increase in size, mass, or speed, the atmosphere has a decreased effect on speed.

To illustrate this concept further, a small handful of sand is sprinkled onto a paper from a height of three feet. The paper, held by its edges, has no problem catching and supporting the sand. Drop a two-pound rock from the same height and it will easily tear through the paper; the paper is representative of Earth's atmosphere—a thin layer of gas and the rock is a surrogate meteorite.

We can see the results of slow-speed impacts of small objects like the Mars Opportunity heat shield in Fig 7.9.5. Observations included a small shallow crater and minimal circular ejecta, less than the diameter of the crater. There is however, a dark ejecta ray clearly indicating the direction of impact.

Continuing our comprehensive examination of craters, Fig 7.9.6 shows four recent Martian impact events. Interestingly, there has been little discussion about the importance of these events or the attributes we are about to discuss. Perhaps this is because scientists have assumed that impact craters are common and comprise nearly all the craters seen on the planets and moons today.

However, the Mars Impacts may actually be very rare. We see some indication of this by comparing the sizes of the craters in Fig 7.9.6. They are all approximately 22 meters across, indicating that the meteorites that made them were about the same size. Moreover, the locations of the impacts are clustered in areas where other impacts took place roughly within a two-year

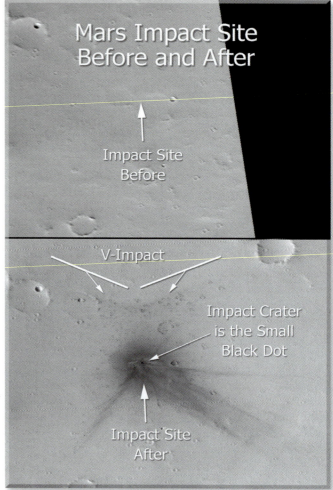

Fig 7.9.4 – The top image was taken April 18, 2003. The lower image of the same area was taken February 7, 2004. Between times, a meteorite created a 19.8 meter (65 foot) crater, located at the approximate center of the dark ejecta blanket. The ejecta material covers an area 20 times the diameter of the crater and ejecta rays extend over 50 times the crater's diameter. Both types of ejecta hold clues as to what high-speed impacts really looks like. At the top of the lower image a shock "V" is evident, indicating the direction from which the meteorite was traveling. V-Impacts result from oblique or low-angled impacts. This impact and several others that occurred around the same time and in the same general area of Mars are the first observations of actual high-speed impacts on a planetary surface. Unfortunately, the details about the craters remain sketchy.

Fig 7.9.6 – (Following Page) The illustration shows four actual impact sites identified in 2006. These are the first observed non-Earth high-speed impact craters. Although the impact event itself was not observed, orbital imaging revealed newly formed craters not visible during previous flyovers. These photos establish criteria for identifying high-speed impact events, including V-Impact ejecta patterns and ejecta blankets that are 20-50 times the diameter of the craters. Although erosional process on weathered planets will eventually destroy the ejecta, non-atmospheric bodies, such as the Moon, provide an excellent laboratory to view crater evidence not spoiled by weathering. The craters are evidence of how the surface of the Moon was shaped. The Mars Impacts ejecta patterns are *very rare* on other celestial bodies; most craters have short ejecta patterns suggesting a hydrocrater origin for craters on bodies throughout the solar system.

CHAPTER 7 THE HYDROPLANET MODEL

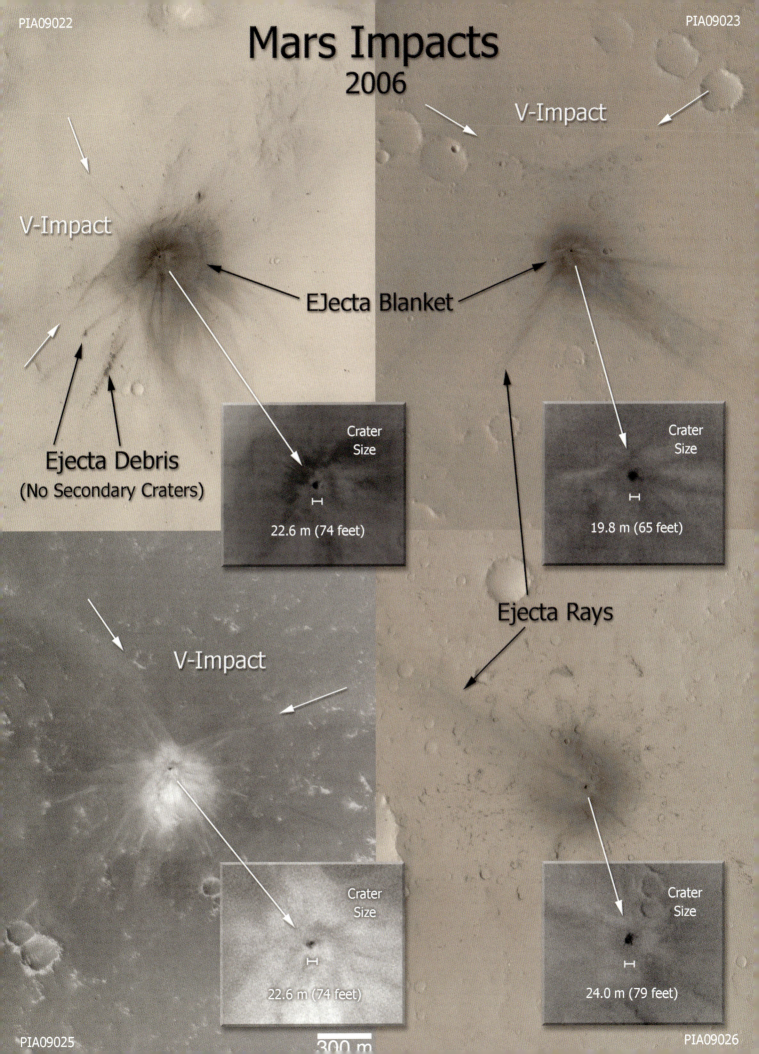

period. This is a strong indication that the meteorites were traveling in the same gravitational orbit. Future photos and examinations may be able to confirm this. Whatever the case, these are the very first high-speed planetary impact sites absolutely known to have come from impact and surprisingly, they have received very little attention.

V-Impact Signature and Ejecta Evidence

One of the primary reasons these impact sites are so important is the unmistakable V-Impact signature seen in three of the four photos in Fig 7.9.6. The **V-Impact Signature** is the unmistakable "V" pointing in the direction the meteorite was traveling. This indicates the impactor was traveling at an oblique angle toward the surface, not straight down or skipping across the surface.

Researchers noticed similar V-Impact "forbidden zones" decades ago on a few of the Moon's craters:

"Impact angle has a far more profound effect upon the ejecta blanket than it does upon the crater itself. The crater rim remains circular down to an impact angle of 10 degrees or less, whereas the ejecta blanket is noticeably modified at angles as high as 60 degrees. As the impact angle decreases from 90 degrees, the ejecta first show a preferential concentration on the downrange side of the crater. At incidence angles less than 45 degrees, **a wedge-shaped 'forbidden zone'** (a term introduced by D. E. Gault and J. A. Wedekind, 1978) develops uprange of the crater." Bib 161 p101

The V-Impact types of ejecta are very *rare* on our Moon laboratory, which is void of atmosphere or weather to change or alter ejecta patterns. The vast majority of hypervelocity impacts would occur at oblique angles, making the V-Impact Signature *the most common* ejecta pattern that **should be** visible on the Moon—yet this is **not** the case. This fact led impact research scientists to draw the following conclusion about V-Impact forbidden zones:

"Little **theoretical understanding** of these ejecta patterns yet **exists**." Bib 161 p101

How such "little theoretical understanding" of the V-Impact Signature can exist with such strong evidence of how they are formed is rather surprising, yet perhaps quite simple to understand, as the next few pages show. The mechanics of how V-Impact Signature patterns are made may be partly understood, but the wisdom of *why* they exist and why they are so rare on the Moon and other planets is not comprehended. Impactologists apparently do not understand nor do they seem interested in even theorizing why. Again, it is because of the notion that the majority of all craters were made by impact.

This impact confusion can be seen by comparing the rays emanating from the Mars impact craters in Fig 7.9.6 with the Moon's rays, radiating from craters seen in Fig 7.9.8. Each of

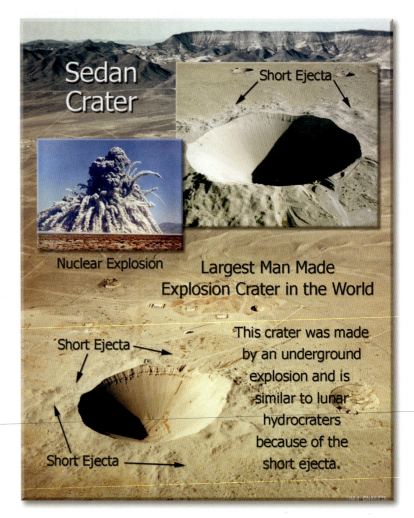

Fig 7.9.7 – Sedan Crater is a manmade subsurface explosion crater; a low-velocity pit with characteristics similar to hydrocraters, which are also underground explosion pits. Seen from a distance, the most distinguishing aspect of an underground explosion crater verses a hypervelocity impact crater is the ejecta. High-speed events have long ejecta and rays, easily discernable in the Mars Impacts diagram, whereas underground explosions produce smaller blankets of material. Another important feature of subsurface explosion ejecta is its circular, non-directional orientation; impact craters exhibit ejecta that reveal directionally oriented formation; evidence of the trajectory of the impactor.

the Mars impacts boast obvious ray systems, yet photos of the Moon are noticeably absent of craters that have ray systems, excepting those on some large craters. Why is this? The Hydroplanet Model has a simple answer—the majority of the craters are low velocity, non-impact hydrocraters, which is why they do not have long rays.

Because impactologists "believe" that almost all craters are of impact origin, they are forced into contriving fanciful theories of how the rays were obliterated:

"It is **believed** that the micrometeoroid-induced 'gardening' or overturn of the regolith eventually **obliterates the rays**, although **the precise mechanism of obliteration is unknown**." Bib 161 p108

Accordingly, impactology's theories on fading rays "probably" work, but are by their own admission not well documented:

"The **rays** of small craters **probably** fade faster than those of

Fig 7.9.8 – (Following Page) The Moon's surface is dominated by craters. Scientists have debated the origin of the craters for centuries; impact being the most recent and dominant theory. The recently viewed Martian impact craters shed new light on high-speed impact cratering, and give us clues to look for in the identification of other impact craters. Of the millions of lunar craters, very few exhibit ejecta blankets and rays that qualify them as being impact craters. Because there is no weather on the Moon, impact ejecta patterns and the V-Impact signature would be readily apparent on thousands of craters, if such patterns existed. Moreover, the absence of atmospheric gases to slow even the smallest meteorites would have allowed them to exhibit high-speed ejecta patterns. These would have been readily observable by the astronauts. NASA's vast photographic archives could be further reviewed too. But no such evidence exists, because most craters are not of impact origin!

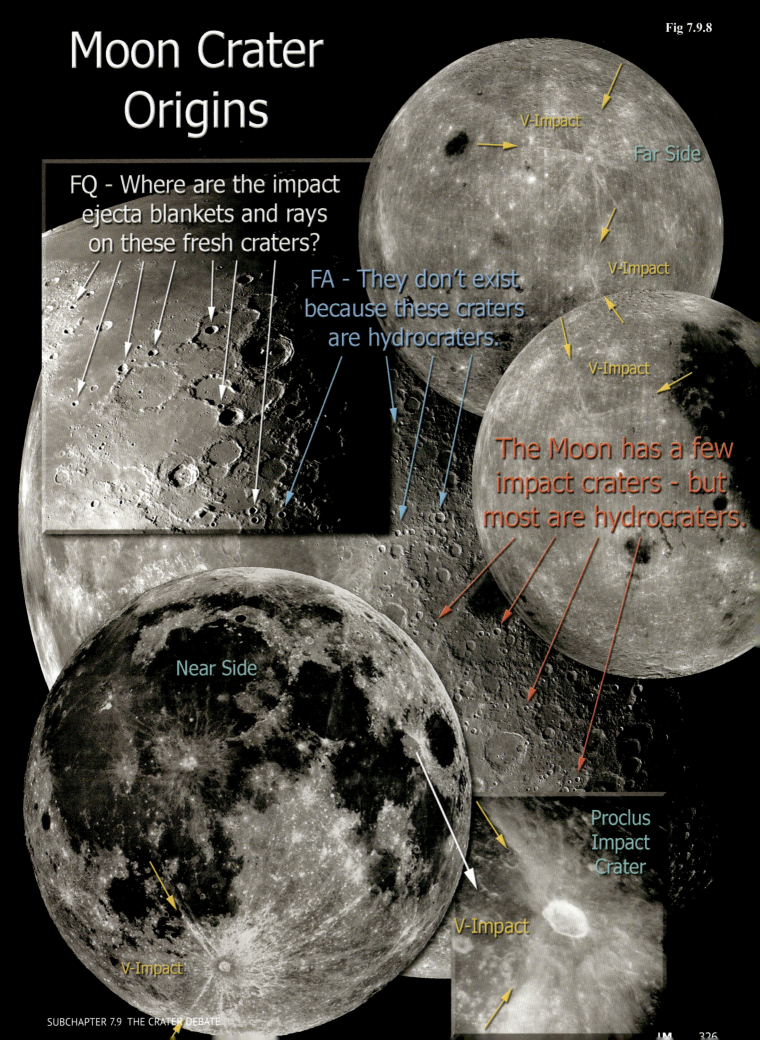

Fig 7.9.8

large craters, **but this cannot be well documented** on the basis of available data." Bib 161 p108-10

Amazingly, laboratory experiments have demonstrated how rays are made. The rays on Mars and the Moon from known impacts are simple to understand—yet impactologists acknowledge:

"The nature of rays **is not well understood**." Bib 161 p110

Why is the nature of rays not well understood? For the same reason the V-Impact Signature patterns are not understood—they *should* be apparent on most 'impact' craters—but most craters do not have them!

Continuing with our observation of the Mars impact craters, compare the diameter of the crater to the diameter of the ejecta blanket and the rays surrounding the craters in Fig 7.9.6. The ejecta blanket is 20 to 50 times the diameter of the crater. In contrast, relatively *small* ejecta blankets surround low-speed impacts and hydrocraters. The Sedan Crater in Fig 7.9.7 is a manmade explosion pit that exhibits a **small ejecta blanket and ray system**. The energy of the explosion came from beneath the surface, defying gravity and the weight of overlying sediment; it was a low-speed explosion. A low-speed explosion cannot propel sediment the same distance a high-speed explosion can, a fact of physics that has apparently been overlooked while studying impact craters.

A hypervelocity impact from above the surface would have gravity on its side. Significantly increased energy would be present, high heat would melt and fuse rock, sediment would be ejected at much higher velocities and larger ejecta patterns would be evident.

The Crater Debate has been chiefly centered on discovering the '*criteria*' that establishes a crater as being formed from impact versus being formed by a volcanic explosion. The contemporary Mars Impacts clearly show very large ejecta patterns relative to crater size, they exhibit rays, and they have V-Impact Signatures. For the first time, scientists have actual criteria for determining if a crater is from an impact event, making these observations very important. On Mars and more especially on Earth, time and weather will destroy most of the surface ejecta material, but on bodies such as the Moon that have no atmosphere, the presence or absence of **known impact evidence** is one way to determine a crater type, even without visiting the crater directly.

The Lunar Crater Origins Evidence

There are three new evidences with which to view the craters on our nearby neighbor, the Moon:

1. Physics of crater formation (low vs. high speed)
2. Actual Mars Impact ejecta and patterns
3. Hydrocraters and hydrocrater formation

With these evidences, we can look at the craters on the Moon in a completely new light—from a Hydroplanet perspective. Our first objective will be to discuss the craters as viewed from a distance, followed by a closer examination of surface Lunar features and craters based on images from the Lunar landings from the late 1900s.

Looking at the surface of the Moon in Fig 7.9.8, there are few "fresh" high-speed impact craters viewed from this distance, based on the absence of large-sized ejecta blankets and ray patterns like those seen on the Mars Impacts. There are four large V-Impact signatures (highlighted with gold arrows), but the majority of the craters do not have them, especially the smaller, newer craters. In fact, of the thousands of Apollo images reviewed, no large-sized impact ejecta blankets or rays were seen on any of the small craters the astronauts imaged while orbiting the moon. This was also confirming evidence that the Mars Impacts of 2006 are a rare phenomenon.

Some investigators turn to micro-impact research and draw the conclusion that micro-impacts eroded away the ejecta blankets and rays over billions of years. However, micro-impacts could not change the general ejecta patterns of *large* craters, only very small craters, on sizes measured in a few centimeters or inches. Micro-impacts are small enough that the human eye cannot see them without the aid of magnification. Relying on micro-impacts to eliminate kilometer sized crater ejecta would be like trying to smooth out the Grand Canyon with very fine sandpaper.

Most of the Moon's large craters seen in Fig 7.9.9 do ***not*** have impact ejecta because they were formed *underwater, as hydrocraters*. Take note how the surface around the craters is smooth, which suggests a water origin and counters the amazing statement by one of the foremost scholars in impact cratering. By completely ignoring the **lack** of ejecta material on the majority of the Moon's craters, he proclaims, "all impact craters are surrounded by a deposit of debris":

"Essentially **all impact craters are surrounded by a deposit of debris ejected from the crater interior**." Bib 161 p89

As we will establish with the following Glass Evidence, one of the longest ongoing debates in science concerns fine Moon dust that covers the Moon. This so-called dust is not from im-

Fig 7.9.9 – These lunar craters are miles across. A close look at the surface of the Moon reveals a *smooth* surface in the low-lying areas, which could not have come from "lava flows," as planetary geologists have proposed. Lava flows produce sinuous ripples or jagged edges as advancing melted rock cools and new lava covers old flows. Where is the "impact ejecta" around the craters in this image? The lack of ejecta is unequivocal evidence that these pristine craters were not made by impact. Instead, they were created in a manner similar to the craters at the bottom of Crater Lake and the ocean—hydrothermal vents. Rather than being absent, the ejecta was dispersed by water under which the eruption occurred.

pact, but from the Moon being a hydroplanet just as the Earth is and that the Moon dust is similar to the fine sediment found on the bottom of the Earth's ocean. The traditional volcanic theory for the surface of the Moon fails because there are no volcanic flows like those that we find on Earth's continents, which have distinct ripple patterns. Nor are there cone volcanoes and lava flows down valleys on the lunar surface as we see all over the Earth.

Some researchers have proposed that impacts *caused* volcanic eruptions on the Moon. Jay H. Melosh of the University of Arizona relates the following in a report titled *Impact-Induced Volcanism: A Geologic Myth*:

"The **idea** that the **impact of a large comet or meteorite may cause a volcanic eruption** is becoming entrenched in the geologic literature. In spite of the apparent appeal of this idea, **there is no evidence that it has ever actually happened, either on the Earth or on any other planet in the solar system**." Note 7.9i

Melosh then goes on to explain why impacts never caused any volcanism on the Moon:

"The total energy required to melt the basalts in many flood basalt provinces greatly exceeds the total energy available from the impacts that have been proposed to form them. Furthermore, study of planets such as the Moon and Mercury that do show signs of antipodal focusing from large impact basins show no trace of volcanism at the antipodes. **There is thus neither evidence for impact-induced volcanism nor a plausible mechanism for creating it. Proponents of an impact origin for flood basalt provinces should probably look for more effective causes**." Note 7.9i

Besides Melosh's point of inadequate evidence for impact volcanism, proponents of 'impact basalt creation' will never find true evidence for basalt formation from a dry impact because basalt is only formed in a hypretherm, a subject we will cover in depth in subchapter 18.7. The basalt forming mechanism that researchers should be looking for is the same for all non-melted minerals—a hyprethermal environment.

The Moon Crater Origins diagram, Fig 7.9.8, also shows that there are only a few large, high-speed impacts exhibiting a large ejecta blanket and ray patterns. How were the rest of the craters formed? There are two new possibilities; phreatic explosion pits (hydrocraters) or high-speed impact in a different environment; one that planetary scientists have probably never given thought to—*a water environment*.

After the solid portion of the Moon crystallized out of water (refer to the Law of Hydroformation, subchapter 7.4), the water evaporated off the surface of the Moon, but this took time. In later stages, submarine basalt and mudflows from hydrofountains filled the maria basins and low-lying areas. This underwater environment left the surface of the Moon looking like the floor of the Earth's oceans, seen previously in Fig 7.8.13 and 14. Most of the craters on the Moon are hydrocraters, especially the newer ones, but it is nearly impossible to be sure simply by looking at them from a distance. Hopefully, future surface expeditions will allow seismic and other observations to clarify this.

Impacts could have occurred in shallow water or muddy terrain. Direct evidence of this will be seen when we examine the Tyco Crater. Impacts into shallow water or mud have a different ejecta pattern as compared with hard surface impacts, such as the 2006 Mars Impacts. High-speed impacts into soft or wet material exhibit characteristics similar to hydrocraters, one reason it is hard to distinguish the origin of older lunar craters.

However, the overriding point of our discussion is that cratering is the result of either impact into a soft, water saturated surface or hydrocratering from hydrofountain activity. The larger point to be made is that the Moon is a **Hydromoon**, not a melted, magmaplanet moon.

The 'Smoking-Gun' Glass Evidence

It was in the Magma Pseudotheory chapter where evidence was first presented, establishing that the Moon did not come from a melt. The surface of the Moon is certainly not glass, which it would have been had it cooled from a molten ball of magma. We can look for Glass Evidence and know with a surety that the Moon was not created by impact as Shoemaker theorized because its surface is not littered with broken glass-like rock.

This 'smoking gun' has been overlooked by impactologists. High velocity impact creates **glass** and since the most widely accepted notion of crater origin is high-velocity impact, the astronauts should have found broken glass all over the surface of the Moon. Why?—the answer is simple:

> A high velocity impact on a solid surface planet produces heat and glass—lots of it.

Understanding Hypervelocity

When we speak of hypervelocity or high-speed impacts, we are speaking of the speed that planets travel around the Sun in the solar system. Because lunar craters are said to have been created by rocks or meteorites traveling around the Sun, we must account for their high-velocities in experiments if we are to reproduce an impact explosion similar to what might happen in nature.

The Earth travels around the Sun at about 30 km/sec, Mars averages approximately 24 km/sec, and Jupiter runs at 13 km/sec. Many meteoroids are thought to come from the asteroid belt that lies between Jupiter and Mars and velocities can be

Fig 7.9.10 – Projectiles launched at speeds up to 17,000 mph (7.6 km/s) impact objects at the NASA Ames Researcher Center.

expected to fall between 13 and 24 kilometers per second for those bodies. If two such bodies were traveling towards each other, the speed would be considerably higher. The 1994 Jovian impact of comet Shoemaker/Levy was reported by NASA to have been 60 km/sec. Note 7.9j

Once we realize how fast the impact bodies are traveling, we can understand that the resulting explosions will be extremely hot. It is *impossible to have low-temperatures* with high-velocity impacts:

"In shock-wave compression, pressure increases as a wave front passes through the medium as a result **of high-velocity impact**... Temperature rises steeply with pressure in such dynamic compression experiments. The dynamic compression can provide simultaneous high P-T conditions, **but not** high-pressure, **low-temperature conditions**." Note 7.9k

The high temperatures associated with high-velocity impacts can be thousands of degrees. Knowing this is extraordinarily important because temperatures of thousands of degrees will melt surface rock, turning it into a *glass*. A paper published in the *Journal of Geophysical Research* reported a temperature of 5,000 K for an impact with a speed of only 5.5 km/sec, well below the typical high-speed planetary impact:

"This figure shows that the initial temperature of the cloud of the vertical and oblique impacts is about **5,000 K**." Note 7.9l

Five thousand degrees Kelvin is approximately equal to 4,727 °C (Celsius) or 8,540° F (Fahrenheit). Since quartz melts at about 1,700 °C, it is evident that the temperature present in an impact at the low-end velocity of 5.5 km/sec are well above melting and vaporizing both the impactor and the surface material, a fact that researchers seem to have overlooked.

Hypervelocity Laboratory Impact Studies

Researchers have known for some time that melting and vaporization during impact cratering events are *important*:

"The production of **melt and vapor is an important process** in impact cratering events." Note 7.9m

The University of Arizona scientists go on to say:

"Because **significant melting and vaporization do not occur** in impacts at velocities currently achievable **in the laboratory**, a detailed study of the production of melt and vapor in planetary impact events is carried out with hydrocode simulations." Note 7.9m

This was the first paragraph of a 1997 journal article titled, *A Reevaluation of Impact Melt Production* in which it stated that "significant melting and vaporization do *not* occur" in the laboratory at velocities currently achievable.

*What if the scientific community, as a whole had simply **overlooked evidence of melt** in all previous experiments?*

What if *glass* or *melt* was the common denominator in all previous hypervelocity experiments, and scientists had just missed that fact? The UM's extraordinary claim is that *large* amounts of glass or melt are present in hypervelocity impacts in the laboratory; so let's examine the extraordinary evidence. We will review five hypervelocity-impact journal articles reporting the results of experiments conducted prior to the publication of the 1997 journal article, *A Reevaluation of Impact Melt Production*.

The **first article** was published in 1983 in the *Lunar and Planetary Science* journal. It discussed the impact of quartz sand targets with aluminum spheres at a velocity of 5.9-6.9 km/sec. The experimenters reported:

"Hypervelocity impact of al-projectiles into quartz sand results in the **complete melting of the projectile, partial melting of quartz sand**, and chemical reactions between aluminum and silica. The liquid projectile is spread over the dense, shock lithified target sand in the **center of the crater**." Note 7.9n

As we can see, there was significant melting observed. This experiment was a vertical impact, perpendicular to the impact surface, producing glass in the center of the crater. An impact from an incoming body at an angle—called an oblique impact—will cause the melt or impactor to be off-center of the crater, based on the direction of travel.

Later in the report, the researchers mention a very important observation that we will want to remember for the next subchapter, the Meteorite Model. They note, "it should be possible to detect remnants of the meteorite projectile" and that the remnants should be "fused metal" and "silicate melt" found on regolith (loose heterogeneous rocks):

"The results of this study bear on the understanding of regolith-forming processes on planetary surfaces. In agglutinate particles of natural regoliths (e.g. lunar, asteroidal) it **should be possible to detect remnants of the meteoritic projectile** material which was not mixed with the target material. The projectile **should be present as fused metal** (iron meteorites) **or silicate melt** (stony meteorites) adhering to shock fussed and lithified agglutinates which originate from small scale cratering events in the regolith of all planetary bodies without atmosphere." Note 7.9n

Perhaps you might have already noticed the words "should be possible" because this (fused metal and silicate melt) is what should be present in the so-called impact craters of the Moon and Earth. Of course, it is not and we will be giving detailed examples as we go along.

The **second scientific article,** also published in 1983 appeared in the Journal of Geophysical Research. It involved impact craters that were reportedly covered with "copious amounts" of glass:

"Small-scale impact craters (5-7 mm in diameter) were produced with a light gas gun in high purity Au and Cu targets using soda lime glass (SL) and man-made basalt glass (BG) as projectiles. Maximum impact velocity was 6.4 km/s resulting in peak pressures of approximately 120-150 Gpa. **Copious amounts of projectile melts are preserved as thin glass liners draping the entire crater cavity**; some of this liner may be lost by spallation, however." Note 7.9o

Fig 7.9.11 is a photo of one of these impact craters, covered with "projectile melt" easily seen with the naked eye. It is certainly obvious from the example that impactologists may have missed the fact that large amounts of glass were produced in this hypervelocity impact, because of the high temperatures.

The **third example** comes from the 1990 publication of the *International Journal of Impact Engineering* and covers a German experiment with impact velocities between 6 and 17 km/sec:

"Figure 3b shows the impact at a velocity of 6 km/s. Residue of the projectiles **can be seen at the walls of the crater**. Figure 3c shows an impact at a velocity of 8.4 km/s. The impacting **projectile is vaporized** and ionized. **Condensed droplets of**

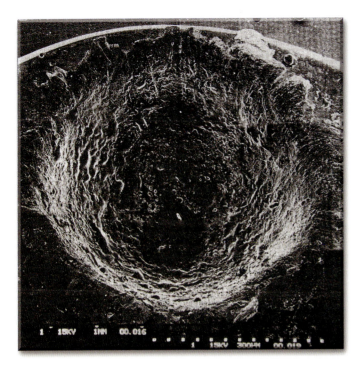

Fig 7.9.11 – A basalt-glass projectile with a velocity of 5.4 km/sec formed this 6.1 mm crater in aluminum. The experimenters described the crater: "Note that the **entire crater** is relatively **evenly lined with projectile melt**". The researchers go on to say that the melt is, "**easily recognized even by the naked eye**". This photo is from the second of five journal articles describing hypervelocity melt.

the projectile can be seen on the bottom of the crater." Note 7.9p

Once again, the hypervelocity impact event produced comparatively large quantities of melt and vaporized material. The investigators also identified a *scaling law* relating a ratio of crater depth to projectile diameter that *held for the entire velocity range*:

"The experimental results discussed in this paper indicate, that for a spherical glass projectile impacting on gold, tungsten, aluminum and iron, no significant difference can be found between the velocity regime up to 8 km/s and the velocity regime beyond 8 km/s up to 17 km/s.

The crater volume seems to be a linear function of the kinetic energy of the impacting projectile and the $V^{2/3}$ scaling law for the **ratio of the crater depth to the projectile diameter applies for the entire velocity regime of 2 – 17 km/s** for the glass-aluminum combination, where projectile and target have the same density." Note 7.9p p280

This scaling law becomes important as we compare experiments. It demonstrates that the larger the projectile is, the deeper the crater it will excavate. This is very interesting when we look at the Moon's surface and see large craters with very shallow depths, an observation not supported by impact evidence, but well supported by hydrocrater formation evidence.

The **fourth article** comes from the 1993 journal, *Nature*. Experimenters here used aluminum projectiles to impact granite at 4 km/second. The researchers report:

"Rock embedded in the Al [aluminum] flyer plate is obvious-ly shocked, **containing abundant diaplectic glass**." Note 7.9q

The researchers observed abundant glass and mentioned that:

"…our **experiments can be scaled directly to larger events** by using numerical codes. These calculations show that crater features scale linearly with projectile size…" Note 7.9q

Being able to scale such an experiment to larger events is noteworthy because science has *not* been able to directly observe large, naturally occurring hypervelocity events and their resulting craters to see how much glass is actually present. For this reason, the abundance—or lack thereof— of glass has been overlooked. Through scaling, we can ascertain that larger projectiles should exhibit more melt and that glass ***should be*** observable on a large scale at impact craters here on Earth.

The **fifth hypervelocity experiment** report can be found in the 1993 *International Journal of Impact Engineering*. Iron micro-spheres were used as projectiles on various materials, including a quartz target. The experiment is judged the best of the five, employing the most realistic type of projectile and target as compared to natural meteorite impacts into a silicate crust. Moreover, a velocity of 3-30 km/sec was used, much more representative of the speeds found in nature. Here were the results:

"Quartz forms **a definite melt region around the crater** in the middle of a larger spallation zone." Note 7.9r

Like the four preceding experiments, this experiment unmistakably produced a "melt region" or large glass area as compared to the crater.

Thus, we can conclude that the 1997 article, *A Reevaluation of Impact Melt Production* was grossly mistaken in stating that:

"…significant melting and vaporization **do not occur** in impacts at velocities currently achievable in the laboratory…" Note 7.9s

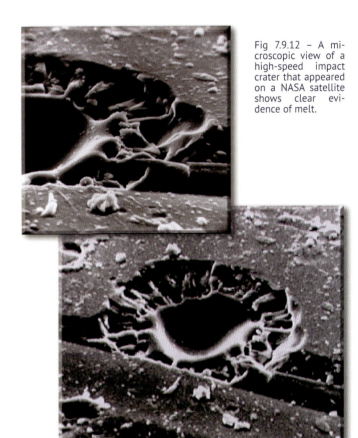

Fig 7.9.12 – A microscopic view of a high-speed impact crater that appeared on a NASA satellite shows clear evidence of melt.

> "Copious amounts of projectile melts are preserved as thin glass liners draping the entire crater cavity…"

Significant melting and vaporization have been produced in the laboratory. This significant fact has been **overlooked** and omitted from the list of *critical criteria* used in determining whether a crater was made by a high-speed impact. A high-speed impact on a hard surface would create an enormous amount of heat and melt—it is just that simple.

Larger Impact Crater Equals More Glass

The one thing these researchers did get right was that the larger the meteorite, the greater the energy that would be produced, which means more melt would be generated:

"We conclude that, for planetary-type impact events where the impact velocities exceed about 15 km/sec, **the amount of melt and vapor can be scaled by the energy of the impactor.**" Note 7.9s p421

Another group of researchers from Canada confirmed this statement:

"Both observations and model calculations indicate that the **relative volume of impact melt increases with crater size.**" Note 7.9t

The final summation is that every high-speed impact crater should contain large amounts of melted material, or glass, readily abundant and easily observable. If lunar craters were primarily impact craters, astronauts should have collected this abundant material as they walked the craters.

Until the 1980s, scientists did not have the technology to produce high-velocity (greater than 10 km/sec) impacts in the laboratory. It was premature for Shoemaker and others to theorize in 1976, that so many of Earth's craters and others in the solar system were from impact, without ever having observed a 10km/sec impact event. Because most meteorites enter our atmosphere at speeds well above 20 km/sec, extraordinarily large quantities of glass should be abundant at the impact sites, and the Moon should be covered with glass. Broken glassy rocks should be everywhere, but they are not, leaving us to speculate; how did the impact notion become so well accepted?

The Trinity Glass Evidence

Because there are no large, naturally occurring hypervelocity impact craters where the impact was directly observed, we are left with lab-produced micro-impacts to determine how much melt is produced from high-speed impacts. Fortunately, there is one large artificial hypervelocity impact event to verify the observations made in the lab.

On July 16, 1945, the world watched as the nuclear age was ushered in, in the deserts of New Mexico. The world's first atomic bomb, known as Trinity, was detonated, creating a large hypervelocity impact crater. The story is told in the book, *Day of Trinity*, where we read an excellent account of the crater created by the detonation of the atom bomb, which had been perched a hundred feet above the desert sand:

"Anderson was unprepared for the sight that greeted him. A half-mile away he saw what looked like **a great jade blossom** amid the coppery sands of the desert. Where the shot tower had once stood, **a crater of green ceramic-like glass glistened in the sun**. The fireball had sucked up the dirt, fused it virescent with its incredible heat, then dumped the congealing particles back on the explosion point. **They lay there inside a 1200-foot-wide saucer some twenty-five feet deep at the center**. The bomb, even from 100 feet up, had so pulverized the earth that the tower's concrete stumps, which once stood above ground, had been crushed to a depth of seven feet beneath the sand, The tower itself had completely evaporated. Within a mile of the crater there was no sign of life or vegetation." Note 7.9u

Fig 7.9.13 shows the tower, the mushroom cloud from the explosion, an aerial view of the darkened, glass covered impact crater and an up-close view of the glass on the ground near the crater. The glass was inches thick, "green ceramic-like glass" looking like a "great jade blossom."

Fig 7.9.14 shows a melt experiment attempting to produce similar green glass. Desert soil was placed in a metal container where it was melted using an acetylene torch. The red-hot melted soil cooled into a 'green jade blossom' similar to that of the Trinity event. Obviously, the heat from the torch melted the sand and pebbles into glass, but what about the trinity nuclear explosion, where did that heat come from? The following statement made by a geologist and accepted, as general wisdom by most people including scientists, is this:

"The nuclear explosion **melts and vaporizes** the surrounding rock…" Note 7.9v

It was not until photos of researchers standing inside the cavity of an underground nuclear explosion were viewed, that the realization that the nuclear explosion itself was not the source of heat that melted and vaporized rock. The underground cavity showed no signs of melted rock.

Heat melts rock, and the heat from these types of explosions comes from *friction*. It was hypervelocity frictional heat; soil moving against soil on the earth below the Trinity tower that generated heat sufficient to leave behind the glass jade blossom in the crater. If it would have been the heat from the mushroom cloud, airborne sand would have fallen to the ground as glass droplets or spheres, but this was not reported. The Trinity aerial photo also shows dark rays of glass emanating from the central blast point.

A similar sized blast of TNT would not have had the same effect on the soil as the Trinity blast because TNT does *not* produce a hypervelocity explosion. Only high-speed explosions—nuclear being about 9 km/sec—have the ability to create enough frictional heat to melt rock and make glass. This is further evidence why we should expect glass melt at ***all*** high-speed, natural impact craters that occur on a relatively hard planetary surface.

The inclusion of *glass as criteria* for determining an impact event has been excluded. Some of the 'so-called' hypervelocity impact craters have little or no glass, which countermands the physics of frictional heating.

On the other hand, hydrocraters can have it either way. They can be void of glass because they are not formed by hypervelocity explosion, however, underground frictional heating can produce steam and melted rock, both of which can come to the surface and be ejected. Of course, this confuses the investigators and keeps the crater debate alive.

Hydrocraters are misunderstood, as are actual hypervelocity impact craters, resulting in their being thrown in together. To this day, correct and definitive criterion for hypervelocity impact craters has yet to be established. To that end, however, the UM has shown that high velocity impact craters have:

1. Large ejecta blankets and rays
2. Large quantities of glass

Fig 7.9.13 – The first atomic bomb explosion occurred on July 16, 1945 at what is known today as the Trinity Site, located in New Mexico, USA. The nuclear bomb was suspended 100 feet in the air on the Trinity Tower. The blast created a 1200-foot wide crater 25 feet deep. The aerial view shows a darkened area of the blast, later described by those that came upon it as a **"green ceramic-like glass."** Many desert sands turn green or black when melted, just like the sand at the Trinity Site. Trinity glass is important because it is the only example of a large-scale high-speed impact event, and it clearly demonstrated the production of a significant amount of melted glass. Above-surface nuclear explosions can reach speeds of up to 9 km/sec, similar to laboratory hypervelocity experiments and some actual cosmic impacts.

Both of these are critical and definitive criteria for identification of hypervelocity impact craters, yet *neither* is recognized as such by impact geophysicists.

The Nuclear Crater Evidence

The Arizona Meteor Crater is also known as the Barringer Crater after D. Moreau Barringer, who purchased the site in 1903. The Barringer Crater Company operates the crater as a commercial tourist site and states this on its website:

"In **1963**, geologist **Eugene Shoemaker** published his landmark paper analyzing the similarities between the Barringer crater and craters created by nuclear test explosions in Nevada. Carefully mapping the sequence of layers of the underlying rock, and the layers of the ejecta blanket, where those rocks were deposited in reverse order, **he demonstrated that the nuclear craters and the Barringer crater were structurally similar in nearly all respects**. His paper provided the clinching arguments in favor of an impact, **finally convincing the last doubters**." Note 7.9w

Unfortunately, Shoemaker did *not* actually "demonstrate" that nuclear craters and Barringer crater were structurally similar and he certainly did *not* convince the "last doubters." Two prominent crater geologists at the time, W. H. Bucher and G. J. H. McCall, were absolutely unconvinced and in 1964, (a year after Shoemaker presumably convinced the "last

Fig 7.9.14 – Desert soil heated and melted by an oxyacetylene torch. The melted soil looks similar to the description of the Trinity glass because the iron minerals in the soil turn the melt green. The 1200-foot wide crater was described as a "great jade blossom" because of the melted and fused desert sand. This is what most rocks look like when they melt—they become glass! This fact remains unrecognized in the quest to pinpoint unambiguous criteria for the identification of high-speed impact craters.

7.9 THE CRATER DEBATE

Nuclear Crater Evidence

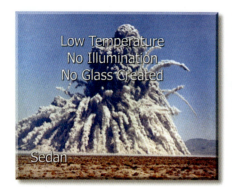

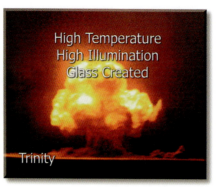

Slow-Speed Explosion (Underground)   **High-Speed Explosion (Above ground)**

Fig 7.9.15 – This is a comparison of the low-speed, subsurface Sedan nuclear explosion with the high-speed, above ground Trinity nuclear explosion. The low-speed Sedan explosion had a low temperature, no illumination and created no glass. On the other hand, the Trinity explosion was a high temperature, high illumination and left the entire crater covered with glass. Although both explosions were nuclear, the difference between them identifies the difference between impact-type craters and phreatic or subsurface explosions.

doubters"). McCall wrote in the journal of *Nature*:

"In the case of certain other craters, cryptovolcanic and plutonic igneous complexes, **far from reasonable and even misleading arguments have been advanced proposing impact explosion as an alternative to widely accepted geological processes**. The evidence cited is tenuous in the extreme, and much of the verbiage is simply conjecture: evidence, often far more weighty, cited by **earlier workers who have rejected impact explosion, has been ignored**." Note 7.9x

We will dive deeper into this later, but this example clearly shows how misleading some 'scientific' statements are and how uninformed the general scientific community is, even today, about the Crater Debate.

As it turns out, the Barringer Crater is not structurally similar to nuclear craters at all; it shares few similarities with the hypervelocity Trinity crater. Comparing Barringer to the *low-velocity* Sedan nuclear crater, many differences stand out. Sedan does not have a large flat floor like the Barringer Crater; Sedan is much more circular than Barringer; Sedan does not have a major fault line running through the crater; Sedan's ejecta material is homogeneous, not made up of completely different materials like Barringer Crater. What Shoemaker **did demonstrate** is that both the Arizona Crater and the Sedan Crater were formed by a **low-speed explosion**, but that was all that Shoemaker's physical evidence demonstrated.

Fig 7.9.15 compares two nuclear explosions, the Sedan and the Trinity. Although both are nuclear, a major difference is the blast velocity and resulting cratering. Trinity was an aboveground blast with high-speed dynamics, interacting with matter over a great distance. The high-speed nature of the blast caused high temperatures, high illumination and large amounts of glass, *all* characteristics of any hypervelocity impact. Conversely, Sedan was an underground explosion and exhibited *low-speed* dynamics, pushing sediment out of the way to form a crater. The low-speed of the Sedan explosion was similar to that of a hydrocrater. Because it came from underground, it resulted in a much smaller ejecta blanket and an absence of rays.

The Wabar Impact Crater Evidence

The Wabar impact crater site, first reported in 1932 by a British explorer is in a very remote area of Saudi Arabia. In 1995, E. M. Shoemaker and J. C. Wynn performed the first "detailed investigation of the geology" of the Wabar Impact Craters. The craters were from a reportedly recent fall and promised to hold significant answers in the study of meteorite impact sites. Fig 7.9.16 gives a broad view of the 116-meter diameter crater, the largest of the three craters. The smaller craters measured 64 and 11 meters respectively. Surprisingly, this important impact site has only been visited and reported on by one team of geologists and very few photos are available for study. The photos in Fig 7.6.16 were taken in 1965 by James Mandeville; they remain the best photos on record.

The Wabar impact is significant for several reasons. First, they can be eliminated from the hydrocrater category because there is no deep-earth water source, no diatreme and there are no directly associated faults. Next, the Wabar site demonstrates that with a comparatively small amount—still several tons—of iron meteorite material, an enormous amount of glass is formed, even though the speed of the impactor was at the low end of hypervelocity and with the sand absorbing much of the energy. The glass fallout contained a significant amount of the melted iron meteorite, a typical expectation of a hypervelocity impact event, due to the enormous amount of heat produced. Finally, the impact glass showed almost no sand abrasion, indicative of the young age of the fall and an aid in evaluating the characteristics of real impact craters.

The bottom left inset photos of the Wabar diagram show two of the primary iron meteorites recovered from the site, each representing several tons. Reportedly, only a small number of additional meteorite fragments were found, giving us a rather good indication that less than 10 tons of iron-nickel material found its way to the surface of the Earth. This smallish 10-ton meteorite (as compared to the speculative size of the Arizona Meteor Crater meteorite, which is said to have been 300,000 tons), was slowed significantly by the atmosphere, but remained speedy enough to melt a significant amount of sand and meteorite material. This is a critical point to remember, because a *larger* meteorite that remains intact as it enters the atmosphere will retain most of its speed, resulting in a higher energy impact and creating significantly *more glass* than that found at Wabar. We read of the glass at Wabar in the 1995 report by Shoemaker and Wynn:

"The exposed rims of the two largest craters **are mantled with fragments of black, slaggy glass called impactite (consisting of about 10 percent iron-nickel from the shock-melted meteorite and 90 percent local sand**) and chunks of instant rock that look like bleached sandstone." Note 7.9y

CHAPTER 7 THE HYDROPLANET MODEL

Fig 7.9.16 – The Wabar Impact Crater was discovered in Saudi Arabia in 1932 by Harry St. John Philby. Wabar is a recent meteorite fall showing very little aeolian sand abrasion on the impactite. It is one of the few impact sites not associated with any volcanic activity. Large meteorite fragments were recovered and the crater had copious amounts of impactite glass. The crater does not have any of the features that would identify it as a hydrocrater, faulting, a water source, or a diatreme. Wabar is important because it demonstrates how much glass would be created by a several-ton iron meteorite impact. The impactor velocity had to have been several km/sec to generate the heat necessary to melt the surface materials, but it could not have exceeded several dozen km/sec or it would have melted and vaporized the meteorites. A small amount of iron/nickel meteorite material, reportedly about 10% of the glass impactite, was mixed in with the ejecta. The Wabar Impact Crater evidence is referred to a number of times to establish that other 'so-called' impact craters are in reality hydrocraters. Photos are courtesy of James Mandeville who visited the site in 1965.

You can see this black layer (from the iron-nickel meteorite) of glass surrounding several of the pieces of impactite in Fig 7.9.16. This type makes up 10 percent of the glass at the site.

All the evidence from the lab-produced impact experiments are confirmed at the Wabar site. Shoemaker saw this but still let his prejudice convince him that other craters that did not exhibit these characteristics were still of impact origin. So why make a big deal about impact glass? If all known hypervelocity impact events, in the laboratory or in the field exhibit these glass characteristics, then craters that do not have them must **not** be impact craters, a fact that investigators have overlooked.

So much glass was found at Wabar, that the glass literally defines the craters and ejecta metamorphically. In other words, the sand dunes would have obliterated everything from the impact long ago if not for the enormous amount of impactite glass and "instant rock" created by the explosion and heat. These events formed a backbone, or shell of the crater structures, allowing the craters to survive to this day:

"The exposed rims of the two largest craters are **mantled with bombs and lapilli of black and white slaggy impactite glass and with large and small clasts of instant rock**. At the 11-m crater, the ejecta rim has been entirely eroded away; only the breccia lining the lower crater walls is exposed. **Survival of the craters** in the shifting sands of the Rub' Al-Khali **is due solely to the presence of the impactite glass and instant rock** that mantles the crater rims and resists transport by the wind and to the resistant breccia lenses lining the crater walls." Note 7.9y

The meteorites seen in the photos are no more than 2 meters long, yet an area 50 x 130 meters of glass material was made:

"The impactite-mantled rim deposit was exposed over an area about 50 m by 130 m. A fallout deposit of glass bombs and clasts of **instant rocks** with scattered **oxidized and metallic meteorite fragments** rests on an ejecta blanket of sand that is about 1 m thick at the crater rim. The rim has been eroded back about 6 m from the preserved 0.5-m thick breccia lining the crater wall, leaving a topographic bench where deformed pre-crater sand is exposed." Note 7.9z

The researchers reported that:

"In effect, **it rained molten glass for more than a half kilometer from the impact**." Note 7.9aa

A truly amazing aspect of the Wabar observations is that the researchers go on to say that the Wabar Impact Craters "resembles" several other "meteorite craters" including Meteor Crater in Arizona:

"Hence the structure of this crater in sand **resembles that of Meteor Crater and of several other meteorite craters** in bedrock mapped by the Shoemakers in Australia." Note 7.9ab

As we will shortly see, these other "impact" craters have almost **no glass**, especially material from a "molten glass" rain. It is unscientific to say that an iron-meteorite traveling at hypervelocity would be vaporized upon impact, without leaving an enormous amount of glass. Surely, some of the vaporized meteorite would end up encapsulated as glass on the surface of the crater and surrounding ejecta blanket.

The Lunar Glass Evidence That Isn't There

If we were to put into one sentence what has so far been established about impact craters, geologist G. J. H. McCall said it best back in 1964:

> "It is not in dispute that impact explosions may have had some role in geology, though it must appear to most geologists, weighing the geological record without sentiment, that this role was an **insignificant** one."
>
> G. J. H. McCall, *Nature*, Vol 201, January 18, 1964, p252

It would probably be difficult to find a professional geologist that would agree with this statement today—yet McCall was right on the mark! It is true that impacts *can* and do play a role in geology—but the evidence supporting many natural hypervelocity impacts is just not there. The Jupiter impact of 1994 and the 1996 Martian impacts previously mentioned, along with a few impacts here on Earth, demonstrate that hypervelocity impacts are a rare occurrence. Determining via empirical evidence that *impacts play a relatively minor role* in geology is a rather simple endeavor.

One of the best places to prove that impact craters play an insignificant role in geology is the surface of the Moon. Without an atmosphere, the Moon affords us an exceptional historical account of the aftermath of crater explosions. The first priority for science is to determine if the explosions came from above as a high-speed impact or from beneath as a low-speed eruption. Compelling, direct evidence that the vast majority of the craters were formed by subsurface eruptions is the absence of discernable ejecta blankets or ray patterns.

Evidence of high speed impacts have been recorded on the Moon. Fig 7.9.17 shows just such an event. The photomicrograph is an actual lunar rock sample enlarged 270 times to show an impact of a high-speed micrometeorite. Like all other hypervelocity impacts, there is an obvious melt region throughout the crater, consistent with reports from researchers involved in hypervelocity impact experiments. Thanks to the efforts of NASA, we can conduct a close-up survey of the lunar surface.

Thousands of lunar photos have been scoured, looking for impact craters that exhibit a glassy or smooth appearance indicative of high-speed impact events. Thus far, they have eluded discovery. Because there are so many craters on the surface of the moon, the lunar surface **should be** littered with countless millions of melted rock fragments (See Fig 7.9.18); but where are they? Moreover, the glassy melted rocks should exhibit vesicles, evidence of escaping volatiles, which are gases in the rocks, such as water from the meteorites.

Fig 7.9.17 – This is a photomicrograph of a micro-impact on lunar material returned to Earth and enlarged 270 times. Does this high-speed impact crater look like any of the craters on the Moon's surface? Because high-speed impact produces abundant melt and glass, and because so few of the lunar craters have this obvious evidence, we can conclude that the majority of the lunar craters are ***not*** impact craters. NASA photo S70-20416 taken 1.06.70.

The Evidence of missing Lunar Glass has stumped planetary scientists for several reasons, but the primary source of confusion is rooted in the twin Pseudotheories of Magma and the Rock Cycle. Because the difference between different types of silica rocks such as quartz (a crystalline mineral) or obsidian (a glassy rock) is not truly understood, the missing lunar glass mystery remains largely unnoticed.

In 1978, astute researchers observed phenomena of impacting granite, quartz sand, and pumice powder at velocities ranging from .05-7.2 km/sec. Notice that a "reservoir" or "puddle" of melt was present after impact:

> "However, as the craters grow in size and the ejection velocities decay, **melt is ejected** throughout the 360° azimuthal range **as it flows up from a 'puddle' or reservoir of melt that lines the bottoms of craters** at the steepest angles of incidence."
> Note 7.9ac

Interestingly, the researchers go on to comment how lunar craters have "lavalike materials" that "flowed or ponded in a fluid state." Two failed theories are employed to explain mysterious once-flowing material. The first of the two theories—volcanic flow—is dismissed, leaving the second as the presumed final answer:

> "The **lavalike materials** appear to have been emplaced and **flowed or ponded in a fluid state and occur around many, generally large, lunar craters**. Although first interpreted to have a volcanic origin (e.g., Strom and Fielder, 1970) Shoemaker *et al.*, (1968) and Guest (1973) suggested that such materials around, respectively, Tycho and Aristarchus, are **melt produced by shock-heating during the impact event that produced the craters**." Note 7.9ac p3862

As seen in Fig 7.9.18, the vast majority of lunar surface rocks do *not* look like Apollo 16 sample 64435, shown at the lower right, which is a specimen of melted rock. The vast majority of lunar rock is not lava from a volcanic flow, nor is the rock from an impact-induced melt-flow. To see scores of lunar surface photos, go to the website, http://www.apolloarchive.com/apollo_gallery.html and view the *Apollo Image Gallery*. Melted rock from volcanoes, high-speed impact melt, and lab or furnace-induced melt has been extensively studied by scientists. There is little physical evidence or photographic documentation that supports the idea that lunar rock could have its origin in these types of *melts*. Basalts on the Moon share similarities with the Earth's basalt; they are not glassy and they did not originate from a dry surface melt. Without physical or photographic evidence, questions about lunar rock origin remain unanswered as does the question—where is the impact melt?

The Basalt Mystery in the Rock Cycle Pseudotheory chap-

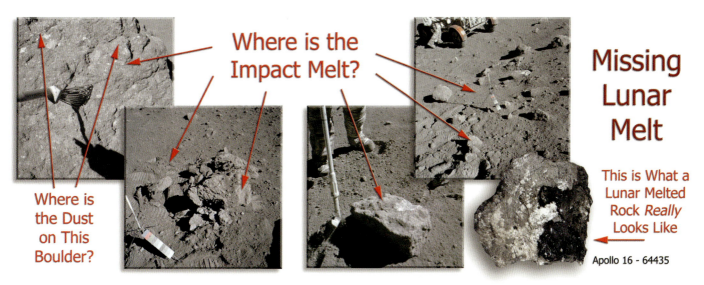

Fig 7.9.18 – These NASA photos from the Apollo 16 and 17 missions are typical of the thousands of detailed photos taken of the Moon's surface. Few of them show glass-like melted rocks similar to the Wabar glass or the Lunar Sample 64435 above (that actually did show a melted edge). This fact is part of the reason the **volcanic-impact crater debate** has gone on for decades, and would continue indefinitely without the new evidence of the Hydroplanet Model. On the Moon, there are no volcanoes with lava flows like those on Earth and impacts are very rare, thus, neither theory can adequately explain the origin of the lunar craters, or the rest of the Moon. Furthermore, where is the impact dust on the boulder in the photo on the left? Many boulders have no dust or sediment that would be present if numerous impacts had occurred, as thought by impactologists. In the Hydroplanet Model, water present during the final stages of the Moon's formation could have removed the dust and small sediment.

ter addressed how geologists have mistakenly called lunar basalt samples "glassy" when in fact they are made of materials denser than quartz and are *not* glass. From the book, *Magmas and Magmatic Rocks*:

"Many of the VLT **basalts are glassy**, but the phaneritic samples usually have subophitic textures, and contain approximately 62 per cent **proxene**, 32 per cent **plagioclase**, 5 per cent **olivine** and only 1 per cent Fe-Ti phases." Bib 87 p109

The minerals in these lunar basalts are not glass, nor have they been melted. The origin of the lunar rocks share similarities with the origin of Earth's minerals; they are formed in a hypretherm. The New Geology subchapter showed that an environment consisting of pressurized and heated *water* was essential to mineral and crystal formation, which are crucial processes in the origin and formation of planets. Since modern geology does not understand how basalt is formed and how *little effect* impact melting has *actually* affected the Moon, it should come as no surprise they do not have a true understanding of lunar basaltic rocks. The latest nature documentary or textbook may *imply* a correct understanding, but deeper research indicates that it is *really* "not clear" at all:

"At present **it is not clear whether these basaltic rocks are primary igneous rocks or pseudo-igneous rocks produced by impact melting**." Bib 87 p110

Astonishingly, if geologists had actually melted basaltic rocks, they would see that they are no longer basalt but a glassy melt! See Fig 5.7.9 for what melted rocks look like. On page 108 of the book, *Magmas and Magmatic Rocks* (Bib 87), it is reported that the number one mineral constituent of various basaltic rocks from the Moon is SiO_2. The UM has shown irrefutably, that quartz and glass, both SiO_2 rocks, have completely different physical properties. Given this fact, it *should be* crystal-clear whether basaltic lunar rocks were "produced by impact melting" or not. The SiO_2 in the rock need simply be examined to see whether it is quartz or glass. If it is a "basalt" rock, the SiO_2 will be present as quartz, not glass, proving the rock was not "produced by impact melting."

There is another reason that lunar rock researchers have been puzzled about the lunar rocks:

"An important characteristic of the intermediate-K Fra Mauro **basalts**, and one that is not found in the low-K Fra Mauro basalts, **is their lack of contaminants derived from impacting meteorites**." Bib 87 p110-111

Impactologists claim the entire surface of the Moon is made up of impact debris; therefore, there should be no "lack of contaminants derived from impacting meteorites." Since most meteorites contain the uncommon element nickel, impactite, or glass with nickel contaminates infused into the rock should be ubiquitous on the lunar surface, but it is not. In comparison, the historic Wabar impact site in the Saudi Desert showed that approximately 10% of the impactor was infused with the impactite glass.

A Major Revision of Our Understanding of the Moon

Of the two popular crater-formation theories, the volcanic theory was quickly dismissed when astronauts went to the Moon because geologists had not come to realize that many of the minerals on Earth's now dry continents were formed underwater, in a former ocean. During the 1960s and 70s, astronauts were trained in surface geology—deep ocean environments and their associated geology would remain generally unknown until the 1990s. Because of their training, they expected to see some evidence of volcanism. In the book, *The Once and Future Moon,* Paul D. Spudis comments on their discovery:

"The skilled crew members were surprised during their moonwalks—**where were all the volcanic rocks they expected**?" Bib 129 p73-4

Of course, the "expected" volcanic rock they were looking for is the type of rock seen coming from *volcanoes* on Earth. However, on the Earth, many volcanic-like deposits exist which have never been observed coming from any type of volcano. Fig 7.9.19 shows many of the "volcanic" hills that exist in the

Phoenix, Arizona landscape, but there are no actual volcanoes! These types of volcanic rocks did not form on land; they formed under an ocean in a hypretherm environment. Underwater technology has just begun to allow us to observe these environments and new discoveries abound. In the Basalt Mystery, subchapter 6.5, we made the claim that no one has observed basalt coming from a land-based volcano, even though basalt is recognized as the most widely distributed volcanic rock on land. Active basalt formation has not been seen because it is formed underwater, a fact unknown to the Apollo team of scientists.

Because the volcanic theory was implausible based on their expectations, the only other theory planetary scientists had, was the impact theory:

"Although ancient **volcanic rocks were expected**, the mission returned **impact-processed rocks instead**. **This finding led to a major revision of our understanding of the Moon.**" Bib 129 p75

So it was, that without lunar volcanic rocks like the dry surface volcanic rocks of Earth, a *major revision* in planetary science took place and the impact pseudotheory became the dominant player in every scientific circle.

Researching lunar rocks is not an easy task. Most articles use authoritative sounding words such as "impact-processed" rocks, but the real fundamental question is whether the majority of lunar rocks were melted or not.

FQ: Were the lunar rocks collected from the Moon melted or not?

Lunar rocks are of course priceless, and there is no centralized location to view them all. They are scattered around the world and unfortunately, good photos are rare. We can get a generalized feel of the researcher's mindset from the NASA site listing the *Top Ten Scientific Discoveries Made During the Apollo Exploration of the Moon*. There, the three primary types of lunar rocks are listed: basalts, anorthosite and breccia.

The lunar basalt of the dark mare basins is similar to the Earth's dark basalt. It is no coincidence that the Earth's oceanic crust is also primarily basalt, rock formed at the bottom of the ocean. Anorthosite comes from the lighter colored lunar highlands, and *breccia* is supposed to be the material remains of impact activities. **Breccia** is defined as being angular fragments of rock cemented together in a matrix. Presumably, the only method by which cementing could occur on the Moon, based on current ideology, is melting, by impact or otherwise. Again, we should expect to see the words 'melt' or 'glass' in the official explanation of how lunar breccia formed from impact:

"**Breccias** are composite rocks formed from all other rock types through **crushing, mixing, and sintering during meteorite impacts**." Note 7.9ad

Why would 'melting' be excluded from the definition of breccias if the lunar samples supported the notion of impact? If impact was the primary source of so common a lunar rock, evidence of melting should have been everywhere the astronauts collected rocks.

NASA's list of the Top Ten Scientific Discoveries of the Apollo mission lists the Moon's "magma ocean" as top discovery #7, and lists #6 as:

"**All Moon rocks originated through high-temperature processes with little or no involvement with water**." Note 7.9ad

The Impact Pseudotheory came about in part because scientific investigators had *no other paradigm in which to frame their work* concerning the Moon's geology, especially when they found no traditional volcanoes or lava flows like those seen on Earth. Without the Hydroplanet Model, researchers succumbed to the "charms of the deductive approach" one lunar scholar warned about:

"'Workers on these samples, which have **endured countless meteorite impacts**, must guard against the **charms of the deductive approach**, deriving the details from **a general theory** in the manner of medieval scholars…'" Bib 87 p110

Deductive reasoning is only as good as the general theory upon which it is based; in this case, the Impact Theory is based upon magma theory. The lack of traditional dry-land volcanic rocks confused investigators because they did not consider that lunar craters could have been the remains of hydrofountains. Thus, a "major revision" of our understanding of the Moon is needed, a revision towards hydrovolcanism and away from impacts.

The Shorty Lunar Hydrocrater Evidence

Although access to all of the lunar photos and research is not readily available, some surprising details are to be had from those that are. A good example of this is the somewhat famous orange-glass soil found at the rim of Shorty Crater, seen in Fig 7.9.20. This soil and other common glassy materials are not recognized as impact glass, but as a volcanic glass "from great depths" in the Moon:

"Picritic **glass beads** are ubiquitous at the Apollo and Luna landing sites and are thought to represent some of the most primitive **magmatism on the Moon**.

"Furthermore, many of these glasses may **come from great depths in the Moon**, some even in excess of 700 km." Note 7.9ae

Paul D. Spudis, describes a "huge fountain of

Fig 7.9.19 – Many of the hills surrounding the Phoenix, Arizona area are thought by geologists to be "volcanic" but not volcanoes. The material they are made of is not lava and there are no noticeable craters at their summits. Where did they come from and how were they formed? They are a geologic enigma just as lunar rocks are. No one has ever seen mountains or hills like them being formed on land. They were not formed on land, but underwater, in a hypretherm. Scientists have recently begun to study these types of formations at the bottom of the ocean.

CHAPTER 7 THE HYDROPLANET MODEL

Fig 7.9.20 – Apollo 17 astronauts visited this crater and found several areas of orange glassy soil. Similar glassy material is ejected from Earthly hydrofountains, of which several examples will be shown in the Universal Flood Model chapter. This glassy material is not impact-type glass. No impact melt-rocks were observed. Because the Impact Pseudo-theory is so persistent, researchers incorrectly assumed the orange soil from Shorty Crater was impact melt, but the evidence clearly shows otherwise.

> "All Moon rocks originated through high-temperature processes with little or no involvement with water."
>
> NASA

liquid rock" that sprayed the orange soil onto the rims of the craters:

"The **orange soil** discovered at Shorty crater turned out to be an **unusual black-and-orange glass**. Like the green glass from *Apollo 15*, **these glasses are volcanic ash**, the product of **a huge fountain of liquid rock sprayed out onto the surface**." Bib 129 p77

What Spudis did not realize was that the "huge fountain" he envisioned was a *hydrofountain*, leaving behind a crater known as Shorty Crater—more aptly named **Shorty Hydrocrater**—from which the glass was taken! All volcanic eruptions are to some degree the result of steam expansion and subsequent explosion. We know this because of the water content in all volcanic materials on the Earth and because water is *the most commonly known liquid in the universe*. Superheated water becomes steam and supplies the explosive force required to form craters. Lunar rocks were formed in a lower gravity environment and thus do not contain the same water content Earth's volcanic rocks do.

The orange and green-black lunar glass described here is direct physical evidence that the Moon is a Hydromoon and that most of the craters are the remnants of hydrofountains.

Typically, many scientists do not know that the orange glass seen in fig 7.9.20 is *not* from an impact melt but is actually from a hydrocrater. One interpretation describes the orange lunar soil thus:

"The **orange lunar soil** discovered by Apollo 17 astronauts, as seen through a microscope. Some particles are spherical, indicating they were formed as molten drops **ejected from an impact**. Others are irregular, formed by mechanical fractures." Note 7.9af

The evidence that the orange glass is not likely to be from impact melt is that there is no record of infused meteorite material within the glass, a condition likely to be present if the glass came from a meteorite-impact melt. Further, impact melt would be present in many different sizes, not just "soil" sized material. Finally, the orange material is not found in the shallow layers of the Moon's regolith in the surface sediment layers. This has led to the suggestion that the orange materials came from deep inside the Moon via a "huge fountain of liquid rock," as noted by one investigator.

Shorty Hydrocrater and many other lunar hydrocraters actually did spray "onto the surface," *large amounts* of sediment including the fine grey dust encountered by the astronauts.

This is contrary to the impactologists' theory that the Moon dust came from micrometeorites, but we'll talk more about this later.

Satellite Impact Evidence—"Molten Craters"

The Long Duration Exposure Facility or LDEF designed in 1970 intended to provide long-term data on meteoroid exposure in the space environment. In 1984, a cylindrical structure 30 feet by 14 feet, containing 86 trays was carried into space by the shuttle, *Challenger*. After 5.7 years in space, the trays were brought back to Earth and closely examined. Researchers observed remnants of high-speed micrometeoroid impacts, craters with signs of melt:

"The large majority of the craters appeared to be the result of a hypervelocity impacts (>6 or 7 km/s) and were generally symmetric and hemispheric in shape, exhibiting the typical lip or rim around the crater. **The crater interior surfaces generally displayed evidence of being molten during the crater formation process**." Note 7.9ag

Similar to high-speed impact observations of lunar material, 'bowel-shaped' craters in the LDEF trays and the interior surfaces of the crater exhibited evidence of being "molten during the crater formation process." These micro-scale impact observations are consistent with the macro-scale Trinity hypervelocity explosion that formed a 'glass bowel' depression.

Another critical observation was that of the impactor. The meteoroids did not vaporize as is commonly suggested by many investigators studying Earth's so-called impact craters. The LDEF researchers reported:

"The majority of the natural meteoroid impactors can be categorized into one of the following groups: (1) Chondritic - largely made up of relatively **well-mixed and homogenized fine-grained matrices**, (2) Monomineralic silicates - characterized by high concentrations of Si, Mg, and Fe and **generally found in molten form**, and (3) Fe-Ni-sulfide rich particles - **also generally found as melts**." Note 7.9ag

Here is specific evidence that a hypervelocity impactor *leaves its mark within the melt in the crater*. It does not simply vaporize without leaving any trace. The impactor's residual melt evidence allowed researchers to determine the specific composition of the meteoroid. They also reported the impact material was "found in molten form." Investigators were even able to distinguish between natural impactors and manmade material orbiting the Earth. This will be important later in the Arizona Hydrocrater and Impact-to-Hydrocrater Evidences.

The Flawed Impact Criteria

Over the last century, an obstacle in the identification of impact craters was the lack of knowledge concerning *correct impact criteria*. This is due in part to the ignorance geophysicists have about hydrocraters and hypervelocity impact events. Numerous "definitive" criteria have been proposed as evidence for identifying hypervelocity meteorite impacts. One by one, these criteria have failed to stand the test of time. The University of New Brunswick hosts a website, http://www.unb.ca/passc/ImpactDatabase, which contains the *Earth Impact Database*, a clearinghouse for terrestrial impact craters and the research on the craters. The site defines the criteria for identification:

"**The principal criteria for determining if a geological feature is an impact structure formed by the hypervelocity impact of a meteorite or comet are listed below**." Note 7.9ah

Jason Hines, the data manager, sets forth six criteria, of which only the first three are deemed "definitive":

"In terms of relative importance, it is generally considered that criteria 1-3 are **definitive** (they all relate to the passage of a shock wave through rock and resulting modification processes)…"
Note 7.9ah

Here are the three "definitive criteria":
1. Presence of **shatter cones** that are *in situ* (macroscopic evidence).
2. Presence of multiple planar deformation features (**PDFs**) in minerals within *in situ* lithologies (microscopic evidence).
3. Presence of high pressure mineral polymorphs [**coesite, stishovite**] within *in situ* lithologies (microscopic evidence and requiring proof via X-ray diffraction, etc.). Note 7.9ah

The UM will demonstrate, one by one why each of the three 'impact criteria' (shatter cones, PDF's

> "Astronomers seem to invoke collisions, however improbable, when they cannot explain odd features of the solar system."
>
> J R Minkel – Scientific American, June 2006

and coesite) are flawed criteria. It is astonishing that the entire scientific world has access to the *Earth Impact Database* criteria and research, yet no one seems to have recognized the serious flaws of the *definitive* criteria.

This is probably due to the same reason so many other extraordinary UM claims and evidences remain unnoticed in science today—modern science is in a paradigm of its own making. There are so many pseudotheories in today's science that scientists are restricted in their vision and in their ability to see outside their paradigm. As Kuhn noted when discussing scientific revolution in Chapter 6, scientists construct "elaborate equipment, the development of an esoteric vocabulary and skills, and a refinement of concepts that increasingly lessens their resemblance to their usual common-sense prototypes." This leads to "an **immense restriction of the scientist's vision** and to a **considerable resistance to paradigm change**." Note 19.9ai

Fig 7.9.21 – There are three "definitive criteria" impactologists seek to determine that a crater was made by impact. To substantiate the claim, researchers have declared that shatter cones, PDFs and coesite are *only* found in impact craters. Unfortunately, this is not true, and the UM presents scientific evidence showing that each so-called definitive criteria are also found in natural geological landforms. Shatter cones, PDFs and coesite are found at **both** impact craters and hydrocraters, demonstrating both events are **surface explosions**. Image is of Victoria Crater on Mars.

Without the ability to change, science cannot progress; scientific revolutions require a paradigm shift. Showing that the current Impact Determination Criteria are flawed is one step closer to the paradigm shift that must occur. As Kuhn points out, "when paradigms change, the world itself changes with them."
Note 19.9ai Chapter 5

The 'impact criteria' are flawed because **both** impact craters and hydrocraters are surface **explosions**. In the past, geologists didn't think steam was capable of generating such large explosions. As it turns out, the largest explosions on Earth, volcanic eruptions are essentially *steam explosions*. Because these surface explosions are associated with hydrofountains and hydrocraters, the definitive impact criteria show up in both impact craters **and** hydrocraters. Impactologists are not familiar with hydrocraters and their associated facts and observations because hydrocraters have not been a part of the impactologist's paradigm.

Cone-in-Cone Structures

"Resemble Impact-Generated Shatter Cones"

Fig 7.9.22 – Natural Cone-in-Cone structures are found in a variety of geological settings, often mistaken for "shatter cones" said to be formed during an impact event. Cones like these have been discovered in locations previously believed to be impact sites that are now known to be hydrocraters, the Bushveld Complex in South Africa being one such example. Despite this, shatter cones are still held as absolute criteria for determining a crater's impact origin by modern science.

> Impactologists must acknowldge that if the criterion for identifying impact craters exists in non-impact craters, the criterion can no longer be held as "definitive" and should not be used to establish a crater's impact origin.

Shatter Cone and PDF Impact Criterion Myths

In the book *Impact Cratering*, the reader is introduced to the "first reliable geological criterion" for identifying impact craters:

"The **first reliable geological criterion** for the recognition of impact structures in the absence of meteorites was established by R. S. Dietz as a result of his studies of the Kentland structure in 1947. At Kentland Dietz found peculiar fractures in the rock that caused it to break into striated cones. The apex of the cones generally pointed toward the center of the crater, indicating the source of the shock that fractured the rock. **Shatter cones** had first been recognized by Branca and Faas at Steinheim in 1905, but **Dietz argued that they occur only in impact craters**, a proposal supported by an experiment in which E. M. Shoemaker and others (1961) produced shatter cones around small-scale impacts in dolomite." Bib 161 p8

Curiously, Shatter Cones are still identified by the experts as one of the 'criteria' for identifying high-speed impact craters. The Arizona Meteor Crater is the first 'discovered,' most well studied, and best documented 'impact' crater, yet it **has no shatter cones**. This should have raised a red flag, but it seems to have been ignored. Moreover, Shoemaker's shatter cone experiment showed that they could be created with high-pressure explosions, *including steam explosions*.

Other geologists, such as G.C. Amstutz, noted in a 1964 publication that the term "shatter cones," which implied impact, were actually diagenetic (sedimentary-process related) cone-in-cone features:

"Another criterion is also rather one-sided and superficial. This is the attempt to prove the impact nature of these structures by the presence of cones which have been given by some the interpretative name '**shatter cones.**' A detailed morphometric study showed some of these cones are likely to be diagenetic cone-in-cone features; **whereas others may have been produced by the breakage and pressure during tectonic movements**. To the knowledge of this author **no shatter cones have ever been found in meteorites themselves**. This, of course, is a rather striking evidence." Note 7.9aj

Another geologist, G. McCall, contemporary with Shoemaker noted:

"I believe that the lesson to be learnt from Vredefort **is the converse of that suggest by Dietz**. It is that **shattercones, erroneously regarded as an indication of impact explosion**, can stem from endogenous geological activity [hydrofountains]. The recognition of random-oriented shatter-cones at Steinheim weakened Dietz's original hypothesis, and there is, in fact, **no** real evidence of any connection between these structures and meteorites." Note 7.9ak

Voices like these became clouded in the Dark Age of Science, as theories continued to become the end-points of science. Recently, younger open-minded investigators recognized that there had to be a reason for so many unanswered questions in impactology. *Impact Tectonics*, a 2005 publication containing an article about cone-in-cone structures in Morocco, introduces direct evidence of the cone-in-cone structures McCall wrote about:

"Several geological features, including sedimentary **cone-in-cone structures** and percussion marks, **may resemble impact-generated shatter cones**. Especially inexperienced workers **may mistake such features for impact deformation**. In 1997, our group investigated an **alleged occurrence of shatter cones** in the Hamada area of southeastern Morocco and found that these are **actually cone-in-cone structures**, probably from the Lower Visean Merdani Formation." Note 7.9al

The mistaken identification and "alleged occurrence of shatter cones" is common, as the authors of *Impact Tectonics* go on to say:

"Shatter cones are mesoscopic fracture phenomenon that is **widely considered** a macroscopic shock metamorphic feature and, as such, a **diagnostic indicator of impact structures** (e.g., French 1998). Indeed, many impact structures have first been noted because of observations of shatter cones. However, **there are several geological phenomena that either resemble shatter cones closely**, including percussion marks/cones, such as those described by Reimold and Minnitt (1996) from the eastern environs of the **Bushveld Complex** in South Africa, or **cone-in-cone structures** that have also, in the past, been occasionally **mistaken for shatter cones** (Amstutz 1965)." Note 7.9al p82

This is additional direct evidence that 'impact-craters' such as the Bushveld Complex, were erroneously deemed to be of impact origin because of the incorrect identification of diatonic shatter cone features. This leads us to ask the FQ:

FQ: What is the origin of 'shatter' cones?

The *origin of the shatter cone feature*, like so many other geological features, remains "unresolved":

"Cone-in-cone structures are among the most spectacular and enigmatic geological features. **Their origin and mechanism of formation have been debated for more than a century and are still essentially unresolved**." Note 7.9al p82

One way to determine if silica-based shatter cones are of impact origin is to examine the silica for evidence of *glass*. Very high pressure and high temperature would produce silica-glass cones. If the silica cone is *quartz*-based, then the cone formed in a hypretherm. Due to their larger, macroscopic (visible to the unaided eye) size, it would have been formed in a diatreme. The time frame associated with an impact event is far too short (less than one second) to grow large, quartz-cones. Macroscopic cones formed over longer periods, in hyprethermal conditions,

the environment present in hydrofountains.

One scientist, John Luczaj took on the challenge of disproving the erroneous notion that a string of about eight crater structures along a 700-km linear array stretching from Kansas through Missouri into Illinois were of impact origin. (see Fig 7.9.23). Luczaj states that these craters **cannot** be impact craters as was previously assumed, but are instead, structures produced by explosive "igneous activity." He reported his views in the journal of *Geology*:

"There has been a long-lived controversy over the origin of a well-defined linear set of eight cryptoexplosion structures [hydrocraters] in the southern mid-continent, United States. The linearity suggests that these structures must be related to either (1) impact of a string of bolides derived from breakup of a large parent bolide, or (2) volatile-rich, explosive ultramafic magmatism associated with structural weaknesses in the crust. **That these structures are of various ages, are at or near intersections of major regional tectonic features, and have closely associated ultramafic magmatism**, rules out the impact-string hypothesis. The **only remaining hypothesis** is that these cryptoexplosion structures were produced by explosive ultramafic volcanic activity. **An important corollary to this hypothesis is that some shock-metamorphic features, including shatter cones and shocked quartz, can be produced by explosive ultramafic igneous activity**." Note 7.9am

The "important corollary" Luczaj refers to establishes the fact that previously held 'impact' structures based on the presence

> "Several geological features, including sedimentary **cone-in-cone structures** and percussion marks, **may resemble impact-generated shatter cones**. Especially inexperienced workers **may mistake such features for impact deformation**."

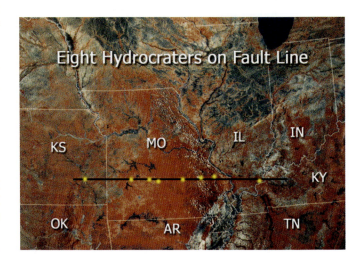

Fig 7.9.23 – The eight yellow dots represent eight "mysterious explosions" that occurred along a 700-km long fault line stretching across four states in the U.S. Although impactologists have tried to link these craters to impact (there are still two on the Earth Impact Database), John Luczaj calculates that the chance alignment of impact structures like these is less than one in a billion. Because the craters are not all of the same age and are related in other regional tectonic features, Luczaj concludes that they are *not* of impact origin but of volcanic origin. An important corollary to this conclusion is that the *shatter cones and shocked quartz* found at some of the craters, which are also of volcanic origin, and thus they *cannot be used as impact criterion*.

of shatter cones and shocked quartz must be completely reevaluated:

"**An igneous rather than an impact origin** for the linear array of cryptoexplosion structures presents an **important corollary to the problem of cryptoexplosion structures in general. If explosive volcanism in these eight structures produced shock-metamorphic features such as shatter cones and shocked quartz, the same explosive igneous processes in other regions are therefore capable of producing these shock-metamorphic features**. The presence of shock-metamorphic features is not inconsistent with a terrestrial origin. Recent work suggests that shatter cones, pseudotachylite, and planar lamellar structures in quartz **can form at lower stresses and lower strain rates than previously thought** (e.g., Shevchenko, 1996; Reimold, 1995; Lyons et al., 1993). Nicolaysen (1972) and Nicolaysen and Ferguson (1990) discuss possible mechanisms for the formation of some shock-metamorphic features in cryptoexplosion structures [hydrofountains] resulting from **violent release of confining pressures on magmatic volatiles**." Note 7.9am p297

Interestingly enough, two of the eight craters are still listed on the 'Earth Impact Database.' Researchers calculated that there is a billion to one chance that these eight structures were aligned by chance:

"Rampino and Volk (1996) calculated an **extremely remote probability** (<10^9) that the alignment of these eight structures was a matter of chance." Note 7.9am p295

When the empirical evidence does not fit the theory, some scientists become philosophers, as Melosh did when he responded to Luczaj and others' extremely remote probability:

"Although the probability of a coincidence between two bona fide impact craters and a pre-existing line of tectonic structures may seem low, **it appears that just such a coincidence must have occurred**." Note 7.9an

Dismissing the possibility that the two impact craters on this line of tectonic structures may *not* have been "bona fide impact craters," Melosh attributes their alignment to mere "coincidence."

The two craters (Crooked Creek and Decaturville) that fall on the 700-km fault line should be removed from the Impact Database and they should no longer be considered impact craters. Luczaj goes on to question *other mid-continent craters* that have been erroneously "interpreted as impact craters" based on incorrect shock features:

"The **origin of similar structures in the midcontinent** that were **interpreted** as **impact craters** on the basis of these **shock metamorphic features is now in question**." Note 7.9ao

As you might imagine, suggesting the need to 'question' other 'established' impact craters based on a reevaluation of shock features (shatter cones and shocked quartz), does not sit well with deeply entrenched impactologists. Luczaj had opened a can of worms and was criticized for simply reporting his scientific observations and their implications. (See H. J. Melosh, *Nature*, Vol. 394, 16 July 1998, p223). Of course, it was just a matter of time before this would happen. The facts have been swept under the rug by impactologists for many decades, and now it is time to look at the data for what it really is.

PDF is an acronym for Planar Deformation Feature, a twenty-dollar word for 'shocked rock.' Recent studies have shown that PDFs, or shocked rock features are found in non-impact structures, which negates their use as a *definitive* impact criteria. Impactologists have long thought that not enough pressure could be generated to produce shock metamorphism, however, Luczaj noted and many researchers have since demonstrated that quartz "can form at lower stresses and lower strain rates than previously thought." Ibid. p297 Moreover, recent volcanic explosions like that of Mt. Saint Helens have provided actual examples of the tremendous pressure that is generated in a volcanic explosion, a fact not previously known. Shocked rocks, formed by phreatic explosion have been observed, although very little has been reported about them in scientific literature, comparing them with supposed 'impact' shocked rock. Because the current paradigm is completely 'impact' in nature, alternate explanations are often summarily dismissed.

The Coesite Impact Criterion Myth

In the Magma Pseudotheory, subchapter, 5.7 Quartz is Not Glass, coesite and its closely related sister, stishovite were established as being high temperature, high pressure forms of silica. Early impact researchers E. M. Shoemaker and Robert S. Dietz were the primary forces behind the establishment of coesite as a defining criterion of impact cratering. They stated that coesite could "*only* be created by meteoritic impact." In 1963, Dietz claimed this in the *American Journal of Science*:

"**All** seem to agree that **shatter cones and coesite validly indicate** intense shock; the writer believes such shock **can only be created by meteoritic impact**." Note 7.9ap

Not everyone agreed with Dietz, including two of the most renowned geologists of the time, W. H. Bucher and G. J. H. McCall. McCall replied directly to Dietz in the journal of *Nature* in January 1964:

"**Coesite**, the other **supposed indicator of impact**, has, as Bucher states, been recognized at the Rieskesse, south Germany, and the Bosumtwi crater, Ashanti. This dense polymorph of silica seems, considering Bucher's devastating arguments, to have been **erroneously given the virtual status of an impact explosion indicator**." Note 7.9aq

McCall hit the nail on the head with his assessment. Impactologists had begun to see every crater with coesite as an impact crater. The problem with this is that non-impact craters should have no coesite. We found no evidence for such a study, and as of this writing, views opposing the Impact Pseudotheory are given little credence in modern impact literature, regardless of hard evidence. Looking at the other end of the coesite mystery—the formation of coesite—researchers were unsuccessful at locating coesite at nuclear blast sites, which impactologists had predicted would be there:

"**Coesite** was also searched for, **unsuccessfully**, at sites of **nuclear explosions**." Bib 119 p115

Researchers tried to make coesite, or "shocked quartz," in impact experiments. Yet, as Shoemaker reported, he and others failed to identify coesite in their high-speed impact experiment conducted in 1963 using sandstone from the Arizona crater:

"Close petrographic examination and X-ray diffraction of the crushed sandstone aggregates **failed to yield any conclusive evidence of the presence of the high-pressure silica polymorph coesite**, such as has been found in the natural crushed sandstone from Meteor Crater." Note 7.9ar

This has not stopped impactologists or most other scientists from the continued use of coesite as definitive geological criteria for impact cratering. In the 1989 book, *Impact Cratering*, one of the foremost impact cratering scholars amazingly stated:

"The discovery of **Coesite**, a high pressure phase of quartz, in 1953 and a still higher pressure phase, Stishovite, in 1961, provide **new geological criteria** for recognizing the high shock pressures that accompany the impact of high speed meteorites. **The transformation pressures of Coesite and Stishovite are so high that they have never been observed to form in any volcanic eruption or by any natural process other than impact.**" Bib 161 p8

Why this is amazing is the contradictory observations made in 1977, over ten years before *Impact Cratering* was published. Joseph R. Smyth of the Los Alamos Scientific Laboratory identified precisely what impact scholars have said for many years *could not exist*—coesite from a natural volcanic process:

"Primary crystals of **coesite** and high sanidine ($Or_{98}Ab_2$) up to 3 mm in greatest dimension **have been identified** in a grospydite inclusion from the Roberts Victor **kimberlite pipe, South Africa**." Note 7.9as

This 1977 report of *coesite from natural non-impact processes* has either been overlooked or blatantly ignored by the scientific community. Kimberlite pipes are evidence of hydrofountains, diatreme pipes that open onto the surface and contain the highest concentration of diamonds anywhere in the world. Kimberlite diatremes were naturally occurring hypretherm environments that facilitated the growth of high pressure-high temperature minerals such as diamonds—and coesite crystals. Smyth continues his 1977 report:

"**Coesite, previously considered diagnostic of impact processes**, is reported for **the first time** from a **natural static pressure environment**." Note 7.9as

The discovery of coesite was not a one-time event. Wohletz and Smyth report in 1984 that a xenolith or nodule (seen in Fig 7.9.24) contained 6% coesite, a surprisingly large amount. Bib 161

It seemed that all one need do is go looking and coesite would turn up. As of this writing, many experts are still saying coesite is an absolute indicator of impact. Why they continue to think this is not clear, except that perhaps there are many who simply trust the impact 'experts.'

In addition to hydrofountain diatremes, coesite was found in quartzite outcrops that have been attributed to lightning:

"In the Diamantina region of Minas Gerais, Brazil, **shattered quartzite outcrops have been found which contain coesite**. The origin of these shattered outcrops has been attributed to the effects of lightning strikes. Microfractures in these quartzites are saturated with water during the rainy season. When struck by lightning, the water temperature in these confined systems rises to over 1500 °C, resulting in an explosive expansion, which can blow rocks apart, and produce estimated pressures >3.5Gpa, indicated by the presence of coesite (**identified by RMP analyses**)." Note 7.9at

The lightning itself did not make the coesite; it superheated water in the rock producing a phreatic explosion that generated intense pressure. Again, the fact that coesite was found in the natural environment with no association to impact refutes the claim that coesite can be an absolute indicator of impact.

An important aspect of coesite is its size. The coesite crystals found at the Wabar impact crater are microscopic and cannot be seen with the naked eye. However, some of the coesite crystals found in the African kimberlite mine were 3 mm in length, the size of a small pebble, easily seen with the naked eye. Crystal size is determined by time. Longer growth periods allow for larger crystal growth. Thus, the size of the coesite crystals can

> "The transformation pressures of **Coesite** and Stishovite are so high that they have **never been observed to form in any volcanic eruption or by any natural process other than impact.**"

> "Coesite, previously considered diagnostic of impact processes, is reported for the first time from a natural static pressure environment."
> Joseph R. Smyth
> Earth and Planetary Science Letters, 34 (1977), p284

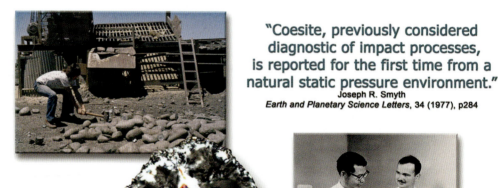

The Coesite Impact Critera Myth

Fig 7.9.24 – Coesite is a high pressure, high temperature silica mineral, found at the Arizona Meteor Crater. This inset photo from 1960 is Ed Chao and Gene Shoemaker discussing their *belief* that coesite is *only* found in impact craters. In the 1989 book *Impact Cratering*, considered the authoritative guide to high-speed impacts, the belief that coesite is only found in impact sites persisted and was again heralded as one of **the** geological "criteria" for determining impact. The beautiful crystal shown above is a large coesite crystal discovered in 1977 by Joseph R. Smyth in a South Africa kimberlite diatreme (a hydrofountain). The upper-left photo shows nodules from the area being broken apart for study. The discovery of the South African coesite totally negates the idea that coesite is definitive impact criteria, because it has been found in "natural process." Impact scientists have completely ignored, or overlooked the fact that coesite occurs in natural settings.

CHAPTER 7 THE HYDROPLANET MODEL

serve as an indicator for how long the high pressures existed in which the crystals could grow. During an impact event, the amount of time that high pressure and high temperature existed is extremely limited, a period measured in nanoseconds. Therefore, only microscopic crystals of coesite would be evident in impact craters, whereas large, 1mm+ coesite crystals would indicate high pressure and temperature was present for a longer period, such as would be present during hydrocrater formation.

The Wabar impact crater established a minimum size in which impact craters in siliceous rocks "should contain coesite":

"Primary crystals of **coesite ... have been identified** in a grospydite inclusion from the Roberts Victor **kimberlite pipe, South Africa**."

"Because the size of a crater increases with the total energy released by impact, the presence of coesite at the Wabar crater suggests that, in siliceous rocks, impact craters of this or greater size should contain coesite." Note 7.9au

This was noted by Chao and others in the journal of *Science* as far back as 1961, however, the absence of coesite in much larger craters has been ignored by those who endorse the Impact Pseudotheory.

In Summary, all three of the so-called "definitive" criteria used to identify impact craters have been shown to be invalid. All of these impact criteria are found in natural non-impact geological settings. Impactologists have been so dogged in their attempt to locate the next "impact" crater, that they have missed many important clues to the real process of crater formation. In the future, many, perhaps even most of the impact craters will come to be known as hydrocraters, based on evidence seen here in the Hydroplanet Model.

The Tektite and Libyan Desert Glass Evidence

Fig 7.9.25 is a photo of tektites, troubling anomalies for geologists. In 1994, O. Richard Norton wrote *Rocks from Space*, and said this about tektites:

"**Tektites** are found in abundance around known impact sites and, like **shatter cones and shocked quartz**, **are considered diagnostic for impact features**." Bib 29 p392

We have already discussed how shatter cones and shocked quartz have proven *not to be diagnostic* of impact features because they are also present in other natural settings, like hydrocraters. Tektites are another example of a "diagnostic" feature that has unsatisfactorily been used to denote evidence of impact cratering. If they were truly diagnostic, tektites should be present at most silica impact sites. In his discussion about tektites, the author of *Rocks from Space* comments on four examples of tektites (shown in images in his book) from Thailand, The Philippines, Australia, and Ries Crater in Germany, saying that:

"**Only one** of the four shown is associated with a **known impact crater**." Bib 29 p390

What happened to the other three impact craters presumably responsible for the fields of tektites? The one "known impact crater," Ries Crater, lies directly on a fault line associated with other volcanic features clearly *not* of impact origin. We will talk more about Ries Crater shortly.

Although books like *Rocks from Space* and popular science programs portray tektites as being evidence for the impact pseudotheory, the truth lies in the details of the scientific journals and publications, in which we see considerable controversy. One scientist, John O'Keefe, noted in the preface of his 1976 book, *Tektites and Their Origin*, that over 900 scientific papers had been written about tektites and at the time of his study, three quarters of the lunar scientists surveyed (there were forty scientists surveyed) thought that tektites came from the Earth. The rest of the lunar scientists believed tektites were of lunar origin.

Referring to a publication in 1984 by the Mineralogical Society of America, note that tektites were dubbed "strange" deposits with all sorts of theories on their origin being expressed:

"Tektites are **strange** glassy blebs and masses found in certain areas of the Earth (Czechoslovakia, Philippines, southeast Asia, Australia, etc.). **Considerable con-**

Tektite and Libyan Desert Glass

Fig 7.9.25 – These glass-like, pitted, dark stones are **tektites** that look somewhat like obsidian but were formed when glass was thrown into the air and cooled. Thin pieces (like the one being held) often show a green color. Impactologists have tried to associate tektites with impact craters or to imagine that the glass came from the Moon. However, evidence clearly shows that neither of these scenarios is correct. Only a large hydrofountain, as part of the hydroplanet model can explain all of the characteristics. The larger specimen in the center of the photo is Libyan Desert glass. It is found in areas that have no evidence of volcanism or impact structures. Sometimes, one side of the specimen is rough, indicating it had been partially buried, the top side sandblasted smooth by relentless dust storms in the area. The evidence also indicates that Libyan glass is not very old. The origin of this piece of Nature's Puzzle can be found in the next chapter—the Universal Flood.

Fig 7.9.26 – These unique tektites are from Australia. These rare specimens exhibit delicate and beautifully shaped geometric forms. Where are such tektites forming today? They defy explanation and refute the notion of formation by meteorite impact and the myth of uniformity.

troversy exists about their place of origin. Various theories have been proposed, the most widely accepted being that they were **formed by large meteorite impacts** on the surface of the Earth that fused surface materials and blasted the melt into high Earth trajectories. A **cometary origin** has also been considered, and argued that tektites are **ejecta from lunar volcanoes**." Bib 155 p558

Tektites are "strange" because their origin does not fit well with *any of the current theories*. Certainly, there would be less controversy if their origins were understood. Fortunately, there is a smoking gun, giving us solid clues as to where the origin of tektites might be found.

First, the Uniformity Myth that the present is the key to the past must be abandoned. Mankind has never observed or recorded natural tektites being formed. Tektites exist in distributions called strewn-fields, vast areas of land and water that take in several countries, indicating the event that produced them was truly colossal. They are not randomly distributed on the Earth; there are heavy concentrations in some areas—Asia— while in other areas—Germany—none have ever been found. This fact alone should open our eyes to possibilities outside of the old uniformity box.

There is a telltale signature to determine whether tektites came from meteorite impacts or not. It is the *nickel-iron* component of meteorites. As high-speed impacts vaporize the meteorite and melt the surrounding rock, the vaporized meteorite, most of which contain at least some nickel-iron components, would mix with the silica of which the tektites are made. Most, if not all tektites **should have traces of nickel-iron** within their glassy matrix—but they do not. A careful reading of a report on tektites from the *American Mineralogist* shows that although nickel was found in metallic spheres *within* some tektites, the tektite glass itself contained no nickel. This is in contrast to the Wabar impact glass, which showed clear signs of nickel throughout the blackened glass." Note 7.9av

The same would be true of cometary impacts and lunar volcanic ejecta. The tektites could not have been from a comet or the Moon because the tektites have tiny inclusions of 'air' in them, which of course does not exist on either of the two possible parent bodies, leading to the supposition of inclusion after formation:

"Muller and Gentner (1968) used high sensitivity gas chromatography to analyze the gases from individual vesicles in various **tektites** and other natural glasses. They found empty vesicles adjacent to others with N/O ratios **like air**, and suggested that the air was trapped '…during the fusion of the tektite forming material or during the flight of the tektite **through the lower atmosphere**…'" Bib 155 p560

Each tektite has a specific mineral signature that, like the residue found on a shooter after firing a weapon, should be present in the tektite. The absence of this residue is like a veritable smoking gun, showing that **tektites are not** of a celestial origin.

Furthermore, other well-known natural surface glasses, such as **Libyan Desert Glass** (the large centerpiece in Fig 7.9.25) are "enigmatic glasses" Bib 155 p560 because they "contained CO_2 and air" and are therefore terrestrial. Curiously, no volcanoes lay in the areas where these surface glasses are found.

"The origin of tektites, on the other hand, is not well understood and the conflict of theories **remains puzzling and unresolved**. In many ways, the large chunks of natural glasses such as **Libyan Desert Glass** and Muong Nong tektites **are even more puzzling** than the more traditionally recognized aerodynamic tektite forms." Note 7.9aw

All of the natural surface glasses, including tektites, must then have formed in some unique way heretofore unknown and unrecorded, yet it must not have been in the too distant past because the glass would be much more decomposed and weathered.

Researchers recognize this when they consider the "homogeneous and relatively bubble free liquid" not found anywhere else on Earth today:

"But, the central issue in determining the impact origin of tektites remains, that is, **how to transform a mass of crushed rock into a homogeneous and relatively bubble free liquid which rapidly cools to a glass**. Even the commercial production of glass takes many hours to relieve the melt of its volatile components. No partially melted material, or target rock inclusions, have **ever been found** in Libyan Desert Glass." Note 7.9aw

Tektites have a very low water content and must have experienced very high temperatures and pressures at a level not normally associated with volcanic activity. These characteristics would have been present in colossal-sized hydrofountains, which would also explain the distribution of the tektite fields that cover such vast areas.

The Crater Depth Evidence

Direct evidence that craters on the Moon and other celestial bodies are hydrocraters rather than impact craters comes from examining the *depth of simple craters* (craters without central uplifts or rings). Meteorites travel at speeds of kilometers-per-second. As they approach a celestial body, a planet or moon, the gravitational force of the object exerts a pull on the meteorite. The added tug is relatively inconsequential, comparable to the weight change of a human after taking a breath of air, and **does not significantly affect the speed or depth of a crater.** If this were not the case, *larger* planets with higher gravity would exhibit greater crater depths.

Has research shown a correlation between crater depth and the size of celestial bodies, and if so, do the larger bodies have deeper craters?

Researchers have identified *a direct but **inverse** correlation* between the depth of simple craters and celestial body size—that is, the *smallest* sized bodies exhibit the *deepest* craters:

"The **depth of the deepest simple crater** at the onset of the simple-complex transition is also a **function of 1/g**. Thus, the deepest simple crater on the **moon** is about **3 km** deep, whereas on **Mercury, Mars**, and the **earth** the deepest simple crater is **1.7 km, 1 km**, and **0.5 km** deep, respectively." Bib 161 p131

This means craters on smaller planets are deeper than craters on larger planets, and crater depth is predictable based on the planet size. This is not compatible with crater origin by impact.

However, if we take a different view of the matter and assume these craters are hydrocraters, the deeper craters are explained quite easily. Imagine that a bullet is shot straight **down** from height (as in impact) into a bucket of soil, a city-sized plot of land, or even a moon, assuming the soil is all the same, the resulting crater would be roughly the same depth. However, if the bullet was shot **up** from beneath the surface, Earth's gravity would slow the bullet and ejecta more than the Moon's gravity would, which has a gravitational pull 1/6th that of the Earth. The increased gravity on larger bodies would result in more ejected material settling back into the crater after a hydrofountain explosion than smaller bodies because the gravitational tug would pull it back in. Thus, craters of the same diameter on the Moon and other smaller planetoids are *deeper* than those present on the Earth or Mars. Smaller celestial bodies have less gravity, which allows the same *upward* explosion (as in hydrofountain) to expel more ejecta than would be expelled from the same sized explosion on larger planets.

This fact is shown quantitatively in the previous quote from *Impact Cratering* but its significance was not realized. The data identifies the deepest simple crater on the Earth is 0.5 km whereas the deepest crater on the Moon is 3 km deep, six times deeper than the Earth's deepest crater, an intriguing correlation between the Earth's and Moon's gravity.

Crater Doublet Evidence

Three independent research groups found unmistakable "non-random" evidence of double craters on the Moon and on Mars. Impacts are random events that should exhibit no specific or repeatable patterns on the celestial bodies they strike. On the other hand, hydrocraters are not random, often associated with volcanic events, forming predictable patterns based on identifiable characteristics. Their interrelated formation confirms that they are of hydrofountain origin. **Crater Doublets** are simply

Crater Doublet Evidence

Crater Doublet Rule:
Smaller craters always intrude onto larger craters.

"Lunar Craters are Nonrandomly Destributed"
Journal of Geophysical Research

How could anyone believe that a simultaneous impact created a second crater right on the edge of all five of these craters?

Where are the crater walls like this found on the Moon?

Fig 7.9.27 – The two black and white photos show examples of Crater Doublets on the surface of the Moon. Crater Doublets are two or more craters that bisect each other. The smaller crater **always** intrudes upon the larger crater, which would not happen through random impact cratering. Multiple eruptions of hydrocraters easily explain the multiple, smaller crater phenomenon. Simultaneous high-speed impacts created in the laboratory produced a crater doublet with a high wall, represented in the color photo above. The wall formed because the energy released from the simultaneous impact collided and dispersed ejecta laterally. The experimentally produced crater doublet with distinct walls has no analogue on the lunar surface.

double craters (or sometimes multiple craters), where a larger crater is overlaid with a smaller crater (or craters), often notably smaller. See Fig 7.9.27 for several examples of Crater Doublets. The Moon has many of these and they can be seen everywhere.

The Lunar Doublets study is the first of three evidences that craters are of hydrofountain origin. The study included 3200 pairs of lunar crater doublets reported in the *Journal of Geophysical Research* in 1971. Investigators determined that the "lunar craters are **nonrandomly** distributed" Note 7.9bc by comparing the azimuths or angles at intersecting craters:

"Azimuths were plotted between all intersecting lunar craters > 10 km in diameter, about 3200 pairs, directed from the center of the older crater to the center of the younger. **The results are statistically nonrandom above the 90 percentile**." Note 7.9bc

These investigators establish that more than 90% of the time, intersecting lunar craters overlapped in a *similar pattern.* This pattern can be easily seen in Fig 7.9.27, inset right, where at least five craters have a smaller crater on the crater rim. Each of these crater doublets are similarly proportioned, as compared with other doublet pairs. This is nearly impossible to have been the result of meteorite impact. Additionally, four of five of the small doublet craters are on the southeast rim of the larger crater—also not a coincidence. This is what the researchers referred to as being statistically nonrandom to the 90th percentile. The Lunar Doublets occur at far too high a frequency to have occurred at different times. They had to have been formed at the **same** time, a *simultaneous 'impact'* if they were of impact origin.

The researchers from the Department of Geology at the University of New Mexico who studied the 3,200 doublet pairs all suggested the origin of these lunar craters was "endogenic," having a subsurface origin. Of course, this is how a hydrocrater is formed:

"The data presented here **suggest endogenic [hydrocraters] origin of most large lunar craters**." Note 7.9ax

Hydrofountain crater formation easily explains the Crater Doublets because hydrofountains can have *multiple* eruptions, where each successive eruption is usually smaller than the preceding one, leaving a smaller hydrocrater behind.

The second Crater Doublet evidence comes from Mars Doublets. This research was conducted and reported by different investigators but in the same *Journal of Geophysical Research.* It included a study of clustered craters on Mars, similar to the Lunar Doublets research reported by Elston and others. Researchers found *nonrandom* Crater Doublets on the red planet as well, which led these investigators to consider the possibility that non-impact forces created them:

"**A large number of Mars craters** are nearly tangential to other craters. They occur in clusters or as isolated **crater doublets**. Results of probability calculations and a Monte Carlo cratering simulation model show **conclusively that many of the Mars craters could not have resulted from random single-body impact**. The possibility that these craters are **calderas** is considered, but they could be **only if calderas on Mars form by mechanisms different from those on earth**." Note 7.9ay

The researchers acknowledge that the Mars Doublets appear to be "calderas" but step back from this obvious conclusion because hydrocrater doublets have not been found on Earth and no clear mechanism exists to explain these phenomena. This reluctance may be the result of a Uniformity Myth bias that should be abandoned thanks to events like the 1980s Mount Saint Helen's eruption.

Researchers attempted to reproduce the Mars Crater Doublets by creating simultaneous impacts. The results of the experiment were deemed unsuccessful because of the creation of a "common crater wall" that was "straight and very high," Note 7.9ay p2430 with no analogue in nature. The wall was created when two impact craters were formed at the same time. No examples of walled Martian crater doublets are known.

Again, reconsidering the data and interpreting the crater doublets as being formed as phreatic explosion pits, one after another, no crater wall is formed and successively smaller craters are formed, each on top of the preceding one.

In a simultaneous impact event, a wall is created *every time* two impactors strike close to one another due to the energy from the explosion dispersing the force laterally, creating a wall extending beyond the crater boundary as seen in Fig 7.9.27, inset lower left.

The third evidence that most lunar craters are of hydrofountain origin is the Crater Doublet Rule, a rule first identified by a

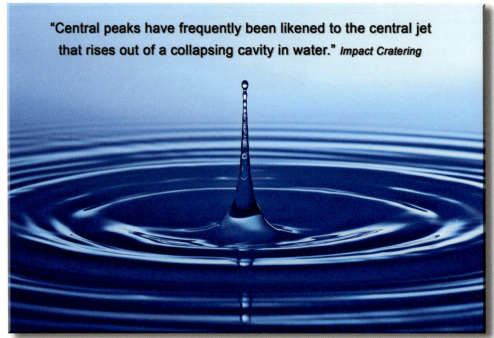

Fig 7.9.28 – Water dropped onto a water surface creates a central peak that is often associated with lunar crater formation. But the Moon neither is nor was liquid rock. If it had been, the Moon would be glass. Moreover, a liquid does not retain this shape, as some geologists have proposed.

long-time lunar expert and author, Paul Moore:

"**When one crater breaks into another it is the rule that it is the smaller crater which is the intruder**. Look for example at Thebit near the edge of the Mare Nubium. Its wall is broken by a smaller crater, Thebit A, which is itself broken by a yet smaller structure, Thebit F. This is to be expected on the volcanic theory; the oldest eruption would be the most violent. **The rule holds good in well over 99 per cent of cases**." Bib 163 p91-2

Amazingly, these three direct evidences of hydrocraters; Lunar Doublets, Mars Doublets and the Crater Doublet Rule were all recognized *over three decades ago,* but the research and the evidence have either been ignored, or in the case of the Doublet Rule—observation was trumped by impact dogma that seemed to be reinforced by "space research."

> FQ: Why do smaller craters over-lie larger craters in over 99% of all crater doublets?

For decades, Patrick Moore used the Crater Doublet Rule to explain how smaller craters overlap larger craters at a rate greater than 99%, a rate *well above any random impact possibility.* Unfortunately, he caved to the pressure of his colleagues saying:

"All this sounds reasonable enough. Only now, with the development of **space research**, have we had to **abandon the caldera hypothesis**." Bib 163 p92

There will be more on Moore's abandonment of his volcanic theory of crater formation as we discuss how scientists have embraced the Impact Pseudotheory, a theory that like so many other modern pseudotheories is held sacrosanct without any *real* scientific evidence. Although Moore ultimately abandoned his caldera hypothesis, neither he nor any scientist since has presented compelling scientific data to explain Lunar or Martian Doublet formation. Neither has there been any "space research" that satisfactorily explains how the Crater Doublet Rule can be modified to account for impact cratering. In fine, the Crater Doublet Evidence is extraordinary evidence that most of the craters are hydrocraters and not impact craters.

The Crater Peaks Evidence

Complex craters afford us compelling evidence that they were not formed by impact. Differing from the simple bowl-shaped crater, "complex" craters exhibit a variety of features such as central peaks, pits, rings and flat floors. Many of the craters on planetary bodies in the solar system have peaks; mountains located inside the bowl of the crater, lending reason to their name, central peaks. How these mountains form, and their location, is important to our understanding of crater formation. The impactologist imagines that an incoming meteorite body explodes at impact on the **hard** surface, creating a crater and leaving behind mountains in the crater. The mountains, commonly called *central peaks* or *uplifts* in scientific literature, occur only in larger craters, many kilometers wide, formed as the ground ostensibly liquefied during impact.

To explain the formation of central peaks, scientists have imagined the material under the crater must have become fluid, i.e. melted and have relied on the physics of water, seen in Fig 7.9.28 to demonstrate this idea:

"**Central peaks have frequently been likened to the central jet that rises out of a collapsing cavity in water**. If central peaks in impact craters actually do form by a hydrodynamic mechanism of this kind, the rock debris beneath the crater **must behave as a fluid for at least a short period of time**." Bib 161 p147

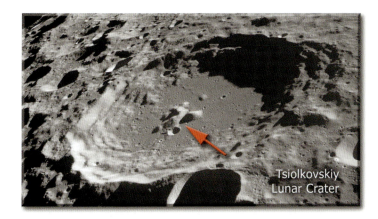

Fig 7.9.29 - Many craters have peaks rising from the crater floors. They are commonly referred to as central peaks but are often not central at all. Sometimes, the peaks may consist of a mountain range, other times there are just one or two peaks. Like their Earthly counterparts, lunar craters with central peaks owe their existence to water and hydrofountain action.
In these images of craters with peaks, note how each is not centrally located. Crater Lake is a known non-impact crater whereas the two lunar craters are believed by scientists to be of impact origin. According to their theory, if the central peaks are the result of impact rebound, they should be right in the center of the crater. Landforms like these, coupled with other hydrocrater evidence testify of Earth's hydroplanet origin.

"Central" Peaks That Are Not Central

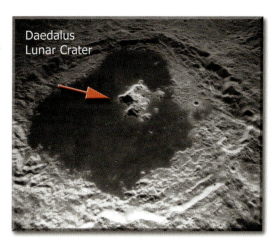

Complex Crater Evidence

"...there are many areas of uncertainty in how such features as central peaks, pits, or rings originate." *Impact Cratering p126*

Why would Impact Occur Here and Not Here?

CHAPTER 7 THE HYDROPLANET MODEL

There are many reasons why the water example fails to adequately explain the central peak phenomenon. The so-called central peaks suggest they are located in the center of the crater, which they should be, if their formation was as the impactologists suggest. However, most peaks *are not located* at the center of the crater and there are often multiple peaks on the floors of the large craters. This can be seen at lower left in Fig 7.9.30. Rather than verifying the traditional impact theory of crater formation, off-center peaks refute it.

The next problem complex craters pose for impact theorists is that experiments performed on *hard* surfaces result in a bowl being formed, not a flat floor with peaks and crater pits. Lunar craters are said to have formed from impact, presumably on the hard, brittle surface of the Moon, without subsurface water. To behave as a fluid implies the material was melted, or semi-melted, which would indicate that glassy crater peaks and floors would be present. Liquefaction—loose sediment acting like a fluid when under force—is seen by some as an explanation for central peak formation, however, liquefaction usually involves water, which we will discuss shortly. The Snowball Crater experiment on Earth demonstrated that a peak of sediment remained in the center of a crater excavated by a 500-ton TNT blast. However, this occurred because the water table allowed the surrounding sediment to be pushed out of the crater. Some of the craters on the Moon may indeed, have been formed by impact where peaks were formed by this mechanism, but only if significant water was present.

One additional hurdle impactologists have is that peaks are formed only in very large craters. If crater peaks were formed by 'rebounding' melt, then both *small and large* craters would exhibit crater peaks. Since they do not, researchers have concluded that the "rebound" process probably "cannot explain the central peaks of complex craters":

"R. B. Baldwin (1963) suggested that central peaks originate by a **'rebound' process** in which rocks below the crater are strongly compressed by the force of the impact and then spring back elastically when the stress is relieved, buckling the crater floor upward into a structural dome. Although this process has no obvious dependence on gravity or crater size, and so probably **cannot explain the central peaks of complex craters**..." Bib 161 p141

The bottom line is that experts in the field of impactology must acknowledge:

"Since the collapse of large craters **has never been observed directly**, there are **many areas of uncertainty** in how such features as **central peaks, pits, or rings originate**." Bib 161 p126

Key words, "many areas of uncertainty," echo the sentiment of many who study craters. Examining nuclear explosion craters formed on hard dry ground, we found *no* complex crater features. Additionally, high-speed impact studies showed *no evidence of central uplift formation* from a single projectile in a single impact event.

Some experiments employed the use of thin materials, such as aluminum foil to simulate an atmosphere. When the projectile struck this membrane prior to impact on the surface, the impactor was seen to break apart, and there was evidence of peak formation. However, these experiments are not applicable on bodies such as the Moon that have no atmosphere. Since the Moon has no atmosphere, experiments simulating impact on its surface must have no obstruction to impede the projectile before its impact. Scientists have been unable to duplicate complex crater features despite having a good idea of the speed, size, and mass of the meteoroids presumed to have been involved in the impact, and the geology of the impact sites.

Fig 7.9.31 – The Bull's-eye double crater on Earth's Moon is an almost impossible impact crater. There is a noticeable lack of impact ejecta on these types of craters, yet most researchers still assume they were made by meteorites. The Hydroplanet Model has a new origin for such craters. Courtesy of NASA (AS15-93-12640)

The Double Crater Evidence

Fig 7.9.31 is the Bull's-eye on Earth's Moon. This crater is located on the floor of the much larger Humboldt Crater. Little has been published about these double craters, which are found on planets and moons throughout the solar system. There isn't a shred of evidence to suggest these types of craters are of impact origin, but some have still suggested they are. To consider the impact premise objectively would be to accept that a smaller, secondary meteorite landed in the exact same spot as a previous meteorite. This meteorite would have had to be small enough not to obliterate the first crater, and not leave any ejecta evidence behind; an extraordinary coincidence, and perhaps possible.

FQ: How were double
(and triple) craters formed?

The universe is filled with bizarre and rare occurrences, but if there were many of these types of craters, no right-minded person could accept that such craters were formed by the impact of meteorites that struck the exact same spot. Fig 7.9.32 shows the surface of Jupiter's moon, Ganymede. There are at least eight double craters in this one area of the moon, far too many of them to consider the notion of impact, leaving much speculatation about their origin:

"Like several other large craters in this scene, the rayed one has a central pit, whose **origins remain speculative**..." Note 7.9az

There are plenty of clues to the crater's origins, right here on the Earth, but those clues only make sense with the Hydroplanet Model. Double craters are formed by *multiple* hydrous

Fig 7.9.30 – On the previous page black arrows point to small lunar "pit" craters on the flat floors of large craters; red arrows show that at higher elevations, there are no pit craters. Notice that non-cratered areas lay right next to cratered areas. Researchers have had to acknowledge that, "...the origin of pit craters remains, for the present, mysterious." Bib 161 p133 The reason "pit" craters are mysterious is that they are still considered impact craters. Why would impacts occur on the floors of large craters but not on the mountains? The formation of hydrocraters on a hydromoon surface holds that secondary smaller eruptions are typical and frequent, as can be seen on Earth in a small scale at geyser basins and in mud craters. This happens because earthquake aftershocks are usually smaller, creating less frictional heat.

Fig 7.9.32 – This image of Jupiter's moon Ganymede shows multiple double craters, both primary and secondary craters are remarkably similar in size. It is statistically impossible for so many craters to have formed by impact with two meteorites hitting the exact same spot. However, double craters are common and are caused by multiple hydrous eruptions. These are common hydrocrater phenomenon. Image only courtesy of NASA (PIA00334).

eruptions that may occur with only minimal ejecta sediment. A similar process can be seen, albeit on a tiny scale, when the humble sand crab expels water from its sand beach burrow. (An example of this can be seen in Fig 7.13.10)

If one will but look for hydrocrater evidence, it can be found everywhere on Earth and in the solar system. Then, simple answers to the difficult questions about planet and moon formation will be uncovered.

The Flat Crater Floor Evidence of Water

Rather than being cup-shaped craters, most lunar craters are shaped more like saucers; shallow pits with flat floors. The mystery intensifies with the recognition that the floors of largest craters on the Moon are not just flat, but actually follow the curvature of the Moon, rising above the surrounding raised rim of the crater. Impact cannot account for the massive uplift of the interior of craters, more than a thousand kilometers across in the largest craters.

Lunar scientists have attempted to explain these vast plains, the lunar maria, as coming from a flow of basalt. The same sci-bi is used to explain the basalt deposits on Earth, but as we learned in the Rock Cycle Pseudotheory chapter, in the Basalt Mystery subchapter, basalt is not simply a dry surface melt. You will recall that there have been no historical basalt flows on the Earth's continents. Glassy-lava flows yes, but no basalt flows. The same must hold for the Moon. Basalt flows could not occur as an impact 'melt' nor could the basalt flows come from magma! If the vast plains on the Moon were from melted rock flowing on the dry surface, glass-like rocks would be abundant, and as we have already discussed, almost no glass-like rocks have been found on the Moon.

Another long-held theory is that the flat-floored craters were formed long ago when the Moon was still molten. However, there are flat floors on the newest craters too, like Copernicus, which invalidates this idea. The dog-magma theories are dead-end ideas that cannot account for all of the data. Recognizing this, some investigators began looking in new directions for answers.

In 1980, Canadian researchers conducted three crater-forming experiments using a 500-pound TNT explosive to excavate each crater. At first glance, it appears the process of how "flat-floored craters on the Moon with both central uplifts and multi-ring structures" were formed:

"Three explosion experiments, called Snowball, Prairie Flat, and Dial Pack, formed large **flat-floored craters with both central uplifts and multi-ring structures**. The surface morphologies of the craters also displayed either prominent central mounds or multirings. In all three cratering trials, the charge type, charge energy, and target media were the same." Note 7.9ba

However, there are at least three disparities in the comparison to lunar craters. First, the 500-ton detonations were performed directly *on the surface*, much different from that expected in high-speed impacts. Second, the craters were excavated by a TNT blast, a low-speed explosion that creates a much different crater structure than a high-speed impact would. The third, and probably the most important discrepancy is the fact that these craters were formed above a *shallow water table* of approximately 7 meters.

Only the Snowball Crater exhibited a central 'uplift' pattern. Rather than being simply uplift, it was the result of **hydrologic** compression and rebound of non-ejected sediment. Shortly after the explosion, water began to seep into the crater by way of micro-hydrofountains, eventually covering the floor.

What the experiments successfully showed was that water played a significant role in the crater formation, a fact impactologists have had to downplay by associating the fluidity with magma. After all, doesn't everyone "know" that the Moon has **no water** on or near the surface? It certainly did not have any during the theoretical period of the 'Great Bombardment.'

Earlier in this subchapter, we introduced evidence that a small percentage of the craters on the Moon are impact craters. Not mentioned at the time was the fact that the impacts occurred primarily in shallow water or mud plains. A good description of the materials involved in the formation of peaks, rings and pits of complex craters comes from the book *Impact Cratering*. Of course, melted rock has been at the top of the list for a long time in most scientific circles, but careful research can bear out the truth. Melosh noted this about the volume of fluidization from melt in complex crater formation:

"One idea that was proposed early in the study of crater collapse is that **the rock debris beneath the crater is fluidized by impact melt**. The debris flows briefly as a melt-solid slurry until it cools and solidifies. Some impact melt is found on the floors of complex craters, **but it is rare in the central uplifts or in the zone of stratigraphic uplift beneath the crater, where fluidization is required**. The volume of impact melt surrounding known complex craters **falls far short of the volume required to fluidize much of the surrounding rock debris**." Bib 161 p151

Plastic materials are tagged as the reason complex craters have a flat floor:

"The **fact** that craters in a **plastic material** collapse back to a constant depth **also accounts** for the development of a **broad flat floor** in the interior of large complex craters." Bib 161 p147

Since impact melt cannot account for the fluidization, what else could contribute to the plasticity of the material?:

"Since conventional rock and debris strength properties **cannot explain crater collapse**, the problem can be turned around to determine **what strength properties the rock must possess to explain the observed collapse**... Perfectly plastic materials lack internal friction: the cohesion is independent of overburden pressure. This failure law provides a fair description of metals and **water-saturated clays**. It is *not* a good description of the strength properties of either **intact rock or broken rock debris**." Bib 161 p145

Both experimentally and in nature, "broken rock debris" does not form complex crater features, but "clay slurries" do:

"**Experimental impacts** into Bingham fluid targets (**clay slurries**) also yield crater forms similar to those observed in complex craters." Bib 161 p149

It is interesting that one of the most ardent supporters of impact cratering, Melosh, demonstrates in his own book the seminal role water plays in cratering; fluidization of rock beneath the crater, water saturated clays describing the correct plasticity of the rock, and experimental impacts into clay slurries that yield cratering results that mirror nature. It seems that *water* related evidence appears throughout the solar system, including Mars:

"The exaggerated development of central peaks observed in many Martian craters may also be **due to near-surface liquid water**, whose presence is suggested by the **fluidized appearance of Martian ejecta blankets**." Bib 161 p151

Amazingly, the researcher can see Martian ejecta blankets as having the "appearance" of fluidization—liquid water—but does not recognize the same fluidization evidence on other celestial bodies' craters like those on the Moon and Mercury:

"Another idea that is irrelevant for lunar or Mercurian craters but which may apply elsewhere is that the debris surrounding the crater is fluidized by the presence of some **interstitial fluid, usually liquid water. Explosion craters on earth are strongly influenced by the presence of subsurface water**." Bib 161 p151

In every case observed thus far, the interstitial fluid is *always* liquid water or steam and the explosion craters on Earth are not only "strongly influenced" by the presence of subsurface water, they are **created** *by the subsurface water*!

Perhaps being this technical about the wording seems trivial, but it is very important to reemphasize the role steam explosion craters play in modern geology. The lack of experience and understanding of hydrofountains, hydrocraters and the Hydroplanet Model have left the modern geologists with an incorrect perception of nature, and that must be corrected.

The Pit Crater Evidences

Presumably, impacts are random in their placement, therefore, why are large crater floors littered with **numerous** small craters, while higher elevations and rims have noticeably *fewer of them*? (See Fig 7.9.30)

FQ: Why are there many craters on the floor of large craters but few on their slopes and rims?

The obvious pattern where many craters are on the floors of large craters but not on the slopes and rims illustrates what some scientists have recognized as being a statistical impossibility. There is no weatherization or erosion on the Moon, so slope and rim impact craters would be clearly visible. (There are some, but relatively few as compared with those on the floor). The impact pseudotheory falls short at explaining why craters are concentrated on the floors of larger craters, but not on the rims. Turning, again to the book, *Impact Cratering*:

"The pits are **not** mere depressions in the crater floor, but are **frequently rimmed and in larger craters show upbowed floors**..." Bib 161 p133

How were they formed?:

"A **number** of explanations have been offered for the formation of central pits." Bib 161 p133

Because of the magmaplanet paradigm and the idea that all craters are from impact, solid answers have eluded scientists:

"**None**, however, has yet reached the point of general acceptance and **the origin of pit craters remains, for the present, mysterious**." Bib 161 p133

The only reason for the mystery is the ignorance of water's role. The material presented here establishes that there is no evidence that complex crater features; crater peaks, flat floors, rings, and pits are formed by meteoroidal impact on hard surfaces. However, the Hydrocrater Model, subchapter 7.8, has no trouble explaining these features. They are hydrocraters and exist here on Earth, the Moon, and many other celestial bodies. Moreover, the formation of these features can be from impact cratering, if the ground is saturated with sufficient water. The bottom line in either case is that the presence of *water is the most important and dominant element* in the formation of complex crater features, and both are a part of a **hydro**-planet environment.

The Crater Chains and Channel Evidence

Channels and rows of multiple craters, some linear, some gracefully sinuous are commonly seen in the solar system:

"Many of the so-called rills are made up basically of small craters which have run together, often with the **loss of their divid-

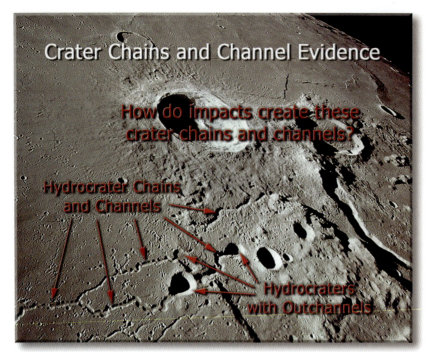

Fig 7.9.33 – Here we see Lunar Crater Chains and Channels without an origin in modern science. It is hard to find more definitive evidence of the Hydroplanet Model than hydrocraters and channels on the Moon. The Impact Pseudotheory fails miserably at explaining such phenomena, yet with the new discoveries in this chapter and the recognition that water is responsible for all mineral growth, such complex crater hydro-features become easy to understand.

tions about Crater Chain and Channel formation follow the Hard Question Principle—hard questions are easy, when we know the answer.

The Crater Chemical Composition Evidence

Another oft overlooked but critical fact about Moon craters is their chemical composition. The so-called impact melt-rocks retrieved from lunar craters have been carefully analyzed, uncovering a mystery:

"One of the principle **mysteries of impact melts** on the Moon is the **chemical composition of their matrices**..." Bib 129 p138

Paul D. Spudis, in his book, *The Once and Future Moon*, discussed this mystery and then posed two *significant* questions:

"How can an impact melt be made that has a **composition completely different from that of the target rocks in which the craters formed? Are we wrong about how impact melts are created**?" Bib 129 p139

ing walls. Others, such as the Hyginus Cleft, are crater-chains in part, though it is true that there are plenty of rills which show no trace of crater-like enlargements. **Crater-chains are very common indeed**..." Bib 17 p117

Examples of Crater Chains and Channels can be seen in Fig 7.9.33. Lunar Crater Chains and Channels, also known as rills are a real mystery to the impactologist. The chains of craters are not aligned and therefore could not have been made by multiple fragmental impacts of a larger impactor. In 1994, the comet Shoemaker-Levy broke apart and crashed into Jupiter. Although scientists had no way of viewing the results of impact on the Jovian surface, many of them supposed this event supported the formation of crater chains by multiple, simultaneous impact events. If one looks at the string of comet fragments, (See Fig 7.16.14) it is apparent that they are not all aligned, and would not have formed a linear chain of craters.

Based on actual observations of lava flows on Earth, flowing, molten rock does create deep channels, like the ones seen coming from the craters on Fig 7.9.33. Moreover, the channel size and length far exceed any comprehensible amount of melt from an impact event, certainly not enough to flow *up* and over the rim of the crater. Clearly, the impact pseudotheory falls short at answering such critical questions.

FQ: How are crater chains and channels formed?

In Fig 7.9.34, we see how pressurized water easily flows out of hydrocraters in channels that are commonly found at hydrothermal sites here on Earth. The crater chains are formed along weak tectonic fault lines where water has risen through the crust. In many cases, as water is heated to steam, explosions form hydrocraters. Ques-

> "Are we wrong about how impact melts are created?"
>
> The Once and Future Moon, Paul D. Spudis

The answer to his first, perhaps rhetorical question is—it cannot. The second answer is much more ominous for the impactologist engaged in the study of the Moon's craters. Scientists have been wrong not only about how impact melts are created, but that they are not impact melts at all. When the rocks harvested from a crater differ from both the surrounding surface and from known meteorites, it is time to consider an alternative to impact origin. If the rocks originated from deep below the surface, not of magmatic origin, but carried to the surface via hydrofountain activity, it is not difficult to understand why they would be of different composition than the surrounding surface rocks.

Fig 7.9.34 – Hydrocraters like this are common in hydrothermal areas around the globe. A mini-hydrofountain formed this crater; it boasts an outwash channel leading away from the crater, like those seen on the Moon.

Fig 7.9.35 - This image of Mud Volcanoes and hydrocraters comes from the Gulf of Mexico, in the Green Canyon Lease Area. Water depth here is approximately 2250 ft. The field of view is about 4-5 m (15 feet) across. Fig 8.8.13 & 14 show similar concentrations of hydrocraters, arranged in non-random orientation. The location of the craters has a direct correlation to hyprethermal water channels running from beneath the surface to the ocean floor. Image Courtesy of Harry Roberts.

Lunar Crater Answers from Oceanic Hydrocraters

Many questions about the crater mysteries, such as crater doublets, flat-floored craters, and channels on the Moon or other planets have no logical answers in the impact paradigm. As we look to what some consider an unlikely place—the Earth's oceans—for answers, new possibilities began to unfold. Oceanic hydrocraters on the sea floor, as seen in Fig 7.9.35, hold important clues to the origin of lunar craters. From *Marine Geology* we read how craters on the ocean floor are "draped with sediment," similar in appearance to what is found on the Moon:

"Some pockmarks [hydrocraters] appear to be **draped with sediment**, making them **smoother and shallower**, and others have **sharp rims and are deeper**. This suggests that there are several generations of pockmarks or that some are younger or active and others are older, having formed long enough ago to be **partly buried**." Note 7.9bb

Here we see possibilities that both smooth and sharp-rimmed craters on the lunar landscape can be explained by studying oceanic hydrocraters. Furthermore, the process that partially buried craters on the lunar surface can be *easily* explained by sediment flow from surrounding hydrocraters and mudflows, whereas partial burial by impact ejecta has always been an inadequate explanation.

In answer to the earlier FQ of why there are more craters on the floors of large craters than there are on the rims, researchers explain a similar phenomenon under the ocean of how pockmarks form "pathways" that "may branch off" for various reasons, forming new craters that overlap previous pockmarks:

"They form overlapping or **superposed pockmarks which seem to have developed first as large overlapping pockmarks** over 300 m in diameter, **with small ones about 100m across formed within them**.

"Near the seafloor, additional pathways may branch off either because some routes are sealed… or because new routes are easier to find in the unconsolidated upper sediments. This could explain the occurrence of **groups of pockmarks** in the same area or overlapping pockmarks." Note 7.9bb p268, 274

Ocean floor hydrocraters answer another question previously asked concerning flat crater floors—why do large lunar craters have them? They have them for the *same reason ocean-floor*

hydrocraters do. From *Marine Geology*:

"…some are V- or U-shaped in cross-section **while other bigger ones have flat floors**…" Note 7.9bb

Actual observation of ocean floor hydrocraters has shown larger craters are filled with sediment from active diatremes, causing a flattening of the floors of larger craters. With direct physical evidence of flat-bottomed craters on the ocean floor, we can begin to understand how the Moon or other planets can have flat-floored craters as well.

We can see similar events that have happened on the floors of inland hydrocraters like Crater Lake, seen earlier in this chapter, in Fig 7.8.10. Often, the initial explosion that resulted in a large crater is followed by smaller secondary hydrofountain eruptions, which may form mounds and/or additional craters. Evidence of this is seen on the floor of Crater Lake as many secondary mounds and craters are found there. Secondary eruptions layer the bottom of the crater, flattening its floor.

What about the Crater Channel Mystery on the Moon; do undersea observations have anything to add to our understanding? The oceanic researchers studying the pockmarks provide another interesting and valuable observation:

"…some suspected pockmarks [hydrocraters] appear to **have channels leading away from them as if they have had outflow of dense fluids**, i.e. were once brine-filled but have emptied." Note 7.9bb p268

With a shift in one's paradigm toward hydrocraters, it becomes rather easy to comprehend the wisdom of *how* these channels were formed.

Enceladus' Large Hydrofountain and Crater Evidence

No doubt, one objection to the hydrocrater model is the lack of observable terrestrial hydrofountains of sufficient magnitude to explain the phenomena. The formation of large hydrocraters is uncommon, thankfully, an event most of us will never see personally. The Mt. Saint Helens eruption, its subsequent hydrocraters, and the ocean floor hydrocraters are recent examples found here on Earth that testify of this ongoing process and they provide unequivocal evidence of phreatic—steam—crater formation. Other examples exist in faraway places in our solar system. Saturn's moon, Enceladus is one such place:

"As it swooped past the south pole of Saturn's moon Enceladus on July 14, 2005, Cassini acquired high resolution views of this **puzzling ice world**. From afar, Enceladus exhibits a **bizarre mixture of softened craters and complex, fractured terrains**." Note 7.9bc

What makes Enceladus' "softened" craters so bizarre? We touched briefly on this in our previous discussion about hydrofountains (see Fig 7.7.13), and now, as we examine the craters in Fig 7.9.36, no patterns of impact ejecta are evident. The reason for this is the makeup of Enceladus' surface—frozen water—and gravitational heating. Frictional heat caused by gravitational flexing of the crust superheated subsurface water. Rising pressure powered spectacular hydrofountains and melting ice smoothed craters. All of these terrain-forming processes were made possible with water, frozen *water*, and lots of it.

Shoemaker never saw the cratered and scarred surface of Enceladus in Fig 7.9.36, but he would have assumed the craters were of impact origin. Shoemaker's and other's statements about impact have influenced scientists for decades, causing researchers to force-fit the impact theory into explaining the cra-

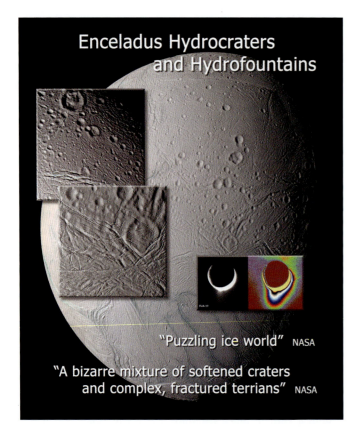

Fig 7.9.36 – Saturn's moon, Enceladus, has caused geoscientists to acknowledge that catastrophic hydrofountains exist! Enceladus clearly demonstrates that huge water fountains, the size of the entire moon can happen. These are powered by heat from tidal friction. Hydrofountains of this scale can produce large hydro-valleys, hydro-channels and hydrocraters. Images courtesy of NASA.

ter's origins and "puzzling" about this "bizarre ice world."

The following discussion from NASA describes the background image and two inset left photos in Fig 7.9.36. This occurred before the discovery of Enceladus' giant hydrofountain eruption in 2005:

"This dramatic scene from Cassini illustrates an array of processes on Saturn's moon **Enceladus**, a once geologically active world. Most of the larger craters appear to have softened from their original, presumably crisp appearance, and are cross-cut here by numerous faults.

"Cassini acquired this high-resolution view of Enceladus during its closest encounter yet with any moon of Saturn.

"Toward the bottom of the scene, terrain containing fractured and softened craters gives way to essentially non-cratered terrain consisting of tectonic faults.

"The softened craters, fractured plains and wrinkled terrain on Enceladus **suggest geologic activity** has taken place in several episodes during the satellite's history. This activity **might continue into the present time**, although imaging team scientists have seen no evidence for current activity on the moon." Note 7.9bd

Of course, we *know* that "geological activity" is continuing on this moon at present because it has been observed. Geologists cannot deny that hydrofountains exist in the solar system, and that they would produce large hydrocraters. The size of Enceladus' present-day hydrofountain defies comprehension and scientists have yet to recognize the importance of the Enceladus Hydrocrater—until now.

The Europa Hydrocrater Features Evidence

The smallest of Jupiter's four Galilean moons, Europa, is the subject of Fig 7.9.37. Close-up images of Europa's icy surface reveal features that researchers now accept as being hydrofeatures—subsurface ice "diapirs" (fluidized pipes of material brought to the surface):

"Here we report on the morphology and geological interpretation of distinct surface features—**pits, domes and spots**—discovered in high-resolution images of Europa obtained by the Galileo spacecraft. The features are interpreted as the **surface manifestation of diapirs, relatively warm localized ice masses that have risen buoyantly through the subsurface**." Note 7.9be

For decades, the notion of 'impact craters' has been so deeply ingrained that researchers substitute the word 'pit' for 'crater' in the description of Europa's hydrofeatures. There are many interesting aspects of Europa's moon to consider. With a paradigm change from impact crater to hydrocrater, new ideas will surface and truths will be unlocked.

The Callisto Hydrocraters Evidence

Another of Jupiter's hydromoons, Callisto, seen in Fig 7.9.38 seems to redefine the origin of many large craters as "upwelling

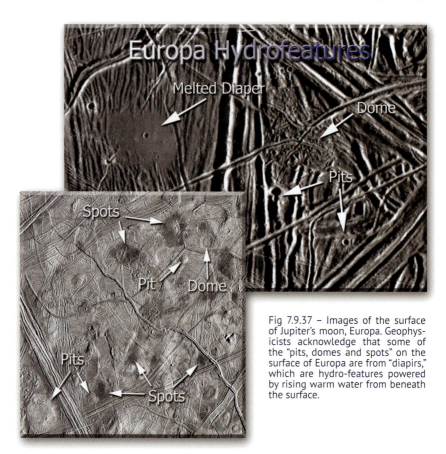

Fig 7.9.37 – Images of the surface of Jupiter's moon, Europa. Geophysicists acknowledge that some of the "pits, domes and spots" on the surface of Europa are from "diapirs," which are hydro-features powered by rising warm water from beneath the surface.

CHAPTER 7 THE HYDROPLANET MODEL

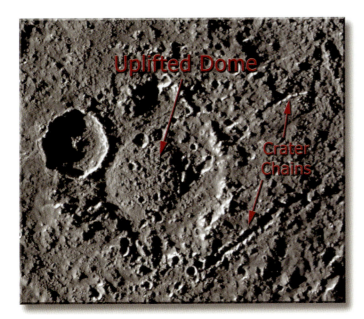

Fig 7.9.38 – Callisto, the second largest satellite of Jupiter is a hydromoon. There are large craters on its surface that could not have been formed by impact. One such crater, Har, is a hydrocrater with a large uplifted dome. Researchers have tried in vain to discover crater dome formation with impact theory. NASA photo PIA01054.

ice" forms a dome in the center of the crater:

"An older, eroded crater called Har has a large, rounded **mound on its floor that was probably caused by upwelling ice**." Note 7.9bf

Note how the mound appears similar to Europa's domes, which were attributed to a "manifestation of diapirs." Relatively warm water pushed up through a diatreme that formed the original crater. There are many similarities to the lunar craters we studied earlier.

The Hyperion Hydrocrater Evidence

When scientists saw Saturn's moon Hyperion for the first time, they were astonished to say the least. The surface displayed an extraordinary density of craters, but the most incredible feature was an apparent crater nearly the size of the moon itself! The image of Hyperion in Fig 7.9.39 affords some interesting observations. For one, there is a distinctive dome in the center of the largest crater. Note also, the apparent orientation of the craters along the rim of the large crater. Instead of being circular shaped, as one would expect if they were high-speed impact craters, they are shaped in such a way as to give them the appearance of radiating outward from the large crater.

How could Hyperion's surface be shaped by impact? NASA researchers, steeped in impact theory think so:

"It is likely that Hyperion **may** have been **bombarded with meteors**, which blew part of this moon away and caused its **highly irregular shape**. The irregular shape of Hyperion and **evidence** of bombardment by meteors makes it appear to be the oldest surface in the Saturn system." Note 7.9bg

No "evidence of bombardment" was given other than the assumption that since there are craters they must be from impact. To explain the deep, dark floored craters on Hyperion, scientists have proposed a theory called "thermal erosion," a process whereby sun-warmed material melts its way into the surface of the moon:

"This up-close view shows a low-density body **blasted by impacts** over eons. Scientists **believe** that the spongy appearance of Hyperion is caused by a phenomenon called **thermal erosion**, in which dark materials accumulating on crater floors **are warmed by sunlight** and melt deeper into the surface, allowing surrounding ice to vaporize away." Note 7.9bh

There are some glaring flaws in the theory that attempts to address the dark-floored phenomenon while adhering to the favored impact origin theory. Saturn is ten times further from the Sun than the Earth, far too distant for the Sun's heat to make an appreciable difference. After all, the ice moons of Jupiter are significantly closer to the Sun yet they exhibit no such phenomena.

"Hyperion is small—on average only 165 miles across. It seems **to be made mostly of ice**, with a dark, dusty coating. Bright white rims surround some craters within craters, suggesting that the dark material is not very deep. Its interior appears to be a series of **voids**, not unlike a loosely compacted snowball." Note 7.9bi

If Hyperion's interior is a series of voids, how did it acquire them? Assumptions that Hyperion is a fragment of a larger moon are founded on the magmaplanet theory and the impact theory, neither of which answer the question of how the moon acquired its interior voids:

"**The origin of the moon is also a mystery**. Its peculiar non-spherical shape suggests that it could be a fragment left over from a **giant collision** or bits and pieces from a bigger moon, says Spilker." Note 7.9bi

The Hydroplanet Model easily explains all of these phenom-

Fig 7.9.39 – Hyperion is one of the most unique moons in the solar system with a look all its own. Like the individuality of a snowflake, each icy hydromoon is unique and suggests a celestial water origin, which is nowhere to be found in modern science.

ena, and while the actual origin of the moon is not yet known, the processes that shaped its surface can be.

Hyperion tumbles unpredictably in its orbit around Saturn, subject to the gravitational torque of Titan, Saturn's largest moon. Tidal forces and gravitational-friction heat water deep beneath the surface, creating intense pressure. Rising steam deforms the crust, creating dome-shaped landforms. Access to the surface through faults or previous waterways gives way to phreatic eruptions, blasting material away, forming hydrocraters. The majority of the material escapes into space but denser material would fall back into the newly dormant diatreme at the base of the craters.

In 2005, NASA's scientists were unaware of the Hydroplanet Model and its numerous evidences and therefore, as Linda Spilker, deputy project scientist at NASA's Jet Propulsion Laboratory in Pasadena, California stated, Hyperion is truly a "mystery."

Voices of Reason Crying From the Dust

Early researchers saw the flaws in the impact pseudotheory but did not have the luxury to observe the actual Martian impacts we have seen today. They did not have high-speed impact studies to review, or new satellite technology that demonstrated how glass is formed in high-speed impacts; their voices of reason were drowned by modern theorists. If the discovery of natural law using the Universal Scientific Method had been the ultimate goal of science instead of theories being the end-point, there might have been a different conclusion.

Walter H. Bucher, a pioneering researcher of the 1950s and 1960s identified geological evidence refuting impact crater origins. Bucher found that many of the so-called 'impact' craters were located on known volcanic fault lines and were "not randomly distributed":

"If meteorites caused these structures [craters], they **must be distributed randomly**, i.e., they must not bear a systematic relation to structures of terrestrial origin nor to magmatic activity. In this paper this test is applied to the largest cryptoexplosion [hydrocraters] structures in Europe, North America, and Africa.

"These test cases are decidedly not randomly distributed, neither in space nor in time. **The meteorite impact hypothesis is therefore rejected.**" Note 7.9bj

Bucher drew the following conclusion of the geologic origin of these craters:

"The writer suggests the release of vast quantities of water vapor through sudden crystallization of supercooled molten rock near the base of the crust as source of energy, carried rapidly into porous rocks near the earth's surface under an impervious cover. **Rapid arrival of the vapor results in an explosion crater...**" Note 7.9bj

Bucher was right on target and his evidence has yet to be disproved, leaving one to wonder why the concept of hydrocraters did not catch on. Perhaps there were several reasons it did not catch on, none of which was verification through actual empirical evidence that craters were formed by impact. Bucher and others who were on the frontline of the crater origin debate were fighting against the popular 1960s space-craze where it seemed everything came from space.

Other researchers such as G. C. Amstutz noted with alarm in 1964 how far the science of the day had departed from the scientific method:

"**The fact that scientists of our generation can depart so much from the scientific method of working with congruent analogues**, or, in other words, the **fact** that the symmetry relationship between the polygonal structures [hydrocraters] and the regional geological pattern can be ignored, **is alarming**.

"This fact may be interpreted historically as one of the **severest consequences of the dissociation between the many fields of human interest**. Not only do the branches of the sciences **not know any more what is going on in the neighboring branch**, but even more so, the sciences do not integrate their basic results any more with those of the humanities, of philosophy and psychology. In other words: **The average natural scientist does not know enough about the basic nature and limitations of a scientific theory. The confusion between facts, observations and interpretation is widespread…**" Note 7.9bk

Bucher, Amstutz, and others like them are literally Voices of Reason crying from the dust as the scientific world gradually confused theory with fact. Today, the typical scientist is trained in the establishment where theory is king, and specialization is the norm, a stark contrast to the universal nature of the UM.

The Crater Debate Summary

For decades, Patrick Moore, well-known lunar scholar, authored many books on the Moon. In his 1976 book *New Guide to the Moon* he did not even mention the word "impact" as being responsible for cratering. Moore was one of the last holdouts of the 'volcanic' theory. Sadly, in his new 2005 book, *Patrick More on the Moon,* he abandons his volcanic theory of lunar crater formation, turning wholly to the Impact Pseudotheory:

"Craters are almost always associated with volcanoes: Vesuvius, Etna, Stromboli and many others. Some are violently explosive, while others are gentler. Go to Halemaumau, atop Mauna Loa in Hawaii, and it is (usually) quite safe to look down into the lava-lake; believe me, it is a fascinating sight. Obviously, then, it seemed logical to assume that the lunar craters were of volcanic origin, similar to our calderae, and for many years most people believed this. **I certainly did**, for the **reasons given in the earlier editions of this book. Sadly, I now have to accept that I was wrong**. There has been extensive volcanism on the Moon, but the craters were produced by meteorites which rained down on the Moon thousand of millions of years ago, during the period of what has become known as the Great Bombardment." Bib 163 p86

Was Moore wrong? His former theory of crater formation was based on active and dormant volcanoes viewable on Earth and it could not have fully explained the lunar surface, but as Moore admits, the volcanic theory does explain *parts* of lunar cratering that impact cannot. Moore and other researchers recognized only a small part of the crater formation process; however, the impact theory has also failed to account for *many* anomalies in the cratering process. Neither Moore nor any other researcher has been able to provide evidence proving that a "Great Bombardment" took place—and they never will because it never happened.

For the first time in science, the UM has brought together real criteria for determining if a crater is of impact origin. It has also presented significant and extraordinary evidence in this subchapter that the vast majority of planetary and lunar craters are hydrocraters.

It is difficult to encapsulate all of the extraordinary claims and discoveries in this subchapter; the Crater Debate is backed with extraordinary evidence. The following sub-subchapter headings are a partial list of the discoveries and evidences covered, and are given here to help incorporate this vast amount of information. The list includes evidences for the few craters that are of impact origin, and a large amount of evidence indicating that most craters are not of impact origin:

- V-Impact Signature and Ejecta Evidence
- Evidence of Glass
- Evidence of Hypervelocity Melt
- Larger Impacts Equal More Glass
- Trinity Glass Evidence
- Nuclear Crater Evidence
- Wabar Impact Crater Evidence
- The Absence of Lunar Glass Evidence
- Shorty Lunar Hydrocrater Evidence
- Satellite Impact Evidence—"Molten Cratering"
- Flawed Impact Criteria
- Shatter Cone and PDF Criterion Myths
- Coesite Impact Criterion Myth
- Tektite and Libyan Desert Glass Evidence
- Crater Depth Evidence
- Crater Doublet Evidence
- Crater Peaks Evidence
- Flat-Floored Crater Evidence of Water
- Pit-Crater Evidence
- Crater Chains and Channel Evidence
- Crater Chemical Composition Evidence
- Enceladus' Large Hydrofountain and Crater Evidence
- Europa Hydrocrater Features Evidence
- Callisto Hydrocraters Evidence
- Hyperion Hydrocraters Evidence
- Voices of Reason Crying From the Dust

One of the goals of this chapter was to present new evidence of hydrocraters and to end the Crater Debate. Impactologists might have assumed the debate was over, but it was not resolved. Of course, there are impact craters and meteorite impacts do occur, but scientists cannot naïvely continue to assume all craters are of impact origin. Eventually, time and truth will win the day. So many people want to know the truth that when it is finally made available, there will no longer be a debate about crater origins.

This has been one of the longest subchapters in the book, yet the evidence presented here barely scratches the surface of the hydrocrater evidence. Space constraints have allowed only this small portion to be included, but there is no doubt that Millennial Science will foster many books on the subject.

New technology and new observations have provided evidence and data for those seeking to understand the real origins for the majority of the craters. Upon what is this evidence based? It is based on the hydroplanet model, a paradigm where the Earth's and the Moon's surfaces were covered with water during their formation. We can observe similar formations and processes occurring right now, at the bottom of Earth's oceans. All one has to do is look with an open mind.

Moreover, during the catastrophic periods of former times, large-scale volcanism unlike anything witnessed in modern times, caused the hydroplanet Earth to spew vast amounts of water and sediment onto the surface through hydrofountains, of which remnant hydrocraters are found all around the Earth, oft times being labeled as 'anomalies' or 'enigmas.' There will be a great deal more discussion about this in the following chapter—the Universal Flood Model.

In the next subchapter, we explore a topic directly related to the craters, the *real* origin of—***meteorites***.

Ending the Crater Debate

7.10 The Meteorite Model

The intellectual history of Earth's meteorites has been recorded in many books, demonstrating that science has been slow to accept the idea that rocks can fall from outer space. O*bserved* meteorite falls are quite rare and when common folk reported rocks falling from the sky, scientists were at first skeptical. Over time, the idea that rocks of extraterrestrial origin fell from the sky became widely accepted and is today, a fascinating and important part of both astronomy and geology, bringing these fields together in ways never before imagined.

Are all the so-called meteorites really of extraterrestrial origin? If large rocks do not fall from space as frequently as is now supposed, and more importantly, if the *true* origin of most meteorites proved *not* to be of extraterrestrial origin, this would certainly have a far-reaching effect on science today. These questions and many others centering on the real origin of meteorites are the subject matter of this subchapter.

From Meteor to Meteorite Pseudotheory

There are dozens of accounts of actual meteors streaking through the sky and landing on the Earth. Some have famously crashed through roofs, damaged cars, and in one case—the only one ever recorded—an 8-pound meteorite struck an Alabama woman's hip in 1954; luckily, she suffered only a nasty bruise. The romance and allure of owning a rock from space continues to drive sales of meteorites; some can command prices of hundreds, even thousands of dollars per gram.

With the help of meteorite collectors around the world, scientists have collected and studied hundreds of meteorites. Based on the samples they examined, three classifications of meteorites have been established. If rocks found by meteorite hunters and lucky enthusiasts fit into one of three classifications, they are deemed 'meteorites.'

The practice of the UM is to question such notions with an open mind. Questioning the authenticity of meteorites today may seem odd, based on the research that has already been conducted, but there is sufficient evidence to suggest scientists have reached wrong conclusions.

The Meteor Model in Chapter 26 will establish that meteor showers, or shooting stars, are not simply bits of sand or dust raining down on Earth and burning up as they enter the atmo-

The Meteorite Model

sphere; a decades-old modern science era idea. Meteorites and impact craters are directly connected, since it obviously requires a meteorite to make an impact crater. Since we just concluded a survey of Hydrocraters and the Crater Debate, this poses an interesting FQ:

> If there are fewer impact craters than previously thought, would there not also be fewer meteorites?

In the Hydrocrater Model, the idea was presented that the vast majority of lunar craters should be recognized as hydrocraters, not impact craters. Therefore, there is a real possibility that the frequency of meteorites on the lunar surface and on our Earth is far lower than previously estimated—fewer impacts mean *fewer meteorites*.

The years of UM scientific research revealed more mystery and enigma surrounding craters and meteorites than any other geological phenomenon. Of course, there really are rocks that fall from space, true meteorites, but this subchapter **challenges the assumption** that rocks fitting modern sciences' meteorite characteristics are actually meteorites.

Science's modern-day meteorite model mingles truth with error. Unanswered questions and claims based on faulty assumptions, such as the presumed abundance of impact craters, have led to an incorrect assessment of meteorite-like rocks. Truth is established through observation and verification of claims, but as we have seen in other subjects, ignored or unanswered FQs lead to unsubstantiated claims, which lead to false theories. The acceptance and proliferation of the modern-day meteorite theory earns it the distinction of a pseudotheory—a false theory taught as fact.

What this subchapter will establish is real physical evidence that challenges the accepted dogma of meteorites and both their literal and imagined impact on celestial objects. The error is confirmed by asking fundamental questions that the theory cannot clearly explain. It is important to recognize that the UM Meteorite Model *does not claim that rocks **cannot** fall from space or that there are no impact craters,* because there are real meteorites and actual impact craters. The Meteorite Model will answer the three following Fundamental Questions:

FQ1: Do meteorites fall from space as frequently as modern science presently believes?
FQ2: Do all the rocks presently called 'meteorites' come from space?
FQ3: What is the true origin of meteorite-like rocks?

Meteorites Defined

The science and study of meteorites is called **meteoritics**:

"**Meteoritics** is a science that deals with meteorites and other extraterrestrial materials that further our understanding of the origin and history of the Solar System. A specialist who studies meteoritics is known as a *meteoriticist.*" Note 7.10a

Meteorites are generally defined as rocks that have fallen naturally to the Earth's surface from space. They are divided into three primary classifications; **stone**, **iron** or **stony-iron**, and then further divided into sub-classes based on specific characteristics illustrated in Fig 7.10.1.

The stony meteorites look rather like common terrestrial rocks, bland and unassuming; they make up more than 85% of all meteorites. Iron meteorites are dense and are the type most commonly associated as typical meteorites. The remaining category is the stony-iron meteorite; they are a minority and are rare. Each type of 'meteorite' holds its own extraordinary evidence, which will be the source of some extraordinary claims later on.

Why Are Meteorites Important?

Richard Norton authored a book on meteorites and recorded his thoughts in the preface:

"Who would have guessed that the study of meteorites would involve the geology of glaciers, the orbits of comets and asteroids, the U-2 spy planes, a spacecraft to Jupiter, and the death of the dinosaurs? This is only half the story. Science, with all of its complex ramifications, is a human activity. The history of our pursuit of meteorites is littered with not-too-admirable human traits involving theft, hoaxes, and courtroom trials driven by personal passions, scientific one-upmanship, and personal monetary gain. In contrast, **the study of meteorites is intimately involved with the search for our beginnings**." Bib 29 Preface

Surely, meteorites have become an integral part of society and science in general. Geoscientists look to meteorites and their impact craters as the *critical foundation* for current theories on planet formation. From the *McGraw-Hill Concise Encyclopedia of Science and Technology* is the theory that:

"Meteorites represent samples of the **primitive interplanetary rubble** that orbits the Sun but did not aggregate into a planetary-sized object," Bib 12 p1199

Over the years, this simple theory has grown in scope in an attempt to answer and explain a host of questions, including the origin of life:

"**Meteorites offer important insights** into processes in stars

CHAPTER 7 THE HYDROPLANET MODEL

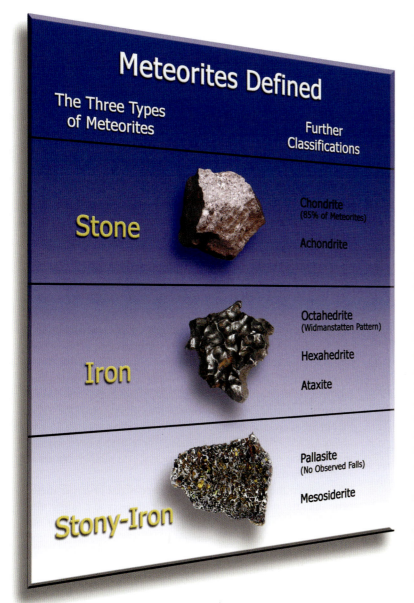

Fig 7.10.1 – Meteorites can be defined or classified into three types: stone, iron and stony-iron, each classification exhibiting markedly different characteristics. The classes are further divided based on other criteria. Stony meteorites make up the majority of all meteorites but iron meteorites are more commonly recognized because of their heavily sculpted appearance. Stony-iron meteorites are the rarest and often the most beautiful meteorites. They are highly prized by collectors and command very high prices on the open market.

and in interstellar regions, the birth of our solar system, the formation and evolution of planets and smaller bodies, and **the origin of life**." Bib 166 Introduction

The modern science paradigm and the gradual expansion of the meteorite theory have influenced many fields of science—astronomy, geology, biology, paleontology and others. For decades, each of their respective theories have in part, been formulated on the Meteorite Pseudotheory. Meteorites are important because they hold foundational evidence of Earth's formative periods, and once known, will aright the fields of science from their pseudo-theoretical paradigms.

The Overblown Meteorite Number

> FQ1: Do meteorites fall from space as frequently as modern science presently believes?

We will investigate this important question. Estimates of the numbers vary widely. In one article published in the *Monthly Notices of the Royal Astronomical Society*, author P. A. Bland estimates a number that translates to 18,000 to 84,000 meteorites bigger than 10 grams per year. The Natural History Museum in London posted an article on its web site titled, *Falls and finds*, which says this about the annual rain of meteorites:

> "**Every year between 30,000 and 80,000 meteorites larger than 20 grams in mass fall from space to Earth**. These can be spectacular events." Note 7.10b

A twenty gram mass is a little less than an American silver dollar. Later on, we will discover that the vast majority of meteorites weighing more than 20 grams lose almost none of their mass as they fall to the Earth, therefore, these "spectacular events" should include meteorites that actually land on the surface to be found by lucky hunters.

If we average the meteorites from the last quote—50,000 meteorites per/year—and multiply that by 100,000 years, (Arizona Meteorite Crater is estimated to be young at 50,000 years) there would be five *billion* meteorites on the Earth! Two-thirds of these would fall into ocean waters leaving over ***1,600,000,000*** *meteorites* for collectors to share. Today, meteorites can sometimes be worth more than gold, and countless meteorite hunters scour the planet looking for these treasures; just how many have been found? One 1998 encyclopedia states:

> "Meteorites consist mostly of metal and silicate material. They can be classified as irons, stony irons, or stones, depending on the relative proportions of the iron and the silicates. **There are only about 2500 meteorites known**." Bib 12 p1199

In Richard Norton's 1994 book, Rocks from Space, Norton says:

> "Stony-iron meteorites are mixtures of nickel-iron and silicate minerals that **seem** to match the mantle in our model of a differentiated parent body. They must be very rare in space; **of the more than 2,600 meteorites known, only 73 are stony-irons**." Bib 29 p235

Since the 1990s, the number of known meteorites has nearly doubled with the recent discovery of a couple thousand found in Antarctica, but even then, the total known meteorites equates to one ten-thousandths of a percent—0.0001%—a very small percentage indeed! Even more surprising is the number of *observed falls* that have been recorded over the past few decades. Three camera networks, one in the Central United States, another in Europe and a third in Canada, have taken tens of thousands of photographs of meteors since the 1950s. Every one of the 50,000 20-gram or larger meteorites that presumably fall to the Earth every year should have been recorded as a bright, spectacular meteor. Yet During the 25 years that the Czechoslovak stations were photographing, only 18 fireballs or bright meteors were recorded using 50,000 photographic plates. Note 7.10c

The overblown number of meteorites becomes even more suspect when we learn how many meteorites were recovered from 25 years of photographic research by these three camera

Fig 7.10.2 – Meteorite lovers find thousands of meteorites for sale, like these at the world-famous Rock and Gem Show in Tucson, Arizona. However, most of these are fragments of larger stones, collected from only a few sites worldwide. In some cases, large rocks are subject to the extreme cold of a nitrogen bath, and then shattered into hundreds of smaller, more marketable pieces. Others are sliced and sold at premium prices.

networks; by 2003, a whopping total of *four*:

"Photographic observations of meteoroids passing through the atmosphere provide information about the population of interplanetary bodies in the Earth's vicinity in the size range from 0.1 m to several metres. **It is extremely rare** that any of these meteoroids survives atmospheric entry **to be recovered as a meteorite on the ground**. Pribram was the first meteorite (an ordinary chondrite) with a photographically determined orbit; it fell on 7 April 1959. **Here we report the fourth meteorite fall to be captured by camera networks**." Note 7.10d

Apparently, it is rare for a photographed meteor to be found on the ground as a meteorite, but it is not because these rocks "burn up" during atmospheric entry. The Meteor Model, earlier in the book explained that atmospheric electrical charges account for many of the shooting stars. These occur at too high an altitude to account for atmospheric frictional heating; they are caused by electrical discharges.

The Overblown Meteorite Number contributes to other problems as science makes assumptions and extrapolates the numbers. Nothing more than mere guesses, they serve to compound mistakes and bolster a paradigm with no basis in fact. This can be seen in the following example taken from the popular book, *The Once and Future Moon*:

"The Moon is struck constantly by the very tiny, less often by the moderately small, and very rarely by the big. The meteoroid flux is dominated by dustlike particles; **objects about the size of a car hit the Moon about once every 100 years**." Bib 129 p84

Estimates like this are commonly made, but where does the number "once every 100 years" come from? No impact event the size of a car, or for that matter, even the size of a dust particle, has ever been observed and documented on the Moon. The largest meteorite ever recovered from an observed fall on Earth is probably the Kirin Meteorite that fell in China on March 8, 1976. This rock was only about a meter in diameter, so the falling of a car-sized rock from space has *never been observed* on the Moon *or* the Earth.

Another problem arises when numbers like this are extrapolated. If we were to assume that car-sized impacts occurred on the Moon at this rate for 4 billion years, then there should be *4 million large impact craters* (from car-sized meteorites only) on the Moon, each with melted and fused impact glass. Scientists are still looking for evidence that *even one* of these occurred recently, within the last couple thousand years or so.

When we see a crater—whether here on the Earth, on the Moon or other celestial bodies, we should not jump to the conclusion that it is impact. Let's do a little more research and question the meteorite theory with an open mind.

Are All Classified 'Meteorites' Really From Space?

Perhaps the first thought that comes to mind when asked if meteorites really come from space is, where else would they come from?

FQ 2: Do all the rocks presently called Meteorites come from space?

We have grown up in a generation that believed there was only one answer for the origin of the meteorites, but that is about to change.

We begin by first looking at how the field of meteoritics has painted itself into a corner with its contradictory concepts of meteorite origins. Astrogeologists have advocated the false theory of accretion to explain how space dust accumulates to create rocks, and eventually planets. There hasn't been any actual observations of this alleged process; it is simply theorist's imaginative modeling.

Moreover, there is no evidence of sand-sized or larger meteorite material crashing into each other to become a larger meteorite. How do we know this? *Meteorites show none of the critical characteristics of having been melted.* Just like the vast majority of rocks on Earth, meteorites were **never melted**. They exhibit a *crystalline structure,* which indicates they were

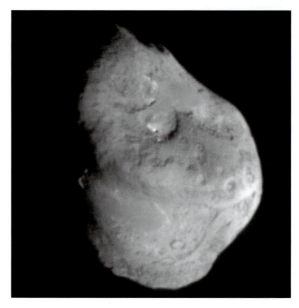

Fig 7.10.3 - Comet Tempel 1 as imaged by the *Deep Impact* mission looks nothing like a meteorite or a fragment of another planetary body. Comets have long served as modern science's primary origin for meteorites, but recent visits have researchers beginning to wonder. This comet has a density less than that of water—far less than the density of any known meteorite rock.

formed in a hypretherm environment.

Another modern science theory of meteorite origin has rocks being torn from parent bodies in our solar system. However, the stone, iron, and stony-iron meteorites are composed of different minerals than those found on the Moon, Mars or any other planet or moon we have so far investigated first hand. For the meteoriticist, there are only two other possible meteorite origins—comets and asteroids. But there are problems with these theories too.

Despite the still-current idea that meteorites come from comets, researchers all but dismissed the idea in 1975. The following quotation comes from the *Journal of Geophysical Research*:

"There is serious doubt, however, that even a single cometary meteoroid has survived atmospheric entry and become a recovered meteorite. Although a cometary origin cannot be unambiguously ruled out, there is good evidence from studies of implanted solar wind, solar flare tracks, and mineralogy that **meteorites do not come from comets and probably have an asteroidal origin**." Note 7.10e

The researchers' use of the word "probably" in explaining how meteorites come *from asteroids* is not very sure, but then again, they use the word "probably" to imply meteorites *come from comets,* all in the same article:

"The majority of meteoric bodies **probably** have **cometary origins**…" Note 7.10e

Why do these last two statements seem so contradictory? Because they are. Meteorite origins have mystified researchers for decades because observations contradict the theories. In fact, new up-close observations of comets and asteroids have only served to make the theory that meteorites come from comets or asteroids even more confusing. One encyclopedia explains meteors "probably" come from comets but admits, "there is *no* compelling evidence that comets are the source" of meteorites:

"Comets are another possible source of meteorites. Cometary debris is **probably** the source of many **meteors**, including the more spectacular meteor showers, but **there is no compelling evidence that comets are the source of any known meteorites**." Bib12 p1199

Comets had seemed the best bet so far, but the more we learn about comets—especially during recent years as we visited them with space probes—the more the evidence shows they are *not* the source of the meteorites.

Similar problems have arisen with the asteroid-origin theory. For example, the largest known asteroid, Ceres, has a density only twice that of water, far below the density of any known meteorite. Geoscientists have *implied* that asteroid and meteorite spectra are similar by comparing the reflected light from their surfaces, but this has yet to be confirmed. After visiting several asteroids with spacecraft, the prospects look even worse. For example, high density, metallic iron meteorites are not representative of any asteroid found to date.

Perhaps future research will shed new light on meteorite origins from other planets, and the door is certainly not closed on the subject. There is much more to discuss about asteroids and comets; two whole sub-chapters are devoted to the subject. Our goal here is to present the facts as they are and to open the mind to the possibility that there are other potential origins for meteorites, and that they did not all fall from the heavens.

FQ3: What is the true origin of meteorite-like rocks?

Even if we accept the assertion that meteorites came from another celestial body, that body could not have been melted, whether by a radioactive heat, impact, or any other means because, like all other minerals, meteorites did not come from a melt or magma. Therefore, meteorites *must have a **new**, heretofore unknown origin*.

The Meteorite Mineral Enigma

All the elements on the Earth and in the meteorites are believed by modern science to have come from a proto-solar cloud that "accreted" to become our solar system. The Accretion Pseudotheory has already been shown to have no basis in fact, and now we examine another evidence supporting its falsity—the Meteorite Mineral Enigma.

Because the minerals in the Earth and meteorites are said to have had the same cosmic origin (the proto-solar system cloud), meteorites, like the minerals on the Earth *should* represent "a variety of compositions." One meteoriticist reiterates this line of reasoning:

"A **cosmic origin** would suggest that **stones [stony meteorites] can come from all over the solar system**; that is, they would have **several origins and therefore a variety of compositions**." Bib 29 p41

Let's compare the variety of Earth's mineral composition with the meteorites. We establish the variety of minerals found on the Earth from the *Complete Guide to Rocks and Minerals*:

Fig 7.10.4 – Some minerals such as this opal do not exist in meteorites. If meteorites and the Earth share the same cosmic origin, why not? Meteorites contain few minerals as compared with all known minerals; and certain elements, like nickel, exist in all meteorites, yet nickel is relatively rare among the rocks found on Earth. Moreover, many meteorites contain certain ratios of elements suggesting that meteorites have only a few places of origin instead of a cosmic wide origin as suggested by modern theory.

"There are **4,000 to 5,000 different minerals** in the Earth's crust." Bib151 p42

With this information, we can ask the logical question, do meteorites share this diversity?

FQ: How many minerals are in the meteorites?

Obviously, there should be several thousand minerals in the meteorites, but that is not the case:

"Before 1962, about 40 minerals were recognized in meteorites with 5 unique to meteorites. In 1972, an additional 47 minerals were recorded, 20 of which are unique. Today, **nearly 300 minerals are found in meteorites** with about 40 found exclusively in meteorites." Note 7.10f

Keep in mind the number of "nearly 300" minerals come from the meteorites, which are *assumed* to have come from space. What if the meteorites did not come from space, but came from an entirely different origin? Because meteorites are made up of **very few minerals**, they contradict the theory of their "cosmic origin."

In fact, the low mineral count in meteorites is just the tip of the iceberg. Another tall hurdle for the meteoriticist to account for is *the presence of nickel in nearly **all** meteorites*:

"Today we know that **iron in meteorites is never found without the presence of nickel and that virtually all meteorites contain nickel in some form.**" Bib 29 p40

This fact is based on actual observations of the meteorites, yet nowhere do we see scientists questioning the absence of nickel in the Earthly mineral assemblages. If virtually every meteorite contains nickel in some form, and if the meteorites and the Earth shared the same cosmic origin, *why* is nickel so rare among the Earth's surface rocks? Pick up a rock from any of the Earth's riverbeds and you would be lucky to find *any* nickel in it, yet *every* meteorite has nickel. How can this be if they share the same origins?

Of course, the modern science answer is that a *magmatic* melt allowed heavier elements to be separated where they sunk to the core. Even though there is *no evidence for magma*, if this was the case, and ancient planetesimals had been ripped apart, **all** the layers would be represented in the meteorites, including the non-nickel layers—but they are not. The enigma of meteoritic nickel is very real and confirms that modern science has not identified the true genesis of most meteorites.

Another enigmatic element for the meteoriticist is carbon. One sub-class of stony meteorites is the carbonaceous chondrites:

"The interest in carbonaceous chondrites stems in part from their relatively **high carbon content**; thus, the name 'carbonaceous.' Carbon compounds give these meteorites a dark-grey color. **They are not the only meteorites containing carbon**, however. **Many contain carbon** in the form of carbides, graphite (amorphous carbon), or even diamonds." Bib 29 p196

If there is so much carbon in some of the meteorites, then there should be ample evidence of this on the Moon, another interesting FQ:

Many meteorites contain carbon, how much carbon is found on the Moon?

The surprise answer is that there is none, at least not enough to measure it at parts-per-million from the rocks collected during the Apollo missions. Bib 162 p140-141 This fact confirms both the Meteorite Mineral Enigma and the Hydrocrater Model; the lunar craters are primarily hydrocraters, not impact craters.

Where did the carbonaceous rocks come from, that do not match any rocks from the Moon? One researcher sees a clue in the oxides—oxygen rich minerals in carbonaceous chondrites:

"Carbonaceous chondrites **must have formed in an oxygen-rich environment**. Most of their metal is either locked in the silicon tetrahedron or **combined with oxygen to form oxides**—magnetite for example—or combined with sulfur to form sulfides." Bib 29 p191

This raises another FQ:

Where in the solar system is the **only known** "oxygen-rich environment"?

Perhaps this will open the mind to another possible origin of the 'meteorites.' From discussions we will find in the Chemical Model Chapter and in the New Geology section earlier in this chapter, we learn that all oxides obtain their oxygen from water. Thus the carbonaceous meteorites were, like all minerals, formed in water.

Meteorite minerals have indeed been an enigma for a variety of reasons: the rarity of minerals in meteorites as compared with those on Earth, the abundance of nickel in meteorites vs. Earth rocks, and the carbon in meteorites but not on the Moon. None of these conditions can be accounted for if meteorites came "*from all over the solar system.*"

All of these conditions **can be accounted for** if the meteorites came from *only a few sources*. It turns out that there is ample evidence for a new theory involving only a few sources for meteorites. If we study the inter-element relationships between the most common stony-meteorites, a remarkable relationship between Ca (calcium) and Al (aluminum) is revealed:

"The **relationship between Ca and Al** in chondrites and basaltic achondrites has been examined and is shown to be **remarkably close.**" Note 7.10g

What does it mean to be remarkably close? When the ratio of Ca and Al was graphed, 16 meteorites, those being from the most common group of meteorites, exhibited nearly identical graphs. Most of the basaltic achondrite meteorites contained over 7% calcium and just under 6% aluminum. The researchers also compared the calcium/strontium ratio and found a similarly close relationship there too, even though strontium measured only a several parts-per-million abundance.

These inter-element relationships demonstrate that the rocks must have shared a **very similar** origin—from a limited number of sources. Moreover, it suggests that they may not have had a celestial origin at all. Paragenesis, a term used in geology to identify groups of minerals that are commonly found together was presented earlier in this chapter; that inter-mineral relationship comes from the unique environment in which the minerals grew, here on Earth. Understanding this process and its applicability to the meteorite mineral enigma is important. More will be presented about this process and environment later on in the book.

"A cosmic origin would suggest ... a **variety of compositions**." Bib 29 p41

However, we find just the opposite in meteorites—a **very select group** of mineral compounds.

Large Meteorites Missing Craters Evidence

In the Hydrocrater Model subchapter, we learned that small hand-sized meteorites are slowed to terminal velocity by atmospheric friction, while large, house-sized (or larger) meteorites are slowed very little, impacting the Earth at cosmic hypervelocity. The lesser, low-speed meteorites leave small nondescript craters, if any at all. On the other hand, very large rocks produce huge craters and large amounts of glass melt. Now, let's look at the in-between sizes of meteorites, from say 1 to 100 tons. These end up reaching the Earth's surface at speeds between the slow and hypervelocity meteorites and therefore produce medium-sized craters and lesser amounts of melt. These facts have been gathered from actual, observed falls of medium-sized meteorites, such as the Sikhote-Alin fall in Russia and the Kirin fall in China.

The **Sikhote-Alin fall** occurred February 12th, 1947. Over 200 craters were created by the fall with the largest being 26.5 meters (87 feet) in diameter. Several thousand iron-nickel fragments were strewn in an elliptically shaped strewn field over a kilometer in length in rural mountainous terrain; the largest recovered being 1,745 kilograms (2 tons). Many of the craters contained a number of meteorite fragments that arrived at the same time. Small, single meteorite fragments fell nearby; they were found lying on the surface with little or no cratering. Bib 167

On March 8, 1976, the **Kirin fall** occurred in China, along an elongated ellipse 72 kilometers (45 miles) in length. Hundreds of meteorites were found at this observed fall, one of which was the largest stony meteorite in the world, weighing in at 1,770 kilograms (2 tons). This meteorite penetrated 6.5 meters (21 feet)

Fig 7.10.5 – The Hoba meteorite in Namibia, Africa, estimated at over 60 tons, is the world's largest known meteorite. Identified in 1920 by scientific explorers, it is a National Monument that draws many visitors each year. There are however, several mysteries surrounding this famous rock. Observed falls of meteorites much smaller than this one are known to form craters, yet the Hoba and other large alleged meteorites apparently did not form craters—why? Additionally, why is there no evidence of significant rusting or sloughed off rust from this iron-nickel rock?

Fig 7.10.6 – The United States' the largest 'meteorite' is the Willamette meteorite, found ironically on land owned by the Oregon Iron and Steel Company. Like the Hoba discovery in Africa, no crater was associated with this giant iron rock, an enigma if the meteorite had really impacted the ground.

into the soil and produced a crater several meters in diameter with a rim .65 meters (2.1 feet) high.

Both of these falls had meteorite fragments weighing approximately 2 tons that made deep *craters measuring several meters across*. With this information, let us ask another FQ:

FQ: In what sized crater should we find a 15, 30, or 60- ton meteorite?

Obviously, we should expect to find them in craters much larger than those associated with the 2-ton meteorites; yet this is not what has been reported. Fig 7.10.5 is a photo of the **Hoba meteorite**, the *largest* meteorite in the world. It weighs over 60 tons. Fig 7.10.6 shows the **Willamette meteorite**, the United States' largest, over 15 tons, found in 1902 in Willamette Valley, Oregon. Neither the Hoba meteorite, located in Namibia, Africa nor the Willamette specimen, was found associated with *any* crater at all!

Based on actual, observed falls of large meteorites, they recognized that "such large irons **would** produce deep craters in the ground":

"Even the greatest of them all, **Hoba** West, has actually penetrated the Kalahari superficial limestone to **only its own thickness**, and lies with its **flat base level with the ground**. But the minimum (computed) penetrations into clay **show that such large irons would produce deep craters in 'the ground'**; not so large, however, but that in the course of centuries the holes would be effaced by the wash of the weather.

"Herein may lie an explanation of the **well-known fact, that the large iron meteorites are so generally found in rock uplands**." Note 7.10h

The investigator, an astronomer, made this comment in 1933 but appears to have missed an important point; in general, larger meteorites have higher speeds and make bigger craters. If the Hoba specimen was in fact, a meteorite that fell from space, it should have created a significant crater, hundreds of meters across with a rim tens of meters high. There was no discernable crater at the Hoba site.

The perplexing fact is that the world's largest 'meteorites' have no impact craters associated with them, leading to another FQ:

> Why do the world's largest 'meteorites' have no craters?

Today's meteoriticist has no valid explanation. Some have thought glaciers moved the Willamette meteorite, but the Hoba—the world's largest—is on the flat, desert plain of South Africa, where glaciers never visited. Moreover, the Willamette specimen shows no striations or erosion marks that would be present if it had been moved any significant distance by a glacier.

There must be another explanation. Why do such large meteorites have no impact craters; could it be they have no crater because the so-called meteorite did not fall from space? This would answer a host of questions, but would obviously bring up a few more.

What is the Real Origin of Meteorites?

Two important, fundamental questions to be answered involve the origin of meteorites:

FQ: **Where** were meteorites made?
FQ: **How** were meteorites made?

By 1985, researchers had come to *accept* that meteorites came from asteroids:

"It is now **accepted** that **most of the meteorites**, and particularly the carbonaceous chondrites, **are fragments of asteroidal material**." Bib 87 p181

This acceptance is founded primarily on the notion that there is no better alternative, not that it had been demonstrated. Once again, real science cannot progress if scientists 'accept' an idea simply because there is no other alternative. It may be hard for scientists to admit publicly that they "do not know," yet science can only progress when we acknowledge just that, when there is no empirical evidence. If a theory is accepted without actual data and without using the Universal Scientific Method (USM), it will result in minds closed to the real scientific truth—because we already "know."

The book *Rocks From Space* states that it is a "fact" that meteorites come from asteroids, yet the author admits it was "hard to imagine" how asteroids could retain sufficient heat to *melt* the meteorites:

"Little more than thirty years ago, a few courageous scientists put forth the argument that meteorites come from asteroids. This **fact** seems obvious to us today, but it was a long 'rocky' road to establish that **idea**. Asteroids were known to be small bodies, too small to have had thermal histories like those of the planets. It was **hard to imagine** that even the largest asteroid was large enough to **retain sufficient heat to melt and differentiate**. Most scientists agreed that asteroids were primitive bodies, unchanged since their formation, and because **most meteorites showed a history of melting**, they must have encountered very different conditions." Bib 29 p345

The ***real problem*** is not *where* the heat came from to melt the meteorites, but the idea that the meteorites were even melted at all. They, like most other rocks, did not come from a melt! We see in Fig 7.10.7 what iron meteorites really look like when they are melted. Iron beads, commonly seen by welders and foundry workers are formed because metals tend to 'bead' when liquefied. Researchers have mistakenly *assumed* that recrystallization follows melting:

"Like achondrites and irons, **stony-irons have suffered melting and recrystallization**." Bib 29 p235

But this is not the case. Meteorites, like the vast majority of other rocks we find on Earth are crystalline, demonstrating that they grew in a hypretherm described in the Hydroplanet Model. *If* the meteorites had indeed been melted, they would not exhibit an orderly crystalline structure, and they would contain *glass*.

Are there natural meteorites that exhibit melt characteristics like the melted portion of the meteorite in Fig 7.10.7—probably not. The vast majority of meteorites show no real evidence of melting even though scholars claim that nearly all meteorites have a "history of melting." This misunderstanding is based

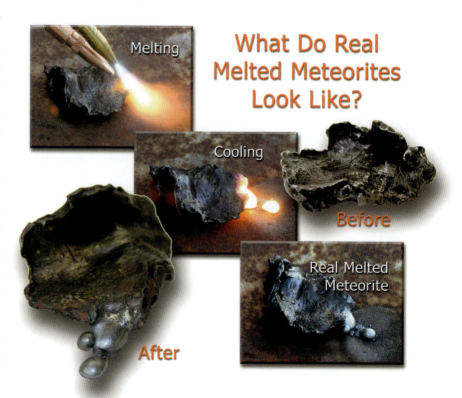

Fig 7.10.7 – This simple melting exercise demonstrates one reason why meteorites did not come from a magmaplanet core. It also shows what meteorites melted during an impact would really look like. Anyone with welding experience can relate to the iron 'beads' that form from melted iron and melted iron meteorites are the same. This photo group includes before and after images of a piece of the Sikhote-Alin meteorite. The melted metal takes on a completely different appearance. The widely accepted origin of meteorites attributes them to the asteroid belt, which is assumed to be the remnant of a magmaplanet core. Most meteorites show no evidence of melting, whether by impact or by being a magmaplanet core remnant.

CHAPTER 7 THE HYDROPLANET MODEL

on dogmagma and the idea that rocks and minerals formed from a melt, when they actually formed naturally out of water.

> Because of dogmagma, scientists believe "most meteorites showed a history of melting".

Widmanstätten Crystalline Pattern Evidence

The largest group of *iron* meteorites is the octahedrites, comprising 85% of the category. These are easily recognized by their unique crystalline pattern, which is the **Widmanstätten pattern** seen in Fig 7.10.8. The Widmanstätten pattern is said to be "diagnostic for iron meteorites in general":

"Since **Widmanstätten figures do not appear in terrestrial iron ores** and since about 85 percent of iron meteorites are octahedrites, **the presence of Widmanstätten figures is diagnostic for iron meteorites in general**."
Bib 29 p221

Meteoritic scholars commonly hold that the Widmanstätten pattern is caused when the metal cools from a high temperature melt:

"**As meteorites cool from high temperatures**, the precipitation of kamacite (α) in a matrix of taenite (γ) occurs, **forming the Widmanstätten pattern**." Note 7.10i

This statement was taken from the journal of *Meteoritics & Planetary Science*, the authoritative publication on meteorite research. However, the same journal published reports by other researchers who determined from their actual experiments that *assumptions* about Widmanstätten pattern formation must be reexamined:

"Recent laboratory experiments in two areas highlight the **need for a reexamination of fundamental assumptions about the meteoritic Widmanstätten structure**." Note 7.10j

Why the need for a "reexamination of *fundamental assumptions*" of Widmanstätten pattern formation if geoscientists already know how to duplicate them? As it turns out, they did *not* know how to recreate the Widmanstätten crystalline patterns from "high temperature" melt precipitates. The researchers further note that a primary mineral in iron meteorites, kamacite, "cannot be reconstructed" using thermal techniques:

"The **thermal history** of body-centered cubic low nickel-iron (alpha or delta) **cannot be determined by metallographic techniques**. Therefore, **the thermal history or mechanism of kamacite formation cannot be reconstructed** based solely on crystal structure." Note 7.10j

The primary reason researchers failed in their attempt to recreate the kamacite mineral and the Widmanstätten pattern was that their mechanism for growth was *thermal* only, and not hyprethermal.

The Widmanstätten Pattern

Fig 7.10.8 – Slices of iron meteorites show the crystalline structures of the minerals. The pattern is called the Widmanstätten Pattern, and it provides definitive proof that iron meteorites did *not* come from a melt. Temperatures exceeding 800° C (well below the melting point), ***destroys the Widmanstätten pattern***. Although meteorite scholars claim that "meteorites cool from high temperatures," the Widmanstätten pattern has never been reproduced in the laboratory, prompting some researchers to state that there is a "need for a re-examination of fundamental assumptions about the meteoritic Widmanstätten structure."

> "... the **thermal** history or mechanism of kamacite formation **cannot be reconstructed**..."

Except for a few of Earth's rocks such as those associated with the intense heat generated by friction at fault lines, all natural minerals do not come from a melt. It was therefore no surprise to find researchers reporting in 2001 that new experiments showed lower temperatures (below 400 °C) provided a "better fit" for growing kamacite:

"**A major revision of the current** Saikumar and Goldstein (1988) **cooling rate computer model for kamacite growth is presented**. This revision incorporates a better fit... particularly **below the monotectoid temperature of 400° C**." Note 7.10k

Recognizing that these crystalline meteorite patterns do not come from high melt temperatures, and that temperatures more commensurate with a hyprethermal environment provide a better fit, we can now evaluate another common sci-bi. Unable to produce observable results, modern science often falls back on the 'millions-of-years' time frame to justify various pseudo-theory statements. Such is the case of meteorite growth theory. Although most crystals grow naturally in a hypretherm in a matter of hours or days, modern geological pseudotheory has locked investigators into an imaginary deep-time period. Recently however, researchers have had to attempt to overcome this hurdle and acknowledge that a "*re*examination" of the millions-of-years time frame is warranted. Finding instead that "a relatively short time frame" is more "consistent with modern metallurgical understanding" in the creation of Widmanstätten structures:

"Thus, the concept that the meteoritic **Widmanstätten structure is the product of a solid state phase transformation over millions of years requires reexamination**. An alternate model [theory], **consistent with modern metallurgical understanding**, is that the meteoritic Widmanstätten structure was produced directly from a melt, logically under microgravity conditions, in a **relatively short time frame**." Note 7.10l

> "Thus, the concept that the meteoritic **Widmanstätten structure** is the product of a solid state phase transformation over millions of years **requires reexamination**."
>
> P. Z. Budka1 and J.R.M. Viertl

Keep in mind that the crystalline meteorite patterns have *not* actually been reproduced, but the investigators are getting closer by recognizing that lower temperatures and short time frames are realistic possibilities for natural growth patterns. Both of these are characteristic of the hyprethermal environment. Although the meteoritic science community frequently states that iron "meteorites cool from high temperatures," the UM position, based on the important Widmanstätten crystalline structure, proves that this statement is not true. The application of a "short time frame" and low temperatures has researchers taking a step in the right direction, toward the hyprethermal process.

The Widmanstätten pattern provides compelling *testimony* that meteorites were never from the **melted** core of a planet because the pattern is erased when heated above 800 °C (1472 °F):

> The **Widmanstätten structure is destroyed** in a short amount of time when heated to just **800° C**.

This fact is recorded in the *Journal of Meteoritics*:

"The statement frequently appears in the literature of meteoritics that the **Widmanstätten structure** of the octahedritic class of nickel-iron meteorites will be **completely destroyed if such a meteorite is heated to approximately 800 °C**." Note 7.10m

Amazingly, this was printed in the *Journal of Meteoritics* in 1954. Since that time, meteorite scholars have been equipped with the knowledge that meteorites that have the curious Widmanstätten pattern could not have come from a melt, but have apparently missed this fact because of dogmagma. Geosciences claim to "know" that the Earth and other planetoids' have iron-nickel-magma cores, making them the source of the iron meteorites. To arrive at this conclusion, some truths have simply been ignored. We asked one meteorite expert how he thought iron meteorites could have come from a melt, knowing that the crystalline structure is destroyed in the melt process; he did not have a defendable answer. His expert response was the platitude of modern science; the "millions-of-years" sci-bi that the crystal pattern grew slowly over time, so of course it was not observable. It did not seem to matter that the **Widmanstätten pattern is destroyed at elevated temperatures, regardless of the time frame**.

Researchers determined that the Widmanstätten pattern could be destroyed in as little as 30 minutes at 800° C. High-speed impacts produce temperatures well above the level needed to destroy the Widmanstätten pattern, but the impact event lasts for only a fraction of a second. Residual heat however, can last for *days*, as witnessed during the impact of Jupiter by the Shoemaker-Levy comet in 1994. Large meteorites entering the Earth's atmosphere at high speed would easily produce enough high-heat energy to last far more than 30 minutes, especially for those meteorites that were mostly vaporized; the theoretical meteorite that produced Arizona Meteor Crater being one example.

In summary, the Widmanstätten patterned meteorites demonstrate three things:

1. Iron-nickel meteorites did *not* come from a melt because heat above 800 °C erases the Widmanstätten pattern.

2. It *cannot* be assumed that the cores of **melted** planets are of the same minerals that Widmanstätten patterned meteorites are, because the Widmanstätten pattern could not withstand the high heat.

3. Large, high-speed impacts produce more than enough heat, which persists long enough for the Widmanstätten pattern of the impactor to be destroyed. Therefore, it can be deduced that meteorites with a Widmanstätten pattern did *not* come from a large, high-speed impactor.

The last item casts serious doubt on the origin of the meteorites that presumably came from an impact event that created the Arizona Meteor Crater. Known among meteoriticists and collectors by the former name of the crater, the Canyon Diablo meteorites exhibit the Widmanstätten pattern; how could they be from an impact event powerful enough to have *vaporized* most of the 300-thousand-ton impactor?

> FQ: Did the Canyon Diablo meteorites from the Arizona Crater, which display the Widmanstätten pattern, come from a large, high-speed impact?

"No Terrestrial Iron-Nickel Metal" Assumption

There are two reasons today's researchers remain unsettled about native iron-nickel metal deposits. First, they believe meteorites are special; *formed somewhere in space, they conceive of them as being the cores of past planetary members of the solar system*. Secondly, the origin of iron-nickel ore, actually, the origin of *all metals and ores* remain a mystery in geology, first documented in the Rock Cycle Pseudotheory chapter.

One erroneous claim from the meteorite experts is the notion that meteorites are the only source of natural nickel-iron alloy; that there is no naturally occurring terrestrial iron-nickel metal found on Earth.

One example of this claim comes from St. Louis' Washington University Department of Earth and Planetary Sciences' website. This informative web page goes on to detail how meteorites and non-meteorite rocks can be distinguished, and it states that metal found in meteorites is *not* found on Earth's surface:

"**No naturally occurring terrestrial rock contains iron metal or iron-nickel metal**. There are two reasons. **First, early in Earth's history** the iron-nickel metal sank to form the **Earth's core**. Second, any metal that did not sink has oxidized (rusted) over **Earth's long history**. The Earth's environment is far more oxidizing (oxygen atmosphere and water) than space, where meteorites originate. Earth rocks *do* contain iron and nickel, **but only in oxidized (non-metallic) form**. Therefore, if you find a rock that contains iron-nickel *metal*, it's **probably a meteorite**." Note 7.10n

"No naturally occurring terrestrial rock contains iron metal or iron-nickel metal... Therefore, if you find a rock that contains iron-nickel **metal**, it's probably a meteorite."

occurred "at the same place":

"The discovery at Ovifak [Disko, Greenland] is remarkable, not only as the **largest discovery of meteoric iron hitherto known** to have been made... Add to this, the **remarkable circumstance**, that

The website's statement, posted prior to 2007, was simply wrong, and by the end of 2007, it was changed in part; "No naturally occurring…" became "With a few rare and well known exceptions." The "exceptions" were apparently not that "well known" when the site was first posted.

The former declaration that, "no naturally occurring terrestrial rock contains iron-nickel metal" is still commonly referred to by meteorite scholars but can be soundly falsified because of the existence of several different terrestrial iron-nickel deposits. Naturally occurring nickel-iron *is* found on Earth and therefore, the metal found in meteorites is not limited only to rocks from space. Fig 7.10.9 is *not* a meteorite; it is a terrestrial iron-nickel specimen from Russia.

Although meteoriticists would like to believe that nickel-iron metal is not found on Earth, worldwide deposits of elemental iron-nickel have been known since at least the 1800s. These include:

1. Disko Island, Greenland
2. Josephinite, USA
3. Kola Peninsula, Russia
4. Kassel, Germany
5. Gorge River, New Zealand

Note 7.10p

Greenland Native Iron-Nickel Evidence

The world's third largest so-called meteorite, weighing in at 31-tons was found in Greenland along with many other large ironstones, including a 20-ton specimen. But, Greenland is also listed among the world's known deposits of native iron-nickel. The tale of Greenland's Native Iron-Nickel and her meteorites is an amazing story, one that has never been fully told. From the little information readily available, we will piece together compelling evidence of the terrestrial origin of Greenland's so-called 'meteorites.'

For centuries, native Greenlanders have gathered iron-nickel nodules along the west coast of the island. They used this material to make various tools and weapons, surprising the Western European explorers who first encountered them. News of the unique iron-alloyed tools gradually spread, attracting scientists and explorers. By 1870, Professor A. E. Nordenskiöld had discovered the largest deposit of native iron-nickel in the world, near the shores of Disko, Grenland. Nordenskiöld had been searching for iron meteorites that had fallen from the sky. From his personal account of the 1870 expedition, we read of his discovery, that by a "remarkable circumstance" *both* native **and** meteoric iron-nickel

Fig 7.10.9 – Natural iron-nickel metal from Jungtukun, Russia. Meteoritic scholars originally thought (and as of this writing, some still do) that iron-nickel metal is **only found** in meteorites. However, this is not the case as there are several known natural deposits of iron-nickel minerals worldwide. Curiously, the natural deposits are located in the same vicinity as the meteorites, raising some interesting questions.

lenticular and discoidal **pieces of native iron occur *at the same place* in the underlying basalt**, that basalt pieces of considerable size, in numerous spots, form a crust on the larger **meteorites**, and even sometimes has been driven through the surface into the iron. Nevertheless, in spite of this, it appears to me that there **cannot be a doubt of the really meteoric origin of the large masses**. Their form, their composition, their appearance, sufficiently indicate this." Note 7.10o

Note that today, there is *no question* among the experts that **native** iron-nickel exists in Disko. Recently, scientists have tried to understand how this metal came up from the Earth's 'iron core' and many research papers have been written on the subject. Some meteorite scholars have declared the coexistence of both native and meteoritic iron-nickel as merely a 'coincidence.' Nordenskiöld *believed* that some of the pieces of the native iron-nickel metal he found fell from the sky as meteorites because he **believed,** like most others in 1870 that the Widmanstätten structure was only found in meteorites. The Widmanstätten Pseudotheory has continued to the present day; a meteorite book in 2004 stated that the "Widmanstätten structure" is "***unknown*** in the rare occurrences of *terrestrial*" iron-nickel metal:

"In most cases etching with acid will show the presence of the **Widmanstätten structure which is unknown in the rare occurrences of terrestrial Fe,Ni metal**." Note 7.10p

Nordenskiöld and other modern researchers failed to realize the indisputable fact that native Greenland iron-nickel has the Widmanstätten pattern that some meteorites have.

Nordenskiöld recorded taking iron-nickel metal from an obvious "basalt ridge" that "exhibited fine *Widmanstädt's figures*":

"Sixteen metres from the largest **iron block a basalt ridge** of a foot high rose from the detritus on the strand, and could be followed for a distance of four metres, and was probably part of the rock. Parallel with this and nearer to the strand ran another similar ridge, also about four metres long. *The former contained lenticular and disk-shaped **blocks** of nickel iron, in external appearance, chemical nature, and relation to the atmosphere (weathering), like meteoritic iron.* On being polished and etched this iron **exhibited fine Widmanstädt's figures**." Note 7.10q

Nowhere is there a better account of the Greenland Native Iron-Nickel Evidence than is found in the 1911 Encyclopedia Britannica. Its authors were open-minded about the

Fig 7.10.10 – Workmen recovering the ½-ton Greenland iron located near Cape York, on the western shores of Greenland. Dubbed the Dog, this meteorite was considerably smaller than others found nearby. The largest of the Cape York group weighed in at 31 tons; it is one of the world's largest meteorites. No impact craters were evident at any of the Cape York meteorite fragment locations, but many nodules of native iron-nickel were identified in the basalt deposits near Disko, also along the western shore of Greenland. Iron-nickel meteorites without any associated craters are found at the same place nodules of native iron-nickel are found—coincidence or clues to a non-meteorite origin?

iron-nickel metal's origin, and challenged the claim of meteoric origins for not only the native iron-nickel metal at **Disko Island** off the northwestern coast of Greenland, but also discoveries of other 'meteorite falls' found in 1895 along the southwestern Greenland coast by **Cape York**. Note that "native iron" is found not only at Cape York, but that these newly discovered deposits of metal also contained the Widmanstätten crystal pattern, and that "no evidence hitherto given seems to prove decisively that it cannot be telluric" (telluric is Latin for Earth):

"*Native* iron was found by Nordenskiold at Ovifak, on Disco Island, in 1870, and brought to Sweden (1871) **as meteorites**. The heaviest nodule weighed **over 20 tons**. Similar native iron has later been found by K. J. V. Steenstrup in several places on the west coast enclosed as smaller or larger nodules in the basalt. **This iron has very often beautiful Widmanstätten figures like those of iron meteorites, but it is obviously of telluric origin**. In 1895, Peary **found native iron at Cape York**; since John Ross's voyage in 1818 it has been known to exist there, and from it the Eskimo got iron for their weapons. In 1897 Peary brought the largest nodule to New York; it was estimated to weigh nearly 100 tons. This iron is considered by several of the first authorities on the subject to be of **meteoric origin, but no evidence hitherto given seems to prove decisively that it cannot be telluric**. That the nodules found were lying on gneissic rock, with no basaltic rocks in the neighbourhood, **does not prove that the iron may not originate from basalt**, for the nodules may have been transported by the glaciers, like other erratic blocks, and

will stand erosion much longer than the basalt, which may long ago have disappeared." Note 7.10r

Several years later, in 1919, Hugh M. Roberts, an experienced geologist and mining engineer noted that Widmanstätten patterns were found terrestrially:

"Prof. A. E. Nordenskiold and K. J. V. Steenstrup, between the years 1870 and 1884, discovered large boulders of native nickel-iron, weighing as much as twenty tons, occurring in connection with basalt flows on Mt. Ovifak, near the Island of **Disco**, off the west coast of **Greenland**. Lenticular and disc-like pieces of the metal were also found embedded in the rock. **These fragments had the pitted appearance characteristic of meteorites and were long thought to be of meteoric origin, for the iron gave the Widmanstätten figures**. Steenstrup found a thickness of fifty feet of basalt filled from top to bottom with iron grains as a constituent part of the rock. **These grains showed Widmanstätten figures**." Bib 126 p38

This last quote is taken from H. H. Nininger's book, *Arizona's Meteorite Crater*. Nininger, a researcher famous for his work on the Arizona Crater he lived near for decades, was an expert on Widmanstätten figures. Nininger however had nothing to say about the Widmanstätten figures in Greenland's *native* iron-nickel, choosing instead to ignore it.

In summary, the native iron-nickel from Disko Island provides the following evidence:

1. Native iron-nickel exhibiting the Widmanstätten pattern is found in association with basalt at Disko. Therefore, the so-called meteorite iron-nickel from the same site cannot be *assumed* to have come from space.

2. No impact craters have ever been identified at the Disko site.

No doubt, there is a great deal more to learn about Greenland's native iron-nickel and its Widmanstätten patterns; a research project for future Millennial Scientists.

The 31-ton Disko Island meteorite, third largest iron-nickel meteorite in the world, the world's largest meteorite, Africa's Hoba, *and* the United State's largest meteorite, the Willamette, all share one thing in common—*no impact crater*. In fact, there are no impact craters for *any* of the Greenland meteorites; this is certainly more than just a mere coincidence.

Josephinite Native Iron-Nickel Evidence

Another occurrence of native iron-nickel is in Josephine County, Oregon, USA, along Josephine Creek. Scientists originally thought the metal "blebs" found there were from space:

"Some investigators have thought that perhaps blebs of molten metal, showered from a large meteorite, might have been responsi-

Fig 7.10.11 – **Iron metal nodules** from China sold as 'meteorites' have a strong magnetic attraction but do not have nickel, so are not considered to be meteorites by the experts. However, their origin is still as mysterious to geoscientists as meteorites are because neither can be easily explained by modern science. Like meteorites, they are found all over the world, sitting on the ground as if they fell out of the sky.

CHAPTER 7 THE HYDROPLANET MODEL

ble for its occurrence; others have reasoned that part of a large meteorite, perhaps the missing Port Orford, Curry County, Oregon, pallasite disintegrated, covering the area with thousands of small metallic fragments, later to be washed down and to be deposited in the Josephine gravels." Note 7.10s

The Josephine specimens were *nodules* with no known source. During this investigator's research in 1949, he collected 75 specimens, the largest being 31.4 mm (1¼ inches). Other collectors reportedly held specimens that were much larger. Later research, from the 1980s revealed that the iron-nickel metal had Widmanstätten patterns:

"Josephinite, a complex, metal-bearing rock from the region of the Josephine Peridotite in southwest Oregon, contains FeNiCo metal alloy phases having exsolution textures. Scanning and transmission electron microscopy **have revealed Widmanstätten patterns**…" Note 7.10t

The native iron-nickel metal spoken of is called Josephinite. Its Widmanstätten crystal pattern is microscopic due to a high percentage of nickel (74%). Modern science has no answer for the origin of the Josephinite nodules and many other iron 'meteorites.'

Fig 7.10.11 shows naturally occurring Chinese iron nodules that were sold as 'meteorites.' Because nickel is absent in the nodules they have not been confirmed meteoric by the experts. Their origin, like most meteorites, remains mysterious. The reason; many meteorites share a common origin with the iron nodules, an idea far from the minds of the professionals.

Other Native Iron "Coincidences"

> FQ: Is it a coincidence that the United States' largest meteorite, the **Willamette**, came from an iron producing area?

The Willamette native iron 'meteorite' was discovered on land owned by the Oregon Iron and Steel Company, near Portland, Oregon. The Willamette iron was stolen, removed from the steel company's land, forcing the company to go to court to retrieve its 15-ton piece of iron. One researcher remarked:

"The problem was that the land it lay on was owned (**ironically**) by the **Oregon Iron and Steel Company**." Note 7.10u

Can this be dismissed as simply another coincidence? Considering the relative scarcity of iron outcrops, it seems odd there would be so much coincidence among meteorites and their surroundings:

"It all started with a man named after bears, one Yakov Medvedev. In Russian, Medvedev means 'Son of Bear.' The year was 1749. The retired Cossack, also a blacksmith, hunter, and collector of valuable metals, was climbing the heavily wooded slopes of south central Siberia's 900-meter-high Mt. Bolshoi Imir in search of gold, red elk, or anything else of value. When an **outcrop of iron ore** beside a huge rock caught his eye, he most likely hefted and examined the lumps of magnetite lying about, dreamed of riches, and made a mental note of the outcrop's location." Note 7.10v

Medvedev later returned with German mining engineer Johan Mettich to evaluate the outcrop of iron ore. They made an astonishing discovery:

"**Some 300 meters from the outcrop** they were startled to see an irregular sphere-shaped object some 70 centimeters [2½ feet] in diameter **just sitting there on the forest floor as if gently lowered by a crane**. When struck the object gave off a pleasant ringing sound. At first supposing it was an iron nugget of terrestrial origin, Mettich soon became much puzzled when he was unable to find among rock specimens along the ridge any that resembled the mysterious nugget." Note 7.10v

The object became famously known as the **Pallas Iron** meteorite, after which all modern pallasites are named, they being the largest category of stony-iron meteorites. Once again, is it coincidental that the largest pallasite meteorite in the world *just happened to fall* next to an iron outcrop; and is it coincidental that there was no associated crater?

The False "Alien to Earth" Claims

Rocks from space have an aura everyone seems to relate to; mysterious and rare, they represent in our minds, places in the solar system we have never been. Sometimes, the excitement of finding a 'space rock' leads to false claims of "cosmic metal," even among scholars. Meteoriticists have declared certain minerals, such as **schreibersite** and troilite occur in all meteorites but are "alien to Earth":

"**Schreibersite** is a truly cosmic mineral, **alien to Earth**. Like troilite, it occurs in all meteorites." Bib 29 p230

Of course, claims like this—certain minerals are found only in cosmic rocks—make meteorites more extraordinary and desirable, which in turn produce higher selling prices. Not necessarily conspiratorial, the claims are simply incorrect. Schreibersite, for example has been found in the basaltic dike of Disko, on the west coast of Greenland:

"The lower iron-rich half of the lens formed by crystallization of iron, cohenite, troilite, **schreibersite** and wüstite (in that or-

Fig 7.10.12 – The dark Fusion Crust on this ordinary, light-colored chondrite meteorite can be easily seen because of the stark contrast between the rock and the dark, glassy, paper-thin melted Fusion Crust. This fragment was found in 1860 in New Concord, Ohio, USA. It is part of the University of Arizona's collection. The thumbprint-like impressions on the top area of the Fusion Crust are called regmaglypts; they are erroneously believed to form by heat generated during entry into Earth's atmosphere, at the same time as the paper-thin Fusion Crust formed.

der), with trace amounts of the phases found in the upper half of the lens." Note 7.10w

The discovery of native schreibersite in Greenland has been known at least since the publication of the above article, published in the *Journal of Petrology* in 1985. It predated the claim in the popular meteorite book, that schreibersite is exclusively of cosmic origin, by nearly a decade. Because those who 'know' the most about meteorites have overlooked this type of evidence, considerable confusion about the real origin of meteorites remains.

The Fusion Crust Enigma

Fig 7.10.12 shows a light-colored stony meteorite with a dark, paper-thin crust along the unbroken outer surface of the meteorite. This dark coating is called a **fusion crust** and is defined as a "thin glassy coating" resulting from the melting of the meteorite surface as it passes through the atmosphere. Meteorite scholars have observed that the fusion crust is "seldom more than 1 or 2 mm thick":

"Meteors enter the atmosphere at speeds of many miles per second. At those tremendous speeds, the friction of the atmosphere causes the exterior of stony meteors to melt. The melted portion is ablated (it sloughs off) and new material is melted underneath. A meteor can lose most of its mass as it passes through the atmosphere. When it slows down to the point where no melting occurs, the last melt to have formed cools to form a **thin glassy coating called a fusion crust**. Fusion crusts are seldom more than 1 or 2 mm thick." Note 7.10x

The statement indicates that a meteor "can lose most of its mass as it passes through the atmosphere," through a process called ablation. This unobserved process would only really happen on pea-sized rocks, if it happens at all. A larger meteorite has only a small fraction of its surface melted as compared with its entire mass. A fist-sized meteorite, for example would have a mass thousands of times greater than the melted fusion crust. Moreover, there are not any good examples of pea-sized meteorites with a fusion crust, perhaps because smaller meteorites slow down much faster in the upper atmosphere and have no time to heat up enough to create a fusion crust.

In 1974, several researchers reported their findings in the *Journal of Geophysical Research* on experiments that included environmental conditions simulating an Earth-bound meteor traveling at about 12 km/sec from an altitude of 70 kilometers. Note 7.10aa The experiments conclusively showed that only a paper-thin fusion crust is created on ablated samples. This is easily understood because most meteorites traveling at an average rate of 20 km/sec pass through less than 100 km of atmosphere, heated in the upper atmosphere *for only a few seconds* before slowing to terminal velocity, (about .5 km/sec). Thus, there is only a narrow timeframe for fusion crust formation. The few seconds of heating forms a paper-thin fusion crust on stony meteorites and even less on iron meteorites. This has created a bit of an enigma.

> Fusion Crust Enigma: A fusion crust does **not account for regmaglypts**, despite modern science's sanction.

The Regmaglypt Pseudotheory

The word **regmaglypt** is an uncommon word, unless you collect or study meteorites. **Regmaglypts** are the thumbprint like impressions on the surface of larger meteorites. Fig 7.10.13 shows the Willamette meteorite and Campo del Cielo fragments displaying many **regmaglypts.** Meteoriticists state explicitly that regmaglypts are "produced by ablation," as explained in the book, *Rocks from Space*:

"A deep pit or cavity on the exterior of meteorites **produced by ablation** of certain minerals in the meteorite as it passes through Earth's atmosphere." Bib 29 p430

Another authority on meteorites, the Department of Earth and Planetary Sciences at Washington University in St. Louis, claims:

"**Regmaglypts** are thumbprint like impressions on the surface of larger meteorites that are **formed by ablation** of material from the surface as a meteor passes through the earth's atmosphere." Note 7.10y

The UM's extraordinary claim is that modern science's origin of regmaglypts is a *pseudotheory;* a false theory taught as fact to the entire world. The fusion-crust formation from actual experiments and on observed falls was less than 2 mm, about

Fig 7.10.13 – The large iron in the background is the 16-ton **Willamette** specimen; the smaller foreground irons are **Campo del Cielo** iron from Argentina. These exhibit surface indentations called **regmaglypts**. Note the unusual depth of the regmaglypts on the large Willamette iron; some appear deep enough to cut through the entire structure. How could ablation during entry into Earth's atmosphere create such deep cavities when observations only show a paper-thin fusion crust from melting occurs? Meteorite scholars are not asking this fundamental question and without the answer to it, we will never know the true origin of these meteorites.

Fig 7.10.14 – Debris from the Columbia spacecraft disaster is laid out for research and investigation. Although these objects were traveling at km/sec, none had regmaglypts on their surfaces. Much of the debris survived the fall because small objects quickly lose speed as they are slowed by the atmosphere. How were inch-deep regmaglypts on fist-sized iron meteorites formed by ablation while passing through the atmosphere? The Stardust Satellite (inset top) entered the atmosphere at a record 13 km/sec, but also showed no signs of regmaglypt formation. Despite the evidence, scientists still see regmaglypts as coming from atmospheric ablation.

the thickness of a dime. The heat that produces a fusion crust is simply not sufficient to produce deep regmaglypts.

FQ: Are regmaglypts formed by ablation while passing through Earth's atmosphere if heat is only suffiecient to form a thin fusion crust?

The answer is that ablation **did not, nor can it** form the large regmaglypts found on meteorite rocks. There is extraordinary evidence that has been disregarded. Fusion crust experiments prove that deep regmaglypts are not from ablation, a fact ignored because of the paradigm that all rocks with regmaglypts are meteorites from space. Ignoring the obvious however, does not change the truth; a new origin—one that can be demonstrated scientifically—is needed to explain how regmaglypts are formed.

Other evidence that regmaglypts did not form through atmospheric ablation comes to us through the study of the debris from the Space Shuttle Columbia, destroyed during re-entry into Earth's atmosphere February 1, 2003. Fig 7.10.14 shows all sorts of spacecraft debris from the Columbia tragedy, which traveled through space and returned through the atmosphere at high-speed. We were unable to identify *any evidence* that regmaglypts formed on any of the metal parts, but why should there be when only a paper-thin layer of melt forms on metal objects entering Earth's atmosphere.

The Willamette iron in Fig 7.10.13 is an excellent example of regmaglypts and large cavities. This particular specimen boasts holes so big they have nearly cut through the iron. A common sci-bi accounting for these holes is rusting, or oxidation. However, upon closer examination, it was confirmed that very little rusting took place at the Willamette site in Oregon where the iron was found.

In 1948, Russell A. Morley investigated the manmade crater excavated to remove the iron and only found "8,500 grams of oxide, or a little over 15 ¼ pounds." Note 7.10z

This was nowhere near the thousands of kilograms that should have been present had the cavities formed from rusting. Although many iron meteorites can and do rust, especially in wet environments, there is no evidence to support the idea that regmaglypts or cavities are from rusting, nor has sufficient oxide waste been identified at the find sites to support the claim. Note too, that the Willamette iron in Fig 7.10.13 shows evidence of regmaglypts *inside* the large cavities.

When rusting does occur, it follows the crystalline structure of the meteorite, not the circular pattern of the Willamette pits, thus further confirming that the circular-shaped pits were **not** formed from rusting. Fig 7.10.15 is a photo of a cut fragment of an iron meteorite that was kept in water and allowed to rust for a period exceeding two years. Just as expected, the rusting followed the sharp crystalline pattern, not a circular one.

Other large irons, such as the **Hoba** specimen, and the **Campo del Cielo** specimens do not have deep circular rusting patterns at all, which should *be present in some degree* since they are quite similar to the Willamette iron. The Willamette's deep pits have sharp edges while other surface areas are not affected at all; another unexplained oddity.

Most iron meteorites weather like a hammerhead left out in the rain to rust; the entire surface oxidizes and eventually crumbles. The 37-ton Campo del Cielo iron and thousands of smaller Campo irons show surface rust, but nothing like the Willamette pitting. When we spoke to those who actually search for Campo del Cielo meteorites, they reported that rusting is minimal and steel brushes easily remove any rust.

There are other theories for regmaglypt formation; one suggests that regmaglypts are formed on impact. However, impact experiments at both high and low-speeds prove unequivocally that this does not happen. Regmaglypt formation has not been reproduced in laboratory impacts nor have any signs of heavy melting in the cavities been identified.

One observed fall, the Russian Sikhote-Alin event of 1947, contained *thousands* of meteorite fragments with regmaglypts. During the fall, some fragments exploded turning the meteorite into shrapnel upon impact with no new regmaglypts forming. The regmaglypts were clearly *already* formed on the meteorite surface *before* they hit the ground.

According to one current theory of meteorite origin, iron meteorites are fragments *broken off from a larger parent body*. To understand what a meteorite fragment should look like if broken in the frozen reaches of space, we turn to the modern meteorite merchant. Few people, except avid collectors and institutions, are interested in purchasing large unwieldy specimens of meteorites and smaller pieces can usually command a higher price per gram. The meteorite merchants know this, and to accommodate their customers and to maximize their profits, they break large specimens into smaller pieces. They do this by freezing a larger rock with liquid nitrogen and

Fig 7.10.15 – This piece of iron meteorite was stored in water for more than two years and had developed a considerable amount of rust accumulation. After washing, the crystalline structure was strongly evident. There is no indication of a rusting pattern suggestive of regmaglypt formation; another incorrect origin proposal.

then crushing it with a 60-ton press. The results can be seen in Fig 7.10.16. The broken fragments do not have regmaglypts or cavities.

In summary, regmaglypts are *not* formed by:
1. A melt on another planet in space.
2. Impact.
3. Atmospheric ablation.
4. Rusting, or erosion.

Since we tend to ignore that which we do not understand, meteorites with regmaglypts have become merely an expensive curiosity for meteorite enthusiasts. The regmaglypts are, however, a *critical* part of the Meteorite Model. If we learn the true origin of regmaglypts, we will know the true origin of the meteorites themselves—these surface 'thumbprints' testify to both how and where the meteorites were made.

The observed 1947 Sikhote-Alin fall in Russia provides us with two important clues as to how and where meteorites are made. Perhaps no other fall has offered more information about iron meteorites than this one. Thousands of meteorite fragments and hundreds of craters have helped establish important criteria regarding meteorites and their associated impact craters. Over the decades, researchers have closely examined many specimens, categorizing meteorites that fell intact on soft ground as 'whole or single specimens' and those that exploded on impact or misshaped by rocky surface impact as shrapnel. The first clue came as the researchers identified a "remarkable morphological feature" in "practically every specimen." This feature was the octahedral composition or *crystalline structure* of the specimen:

"The most **remarkable morphological feature of single specimens** is that practically every specimen shows the elements of their **octahedral composition**, in some cases very marked." Bib 167 p350, see Fig 245, specimen No. 1434

Photos in the book, *Giant Meteorites* illustrated this phenomenon showing whole specimens with "the *form* of a union of two beams" or intersecting crystals. The researchers noted the axis of the individual elements:

"…bisect each other at angles of 60°, 90° and 120°. On nearly every sample isolated, almost completely flat surfaces may be found that intersect at the same angles, usually forming sharply-marked planes and ribs." Bib 167 p350

These whole meteorites could not have maintained their octahedral crystalline structure if they had been melted, and as seen in previous photos, meteorite specimens that were either melted or broken do not look at all like the regmaglypt-included Sikhote-Alin specimens. The meteorites achieved their form and shape because they grew that way.

Yes, *grew.* Like all other natural crystalline structures, the Widmanstätten pattern identified in all the meteorite specimens we have been discussing is an octahedral *crystalline* structure, formed in a hypretherm.

The second significant clue that verifies that these crystalline-structured meteorites formed in a hypretherm and not a melt in space is the *relationship* between the size of the regmaglypts and the size of the meteorite:

"The specimens of the Sikhote-Aline meteorite shower are distinguished by their sharply-marked, well-developed regmaglyptic relief. Study of this relief established the **important fact that the size of the regmaglypts is closely connected with the size of the meteorite**, more precisely with its cross-section at right angles to the **direction of motion**. It was established that, on an average, **the diameter of the regmaglypts is approximately 0.09 of the meteorite's cross-section**." Bib 167 p354

Evidence that the regmaglypts size is directly related to the cross-section size of the meteorite clears up the puzzle as to why larger meteorites have larger regmaglypts. It also refutes the idea that regmaglypts were formed in seconds following entry into the atmosphere. If the indentations had been formed by ablation, regmaglypts would all be of similar size. The fact that regmaglypts—a surface morphological feature—are directly related to the size of the specimen, makes them similar to thousands of other minerals grown in the Earth whose surface features are also directly associated with specimen size.

The "direction of motion" referred to in the previous statement has to do with regmaglypts that appear to flow in a single direction. They are thought to have been formed by atmospheric friction, but as we have discussed, regmaglypt formation from melting has not been proven, certainly not to the degree that large (centimeters or larger) flows of melted material could have occurred in the short time meteorites pass through the atmosphere. However, as the connection between meteorite origins and hydrofountains is explained, directional regmaglypt formation is easy to understand. Water chemistry and pH changes to water flowing past meteoritic material can facilitate dissolution and creation of ablation pits. In subchapter 8.15, the Dissolved Surface Mark, there is an actual example showing how regmaglypt-like indentations and directional flow patterns form on minerals because of chemical/gas activity.

The Regmaglypt Pseudotheory, like the obsidian pseudotheo-

Real Broken Meteorites Like These, Do Not Exhibit Regmaglypts

Fig 7.10.16 – These fragmented Campo del Cielo irons were made when a larger specimen was crushed under a 60-ton press after being cooled with liquid nitrogen. A similar result would be expected in the cold outer space environment, showing that regmaglypts are not impact related. Courtesy Bud Eisler at Cosmic Cutlery.

ry, or the basalt pseudotheory, are false theories taught as fact, based on imagined solutions to basic questions. These pseudotheories remained unchallenged because modern scientists simply did not question *everything* with an open mind—the first principle in the UM.

The Olivine Crystals Evidence

The most beautiful and well-known type of stony-iron meteorite is the pallasite. The **pallasite** is unique among the meteorites because it contains the iron-nickel metals and a large amount of the greenish-yellow crystal, olivine. Fig 7.10.17 illustrates the internal structure of the Fukang pallasite, typical of many pallasites. Transparent olivine crystals are clearly discernable and make the pallasite a highly attractive meteorite.

> "The specimens of the Sikhote-Aline meteorite shower are distinguished by their sharply-marked, well-developed regmaglyptic relief. Study of this relief established the **important fact that the size of the regmaglypts is closely connected with the size of the meteorite...**"
>
> Giant Meteorites, E. L. Krinov, p354

The Olivine Crystals Evidence story begins with a discussion between the UM author and a renowned meteorite expert. During the course of that meeting, the following fundamental question was asked:

FQ: How can pallasite meteorites with olivine crystals come from a melt if olivine crystals are destroyed in a melt?

The meteorite expert, Dr. Moore answered:

"I agree, the olivine and iron-nickel mineral combination in pallasites are a great difficulty for scientists." Note 7.10aa

Let's look at the pallasite in Fig 7.10.18 and see why this mineral combination is such a "great difficulty" for scientists. The beautiful, transparent-green olivine crystals look quite pure, but olivine actually contains two metal elements, magnesium and iron. When heated, the green crystals turn a brownish-grey color as they melt (see Fig 7.10.18).

During the 1970s, researchers discovered that heating during the ablation process destroyed olivine crystals and produced the iron mineral magnetite:

"The melting and oxidation of olivine during meteor ablation, **which produce magnetite and an Fe-deficient olivine**, are indicative of the ablation environment and **may be used as criteria for the identification of ablation debris**." Note 7.10ab

Such evidence confirms that pallasite specimens did not originate from a melt. Furthermore, if pallasite meteorites fell from the sky, they should have a fusion crust that would consist of melted olivine crystals, forming a thin magnetite shell. (See melted olivine crystals in Fig 7.10.18). To validate, or refute the theory that pallasites are of cosmic origin that found their way to Earth, we can ask this FQ:

FQ: Where is the fusion crust on the pallasite meteorites?

Melted olivine bears a similarity to glass and is resistant to weathering. If they actually came through the atmosphere, we should find evidence of such on most finds. There are several dozen pallasite finds with thousands of kilograms of recovered specimens; none exhibit evidence of olivine melting. Our search for pallasites with fusion crust proved unsuccessful. No mention of fusion crust was found in journal articles, no evidence in any pallasite photographs, and no fusion crust on *any* specimens observed first hand. An entire class of meteorites—the pallasites—exists *without a fusion crust.*

There are only four observed pallasite falls recorded. Note 7.10ac Three of these falls occurred prior to 1902, but were not scientifically documented. The most recent event occurred May 16, 1981 in Omolon, Russia, but only the observance of a fireball was recorded; the Omolon pallasite was not collected until two years later. It is equally likely that this specimen was already on the ground, found because of the concentrated effort to locate the meteorite associated with the previously observed fireball. None of the observed pallasite falls holds up under scrutiny; their incidences based only on circumstantial evidence.

A similar story is told about the Russian **Shirokovsky pallasite,** said to have fallen through the ice February 1, 1956 following a bright fireball; then 46 years later, in 2002, divers recovered about 150 kg of pallasite-like samples. Note 7.10ad The pallasite however, did not "fit" the accepted chemical composition of other pallasites; the Shirokovsky specimen was nearly absent

Fig 7.10.17 – The **Fukang Pallasite** meteorite was found in China in 2000. It is nearly a meter long and weighed in at just over 1000 kg (2205 pounds). It is one of the largest pallasites ever found. Pallasites are a subgroup of the stony-iron meteorite class of meteorites and contain beautiful olivine crystals. The yellow-green crystals embedded in nickel-iron metal have been both a fascination and a mystery to researchers ever since German naturalist, Peter Simon Pallas first discovered them in 1772. Because they are believed to have come from a melt, laboratory experiments have proven unsuccessful at reproducing them. Olivine is a common silica-based mineral that like quartz, is grown only in a hypretherm. Thus, pallasites with olivine *sealed* in a nickel-iron matrix demonstrate that *both* the silica-based olivine crystal and the nickel-iron metal in these types of specimens both grew in a hypretherm.

7.10 THE METEORITE MODEL

Fig 7.10.18 – What does olivine look like when it is melted? As seen above, magnesium and iron metals locked in the green olivine crystals come out when melted, **totally changing the appearance and properties** of the melted olivine. Interestingly, meteorite books and research papers appear totally void of practical tests and observations such as these. Because olivine is a primary mineral in the Pallasite meteorites, why are there no melted olivine-like metals on the surfaces of the pallasites, like the Fukang specimen, if atmospheric ablation (melting) really occurred? Meteoriticists have been unable to satisfactorily explain how metals and silicates could have formed together in a melt—because olivine obviously did not grow from a melt.

of iron but had a high nickel content (20-47% by weight)—more than most pallasites—and about 50% olivine; the crystals not being in a typical form. The recovered material is currently dubbed a **pseudometeorite** because the Meteorite Nomenclature Committee concluded:

"…the petrology and geochemistry of this object strongly suggest that it has a **terrestrial origin**." Note 7.10ae

This is important! If this pallasite-like specimen (metal and olivine mixture) *could be terrestrial*, then why couldn't other pallasites be terrestrial? No pallasites have been recovered from scientifically documented falls; no one has observed an actual pallasite falling and recovered the rock. Thus, the possibility that meteorite-like rocks could have been made right here on (or in) the Earth must be considered. Ponder this for a time; we will come back to it later.

The **Huckitta pallasite** was found in Australia in 1937. Investigators have assumed that this 1411 kg pallasite, along with many other metallic minerals we will discuss in later sections weathered (rusted), leaving behind byproducts of hematite and magnetite:

"The mass was surrounded by 900 kg of **iron shale**. The majority of the material had been residing at depth and has been **weathered** to the point that the nickel-iron matrix is **transformed into hematite and magnetite**." Note 7.10af

Nickel-iron can rust, but it **cannot** be "transformed into hematite and magnetite." Neither hematite nor magnetite contains nickel, nor are they 'rust.' These two iron minerals are formed in water, most specimens being grown in a hypretherm. The Huckitta pallasite does not "fit" the profile of other pallasites; it contains high amounts of unique elements, the presence of which is not easily explained. In the next chapter, we will learn more about biogenic minerals and how a hyprethermal, biologic environment is responsible for many such minerals found on the Earth today.

In the Magma Pseudotheory chapter, we refuted modern science's idea that the Earth has an iron core and the corollary of that theory; that meteorites are the iron cores of other planets. The Olivine Crystals Evidence also refutes modern science's position and their long-held *assumption* that Earth's magma interior is made of pallasite-like meteorite material, without ever having had any real evidence. We see this idea stated in the book *Rocks From Space*:

"**If** researchers could sample the core-mantle boundary deep within Earth, they would find it **similar in composition to pallasites**." Bib 29 p237

This is a byproduct of *not* using the Universal Scientific Method, the building of theory upon theory. Multiple experiments, including the drilling of many deep boreholes, produced no real evidence or explanation of how the pallasite nickel-iron-olivine minerals formed. Olivine is rich in *magnesium*, a metal much less dense than nickel or iron; gravitational separation based on their individual densities would have kept them from combining if a "magma soup" had ever existed.

The "Enigma of Chondrules"

Of all the meteorite types, Chondrites are the most common. Among the ingredients of these meteorites, chondrules are spherical inclusions that are an "enigma" to modern geology. One example can be seen in Fig 7.10.19:

"In spite of extensive studies over a long period of time, the **enigma of chondrules in meteorites is still with us**. Many theories have been proposed to explain the origin of these millimeter sized spheroids and the manner in which they have been incorporated in the chondritic meteorites, **yet many questions remain unanswered**." Bib 155 p561

Fig 7.10.19 – This slice of a chondrite meteorite shows its **chondrules**, which are spherical mineral inclusions. The existence of the crystals along with the presence of other minerals such as calcite and gypsum has proven to be an "enigma" for geoscientists. Speculating about their origin has led to many complex but incorrect theories of their origin.

Among the questions that remain unanswered is; how do calcite and gypsum, two minerals that form in "low temperature hydrous" environments (think ocean water with microbes) get into the matrix of rocks that came from space? (Bib 155 p562, Murchison carbonaceous chondrite) Presumably, these space rocks were made in the vacuum of space, in an environment in which they were melted. But the physics of outer space does not allow for volatiles (water) to remain in small melted bodies like meteorites, nor does it account for the formation and presence of minerals in the melt like calcite and gypsum. The scholar, Edwin Roedder wrote *Fluid Inclusions*, the definitive work on liquid inclusions in minerals. He realized that many questions remain unanswered with regard to chondrite meteorites:

"The **low-temperature hydrous** part of the sequence of environments or events that yielded this obviously bimodal distribution [calcite, gypsum and other minerals] **is itself a question of considerable consequence in understanding the origin of the solar system**…" Bib 155 p562

Modern science has painted itself into a corner with its origin of the solar system by accretion theory; a theory with extremely hot temperatures and almost no water and no mechanism for minerals like calcite and gypsum to form in rocks from space. Truthfully, the minerals in meteorites do pose "a question of *considerable consequence* in understanding the origin of the solar system," the consequence being the utter abandonment of the old accretion theory to be replaced by a model with *water at its foundation*. Whether these rocks came from space or right here on the Earth, the Hydroplanet Model can account for their formation.

Meteorites Are Ejectites

Astronomers and meteoriticists have been of a mind that meteorites were all formed by some sort of melt or melt related event, and so, have not ask this basic question:

FQ: From where did meteorites come?

Answers to this question are found in the Crystallization Process we examined earlier in this chapter. Like many common Earth rocks, most meteorites contain quartz, which means they were grown in a *hypretherm*. The hypretherm, a pressure, heated, aqueous environment facilitated the growth and crystallization of rocks later ejected through powerful hydrofountains. These rocks are **ejectites,** a new UM rock classification defined as being any material ejected from a hydrofountain.

Ejectites that are expelled with sufficient force to escape the planet or moon's gravitational hold and later, perhaps hundreds or thousands of years later, fall to the Earth (or other celestial body) are **meteorites**.

Ejectites are often found near hydrocraters, and though they have been erroneously dubbed 'meteorites,' most never fell from space. Instead, they were blown out through the hydrocrater and lacking the energy to achieve escape velocity, they returned to the Earth.

Most ***meteorites are ejectites*** from hydrofountains on the Earth or other celestial body. Some meteorites *might* be 'impact' debris from a violent collusion in space, but this has yet to be documented.

In the not-too-distant past, the *only* origin for meteorite-like rocks was either comets or asteroids. Recently, scientists gave meteorites another possible origin—the inner *planets*, though no idea for how the rocks left the planet was given. We read of this from the book *Meteorites and Their Parent Planets*:

"The hypothesis that **meteorites might be derived from a planet** such as Mars, Venus, or Mercury is a **recent suggestion**. The difficulties encountered in extracting a rock from a planetary gravitational field are formidable, but apparently **not prohibitive**. So far, only meteorites thought to be from Mars have actually been recognized." Bib 166 p36

Researchers were coming to the realization that comets and asteroids were not the best candidates for meteorites and so have looked to the next logical source—the planets and their moons. Surprisingly, the Earth was left off the list of planets that produce meteorites, even though evidence of ejectite material is abundant.

Fig 7.10.20 is an illustration of hyprethermal formation and expulsion of ejectites. The power of steam explosions should not be underestimated. They are capable of releasing more energy than any atomic bomb man has ever detonated. Even small moons, such as Saturn's Enceladus are capable of ejecting steam, water and ice rocks far into space, through hydrofountains powered by heating from gravitational friction.

Not all of the ejected material would escape the planet or moons tug and some of the ejectites would eventually return to the surface. Ancient hydrofountains were far more powerful than any seen or recorded by modern man, and the ejectites from them were more diverse. We will uncover empirical evidence of where, when and how these colossal hydrofountains were active in the following chapter.

Many small and medium eruptions have occurred in modern times, including hydrofountains that produced no volcanic materials. As shown in Fig 7.10.20, other minerals formed in the hypretherm can become ejectites during a steam explosion. Different subsurface materials, temperatures and pressures determine the type and magnitude of mineral growth. These variables also influence the magnitude of the explosions, some of which are underwater eruptions. Metallic, meteorite-like ejectites can be produced in some cases, but not all. Other terrestrial minerals such as iron shale, hematite, magnetite, and rocks like the Huckitta pallasite are all the result of similar formative processes.

Geologists recognize that a great majority of Earth's craters are hydrocraters; craters from steam explosions (see the Hydrocrater Model subchapter). Couple this fact with the Crystallization Process, the presence of a hypretherm, and hydrofountains, and the origin of metal nodules that are sometimes formed and ejected and sometimes not becomes clear.

Let's return to the modern science 'impact' crater story. According to The Earth Impact Database, November 2010, Note 7.10ag there are 188 "confirmed" listed impact sites. The great majority of them have no meteorite fragments associated with them. Some of those that list related meteorites do so based on circumstantial evidence. If these impact structures were actually of impact origin, the laws of physics imply that there should be ample remnant meteorite fragments. Because they are not present, the idea that they weather away is invoked, as seen in the textbook *Introduction to Planetary Geology*:

"**Meteorite fragments** are often found in association with young impact craters and serve as **the best criteria for recognition of meteorite impact**. However, meteorite fragments **weather rather quickly**, and only about 13 known impact craters have meteorite fragments associated with them. **In every case where meteorite fragments are found, the fragments indicate that the crater was caused by an iron or stony-iron meteorite.**" Note 7.10ah

The weathering sci-bi is amazing. Over 85% of meteorites are chondrites, which are similar in many respects to ordinary rocks like granite and are resistant to weathering and erosion. There may be some decomposition over time, but most of the craters exhibit characteristics of a youthful age.

Moreover, high-speed impact sites should have abundant impact melt with infused iron-nickel from the meteorite. However, impact glass exists at only a few of the so-called impact sites. This fact, coupled with the lack of meteoritic fragments, should have raised concerns about the validity of the impact claims.

Figs 7.10.21, 22 & 23, show three rocks that are considered geologically old, yet all show little or no weathering. None of these minerals came from volcanoes. They are not glassy but are nodule-like, have *convex* or *concave* surfaces and so could not come from a magma melt. These are rocks formed in a hypretherm; likely ejectites expelled through a hydrofountain judging by the sediment in which they were found. These examples exhibit similarities to various meteoritic characteristics, chondrule-like nodules and regmaglypt-like pits. The hydroplanet model accounts for the origin of hundreds of types of nodules found worldwide, formed through a demonstrable hypretherm growth mechanism.

Fig 7.10.20 – Meteorites, like most other minerals did not originate from a melt—they grew in a hypretherm. This is how the regmaglypt surface features formed and why meteorites are not glass-like melt materials. Knowledge of this new meteorite origin is possible because of the knowledge contained in the Magma Pseudotheory Chapter and in the Hydroplanet Model Chapter. Meteorites were first thrown into space as **ejectites** after being formed in a hypretherm as illustrated above.

The Hydrofountain Origin of Meteorites

Meteorites are really ejectites born of a *water* environment in diatremes and hydrofountains. As verification of this, scientists have found *water-filled inclusions* in meteorites, contrary to every current meteorite origin theory. Earlier in this chapter, we quoted from the book, *Fluid Inclusions*, by Edwin Roedder, a well-known scholar in the field. He wrote:

"After the investigations of the lunar samples, inclusions in meteorites came under study, in part as a result of an hypothesis that some meteorites may have originated on the Moon. In addition to the **well-known and expectable glass inclusions, inclusions of aqueous fluids have been found in a series of meteorites** (Flieni et al., 1978). **This discovery was completely unexpected and still remains thoroughly enigmatic after several years of study**.

The current status of research on inclusions in extraterrestrial materials is not very satisfactory. As will be evident in the tentative interpretation in this chapter, **too many inexplicable aspects remain**, but the field is young. **Perhaps a second decade of research will provide a more cohesive and satisfying body of knowledge." Bib 155, p534

Almost three decades have passed since Roedder's 1984 book was published, but the "body of knowledge" remains egregiously thin explaining the "inclusions of aqueous fluids." The fact that the discovery of water in meteorites was so "completely unexpected" and that it "remains thoroughly enigmatic" is understandable, given that the old paradigm with its origin of meteorites is incorrect.

Comparing the amount of water (by weight) locked inside the microstructure of the meteorite shown in Fig 7.10.25, we begin to appreciate the significance of the dilemma facing investigators stuck within the dogmagma paradigm!

How could meteorites have come from a melt and have so much water locked within them?

Meteorites have some of the highest water content of all minerals, a mystery for the modern meteoriticist trying to make sense of the meteorite-from-melt origin:

"**The water content (by weight) of the meteorites is about 11 percent** for type 1 chondrites, about **9 percent** for type 2, and **2 percent** of or less for type 3." Note 7.10ai

The Hydrofountain Origin of Meteorites explains *why* the largest 'meteorites' found on Earth have no craters. Rocks ejected from an extraordinarily powerful hydrofountain *can* escape Earth's grasp and make it into space, orbiting the Earth in unstable, decaying orbits. As the ejectites return to the surface of the Earth or another planet, they become meteorites. Of course, most of the rocks ejected from hydrofountains never reach space. Trapped by the inexorable tug of the Earth, heavier and less-speedy stones, unable to reach escape velocity, fall to the Earth at terminal velocity or slower. Regardless of their size, they produce minimal, if any cratering. Probably, these ejectites would land in relative proximi-

Fig 7.10.21 – This *convex* mineral specimen is a **hematite nodule.** It formed naturally in a hypretherm. Hematite is an iron mineral with a density similar to meteorites. Notice how the iron crystals formed in spherical shapes, even on the surface. When broken, these nodules, like meteorites have sharp edges, seen on the left side of the nodule. This indicates a surface structure that was not broken or melted as it was formed.

Fig 7.10.22 – This *convex* non-iron rock was found with others of various sizes in a sandy hillside near Young, Arizona, USA. The sediment these rocks were found in was of light brown sand and clay, quite different from this specimen. The nodule-like surface resembles the regmaglypts of meteorites, which is a form of crystallization. In rocks like these, and with meteorites, broken, sharp edges are uncommon and erosion is minor. In this case, no surface erosion was noted.

Fig 7.10.23 – **Quartz nodules** with pitted surfaces that appear similar to some regmaglypts. *Concave* nodules such as these are formed in a hyprethermal environment when quartz fills cavities where other mineral and rock formation is occurring.

ty to the hydrofountain from which they came, explaining why iron ejectites are sometimes found near iron ore deposits. More examples of hydrofountains and their various types of ejectite minerals will be examined in the following chapter.

The Meteorite Enhydro Evidence

As we consider the modern science environment where meteorites were *supposed* to have come from—an outer space melt, we see the corner into which meteoritics has painted itself. The scenario where rocks are formed from anhydrous melts (melt without water) under no pressure cannot account for enhydros—encapsulated water—and "moving bubbles" in meteorites, but such have been reported:

"Most **meteorites**, and chondrules in particular, **seem** to have formed at **high temperatures**, from **anhydrous melts**, out in space at presumably near-zero pressures. As a result, silicate-melt inclusions are expectable and apparently ubiquitous, but the presence of **actual liquid inclusions (i.e., with moving bubbles at room temperature) would seem almost impossible. However, a few have been reported**." Bib 155 p566

Within the known laws of physics, water in a melt is a volatile, like air bubbles in water—the volatiles would escape to the surface and vaporize. If meteorites came from a melt, they should *not have water,* but they do, and lots of it—as seen in Fig 7.10.25. This would seem to prove "impossible" for meteorites to have come from a melt, but the undeniable facts of water-versus-magma get even better; the experts in mineralogy and inclusions go on to report:

"Liquid inclusion in chondrites. As a diogenite is, in effect, an igneous rock, the presence of aqueous liquid inclusions, trapped during late stages of crystallization of a 'magma chamber' of some unknown (but presumably small) size, is **not too difficult to ac-**

cept. However, **chondrules** almost certainly must have formed at very high temperatures, and suspended in space rather than in a magma body, making **the possibility of liquid inclusions that much more remote. Nevertheless, Warner et al. (1983) reported liquid inclusions, with moving bubbles, in chondrites** Bjurbole (L4), Faith (H5), Holbrook (L6), and Jilin (H5)." Bib 155 p568

In writing the comments, Roedder, the fluid inclusion expert, sheds light on the matter and candidly lays the facts on the line, stating things as they *really* are:

"Interpretation of liquid inclusion in meteorites. **These liquid inclusions present us with an enigma**. Bib 155 p569

One enigma he goes on to identify is the coexistence of fluid in olivine rock:

"... **how** aqueous fluids could be condensed, and **how** they could be trapped as inclusion in the olivine of apparently unmetamorphosed chondrules, without reaction, **is paradoxical**. **The origin of these fluid inclusions is unknown**." Bib 155 p569

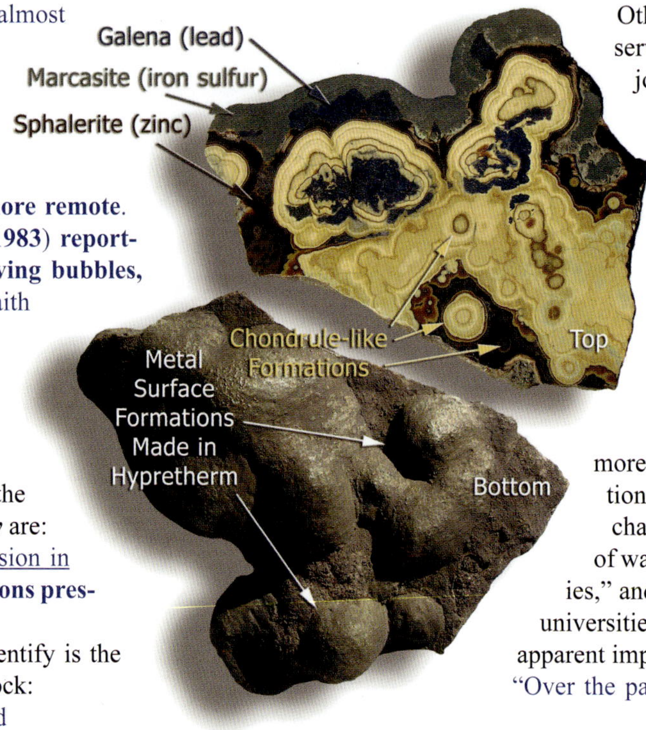

Amazingly, Roedder seems to have missed the answer he had already given himself about the origin of the fluid inclusions; it was in the *opening sentence* of his lengthy 600-page book on fluid inclusions, published in **1984**:

"...**all crystals** in all terrestrial and extraterrestrial samples **have grown from some kind of fluid**." Bib 155 p1

Toward the beginning of this chapter, we asked the question; without magma, what is the only other liquid in space? (See fig 7.1.3) By now, it is probably obvious that scholars, like Roedder have not been able to escape the modern science magma straightjacket. There had never been an alternative, and so, as science philosopher Thomas Kuhn recognized that "normal science" goes on thinking it is progressing within its current paradigm, but all it is really doing is furthering the crisis that already exists. You may have noticed how often the words mystery, enigma, or paradoxical are used by scientists; all spawned from the many observations of **water in rocks**, including meteorites. Years of dogmagma continue to influence researcher's interpretations, but there are moments of coherent reasoning.

In **1995**, French researchers recognized that:

"From the present results, it seems clear that **water, in at least two meteorites, can no longer be considered as a simple product of condensation** under thermodynamical equilibrium conditions." Note 7.10aj

Fig 7.10.24 – This mineral specimen from Olkusz, Poland contains at least three metal minerals. The sliced and polished specimen shows chondrule-like spheres toward the top and an iron metal surface on the bottom. Observing mineral growths like these helps establish how spherical chondrules and regmaglypts were grown on ejectites and meteorites. They are all grown in hot water under great pressure, but not from a melt.

"...inclusions of **aqueous fluids** have been found in a series of meteorites. This discovery was **completely unexpected** and **still remains thoroughly enigmatic after several years of study**."

Other investigators reported new observations about meteorites in a **1996** journal article in *Nature*. Their findings:

"...**clearly establishes aqueous alteration** as one of the **earliest processes in the chemical evolution of the Solar System**." Note 7.10ak

Modern science's Accretion Theory, reviewed in the Magma Pseudotheory, does not account for water and it had been left out of the planet creation process for more than a century. Too many observations, including meteoritic water, have changed the "fundamental importance of water" in "primitive solar system bodies," and in 1999, researchers from various universities released a report documenting the apparent importance of water:

"Over the past three decades, we have become **increasingly aware** of the **fundamental importance of water and aqueous alteration on primitive solar system bodies**." Note 7.10al

Meteoriticists may be "increasingly aware" but remain almost completely ignorant of water's fundamental importance to meteorites—and to all crystallized natural minerals. The meteorites did not acquire water in their matrix by happenstance; following the Law of Hydroformation, they were formed in a hypretherm. The origin of all natural crystallized minerals, including meteorites can only be understood and comprehended with this law in mind. These researchers, lacking knowledge of the hyprethermal process, acknowledge they simply *do not know* the nature of the aqueous fluids:

"Nevertheless, **we do not know the location and timing of the aqueous alternation or the nature of the aqueous fluid itself**." Note 7.10al

Because the actual role of water in mineralization is not understood, unanswered questions like this are expected. With the Hydroplanet Model, we can draw many new conclusions to answer numerous questions that have heretofore been unanswered. For instance, we know that meteorites did not come from a melted planet or asteroid *because minerals do not crystallize in a melt*.

As Kuhn said, it takes extraordinary claims and demonstrable evidence to spur a scientific revolution, and that is what the UM is all about. Next, we will look at more compelling evidence of the hydroplanet model; meteorites not just coming from water, but from *terrestrial seawater*.

The Seawater Evidence

Several decades ago, Roedder noted in his book, *Fluid Inclusions* that "brine" was found in meteorites:

> "The most reasonable interpretation of these data is that **the liquid inclusions contain a highly concentrated brine**. This interpretation would explain the apparent viscosity and the behavior on cooling." Bib 155 p568

Without a way to fit such an observation into the worldview of meteoritic thought, it remained shelved. A chance event occurred March 22nd, 1998 adding new evidence of this oddity when seven boys witnessed the fall of a meteorite in Monahans, Texas. The

> **Science has looked to the skies to find the origin of rocks and minerals— but the answer lies at the bottom of the sea.**

Why has this amount of water in this meteorite been almost totally ignored by researchers?

Fig 7.10.25 – This ordinary chondrite meteorite weighs 725 grams. Researchers discovered that the water content of these types of meteorites is approximately 11% by weight, which translates to 80 grams (1 ml of water ≈ 1 gram) of water, represented in this beaker. This is a huge amount of water for any type of rock, but for a rock presumably from a once-melted planet—this defies all reason and logic. Astonishingly, the water content of meteorites has been overlooked or ignored by almost all meteorite researchers.

stone was recovered immediately and taken to Johnson Space Center to a clean room where it could be examined. Because of the meteorite's rapid recovery, chances of contamination to the interior of the meteorite had been effectively eliminated. Upon opening the first sample, scientists were surprised at what they found:

> "After breaking open the first sample, we noted that the gray matrix contains locally abundant aggregates of **purple halite (NaCl) crystals, measuring up to 3 mm in diameter**. To the best of our knowledge, megascopic halite has not been previously seen in any extraterrestrial sample, although microscopic halite has been reported in several meteorites. Backscattered electron imaging of Monahans reveals that crystals of sylvite (KCl) are present within the halite crystals, **similar to their occurrence in terrestrial evaporites**." Note 7.10al p1378

Large salt crystals similar to terrestrial evaporites are compelling evidence that some of the Earth's meteorites were originally ejected from hydrofountains **right here on Earth**. Once such ejectites had been blasted into space, their orbits slowly decayed until they found their way back home to the Earth.

Scientists who study meteorites have determined that certain chemical compositions are the key to identifying which rocks are 'meteorites.' Of those that qualify, the possibility of terrestrial origin has been ruled out, leaving only off-world origins to try to explain observations of terrestrial-like brine inclusions. As a result, modern researchers have chosen to believe that the brine is evidence of "early solar system material," present at the formation of the Monahans meteorite:

> "The **most important conclusion of this study** is that an actual sample of the aqueous fluid **presumably** responsible for aqueous alteration of **early solar system material** has been identified and characterized, and this fluid is **a brine**." Note 7.10al p1379

Of course, brine-infused early solar system material also supports the Hydroplanet Model but it is much more probable that this meteorite was made here on Earth. After all, the ≈3 mm salt crystals were found *inside* the matrix of the meteorite, proving the rock formed from a briny-water solution.

NaCl and KCl salts are present in abundance in the Earth's oceans but are unknown elsewhere in the solar system. This probably led some investigators to ask whether the salts found in the meteorite matched the salts in Earth's oceans. Sea salt types are relatively consistent in their distribution, contiguity, and relative magnitude. If evidence of similar salt distribution in meteorites were found, this would indeed be *very* compelling.

According to an article in the 2000 *Meteoritics & Planetary Science* journal, this was **exactly** what researchers found while

> "...**all crystals** in all terrestrial and extraterrestrial samples **have grown from some kind of fluid.**"
>
> Edwin Roedder, *Fluid Inclusions*, 1984, p1

examining the Nakhla meteorite:

> "The primary observation is that the **suite of species [minerals] found in Nakhla is similar to the most common ions present in contemporary terrestrial seawater**... In addition, **the relative magnitude** of the species [minerals] is **similar to that of seawater**..." Note 7.10am

It seems that around every corner there is compelling evidence that there is water; water everywhere, encompassing the origin of all natural substances.

> The Nakhla meteorite salts and their relative magnitudes are **the same** as Earth's seawater. How could the Nakhla meteorite have come from anywhere else but the Earth?

The Lunar Salt Evidence

Having firmly established the fact that various *sea salts* are present in some meteorites, we will explore the importance of this observation and how it relates to the Moon.

According to modern science, the surface of the moon has been bombarded by meteorites, especially so during the Late Heavy Bombardment period, which they figure happened 4.1 to 3.8 billion years ago. Modern science arrives at this conclusion based on evidence from lunar samples collected during the Apollo missions. Since salt (NaCl) is found in some of the meteorites on Earth—the Monahans meteorite for example—and if the presence of salt is tied to early solar system material, *why didn't any of the rocks brought from Moon have **any** salt or chlorine* in them?

> Where is the Salt on the Moon?

Salt is a ubiquitous and abundant mineral on Earth. Most theories about lunar origins postulate that the Earth and Moon originated from the same Solar Nebula or even the same celestial body, therefore, one would expect salt to be evident in lunar samples, perhaps even in great quantities. The fact that it is not should raise important questions. However, if salt comes from a biologic source (we discuss this in the following chapter), the absence of salt on the Moon finally makes sense.

The Tip of the Meteorite Iceberg

In summary, this subchapter has defined meteorites as never before. A new meteorite origin comes from asking unconventional, but fundamental questions that have yielded surprising answers.

Some meteorites exhibited a crystalline structure—the Widmanstätten pattern—revelatory of its origin; it could not have come from a melt, nor could they have been subjected to significant melting while passing through Earth's atmosphere because 800 °C heat destroys the crystalline structure.

There are relatively few known meteorites, perhaps between 3,000 and 5,000 worldwide, including all observed falls, yet scientists have told us tens-of-thousands of meteorites larger than 10 grams fall to Earth each year, a truly superfluous claim.

Over the last few pages, a number of critical observations have been discussed: crystalline olivine in pallasites that could not have come from a melt; the world's largest meteorites have no crater associated with them; the world's largest craters often do not have meteorites; the enigma of chondrules—spherical inclusions—in meteorites; saltwater enhydros (trapped saltwater) meteorites; and the coincidental existence of very rare native nickel-iron deposits where meteorites of similar composition are found.

Taken alone, each of these observations are unexplainable anomalies, left on the shelf by scientists who hope future research will provide answers. Together, they are the evidence that many meteorites have a completely different origin than 'outer-space.' The discovery of native nickel-iron deposits in Earth's crust, which are similar in composition to meteorites found nearby, is firsthand evidence of their hydro-fountain origin.

More correctly termed ejectites; the extraordinary evidence presented here is just the tip of the meteorite iceberg and there are still many unanswered questions. The subject of meteorites and craters affect many scientific fields, and it is one of the more complicated subjects covered in the UM because there is so much pseudotheory entwined with actual observation.

Our knowledge of both ejectites and meteorites will no doubt increase considerably, as Millennial

Fig 7.10.26 - Astronauts probably never imagined that there was no real evidence that meteorites created the lunar craters as they were taught. Hundreds of rocks samples were collected from across the Moon and brought back to Earth for intensive study, which revealed that **no salt or even elemental chlorine** was found in any of the soils or samples collected (Bib 166 p140-1). However, salt was identified in some of the meteorites found on Earth. This means either the 'meteorites' with salt found on Earth are not from space, or meteorites with salt have not impacted the Moon—actually, both of these statements are true.

Scientists build on the extraordinary claims in the Meteorite Model. Many new books will be written to cover this one small subchapter, as investigators employ the use of the foundational UM principle; question everything with an open mind.

The last two subchapters of the Hydroplanet Model have prepared a scientific foundation for the next subchapter—The Arizona Hydrocrater. Using information gleaned from our previous study and discussion about both craters and meteorites, we will consider the most famous, and one of the most controversial craters in history. One thing you can be sure of; this road of discovery will take us through completely new territory.

7.11   The Arizona Hydrocrater

From Meteor Crater to Hydrocrater

Anyone who has watched history or science programs on television, or who has read popular science literature has surely found informative and graphically captivating presentations on asteroids, meteorites and impact craters. No doubt, there have been many discussions about the Arizona Meteor Crater and the words of Eugene Shoemaker, where he "confirmed beyond any doubt" that the Arizona crater was an impact crater. Why would anyone question something so seemingly absolute? If we have learned anything from the UM, it is that we should *question everything with an open mind*. If we are observant and if our knowledge base is built on truth, this attitude will lead us to new discovery.

The Arizona Crater is particularly special and has been given a subchapter all its own, partly because it was the first 'Meteor Crater' to have been "confirmed beyond any doubt," and because it is probably the most studied of all impact craters; it is touted as *the seminal example* of what an impact crater should look like.

Harvey H. Nininger, one of the first to have studied the crater, and a long-time scholar of craters and meteorites said this:

"The **impact** that the concept of meteorite craters must eventually have on **the stream of geological thought** is scarcely exceeded in magnitude by the physical impact which produced the crater itself." Bib 126 p145

He went on to note the status he thought the Arizona Crater ought to occupy in scientific history:

"The **place which the Arizona crater occupies** in these considerations is, and of necessity **must always be, most important. The birth and growth of the entire impact-crater idea** has coincided with the discovery and explorations of this great crater." Bib 126 p145

The Arizona crater has been considered *the* example of impact craters for *over half a century*. Those meteoriticists who studied the crater were absolutely sure of its meteoritic origin, prompting them to remark as Frederick C. Leonard of the University of California at Los Angeles did in 1945:

"'The amazing thing is, however, that **any geologist should still disbelieve in the meteoritic origin** of what is **undoubtedly the finest example of** a meteorite crater known on the earth!'" Bib 165 p334

It may seem preposterous to challenge such a long-held and universally accepted tenet; that the Arizona Crater is an impact crater. Nevertheless, the UM does, and for good reason!

The Arizona Crater is not an impact crater, it **is a hydrocrater**.

Others might also have come to this conclusion if they had had the influence of the Universal Model. However, they have not, and as we will see in later chapters, the well-established belief that this crater is an impact crater is a foundational pillar of modern science and a major underpinning for a variety of pseudotheories discussed in the Age Model, the Fossil Model, and the Evolution Pseudotheory Chapters. There is much at stake here, and modern science has either not seen the evidence or has purposefully swept away any conflicting evidence associated with this highly favored "impact" crater.

Having completed the Hydrocrater Model, the Crater Debate, and the Meteorite Model, we can bring the new evidence from those three subchapters together to re-examine the Arizona Crater, and prove whether it is *an impact crater or a hydrocrater*.

Throughout this chapter, the UM has shown that since the 1990's, astronomers and geologists have amazingly come to a consensus that almost *all* planetary craters, lunar craters, and many of Earth's craters are from "meteor impacts." This consensus has come without actually observing firsthand, an impact crater created by an actual high-speed impact event, although Hollywood has portrayed otherwise. If it is as the UM claims, and the Arizona Crater proves *not* to be an impact crater, *imagine the impact* this single UM claim will have on all of science.

Much of the basic information concerning the Arizona Impact Crater Pseudotheory has been gathered from www.barringer-crater.com, a website represented by the current owner of the crater. Many professional impactologists and other scientists have contributed to the content, and they continue to advise the owner of the crater. The website is a repository of that information and helped generate the following statistics, which can be found there.

Arizona Crater Brief History

The Arizona Crater is a mile-wide (1100 meters), 570 foot (174 meters) deep hole in the Arizona desert, 35 miles east of Flagstaff; its rim rising an average of 150 feet above the surrounding plane. As of this writing, scientists 'believe' a 150-foot wide, 300,000-ton iron-nickel meteorite traveling 40,000 miles per hour (18 km/sec) impacted the Earth some 50,000 years ago. The impactor vaporized, leaving, at most, 30 tons of iron-nickel that has been collected near the crater. Thus, 299,970 tons or 99.9% of the meteorite vanished, and no one has been able to explain where it went.

To understand the Arizona Crater, the most debated crater in the history of the world, and to discover its origin, we must begin by learning some of its recent history. We begin with the early scientific pioneers, many of whom spent a great portion of their lives studying and trying to comprehend the origin of this remarkable landform. Here are some of the primary historical characters that figure prominently in the crater's history; they will be introduced more fully as the Arizona Hydrocrater Model story unfolds:

Grove K. Gilbert – Chief geologist for USGS. He provided the first scientific study of the crater, (1890s).

Daniel M. Barringer – Mining engineer and businessman. He obtained mineral rights to the crater (early 1900s).

Harvey H. Nininger – Conducted the most intensive study of the crater and provided a museum for the crater (1942-60).

Eugene M. Shoemaker – Astrogeologist for the USGS. He was the primary researcher to champion the impact theory of the Arizona Crater (1960s-80s).

Originally, the Arizona Crater was known as Coon Mountain, then later as the **Canyon Diablo Crater.** Then in 1903, a mining engineer named Daniel Moreau Barringer suggested the crater had been made by a meteorite impact. Intent on finding his fortune, Barringer filed a mining claim on the crater site and formed the Standard Iron Company. He spent 27 years trying to find the large iron meteorite he believed had made the crater, and that must be lying beneath the crater floor. Barringer and his investors spent $600,000 (equivalent to more than $10 million in modern times), painstakingly drilling and searching for the valuable chunk of rock. With his funds exhausted, exploration ceased and a recent scientific report concluded that the meteorite would have vaporized upon impact. Within weeks, Barringer suffered a heart attack and died.

Today, most scientists see Barringer as a heroic pioneer in the trendy new field of impactology and still often cite the crater as the **Barringer Crater**, even though it is now known officially as **Meteor Crater** and is designated as such in most texts and presentations. Both of these names will be seen in the historical statements made about the crater, but will be commonly referred to as the **Arizona Crater** or the **Arizona Hydrocrater** in the UM text.

"Science" Backs Meteor Crater

No scientist has studied the Arizona Crater more than H. H. Nininger, who wrote the book, *Arizona's Meteorite Crater* in 1956. He spent two decades studying and living beside the crater. A good amount of Nininger's research is referred to in this subchapter. In his book, Nininger correctly recognized a fact also demonstrated in the UM Hydrocrater Model subchapter, that is; *no criteria exists* in modern science as to what characteristics are "considered final proof" of an impact crater:

"Even today, however, **no general agreement among scientists has been reached** as to just what characteristics **should be considered final proof** that a given feature has resulted from a collision of cosmic matter with our planet." Bib 126 p4

Nininger makes this statement right near the beginning of his book; even though he 'believed' the Arizona Crater was an impact crater. Nininger knew there were too many unanswered questions surrounding the crater to say that it was *unquestionably* of meteoritic origin.

Three decades earlier, in 1924, Daniel Barringer had made statements that were much more absolute (see Fig 7.11.1). These were totally unfounded and demonstrated Barringer's commercial mining interests outweighed his scientific views of the crater. This is further confirmed in a 1926 paper Barringer published in the *Engineering and Mining Journal*, wherein he asserts "the origin of the crater having been proved beyond all

Fig 7.11.1 – On page 10 of the July 1924 edition of *Scientific American*, Daniel M. Barringer stated the following: "Owing to the fact of my having been privileged to be the discoverer of the *origin of the Meteor Crater* in Arizona and to be able to **prove** that it was produced by the impact of a large mass of meteoric iron, very probably, as I have stated, in the form of a compact cluster of iron meteorites..." The editors of *Scientific American* added the following commentary: "For many years the consensus of scientific opinion rejected the suggestion that this was an "impact crater." Mr. Barringer, its owner, has now **proved incontestably** that it is such and seeks to extend the impact hypothesis to the moon's many craters." Of course, true science *must always be contestable* to demonstrate its truthfulness. The UM contests the impact theory of the Arizona Crater, and for the first time ever, introduces empirical evidence that will confirm that there was no impact and that the crater was formed by a steam explosion. The new Arizona Crater science will radically change how modern science views cratering processes on the Earth and elsewhere.

The Arizona Crater

> ### Science Backs Meteor Crater
>
> Because certain people, reluctant to believe the unprecedented, regard as sensational the theory that Meteor Crater was formed by the impact of a giant meteor which struck the earth, we have obtained the following definite statements from two well-known scientists:
>
> "I am perfectly willing to make a strong affirmative statement in support of Mr. Barringer's article," writes Dr. W. F. Magie of the Palmer Physical Laboratory, Princeton University, "but there ought to be no need for it. There is no reasonable doubt that the Crater was formed by the fall of a meteor and that this meteor is buried in it."
>
> Dr. Elihu Thomson, Director of the Thomson Laboratory of the General Electric Company, writes, "I am very willing to be quoted as follows: 'There can be no question of the Crater being made by masses of meteoric iron, and that an enormous mass of such iron remains buried under the south wall of the Crater.'"
>
> *The Editor.*

Fig 7.11.2 – This note from editor on p52 of the 1927 Vol. 137 edition of *Scientific American* is a perfect example of how modern science has tried to exert 'authority' to convince the public of its theories instead of backing claims with empirical evidence. This box should have been titled *Scientists Back Meteor Crater*—not "Science" and should stand as a reminder that scientists have gotten it wrong many times. We should always look to the evidence to back the theories and models, not just a give way to popular opinion. This subchapter will demonstrate that real science does not back the "Meteor Crater" claim. The two scientists in this article were wrong in thinking a large meteorite was buried beneath the crater and they were wrong that the crater was formed by the fall of a meteorite. Of course it took years to accept the truth that there was no meteorite buried in the crater just as it will take time for scientists to accept the UM evidence proving the crater was not made by impact at all, but by a hydrofountain.

question…" He went on to say:

"The impacting mass of meteoric iron, as has been stated in my final scientific paper on the subject, **has been located just where I predicted in a paper**, published in 1909, that it would be found; that is, **under the southern wall of the crater**." Note 7.11a

More concerned with proving his 'theory' than with finding the truth, Barringer, like many theoreticians, fell into a trap. Ultimately, he failed miserably and neither he nor anyone else ever found the large "impacting mass of meteoric iron" under the "southern wall of the crater" or anywhere else, as he had claimed.

There have been others who have expressed confidence that the Arizona Crater is of meteoritic origin. Robert. S. Dietz, one of the foremost proponents of the meteor crater notion, published the following in the *American Journal of Science*:

"The writer considers that the Barringer Crater (Meteor Crater) is **unquestionably of meteoritic origin** and that **criteria for astroblemes [impact craters] may validly be extrapolated from it**." Note 7.11b

Dietz proclaimed his absolute position, stating the crater is "unquestionably" of meteorite origin. His statement is a perfect example of how reason has been hindered and why modern science has made so many mistakes over the past 100 years. Perhaps scientists make statements like this for popularity or to appear more knowledgeable than they really are; whatever the reason, they confuse and *mislead* the public and other scientists, as well as science in general.

Later, in the ***same*** **journal article**, Dietz concludes his remarks about the 'meteor crater' by admitting, "The case is *not* yet proven":

"I concur with Bucher that traditional thinking should not be overlooked; we should not forgo the mundane in favor of the esoteric…I **believe** we stand today where we stood in 1930 relative to the acceptance of the **Barringer Crater as a bona fide meteorite crater**, following the writings of Barringer, Rogers, Spencer, and Blackwater. **The case is not yet proven** but the scales are tipped in its direction." Note 7.11b p663

If science is based on 'belief' it would be a religion; real science is about 'knowing' through reproducible and repeatable observations, things that are predictable. Dietz knew there were "many puzzling and unresolved aspects" of impact craters and that he was far from understanding "their true origin" as he continues:

"The writer agrees with Bucher that there are **many puzzling and unresolved aspects of cryptoexplosion structures [impact craters]** and that we are **still far from a definitive understanding of their true origin**." Note 7.11b

Both Shoemaker and Dietz were popular impactologists during the 1960s and exerted significant influence on the science of impactology by using their strong personalities and outspoken statements regarding impacts, particularly the Arizona 'impact'.

The Arizona Crater website states that it was Shoemaker's "landmark paper" that finally convinced the last doubters:

"In 1963, geologist Eugene Shoemaker published his landmark paper… His paper provided the clinching arguments in favor of an impact, **finally convincing the last doubters**." Note 7.11c

With the worldview apparently settled, science moved on to build whole theories and models based on the impact origin of the Arizona Crater. But there are still some threads of doubt, cracks in the foundation, as it were. In this chapter, direct evidence, some from Shoemaker's own papers, will show a very different picture. The science and the data, when viewed with an open mind, reveals a whole new Arizona Crater paradigm, complete with empirical evidence explaining why the Arizona Crater is *not* an impact crater, and establishing that it *is* a hydrocrater.

Evidences Against the Arizona Crater Impact Theory

The following is a partial list of the evidences against the Arizona Crater being formed by impact; each will be treated separately, later on:

1. Lack of impact glass.
2. Lack of melt-evident meteorites.
3. Lack of residual 'vaporized material.'
4. Presence of Widmanstätten pattern in meteorites.
5. Lack of shrapnel fragments.
6. Lack of crater imbedded non-vaporized meteorites.
7. Multiple iron sources require multiple impactors—and multiple craters.
8. No oblique strewn meteorite field.

9. The limestone in the crater unheated by high-speed impactor.
10. Absence of shatter cones.
11. The amount of iron at the crater is insufficient to produce the crater itself.

We will discuss how these evidences relate to *both* high-speed and low-speed impact theories. Crater investigators have never found compelling evidence to completely satisfy either of the two impact theories, and often jump back and forth between them; hence the reason to address both.

1. Lack of impact glass.

The Hydrocrater Model subchapter set forth the information needed to show that large meteorites retain most of their cosmic velocity. Geoscientists have known for a long time that the atmosphere slows down very few meteorites, especially those that weigh thousands of tons:

"Meteorites that weigh **hundreds or thousands of tons** are **not** retarded much by the atmosphere, **retaining a large fraction of their cosmic velocities.**" Bib 29 p53

The Arizona meteorite was estimated to be 300,000 tons, leaving no question that it fits the 'large' category; thus it would have retained its cosmic high-speed upon impact. One distinguished impact-cratering expert, Jay Melosh, explains how melted rock lines the interior crater of high-speed impacts in his book *Impact Cratering*:

"As the crater expands during excavation the shocked and **melted rocks line the interior of the growing crater cavity.**" Bib 161 p128

The melted rocks and glass **lining the interior of the crater** then "slide onto the floor":

"The last material to **slide onto the floor is the clast-rich melt rock that originally lined the transient crater rim.**" Bib 161 p128

In a mile-wide crater like the Arizona Crater, there should be ample melt-rocks and glass; at least thousands of tons! Where is it?

The Arizona Crater has no impact-glass.

Researchers know it should exist, and many Arizona Crater publications *imply* or state absolutely that there is, or was a large amount of glass at the crater. The trouble is it doesn't exist. There are no written accounts, no photographs or drill samples making evident the *meters of glass* that should exist, especially since 99.9% of the meteorite mass presumably vaporized. During an excursion to the crater, we asked the management where we could view glass found at the crater. There wasn't any, and the guide who had spent 25 years working and living at the crater had never seen any glass himself.

Handfuls of lechatelierite, a type of silica glass was found in a few scattered areas. These samples were cited as 'support' for the glass that researchers knew should be present. However, lechatelierite is also formed by lightning strikes, and it can be attributed to "old explosive volcanoes…" which it was in one researcher's account in the book, *Coon Mountain Controversies.* Bib 165 p337

The large amounts of impact glass are missing at the Arizona Crater, but great quantities of it are present at impact craters such as the Wabar site in Saudi Arabia, and at experimental high-speed impact test sites; both Wabar and the experimental

"The writer considers that the Barringer Crater (Meteor Crater) is **unquestionably of meteoritic origin…**" Then, the writer says in the same journal article: **"The case is not yet proven…"**

Robert S. Dietz

high-speed impact craters are *lined with surface glass.* Impactite—impact glass—***should*** be present *everywhere in the crater* if a high-speed impact had occurred, and it shouldn't be hard to miss.

But the researchers have missed this unassuming fact, and they have overlooked another simple observation; one that Melosh did note:

"**Solidified melt pools are not found at Meteor Crater** or Lonar Lake…" Bib 161 p127

Melosh reiterates further, the lack of melt at Meteor Crater in a 2005 article in *Nature*:

"This result may explain the **old observation that there appears to be much less melt in Meteor Crater than would be expected**…" Note 7.11d

Actually, there is almost no evidence of melt. To account for this, various sci-bis have been put forth, but none comes close to the real explanation. One theory is that "subsurface water" contributed to the absence of melt:

"**The absence of melt in these craters** may have been the result of a **subsurface water table** which **dispersed the melt by steam explosion.**" Bib 161 p127

H. J. Melosh was not the only one to imagine a "steam explosion" to account for the missing glass. William Graves Hoyt in the book, *Coon Mountain Controversies* states:

"As the meteorite penetrated deeper, it progressively lost energy and, reaching the Coconino sandstone and the water table, generated superheated **steam which exploded. The relative**

scarcity of silica glass, or lechatelierite, in the Coconino then was "due to the **reduced energy** of the meteorite and premature explosion, together with the cooking action of water volatilization." The absence of fusion products in the Moenkopi sandstone, above the Kaibab and on the surface, was due to the meteorite's **greater energy** at the moment of impact, which had been sufficient to volatilize rather than merely fuse the rock." Bib 165 p342-3

Unfortunately, Hoyt, like many impactologists, left reason lying in the wastebasket, trying to create logic where none existed. A massive, split-second explosion had presumably "reduced energy" while *at the same time* had "greater energy…sufficient to volatilize" the rock rather than melt it. Hoyt did get two things right, although they aren't connected in his account; first there is a "scarcity of glass" and second, heated water caused a steam explosion at the Arizona Crater.

If the melt had indeed been "dispersed" by *steam explosion,* water infused glass—enhydro impactite—should be evident, but they *are not*. The absence of melt is a perplexing characteristic of many other so-called terrestrial 'impact' craters. Did they all have subsurface water to 'disperse' the melt?

The newest explanation to account for the absence of melt comes from an article in *Nature* entitled, *Meteor Crater formed by low-velocity impact* (H. J. Melosh, G. S. Collins, *Nature*, Vol 434, 10 March 2005, p157). Surprisingly, researchers merely conclude that a computer modeled lower velocity of *12 km/sec* (8 miles/sec) instead of the previously assumed rate of *15 km/sec* (9 miles/sec) would account for absence of impact melt. Amazingly, the investigators neglect the fact that 12 km/sec is still hypervelocity, *well above* the velocities of the five laboratory tests presented in the Hydrocrater Model subchapter. All five of those tests showed large amounts of melt were produced at speeds slower than 12 km/sec. Reducing the velocity by 3 km/sec, from *15 to 12km/sec* does not account for the lack of melt.

Moreover, impactologists have said that 99.9% of the meteorite that created the crater "vaporized," leaving only 30 tons of the estimated 300,000-ton meteorite to be found.

> FQ: How does nearly 300,000 tons of iron-nickelmeteorite vaporize without melting a tremendous amount of contact rock?

The obvious answer to this FQ has led some impactologists to fall back to the position of a *low*-speed impact instead of a high-speed impact. But the physics *does not allow for this either*. The vaporization of almost 300,000 tons of metal **requires** high-speed, and the orbital dynamics of the solar system *require high speed too*. Large objects enter Earth's atmosphere at *a minimum velocity of 11.2 km/sec* and a maximum of 72.8 km/sec:

"The minimum encounter velocity between a planet and a meteoroid is the planet's escape velocity. The maximum is a combination of its escape velocity, heliocentric orbital velocity, and the velocity of an object just barely bound to the sun at the planet's orbital position. **For the earth the minimum velocity is 11.2 km/second and the maximum is 72.8 km/second.**" Bib 161 p205

How do you slow a massive meteorite down? Since the so-called meteorite did not break-up during its passage through the atmosphere (there would have been multiple craters if it had), it could have only been slowed by the atmosphere. Regrettably, this also fails the physics test. From the book *Impact Cratering*:

"When a meteoroid collides with the earth or some other planet with a substantial atmosphere it thus encounters atmospheric gases at high speed. Whether it will retain its high velocity after traversing the atmosphere depends upon its **mass or diameter and the thickness of the atmosphere**." Bib 16 p205

Impact Cratering gives the **minimum** diameter that a projectile would need to be to *avoid being slowed* by the Earth's atmosphere:

"…the **minimum diameter projectile** that can penetrate an atmosphere…" for an iron meteorite traveling through the Earth's atmosphere is, "**20.0 meters**". Bib 16 p205-6

The Meteorite that allegedly formed the Arizona Crater is said to have been *46 meters* (150 feet - barringercrater.com); over double the size of a meteorite that could be slowed by the Earth's atmosphere. Thus, the evidence shows, both empirically and quantitatively that *if* the Arizona Crater was made by an impact—it would have had to be a high-speed impact event, which would have formed a tremendous amount of melted impact glass; glass that does not exist.

2. Lack of melt-evident meteorites.

While at the Wabar crater, Shoemaker noted that about "**10 percent iron-nickel from the shock-melted meteorite** and 90

> Where are the **melted** rocks that should "line the interior of the…crater cavity" of the Arizona Crater?

SUBCHAPTER 7.11 THE ARIZONA HYDROCRATER

Where are the melted meteorites at the Arizona Crater?

Fig 7.11.3 - Researchers have painted themselves into a corner when it comes to the disappearance of the Arizona "meteorite." Over 99.9% of the iron that should be present from the meteorite that was supposed to have made the crater does not exist. If the high-speed impact had vaporized the meteorite, *most* of the remaining iron fragments should have been melted, or shown signs of melting, but such evidence *does not exist*. Seen above, an actual Arizona Crater iron is melted, producing round beads of metal melt of a type that should be evident.

percent local sand" Note 7.11e made up the large amount of glass melt at the impact crater in Saudi Arabia. Back at the Arizona Meteor Crater, the huge amount of melted indigenous rock that should be there is missing, and so are the *thousands of tons of melted meteorite* that should have been *in* the melted rock!

Shoemaker, conducting a modest 4-km/sec impact experiment using steel projectiles into sandstone surfaces, demonstrated that "frictional heating" offered the best explanation for the "fused steel" *on the chips* of the projectile:

"Thus, **frictional heating** appears to be adequate to account for the fused steel recovered in the experiment independently of the possible contribution of heat from the target sandstone. In particular, **frictional heating appears to offer the best explanation** for the local occurrence of **fused steel on the surfaces of the larger recovered chips of the projectile.**" Note 7.11f

Shoemaker's impact experiment also proved that melted iron "droplets" and "spheres" produced surfaces "like a welding bead":

"**Droplets and rivulets** occur locally on the striated surfaces. **Isolated spheres** generally have a smooth specular surface **like a welding bead**." Note 7.11f p675

Shoemaker's impact study produced fused steel as melted droplets on the surfaces of the steel projectile, all at much lower velocities than the assumptive rate of the Arizona Crater meteorite. Obviously, a higher velocity would have been required to vaporize and melt the Arizona Crater irons, and additional heat would have cause *even more meteorite melting*.

FQ: Where are the melted or partially melted Arizona Crater meteorites?

Thousands of iron-ejecta rocks have been recovered from the crater; none have been reported as being melted. By 'melted' we mean large iron beads, similar to those in Fig 7.11.3, of the type a welder might see when welding steel.

The widely read book, *Rocks From Space* cites the size a meteorite must be to survive impact:

"**Today we know** that meteorites weighing more than 100 tons cannot survive impact with Earth. Pieces may fragment off, but the main mass **will be vaporized**." Bib 29 p126

The theoretical Arizona Crater meteorite was supposed to be thousands of times larger than 100 tons; the physics of high-speed impact presumes vaporization of most of the meteorite. If 99.9% of the meteorite was vaporized, then certainly most, if not all the remaining meteorite fragments would show evidence of melting—however, they do not.

3. Lack of residual 'vaporized material.'

Over the last several decades, most researchers have come to recognize, because of impact physics and nuclear test explosions, that a crater the size of the Arizona crater could not have been created by a low-speed impact. To create such a hole, vaporization of

Fig 7.11.4 - The **Volcanic Spheroids** from the Arizona Crater are about a millimeter or two and smaller. Crater expert, H. Nininger estimated that between 4,000 and 8,000 tons of iron-nickel spheroids exist at the crater, the alleged remnants of the vaporized impactor. In one experiment, they were immersed and stored in distilled water for several years without *showing any significant rusting or deterioration*. Like other ejected volcanic material, these spheroids are resistant to erosion. This simple experiment demonstrates that these little Arizona Crater irons are not deteriorating at a rate that would account for the tens of thousands of missing tons of vaporized meteorite. Moreover, Nininger noted the spheroids were not found in random distribution at the crater but were conspicuously nonrandom, indicating they were not from a single impact explosion, but from multiple eruptions. A volcanic origin accounts for the nonrandom, multiple eruption distribution as well as the mysterious phosphorous and sulfur present in the spheroids. Volcanic does not necessarily imply melted, in fact, the spheroids' absence of rusting suggests they were not melted at all.

nearly the whole meteorite body is expected, resulting in hundreds-of-thousands of tons of vaporized meteorite. Where did it all go?

FQ: Where are the 299,970 tons of vaporized meteorite?

Vaporized metal cannot simply disappear like water vapor. Some scientists have come to believe that the 'vapor' from the meteorite coalesced into tiny droplets that fell to the ground. At first, this view seems to be supported because small, iron-nickel particles are found around the crater, but at nowhere near the volume to account for the vaporization of 300,000 tons of meteorite. The "spheroids" as Nininger called them can be seen in Fig 7.11.4. They are quite resistant to rusting and deterioration.

The evidence of this UM claim is that the numbers do not add up. Note the volume of the coalesced iron-nickel material, estimated by Nininger from data collected at 60 locations around the crater:

"**The measurement made in 60 locations** indicate that from **4,000 to 8,000 tons** of the **little spheroids** are now present in the upper four inches of soil within an average radius of two and one-half miles from the rim of the crater. Their presence has been demonstrated also at distances of six to seven miles north and northeast of the crater but in too small a quantity for satisfactory measurement." Bib 126 p91-3

Four to eight thousand tons is a far cry from the 300,000 tons of presumably vaporized meteorite. This important fact has been regularly dismissed by modern day researchers.

Some researchers have proposed that "erosion" carried most of it away. If so, to where was it carried? Iron sediment is twice as heavy as the surrounding silica sediment and it tends to clump together in streams (noted in the Rock Cycle Pseudotheory), leaving behind identifiable erosional patterns. Inspection of surrounding wash beds and the nearby Diablo Canyon, west of the crater produced no iron-nickel spheroids. Answers to account for the large amount of vaporized material have yet to be produced.

In 1999, computer modeling convinced investigators that *over* 80% of the impactor had been melted—vaporized too finely to be recovered, as they "infer":

"Although the impact modeling calculations for an impact velocity of 20 km s-1 show that **~80% of the Fe projectile melted**, the estimated mass of the Canyon Diablo **spheroids, 4000 to 7500 metric tons, accounts for less than 10% of the assumed meteoroid mass, ~100,000 metric tons**. We **infer** that the missing molten impactor material **became too finely dispersed to recover** during the expansion of shocked, volatile-rich target material (limestone) upon release from high pressure." Note 7.11g

Of course, science should be about proving—not inferring, but something had to account for the missing "molten impactor material;" it could not simply 'disappear.'

There are indications that the spheroids found at the crater are not from a *single* impact explosion. Nininger notes that the magnetic spheroid samples he collected from a 100 square-mile area produced a fallout pattern that was "by no means simple and uniform:

"I now began to inspect all of the **magnetic samples** and to gather many more. A lot of field work was necessary to establish their **distribution**, which I was eventually to learn **was by no means simple and uniform as was the distribution of volcanic ash**..." Bib 126 p89

The point Nininger said he eventually learned; that volcanic ash distribution was uniform, was not entirely correct. Like the spheroids, volcanic ash is *not* always subject to simple and uniform distribution. This would lead Nininger to a number of misconceptions. Impacts create a *single* explosion, a point in time, whereas volcanic vents and hydrofountains can have multiple eruptions from multiple vents, each producing a variety of ejecta, including sprays of material that leave discernable patterns and rays around the perimeter of the crater.

The ash around the crater actually renders some very interesting clues. Fig 7.11.5 contains ash collected near the crater, showing that the size of the ash decreases as distance increases. This is similar to the ash distribution from Mt. St. Helens, also shown in the diagram. The orientation of the ash in relation to the crater, and the fact that the ash contains nickel, identified

Fig 7.11.5 – These are photos of volcanic ash found near the Arizona Crater. Both piles of ash particles were collected northwest of the crater at 3 and 5 miles as noted. The ash obviously had high iron content, easily detectable with a magnet. Early researchers including Nininger noted that the ash *appeared* to be "like volcanic ash." This is because *it was* volcanic ash! Size and distribution of the particles follow a pattern similar to that of Mt. St. Helens, the greater the distance, the finer the particulate matter. Nininger had the Arizona Crater ash tested for content, which showed the material contained both iron and nickel. Although this seemed to confirm the ash included vaporized meteorite material, it could also mean the so-called meteorites came from a volcanic source. But that possibility was ignored, there seemed to be no reason to pursue it under the impact paradigm.

by independent lab testing, which suggest the crater is at least associated with a volcanic origin.

Nininger, like others in the scientific community knew there were no large "molten drops or bombs" around Arizona Crater and he wondered why:

"After Spencer's announcement of the occurrence of silica-glass in connection with the Australian [Henbury] and Arabian [Wabar] craters, the question arose among American scientists as to **why no similar scattering of molten drops or bombs had occurred at the Arizona crater**, and if it had, **why no specimens had been recovered**." Bib 126 p118

It turns out that the reason why no specimens had been found is because they were *so small* and they were *buried*. Nininger made a complete inspection of the crater and found what he called "clinkers" and other small pieces of nickel-infused material that to him seemed to be "volcanic cinders":

"Next morning an early trip was made to the southeastern rim where the '**clinkers**' were collected and inspected. Some of these were without doubt the same as the particles found in the **gravel pit**, but others seemed **just as certainly volcanic cinders**." Bib 126 p120

Nininger thought these particles were volcanic because they looked like volcanic cinders and had all the properties of volcanic material—yet they had nickel *infused* in them. This seemed to prove the impact origin of the crater, which Nininger had been looking for. Indeed, nearly all the miniature clinkers tested included nickel, but he was not acquainted with the terrestrial occurrences of nickel ore, and had no reason to suspect that these were of any other origin than impact, yet Nininger "could not escape the feeling that perhaps some of what "appeared like volcanic ash" was "of purely terrestrial origin":

"In spite of this very strong evidence, some of the specimens we picked up on the rim were not bomb-like. They were merely vesicular fragments that in all of their outward features appeared like **volcanic ash. I could not escape the feeling** that perhaps some of these **were of purely terrestrial origin**." Bib 126 p121

Instead of the unproven vaporized-meteorite theory, there is another explanation that can easily account for the presence of the small iron-nickel particulates. A hydrofountain could have easily carried both small and great iron-nickel ejectites to the surface. There are many examples of hydrocraters in the area surrounding the Arizona Crater that exhibit similar ejecta material patterns. Moreover, there is no minimum or maximum amount of material to account for from a hydrofountain, as there would be with a meteorite impact scenario. Multiple eruptions account for the *uneven and nonrandom manner* in which the spheroids are found.

Another observational fact Nininger overlooked was the *chemical difference between* the spheroids and the irons distributed around the crater. Although he noted that the spheroids contained 0.56% phosphorus and 0.96% sulfur, Nininger neglected to emphasize that the large **irons at the crater did not** contain these elements, even though he included the makeup of both the spheroids and irons on page 88 of his book, *Arizona Meteorite Crater* (Bib 126). This would prove to be a major oversight.

FQ: Why did the spheroids have phosphorus and sulfur?

Sulfur is commonly found in *stony* meteorites, yet Nininger reports that the *crater's* **irons have no sulfur or phosphorus**, even down to the level of *parts-per-million*. Thus, if the spheroids did not get these elements from the iron 'meteorite,' where did they come from? The crater areas' sandstone and limestone have these elements in such minor quantity that no more than 1.5% of the spheroids could have derived it from there.

The only way the spheroids could contain phosphorus and sulfur, and the irons not, is if they came from a different source. The source might have been similar, but certainly could not have been identical.

The source was actually various vents in the diatreme of the crater. Volcanic craters have long been known to eject both phosphoric and sulfuric emissions and minerals, recognizable as a rotten-egg smell present at volcanically active centers, such as Yellowstone National Park. Both phosphorus and sulfur are byproducts of microbial waste and are directly related to many of the metallic ores, something we will learn more about later.

There is one additional piece of evidence pertaining to the origin of the spheroids, their silica content. If the spheroids precipitated from the post-impact vapor cloud, they would be made of the material vaporized during the explosion, that is, metal from the meteorite and silica from the sandstone it impacted.

FQ: How much silica do the spheroids contain?

According to Nininger, the spheroids have no silica! It seems impossible that a high-speed impact would vaporize thousands of tons of meteorite without vaporizing the silica sandstone too. Therefore, we see, the Arizona Crater has no glass, a characteristic of high-speed impact, and the spheroids present have no silica.

There are four compelling reasons that the crater's spheroids are of a hydrocrater origin rather than precipitates of an impact vapor cloud:

1. Over 90% of the spheroids are missing if they came from the vaporization of a meteorite impactor.
2. The distribution of the spheroids is not uniform, as if from an impact explosion, but is uneven and nonrandom because of multiple hydrocrater eruptions.
3. The sulfur and phosphorus in the spheroids did not come from the irons or the surface rock, but rather from hydro-volcanic sources.
4. The spheroids are devoid of silica, highly improbable if they had precipitated from an impact vaporization explosion.

4. Presence of Widmanstätten pattern in meteorites.

The Meteorite Model of the previous subchapter verified that

> "...in all of their outward features appeared like **volcanic ash. I could not escape the feeling** that perhaps some of these **were of purely terrestrial origin**."

If Arizona Crater irons like this specimen were involved in a high-speed impact they would have been subjected to heat over 800° C, which would have destroyed the Widmanstätten crystalline pattern.

Fig 7.11.6 – This Arizona Crater iron (Meteor Crater, or Canyon Diablo Meteorite) has been sliced and etched to reveal its interior crystalline Widmanstätten pattern. If irons like these were really fragments of a meteorite that had been 99.9% vaporized, *the vast majority of the fragments* would have been subjected to tremendous heat, well above 800° C, which would have *destroyed* the Widmanstätten pattern. This sample and thousands of others like it show very little evidence of heat alteration to the crystal pattern.

the Widmanstätten pattern disappears at only 800 degrees Celsius, whereas it takes several thousand degrees to vaporize the metal. If 99.9% of the Arizona meteorite was vaporized, then the remaining pieces of the meteorite should have been melted, or at the very least, the Widmanstätten pattern should have been destroyed.

The Arizona crater impact would have been so violent that the surrounding rock would have been melted for hours and extraordinarily hot for many days. The greater the mass of the heated object, the longer it would take to cool down.

Even smaller pieces of ejected iron would have been subjected to substantial heat over a period of time sufficient to destroy the Widmanstätten pattern—yet no evidence of this has been presented. The crystalline Widmanstätten pattern would certainly be affected on smaller, unbroken specimens; complete with regmaglypts, but this too, is not the case. Nininger comments on another researcher's (Moulton) theory:

"It is **inconceivable that a swarm** such as he depicted **could have given us the several concentrations** of large and small irons two to four miles from the crater, **none of which show heat alterations**.

"Again, the explosion which Moulton believed formed the crater and scattered the fragments over the plains **should have** deposited only heat-altered fragments. Especially small masses could not possibly have come through this ordeal without being heated above the gamma-alpha transformation temperature. But, as we have pointed out, the irons found on the plains away from the crater, whether large or small, show good Widmanstätten figures, which means that they have not been subjected to a temperature much above 800° C." Bib 126 p70

Widmanstätten patterns are crystalline in nature, unequivocal evidence of a hyprethermal growth environment, as explained in the Meteorite Model. The Arizona Crater 'meteorites' are simply iron ejectites that grew deep in the pressurized environment of the hydrocrater diatreme; ejected from the crater by a phreatic explosion.

Further evidence discounting the high-speed, high temperature impact-vaporization event was shown by Nininger's own laboratory experiments; artificially heated irons would "*entirely disintegrate within a few years*":

"Laboratory experiments have **proven** that application of heat to surviving Canyon Diablo iron **speeds up the process of oxidation** so that some which have survived the weather through thousands of years since landing may **entirely disintegrate within a few years after being artificially heated to 700° to 800° C**." Bib 126 p78

Since nearly all of the recovered irons were found either on or near the surface, and since they were exposed to weathering, there should have been a great many disintegrated specimens.

FQ: How many disintegrated or partly disintegrated irons were found at the Arizona Crater?

There were none, which demonstrates that there was no high-speed impact. The intense heat of the vaporization impact would have destroyed the Widmanstätten pattern in *thousands* of the small irons found around the crater, and *thousands* of disintegrated, or partly disintegrated irons would have been recovered at the crater site.

5. Lack of shrapnel fragments.

To account for the lack of melt, some impactologists have theorized that the Arizona Crater impact was *not* a high-speed impact (as Melosh and Collins claim in *Nature* 10 March 2005). This however, is quite impossible because of the crater's size; it is far too large to have been formed by the small amount of iron material found outside the crater. For a good comparison, the amount of iron found at the Crater is analogous to shooting a rifle into the ground to create a swimming pool—a literal impossibility. (For the curious, a rifle-bullet will produce 2" crater in dry, consolidated sediment.)

Because the physics and the geology at the Arizona Crater disagree with accepted theories, researchers must return to the drawing board. Only high-speed impact could account for 99.9% vaporization of the impactor and a crater of this size, but there is insufficient impactite melt, there are no melted meteorites, and the Widmanstätten pattern still manifests in existing meteorites when it should have been erased by the intense heat. These facts clearly indicate that there was no high-speed impact.

However, if researchers abandon a high-speed impact in favor of a *slower* speed impact, they must account for the *hundreds of thousands of tons of missing iron* and they must account for the absence *of fragmented meteorite shrapnel*.

The Russian Sikhote-Alin fall of 1947 showed how much metal remained following a low velocity impact. The observed Sikhote-Alin meteorite impact and cratering proved to be a productive field for studying lower speed impacts of iron-nickel meteorites in the 50-100 ton range. The evidence proved that slow-speed impacts of this mass range do not produce high heat

A slow-speed impact would have produced many broken fragments and shrapnel, like these Sihkote-Alin specimens.

Fig 7.11.7 - These specimens are shrapnel fragments of the Russian Sikhote-Alin meteorite that fell in 1947. If the irons from the Arizona Crater were from a low-speed impact, there would be **many** that exhibited this type of shrapnel appearance—but there are not.

nor melt. The energy of the impact reduced the meteorites to shrapnel when it neared the ground because the mass was insufficient to keep the incoming speed above terminal velocity. The larger mass of the Arizona Crater meteorite would have experienced much less atmospheric braking and the higher velocity would have yielded more fragments, but a century of investigation has failed to support this notion.

Early in the investigation of the crater, researchers such as Edwin E. Howell noted that the iron rocks found in proximity of the crater were *not* broken fragments:

"Howell had examined many of the meteoritic irons, concluding that each was a 'complete individual' and that, while all were found together in a relatively small area, **apparently none had been broken off from a larger iron mass**." Bib 165 p52

Anyone who studies Arizona Crater irons will notice the apparent rarity of broken irons. If the irons were neither *broken* nor melted, how could they be fragments "from a larger iron mass"? Perhaps researchers are confused because the impact theory was not questioned, as it should have been. The original account of the observation comes from G. K. Gilbert's 1896 report in which he states:

"Our fellow-member, Mr. Edwin E. Howell, through whose hands much of the meteoric iron has passed, points out that **each of the iron masses, great and small, is in itself a complete individual**. They have **none** of the characters that would be found if they had been **broken one from another**, and yet, as they are all of one type and all reached the earth within a small district, it **must be supposed that they were originally connected in some way**." Note 7.11h

The assumption made by Gilbert (that the irons were connected) over a century ago is widely accepted today. However, as the Meteor Model demonstrated, there is no evidence to support this assumption, and in fact, it has been verified that all the surface regmaglypts on the recovered irons could *not* have been made by impact or by passage through the atmosphere.

It is understandable that over a hundred years ago Gilbert was not familiar with the variety of **nodules** that exist in nature, but modern science has added much to the subject; observations that include the discovery of vast quantities of manganese nodules on the seafloor. These nodules were not "originally connected" to a parent metal-body, but **grew from solution** as all nodules do. In the Meteor Model, we saw metal and quartz nodules whose formation has never been explained by geology, unknown because the geological past was not like the present, contrary to the uniformity myth. These nodules, and the Arizona Crater irons share commonality in their origin, the knowledge of which will be greatly enhanced by the end of this and the next chapter, as the Universal Flood event is established.

6. Lack of crater embedded non-vaporized meteorites.

There are a number of reasons the Arizona Crater could not be the result of a low-speed impact. One of those reasons is that meteorite fragments were not found *in the crater itself*. The Sikhote-Alin impact in Russia and the Kirin impact in China were both impacts from meteorites of slower velocities, both had a large number of meteorites, both impacts left craters, but exhibited neither surface melt nor significant melting of the meteorites.

FQ: How many meteorites were recovered **inside** the Arizona Crater?

Answer: None.

However, the data from the Russian and Chinese observed falls confirmed that a large portion of the recovered meteorite material came from **within** the craters, contrary to that found at the Arizona Crater. Thousands of irons have been found *outside* the Arizona crater, but there is *no* documentation of *any* irons being recovered **within** the Arizona Crater.

Observed low-speed impact events have numerous meteorites embedded in their respective craters, but the Arizona Crater has none, reported in the first scientific paper on the Arizona Crater from 1896:

"**No iron has been found within the crater**, but a great number of fragments were obtained from the outer slopes where they rested on the mantle of loose blocks. Many others were obtained from the plain within the region of scattered debris, and others, though a smaller number, from the outer plain. One **large piece** was discovered **eight miles east of the crater, or almost twice as distant as any fragments of the ejected limestone**." Note 7.11h p8

Fig 7.11.8 illustrates the iron-ejecta are **not** *randomly distributed* around the crater. If the Arizona Crater had been from impact, a single explosion and a large quantity of ejecta should have produced a randomly distributed ejecta pattern, where generally heavier rocks would fall closer to the crater than smaller rocks. Curiously, the largest recovered piece of meteorite was discovered eight miles from the crater, farther than any of the smaller pieces. There was also no 'ray' of ejecta material in the direction of this specimen's point of discovery.

Although the impact theory cannot account for this anomaly, volcanoes and hydrofountains can eject large rocks many miles from the mouth of the explosion. Such would be the rocks being shot from the diatreme during a pressurized steam eruption.

The non-random pattern on Holsinger's map confirms that the Arizona Crater was not formed from a single, random impact explosion, and the absence of meteorites *in* the crater confirms that the crater was not a low-speed impact crater.

If the impact theory were expanded to include an event with multiple meteorite impactors, we would soon discover that for

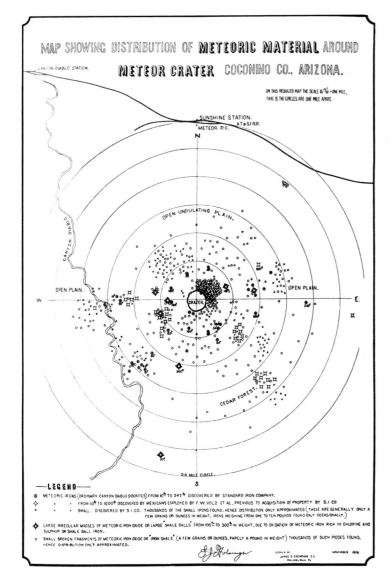

Fig 7.11.8 – The Holsinger Map was created in 1908 and remains the best document for showing the distribution of irons around the Arizona Crater. There is clearly **no elliptical strewn field** that would have existed had a giant meteorite impacted the area. The regmaglypt evidence presented in The Meteorite Model also shows that the Arizona irons did not come from a single body nor were they broken off or melted from another meteorite body. This leaves only one plausible explanation—the irons were formed individually, beneath the crater in a diatreme, and then later blown out of the crater. This explains why no irons were found in the crater. **Multiple** eruptions explain, for the first time why the irons cluster in groups around the crater in a radial pattern. Only a hydrocrater origin can account for all of the features of the crater and its iron ejecta.

this to be valid, there would need to be an elliptical debris pattern, not a circular pattern. Further, multiple meteorites would have produced multiple craters. Neither of these is present at the Arizona Crater site.

High-speed impact or low-speed impact; neither works. If it had been a low-speed event, nearly all of the meteorite fragments would have been found *in* the crater. If it had been a high-speed impact, some of the meteorite would have fused with the surrounding rock. Melt-rock, like that at the Wabar impact site would be abundant; but it is not, because it is not an impact crater at all!

7. Multiple iron sources require multiple impactors—and multiple craters.

From Barringer's day to the present, every Arizona Crater researcher has had to acknowledge there are "problems" with the distribution of irons around the Arizona Crater. Nininger in his book, *Arizona Meteorite Crater,* discusses this under the heading, *Problems of the Northeast Rim Concentration*:

"The Holsinger map published by Barringer in 1909 [Fig 7.11.8] had shown a **great concentration of small meteorites on the northeast rim of the crater which had never been accounted for**. We felt that there **must be** some explanation of this concentration." Bib 126 p62

Holsinger's map in Fig 7.11.8 does show a large concentration of irons on the northeast rim, just as Nininger noted. The ejecta pattern apparent on the map poses problems for a high-speed impact explosion because the ejecta should be *randomly* distributed around the crater. Regardless, Holsinger's map clearly shows a *non-random pattern,* with pockets of irons distributed at various locales around the crater. Particularly troubling to researchers was the large concentration near the northeast corner of the rim. Nininger set out to find an explanation for this but only found more trouble. He noted that treasure hunters had exploited the area for many years. Despite the fact that the area had been well picked over, some 6,000 specimens were collected during 1947 and 1948 from northeast area of the crater. The recovered irons weighed up to 17 pounds with the average being approximately four ounces.

After cutting and analyzing a number of the specimens, Nininger made the following comment:

"No one who has had considerable experience in etching and studying meteorites **can possibly conceive of these irons as having come from the same parental mass**, any more than one could conceive of an elm leaf and an oak leaf coming from the same tree." Bib 126 p64

It was an insightful comment. Not only did Nininger recognize the 'distribution problem' of irons around the crater, a problem he was never able to solve, he realized different "parental" masses had birthed different pockets of irons, located randomly around the crater. So now we ask the following FQ:

Are there multiple craters at the Arizona Crater site?

There has never been a simple or clear answer for these two problems, but one fact was clear, the various clusters of irons were *not associated with* **multiple** *impact craters*! The Russian Sikhote-Alin and other observed falls had produced *multiple impact craters* with clusters of meteorites associated with each crater. Nothing similar existed at the Arizona Crater, but that did not prevent Nininger and others from making the claim that several meteorites fell, although he does so without convincing evidence:

"It then seems fair and reasonable to **assume** that a group of several meteorites encountered the earth at the time the crater was formed. In other words, this has been a *composite* fall consisting of an *undetermined number* of *components*." Bib 126 p65

As we have already established, the thousands of irons found around the crater generally show no signs of being melted or broken, and the original researchers identified evidence that the irons were of different original masses. Having determined that they did not come from a parent body does not necessarily establish that there were multiple impactors.

One of the primary facts that refutes the multiple impactor theory is the absence of other craters; swarms of individual meteorites do not all land in one crater. The atmosphere would have perturbed their trajectory, spreading them to create multiple craters. From the earliest investigations of the crater site, Barringer and others found the irons were always on or very near the surface:

"On the distribution of meteoritic material, Hager quoted Barringer's 1909 paper, italicizing passages pertinent to his point that the ordinary Canyon Diablo irons, at least, were **always found 'on or very near the surface,' a fact that Barringer had been unable to explain**." Bib 165 p338

Lacking multiple craters, a low-speed impact had no validity and Barringer knew it. Because there was little glass found at the crater, modern researchers turned to a "low-velocity" impact theory to account for the lack of melt. However, this idea also fails when we realize that this theory requires a "swarm" of meteorites as Nininger had long ago theorized but failed to account for. Melosh notes in his 2005 paper on the crater:

"Our proposal that Meteor Crater was created by a dispersed **swarm of low-velocity iron fragments** (many of which were dispersed beyond the central debris cloud) is also consistent with the recovery of **large numbers of small, unmelted iron-meteorite fragments near the crater**." Note 7.11i

This theory does not account for:
1. The lack of *multiple* craters if there had been multiple meteorites.
2. The lack of iron meteorites *in* the crater that have been observed at contemporaneous falls.
3. The mechanism to account for the *non*-random distribution of the irons.
4. The absence of a *strewn field* (discussed next) from a meteorite swarm.

Viewing the Arizona Crater as a Hydrocrater *answers all of these questions* and many more. By simply understanding the process of mineral growth, including iron ejectites in a hypretherm, and by understanding that enormous quantities of material was ejected from multiple hydrofountains, the evidence at the Arizona Hydrocrater suddenly makes sense.

8. No elliptical meteorite strewn field.

The Homestead Meteorite broke the placid evening skies in Iowa in February of 1875, leaving behind the Homestead Meteorite Strewn Field illustrated in Fig 7.11.10. A strewn field is an elliptical pattern of debris left behind on the surface when an object or a number of objects traveling together in the atmosphere break up prior to impact. All large *observed* mete-

Fig 7.11.9 – The Holsinger iron, the largest Arizona Crater specimen is about a meter in length. It was found several miles from the crater and is housed today at the onsite Arizona Crater museum. All ***observed*** falls of similarly sized meteorites produced deep craters, usually several meters wide. How did this piece and other large irons like it come from space without forming craters? The logical answer is that they did not come from space at all, but came from within the Earth, ejected during a low-speed, steam explosion event.

orite falls have produced a strewn field similar to the Homestead strewn field. If the Arizona irons came from a low-speed meteorite, they too would have produced a similar strewn field. However, Nininger identifies no such field at the crater:

"First, we should **definitely expect such a swarm** to lose its outermost members to aerial friction before it struck and such of these as were not consumed **would form an extension of the strewn field** in the direction from which the meteorite approached. **Actually, no such extension of the strewn field has been found on any side of the pit**. Remember, we are here considering the plains specimens beyond the rim." Bib 126 p69

Another modern meteoritic scholar confirms "the typical elliptical strewn field expected for multiple falls" was not observed at this site:

"Other studies concentrated on the distribution of meteorites on the plains surrounding the crater, **the typical elliptical strewn field expected for multiple falls was not seen here**." Bib 29 p123

FQ: Is there a strewn field at the Arizona Crater?

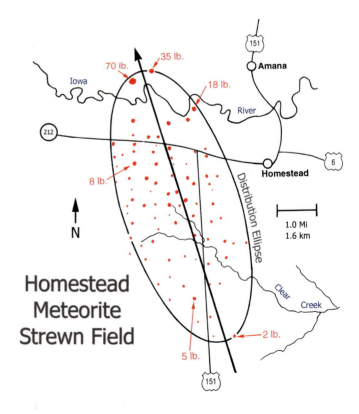

Fig 7.11.10 – This is an example of a strewn-field, an elliptical distribution of impact debris left behind when large meteorites land on Earth. Every large-sized *observed meteorite fall* has left behind similar distributions of materials, which follows the laws of physics. If there were multiple meteorites in the Arizona desert, as most investigators believe, why is there no strewn-field at the Arizona Crater? Adapted from *Rocks From Space*, Bib 29, p71.

CHAPTER 7 THE HYDROPLANET MODEL

Although there is no strewn field at the Arizona Crater, Nininger noted there was clear evidence of *multiple* iron bodies! Nininger wrestled with this conundrum for a long time, giving much thought to how an impactor might leave behind a crater such as this while accounting for all the physical observations of the crater and the surrounding irons. He articulated his own impact theory where he believed: "A large nickel-iron mass of a few hundred thousand tons" was accompanied "by a small number of lesser bodies…" Bib 126 p74 He envisioned that these meteorites were traveling between "5 to 10 miles per second" Bib 126 p75 and described the explosion as follows:

"As the body of the meteorite bored into the earth, the escaping compressed air, steam, and other gases **torched away the surface of the mass** in a sheath of liquid and vaporous flame. At the same time, shock waves, racing forward into the "immovable object" and backward through the "irresistible force," were building up to a grand finale in the form of a dual explosion. As it was forced to stop, the fatally decelerated invader from space and the compressed sediments with their trapped **super-heated steam exploded** with the violence of a **hydrogen bomb**." Bib 126 p76

Nininger's theory included a *steam explosion*, a remarkable step in the right direction, one that some modern impact researchers have continued to support. Unfortunately, his other

Limestone on the Arizona Crater Rim

Iron Nodules

Quartz Geodes

Instead of Limestone powder or glass caused from impact heating, iron nodules and quartz geodes line the limestone at the crater.

Fig 7.11.11 – The lighter colored deposits seen here are from the limestone layer that existed on the surface of the Arizona plain before the crater explosion took place. Compare their appearance to the heated limestone in Fig 7.11.12. If this limestone had been subjected to the intense heat of a hypervelocity impact, there would be massive amounts of powdered limestone and glassy deposits would abound. A hike around the crater's rim revealed no evidence of limestone heating *anywhere*. Instead of finding heated or melted rocks, there was ample direct evidence that the limestone and sandstone surface rocks at the Arizona Crater had been shaped by water. They compared favorably to the rocks at hydrocraters such as the Colorado and Montezuma craters, which were also unheated sandstone and limestone. Iron nodules and quartz geodes were also found, which are evidence of an earlier hypretherm, preceding the steam explosion. This event will be discussed in the next chapter.

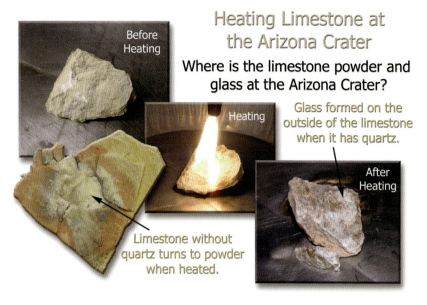

Heating Limestone at the Arizona Crater

Where is the limestone powder and glass at the Arizona Crater?

Before Heating

Heating

Glass formed on the outside of the limestone when it has quartz.

After Heating

Limestone without quartz turns to powder when heated.

Fig 7.11.12 – Arizona Crater researcher H. Nininger observed that when ordinary limestone is heated, it crumbles and turns to powder. If the limestone has quartz (silica) mixed in with it, during heating, the quartz turns to glass forming a protective layer, resistant to erosion. Both of these observations are reproduced here. Neither limestone powder nor glass encased limestone is found at the Arizona Crater, refuting the claim that the crater was made by high-speed impact.

ideas were still tied to impact. Nininger's multiple impactor theory never explained why the Arizona Crater had no multiple craters, a characteristic of all large *observed* falls with multiple meteorites.

Nininger's speculated velocity of 7.5 miles-per-second (12 km/sec) is high-speed, and his further reference to a "hydrogen bomb" and vaporous flames torching "away the surface of the mass," are fanciful allusions of tremendous heat and melting, of which there is little evidence.

9. Limestone at the crater has not been heated.

The majority of the sediments that comprise the Arizona Crater strata are sandstones, but the top layer is the Kaibab formation, a layer of limestone and dolomite approximately 270 feet thick, represented in Fig 7.11.11.

When limestone is heated above approximately 1000° C, it turns to *powder*, seen in Fig 7.11.12. Nininger performed this simple test himself and found that the Kaibab formation limestone "began to crumble and fall apart" within only a few days after heating:

"**Fragments of the Kaibab formation** were subjected to an **oxyacetylene flame**" for "120 seconds... in a few days these products began to **crumble and fall apart**." Bib 126 p129

FQ: Is the limestone at the Arizona Crater powdered?

Some of the limestone around the crater is *siliceous* limestone, carbonate layers formed when ocean waters covered the area, comprised of varying amounts of limestone and silica, or quartz. Fig 7.11.12 is siliceous limestone from the crater being subjected to heat. Silica in the limestone becomes glassy, forming a glaze that holds the limestone together. Non-siliceous limestone turns to powder when heated, also seen in Fig 7.11.12.

FQ: Where is the glazed siliceous limestone, glazed from high temperatures at the Arizona Crater?

10. Absence of shatter cones.

Shatter cones were first introduced in the Hydrocrater Model subchapter. There we explained that researchers have "argued that they occur **only** in impact craters":

"The **first reliable geological criterion** for the recognition of impact structures in the absence of meteorites was established by R. S. Dietz... **Shatter cones** had first been recognized by Branca and Faas at Steinheim in 1905, but **Dietz argued that they occur only in impact craters**..." Bib 161 p8

The emphasis being on "only" because an impact crater can only be identified if there is "geological criterion" that occurs **only** in impact craters. Shatter cones are supposed to be one of those criteria. The author of the book *Rocks from Space* tones down the absoluteness of this criterion by saying the presence of shatter cones "**helps authenticate** meteorite craters":

"There is one more shock-related feature that **helps authenticate meteoritic craters**. Curious cone-shaped rocks called **shatter cones** are **frequently found** around many old and weathered impact craters." Bib 29 p128

Why tone down the importance of shatter cones if, as Dietz claims, "they occur only in impact craters"? The answer comes from the next page in the book, *Rocks from Space*:

"**Curiously, no** well-formed **shatter cones** have been found at **Meteor Crater**." Bib 29 p129

The icon of the world's impact craters, Meteor Crater, is lacking a significant impact crater indicator. Indeed, this is curious, and is a fundamental question that needs an answer!

FQ: Why are there **no shatter cones** at the Arizona Crater?

The "best preserved example of a meteorite impact in the world" is missing "the first reliable geological criterion." There is no good answer to this FQ, except to simply recognize that the crater was not made by impact, but nobody wants to admit that!

Another significant criterion—coesite—was not found where it should have been in the sediment of the Arizona Crater, but coesite was discovered in 1961 "as an inclusion in the diamonds" Bib 119 p128 in the iron ejectites (meteorites) at the crater. However, just as we established in the Hydrocrater Model subchapter that shatter cones are found in natural geological settings (diatremes) not related to impact events, so too, is

---

"The first reliable geological criterion for the recognition of impact structures..." was **shatter cones**.

"...they occur **only** in impact craters..."

Then why does the Arizona Crater have **no** shatter cones?

CHAPTER 7 THE HYDROPLANET MODEL

coesite. The **absence** of coesite in the crater sediments solidly refutes the impact theory whereas the **presence** of coesite in the diamonds in the iron ejectites reliably substantiates a diatreme-hydrocrater origin. Coesite is now widely documented in diamonds found in hydrocrater diatremes around the world.

11. The amount of iron at the crater is insufficient to produce the crater itself.

In 1963, Eugene Shoemaker performed a high-speed steel-projectile experiment into the Coconino sandstone of the Arizona crater. More than half of the mass of the projectile was easily recovered from the debris:

"Slightly **more than half of the original mass of the projectile was recovered** from the coarse fractions of the debris by extraction with a hand magnet. The bulk of the remainder is present partly as free steel particles in the fine fractions and partly as impregnations in crushed sandstone aggregates." Note 7.11j

This statement is probably the source of the claim shown on the diagram at the crater's museum, seen in Fig 7.11.13. The caption states that about half of the 'meteorite' was "blasted out upon impact, raining tiny fragments on the rim of the crater and the surrounding plain." Left unsaid is the fact that only *several* thousand tons have been recovered of the *tens-of-thousands* of tons that should exist. The missing material has yet to be accounted for, and has actually never been scientifically proven to exist.

The *other half* is supposed to be "present in very small to microscopic iron-nickel spherules and fragments in the breccia lens beneath the crater floor to a depth of **3,000 feet**".

One reason modern science needs to be replaced with millennial science is terrific statements like this, common in modern science. This should never have been made because it is based wholly on inference—not statements of fact. The deepest drilling project penetrated only 800 feet beneath the crater floor, making it impossible for researchers to know if "microscopic iron-nickel fragments" exist below this level.

Museum claims should be based on demonstrable facts and reliable data. We have been misled by modern science far too long.

The Arizona Hydrocrater Evidences

For over a century, researchers have searched in vain for the 'smoking gun' that would prove a craters' impact origin. The insurmountable problem for researchers is that *no one* has ever directly observed a **large-sized** natural occurring high-speed impact, and so, researchers cannot be sure what one should look like or what it should consist of. Whether an explosion that forms large craters occurs below ground or above ground, the resulting crater is quite similar, and this has been the cause for a great deal of confusion over the last 100 years.

Another cause for error came from the space exploration program and the related enthusiasm that raged during the 1960s and 1970s. Many scientists had a great desire to find something from space right here on Earth and since rocks had been known to fall from space, it was only natural to expect excitement over each new fall and every impact crater that might be the result of a fall, whether new or of ancient date. The excitement should not have negated the scientific scrutiny that needed to take place, but it was so by many impactologists.

Errors in the classifying of many craters as impact craters rather than hydrocraters have occurred primarily because geology does not correctly understand the Hydroplanet Earth. Even though there are mountains of evidence that the Arizona Crater is not an impact crater, the most compelling evidence, strong enough to convince even the most skeptical reader is yet to come. This empirical evidence is not new and is taken from the pages of geological journals, facts already known, *yet blatantly ignored*. It is the evidence of the Arizona **Hydrocrater**.

Understanding why crater investigators have been stumped for so long may be difficult to comprehend until we realize that without the evidence presented in the Magma Pseudotheory, science remained unaware of how heat is generated inside the Earth. The significance of friction in fault lines and water under pressure does not seem important until we read of the Crystallization Process and begin to comprehend how minerals are made. It is with these new Natural Laws that we can comprehend how the diatreme hypretherm is essential in the formation of ejectites, some of which are known as meteorites.

At the Arizona Meteor Crater museum, you will hear all about impacts and multimedia diagrams will immerse you in the power of a giant impact. But, there is another crater of similar size and shape of the Arizona Meteor

> "Curiously, **no** well-formed **shatter cones** have been found at **Meteor Crater**."
>
> O. Richard Norton

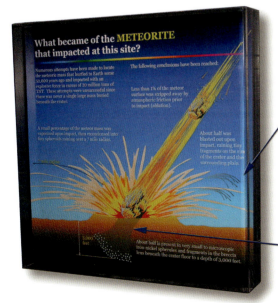

How do they know half of the meteorite was blasted onto the rim and surrounding plain if the material has never been recovered?

How can anyone know meteorite fragments lie 3,000 feet below the crater floor if drilling has only reached 800 feet?

Fig 7.11.13 – This graphic that hangs in the Arizona Crater museum explains what became of the 'meteorite' that made the crater. The interpretative graphic states that half of the 'meteorite' was "blasted out upon impact" while the other half remains buried "beneath the crater floor to a depth of 3,000 feet." Despite the absoluteness of the claim, there is no geological evidence to support it and none is offered, because it does not exist.

Crater that was observed in 1955; it was formed by a hydrofountain explosion. As of this writing, is *not discussed* in the museum's presentation. To be objective, the museum should probably discuss both impact and steam explosion theories, even as it explains why the Arizona Crater is now said to be an impact crater.

The Nilahue Hydrocrater in Chile is 1,100 meters in diameter and 350 meters deep, nearly the *same* size, depth and shape as the Arizona Crater. It deserves recognition among crater researchers if only for the historical sake of the crater debate. Unfortunately, the Nilahue Hydrocrater and others like it get almost no attention in research journals or in the scientific literature in comparison with their famous cousins, the 'impact' craters.

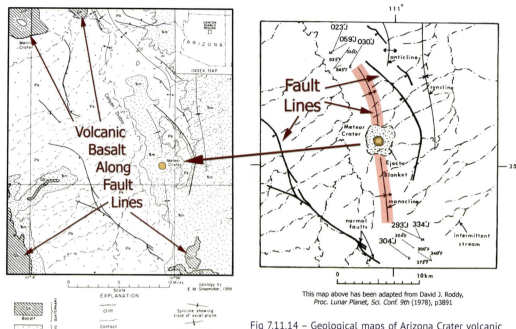

This map has been adapted from the USGS, Eugene M. Shoemaker, International Geological Congress XXI Session 18, p419, 1960

This map above has been adapted from David J. Roddy, Proc. Lunar Planet, Sci. Conf. 9th (1978), p3891

Fig 7.11.14 – Geological maps of Arizona Crater volcanic area from both Shoemaker and Roddy have been adapted to illustrate the relationship between the crater (in yellow), the fault it lies on (shaded red), and other faults in the area (solid lines). The hatched areas indicate basalt flows. These relationships have been overlooked because they do not support impact theory.

We will give them their due credit here. The Hydrocrater Model demonstrated there are three criteria to identify every crater that is formed by a steam explosion:
1. Water source
2. Faults
3. Diatreme

These are definitive criteria and as it turns out, the Arizona Crater has all three—and a lot more. Here is a list of the Arizona Hydrocrater Evidences we will discuss:

1. **Water Source Evidence.**
2. **Bisecting Fault Evidence.**
3. **A Diatreme—the 'smoking gun.'**
4. **Volcanic District Evidence.**
5. **Shale Ball Evidence—they are not meteorites.**
6. **Diamonds—known to form only in diatremes.**
7. **Pure Silica—the second 'smoking gun'.**

1. Water Source Evidence

The power behind all hydrofountains and their remnant hydrocraters is water; evidence of it is crucial in the identification of a hydrofountain. In the case of the Arizona Crater, water was a significant hindrance; Barringer discovered this as he tried to penetrate below the surface of the crater over a century ago:

"In passing I should say that it was found impossible to sink our main shaft more than 222 feet, or about 25 feet below water level, because of the peculiar **quicksand, formed by the water saturated silica**, which we encountered; also that, in my judgment, it will be impossible to properly explore the depths of the central portion of the crater until we have removed the **water which we now know fills its lower portion**." Bib 126 p29

The Crater's water source persists and even to this day continues to meet the water needs of the visitor center through on-site wells. We will learn shortly that water flows through faults and joints, plentiful in this area and made evident by surface vegetation that has exploited this bounteous resource through deep aquifer-penetrating roots.

2. Bisecting Fault Evidence

Amazingly, crater researchers apparently dismissed the presence of faults "in the region of the crater" and directly beneath the crater itself as being unimportant. Of course, the relationship of faults with the Gravitational-Friction Law has only become evident here in the UM; frictional heating along fault lines with subsequent steam explosions is a totally new concept for modern geology.

Shoemaker notes evidence of faults in the area surrounding the crater (seen in Fig 7.11.14) in the following statement:

"The crater lies near the anticlinal bend of a gentle monoclinal fold, a type of structure characteristic of this region. The strata are broken by wide-spaced **northwest-trending normal faults**, generally many miles in length but with only a few tens of feet to about a hundred feet of displacement. Two mutually perpendicular sets of vertical joints of uniform strike **occur in the region of the crater**." Note 7.11k

Present-day researchers have added to the documented fault population that Shoemaker first identified. These can be seen in the Diagonal Regional Joint Trends, Fig 7.11.16. Details of these faults are revealed in the book, *Impact Cratering*:

"**Joints, faults, or planes of weakness in the target rock** also play a role in affecting the **final crater form**... One of the most striking examples of structural control, however, is Meteor Crater, Arizona. Seen in plan view, the crater is actually **more square than circular**. Small tear faults occur at each 'corner' of the rim. These faults formed when adjacent plates of rock were uplifted different distances by the cratering flow. **Meteor Crater's peculiar shape is caused by two orthogonal sets of vertical joints traversing the sedimentary rocks in which it formed**. The joint directions form approximate '**diagonals' across the crater**. The cratering flow exploited these planes of weakness and peeled back the surface rocks much as the petals of a flower open, tearing along the joint directions." Bib 161 p84

The Fault Evidence supporting the Crater's hydrofountain origin could hardly be clearer, having "two orthogonal sets of vertical joints" traversing the crater. This relatively rare geologic phenomenon is also evident in Yellowstone National Park in Wyoming, USA, where numerous bisecting faults are present, creating one of the hottest hydrothermal locations in the world (refer back to Fig 5.3.15). The Yellowstone faults verify the steam producing mechanism of frictional heating and phreatic eruption. The bisecting faults at the Arizona Crater along with abundant water are two of three necessary components of a hydrofountain system.

3. A Diatreme—the Smoking Gun

The Geosciences community well knows that both aboveground and underground explosions can form craters. The similarity of the two, as described in the Hydrocrater Model has created a great deal of confusion in the recognition or distinguishing of impact craters from hydrocraters. The question scientists have needed to ask is:

> FQ: What definitive proof can distinguish an impact crater from a steam-eruption crater?

Instead, the emphasis has been only on the identifying of characteristics that are purported to be present at an impact crater—shatter cones, coesite, impact-glass and others. If the right question had been asked, the true origin of Arizona's famous crater would likely have been discovered long ago. One could *infer* that the Crater is not an impact crater simply by the lack of impact-qualifying evidence, but there is a more definitive way to prove **absolutely** how the crater was formed; whether by impact or by subterranean eruption—a *diatreme*.

Lying beneath the Crater is the smoking gun that the Arizona Crater is indeed, a hydrocrater. A diatreme is a funnel-shaped pipe below the crater, through which superheated subterranean water traveled explosively toward the surface. Without this proof, the truth of the crater's origin might be forever mired in uncertainty. The presence of such a diatreme establishes that the explosive force came from below; it is that simple!

Drilling into "Undisturbed Sediments"

Probably, one of the most difficult aspects of a scientist's job is to remain objective and not allow one's personal agenda to overshadow his or her professional work. Financial interests compound the challenge of remaining objective and Barringer's commercial mining interests certainly clouded his investigation of the crater and the subsequent release of the drilling data he obtained. Barringer was a very influential person and was able to convince others to part with their money based on his ideas and purported observations, one of which was that the sediment below the crater was "undisturbed." However, truth and time, as we will see, are on the same side. Sometimes, as was the case in this instance, it takes a century to uncover the truth.

From the 1933 *Geographical Journal*, a prominent geologist gave the following statement to his colleagues with respect to drilling at the crater:

"Mining claims were taken out by the Standard Iron Company in 1903 and many trial shafts (six) and bore-

Fig 7.11.15 – George Herman with CBS interviews David J. Roddy at the Arizona Crater. Roddy along with Shoemaker became the two prominent geologists to report on the Arizona Crater during the 1960s-70s.

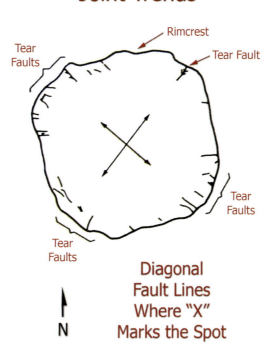

Fig 7.11.16 – The outline of the Arizona Crater's rim reveals that it is slightly **square rather than circular**. The data for this diagram comes originally from Shoemaker and is included in the book, *Impact Cratering*, Bib 161 p84. The "Regional joint trends" referred to are "vertical joints" that traverse **the crater diagonally**. Although such fault lines are well known among impact researchers, without the Lava-Friction Law established in the Magma chapter, the importance of the diagonally bisecting fault lines, which is the mechanism for frictional heating and later steam explosion remained largely unknown. This is further evidence that the crater is not an impact crater.

holes (twenty-three) were put down at considerable expense, but nothing of value was found. After passing through crushed and metamorphosed sandstone and abundant rock-flour, **undisturbed sandstone was met at a depth of 620 feet**." Note 7.111

This geologist made it clear that the sandstone beneath the crater was "undisturbed." Later in the same journal, another professor of geology from the University of Cambridge corroborates his colleague's claim that the underlying sediments were "undisturbed" and takes this as proof that the crater was formed from above:

"The interior of that crater has been bored and the underlying sediments were found to be pulverized to nearly 1000 feet below the plain. **Below that level the strata are still undisturbed**. These investigations **seem** to **dispose of any idea that the crater is blown up from a depth**. Consequently it **seems certain** that the crater was caused by something which **penetrated from above**." Note 7.111 p246

It all sounded pretty certain. A few years later, 'the' *Textbook of Geology*, published in 1939, proclaims as "fact" that the strata beneath the pit is "completely intact and undisturbed":

"Exploration for the meteorite by **drilling** and otherwise has established **the fact** that the **strata beneath the pit are completely intact and undisturbed, thereby proving that the pit is of nonvolcanic origin**." Bib 125 p290

Apparently, Barringer had 'convinced' the experts. Even the most dedicated observer of the crater, Harvey Nininger was convinced, and he said so in his comprehensive 1956 book, *Arizona's Meteorite Crater*. He believed that beneath the crater lay "undisturbed sediments":

"Below this pulverized bed of silica the **drill passed gradually into undisturbed sediments**." Bib 126 p25

In 1960, Eugene Shoemaker made it 'official,' by stating before the International Geological Congress that the siltstone and sandstone beneath the crater, at depths of "700 feet and deeper" were "ordinary":

"Cores of **ordinary siltstone and sandstone** of the Supai **were obtained at depths of 700 feet and deeper**." Note 7.11m

Shoemaker's prestige put to rest any doubts of volcanic origin because "ordinary siltstone and sandstone" existed at 700 feet and below (see Fig 7.11.17). The only problem was—it was not true.

Nininger came on the scene after Barringer's death in the 1930s and was in the unique position of knowing more about the Arizona Crater and its history than anyone else. Nininger operated the American Meteorite Museum near the crater site from 1942 to 1953 and then from Sedona, Arizona from 1953 through 1960 during which time he amassed the largest meteorite collection in the world. Most of this collection ended up at Arizona State University's Center for Meteorite Studies, which is currently the largest university-based repository of meteorites in the world.

Nininger turned his attention to the drilling logs and took issue with the interpretations of the logs and the prospects derived from it. The financial *bias* Barringer had in finding a large iron-nickel body below the crater had probably influenced the interpretation of the log. Nininger states clearly that some respected geologists had legitimate reasons to flatly reject the drilling log:

"As is always the case in **prospecting of such a nature**, there are **different interpretations** to be placed upon the findings made. The **log of this drilling** has been studied by men who agree with the Barringers' interpretation of it, but also by others **who strongly disagree**. Some highly respected geologists **flatly reject the entire log on grounds that it shows careless handling on the part of the driller**." Bib 126 p43

The handling of the drilling logs may have been more than just careless as they were not necessarily in support of Barringer's views. It is a sad chapter of the Arizona Crater story where facts from drilling were suppressed to keep the 'impact' theory alive and well.

Claims of "undisturbed sediment" persist even today. Person-

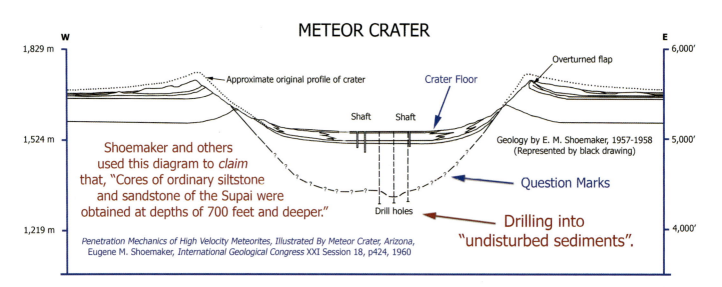

Fig 7.11.17 – Another of Shoemaker's illustrations, his Question-Mark Diagram, shows a cross section of the Arizona Crater (Shoemaker's original drawings are black). Almost every scientific report about this crater uses or refers to this diagram. The names of the layers have been omitted but Shoemaker's bowl-shaped inference, aptly drawn using question marks, rightly demonstrates that lack of knowledge about the crater and what lies beneath it. Impactologists have claimed repeatedly that Barringer's drillers found "undisturbed sediments" below the crater. However, modern seismic and magnetic evidence have proven that all of the previous researchers have been mistaken about this. The Arizona Crater's true shape clearly identifies it as having a diatreme, which proves unequivocally that it did not come from impact.

al conversation with present-day tour guides at the Crater and with other consulting researchers produced assurances that Barringer's drilling efforts identified "undisturbed sediments." However, there were no new borings or underground research conducted from Barringer's early 1900s era drilling through the 1970s.

New information was published in 1975 by a USGS research team detailing a seismic study (sound wave mapping) of sub-surface sediments below the crater. That report found that Barringer's team had *not* passed the fractured zone as he had stated:

"**Drilling** by Barringer and his associates (1905-1914) did **not pass completely through the fractured zone**."

"**Drilling** by Barringer and his associates (1905-1914) did **not pass completely through the fractured zone**." Note 7.11n

Unequivocal Seismic Evidence of a Diatreme

One should be careful when using the word unequivocal, for it fosters the idea that there is no doubt; that such facts are unambiguous. This describes *exactly*, the Seismic Evidence of a Diatreme beneath the Arizona Crater. It is, because it could be no other way; it is the veritable smoking gun of the Crater's origin.

In 1975, three USGS geologists reported results from the latest seismic survey of the crater, gathered from dynamite-induced sound waves produced in 'shot holes' at various locations around the Crater. Their findings are recorded in the *Journal of Geophysical Research*.

Surprisingly, instead of merely showing what the seismic data revealed, the researchers employed the use of the image in Fig 7.11.19, which is very similar to the diagrams by Shoemaker and the modern CTH impact models depicting a *bowl-shaped convex* fractured zone. The bowl-shaped fractured area is inferred because it is the shape expected from an actual impact.

The researchers state the bowl-shaped area of fractured rock is:

"…based on **equations** of Beals et al." Note 7.11n p767

The researchers compiled their own seismic model (Fig 7.11.20) showing the fractured zone under the crater from *actual seismic data,* describing the detailed diagram in Fig 7.11.20 in these terms:

"Cross section of Meteor crater showing the approximate configuration of the brecciated, debris, and **fractured zones as interpreted from seismic data**." Note 7.11n p774

Fig 7.11.20 is *the actual diagram from the journal article, configured from the seismic data* (taken from TNT dynamite explosions made), and colored-shaded red here for easier recognition in the diagram. The investigators went on to say:

"Rocks around the crater **are fractured** as far as 900 m from the rim crest and to a **depth of at least 800 m [2,625 feet] beneath the crater floor**."
Note 7.11n p774

Compare Fig 7.11.19 with Fig 7.11.20 to see how different they are. Despite the obvious differences, Shoemaker and others continue to assert the *1960 claim* that **ordinary** (non-fractured) siltstone and sandstone were obtained at depths of *700 feet and deeper*:

"…**ordinary siltstone and sandstone… were obtained at depths of 700 feet and deeper**." Note 7.11o

You don't have to be a scientist to recognize the substantial difference between 700 feet and **at least 2,625 feet** beneath the crater floor! The seismic study found that the fractured zone extends more than three times the depth Shoemaker, Nininger and Barringer

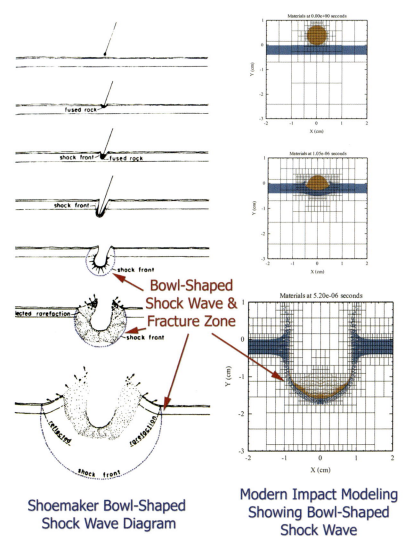

Fig 7.11.18 – These two diagrams are adapted from Shoemaker and from the computerized impact-modeling program, CTH, produced by Sandia National Laboratories. In this illustration, CTH models the shock wave and physics of an impact event, clearly showing the 'bowl' shaped shock wave traveling through the target material. The bowl-shaped fractured pattern has been verified in numerous experiments involving the actual impact of certain objects. This pattern is evident in Shoemaker's 1960 diagram on the left. Is there a similar bowl-shaped fracture zone at the Arizona Crater? The physical data produced from the Arizona Crater *does not show a bowl-shaped fracture zone.* This would have to exist if an actual impact had occurred.
Shoemaker Diagram from *International Geological Congress* XXI Session 18, p424, 1960.

previously claimed, and was *funnel*-shaped, not *bowl*-shaped as had been inferred.

Here was solid evidence that the Crater was formed not by impact, but by an underground steam eruption, yet the researchers and their reviewers chose to ignore the *obvious*. In their Summary and Conclusion, they completely dismissed the seismic data showing evidence of the diatreme and neglected it with the inferred bowel-shaped impact fractured zone.

Shoemaker's claim:

"... **ordinary** siltstone and sandstone ... were obtained at depths of **700 feet** and deeper."

E. M. Shoemaker

Actual seismic study proved:

"Rocks ... are **fractured** ... to a depth of **at least 800 m [2,625 feet]** beneath the crater floor."

H. D. Akermann, et al.,

could have produced the high pressures required to shock metamorphose the Coconino sandstone." Note 7.11p p765

The truth is, coesite, like diamonds is a naturally occurring mineral that forms in diatremes where high pressure and high temperature *do exist*. Two years after the 1975 seismic study, geologist Joseph R. Smyth would discover coesite in an African diatreme, falsifying the notion that coesite is formed only during a high-speed impact.

In the Magma Pseudotheory, the Silica Phase Diagram, seen in Fig 5.7.6 identifies that coesite is known to form under as little as *20 kbar* pressure versus the overstated pressure of *100 kbar*.

Why have geologists chosen to ignore the Unequivocal Seismic Evidence of a Diatreme?

The opening paragraph of the 1975 Arizona Crater seismic study, published in the *Journal of Geophysical Research*, answers this question in part because of the supposed settlement of the Gilbert (steam explosion) and Barringer (impact) controversy 15 years earlier by the discovery of coesite and stishovite at the crater:

"The Gilbert-Barringer controversy was **finally conclusively settled** when coesite and stishovite, both high-pressure polymorphs of quartz, were found in crater breccia derived from the Coconino sandstone [Chao et al., 1960]. Coesite and stishovite require formation pressures of **more than 100 kbar**, far greater than the maximum pressures that can be generated by a volcanic explosion [Roddy, 1968]. **Only** a hypervelocity impact

The scientists of 35 years ago falsely believed that coesite was only producible at these high pressures and only naturally in impacts. Remarkably, these same false theories are still propagated today and the Arizona Hydrocrater is still viewed as an impact crater.

The researchers had fallen into the same trap many modern scientists fall prey to—they failed to question everything with an open mind. We begin to think that scientific knowledge is absolute—but it is not; we must constantly apply the UM Question Principle, being in the paradigm that we must question everything with an open mind, that by so doing, our daily experience can confirm scientific truth, if it is actually true. Truth does not change, but our knowledge of it does. If truth is our goal, and we are not afraid to challenge current thinking—whatever the discovery, it will serve to move us closer to truth.

The seismic study researchers were unaware that their data "finally conclusively settled" the question of the Arizona Hydrocrater by identifying the Arizona Crater Diatreme!

There was ample circumstantial evidence of this fact; the champion of impact cratering, E. Shoemaker documented many "diatremes, or explosion pipes" that occurred in the vicinity of the Arizona Crater. During the 1950s, Shoemaker became well acquainted with "**maars**," which are simply *hydrocraters*:

"**Shoemaker** was well acquainted with regions in the southwestern United States where **maars** occur. In 1956 he described the geologic setting of uranium deposits which he had discovered in diatremes on the Navajo and Hopi reservations in Arizona, New Mexico, and Utah. This work involved careful study of the **diatremes, or 'explosion pipes,'** which, as we have seen, culminate at the ground surface

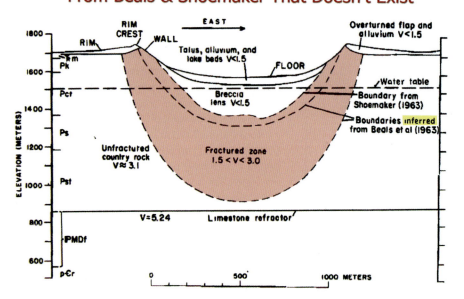

Fig 7.11.19 – During 1975, seismic study of the fractured zone area of the Arizona Crater was conducted. This diagram was included in the report as representing the *theoretical 'bowl' shaped fractured zone* evident at other actual impact events, and *was expected* at the Arizona Crater. The shaded area is the expected bowl-shaped fractured zone area beneath the crater as it should exist, *if* the crater had indeed been formed by impact. The observations did not fit the expectations, and as Fig 7.11.20 clearly shows, a *convex, funnel-shaped* **diatreme** lay below the crater. This provided the unequivocal evidence that the crater was not of impact origin.
Diagram adapted from: H. D. Akermann and R. H. Godson, J. S. Watkins, *Journal of Geophysical Research*, February 10, 1975, p767.

in maar-type volcanic craters. At least 250 maars are well exposed in the Hopi Buttes region, more than are known in any other area of comparable size."
Bib 119 p142-3

The Hopi Buttes volcanic hydrocrater region lies just north of the Arizona Crater. In this story, the historical researcher-author sees the Shoemaker-scientist as such an authority that no one would doubt the work he had done. Instead of focusing on the data to determine whether it was factual or not, the popular expert was taken at face value. This will prove a blow to impactology as decades of false information is uncovered.

The Magnetic Diatreme Evidence

The Seismic Evidence of a Diatreme was identified as a 'smoking gun' because of its simplicity. The Magnetic Diatreme Evidence parallels the seismic testimony and testifies even further of the crater's diatreme. This magnetic evidence comes from a survey that was taken at approximately the same time as the seismic survey of 1975, but by different researchers from the USGS. These two excellent underground surveys have been shoved onto the back shelf because of 'conflicting evidence.'

In subchapter 5.12 of the Magma Pseudotheory, we discussed the Earth's energy field and referred to the familiar compass with its floating magnetic needle pointing in the direction of

> **Without the UM Question Principle—question everything with an open mind, we begin to think that our scientific knowledge is absolute—and it is not.**

the magnetic North Pole. In our experiment, changes in the field were noted when the compass was moved slowly toward a large metal object; the compass' needle began to turn towards the metal object. This response is called a "positive magnetic anomaly" whereas a negative anomaly is referred to as a "magnetic minimum." Therefore, if we measure the Earth's magnetic field, and then place an iron meteorite underneath the sensor, a *positive* magnet anomaly is "anticipated":

> "…**positive magnetic anomalies are anticipated over iron meteorites** because of their intense magnetic polarization…" Note 7.11q

If a large iron meteorite were to be buried under the crater, or if there were several hundred thousand tons of iron-nickel material that had settled around the crater, the magnetic field in the area of the crater versus the out-lying plains should be expected to produce a positive magnetic anomaly, but instead, it was **negative**:

> A magnetic **minimum** was found over the crater instead of the anticipated positive magnetic anomaly.

This caused the researchers to conclude:

"There is **no** evidence for a **buried massive meteoritic body**, and there is **no** gross asymmetry to the subcrater structure that

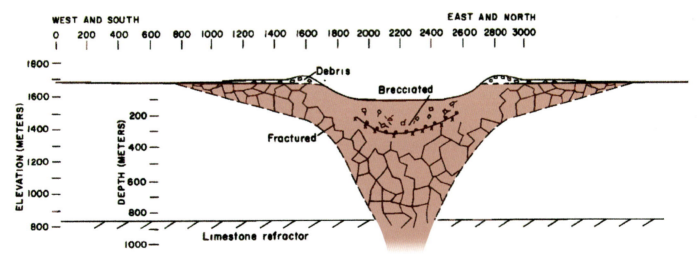

A Seismic Refraction Technique Used for Subsurface Investigations at Meteor Crater, Arizona, H. D. Ackermann and R. H. Godson, J. S. Watkins, Journal of Geophysical Research, Vol. 80 No. 5, February 10, 1975, p774

Fig 7.11.20 – This **Seismic Diatreme Evidence Diagram** represents the first 'smoking gun' of the Arizona Hydrocrater Model. It is the **actual** diagram (area shaded red for clarification) taken from the seismic study at the Arizona Crater as reported in the *Journal of Geophysical Research*. The area shaded red stands in **stark opposition** to the proposed crater shapes shown in Fig 7.11.18 &19, which show concave *bowl-shaped* fractured areas. This diagram, constructed using actual data of seismic sound waves, unambiguously shows the *convex funnel-shaped diatreme* beneath the crater. Astonishingly, the researchers on this project still concluded the above fracture zone was made by impact! It is anyone's guess why the researchers totally ignored the facts that were before them. However, as they said at the beginning of their paper, the controversy between the crater being an impact versus a steam explosion crater, "was finally conclusively settled" because "coesite" had been found at the crater. Nevertheless, this is a sign of their biased selection towards impact because coesite has now been shown to come from *diatremes*. Amazingly, this data has existed for at least four decades, yet it remains hidden and ignored by the scientific community. The crater's diatreme is the 'smoking gun' of the Arizona Hydrocrater.

The Shoemaker Impact Continues

Scientific Knowledge is not Absolute

Fig 7.11.21 – E. M. Shoemaker points out 'impact' evidences of the Arizona Crater to a crowd gathered a few decades ago. With the correct scientific principles in the UM and with the pseudotheories of the past now revealed, we stand poised to escape the grip of the Scientific Dark Age.

would **indicate the trajectory of the impacting meteorite**. However, several **interesting observations** have been made from the gravity and magnetic studies of this young well-preserved meteorite crater." Note 7.11q p788

There was no evidence of Barringer's alleged large meteorite body nor was there evidence of an accumulated mass of small meteorites. This was an indication that there was not a low-speed impact either, because, as the Sikhote-Alin fall demonstrated, the iron meteorites would have been found *inside* the crater.

Something was causing the magnetic field around the crater to be at minimum, weaker than the surrounding area, but what could it be? The terrain of the crater itself was taken into account but no explanation was apparent, leaving investigators to conclude the source for the low reading was a "subcrater source":

"*Source of anomaly.* **The origin of the magnetic minimum may be a subcrater source totally unrelated to the meteorite impact** and its effect on the rocks surrounding the crater. However, the **coincidence of the magnetic minimum with the crater suggests a causal relationship** that must be investigated." Note 7.11q p786

Perhaps these geoscientists were more open-minded than their seismic-study colleagues were. They acknowledged that the magnetic minimum might be from a source "totally unrelated to the meteorite impact" and noted that whatever caused the crater likely caused the magnetic minimum, and should be investigated further. Unfortunately, dogmagma has obscured the real source of Earth's energy field, which is directly related to the magnetic minimum at the crater. We'll go into the details of Earth's energy field later, in the Weather Model.

We can be certain that this magnetic minimum does not come from a magmatic source deep inside the Earth, but from the crystalline rocks in the crust itself. The crust's crystalline structure was disrupted by the violence of the steam explosion from the diatreme, fracturing the country rock and impairing its ability to produce electrical energy (the piezoelectric effect discussed in subchapter 5.8). This is borne out by the fact that the negative magnetic anomaly is not just over the bowl of the crater, but extends beyond the crater rim to a point approximately equal to the edge of the diatreme visible in Fig 7.11.20:

"…the observed negative anomaly **extends out from the rim crest** for a distance approximately equal to the crater radius… **Attempts to model** the observed magnetic anomaly have been **unsuccessful**." Note 7.11q p787-788

The negative magnetic anomaly extends beyond the rim crest about one crater-length radius because of the underlying *fractured rock,* making the magnetic minimum another direct evidence of the crater's diatreme. The investigators attempts at modeling the magnetic anomaly beneath the crater likely failed because they were based on incorrect theories of how the fractured rocks interact with the Earth's energy field.

The magnetic study and the previously discussed seismic study were published at the same time and in the same journal. Although the researchers corresponded with one another, a major difference is apparent. Those involved with the seismic research totally ignored the data, concluding instead that the *actual* diatreme shape and the *inferred* bowl-shaped morphology "approximately agree," whereas the magnetic survey team recognized that *no model was available* or could be made to explain the magnetic minimum found at the crater. Why the disparity?

The answers can be found at the beginning of each of the journal articles, in the opening paragraphs. Here, the open-mindedness of the two research groups *before* they performed their study can be seen. The seismic group noted that the crater "controversy was **finally conclusively** settled" in favor of an impact crater, whereas the magnetic survey group was not so settled:

"On the earth, circular depressions similar to those observed on other planetary bodies occur primarily as the result of **two natural phenomena, volcanic explosion or collapse and meteorite impact. Because both these phenomena can produce similar-appearing craters**, the problem of identifying their origin is often **quite difficult**, as is evidenced by the scientific discussions on the origin of lunar craters. Indeed, a controversy has continued for more than 70 years over the origin of one of the earth's most famous and better known craters, Meteor Crater, Arizona, although an **impact origin for this crater is now generally accepted**." Note 7.11q p776

The objective approach taken by the magnetic survey team led to a more accurate result. This example will probably remain a classic Millennial Science example of the open-mined approach to scientific learning. It is a fine example of how questioning everything with an open mind can lead to new discovery.

4. Volcanic District Evidence

One of the simplest evidences geologist researching the Arizona Crater must consider, is the fact that the crater lies smack in the middle of a "volcanic district," a fact many meteoriticist seem to have overlooked. When the U.S. Geological Survey was founded, Grove K. Gilbert was appointed as Senior Geologist, a position he held until his death. He was one of the most influential American geologists of all time and one of the first scientists to study the Arizona Crater. He addressed the Geological Society of Washington in 1895 on the subject of the Arizona Crater. He saw a relationship between the Arizona Crater and the crater from the Japanese Ko-Bandai volcano that erupted July 15, 1888. The old volcano had experienced a major phreatic (steam) explosion, entirely changing the face of the landscape. Gilbert goes on to compare the two craters:

"The competency of volcanic steam for the production of a crater is thus shown by a **parallel instance**, and the only conspicuous difference between the **Japanese** case and the **Arizonan** lies in the fact that in the one the disrupted rock was volcanic and in the other it was not. **This difference seems unessential**, for in neither case was there an eruption of liquid rock; the ancient lavas of Kobandai had been cold for ages, and their relation to the catastrophe was wholly passive. Moreover, the manifestation of volcanic energy is no more exceptional on the Arizona plateau than in the Bandai district. **The little limestone crater is in the midst of a great volcanic district. The nearest volcanic crater is but ten miles distant, and within a radius of fifty miles are hundreds of vents from which lava has issued during the later geologic period**s." Note 7.11r

Fig 7.11.22 is an aerial photo showing the Arizona Crater Volcanic District. Major fault lines run from the northwest to the southeast throughout the district and there are obvious volcanic forms in both areas. Had impactologists studied the volcanic district more carefully, they would have discovered many minerals—like the iron ejectites in the Arizona Crater—that are not accounted for by the magma theory.

R. S. Dies, second only to Shoemaker in the promotion of Meteor Crater during the 1960s, casually noted that the crater lies among volcanic effects, declaring it "a most confusing thing for a meteorite to do":

"Although this **crater lies in the midst of a great volcanic district**—the volcanic San Francisco Peaks to the northwest; the White Mountains, peppered with volcanic craterlets, to the southeast; and a world-famous group of **diatremes**, the Hopi Buttes, to the northeast—he [Dietz] interpreted its location as an example of the **random** placement of meteorite craters. He [Dietz] remarked that 'landing amidst this full span of volcanic effects was **a most confusing thing for a meteorite to do** but, with the perversity of nature, it **apparently** did so anyway.'" Bib 119 p151

Interestingly, he *knew* then that neither he nor anyone else had "final proof" of impact:

"...Dietz admitted that **no final proof had yet been achieved**..." Bib 119 p151

Sadly, Dietz, Shoemaker and others of the 1960s era were *only looking for proof of impact*. Had they been motivated to look for a *diatreme*, they would have found it. And so it is that Arizona Crater visitors today are given a handout explaining the Crater's brief history saying:

"Dr. Eugene Shoemaker, former Chief of the Branch of Astrogeology of the U.S. Geological Survey in Flagstaff, **proved in 1960, beyond any doubt** that Meteor Crater was indeed the

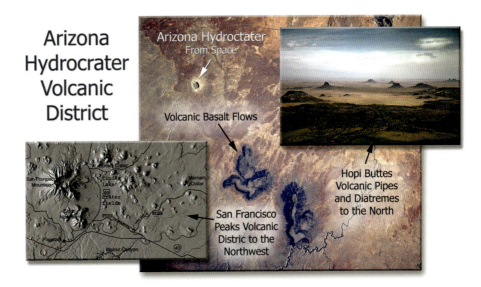

Fig 7.11.22 – This aerial photo of north central Arizona shows the Arizona Crater Volcanic District. Regional lines of faulting run generally northwest to southeast from the Grand Canyon down through the Arizona Crater area along the Arizona Plateau. To the northwest of the crater lies the San Francisco Peak Volcanic District, which includes many hydrocraters. To the north of the crater, in an area called Hopi Buttes, hundreds of volcanic pipes and diatremes are seen. Contrary to Dietz's supposition that this is a "most confusing thing," there is no confusion that the scientific evidence has always pointed to a volcanic origin of the Arizona Hydrocrater.

product of a giant impact event." Note 7.11s

Whether this is true or not behooves the reader to be the judge. For his proof, Shoemaker presented a paper on the Arizona Crater to the International Geological Congress in 1960, which included the following:

1. A diagram of the crater being on an "anticlinal bend" (an uplifted fault) in a *volcanic* region.
2. A diagram of the impact showing an inferred "*bowl*" shaped shock front.
3. A cross-sectional diagram of the crater with a line of *question marks* marking a bowl-shaped bottom of the crater.
4. The statement: "The **mechanics** of **typical maars or diatremes** lead to **collapse** of material from the crater walls into the vent, generally leaving the **surrounding country rock undisturbed**." Note 7.11t

Rather than proving "beyond any doubt" the Arizona Crater was from impact, these four points establish just the opposite. Shoemaker was supposed to be *the* expert at distinguishing impact craters from steam explosion craters, yet his own statement reveals that he was unaware of the mechanics of steam explosions and obviously did not know about hydrocraters.

Maars or diatremes are formed by steam explosions. Collapse craters, a different type of hydrocrater altogether, are rare in comparison. The Southwestern USA Hydrocrater Survey, illustrated in Fig 7.8.5, includes 11 examples of hydrocraters, a small sampling of the dozens that were surveyed. Of these, only the Chevelon Hydrocrater, and possibly the Montezuma Hydrocrater are collapse craters. The rest are *steam explosion* craters. At all of these craters, the surrounding country rock was significantly disturbed because of the steam explosion and subsequent fallout.

Nininger said it best when he quoted one of his former professors, G. W. Stevens saying:

"There are no authorities in science. Only nature is Authority." Note 7.11u

5. Shale Ball Evidence

Shale balls look very much like an ordinary rock on the outside. They do not resemble typical meteorites and they do not have the traditional regmaglypts of other iron meteorites. Fig 7.11.23 shows an example of a shale ball from Wolfe Creek Crater, which is similar to some iron-rock specimens collected at the Arizona Crater. When opened, these rocks reveal layered laminations over a round or oblate core. This similarity to shale is the reason for their name. Some large specimens were reportedly over three feet long and weighed upwards of 300 pounds. According to Nininger, Barringer attached "great importance" to them:

"In addition to the metallic fragments, all collectors had found masses of **iron oxide** associated with the crater. Most of these were in the form of **laminated chips or blocks**, the larger of which showed more or less **curvature as to form**. Barringer justifiably attached **great importance** to these. He called attention to the fact that they had evidently been detached from more or less **rounded masses** of which he found many and which were called *shale balls*." Bib 126 p30

Nininger conducted a detailed examination of the shale balls and noted the roundness was in "striking contrast" to the irregular shapes of the irons:

"Attention was also called to the fact that the tendency to **roundness in the shale balls was in striking contrast to the varied and irregular shapes of the irons** and to the further fact that in their distribution the irons showed a **strong tendency to be associated with the oxide chips**." Bib 126 p30

What Nininger meant by a "strong tendency" for the irons to be "associated with the oxide chips" was that the oxide chips, which were outer fragments of the shale balls, were of different composition than the irons but shared proximity. Nininger recognized the difference in composition but failed to realize the significance; the "chips" were *not* oxidized fragments of the nickel-iron meteorites. What was the origin of the iron oxide chips? Fig 7.11.11 shows the iron nodules on the crater's rim that are embedded in the limestone; they are similar to the iron chips in our discussion.

The Shale Ball Evidence is apparently not important to modern geologists; we contacted the Arizona State University Meteorite department to inspect shale balls from their collection, but the curator did not even know what they were.

Another crucial piece of evidence that refutes the meteorite theory comes from a particular element more abundant in the shale ball cores than in the irons:

"Some of these were described as having **iron cores**, and chemical tests by the U.S. National Museum showed the cores to be slightly **different** in composition from the unoxidized metallic masses which, since 1891, had been known as the **Canyon Diablo irons**. The iron cores were shown to contain **chlorine** in greater amount than did the regular Canyon Diablo irons." Bib 126 p30

The chlorine Nininger refers to is a component of salt, and salt is direct evidence of the terrestrial origin of the shale balls. Subterranean saline water is a ready source of the elemental material in the shale balls and the oxide chips.

It appears that almost no modern research has been conducted on the shale balls, but Barringer realized their importance, and

Fig 7.11.23 – This shale ball from Wolfe Creek Crater, Australia is similar to the shale balls found at the Arizona Crater. Shale Balls look like typical rocks on the outside, but when cut, they reveal a metallic iron interior. These types of geological specimens are often ignored by meteoritics scholars because they seem to have no connection to the meteorites. In fact, they can think of no reason why they should exist. They do not fit into any of the 'meteorite' categories, being more like terrestrial iron oxide rocks, yet they are often found in association with other traditional iron 'meteorites'. Fig 7.12.16 contains more shale ball examples.

came surprisingly close to identifying their origin:

"On the other hand, the shale-balls invariably are a rounded shape. This immediately suggests the shape of **cobblestones in a creek-bed**. The **only** two ways in which they could have attained this shape are by being **abraded from without, or by being fused and 'dripped,'** as molten lead is dripped to form shot. We know that they could **not** have been molten, for all the pieces of iron and even some of the shale **show the Widmanstätten** crystallization figures, which **completely disappear when the iron is heated** to 1500 degrees, Fahrenheit. Therefore they **must have been bumped and rubbed into their rounded shape**—and this could only have been done by other pieces of iron, floating near them in space." Bib 126 p30

Barringer was dead on when he supposed they were shaped by some form of erosion, but was unaware of the physics of space. He was still trying to fit his observations into the meteorite-from-space theory. Of course, our experience with space today reveals that floating balls of iron in zero gravity are not simply "bumped and rubbed" together to become rounded.

However, it *does* happen on Earth. Both on the surface and under the Earth's crust, rocks are moved against each other in rivers, underground aquifers, and *diatremes*. Additional evidence that Barringer's rubbed and rounded space-rocks theory was wrong, and can be had by answering this FQ:

FQ: Have **observed** meteorite falls ever produced shale balls?

Because shale balls are never found associated with observed falls, one could suppose that the "shale" did not come from space, leaving us to find an alternative origin. That origin is a diatreme.

Finally, the shale balls provide *direct* evidence of the *actual origin* of the Arizona Crater irons. Some of the iron ejectites were found encapsulated inside the shale balls; nickel-iron "*nodules*" that resemble the Canyon Diablo irons:

"…some of the **shale balls**, upon being broken apart, showed instead of a metallic core **several nodules of nickel-iron resembling in shape the normal Canyon Diablo iron**." Bib 126 p30

Nininger's use of the term "*nodules*" was another close call in identifying the real origin of the irons. Like other terrestrial nodules, the Arizona Crater irons share a unique origin; they were formed in a hypretherm. Further discussion on the general origin of nodules will take place toward the end of subchapter 8.14.

Hugh M. Roberts, an experienced geologist and mining engineer, was hired to inspect the crater and provide the known scientific facts concerning the crater. This was in 1919, before the scientific community had *assumed* the crater was from impact. He provided an excellent report along with a challenge to learn whether the irons at the crater were terrestrial or meteoric, and that a rational theory must account *for all of the facts*:

"Since native nickel-iron is generally **assumed** to be of meteoric origin, it has been concluded **regardless of the obvious geologic relations** that the crater was formed by the fall of a large meteor. **If, in reality, this iron is not meteoric but terrestrial, and if a rational theory of its origin in connection with the formation of the crater can be maintained**, we will arrive at an explanation which **accounts for all of the facts**." Bib 126 p37

The Arizona Hydrocrater Model does just that—it is a rational theory that "accounts for all of the facts."

6. Diamond Evidence—Diatreme Formation

For over a century, investigators have tried to prove beyond a reasonable doubt that the Arizona Crater was formed by impact. From time-to-time, new "impact criteria" surfaces with which to set the record straight. However, each new impact criterion usually fails to accomplish its objective, and instead, ends up supporting the hydrocrater origin. One new criterion is diamonds, but as we shall see, diamonds have the same origin as the Arizona Crater—a *diatreme*, *not an impact*.

Diamonds were first discovered in 'meteorites' at the Arizona Crater and it did not take long for diamonds to become a hot topic amongst the researchers. Following the discovery of diamonds at the crater, two investigators, heavily steeped in the

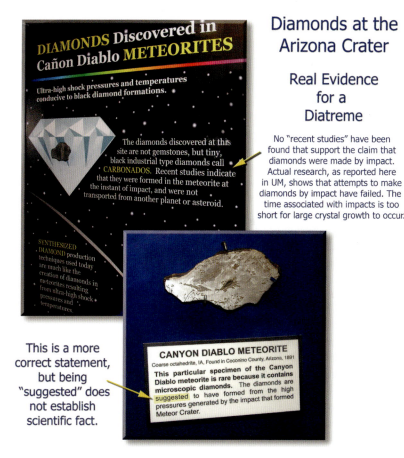

Fig 7.11.24 – Diamonds have always held a fascination with mankind. They are the hardest gemstone and one of the most beautiful when cut. Diamonds were discovered in the irons at the crater, and researchers have tried diligently but unsuccessfully to reproduce them, even though the crater museum display (top photo) indicates that modern diamond production techniques can. Journal articles included here in the UM show that diamond production by the environment sustained during the instant of impact is not possible. The lower right photo of a meteorite sample from the ASU Meteoritic Museum more accurately expresses the truth. Diamonds were not formed by impact shock they were formed in diatremes. Millimeter-sized diamonds come from natural diatremes, another evidence of the incredible pressure, temperature, and water environment that once existed, an environment most researchers have ignored.

magma paradigm, Wentorf and Bovenkerk, suggested that laboratory-grown diamonds, produced under high heat and pressure could explain how the Arizona Crater diamonds were formed:

> "A few months later, R. N. Wentorf, Jr., and H. P. Bovenkerk suggested that there had been enough **experience in laboratory production of diamond for useful comparisons to be made** between man-made and natural ones. Small black diamonds very similar to those found in meteorites had been produced in the laboratory from carbon in the presence of **molten metal**, or from a mixture of the two, under conditions of relatively low temperature and moderately high pressure." Bib 119 p175

However, it has already been demonstrated that the iron meteorites did not come from a melt; therefore, the diamonds could not have been formed in the presence of molten metal. A bigger problem arose when two other researchers, Lipschutz and Anders theorized that the diamonds in the meteorites were formed from the *shock* of the impact. Since the scientists saw diamonds *in* the meteorites, they 'knew' they came from space, but they also knew they were not created in space. Therefore, they concluded the diamonds must have been formed at impact. The shock-origin theory of meteoritic diamonds was born.

Their colleagues disagreed, stating the **time needed** for diamond crystals to grow in the Canyon Diablo (Arizona Crater) meteorites did not exist in the shock-origin theory:

> "With respect to Canyon Diablo diamonds, they found the arguments of Lipschutz and Anders convincing in many ways, except that the average diameter of the tiny diamonds found in meteorites was known to be about 0.01 millimeter. **Lumps of diamond with diameters of more than 1 millimeter—lumps a million times larger—had been observed in Canyon Diablo iron.** They felt that this provided a strong argument **against impact origin** because, judging by their experiments, a time of **at least one second would have been required** for the growth of such relatively **large diamonds**; but the duration of the required pressures at the moment of impact had been calculated at **only one tenth of a second.**" Bib 119 p175

Still today, most scientists do not understand the time required for rocks, minerals, and crystals to grow, partly because so few scientists actually *grow rocks*. Minerals, including 1-millimeter diamonds, do *not* take millions of years to grow, nor do they form in a *fraction of a second*. Typically, a 1-millimeter crystalline mineral will take approximately an hour up to several hours to form. This has been confirmed by our own *actual* rock-growing experiments and from other research involving synthetic mineral growth. Bib 104

Lipschutz and Anders tried to come up with their own experimental tests to make diamonds from the "shock of impact" but had no success:

> "In the same year, Lipschutz and Anders made experimental tests of several ways in which they **thought** that diamonds might have formed in the Canyon Diablo iron during the **shock of impact. None of their experiments produce diamond...**" Bib 119 p176

Lipschutz and Anders failed to show that diamond crystals could be produced from shock. Other researchers, such as Carter and Kennedy identified other flaws in the shock-origin theory:

> "We have discussed, in the light of our observations, several of the more important criteria from which Nininger [1956] suggested and Lipschutz and Anders [1961a] **deduced a shock origin for the diamonds in the Canyon Diablo meteorite** and the ureilites. We have found that these criteria do not survive a careful scrutiny and that the physical and textural features of the diamonds themselves almost conclusively **prove that they were not formed by shock**." Note 7.11v

Lipschutz and Anders' shock-theory failed, but they were successful in showing that their colleague's (Carter and Kennedy's) diamond-from-melt theory was also flawed. :

> "**If** the diamonds had **formed from a melt**, they would have floated to one end of the nodule, **which is contrary to observation**." Note 7.11w

In the end, Lipschutz, Anders, Carter, and Kennedy were all correct in proving each other's diamond-origin theories were wrong, and the Canyon Diablo diamonds remained a matter of debate for many years:

> "Although examples of shock metamorphism were becoming more common, the origin of the Canyon Diablo **diamonds remained a matter of debate**." Bib 119 p177

Over the next several decades, research on coesite and other shock-made materials would continue, as did the search to identify an absolute test for determining impact. So far, **none** of the impact criterion have been successful simply because all can be created by both explosions *from above* (meteorite) *and by explosions from below* (hydrofountain). Trying to find evidence that comes only from explosion will probably *never* be conclusive.

Rather than conclusively ending the debate, and despite the absence of any proof that diamonds are formed by the shock of impact, researchers conceded and the controversy simply died away:

> "And eventually, **perhaps** because of so many dramatic demonstrations of the transformation of material by shock, the diamonds in the Canyon Diablo iron were **generally conceded** to have been formed by the shock of impact; and the **diamond controversy died away**." Bib 119 p178

The concession certainly did not solve the puzzle, and any rational theory should account for all of the facts, therefore, this challenge is extended to the supporters of the shock-origin theory:

Arizona Crater Diamond Challenge
Create a 1mm diamond similar to those in the Arizona Crater irons within one tenth of a second or less.

The Barringer Crater Museum continues to promote diamonds in the meteorites as being "evidence" of its impact origin. However, there is no explanation as to *why* the diamonds were evidence of impact. The ultimate origins of all natural diamonds (larger than 1 millimeter) are *diatremes*:

> "To this day, Canyon Diablo meteorites hold the record for the **most diamonds recovered**." Bib 29 p232

7. Pure Silica—The Second 'Smoking Gun'

If the first Arizona Hydrocrater's smoking gun—the diatreme beneath the crater—was not enough to convince the most ardent meteoriticist, there is another; the large *pure silica deposit* on the south rim of the crater.

What makes this 'smoking gun' so astonishing is that it has been right under the nose of every scientist that has investigated the crater—and yet they missed it. One reason the pure silica evidence stands out is the sediment-origin questions addressed in the UM and the recognition that the vast majority of the Earth's sediment is not from erosion. We've already discussed this in part, earlier in this chapter, and there is much more in the next chapter, detailing how sand and clay were formed in a hypretherm and ejected through hydrofountains.

Back in Fig 7.7.14, Hydrosand Fountains are evidence of the connection between sediments and earthquakes, where subterranean sand is ejected onto the surface by earthquakes and water. A simplistic example of the diversity of sediments that can be brought to the surface through hydrofountains can be seen in Fig 7.11.25. In this image, beach crabs have buried themselves at different depths, each ejecting the sand from their burrow onto the surface.

Now imagine the events of a single impact explosion; sediments of many layers; red, brown, white; all thrown violently into the sky, vaporizing some of the material, and *indiscriminately mixing the rest, producing a homogenous blend of brownish-red ejecta, to be deposited around the crater rim*. The explosion would have been so powerful that debris would have rained down for some time. So, how did a large deposit of pure white silica come to rest on the rim of the crater?

> FQ: How did the large deposit of pure white silica come to rest on the rim of the Arizona Crater?

Obviously, not from an impact! The silica deposit seen in Fig 7.11.26 is exceptionally large, stretching about a half mile on the south side of the crater. Previous investigators noted the following:

"**Vast quantities of powdered rock** (which they called '**silica**') were found in both the rim and the crater. Under the microscope this powdered rock, in some cases so fine that no grit could be discerned when it was placed between the teeth, was seen to be made up of broken **grains of quartz sand** which looked like angular pieces of ice." Bib 119 p32

It has only been recently that geologist have come to understand that sediment, deep in the Earth *could* be brought to the surface by water during earthquakes through liquefaction. They still do not recognize that many of the Earth's vast surficial deposits occurred this way, made below the surface in a hypretherm.

Researchers observed the following:

"In the rim, on the contrary, **boulders and rock fragments and pieces of meteoritic iron and fine debris were indiscriminately mixed**." Bib 119 p32

Looking at Fig 7.11.26, do the deposits on the rim look as if they are "fine debris" mixed "indiscriminately" with the red and brown sand from the crater? Instead, large red sandstone boulders are lying on the pure white silica sand, as if they had come from an outburst after the silica was deposited, despite the claims of some researchers:

"In addition, if there had been sufficient heat and underground water to create a steam explosion of such magnitude, there would probably have been a series of events. **Nothing**, however, was found to suggest **more than a single outburst**." Bib 119 p32

> FQ: By what means did large boulders fall **after** the deposition of the fine white silica powder?

Large rocks, some up to a meter long, are on or near the surface, imbedded into the white silica. A simple experiment will demonstrate that powdered material would remain airborne longer than large rocks; which means we should see larger rocks on the bottom of the white deposit—but this is not the case at the crater.

The ***only*** explanation that accounts for the larger rocks being on the surface of the white silica and not below it is that they were ejected later, ***after*** the white silica had been deposited. The evidence of multiple outbursts is well supported by a hydrocrater theory, but not so by the impact theory.

To be fair, the geologists of the 1900s were unaware of hydrocraters, even though there were three excellent examples nearby, such as the Colton, the Buell and the Zuni hydrocraters, all within an hour's drive or so from the Arizona Crater. From researchers circa 1909:

Fig 7.11.25 – Tiny little craters and their ejecta were created by crabs buried in the sand at the seashore. The crabs burrow to different depths, expelling sand and debris from their homes during low tide. Different kinds of sediment exist at different depths beneath the tiny craters just as it does beneath giant ones. This example demonstrates how hydrocraters can eject different types of sediment even at the same location. This can happen because of multiple vents, successive eruptions, or changes in the subterranean hydrologic flow.

"Another interested person was H. L. Fairchild, of the University of Rochester, who also visited the spot. He agreed with Branner that **if the crater was not of meteoritic origin, it was the most interesting geological puzzle of the present time**." Bib 119 p34

Driving to nearby hydrocraters might not have been possible in 1909, but within a few decades, automobiles shuttled investigators and geologists around, and they were able to see much more of the landscape. By 1942, geologists had found and identified diatremes that contained matter other than volcanic material, "silica" of "hydrothermal origin":

"A few **diatremes contain** gypsum, travertine, or **silica**, minerals which may be of **hydrothermal origin**. Most diatremes contain tuff, which is altered to or cemented with calcium carbonate. These minerals **must indicate that the vents contained circulating waters**, which in part may have been juvenile." Note 7.11x

The geologists knew in the 1960s of the hydrothermal origin of silica, but their minds were closed to this fact at the Arizona

Fig 7.11.27 – These are two of the rare specimens of 'glass' taken from the Arizona Crater. They are a part of the ASU Meteoritic Museum collection. The mineral, Lechatelierite, is commonly formed naturally by high temperature fusing of quartz sand, such as during a lighting strike, or high-speed impact. One distinctive feature of these specimens is that they have no meteorite material in them at all, which is totally unlike the Wabar impact glass, which included black meteoritic material in every piece. Obviously, researchers have had a difficult time connecting the pure white silica at the crater to an impact event. As Crater researcher H. Nininger noted, they are "not unlike volcanic pumice" and labeled them as being associated with "steam." Like other anomalous Arizona Crater evidence, lechatelierite can be formed by hydrocrater activities.

a violent phreatic explosion. Some of the silica was fused in the "volcanic-like pumice," the lechatelierite that is found on the rim of the Arizona crater, similar to the material at the Ubehebe crater in Death Valley.

"We Have Found Very Little True Silica-Glass"

The Smoking Gun evidence of the pure white silica is important for another reason; there is an absence of the silica-glass, ***supposedly*** melted from the impact as the Coconino sandstone beds were penetrated, as Nininger himself explains:

"Even though the Coconino sandstone is nearly pure silica and constitutes about two-thirds of the sediments penetrated by the meteorite, **we have found very little true silica-glass**. The lechatelierite found by Barringer constitutes **almost all of the fusion product derived from that formation** that has so far been recovered. Its quantity has **never been determined**, but the Barringer report indicated that at several locations the drill **penetrated more or less of it**." Bib 126 p128

Only a couple of handfuls of glass from the drilling project exist, and there are *no mass quantities of glass on the surface*. The small amount of white silica-glass that was encountered during drilling operations beneath the surface, *in the diatreme,* is precisely what would be predicted from a steam explosion, which would have diffused the Coconino sandstone by pressurized steam with the hottest areas in the diatreme being closest to the fault lines. Enough heat could have been produced to fuse *some* of the silica. The steam-powered eruption could have separated and purified the sandstone in a differentiation process similar to the distilling process of water, where impurities are removed by pressurized steam.

Being in possession of the aforementioned evidence, we can begin to comprehend *why* the FQs asked by Nininger have never been answered. Here are three of Nininger's questions he posed in the *American Journal of Science* in 1971:

1. "In view of the abundant evidence of extreme temperatures produced by this impact, why do the exposed formations of the inner crater wall **show no glazed surfaces**?
2. "Since the principal formation (Coconino sandstone) invaded by the meteorite is almost pure silica, **why do we not find this mineral** strongly dominant in the fused product?
3. "Why is **no nickel-iron found in lechatelierite** [glass]? Note 7.11y

These three simple, but fundamental questions Nininger raised are among the questions asked in the UM. The glass and minerals puzzled over by Nininger as well as the answer to these questions can be realized when we understand that:

Crater. Why? Because everyone had 'agreed' on an impact origin—why look for anything else?

Steam-Produced Pumice Evidence

Nearly 600 feet below the crater floor, a deposit of pure silica (rock flour) was encountered; presumably transformed Coconino Sandstone. This contained a mineral later called **Lechatelierite** (Fig 7.11.27). Nininger was one of the few individuals to observe this material firsthand and recognized its importance and its similarity to volcanic pumice:

"Variety 'B' was even more impressive. In it nearly all of the sand grains had lost their identity, evidently having been reduced to **a liquid condition in the presence of steam**, yielding a product **not unlike volcanic pumice but mineralogically consisting of almost pure quartz**." Bib 126 p29

Nininger rightly attributes the "not unlike volcanic pumice" material's origin to steam as seen in Fig 7.11.27. The steam, however, did not come from an impact into the upper red and brown layers, which would have mixed with the lower white layers of Coconino sandstone. In addition, the white silica deposit did not contain meteoritic material.

Instead, water from underneath the crater, a hypretherm, possibly separated the silica from the white Coconino sandstone or more likely derived the silica from an even deeper source, carrying it to the surface along cracks and faults, finally erupting in

Fig 7.11.26 – **The Pure Silica Evidence** exhibited on the previous page is compelling evidence that a steam explosion is the only logical origin of the Arizona Crater. The sizeable deposit of white silica flour, easily seen on the southern rim of the crater in the aerial photo, is remarkably pure. The sample in the vial illustrates its purity. Pure silica like this has only been documented to form in a hydrothermal environment. The origin of the white silica lays hundreds of feet below layers of red and brown sandstones, yet it was *not* mixed with material from those layers when it was deposited on the rim. Drill tailings in the bottom of the crater, left inset, are remnants of Barringer's drilling activities when he was searching for the 'big meteorite.' He never found it. Instead, he unknowingly discovered the Pure Silica Evidence that this crater is a hydrocrater. The center inset photo shows a TNT explosion and indiscriminate *mixing* of the disturbed layers into the ejecta outside the crater, similar to what would be expected during an impact explosion. This is not what is found at the Arizona Crater. Multiple eruptions, establishing the crater as being a hydrocrater, are confirmed by the pure white silica dikes on the crater's edge that has no meteoritic material, and the large sandstone blocks embedded in the top layer of the white silica, evidence of at least two different eruption cycles.

The crater was **not** made by a meteorite impact.

Once again, the Hard Question Principle is evident; previously hard to answer questions are now easy, once we know the answer, proving that in Nature, the simple truth is that truth is simple.

In Nininger's final paragraphs, he outlined the "Need For Extensive Research At The Arizona Crater," which is needed more now than ever. [17.11y p658] Until now, simple clear answers to Nininger's questions were unavailable because of the impact paradigm Nininger and those who followed him are in.

The Ubehebe Silica-Dike Evidence

It so happens that there is another hydrocrater, which exhibits a concentration of white silica not unlike that of the Arizona Hydrocrater. Roughly 400 miles northwest, in Death Valley, California, lays Ubehebe Crater, which modern geologist agree is undisputedly the result of a phreatic explosion. We discussed this crater earlier in the Hydrocrater Survey, seen in Fig 7.8.5 & 6. At Ubehebe, there is direct evidence of multiple explosion events and a prominent, white silica dike that can be seen in Fig 7.11.28. The silica dike is very similar to the one shown in the Pure Silica Evidence montage, Fig 7.11.26; compelling evidence that both are hydro-volcanic structures created by steam explosions.

They both would have experienced multiple eruptions or the ejecta and related debris would have been of a more homogeneous nature—all of the materials mixed together. Instead, there is deposit of almost pure silica that verifies that there must have been multiple eruptions of various materials from diverse layers beneath the crater. Modern geologists have apparently not seen the 'silica-dike connection' between the Ubehebe and Arizona Hydrocraters because the Arizona crater's origin has been assumed to be from impact.

The Mars Silica Evidence

When stuck in an old paradigm, such as the Impact Pseudotheory, it is easy to ignore the evidence right in front of your nose, like the pure white silica of the Arizona Crater, and its similarity to the silica deposits of other craters. However, there is even more evidence. Stepping away from the impact pseudo-theory increases the possibility of seeing the real significance of the pure silica deposits.

Recently, on Mars, a discovery of "nearly pure silica" was made (Fig 7.11.29). In the news release issued by NASA about the discovery, no mention of 'impact' was made; instead, researchers speculated only about *water related origins*:

"Researchers using NASA's twin Mars rovers are sorting out two possible origins for one of Spirit's most important discoveries, while also getting Spirit to a favorable spot for surviving the next Martian winter.

The puzzle is what produced a patch of **nearly pure silica** -- the main ingredient of window glass -- that Spirit found last May. **It could have come from either a hot-spring environment or an environment called a fumarole**, in which acidic

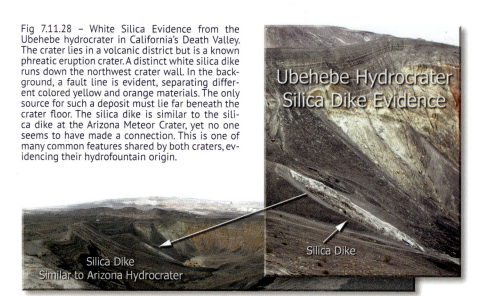

Fig 7.11.28 – White Silica Evidence from the Ubehebe hydrocrater in California's Death Valley. The crater lies in a volcanic district but is a known phreatic eruption crater. A distinct white silica dike runs down the northwest crater wall. In the background, a fault line is evident, separating different colored yellow and orange materials. The only source for such a deposit must lie far beneath the crater floor. The silica dike is similar to the silica dike at the Arizona Meteor Crater, yet no one seems to have made a connection. This is one of many common features shared by both craters, evidencing their hydrofountain origin.

steam rises through cracks." Note 7.11z

The researchers were right in envisioning a "hot-spring environment" origin of the white sand, but it is still a "puzzle" to them. Hot springs and fumaroles on Earth typically do not produce pure silica **sediment**. They do produce a silica mineral called **geyserite** that forms around the openings of hot springs. Geyserite grows in layers around the mouth, or crater of hot springs as siliceous water wells up from below the crust. As it rises and cools slightly, the silica falls out of solution. It does not form silt-sized or sand-sized sediment. Erosion of the geyserite cannot explain the existence of large homogeneous sediment deposits because erosion does not produce only one size of material. Moreover, geyserite consists of several minerals other than silica, which would also erode into various sizes of sediment.

Fig 7.11.29 – This photo is of the surface of the planet Mars. The rover space probe revealed white silica as it disturbed the soil. Researchers speculate this "nearly pure silica" deposit must have come from a "hot spring environment".

CHAPTER 7 THE HYDROPLANET MODEL

Pure silica sediment deposits of large, similarly sized grains consist of individual crystals formed in a hypretherm, which required high temperature *and high pressure*. Present day hot springs are not under sufficient pressure to supersaturate the heated water with the sufficient silica required for large, individual sand-grain crystal growth.

NASA researchers tried to explain the formation of the silica sand by simply heating and cooling the water without the knowledge that *high pressure is required* to grow the silica grains:

"'This stuff is more than 90 percent silica,' Squyres said. '**There aren't many ways to explain a concentration so high**.' One way is to selectively remove silica from the native volcanic rocks and concentrate it in the deposits Spirit found. **Hot springs** can do that, **dissolving silica at high heat and then dropping it out of solution as the water cools**. Another way is to selectively remove almost everything else and leave the silica behind. Acidic steam at fumaroles can do that." Note 7.11z

The acid sci-bi is not adequate at explaining the selective removal of other minerals for several reasons. First, it has not been demonstrated on Earth; secondly, highly acidic hot springs are rare. A similarly false acid sci-bi has been rationalized as being responsible for the formation of limestone caves, a theory demonstrated to be untrue in the Rock Cycle Pseudotheory chapter.

If the hot springs environment was acidic enough to dissolve some minerals, it does not mean that it would dissolve *all the other minerals*, leaving only quartz. As far as we have been able to find, there has been no demonstration of this theory.

The Scientific Investigator versus the Theorist

After a careful reading of the Hydroplanet Model, one might be tempted to ask; how could so much physical evidence of the Arizona Crater's hydrofountain origin have been overlooked? One answer comes from a quote G. K. Gilbert made in 1896 when commenting on "the prime difference between the investigator and the theorist":

"In the testing of hypotheses lies **the prime difference between the investigator and the theorist**. The one **seeks diligently for the facts** which may overthrow his tentative theory. The other **closes his eyes to these**, and searches only for those which will sustain it." Grove Karl Gilbert, Bib 119 p29

Gilbert was ridiculed for his view that the Arizona Crater came from a steam explosion. Kathleen Mark, author of the well-received book, *Meteorite Craters*, said:

"…Grove Karl Gilbert said that the Coon Butte [Meteor Crater] had been created **by a steam explosion**." Bib 119 p30

Mark went on to explain how detrimental she thought "Gilbert's mistake" was:

"**A mistake, like the evil that men do, lives after them**. Good scientific work, while not exactly interred with their bones, is often so completely assimilated that it becomes an anonymous part of the discipline they have helped to develop." Bib 119 p30

Was Gilbert really mistaken about the Arizona Crater's steam explosion origin? Time will tell. Gilbert was certainly capable of making mistakes and of missing essential evidence, and so too is the UM. Nevertheless, Gilbert exposed the *real evil that takes place in science*, which occurs when a theorist closes his or her eyes to the facts instead of being open minded to wherever the truth leads. The legacy Grove Karl Gilbert left us with is the definition of a **true scientific investigator**:

> One who seeks diligently for the facts, which may support or overthrow the tentative theory.

A theorist, D. M. Barringer's life was intertwined with the Arizona Crater like the veins in his body. Barringer had put literally everything he had on the table, betting the hole in the Arizona desert had come from a giant space-rock; that a meteorite lay buried in the crater. In the end, Barringer could not accept the truth that the physicist's calculations showed; *if* the crater had been made by impact, the heat and energy from the impact would have vaporized most of the meteorite.

Barringer's commercial mining interests and his reputation had blinded him to the facts. He passed away just weeks after hearing new scientific reports on the crater that said he would not find his coveted meteorite. Unfortunately, most researchers and the scientific establishment are not much better off than Barringer. They have financial incentive and reputations to uphold and they close their eyes, as Gilbert said, to the facts, including those presented here in the UM. It is a grave disservice to students and to the scientific community as a whole.

In this respect, the UM's goal is somewhat unique. Being a private venture, there is neither financial incentive nor reputation to support. If there are errors or wrong assertions, so be it. The purpose of the UM and of Millennial Science is to build on demonstrable, observable and reproducible facts and processes, and to discard improvable theories.

By the middle of the last century, modern science had become accustomed to accepting "theories" as fact, and many new pseudotheories were formulated; built around the erroneous idea that the Arizona Crater was an impact crater. The impact of E. Shoemaker (see subchapter 7.9, The Crater Debate) on the geo-planetary science fields was based on observations and interpretations of the Arizona Crater and resulted in theories built on very little truth. One meteoritic author records:

"Many surveys and studies have been conducted at the crater since Barringer's day. Some of these studies confirmed his conclusions, some did not…

"Most of these studies were not taken seriously by geologists, and **by the mid-1950s few believed the crater had been caused by impact**. Then in 1957 a young geology graduate student, **Eugene Shoemaker**, began investigating the crater anew" Bib 29 p122-3

What changed in the minds of geologists that caused them to think the crater had an impact origin? Throughout the history of science, the influence of the human factor cannot be underestimated. Shoemaker was a very persuasive impactologist; single-handedly changing most people's beliefs during the 1960s. His Impact Statement gradually changed the theory of planet formation. His new observations would someday be a part of the empirical evidence that would show the Arizona Crater is a hydrocrater, but not during his time. In spite of the physical facts that would eventually disprove the impact theory, Shoemaker's persistence swayed the minds of geologists and others:

"It was **Shoemaker's persistence** that eventually **convinced geologists** that Barringer Crater was the youngest and best preserved impact crater on Earth." Bib 29 p123

Investigators had failed to recognize the Arizona Hydrocrater for what it was because of the dearth of hydrocrater research.

Possibly 99% of the time and money spent on crater research has been orientated toward impact, probably due to the excitement of rocks that fall from space. That excitement can easily lead to imagination without merit. Nininger gives us an example of this in 1971:

"Our first lunar explorers will want to land their ship in one of the deep craters of small diameter **for safety**. Exposure on the moon's surface outside such a protective feature will be incomparably more dangerous than will the flight between us and the moon so far as **meteorites are concerned**." Note 7.11aa

Nininger's "safety" concern never materialized. The allure of space exploration and the connection to Earth through meteorite impacts continues to propagate imaginative theories which have little data to support them. Nininger, the quintessential Arizona Crater researcher commented on the crater "pit" and its dissimilarity to known volcanoes:

"**This pit is unlike known volcanoes**. It is unlike known sinkholes. **It lacks certain essential features of steam or gas blowouts**. There are **few other rimmed pits similar in structure** and in each case where they have been adequately explored **meteorites have been found associated**." Bib 165 p339

Unfortunately, Nininger ignored or was ignorant of the Buell and Zuni Hydrocraters, directly east of the Arizona Crater, near the Arizona-New Mexico border. These craters are also not considered typical 'volcanic' structures, yet are indisputably formed by volcanic action. Interestingly, both craters are "similar in structure," large craters with upturned rims, having all the "essential features of steam or gas blowouts" (water, faults and a diatreme), just like the Arizona Hydrocrater. There are hundreds of hydrocrater examples worldwide, but Nininger and others have apparently chosen not to study these hydrocraters. This is *why* impact craters were never really understood; most researchers ignored the obvious hydrocraters and studied only what they 'thought' were impact craters.

Summarizing the Arizona Hydrocrater

A trip to the Arizona Crater's museum or the company's website shows immediately the strength of false tradition. Before

> "In the testing of hypotheses lies **the prime difference between the investigator and the theorist**. The one **seeks diligently for the facts** which may overthrow his tentative theory. The other **closes his eyes to these**, and searches only for those which will sustain it."
>
> Grove Karl Gilbert,

the UM—the entire world believed that this crater was made by impact—yet it was not. This is one of the most important, extraordinary discoveries of the UM.

Here is a partial list of the evidence that the crater is a **Hydrocrater**:

1. There is no impact glass from a high-speed impactor.
2. There are no meteorites showing evidence of melting.
3. There is insufficient residual material if the meteorite actually vaporized (no meteorite-infused glass particles).
4. The Widmanstätten pattern establishes that the irons near the crater were formed at a low, non-melt temperature as compared to the supposed temperature of impact.
5. There are no shrapnel meteorite fragments from a low-speed impact and disintegration of a large impact body.
6. No embedded meteorites were found in the crater.
7. Two different forms of irons were found at the crater, meaning that there would have had to be multiple impactors and multiple craters. This is not supported.
8. The strewn-field of iron fragments is not elliptical as it is with known impact events.
9. Limestone at the crater shows no evidence of heating, which should be evident from a high-speed impact.
10. No shatter cones were found.
11. The amount of iron found at the crater is far less than the iron necessary to form a crater of this size.
12. There is evidence of subterranean water.
13. Bisecting faults lie beneath the crater.
14. The geomorphology below the crater is in the shape of a diatreme, not an impact bowl—the Crater's Smoking Gun.
15. The Crater lies in a volcanic district.
16. Shale Balls are not meteorites; they are a form of iron ore and are found at the Crater.
17. Diamonds are present, which are known to form only in diatremes.
18. A significant deposit of pure white silica on the rim and in drilling remnants at the base of the Crater attests to multiple eruptions of subsurface waters. This is the Crater's second smoking gun.

There are many more points to be made, and many yet to be discovered. The truth will become evident as the shift from an impact origin paradigm to a hydrocrater origin begins to happen.

The Arizona Crater, known today by the misapplied name of Meteor Crater has endured a colorful life of heated controversy, unscrupulous businessmen, misinformed geologists, and false science. Today, the world's most famous crater has a true story to tell, and this gem of the desert will outlive these errors and misconceptions as it becomes known as the **Arizona Hydrocrater**, the world's most famous *Hydrocrater*.

7.12 The Impact to Hydrocrater Evidence

Impact From the Deep

During the influential 1960s, geologist K. L. Currie of the Geological Survey of Canada spent a great deal of time investigating many of his country's craters. He came to a profound conclusion:

"**None** of these craters can be explained either by 'classical' impact theory or by analogy with known volcanic craters. **New ideas are needed**." Bib 119 p156

New ideas were slow to come, until now. New ideas to challenging old problems is the focus if the UM, but unlike almost every other new idea or theory that has come along, the ideas expressed here, in the UM, are not unsupported musings left to stand among other hundreds of unsupported theories. Instead, each new concept is buttressed by other connected observations, predictions and natural laws that are all a part of nature's puzzle. When the pieces of any puzzle are arranged in their one ***true*** order—it takes little effort for them to 'fit.'

Several decades ago, paleontologists recognized evidence of mass extinction events. Worldwide, the fossil record showed that tens-of thousands, even millions of plants and animals died, entombed in the crust of the Earth. It was obvious that many had experienced a quick burial by a global catastrophe. However, the Uniformity Myth prevented acceptance of this notion and investigators continued to publish papers espousing a slow, *gradual* process, perhaps driven by climate change that eventually killed off many of the world's species. This persisted until the 1980s when a popular new theory appeared that said an asteroid had struck the Earth, killing the dinosaurs and other life. The asteroid landed just off the Yucatan peninsula and left behind the Chicxulub Crater, seen in Fig 7.12.1. The demise of the dinosaurs was settled!

In a move away from uniformity, impact catastrophe theories became the rage, and geologist began to sidestep the geological evidence of hydrocraters in favor of impact craters. Nevertheless, time (sometimes a very long time) has a way of revealing concealed facts and clearing up muddy waters.

A 2006 *Scientific American* article titled *Impact From the*

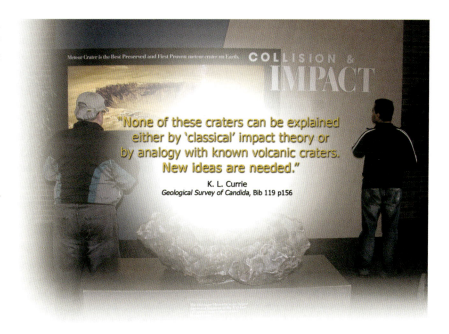

Deep included the following subtitle:

"**Strangling heat and gases emanating from the earth and sea, not asteroids, most likely caused several ancient mass extinctions.**" Note 7.12a

Peter D. Ward, a professor from the University of Washington's biology department found new geochemical markers amongst the world's fossil remains that has encouraged some scientists to reconsider the notion of impact induced mass extinctions. In his article, *Impact From the Deep*, Ward reexamines the so-called 'unquestionable' evidences of impacts and even expresses doubt about "*any*" impacts at two of the five supposed mass extinction eras:

"Indeed, further investigation of the evidence **has called into question the likelihood of *any* impacts at those two times**. No other research groups have replicated the original finding of buckyballs containing extraterrestrial gas at the end Permian boundary. **A discovery of shocked quartz from that period has also been recanted, and geologists cannot agree whether purported impact craters from the event in the deep ocean near Australia and under ice in Antarctica are actually craters or just natural rock formations**. For the

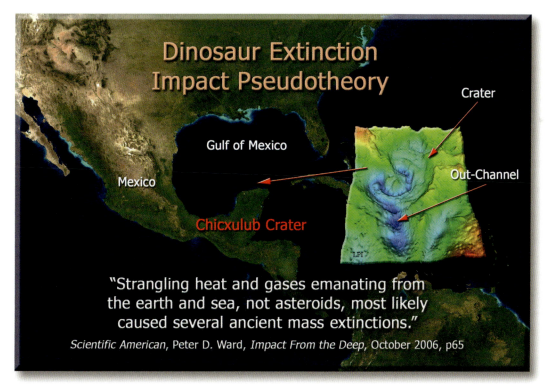

Fig 7.12.1 – The Dinosaur Extinction Impact Pseudotheory has no basis in fact, and recently several researchers have found further evidence against this impact theory. The supposed impact crater, located in Mexico, is shown above in enhanced 3-D relief. The image clearly shows a large ridge extending inside the crater and beyond its rim (upper portion of the crater). This same type of ridge is evident at the Buell Hydrocrater seen Fig 7.8.7. Impact crater theory cannot account for this type of ridge formation, but hydrocraters can. In this image, blue represents the lowest elevations, which shows a large out-channel (bottom of the inset image). The Upheaval Dome Hydrocrater, seen in Fig 7.8.5 and later in this subchapter in Fig 7.12.18, shows the same type of out-channels. These were created by the outflow of water and sediment, probably *underwater*. New discoveries of superheated gases coming from the Earth support hydrofountains and hydrocraters, both of which will be seen in even clearer light in the following chapter—The Universal Flood.

7.12 THE IMPACT TO HYDROCRATER EVIDENCE

end Triassic, the iridium found is in such low concentrations that it might reflect a small asteroid impact, but nothing of the planet-killing scale seen at the K/T boundary. **If impacts are not supported as the cause of these mass extinctions, however, then what did trigger the great die-offs? A new type of evidence reveals that the earth itself can, and probably did, exterminate its own inhabitants.**" Note 7.12a p68

We will continue to explore this concept in the coming pages, and especially, in the next chapter where we will investigate new biomarker evidence suggesting the great dying that took place in the Earth's past was caused from *within* the Earth's crust, just as Ward suggests.

The Impact Fad

It is hard to say whether the science of impactology has driven demand for "meteorites" or whether the commercialization of meteorite specimens has increased interest in impacts; whatever the cause, both have certainly affected the 'Impact Fad.' The scientific community shares responsibility in perpetuating the fad by routinely repeating basic mistakes in classroom texts. From the textbook, *Introduction to Planetary Geology,* we read the following:

"A **circular outline and raised rim** of a crater suggest that the structure **was formed by impact**, especially when the rim is underlain by an anticline or associated with an **overturned flap**. Also **indicative of an impact origin is an ejecta blanket** with reversed layering which may have secondary craters superimposed upon it." Note 7.12b

You don't have to be a geologist to recognize the error of this statement. Hydrocraters *also* have "a circular outline and raised rim" and an "overturned flap." Moreover, "ejecta blankets with reversed layering" are also present on most hydrocraters. Shoemaker focused attention on these characteristics when trying to convince the scientific community back in the 1960s about the impact nature of the crater. Rather than clearing up the subject, Shoemaker added to the confusion because he also did not recognize and identify the *difference between hydrocraters and impact craters*. As noted in the Hydrocrater Survey, for every so-called "impact crater" there are at least 99 hydrocraters. Had the author of the previous quote from the *Introduction to Planetary Geology* been more aware of Earth's most abundant type of crater, it is doubtful he would have made such a statement.

Without studying both types of craters—impact and hydrocrater, one could not expect to know the difference between them. Research into steam-explosion craters is minimal at best; on the other hand, the web-based *Impact Database* notes that there has been "exponential growth in publications relating to impact":

"The Earth Impact Database represents a compilation of information from around the world. Maintaining this site is both a formidable task and a formidable responsibility. Moreover, it is a task that is **growing because over the last 25 years the impact process has been increasingly appreciated by the Earth Sciences community as an important planet-building and planet-modifying process**. A consequence of this is the **exponential growth in publications relating to impact**." Note 7.12c

Appreciating a theory does not make the theory true. Because most craters in our solar system are not from impact, which has become widely accepted over the past 25 years or so, the simple facts of cratering become complicated as we try to match them to a process that did not happen. In 1989, one of the most respected scholars in his field, May Melosh, wrote a comprehensive book called *Impact Cratering*. On the first page, he comments how ignorant the scientific community is on the cratering process:

"In spite of these major strides forward in the understanding of impact cratering, **many geologists, and even planetary scientists, remain ignorant of vital facts about the cratering process and its resulting craters**." Bib 161 Preface

Melosh goes on to express his opinion as to why there is so much "widespread ignorance":

"I have long felt that the reason for this **widespread ignorance** is the **lack of any one place** where the major facts about impact cratering have been collected together and presented in any accessibly way." Bib 161 Preface

Melosh was only partly right; a single authoritative book on the subject has been missing, but that was only a minor part of the crater origin puzzle. A much bigger issue is the fact that cratering is a *multi-field* scientific issue. Few researchers follow an extensive course of study in astronomy, geology, physics, and chemistry. Without a solid understanding in each of these fields, and a detailed background in the historical papers and books written on lunar and terrestrial craters, the real issue cannot be addressed. Even then, with all this knowledge, there have been so many incorrect assumptions about the cratering process, both present and past, that it is hard for anyone to find the truth.

Melosh's *Impact Cratering* book helped to clear up some of the cratering misconceptions, but the author's entire approach to the question of how craters are formed neglected to **even consider** the formative processes of 99% of Earth's craters. The book approaches the cratering process from a physics standpoint, and deals heavily with lunar craters, which are *all assumed to be of impact* origin. Ask the author about the Crater Debate and you would almost certainly hear, "what crater debate?" For him, there is no debate:

"It is **now recognized** that the cratered landscapes of the

Fig 7.12.2 – Meteorites and Impact craters became a fad in the late 1900s and many parts of modern culture were literally *impacted* by the science of impact. However, as the Hydroplanet Model has demonstrated, the correct origins of both meteorites and the majority of the craters found throughout the solar system have been unknown, until now. The Universal Model has combined observations from all of the fields of natural science to discover the real impact story. There are many hydrocraters with new stories to tell; we will explore some of them here.

moon, mercury, Mars, and many of the solar system's satellites are sculptured **predominantly by repeated impacts of all sizes.**" Bib 161 Preface

Why then, does Melosh acknowledge that "many aspects" of common complex craters are "either unclear or controversial" and go on to say that the "origin of the prominent central peaks is still argued heatedly":

"**Complex craters** constitute one of the **most common** large-scale surface features found on airless bodies through the solar system. They have been studied intensively since before the era of spacecraft exploration of the moon and planets. Several examples of complex craters are even known on the earth in various states of exposure and preservation. Nevertheless, **many aspects of their structure and formation are either unclear or controversial**. The **nature and origin** of the prominent central peaks is still **argued heatedly**. The collapse that seems obviously necessary to account for complex crater formation **violates the standard notions of material strength of geologic materials**. The origin of multiple rings, both internal and external, **has nearly as many interpretations as there are investigators**." Bib 161 p131

Such wide-ranging disagreements would likely not happen if researchers were working within the correct paradigm. As the UM has continually documented, major, unresolved issues like these exist in many fields of scientific study, which is a clear indication that researchers have not understood the empirical evidence and how it applies to natures puzzle. Once open-minded individuals do, progress will be made and the wisdom that is lacking in so many scientific endeavors will increase.

There are a number of other craters allegedly of impact origin, but when we look for evidence of steam-explosion origin, the evidence is sometimes overwhelming. All we need do is simply look. With the background of the Crater Model, the Meteorite Model, and Arizona Hydrocrater Model, it will be easier to discuss and understand the upcoming craters, which are still thought by modern science to be impact craters.

The Arizona Crater may have been the *best* example of an 'impact crater' turned hydrocrater, but there are other craters that provide clues and a broader understanding of the significance of hydrofountain cratering activity.

The Wolfe Creek Crater

The second largest crater in the world with iron ejecta materials similar to the Arizona Crater is the Wolfe Creek Crater (Fig 7.12.3). The Western Australia crater, originally discovered by airplane in 1947 is 880 meters (2,887 feet) in diameter; about three-quarters the size of the Arizona Hydrocrater.

In 2004, the Wolfe Creek Crater National Park web site explained the crater's impact origin:

"Scientists have made an **intensive study** of the Wolfe Creek meteorite crater… It would have weighed more than **50,000 tonnes** and is thought to have been traveling at **15 kilometers a second**, a speed which would have taken it across Australia in five minutes." Note 7.12d

As we investigated the
, we discovered the "intensive study" wasn't all that intense. In many respects, the Wolfe Creek Crater was less qualified as an impact crater than the Arizona Crater. A 50,000-tonnes meteorite with a velocity of 15 km/sec should be amply evident during a thorough investigation of the crater, but there was no such evidence. Despite this, impactologists still believe as firmly as researchers at the Arizona Crater do, that a meteorite impacted the Australian countryside. Within three years of its discovery, Australian geophysicists from the Bureau of Mineral Resources reported that the crater was without doubt, formed by impact. Three decades later, in 1979, a geophysicist from the Smithsonian Institution, listed "definitive evidence" as proof of its impact origin:

"The association of **meteoritic material**, the crater's indisputably **explosive origin**, and the **lack of any suggestion whatsoever of volcanic activity** are almost universally regarded as **definitive evidence of Wolfe Creek Crater's impact origin**. No doubt this is the correct interpretation." Note 7.12e

R. F. Fudali in the *Journal of Geology*, lists three specific evidences of the crater's origin. The statement shows how little geology knows about craters. It is rather common knowledge that more than 99% of all Earth's craters are volcanic or steam-*explosion* craters, yet the geophysicist claims "explosive origin" as evidence for impact. The lack of volcanic activity as evidence of impact was also dispelled in the Hydrocrater Model subchapter; volcanic material need not be present or exposed during a phreatic explosion. Steam does not leave much of a fingerprint and as the Magma Pseudotheory confirmed, *volcanic activity* has not been fully comprehended by geologists. Thus, these two so-called *definitive evidences* should be dismissed.

The third and final evidence is the least supporting of all—the association of so-called "meteoritic material" at the crater. In 1954, William A. Cassidy reported on one of the first scientific expeditions to the crater:

"In spite of diligent search, **no wholly metallic meteorites were found**…" Note 7.12f

Not one "meteorite" was found at the crater. The area was searched for decades without success. There were a few possible meteorites found 3.9 km (2.4 miles) away, but in exceptionally small quantity, totaling only 1,343 grams (47 Ounces) in weight:

"At Wolfe Creek iron meteorites have been found in the neighborhood of the crater, **although in small numbers**. In 1965, P. Kolbe and E. Pederson found 1,343 grams of iron meteorites 3.9 km south-west of the crater in an **elliptical area 30 by 20 meters**." Bib 168 p8

Given the distance, the amount of material collected, and the concentration of the collected material, it is unlikely that the meteorite material is directly connected to the Wolfe Creek Crater, even though it is in the "neighborhood."

However, there were 1,400 pounds of iron "shale balls" found just beyond the rim of the crater, but these are not meteorite material; they are similar to those of the Arizona Crater. Even if the shale balls were to be counted as part of the meteorite mass, it is a long way from the presumably 50,000 tonnes of iron!

If a 50,000 tonne meteorite formed the Wolfe Creek Crater, where is the remnant meteorite material?

Erosion is so scant in Western Australia that the researchers were of the opinion that the crater had experienced little erosion:

"The **erosion of the crater is slight**, and signs of erosion on the **steep walls of the crater** are **not** well marked." Note 7.12g

Even massive erosion could not easily account for the disap-

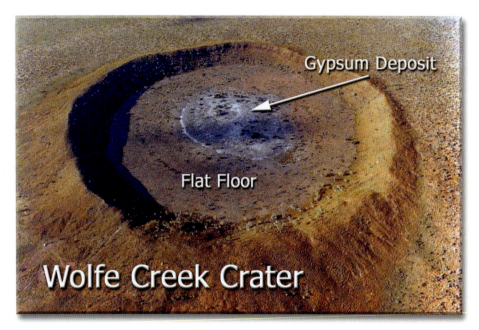

Evidences Against Impact

1. No Meteorites
2. No 50,000 Tonnes of Iron Material
3. No Glass of Any Significance
4. No Shatter Cones

Evidences For Hydrocrater

1. Fault Lines Found Along Sink Holes
2. Water Presence From Gypsum Deposit
3. Flat Floor From Diatreme
4. Iron Shale From Diatreme
5. Uranium From Diatreme
6. Magnetic Pipe From Diatreme

Fig 7.12.3 – The Wolfe Creek Crater in Western Australia is similar to the Arizona Crater and has many direct evidences that dismiss impact origin in favor of a hydrocrater origin. Once we take an objective approach to how these craters were formed instead of being biased toward impact, the facts reveal some amazing clues, and testify that most craters are hydrocraters.

The hydrofountain processes were not known during the crater's investigation, and so, as one science writer wrote, crater researchers were left to choose between two inferior possibilities:

"As Boon and Albritton remarked, one hardly knew which was better—a meteorite hypothesis **without meteorites** or a volcanic hypothesis **without volcanics**." Bib 119 p73

A meteorite theory without any meteorites is not easy to sell, and though the volcanic hypothesis includes water and steam, it was overlooked because the Hydroplanet Model did not yet exist to account for the origin of water and heat.

Other direct evidence that the Wolfe Creek Crater is not from a high-speed impact is the absence of *glass*:

"**No silica glass or sintered rock has been discovered in the area**, but there is every probability that further work will disclose the presence of material of this character." Note 7.12g p34

As we have already learned, there should have been truckloads of glass present on the surface if the crater was really of impact origin. Except for a photo of a couple of small pieces of nondescript "tektite-like" melt (in Bib 168), none have been reported.

Other missing evidences that contradict the impactologists "impact criterion" are shatter cones. Neither the Arizona Hydrocrater nor the Wolfe Creek Hydrocraters have them:

"Shatter-cones have **not** been found at Wolfe Creek…" Bib 168 p19

It isn't just the missing evidence to support claims of Wolfe Creek Crater's impact origin that should be weighed; it is the *presence of hydrocrater evidences*. These have been overlooked or ignored.

Wolfe Creek Hydrocrater Evidences

Once in a while, researchers ask the right questions about crater origins:

"The question is, how did these craters and circular structures form? It is clear that all were formed by an explosion, but was the source of the explosion **from within the Earth's crust, or was it from an extraterrestrial source**, such as a giant meteorite hitting the Earth?" Bib 168 p4

These questions were posed by Australian scientists who realized that *objective* crater research must account for all possibilities, not just "impact criterion." Unfortunately, without the Hydroplanet Model, the researchers unknowingly contradict themselves by 'thinking' they understand the geological pro-

pearance of 50,000 tonnes of iron material. Remnants would certainly be found in the washes and creeks—but there was none reported, because it never existed. Inside the crater there was nowhere for the meteorite material to go, except onto the crater floor. Moreover, if iron meteorite material had been present for 'hundreds of thousands of years,' iron oxide from rusting would be evident in the landscape and would *still be present in the soil*. But like the Arizona Crater—it is not. These facts should have raised the impactologists doubts long ago.

A steam-explosion hydrocrater origin has no such requirements. The hydrocrater origin follows the Simple Truth Principle of the UM; in nature, the simple truth is that truth is simple.

cess of craters and crater forming events, yet they acknowledge the origin as being "clearly enigmatic" and that the "cause of the explosion is unknown":

"There are **thousands** of circular structures on the Earth's land surface and many of these can be explained by the action of **well understood geological process**—such as volcanism. A number of these structures do **not** occur in volcanic terrains, **nor** are they associated with volcanic material. **Their origin is clearly enigmatic** and some scientists have described them as 'cryptovolcanic' or 'cryptoexplosion' structures, suggesting that they are the result of **explosive eruptive activity** or that **the cause of the explosion is unknown**." Bib 168 p16

The next excerpt, from the same book, *Australia's Meteorite Craters*, records researcher's observations that establish the presence of the first two requirements of a hydrocrater—faults and water. The "sink holes" lying along "two intersecting lines" in the crater is quite similar to the bisecting faults found in the Arizona Crater (Fig 7.11.15). These fault lines would have been the source of subterranean heat and water, and an avenue for the gypsum identified in the crater, seen as gray in the center of the crater in Fig 7.12.4. Gypsum minerals are commonly associated with water.

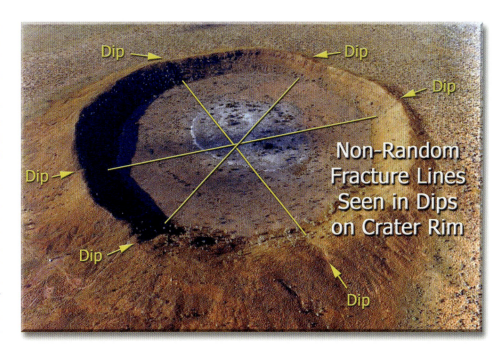

Fig 7.12.4 – The dips around the rim of Wolfe Creek Crater correspond to non-random fracture lines across the crater. Laboratory experiments have shown that impacts randomly fracture the material at and below the surface, paying little attention to preexisting non-random fault lines. Steam explosions emanating from beneath the surface are necessarily tied to lines of faulting and break the surface in an orderly way, like pealing a banana instead of smashing it. Faults are evident in many of the world's craters.

The crater also exhibits a relatively flat floor, with a slight rise, typical of what would be expected of a hydrocrater:

"The current floor of the crater is **flat**, except for a slight **rise in the centre**, a feature found in **many such craters**. It is largely sand covered and, depending on seasonal variations, supports a sparse growth of spinifex and scattered trees. The central area, made of porous **gypsum**, supports the densest vegetation and is pierced by a number of sink holes. Trees of unusually large size for the area grow here, no doubt drawing on **reserves of water** trapped after summer rains. The **sink holes lie along two intersecting lines** and probably reflect the position of stress fractures formed by the explosive excavation of the crater." Bib 168 p8

The distance from the rim to the floor of Wolfe Creek Crater is only 29 meters (95 feet) compared to 174 meters (570 feet) of the Arizona Crater. Why such a difference in height when the diameters are so similar? Both craters lie in flat, red sandstone desert environments, and geologists have no explanation to account for this in the impact theory. Look at Fig 7.12.5 and think about how such a large flat floor could be the result of an impact. No impact experiments have produced anything like it, but in the Hydroplanet Model, sediment-laden water flowing upward easily accounts for the flat-floored craters.

As we gaze at Fig 7.12.6, a question may come to mind about the large area of tree growth at the center of the crater. There, gypsum and other hydrated calcium minerals are concentrated and water is more abundant. What impact process could be responsible for concentrating hydrous minerals, cause a slight rise in the crater floor, and cause water to be in greater abundance there? Instead, think of the alternative; hydrofountain processes easily account for *all of the phenomenon* present at Wolfe Creek Crater. Upwelling mineral rich water would carry sediment to the surface, flatten the crater floor of a previous phreatic erup-

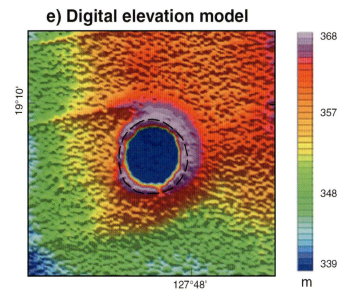

Fig 7.12.5 – A digital elevation model of Wolfe Creek Crater performed by the GSWA of the Australian Government shows a high elevation of 368 meters at the rim and a low of 339 meters on the crater floor. The mean difference is a distance of 29 meters (95 feet), which is only 16% of the Arizona Crater's rim to floor distance of 174 meters (570 feet). How do impactologists account for this if both craters where from impact? The Arizona Crater is only slightly smaller and both lay in similar sandstone sediments. The easy, but ignored answer is that excess sediment from the Wolfe Creek Crater diatreme filled the crater.
Image courtesy of gov.au Department of Industry and Resources, GSWA Record 2003/10.

SUBCHAPTER 7.1 MAGMAPLANET TO HYDROPLANET

tion, swell directly above the now-backfilled diatreme, and be an ever-ready source of water, for use by the vegetation.

There is an absence of meteorites at Wolfe Creek 'Meteorite' Crater, but the crater is still considered to have "meteoritic material." The reason is that three large iron "shale balls" were found on the rim of the crater:

"This paper gives a brief preliminary report on the expedition to the Wolfe Creek Crater, which recovered in excess of **1400 pounds** of more-or-less completely oxidized meteoritic material, including **3 large**, unusually dense, masses weighing 352, 336, and 324 pounds, respectively. Each of these big masses exceeds by nearly 100 times the largest specimens heretofore found at the Wolfe Creek Crater. The smaller, less dense, specimens recovered appear to be **much like the so-called "shale balls"** found at the Barringer and Odessa Meteorite Craters." Note 7.12h

These shale balls were never considered 'meteorites' and they are not rusting. Typically, meteorite irons rust from the outside in; the inside, unexposed to water and weather remains pristine with the iron-nickel crystalline pattern easily seen when cut and etched. This is not the case with the shale balls. They contain almost no iron-nickel metallic minerals and are so named because of their appearance—they look like shale. The specimens look like shale because they contain organic elements, which is an "important clue" to their origin and to the "the crater itself":

"Sometimes these shale-balls occur in clusters, and often they are actually **welded to the surface of the laterite**. Some shale-balls have yielded two minerals previously unknown to science: *reevesite*, a nickel-iron **carbonate**, and cassidyite, a **calcium**-nickel-magnesium **phosphate**. These rare nickel-bearing minerals provide us with an **important clue to the origin of the shale-balls, and the crater itself**." Bib 168 p8

The researchers never reveal the "clue" they think the shale balls contain; perhaps they were completely unaware of it. The clue is that carbonate, calcium and phosphate are all common biogenic materials commonly associated with ores formation, and many ores are formed in diatremes. The details of such ore formation will be discussed in the next chapter. The mineral **laterite** is important because it is associated directly with the iron shale, which is connected to its formation. The "welding" mentioned is not a melt-weld, but a bond between the laterite and the iron shale that formed as they 'grew' together in the diatreme beneath the crater. It turns out that laterite only forms "beneath the surface" where "interrelations of ground water" are favorable:

"**Laterite develops beneath the surface** in soil zones, unconsolidated sediments, or decomposed rocks where **interrelations of ground water, soil/water table, and topography are favorable**." Note 7.12i

If this does not sound like a diatreme environment, consider that laterite commonly consists of "pellets of iron" ores:

"**A common type of laterite** consists of closely packed, **round pellets of iron**, manganese, or aluminum ores." Note 7.12i

Another factor testifying to the origin of the large iron shale specimens is their conical shape:

"In shape, the two principal masses, M_1 and M_2, are low, **elliptical conoids** with vertices somewhat rounded on M_1 and very much rounded on M_2. The dimensions of the roughly **elliptical bases of the cones**..." Note 7.12j

Cone shapes are *not* a common shape formed by random impact fracturing. Similar conical-shaped iron ejectites and meteorites are found around the world, the Willamette being one of the largest and most famous. The shape is commonly associated with meteoritic materials, allegedly defining the atmospheric "entry direction," of the mass. However, as we noted in the Hydrocrater Model, surface ablation of an actual in-falling meteorite body is paper thin, and does not account for the reshaping of large meteorites. Therefore, there must be an alternate origin for the Wolfe Creek Crater specimen's conical shape.

Again, the hydrofountain process provides a simple explanation for the iron specimens; they are grown in a high-pressure hypretherm with water flowing in a single direction. This would promote crystal growth at the down-stream end of the specimen, creating the cone-shaped structure.

There are other examples of hyprethermal environment iron-materials at Wolfe Creek Crater. 'Meteoritic iron' was found to have hydrated iron oxides with silicate minerals and chalcedony, all of which are formed in a hypretherm diatreme environment:

"Samples of **meteoritic iron** were submitted to the Western Australian Government Chemical Laboratories. The following information is taken from their report. The samples consisted of two fragments, A and B. A weighed approximately 300 gm. And B approximately 500 gm. The material is sufficiently magnetic that fragments of pea size may be lifted by a bar magnet, and it consists mainly of **iron oxides**, **hydrated in part**, **with some silicate minerals** too highly impregnated with iron oxides to be identifiable, and a little **chalcedony**. After fine grinding, specimen A yielded a **very small amount (.06 per cent) of metallic iron** which was retained on a 90-mesh screen." Note 7.12k

Only a very small amount of metallic iron was found in the iron samples, but a much larger amount of another very important mineral was discovered.

The Wolfe Creek Nickel-Uranium Evidence

The important mineral, **bunsenite** has the chemical formula NiO–nickel oxide. Bunsenite is very rare in nature, but here, the iron samples were 1.9% NiO:

"Fragments of various size of heavy metallic material were found around the rim of the crater, particularly along the southern sector. R. O. Chambers, curator of minerals of the Australian Museum, has advised that 'the specimen contains 1.9 per cent of **NiO which is far in excess of what would be expected in terrestrial rock**.'" Note 7.11k p34

Why is the presence of bunsenite at the crater so important? It is literally undisputable evidence that the crater at Wolfe Creek is a Hydrocrater. Bunsenite is connected to another rare element—uranium. The natural environment that bunsenite is found in is a "hydrothermal Ni-U vein":

Fig 7.12.6

"Environment: In a **hydrothermal Ni-U vein**." Note 7.12l

Ni-U is a *nickel-uranium* ore, and it just so happens that the Wolfe Creek Crater eruption spewed "radioactive elements," *including uranium* from its vent, away from the crater itself:

"The most striking feature here is the **concentration of the radioactive elements (Uranium, Thorium and Potassium) around the crater rim together with a south west-to-west concentration away from the crater itself**." Note 7.12m

Significant levels of both nickel and uranium have been documented, and so has the existence of a hydrothermal Ni-U vein. This is significant because both elements are rare in the Earth's crust. On average, Uranium exists at only 1.4 ppm in the crust and even more rarely in meteorites at only .008 ppm. The high concentration of uranium at the crater can be seen in Fig 7.12.7, represented by the white color. This phenomenon is inexplicable with the impact pseudotheory and geologists have declared, "It is difficult to explain…"

"**It is difficult to explain** the anomalous concentration of **radioactive uranium**, thorium and potassium as the result of the catastrophic melting of the Wolfe Creek Meteorite itself on impact, given the **extremely low abundance of uranium in meteorites**." Note 7.12m

One Australian scientist, a consulting diamond geologist, had direct experience with diatremes during his diamond geological surveys. Because of this, the geologist was able to recognize that elevated radiometric uranium counts are often associated with the mineral laterite, also connected to ores in a hydrothermal environment. The ores were brought to the surface by **lamproites**:

"Experience elsewhere in this region shows that **elevated radiometric counts are often associated with surface laterite deposits**, one such example occurring on NookanBah Station hundreds of kilometers to the west where a similar geophysical survey was conducted by the author on behalf of a client over a tenement hosting **igneous intrusions known as lamproites** during 2001." Note 7.12m

Lamproites are a form of hydrofountain processes, "diatreme" or "cone edifices":

"**Lamproite** volcanology is varied, with **both diatreme styles and cinder cone or cone edifices known**." Note 7.12n

Diatremes and cone edifices are water-driven processes, directly involved with hydrocrater formation.

Rather than being a good example of an impact crater, the Wolfe Creek Crater and its nickel-uranium content establishes hypretherm ore formation and subsequent surface deposition through subterranean pipes and diatremes, and exposure by phreatic eruption.

A diamond geologist, Louis A. G. Hissink, recognized that the uranium radiometric anomaly on the crater's rim was "not derived from the meteorite itself":

"All that could be said is that the **radiometric anomaly is associated with the crater rim**, but is **not derived from the meteorite itself** since these meteorites **don't have enough uranium in the first instance to create the measured anomaly**." Note 7.12o

This was probably the only geologist to actually go on the record, questioning the meteorite impact origin of the crater. The radiometric readings at this crater and others can provide evidence of hydrocrater origin. A simple aerial survey would be one way of confirming this possible connection at other craters sites.

The Wolfe Creek Magnetic Survey Evidence

Magnetic surveys are commonly conducted by geologists to assist in the determination of subsurface materials. They are usually performed aerially. Some minerals, particularly ores, show a higher magnetic reading, helping the mining industry or gemstone firms find diatremes that contain economically valuable products.

Fig 7.12.8 shows the results of a magnetic survey of the Wolfe Creek area taken in 2003. The "bull's eye" stood out when the

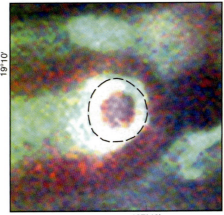

f) Ternary radiometric image

Fig 7.12.7 – Radioactive Uranium, showing white in the above image of Wolfe Creek Crater, has geologists scratching their heads, finding it "difficult to explain" such excess uranium around a meteorite crater. Uranium is typically very low in meteorites; if uranium was present in the meteorite, it should have been found inside the center. Alternatively, uranium ejected from a diatreme makes sense, and there is analog in Western Australia, were uranium ore is a common ejectite ore.
Image courtesy of gov.au Department of Industry and Resources, GSWA Record 2003/10.

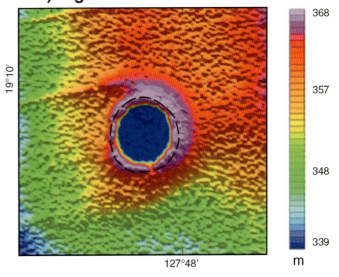

e) Digital elevation model

Fig 7.12.8 – A magnetic survey of the Wolfe Creek Crater gave clear indication of elevated magnetism on the crater's rim and center "bull's eye." The surrounding terrain registered a gradual change in magnetic readings, leaving the question of what was causing the high readings at the bull's eye and rim unanswered. A high-speed impact is generally thought to indicate vaporization, so the bull's eye reading is not thought to be of meteoritic origin, even though meteoric iron metal would cause elevated magnetic readings. If we evaluate the data based on the Hydrocrater Model, then the bull's eye is merely a diatreme pipe that carried magnetic material, such as that found to the west of the crater, to the surface.
Image courtesy of gov.au Department of Industry and Resources, GSWA Record 2003/10.

geologist read the data:

"The Magnetic response of the crater is fairly weak, with a thin annulus **magnetic high corresponding to the crater rim**, and a very small **high in the centre of the bull's eye in the crater**. The regional magnetic field slopes from west to east." Note 7.12o

No explanation for the bull's eye was given, but if we look to the Hydrocrater Model for that explanation, it is easily understood. The magnetic high at the craters center is merely the pipe below the diatreme, where increased magnetic material was carried to the surface. The pipe below ground might very well look like the one back in Fig 7.7.16, a vertically oriented fossil hydrofountain.

The Odessa Impact Crater Myth

Southwest of the city of Odessa, Texas, USA, lies a 168 meter (551 foot) wide crater that is *only* 5 meters (15 feet) deep at its lowest point. This wide, flat crater is shaped similar to a dinner plate. How did this dish-shaped crater, with an elevated rim around its perimeter become filled in with sediment? Odessa Crater is in a flat, desert plane with no nearby rivers or evidence of heavy erosion. This unanswered question went right over the heads of the first explorers of the crater.

Brandon Barringer, brother of D. Moreau Barringer, Jr. reports his brother's findings from a letter written to his father, who at that time was trying to convince investors that the Arizona Crater contained a large amount of profitable iron. The young Barringer's letter said:

"It is a meteor crater, beyond a shadow of a doubt." Note 7.12p

The Barringers paid for drilling at the Odessa Crater to find their elusive iron fortune, like they thought existed at the Arizona Crater. Unfortunately, their efforts mirrored the results of the Arizona Crater:

"… its primary objective was **never** realized, for the **expected crater-making meteorite was not found**." Bib 169 p11

In fact, no iron was ever found within the crater. But that wouldn't be known for some time. Meanwhile, Barringer's son said in his letter to his father that he "knew" what kind of crater it was *before he even got to the crater*:

"As we came up to the rim, **I knew what it was**… The whole thing is so absurdly like M. C. [Meteor Crater] that it doesn't seem possible." Note 7.12p p162-3

Here again we see the first principle of the UM—question everything with an open mind—being ignored. As we will see shortly, the scientific inferences made at this crater will also fail the test of time.

The Odessa Crater was one of the first craters after the Arizona Crater to elicit serious scientific study. Drilling operations, trenching, and the digging of a main shaft in the center of the crater were carried out to allow for the inspection of the sediment beneath the crater. With so much effort and geological work, one would think geologists would have gained a firm understanding of how this crater was formed—but they did not. Here again, the primary reason confidence waned was that observational facts did not align with impact theory. Investigators focused their energy on the impact 'evidence' but totally ignored the abundant hydrocrater evidence.

As before, the geologists did not have the UM, the hydroplanet model, or the hydrofountain model, but steam-explosion craters were a well-known fact of the day. Nevertheless, the research shows a mindset consistently directed towards impact origins. In one example, the driller "encountered an impenetra-

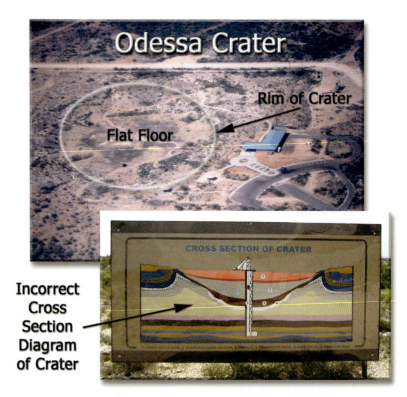

Evidences Against Impact
1. No Meteorites in Main Crater
2. No Melted Meteorites
3. No Shrapnel Meteorites
4. Not Enough Meteorite Material
5. No Glass or Coesite
6. No Shatter Cones or Shocked Rocks

Evidences For Hydrocrater
1. Lateral Fractures Below Crater Found
2. Water Presence at 200' Below Crater
3. Flat Floor From Diatreme
4. Iron Shale Balls From Diatreme
5. White Rock Flour From Diatreme
6. Oil/Salt Diatreme District
7. "Inverted Cone" Diatreme

Fig 7.12.9 - The Odessa Crater, located near the city of Odessa, Texas, USA, is the less famous 'impact crater' cousin to the Arizona Crater. The evidence outlined above and in the narration contradicts the impact theory. Just like the Wolfe Creek crater and the Arizona Hydrocrater, the evidence is strongly supportive of a hydrocrater origin. The Cross Section diagram of this crater, as shown at the crater site is incorrect. The details are discussed in the text.

ble object at a depth of 164 feet," with joy, they assumed they had found the meteorite:

"The drill holes in the crater have not been extended beyond the level of the **meteor encountered at 164 feet**, but the strata below the meteor **presumably** are **undisturbed**." Note 7.12q

Drilling at the Arizona and Odessa craters was not 'scientific' because the drill rigs were designed to drill water wells. They did not produce core samples and provided little reliable data for geological study. With assurance from the driller, drilling continued to a depth of 165.5 feet where the truth about the 'meteorite' was revealed:

"The driller, who had had much experience drilling water wells in the area with the same rig, **assured us** that he had never before encountered 'a rock' that **he couldn't drill through**.

"Despite all of this, the impenetrable rock turned out to be just that—a **highly silicified segment of conglomerate not noticeably altered**, and in its normal stratigraphic position." Bib 169 p13

Surprisingly, the failure to locate a meteorite did not alter the researchers' conclusion that the crater was impact-derived. *If* the crater was made by a meteorite, the smallish size of the crater, (168 meters) meant that the meteorite would have been slowed by the atmosphere and would therefore not have been traveling at high speed. Because the crater was *not* large enough to have been formed by a high-speed impact, the possibility of a vaporized meteorite was dismissed.

Researchers had to ignore direct evidence from the 1960s Sikhote-Alin meteorite impact in Russia. Craters from that observed fall were similar in size to Odessa. The Russian event *demonstrated* that smaller impact craters would *contain meteorite fragments*. Still, researchers chose to hold to the belief that the Odessa Crater was of impact origin.

Questions as to why the meteorites found *outside* the crater were not in their own crater, as they should have been had they 'really' come from space—remained unanswered.

A 1967 report further documents that *neither glass nor*

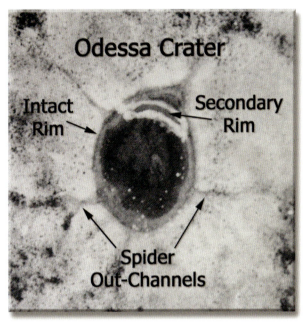

Fig 7.12.10 – This high-altitude aerial photo of the Odessa Crater was taken during the 1930s *before trenching* and excavation work had begun. Notice the crater's non-circular shape and intact rim around the crater. The primary crater had been nearly filled and a second crater was discovered nearby that was completely filled in. No clear geological explanation has been put forth to explain how these craters were filled if they had been formed by impact. There was also no explanation for the secondary rim on the north end of the crater or the 'spider out-channels' leading away from the crater. None of these features fit with the impact theory, but they fit well with hydrocrater formation. Buell Hydrocrater is another example (see Fig 7.8.7). Courtesy of G. L. Evans and C. E. Mear (2000).

Fig 7.12.11 – Iron ejecta found outside the Odessa Crater contains the familiar Widmanstätten pattern, which was 'thought' to exist only in meteorites from space. This contributed to the belief that the crater was an "impact" crater. However, no iron specimens were ever found **in the main crater**, instead they were **ejected** out of the crater. Other 'meteorites' found in Crater No. 2 looked nothing like these irons, and were made of minerals such as limonite, minerals associated directly with hydrothermal environments.

the other two 'impact criterion,' shocked rock and coesite, existed at the Odessa Crater:

"…**no other evidences of explosion**, such as **coesite, glass, shocked rock**, or **resolidified droplets have been found**." Note 7.12r

No evidence of either low-speed or high-speed impact was found, yet researchers tenaciously clung to the belief that Odessa was an impact crater! Because iron-nickel nodules found outside the crater had no analogous origin in modern science, they were deemed "meteorites" from space. Investigators were not asking why they were not melted or why they were not curled into shrapnel, like other documented falls—they simply accepted that they were meteorites, period.

In a 2000 report on the Odessa Crater, the researchers are clearly not aware of the requirements of a hypervelocity impact or of hydrocrater formation evidence:

"Among the unique changes produced in rocks commonly associated with meteoroid explosion craters are several kinds of shock metamorphism retained in the impacted rocks. Perhaps the best known and **most definitive** of these features are known as **shatter cones, coesite, and rock flour**, all of which are formed instantaneously by **hypervelocity** shock waves from the explosion of crater-making bolides. **Rock flour and shatter cones are present at the Odessa Main Crater**." Bib 169 p25

The researchers provide no evidence of a hypervelocity impact at the Odessa Crater; glass and vaporized meteorite material was totally absent. Because of the smallness of the crater, coesite, or shatter cones from impact are unlikely to be encountered. One interesting characteristic of the crater, rock flour, has never been explained or proven how it comes from impact. Like the shatter cones, *it can come from hydrothermal vents*. The researchers reported *one possibility* of "a shatter cone," but shatter cones are almost always naturally cone-in-cone structures, found in various non-impact geological settings.

The researchers acknowledged that other shatter cones would have to exist if they are to be used as 'evidence' for impact, but they assumed they were "probably" destroyed:

"Nearly all of the naturally-exposed limestone boulders in the ejecta rim have been reduced greatly and modified by erosion, which **probably destroyed** or obscured any segments of **shatter cones** that may have been present in them." Bib 169 p25

Of course, this completely ignores the fact that shatter cones should have been evident in the shaft where they would not have been subject to erosional modification. In summary, the evidence of impact origin, the meteoritic material, the glass, the shatter cones, coesite, and shocked rock and the rock flour, or more correctly the lack of these evidences, clearly do not support the impact origin of the crater. The evidence *for hydrocrater origin* is even more compelling.

The Odessa Hydrocrater Evidences

The 165-foot shaft dug at the crater is the only known example found in which researchers were actually able to go beneath the crater and observe a portion of the diatreme. Vandalism caused the shaft to be sealed years ago, but perhaps it can be reopened in the future, since all of the observations were made with an 'impact paradigm' bias. We can still pull many facts from the written papers that support a hydrocrater origin. The first of these are the east-west lateral "fractures":

"In the still intact part of the original unit of rock flour exposed in the walls of the shaft were a number of **east-west fractures** and minor wavy features somewhat **similar to those seen in alluvium in areas of active earthquakes**. Moreover, its noticeably undulating contact with the immediately underlying stratum of basement sandstone was suggestive of some **back and forth lateral movement**." Bib 169 p26

An impact would probably have produced *random* fracturing beneath the crater, not evidence of "back and forth lateral movement" in fractures "similar to those seen" in areas "of active earthquakes." This establishes the first evidence of hydrocrater origin; fault lines or **fractured earthquake features**. The next evidence of hydrocrater origin is the presence of **water** in the vicinity of the crater. Researchers reported that water was found 200 feet below the crater in a well drilled near the main crater:

"A water well drilled near the Main Crater in 1940 encountered the **water table at about 200 feet** beneath the surface." Bib 169 p38

The smoking-gun evidence that Odessa is a hydrocrater is the presence of a **diatreme**. Researchers during the 1967 "exhaustive drilling of the Odessa Crater noted that the crater had the "shape of a bowl":

"The greatest puzzle, however, lies in the fact that the **exhaustive drilling** of Odessa indicates that the original crater had the **shape of a bowl, the sides of which apparently** had **not** been penetrated by the meteorites. This is what one would expect from a **true explosion crater**..." Note 7.12r

Brandon Barringer was correct in his assessment that a "true" **impact** "explosion crater" would have the "shape of a bowl." But the "exhaustive drilling," as seen in Fig 7.12.12 clearly shows that drilling was *never conducted on the sides of the crater,* as represented by the drilling markers, which means it was not so exhaustive after all! Once again, Barringer ignored the direct evidence from the trenches that *were dug at the edges* of the crater, where the **true shape** of the original crater was revealed.

Fig 7.12.13, made by Evans and Mear in 2000 from the trench excavation observations, verifies the edges of the crater are of a thin **convex form, matching the shape of a diatreme,** opposite the concave bowl shape Barringer had presumed. In fact, Fig 7.12.13 is quite similar to the left edge of the Arizona Crater diatreme, seen in Fig 7.11.19.

Barringer apparently missed the 1941 research by Sellers and Evans, who were impact supporters themselves. They described the filled area under the crater from drilling and trenching evidence as an "**inverted cone**":

"From the center of No. 1 the rock was expelled, as previously described, to a maximum depth of at least 90 feet and was shattered to at least depth 103 feet, **probably deeper**. In addition to expulsion, the rock was thrust or shoved laterally. The resulting space from

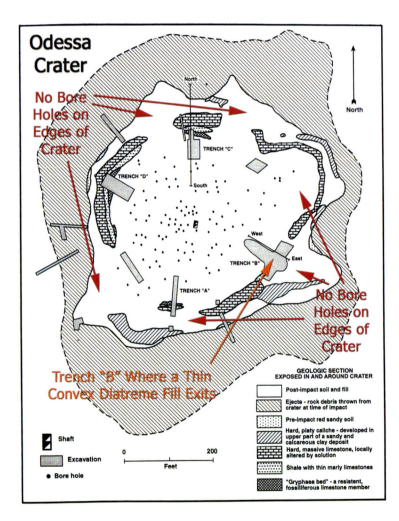

Fig 7.12.12 - This geological map of the Odessa Main Crater shows the location of the boreholes in the crater. Brandon Barringer had indicated that the drilling was "exhaustive," stating that the ground beneath the crater was in the "shape of a bowl." The red arrows in the above map indicate the absence of drilling activities around the edges of the crater, which meant Barringer's determination of the crater's underground shape was faulty because he had not seen the whole picture. It was after the trenches (shown in the diagram) were excavated on the crater's rim that a **convex-shaped diatreme-like fill** was revealed. Fig 7.12.13 is the photo and diagram drawn by the original investigators showing the diatreme-shaped edge of the Odessa Crater (Trench "B" above).
This map is courtesy of G. L. Evans and C. E. Mear (2000), adapted by the author.

The cross section of the east rim of Odessa Crater does not exhibit the steep slope of a bowel-shaped crater.

How did scientists miss such obvious evidence? What other 'impact craters' have been misidentified?

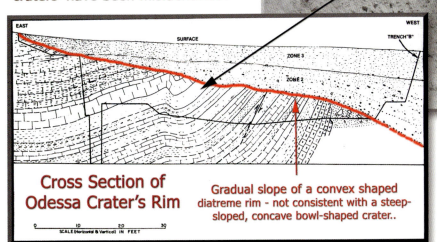

None of the bore/drill holes penetrated the edges of the crater and it was not until the trenching as seen in this photo was done that the thin diatreme edges of the crater were found.

Cross Section of Odessa Crater's Rim

Gradual slope of a convex shaped diatreme rim - not consistent with a steep-sloped, concave bowl-shaped crater..

Fig 7.12.13 - This diagram and photo represent a cross-section of the primary Odessa Crater through its East rim. Like the Arizona Crater, the rim of this crater is *very revealing*. If the crater was formed by impact, the rim would exhibit the **steep, downward trend** of a bowl. The Odessa crater and the Arizona Hydrocrater both have a **gradually sloping, convex rim**. It seems amazing that investigators missed important evidence revealed in the trenches that the Odessa Crater had a diatreme beneath it. Because iron 'meteorites' were found outside of the crater, impactologists apparently decided that the crater was made by an impact and completely missed the evidence of the diatreme. Images courtesy of G. L. Evans and C. E. Mear (2000).

which rock was expelled is of the form of an **inverted cone**. The width of the **cone** at plains level is about 500 feet; the depth is at least 90 feet or, including the 'rock flour', **at least 103 feet**." Note 7.12s

Even in 1941, researchers realized the shape of the sediment below the crater was not in the "shape of a bowl," but was more like an inverted cone. Researchers throughout the history of this crater have struggled with what they saw, but they diligently tried to make it 'fit' the impact theory. A 2000 report on Odessa Crater states:

"About 70 feet of **relatively undisturbed** strata that underlie the base of the crater were observed to a depth of 165 feet in the main shaft." Bib 169 p22

What does "relatively undisturbed" mean? Two pages later, the researchers note the fact that this "relatively undisturbed strata" actually contains "several zones" that are fractured and faulted:

"The section from 96.8 to 165.5 feet consists of sandstone and gravel as described in Table 1, but **several zones are fractured and faulted**, probably as a result of the meteor impact." Bib 169 p24

Unfortunately, we have been unable to find clear diagrams of the fractured zones and do not know how far down the 'disturbed' sediment goes.

Deeper fault zones are the evidence supporting frictional earthquake heating, which is necessary for a steam explosion, and they confirm one other thing about the diagram posted at the crater site—the Odessa Crater cross-section in the visitor's center is not correctly drawn. That diagram, the colored version in Fig 7.12.14, shows disturbed layers of *only* the first 95 feet below the crater floor. The diagram is *very* misleading for several reasons; First, is does not correctly show the crater rim as being *convex* shaped, as shown in Fig 7.12.13 and redrawn on Fig 7.12.14 in red. Second, the diagram does not show the fractured and faulted area below the crater, which along with the convex rims establishes the "inverted cone" shape of a diatreme.

Finally, Odessa Crater researchers failed to account for the "bowl shaped" fractured area of rock, noted in blue in Fig 7.12.14, that should be present if the crater was formed from impact.

A satisfactory method to clear up the confusion of the Odessa diatreme would be to conduct a seismic study, similar to the study at the Arizona Crater. This would establish that the "meteorites" found near the crater were actually iron ejectites blown out during a phreatic explosion. The study will aid in the understanding of other hydrocraters and the types of geological environments in which they occur.

The Odessa Flat Floor Evidence

Even though the Odessa Crater is over 550 feet in diameter, it is currently only 15 feet (5 meters) deep at its lowest point. As Fig 7.12.14 shows, it is almost completely filled with sediment materials. How did the sediment get there? Laboratory impacts and near-surface nuclear explosions simulating impact excavate 'bowl shaped' craters with a minimal amount of fall-back into the crater. The 'excess fill' in the crater is a huge problem for most of the large so-called impact craters on the Moon where the erosion sci-bi does not even come into play.

Impact theorists at the Odessa Crater have only one explanation for the excess fill, and that is erosion. However, the evidence dismisses any possibility that the crater was filled by erosion. First, let's consider the relatively 'short time' of the crater's existence. Crater scientists have determined the crater is at most, only 25,000 years old:

"Actually, Shoemaker puts the age of Barringer from the geology at about 20,000 years, and Evans gives **25,000 years for Odessa**. Goel (1962), from a study of the Carbon 14 content, finds **11,000 plus or minus 4,000 years for Odessa**, which could well mean 19,000 or more, if the actual error is twice the

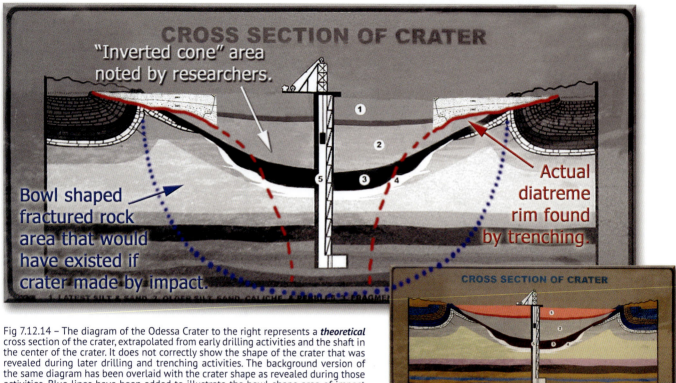

Fig 7.12.14 – The diagram of the Odessa Crater to the right represents a *theoretical* cross section of the crater, extrapolated from early drilling activities and the shaft in the center of the crater. It does not correctly show the shape of the crater that was revealed during later drilling and trenching activities. The background version of the same diagram has been overlaid with the crater shape as revealed during those activities. Blue lines have been added to illustrate the bowl-shape area of impact fractured rock that would exist if impact had really occurred. The red lines show the actual "inverted cone" shape of a diatreme identified by investigators during excavation of trenches on the crater's rim.

probable." Note 7.12t

Accordingly, the geologist cannot assume a 'shallow sea' or a glacier covered the crater, a common deference to explain such geological enigmas. Scientists also claim the environment throughout the craters existence was "semiarid":

"Based on studies of Late Quaternary deposits of the High Plains it appears that **the present semiarid climate has existed for most of the Holocene, and the plains have been some type of grassland for much of the past 20,000 years** (Holliday, 1995)." Bib 169 p15

Therefore, a *dry non-changing environment* existed for most of the craters life. As was noted in Fig 7.12.10, when the crater was first studied, the rim of the crater was intact. Formed of limestone, it did not allow any outside erosional forces, besides the occasional rain and wind into the crater. This leads us to ask another FQ:

FQ: How do erosional forces outside the impact rim fill the crater with bedded layers of **different colored clays and gravel?**

Of course, they could not. The crater was never filled by surface erosion because these types of clays and gravels do not exist on the surface near the crater. The 'excess fill' could have come *only* from within the crater, from a diatreme below. This explains the *different* colored clay and gravel bedding, all from multiple eruptive events, as different materials were carried from depth during successive eruptions. Moreover, "manganese fills joints in the clay"—a process that can only occur in a hydrothermal environment:

"Overlying the basal fall-back material, from 64 to 77 feet in Zone 2, are **thin beds of gravel**, one-half to four inches thick interbedded with thicker beds of laminated silty, light **brown-ish-red and yellow clay. Manganese fills joints in the clay. The gravel, clay, and silt are distinctly bedded, and lie almost perfectly flat.** Bib 169 p.26-27

From a survey around the main crater, researchers identified "two unexplained anomalies." The first was that iron ejectites, cast from the main crater were found "at shallow depth." This is what one would expect if it had come from a steam explosion. It is anomalous because the observations were the *opposite* of what would have been expected had they been actual meteorites traveling together en masse. The other 'meteorites' would all have each formed deep craters themselves—such was not the case.

One other crater was found, called simply Crater No. 2, which was the other anomaly:

"In connection with the magnetometer survey of the large crater area **two unexplained anomalies were noted**. One of these was shown by subsequent excavation to be due to some meteorites in the soil **at shallow depth**, and the **other**, upon excavation, proved to be a small meteor crater, called **Crater No. 2**." Bib 169 p33

Why was Crater No. 2 an anomaly? Because like the main crater, it was filled with a "clay-like deposit" with no origin from the surface:

"This small crater is approximately circular in outline, about 70 feet across and 17 feet deep. It had become **completely filled**... was formed in relatively incoherent materials, chiefly a **clay-like deposit, probably largely solution residue from the Cretaceous limestone**." Bib 169 p33

The geologists had no mechanism to account for the completely filled Crater No. 2. How could surface limestone have eroded to fill the crater with a "clay-like deposit," and leave no evidence of erosion? Limestone is not silica-based like the clay sediment and there is no method to explain how rainwater

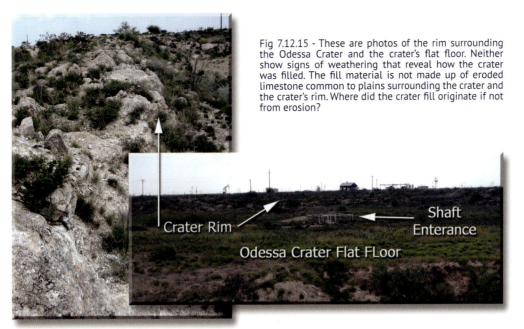

Fig 7.12.15 - These are photos of the rim surrounding the Odessa Crater and the crater's flat floor. Neither show signs of weathering that reveal how the crater was filled. The fill material is not made up of eroded limestone common to plains surrounding the crater and the crater's rim. Where did the crater fill originate if not from erosion?

overflowed the crater's rim from outside the crater.

Researchers state that Crater No. 2 was filled primarily with "post-impact **pond** deposits", [Bib 169 p22] however, they give no explanation where the "pond" fill came from. As with the main crater, it could not have come from outside the rim—it could have only come from below.

This filled second crater was truly an *anomaly* for the researchers, because it could not have been filled by an impact and there was not sufficient time for it to be filled by erosion.

The impact theory meant there was only one explosion at Odessa, and no mechanism to account for the infill of either crater. With the Hydrocrater Model, multiple eruptions are common, and the interbedding of silt, clay, gravel and sand is easily understood. The layering provides us a *history* of the crater's existence.

The Odessa hydrocrater and many others like it have evidence of water; some being filled with water, others, like the Wolfe Creek Hydrocrater, have a relatively high water table that supports thicker vegetation. The availability of water in such places would be oases for many animals, both the hunters and the hunted. Kills at the water hole would have been common, and subsequent eruptions would bury the bones and remains of these animals. Trenching and shaft diggings at the crater did reveal many poorly fossilized bones:

"Fragmentary, but unidentifiable, fossilized bones were recovered from Zone 3 in the Main Crater shaft at depths ranging from 8 to 20 feet below the surface. Apparently, the deposition of sediments in this upper part of the section was at such a slow rate that the bones were **heavily weathered**. **Badly** decomposed jaws of the **extinct horse**, *Equus* sp. were found in Zone 3 between a depth of 9.5 feet to 11 feet. A fragmented **elephant tooth** was collected from the shaft at a depth of 17 feet." [Bib 169 p27-28]

The fossils found at the crater were weathered and decomposed because they were not subjected to high enough pressure and temperature to have been transformed fully into siliceous rocks. Details on this will be given in the Fossil Model chapter.

Other than the incidental reporting of the fossils in the crater, researchers have been surprisingly silent. One wonders if it is because the fossils do not support the impact hypothesis; one more pesky anomaly to account for. Normally, erosion could explain the bedded layers and burial of bones, but here, erosional materials could only come from the crater's rim—and that doesn't work.

The trenches in the main crater revealed no evidence that the eroded sedimentary material came from outside the crater, but excavations at Crater No. 2 did produce some iron specimens in the crater.

The Odessa Iron Ejectite Evidence

Although all of the known iron ejecta were thrown clear of the Main Odessa Crater, Crater No. 2 was found because iron specimens were buried near the center of the crater causing elevated magnetic readings. It is possible that there are other craters in the surrounding area that have no iron; future seismic readings may yet identify them.

The main crater, Crater No. 2 and probably other craters in the area were filled with subterranean sediment from multiple eruptions, much as old geysers and mud craters are today at Yellowstone Park in Wyoming, USA. The size of the crater is indicative of the size of the explosion, and hence the ability to eject denser materials. The primary crater at Odessa is 550 feet in diameter, significantly larger than Crater No. 2, which is only 70 feet in diameter. Apparently, the explosion that formed it was not powerful enough to have ejected the iron masses; the iron may have been formed far below the surface much the same way continental ore deposits were— formed in diatremes.

What is known is that Crater No. 2 is not "bowl shaped" but has the shape of a diatreme, much like the main crater. [Bib 169 p22 Fig. 14]

The iron masses taken from Crater No. 2 are in the form of "shale balls." Two examples can be seen in Fig 7.12.16. Notably, they look nothing like the other 'meteorites' from the area around the main crater, shown previously in Fig 7.12.11. How could the two irons be so different if they came from the same meteorite? This is not an easy question to answer if the iron's origin was an impacting meteorite.

Diverse environments below the crater can produce irons of varying composition, and may be different in separate hydrocraters, even though the craters are in close proximity to one another.

According to researchers, iron found in Crater No. 2 had "oxidized to a kind of limonite":

"All of the estimated **six or seven tons** of meteorites found buried in bedrock beneath the crater [#2] base have been completely **oxidized to a kind of limonite**." [Bib 169 p19]

The dilemma facing the researchers is that real meteorites do not *oxidize into limonite*. They can rust, but limonite is not a form of rust, it is an *ore*. The iron ore limonite is a minor source of iron ore but is found worldwide; it, along with the

Fig 7.12.16 – Odessa "shale balls," also called oxidized limonite meteorites are found at the Odessa Crater site at Crater No. 2. Notice that the cut specimen on the left does not exhibit the crystalline structure typical of 'meteorites.' The cavity actually reveals a connection to hydrothermal minerals, which do not come from outer space. These iron specimens are not 'rusting,' nor do they exhibit regmaglypts seen on the iron ejectites in Fig 7.12.11. Why are they different from the iron specimens found outside the main crater if they all came from the same meteorite? Impactologists have no idea how two remarkably different types of iron meteorites—"shale balls" and irons—can come from the same meteorite. These specimens are an excellent example of the different kinds of iron ejectites that can form in hydrofountains and diatremes. They could have been formed at different times and at different depths, a common phenomenon of hydrocraters.

more abundant ore hematite, is created in a hypretherm, not from rusting. Geologists saw similarities between the Crater No. 2 irons and other *known* mineral deposits, commenting on the "close similarity to septarian concretions":

"Numerous specimens, when cleaned, **had a close similarity to septarian concretions**, and would break into angular fragments when removed from the earth." Bib 169 p19

The ***close similarity*** to septarian concretions is due in part to their common origin, but science does not understand how meteorites or septarian concretions are formed. This is noted in an encyclopedia entry on **septarian concretions**:

"The process which created the septaria, which characterize septarian concretions, **remains a mystery**." Note 7.12u

The next chapter will more fully reveal the answer to the septarian concretion mystery, but for now, it is important just to recognize that these concretions are a mystery, and that they are only found near the surface of the Earth—like the iron ejectites are.

The Odessa Rock Flour Evidence

Like the Arizona Hydrocrater, the Odessa crater has a similar 'smoking gun'—pure rock flour. We have already addressed the significance of the pure white silica and how it could not have come from impact, but the Odessa crater adds a new element to the silica 'smoking gun'—*rock flour blocks*:

"The rock flour at the Odessa Main Crater is in a **dish-shaped unit**, about 250 feet in diameter, and roughly **circular in outline**. It lies immediately beneath the post impact crater fill, and is estimated to occupy about 5,000 square yards, with its center in the immediate vicinity of the shaft. Its thickness ranges from less than one foot at the margins to some five or six feet at the center. Clearly, however, its original thickness was momentarily much greater, at least in the central area, for a number of **large intact blocks of rock flour and numerous smaller ones are seen in a jumbled disarray in the basal fallback unit of crater fill**." Bib 169 p25-26

Try to imagine an impacting meteorite changing various multicolored rock layers instantly into a single layer of white silica fine flour. And then imagine how the supposedly instantaneously created white flour became *solid blocks* that were thrown back into the air, coming to rest on the bottom of the crater.

The Odessa crater is not large enough to have been a hypervelocity impact event, and therefore has no mechanism to account for the homogenization of the multiple rock layers into white silica flour.

The Russian Sikhote-Alin fall contained significantly more iron than Odessa, and all else being equal, would have experienced a higher speed of impact. Yet there was no rock flour there.

Although we had fun imagining how the rock flour and rock-flour blocks were formed according to science, the very idea that rock flour could have been made from a slow-speed impact is wrong, and it is serious business for researchers to have totally missed the mark at Odessa in their interpretation of the data.

While investigators have not been able to provide an adequate explanation of how impact forms rock flour, researchers at Odessa speculated about one method actually quite close to the truth—a "steam explosion":

"Evidently **these blocks were broken free and thrown upward** by a sharp rebound of compressed bedrock—perhaps aided by a localized **steam explosion of super-heated ground water in the underlying aquifer**." Bib 169 p26

The Odessa Oil/Salt Diatreme Evidence

One cannot visit the Odessa Crater without taking notice of the nearby oil wells (Fig 7.12.17). Some are located just beyond the rim of the crater. Amazingly, crater researchers do not mention this fact in their research, but they did acknowledge that other 'meteorite impact' features "produced large quantities of oil":

"Some of the buried **complex meteorite impact features** (astroblemes) that have been discovered during exploratory drilling for oil and gas **have produced large quantities of oil**." Bib 169 p47

The scientists saw a *direct* relationship *between craters and oil*—yet they *still think* the craters are made by rocks from space! A 2003 geological report from *The Leading Edge* states, "meteorite im-

> "Evidently these blocks were broken free and thrown upward by a sharp rebound of compressed bedrock— perhaps aided by a localized **steam explosion of super-heated ground water in the underlying aquifer**."

pact structures/craters *can be very good petroleum prospects*":

"The **buried meteorite impact structures/craters can be very good petroleum prospects**. Steven River in Alberta, Viewfield in Saskatchewan, Ames Hole in Oklahoma, and Red Wing Creek in North Dakota are **examples of oil-and-gas fields producing from buried impact structures**." Note 7.12v

Did they actually say that? Apparently, rocks from space have a "very good" track record for predicting "petroleum prospects" and maybe the next documented fall will spur a rush of anxious oil drillers to the impact site in a quest for oil!

This may sound strange, but for the moment, as of this writing, the entire world thinks the Odessa Crater was made by impact, along with a whole crop of oil producing impact structures.

The Odessa Kaolinite/Mercury Evidence

One final direct evidence of the Odessa Hydrocrater diatreme is the **Kaolinite/Mercury Evidence**. This evidence is as simple and as powerful as the Rock Flour evidence reviewed earlier, and it should give researchers the basis of a method for determining if a crater is of hydrofountain origin. All they need to do is to analyze the type and quantity of elements and minerals inside the crater and compare them to those outside the crater. Just as was discovered during radiometric testing at Wolfe Creek Crater, many hydrocraters have elements and minerals in a diatreme pipe that do not exist beneath the ejecta field surrounding the crater. Other craters' diatreme structures may share similarities with the surrounding plain, but at elevated levels. In either case, it is because the diatreme is the remnant of a watery environment very different from the surrounding plain. There, heat, pressure, and diverse microbes present in a unique hypretherm formed a host of new minerals.

One such mineral is **Kaolinite**, a very distinct aluminum silicate mineral used commercially in ceramics, medicine, light bulbs and even toothpaste. The diatreme beneath the Odessa Hydrocrater was once an environment not unlike the hyperthermal environment that sometimes exists at the bottom of the ocean, a place inhabited by unique microorganisms that thrive in a high pressure, high temperature environment. The byproducts of these organisms are biogenic ore minerals. Researchers are only now beginning to understand this mineral hydroformation process.

At the Odessa Crater, the rock flour inside the crater consisted of a "high percentage" of kaolinite. By comparison, it was not found outside the crater area at all. To explain the apparent oddity, researchers reasoned that it must have come after the impact and they employed a favored sci-bi, "weathering," to explain its existence:

Fig 7.12.17 - This view from the rim of the Odessa Hydrocrater shows how close oil rigs are to the crater—the Odessa Hydrocrater lies in the middle of a rich oil field, also known for its abundant salt diapirs. Scientists have actually said "The buried meteorite impact structures/craters can be very good petroleum prospects." One wonders how a rock from space knows where the oil deposits are.

"Petrographic and x-ray investigations (not now available) revealed a **high percentage of kaolinite present in the rock flour** along with angular particles of shattered quartz sand. **By comparison of the rock flour with samples of rock from the same stratigraphic zone outside of the central crater area**, it seems evident that most or all of the kaolin is an **alteration product** resulting from post-impact weathering." Bib 169 p26

The weathering sci-bi is unsubstantiated for several reasons; the crater was not filled with erosional materials, as already discussed, and no evidence of kaolinite-by-weathering outside of the crater was shown. The implication that the kaolinite is an "alteration product" from rock included in the "stratigraphic zone" outside the crater is unsupported because of the presence, or rather absence of mercury.

"Roach *et al.* (1965) assayed rock samples to a depth of about 130 feet below the surface of the Main Crater and found that the rocks have **significantly less mercury than stratigraphically equivalent samples collected 1.1 miles west of the Main Crater**." Bib 169 p26

Impactologists have had a difficult time accounting for both of these chemical anomalies at the Odessa Crater. Their research continues to point them to water, but the dry, desert environment has left them with leaching or weathering, and little else.

The Upheaval Dome Impact Myth

Located on the Colorado Plateau in Southeastern Utah, Upheaval Dome Crater (Fig 7.12.18) is another well-studied crater, probably second only to the Arizona Crater. Like the Arizona Crater, it has generated significant controversy in the scientific community. Like the Odessa Crater, there are oil wells located northeast of the crater in an area known as the Big Flat Oil fields. 1000 meters below the Upheaval Dome lays the thickly layered Paradox Formation, which includes at least five layers of salt, one being nearly 100 meters thick.

It seems meteorites are not only useful for identifying "good petroleum prospects" with the impact craters they create; they are good at locating salt deposits too!

Since the 1920s and through the 1990s, the plasticity of salt was the sci-bi that explained how trillions of tons of material were pushed up into a crater or dome complex. We can't explain this bizarre pseudotheory here, but suffice it to say that it held the number one spot until the 1980s, when Shoemaker came on the scene. As impact-mania gripped the world, many other craters were shuffled into the impact basket and Upheaval Dome was no exception.

Although Upheaval Dome is one of the *most obviously* identifiable hydrocraters, the science of meteoritics went out of their

Upheaval Dome Crater

Evidences Against Impact
1. No Meteorites
2. No Glass or Melted Rocks
3. No Coesite/Stishovite
4. No Shatter Cones
5. No PDFs

Evidences For Hydrocrater
1. Hydrofountains/Dikes
2. Pure White Silica Deposits
3. Quartz Nodules
4. Green Mountain Deposit
5. Dome & Out-Channel

Fig 7.12.18 – The Upheaval Dome Hydrocrater has astonishingly been called the "best exposed impact crater on earth," yet the evidence, listed to the right, is so easily observed and so simply understood, it is baffling how scientists could have come to such a conclusion. Actual dome formation by steam explosions is understood; the Panum Hydrocrater in California is a good example of a hydrocrater dome. Impactologists have overlooked other Upheaval Dome evidence, such as the massive Out-Channel coming from the crater.

in 1999, one influential planetary scientist summed up their thinking:

"**Upheaval Dome** in the northern part of Canyonlands National Park is the **best exposed impact crater on the earth**." Note 7.12w

Sometimes it is good just to see what our tax dollars are paying for. After the geologist opened his article in the Utah Geological Association's publication with the statement that Upheaval Dome Crater was the "best exposed impact crater on the earth," *he went on to admit*:

"**No fragments of the impactor, no highly shocked target rocks, and no melt rocks produced by the impact have been identified**." Note 7.12w

So, definitive impact criteria such as a meteorite body, shocked rock, and melted rock don't exist, but it doesn't end there. Continuing to the final paragraph of the geologist's report on Upheaval Dome Crater, we discover the absence of even more impact criteria:

"At this writing Upheaval Dome has **not** been conclusively proven to be an impact crater to the satisfaction of the last skeptic because the 'smoking gun' in the form of an **impactor fragment, true shattercone, impactite, melt rock, planar deformation feature, coesite, stishovite**, or some such definite feature **remains to be discovered**." Note 7.12w p9

The researcher eliminated every major qualifier of impact, so what did he say was the "primary evidence" supporting the impact origin of the best-exposed impact crater?

"**The primary evidence that Upheaval Dome is an impact structure includes**: (1) a morphology that is consistent with

way trying to link the crater to impact, making mistake after mistake in defining its origin. Following the Lunar and Planetary Science Conference

"**Upheaval Dome** in the northern part of Canyonlands National Park is the **best exposed impact crater on the earth**."

proven impacts, and (2) the presence of subsidiary structures having sequencing, forms and stress indicators expected in an impact." Note 7.12w p5

That was it. The "primary" evidence is a consistent morphology (same form) of "proven impacts." Note that the three best "proven impacts" *were* the Arizona, Wolfe Creek and Odessa Hydrocraters. Upheaval Dome is practically off the scale in dissimilarities to these three craters. To begin with, no hard surface impact study (without significant underground water) has ever produced a central uplift. Impacts create 'bowls,' not 'domes.'

This researcher was not the only scientist to stake the claim of 'impact origin' for Upheaval Dome. In 1999, seven geologists jumped on the bandwagon and summarized their position at that year's Lunar and Planetary Science Conference:

"Stratigraphic uplift observed in the center of **Upheaval Dome** is the result of convergent displacement of the wall of a transient cavity **formed by hypervelocity impact**, not Paradox salt diapirism." Note 7.12x

The evidence they presented that the crater *was made* by impact was that Upheaval Dome had "faults," "dikes" and "no exposures of salt."

Where were the "definitive impact criteria" asserted by earlier scientists—shatter cones, PDFs and coesite? A crater of this size should have all of these criteria present, in abundance, and physical evidence of the impactor (meteorite) should be discernable, if Upheaval Dome was *truly* an impact structure. It seems impact science could not have drifted any further from the truth. Without the Hydroplanet Model, modern science researchers look further and further into the abyss of the Dark Age of science.

Upheaval Dome's Hydrocrater Evidences

There are many and specific evidences that Upheaval Dome is a hydrocrater. In 1995, Shoemaker and his colleague attempted to link a fault line called Robert's Rift to the impact of Upheaval Dome, but cited no specific, hard evidence of how impact caused the feature. Re-evaluating the report, we find the evidence well supports hydrofountains in the Upheaval Dome area, including Robert's Rift, a canyon wall nearby that included limestone blocks transported from "1,000 meters below":

"Hite found that **Roberts rift** contains breccias that are comprised of **upwardly transported clasts**. **Most startling** was his discovery of fossiliferous limestone blocks up to 0.3 m across which were demonstrably derived from the underlying Pennsylvanian Honaker Trail Formation some **1,000 m below**. Black shale fragments in the breccia could have come from as low as the underlying Paradox Formation." Note 7.12y

This was unrecognized hydrofountain activity and therefore "most startling" to geologists because they did not know about the Hydroplanet Model. Because researchers have been so focused on impact, they have missed some of the best geological evidence of hydrocraters, examples being the ***white silica deposits*** at the Arizona and the Odessa Hydrocraters.

In the Rock Cycle Pseudotheory, we revealed how erosion does not generate pure silica sediment. This type of granular sediment grows only in a hypretherm. The same holds true for impact explosions—no scientist has ever shown how an impact can produce a large homogenous body of pure white silica from an impact into red and brown sediment; an idea almost beyond imagination. Upheaval Dome also contains *dikes of white sandstone,* at the center of the crater, where fluxionary waters from the diatreme would have continued after the eruption, bringing the pure white sediment to the surface:

"The **clastic dikes** are composed of **orange, red, or white quartzose sandstone. White dikes are found mainly in the center of the dome, and the orange and red dikes occur near the center to the perimeter of the dome**. Orange-colored dikes in the Chinle Formation have in some cases been physically traced to sources at the base of the Wingate Sandstone, but in general, protoliths for the clastic dikes are **inferred** on the basis of color, mineralogy, and the **assumption** that they are not far traveled." Note 7.12z

Because they know of no practical origin based on impact theory, researchers are left to 'infer' and 'assume' where the sandstones, including the white quartzose (silica) sandstone originates. Fig 7.12.19 is one example showing the dikes and the evidence of the pipes that brought sediment of different colors to the surface. Note also, the open, fossil hydrofountain pipe.

Upheaval Dome Quartz Nodule Evidence

Hydrorock and Hydrofountain Sediment were first introduced in Fig 7.7.17, which shows cobblestones and large boulders that were carried to the surface and solidified in a sandy/clayey matrix. In other words, it was not only clay and sand sized particles that steam and water pushed to the surface—it was all types and sizes of sediment and minerals.

Upheaval Dome is a great source of *quartz nodules* deposited on the surface by hydrofountains. Not all researchers were caught up in the 'impact wave,' and by taking a contrarian view of the crater, they interpreted the quartz nodules and other materials as being low-temperature, possibly hydrothermal in origin. The evidence gave "no indication of being impact-derived":

"**Upheaval Dome**, in Canyonlands National Park, Utah, USA, is a unique structure on the Colorado Plateau. It has earlier been interpreted as an impact structure or as a pinched-off salt diapir. Some **subrounded quartzose fragments** were found in a ring depression near the eastern margin of the structure and, based on vesicularity and apparent flow structure, the fragments were interpreted by early researchers as 'impactites.' Our petrographic studies show no indication of a high-temperature history and are in agreement with a slow, low-temperature formation of the **quartz nodules**. Compositionally, the lag deposit samples are almost pure SiO_2. They show **no chemical similarity** to any of the possible target rocks (*e.g.,* Navajo Sandstone), from which they **should have formed by melting if they were impactites**. Instead, the samples have relatively high contents of elements that indicate **fluid interaction** (*e.g.,* **hydrothermal growth**)... Thus, we interpret the 'lag deposit samples' as normal low-temperature (hydrothermally-grown?) quartz that **show no indication of being impact-derived**. In addition, a petrographic and geochemical analysis of a series of dike samples yielded **no evidence for shock metamorphism or a meteoritic component**." Note 7.12ae

The nodules are much more significant than modern geology has recognized. They are found everywhere on the Colorado Plateau and other geologically similar environments worldwide. Until now, there had been no explanation for their origin.

Once we understand the magnitude of sediment and water movement beneath the Hydroplanet Earth's crust, far more than has ever been imagined, we can begin to see how cobblestones and nodules can be formed and eroded before they are brought to the surface in a steam explosion, the likes of which are rarely experienced today. In the next chapter, the Universal Flood, the

Fig 7.12.19 – The vertical rock structure located in the heart of the Upheaval Dome Hydrocrater is a dike, or fossil hydrofountain. On the left are vertical layers of red sandstone, a white sandstone pipe, and an open vertical pipe where water and perhaps nodules were carried to the surface. Hydrofountain evidence like this covers a surprisingly large area of the surface near the dome, attesting to the naturally eruptive nature of the crater, and contrary to the supposed downward trend of an impact explosion.

widespread occurrence of nodules will be examined, along with the heretofore-unexplained dark desert varnish on the rocks at Upheaval Dome:

"The nodules are typically 5 to 15 cm in size, have rounded to angular shapes, and are mostly coated with thin **desert varnish**." Note 7.12aa p862

The researchers reasoned that the nodules must have come from deep below, because they are not found in any of the exposed geological layers. To them, the crystalline microstructure of the nodules revealed material that was obviously "the result of hydrothermal growth processes":

"The nodule ('lag deposit') samples are quartzose rocks, with only minor amounts of other minerals… Some larger crystals in such material are **twinned, obviously the result of hydrothermal growth processes**." Note 7.12aa p863

The composition of the nodules and the dikes in the dome formation provided "no evidence for an extraterrestrial component":

"Thus, **no evidence for an extraterrestrial component** in the quartzose nodules or the dike rock samples **was found**." Note 7.12aa p867

The geological components of Upheaval Dome were clearly *not* of extraterrestrial impact origin, but from directly beneath the crater, the culmination of high-pressure steam and subterranean sediments. Despite the overwhelming scientific evidence that Upheaval Dome Crater is a hydrocrater, the worldview of the crater's origin is *still* one of impact, as noted at wikipedia.org:

"**Upheaval Dome is an impact structure, the deply eroded remnants of an impact crater**, in Canyonlands National Park southwest of the city Moab, Utah, in the United States." Note 7.12ab

The impact-theory juggernaut continues to influence all fields of scientific thought, attributing the primary planet-shaping force in the solar system to impact cratering without satisfactorily providing the physical evidence to support it. The impact pseudotheory has not been sustained by truth, but by falsehoods, creating a crisis, which, as Kuhn said (in chapter 4), has laid the foundation for a scientific revolution.

Green Sediment and Water Ripple Erosion Evidence

In addition to the physical makeup of a landform or deposit, one of the geologist's primary objectives is to establish its age. In the case of Upheaval Dome Crater, more than 80 research papers and articles have been published since 1927, with no definitive age being determined:

"The **age of Upheaval crater** has **not** been determined." Note 7.12ac

With the arsenal of dating methods modern geologists use, surely some indication of age would have been derived, but the scientific community is silent on the crater's age, probably because the geological phenomenon is not accounted for by any theoretical impact-based age. This should have caused researchers to hold their opinions (theories) on the crater's origin until a new model was discovered.

It was Shoemaker who stated in 1984 that the Upheaval Dome Crater was 'actually' the bottom of an impact crater, and that "at least 1 km of strata" had been removed by erosion:

"**At least 1 km of strata** has been removed by **erosion** since the dome was formed." Note 7.12ad

Shoemaker made no mention of what erosional force could have stripped away that kilometer of strata, two-thirds of a mile, or over 3,000 feet. The arid environment of the crater

Fig 7.12.20 – Upheaval Dome has a mountain of green sediment surrounding the central dome, similar in consistency and color to the green deposits at the Buell Hydrocrater and at Green knobs in Arizona (Fig 7.8.7 & 8) as well as other places. Instead of eroding from other deposits, this green sediment and other colored material are biologically formed ore sediments formed deep in the Earth and brought to the surface through diatremes and hydrofountains. Such a concept is new to geology because the earthquakes required to produce such massive structures rarely occur today, but this does not mean they did not occur in the past, contrary to the modern science Uniformity Principle.

Where is the evidence of millions of years of erosion on these uplifted blocks of sediment?

This green colored formation is a mountain of sediment deposited by hydrofountain activity after the hydrocrater and out-channel was formed. Minimal erosion has taken place since.

Upheaval Dome Crater

Satellite View

Green Sediment is absent in out-channel

CHAPTER 7 THE HYDROPLANET MODEL

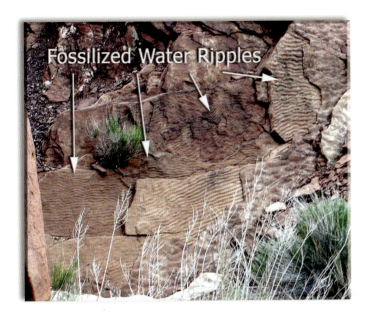

Fig 7.12.21 – These vertically oriented sandstone layers showing fossilized water ripples are found on the dome of the Upheaval Dome Hydrocrater. Why these fossilized ripples were not shattered by the meteorite responsible for the crater is a mystery to researchers. Hard questions like these become easy when we know the answer, and answers about these ripples will be set forth in the following chapters.

Flood event. This is apparent when examining the green sediment both up close and in the aerial photos in Fig 7.12.20. It is apparent that no significant erosion has taken place. Had the erosion taken place as Shoemaker imagined, gullies and desiccated rivers of green sediment stemming from the crater would be evident, but this is **not** the case.

Another piece of the erosion puzzle exists around the sides of the dome in the crater. Fig 7.12.21 shows sandstone slabs that were pushed up by hydraulic steam pressure. The slabs show layers of fossilized Water Ripples whose origin is unexplained by paleontologists. These are a huge mystery to the impactologist.

Why is the water-rippled sandstone **unshattered** if it was at the base of the impact crater?

The sandstone would certainly have been shattered by an impacting meteorite and erosion would have dealt a heavy hand had they been lying around for millions of years.

Upheaval Dome Out-Channel Evidence

One final and compelling piece of evidence is the obvious channel extending from the crater toward the Green River, seen in the distance of Fig 7.12.22. The Upheaval Dome Out-Channel canyon winds northwest, cutting through the outer rings of the crater and opening into a larger canyon as it makes its way to the Green River. There were no discussions among the impact theorist's reports concerning this huge canyon; moreover, no observed impact ever produced an out-channel.

The Out-Channel Evidence of Upheaval Dome is not unique to this crater; out-channels are a common feature of hydrocraters, many of which exhibit them in some form. Other hydrocraters, such as the Richat Hydrocrater have out-channels that provide evidence of sediment flow from beneath the crater. Out-channels exist on the Moon and on other planets, each one unequivocal evidence of their hydrofountain origin.

The Richat Hydrocrater Evidence

A composite of several satellite and 3D images of the Richat Hydrocrater in northwestern Africa are seen in Fig 7.12.23. Other circular, crater-like features can be seen nearby. Because

would seem to require millions of years of erosion for this to happen, if it could happen at all.

The reality of the crater's erosion is laid clear as the impact pseudotheory is abandoned and replaced by the Hydroplanet Model. That model, along with the Universal Flood, clearly establishes that such a crater was formed only a few thousand years ago, and it answers many puzzling geologic anomalies.

One of those anomalies is the mountain of green sediment seen in Fig 7.12.20. This fine-grained sedimentary material is located around the dome of the crater. Scientists have made little mention of this significant deposit and have been unable to define a true origin for the subaerial multi-colored clay and sand. They are a form of hydrosediment that was brought to the surface by hydrofountain pipes or, as in the case of Upheaval Dome, through a diatreme during the end of the Universal

Fig 7.12.22 – This aerial view of the Upheaval Dome's inner ring shows the Out-Channel that runs northwest into the Green river that can be seen in the background. No explanation how the Out-Channel canyon formed from impact is included in any known scientific literature. It is by far the largest single structure connected to the crater, affecting all the rings and even the dome area, yet no one is talking about it. Impact theory cannot account for the outflow of material that occurred when the sediment was in a softer stage, and no theories of erosion account for the flat plains surrounding the crater. The plains have almost no erosion as compared with the rims of the crater, which are well rounded. Such questions remain unanswered in the modern impact theory.

7.12 THE IMPACT TO HYDROCRATER EVIDENCE

modern geology has never seriously considered the idea that the Earth is a hydroplanet, many obvious hydroplanet features have gone unnoticed, been ignored, or dismissed as being unusual or anomalous. The body of scientific hydroplanet evidence is overwhelming, and coupled with the evidence of the Universal Flood in the following chapter; the origin of the world's hydrofountains and hydrocraters becomes irrefutable.

Crater out-channels should have long ago convinced planetary scientists that many craters were of hydrofountain origin. That evidence, together with the lack of impact-glass and other missing impact evidence, and with the evidence of extended sediment transport, demonstrates the close relationship of water and crater origins.

Next to Shoemaker, Robert Dietz was one of the foremost impactologists in 1969 when he made the following statement about Richat Hydrocrater. This demonstrated how hydrogeological features were *foreign* to him and his colleagues:

"An endogenous origin [hydrofountain] for a large, highly circular dome requiring the application of purely vertical forces, in an area **conspicuously lacking in any other tectonic features, is unusual, but is not impossible**." Note 7.12ae

Dietz and his colleagues were unable to grasp the true origin of minerals due to the dogmagma they were in. Not only were hydrocraters "unusual," but simple geological features such as the "central nest of chert" (a quartz based conglomerate of broken rock cemented together in the hypretherm) were considered the "Richat problem":

"The **central nest of chert breccia** is a remarkable aspect of Richat, and its **origin** would seem to be rather critical to the **Richat problem**." Note 7.12ae p1369

It was not only the origin of the chert that was a problem; to the impact theorists, the crater looked like many other 'impact craters' on the Earth, Moon and other planets. Richat Crater *was* once widely regarded as an impact crater even without physical evidence of the claim:

"The 38-km-diameter **Richat Dome**, in central Mauritania, **has been widely regarded as a possible astrobleme [impact crater]**." Note 7.12ae p1367

Nevertheless, Dietz and his colleagues concluded:

"We conclude that the total lack of shock metamorphic effects is significant and that **Richat is *not* an astrobleme, but is endogenous [from within the Earth] in origin**." Note 7.12ae p1370

Dietz drew his conclusions in 1969, conclusions that should have been reached about *most* of the supposed impact craters. Shatter cones, coesite and shocked rock are all false evidences of impact that convinced researchers the majority of craters were made by impact. With the UM's dismissal of the so-called indicators of impact and proof that the long-held criteria are *not* absolute, the entire *Earth Impact Database* should be reevaluated and the vast majority of the craters removed.

Bushveld Complex Pseudotheory

One mineral-rich province in South Africa, The Bushveld Complex, seen in Fig 7.12.24, is famous for its abundant mineral ores. It serves as a definite example of how two competing pseudotheories have kept geology in the Dark Age. For most of its history, the Bushveld circular structures were thought to have come from deep-Earth magma; however, impact theorists deemed it the result of an impact. In a 2006 book titled *A Complete Guide to Rocks & Minerals*, researchers reconciled their ideas, combining both magma and impact pseudotheories into a joint origin:

"South Africa's Bushveld complex in the former Transvaal is one of the **world's great geological wonders**. It is by far the largest layered intrusion, covering up to 65,000km²/25,097 sq. miles and reaching up to 8km/5 miles thick. **It is incredibly rich in minerals, containing most of the world's chromium, platinum and vanadium resources, as well as a great deal of iron, titanium, copper and nickel**.

"Some geologists thought it was created by the hotspot above a mantle plume; others, noting the coincidence of dates with the nearby **Vredefort meteor crater, thought it was a meteorite impact feature**. A recent theory **uses both ideas**, sug-

Fig 7.12.23 - The Richat Hydrocrater is a typical example of a large, multi-ring hydrocrater (38 km/24 miles) with an Out-Channel. Such features clearly establish this crater as volcanic in origin, however, because there are no other "volcanic features," researchers struggle to understand how such craters are formed. They struggle because they do not have the Hydroplanet Model, but there are faint signs of a shift in science as some 'impact craters' are now being recognized as "endogenous" or of hydrocrater origin.

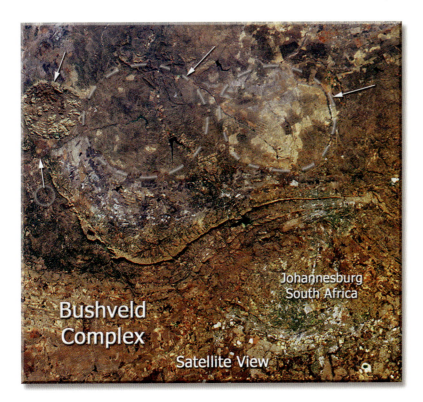

Fig 7.12.24 - The Bushveld Complex is one of the largest ore producing areas of the world and contains several circular features as noted above. Even though ores are being pulled out of many circular diatremes around the world, impact dogma has caused researchers to still see rocks from space as being the origin for such circular structures.

gesting a meteorite impact triggered off a mantle plume to fountain huge amounts of magma up from the mantle—almost as if the meteorite had burst the earth's crust." Bib 151 p91

We recall from the Rock Cycle Pseudotheory that ore deposits were designated "freaks of nature" by the geologists who study them. Just as we identified there, science can never hope to discover the true origin of ore deposits as long as it bases its ideas on the Magma and the Impact Pseudotheories. Ores come from the same natural process that all minerals come from—a heated, pressurized, hydrous environment, the hypretherm!

The Sudbury Impact Pseudotheory

An extreme example of the impact fad is the Sudbury Basin, a 62 by 30 kilometer basin, 15 km deep; a geological wonderment and North-American counterpart to the African Bushveld Complex. Geologists studying the Sudbury Basin could not even agree on what they were actually looking at:

"In 1883 Ore was discovered near Sudbury, Ontario (Canada). Since that time several billions of dollars worth of nickel, copper, iron, and other heavy metals have been produced by mines in that region. Geological studies were undertaken in order to understand the structure and emplacement of the **ore bodies**, but although the investigators were competent geologists, there was much disagreement among them. They **disagreed not only** about the interpretation of what they saw; **they disagreed about what actually had been seen**. Evidence at one location contradicted evidence observed somewhere else, and attempts to understand the ore bodies became a tangle of **confusion and frustration**." Bib 119 p205

Frustrated and confused, geoscientists turned to space to get their answers because they have had such a poor understanding of how ores are made. In another combination of two competing theories, one geologist, J. Guy Bray, stated in 1972:

"I consider that the **Sudbury Basin structure** was initiated by the explosive impact of a **large meteorite**; I do not believe the ores came from the bolide; but I do think its impact acted as a trigger for endogenic magmatism that produced the Sudbury Nickel Irruptive, of which the ores are a part." Bib 119 p224

Dietz was not content to share the credit for the Sudbury nickel ore and gave more credit to the theoretical meteorite:

"Nevertheless, **Dietz maintained that the nickel in the Sudbury ores**, which were unique in their grand scale and apparently underlain by no magma chamber, **must have been derived in some manner from the meteorite**." Bib 119 p224

Dietz was compelled to acknowledge there were still a lot of unanswered questions:

"Of course, the mechanics of impact on the scale envisioned at **Sudbury remains poorly known so that almost any scheme remains possible. Many surprises await**..." Bib 119 p225

It wasn't long before large ore bodies were found at Wanapitei Lake, twenty-five miles northeast of the Sudbury Basin, and it wasn't long before it too, was dubbed an 'impact structure.' The "improbable coincidence" had struck again; meteorites landed directly on a rich ore deposit. Again, we see that if the true origin of the ores were comprehended, the crater structures would not be attributed to impact. Geologists would realize that many ore deposits lie in diatremes formed not too long ago, when the Earth was covered with water. Ores are of a biogenic origin; they come from microbes in a hypretherm environment. This explains why they are found relatively near the surface and in great abundance near crater features.

Without the Hydroplanet Model, investigators have been lost in a vast sea of data with no rudder to steer them in the right direction. Large multi-ring craters are another baffling landform with no known formative mechanism, yet they are common. One research group evaluated the Sudbury structure and noted "contention" among the investigators:

"It is generally accepted that **multi-ring basins** are the consequence of **very large impacts, but the mechanism** by which they form **is still a matter of contention**." Note 7.12af

One observation not easily dismissed is the melted rock found at *very specific* locations in the Sudbury complex. Large mega-impacts would have all been high-speed, melt-producing events, with copious amounts of glass laid down indiscriminately in and around the crater. However, when the investigators searched the crater, they found "four rings," each containing the byproducts of "frictional melting":

"Our **evidence** indicates that the majority of the Sudbury pseudotachylytes [frictional melt], and especially the large **dyke-like bodies, are the products of slip-induced frictional melting and are fault-generated**." Note 7.12af p131

The melted rocks were found precisely where lava would be generated—in faults, not on the surface of the crater where impact heat would have been generated. The melt-rock occurred in four distinct circular, dyke-like forms corresponding with lines of faulting. These surround the ore body that was produced by the upwelling of heated subterranean waters.

The Haughton Hydrocrater

Scientists recently studied Haughton Crater, which lies on Devon Island in northern Canada (Fig 6.12.25). Researchers were excited about its similarities with Martian geology. In the journal of *Science,* the following statement beautifully illustrates the modern science 'impact paradigm':

"**All planets have one thing in common: a relentless pounding by renegade cosmic debris**. About 23 million years ago, one such wanderer **slammed into** the Canadian High Arctic, blasting a 20-kilometer-wide scar. Now called **Haughton crater**, the impact has researchers seeing red—not just in the minerals at the **crater's dead hydrothermal pipes**, but by comparison to Mars." Note 7.12ag

The "relentless pounding" pseudotheory given to modern science by Shoemaker's Impact Statement has completely warped the thinking of all planetary scientists' ideas about planet formation. However, there is a morsel of truth in the above quote; physical evidence of the crater's true formation—"dead hydrothermal pipes"—was right in front of the researchers, but their importance was missed. Haughton crater has several deep out-channels that defy impact-origin explanation, but are a standard of hydrocrater formation.

"Dome Crater" Pseudotheory

Fig 7.12.26 illustrates just how far modern planetary scientists have had to stretch their imagination to stay within the lines of the impact paradigm. Craters and domes are of course exact opposites; but they often occur together, another anomaly that does not fit the Impact Pseudotheory:

"Smaller impacts have smashed into Callisto after the forma-

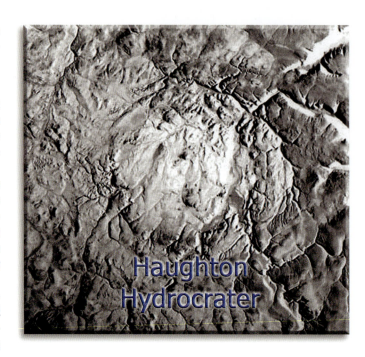

Fig 7.12.25 – The Haughton Hydrocrater, located in northern Canada is 23 km (14 miles) in diameter. Like many other craters around the world, it is considered an 'impact' crater by modern science even though it does not have any clearly defined impact evidences. Because of the impact dogma that exists throughout all the natural sciences today, geologists have been blinded to empirical, crater-origin evidence right in front of them. Haughton Crater is a good example, having a number of "dead hydrothermal pipes" that are totally ignored.

tion of Asgard. The young, bright-rayed crater Burr located on the northern part of Asgard is about 75 km (46 mi) across. Galileo images show **a third type of impact crater** in this image, a **dome crater** named Doh, located in the bright central plains of Asgard. Doh (left image) is about 55 km (34 mi) in diameter, while the dome is about 25 km (15 mi) across. **Dome craters contain a central mound instead of a bowl shaped depression** or central mountain (peak) typically seen in larger impact craters." Note 7.12ah

Lab conducted high-speed impact tests have never produced a crater with a 'dome.' Instead of a central peak, the domes are smooth rounded structures formed by expanding underground water-steam pressure, with heat derived according to the Gravitational-Friction Law.

Earth Impact Database Pseudotheory

During the 1970s, science decided that several hundred large impact craters should be found on the Earth's surface:

"B. M. French pointed out in 1970 that **calculations from astronomical and geological data** indicated that during the past two billion years meteoritic impact in the land areas of the world had **probably** created about twenty craters the size of Sudbury or Vredefort, about 6,000 craters larger than three miles in diameter, and approximately 100,000 craters larger than Meteor Crater, Arizona. In his opinion 'astronomical and geological data both support the present existence of perhaps **several hundred large impact craters on the land areas of the earth**.'" Bib 119 p227

On the surface, this seemed like a useful prediction,

Fig 7.12.26 – In this photo is the Doh crater feature, which is located in the large crater of Asgard on Callisto, one of Jupiter's moons. Amazingly, researchers refer to Doh as being a **"dome crater"** structure. Of course, a crater and a dome are not the same, and it requires an imaginative stretch to explain how swelling domes are formed inside impact craters. Trying to force domes into the Impact Pseudotheory illustrates how far modern astronomy and geology have strayed from observable natural laws to theoretical conjecture. NASA image PIA01648.

and it looked like it was proven by the Earth Impact Database. The problem is that most of the structures on the database were *not formed by impact*. The mistake stems from the **assumption** that most of the craters on the lunar surface and other planets are impact craters too, but as the Hydrocrater Model subchapter has thus far demonstrated, this is not the case. French's *theory* that "several hundred large impact craters" exist on Earth was based on inaccurate data.

> Not learning from correct data is worse than using false data.

A good example of how original researchers missed the mark occurs at the Pretoria Salt Pan in South Africa:

"In the opinion of various investigators, for example, the Pretoria Salt Pan in South Africa was probably a meteorite crater; its circular shape, raised rim, and **apparent absence of volcanic materials** made an impact origin seem **highly likely**. It was felt that conclusive evidence would eventually be obtained by drilling through the lake beds on the crater floor. **Contrary to expectations, no positive indications supporting an impact origin were found** during investigations in 1973. The conclusion that the Pretoria Salt Pan was **not an impact crater was supported when drilling revealed** that soft saline sediments existed to a depth of more than 500 feet below the floor of the structure." Bib 119 p228

What the researchers found was a diatreme! There are *two very important points* to take from the Pretoria Salt Pan experience:

1. No volcanic materials were found at the Pretoria Salt Pan crater, now known as a hydrocrater.
2. The crater was believed to be impact until drilling revealed a diatreme.

One of the initial reasons scientists thought the Arizona Crater was an impact crater was that there was no volcanic material at the crater. The Pretoria Salt Pan showed that hydrocraters do not always eject melted material. Drilling confirmed the presence of a diatreme and changed ideas about the origin of this crater, once "highly likely" an impact crater. Had geologists done the same at other supposed impact structures and correctly interpreted the data, the number of craters listed on the 'Impact' Database would be significantly reduced.

The Earth Impact Database lists the false impact *criteria* (shatter cones, PDFs and coesite) addressed in the Hydrocrater Model subchapter. Because these criteria are invalid, no inquiry into the origin of any crater should include *only* those supposed "impact" criteria. Unfortunately, the site lists no criterion that would identify a hydrocrater, and makes no mention of diatremes, the one clear evidence that a crater is not an impact crater. Impact theorists have been so one-sided, that *no credible* database of the Earth's craters exists. A Millennial Science priority will be to create one.

Four of the craters on today's "Earth Impact Database" were investigated almost half a century ago. K. L. Currie, an objec-

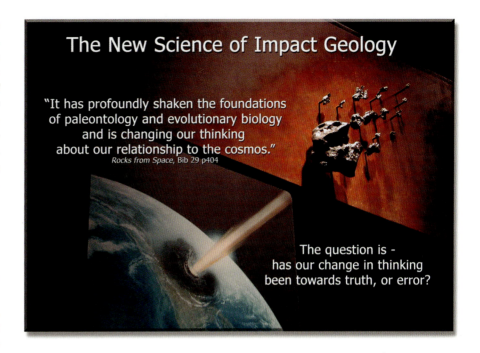

The New Science of Impact Geology

"It has profoundly shaken the foundations of paleontology and evolutionary biology and is changing our thinking about our relationship to the cosmos."
Rocks from Space, Bib 29 p404

The question is - has our change in thinking been towards truth, or error?

tive scientist looked at the geological evidence and outlined what meteoritics and the geosciences needed to do decades ago—to recognize that a new idea was needed to explain some of the craters:

"The **four craters, New Quebec, Clearwater Lakes (two craters) and Manicouagan are compared**. New Quebec shows evidence of slow uplift accompanied by hydrothermal activity, possibly culminating in a minor volcanic outburst. Clearwater East is interpreted as the result of regional uplift followed by local collapse. Clearwater West shows similar features accompanied by a central peak and considerable volcanism. The latter two features are believed to be related. Manicouagan has a well developed central peak and an extensive volcanic pile. The characteristic cycle; upwarp, collapse, volcanism is beautifully shown. **None of these craters can be explained either by 'classical' impact theory, or by analogy with known volcanic areas**. **New ideas are needed**." Note 7.12ai

No "new ideas" were presented and so it was, the science of cratering and impactology was to remain in the Dark Age for another half century.

Impact Geology Gone Awry

The Meteor Model, the Meteorite Model, the Crater Model, and the Arizona Hydrocrater all contain evidence and comprehensive explanations about the Earth's astronomical and geological features—the craters; extraordinary claims that directly challenge prevailing scientific theory. There is one thing modern science and the UM are in agreement with; the science of impact geology has "profoundly shaken the foundations of paleontology and evolutionary biology…and…our relationship to the cosmos":

"The latter half of the twentieth century has seen the rise of the **new science of impact geology**. It is truly a hybrid science requiring a free exchange of ideas from a diverse group of established sciences. **It has profoundly shaken the foundations of paleontology and evolutionary biology and is changing our thinking about our relationship to the cosmos**. All this started so modestly. Flashing trails of light in the sky. Stones pelting Earth from space. A gaping crater in the northern Arizona desert." Bib 29 p404

The question is—has the change in
thinking been towards truth, or error?

The subchapters have pointed out many errors in the logic and observations of meteorites and impact craters. These errors exist because of the absence of a Hydroplanet Model and its accompanying interpretive opportunities. This new model literally creates a *whole new geology*, which will completely change paleontology, evolutionary biology, our thinking about our relationship to the cosmos, and a host of other fields of science. By placing water at the center of geology and astronomy, we have an opportunity to comprehend and understand the wisdom and knowledge of our *real* relationship to the cosmos.

The Impact Paradigm Shift

Time is the test of truth. More and more revelations of craters and pits that do not fit the Impact Pseudotheory are being made, terrestrially and celestially. A shift from impact has a long way to go, and there are indications such a shift is about to happen. The unknown origin of some of the craters on Jupiter's moon, Callisto in Fig 7.12.27 has spurred some scientists to consider an alternative to impact. They are considering a subsurface process:

"This image of Jupiter's second largest moon, Callisto, presents one of the **mysteries** discovered by NASA's Galileo spacecraft. In the upper left corner of the image, what appear to be very small craters are visible (See enlargement.) on the floors of some larger craters as well as in the area immediately adjacent to the larger craters. Some [of] these smaller craters are not entirely circular. They are very similar to a population of unclassified '**pits**' seen in one Callisto mosaic from Galileo's ninth orbit. One possible explanation for the pits is that they represent a class of **previously unseen endogenic (formed by some surface or subsurface process, rather than an impact) features. Another explanation** is that they are partially eroded **secondary craters**." Note 7.12aj

Of course, the other "explanation" is just the same old sci-bi; partially eroded, apparently impact formed craters. Mountains of evidence and decades of observations testify that most of the craters could not have been made by impact. The physical attributes of the craters, the residual evidence at the craters, impact physics, and observations of actual cratering events; both high-speed and low-speed, show how poorly the scientific community has interpreted these ubiquitous landforms. A reevaluation of the data based on the Hydroplanet Model will revolutionize the science of cratering and even bring about change to current planet formation theory.

In the next subchapter, we bring the evidence of hydrocraters together with the hydroplanet evidence to take a new look at the Moon. If the Earth is a hydroplanet, then why not the Moon?

7.13 The Hydromoon Evidence

The Anhydrous Moon Myth

Like many of the Hydroplanet subchapters, the Hydromoon Evidence subchapter could easily be expanded into several volumes; the amount of evidence is astounding. It is an affirmation of the truth and beauty of Nature's Puzzle. This subchapter will focus on two primary points—information concerning the Moon's water origin and the absence of evidence proving impact.

We have already seen the word **hydromoon** earlier in the chapter, while exploring other moons in the solar system. A **hydromoon** is simply a moon formed out of water, whereas an **anhydrous moon** is just the opposite, meaning it was formed without water. Having already established that all of Earth's rocks contain some water, and having learned something about how they grew out of a water solution, it is natural to extend this search to the Moon's rocks. After all, they shouldn't be much different from Earth rocks, especially if the Moon once was part of the Earth, as most present day researchers believe.

Lunar scientists who were looking for water in rocks brought back from the Moon found both similarities and differences when comparing terrestrial and lunar minerals:

"At first glance, a section of a **lunar mare basalt resembles many terrestrial basalts**, but some of the differences are significant. I joined the group studying the samples brought back by Apollo 11 **in the hopes of finding H_2O or CO_2 inclusions**." Bib 155 p534

An important distinction to remember is that the researchers noted that lunar and terrestrial basalts were similar; the lunar rocks were *not glass*! The Basalt Mystery in the Rock Cycle Pseudotheory chapter described the crystalline form of Earth's basalts and the fact that they are not glass; this holds true with the lunar basalts. This means the lunar basalts did not flow out of a volcano on a dry surface. Commonly, lunar geologists believe the lunar basalts are dry lava flows, but this cannot be. There has never been a documented, dry basalt flow observed on the Earth and there is no reason to think any different about the Moon.

Returning to the search for water in lunar rocks, the researchers reported:

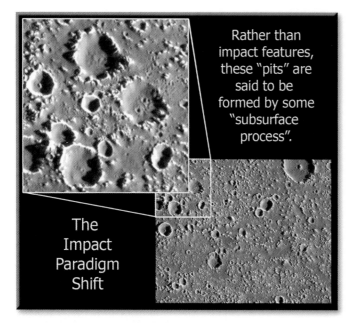

Fig 7.12.27 – Here we see craters described by NASA on Jupiter's second largest moon, Callisto, that do not fit the "impact" pseudotheory. Instead of impact features, NASA says these "pits", "represent a class of previously unseen... subsurface process." A subsurface process that makes a surface crater is a **hydrocrater**! NASA image PIA01630.

CHAPTER 7 THE HYDROPLANET MODEL

"Unfortunately, I was **unable to find even a single minute H_2O (or CO_2) inclusion in these samples**. They show none of the late-stage alteration features so common in terrestrial rocks (particularly in association with ore deposits); serpentine, actinolite, epidote, chlorite, sericite, **secondary aqueous fluid inclusions (and the ore deposits) are missing completely. We know now that the lunar samples are the driest rocks known to man**." Bib 155 p534-5

The absence of water came as a complete surprise to everyone, and shaped the opinions of planetary geologists for over 30 years. Because Moon rocks were anhydrous (without water), researchers assumed water must *not* have been a part of the Moon's formation, but they evidently did not follow this idea to its natural conclusion; how were the Earth and Moon formed together if one has water in all of its rocks and the other none?

FQ: How were the Earth and Moon formed together if one has water in all of its rocks and not the other?

The Anhydrous Moon should have remained nothing more than a theory, but dogmagma convinced lunar geologists that there wasn't any possibility water played a major role in the formation of the Moon. However, there were telltale signs that lunar rocks were *not completely dry*:

"…although the rocks are indeed exceedingly dry, **they do contain a few parts per million of** CO_2 **and** H_2O **or H**." Bib 155 p540

The Anhydrous Moon paradigm kept early researchers from seeing the possibility of water on the lunar surface, but recently, new high-tech instruments helped astronomers identify frozen water at the Moon's poles; an idea still hotly debated because of the dogmagma mindset.

Hydromoon Fundamental Questions

With the UM's New Geology we can answer the question of why there is so much less water in lunar minerals versus the terrestrial minerals; an answer not found in modern planetary geology. To find such an answer, we need merely ask the right question, within the Hydroplanet paradigm:

Why is water not a part of the crystalline structure of some lunar rocks?

The answer to this question comes after answering another simple question:

What caused the Moon to lose its oceans and atmosphere?

It is a matter of gravity. The gravitational tug on the Moon is only 1/6 that of the Earth. In simple terms, a 200-pound man will register only 33 pounds on the Moon. Likewise, when water is vaporized, the Moon's gravity cannot hold it and it drifts into space, easily overcoming the Moon's low escape velocity. Said one author about the vapor of meteorites that strike the Moon:

"This vapor is usually driven off the Moon (the lunar escape

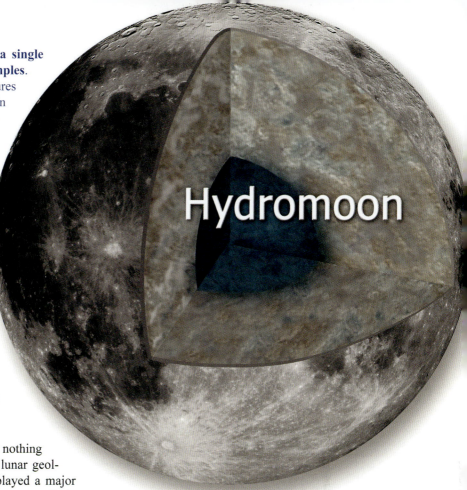

Fig 7.13.1 – The Earth's lunar companion is a Hydromoon because of its formation in water, its rock-water mantle, and its liquid-water core. Most of the scientific evidence of the Moon's water has been around since the Apollo mission during the 1960s when we went to the Moon, but dogmagma has kept it buried. Keeping such information out of the classroom has contributed to the Dark Age of Science and has stifled new research on the Moon.

velocity is only **2.5 km/sec** a low value compared with Earth's **12 km/sec** escape velocity)… **This water is mostly lost from the Moon, boiled off during the heat of the lunar day**." Bib 129 p93

Although his comments were directed at the water lost from alleged vaporized meteorites, why not the whole lunar surface?

A Dehydrated Moon

Taking the statement in the previous quote at face value, the words take on a new dimension. The authors comment that the "…water is mostly lost from the Moon, boiled off during the heat of the lunar day" has application to the lunar seas present on the Moon's surface when it was formed. It would not take long for surface water to evaporate or sublimate under the Moon's wide temperature swings, especially without the protection of an atmosphere.

There is physical evidence for such a theory. Water vapor does exist in interplanetary space in the solar system. Perhaps you will recall the SOHO satellite discussion (see Fig 7.2.6) at the beginning of this chapter; water severely affected the telescope's instruments until a fateful tumble allowed the Sun to melt the previously undetected ice.

Although it is possible that the vapor cloud surrounding the Earth could have been the source of some of the interplanetary water, the Earth's strong gravitational pull would have had to be overcome. The solar wind could have stripped and redistributed

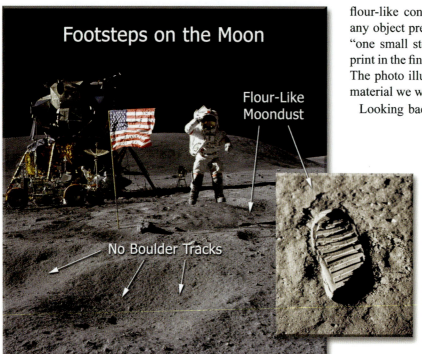

Fig 7.13.2 – Human boot prints left on the Moon decades ago will look the same decades into the future because there is no atmosphere or weather on the Moon. Flour-like Moondust covers most of the lunar surface, but how did it get there without erosion? Impact theorists would like to think Moondust is from 'micrometeorite bombardment'—yet this would mean the surface of the Moon would have lots of melted surface rock along with micrometeorite material. The absence of such material and the absence of 'Boulder Tracks' (See also Fig 7.13.3) is direct evidence against a massive meteoritic bombardment of the Moon's surface.

flour-like consistency that forms near perfect impressions of any object pressed against it. Neil Armstrong's famous words, "one small step for man…" was memorialized with his boot print in the fine, powdery lunar surface, inset right of Fig 7.13.2. The photo illustrates the impressibility of the lunar regolith, a material we will refer to as **moondust**.

Looking back to the question as to why water is not a part of the crystalline structure of lunar rocks, it can be understood that because lower gravity did not hold on to the surface water and evaporated water vapor, the dehydrated environment would have affected the crystalline growth process. Crystal growth experiments performed aboard the Space Station's microgravity produced results significantly different from those grown on Earth. Thus, lower gravity and environmental conditions on the Moon affected the crystallization process helping us to understand better the environment where celestial bodies were formed. The abundance of water in meteorites confirms that the meteorites did not come from an asteroid or comet, but from a larger celestial body or from the Earth itself, as an ejectite.

The Boulder Track Evidence

The fine, grey regolith is more than just a good medium for recording the astronaut's boot prints. Moondust covers almost the entire surface of the Moon, and the accompanying photos show it to be several centimeters thick or more. The fine, powdery material is easily imprinted, leaving behind a record of the smallest degree.

If impacts had really happened on the Moon as frequently as meteoriticists claim, boulders and chunks of rock would have been thrown skyward, returning to the surface both in and out of the craters. Proponents acknowledge this because the Arizona Meteor Crater has ejecta boulders that fell on the inside and the outside of the crater.

> FQ: What would large rocks and boulders do to the Moon's silty surface as they fell and rolled to a stop?

The answer to this FQ seems to have been completely overlooked by impact theorists. We should expect to see **Boulder tracks,** and in fact, we do, but they are unusually rare. The collection of images in Fig 7.13.3 includes one of these rare boulder tracks as well as two photos of snowball tracks on Earth to illustrate how tracks form in soft sediment as they roll down the hillside. The rarity of the tracks is simple evidence of a hydromoon:

> There **should be** many boulder tracks down the lunar hillsides and crater rims—**but there are not**.

Most scientists believe the rocks that are strewn around the lunar surface are ejecta from impacts that happened over millions of years. If this were true, there would perhaps be hundreds of photos, of the thousands taken during the Apollo program and since, showing boulder tracks. Because there is no weather, they would have never been covered, left to remain like the astronaut's boot prints, frozen in time.

some of that water vapor too. However, evaporated water from the Earth's moon, other moons and smaller planets could have played a major role in the water that resides in interplanetary space.

The point to be made is that we need to take a new look at the Moon, a paradigm shift in our evaluation of the observations that have been made. Instead of the dry, desert mentality; look at the surface of the Moon as if it were underwater. Compare Fig 7.8.13 and 7.9.33, showing pockmarks and surface forms beneath the ocean with the pitted and smoothed lunar surface seen in 7.9.30 and 7.9.31. The underwater images show that a fine silty material covers everything, softening the landscape. Some of the silt is marine snow, the detritus of dead animals sinking into the abyss, but much of the material came from mud volcanoes and hydrofountains that left hydrocraters behind as evidence of their existence. Greater details of the origin of the silt and sand will be given in the Sand Mark of the next chapter, where the Sand Mystery, introduced in the Rock Cycle Pseudotheory, will be resolved. The vast majority of Earth's sedimentary material did not come from erosion on the surface, but from underground. Of course, these ideas challenge all that modern geology teaches, but they are strongly supported by evidence such as the Sediment Size Table (Fig 6.10.5) that outlines the missing sizes of pebbles, sand and silt in the natural erosion record.

Not only does the lunar surface look like silt on the bottom of the ocean, it acts like it. With diminished gravity and no atmosphere, the Moon's oceans evaporated leaving sediment of

Boulder Track Evidence

No Lunar Boulder Tracks in Process of Being Coverd Up

Rare Lunar Boulder Tracks

Snowball Track Examples

Micrometeorite impacts can't explain missing Boulder Tracks all over the moon.

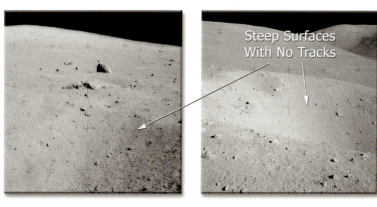

Steep Surfaces With No Tracks

Where are the boulder tracks on the Moon's surface that should exist on thousands of photos like these?

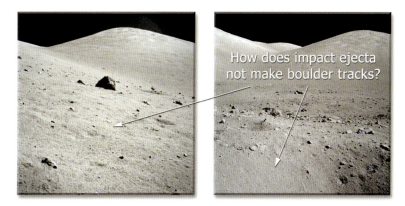

How does impact ejecta not make boulder tracks?

Fig 7.13.3 – Boulder Tracks are a rare phenomenon on the Moon, but hold important clues. Boulder Tracks on the lunar surface are similar to snowball tracks on Earth, made as a snowball rolls downhill on fresh snow as seen above. The soft Moondust covering the lunar surface makes for an ideal observatory for examining this evidence. If impacts **actually** dominate the lunar surface, the ejecta from such impacts would sometimes fall on steep slopes and roll downhill, leaving behind Boulder Tracks as seen in the 'Rare Lunar Boulder Tracks' inset photo above. The tracks in this photo were the result of moonquakes. Thousands of lunar images like those above right were studied, but revealed only a handful of Boulder Tracks. Where are the expected thousands from the impact ejecta? One common sci-bi is that micrometeorite bombardment erased such tracks, but although evidence indicates that micrometeorites actually do occur, there is no evidence of massive microscopic bombardment. Moreover, the Moondust appears to be void of the high quantities of melt-rock and glass from a micrometeorite bombardment. Such a theory might be able to explain the destruction of *some* of the smaller 'boulder tracks,' but cannot account for the absence of large tracks.

Bombardment by **micrometeorites** is one sci-bi scientists fall back on to explain the Moon's missing boulder tracks: Billions of years of micrometeorite impact presumably smoothed the surface and erased all evidence of the tracks. At first glance, this seems like a reasonable idea; however, the explanation falls short once we take into account the whole picture. Some tracks would be protected by outcrops and bigger boulders. The microscopic size of the micrometeorites would have the impossible task of erasing tracks of large boulders. Evidence of newer tracks would exist, and finally, there would be tracks with micrometeorite erosion **in process**. A detailed examination of more than a thousand high-resolution photos of the Moon revealed *no partial tracks*.

The reason the astronauts identified the boulder tracks at all were the periodic **moonquakes** that dislodged unstable boulders and sent them tumbling downhill.

Furthermore, micrometeorites produce micro glass-lined craters. These do exist, but *by no means do they extensively cover the surface of the Moon*, and therefore cannot account for covered boulder tracks. Craters larger than 4 mm are even rarer than smaller, glass-lined micro craters:

"Craters or pits smaller than a millimeter have been observed on the rocks returned to the earth by the Apollo 11 mission. **Craters smaller than 4 mm in diameter commonly are glass-lined** whereas the **craters larger than 4 mm are only rarely glass-lined and are commonly irregular in form**." Bib 171 p453,

The "glass-lined" micro craters were seen on lunar rocks by Shoemaker and others back in 1970. Their examination did not reveal evidence of micro cratering sufficient to cover a single rock, and certainly not enough to erase all trace of boulder tracks.

What Shoemaker and others missed was that the glass-lined micro impacts on the Moon rocks and in the laboratory both testify that all sizes of high-speed impact produce *glass-lined craters*, not just those less than 4 mm. Larger impacts produce higher energy, higher temperatures, and create more melt in their craters.

Scientific equipment orbiting the Earth in outer space has confirmed what the Moon rocks already showed; micrometeorites are a real phenomenon that produces glass-lined craters. However, the frequency of such impacts is overestimated, with no evidence showing they could erase the anticipated Boulder Tracks. The glass-lined microcraters on Moon rocks are too sparsely distributed to support a micro-impact origin of moondust.

The Hydromoon Model explains why there are so few Boulder Tracks on the Moon and how the lunar surface came to be coated with fine-grained moondust. Unlike the impact theory's millions-of-years of micro-impact on a dry surface, the hydro-moon surface was shaped in a short period of time on a very wet surface. At the end of the hydrocratering period, water began to disperse. Silt and colloidal minerals in the water began to settle, blanketing the lower lying areas of the Moon. Any boulder tracks that had existed were covered and smoothed over, much as we see at the bottom of Earth's oceans today.

Occasionally impacts do occur on the Moon and some do create Boulder Tracks, but their scarcity is a validation of the Hydrocrater Model.

> It is astonishing that the simple mystery of missing boulder tracks has been so neglected!

It is astonishing that researchers came so near the subject, but were stopped short because of their inability to conceive of the possibility of a hydromoon.

Impact Boulder Mystery and Answer

In 2002, planetary scientists studied new images of the asteroid Eros and noticed that among craters of similar size, some had many boulders inside them while others had few. This was puzzling because 'impact' craters of roughly the same size and degradation *should* have roughly the same amount of boulders. The researchers turned to the best high-resolution images of the Moon available—the Lunar Orbiter images—for further study.

Using 12 of these images, they identified 213 craters greater than 100 meters, over a 124 km² area. Did they find what the impact theory predicts—craters of the same size and same degradation in close proximity with roughly the *same* boulder populations?

"Mapped crater diameters range from 100 to 1700 m, with an average diameter of 185 m. We find it **significant** that **over half of the mapped craters are boulderless** (H) and that **88% have no boulders to just a few boulders**... **Craters with abundant boulders are rare** (3%, Classes A and B). We find it **significant that craters of the same size and same degradation state can be found in close spatial proximity with vastly different boulder populations**." Note 7.13a

The investigators found *the opposite* of the impact theory postulate and recognized this observation as being "significant." They did not explain the differences in boulder populations between similar craters.

In their closing summary, they merely said the Eros boulder distribution was "not unexpected" Note 7.13a because they found the same condition on the lunar surface. This of course, did not explain the 'boulder population mystery' on either surface, or any other celestial surfaces for that matter. Without the Hydroplanet Model, scientists would probably continue with such observations indefinitely, without ever perceiving the truth. Seeing the craters as hydrocraters makes the answers easy—some craters have boulder-sized sediment carried from the depths whereas others had finer sediments.

The KREEP Evidence

Another puzzling mystery that has evaded researchers is the so-called KREEP Evidence. From the book, *Planetary Science: A Lunar Perspective* we read:

"It is probably fair to say that **no topic has caused more confusion in our understanding of lunar petrogenesis than has KREEP**." Bib 162 p214

KREEP is an acronym for **K** (potassium) **R**are **E**arth **E**lements and **P** (phosphorus). These elements are found in some lunar rocks. The KREEP Evidence shows that heavy, rare Earth minerals along with lighter potassium and phosphorus elements are found together in Moon rocks, a fact "puzzling" to researchers because the heavier elements *should have* sunk to the center of the Moon when it was molten. Yet they were found on the lunar surface. This makes KREEP rocks "incompatible" with "magma crystallization":

"The last major rock type of the crust is one of the **most puzzling** in that it was first recognized as a chemical component and was found as a rock type only after considerable study. It was noted that breccias from the *Apollo 12 and 14* landings sites were particularly enriched in a group of the elements [KREEP] that include **samarium, uranium and thorium** (making this component radioactive). These elements do **not** enter into the crystal structure of the common minerals **during magma crystallization because of their size [weight]** or charge and are thus termed '**incompatible**' trace elements." Bib 129 p150

The KREEP elements are "incompatible" in the same way metal ores found on the Earth's surface are considered by some geologists to be "freaks of nature." They too *should have* sunk into the Earth's 'magma ocean.' Notwithstanding their similarities, KREEP is different from the Earth's ores because KREEP covers large areas of the lunar surface. Researchers also recognized that KREEP comes "from a common source":

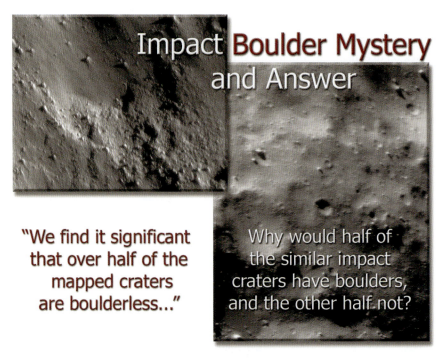

"We find it significant that over half of the mapped craters are boulderless..."

Why would half of the similar impact craters have boulders, and the other half not?

Fig 7.13.4 – Two images of the surface of the asteroid Eros; the overlapping image on the left is a close-up. Researchers found it "significant" that similar sized craters in the same vicinity contained completely different populations of boulders. How does impact theory account for this? Hydrocraters provide an easy answer for this phenomenon, multiple eruptions from different sublevels, each producing different populations of boulders in different craters.

"Because **KREEP** is so widespread, it **cannot be dismissed as a small or trivial volume of residual** melt produced from crystallization of a local intrusion. In magma **ocean models**, **it represents perhaps the final 2% residue**. Studies of Sm/Nd isotopic systematics indicate that the samples measured **come from a common source**." Bib 162 p209

This common source of KREEP creates "confusion" among geologists because the source is obviously not from meteorites, nor should it be on the surface because of its density. However, the Hydroplanet Model would have the "final 2% residue" coming from an ocean—just not a magma ocean. The residue was indeed "brought to the surface" by hydrofountains capable of carrying rock from deep within the Moon. From the book, *The Once and Future Moon*:

"The occurrence of KREEP in lunar rocks suggests that it resides in the deep crust and is **brought to the surface** by basin-scale impact **or by volcanism**, in which lavas moving through the lower crust can **bring the rock up from depth**." Bib 129 p152

Lunar scientists missed the true origin of KREEP and the majority of other lunar surface features because mysteries like the Basalt Mystery, in the Rock Cycle Pseudotheory chapter, remain unanswered. Basalt on the Moon is similar to Earth's basalt in that it is not melted lava glass. Basalts have a crystalline structure and therefore do not come from dry volcanoes on the Earth's continents; nor do such lava *flows* on the Moon exist. The next chapter will describe in some detail how basalts were also formed in a hypretherm, with hot water and intense pressure.

The 'molten Moon' theory will never work because water has been left out of the equation, therefore the true chemical composition of both the Earth and Moon remain elusive:

"Even if the **Moon was largely molten** at any given time, a chilled crust would have quickly formed as insulation against the cold vacuum of space. This chilled rind would have had a **chemical composition identical to that of the bulk of the Moon**, and we have searched the sample collection for pieces of it, although **none have been identified**." Bib 129 p152-3

Again, the simple answer for the origin of KREEP is the same for the origin of Moon dust and most of the Moon's craters—hydrofountains. In this way, minerals were concentrated in specific areas.

A corollary to the KREEP mineral mystery is the reduced abundance of mineral types on the Moon. On the Earth, many of the ore minerals in the crust are known to be of biogenetically related origin. In other words, organisms played a critical role during the hyprethermal concentration and formation processes.

The Moon is void of life and therefore is expected to have fewer minerals than the Earth. Scientists recognize this fact, but unfortunately attribute it simply to the lack of water:

"The **lack of water** on the Moon **has greatly reduced the number of minerals found** on that planet as compared to the Earth." Bib 87 p109 from Magmas and Magmatic Rocks

Of course, this statement is partially correct. The lack of water reduced the number of minerals found on the Moon, because it reduced the ability of life to produce them. The law of hydroformation holds that all natural crystalline minerals were originally formed from water. The Law of Hydroformation holds that all natural crystalline minerals formed originally out of water. The Law of Biotrans in Chapter 24, the Chemical Model states that chemical elements can be created or transformed into other elements by biological organisms. These natural laws are confirmed by the diversity of mineral types on the Earth as compared with the Moon.

The Lunar Core Evidence

The Moonquake Tidal Evidence in the Magma Pseudotheory chapter provided important confirmation of the Gravitational-Friction Law. On Earth, gravitational friction from the Moon can cause earthquakes. During the Apollo missions, a discovery was made about the Moon's core because of moonquakes. Like Earth, shear waves do not travel through the liquid core of the body (see Fig 7.13.5).

The speculative liquid for those living in the magma paradigm was molten rock, as old time lunar expert Patrick Moore noted in his book, *Mission to the Planets*:

"Before the Apollo missions, we did not know anything definite about the Moon's interior. Some astronomers believed it to be cold, so that the globe would be solid and inert all the way through, while others thought that it was more likely to be hot. It now seems that below the regolith there is a half-mile (1-km)-thick layer of shattered bedrock, while underneath this comes a layer of more solid rock which goes down to around 16 miles (25km) in the mare regions. Lower down are various other layers, **while the actual core, from 600 to 930 miles (1,000 to 1,500km) in diameter, is hot enough to be molten**." Bib 149 p36

The liquid nature of the lunar core is well documented by numerous seismic recordings, but why assume the liquid is molten

Fig 7.13.5 – Although planetary geologists have tried to link the Moon's core with the Earth's 'magma' core, the seismic evidence about the lunar core confirms there is a liquid **water** core. The above Hydromoon diagram agrees with research from the 1970s and recent seismic studies in 2005, all of which contributes knowledge about the Earth's own Hydrocore.

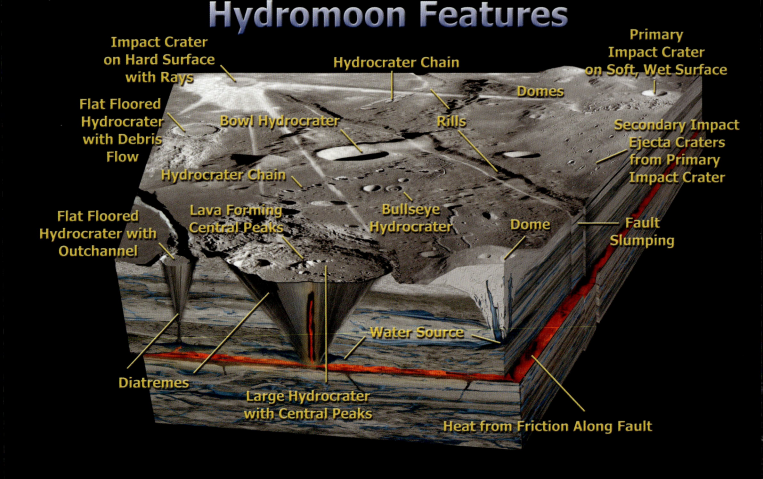

Fig 7.13.6 – This diagram illustrates the Hydromoon Features, many of which are discussed in this subchapter. Without a Hydromoon paradigm, most of the geological features on the Moon are anomalous or enigmatic. Researchers have been unable to rationally explain how features such as crater rays, flat floored craters, bull's-eye craters, domes, crater chains, and secondary impact craters are formed in the traditional magma/impact paradigm. For the first time, these features are comprehensible using common sense and a new paradigm—the Hydromoon.

rock? Scientists acknowledge that the core is too small to have retained its heat in the accretion pseudotheory. Probably dogmagma has kept the hot moon core theory alive in spite of the fact that in 1973, during the Proceedings of the Fourth Lunar Science Conference, researchers reported that a "molten core" did not work after analyzing the moonquake data:

"A striking contrast has been found between signals that originate on the near-side of the moon and those from far-side sources. Direct shear waves, normally prominent in signals from near-side moonquakes, and weakly defined in near-side meteoroid impact signals, cannot be identified in the seismograms recorded at several of the seismic stations from far-side events...

"Although available data are not sufficient to derive a detailed seismic velocity model for the deep interior, these observations **can be explained by introducing a 'core' with a radius of between 600 and 800 km, in which shear waves do not propagate** or are highly attenuated...

"**A completely molten core of the size indicated, however, is not likely** because a decrease of the compressional wave velocity exceeding the value of 0.3 km/sec obtained by our preliminary analysis would be expected. **Other possibilities, such as increased volatiles [water] in the deep interior of the moon, however, cannot be ruled out at present**.

"The core radius of about 700 km, inferred from the seismic data, is inconsistent with a high-density molten metallic core material** both by moment-of-inertia and seismic wave velocity considerations.

"**From the foregoing, it appears that the transition at about 1000 km from an outer lithosphere to an inner asthenosphere may be similar to, though much deeper than, its terrestrial analog.** The lunar lithosphere is a relatively rigid outer shell, while the less rigid asthenosphere of the moon has properties similar to those of the asthenosphere (low-velocity zone) of the earth." Note 7.13b

So the Moon's liquid core was identified, but molten material was inconsistent with other scientific findings. Instead, increased volatiles—water—could not "be ruled out at present." This keen observation was made in the 1970s before any evidence of water was found on the Moon. The geophysicists had the truth right in front of them; they knew that the moonquakes were caused by tidal energy (gravitational pull of the Earth) but they still turned to Earth's so-called "internal heat energy" as the source for the Earth's quakes:

"**Tidal energy appears to be an important, if not the dominant, source of energy released as moonquakes... In contrast, earthquake activity** is concentrated at shallow depth, caused by movements of a relatively thin, mobile lithosphere, **presumably driven by internal heat energy**." Note 7.13b p2526

One of the researchers from the article, Yosio Nakamura, reanalyzed the Apollo moonquake data using newer technology. He reported:

"This effort was highly successful, and increased the number of positively identified deep moonquakes by more than fivefold, from 1360 to 7245, and about 250 new nests were discovered." Note 7.13c

Although the data did not definitively confirm an aqueous lunar core, evidence pointed to a hydrocore similar to that of the Earth. Research in 2005, using the Lunar Prospector (satellite) and 35 years of lunar laser ranging data identified additional evidence of a "fluid core" from the orbital motion of the Moon:

"Recent progress in the determination of CMB flattening has improved and now seems significant, which together with the present results strengthens the case for a **fluid core**." Note 7.13d

The researchers of course assumed the liquid core was of molten iron, but there is no evidence of either an iron core or a hot core. Neither was discussed in any detail. The valid point is that lunar seismic waves do not travel through a liquid core.

Hydromoon Features

Fig 7.13.6 illustrates a number of Hydromoon Features. These typical lunar features have no rational explanation from the magma or impact pseudotheories. This will be shown using quotes from scholars and other observational evidence. On the other hand, the Hydromoon model provides an opportunity to make sense of these unique features and to comprehend how they formed.

Australia's Wolfe Creek Hydrocrater showed how craters with "shallow flat floors" are back-filled diatremes—a correct and simple answer to an old mystery. Contrast this with any of the impact theory answers; they will likely be as complicated as they are incorrect.

Thousands of lunar craters exhibit "anomalously shallow flat floors" and "central humps," as researchers noted during the return to the Moon in 1970:

"Near the lunar module the surface of the regolith is pockmarked by abundant craters ranging in diameter from less than 2 cm to several tens of meters. A number of craters in the area, in the size range from 30 to 55 m, have sharply formed raised rims and **anomalously shallow flat floors or floors with central humps**." Bib 171 p452

The Hydromoon Features shown in Fig 7.13.6 include several flat-floored craters on the left and a dome on lower right of the diagram. The Hydromoon Model easily explains all of these moon-wide features.

Fig 7.13.7 – Real melted basalt does not look like the Earth's or lunar basalt. This basalt was melted using an acetylene torch, showing that it was obviously not a 'lava flow' on the Earth or the Moon, as geologists have thought for decades.

Paul D. Spudis, well-known lunar scholar and an outspoken scientist describes many of the Moon's features as being strange—and for good reason. The "strange landforms" do not conform to the pseudotheories of "impact or volcanism":

"Across the Moon, both in highlands and in maria, we find **strange landforms that do not conform to our notions or understanding of lunar processes**. Oddly shaped pits, depressions, and irregular dimples **do not resemble the known landforms created by either impact or volcanism**, the two dominant surface processes. Sometimes these strange landforms are associated with bright or dark patches. Such areas may be places **where pockets of gas have escaped out of the deep interior**." Bib 129 p42-45

Where did Spudis think the strange landforms came from? He concluded, "pockets of gas...escaped out of the deep interior." Steam is, of course, the most common type of gas.

The Maria Basalt Hydroevidence

Seventeenth century astronomer, Galileo saw the dark areas covering vast expanses of the Moon and mistook them for oceans. Today, science sees them as basaltic plains, as explained in this 1985 text:

"**We now know** that the **maria are seas** of concealed **basaltic lava**, and that the Moon as a whole is **remarkably free of water**." Bib 87 p107

Since then, water has been found in the frozen polar regions of the Moon. Water is also at the very heart of the lunar basalt formation process. Geology is still completely unaware that basalt is not lava that flowed across a dry landscape, which is so commonly stated throughout the scientific literature. We "know" lunar basalt did not flow across the dry landscape because the lunar basalt is crystalline SiO_2 (quartz), not glass Bib 162 p286. Almost half of the basalts brought back from the moon had a chemical composition of SiO_2, with much of it being olivine, the familiar yellow-green crystal in pallasite meteorites. The Crystallization Process earlier in this chapter explained how quartz and olivine are hyprethermal minerals—grown *only* in a hot water, high-pressure environment.

We can see what basalt would look like if it had actually been from a melt *by simply melting it*. Fig 7.13.7 shows melted terrestrial basalt. The shiny melted portions of the basalt specimen look entirely different from the unmelted portion. This simple demonstration is reproducible by almost anyone, yet it is rarely done. The Basalt Mystery (subchapter 6.5) explained that basalt

> "Across the Moon, both in highlands and in maria, we find strange landforms that do not conform to our notions or understanding of lunar processes."

7.13 THE HYDROMOON EVIDENCE

has ***never*** been observed flowing from volcanoes on the continents, even though this sci-bi is commonly taught. ***Basalt was not made on a dry Moon***.

In addition to the micro-melting evidence of basalt, the macro-basalt flow evidence holds clues. Fig 7.13.8 shows The Lunar Basalt Evidence illustrating the difference between lava flows on Earth and the basalt flows on the Moon. When rock is melted on the surface of a planet (especially a cold moon), it would flow for only a short period of time before it cooled, creating irregular and jagged ripples, like Earth's lava flows.

Compare the actual lava flows on Earth with the lunar surface photos. The stark contrast between the two made it hard to accept that there were lava flows on the Moon:

"The vast plains, which cover 17% of the surface of the Moon, **contrast with many terrestrial volcanic landforms**, a difference which **delayed their acceptance as lava flows**. The relief on the mare surfaces is nowhere great, and they are, in fact, exceedingly smooth. Slopes of 1:500 to 1:2000, and differences in elevation of less than 150 m over distances of 500 km are common. This smoothness of the maria is **due to the low viscosity of the mare lavas**." Bib 162 p263-4

One wonders why the Moon's so-called "lava flows" were *ever* accepted as being similar to Earth's lava flows. The claim that the smoothness of the maria is due to the "low viscosity of the mare lavas" is incorrect speculation. The maria basalt is not a glassy lava. Even a thin, low viscosity flow of melted rock must still cool at its snout, which would be evident as ripples and flow lines in the basalt, but such forms are *not* visible in all of the basalt filled marias:

"For the most part, **the maria appear to form a smooth, nondescript surface, and discrete flows are not visible**". Bib 129 p114

Basalt on the Moon is truly a mystery, but with the Hydroplanet Model, which includes Hydromoons, the simplest questions about the Moon's surface features can be answered. One such question centers around the dark basalt patches seen on the Earth-facing side of the Moon, but not on the other.

The Lunar Mare Basin Evidence

The Moon is tidally locked in its orbit around the Earth, which keeps the same hemisphere facing the Earth all the time. Man had never seen the far side of the Moon until the advent of the rocket allowed him to go there and photograph it. Of the two hemispheres shown in Fig 7.13.9, the near side is the view common to the Earth during each full Moon. The long-unseen far side of the Moon is remarkably different from the familiar Earth facing side when it was revealed to the world in 1959. This proved to be one "of the biggest surprises of the space age":

"Galileo Galilei, the 17th-century astronomer, called the lowlands ***maria***—**Latin for 'seas'**—because of their smooth, dark appearance. **One of the biggest surprises of the space age came in 1959, when the Soviet spacecraft Luna 3 photographed the moon's far side, which had never been seen before because it is always turned away from Earth. The photographs showed that it almost completely lacks the dark maria that are so dominant on the near side. Although scientists now have some theories that could explain this dichotomy of terrain, it remains an unsolved puzzle**." Note 7.13e

The surprise was that the far side of the Moon had almost none of the dark maria, which leads us to ask this FQ:

FQ: Why are the dark, lunar mare basins concentrated on the Moon's nearside?

The reason for the absence of the maria on the far side of the Moon should not be too difficult to understand, except that science has never truly understood the origin of basalt on the Earth or the Moon. As we have learned, basalt is not a dry-melt rock, and Galileo was more right than he was given credit; the seas on the Moon were *really* seas, that is, wet basalt seas. The "unsolved puzzle" only remains so because of the dogmagma modern science is in. The wisdom of the Hydroplanet Model makes clear how the Moon's hyprethermal formative period saw the vaporization and dissipation of the Moon's water. The

Fig 7.13.8 – Why do so-called lava flows on the Moon look nothing like lava flows on Earth? Geoscientists have been unable to answer this question with any degree of certainty. Note how formerly molten, liquid rock on the Earth forms 'flows' and ripples. No landforms of this sort have ever been seen on the lunar surface. Could the Moon have had a watery origin? If so, this would answer enigmatic lunar mysteries that have persisted for more than a century.

CHAPTER 7 THE HYDROPLANET MODEL

Fig 7.13.9 – Because the Moon does not spin on its axis, the same side faces the Earth all the time. The dark mare basins are the common and dominant feature of the Moon's nearside, and it was presumed the far side was similar. However, satellite imagery from orbiting spacecraft revealed one of biggest surprises to come from the early days of space exploration; the lunar far side was almost completely void of such maria. This grand mystery remains one of the many "unsolved puzzles" of the Moon.

Earth's gravity pulled residual water towards the Earth-facing hemisphere, just as the Moon pulls the Earth's oceans toward it today.

Evidence for this can be found in the fact that the dry maria basins have some of the lowest elevations on the Moon. These are the areas where puddled water last accumulated, and the reason they are the smoothest, darkest and least cratered areas of the Moon. Other direct evidence comes from the Earth's own basaltic oceanic crust. This crust was also formed under the ocean, in a hypretherm. Pressure, heat, and mineralized seawater were all a part of the basalt formative period. The following chapter will reveal more details about how this happened on the Earth.

That the mare basins were indeed 'seas' solves a sticky problem for astronomers. If the mare craters were impact craters, there would have to be an explanation why these massive young crater basins were concentrated on the side of the Moon facing the Earth. The nearside hemisphere is the most unlikely location for such a concentration because the combined Earth-Moon gravitational field would have blocked or deflected incoming meteors of such great size. Had the mare basins been on the far side, it would not be such a puzzling mystery.

Moreover, there is no supporting evidence that impact can cause lava (especially not basaltic lava) to come to the surface and fill the crater basins. The author of the book, *Impact Cratering*, was quoted in the Hydrocrater Model, confirming the lack of evidence and stating the physics that refutes the theory that lava or basalt comes from impact.

With the impact theory being the most unlikely explanation for the origin of the basaltic mare basins on the Moon, we turn to the Hydroplanet Model for answers. In this model, a number of mysteries are made plain, as we have already seen. The mare basins include countless hydrocraters where lava flowed through diatremes subsequent to primary steam eruptions, forming hills, mountains, and other landforms throughout the hydrocrater basin. The gravitational environment on the Moon is markedly different from the Earth due to its smaller size and reduced mass. This difference aids the flow of liquid rock under hyprethermal conditions, such as the basalt that filled the mare basins.

The lunar maria evidence is an example where scientific truth is beautifully simple and simply beautiful. Without the Hydromoon Model, there is no clear answer, but knowing the lunar landscape formed under water allows us to learn lessons from the Earth's *tides and oceans*. Just as the Moon's gravity pulls the Earth's oceans towards itself today, the Earth, with a gravitational tug six times greater than the Moon, pulled the lunar oceans Earthward until the last remaining seas evaporated, leaving behind the vast, basalt-filled mare basins as testimony of its once hydrous youth.

From Lunar Impact Crater to Hydrocrater

How did astronomers 'discover' that lunar craters were from impact? Initially not all thought the craters *were* impact craters. In fact, most did not; but eventually, they came to agree, a sentiment reflected in the preface of the book *Meteorite Craters*:

"It is at last **agreed, after much controversy**, that most of the craters on the moon were caused not by volcanism but by the impact of meteorites." Bib 119 Preface

'Agreement' of this sort seems to have taken place each time we find a pseudotheory being taught as fact. In such cases, empirical evidence is not evident and controversy often rages for decades or even centuries before one side finally capitulates, and by common 'agreement,' scientific fact is claimed. Although consensus may be reached, a *scientific truth* does not change over time and a consensus based on faulty assumption must eventually dissolve.

Impact cratering was not at first understood, but it eventually came to be the accepted method of lunar cratering:

"Three centuries passed before the process of impact cratering was **properly understood** and it took another half century before **most of the doubters were convinced that the moon's craters are caused by the impact of large meteorites**." Bib 161 p3

Because the Moon is inaccessible for first hand observations, other than the few rocks brought back to Earth during the Apollo program, scientists look to terrestrial craters to gain understanding. From the book, *The Once and Future Moon*, we read how "scientists typically extrapolate results" to arrive at their conclusions:

"**Scientists typically extrapolate results** from study of **terrestrial impact melts to lunar rocks**. Indeed, our **fundamental understanding of the process of impact melting comes from the use of craters on Earth as a guide**." Bib 129 p138

If the "fundamental understanding" of impact cratering on Earth is flawed, the understanding of lunar 'impact craters' would also be flawed. If the Earth's most celebrated and most studied crater, Arizona's Meteor Crater, accepted by the world today as being an impact crater, was not from impact, what then of the Moon's craters?

There are perhaps hundreds of examples where it can be seen that there is a fundamental ***mis***understanding of the lunar cratering process. The Moon's Orientale basin is one such example.

Fig 7.13.10 shows the 578 mile (930 km) multi-ring Mare Orientale, the dominant feature in the image of the Moon. Orientale is on the extreme western edge of the nearside hemisphere of the moon, making it difficult to observe from the Earth. The vast majority of astronomers and geologists today believe the Orientale basin complex and its associated rings were made by impact:

"**Orientale** provides an example of a **fresh**, multiring basin. Because it is the youngest basin, it is the only nearly unmodified example of its type on the Moon. Like complex craters, **basins are surrounded by deposits that were laid down at the time of impact**." Bib 129 p30

The reference to "fresh" is curious since it is believed to be billions of years old, but just where is the impact ejecta or rays presumably surrounding the crater? Multi-ring craters also appear on Mars and the Earth, but they have never been proven to be formed by impact. The Lowell Crater on Mars, right inset in Fig 7.13.10, is blanketed with frost. Scientists now think water was a major force in the shaping and formation of many geological features of the red planet because of the recent discovery of abundant water. In the tiny world of beach sand on the shores of the Earth's oceans, we find a clue to multi-ring crater formation; a double ringed crater in the sand (Fig 7.13.10 center inset), marking the home of a sand crab. In miniature scale, he has reproduced the hydrocratering process of multi-ring crater formation.

The writer of the above statement about the Orientale Basin leaves the reader with the assumption the basin is of impact origin, but several pages later, *acknowledges* that multi-ring basins are a mystery, seeing a "deep seated structure" where material is carried "from great depth" through a fault "pipeline" to the surface:

"**The origin of basin rings is still something of a mystery**. It looks like **a very deep seated structure is associated with the basin rings** because they are often **the sites of eruption** of volcanic lava **from great depth**. This association might indicate that the rings are giant fault scarps and that the fracture along which the scarp formed served as **a pipeline for the movement of lava to the surface**." Bib 129 p37

The impact mindset has scientific researchers looking for

Fig 7.13.10 – Orientale Crater is one of the Moon's best-preserved, "fresh" multiring craters, a classic lunar "mystery" with no solid evidence of impact or impact ejecta. No experimental evidence has been produced and no logical argument exists to support the formation of multiring or basin ringed craters by impact. However, multiring craters and ringed basins are easily explained by hydrocratering. Nature supplies a simple analog of this with the small double-ringed crater made by a common crab in ordinary beach sand (inset center). The Lowell Crater on Mars is covered with white "frost" in this (inset right) image. It is another example of a multiring crater or ringed basins that does not fit the impact paradigm. Note also the smaller craters around the larger Lowell Crater; none of them shows evidence of impact ejecta.

"The origin of basin rings is still something of a mystery. It looks like a very deep seated structure is associated with the basin rings because they are often the sites of eruption of volcanic lava from great depth." Bib 129 p37

Orientale Crater Mystery

Orientale Crater

Double Ringed Sand Crab Crater

Lowell Crater of Mars with "Frost"

ways to explain the features as they relate only to impact. That influence apparently bars them from making the connection to hydrocraters, a relationship they were so close to seeing with giant fault scarps, pipelines for carrying materials from great depths, even the inner-basin melt rock. Most scientists are searching for truth, but sadly, consensus too often outweighs intuition.

Enceladus' Water Fountain Evidence

Today, there is hard evidence documenting that a vast amount of water can be spewed into space from a moon-sized body, escaping the surface and atmospheric hold of the parent body forever.

"Enceladus has joined the small but select band of moons known to have an atmosphere. The Cassini spacecraft, currently orbiting Saturn, has found **a layer of water vapour surrounding the icy moon**, which is likely to be issuing from its surface or interior...

"'It was a **complete surprise** to find these signals at Enceladus,' says Michele Dougherty of Imperial College London, who leads the magnetometer team.

"At just 500 kilometers across, the moon's gravity is **insufficient to keep hold of the water vapour for long. This means that a strong flow of water must continually replenish the atmosphere, suggesting that Enceladus may be volcanically active or possess steamy geysers.**" Note 7.13f

Volcanically active or steamy geysers from a small icy moon perhaps billions of years old—better evidence than this could not have been fabricated! Fig 7.13.11 shows the Hydromoon Enceladus' erupting water fountain spewing water into space.

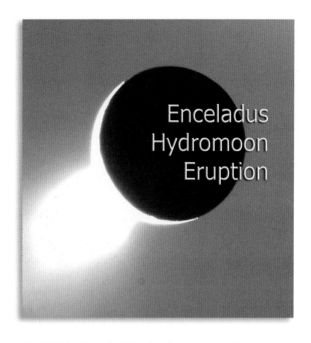

Fig 7.13.11 – Saturn's 300-mile diameter moon, Enceladus, is a near-perfect example of a Hydromoon. Comprised almost totally of water and ice, it has an active hydrofountain that is enormous in both width and height. This eruption is powered by tidal forces as explained by the Gravitational-Friction Law. This should have been one of the most paradigm-challenging photos ever taken, yet few have even seen it.

There was no impact and there is no magma. Enceladus is a hydromoon with clues of its age in the rate at which the water is escaping its surface. This dramatic discovery should have spurred a renewed discussion about our own Moon, but unfortunately, it has not.

The Secondary Impact Evidence

The Copernicus lunar crater is a well-known and easily identifiable nearside crater with a fabulous ray system. Copernicus can be seen near the top of the Apollo 17 photo in Fig 7.13.12.

Secondary Impacts are another fascinating physical evidence of the Hydromoon Model and are formed when material ejected from the primary crater lands nearby, leaving an impression on the surface around the primary crater. Although there may be disagreement on the origin of the Copernicus crater, the Secondary Impact craters are unarguably caused from ejected matter that would ***not*** have had the high velocity of an incoming meteorite. Ejecta material thrown from either impact or hydrocrater eruption would have only the force of gravity to pull it back to the surface, and on the Moon, that force is only 1/6 that of the Earth,

In the authoritative book, *Impact Cra-*

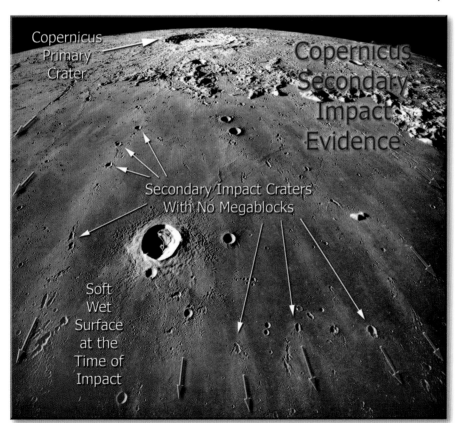

Fig 7.13.12 - This fabulous shot of the Copernicus Crater (background) and secondary impacts (foreground) was taken during the Apollo 17 mission. Secondary impacts are low velocity events requiring exceptionally large blocks to form craters on dry ground. Why are no such "megablocks" seen in any of the secondary craters? On a wet surface, secondary craters can be formed by smaller projectiles, and projectiles of any size can conceivably be swallowed by the liquid mud. Courtesy of NASA (AS17-2444).

tering, researchers report that Copernicus's secondary ejecta blocks were traveling an average of ".5 km/second" Bib 161 p105 (500 m/sec), or at about half the speed of a rifle bullet. With the modern science Dry Moon Pseudotheory idea in mind, we ask this FQ:

FQ: How were Copernicus' low-speed secondary impact craters formed without leaving any observable trace of large impactor blocks?

In a dry environment, the cratering effect of falling sand and small rocks is negligible, even in mass quantity, especially under the influence of the Moons' diminished gravity. Furthermore, the breakup and disintegration of large blocks is not to be expected during the slower speed impact of secondary ejecta. The fact that Copernicus' secondary craters exist necessitates an explanation, for such cratering requires reasonably large, *intact blocks* of rock.

This concept is illustrated in Fig 7.13.13 where a Secondary Impact crater is formed by an impactor on dry, loosely packed volcanic material. In this example, the 1-meter impactor is intact in the center of the crater. Although impactologists know intuitively that giant "megablocks" are needed to form Copernicus' secondary impact craters, the FQ of where those blocks went remains unanswered.

Turning to the Earth for analog, Melosh's book, *Impact Cratering*, records an instance of megablocks on Earth, limestone blocks of up to 25 meters, blown from a 22-km crater in Germany. The blocks, traveling about the same velocity as the impactors that would have formed Copernicus' secondary craters, were seen as far away as 35 km from the primary crater. With this, we gain some understanding of what secondary lunar craters and impactors should look like—yet, in the Dry Moon Pseudotheory, nothing seems to add up, and researchers are left to *infer* block sizes of extraterrestrial (lunar) crater impactors:

"**Intact 'megablocks'** of Malm limestone have been found great distances from the 22-km diameter **Ries Crater in Germany**. These blocks, all from the uppermost geologic unit in the target area, range from one block 1000 m in diameter that was thrown at least 7 km from the transient rim down to blocks 25 m in diameter at ranges of 35 km, corresponding to an ejection velocity of **600 m/second**. The existence of such large ejected blocks on earth **suggests** that the **block sizes inferred** from extraterrestrial secondary craters **are real**." Bib 161 p105

Of course, the reason the extraterrestrial secondary block sizes are inferred is because they are not visible, and it was only because blocks were found from similar cratering events on Earth that researchers accepted their inferences as being real.

There is surely a solution to this puzzle, and learning it requires asking questions, the right questions, with an open mind, no matter how improbable the question may seem:

FQ: Was the Moon **dry** when Copernicus' Secondary Impact Craters where formed?

Because the secondary craters were formed by impactors that were not traveling at hypervelocity, the "megablocks" that formed them could not have vaporized. There are no atmospheric erosional forces at work on the Moon that account for the complete obliteration of the impactor. Bombardment by micrometeorite has been claimed by some to be the only erosional force at work, but that process, if true at all, falls far short of

Fig 7.13.13 – This secondary crater was created when a primary crater's ejecta, travelling at low-speed, landed and formed this crater. This example is a 6-foot (2-meter) crater formed in soft, tuff sediment where the impactor (seen in the center of the crater) landed. On hard, dry material, very little cratering is evident. A Secondary Dry Impact is a low-velocity event where the impactor is not destroyed and can be found either in the crater or nearby.

explaining the disappearance of the megablock impactors.

However, if the lunar surface was soft and muddy, solid ejecta could have been easily swallowed by the lunar mud. The ejecta could also have been semi-soft blobs of material that was absorbed into the surrounding plain. Either explains the absence of the impactors.

How soft was the muddy lunar surface? The speed of the secondary impacts on the Moon would have been similar to the speed satellites fall to Earth after being slowed by the atmosphere. The satellites barely dented the hard, dry surface of the Earth, depicted earlier in the Hydrocrater Model (see the impact of the Genesis satellite in Fig 7.9.5). Thus, the empirical evidence on the Moon's surface attests to its plasticity during the final days of its formation.

A rock's softness or malleability is achieved in one of two ways, either by heating, or by hydrating with an aqueous solution. The Magma Pseudotheory demonstrated that the Moon's surface could *not have been melted* because of the absence of glass or glassy rock. Thus, the alternative is that the Moon's surface was at some time wet and muddy, which is supported by Copernicus' Secondary Impact Crater evidence, which in turn supports the Hydromoon and Hydroplanet Models.

We know that hydrocraters are formed when water inside a planet is heated by friction along lines of faulting. This is only possible when rigid material grinds together, and so, not all of the Moon was water and soft materials. In addition to the fine silt suspended in the water contributory to the Moon's final form, the lunar surface consists of brecciated rock (angular pieces of rock cemented together by hyprethermal processes). Some of the lunar breccia is a form of volcanic (not impact) glass, containing broken fragments of other rocks. It was formed in underground volcanic vents or diatremes and blasted to the surface along with rocks, water and other debris through hydrofountains, and altered to its final form by successive events.

"Testifying to Their Common Origin"

Before the Impact Pseudotheory basked in its popularity in the 1960s, lunar scientists were open to alternative theories of the Moon's origin. However, the Apollo astronauts found no

Fig 7.13.14 – The Humboldt Hydrocrater is a classic example of a flat lunar hydrocrater. Its surface exhibits a number of hydrocrater phenomena: diatreme mountains, domes, circular faults, a bull's-eye crater, a crater chain, scarps, and clefts testifying of their common origin, the Hydromoon model, and tectonic weakness. The scientists who recognized these volcanic structures during the 1950s and 60s were ignored while scientists working on the Apollo missions paved the way for the Impact Pseudotheory based on a dry Moon.

water on the moon and the rocks they brought back were surprisingly dry. Planetary scientists began to assume there had never been much water on the Moon. Today, the tables have turned as researchers have discovered large amounts of water beneath the Moon's surface. From an article in *Scientific American, The New Moon*:

"Current estimates indicate that more than **10 billion tons of ice** exist within the upper foot or so of the surface at both poles." Note 7.13g

Before researchers rejected Hydromoon ideas, geologist G. C. Amstutz wrote of a "regional framework" of structures on the Moon that were similar to the calderas and "explosion diatremes" of volcanic areas on the Earth. Amstutz's 1964 article identifies craters on the Earth and Moon as "polygonal structures" because of the "parallel sets of polygonal faults and parallel folds":

"The shapes of the *Polygonal structures on the moon* are more closely comparable to the polygonal structures of the Middle West than to impact features such as bomb craters. One of the typically polygonal structure areas of the **moon** is the Ptolemaeus region. Here, the walls of the polygons are **clearly parallel to what may be considered as regional fault patterns**. One feature is very important for both, the moon structures **and the polygonal structures, and absent at the impact holes: their distinct association with regional structural trends**. This is perhaps the most important observation and applies also to the volcanic and ring complexes. **All fit into the regional framework and are an integral part of it**. Therefore, other features need to be compared to see whether more analogous ones occur.

"The calderas of Hawaii, of Iceland, of the Azores, of Crater Lake and of other similar structures fit into their regional volcanic belt. The numerous dikes, **explosion diatremes** and tuffisite pipes of the Middle West also form a structural pattern and the Wells-Creek basin, and Higgs dome fit well into this pattern." Note 7.13h

Amstutz further noted the impossibility of a meteorite's ability to find a "regionally favorable" place to impact:

"Anyone examining the polygonal and ring structures and their geologic environment **carefully** can hardly fail to dismiss the assumption that a meteor would fall on the earth like a guided missile, and **sense exactly where there is a regionally favorable place for punching a hole into the earth**." Note 7.13g

Another careful observer, V. A. Firsoff published the *Moon Atlas* in 1961 after a year of detailed lunar observations. After carefully mapping the features of the lunar surface, Firsoff identified lines of "tectonic weakness" that stretched around the Moon "testifying to their common origin":

"A grid line is essentially **a line of tectonic weakness**, along which land movements have occurred and which may manifest itself as **fracture, cleft, fault, thrust line, crater chain, ridge or scarp**. **All these types of formation are often present in one and the same line, thus testifying to their common origin**." Bib 111 p32

Evidence of these features shows up in Fig 7.13.15, the Lunar Out-Channels diagram. The Hydrocraters in this image show out-channels that flow from the craters onto the surrounding plain. These formations clearly show that water flowed during the late stages of lunar surface formation. These are not lava flows and do not exhibit the characteristic rippling of rapidly cooling lava on a dry surface. Instead, they consist of basalt and other minerals crystallized while in a hypretherm, under water.

Similar landforms exist today under the Earth's oceans. In most cases, they are not crystallized unless the temperature of the water, the pressure, and the chemistry are such that a hyprethermal environment conducive to crystal growth existed. The similarities do consist of large, flat plains and shallow depressions. When evaporated, the surface is quite similar to the Moon, as Firsoff noted:

"Characteristically, **lunabase** is lacking in vertical relief, and **large lunabase areas are plains**, sometimes smooth, but often knurled or scooped out into **innumerable shallow depressions**, so as to resemble the surface on a *névée*, **produced by evaporation**." Bib 111 p8

Firsoff realized that the geological processes present on the Earth today cannot wholly account for the formation of the lunar surface features, but he realized that many of the features he saw were "like our geysers, or perhaps mud volcanoes" consisting "mainly of emission of gas, which was primarily steam and may occasionally have turned into liquid water":

"The meteoritic hypothesis has its points, but it fails to account adequately for the varied features of the lunar surface and vulcanism has constantly to be brought in by the back door, as it were, to supply the deficiencies of meteoritic interpretation. Yet the volcanic alternative, too, has some awkward hurdles to take and may require further modification.

"Its supporters incline to talk of fire and smoke, ash and lava; and these are not very much in evidence on the Moon, while such evidence as there is may be questioned. The **lunavoes** would seem to have been **more like our geysers, or perhaps mud volcanoes**, than like Cotopaxi, Etna, or Krakatoa. **Their activity consisted mainly of emission of gas, which was primarily steam and may occasionally have turned into liquid water**." Note 7.13i

This was a remarkable observation and showed that Firsoff had gained a correct insight into the true formative processes of the lunar surface and his meticulous study of the Moon's surface.

Few scientists have taken the time to make a careful study of the lunar surface. For Firsoff and those with him, the lunar surface was littered with hydrofeatures analogous to those found on Earth, and it was a simple conclusion that the lunar surface had once been covered with water and that the mare basins were at one time water-filled.

The compelling evidence was eventually dismissed and by 1995, lunar experts thought it "strange" as they recalled leading scientists once thought "there was absolutely no doubt that the lunar seas had once been water-filled":

"Moreover, there was no trace of any hydrated materials – that is to say, materials involving the past presence of water. **It was strange to recall** that at a meeting of the International Astronomical Union as recently as 1966, Dr Harold Urey, one of the world's leading planetary geologists, had told me that in his **view there was absolutely no doubt that the lunar seas had once been water-filled.**" Bib 149 p32

Someday it will be probably be equally "strange to recall" how scientists dismissed the overwhelming evidence of the Hydroplanet Model and actually thought the Moon, Earth and other planets were once *melted*.

The Moon's Gravitational "Anomalies"

With the introduction of the Hydroplanet Earth, in subchapter 7.6, and the empirical evidence that it supports, we became acquainted with three reoccurring components of the Hydroplanet Model. These are hydrocraters, hydrofountains and diatremes. These elements appear throughout the model and are critical parts of the formation process of all the celestial bodies discussed thus far. Mainstream geology has pretty much ignored the hydrofeatures over the last 50 years and has failed to recognize their importance because the dogma of magma has had such a firm grip on the minds of the investigators and scientific leaders.

It may seem a simple thought that the origin of rocks should be left open to debate since modern science has yet to reproduce most rock types. However, the Age Pseudotheory has the establishment thinking rocks take millions of years to form and so, the idea that rocks can form almost instantly through the hyprethermal crystallization process has gone completely unnoticed.

Although we have not penetrated the crust of the Moon more than a couple feet, recent gravity measurements of the lunar surface were obtained by orbiting spacecraft that revealed evidence of the features that lie beneath the surface of the Moon. This has enabled a detailed map of the Moon's relative surface density, which can be seen in Fig 7.13.16. These maps provide additional supporting evidence of the Hydroplanet Model.

To begin, we refer to a quote from the 1989 *Impact Cratering*, the authoritative book on the science of impact cratering:

"The grooved and lineated ejecta blanket surrounding all lunar basins, the associated secondary craters, and samples returned from the moon's surface **have convinced all but the most stubborn planetary geologists that lunar basins are of impact origin**." Bib 161 p163

This statement expresses the sentiment felt by nearly all scientists at the later part of the 20th century; that lunar basins are of "impact origin." By 1959, the soviet space probes had imaged the far side of the Moon, revealing the surprise that there were none of the great mare basins on the nethermost side. Then in 1968, new surprises had come when gravitational 'anomalies' (high gravity readings) were discovered. On Earth, the higher elevations such as those with mountain ranges are generally associated with higher gravitational readings, not the low-lying valleys. On the Moon, the basins have the lowest elevations. Less rock should translate to lower gravity readings, but this was not the case; in fact, the majority of the "mass concentrations" were located *in the basins* (see Fig 7.13.16):

"**Mascons,** an acronym for 'mass concentrations,' were first discovered in 1968 when large perturbations were noted in the orbits of the Lunar Orbiter spacecraft series. **Mascons are anomalous mass excesses associated with nearside lunar basins.**" Bib 161 p180

The anomalous "Mascons" remained mysterious until 1994 when NASA sent the Clementine spacecraft orbiter to the Moon to collect detailed data. That mission was followed by the 1998 Lunar Prospector orbiter mission, which confirmed that the mascons were *actual underground structures* on the Moon. One significant problem; impacts do not account for the mascons.

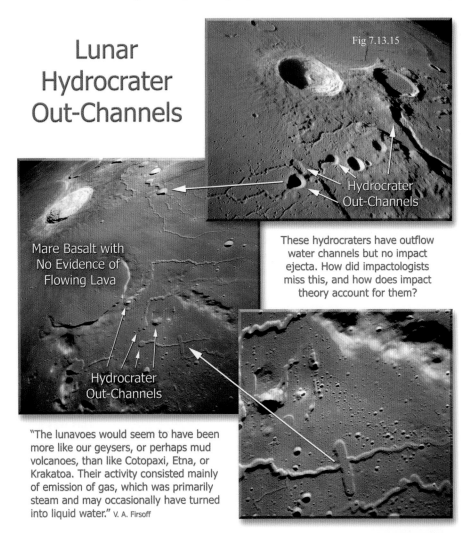

Lunar Hydrocrater Out-Channels

Fig 7.13.15

These hydrocraters have outflow water channels but no impact ejecta. How did impactologists miss this, and how does impact theory account for them?

"The lunavoes would seem to have been more like our geysers, or perhaps mud volcanoes, than like Cotopaxi, Etna, or Krakatoa. Their activity consisted mainly of emission of gas, which was primarily steam and may occasionally have turned into liquid water." V. A. Firsoff

Without a logical explanation for the excessive mass readings, scientists were left to speculate how such masses got there. Still in the impact-cratering paradigm, scientists attempt to explain the phenomenon with basalt filled basins:

"After several years of **speculation and study** it became clear the excess mass is due to the mare basalts filling the basins." Bib 161 p180

Years of "speculation and study" have yet to clear this mystery and there are a number of problems with their assumption. The basins are *depressions,* and have the lowest elevation of the surrounding areas. Even with the obviously vast quantities of basalt, there is no justification that the areas of lesser material would have a higher gravitational reading based on the basalt materials alone. The basalt is not of significantly denser material than the surrounding rock to account for the difference.

More importantly, researchers understand the basalt in the basins was not from impact melting. Jay Melosh, the author of *Impact Cratering* also wrote an article in 2001 entitled *Impact-Induced Volcanism: A Geologic Myth* that was published in the journal *Planetary Geology*. He refutes the notion that impact could "cause a volcanic eruption" as being unfounded and lacking "evidence":

"The **idea** that the **impact of a large comet or meteorite may cause a volcanic eruption** is becoming entrenched in the geologic literature. In spite of the apparent appeal of this idea, **there is no evidence that it has ever actually happened, either on the Earth or on any other planet in the solar system.**" Note 7.13j

Melosh continued to explain why impacts *never* caused volcanism on the Moon:

"The total energy required to melt the basalts in many flood basalt provinces greatly exceeds the total energy available from the impacts that have been proposed to form them. Furthermore, study of planets such as the Moon and Mercury that do show signs of antipodal focusing from large impact basins show no trace of volcanism at the antipodes. **There is thus neither evidence for impact-induced volcanism nor a plausible mechanism for creating it. Proponents of an impact origin for flood basalt provinces should probably look for more effective causes**." Note 7.13j

In summary, we can draw these two conclusions from the foregoing statements:

1. Lunar basins "are of impact origin" and are flooded with volcanic basalt.
2. Impacts cannot "cause a volcanic eruption" and fill these basins with basalt.

The two contradictory statements cannot both be true, which should have been the catalyst to look in new directions; it is obvious that there was no impact, leaving us with the question of the basalt origin:

FQ: How did basalt get into the crater basins?

Clues to the origin of the basalt and the mass concentrations in the lunar basins are within the pages of the book *Impact Cratering*:

"Recent studies show that **the center of each basin is underlain by a plug of dense lunar mantle material that intrudes through the less dense lunar crustal rocks** that surround the basin and underlie the ring structures." Bib 161 p167

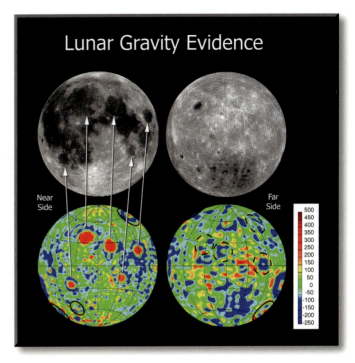

Fig 7.13.16 – The two colorized images indicate the strength of gravity on both near and far sides of Moon as taken during the NASA Lunar Prospector mission. Red areas are indicative of the highest gravitational influence and blue the least. Planetary geologists have presumed that the dark, circular mare basins are from impact. The basins are some of the lowest points on the lunar surface, which *should* correspond to *lower* gravitational levels. An impact event should have left fractured and melted debris in the crater, which is also less dense, contributing further to *lower* gravity readings. Yet, as the maps clearly show, many of the basins have a **higher** (areas in red) gravity reading—opposite that of impact expectations. The only explanation for such high gravitational readings in the flat mare basins are massive diatremes beneath the crater basins. These diatremes are filled with dense crystalline rock from depths. Researchers choose to call these diatremes "plugs," but still cling to an impact origin for the large mare basins.

Recognizing that "the center of each basin is underlain by a plug" should have been a moment of serendipity for the impactologist, but it was not to be. The identification of the plug was strong evidence of the presence of a diatreme—empirical evidence of a hydrocrater!

Another problem for the 'impact' researchers was that the craters with elevated gravitational fields, identified as red in the previous diagram, are located near craters of similar age that did not have elevated levels. How did impact cause such a phenomenon? The experts did not know, but if we reassess the evidence, the diatreme "plugs" are remnants of massive eruptions where matter from various depths with different densities was carried to the surface; some of which were dense enough to account for the elevated gravitational readings.

The discovery of the dense plugs, or diatremes was not a new concept. In 1735, French mathematician Pierre Bouguer observed and named the phenomenon the Bouguer anomaly. His ideas are still shared by many modern planetary scientists, who still include the plug explanation as part of the impact pseudo-theory:

"One **possible explanation is that plugs** of dense rock from the lunar mantle [magma] may have **risen toward the surface** of the basins after impact." Note 7.13k

To continue to view such craters as "impact" requires the diligent exercise of the imagination and the dismissal of part of the evidence. As Melosh said in his book, *Impact Cratering*,

7.13 THE HYDROMOON EVIDENCE

there has *never been* an observation or plausible mechanism for *volcanism caused by impact*.

One does not have to 'imagine' how the high gravity areas on the Moon formed; we can examine similar features here on the Earth. There are countless examples where minerals were formed in diatremes and pipes, many of which are exposed at the surface. Mining companies commonly conduct flyover surveys using sensitive gravitational equipment that identify circular formations which exhibit both high and low readings.

Like those on the Moon, the gravitational anomalies are diatreme pipes or 'plugs' that are remnants of past hydrofountain activity, having densities based on the materials that were carried from the depths.

The Hydromoon Summary

In summary, proponents of impact derived basin basalt will never find scientific evidence that such basalt formed from a 'dry' impact because crystalline basalt only forms in a hypretherm. The basalt-forming mechanism researchers must understand is the water-based hypretherm. This hyperthermal environment is the same for all non-melted minerals. The lunar gravity evidence in part with other Hydroplanet Model evidence further confirms that our lunar neighbor is a hydromoon.

The Hydromoon Evidence subchapter tells the story of the relatively dry Moon. It did not begin as an anhydrous dry, melted body, contrary to the belief of nearly all modern planetary geologists; but it did become dry, through dehydration. This comprehensive subchapter introduces a tremendous amount of new material and old observations, presented as part of a completely new model. Missing boulder tracks, the overemphasized effects of micrometeorites, and boulderless craters each provide a connecting link to the solution of one of nature's puzzles of how the Moon was formed. The abundant Hydromoon features; dome hydrocraters, flat-floored hydrocraters with out-channels, central peaks and landforms in complex craters, crater chains, and a host of other hydro-related features further testify of the Moon's water-born origin.

Add to all of this, the gravitational anomalies and evidence of the tidal effects of Earth's gravity on the lunar mare basins, its ancient seas and crystalline basalts, and a clear picture begins to emerge; the Earth's moon was indeed, a once wet hydrobody, beautifully preserved for human discovery and exploration.

We are now poised to learn about other tremendously important hydrobodies that continue to confirm the Hydroplanet model—the Hydrocomets, infrequent denizens of the sky.

7.14 The Hydrocomet Evidence

There is probably no night-sky object that has evoked more fear or reverence among Earth's human inhabitants than the comet. In ancient times, comets heralded bad omens and God's blessings alike, but it was not just the ancients whose lives were affected. The Daylight Comet of 1910 created panic as people in Chicago sealed their windows to protect themselves from the comet's poisonous tail. Others bought "Comet Protecting Umbrellas," and anti-comet pills.

Comets, or rather Hydrocomets, are another piece of the Hydroplanet Model, and, like so many other pieces of the puzzle, remain mysterious and misunderstood. In this subchapter, we will dispel some of the mystery of one of nature's most beautiful signs of celestial water.

The Origin of Comets

People did not always know what comets were. In ancient times, people thought comets were the 'power rays' of supernatural beings. They thought comets contained fire because they were so bright in the sky. In the 4th century BC, Aristotle imparted his belief that comets were atmospheric phenomenon; hot, dry exhalations that occasionally burst into flame. This idea persisted for two thousand years.

Today, it is known that comets are small members of our solar system, most of which are thought to come from just beyond the furthest reaches of the Sun's influence, in an area called the Kuiper Belt. A popular NASA web site posted an animation describing how a "comet is born":

"As the animation zooms into the disk, micron-size particles of dust can be seen sticking together to form centimeter- and millimeter-sized rocks. As the rocks become more massive, gravity takes over, forcing other surrounding pebbles and dust particles to collide with the larger rocks. The process continues until a comet is born." Note 7.14a

The animation sequence and accompanying description is a classic example of what is wrong in modern science today. Although the description of Fig 7.14.1 that goes along with this comet animation states that this is "one of the most widely accepted theories pertaining to the origin of comets," it is taught as though it is fact.

No hard evidence exists that comets or any other planetary body came from the collusion of rocks in space (accretion). In the Magma Pseudotheory, we looked at the accretion theory and discussed how high-speed impacts create melt, which cools as *glass*. Comets are not made up of glass. As one researcher noted in *Science*, the interstellar dust (between stars) appears to con-

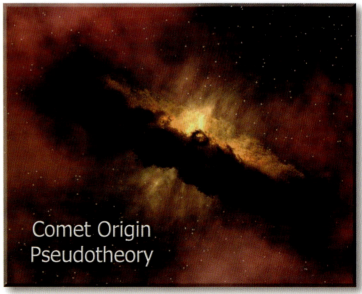

Fig 7.14.1 – Genesis of a Comet by NASA. This artist's concept of comet formation includes accretion and impact of dust particles that become rocky comets. However, this pseudotheory does not take into account how all rocks originally formed—from water. How do we know comets did not come from impact? Because they would be made of glass—and they are not.
(PIA02107 from: photojournal.jpl.nasa.gov).

tain amorphous or *non*-crystalline silica, which is glass—**but** the dust from comets is a "crystalline silicate":

"As in the interstellar medium, much of the dust from comets consists of **silicate minerals**, **but** despite the similarities, there are **puzzling differences**. For example, interstellar dust shows the absorption signature of **amorphous [glass] particles** with a silicate composition, whereas **Hale-Bopp and other comets have crystalline silicate**, probably in the form of Mg-rich olivine." Note 7.14b

Unfortunately, modern science is no closer to understanding the origin of comets than it was 100 years ago, even after visiting them with spacecraft. Publicly, the theory that comets and planets are accretions of material from the first 100 Myr (million years) of the solar system is portrayed as fact through handsome NASA animations and science programs shown on television. A search through science journal articles, however, reveals just how unclear this process really is:

"**Comets** are **thought to be** the best-preserved relics of the first 100 Myr of the proto-solar nebula, when interstellar medium (ISM) dust particles of size ~0.1 μm merged and accreted to form planetesimals of size ~1 km… **the details of the mechanism of accretion are quite unclear. Without a detailed understanding of the mechanisms driving this process, it is impossible to understand and predict how planetary cores are built and how they evolve**…" Note 7.14c

The truth is the mechanism of accretion will remain "quite unclear…impossible to understand and predict how planetary cores are built and how they evolve" until the current pseudotheory is replaced.

Planets and moons are not the only celestial bodies to show *direct evidence* of the Hydroplanet model; asteroids and comets were also formed out of water, and we will examine the evidence that supports this claim. First, we need to define what a comet is.

Comets Defined

The word "comet" comes from the Greek word for "hair." Our ancestors once thought comets were stars with what looked like flowing hair trailing behind. Advances in technology and math have made it known that comets are neither atmospheric phenomenon nor "hairy stars," but are a part of our own solar system. By the end of the nineteenth century, comets were believed to be a loose pile of rocks coated with an icy layer.

The planets and asteroids in the solar system orbit the Sun in elliptical orbits along a generally flat plane called the ecliptic. Other celestial bodies that are bound to our solar system include dwarf planets and comets that follow highly elliptical orbits, noticeably unrelated to the ecliptic followed by the planets. The long elliptical orbits of comets makes them unique in the solar system, and it contributes to the special feature they are known for—their tail.

We cannot see asteroids with the naked eye from Earth because of their small size, but sometimes, comets become a visual delight in the night sky because of their long graceful tails. It was thought that the tails came from melting surface ice as the comet approached and passed the Sun. However, the theory *failed to explain* how a body that contained only a little ice on its surface could continue to put on such a brilliant display after several passes around the Sun.

If not surface ice, how do comet tails form and what are they

Fig 7.14.2 – This sunset scene includes the Hale-Bopp Comet as it appeared near Pazin, Istria/Croatia. Although comets have been observed since mankind looked to the sky, only recently have direct observations from spacecraft orbiting them given us the knowledge and wisdom of what they are and how they were made. Courtesy of Philipp Salzgeber (http://salzgeber.at).

made of? Scientists have long known the "most abundant" component of the comet's tail and luminescent head—the coma—is water:

"Being very small and dark, the actual nucleus of the comet is extremely difficult to observe. Thus, our best knowledge about the size and activity of a comet is obtained through observations of **cometary water production, ordinary water being the most abundant volatile component in the coma**." Note 7.14d

Comet West in Fig 7.14.3 shows the beautiful watery blue tail typical of many comets. Earth-bound telescopes are not able to resolve the comet itself, but instead reveal the coma and the tail, which is made primarily of water, a material we are interested in studying. In fact, there is lots of it, far more than exists in a surface ice sheet, a fact that has confused planetary scientists.

The abundance of water led scientists to theorize during the 1950s that comets were "dirty snowballs" or icy worlds of loose piles of rocks and dust originating far out in the solar system.

Then in 1986, 1996 and 1997, space probes were sent to observe respectively, Halley's Comet, and Comets Hyakutake and Hale-Bopp. From those missions came direct observational evidence of the comets mysterious nature. The first close-up image of a comet was Halley's Comet, in Fig 7.14.4, which revealed that the coma and tail was more than just melted surface ice. There were active hydrofountains emitting jets of steam and enormous amounts of subsurface water and sediment, dust seen in the comets' tails; the same type of material exists on the surface of other celestial objects like asteroids and moons. Spectral

Fig 7.14.3 – This is Comet West showing off its beautiful tail, which always points away from the Sun. As comets travel close to the Sun, solar radiation and the increased gravitational effects of the Sun and nearby planets cause water in the comet to be jettisoned, forming tails often visible with the naked eye. Courtesy of J. W. Young, NASA.

Their comment that "somewhat more" water lies beneath the Earth than is contained in Earth's oceans is vastly understated and reveals the researchers ignorance of the Hydroplanet Model evidence. This carries into their observations and interpretations of the lesser solar system bodies, although they recognize that the small bodies hold important "clues" to all of the celestial body formation events in our solar system.

In June 2004, photos of Comet Wild 2, in Fig 7.14.5, provided the clearest evidence yet that comets were more like asteroids and the Earth's Moon than loose *piles of rubble*, as astronomers had theorized:

"Yet Wild 2 is **not a fractured pile of rubble that would all fly apart when hit**, **as some astronomers expected**. Brownlee: 'We're sure this is a rigid material because it can support cliffs and spires.'

"**What sort of material can crumble under impact, leave sheer walls and allow its parent body to remain intact**?" Note 7.14f

This very good question has *not* been answered by modern science's impact theories. Every known celestial body of a size similar to Comet Wild 2 with equally large craters would have been smashed to pieces by such a large impactor. With mounting physical evidence, even planetary geologists well heeled in impact theory are questioning the impact conception.

This is important because as we study the craters on the comets and on the Moon, we find the same hydrofeatures on both of them. If researchers would question the impact theory with an open mind, new ideas from already existing observations would begin to take shape. There are other types of *explosions* that can form craters on the comets. This is alluded to in the following quote where it was said that "other forces" might be responsible

analyses of the material streaming behind the comet had indicated that the primary substance in the tails of the comets was indeed water, and now the probes had confirmed it, ending *part* of the speculation.

Water was not coming 'off' the comet, but *out of the comet*, which is why we call them **hydrocomets**. All known comets have hydrofountains that spew jets of water and sediment from beneath their surfaces as they approach the Sun.

Researchers have long thought it was only heat from the Sun that caused vaporization of water on the comet's surface, but they have been forced to rethink that idea because the water is coming from *inside* the comet body. The vaporization is not caused only by the Sun's heat, but from the effects of gravity, as outlined in The Gravitational Friction Law. Gravitational friction generates heat below the surface of the celestial body as it is acted upon by the tidal stresses of the Sun as it approaches perihelion.

Hydrocomet Evidence

With the evidence uncovered by the comet probes, the significance of water became more apparent and a *Science News* article in 2002 reported that comets are "about half ice":

"Beyond these planets, **water condensed in large quantities and formed comets, which are about half ice**. Compared with these icy objects, **Earth contains little water**. Only about 0.02 percent of its mass is in its oceans, and **somewhat more** water sits beneath the surface." Note 7.14e

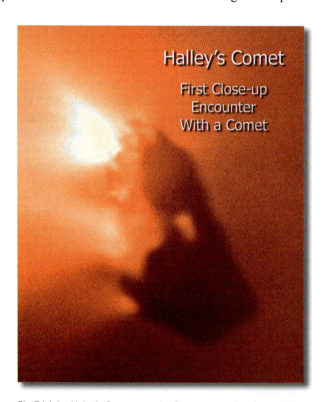

Fig 7.14.4 – Haley's Comet was the first comet to be observed up close by spacecraft. This revealed clues about the inner workings of the coma. Ice was not melting on the surface of the comet, but rather hydrofountains of steam, water ice, and dust were being ejected from identifiable hydrofountains. Courtesy of NASA.

Fig 7.14.5 – These are different images of the same comet, Wild 2 as it was observed up close in 2004. This proved that comets were not piles of rubble from impact. Instead, these hydrous bodies emit "steam explosions" just as the Hydroplanet Model predicts. Planetary scientists realize that if craters of the size observed on this comet were from impact, they would have broken the body apart. However, they still have not been able to recognize them as being hydrocraters. Courtesy of NASA.

for the cratered surface features:

"That all **assumes** the 'footprints' and other depressions on Wild 2 **are in fact impact craters. Other forces could be at work**. 'Comets do blow up unexpectedly,' Brownlee pointed out, adding that **built-up internal pressure and 'steam explosions' might be** responsible for some of the surface features." Note 7.14f

Of course, the "other forces" *are* steam explosions. Every mission science has taken to learn about the comets has provided additional evidence in support of the Hydroplanet Model, and that tradition continues with each new mission. Three years after Comet Wild 2 yielded her secrets, Comet Tempel 1 opened a window to the inside of the comets.

Tempel 1 Reveals More Hydrocomet Evidence

One of the most recent comets to be studied was Comet Tempel 1. A mission known as Deep Impact was designed to crash a projectile directly into the comet, to learn more about its makeup. On 4 July 2005, that mission was carried out as Tempel 1 was impacted by a 370-kilogram copper impactor-spacecraft. The comet turned out not to be solid as was previously assumed:

"The observations bolster theories that comets may have seeded Earth with the raw materials for life and suggest they may be **sponge-like – rather than hardened – at their cores**." Note 7.14g

Here was more evidence of the Hydroplanet Model. Rather than solid, hardened cores, comets were porous, sponge-like celestial bodies, a lot like the hydroplanet Earth. One primary difference between the Earth and a comet, other than the obvious size disparity was that the comet would travel close to the Sun and have its exterior and interior water baked out. Researchers now realizing this have had to revise their theories about comets and the water they contain. From the 2005 journal, *Nature*:

"Because the comet passes close to the Sun on each orbit, solar heating could have built up gas pressure inside, releasing a fierce blast when its crust was punctured. Instead, the Sun's heat seems to have **baked out volatile materials** from the outermost metres, leaving a fine dust of minerals and organics. '**Theories about the volatile layers below the surface of short-period comets are going to have to be revised**,' says Charlie Qi, an astronomer at the Harvard Smithsonian Center for Astrophysics in Cambridge, Massachusetts." Note 7.14h

Why do the theories of comet interiors need "to be revised?" Because they are partially hollow!

"The wider view shows debris extending about 720 kilometres from the nucleus. Rough calculations from the way it has scattered show **the comet is much less dense than water ice, and is extremely porous**." Note 7.14h p158

Sciences original theories have themselves become void. Hydrocomets had water in them originally, when they formed, but heating by the Sun and by gravitational friction caused the interior water to escape through hydrofountains leaving behind this

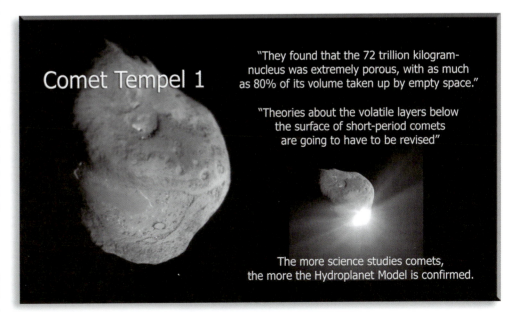

Fig 7.14.7 – Comet Tempel 1 was impacted on July 4th, 2005 by the washing-machine sized probe, Deep Impact. The smaller image was taken after the 10-km/sec impact showing the heat and dust generated from the impact. Instead of meteorite material, the comet proved to consist of at least 80% water that was being ejected by steam jets. These were hydrofountains, a surprise for the astronomers and geologists. Comet "theories" were going to have to be completely revised.

"empty space":

"They found that the 72 trillion kilogram-nucleus was **extremely porous, with as much as 80% of its volume taken up by empty space**.

"'That tells me there is **no solid layer all the way down to the centre**,' says Mike A'Hearn, the mission's principle investigator at the University of Maryland in College Park, US…

"'**It's like a sponge, with a lot of cavities**,' agrees Horst Uwe Keller, an astronomer at the Max-Planck Institute for Solar System Research in Germany. He observed the event with Europe's Rosetta spacecraft and says the discovery confirms previous observation suggesting **other comets are also porous**. 'When you touch it, it just crumbles under your hands.'" Note 7.14i

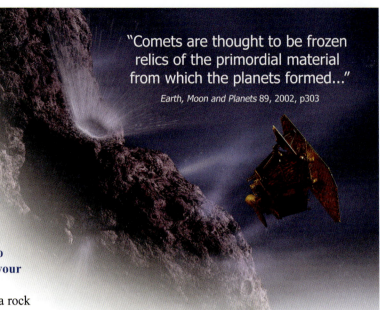

"Comets are thought to be frozen relics of the primordial material from which the planets formed…"
Earth, Moon and Planets 89, 2002, p303

This was simply an amazing discovery! On the Earth, a rock that is 80% hollow but with a solid exterior crust is called a geode. Tempel 1 and many comets like it are essentially nothing more than large geodes! All known geodes are created in the same way, which is why comets have or have had water inside them—they were all made in a liquid water environment. The words 'impact crater' are absent in the *entire* articles these quotes are taken from. A sponge-like planetoid that crumbles when touched would definitely not hold if struck by another rock at hypervelocity, and could not have left a crater half the diameter of the planetoid itself. Such giant craters show up on many asteroids, comets, moons, and planets. That craters of such immense proportions could occur from impact without completely obliterating the body where the crater occurs is almost inconceivable. However, the "*steam explosions*" mentioned by the Comet Wild 2 investigators *are* conceivable, and they fit the data.

Rewriting the Textbooks Again

After NASA's Deep Impact spacecraft crashed into Comet Tempel 1, researchers discovered that there was more than water and empty space inside this comet. One science writer noted that astronomers threw their hands in the air when trying to understand how clay and carbonates got into the comet:

"Clay and carbonates are thought to **form in liquid water**…
"'**How did clay and carbonates form in frozen comets**?'" Note 7.14j

Only when the true origin of clay and carbonates is understood can their presence on hydrocomets be explained. That origin will be established in the next chapter. Nevertheless, the existence of clay and carbonates and the presence of water instead of magma is a clear sign that the 'standard theory' of coalescing gas and dust from the solar nebula as a foundation for planetary formation is in trouble:

"**This argues against the standard theory** that dust and gas in a disc around the Sun simply clumped together with other material at the same distance from the star." Note 7.14j

The astronomers could see the writing on the wall:

"The discovery of carbonates '**creates problems one way or another**,' he says." Note 7.14j

The presence of the carbonates poses no problem once the comprehension of how hydrocomets are formed is known. The problem lies with the magma theory of comet formation. Look past the rhetoric in the scientific publications, and ask what modern scientists *really* know—in many cases, apparently very little. The fields are fraught with controversy and contradiction at even the basic level. For example, notice the contradiction in the entry on comets from a popular encyclopedia:

"**Comets are another possible source of meteorites**. Cometary debris is probably the source of many **meteors**, including the more spectacular meteor showers, but **there is no compelling evidence that comets are the source of any known meteorites**." Bib 12 p1199

Because the pseudotheories of modern science are wrong on such a fundamental level, there is no reason to expect anything other than "problems" and contradictions.

Another example that carries this point further are the comments by the influential planetary scientist Gene Shoemaker as he extrapolates false ideas to their utter end by saying, in his words, "it can be plausibly argued…":

Spacecraft are providing more hydrocomet evidence every time they fly by.

Fig 7.14.6 – In this artist's rendering of a comet-seeking spacecraft, we are reminded that the real impact on planetary science has not been by meteorites, but by hydroplanet discoveries. Each mission to the smaller bodies of the Solar System uncovers new evidence of hydrofountains.

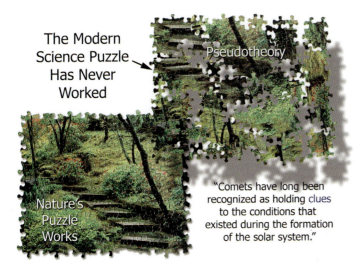

The Modern Science Puzzle Has Never Worked — Pseudotheory

Nature's Puzzle Works

"Comets have long been recognized as holding clues to the conditions that existed during the formation of the solar system."

"It can **plausibly be argued** not only that **we are the direct descendants of cometary material**, but that the rise of our species is one ultimate consequence of the rapid increase in the variety of mammals inhabiting Earth following the mass extinction of 65 million years ago. The proposition, in other words, is that **we are the progeny of comets and our own species has arisen because the course of evolution was fundamentally altered by a comet impact.**" Note 7.14k

How plausible could any such idea be that is based on assumption after assumption? Such 'plausible arguments,' and so many like it are seen throughout the natural sciences; completely unfounded, they are not science at all. That comets are the deliverers of Earth's life-giving molecules and substances—unsubstantiated theory. Mass extinction by comet impact 65 million years ago—more theory, filled with holes even by the scientists who embrace it. That humans descended through ever branching cross speciation events from slime—not a shred of evidence. The whole basis of Shoemaker's argument is based not on one, *but a handful of false theories*. Even the idea that comet water is related to the Earth's water is fallacious:

"…recent analysis of comet water has shown that **comet water is significantly different from typical ocean water on Earth.**" Note 7.14l

Modern science seeks to reconcile the "argument" that life came to the Earth aboard a comet as it does with most of the pieces of the modern science puzzle—force the pieces together, whether they fit or not.

Comets hold many "clues" to the conditions that existed when the solar system was formed. All one has to do is be willing to look at the truth that in some cases, the researchers have identified but found hard to accept—that water is "the primordial material from which the planets formed":

"In addition to being fascinating objects in and of themselves, **comets** have long been recognized as holding **clues to the conditions that existed during the formation of the solar system. Comets are thought to be frozen relics of the primordial material from which the planets formed as they have probably undergone little processing since their formation.**" Note 7.14m

7.15 The Hydroid Evidence

One of the most popular video games of the arcade game era was *Asteroids*, released in 1979 by Atari Inc. The player's objective was to shoot asteroids without being hit by the oncoming asteroids or the speedy fragments of previously blasted asteroids. The game became Atari's best selling game of all time.

Asteroids have always been a source of great mystery and awe, especially as teachers explained to young people how there was once a planet between Mars and Jupiter, and that it had been crushed and destroyed, leaving behind only rocky fragments. How many wondered, mouth agape, could this happen to the Earth?

How do we know what we know about asteroids? Could they share commonality with the comets and some of the moons in our solar system?

Hydroid Defined

A **Hydroid** is a hydro-asteroid or an asteroid that formed from water. Asteroids are a group of objects that inhabit the solar system in the general area between Jupiter and Mars, but are not limited only to that area. They are usually defined as minor planets that do not have a watery coma or tail like comets do. Allegedly, they are the origin of the meteorites, but as we have seen earlier, this is not necessarily true.

The meteorites are divided into three basic groups: stones, irons, and stony irons. How many asteroids have scientists observed, up close and firsthand that match any of the meteorite categories? Surprisingly, none of them do. After reading the Meteorite Pseudotheory chapter, it probably does not really come as much of a surprise.

The story is similar to the comet story; the theories based on dogmagma and impact have grossly misconstrued what asteroids should look like and how they were formed. In fact, researchers realize that the distinction between asteroids and comets is "ambiguous":

"The **distinction between asteroids and comets** is sometimes **ambiguous**: comets are typically more volatile-rich and form farther from the sun." Note 7.15a

By the end of this subchapter, we will see why this is so, and the similarities between hydroids and hydrocomets will be evident.

The Asteroids That Never Existed

Because asteroids are so small and so far from the Sun, we never had a clear picture of them until we sent spacecraft to investigate. The first target was the asteroid Gaspra in 1991, followed by the asteroid Ida and its moon Dactyl in 1993. Before that time, planetary scientists 'knew' that they looked like "fragments":

"…asteroids look very much like **fragments**…" Note 7.15b

This avowal is from the journal, *ICARUS* where eleven scientists reported their findings from numerous hypervelocity impact experiments using a 1-gram impactor and 10,000-20,000-gram objects of impact. Their experiments produced surprising results. They reported that impacted rocks produced only sharp "fragments" that were visible to the eye and that *no craters were made.*

Why would investigators expect anything different? In 1986, the small celestial bodies—the asteroids—were expected to be heavy, like meteorites, colored like meteorites, and composed of sharp fragments from impact.

Fig 7.15.1 symbolizes the asteroid that never existed; it is a photo of a rock fragment representing what asteroids *should look like*. The images of real asteroids in Fig 7.15.2 are remarkably different. The grey, smooth-edged shapes of actual asteroids look nothing like impact-melted or fractured rocks. They look more like the surface of the Earth's Hydromoon!

Incredibly, many asteroid scholars have been so prejudiced by the impact and accretion pseudotheories that even after looking at photos of Gaspra and Ida and their smooth, rounded shape and lunar-like surfaces, they still choose to describe the asteroids thus:

"We have already seen that **all** the asteroids visited by spacecraft **have angular shapes, suggesting that they are collisional fragments**." Bib 164 p107

It is unclear how the author of this article can view the asteroids as being "collisional fragments" and having "angular shapes," but it illustrates the power of how old pseudotheories influence even the best minds in science.

In the Hydrocrater Model, high-speed impact experiments produced impact glass, and none of the asteroids appear to have anything like that on their surfaces or in their craters.

How else could such smooth-surfaced celestial bodies form? In the Hydrospheres subchapter, 7.3, the concept was shared that large members of the solar system formed spheres because of their greater internal gravitational force. During the crystallization from water process, smaller bodies were deformed by external gravitational forces. Most of the asteroids we have images of, like those in Fig 7.15.2, indicate a single body formation and do not show any evidence of impact ejecta or boulder tracks. The asteroid Ida even has its own moon, Dactyl. Where are the boulders that should be laying on the surface of Ida? If the asteroid was formed because of an impact, where are the other fragments? Such nagging questions remain unanswered despite increased frequency of asteroid visits by space probes.

By the year 2000, scientists' predictions had been overturned

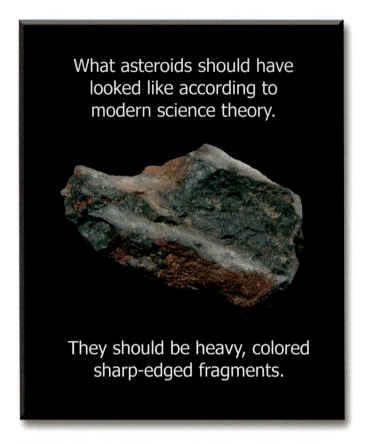

Fig 7.15.1 – The asteroid that never existed. This is what hypothetical asteroids *should have looked like* if their origin had been from impact—sharp broken fragments of heavy, colored rock. However, no asteroid that looks like a "fragment" has been observed. Unbroken and less dense than iron meteorites, asteroids are light, grey and smooth as if they had been shaped by water.

Fig 7.15.2 – These are the first close up images of actual asteroids ever obtained. Gaspra was photographed in 1991, Ida was reached by spacecraft and imaged in 1993 and Eros in 2000. Ida was unique because it has its own moon, Dactyl. These asteroids do not look like broken fragments of rock. Planetary scientists were surprised; no impact theory or magmaplanet theory has been able to explain how they were formed. Images courtesy of NASA.

by the close-ups of the asteroids:

"Egged on by this nervous curiosity, we are entering the golden age of comet and asteroid exploration. Over a dozen have been imaged, and each new member of the menagerie is welcomed with delight and perplexity. **They are not what we expected. Small asteroids were predicted to be hard and rocky**, as any loose surface material (called regolith) generated by impacts was expected to escape their weak gravity. **Aggregate** small bodies **were not thought to exist**, because the slightest sustained relative motion **would cause them to separate**." Note 7.15c

Amazingly, this report was based on the best high-resolution images available. Do the asteroids in Fig 7.15.2 appear to be an aggregate of several rocks or pieces of former planet's asteroids? Maybe this is why the planetary scientist cited in the previous quote made the following statement:

> "**Asteroids have become notorious menaces** but are best appreciated in a positive light, as surreal worlds **bearing testimony to the origin of the planets**."
>
> *Scientific American*, May 2000, p46

Asteroids actually do bear "testimony to the origin of the planets," and the reason they have become such "notorious menaces" is simply because they disprove the modern science Solar System Origin Pseudotheory!

The Asteroid Impact Menaces

One of the biggest problems researchers have with asteroid craters is that some of them must have been from "colossal impacts":

"**Most of them [asteroids] have one or more extraordinarily large craters, some of which are wider than the mean radius of the whole body. Such colossal impacts would not just gouge out a crater—they would break any monolithic body into pieces**. Note 7.15c

Fig 7.5.3 illustrates the magnitude of the problem. The asteroid, Mathilde boasts a crater so large it baffles the imagination! Impactologists are well aware of the expected outcome if an impactor is of such large size. Scientific experiments from the 1970s had scientists saying:

"**Experiments** suggest that almost all rocks **are destroyed** if they are hit by a particle that would have formed **a crater equal in diameter to the rock**." Bib 171 p453

Why then, do researchers continue to believe that the large craters on asteroids are impact craters? They have no other cratering mechanism to turn to, so they continue to rely on their old sci-bi, and the asteroids continue to "menace" them. The only way to explain this obvious flaw is to stake the claim that the small bodies must be made up of "aggregates," or fragments of rocks. Somehow, this loose pile of rocks came together in space and survived the onslaught high-speed impacts that formed craters of enormous proportions.

Of course, planetary scientists had no choice but to imagine something like the "loose rubble" pile theory because of the density issue. Mathilde has a density of 1.3 g/cm³, only barely above that of water! This "improbably low" density forced researchers toward the rubble pile theory:

"**Evidence of fragmentation also comes from the available measurements for asteroid bulk density. The values are improbably low**, indicating that **these bodies are threaded with voids of unknown size**. In short, asteroids larger than a kilometer across may look like nuggets of hard rock but are **more likely to be aggregate assemblages**—or even **piles of loose rubble** so pervasively fragmented that no solid bedrock is left." Note 7.15c

To exacerbate the problem further, even the rubble pile theory cannot account for the low density of 1.3 g/cm³. Geoscience literally has no answer for how Mathilde formed. But the Hydroplanet Model does. Mathilde can be easily explained as a classic **geode** that was filled with water before being heated by gravitational friction, which released the water, leaving the interior with ample "open space":

"**Up to 50% of the interior volume of 253 Mathilde consists of open space**. However, the existence of a 20-km-long scarp may indicate that the asteroid does have some structural strength, so it could contain some large internal components." Note 7.15d

It does not take a planetary scientist to see that Mathilde in Fig 7.15.3 is not a loose "rubble pile." The researchers that suggested this dreamed-up theory had to abandon the "elementary" physics they learned in school:

"From a physics perspective, the simple **billiard-ball analogy** of collisions between rocks knocking each other around the inner Solar System **does not stand up to elementary scrutiny**. It is no easier to 'bump' icy, rocky or even metallic objects, with finite material strengths, from the asteroid belt into Earth-crossing orbits **than it is to hit eggs around the fairways with a golf club**." Note 7.15e

Golfing With Eggs, Fig 7.15.4, dramatically smashes the billiard-ball analogy that has been included in science textbooks

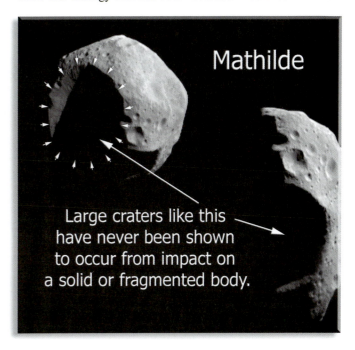

Fig 7.15.3 – The Mathilde asteroid was visited by spacecraft in 1997, which revealed some of the largest craters seen on any small body. This was not the biggest surprise Mathilde had to offer. This solid looking rock was anything but solid. Mathilde's density proved to be barely above that of water (1.3 g/cm³)! Images courtesy of NASA.

Golfing With Eggs

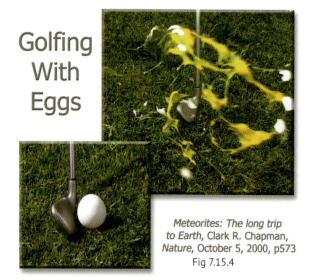

Meteorites: The long trip to Earth, Clark R. Chapman, Nature, October 5, 2000, p573
Fig 7.15.4

"From a physics perspective, the simple billiard-ball analogy of collisions between rocks knocking each other around the inner Solar System does not stand up to elementary scrutiny. It is no easier to 'bump' icy, rocky or even metallic objects, with finite material strengths, from the asteroid belt into Earth-crossing orbits than it is to hit eggs around the fair ways with a golf club."

for years. No scientist has ever demonstrated or explained how large craters, like Mathilde's, could have been made by impact, whether on a solid or a fragmented body—and they cannot because they are not impact craters.

The Itokawa Evidence

The Itokawa asteroid is another example of Hydroplanet Model evidence. Dissimilar in most aspects to the Mathilde asteroid, Itokawa revealed a completely new world as the Japanese Hayabusa spacecraft came in close contact with the asteroid in late 2005. A variety of new data, including the stunning photos shown in Fig 7.15.5, was obtained. Researchers were again surprised at what they saw, and immediately saw a "complicated" geology:

"The origin of the boulders on the smooth terrains seem to be **complicated**: some boulders are surrounded by shallow depressions, whereas most others are not." Note 7.15f

It was "complicated" because the observations "cannot be explained" by the existing impact pseudotheories:

"The existence of decameter-sized boulders on Itokawa, as well as the abundance of meter-sized boulders, **cannot be explained by simple impact-cratering processes**." Note 7.15f

Again, the "impact-cratering processes" as envisioned by modern science cannot explain the Itokawa landscape. Researchers theorized that some type of catastrophic disruption might have taken place, but that still led them back to an impact theory relic, the "possible rubble-pile":

"Thus, the boulders might have been produced when Itokawa was generated by a **catastrophic disruption**, which is consistent with the **possible rubble-pile** structure of Itokawa." Note 7.15f

Here we have a new problem with the rubble-pile impact theory. No other observed asteroids share a rough bouldered surface like Itokawa. If this was an example of how asteroids are made by impact, then other asteroids should exhibit somewhat similar morphologies—but they do not.

Another issue with the rubble-pile impact theory is the non-spherical potato shape of the celestial body. The accretion theory of rocks randomly impacting one another does not work for a number of reasons; glass would be common from the impact melt generated from impact heat; the revolving body of the asteroid would have attracted billions of rocks from all directions, contributing to a much more spherical shape. Some have suggested that Itokawa is made up of only *two* large fragments and many small ones. If this is so, the two larger lobes appear to be well *rounded and without craters!* How does this happen from a high-speed impact?

A third problem investigators identified was Itokawa's heterogeneous color and brightness. In other words, the surface of this asteroid is ***not*** generally the same color everywhere. The photos in Fig 7.15.5 do not show the color discrepancy, but certain minerals exist on the surface of Itokawa in diverse areas. Random impact on the surface *should have mixed* and homogenized the surface. Marked differences in surface color and reflectivity (albedo) became very apparent and were "most intriguing":

"Among the **most intriguing characteristics** of Itokawa are the **heterogeneities in color and albedo**, which are unusual

Fig 7.15.5 – The Itokawa asteroid (535 meters across) was imaged in 2005 by Japan's Hayabusa spacecraft. The large boulders on its surface make Itokawa surprisingly different from other asteroids previously visited. There are no signs of boulder tracks or craters of any significance. How could this be if such solar system bodies were created by impact? Images courtesy of JAXA.

because no previously observed asteroid bodies show large variations in both of these characteristics. We found a variation of more than 30% in v-band albedo…" Note 7.15g

The Hydrocrater Model, subchapter 7.8, demonstrated how different minerals are deposited on the surface of the Earth when successive eruptions draw material from different depths beneath the surface. These hydrofountain eruptions often have very *different* compositions.

The Ceres Hydroid Evidence

Ceres is the smallest identified dwarf planet in the solar system and by far the largest body in the asteroid belt. Ceres is about 590 miles (950 kilometers) in diameter and, as shown in Fig 7.15.6, has a spherical shape. We have yet to send a satellite to Ceres, so we have to settle with this photo from the Hubble Space Telescope. Although the images are indistinct, astronomers have been able to calculate that the density of Ceres is only twice that of water. It also has a bulge at the equator like the Earth. These new details lead investigators to conclude that Ceres may have an "icy" interior:

"**An icy interior for Ceres**? Observations indicate the largest main-belt asteroid **may have an icy mantle beneath its surface**." Note 7.15h

Ceres, the largest member of the asteroid belt is a fine example of a Hydroid. Astronomers have calculated that this celestial body is probably one-quarter *water*:

"The team developed computer models of Ceres' interior using available data on the asteroid. Their findings: **Ceres has a rocky core and could have an icy mantle as thick as 77miles (124 km), amounting to about one quarter of its mass.**" Note 7.15h

Then in 2014, the Herschel spacecraft confirmed that water was being emitted from Ceres:

"ESA's Herschel space observatory has discovered water vapour around Ceres, **the first unambiguous detection of water vapour around an object in the asteroid belt**." Note 7.15i

No longer computer model generated, the first actual detection of water on Ceres proves that the largest asteroid in our solar system is also connected to a water origin:

"'**This is the first time that water has been detected in the asteroid belt, and provides proof that Ceres has an icy surface and an atmosphere**,' says Michael Küppers of ESA's European Space Astronomy Centre in Spain, lead author of the paper published in *Nature*." Note 7.15i

In 2015, NASA's Dawn mission is set to arrive and orbit Ceres and produce even more Hydroid Evidence.

If we set aside the many scientific theories that have been taught for so long, and ponder the magnitude of water evidence that exists in the solar system, the galaxies, and in the universe, how could such liquid spheres, including our own Earth, have ever formed without water?

Fig 7.15.6 – The latest image of Ceres, the largest near-Earth asteroid (590 miles/950 km diameter) traveling around the Sun. Scientists now estimate Ceres has at least a 77 mile/124 km mantle of ice that represents one quarter of its mass. Ceres is truly a hydroid by definition. Courtesy of NASA, HST.

Small Hydrobodies of the Solar System

As we come full circle in our discussion about hydrocomets and hydroids, there is one last observation to consider—both of these types of celestial bodies are the *same thing*—Small Hydrobodies. In the April 2000 issue of Nature, this similarity was noted:

"Discoveries of **comets that behave like asteroids and asteroids that behave like comets** are making us **reassess our view of Earth's smallest neighbors**." Note 7.15j

In the past, the orbital behavior of the different bodies causes scientists to draw distinctions between the comets and the asteroids. Not too long ago, astronomers never dreamed that asteroids were like the comets with abundant water and with the ability to emit jets of that water to create tails just like their cousins, the comets—yet this is exactly what has happened:

"We now have **comets in asteroid-like orbits and asteroids in comet-like orbits**. Both comets and asteroids can evolve from the Oort cloud into highly inclined, even retrograde, orbits about the Sun, so orbital behavior is no better than physical behavior for telling them apart. Our attempts to sort comets and asteroids into separate boxes **have failed** and astronomers should now consider these objects as members of a highly diverse family—**the small bodies of the Solar System**." Note 7.15j p830

Why have so many of the theories failed to provide answers? The solar system was not made from a collapsing cloud of dust; these "small bodies of the solar system" are much more connected to the watery beginnings of the solar system than ever imagined. Because of their orbit in the ecliptic plane, the asteroids were thought to be formed closer to the Sun and so, did not have water in their makeup, like comets. However, recent observations have changed this thinking and asteroids are known to "exhibit a long dust tail typical of icy comets" if they are heated sufficiently, as comets are when they approach other celestial bodies because of the effects of the Gravitational Friction Law. Small Hydrobodies do not require the Sun's heat to exhibit this behavior. Frictional heat by changes in gravity alone can heat the water in the hydrobodies. Because of the low pressure that exists in outer space, even a very small amount of heat-energy can cause water to boil and vaporize. Thus only minor heating is required for comets and asteroids or hydrobodies to spew water vapor.

It was a new observation of comet 133/Elst-Pizarro that brought the similarities of the solar system's Small Hydrobodies to light:

"In both 1996 and 2002, the 'original' main-belt comet, 133P/Elst-Pizarro (named after its two discoverers), **was seen to exhibit a long dust tail typical of icy comets**, despite having the flat, circular orbit typical of presumably dry, rocky asteroids. As the only main-belt object ever observed to take on a cometary appearance, however, 133P/Elst-Pizarro's true nature remained controversial. **Until now**.

"'The discovery of the **other** main-belt comets shows that

133P/Elst-Pizarro **is not alone in the asteroid belt**,' Jewitt said. 'Therefore, it is probably an ordinary **(although icy) asteroid**, and not a comet from the outer solar system that has somehow had its comet-like orbit transformed into an asteroid-like one. **This means that other asteroids could have ice as well**.'"
Note 7.15k

Never before had scientists thought that comets and asteroids might share such similarities. A consequence of the new observations has been a reevaluation of how our blue planet got its water. With the magmaplanet Earth paradigm, geoscientists theorized for decades that the Earth got its water from comets—until now:

"**The Earth is believed to have formed hot and dry, meaning that its current water content must have been delivered after the planet cooled**. Possible candidates for supplying this water are colliding comets and asteroids. Because of their large ice content, comets were leading candidates for many years, **but recent analysis of comet water has shown that comet water is significantly different from typical ocean water on Earth**." Note 7.15k

The University of Hawaii Institute for Astronomy researchers, like almost all astronomers and planetary geologists, has a serious impediment to the truth, being in the magmaplanet paradigm. They say the Earth is "*believed*" to have formed hot and dry," and the almost daily eruption of Hawaii's volcanoes seems to support their belief. Yet they recognize that their one source of water for the Earth (from comets) does not work. Therefore, they are back to the drawing board:

"Essentially, **they [comets] should be just rocks**. With the discovery of the **main-belt comets**, we now know this is not the case, and that, in general, **the conventional definition of comets and asteroids are in need of refinement**." Note 7.15k

In the next chapter, the reason water on Earth is so different from other celestial bodies is something few geologists and astronomers study—organisms. As the Chemical and Water Models demonstrated, when life is introduced to water and minerals—the elements and molecules are changed in ways never anticipated.

7.16   More Hydroplanet Evidence

There are an almost infinite number of evidences of the Hydroplanet Model in the solar system. They of course, and to the relief of the reader we are sure, are not all included here. Nevertheless, a few other important pieces of evidence that come from various places around the solar system should be pointed out. These evidences from Venus, Mars and a couple of moons add an important dimension to the picture.

The Venus Hydroplanet

The new Hydroplanet paradigm is like having a microscope or telescope for the first time; everywhere we look, there is something new. Those who choose to look with an open mind will find a universe full of water evidence.

Turning to the planet Venus, Earth's shrouded sister planet, the evidence of a water-planet shows up again. The Aurelia Crater on Venus' surface is a clearly manifested example of a hydrocrater that also contradicts the meteorite impact theory. Note the dramatic contrast between the white ejecta and black background in Fig 7.16.1. Meteorites and impact ejecta cannot

Fig 7.16.1 – The Aurelia Hydrocrater on Venus has a diameter of 20 miles (32 km). It is a 'fresh' crater with surges of white sediment flows surrounding the crater. Surges like this are not associated with impact, but commonly occur in a water environment.
Courtesy of NASA (PIA00239).

make the type of ejecta patterns seen in this image and no impact has ever been known to create a flow of white sediment. As we have seen on Earth and elsewhere, many sedimentary materials that are ejected from hydrocraters are white. Note also, the flow patterns of the material and the evidence of *flow* down valleys and channels toward the bottom right of the photograph.

A similar flow of material is seen around Venus' Dickinson Hydrocrater in Fig 7.16.2. This photo was taken by the Magellan space probe in 1990; it is a dramatic example of the Venus Hydro-

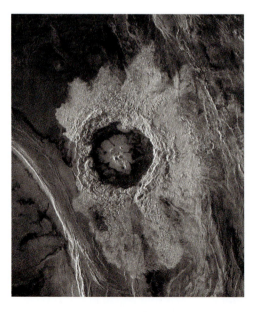

Fig 7.16.2 – The Dickinson Hydrocrater on Venus has a diameter of 43 miles (69 km). Even with obvious water surges at this crater, planetary scientists still turn to impact pseudotheory and claim the sediment surge is impact melt or a volcanic flow from the impact. Courtesy of NASA (PIA00479).

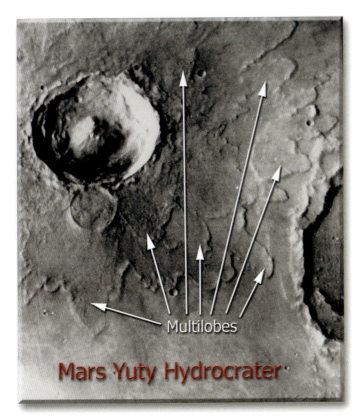

Fig 7.16.3 – The Yuty Hydrocrater on Mars is an excellent example of a multilobe flow, or multiple surges of liquid water and sediment. Multiple eruptions or surges can only come from a hydrocrater, not impact. These are direct evidences of large craters the erupted non-melted sediment in a manner not directly observed before. This hydrocrater evidence is one of thousands that exposes the Uniformity Myth, where activities in the present cannot explain the past.

water driven are forced into untenable impact scenarios:

"As understanding of the differences between impact and volcanic craters developed, one class of volcanic craters (maars and tuff rings) were found to have **similar profiles as impact craters**. Since these volcanic craters have ejecta blankets consisting of base surge materials, workers considered the possibility that impact craters also produced **base surges**. However, at that time volcanic base surges were thought to require steam. **The anhydrous [without water] nature of the lunar surface seemed to preclude base surge formation. This debate still continues to this day, with laboratory simulations of impact cratering showing ejecta moving in orderly ballistic trajectories. But, Viking observations of Martian craters showed ejecta deposits with structures typical of lateral flow, difficult to explain by ballistic transport.** Since the surface of **Mars** was likely **wet** during much of its impact history, Wohletz and Sheridan (1983) suggested that base surge can be a part of impact cratering." Note 7.16a

"Difficult to explain by ballistic transport" means that the impact theory cannot explain these craters. Impact experiments have not been able to reproduce ejecta with surge formations. The Dickinson Hydrocrater in Fig 7.16.2 shows a flower shaped surge of white sediment material. Both of the Venus hydrocraters shown here are void of smaller 'impact' craters that are more prevalent than large craters.

Multilobed Ejecta Evidence

Besides the Base Surge Evidence showing the flow of massive amounts of water and sediment (more than a dozen miles wide and several thousand feet high), there are **multiple flow lobes** or 'multilobes' in these surge areas demonstrating that *multiple* eruptions took place. The Yuty Hydrocrater on Mars in Fig 7.16.3 is one such place. Impact-cratering expert Jay Melosh states that the "most straightforward" (and only scientific) explanation for multilobed ejecta patterns is that they formed from "liquid water":

"In summary, the most straightforward explanation of the Martian **multilobed ejecta patterns is that liquid water is incorporated into the ejected debris.**" Bib 161 p100

Melosh goes on to say that "Much research must be done…" in this area, however, as noted in previous statements, laboratory simulations show ballistic ejecta moving in orderly *single* patterns.

Researchers know that multiple flows mean multiple eruptions—not multiple impacts! However, because impact dogma dominates modern science today, this fact is extraordinary evidence that has been swept under the scientific rug.

Melosh continues stating how many Martian craters' central peaks are not from impact, but from "some fluidizing agent peculiar to Mars".

"Many Martian craters have **abnormally large central peaks** and other internal

planet. Clearly not fall-out from an impact explosion, this white, flowing material is similar to the mudflows of the Mt. Saint Helens eruption.

The central uplift in this crater is very large, being a third of the diameter of the entire crater.

The Aurelia Crater seems to be a fairly recent crater that shows little erosion. Moreover, Aurelia is a large hydrocrater, 19.8 miles (31.9 km) in diameter; if it had been formed by impact, it would have had ejecta spewed to a distance several times the diameter of the crater, significantly further than the material seen in the image.

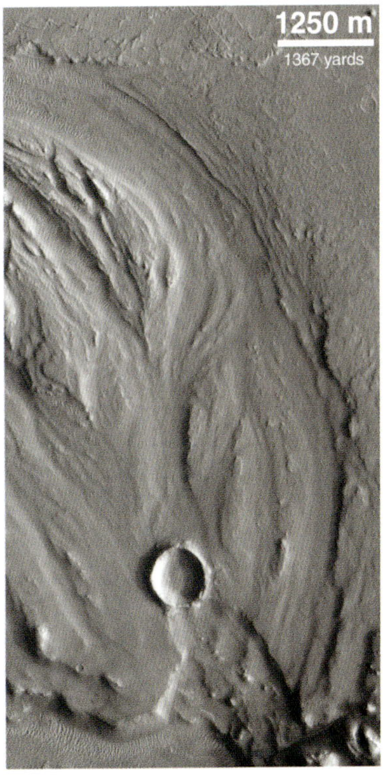

The Hydrocrater Base Surge Evidence

The geosciences continue to struggle with the new hydroplanet evidences that appear with each new space mission. In the following statement from the 1998 book *Magma to Tephra*, we see that researchers have looked at *all craters* through 'impact glasses' instead of realizing that impact cratering is only a small percentage of the overall cratering process. Without the knowledge that there was once water on the Moon, lunar base surges that are visibly

Fig 7.16.4 – The Mars Hrad Vallis water channel and hydrocrater are examples of underground water coming to the surface seen only very rarely today on the Earth. Courtesy of NASA (PIA02076).

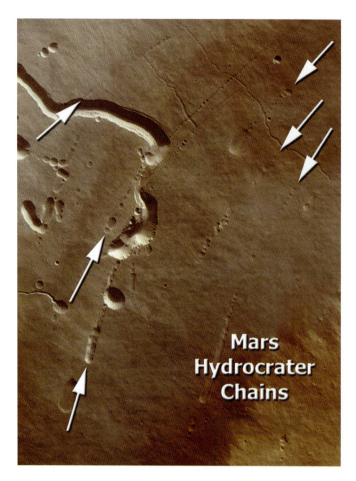

Fig 7.16.5 – This is a close up photo of Pavonis Mons area taken from the Mars Express satellite. It shows some of the typical **Hydrocrater Chains** on Mars. They follow a large circular path around the Pavonis Mons summit. These are collapsed hydrocraters and hydrovalleys that formed when underground water eroded channels along fault lines around the volcanic summit. Image only courtesy of ESA.

"Where did the water come from?" Throughout this chapter we have learned that most of the water in the celestial bodies comes from inside. Hydrocomets are able to spew out water for thousands of years (not millions by the way)—because there is so much water within them.

When the water in the crust is turned to steam by the radiant heat of the Sun or by gravitational frictional heating, it escapes through hydrocraters and fault lines, which create hydrovalleys, if the water is liquid during the outburst.

Fig 7.16.4 is a close up of the Mars Hrad Vallis water channel. The significant feature in this image is the hydrocrater at the bottom of the image. Notice the distinct water channel flowing directly *out of the crater* toward the lower elevations below.

The out-flow channel, and the lack of ejecta is a clear sign the crater is not of impact origin. Hydrofountains can create craters and eject only water, leaving behind little trace of ejecta material. Hydrocraters can be formed as underground aquifers remove sediment and allow overlying sediment to collapse. This can form craters, chains of craters and collapsed valleys.

Only a decade ago, planetary scientists describing craters on Mars as being hydrocraters and not impact was unheard of. The idea of outflow channels and subsurface water being "catastrophically released" was a bold new move and it happened in 2005:

"Aram Chaos, a 280-kilometer-diameter crater, has an outflow channel and is filled with layered rocks that contain hematite. Gigantic blocks of rock litter the crater floor. **It looks as though a torrent of subsurface water was catastrophically released**, **causing the overlying terrain to collapse**. Some of the water ponded in the crater, forming the layers of hematite-bearing sediments." Note 7.16c

Mars Hydrocrater Chain Evidence

The Mars Express photos taken by the ESA (European Space Agency) show surface phenomena on Mars in amazing detail.

collapse structures compared with lunar or Mercurian craters, also suggesting the presence of **some fluidizing agent peculiar to Mars**." Bib 161 p98

That "fluidizing agent" is of course, water. Why is it so difficult for scientists to accept the concept of water being the primary agent in the formation and erosion of planetary surfaces?

The Mars Hydroplanet

The successful deployment of several satellites and rovers to Mars has resulted in an abundance of new information and data. Evidence that there was once plentiful water on Mars has come to light. One research geophysicist at the USGS states the "most important advance" in scientific exploration:

"**The most important advance in the last 25 years**: In January 2004 the Mars Exploration Rover Opportunity landed in a small crater and discovered sedimentary strata showing that open, **shallow seas once existed there**." Note 7.16b

Near the beginning of this chapter, we discussed the shock researchers had with so much evidence for large volumes of water on Mars. Now that the Mars rovers and the Phoenix Mars Lander have confirmed the abundant nature of water on Mars, the next question is to ask,

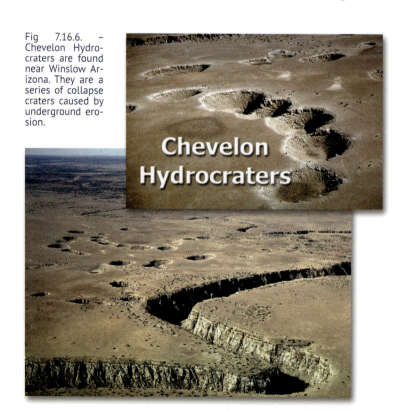

Fig 7.16.6. – Chevelon Hydrocraters are found near Winslow Arizona. They are a series of collapse craters caused by underground erosion.

Researchers look at almost every crater on celestial bodies as being "impact" craters, images like the one in Fig 7.16.5 are confusing to the impactologists because, although the circular depressions look like other so-called impact craters at different locations on Mars and other celestial bodies, the craters and depressions are *aligned in chains around mountain summits*. They are obviously *not* from impacts. Perhaps to differentiate themselves, the researchers refer to these depression "craters" as "pit chains":

"**Pit chains**, strings of circular depressions **thought to form as the result of collapse of the surface**, are also visible within the colour image." Note 7.16d

The word "pit" is generally used to describe small depressions, but these "pits" can be many miles across, therefore the word "pit" is a poor description of the depression and it does not correctly describe the nature of the depressions. Faulting alone has never been shown to create such chains of depressions. Hydrocrater chains are formed by the underground erosion of water along fault lines. Examples can be seen in Fig 7.8.12 (Mt. Saint Helens hydrocraters) and Fig 7.8.13 (pockmark craters) where surface material has collapsed into circular depressions along fault lines. Therefore, these types of structures are more properly named **Hydrocrater Chains**.

Chevelon Hydrocraters

30 miles from the Arizona Hydrocrater (formerly Meteor Crater) are a chain of craters called the Chevelon Hydrocraters (Fig 7.16.6). These depressions are called "pot holes" on local maps.

The Chevelon Hydrocraters, named after a nearby canyon, are similar to many of the collapsed hydrocraters seen on Mars. The geological setting of the Chevelon Hydrocraters is very similar to the Arizona Crater with similar layers of sediment and even iron nodules like those found at the Arizona Crater. The Chevelon craters do not show evidence of a steam explosion or of ejectites. Both craters have underground aquifers and nearby canyons. Notice how little erosion there is on the surrounding plains in Fig 6.16.6. Like the nearby Grand Canyon, this canyon was not created by surface erosion such as we see happening today—but by a catastrophic geological event of the past.

An excellent example of this type of event can be seen earlier in this chapter, in Fig 7.8.12 where the Mt. Saint Helens eruption created a massive mudflow with water that came *from* the volcano. After the mudflow, underground water channels began eroding away the soft volcanic sediment, which at first

Fig 7.16.7 – The Mars Express spacecraft took this photo of mysterious channels and valleys running down the slope of Pavonis Mons in 2004. This photo takes in about 26 km (16 miles) across. It caused many questions for modern geologists. Although researchers thought these structures were collapsed lava tubes, six items discussed here, in the text explain why these channels are Hydrovalleys and not collapsed lava tubes. Only when we can come to understand that Mars, like the Earth, is a hydroplanet, can we begin to comprehend how such structures were formed. Courtesy of ESA.

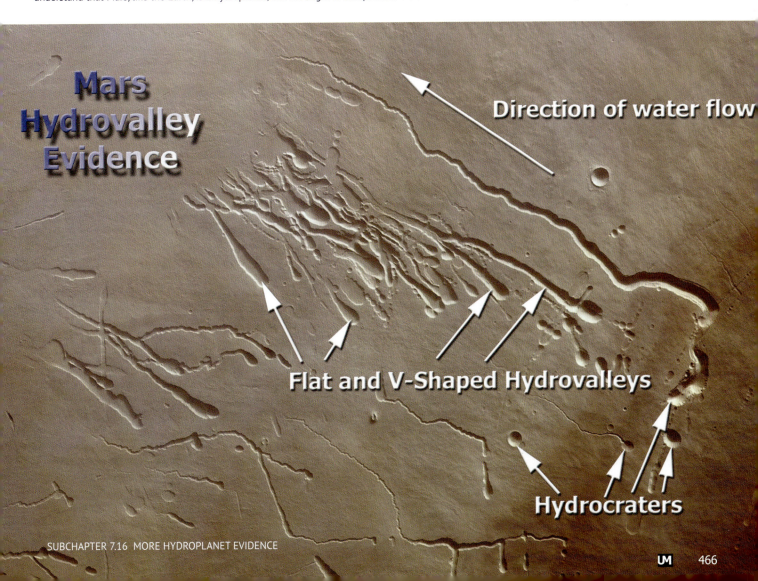

created a number of hydrocraters seen in Fig 7.8.12. Both steam eruptions and collapse craters resulted. The hydrocraters are not visible today because rainstorms, river flooding and subsequent mudflows destroyed all evidence of the depressions.

On the smaller celestial bodies of the Moon and Mars, surface waters quickly disappear because of reduced gravity. Weather on Mars is not as severe as it is here on the Earth and the surface depressions are not erased as quickly. It is important to note that the massive erosion at Mt. Saint Helens happened in a just few short years, accelerated because the sediment was not hard, glassy lava but mostly soft, muddy sediment. This soft sediment is very different from the typical Hawaiian lava flows that harden as they cool.

Most planetary geologists envision hard volcanic deposits when they speak of volcanic material blanketing the Moon and Mars. Instead, it was soft sediment with occasional hard, rocky deposits, very much like the Mt. Saint Helens eruption.

If we were to see Mt Saint Helens underwater and watched the explosions, mudflows, and cratering form, along with the gradual disappearance of the overlying water, we would be left with a surface that looks like many areas of the Moon and Mars.

The Mars Hydrovalley Evidence

The Tharis region of the planet Mars includes the largest mountains in the solar system. The smallest of the three Tharis Montes volcanoes, Pavonis Mons—Latin for "Peacock Mountain"—rises 8.7 miles (14 km) above the mean surface level, one and a half times taller than Earth's Mount Everest. On its flank lie hundreds of landforms that are stunning testimony of Mars' wet history. Fig 7.16.7 is a high-resolution view of an area on the slopes of Pavonis Mons showing many hydrocraters and water-formed channels called hydrovalleys. A **Hydrovalley** is a channel that formed quickly by the movement of a large volume of sediment-laden water. In this image, hydro valleys both large and small can be seen along with a number of hydrocraters.

Instead of recognizing that these channels were formed by water, geologists hold the opinion that they are **collapsed lava tubes:**

"Researchers believe these are **lava tubes**, channels originally formed by hot, flowing lava that forms a crust as the surface cools. Lava continues to flow beneath this hardened surface, but when the lava production ends and the tunnels empty, the surface **collapses**, forming elongated depressions. **Similar tubes are well known on Earth and the Moon**." Note 7.16d

The last part of this quote is both true and false. "Similar tubes are well known on…the Moon" is true in part because we do see the same landforms, but these are not found on the Earth. Granted, there are lava tubes on the Earth, but they are tiny in comparison to the landforms on Mars. There are six reasons these channels and valleys are *not* collapsed lava tubes.

First, there are no comparable collapsed lava tubes on the Earth, at least not in size. Nothing on Earth even comes close to the size of the Martian channels, some being *several* kilometers wide and extending for many kilometers. The largest collapsed lava tubes on the Earth's surface are about one hundred times smaller than the Martian valleys (see Fig 7.16.8 for examples). There is simply no evidence that lava tubes are capable of being

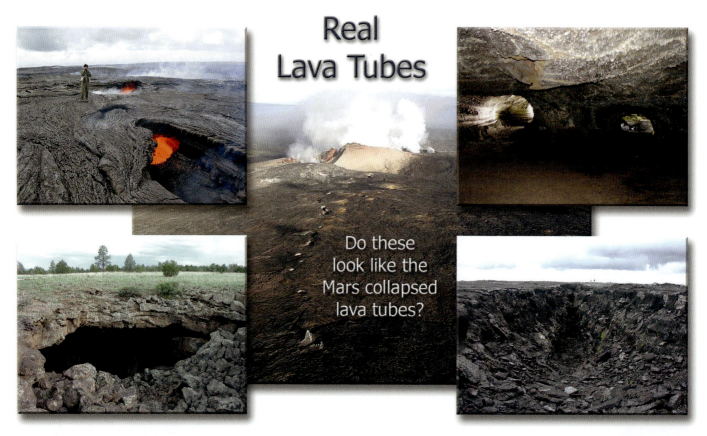

Fig 7.16.8 – These actual Lava tubes are in Hawaii and New Mexico, USA. Notice how little resemblance they have to the so-called collapsed lava tubes on Mars in Fig 7.16.7. Without the Hydroplanet Model, science is forced to think in the Magma Pseudotheory paradigm, which introduces errors. Only a huge water event on Mars could have formed the large hydrocraters and hydrovalleys that exist there.

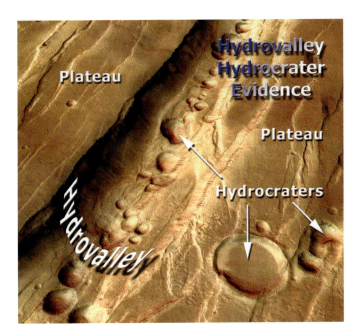

Fig 7.16.9 – This 3D enhanced photo of Mars' Phlegethon Cantina Hydrovalley and the surrounding plateau and hydrocraters covers an area approximately 10 km (6 miles) across. It dramatically illustrates the type of channels and craters that are created in a water-steam environment, similar to some of the events that ocurred during Earth's Mt Saint Helens eruption. The valley could not be a collapsed lava tube and there is no debris in the valley to support such an idea. Additionally, the craters show no evidence of lava flows coming out of them. If these landforms are seen as collapsed and/or eroded hydrovalleys and hydrocraters, the telltale evidence of flow and erosion suddenly makes sense. Similar examples exist under the Earth's oceans. Image only courtesy of ESA.

so large.

Secondly, such enormous tubes would suggest that when the necessarily thick lava tube ceilings collapsed, enormous boulders and blocks of lava would have been left behind in the valleys. However, no evidence of this is seen in any of the high-resolution photographs.

Third, the kilometer-long channels and valleys gradually taper off, decreasing in both width and depth. This is not a characteristic shared with Earth's collapsed lava tubes, but it is seen in valleys eroded by water. Additionally, most of the Martian valleys run continuously along the surface, showing that they were formed *on* the surface, whereas lava tubes on Earth originate underground and do not always flow parallel with the surface. Thus, many of Earth's lava tubes have col-

Fig 7.16.10 – This is a hydrocrater chain inside a rill or hydrovalley on Mars. The chain clearly illustrates the non-impact nature of the crater structure. These craters are unique in that they are elliptical and have some rim structure. Not found to be occurring today, these features refute the Uniformity Myth. Courtesy of NASA (PIA01686).

lapsed sections when they are near the surface and uncollapsed sections that are deeper. The Martian channels exhibit relative consistency the entire length of the channel. This is indicative of surface water erosion.

Fourth, Actual collapsed lava tubes on Earth exhibit rough, flat-bottomed channels whereas the Mars channels exhibit *both*

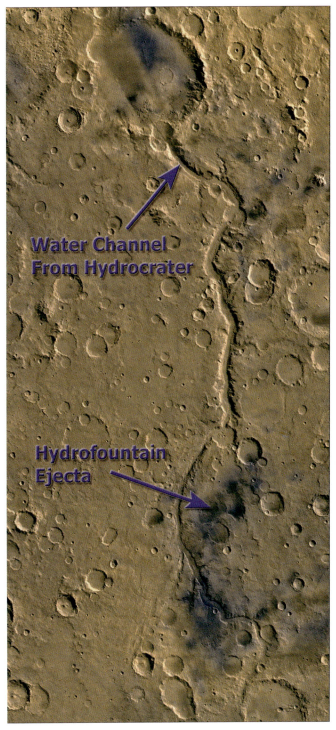

Fig 7.16.11 – The Mars Ma'adim Vallis water channel and hydrocrater. This channel is a canyon that is larger than the Grand Canyon on Earth. It had to have formed quickly. Its flat-floored craters and valleys have no nearby mountains that could have supplied the needed rivers of water. Modern geology today does not accept that hydromountains and hydrocanyons are common landforms on the Earth's surface, or that such features could have formed in a short time period, yet this is exactly what the surface of Mars suggests happened. Courtesy of NASA.

7.16 MORE HYDROPLANET EVIDENCE

smooth, flat-bottomed channels and V-shaped valleys, which cannot be accounted for with a collapse theory. However, both are accounted for with a hydrovalley model; flat-bottomed channels form when water heavily laden with mud and thick sediment flowed down the valleys, whereas V-shaped valleys would form with high energy water movement during periods of high water flow or down steeper flow gradients, both of which account for V-shaped channels.

Fifth, most lava flows on Earth are dark, mostly black in color, as seen in Fig 7.16.8. The iron in Earth's soil contributes to the characteristically dark color. The Martian surface has lots of iron, but the valleys seen in the Mars Express photos are clearly not black. There is comparatively little wind and weather at the elevation where these valleys lie, so it is unlikely that they were covered by wind born sediment. In any case, the terrain is uncannily uniform and no black lava boulders are lying in the valleys. Mars does have plenty of such dark lava-like boulders, but they are found elsewhere.

The **Sixth** and most compelling evidence that the giant Martian channels are hydrovalleys lies inside the valleys themselves. The channel in Fig 7.16.9 is lined with craters along its bottom, which can also be seen in Fig 7.16.10. Such craters obviously did not come from impact, and it is equally obvious that the craters are not from a collapsed lava tube.

The chains of horizontal depressions on a planet's surface form along lines of tectonic weaknesses as discussed in the Hydromoon Evidence subchapter. These fault lines intersect deep aquifers and provide channels for water movement. Gravitational friction superheats the water, which rapidly expands toward the surface where it explodes, blowing overlying sediment away. Following the initial burst, rapidly moving water continues to flow from the newly created hydrocrater, washing away sediment and forming outwash channels. The removal of so much sediment along fault scarps reduces overburden pressure furthering additional phreatic eruptions.

To summarize the reasons the Martian channels are *not* collapsed lava tubes, but are instead Hydrovalley channel structures:

1. No similar-sized collapsed lava tubes are known to exist on Earth.
2. There is a noticeable absence of collapse debris in the valleys.
3. The valleys are long and continuous, unlike any known lava tubes.
4. The channel valleys are both flat and V-shaped and erosion appears too fine to be collapsed lava tubes.
5. Almost all lava tubes observed on Earth are black and lava-like, whereas the Martian valleys are not.
6. Mysterious craters in the channel bottom do not fit the lava tube theory, but it is easily explained by hydrocratering.

All of this empirical evidence leads to the inevitable conclusion that Mars is a hydroplanet with hydrocraters and hydrovalleys. Another very important aspect of the Martian hydro-evidence is the amount of time it took for them to form. Like most of Mars' hydrofeatures, those seen in Fig 7.16.11 were not created over millions of years as many geologists think.

On the Earth, weather and the infamous sci-bi of geologic time along with the pseudotheories of uplift and plate tectonics are invoked to explain the wearing down of geological formations. Geologists turned to the pseudotheories to account for the formations of the Earth's continental landmasses, known to have come from oceans. They incorrectly believe that the continents moved up and down allowing "shallow seas" to come and go.

On Mars however, there appears to be little, if any plate tectonic activity and atmospheric water is almost non-existent. Therefore, the erosion on Mars could not have been caused by these factors, yet Fig 7.16.11 clearly shows *flat-floored* craters and a *flat-floor* valley that is evidence of voluminous, sediment-laden water.

Ma'adim Vallis is over 20 km (12 miles) wide, attesting to

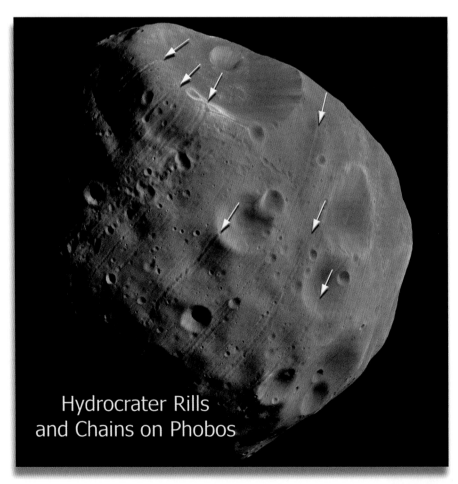

Fig 7.16.12. – Mars' largest moon, Phobos, shows off its hydrocrater chain, evidence that it is a hydromoon. Notice that some crater chains continue over the rim of the large crater. It is perplexing that planetary scientists still view these craters as "impact craters." A complete paradigm shift is needed before answers to some of the most basic geological questions about this moon can be given.
Image only courtesy of NASA

Fig 7.16.13 - A crater chain on Jupiter's largest moon Ganymede reveals light colored ejecta material only on the upper half of the crater chain. These are not a chain of impact craters, but are a line of hydrocraters, occurring along a line of tectonic weakness. The light colored ejecta material would have come from layers beneath the crater. The smooth plain includes other small hydrocraters with white ejecta material around them as well. The impact theory cannot clearly account for many such formations, but the Hydromoon Model can. There are simple answers for how these features formed and why they look the way they do. Courtesy of NASA (PIA01610).

the presence of truly enormous sediment-laden water volume.

Planetary scientists long aware of these apparent contradictions are left to "puzzle" over how such a dry world could have once had so much water:

"Lakes and riverlike networks do exist, but **water may have flowed through them only briefly**. It is possible that water remained frozen for most of the time, was occasionally released and quickly refroze. Still, **planetary scientists puzzle over how a world that was so arid in general could have been so watery at certain places and times**." Note 7.16e

The puzzle is made clearer as one comes to realize that Mars was *formed out of water*. Mars' smaller size and reduced gravity allowed evaporated water to escape the planet. There is real evidence that large amounts of water once existed on Mars. Some idea of how much can be gained by noting how much water is still on the arid planet's surface:

"The second is the discovery that **Mars has colossal reservoirs of frozen water that migrate around the planet as its climate changes**. To begin with, **both poles have deposits of ice or ice-rich sediments that are up to several kilometers thick over a combined area nearly twice the size of Arizona**." Note 7.16e

Phobos Hydromoon Evidence

Mars' largest moon, Phobos, looks very similar to the hydrocomets or hydroids, except that Phobos exhibits some fantastic hydrocrater chains. Phobos' crater chains seen in Fig 7.16.12 are important evidence of hydrocrater activity in parallel rills and valleys that occur on the outside and inside of Phobos' most massive crater. Note that some crater rows appear to transverse the crater rim. Amazingly, researchers still refer to Phobos' craters as 'impact' craters, even the largest one. Such an impactor would have pulverized this small moon. On the moon's surface, no ejecta is visible, but there are many hydrocrater chains! Impactologists cannot explain how so many parallel 'impact chains' could have formed, especially by different meteorites at different times.

The low density of Phobos, only 1.9 g/cm^3, is barely twice that of water, and testifies of its watery origin. Knowing this, researchers have suggested:

"Phobos's density is too low to be solid rock, and it is known to have significant porosity. These results led to the suggestion that Phobos might contain a **substantial reservoir of ice**." Note 7.16f

The Ganymede Hydromoon Evidence

Because of its density, Jupiter's largest moon, Ganymede, is estimated to be composed of rocky material and...water:

"The average density of Ganymede, 1.936 g/cm^3, suggests a composition of approximately **equal parts rocky material and water**, which is mainly in the form of ice." Note 7.16g

Most planetary scientists agree that there is considerable water on this moon that is only slightly larger than the planet Mercury.

It is interesting to see the hydro-paradigm shift occurring as mankind continues to discover watery worlds in our solar system and beyond. But, it has a long way to go to overcome the age-old impact pseudotheory.

Fig 7.16.13 shows a chain of craters on Ganymede that is still referred to as an impact crater chain in every science publication we found, but how can we know that meteorites did not form this crater chain? The 1994 Jupiter Impact of comet Shoemaker-Levy (Fig 7.9.3) demonstrated that an impactor can break into pieces creating multiple impactors, but it also demonstrated that they do not occur in a straight line. Fig 7.16.14 shows the row of broken fragments as they approached Jupiter. The fragments are not in a straight line and are not of equal spac-

Fig 7.16.14 - Comet Shoemaker-Levy is looked at as providing support for the origin of "impact crater chains." The comet broke apart as it approached Jupiter where it met its end and as it did, astronomers anxiously watched and recorded every detail. This image captures the line of comet fragments just before impact. Note that the impactors are not on a straight line and would therefore not create a straight line of impact craters. Note also, the distance between each impactor. This difference would translate into significant distance between craters. Both of these facts demonstrate the near impossibility that crater chains like those in Fig 7.16.13 are made by impact.

ing, and would therefore not form a chain of craters like Ganymede's in Fig 7.16.13.

Final proof came from observations of the actual Shoemaker-Levy impact event that clearly showed there was no impactor alignment. Instead, the impact debris that entered Jupiter's atmosphere was widespread and very much non-linear.

There are countless evidences of the Hydroplanet Model throughout the universe. This subchapter has included only a few. Hopefully, the examples given will ignite a search for truth in future researchers to seek out new and different evidences, or to explore those presented here in great detail.

7.17 The Hydroplanet Frontier

Summary of Hydroplanet Model

The Hydroplanet Model is the longest chapter of the Universal Model project and can be seen as being the 'core' of the book. In a way, all of science rests on the foundation of geology. The modern science geology is a cobbled together group of theoretical magmaplanet ideas, which are consistently unable to account for the real origins of Earth's structures.

The Earth, the Moon and other celestial bodies in our solar system are hydroplanets, not molten magma-planets. There is ample evidence of this already and it is only a matter of time before a flood of evidence becomes available; we have only scratched the surface. Evidence of the Hydroplanet Model was found in almost every field of science we researched; all we had to do is look for it. Geology's hard questions became easy with the Hydroplanet Model as a guide. Of course, one book contains a finite amount of space and it was difficult to decide what to leave out, because everything is so connected. It will be the job of Millennial Science to continue the quest for the discovery of scientific truth.

Direct Versus Indirect Evidence of Hydroplanets

This chapter has amassed a large amount of evidence to establish the reality of the Hydroplanet Model. There is so much new information that it is difficult to review and compartmentalize the New Geology that is a part of the Hydroplanet Model—it will simply take time.

Many scientists may have a difficult time accepting the Hydroplanet Model because the tradition of a hot, molten, magma interior is deeply entrenched in our psyche, and it will remain so until research begins with the new UM paradigm. Time and technology have frequently changed the course of history, and it will do so here as well. That's the nice thing about searching for truth instead of error—time is on our side.

Although some of the empirical evidence in this chapter is indirect, most is not. There are many direct evidences such as hydrocraters, hydromoons, hydrocomets, hydroids and the spherical shape of the planets that undeniably testify that planets were made from water. Huge amounts of water have been observed spewing out of comets, and apparently, this has been occurring for thousands of years. Throughout the solar system, only a very small amount of lava exists. The **only** other liquid in the universe, and the most abundant substance in the universe, which is directly responsible for the formation of our quartz-based planet, *is water*.

The beautiful, spherical shape of every planet and moon bears testimony to its water origin. The amazing part of this truth is that it has been right in front of our eyes all along. To really appreciate the matter we must explore the "final frontier" right here on Earth!

The Final Earth Frontier

In 2004, Joel Achenbach, a science writer wrote in *The Science of Things* section of the *National Geographic* an article titled, *To the center of the Earth*. Achenbach makes an excellent observation about a "nagging problem for humans":

"A nagging problem for humans, a species that likes to brag about all the distant planets and moons it has surveyed, is that

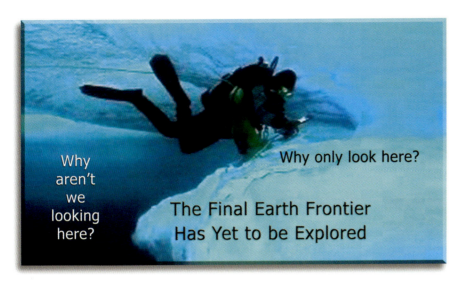

Fig 7.17.1 – Fewer people have been to the bottom of Earth's oceans than have set foot on the Moon, leaving it relatively unexplored. What little we have seen of it has revealed bizarre life forms, extraordinary environments and unsolved riddles. In the strange and hostile abyss lie more clues to Earth's origin and the processes that have shaped it. To truly understand these processes, we must look deeper, toward Earth's final frontier, to places and environs never before seen, or imagined by humankind. Secrets untold await the courageous and open minded souls who do.

we've never taken a good look right under our noses. The interior of the Earth is tantalizingly close, by cosmic standards, but how do you get there?" Note 7.17a

Achenbach's question is intriguing, but he posed it while under the umbrella of the magma paradigm. The Hydroplanet Model didn't exist when the article was written, yet Achenbach has the instinctive feeling that the magma planet idea he proposed wouldn't work:

"What little we know about the interior of the Earth (like the fact there's a crust, a mantle, and a core, or that there aren't mole people down there) comes from indirect evidence, such as the analysis of earthquakes.

"So maybe it's time for a radical new approach to exploring Earth's interior. Caltech planetary scientist David Stevenson says we should forget about drilling holes. Instead we should open a crack.

"Stevenson proposes digging a crack about half mile long, a yard wide, and a half mile deep (not with a shovel—imagine the back strain) but with an explosion more on the scale of a nuclear bomb. Next, he'd pour a few hundred thousand tons of molten iron into the crack, along with a robotic probe. The iron, denser than the surrounding crust, would migrate downward at about 16 feet per second, carrying the probe with it and opening the crack deeper and deeper. The iron blob would drop for about a week and 2,000 miles to the outer edge of the Earth's core, the probe beaming data to the surface.

"Stevenson compares his idea to space exploration. 'We're going somewhere we haven't been before,' he says. 'In all likelihood, there will be surprises.'

"This proposal can probably be filed in the drawer marked Ain't Gonna Happen.'" Note 7.17a

Such ideas are probably based on the Hollywood interpretation of geology. Hollywood has entertained us with science's pseudotheories, entertainment that is repeated continuously on documentary cable channels, whose funding comes from the institutions most interested in the status quo. Achenbach closed his article with these words:

"Great things can come from what seem like crackpot notions." Note 7.17a

The UM may be considered by many to be just such a "crackpot notion," however, as Achenbach said, great things can come from what *seem* like crackpot notions.

How can any great scientific discovery ever come from something that is not true?

Some balk at the idea of going to the center of the Earth because it has never really been seriously considered. Imagine the scientific data that would be gained if we could get there. An entire new field of science, titled here in the UM as Veology, would be created. **Veology** is the study of Earth's interior through physical contact.

How could such a scientific field come about? In the past, the focus was how to create probes that could withstand the heat of magma. But looking in that direction was wrong. There are deep ocean trenches that have never been penetrated, and they lie waiting for the human spirit to carry them towards the center of an unknown world. This new field of science—Veology, will be based on actual observations of water channels that penetrate deeper and deeper into the Earth. Instead of heat being the primary technological hurdle, it will be pressure.

Overcoming high pressure is not new, and there are already submarines and probes that have explored many remote areas of the ocean depths. Creating high-pressure vessels would be very similar to building space ships that leave the Earth's surface to travel to the outermost regions of the solar system. Just as we first sent probes to the Moon, probes will lead the exploration into the inner Earth, probably through water channels on the bottom of the ocean. Multiple probes could be left along paths toward the interior relaying information to the surface from an ever-deeper array of probes, similar to how we use satellites around the Earth today.

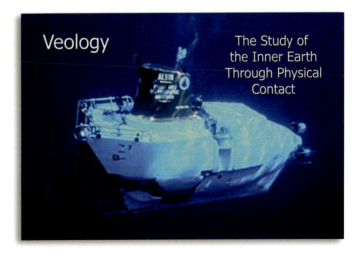

Fig 7.17.2 – Veology is a new field of science that seeks to study the interior of the Earth through physical contact and observation. Such a notion may be considered impossible under the current magma mindset, but a change in paradigm will open new doors of incredible discovery.

Fig 7.17.3 – Just how will the interior of the Earth look when we see it? There are many indications that the Earth is a giant geode, with crystalline rocks bridging giant water-filled chasms. The area beneath the crust will surely reveal much. This fanciful image is an actual geode, polished to show its remarkable texture and lighted from beneath to reveal its inner secret; a bubble (the round shadowed area in the center) floating above water. Like the Earth, this geode is filled with water.

Eventually, manned probes will carry humans down in the ultimate cave exploration project of all time to explore and add new dimension to the world beneath our feet, making discoveries of never before imagined minerals and organisms.

The first step toward the Moon was to be able to see it as it really was. Galileo did this with his telescope, which opened a new vision for mankind. Perhaps the Hydroplanet Model, with its many new observations and discoveries, many of which were never before known or seen—will open the minds and hearts of the world's researchers to explore the inside of our planet in a new way. New microbes, new biology, new rocks, new geology and a new water frontier.

The Universal Flood Evidence

One of the most compelling direct evidences that the Earth is a hydroplanet is the Universal Flood. With the Hydroplanet Model behind us, we are prepared to discover one of the most interesting and empirically documented major events in Earth's history, an event that appeared to be soundly dismissed while the world's centers of education were locked in the throes of the Magma Pseudotheory paradigm. This astonishing event—the Universal Flood—was the most crucial geologic event ever to occur on the Earth since its creation. And it wouldn't have happened if the Earth wasn't a Hydroplanet. **The Universal Flood Model** is the subject of the next chapter.

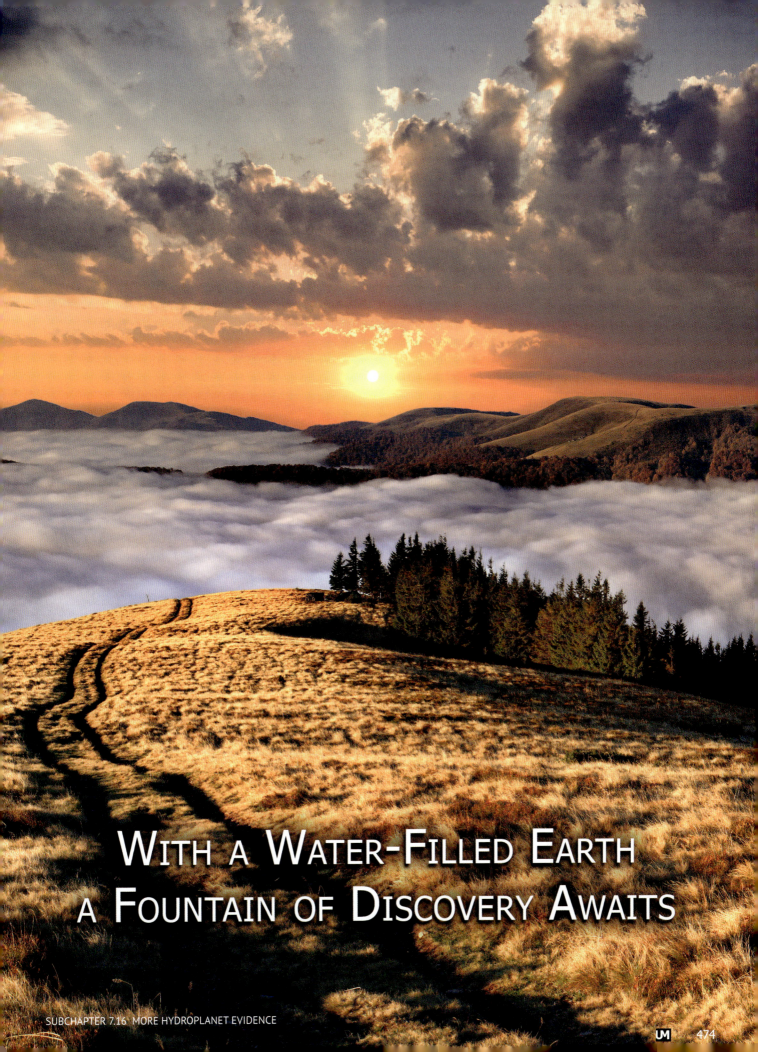

WITH A WATER-FILLED EARTH
A FOUNTAIN OF DISCOVERY AWAITS

SUBCHAPTER 7.16 MORE HYDROPLANET EVIDENCE

8

The Universal Flood Model

8.1   The Universal Flood History
8.2   The Acknowledged Flood
8.3   The Universal Flood Mechanisms
8.4   The Hydrofountain Mark
8.5   The Sand Mark
8.6   The Erosion Mark
8.7   The Depth Mark
8.8   The Carbonate Mark
8.9   The Salt Mark
8.10 The Oil and Gas Mark
8.11 The Coal Mark
8.12 The Pyrite Mark
8.13 The Ore Mark
8.14 The Surface Mark
8.15 The Diamond Mark
8.16 The Inclusion Mark
8.17 The UF Summary

"The present condition of the earth can not be assumed to be the only 'normal' one in earth history."

W. G. Woolnough, Geologist

The idea of a worldwide, all-encompassing deluge of water is not a new idea. Stories from nearly all cultures tell of a Flood sent to destroy civilization. Although the most familiar and well-known account is the Hebrew account of Noah as told in Genesis, all Abrahamic religions have variations of the story, including the narrative found in the Qur'an. Flood stories are recounted in the Mesopotamian Epic of Gilgamesh, the Popol Vuh of the ancient Maya of Central America, and in accounts from Asia, Australia, Africa, and India. Such written words once had great worth for the people who read them but are dismissed today as being myth and legend. Words committed to writing do not make them true, but truth may be locked within them. Finding the truth in anything written is a challenge we all share—especially if one does not know what truth is.

The cultural, religious, and familial backgrounds we come from are powerful forces in the shaping of our ideas and belief systems, so it is natural to prejudge the content of this chapter. To get the most out of it, it is suggested that you place your preconceived notions concerning the worldwide flood temporarily on the shelf and look for just one thing—the scientific truth. If the Deluge, the largest and most cataclysmic event ever recorded by man *actually happened*, one should expect to find scientific evidence of it literally everywhere and in every corner of the world. This is precisely what this chapter aims to do.

Throughout this chapter, the epic event known as the Universal Flood (frequently referred to as the 'UF' for convenience), will take on a profoundly new meaning as its importance to our modern world and to human history becomes evident. No event since the Earth was first formed had a more wide-reaching effect on the physical world we live in than the Flood, and the marks of that event are everywhere.

The subchapters herein address a variety of geological evidences; 'Marks' left behind as evidence of the Universal Flood that occurred only a few thousand years ago. Some of these marks are new observations and some are old evidence looked at in a new light.

8.1   The Universal Flood (UF) History

This subchapter includes a brief history and definition of the UF, a discussion of the Dark Age of Geology and its view on the Flood Myth, as well as an introduction to the Universal Flood Model.

No Scientific Evidence for a Universal Flood

The UM saw its beginnings in 1990; the result of an archaeological research project. People often asked what sort of "scientific research" was being conducted, which was not easy to explain because today, scientists generally focus on only one field, whereas the UM involves a remarkably wide range of research, touching all fields of science. The Scientific Dark Age needed to be exposed and pseudotheories uncovered, and there were new scientific discoveries to uncover in every major field. Interestingly, this cross-field approach was one of the keys behind many of the new discoveries in the UM; the answer to a question in one scientific field was *only* to be found in a completely different field of study. To find the truth in archaeology required knowing the truth in other fields such as geology and physics. It did not take long to discover that the scientific theories from those fields did not match the data and the observations of what lay buried in the ground. And the truth was not to be found in the scientific, geologic, or archaeological texts that were available. Then in 1993, an event occurred in the office of a well-respected archaeologist, a professor at an acclaimed university. That event started to snowball, eventually becoming an avalanche of ideas and possibilities.

I had stopped by the professor's office to discuss certain archaeological research I had been involved in over the past several years. After a few basic questions, it quickly became evident that he viewed me as no more than a pesky fly on his computer monitor. He treated each question with contempt. Having never experienced such disdain from other professors, and after receiving peculiarly ambiguous answers, I decided to ask one last question: *if* flood waters from a worldwide flood event so great as to have covered all of the mountains, and then to have drained away after only a year's time, as recorded in world history, there should be unequivocal geological evidence of such an event, and the entire surface of the Earth would show that evidence. Had he seen this evidence? The professor stopped me almost mid-sentence, and said:

"There will **never be**—*any scientific evidence*—for a Universal Flood."

I was momentarily stunned.

I thought perhaps I had misstated the question or maybe he had misheard it. I realized that this was not the case, and it was at that moment I recognized how deep the roots of the modern scientific establishment ran.

Here was a proverbial tree that had grown to overshadow the entire educational world with roots that seemed to reach to the very center of the Earth. The scientific community had attached a stigma to anything hinting of religion; the subject of a Noachian flood was taboo. It also meant that if the scientific community would not objectively consider the idea of a universal flood, then perhaps many of the ideas held in other various scientific fields could be wrong too. If a stealthily clad modern science agenda was barring the discovery of truth, then a *revolution in science* would *have* to take place before any change would occur, change necessary to reignite the scientific discovery of truth.

For various reasons, I ***knew*** as I walked out of the professor's

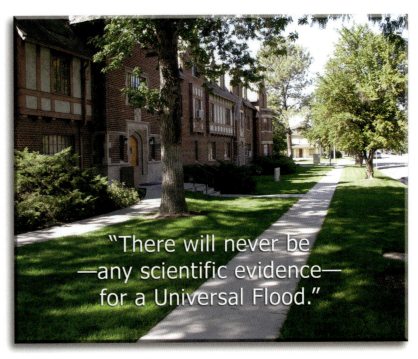

Fig 8.1.1 – Within the walls of this building, a professor once said, "There will never be—any scientific evidence—for a Universal Flood." This event put the gears of a New Millennial Science in motion that is changing the way the world looks at the Universe and man's role in Nature. It was on this sidewalk, in 1993 that what you are reading—The Universal Model—was first envisioned.

office that there *was* evidence; powerful, provable scientific evidence of a Universal Flood—and that the archeologist I had spoken with was *wrong*. I was only a fledgling scientist at the time, but knew I would find the truth, and ***knew right then*** that the material you are now reading would come to be.

Universal Flood Defined

For clarification, The Universal Flood Model in the UM is a **global-scale** event described as follows:

1. Floodwaters covered the entire surface of the Earth, including all mountains, for about one year's time.

2. The Universal Flood occurred on Earth about 4,000 years ago, as recorded by mankind.

3. Natural geological, chemical, and biological evidences confirm that this event took place.

The words of the distinguished geologist, W. G. Woolnough appear on the title page of this chapter. They are drawn from the following statement:

"**The present condition of the earth can not be assumed to be the only 'normal' one in earth history**. In cases where the characteristics of geological formations require explanations **other than** those afforded by the strict application of the Doctrine of Uniformity it is **legitimate and necessary to invoke cases not now observed in operation**, provided always that *objective evidence*, not conformable with conventional explanations, can be adduced." Note 8.1a

Woolnough was one of only a few mainstream geologists to ever acknowledge that the present condition of the Earth cannot be assumed to be the only "normal" condition. He understood that geology's doctrine of uniformity could not explain the origin of major geological resources such as coal, salt, sediment, and oil deposits, all of which will be discussed throughout this chapter. Although such statements could result in the branding one as a heretic, Woolnough's position was as logical and reasonable as any other well-known geologist's. Today, his statements have still **never** been disproven.

Moreover, Woolnough noted that we must always provide "*objective evidence*," and though it may not be conformable to conventional explanations, it must be adduced by anyone. This is especially important as we provide explanations for all these geological formations that cannot be presently explained. Employing "objective evidence" means to reach beyond being merely 'skeptical' or 'critical' of current scientific theory—we must find empirical evidence and question observations and interpretations **open-mindedly**. This is the process promoted throughout the UM, and with the Universal Scientific Method, a process you will be able to examine in this chapter. Instead of theories without prediction, or observation without testing, both of which are found in modern, twentieth century geology—actual evidence of the universal flood will be revealed. Answers to the geological mysteries presented in the Rock Cycle chapter will be explained and verified. Such wisdom would not be possible without the Universal Flood Model.

The Dark Age of Geology

Unfortunately, modern science has chosen to deny *the most important geological event in the Earth's history*. Such denial would be one of the biggest contributing factors to the present Dark Age of Science. The decision by modern science to disregard such a large body of scientific evidence and to interpret the observations of this great historical event with such obvious bias led to the period is now called the **Dark Age of Geology**.

The Dark Age is a time where the lack of scientific light and truth brought on a dearth of geological knowledge and wisdom for a period of more than a century. Thousands of geological and biological questions remained unanswered. This intellectual drought was the stimulus for the Hydroplanet Model, to bring about a revolution in geological thought and to identify the crisis in geology outlined in the Magma and Rock Cycle chapters. Next to the Hydroplanet Model, no other chapter in this book is as long—or as important; the quantity and quality of physical evidence of the Universal Flood is truly astounding. Even then, this chapter barely touches the surface of the vast ocean of evidence.

To understand the magnitude of the evidence that exists, we'll draw an analogy. Suppose you are a juror in a case against a person accused of robbery. The evidence includes the following: 1) The police find the suspect's fingerprints all over the crime scene; 2) There were four eye-witnesses that observed the suspect at the crime scene placing the stolen jewelry into his pocket; 3) The suspect's DNA was identified at the crime scene; 4) There was video surveillance clearly showing the suspect stealing the jewelry; and finally 5) The fleeing suspect was apprehended and the stolen jewelry was in his possession. With

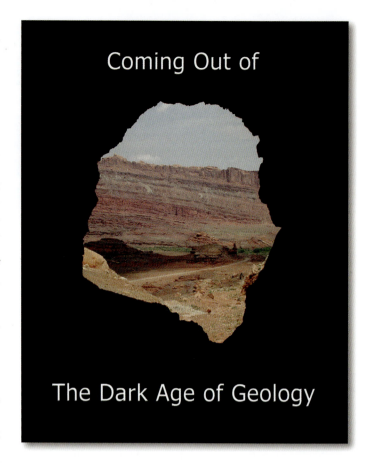

Coming Out of

The Dark Age of Geology

evidence like this, no jury would have a difficult time deciding the suspect's fate. With sufficient evidence, the ability to make a judgment *beyond a reasonable doubt* becomes easy.

The Universal Flood Model chapter includes more than enough evidence of the Flood to convince any unbiased jury. Covered in this chapter, there are 13 physical '**Marks**' the Universal Flood left behind for us to *observe and investigate*. Each piece of evidence in the case of the robbery was observable and testable; the physical Marks of the Flood are observable, testable, and can be examined firsthand by you, the juror. You can decide on the validity of the evidence for yourself.

It was a difficult process to narrow the selection of Marks to only thirteen. The deeper we looked, the more Marks we found, but in the end, those that were the most readily observable or that seemed the most obvious were chosen for inclusion in this chapter. There must be no doubt that the Universal Flood was a real event. Why? Because the UF not only affects many other fields of science, *it completely changes them*.

Each Mark is its own subchapter. Within each subchapter are *many* sub-subchapters, which are in a sense, individual Marks themselves. Thus, there are over a hundred physical testimonies of the validity of the Universal Flood included in this chapter.

Even though the Universal Flood will be shown to have occurred during recorded history, the fact that the history was written does not necessarily *prove* written history is true. To validate the historical nature of the Flood and its place in time, the Marks of the UF must be of a physical nature so that they can allow the scientific testing of each extraordinary claim that the UF actually happened. For the most part, the scientific tests and observations are readily accessible and simple enough for the general public to repeat.

Modern Science and the Universal Flood 'Myth'

There are few subjects on which modern scientists more readily agree than this—that the Universal Flood is a myth; a cultural bedtime story told by an ancient uneducated race of humans. This belief is so widely held by today's science professionals that we need not even quote the dozens and dozens of books that scoff at and hold in contempt the Universal Flood event. For reference purposes, here are two web sources with books on ancient cultures both for and against Noachian Flood accounts:

Man Remembers the Flood: Non-Biblical Flood Stories and Legends - http://library.puc.edu/heritage/bib-ManRemF.html

Flood Stories from Around the World - http://www.talkorigins.org/faqs/flood-myths.html

It is clear in the World History Model chapter why so many ancient cultures have stories of the Great Deluge. It is not by chance nor is it a coincidence that so many people, who speak countless languages and exist isolated from one another, share this common story. If there had been such a cataclysmic event in Nature, even if it was long ago, it would be very hard to hide the evidence, and stories would be handed down for generations. Indeed, the Universal Flood has two distinctive features; first, the UF is the *largest* natural disaster ever to occur in the history of mankind, second, it was the most *universally widespread*. Geologically speaking, it should be extraordinarily *easy* to prove the reality of a Universal Flood event, if it actually occurred.

Why Does Modern Science Refuse to Consider the Possibility of a Universal Flood?

There seem to be two primary reasons modern science refuses the possibility of a worldwide flood. One of those reasons will be covered in the Evolution chapter, where we uncover an agenda in modern science, which directly counters any physical evidence, no matter how scientific or solid, supporting Biblical events. The serious nature of this agenda, the controversy that will likely ensue as scientists try to deny the agenda, and a detailed, scientifically laid out case of the agenda is set forth in the Evolution chapter.

The second reason that modern science cannot seriously consider a universal flood event is the foundation upon which all modern geology is built. Lyle and Hutton, the founders of modern geology, built the modern geologic edifice on the Doctrine of Uniformity and on the theory of an ancient, hot molten Earth. Leaving this foundation behind today would mean changing much of modern geology.

Nevertheless, some astute geologists, like W. G. Woolnough, who had both experience in the field and in the research of scientific literature, made this observation:

"Most geologists will exclaim against the suggestion **that they are blind to the possibility of limitations in the application of the Doctrine of Uniformity**, and will claim, always in perfectly good faith, that their minds are open to all the implications of such limitations. **Yet, in almost any paper which one selects at random, critical analysis will reveal explicit or implicit re-statement of the limited Doctrine of Uniformity.**" Note 8.1b

In the last chapter, several examples of this doctrine were encountered. Throughout the world today, scientific textbooks and movies are based on the Doctrine of Uniformity. To this, Woolnough quotes a geologist; "...the fundamental conception that **all** geological processes in the past are **not fundamentally different** from those which are still active at the present day...", and then adds his own commentary:

"Such statements, perfectly correct in their immediate context, represent the explicit statement of the limited Doctrine of Uniformity, and **do not suggest a complete admission of the possible existence, in past geological periods, of sets of conditions vastly and fundamentally different from those prevailing at the present time.**" Note 8.1c

This chapter is about one of those past periods in which a set of conditions "vastly and fundamentally different from those prevailing at the present time" existed. It was the time of the **Universal Flood**.

Of course, for scientists to accept this evidence, they must first abandon the uniformity doctrine completely. Because it has been taught to them in every science class, they naturally view the world through uniformity glasses. If a person with a scientific mind expects to progress, this new material must be approached with an open mind.

The Universal Flood Chapter Outline

The UF chapter consists of four basic premises. These will be addressed and presented using scientific, empirical evidence as follows:

1. A Universal Flood has been indirectly acknowledged.
2. A specific mechanism caused the Universal Flood.
3. A catastrophic event caused the Universal Flood.
4. There are undeniable Marks of the Universal Flood.

8.2 The Acknowledged Flood

This subchapter includes evidence that modern science acknowledges that all of the continents were once simultaneously flooded and that the dinosaurs, along with many other organisms perished during the "Great Dying." Evidence that all of the Earth's mountains were once covered by a sea will be presented along with evidence that the subduction of the continents did *not* take place. A Martian Deluge and Earth's Mega-floods, all acknowledged by modern science, provide another perspective of the worldwide catastrophic nature of the Universal Flood.

An Acknowledgment that All Continents Were Flooded

What is to be found in the scientific literature concerning the covering of Earth's continents by floodwaters? Not too surprisingly, modern scientists acknowledge that *all* of the continents were once covered by oceanic waters because of the marine fossils found on or near the surface of every one of them. However connected this may seem, researchers study only one area of the Earth at a time and seem to miss seeing the possibility that all other parts of the world were flooded at the same time.

The following example is typical of the concepts taught

to young scholars about the flooding of the interior of North America:

"In **fact**, the **interior of North America was flooded**, all across the Great Plains, from the Gulf of Mexico northward through Canada. **Sea level was several hundred feet higher than it is today.**" Note 8.2a

With this very important "fact" established, we can ask an FQ that is easily overlooked by geology today:

> FQ: If the sea level rose several hundred feet, flooding the interior of North America, what was happening **to all the other continents?**

Although geologists see similar evidence of flooding on all continents, modern geology does *not* embrace the idea that a recent Universal Flood ever took place for several reasons:

1. Flooding was regional, taking place at different times.
2. Global Flooding is generally attributed to melting glaciers.
3. The Uplift Pseudotheory, which proposes once-flooded continental areas were raised to the mountains we see today.

These three *assumptions* have been taught as though they were fact in the modern science classroom for over a century, although none of them have been proven with empirical evidence, and all have flaws. The first assumption, the idea that flooding was regional, does not fit with a rising sea level. If the world's oceans were rising, it would have affected *all of the continents*; there is no real mechanism in *modern* science to explain how this could happen, although glaciers and glacial melt attempt to try.

The second assumption is that the melting of glaciers caused regional flooding—but other than very restricted areas of flooding, this idea is not logical because, on a continental scale, melting glaciers would actually cause the continents to rise! Global warming advocates are incorrect in their assumptions that melting glaciers and large ice masses at the poles would cause sea levels to rise on the continents. No evidence conclusively shows that the sea levels have risen *at all* from glacial ice melting. In fact, as noted in the Weather Model (subchapter 9.9), the glacial retreat in Alaska has caused just the opposite to happen—the reduced weight has allowed the land to *rise*, effectively lowering the sea level. In that chapter, we further explain why all the ice in the world could melt and ocean levels would *still not rise*.

The third assumption of continental uplift was proven a false theory in the Magma Pseudotheory Chapter. Modern science has never shown any evidence for the slow rising of large landmasses, although many claim they have.

Continuing with our focus in this subchapter, modern scientists acknowledge that all the continents were flooded by "global sea level fluctuations," as demonstrated in the following quote from Annabelle Foos, professor of geology at the University of Akron. Foos describes the flooding of the Grand Canyon craton (a **craton** is considered an old and stable portion of a continental landmass not subjected to major deformation for a prolonged period):

"The craton experienced a series of major transgressions and regressions where sea levels rose, and flooded the continent, followed by a lowering of sea level...

"The sequence boundaries are **not restricted to North America and can [be] correlated with sequences on other continents, suggesting they represent eustatic or global sea level fluctuations**." Note 8.2b

The conclusion to be drawn here, which is made abundantly clear in the scientific literature, is that modern science *already acknowledges simultaneous flooding of the continents* by at least several hundred feet, as it accepts the physical evidence for "global sea level fluctuations." This, of course, is *universal flooding*, albeit by only several hundred feet. But this is only the first step in our journey.

UF Proven to be a Single, Global Event

The 'Marks' later in this chapter demonstrate that there was a Universal Flood covering the entire Earth, and they prove that it was a one-time, global event. The average person will easily recognize this fact because the flood 'Marks' could have only been made when the entire world was covered with water at the same time, not just during *localized flood events that happened at different times*. This one-time global event will be verified by the observation that the flood marks happened together, at a single point in time, and it will also be shown that many of those marks are not being formed today.

Paleontologists have been puzzled for a long time as to why so many of the Earth's species disappeared at the same time. In a *Nature* August 2003 article *Boiling seas linked to mass extinction*, one Harvard paleontologist concluded that "there's no consensus as to what happened:"

"Up to **95% of Earth's marine species** disappeared at the end of the Permian period. Some **70% of land species**, including plants, insects and vertebrates, also perished. 'It's arguably the single most important event in biology but there's no consensus as to what happened,' says paleontologist Andrew Knoll of Harvard University in Cambridge, Massachusetts." Note 8.2c

The article does identify some interesting points that paleontologists *do know* about the Earth's mass extinction; points on which they all seem to agree:

1. 95% of Earth's *marine* species disappeared.
2. 70% of Earth's *land* species disappeared.
3. The event was the *single most important event* in biology.
4. There's *no consensus* as to what happened.

The researchers go on to add their observations to the list by identifying that **Boiling seas** were linked to mass extinction. These points lead to a fundamental question:

> FQ: What catastrophic event included the following; 95% of marine species disappeared, 70% of land species disappeared, was the single most important event in biology and was linked to boiling seas?

Most people are unaware of such unambiguous empirical evidence, and because modern science is in a state of outright denial about the Universal Flood, it is unlikely that a consensus as to what happened will ever be reached on the matter; at least not under the current modern science mindset.

By their own words, the catastrophic disappearance of a majority of the Earth's species was "the *single most important event* in biology." Such an "event" was not a series of separate flood events spanning millions of years. The researchers recognize the event was global, and furthermore, they recognize the implications of this event and its *importance* to all of modern science. Sadly, the mere suggestion of a Universal Flood could get you ejected from premier universities, or at the very least, denied access to revered publishers. Discoveries that point to a Universal Flood have been rejected by mainstream scientific journals for more than a century.

What Really Happened to the Dinosaurs?

One of the hottest scientific topics linking science with the public is the great mystery of the extinction of the dinosaurs. A hundred years of geological and fossil evidence piled up before scientists were forced to begin questioning the coveted uniformitarian principle, reluctantly admitting that massive cataclysms had occurred in the past. If anyone will look with an open mind, they can find incredible evidence that there was a sudden end to huge numbers of life forms right on the surface of the Earth and into the ocean depths.

In the previous chapter, Fig 7.12.1 included recent evidence that paleontologists really don't know how the dinosaurs died. That section of the book questioned the popular Dinosaur Extinction Impact Pseudotheory that researchers claimed in a 2003 *Scientific American* article was common knowledge:

"By now it is **common knowledge** that the impact of an asteroid or comet brought the age of the dinosaurs to an abrupt end." Note 8.2d

The researchers had proposed a new theory where *fire* was a part of the death-mechanism following the impact catastrophe:

"In sediments deposited in what is now Colorado and Montana, Iain Gilmour and his colleagues at the Open University in England have found chemical and isotopic fingerprints of methane-oxidizing bacteria—**a sign that the loss of so much life may have temporarily created anoxic, or oxygen-starved, conditions in small freshwater ecosystems**. Although the success of these bacteria is **not a signature of fire per se, it does indicate the pervasiveness and abruptness of death, which requires a mechanism such as a global conflagration to explain**." Note 8.2d p104

As they point out, there was no actual "signature of fire per se," but there was a complete disruption of natural *water systems* that came into existence, after the killing of many of Earth's species. This "abruptness of death" across the Earth certainly did create an environment where an explosion of "oxygen-starved" bacteria left a "signature" for us to learn more about, which we will do later in this chapter.

In an accompanying photo in the article, one of the authors was shown standing next to a sedimentary layer of sandstone that had obviously been laid down in *water*—and it showed no indication of the supposed fire. He was pointing to what was identified as the "fingerprints" that required a "global" mechanism, the implication being that the "fingerprints" showed up around the world. The fingerprints to which he was referring was explained later in the article:

"Mass grave left by the Chicxulub cataclysm has been preserved as a **light-colored layer of clay**..." Note 8.2d p104

Clay is a silica-based mineral. It is not formed by fire, but the researchers did identify an environment that was evidently void of oxygen, leading them to suppose that this clay layer was a signature of a global fire. The researchers had to look past the uniformity myth to accept the "abruptness of death" that occurred on such *a global scale and in a single event*. What they could not see because of their training was that such an event was caused by water on a worldwide scale never before imagined, not by fire.

To uncover the evidence for what really happened to the dinosaurs we need to review the fieldwork of the scientists who work with the remains of dinosaurs, instead of the meteorite impact specialists. In the book, *Digging Dinosaurs*, investigators describe the unusual position of the bones and speculate on the death the animals possibly experienced:

"The bones were in awful condition. Some even looked as if they had been sheared lengthwise. However, right next to a badly damaged bone would be one that was untouched. Furthermore, we found some of the bones standing upright. **Bones caught in water or lying on the ground and buried by sediment don't stand up vertically**. But if the creatures had been caught in a **mud slide**, they might have just been bashed to pieces and left in these odd positions as the mud settled." Note 8.2e

"Sea level was several hundred feet higher than it is today".

Massive worldwide mudslides have occurred but often go unrecognized because scientists have not looked for evidence of flooding of this scale before. We walk on solid ground today never realizing that we are likely walking on the remains of a massive mudflow from the Universal Flood. In this particular case, the mudflow was identified because the fossilized bones of dinosaurs caused a thorough investigation of the entire area. In 1978, the first Maiasaur dinosaur fossils were found in western Montana, USA, which would eventually result in the creation of the Museum of the Rockies in Bozeman Montana. Note the type of mudflow that was interpreted at the excavation:

"...John Lorenz, who was working on the dig in 1980, thought the animals might have been caught as a group in some kind of **catastrophic mud flow**—a flood, but of mud instead of plain water." Note 8.2e

Perhaps it is important to mention here that as we review the evidence of the Universal Flood, the UF waters were not merely rain or meteoritic water. The water included water from several sources with a variety of compositions including mud, alkaline or acidic water, briny water, and others; all a part of the Flood event. It is hard to comprehend the enormous forces involved in this event or the sheer volume of water and mud, because no one has ever experienced anything like it. There would have been so much water and sediment moving at such rapid velocity, that life forms of immense size would have been tossed about like chaff in the wind. This period of burial contributed to the demise of all kinds of life and left us most of the fossils we find today. Note that the researchers have difficulty conceiving such a catastrophic event themselves:

> "How could any mud slide, no matter how catastrophic, have the force to take a two- or three-ton animal that had just died and smash it around so much that its femur—still embedded in the flesh of its thigh—split lengthwise?" Note 8.2e p122-3

The researchers had an interesting experience as they went about setting up camp near their anticipated dinosaur fossil dig site. Each time they would attempt to drive a tent stake into the ground, they would encounter more bones. After finally getting the tents set up, sleeping became a real challenge because there were so many "bones poking" them. Their "extreme discomfort" caused them to move their tents and begin digging right in camp! What they found was almost unbelievable:

> "**There was no question anymore**. We had one huge bed of maiasaur bones—**and nothing but maiasaur bones**—stretching a mile and a quarter east to west and a quarter-mile north to south. Judging from the concentration of bones in various pits, there were up to **30 million fossil fragments in that area**. **At a conservative estimate, we had discovered the tomb of 10,000 dinosaurs**." Note 8.2e p128

Just imagine *10,000* large dinosaurs being smashed together and preserved for our study, yet their demise is just the tip of the iceberg when it comes to the enormity of the Flood catastrophe. The researchers go on to ask an important question and to make several key points:

> "What could such a deposit represent? **None of the bones we found had been chewed by predators**. But most of the bones were in a poor condition. They were either broken or damaged some other way, some broken in half, some apparently sheared lengthwise. **They were all oriented from east to west, which was the long dimension of the deposit**. Smaller bones, like hand and toe bones, skull elements, small ribs and neural arches of vertebrae, were rare in most of the deposit. At the easternmost edge of the deposit, however, these bones were the most common elements. All the bones were from individuals ranging from 9 feet long to 23 feet long. **There wasn't one baby in the whole deposit**. The bone bed was, without question, an **extraordinary puzzle**." Note 8.2e p129

The investigators knew they had one of the biggest puzzles ever identified on their hands—and some of the fundamental questions they asked should have been easy to answer, but they never were. For example, where were the baby dinosaurs that must surely have been with the herd? Mothers would not just leave their young, unless their life was threatened and unless they quickly traversed great distances, making it impossible for their offspring to keep up. Once again, this points to a catastrophe larger than any volcanic mudslide ever witnessed by science. Moreover, if the burial of the animals was only a local phenomenon that happened under typical conditions, there should have been other fossils of plants and animals that shared the plains and valleys where the dinosaurs roamed. They should have been buried alongside the dinosaurs, but there were none.

Strangely, none of the bones showed evidence of being chewed by scavengers. Such puzzling facts are easily answerable by a flood event that was *universal* in Nature, and that took place over an *extended* period. Landforms of new sediment laid on the surface, deposited by hydrofountains. They created islands where animals congregated in an environment completely different from the one we are used to seeing today. Evidence of these new geological and paleontological discoveries will continue to be introduced in later subchapters.

The isolated animals were only able to congregate for a time, ultimately destined to perish in a flood of incomprehensible proportions; "no ordinary spring flood" concluded the researchers:

> "This was **no ordinary spring flood** from one of the streams in the area, but a **catastrophic inundation**." Note 8.2e p131

"The Great Dying"

In the first decade of the 2000s, paleontologists finally jumped on the catastrophe band-wagon, admitting that fossils proved there was a time of "Great Dying":

What Really Happened to the Dinosaurs?

"This was no ordinary spring flood from one of the streams in the area, but a catastrophic inundation".

"Life was flourishing on the Earth about 250 million years ago, then during a brief window of geologic time **nearly all of it was wiped out**...

"The terrible event had been lost in the amnesia of time for eons. **It was only recently** that paleontologists, like hikers stumbling upon an unmarked grave in the woods, **noticed a startling pattern in the fossil record: Below a certain point in the accumulated layers of earth, the rock shows signs of an ancient world teeming with life. In more recent layers just above that point, signs of life all but vanish.**

"Somehow, most of the life on Earth perished... Scientists call it... '**the Great Dying**.'" Note 8.2f

Sadly, paleontology has known about the epic story of the Great Dying since fossils were first dug from the ground. At a place "just above that point" in the layers of the Earth's crust where an "ancient world teeming with life" existed, "signs of life all but vanish."

When paleontology first became a recognized science in the late 1800s, its practitioners had no interest in any evidence of catastrophism. In fact, there was such a marked disdain against the 'Great Flood' that any evidence of catastrophe was summarily dismissed or denied. Nevertheless, with over 200 years of scientific evidence mounting so high, catastrophic events had to be acknowledged. But it was assumed that they happened over 'vast amounts of time.' Events of the last several decades, such as the Jupiter impact by comet Shoemaker-Levy and the catastrophic eruption of Mount Saint Helens, compelled science to finally recognize what the geological and paleontological evidence had shown all along; an "extremely abrupt" catastrophe had occurred:

"'I think paleontologists are now coming full circle and leading the way, saying that the extinction was **extremely abrupt**,' Becker notes. 'Life vanished quickly on the scale of geologic time, and it takes something **catastrophic to do that**.'" Note 8.2g

Throughout the last century, the truth had been accumulating, to the point of boiling over, that a single catastrophic event, unlike anything ever imagined and "not anywhere operative at the present day" had occurred. It was reported in the *Bulletin of the American Association of Petroleum Geologists*:

"The most essential point to be remembered is that, *under normal marine conditions*, the assemblage of organisms at any one time and place is a BALANCED ASSEMBLAGE, in strict equilibrium with the environment. Any disturbance of this delicately poised equilibrium is followed immediately by an alteration in the nature of the assemblage, more or less conspicuous in direct ratio to the character and extent of the variation of the dominant factors of environment. If, then, the impact of environmental factors is adequately known, **it is possible to trace the nature of the consequential changes, even of very ancient formations, and deduce many vital factors in relation to paleogeography**.

"The existence at any time or place of the *unbalanced assemblage* of organic forms **is proof positive of serious disturbance of equilibrium, points to definite abnormality of conditions, and demands explanation**. Such explanation may require the postulation of permutations and combinations of **factors not anywhere operative at the present day**." Note 8.2h

In addition to the obvious evidences cited in this chapter, the Fossil Model, World History Model, and Clovis Model Chapters all provide more evidence showing conclusively that the UF was a colossal one-time event.

All Mountains Covered by the Sea

The pressure required to form quartz-based crystalline rocks found on the land's surface sheds significant light about the depth of the oceans during the Flood; the deeper the water, the higher the pressure. The Hydroplanet Model Chapter revealed

The Great Dying

SUBCHAPTER 8.2 THE ACKNOWLEDGED FLOOD

the amount of pressure that must have been present during quartz crystallization and fossilization. Later in this chapter we will discuss the height of the water column needed to form other rocks that are on the continental surface.

There are actually a number of reasons geologists already acknowledge that oceans once covered the entire world, *including all the mountains*. We will discuss three of those reasons here:

1. **Pillow lava**
2. **Sea shells**
3. **Limestone**

All three of these rock types can be found at the tops of the world's highest mountains today, and all three have their origin in a former ocean. In fact, there are so many of these rock types high in the mountains that it almost defies understanding. Modern science cannot merely say that the oceans were once only a few hundred feet because the mountains are obviously many times taller than this! Of course, modern science has side-stepped the physical evidence of the UF by asserting that the mountains were uplifted slowly, but that pseudotheory was exposed in chapter 5.13. There is **no** empirical evidence of whole continental areas uplifting.

Without the uplift and subduction pseudotheory, geologists have no way of explaining how the Colorado Plateau could have been lowered so that seas could cover it. In their minds, this uplift and subsidence occurred no less than five times, each time to account for the massive layers of limestone that exist between the layers of sandstone exposed in the Grand Canyon series. Although geologists should be able to date the theoret-

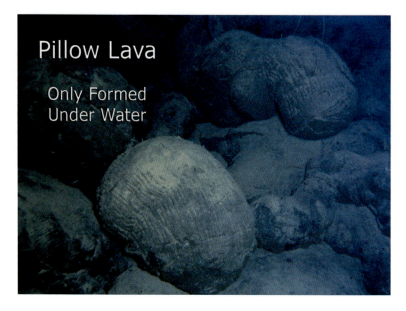

Fig 8.2.1 – These rounded objects are actually pillow lava that formed off the Hawaiian coast. This type of lava forms in shallow water, demonstrating that the tallest mountains in the world were once covered with water because they have pillow lava deposits high on their slopes. Courtesy of NOAA.

ical uplifts, they are unable to resolve even one of those dates, as recorded in *Arizona Geology*:

> "And, if the Colorado Plateau was not uplifted during the past 6 million years, when was it uplifted? Various other types of **evidence for age of uplift are contradictory, and this issue remains unresolved**." Note 8.2i

One thing the Arizona geologists generally agree on is that the development of the Grand Canyon involved "*catastrophic flooding* and rapid erosion":

> "Details of the spillover and development of the Colorado River through the Grand Canyon are not known, but we suspect that it **involved catastrophic flooding and rapid erosion**." Note 8.2j

We will return to the Grand Canyon and the Colorado Plateau later on.

Pillow Lava Mountain Formations

When lava is released under water, it cools rather quickly. Pressurized lava continues to be extruded, pumping out in waves, creating rippled **pillow lava**. Fig 8.2.1 illustrates the spherical looking 'pillows' formed as lava flows under deep water. Pillow lava is commonly found at active marine volcanic chains and along the plate boundaries of the mid-ocean ridges.

What makes pillow lava so important isn't that it is forming right now in Hawaii, but rather—that it formed "in the Alps":

> "Pillow lava is a familiar phenomenon. It can be seen in shallow water off Hawaii and even **in the Alps**…" Bib 112 p46

Pillow lava is present on many other mountains as well; unequivocal evidence that the tallest mountains on Earth were once covered by ocean water during a time when lava was actively flowing.

Fig 8.2.2 – Fossilized seashells are found near the summits of Mt. Timpanogos, Utah, USA and Mt. Cloudcroft, New Mexico, USA, both over 9,000 feet (2,700 meters) in elevation with no evidence they are rising. Since such fossil formations could have only formed in an ocean, it is clear that water once covered these and other mountains around the globe. The real question is *how were they covered?* Geologists have relied upon uniformitarianism—but this theory has failed, leaving modern geology with no real mechanism to answer this question. The Universal Flood Model finally identifies the mechanism explaining how a universal flood covered the entire surface of the Earth, including all of its mountains.

Sea Shells and Limestone on Mountain Tops

Anyone who hikes the world's tallest mountains will find shells from an ancient sea and many other types of oceanic fossils (Fig 8.2.2). Such fossils are well known, commented on here, in the book titled *The Miracle Planet*:

"At Kathmandu, Nepal, nearly a mile above sea level, **fossilized sea shells** are sold as souvenirs in open-air markets. They come not from the coast, 500 miles away, but **from the Himalaya Range to the north**." Bib 112 p50

The book informs us about the "countless marine organisms" lying on top of the tallest mountain in the world:

"Even the summit of **Mount Everest** displays yellow bands of limestone that were formed underwater out of the remains of **countless marine organisms**." Bib 112 p55

Fig 8.2.3 – The Dolomites are a series of mountains in northeastern Italy that get their name from the dolomite mineral of which they are comprised. Dolomite is a form of limestone known to come from oceans. Although fossils such as starfish, center inset above, are found in these mountains, no evidence of gradual uplift or subsidence over 'geologic time' is found. To understand how the Dolomites formed and how they came to include oceanic fossils requires knowledge of a previously unknown mechanism of seawater inundation capable of covering whole mountain ranges.

From the book, *Earth, The Making, Shaping and Working of a Planet,* we see another example of how high the oceans once were:

"Mount Makalau on the border of Nepal and Tibet is nearly 27,800 feet (8,500 m) high, only marginally lower than the tallest mountain in the world – Mount Everest – which stands to the west of it. **On its crest lie the youngest rocks and the ancient seabed of the primeval Tethys Ocean**." Bib 113 p45

Now that we have established the observed fact that ancient sea beds exist on top of the tallest mountains in the world, we can ask some simple, fundamental questions. According to modern geology, these ancient sea beds were "uplifted millions of years ago." If so,

FQ: Where is the evidence of millions of years of erosion of sediments and fossils at the tops of the mountains?

In the tops of the mountains, the rocks in which fossils are found are not smoothed and rounded, they have sharp edges and a crisp angular shape, testifying to a recent event that created them. The Dolomites of Italy, as seen in Fig 8.2.3, are a classic example of massive mountains that contain thick layers of ancient ocean sediment and fossils showing very little erosion from millions of years of "uplifting."

Because of the presence of pillow-lava, seashells, and marine sedimentary rocks found at the tops of the world's highest mountains, modern science recognizes that ancient seas once covered *the entire surface of the Earth*. Unable to discern the real mechanism of how these ancient marine sediments were covered with water, modern geology turned to the sci-bi of Plate Tectonics. However, Plate Tectonics employed a theoretical vertical movement of the crust that fails for the same reason the magma theory fails; neither theory has been proven with actual, observable evidence. Of course, there have been many observations, which have been cobbled together to *infer support* for the two theories. Nevertheless, it turns out the observable evidence actually *contradicts the theories*.

Moreover, we can demonstrate that Plate Tectonics did not take place over hundreds of millions of years by identifying **when** worldwide ocean waters covered the mountains.

The Plate Tectonic Uplift Deception

Although the public may be generally unaware of the evidence showing that all the mountains were once covered by ocean water, most geologists are. To make sense of this evidence, which includes pillow lava, seashells and limestone, they turn to the sci-bi of plate tectonics:

"Most of the fossils are not new discoveries. Early in the nineteenth century, their presence in these mountains was cited as proof of Noah's flood, **but a more persuasive explanation lies in the processes of plate tectonics**." Bib 112 p55

The *horizontal* movement of the Earth's crustal plates is both observable and quantifiable, but geologists have mingled this truth with the Uplift Deception. It is a 'deception' because of how it came to be a part of the modern science theology. Early researchers, including James Hutton and Charles Lyell, first proposed the uniformitarian concept of continental uplift over long periods of time two centuries ago. Since that time geologists have *continued* to teach this doctrine as though it was fact, which deceives the public, because there are no scientific observations to back it up.

In the Magma Pseudotheory, real scientific evidence established the fact that modern geology has never had any empirical proof for the vertical uplifting of continents. If anyone says they do, just ask to see the evidence; they will come up empty every time. Remember we are talking about evidence of entire continents rising and falling over long stretches of time, not just individual mountains or valleys here and there. We know by observation that these can rise and fall rapidly, during regional earthquakes.

Two astute researchers, after spending considerable time researching the mechanics of mountain building, wrote a book called *The Origin of Mountains*. The Australian researchers began to ask fundamental questions outside the dogma so prevalent in geology today, and found that the uplift of mountains occurred over *a relatively short and distinct time*, a *deviation* from uniformitarianism:

"**Uplift occurred over a relatively short and distinct time**. Some earth process switched on and created mountains after a period **with little or no significant uplift. This is a deviation from uniformitarianism**." Bib 141 p303

Additionally, the researchers noted that, "Some earth process switched on…" to create mountains, but what was the earth-wide process that "switched on" the mountain building process and is it taking place today?

To the credit of the researchers, they noted that no one in the scientific community, including themselves knows what caused this "short, sharp period of uplift," and they also explained the notion that must be abandoned to find out:

"**We do not yet know** what causes this short, sharp period of uplift, but at least **the abandonment of naive mountain building hypotheses might lead to further realistic explanations**." Bib 141 p303

Wow! The "naive mountain building hypotheses" ***must be abandoned*** to know the cause of the "short, sharp period of uplift." Dedicated research and time had uncovered a solid geologic truth. The evidence provided by GPS satellites as given in the Uplift Pseudotheory (chapter 5.14) is the proverbial nail in the coffin of long-term continental uplift. There is simply no long term uplifting or subduction of large areas observed anywhere today!

The Sinking Subduction Myth

There is clear evidence of worldwide flooding, a fact not disputed by modern science because marine fossils, pillow lava, and limestone are at the tops of the world's mountains. How did they get there if not by continental uplift? Science devised another explanation—a sci-bi—that envisions the lowering of whole landmasses and mountain ranges through a process called **subduction**. Through subduction, the land was lowered and the seas rose above the mountains; a simple concept if it was real—but it is not.

Researchers from the University of Western Australia and Geoscience Australia realized that uplift was not happening continuously, but "over a relatively short and distinct time," and they appeared to be right on the mark in their discussion about subduction:

"Plate tectonics has no real explanation for mountains on passive margins. Uplift is real, **but there is no possibility of subduction**. Ad hoc explanations have to be proposed, such as thermal effects or diapirs." Bib 141 p272

Throughout their discourse on subduction, they point out a significant number of inconsistencies in the subduction theory. While they conceded that occasionally, a mountain range might fit the theory, there were far too many contradictions to accept it:

"**The real problem with subduction is that it can do everything**. Plate collision may be invoked 'to explain uplift (making mountains), or subsidence (making deep trenches). It may make folds by compression, but makes backarc basins by tension. **The fact that the subduction hypothesis can account for both uplift and subsidence, compression and tension, means that it has too many degrees of freedom. It can account for opposite effects and is not testable**.'" Bib 141 p300

As the geologists looked outside of geology's dark box of uniformity, they had confirmed that subduction was a pseudotheory. They recognized, as the UM has confirmed, that: "**Nobody** has observed subduction…" Bib 141 p308

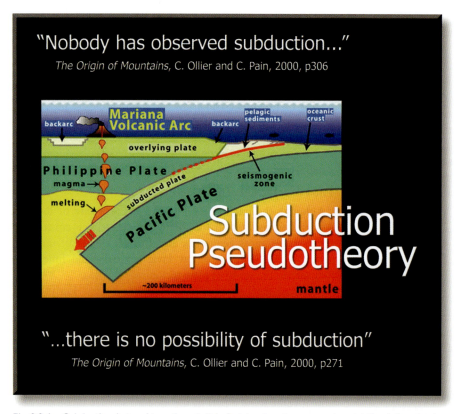

Fig 8.2.4 – Subduction is taught as though it is fact in almost every geology textbook in schools today. However, researchers have shown that the subduction hypothesis "is not testable" and therefore not proven. Without subduction, modern geology has no explanation for most of the acknowledged worldwide floods. In the colored NOAA diagram above, melted and rising magma is shown as though it is generated when the subducted plate sinks into the hot mantle of the Earth. In actuality, melted rock that shows up in volcanoes is generated by *friction*. Friction-induced lava (extrusive lava) has become more accepted in some geological circles because researchers finally realized there are earthquake generating faults below all volcanoes. Those faults move rock that is under great pressure, melting it where it can rise to the surface as lava.

It couldn't be any clearer; no subduction—no shallow seas, and without subduction, geology has no mechanism to explain *multiple* layers of limestone on the mountains.

Why has modern science so willingly 'believed' in uplift and subduction without empirical evidence to support it? Is this simply an example of ignorance or has this been a deliberate attempt to deceive the public with false pseudotheories? The answers to these important questions will be found in later chapters, but first we must continue with the framework that supports the answers.

Megaflood Evidence on Mars

On July 4th 1997, scientists at the Jet Propulsion Laboratory in Pasadena, California were not waiting for the evening fireworks show; instead they were anxiously watching their monitors for images from the Mars Pathfinder mission. The Pathfinder spacecraft had recently landed on Mars and had caught scientists by surprise as images showed evidence of a deluge far larger than anything ever conceived of on Earth:

"The Pathfinder spacecraft's spectacular basketball landing on Mars last summer brought it to a stop on a boneless dry flat, but its cameras revealed **clear signs of a spectacular flood that raged across the surface** more than a billion years earlier. Had you been standing on the 60-mile-wide Arés Vallis floodplain at the time, you would have seen **a torrent hundreds of feet high racing toward you at breakneck speed. The deluge would have made the combined flow of the Mississippi, Amazon, and Nile rivers look like a trickle, and could have filled the Mediterranean Sea overnight**. Pathfinder's Sojourner rover trundled past unmistakable vestiges of this cataclysm: water-worn stones and rocks piled up like cars in a series of rear-end collisions.

"Scientists have a hard time conjuring up a source for so much water. One remote possibility is volcanic activity, which might have rapidly melted Martian glaciers and triggered the flooding. Liquid water might also have pooled in aquifers beneath surface permafrost, only to be suddenly released when a tremor—a Marsquake—or cometary impact cracked the seal." Note 8.2k

This speculation, when looked at closely is quite remarkable. First, the physical evidence revealed a flood so "spectacular" and so large that nothing geologists had ever observed on Earth could compare. Second, the size of the Flood proved that large floods happen in the solar system, even on a planet with no visible water! Think of that! Scientists of the past have had a hard time "conjuring up a source" of water to explain Earth's Universal Flood, and so have rejected it. Meanwhile Mars, a planet with *no* surface *oceans,* had evidently experienced colossal flooding greater than anything before known on Earth.

Moreover, it was exciting to see the ideas expressed by the researcher who was not held back by the restraints of scientific dogma of a global flood event here on Earth. His musings led him to speculate that the water responsible for the massive deluge on Mars possibly came from "aquifers beneath" the Martian surface, perhaps released when a marsquake broke the seal and released the flood waters. The next subchapter will reveal a similar mechanism to account for the Earth's global flood event; the same thing that caused the Earth's Flood, probably caused Mars's Flood too.

> "Uplift occurred **over a relatively short and distinct time**... This is a deviation from uniformitarianism."

Notice that the Martian Deluge, which shared similarities with the Earth's Deluge, lasted only a short time, perhaps "only weeks or months":

"Geologic violence on Mars also carved the huge channel of the northern Kasei Valles. Despite the feature's great size— some 500 kilometers wide and nearly 3,000 kilometers long— some scientists think the catastrophic releases of groundwater that carved the valley may have lasted **only weeks or months**. Others believe that the water was confined underground by ice. **When the ice ruptured, possibly due to volcanic activity, water burst forth violently**. The water flow rate through Kasei could have been a billion cubic meters per second, dwarfing the capacity of channels on Earth such as the Amazon River and Strait of Gibraltar." Note 8.2l

There was plenty of flood evidence, including flood related sediments:

"Scientists were right about the flood. The hallmarks of an ancient deluge appear as far as Pathfinder's imager can see. Boulders are angled and stacked in the same direction. Evidence of layered sedimentation shows up both in nearby rocks and the Twin Peaks." Note 8.2m

If the Mars Flood lasted only weeks or months, there should have been no time to produce rounded pebbles and cobblestones, but because modern geology has not been in the hydroplanet paradigm, they have not understood how the model works. The rounded pebbles and cobblestones on both Mars and Earth can be explained by Hydroplanet Model mechanisms:

"**Intriguingly**, not all the rocks appear to be volcanic. Some have layers like those in terrestrial sedimentary rocks, which form by deposition of smaller fragments of rocks in water. Indeed, rover images show **many rounded pebbles and cobbles**

Fig 8.2.5 – This panoramic image from the Mars Pathfinder spacecraft was taken in 1997. Pathfinder landed in Ares Vallis, which from space, had the appearance of a large flood plain. As it turned out, "Scientists were right about the flood. The hallmarks of an ancient deluge appear as far as Pathfinder's imager can see." The large boulders on the flat plain were direct evidence that floodwaters had passed through there. The Martian landscape preserves a number of evidences of its own Universal Flood, which was very similar to the Earth's. However, on Earth, vegetation, a thicker, more active atmosphere, and more extensive erosion forces have concealed some of the Earth's own Flood evidences. According to researchers, the Martian Flood was so immense that "the deluge would have made the combined flow of the Mississippi, Amazon, and Nile rivers look like a trickle, and could have filled the Mediterranean Sea overnight."

on the ground. In addition, some larger rocks have what look like embedded pebbles and shiny indentations, where it looks as though rounded pebbles that were pressed into the rock during its formation have fallen out, leaving holes. These rocks may be conglomerates formed by flowing liquid water." Note 8.2n

In the Hydroplanet Model, a great majority of the hydroplanet's erosion takes place *underground* in a process called **endoerosion**. **Endoerosion** takes place where most of the hydroplanet's water is located, in aquifers beneath the crust. The suboceans and aquifers of hydroplanets keep rocks constantly moving, rounding pebbles and cobblestones of all sizes. Hydrofountains bring these rocks to the surface. This is likely how most of Mars' pebbles and cobblestones were formed. On Mars, surface water evaporated into space because of Mars' lower gravity, eliminating most surface water erosional processes seen on Earth.

The discussion about the UF Mechanisms in the next subchapter reveals evidence for how the Flood took place on Earth. It is quite possible that the flooding on Mars happened about the same time because the event that perturbed the Earth—a comet-like celestial body—may have also affected Mars, causing the suboceans there to burst onto the surface.

Global Evidence of Water on Mars

Although the evidence for the Mars Flood is overwhelming and the deluge is now widely acknowledged, some still think to downplay the size of the Flood, looking at it as a localized event. Some researchers who have studied the Martian landscape for decades suggest that Mars "had lakes all over its surface at one time":

"Actually Malin, 50, had been fretting about this issue for decades and even predicted elements of the hypothesis in his thesis written in 1975, based on Mariner 9 data, with its thousand times poorer resolution. 'Nobody bought my story,' he said, 'so I stopped harping on it.'

"Until now. 'If I had to bet, I probably would say these layers represent lakes that occurred on Mars very, very early in its history. I probably would say that **Mars had lakes all over its surface at one time**, and materials were being transported in these lakes.'" Note 8.2o

The fact that erosion from the Martian Flood is similar across the planet's surface testifies to its universal nature and to its "short-lived" time frame. With each new picture of Mars' surface transmitted from the rovers, Spirit and Opportunity, the evidence of a flooded Mars mounted:

"The rovers, Spirit and Opportunity, have found **direct and convincing evidence that water sometimes sloshed across Mars**, almost certainly during the planet's earliest epoch well over three billion years ago. Perhaps only puddles, streams, and flash floods came and went, drying repeatedly. But it's looking more and more likely that **liquid water was once abundant, though short-lived**." Note 8.2p

These new observations of water on Mars spurred new fundamental questions, some of which were posed by investigators in 2005:

"**How could the water have disappeared so fast? Where did it go?** Maybe some was lost for good when it evaporated and rose high in the atmosphere, where radiation split it into atoms that escaped to space. Some of it remains on Mars, frozen at the poles. And **much may be hiding underground, possibly in glaciers buried beneath the dust**." Note 8.2q

If water disappeared rapidly off Mars, with one potential place of its disappearance being underground, we add another FQ:

If vast areas of Mars were once covered by water that disappeared from the surface, why could the same not happen on Earth?

In the old modern science paradigm, each new observation yields more questions than answers. Astronomers back in 1997 had struggled with these same questions:

"At least for the moment, Pathfinder's images and data **pose more questions than answers**. Where did all this water come from? Where did it go?" Note 8.2r

One exciting aspect of the UM is learning by combining new wisdom with new knowledge, thereby growing in our intelligence; it is the Learning Process in action. Where science has been unable to understand processes that drive a magmaplanet, the hydroplanet model exhibits processes that make sense and that are easily explainable:

"Where there is evidence that water once flowed, **scientists don't understand what processes drove it**." Note 8.2s

By studying other planets, researchers are beginning to realize that to reach the truth they must *give up the old Uniformitarian Principle* James Hutton gave them:

"Perhaps the layers in the Surveyor images represent the only record of the erosion of landscapes long gone **because the processes that created them no longer operate on Mars**. 'Craters the size of Washington, D. C., were completely filled and then exhumed,' says Edgett. 'Unbelievable amounts of material were moved around **in ways that just don't add up**.'" Note 8.2s p38

Amazing isn't it?

Earth's Megaflood Evidence

Most scientists really want to know the scientific truth. If the Universal Flood really occurred, it was only a matter of time before independent research would find evidence of catastrophic flooding. New technology would help. By launching satellites into space and by probing the ocean depths with video, we are seeing things never before seen. Even though the majority of researchers have been taught the anti-universal flood orthodoxy of modern science, a few have published papers in scientific journals outlining a number of Earth's "megafloods."

The now known super-sized floods are still considered localized events that required an enormous amount of water to be released from higher elevations by some natural mechanism. The most often cited mechanism driving the localized megaflood event is glacial ice-dam breakage. This presumes the existence of the ice age, a pseudotheory discussed in the Global Warming Pseudotheory subchapter of the Weather Model Chapter. There, we discover the so-called ice-age *'evidence'* can be found all the way to the equator, which of course creates real problems

> "Perhaps the layers in the Surveyor images represent the only record of the erosion of landscapes long gone **because the processes that created them no longer operate on Mars**."

for life if there was Pole-to-Pole ice.

Here we will discuss two of Earth's known Megafloods, both of which are acknowledged by modern geology as being catastrophic events. The English Channel Megaflood in Europe was a catastrophic event only recently discovered; it has been well received but the Channeled Scabland Flood in northwestern USA, first proposed in 1923, was bitterly resisted by geologists for over half a century.

English Channel Megaflood Evidence

The following statement comes from a news report on a recent study in the journal *Nature*, which cited evidence supporting a catastrophic flood of such proportions that it separated the United Kingdom from France (Fig 8.2.6):

"The theory that **Britain became an island** during a **catastrophic flood—rather than through the course of normal erosion**—was first proposed in the 1980s. The new study, outlined in the scientific journal Nature, used high [higher]-resolution sonar data that were previously unavailable to produce three-dimensional, high-quality imagery of the region.

"In a commentary in the journal, Philip Gibbard, a University of Cambridge geologist who was not involved in the study, praised the research. 'It is no exaggeration to say that this Channel flood was probably ... **one of the largest ever identified** ... (and) it had **profound long-term geographical consequences**,' he wrote." Note 8.2t

Another source revealed just how large this catastrophic flood was:

"'The **first** was probably **100 times greater than the average discharge of the Mississippi River**,' said Sanjeev Gupta, a geologist at Imperial College London and co-author of the study. 'But that's a conservative estimate—**it could have been much larger.**'" Note 8.2u

With a discharge rate estimated at being at least 100 times that of the Mississippi, we can know that no modern human had ever seen anything like this in Nature before. This is an important point for those who think that because nobody has experienced a universal-sized flood in modern times, it could not have happened in the past.

The author of the previous article included the word "first," suggesting there was more than one flood event. We'll return to this shortly, but first, see why researchers think only a giant flood could be responsible for the carving out of the channels, and not erosion over time:

"He explained that erosion by river or ocean also **can't account for the underwater valley**, because it is too wide and has structures characteristic of a major flood.

"'The valley cuts across a large number of rock types, simply ignores the different layers,' he said, explaining that **only a rapid, enormous and powerful flood can account for such bedrock-scouring features**." Note 8.2u

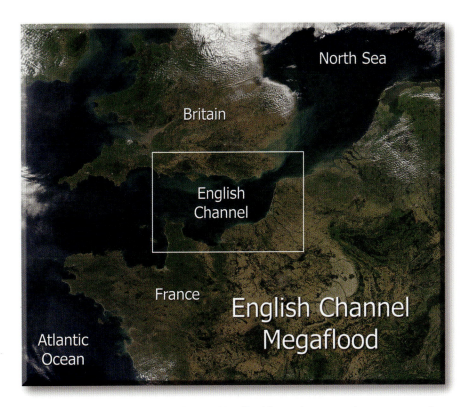

Fig 8.2.6 – Evidence of the English Channel Megaflood lies under present-day ocean waters between Britain and France. Modern science has never seen or recorded a megaflood of this magnitude, demonstrating that although geological evidence exists, the flood event was unknown to mankind. The evidence for the catastrophic flood that carved out the English Channel did not become available until 2007 when the channel topography was revealed using new underwater mapping technology.

The megaflood evidence the researcher cites—cutting across a large number of rock types—is a good flood indicator that can be used elsewhere in the world. The researchers note that the English Channel valley network is too flat to have been carved over long periods of time; they also explain that other slow erosion factors are not apparent:

"Our study provides the **first direct evidence** that a megaflood event was responsible for carving the English Channel valley network. **Normal fluvial processes cannot explain** erosion of the Northern Palaeovalley, because **before flooding no significant river was sourced** from the Weald-Artois ridge to the east. Tidal scour is not a viable mechanism because the superposed dendritic drainages indicate subaerial exposure of the valley floor after incision. **Erosion by glaciers is untenable**, because there is no evidence that these advanced into the English Channel. Our observations are consistent with erosion by high-magnitude flood flows, **as in the Channeled Scabland**, in which analogous landforms were indisputably formed by catastrophic drainage of the glacial lake Missoula." Note 8.2v

With this evidence, the researchers devised a theory that involved a *single* rock dam at the Dover Strait that eventually failed and released the floodwaters:

"Our observations support the megaflood model, in which **breaching of a rock dam at the Dover Strait instigated catastrophic drainage** of a large pro-glacial lake in the southern North Sea basin." Note 8.2v p342

Although the researchers found important direct evidence of the megaflooded area, there are two reasons that the English Channel flood event was *not* an independent event, but part of a much larger universal flood event involving water from multiple sources at different times. One reason is the evidence that "*at least two episodes* of flooding" eroded the valley:

> "The presence of a bedrock bench at the valley margin indicates that **at least two episodes of flooding eroded the valley**. The preservation of small truncated and beheaded channels on the upper surface of island M1 is evidence that an interlude of normal fluvial processes operated on the valley floor following the initial episode of flooding but before final valley incision by a second flood episode. We **cannot** resolve the absolute timing of the flooding events." Note 8.2v p344

Here we see researchers acknowledging multiple flood episodes, but giving no answer as to how multiple rock dams were created and breached allowing the multiple episodes of flooding. The separate flood episodes had to have been very close together in time or significant secondary erosion would have been evident after the earlier event. Where is the evidence in Nature that large natural rock dams form during short time intervals, break loose and are completely reconstructed to be breached again? This simply cannot happen with normal sedimentation and erosion, and certainly not following a megaflood event. With **no** mechanism to account for the formation of *multiple*, extremely large, *rock dams* at the same site, researchers should have been clued to look for another source of water.

Although the actual source of the water has eluded the researchers, evidence of it can be seen in Fig 8.2.7, which is the diagram researchers constructed and published in the journal *Nature*. The diagram shows the carved-out valleys and channels that were formed during several megaflood flow events, and it also shows **Secondary Flow Channels** from higher elevations (shown by white arrows). If the identified flood channels were really only created by the breaching of the theoretical Dover rock dam on the east end of the English Channel, **only** channels running from the Dover Strait dam site to the western Hurd Deep would exist. This is clearly not the case.

The white arrows identify a number of separate channels that run from many *higher* elevations *on both the north and south sides of the channel*. These are not connected with the theorized Dover Strait rock dam. Water obeys gravity and will not flow rapidly down a valley then back up a different valley at a higher elevation while speedily carving channels along a 75 kilometer (47 miles) stretch. This is what would have had to happen in the Dover Strait dam theory. In fact, the single *widest* flood channel is not even the English Channel! It is the Seine River Channel on the north of France, where the Seine River empties into the English Channel today (see fig 8.2.7).

Obviously, the researchers did not see megafloods along the north and south sides of the English Channel creating secondary valleys leading to the center of the English Channel and on to the lower elevation Hurd Deep. Why did the researchers fail to consider all the megaflood evidence that carved the north-south flowing channels that could not have come from the eastern Dover rock dam? It was certainly not because of the lack of 'direct evidence' now available, thanks to new high-quality imagery of the region. Doubtless, it was because modern science was not prepared for the reality of a flood of such epic proportions.

> The water required for megaflood events to occur during a common time over such a large area has never been seen or imagined in modern times.

Such a water source is far beyond the paradigm of mainstream scientists today. Until the publishing of the Universal Model, the scientific evidence needed to demonstrate the reality of a worldwide universal flood was unknown.

Channeled Scabland Megaflood Controversy

A controversy began in the 1920s, centered on the eastern

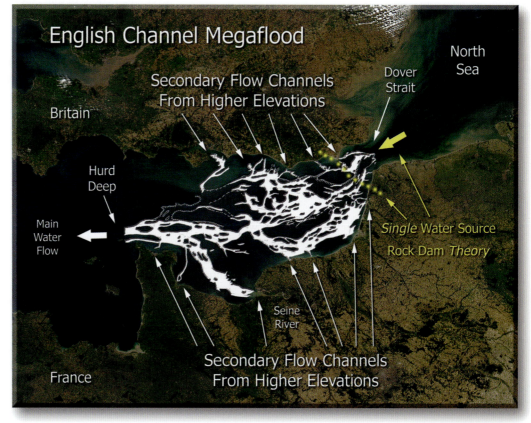

Fig 8.2.7 – This diagram illustrates the "palaeovalleys" and channeled areas shown in white. These channels were formed during a megaflood that researchers think was the result of a dam breach at Dover Strait (dotted yellow line on the right), which produced a single flow of water. However, the Secondary Flow Channels identified with white arrows clearly demonstrate that the flooded channels were not produced by a single water flow—but by *multiple sources from higher elevations*. Because the flood channels formed in a short time frame, and because the water flowed from so many sources—a completely new model is needed to explain them. The *Universal Flood Model* answers such questions and many more.

The palaeovalleys and channels shown in white in this diagram were originally drawn by the UK Hydrographic Office and adapted from a journal article in *Nature*, 19 July 2007, p342.

side of the state of Washington, USA. This controversy should become a permanent historical reminder of how easily pseudotheory can stop the progression of true science. It began when geologist J. Harlen Bretz shocked the geological community when he proposed an ancient flood of gigantic proportions had occurred right in the middle of the eastern side of Washington State. He called the channeled area carved out by this flood a "Scabland." The flow rate he estimated was ten times the combined flow of *all the world's rivers*; the Amazon River alone flows at six million cubic feet per second.

The uniformitarian foundation of modern geology along with the magmaplanet pseudotheory doesn't make it hard to imagine the reaction of the scientific community. Victor R. Baker explains in *Benchmark Papers in Geology* (1981) that for over half a century, men who "thought they were upholding the very framework of geology" twisted the evidence that Bretz presented:

"The recounting of the Spokane Flood controversy is thus something more than a discourse on the topic of catastrophic flooding. The papers by Bretz and others represent a marvelous exposition of the scientific method in action. **One cannot but be amazed at the spectacle of otherwise objective scientists twisting hypotheses to give a uniformitarian explanation to the Channeled Scabland.** Undoubtedly **these men thought they were upholding the very framework of geology** as it had been established in the writings of Hutton, Lyell, and Agassiz. The irony may be that Bretz's critics never really appreciated the scientific implication of Agassiz's famous dictum, 'Study nature, not books.' Perhaps no geologist has understood and lived the spirit of those words more enthusiastically than J Harlen Bretz.

"The most amazing twist to the students of the Channeled Scabland is the most recent. As the Viking spacecraft were orbiting Mars in the summer of 1976, their cameras were trained on the great Martian channel systems. The photographs revealed uplands streamlined by fluid flow, eroded scabland on the channel floor, and many other features that **we now know to be diagnostic of bedrock erosion by catastrophic flooding**. **Fifty years** after J Harlen Bretz's theory of scabland erosion on the Columbia plateau was denounced at an infamous meeting of the Washington Academy of Science, **astrogeologists were using Bretz's well-documented studies of the Channeled Scabland as the major earth-analog to Martian channeled erosion**. The origin of the great Martian outflow channels is a new controversy of which Bretz is not part. Nevertheless, the **fact** remains that few geological concepts have continued to have such relevance to modern geology as Bretz's theory of catastrophic flood origin of the Channeled Scabland, **born amid bitter controversy over a half century ago**."
Bib 157 pix

There was a "bitter controversy" between the geologists because the Doctrine of Uniformity was under attack as well as the introduction of catastrophic flooding that had never before been seen. Baker, Bretz, and others recognized that uniformitarianism became the "alternative geological explanation to assumptions of special creations and interferences by Divine Providence":

"As a hypothesis, substantive **uniformitarianism evolved as an alternative geological explanation to assumptions of special creations and interferences by Divine Providence**."
Bib 157 p4

The evidence will continue to accumulate as we move through this and other chapters, showing that modern science created many "hypotheses" as an *alternative to anything* that supports or suggests the hand of Divine Providence, even at the cost of the truth. This is *why* scientists berated Bretz's empirical evidence as an "outrageous hypothesis":

"The debates over that hypothesis were not always marked by scientific objectivity, but their recounting here provides a fascinating example of the triumph of an **outrageous hypothesis** (Baker 1978). **Only in the last two decades has the flood hypothesis gained general acceptance**." Bib 157 p1

Victor R. Baker edited CATASTROPHIC FLOODING The Origin of the Channeled Scabland, writing an excellent summary in the Forward of the book outlining the Scabland Megaflood as an example of Kuhn's "normal science," introduced in Chapter 4 of the UM, which explained why scientists were unprepared for the paradigm shift required to understand the Scabland Flood:

"The concept portrayed in scientific textbooks is that knowledge is gained by a steady, cumulative process. One philosophical view, however, holds that science is characterized by **long periods of conformity to a commonly held theoretical framework (Kuhn 1962)**. This framework, which Kuhn calls the 'paradigm,' is the tradition in which new investigations proceed. The science that occurs during paradigm-governed periods is '**normal science**' and mainly consists of solving puzzles generated by the paradigm. It is 'normal science' that appears in textbooks.

"A corollary to Kuhn's paradigm concept is that the new hypotheses generated to explain phenomena in a particular scientific field will be most often conditioned by the paradigm. In essence, a set of unstated 'ground rules' dictates the manner in which science will identify puzzles for its study. **These rules will also act to exclude many possible explanations for natural phenomena if such explanations lie outside the bounds of the accepted or 'normal' science of the day**.

"Kuhn's thesis holds that normal science eventually turns up **puzzles that cannot be solved within the existing theoretical framework**. Such anomalies are ignored, and attempts may even be made to suppress their study in order to prevent interference with the rapid progress that appears to occur in normal science. However, these anomalies hold the key to a superior kind of scientific progress. **When their importance can no longer be ignored, a scientific field may experience crisis, pronounced change, and eventually a completely new synthesis. This is the 'scientific revolution.'**" Bib 157 p2

> "As a hypothesis, substantive **uniformitarianism evolved as an alternative geological explanation to assumptions of special creations and interferences by Divine Providence**."

This marvelous summary of Chapter 4, Scientific Revolutions and the Scablands incident are a classic example of Kuhn's paradigm shift taking place. Normal science has indeed turned up "puzzles that cannot be solved within the existing theoretical framework" as we have seen with Bretz's research on the Scablands. This created a crisis, which required a change; a completely new synthesis as Baker stated; eventually a scientific revolution. The Hydroplanet and Universal Flood Models will stimulate the geological revolution Kuhn envisioned as "normal" science has stayed within its theoretical framework for a long time without discovering new natural law.

Baker goes on to say that a "completely new frame of reference was required" for normal science to finally see the truth about the Channeled Scabland Megaflood:

"J.H. Mackin has been quoted as saying, '**to understand the scabland, one must throw away textbook treatments of river work**' (Bretz and other, 1956, p960). Certainly a failing of Bretz' critics in the Spokane Flood debate was their **insistence that the Channeled Scabland conform to 'established' geomorphic processes. The scale of the problem was key, and a completely new frame of reference was required**. Bretz provided the required viewpoint: 'Channeled Scabland is **river bottom topography** magnified to the proportion of **river-valley topography**.'" Bib 157 p276

We cannot begin to understand the science presented in the Universal Flood Model unless we start with a completely new frame of reference, just as the Channeled Scabland required. We *begin* by acknowledging that large flood events occurred that were not formed by a continuous progression of processes that we see today, as Baker states:

"The Channeled Scabland **emphatically** did not form by a continuous progression of processes that were similar to those we observe in action today." Bib 157 p5

The Catastrophic Nature of the UF

Periodic flooding is experienced around the world every year. Occasionally, dramatic flooding occurs, termed 100-year flood events, calculated by governmental agencies for construction purposes when real estate or roads are to be built in flood prone areas. What this means is that every 100 years or so, a potential flood zone can expect to experience a dramatic flood event. Beyond this, a **catastrophic flood** is stated as being merely a flood

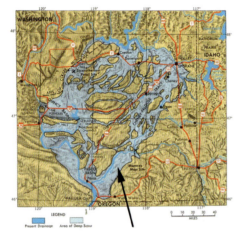

Scabland Flooding

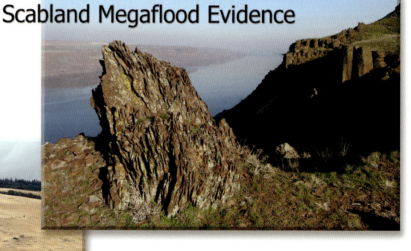

Scabland Megaflood Evidence

"One cannot but be amazed at the spectacle of otherwise objective scientists twisting hypotheses to give a uniformitarian explanation to the Channeled Scabland."

"Only in the last two decades has the flood hypothesis gained general acceptance."

Victor R. Baker, 1981
*Catastrophic Flooding
The Origin of the Channeled Scabland*

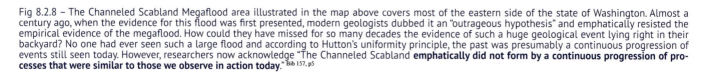

Fig 8.2.8 – The Channeled Scabland Megaflood area illustrated in the map above covers most of the eastern side of the state of Washington. Almost a century ago, when the evidence for this flood was first presented, modern geologists dubbed it an "outrageous hypothesis" and emphatically resisted the empirical evidence of the megaflood. How could they have missed for so many decades the evidence of such a huge geological event lying right in their backyard? No one had ever seen such a large flood and according to Hutton's uniformity principle, the past was presumably a continuous progression of events still seen today. However, researchers now acknowledge "The Channeled Scabland **emphatically did not form by a continuous progression of processes that were similar to those we observe in action today.**" Bib 157, p5

that changed the surface of the Earth over a *large* but localized area. However, no government agency or scientific establishment has considered or conceived of a worldwide Universal Flood.

On the catastrophic scale, just how large is large? Fig 8.2.9 is a diagram of the Colorado Plateau Swift Erosion process, which illustrates the type of catastrophic flooding that occurred around the Lake Powell area of Utah and Arizona. These four plates encompass an area extending from Northern Arizona into Southern Utah, and show the successive events that shaped this portion of the Colorado Plateau. This flood was similar to the English Channel Megaflood and the Channeled Scabland Megaflood in that they are all excellent examples of catastrophic flooding that occurred in the not too distant past—but were *not observed by mankind*. They are examples where the Uniformity Myth fails, and the Catastrophic Principle, a significant element of the New Geology presented in the Hydroplanet Model Chapter, succeeds at explaining the events.

Yet, even in the realm of catastrophic floods, the **Universal Flood** *is completely unique*. For example, a hurricane or a tsunami can cause what is normally considered catastrophic flooding. Such an event can force a tremendous amount of ocean water inland, flooding vast tracts of land. Although destructive, relatively little sediment is left behind from the flooding. In this chapter we will discover that the UF was not just limited to an inconceivable amount of water. A tremendous quantity of new sediment; both hydrosediment and erosionary sediment were created and deposited around the world, completely changing the landscape from its pre-flood state.

Another distinguishing feature of the megafloods is that large areas of the Earth's surface were carved and changed in such a way that only a megaflood could account for the magnitude of the change. Notwithstanding this erosional evidence, the Universal Flood includes many other unique, worldwide environmental conditions that will demonstrate the extent and magnitude of the Deluge. Such environmental changes to the Earth's surface will become clear as we learn about the Universal Flood Mechanisms.

8.3 The Universal Flood Mechanisms

The impact that the Hydroplanet Earth had on the Universal Flood will be extraordinarily clear in this subchapter. The various mechanisms that started the Flood, events that happened during the Flood and the catastrophic effect on the Earth's crust will be presented along with a few geological events that support the UF Mechanisms. We will also discuss the environmental conditions that prevailed during the Universal Flood event.

Origin of Floodwaters

The largest obstacle barring a scientific explanation of the Universal Flood has been a question of the origin of the floodwaters. *Where did all the water come from*? In the historical

Colorado Plateau Swift Erosion

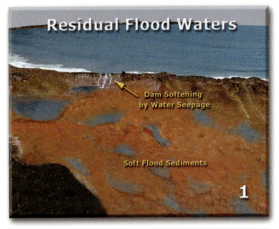

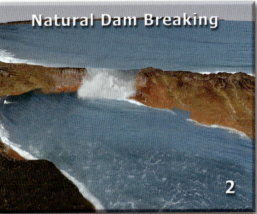

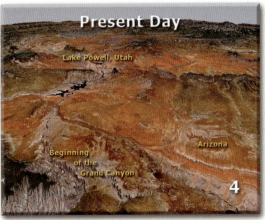

Fig 8.2.9 – Four sequential diagrams illustrate a likely scenario of how the area around present-day Lake Powel and the upper Grand Canyon formed. Residual UF floodwaters would have formed large continental lakes and natural dams. Eventually, dams broke causing rapid erosion over a very short period. The Scablands of eastern Washington, USA, and English Channel Megafloods are areas already acknowledged by researchers as being formed by a massive flood event. The modern day Mt. Saint Helens eruption provided another example of rapid erosion and deposition of many layers of differing sediment, which geologists had previously thought impossible over such a short period of time.

Biblical account, water covered *all* of the mountains. Even if the mountains were not as high as they are today, that was still a lot of water.

Scholars are well aware the rain recorded in the historical account lasted only forty days:

"And the rain was upon the earth forty days and forty nights."
Genesis 7:12, KJV Bible

This caused most researchers to scoff at the idea that the Earth's mountains could have been covered by the water of only forty days rain—and for good reason. If all the moisture trapped in the atmosphere fell as rain, it would not amount to much:

"If all of it [the water vapor in the atmosphere] abruptly fell as rain, the 3,100 cubic miles of water would cover the earth with **barely an inch**." Bib 13 p39

Trying to explain how water covered the mountains of the Earth with an inch (2.54 cm) of rain just isn't going to cut it. Some researchers propose that comets or meteorites are the source of the Earth's ocean water, but the evidence presented in Subchapter 7.6 showed the amount of deuterium in the oceans and xenon in the atmosphere would exist in greater abundance than now exists if this was the case. Therefore, there is no known external source for the universal floodwaters, let alone an amount needed to cover the mountains. End of story? Not quite.

If we go back to the historical record, we discover the floodwaters did not come *from rain*:

"In the six hundredth year of Noah's life, in the second month, the seventeenth day of the month, the same day were **all the fountains of the great deep broken up**, and the windows of heaven were opened." Genesis 7:11, KJV Bible

What were the "fountains of the great deep" and how do they fit into the picture? These "fountains" are the hydrofountains first introduced in the Hydroplanet Model. In an upcoming subchapter, the Hydrofountain Mark, detailed evidence of these worldwide fountains will be introduced.

Hydroplanet Earth Origin

Knowing the floodwaters did not come from the Earth's atmosphere or from space, we can turn to the **one** other source already known to exist. Amazingly, this source has been right under our feet all along, as scientists are well aware:

"In most parts of the world, any hole dug deep enough will yield water—with or without the aid of the diviner's forked hazel stick. Water is plentiful in New England, and at least two to five gallons of water per minute can be produced from almost any hole deeper than 20 feet. Near Miami, a yield of 1,000 to 1,500 gallons per minute is considered commonplace from wells averaging 50 feet in depth. The sand underlying Tallulah, Louisiana, yields 7,000 gallons per minute. Even around Tucson, drillers count on striking a dependable household supply at several hundred feet.

"Potable water exists in the ground—in some quantity, in some form, at some depth—**nearly everywhere on earth**. The Sahara itself, a synonym for total aridity, is underlain by water: an estimated 150,000 cubic miles spreading over 2.5 million square miles of land area." Bib 13 p55

Since flowing water is found "nearly everywhere on earth," one would think that the quantity estimates of underground water would be high, yet from a widely used 1998 geology textbook (Fig 7.6.6) we find that **only 1%** of the water in the Earth is reported to exist underground. In contrast, the oceans are reported to contain 96% of the Earth's water.

However, since this particular geology textbook was written, technology has improved observations of Earth's subsurface and as subchapter 7.6 documented, researchers now acknowledge underground oceans of 10, 30, even 100 times as large as the oceans on the Earth's surface! These new discoveries completely change our knowledge of the amount of water below the surface. This knowledge has come about only during the last decade. In Fig 7.6.6, in the Hydroplanet Model chapter, evidence of an enormous amount of salt water in the lower crust is shown. There is so much water in the crust alone, that if released to the surface, it would certainly cover the highest mountains.

Why haven't such important new discoveries become well known? Because these newly discovered scientific truths don't fit the current paradigm and especially do not bode well for the general scientific community who has preached against the hydroplanet Earth for more than a century, they are given little attention. *Any* model of the Earth that accounts for enough water to support a Universal Flood paradigm goes directly against the current scientific agenda.

It is significant in the history of science that for the first time, real quantitative empirical evidence has been presented in the Hydroplanet Model that verifies the large amount of water in the Earth's crust, sufficient to account for the source of water for the Universal Flood.

Universal Flood Mechanism Diagram

Fig 8.3.1 identifies the eight steps presented in the Universal Flood Mechanism Diagram. No internal mechanism that would facilitate the quick release of considerable subcrustal water has ever been recognized. Throughout time, large-scale planetary actions and movements have been accomplished through gravitational movement, which means another planetary body must have come close to the Earth in the past, enabling the gravitational forces of both the Earth and the approaching body to act upon each other.

The uniformitarianism pseudotheory has kept knowledge of catastrophic events of the past at bay for centuries, but new direct observations such as the recent Shoemaker-Levy Comet impact dispelled much of this false doctrine, leaving researchers to acknowledge the periodic close interaction of celestial bodies. The Earth's close interaction with a comet in its recent past and the affect this comet had will be discussed following the Universal Flood Mechanism Diagram.

1. Two Forces Hold Earth's Crust in Equilibrium

Subchapter 25.7 identifies evidence of the Revolutionary Universe, which includes the revolutions the planets make around the Sun and their axial spin. The Universal Energy Laws in the Universe System will explain how all matter in the universe is directed by a Central Universal Energy source that affects the rotation of the Earth on its axis, which keeps the Earth spinning at its constant rate. Without an external energy source, tidal friction and friction from the solar wind would slow the planet's axial spin, eventually stopping it. It would also affect its revolutions around the Sun. The Principle of Resonance directs Universal Energy through all matter, from atoms to the

Universal Flood Mechanisms

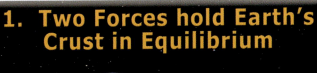
1. Two Forces hold Earth's Crust in Equilibrium

2. Comet Passes Close to the Earth
Earth's Motion Disrupted

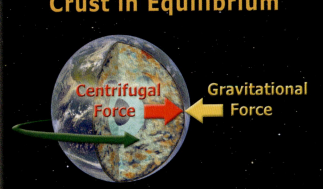
3. Earth's Rotation Rate is Reduced
Less Centrifugal Force — Same Gravitational Force

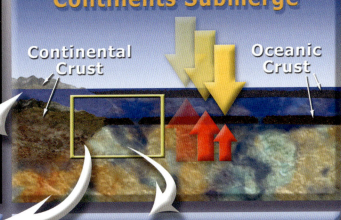
4. Crusts Collapse, Continents Submerge
Continental Crust — Oceanic Crust

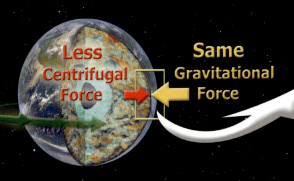
5. Hydrofountains at Plate Boundaries
Water from below is forced through cracks in the crusts
Continental Crust — Oceanic Crust

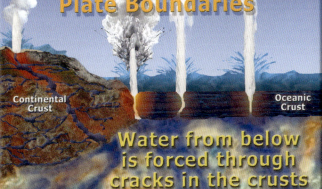
6. Hyprethermal Conditions
Water Pressure 30,000 ft. = 13,395 psi
Frictional Heat 350°–400°C (650°–750°F)
Continental Crust — Oceanic Crust

7. Rotation Rate Increases, Flood Waters Recede
Increasing Centrifugal Force — Same Gravitational Force

8. Two Forces Came Back into Equilibrium
Centrifugal Force — Gravitational Force

Fig 8.3.1

Earth on its axis can be altered. Just like a spinning top, it can be 'bumped' into a wobble. The Earth is spinning like a top and it has a relatively thin crust floating on its internal water. Fig 8.3.2 shows a metal paperclip 'floating' on the surface of the water because of water's surface tension. To understand the thinness of the Earth's crust, imagine shrinking the 8,000-mile-diameter Earth to the size of a basketball; the crust can be represented by a single sheet of paper. This crust is kept floating on the surface by surface tension and by the centrifugal force created by the Earth's spin. Centrifugal force is the same force that holds a rock in a sling as we swing it about our head. Stop the motion of the sling and the rock will fall.

Centrifugal force created by the Earth's spin is *thrusting* the crust of the Earth away *from* the center of the Earth. However, this force is balanced by the Earth's gravity, which is *pulling* the crust *toward* the Earth's center. This gravitational force keeps us stuck to the Earth's surface because we are also pulled toward its center. If this force were to suddenly vanish, we would fly off the planet, just like a rock released from a swinging sling. The crust of the Earth is held in *remarkable equilibrium* by the opposing forces of centrifugal motion and gravity—but this equilibrium can be disrupted.

2. UF Comet Passes Close to the Earth

A toy top spins until slowed by friction and it topples over. A magnetic top seems to spin effortlessly, floating in the air, but its motion can be disrupted by the influence of another magnet that is moved into its proximity. In like manner, the gravitational force of a roving celestial body can affect the spin of the Earth if it were to come within proximity of the planet. In the era of the UF event, the spin of the hydroplanet Earth was disrupted by the passing of a comet or other celestial object near the Earth.

Science was slow to accept the periodic passage of comets near the Earth and around the Sun. It was Edmund Halley (1656-1742) of whom Halley's Comet is named, that predicted the return of his namesake comet in 1759, 77 years from its most recent appearance in 1682. The periodicity of comets was later championed by William Whiston (1667-1752) who became an "object of ridicule by the entire world" because of his calculations and measurements of comets and their trajectories that lead him and others to the "idea" that comets would return:

"But the very idea of a **periodicity of comets**, gleaned by **Whiston from Halley**, was not yet accepted. In 1744 a German author wrote: 'It is well known that Whiston and others like him who wish to predict the comings and goings of comets, **deceive themselves, and have become an object of ridicule by the entire world**.'" Note 8.3a

Whiston, who was chosen by

Fig 8.3.2 – Although this paperclip is several times denser than water, it is floating because of surface tension, which causes the water to behave like an elastic sheet, holding the heavier metal on its surface. The force of surface tension and centrifugal force, which is caused by the Earth's rotational spin, cause the Earth's crust to 'float' on the surface.

Isaac Newton to take his mathematics chair at Cambridge University, advocated the idea that a comet was the cause of the Earth's Universal Flood. Unfortunately, neither Halley nor Whiston lived to see the predicted return of Halley's Comet, but to the credit of science today, both men are now recognized for their achievements in comet research. There are other Enlightenment era scientists that advocated the Deluge as being a real event; the UM evidence will bring some of their names and ideas back into favor as the truth becomes known.

How real is the possibility of a Universal Flood era comet? If the historical records are correct, such a comet must have approached the Earth about 4,300 years ago. Initially, comet periodicity seemed to be short-lived; Halley's was only a 75-year return cycle. The orbital cycle of Pluto, the furthest planetoid to be discovered in early 1930, is only 248 years. The question is; are there long-period celestial bodies in highly elliptical orbits around the sun that can potentially intersect Earth's orbit?

During recent years, new astronomical discoveries have been made using computers that scan old telescopic images. These discoveries make the UF Celestial Event seem a very likely possibility. One example of a very long-period orbit is Sedna, discovered in 2003. Fig 8.3.4 shows the approximate orbit of Sedna in relation to the solar system, as well as a relative size comparison with the Earth. Note 8.3b Sedna's orbital period is 11,400 years long!

Sedna is just one of over 3,000 known comet-like objects and was detected because it is now near its perihelion; the point along its orbit that is closest to the Sun. Sedna's orbital period of over 11,000 years

Fig 8.3.3 –William Whiston (1667-1752) advocated the existence of comets and believed they were predictable objects that orbited the sun. Whiston was also a noted historian and scientist and explained how a comet could have been the cause of the Universal Flood. Although ridiculed for his beliefs, the scientific evidence of both comets and the Universal Flood is a reality today.

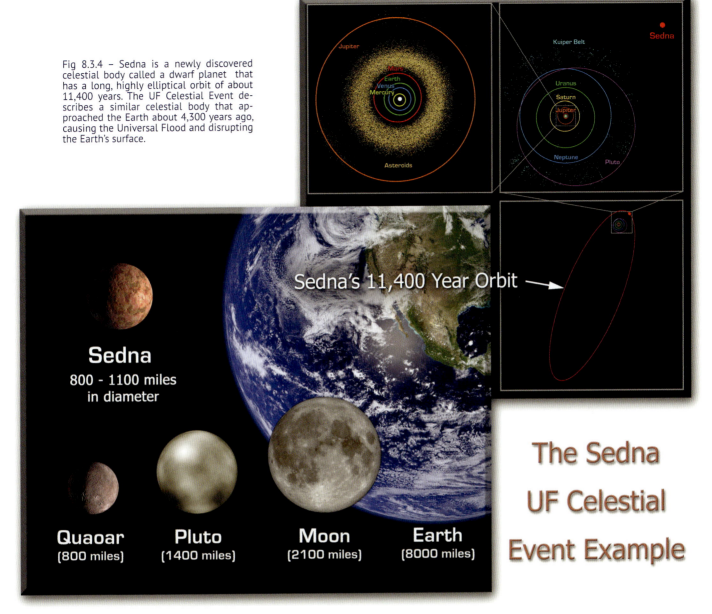

Fig 8.3.4 – Sedna is a newly discovered celestial body called a dwarf planet that has a long, highly elliptical orbit of about 11,400 years. The UF Celestial Event describes a similar celestial body that approached the Earth about 4,300 years ago, causing the Universal Flood and disrupting the Earth's surface.

made it one of the most distant solar system objects yet observed, with many new long-period comets being discovered every year.

For example, comet McNaugt has an orbit of roughly 92,000 years and comet West has an estimated orbital period of as much as 6 million years. These long-period comets have highly excentric orbits and periods that just a few years ago were unknown and support the UF comet as proposed here in the Universal Flood Model. Note 8.3c

Although Sedna has a highly elliptical orbit that is still not yet precisely known, it satisfies the question that there are celestial bodies that could have passed very close to the Earth over 4,000 years ago. It is also important to note that the comets' current orbits could have been altered by interactions with other celestial bodies over thousands of years, meaning their current orbit might not be reflective of past orbits. Regardless, we have direct evidence that a comet or comets, in the solar system could have passed by the Earth several thousand years ago. Such a pass would cause a major disruption to the Earth's axial spin.

The UF comet would have passed by so long ago that it is, no doubt, deep into its long run, far out into space, beyond the solar system, perhaps even beyond the reach of modern telescopes. No doubt, the coming years will reveal many more solar system objects that have long elliptical orbits, perhaps even *the* UF Comet itself.

3. Earth's Rotation Rate Reduced

The third step in the Universal Flood Mechanism Diagram is the reduction of Earth's rate of rotation. The slowing of the Earth's spin disrupted the delicate balance between centrifugal force and the force of gravity, seen in step 3 of Fig 8.3.1. Gravity remained constant, but the centrifugal force, the force that holds the Earth's crust above the suboceans, decreased.

Another aspect of the UF Comet's disruptive approach is its effect on the Earth-Moon planetary system. The Moon is also subject to gravity and centrifugal forces. As the UF comet drew near the Earth, it would have effected a change to the *entire planetary system*, temporarily altering their orbits. That change would cause the Moon to periodically swing in closer to the Earth. The author of the book, *The Restless Earth* wrote of the effects of such an approach:

Crusts Collapse and Continents Submerge

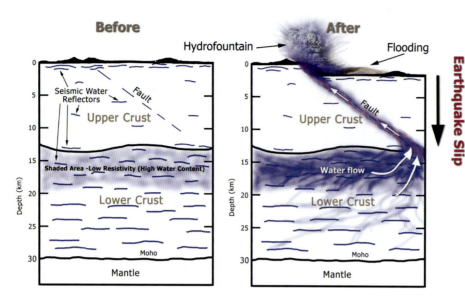

Fig 8.3.5 – This is an expanded view of the events that occurred during step 4 of the Universal Flood Mechanisms. The diagram on the left represents the Earth prior to the UF event, as adapted from D. Ian Gough's journal article in Nature, discussed previously in the Hydroplanet Model Chapter 7.6. Seismicity, Resistivity and actual borehole observations have established water's presence in the continental crust approximately as shown in the left illustration. When the Earth's crusts began to collapse as the rotational rate slowed, water flowed along fault lines as shown in the diagram on the right. This stylized diagram illustrates the events that took place during that tumultuous period, showing rapid movement of pressurized fluids from the Lower Crust, heated as it moved along fault lines toward the surface, flooding the Earth. Because so much water resides in the lower crust, only minor slippage need occur relative to the crust's total thickness for a period of universal flooding to take place.

"But did the Moon ever approach very close to the Earth? If so, huge tides, not just in the Earth's oceans but in the solid rock of both bodies, could have caused dramatic events in Earth and Moon. Even today, there are moonquakes every month, caused by the Earth." Bib 127 p97

As we continue through the chapter we will see geological evidence for the "dramatic events" the author described. As he noted, the huge tides would not be limited to the oceans; the Earth's land would experience large tidal movements. The gravitational forces of the Moon, the UF comet, and the change in Earth's axial spin along with the rising and falling of continental and oceanic crusts would facilitate a complete collapse of the Earth's crust.

4. Crusts Collapse, Continents Submerged

In the last chapter we discussed the equatorial bulge that makes the Earth's diameter 27 miles (43 km) larger at the equator than it is from pole-to-pole (Fig 7.6.5). Scientists already know this is due to the centrifugal force of the rotating Earth, which moves close to 1,000 feet per second at the equator. If the Earth was more solid, the bulge would be much smaller, but because of water under the crust, the planet's bulge is large. A slowing of the Earth's rotation would trigger a tremendous outward flood of subsurface ocean water.

This is illustrated in Fig 8.3.5. The left diagram represents the crust in its pre-collapsed position; the right illustrates a continental scale fault slip with a range of several kilometers. The water of the upper portion of the Lower Crust would be forced to the surface through fault lines as the dense Upper Crust collapsed onto the weaker and less dense Lower Crust.

To put this into perspective, the present-day continental landmass that is above sea level is only about a third of the surface

Fig 8.3.6 – During the fourth and fifth steps of the UF Mechanism, the crust collapses, continents submerge and hydrofountains erupt. In this image, we see a tiny version of actual crustal collapse and hydrofountain activity from the 1994 and 2003 Hokkaido Toho-Oki and Tokachi-Oki earthquakes in Japan. Although the quakes were considerably smaller than those of the UF, they help us understand the processes of heaving, sinking, hydrofountains and hydro-sand boils. During the UF, these processes were thousands of times more powerful, almost incomprehensible to mankind.

CHAPTER 8 THE UNIVERSAL FLOOD MODEL

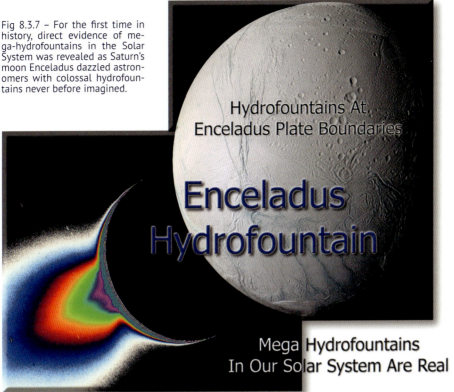

Fig 8.3.7 – For the first time in history, direct evidence of mega-hydrofountains in the Solar System was revealed as Saturn's moon Enceladus dazzled astronomers with colossal hydrofountains never before imagined.

area of the planet. Furthermore, if the land area's elevation was averaged, it would stand only 410 feet above sea level (see Fig 6.1.2 for illustration). In contrast, the much larger, two-thirds surface area of the oceans is on average 3700 meters (12,100 feet) deep! (http://en.wikipedia.org/wiki/Ocean – Accessed 4.6.15) This means that there is more than 30 times the volume of *water below sea* level than there is *land above sea level*. Even this significantly larger amount of ocean water, as compared with the dry land, is only a *small* fraction of the total water in the Earth's suboceans.

Estimates from researchers cited earlier in the book recognized that there may be up to *100 times* more water beneath the Earth's crust than is in its oceans; and that number continues to climb. There is no question that more water will be found beneath the crust as technology improves. That will further justify the already easily explainable collapse and submergence of the continental crust.

We can actually see, on a small scale, how the crust might have appeared during this period of collapse. The aftermath of two Japanese earthquakes are shown in Fig 8.3.6. The earthquakes caused entire land areas to sink, leaving incidentally supported infrastructure, such as the sewer system's manhole intact. The process, called **liquefaction**, has been shown to be responsible for the subsidence of heavier sediment and the rising of water and lighter sediment during the shaking of earthquakes. This process and the release of superheated water and steam by hydrofountains brought the hydrosand seen in Fig 8.3.6 to the surface.

5. Hydrofountains at Plate Boundaries

We can see how big mega-hydrofountains can be in Fig 8.3.7, which is the recently imaged eruption of the Mega Hydrofountain on the moon Enceladus. This hydrofountain erupted with such force that it extended beyond the diameter of the moon itself! Liquid water expands very little when heated, until it experiences a change in phase from a liquid to a gas. On small planets and moons, lower gravity and little or no atmosphere creates an environment of low pressure on their surfaces. Under lower pressure, water can turn to steam at a lower temperature, easily erupting and breaking the surface of small planetoids.

Although most of the Universal Flood hydrofountains were probably not as large relative to the Earth as the Enceladus hydrofountain is, they were certainly much larger than anything ever experienced in modern times. When water turns to steam at sea level, it expands 1,700 times its liquid volume almost instantly. During the UF event, subsurface waters would have been heated by forces explained in the Gravitational-Friction Law at plate boundaries and along fault lines around the globe. Fig 8.3.8 is an artistic conception of giant hydrofountains along Earth's plate boundaries during the UF event.

6. Hyprethermal Conditions

Step six in Fig 8.3.1 illustrates the Hyprethermal Conditions present during the Flood. These conditions are difficult to visualize because the Earth is so large but

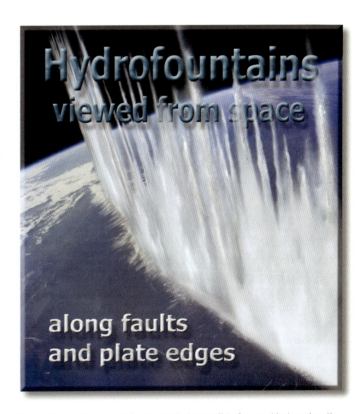

Fig 8.3.8 – Until recently it was nearly impossible for mankind to visualize the scope of the catastrophes that impacted the whole world. Only since the 1960's, during the advent of the era of space exploration have we been able to view what the whole globe looks like. Perhaps the hydrofountains of the UF event would have looked like those shown in the artistic rendering above. The dramatic effect hydrofountains had on the surface of our planet will be outlined in this chapter.

it was very similar to what exists today in certain areas on the bottom of the ocean.

To comprehend the depth of water and the pressure on the surface of the collapsed continents, we have to use an analogy. If you have ever been on a commercial airplane, you have probably heard the pilot say, "We have reached our cruising altitude of 30,000 feet." At this altitude, the giant jumbo jet appears as a tiny, barely discernible aircraft to an observer on the ground. This is the estimated height of the ocean waters during the peak of the UF event. 30,000 feet of water equates to over 13,000 psi pressure, which coincidentally, is the precise amount of pressure needed to grow rocks in a hypretherm. Such incredible pressure, along with estimated temperatures of between 350ºC and 400ºC, which resulted from frictional heating from the collapse and flexure of the Earth's crust, created the necessary environment for massive and rapid mineral and rock growth worldwide. It was this unique hyprethermal environment that existed during the Universal Flood that gave rise to dozens of unsolved mysteries and enigmas in modern science. Many of these were introduced in the Magma and Rock Cycle chapters.

At first, thirty thousand feet (5.7 miles/9.1 km) of water seems to be an awful lot of water as one imagines looking down from an airplane at cruising altitude. But in comparison to the Earth's overall diameter of nearly 8,000 miles (13,000 km), it is actually quite small; less than 1/1,000 of the diameter of the Earth, as thin as a single sheet of paper on a basketball-sized Earth. Although the catastrophic collapse of the crust was the largest cataclysm in history, only a small fraction of the Earth's outer area changed.

Picked apples shrink as the water inside evaporates. Just as it is difficult to see the shrinking of the apple skin (it takes place over several days), an astronaut viewing the universal flood from the Moon would have seen the green and brown colors of the Earth's continents change to white and blue as floodwaters enveloped them, but changes in the crust as it collapsed would have been nearly indiscernible.

7. Rotation Rate Increases, Flood Waters Recede

There are two practical questions raised when discussing the reality of the flood; where did all the water come from and where did it all go when it was over. Having answered the question of the water's origin, we turn to the other fundamental question:

FQ: Where Did All the Water Go After the Flood?

This question is easily answered when we understand the forces at work. As the comet body passed by the Earth and began to recede back into space, the Resonance Energy (described in the Universe Model, Chapter 25) from the Sun and the CUE (Central Universal Energy) began restoring the Earth to its original rotational speed.

The Earth's Moon was extremely influential in the axial spin stabilization period the Earth experienced after the passing of the UF comet. The resonant orbit of the Moon around the Earth would have acted like a flywheel, storing inertial energy during the time of the disruption, allowing it to aid in the stabilization of the Earth's rotation afterward.

The restored axial spin increased the centrifugal forces acting on the Earth's thin crustal plates. As the crust lifted, the floodwaters on the surface began to subside. The restoration of the crust to its antediluvian elevation was not a gentle event, and monumental heaving and thrusting occurred. The crust was cracked and buckled, stretched and crushed, as landmasses rose and crashed together. One of the major geological events was the recession of flood waters through open cavities in the rising crust. The geological evidence of these events is awaiting further discovery and can be easily seen, if the Universal Flood Mechanism is the paradigm through which one views the landscape.

8. Two Forces Return to Equilibrium

As the Earth returned to its original rotational rate, the opposing centripetal (the inward pull of gravity) and centrifugal (the outward pull from spin) forces returned to equilibrium. This brought order back to a chaotic planet that now looked very different on its surface. Some continental areas had high, mountainous elevations, completely changed by high velocity floodwaters, other areas escaped with relatively little change, except for the flood sediment left behind, which varied from only a few inches to several hundred, even thousands of feet. In some locales, the drainage valleys exhibit sediments that are miles deep, sediments that defy conventional uniformitarian time periods as outlined in the Rock Cycle Pseudotheory Chapter.

The wide-ranging effects of the Earth's crust's return to equilibrium included the colossal breakup of the Pangaea super-continent. There is no doubt among scientists that present day continents were once conjoined; the empirical evidence proves this fact, but the true mechanism that separated the Earth's continents has eluded modern scientists. A stylized image of the pre-flood Pangaea is shown in Fig 8.3.9. The collapse of the crust during the Flood and especially its resurgence as equilibrium returned broke the crust into giant pieces. These pieces, or plates as they are known today, began drifting apart quite rapidly during the few tens or hundreds of years following the Flood, eventually arriving at their current location. Today, they move only a few centimeters per a year.

The drifting and breakup of Pangaea during the Flood is the explanation for the Earth Crust Mystery of the Rock Cycle Pseudotheory. In Chapter 6.12, we learned that the Earth's oceanic crust is six times thinner than the continental crust, and that it is made primarily of basalt, whereas the continental crust is predominantly granite. Modern geology has no method to account for the differing thicknesses or composition of the two crust types. However, the Universal Flood Model has simple answers for these seeming anomalies.

The primary reason the two crusts are so different in composition is that they were formed at markedly different times and in very different environments. All minerals are entirely dependent upon the hyprethermal environment in which they were formed, as they crystallized into the different types of minerals and rocks we find today. The 36km-thick continental crust was formed during the Earth's original formative period, before organisms were present, which allowed the predominantly granitic continental crust to grow without the influence of biogenic minerals. On the other hand, the oceanic crust is thin because it formed more quickly. The heat necessary for oceanic basalt growth was produced by friction between slowly moving masses of underwater rock. The heat was coupled with pressure, making a hypretherm which lasted until the Earth's axial velocity increased during its return to equilibrium. The oceanic rocks between the continents are dark basalt for the

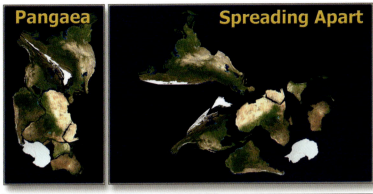

Fig 8.3.9 – Geology has never had a correct mechanism to explain the breakup of the Pangaea supercontinent into the drifting continents of today (No magma—no mechanism for breakup). On the other hand, the UF Mechanism easily accounts for the breakup of Pangaea, the mechanisms for the creation of the basaltic oceanic crust, and a host of other anomalous phenomena.

trifugal force of the spinning Earth. Furthermore, the understanding that the formation of oceanic basalt came from frictionally melted rock deep underwater, high pressure, natural mineralizers, and a huge concentration of floodwater microbes combined with silica and other minerals to form the varied forms of basalt.

Geological Events Supporting UF Mechanisms

There are three areas that help further explain the 4th, 5th, and 6th steps of the UF Mechanism:

1. The Juan de Fuca Ridge Events
2. The Grand Canyon's Earthquake Origin
3. The Waimea Canyon Evidence

These landforms contain geological evidence supporting the collapsing and rising of the Earth's crust over a short period of time, by earthquakes the size of which have never before been seen. Additionally, there is empirical evidence that ocean water can be heated by earthquakes, supporting hyprethermal conditions that existed during the Flood.

The Juan de Fuca Ridge Events

The first supporting evidence of the UF Mechanisms comes from the Juan de Fuca Ridge, an underwater mountain range off the northwest coast of the United States. This landform was first introduced in the Magma Pseudotheory chapter, and can be seen in Fig 5.3.12.

same reason black smokers emit black sediment, forming dark minerals at the bottom of the ocean today; they are rich in iron, manganese and other mafic minerals, which come from the *microbes* that live in the hypretherm environment.

Researchers have recently discovered evidence that the dark minerals found on rocks are indeed, biologically formed. More details of this discovery and how it connects to other mineral formations will be found later on, in the Pyrite and Ore Mark subchapters (subchapters 8.12 & 13).

With the Universal Flood Model we can finally explain how the continents drifted around the world, due in part to the cen-

A minor earthquake swarm struck the ridge on March 1, 1997, causing a significant increase in fluid vent temperatures at depths of around 2,500 meters. Later, on June 8, 1999 as temperature observations of the thermal vent systems continued, a small 4.5 earthquake occurred. What made this event important was that the earthquake was attributed to tectonic activity that was known to be unrelated to a magmatic event:

"Hydrothermal vents on mid-ocean ridges of the northeast Pacific Ocean are known to respond to seismic disturbances, with **observed changes in vent temperature**. But these disturbances resulted from submarine volcanic activity; **until now,**

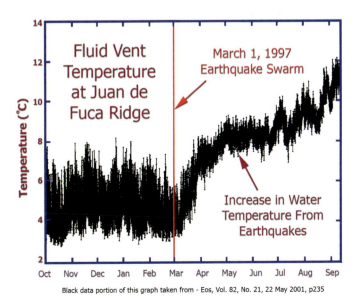

Black data portion of this graph taken from - Eos, Vol. 82, No. 21, 22 May 2001, p235

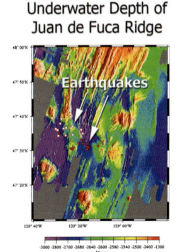

Fig 8.3.10 - The graph at left represents the earthquake-induced increase in fluid vent temperature of the Juan de Fuca Ridge off the NW USA coast. The location of the earthquakes is identified by red dots at right. This was the first time researchers recognized that an earthquake swarm was responsible for the increase in ocean water temperature instead of a magma source. This is direct evidence of the UF Mechanisms we have been discussing, where large earthquakes are able to increase water temperatures and create hyprethermal conditions. During the UF, these happened on both oceanic and continental crust surfaces, because of the enormous volume of water.

SUBCHAPTER 8.3 THE UNIVERSAL FLOOD MECHANISMS

there have been no observations of the response of a vent system to non-magmatic, tectonic events." Note 8.3d

The researchers were contemplating the seismic events from within their magma-planet paradigm. They, like most geologists, probably saw that magmatic events drove all earthquake activity, unaware of the effects of gravitational frictional heating. However, the physical evidence their research showed convinced them that *non-magmatic earthquakes exist*:

"...earthquakes generated by **non-magmatic, tectonic movements** along normal or strike-slip faults are also **frequent** in the northeast Pacific..." Note 8.3d p174

This is a significant mindset change—"frequent" earthquakes caused by "tectonic movements" along fault lines. Moreover, in the May 22, 2001 edition of *Eos*, a geophysical journal, the same researchers published an article, *Earthquakes' Impact on Hydrothermal Systems May Be Far-reaching*. Their "far-reaching" observation was still a remarkable understatement after realizing *the extent of water's influence on the formation of crustal minerals*. The investigators did recognize an "association between earthquakes and hydrothermal circulation":

"Examination of earthquake activity and time-series vent fluid temperature from historical records on the Juan de Fuca Ridge **have added new weight to the proposed association between earthquakes and hydrothermal circulation**." Note 8.3e

Such deep ocean observations have begun to change the way geologists look at the ocean and its connection to mineral creation. Instead of forming out of imaginary magma deep inside the Earth billions of years ago, minerals are forming right now, at these and other hyprethermal vents, in "chemical processes associated with *crustal formation*":

"The proposed **earthquake-induced thermal flux changes appear to be substantial**, and if they occur over a considerable portion of a mid-ocean ridge axial valley, they would **profoundly impact the physical and chemical processes associated with crustal formation**." Note 8.3e p236

Having direct evidence that heated water comes from earthquake-induced thermal vents establishes this necessary component of crustal mineral formation. In order to account for whole areas of new crust formation in a hypretherm, we must identify evidence of a lot of *big* earthquakes.

The Juan de Fuca Ridge is one example where researchers identified evidence of earthquake-induced "hydrothermal circulation," which under deep water facilitates hypretherm conditions. What about the dry surface of the continents—could minerals found on the desert floor today have come from earthquake-heated deep water? The answer is actually quite simple. Modern geology already ac-

Fig 8.3.11 – The Seismicity of Arizona is revealed as red dots mark earthquake locations. Notice that the majority of them are located near the Grand Canyon. USGS source.

knowledges that all of the deserts were once covered by seas; there are just conflicts as to the timing, depth and chemical composition of those seas. In this chapter, dozens of geological evidences (not just one or two) unequivocally demonstrate that the surfaces of the present-day continents contain minerals that formed only a few thousand years ago in frictionally (from earthquakes) heated ocean water.

Grand Canyon's Earthquake Origin

The Grand Canyon evidence is one of the finest examples of the Universal Flood after-marks. Its physical characteristics and the geology surrounding the canyon confirm that a catastrophic event occurred several thousand years ago on an unprecedented scale. Using the Flood Mechanisms previously discussed, we will explore how the Colorado Plateau and the Grand Canyon could have been formed, and then investigate the geological evidence behind that claim.

As the Earth's rotation slowed, the thin crust began to collapse, cracking and buckling the crust in weak or previously faulted areas. In some areas, friction from continued crustal plate movement heated deep ocean water, creating a hypretherm. Abundant biological sediment (limestone) and new clay or

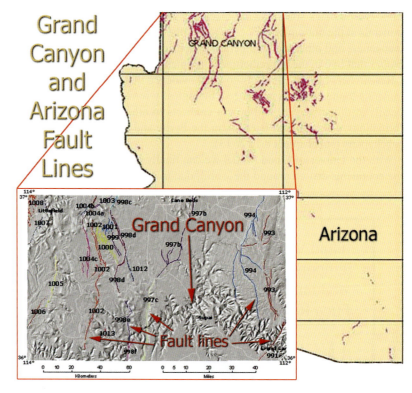

Fig 8.3.12 – These maps from the USGS show Arizona's fault lines; however, Arizona's largest fault of all has been overlooked. It is the Grand Canyon itself! The direct evidence of this massive fault is the north and south canyon rims, which are respectively 8,000 and 6,800 feet. The matching vertical patterns in the layers revealed in both sides of the canyon prove they were once connected. However, the South canyon rim is now lower than the North rim by 1,200 feet, and the only thing that could have done that was a massive Grand Canyon Earthquake. Diagram adapted from http://earthquakes.usgs.gov/qfaults/az/index.html.

sand deposits were formed. Some of these materials were cemented together in the hypretherm, forming the multiple sandstone, limestone and claystone layers of the Colorado Plateau in Northern America.

Because the layers of the Colorado Plateau were formed during an extremely short period, about a year's time, we should *not expect* to find certain Missing Layers in the Grand Canyon series discussed in Chapter 6.12.3. There are no 'layers' of oil, soil, coal, lava flows, coral reefs or salt deposits, all of which *would have been present* among the Canyon's multiple layers if the entire plateau had been raised and lowered five times, as modern geology holds. The fact that these layers are missing proves that millions of years were not involved in the making of thousands of feet of sediment on the Colorado Plateau. The massive sandstone layers are homogeneous sand crystals of extraordinarily uniform grain size. This establishes that the layers were not formed by traditional aeolian (wind) or fluvial (river) erosion processes; they too, were formed in the Flood's hypretherm.

After the UF deposited the sedimentary materials on the plateau, the observable geological evidence demonstrates that as the Earth's rotational velocity began to return to its antediluvian rate, giant cracks, like the Grand Canyon in Arizona showed up in the newly formed Flood sediments. These cracks became the deep canyons we see today. Contrary to modern geology, it did not take millions of years to form such a deep canyon. Like most of the world's canyons, the Grand Canyon was formed rather suddenly. All it took was a *big earthquake*.

Our next step will be to provide scientific evidence of such an event; the mega-earthquake that created the Grand Canyon. By carefully observing geological formations in the canyon we can see, from the following evidences, how the Grand Canyon was formed. Each piece of evidence provides an important part of the picture that establishes the Grand Canyon's Earthquake Origin.

1. The Grand Canyon is Arizona's most active earthquake area.
2. Dozens of faults lie within and just outside the Canyon.
3. There is a massive 1,200-foot elevation difference between the South Rim and the North Rim of the Canyon.
4. There were two primary earthquake events, first a lowering event and then a raising event.
5. The Cardenas Lava lies at the base of the Grand Canyon.
6. The Canyon was not formed by erosion.

1. The Grand Canyon is the most seismically active area in Arizona, which can be seen in Figure 8.3.11. The largest earthquakes in Arizona between 1990 and 2001 are represented by the red circles in this diagram. There is clear evidence of their clustering near the Grand Canyon. Since the USGS began monitoring earthquakes in Arizona, the Grand Canyon area consistently ranks highest for both magnitude and frequency of quakes. The earthquakes occurring today fall on faults that would have also experienced slippage in the past.

2. There are dozens of faults within and just outside of the Grand Canyon area. Some of the faults like the Butte Fault actually run along the Colorado River:

"The Butte Fault parallels the Colorado River…" Note 8.3f

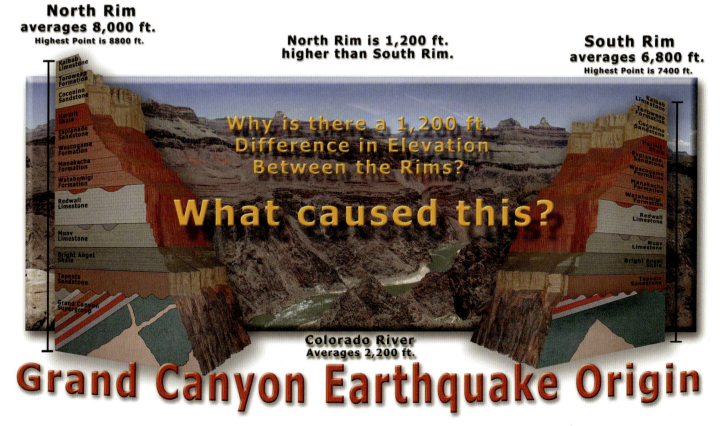

8.3.13 – The Grand Canyon Earthquake Origin diagram illustrates the 1200-foot difference in the two rims of the world's most famous canyon. Geologists have apparently overlooked the physical evidence of the Grand Canyon's origin. They have never seen an earthquake shift anywhere near the size of the shift produced by the Grand Canyon Earthquake. Grand Canyon data - http://www.nps.gov/grca/parkmgmt/statistics.htm.

SUBCHAPTER 8.3 THE UNIVERSAL FLOOD MECHANISMS

Although the Butte fault parallels the Canyon's river, most of the faults run *perpendicular* to the river, which can be seen on the USGS map in Fig 8.3.12. The fact that the Colorado River runs *perpendicular* to most of the faults in the area, is evidence that the Canyon was *not* made over a long period of time. Cracks in the surface form natural channels for water to follow, which would have affected the formation of side canyons, had the Canyon really been formed by erosion. The fact that these perpendicular faults are not present-day canyons is evident on the USGS Quaternary Fault and Fold Database for Arizona in Fig 8.3.12.

Next, we will examine why only massive movement of the crust in this area, on a scale greater than anything ever witnessed by mankind, could cause such a deformation in the Earth's crust.

3. Few people are aware of the substantial 1,200-foot difference in elevation between the North and South Rims of the Grand Canyon. The North Rim's average elevation above sea level is 8,000 feet, whereas the South Rim averages only 6,800 feet, a remarkable difference. It is indisputable that this difference is due to a colossal *earthquake*, even among scientists who still think the Canyon was formed by erosion. There is no controversy among geologists that the two sides of the canyon were once at the same elevation because the exact layers are evident on both sides of the Canyon, fitting together like pieces of a puzzle.

FQ: What is the only known mechanism for large areas of the Earth's crust to drop quickly?

The answer is of course—earthquakes! To put the Grand Canyon's 1,200 foot (366 meter) drop in perspective, compare it to the movement that occurred during one of the largest earthquakes in modern recorded history; the December 26, 2004 Sumatra Earthquake and Tsunami in the Indian ocean that claimed more than 225,000 lives. Researchers observed:

"Yet later analysis revealed that the magnitude 9 shock raised a 1,200-kilometer stretch of seafloor by **as much as eight meters** in some places..." Note 8.3g

The **26-foot (8 meter)** drop in the Indian Ocean floor generated a 9.3 earthquake. The earthquake from a 1,200 foot (366 meter) earth movement at the Grand Canyon is unfathomable, clearly beyond anything man has ever seen. Imagine a sudden jolt, or a series of sequential jolts that dropped the crust *46 times* greater than the largest earthquake ever measured.

This may be difficult to accept for some scholars, but like the Scabland Megaflood evidence, nothing like it had ever been seen before, but the proof is there and cannot be ignored.

The Grand Canyon Earthquake produced a massive crack, shaping the present-day canyon.

As we look further, there are additional facts that attest to the earthquake origin of the Grand Canyon. Notice that the walls of the canyon are sharply defined, not smoothed by millions of years of erosion. Fig 8.3.14 is an image of the Little Colorado River Canyon, a tributary canyon of the Grand Canyon. It is immediately obvious that the sides of the Little Colorado Canyon are sharp and angular, with very little erosion evident on the surrounding plains leading *to the canyon!* This phenomenon has raised questions as to how such flat plains surrounding the Grand Canyon and canyons like them around the world were formed, with so little erosion on the plains themselves. This is an example of the Planation Mystery identified in the Rock Cycle Pseudotheory Chapter, seen in Fig's 6.11.9, through 6.11.12. Although a mystery to modern science, a worldwide flood, laden with sediment that settled to the bottom of the entire planet's flood basins, accounts for what we see today.

The Earthquake that formed the Grand Canyon must have occurred toward the end of the UF or sometime thereafter because there is so *little* fluvial erosion along the canyon's walls and upper rim. This is further evident when comparing the crisp walls of the Grand Canyon to an area northeast of the Grand Canyon in the area of the current man-made Lake Powell reservoir in Arizona and Utah. Fig 8.3.15 contrasts the *smooth, flood-worn lower* canyon with the crisp, angular upper canyon walls. The lower areas were scoured smooth during late stage UF floodwater recession. Smoothed walls like these are found where massive quantities of rock and boulders were carried by high velocity megaflood waters across great distances. More examples of megafloods and the scouring action that accompanies them are in the upcoming Erosion Mark subchapter.

An example of floodwater eroded sandstone is the Antelope Canyon in Arizona, world famous for its beautiful slot canyons, seen in Fig 8.3.16. This remarkable canyon was shaped

Fig 8.3.14 – From the air, this section of the Grand Canyon, the Little Colorado River Canyon looks like a crack in the Earth's crust—because it is. Notice the two small hills in the background; these are volcanic, formed by earthquakes. The edges of this canyon are sharp, indicative of only a few thousand years of erosion, not millions of years as taught in geological textbooks.

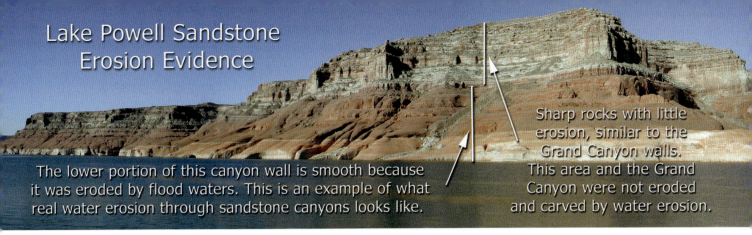

Lake Powell Sandstone Erosion Evidence

The lower portion of this canyon wall is smooth because it was eroded by flood waters. This is an example of what real water erosion through sandstone canyons looks like.

Sharp rocks with little erosion, similar to the Grand Canyon walls. This area and the Grand Canyon were not eroded and carved by water erosion.

Fig 8.3.15 – The mountains surrounding Lake Powell on the Arizona/Utah border show a clear distinction between layers exposed to heavy erosion and layers showing almost no erosion. The smoothed and well worn lower areas are strong evidence of a megaflood in this area. It would have been similar to the Channeled Scabland Megaflood in Washington and the English Channel Flood, yet there is no mention of such an event in the geological literature. The modern man-made Lake Powell's water level does not cover the smoothed sandstone rocks. Because so little erosion has occurred since the walls were smoothed, the megaflood event that shaped them must have been within the last several thousand years. There are many areas that show similar flood-worn rocks lying in areas where little water exists today. When the significance of the UF event is realized, the origin of places like this will finally be known.

by high velocity floodwaters laded with large rocks and boulders, smoothed by the abrasive action of the water's movement. No such canyons exist anywhere in the Grand Canyon system, because the Canyon was formed quickly, the result of an epic earthquake.

4. The Grand Canyon Earthquake Origin theory presented here includes the extraordinary claim that the Earth's crustal plates *dropped* when the spin of the Earth was slowed. Floodwaters and sediment from inside the Earth covered the crust, and finally, as rotational speed increased, the plates *moved back up*, causing additional monumental earthquakes. As we explored the evidence of this extraordinary claim, we were surprised to find investigators describing a similar *dropping and rising of the Grand Canyon layers:*

"Normal and reverse movement along Butte Fault is thought to **have first dropped the Supergroup by about 5000 feet** on the fault's west side prior to its burial by the Paleozoic column. **This fault was later reactivated with the block on the west side moving back up about 2700 feet**, warping the overlying Paleozoic column and creating the East Kaibab monocline in the process." Note 8.3h

Such earth movements account for the dropping and rising of the Canyon's rims; they also account for the difference between the initial 5,000 foot drop and the subsequent 2,700 foot rise.

Another significant fact that cannot be overlooked is the elevation difference between the South Rim and the North Rim. It isn't just the South Rim that is 1,200 feet lower than the North Rim; the entire southern plateau stretching across the northwestern part of the state of Arizona is lower by 1,200 feet. Never has mankind, through the eyes of modern science, witnessed such an event.

The Catastrophic Principle is confirmed by this geological evidence that *scientists already acknowledge*, but have yet to recognized the significance.

5. Additional strong evidence that the Grand Canyon sediment layers were deposited during the UF event and later broken by the Grand Canyon Earthquake lies in the bottom of the Canyon. The colossal break that caused the South Rim and the whole southern plateau to drop also produced friction along the lines of faulting. A rapid displacement would be evident by the presence of melted rock along the fault lines at the base of the Grand Canyon Series between the bottommost sediment layer and the original continental rock, where the slippage took place.

Although there is almost no lava found between the other layers of the Canyon Series, (as seen in Fig 6.12.3) the Cardenas Lava is a melted layer that exists near the bottom of the Grand Canyon. This slippage area, called the Cardenas Lava flow, is found at the base of the Canyon among the layers identified as the Supergroup. It is exposed in the eastern Grand Canyon and ranges between about 800 and 1,000 feet thick. Further evidence that this Lava flow is related to the UF event comes from a researcher well acquainted with the Canyon:

"Lower and upper members can be identified in the **Cardenas** based on their topographic expression... The chemistry, nodular character and glassy texture of the lower member suggest the basalt may have been **quenched in brackish or marine water as it extruded**." Note 8.3h

The evidence that marine water is associated with the Cardenas lava flow and the actual presence of the lava flow are both strong indicators of the UF event, but there is still more. Lava flowing on the surface of the dry continents cools before sediment of any consequence intrudes the flow. During the UF, water and sand flowed almost continuously, especially through the newly formed Grand Canyon, a natural drainage channel to the ocean near the end of the UF. With this in mind, it would not be surprising to find flood sediments comingled with the Cardenas lavas at the bottom of the Grand Canyon, a characteristic rarely found in most basalts:

"Local features and mixing at the contact suggest that **the basalt overflowed wet**, unconsolidated Dox **sand and sediments**. The Cardenas is characterized by **recurring horizons and discontinuous lenses of interbedded sandstone**..." Note 8.3h

Lenses of sandstone are horizontally concentrated deposits, thick in the middle and thinning at the ends. For such interbedding to occur in the melted rock, significant sand-laden water must have been flowing across the melt-rock as it was extruded. Although there is no *present day* analogue of the Cardenas Lava Flow where 'lenses' of sandstone inclusions are encapsulated in observed lava flows, this phenomenon is expected during the Universal Flood.

6. Theories on the origin of the Grand Canyon have varied

widely. Once, the Canyon was believed to have been carved by the simple erosional forces of a small river over a very long time. As scientific data filtered in, the age of the Canyon was dropped from being a hundred million years old to just 6 million years, a figure that has stuck the longest, although two newer theories have proposed ages of 17 million years, or 55 million years. Radiometric dating has never been conclusive, and so there is no consensus as to the Canyon's age.

One glaring shortfall of all erosion theories is the missing, enormous delta at the end of the Colorado River; it *does not exist*. Where did the washed away sand and eroded material go? We found no explanation for the disappearance of the sediment, which would have been one of the largest sediment transport events on the planet. Eroded red sandstone does not exist along the river or in the Sea of Cortez, except in incidental quantities related to current river flow activity.

The challenges related to the missing sediment and the time variables spawned another theory of Canyon erosion. The theory, proposed by geologists, included a breached-dam element.

The breached dam theory essentially proposes that a large body of water held northeast of the Grand Canyon by a natural dam was breached, allowing for a more rapid canyon-carving event. Such a theory eliminates the long-time sci-bi of the Canyon's origin by accepting the Catastrophic Principle. However, while the breached dam theory includes a large amount of high velocity water flowing through the canyon for a short time (which could also have happened at the end of the UF event), it ignores the earthquake evidence and does not account for the deep, side-canyon cracks and fissures, north and south of the main canyon (Fig 8.3.14 is an example). Some of these tributary canyons are many miles long and *would not have been eroded by the water from a breached dam that flowed through the main canyon*.

The simple fact is that these side-canyon cracks could not have been formed by a breached dam, just like the breached dam did not form the secondary flow channels of the English Channel. The presence of the deep, side-canyons confirms that a single dam break could not have created the Canyon by itself. Furthermore, the dam-break theory offers no explanation for the 1,200-foot elevation differences of the Canyon's rims!

In summary, the six evidences herein show that an earthquake of mammoth proportions formed the Grand Canyon after the UF, not during millions of years of river erosion. Only a massive earthquake, coupled with voluminous receding floodwaters, which cleared the bottom of the Canyon, can account for all of the Canyon geology as it exists today. With new UM insights, there will surely be many new books written about this incredible Canyon.

The Waimea Canyon Evidence

There are many canyons that hold clues to the UF event, and like the Grand Canyon, there are other large canyons that were formed primarily by earthquakes, not erosion. Another outstanding example is found on the island of Kauai, Hawaii; a canyon that has come to be known as the Grand Canyon of the Pacific—**Waimea Canyon** (Fig 8.3.17). In contrast to the multi-colored, dry desert Grand Canyon in Arizona, Waimea Canyon is graced with a lush and beautiful forest.

It doesn't take much to see what these two Grand Canyons have in common besides their name. A popular encyclopedia's description of Waimea Canyon includes the typical sci-bi of nearly all canyon formation—fluvial (river) erosion:

"Waimea Canyon, also known as the Grand Canyon of the Pacific, is a large canyon, approximately ten miles (16 km) long and up to 3,000 feet (900 m) deep, located on the western side of Kaua'i in the Hawaiian Islands of the United States. **The canyon was formed by a deep incision of the Waimea River** arising from the extreme rainfall on the island's central peak, Mount Wai'ale'ale, among the wettest places on earth." Note 8.3i

However, as we read along a little further, the description of Waimea Canyon unveils a clue to its real origin—"an enormous fault along which a

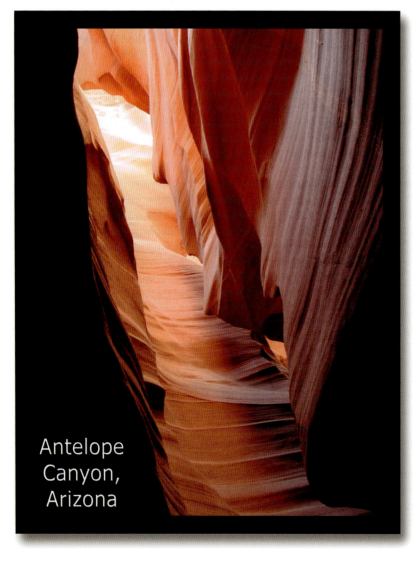

Antelope Canyon, Arizona

Fig 8.3.16 – One of the most beautiful slot canyons found on Earth is Antelope Canyon in Arizona. This canyon was not carved by geological time, but by flooding. As we learn throughout this chapter, floods are as much a creative process as they are a destructive process. This slot canyon's smooth walls represent how sandstone canyon walls really look after being eroded by fast-moving floodwater. The rock debris was carried quickly away, leaving no trace of it.

large part of the island moved downwards in a big collapse."

"Geologically the canyon is carved into the tholeiitic and post-shield calc-alkaline lavas of the canyon basalt. The lavas of the canyon provide evidence for **massive faulting and collapse in the early history of the island**. The west side of the canyon is all thin, west-dipping lavas of the Napali Member, while the east side is very thick, flat-lying lavas of the Olokele and Makaweli Members. **The two sides are separated by an enormous fault along which a large part of the island moved downwards in a big collapse.**" Note 8.3i

Scientific observation of the faults and the obvious earthquake activity are pieces of the puzzle previously unknown. With the new paradigm of the UM, we are able to see how the pieces of Nature's Puzzle can only fit one way, and when properly placed, compose a beautiful picture that is understandable and easy to comprehend.

Both canyons have massive faults running through them and both show evidence of massive earthquake activity. Neither canyon was formed by erosion, they were formed by *earthquakes*. And they were not small earthquakes either! They were of such size as to defy imagination, far beyond anything mankind has ever seen.

UF Environmental Conditions

The eight Universal Flood Mechanism steps and some examples of the physical evidence supporting them have been presented. Of course, it is impossible that one chapter can do justice to the complexity and wide-ranging details of the UF Mechanisms. Our purpose here is to outline the form of a model wherein some of the indisputable marks of the UF can be presented and summarized. The Universal Flood Model will also provide documentation for five unique and important environmental changes the entire Earth experienced during the UF event. These changes had never before occurred nor have they since the Universal Flood. The five unique environmental conditions were:

1. **Global hyprethermal conditions on the Earth's surface**
2. **Massive worldwide bacterial and algal blooms**
3. **Unprecedented global landform creation, including sand and rock formations, and mountain building**
4. **Massive, global ore and fossil fuel creation and deposition**
5. **Worldwide, short-time-period erosional events**

All five of these geological environmental events will be described in detail throughout the chapter, in the Marks of the Flood. The unique environmental conditions of the UF can be understood because of the Hydroplanet Model, which shows us how it is possible to have *the amount of water necessary* for a Universal Flood.

The Magma Pseudotheory Chapter revealed the mechanism necessary to account for the thermal requirements of the Flood through the Frictional-Heat Law. All the components of a global hypretherm were present during the Flood; temperatures of roughly 300-400°C, pressure of approximately 13,000 psi was achieved from floodwaters that were approximately 30,000 feet (8,850 m) deep, massive bacterial and algal blooms, both basic and acidic waters, and mineralizers all contributed to the UF hypretherm that made mineral and rock growth possible. This along with new biogenic sand and rock formations, and mountain building events occurred globally during the Flood.

We begin the journey with the first UF Mark, the Hydrofountain Mark. Although the concept of the hydrofountain has already been introduced, the enormous scope and magnitude of the fountains during the UF cannot be fully realized without the following empirical evidence. The amazing thing about this mark is that the hydrofountains have been overlooked by nearly all of geology for so long.

8.4 The Hydrofountain Mark

For the Universal Flood to have taken place, the bowels of the Earth were opened, releasing enough water to cover all the mountains. The Earth's surface cracked and buckled from the gravitational pull of the passing celestial visitor and the crustal plates collapsed from the decreased spin rate. Enormous amounts of water were ejected onto the continents and below the oceans; water came up wherever the least resistance was found. Water heated by gravitational friction flowed explosively through newly formed cracks bearing all types of sediment. These were hydrofountains.

Hydrofountains were introduced in the last chapter as being an important part of our planetary system's history. They have been largely overlooked, especially right here on Earth. There were several varieties of hydrofountains that changed the surface of the Earth during the UF, and many of them will be discussed in this subchapter. We will also discuss the Spouting Processes of fountains, hydrorock fountains, and the Skipperock Origin along with the processes by which rock pillars are formed. Mudfountains and their worldwide significance will be examined, including examples that have not been recognized as such before.

Fig 8.3.17 – Waimea Canyon on the island of Kauai, Hawaii, U.S.A., is known as the Grand Canyon of the Pacific. The canyon's acknowledged "enormous fault" and a "big collapse" of part of the island is anecdotal evidence further confirming the earthquake origin of this canyon and Arizona's Grand Canyon.

Hydrofountain History

The Hydrofountain Mark is a continuation of the hydrofountain concept first introduced in the Hydroplanet Model subchapter 7.3 and later defined in 7.7. In the Hydroplanet Model subchapter, we learned that when liquid water is converted to steam, it can expand 1,700 times its original volume. Liquid water in the crust, heated by gravitational friction may experience such an expansion, causing it to escape through the crust in pipes and diatremes. As we will see momentarily, hydrofountains under deep water can build solid rock forms, even mountains as the temperature cools or as pressure is reduced. Sometimes the superheated steam explodes through the crust leaving only a hydrocrater behind, but usually, the explosion is accompanied by rocks, sediment, and minerals, all flowing out and onto the surface. Today, as land-based volcanoes, such as Kilauea in Hawaii, erupt they form igneothermic minerals which are minerals changed to glass from heat with minimal water. However, during the UF event the vast majority of mineralization occurred underwater, forming hyprethermal and hyprethermic minerals. These are the rock types most commonly found on the continental surface today.

Our first discussion in this subchapter will be the geyser hydrofountains that eject primarily hot water and steam. Then we'll take a look at fossil hydrofountains; the preserved pipes and dikes of ancient hydrofountains that can still be observed today. They are comparatively rare. The most abundant forms of hydrofountains are the mud and sand hydrofountains that left their mark in the form of fine sediment. This sediment, contrary to conventional geological theory, did not come from the surficial erosion of mountains. Following that, the geological evidence of rock hydrofountains and hydro-mountains will be introduced.

The Hydrofountain Mark includes physical evidence that massive hydrofountains once erupted worldwide, and the 'Marks' they left behind provide evidence of *when* they erupted. For the first time in history, knowledge about the origin of many of Earth's sedimentary materials will be revealed, prior to which modern geology could only list as 'anomalies.'

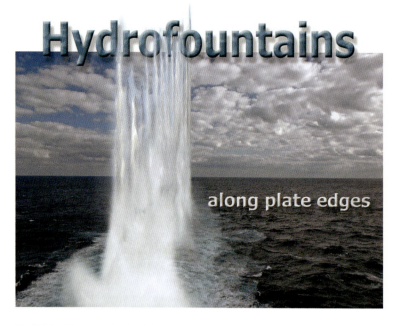

Fig 8.4.1 – The magnitude of the Universal Flood Hydrofountains that occurred on Earth can be hard to visualize, even after seeing similar hydrofountains in other areas of the solar system (Enceladus). This conceptualized diagram illustrates what fountains erupting along crustal plate edges might have looked like. Pressurized subterranean water created a sight never before seen as it was expelled through colossal fountains thousands of years ago. Today, the physical evidence of these events exists throughout the geologic record, allowing us to observe how these fountains shaped the landscape. This subchapter has been written to identify the Mark of the UF Hydrofountain.

Geyser Hydrofountains

The images in Fig 8.4.2 are representative of the geyser hydrofountains at Yellowstone Park, USA; they expel hot water and steam, but very little sediment. Geologists believe that geyser water is heated by a great ocean of magma that resides not too far below the surface. However, since magma is an unproven theory, a pseudotheory, we should expect to find an alternative source for the heat that drives the geysers. And we should find such a heat source at all of the world's geyser sites. This is exactly what we do find—earthquakes. Some are quite small and may often occur in swarms, but in the geyser basins of Yellowstone, earthquake activity is continuously and carefully monitored because of the link to geo-

The Geyserite Evidence

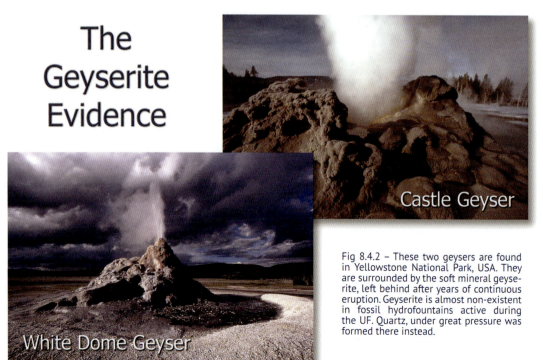

Fig 8.4.2 – These two geysers are found in Yellowstone National Park, USA. They are surrounded by the soft mineral geyserite, left behind after years of continuous eruption. Geyserite is almost non-existent in fossil hydrofountains active during the UF. Quartz, under great pressure was formed there instead.

Hydrofountain Formations

Fig 8.4.3 – Some hydrofountain formations occurred on dry land spewing water and sediment onto the surface, forming a wide variety of landforms still visible today. Large craters are the result of a massive eruption; vast quantities of underground sediment, previously crystallized in hyprethermal conditions were ejected. One type of present-day hydrofountain—geysers—are common, but are very small in comparison to the large eruptions in the past.

thermal activity and because geologists think an event far larger than the 1980 eruption of Mount Saint Helens is imminent.

An interesting characteristic of geysers is the mineral geyserite. Geyserite is a soft, opaline silica mineral that combines with microorganisms in the hot pools to form the beautiful rainbow of colors seen near many geysers.

What do present day geysers and the minerals formed by them have to do with the Universal Flood? Decades, even centuries of geyser activity have left geyserite deposited around present day geysers; relatively small mounds of the mineral can be seen at Yellowstone. Elsewhere in the world, there is evidence of pipes and fossil hydrofountains far more massive than is evident in the world's geyser basins. The fossil hydrofountains seen in Fig 8.4.4 are pillars of sediment that are obviously much older than the geysers in Yellowstone, they are *not* made of geyserite, and they have no water coming from them today. Instead, these fossil hydrofountains are composed of quartzose clay, sand or larger sediments, lithified with a quartziferous cement. This means they were formed in a *hypretherm*—not from unpressurized hydrothermal waters like geyserite is formed in.

Fossil hydrofountains like those in Fig 8.4.4 were not formed like the smallish geysers we observe today. They were formed ⌐er eruptions were being driven by massive ⌐rth's crust as it buckled and folded during the ⌐ why these old hydrofountains do not contain ⌐y there are generally no remnants of ancient ⌐ite in the rock record.

⌐ing the hydrothermal mineral *geyserite* in the ⌐tains at Kodachrome Basin, we find *quartz*, a ⌐neral *requiring high pressure* and we find the ⌐between the sand particles is also of hyprether- ⌐s the remarkable difference between geyserite, ⌐rmed on the surface today and the quartz hy- ⌐o formed *on the same surface*, but during the ⌐ressure. Pressure of such magnitude that came ⌐ five miles of water.

The Kodachrome Fossil Hydrofountain Evidence

As sediment laden water travels upward in the hydrofountain pipe, the column itself can become a hyprethermal environment, complete with high pressure and heat, allowing the sediment in the pipe to become solidified into rock. If the elevated temperature and pressure within the pipe did not extend to the area outside the pipe, the column of solidified sediment would become harder than the surrounding sediment. As the UF floodwaters began to recede, tremendous water volume carried unconsolidated sediment to lower elevations, exposing the hard pipes of sediment, leaving them standing as testaments to the nature of the Universal Flood.

Fig 8.4.4 – The pillars in Kodachrome Basin, Utah, USA are excellent examples of how sandstone across the entire Colorado Plateau was cemented. Instead of just geological time (which never cemented anything), these pillars were conduits that brought frictionally heated water from below while the Earth was being covered with Flood waters. Notice the pillars are generally formed of a sandstone whiter than the surrounding rock. This is because the material that formed them was from a different location. See also Fig 8.4.5 for a diagram of how these rock pillars formed.

SUBCHAPTER 8.4 THE HYDROFOUNTAIN MARK

Fig 8.4.4 includes several examples of the fossil hydrofountains at Kodachrome Basin, Utah. These well preserved hydrofountains have been left intact; normal erosion would have tumbled them long ago. The Pedestal Mystery in Fig 6.11.14 documented many such rock pedestals that have no clear origin in modern geology, but are easily explained through the UF hydrofountain processes. Fig 8.4.5 explains how rock pillar formations are created using the science of the Hypretherm, Hydrofountains and Universal Flood.

The "Fluidization" Evidence

Many pipes contain sharp, angular rocks in the column sediment, cemented together with quartziferous sandstone that originated deep in the Earth. These are called "breccia pipes" in current geological literature. The process by which they were made was hotly debated for many decades and is still not completely accepted in geological circles for one reason—they are not being formed today. The term used to explain the geologic process that formed the rock pipes is "**fluidization**," an obvious allusion to the formative process's connection to *water*.

Although the knowledge that hydrothermal water exists in the cracks and crevices of rocks and that water is connected to the formation of many minerals, the Crystallization Process described in chapter 7.4 has not been conceptualized by modern geology. The recognition that pressure, along with heat and water are essential for crystal growth has eluded geologists. An example of this can be seen in the following statement taken from an article titled, *Experimental Evidence for Fluidization Process in Breccia Pipe Formation*:

"The significance of the contribution provided by Reynolds (1954) in focusing on **fluidization as a possible geologic process** is evidenced by the number of workers who have utilized the concept during the past three decades. Several features typically observed in many breccia pipes **defied viable interpretation until the fluidization mechanism was invoked**. Explaining the presence of abundant well-rounded fragments was particulary troublesome, especially those found so near their source that the transport distance would have been inadequate to account for their degree of abrasion. No evidence for thermal spalling has been found in the majority of these systems and some rounded mineral fragments exhibit surface polish, pitting, and striation, properites typical of an abrasional process. Formation of abundant rock flour matrix and through mixing of fragments of unlike compositions would be difficult to achieve through simple explosive, collaspe, or chemical brecciation processes. **Furthermore, no other reasonable mechansim has been advanced to account for the presence, in a single small block of breccia, of fragments derived from well below and well above**. This mixing phenonenon along with differences between central and marginal breccia phases in some **pipes suggests a movement pattern analogous to spouting**. Size sorting also may result from the settling of blocks that are too large to be fluidized and which tend to concentrate along pipe margins and at the bottom of the system, becoming part of the fixed bed through which gas continues to rise. Prolonged fluidization, particularly under high-velocity flow conditions, could lead to reduction in block size by abrasion and eventually these blocks could be incorporated within the fluidized bed."
Note 8.4a

Scientists found "no other reasonable mechanism" besides fluidization (hydrofountain) to account for large blocks of breccia rock being brought through the pipes from lower layers. Another well-made point investigators brought to light was the "spouting" mechanism, which explains the pressure-driven eruption as a reoccurring event; spouting, pressure release, pressure builds and spouting again. Modern day geysers repeat the ***spouting process*** during their cyclical eruptions. This process, which explains how rocks can be brought to the surface from deep within the Earth, ***and*** how rocks and other surface debris such as trees *can be sucked back into the pipe* when the pressure is reduced, will be discussed in detail shortly.

Rock Pipes Found Worldwide

The Kodachrome Basin Fossil Hydro-columns tell an amazing story, but skeptics dismiss them as being nothing more than an 'anomaly,' a 'classical geological mystery.' However, there are far too many to dismiss so casually! The book, *Neglected Geological Anomalies*, includes some of the cylindrical structures or 'pipes' found **around the world**. In this sample, the "pipes" around New York State and Ontario are mentioned:

Rock Pillar Formation

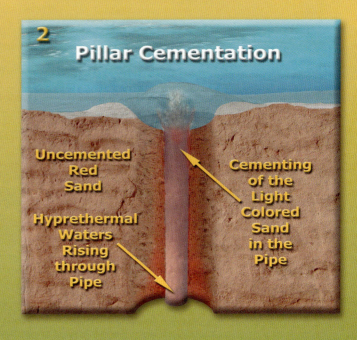

Fig 8.4.5 – This diagram illustrates how sandstone rock pillars were formed in a hypretherm during the UF. Sand was pumped up from below as crusts shifted and sank. This continued as floodwaters inundated the area, piling layer upon layer of sand and sediment. An underwater eruption of white sand from lower layers heated by frictional heating created a hypretherm, in which the column of white sand became hardened from the heat and pressure of the hypretherm that existed in the column. Underwater flood erosion carried away loose sediment layers exposing the hardened column of white sandstone and the semi-hardened underlying red sandstone. The lower layers of sand were hardened as hyprethermal waters from faults allowed the cementation and growth of interstitial quartz crystals. Plate 5 is an actual pillar in Kodachrome Basin, Utah, U.S.A. today.

Without the UF hydrofountain and hypretherm mechanisms, modern geology has been left with only erosion to explain such landforms, despite its inability to account for how such tall, fragile columns withstood millions of years of weathering. Neither can it explain why there are no present-day processes forming any new columns.

Monument Valley Arizona Rock Pipes
"Neglected Geological Anomalies"

Fig 8.4.6 – For the first time, a real scientific mechanism can explain the formation of Rock Pipes—the hypretherm of the Universal Flood. They are connected to the worldwide event that formed them under increased pressure and temperature that does not exist anywhere on the continents today.

"**Cylindrical structures, often called 'pipes', are not uncommon** in sandstone. Geologists have proposed a wide variety of mechanisms to explain them.

"As long ago as 1850, F.B. Hough described cylinders appearing in the Paleozoic Potsdam sandstone, which occurs in northern New York State and adjacent Ontario.

"A frequent and very interesting phenomenon is observed in the Potsdam sandstone of **Jefferson and St. Lawrence counties, N.Y.**, and consists in the occurrence of vertical cylinders, often of great size, and of undetermined length, having a concentric stratification, but not differing in color or texture from the surrounding rock.

"The diameters of these masses vary **from two inches to twenty feet and upward**." Bib 137 p7

The book continues listing fossil pipes in Canada:

"A later, more detailed study of these same cylinders was conducted by J.E. Hawley and R.C. Hart in the 1930's. We present here their summary and one of the their illustrations.

"'**Cylindrical, tree-like structures** in the basal Paleozoic sandstone near **Kingston (Ontario)** are described in detail. From an unexposed horizon these rise vertically through nearly horizontal strata. They are shown to be **composed of sand grains** similar in characteristics and in general size and assortment to those of the enclosing beds, and considered to have formed during, or after, deposition of the sand, but before final cementation. The concentric, cylindrical color banding is believed to be concretionary in nature, localized by the presence of vertical cylinders of uncemented and not excessively disturbed sand. **The formation of these is attributed to the action of currents of water, rising vertically through the strata from, possibly, a buried fault line or other controlling structure, destroying the original bedding throughout the cylinders, forming quicksand therein, and appearing at the surface as submarine springs**.'" Bib 137 p8

Next, it lists fossil pipes in Syria:

"**Syria**. A thick deposit of calcareous sandstone, extending upwards of 300 miles along the Mediterranean coast, is penetrated by many sand-filled cylindrical cavities. 'These holes are always vertical, from 10 to 20 inches in diameter and from 5 to 15 feet deep, spreading out funnel-like at the surface, and tapering like a cigar at the bottom.'

"The author of this paper, E.A. Day, remarked that these holes were sometimes used as tombs. As for the origin of the holes, he doubted that a percolating water mechanism, as profferred for the origin the similar pipes in English chalk could make such smooth, round, regular holes." Bib 137 p8

Then, pipes from Arkansas USA:

"The **sandstone pipes** in **Arkansas** were first noted and described by A.H. Purdue and me, in the Eureka Springs and the Harrison quadrangles. **They range in diameter from 2 ½ inches to 150 feet, stand vertical**, and pass through the full thickness of the Newton and the Kings River sandstone members of the Everton formation." Bib 137 p9

Clearly, these pipes are not limited to a few insignificant corners of the natural world. The geological community does not understand their significance because of the Rock Cycle Pseudotheory paradigm and the Uniformity Myth. This led to a neglect of these types of formations, and there is still today, *no real origin* among modern geologists for these magnificent structures.

One researcher describes having studied pipes from around the world, including some in Arizona, Colorado, Nevada, Montana, Utah, Chile, the Philippines and Australia. He describes and performs some excellent experimentation using pressurized air to show how water and steam could have created the pipes by forcing sediments to the surface, Note 8.4b but in the journal report, he stops short of explaining *how the pipes were solidified*.

So far, our discussion has been limited to hydrofountain evidence on dry land, but the UF fountain mechanism claims that most of these fountains actually occurred while an ocean covered the entire earth. Fountains that consistently extrude water and sediment to the surface might behave like the volcanoes in Hawaii, which slowly build broad expanses of lava. Do such fountains exist in the oceans today?

Mudfountains Are a "World-Wide Phenomenon"

The word "volcano" conjures images of red-hot lava flowing from cone-shaped mountains, but there are many other types of volcanoes, including 'mud' volcanoes. Science has known about mud volcanoes for several hundred years, yet little is understood about them. Back in Fig 7.7.9 and Fig 7.7.10, the well-known Mount Saint Helens volcano was shown to have expelled enormous quantities of mud, more than almost anything else, making it, in fact, a mudfountain. Fig 8.4.7 shows the type of mudflow that took place. These facts demand an entirely new perspective toward 'lava' volcanoes and traditional lava eruptions taught in modern geology textbooks. As will become abundantly clear throughout the next several subchapters, the Earth's continental crust is covered with voluminous sediment spewed from hydrofountains, many of which are **mudfountains**.

Mudfountains are truly a "world-wide phenomenon" and only recently, as noted in this 2003 journal article, have researchers begun to recognize their "significance":

"**Mud volcanism is a world-wide phenomenon, both on the seafloor and on land**. Since its discovery on Java in the early 19th century (Goad 1816), it has been described by numerous workers (e.g., Abich 1863), **but until recent improvements of marine geophysical data acquisition its significance has not been fully acknowledged**. As a result of the tremendous efforts and submarine drilling and sampling during the last few decades, however, some light has been shed on the mechanism of mud extrusion as well as the source of the components involved." Note 8.4c

Researchers defined mud volcanoes by what they saw being extruded onto the surface:

"In **mud volcanoes**, plastic, **clayey materials or matrix** from deep source strata are extruded to **the sea floor** by various driving forces." Note 8.4d

In defining how far down the natural supply pipes of the mudfountains go, researchers said this:

"Mud volcanoes can root **several kilometers below the seafloor**." Note 8.4e

The figure of several kilometers is actually only the beginning of what new technology has revealed. Because it is a new field of study, technology is rapidly increasing the depth which can be analyzed. One researcher, using the CMP profile reported the depth of a mudfountain in the South Caspian Basin :

"… penetrates to the depth of **9 km**…" Note 8.4f

Nine km is deeper than all the boreholes we humans have ever drilled, except for one. This establishes that a tremendous amount of sediment was displaced by the hydrofountain. It also helps establish the enormity of the sediment load that can be expelled onto the surface by these massive fountains.

If we recall that geology already acknowledges all of the continents were covered by water, at least once, and if we recognize that mudfountains are a world-wide phenomenon, it shouldn't be too much of a stretch to imagine the sediment on the continents and mudfountains are connected.

Marine scientists have explored only a very small fraction of the ocean floors, yet estimates of the number of submarine mud volcanoes run high:

"**Submarine mud volcanoes** are distributed on the Earth more extensively than their subaerial [on land] analogs. The estimated **total number** of known and inferred deep-water **mud volcanoes is 10^3-10^5**." Note 8.4g

Although their estimates are as high as 100,000, the number is actually much larger as we look for them through a new Universal Flood paradigm.

Because of *dogmagma*, some researchers theorized that mudfountains were driven by mantle magma from deep within the Earth. However, new data has documented that this is not the case:

"Based on published data and 65 new determinations of He isotopes in gases from mud volcanoes of the same regions, Lavrushin *et al*. (1996) concluded that **mud volcanism is independent from mantle magmatism** and the absence of mantle-derived helium in natural gases… **unambiguously implies the crustal source for hydrocarbons and all other components**. The exception is provided by methane exhalations of mud volcanoes of Georgia and Sakhalin where the presence of a

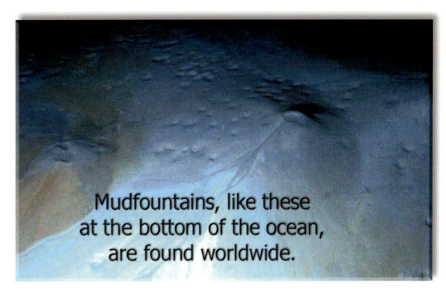

Fig 8.4.8 – Mudfountains are very important element of the UF because much of the sediment on the continents that geologists thought came from eroded mountains actually came from mudfountains, sandfountains, and rockfountains. Mudfountains are a truly worldwide phenomenon.

Mt. Saint Helens Mudfountain

Fig 8.4.7 – The May 18, 1980 mudflow from Mt. Saint Helens not only buried or destroyed everything in its path like this bus, the water and mud from this mudfountain moved large objects like this boulder.

mantle component is possible." Note 8.4h

The "crustal source" of heat and pressure that brings sediment to the surface comes from forces described by the Frictional Heat Law. Mud volcanoes—mudfountains—have many of the same features as the hydrocraters that were discussed in subchapter 7.8, The Hydrocrater Evidence. They are evidence for some of the mountain formations on the Moon. Just like the Hydrocrater identification criteria in the last chapter, Mudfountains have the three identifiable characteristics: a water source, a fault, and a diatreme. As one researcher noted:

"The second (and more common) mechanism is the formation of mud volcanoes as a result **of the rise of fluidized mud along faults and fractures**…" Note 8.4i

Another Russian geologist, V. N. Kholodov, did extensive research on mudfountains, detailing their connection with earthquakes. Note his observation that "chains" of craters exist in the submarine environment, just like they do on the Moon and on other planetary bodies, and that "intersections" along lines of faulting, like those below the Arizona Hydrocrater, create larger specimens of mud volcanoes:

"The spatial localization of mud volcanoes is very often determined by the distribution of tectonic fractures. This peculiarity was first noted by Abikh (1863, 1873) who studied mud volcanoes of Azerbaijan and emphasized that they are grouped in chains extending in two perpendicular (southeast-northwest and southwest-northeast) directions. The largest volcanic edifices are usually localized at intersections of these two main directions. He stressed that the distribution of mud volcanoes is controlled by a network of normal faults, which are marked in the southeastern part of the Caucasus by recent tectonic movements and high seismicity." Note 8.4j

Kholodov and others also documented the cyclical nature of mudfountain eruption, which follows a cycle of earthquakes long discounted by traditional geology:

"It appears that mud volcanoes of Azerbaijan erupted diachronously in the wave-like manner from the north to south. During the considered period of 160 yr, the region witnessed three or four such waves, which initiated on the Apsheron Peninsula and gradually migrated toward the Kura Depression where they terminated to start again in the first area during the next time interval."
Note 8.4j p303

In the Azerbaijan, Kerch-Taman area near Russia, over 40 billion cubic meters of mud are estimated to have come to the surface. This of course answers the questions of why some of the region collapsed:

"Such a significant removal of material from deep layers should be compensated by subsidence and collapse. This explains the origin of impressed synclines of the Kerch Peninsula." Note 8.4j p305

Moreover, this process explains the existence of rills, depressed valleys and channels on the lunar surface, which has no explanation from the impact pseudotheory of the Moon. Lunar hydrodynamics can easily account for erupted mud and sediment on the surface that caused both the collapse of rills and the building of lunar domes. Some domes are the result of extruded sediment while others are areas swollen from rising sediment but without enough pressure to erupt.

Lusi, the Java Mudfountain

As of the writing of this book, the Indonesian island of Java experienced the relentless power of a modern day mudfountain that unleashed its furry on the countryside (see Fig 8.4.9). May 2006 saw the beginning of a mudflow dubbed **Lusi** that began erupting *heated mud and steam* near a gas exploration well. Just two days earlier, an earthquake had struck the area. The January 2008 *National Geographic* did an article titled, *Drowning in Mud,* and showed thousands of homes covered in mud:

"Lusi, as Indonesians call the mudflow, is one of the **more bizarre expressions** of Indonesia's geologic turmoil. Since May 2006, it has spewed millions of barrels of heated sludge, blanketing an area twice the size of New York City's Central Park. Villages have disappeared under the mud, 60 feet deep in places, and 10,000 families have been forced from their homes." Note 8.4k

Geologists called the mudflow a "bizarre expression" of geological turmoil because they are so uncommon. Worldwide mud deposits have often been *incorrectly classified* as "volcanic ash" instead of mudfountain deposits simply because of the magma pseudotheory. However, there was a time when they were far more violent and widespread, a time in the not-too-distant past. Thousands of examples exist where mudfountain deposits like the Badlands of South Dakota, USA, demonstrate evidence of a Universal Flood origin rather than a volcanic ash or erosion origin.

The Badlands Mud Evidence

Until the last couple of decades little had been written about mud volcanoes because they did not align with modern geological theories. Deep-ocean submarines had not yet documented their profusion on the ocean floor. Having established at the beginning of this chapter that all the world's continents were once covered by water, we have to ask this FQ:

> FQ: If the present-day ocean floor is littered with mudfountains, why wouldn't the continents have had mudfountains when they were covered with oceans?

The ground truth in geology today is that many mudfountain deposits on dry land are not recognized as mudfountains. To explain them, geology has turned to the Uplift Pseudotheory and to the erosion sci-bi. One such example is the Badlands of South Dakota (see Fig 8.4.10). The origin of the Badlands is not

Fig 8.4.9 – Lusi, the modern mudfountain on the island of Java in Indonesia was responsible for displacing more than 10,000 families over the past several years because of its unstoppable heated mudflow and steam eruptions. These types of geological catastrophes appear "bizarre" because we are not used to seeing them. However, they were prevalent in the past, during the UF, leaving behind their distinctive Mark of the Flood.

described in detail or with clarity anywhere we searched. The National Park Service's Badlands website was one of the few places that recounted how the Badlands area was purportedly formed, but its story is wholly inadequate.

First, we are told that the light colored sedimentary formations seen in the Badlands Mudfountain Diagram (Fig 8.4.10) are accumulated from millions of years of erosion from the Rocky Mountains that "reared up" spreading sediment:

"A geologic story is written in the rocks of **Badlands National Park**, every bit as fascinating and colorful as their outward appearance. It is an account of **75 million years** of accumulation with intermittent periods of erosion that began when the Rocky Mountains **reared up** in the West and spread sediments over vast expanses of the plains. **The sand, silt, and clay, mixed and interbedded with volcanic ash, stacked up, layer upon flat-lying layer, until the pile was thousands of feet deep.**" Note 8.41

The scientific evidence that the Rocky Mountains rose 75 million years ago is lacking, but more importantly, the light colored sediments of the Badlands did *not* come from the Rocky Mountains. There is no trail of sediment from the western Rocky Mountains to the Badlands, and there are *no volcanoes in the area* to account for the supposed "volcanic ash." Ash can usually be traced back to its source by examining the fallout path to its volcanic origin—but *there is no ash fallout source* for the Badlands sediment. It is incredibly unscientific to have ever designated erosion or ash fallout theories as an origin for Badlands sediment.

Moreover, the Badlands mud sediment is of a completely different composition than the Rocky Mountains or the Black Hills. The Badlands' multicolored layers were laid down similar to the colored layers identified beneath the Odessa Crater (Fig 7.12.9), which were formed by hyprethermal waters in the diatreme. Aeolian sediment blown in from the Rocky Mountains would have traveled a great distance before coming to rest on the Badlands area, and would been a heterogeneous mixture of many different minerals, not graded into distinctively different colored layers. As we will learn about shortly, the colored layers are caused by different bacteria that lived in the heated waters and from different underground sources.

But first, let's look at the fact that the Badlands layers are at most, only several thousand feet thick, and let's look at the length of time erosion has really been at work there. According to the park service website:

"Erosion is so rapid that the landforms can change perceptibly overnight as a result of a single thunderstorm." Note 8.41

Most of the hills consist of flour-like sediment, and like similar hills on the Colorado Plateau (Painted Desert, for example), they experience remarkably rapid changes after each thunderstorm, enough to make the difference discernible in photographs of the area. The Park's website posts:

"On average, the White River Badlands of South Dakota **erode one inch per year.**" Note 8.41

Add to this the time element since the era that supposedly began the erosion processes of the Badlands district (Note that this event was an "uplift" event):

"**Broad regional uplift** raised the land about **5 million years ago and initiated the erosion that created the Badlands.**" Note 8.41

If we take the conservative one-inch-per-year erosion rate and multiply it by the number of years from the last "broad regional uplift," presumably "5 million years ago," the event that *supposedly* created the Badlands, what do we get? *5 million inches; sediment **79 miles (127 kilometers) deep**!*

FQ: Where did 79 miles of sediment go if the erosion rate is 1" annually?

It is apparent that the geologists working the site did not think through the numbers. Nowhere on Earth is the crust *79 miles thick*, especially not eroded sediment. Alternatively, if the thousand-foot thick Badlands sediment came from mudflows about 4,000 years ago, during the UF event, the total erosion would be about 4,000 inches, or just over 300 feet—an amount very close to the actual eroded landscape that exists at the Badlands today. The Age Model chapter demonstrates further that the millions-of-years geological time frame has no basis in fact and that many scientists have deceived themselves and the public into thinking it is a real timeframe.

Science is so much fun when truth is on your side—if a theory is correct, the Universal Scientific Method will prove it. A careful inspection of the Badlands or other similarly formed mudfountain areas around the Colorado Plateau will reveal lines of **faulting in the mud deposits.** One example can be seen in Fig 8.4.10.

The vertical or semi-vertical faults are filled with hardened silica, formed when Flood related hyprethermal waters flowed through cracks in the mud after it had been deposited. There is no explanation how such vertical faults could have formed "over geological time" with erosion or airborne ash as the only operative process; such sediments are only capable of horizontal sediment placement.

Another interesting fact is that the silicified fault lines were formed in the same hyprethermal waters that fossilized the plant and animal remains found in the Badlands sediment, also seen in Fig 8.4.10. The park service's explanation for the preservation of animal remains acknowledges the animals "**died in floods**":

"During the Late Eocene and Oligocene Epochs 37 to 28 million years ago, the region that is now the **White River Badlands** supported many kinds of animals. The land was then lush, well watered, and much warmer than now. The animals, mostly mammals, roamed the floodplains; **many died in floods and were quickly buried in river sediments**. Conditions for preservation were **excellent**; the Oligocene beds are one of the world's richest vertebrate fossil sites, though they represent only a short segment of Earth history." Note 8.41

It is illogical to assume that animals can be buried in warmer waters and not rot away, regardless of the speed at which they were buried. The truth is, animals buried in a floodplain like the Badlands today do not turn into rocks—yet there are millions of animal fossils in the Badlands mud. Those fossils testify to the events that formed the Badlands, in fact, the connection between the silicified fault lines and silicified fossils found in the mudfountain sediment are both made of the same minerals and are often found together in the same mounds.

Across the Colorado Plateau, the most common fossil, also frequently found in mudfountain deposits, is petrified wood, made of silica. Fig 8.4.11 includes three examples among hundreds that have been personally observed by the author from

Fig 8.4.10

Fig 8.4.11 – These Colorado Plateau Mudfountains are examples of mudfountains found around the world—but they are not erupting today. Such mudfountains are not erupting for the same reason we do not find buried wood turning into rock today. Both of these worldwide phenomenon required tremendous earthquakes unlike any ever seen by mankind and the creation of hyprethermal minerals and fossils through processes also unknown. Many mudfountains exhibit minerals similar to the kimberlite minerals found in Buell Hydrocrater, which are found only on or near the surface, a testament to the Universal Flood event.

many Western States of the USA.

Kimberlite Hydrofountains

Another type of hydrofountain is a kimberlite formation. Kimberlite is the name of a rock type first categorized over a hundred years ago and based on descriptions of the diamond-bearing diatreme pipes found in Kimberley, South Africa. Kimberlite Diatremes consist of green and gray minerals, which are formed in the hyprethermal vent pipe from deep underground. If the vent had enough pressure to explode at the surface, it formed a hydrocrater. Kimberlite is commonly found in fields or in clusters of diatremes worldwide. It is the mineral associated with the origin of most diamonds:

"One of the most valuable minerals, diamond, occurs chiefly in ultramafic igneous rocks called **kimberlites**, named for Kimberley, South Africa, where they are found in relative abundance. These rocks were forcefully intruded to the surface from deep in the crust and upper mantle in the form of long, narrow pipes. **We know** these diamond-bearing kimberlites originate at great depths because diamonds and other minerals found in them can be formed **only** under the conditions of extremely high pressure that exist in the upper mantle. **Kimberlites erupt to the surface at high speed, propelled by pressurized volatiles such as H_2O and CO_2. No one has ever seen a kimberlite eruption.**" Bib 59 p596

Several points are to be made about this quote from the geology textbook - *Understanding Earth*. The first, geology does *not* "know" that the diamond-bearing kimberlite originates only from the mantle. The origin of diamonds is still a mystery to geologists because the great pressure required to make diamonds did not come from magma. Our research has not identified *any* magma or lava flows that are associated with Kimberlite Fountains, instead researchers admit, as seen in the former quote, that the sediments erupted to the surface at high speed, "propelled by pressurized volatiles such as $H2O$…"

Furthermore, Kimberlite Fountains are another important example of modern geology's uniformity principle gone awry. As noted in the last quote, "*No one* has ever seen a kimberlite eruption." Thus, Hutton and Lyell's uniformity principle fails because no historical observation can account for the kimberlite diatremes. Of course, many past events have never been observed by modern geology, some of which, including the UF, are the real keys to the past. This helps to clarify why today's geology has never been able to explain some of the most fundamental geological questions.

The truth of diamond formation and the making of the kimberlite pipes can be seen in the collapsing plates of the Universal Flood Mechanism. Rapid shifting of the plates and concurrent heating produced magnificent heat and pressure, all sufficient enough to account for diamond formation; more on this later in the Diamond Mark.

Buell Hydrofountain Evidence

In the area east of the Grand Canyon, in Arizona, several kimberlite pipes or diatremes are to be found. The Garnet Ridge area and the Buell Hydrocrater, discussed previously in Fig 7.8.7, are two examples. Researchers examining Garnet Ridge discovered that the hydrofountains in the area had expelled "xenoliths" or foreign minerals ranging in size from sand grains to *boulders* from great depth; materials definitely *not* from the surrounding mountains. The xenoliths showed no indication of melting by magma, but instead were smoothed, polished and even striated (grooved):

"The surfaces of xenoliths in contact with serpentinite **are smoothed or polished and, in some cases, striated**. The inclusions of fossiliferous Paleozoic limestone **show no evidence of metamorphism.**" Note 8.4m

Such rocks could have only come from hydrofountains that have pipes that run very deep. The researchers continue:

"The **xenoliths** within the pipes include abundant blocks of Jurassic sedimentary rocks, **many of which are extremely large…the larger inclusions, however, are well rounded and reach diameters of 10 feet**. The fragments derived from the igneous and metamorphic basement **have ascended at least 4000 feet.**" Note 8.4m

Imagine boulders 10 feet in diameter being carried from depths of at least 4000 feet below the surface! No one has ever witnessed anything of such magnitude, yet such diatremes exist around the globe—***by the thousands***. What powered the hydrofountains seen on the surface of the Earth everywhere, all demonstrating a similar eruptive time frame? Hydrocraters and hydrofountains like these are direct evidence of the Flood, especially if the crater or rim of the crater was washed away leaving only the diatreme.

The hydrofountain that formed the Buell Hydrocrater had several pipes. One of them expelled an "olive-green" mud that can be seen in Fig 8.4.11. This green mud is not "kimberlite tuff," it was not glass, nor is it melted. Instead, it is "quartzite" formed in the Buell hypretherm:

"The **kimberlite tuff** is a purplish-brown or **olive-green** rock consisting of a fine-grained matrix containing innumerable fragments that are mainly light gray fine grained **quartzite**." Note 8.4n

The Buell Hydrocrater provides evidence for many of the UM chapters because geologists think that the quartzite in the crater came from magma deep within the Earth. However, *quartzite is not glass* and did not come from a melt or from magma. Researchers observed that the kimberlite in the Buell Hydrocrater contained significant water! Four samples were tested and contained from 7.25% to 13.41% water, which would not be present if the kimberlite had been melted at the surface of the crater. Note 8.4n p267

What really formed the crater and fountain if not magma? It is so obvious, yet even the most renowned geologists, such as E. Shoemaker could not envision the abundance of liquid water below the surface. All they could do was "infer" fluidization:

"Shoemaker, Roach, and Byers **infer** that the rock fragments spalled from the walls of the **boiling magma column** would become entrained in the moving gas and liquid to produce a **complex** fluidized system. As they point out, 'Intricate mixing, **rounding, and polishing of debris derived from depth**, particularly in some of the kimberlite-filled diatremes in the northern part of the Navajo Reservation... **suggest fluidization**.'" Note 8.4n p268

Magma melts rocks, *water* rounds and polishes rocks; it couldn't be simpler. But to the modern geologist, the fluidized system is "complex" because of their magma and rock cycle paradigm. As a final point of confusion, the age of the rocks presumably derived from "boiling magma" was indeterminate, although they should have been easily dated based on the geologists 'geological time' scale:

"... kimberlite has **not been established** with narrow limits." Note 8.4n p269

This simply meant that the 'dates' were all over the board, so no dates were reported. However, because the Buell Hydrocrater has very little surface erosion, we know the age of the minerals must be only a few thousand years since they were formed in the crater.

> **Many past events have never been observed by modern geology, some of which, including the UF, are the real key to the past.**

Fig 8.4.12 – John Wesley Huddleston in 1906, about the time he purchased a 160-acre farm in Arkansas, USA. Huddleston discovered diamonds right on the surface of his farm, the first outside South Africa. His discovery eventually led to America's own diamond rush.

It turns out that kimberlite formations have been looked at closely by researchers, not for their fountain geology, but for their economic value in *diamonds*. Kimberlite fountains happen to be where almost all natural diamonds are found. This is a classic example of how scientific truth comes forth because of economic reasons, not because of scientific inquiry. In fact, the overall importance of kimberlite fountains was still completely overlooked.

America's Diamond Rush

January 24th, 1848, Gold was discovered in the American River, at Sutter's Mill, in Coloma, California. Tens-of-thousands rushed to the gold fields in what would become one of the world's most well-known gold rushes. Although not as well known, and certainly not as rich in terms of quantity, America had its own Diamond Rush early in the 1900s, in the state of Arkansas. The story behind the diamond diatreme, where visitors still hunt for diamonds today, is interesting for its historical significance and because it holds direct evidence of the Universal Flood. The story begins:

"...in early 1906, an illiterate pig farmer named John Wesley Huddleston made a deal to buy the 160 acre 'Old McBray place' for $1,000. 'He didn't have the hundred dollars the owners wanted for a down payment, so he offered a mule and they took it. He made his X on a note to take care of the rest of his debt.' His purchase included most of the land on which the kimberlite lay. Mr. Huddleston had hardly made his first payment on the land when, on August 8, 1906, **he found two, lustrous stones on his property**." Bib 175 p5

The stones were diamonds. Not many people know what a diamond in the rough looks like, and the Arkansas diamonds look a lot like quartz. Amazingly, it wasn't a geologist or a mineralogist that first recognized the diamonds, it was John Huddleston himself:

"History has portrayed John Huddleston as somewhat of a backwoods hillbilly. Maybe his lack of educational and vocational opportunities did leave him to be labeled as 'an illiterate pig farmer in rural Arkansas.' But that does not mean Mr. Huddleston was not an intelligent man. **He was able to outsmart college degreed geologists by being the first to discover diamonds in their matrix in the United States**." Bib 175 p6

Huddleston may not have been able to read or write, but he proved that it did not take formal education to be successful. With an uncommon sense about him, he sold his diamond-bearing property and the diamonds he had collected at a substantial gain:

"Their first order of business was for Reyburn and Cohn to travel to Murfreesboro where they negotiated to buy John Huddleston's farm from him for **$36,000**. Even after paying off the thousand-dollar balance on the farm, Huddleston had amassed a virtual fortune when you consider what money was worth nearly a hundred years ago. It has also been reported that John walked away with twenty-one diamonds that he had found during the interim period before the deal was closed. Later he told people that he had sold the tobacco sack full of diamonds for **$40,000**." Bib 175 p7

News of the Arkansas diamond find spread quickly and fortune hunters poured in from all over, each with the hope of finding the diamond that would make them rich. Thus began America's Diamond Rush. Guards were posted at the property no longer owned by Huddleston, but an enterprising neighbor capitalized on the crowds by letting them hunt on his property for fifty cents apiece. One early searcher picked a 13.25-carat diamond right off the surface, near where the Mauney mine would later be located. A few eager fortune hunters and early diamond prospectors are seen in Fig 8.4.13 along with some of the diamonds found at the site.

In 1972, the State of Arkansas purchased the property and turned it into the Crater of Diamonds State Park; allowing anyone to search for their treasure for a nominal entrance fee.

Arkansas Diamond Diatreme Wood Evidence

The Arkansas Diamond Diatreme also provides direct evidence of a UF mudfountain. The prestigious geologic journals were surprisingly quiet on the subject, but Glenn W. Worthington, an amateur geologist wrote the book *A Thorough and Accurate History of Genuine Diamonds in Arkansas* after working the site for many years. Worthington recorded an interesting discovery from one of the commercial miners working the area. Carbonized wood had been found in lamproite, a grey mudstone similar to kimberlite (see Fig 8.4.14). He described the event as follows:

"When Star excavated their American Mine, they discovered that an **unexpected, curious ingredient** had been mixed into their lamproite when it was intruded. They found that it contained **large pieces of black, carbonized (or coalified) wood**. These were not pieces of petrified wood in which the wood fibers had actually been replaced by minerals... Some pieces were **covered with beautiful, golden marcasite (like pyrite) that had grown on the wood** over past millennia.

"But **how did huge chunks of trees get mixed into the ore seventy feet deep in the ground at The American Mine**? And, at Black Lick, geologist Robert Allen found smaller pieces of wood in drill core taken from **a depth of two hundred feet**." Bib 175 p128

FQ: How were trees buried and preserved hundreds of feet below the surface in diamond-bearing mud?

This is an interesting question—but like many of the questions we asked, there were no answers in modern geology. Worthington identified gravel deposits above the diatremes and wondered where the rounded stones came from. Worthington would search for years to find answers to his questions:

"When examining the occurrence of coalified wood, we should first consider that there is twenty **feet of tokio gravel** and tokio conglomerate on top of these lamproite bodies. Tokio gravel consists of **well-rounded, former beach pebbles** of novaculite, jasper, flint, and other cryptocrystalline quartz. Some of the pebbles are **cemented together by iron** to form a natural concrete called conglomerate. For two decades I had been told that the conglomerate was here *before* the volcanic intrusions. If so, are we to believe that the lamproite did bust through 125 miles of other rock during its ascent; but when it reached the conglomerate at near surface, it was unable to break through the natural layer of concrete?

"The American, Black Lick, and Kimberlite Mines are all 'crest of ridge' occurrences. In other words there is a definite, topographical high (or bulge) in the earth where these three lamproite bodies lie buried. Standing there you can almost picture the intrusion coming near surface and then being diverted horizontally as if a giant mole had made a tunnel along the ground. But if that were the case, **how did trees get mixed into the ore**? Everyone knows that trees always grow on top of the ground instead of under twenty feet of concrete. Between The American and Black Lick outcroppings there is a valley. **Did the volcanic material come to surface, swallow trees**

American Diamond Rush

Star of Arkansas

Fig 8.4.13 – The Crater of Diamonds State Park in Arkansas was created in 1972 after nearly seven decades of commercial mining operations. Early diamond hunters rushed to the area in 1906 seeking their fortune. Some found it. In 1956 the Star of Arkansas, at 8.27 carats was cut from a stone weighing originally 15.31 carats. The upper right inset shows diamonds typically found at the park today, of which about 600 are found each year.

in its molten slush and then go back down underground? No, I cannot conceive that anything forceful enough to break through 125 miles of rock to come to the surface would have turned around and crawled back under a layer of rock.

"**For years I had been told incorrect geology** about the conglomerate forming *before* the volcanic activity. What I did not know for a long time was that Miser and Purdue had published in a 1929 report that the lamproites intruded **all of the way to the surface. After that, way back when Arkansas was still beachfront property, the tides washed the cobbles over the top of the lamproites and conglomerates were formed over them.**" Bib 175 p128

One can see the confusion that comes from the magma paradigm. It seemed incredulous that wood could be deposited so far below the surface in a deposit that supposedly came from magma. And why were there rounded beach rocks laying over the diatreme, which was higher than all of the surrounding area? This was all a part of the mystery Worthington was trying to solve. To learn what species of trees were buried in the diatreme, he sent samples of the coalified wood to various experts. He learned that the trees did not grow in water, especially not salt water, and that they had grown in a seasonal environment:

"One paleobotanist told me the tree specimen was definitely from an **angiosperm (flowering tree)**. The others determined that the trees had been **conifers—cone-bearers**, the Cretaceous equivalent of today's pine trees. I also learned that if these were living trees at the time of the intrusion, **the lamproite had not been intruded into a shallow sea**, as previously suspected. **These trees had not grown *in* water**, but actually had fibrous trunks that carried water up the trees. The wood samples also showed distinct growth circles, which indicated there were **seasons**—periods of hot and cold, growth and dormancy—while the trees were living." Bib 175 p129

Seashells found in the same area added to the confusion because they were the ***evidence of an ocean that covered the area*** around the same time:

"In a nearby creek bed there are conglomerates of seashells as **evidence of the years this land was submerged in a shallow sea**." Bib 175 p129

The scientific literature revealed *no explanation* for many of the questions. Had modern geology skipped over the important geological features of the Arkansas Diamond deposit? In a surprising find, the *Nashville News* published a story about the Kimberlite Diatreme around 1912. It was amazingly insightful, accurate, and was based on common logic—refreshing considering it was made during an era already steeped in dogmagma:

"It is now practically certain that the areas of this **diamond bearing Kimberlite**... represent two or more 'Pipes' or 'Vents' filled from the great depth of the earth's interior by volcanic action, somewhat resembling a '**mud volcano**'; in as much as little or **no metamorphic effects are observable on the inclusions of the soft shales and other minerals that would naturally be partially or wholly** destroyed if the great heat of flaming gases and molten lava characteristic of the volcanoes of the ordinary kind had been present. **From this we reason**, that while the explosive vapors undoubtedly accompanied the eruption, the action was due **to hydrothermal rather than igneous agencies**.

"Such a theory is borne out by a close examination of all the Kimberlite vents of Africa, in many of which mining has now reached depths of several thousand feet." Note 8.4o

In the final analysis, it was "hydrothermal rather than igneous agencies" that formed the Arkansas Diamond Diatreme! As it turns out, this was not the only location that organic matter had been buried deep within the diatremes.

The Spouting Process Evidence

A modern-day mineral rush began in Canada in 1991 with the discovery of diamonds in Canada's Northwest Territories. Canada's first commercial diamond mine, the Ekati mine, began mining the diamond diatremes in 1998. By 2004, over 26 million carats had been produced at the mine.

Geology is still relatively unaware of this type of mudfountain diatreme deposit because their origin is seen as being:

"...from **very deep mantle**-derived sources." Note 8.4p

Fig 8.4.14 – How does wood become buried 200 feet in a diamond bearing diatreme, which supposedly came from magma deep in the upper mantle? Obviously, the wood did not come from magma, but neither did the diamonds or the marcasite minerals found with the wood (below center). Glenn Worthington worked the commercial mining operations at the site and is shown holding some of the coalified wood found in the grey lamproite deposit being excavated. Neither the wood, nor the marcasite, nor the overlying tokio gravel come from typical volcanoes, leaving the Arkansas diamond diatreme, like the diamond diatremes in Africa, an unsolved mystery for modern geology but direct evidence of the Universal Flood.

Finding Wood in the Arkansas Diamond Diatreme

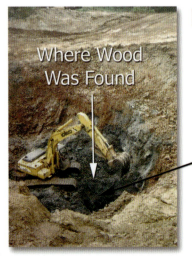

CHAPTER 8 THE UNIVERSAL FLOOD MODEL

However, there were other mysterious inclusions in the diamond bearing pipes:

"During early diamond drilling exploration abundant... **fossils were being recovered... From a few of the diamond pipes rare leaves, turtle bones and fish parts have been recovered**." Note 8.4q

Leaves, turtle and fish parts? How could organic matter withstand the heat of magma from the "very deep mantle?" Some of the specimens appeared to be "normal cedar logs" that had "not been exposed to any burning":

"Some specimens have **not been exposed to any burning and look like normal cedar logs**." Note 8.4q

The researcher who documented the buried plant and animal remains asked a fundamental question that was asked earlier in the UM:

"**How did the wood and other fossils get buried 400 meters (1300 feet) and deeper within the pipe?**" Note 8.4q

Thirteen hundred feet is an incredible depth for animals and plants to have been buried and preserved. Such examples identify several important points about the diamond pipes:

1. The pipe material was not molten or melted as current theory suggests.
2. The process that formed the pipes is a hydrofountain process.
3. The fossilized organic matter was preserved in a hyprethermal environment.
4. The organic matter was buried during a "spouting process," from multiple eruptions.

The eruptions were of a recent date because organic material *from the surface was found near the surface,* in the pipe rather than being eroded away. This demonstrates that the pipes are surface phenomena. In many cases, the funnel-shaped opening at the surface is still evident, demonstrating that the pipes are a recent surface-geology event.

Often the eruption of large-diameter pipes provides direct evidence of surface fragments being buried as deep as 1,500 meters (4,921 feet)! Buried surface material, as described in *Economic Geology,* is the result of the "spouting process" of erupting hydrofountains, where geysers explode then collapse as noted in the fourth item above:

"The presence of fragments of **surface or near-surface materials** such as the **large slab of Mancos Shale** that occurs in the Mule Ear diatreme, Utah, **1,500 m below its original stratigraphic horizon** (Stuart-Alexander et al., 1972), **charcoal fragments at 650 m below the present surface** in the Philippines Balatoc pipe (Sawkins et al., 1979, p. 1421), and **a carbonized tree trunk at the 244 m level in the Cripple Creek breccia pipe**, Colorado (Lindgren and Ransome, 1906, p. 31), **support the spouting process**." Note 8.4r

The "spouting process" was described as multiple eruptions in the Hydroplanet Model Chapter. Geology has left water out of their dry Earth theories all too often, but research conducted by geologists like M. E. McCallium, identified the importance of water "fluidization" in 1985. This was apparent not just in kimberlite diamond pipes—but in all "volcanic pipes":

"The probable **importance of fluidization in the formation of kimberlite pipes** has been emphasized by Dawson (1962, 1971), McGetchin (1968), Woolsey et al. (1973, 1975), Clement (1975), McCallum (1976), and others, and a host of workers either stress or allude to **the role of a fluidization mechanism in the genesis of a variety of diatreme and volcanic pipe structures** (e.g., Shoemaker and Moore...) There is **little evidence to suggest that similar features observed in many ore-bearing breccia pipes and in various nonmineralized diatremes and volcanic pipes were not generated by essentially similar processes**." Note 8.4r p1527-8

Here the hydrofountain process is described as *the process for volcanic pipe generation*. The question to ask is:

Why are so many hydrofountains found on the **surface** of the Earth today, unless they all occurred at the same time several thousand years ago?

Historically there are no records of any eruptions over the last several thousand years large enough to account for large diatreme pipes. Beside the occasional eruption of traditional volcanoes or mud volcanoes, there is no record of *numerous* hydrofountains erupting on the continental areas.

There are a couple more examples of the mudfountain and hydrofountain processes to talk about. Whether mud, cob-

"How did the wood and other fossils get buried 400 meters (1300 feet) and deeper within the pipe?"

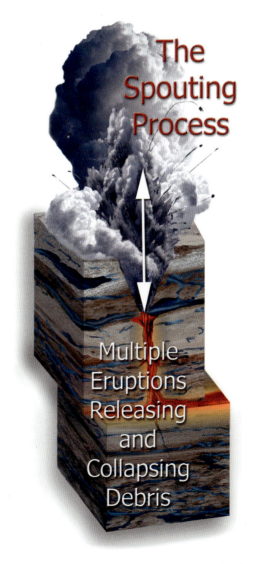

The Spouting Process

Multiple Eruptions Releasing and Collapsing Debris

blestones, or boulders—tens-of-thousands of hydrofountains erupted *recently*. This time frame is verifiable in each case because erosion has not swept the hydrofountain debris away and no erosional patterns exist to suggest differently.

Artists Pallet Evidence

The Artists Pallet is a beautiful example of the Hydrofountain Spouting Process forming an array of colorful clay and ore mineral deposits toward the end of the Universal Flood event. This mineral complex is found in Death Valley, California, USA. Fig 8.4.15 illustrates the extent of the colorful minerals, clearly the handiwork of hydrofountains. These hydrofountains mixed lighter material with darker volcanic material that also came from beneath the surface.

Death Valley mountain ranges have many areas of dark volcanic eruptions and many areas of light, sand hydrofountain materials that we will explore further in the Sand Mark subchapter. The long-held origin sci-bi for the white sands of the Valley has been erosion—but the surrounding mountains are not white. The erosion evident at the Artist Pallet formation is clearly a *very recent* phenomenon due to the lack of blue, pink and white colored clay and mineral debris in the gullies, washes and flat areas of the Valley. The orange mineral, seen to the right of the background photo in Fig 8.4.15 has experienced the greatest amount of erosion, but it is a much larger and softer deposit, which allowed it to spread as far as the road. Area rainstorms erode the sediment noticeably; millions of rainstorms (being millions of years old) would have eroded the orange material so significantly that it would have easily stretched *much further into the valley*—but it has not.

Hydrorock Fountains Evidence

In subchapter 7.7 the Chi-Chi Hydrorock Fountain was introduced to demonstrate the reality of Hydrorock Fountains. The Chi-Chi Fountain established that Hydrorock Fountains are an event that must be considered in the geological sediment survey of any area. The FQ to ask is this:

FQ: When did Hydrorock Fountain eruptions occur in the past and how widespread were they?

Present-day geological conditions are rarely suitable for modern day Hydrorock Fountain eruptions. The Chi-Chi example is uncommon because of the unique combination of characteristics required for such an event. For this to happen, a large earthquake must occur near the surface, heating a tremendous amount of water, sufficient to produce enough pressure to expel large rocks forcefully onto the surface.

It was pointed out earlier in the Hydroplanet Model that the largest amount of erosion by far, takes place below the surface of the Earth. Modern geology is unaware of this fact. The vast majority of the Earth's moving water lies beneath the surface, and it is there where most rocks become rounded, which are then brought to the surface through hydrofountains. This process, called **endoerosion** (underground erosion in aquifers) and the hydrofountain paradigm must completely replace the incomplete aeolian (wind) and fluvial (surface river) erosion paradigm that has heretofore been the primary erosion mechanism. Because endoerosion and hydrofountains are the primary erosional factors that shaped the Earth, modern geology textbooks will have to be rewritten.

Traditions caused few researchers to understand what they were looking at as they studied Hydrorock Fountain events because these types of fountains erupt so rarely in modern times. However, one researcher, M.E. McCallium described how channeling or gas-streaming was the origin of "pebble dikes" found *worldwide*:

"More abundant features of probable channeling or **gas-streaming origin are the pebble dikes that have been described from numerous breccia pipe localities** (e.g. Tintic, Nevada, Farmin, 1934; and Lovering, 1949; Braden pipe Chile, Howell and Molloy, 1960; New South Wales, Wilshire, 1961; Mount Morgan pipe, Queensland, Cornelius, 1967, Bisbee, Arizona, Bryant, 1968, 1974; El Salvador, Chile, Gustafson and Hunt, 1975; Los Bronces, Chile, Sillitoe and Sawkins, 1971; Engineer Pass pipe, Colorado, Maher, 1983).

Fig 8.4.15 – A hydrofountain deposit known as Artists Palette in Death Valley National Park, California, U.S.A. Very little erosion at this site verifies that this hydrofountain sediment, along with many others in the valley is of recent origin. Eruptions that formed them are not seen in modern times because they have, for the most part been dormant since the UF event.

These pebble dikes exhibit a close spatial relationship with ore-bearing breccias and generally are characterized by angular to well-rounded fragments in a fine-grained, commonly flow-banded matrix. **Well-rounded fragments typically reflect considerable transport distance from source whereas angular fragments generally appear to have been derived from the enclosing wall rock**. Matrix material may comprise in excess of 70 percent of dike volume at some localities and generally consists of clay to sand-sized particles." Note 8.4s

McCallium correctly accounts for the "transport distance" factor in well-rounded pebbles and the "close spatial relationship with ore-bearing breccias" in subsurface pipes. Still unrecognized is the importance of heated water and its contribution to pressure and eruptive forces, or the necessity of it (the hot water) for the environment where biogenic minerals grow to form ores. This will be dealt with later in subchapter 8.13, the Ore Mark.

A road cut on California's Highway 395, 23 miles northwest of Bishop, California, USA, gives us a unique opportunity to see Hydrorock Fountain geology. The 1957 road cut, seen in Fig 8.4.16, includes many Hydrorock Fountain features.

A thin 'cap rock' layer of sediment on the top of the hill is sediment expelled by hydrorock fountains through diatremes seen in the road cut. All over the world, endoerosion sediment has been spewed onto the surface, but modern geology still sees it as mountain erosion instead.

The *underground water action* involved in the formation of another feature of the hydrorock fountain, called a "clastic dike," remains unknown. The 1997 edition of the popular book series titled *Geology Underfoot*, discusses the clastic dike (hydrorock fountain) shown in Fig 8.4.16:

"**Most geological dikes are of igneous origin**, formed by intrusion of molten rock along fractures in older rocks. These dikes, however, consist primarily of pumice fragments including some **sand, pebbles, and smooth, rounded cobbles like those in the fluvial gravel that caps the cut**. Geologists call this type of dike a clastic dike. Most clastic dikes form when mobile material, typically **slurries of sand and rock fragments, intrude from the side or from below**." Bib 128 p231

The word "water" doesn't appear in this statement or anywhere else in the chapter on this subject. The closest the authors come to acknowledging the effects of water is the comment "slurries of sand and rock," but no discussion of where the water in the slurry might have originated. In the minds of geologists dogmagma has kept the Earth a dry, desert interior, leaving many clastic dikes and other hydrofountain features unrecognized.

Notwithstanding the literal drought of knowledge that has taken place, a flood of new data is beginning to change the geological landscape. In a 2006 study of 250 clastic dikes near the Dead Sea in Israel, five researchers acknowledge, as the Rock Cycle Pseudotheory previously stated, that the mechanism for

Fig 8.4.16 – This road cut exposed a Hydrorock Fountain on Highway 395 in California. There, cobblestones identify the pipe and the filled diatreme at the top of the hill. Ejected hydrorock forms the cap rock on the surface of the hill. Similar cap rock lies atop many desert plateaus around the world, all made by the same process.

Clastic Dike Hydrofountains Near the Dead Sea

Figs 8.4.17 – These Dead Sea dikes are examples of the hydrofountain process, which involves water fluidization, hyprethermal conditions, and strong earthquakes.

clastic dike formation is "poorly understood":

"**The mechanism of clastic dike formation is poorly understood**, and interpretation of field observations is commonly ambiguous." Note 8.4t

After studying several hundred clastic dikes in the Dead Sea area, the geologists made two profound points that confirm the hydrofountain process:
1. The clastic dikes (hydrofountains) were **formed by water fluidization**, and
2. The clastic dike fluidization was triggered by **strong earthquakes**:

"Our AMS analysis, field observations, and interpretations demonstrate that **the formation of most clastic dikes in the Amil'az Plain is associated with fluidization triggered by strong earthquakes** along the Dead Sea transform after deposition of the Lisan Formation." Note 8.4t p72

It is an interesting and important side note that many of the dikes were accompanied by green clayey sediment similar to that which we discussed in Fig 8.4.11, which is common on the Colorado Plateau:

"**Most abundant** are **dikes composed of green clay, silty quartz, and aragonite**, with a composition similar to that of the lower layers of the Lisan Formation." Note 8.4t p70

The colored clay sediments on the surface of the Earth all come from hydrofountains. How much of the clay, silt and sand sediments were formed underground and deposited on the surface during the UF will be covered in the next subchapter.

Having a simple explanation for the formative mechanism of the Dead Sea dikes in Israel, seen in Fig 8.4.17, the Dike Mystery introduced in Fig 6.4.8 can now easily be solved. Instead of an intrusion of lava or melted rock as once thought, it is a hydrofountain channel, formed in a hyprethermal environment; exactly the environment in which these mineral types grow.

Alternatively, no clear explanation for how magma forms "tabular" intrusions in solid rock has been found. There is no mechanism to force the magma towards the surface, let alone for it to form tabular intrusions. Anyone who has ever taken a water hose with a spray nozzle attached and pointed at the ground has experienced the force of water under pressure. When water, and especially steam, is under high pressure, it can easily cut through rock and, at the same time, form hydrothermal minerals.

The actual formation of a clastic dike is something modern geology has never seen, yet there are probably millions of them *all over the surface of the Earth today*. A worldwide hypretherm event is the only explanation for this wide spread phenomenon.

Skipperock Hydrofountain Evidence

The Skipperock Mystery is a favorite geological anomaly of the UM and one of the unique features of the endoerosion processes at work during the Flood. Today, Skipperocks are found along California's Pacific coast in sandy sediment, and further inland. These smooth, flat, and similar sized rocks, first introduced in the Rock Cycle Pseudotheory chapter in Fig 6.11.7, have no explanation for their origin in modern geology. Where did they come from and how were they made? Looking at the surface erosional processes at work today, the typical stream-

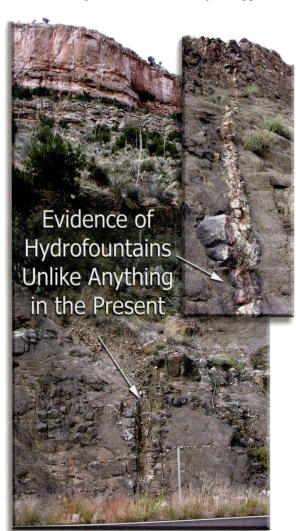

Fig 8.4.18 – Evidence of hydrofountains, like these fossil pipes along HWY 60-70 near Show Low, Arizona, U.S.A., are found right on the surface, yet are not being formed today. If they were formed millions of years ago, why have they not eroded away? When did they really form?

bed has tumbled rocks of all sizes, irregular shapes and a number of broken rocks. The ocean's well rounded rocks are usually not flat. A Skipperock streambed or Skipperock-forming ocean beach does not exist!

The Skipperock Origin Diagram in Fig 8.4.19 establishes how these unique rocks are made. Just as rivers and streams sort rocks depending on velocity and depth of water, hydrofountains also sort sediments. In some instances, a vertically oriented hydrofountain containing sorted material will have a continuous upward flow of water of modest velocity. Gravity and overlying pressure holds rocks of particular size back, restricting their ability to tumble while the upward movement of water vibrates the stones. Smaller pebbles and sand are ejected, but the larger rocks, too heavy to be ejected, continue to be eroded; held in place by the force of gravity, they take on the flattened, but rounded shape of the Skipperock.

This process, the only explanation for skipperock formation we know of, can be seen in Fig 8.4.19. There is evidence of this action in natural pipes, where a bed of particles could be "balanced by the upward drag force on the particles created by an upwardly moving fluid" as reported in *Mineralium Deposita*, 2003:

"**Fluidization** can simply be viewed as an equilibrium state wherein the buoyant weight of a bed of particles is balanced by the upward drag force on the particles created by an upwardly moving fluid (Wilson 1984; Davidson et al. 1985). Fluidized bed technology has been extensively reviewed in the **engineering** literature, **but little research has been conducted on fluidization in natural geologic processes**." Note 8.4u

As previously noted extensively in the Hydroplanet Model and throughout the UM, technologists have far outpaced the geologists in many areas; pipe fluidization is no exception. Nevertheless, four geological researchers examined a natural 10-meter diameter pipe in the famous Bushveld Complex ore deposit in South Africa, commenting on this very subject:

"We re-examine a breccia pipe in the Bushveld Complex to investigate the role of volatile fluid over-pressure in its emplacement. **This 10-m-diameter breccia pipe was emplaced vertically** in the anorthosite of the Upper Critical Zone of the Bushveld Complex and contains blocks of Bushveld anorthosite, norite, and pyroxenite, apparently derived from lower in the stratigraphy." Note 8.4u p356

Here the researchers looked at water flow mechanics and their impact on the rocks in vertical pipes. Their research dealt specifically with velocity and the volume of water that kept the long axis of the rocks vertically aligned; in this case, it was the *excessive speed* of the water. In the case of the Skipperocks, a more modest flow of water allowed the expulsion of smaller sediment, but was not sufficient to align the rocks vertically. In the Bushveld Complex's 10-meter hydrofountain, the speed of water actually balanced the *vertical* position of rocks during fluid flow:

"Blocks with high aspect ratios are generally **oriented vertically**, with their long axis perpendicular to the general stratigraphy of the complex and **parallel to the long axis of the pipe**, **indicating vertical fluid movement**. Seeing that the blocks maintain this orientation, one can infer that there was a significant solid fraction to the material which **moved up the pipe during emplacement, such that the blocks did not signifi-**

Fig 8.4.19 – The Skipperock Mystery is finally solved, as shown here; the Skipperock Origin Diagram. Skipperocks are made in a vertical hydrorock fountain in which the rocks are held in place with gravity, but are vibrated and rotated about their flat axis in upward, low velocity flow of water.

cantly shift during cessation of fluid flow." Note 8.4u p363

It is important to note that in both the skipperock and Bushveld examples, the fluid flow velocity would have remained constant for an extended period—otherwise, the change in velocity would allow the rocks to vary their horizontal or vertical orientation. How much time was involved? In the case of the skipperocks, enough time for the endoerosion process to shape them. Anyone familiar with rock tumbling and polishing knows the process only takes weeks or months as rocks are in constant movement, vibrating against each other. The Bushveld pipe requires an assessment of the velocity of the water. The minimum fluidization velocity was surprisingly fast:

"The **minimum fluidization velocity which would be required** to emplace blocks of the size and density found within the breccia pipe is found to be **10-110 m/s [22-246 mi/hr]**." Note 8.4u p356

That is, the largest blocks required a fluid movement of at least 246 miles per hour; no less than the speed at which geysers erupt. Now we have to ask another FQ:

> FQ: How were the pipe's contents consolidated and cemented while erupting at a velocity of up to 246 miles per hour?

It is interesting that the researchers did not even consider this question or at the very least, failed to mention it in their report. The reason the answer to this question is so fundamental is that as soon as the flow of the water was reduced at all, the vertically oriented rocks would become reoriented, just as if they were dropped to the ground.

Without the UF event and the ensuing Hypretherm, there is no good answer. Not aware of these processes, the researchers assumed there must have been a "significant solid fraction" to the material in the pipe. Nevertheless, the hypretherm environment present in the hydrofountain pipes included high pressure and high temperature, and the slightest change can trigger a sudden crystallization of the silica in the water, which causes a thickening of the sediment, locking the rocks in their *vertical* position as velocity is reduced. As the flow continues to slow, the pipe becomes cemented with the remaining silica. Even though clay or mud could also have thickened the fluid in the pipe, thickening alone does not account for the formation in the Bushveld pipe because clay and mud do *not become* **cemented** with silica outside a hypretherm.

Capitol Reef Basalt Hydrofountain Evidence

Back in Fig 6.5.7, dark basalt rocks are laying on the surface across hundreds of square miles around Yuma, Arizona, USA. The origin of these rocks is a mystery, a part of the Basalt Mystery in that subchapter, which also demonstrated that basalt did not come from land-based volcanoes as modern geology has assumed because the basalt is not melted. Fig 8.4.21 shows basalt cobblestones and boulders on the surface in Capitol Reef National Park, Utah. The same missing origin applies here. In the *Capitol Reef, Black Boulders* Bulletin of June 2006, Richard Waitt of the USGS describes the basalt boulders as being "strikingly out of place":

"**Large black boulders** are strewn along several valleys that cross Capital Reef National Park. In the Fremont River Valley they cover Johnson Mesa and scatter the hillsides of Fruita. **The boulders are strikingly out of place** among the tilted red and white bands of sandstone and shale that form the Waterpocket Fold. Igneous rocks, **consisting of basalt and andesite** that followed over ten million years ago, now form cliffs that edge Boulder Mountain and the Thousand Lakes Mountain plateaus west of the park." Note 8.4v

The first place geologists usually go to explain "out of place" boulders is ice-age glaciers—however, the theoretical ice-age did not reach into Southern Utah. Still, the boulders were in many instances too large to be moved easily, and many are "angular in shape" verifying that they had ***not*** been rolling along for ten million years:

"**Geologists long thought** the boulders had moved **from Boulder Mountain in Ice-Age glaciers and streams** that carried the rocks down valley. Studies show that the glaciers were **small** and the streams **lacked the power to move boulders nine feet or more in diameter** such as those found around Fruita. Many of the boulders are **angular in shape, whereas rocks rolled by streams become rounded**." Note 8.4v.

In fact, the geologists have identified no source for the Boulder Mountain basalt boulders or the boulders in Capital Reef Park. Instead, the boulders in Capital Reef are unscientifically theorized to have traveled many miles in "landslides" of dense flow, "rafting" the boulders to their present location. If this had been the case, *there would be a trail of boulders* from Boulder Mountain to Capital Reef Park; such a trail doesn't exist. Moreover, the basalt boulders in the Park are not mixed with other country rock in most locations, indicating that they came from a *single, unique source*—a hydrofountain. Note that the layer of black boulders in the bottom center and right images consists of a single layer laid down independent of the other layers. We would expect multiple layers if 10 million years of erosion and boulders 'rafting along landslides' were true. However, geologists do acknowledge that a flood took place at this very location:

"East of the park, black boulders form flat benches where **ancient floods emerged at the mouths of Pleasant Creek** and Oak Creek Canyons, i.e.: Notom Bench." Note 8.4v

Imagine a time when a flood laid boulders ***on the top of hills but not in the valleys,*** yet this is exactly what is seen in the cen-

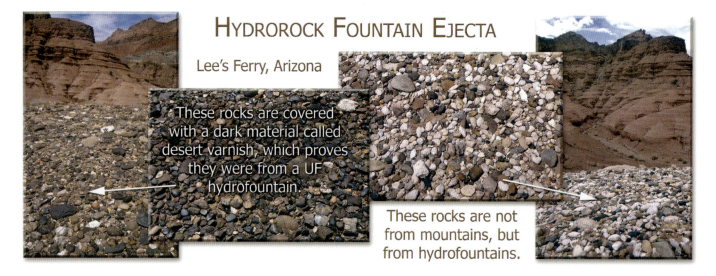

Fig 8.4.20 – Hydrorock Fountain pipes are hard to identify because they are usually underground, but the ejecta brought to the surface by them is easily observed, especially in areas where there is no mountainous source for the rocks. In this example, piles of fluvially eroded rock line the tops of mesas at Lee's Ferry, near the Grand Canyon, Arizona, USA. Some piles show off the lighter color of the natural minerals, other deposits are darkened by desert varnish, a biogenic manganese/iron coating that grew on the exposed surfaces of the rocks after being ejected from a hydrofountain. Desert varnish is another key piece of evidence of the Universal Flood.

Fig 8.4.21 - These basalt boulders are "strikingly out of place" as researchers have indicated. Although ice-age glaciers were once used to explain their out of place location and origin, this theory no longer explains the boulder's lack of erosion and their location on the tops of hills and not being in valleys. Only the UF Hydrofountain can succinctly explain their origin.

ter image of Fig 8.4.21, at the mouth of Pleasant Creek! Investigators are mum on this except to say that boulders are found "600 feet above" the present river level:

"Black basalt boulders east of the park. Distant mesa **capped by boulders is 600 feet above present river level**." Note 8.4v

The reason the non-native, rounded boulders are found on hilltops but not in the valleys of Capitol Reef and many other places, is that they were expelled through hydrorock fountains that erupted during the Universal Flood. This statement may sound somewhat bold, but keep in mind that we have just begun presenting the UF Evidences, of which there are hundreds. This claim, along with many to come, is made with the whole picture and all of the evidences in mind.

Mars Geyser-Hydrofountain Evidence

There are numerous examples of hydrofountains on other planets. The Enceladus hydrofountain in Fig 7.7.13 clearly demonstrates the enormous influence the Gravitational-Friction Law has on hydrofountains in the solar system. Hydrofountains and hydrocraters exist on almost all of the celestial objects in our solar system, but we will take the time to illustrate one new evidence on our neighboring planet—Mars.

Having sent several orbiting satellites and land rovers to Mars, it has become an explorer's haven. One recent discovery, shown in Fig 8.4.22, revealed a number of dark ejecta deposits, windblown toward the northeast. This method of sediment deposition on the surface and its effects on entire landscapes has been underappreciated in modern geology. Researchers describe how the "dark dune spots" were created:

"The seasonal frosting and defrosting of ice results in the appearance of a number of features, such dark dune spots with spider-like rilles or channels below the ice, where spider-like radial channels are carved between the ground and carbon dioxide ice, giving it an appearance of spider webs, then, **pressure accumulating in their interior ejects gas and dark basaltic sand or dust, which is deposited on the ice surface and thus, forming dark dune spots**. This process is **rapid**, observed happening in the space of a few days, weeks or months, a growth rate rather unusual in geology - especially for Mars." Note 8.4w

The eruption process was rapid and did not take 'geological time' to occur; in fact, the recent nature of the debris is easily established. Although CO_2 plays an active part in Martian geology, water is also interwoven into the eruption recipe:

"Data obtained by the Mars Express satellite, made it possible in 2004 to confirm that the southern polar cap has an average of 3 kilometres (1.9 mi) thick slab of CO_2 ice with varying contents of frozen water, depending on its latitude: the bright polar cap itself, is a mixture of 85% CO_2 ice and 15% water ice. The second part comprises steep slopes known as 'scarps', made almost entirely of water ice, that fall away from the polar cap to the surrounding plains. This transition area between the scarps and the permafrost is the 'cryptic region', where **clusters of geysers are located**.

This model explores the possibility of **active water-driven erosive structures, where soil and water derived from the shallow sub-surface layer is expelled up by CO_2 gas through fissures eroding joints to create spider-like radiating tributaries capped with mud-like material and/or ice**." Note 8.4w

As time marches on, the hydrofountain evidence continues

Fig 8.4.22 – The upper left image is of the Martian surface as seen from a satellite. It covers an area about 1 km (.6 mile across) and shows a dramatic array of geyser-hydrocraters. The image on the right is an artist's impression of how the geysers might look from the ground. They consist primarily of CO_2 and water vapor, forming rapidly in a matter of months or even just a few days. These hydrocraters, like many others on the planet are not geologically old, but fresh.

to build, not only on this planet, but on many other planets and moons. Mars is unique because of its proximity to the Earth, which may have allowed it to be influenced by the passing of the UF Comet. If this was the case, then many of the large, water-formed features on its surface happened in an event that occurred not long ago; an event shared with the Earth.

The Hydrofountain-Hydrocrater Connection

In this subchapter we examined many types of hydrofountains and hydrocrater processes resulting from the Universal Flood. Some of these include: geysers, fossil hydrofountains, worldwide rock-pipes, mudfountains, the Arkansas diamond diatreme, the spouting process, the skipperock origin, and hydrorock fountains. Hydrofountains are the mechanism behind the creation of hydrocraters, which was discussed in detail in the Hydrocrater and Crater Models, subchapter 7.8 & 7.9. These new models will completely change our understanding of all of the craters and the cratering processes in the solar system and by default, will also change the way we view planet formation. But the hydrocratering process can *only* be comprehended if we know how they are formed, which can be easily explained and understood. It is for this reason that the Hydrofountain-Hydrocrater connection is so important.

Could the thousands of hydrocraters and hydrofountains that exist primarily on the surface of the Earth not have come from a worldwide earthquake-water event?

Since geology has not recognized the abundance, the origin, or the importance of the hydrofountains and hydrocraters on the Earth, this question is a key concern for future research. We recognize how little emphasis is put on hydrocraters in the March 2008 *National Geographic* magazine article on Iceland. Iceland is located in a highly active earthquake zone because of the underlying crustal plates that meet at this area in the Atlantic Ocean. As a result, geothermal activity abounds, and many craters dot the landscape. In the *National Geographic* article, the authors chose to equivocally identify the steam explosion craters thus:

"Molten lava flowing across wetlands and into the cold waters of Lake Myvatn set off **steam explosions** that created a chain of **pseudo craters** more than 2,000 years ago—a landscape that draws sightseers today." Note 8.4x

"Pseudo" means 'false,' 'pretend,' or 'unreal,' a point firmly made in the UM regarding the 'pseudotheories' of modern science. They are false theories taught as fact. But what is false or unreal about the steam explosion craters in Iceland? Nothing—and that is the point. The true importance of "steam explosion" craters like these has never been fully understood, until now.

Science may choose to think of hydrocraters and the hydrofountains that made them as just an anomaly—or something to 'pretend' is not real, but this is about to change, because nothing could be further from the truth.

8.5 The Sand Mark

This subchapter is a follow up to the Sand Mystery introduced in Chapter 6.3. Here we explore the origin of most of the Earth's sand and other sediments, and how the hypretherm present during the Universal Flood is connected. Specific examples of how individual grains of sand are born and how they formed large deposits are included. Finally, the key to the understanding of how the Earth's vast sandstone deposits were made will be shown by experiment, which will reveal how sand becomes sandstone. Sand is clearly an unequivocal Mark of the Universal Flood.

The Sand FQs lead to FAs

In The Sand Mystery of 6.3, we cited a Scientific American article where researchers stated that "the little grains of sand have been somewhat neglected by geology..." Also, in an article in the *Journal of Geology* from 2003, one researcher noted that "the origin of pure quartz sheet sandstones" was considered a "major puzzle" and when coupled with the origins of dolomite, red beds, and other sedimentary deposits created "seemingly intractable geological problems." Each such statement preceded some new interpretation or conjectural advancement in the researcher's field intended to contribute to the *overall understanding of how Nature works*. However, the Magma and Rock Cycle Pseudotheory chapters showed that much of the research is based on false premises resulting in faulty interpretation.

One primary reason the modern geological rock cycle is incorrect is that the actual origin of most sedimentary rocks on the Earth is unknown. It is out of the small and simple things that many great things come, including the answers to major geological problems. For this reason, it is essential to learn the true origin of sand and sandstone, an idea not lost on modern geologists:

"...geologists are realizing that **sandstones and sands hold the key to many questions still unanswered**." Note 8.5a

In 1957, F. J. Pettijohn opened his book, *Sedimentary Rocks*, considered by some to be one of the most comprehensive geological works on the sedimentary layer of the crust, with this statement concerning the origin of sediments:

"**The origin and accumulation of sedimentary rocks might, at first thought, seem relatively simple**. Sands and muds are seen to form and be carried by the rivers from the continents into the sea. The origin of sedimentary rocks, unlike that of many igneous and all metamorphic rocks, is apparently open to inspection and study. **Unfortunately the matter is not so simple**. Not all of the formative processes can be seen. The diagenetic changes, in particular, which include intrastratal solution, **cementation, formation of concretions**, and so forth **cannot be readily observed. Neither can the turbidity currents responsible for the transport, deposition, and structures of many marine sediments. The formation of many chemical sediments has never been seen**. And so, as in the case of most other rocks, the history must be reconstructed from the geologic record—the effects produced by processes **no longer operative**." Bib 159 p2

Pettijohn hit the nail squarely on the head with his statement that the cementation of sandstone and the currents responsible for making much of the present-day hardened marine sediments "cannot be readily observed." He also correctly noted that the

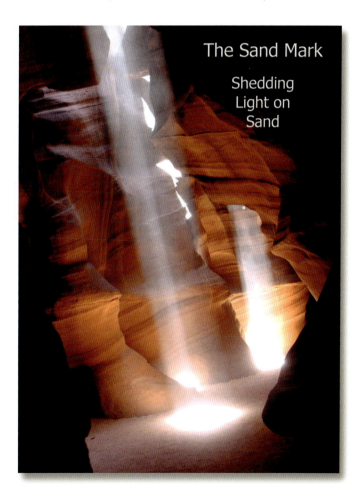

The Sand Mark
Shedding Light on Sand

"formation of many chemical sediments has never been seen."

> FQ: Why are "the origin and accumulation of sedimentary rock...processes no longer operative?"

Clearly, such observations unequivocally demonstrate the falsity of the theory of uniformitarianism as set forth by Hutton and Lyell. It is this false theory that contributed to the last century of the Geological Dark Age. Recognizing that the geological processes that formed much of the Earth's surface are "no longer operative" opens the door for new discovery through new models, such as the Universal Flood Model, which will reveal how once-operative geological processes of sedimentation, cementation, and formation worked.

No hypotheses *under the current Rock Cycle paradigm* can account for the missing sizes of sediments described in subchapter 6.10. Pettijohn explains this in *Sedimentary Rocks,* concerning the two scarce or missing layers of sediment:

"**It is difficult to see, under any of the hypotheses** accounting for the apparent deficiencies of certain sizes, if these sizes were produced by weathering or abrasion, **what has become of them.**" Bib 159 p50

Recall that one of the missing sizes of sediment is precisely the size of the loess grains, one of the world's most important and abundant sediments. The origin of this particular sediment remains a complete mystery to modern geology. The loess sediment is a part of the Carbonate Mark in subchapter 8.8.

The FQs about the origin of pure quartz sand, cemented sandstone deposits, and missing sediment sizes have set the stage for a fundamental change in our understanding of sedimentology, and through the Universal Flood Model, "processes no longer operative" can be explained.

The Origin of Sand and Other Sediments

The most rudimentary explanation of sedimentary rock presently accepted is that:

"Sedimentary rocks are secondary rocks. **They originate on the Earth's surface** from the weathered debris of other rocks..." Bib 15 p260

Unfortunately, this widely taught tenet of modern geology is incorrect. Sedimentary material that originated from the weathering of the Earth's surface constitutes approximately 1% or less of all sedimentary material. There are actually two very different types of sediment:

1. **Erosionary Sediment** is sediment from surface weathering processes—conventional surface erosion, or sediment from subsurface processes—endoerosion.
2. **Hydrosediment** is sediment formed authigenically, which may be carried to its final location through aquifers and hydrofountains.

What distinguishes the two types of sediments is *how* they were formed. Detrital, or surface erosionary sediment is formed when fragments of rock are dislodged from their parent by conventional erosional forces, such as wind abrasion; it is the result of surface erosion. Endoerosion is sediment from subsurface processes. Hydrosediment consists of authigenic sediment, which was formed in-place, although it may be carried and dispersed elsewhere after its formation. So then, there are two ways in which detrital or erosionary sediment is formed (as defined in the New Geology subchapter 7.5):

1. **Surface Erosion** – conventional weathering of surface rocks via wind abrasion, water, or chemical processes.
2. **Subsurface Erosion** – erosion of subsurface rocks by water motion beneath the crust, also called endoerosion.

Modern science recognizes surface erosional processes. Investing heavily in research to understand them, geologists have assumed that nearly all of the Earth's sediment comes from such processes. A review of the available research on the subject reveals almost total ignorance of the processes of endoerosion and authigenic formation. The estimation that surface erosion processes account for approximately 1% or less of the Earth's sediment comes from the author's own research on a collection of hundreds of samples of clay and sand collected from around the world. However, the estimated figure is supported by observations in the Sand Mystery subchapter 6.3, which showed that further erosion of small sediment by wind or water rarely occurred in laboratory experiments. Clay, silt and most sand particles were simply too small for wind and water to propel the particles together with enough force to break the particles down any further.

Prior to the discovery of the Hydroplanet Model, no one realized how much water was right beneath our feet—except for perhaps the well drillers. It was interesting to talk to those who drill for water as a profession. They would describe the vast "lakes of underground water" they had encountered; water most of us are completely unaware of.

In the Hydroplanet Model, the Earth's hydroplumbing system is redefined, which is supported by piles of evidence that there is easily 100 times more water *below* the Earth's crust than there is on its surface. Most of that water is in constant flux, carrying sediment as it moves throughout the crust, being subject to endoerosion (underground erosion) processes. This has a greater quantitative influence than does weathering on the surface. Thus, most of the erosionary sediment material on our planet is not surficial, but from deep within abundant underground aquifers.

During the UF, active hydrofountains expelled copious sediment around the globe, forming the mysterious, homogenous sand, silt, and clay formations that lie plentifully across the world's landscape. Less common are heterogeneous sand and sediment deposits, which were formed during the last four-thousand years or so, since the UF took place; they are only a relatively minor constituent of the sedimentary material on the Earth's surface.

Among the world's vast, homogenous sand deposits, the majority are predominantly instances of *quartz* crystals. Where quartz of such purity originates has been a subject of much debate among modern geologists. There are no pure quartz mountains on the Earth and there is no pure quartz 'mountain' underground either, so the conventional explanation of sand's origin through erosion fails, and that suggests another FQ:

> If quartz sand did not come from **erosion** on the surface or beneath the crust, what was the source of this truly abundant sediment?

We already know from the discussion in Chapter 7.4 how

quartz crystals are made; in a *hypretherm*. Fig 8.5.1 illustrates the **UF Hypretherm** that existed during the Flood. It was the largest, most extensive hypretherm to have occurred since the Earth was formed, and it was during the time of the UF Hypretherm that a great majority of the Earth's sediments were formed. The new mineral classification of **Hydrosediment** presented in 7.5 defines the Earth's most abundant sediment; authigenic minerals crystallized in the supersaturated waters of the UF. When the erosionary sediments formed from surface wind and river action are compared with the vast deposits of hydrosediment materials of both loose and lithified structure, the former are almost inconsequential.

The Hyprethermal Sand and Sediment Formation Mechanism

There are many conditions necessary for mineral growth in the hypretherm, including an important biogenic factor. The role of microbes in mineral growth will be discussed throughout this chapter, including bacteria that dissolves and produces silica for mineral growth. Water, pressure, heat, and natural mineralizers were all present on the submerged surface of the Earth and beneath the crust all across the globe during the Universal Flood (Fig 8.5.1). There are places at the bottom of the ocean today (examples given later) where these condition exist, but on a tiny scale in comparison to the enormity of the UF event.

There were two locations where hyprethermal sand and other sediments formed:

1. **Subcrustal Hypretherm**–sediment formed underground is carried to the surface through hydrofountains where it is deposited.
2. **Surface Hypretherm** – sediment forms in hot, turbid ocean waters above the surface where it is drifts down, being shaped by the influence of active water movement.

Fig 8.5.1 shows the hyprethermal conditions that existed on the surface and below it. Pressure is created by water above the surface, heated by earthquake frictional heating. The environment that existed in the UF hypretherm was very similar to the experiment presented in Fig 7.4.13, which included the elements of heat, pressure, and water in an experiment to reproduce hyprethermal crystal growth.

Most of the sand and clay sediments on the Earth have a hyprethermal origin (see Fig 8.5.2). This means that they are crystals that precipitated and grew in a pressurized, hot-water environment. ***This is the mechanism*** *in which pure quartz sand, silt or clay deposits form.*

The **prethermation process** described in The Crystallization Process in subchapter 7.4 accounts for crystallization in hyprethermal environments in the following three ways:

1. **Temperature lowering**
2. **Pressure lowering**
3. **Pressure increasing**

With a change to at least one of these attributes, changes in the silicate water solute takes place. Even if no seed crystals are present, crystals can form:

"Crystallization in these hydrous systems can be promoted by **temperature lowering, pressure lowering, or pressure increase**." Note 8.5b

It is quite amazing that geology has puzzled over the Sand Mystery for decades, especially since researchers have known since the 1950s that quartz crystallization can occur by lowering the temperature or altering pressure in a hypretherm. Unfortunately, the two pieces of Nature's Puzzle, quartz crystallization and the prethermation processes have never been put together before, and since modern geology does not consider the possibility that the UF actually occurred, processes that form sand and other quartz sediment have been completely overlooked.

In the hyprethermal quartz growth experiments, crystals were grown in high-pressure autoclaves with an elevated temperature at the bottom of the system and slightly decreased temperature at the top. This created natural convection currents that facilitated crystal growth as the silica saturated solution experienced a temperature reduction. Crystallization of fine sediment during the UF would have been affected by several factors (see Fig 8.5.2). In one scenario, microscopic silica seed crystals would begin to grow from the silica-rich waters formed as quartz rocks dissolved in hotter areas of

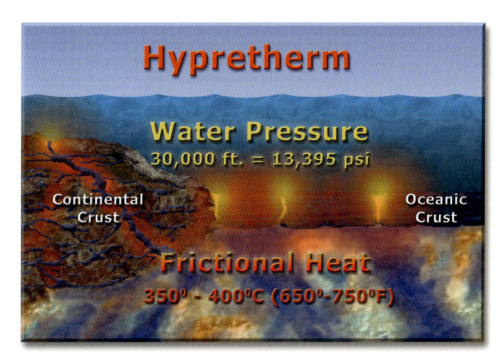

Fig 8.5.1 – The Hypretherm is created when water is under high pressure and high temperature. Today, hypretherms exist at the bottom of the ocean in areas where frictional heating supplies the necessary temperature, in places such as plate boundaries. The most extensive Hypretherm since the Earth's formation was the UF Hypretherm, when water covered entire continents to great depths, perhaps exceeding 30,000 ft (9000 Meters). Great land movements generated tremendous frictional heat needed for the Hypretherm environment.

Hyprethermal Sand Origin

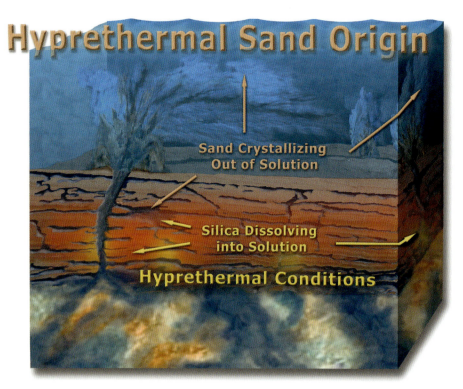

Fig 8.5.2 – This diagram depicts the Hyprethermal Sand Origin, which is the origin of much of the Earth's sand. During the UF, the entire surface of the Earth was covered with water heated by frictional earthquake heating; areas on or within the crust of sufficiently high heat and pressure experienced hyprethermal conditions. Dissolved preexisting silica from quartz-based rocks provided the material required to start the crystallization process of silica sediment. Some of the sediment formed in the water above the crust and fell to the ocean floor of the Flood, whereas the quartz sand crystallized beneath the surface and was ejected through hydrofountains over vast areas, such as the Badlands in South Dakota, USA.

the crust. As the crystals grew, becoming too large to remain suspended in ocean waters, they fell towards the ocean floor. Many seed crystals grew from oceanic microbes, which we will discuss later. Sediment crystals of clay, silt, and sand formed in the hyprethermal waters due to an *increase in pressure* as particles fell deeper and deeper until finally resting on the ocean floor. This sand-forming mechanism explains why the mineral content of sand crystals is so pure, and why there is such similarity in size.

Sediments could have crystallized in underground pipes or hydrofountains as *pressure was increased or decreased* by temperature changes brought on by frictional earthquake heating. The crystallized sediment could then have been ejected onto the surface, as seen on the bottom of the ocean today where mudfountains spew sediment at tectonically active zones. Another common sediment crystallization process occurs when *water temperature undergoes a change* in an upward flowing 'pipe' filled with micro-crystals in suspension. In each of these cases, the prethermation process forms different types of sediment, based on the minerals dissolved in the solution and the environmental changes that take place in the hyprethermal waters. In this way, clay, silt, loess, sand, and other sediment grew out of solution to be deposited in the ocean and on submerged continental surfaces everywhere, during the UF.

The Hyprethermal Sediment Mechanism explains how miles of various homogenous sediments were deposited across the surface of the Earth's crust. This is the same mechanism that cemented the various sediments, turning them into stone!

Although geologists do not realize that prethermation processes formed much of the Earth's sediment, some scientists have recognized a similar process in sugar crystal formation. A "seeding" process is used to crystallize brown sugar to form crystals of uniform size, just like we find with pure sand crystals. From *Nature* 2004:

"The generation of a solid from a solution through crystallization might sound such a simple, familiar process that one could be forgiven for thinking that it is fully understood. But despite the work of Wilhelm Ostwald on crystal nucleation, and the development of classical nucleation theory, **this is not so**. Take an everyday example such as brown sugar, millions of tonnes of which are crystallized annually, to be dissolved in tea and coffee. **The crystals are all of uniform size—no really big ones, no really small ones—and all are nicely faceted. This is no accident: the effect is achieved by a process known as 'seeding', in which small crystals of pulverized sugar are introduced in the solution to act as seeds on which crystal growth can start**. This seeding, however, does not follow the rules of the garden, with one new plant (or in this case, crystal) per seed. In fact, the amount of seed needed to catalyze the whole crystallization may be less than 1% of the mass of the final product." Note 8.5c

It is an interesting side note that while quartz growth experiments were being conducted as part of the UM research, the seeding process was observed to occur naturally during some of the test runs. When experiments included organic material during the fossil making process, a seeding of the entire autoclave reactor occurred producing some very interesting results. These will be discussed later in the Surface Mark subchapter and in the Fossil Model chapter.

Before we continue with the Sedimentation Model and the explanation of how the majority of crustal sediment was formed, a short discussion about the Double Terminated Crystal Mystery in 6.4 (Fig 6.4.15-16) will help conceptualize the growth process of sand crystals. Although it requires a microscope to clearly distinguish individual sand grains, some quartz crystals, like those seen in Fig 8.5.3 can be several inches long, making them easy to examine and compare with smaller sand grains.

The Hyprethermal Sedimentation Model

Having set forth the Hyprethermal Sand and Sediment Formation Mechanism, we are in a position to solve the Sand and Clay Mystery from the Rock Cycle Pseudotheory Chapter. Although the typical geology textbook claims the erosion sci-bi is the origin of sand, silt and clay, the Loess Mystery of 6.10 cited investigators from the field of sedimentology acknowledging that the small particles are not now, nor have they been formed by "normal stream abrasion." Additionally, the crushing process typical of weathering does not create these particularly small particles:

"The general ineffectiveness of normal stream abrasion on particles smaller than 1 to 2 mm **precludes the formation of silt and clay by continued size reduction of coarse materials**. Furthermore, lognormal distributions are **not** formed by single-stage **crushing processes**." Note 8.5d

Even back in 1957, after technologists had learned how to grow quartz crystals in the lab, some geologists realized that there was a *real possibility* that an "appreciable volume" of crustal sediment was *not* "produced by rock disintegration":

"Either they have been **produced by rock disintegration** but are mechanically unstable **or they never were formed in appreciable volume**." Note 8.5e

However, this bite was simply too big for geology in general to chew. Everyone 'knew' sediment came from the disintegration of rocks, and no scientist cared to look further. Even so, the Sand Mystery questions were never answered. Some, like Francis John Pettijohn, who was familiar with the quartz crystallization process technologists had perfected, saw that clay could also be "precipitated from solution":

"'In other cases, however, the **clay is coarsely crystalline** and may possibly be introduced and **precipitated from solution**.'" Note 8.5f

This was a very important observation because it identified the *only* way that almost all of the clay and claystone deposits formed—"precipitated from solution." Clay is far too small to erode from larger sediment as discussed in the Rock Cycle Sand Mystery, and is generally homogeneously pure quartz, too pure to have originated from granite or other parent sources via erosion. Clay is "coarsely crystalline," not eroded smooth because it was formed in the Universal Flood Hypretherm.

From the former discussion on the origin of sediment and the mechanism by which it is made, we can now identify the **Hyprethermal Sedimentation Model:**

1. Surficial Erosion did not form the majority of sand, silt and clay sediment.
2. The Missing Pebble and Sand sizes were not formed in the first place.
3. The majority of all sediments prethermated from a hyprethermal solution.

As with any new scientific model, it must stand up to scrutiny, and to be objective we must ask:

FQ: What scientific proof is there of a Hyprethermal Sand Origin?

To demonstrate the reality of the Hyprethermal Sedimentation Model we must turn to the field of sedimentology for evidence that the majority of the world's sediments were formed authigenically—formed 'in-place' instead of being detrital or from conventional erosion. No one would dispute that this is an extraordinary claim, and as such, must be backed by some extraordinary evidence.

Double Terminated Quartz Mystery Solved

The mystery of the double terminated quartz crystals first appears in The Rock Cycle Pseudotheory in Fig 6.4.15. Although most quartz crystals grow on a surface such as the inside of a geode or in cracks and veins, double terminated quartz crystals are terminated naturally at both ends of the crystal, which means that the crystal remained detached during the growth process; it

Fig 8.5.3 – These naturally double terminated quartz crystals from China average about 1cm. They were formed in a hypretherm while suspended in solution in a hydrofountain with a flow great enough to suspend the crystals during their growth period. Once they were too large to remain buoyant, they fell out of the precise crystal-forming hypretherm. Many quartz sand deposits were formed in a similar prethermation process during the UF.

was not 'touching' the surrounding rock when it formed.

FQ: How did double terminated crystals form?

Modern geology has no idea how these double terminated crystals, sometimes called "floaters" among mineral collectors, formed. In some locales, they are relatively common among various surface sediments.

If a crystal shows no attachment point on any of its faces, it must have formed while ***suspended in solution***. Small clay-sized or silt-sized crystals take only minutes to grow under the right conditions and are light enough to remain suspended for long periods. Quartz crystals are two and a half times heavier than water, so larger crystals sink quickly, which indicates that large double terminated crystals formed in a solution that was *moving upward* to offset the force of gravity.

An upflow of hyprethermal water in a hydrofountain lasting hours or days would allow crystals to grow from microscopic to several centimeters in size without being attached to any surface. Larger crystals required more time, and a continually replenished source of supersaturated hyprethermal fluid. Once again, the understanding of the *hydrofountain* concept is critical to comprehend how these beautiful jewels of Nature were formed.

Significance of Double Termination

Why are double terminated crystals so important? The answer to this question comes by understanding the environment that had to be present for the crystals to grow. First and foremost, hyprethermal conditions must be present. Secondly, a constant source of upward moving, silica-rich water must be present. Since we know this is how these crystals formed, we can ask another FQ:

Are double terminated crystals being formed today?

There are no known instances in modern times or in recorded history that such crystals are being formed naturally.

Double terminated crystals are found in sedimentary material usually ejected by hydrofountains or in cavities in various locations worldwide. One popular place is Herkimer, New York, USA, where Herkimer "Diamonds" (double terminated and single terminated quartz crystals) are found in pockets in dolomite, which appears to have hardened after the flow of water ceased. This formation is found on the surface (Fig 8.5.4). Another location where pockets of so-called 'diamonds' in dolomite are found is Diamond Point, near Payson, Arizona. There, perfectly clear quartz crystals, including many double terminated specimens can be found in the surrounding sediment see Fig 8.5.5.

It isn't too surprising that little research exists on double terminated quartz crystals even though they are a popular collector's item readily available at the world's mineral shows. Until now, little geological importance was placed on them. But now, their presence is significant proof of the hyprethermal conditions that once existed, and they set the stage for the Hyprethermal Origin of Sand.

Another important aspect of the double terminated crystal mystery is the fact that the crystals show little or no weathering:

> FQ: Why do most of the crystals show no signs of weathering?

Such questions are difficult to answer by studying modern geology's origin of rocks. The obviously untransported crystals must have formed where they were found. Reviewing available research from continental borehole drillers revealed no evidence that a hyprethermal environment of sufficient temperature, pressure, and mineralizer has ever been encountered, where quartz crystals of this type are growing. However, exploration of the ocean floor has revealed that there are places with just such an environment, one being Tag Mound, located off the Pacific coast of North America. There lies conclusive proof of the hyprethermal conditions required for most crystal formations found on the Earth's surface, both under the ocean and on the continents. The reason scientists have not found crystals like the Herkimer Diamonds forming anywhere on the continents today is that they, like the vast majority of all rocks and minerals, whether on the mountains, the plains, or the valleys, formed in a *deep, hot ocean*. The evidence supporting this new geological discovery is enormous, and surprisingly intuitive.

Herkimer "Diamonds"
(Quartz Crystals)

Fig 8.5.4 – Herkimer Diamonds are beautiful quartz crystals from New York. In the background image, a diamond is shown in situ, as it was found in a pocket, partially attached to the surrounding rock. In the foreground, a double terminated crystal with a water enhydro is shown.

Diamond Point
Quartz Crystals

Fig 8.5.5 – Diamond Point quartz crystals seen here on dolomite matrix. These crystals formed in place, on the rock in silica saturated water. Double terminated crystals must have formed in an environment of upwelling water keeping the crystals floating while they grew unattached. The natural method they grew in has not been observed, making them a geologic mystery. The hypretherm and other processes active during the UF solve the mystery of the double terminated 'floaters.'

The answer to the previous FQ, why there is almost no weathering evident on these crystals, is that *they are not very old*. The lack of erosion proves that the crystals were not transported after they were formed, they were simply buried. This of course, is not easily explained using the millions-of-years uplift/erosion/subduction theory. Their formation and subsequent burial is easily understood as active fluidization of heated, silica-saturated water, under specific pressure

flowed for a period of only a few days or weeks. Changes to the system; temperature, pressure, mineralizer, fluid velocity, etc, ceased crystal formation. However, heat, pressure, and mineral saturated water was still present, but in a not-so-perfect balance as to form specimen quality crystals. Many of the crystals exhibit clues to the concurrent crystallization of dolomite and calcite, and they exhibit a distinct and direct connection to the formation of hydrocarbons. Because some crystals actually have oil inclusions in them, there is no doubt that the crystallization process can be associated with biogenic activity.

The Biogenic Opal Evidence

It has been only recently that researchers identified unequivocal proof of the biological origin of some minerals. From the 2003 *Earth-Science Reviews*:

"There is an **increasing awareness** of the role of biota [organisms] in the regolith [sediment] generally. Examples include the **precipitation of iron and manganese** (Skinner and Fitzpatrick, 1991) and **interaction between bacteria and minerals** (McIntosh and Groat, 1997)." Note 8.5g

Indeed, bacteria in the sediment have been found to dissolve silicates and store it in their cells:

"Many groups of regolith bacteria are known to **dissolve silicates through enzyme and organic acid secretion**. Some, such as *Proteus mirabilis*, also **store the silica within their cells and in slime layers** as monomeric silica." Note 8.5g p181

The mineraloid opal is an amorphous (non-crystalline) form of silica, very similar to quartz, but much softer because it is not made under high temperature and high pressure like quartz. Silica-loving bacteria and an increase in the dissolved silica in natural water solutions make more silica available to crystallize and grow into sediment, given the right hyprethermal environment. The dissolved silica can also be used by other organisms to grow **Biogenic Opal**. In the past, biogenic opal was once considered an insignificant component of Earth's regolith (surface sediment), but is now known to be "a significant component of many soils":

"Wilding et al. (1989) regarded most soils as containing up to 3% biogenic opal and cited examples where up to 20% was present. **These values mean that opal comprises a significant component of many soils** and can commonly exceed the abundances of potassium, calcium, magnesium, sodium and phosphate. However, **biogenic silica is normally regarded as a minor component and the potential role of organisms in the large-scale deposition of silica in the regolith is largely ignored**.

"Despite this, there is considerable evidence that opal from phytoliths, diatoms, sponge spicules, and other organisms are often abundant in the regolith in many localities. **Widespread and common accumulations >2% by grain abundance are known from every continent except Antarctica in environments as diverse as coastal and inland swamps, forests, grasslands, and flood plains**. They are present in the tropical, temperate, and semi arid regions." Note 8.5g p175-6

As we can see, researchers have identified opalized organic *sediment* the world over; however, *living* examples of organic opal biota are not present in all of these areas. Why is this? The answer in *Earth-Science Reviews* comes from another very important observation; siliceous sponges normally comprise "only a minor component" of certain ecosystems, yet they can be prolific. One example; "*100% sponge spicules, form the land surface over large areas of southern Australia*":

"Siliceous sponges are **normally only a minor component of marine and terrestrial ecosystems**. Their presence in sediments and soils are similarly normally minor. Under conditions **not fully understood** they can, however, **proliferate and even dominate**. Late Eocene specular marine and marginal marine sediments, locally with up to **100% sponge spicules, form the land surface over large areas of southern Australia**. The reasons for this proliferation **during a narrow time interval are not fully known**, but probably related to unique confluence of runoff, turbidity, and nutrient conditions." Note 8.5g p180

The worldwide distribution of biogenic sediment has been documented with some continents being covered by as much as 100% of the "*land surface over large areas.*" This happened because of "*flooding*," which allowed a "*sponge bloom*" of a magnitude great enough to be a significant component of the sediment":

"The common feature in all of these environments is seasonal to permanent **flooding**. Under such conditions the diatoms and sponge **bloom in sufficient quantities so as to make a significant component of the sediment**." Note 8.5g p186

Floodwaters bring various colloidal and fine sediments that provide the necessary nourishment for the microbes to bloom in

Fig 8.5.6 – Diatoms are microscopic organisms living in the seas. They are composed of opalized SiO_2 or quartz. In this Homegrown Authigenic Quartz mechanism, dissolved silica attaches to and fills the micro-shells in profusion during the UF. Both quartz and pyrite formed in the same hyprethermal; the microorganisms living and dying in the Flood waters provided the **seed** material for crystallization and for most of the Earth's clay, silt, and sand sediments.

extraordinarily large numbers. Present-day blooms still occur, but not on the scale recorded in the Earth's soil. We'll visit the subject of biological blooms again because of their importance to the geological record and the Universal Flood.

Homegrown Authigenic Quartz

Knowing that the extent of the biogenic silica is worldwide and that some areas are comprised of up to 100% biota, we can now understand another recently discovered mechanism describing the creation of *quartz* sediment grains. The significance of this discovery has important implications for the Universal Model but remains largely unknown because it does not fall within modern geology's Rock Cycle paradigm.

Research has revealed that a large percentage of the *quartz* sand and clay on Earth's crust is *not* detrital (from surface erosion), but *is formed in-place*. This material is called **authigenic** or 'homegrown' quartz.

Jürgen Schieber of Indiana University notes in an article in the journal, *Nature*, that geologists have long considered the most abundant sedimentary rock type as being "detrital," eroded from "continental crust":

"Mudstones are the most abundant sedimentary rock type, containing most of the quartz in the sedimentary record. This quartz is largely in the silt size range and **has long been considered detrital, derived from continental crust**." Note 8.5h

However, when Schieber reexamined the mudstones, he found them to be *quartz filled algal cysts*:

"Petrographic reexamination (by light microscope) of Late Devonian mudstones from **across North America reveals abundant quartz silt grains that are texturally identical to those identified here as authigenic quartz infills of algal cysts**." Note 8.5h

Fig 8.5.7 includes images of microscopic sediment showing quartz and pyrite filled algal cysts that are the skeletal remains of a once prolific algal bloom that occurred in the sea. Instead of mudstone and sandstone originating from surficial erosion as once thought, *it formed in place*, prethermating from seawater as temperature decreased or pressure changed (increase or decrease). The homegrown authigenic quartz also fills large, sand size (0.1-1.0 mm) algal cysts:

"Very early diagenetic **quartz also fills large, sand size (0.1-1.0 mm) algal cysts** in these rocks. These cysts, remains of planktonic green algae, have walls that consist of complex lipoid-like substances that are highly resistant to chemical breakdown and bacterial degradation." Note 8.5h

> "…as much as 50% of the quartz "grains" in the rock record (and I am helped here by the preponderance of mudrocks) are of authigenic origin."
>
> Jürgen Schieber

"I would like to know that one myself, but I would not be surprised if **as much as 50% of the quartz 'grains' in the rock record** (and I am helped here by the preponderance of mudrocks) **are of authigenic origin**." Note 8.5i

How does such a discovery impact the interpretation of the 'geological record'?

> What would happen with the interpretation of the 'geological record' if much of the Earth's sediment did **not** erode from mountains over millions of years?

Schieber himself realized:

"If authigenic/intrabasinal quartz silt is widespread, **a large portion of the sedimentary record may have been misinterpreted, with important implications in a variety of research areas**. Although the identification of authigenic quartz silt is not a trivial matter, its recognition offers a variety of **new opportunities to better understand the geologic record**." Note 8.5j

After reading Schieber's research papers on authigenic quartz, we asked him how widespread he thought authigenic quartz grains were as compared to the distribution of detrital quartz grains. Schieber responded:

Fig 8.5.7 - These photomicrograph slices of small pieces of sediment are some of the most important images ever taken of micro minerals. The slices of sediment show they grew from microorganisms in the ocean in supersaturated hyprethermal mineral waters. They represent solid evidence that the majority of the Earth's sediment is authigenic hydrosediment (clay formed in place during the UF Hypretherm, not by erosion), not detrital sediment. The top right image is an organic quartz grain cyst in a sandstone bed. The outer rim 'R' is chalcedony. The grey bar in the image provides scale, it is .05 mm. To the left is a pyrite Ooid, taken from Chattanooga Shale. The white bar is larger, at .5 mm. The bottom right image is of a spherical cyst in a dolomite shale bed showing the radial fibrous texture of chalcedony that filled the organic cyst. Here the scale bar is .1 mm. These images are courtesy of Jüergen Schieber (Indiana University), who established the prevalence of organically grown sediments in the geological record is "as much as 50% of the quartz grains in the rock record." The origin of this common, authigenic sedimentary material has never been explained with much clarity, even though it represents such a significant portion of the world's sediment.

CHAPTER 8 THE UNIVERSAL FLOOD MODEL

Fig 8.5.8 – This silica sand mine near Overton, Nevada, USA produces some nearly pure, clear quartz sand (inset photo). White deposits of sand lay right next to red deposits, as can be seen in the image. Such sand deposits cannot be explained by erosion processes, and are only comprehended in the light of the UF Hypretherm. Notice the dark iron minerals lying on the surface of the rocks in the foreground; these iron minerals are the same as those found on the rim of the Arizona Hydrocrater. They can be found throughout the Colorado Plateau area, only on the surface. The iron minerals are additional evidence of the UF but will be discussed later in the Surface Mark subchapter.

face of the Earth, which has seen little change since its formation and deposition.

2) The environment included silica-rich, deep, hot (at least warm) sea water. Only a hyprethermal environment supersaturated with silica can account for crystal growth large enough to fill the algal cysts.

Even though research on homegrown quartz is straightforward and uncontested, and though it changes the entire geological-time paradigm, modern geology remains essentially unaware of it. Schieber has only just begun research on other sediments, which he thought might also have an authigenic, or homegrown origin. This is an area holding great research potential for future Millennial Scientists!

The Clay Evidence

Clay is a very important constituent of the sediment found on the Earth. The book, *Rocks and Minerals & Gemstones* states this:

"**Clay and claystone** are the **most widely distributed of all sedimentary rocks**." Bib 15 p276

There is an obvious reason for clay be-

In what might be considered the scientific understatement of the year, Schieber stated:

"...a **large** portion of the sedimentary record may have been **misinterpreted**..."

Schieber's discovery of Authigenic Quartz ran through the scientific community. In an article in *Science* titled, *Homegrown Quartz Muddies the Water*, the author states:

"Mudstones, now commonly exposed in cliffs and roadcuts, were formed in the sea by clay that was washed from land. The clay contains fine grains of quartz. The size and distribution of these grains, **it's believed**, can reveal how far the sediments traveled from shore or even whether they took an airborne journey from a desert. Such **inferences assume** that quartz silt, like mudstone clay, **started out on land. However**..." Note 8.5k

The article goes on to say that instead of land-surface erosion, "the silt had grown in place":

"The mudstone quartz grains were more similar to **quartz precipitated under the sea than quartz from mountains**. The local quartz was surprisingly common, too. In some samples, Schieber found that **all the silt had grown in place rather than eroding from land**." Note 8.5k

The scientific community's reaction to the discovery of homegrown quartz is apparent in the article as one geologist commented; "'The finding "makes life more complicated..."'" Note 8.5k The answer to why a geologist's life would be more complicated is that instead of supporting the Rock Cycle and Uplift Pseudotheories, the new discovery suggests an environment that included:

1) A sea that covered the entire Earth over a short period of time, not too long ago. This is apparent because of the widespread distribution of authigenic sand, silt and clay on the *sur-*

Fig 8.5.9 – This view of sand from Carlsbad Beach in San Diego, California, USA, shows a mix of mostly pure quartz and iron oxide crystals. The black iron particles are seen attached to a magnet at the top of the photo. This sand is similar to the Imperial Dune sand we examined in the Rock Cycle Pseudotheory chapter. As we already pointed out, there are no inland mountains of only quartz and iron where this sand could have eroded. With the UF paradigm, we can now understand how these types of quartz sands crystallized in a hypretherm along with a variety of iron reducing bacteria, which formed the iron oxide crystals. This is why both grains are of similar size and why there is such a concentration of these minerals.

SUBCHAPTER 8.5 THE SAND MARK

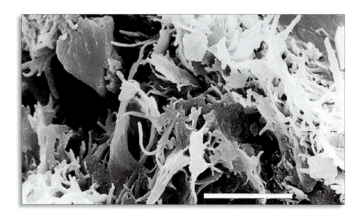

Fig 8.5.10 – Like all other sediment, clay was believed to be sedimentary material from erosional processes. But as can be seen in this scanning electron micrograph (SEM), the rounded characteristics of erosion are noticeably absent. Modern science has since turned to "chemical weathering" by acidic water as the primary mechanism for clay's origin, but as the Rock Cycle chapter demonstrated, natural waters are almost all alkaline—not acidic, thus refuting their theory. Instead of erosion or weathering, the stringy structures of the clay are a form of crystallization, a process that takes place in water.

ing the "most widely distributed" sediment, but the geological community, staunchly loyal to the erosion sci-bi, is unaware of it. Advances in technology have made it possible to view clay like never before, as seen in Fig 8.5.10. In this scanning electron micrograph of clay, it is apparent that there is no evidence of erosion. Research into the Loess Mystery introduced in Chapter 6.10 revealed that clay particles are far too small to originate from any type of mechanical erosion. The realization of this fact led to the introduction of another unfounded "chemical weathering" process, best described by *Wikipedia*:

> "**Clay minerals** typically form over long periods of time **by the gradual chemical weathering of rocks,** usually silicate-bearing, **by low concentrations of carbonic acid** and other diluted solvents. These solvents, usually acidic, migrate through the weathering rock after leaching through upper weathered layers." Note 8.51

Once again, geology was misled down the back alley of erosional "weathering." The Carbonic Acid Cave Formation Pseudotheory, mid-chapter 6.9, showed that the vast majority of crustal sediment waters do not have "low concentrations of carbonic acid," instead, have just the opposite—*they are slightly alkaline*. Caves were not formed by "acidic" solvents and neither was clay sediment. Researchers seemed to know this but were apparently unwilling, or unable to accept where the evidence was leading; that clay is grown in water. In the end, *Wikipedia* concedes that "some" clay sediment comes from "hydrothermal activity":

> "In addition to the weathering process, **some clay minerals are formed through hydrothermal activity**."
> Note 8.51

The real source of the most widely distributed of all sedimentary minerals is silica crystallized in water. Although extremely small, single silica crystals have been seen to grow in experiments at room temperature, they have never been shown to produce ***deposits*** of clay, even over long periods of time. This is because there is not enough dissolved silica in natural streams, lakes, or the ocean, and because the temperature and pressure is insufficient. In other words, a *hypretherm* is not present in the environment today, and large, quartziferous (clay) rock deposits cannot form. Earlier in the chapter, we showed that warm silica water under low pressure does not produce quartz sand or clay, it produces geyserite. But what does this all mean?:

> If clay is formed in a hypretherm and is the "most widely distributed" of all of the Earth's sediments, the *entire surface of the Earth* **must have at one time been in a hydrothermal environment!**

Researcher's assumptions that clay and claystone are byproducts of chemical weathering or low-pressure hydrothermal activity suggest that large deposits of clay should be abundant in these rare environments—but they are not. Such false assumptions have also led to other incorrect ideas, like the notion that:

> "Clay rocks are **always** stratified." Bib 15 p276

This statement comes from one of the most popular rock and mineral books in the world, and demonstrates how common misstatements like this are. Fig 8.5.11 is an image of beautifully colored claystone from the Colorado Plateau. Similar specimens can be found around the world. Obviously, they are not "stratified," but are instead, a testament to the formative environment of the hypretherm. These specimens show that they did not form simply by the weathering of other rocks. The brightly colored bands in the specimens in Fig 8.5.11 (and in the Rock Cycle chapter Fig's 6.3.17, 18 and 19) are indicators of temperature changes to the water in which they formed and of diverse bacteria that thrive in those varied temperatures.

Even today, at hot water geysers, water of different temperature supports different species of bacteria, but no quartz clay or sandstone is being formed because the geysers, like those at Yellowstone Park, are low-pressure surface features.

For the first time, there is a clear and concise explanation for the Clay, Mudstone, and Zebra Rock Mysteries identified in the Rock Cycle Pseudotheory chapter. These are also additional testimony of the Universal Flood event.

Death Valley Sand Dune and Clay Evidence

Look at the top and right images of Fig 8.5.12 and ask your-

Fig 8.5.11 – Layered sandstone and claystone, like these specimens from the Colorado Plateau, USA have not had an explanation for their origin in modern geology. However, the mystery can be answered with the UF Model. For lithification to take place and for the cementation of sand grains with silica to occur, hot water saturated with a mineralizer under high pressure is necessary. The UF model provides the hypretherm, which explains the origin of beautiful deposits of nature like these.

self; did the light-colored sand come from the dark Cottonwood Mountains in the background?:

"The primary source of the dune sands is **probably** the Cottonwood Mountains which lie to the north and northwest. The tiny grains of quartz and feldspar that form the sinuous sculptures that make up this dune field began as much larger pieces of solid rock." Note 8.5m

It is astonishing that key facts about sand formation have been overlooked for so long. The erosion sci-bi evident in the above quote is a foundational pillar of the Rock Cycle Pseudotheory and has been so for more than a century. Because of error, and a subversive agenda, modern geology threw out any possibility of recognizing the Universal Flood event, and as a result, its search for the true origin of sand deposits, like those in Death Valley has proven fruitless.

The false-color Landsat image in Fig 8.5.12 reveals different mineral deposits, with no evidence that the mountains surrounding the Racetrack Playa clay deposit or the Mesquite Sand Dunes are of the same composition. This is strong evidence that the deposits did not come from *any* conventional weathering process.

One feature abundant throughout Death Valley, but uncommonly noticed are hydrofountains. They deposited sediments of all types in the Valley; clay, sand, cobblestones and even larger sediment. One reason this happened is that the area is criss-crossed with faults, which produced frictional heat and high-energy steam eruptions:

"The valley is bisected by a right lateral strike slip fault system, represented by the **Death Valley Fault** and the **Furnace Creek Fault**. The eastern end of the left lateral **Garlock Fault intersects the Death Valley Fault**." Note 8.5n

In the final stages of the UF, soft clay and sand deposits of great variety beautifully grace the flanks of now-dormant hydrofountains, long since inactive. It is curious how anyone could have thought sediments, like those in Fig 8.5.13 could have been deposited by weathering.

The Origin of Sandstone

Sandstone is comprised primarily of individual grains of quartz (most common sandstones), but what is the glue that cements sand grains into sandstone?:

"**Quartz** is the most common **cement** in sandstones." Note 8.5o

The seemingly simple chemistry and crystallography of quartz is a "source of many conflicts and disagreements" among scientists, according to a conference of the International Association of Sedimentologists, in 2000:

"**Quartz** is a major porosity-destroying cement in many **sandstones**. Despite its simple chemistry and crystallography, it is the **source of many conflicts and disagreements** within the petrographic community about practically **every aspect of its genesis**." Note 8.5p

The reason the cementing agent of sandstone caused such a division amongst scholars is because *no one knows **how** or **when*** quartz became the cementing material. Even back in 1957, Pettijohn declared that the most fundamental questions of "how and when" sands become cemented, and the source of the cementing material was "unresolved":

"The problems of *how* and *when* sands become cemented and the source of the cementing material **are still unresolved**. There has been renewed interest in these problems in recent years.

"It was long supposed that meteoric or artesian circulation carried the cementing materials into the sandstone and there deposited their silica or carbonate. It is known that ground

Fig 8.5.12 – In the false color Landsat satellite image of Death Valley, California, a variety of mineral deposits are apparent. Notice the two patches of white, which are Racetrack Playa Clay and Mesquite Sand Dunes. There are no 'white' mountains around them from which weathered or eroded material appears to have come. Apparently, this was overlooked by geologists, which is understandable since they had no knowledge of hydrofountains and their landscape-changing role, especially during the Flood. Fault lines run right through this area, evident by the lava extrusions in the vicinity, which testify to the earthquake friction that was generated here, which caused hydrofountains to erupt and spew clay and sand onto the surrounding surface.

SUBCHAPTER 8.5 THE SAND MARK

Fig 18.5.13 – Colorful clay-sand mounds near Caineville, Utah, USA provide convincing testimony that the foundation of sedimentary geology, uniformitarian-ism—is false. With the new UF paradigm, these types of deposits, never before recognized as Flood deposits, are clearly understood. Across the landscape there are thousands of such examples, making it a wonder why we have been unable to see them for what they are.

waters do carry these materials in solution, that artesian flow does occur, and that materials may be precipitated from solution. Van Hise (1904) reviewed the problem at some length. He supposed that the silica (or other cementing material) was dissolved in the belt of weathering and redeposited in the belt of cementation. Downward moving waters would dissolve as they were warmed up, and upward moving waters would precipitate as they moved to cooler regions. Van Hise pointed out, **however, that the silica content of ground water was very low**. On the average there is only 1 part of silica to 50,000 of water. **To cement a cubic mile of sand** (with average porosity of 26 per cent) **would require 130,000 cubic miles of average ground water**." Note 8.5q

It is hard to imagine *130,000 cubic miles of ground water*. The highest mountain in the world is not even six miles above sea level. The amount of ground water purported to have flowed through one cubic mile of unconsolidated sand to cement it into sandstone is 130,000 cubic miles—a ratio of 130,000 to 1, a number beyond comprehension. This process is not happening anywhere in the world today, which is why forthright geologists must acknowledge they *do not know* "how or when" the world's sandstone deposits formed.

A more recent publication identifying the **significant lack in the understanding** of the diagenetic process (hardening of sediment after deposition) can be found in the publication: *Sandstone Diagenesis Recent and Ancient, a Reprint Series of Volume 4 of the International Association of Sedimentologists,* 2003, p36. In it, they say:

"**Significant gaps remain in our fundamental understanding of diagenetic processes** and mineral properties that require research focused on **low-temperature geochemistry for their resolution**." Note 8.5r

"Low-temperature geochemistry" (at or around room temperature) will never reveal how sandstone is formed from sand because large natural sandstone formations did not form in low temperatures. This is why there are no large natural sand deposits that are turning into sandstone formations today.

There are many reasons the groundwater cementing process of sandstone does not work today; for one, miles of heterogeneous, loose sand from surface erosion is rare. If geological time was real, heterogeneous sand would account for the vast majority of the world's sand. However many of the Earth's enormous sandstone formations are *homogeneous*, meaning they are made of nearly identical material, without major changes in color or sediment type and exhibit very little layering (see Monument Valley, Arizona, USA in Fig 8.5.14).

If the sand had been deposited over a long period of time, climate change, wind, and water would dictate different patterns, leaving behind layered and impure sand. Yet Monument Valley sandstone deposits, like many of the World's sandstones, testify that a single, thick layer of homogeneous sand was cemented by quartz, *at one time*.

Moreover, if flowing groundwater had cemented the sand grains together, a significant *cementation gradient* would be evident. A typical desert may experience around six inches of rainfall per year, which means that most rainstorms produce only a fraction of an inch of precipitation, wetting only the surface. Larger storms would moisten grains several inches below the surface but occur less frequently. Even during periodic downpours, penetration of rainwater would not be expected to exceed several dozen feet. In all, the upper surface layer would experience considerably more evaporation and silica deposition than deeper layers—which would contribute to a cementation gradient, but this is not seen in Nature.

In fact, it appears that ambient temperature ground water releases very little silica, far less than that required to crystallize in the interstitial (pore) space between the sand grains. Even if we discount the fact that there is little evidence of silica from surface water flow, sandstones cemented by silica from a supersaturated solution, do not remain cemented when exposed again to water—*they disintegrate*.

Pettijohn revealed another point why meteoric water could not have been involved in the cementation of sandstone:

"Owing to the fact that some sandstones, especially those in the deeper basins, **are filled with salt waters**, many have supposed that **no meteoric waters have ever circulated through them, and therefore another source for their cement must be sought**." Note 8.5s

And so it was that for 50 years, sedimentologists have been searching for a new source. According to the International Association of Sedimentologists conference in 2003:

"**Numerous sources** of quartz cement have been advocated…" Note 8.5t

The result of numerous attempts to identify a source of the quartz cement yielded some fruit; though sedimentologists have yet to figure out that high pressure is needed for cementation, they became aware of the high temperature required, identifying it as a "thermodynamic driving force":

"The occurrence of quartz cement in sandstones depends on two key controls. For quartz cementation to occur at all, there **must be a thermodynamic driving force**…" Note 8.5u

In another statement, made by sedimentologists at the 2003

Fig 8.5.14 – Monument Valley, Arizona, USA, is famous for its red sandstone spires that rise majestically into the sky. These landforms are comprised of a continuous series of layers of homogeneous sand unsullied by sediment and materials from rivers or wind-borne weather phenomena. Moreover, there is simply no mountain source from which the sand could have eroded. The true source of the sandstone is the UF hypretherm.

conference, researchers realized that it was a "dangerous assumption" to use temperatures *higher* than *exist today* in sand deposits, and to then "extrapolate" the data back to lower temperatures, acknowledging that such extrapolation is "risky business":

"Experimental approaches have been used successfully in metamorphic petrology but the relatively low temperatures encountered in diagenesis mean that reactions **cannot** easily be synthesized under realistic conditions in a laboratory. The usual route is to **increase the temperature** (beyond the diagenetic realm) and then **extrapolate rate data back to lower temperatures. This requires the dangerous assumption that the mechanism is identical at high and low temperatures. Over and above the mechanism issue, extrapolation is always a risky business with ever greater uncertainty away from the control region**." Note 8.5v

Why were higher temperatures employed at all, instead of the temperatures that exist at sand dunes and on the bottom of the ocean today? Because low temperatures do not work, this is one source of the confusion surrounding the current quartz cementation theory. In fact, the temperatures required are elevated considerably above the normal ambient range:

"**Quartz cement** may provide similar opportunities for finding **inclusions that can delineate the environment of sandstone digenesis**. R.C. Nelson (pers. Com., 1975) has studied quartz overgrowths in the Crystal Mountain and Blakely Sandstones (Lower and Middle Ordovician) from central Arkansas, USA. Most of the inclusions are trapped at the original grain boundaries of the detrital quartz grains, but some occur within the overgrowth, and considerable care must be used to avoid confusion with the inclusions of the original detrital grains. Th [temperature of the water solution] for inclusions in the overgrowths ranged from **97.5 to >150°C (possibly as high as 175°C)**." Note 8.5w

Researchers were careful at distinguishing the difference between what they supposed to be detrital sand grains and pore-space quartz, believing water solution temperatures possibly as high as 175°C (347°F) would have been required for the cementation processes to occur. But from where did the water come, and how was it heated?

In addition to the high temperature problem researchers struggled with, the first report presented at the International Association of Sedimentologists in 2000 was on *Quartz Cementation in Sandstones*. The authors indicated that theories continued to be proposed even though many problems existed and "direct evidence" was lacking:

"However, sources external to a given sandbody are still proposed, **despite the lack of direct evidence and problems of the low solubility of silica and the vast quantities of water required to accomplish quartz cementation**." Note 8.5x

Researchers began to consider the possibility of *a short period* cementation process, and the essential factor of *pressure*:

"It is not yet proven whether quartz cement forms **continuously at slow rates or rapidly over short periods. Pressure** (effective stress) is potentially **an important control on quartz cementation**…" Note 8.5y

Because "coarsely crystalline" clay was present in the sandstone, sedimentologists also considered the possibility that it was "precipitated from solution":

"Some sandstones contain clay. Some of this seems to be material entrapped at the time of deposition, and is therefore a detritial component. In other cases, however, **the clay is coarsely crystalline and may possibly be introduced and precipitated from solution**. Note 8.5z

Another cause for confusion in the cementing processes of sandstones is that quartz is not the only cementing agent:

"Some sandstones have more than one species of cementing mineral. If more than one such mineral is present, it is important to determine the **paragenesis or relative age of the cementing minerals**. Waldschmidt (1941) studied the sandstones of the Rocky Mountain area, ranging in age from Pennsylvanian to Cretaceous, and concluded that a definite order of precipitation could be established. According to him this order is **quartz followed by calcite**, or if there are **three cementing minerals, quartz, followed by dolomite, in turn followed by calcite**, or if there are four cementing minerals, the three last named would be followed by **anhydrite**." Note 8.5aa

This sequential process of cementation is important because it must have happened at essentially the same time, and there is only one environment in which this could have taken place, on a global scale—during the UF Hypretherm. The Rock Cycle Pseudotheory chapter demonstrated that anhydrite (a salt) is not formed by precipitation through evaporation. Shortly ahead, in the Salt Mark subchapter, it will be shown that anhydrite, like quartz, calcite, and dolomite prethermate in a hyprethermal environment.

For the modern sedimentologists, sandstones with more than one type of cementing mineral continue to create problems:

"A sandstone with two or more cementing minerals is **further a problem** and **none of the theories of sandstone cementation has adequately dealt with this problem**. Most authors have been content to establish the relative age of the several cementing minerals **without explaining why** under their theory of cementation, **there should be two cementing materials or why they are deposited in a particular order**." Note 8.5ab

Theories that have come the closest to the real processes are those that see a cementation process occurring at the "bottom of the sea," however, the "low concentration of silica in the sea water and the absence of present-day cementation" have nixed even these theories:

"Krynine says that the precipitation of the silica cement 'takes place at the bottom of the sea immediately following the deposition of the sand grains.' Krynine does not present the evidence to support this conclusion, and the low concentration of silica in sea water and the absence of present-day cementation of marine sands contemporaneous with their formation make the hypothesis **untenable**." Note 8.5ac

Some researchers acknowledged the real lack of understanding that exists in the field:

"The analysis suggests that **the pattern of quartz cementation** must either arise from redistribution of material already present within the reservoir or from **a completely new process that we do not understand**" Note 8.5ad

Being unable to identify the source of sufficient silica "material already present" in natural reservoirs, there is only one other alternative:

> The process of natural quartz cementation of sandstone is "**a completely new process that we do not understand.**"

Although the process of sandstone cementation is not understood in modern geology, it is recognized as a fundamental, worldwide process, and it should "be possible to predictively *model* quartz cementation":

> "**Quartz cementation** follows similar patterns **worldwide**. This suggests that quartz cementation is controlled by fundamental processes that are common to all basins. It follows that **it should be possible to predicatively model quartz cementation**." Note 8.5ae

And in one sense, it is. To model it, all we need to do is to **make** sandstone.

How to Make Sandstone

What better way to understand how natural sandstone is made than to make it, the same way Nature makes it? With the shorter time frame of the Universal Model and with the knowledge of how rocks are made, we will approach the sandstone making process from a different perspective. The Identical Principle will be employed to produce **Synthetic Sandstone** by reproducing natural environmental conditions not extant today, but were in existence in the past.

To begin, we must understand that sandstone *is not formed* when sand is buried and subjected to the pressure of overlying sediment, with normal-temperature rainwater percolating through it. Present-day sand dunes, no matter how large, are not forming sandstone. It's that simple.

Additionally, laboratory experiments have shown that water at ambient surface conditions (average surface temperature and normal atmospheric pressure) will **not** dissolve eroded quartz material and become supersaturated with silica. Without a supersaturated silica solution, natural waters cannot grow quartz crystal sand grains nor can it cement the grains together. Thus, the conditions present in today's aquifers generally cannot account for sandstone formation from sand.

The following quote from the journal, *Science*, represents researchers' knowledge about the reaction between 25° C seawater and ground quartz:

> "During the course of reaction with ground quartz, the silica concentration of the seawater rose to about 3 ppm in 1 month and to 4.4 ppm within 1 year. **No large degree of supersaturation of the seawater with respect to quartz was observed during the course of the experiment.**" Note 8.5af

The researcher's conclusion was that quartz reached a *maximum* silica solubility of 6 ppm at 25° C and 1 atmosphere. The six-parts-per-million silica solution produced micrometer crystals, less than millionth of a meter in diameter. Even over extended periods, the micro crystals were only visible with the aid of an electron microscope. Such microcrystals are far too small to cement sand into sandstone. In contrast, the natural crystals that formed during the cementation process are often as large as the sand crystals they cement. In fact, higher silica saturation can facilitate the formation of agates and geodes in open cavities. We will discuss this further in the Surface Mark subchapter.

How does one go about growing quartz crystals large enough to cement sand grains together the way Nature does? The first step is to obtain a highly saturated solution of silica, which can be easily accomplished by increasing the temperature and pressure of a solution including water, un-dissolved silica and a natural mineralizer (NaOH). The high pressure and temperatures of the autoclave apparatus in Fig 8.5.15 produced a supersaturated silica solution, which was poured into a glass vial partially filled with natural sand and allowed to evaporate over several days in a warm environment. The relatively hard sandstone that formed can be seen in Fig 8.5.16. The glass vial had to be broken to release the newly formed sandstone.

This method of adding supersaturated silica water to sand created synthetic sandstone that *appeared* indistinguishable from natural sandstone. In Fig 8.5.17, two pieces of sandstone, one natural and one synthetic, made using the method just described are shown. However, the synthetic sandstones in Fig 8.5.16 and 17 were formed under *ambient temperature and pressure*, and as such, degraded easily, crumbling when soaked in water and broke when dropped. Some natural sandstone rocks share this characteristic, crumbling easily when subjected to water. The sandstone rock collected near the lakeshore in Fig 8.5.18 is easily crushed and returned to lose sand with little effort because it was formed at low temperature and pressure.

The next step was to reproduce sandstone similar to the vast majority of the massive natural sandstone deposits that withstand weathering and erosion, that won't dissolve when soaked

The Sandstone Mark

in water. This process involved the *cementing* of individual sand grains by relatively large quartz crystals. Such crystals grow only at high temperatures and pressures, in a hypretherm. The UF Hypretherm that existed worldwide can be simulated in the autoclave. There, the high temperature and pressures necessary for quartz crystal growth will stimulate interstitial crystallization, cementing the sand grains together in a much more permanent fashion.

UF Sandstone Experiment

The UF Sandstone Experiment, illustrated in Fig 8.5.19, consists of two sandstone experiments—one using fine, unconsolidated sand and the other using coarse sand. The purpose of these experiments was to demonstrate how the world's natural sandstone is formed by reproducing the same hypretherm environment that existed during the Flood. According to the steps of the Universal Scientific Method subchapter 2.5, a **theory** is proposed that the natural sandstone deposits found worldwide are the result of the Universal Flood hypretherm. The **prediction** is that synthetic sandstone can be produced in the same fashion as natural sandstone by recreating the hypretherm environment that existed during the Flood. We **test** our theory and our prediction by **observing** the results of the experiment and **evaluating** the outcome. This experiment, and others in the UM have resulted in the discovery of new geological **laws**, found herein.

The autoclave in Fig 8.5.20 was heated using a temperature gradient of 350° - 400°C, which allowed for convective water currents to flow. Similar currents have been observed deep in the ocean today, near plate boundaries, where frictional heating occurs. The pressure in the autoclave ranged from 10,000-17,000 psi, which is the pressure range continental sand would have been subjected to during the Flood, if the world's tallest mountains were covered with water.

Brass tubes with screens at each end held the sand in place while it was subjected to superheated, pressurized water, moving through the sand. To ensure crystal growth occurred, quartz seed crystals were placed on the holding wire with the tubes of sand. The Fine Sandstone Experiment lasted for 60 hours; at the conclusion, the seed crystals were larger and small quartz crystals were observed on the screen of the tubes (Fig 8.5.19). This confirmed the flow of silica saturated water through the pore spaces of the sand.

The Coarse Sandstone Experiment lasted 48.5 hours. The seed crystals also exhibited significant growth, suggesting promising results for sandstone formation. The newly formed sandstone was very hard—we were able to cut the brass tube from the sandstone without breaking the sandstone—and it did not dissolve when soaked in water. In fact, *they did not dissolve* after being soaked in water for several years. This had been one of the significant tests to compare synthetic sandstone to natural sandstone—would it hold up under water.

At the conclusion of the experiment, the prediction held true; synthetic sandstone was produced that was similar to natural sandstone. Other observations included the curious grey-green color of the fine sandstone, which happened because some of the metal from the tube casing and the end screens dissolved in the solution, rendering coloring similar to natural sandstones. The large, coarse sand from the Lake Tahoe, Nevada, region was well cemented, demonstrating the importance of the hypretherm to crystal growth and the cementation process. The sandstone experiments provided direct evidence that synthetic

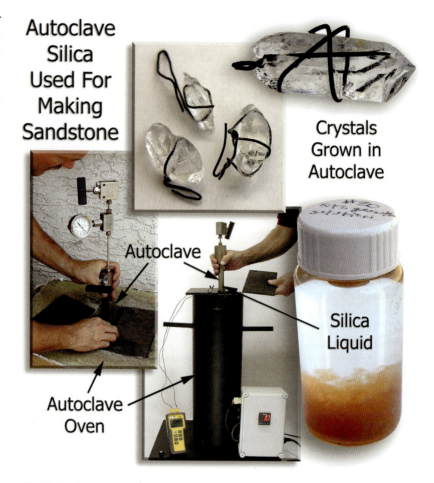

Fig 8.5.15 – Dissolved silica for making sandstone was obtained using a high temperature autoclave. The supersaturated silica solution was removed and transferred to a glass vial, partially filled with natural sand. The silica solution was allowed to set and evaporate for a couple days. Fig 8.5.17 shows the resulting sandstone cemented by the silica solution.

Fig 8.5.16 – This sandstone was formed in a glass vial that had to be broken to release the newly formed sandstone. It was made as a supersaturated silica solution taken from an autoclave reactor was poured over loose sand, which was allowed to evaporate at room pressure and temperature. The hardness of the sandstone is evident from its integrity after the glass vial was broken away to retrieve the sandstone for testing. Unfortunately, because this type of sandstone was **not** formed under pressure, it did **not** hold up when soaked in water. See Fig 8.5.19 for a natural analog to this type of sandstone.

sandstone, which exhibited visual and dissolvability characteristics indistinguishable from natural sandstone, was formed in a hyprethermal environment similar to that of natural sandstone.

Another Sandstone Mystery Solved

In the Rock Cycle Pseudotheory, sandstone is considered a 'sedimentary' rock. Scientists have always assumed that sandstone formed in 'layers' as fine particles of eroded material from other rocks was laid down:

"Sandstones are **always** layered."
Bib 15 p272

However, there are many types of sandstone that clearly *did not* form in 'layers.' There are many variants of sandstone, claystone, and other sedimentary material that exhibit non-layered features, some of which can be seen in Fig 8.5.20. Because modern geology is so intimately tied to the uniformity myth and to geological time, it has not been able to explain how sandstones and other formations were formed.

It is true that with only two forces at work on sandstone formation—wind and water erosion and deposition—all sedimentary formations *should be in layers*, but since they are not, other forces must have been involved. The hypretherm process makes it possible to understand how sedimentary materials were cemented without indication of layers, in all kinds of unique formations. Before this, the origin of 'wonderstone,' 'zebra rock,' and many other mysteriously colored sandstone and claystone forms was not known.

We learn that gravity was not the only force holding mineral

Fig 8.5.18 – Some natural sandstone crumble easily when exposed to water. This indicates the rock was formed in a low pressure/temperature environment. Silica cemented the sand grains together, but lacking sufficient pressure and heat, it did not crystallize, and therefore did not harden. Most sandstones are more resistant to the corrosive effects of water, as seen in the Sandstone Experiments, because they formed in the high pressure/temperature environment of the hypretherm.

forms together as they hardened. The great pressure exerted by water during the UF Hypretherm immobilized sand grains while minerals from bacterial reduction reactions combined with the quartz to form many unique colors and circular structures. Some of the circular structures in Fig 8.5.20 are actually caused by micro-bacterial mats that grew in favorable temperatures as nutrients became available in the water.

Vast blooms of microbes occurred during the UF, creating both acidic and alkaline waters as a variety of microbes grew and died. Chemically different Floodwaters were a major factor in the formation of uniquely colored sandstone, which could never have come from ordinary weathering processes.

Fig 8.5.21 illustrates the important role microbes had on sand and sandstone formation. The jar contains beach '**biosand**' collected from a Mexican lagoon and stored in a dark environment for several months. Abundant microbial growth can be seen in the jar on the left, after only a few months. Over time, the microbes—dark colored areas—experienced a continual change in appearance and caused a significant shift in chemistry as compared with their surroundings.

If this biologically active sand had been in a hypretherm, the multicolored sand would have become *non-layered sandstone*, similar to that found in Nature, and very different from when it was first deposited by ocean water. Thus, microbes, short time periods, and the hypretherm dictate the shapes and colors of the sandstone, not millions of years of surface erosion.

The Sandstone and Limestone Connection

As we prepare to leave the Sand Mark and move on to other Marks, there is one mystery that deserves some attention. It has to do with sandstone or limestone deposits that lie on top of each other, in an order seemingly out of sequence. This question has perplexed geologists for ages, but can now be answered.

> FQ: How does erosion deposit pure, homogeneous layers of sandstone on top of limestone (or limestone on top of sandstone)?

Such a deposit has no rational answer in modern geoscience because there is just no way surface erosion would form a pure layer of *anything*. Furthermore, detrital erosion cannot explain how one layer transitions into a completely different layer, *instantly* without any gradient! If an ocean had once covered the desert, receding over millions of years as geology claims, an instantaneous transition from pure sandstone to pure limestone would be *impossible*.

Having answered how the large deposits of homogeneous sand were formed, in a hypretherm during the UF, we can

Fig 8.5.17 – One of these sandstones specimens is natural. The other was made from dissolved silica from the autoclave reactor and natural sand. When asked to pick the synthetic rock, most selected the one on the left, which was incorrect—it's the natural one. But looks can be deceiving! Since the synthetic rock was formed under normal atmospheric pressure, it did not survive long and was easily broken when it was dropped. The experiment revealed important clues about pressure.

Fine Sandstone Experiment

Coarse Sandstone Experiment

Fig 8.5.19 – Two sandstone-growing experiments involving both fine and coarse sand were conducted to demonstrate the hypretherm processes associated with sandstone formation. In both experiments, natural sand was placed in brass tubes and held in place by screens at each end. This allowed pressurized, superheated water to pass through the pore spaces of the sand. The sandstone that formed was very hard, similar to natural sandstone. It was then soaked in water for several years, where it showed no sign of being dissolved.

apply the same process to the carbonates. The upcoming Carbonate Mark details the origin of limestone, but understanding that both sand and carbonates precipitate from ocean waters at different times to form completely different deposits, one on top of the other, with no discernible gradient, answers this old geologic mystery.

The Fundamental Answer comes from an observation most magma researchers have shown little interest in—pH values. Traditionally, pH values are associated with water, and since rocks and water are not thought to be directly connected, certain

Fig 8.5.20 – These pieces of sand and claystone are from across the Colorado Plateau, USA. The UF Model explains how such rocks formed. In the hypretherm, biota played a major role in the formation of these and other curiosities. These bioclay and biosand forms do not readily dissolve in water.

clues have escaped notice. However, as the temperature of the UF waters increased, the pH, or acid/base relationship of the water decreased:

"**All pH values**, both calculated and directly measured, either *in-situ* or at room temperature, **decrease with an increase in reaction temperature**. The pH of saturated hot water-granite interactions at 100°C is 8.2, but shifts to a near-neutral state [7.0] as the solution temperature increases." Note 8.5ag

The change in pH is important to recognize, because in the hypretherm, as temperatures increase, *more* quartz (sandstone) dissolves into solution with which to make sand, and a near-neutral pH solution allows for different microbes to form, which creates various carbonate (limestone) deposits to also form at different times. Thus, as the temperature fluctuates, it passes through various critical points, which along with pressure and pH levels, can trigger sand or limestone to start or stop growing *instantly*. The pH level plays a critical role in biological growth processes. Changes in the pH levels of the UF waters were directly responsible for a variety of massive algal blooms and other proliferation of microorganisms that produced materials necessary for many of the minerals we find on (and in) the Earth's crust today.

The Old Geology of erosion and shallow, cool seawater, alone were never able to explain the origins of sandstone and limestone. Conversely, during the UF Hypretherm, all of the conditions for instant sandstone and limestone growth were present and can be scientifically confirmed.

8.6 The Erosion Mark

The Erosion Mystery, subchapter 6.11, documented a number of erosion mysteries not clearly explained by the uniformitarian view of modern geology. This subchapter contains simple explanations for those mysteries, which are each 'Marks' of the Universal Flood. Additional evidence is included, establishing the reality of a massive erosion event that took place only several thousand years ago.

"Extreme Events Shape the Landscape"

Fig 8.6.1 shows a large amount of debris including large boulders that were rapidly displaced in a short amount of time by water. Events like this led geologists to conclude that 99.9% of the time, only minor changes to the landscape occur:

"Humans tend to perceive the landscape as unchanging, however, because so little change typically occurs during our lifetimes. This recent debris flow activity helps bridge this gap in perception. **Only minor changes occur in the landscape 99.9% of the time, but a tremendous amount of change can occur in the span of a few hours**." Note 8.6a

This statement was made in a 2006 article titled, *"Extreme events shape the landscape"* in the Arizona Geological Survey. Even so, most geologists don't realize the truefulness of this statement. Without the Universal Flood paradigm as presented here in the UM, questions about how the geological landscape was eroded will remain unanswered. Modern geology cannot answer the unanswerable because of its refusal to accept a Universal Flood model and because of its tenacious adherence to geological time. However, the false theories of modern science will ultimately fail in their attempt to stop mankind from discovering the truth about what really happened just over 4,000 years ago; an erosion event of monumental proportion.

Answering the Unanswerable

In Chapter 6.11, Fundamental Questions about a variety of erosion events were posed, and shown to be not answered by modern geology. There were seven specific examples selected because of their widespread occurrence. These will now be discussed with the UF paradigm in mind;

Before   After

Fig 8.5.21 – The above images are of the same jar that changed over time. The jar on the left contains wet biosand taken from an ocean beach on an inlet where sand had accumulated along with other living and dead organisms. The jar was stored in the dark for several months, during which time many areas of biologic activity became evident (dark areas). The dark matter in the right jar indicates a biologically active biosand, which has a completely different chemistry around the sand grains. Had this biosand been subjected to a hypretherm instead of just lying on the beach, it would have become sandstone, stained with the colors of the biota and the new minerals present. This example of '**biosand**' helps to illustrate why microorganisms are so important in sandstone formation and coloration.

1. Granite Boulders   5. Planation
2. Arch Formation   6. Pedestal Formations
3. Soil Formation   7. Alluvial Fans
4. Skipperocks

1. The Granite Boulder Evidence

Among the Erosion Mystery Examples in chapter 6.11, the

Granite Boulders were the first to be discussed. Large granite boulders, like the ones in Fig 8.6.2, can be found all over the world, from deserts to high plains to the tops of the Sierra Mountains in California. Obviously, the well rounded boulders did not break away from a mountain already rounded—they were shaped by erosion. Granite boulders and other types of boulders do experience weathering, where pieces are broken off as they encounter the relentlessness of wind and water, freezing and thawing. The pieces are not small enough to be blown away by the wind, and so we are left with water as the only means to carry away the eroded materials. Almost always, there are no streams or water channels to indicate where such material went, if it had been there.

To answer the puzzling mystery, geologists assume that some granite boulders were carried to their final resting spot by glacial transport. However, the boulders in Fig 8.6.2 are on the tops of mountains *and in deserts*; no recent ice movement here! It is also apparent that the boulders could not have been there for millions of years because observed decomposition over a couple hundred years has rendered an appreciable amount of boulder breakdown, establishing an age of not more than several thousand years.

What are we left with? These boulders are not being formed in place, they are not formed quickly today, yet they are found worldwide. What is the explanation of their origin? It is true that hydrofountains are known to have expelled enormous boulders, but most of the boulders in Fig 8.6.2 are part of a massive, well-rounded outcrop of granite. The granite outcrops exhibit the same erosional patterns as the boulders, which also include very little present-day erosion, and so could not have been 'expelled' during a hydrofountain eruption. Therefore, the

Fig 8.6.1 – Modern geology textbooks commonly expound on the 'vast geological time' that shaped the landscape we observe around us today. The reality is that, "Only minor changes occur in the landscape 99.9% of the time, but a **tremendous amount of change can occur in the span of a few hours**." In fact, this is how the vast majority of real erosion occurs. The Erosion Mark explains how erosion like this has taken place only recently, over a few thousand years.

rounded granite boulders and the well-rounded outcrops, many of which cover hundreds of square miles, must have been eroded quickly, in a common event, not too long ago.

Remember that during the UF event, the Earth's plates fell several miles allowing subsurface water to rush to the surface as the crust collapsed. Rushing water *several thousand feet high, filled with of all kinds of sediment* moving quickly across

Granite Boulder Evidence

There are no piles of decomposed granite under these boulders because they were not shaped by surface erosion.

Fig 8.6.2 – Granite boulders like these are found all around the world in all sorts of environments. They did not come from the rafting of glaciers, yet they were placed where they are today only very recently which is clear because of the lack of erosion debris beneath the boulders. Without ice movement, there simply is no other mode of transportation outside the mechanisms active during the Universal Flood that carried giant boulders even to the tops of the mountains.

the surface, scouring the landscape can be envisioned. Such massive water movement, coupled with giant hydrofountains of mud and water, stripped away sediment, leaving behind a smoothed landscape, littered with rounded boulders. The Turbidity Model, a few pages later in this chapter will detail additional characteristics of the catastrophic erosion that occured; details that have been documented in the field and in the laboratory, yet have not been known to occur in modern times.

2. The Arch Formation Evidence

The Arch Formation Mystery in chapter 6.11 included the following question:

> Where is the debris that fell from Double Arch or Delicate Arch in Arches National Park?

The Arch Formation Evidence diagram (Fig 8.6.3) shows there is almost no debris below any of the natural arches, excluding recent, historical falls. Where did it go? A closer examination of the photos in the diagram reveals no debris below the arches, *no trail of debris from the area, and no rivers or channels* to explain how 'millions' of years of water carried the arch detritus away. And yet, it seems no scientist is asking this very simple question of where all the debris went.

Fig 8.6.3 includes a photo of Landscape Arch from the ground and one from the air. Landscape Arch is the longest arch in Arches National Park in the USA. Since 1991, three separate falls have been documented; the boulders from the fall can be seen in the foreground. In an even more recent fall, Wall Arch suffered a complete failure in 2008. In both cases the boulder debris from the fall lies on the park floor, beneath the arches, where it will stay for a long time. But what geologists have failed to ask is; where did the boulders from all the previous falls over 'millions' of years go? The arches lay in the desert, there are no rivers of water to carry the boulders away, and the sandstone boulders aren't going to dissolve away any faster than the sandstone arch they fell from.

There are some 2,000 arches that exist today. If a modern erosion process was still forming arches, the process should be evident by the existence of countless new arches in various stages of formation and collapse. If they were formed over 'geological time' there would be many *thousands in process*, with ample debris beneath them. There should also be *many thousands of fallen* arches, but these are noticeably absent.

The vast majority of the arches in Arches National Park appear 'fresh' with very little debris beneath them, testifying of their recent formation just a few thousand years ago. The boulders that fell from the arches were swept away in the UF as high speed water eroded rock, punching through the fins and thin walls of the weakened sandstone. Without the Universal Flood model, arch formation must forever remain a mystery. It is the flood mechanism of the UF that makes this understanding possible.

3. The Soil Formation Evidence

The Soil Mystery in chapter 6.11 established another very important geological fact that has been overlooked in modern geology. Fig 6.11.5 showed examples of hundreds of feet of sediment which does *not* have multiple layers of *organic soil*. Fig 8.6.4 shows additional images of road cuts where only a single layer of organic soil is found. Where are the missing layers of organic soil? These images are typical of the vast majority of Earth's continental surface, wherein only one layer of organic soil exists at or near the surface. How could this be possible if millions of years have gone by with ever-changing landscapes? Why do *we not observe multiple layers* of organic soil that should exist if the land had been 'subducted' and covered by oceans, covered again with new sediment, and then, as modern geology claims, 'uplifted' to have another organic layer on the new surface?

The answer is simple, and just as we discovered in the Granite Boulder and Arch Formation Evidence, these 'other' organic layers *never existed*. The only explanation that fits *all three* of these erosion marks is the Universal Flood. During the Flood, floodwaters swept away sediment from eroding boulders, detritus from beneath the arches and other rock formations, and deposited the massive layers of sediment we find on the surface today, in a worldwide event some *4,000 years ago*. This short time frame would only allow a single, relatively thin layer of organic soil on the Earth's surface.

Throughout this chapter, the approximate figure of 4,000 years is referred to quite consistently because multiple lines of evidence that support that approximate age. For example, H. H. Bennett, the chief in charge of the Bureau of Chemistry and Soils, U.S. Department of Agriculture gave an important report in 1932 on soil and erosion titled: *Lectures on Soil Erosion: Its Extent and Meaning and Necessary Measures of Control*. In this report Bennett stated an overlooked fact about the *time* it takes Nature to make one inch of topsoil:

"According to some of the **quantitative measurements** made at the erosion stations, nature requires **not less than 400 years to build one single inch of the topsoil** of some of our important types of farm land. This appears to be true, for example, of the very extensive soil, the Shelby loam, occurring over the rolling parts of the Corn Belt, in Northern Missouri and Southern Iowa." Note 8.6b

After identifying his "quantitative measurements," he goes on to discuss the average depth of topsoil in the area of Northern Missouri and Southern Iowa:

"Many people have the idea that the soil (as distinguished from the subsoil) is much deeper than it really is. **On examining 172 soil samples collected from 34 states** and representing, very largely, important upland types it was found that the soil depth as recorded **averaged only 9 inches**. Many of our most important types of farm land range from only about 3 to 7 inches in depth of topsoil." Note 8.6b

A little math and Bennett's "**not less** than 400 years" to produce one inch of topsoil derives some interesting figures. We'll use his minimum rate of 400 years-per-inch and then compare that with a rate of 500 years-per-inch to arrive at the approximate time necessary to produce 9 inches of top soil:

400 years/inch x 9 inches = **3,600 years**
500 years/inch x 9 inches = **4,500 years**

A period of 4,000 years falls roughly between these two possibilities, which is supported by Bennett's Corn Belt topsoil analysis of the central USA. In areas that are dryer, one inch of topsoil takes much longer to form. Consequently, topsoil depth is much less in more arid climates. In either case, the time frame of several thousand years for the single layer of topsoil is a tremendously simple, yet *significant* Mark of the Universal Flood.

Arch Formation Evidence

Other than recent falls, where is all the debris?

Landscape Arch From the Air

Where did the debris from under the arch go?

Landscape Arch

Piles of boulder debris never existed here.

Boulder debris from 1991 fall and later.

Wall Arch After Fall

Boulder debris from 2008 fall.

The UF waters carried the arch debris away.

No Debris and no present-day water to wash it away.

Wall Arch Before Fall 2008

Boulders beneath the arches did not just disappear, they were carried away in a huge flood.

Fig 8.6.3

CHAPTER 8 THE UNIVERSAL FLOOD MODEL

Soil Formation Evidence

*Only one layer of organic soil on surfaces worldwide.
No multiple layers created over 'millions' of years.
Quanitative measurements indicate the soil formed ~4,000 years ago.
Does this look like Universal Flood evidence to you?*

Fig 8.6.4 – Here the cross sections of topsoil layers from different climate environments from around the world show a *single* *layer of organic soil* at the top of the section. This defines the vast majority of the surface area on the continents and testifies of two things; first, the continents were not subducted and uplifted multiple times as modern geology claims, and secondly, the thickness of the organic soil layer on the surface identifies the time each layer took to form. Because soil formation times can be generally determined, such soil layers indicate a worldwide event took place only several thousand years ago, depositing the sediment beneath the soil layer.

4. The Skipperocks Evidence

The Skipperock Mystery was introduced in Fig 6.11.7, which showed a deposit in Carlsbad Beach, California, USA, that consisted of a variety of rocks that were primarily flat, oval-shaped rocks. Streambeds with similar rock and mineral types show a variety of shapes and sizes, not a majority of flat, oval shaped rocks. The origin of the 'skipperocks' was outlined previously, in the Skipperock Origin Diagram (Fig 8.4.19). It is shown again in Fig 8.6.5. The presence of hydrofountains below the surface allows for the accumulation of rocks that are held in place by gravity, but vibrated as water flowed upward slowly; allowing the rocks to rotate horizontally but not enough to flip vertically. The underground hydrofountain erosion origin explains what no surface river could do.

5. The Planation Evidence

The Planation Mystery in subchapter 6.11 described the challenge geologists are faced with when trying to understand the world's vast, flat plains. One researcher, who spent a lifetime observing the plains and mountains concluded; "*The remarkable thing is that plains of great perfection are ever made.*" One obvious complication is at present, there are no processes at work that are forming such great expanses of continental flatlands.

Researchers must also deal with the contradictory problem of geological time. Clearly, plain sediment was deposited in a water covered environment, and if you add uplift and millions of years to the equation, deeply eroded valleys should be evident everywhere. However, the plains surrounding Ayers Rock in Australia, seen in Fig 8.6.6, are strikingly flat with *no deep valleys*, a characteristic shared by many of the world's great plains. This is a "mystery" to modern scientists, as recorded in the book, *The Origin of Mountains*:

"At present, the cause of the observed high rate of planation **remains a mystery**.

"**It is even more difficult to make a planation surface if the land is rising tectonically**, yet the planation surfaces are there." Bib 141 p302

Plains-making by uplift over geological time just does not work. The modern geologist sees a "global tectonic quiet period" (no raising or lowering of plains), which also makes no sense to researchers. Ollier and Pain in *The Origin of Mountains* state:

"This suggests **tectonic quiet** in many different places. **It is virtually a global tectonic quiet period. Why should this be?**" Bib 141 p302

The way in which Nature *really works* is that uplift and subduction do happen, but on a global scale, over a short period, not *gradually*. A very large subduction event was followed by an energetic uplift period during the Universal Flood, which can easily answer researchers' questions:

"And **why should a period of tectonic quiet be followed so rapidly by a period of great uplift?**" Bib 141 p302

The "great uplift' was merely the Earth's crust returning to its pre-flood level as the Earth's rotation increased back to its normal rate. The period of tectonic "quiet" occurred while massive areas were covered with sediment generated during the UF event. All of the observations recorded by these researchers can be accounted for by events at the onset of the Flood, as the UF

Skipperock Evidence

Skipperock and clay deposits for the first time have a real origin in the UF paradigm.

Fig 8.6.5 – The Skipperock Origin and the clay deposit in which the rocks are found has a simple and easy to understand explanation in the paradigm of the Universal Flood. Erosion on the continental surface cannot explain deposits like these, but a hypretherm and a hydrofountain activity can. Skipper rocks were originally deposited in horizontal layers; although irregular in shape, they were sorted roughly by size in flood surface erosion processes. Underground aquifers loosened the rock layer as pressurized water moved past and shaped the rocks. Hydrofountains carried the rocks to the surface after spending time held in place in the diatreme pipe, seen in the above right diagram.

Comet passed by, or by the events during and shortly after the Flood.

The period of "high tectonic activity just before the planation" that researchers observed was the breakup of the crust during the early stages of the event:

"Furthermore, in many regions the planation surfaces cut structures that indicate **high tectonic activity just before the planation**." Bib 141 p302

The "high tectonic activity" included frictional heating that occurred as the crust was jostled about right before massive floodwater-created sediments were spread out over the Earth's surface.

6. The Pedestal Evidence

Many people are familiar with the strange pedestal landforms that are found around the world. They remain mysterious to both the general public and to modern geology, and there is no clear explanation of how they hardened or why the lack of erosion beneath them. However, the Rock Pillar Formation diagram in 8.4.5 & 6 explains how the pillars and pedestals were made. More examples of Pedestal Forms can be seen in Fig 8.6.7.

Fig 8.6.6 – Ayers Rock, a fossil hydrofountain, stands in the middle of a massive flat plain that has no origin. No other explanation other than the UF can document how the vast plains were formed.

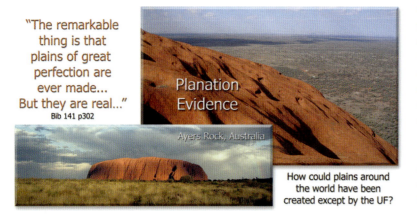

"The remarkable thing is that plains of great perfection are ever made... But they are real..." Bib 141 p302

Planation Evidence

Ayers Rock, Australia

How could plains around the world have been created except by the UF?

To comprehend rock pillar and pedestal formation, we must first have a clear understanding of the hydroplanet Earth. The new discovery of water in the planet's interior and the knowledge that the water is heated by gravitational friction contributes to our understanding of the vertical structure of hydrofountains. Add to this the knowledge of the hypretherm environment and the mechanism for pedestal formation comes clear. To complete the picture, this all happened during the UF event because the Flood is the only mechanism that could possibly have removed all the sediment from around the pedestals, without leaving hardly a trace!

The removal of the sediment by normal weathering over geological time would have toppled many of the pedestals and pillars long ago, which is why they are such a mystery. In fact, some of the pedestals that *have* fallen, just like the previously mentioned arches, still have the boulder debris right beneath the pedestal. The debris doesn't go anywhere, and like the arches, there are few toppled pedestals. In both cases, Floodwaters swept away the debris of the flood-created landforms.

7. The Alluvial Fan Hydrofountain Evidence

In the Hydrofountain Mark (subchapter 8.4) we introduced empirical evidence of hydrofountains ejecting tremendous amounts of clay, sand, and even boulders. The continuation of this concept is that as the hydrofountain continues ejecting copious sediment and rock over an extended period, it eventually begins to build up, forming a mountain of discharged material. Eventually, the water will drain away, but large rocks are not easily moved unless a river of sufficient size is present, which is a rare event on desert mountains. But this is exactly how the spectacular alluvial fans in Death Valley, California came to be, as seen back in Fig 6.11.15.

Fig 8.6.8 shows additional images of the Death Valley alluvial fans and also two images of rock flows associated with the eruption of Mt Saint Helens in 1980. Few people are aware that no red-hot flowing lava came from this eruption, but there

Pedestal Formation

Fig 8.6.7 – The Pedestal Mystery of the Rock Cycle Pseudotheory chapter becomes the Pedestal Formation evidence of the UF because of their hydrofountain origin. The Hypretherm explains how hardened pedestals and pillars are formed as high temperature silica and calcite rich waters seeped up through sediment under high pressure, forming the ubiquitous pedestals. Hydrofountains created the vertical structures above, some of which show a hardened crust on the top, indicating that they are of recent formation without much erosion. Some pillars even exhibit open fountainheads, clearly establishing that they are Hydrofountains. The curious absence of erosional debris beneath the pillars is indicative of the scouring action of water after they were formed, and also their youthful age, being only several thousand years old. Clockwise from upper left, these pedestals are in Cappadocia Turkey, Balanced Rock, Utah, Nambung Australia, Grand Canyon area (2 images) and a tall pillar from Kodachrome Basin, Utah.

were terrific mudflows and hydrorock flows. In the images of the hydrorock flows, large rounded rocks and boulders are evident, with very little small sediment. These flows are very similar to the alluvial fan-shaped rock flows in Death Valley and in other alluvial fans around the world. Of course, the actual rock type differs, and the Mt Saint Helens flow was far smaller and of significantly shorter duration than the Death Valley event.

In the case of Death Valley, imagine the effect of an event much larger than the eruption of Mt Saint Helens. The expulsion of water and rocks would quickly form a fan of epic size, especially if a tremendous volume of water was flowing at high velocity. A similar effect can be seen, but with much smaller sediment (silt) in river delta fans as rivers unload their sediment when water velocity slows. A mini alluvial fan is shown in Fig 8.6.9, where sand-grain sized sediment and small pebbles were carried by a temporary flow of water. It is obvious that the size of the alluvial fan sediment is dictated by the volume of water and its flow velocity. Not surprisingly, we found no such calculations for Death Valley alluvial fan formation, probably because the *enormous quantity of water necessary for the creation of such giant fans has never been seen and is inconceivable to Death Valley researchers.* This is a worthy project for Millennial Science to pursue sometime in the future.

When the sediment expelled from a hydrofountain is large enough, as in the Death Valley alluvial fan sediment, traditional surface erosion and weathering *cannot easily wash the rocks down* the mountain. Thus, their existence is literally a geological testament of hydrorock fountains erupting near the end of the UF event, forming steep-gradient alluvial fans of large rocks in many worldwide locales. It is amazing how the UF paradigm can make the formation of such incredible landforms so comprehensible after being neglected by modern geology for so long.

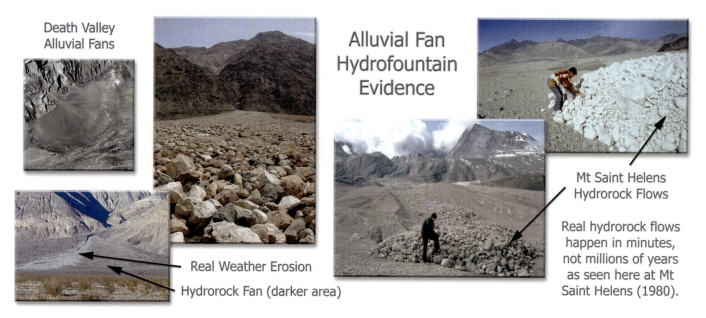

Fig 8.6.8 – The three images on the left are Death Valley alluvial fans. Actual recent weather erosion is evident by the lighter areas whereas the original hydrorock fan is darker. The dark hydrorock fan was not formed by any weather process known in modern times, and these types of alluvial fans are not seen forming today because the vast amount of water required to move such a volume of material does not exist. There are occasional events that give us clues though. The eruption of the Mt. Saint Helens hydrofountain in 1980 (two images on the right), produced a large flow of rock transported to their final location by a large pulse of water. Similarly, the rocks at Death Valley were ejected by a hydrofountain that included a brief, high volume surge of water. If this had been otherwise, the flow would have produced more pronounced river valleys and the rock flows would exhibit a different sediment composition.

The Kaolinite Evidence

We find the kaolinite mineral throughout the world:

"**Kaolinite** is one of the **most common minerals**; it is mined, as kaolin, in Pakistan, Vietnam, Brazil, Bulgaria, France, United Kingdom, Germany, India, Australia, Korea, the People's Republic of China, the Czech republic, Spain, and the United States." Note 8.6c

Kaolinite is a white clay mineral supposedly produced by a "chemical weathering" process:

"It is a soft, earthy, usually white mineral (dioctahedral phyllosilicate **clay**), **produced by the chemical weathering of aluminium silicate minerals like feldspar.**" Note 8.6c

Chemical weathering generally means the breakdown of rocks and minerals by chemical reactions in the atmosphere or the soil. Feldspar and other minerals do not exist by themselves; other eroded minerals must also be present and mixed in with the white 'chemically weathered' kaolinite material, yielding any number of colors besides white.

The presence of Kaolinite confirms the Hydrofountain, Sand and Erosion Marks of the Flood because surface erosion could **not** produce the kaolinite deposit being mined in Fig 8.6.10. The only *proven* way kaolinite is made is in a *hypretherm*. From *Chemical Geology* (1999), several European investigators produced kaolinite using hypretherm environmental conditions in a matter of hours, similar to the timeframe previously associated with quartz crystallization:

"**Kaolinite was hydrothermally precipitated** starting from amorphous aluminosilicates, with Si/Al ratio from 1.8 to 0.76, at **temperatures of 150, 175, 200, and 250°C**, at time periods varying from **6 h to 60 days**." Note 8.6d

The researchers did not mention the *pressure* involved in the experiments in the journal abstract, but did later in the body of the article:

"The gels (2.5 g) were **hydrothermally aged** in KOH 0.1 M **solution** (10ml), in Teflon-lined **reactors**, by heating in an oven at the pre-fixed temperatures (150, 175, 200, 250°C). The **pressure inside the reactors** was at the corresponding water vapour pressure: 4.7, 8.8, 15.3, 25.2 and 39.8 atm [39.8 atm equals 858 psi], respectively." Note 8.6d p174-5

In times past, the pressure component critical for growing rocks was neglected. This is evident here where the researchers refer only to the "hydrothermal" environment, even though the high pressure "inside the reactors" was present. **Hypretherm**, which includes pressure as a *critical* component of the heat-pressure-water rock formation process, was discussed previously in the Hydroplanet Model.

Fig 8.6.9 – A mini alluvial fan (about four feet across), shows how sediment is deposited when fluid flow and sediment size are matched. This fan will not last long because the next storm that comes along with a greater flow rate will wash it away. As flow rate slows, or during subsequent low-flow events, lighter sediment is moved and channels are formed. This is reproduced in large scale on the lower left fan in Fig 8.6.8.

CHAPTER 8 THE UNIVERSAL FLOOD MODEL

Kaolinite Deposit, Bulgaria

Does this white kaolinite look like it 'chemically weathered' from the brown layer?

Fig 8.6.10 – Kaolinite is a white clay mineral containing silica and aluminum. Researchers in the Rock Cycle Pseudotheory paradigm think that kaolinite is "produced by the *chemical weathering* of alumininium silicate minerals like feldspar." In the above image from Bulgaria, the white clay was obviously laid down before the brown deposit. How did weathering and erosion from a mountain source produce a pure white deposit that was immediately afterward covered by a brown deposit? How did the kaolinite erode or 'chemically weather' from the brown deposit? One can see similarities with the Arizona Hydrocraters' white and brown deposits. Clearly, modern geology has no good answer to these questions. Conversely, the Hydroplanet and UF Models easily explain how the deposits were formed and from where the material came. Actual experimental studies reported in the 1999 journal, *Chemical Geology*, verify that kaolinite clay was made in a *hypretherm*. Furthermore, no observations of 'chemical weathering' producing kaolinite deposits have been made. Why? Because the only time kaolinite, "one of the most common minerals" on the face of the Earth was formed, was in the Universal Flood hypretherm.

Notice the relationship between high temperature, high pressure, and time and how it factors into the making of crystalline kaolinite:

"Kaolinite was the only crystalline phase found in the products of the runs and its formation and **crystallinity depended on time, temperature, and Si/Al ratio of the starting material**. Products synthesized at **higher temperature contain more kaolinite which was more crystalline than in the experiments carried out at lower temperature**." Note 8.6d p171

When compared with larger quartz crystal growth, which occurs at higher temperatures and pressures, kaolinite crystallized at temperatures of under 250°C, relatively low for crystallization. Although today there are no known large bodies of water on the Earth with temperatures anywhere near 250°C (482°F), they were common during the UF event. This is direct evidence of the Universal Flood event, which continues to show up as we explore the Earth's crust while applying the Question Principle—to question everything with an open mind.

One surprising omission from the kaolinite experiment regarding the world's natural kaolinite deposits was the imputed origin of the high temperature *and high pressure* required in the production of the kaolinite. In fact, the pressure, although recorded, was not mentioned as a significant element of the crystallization process. Should we be surprised? Probably not. Most of today's geologic researchers came through an academic environment alienated from any possibility of recognizing physical evidence that points to a Universal Flood.

The Turbidity Model

The Sand and Loess Mysteries in the Rock Cycle Pseudotheory chapter exhibited many examples of very fine sediment, some of which is now stone. Besides the fine sediment, *larger rocks,* from pebble-sized to boulder-sized are seen, sorted and layered in the mudstones, claystones, and siltstones that make up the majority of the sediment on the Earth's surface.

In Fig 8.6.11 we see an example of interrelated laminations of fine and course sediment not typically highlighted in geology textbooks because their origin is not really known. If what we are presenting was shown in modern textbooks, the authors would have had to acknowledge that the mechanism responsible for such formations has not been observed and has no analog in Nature today, which is another refutation of the uniformitarian philosophy, one of the great pillars of modern geology.

Some fifty years ago, a model explaining the deposition of these mysterious sedimentary layers *was* presented in the *Journal of Geology* (1950). P. H. Kuenen (who performed the loess experiments described in the Loess Mystery subchapter) and C. I. Migliorini independently concluded, through field observations and laboratory experiments, that the primary mechanism for the deposition of most sedimentary material, including the *sorting of larger rocks* within the *finer sediment*, were turbidity currents. In the UM, we will refer to this as the **Turbidity Model**.

Turbidity is defined as the cloudiness or muddiness of water, but in a geological sense, turbidity is used to describe dense *currents of water* responsible for laying down both fine and coarse sedimentary material, such as that seen in Fig 8.6.11. A present-day turbidity current off the California coastline is shown in Fig 8.6.12. It is apparent that turbidity currents of the past were far more powerful than anything observed today, driven by a natural disaster of unimaginable size with the energy to move massive amounts of sediment across great distances and out onto flat oceanic plains.

Kuenen and Migliorini describe how they formulated the Turbidity Model:

"The present authors **arrived independently at the conclusion that the most important types of graded bedding appear to have been produced by the action of turbidity currents of high density on the sea floor**. But, whereas one of us (Migliorini) was engaged in the investigation of graded rocks encountered **in the field**, the other (Kuenen) had studied artificial turbidity currents of high density **in the laboratory**. When they found how closely their conclusions tallied, although arrived at from opposite starting points, they decided to present their results jointly to a wider pubic." Note 8.6e

With observations from *both* laboratory and fieldwork, the researchers found the following:

"In **typical thick series of graded beds** the areal extent of

Turbidity Deposits

Fig 8.6.11 – Sedimentary deposits like the one seen here have no clear explanation in modern geology, although they are excellent examples of the UM Turbidity Model. Earthquakes and storm water, more powerful than any observed in modern times, moved massive amounts of sediment across flat areas inundated by seawater. During this event, turbidity currents carried thick mud across vast plains, laying down clay, silt, and stones in deposits like those seen above.

These kinds of deposits exist over vast areas, yet they are not being formed today.

each layer tends to be so large that it **cannot represent the accumulation on a river bed**. In many cases **marine fossils also prove that the deposition took place on the sea floor**. Here the only currents of sufficient strength to move coarse sand are of **tidal nature**. These do not attain great velocities over wide areas but only in **restricted channels**. Where they do occur, scour tends to prevail, and ripple marks are common." Note 8.6e p120

The large areas of sedimentation precluded river transport and deposition, whereas "…marine fossils also *prove* that the deposition took place on the sea floor." Such large-scale events suggest that *very large tides* were present to move larger sediments and deposits over such vast areas.

Next, we read that the massive quantity of loose material making up the deposits must have "accumulated elsewhere" and that turbid waters must have been dense enough to carry pebbles and rocks in suspension instead of rolling along the bottom:

"The volume of the thicker beds is **too great to be accounted for by deposition in one season**. Hence the material must have **first accumulated elsewhere over a long period**, then been **transported and finally redeposited in a graded bed** at the present site. This transportation **must have occurred without erosion of the former unconsolidated and fine-grained top layer** of the preceding bed. Yet coarse sand and **even pebbles are found at the bottom of coarser members**. These larger particles were **not rolled along the bottom but were carried in suspension**: otherwise cross-bedding would inevitably have developed." Note 8.6e p120

The researchers also discovered that the processes involved in laying down these large worldwide deposits of **turbidite** (mudstone, claystone and siltstone) were not active over geological time—but occurred "in a sudden brief act":

"The sedimentary matter, accumulated over a long period of time on the upper slopes of a basin, can be raised and carried to the deeper areas **in a sudden brief act**." Note 8.6e p122

The final step of the Turbidity Model includes the fact that processes that occurred around the world in a sudden brief act, "*have never been observed to date*":

"There may be some **doubt** that turbidity currents of high density are ever developed in nature **because they have never been observed to date.** But it should be borne in mind that by their very nature they must always occur where they cannot be directly observed. Also, (1) they cannot be very common; (2) they flow below standing water and must soon descend to considerable depths below the surface; (3) they belong in turbid waters, otherwise there is not sufficient lutite available; and (4) **either storms or earthquakes may initiate them when the observer is not in a favorable position to note any minor features that might be visible at the surface**." Note 8.6e p125

The researchers included indications of what might cause

Fig 8.6.12 – This example of turbidity currents formed off the coast of California, USA, after a storm. It is probably a thousand times *smaller* than the colossal turbidity currents that carried mud, boulders, and other debris from higher elevations across the world's continents "in a sudden brief act" of the past. Such catastrophic flows have never been seen in modern times, but they transpired across the continents during the Universal Flood.

CHAPTER 8 THE UNIVERSAL FLOOD MODEL

these types of underwater mudslides—"storms or earthquakes"—but did not elaborate on how large, or why such storms and earthquakes might have occurred. With the added knowledge of the Hydroplanet Model and Universal Flood Model, we are in a better position to comprehend the massive nature of never-before-seen storms and earthquakes, and to understand how sediment that had "accumulated elsewhere" came to be. The UF Celestial Event explains the gravitational force necessary to produce incredibly large storms and earthquakes, while the Universal Flood Mechanisms created the hypretherm environment and powered the hydrofountains that necessarily supplied both fine and coarse sediment.

The Turbidity Model is summarized as follows:

The Turbidity Model

1. **Turbidites** are sorted fine and/or coarse sediments transported and deposited by turbidity currents.
2. Most mudstone, claystone and siltstone are turbidites.
3. Worldwide turbidites were formed in oceanic flood waters that covered all of the continents.
4. Turbidite sediment accumulated in hydrofountains that erupted onto the surface.
5. Turbidite deposits were formed in a sudden, brief act.
6. Turbidites in large quantity are not observed forming in modern times.

The Turbidity Model is the answer to the Classic Geologic Anomaly illustrated in Figs 6.3.12, 13, 14, and 15. Finally, there is a simple scientific explanation for how mysterious geological anomalies of clay, mud, and siltstone were created. In fact, Kuenen's and Migliorini's experiments (*The Journal of Geology*, Vol. 58, No.2, March 1950, Plate 2B) showed mudstones being formed similar to those in Fig 6.3.17, the primary difference being their hardness because they had not been subjected to a hypretherm.

Perhaps surprisingly, Kuenen and Migliorini's *overall* model has not been rejected, probably because no one has produced any other viable theory to explain the possible origin of mudstone, claystone and siltstone rocks. Today, the term *turbidite* is commonly used to describe sediment deposited by turbidity currents.

Previously, in the Hydroplanet Model, we established the necessity of the hypretherm to cement mud, clay and silt into stone. Also, in the Biogenic Opal and Authigenic Quartz sub-subchapters, (see Fig 8.5.7), we recognized that many of the individual sediment grains grew from biogenic sources in the UF ocean waters. This, together with the Turbidity Model, provides clear evidence of how mudstone, claystone and siltstone came to be deposited and cemented in thick laminations, formed with sorted pebble-through-boulder sized rocks, on all the world's continents.

To be fair, the subject of turbidity is not without controversy. For over 50 years, the "*properties* of turbidity currents" has been debated:

"The **turbidite paradigm**, which is based on the tenet that a vast majority of deep-water sediment is composed of turbidites (i.e., deposits of turbidity currents), has been the subject of **controversy for nearly 50 years**. The crux of the controversy revolves around disagreements concerning the **hydrodynamic properties of turbidity currents** and their deposits." Note 8.6f

However, the controversy was primarily due to "semantic difference in definitions of sediment":

"...recent **controversy** concerning distinction between **turbidity currents and debris flows is due to semantic differences in definitions of sediment**—gravity flows." Note 8.6f p312

Although many researchers have added understanding to the subject of the turbidite, Kuenen and Migliorini were not properly credited for being the first to recognize the general mechanism of turbidite formation. Because volumes of mudstones, claystones, and siltstones are not forming today, anywhere we can observe, researchers will continue to argue the details, but Kuenen and Migliorini should at least be given credit for the Turbidity Model they defined.

There is good reason the "turbidity paradigm" is an important aspect of the whole UF story. Not only is it well established in modern geology as the mechanism responsible for the majority of the Earth's lithified sediment, it happened underwater. But it didn't happen over a long geologic time period; it happened worldwide, it was under ocean water, it was a sudden brief act, and is not happening today.

Colorado Plateau Liquefaction Evidence

In the Sand Mark, the UF Hypretherm and hydrofountains were shown to be responsible for bringing much of the Earth's sediment to the surface. Firsthand accounts of this type of activity are rare because modern earthquakes are not intense enough to produce the adequate shaking and heating of underground water present in times past. The process of **liquefaction**, where loose sediment acts like a liquid during intense earthquake shaking, was responsible for bringing to the surface much of the Earth's sediment. Two relatively modern examples show this process at work. The first occurred on December 16, 1811 when the Realfoot fault beneath New Madrid, Missouri released its pent up energy in an earthquake that brought hundreds of acres of sand to the surface of the Central United States. One source noted the "unfamiliar phenomenon":

"Survivors of the **New Madrid earthquakes** reported not only intense ground shaking and land movement, as would be expected during an earthquake, but also an **unfamiliar phenomenon: water and sand spouting up through fissures, or cracks, in the Earth's surface.**" Note 8.6g

The other example, as reported by *Earthquake Spectra* occurred on January 26, 2001 in Bhuj, India. This large earthquake initiated liquefaction, triggering hydrofountains that spouted sand over an area **greater than 15,000 km² (9,320 square miles)** in India. Note 8.6h

Soil liquefaction occurs as water and loose sediment combine and rise to the surface during seismic shaking:

"**Liquefaction** is more likely to occur in loose to moderately saturated granular soils with poor drainage, such as silty sands or sands and gravels capped or containing seams of impermeable sediments." Note 8.6i

Liquefaction is not a new concept, but has only recently caught the attention of engineers:

"Although the effects of **liquefaction** have been long understood, it was **more thoroughly brought to the attention of engineers** in the 1964 Niigata earthquake and 1964 Alaska earthquake. It was also a major factor in the destruction in San Francisco's Marina District during the 1989 Loma Prieta earthquake..." Note 8.6i

Fig 8.6.13 is a satellite image showing sediments of the Colorado Plateau. Where did the red and orange sediments originate? The long-held modern science answer is that they eroded from far away mountains, over millions of years; but there is no evidence of such mountains, because they don't exist.

The erosion theory is flawed because it does not explain the homogenous nature of the sand grains which make up vast areas of the Plateau. This can only be understood in the light of the Hydroplanet Model. We pointed out in the Rock Cycle Pseudotheory chapter that there are no pure-quartz mountains today and none have been shown to exist in the past. Because the grain sizes of the red and orange quartz sand are similar, there must be an origin other than surface erosion.

Modern geology continues to ignore this important fact because geologists are *only* looking at *surface* erosion—not the largest source of Earth's erosion—sub-surface erosion. The modern examples of sedimentation through hydrofountain expulsion, such as witnessed in Bhuj, India, establish how sediments can come from beneath the Earth's surface instead of the traditional system of river or aerial erosion.

Combined with the hypretherm model, we have a mechanism for homogeneous, similar-sized sand grain formation, expulsion through hydrofountains, deposition, and coloration from underground iron oxide and other biogenic sources. Sand, once formed and brought to the surface through hydrofountains, stained red, orange, or yellow by a variety of influences, was made into sandstone via the Hyprethermal Sedimentation Model presented earlier in the Sand Mark. The one unmistakable factor is the vastness of the Colorado Plateau. There are many places in the world today with similar characteristics, covering thousands of square miles with colored sandstones and other sediments. All of them share a common origin; a worldwide hypretherm event that occurred several thousand years ago, as attested by the lack of erosion on the plateaus.

Agathla Peak Hydromountain Evidence

There are thousands of examples of *false erosion claims* in the scientific literature due in part to the pseudotheories of geological time and continental uplift. They cannot all be addressed in this book, but there are a couple of interesting examples that have a story to tell. One is the Agathla Peak Hydromountain located in northern Arizona (see Fig 8.6.14).

"**Agathla Peak is an eroded volcanic plug** consisting of volcanic breccia cut by dikes of an unusual igneous rock called minette. It is one of many such **volcanic diatremes** that are found in Navajo country of northeast Arizona and northwest New Mexico. **Agathla Peak** and Shiprock in New Mexico are the most prominent. These rocks are part of the Navajo Volcanic Field, in the southern Colorado Plateau. Ages of these minettes and associated more unusual igneous rocks cluster near **25 million years**." Note 8.6j

Along with the theory that Agathla Peak is 25 million years old, volcanic plugs like Agathla are said to be *exposed through erosion* as surrounding sediment and rock is weathered away over 'geological time':

"If **a plug is preserved, erosion may remove the surrounding rock while the erosion-resistant plug remains**, producing a distinctive upstanding landform." Note 8.6k

The problems with such 'theories' are manifold and can be understood by evaluating the *actual* erosion and the geological evidence of the mountain itself.

Looking at Fig 8.6.14, there are no signs of significant erosion around Agathla Peak. Incidental surface erosion is evident, but there are no deep valleys and riverbeds that presumably carried off the billion-plus cubic feet of sediment that once surrounded Agathla. Could rain and snow erode over 1,500 feet (457 meters) of sediment, leaving behind such a flat plain? Actual erosion seen in Fig 8.6.14 indicates that weathering processes could have only been at work here for no more than several thousand years.

The icing on the cake is the dark mineral, minette that erupted out of the open hydrofountain seen at the top of the mountain. It surrounds the mountain today, but *if massive erosion had really been at work here*, the minette debris on the surface in the center panoramic image would certainly have been swept away long ago, along with 1,500 feet of overlying debris that presumably concealed the erosion-resistant plug. Clearly, this was not the case.

The actual events that formed Agathla were liquefaction, rap-

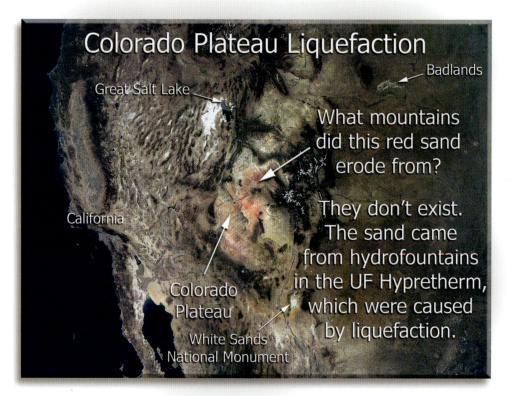

Fig 8.6.13 – This NASA image of the Western United States shows the results of the Colorado Plateau Liquefaction processes that took place during the UF. There are no red or orange mountains from which the red and orange sand and clay eroded, a fact apparently missed by modern geology. Modern day liquefaction examples prove that large earthquakes can expel tremendous amounts of clay and sand onto the surface in a manner analogous to the hydrofountains of the Universal Flood. Liquefaction was the driving force behind the final shaping of the Colorado Plateau and its characteristic plateau and valley morphology.

Agathla Peak Hydromountain

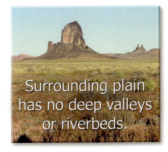
Surrounding plain has no deep valleys or riverbeds.

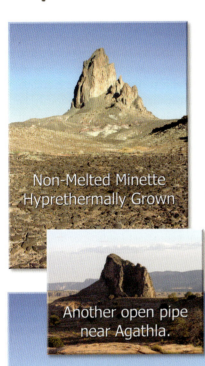
Non-Melted Minette Hyprethermally Grown

Hydrofountain Opening

Erupted Minette Debris

Agathla minette was not melted when it erupted on to the surface, as proven by this artificially melted minette specimen.

Another open pipe near Agathla.

How was Agathla covered by sediment that eroded away leaving only the minette mineral behind?

Colored clay deposits at base confirms hydrofountain origin.

Non-eroded brown sandstone at base verses red sand in surrounding plain coinfirms hyrofountain.

Petrified wood showing no sign of weathering among minette debris confirms hypretherm origin.

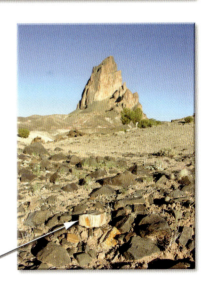

Fig 8.6.14 – Agathla Peak is a hydromountain in Monument Valley, Arizona, USA. It has an open hydrofountain pipe still evident at its peak that ejected the dark mineral minette seen surrounding the 1,500-foot high mountain. The mineral was not melted and extruded onto the surface like lava. When the minette is melted, it turns into a glass (see the melted rock in the center, above). The minette mineral was grown in a hypretherm and ejected after Agathla punched through the surface. It was deposited along with brown, lithified sand and colored clay as seen in the images. The sand and clay did not erode from the mountain or anywhere else in the area since the surrounding plain is made of red sandstone. Another evidence that this is a hydromountain formed in the UF Hypretherm is the petrified wood in the area. It is not eroded, and thus not transported, but is found with the minette mineral all around the mountain. Although modern geology claims 1,500 feet of sediment covering Agathla and other "volcanic plugs" in the area was eroded away over the last 25 million years, no deep valleys and no evidence of such extreme erosion exists around Agathla Peak to support this pseudotheory.

SUBCHAPTER 8.6 THE EROSION MARK

id up-thrusting, and hydrofountain activity, which expelled the minette minerals in a hypretherm. How do we know this?

First, there is a brown, sandstone pipe deposit, easily seen in the diagram; it could have only hardened in a hypretherm. The sand that formed this pipe must have come from great depths because the sand on the surrounding plain is *red*—not brown.

The multicolored clay at the base of Agathla is unmixed, which, as established in the Sand Mark, came from a hydrofountain. And finally, *petrified wood* is found among the minette minerals and the multicolored clay. Because the petrified wood is a quartz-based crystalline rock, it had to have been formed in a hyprethermal environment, right along with the minette. The petrified wood shows no signs of erosion or rounding due to transport, testifying that it was made right where it is found only several thousand years ago.

It should be no surprise to discover that the minette mineral is similar to kimberlite or to lamproite, which are both known to contain diamonds. It is also similar to the Arkansas diamond diatreme, seen in Fig 8.4.14, in that it includes hyprethermally altered wood and minerals. One of the major elements of petrified wood and fossil formation is their direct association with multicolored clay deposits and hydrofountains. Surprisingly, this essential element is almost wholly overlooked in both geology and paleontology.

One final piece of evidence comes from a simple melting experiment; a piece of the minette mineral taken from the slopes of Agathla was melted, proving that this mineral was ***not*** melted when it erupted on the surface. It is similar to non-vesicular basalt, also formed in a hypretherm. We will elaborate on this further, in the next subchapter.

Mountain Liquefaction Evidence

Modern geologists know of dozens of "volcanic diatremes" across the Colorado Plateau, such as Agathla Peak, but *obvious similarities* between the dark mineral pipes and the red sandstone structures, like those in Monument Valley, Arizona (seen in Fig 8.6.15) have eluded them. To explain how the red sandstone monoliths in Monument Valley were formed, geologists turn to the same old erosion sci-bi used to explain the origin of Agathla Peak. But the theory again fails because there are **no deep gullies or river-formed valleys** *between the sandstone structures*, although they should be there if there really was millions of years of erosion.

The absence of erosion is because the sandstone mountains were ***not*** left standing after the removal of sediment by water—but were exposed by **Mountain Liquefaction**. A modern analogy of this process is seen in Fig 8.6.15 & 16 where rigid infrastructure, such as sewer pipe risers were raised above the street surface during an earthquake in Japan. In this way, colossal earthquakes raised hardened sandstone pipes and solidified landforms during the UF hypretherm, while the surrounding plains dropped. Modern geology has overlooked specific connections between volcanic diatremes, sandstone pipes, and liquefaction brought on by very large earthquakes, leaving the scientists in the dark about major landscape-altering forces affecting areas as large as the Colorado Plataea.

Mountain Liquefaction is defined as the (relatively rapid) raising of hyprethermally hardened crust by large earthquakes. With this process, we can, for the first time understand how the extraordinary mountains in Monument Valley formed, and not over millions of years, as plainly manifested by the absence of deep gullies and valleys betweens the structures.

The new hydrofountain paradigm is a crucial factor in the discovery of how these fascinating pillars were formed. Coupled with the study of recent, large earthquakes and other new discoveries, we can begin to understand the magnitude of the liquefaction process. The modern analog of raised sewer pipes in Fig 8.6.15 & 16 provide a visual example of the vertical aspect of this process, but there are also *horizontal* landforms formed during the UF Hypretherm.

During the Universal Flood, exceptionally large earthquakes raised whole sections of crust that had been previously hardened in the hypretherm. These horizontal landforms were raised above the surrounding plains by mountain liquefaction to become the great plateaus of the desert Southwest in the United States.

Endless UF Erosion Evidence

Flood erosion evidence of the UF is seemingly endless. Because the Universal Flood was responsible for the creation of most of the Earth's current geological landscape, the evidence for the flood can be seen worldwide, in every corner of the countryside. A few examples among millions, stand out. One of those is a curious granite formation in Western Australia. Known locally as Wave Rock or simply the Wave, the landform has generated heated controversy among geologists for as long as it has been known:

"**Opinions vary and arguments rage between geologists over this**. Some have suggested that the exposed rock surface of the lip of the **Wave** is harder than that below. Hence water eroded the softer rock below into a wave shape." Note 8.61

Such controversy exists because desert water alone could not shape this rock. Elsewhere in the world, landforms such as this lay near rivers where ideas of floodwater erosion are easy to accept. Here in the desert, boulder-laden floodwaters coursed by, rapidly grinding away at the tough granite rock. Today, large rounded boulders lie on the ground just 100 meters west of Wave Rock.

On the following pages, several examples of UF sandstone formation and erosion are shown. Liquefaction, UF landscape erosion, and UF gravel deposits are shown. Each of these landscape types *cannot* be explained by the mythical theory of uniformity, where weather and time combine to imperceptibly shape the landscape we observe today. In fact, if we look at the type of weathering going on right now, in almost every case, the observed erosion accounts for only several thousand years of activity.

The Colorado Plateau exhibits deep primary valleys but no secondary valleys or gullies that should be evident across the large flat plains if time and weather had been the maker of the deep valleys. We find similar evidence at the Valley of Fire in Nevada. There are no erosional patterns supporting the hypothesis of traditional erosion. There is little erosion evident at the base of the hardened red sandstone islands because they were not formed by modern science's traditional erosional processes.

Similar observations can be made at all the locations shown in the panoramic images depicting UF Landscape Erosion. In each location, day to day weathering is not carving away the landscape, a fact known by many geologists already (from *Arizona Geology, Arizona Geological Survey*):

Mountain Liquefaction Mechanism

Liquefaction occurs when the landscape settles during large earthquakes. The mountains in the background remained elevated as the surrounding plains sunk, similar to the manhole in the above image.

Large earthquakes bring water and sediment to the surface during the shaking, thrusting some rigid structures upward while the surrounding landscape sinks.

Fig 8.6.15 – The Mountain Liquefaction Mechanism explains how many mountains like those in Monument Valley, Arizona, USA, were formed. Large earthquakes shaking for extended periods of time along with underground systems of hardened rock, cemented during the UF hypretherm, rose to the surface as fossil hydrofountains, in a manner similar to the sewer line during a recent Japanese earthquake. The old geological sci-bi of erosion over millions of years falls short in explaining such mountain ranges because of the lack of deep, erosion-carved valleys that *should* exist on the plains between the mountains.

"Only minor changes occur in the landscape 99.9% of the time, but a tremendous amount of change can occur in the span of a few hours." Note 8.6m

Their problem is that this fact has not been applied to the *all of the surrounding* landscape. **When** did the "tremendous amount of change" occur spoken of? The answer can be seen in *actual* weathering erosion, which only accounts for a couple thousand years. Actual erosion by weathering is evident by the few pieces of granite below well-rounded boulders; it is the few fallen blocks below the still-standing delicate sandstone arches, or the thin layer of soil formed only at the surface. It is the handful of rock fragments below pedestals, the few feet of blow sand around desert vegetation, or the small dry wash down a huge alluvial fan. There are hundreds of examples that could have been added, and all share one common characteristic when viewed in the light of the Universal Model; present-day sediment deposits were not laid down or created over geological time. It is only in light of the Universal Flood that the Erosion Marks and erosionary features across the landscape can be clearly and succinctly explained and comprehended.

8.7   The Depth Mark

This subchapter provides the evidence for the extraordinary claim made in the Basalt Mystery and the Obsidian Mystery in subchapters 6.5 and 6.6, which identified that the definition of these minerals is incorrect and that:

> No one has observed fine-grained basalt (without vesicles) or obsidian from a present-day continental lava flow.

Unequivocal evidence for this claim resides in one particular physical constraint required for all non-vesicular basalt and obsidian—pressure. As we will show, the vast majority of basalt and obsidian deposits on land had to have been subjected to the pressure equivalent to being buried by at least 3,000 meters (9,842 feet) of water.

The Basalt and Obsidian Mysteries Are Solved

In the Rock Cycle Pseudotheory chapter, we discovered that

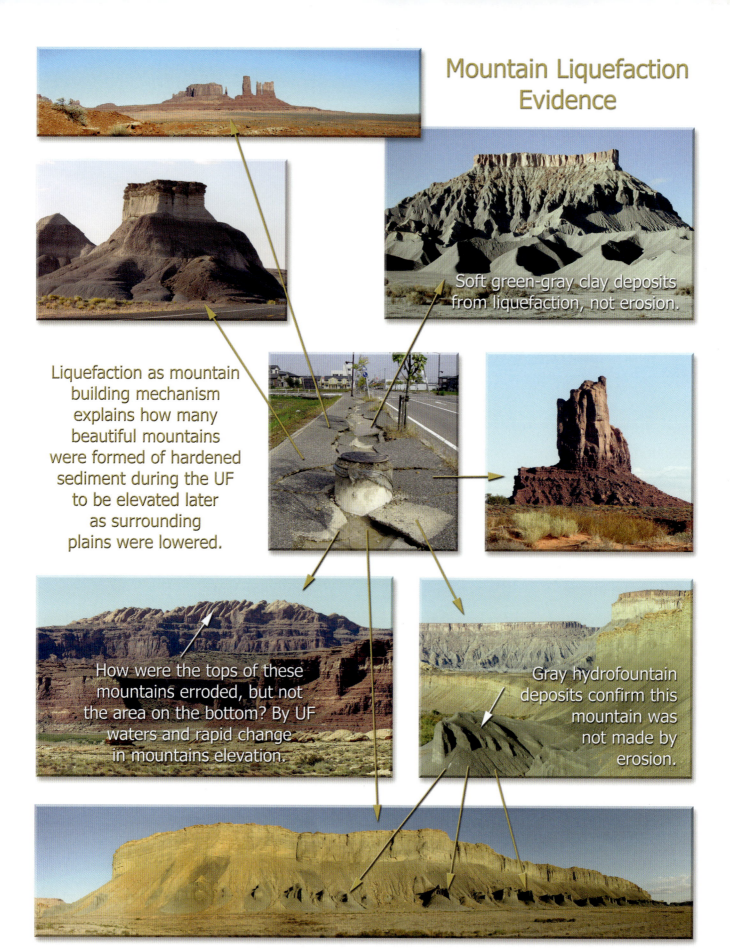

Fig 8.6.16 – Countless examples of Mountain Liquefaction Evidence, like the examples above, show up all over the Colorado Plateau. For the first time, a logical explanation describing the origin of these mountains and their features is available with the UF and liquefaction processes. The erosion sci-bi does not explain the lack of large-scale erosion between the mountains, nor does it account for the soft clayey hydrofountain sediment at their bases that would have surely washed away long ago, if traditional weathering and erosion was the real process.

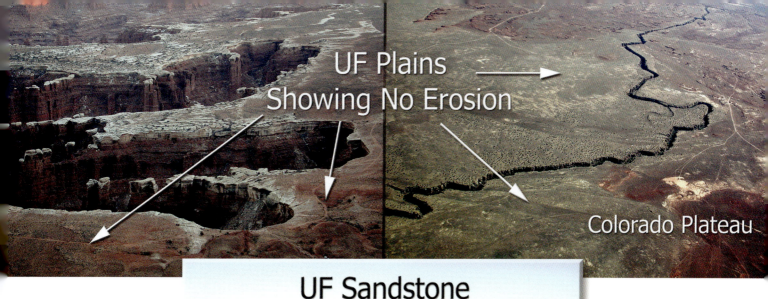

UF Plains Showing No Erosion

Colorado Plateau

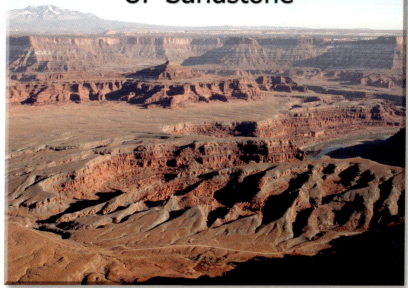

UF Sandstone

The Universal Flood impacted the entire surface of the Earth.

The Erosion Mark exists all over the world, all we have to do is look.

UF Hydrofountain Sandstone

Sediment thousands of feet deep showing no erosional patterns.

Where is the eroded red sand?
Only the UF has the answers.

Valley of Fire, Nevada

Universal Flood Landscape Erosion

UF Gravel Deposits in Montana, USA

This entire valley had to have been covered by fast moving water to deposit these cobblestones.

Large gravel deposits like these can be found worldwide and were deposited by the profuse floodwaters of the UF. Even the largest mudflows observed at Mt Saint Helens did not sort cobblestones like we see in these deposits.

Actual Hydromountain Erosion at Mt Saint Helens

These hydrothermal and other deposits were laid down in days, and are similar to many of the UF deposits.

SUBCHAPTER 8.6 THE EROSION MARK

neither fine-grained basalt nor obsidian has ever been observed coming from land-based volcanoes. Although both types of rocks are classified as 'volcanic' in every geology textbook, the mystery of how these minerals formed is solved once we stop looking to volcanoes on land and start looking at how minerals form *deep in the ocean*.

It is their *lack of vesicles*–holes formed by escaping gases– that provides a critical clue in the understanding of how such minerals formed. Basalt and obsidian have never been observed coming from volcanoes because at the atmospheric pressure on the surface of the Earth, water vaporizes and easily escapes the cooling, melted rock. This produces lava rocks typical of what most of us associate with volcanic rock; sharp, jagged edges, filled with holes, like those in Fig 8.7.1.

Most fine-grained basalt and obsidian formed under great pressure, *in a deep ocean* so that as the mineral melt cooled, the volatiles (water and water vapor) trapped in the melt were unable to escape, leaving the hardened minerals without vesicles.

Remember that all rocks contain water because our planet was formed of water. When rocks are heated, or melted, that water will quickly escape at the surface, leaving behind vesicles because outside ambient pressure is low. If the lava that was exposed to atmospheric pressure contains silica, it will always be a *glassy* form of silica versus the crystalline form of quartz. The crystalline form is found in many basalts, proving that the basalt was grown in a hypretherm.

The Origin of Obsidian

In subchapter 6.6, the Obsidian Mystery, we posed the following question—does modern geology really know how obsidian is formed? To establish generally what most geologists think, we take the following from the *Origins of Igneous Rocks*:

"**Obsidian is commonly formed** by silicic **melts** that are quenched **in air**." Note 8.7a

Dictionary.com includes the following description of the origin for obsidian in science:

"A shiny, usually black, volcanic glass. **Obsidian forms above ground from lava** that is similar in composition to the magma from which granite forms underground, but cools so quickly that minerals do not have a chance to form within it." Note 8.7b

These false definitions have impacted the technological world, as read in the *Encyclopedia of Glass, Ceramics, and Cement*:

"Obsidian, for instance, is a naturally occurring combination of oxides fused by intense volcanic heat and vitrified (made into a glass) **by rapid air cooling**." Note 8.7c

These ideas are simply incorrect. If we carefully observe *historical* land-based lava flows, **no newly formed obsidian** will be found. Although natural obsidian (without vesicles) is found in relative abundance around the world, and in large quantities like Obsidian Mountain, California, USA, seen in Fig 8.7.2, typical non-vesicular obsidian has *historically* **never** been observed forming in any significant quantity. In one case, in 1957, there was a claim that obsidian had formed in association with rhyolite when the Tuluman Islands were formed near Papua New Guinea. Researchers reported:

"Rock types were silica-rich and included **highly vesiculated** rhyolite and pumice, commonly **with bands of black obsidian**." Note 8.7d

Fig 8.7.1 – This is the Sunset Crater field in Flagstaff, Arizona, USA. It is a good example of volcanoes and lava formed in the *atmosphere*, on the surface. There are no fine-grained or columnar basalts. The inset photo shows lava and the vesicles, or holes that formed when gases escaped the cooling lava—a sure sign the lava formed while exposed to the air and not under the pressure of deep water.

But as it turned out, the obsidian was not solid, it was "*vesicular*":

"Later inspection showed that the flows consisted of hard, slightly **vesicular obsidian** and dark grey vesicular rhyolite." Note 8.7d p295

One additional key to this rhyolite-obsidian instance has to do with the environment in which they formed. The *minerals were not formed from a dry continental flow, but were in a pressurized, heated* **submarine** *habitat*:

"Initial activity was **submarine**, but later the volcano gradually built up to sea level." Note 8.7d

The Tuluman Islands obsidian was formed amid constant steam eruptions, which included initial high pressure and water, both important in its formation. The high-pressure aqueous environment was also present at other non-historic obsidian formations. Obsidian Dome, located in the Mammoth Lakes region of California also consists of brown rhyolite and black obsidian, seen in Fig 8.7.3. There, low-grade obsidian is riddled with bands of highly vesicular obsidian along with more solid segments. We know that pressurized hot water played an important role along the fault associated with the dome because the Inyo Hydrocraters are located on the same fault. Three large steam-explosion craters were formed when a great deal of superheated water blasted away the overlying rock strata. Panum Crater lies on the same fault northeast, near Mono Lake. It is a hydrocrater as well, with a rhyolite and black obsidian dome in its central uplift area.

The extremely rare occurrence of obsidian and fine-grained basalt in historical times is difficult for modern geology to understand. Of all the historic continental lava flows studied, almost none occurred in *pressurized water environments*, which is why so few of them have obsidian or basalt flows. Obsidian Dome and Panum Crater formed small amounts of poor quality obsidian within their subterranean high-pressure aqueous environments before they broke the surface. Conversely, the largest and best quality obsidian deposits, like Obsidian Mountain in California, were formed when the entire mountain range of obsidian cooled on the surface, **under** thousands of feet of water.

For such a quantity of molten rock to have created a mountain-sized deposit of obsidian on the surface, it had to have been beneath enough water, in a **subocean hypretherm**. That is the only environment with sufficient pressure in which high quality obsidian could have formed; and it was only during the Universal Flood that such an environment existed on the continental surface.

A corrected definition of obsidian's origin must include the specifics about the environment and the mechanism where it forms:

Obsidian is glass that formed
from a silica melt in a **hypretherm**.

The Origin of Basalt

The solution to the Basalt Mystery is very similar to the Obsidian mystery in that fine-grained basalt also requires a hypretherm for growth. However, basalt is different than obsidian in two important ways. First, its composition includes less silica than obsidian and second, the cooling time was longer than for obsidian.

There are several descriptions of basalt and its habits. One distinctive feature of basalt is identified in this definition from Dictionary.com:

"…the **dark, dense igneous rock of a lava flow** or minor intrusion, composed essentially of labradorite and pyroxene and often displaying a **columnar structure**." Note 8.7e

The "columnar structure" is not found in quickly cooled glassy rock like obsidian. Fig 8.7.4 shows re-melted basalt, cooled relatively quickly in air. The resulting glassy appearance shows what air-cooled basalt really looks like. Moreover, many basalt deposits contain crystalline quartz, which, as has been established, requires a hypretherm to grow. We will return to the significance of basalt's columnar habit shortly, but first, we are in a position to answer the Earth Crust Mystery described in Fig 6.12.1. *The American Heritage Science Dictionary* explains where the largest amount of **basalt** is found on Earth:

"A dark, **fine-grained**, igneous rock consisting mostly of plagioclase feldspar and pyroxene, and sometimes olivine. **Basalt makes up most of the ocean floor and is the most common type of lava**. It sometimes cools into characteristic hexagonal **columns**, as in the Giant's Causeway in Anterim, Northern Island. It is the fine-grained equivalent of gabbro." Note 8.7e p2

The fact that basalt makes up the majority of the "ocean floor" is not by chance. It is where the largest hypretherm in the world still takes place. Modern geology hasn't been able to provide a sound answer as to *why* the continental crust is so much thicker than the oceanic crust, or why the continental crust is comprised mostly of granite whereas the oceanic crust is predominantly dark basalt. Compared with all other minerals, basalt is found only rarely on the continental crusts because the hypretherm that existed for continental basalt formation lasted for only *a short period* during the Deluge.

Fig 8.7.2 – Obsidian is incorrectly defined as a silica melt that was quenched quickly in air. If this were true, it would be easily reproducible by melting and re-cooling obsidian. however, as soon as a torch is put to the mineral, it explodes as the internal pressure of heated water becomes steam. Obsidian must be formed in a hypretherm; it cannot be formed in air. Moreover, obsidian was formed in a pressurized, alkaline water environment. Evidence for this comes from obsidian specimens like the one in the top right corner of this figure. The lime rocks imbedded into the obsidian are partially dissolved, which occurred after the obsidian was formed as the waters became acidic. The dissolving action and acidic environment will be discussed in detail in the Surface Mark, later in this chapter.

Obsidian Dome

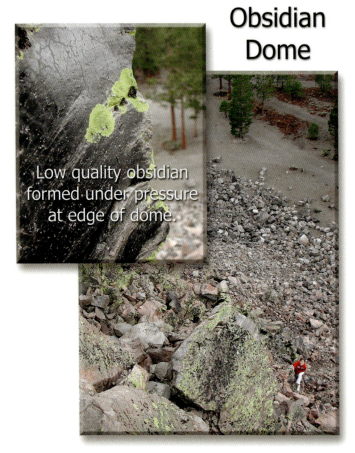

Fig 8.7.3 – Obsidian Dome, located in California, USA, is a rhyolite deposit accompanied with a small amount of poor quality obsidian. The black obsidian is located only along the edges of the dome where pressure was the highest when the volcanic structure was formed by earthquakes and water pressure. Thus, a hypretherm was involved in these smaller deposits of obsidian too.

However after the Flood, the now cracked Pangaea supercontinent began to separate as the Earth returned to its original rotational speed. The Universal Flood mechanism explains both the Earth Crust Mystery and the Pangaea Mystery in Fig 6.12.2. These closely linked mysteries account for the spreading of the continents, a concept modern geology recognizes but has not had a mechanism for how it occurred. This ongoing process is still forming new oceanic crust today, although on a much smaller scale.

Over time, perhaps several hundred years, the Earth's rotational speed slowly returned to its former velocity due to the Resonance Energy as explained in the text associated with Fig 8.3.9. The rising rotational speed increased the centrifugal force acting on the broken fragments of the Pangaea supercontinent, pulling them apart just like the riders in the swinging chairs of an amusement park ride in Fig 8.7.5. Friction from moving oceanic crust melted crustal rocks along the mid-ocean ridges forming heavy, dark basaltic minerals. The darker color of basalt rocks comes from magnetite and other heavy mineral producing microbes that are present on the ocean floors. A key factor associated with the formation of basaltic flows is the depth at which the flow took place. The depth of the basalt flow has a direct relationship to the size and distribution of the vesicles imbedded in the basalt, because depth affects pressure.

The Key—Vesicles

As we explore the key factor of basalt formation; the formation, size, and abundance of vesicles in the rock, we learn more about the environment in which both obsidian and basalt formed. Researchers have just begun to realize just how significant these vesicles are. From a 2002 *Lunar and Planetary Science* article:

"The **significance of vesicles** for understanding **lava flow emplacement** has become increasingly recognized over the past decade." Note 8.7f

"Lava flow emplacement" is simply the *environment* where the lava *cooled*. The presence (or absence) of vesicles help quantify the environment in which basalt and obsidian formed here on Earth, and they can help establish the environment that existed on other planetary bodies exhibiting similar minerals. The article from which the last quote was taken, *Vesicles: A Fundamental Characteristic of Planetary Surface Rocks*, showed that researchers now realize that these small holes hold big answers to planetary formation mysteries.

The reason for their importance is that vesicles are produced by escaping gases, which is affected by pressure surrounding the cooling rock. If the vesicles are small, the rock cooled under pressure restricting the ability of volatile liquid (water) to form large bubbles. No vesicles indicate even higher pressure, and when non-vesicular glassy rocks form on the surface of the Earth, they did so in only one environment, under an ocean of water!

The Critical Depth Fundamental Answer

Research has shown that below the critical depth of approximately **3,000 meters** in the ocean, an increase in temperature or pressure affects the liquids and gases in rocks such that they cannot expand as they do on the surface. Therefore, rocks melted below this critical depth exhibit different properties than they would if they had been melted at the surface. This is why

Fig 8.7.4 – Fine-grained basalt without vesicles does *not* come from continental lava flows, although geology has assumed it does. It would be glass, as seen in the above melted specimens of basalt. Instead, basalt must form in a hypretherm, if it has no vesicles or if it contains crystalline quartz. A longer time, measured only in days, is required for fine-grained basalt to form, as compared with obsidian, because it is normally found in a crystalline or columnar structure, observable in the way it fractures. Faster cooling times and lower pressure produce lower quality basalts, with increasingly larger vesicles.

Fig 8.7.5 – The centrifugal force pulling these amusement ride chairs apart is increased as the rotational speed of the ride is increased. The same force pulled the plates of the Earth apart, forming new basaltic crust under the ocean after the UF, as the rotational speed of the Earth returned to its former velocity.

Increased rotational speed causes these chairs to move apart just as the continents moved apart when the Earth's rotational speed increased.

land-based lava flowing from a volcano will never become fine grained basalt or obsidian. The details are outlined in *Volcanoes, A Planetary Perspective*:

"Beyond a certain pressure, known as the *critical point*, highly compressed steam behaves in the same way as liquid water: it no longer decreases in volume with further increased pressure. For sea-water, the critical point is at 315 bars, equivalent to a depth of about **3000 metres [9,842 feet]**. From a volcanological point of view it follows that below the critical depth, explosive fragmentation of magma is impossible, no matter how volatile rich it is, or how high the temperature. In volatile-rich magmas it is possible for vesicles [cavities in rocks] to nucleate, and indeed deep-water vesicular basalts are common. But the vesicles can *never* expand and grow. Thus, the volcanology of the abyssal ocean floors is essentially that of lavas. There are no pyroclastic rocks. Similar arguments apply to silicic lavas erupted underwater on continental margins. At depths of about 1000 m, dredge samples of volatile-rich dacitic [silica] lavas sometimes appear remarkably vesicular**; at depths below 3000 m they are dense, dark, and glassy**." Note 8.7g

This observation is direct evidence of the true origin of basalt and obsidian, and it explains *why* fine-grained basalt and obsidian are *not* forming on continental volcanoes today; they aren't under at least 3,000 meters (about 2 miles) of ocean! The 3,000 meter Critical Depth element identifies the necessary pressure to prevent water in the basalt and obsidian melts from expanding and experiencing a gaseous phase change of heated water. This is further in a 2003 article from the American Geophysical Union titled, *Explosive Subaqueous Volcanism*:

"Water differs from air in many ways, and we recognize four major roles for water in affecting the characteristics of subaqueous eruptions. (1) The role of steam. Water will boil and expand dramatically in shallow water, where explosive eruptions are well known. **This process is damped with increased depth, and there is no phase change during expansion below roughly 3 kilometers' depth in the oceans**..." Note 8.7h

Subaqueous eruptions are directly affected by the type of melted crustal rock, the depth at which the eruption occurs, and the pressure of the overlying water, all important factors contributing to the physical structure of the cooling mineral (see Fig 8.7.6).

The constitution of the seawater is another important factor contributing to the structure and nature of the subaqueous cooling lava. During the Flood for example, the oceans salinity and chemical composition varied, and thus the depth at which non-vesicular basalt or obsidian varied too:

"The critical depth (pressure) is **2.2 km for pure water** and approximately **3.1 km for sea water**." Note 8.7i

Thus, the primary factor involved in vesicle formation is the subaqueous depth in which melted rock cooled.

Puna Ridge Subaqueous Basalt Columns

The Puna Ridge is an underwater area located along the Hawaiian volcanic chain. As researchers have studied the morphology and physical structure of subaqueous cooled lava in this area, they identified *water depth* as being one of the "simple trends" associated with the structures found there.

"A **few simple trends** in overall morphology are **a function of water depth**. Only two fissure vents are identified, both on the shallow ridge crest (<1100 m depth). Large central-vent eruptions dominate farther down the ridge crest. Large cratered cones are observed only at crustal depths between 600-2200 m..." Note 8.7j

Fig 8.7.7 shows the columns of subaqueous basalt at the Puna Ridge site, described as follows by researchers:

"On this cone rocks with **well-formed columnar joints** were photographed near a breach on the northeast side of the crater, and **columns a few meters long have tumbled down the northeast flank**. The columns indicate that flows at least a few meters thick cooled within the crater." Note 8.7j p134

These columns are only estimated to be from hundreds to several thousand years old:

"**Submarine lavas** from the crest and flanks of the **Puna Ridge appear to be older than most subaerial lavas**. Holcomb (1987) estimated that 70% of the subaerial portion of Kilauea is younger than ~500 years. Based on palagonite thicknesses, Clague et al. (1995) estimated that dredged lavas from the Puna Ridge are 700-24,000 years old, mostly 2000-7000 years old." Note 8.7j p126

If basalt columns were formed here just several hundred or perhaps a few thousand years ago, why are there not similar columns of basalt formed historically on dry land and recorded by modern man?

Were all continental basalt columns formed subaqueously—under an ocean?

Devils Tower Mark

In the Basalt Mystery, subchapter 6.5, we discussed the debate among geologists who cannot agree on how Devils Tower in Wyoming was formed. The tower is an obvious 'intrusion' onto the surrounding plain; it is a volcanic rock that was pushed through the surface. But how it formed can never be understood within the magma-rock cycle paradigm. None of the materials

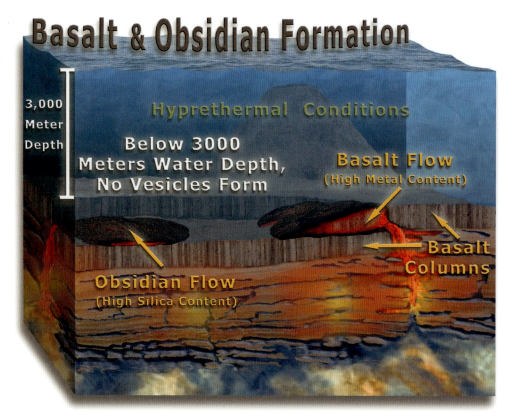

Fig 8.7.6 – This is the Basalt and Obsidian Formation Diagram, illustrating the mechanisms involved in the formation of both mineral types. To prevent the formation of vesicles in the rocks made by escaping gases, the minerals were subjected to extreme pressure, enough to prevent water from becoming gas and expanding. Most researchers recognize 3,000 meters (1.9 miles) of ocean water as the critical depth below which vesicles do not form.

hydrofountain activity, what evidence is there that the Tower was at the bottom of an ocean? The Tower's large crystalline columns indicate they were formed under pressure during the Flood, but there must be other empirical evidence of the Universal Flood in the area. The Devils Tower National Monument web site page states the following concerning the Tower's geology:

"The oldest rocks visible in Devils Tower National Monument were laid down in a **shallow sea**…

"Seas retreated and returned. Climates changed and changed again. **Gray-green shales** (deposited in low-oxygen environments such as marshes) were interbedded with fine-grained **sandstones**, **limestones**, and sometimes thin beds of red **mudstone**." Note 8.7l

In the Sand Mark, we noted that much of the world's homogeneous sandstones and mudstones are of authigenic origin, meaning they were made in place during the Universal Flood and are not from traditional weather erosion. Later in this chapter, in the Carbonate Mark and in the Ore Mark, we will show how microorganisms from inside the surrounding the tower show any evidence of being *melted*, and according the old rock-cycle thinking, volcanic structures imply melting. Moreover, there is *very little* erosion of the Tower itself; the columns are not rounded by weathering processes. The utter lack of any relevant erosion evidence should be cause enough to dismiss the millions-of-years erosion theory. So just how did the Tower form?

Having previously established the reality of hydrocraters, hydrofountains and hydromountains, new light can be shed on the origins of Devils Tower. The first step is to look for identifiable geological features suggestive of a diatreme. On the official Devils Tower National Park web page we read:

"Devils Tower is near the middle of the **collapsed dome**." Note 8.7k

A "collapsed dome" is a 'crater' in what the park service further identifies as being a "Shallow structural basin." The first clue that Devils Tower is associated with a hydrocrater-mountain is then, evident. But there are three geological features necessary to establish a hydrocrater-mountain origin. They are a diatreme, a fault, and the presence of a water source. The tower 'intrusion' itself is evidence of a diatreme, so the first feature is a given. Next, a hydrocrater-mountain must include *faults* for heat generation and to allow passage for underlying material to come to the surface. The Devils Tower National Park web page continues:

"**Three faults** were observed in the area of the National Monument." Note 8.7k

Finally, there must be adequate water for the hydromountain-hydrocrater system to work. In this case, Devils Tower sits alongside the Bell Fourche River. Beyond the evidences of

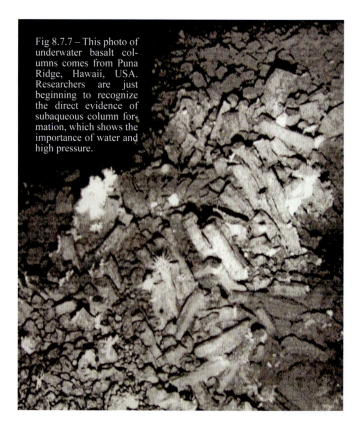

Fig 8.7.7 – This photo of underwater basalt columns comes from Puna Ridge, Hawaii, USA. Researchers are just beginning to recognize the direct evidence of subaqueous column formation, which shows the importance of water and high pressure.

CHAPTER 8 THE UNIVERSAL FLOOD MODEL

Fig 8.7.8 – For the first time, we have a mechanism that can explain how Devils Tower, in Wyoming, USA, was formed. Investigators of this magnificent Tower have identified all the geological features needed to explain the landform as a Hydromountain. The Tower lies in a crater, has a number of faults, and a source of water. A diatreme must also be present for this mountain to be an intrusive feature of the landscape.

Earth were responsible for the production of gray-green shales and limestones, like those found at the Tower. It was no "shallow sea" that produced any of these sedimentary materials! Nowhere in the world are sandstone, mudstone, shale, or limestone being formed to the degree we find these materials on Earth's present-day surface.

The same goes for Devils Tower; there are no geological processes at work today that are capable of producing such a structure. It is only through the light of the UF that such deposits and structures can be understood and comprehended. Devils Tower's minerals formed, like all fine-grained, crystalline basalt, in a hypretherm. This required a deep ocean covering the entire continent, and thus the entire world. It also required the action of extremely large earthquakes, the likes of which modern man has never seen.

Potato Starch Column Experiment

Although no one has ever seen basalt columns forming in Nature, a simple experiment can illustrate how these types of columns are made. The evidence of columnar formation can be seen in Fig 8.7.9 where potato starch was used as a substitute for basalt to create columns of a form similar to columnar basalt. Having established that pressure is required to prevent gases from escaping and forming vesicles in the material, and having established the necessary pressure came from a deep ocean, we can use a Potato Starch Column Experiment to understand column formation *in water*.

Fig 8.7.9 shows the result of the experiment where potato starch was mixed with water in a container to its normal consistency and placed under a hot lamp to evaporate the water. The distance of the lamp had to be varied such that the starch mixture could dry in about a day; too short or too long of a drying period changed the result.

Although this is a simple kitchen experiment, it is not to be taken lightly, even by skeptics. The *Journal of Volcanology and Geothermal Research*, published an article describing just such an experiment, stating:

"**Basalt cooling and starch desiccation are similar processes**, because they are diffusive. In both cases the resulting contraction is strong enough that contraction stresses exceed the material strength. As a consequence, **the crack systems in both media are basically very similar**, in spite of extreme differences in microstructure and elastic properties." Note 8.7m

This starch column experiment identified that *water is an essential* element in the formation of the columns, and so too, it is with the basalt columns. It turns out that other starches such as corn or rice starch work well, and it also turns out that other minerals, such as feldspar exhibit columnar geology. Fig 6.5.6 in the Basalt Mystery subchapter illustrates *quartzite columns* that formed a crystalline morphology similar to the Devil's Postpile in Fig 8.7.10. There, the crystal structure formed both vertically and horizontally. All columnar structures did not come from a melt on the *dry* continental surface, but from a subaqueous hypretherm.

The Ignored Historical Basalt Evidence

The evidence of the submarine origin of basalt has been right in front of the researcher's eyes, yet generally it was ignored. One geologist, Haraldur Sigurdsson, wrote *Melting the Earth* in 1999, noting that the evidence of submarine volcanism was and has always been very real:

"One of the few scholars who studied active volcanoes in this period was Guy Tancrede de Dolomieiu, a Knight of Malta, adventurer, French army officer, and professor in the Ecole des Mines in Paris. The memory of this great geologist is preserved in the calcareous Alpine mountains that bear his name: the Dolomites. As noted previously, during his travels in Italy and Sicily he was the first to recognize volcanic ash deposits interbedded with marine sediments, his most important contribution to volcanology. **In 1784** he showed that marine limestones in Sicily contain numerous layers of dark volcanic ashes **and basalts**, and proposed that **submarine eruptions were discharging volcanic products at the time when the limestones were accumulating on the ocean floor. He was the first to demonstrate that basalts sandwiched between sediments need not be derived solely from igneous intrusion, as proposed by**

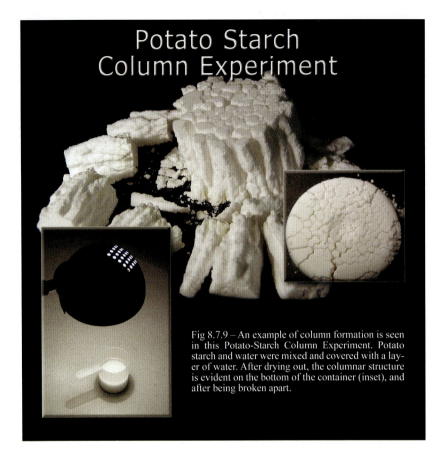

Potato Starch Column Experiment

Fig 8.7.9 – An example of column formation is seen in this Potato-Starch Column Experiment. Potato starch and water were mixed and covered with a layer of water. After drying out, the columnar structure is evident on the bottom of the container (inset), and after being broken apart.

Hutton (who referred to them as 'unerupted lava'), **but could equally be the products of submarine volcanism**." Bib 136 p137

Sandwiched between ocean sediments, basalt essentially proves "submarine volcanism" took place. Moreover, Sigurdsson noted that in 1802, John Murray (a Neptunist who correctly believed the Earth's minerals crystallized from ocean waters) proposed that basalt and granite formed from "a very hot chemical soup" ocean, based on his observations of geysers in Iceland:

"By the early nineteenth century, the Neptunist theory had become severely weakened and was encountering increasing opposition, especially when it was shown that the silicate minerals that compose crystalline rocks such as basalt and granite are **insoluble** in aqueous solutions **at normal temperature**. Few adherents remained but John Murray (1802) attempted to rescue the theory from this trap by pointing out that **silica is found in solution and precipitated from the high-temperature waters and exhalations of the geysers of Iceland**. He also proposed that the primordial **ocean was a very hot chemical soup that dissolved alkalis and silica, and was filled with the 'saline, early and metallic matters.'**"
Bib 136 p123

In reality, the Neptunists were on the right track; *all of the evidence visible to the geologists* testified of basalt's submarine origin. *Melting the Earth* continues:

"The depths of Raspe's early understanding of the volcanic origin of basalts are also well illustrated in his 1769 correspondence with **Sir William Hamilton**, the patient observer of the eruptions of Vesuvius **who had examined more lava flows than any other student of volcanoes at the time**. In a letter to Hamilton, Raspe inquired, 'I would like to know from your Excellency, if you have ever found among the new lavas of Vesuvius, anything which confirms the interpretation of Mr. Desmarest; in other words, **do any of the cooled lava flows of the Vesuvius display toward their end something similar to prismatic basalt**?' Hamilton replied, '**Concerning the basalt, I have, since investigating lavas, always considered it as a type of lava. However, I have never found any kind which is columnar or polygonal like basalt, among the numerous types of lavas of the Vesuvius, Sicily and the island of Ischia**.'

"From his own observations and those of Hamilton, Raspe concluded that while columnar basalt was undoubtedly volcanic and formed when lavas erupted on the sea floor, not as the product of subaerial [on surface in air] eruptions, 'because its mass cooled faster.'" Bib 136 p142

It couldn't be any clearer, but the *modern* geologist has generally *ignored* these observations and so remains unaware that the vast basalt deposits on land are a submarine or subaqueous mineral!

Why have researchers chosen to ignore the submarine origin of continental basalt for more than a century?

For one simple reason: continental basalt was ***formed only recently***. The dearth of erosion is evidence of the basalt's several-thousand-years young age. This simple fact completely changes the origin of the world's "most common lava" and has

Fig 8.7.10 – Devils Postpile National Monument lies in the High Sierra backcountry in California, USA. It is one of the world's best examples of columnar basalt. Long ago, mankind looked to these 'strange' formations with awe, and geologists still do today. Whether a postpile or a tower, their true magnificence is only evident with the Universal Flood paradigm, because they tell a story about one of Earth's greatest formative periods.

CHAPTER 8 THE UNIVERSAL FLOOD MODEL

revolutionary consequences.

The Ignored Modern Day Empirical Evidence

It is interesting that the answer to the Basalt Mystery continues to pop up directly in front of the researchers. Several decades ago, geologists from the Australian National University had this to say:

"Hydration of **modern basaltic** crust occurs at the mid-oceanic ridges. Here magma chambers [actually frictional heat at faults] below the ridges act as heat sources for convective systems **which cycle hot sea-water through fractures in the crust**, resulting in deep and wide-spread hydration of **sea-floor basalts**. The process is **effective because the heat source is overlain by water**." Note 8.7n

The question arises that if geologists knew "modern basalt" was formed in association with "hot sea-water" "overlain by water" of the deep ocean, why didn't other researchers make the connection with the formative process of all the 'old basalts' on the dry continents? Perhaps it is that geologists are unwilling to admit that the continents of the world were *not* covered by "shallow seas," but by deep, hot ocean water. Such a notion would be difficult to fit in the geologist's world view, especially because the continental manifestations of basalt and obsidian are dated as being thousands, not millions of years old. Contrary to the 'general consensus' of the geological community today, scientists are beginning to recognize the role of water in the origin of many minerals, including basalt. Even without the big picture of the Hydroplanet Model and Universal Flood Model, they are uncovering geological truths that are challenging many long held beliefs.

The Biogenic Origin of the Earth's Oceanic Crust

In the Earth Crust Mystery (subchapter 6.12), we asked two important questions about the Earth's crust, restated simply here:

Why is the oceanic crust material so different and so much thinner than the continental crust?

If both crusts cooled from magma, there is no simple answer to this question in modern geology. Additionally, without magma, there appears to be no mechanism for the separating of the Pangaea Supercontinent that once existed. Without a magmaplanet, tectonic plate movement doesn't make sense to the modern geologist—but there are alternative answers to these questions, if we look outside of the modern geology box.

The Depth Mark establishes the hyprethermal nature (3,000 meter ocean depths) in which most (non-vesicular) basalt and obsidian formed. This was the deep, hot, water environment necessary for the formation of fine-grained basalt, which is also the reason this type of basalt is not seen coming from land volcanoes today. Floodwaters on the earthquake-heated continental crust lasted only a relatively short time. This explains why the continents are *not* entirely covered with basalt, whereas the oceanic crust *is* almost entirely basaltic rock.

Although most Floodwaters were evacuated from the continents after perhaps a year's time, the breaking apart of the Pangaea Continent into newly formed and separating continental plates took place over a much longer period of time, coinciding with the return of the Earth's rotational velocity and a return to equilibrium of the system. The new continents began to slowly migrate around the Earth, driven by centrifugal forces and changes in orbital and rotational dynamics. The ocean floors between the continents were pulled apart and melted in a hypretherm by the heat generated from tectonic plate movement and earthquake friction. It was then that most of the mafic rocks and dark minerals making up the ocean floor formed and why they are so much younger than the continental rocks, grown when the Earth was first created.

Later, in The Ore Mark subchapter (8.13), we explain the

Giants Causeway Ireland

biomineralization that accompanied the creation of the oceanic basalts and obsidian. The darkness of these minerals comes from their rich iron and magnesium content, trapped within each mineral's structure as it formed at the bottom of the UF ocean, in the hypretherm. Where did these metallic minerals come from—Microbes! Ocean water is teaming with life and when the temperature and proper nutrients are available, microbial life-forms explode into great ocean-wide blooms. Such an event occurred during and shortly after the Flood, impacting the creation of the thin, newly forming oceanic crust. This explains why the oceanic crust is thin and completely different than the thicker, granitic continental crust.

Biomineralization was active during the formation of deep-ocean basalt, whereas the formation of original granitic continental crust was formed of other hyprethermal minerals, and no biomineralization—it was not made in the presence of microbes!

Taking this further, the oceanic and continental crusts are so different because no life existed on the Earth when it was first formed. The Earth's basalt is a biogenic mineral, granite is not, which is why there are no fossils found in granite; granite was formed when no life was on the Earth.

Direct evidence of this mode of oceanic crust formation is seen today on the ocean floor, at plate boundaries where friction is heating and melting underlying rock, stimulating the creation of biogenic basalt. Early on, rapidly spreading crustal plates eventually bumped against other plates until the Earth's spin rate finally reached equilibrium. This slowed the movement of the crustal plates, almost to the point of stopping, which is evident today with horizontal movement being at most a few centimeters a year.

The Basalt-Carbonate Connection

The Depth Mark redefined the origin of both obsidian and basalt, based on empirical evidence that a hypretherm is required to form either mineral.

A corrected definition of Earth's **Basalt** is as follows:

> **Basalt**: a dark crystalline mineral formed in a submarine hypretherm in the presence of biomineralization.

Adding the *hypretherm* process in the description of how obsidian and basalt forms, completely changes the way in which large portions of the Earth's crust formed. Instead of dry continents covered by cool, shallow seas millions of years ago, the surface was submerged under *at least* three kilometers (1.9 miles) of water, as evidenced by the fine-grained basalt on the continental surface; basalt showing almost no erosion because of its recent deposition.

Another interesting factor is the origin of lunar basalt. Lunar basalt is quite different from the Earth's basalt, containing, for example, large amounts of titanium. Despite this, both basalts share many properties. Did lunar basalt include the element of microbes? We do not yet know, but we must be open to this possibility. Until we return to the Moon for further investigation, many lunar questions will remain unanswered. It is surprising how little we actually know about our own Moon, even as close and accessible as it seems to be.

The UF paradigm reveals a *direct* connection between basalt and the carbonates; a connection not made in the basalt research literature we read. We'll call it the Basalt-Carbonate Connection. Fig 8.7.13 shows a layer of basalt on top of layers of clay. The basalt is the darker grey material. On the faces of the basalt and in the cracks, a *white mineral* can be seen. This material is calcium carbonate, which will be explained in detail in the next subchapter, the Carbonate Mark. The association between these two minerals is not coincidental. The vast majority of continental basalt exhibits carbonate (limestone) in the cracks, on the surface, and also *inside* the basalt itself! Many samples of basalt have been broken, revealing carbonate-filled voids. Surprisingly, the origin of the carbonate inclusions is *known* among geologists—it is from bacterial algae grown in oceanic waters!

Here again we have direct evidence that basalt formed in an ocean. The Basalt-Carbonate Connection can be observed in almost *every basalt deposit* on every continent worldwide. So strong is the evidence of a Universal Flood that it is hard not to see it once these connections are known. One powerful example comes from Yellowstone National Park in Wyoming. Two distinct layers of columnar basalt exist between layers

Fig 8.7.12 – Modern geology has no explanation to account for the differences between continental and oceanic crusts. However, the UF makes it possible to comprehend the biogenic nature and rapid formation of oceanic basalt crust. As floodwaters drained quickly off the continental landmass, very little basalt had formed on it. As the Pangaea supercontinent broke into several large landmasses, each moved rapidly apart, creating frictional heat and hyprethermal conditions at the quickly spreading plate boundaries. This stimulated prolific biomineralization in the deep ocean, forming the Oceanic Basaltic Crust. In contrast, the *original* (pre-Flood) continental crust was formed during Earth's primeval watery hypretherm. This occurred prior to life's arrival, so it did not include biogenic processes.

Fig 8.7.13 – What is this white carbonate material on, around and inside this basalt deposit? It is limestone that modern geology already acknowledges comes from oceanic waters. The Basalt-Carbonate Connection is direct evidence that basalt deposits on the continents were formed in a submarine environment, and it supports the new UM definition for basalt.

of sediment, seen in Fig 8.7.14. Large, rounded cobblestones are in the layer just below the upper basalt columns. These are spread across a vast flat plain, but have no origin in the traditional geologically based erosion time scale. Below the lower layer of basalt are sediments exhibiting ore-like characteristics. These types of deposits along with the ore deposits found in almost every country verify that a worldwide catastrophic event occurred recently across every continent, under a deep ocean with some areas being extraordinarily hot.

A Final Look at the Depth Mark

The unambiguous fact that fine-grained basalt and obsidian formed in a deep hot ocean has not precluded geologists from speculating wildly about how such minerals formed. One key piece of evidence is the lack of vesicles, or the smallness of their size in some specimens, but there is much more, such as the biogenic origin of the mafic minerals, the columnar habit of basalt, and the carbonate connection. Another curious bit of evidence comes from the book, *Petrified Wood* by Frank J. Daniels:

"The association between volcanism and petrification is quite apparent when **petrified logs are dug directly from basalt** lava flows in Washington". Bib 194 p120

Certainly, hot lava rolling through the forest would consume the forest, but the Universal Flood Hypretherm offers an explanation for how both the basalt and petrified wood formed—under deep water, high pressure, and high temperature.

8.8 The Carbonate Mark

The Carbonate Mark subchapter is an in-depth look at the carbon sediments in the Earth's crust, whose origins, at best, are unclear or muddy in modern geology. The true origin of carbon will be explained in the Chemical Model chapter by a biological transformation (biotrans) process, which along with a formerly unknown source of carbon, explains many of the Carbon Mysteries in the Rock Cycle Pseudotheory chapter. In this subchapter, the connection between quartz and carbonates is explained, which also helps identify the real origin of loess.

The dolomite problem is also addressed along with the actual processes of how carbonate caves and carbonate pinnacles are formed.

The Origin of Carbonates

The Carbon Cycle Pseudotheory in Chapter 24, the Chemical Model, will explain the origin of carbon that is discussed here in the Carbonate Mark. The theory that modern science uses in the globalized carbon cycle will be shown to be incorrect because the origin of over 99% of global carbon is unaccounted for with the inclusion of carbonate sediments. For many decades the carbon cycle diagram in Fig 8.8.1 has been taught as part of the theoretical carbon cycle as though it were fact, making it a pseudotheory.

In summary, the diagram of the Carbon Cycle is in error because it excludes the 78,000,000 units of sedimentary carbon stored in the world's limestone and dolomite deposits. If we add up the carbon contained in the atmosphere, the vegetation, soils, fossils fuels, rivers, surface oceans, marine and dissolved carbon, and the deep-ocean carbon identified in this NASA diagram, the total comes to about 47,000 carbon units. The figure of 150 units of sedimentary carbon near the bottom of the diagram is vastly short of the correct 78-million figure. Comparing the 47,000 carbon units representing all the carbon

Fig 8.7.14 - Two sets of columnar rock deposits are located near the Yellowstone Tower Falls in Wyoming, USA. Located beneath the upper columnar sill are sedimentary layers that contain large rounded rocks on what was once a flat plain. What process formed large flat plains of heavy sediment as well as the many different layers of colored ore minerals above the lower column of rocks? No erosion or process that is observed today can explain these formations. Only the UF Hypretherm can explain how geological formations like this occurred.

cycling throughout the environment with the actual carbon contained in the carbonate sediments, we see that over 99% of the 78 million carbon units in the Earth's crust are unaccounted for.

Obviously, a theory that misses this much of the data cannot be correct. But scientists are reluctant to include the large amount of carbon in the sediment for one reason—there is no known origin for it. As the Carbon Cycle Pseudotheory (Chapter 24) will demonstrate, there is almost no carbon in the rocks the Earth was originally made of. This is apparent in the elemental composition of igneous rocks, shown in the table in Fig 8.8.1. The Carbon Model will present experiments demonstrating the process of biotrans—biological transformation. In this newly identified biological transformation model, the Law of Biotrans completely changes the paradigm we have with respect to all forms of life. The creation of carbon by microbes transcends all of modern biology, which places us in a unique position. Armed with new scientific knowledge and wisdom, we are prepared for the first time, to discover how such massive amounts of carbon sediment were created on the Earth's surface.

Furthermore, this subchapter will answer many of the carbonate mysteries outlined in the Carbonate Mystery subchapter, 6.9. These mysteries exist in part because calcium carbonate, the hard, white residue left behind on your drinking glass, your dishwasher, and on your showerhead is still believed to be a *chemical* process. In truth, however, it is a *biological* process.

Another reason the Carbonate Mystery exists in geology today is that geologists have not yet devised a mechanism to explain how immense global carbonate deposits came to be and in so short a time. Global surface carbonate deposits are far too pure and show too little erosion to have been laid down on a geological time scale. It is clear they were formed and deposited quickly, a fact that has many geologists proposing incorrect theories that attempt to explain actual observations. One such example is the Snowball Earth Theory.

Cap Rocks from "Snowball Earth" Theory

Cap Rocks, shown in Fig 8.8.2, are layers of hardened sediment that "cap" other layers of sediment. There are many types of Cap Rocks, but a large number consist of the yellow, carbonate-based layers seen in Fig 8.8.2. The *global* relationship of the Cap Rock material has been largely overlooked by researchers, but must be accounted for in any theory attempting to explain the Cap Rock origin. One group of researchers devised a theory to explain the origin of Cap Rocks, which they reported in the journal of *Science*. Their theory was dubbed the "snowball Earth" theory:

"Negative carbon isotope anomalies in carbonate rocks bracketing Neoproterozoic glacial deposits in Namibia, combined with estimates of thermal subsidence history, suggest that biological productivity in the surface ocean collapsed for millions of years. This collapse can be explained by **a global glaciation** (that is, a **snowball Earth**), which **ended abruptly when subaerial volcanic outgassing raised atmospheric carbon dioxide to about 350 times the modern level**. The rapid termination would have resulted in a warming of the snowball Earth to extreme greenhouse conditions. **The transfer of atmospheric carbon dioxide to the ocean would result in the rapid precipitation of calcium carbonate in warm surface waters, producing the cap carbonate rocks observed globally**." Note 8.8a

This technical description of the snowball Earth theory can be broken down into a few simple steps that attempt to describe

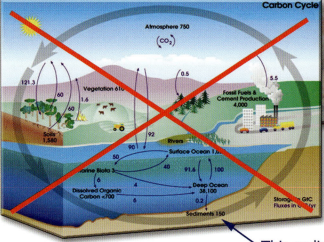
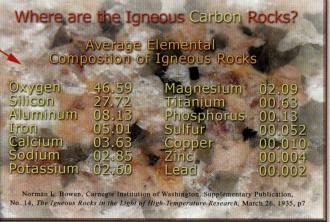

Fig 8.8.1 – The Carbon Cycle cannot be properly understood with the current cycling theory. As will be further established in the Chemical Model, the massive carbon sediments (limestone/dolomite) on the surface of the Earth have never been properly accounted for in modern geology. Despite the widespread acceptance of this cycle, it still does not account for the actual origin of the elemental carbon on the Earth. The Earth's vast carbonate deposits and the true carbon cycle can be understood only after possessing the knowledge of the Earth's Deep Biosphere, and the Universal Flood with its accompanying processes.

how the world's oceans produced the "cap carbonate rocks observed globally." First, a period of "global glaciation" plunges the Earth into a deep freeze—the snowball Earth period. No mechanism is given to explain the onset of this *cold spell*, or how the Earth's organisms survived it. Secondly, the global glaciation abruptly ended with a sudden *worldwide heating* coupled with "volcanic outgassing" of carbon dioxide, but again, absolutely no mechanism to trigger such a change is offered.

Thirdly, no explanation for how the Cap Rock became hard is given. The bottom line is snowball Earth theory is like so many other modern science theories—no basis in fact, the primary reason we have been through a Dark Age of Science.

One aspect the snowball-Earth theory does reveal is that the researchers **knew** the global carbonate cap-rocks were associated with a ***global ocean,*** a subtle witness of the Acknowledged Flood in subchapter 8.2. Snowball-Earth researchers may have

Fig 8.8.2 – Examples of Cap Rocks, seen below, are typical of a global phenomenon unexplained by modern geology with any degree of certainty. Carbonate cap rocks that exist globally on the surface today were created in an ocean, yet the type of ocean they occurred in does not exist today.

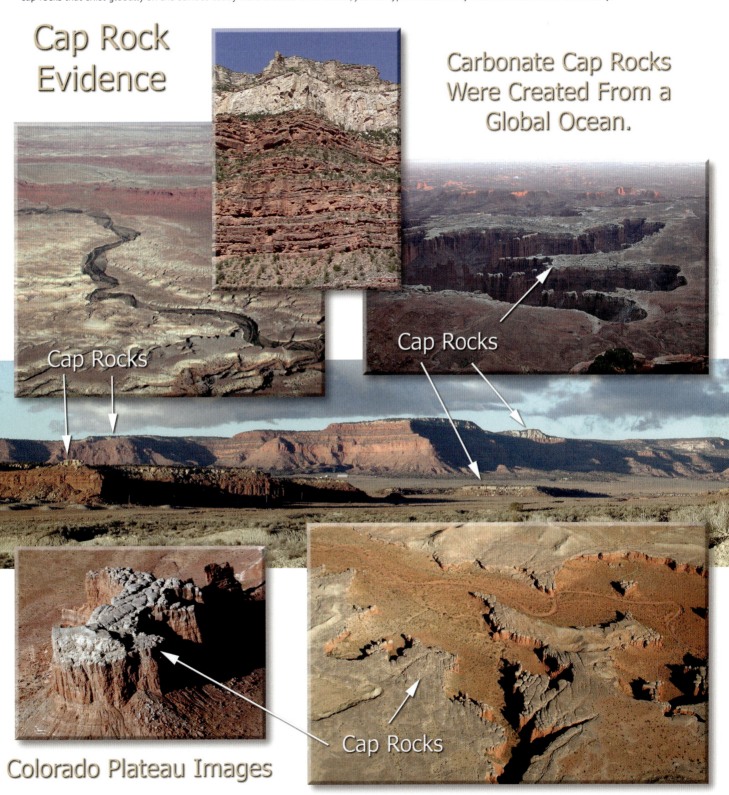

Cap Rock Evidence

Carbonate Cap Rocks Were Created From a Global Ocean.

Colorado Plateau Images

SUBCHAPTER 8.8 THE CARBONATE MARK

realized the important fact that the event was "global" and "rapid," but they failed to identify the mechanisms that started or ended the event. This of course is where the Universal Flood comes in.

Microbe Dogma—"Was Wrong"

Because the Magma and Rock Cycle Pseudotheories are so deeply ingrained in modern science, the thought that anything could be living deep inside the Earth, amongst the melted and metamorphosed rocks seemed an utter impossibility. That dogma held until the mid-1990s when researchers discovered they were "wrong," as they reported in 1996:

"With countless novel uses still awaiting discovery, biologists continue to scour the surface of the earth in search of microbes that might prove valuable in formulating new drugs or improving industrial processes. But until recently, **few such bio-prospectors thought to look deep inside the earth. Long-standing scientific dogma held that this realm was essentially sterile. But that belief, as it turns out, was wrong**." Note 8.8b

It was *dogmagma* that caused the "long-standing scientific dogma" that the Earth deep inside was "sterile."

The Endobiosphere Evidence

Endobiosphere is a new term in the UM that describes the *biosphere within the Earth*. This includes the microbial environment that exists from the crust to the core. It has only been during the last decade that researchers have come to realize the "hordes of unknown microbes" that exist inside the Earth "in environments presumed impossible" to support life:

"Not long ago, the very idea of finding life at such depths was considered less likely than discovering it on other planets. Life was thought to peter out a little past the depth of a grave. A billion bacteria may exist in a pinch of topsoil, but biologists found that the numbers dropped to the millions, then the thousands, then the hundreds per sample as they dug down farther from the sun, air, and sources of food. Then in the last decade, researchers digging deeper and using new analysis techniques **were astonished** to find **hordes of unknown microbes**. Isolated from the surface for eons, these organisms have been living in environments **presumed impossible**—oil wells, aquifers, and **deep rock**." Note 8.8c

The researchers were utterly "astonished" to find "hordes of unknown microbes" at great depths because their wisdom held that without sunlight, air or carbon, life could not be supported. However, the Law of Biotrans is unknown to modern science. With the processes of transformation and carbosynthesis, found in the Chemical Model, we will be able to understand the mechanism that drove these events and we recognize the mighty role microbes played in the geology of the Earth.

Researchers have long had their microscopes, oil wells and aquifers in which to see these microbes, but they have been missed because their paradigm was wrong. Researchers from the University of Washington found:

"Seafloor crust at mid-ocean spreading centers harbors a substantial microbial infauna, supported in part by nutrient fluxes released during crustal accretionary episodes. Evidence for this subsurface microbial community is compelling. All eight submarine diking-eruptive events examined recently resulted in massive effusions of microbial floc, including very high densities of live thermophilic bacteria. **The precise source of this material is not certain**. Perturbations in established hydrothermal flows (**triggered by volcanism**) may have **flushed existing microbial products from pores and cracks in the sub-seafloor**. Alternatively, fracturing and magmatic intrusion may have enhanced nutrient fluxes, resulting **in microbial blooms carried upwards to the sea floor by upwelling fluid**. A significant, but unexplored, fraction of the oceanic crust seems to be a volcanically-supported sub-seafloor biosphere." Note 8.8d

These researchers are "not certain" of the source of the microbes because they do not recognize the existence of the endobiosphere. The "microbial blooms" were not 'born' inside dry rocks or molten magma; all life needs water, and the microbes needed a path to the surface. The "pores and cracks" between crustal rocks play an important role in the formation of biogenetic rocks and the Law of Biotrans. How deep and how much biogenetic rock lies beneath the surface? Answer this fundamental question and the mystery of the world's thick limestone and dolomite beds will be solved.

We return to the second deepest and most studied borehole in the world, the KTB well in Germany. The rock cores drawn from the KTB borehole revealed an astonishing fact apparently overlooked by geologists. Among the numerous rock minerals encountered in the borehole cores, the most common was *not* an inorganic mineral; it was *calcite*, which is calcium carbonate. Note 8.8e In addition to being the most abundant mineral encountered, calcite was found as the drill bit reached **over 9,000 meters into the Earth**. The direct evidence of carbonate rocks as deep as 9 km created a problem; how did the carbonate rocks get there?

Some investigators argued that the carbonates found deep in the crust came not from microbes but from an abiogenic or non-biological mechanism. They claim the carbonate rocks must have come from carbon in the igneous rocks present when the Earth was created. However, Fig 8.8.1 shows the constitution of Earth's igneous rocks, indicating that carbon is almost totally absent in them.

Unequivocal evidence that almost all hydrocarbons come from microbes came from an important study published in *Nature* in 2002. The study examined isotopic signatures to determine if fossil fuel hydrocarbons originated from igneous or magmatic rocks. Researchers ruled out any significant non-biological source for the hydrocarbons:

"...**we can now rule out** the presence of a globally significant abiogenic [non-biological] source of hydrocarbons." Note 8.8f

What this means is that the Earth's hydrocarbons, such as petroleum, did not come from rocks. Where then, did all of the Earth's carbon originate? Without the Law of Transformation, researchers turn to the magma pseudotheory:

"These abiogenic [non-biological] hydrocarbons are generally formed by the **reduction of carbon dioxide**, a process which is **thought** to occur during **magma cooling**…" Note 8.8f

Here we see again that magma cooling is modern science's best option, and without it, they have no real origin for much of the Earth's carbon. Why is this important and why the emphasis on the Earth's Endobiosphere? Because of the role it played during the UF:

When UF hydrofountains burst, they released far more than just water; a diverse concoction of microbes was spewed out, altering the pH of the water and creating diverse, new environments no longer in existence today. Here were formed the biogenic minerals found in such abundance today; limestone, dolomite, loess, salts and the ore minerals.

Here for the first time, the real origin of many mysterious minerals discussed in the Rock Cycle Pseudotheory is revealed. These minerals came from microbes, but not just any microbes. They were special microbes, indigenous to the Universal Flood. One reason modern science continues in its ignorance toward the microbes of the Endobiosphere is that evidence of their existence is relatively new, leaving most researchers with the idea that most life is fueled by "sunlight":

"**Most life** on earth is fueled directly or indirectly by **sunlight**." Note 8.8g

However, with the revelation of the Hydroplanet Model and its recognition of abundant water inside the Earth along with the knowledge that all natural waters contain microbes, science is just beginning to learn how much living carbon there is inside our dark Earth. As more and more microbial rich water is identified inside the planet, the importance of sunlight for many organisms is diminished, and science will slowly begin to comprehend the Earth's ultimate carbon source—the Endobiosphere.

In the next section, we will see that anaerobic (without oxygen) bacteria are major contributors to the carbonates in the Earth's crust, and that this bacterium also happens to be found inside the Earth.

Carbonates Produced Experimentally by Microbes

In the Chemical Model we will discuss how photosynthesis and chemosynthesis use microbes (certain bacteria and fungi) and water to break down minerals in the Earth and transform them into carbon. They do this instead of using airborne carbon dioxide, a long-held fundamental assumption of the photosynthesis process. Neither of these processes—photosynthesis or chemosynthesis—can account for all the limestone deposits in the Earth. The Chemical Model introduces a new carbon producing process called carbosynthesis. The biochemical transformation process, or biotrans, is a process whereby water and other minerals are transformed into carbon and other elements. This process, which was demonstrated with the biotrans experiment (biojar system), accounts for the carbon in the Earth's vast geological deposits.

Unfortunately, modern science researchers are unaware of this process because carbosynthesis is new in the UM. For the scientists, the calcium carbonate in the ocean is still thought to come from soluble limestone dissolved by acidic waters even though the Rock Cycle Pseudotheory chapter demonstrated this to be false. During the past few decades, laboratory tests have shown that calcium carbonate ($CaCO_3$) comes from a biogenic process instead of an inorganic process, but this information is still not widely known. Calcium carbonate is defined in a current popular encyclopedia in this way:

"**Calcium carbonate is a chemical compound** with the chemical formula $CaCO_3$. It is a common substance found as rock in all parts of the world, and is the main component of shells of marine organisms, snails, and eggshells. Calcium carbonate is the active ingredient in agricultural lime, and is usually the principal cause of hard water. It is commonly used medicinally as a calcium supplement or as an antacid, but high consumption can be hazardous." Note 8.8h

Notice that the definition states calcium carbonate as being a "chemical compound," explaining where it is found, but containing **no origin** for the $CaCO_3$. However, asking the right questions led many investigators to the important knowledge that natural $CaCO_3$ *is from microbes*. Back in 1984, a number of researchers confirmed an important previous discovery by observing the calcification of mats in Solar Lake, Sinai:

"**The major question is** whether most of this authigenic $CaCO_3$ is produced via evaporitic (i.e., **inorganic**) or bacterial (i.e., **biogenic**) processes." Note 8.8i

By asking this question, investigators were able to answer whether natural carbonates came from an inorganic or biogenic process because they grew bacteria in the laboratory:

"Pore-water data **confirm the observations** of Krumbein and co-workers that **the oxidation of the cyanobacterial mat by anaerobic bacteria is a major process leading to the precipitation of $CaCO_3$.**" Note 8.8i p626

Natural calcium carbonate is indeed a ***biogenic process!*** Nevertheless, this research was preliminary, involving only small amounts of cyanobacteria on mats just below the surface of the water in the lake, far short of the massive carbonate deposits on Earth. But then in 1998, several Floridian researchers stated their investigations "require reevaluation" of the contribution of microbial carbonates to carbonate sediment:

"The contribution of **microbial carbonates to carbonate sediment budgets has generally been overlooked**. However, this research indicates that the magnitude of this process is such that these budgets may **require reevaluation** as new data on microbial calcification potential is acquired." Note 8.8j

The role of *microbial carbonates* in the carbonate sediment budgets (massive limestone/dolomite layers) truly "*has been generally overlooked*" because the carbonate deposits have always thought to have been made up of dead sea creatures that had taken in *inorganic* carbon from ocean waters. Remarkably, the Florida research team found that in the control samples that had *no bacteria, no precipitation* of calcite occurred, while those samples that included bacteria did show precipitation of calcite, leading them to state; "cells are required for precipitation of calcite":

"While experimental cultures and controls were supersaturated (SI = 91.4) with respect to calcite, **controls showed no mineral precipitation** within the incubation period of 4 h. Periodic microscopic examination of semi-continuous stock cultures of *N. atomus* in undersaturated conditions frequently showed mineral precipitation indicating that cells are capable of inducing precipitation in undersatu-

Why were researchers utterly "astonished" to find "hordes of unknown microbes" at great depths? Wisdom reveals the answer—dogmagma replaced with a hydroplanet creates a new environment microbes could live in.

rated media, and that cellular metabolism likely plays a role in generating a microenvironment near the cell that is conductive to calcification. Results of calcium carbonate precipitation experiments indicate that **the presence of live *Nannochloris atomus* cells are required for precipitation of calcite** in BG11 at the experimental parameters described above." Note 8.8j p1506

Not only were bacteria "required" for calcite precipitation, the environment showed clues important to discuss. The pH level was kept at a constant 8.5 and the temperature remained at 33° C (91° F), both of which are slightly higher than is found in the ocean today. Fig 8.8.5 shows the world's sea-surface temperatures peaking at only 30° C, a full 3° C (10° F) lower than the experimentally grown carbonate. This relatively small temperature difference is very important to many bacteria species. Furthermore, the ocean pH averages only about 8.1 compared with the elevated experimental pH of 8.5. Note 8.8k Higher pH levels dictate the quantity and type of microbial mass to be produced.

This is one reason so little natural calcium carbonate is found forming in the oceans today. In contrast, the hyprethermal ocean environment included higher pH levels and elevated temperatures, dramatically affecting microbe colonies responsible for producing the massive layers of carbonate sediment found on the surface of the continents today.

Another example of the "greatly underestimated" contribution of bacteria to the terrestrial carbon budget comes from an article in the August 2002 journal of *Naturwissenschaften* titled, *Is the contribution of bacteria to terrestrial carbon budget greatly underestimated?* In this article, researchers reported that calcium carbonate production by bacteria is a "common and rapid process":

"Calcium carbonate production by bacteria from low molecular weight organic acids seems to be a **common and rapid process**. On acetate- and citrate-containing media, **results clearly show the influence of bacteria on the mineralogical nature and shape of $CaCO_3$ crystals** produced on similar media. **Precipitation of biogenic calcite crystals is quite common.**" Note 8.8l

These Swiss researchers also reported that the pH "rapidly increases during the exponential growth of the colony":

"The growth of R. *eutropha* and X. *autorophicus* in the liquid Schlegel medium containing potassium oxalate showed a rapid consumption of oxalate associated with a continuous increase in pH. **The pH rapidly increases during the exponential growth of the colony** and continues to increase at a lower rate after the stationary phase has been reached. This can be explained by the release of metabolites during lysis of dead cells, which are probably quickly reused by living cells, after the oxalate source has been totally exhausted. Final pH after 7 days of incubation was **>9.5 in each case**. Calculations show that the theoretical pH should reach a value of 9.55±0.05 after total oxalate consumption, **emphasizing the role of bacteria in providing the required conditions for calcium carbonate precipitation.**" Note 8.8l

The exponential growth of such microbes is halted in today's oceans because rarely if ever does a 9.5 pH level exist over a widespread area of seawater (seawater typically has a pH of about 8.1). But during the UF event, this was exactly the environment, as a substantial volume of microbes from the endobiosphere were ejected onto the surface through hydrofountains.

In 2003, another group of Swiss experimenters discovered that when calcite crystals precipitated, they contained "marks and holes of the same shape and size" as the bacteria used in the experiment:

"The results of these initial experiments are encouraging and **demonstrates by direct measurements the potential of cyanobacteria Syneccococus to precipitate calcite**. The amount of the precipitated calcite varied in experiments using solutions with a different ratio of dissolved inorganic carbon and calcium. **Remarkably more calcite precipitated in the solution with higher carbon but lower calcium concentrations**. The results of the experiments provide some evidence that the cell walls of the cyanobacteria acted as a substrate of nucleation of $CaCO_3$. The unicellular cyanobacteria *Synechococcus* induced the precipitation of the rhombohedric calcite crystals under both conditions. Ion-selective electrodes were shown to be a useful tool for precipitation experiments. **The precipitated crystals contain marks and holes of the same shape and size as the *Synechococcus* cells.**" Note 8.8m

The results of the experiment produced a phenomenon very similar to the silica cysts observed in Homegrown Quartz (in wide-spread deposits of claystone) seen in Fig 8.5.6, except that the silica cysts are *quartz* whereas the carbonate bacterium-formed cysts are *calcite*. This is significant because these two minerals comprise the vast majority of all natural cement in *all the sedimentary rocks* found in the Earth's crust.

The increase in precipitated calcite despite having higher carbon concentration may yet be another confirmation of carbosynthesis and the Law of Biotrans. Although the ability of bacteria to transform non-carbon bearing minerals into elemental carbon is a recent UM discovery, evidence should be widespread and apparent, once researchers begin looking for it.

Endobiosphere Environment Discovered

During the last decades of the 20th century, technology supplied the wherewithal to study the bottom of the ocean, opening a completely new field of research. In a classic article titled *Floor Show*, John R. Delaney with the New York Academy of Sciences describes what researchers found in 1977 while ex-

Fig 8.8.3 – Black smokers were first observed along the mid-oceanic ridge where the oceanic plates are in constant motion, producing frictional heat that supports an endobiosphere thriving with microbes dependent on heat, pressure, and unique chemistry to survive.

ploring earthquake-prone areas in the seafloor. They discovered evidence of the endobiosphere, "a previously unsuspected biosphere below the seafloor":

"As it turned out, vent fluids from the area near the bacterial plume found by Embley were carrying hyperthermophilic Archaea. Indeed, in culture, they would **not** begin to multiply until they were in **oxygen-free conditions and at temperatures between about 130 and 194 degrees F.**, well above the sixty-six-degree F. temperature of the effluent from which they were sampled. In the laboratory we found that, under optimum conditions, **the doubling time for the populations of such microbial communities can be less than an hour**. But no one knows what gave rise to the outpouring of microbial material from below the seafloor. Had the eruption released nutrients that triggered an explosive bloom of microorganisms below ground? Or had the eruption simply flushed out what was already there? Either way, the effluent, luxuriant with microorganisms, pointed toward **the existence of a previously unsuspected biosphere below the seafloor, supported by the output of volcanic volatiles**." Note 8.8n

Why was this newly discovered biosphere "unsuspected"? Because modern science believed that all life got its energy from the Sun, and the ocean depths were impenetrable by the Sun's rays. The submersible Alvin revealed black smokers (Fig 8.8.3) and mats of bacteria in anaerobic or oxygen-free environments, completely changing the perception of life in the deep ocean and beneath the crust of the Earth. Researchers have been especially interested in the new endobiosphere environment to help understand how life could exist on other planets. The UM has used the same information to help learn how the deep Earth biosphere impacted the crust during the UF and how the endobiosphere affects the environment on the surface where we live today—including the weather.

This becomes more apparent in the next chapter, the Weather Model, as we explore the cyclical nature of earthquakes and the heat they provide, which is critical to the Earth's weather cycles. Since researchers are generally not yet aware of the cyclical nature of earthquakes and the heat they generate beneath the surface, they missed the significance of events like the "volcanic springtime on the seafloor":

"**Bacterial blooms signal a kind of volcanic springtime on the seafloor**. Here seasons are not determined by the tilt of the earth's axis or by the orbital relations of the earth and the sun. **They depend only on episodic releases of molten rock or gases from deep within the ocean crust**. In that dark habitat the microorganisms are the primary producers—they are the grass or phytoplankton of the deep seafloor. When the blooms take place, opportunistic foragers move into the local area to graze on the vastly increased food supply. The first wave of spider crabs, fifteen inches across, may be followed by highly mobile predators and scavengers, such as octopods or fish." Note 8.8n p30

Obviously, we haven't enough knowledge of the ocean floor nor observations to determine the real cycles of Nature when we read claims like the following from Delaney:

"A submarine diking event—**whereby magma is injected into rocks and cracks from a reservoir below**—is one of the **most basic of geologic processes**, the irreducible event that has formed the seafloor throughout more than four billion years of geologic history." Note 8.8n p31

Only after the realization that the "most basic of geologic processes", such as magma dikes, are based on modern science pseudotheory, will investigators be able to look beyond the dogmagma and begin to comprehend how the basaltic sea floor was really created through tectonic activity. Millennial Science will find many new and exciting weather patterns evident on the seafloor, patterns and events never before observed. Springtime on the seafloor will become a surprising reality.

Organic Eruptions Explained

The Hydroplanet Model previously outlined volcanic emission rates showing that over 94% of the gases from observed volcanoes is water in the form of steam. This helps establish the presence of a considerable amount of water beneath the crust. But the other gases that come from deep inside the crust have long perplexed Geologists as to their origin. Substantial carbon and sulfur gases are released during continental volcanic eruptions; where did all the carbon and sulfur come from since

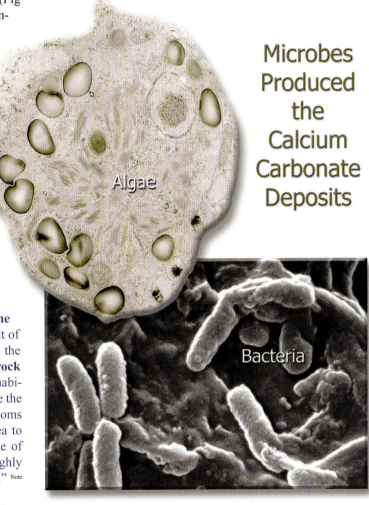

Fig 8.8.4 – Science's long-held belief is that life on Earth is fueled by sunlight, but new discoveries of microbes deep beneath the surface are changing this view. Without sunlight microbes can flourish, and it is they who are responsible for creating the massive carbonate deposits that contain over 99% of the carbon on Earth.

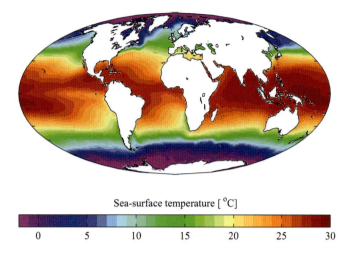

Fig 8.8.5 – This map shows the surface temperatures of the world's oceans in 2001 according to NOAA. The hottest temperatures, shown in red exist near the equator. However, even the warmest temperatures are not high enough to facilitate the growth of the microbes responsible for the Earth's calcium carbonate deposits. When conditions were right, they grew exponentially over a short period. Hydrofountains in the ocean produced the necessary microbes, the increased pH levels, and the nutrients from the endobiosphere, fueling massive microbial blooms during the UF.

these elements are almost non-existent in the igneous rocks presumably of which the Earth was formed? The biomass in the endobiosphere now explains this, showing that these gases came from microbes.

If we turn our attention to the underwater volcanoes, we find in an article titled *Where Biosphere Meets Geosphere* (2002 *Scientific American*), where Delaney once again discusses the "massive effusion of biomaterial" that erupts from underwater volcanoes:

"At this past December's meeting of the American Geophysical Union (AGU), oceanographer John Delaney of the University of Washington at Seattle, a major contributor to hydrothermal vent ecology research, spoke of 'conceptual shifts' regarding microbial biospheres. '**In 11 of 11 cases, underwater volcanoes have been found to have a massive effusion of biomaterial**...'" Note 8.80

The newly discovered bacterium from beneath the Earth's crust is only the first step in understanding the origin of the material forming the Earth's massive limestone and dolomite layers. The next step is to recognize the ocean's invisible forest.

The Ocean's Invisible Forest

In an August 2002 *Scientific American* article called *The Ocean's Invisible Forest*, new discoveries of phytoplankton are reported, moving us closer to understanding the origin of the carbonates and the mechanism needed to produce the huge quantities of organics found in the Earth's crust today. The article starts out:

"**Every drop of water** in the top 100 meters of the ocean contains thousands of free-floating, microscopic flora called phytoplankton." Bib116 p56

Each of the phytoplankton is an organism; algae and diatoms, seeds that given the right environmental conditions, can *bloom* into colonies of organic matter of enormous magnitude (Fig 8.8.7). Only since the 1997 launch of the NASA Sea WiFS satellite have researchers discovered how sensitive these organisms are to temperature and nutrients in the sea:

"New satellite observations and extensive oceanographic research projects are **finally revealing how sensitive these organisms are to changes in global temperatures, ocean circulation and nutrient availability**." Bib 116 p56

Anyone who has ever had a pond or stream knows that in the summer, as the temperature rises, algae growth increases dramatically. But there is another important factor that enables phytoplankton to bloom so profusely—nutrients, three in particular.

Researchers have long known that the phytoplankton need nitrogen and phosphorus to spread quickly, and assumed that it was the lack of phosphorus in the ocean that kept the plankton in check. Later oceanographers came to realize that it was actually nitrogen that was the most crucial, and that the nitrogen did not come from the air, but from a particular source. It is not a coincidence that the source is from our little friends that are now known to exist in vast quantities *inside the Earth*—bacteria from the endobiosphere. It is not living bacteria, but decaying bacteria that release the nitrogen:

"The vast majority of **nitrogen** is fixed by small subsets of **bacteria and cyanobacteria** that convert N_2 to ammonium, which is released into seawater as the organisms **die and decay**." Bib 116 p59

With two of the three nutrients needed for the plankton blooms identified—nitrogen and phosphorus—what is the third nutrient?

In the 1980s, chemist John Martin theorized that plankton blooms would increase in the ocean if **iron** were added to the water. Investigators had noted that iron was needed as a *catalyst* for the algae to absorb the other nutrients, but thought there was plenty in the ocean already. However, this assumption proved to be incorrect:

"Using extremely sensitive methods to measure the metal [iron], he discovered that its concentration in the equatorial Pacific, the northeastern Pacific and the Southern Ocean **is so low that phosphorus and nitrogen in these surface waters are never used up**." Bib 116 p59

Martin also noted that the iron was being carried out onto the open ocean primarily by the wind:

Fig 8.8.6 – The submersible Alvin was one of the first to visit and observe the deep seafloor and the impact of the endobiosphere. Hot springs originating from beneath the seafloor exhibit a variety of pH levels and temperatures, each responsible for new environments and marine life never before seen.

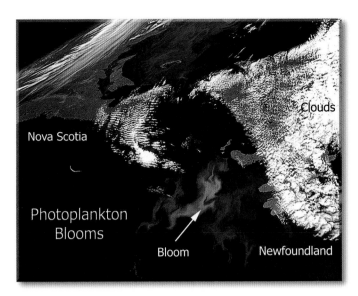

Fig 8.8.7 – Phytoplankton blooms seen from satellite are called coccolithophore swirls caused by warm waters and nutrients welling up from the ocean depths. They can occur in days but are relatively lightly concentrated in modern times. In contrast, warmer waters and increased iron available during the UF fostered much larger and denser blooms, which provided the carbonate material needed for the vast carbonate deposits found globally today.

"...Martin and his coworkers pointed out that practically the only way **iron** reaches the surface waters of the open ocean is via **windblown dust**." Bib 116 p59

In 1995, Martin and other investigators set off to test his theory. They slowly spread several hundred kilograms of iron over a 50-square kilometer patch of ocean. The answer was apparent:

"...the results were clear: the additional iron dramatically increased phytoplankton photosynthesis, **leading to a bloom of organisms that colored the waters green**." Bib 116 p60

Three other groups, from New Zealand, Germany and the U.S., have demonstrated unequivocally that dramatic blooms can be achieved with the addition of iron to the seawater:

"Preliminary results indicate that one ton of **iron** solution released over about 300 kilometers **resulted in a 10-fold increase** in primary productivity in eight weeks' time." Bib 116 p60

The third ingredient is such an important component of the UF because of its relative abundance on the surface today. It is the metal that turns claystone, sandstone and deserts red, and is the most widely spread metal in the crust of the Earth. It is the black silty grains that stick to your magnet when you run it through sand or soil. It is *iron*.

It is this metal that was spread around the world, throughout the seas; expelled with muddy water erupting from the hydrofountains active during the Universal Flood.

Because the UF really happened, there is evidence of increased mineral content in the ocean in the past, including iron. Evidence for this can be found in the ice cores drawn from the *frozen poles*:

"**Historical evidence** buried in layers of ice from Antarctica also supported Martin's hypothesis. The Vostok ice core..." showed that in the past, "**the amount of iron was much higher** and the average size of the dust particles was significantly larger..." than in more recent times. Bib116 p60

"Enigmatic" Blooms "Confounding Scientists"

Phytoplankton blooms are usually green due to the chlorophyll within their cells. However, blooms can come in many colors, one being black:

"An **enigmatic mass of black water more than twice the size of London** has appeared this year in waters off Florida, alarming fishermen and **confounding scientists**.

"The first reports of the black water came from fishermen in the region in January. Archived satellite images revealed that the black patch emerged off south-west Florida late last November and eventually spread to cover about 3400 square kilometers of water, including a large portion of the Florida Keys National Marine Sanctuary." Note 8.8p

This particular bloom (Fig 8.8.8) occurred off the coast of Florida in 2002, confounding scientists because of its rarity. We are just beginning to understand the many types of blooms that can occur, many of which have never been seen before, each responding to a particular environment including temperature and nutrients generated from eruptions on the seafloor or from mainland flooding. Both of these scenarios incited blooms during the UF, creating similarly rare blooms then, only on a much grander scale than anything seen today.

The black bloom in Florida confused researchers. At one point they looked to "diatoms" to explain the bloom. But, diatoms are *clear*, making it difficult to understand why they would ever be chosen:

"But researchers who met in the region last week agreed that there's a more likely explanation: a bloom of another type of **algae called diatoms** could be darkening the water." Note 8.8p

A better explanation came by looking to the seafloor—bacterial mats. These mats can double in size in hours when conditions are right:

"Another hypothesis implicates an algal bloom as an indirect cause of the black water. **Blooms can lower oxygen levels in water**, causing plants on the sea floor to die. These conditions would promote the growth of **bacterial mats** that could have darkened the water as they broke up." Note 8.8p

However, another report identified "oxygen in the water column—it was not a "dead zone," and thus, the black water did not come from dead bacterial mats. The bottom line—researchers do not know what caused the enigmatic bloom because we have not studied these types of biological phenomenon enough. This is the primary reason investigators do not realize the importance of identifying the origin of over 99% of the carbon (carbonate deposits) locked up in the Earth's carbonate rocks.

Plankton Blooms Decline as Fast as Created

Having established at least one of the origins of the carbonates that form worldwide limestone and dolomite formations—microbe blooms—how long would it take for the limestone layers to form; millions of years? Present-day blooms demonstrate, through experimentation and observation of natural phenomenon that it takes only days for massive microorganism growth to occur, when the proper nutrients are available and the ocean water conditions are right. Experimenters conducted further experiments to determine how long an iron-stimulated bloom would last:

"Here we **report on the decline and fate of an iron-stimulated diatom bloom** in the Gulf of Alaska. **The bloom terminated on day 18**, following the depletion of iron and then silicic acid, after which mixed-layer particulate organic carbon (POC) concentrations declined over six days. Increased partic-

Fig 8.8.8 – A Blackwater Bloom appeared in 2002, shown in this satellite image off the southern tip of Florida, USA. The bloom confounded scientists because, like many types of the rare blooms seen off the southwestern tip of Florida, they are little understood and rarely studied in detail. The unusual conditions that promote these rare blooms also existed during the Flood, but this too is not recognized despite the knowledge that the vast continental carbonate deposits came from ocean organics—which were primarily massive blooms. Future research into these types of blooms will reveal many new details about how the carbonate minerals were formed.

ulate silica export via sinking diatoms was recorded in sediment traps at depths between 50 and 125 m from day 21, yet increased POC export was not evident until **day 24**." Note 8.8q

Once the nutrients were depleted, the bloom died out; and *it only took days*, just like it took only days for exponential growth. We now have, for the first time, scientific evidence showing that world-wide oceanic blooms can grow exponentially, in a matter of days, when the iron nutrients are in place, and the water environment is right. Furthermore, when the blooms die, they form layers of sediment, also in days. With the new Carbonate Mark evidence, we should ask:

Why is it critical that microbial blooms end quickly?

The reason is simple; recall how the carbonate deposits are formed, there is only one way dozens-of-feet thick layers of carbonate can form *without any other* layers of sediment—quickly! If any amount of extended time is involved, even a year or two, the cycles of weather and storms would have carried other sediment in and mixed it with or covered the carbonate layers as they were being formed. Thus, the UF Hypretherm had the ability to form thick, relatively pure layers of sediment by first growing the microbes, and then by killing them quickly through iron and nutrient depletion, temperature change, and alteration of the pH factor. This is the only way in which thick layers of pure carbonate sediments could have formed without being interbedded with layers of other sediment that would have to be present if geological time had been involved.

Oceanic Crust Earthquakes Producing Microbes Today

There is direct evidence that earthquakes under the ocean today are causing a tremendous rise in local microorganism fauna. If these events are happening today, why would they not happen in the past? From the *Science* article *titled, Stirring the Oceanic Incubator*:

"After the June 1999 Juan de Fuca Ridge earthquake swarm, **observations of dramatically increased biological activity** were the first indication **that something unusual had occurred**. Evidence included the recent formation of large bacterial mats, abundant water-suspended bacterial floc and particles, and vent-specific animal communities that were **visibly expanded in size since the previous year**.

"The microorganisms that reside in the porous upper crustal rocks are also affected by earthquakes. This subseafloor microbial community is extremely sensitive to temperature, oxygen content, pH, and other environmental variables. It thrives in subsurface zones where the hot hydrothermal fluid mixes with entrained seawater." Note 8.8r

Perhaps it is "unusual" to see such "dramatically increased biological activity" occurring under such favorable conditions, but these conditions included being under the ocean, where the necessary minerals are present, and where earthquake swarms produce the necessary heat for the growth of "large bacterial mats" in the water. These conditions do not exist in the world today, except in isolated places. But they had to have happened one time, at the same time, to form the tremendously thick, *single layered* limestone and dolomite deposits on a global basis, near the continental surface where they exist today.

Carbonates do not simply become rock after precipitating out of ocean water, even after a long time. Even man-made cement, which is made from carbonate material, must be heated to over

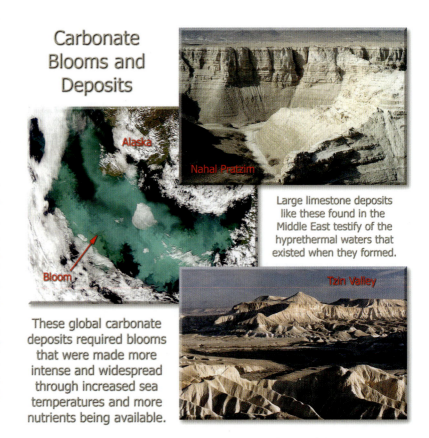

Carbonate Blooms and Deposits

Large limestone deposits like these found in the Middle East testify of the hyprethermal waters that existed when they formed.

These global carbonate deposits required blooms that were made more intense and widespread through increased sea temperatures and more nutrients being available.

CHAPTER 8 THE UNIVERSAL FLOOD MODEL

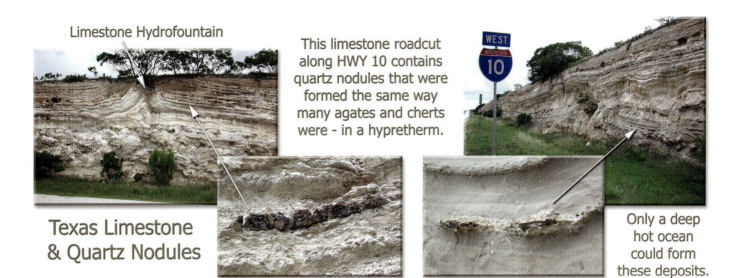

Fig 8.8.9 – Many limestone deposits, like this one in Texas, USA, contains quartz inclusions that were formed along with the limestone. The Crystallization Process in the Hydroplanet Model demonstrated that the quartz nodules seen in roadcut were formed in a hypretherm, which indicates the environment wherein the limestone was formed. Deposits like these, which are seen on the surface around the world, are a witness of the Universal Flood.

1,000° F before it will react with water to become solid. Naturally hardened carbonate rocks were subjected to the same hypretherm the sandstone was; it being often interbedded within the carbonate layers. The high pressure, high temperature environment of the hypretherm existed all across the Earth at only one time—during the Deluge.

Direct evidence of the UF carbonate hypretherm can be seen at a road-cut along Interstate 10 in Texas, Fig 8.8.9, where quartz nodules are encased in a limestone deposit. Formations like this exist worldwide and *require a hypretherm* for *quartz formation*. Researchers know:

"Chert nodules **are common** in limestone layers…" Note 8.8s

This common association of quartz and carbonate deposits, where quartz nodules are cemented throughout limestone or dolomite material demonstrates the prevalence of the UF Hypretherm during deposition and lithification.

Where Mineral Deposits Get Their Colors From

Pure quartz, like pure diamond is clear, but most minerals have their own distinctive color; rubies are deep red, pyrite is yellow, emeralds are green. Many mineral deposits and probably most mineral specimens exhibit a variety of colors and banding (see the agates in Fig 8.8.10). The traditional explanation for the different colors within an ore deposit or within a specific mineral specimen involves the identification of specific chemical formulas of the minerals. It is assumed that the added elements just happened to be in the solution when the mineral crystallized. However, this is only part of the story. *Microorganisms* contributed much to the colorization and banding of Nature's handiwork. We will return to this concept again, in the Ore Mark, where microbes such as cyanobacteria and other organisms lived and died in the water solutions, their remains becoming a part of the newly formed mineral.

To help understand how each microbial species contained specific elements that create color in minerals, we reviewed cyanobacteria research that previously established bacterial reaction to high temperatures. From the research on the geysers at Yellowstone comes the following:

"You can tell the approximate **temperature of a stream by the color of its cyanobacteria**. If there are no cyanobacteria, the temperature must be greater than 167°F (75°C). If the bacterial mat is bright **yellow**, the temperature is around 160°F (71°C); brilliant *orange*, about 130°F (57°C); and pure **green** shows up at around 120°F (50°C) and below." Bib 134 p19

Examples of a high temperature environment and its effect on various colored bacteria are illustrated in Fig 8.8.11. Most lakes and rivers do not have multi-colored microbes because they only grow in high temperature water. However, during the UF, global water temperatures increased boosting the numbers of many microorganisms that today remain relatively scarce. Changes in pH, temperature, and other factors caused a microbial bloom of epic proportion, many becoming entombed on the surface of the continents, in large bio-mats of dead organisms. From these came the vivid colors locked in the carbonate and silicate rock layers.

The Limestone-Dolomite-Loess Connection

In the Rock Cycle Pseudotheory chapter, under the carbonate mystery, we asked many fundamental questions with respect to the carbonate minerals that have no answers in modern geology. One myth responsible for holding back the discovery of truth about this mystery is the Carbonic Acid Pseudotheory. Amazingly, carbonic acid, which is not seen dissolving or forming caves or speleothems, and which is not found in any abundance *anywhere on the planet*, is said to be responsible for dissolving and re-depositing carbonate cave minerals.

The true origin of the mysterious carbonate deposits is revealed here. The endobiosphere and microbial blooms are the source of the carbonate deposits. This new evidence accounts for limestones, the dolomites, and the loess deposits, *connecting all three* in ways never before known. Each of these deposits formed in a different hydrothermal or hyperthermal environment, unique to each of their morphology and composition. Yet all three have an important *biogenic* component that was discussed in the Crystallization Process of the Hydroplanet Model. The vast majority of all biogenic deposits on the Earth

Fig 8.8.10 – Banded agate specimens show off some of nature's most unique and beautiful designs, yet the answer as to how they were formed or from where they got their colors has eluded investigators, until now. These agates are primarily cryptocrystalline quartz, and were thus formed in a hypretherm; the majority found in carbonate deposits, such as the Texas Limestone deposit in the previous diagram. Their striking colors come from a combination of microbes that left their mark by leaving behind various minerals when they died. Each species of microbe is unique to its environment. This fact, coupled with their existence only in surface deposits bears witness of their creation during the Universal Flood.

today do not fit the uniformity myth and are not forming today—they are each a testament of the Universal Flood.

The most glaring, fatal flaw of the Carbonic Acid Pseudotheory is its inability to explain how *dolomite*, a carbonate mineral making up large areas of the Earth's surface, was formed in a manner *similar to the limestone deposits*, yet was *not eroded* by carbonic acid. Dolomite, which has the same chemical formula as calcite except for the added element of magnesium, is highly resistant to acids, rendering the acidic theories invalid at explaining dolomite.

Again, it is the UF microbial blooms that connect the origin of limestone, dolomite and loess. Until fairly recently, research documenting dolomite's origin was unavailable, but now, with the introduction of the Hydroplanet Model and the UF Model, we have a framework within which to work.

The Real Origin of Dolomite Deposits

The "Dolomite Problem," as named by modern geology, was discussed in Fig 6.9.19, where we showed the truth about the massive dolomite deposits:

"**Dolomite at present time, does not form**

Fig 8.8.11 – Colorful species of algae and bacteria as seen on the left are not common in most rivers or lakes, but they are frequent inhabitants of hot springs. Brightly colored yellow hot spring in Yellowstone National Park, and the other hot springs near the base of Hoover Dam on the Colorado River in Arizona, USA show colors based on the microbes living there. The temperature and chemical makeup of the hot springs differ, each supporting a different variety of life. The minerals from the waste products and the remains of dead microbes during the UF left behind the great variety of colors we see in the minerals represented in Fig 8.8.10 above. During the UF Hypretherm, unique microbes processed existing minerals, and formed many new ones.

CHAPTER 8 THE UNIVERSAL FLOOD MODEL

Fig 8.8.12 – The beautiful Havasupai Falls in the Grand Canyon flow over carbonate deposits that are enigmatic to geologists. For the first time, the color of the red sandstone and limestone layers can be explained by iron producing microbes present long ago.

on the surface of the earth; yet massive layers of dolomite can be found in ancient rocks. That is quite a problem for sedimentologists..." Note 8.8t

The primary reason the origin of dolomite remains such a problem is that dolomite formation, just like limestone, was long thought to be an inorganic process. But as we have been discussing, researchers were recently able to confirm the "magic ingredient"—microbes:

"**Microbiology** has apparently solved at least one geological mystery to date: that of **dolomite**—a soft, white, vitreous mineral composed of calcium, magnesium and carbonate crystals. **By all accounts, the world should be covered with the stuff**. Huge dolomite deposits do form sedimentary rock in the American Midwest, as well as in parts of Europe and Mexico. But in many places where you would expect to find the mineral—**marine waters**, for example, usually contain an overabundance of the necessary ingredients—**it is entirely missing**.

"'Dolomite shows up in the rock record all over the place, but it rarely occurs in modern low-temperature environments,' Rogers says. 'There are some very special environments, hypersaline environments that do have dolomite precipitation,' such as Rio de Janeiro. **But it's hard to find a place where dolomite is forming now, never mind at rates that explain large deposits in the past.**

"In 1995, Crisogono Vasconcelos of the Swiss ETH-Central Geological Institute and co-workers seemed to solve the puzzle of the missing dolomite deposits, **publishing a landmark paper in *Nature* that documented just how the mineral forms in a laboratory setting. The magic ingredient? Microbes**." Note 8.8u

Once investigators realized that microbes were involved with the formation of dolomite, they encountered a second problem; the environment in which the lab-grown dolomite was created is almost totally non-existent today. These researchers reported identifying the magic ingredient in their *Nature* article, that the dolomite was actually produced in a "low-temperature *anoxic*" environment:

"Together, our analyses provide **conclusive evidence that the bacterial production of dolomite can be achieved in low-temperature anoxic conditions in a relatively short-time**." Note 8.8v

"Anoxic" means to be *without oxygen*, a rare environment on the *surface* of the Earth. Therefore, the scenario is not realistic for growing conditions today, which helps explain why dolomite is not found forming in large areas on the surface of the Earth. But the non-oxygen environment existed in many areas during the Universal Flood, because of the abundant anoxic water coming from below ground. Moreover, we know how long it took for the bacterial production of dolomite. It was simply a matter of days. Bacteria are one of the fastest reproducing organisms alive, researchers noting they had the ability to grow in a "relatively short-time."

Laboratory research has led scientists to acknowledge that the "environments where dolomite formed in the geologic past differ significantly from those where it forms today":

"Much modern dolomite differs significantly from the bulk of the dolomite found in the rock record, leading researchers to speculate that **environments where dolomite formed in the geologic past differ significantly from those where it forms today**." Note 8.8w

The significance of the differences can be found in the Dolomite Mark of the Flood.

The Dolomite Mark

With all the pieces in place, we can resolve the "dolomite problem." The extremely thick dolomite (and limestone) formations from around the world could not have formed out of today's oceans, contrary to the claim in most textbooks. No one has seen this process in action. The carbonate deposits grew from massive bacterial blooms that are not in process today. These blooms happened *spontaneously,* and rapidly, forming pure deposits void of contamination from erosion. This is in direct opposition to the carbonate pseudotheory that states these deposits were laid down over millions of years.

One apparent clue that dolomite is a mark of the Flood is that the 'stones' are cemented. As we previously discussed, the hypretherm is the only mechanism by which the diagenesis or the cementing together of these carbonate clays or mud can take place. One way to confirm this is through the study of water-included rocks, with micro enhydros inside the carbonates. More details about inclusions will be had near the end of this chapter, but for now, research into the fluid inclusions confirms the hypretherm environment that natural dolomite grew in. From the book, *Fluid Inclusions*, researchers found:

"**Fluid inclusions** may help clarify the '**dolomite problem**,'

Fig 8.8.13 – This is a carbonate cave in Alabama, USA. Scientists know that "environments where dolomite formed in the geologic past *differ significantly* from those where it forms today."

i.e., when and in what sequence did various types of dolomite form?" Note 8.8x

The investigators noted that dolomite formed in a solution that was both "hotter and saltier" than calcite:

"Freeman (1973) has reported that the fluid inclusion in some epigenetic dolomite-calcite cement show that **the fluids forming the dolomite were both hotter and saltier than those forming the associated calcite cement**." Note 8.8x p321

The environment in which the inclusions grew is again confirmed in the book, *Fluid Inclusions*, as prepared by the Mineralogical Society of America:

"Nahnybida et al. (1982) were able to show that **calcite, dolomite, and anhydrite cements in upper Devonian dolostones formed at 140-155°C and >2.5 km depth, from 2.4-3.7 molar solutions (NaCl equivalent)**." Note 8.8x

This research confirms the "dolomite problem" can be solved once we realize that dolomite, calcite, and anhydrite, three of the most abundant sedimentary deposits on the Earth were formed in a *hypretherm*. It is *only* with high temperatures (140-155°C or 284-311°F), ocean waters of considerable depth (>2.5 km or 1.5 Miles), and a mineralizer (NaCl ocean water as an example) that such natural mineral deposits could have formed. The fluid inclusions are the 'fingerprints' left behind telling us exactly the type of water solution the minerals formed in.

In the journal *Sedimentology*, a more recent study reported the results of a comparison between synthetic dolomite and natural dolomite; they were similar in crystal structure, even when observed with a scanning electron microscope. The key is that the synthetic dolomite crystals were formed *only* under high pressure at a temperature of 200°C:

"Examination with scanning electron microscopy (SEM) and scanning force microscopy (SFM) revealed etch pits, layers and islands on dolomite crystal faces synthesized from calcite in Ca-Mg-Cl solutions at **200°C and a wide variety of natural dolomites**." Note 8.8y

Obviously, anoxic (oxygen-free) environments of high pressures and high temperatures covering vast expanses do not exist on the surface of the Earth today, leaving scientists with the "dolomite problem" because at the present time, dolomite "does not form on the surface of the earth." This simply means that "environments where dolomite formed in the geologic past *differ significantly*" from the environment that exists today.

The Loess Mark

In the Rock Cycle Pseudotheory Subchapter 6.10, a number of FQs were posed regarding the ubiquitous sediment, Loess, along with the following six reasons Loess remains a mystery:

1. **No clearly defined origin for loess can be found in modern geology.**
2. **No verifiable mechanism of how loess was deposited exists.**
3. **No comprehensive study of loess has been performed.**
4. **A complete worldwide map of loess deposits has not been made.**
5. **Loess is a part of the Untold Story of Sedimentology.**
6. **Loess is a part of the Missing Sediment Mystery.**

Right before the Loess Mystery in the Rock Cycle Pseudotheory chapter, the Carbonate Mystery revealed that loess is a worldwide phenomenon. This can be seen in Fig 8.8.14. There were also four geological facts concerning loess that have been overlooked; first, that loess exhibits a vertical columnar habit and not one of horizontal layering; second, that the unique combination of loess minerals is not linked to the minerals of neighboring mountains; third, the particle sizes of which loess is formed must have come about by some means other than erosion, and finally, there is little variation in the diameter of loess particles worldwide.

There had to be *some mechanism other than wind or glaciation* that formed this sediment; something that created and deposited it across the globe, and since loess deposits are only found on the surface, the depositional event must have been recent and unique.

Diligent researchers performing water and wind abrasion experiments recognized that wind did not form loess particles, and that "loess must have been formed in some other way":

"**Curiously** one does **not** find the tiny sand chips in the loess laid down by ancient dust storms. The bulk of loess is quartz, but **its particles fall outside the size range produced by wind action on sand. This shows that the loess must have been formed in some other way**..." Note 8.8z

The Universal Flood Model finally gives us an opportunity to solve the Loess Mystery. One key to the solution is the presence of a large percentage of carbonate material, giving it its distinctive yellowish color when exposed to the air. The following question posed by researchers illustrates the difficulty of the Loess Mystery solution, and that carbonates hold clues to the provenance of the loess sediments:

"The studied fine-grained materials [loess] consist mainly of quartz exhibiting medium to low degrees of roundness. In minor amounts—but sometimes up to 30%—carbonates and micas are present, too. **Especially the carbonate minerals may give some information about the provenance of the silts**. The drainage channels of the Wadi Firan / Wadi esh Sheikh system are incised in the crystalline basement; the tributaries and the Wadi Firan **do not cross or reach any region where carbonaceous rocks are outcropping. Where does the carbonate come from?**" Note 8.8aa

Fig 8.8.14 illustrates the globalized scope of loess sediments, which, as the researchers previously noted, does not match local outcrops of other minerals, leaving the International Society of Soil Science authors to ask this simple question; "Where does the carbonate come from?"

Although they asked, they did not have an answer. Further complicating the matter, the particular Egyptian loess deposit they were studying had a substantial dolomite constituent, but it was *80 km to the nearest dolomite rocks!* There was simply no rational scientific explanation using traditional aeolian (wind) or fluvial (river) erosional processes; there was no 'debris trail' from the supposed source outcrop.

Without the benefit of the UF paradigm, researchers explored an even more improbable explanation of loess' origin. A research group from the United Kingdom offered in their report, three alternatives to the failed aeolian and fluvial theories:

1. Grain fracture and release during **salt** weathering.
2. Grain fracture and release during **frost** weathering.
3. Grain fracture and release during deep weathering of **plutonic rocks**. Note 8.8ab

However, they appended their ideas in the 2002 report, "origins of desert loess" with this:

"Through studies of these mechanisms it is **hoped** that grain fracture **may be better understood and that viable origins can be identified for quartz silt in and around hot deserts**."
Note 8.8ab

The article was an extensive study of other studies leading to no conclusions of any significance. Because loess is known to exist in many desert environs, the glacier theories are left lacking, with the authors *hoping* to identify a "viable origin." But they offered *no experimental evidence* for their salt, frost or plutonic rock weathering theories.

The Argentine Loess Evidence

One year prior to the UK report, another article, published in *Elsevier*, demonstrated that multiple layers of loess were laid down in a "remarkably homogeneous" deposit that contained multiple layers of loess that *cycled back and forth in pH* levels, sandwiched between other layers of soil in an Argentinean deposit. This type of direct evidence refutes the UK report's theories of salt, frost or plutonic rock fracture and release, but was unfortunately excluded from the UK report.

The Argentinean report on loess clearly demonstrates that no theory explains observations made during the study, with one exception; the Universal Flood's sediment making processes. Observations of a 42-meter thick Argentinean loess deposit reveal 24 layers of loess separated by 28 other layers of soil. The loess layers were described as follows:

"Morphologically, **loess layers are distinguishable from the buried soil horizons** because of light colors, massive structure in large polyhedrons, high silt content, **effervescence to HCL** and absence of organo-clay cutans." Note 8.8ac

On the other hand, the soil layers (paleosols) were described as such:

"Compared to the loess layers, the **paleosols have, among other distinctive characteristics**, less thickness, slightly darker colours, prismatic structure, organo-clay cutans on the surface of the structural faces and **no effervescence to HCl**."
Note 8.8ac p2

Fig 8.8.14 – Because little global research on loess is available, this is the most complete map of Loess and Limestone Deposits. It is still incomplete. However, the vast tracts of land covered by these carbonate deposits is readily apparent. For a long time, it was thought that Loess only existed in glacial areas, but as the map shows, many desert areas that were never covered with ice *also* contain loess. This forced researchers to revise their theories again, although there is still no clear origin for the sediment in modern geology. Loess, like limestone and dolomite deposits cannot be explained using current geological models. Both the carbonate and the silica components of loess are troubling enigmas in modern science, but can be logically and clearly accounted for, *only* through Flood Geology.

Understanding the difference between these two layers is essential if we are to comprehend how they were laid down. There were differences in their colors, their polyhedron crystal structure (recall the basalt columns—starch experiment), silt (quartz) content, organic content and most importantly, the comment of effervescence to HCL. Effervescence to HCL simply means that when subjected to hydrochloric acid (HCL), the loess sediments would fizz, whereas the soil layer material would not. The effervescent (fizz) reaction occurs if calcium carbonate is present in the layer, and this gives us clues as to how these layers were made.

The researchers emphasized the remarkably systematic repetition of the cycles:

"Thus, the same conditions prevailing during the **loess deposition, on the one hand, and during the subsequent soil formation, on the other, have systematically repeated**, with only minor modifications, during **at least 24 cycles**... **This cyclicity is probably the most remarkable feature of the sequence**." Note 8.8ac p3

The *cycles* between loess and soil deposition are definitely "the *most remarkable* feature of the sequence," but unfortunately, the researchers had nothing of any consequence to say about it. Does any modern geology model explain how layers of soil and loess can cycle back and forth 24 times while accounting for different pH levels?

Can ordinary erosion processes create cyclical pH levels in soil/loess layers?

We have found no evidence where *ordinary erosion* can cause a cyclical repetition of pH changes in loess/soil layers, as observed in the Argentine study:

"The **soil pH** is usually **0.5 to 1 unit less** than the pH of the associated **loess layer**." Note 8.8ac p3

Ordinary surface streams do not experience significant changes in pH like this (aside from yearly seasonal cycles). In this case, researchers believe the whole 42-meter sequence is 10,000 years old, making each layer 400+ years old. There is simply no way to explain how the environment changed back and forth every 400 years, depositing alternatively higher and lower pH soil along with the other differences noted. Hard questions become simple when we know the answer, and there is a simple, comprehensible answer, which explains how water laid down layers of sediment of alternating pH and other factors.

A simple answer is connected to our previous discussion on microbe blooms. Microbial blooms are produced by a variety of organisms in a highly variable ocean. As conditions changed, one species of microbial growth exploded while the previous one died out, depositing massive layers of organic detritus. Heated hyprethermal waters further encouraged microcrystalline growth and additional bioactivity. As the turbid waters shed their silty sediment load, massive layers of loess sediment accumulated on the ocean floor.

This is the origin of loess.

Further direct evidence supporting this process is the "nodules" of calcium carbonate found in the Argentine loess layers:

"**Calcium carbonate** is present as **nodules**, from frequent to abundant, and as powdery coatings only in the upper and lower parts of the sequence." Note 8.8ac p2

The limey "nodules" have no origin in the aeolian, fluvial, salt, frost or plutonic rock weathering theories, but they are a direct link to the microbe bloom events. Ocean currents on the seafloor create eddies that sort and accumulate sediment of similar density in specific areas and around objects lying on the ocean floor. This is one reason why white calcium carbonate is associated with many fossils, like petrified wood, of which we will read more about in the Fossil Model. These accumulations of carbonate or silica crystallized and lithified as temperature and pressure (water height) increased, forming nodules.

Other direct evidence of loess deposition in a mild hyprethermal environment are thin "calcite sheets" found within loess deposits:

"Polyhedrons are separated by small cracks, 2-3 mm wide and filled with **calcite sheets**." Note 8.8ac p1

The Calcite Mystery revealed that such calcite sheets or veins are not observed growing in typical surficial waters, but only in warm, bacteria filled waters in the laboratory.

With such dramatic changes in pH levels, prismatic (columnar) structure, and color of the sediment, one might suppose the investigators would argue for changes in climatic conditions as being the contributing factor in the loess/soil layer deposition, which in fact, they did:

"The **rhythmic repetition of 24 loess layers**, interbedded with 28 Bt horizons featuring clay illuviation and pedogenic clay formation, **required the recurrence of alternating climatic conditions different from the current ones**." Note 8.8ac p6

How were the "alternating climatic" changes driven? No explanation was given.

The investigators did comment how "remarkably homogeneous" the whole loess section was, but they contradicted their "required" recurrence of alternating climatic conditions by stating that "the environmental conditions controlling the loess deposition *did not change* over a notably long time span":

"The loess material remains **remarkably homogeneous along the whole section**. From top to bottom, the sand fraction of the C horizons slightly decreases from 35% to 31% and the clay fraction slightly increases from 6% to 11%. But the silt fraction remains stable around 58-59%. Thus, even if the depo-

Fig 8.8.15 – A close study of loess deposits within the UF paradigm reveals a simple answer to the origin of loess; long mysterious and the subject of many unfruitful theories. Loess' silica, carbonate and unique chemistry reveal a sediment formed by microbes and hyprethermal water, in an event that spanned the globe only a few thousand years ago.

Question: How much erosion evidence was observed in the 42-meters (138 feet) deposit of fine loess silt supposedly laid down over 10,000 years?

Answer: "No erosion evidence...was found."

Alfred J. Zinck, Jose Manuel Sayago

Only during the Dark Age of Science would the importance of such simple evidence be overlooked.

sition periods have shortened from the beginning to the end of sequence, **the environmental conditions controlling the loess deposition did not change over a notably long time span."** Note 8.8ac p3

Perhaps this sounds like the scientists were confused, however, contradictions like these are common in modern science literature because of the Dark Age of Science. Researchers did recognize the lack of change over time to the remarkably homogeneous loess deposit of over 42 meters (138 feet), recording the *utter lack of "erosion evidence"*:

"Surficial A [**organic topsoil**] and subsurficial E horizons **are absent** from the whole sequence. An obvious **assumption** is that any existing A horizon was removed by wind or water erosion, before a new loess layer buried the remaining Bt horizon. **No erosion evidence**, such as incision unconformities, angular truncation of strata, buried rill and pond microtopography, local pavements of coarse fragments and the like, **was found**. Paleosol horizons and loess layers are **strictly parallel** and have concordant dip and strike values. Moreover, it is **quite unlikely that all A horizons, if they originally existed, would have been fully removed** in all 24 Bt/C cycles of the sequence. Thus, **A horizons might have never existed as such**." Note 8.8ac p4

The "strictly parallel" horizon, evident on all layers meant they were flat and relatively void of erosion during deposition, which is *unequivocal evidence* that the 42 meters of loess and soil sediment was *not* laid down over an extended geological time, or in an environment that exists on the Earth today. The complete absence of topsoil in *any* of the layers is further direct support of the Soil Formation Evidence documented previously in the Erosion Mark (see Fig 8.6.4).

The Loess Model

The Loess Model can be summarized as follows:
1. Loess deposits were formed in UF hyperthermal waters from microbial blooms.
2. Microbial blooms grew and died quickly, along with siliceous organisms, which then fell to the ocean floor where pressure and higher temperatures existed.
3. The Hypretherm in the deep ocean allowed for microcrystalline quartz and other crystals to grow, forming silt that mixed with the carbonate detritus of dead microbes as it fell to the ocean floor.
4. Loess deposits were created in similar fashion to the sand and carbonate deposits in the UF, except that they generally exhibit both silica and calcium carbonate constituents combined in thinner layers.
5. Loess helps explain the Untold Story of Sedimentology because it is not formed by erosion, but grown authigenically.
6. Loess, and other homogeneous silt and sand deposits have an authigenic origin, which explains the two "missing sections" in the Sediment Size Table. (They were not missing, *they were never really there*.) **The vast majority of the Earth's sediment is not detrital material from traditional erosion, but is of authigenic (made in place) origin, formed during the UF Hypretherm.**

It isn't any wonder why loess has remained such a mystery in modern geology for so long. The Loess Model is a part of the Universal Flood Model, which is a subject markedly taboo in modern science. However, truth is independent of its origin and because it does not change—it never goes away; it awaits discovery by those who seek it, open-mindedly.

The Carbonic Acid Cave Pseudotheory Debunked

Most who have taken a tour through any of the many public caves have heard something like this:

"A speleothem called a **stalactite** begins with a drop of water on a cave ceiling. The drop **contains dissolved limestone that it picked up as it moved through the layers of rock above the cave**." Note 8.8ad

Unfortunately, this commonly told story taken from the book, *Caves*, is not true. The Carbonic Acid Cave Formation Pseudotheory in chapter 6.9 showed that natural water in almost all caves is not acidic, and thus it is not chemically dissolving the caves. The modern geological origin of caves is also incorrect. The first line of direct evidence comes from the observation that no solid limestone deposits are being formed anywhere in the world today. Then, there is the problem of the missing layers of sand and silt that would be present between the layers of limestone if the depositional events had occurred over long periods of geological time, along with the rising and falling of continents. Finally, there is no evidence that acid dissolved away holes in the limestone to form the caves.

If the holes in the limestone were not eroded away by acid over long periods of time, how else were they created? Clearly, they had to have been formed when the limestone deposit was first laid down.

It is surprising that the Carbonic Acid Pseudotheory has been able to remain in favor for so long, despite the fact that the caves are not eroding today—but growing! Under the current geological time 'theory,' the limestone layers where caves are found had only one way of forming—by slow deposition. Through this process, the layers would have been solid, and the caverns would have had to be eroded, but this never happened.

Real Origin of Caves

Carbonate caves were formed shortly after the sediment was laid down during the UF, by aquifers that ran through the sediment. Then the carbonate layers hardened during the UF hypretherm, preserving the voids in the rocks. The beautiful formations then grew from microbes, not by the redeposit of dissolved carbonate.

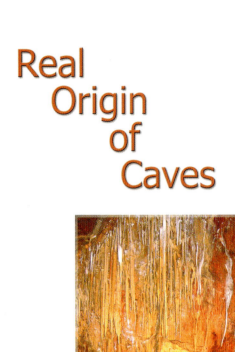

Microbes, not acids created the cave formation.

Carbonic caves are growing, not eroding today.

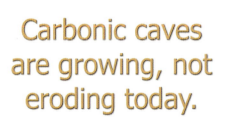

CHAPTER 8 THE UNIVERSAL FLOOD MODEL

Instead, caves are merely large geodes, created the same way small geodes were—in a hypretherm.

The Real Origin of Carbonate Caves

Carbonate caves were formed at the same time as the carbonate layers were being deposited. They were shaped *by underground and underwater streams* able to erode the caverns while sediment was still soft and pliable, before the hypretherm hardened the carbonate sediments. This actually happens today, in unconsolidated and uncemented sediment. Underground aquifers erode cavern-like voids all over the world. What we do not see is underground *carbonic acid* aquifers.

Having already identified the true origin of the carbonates—microbes—what should we expect to find as the true origin for the beautiful speleothems in caves? It should come as no surprise that cave crystals show evidence of being a "consequence of biologic activity":

"In this article, **we present a previously unreported morphology of bacterial precipitated calcite** (determined using XRD, FTIR, and SAED) occurring subaqueously in Weebubbie Cave. Observations using FESEM and TEM revealed spindle-shaped crystals with curved {hk.0} faces lying parallel to the c-axis. Calcite precipitated under conditions designed to mimic the inorganic solution chemistry of the cave **revealed a different morphology. These differences between the crystals suggest that the formation of the cave crystals is a consequence of biological activity.**" Note 8.8ae

This article, from the 2001 *Geomicrobiology Journal*, conclusively demonstrates that cave speleothems and formations are associated with microbes. Another article in the same journal noted that microbes "play a direct and active role in calcite precipitation":

"**Calcified microbes contribute directly to the growth of speleothems. The microbes are important because they act as nucleation sites for calcite precipitation and in many cases appear to control the types of crystals that form**. This notion is supported by the fact that some substrates have calcified microbes lying close to other microbes that have not been calcified. Observations like this led to the suggestion that crystal morphology and calcification style are taxa specific. This suggestion assumes that the microbes, through their metabolic processes, **play a direct and active role in calcite precipitation**." Note 8.8af

Here we see that the type of microbe can determine the type of calcite crystal that will form! The investigators also noted that the water in Weebubbie Cave was not acidic—but alkaline. Moreover, laboratory analysis of the water and the calcite crystals showed that supersaturated solution alone did not form the cave crystals:

"All analyses of the cave crystals confirmed them to be calcite. The presence of these calcite crystals **cannot be explained purely by precipitation from a supersaturated solution**." Note 8.8ag

These observations and experiments show that caves are growing by "bacterial precipitation of calcium carbonate," not from a mineral precipitate of acidic waters. These discoveries directly challenge the core of modern cave formation theory. Although these recent discoveries confirm the extraordinary UM claim that stalagmites and stalactites in caves are growing from microbes, not from carbonate solution dissolved by acidic waters, investigators have yet to see the significance of this discovery.

Even today, geologists are still convinced that carbonic acid is flowing through cracks and crevices in the limestone strata surrounding the caves, eroding them further, even though they have no evidence to support their pseudotheory. In reality, the very opposite is true—microbial rich alkaline water is flowing through limestone caves and causing them to grow! Later, in the Age Model Chapter we will discuss how modern science has applied great age to the caves, but now, with careful measurements, they prove to be only a couple of thousand years old. Should this come as a surprise? The facts have been overlooked for years because the Universal Flood Model did not exist to tie all these loose observations together.

Present-day stalagmites are growing every year; observable by the naked eye, a fact the does not fit within the typical geological age sci-bi that caves are millions of years old. With the UF paradigm, the origin and the age of carbonate caves can be established and understood.

The Tufa Pinnacle Evidence

In Fig 8.8.16, deep 'pipes' in carbonate deposits in Yucatan, Mexico are similar to other 'pipes' found worldwide. The Carbonic Acid sci-bi fails at explaining the formation of *cylindrical pipes* in carbonate layers. However, the hydrofountain easily explains these formations. The Hydrofountain Mark included examples of fossilized silica hydrofountains formed during the UF. Fig 8.8.17 illustrates similar fossil hydrofountains, except they are formed of calcium carbonate. These carbonate pillars are called **Tufa Pinnacles** and are similar to the stalagmites in caves in that they are formed by microbes, except that the tufa pinnacle was formed in water and the microbes came from a hydrofountain below the lake bottom's surface.

Not surprisingly, researchers acting under the Carbonic Acid Pseudotheory umbrella still look to the "slightly acidic" trickle down water theory to describe the origin of tufas like the Pinnacles Desert in Nambung National Park, Australia, seen in Fig 8.8.17. The official Nambung National Park web site records the following:

"The raw material for the limestone of the pinnacles **came from sea shells** in an earlier epoch rich in marine life. These shells were broken down into lime-rich sands which were brought ashore by waves and then carried inland by the wind to form high, mobile dunes. Three old systems of sand dunes run parallel to the WA [Western Australia] coast, marking ancient shorelines.

"The oldest of these, known as the Spearwood dune system, is characterised by yellow or brownish sands. In winter, **rain, which is slightly acidic, dissolves small amounts of calcium carbonate as it percolates down through the sand**. As the dune dries out during summer, this is **precipitated as a cement around grains of sand** in the lower levels of the dunes, binding them together and eventually producing a hard limestone rock, known as Tamala Limestone." Note 8.8ah

Some of the formations at the Pinnacles Desert exhibit openings near their tops, clearly indicating the hydrofountain nature of the structures, but this seems to have gone completely unnoticed. The connection between the carbonate pinnacles currently forming in Lost City on the floor of the *Atlantic Ocean* (Fig 8.8.17) and the tufa pinnacles on the continental crust has also been missed. Instead, we read of imaginative theories like the one from the Nambung National Park web site:

"The acidic soil accelerated the leaching process, and a hard layer of calcrete formed over the softer limestone below. Cracks which formed in the calcrete layer were exploited by plant roots. When water seeped down along these channels, the softer limestone beneath was slowly leached away and the channels gradually filled with quartz sand. This subsurface erosion continued until only the most resilient columns remained. The Pinnacles, then, are the eroded remnants of the formerly thick bed of limestone.

"As bush fires denuded the higher areas, south-westerly winds carried away the loose quartz sands and **left these limestone pillars standing up** to three and a half metres high." Note 8.8ah

Descriptions like this come about because of the lack of thorough investigation by scientists and because of the taint of modern science pseudotheories. Geology remains unaware of how most of the world's quartz sand and carbonates grew, completely ignorant of the UF processes, and therefore unable to visualize the true origin of Tufa Pinnacles.

Until the Lost City tufa pinnacles on the Atlantic floor were found in 2000, researchers had looked to the sun and to algae in a *shallow cool sea* as a possible source of the carbon in Tufa formations like those found at Mono Lake and Searles Lake in California (Fig 8.8.17).

Running against the grain of the prevailing inorganic formation theories, David E. Scholl proposed in 1960 that the pinnacles in the California desert at Searles Lake were from microbes:

"It is proposed that **the pinnacles were precipitated by algae about the orifices of sublacustrine springs issuing along faults**..." Note 8.8ai

Scholl made important discoveries when he noticed the microbial cells in these pinnacles and without fully realizing it, documented the springs that fed the pinnacles were on fault lines. Back in 1885, I. C. Russell actually observed spring water rising through the tubular porous centers of submerged pinnacles at Mono Lake. What the early researchers did not know is that the faults had heated the water, creating an environment for the microbial growth of the calcareous pinnacles. What they got right was the correct origin of the carbonate pinnacles—microbes, despite the prevailing view:

"The prevailing view is that the **tufa deposition at Mono Lake**, east of central California, **is dominated by inorganic processes**. This paper asserts that **algae are important contributors to the formation of lithoid tufa at the lake, and probably also of other varieties**." Note 8.8aj

Unfortunately, without the knowledge of earthquake heated water, investigators have failed to understand the complete

Hydrofountain Caves

Fig 8.8.16 – Many deep pits exist in limestone deposits worldwide. Like these in Yucatan, Mexico, most are curiously ***round***. Some are so deep that thrill seekers base-jump into them with parachutes. How were they made? Carbonic acid is *not* appreciably eroding them today, and long-term erosion cannot account for the cylindrical form. These pipes are ancient hydrofountains, created during the UF, a testament of the Universal Flood that modern geology has completely missed.

mechanism of tufa formation:

"Although substantial research has been done on the subject, particularly at Mono Lake, **the specific mechanisms for tufa formation are still not fully understood**." Note 8.8ak

Again, the primary reason the formation mechanism of the tufa pinnacles at Mono Lake and Searles Lake are not understood is merely because no one has seen them growing, because the conditions in which they grew are not active today. Does this sound familiar? The uniformity myth fails again as we see no tufa growth in these lakes.

Where do we find carbonate tufa growth today? Deep in the Atlantic Ocean at Lost City, where temperatures range from 40° to 90° C (104-194°F) and with pH levels from 9 to 11. Some of the pillars have been called "white smokers" because of the prolific spewing of microbes. This is a *hypretherm*!

The previous article cited was from a 1992 *California Geology* article, published eight years before the carbonate pinnacles were found on the ocean floor. Since then, the connection between the examples of the pinnacles on land and in the oceanic pinnacles shown in Fig 8.8.17 would seem rather obvious, but it has been difficult for investigators to make the leap because of the notion that *shallow cool seas* covered the continents.

Grasping the importance of the abundant evidence this chapter provides of deep, *hot seas* and *alkaline water* is critical for the comprehension of how carbonate pinnacles and other geological structures formed. The Depth Mark established some of this evidence and the Quartz Mark, the Surface Mark, and the Ore Mark will continue to build on this foundation.

The Hyprethermal Marble Evidence

In the past, geologists thought the only way vast areas of limestone could have been transformed into marble is through the metamorphic sci-bi of the Rock Cycle. One troubling problem with the theory of metamorphism is that there has never been any way to observe or to prove that it really happens; no uplift or subsidence of large, continental land masses has ever been witnessed. Furthermore, there is no proof that metamorphism by only heat and pressure exists. What is observed with relative frequency, through deep-sea submersible missions, is the hypretherm in action. There, carbonate, quartz, and ore minerals like those found on the surface of the continents today are forming. The new minerals are being created in deep, hot water. This is precisely how marble was formed. Strong evidence that marble was created in the UF Hypretherm is the occasional quartz cluster, or vein within the marble. Such inclusions demonstrate a specific temperature, pressure, mineralizer, and pH of the water system were present. There was only one time all these factors existed globally on a scale grand enough to form all the world's marble deposits—the UF hypretherm.

The Carbonate Mark Summary

It is time to bring the Carbonate Mark to a close, where the pieces of the carbonate puzzle can come together to establish how the various carbonate materials found on the surface of the Earth actually formed. Here is a quick review:

1. The origin of limestone and dolomite is now established as coming through the carbosynthesis of endobiosphere microorganisms.
2. Massive plankton blooms formed and died quickly, forming single, thick layers of dolomite and limestone.
3. The underground microbe evidence provides the first step needed to understand the production of extensive layers of carbonate material.
4. Massive algal blooms occur when three conditions are met; warm water temperatures, sufficient dead bacteria, and the presence of an iron catalyst.
5. The origin of loess is established by quartz precipitation in the UF environment with microorganisms forming a carbonate sediment and biogenic cementing agents.
6. Carbonate Caves and Pinnacles were created by microbes and flowing water, in a hypretherm, not carbon-

Pinnacle Desert, Australia

Fig 8.8.17 – Examples of Carbonate Pinnacles from Australia and California defy explanation by modern geology because they are not being forming on land today. Although they exist globally, the environment they grew in was a hypretherm, an environment that has not existed on the continents for many thousands of years. When the Hercules submersible visited the bottom of the Atlantic Ocean, similar formations were found, but these had apparently been formed more recently, and in some cases, were still growing. They grow in microbially rich, hot ocean waters, deep in the Atlantic.

Searles Lake, California

Tufa Pinnacle Evidence

Lost City Carbonate Pinnacles

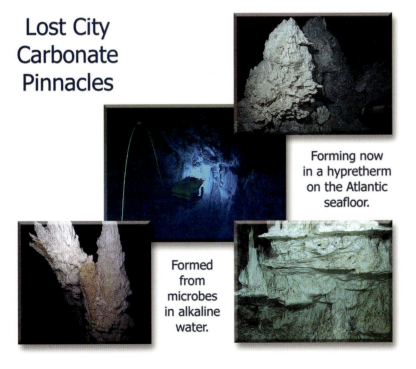

Forming now in a hypretherm on the Atlantic seafloor.

Formed from microbes in alkaline water.

Mono Lake, California

SUBCHAPTER 8.8 THE CARBONATE MARK

ic acid.
7. The UF produced the only event in which all the components were present allowing all of the conditions to be met to produce immense carbonate deposits, caves and pinnacles worldwide.
8. Similar UF environments exist today in limited, deep-sea areas, proving how microorganisms lived, died and formed the carbonate structures of the UF. These carbonate formations stand today as a testament of the UF.

The Carbonate Mark will be further discussed in the Chemical Model, with the Carbon Cycle Pseudotheory. Modern geology states that igneous rocks formed the Earth. However, they contain no carbon. Amazingly, modern science has no origin for the Earth's carbon. The newly discovered process of carbosynthesis and the endobiosphere is this origin. Many researchers now acknowledge that the old dogma of no life inside the Earth…was wrong.

With the new perspective that there is more life inside the Earth than on its surface, we can begin to understand how 99% of the carbon locked in limestone/dolomite deposits came from deep-Earth microbes. Hydrofountains brought massive quantities of bacteria to the surface, along with iron sediment, which provided the necessary nutrients for large-scale algae blooms that in turn facilitated the Earth's great carbonate deposits.

The UF process explains how carbonate cap rocks formed and, for the first time, solved the dolomite problem with microbes. With the UF Hypretherm, the quartz–carbonate connection is apparent in the extensive loess deposits and imbedded quartz nodules in limestone, which can now be understood. And finally, for the first time in modern history, the origin of carbonate caves and pinnacles can be articulated with clarity.

8.9 The Salt Mark

The Rock Cycle Pseudotheory chapter opened with interesting stories that revealed the enigma and mystery of the most abundant mineral on the surface of the Earth—salt. In the Salt Mark, we will answer the mysteries by learning about acid-base neutralization and the Acid-Base Biosalt Origin. These new concepts are the core of the UF Salt Model and are supported by several independent salt experiments, which, along with observations of natural salt deposits reveal how large ancient salt formations were created.

The Few That Knew

The Salt Mark is one of the most important Marks of the Flood in part because of its simplicity; salts are simple minerals and are sometimes visible in dramatic expressions in the geological record.

The study of salt in the Rock Cycle Pseudotheory chapter began by demonstrating that there is more salt in the oceans than there is land above sea level! This is a spectacular fact very few people, including geologists, are aware of.

Moreover, vast continental salt deposits, spread peculiarly across the Earth's surface remain a complete mystery in modern geology, with few scientists ever even recognizing the enigma. The Zechstein deposit in Germany is 8,000 meters (5 miles) deep, as large as Mt Everest is high! This is baffling for the geologists who learned that the igneous rocks that formed the Earth have no chlorine with which to form such huge salt formations. *Salt*, like carbon, *has no origin* in modern geology.

A few dedicated, open-minded geologists, such as Walter Woolnough, studied both the scientific literature and the physical landscape, receiving the Clarke Medal from the Royal Society of New South Wales for distinguished work in the natural sciences. Woolnough was one of *the few that knew*:

"**The writer was engaged for a number of years in an intensive investigation of salt deposits**, in the course of which he **studied the literature carefully**, and visited arid regions where, if anywhere, such deposits should have been observed in course of formation. **In no instance were conditions encountered which could conceivably have produced any of the *major* primary salt deposits of the geological past. *Some* circumstances or set of circumstances, entirely lacking under present day conditions, *must* have been operative when such major salt deposits were generated.**" Note 8.9a

His words speak for themselves still today and remain a *powerful reminder* that truth does not change.

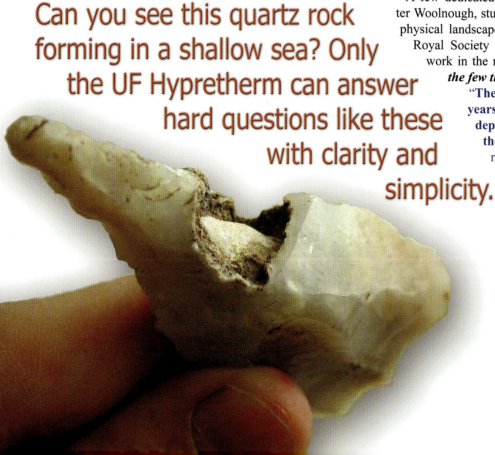

How did this carbonate form inside this quartz arrowhead? Can you see this quartz rock forming in a shallow sea? Only the UF Hypretherm can answer hard questions like these with clarity and simplicity.

The most abundant mineral on the entire surface of the Earth was created in "circumstances, entirely lacking under present day conditions." In the UM, we will learn what those circumstances were.

Universal Flood Salt Model

Woolnough was in a unique position to have studied many salt deposits and the early scientific literature surrounding them. His careful observations and straightforward approach verifies the Salt Mark's extraordinary claim that current conditions on the Earth's present-day surface could not "*conceivably have produced any* of the major primary salt deposits." Woolnough was right on the mark, proclaiming that the circumstances "*entirely lacking* under present day conditions *must have been operative* when such major salt deposits were generated."

For these facts to be accounted for, a completely new model was necessary, capable of explaining *simply and logically* how large salt deposits were formed. The model had to account for and explain salt in the seas and in other major briny lakes, clearly identifying where the salt actually came from. The new model is the **Universal Flood Salt Model**, comprised of three primary aspects.

UF Salt Model
1. Acid-Base Biosalt Origin
2. Hyprethermal Environment
3. Universal Flood Event

These three essential components will be examined with the intent of demonstrating the critical nature of each in the overall salt-origin model. The first of these is the **Acid-Base Biosalt Origin**, which explains how natural salts originated from the neutralization of chemical acids and bases created by the proliferation of biological waste.

The second component of the UF Salt Model is the **Hyprethermal Environment** needed to *prethermate* the massive salt deposits found throughout the Earth's upper crust. Large salt deposits form only in supercritical conditions, when salt-water is at an elevated temperature and pressure, like those found in the depths of the ocean.

The Universal Flood Event is the third component of the UF Salt Model. It describes the only time in world history where both the biosalt origin of salt and a global hypretherm existed, creating kilometer-thick deposits of pure salt throughout the upper crust of the Earth.

Researchers That Came Close

In the Rock Cycle Pseudotheory chapter, we read modern geology's acknowledgment that salt dome theory had "evolved from 100 years of speculation" and that salt dome origins remain "highly complicated and not wholly understood." The Salt Mark will change the speculation by replacing theoretical origins with actual facts that bear out salt's true origin.

Although the *origin* of the ocean's salt continues to elude geology as it has since its beginning, some turn-of-the-century researchers came awfully close to identifying the mechanisms of salt dome formation. Harris (1906), Campbell (1911) and DeGolyer (1926) were all on the right track, noting the following in their work:

"**Artesian waters**, which had entered outcropping formations north of the general salt dome areas, having descended to great depths along pervious beds, becoming heated by the high earth temperatures existing at such depths... **rose under hydrostatic pressure at points of weakness, occurring mainly at the crossing of faults** in the pre-Tertiary formations. Deposition of salt by the cooling of this hot saline solution resulted in the formation of a slender pencil-like cone of rock salt." Bib 144 p29

Instead of the false modern theory that evaporation is responsible for surface salt deposition, where salt beds would eventually be covered with sediment and later pushed to the surface, these investigators took a novel approach, including a "hot saline solution" that presumably occurred at the "crossing of faults," which produced a "slender pencil-like cone of rock salt." Another early researcher, Hagar, recognized the process that apparently caused the "slender pencil-like cone" of salt domes or plugs to form:

"Hager (1904) presented yet another approach where **heated waters carrying salt and other minerals vented to the surface along channels previously formed by volcanic gases. The salt was precipitated from saline waters into necklike masses.**" Bib 144 p29

Like many areas of scientific investigation, the work of these early investigators was sidetracked by the powerful pseudotheories that had taken over modern science. The twentieth century had begun its descent into a Dark Age as scientific progress was halted; speculation and theoretical ideas became more important than finding scientific truth.

Although at the time, these turn of the century geologists were unaware of the hydroplanet Earth and the many types of hydrofountains, they were able to see that "heated waters carrying salt and other minerals vented to the surface along channels previously formed by volcanic gases," a remarkable observation. The Fountain Mark demonstrated that there are many filled diapirs, such as the kimberlite pipes that contain diamonds. These worldwide land forms run very deep, just like salt diapirs do.

We will expand on the salt dome/plug investigation later in the subchapter, after learning where the salt actually came from:

What is the origin of salt?
Na + Cl Does Not Equal NaCl

To comprehend how the Earth's most abundant single mineral deposits were made, we must know from where natural salts

---

"In no instance were conditions encountered which could conceivably have produced any of the major primary salt deposits of the geological past. Some circumstances or set of circumstances, entirely lacking under present day conditions, must have been operative when such major salt deposits were generated."

W. G. Woolnough

come. Like the carbon discussed in the Carbonate Mark, modern geology has long **assumed** that salt came from "igneous rocks." From a 1957 book titled, *Sedimentary Rocks*:

"…we **assume** that all the **sodium in the ocean** is derived by leaching of primitive **igneous rocks**…" Note 8.9b

However, the sodium in the "primitive igneous rocks" is not sufficiently abundant to account for ocean's salt content. Furthermore, the notion that sodium is derived by "leaching" of igneous rocks is absurd because sodium does not exist in nature as an isolated element.

The Rock Cycle Pseudotheory chapters confirmed and the Chemical Model will show that like carbon, there is virtually *no chlorine* in the igneous rocks that presumably formed the Earth. Without chlorine, the most abundant of the salts—NaCl—would not exist. We therefore, must identify the actual source of both the sodium and the chlorine from somewhere other than the igneous rocks. That source is the biota of a now largely dormant event.

Chapel of the Blessed Kinga Poland

Wieliczka Salt Mine

Fig 8.9.1 – The Wieliczka Salt Mine in southern Poland has been an important source of rock salt since the late 13th century. There are over 200 km (124 miles) of underground passages on nine levels, some reaching depths of 327 meters below the surface, connecting over 2,000 excavated salt chambers. Miners began carving sculptures out of the native rock salt, carving fittings and fixtures, walls and alters. These images are of the Chapel of the Blessed Kinga Poland, some 300 feet below ground in the old salt mine. Everything in the room, even the chandeliers are made of salt. This salt deposit is now a museum receiving over a million visitors per year, many of whom ask—how did all this salt get here?

The **key** to the origin of salt is that salt is a *biogenic* mineral—formed from microbes!

To understand how microbes create salt we must first examine the differences between salts and non-salt minerals.

To Dissolve—Or Not to Dissolve

That is the question; in fact, there are two FQs that we must consider:

Why do rocks and minerals such as quartz, mica, or pyrite **not** dissolve in pH **neutral** water?

Why **do** solid mineral salts such as halite (NaCl) dissolve in pH neutral water?

The answer to these questions begins back in the New Geology subchapter, 7.5. There we learned that many minerals, including those that are quartz-based, dissolve in *alkaline water* under increased pressure and temperature. Since most rocks dissolve in alkaline water—they will also *grow* or re-crystallize in alkaline water as the temperature and/or pressure is altered. This is how the majority of the rocks we walk on were formed.

We can answer the first question by simply recognizing that typical quartz-based rocks do not dissolve in neutral pH water **because they did not grow in pH neutral water**. We can also answer the second question by recognizing that salt rocks **do** dissolve in pH neutral water **because they grew out of pH neutral water**. Whereas most minerals grew out of alkaline water, they will therefore *not* revert back to their dissolved state in neutral water.

To create new forms of quartz-based minerals, we must *start* with quartz-based igneous rocks and dissolve the minerals in high pressure/temperature alkaline water. This cannot be done with salt minerals, why—because the igneous rocks that presumably first formed the Earth, do not contain all of the elements necessary to form salt minerals. Very little sodium and almost no chlorine exists in igneous rocks wherewith to create sodium chloride (halite) salt. The Universal Flood Model origin for both sodium and chlorine of which salt is comprised is the *same* origin responsible for the carbon in the carbonate rocks—biological transformation.

Just as the carbon in carbonate rocks was formed through biological transformation, a process to be detailed in the Chemical Model Chapter, the world's massive salt deposits were formed as microbial blooms of incomprehensible size which produced the necessary compounds. Moreover, the method by which microorganisms transformed original rocks and water into other minerals and recombined them is the Acid-Base Neutralization Process.

Acid-Base Neutralization Process

The natural process of forming salts from scratch involves the neutralization of acids and bases. In chemistry, this is a rather simple and commonly taught practice, however in geology, especially as it relates to rock formation, the Acid-Base Neutralization of *natural salt deposits* is a completely new concept.

To see several examples of how salts can be made from common acids and bases found in Nature, refer to Fig 8.9.3. The formulas of both halite (common table salt) and gypsum (the

Fig 8.9.2 – This salt mine in Michigan, USA, is one of many around the world made possible because of salt created during the Universal Flood. Modern science has long denied the occurrence of a global flood, but the Salt Mark demonstrates with empirical evidence, that such salt formations could have formed only in a deep ocean in hyprethermal conditions. This is why there is no present-day analog to these ancient deposits forming today, and why geologists cannot logically explain their origin through evaporative processes.

second most common salt) are shown there.

Long ago, cement manufacturers learned that they could use either natural or synthetic (manmade) gypsum salt in the cement manufacturing process. Where does "synthetic gypsum" come from? From "acid neutralization":

"The second source of **synthetic gypsum is acid neutralization**. The sulfate production of TiO2, used as a white colorant in many retail products, **yields byproduct gypsum** in the process of **neutralizing acidic waste**. The chemical gypsum produced is of high quality…" Note 8.9c

The synthetic gypsum is chemically the same as natural gypsum because they were both formed in a similar process involving the neutralization of acids and bases. Although the Acid-Base Neutralization of natural salt deposits is the first step in understanding the Acid-Base Biosalt Origin, we recognize that there are other chemical-exchange possibilities in salt prethermation. In Fig 8.9.3, the last formula includes algae, decaying matter, and water to create gypsum salt and methane gas, even though it is not a strong acid-base relationship. Methane gas is a natural gas containing large amounts of H_2S (hydrogen sulphide), which demonstrates a definite relationship with salt's origin.

"Hydrogen sulphide often results from the Prokaryotic breakdown of organic matter in the **absence of oxygen gas**, such as in swamps and sewers; this process is commonly known as anaerobic digestion. H_2S also occurs in **volcanic gases, natural gas and some well waters**." Note 8.9d

These biosalts are related to anaerobic bacteria from deep within the crust of the Earth that requires no oxygen.

Such microbes are just beginning to be understood by modern science; like algae in the surface oceans, they can bloom quickly and in great profusion when conditions are right. Modern examples of the right hyprethermal conditions are present at the black and white smokers on the ocean floor today (Fig 8.9.4). The Lost City Carbonates deep in the Atlantic Ocean are also a good example (see Fig 8.8.17). There, warm, highly alkaline water was being released from fractures with a pH of 11:

"The reaction releases heat and dissolves some of the minerals in the rock to form warm, **alkaline water that rises from fractures in the sea floor. The water can reach 90°C and pH 11**. When it meets cooler sea water, **calcium carbonate precipitates**, which gradually builds up the chalky chimneys." Note 8.9e

The highly alkaline pH of the water at Lost City is not indicative of all of the water along the mid-ocean ridges; water with an acidity of "2.0 or less" was reported in many vent fields, bathed in an acidic sulfide-rich environment. When various acids and bases combine in the Earth's underground aquifers and caverns, *salt will prethermate* as the temperature is lowered or the pressure is changed.

Acid-Base Biosalt Origin

In Fig 8.9.3, the Acid-Base Biosalt Origin diagram illustrates several chemical formulas that result in halite and gypsum salts. Other anciently deposited, less-common salts are a prethermate, formed when specific microbial waters were heated and combined. The process of combining the acids and bases of microbial fluids to form salts is called **biosalt synthesis**.

Modern geology has no simple, rational, or logical way to explain how miles-thick layers of NaCl salt precipitated through evaporation of seawater. Moreover, it is impossible to conceive when considering the variety of other salts in the ocean. Evap-

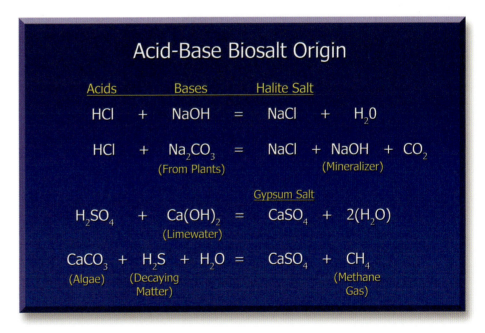

Fig 8.9.3 – The origin of the Earth's large salt deposits can now be comprehended with the Acid-Base Biosalt Origin. Biosalts are salts formed from biogenic sources, which processes are observable today on the ocean floor around active hydrofountains. White smokers produce a discharge more acidic than the black smokers' alkaline fluids (see Fig 8.9.4). Most ancient salt deposits can be shown to have prethermated from an acid-base solution that originated from microbes living within the crust of the Earth. In the diagram above, several possible chemical formulas for the production of halite and gypsum are suggested. There are many other salt formulas involving a wide range of chemicals; a field ripe for millennial science as the true nature of salt origins are understood.

oration can only account for the formation of several inches of salt. However, the UF Hypretherm reveals an easy explanation for thick salt deposits.

Changes in temperature and pressure (P-T-X) of the hypretherm causes different salts to prethermate at different times. (see subchapter 7.4) In salt-water systems, "fluid separates" into low salt vapor and high salt liquids:

"The topology of **salt-water systems**, such as $NaCl-H_2O$ shows that over a wide range of P-T-X conditions, **fluid separates into a low-salinity vapor and a high-salinity liquid (brine)**." Note 8.9f

With high pressures and temperatures, it becomes possible to see how ancient, massive deposits of pure NaCl or anhydrite could form. As the temperature/pressure in the Earth's crust increased from the extraordinary fall and rise of the crust during the UF and its related earthquakes, changes in underground aquifers caused vapors and fluids within the pipes and hydrofountains to "separate" as researchers observed. The physical characteristics of the brines became as different as oil and water, allowing the accumulation of substantial quantities of pure brine in some locations. The brines then prethermate at different times, facilitating the growth of pure salt minerals, limited in thickness to the volume of the accumulated brine, forming massive layers seen today as ancient salt domes, plugs, and other formations.

The Acid-Base Biosalt Origin also explains why these kinds of salt formations are not happening today—deep hot oceans are not covering the continents! Thick layers of pure salt do not form from evaporation; the thick layers in Nature today lack horizontal bedding of sediment that should be evident as storms would have driven all types of sediment and plant matter into the supposed salt-forming bays and lagoons.

The Acid-Base Biosalt Origin diagram in Fig 8.9.3 is lacking many details of the biosalt synthesis process, but time and Millennial Science research will work the details out. Today, however, we can look at several of the chemicals in the biosalt synthesis process and discuss where they are being made in Nature today and how they would have been produced even more abundantly during the UF.

Na_2CO_3 (sodium carbonate) or soda ash is extracted from the ashes of many plants. When this base is combined with HCl (hydrochloric acid), which can be found at some of the ocean floor hyprethermal vents, the products of halite (NaCl), sodium hydroxide (NaOH) and carbon dioxide (CO_2) are formed. NaOH is an important base in the biosalt synthesis process because it is the mineralizer that will dissolve quartz when under high pressure and temperatures. NaOH is the mineralizer used in the Crystallization Process in subchapter 7.4 to grow quartz crystals. The natural biosalt synthesis process was essential in the production of many quartz formations during the UF, a subject discussed in detail in the upcoming Quartz Mark.

When H_2SO_4 (sulfuric acid) is combined with limewater, gypsum salt is formed. Another way in which gypsum or anhydrite may be formed, occurs when algae (from calcium carbonate) combines with byproducts (H_2S hydrogen sulphide) of decaying organic matter. These formulae give us clues as to why carbonate caps and methane gas (CH_4) are associated with salt deposits.

Changing the origin of salts from an 'evaporite' to a 'prethermite' is an *enormous change* for modern geology and such change is not likely to come easy.

Ancient Salt Deposits Are Prethermites Not Evaporites

Over 100 years ago, well-respected geologist Van't Hoff and his associates conducted an extensive array of evaporative experiments which suggested that geological salt deposits "are *not* formed by simple evaporation of sea-water":

"Between 1896 and 1908, Van't Hoff and his co-workers carried out a massive experimental program designed to elucidate marine evaporite deposits…Van't Hoff was keenly aware of major discrepancies between prediction and observation. He concluded that '**These results correlate with natural relationships qualitatively, but not at all quantitatively, which indicates that salt deposits are not formed by simple evaporation of sea-water**.'" Note 8.9g

Van't Hoff is extensively cited in technical literature, but the results and conclusions from his experiments have not been included in modern geology. The reason is simple. No theory, including evaporation has ever been proposed that can explain the origin of the world's salt deposits. Thus, modern geological textbooks continue to teach that salt deposits are "evaporites" formed as seawater evaporated, despite having been proved otherwise.

Toward the end of subchapter 6.1 in the Rock Cycle, it was stated that—salt is formed by *methods other* than evaporation. The Crystallization Process in subchapter 7.4 identified this new process as **prethermation**. Prethermation is a process by

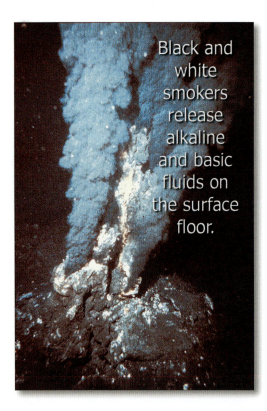

Fig 8.9.4 – The discovery of hot smokers on the seafloor is completely changing the way science looks at mineral crystallization. We can observe salt formation 'in process' that reveals how geological salts and ore deposits were formed. It is only with acidic and alkaline fluids from microbial wastes, in a hypretherm environment that salt minerals form. There are many unknown aspects of the smokers that are awaiting millennial scientist's discovery.

which common quartz rocks are made, and it is the method by which almost all of the large salt deposits were created. This is due to the fact that both quartz and salt crystals formed when the pressure or temperature of the mineral solution in the hypretherm changed.

> Prethermation replaces evaporation as the process that formed the world's thick salt deposits.

Only during the last two decades have investigators begun to take a hard look at methods of salt precipitation other than evaporation. Lawrence A. Hardie, a maverick researcher took a novel approach to the subject, suggesting caution to his 1991 audience:

"**The reader is warned** that some of the views presented here are **by no means widely held** among current workers in the **field of evaporites**." Note 8.9h

Hardie's research turned out to be cutting-edge ideas of the new geology here in the Universal Flood Model. Hardie explained that we "no longer can simply assume" that "evaporites were formed by evaporation of ancient seawater":

"All this information on Quaternary nonmarine saline lakes and evaporites means that *we no longer can simply assume that all ancient gypsum-anhydrite-halite-potash evaporites were formed by evaporation of ancient seawater*." Note 8.9h p139

Geologists have long-held the "basic assumption" that ancient large salt deposits "were precipitated from seawater," but Hardie makes it *very* clear—this assumption results in a dilemma that "cannot be resolved":

"In the past this has not been considered a problem because, **almost without exception**, studies of evaporites have started with **the basic assumption** that all ancient gypsum-anhydrite-halite deposits and all potash deposits **were precipitated from seawater**. As stressed above, **this can no longer be considered as a valid assumption**. This approach results in a dilemma that at this point **cannot be resolved**." Note 8.9h p140

Having asked the question, "Where, then, does all this leave us?" Hardie was able to move outside of the scientific black box by recognizing the contribution of *hydrothermal brines* thermally convecting below the seafloor.

"**Hot (300°C) hydrothermal brines**, rich in Ca and K and almost devoid of Mg and SO_4, **are thermally convecting at depths as shallow as 1.5 km beneath the rifted floor of the Salton Sea trough in southern California**..." Note 8.9h p150

For the first time, a mechanism besides evaporation was being explored to explain the separation of salts and their apparent "hydrothermal origin." Hardie observed the "*thermal event*" was not only creating the upwelling brines but also "other metallic ores":

"It is possible, indeed probable, that many present-day deep basinal brines **had a similar hydrothermal origin** and are

Fig 8.9.5 – These examples of Natural Biosalt are common if one looks in the right places, and at the right time. Sometimes after a rainstorm, the desert soil blooms with microbes, forming white patches of a variety of salts in a natural biosalt synthesis process. Hot springs, like those flowing into the Colorado River, exhibit different temperatures of spring water, each with a unique microbial community. When the byproducts of the microbes combine, they can form the white, salt residue seen attached to some of the rocks in the spring. In a similar fashion, in present-day, deep ocean hyprethermal conditions, microbes are making salts.

now **simply cooled down hydrothermal brines** that are relatively static and in conductive thermal equilibrium with their host bedrocks. **As such they would have originated during a particular** *thermal event* when they not only acquired their $CaCl_2$ chemical signature (together with increased heavy metal concentrations as per the Salton Sea system) but were convected upward as a result of thermally induced density instabilities. **As the upwelling brines began cooling in the upper reaches of the convection cells, they may have deposited Pb, Zn, Cu, and other metallic ores, together with quartz, calcite, dolomite, and anhydrite cements**. With a favorable plumbing system some of the brines may have reached the surface as brine springs, where they would have acted as source waters for magnesium sulfate-poor evaporites, as outlined earlier. **When the thermal event eventually decayed...the hot brines cooled down and migrated gravitationally to the deeper parts of the basin to become static Na-Ca-Cl basinal brines**." Note 8.9h p160

Salts that are from "simply cooled down hydrothermal brines" are *prethermites*. Hardie only missed the critical element of pressure in his observations, but he did note connections between carbonates, oil, gas, and ores:

"This is all the more important because the **evaporite-hydrothermal brine-basinal brine cycle** includes stages that may involve the (a) the acquisition of heavy metals and **their deposition as hydrothermal ores**, (b) the metasomatism and burial diagenesis of basinal sediments (**cementation, dolomitization, calcitization, etc**), and perhaps also (c) **the migration of oil and gas**." Note 8.9h p163

Each of these are critical and specific marks of the UF event. The Carbonate Mark has already been discussed and the Salt

Mark, the Oil and Gas Mark, and the Ore Mark are in the pages ahead. Each of them are the result of or impacted by hydrothermal "upwelling brines" from below the ocean floor, created by a thermal event.

Hardie summarizes his article by noting that "evaporites" remain a "dominant paradigm in evaporitology," but does not identify specifically a name for the newly identified process he suggested. He does however, state that "the time seems ripe for a reassessment" of the evaporites and recognized the likelihood that many evaporites must be reinterpreted to be of "*nonmarine or hybrid origin*":

"**The assumption** that almost all ancient evaporites are seawater residues that were deposited in marine environments located in the subtropical high-pressure climatic belts **remains the dominant paradigm in evaporitology**. It is hoped in the light of what has been discussed in this paper that **the validity of this paradigm will be given a second look by Earth scientists of all kinds. The time seems ripe for a reassessment of the geochemical significance of evaporites and for students of evaporites to take into consideration the possibility, indeed the likelihood, that many evaporites interpreted in the past to be marine are, in chemical terms, of nonmarine or hybrid origin**." Note 8.9h p163

Hardie was on the right track when he first questioned the evaporite pseudotheory with an open mind and was rewarded by identifying a new source for salt deposits—hydrothermal vents on the bottom of a hot ocean.

What is needed now are laboratory experiments that take both high temperature and high pressure into account. The results must then be compared to field observations. In this way, the prediction that the UF Hypretherm produced the ancient salt deposits can be tested and confirmed.

The Norwegian Salt Formation Evidence

In the Crystallization Process of the Hydroplanet Model, we introduced a study from 2006 in support of the Salt Deposit Model. Five Norwegian researchers used *laboratory experiments and field observations* which help to demonstrate that the large salt deposits formed in a *hypretherm*. The new model proposed by the researchers employed high temperatures and high pressures as an "outsalting mechanism" instead of evaporation. The article in The *Oil & Gas Journal* reported:

"Geologists, whose current model for salt deposition and accumulation relies **only on solar evaporation of seawater, have overlooked this novel hydrothermal outsalting mechanism**.

The **main beauty of the new model** is its lack of demand for large ocean basins (such as the Mediterranean Sea) to evaporate up to 10 times for salt several kilometers thick to accumulate." Note 8.9i

The beauty of the Norwegian salt model, says the author, is that the old problems associated with salt evaporation disappear with salt precipitation from temperature and pressure changes. The Norwegian researchers noted that a cool, shallow sea did not create the large salt formations, but a hot, deep ocean, greater than 3 km (1.9 miles) and hotter than 430°C (806°F) must have existed for "spontaneous precipitation of salt" to occur:

"**Laboratory observations** and molecular modeling **confirm that a spontaneous precipitation of salt occur when brines or seawater attain supercritical conditions. These conditions are attained in the sub-surface at >3km depths (300 bars) and at temperatures above 430°C.**

"A simple laboratory experiment confirms that salts (anhydrite and halite) precipitate when seawater boils in submerged porous sand, and that the salt is displacing the sand in contact with the heating element; thus forming a 'pure' salt body. Based on published information from the Atlantis II Deep, Red Sea, and the Lake Asale area, Ethiopia, it is suggested that **most of the salts deposited in these geological settings may have formed by the various modes of hydrothermally associated salt forming processes referred to above**." Note 8.9j

The simple answer as to how the two most common salts (anhydrite and halite) became separated during precipitation was revealed in a laboratory experiment that heated seawater to ***430°C*** (806°F) with ***pressures greater than 300 bars (4351 psi or equivalent to 3 km (10,000 ft) ocean depth)***. These con-

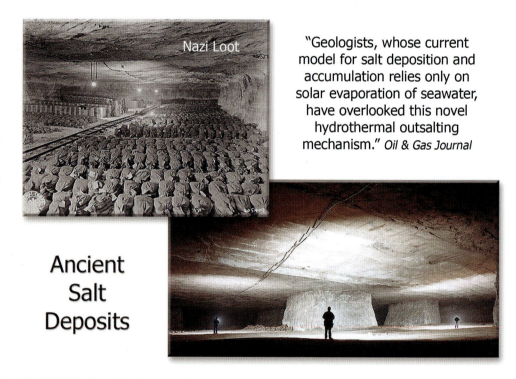

"Geologists, whose current model for salt deposition and accumulation relies only on solar evaporation of seawater, have overlooked this novel hydrothermal outsalting mechanism." *Oil & Gas Journal*

Ancient Salt Deposits

Fig 8.9.6 – The sheer size of the world's underground salt deposits is hard to comprehend, unless you've actually been fortunate to see one first hand. Mines stretch for miles, sometimes underlying entire cities. They may have multiple layers, some of which have chambers the size of large warehouses. During World War II, Nazi Germany stored vast wartime supplies, including an enormous horde of gold. Norwegian scientists developed a model of ancient salt formation very similar to the UM salt model, except that they identified no mechanism for continental deposition. The Universal Flood provides the mechanism for salt origin, deposition, and for minerals commonly associated with the salt deposits, all of which were a part of the UF.

ditions are the hyprethermal conditions present during the UF, with the depth being the same necessary for obsidian and basalt growth as identified in the Depth Mark.

Sadly, new research like this and many other important and valuable new discoveries get relatively little attention in modern science because of dogmagma and the rock cycle pseudotheory. Within the context of the UM we can fully appreciate the critical nature of this salt research. The Rock Cycle Pseudotheory chapter established that the largest single mineral-rock in the world is a salt dome, some being over seven miles deep. It was also noted that there is more salt in the ocean than on all the land above sea level! The realization that there is enough salt in the ocean to create a pile over 500 feet thick across every inch of the continents makes salt the most abundant mineral on the entire surface of the Earth, and yet, it has apparently been one of the least studied minerals in geology.

The Norwegian salt experiments and the results of their research show that the great halite and anhydrite deposits came about due to "heat sources" at the bottom of a world-wide ocean, "refining" the salt water into various pure salt deposits:

"Salt production will, in large, follow the same pattern as that described for the Atlantis II Deep. **Anhydrite** will precipitate in the recharge zones, while **halite and other salts precipitate in the supercritical zone nearer to the heat sources**. Salt will also precipitate at the surface by solar evaporation in shallow pools (Schreiber, 1988). In order to account for the anomalous high-magnesium salts that precipitate, we theorize this to be a result of 'seawater starvation'. Because the Lake Asale region is cut off from the sea, there is likely very little new seawater being drawn into the system. **The circulating water will preferentially leach the most soluble salts, thus 'refining' the primary salt. This preferential leaching of potassium and magnesium is a manifestation of processes that causes the 'Potassium dilemma', i.e., the under-representation of potassium and magnesium in common rock salt compared to the composition of seawater**." Note 8.9j p865

The researchers had discovered that the halite, gypsum, anhydrite, potash, and other pure salt deposits came *from a hyprethermal process*, there being no other physical or chemical way for them to form. The simple fact is that the UF Hypretherm that created quartz sand across the surface of the Earth several thousand years ago also created the largest and most abundant mineral deposits in the world—salt.

The "potassium dilemma" was one of the salt mysteries identified in the Sea Salts Mystery in Fig 6.1.15, which is linked to the uniformitarian myth that evaporation formed the ancient salt deposits. The "potassium dilemma' is solved in the UF Hypretherm.

Hyprethermal Potash Evidence

Potash, (KCl) or potassium chloride, is one of the minor salts. Also known by its mineral name **sylvite**, it is found worldwide and is an important ingredient in fertilizers. Writing an article

> "The time seems ripe for a reassessment of the geochemical significance of evaporites and for students of evaporites to take into consideration the possibility, indeed the likelihood, that many evaporites interpreted in the past to be marine are, in chemical terms, of **nonmarine or hybrid origin**."
>
> Lawrence A. Hardie

on the *Origin of Ancient Potash Evaporites* in the journal of *Science*, researchers mention that "potash salts fail to match" *evaporated* seawater predictions:

"One major problem, however, is that ancient evaporites containing soluble **potash salts fail to match** the mineralogical sequences **predicted in the evaporation of modern seawater**." Note 8.9k

This should be familiar by now; predictions based on modern science theory not matching observations. Realizing the evaporation sequence failed to match geological potash deposits, researchers went searching in western China for more answers. They found evidence that salt deposits came from underground, *non-marine* sources:

"These so-called anomalous potash deposits **may have formed from non-marine parent waters**." Note 8.9k p1092

With the paradigm change that ancient salts like potash did not evaporate from seawater, it is easier to look to other sources for answers. Hyprethermal conditions and biogenic vents on the seafloor, which are heated pipes where microbes live and die, produce the acidic and alkaline waters needed for salt formation.

The challenge facing researchers is discovering how salts like potash can precipitate or crystallize in large quantities, from a brine solution, independent of other salts in the mix. Evaporation simply doesn't work, but laboratory experiments show precipitation is possible with *increased temperature*:

"Where brine is mutually saturated with KCl and NaCl, the solubility of NaCl decreases slightly with **increasing temperature up to 90°C**, above which the solubility of NaCl increases with temperature. Since the solubility of KCl increases dramatically with increasing temperature up to **90°C**, a warm brine would dissolve much less NaCl if there were sufficient sylvite available to saturate the solution with respect to KCl. **When the brine is cooled, sylvite will precipitate more rapidly than halite**, and at certain temperatures, a solution which is saturated with respect to both KCl and NaCl **will precipitate only sylvite**, and NaCl will remain in solution." Note 8.9l

So far as is known, increased temperature is the only way scientists have identified to precipitate sylvite (potash) by itself; temperatures far above those in the ocean today. In addition to the lab-controlled experiments, fluid inclusions in actual ancient salt deposits were analyzed. The results showed that sylvite and the parent salt, halite, were formed at *increased hydrothermal temperatures*:

"**Sylvite daughter crystals** from fluid inclusions in halite dissolve at moderately high temperatures (avg of **63°C** for Rhine Graben, **71°C** for Salado, and **39°C** in Prairie). **These results indicate that halite crystallized from warm surface brines** rich in KCL. Cooling of these brines at or just below the surface **is a major mechanism by which to achieve supersaturation with respect to sylvite**." Note 8.9m

After a century of evaporite experiments, which failed to match the mineralogical sequence of salts, researchers are beginning to realize that a hydrothermal environment was required to produce the ancient salt deposits.

Gypsum/Anhydrite Evidence

Gypsum and its near cousin, anhydrite, together are the second most abundant salts. We first encountered the Gypsum and Anhydrite Mysteries in the Rock Cycle Pseudotheory, subchapter 6.1. Here we give reasons for why the world's massive gypsum and anhydrite formations remain inadequately explained in modern geology. As a reminder, the two minerals share nearly the same chemical formula, but gypsum is almos*t twice as soft* as anhydrite and contains a good portion of water in its crystal structure. Anhydrite, which contains very little water, is not located in surface soils and in dunes like gypsum is. On the continents, anhydrite is found in primarily large deposits, in diatremes, and is commonly associated with halite salt in salt domes.

In the scientific literature and in books written to the interested public, like the book, *The Complete Guide to Rocks and Minerals* (2006), we find the following description:

"Like **gypsum**, **anhydrite** is a white powdery mineral that typically forms in **thick beds when water evaporates**." Bib 151 p181

Having documented that mainstream scientific literature on the subject is ***incorrect***; that neither gypsum nor anhydrite is deposited in any significant degree by evaporation of seawater, how do these two minerals, chemically the same, differ so radically in hardness, if both formed merely by evaporation? Anhydrite is harder than gypsum for the same reason diamonds are harder than graphite. Although diamonds and graphite share the same chemical nature (both are carbon), diamonds *were formed under tremendous pressure*, making them much harder. *This is the same reason* anhydrite is so much more dense than gypsum. Being formed under pressure (the prethermate process) allowed water to be squeezed out of the crystal matrix of anhydrite, but not gypsum.

We saw similar characteristics with vesicle-free basalt and obsidian that were formed below 3km of ocean water. The high pressure from deep water creates different forms of minerals, even though the chemical properties may be the same. With these new insights we can explain why gypsum is found primarily on the surface (see Fig 8.9.7) while anhydrite can be found at depth, one instance being the series of boreholes near Humble Arizona, where anhydrite was identified at 6,000 feet deep.

Elsewhere, gypsum is found on the Colorado Plateau, as in Fig 8.9.7 in homogeneous, colored clay hydrofountain deposits. In Flood deposits of the same type of clay, calcite and silica fossils can be found. It is not a coincidence that all of these minerals and fossils are found in Flood hydrosediment.

When the team of Norwegian scientists applied high temperatures and pressure, they were able to see anhydrite form separately from brine water. Interestingly, as a consequence of removing some of the laboratory equipment, the researchers found a patch of white mineral on the floor:

"When the tank was removed from the pool, **a 30 cm di-**

Gypsum Crystal Deposit

These gypsum crystals did not grow from evaporation, they could have only grown from UF hydrothermal water.

Nevada, USA

Similar clay deposits that formed in the UF contain quartz & calcite fossils.

Fig 8.9.7 – Large gypsum crystals like the ones shown above are not found in evaporation ponds and they don't form when seawater evaporates, even though scientific literature states this to be the case. They grew in hydrothermal or hyperthermal waters in soft clay deposits. The clay deposits are the same type of material in which petrified wood and other fossils can be found; geological flood formations created in the hypretherm of the Universal Flood. Most colored clay mounds are hydrofountain sediments from thermal water vents and veins active during the flood event. When conditions and mineral content were just right, gypsum crystals grew rapidly, sometimes in profusion, and sometimes to extreme sizes.

CHAPTER 8 THE UNIVERSAL FLOOD MODEL

ameter white patch appeared on the concrete floor underneath the tank. ESEM (Environmental Scanning Electron Microscope) analysis showed it to consist of **gypsum** ($CaSO_4·2H_2O$)." Note 8.9n

The fine grained gypsum formed because it was not under pressure but was separated from the high temperature and pressurized apparatus that made the anhydrite. However, large gypsum crystals require elevated temperatures and possibly some elevated pressure. The largest known gypsum crystals in the world were found in the Naica mine located in Mexico. There, gypsum crystals of up to 12 meters (39 feet) inside a large underground cavity were revealed after pumping out all of the water (http://giantcrystals.strahlen.org/america/naica.htm). The huge gypsum crystals obviously did not grow out of evaporated water, but precipitated in a hyprethermal process, with a change in temperature or pressure.

Another example of giant gypsum crystals comes from Spain, where a huge geode was recently discovered, with a cavity large enough to seat 10 people. The crystals in the geode (the largest were up to 2 meters) were apparently formed first in *freshwater* and finished with *marine water*:

"Fluid inclusions and stable isotopes of **giant transparent gypsum crystals in a huge geode** recently discovered in Pulpí (SE Spain) are useful tools for explaining its geological formation. Fluid inclusions suggest a mixture of fluids, **starting with freshwater and including seawater in more advanced stages** of the gypsum crystal growth. The isotopic composition of the hydration water in the gypsum crystals agrees with a meteoric fluid. The gypsum is enriched in ^{34}S, which **denotes genetic links with marine sulphates via freshwater dissolution-recrystallization of earlier marine evaporites**." Note 8.9o

There aren't many places where hydrothermal freshwater could become hydrothermal seawater today, yet during the UF Hypretherm, freshwater aquifers and hydrofountains would have been common as underground continental fresh waters were heated by earthquake friction, but would have quickly been saturated with seawater as the Flood waters spread across the continents.

Other direct scientific evidence that anhydrite is of a hyprethermal origin are the instances where anhydrite grew together with other hyprethermal minerals, even being included inside natural quartz crystals.

Decades ago, researchers identified the reason anhydrite precipitates at a different time than other salts—instead of becoming *more* soluble at higher temperatures, like halite, "it becomes *less* soluble with higher temperature":

"It is well known that seawater easily becomes supersaturated with respect to **anhydrite** ($CaSO_4$) **when heated beyond 130 °C** (Bischoff and Seyfried, 1978). This happens because anhydrite has a retrograde solubility in water, i.e., **it becomes less soluble with higher temperature**." Note 8.9p

Fig 8.9.8 – Sand "roses" are gypsum crystals found near the surface in many of the World's deserts, but they are not known to be forming anywhere in the world today. The large crystal conglomerations formed in hyprethermal conditions, which is direct evidence of the Universal Flood event.

The unique physical property of anhydrite salt is *only* observed, "when it is heated beyond 130 °C," which is nearing hyprethermal conditions. We have found only one place in Nature where anhydrite is currently being formed—and that's at the bottom of the ocean.

The Sand Crystal Mystery Answered

Large Sand Crystals made of gypsum, Fig 8.9.8, were first introduced in the Rock Cycle Pseudotheory chapter. Like so many mysterious rock specimens, the question of how they were formed remains unanswered. The uniformity myth fails because sand crystals like these are apparently not forming today; at least no one has seen them being formed, either by evaporation or in "shallow cool" seas, like those that supposedly once covered all the continents.

Just like large calcite and quartz crystals, gypsum crystals require deep, hot seas to precipitate. There must also be a constant flow of dissolved gypsum from the combination of acidic and basic solutions provided by bacteria and other microbes.

Because of their widespread distribution, a *large area* of frictional heating must have been present, not just along crustal plate boundaries. Today, most gypsum crystals are found commonly in the center of the continents, not necessarily associated with plate boundaries or nearby volcanic activity. Moreover, they are found generally on or near the *surface*. All of these parameters require a global hypretherm of a recent date.

White Sands Gypsum Evidence

Turning our attention to the White Sands Gypsum Salt Mystery, also discussed in the Rock Cycle chapter, how did this unique deposit of pure gypsum sand crystals form? Fig 8.9.9 reveals two fundamental questions and answers to the origin of the gypsum sand. The answers come not from evaporation and erosion, but from precipitation in thermal ocean waters.

In the 2006 *Marine and Petroleum Geology* journal, researchers documented having found that salt "venting fountains" on the seafloor "will precipitate and accumulate on the seafloor":

"Finally, salt may form within the brine-pools on the seafloor, **fed by the brine venting fountains** (Schreiber, 1988). When warm and high-density brines enter these pools, they start cooling, and many of the common sea salts (e.g. halite) **will precipitate and accumulate on the seafloor**." Note 8.9q

The "venting fountains" they observed are actually hydrofountains offering additional direct evidence of both the Sand and Salt Marks. Depending on depth, temperature, and dissolved minerals, various clays and sands of salt and other minerals, particularly quartz, iron, and pyrite will be produced. Observations like these will continue to force modern science to rewrite its understanding of continental sediment layers.

The gypsum sands at the White Sands National Park were formed as individual crystals, much like double terminated quartz crystals were formed, in free-floating water (see Fig 8.5.3). The gypsum sand crystals precipitated from the acid-base neutralization of microbial wastes and were ejected from hydrofountains or fell out of solution on the ocean floor, which is now the desert floor of New Mexico, USA.

Hyprethermal Salt Plug Evidence

Of the hundreds of scientific evidences that prove the Universal Flood, salt plugs (or pipes) have to be near the top of the list; extraordinary examples of the Flood event. The Rock Cycle Pseudotheory found salt domes to be geological structures that are "highly complicated and not wholly understood." However, hard questions are easy, when we know the answer.

The salt dome expert cited in the earlier chapter made an excellent point about gathering data, but analyzing it with the wrong theory:

"Perhaps if there is one phase of geology on which we have accumulated more data, yet have more to learn about, it is salt dome geology... **Ideas on origin and mode of development have not progressed significantly from the stage of theory, despite widespread acceptance today of a few basic concepts.**" Bib 144 p1

In other words, salt plugs/domes have completely baffled the geologists. The reason is simple—*hydrofountains*. Without the Hydroplanet Model and an understanding of the essential role of hydrofountains, salt domes, plugs, or pipes will never be understood.

A salt plug is simply a hydrofountain pipe filled with salt.

The weather resistant 'dome' or a cap rock on top of a deep rooted pillar of salt is responsible for the these features commonly being referred to as salt 'domes.' Looking beneath the surface, the salt 'dome' takes on a whole new look. Seen in Fig 8.9.10, the feature's shape is more correctly a 'salt plug,' thinner at its base, wider at its top, as it nears the surface. The term 'diapir' is often used to describe salt domes and plugs, but a diapir is usually defined as an intrusion of soft, deformable material that was forced to the surface. We find no physical or geological evidence that this ever took place. Soft, pliable material under pressure and over a long period may flow upward along fault lines, but it would *not* form vertical, cylindrically shaped pipes! The term diapir must either be redefined or not used to define the salt pipes.

The Big Hill Salt Dome in Fig 8.9.10 was imaged using gravitational data from GETECH's 3-D gravity modeling program, a method used in modeling salt domes from Matagorda County, Texas to the Mississippi River in Louisiana. Geological studies of the plug have found that it is sound enough to house large petroleum reserves in mile-deep caverns. If the plug had really been pushed up from a salt bed, fracturing would be evident, and contamination of the salt body with surrounding sediment would be apparent, even if the cracks and fractures were sealed with water or pressure at a later time. There appears to be no such evidence of cracking.

The reason salt dome geology's "ideas on origin and mode of development have *not progressed significantly from the stage of theory*" is because all of their theories of evaporite deposition and subsequent plastic extrusion are simply incorrect. Furthermore, salt plugs have no fossils, a difficult fact for researchers to reconcile who believed the salt body came from an ancient evaporated seawater basin. One of the most diverse biospheres on the planet is the shallow ocean near the shore, presumably where the evaporite salt formed. The only possibility a nearly pure halite salt body could have formed is the acid-base, Universal Flood, Hydrofountain pipe origin of salt plugs.

Seeing the hydrofountain origin of salt plugs, *all* of the previously unanswered mysteries are solved and the questions are answered. The salt deposit is a *biogenic mineral*, formed from acidic and alkaline wastes of microbes. The biosalt synthesis

FQ - Where are the rivers transporting the gypsum sand from the mountains?

FA - They don't exist because the salt crystals came from a hyprethermal mineral deposit created in the Flood.

FQ - Rain over millions of years would have dissolved the gypsum sand crystals. Why do they still exist?

FA - Because the gypsum sand formed only several thousand years ago in the Flood.

White Sands National Monument

Fig 8.9.9 – The deposit of gypsum sand in the White Sands National Park in New Mexico, USA, was discussed in the Rock Cycle Pseudotheory chapter. Now, with the Universal Flood model, we can answer FQs about that gypsum deposit. The gypsum sand crystals did not form from evaporating seawater as modern geology has claimed; instead, they precipitated out of biologically active hydrothermal waters during the UF.

process and the Law of Biotrans reveals the true origin, not just of salt plugs and domes, but of over 99% of all of the Earth's salt deposits.

There is another piece of evidence that shows the hydrofountain/hyprethermal origin of the salt plug, and it has to do with the cap of the plug. Cap rocks usually contain anhydrite salt. The Norwegian research previously revealed that anhydrite is formed in a *hypretherm*, not from evaporation. Through the hypretherm process, a change in the pressure or temperature inside the pipe can cause the precipitation of anhydrite salt, independent of other salt precipitation. Because the anhydrite becomes less soluble as temperature increases, the mechanism for large, homogenous anhydrite and halite deposits, in direct contact with each other is revealed.

The process involving the acid-base hyprethermal mechanism that was active during the UF is confirmed today by an ***active*** salt diatreme on the bottom of the ocean at the TAG Hyprethermal Mound.

The reason sediments near the top of the salt plug are "pushed up" is because of the immense pressure exerted by the hydrofountain that formed the diatreme. Although not as powerful near the end of the hydrofountain's life, the continued upwelling water through the slowly dying fountain pushed newly formed salts and sediment upward.

In traditional geology, such up-thrusted sediments confused investigators because there has been no way to explain the tubular shape of the pipes and the purity of salt. Softer sediments under pressure follow the path of least resistance, extruding along fault lines and aquifers—not in vertical pipes that seem to defy the surrounding strata.

The vertical pipes and diatremes are easy to explain with the Hydroplanet and Universal Flood Model because the sediment that the salt pipes were formed in was originally *soft and unconsolidated*. This allowed the formation and upward thrust of vertically oriented, cylindrical pipes, which are known to exist all over the world. As the Flood event progressed, many of the sediments became hardened; lithified in hyprethermal conditions as sandstone and limestone layers atop fossil pipes we see today.

Another interesting evidence of the UF origin of salt comes from *salt glaciers* in Iran. Seen in Fig 8.9.11, geologists claim the dark salt formations, which contain clay, are the result of "geologic time":

"Thick layers of minerals such as **halite (common table salt)** typically accumulate in closed basins during alternating wet and dry climatic conditions. **Over geologic time**, these layers of salt are buried under younger layers of rock. The pressure from overlying rock layers causes the lower-density salt to flow upwards, bending the overlying rock layers and creating a **dome-like structure**. **Erosion has spectacularly revealed** the uplifted tan and brown rock layers surrounding the white Kuh-e-Namak to the northwest and southeast (center of image)." Note 8.9r

There is a simple FQ that goes along with the illustration in Fig 8.9.11:

Why are the salt glaciers not dissolved and eroded far more than the surrounding mountains?

Some of the salt glacier should dissolve each time it rains, but the mountains would not. We should see ***significantly*** more erosion of the salt glaciers and the Kuh-e-Namak salt mound than the surrounding landscape, if "geological time" were responsible for these deposits—but we do not!

Neither image shows evidence of significant salt erosion from the white halite salt dome or the dark salt glaciers.

How could such salt deposits be older than a few thousand years?

The erosion is indicative of only several *thousand* years, not millions.

Before moving on to the Hyprethermal Tag Mound Evidence, we need to explain a salt dome term first introduced in the Rock Cycle chapter. It is called "false cap rock.":

"The sands of the **false cap rock** are usually **cemented by calcite or other water insoluble minerals such as pyrite.**" Bib 144 p45

Since pyrite does not crystallize out of evaporating seawater, geologist have looked at this apparent unrelated cap rock as being a hard,

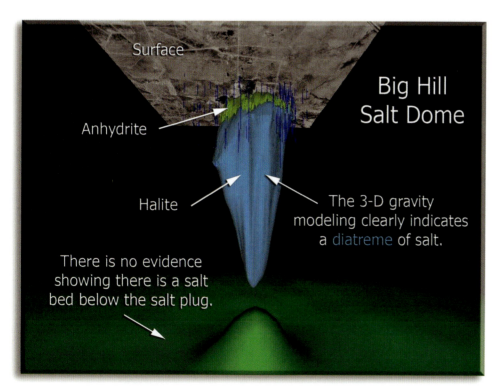

Fig 8.9.10 – The Big Hill Salt Dome in Texas, USA, is a typical *salt diatreme* that should more correctly be called a **salt plug** because the diatreme became 'plugged' with salt after pressure and temperature changed during its hyprethermal growth period. In this 3-D modeled image, the base of the plug takes on a point shape, not indicative of a "bed of salt" as modern geology had assumed. The smoking gun of this prethermite salt plug deposit is the anhydrite salt cap, shown in green at the top of the plug. Anhydrite salt is a gypsum salt that forms *only in a hypretherm*, indicating that the entire plug experienced hypretherm conditions. Moreover, the salt deposit shows no evidence of *layering or that it originated from a deep bed of salt*.

but 'false' cap over the underlying salt structure. Now, as we examine the relationship between pyrite on TAG Mound and other mineral features we have been discussing, we will see the geologists were wrong yet again.

TAG Hyprethermal Mound Evidence

The UF Salt Mark evidence thus far has been simple and perhaps refreshingly new. Unfortunately, or rather fortunately, there has been no global flood recently from which we can examine salt diatremes and other UF related phenomenon forming on the continents. However, the *ocean floor* has places where earthquakes and hydrothermal vents are still active. Is it possible the right combination of microorganisms, pressure and temperatures are present where some of the hypretherm events can be viewed?

The Hydroplanet Model revealed the existence of hydrocraters *on the seafloor*. Researchers, just beginning to study them in detail call many of them "pockmarks." Hydrocraters have diatremes, which are on fault lines, so let's look at what some research has revealed about one particular pockmark hydrocrater below the Mediterranean Ocean:

"Dive observations confirmed the presence of a **brine-filled** pockmark about 250-300 m across and at least 8-10 m deep, now known as the Nadir Brine Lake." Note 8.9s

The Mediterranean Ocean is already salty, so a "brine-filled" crater would have to be in an area of upwelling warmer water. It just so happens that the Nadir Brine crater lake lies not only in a mud volcano field, it is located directly on two bisecting fault lines:

"The **Nadir Brine Lake** lies at the conjunction of a NW-SE and two N-S **faults**, to the northwest of Maidstone mud volcano." Note 8.9s

A number of "brine-filled" hydrocraters found in selected fields are similar to what we find in salt diapirs/diatremes on the continents today. Nevertheless, it would be even more convincing if we could *actually observe salt being deposited today* inside these hydrocrater diatremes on the ocean floor. Could there be hydrocraters that have fluid mixes from black smokers and white smokers that are considerably different in pH and could they create salt deposits? To find out would take a tremendous amount of technical diving and drilling equipment, money and planning.

In 1994, along the Mid-Atlantic Ridge, 3,700 meters below the ocean surface this is exactly what took place. The **TAG hyprethermal mound** was the first "active ore deposit" to be drilled into. Were there any surprises? It was "large amounts of anhydrite within the mound" that "strongly influenced" the researchers' understanding of salt deposits:

"In 1994, the ODP [Ocean Drilling Program] drilled through the active **TAG mound** to determine **for the first time the internal structure of an active ore deposit**. Although **anhydrite precipitation** had been predicted from fluid studies, the occurrence and location of **large amounts of anhydrite within the mound have strongly influenced our understanding** of the formation of sulfide structures at the seafloor." Note 8.9t

Fig 8.9.12 is an illustration of the TAG Hyprethermal Mound showing the hydrocrater and diatreme that is filled with basalt breccia. Hyprethermal fluids flowing up through the diatreme

Fig 8.9.11 – There are over 200 salt domes and glaciers in the Zagros Mountains in southwestern Iran. Modern geologists believe the mountains and salt diapirs were formed by evaporite deposition over geological time. However, halite–NaCl salt–dissolves and erodes thousands of times faster than the surrounding quartz-based rock. Therefore, the salt deposits of the Kuh-e-Namak salt dome and the dark salt glaciers should appear significantly more eroded, perhaps thousands of times more, than the surrounding landscape—but they are not. In reality, these salt deposits are showing off their youth, direct evidence of their formation during the UF. Satellite images courtesy of NASA.

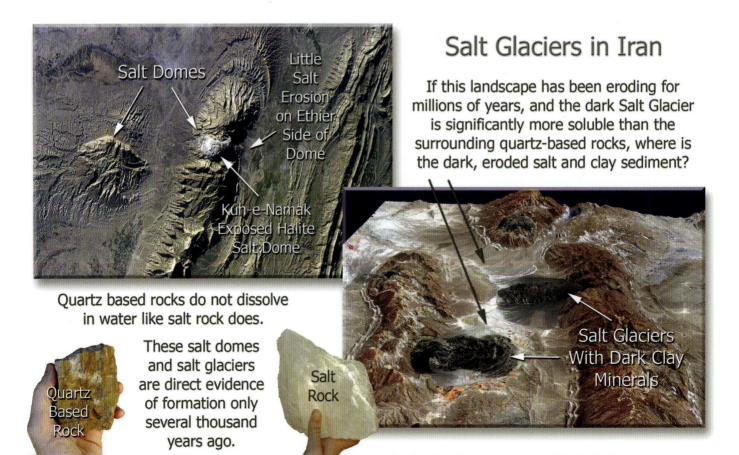

mix with seawater creating environments with different temperatures within the mound. This in turn fostered unique ecosystems of various microbes, some acidic and some alkaline, manifesting as white or black smokers:

> "**White smoker fluids** exiting the Kremlin area are approximately **60° to 90° cooler than black smoker fluids, and are considerably more acidic**, and less Fe- and H2S- rich; these differences are attributed to the white smoker fluid being a mix of roughly five parts black smoker fluid and one part seawater, with precipitation of Fe-sulfide within the mound accounting for the low Fe, H$_2$S, and pH." Note 8.9u

Here is direct evidence of the Acid/Base Biosalt Origin. Although modern day Earthtides create frictional earthquake heat, it is relatively tame when compared with the UF Earthquake activity. The Tag hyprethermal mound is only about 200 meters across, but has produced a considerable amount of anhydrite salt:

> "The significance of seawater entrapment within the TAG active mound is underscored by the presence of **a large volume of anhydrite, estimated to be on the order of 2 x 10^4 m^3**".
> Note 8.9u p188

This amounts to 20,000 cubic meters of salt on just one small mound with relatively minor faults! Imagine the scale of the Acid/Base Biosalt process during a global event the size of the Universal Flood; enormous formation of salt diapirs all across the continents.

Another important factor of the TAG mound anhydrite formation was the length of time for deposition. The Crystallization Process laid out previously in the Hydroplanet Model revealed that the Earth's mineral deposits did not need 'geological time' in which to form. This is confirmed at the TAG hydrocrater mound, where a relatively modest flow of fluid took only a short time to form the anhydrite salt deposit:

> "The **length of time** required to deposit this volume of anhydrite, assuming flow rates of 1.2 to 12 kg/s, is **80 to 800 yr**." Note 8.9u p188

Even this length of time is a long period of salt prethermation; laboratory results show that in a matter of days, large quantities of salt *can* accumulate in a hypretherm. Larger quantities of salt simply require a larger hypretherm environment, which is exactly the environment the global flood produced.

We want to recognize the TAG hyprethermal mound not only for the large amount of anhydrite, but also for tremendous amount of pyrite and quartz found during explorative drilling. These can be seen in Fig 8.9.12, and will be referred to again in the Pyrite and Quartz Marks later in this chapter.

Setting the Salt Record Straight

This subchapter included considerable discussion on miles-thick salt deposits. Though uncommonly seen, the salt plugs and huge underground salt deposits are an important part of geology and the entire history of the Earth. Few people know much about the giant salt bodies, and even geologists place little emphasis on them.

Conversely, the salt deposits most familiar to the majority of people are salt flats, like those near the Great Salt Lake in Utah, USA (Fig 8.9.13). Generally speaking, the salt flats are not the geological salt deposits referenced in this subchapter because the flats are very thin laminations, truly the result of evaporated seawater. Many salt flats are the mineral residue of bounded post-Flood lakes that eventually dried up. In fact, digging in one area of the Great Salt Lake flats, clay was found only a foot (.3 meter) beneath the salt surface. Great Salt Lake authority, Dr. David Miller is shown drinking from a freshwater spring with a source only several meters below the surface (in Fig 8.9.13).

Researchers continue to be misguided in their research into salt deposits because of the assumption that evaporation was the only mechanism for their origin. There are *three kinds of deposits*, the **first** and by far the most prevalent are the thick salt diapirs and massive underground beds that span large areas of the continents. These were created in the UF Hypretherm. The **second** most common salt deposits are the thin surface laminations in places like the Great Salt Lake salt flats. These are the result of evaporative processes; they contain several different salts, all of which are represented in the ocean. The **third** type of salt deposit is the *salt sand*, like the gypsum sand from

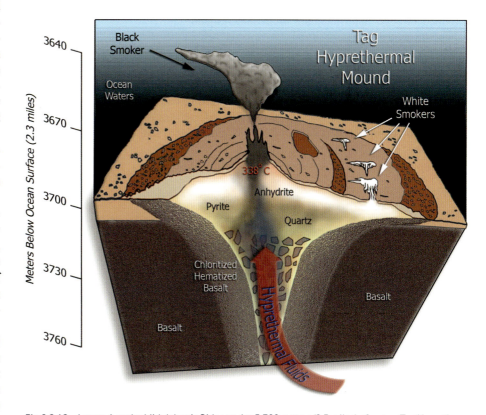

Fig 8.9.12 – Located on the Mid-Atlantic Ridge under 3,700 meters (2.3 miles) of water, Tag Hyprethermal Mound is an "active ore deposit," the first of its type to be explored. In hyprethermal fluids of over 338° C, black and white 'smokers' exhibit a variety of fluids of different pH and biogenetic diversity. Drilling revealed a hydrocrater diatreme beneath the mound that is filled with basalt breccia. Quartz, pyrite and anhydrite are common biogenetically formed minerals, being actively formed in the hyprethermal environment. The Tag Mound is an important example of the processes involved with mineral growth and formation, the origin of the Earth's ores, and true source of the majority of the world's salts.

This fresh water artesian well sits on the edge of the Great Salt Lake in Utah, USA. It gets its water from a fresh water aquifer that is only several meters below the surface.

Fig 8.9.13 – Evaporative saltpans or salt flats exist all around the Great Salt Lake. However, thin layers of evaporative minerals represent only a tiny fraction of the salt deposited on and throughout the crust of the Earth. The flats surrounding the Great Salt Lake are less than a meter thick as attested by the image above, showing an artesian well whose source is only several meters deep.
Photo courtesy of Dr. David E. Miller, the man drinking from the well.

White Sands National Park in New Mexico, USA. Salt sands are uncommon, created in the same way most of the quartz sands were, in a hypretherm environment. They either precipitated out of hyprethermal waters above the ocean floor or were expelled through hydrofountains from underground aquifers where the crystallization occurred. The wisdom behind the origin of each of these three kinds of salt deposits comes because of the knowledge of the UF Hypretherm.

Although salt has long been an essential preservative and an important commodity to mankind, it is significant that its biogenic origin has eluded mankind for so long. Moreover, the mechanism behind the deposition of 99% of the world's salt formations—the UF Hypretherm, may prove to be one of the most important geological discoveries ever made.

The direct result of the knowledge of the little known fact that salt is the most abundant mineral on the **entire** surface of the Earth may be one the most important geological discoveries ever made.

The Salt Mark Summary

Summarizing the findings of the Salt Mark of the Universal Flood begins with this quote from geologist W. G. Woolnough:

"**In no instance were conditions encountered which could conceivably have produced any of the *major* primary salt deposits of the geological past. *Some* circumstances or set of circumstances, entirely lacking under present day conditions, *must* have been operative when such major salt deposits were generated.**" Note 8.9v

Woolnough wrote these words in 1937 after having studied many hundreds of the Earth's salt deposits and the scientific literature surrounding them. Woolnough's words echo the empirical evidences in the Salt Mark, verifying the extraordinary claim that current conditions on the Earth's surface could, in no instance, "*conceivbly have produced **any** of the major primary salt deposits.*" Woolnough further noted that some set of circumstances, "*entirely lacking under present day conditions, **must have been operative** when such major salt deposits were generated.*"

This required a complete revision of the geological salt model to explain *simply and logically* why there are so many massive ancient salt deposits and how these deposits were formed. The **Universal Flood Salt Model** was created to answer that challenge. It has three primary parts.

UF Salt Model
1. Acid-Base Biosalt Origin
2. Hypretherm Conditions
3. Universal Flood Event

The first aspect of the model is the **Acid-Base Biosalt Origin**, which explained that natural salts originated from the neutralization of chemical acids and bases that were created from biological wastes.

The second part of the UF Salt Model is the **Hypretherm. Hyprethermal Conditions** were necessary to *prethermate* massive salt deposits. These conditions existed when the salt water was at elevated temperatures and pressures deep in the (continent covering) ocean.

The **Universal Flood Event** is the third feature of the UF Salt Model. It was the only time in world history when both the biosalt origin of salt and hyprethermal conditions existed on a global scale, making it possible for multi-kilometer thick salt deposits to form in the crust.

The Universal Flood Salt Model utilized new discoveries from the present-day floor of the ocean that revealed black and white smokers emitting acidic and alkaline fluids. These environments allowed certain microbes to thrive, deep in the ocean, and even beneath the ocean floor. The Acid-Base Biosalt Origin then explained how a variety of microbially rich acid-base waters were combined and subjected to hyprethermal conditions

Blue Salt

Fig 8.9.14 – If the Salt Mark subchapter was successful, the connection between salt and the Universal Flood should be apparent. These unique specimens of blue salt are a form of halite from the Klodawa Salt Mine in central Poland. When dissolved, the blue color disappears with no trace of blue in the water. The blue color is the result of a unique crystal structure that refracts light and from a small amount of KCL (potash salt). It was formed in specific Hyprethermal conditions.

facilitating the prethermation of large volumes of salt in the crust.

New hyprethermal experiments with salt brines performed by Norwegian scientists, along with the direct evidence from potash and salt plugs help support the UF Salt Model. Further confirmation of the UF Salt Model comes from the gypsum/anhydrite evidences, salt glaciers in Iran, and new evidence from the TAG mound expeditions.

With all of this refreshingly new scientific knowledge and wisdom, Millennial Scientists can expect to fill in even more of the gaps in Nature's Puzzle as time reveals new truths to those who seek with an open and inquiring mind, but from within the correct paradigm.

8.10 The Oil and Gas Mark

Although students have been taught for decades that oil and dinosaurs are somehow connected, the real connection is between oil and salt. Moreover, the marriage between petroleum and natural gas is still debated to this day. In this subchapter, we will discuss the two sides of the debate, showing that neither has been able to sufficiently explain major deficiencies in their theories. The Oil and Gas Model will be presented, the salt connection, and the UF Hypretherm will reveal that 'geological time' is not needed to explain the oil and gas deposits found in the crust of the Earth today.

The Oil/Gas Debate

Traditionally, petroleum and gas are considered "ancient *organic* materials":

"Geologists view **crude oil and natural gas** as the product of compression and heating of **ancient organic materials** (i.e. kerogen) over geological time." Note 8.10a

It is considered by most geologists to be the prehistoric remains of "zooplankton and algae":

"Today's **oil formed from the preserved remains of prehistoric zooplankton and algae**, which had settled to a sea or lake bottom in large quantities under **anoxic conditions**." Note 8.10a

However, not everyone is aboard this science bandwagon. A number of Russian geologists and a westerner, Thomas Gold, maintain that hydrocarbons of purely inorganic origin exist within the Earth and that they are responsible for petroleum and gas deposits. This claim led to a debate over whether or not oil and gas are biogenic in origin. The Oil/Gas Debate is by no means a trivial matter in the petroleum industry. The entire basis upon which future oil reserves are based—depends on which side of the debate is correct.

Abiogenic Oil Verses Biogenic Oil

The non-organic or Abiogenic Oil Theory began in modern times during the 1950s in Russia and Ukraine. In 1985, Thomas Gold wrote a noteworthy opposing viewpoint of the abiogenic oil theorists in the *Annual Review of Energy* concerning the origin of natural gas and petroleum:

"**Why should we doubt the biological origin of the earth's hydrocarbons**? Has not an immense amount of work demonstrated that the locations of hydrocarbon deposits, the chemical natures and contents of the oils, and the isotopic composition of the carbon, all fit the biogenic theory? Has the biogenic theory not provided the basis for the exploration which has been so successful in the past?

"This is what is generally said, **but when we look at the various points in detail the situation turns out to be not nearly so clear-cut**. Several leading investigators, in both geology and chemistry, have expressed doubts. Hollis Hedberg, a well-known petroleum geologist, wrote, '**It is remarkable that in spite of its widespread occurrence, its great economic importance, and the immense amount of fine research devoted to it, there perhaps still remain more uncertainties concerning the origin of petroleum than that of any other commonly occurring natural substance.**'" Note 8.10b

Indeed, the abiogenic theorists have a point when it comes to the "uncertainties concerning the origin of petroleum." In one example, the biogenic theory of oil states that petroleum comes from organic remains, which "settled to a sea or lake bottom *in large quantities* under anoxic conditions." If this were true, there should be evidence of it today:

FQ: Where are "large quantities" of organic material resting on the bottoms of lakes or seas today buried under sediment, being 'pressure cooked' into petroleum?

Oil drillers have not recorded borings of graduated organics, where modern organic material has gradually changed to crude oil as depth increases. Drillers typically transition from inor-

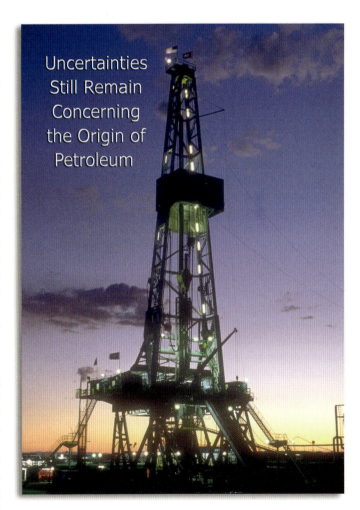

Fig 8.10.1 – Why would there still be uncertainties concerning the origin of petroleum after spending so much research, time and money pumping it out of the ground? Today, there are multiple theories of the origin of oil and gas, but are they correct?

ganic clay, silt or sandstone directly into oil reserves.

Organic chemistry research has clearly shown that high pressure and temperatures are required to convert organic material into oil. If this occurred over 'geological time,' all stages of oil should exist in the deposits; from modern, partially processed oil to fully developed ancient oil. However, oil in various stages of formation does not exist. Furthermore, if "source rocks" was really the source of the oil, there would be *observable paths* toward the oil pockets—but these paths do not exist. Why not?

Let's look at some of the facts the abiogenic theorists point out, which refute the Biogenic Theory. First on the list are the assumptions held by the organic theorists, as expressed by Ukrainian SSR geochemist Dr. Porfir'ev:

"The idea that **petroleum** is extraordinarily stable and that it **can preserve all its properties for hundreds of millions of years also is a necessary assumption of organic theory**.

"**Another assumption** is that the **processes of natural petroleum formation**, migration, and accumulation are a **permanent process, taking place from Proterozoic time to the present**." Note 8.10c

Fig 8.10.2 – The "petroleum process" is not a process understood in modern geology today, simply because it is not a process that can be observed.

The Soviet geologists and geochemists clearly saw a problem with the "assumption" that the petroleum process is a "permanent process" because it has not been observed rendering it "unproved":

"These assumptions, unproved, lead to such an impasse that further investigation of dispersed hydrocarbon theory seem hopeless." Note 8.10c

Abiogenic theorists point to several boreholes that appear to have produced petroleum in crystalline basement rocks, which to them was "direct and most convincing proof":

"The **most direct and most convincing proof of the inorganic theory** is the existence of commercial **petroleum** accumulations in **crystalline and metamorphic basement rocks**." Note 8.10c p14

Drilling in this type of rock is expensive and there are relatively few boreholes from which to study. Moreover, the data gathered from them is debatable. What no one is debating is that the more we drill, the more microbes we find.

Thomas Gold, a maverick, who in his early years, took on the radio astronomer's claim that the first celestial radio signals were stars. He championed the idea that the radio sources were actually distant galaxies, and after a long dispute and harsh ridicule, he was eventually proven right. He ran afoul again with the discovery of neutron stars; colleagues thought Gold's ideas were outrageous, although today, his ideas are widely taught. Gold, along with Fred Hoyle and Hermann Bondi, developed the Steady State Theory of the universe, which was abandoned with the advent of the Big Bang Theory of the 1970s. (The Universe Model chapter will be demonstrating that Gold's Steady State Theory was much closer to the truth than the pseudotheoretical Big Bang theory).

These examples show Gold's broad-minded approach toward research, and his penchant for looking outside the box. He was able to integrate his research across several fields of science, including geochemistry, and he was a strong proponent of the Abiogenic Theory of oil. Although the theory has not been proven, Gold's observations are factually interesting and in *direct contradiction* of the Biogenic Theory of oil-over-geological-time.

In a Professional USGS Paper published in 1993, Gold made points about the geology of oil and that a "larger scale phenomenon for the cause of the oil supply" must be considered:

"Everyone now thinks of Arabia, the Persian Gulf, Iran and Iraq as being the oil region of the world. It is indeed one connected large patch that is oil-rich, stretching for 2,700 km from the mountains of Eastern Turkey down through the Tigris Valley of Iraq and through the Zagros Mountains of Iran into the Persian Gulf, into Saudi Arabia and further south into Oman (Figure 2). There is no feature that the geology or the topography of this entire large region has in common, and that would give any hint why it would all be oil and gas rich. **The various oil deposits are in different types of rock, in rocks of quite different ages, and they are overlaid by quite different caprocks**. They are in a topography of folded mountains in Turkey and the high Zagros mountains of Iran, in the river valley of the Tigris in Iraq, in the Persian Gulf itself, in the flat plains of Arabia and in the mountainous regions of Oman. **It cannot have been a matter of chance that this connected region had so prolific a supply of oil and gas, but resulting from totally different circumstances in different parts of the region. These hydrocarbon-bearing formations represent times so different from each other that there would have been no similarity in the climate or in the types of vegetation that existed there during deposition, just as there is no similarity in the reservoir rocks or in the caprocks of the different regions now**. Yet it is a striking **fact** that the detailed chemistry of these oils is **similar over the whole of this large region** (Kent and Warman, 1972). Surely this is an example of the need to invoke a **larger scale phenomenon for the cause of the oil supply than any scale we can see in the geology of the outer crust**." Note 8.10d

These observations go hand-in-hand with Dr. Porfir'ev's view that there is no "permanent process" of natural petroleum production. Toward the end of Gold's paper, he stresses that "a range of new processes will have to be investigated":

"**A range of new processes will have to be investigated for the understanding of mineralization in the crust**, and the search for **hydrocarbons** may become associated with the search for certain minerals. The microbiology in the ground which is fed by hydrocarbons may have contributed to highly selective processes; just as we saw concentrated magnetite in the boreholes in Sweden, apparently concentrated by microbial action, so perhaps all the large magnetite deposits of Sweden have a similar origin. Judging from the quantities of microbial material that have been identified in hydrocarbon regions (Ourisson and others, 1984), **microbial processing may have been of major importance in the evolution of the crust**." Note 8.10d

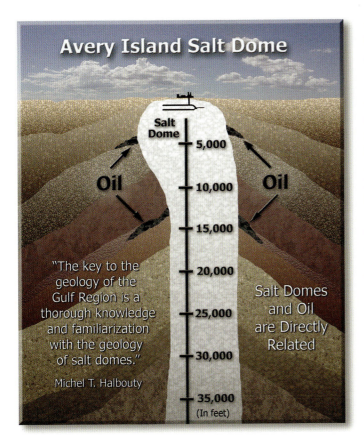

Fig 8.10.3 – There is a direct relationship between crude oil and salt domes, but the relationship's origin is largely misunderstood. Why are there so many oil deposits found on the sides of salt domes but not so often in other areas? If salt domes and oil deposits were really made on flat plains, why were they not mixed together as the salt was "pushed up"? As investigators have said, simple questions must remain unanswered until a "new process" is identified to explain their creation. Here for the first time, a new process is presented combining hydrofountains and the Salt and Oil Models of the UF. In this model, the biogenic origin of both salt and petroleum is explained.

Both the Carbonate Mark and the Salt Mark held evidence of vast, unprecedented microbial activity. Thomas Gold's comment that "microbial processing may have been of major importance in the evolution of the crust" is right on the mark. Each of the next three subchapters will add further corroboration to this view.

The Oil/Gas Debate has never been settled because both sides are in possession of facts that support one theory and refute the other, which leads one to suppose that **portions** of both theories are correct. The abiogenic scientists point to the fact that no large-scale depositions of "zooplankton and algae" are taking place right now and that the "petroleum process" requires heat and pressure, which is also not observable in present-day Nature. On the other hand, geochemists and geologists in support of biogenic theory point to organic markers clearly showing that crude oil is biogenic. In a 1984 article in *Scientific American*, *The Microbial Origin of Fossil Fuels*, we read:

"We have found a striking similarity in hundreds of sediment samples from throughout the world: they all seem to be made up principally of **microbial cell debris**. The compounds in **petroleum are derived from precursors found in the cell membrane of unicellular plankton and of bacteria and other microorganisms that inhabit the sea floor**." Note 8.10e

The UM enters the petroleum origin debate with a completely different twist. By simply taking the actual observations and facts from both sides of the debate, a new picture of Nature's oil puzzle begins to emerge.

First, from the Biogenic Oil Theory side, petroleum deposits exhibit undeniable biologic compounds that are of a deep ocean origin. However, Instead of trying to identify microbes coming from present-day lakes and oceans, which has thus-far proven fruitless, we turn to observations coming from deep oil and gas boreholes. Those observations indicate the petroleum microbes are primarily of a deep-crust origin. Furthermore, the 'petroleum process' is *not active today,* except, perhaps in small, isolated areas, yet unknown. It was during the Universal Flood Hypretherm that conditions were exactly right; high pressure, high temperatures, and the expulsion and proliferation of deep-crust microbial life forms. The convergence of all the aspects of the UF made the petroleum process possible on a grand scale. The UM Oil and Gas Model can be summarized as follows:

<u>The Oil and Gas Model</u>
1. Oil and Gas are microbially derived.
2. The microbes came from deep in the Earth's crust.
3. Petroleum deposits formed during the UF Hypretherm.

There are two other unique observations that support the UM Oil and Gas Model. The first is the relationship between oil, gas, and salt. The other is an observation of an "Instantaneous" Geological Time Frame.

The Oil/Gas and Salt Relationship

Oil, natural gas, and 99% of the world's salt deposits are *directly related*. In fact, petroleum geologists have known since petroleum was first discovered in large quantity that the chances of finding oil were greatly improved if salt was encountered during exploration. The same held true for the discovery of natural gas.

FQ: Why are oil, gas and salt related?

The biogenic origin of salt deposits shares commonality with the oil and gas deposits, and without the understanding revealed in the Salt Mark, this question *could not be answered correctly*. There are actually two keys necessary to comprehend salt's origin:

1. Salt's biogenic origin
2. The UF Hypretherm environment

Michel T. Halbouty, a distinguished geologist and petroleum engineer provided some of the best research and field exploration in the oil, gas, and salt fields during his seven decades in the field. In 1950, Halbouty and his partners drilled 29 wells with only two dry holes, then six years later, he drilled 14 exploratory wells of which 12 were successful. With over 400 articles on petroleum geology and engineering to his name, Halbouty was in a unique position that allowed him to see an important relationship between oil, gas, and salt domes:

"The **key** to the geology of the Gulf Region is a thorough knowledge and familiarization with the geology of **salt domes. More than four-fifths of all oil and gas accumulations (fields) in the Gulf Coast province have been geologically affected by the growth of many domes**…" Bib 144 pxiv

Halbouty was absolutely correct; the key to understanding the geology of petroleum is to understand salt domes!

FQ: Why is an understanding of salt domes necessary to understand oil and gas accumulations?

Answer: Petroleum and salt share a common origin: microbial waste.

The diagram showing the rock layers beneath the Avery Island Salt Dome is seen in Fig 8.10.3. This illustration was first discussed in the Rock Cycle Pseudotheory chapter. Salt domes like this one are found on land and under ocean floor sediment. Accumulations of oil or gas are often situated *directly against* the salt diapir in a curiously *predictable pattern*, not simply because they happened to migrate there.

Oil rigs along the US Gulf Coast seen in Fig 8.10.4 are associated with salt. The drilling process nearly always encounters salt formations, and drillers know the best method for finding oil; "we find out where the salt is, and know that the oil is next door."

In one drilling waste pipe (in Fig 8.10.4), salt water flows into a holding pond from a nearby borehole. Experienced rig operators know how deep and how far from the main salt body they should be to have the best chance of striking oil. For them, the oil and salt relationship is an undeniable fact.

But there was a time when geologists thought it was impossible to find oil in such locations. Halbouty explains this in his history of the most famous oil discovery of all time—Spindletop:

"**The discovery of oil at Spindletop changed the thinking of geologists, engineers, chemists, and economists the world over** because it proved that vast quantities of oil could be produced in a short time from a single source. The age of liquid fuel was born at Spindletop. The United States, as a result of the discovery, rapidly surpassed Russia, the largest oil country of that day, in oil production and in proven reserves. Spindletop was a symbol of the economic might of a fledgling industry—portrayed by a towering gusher of oil. Pattillo Higgins, the man of the hour at Spindletop, with his associate Anthony F. Lucas, **discovered oil, in a place where the best geologists said it was impossible**. A new era of geological thinking was born which, once born, remained unchanged for three decades." Bib 144 pxv

The Lucas Gusher of 1901, Fig 8.10.5, lasted for nine days, blowing out over four million gallons of oil before being brought under control. As with most paradigm-challenging ideas, there were plenty of skeptics around, many of who were the "learned professionals" who looked at the drillers as being "foolish," "destined for certain failure," as Halbouty continues:

"Skeptics were plentiful at the time Columbus 'Dad' Joiner began drilling for oil on a non-existent geological feature which he termed 'the Overton anticline.' In reality the Overton anticline was a nondescript topographic feature devoid of subsurface expression. **Learned professionals surmised and believed that oil was to be found only on structural highs. Scientists accepted the anticlinal theory of accumulation as dogma and they publicly stated that any deviation from its principle was not only foolish but destined for certain failure**. Fortunately, Joiner did not heed the advice of the learned explorationists of his day. It is now evident that he did not know the correct geologic picture but he believed strongly that his efforts were right. This he proved by the discovery of the **East Texas field, the largest oil field on the continent** and one of the largest purely stratigraphic oil accumulations yet found in the world." Bib 144 pxv

What lessons can we learn from the past? First and foremost, we should be willing to question everything in science with an open mind, realizing that modern science has been in a Dark Age for the last century. We should also be able to observe processes in Nature such as the natural production of oil, *if* those processes are actually occurring. If the theories are true—the test of time will bear them out, but this has not been the case with either oil or salt deposits.

Both are undeniably connected in their origin because of the fact that they are often in such proximity. Salt plugs are simply remnant hydrofountains filled with salt prethermate from mi-

"More than four-fifths of all oil and gas accumulations (fields) in the Gulf Coast province have been geologically afftected by the growth of many [salt] domes..."
Michel T. Halbouty

Oil and Salt Deposits are Directly Connected by their Similar Origin

Fig 8.10.4 – Oil derricks and drill rigs like these are found along the Gulf Coast in the United States. Seasoned drillers look at salt as strong evidence for oil when exploring for productive deposits. The drill operator said that they knew they were near the oil when they started drilling through the salt. The oil and salt connection is an interesting method of predicting where the best drill sites for oil will be, but this important connection has never been fully evaluated in modern geology because geologists lack the knowledge of the biogenic origin of **both** salt and petroleum. Note the salt water being discharged from a pipe carrying waste pumped from a nearby borehole.

CHAPTER 8 THE UNIVERSAL FLOOD MODEL

crobial waste, why not the same for petroleum?

Instead of millions of years of burial as the textbooks claim, could there be a demonstrable natural process of petroleum production that takes place over a short time frame? All one need do is look with the new paradigm of the UF.

An "Instantaneous" Geological Time Frame

The uniformity based geological time sci-bi is the most called upon 'crutch' of modern science, especially when it comes to explaining the unexplainable. It is almost impossible to find scientific literature not beholden to this crutch, and it is no surprise to see petroleum research routinely refer back to geological time. However, just as we saw with salt and as we will see shortly with coal, long periods of time are not required for the formation of oil and gas by natural means. Today, deposits of *in-process* crude oil in different stages of formation are not found anywhere, but laboratory production experiments have produced some interesting results. In one case, in 1984, oil and gas were produced in the laboratory in a matter of a few years, by subjecting source material to high temperatures:

"Consequently, we have heated potential source material from **100 to 400 °C over six years**, increasing the temperature by 1 °C per week." Note 8.10f

The resultant crude oil appeared to be "indistinguishable" from some natural crude oil:

"After four years, a product **indistinguishable from a paraffinic crude oil was generated** from a torbanite, while a brown coal gave a product distribution that could be related to **a wet natural gas**." Note 8.10f

As time passed, researchers began to realize "the importance of *water* in laboratory experiments designed to simulate natural processes." One such researcher was M. D. Lewan, who in 1993 wrote an important article titled, *Laboratory Simulation of Petroleum Formation*. In that article, Lewan introduced the subject by noting the importance water had played in "granite melts," "coal formation," and various technological oil processes, and the following:

"Prior to 1979, organic geochemists **inadvertently ignored** these observations and **the ubiquity of water in sedimentary basins when considering the natural process of petroleum generation**. A notable exception is the work reported by Jurg and Eisma in 1964. Noting differences in the thermal decomposition of behenic acid in the presence and absence of liquid water, these investigators suggested that **water played an important role in petroleum generation**." Note 8.10g

Most of the experiments Lewan observed were conducted at 350 °C, over 72 hours. He noted that water in liquid phase in contact with heated source rocks produced "a free-flowing liquid oil similar in composition to natural crude oil":

"Hydrous pyrolysis is a laboratory technique used to **simulate petroleum formation**. The technique **maintains a liquid water phase in contact with potential source rocks while they are heated at subcritical water temperatures**. If the proper experimental time and temperature conditions are employed, **a free-flowing liquid oil similar in composition to natural crude oil is generated and expelled from the rock**. The expelled oil accumulates on the surface of the water, where it may be quantitatively collected at the end of an experiment." Note 8.10g p439

The temperature, pressure, and water that produced free-flowing oil similar to natural crude are characteristic of a **hypretherm!** One essential problem with both the biogenic and abiogenic oil origin theories is that neither acknowledges the importance of water in the process. Like many areas of modern geology, ignorance of the Hydroplanet Model and its abundant crustal and sub-crustal water prevents an accurate understanding of biogenic and abiogenic geological deposits formed during the Flood.

Lewan finds himself in this unfortunate spot as he explains the "inability" of anhydrous experiments to produce oil "in a manner analogous to that in the natural system":

"Petroleum formation may be defined as hydrocarbon generation within and expulsion from a source rock. **The major problem** with anhydrous pyrolysis experiments (i.e., **no added water**) is their **inability to generate and expel an oil-like pyrolysate in a manner analogous to that in the natural system**." Note 8.10g p421

The result of the laboratory experiments showed that geological *time* **is not** necessary, but that water **is** needed to produce oil. A recent discovery off the coast of California was reported by Bernd Simoneit, a geochemist in an article titled, *Hydrothermal Alteration of Organic Matter in Marine and Terrestrial Systems*. In the article, he said the conventional geologically slow process of petroleum formation is challenged with a new natural process creating an "instantaneous" geologic time frame:

"**Conventional petroleum formation is a geologically slow process** tied to basin subsidence and organic matter maturation. In the case of **hydrothermal systems**, organic matter maturation, petroleum generation, expulsion, and migration

Fig 8.10.5 – On January 10, 1901 the Lucas Gusher at Spindletop spewed 100,000 barrels of oil a day for nine days before being controlled. This well was drilled in an area where "the best geologists said it was impossible" to find oil. Most geologists then, as now, were constrained by their inability to escape the dogma of modern science, and so remain ignorant of the true origin of the world's petroleum.

are compressed into an **'instantaneous' geologic time frame**. At seafloor spreading axes, hydrothermal systems active under sedimentary cover (e.g., Guaymas Basin and Escanaba Trough) generate petroleum from generally immature organic matter in the sediments. **The hydrothermal petroleum migrates rapidly away** from the high-temperature zone, usually upward, and leaves behind a **spent carbonaceous residue**." Note 8.10g p414

Simoneit calls this newly identified process that compresses petroleum formation into an "instantaneous geologic time frame" a "hydrothermal oil generation process," which is analogous to the hyprethermal process active in underground aquifers during the Flood. The word 'hyprethermal' is an important UM addition because '*hydro*thermal' neglects to account for the *pressure* that was present on or under the ocean floor. It was *pressure*, along with the other hypetherm factors, that caused the "spent carbonaceous residue" Simoneit observed to remain after the petroleum migrated away. This residue is similar to the carbonate deposits found today on the continents. Simoneit continues:

"In general, **hydrothermal oil generation processes differ significantly from the conventionally accepted scenario for petroleum formation** in sedimentary basins, where organic matter input, subsidence, geothermal maturation, oil generation, and oil migration are discrete successive steps that occur **over a long period of geologic time**. In **hydrothermal petroleum formation**, several of the steps of oil generation occur **almost simultaneously**, and the oil-generating process has been shown to be completed in **short periods of geological time**." Note 8.10g p414

Here is *direct empirical evidence* that petroleum can be formed in "short periods of geological time" when subjected to a "hydrothermal oil generation process" instead of "the conventionally accepted scenario for petroleum formation" that presumably happened over a long period of geologic time.

The investigators had stumbled upon a small pocket of present-day hyprethermal activity that once happened on a global scale, during the Universal Flood. Although they may not have understood the significance of their discovery, or what it would later confirm, they recognized that "hydrothermal oil generation should be evaluated thoroughly":

"**Hydrothermal oil generation should be evaluated thoroughly** to document its occurrence and implications, and to obtain a better understanding of the geological constraints for the process. **This approach would open potentially new and, up to now, unconventional oil exploration targets**." Note 8.10g p414

The Oil and Gas Mark is only a small portion of the paradigm change geology will experience when it comes to the realization that the Universal Flood was an actual event. When it does, many fields will be forever changed, including the exploration of oil and gas. When the importance of the 'hydrothermal' discoveries can no longer be ignored, science will experience a crisis which will be the catalyst of a scientific revolution.

8.11 The Coal Mark

Coal, like petroleum, plays a critical role in modern society, primarily as an energy source. Unfortunately, modern geology is unable to adequately answer the most basic questions concerning its origin. In a manner similar to synthetic quartz production, it was the *technologists* who figured out how coal was

Hyprethermal Oil Experiment

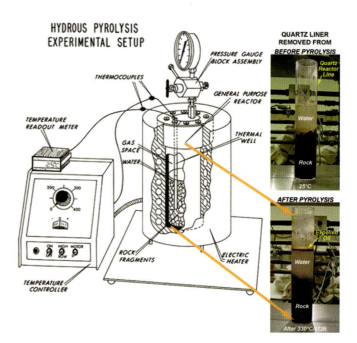

Fig 8.10.6 – Oil production experiments conducted by the USGS involving high pressure and high temperature resulted in oil similar to natural crude oil found in the ground. The oil was obtained from source rocks after only 72 hours under pressure, at 330° C in water. Researchers note that this new type of hydrothermal (hyprethermal) "oil generation should be evaluated thoroughly." (http://energy.er.usgs.gov/gg/research/)

really formed, yet this discovery remains seemingly unnoticed in geological circles because it does not fit the deeply ingrained coal theory dogma. It was discovered that coal is formed at an elevated temperature in an acidic environment that exists nowhere today, except in incidental settings. Now, however, the UF environment provides clarity and wisdom to answer the mysteries first laid out back in the Rock Cycle Pseudotheory chapter, and to show how and why coal was formed only in the upper crust of the Earth.

The Coal Mystery Unveiled

Those of us that grew up in the 1960s may remember *Jonny Quest*, the popular television cartoon adventurer. He shaped children's thoughts about dinosaurs dying millions of years ago in murky swamps to become coal. Even today, dinosaur artists and movies depict dinosaurs dying in swamps that will eventually turn to coal. In reality, we know that coal is comprised almost exclusively of plants due to the abundance of fossilized plants found within many deposits. See Fig 8.11.1 for an example. But what is the process scientists suppose happens as dead plants become coal?:

"Coal is formed from the remains of plants that have undergone a series of far-reaching changes, turning into a substance called **peat**, which subsequently was buried. Through **millions of years**, the Earth's crust buckled and folded, subjecting the peat deposits to **very high pressure** and changing the deposits into coal." Bib 130 p157

This answer comes from *The Handy Science Answer Book*, compiled by the Science and Technology Department of the Carnegie Library of Pittsburgh, but it is typical of the general idea, where "millions of years" and "very high pressure" are required to change peat deposits into coal. However, this is clearly not the case, as we will now discover.

The Coal Mystery in the Rock Cycle Pseudotheory chapter established that:

1. Coal formation is not in process today.
2. The Four-Stage Coal Formation process is a pseudotheory.
3. The origin of peat remains a mystery.
4. The "Knife-Sharp" Sediment Contact between coal and other rock is a mystery in geology.
5. Many other coal mysteries exist in geology.

To solve the Coal Mystery, we must look beyond the current geological time sci-bi. We begin by conducting experiments in the laboratory.

The Real Origin of Coal

FQ: If coal takes millions of years to form, then coal, **indistinguishable from the real thing,** could **not** be made in the laboratory, right?

From the scientific journal of *Nature*:

"...Winans and his colleagues at Argonne National Laboratory have taken **less than one year** to prepare a thoroughly characterized synthetic coal. **The material they produce is indistinguishable from the real thing** by all the techniques so far applied to it and its synthesis **raises many interesting questions in coal chemistry**." Note 8.11a

This discovery certainly does raise many interesting questions, not just regarding coal chemistry, but questions about modern science in general. If coal "indistinguishable from the real thing" can be produced in "less than one year," then the Identity Principle, which is that identical results come from duplicating natural processes, would suggest there is a similar *process that produced all of the world's coal*. In this instance, the synthetic coal was subjected to a number of analytical procedures to determine how close it was to the real thing:

"Fourier-transform IR, NMR, electron-spin resonance, gas chromatography, mass spectrometry and oxidative degradation were used to analyze the product. **The changes that were observed are similar to those that occur in natural coal production.**" Note 8.11b

For more than a century, numerous attempts at duplicating the coalification process in the laboratory produced varied results. In one experiment back in 1913, researchers produced a coal-like product in *hypretherm* conditions:

"When subjecting cellulose to hydrothermal carbonification (**heat with water under pressure**), these workers [F. Bergius and H. Specht, 1913] **obtained a black, coal-like product**." Note 8.11c

In the 1950s, more comprehensive *hyprethermal* experiments were performed:

"**Natural Coalification was imitated** by subjected original plant materials (wood and some of its components) and a partially transformed substance (peat) to a **hydrothermal treatment**. This was done by heating the basic materials in a vacuum or a nitrogen atmosphere, **in the presence of water in a rotary autoclave**." Note 8.11c p224

Each of these hyprethermal experiments revealed a very important observation:

"If the examination is carried out in water (without additives)—as was done in the experiments discussed above—**the medium rapidly turns acid**." Note 8.11c p230

Researchers had recognized that both artificial and natural coalification processes required an *acidic* medium:

"The investigations into the effect of the hydrogen ion concentration **have shown that the normal coalification has in all probability taken place in an acid medium. In natural, and also in artificial, Coalification such a medium exists owing to the presence of acid decomposition products formed in the primary stage**." Note 8.11c p234

This leads us to ask two simple FQs:

FQ: Where today can we find large **acidic swamps**?

FQ: Where do we find an **acid medium** in boreholes into the Earth's crust?

Fig 8.11.1 - This is a cross section of a core of coal from Bowen Basin, Australia. Fern leafs can be seen in this coal fossil. Many other plants have been identified in coal deposits, including pine cones and large tree logs.

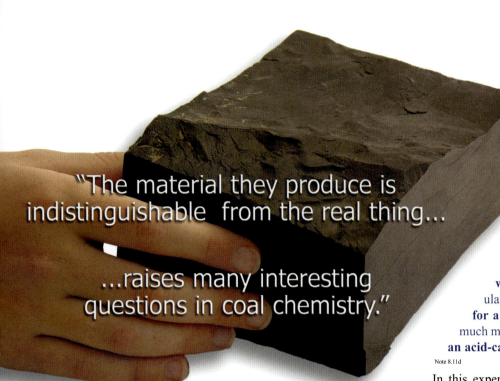

"The material they produce is indistinguishable from the real thing...

...raises many interesting questions in coal chemistry."

The answer to both questions is that we don't. However the Universal Flood event included both acidic and basic environments, which facilitated both salt formation and the coalification process in specific areas of the crust where the acid medium and hyptherm conditions prevailed. The Rock Cycle Pseudotheory explained that modern geology's Four-Stage Coal Formation process has never been observed and has no basis in fact. In that hypothetical process, peat becomes lignite, which becomes bituminous coal, which eventually becomes anthracite. There are three types of coal, but no evidence to support the notion that there is a coalification gradient at work. Instead, there must be other solutions to solve the mystery of coal's formation, and as it turns out, there is; different recipes create different coal types.

One part of the coal-making process involves the separation of the woody materials; the cellulose and lignin. Cellulose is the fibrous substance in wood used to make paper, lignin is the dark oily resin or sap sometimes seen oozing from pine trees. Lignin is easily removed by soaking the wood in a basic solution of sodium hydroxide (NaOH), which, if you recall from the Salt Mark, was a major component of the UF essential in the forming the Earth's salt deposits. NaOH also appears in the Quartz Mark as an indispensable part of the quartz crystal growing process. It is one of the crucial mineralizers associated with processes that created most of the minerals on the surface of the Earth today. But those processes are not happening around the world today, thus, the wide variety of deposits associated with NaOH are each a testament to the authenticity of the UF event.

Once the lignin was removed from vegetation floating in the NaOH rich floodwaters, the lignin did not need pressure to become coal. All that was required was heat and clay! Today's swamps are obviously not covered with great quantities of clay, nor are they heated to over 150°C (302°F), but this is exactly the condition in some of the flooded environments near the end of the UF destruction, as crustal plates began to return to their former heights when acted upon by the outward pull of centrifugal force caused by the Earth's stabilized axial spin. Receding waters coupled with copious amounts of heat generated by the friction of the moving crust accelerated the coalification process.

Coalification during the UF took only several months, and in 1985, in laboratory experiments using the same type of processes, synthetic coal indistinguishable from the real thing was produced:

"**The synthetic coal is produced by warming lignins** (highly aromatic molecular components of woody tissue) at **150°C for a few months** in the presence of twice as much montmorillonite clay, which seems to **serve an acid-catalytic role**." Note 8.11d

In this experiment, the clay played an acid-catalytic role, which was critical at the 150°C temperature, but if lignin had been buried by Flood sediment that was deprived of oxygen while the temperature rose to 350°C, clay would no longer have been required for the lignin to become coal. Further testing revealed high temperature pyrolysis, which is the chemical decomposition of organic material without oxygen, can produce coal products:

"In contrast to the low temperature clay catalyzed reaction, **high temperature pyrolysis of lignin at 350-400°C in the absence of clay did not affect its transformation into coal-like material corresponding to lignite or vitrinite**. A similar observation has been previously made from hydrothermal carbonification of lignin (Schujmacher *et al.*, 1960)." Note 8.11e

All chefs know that when the temperature is raised, it takes less time to cook, and that is just what researchers discovered. At temperatures of 300°C and higher, it did not take millions of years, or thousands of years, or even hundreds of days, it took only about *one hour*:

"Interestingly, pyrolysis of **pine wood at 300°C for one hour produced coal-like materials**, which contrasts with Winan's observation using lignin as a starting material. According to H. Retcofsky, who is completing a study of the Bureau of Mine's samples, **pyrolysis for eleven years at 200°C is roughly equivalent to one hour at 290°C**." Note 8.11f

Never before has science considered the possibility that the Earth's crust was subjected to high temperatures for only a limited amount of time, (perhaps only hours or days) which allowed many new geological formations! For the first time, with the light of the UF, we can see how large areas of vegetation were rapidly buried and subjected to elevated temperatures, which formed a variety of coal types. Furthermore, acidic environments with temperatures of less than 150°C, the lignin that was separated from woody tissues accumulated in vast areas of the Earth, becoming low rank coals.

Now contrast this simple explanation of coal formation with

CHAPTER 8 THE UNIVERSAL FLOOD MODEL

modern geology's explanation. Note how the "evidence" does not support "any particular theory of coalification" because the "evidence" did not follow the assumed "natural evolutionary pathway" of coalification:

"However, to the present day, just as was pointed out earlier by Dryden (1956), **there has been no incontrovertible evidence to support any particular theory of coalification**. During the past decades, attempts to reproduce the natural coalification processes under laboratory conditions have been carried out using lignin, cellulose, wood, spores, and very low rank coals. However, most of these early coalification studies **were unable either to demonstrate a relationship to natural evolutionary pathway** or to elucidate the detailed physical and chemical nature of the products." Note 8.11g

Additional researchers also noted that although the traditional coal theory "has never been seriously challenged for lack of contradictory evidence," the evidence used to support the theory requires that humic acids condense within the coal over time. As with other, similar coalification theories, this has never been proven:

"Although it is **difficult to explain the oxidation of lignin under the anaerobic [absence of air] conditions that exist during coal formation, this theory has never been seriously challenged for lack of contradictory evidence**. Unfortunately, the evidence used to support the theory, namely the decreasing abundance of humic acids in coal of higher ranks as compared with that in peat, does not unambiguously indicate that the humic acids are condensing with time."
Note 8.11h

The "natural" pathway to coal is not 'evolutionary,' because low rank coals do not 'evolve' into higher-ranking coals. Even with empirical evidence to the contrary, and the fact that the evolutionary pathway has never been seen or demonstrated, science will continue to hold fast to the pseudotheory taught for over a century because no other explanation with demonstrable evidence has been presented. This same situation occurred with the Evolution Pseudotheory in Chapter 12. Now however in both cases, new explanations supported by actual scientific observations are emerging.

These new observations provide powerful evidence for new explanations, wisdom reveals that truth can be no other way than the way it is. For years, investigators struggled to make coal using cooler temperatures that correspond to current theory, but cooler temperatures failed to produce satisfactory results, like the baker who tries to bake cookies at low temperature; even years fail to yield good cookies. The recipe is critical and so is the temperature. The minimum temperature for coal-making lies well above typical surface temperatures, or the temperature found in today's coal mines.

Furthermore, heat speeds decomposition of organic substances. It is doubtful that buried vegetation subjected to heat of perhaps 200°C for an extended period of time would hold up long enough to produce coal. Amazingly little research on critical questions like these shows up in the journals, probably due to the uniformity paradigm, which is difficult to see past without a completely new approach.

In summary, the realization that coal, "indistinguishable from the real thing" can be formed, we need ask what *should* such coalification experiments lead to? One researcher in *Nature* noted:

"The production of artificial coal **will open a new era of the study of coal formation** by allowing the direct experimental test of hypotheses." Note 8.11i

Incredibly, this statement was made several decades ago, in 1985, yet since that time, not a single study has been found developing this new discovery. The scientist (John Larsen) involved in the discovery responded to questions about the lack of further study stating simply that interest in the field of coalification has waned. He knew of *no further* coalification experiments.

Here was a process about how to make coal indistinguishable from the real thing, yet *no one* has expressed interest in further discovery about how Nature created the most abundant energy

Fig 8.11.2 – Sand and limestone in coal deposits have never made much sense; after all, how did **white sand**, which has no source, become deposited next to coal without any pebbles or river rock? Why was limestone, which presumably came from a sea, left behind without a large amount of salt?

Coal Deposits Only Become Clear in a UF Paradigm

deposit in the United States. One must ask why?

> Could it be that no theory acceptable to modern science fits the actual laboratory evidence?

The Coal/Sphalerite Inclusion Evidence

Thus far, any scientifically minded person would find it hard to deny the evidence produced in the laboratory supporting the idea that coal formation occurred in a hydrothermal environment. Such an environment was present when the waters of the Flood were receding. Nevertheless, if coal came from the UF and not from peat swamps, conclusive proof that *natural coal deposits* are of hydrothermal saline origin should be evident. Today, there are no peat swamps in hot saltwater, but there is a transparent mineral called sphalerite imbedded in some coals. The mineral contained saltwater inclusions, letting researchers know the water temperatures and the chemical makeup of the water when the coal and sphalerite formed. Imagine the head-scratching this must have caused!

> "In some **coal**, however, **sphalerite** is a significant constituent, and as it is a transparent mineral, fluid-inclusion studies on it may yield information on the conditions of its origin. Leach (1973) studied inclusions in sphalerite from coal mines in central **Missouri**, USA, and found Th [upper temperatures] = **80-110°C and strongly saline brines** (>22 wt % NaCl equivalent). As the sphalerite from the adjacent northern **Arkansas**, USA, **zinc district** had similar fluid inclusions (**83-132°C** and > 22 wt % NaCl equivalent). Leach suggested that **these two mineralizations formed from a single episode of fluid flow.**"
> Bib155 p330

Besides being formed in a saline hydrothermal environment, the two formations—from Missouri to Arkansas—were apparently *"formed from a single episode of fluid flow."* What would cause an accumulation of dead vegetation of such magnitude to become a coal deposit from a single episode of saline hydrothermal fluid flow? The mineral sphalerite is throughout the coal, and:

> "Fragments of coal are found **embedded in the sphalerite**."
> Bib155 p330

Thus, sphalerite and coal formed in the *same* environment, evidenced by the fragments of coal embedded in the sphalerite mineral, which was itself an inclusion in the coal. The Coal/Sphalerite Inclusion Evidence is not limited to the coal deposits. Research on *zinc ore deposits* 150 km to the north revealed the same sphalerite saltwater inclusions, indicating a "basin-wide event, making it evident that large-scale fluid movements must be involved":

> "A **connection** might exist between the fluids depositing this sphalerite and those forming the sphalerite in the Upper Mississippi Valley Zn [**zinc**] deposits ~**150km to the north**. The Th [upper temperatures] values fond in the Upper Mississippi Valley Zn [zinc] deposits by earlier workers range from **75-121°C**... These fluids were therefore in the same temperature range but somewhat **more saline** than those in the coal beds. As pointed out by Hatch et al. (1976), from the color banding, the deposition of this sphalerite was **basinwide event, making it evident that large-scale fluid movements must be involved.**" Bib155 p330

This certainly seems to describe the UF event! Basins are *surface* depressions, so how does the geologist account for rock or ore formation in an *underground* 'basin'? Although researchers noted "large-scale fluid movements must be involved," they had no idea the scale of the event, and did not know where the saltwater and heat originated. The UF model explains all of this.

From Peat Mystery to UF Peat

Another of the Rock Cycle Pseudotheory chapter mysteries was the Peat Mystery, in part because it has been assumed by many scientists to be the source of coal. Here we will discuss a few interesting facts about peat and why it could not have been the predecessor of the world's coal deposits. One of the best examples comes from Ireland, where peat has been used for hundreds of years for both domestic and commercial purposes. The Irish peat does not exhibit lush swamp-like plants traditionally associated with coal and peat:

> "There are three peat types found. Woody fen peat with the remains of **pine and birch wood**, peat with abundant remains of **bog bean**, and **reed peat**."
> Note 8.11j

Pine and birch don't normally grow well in swampy areas. Moreover, coal is quite commonly found in seams with flat horizons both above and below the coal. If peat bogs were the antecedent of the coal deposits, they should exhibit similar characteristics:

> "Interestingly, pyrolysis of pine wood at 300°C for one hour produced coal-like materials..."
> *From lignin to coal in a year, Nature, Vol. 314, 28 March 1985, p316*

FQ: Since peat is supposed to be the predecessor of coal, which occurs in flat seams, do 'peat deposits' exhibit the same flatness of layers?

"The ground **beneath the peat mass is not flat. There are depressions and ridges**. The sub-peat mineral soils comprise silty clay soils, silt, weathered and unweathered **glacial till soils, gravel and lake marl**." Note 8.11j

The peat deposits are not flat like the coal deposits and they lie above "till soils, gravel and lake marl," alluvial deposits rarely found in coal deposits.

The Coal Mystery also revealed the fallacy of the theory that lower ranked (soft) coal becomes higher ranked (hard) coal as depth increases. Although this is *not* observed in coal, it does show up in peat layers in the midlands of Ireland:

"Turf cutters divide a raised bog into **four distinct layers**. The top layer of vegetation and peat was known as **top scraw**. This was followed by a layer of **fibrous white turf**, which gave way to the underlying **brown turf**. The bottom layer or **black turf** is the most valuable as it is a long burning fuel." Note 8.11j

The progression of lower energy peat to higher energy peat as depth increases is not evident in coal deposits. Coal requires UF environmental conditions to form; no Universal Flood, no coal. Many examples bear this out, and if we take the time to look, with the UM paradigm, it is clear that millions-of-years old peat deposits are not the predecessors of today's coal.

The Universal Flood model accounts for the "accumulations of vegetable matter" thousands of feet thick and renders the subsequent coalification easily comprehensible. During the Flood, vegetated landscapes in their entirety were washed into deep valleys and buried by massive floodwater sediments and microbial detritus, then heated and baked by earthquake frictional-heating, producing the coal deposits we see today.

Ignoring the Coal Reality

The Coal Mark established that geology has not shown that peat bogs, buried over geological time ever produced any type of coal. Instead, research has revealed that coal more likely came from plant remains that were buried quickly by water and sediment and subsequently heated to high temperatures. The evidence for this, although commonly ignored by researchers, comes directly from the coal deposits and mines, where vast forests that were buried rapidly and *preserved* were uncovered. This only happens in deep water, free of oxygen.

In one example, in Vermilion County, Illinois, USA, coal miners discovered large tree trunks up to 33 meters (108 feet) in length, well-preserved in an ancient forest buried in place by "catastrophic drowning," as reported in the journal of *Geology*:

"The contact between the Herrin Coal and the Energy Shale is everywhere sharp in the central area, **suggesting that this segment of mire forest was abruptly drowned when one or more earthquakes resulted in several meters of subsidence and the formation of the estuary**. A recent analogue may be the New Madrid earthquakes of 1811–1812 that drowned a lowland forest and formed Reelfoot Lake, Missouri (Penick, 1981). **Catastrophic drowning** is supported by widespread and abundant *Lepidophloios* and *Cordaites* crown branches, which show full leaf arrays; both groups commonly shed leaves during natural senescence. Spring-neap cycles suggest a typical sedimentation rate of 2 cm/mo away from tidal channels, similar to those calculated for adjacent sites (Feldman et al., 1993). Although such features are not seen in the thin, plant-rich siltstone beds, **the entire fossil forest assemblage was likely buried within less than two months**." Note 8.11k

Although the researchers correctly note the entire forest assemblage was buried quickly in a catastrophic event, they failed to recognize that a several-meters drop in elevation does not explain the absence of oxygen and microbes, which is evident from the highly preserved fossils. Some fossils retain even very small branches and delicate leaves. The burial would have been rapid, very deep, and associated with high temperatures to enable the conversion of plant material into coal, a process foreign to the imaginary world of uniformity that most investigators are trained in.

Furthermore, the *marine deposits* in coal that have no explanation in the swamp theory, fit well within the UF Model. The *salt content* in the Wisła River in Poland increased significantly during this half of the century because of the increase in coal mining activity. There, coal seams are heavily impregnated with *salt*, and as the salt is washed out of the coal deposits, the salinity of the river is raised.

Another example showing the extent of salt in coal comes from mines in the Collie area and in the Perth Basin of Australia. The brown coal found there is so high in salt content that currently, it is economically prohibitive to mine.

Modern geology attempts to explain why salt and other marine deposits are found in hundreds of areas where coal is located by attributing it to local flooding brought on by earthquake activity. They reason that the coal is relatively near the surface and that in recent times, subsidence caused the forests to be buried. But this does not explain the enormous depth of the coal mines, the global instances of buried forests, the large amount of water necessary to gather massive quantities of foliage covering hundreds of square miles, or the elevated water temperatures, all of which contributed to the vast deposits of worldwide coal.

Coal Inclusion Evidence

Coal in Missouri, USA, had inclusions showing the coal was made between "80-110°C" (176-230°F) and in "strongly saline brine".

The funny thing is, scientists *know how to make coal*! Since this process is known, one would think the debate over coalification is over:

> FQ: Because scientists know how to make coal indistinguishable from the real thing, is the debate over natural coalification over?

Unfortunately, no; from the 1983 edition of *Chemical and Engineering News*, we read from an article titled, *Argonne Scientists Make Artificial Coal*:

"The most widely held **view** of natural coalification—the sum of the geological processes that produced coal—is that organic plant material was transformed microbially to humic materials. These, in turn, were transformed by abiotic thermal processes to lignite, bituminous coal, and anthracite.

"Randall E. Winans, who leads the group from Argonne's chemical division investigating coal structure, says that **as appealing as this view may be, it is no more firmly based than other views. In fact, he maintains, there is no incontrovertible evidence to support any particular theory of coalification. How coal was formed is still under debate**." Note 8.11l

Why would there be *a debate* if the scientists themselves admit "The material they produce is indistinguishable from the real thing…"? In fact, *there appears to be* **no debate** *among modern scientists about how coal formed today*, but it is not because they learned *how* to make coal. Instead, they choose to accept the old uniformity theory of coalification-over-geologic-time, dismissing observable evidence by claiming "How coal was formed is still under debate."

Modern geology knows most coal deposits were buried in oxygen-free, hot, saline waters, but because this does not fit the 'theoretical paradigm' of modern science—the answer has been ignored.

"How coal was formed is still under debate."

The Coal Reality

"The entire fossil forest assemblage was likely buried within less than two months" by "catastrophic drowning" from "one or more earthquakes".

The Sand, Carbonate, Salt, Oil and Coal Connection

Reaching the end of the coal discussion and looking back on the previous Marks of the Flood one can see a clear biological connection between most of these geological deposits. Not only did these vast geological formations come from living organisms, there is no clear evidence that any of them are actively forming today. The geological researcher and scholar, W. G. Woolnough, introduced previously, was one of a few scientists that recognized a portion of the truth regarding these biogenically formed geological deposits. Although he was *incorrect* in his thinking at the time that uniformitarianism would "always form the basis of geological theory," Woolnough conceded that the formation of such vast deposits has "not been observed at all":

"While the DOCTRINE OF UNIFORMITY must **always form the basis of geological theory**, it is **necessary to admit** phenomena have occurred during geological time which have **not been observed at all,** or not in their entirety, during the very limited period of modern scientific investigation.

"Reasons are given for the belief that **major coal measures, major primary salt deposits, major fresh-water series of sediments, and sources of oil deposits are not now being deposited anywhere on the earth**." Note 8.11m

The answer as to why these processes have not been observed should be abundantly clear by now:

> The world's deposits of Sand, Carbonates, Salt, Oil, and Coal were formed in the Universal Flood.

Turning our attention to new UF Geological Maps will help explain why such deposits are not being laid down today, and will show the relationship between them. The maps show a series of integrated pieces of the geological puzzle that fit together surprisingly well.

The UF Geological Map Evidence

On the next pages, a series of geological maps of the United States show relationships never before contemplated. They have been interpreted with a Universal Flood perspective. Each map was adapted from a variety of professional geological sources, which are detailed in the notes section, Note 8.11n.

The UF Geological Maps (Fig. 8.11.3) compare major geological mineral deposits on or near the surface of the Earth. Although they show only the United States, similar associations occur worldwide. Surprisingly, we found no research seeking to compare these important deposits, which comprise the majority of the sedimentary material found on the Earth's surface. Why this is the case is perplexing; if loess had really eroded from larger sand deposits, shouldn't scientists be comparing loess to the sand deposits to better understand the origin of loess? An old 1951 map reference comparing loess and sand was the best resource available for the comparison. The other comparisons did not exist at all, leaving it up to the UM to create overlays to show the relationships.

Because modern science has become so specialized over the last century, researchers rarely step back to look at the 'big picture' and ask the big questions. Most scientists are content to assume the big, foundational questions were answered long ago. This is obviously not the case. As the Rock Cycle Pseudotheory showed, some of the most basic questions about the carbonates, salt, sand, loess, and coal, had no clear or simple answers. The introductory chapters of Universal Model identified why this happened; a Dark Age of Geology instituted a period of ignorance about the Universal Flood and its evidences.

As we compare the UF Geological Maps, the carbonates occupy completely different areas than coal, so much so that the carbonates, identified in yellow, *surround* huge coal reserves with little overlap. Did the ocean waters that made the carbonates just happen to go *around* the coal deposits that were presumably formed during a completely different time? This would require different elevations of the carbonates and the coal deposits—but this is not the case.

The salt and carbonate deposits also display a curious relationship. Supposedly they both formed out of ocean waters at about the same time, so why are they found in completely different locations? Notice that the coal and salt deposits fit next to the carbonate deposits like puzzle pieces.

Sand is another piece of the puzzle that shares a relationship with the carbonate formations. Some of the microbes that form carbonates along the Gulf Coast are in the ocean along with sand today—so why are the ancient sand and carbonate deposits not formed together? The UF Sand Map shows how well sand deposits fit next to the yellow carbonate deposits. Once again, these ancient geological deposits should be more intertwined if they had formed together in a shallow seas over millions of years of geological time. The facts obviously paint a different picture.

If we study the underground rivers and aquifers and consider the UF Hypretherm, we find the answer. A variety of temperatures, pressures, and microbes existed at different locations throughout the great Deluge. These variables created a host of different water environments, each with distinct pH levels, and each supporting a particular type of mineral growth. Some environments encouraged massive algal and bacterial blooms that formed the carbonates, other environments preserved the forests and jungles as coal, or were combined to form the different salt deposits or to precipitate ores in the diatremes. Sand and clay precipitated from silica rich hyprethermal waters, while loess was formed toward the end of the Flood as microbial and silica-saturated waters combined under minimal pressure, generating copious amount of fine loess silt. We find direct evidence of the water environments that existed during the UF by examining the many aquifers that still exist today. These are often aligned with the various mineral deposits shown on the UF Geological Maps. These ancient water systems were responsible for delivering the types of sediment that formed in certain areas of the crust.

One reason simple questions are not being asked about the obvious relationships seen in the UF Geological Maps is that such questions do not fit the *imaginary* geological timescales associated with the false theories of uplift and subduction so deeply ingrained in the minds of the geologists. Such *theoretical* events and long-time scales will be dealt with in detail in the upcoming Age Model chapter, but one thing the reader can be sure of is that there has never been *any* empirical evidence to support them. The imaginary geological time scale was envisioned back in the 1800s, long before there was any 'absolute dating method' so contrived by modern science. But their 'absolute dating method' is based wholly on sediments whose origins remain obviously unknown.

> FQ: Can sediments be properly dated if the origin of **how** the sediments were formed is unknown?

This amazingly simple question is one every geologist must ask themselves. Numerous geological mysteries brought out in the Rock Cycle Pseudotheory chapter can be examined under the new UF paradigm, which sheds new light on these old mysteries. The Marks of Sand, Erosion, Depth, Carbonate, Salt, Oil, Gas and Coal reveal new interpretations of old discoveries and observations that *answer* the mysteries and align **perfectly** with all the other Marks of the Universal Flood, of which there are still five to come.

8.12 The Pyrite Mark

The shiny gold mineral pyrite, long known as the gold of fools because of its deceiving wink and luster is made up of iron and sulfur (FeS_2). This iron and sulfur mineral is the world's most common sulfide mineral. In this subchapter, we will explore the biogenic origin and hyprethermal growing conditions of this mineral, which is another mark of the UF Hypretherm. It is no coincidence that iron grains (easily found by moving a magnet through most sediment) have no origin in modern geology. Their origin is not from larger eroded rocks, but from microbes that lived in elevated temperatures, when pressure allowed the growth of larger grains. Biogenic meteorite evidence and hyprethermal sulfur evidence is included before the chapter concludes, documenting small pyrite framboids and spheres found globally in the sediment along with biomarker evidence and pyrite sun evidence that tie pyrite directly to a UF Hyprethermal origin.

Iron, Iron, Everywhere

Before we begin the pyrite discussion, a short exposé on iron is in order. Pyrite is made up of iron and sulfur (FeS_2). Chapter 6.7 included several fundamental questions about pyrite and iron deposits from around the world. Fig 8.12.1 shows iron magnetite sand, discussed previously and also found worldwide. The Iron Soil Experiment (Fig 6.7.5) demonstrated that iron magnetite sand did ***not*** come from erosion, as is commonly thought, because the surrounding rocks did ***not*** produce magnetite sand when the rocks were crushed and tested with a magnet.

Why do quartz-based sedimentary materials such as sand and clay usually have large amounts of iron (hematite and magnetite) minerals, yet very few larger rocks, from which the sediment was supposedly eroded, contain these iron minerals? This FQ can finally be answered, through the Universal Flood Model, which explains the iron grain deposits (silt, sand or larger) with the three following observations:

1. **Iron grain minerals are a worldwide phenomenon.**
2. **Iron grain minerals have a biogenic origin.**
3. **Iron grain minerals grow only in a hypretherm.**

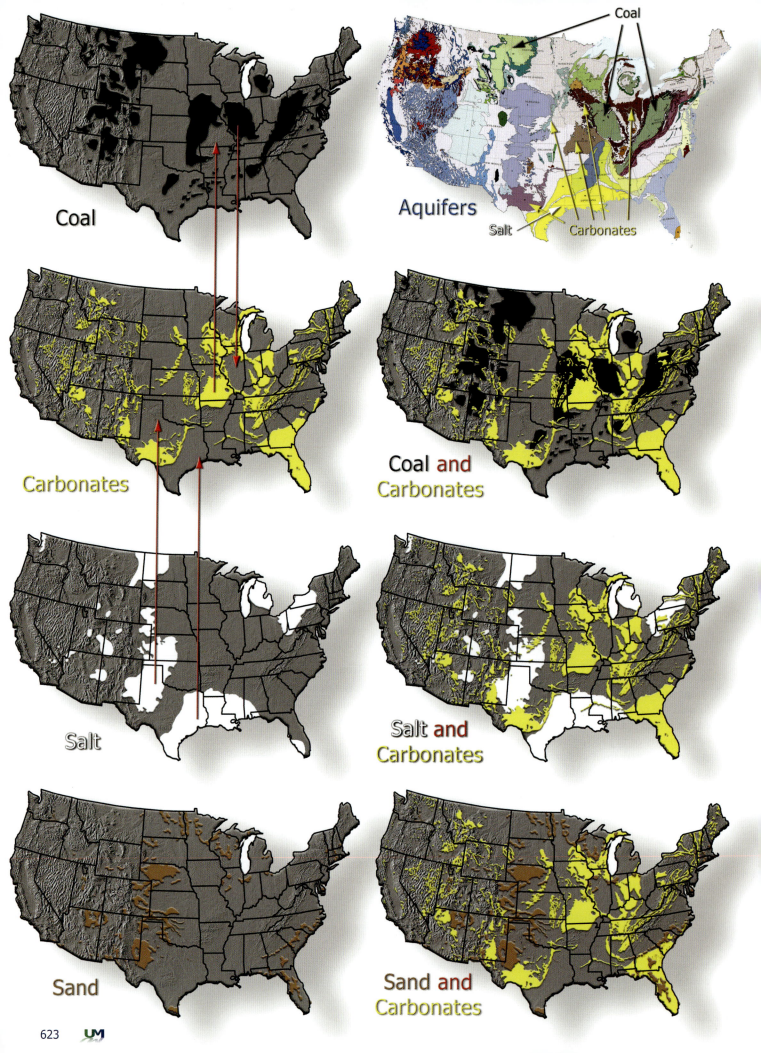

UF Geological Maps

Loess

Sand

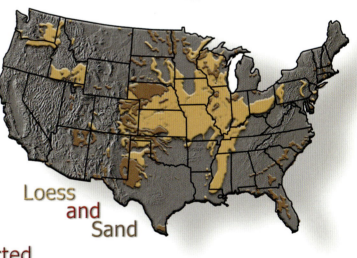
Loess and Sand

Are coal, carbonate, salt, sand, loess and aquifers direct evidence of the UF?

Were the deposits laid down *randomly* at different times and places, over millions of years, or were they laid down *at once*, together, like pieces of a puzzle?

The evidence speaks for itself; maps of different deposits reveal *extraordianry relationships*.

Why are ocean carbonates connected with coal deposits? Why are carbonates and salt in seperate locations if they are both evaporites from shallow seas? Why are sand and carbonate in distinctly different areas when both are found together on present-day seashores? If loess is the product of erosion from larger sand grains, why are loess and sand found in different places?

The UF Model explains how different deposits *are formed independently* in Flood waters, through underground aquifers that exist today.

During the 1980s and 90s, mineralogists made an amazing discovery concerning iron minerals (primarily hematite and magnetite, the two *most common* metal minerals found in the crust). What was this 'amazing' discovery?

Iron grains found in sediment are biogenic—
or in other words, made from bacteria.

This came as a surprise; the black sand children find by running a magnet through loose soil came from microbial organisms! The iron sand grains attracted by the magnet is magnetite; a chief ore of our modern iron world. Just imagine, the iron in the car you drive came from microbes!

Several decades ago investigators made important discoveries about the role of organics in iron sediment. **Biomineralization**, is a new term used to describe the biological mineralization process. Although in its infancy, it promises to yield many important insights, some of which we will explore in the next subchapter, the Ore Mark.

The 1998 edition of the journal, *American Mineralogist* reported that there are two basic ways by which microorganisms make minerals. They either use their body structure in a biologically *controlled* mineralization (BCM) process, like we saw in Fig 8.5.6 as quartz sediment was formed when organic cysts were filled by dissolved silica in a hypretherm; or through a biologically *induced* mineralization (BIM) process, where "metabolic byproducts" are exported "into the surrounding environment":

"Many microorganisms facilitate the deposition or dissolution of minerals. Minerals formed by bacteria are known to be synthesized in two fundamentally different modes of mineralization. In **biologically induced mineralization (BIM)**, the biomineralization processes are **not** controlled by the organisms; mineral particles produced in this manner are **formed extracellularly, have a broad size distribution, and lack a consistent, defined morphology**. **Magnetite (Fe_3O_4) can be formed through BIM** by iron reducing bacteria. These microorganisms respire with ferric iron... under anaerobic conditions to form magnetite. Thus, this uncontrolled **extracellular magnetite formation results from the export of metabolic byproducts into the surrounding environment**. Therefore, external environmental parameters such as the pH and Eh can greatly affect mineral formation." Note 8.12a

These "metabolic byproducts" are the fundamental answer to what created red sand in the sandstone found ubiquitously throughout the world's deserts, as seen in the Sand Mystery of chapter 6. Fig 8.12.2 shows several specimens of the beautiful iron-laced sandstone rock called 'Wonderstone' from the area of Kanab, Utah, USA. This sandstone was discussed in Fig 6.3.19. Geologists have never understood how this rock formed, but now, with the UF model, its origin is clear. Laboratory tests revealed the conditions in which certain iron-producing bacteria grow, but there had been no way to account for hot, pressurized, acidic or alkaline water that apparently covered whole areas of the continents. This bedded sandstone testifies of exactly the conditions that existed when it formed; an environment that generated biologically induced mineralization.

Iron byproducts released into the hyprethermal environment during the Flood also answer another modern geology mystery —how were natural iron crystals like those found in Fig 8.12.3 and 4 formed? Both types of crystals were recovered from soft sediment but showed no sign of having **ever been eroded**, testifying that they are of a *young age and were created in situ—where they were found*. Each of these biologically grown minerals is direct empirical evidence of the Universal Flood and its accompanying hypretherm. It is only under high pressure, high temperature, and in a proper pH environment that supports particular microbes, that large hematite crystals grow.

The second type of biologically controlled mineralization (BCM) produced the *smaller magnetite sand crystals* seen in Fig 8.12.1 (iron sand on the desert floor). We read in the *American Mineralogist* that when the mineralization takes place on or inside the organic vesicle, a "high degree of control over the composition, size, habit, and intracellular location of the mineral" takes place. This produces "*well-ordered crystals*" that have a "narrow size distribution, and well-defined, consistent morphologies":

"Magnetite can also form through **biologically controlled mineralization (BCM)**. In BCM, minerals are deposited **on or within organic matrices or vesicles inside the cell**, allowing the organism to exert a **high degree of control over the composition, size, habit, and intracellular location of the mineral**. Because the intracellular pH and Eh are strongly controlled by the organism, mineral formation is not as affected by

The Biomineralizaton Origin of Iron Sand

Fig 8.12.1 – This photo includes an area of about one square meter of a dry riverbed in Arizona, USA. It contains a large amount of iron magnetite sand. Iron sand has a biogenetic origin and is being formed today only in **hyprethermal environments** deep in the ocean. Magnetite sand is distributed worldwide and is found in many layers of sediment, but how did it get there? The cold shallow sea theory that covered the continents, as modern geologists imagine does not work. Only the deep-hot oceans of the UF Flood provide the answer for the formation and widespread distribution of iron sands.

Fig 8.12.2 – These samples of sandstone come from Kanab, Utah, USA. They are colored beautifully by the iron bacteria that exported "metabolic byproducts" from the surrounding environment. The dark lines are hematite and the red coloring was produced by what researchers call Biologically Induced Mineralization (BIM). Geologists do not recognize that the pure, clear quartz matrix of which the sandstone itself is made was formed at the same time as the iron byproducts, in the UF hypretherm. There has not been another known process which could have produced these global surface sandstone deposits.

external environmental parameters as in BIM. **Magnetite particles** formed through BCM by the magnetotactic bacteria (the subject of this paper), are **produced intracellularly, occur as well-ordered crystals, have a narrow size distribution, and have well-defined, consistent morphologies**." Note 8.12a

We now have a simple answer for the origin of the magnetite in the sand and silt, and the reason it did not erode from larger iron rocks—it was made by iron producing bacteria in the UF Hypretherm. Just like other crystal growth shows, the larger the magnetite crystal, the longer it was growing in the hypretherm. Smaller, colloidal iron may not require pressure, but the size of the crystal is limited to the level of increased pressure. This simple explanation also accounts for why there is "a narrow size distribution" of the sand itself and why the shapes are well-defined and consistent.

Although the biogenic magnetite growth seen by these researchers was quite small, the fact that they noticed its occurrence is tremendously important. Microbes are able to generate magnetite quickly and under the right hyprethermal conditions, the magnetite can be dissolved and grown into sand-sized grains in only a few days. A massive bloom of magnetite-producing microbes could produce an enormous surge of magnetite. This is evident in the global deposits of quartz sand, which contains primarily quartz and an unusual amount of magnetite (see Fig 8.12.5 for an example).

These new discoveries of biogenic iron drastically change the iron-from-chemicals pseudotheory taught since modern geology came to be. In fact, observations showing the role of microorganisms in the deposition of iron and other metals require that *all* "current genetic models *need to be revised*":

"New concepts of **the role of microorganisms in the deposition of iron** and other metals in ironformations and other metalliferous sediments throughout geologic time **suggest that current genetic models need to be revised**. Recent investigations of bacterial activity in surface and subsurface waters have demonstrated the ability of **bacteria to mediate many geochemical reactions**." Note 8.12b

There is *only one* biogenetic model that accounts for both the biogenic magnetite and quartz minerals seen in Fig 8.12.5—the UF Hypretherm model. As was explained in the Rock Cycle chapter, erosion cannot account for the presence of *only* these two minerals in the sand deposit. The magnetite is twice as dense as the quartz grains; other minerals have densities between that of quartz and magnetite and would have been present if the sand had actually come from erosion. The only explanation for these types of worldwide deposits is that they are authigenic—grown in place, and since we know exactly what the environment was for the particular size of quartz grain growth, we can calculate the environment wherein the magnetite grew. Thus, the new biogenetic iron forming model includes the hypretherm.

Recent research helps establish the presence of the hypretherm in which iron sand grains grew. The previous quote and the next four all come from an important summary paper *The Impact of Bacteria on the Deposition of Iron formations* that was published in the *Canadian Society of Exploration Geophysicists* in 2000 by D. Ann Brown and Gordon A. Gross. Through experiments, the researchers made observations confirming the extraordinary claim in subchapter 7.5, A New Geology, that biogenic minerals are responsible for iron deposits, which required the inclusion of a new classification of rock type. This refuted the long-standing theory that iron deposits are "solely a chemical process." In the Magma Pseudotheory paradigm, geology is unable to visualize the massive endobiosphere that exists inside the Earth,

Fig 8.12.3 – This iron ore hematite crystal is 5 cm (2 inches) long. It is called a pseudomorph, and according to modern geology, it formed at the "bottom of a cold sea." Actually, large hematite crystals like this were not created at the bottom of "a cold sea" at all, but are biologically induced minerals created in the UF Hypretherm. Both laboratory controlled experiments and observations at heated tectonic ridges have established the actual environment where this type of mineral grew.

Fig 8.12.4 - These very large, exceptionally geometrical magnetite crystals form *only* in a hypretherm. Geologists have never seen crystals like these being formed anywhere on the Earth today because the very specific growing conditions they require are dormant. Even the limited deep-ocean expeditions have not revealed new crystals being formed. These minerals demonstrate the uniqueness of the UF Hypretherm and invalidate the false uniformity principle.

but as more and more Hydroplanet evidence is revealed, the reality of a living interior Earth is coming to light.

These scientists now recognize that iron minerals come from *biogenic* sources—and that no known *chemical* processes accounting for the production of natural iron has been "demonstrated":

"The reduction of metal ions in the natural environment has been **thought** to be solely a chemical process, but while bacterial reduction of iron by specific reactions has been clearly demonstrated, **geologically significant abiotic [absence of organisms] chemical reduction of iron has yet to be demonstrated**." Note 8.12b

The banding, lamination and interlayering of iron and silicate minerals (see Fig 8.12.2) are now recognized as having a "biogenic origin":

"**Evidence** of such biogenic processes in the genesis of these ferruginous rocks is shown by various microfossils, some stromatolite-like forms, and by relics of primary sedimentary features thought **to be of biogenic origin** such as ovoids, **granules**, oolites, **nodules** and **microbanding. The delicate concentric banding, lamination and interlayering of iron and silicate minerals and the varied shapes of these features are attributed to the important role played by microbial mats and stromatolites in their development**." Note 8.12b p2

Obviously, researchers look to find these types of minerals forming today, and in so doing, recognize that volcanic systems on the seafloor present similar features to the ancient rocks on the continents that contain quartz. These features are formed in a hypretherm, whereas the processes occurring in Iceland and at Yellowstone geysers produce only geyserite, *not* quartz because these systems are only *hydro*thermal, not *hypre*thermal:

"Although these features occur in ancient rocks, **they are also found in rocks that are forming today in Iceland, basins in the Red Sea, and at Santorini and other places in the Mediterranean Sea**; as well as along the tectonic ridges and volcanic arc systems that encircle the globe.

"Furthermore the laminated sediments of Baja California bear a striking resemblance to those of the Archean Swaziland System. Whereas most stromatolite forming today are in carbonate environments, siliceous stromatolites, comparable to those in the **Gunflint Formation chert, are presently being formed in geyserite at Yellowstone National Park, U.S.A.**" Note 8.12b p3

Future research will likely show that some varieties of microbial species require elevated temperatures and elevated pressures to exist, and are therefore found only in the deep sea near seismically active zones in hypretherm environments. It was similar iron reducing bacteria active during the Flood that were responsible for much of the crustal iron deposits and the red coloration of worldwide sediment today.

The investigators also noted that the iron formations were "widely distributed throughout the world" and were found ver-

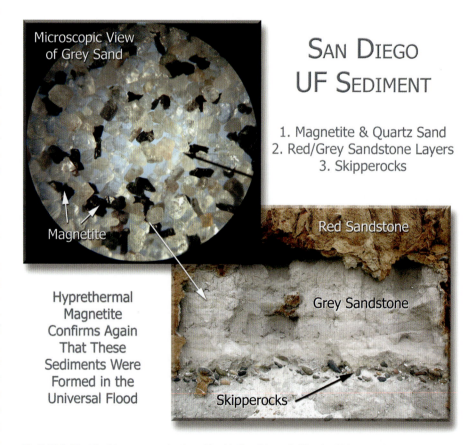

San Diego UF Sediment

1. Magnetite & Quartz Sand
2. Red/Grey Sandstone Layers
3. Skipperocks

Fig 8.12.5 - The black iron magnetite found in this San Diego, California, USA, sand deposit confirms that all of the sediments seen in the lower right photo were formed during the UF. Researchers found that "iron formations commonly occur as **distinct grains** in a cherty or granular quartz matrix" and that today, "many sites of hydrothermal discharge **on the ocean floor** all bear evidence of biological activity in their development." In this example, the iron magnetite grains were formed in a hypretherm, as was the red and grey sandstone. The skipperocks required hydrofountain action to shape them, and to deliver them to their final place of deposition. No geologic explanation other than the Universal Flood accounts for all of these sediments.

CHAPTER 8 THE UNIVERSAL FLOOD MODEL

tically throughout the geological record, occurring as "*distinct grains* in a cherty or *granular quartz matrix.*" More importantly, they noted that today "many sites of hydrothermal discharge on the ocean floor" are biogenic:

"On the other hand, the ironformations and iron-rich stratafer sediments are **widely distributed throughout the world and range in age from Early Archean to Recent**. The oxide, silicate, carbonate or sulphides iron minerals in the ironformations **commonly occur as distinct grains in a cherty or granular quartz matrix. Those forming today at many sites of hydrothermal discharge on the ocean floor all bear evidence of biological activity in their development**." Note 8.12b p3

In summary, it is difficult not to find iron throughout the geological column. Iron deposits of "distinct grains" are being made right now in deep-ocean biogenic hyprethermal areas—in fact, they are being formed *only in a biogenic hypretherm.*

An example of how deep iron reducing bacteria have been found, we turn to a drilling expedition in 1993, where 12 tons of "fine-grained magnetite" was pumped from a Swedish borehole that had been drilled through 6,779 meters of granite:

"**Fine-grained magnetite** was found in large amounts in the borehole: some **12 tons** were pumped up. The formation of such small magnetite grains has previously been associated with **the action of bacteria**..." Note 8.12c

Was the entire sedimentary column associated with these iron grains formed in a biogenic hypretherm?

The Biogenic Meteorite Evidence

Where do meteorites really come from? Just such a question was posed in subchapter 7.10, the Meteorite Model, where fundamental questions about the origin of meteorites and iron ejectites were answered. The true origin of most meteorites has long been an unanswered 'mystery.' Because of modern science's foray into the Dark Age of Science, it was not even aware of the hydrofountain origin of such minerals.

Iron formations within the Earth's crust have already been established as having a *biological origin rather than a chemical origin*, so why would this not also apply to most of the 'meteorites'? In the Meteorite Model, these were shown to be iron ejectites with a biological origin.

Carbonaceous chondrites make up about 3% of the meteorites. Researchers have deduced that these meteorites were formed in water at what they believe were "low" temperatures because they have the same salts in their matrix that is found in the ocean; "epsomite, gypsum, and calcite," as recorded in the book, *Soils of the Past*:

"... **carbonaceous chondrites** [meteorites] also have a matrix of minerals that formed at very low temperatures (0-58°C) **in the presence of water**. This includes a variety of iron-rich smectitic and serpentinelike clays and organic molecules. **Other evidence of hydrous alteration are framboids and plaquettes of magnetite and veins of epsomite, gypsum, and calcite**." Bib 118 p318

The Salt Mark demonstrated that most salts were created in a hypretherm, but researchers were not aware of the higher temperatures and pressures required to make these salts. Moreover, magnetite's biological origin is connected to the origin of framboids, which formed in a hypretherm too—not at "low" temperatures. All of the evidences found in carbonaceous chondrites point right back to the hydrocrater diatreme, which was a natural hypretherm right before it exploded, sending iron ejecta high into the air, perhaps even high enough to break free of the Earth's gravitational tug; orbiting the Earth in unstable, decaying orbits, destined to eventually return to the Earth from whence they came.

No comet or asteroid is believed to have an ocean with salts similar to that of the Earth, yet this was the environment that seemed to birth the carbonaceous chondrite meteorites. Scientists found direct evidence pointing to the fact that these types of meteorites were formed in "soils" with:

"...an overall alkaline pH **like that of desert soils on Earth**." Bib 118 p320

The researcher came surprisingly close, but no known asteroid or comet has soils like the Earth.

The 'meteorites-are-from-space' paradigm makes the researcher unable to see one obvious potential of the chondrite's origin—*that meteorites are rocks that formed in the Earth's soils along with so many other types of iron rocks*. Knowledge of hyprethermal rock and crystal-forming processes and the wisdom of the Universal Flood mechanism is the solution to the meteorite mystery.

Researchers that study all types of rocks and meteorites, found convincing evidence that carbonaceous chondrite meteorites did not come from gaseous nebular dust (accreted solar system) as is popularly theorized, but from "water in a soil":

"These [structures in carbonaceous chondrites meteorites] are **all more like alteration of material by water in a soil than individual reactions between dispersed gas, liquid, and dust**." Bib 118 p319 Note 8.12d

The Hyprethermal Sulfur Evidence

With evidence that natural iron sediment was formed in a biogenic process and not a purely chemical process as was once thought, we turn our attention to the other element in pyrite, to discover the processes involved in producing it. Natural sulfur compounds give off an unpleasant odor reminiscent of rotten eggs, garlic, or skunk odor. Like iron, sulfur is of organic origin and is dependent on microbial processes far more than is currently realized by modern science. Sulfur plays a critical role in all living organisms, and as will be discovered in the Human Model chapter, it is essential to human life.

The mineral form of elemental sulfur of any quantity is found *almost exclusively* in one of two natural environments:

"Elemental sulfur can be found near **hot springs and volcanic regions in many parts of the world, especially along the Pacific Ring of Fire.** Such volcanic deposits are currently mined in Indonesia, Chile, and Japan... Sicily was a large source of sulfur in the Industrial Revolution.

"...**Significant deposits in salt domes occur along the coast of the Gulf of Mexico**, and in evaporites in eastern Europe and western Asia." Note 8.12e

Significantly, both sources share environmental similarities—hydrothermal or hyprethermal conditions. This is not by chance, because sulfur-producing microbes thrive *only* in these environments. The Salt Mark established the hyprethermal biogenic origin of salt, which is also the origin of the sulfur mineral deposits on the cap rocks of the domes and plugs.

A connection between the Hydroplanet Model and the Uni-

Carbonaceous Chondrite "Meteorites"

Evidences For Earth Origin

1. Formed in Water
2. Framboids
3. Magnetite
4. Gypsum
5. Calcite
6. Alkaline pH

Fig 8.12.6 - This carbonaceous chondrite meteorite fell in Chihuahua, Mexico on February 8, 1969. Planetary science assumes all meteorites that fall from the sky are **not** from the Earth, but the physical evidence suggests otherwise. Rocks of this type are found near hydrofountains, and like the meteorites documented to have fallen from space, are known to have formed in a watery environment. They may contain pyrite framboids, magnetite, gypsum, and calcite and they may have an "alkaline pH like that of desert soils on Earth."

versal Flood Model is *hyprethermal sulfur*. Mount Pinatubo's eruption in 1991 was the largest volcanic explosion of the past century. Like the Mount Saint Helens eruption, it produced far more water than researchers realized because it was a hydrovolcano. The third most abundant gas behind carbon dioxide and steam was sulfur dioxide (SO_2), which provided what USGS geologists termed 'excess sulfur':

"The climactic June 15, 1991, eruption of Mount Pinatubo injected a minimum of 17 Mt (megatons) of SO_2 into the stratosphere--**the largest stratospheric SO_2 cloud ever observed**. This study is an investigation of the immediate source of the sulfur for the giant SO_2 cloud. Approximately 100 electron microprobe analyses show no significant differences, at the 95 percent confidence level, in S or Cl contents between glass inclusions and matrix glasses of the erupted dacite. These results indicate that there was no significant degassing of S or Cl from melt during ascent and eruption. Furthermore, the 17-Mt SO_2 **cloud contained over an order of magnitude more sulfur than could have been dissolved in the quantity of erupted silicate melt at the pre-eruption conditions. A major source of 'excess sulfur' is therefore required to account for the SO_2 cloud. Degassing of melt in non-erupted dacite as a source of the excess sulfur implies volumes of non-erupted dacite larger than the estimated volume of the magma reservoir beneath the Mount Pinatubo region.**" Note 8.12f

This article was published in the 1990s, just as researchers were beginning to discover present-day hyprethermal environments on the bottom of the oceans. These thriving ecosystems would later be the key to the excess sulfur, but before that, researchers did not know there was *not* a "magma reservoir" below Mount Pinatubo. They assumed the heat was from magma, but reached a surprising conclusion about the eruption through indirect calculations of sulfur dioxide and heated water inside the volcano:

"It is proposed that the dacite erupted on June 15 was **vapor-saturated at depth prior to eruption**, and that an accumulated **vapor phase in the dacite provided the immediate source of excess sulfur** for the 17-Mt SO_2 cloud. Investigations based on exploration drilling for geothermal energy suggest that magmatic volatiles were discharged into **the Pinatubo hydrothermal system** from the vapor-saturated dacite prior to the 1991 eruption. Experimental studies, geobarometer results, and the H_2O and CO_2 contents of glass inclusions **indicate that the Pinatubo dacite was saturated with water-rich vapor before ascent and eruption.**" Note 8.12f

Indeed, the "Pinatubo hydrothermal system" was responsible not only for the largest explosion in modern human history, but for hosting one of the largest colonies of sulfurous microbes ever known to be ejected into the atmosphere. The researchers noted that the volcanic rock, dacite did not contain enough sulfur to account for the huge cloud that was spewed into the air, but they were not aware of the sulfur producing hyprethermal system inside Pinatubo. That system is analogous to the hot water vents on the ocean floor, like those at TAG Mound. Deep inside the volcano, water, heat and pressure combine to create a hypretherm enabling sulfur and carbon based microbial life, which generate the various gases (SO_2 and CO_2) present during the eruption.

In Nature, the most widely distributed sulfurous mineral is **pyrite**. Commonly known as fool's gold, it was first introduced in the Iron Mystery of chapter 6. There, Fig 6.7.1 discusses the sparkling gold-like metal found in many streams and sediments. Pyrite is scattered abundantly across the Earth in diverse forms, and sometimes is found comprising a whole fossil.

There are two steps in comprehending how pyrite was formed in so many geological settings. The first involves the environment necessary for the formation of microcrystal pyrite; the second is discovering the setting required for large pyrite crystal growth.

The Framboid Pyrite Evidence

The smallest recognizable pyrite crystals cluster in a framboidal shape, seen in Fig 8.12.7, a photomicrograph of tiny microcrystals measuring around 10 microns (1/100,000 of a meter) in size. According to the book *Soils of the Past*, framboids grow in an environment of "waterlogged soils":

"Another product of microbial metabolism is the distinctive sand-sized aggregates of tiny balls of pyrite known as **framboids. These are produced in waterlogged soils**, by reduction of ferric iron, sulphate, and sulfur, coupled with the breakdown of organic matter, by microbes such as *Desulfovibrio*." Bib 118 p183

An in depth article titled *Experimental syntheses of framboids—a review* (Hiroaki Ohfuji, David Rickard) published in 2005, reported on the results of eleven tests from different investigators over a 30-year period. The following quotes are taken from this study, the first of which identifies what framboids are and reveals clues as to why they are important in geology and especially to the UF:

"**Framboids** are microscopic spheroidal aggregates of microcrystals with a distinct internal microarchitecture. They are most commonly developed in pyrite. Indeed, **framboidal pyrite represents one of the most dominant pyrite textures observed in nature** such as in sedimentary rocks..." Bib 147 p147

The study of framboids is critical because as noted, they rep-

CHAPTER 8 THE UNIVERSAL FLOOD MODEL

Natural Sulfur

Natural sulfur has two primary sources - volcanic regions and salt plug cap rocks, supporting a biogenic hydrofountain origin.

resent "one of the most dominant pyrite textures observed in nature." Framboids are not only common in sedimentary rocks, but their *origin* represents the **global environment** wherein many types of sediments were simultaneously created.

Unfortunately, without the UF paradigm, geology has been unable to see the world submerged in hydrothermal waters. Researchers spent 30 years trying to create framboids in alternative environments at *ambient* (room) temperature instead of hydrothermal conditions, leading to 'problems' in framboid research:

"The central **problem** about framboidal pyrite **concerns its origin**." Bib 148 p147

Applying heat revealed new evidence:

"**The aggregate size [of the crystals] was observed to increase in proportion to the heating temperatures and experimental duration**..." Bib 147 p157

Indeed, the largest synthetic framboids ever produced (Graham and Ohmoto 1994) were grown during periods of a few days at the highest temperatures (~350°C).

However, pyrite crystals of this form found near the surface of the Earth are assumed to have been formed at ambient temperatures (about 23°C), because they are so widespread and because there has never been an explanation for how most of the Earth's surface was exposed to elevated temperatures while underwater. Nonetheless, the researchers found the "clearest" correlation between experiment and Nature to be "elevated temperature":

"One of the **clearest and perhaps surprising correlations** between experimental conditions and framboid synthesis is **temperature**." Bib 147 p164

The report continues:

"…the syntheses of large, well-developed framboids **analogous to natural equivalents have been performed consistently at elevated temperature**. Indeed, the **only** synthetic framboids with an organized internal microarchitecture were synthesized by Graham and Ohmoto (1994) **at 150-350°C**. Sunagawa et al. (1971) synthesized large framboids at **200-300°C**. Sweeney and Kaplan (1973) used the lowest temperatures **60-85°C**. Framboid syntheses **at ambient temperatures** have generally resulted in the formation of **pseudo-framboids** or small framboid-like spherulites.

"The role of temperature is generally assumed to increase the rate of reaction, particularly with respect to pyrite formation. The relationship of this process to framboid formation remains **unclear**. It may imply that **framboid formation requires rapid reaction kinetics [high temperature] and that framboids develop naturally where reaction rates are high**." Bib 147 p164

By changing our paradigm to *expect* higher temperatures and reaction rates, framboid formation goes from "unclear" to crystal clear. Furthermore, the best and largest framboids were produced at temperatures over 100°C (300-350°F), which *required elevated pressures* to prevent the water from boiling and disrupting the growing environment. Therefore, the Framboid Pyrite Evidence is direct empirical evidence of the UF Hypretherm for these reasons:

1. Framboids are a global phenomenon.
2. Pyrite framboid crystals require high temperature, pressures, and water to form.
3. Framboids are found throughout the 'so-called' geologic column.

Although this is self-evident, it must be made clear that because the Universal Flood took place within the past several thousand years, the entire geological rock time scale must be revised to include the many empirical evidences shown in this chapter, including the framboid testimony.

Framboids produced in the laboratory were comparable to those found "in natural sediments" following the UM Identity Principle—identical results come from duplicating natural processes found in Nature. The researchers said:

"These textural and structural attributes of the **synthetic framboids are comparable to those displayed by common pyrite framboids observed in natural sediments**." Bib 147 p158

If framboids, equivalent to natural forms, *did not form at ambient temperatures*, but only in conditions equivalent to the UF Hypretherm, one should come to expect problems trying to form framboids *at* ambient temperatures:

"In the meantime, the formation of large framboids **equiv-**

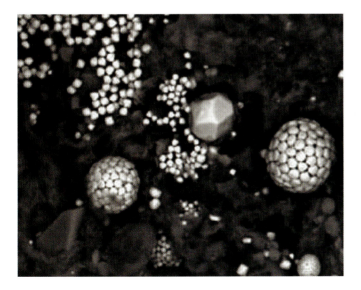

Fig 8.12.7 – The smallest crystals in this SEM photomicrograph are pyrite crystals which join to form framboids. The larger crystal form in the center is also pyrite and is called a pyriteohedron. Pyrite deposits are found worldwide in a variety of forms. It is replicated in the laboratory setting using high temperatures and pressures not found on the Earth's surface today. Their abundance and global distribution testify of their formation during the Universal Flood.

alent to natural forms at ambient temperatures remains a challenge to the experimentalist." Bib 147 p164

This is an example of the UM Hard to Prove Principle—where trying to force 2 x 3 = 7.259 will always result in 'challenge' to the research. With the UF, the answer '6' is simple.

As you have been reading the Universal Flood chapter, you may have noticed the Marks of the Flood are remarkably inter-related. This is the way both truth and arithmetic work. For example, if 2 x 3 = 6, then 6 divided by 2 will always equal 3. The subchapters in this book each reveal a piece of Nature's puzzle that interconnects with other pieces, providing the most complete picture of Nature yet seen by science. The Pyrite Mark is one of those intriguing pieces. The framboid researchers found further confirmation when hyprethermal conditions (300°C) produced magnetite, anhydrite and elemental sulfur by simply adding calcium carbonate ($CaCO_3$) to the solution:

"**Framboidal pyrite was successfully synthesized** in the experiments where **$CaCO_3$ was added** to the initial solution to maintain the pH at 6-6.5. The solid reaction products from these runs were composed of pyrite **magnetite, anhydrite (or gypsum)** and unreacted **elemental sulphur**." Bib 147 p156

When $CaCO_3$ was excluded from the experimental processes, completely different minerals were formed:

"The experiments, performed **without the addition of $CaCO_3$**, produced **pyrite**, **marcasite** and **hematite** displaying rose-flower like aggregate shape." Bib 147 p157

In all cases, hyprethermal conditions were essential. Because there are so many possible combinations of elements, an important question is to ask why certain minerals are so common and widespread:

> FQ: Why are certain minerals, like pyrite, magnetite, and hematite so common and widespread throughout the sediment layers?

Iron minerals like pyrite, magnetite and hematite are common in the crust because they are primarily *biogenic* in origin and are created by only certain species of bacteria. They are commonly found together, lending support for the Law of Paragenesis set forth in subchapter 7.5. The Law of Paragenesis identifies one reason why so many mineral groups occur *only* with certain other minerals. The microbial species responsible for certain metallic minerals *require* specific hydrothermal or hyprethermal environments to thrive, which is why we do not find them actively growing today on a global basis, even though the Earth's sediments *contain them in great abundance*. This fact certifies that they were all produced globally and concurrently in the not too distant past (see Fig 8.12.8).

The Carbonate Mark showed that common algae blooms could produce the $CaCO_3$ (calcite) in the limestone layers when in the presence of a hypretherm. The $CaCO_3$ producing algae and other bacteria were obviously mixed throughout the UF waters, stimulating the growth of framboids, magnetite, anhydrite, elemental sulfur, and a host of other minerals during the UF Hypretherm.

The Pyrite Fossil Sphere Evidence

In addition to pyrite framboids, pyrite spheres and other shapes are common in many sediments. The following discussion about **pyrite fossil spheres** is a continuation of the research presented in Fig 8.5.6, where Homegrown Quartz was demonstrated to have crystallized inside organic cysts, which is responsible for the origin of at least half of the sediment in the rock record. This discovery should have led to a monumental shift in the geological paradigm, yet it gained little attention because the discoverers didn't realize the environment quartz grows in. Instead of shallow cool seas, deep, hot water was needed, not only for quartz crystallization, but for the growth of pyrite and most other minerals. This can be seen at the TAG Mound, previously discussed in the Salt Mark. There, pyrite, quartz and anhydrite are all formed together in a present-day hypretherm.

The research on pyrite spheres was conducted several years before the discovery of framboids and recent investigation that showed high temperature and pressure was critical to pyrite growth, especially where larger crystals were formed. Thus, the pyrite fossil sphere researchers were unaware of the hypretherm environment in which the pyrite spheres grew—however, their logical observations of pyrite spheres and minerals associated with them produced several excellent examples of UF Hypretherm minerals. *From the Journal of Sedimentary Research*:

"**Direct observation of cyst cuticle in association with pyrite spheres** suggests that localized bacterial sulfate reduction in *Tasmanites* interior voids led to formation of localized pyrite deposition, in a manner **similar to that described from certain ammonoid chamber settings**." Note 8.12g

Fig 8.12.9 is a fossil ammonite shell that was transformed into pyrite. Just like the smaller cysts, some of the outside of the ammonite shell is still present while most of the interior has been replaced with pyrite. Researchers discovered small grains of pyrite that had formed inside the cysts that held various microbes.

The investigators found clues about the environment in which these pyrite sphere cysts formed by observing and recording other materials in the shale (hardened mud) deposit:

"In the Chattanooga and New Albany shales, enrichments of **pyrite spheres and half-spheres** are typically found in **lag deposits**, where they are mixed to various proportions with **quartz grains, fish bone fragments, conodonts, glauconite grains, and carbonized wood debris**." Note 8.12g p159

Obviously, the pyrite spheres found on continents today were formed in an ocean—the question is, at what temperature and depth were they formed, and when? Fig 8.12.10 shows even larger specimens of pyrite along with plant and animal fossils found outside the city of Dallas, Texas, USA. These were formed together under a hyprethermal ocean. Pyrite is a unique, but common fossil preserver that deserves further comment, which we will do in greater detail, later in the Fossil Model.

Researchers looking at today's oceans are compelled, because of their uniformitarian mindset, to see only one answer to this question about temperature and depth—they "appear" to form in shallow cold waters:

"…it **appears** therefore quite possible that **pyretic lags with pyrite spheres formed in shallow water** during seafloor reworking **by storms**, and did not require anoxic bottom waters and internal waves for their formation. Considering that these grains are **as large as 0.9 mm**, twice as dense as quartz, and associated with bone fragments up to 20 mm in size, **the flow**

Fig 8.12.8 - This specimen contains pyrite, quartz, and sphalerite. The UF Biogenic Hypretherm supports the Law of Paragenesis (subChapter 7.5) where specific microbes grow together, producing byproducts that precipitate into crystals like this specimen in a hypretherm. This is one reason why some minerals appear in regular, predictable associations and others do not.

velocities at the seabed should have been on the order of 1 m/s or more." Note 8.12g p163

Although the fossil pyrite spheres are a very common global phenomenon, if they were made in shallow cold seas we would expect to find them in "modern basins":

"We **believe** therefore that the described process of pyrite sphere formation is **uniformitarian** and **will eventually be identified in modern basins**." Note 8.12g p158

Merely *believing*, is of course, unscientific. The Pyrite Framboid Evidence, TAG Mound observations and many other examples *demonstrate* that pyrite (especially larger >1-mm or more) does ***not*** form naturally in ambient temperature waters. Cold water pyrite sphere formation remains an unfounded theory.

The investigators actually had some of the clearest evidence of hyprethermal pyrite spheres growth when they documented the "intergrowth and coprecipitation of ***silica*** *and pyrite* in these cysts":

"A close temporal coincidence or **overlap between pyrite and silica deposition** in *Tasmanites* cysts is also indicated by **intergrowth and coprecipitation of silica and pyrite in these cysts**, and by silica cement within pyretic cyst fills described in this study." Note 8.12g p162

Quartz is the most common form of silica, and as the Magma Pseudotheory chapter illustrated with the Silica Phase Diagram, quartz crystallization larger than microscopic size occurs ***only*** in a hypretherm. Therefore, the fact that both silica quartz and pyrite are found in the cysts *establishes that both minerals* were formed in a hypretherm.

Another important fact the researchers know is that pyrite grains are found throughout the geological sediment column—from the Precambrian to the Holocene:

"The **fact** that comparable **pyrite grains can be found in sediments that range in age from the Precambrian to the Holocene** suggests that early diagenetic pyrite spheres as described here **are much more common in the sedimentary record than currently presumed**." Note 8.12h

Because the pyrite grains are so "much more common" than previously known, and because pyrite is known to grow in a hypretherm, we have a simple FQ:

FQ: Were pyrite spheres formed over billions of years in multiple, intermittent global hypretherms, or were they formed in the UF hypretherm, when most of the sedimentary layers were formed?

The Simplicity Principle dictates that the simplest, correct explanation is the best, and when we recognize that sediment with well-defined pyrite crystals required elevated temperatures and pressures, which are only found on heated fault lines on the ocean floor today, we can see why the UF Hypretherm is not only the simplest, but likely the *only* correct explanation. Modern science has refused to consider a world covered with a hot ocean even once, and it certainly does not conceive of multiple episodes over billions of years.

The Biomarker Evidence

Following several lines of evidences, the Pyrite Mark reveals that most of the iron, sulfur and pyrite minerals in crustal sediments were formed in hyprethermal conditions. Sulfur is a primary ingredient in pyrite, but where did the sulfur originate? We find answers to this question by looking at **biomarkers**, which are "tough organic molecules" that survive the decay of an organism's body to become fossils themselves in sedimentary rocks. Biomarkers were first discussed in the Hydroplanet Model under Impact-to-Hydrocrater Evidence. Peter D. Ward

Fig 8.12.9 – An Ammonite, a nautiloid shelled creature that once lived in the sea became a pyritized fossil. These beautiful specimens are abundant in Madagascar, but are a perplexing mystery to scientists today because there are no fossils becoming pyritized anywhere in the world today. The chambers of this ammonite reveal details similar to the spherical pyrite cysts found globally in most sediment. They are not forming in ambient shallow water conditions today. Laboratory tests have shown that hyprethermal conditions are necessary for the formation of pyritized specimens. How were pyrite framboids and cysts deposited in the sediment layers of the Earth? Pyrite crystallization and sediment deposition are related UF events, directly associated with the UF Hypretherm.

Hyprethermal Pyrite and Fossil Deposit

Fig 8.12.10 – Petrified wood, coral and other fossils lie next to pyrite secretions in clay just outside of Dallas, Texas, USA. Laboratory experiments confirm hyprethermal conditions are required to form pyrite specimens, which is supporting evidence that the same hypretherm conditions that formed the petrified coral, wood, and quartz also formed pyrite. These fossils and other marine fossils like coral and shark teeth were silicified in a hypretherm at the same time, evident by their colocation and shared characteristics. The event that did this was not a mere common occurrence, and it doesn't happen today.

in his article, *Impact From the Deep,* saw them as evidence that mass extinction could be explained by heat and escaping gases *from inside the Earth* instead of impact by asteroids:

"Strangling heat and gases **emanating from the earth and sea**, **not asteroids**, most likely caused several ancient mass extinctions." Note 8.12i

For several years now, Ward and other researchers have been using "biomarkers" to help understand the Earth's history:

"About half a decade ago small groups of geologists began to team up with organic chemists to study environmental conditions at critical times in the earth's history. Their work involved extracting organic residues from ancient strata in search of chemical 'fossils' known as **biomarkers**. Some organisms leave behind tough organic molecules that survive the decay of their bodies and become entombed in sedimentary rocks. **These biomarkers can serve as evidence of long-dead life-forms that usually do not leave any skeletal fossils**." Note 8.12i p68

Biomarkers are important because they help establish past environmental conditions. The biomarkers demonstrate that certain conditions, including temperature and pH were present when the biomarker's parent bacteria were living. Furthermore, the biomarkers establish the enormity of the bacterial event and range, or region in which the bacteria grew. Because biomarkers are found worldwide, it is apparent that the extinction event they are connected with was global in Nature.

From the biomarkers, we discover that sulfur bacteria produced the sulfur that was used in the formation of many sulfide minerals that were formed in anoxic marine environments:

"Among the **biomarkers** uncovered were the remains of large numbers of tiny photosynthetic green sulfur bacteria. Today these microbes are found, along with their cousins, photosynthetic purple sulfur bacteria, living in **anoxic marine environments such as the depths of stagnant lakes and the Black Sea**, and they are pretty noxious characters. For energy, they oxidize hydrogen sulfide (H_2S) gas, a poison to most other forms of life, and **convert it into sulfur**. Thus, **their abundance at the extinction boundaries opened the way for a new interpretation of the cause of mass extinctions**."
Note 8.12i p68

Biomarkers certainly have opened a way for "a new interpretation of the cause of mass extinctions," leaving researchers to scramble for new sources of sulfur and mechanisms to account for the biomarkers. As one might expect, the Universal Flood Model provides the precise environment to explain various areas of the Earth where biomarkers and fossils testify that a once harsh environment existed, when large portions of the oceans were heated. These hot areas contained both acidic and alkaline waters, providing perfect conditions for ocean-wide blooms of sulfur-loving bacteria:

"But the **biomarkers in the oceanic sediments from the latest part of the Permian, and from the latest Triassic rocks as well, yielded chemical evidence of an ocean-wide bloom of the H_2S-consuming bacteria**. Because these microbes can live **only in an oxygen-free environment** but need sunlight for their photosynthesis,

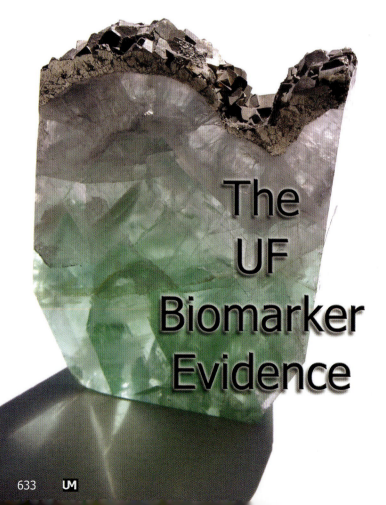

The UF Biomarker Evidence

their presence in strata representing shallow marine settings **is itself a marker** indicating that even the surface of the oceans at the end of the Permian **was without oxygen but was enriched in H$_2$S**." Note 8.12i p70

Meteor impact theories cannot account for oxygen depletion over large areas, whereas the UF model doesn't need to explain surface oxygen deprivation, because bacteria came from anoxic areas of the endobiosphere (beneath the crust) and were released in the Flood. This release contributed to the "ocean-wide blooms of the H$_2$S-consuming bacteria" that is evident in the sediment record.

Another important factor of the "mass extinctions" event is that meteorites do not cause large lava flows. The calamitous events of the Universal Flood included large earthquakes and massive lava flows, and these are evident in the geological record. Researchers see this in the sediments they investigated:

"**Around the time of multiple mass extinctions, major volcanic events** are known to have **extruded thousands of square kilometers of lava onto the land or the seafloor**." Note 8.12i p71

Because the scale of the Universal Flood was so large, major volcanic events the like of which have never been recorded by modern man, along with the apparent mass extinction, would have left obvious marks throughout the Earth's sedimentary layers because they *were all formed during the Deluge*. This is just what researchers have found; biomarkers in more than just the Permian layer:

"**Biomarkers and geologic evidence of anoxic oceans** suggest that is also what may have occurred at the end **Triassic**, middle **Cretaceous** and late **Devonian**, making such extreme greenhouse-effect extinctions **possibly a recurring** phenomenon in the earth's history." Note 8.12i p71

A recurring phenomenon? Researchers suggested no mechanism to explain how large areas of the Earth's oceans and continents became **hot, oxygen-starved, sulfurous cesspools**. In fact, there is no clear explanation that can explain *why* such a harsh environment in which few living things could survive would ever *reoccur* over large areas of the Earth.

By realizing that most of the geological sedimentary columns were formed in the UF, and that the Flood event is recorded by the biomarkers *and fossils* in these layers, we can comprehend why such a harsh environment existed that is recorded throughout the geological column.

The Pyrite Sun UF Evidence

Fig 8.12.11 shows a Pyrite Sun encased in black slate. Because larger pyrite crystals like these are known to grow only in a hypretherm, they help explain how the slate and coal associated with this specimen formed. Pyrite cubes, like pyrite suns, require a unique environment, near the surface that exists nowhere today because it was a one-time UF event.

As each subchapter Mark of the Universal Flood is read, it becomes clear that all of the Marks are physically interrelated. They all exhibit some combination of heat, pressure, water, and an explosion of microbial life, creating a Universal Flood model of a global geological disaster that completely changed the surface of the Earth, not long ago.

As the facts are presented and the evidence weighed, it is astounding to think that modern geology has never even had a word to describe the hypretherm environment of high pressure, high temperature, and water where most of the minerals on the surface of the Earth were formed. The word

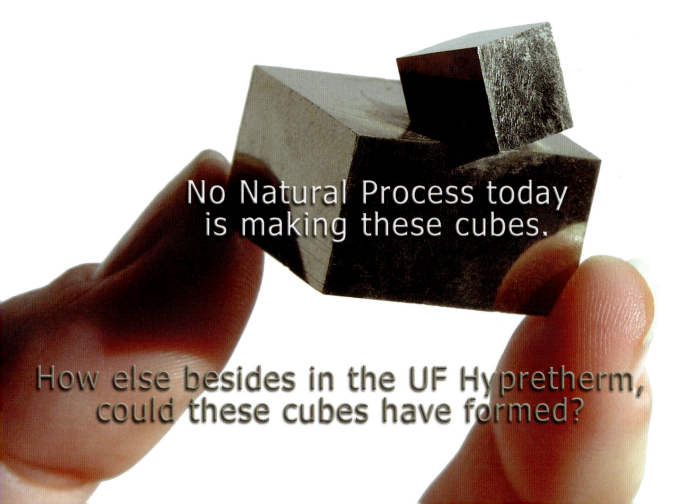

No Natural Process today is making these cubes.

How else besides in the UF Hypretherm, could these cubes have formed?

"hypretherm" should become as common as the word salt—because no rock on the surface of the Earth today escaped the effects of this worldwide environment.

8.13 The Ore Mark

The Ore Mark is the last of six subchapters documenting the *biogenic* hyprethermal nature and formation of the majority of the Earth's minerals. These included the Carbonate Mark, Salt Mark, Oil and Gas Mark, Coal Mark, Pyrite Mark, and finally now the Ore Mark. This subchapter will show how researchers have shifted from the long-held dogmagma origin of ore to a new origin that confirms the reality of the UF.

Biomineralization, a term used to describe biologically formed minerals, explains the origins of uranium ore and its hydrofountain deposition. Fossils found among ore deposits in association with the famous K-T Boundary provide more evidence of the Ore Mark of the UF. These clues also help explain the origin of the oceanic crust.

From Dogmagma Ore to Hyprethermal Ore

In the Rock Cycle Pseudotheory, chapter 6.8, the mystery of certain ore formations had caused some geophysicists to think of ore deposits as being "freaks of nature":

"Minable ore deposits are **really freaks of nature** created by the combination of exceptional factors." Bib 120 p146

The primary reason *why* ore deposits are such an enigma to geology is that their origin has been based in dogmagma:

Fig 8.12.11 – Pyrite Suns were introduced in Fig 6.7.3. They are common in Illinois, USA coal mines. The Coal Mark demonstrated that coal is formed easily in a hypretherm, which is further confirmed by the presence of pyrite, because large agglomorations of pyrite also grew in a hypretherm. The Pyrite Mark shows that pyrite, coal, and the dark slate encasing the Pyrite Sun above all formed in the same environment, at the same time, in a hypretherm.

"Many important **ore deposits** are linked to **magma chamber**, where melted rock collects. As the magma cools and begins to solidify, **heavier minerals begin to sink to the bottom of the chamber**." Bib 151 p52

Despite decade-old direct observations of "ore factories" forming ore through biomineralization in a hypretherm, *not* in "magma chambers," this statement, from the 2006 edition of *The Complete Guide to Rocks and Minerals,* is typical of the modern geology mindset. Why are many scientists and the general public not aware of this? New discoveries in science, *if not deemed significant*, may take many years, if not decades before they trickle into the textbooks and scientific literature. This might seem surprising, especially with the way information is disseminated so quickly today. The problem may lie in the challenge the new discovery poses to long-held scientific paradigms; what then, is science to do?

Being in the Dark Age of Science, geologists have not seen the significance of this new origin of ore, only recently discovered in the 1990s. Probably the only reason some attention has been given to the *new origin of ore* is the economic importance of ore minerals. The discovery and recording of new, natural scientific data (knowledge only) over the last century has been astonishing, driven by the economic and technological pursuits of mankind. Whereas the advancement in wisdom from 'pure science' that would allow us to comprehend how and why Nature *is*, has fallen remarkably short of the mark.

Mineralogists have kept the faith, steadfastly believing that magma exists, and that theoretical 'magma plumes' continue to rise, but the old sci-bi that "heavier minerals" (ore deposits) were brought from the bottom of a magma chamber—is fading. Magma plumes are an appendage of the passing theory of magma, seismic evidence clearly showing that they do not exist (see Fig 5.14.2).

Ore formations, like most other rocks discussed in the Hydroplanet Model, are associated with a *Hyprethermal Process*, and most researchers can see the hydrothermal writing on the wall, having no other way to explain ore's origins. The geophysics who said, "ore deposits are really freaks of nature," also said this:

"**The hydrothermal activity of [oceanic] ridges is a fundamental phenomenon**, the full implications of which are not yet fully understood. It is at the **origin of certain mineral deposits**, mostly of copper, which are **precipitated at the vent of the hot springs [at the bottom of the ocean] and are deposited onto the ridge basalts**." Bib 120 p81

One reason hydrothermal activity at the origin of the ore deposit remains "not yet fully understood" is that ores are a surface phenomenon, in the upper part of the Earth's crust. This indicates the ores were formed in *a short period of time*, which further exacerbates the modern geologic view expressed in the theoretical 'Rock Cycle' that the ores should have been 'pushed' downward as they were covered with millions of years of sediment. This would show layers of ores throughout the geological column—but multiple ore layers extending deep into the crust *do not exist*. Thus, crustal ores were formed in a short period of time, not long ago because they are only found in the upper surface of continental and oceanic crusts.

The discovery of hot vents spewing a chemically rich soup deep in the ocean was an essential step in the recognition of this newfound process. The importance of the hot spring vents on the ocean floor known as 'black smokers' is expressed by one researcher in a 1993 journal article:

"Each year, we learn more about the nature of volcanism beneath the oceans, from side-scan sonar and submersibles. **It is hardly likely, though that any future discovery will match in importance that of 'black smokers'**, plumes of hot, mineral-rich water jetting from vents along ridge axis. **These have helped to explain the origin of some major economic mineral deposits**, such as the massive sulphide deposits of Cyprus, sources of the **copper** which gave Cyprus its name." Note 8.13a

The reason the discovery of "black smokers" was so important was that it "helped" to *explain* the origin of mineral deposits. For the first time, modern geology had the correct answer to the origin of many of the minerals in the Earth's crust, and it wasn't from magma—it was from water.

Another reason ores do not fit the Rock Cycle Pseudotheory is that they are intimately connected with living organisms. However, living things do not do well in magma; fitting the essential ingredient of life in the formation of ore has been a huge stumbling block for geologists. The only way to reconcile this has been for geologists to assume some kind of "ore-forming hydrothermal fluid" is coming from a "magma chamber":

"Copper (Cu), gold (Au), and iron (Fe) are first concentrated in a sulfide **melt during magmatic evolution** and then released to an **ore-forming hydrothermal fluid** exsolved late in the history of **the magma chamber**." Note 19.13b

The general idea that magma, and "hydrothermal fluid" must occur together to form ore has gained momentum since this statement was published in the journal of *Science* in 2002. Nevertheless, complete comprehension of how the ores of the crust were made will only come with the discovery of the true source of heat and water that made them.

The Hyprethermal Origin of Ore

The technology of the 1980s deep ocean submersibles, described in the Carbonate Mark, began to completely change geology's mineral origin paradigm. In the journal of *Science*, March 27th of 1998 an article, *Origins of Hydrothermal Ores*, begins with the following:

"The **dominance of hydrothermal deposits as major industrial sources of many elements** has stimulated intense study of their genesis for over a century. **Their origin is unequivocally by precipitation from aqueous solutions within the upper several kilometers of the crust.**" Note 8.13c

By 1998, geochemists saw ores as being of "hydrothermal" origin. Although they did not include the critical factor of pressure, they recognized the importance of the **high temperature water environment** necessary for the production of the "major industrial sources of many elements."

This was an unprecedented departure from the dry molten magma theory (without water) James Hutton and thousands of later uniformitarian geologists envisioned.

Even though researchers had not yet recognized the importance of pressure in ore creation, they deserve partial credit for recognizing that the origin of ore is "*unequivocally* precipitation from aqueous solutions." The Hydroplanet Model previ-

Fig 8.13.1 - The colorful layers of metal ore waste can be found in copper mines around the world. They represent powerful confirmation of the Universal Flood. Due to recent observation of "ore factories" on ocean floors, the mineral deposits on land are now considered by modern geology to be of an unusual origin; to have formed in the same environment extant during the UF event—a hypretherm.

ously established that the precipitation process is not an evaporative process, but a process involving temperature lowering or pressure change. It is a process called **prethermation**.

Researchers used temperatures of "at least 350°C" when creating lab-grown ores and that Cl (chlorine) from salt and organic sulfur were involved in ore deposition:

"Many heavy metals occur in ores as sulfides or oxides that are insoluble under most conditions. **Laboratory experiments at hydrothermal temperatures, in combination with chemical modeling, show that chemical complexing with Cl and S and favorable pH and redox state are the main controls on solubility**. Solubilities can now be calculated thermodynamically for most important ore metals at most depositing and transporting conditions to **at least 350°C**." Note 8.13c

Later in the Inclusion Mark, we will see specific temperature/pressure relationships in the formation of many minerals we find on the surface, but here, the presence of fluid inclusions in the ores, establish both the temperature, pressure, and chemical components present in ocean salts when the ores were made:

"**Fluid inclusions** offer samples of the fluids passing through the deposit and, often, approximations of the **associated temperatures and pressures**... Chemical analyses of bulk solutions freed by laboratory crushing of hundreds of inclusions show that **Na, Cl with lesser Ca, Mg, K, and sometimes major CO_2** are the major solutes." Note 8.13c

It is apparent that ancient sea salts present when the ores were made were not much different than they are in today's oceans. The inclusions further confirm the hyprethermal origin of the ore and its association with hyprethermal salt deposits. Recently, five Norwegian researchers made this important connection themselves, in a 2006 *Marine and Petroleum Geology* journal article:

"The DSDP [Deep Sea Drilling Project] drilling experience also **clearly demonstrates** that the sediments within the Atlantis II Deep **are not 'normal' marine sediments**, and that we are dealing with **not only a hydrothermal salt producing system, but also a hydrothermal ore-forming system**." Note 8.13d

The Deep Sea Drilling Project was an important study of existing hyprethermal conditions in present-day deep seas. Those conditions simulate the environment of the UF Hypretherm. As the researchers noted, the drilling experience "clearly demonstrates" that salt *and* ore-forming sediments are *not* formed in typical 'cold shallow sea' conditions so commonly described in today's geology textbooks. The low temperature, low pressure environment of cold shallow seas does not produce salt or ore deposits. It's that simple. The Earth's vast ore and salt deposits could have *only been formed in a hypretherm*.

If one ponders the hyprethermal nature of the Earth's ore and salt deposits for a moment, it becomes obvious that a *monumental change* in modern geology is needed, but the biggest challenge geologists face is coming to terms with the global event that created deep, hot oceans.

Figure 8.13.2 – This is a fragment of a sulfide chimney from Mystic Mound at Explorer Ridge off the coast of Canada. Recent discoveries on the bottom of the ocean confirm the hyprethermal origin of ore deposits, which means ores were made in **hot deep** areas of the ocean, not "shallow cool seas." Note the coin for scale.

FQ: **How** were massive ore deposits on the now-dry continents formed in a **deep, hot** ocean?

Shallow Cold Sea Pseudotheory Overturned

Along with the pseudotheories of Magma, Continental Uplift, and Uniformity, the progressive theory that has kept modern geology in the 20th century Dark Age of Science is the **Cold, Shallow Sea Pseudotheory**. The fifteen Marks of the Flood included in this chapter provide dozens of geological proofs that a deep, high temperature ocean existed in many areas, covering whole continents. The world's vast sandstone deposits alone confirm the presence of a global hypretherm, because it was in such an environment that the sand was cemented into sandstone.

Thanks to the TAG Mound expedition, researchers now recognize that "ore factories" on the ocean floor show how continental ore deposits were originally formed. There, beneath more than 3,000 meters of water, fluids of over 338°C are flowing from constantly active pipes. Scientists say this "represents a *significant* step forward in our understanding" of ore deposits:

"'The results of drilling at the **TAG mound represent a significant step forward in our understanding of how some of the large base metal ore deposits on land were formed at the seafloor** millions of years ago,' says Herzig with the Institut fur Mineralogie und Lagerstattehre der Rheinisch Westfallschen in Germany. '**We have learned** how complex these seafloor mineral deposits are and we begin to realize that the activity of those **ore factories in the deep sea must have a considerable impact on the composition of both seawater**

Fig 8.13.3 - Evidence of ore deposits can be found along many highways across the Western United States. They are often exposed in road cuts that reveal the vertical nature of the fossil hydrofountains associated with them. The upper image is from the Death Valley area in California; the lower image comes from central Arizona. After reading this chapter, the road cuts will take on a whole new meaning. With this paradigm-shift, many readers report enlightening experiences on their road trips as they recognized that these deposits **became rock by microbial activity in a deep, hot ocean**. As one views these deposits, glancing up at the commercial airliner passing high overhead, cruising at an altitude approximately equal to the depth of the ocean during the Flood, the experience inspires awe, affecting the way in which we look at the entire world.

Road-Cut Ore Deposits

Ore deposits can be seen at road cuts in many areas of the Colorado Plateau.

and the oceanic crust.'" Note 8.13e

Researchers had finally observed *how* ore deposits were made, but *not when*. Apparently, it did *not* take millions of years to deposit the ore, and the scientists found no direct evidence supporting a deposition model spanning 'millions-of-years.'

In addition to the evidence that crustal minerals are formed in hyprethermal vents on the ocean floor, ore geologists now realize many minerals that were once thought to come from molten rock actually came from a "highly saline fluid." A copper mine in central Chile afforded researchers evidence of the temperature and salinity of the ocean fluid associated with the ore deposited in the copper mine:

"Temperatures of crystallization of these minerals, as determined by the highest homogenization temperatures of **highly saline fluid inclusions, range from 400 to >690°C**." Note 8.13f

When water is heated above 100°C, it cannot remain in a liquid state unless it is under pressure. To maintain liquidity at such elevated temperatures, 400°C to greater than 690°C, there had to have been a great deal of pressure, which was necessary for crystallization to take place. This completely refutes the cold, shallow sea notion.

Further supporting the deep, hot sea origin of the central Chilean mine's copper ore is the wide variety of other minerals which are known to form in a hyprethern:

"It is matrix supported, with between 5 and 25% of the total rock volume consisting of breccia-matrix minerals, which include **tourmaline, quartz, chalcopyrite, pyrite, specularite, and lesser amounts of bornite and anhydrite**. An open pit mine, centered on this breccia pipe, has a current production of **50,000 tonnes of ore per day** at an average grade of 1.2% copper, and copper grade in the breccia matrix is significantly higher." Note 8.13f

The daily production of 50,000 tons of ore at this mine focuses on the breccia pipe because this was where mineral rich hot fluid flowed and where biomineralization and copper ore accumulation occurred. The breccia pipes contain rounded sand, pebbles and rock, cemented and usually bounded by siliceous (quartz) material. This rounded material verifies the violent frictional action that took place in these hydrofountains before the sediment was cemented by the hyprethermal waters.

Another large copper mine just south of Superior, Arizona, reveals the immense, almost overwhelming scale of the once-active ore factory process.

One discovery that continues to elude ore geologists in their research into the 'ore factories' at the bottom of today's oceans is that there are no known active 'ore factories' anywhere near the size of the enormous ore deposits on the continents.

There is good reason why large ore diatremes are not likely to be found on the ocean floor today; the constant pressure of overlying water, which restricts the rapid and massive eruptions that were possible on land, prevents such large scale explosions. A diatreme several miles across may reach many miles deep into the crust of the Earth. The diatreme is generally a funnel-shaped, breccia-filled, subsurface landform, created by the hyper-expansion of extremely hot upwelling water. The resulting enormous explosion is possible on land because of rapid decompression of the water as it nears the crustal surface in rather shallow UF-waters. The process was kept in check in the deep ocean because of the elevated overlying pressure of miles of ocean water and the added miles of UF water above the former sea floor.

The continental diatreme-pipe ore deposits were then first formed when the continental surface was under relatively low water pressure, the explosion evacuating enormous hydrocraters of debris. Following these first effects of the oncoming deluge, the entire landscape began to sink, becoming submerged in increasingly deep water, heated by constant and violent earthquakes. Torrential biogenic fluids gushed from hundreds of thousands of wounds in the Earth's crust, which were the pipes connecting the subterranean biosphere to the new continental ocean floor. Trillions upon trillions of piezophilic and thermophilic microbes inundated the submerged continent, thriving in the UF hyprethern. Some produced the massive ore deposits left behind in the pipes and diatremes on the continents today.

Modern geology has not been able to describe how such an event happened, or the mechanism that might have driven it, or even that such an event was possible. But there is no problem with such a scenario in the Hydroplanet Model, with hydrofountains and the Universal Flood being the force behind it.

In Summary, the Shallow Cold Sea Pseudotheory is easily re-

"The results of drilling at the TAG mound represent a significant step forward in our understanding of how some of the large base metal ore deposits on land were formed at the seafloor..."

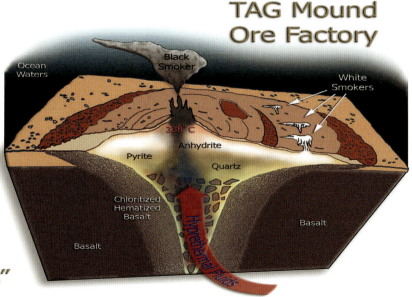

TAG Mound Ore Factory

futed by the hyprethermal ore making process and a number of other hyprethermal evidences presented in this chapter.

The Biological Origin of Ore

Amazingly, the true origin of most ores comes down to *one element*, directly in front of the researchers, but apparently remarkably underappreciated. The element is sulfur (spelled sulphur in some circles). In the Pyrite Mark, sulfur was noted in the hypretherm and in biogenic environments, reviewed here in the 1939 edition, *Textbook of Geology*:

"In the primary ores of copper, lead, zinc, and silver, the metals were originally deposited in compounds in which they are united with **sulphur—in short, sulphides**. Galena (Pb**S**) is the ultimate source of all the world's lead; sphalerite (Zn**S**) is the chief source of all our zinc; chalcopyrite (CuFeS_2) and other copper-bearing sulphides are the ultimate sources of our copper. Gold, it is true, is deposited as a native metal, but it is almost invariably associated with pyrite (FeS_2), that most ubiquitous of **all sulphides**." Bib 125 p479-80

Sulfur is a rare element in igneous rocks (refer back to Fig 6.1.3), found typically to contain only .05% sulfur. Why then, is sulfur associated so abundantly with *almost all ores*, and what makes this an important fact? The Law of Paragenesis (subchapters 7.5 & 8.12) holds the answer key to this question; Sulfurous ores formed as the Law of Paragenesis states, in hyprethermal deposits at the same general time and in chronological order. For the ores on land, this was during the UF. The sulfide ores are connected directly to microbes, in particular, bacterial byproducts. Present-day analog of this is seen at active hyprethermal vents on the ocean floor.

Researchers noticed, for example, that iron (Fe) ores are produced with sulfur in very low pH conditions, where the bacteria "*T. ferrooxidans*" could grow:

"Recent work on the **bacterial catalysis of Fe(III) reduction with sulphur** has raised several questions. An Fe(III) reduction system with sulphur as an electron donor was found to have an **optimum pH of 2.8-3.8** in *T. ferrooxidans* with **no reduction observed below pH 1.4 or above 4.8**." Note 8.13g

The same bacteria were also used successfully in the extraction of gold from sulphide minerals:

"***T. ferrooxidans* has been used successfully in the extraction of gold** from refractory **sulphides** concentrates, and there are no reports of the bacterial activity in the process being inhibited by the gold." Note 8.13g p106

Everywhere researchers turned, it seemed, microbes were involved in mineral production and crystal growth—even "gold crystals":

"For instance, Southern and Beveridge (1994) have shown that fine-grained gold colloids immobilized **within bacteria** can be altered during diagenesis at 60°C to coalesce and form **gold crystals**." Note 8.13h

Some of the top researchers in the field made a simple but *profound statement* when they identified the cause of the large

Copper Ore From Superior Mine

Superior Arizona Ore Mine

Natural ore producing factories with diatremes this large are not found on the ocean floor today.

deposits of metal (iron):

"**The biological accumulation of large amounts of Fe [Iron] in the environment is the result of biomineralization.**" Note 8.13h

Over the past decade, prior to the publishing of the UM, the concept and processes of biomineralization was discussed with hundreds of people. Yet, none of the laypeople and few scientists knew that ores and metals were the byproducts of microbes! This was a completely new concept; for most, the ores and metals had presumably come from melted rock, but now, direct observation has proven that they came from ***microbes in heated water!***

The Uranium Ore Diatreme Evidence

An important example of the interaction between heavy metals and biomass comes from uranium:

"**Uranium removal has been demonstrated using biomass particles** immobilized using various polymeric membranes and such preparations may capable of **>99% metal removal from dilute solutions**." Note 8.13i

Today, microbes have become extremely important in the cleaning up of highly toxic, heavy metal contaminated sites, and they do so naturally and economically as the bacteria involved actually absorb the noxious metals into their cell structure:

"**Precipitation within or on cell-walls** may be particularly evident with radionuclides such as **uranium and thorium**. In *S. cerevisiae*, **uranium was deposited as a layer of needle-like fibrils on cell-walls reaching up to 50% of the biomass dry weight**. That such a large amount was bound by the cells implied that additional uranium had 'crystallized' on already bound molecules." Note 8.13i p21

The reason these microbes absorb and precipitate certain metals is that the metals are an important and integral part of their existence. Just as human beings depend on calcium for their bone structure, some microbes depend on specific metals in their cells to function. When the bacteria die, the concentrated metals became dissolved in the hyprethermal waters just as soil microbes dissolve bones once they are buried in the soil.

Uranium ore is only found in very specific areas of the crust because the microbes that formed the ore live only in a very select environment.

Eugene Shoemaker, the impactologist who convinced the scientific community and the world that the Arizona Crater was an impact crater (see subchapter 7.11), had ironically studied many hydrocraters and diatremes that had become filled with hardened mineral ore in the Four Corners area of the Southwestern USA. There are at least 250 diatremes scattered across this area of the United States, and Shoemaker described the diatremes, ignorant of the fact that they were all filled with UF sediments:

"Many kinds of material fill the diatremes. In its upper part a mature diatreme is filled with bedded tuff and **limestone** and locally with thinly laminated **clay** and **silt**, evaporites [**salts**], and some **bedded chert**. Lower in the vent these sediments give way to more massive tuff, breccia, huge country rock, and finally agglomerate and solid igneous rock. Much of the tuff of the filling, which in some vents is at least 2,000 feet thick, is well bedded and **shows many features of fluviatile deposition**, such as cross lamination, channels, and erosion surfaces. Limestone interbedded with the tuff ranges from massive to thinly laminated, and much of it is argillaceous though a few beds are composed of relatively **pure carbonate**." Note 8.13j

The clay, silt, salt, chert, and limestone deposits in the diatremes were all previously shown to have formed in the UF hypretherm, and are here being *interconnected* once again. The force evident in the diatremes is greater than anything modern geology has ever seen, with "blocks over 100 feet long" incorporated in the diatreme pipes:

"Fragments of rocks from all parts of the sedimentary column and, in places, from the crystalline Precambrian basement are generally incorporated throughout the tuffs. The fragments range from minute grains to **blocks over 100 feet long**."
Note 8.13j p180

These diatremes contained another important mineral, an ingredient critical to the nuclear power plants industry—uranium:

"Much of the **limestone** in the Hopi Buttes diatremes contains from **0.001 to as much as 0.02 percent uranium, and nearly every diatreme that contains exposed limestone is conspicuously more radioactive than the surrounding rocks**. Concentrations of uranium higher than 0.02 percent are found in some limestones but more commonly in **siltstones and claystones**."
Note 8.13j p184

Until now, geological interest in uranium primarily rested with the *technologists* focused on mining and enriching uranium for use in nuclear reactors. Even recent biomineralization of uranium in well-sites focuses primarily on cleaning up uranium mine waste. There is an important *geological* reason as to why this "conspicuously more radioactive" uranium mineral is found in diatremes; the hyprethermal environment and a very specific microbial component.

There were microbes throughout the Floodwaters—***all** crustal rocks formed in the Flood were impacted by microbes, which include the radioactive minerals in the diatremes*, but why is this important?

The dating of rocks is based on the **assumption** that microbes were **not** involved in the making of the rocks.

Natural Gold Deposits

All the ore metals including gold can be traced to biomineralization.

Uranium Ore Diatreme Evidence

Typical Uranium Ore Diatreme

1904 Oyler Mine, Southern Utah

Mine Located in UF Deposits

Over 250 hydrofountain diatremes are known in the Colorado Plateau Grand Canyon area.

"Nearly every diatreme that contains exposed limestone is conspicuously more radioactive [contains uranium] than the surrounding rocks."

Uranium Diatremes also contain: pyrite, quartz, hematite & carbonate minerals proving the hyprethermal connection between these minerals.

Pitchblende Uranium Ore

Grand Canyon Rim Uranium Mine

CHAPTER 8 THE UNIVERSAL FLOOD MODEL

Biomineralization completely changes at least six decades of assumptive geological dating that was based on the radioactive decay dating method, where scientists presumably 'know' how much of a particular radioactive element was present when the rock was formed. Because microbes were involved in uranium and lead mineral formation, the quantitative amount of uranium or lead from previously dated sediments must be completely reanalyzed.

More specific details of rock dating will be dealt with in the Age Model; but for now, we must stress the importance of biomineralization, radioactive minerals, and dating.

During the 1970s, much work was done on the world stage of uranium because of the "energy crises" and the scare tactic that the world was going to run out of petroleum within the next couple of decades. One important report titled *Hydrothermal Uranium Deposits* was completed in 1977 and noted the shallowness of the diatreme pipe:

"**Most hydrothermal uranium deposits are quite shallow**, rarely extending to **depths greater than 300 m**; however, mining at the Schwartzwalder mine, Colorado, and the Beaverlodge district, Saskatchewan, has already proceeded to **depths of 700 and 1,500 m respectively**." Note 8.13k

These researchers saw a 'problem' with the "shallow" depths of the uranium deposits:

"**What is clear**, however, is that at least some hydrothermal uranium deposits, such as the Fay-Verna mine in the Beaverlodge district, were demonstrably formed over a depth range of more than 1000 m, and that **others were formed at temperatures which are difficult to reconcile with a very shallow origin except in regions of exceptionally strong geothermal gradients**." Note 8.13k p70

The Magma Pseudotheory (chapter 5) demonstrated that the "geothermal gradients" (high temperatures) needed to produce uranium ore should not have been found near the surface where the ores were produced! Presumably, the heat should have come from deep inside the magmaplanet, but this made no sense because most of the diatremes were north or south of the Grand Canyon, but not *in the Canyon*. The uranium ore was in an area not bounded by a continental plate, but was far inland, in an area difficult for geologists to explain how magma could have traveled through the thick continental crust to the surface. This makes it "difficult to reconcile" the shallowness of the ores.

In some cases, the formative temperature of the uranium ore (pitchblende) diatremes was as high as 350°C. Moreover, there was an abundance of NaCl:

"During post-**pitchblende vein formation** at Limouzat, the P-T-X conditions of the hydrothermal fluids were quite variable. Inclusion filling temperatures range from **25 to 350°C**, and inclusion fluids contain between **1.7-23 equivalent weight % NaCl** and 0 to **7.7 % CO_2**". Note 8.13k p28

When this report was published in 1977, no one knew there were active hyprethermal vents on the ocean floor, and researchers must have wondered from where so much NaCl had come. Today, observations from the ocean's abyss and the Universal Flood Model make it easy to understand how the global hypretherm created large salt deposits on land and why some of it was left behind in the ore diatremes.

Another key observation researchers made was the contact zone between the rock walls and the brecciated pipe. A similar phenomenon was shown in the Rock Cycle (chapter 6) where intruding dike walls showed no signs of melt, although they are claimed to have been hot magmatic intrusions. Investigating the rock walls adjacent to the uranium vein, it was reported:

"If the flow of hydrothermal fluids is **too rapid for fluid-wall rock thermal equilibration, the temperature of ascending solutions should be higher than that of the adjacent wall rocks; therefore alterations sequences should normally develop in a negative temperature gradient outward from veins**. The reverse would be true for descending solutions. Unfortunately, **wall rocks surrounding hydrothermal uranium veins are usually only weakly altered**, and where alteration is strong the means for determining the direction of temperature gradients are not yet available." Note 8.13k p70

Pressurized high temperature fluids flowing *for long periods of time* would completely alter the walls of the diatreme, so there was no period of 'geological time' here. The event that occurred here must have taken place quickly.

Uranium accumulations formed *quickly*, so other minerals deposited with the uranium must have formed *quickly* too, inside these hyprethermal diatremes:

"A study of paragenetic diagrams for **41 of the uranium deposits**... showed the presence of the following minerals in more than half of the deposits: **pyrite/marcasite** (100%), base metal **sulfides** (88%), **quartz** (81%), **hematite** (71%), and **carbonate** minerals (55%)." Note 8.13k p11

Where else have all of these hyprethermal-diatreme minerals been found? There is no coincidence here; the fact that they have been found all over the Earth establishes the environment that once existed, and the time in which these UF minerals were formed.

The Ore-Fossil Evidence

The uranium ore deposits in the sandstone of the Colorado Plateau have yielded one more incredible clue about how and when they were formed:

"In these deposits [Colorado Plateau sandstone] the **uranium-vanadium minerals fill the voids around the sand grains**. Locally they may **replace fossil plant materials, including large logs, and may even occur as replacement of ancient bones**." Note 8.13l

Having established with empirical evidence, that these uranium ore deposits were formed in the hot vents of a deep ocean hypretherm, we ask another FQ:

> How were plants, large logs, and terrestrial animal bones buried and preserved at the bottom of the deep hot ocean?

Is this happening today? Modern day adventurers scouting

Fig 8.13.4 - On the previous page, two uranium ore diatreme pipes are represented; one from southern Utah and the other from the south side of the Grand Canyon, Arizona. Uranium ore pipes, like all other ores, were formed in a hypretherm. They help establish the UF hydrofountain activity. The biologic and radioactive nature of the ores raises new questions about all crustal rocks formed in the UF Hypretherm. Modern rock dating does not account for microbial contamination present during the crystallization process of most rocks. This new revelation challenges all modern geological rock dating techniques. The diagram illustrating a Typical Uranium Ore Diatreme comes from the USGS.

the oceans depths have not seen it, and there is no reason to suspect it. There are a number of debunked fallacies that paleontologists turn to in an attempt to explain the presence of the fossils. One might say the land fossils became lithified on land prior to the submersion of the continent, were deposited in the uranium ore pipe, and then uplifted to where they are today. There is no evidence of slow continental subsidence or uplift taking place today or in the past, and, to become hardened quartz rocks, the fossils had to have already been subjected to a deep, hot oceanic hypretherm themselves! (This is a sneak-peak into the upcoming Fossil Model).

A related example comes from the Kennecott Copper Mine in Bingham Canyon, Utah, the largest excavation site in the world. When mining operations first began, fossilized bone fragments and teeth were found 70 feet below the surface, shown in Fig 8.13.5.

The copper ore body was formed in a growing mountain under an ocean of increasingly deep water. A vertical diatreme of sediment-laden hyprethermal fluid carried material to the surface from miles beneath the crust. As the frenetic flow quieted, the diatreme collapsed, backfilling as suspended and floating debris settled. Some of that debris was the remains of animals or plants, and as they sunk and were subjected to hypretherm conditions, they became fossilized. A similar series of events was revealed in Fig 8.4.14.

The Mystery of Ore Revealed

We noted earlier that some geophysicists stuck in the magma paradigm believe "Mineral ore deposits are really freaks of nature…" Bib 120 p146 Indeed, if we are to assume that all rocks and minerals came from a melt, ores would seem freakish for several reasons. One is the ore's crystalline nature, which appears completely different from ores coming from a high temperature melt. Another is the location of ore bodies; they are typically found only near the surface or in diatremes that extend far below the surface. Finally, there are no ore bodies of any significance being formed anywhere on Earth today.

The "freaks of nature" quote comes from a 1995 geophysical book, *The Mid-Atlantic Ridge*. It lends perspective to the attitude held by geologists for some time. A few years later, we see the mindset of some geological researchers beginning to change. Truth and time are on the same side. Physical evidence continues to mount and cannot be ignored as geology comes to a literal crossroad of thought. An August 1999 *Science* article, *Mineralogy at a Crossroads*, hints at the change:

"The materials of the planet's interior exhibit physical and chemical properties that can be far different from what is observed near the surface. For example, the discovery that hydrogen [water] **can be bound in dense silicates and metals has given rise to the possibility of oceans of water locked up within Earth's interior**" Note 8.13m

"Oceans of water" in the Earth's interior is more than just a possibility; the water trapped in rocks, the incredible abundance of microbial life, especially the apparently now-dormant piezophile and thermophile bacterium, and the biological origin of the world's ores all testify that the Earth is actually a hydroplanet. Biology and geology actually become one in "biogeochemical cycles" according to the *Science* article:

"Issues of complexity are leading directly to the boundary with biology. **It is now recognized that microbes control many geological processes, such as ore deposition and crucial biogeochemical cycles, echoing the connection between biology and geology** advanced by Vernadsky." Note 8.13m p1027

Vladimir Vernadsky had advanced the crucial role of biology in geology back in 1922, but there wasn't enough scientific evidence concerning the biogenic nature of minerals at the time, and the Magma Pseudotheory was far too strong to overcome without extraordinary evidence.

The Ore Mark is one of many extraordinary evidences of the UF event. Today, empirical evidence has convinced almost all researchers that ores are formed in a biological hypretherm, but the significance of the recent conclusion has yet to be realized. The Universal Model will change the perception of the natural world as the undeniable evidences of the UF are seen in a new

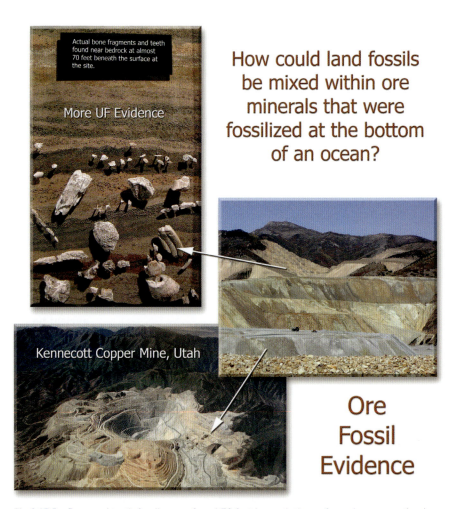

Fig 8.13.5 – Bone and teeth fossils were found 70 feet beneath the surface when excavation began at the Kennecott Copper Mine. The mine lies near the top of a mountain making it difficult to see how land-based animal bones were deposited deep in an ocean where they became fossilized, without the Universal Flood.

light. One final point of the Ore Mark is the unique world-wide iridium mark, known as the K-T Boundary.

The K-T Boundary Evidence

Iridium ore and dinosaurs seem to have a lot in common, according to modern scientists. To understand the connection between the two, we turn to the **K-T Boundary**, the thin band of sediment that separates the Cretaceous (K) period and the Tertiary (T) period in modern geology's theoretical timeline. The sediment between the two zones is quite different and geologists have long thought that some "worldwide catastrophic event" had to have contributed to the change in the layers:

"**Geologists and paleontologists were slow to accept** the theory of a **worldwide catastrophic event**, especially one initiated extraterrestrially. Most scientists feel much safer considering orderly Earthbound processes working gradually and predictably over geologic time. **Catastrophism smacks of biblical ideas like Noah's flood**..." Bib 29 p380

The attitude *against* the Flood, which many researchers have, makes them skeptical of major cataclysmic events, but there is so much evidence of a "worldwide catastrophic event" that members from *all* disciplines of science acknowledge the "event" cheerfully, thanks to the trendy new impact pseudotheory currently in vogue. The trouble is that the vast majority of meteorites are *not* from asteroids or comets and have no real origin in modern science, but their real origin will be explained in the Meteorite Model (subchapter 7.10). That chapter explained how *metal* compounds were formed in hyprethermal diatreme and ejected with extraordinary force, perhaps even sufficient to escape the gravitational hold of the Earth. Although they return to Earth as meteorites, most ejectites didn't leave Earth's atmosphere. Furthermore, the Ore Mark established the biogenic origin of the Earth's ores, which include the agglomerations of metals in the ejectites.

The K-T Boundary layer from the "worldwide catastrophic event" is usually seen as a thin band of sediment relatively near the surface, seen in Fig 8.13.6. During the 1980s, researchers noticed an excess concentration of the element iridium in this thin band of sediment that seemed to appear in worldwide sediment layers. They theorized the excess iridium must have come from the impact of a meteorite:

"When Alvarez and colleagues discovered the excess iridium in the K/T clay, they realized that **something special** must have happened to cause the stuff to be deposited. **So where had it come from? It must have come from space, they argued. The impact of a comet or asteroid could account for the iridium excess**." Note 8.13n

The "something special" found in the K-T Boundary was an important key—but it was not from an "impact of a comet or asteroid" and it wasn't simply *elemental* iridium. In the Chemical Model it will be shown that almost all 'elements' are found in combination with oxygen or some other element, and only very rarely in isolated elemental form. This holds true with the iridium, which is contained in ores such as platiniridium and ruthenosmiridium, which are biogenic ores. It was not an extra-terrestrial impact that created and consolidated the ore; it was Flood-driven microbial activity, which was also responsible for the K-T Boundary and its associated heavy metals. Research in *Geochimica Et Cosmochimica Acta* (1989) found 'meteoric' metal elements but also found "*non*-meteoritic" elements; Arsenic (Sb), Antimony (As), and Zinc (Zn) in the K-T Boundary layers:

"**Three non-meteoritic trace elements (Sb, As and Zn) are strongly enriched at eleven K-T boundary sites**, along with mainly or partly meteoritic elements (Ir, Ni, Cr, Fe and Co). The proportions (As, Sb, Zn/Ir) are remarkably constant over a ~100-fold range in concentration. This correlation persists in sub-layers of boundary clay and even extends to soot (from burned land biomass). Apparently, **all the components, despite their diverse origins, became associated in a single, global component prior to deposition. No wholly satisfactory source is available for As, Sb and Zn**: the trace element pattern in volcanic gases does **not** match that in K-T boundary clay, with ratios to Ir falling short by 1 to 2 orders of magnitude, terrestrial rocks do **not** reach high enough concentrations and (modern) ocean water contains too little Zn..." Note 8.13o

There was no "satisfactory source available" for the three "*non*-meteoritic" metals because the researchers had supposed they might be from a 'meteorite.' Instead, *all of the metals* came from "a single, global component prior to deposition." To understand the origin of the metals in the K-T Boundary, whether meteoritic or *non*-meteoritic metals, we need look no further than the biogenic processes of ore production already known.

This makes the evidence in the Ore Mark essential in comprehending many past geological processes. They key of course, is the microbes. *Science News* hosted an article titled *Microbes complicate the K-T mystery*:

"'**This could have a really important role in deciding what the nature of the catastrophe was**,' says biologist Betsey D. Dyer from Wheaton College in Norton, Mass. She and her colleagues describe their experiments in the November [1989] GEOLOGY. Dyer says the study is the first to **demonstrate that microorganisms can both enhance and erase the iridi-

Fig 8.13.6 – The K-T Boundary is a band of sediment between two ***theoretical*** geological time periods, Cretaceous (K) and Tertiary (T). Geologists have long-known of this worldwide sediment band but have never been able to account for it with uniformitarian dogma. The advent of impactology in the late 1900s brought in a new theory that included the universal destruction of the dinosaurs by an impact event. It was supposed that the elevated concentration of the element iridium in this worldwide layer of sediment came from a meteorite. However, if a meteorite was responsible for the K-T Boundary, the sediment should also exhibit large amounts of nickel and other elements from the meteorites—but it does not. In fact, the K-T layer is enriched with many ***non***-meteoritic elements, such as arsenic and antimony, both rare elements in the Earth's crust. Where did they come from? They all share a common origin—microbes.

um concentration in rock." Note 8.13p

As early as 1989, researchers had learned that microbes could *enhance* iridium concentration in rocks. There was no need to look to a meteorite from space to do it, and if there had been a meteorite, it probably got its iridium from microbes on Earth during the UF. One thing that kept researchers in the dark was the *relationship between nickel*, common in almost all meteorites, *and iridium*. But the one place these two elements occur together naturally—**nickel ores**—from terrestrial mines:

"Naturally occurring iridium alloys include osmiridium and iridiosmium, both of which are mixtures of **iridium** and osmium. It is recovered commercially as a by-product **from nickel mining and processing**." Note 8.13q

Nickel ores in the crust, including the ore in the K-T Boundary sediment, is formed just as the Ore Mark demonstrated—in a *hypretherm*, a high pressure, heated ocean.

What has the research on the K-T Boundary confirmed positively? That the Iridium, Arsenic, Antimony, and Zinc-rich ore layer found *on the surface of the continental crust* occurred *globally*. Summarizing the K-T Boundary Evidence, we find:

1. The K-T Boundary sediment is the result of a global, catastrophic event.
2. The K-T Boundary sediment layer contains ores.
3. The event was recent, occurring on the surface.

If we include in the summary, the knowledge that the ores come from microbes and the microbes grew in hyprethermal water, we can ask this important FQ:

> FQ: What recent, catastrophic event produced a hypretherm capable of producing an Iridium-rich metal ore deposited across the entire surface of the Earth?

Ore Mark Summary

The Ore Mark concludes the series of six subchapters that are all connected by the biologic events of the Universal Flood. Dogmagma once caused many researchers to see ore deposits as "freaks of nature" and that idea still pervades much of science because the hyprethermal nature of ore is only a recent discovery. It has yet to be incorporated into the big picture of geology, but the UF paints a beautiful picture on the canvas of truth that explains, for the first time, the true origin of the so-called freaks of nature.

The Cold, Shallow Sea Pseudotheory was completely overturned and replaced with a model of the real environment that existed not long ago, covering all the continents of Earth under a global, deep, hot ocean.

The biological origin of ore transforms the thinking of where all rocks come from, showing that microbes played an integral part in the production of all metal ores we have come to rely on in modern society. The Uranium Ore Evidence established that even the heaviest metal ores came from hyprethermal biologic activity, opening the door to a necessary reevaluation of long-established ages of rocks.

The ore-fossil connection revealed another fascinating but unanswered modern science mystery. The answer to that mystery, along with the K-T Boundary Evidence firmly establishes the global nature of the Universal Flood.

With this new information, it will be difficult to ever again look at limestone, salt, oil, gas, coal, iron, pyrite, and ore deposits without seeing their UF origin. Many readers comment that after reading this chapter, their eyes have been opened to the beauty and simplicity of many geological features and their origins as they travel around the globe.

8.14 The Surface Mark

Of all the subchapters in the Universal Flood Chapter, the Surface Mark contains some of the most unique and plainly manifested biological and geological evidence of the Flood. Once we learn how these surface features were created, it will be fascinating to go out in Nature and see firsthand the exquisitely formed UF surface rocks.

Seeing for the first time how these special Marks of the Flood were created may be a treasured experience as the Surface Mark broadens the understanding. There are four types of Surface Marks described in this subchapter; quartz, carbonate, varnish, and nodules, each having no previous, logical origin in modern geology. Here, with a UF paradigm, the origin of these 'Marks' will become clear and distinctively obvious.

Surface Marks Defined

This subchapter deals with the Surface Marks of the Flood, which are found globally and primarily on the surface of the world's deserts. These Flood Marks present a unique opportunity to examine mineral deposits that formed directly on the surface during the UF and are still there today.

A common characteristic among most of these marks is their location; they persist in areas where there is little vegetation and rainfall, which allowed for their preservation. Thus, the desert environment is a natural laboratory for investigating the Surface Marks of the Flood, that in many cases still lay where they were at the end of the Flood event. This also affords us the opportunity to study the geological conditions at the end of Deluge.

The following four critical factors confirm the UF as being the ***only*** possible origin of the Surface Marks discussed in this subchapter:

1. They are found globally, on or near the *surface*.
2. They were formed within the last few thousand years, evident because of minimal *erosion*.
3. They are *not forming today*, confirming the Catastrophic Principle.
4. They were formed in a *global hydrothermal or hyprethermal environment* that does not exist today.

The four types of Surface Marks have been organized into the following categories:

1. Quartz Surface Marks
   A) Surface Chalcedony
   B) Geodes
2. Carbonate Surface Marks
   A) Caliche Rocks
   B) Limesurface Rocks
   C) Dissolved Rocks

D) Great Salt Lake Sand
3. Varnish Surface Mark
4. Nodule Surface Mark

Some of these Marks may seem familiar because we have discussed similar phenomenon already. Quartz was discussed first in the Quartz Mystery in subhapter 6.4, but is revisited here because of Surface Chalcedony and Geodes, both fascinating surface evidences of the UF. The world's massive carbonate deposits testify of the Flood, but so do the simple surface marks addressed here. Hopefully, the exploration into their origin will be enlightening and rewarding!

The Surface Chalcedony Mark

Fig 8.14.2 is an example of Surface Chalcedony, a cryptocrystalline quartz rock that normally grows on a solid surface. These were introduced in the Quartz Mystery, in Fig 6.4.5. Chalcedony is most commonly seen in agates and nodules, but in some cases, it did not grow on a solid surface. Surface Chalcedony formed on loose, unconsolidated sedimentary material. Remember the FQ from chapter 6.4:

> How can chalcedony be formed on the *unconsolidated surface* of the Earth and still be resting in its original position if it formed millions of years ago?

The answer is quite simple—it can't. The Surface Chalcedony that grew on unconsolidated or loose sedimentary material was formed only recently.

Some of the most beautiful examples of Peruvian surface chalcedony that grew on an *unconsolidated* surface are seen in Fig 8.14.3. Notice the quartz crystals that formed on the top side of these specimens. The crystals formed on the side facing the water, whereas the bottom side is flat and dull where it was in contact with the loose sediment. The Peruvian Surface Chalcedony rocks are some of Nature's finest creations and are simple testimony of the Universal Flood because they demonstrate the environment on the floor of the continental ocean (the hypretherm) and they prove the *time frame* in which the hot-deep ocean occurred. These rocks show little evidence of weathering because they are only several thousand years old.

Is it not a coincidence that unconsolidated Surface Chalcedony verifies the UF timeframe, which was previously established by the Erosion Mark to be only several thousand years ago. The young age of Surface Chalcedony is also seen at a collecting sight near Safford, Arizona (Fig 8.14.4) where beautiful specimens can be found that exhibit almost no weathering. There, specimens that formed on unconsolidated sediment can be found along with samples that grew on rocks.

Professional Chalcedony Research

Chalcedony is a mineral not widely known to the public, but is very popular among collectors and is well known to min-

Fig 8.14.1 – This agate geode from Argentina shows off its brilliant red interior banding. Geodes are not found throughout the geological column but only on or near the surface. They are an important Surface Mark of the Universal Flood.

eralogists. However, several decades ago the last known experiments on synthetic chalcedony were performed, and at the time of this writing, only one U.S. investigator, Peter J. Heaney, could be found pursuing any research on the rocks. Even then, it comprised only a small portion of his research efforts. It was as if geologists had given up the study of these beautiful mineral specimens, or perhaps they had no further interest. Either way, it was puzzling.

At first, geologists assumed hydrothermal waters were responsible for the formation of chalcedony; that increased temperatures alone would dissolve silica in nearby minerals and redeposit it in geodes and other forms of chalcedony. But quartz does not grow this way. In fact, the *only* way scientists had ever seen chalcedony grow, was, like all other crystalline quartz polymorphs—*in a hypretherm*—under increased temperatures *and pressure*. Heaney reports in a 1993 article, *A proposed mechanism for the growth of chalcedony*:

"**Experimental syntheses of chalcedony** have been cited **as support for precipitation at elevated temperatures**: White and Corwin (1961) **produced chalcedony at 400°C and 340 bar**; Oehier (1976) performed runs at **100-300°C and 3kbar**; and Kastner (1980) obtained chalcedony at **150-240°C**. In addition, Blankenburg and Berger (1981) suggest formation temperatures for agatoid chalcedony **in excess of 375°C** on the basis of crystallite size geothermometer." Note 8.14a

Surface Chalcedony
Attached To a Solid Surface

Fig 8.14.2 – This example of surface chalcedony grew on a solid rock surface. Notice that the chalcedony shows little sign of weathering, indicative of its young age. The environment in which chalcedony, a cryptocrystalline quartz rock grew, was the same environment all quartz grows in—a *hypretherm*.

The pressure noted in this article was 340 to 3,000 bar (4,931 to 43,511 psi), which is hyprethermal pressure. Now let us put ourselves in the shoes of the geologists who realized that besides high temperatures, high pressures was required. That meant natural Surface Chalcedony that grew on unconsolidated silt or sand had **miles** *of ocean water above it*. Inconceivable!

> When in the recent Earth's history, did the entire surface of the Earth have hyprethermal conditions to grow the surface chalcedony we found globally?

Modern researchers have probably never thought of this question because the Flood 'myth' is the farthest thing from their mind. Researchers today learn early on that more than a century ago, modern science disproved the Flood 'myth.' So without a hypretherm to create these Surface Chalcedony specimens found all over the continents, there is no logical way for them to grow and researchers, as noted before, merely lost interest. Heaney is one of the few investigators who has given any serious thought into the matter. He notes that it is *"possible* that silica-rich fluids flow from deeper within the earth," but these fluids do *not* make *chalcedony*—they make "geyser fluids," which make geyserite:

> "It may be possible that silica-rich fluids flow from deeper within the earth into near-surface cavities and become highly viscous upon the **sudden reduction in pressure and temperature**. Such behavior is typical of geyser fluids, which contain dissolved silica in the hundreds of ppm. **However, geyser fluids** precipitate a siliceous sinter that is either amorphous or cristobalitic; as with marine silica gels, **chalcedony is not deposited directly from these fluids**." Note 8.14b p68

Direct deposition by upwelling volcanic water or hot springs was not the answer, and modern researchers had nowhere else to go for answers.

There is, however, a simple key to understand chalcedony growth and it is the *surface* habit of the mineral. Fig 8.14.5 is a dry creek bed in the Fire Agate Surface Chalcedony field near Safford, Arizona. Looking for chalcedony in banks of the eroded creek bed will yield poor, if any specimens. To find the best pieces requires scouring the surface of the surrounding desert countryside; the prized finds are often not buried. Fortunately for the collector, Surface Chalcedony is found on the ground in pristine areas of the desert that has experienced little erosion, with the vast majority of it lying flat side down. Some rockhound books say the Surface Chalcedony (also known as desert roses) may have been completely picked up in recent years, but in time, more of them will appear as the surface is subjected to weathering and erosion. Unfortunately, when the areas were revisited, the hoped for new crop of 'desert roses' was not to be found. Surface Chalcedony specimens are like fossils—they were made in a very special environment, and are not being made anymore on the continental surfaces. Note 8.14c

The best description of chalcedony genesis came from a 2008 report from the Thailand *International Gem & Jewelry Conference*. The researcher was unaware of, or chose not discuss the habit of Surface Chalcedony found on unconsolidated sediment, and, despite not having a global Universal Flood perspective, he rendered a pretty accurate report on chalcedony origins:

> "Chalcedony is a secondary mineral formed from watery silica gel at relatively low temperature. The silica is often released by the weathering process of rocks that are of initially void, for example basalt, and the formation of Chalcedony took place accordingly at **very near to the earth's surface**. Chalcedony is not only found in weathering volcanic rocks, **but also in sedimentary ones**. In igneous or metamorphic rocks, Chalcedony is rather **rare and only form as veins in cracks or cavities that have been percolated by warm rising silica – rich brines**. Occasionally, Chalcedony is found as a **petrifying agent in fossils**. Although Silica can redeposit in various forms, the watery silica gel cannot form to be Chalcedony without the proper condition such as degree of **acidity and basicity, temperature, pressure and it cannot be formed in the interfered places such as the cavities of rocks or the deepest of lakes**." Note 19.14d

The Thai article's author recognized that chalcedony formed only very near the Earth's surface, in sedimentary material, not inside igneous or metamorphic rocks. He also commented on the importance of briny fluids, the correct pH, and the components of temperature and pressure, all parts of the UF Hypretherm.

Besides the physical location of Surface Chalcedony, what other evidence is there that can validate how chalcedony is made? Some of the best evidences are agates and geodes. Fig 8.14.6 shows two chalcedony geodes formed as chalcedony precipitated on rolling geodes. These particular items were found in limestone, *near the surface*, which is an important aspect of geode identification. In an upcoming section, the reason open cavity geodes only grew very near the surface will be revealed.

Field evidence has shown that Surface Chalcedony is not growing today on the *surface of any continent*, not even in geysers with elevated water temperatures. Even with the Surface Chalcedony Mystery seemingly solved, there is still stronger proof, a veritable smoking gun that chalcedony grows *on the surface in a hypretherm*.

The Surface Chalcedony Smoking Gun

Once again, the TAG Mound on the Atlantic seafloor surren-

Peruvian Surface Chalcedony

The top side of these chalcedony specimens show their crystalline faces that were exposed to the hyprethremal water environment.

Fig 8.14.3 – These delicate examples of Surface Chalcedony come from Peru, where they grew and crystallized on **loose unconsolidated** sediment such as silt or sand. They were not attached to a rock but were simply picked up off the surface. The difference between the top side and bottom make it easy to see which side grew on the sediment. Surface Chalcedony can be found around the world, but they require hyprethermal conditions for growth. They often show little signs of erosion, validating their UF Hypretherm origin.

The specimens were formed on unconsolidated material; the bottom side is dull and uncrystallized because it was not directly exposed to the hyprethermal solution.

ders a 'smoking gun,' revealing evidence of chalcedony genesis. As investigators drilled into the mound they found "*no chalcedony*":

"**Samples taken from TAG 1 and 5 from below the anhydrite zone contain no chalcedony. Instead they contain subhedral quartz crystals which show oscillatory zoning in aluminum.**" Note 8.14e

However, chalcedony was found (see Fig 8.14.7) on "*the top and sides of the mound*":

"**Length-fast chalcedony occurs in an outer low temperature envelope across the top and sides of the mound.**" Note 8.14e

The remarkable evidence of the formative environment of Surface Chalcedony confirms its UF origin. This is the *only place in the world* where chalcedony is seen *forming* in Nature:

The only place in the world where chalcedony is forming today is on a frictionally heated fault on the bottom of the ocean under 2.3 miles of water!

Further validity of the UF hypretherm is available by examining many other minerals, such as the Hyalite Surface Opal seen in Fig 8.14.8, but in the interest of space and brevity, only surface chalcedony is discussed here.

The Geode Mark

Geodes and chalcedony are commonly associated, and to understand how they were formed, we first look to modern science's version. From the 2006 book titled *Complete Guide to Rocks & Minerals* we read how chalcedony and agate are believed to have formed:

"Silicate minerals dissolve from the **lava** and filter into the gas bubbles. As the lava cools, these dissolved silicate minerals

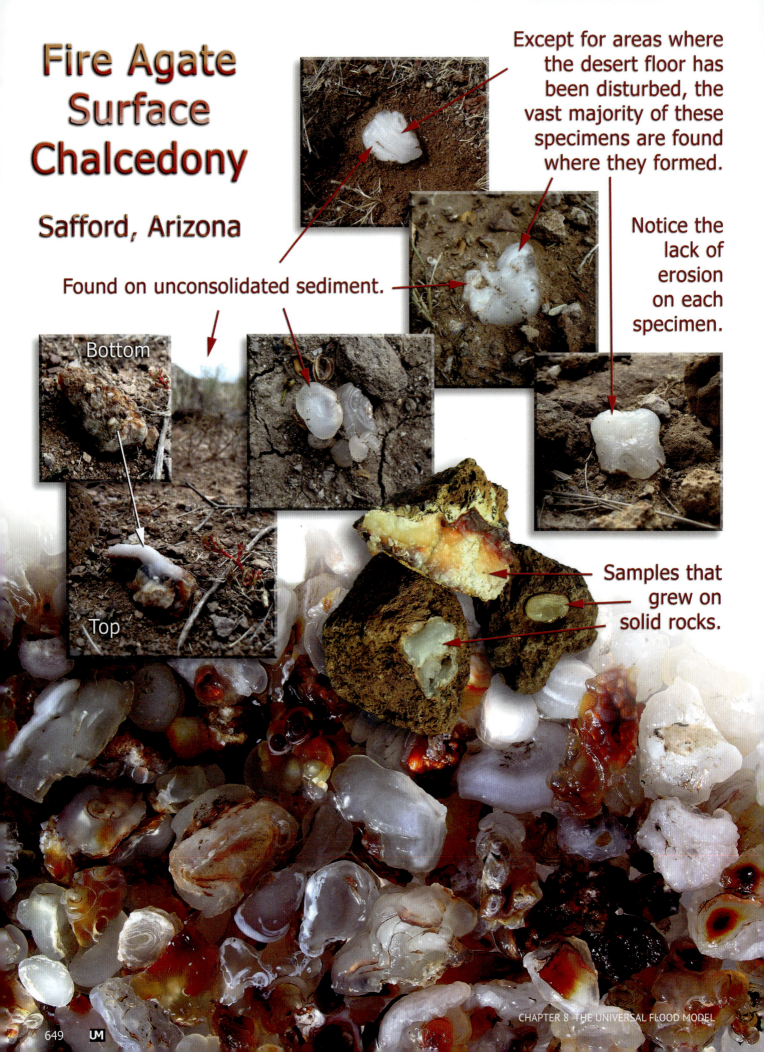

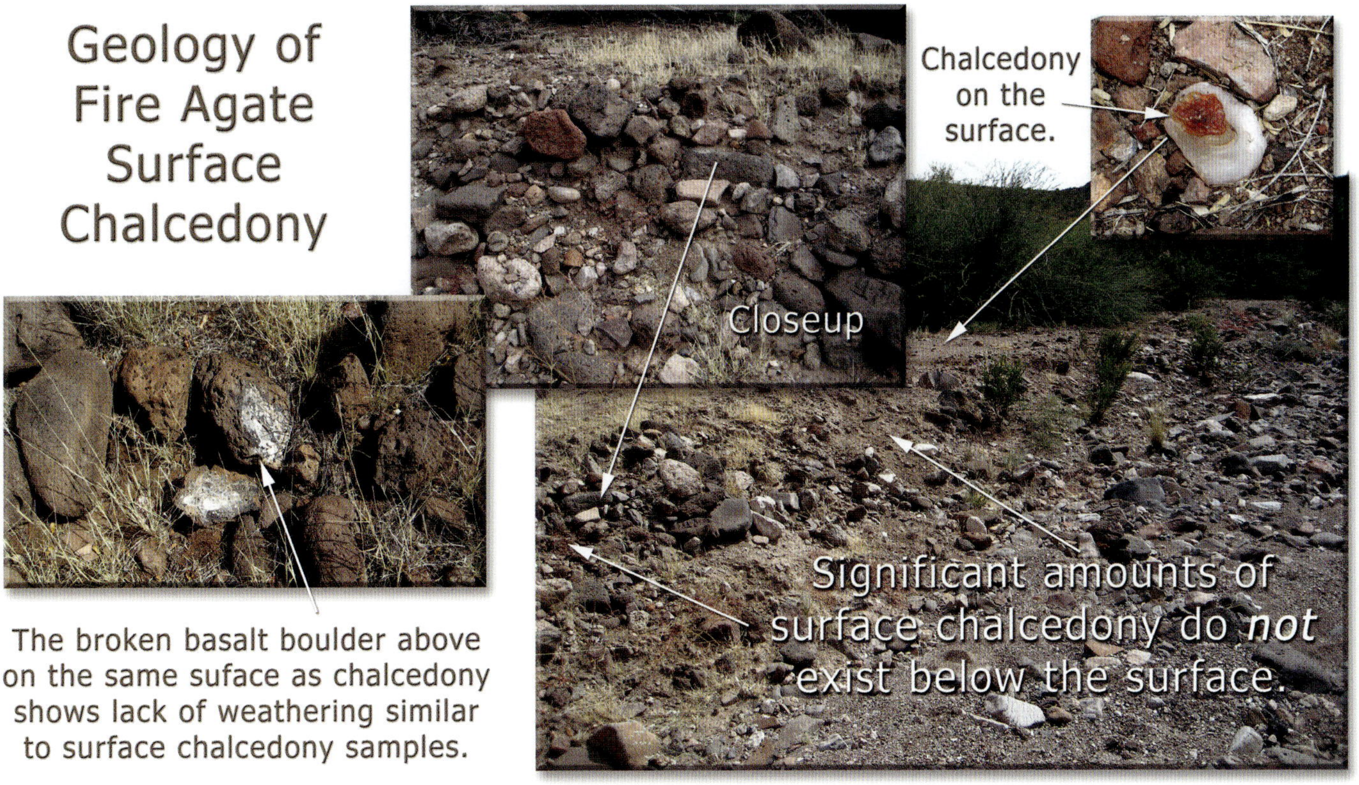

Geology of Fire Agate Surface Chalcedony

The broken basalt boulder above on the same suface as chalcedony shows lack of weathering similar to surface chalcedony samples.

Closeup

Chalcedony on the surface.

Significant amounts of surface chalcedony do not exist below the surface.

Fig 8.14.5 – The Geology of Surface Chalcedony has been long misunderstood. Specimens lie on the surface of all kinds of unconsolidated sediments, not just in 'volcanic areas.' In fact, no one has ever reported seeing chalcedony form in a volcanic setting on land. Of course they wouldn't, because chalcedony is a form of quartz that grows in a hypretherm (under deep hot water). The images above come from the Fire Agates Surface Chalcedony field discussed in Fig 8.14.4, near Safford, Arizona, USA. Notice that the wash has eroded the landscape, revealing lower layers of sediment that contain very few chalcedony specimens. Similar to the chalcedony rocks, many of the basalt boulders in the area show little or no erosion. The broken basalt boulder in the leftmost image contains carbonate, indicative that it formed in an ocean. There are no lava flows or volcanoes in the region, so the source of the basalt nodules and boulders seems unclear. This area is an example, similar to Fig 8.4.21, of UF Hydrofountains bringing many of the rocks to the surface. This is why they do not appear to have a surface origin. The rounding was caused by endoerosion; movement in underground aquifers. They were ejected onto the surface near the end of the Deluge, when the final frictional heating occurred, as continental plates moved back into place.

coagulate as a gel inside the bubbles. Then iron and manganese compounds from the surrounding rock infiltrate the gel, creating layers of iron hydroxide. Eventually the whole bubble hardens and crystallizes, **forming distinctive nodules of banded agate.**" Bib 151 p205

The nodules and agate could not have come from lava or they would have been made of *glass*. It is amazing that modern science's version of agate origins is the absence of the word—*water!* One wonders why geology professionals lack the most basic knowledge of agate and geode formation. Dogmagma—the belief that minerals came primarily from a melt is false. It is extremely rare to find a modern geologist or mineralogist actually growing quartz or silica rocks. Actually, we have yet to meet ***anyone*** who has. As soon as the educational community begins to grow quartz-based rocks the way Nature does, students, teachers, and even the world must experience a complete shift in their paradigm as they begin to see how the Flood impacted the world we live on. The UF Hypretherm challenges literally every aspect of modern geology and sets the world on a new, higher level of understanding as scientists reach for answers to questions never before envisioned.

The Agates and Geode Mystery of chapter 6.4 persists because today's researchers do not understand the hypretherm and hold to the false notion that quartz based minerals including agates and geodes came from lava or hydrothermal fluid *without added pressure*. After having spent over two decades researching and conversing with dozens of geologists, mineralogists, and collectors, it became very apparent where the real knowledge about agates and geodes was—and it was not with the professionals.

The Thunderegg Evidence

The best example to illustrate the superiority of amateur knowledge and wisdom regarding agates and geodes comes from Mr. Robert Colburn, known also as the **Geode Kid**. He

Chalcedony Geodes

Chalcedony Rolling Layers

Fig 8.14.6 - Geodes from Mexico are a good example of chalcedony formation. The outside layers indicate the first layer was originally formed inside a limestone fluid cavity. That covering was subsequently removed. After rolling about, layers of swirly chalcedony grew; quartz crystals seen on the specimen on the right grew in place at that time. Geodes like these are found only near the surface, verifying the surface characteristics of chalcedony.

TAG Mound Surface Chalcedony

Chalcedony grew on the *surface* and sides of the mound, *not* below the surface.

"Samples taken from TAG 1 and 5 *from below* the anhydrite zone *contain no chalcedony.*"

"Length-fast *chalcedony occurs* in an outer low temperature envelope across *the top and sides of the mound.*"

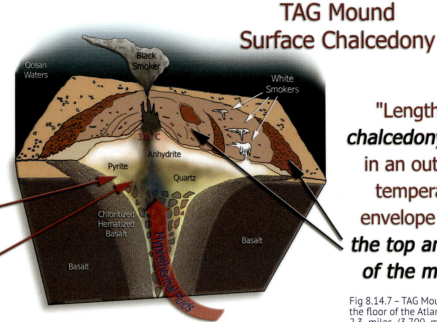

Contributions to Mineralogy and Petrology, Vol. 137 No. 4, December 1999, Abstract

Fig 8.14.7 – TAG Mound is located on the floor of the Atlantic Ocean under 2.3 miles (3,700 meters) of ocean water.

has personally collected and cut tens of thousands of 'thundereggs' over nearly half a century. He recently donated his collection to the Deming Luna Mimbres Museum in downtown Deming, New Mexico, USA. Colburn has collected from over 100 locations, gathering some of the most unusual and beautiful specimens ever found.

One type of geode, a **thunderegg** is a particular type of nodule that exhibits a star-like structure when cut. It often contains various types of quartz such as agate, jasper or chalcedony and many other minerals.

Colburn had interacted with Peter J. Heaney of Penn State University and many other professionals over the years. He had the best 'hands on' experience of any individual we talked with, and he shared much about the geology and circumstances in which these agates and geodes formed. Colburn produced a 26 page report on his research titled, *The Formation of Thundereggs*. In his report, he makes the following important points:

"This last point – of a **thunderegg being filled with silica bearing aqueous solutions** is important in light of several others attempting to develop various theories on how thundereggs form, specifically by any other means than a gas cavity filled by silica solutions. In a **hot molten silica gel accretion theory**, it is instructive and necessary to quote a passage from B.M. Shaub's dissertation published in the February (and continued in the March issue) of the *Lapidary Journal* in 1979, page 2348, he says of all other gas-cavity based theories,

"'All base their theories on the assumption that by some force, mechanism or other culminating processes of nature, large cavities were developed in the **acid magmas**. The cavities are in reality the **only basic starting point for every theory known to the author**. The interiors of gas cavities are in general very smooth. Unless deformed after formation, the cavity walls have no distortions or irregular interior surfaces. The writer has never seen rough, irregular inside walls in scoria or other rocks with gas cavities large enough for the examination of their inside cavity walls...'

"Then he concludes:

"'Hence it must be certain that thundereggs are not cast of gas cavities and all theories based on gas cavity filling must be completely defective...'

"He goes on to state that,

"'...even our renowned petrographers and mineralogists have been led astray on fruitless routes by trying to match the spherical shape of thundereggs with that of the gas cavities...'

"He continues to describe his **theory on thunderegg genesis** which postulates an accretion of silica gel as an immiscible **directly out of molten rock** which then spontaneously solidifies to the entities we find today." Note 8.14f

Hyalite Surface Opal

Fig 8.14.8 – This interesting specimen from Valec, Czech Republic is a form of opal called hyalite. This sample also formed on the surface. It is one example of many among other types of minerals that formed on the surface only recently. These types of minerals are additional Marks of the Flood and, like the sample above, show no erosion. Modern geology apparently missed the fact that chalcedony and other types of minerals like Hyalite Surface Opal are not forming today on the continental surface. They are Marks of the Universal Flood.

Bucking the norm, Colburn recognized that geodes did not come from "molten rock." He continued describing his experiences, stating that thundereggs can be best described as being filled with a silica-rich fluid:

"I must credit this gentleman with piquing my intuition and inspiring me to reevaluate and write about what I have seen in my specimens and how they occur in the field. My first reaction to this 'hot gel' theory was to ask **how a calcite crystal could get into such a hot environment**, become encased and circumscribed with banking of the agate. **Calcite crystals cannot form in the 2000°+ Fahrenheit temperatures in the relatively low pressures of a lava flow**. Such temperatures would reduce calcite to calcium oxide and carbon dioxide. **I have scores of specimens with calcite crystals well inside the agate, as well as other water-based inclusions such as manganese dendrites, zeolites (sagenite) plumes and stalactitic growth**. These growths, which in most cases are the first implacements in thunderegg cavities, have subsequent bandings of **chalcedony and/or quartz** coating and outlining these as well as every other protuberance on the cavity wall itself. This is why the radial fibrous structures are seen on weathered agate cores of **thundereggs which are often mistaken for agate casts of fossils or coral, which we shall be entertained with later**.

"Perhaps we can get closer to understanding thundereggs by approaching them as **gas cavities filled with water-based minerals**, like those of amygdaloids, an idea fairly well accepted by most geologists.

"An amygdaloid is a mineral-filled gas cavity found in basalt and andesite lavas such as the well-known huge deposit of Brazilian agates. These nodules have a round to tear-drop shape, and walls in these lavas are smooth from which amygdaloids are removed or weathered. The lavas that contain these cavities, filled or not, are called amygdaloidal basalt and/or andesite. These lavas are the largest source of igneous-based agate nodules and geodes in the world. **It is intuitively simpler for me to describe a silica-water based filling for gas cavities in thundereggs as well as for amygdaloids. But it only seems intuitively contrary for expanding gases to form box-like asteriated holes**.

"Therefore, it would be necessary to find a natural mechanism that would produce such a cavity. After doing this, **all water-based solution fillings of a gas cavity would then be explained**." Note 8.14f

Colburn makes the case for hyprethermal growth by describing the necessary increased temperature and pressure in a water environment, and he sets the stage of the Surface Mark by discovering an interesting relationship in the formation of the cavities, illustrated in Fig 8.14.11. The Thunderegg Formation Sequence shows a relationship between the size of the cavity and the depth at which it was buried, and it is obvious that the deeper the geode, the smaller the cavity.

Colburn did not recognize the significance of his discovery. From the UM perspective, the geodes demonstrate the existence of a *recent, global hypretherm* that was extant right on the surface we walk on today.

Are geodes like these found in deep layers of the Grand Canyon? No, they are not.

We know geodes are not found in the deep layers of the Grand Canyon because they were formed at only one time in the Earth's *recent* history, and we know exactly when that was. Colburn also mentions that "agate casts of fossils or coral" are often mistaken for thundereggs.

FQ: If agate "casts of fossils" are often mistaken for thundereggs, what does it tell us about fossils that are made of agate?

The Fossil Model chapter utilizes all of these new discoveries to make even more new scientific discoveries, but we're ahead of ourselves. Look at the

Fig 8.14.9 – Geodes and agates are some of the most beautiful creations of nature, and one of the biggest mysteries in modern geology. It is only through the mechanism of the UF that we can comprehend their origin.

Fig 8.14.10 - A particular type of agate or geode nodule that contains star-like or other geometric shapes is often called a thunderegg. They are generally a few inches in diameter and come in every imaginable design and color. Thundereggs, like all geodes hold important clues for geology because they help explain surface mineral formations that were formed with them, in the hypretherm, on or near the floor of a deep, hot, continental ocean. Certain biominerals, then living, are responsible for the thunderegg's beautiful colors.

dinosaur bones in 8.14.12 that were fossilized. Their centers are now filled with an array of beautifully colored agate. Bones contain a soft center, where the marrow carried the lifeblood of the animal. The softer tissues gave way to the solvent action of acidic or alkaline water, leaving behind cavities to be filled with siliceous rock. When the fossilized bones are crosscut, they exhibit the same crystalline pattern toward the center of the bone that many agates and geodes do, because the bones and the geodes contained similar fluids. The fluid in the bones and the fluid in the agates and geodes was heated and pressurized in the UF hypretherm environment on the floor of the continent-covering ocean, which is why they are found only on or near the surface today.

The First Man-made Geode

Fig 8.14.13 illustrates an experiment where we encountered the first man-made Geode. It had been a surprise that the geological community never followed up on the discoveries technologists were making about how quartz is made by applying known hyprethermal techniques to natural geological processes. Man-made sandstone was formed using a hyprethermal reactor, shown in Fig 8.5.17, and the first man-made geode was formed in the same way, a reproduction of the UF Hypretherm.

Geodes are generally defined as rocks with an open cavity with crystals growing on the inside surfaces. Fig 8.14.13 shows thousands of tiny quartz crystals growing on the inside wall of the reactor. This did not happen during normal experimental runs involving just quartz. The nucleation of small quartz crystals occurred only with the introduction of wood into the system! Dozens of trial runs failed to produce petrified wood because of a multitude of factors, too much heat, too rapid heating, and incorrect mixtures, but those runs often created beautiful geodes, profuse with quartz crystals. The dissolved wood particles became the nuclei for crystals growth throughout the reactor. As fascinating as this new discovery was, future runs eventually yielded petrified wood, an exciting story we will visit in subchapter 11.4.

This concludes our brief discussion about the variety of Quartz Surface Mark evidence; surface chalcedony, agates, geodes, and other quartz related surface minerals found at the surface of today's continents, some still lay untouched in dry desert areas, unmoved since they were formed in the UF hypretherm.

Thunderegg Formation Sequence

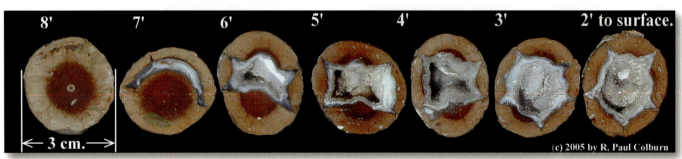

Fig 8.14.11 – This rare sequence of thundereggs shows the typical formation of geode crystals in specimens collected at 8', 7', 6', 5', 4', 3' and 2 feet from the surface. This sequence proves that not only did the formation of thundereggs take place near the present day surface (where they were found), the closer the thunderegg was to the surface, the larger the cavity. Specimens like these are found worldwide, but **only** on or near the surface, so how could they possibly be millions of years old, even *if* the Uplift and Subduction Pseudotheories were correct? Thundereggs are direct evidence of the *recent, global hyprethermal event* known as the Universal Flood. Image courtesy of Robert Colburn.

CHAPTER 8 THE UNIVERSAL FLOOD MODEL

The Carbonate Surface Marks

The Carbonate Mark (subchapter 8.8) established the fact that large, crustal carbonate *deposits* are not being formed today. It also established that the existing large carbonate deposit's origins are not explained with any degree of clarity by modern geology. However, the UF easily accounts for and clearly explains the genesis of carbonate deposits with the UF Hypretherm and with biomineralization.

There are four Carbonate *Surface* Marks (Caliche, Limesurface, Dissolved Rocks, and Great Salt Lake Sand), and several examples of carbonate deposits found only on the *surface* today, not deep in the sedimentary record. The emphasis of the Surface Marks is the surficial nature of the mark, and that it is not apparent deep in the geologic column.

One of the extraordinary claims made in the UM is that these Surface Marks remain unnoticed by the geological community at large. The primary reason the Surface Marks continue to be neglected despite the simple nature of the observations, is that modern science lacks the UF paradigm. With so many Flood evidences already laid out, the reality of the Deluge becomes more apparent than ever. By looking at Nature through new Flood glasses, a whole new geological world opens to our view.

The Caliche Surface Mark

The Caliche Mystery was introduced in Chapter 6 (see Fig 6.9.18). Later in the Basalt Carbonate Connection, (see Fig 8.7.13) a layer of limestone cement just below the surface was identified. When rounded rocks are cemented together with naturally occurring lime, it becomes a **calcareous conglomerate**, which is commonly known as **caliche**. Examples of it are shown in Fig 8.14.14, and similar forms of light colored caliche are found around the world.

Researchers have *assumed* that caliche is formed merely through the leaching of minerals "from the upper layer of the soil":

"*Caliche* generally forms when minerals **leach from the upper layer of the soil** (the A horizon) and accumulate in the next layer (the B horizon), **at depths of approximately 3 to 10 feet under the surface**." Note 8.14g

The problem is that there are many cases where sediment having no relationship to the caliche is lying above the caliche layer, such as the clay layer in Fig 8.14.14. An even more obvious refutation comes from the study of any deep canyon. One should find multiple layers of caliche that presumably formed over 'millions' of years. They just don't exist, but this has not stopped modern science authors from stating that they do!:

"[Caliche] layers vary from a few inches to feet thick, and **multiple layers can exist in a single location**." Note 8.14g

Where are the examples of "multiple layers" of caliche? There are *none!* This is one great example of the "just trust me" mind-set that runs throughout modern science, and it must be replaced with the "show me" attitude of the UM. The simple act of *showing* science students instead of merely teaching them pseudotheories will completely transform modern science into Millennial Science.

FQ: Where do we find caliche being formed today?

We found no instances of modern caliche deposition in any of our research. In fact, there was no discussion on the length of time it takes to form. Because caliche is almost always relatively near the surface, erosion precludes the possibility of a long formative period. One of the ways in which the origin of caliche has been explained involves a process of leaching and subsequent rising of dissolved minerals:

"However, *caliche* also form in other ways. It can form when water rises through capillary action. In an arid region, rainwater will sink into the ground very quickly. Later, as the surface dries out, the water below the surface rises, **carrying up dissolved minerals from lower layers**." Note 8.14g

Despite a number of hypothetical claims, caliche has not been shown to be forming in any soils anywhere around the world today, because the right environment does not exist for the cementing process to take place. This supposed dissolution of minerals that then rise through capillary action is similar to the Carbonic Acid Pseudotheory that attempts to explains cave formation (chapter 8.8). Both theories are fraught with prob-

Fossilized dinosaur bones were changed into agate just as other geodes and agates were, in the UF Hypretherm, only on or near the surface.

Fig 8.14.12 - These fossilized bone specimens are filled with agate. Comprised of the same material as geodes, they were both formed in a hypretherm environment at the same time. Open cavities, whether inside bones or pockets in the sediment, were filled with siliceous fluid that crystallized into agate rocks during the UF Hypretherm.

Dinosaur Fossil Agate

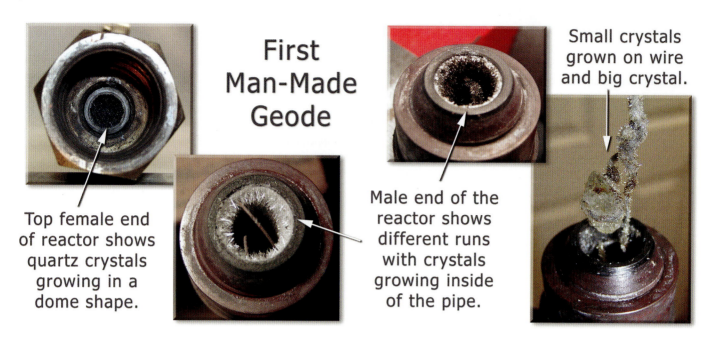

Fig 8.14.13 – These images show the interior of our quartz autoclave reactor after runs that produced geode-like results. The top-left image shows the female end cap with a dome shaped array of quartz crystals that was discovered when the reactor was opened. The two center images are of the open pipe part of the autoclave after two different runs, showing different crystal structures that grew on the inside of the pipe. The large crystal wrapped in wire was covered with tiny quartz crystals, a surprise since this had not happened during previous quartz growth experiments. What caused this nucleation of small crystals to grow throughout the inside of the autoclave? The only thing that changed was the addition of wood. These man-made geodes are the result of wood petrification experiments inside the autoclave. The disintegration of the wood in the reactor created tiny seeds that stimulated the precipitation of the quartz in solution, causing a remarkable crystallization event. More details about these experiments will be discussed in the upcoming Fossil Model Chapter.

lems; first, there is rarely any carbonate above or immediately below the caliche layer that can be dissolved, and secondly, the supposed carbonic acid doesn't exist to dissolve it! Moreover, as the Carbonate Mark subchapter explained, carbonates come from a biomineralization process, not by leaching or mineral dissolution (see Fig 8.14.15). Put simply, caliche **grew** by microbes in a specialized hydrothermal environment toward the end of the Flood. This is why caliche is only found on or near the crust's surface.

There is a great deal of similarity between the Surface Carbonate Marks of caliche and limesurface, our next topic. Caliche grew on the *bottom of surface rocks, sitting in the soil*, whereas a different type of carbonate growth occurs on rocks submerged in a river or lake, which *completely covers the rock* with a carbonate **limesurface** layer.

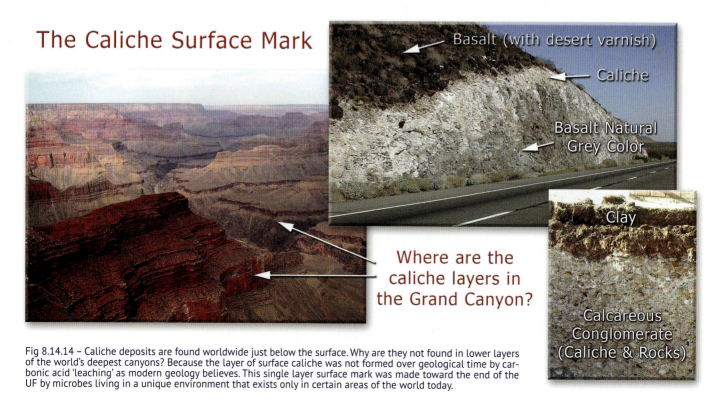

Fig 8.14.14 – Caliche deposits are found worldwide just below the surface. Why are they not found in lower layers of the world's deepest canyons? Because the layer of surface caliche was not formed over geological time by carbonic acid 'leaching' as modern geology believes. This single layer surface mark was made toward the end of the UF by microbes living in a unique environment that exists only in certain areas of the world today.

CHAPTER 8 THE UNIVERSAL FLOOD MODEL

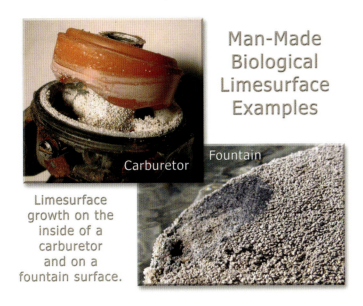

Man-Made Biological Limesurface Examples

Carburetor   Fountain

Limesurface growth on the inside of a carburetor and on a fountain surface.

Fig 8.14.15 - Limesurfaces are not just found in geology but have been documented in many instances, including man-made scenarios. Here we see a carbonaceous surface on the inside of an old carburetor that water and fuel had accumulated in. The picture to the right shows the effects of a man-made fountain. A continuous spray of water coated the surface, which evaporated as the limesurface grew.

The Neglected One-Sided Limesurface Mark

Many of the UF Marks have been disregarded, ignored or not recognized, but the Limesurface Mark has to be one of the simplest and most obvious. It is easy to see, easy to understand, and has been right in front of our noses all along. It was one of the exciting discoveries. The Limesurface Mark certainly provides unequivocal evidence of the UF, but we need to review several images to see what has been missed.

With the biogenic origin of limesurface established, we can discuss these two types of naturally occurring limesurface rocks:

1. Water Limesurface Rocks
2. Soil Limesurface Rocks

Caliche is a macro form of soil limesurface, accumulating in sometimes thick masses over large areas, but there is another type of limesurface rock, specially formed and lying at the heart of our discussion. But first, we need to find out what water limesurface rocks look like. Fig 8.14.16 shows rocks along the Colorado River and at the shore of Lake Mead, just west of the Grand Canyon. In this environment, we find water with just the right mixture of microbes able to deposit the white carbonate limesurface on the rock walls of the canyons and along the river bottom.

Not all natural water bodies leave behind the evidence of microbial action.

Microbes need a host of minerals to live and many require elevated temperatures to reproduce in large enough numbers to form the white carbonate film left behind on some lakes and rivers. When warm water and the right minerals are present, limesurface grows in modern times. Such environments are typical of desert bodies of water. One example is the Pinacate Craters group in northern Mexico, seen in Fig 8.14.17. There, limesurface shows up around both the Elegante and the Colorado Hydrocraters, which are located in a volcanic field that last erupted only several thousand years ago. There are cracks filled with limesurface that came from the hydrothermal waters that escaped from the outer surfaces of the Colorado Hydrocrater. Typically, cracked rocks or soil fissures in the desert do not exhibit limesurface growth because the soil is not saturated with hot, microbial rich water. The thickness and hardness of the limesurface are other attributes to consider. The hotter the water, and the longer the hydrothermal microbial waters existed, the thicker and harder the limesurface.

If the right microbes are present in the water, the temperature does not have to be extremely high. The Fossil Creek limesurface example in Fig 6.9.5 is a rare example of extreme carbonate biomineralization that took place very quickly because of the ideal environment that allowed microbes to reproduce profusely, encouraged perhaps by frequent water level changes.

Another example comes from Rock Canyon, in Provo, Utah, Fig 8.14.18. This dry creek bed contains various types of rocks that were only recently *entirely* covered with limesurface. The

Lake Mead and Colorado River Limesurface

Limesurface

Fig 8.14.16 - These images were taken along the Colorado River and at Lake Mead, Arizona, USA. They clearly show the limesurface on the rock walls of the lake and along the river. Not all lakes and rivers have such white lines, which indicate the growth of a limesurface mineral. A well defined limesurface is usually only evident in warmer climates where the right microbes are present. The limesurface is seen when water levels change. These carbonate surfaces are not created simply because of the evaporation calcium carbonate rich water as was once believed, but are formed by a biomineralization process that does not involve the dissolution of other carbonates.

Pinacate Craters Limesurface Evidence

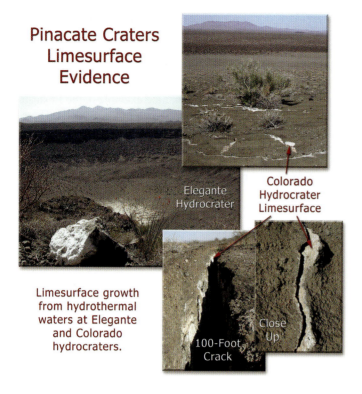

Limesurface growth from hydrothermal waters at Elegante and Colorado hydrocraters.

Fig 8.14.17 - Although limesurface is rather common in desert areas, it does not occur just anywhere. Here we see limesurface in cracks on the desert floor and on a boulder within the two hydrocraters at the Pinacate preserve, Mexico. The carbonate microbes grow in hydrothermal waters when the right mix of minerals was present.

limesurface on these *modern day* rocks is some of the thickest ever observed in over 20 years of field research.

The canyon is an excellent example of what happens when rocks move in a river loaded with limesurface microbes—the ***entire surface*** *of the rocks became coated with carbonate material*. Compare the *Water* Limesurface Rocks of Rock Canyon or Lake Mead with *Soil* Limesurface Rocks in Fig 8.14.21.

According to modern erosion theory, desert surface rocks are moved and turned during periods of erosion, exposing all sides to the soil. Although desert rocks take more time to move than those in the dry riverbed in Fig 8.14.18, modern science sees this happening over millions of years. Based on that, we need to ask an important FQ:

> What should limesurface rocks on the desert soil look like after millions of years?

This question suggests a simple answer—they should look similar to Water Limesurface Rocks, being *entirely* covered with carbonate on all surfaces. But are they?

In Fig 8.14.19, rocks from Capitol Reef Park, Utah show the limesurface covering on only one side of the rocks! This phenomenon is why we refer to the rocks as One-sided Limesurface Rocks. This type of limesurface is not exclusive to rocks from Capitol Reef, they represent a wide-spread phenomenon.

Fig 8.14.21—The Global One-Sided Limesurface Reality diagram shows rocks that have been displaced by erosional processes. It represents One-Sided Limesurface Rocks that are found on the desert surfaces *all across* the Colorado Plateau and the Western United States. Similar samples can be found in deserts throughout the world. Many One-Sided Limesurface rock samples collected over the past decade exhibit limesurface thicknesses of up to an inch (several centimeters). The limesurface on these rocks is much harder than Water Limesurface samples.

There are two extraordinary claims based on the One-Sided Limesurface, both very similar to the extraordinary Surface Chalcedony claims. First, the limesurface was formed on these rocks globally, over a short period of time and in a unique environment that does not exist today. Secondly, these rocks are found only on or near the surface, not in deep sediment because the global event in which they formed took place only several thousand years ago. These extraordinary UM claims come with extraordinary evidences, if we will but look with the new UM paradigm.

Because answers come from questions, let us consider three FQ's to find the evidence for our extraordinary claims:

> If carbonates are being deposited in the soil on rocks today, *where* can we observe this process?
>
> If the rocks have been eroded and moved for millions of years, why is there only *one* limesurface?
>
> Where are the carbonate coated rocks with different planes of limesurface coating?

We find the answers to these questions in Fig 8.14.22. Of the

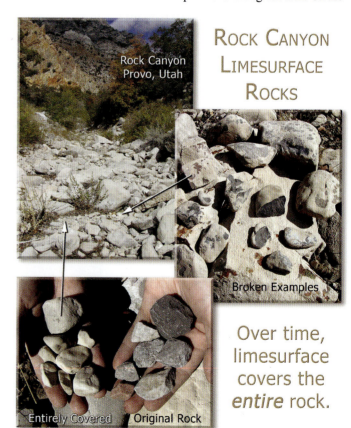

Rock Canyon Limesurface Rocks

Over time, limesurface covers the *entire* rock.

Fig 8.14.18 – because of a lack of limesurface microbes, specific nutrients in the water, and low temperatures, *most* riverbeds do not contain rocks that are entirely covered with limesurface. In this example from Rock Canyon, Utah, USA, however, the rocks are coated with limesurface because conditions were right, allowing the surface coating to grow, ***over time***.

CHAPTER 8 THE UNIVERSAL FLOOD MODEL

One-Sided Limesurface Evidence

Capitol Reef Park Utah, USA

How could these rocks be covered with limesurface on *only* one side if they were really transported long distances over millions of years?

Fig 8.14.19 – Why do these limesurface rocks from Capitol Reef Park, Utah, USA, not look like the limesurface rocks from Rock Canyon in Fig 8.14.18? After millions of years, there has been plenty of time for erosion to turn the rocks so that the limesurface could grow on all sides of the rocks. Why hasn't this occurred? Why is the limesurface on only one side? Answers to these questions have eluded modern geology.

hundreds of desert locations observed over the past decade, it was rare to find any evidence of modern soil limesurface processes. When rocks had been *reburied* as seen in Fig 8.14.22, there was no clearly discernable evidence of new limesurface growth. Over 95% (based on the authors own observations) of the reburied rocks examined, showed no signs of a new limesurface growth.

Occasionally, a rock would be found showing more than one limesurface plane, but the second line rarely corresponded to the orientation in which it was buried. The secondary line on these rare rocks can be attributed to the rocks movement during the short period of active limesurface formation, or the result of uneven soil surface while the carbonate surface was growing. In either case, the occurrences are rare. One-Sided Limesurface rocks can be easily found in most arid deserts, the limesurface corresponding to the partially buried rocks.

This scientific discovery is very important and it is simple and observable, yet it profoundly demonstrates the global surface environment that existed not long ago, and has not been reproduced since. If the *warm, wet, microbe rich soil environment* that occurred toward the end of the Flood had occurred more than once, we would find a *large number of rocks with multi-lined or banded limesurfaces*. But they do not exist.

It is the same reason the thick caliche layer that exists around the world is found only near the present day surface, unless the surface was eroded or moved since its formation. It is quite astonishing how common these rock types are in the world's deserts, yet no serious discussion concerning the one-sided limesurface exists in the scientific literature. Of course we have not looked at every possible discourse on the subject, but thousands of books, journals, articles and web sites revealed nothing of consequence on the subject. The Carbonate Surface Mark has probably been the most neglected and yet is the simplest to understand of the geological evidences of the Universal Flood.

The Dissolved Surface Mark

The third type of Carbonate Surface Mark is the **Dissolved Surface Mark.** It is similar to the Caliche and the Limesurface Marks in that it involves rocks found directly on the surface that have been neglected by modern geology probably because they are such an enigma. The Dissolved Surface rocks directly refute modern erosion theory.

In the desert north of Las Vegas, Nevada, USA, a collection of surface and subsurface rocks, seen in Fig 8.14.23, tell the story of the Dissolved Surface Mark. Rocks at this location are rounded, light colored clay/limestone cobblestones. As is evident in the diagram, the cobblestones that were buried beneath the surface are round and smooth and show no sign of being dissolved at all. However, rocks from the surface are quite different. Some individual rocks are *partially or almost completely dissolved,* as seen in the diagram.

FQ: What caused such a large portion of these rocks to disappear?

The answer to this question is partially revealed by the name given to them, the subject of this section—*dissolved* surface rocks. By **dissolved** we do *not* mean weathered by erosion, but

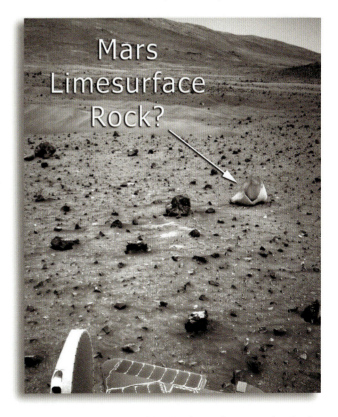

Fig 8.14.20 – This photo of the Martian surface was taken by the Spirit Rover on January 26, 2006. A rock that may be covered by limesurface is a few meters away from the Rover. If limesurface rocks on Mars are confirmed, then Mars was not covered by just an ocean, it was covered by a microbial filled ocean.

Global One-Sided Limesurface Reality

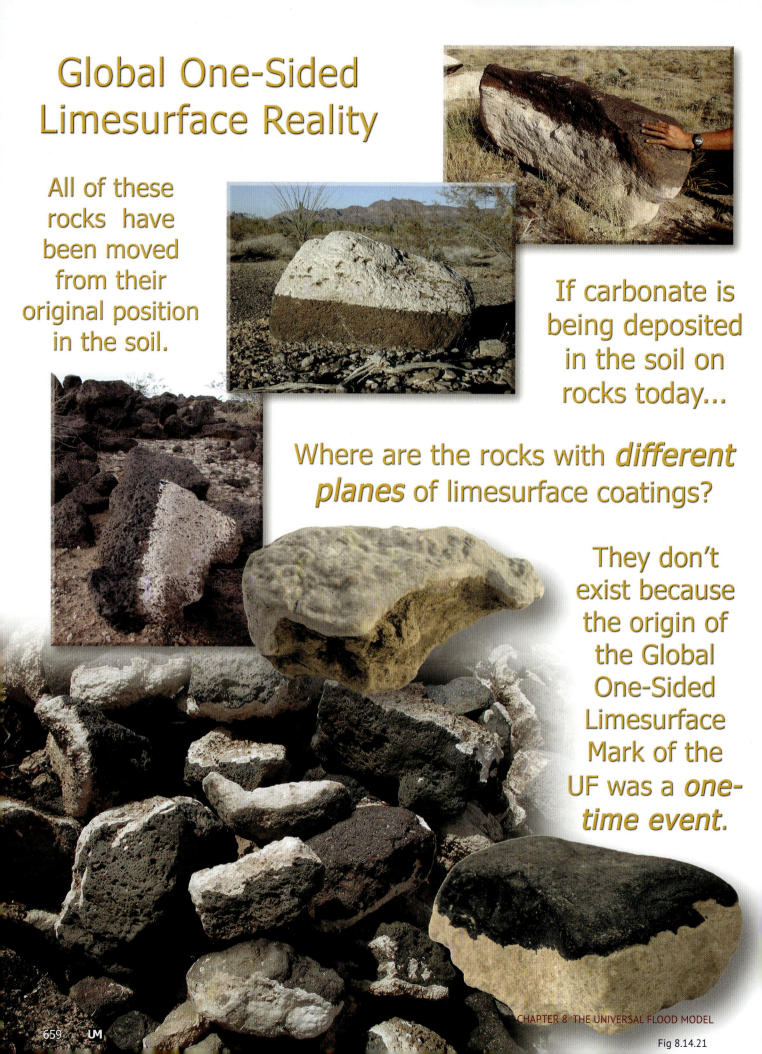

All of these rocks have been moved from their original position in the soil.

If carbonate is being deposited in the soil on rocks today...

Where are the rocks with *different planes* of limesurface coatings?

They don't exist because the origin of the Global One-Sided Limesurface Mark of the UF was a *one-time event*.

CHAPTER 8 THE UNIVERSAL FLOOD MODEL

Fig 8.14.21

Limesurface Reburial Reality

Has there been new limesurface growth on the underside of this repositioned rock? Lifting the rock reveals no new limesurface.

Observation of the rock revealed no evidence of a secondary growth of limesurface.

This Rock has been turned vertically and reburied. Will another horizontal limesurface grow?

Fig 8.14.22 – These examples of Soil Limesurface Reburial were found near Wickenburg, Arizona, USA, but similar examples can be found worldwide. As these rocks confirm, new limesurface is **not** growing on the rocks after they were repositioned and reburied. The rocks showed no evidence of multiple planes of limesurface that should have been evident if the process was ongoing. This indicates there was a single-limesurface event that took place not long ago because the weathering on the limesurface rocks is minimal. Soil Limesurface is the fourth verification (surface chalcedony, geodes, and caliche being the other three) of the unique mineralization process that took place near the end of the UF event on the Earth's surface. Each of these Surface Marks has only one simple clear explanation—the Universal Flood.

exactly what we said—dissolved.

At first glance, investigators may look at these rocks and think that perhaps they are just deeply eroded and cracked because of 'millions of years' in the desert. Nevertheless, careful observation reveals that this is not the case. Other siliceous rocks found in the same location, on the surface or just below the surface are not 'partially dissolved' and more importantly, they do *not* have a *rough and cracked surface* like the heavily dissolved surface samples do. Put simply, if erosion had caused these rocks to look the way they do, they would have been moved and tumbled, making them *smooth* again, or there would be a variety of partially dissolved rocks buried beneath the surface—but this is not the case. Instead, the specimens embedded in the caliche that was beneath the surface are more dissolved than the white carbonate caliche material surrounding them. In fact, the carbonate material surrounding some of the specimens preserved the original size and shape of the clay/limestone cobblestones, seen in Fig 8.14.23.

There is no evidence that rain or weathering is reducing the size of these rocks or making them rough and cracked. The two-page layout, Fig 8.14.24, shows similar Dissolved Surface rocks in the Valley of Fire State Park, Nevada, USA. The unusual consistency of size says something of the depositional environment. Where are the smaller or larger rocks on the white sandstone that should be present if they were placed there by erosion?

The rocks became rough, cracked and dissolved right on the surface because of *acid*. When the cobblestones are put in a hydrochloric (HCl) acid bath, they rapidly dissolve in only a couple of hours. Any attached silica rocks remain unaffected because quartz based minerals do not dissolve in an acidic solution like the carbonate rocks do. Some researchers attempt to explain the dissolution of the rocks by Carbonic Acid; a long held Pseudotheory that does not work for these rocks for the same reason it does not work in limestone caves. First, carbonic acid has *never* been shown to exist in the *current or past environment* from typical weather conditions over large continental areas, areas where these rocks exist. Secondly, if naturally occurring acid rain had been responsible, that would still take 'geological time' which would be offset by the *smoothing action of erosion*.

There is only one simple logical explanation for the geneses of these rocks:

They were dissolved in a short-lived acidic environment not long ago.

This is why Dissolved Surface rocks are only understood with Flood Geology and not modern geology. The TAG mound showed unequivocally and with empirical evidence that at the bottom of a deep ocean, in a hypretherm, *acidic* water is abundant, along with silica minerals. This is the exact environment needed to form the dissolved specimens we have been discussing. Another factor involved is the origin of the cobblestones themselves. There was no nearby source and there was not a local surface deposit. The buried cobblestones were much *too smooth* to be formed by surface erosion. The source of these cobblestones was certainly a hydrofountain.

Dissolved Surface rocks are not as common as limesurface rocks, but are found about as frequently as surface chalcedony is in many arid regions of the Earth. Samples have been collected in desert areas throughout the western United States from high mountain elevations, mid-elevation plateaus, and both in a variety of desert locales. All of them share similar characteristics of acidic dissolution.

The Great Salt Lake Sand Evidence

One final example of the Carbonate Surface Marks of the Flood is the Great Salt Lake sand from around Utah's largest lake. Fig 8.14.25 shows the two common shapes of this unique and beautiful sand: spherical and cylindrical shapes. The two shapes were formed when bicarbonate-cementing agents in the water began combining with tiny bits of sand as they rolled around on the lake floor. The cylindrical shaped sand formed

Fig 8.14.23

Las Vegas Dissolved Rock Evidence

These clay-limestone cobblestones were found below the surface and were *not* dissolved.

Original Cobblestone Size

These dissolved cobblestones are found *only* on the surface.

Before

After

Dissolving

Dissolved Cobblestones

The cobblestones were found to dissolve in HCl acid and leave behind silt or sand residue.

Partially dissolved specimens occur when the rock is covered with a thin layer of sediment.

Some of these cobblestones had quartz fossils embedded in them that showed little erosion, verifying their formation in a recent hypretherm.

CHAPTER 8 THE UNIVERSAL FLOOD MODEL

this way too, but was probably near the shore where wave action rolled the grains in a more uniform direction, forming the long cylindrical shapes. Both formed when the water supported carbonate organisms in an environment much different from today.

How do we know the Great Salt Lake sand formed under conditions different from today? Geology's uniformity myth reveals the answer. Through careful observation of the sand while keeping an open mind, the following FQ presents itself:

Where are the proto-Great Salt Lake
sand grains that are just beginning to form?

Great Salt Lake sand is remarkably uniform in size. *If* the sand had formed over millions of years, and *if* the natural process that formed them is still active today, there would be sand grains orders of magnitude larger and smaller than what is found, especially given the evidence that the sand grains appear to have *grown—not eroded*. When the spherical sand grains are cut in half, they exhibit many layers of growth, somewhat like an onion, which suggests the Great Salt Lake sand should be forming today, with very small grains growing into larger grains. Significantly, a survey of the sands around the Lake revealed an obvious absence of smaller and larger grains of sand.

The uniformity of the existing sand suggests a formative period where the grains were all formed quickly, at a time not long ago when a carbonate-microbe rich environment existed; microbes created the carbonate cement that bound clay and sand grains together. Additional evidence comes from the uniqueness of the sand—there are very few sands in the world with spherical and cylindrical grains. If they were forming today, it is likely that other salt lakes would exhibit similar sand formation. Furthermore, no sands similar to the Great Salt Lake sands are known to exist in the deep layers of exposed canyons, like the Grand Canyon.

The youthful age of the Great Salt Lake Sand is apparent because of the small amount of eroded quartz sediment transported into the lake from fresh water river sources. If eroded silica-based sediment had been flowing into the Great Salt Lake basin for millions of years, a significant amount of it would be mixed in with the sand—but it's not there. Great Salt Lake sand grains are not forming today because they are surface fossils of microbial carbonate and silt that were formed in an environment that does not exist today.

The Great Salt Lake Sand Evidence is one of the four Carbonate Surface Marks which are complimentary to the two Quartz Surface Marks. Remember, the Carbonate and Quartz Surface Marks presented here are in no way a complete list of the Surface Marks that support the Universal Flood. There are many others, two more of which we will discuss: Rock Varnish and Surface Nodule Marks.

Rock Varnish Defined

Rock Varnish is a thin, dark coating on silicate rocks. Because they are ubiquitous in arid deserts, the coating is often called desert varnish. But varnished rocks are not restricted to desert environments, they are found in many climates around the world. In areas where water has not appreciably moved surface rocks for several thousand years, the varnish darkens entire rock landscapes, covering whole plains, hills and mountainsides. The dark color is so complete that one would guess the landscape to be of volcanic origin, but the varnish is deceptive. The thin covering conceals the true color of the rock, which is usually strikingly light, as seen in the following two page spread, Fig 8.14.26.

Ancient petroglyphs (Fig 8.14.27) hundreds or even thousands of years old are the artistry of ancient humans. With rock walls thickly coated in dark varnish as their medium and chipping tools as their brush, they chiseled their life stories into the indelible surface, exposing the natural color of the underlying rock. Today, we ponder the thoughts of the early inhabitants as we gaze upon their art, noticing that the ancient symbols show no sign of being recovered by new layers of desert varnish. That thought is the basis for this FQ:

From where did the dark Rock Varnish
come and how was it formed?

The Varnish Mark is one of the most interesting Marks of the Flood because it is so widespread and so easily understood, but the origin of Rock Varnish has been a mystery for years. Although modern observations have revealed much about the microscopic nature and makeup of the varnish, its origin remains unknown. Now, however, the UM has brought the light of truth to bear on this mystery, uncovering new pieces of Nature's puzzle that fit well and are easy to understand and comprehend.

Rock varnish has fascinated scientists for hundreds of years yet no theory has ever been able to clearly and simply explain how rock varnish formed. Here is one account from the 2006 *Applied and Environmental Microbiology* journal:

"**Rock varnish** (also known as **desert varnish**) is a dark, thin (usually 5 to 500 μm thick), layered veneer composed of clay minerals cemented together by oxides and hydroxides of manganese and iron. Nineteenth century references to rock varnish include those of Humboldt and Darwin. Modern observations of varnish were initiated with the studies of Laudermilk and Engel and Sharp; **however, despite decades of study, the nucleation and growth mechanisms of rock varnish remain a mystery**." Note 8.14h

Fig 8.14.25 – This is a microscopic view of the Great Salt Lake sand from Utah, USA. The sphere and cylindrical shapes form as wave action moves the grains while they grow from carbonate produced by the microbes in the water. If these grains of sand are growing today, where are the smaller and larger grains? To be in-process, much smaller grains should be present in the sand deposits, but they are not. These grains were formed in an environment considerably different from today's environment.

Fig 8.14.24 Valley of Fire Dissolved Rock Surface Mark

The acid needed to dissolve these rocks is not present today. These homogeneous sandstones, basalt boulders and dissolved rocks require the UF to be clearly explained.

These loose rocks are all hydrofountain rocks.

The sharp cracks in this rock are not 'in process' today, but are evidence of a bygone environment.

Rock Varnish Evidence

Desert Varnish Rock Pavement

Dark and Thick

Thin and Lighter

Top Side

Back Side

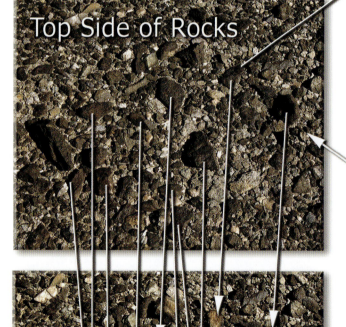

Top Side of Rocks

Back Side of Rocks

Geologists acknowledge that rock varnish is *only* several thousand years old, otherwise these rocks would have moved.

Rock varnish is *only* found on the surface.

Exposed Varnish Vein

Rock varnish forming in vein, showing hydrothermal source.

How could something so simple as the dark coating of rocks found only on the surface, not be understoood?

"However, despite decades of study, the nucleation and growth mechanisms of rock varnish *remain a mystery*."

Applied and Environmental Microbiology, Vol. 72 No. 2, February 2006, p1708

Why don't we find rock varnish in any layers of the sediment below the surface?

Because the varnish was only formed one time at the end of the Universal Flood.

Fig 8.14.26

Why isn't varnish forming here?

Why should rock varnish "remain a mystery" after hundreds of years of observing, even with electron microscopes and all sorts of high-tech equipment. Researchers know the brown or black coating is comprised of iron and manganese oxides, so why can't they tell how it formed? What's the mystery? There is only one answer—and it's the same answer that has kept much of modern science in the Scientific Dark Age over the last century; modern scientists are unable to get outside the magma-uniformity box, and that dark box will remain until it is opened and long held *assumptions* are dismissed. One such assumption is the idea that rock varnish is still forming.

FQ: Is Rock Varnish being formed today?

Rock Varnish Research History

The story of Rock Varnish is interesting and fascinating, and it will go down in history as one of the most neglected evidences of Flood Geology. Modern science's ignorance is evident here in a geological textbook that speaks of hypothetical processes:

"Desert varnish is **hypothesized** to form very slowly from a combination of **dew, chemical weathering that produces the clay minerals and iron and manganese oxides**, and the sticking of windblown **dust** to exposed rock surfaces." Bib 59 p361

Perhaps the windblown dust and dew sci-bi might have possibilities *if* iron and manganese deposits were nearby. There aren't any, except in rare cases. Because the 'hypothesis' lacks evidentiary support, it should be dismissed. Like the rock varnish hypothesis, 'chemical weathering' theories are generally mere speculation and are perpetuated as false ideas are planted in the minds of students who later become teachers. Such speculation impedes the advancement of new scientific discovery. Chemical weathering processes, by definition, break down rocks—it does not build them up, nor does it account for the rock varnish.

Wikipedia relates that most researchers now realize varnish is *not* "drawn out of the rocks it coats." However, we also find the *false assumption* that varnish ingredients came by "wind":

"Originally scientists thought that the **varnish was made from substances drawn out of the rocks it coats**. Microscopic and microchemical observations, however, show that a major part of varnish is **clay**, which could **only arrive by wind**. Clay, then, acts as a substrate to catch additional substances that chemically react together when the rock reaches high temperatures in the desert sun. Wetting by **dew** is also important in the process." Note 8.14i

The Sand Mark, earlier in this chapter, demonstrated that modern geology had misinterpreted the origin of clay. This includes the origin of the clay constituent in rock varnish. Curiously, researchers have yet to acknowledge that some other 'event' must have formed rock varnish because *no one* has been able to replicate the 'dust blown, manganese-clay dew coating process' in the laboratory, despite the assumption that it is happening right now, all over the world. Further exacerbating the mystery is the presumptuous fact that:

Researchers **know** it was formed **only** "over the course of **thousands of years**".

From the September 2006 issue of *Geotimes*:
"Scientists **know** that desert varnish forms slowly, over the course of **thousands of years**, but they have debated its origin." Note 8.14j

How do they know this? Because countless millions of unmoved varnished rocks lie on the surface all over the world. These would surely have been moved and overturned if they had been left for millions-of-years of geological time. Refer to Fig 8.14.26, the two-page Rock Varnish Evidence spread, to see the darkened top side of varnished rocks and the lighter underside after being turned over for the photo. The short time frame of varnish formation is a *primary impediment* for researchers seeking the true origin of rock varnish. Over the last several thousand years—there is no possible global environment known to modern science other than the environment we live in today, thus continuing the rock varnish mystery.

One reason past researchers were unable to comprehend how varnish forms on rocks is because of the association of its former name, *desert* varnish.

"**Rock varnish** was previously called **desert varnish** because the thin coatings **were thought only to form in desert or semi-desert terrains, but** rock varnishes **also occur in many other microsites and environments**." Note 8.14k

The "many other" Rock Varnish environments include tropical areas and deeply frozen areas! Rock varnish has been observed from Hawaii to Antarctica. This one fact refutes the entire 'varnish-from-

Fig 8.14.27 – The rock varnish on this wall has been beautifully decorated with petroglyphs by the ancient inhabitants of North America. The people that lived near this site used art to convey images about their life. The tools they used to chip away the dark surface varnish exposed the natural color of the underlying rock. In the past, scientists attempted to date the rock varnish to establish a time frame in which these people lived, but all dating efforts produced erroneous data. Today scholars recognize that dating rock varnish and petroglyphs produces no usable dates and previous age estimates are no longer used.

air' theory because many environments do *not* contain the clay or other minerals present in the varnish—including the most important mineral; manganese:

"Another **important characteristic** of desert varnish is that it has an **unusually high concentration of manganese**. Manganese is relatively rare in the earth's crust, making up only 0.12% of its weight. In desert varnish, however, **manganese is 50 to 60 times more abundant. This significant enrichment** is thought to be caused by **biochemical processes** (many species of bacteria use manganese)." Note 8.14l

Some sources have observed manganese concentration as being 100 times more abundant in Rock Varnish than in the surrounding environment. There must be a source of manganese other than the surrounding mineral outcrop because of the extraordinary abundance of the mineral. We find the answer earlier in this chapter, in the Ore Mark—biomineralization. The manganese comes from bacteria that produce manganese secretions in their natural life cycle. Actually, this was first documented back in 1981 by one of the foremost experts in the field, Ronald I. Dorn. In the laboratory, he produced a manganese varnish somewhat similar to natural varnish; similar except for its softness, measuring only 2.5 on Mohs hardness scale instead of natural varnish's typical 5. Note 8.14m

More recent technology revealed amino acids and DNA in varnish, but one overarching obstacle has prevented the acceptance of any biomineralization processes; researchers "never saw many bacteria on the surface of the varnished rock":

"One prevailing idea among scientists has been that **bacteria play a role in desert varnish** by producing manganese oxide, the component that gives the coating its dark color. The idea gained credence when **scientists found amino acids and DNA in varnish coatings**. Randall Perry, an organic geochemist at Imperial College in London, says that he, too, always expected a biological mechanism, yet such explanations '**never really made sense since we and others never saw many bacteria on the surface of the varnished rock.**'" Note 8.14n

Another Rock Varnish investigator notes:

"We did **not** observe organisms known to be involved in metal oxidation." Note 8.14o

This proved to be an insurmountable hurdle for the acceptance of the biological origins of varnish, leaving an unanswered FQ:

If microbes made the varnish—where are they now?

The absence of manganese producing microbes on the varnish caused researchers to pursue other directions in their quest to identify varnish origins. Because silicon is the primary element in the varnish, investigators took a non-biological approach, working on the origin of the silica glaze that encompasses the varnish, which gives it its shiny appearance:

"Subsequent analyses using microscopy and spectroscopy techniques suggested that **silicon is the primary element in desert varnish**, as Perry and colleagues reported in the July issue of *Geology*. The team now **thinks** that varnish coatings form through a **nonbiological process when silica dissolves out from minerals on rock surfaces, then gels, bakes and hardens.**" Note 8.14p

The team had identified an important component of Rock Varnish—hardened silica, which we will discuss in detail later. However, "Perry and colleagues" were not able to explain what Dorn had made very clear—no manganese, no Rock Varnish:

"Not all scientists, however, are sold on the silica explanation. **Ron Dorn**, a geographer at Arizona State University in Tempe, says the **silica model falls short** in explaining a number of varnish characteristics, including the **higher concentration of manganese in varnish compared to that in the underlying rock, dust, or nearby soil.**" Note 8.14q

Dorn hits the nail on the head here, showing that other researchers had no origin for the manganese in the varnish. It did not come from the host rock, from dust, or nearby soil as Dorn noted. Neither did Perry or his colleagues explain how the acidic silica solution gelled, baked, or hardened on the tops of natural surface rocks. In the history of Rock Varnish Research, the mystery of the dark varnish continued to be an enigma, and it will remain so until a correct understanding of the silica-making process is recognized.

There are present-day incidences of rock varnish, but they are all associated with active water and the varnish is softer and far less resilient than old varnish. Just like old varnish, new varnish comes from microbes living in water, but most likely of a different species.

The rock varnish process, which includes several key metal elements *and* silica, will shortly be explained, but first a quick history of the Dorn Controversy to set the stage for the discovery of Rock Varnish origins.

The Dorn Controversy

The real history of Rock Varnish research cannot be told without a brief review of the Dorn Controversy. Ronald I. Dorn of ASU, previously noted as being one of the key rock varnish investigators, contributed dozens of articles on the subject over the past several decades, and has been recognized as a leader in the field. As it turns out, another ingredient was discovered in rock varnish; carbon. This inspired Dorn to try to obtain radiometric dates for the rock varnish. Over the years, Dorn sent many samples from petroglyphs to carbon 14 dating laboratories in an attempt to aid archeologists in their dating of the rock art. In 1998, no less than eight researchers from various universities published a paper in the journal of *Science* discussing two interesting substances found by Dorn in his rock varnish samples—*charcoal and coal*:

"An attempt was made to date rock surfaces with accelerator mass spectrometry (AMS) radiocarbon measurements of rock varnishes or rock weathering rinds. In two case studies, samples pretreated in the laboratory of Dr. Ronald Dorn prior to AMS analysis have been found to contain significant quantities of carbon-rich materials of two distinct classes. Type I material resembles **coal**, whereas type II material resembles pyrolized **wood charcoal** fragments. In samples where these type I and type II materials were separated and AMS-radiocarbon dated, they were found to have **widely differing radiocarbon ages**. In these cases, the measurement of the radiocarbon age of the entire sample would yield results that are, **at best, ambiguous**. Neither type I nor type II materials **were found in comparable samples that were independently prepared.**" Note 8.14r

The researchers all recognized a major problem. As noted earlier, rock varnish was only *thousands* of years old, but coal, according to modern science sentiment, is *millions* of years old.

New and Old Rock Varnish Evidence

New varnish with a water source.

Old varnish with no water source today.

Old varnish with no water source today.

How did a varnish layer that supposedly formed by common air/rain interactions on desert rocks have coal in it?

Thinking perhaps, that the material only "resembled" coal, researchers forwarded a sample to a coal expert:

"A specimen of this material, separated from a sample pre-treated by Dorn, **was forwarded to an expert on identification of coal, who identified the specimen as sub-bituminous coal** from a vitrian layer." Note 8.14s

This created an even bigger problem because the researchers had already dated the coal in the rock varnish. These are the dates they reported in *Science*, Vol. 280, No. 5372, 26 June 1998, p2135:

Type 1 (vitrinite coal) – **27,520** to **40,030** years old

Type 2 (charcoal- CWT) – **800** to **4,181** years old

Why were these dates considered "at best, ambiguous"? Was it because the charcoal and coal ranged from 800 to more than 40,000 years? Or was it because the samples of charcoal ranged from 800 to over 4,000 years? Maybe the coal dates were thought to be ambiguous because everyone 'knows' that coal is millions of years old.

As it turned out, the "ambiguous" dates were the last thing on the researchers minds. The ambiguity of almost all modern science dating is a topic we will deal with two chapters ahead in the Age Model. In relation to Dorn's findings, the possibility that the dating method was fatally flawed was not even a consideration, because scientists have no alternative to fall back on.

So the researchers *assumed* that Dorn was "manipulating his samples. How else could "coal" get into the rock varnish? From an article in *Science* titled, *Rock Dates Thrown Into Doubt, Researcher Under Fire*:

"The findings also raise the specter of **scientific misconduct by Dorn**, who says the **researchers are accusing him of manipulating samples**. The *Science* Technical Comment, whose lead author is geoscientist Warren Beck of the University of Arizona, Tucson, **stops short of accusing Dorn of fraud**. 'We do not intend to use the "f" word with any reporters,' says Beck, who, like Dorn, has consulted with attorneys in this matter. But the paper reports that the carbon granules do not appear in samples processed by other researchers. Those close to the case say that the National Science Foundation (NSF) and ASU are reviewing the possibility of misconduct." Note 8.14t

Anyone reading the accuser's 1998 report may find it difficult to accept that following the scathing accusations, Dorn would still be teaching at any university. Those who were not able to reproduce his results thought the idea that *coal* embedded on the tops of rocks around the world was just too hard to believe. To Dorn's credit, he did not flinch. He laid out seven issues in detail in a rebuttal article in *Science,* addressing every point his accusers had posed, including the reason why some researchers were unable to reproduce his coal findings and why others actually had. Dorn restates the "gist of their argument":

"Here is the gist of their argument: **they observed vitrinite and carbonized woody tissue (CWT) of different ages in my rock varnish samples**. Beck *et al.* seem to **imply that such combinations cannot occur naturally in the same sample**, in part because of their failure to find them. In commenting on Beck *et al.*'s presentation at an Australian conference, A. Watchman is reported as stating '**coal and charcoal do not occur together**' and that 'it took "deliberate human action" to bring them together.' **Thus, the critical issue in this controversy is whether vitrinite and CWT naturally co-occur with rock varnish**." Note 8.14u

Dorn was absolutely right in his assessment that the "critical issue in this controversy" was whether vitrinite (coal) and CWT (charcoal) could *naturally co-occur* within rock varnish! And the final answer is—yes they do…, but as one might imagine, this would not sit well with those who recognize the implications of millions-of-years old coal being directly associated with a surface mineral deposit that was clearly only several thousands of years old. The reality of coal embedded inside rock varnish was just too much for some experts, like Alan Watchman, to bear. Watchman stated his views in *Earth-Science Reviews* in 2000:

"Cellulose, lichen, algae, fungi, diatoms and oxalate salts can occur together in a rock varnish, **but it is incongruous for coal or graphite to exist in rock varnishes** together with charcoal, diatoms, oxalate salts, fungi, algae, lichen or plant remains (despite the findings of Beck et al., 1998). **Such an occurrence would represent a contradiction** between the form of carbon and its primary environment of formation and the existing rock varnish setting. **It is unnatural for either coal or graphite to occur in rock varnishes with these other carbon-bearing**

substances because coal and graphite are geologically old and have formed under metamorphic conditions requiring considerable heat and pressure." Note 8.14v

Based on modern science tradition, Watchman was absolutely right; coal (geologically old) and varnish (geologically young) occurring together "represent a contradiction" to *modern geology*. For him, it was "unnatural for either coal or graphite to occur in rock varnishes" around the world because of the dark box of *modern geologic theory*.

The coexistence of coal, charcoal, and varnish are a natural, even expected occurrence in Flood Geology.

Watchman further lamented that "there is no satisfactory explanation" (in modern geology) for coal impregnated rock varnish in Arizona, Hawaii, or the other samples housed at AMS dating laboratories:

"Whereas different carbon-bearing components can coexist in natural rock varnishes, **there is no satisfactory explanation that accounts for large quantities of coal-like and charcoal-like particles in accretions from the volcanic island of Hawaii** (Beck et al., 1998; Dorn, 1998). **Even if the presence of coal can be explained in the Arizona samples, there is no plausible reason for coal or coal-like substances to exist in the Hawaiian varnishes because there are no local sources of coal on Hawaii. It also seems statistically improbable that coal should exist in all of the stored samples of varnish dating experiments at three AMS laboratories**." Note 8.14v p274

Bringing the Dorn Controversy to a close, it is important to recognize Dorn's name was finally "cleared":

"**Dorn** became the subject of a misconduct investigation, **but was cleared by his university of manipulating his samples in September 1999**." Note 8.14w

In the end, the lawyers were paid, and the case against Dorn was closed, but the controversy surrounding coal in rock varnish lives on and will not be settled until the scientific community acknowledges the Universal Flood and the processes connected to it. The reason for this is simple—we cannot change the physical fact that rock varnish and coal are *directly* associated and formed together.

If there was a positive outcome from the Dorn Controversy, it would have to be that the scientific community became "skeptical of rock art dating," Dorn himself agreeing that his own previous publications on varnish dating should now "be viewed with skepticism":

"No matter how the NSF and ASU inquiries turn out, **Dorn agrees that dates included in more than 20 of his publications over the last dozen years must now be viewed with skepticism**. However, **many archaeologists have long been skeptical of rock art dating anyway**, notes archaeologist Benjamin Swartz Jr. of Ball State University in Muncie, Indiana. Thus, although Dorn estimated some southwestern rock art to be more than 15,000 years old--implying an early peopling of the Americas—'**his dates weren't any better than any others**,' Swartz says." Note 8.14x

The acknowledgement that, "dates included in more than 20 of his publications of the last dozen years *must now be viewed with skepticism*" is a preview of the dating challenges addressed in the upcoming Age and Clovis Model chapters, both being directly impacted by rock varnish dating research. In the Age Model, a mountain of direct evidence showing that *all* rock dating methods are flawed will be presented along with documentation of false carbon 14 dating associated with the Clovis culture in North America.

The Varnish Surface Mark of the Flood

Having reviewed the "contradictions" in modern geology's ideas about rock varnish, we are set to discover the Flood Geology origins of rock varnish. The most abundant element in varnish is silica, which gives rock varnish its distinctive "luster, hardness and growth patterns." Understanding silica is an important factor in discovering the genesis of varnish. Notice that rock varnish researchers describe areas other than the world's deserts that contain similar "silica glazes." They also identify the *particular environments* where they find them:

"The centrality of silica to this mechanism for cementing and forming coatings means that **silica glazes and desert varnish should be considered as part of the same class**. Silica glazes from Peru, Oregon, and Hawaii **resemble desert varnish in their luster, hardness, and growth patterns. Siliceous sinter deposits from hot springs**, e.g., the Taupo Volcanic Zone, New Zealand, **are also similar and contain evidence of past life**." Note 8.14y

Hot springs…Finally, the research seems to be getting closer to hitting the target, but it is nowhere near the bull's-eye yet.

SUBCHAPTER 8.14 THE SURFACE MARK

Researchers made a critical observation in an article found in the 2006 journal of *Geology;* they recognized that *silica* was the major component of varnish and that a "similar" silica glaze exists on rocks found in hot springs.

Another important aspect of hot spring water is that many of them are *acidic,* supporting a variety of heat-loving *microbes* in mineral rich water. Research revealed the importance of acidic water during laboratory experiments where *hot acidic water and other ingredients* were necessary to grow synthetic coatings similar to natural varnish:

"Synthetic coatings similar to natural coatings were made in the lab. It is unknown, however, how the addition of **metal salts, UV light, heat, and amino acids** affected the condensation process. **Making the coatings requires dilute silicic acid, as earlier experiments failed when using concentrations higher than those in this experiment**." Note 8.14z

The hot acidic water was important for the silica glaze to form, but how were desert rocks covered in it? Moreover, where are the "metal salts" and "amino acids" on desert rocks? Although this experiment was an important step toward establishing the environment the siliceous varnish formed in, it lacked an explanation for the origin of the most mysterious mineral in the varnish—manganese. It is manganese that gives varnished rock its black coloring.

Perhaps the reason previous researchers missed the true source of the manganese is that rock varnish occurs most plentifully in the **desert** and not in the **ocean**. Research has revealed manganese origins are closely related to bacterial activity in "aquatic environments":

"**Manganese oxides** and manganates form as the result of either chemical or **bacterial oxidation of Mn (II) in a variety of aquatic environments**, although recent evidence indicates that **microbial influence may be the dominant factor in most cases** (Nealson et al., 1988). Microbial formation of oxidized manganese minerals has been observed just above the O_2/H_2S interface in **anoxic basins** (Emerson et al., 1982: Tebo, 1991; Tebo et al., 1984), in fresh water lakes (Chapnick et a;., 1982; Tipping et al., 1985), **hydrothermal vent fields and plumes** (Cowen et al., 1986; Mandernack and Tebo, 1993), **freshwater thermal springs** (Ferris et al., 1987b), and in **estuarine waters and sediments** (Edenborn et al., 1985; Sunda and Huntsman, 1987)." Note 8.14aa

Thermal springs, again…This time, the manganese oxide coatings were precipitated at higher concentrations with higher temperatures in "*seawater*":

"In general **agreement with thermodynamic principles**, Mn (II) oxidizing bacterial spores of a marine *Bacillus* sp. strain SG-1, **precipitated** a variety of lower valence state manganese oxides/oxyhydroxides at **higher Mn (II) concentrations and temperatures** whereas higher oxidation state (>3.10) manganates formed at **lower Mn (II) concentrations and temperatures**, both in HEPES buffer and in **seawater**." Note 8.14ab

Biological oxidation of varying thicknesses is now understandable—hotter springs release higher concentrations of silica and manganese agents, creating thicker varnishes. Furthermore, varnished surfaces in caves, on hillsides, and on the surfaces of large rocks not on the floor are explained by concentrated varnish solutions precipitating onto these surrounding structures.

This explanation accounts for the silica, manganese, coal inclusions, and other important trace elements yet to be discussed; all part of the minerals in the rock varnish.

All of the varnish constituents came from *hot springs* and were *precipitated on rocks* that lay on the surface of the continents at the end of the Flood.

Amazingly, the genesis of rock varnish was discovered as far back as 1977, but researchers apparently did not realize it because the geology of the Universal Flood was then unknown. The first direct observations of the "thin coating of manganese oxide" forming "on the surrounding basalt pillows" was made by the submersible, Alvin (Fig 8.14.30):

"The existence of hydrothermal activity at oceanic spreading centres has long been suspected but it was only four years ago that the **first direct observations** were made. During dives of the **submersible Alvin in Spring 1977**, fields of **hot springs** were discovered at a depth of 2,500 metres on the Galapagos spreading ridge (86°W). The hot spring water had an exit temperature of

Israel Rock Varnish

Some of these dark rocks are varnished and some are extrusive rock from beneath the crust.

Could both of the dark minerals have come from the same source?

Fig 8.14.29 – Modern geology concedes that "varnish remains a mystery," since it hasn't found another source for the varnish minerals. If surface mineral formations like these dark intrusions (see above) came from beneath the surface, why shouldn't we consider a mechanism that could bring rock varnish microbes to the surface to be deposited on the surface rocks?

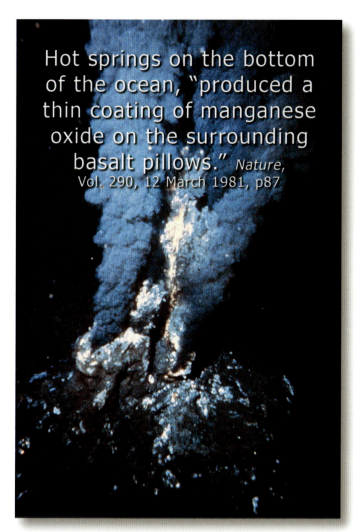

Fig 8.14.30 – Hyprethermal waters containing a variety of biogenic minerals are released in **hot springs** at the bottom of the ocean. In 1977, the submersible Alvin found the "first direct observations" of hyprethermal springs. Although more than three decades have passed since this discovery, modern geology has not seen the connection between manganese rock varnish on land and the manganese varnish being produced today on the floor of the oceans until they gain an understanding of Flood Geology.

15°C above that of the surrounding ocean (~2°C) and had **produced a thin coating of manganese oxide on the surrounding basalt pillows**." Note 8.14ac

An example of rock varnish on dry land and its source can be found in Fig 8.14.31. The lower right image shows rocks darkened from the precipitating silica-manganese solution that came from hydrofountain hot springs. These ancient hot springs are difficult to see today because they lay buried underground. Only the occasional road cut reveals the dikes and diatremes indicative of once active hot spring hydrofountains.

Although research is ongoing on oceanic manganese crusts and varnishes, researchers already know that these types of crusts are "rather common" in the abyssal elevations (3,000 to 6,000 meters) on the seafloor:

"Compared to manganese nodules, only few **manganese crusts** from the nodule belt were analyzed. Though crusts may be **rather common in the close vicinity of abyssal elevations**, their collection is limited mainly to dredge hauls." Note 8.14ad

The Rock Varnish Trace-Element Smoking Gun

The smoking gun evidence for seawater origins of rock varnish (or more appropriately, **Flood Varnish**) is the trace elements found in the varnish. The Rock Varnish Trace-Element evidence is similar to the KREEP evidence in the Hydromoon subchapter, 7.13. Rare-earth elements are *not* concentrated in certain areas by chance. There are very specific mechanisms that account for the high concentrations of rare-earth elements. Just as hydrofountains were shown to be the source of the Moon's rare-earth elements, the Earth's hydrofountains on today's ocean floors and on the flooded continent during the UF produced enriched trace-elements in abundance. The evidence of this shows up on the desert's varnished rocks.

Two researchers from Rice University found the connection between trace elements in rock varnish on land and on the ocean floor. They reported their findings in a 2004 article found in *Earth and Planetary Science Letters*. They note that trace-element abundances "can only be explained" with an "aqueous environment":

"**Varnishes** collected from smooth rock surfaces in the Mojave Desert and Death Valley, California are shown here to have **highly enriched and fractionated trace-element abundances relative to upper continental crust** (UCC). They are highly enriched in Co, Ni, Pb and the rare-earth elements (REEs). In particular, they have anomalously high Ce/La and low Yb/Ho ratios. These features **can only be explained** by preferential scavenging of Co, Ni, Pb and the REEs by Fe–Mn oxyhydroxides **in an aqueous environment**." Note 8.14ae

The researchers were able to correctly explain how the varnishes could have *only* come from an aqueous environment and they were able to partially explain why:

"The **trace-element abundance patterns** described above **require an aqueous origin because the relative enrichments or depletions of each element appear to qualitatively reflect their relative solubilities** and/or their **propensity for being scavenged by Fe–Mn oxyhydroxides**." Note 8.14ae p136

The solubility of rare-earth minerals is only part of the answer. Instead, the biogenic nature of the minerals (discussed in the Ore Mark) is the primary reason why some minerals exist in varnish and some do not. The details are in the microbes. Nevertheless, these scientists did have *a reason* for declaring that the rare-earth mineral combination found in varnish could only come from a water environment; they had a very specific example—from the *South Pacific Ocean Floor*.

Fig 8.14.32 is a chart showing trace elements present in Death Valley and Mohave Desert rock varnish, and a comparison with ocean floor varnishes. The concentration of trace elements in all the samples was remarkably similar, *confirming* that desert rock varnish and ocean varnish shared similar origins.

Because the trace elements are generally *rare*, different locations have a wide range of concentrations. Charting the trace elements on rock surfaces from different hot springs around the United States reveals great diversity in microbial populations, mineral content, and temperature, resulting in completely different concentrations of rare-earth elements. However, charting the trace elements on rock varnish surfaces shows relatively little variation, which is also true of deep oceanic hot springs; seawater is relatively *uniform* in trace element and mineral composition. Trace elements from rock varnish on the desert and ocean floors share this unity because they were both formed in seawater.

Death Valley Varnish Source

Why do almost all broken rocks show no signs of *partial* varnish on the broken face?

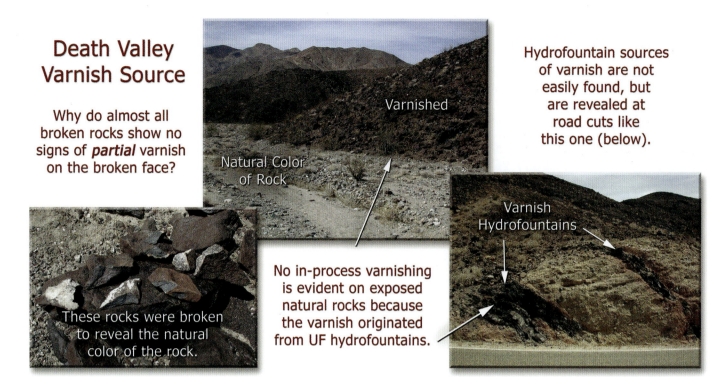

Hydrofountain sources of varnish are not easily found, but are revealed at road cuts like this one (below).

No in-process varnishing is evident on exposed natural rocks because the varnish originated from UF hydrofountains.

Fig 8.14.31 - Death Valley, California, USA, has many spectacular examples of rock varnish. One particular area near Panamint Springs, aptly named because of the hot springs that once existed here near the end of the Flood, emitted the hot silica-manganese solution that precipitated on the surrounding countryside. Possible varnish hydrofountain remains are seen in the road cut, (lower right) where vertical pipes contain dark minerals similar to those of the surrounding area. Finally, we have a real mechanism for the origin and deposition of rock varnish that explains why rock varnish is not occurring today and why previous theories have failed. The UF Rock Varnish Model easily shows why deserts all over the world contain varnish, even though the microbes and acidic hot water needed to form the varnish is absent on the desert rocks today.

Rice University researchers state that rare-earth trace elements and the iron-manganese crusts can "*be explained*" by the direct precipitation of the minerals "*in seawater*," but they stop short of seeing all the deserts of the world covered by seawater several thousand years ago. This leads the researchers to conclude the marine environment is at best a "*very remote* analogue for varnish formation":

"**Fe–Mn [iron and manganese] crusts formed in deep-sea marine environments** may be a **very remote analogue for varnish formation**. It is generally believed that **Fe–Mn crusts form by direct precipitation of dissolved Fe, Mn and metals in seawater**. They have elevated trace metal concentrations, positive Ce anomalies and depletions in Th relative to U. It can be seen in **Fig. [8.14.32]** that Fe–Mn crusts are roughly enriched and depleted **in the same elements**. The enrichments in Pb, Co, Cr, REEs and, especially, Ce **can all be explained** by the tendency for these elements to adsorb onto Fe–Mn oxyhydroxide colloids. The lack of enrichments in HFSEs and Th **can also be explained as a feature inherited from seawater**, which is itself depleted in these largely insoluble elements. Rb and Cs, being highly soluble, are not precipitated with Fe–Mn oxyhydroxides, **explaining** the similarly depleted Rb and Cs contents of Fe–Mn crusts and varnishes." Note 8.14ae p139-40

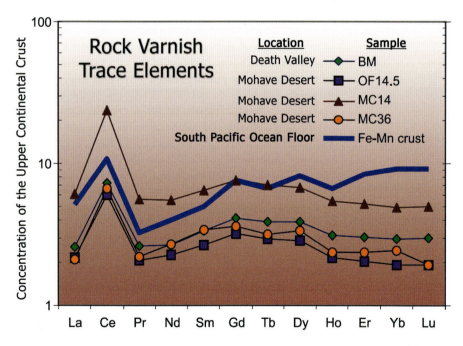

Fig 8.14.32 – This chart compares the concentration of rare-earth trace elements found in rock varnish from the desert floor and from the ocean floor. Normally, there is little correlation between trace elements found in various coatings on rocks that come from a variety of known sources, such as hot springs. Each spring contains different minerals and microbes and thus, the trace element composition bears little resemblance. As we compare the relative abundance between the rock varnish groups on continental deserts and on the ocean floor, a striking similarity appears. The reason for this similarity is seawater. The extraordinarily uniform concentration of minerals and trace elements found in varnish from the ocean floor and on desert floors suggests that they came from the same source—seawater. This diagram is adapted from the chart in *Earth and Planetary Science Letters* 224, 2004, p135.

Amazingly, the researchers actually state three times that the trace elements

CHAPTER 8 THE UNIVERSAL FLOOD MODEL

in *desert* varnish can "be explained" by looking at seawater, as shown in the analysis in Fig 8.14.32, but they can't quite bring themselves to say the varnish **came from seawater**.

Instead, they fall back on the prevailing rock varnish theory; that varnish is caused by wind-blown particles and rain drops on rocks. Although they recognize "overall similarities" between the two varnish types—ocean and desert, they "speculate" that the desert varnish forms in a sub-aerial [in the air near the surface of Earth] environment instead of a marine environment:

"We recognize that the **overall similarities in trace metal abundance patterns between Fe–Mn crusts [ocean floor rock varnish] and varnishes [desert rock varnish] may be entirely fortuitous**. However, **we speculate that even though varnishes form in sub-aerial [near surface of Earth in air] instead of marine environments**, varnishes are essentially **chemical precipitates from aqueous components in the atmosphere**." Note 8.14ae p140

It would be difficult to find a clearer example of modern scientists being blinded by the Dark Age of Science despite having the scientific truth right in front of their eyes. Sadly, our research has uncovered similar happenings in every major field of science today.

Endless Flood Varnish Evidence

There is substantial Flood Varnish Evidence, enough to fill an entire book, but space constraints limit us to just a few examples. One of those examples is the microbial **fossils** found in rock varnish, pegged as being fairly young.

From the 2003 article titled *Microbial Fossils Detected in Desert Varnish*:

"**Desert varnish appears to preserve microbial fossils,** although the duration of their existence is unknown. All observed fossilization appeared to have been located on the surface of the varnish or in natural breaks. Therefore, their formation was probably **fairly recent**. In agreement with the observations of Krinsley our study observed many nodules high in Mn and Fe oxides. Krinsley provided a mechanism for their formation in his 'biodiagenetic model.' In this model, bacterial casts and wastes products are **formed by biomineralization**." Note 8.14af

Like several other UF Marks have shown, biomineralization *commonly takes place in seawater*—yet the biomineralization of manganese fossils on varnished rocks in *rainwater* has never been observed.

Another evidence of the Flood origin of the varnish is the **absence of multiple, crossing layers** of varnish that should exist on transported (moved) varnished rocks. These layers are not expected if the varnish was made at one time. The example in Fig 8.14.33 shows how easily this can be observed.

Those willing to hike deep desert canyons will notice that varnished rocks are absent within the **deeper sedimentary layers** of the canyon. Varnished rocks are on the *exposed surface* of some of those layers, but not within the layers. This is especially interesting in the great valleys like the Grand Canyon. The fact that such observations have been overlooked ought to be astonishing to anyone with a geological background. Of course, not knowing or believing that the Earth was recently covered by a Universal Flood would explain why such evidences are overlooked.

Why No Previous Rock Varnish Theory Works.

For many readers, rock varnish is a new topic and the facts may be a little confusing. The following is a list of the reasons why no previous rock varnish theory works:

1. Unmoved varnished rocks establish a time frame of *only* **thousands of years** from their genesis.
2. The young age indicates that **the present conditions** (desert environment) are a key to the past.
3. Today, rock varnish exists in **many different environments**, globally, with no source for varnish ingredients.
4. No source for the **manganese** in the varnish.
5. Manganese is known to be biogenic—however **manganese microbes are not present** on the varnished rocks today.
6. No source of silica or mechanism for the **silica glaze** in the varnish.
7. No source for **coal** in varnish, and no explanation to reconcile coal and charcoal dates with varnish dates.
8. No source for the traces of rare Earth **elements** in the varnish.
9. No explanation why there are **no multiple or crossing layers** of varnish (similar to limesurface rocks).
10. No explanation for the absence of varnished rocks in the **deeper sedimentary layers**.

Universal Flood Rock Varnish Model Summary

1. Rock varnish was precipitated in the same manner observed today on the bottom of the ocean.
2. Hydrofountains rich in manganese, iron and rare-earth biominerals contributed to rock varnish zones toward the end of the Flood.
3. UF Hydrothermal acidic waters **saturated with silica** created the siliceous "varnish" coating during a **one-time event**.
4. **Coal** particles and other organic matter suspended in UF waters were entombed as fossils in the varnish.
5. The one-time UF event explains why there are no **multiple or crossing layers** of rock varnish.
6. Only the Universal Flood Model **clearly and accurately explains** the instances of global rock varnish that exists today.

Although rock varnish could have been deposited while the receding waters were still quite high (hyprethermal conditions), most of the varnish appears to have been precipitated in shallower waters. This conclusion comes from the fact that the iron precipitate is usually amorphous, or without form and not in the usual crystalline form it takes when it is under high pressures. Additionally, deeper waters would have the tendency to disrupt the final surface of the continents, whereas the last remaining shallower waters might have abated more gradually.

Why the UF Rock Varnish Model Works

1. It explains **why** the Rock Varnish was deposited only several thousand years ago.
2. It explains **why** Rock Varnish is a globally precipitated deposit in environments without nearby manganese deposits.
3. It explains **how** the silica glaze could have been produced in acidic UF hydrothermal waters.
4. It explains **how** coal and other fossils could become en-

No Multiple-Crossing Layers

Fig 8.14.33 - Rock samples with dark, varnished topsides and limesurface encrusted bottom sides number in the millions in rock varnish areas. They each contain a single layer of rock varnish and limesurface, often leaving some of the original rock exposed. If the varnish and limesurface mechanisms were actively occurring today, many rocks would be overturned by erosional forces, and over time, would show multiple-crossing layers of varnish and limesurface. It is very rare however, to find anything but a single layer of varnish and limesurface on the rocks, verifying that a recent, one-time global hydrothermal event created them.

cased in the varnish layer.
5. It explains **why** there are no multiple or crossing varnish layers or varnished rocks within the deep sedimentary layers.

The simplest reason the Flood Varnish model works is because it is true. An unfortunate consequence of the false 'rock-varnish-from-air' theory is that it has been used to obtain false dates for many archeological sites, dates that misrepresent the age of petroglyphs. With the new origin of rock varnish and with the light of the UF, we can put the pieces of Nature's puzzle together for a better look into the archeological mysteries.

Because of the Dorn Controversy, rock varnish research waned during the 1990s, but with the recent data from the Mars' Rovers, rocks which appear to have rock varnish have ignited new interest. Thanks to the new research, the UM model was able to be completed.

The Nodule Surface Mark

The last of the Surface Marks to be presented are nodules. In geology, a **nodule** is a rock that precipitated or formed in or from water. Although generally spherical or rounded, it is not uncommon for irregular shapes to occur (see Fig 8.14.34). An easy way to recognize most nodules is the band of material that wraps around an inner core. This band is important because it signifies that the specimen *grew* to its final shape rather than eroding from a parent rock. Most nodules are small enough to hold in your hand, but they can reach several meters in diameter. We first discussed nodules in Fig 6.4.7, where we examined the Limestone-Quartz Chert Mystery. There we read that, "The origin of chert nodules has **not** been completely explained" Bib 15 p294 This sentiment is shared with almost all nodule types.

The primary reason modern geology is unable to completely explain nodule origins is that they are a Surface Mark, and this is not commonly known among geologists. This is clearly evident in the deep layers of canyon walls because there are no nodules found there. Geodes and thundereggs are nodules that share the same mysterious origin as other nodules—they lack an adequate explanation of their origin because geologists have not figured out how they formed all over the world on the *surface* and not in deeper layers.

Another reason the origin of chert nodules and most natural quartz minerals remains unexplained is that modern geology is not aware of the hyprethermal crystallization process requirements. *All quartz-based minerals required a hypretherm* for growth, including those grown on the continental crust! This means that all of these quartz-based Surface Marks, such as surface chalcedony and most surface nodules grew in water that was generally around 350° C and at least 9km (5.6 miles) deep, under 13,000 psi pressure.

The only time water was this deep, and of such high temperature on the continental surface was during the UF. Chert nodules, like those in Fig 8.14.35 were able to precipitate into rounded silica structures because the hyprethermal conditions of the Flood caused many areas near the surface to be supersaturated with silica. Layers of the soft carbonates, saturated with this silica, sunk to the seafloor, where spherical voids of water and air formed during rapid burial, and were later filled with a siliceous fluid. As the temperature dropped or pressure changed, supersaturated siliceous waters began to crystallize, forming the chert nodules and geodes, lithifying or cementing many of the limestone and sandstone shapes.

This is easy to understand in Flood Geology, but not in modern geology. From the *Essential of Geology* textbook:

"**Chert deposits** are commonly found in one of two situations: as irregularly shaped **nodules** in limestone and as layers of rock. The silica composing many **chert nodules may have been deposited directly from water**. Such nodules have an inorganic origin. However, it is **unlikely that a very large percentage of chert layers was precipitated directly from seawater, because seawater is seldom saturated with silica**." Bib 172 p131

CHAPTER 8 THE UNIVERSAL FLOOD MODEL

Is it possible that chert nodules "may" have been deposited directly from water? No—they *had* to form with water, just as all chert does, which follows the Law of Hydroformation—that all natural crystalline minerals originally formed from water. Furthermore, as the Inclusion Mark will demonstrate, the water was *seawater*, known because of the salts and other minerals found in the natural mineral inclusions. By this, we know certain areas of the Flood seawaters were saturated with silica, which precipitated the chert nodules.

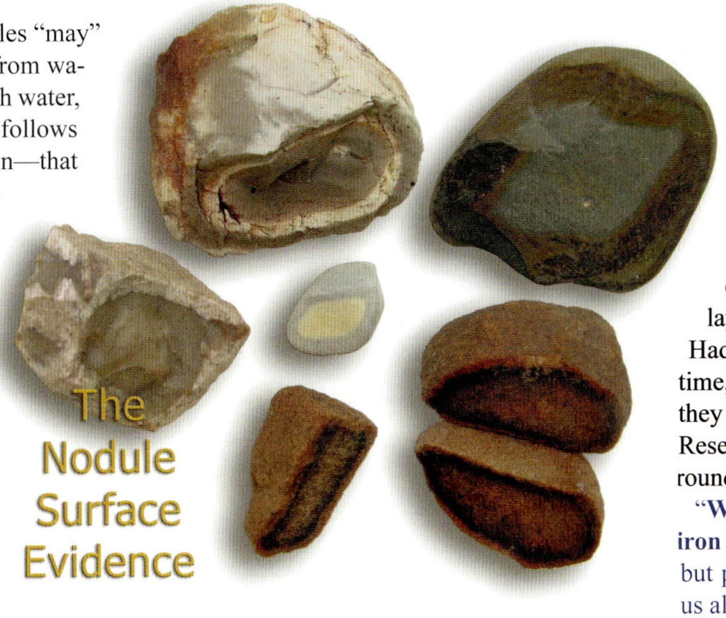

Another name for nodules is concretions. Although **concretions** are typically described as sedimentary rock cemented into spherical or other irregular shapes, many nodules fit this category. As we will see, both nodules and concretions are precipitated out of water, but are not observed forming today, except in rare instances.

Moqui Marble Nodule Evidence

Moqui Marbles is the folk name given to the prolific concretions found in the desert southwest. Seen in Fig 8.14.36, these round nodules are found right on the surface, although some are evidently weathered from surface sandstone layers. We know of no examples of Moqui Marbles in the world's deep sediment layers, including the sediment layers in the Grand Canyon. Had they formed over geological time, as modern geology suggests, they certainly should be there. Researchers are baffled by their roundness:

"**We don't know why some iron concretions are so round**, but perhaps some 'seed' or nucleus alters local chemistry to precipitate iron in a uniform (spherical) manner." Note 8.14ag

One reason modern geology cannot explain the spherical shape of the nodules is the assumption that they formed "underground":

"All of this iron dissolution and transportation **takes place underground**. Even mixing with oxygenated water is a **subsurface process**. The precipitation of the iron in concretions takes place **hundreds of feet or more below ground**." Note 8.14ag

Researchers have never actually observed the nodules forming underground, they simply rely on the old geologic-time

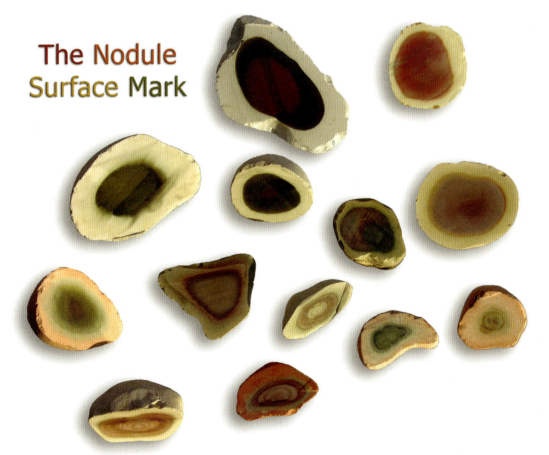

Fig 8.14.34 – This variety of nodules are an important Surface Mark of the Flood. Nodules form only on or near the ground surface under water and are made of both solid minerals and/or sedimentary material. Their origin is not adequately explained by modern geology and without a UF connection, there can be no simple answers for their origin.

Fig 8.14.35 – Colorful chert (silica) nodules surrounded by lighter limestone are found worldwide in limestone deposits. Obviously they grew in a seawater environment because limestone comes from the ocean. Geologists however, declined to say just how this happened because they know the seawater would *have had to be* saturated with silica. But they have also said, "seawater is seldom saturated with silica" (Bib 172 p131). Silica saturated seawater exists only in hyprethermal conditions, i.e. the Universal Flood. Although the origin of the chert nodules escapes modern geology—they are easily explained with Flood Geology.

Fig 8.14.36 – Moqui Marbles are a common nodule on the Colorado Plateau. These are from around Lake Powel, Utah, USA, area. Similar nodules, some attached and some not, share the rounded shape of Moqui Marbles. No rational explanation for the origin of the nodules exists in modern geology because they occur only on or near the surface and because they are not forming today. With Flood Geology, Moqui Marbles are another example of the Surface Mark, corroborated by the Manganese Nodules found in extraordinary numbers on the seafloor today (see Fig 8.14.37).

sci-bi, stating that the nodule's "long history" was shaped by "enormous amounts of water" flowing through the sandstone:

"Over the **long history** of these rocks, **enormous amounts of water** have flushed through this porous sandstone." Note 8.14ag

Today however, the desert environment is void of the enormous amounts of iron rich underground water necessary to support their theory. In fact, the mineral in greatest abundance in the underground water today is calcium carbonate, a mineral not significantly represented in the Moqui Marble Nodules! Although their theory is full of holes, it continues unchallenged. Other investigators hoped to explain the origin of the iron in the nodules through weathering, where eroded minerals from some unknown source mixes with groundwater. But like the previous theories, they invoke unobserved and unworkable processes, such as the following:

"The iron forming these concretions came from the **break down of iron-bearing silicate minerals by weathering** to form iron oxide coatings on other grains. During later diagenesis of the Navajo Sandstone while deeply buried, reducing fluids, likely hydrocarbons, **dissolved these coatings**. When the **reducing fluids containing dissolved iron mixed with oxidizing groundwater**, they and the dissolved iron were oxidized. **This caused the iron to precipitate out as hematite and goethite to form the innumerable concretions found in the Navajo Sandstone**." Note 8.14ah

One correct point in this Moqui marble genesis theory is that *iron did precipitate* out of water. Looking at the nodules more closely reveals that the iron is not holding the nodule's sand grains together—it is quartz! The nodules were cemented in the same way the sandstone around them was—in a hypretherm. They were not formed in a mystical underground process where rainwater dissolved iron from an unknown source, depositing it in neat little round balls in the sandstone and on the surface. One important fact the theory misses is that the iron shell of the nodules is a biogenic, mineral similar to rock varnish! This suggests an altogether different environment from what the researchers envisioned. It just so happens that a great example of the formation of metalliferous nodules exists on the floor of the modern ocean. The nodules, called Manganese Iron Nodules, explain the wisdom of how nodules on the continents are formed.

Manganese Iron Nodule Evidence

Fig 8.14.37 shows manganese nodules on the bottom of the seafloor. These nodules consist of primarily manganese, iron, and silica, the same minerals in the rock varnish on desert rocks. It is no surprise that the manganese crusts discussed earlier are directly connected with the manganese nodules; they are located near hot spring vents (hydrofountains) in the deep ocean.

First discovered over a century ago, the deep ocean nodules proved difficult to obtain and so, had not seen much attention until the 1970s when they were being considered as a possible new mining source. Being in international waters, the question arose as to who owned the new found minerals. Ultimately, legal issues and deteriorating market prices left the nodules on the seafloor undisturbed.

One outcome of the short-lived attention was the recognition that the seafloor nodules were very similar to the Moqui Marble nodules, including their "controversial" origin.

"**Their origin is controversial**. We do **not** know the source of the metals nor how and why they are concentrated in nodules." Note 8.14ai

The geology professor cited, went on to discuss possible sources of the metals in the nodules, concluding the most likely possibility was—"black smokers":

"The metals may be delivered as dissolved species or as sediments from land. **A more likely origin is volcanic activity and hydrothermal alteration of ocean crust at mid-ocean ridges.** Remember the '**black smokers**?' Much of the **suspended particles are Mn and Fe oxides**." Note 8.14ai

The seafloor nodule controversy can be easily solved with a little detective work and the new UM discoveries. A solid clue

CHAPTER 8 THE UNIVERSAL FLOOD MODEL

in the discovery of how manganese nodules and manganese crusts form is their *location* on the seafloor. It turns out that they are primarily found in the "deepest of oceans and near mid-ocean ridges":

"**Manganese nodules** occur as nodules and crusts, primarily in **deepest of oceans and near mid-ocean ridges**." Note 8.14ai

Certainly, pressure from the depth of the water is key, but so is proximity to the volcanically friction-heated mid-ocean ridge zones. Thus, it would appear that the hyprethermal environment is important in the formation of the manganese nodules and crusts. Furthermore, the presence of these metals is abundant on the abyssal surface of all major ocean basins, but the metal-rich nodules and crust occur *only* near the surface:

"The **absence of manganese nodules from deep sea cores** indicates that they do not normally survive burial by younger sediments." Note 8.14aj

This is of particular interest because it suggests the manganese iron nodules, which occur only on the surface of the ocean floor, are a Surface Mark of the Flood, because they need deep, hot water. The geologist's own statements give us a good indication of this, because we know that these ocean nodules are a *worldwide* **seafloor surface** *phenomenon*. To back this up, excerpts from a detailed 1988 report by a group of scientists from around the world said the following:

"The available data on abundance of nodules per unit area show that especially in the Pacific nearly all stations yielding more than 13 kg of nodules per square meter occur at **water depths between 3,700 m and 6,000 m**…" Note 8.14ak

Indeed, water depth and corresponding pressure is critical for seafloor nodule formation, but so is heat. Manganese nodules are not randomly spread across the ocean floor, their appearance coincides with fracture zones:

"The **boundaries** of this [manganese nodule] belt **coincide** approximately with the clarion **Fracture Zone** in the north and the Clipperton **Fracture Zone** in the south." Note 8.14ak p10-12

Fracture zones are the location where earthquakes happen, which produces frictional heat. Heat, water, and pressure are the components necessary for hyprethermal conditions, which is required for the precipitation of the metals into the nodules. Another question to answer; what is the source of the manganese and iron precipitates? Are they coming from the black

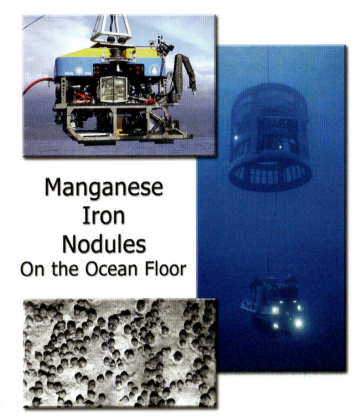

Manganese Iron Nodules
On the Ocean Floor

Fig 8.14.37 – Deep on the floors of world's oceans lies a particular nodule that has mystified researchers for decades. These relatively uniformly sized Manganese Iron Nodules are found right on the surface of the seafloor, but it seems to have eluded modern science as to how they came to be there. Further adding to the mystery is that belts of nodules lay along the seafloor along dormant "fracture zones." Once, when earthquakes heated the water and seamounts spewed abundant biominerals, the growth of these nodules proliferated. But that time has passed, and they lay quietly as a testament of a once violent, heated past.

smokers in the area?

No black smokers were reported in the area where the nodules were located, at least not in numbers of any significance. Instead, researchers noted that the nodules occurred in the "vicinity of seamounts":

"**Nodule** types that occur mainly on the slopes and **in the vicinity of seamounts** accumulate **on the sediment surface**. They generally have **spheroidal to ellipsoidal shapes** and are often in the form of polynodules [irregular forms]. The size of these nodules varies between 0.5 and a maximum of 8 cm. Since these nodules **predominantly lie on the sediment**, their growth is **obviously the result of the precipitation of colloidal particles of hydrated metal oxides**

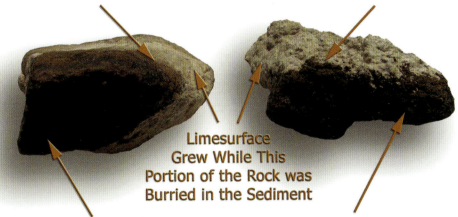

Varnish/Limesurface Equatorial Band

Limesurface Grew While This Portion of the Rock was Burried in the Sediment

Varnish Formed in Hyprethermal Spring Water

Fig 8.14.38 – Occasionally both rock varnish and limesurface can be found on the same specimens, as seen in these two desert stones and on the specimens in Fig 8.14.33. They exhibit a "distinct equatorial band" which marks the position where the dark varnished area was exposed to sea water. The lower, light colored area was buried in the sediment where biogenetic material grew on the stone. The distinctive band documents the UF environment to which the stone was subjected; the single one-time event associated with its creation.

UF Nodule Evidence

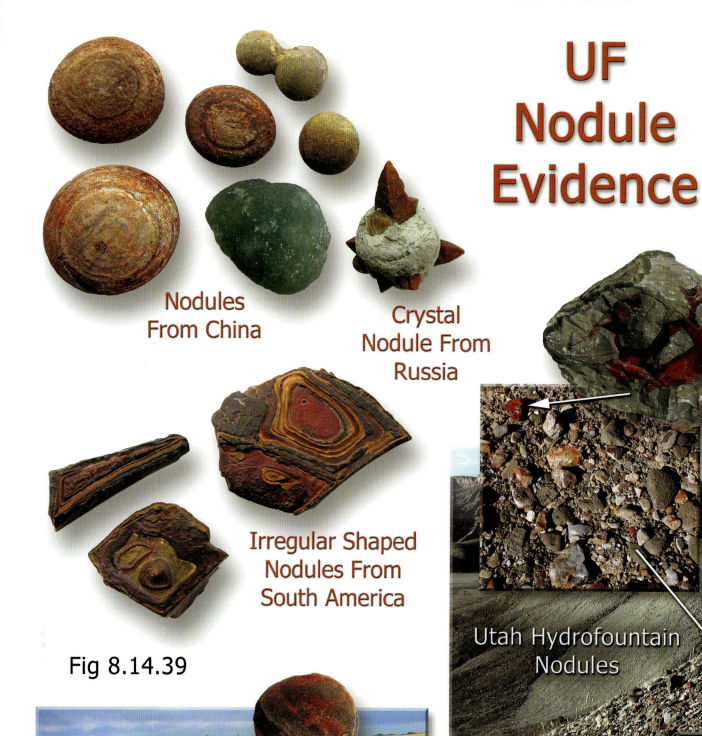

Nodules From China

Crystal Nodule From Russia

Irregular Shaped Nodules From South America

Fig 8.14.39

Utah Hydrofountain Nodules

Silica Mine

Nevada USA Silica Mine

Surface nodules like these are not found in the lower layers of this silica mine.

Lake Powel, Utah

Pyrite Shale Nodules from China

Why are surface nodules like these *not* forming anywhere in the world today?

Because they did not grow in an environment that exists today.

These nodules grew in the Universal Flood.

Surface Nodules From Utah USA

Sandstone Nodules in Situ

Utah Surface Nodules

from the sea water." Note 8.14ak p18

Because observation of the seamounts showed no ongoing eruption or expulsion of hyprethermal fluids at the time, and with the feet of modern geology firmly planted in uniformity cement, there was only one possible outcome of the research; the nodules must have formed "very slowly":

"A further important factor determining the abundance of nodules and encrustations in the deep sea is the **past growth time**. Nodules and crusts **accumulate very slowly, several mm to a few cm per 10^6 years**." Note 8.14ak p16

However, the UM has demonstrated in every case in which millions-of-years supposedly took place, direct empirical evidence that refutes the geologic time assumption.

> If seafloor nodules were formed over millions of years, they would be found throughout the sediment layers, but they are not.

The researchers continue in their report:

"The processes of dissolution and formation of authigenic minerals **seem** to be restricted mainly to the **top several decimeters of the sediment**." Note 8.14ak p124

The record shows quite reliably that the nodules do not merely 'seem' to be restricted to only the top several decimeters (inches) of sediment—*they are only in the upper sediment zone!* The vast majority are right on the surface, proven by dredging and drilling activities. Scientists seem surprised by this because it goes against the protracted uniformitarian view of sedimentation and mineral deposition, resulting in an ongoing controversy concerning ocean nodule formation.

Not all Pacific Ocean nodule belt researchers agree that the nodules "accumulate very slowly." Some noticed that there are no disintegrated nodules beneath the surface, and more importantly, they concluded that *all* the nodules "originated at the *same* time":

"From the **similarity** of the inner part of **buried and surface nodules**, it is concluded that **both originated at the same time**." Note 8.14ak p125

Since that time, researchers determined that there had been a time of inactivity, or as they called it, a "…long period of the **hiatus**". Note 8.14ak p125 Moreover, they recognized the period of inactivity was brought on by a change in the solution surrounding the nodules:

"Nodules seem to develop especially **during hiatuses**, caused either by currents or by solution." Note 8.14ak Forward

Some researchers continue to espouse the idea of the extended hiatus because the nodules are not growing right now—their reason?

> The nodules thin outer lamination is of a different composition than the rest of the nodule.

"Only a **very thin, outer layer of the buried nodules, 0.2-0.5 mm thick, is different in composition and growth structure**. This thin, outer layer has a **laminated** to cuspate growth **structure**…" Note 8.14ak p125

The thin outer lamination could also be identified as varnish, since it consists of the same metals as the rock varnish discussed earlier in this chapter. It was added *after the main nodule formed*, in a completely *different* environment, one rich in metalliferous biogenic minerals, leaving behind a thin layer "different in composition and growth structure" from the original nodule. The hypretherm environment that created this coating was global—because ocean nodules are a global phenomenon.

The Ore Mark documented the significance of biomineralization, citing researchers' observations that all iron minerals come from microbes. Researchers also realize that the "most important factor" affecting the mineral growth on these nodules is the "biogenic matter" in the water:

"The **most important factor** controlling the supply of Mn [manganese] **for nodule accretion** is the sedimentary flux of **biogenic matter**, which is governed by the biological productivity of the ocean surface waters." Note 8.14ak p152

Another group of researchers put it this way:

"**Growth conditions of manganese nodules are mainly determined by the supply of material**." Note 8.14ak p164

An increase in material availability accelerates growth opportunities. When conditions are right, crystal or nodule growth can happen in a matter of days. The small size of the nodules, most of which are less than 8 cm (3 inches), is further proof that they formed during a one-time event. Had they grown over many millennia, some areas would have experienced extended periods of hyprethermal fluid output from active black smokers, which would have produced a wide range of sizes—from microscopic to meters in diameter. Instead of a variety of sizes, the worldwide manganese beds are littered with nodules of relatively uniform size.

What caused the worldwide output of hyprethermal fluids and microbial mineral solution? *Earthquakes* and earthquake friction created heat in which bacteria reproduced exponentially. Because there are no global earthquake swarms occurring, there is little hot water for the microbes to live in, and the once active vents lie dormant across the vast abyssal plain. Not long ago, however, at the end of the Universal Flood event, when the continents moved toward their present-day positions, frictional heat from earthquakes was widespread. Giant seamount hydrofountains were active, expelling supersaturated-hyprethermal water, rich in manganese-iron microbes that coated the ocean floor and the millions of nodules we see today. This environment was very similar to the environment that produced the nodules on the continents; during a time when many parts of the land were still covered by deep, hot ocean waters.

Equatorial Band Evidence

The oceanic manganese-iron nodules are corroborative evidence of two previously mentioned Surface Marks—rock varnish and limesurface. Like the rock varnish and limesurface mark, ocean nodules share a similar characteristic of a *single* band or line around the varnished and limesurface rocks. Researchers describe this band around seafloor nodules as a "distinct equatorial band" or line that separates the top and bottom of the nodules and they explain *how* this band formed:

"**Hydrogenetic nodules** form by direct accretion of colloidal hydrous metal oxide from near-bottom sea water, i.e. the process of precipitation is influenced mainly by the conditions in the water column and controlled by colloidal chemical surface reactions. The mixed-type nodules show a **distinct equatorial band**, which marks the position to which the nodules were formed by **both** hydrogenetic growth and early diagenetic accretion: **The matter of the upper part of these nodules has been supplied from near-bottom sea water, whereas the lower part consists of diagenetic matter removed from the**

peneliquid layer of the underlying sediment." Note 8.14ak p152

Even with their technical explanation, it is an unambiguous explanation of how rocks with both varnish and limesurface formed. Fig 8.14.33 & 38 shows similar rocks from the desert that exhibit the varnish/limesurface equatorial band. Biota from black smokers in deep, hot seawater coated the *top* of the rocks black, while the lower portion of the ocean nodules and desert rocks were coated with a limesurface by calcium carbonate microbial activity.

More UF Nodule Evidence

Nodules and concretions are a fascinating geological topic because of their mysterious origins and their habit of growing near the surface. But it doesn't seem modern geology is even aware of this habit, which is important because it is a key element in the understanding of the environment that nearly all nodules were formed in, and why they remain anomalous. In the following two-page diagram, various nodules from around the world are featured, showing a great variety of mineral types. They all have one thing in common:

None of them are seen forming today.

Despite the millions-of-years rock-age sci-bi, rocks like these nodules often take only hours or days to form. They are not formed deep underground as some have suggested because deep canyons, mines, caves and boreholes show no evidence of nodules like those found on the surface. There are no observations of nodules-in-the-making in caves or mines. Indeed, nodules are truly a Surface Mark, evidence of the one-time environment they were formed in!

There is one place where one can observe nodule growth, but the nodules that are growing don't contain pyrite or other metal minerals and they are not quartz-based nodules because the nodules forming today aren't in a *hypretherm*—they are forming in a *hydrotherm*. They lack the essential component of *pressure* and the necessary ore microbes are absent. What they do have is heated water. **Geyser Nodules**, seen in Fig 8.14.40, consist of the mineral geyserite, which forms in hot springs, precipitating directly out of hot surface water. These rare mineral formations are from Yellowstone National Park; some exhibit the rounded shape associated with nodules because they are constantly moving about in the flowing water. Geyser Nodules are not found lying on the desert floor or in the mountains from long-dead hydrothermal vents, in fact they are extremely rare even in contemporary hot springs, which leads us to ask this FQ:

If nodules require hot, mineral rich water to form, and if quartz and crystalline metallic minerals require high pressure to form, what environment must have existed for the desert surface nodules in Fig 8.14.39 to form?

One final, important nodule fact is that they are often directly connected to many fossils. *Fossil nodules* are common, and as we will discover in the Fossil Model, there is a reason why nodules and fossils are both found *only* on or near the surface.

Why Nodules Remain a Mystery or are Misinterpreted

1. Nodules are thought to **form *only* while buried**.
2. The burial formation process has ***not*** been observed.
3. Nodules are found *only* on or near the surface—***not*** in deep sediments.
4. No mechanism exists to explain the nodule shapes from burial or the extremely **spherical shaped** nodules.
5. The **environment** necessary for the creation of quartz and other minerals found in the nodules was not possible in the inferred burial process.

Universal Flood Nodule Model Summary

1. Nodules follow the Law of Hydroformation; they are **precipitated** out of water.
2. The **UF Hypretherm** and/or hydrofountains provided the necessary minerals and environment for nodule formation.
3. Nodules precipitated out of hyprethermal waters on or near the surface during a **one-time event**.
4. Rolling on the surface underwater formed the **extremely round and oval shaped** nodules—they were not formed in burial.
5. Only the Universal Flood Model **clearly and precisely explains** the global nodule phenomenon on the Earth today.

Why the UF Nodule Model Works

1. The UF model explains **why** nodules were deposited only several thousand years ago.
2. It explains **why** nodules are a global deposit, precipitated in many different environments, but at the same time.
3. It explains **how** the environment was created which formed the quartz and metal minerals in the nodules.
4. It explains **how** nodules could only be on or near the surface as opposed to being in deeper sediments.
5. It explains **why** many nodules are generally of a spherical shape and consistently smallish in size.

Surface Mark Summary

Two primary types of Surface Marks exist—Quartz and Carbonate. Examples of each have been shared, and though it may seem like we have provided exhaustive evidence, just the opposite is true. Surface marks of the UF were found in nearly every geological trip we ever took, and a single book cannot contain descriptions of them all. Still, modern geology's explanation of rainwater seeping through sediment to form the spherical nodules underground as other sediment surrounded them is truly perplexing. The Nodule Surface Mark demonstrates two important aspects of the UF. First, since most of the surface formations required a hypretherm environment, and because they are found on the surface worldwide, a *hypretherm* of global proportions occurred, covering most of today's continental surface. Second, the lack of erosion on and around the surface mark specimens themselves, and the desert environment in which they are found indicates *a young age* of the global hypretherm.

8.15 The Diamond Mark

Everyone knows what a diamond is, but you probably did not know that diamonds come from microbes! According to modern science, diamonds are supposed to have come from 'deep within the hot mantle' where microbes don't exist; one can certainly visualize the "problem" recent diamond discoveries created, which we will review shortly. In the discovery of the origin of diamonds, the inherent flaws and inclusions within the diamonds come to a veritable rescue, providing the necessary clues to decode their true origin.

Yellowstone National Park Geyser Nodules

These are geyserite Nodules.

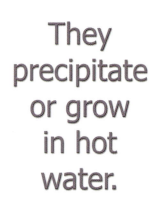

They precipitate or grow in hot water.

These nodules are rare, unlike others shown.

Geyser Nodule Evidence

The Diamond Origin Pseudotheory

How did diamonds form? Because they are Nature's hardest natural mineral, geologists long thought diamonds were formed under exceedingly high pressures and temperatures, deep inside the Earth, "120 kilometers or more." According to modern theory, once the diamonds were formed in the molten 'magma' interior, they came to the surface through volcanic eruptions in "narrow conduits" or pipes:

"According to geologic textbooks, diamonds can grow **only** in Earth's mantle, **at depths of 120 kilometers or more**. It takes the **exceedingly high pressures and temperatures of the mantle**—40,000 atmospheres and 900°C—to squeeze carbon into the ultracompact crystal structure of a diamond. The gems reach the surface when explosive volcanic eruptions force them up **narrow conduits through the mantle and crust**." Note 8.15a

It is no surprise that new discoveries about diamonds "are forcing geologists" to change their thinking about the origin of diamonds. Generally, diamonds are found in *a sedimentary matrix, not in old lava flows* tied to an imaginary deep mantle source. But this is not what is taught in today's geology. Only recently are researchers learning that the origin of Nature's hardest natural mineral doesn't fit the Rock Cycle Pseudotheory.

Rethinking "Cherished Ideas About Earth's Continents"

Science News printed an article titled, *Microscopic diamonds crack geologic mold*:

"Tiny diamond grains, discovered in rocks from southwestern Norway are forcing geologists to rethink cherished ideas about Earth's continents." Note 8.15a

The article continues:

"At only 20 to 80 micrometers in size, the diamonds are too small to see without a microscope. Yet they have dazzled scientists because **they formed within the continental crust, an unlikely birthplace for the world's hardest natural mineral**." Note 8.15a

Diamond experts have since found several locations within the Earth's continental *crust* where diamonds were formed. The "cherished ideas" of a formation deep within a hot mantle is slowly melting away.

"The Norwegian diamonds break the standard mold because they **do not come from volcanic mantle rocks**. Instead, they appear in… **ancient sedimentary deposits at Earth's surface**." Note 8.15a

Diamonds, like ores and most minerals near the surface of the Earth, all share a similar origin—they formed in sedimentary deposits in a hypretherm; they did "not come from volcanic mantle rocks." As it turns out, there has always been a "problem" with science's Diamond Origin Pseudotheory because it is based on the false notion of the up-and-down movements described in the Plate Tectonic Pseudotheory.

"A Tectonic Problem"

The Magma and Rock Cycle Pseudotheory chapters showed that most rocks in the crust, including the sedimentary deposits, have a much lower density than the mantle. Geologists therefore, consider crustal rocks too buoyant to sink into the mantle. Like a piece of cork sinking in water, it just doesn't work. Diamond genesis is only one of the many pieces of Nature's puzzle that plate tectonics cannot account for. Thus, we discover a "problem exists" when trying to find the mechanism to explain how crustal blocks are subducted, exhumed, and brought to the surface. From a journal article in *Geology* discussing microdiamonds we read:

"**A tectonic problem exists** in elucidating the mechanism(s) by which crustal blocks are first subducted to such a depth and then exhumed to the surface." Note 8.15b

The "problem" is that less dense substances don't sink in denser material, and the problem will not be made clear until we change the theory. Some prominent researchers in the field see the new discovery of diamonds in sedimentary deposits as "a wake-up call":

"'This is a spectacular discovery. **This is a wake-up call**,' comments geologist Stephen E. Haggerty, **a diamond expert** at the University of Massachusetts in Amherst." Note 8.15c

From the observations and information gathered by technologists, and by questioning everything with an open mind, including long-held diamond genesis pseudotheories, the "tectonic problem" can be solved:

"Haggerty, however, suggests **that diamonds might have formed without a trip into the mantle**. Industrial researchers, he notes, have learned how to grow extremely thin diamond films **at very low pressures**." Note 8.15c

If the diamond deposits are not from deep in the Earth as previously thought, another geological problem arises. Besides problems with the structure and movement of the crust (the tectonic problem), from where did the increased temperature and pressure come that formed diamonds found *on the surface*?

The hypretherm environment required for the formation of diamonds is the same environment that quartz crystals and other surface minerals required. Therefore, the Diamond Mark, just like the Surface Mark, is evidence of a recent global hypothermal event. Without the hypretherm and the Crystallization Process explained in the Hydroplanet Model, researchers will not be able to understand fully the diamond forming process.

Indeed, if the diamonds were not formed on a trip through the mantle, and if the crust is not being subducted and uplifted as was once thought, diamond expert Stephen Haggerty, notes the obvious:

"In any case, **geologists will have to rewrite some basic**

Kimberlite Diamond Diatreme Example

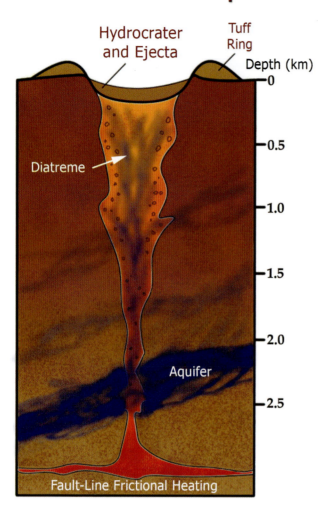

Fig 8.15.1 – Diamond bearing Kimberlite 'pipes' are really just hydrofountains. Some have hydrocraters at the surface. Modern geology has tried to define a mechanism for their formation through the magma paradigm, but to no avail. As we apply the lessons learned from the Crater Model, frictional heating from fault lines across aquifers easily explains their function. They were plentiful during the UF, and there is ample evidence of them on the surface today.

"In any case, **geologists will have to rewrite some basic textbooks**". Note 8.15c

> "In any case, geologists will have to rewrite some basic textbooks."
>
> Stephen E. Haggerty

The Diamond Inclusion Mark

Diamonds are made of carbon—a *single* element. Although it does not sound "complex," researchers think it is:

"**Diamond** is a fascinating and surprisingly **complex** mineral..." Note 8.15d

What would make the diamond mineral "complex"? After numerous careful studies, existing pseudotheories about the origins of the world's diamond leave many questions unanswered. From the research of *Fluid Inclusions in Minerals*:

"In spite of **numerous careful studies**, the available data on the fluid inclusions in diamond **leave many questions unanswered**." Note 8.15d

According to the Rock Cycle Pseudotheory, diamonds were formed so deep inside the Earth that the high temperatures and pressures should have forced all fluid out of the diamonds. However, like all other natural minerals—some diamonds have fluid inclusions:

"A series of recent studies, however, have shown that **fluid inclusions of several types are indeed present in** some diamonds, but their compositions range widely, and present us with some **enigmas**." Note 8.15d

The "enigma" exists because the inclusions are "hydrous" (water based):

"A connection between carbonatites and diamonds has been suggested by many, and Schrauder and Navon (1994) have shown that their studies of the fluid inclusions in fibrous diamonds from Botswana **indicate the presence of both hydrous and carbonatitic fluids during diamond growth**." Note 8.15d

Furthermore, the water inclusions don't contain just water, but a substance certainly not associated with the deep mantle of the Earth—**salt**:

"Chen et al. (1992, 1993) also found relatively **large inclusions in diamonds** from Shandong Province, PRC, they were high in Na and Cl as well as K, and the authors report **actual crystals of halite [salt]**." Note 8.15d p287

Salt from the deep mantle fits nowhere in modern geology, but in Flood Geology salt is an integral part. In fact, with the newest technology, the analysis of very small inclusions within diamonds directly confirms the real origin of diamonds—a bacterial rich ocean!

The Real Origin of Diamonds

New procedures for examining inclusions in diamonds have begun to unlock their origins. In 2003, Daniel Schulze from the University of Toronto, published a paper in *Nature* describing the inclusions, calling them time capsules containing substances and clues about the environment the diamonds were formed in. According to Schulze:

"'**This proves these diamonds have an oceanic heritage**.'" Note 8.15e

Schulze and other researchers studied diamonds from Venezuela, observing that the carbon that formed the diamonds did not come from a carbon source deep in the mantle, but from "sea floor bacteria":

"In addition, these particular diamonds, he says, seem to have 'biogenic' carbon signatures, indicating that some of the **carbon that formed the diamonds came originally from living organisms, such as ancient sea floor bacteria**." Note 8.15e

Despite such strong evidence, the researchers still look to the Earth's mantle and its heat and pressure as the mechanism that created the diamonds! The *dogmagma* paradigm is certainly tenacious. Another interesting fact Schulze recorded was that the mineral *coesite* (discussed in the Crater Model, Fig 7.9.24) was an "abundant" inclusion in the diamonds:

"...**coesite** is unusually **abundant as a diamond inclusion mineral**..." Note 8.15e

Impactologists firmly believe that coesite is only produced on Earth during high-speed impacts. However, because coesite is "abundant" in natural diamonds, the notion that it forms only

Natural Diamond Shapes

Fig 8.15.2 – Prior to cutting, many natural diamonds are in the shape of an octahedron or a cubic shape. These two natural diamonds come from Russia and Congo respectively. For many decades, diamonds were seen as evidence that rocks moved from the mantle to the surface, but new technology has proven they were made on an ocean floor. Then again, if both the Uplift and Subduction Pseudotheories are incorrect, how could diamonds have come from the mantle at all? The answers locked in the diamonds crystalline structure testify of the momentary intense hyprethermal environment that was once on the surface of the Earth.

8.16 The Inclusion Mark

Inclusions are the pockets of liquid and gas trapped within the minerals as they were formed. They hold clues to the environment that existed when the mineral formed. Throughout the book, inclusions have been lightly touched upon, but here we look a little deeper, investigating the constraints the inclusions place upon the environment that must have existed when the mineral formed.

Both organic and inorganic substances are present in water inclusions; these will be examined along with other physical factors required during mineral formation. These include the actual temperature and pressures, now known through the use of new technology capable of extracting and examining the fluid inclusion material. These observations are the final compelling evidence of the Universal Flood Model.

Inclusions Defined

Inclusions hold a special place in the UM. In the Hydroplanet Model, they dramatically established and helped explain the Hypretherm that the minerals grew in and they helped demonstrate the true origin of most of the Earth's meteorites. But they can do even more; they can show us the physical con-

from impact is false. Furthermore, the diamonds come primarily from kimberlite *pipes and diatremes*, which means the coesite inclusions prove the fallacy of the impactologists' claims that craters with coesite are impact craters. Instead, everywhere we look, the evidence confirms that the vast majority of craters are of hydrofountain origin, including the world-famous Arizona Meteor Crater. That crater is clearly a hydrocrater, and it has diamonds in the iron ejectites that came from it.

Diamond Mark Replaces Diamond Pseudotheory

Why have diamond bearing kimberlite pipes "remained a mystery"? From 2007 *Geotimes* article:

"**Kimberlite pipes** form from a **deep** volcanic eruption, but how **remains a mystery**. '**We've never seen them erupt**,' explains Kelly Russell, a volcanologist at the University of British Columbia in Vancouver..." Note 8.15f

Most of the world's diamonds come from Kimberlite pipes. They are a mystery simply because they have never been known to erupt—no one has ever seen one in action. The reason? The pipes are an artifact of the Universal Flood—hydrofountains! If we step outside the box of the Dark Age of modern science, Flood Geology and carbon from "sea floor bacteria" help us make sense of the diamond mystery.

The Diamond Mark touched on the importance of inclusions, but the story goes much deeper. The Inclusion Mark is the last of the marks we will cover in this chapter. It affords additional unequivocal evidence of Flood Geology.

Kimberley Diamond Mine

"This proves these diamonds have an oceanic heritage."

Daniel Schulze

Fig 8.15.13 – Discovered in 1871, the Big Hole in Kimberly, Africa is the largest hand-dug excavation in the world. The mine, located in a geologic structure called a Kimberlite Pipe, of which this was the first discovered, produced over 2,720 kilograms (6,000 lbs.) of diamonds from 1871 to 1914. The black and white image taken during the mid-1870s (inset right) shows the spider web of cables used to service the mining claims. Contrary to modern science pseudotheory, the Big Hole's minerals did not come from a molten magma core and they did not originate from a mass of melted rock. The crystalline diamonds for which this mine is famous formed in hyprethermal conditions created when this UF hydrofountain was active.

ditions of the Hypretherm, including temperatures, pressures, and even allowing us physical evidence of what was actually *in* the UF waters during the Flood event.

"In the case of **growth from solution, solvent often is entrapped**. This mechanism accounts for **the entrapment of liquid inclusions in minerals formed hydrothermally**." Note 8.16a

We have already seen inclusion evidence throughout the chapter. The Sandstone Mark, the Carbonate Mark, the Coal Mark, and the Diamond Mark all had inclusions that were investigated by researchers whose work actually confirms the UF Hypretherm. Inclusions also provide a critical link to UF deposits by allowing the study of the environment that existed when the inclusions were made.

"**Edwin Roedder** and coworkers at the U.S. Geological Survey have **pioneered in the study of inclusions** of liquids and gases in minerals. Such inclusions often contain the hydrothermal mineralizer solutions from which the mineral has crystallized, and so **can provide information about the pressure and temperature conditions of mineral formation**.

"**The mechanism of gas-liquid inclusion is easy to understand**. If crystal growth is very rapid, diffusion becomes important and constitutional supersaturation is likely. Growth protrusions will form rapidly and liquid solvent will be trapped between the protrusions. Such inclusions **provide a sample of the growth environment** by entrapping solution of the density and mineralizer concentration present at the time of crystallization."

Note 8.16a p32

The study by Edwin Roedder (1984) and coworkers, is considered the most complete and authoritative work on the subject of inclusions and is our primary source of scientific quotes in this subchapter. Critics of the UF must address this commonly overlooked geologic phenomenon, that until recently was ignored by most practitioners of geology. Inclusions, along with new technological advances in physics and chemistry have provided us with new information on the environment in which rocks formed.

Roedder explains in the following statement that the fluid "trapped in the crystal as a fluid inclusion" is the fluid out of which the crystal grew:

"Regardless of the process whereby the fluid became saturated with respect to the host mineral for the inclusion—whether by loss of gases, mixing with other fluids, loss of heat, or reaction with wall rocks or earlier vein minerals—**the fluid passing the protruding crystals is the fluid from which that crystal is growing, and if that fluid is trapped in the crystal as a fluid inclusion, that inclusion is a sample of the ore-forming fluid, not a 'final spent fraction**.'" Note 8.16b

Investigators can no longer assert that the fluids and gases trapped inside rocks seeped in from the outside because technology has advanced to the point where we can measure microscopic sized inclusions, some of which contain a vacuum and some of high pressure. Neither would be possible if there was leakage to the atmosphere. Furthermore, observational physics have shown that an accurate determination of temperature and pressure of the liquid at the time of crystallization can be calculated by heating and/or freezing the inclusion while observing the fluid as it changes from a liquid to a solid, or to gas.

Nicolas Steno (1638-1686), a famous geologist who developed the Law of Stratification, observed inclusions in quartz and concluded that the crystal grew from precipitation out of solution and that the liquid inclusion was the same solution wherein the crystal grew. Sadly, dogmagma would hold back the critical importance of this observation until only recently. Now, the new UM paradigm and ample scientific evidence makes it possible to establish the hyprethermal conditions that existed simultaneously across the world's continents just a few thousand years ago.

The Methane Inclusion Evidence

There is a point to be made about the solids or **solutes** that are trapped inside the water inclusions. But first, let's look at what deep-seated rocks found in the Earth's crust do *not* contain.

> FQ: Is salt found in granite or in the quartz, feldspar, or mica that make up the granite rocks?

None that we have found. Related to the Inclusion Mark and the UF event comes this ***extraordinary claim***. If ocean waters covered the continents during the Flood, most surface sedimentary *should contain salt and other organic materials* in their inclusions because they grew out of Flood waters. However, deep mantle materials, that is, rocks that formed before the Deluge such as granite, should ***not*** contain a significant amount of salt or organic waste materials.

Keep in mind that this theory is completely opposite of the claim science makes today, because modern geology holds that whatever is on the surface of the Earth is being 'subducted' and later 'uplifted' to the surface, over and over again, for billions of years through 'mantle convection.' Thus, the salts and organic waste from the ocean

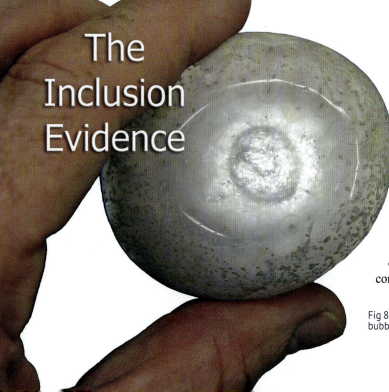

Fig 8.16.1 – This polished agate enhydro from Brazil shows a gas-water (air bubble in water) inclusion in the center of the specimen.

should be mixed within the deep mantle rocks. In other words, mantle materials should have a mix of salt and organic waste within the magmatic solution.

But do salts and other organic materials such as methane, the principle component of the world's abundant natural gas, exist in the inclusions of deep igneous rocks? Roedder explains:

"Fluid inclusion from **mantle materials** of various sorts, as discussed in this paper, consistently show **essentially zero methane**. Several thousand analyses have been made of the gases trapped in fluid inclusions in minerals, including some deep sources (Roedder 1972, 1984). **Methane is commonly found among the gases in inclusions trapped in minerals that grew in an organic-rich sedimentary environment**. When **inclusions in igneous rocks** are examined, however, **many have no detectable methane**, and only small amounts, usually less than 0.1%, are found in others, along with major amounts of H_2, H_2O, and CO_2." Note 8.16c

As we see, igneous inclusions and other deep mantle minerals (like granite) have "essentially zero methane" but methane is "commonly" found inside sedimentary rock inclusions apparently near the surface. These grew in the UF hypretherm.

Modern geology insists that diamonds are evidence of mantle subduction and uplift, and if the diamonds had no methane inclusions, their assertion may have some validity. However, if the diamonds do contain methane in their inclusions, the notion that they formed deep in the Earth ought to be summarily dismissed. Methane inclusions in diamonds would testify that they almost certainly came from UF hydrofountains, not from igneous rocks deep in the mantle. From the *American Mineralogist* we read:

"In summary, our results show that H_2, H_2O, CO_2 and CH_4 [methane] are **persistent molecular compounds** of the gas included in **diamonds**." Note 8.16d

So there it is, methane, a common organic gas is found in diamonds! Obviously, the deep mantle rocks like granite were formed in a completely *different* hypretherm than diamonds and other quartz-based minerals that contain methane and other organic ingredients.

We can refer to the granite-forming environment as the **Creation Hypretherm**, which included the formation of Earth's original minerals at a time before life (including microbes) existed on this planet. Because salt minerals have been determined to be biogenic (Salt Mark), the Creation Hypretherm easily explains what modern geology cannot—why the igneous rocks on the Earth and the lunar rocks don't contain the salts present in the Earth's salty oceans. The Moon does not contain salts because it did not contain the microbes necessary to form constituent molecules, such as HCl acid and NaOH base, which combine to form salt and water.

The Japanese Alps Granite Inclusion Evidence

A great example of the difference between the Creation Hypretherm and the Flood Hypretherm is the Takidani Granodiorite in the Japanese Alps. Granite and other igneous rocks that were formed in the Creation Hypretherm do not typically contain salts that were created and transformed from microbial byproducts in the Flood Hypretherm. However, many of the world's granite mountains were cracked and fractured during the crustal failure of the UF event, allowing hyprethermal waters to flow through the fractures. These fractures were sealed with a mix of greater than 70% quartz minerals that contained fluid inclusions. If UF oceanic waters flowed through the fractures near the tops of the granite mountains, it should be evident that it was hot and salty, which would be apparent in the fluid inclusions. Researchers reported:

"**Fluid inclusion evidence indicates that fractured** Takidani Granodiorite at one time hosted a liquid-dominated, convective hydrothermal system, with **<380°C**, **low-salinity** reservoir fluids at hydrostatic (mesothermal) pressure conditions. 'Healed' microfractures also trapped **>600°C, hypersaline** (~35 wt% $NaCl_{eq}$) fluids of magmatic origin, with inferred **minimum pressures** of formation being ~ 600-750 bar [8,702-10,878 psi], which corresponds to fluid entrapment at ~ **2.4-3.0 km depth**." Note 8.16e

Both temperatures and pressures are in line with the successful quartz growing experiments performed by UM researchers. The salt content of the hypersaline inclusions is equivalent to present-day seawater, whereas low-salinity coincides with freshwater/seawater mixtures, found in lower salinity inclusions. In other words, the UF Hypretherm described at the beginning of this chapter describes the environment that existed around the tops of the Japanese Alps, as identified by independent researchers, at slightly less pressure because of the apparent altitude of the Alps during the Flood event.

The researchers attempt to account for the pressure by asserting a burial of "2.4-3.0 km depth" presumably over some unstated eons of time. This inference is completely lacks empirical evidence. Furthermore, the depth of a couple of kilometers is nowhere near the depth necessary to produce ">600°C" temperatures. In fact, no geologist has demonstrated that such temperatures can *ever be obtained* by depth alone (see the Magma Pseudotheory Chapter). The world's *deepest* boreholes in Germany and in Russia were each *several times* deeper than the 3 km distance suggested by researchers, but never reached temperatures anywhere near the hundreds of degrees Celsius required to grow the quartz-filled fractures!

The Russian Kola hole reached only 50°C at 3 km Note 8.16f, a far cry from the temperatures suggested by the Takidani Gra-

Fig 8.16.2 – Inclusions in enhydros contain more than water; they include the remains of microbial activity in the form of gases. Carbon dioxide and methane are common gases found in inclusions formed on or near the surface of the Earth's crust. Conversely, recycled (uplifted and subducted) igneous surface rocks, *theorized* to have formed deep in the mantle, have as Roedder observed, "essentially zero methane." This is direct evidence that "mantle materials of various sorts" lacking methane, were never subducted or uplifted, but were created in a lifeless (Earth's early creation) hydrothermal environment.

nodiorite researchers. In fact, temperatures of 380-600°C have never been produced simply by 'burial.' There is no direct evidence that the Japanese Alps are uplifting, and so we are left with only 'imagination' of how this all occurred, according to the leading scientists. Conversely, the host of direct evidence of the Universal Flood makes it easy to comprehend how 20,000 feet of UF water rested over the 10,000 foot mountain peaks. With the element of pressure satisfied, the Frictional-Heat Law satisfies the high temperature requirement because massive plate movements would have generated colossal amounts of heat. Anyone living in Japan today knows that earthquakes are a frequent occurrence because of the many fault lines that run through the mountains.

The St. Peter Sandstone Inclusion Evidence

The importance of the inclusions in determining the UF origin of sedimentary deposits is illustrated with the St. Peters Sandstone deposit, stretching from Missouri to Illinois and from Nebraska to South Dakota, in the United States. Up near Lake Michigan, sandstone lies buried between layers of sediment totaling 6 km (3.7 miles). This 99%+ pure quartz sandstone layer is generally 100-350 meters thick around the Lake Michigan area. It has been a source of fascination for geological investigators for decades and is actively mined for its pure quartz qualities.

The simple question of the origin of this sandstone cannot be answered by modern geology and its origin remains "controversial":

"The **origin of authigenic quartz in sandstones** and of **silica-transporting aqueous fluids are controversial**." Note 8.16g

As discussed in the Sand Mark, the genesis of sand deposits like the St. Peter Sandstone have little chance of being discovered by modern geologists simply because they do not comprehend Flood Geology. We get a feel of how frustrating it has been for investigators to work in the modern geology paradigm as we read the following statement from a geological journal article. Note how researchers recognize that the sandstone is not buried deep enough to account for the high pressure and high temperatures necessary to cement the sand grains together. They know that the lower temperatures suggested by other researchers only create "amorphous silica or opal" not the quartz that actually holds the grains together.

"Stratigraphical evidence indicates that the **St. Peter Sandstone** cropping out on the Wisconsin Arch (on the western flank of the Michigan Basin and the northern flank of the Illinois Basin) was buried to depths of less than 1 km, consistent with **the lack of evidence for intergranular pressure solution**. Along a normal conductive thermal gradient, authigenic quartz growth would occur in the temperature range of **20-50°C. The ability of authigenic quartz to precipitate at these low temperatures is controversial**, and both fluid chemistry and ki-

> "Such inclusions **provide a sample of the growth environment** by entrapping solution of the density and mineralizer concentration present at the time of crystallization."

meteoric water in the North Sea **show little evidence of quartz cementation at shallow depth**." Note 8.16g p5104

The confusion amongst researchers grows as they study the "fluid inclusions" in the pore-space cement holding the quartz sandstone together because the calculated temperatures of the sandstone cementing process exceed 200°C (392°F)!

"**The temperature range of quartz cementation** of the deeply-buried **St. Peter Sandstone** in the Michigan Basin may be estimated from the homogenization temperatures of two-phase **fluid inclusions**. Drzewiecki et al. (1994) measured such temperatures for inclusion in quartz overgrowths in the range of 81-153°C; after approximate pressure correction, **quartz crystallization temperatures are calculated to lie in the range 135-205°C**, and this range is adopted in interpreting the isotopic data obtained in this study." Note 8.16g p5104

Why would the high temperatures of quartz cementation be so "*controversial*" to researchers?

> Because this sandstone was supposedly laid down under a **shallow, cool sea**, cemented, then uplifted to its present location.

But quartz cementation cannot happen in shallow, cool seas. The much bigger question of how the rounded, pure quartz grains were formed and deposited in a homogeneous deposit without any other mineral contamination is the proverbial icing on the cake. To the researchers' credit, the conclusion drawn acknowledges that the quartz cements were formed from "brines" at "high temperatures":

"**Quartz cements** in the Michigan Basin **formed from** basinal **brines or formation waters at high temperatures**, with

netic factors may be important. The rate of precipitation **below 70-80°C** may be very low, and SiO_2 may **precipitate as amorphous silica or opal**. Sandstone flushed with

Fig 8.16.3 – The Japanese Alps Granite includes cracks filled and sealed with crystalline quartz minerals that grew in a hypretherm. The temperatures and pressures of the hypretherm that formed the quartz in these fracture zones corresponds to the UF Hypretherm introduced at the beginning of this chapter.

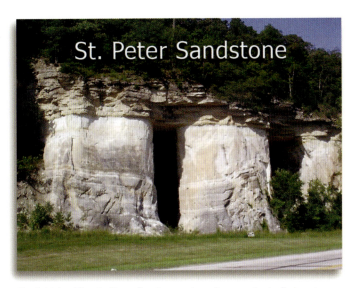

Fig 8.16.4 – The St. Peter Sandstone deposit, comprised of almost pure quartz sandstone, covers a large portion of the Northeastern United States. The origin of the sand grains and the interstitial cementing water remains a complete mystery in modern geology. Investigators studying inclusions in the sandstone cannot account for the high temperatures and pressures present during the cementation of the sandstone. However, Flood Geology and the accompanying hyprethermal environment satisfy all of the requirements of origin prediction.

little or **no meteoric water component**." Note 8.16g p5115

"No meteoric water component" means that there was no water involved from the atmosphere. The brines were merely oceanic waters.

Without Flood Geology, deep hot oceans cannot be envisioned. The physical constraints of liquid inclusions along with the salts and other minerals are evidence of the pressure and temperature that existed when these Flood deposits were laid down. Liquid inclusions in the St. Peter Sandstone show the Self-Evident Principle with respect to Flood origins—it is, because it can be no other way.

Quartz Sediments Formed in a Hot-Deep Ocean

Let us expand our scope from one deposit to all sedimentary deposits *in general* and consider this FQ:

> Is there a consistent environment that can be discovered from inclusions in most sediments?

Edwin Roedder, the expert and author of the book *Fluid Inclusions*, expounds on the subject, identifying salt abundance and salt diversity in sedimentary rock inclusions:

"**The liquid is generally a water solution** with less than 10wt % **solutes**, but concentrations range from more than 50 wt % to practically 0%. The solutes consist of major amounts of **Na, K, Ca, Mg, Cl, and SO_4**, and less amounts of many other ions. Many individual ions in this list may predominate, although **Na and Cl are generally the most abundant**." Note 8.16h

NaCl, or common table salt, is the most abundant mineral on the entire *surface* of the Earth, while all of the other elements listed: K, Ca, Mg and SO_4 are constituent products of other *sea salts*. Simply put, most of the Earth's sedimentary rocks had to have been formed out of ocean water! And it could not have been a *shallow cool ocean*. The liquid inclusions are another unequivocal physical element that proves that the vast majority of the quartz-based sediments could have only formed in a hypretherm.

The Inclusion Prediction

With so many marks of the UF now clearly set forth for all to see, understand, and comprehend, the Learning Process demands that we *apply* what we have learned if we expect to progress and learn more. Otherwise, the "enigmas and anomalies" prevalent in modern geology and paleontology will never go away.

One way to apply the information that we have learned from the Universal Flood, is to use the classic example of the Catastrophic Principle, which supplants the Uniformity Principle set forth by Hutton. The catastrophic nature of the Flood, which was detailed in subchapter 8.3, demands that both organic and inorganic processes take place simultaneously and within a short period of time. This of course, does not take place in the 'geological time scale' invented in the 1800s.

For example, petroleum and quartz geodes found only at the surface are simply not made together in the world of 'geological time,' but the Catastrophic Principle of the UF **predicts** inclusions of *petroleum, inside quartz geodes that were made together on or near the surface!* Let's see how Nature supports the Catastrophic Principle and the UF by providing examples of organic substances, such as petroleum formed together with quartz, in a hyprethrem on the surface of the continents.

The Tyson Creek Geode Inclusion Evidence

William Sanborn, the author of the book, *Oddities of the Mineral World*, provides an intriguing account of an "oddity" scientists once regarded as a "practical joke":

"Now we come to the frosting—or would you believe oil—on the geode cake! I have encountered very knowledgeable mineral collectors and lapidary enthusiasts as well as **professional geologists and paleontologists who have regarded the following data as unreal or some sort of practical joke**.

"It is a **fact** that there is a bed of small-to medium-sized quartz-crystal-lined geodes not more than 50' from a main highway in Illinois that almost has to be seen to be believed! In this totally inconspicuous bed **many of the geodes are filled with either a solid or liquid form of bitumen (or if you prefer, tar or petroleum)**. These are the truly **odd** and unique oil geodes found on a tributary of Tyson Creek about 1 ¼ miles south of Niota, Hancock County, Illinois, on State Highway 96. **They are real mind-bogglers**. Many of the geodes are filled with a hard asphalt like bitumen, and others with a very thick, viscous liquid bitumen. The latter types do not splash when shaken, due to the viscosity of the liquid; however, they definitely constitute the messiest collecting imaginable! To the novice it seems incredible to dig out a hard, grayish, round rock. Ring it with a gentle taps of a rock hammer, and have it pop open, **spilling dark black oil on your hands!**

"Through the courtesy of Dr. S. R. Silverman of Standard Oil of California's California Research Corporation, a very thorough analysis was conducted by their laboratories on bitumen-filled geodes I collected at this site—one with solid—the other with fluid-type bitumen. A series of complicated chemical and isotopic analytical procedures were applied to the specimens, resulting in a lengthy and highly technical report. Some of the more interesting factors are condensed here. First of all, the geodes themselves are of **inorganic origin, while their contents (the bitumen) are of organic origin**. The organic substances in both geodes were completely soluble in carbon disulfide, and an infrared spectral analysis indicated that the solid and fluid bitumens were virtually identical chemically. The analysis also revealed the bitumen-paraffinic composition to be **quite typical of Illinois crude oils**.

"Incidentally, **oil-filled geodes are not limited to this one locality**: the same general region has turned up isolated occurrences, and it is not uncommon to find oil filling small solution cavities in some limestone quarries. An interesting related sidelight is that in certain pockets of fine fluorite crystals from the mines of Rosiclare and nearby Cave-in-Rock, Illinois well-crystallized fluorite **often occurs completely coated with bitumen**." Note 8.16i

Another example, similar to the organically filled Tyson Creek Geodes, are the Brazilian Enhydro Geodes as seen in Fig 8.16.5. Because these geodes are made of quartz, and are found only on or near the surface, they provide physical evidence establishing the presence of a recent hypretherm. All of the minerals with organic inclusions are evidence of the UF Hypretherm, because it is the only answer for such surface-formed minerals.

The Herkimer Quartz Inclusion Evidence

Another example that conflicts with modern geologic theory is "anthraxolite," which is formed within quartz crystals found in New York, USA. In Fig 8.16.6, an example of the famous Herkimer 'Diamond' quartz crystals is shown. This one contains a black mineral called **anthraxolite**, along with other organic substances:

"The quarts crystals from **Herkimer** and Adjacent Montgomery County, New York, USA, are in **vugs** in a crystalline petroliferous dolomite. **Inclusions in them contain, in addition to brine, several types of organic matter: high pressure supercritical mixtures of organic gases**; colorless to yellow, highly fluorescent **oils**; and broken fragments of **anthraxolite, a black organic material having a conchoidal fracture**, which is also found in the vugs with the quartz (Dunn and Fisher, 1954)." Note 8.16j

Why is the black anthraxolite mineral in Herkimer quartz crystals so important?

Because anthraxolite is a type of coal.

"**Anthraxolite** is a hard, black lustrous graphitic **coal**..." Note 8.16k

This is another example of *coal* that is formed on the surface, like the coal particles discussed in the Rock Varnish Surface Mark. Coal formation without burial is in direct opposition to modern geology's coal origin theories, but it is an expected phenomenon of Flood Geology.

The Reason the Inclusion Mark Evidence is so Compelling

The scientific evidence that ties all the Marks of the UF together are the *inclusions*. The inclusions *"provide a sample of the growth environment"* wherein they grew, which is the UF Hypretherm. For the first time in history, all of the UF Marks come together scientifically to form a common sense image of Nature's Puzzle. Just as 2 x 3 is equal to 6, we can deduce that 6 ÷ 3 is equal to 2. In similar manner, each of the UF Marks are interconnected.

We have seen how hydrofountains and their associated diatremes contain minerals like diamonds that form only in a hypretherm. The immense deposits of sand, carbonate, salt, oil/gas, coal, and iron-pyrite contain inclusions confirming their relationship with the hypretherm or hydrothermal environment extant during their formation. Several examples have established the connection to the Universal Flood event, and of the many that exist, we shall discuss one final example.

Leading inclusion expert, Edwin Roedder, in his book *Fluid Inclusions*, discussed the inclusion minerals sphalerite, fluorite, celestite, and quartz in rocks from the Great Lakes area including the states of Ohio, New York, Iowa, Missouri, Illinois, Michigan and Tennessee. Roedder went on to draw a direct connection between the fluid inclusions in these widely distributed surface minerals and another Mark of the Flood recently discussed—ore deposits:

"Many suggestions have been made that these scattered occurrences of **sphalerite, etc., have formed from the same kind of fluids as did the Mississippi Valley ores. The general geologic environments and mineralogy are very similar**; only the $^{87}Sr/^{86}Sr$ values and the volumes of ore minerals are different. Although the **fluid inclusions in these minerals are similar** to those in the Mississippi Valley [ore] **deposits**, in that **they were formed from hot, strongly saline brines, containing much organic matter**, the temperature range and salinity are both a little lower." Note 8.16l

Roedder's realization that the surface minerals **and** the ore deposits spread across this wide area (Mississippi Valley) had indications of being "formed from hot, strongly saline brines,

Fig 8.16.5 – This highly polished Brazilian Enhydro Geode contains a large cavity 50% filled with dark organic fluids and the balance a mixture of gases. The movement of the dark liquid is easily seen inside the cavity. Such fluid-included specimens support the Surface Mark because they are found on or near the surface and contain organic brine fluids that had to originate in a hot, deep ocean environment. The quartz outer shell further establishes the heat and pressure that existed when the geode itself formed.

Herkimer Quartz Crystal With Anthraxolite Coal

Fig 8.16.6 – Herkimer "Diamonds" are remarkably clear quartz crystals from New York, USA. Many are prized for their inclusions. Some crystals contain water and a surprising, unexpected black substance—coal. Since these crystals are found only near the Earth's surface, how did coal end up in this inclusion; unless it was formed during the Universal Flood?

Quite astonishingly, modern science was oblivious to the Dark Age of Science, and continues headlong into mass confusion with more and more 'mysteries,' 'anomalies,' and 'complex' explanations of natural phenomenon popping up each year as new observational facts refute the pseudotheories. Without the knowledge of the UF, modern science has gone on and on, trying to force the complex to explain the simple; to prove, in their minds, that 2 x 3 = 7.159. This is *why* natural science is so complicated for so many researchers and laypeople.

Notice that James Hutton is given the credit for "a completely original theory" of the Earth without even realizing that it is ***only a theory***—without any natural law to back it up or to establish it as scientific truth. In addition, and without even knowing it, Jack Repcheck, a scientific writer, describes exactly why Hutton's theory was incorrect; Hutton *assumed* "erosion" in the past took place just as it does today. He *assumed* that there was uplifting of the continents over time. And finally, he *assumed* the Earth was full of "subterranean heat":

containing much organic matter" in the same "general geological environment" has profound implications!

The UF Hypretherm must replace the old 'shallow cold sea' sci-bi, and by so doing, the genesis of all the geological deposits mentioned in this chapter are easily understood and more thoroughly comprehended.

The Inclusion Mark is beautifully simple and simply beautiful; directly entrapping the former deep once hot ocean water evidence within these surface mineral deposits for each of us to witness. Note 8.16m

8.17 The UF Summary

Geotheoretical to Geological

Beginning with the Magma Pseudotheory chapter and the Rock Cycle Pseudotheory chapter, we uncovered the geo-theoretical model of modern science. Following those chapters, we introduced the geo-*logical* Models of the Hydroplanet and Universal Flood that will necessitate a complete rewrite of geology textbooks and drastic changes to most texts in other fields of modern science.

> If the Universal Flood really happened, how could it be any other way?

The professor, who at the beginning of this chapter said, "There will never be—any scientific evidence—for a universal flood" is, like most other modern scientists, ignorant to the possibility that the 200-year-old modern science establishment could be suddenly turned on its head. What this professor and others fail to realize, is that modern scientists have been stumbling around in a period easily identified in the History of Science Table as—the **Scientific Dark Age**.

The Scientific Dark Age of the 20th century is real, and has left us empty with respect to new natural laws, like those discovered in previous centuries. The should-be laws of the 20th century were replaced by pseudotheories—and the destructive influence on mankind because of this, will continue to become more apparent as we move forward in the UM.

"In addition to giving geology, as Stephen Jay Gould stated, **its most transforming idea—that the earth was ancient**—Hutton devised the first rigorous and unified **theory** of the earth. His theory posited that the earth was constantly restoring itself. He based this concept on a fundamental cycle: **erosion** of the present land, followed by the deposition of eroded grains (or dead ocean organisms) on the sea floor, followed by the consolidation of those loose particles into sedimentary rocks, followed by the **raising of those rocks** to form new land, followed by erosion of the new land, followed by a **complete repeat of the cycle, over and over again**.

"Hutton was also the **first** to recognize the profound importance of **subterranean heat**, the phenomenon that causes volcanoes, and he argued that it was the **key to the uplifting** of formerly submerged land. It was a **completely original theory**." Bib 154 p7-8

So yes, it was a "completely original theory"—but it was also completely wrong. Although Repcheck was unaware of Hutton's errors in 2003 when he made these comments about the man who gave us magma, Hutton will probably always be known as the father of mankind's "*modern* theory of the earth" upon which many other modern theories rest, like those of plate tectonics and the role of the ice ages:

"His ideas were the starting point for the modern theory of the earth, which now includes plate tectonics and the role of the ice ages." Bib 154 p8

But both plate tectonics and the ice age fail the test of time. If the Flood happened 4,400 years ago, any geological evidence of an ice age 10,000 years ago would have been nearly wiped out. Empirical evidence for the 4,400-years-ago time frame of the Deluge is in the upcoming Age Model chapter, which also includes a discussion of how the ice age theory fails.

The Magma Pseudotheory was Hutton's and Lyell's new foundation for all geological discussion in modern science, but if their magma theory was *not* built on a *sure foundation*, and if the center of the Earth is not melted rock, then we should expect a wholly confused 'modern geology,' full of mysteries surrounding even the most basic geological concepts.

That is exactly what the Rock Cycle Pseudotheory demonstrated; dozens of mysteries, many presented with the work of the scientists who investigated them, who admit their observations often did not fit the classic geology paradigm. It is from the geo-*theoretical* world to the geo-*logical* Models of the UM that we are moving towards, with a New Geology based on the most important and ubiquitous substance in the universe—water.

The Hydroplanet Model included empirical, scientific evidence that examined and clearly explained how rocks and minerals are formed, something basic geology acknowledges it has never been able to do. Moreover, the simple processes associated with the formation of the Moon, Earth and other celestial bodies was presented, using water and empirical examples from multiple disciplines.

With the Magma Pseudotheory, the Rock Cycle Pseudotheo-

HISTORY OF SCIENCE TABLE

Year	New Observations	New Technologies	New Theories	New Laws
2000	James D. Watson – DNA double helix Har Gobind Khorana – DNA protein codes Dorothy C. Hodgkin – structure of penicillin, insulin, vitamin B12 Barbara McClintock – gene transposons	Rosalyn S. Yalow – radio immunoassay Rosalind E. Franklin – x-ray diffraction, photography of DNA Frederick Sanger – structure of proteins	Steven Hawking – black holes Tsung-Dao Lee – overthrew parity theory Luis W. Alvarez – bubble chamber theory Richard P. Feynman – nuclear shell theory	**The Scientific Dark Age**
1900's	Margaret Mead – cultural anthropology Enrico Fermi – nuclear reactions Irene Joliot-Curie – radioactive elements Arthur H. Compton – Compton Effect Edwin P. Hubble – red shift of light Selman A. Waksman – microbiology Alexander Fleming – penicillin Nettie M. Stevens – heredity	Charles H. Townes – maser, laser Jacques Y. Cousteau – aqualung, submersibles Shockley, Bardeen, Brattain – transistor Grace B.M. Hopper – supercomputer, COBOL language Enrico Fermi – nuclear chain reaction Robert Watson-Watt – radar Selman A. Waksman – antibiotics	Marie Goeppert-Mayer – nuclear shell theory Werner Heisenberg – quantum theory Linus C. Pauling – resonance, electro negativity theory Edwin P. Hubble – red shift Niels Bohr – quantum theory Albert Einstein – theory of relativity Lise Meitner – nuclear theory	
1800's	Ernest Rutherford – radioactive half-life, atom's nucleus and proton Henrietta S. Leavit – Cepheid stars, period-luminosity Marie S. Curie – polonium and radium George Washington Carver – soil nutrients Joseph J. Thomson – sub-atomic particles Albert A. Michelson – measured speed of light Antoine H. Becquerel – spontaneous radioactivity William Ramsay – discovered argon, krypton, neon and xenon gases Ivan P. Pavlov – reflex action conditioned Luther Burbank – plant hybridization and grafting Wilhelm K. Roentgen – x-rays Dmitri I. Mendeleyev – periodic table Joseph Lister – antiseptic surgery Louis Pasteur – pasteurization Jean Bernard Leon Foucault – speed of light in water, pendulum Matthew F. Maury – oceanography Michael Faraday – electromagnetic induction Jons Jakob Berzelius – cerium, selenium, silicone, thorium Humphry Davy – potassium, sodium, barium, trontium, calcium and magnesium discovered through use of electricity Alexander von Humboldt – explorer, scientific encyclopedia Georges Cuvier – paleontology	John A. Fleming – diode Thomas A. Edison – phonograph, lightbulb, movie projector William H. Perkin – synthetics James C. Maxwell – color photograph Jean Bernard Leon Foucault – gyroscope, silver glass mirror Augusta Ada Byron – binary notation Joseph Henry – electric doorbell, assisted with telegraph Charles Babbage – mechanical calculator, speedometer, ophthalmoscope Michael Faraday – electric generator, motor, transformer Karl F. Gauss – heliotrope, first telegraph	Sigmund Freud – theory of psychoanalysis James C. Maxwell – kinetic theory of gases Friedrich A. Kekule – molecule theory, organic compounds William Thomson – Kelvin temperature Charles R. Darwin – theory of evolution Louis Agassiz – ice age theory **During These Time Periods, Theories Were Tested and Either Became Natural Laws or Were Discarded.**	William Thomson – second law of thermodynamics James C. Maxwell – electromagnetism Gregor Mendel – heredity laws James P. Joule – Joule-Thomson effect, Joule's Law, Law of conservation of energy Jons Jakob Berzelius – law of definite proportions Joseph L. Gay-Lussac – Charles law of gases John Dalton – law of partial pressures, law of definite proportions
1700's	Edward Jenner – vaccinations William Herschel – astronomy Joseph Priestley – carbon dioxide and oxygen Henry Cavendish – hydrogen Carolus Linnaeus – binomial nomenclature Leonard Euler – mathematical nomenclature Benjamin Franklin – lightning is a static discharge Daniel Bernoulli – Bernoulli's effect	Alessandro Volta – electric battery Benjamin Franklin – cast iron stove, lightening rods, bifocal glasses		Antoine L. Lavoisier – law of conservation of mass Benjamin Franklin – static electricity laws
1600's	Edmund Halley – star positions, Halley's comet Isaac Newton – white light contained all colors of the spectrum Robert Hooke – discovered cells Anton van Leeuwenhoek – micro-observation Christian Huygens – astronomy, Huygens' principle Robert Boyle – vacuum pump, chemistry Blaise Pascal – Pascal's principle, mathematics Rene Descartes – mathematics William Harvey – the heart pump's blood	Robert Hooke – weather instruments Anton van Leeuwenhoek – microscope Christian Huygens – pendulum clock Robert Boyle – first match from phosphorus		Isaac Newton – law of gravity, three laws of motion Robert Hooke – law of elasticity Robert Boyle – Boyle's gas law Rene Descartes – law of reflection Johannes Kepler – Kepler's three laws of planetary motion
1500's	Galileo Galilei – astronomy, gravity Andreas Vesalius – human anatomy Nicolaus Copernicus – astronomy	Galileo Galilei – telescope used in astronomy		Galileo Galilei – law of inertia
1000	Hakim ibn-e-Sina (Avicenna) – medicines Galen – human anatomy Eratosthenes – scientific history Euclid – geometry Aristotle – treatises on logic Hippocrates – diseases come by nature, medicine Pythagoras – mathematics, Pythagorean theorem			
500 B.C		Archimedes – Archimedes' screw for raising the level of water, catapult		Archimedes – law of buoyancy, law of simple machines Euclid – geometric laws

Fig 8.17.1 – This is the History of Science Table first introduced in Chapter 2.2. It shows the Scientific Dark Age of the 20th century that came about by the acceptance of unproven theories that are not supported by direct evidence. This caused a lack of research into scientific truth and natural law, which had been supplanted by pseudotheories with underlying agendas.

CHAPTER 8 THE UNIVERSAL FLOOD MODEL

ry, and Hydroplanet Model chapters as a foundation, and with the New Geology, the Universal Flood Model chapter provides the finishing details of Flood Geology.

Eventually, as Flood Geology becomes more widely accepted, the realization that mankind has lost valuable time understanding geological progresses over the last 150 years will become apparent.

The Dark Age of Science remains despite advances in technology that each year prove the magma and uplift theories to be myths. Modern geology cannot let go of these theories for one reason; the scientific facts point to a Hydroplanet Earth, where weather, and Earth's age all support the Universal Flood. If there really was a Universal Flood, *why* are geologists so adamant in their denial of it? The answer to this question can be found in the upcoming Chapter 12, the Evolution Pseudotheory.

The UF Creation

When a person hears the words 'Noah's Flood', typically the first thing that enters the mind is catastrophe. Represented in Fig 8.7.2, The Deluge by Gustave Doré captures the Flood in the most grim of circumstances. The Flood marked the end of a dark period for mankind and the beginning of a new era. There is much to learn from this great event, a time of marvelous creation, of preparation and beautification for future generations of human habitation.

After studying the facts presented in this and previous chapters, truthseekers should comfortably comprehend and understand the reality of the Universal Flood. With this realization, we can begin to step beyond the cataclysmic nature of the Flood and step into the real light this chapter shines on history–Earth's history and human history. The UF was not only a catastrophe—it was a force behind the creation of beautiful mountains, canyons, rivers, and many geological wonders from around the world.

The beauty of the **UF Creation** is unparalleled; however, there is an aspect of the UF Creation that is perhaps, the most profound consequence of the Universal Flood.

> Without the Universal Flood—the modern world as we know it today would not exist.

Why? Because it was the incredible events of the Universal Flood that contributed so much of what we have come to take for granted. The world's coal, the oil and gas, the ore deposits used to smelt the *metals* used in our modern society, the rare earth elements in our digital devices, and an almost endless list of other necessities. Without the energy from the oil, gas, and coal, and the metal products from the ore—modern society and

> "...**the fluid inclusions** in these minerals are similar to those in the Mississippi Valley [**ore**] **deposits,** in that they were **formed from hot, strongly saline brines, containing much organic matter**..."

its appurtenances would not exist.

From this perspective, the UF Creation event was an extraordinary event, something so unique as to stimulate the mind to ponder if there was purpose in this extraordinary event responsible for creating both beautiful landscapes and incredible technological opportunities for mankind to enjoy.

The UF Connection

This Universal Flood chapter is *interconnected* with all other chapters in this book. The UF completely changes the modern science paradigm, restoring the *real* purpose of *science* as defined in *Webster's Unified Dictionary and Encyclopedia*:

"A branch of study which is concerned either with a connected body of **demonstrated truths** or with **observed facts** systematically classified by being brought under **general laws**, and which include trustworthy methods for the discovery of **new truths** within its own domain." Bib 98 p3783

In the Dark Age of Science (Chapter 3), it was explained that this *correct* definition of science was taken from a *1960* Webster's dictionary because "modern" dictionaries and encyclopedias had been diluted, muddying the definition of what science really is—the study of **demonstrated truths** and **natural laws** that describe and explain Nature. Without demonstrated truths and new natural law, science cannot progress.

Where do the new truths of the Universal Flood lead?

Beyond the UF Creation events already outlined, the *origin* of weather must be reconsidered, for the first time, based on the knowledge of the UF and the Hydroplanet Models.

Certainly, the UF completely challenges geological time, which the Age Model chapter takes full advantage of by connecting concepts in the UF with the age of the Earth. The New Geology from the UF must bring about a complete revolution to the geological time scale foundation of the old 'modern' geology. Instead of 'geological time' that was designed without empirical evidence to satisfy the evolution agenda, the following chapters will refer to **flood time** as a reference to the approximate 4,400 years that have elapsed since the Universal Flood events. The Age Model stands on its own pillars of scientific evidence with the UF evidence presented in this chapter being only part of the extraordinary age evidences that establish the extraordinary Age Model claims.

Following the Age Model, the Fossil Model falls into place, being closely associated with the surface rocks that were formed in the UF. Because most fossils are rocks, it makes sense, or at least it **will make sense** that fossils were formed during the Flood. But once again, we take the approach that each new Model must stand on its own scientific evidence and observations.

Since the Universal Flood was a real event, as recorded in multiple world history accounts, most of the animals and plants were killed only several thousand years ago. This of course, completely refutes the validity of the evolution theory. Evolution cannot exist without geological time and the false fossil claims made by modern science. Evolutionists will contend with Universal Flood Model evidence because it weakens their stance.

Per recorded history, the survivors of the Flood landed on Mount Ararat (Turkey area) and would have populated the world we know today; there must be scientific evidence of the spread of human civilization originating from this area of the world. This evidence is explicitly established in Chapter 14, the

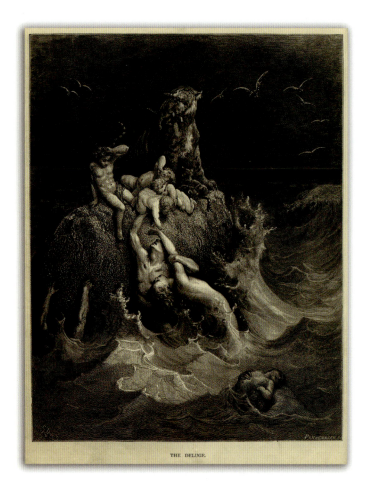

Fig 8.17.2 - The Deluge by Gustave Doré (1832–1883). This painting represents imagery of an era darkened by destruction shared by many when thinking of a world-wide flood scenario. It was a gloomy time for the perished masses and a time of spectacular upheaval, but the Great Flood was also a time of marvelous creation, a time of reshaping the Earth's surface into what we see today—a world of unprecedented beauty in which lay the means for mankind's innovation through the use of fossil fuels, metal ores, and new technology. In the Fossil Model, countless other fossil materials will be shown to have formed as well as hydrofountain landscapes to behold and explore, all created in the magnificent UF Hypretherm.

World History Model.

And what of the human civilization that existed prior to the Flood? If physical evidence were to be established for such a civilization, its existence would further support the UF. Chapter 15, the Clovis Model, was written specifically for this purpose, and it confirms the location where the first humans lived on our planet.

In summary, the Universal Flood is far more than a catastrophic event. It forever changed the world and human history. Knowledge of it turns modern geology into an old geology. It literally provided the means for our modern technological world through the creation of the ores and metals, coal and oil in the crust. But most importantly, the Universal Flood is a part of *our world history*, which tells us where we came from and who we are. We are human beings—as the Living Model and Human Model will also reveal—that have an important purpose.

Because there actually was a Universal Flood, and because modern science has been so determined to deny its existence, modern geology lost 150 years of progress and should be dismissed as an old, antiquated geology.

CHAPTER 8 THE UNIVERSAL FLOOD MODEL

The Universal Flood Chapter connects with all of the chapters and discoveries in this book, and with it comes the restoration of science's purpose: To describe and explain Nature so that we can understand and comprehend it.

The Universal Flood was an extraordinary event responsible for creating both beautiful landscapes and an incredible opportunity for humanity to begin anew.

9
The Weather Model

9.1 The Mystery of Weather
9.2 The Origin of Weather
9.3 The Earthquake Cloud Evidence
9.4 The Global-Weather System Evidence
9.5 The Geofield Model
9.6 The Geofield Evidence
9.7 The Aurora Evidence
9.8 The Ozone-CFC Pseudotheory
9.9 The Global Warming Pseudotheory
9.10 Weather and Geofield Prediction

Fig 9.1.1 – A funnel cloud over the ocean near Puerto Peñasco, Mexico (Rocky Point) in 2008. For as long as humankind has been around, weather has proven to be mysterious. A mystery that continues today with modern meteorologists acknowledging, "forecasting is still an inexact science." Obviously, there is a lot to learn about weather, including the *origin* of weather itself.

Many are surprised to find that the foundation of New Geology in the Hydroplanet Model is also the foundation of the Weather Model. The principles learned there helped define the new model. Earlier in the book, The Magma Pseudotheory chapter did more than identify the Dark Age of Geology—it revealed a veritable Berlin Wall between modern meteorology and the true origin of weather! And so it is, The Dark Age of Science has encouraged 'cloudy meteorology' during the past century. Nevertheless, as the Sun breaks through the clouds of a dark and dreary storm, new light is dawning on a completely new, refreshing, and more complete understanding of the Earth's weather.

Six new natural laws are outlined in the Weather Model, each of which is based on the Universal Scientific Method (USM), including new experiments and observations. The chapter can be divided into two main parts: The origin of planetary weather and the energy field surrounding planets. The interrelatedness of these two topics are documented and explained.

The Earth's Geofield (the magnetic field) will be discussed, including its effect on the Earth's weather, and its protective surrounding of the Earth, shielding us from the Sun's harmful radiation.

"We know for a **fact** that there are **many things we simply don't understand about the complex interactions of our atmosphere with the Earth's surface** and the not-so-empty space surrounding our planet... there is a **huge world of discovery** awaiting the curious."

Walter A. Lyons, Fellow, American Meteorological Society, Bib 66, Introduction

Climatologists say that, "forecasting is still an inexact science," even though satellites can 'see' the weather coming. Why?

Fig 9.1.2 – Although atmospheric science is a scientific field based primarily on observation, investigators increasingly rely on computer models to try to predict future weather events. The science of weather is a field rich in technology, with many new sophisticated satellites, equipped with a multitude of measuring and surveying devices that produce an abundance of detailed and accurate historical and real-time data. However, if there is no wisdom as to how the data fits into nature's complete weather model, all the data in the world will fall short of explaining and predicting future weather events and patterns. We must first understand the *cause* and origin of weather.

9.1 The Mystery of Weather

After defining what is meant by 'weather,' this subchapter looks into the 'art' of weather forecasting. With all their advances in weather technology, we'll discuss how well meteorologists are able to forecast weather events. The Air-Water Model (Chapter 23) revealed new discoveries about gases and the most abundant liquid in nature—water. The crystalline nature of both air and water reveals an amazing relationship among the factors of air, humidity, pressure, and temperature. The Magma Pseudotheory chapter and the Hydroplanet Model set the stage for the discovery of the true origin of weather and the events that effect its changes. We also look at weather from a global perspective to understand the incredible forces behind it. There is indeed, an incredible driving force behind Mother Nature and her weather.

Weather Defined

The *McGraw-Hill Encyclopedia of Science and Technology* defines **weather** as:

"The state of the atmosphere, as determined by the simultaneous occurrence of several meteorological phenomena at a geographical locality or over broad areas of the Earth." Bib 12 p2123

For most of us, we tend to discuss weather in terms of **weather elements**:

"A **weather element** is any individual physical feature of the atmosphere. At a given locality, **at least** seven such elements may be observed at any one time. These are **clouds, precipitation, temperature, humidity, wind, pressure, and visibility**." Bib 12 p2123

These elements of weather are what most of us commonly recognize and discuss as the 'weather.' But there are other elements related to weather, although not necessarily considered an *element* of weather. These are delineated as the encyclopedia continues:

"Certain **optical and electrical phenomena** have long been observed **among** the **weather elements**. These include **lightning, aurora, solar or lunar corona, and halo**." Bib 12 p2123

How are these electrical phenomena connected to the weather?

This important question and its answer will be explored in detail later in this chapter. It will help reveal the real origin of weather.

The encyclopedia also comments on other "weather-related phenomena" including:

"…waves at sea and floods on land." Bib 12 p2123

Consider the previous chapter on the Universal Flood and think about the evidences we identified leading to the cause of it. The role of gravity and its effect on the crust plays a much bigger part than many have realized. It also plays an important role in the Earth's weather. As we move through this chapter, the significance of two invisible factors affecting weather will become very apparent. Those factors are gravity and humidity.

Meteorology and climatology are both rooted in the discipline known as atmospheric science, which includes the study of weather. **Meteorology** deals primarily with the physics of weather whereas **climatology** focuses on the physical geography. This chapter, like most chapters of this book, will address broad aspects of weather, with a few specific areas to help demonstrate newly discovered natural laws of weather.

One amazing aspect of the Weather Model is how a wide range of science disciplines, when brought together, help us understand and comprehend the mysteries of weather. There are many new UM discoveries from the Universe Model and the Earth Systems Models that will open a vision of what controls Earth's weather.

Forecasting—an "Inexact Science"

To forecast the weather today, a meteorologist pursues a course in atmospheric science. The accuracy of forecasting however, is a common target for comedians; after all, we all know how exact "forecasting" is, don't we? A member of the National Weather Service commented after a ferocious storm tore up an afternoon wedding:

"'**Forecasting** is still an **inexact science**,' says Charles McGill, who works in the Burlington office of NOAA's National Weather Service (NWS). A day before the wedding, the office had predicted storms for late in the evening, but not for that afternoon. 'Maybe someday the forecast models will be good enough to **pinpoint when and where individual thunder-**

"Predictability: Does the Flap of a Butterfly's Wings In Brazil Set Off a Tornado in Texas?"
Ed Lorenz

storms will occur, but I think that day is a **long way away**.'"
Note 9.1a

Why, after major technological advances and years of study, is forecasting weather still an "inexact science"?

Even after major technological advances, and years of study, the **origin** of weather is still a mystery.

Before Kepler discovered the Laws of Planetary Motion, the movement of planets in the solar system was a mystery, for the same reason weather remains a mystery today—the natural laws that govern weather are not understood.

Long before the telescope, the planets were known only as points of light in the sky that seemed to wander around, but no one knew why. Until humanity could predict where a planet would appear in the sky, and at what time it would be there, those who studied them had to admit they did not comprehend their movement. Kepler's new laws of planetary motion changed all that. With the Sun in the center of the system, his newly defined laws allowed scientists to visualize the planets in an elliptically oriented solar system which brought understanding to the movement of the planets.

The same holds true for weather. Until we can predict where and when major storms and weather patterns will occur, we cannot say we understand the natural laws of weather. And we cannot make a prediction until we understand where they come from.

The Scientific Dark Age has kept most climatologists from being taught that all of Nature, including weather, is based on natural laws. For example, a forecaster stated:

"**All it takes** to create weather is air and solar heating." Note 9.1a

In a moment, we will show that this statement is *incorrect*. In fact, how can such a statement be made when forecasting is acknowledged as being an "inexact science"? Until we can forecast weather with some degree of *exactness*, how are we any better off than early astronomers attempting to forecast the time and location of the wandering planets before Kepler's laws were revealed?

As with all scientific disciplines, the lack of natural law in the atmospheric sciences has led to many theories that have done little to further the field. For example, one popular theory, we'll call it the Butterfly Weather Theory, continues to influence forecasting:

"Small things can have enormous consequences. That was the thrust of meteorologist Ed Lorenz's influential 1972 lecture, '**Predictability: Does the Flap of a Butterfly's Wings in Brazil Set Off a Tornado in Texas**?' Lorenz argued that complex systems like the Earth's atmosphere can be dramatically altered by seemingly insignificant factors—a puff of wind here, an extra degree of warmth there. That poses a challenge for weather forecasters." Note 9.1a p98

It is a correct principle—that out of small things, great things come to pass. But a butterfly is certainly not going to measurably affect a tornado. 'Butterfly Weather' seems to be an interesting concept and is fun for meteorology professors to teach, but it defies common sense and has inadvertently caused forecasters to lose faith in the possibility that anyone will ever be able to "pinpoint when and where individual thunderstorms will occur."

Of course, meteorologists point to how their predictions have improved over the past decade, but the reason their predictions have improved over the short period is simple; they can literally *see* the weather coming from farther away. Advances in technology, especially **satellites**, give the meteorologist continuous images of global weather phenomenon. The real-time data has made a big impact, says *National Geographic*:

"Thanks mainly to **keener instruments** and **more powerful computers**, forecasters are extending their reach into the uncertain future." Note 9.1a p94

Obtaining more data via new technology only increases knowledge of weather patterns *as they happen*, but it gives little help in predictions made beyond a couple days because the ***origin** of weather is still not understood*. Satellites looking down on clouds moving our way allow meteorologists to make weather predictions in the same way our great-great-great grandparents did, but just a little farther out!

Why is meteorology, with its multitude of weather sensors, computer models, and scientific knowledge not able to predict the weather with more accuracy? Obviously, our knowledge of weather is still very limited.

Magma Pseudotheory and the Weather Model

There is a specific reason that the origin of weather eludes scientific thought. Meteorology is simply missing some important pieces of the natural weather puzzle; the largest piece has been missing for over a century. It is dubbed an "interesting problem of oceanic geography." A 1901 textbook titled *Lessons in Physical Geography*, documents a long-known "fact" about the temperature in the crust of the Earth:

"The **fact** that while the **temperature** of the earth-crust **increases downward**, the **temperature** of the **sea decreases in**

the same direction, constitutes one of the **most interesting problems of oceanic geography**." Bib 142 p252

Within the magma-planet paradigm, this is a *huge* problem!

Heat only moves from hot to cold, so we should see an **increasing** heat gradient from the hot interior of the Earth to the surface in **both** the continental crust **and in ocean waters**— but we do not.

The ocean water temperature *decreases* as depth increases, whereas the thermal gradient in the crust *increases* toward the locale of frictional heating. The decreasing gradient is a "problem" in oceanic geography because if the Earth was a magmaplanet, both the crust and oceans should reveal increasing temperatures, as one approaches the magmatic heat source.

Furthermore, water has a higher heat capacity than crustal rocks, which means water can bring the heat from the Earth's interior through convection and conduction. Thus, we *should* see the temperature of the oceans *rising more rapidly* than the continental crust as ocean depth increases, *if magma was real*. Without the new Frictional-Heat Law in the Magma Pseudotheory chapter, this "problem" could never be solved.

Moreover, the Frictional-Heat Law also sets the stage for the comprehension of how weather works. It is one of the missing puzzle pieces of the Weather Model. Modern science's belief in magma has held back the recognition of the real heat generator in the Earth's crust, which was shown in the Magma Pseudotheory chapter to be earthquakes. In this chapter, we will examine even more proof that earthquakes and earthtide are responsible for the heat below our feet and for the weather above our heads.

Hydroplanet Model and the Weather Model

Before we can ever come to an understanding of weather, we must know where the Earth's internal heat really comes from and how much water is inside the Earth. Intuitively, researchers have seen this coming for a long time. Back in 1970, one insightful investigator noted:

"Man's increasing knowledge of the **hydrologic engine** has answered many of the ancient **riddles of water**, but not all of them. We know where the rain comes from, yet we still **cannot** order its coming. We can explain why water fills the well, yet we still **cannot** always forecast correctly how much water any particular well will furnish. Today's attempts to control **the great global engine** are, like the incantations and dances of other times, merely attempts. **Only new understanding, rising from an intensified study of water**, may finally give man more effective influence over the bountiful compound that is essential to life on earth." Bib 13 p41

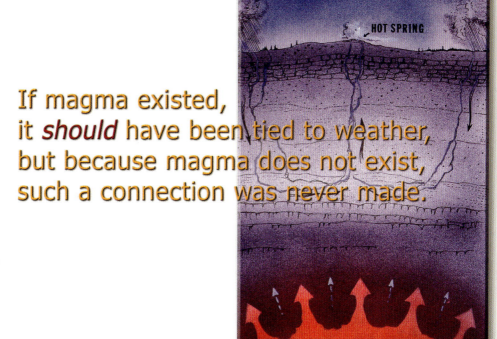

If magma existed, it *should* have been tied to weather, but because magma does not exist, such a connection was never made.

To gain a "new understanding" of water, we cannot simply *intensify* our "study of water" under the current modern science paradigm; it doesn't matter how much one studies a false theory, it is still false and therefore cannot ultimately provide the answers we seek. Without the Hydroplanet Model, the UM Weather Model would not have come to be.

Discovering and Confirming the Origin of Weather

Walter A. Lyons describes the attitude that pervades some scientific disciplines in the book, *The Handy Weather Answer Book*:

"It might seem that all there is to be discovered in science has been discovered. That claim has frequently been made during the last century by some of the more arrogant members of the scientific establishment. Atmospheric scientists are generally more humble—we know for a **fact** that there are **many things we simply don't understand about the complex interactions of our atmosphere with the Earth's surface and the not-so-empty space surrounding our planet**. While many significant advances have been made, there is a **huge world of discovery awaiting the curious**." Bib 66 pxiii

It seems Lyons and other atmospheric scientists are more reserved in proclaiming their theories as though they are fact than scientists in other fields of study—why?

Atmospheric scientists are more cautious about declaring their theories as fact than scientists from other fields, because they know the public can more easily test their theories.

How does the public test the Big Bang Pseudotheory, or the Magma, Geological Time or Evolution Pseudotheories? Even scientists themselves have a difficult time producing hard evidence for the theories. Perhaps atmospheric scientists are held to a higher degree of accountability than other fields of natural

SUBCHAPTER 9.1 THE MYSTERY OF WEATHER

science because what they say is often proved right or wrong in fairly short order whereas other disciplines are so confusing or difficult the average person turns away. One modern science leader, cited in subchapter 2.5, went so far as to say:

"People in general find science **grim** and seem to fear it."
Note 9.1b

Try as it may, modern science has yet to reverse this attitude. Countless modern science societies lament the dearth of upcoming new scientists, wondering what to do to encourage more young people to pursue science. How can science ever expect to reverse the trend and restore confidence without accountability? There is no Universal Scientific Method to test and gauge new scientific discoveries by, and scientists themselves cannot agree on some of the most basic scientific issues in the public's view—like global warming, an issue fraught with tales of deception and intrigue.

One of the purposes of the UM is to bring accountability and leadership to a failing science and to restore the public trust. This will be accomplished by restoring truth in science, where the purpose is the discovery of new natural laws instead of endless theory. This chapter introduces several new weather laws and planetary energy-field laws that are easily understood; laws that can be researched and tested by the general public. Lyons noted in *The Handy Weather Answer Book*, "There is a *huge world of discovery awaiting the curious.*" By explaining the origin of weather, the Weather Model opens up a new world of meteorological discovery for all who are curious.

9.2 The Origin of Weather

The mystery surrounding the origin of weather would remain had we not discovered the origin of earthquakes. Having established the true source of earthquakes in the Magma Pseudo-theory chapter, new meteorological possibilities open up. This subchapter will explore earthquake weather and critical new information about heat from earthquakes and its effect on the water in the Earth's crust. This subchapter also introduces a new Water Cycle as part of the new Weather Model, and three new Weather Laws. These are confirmed through several new weather observations and experiments.

The Missing Factor of Weather

Richard A. Keen, a popular weather book author and meteorologist, wrote:

"**Five basic factors** combine to make it inevitable that the Earth has the kind of weather that we have grown accustomed to. The first, and most obvious factor, is that Earth has an **atmosphere**. Second, Earth is **sunlit**. The third factor is **Earth's rotation**. The next factor—and one unique to Earth—is our planet's vast supply of liquid **water**. And finally, there is **geography**—the variety of surfaces, from oceans to continents to ice sheets, that cover Earth." Note 9.2a

Which of these "five basic factors" is the *cause* of weather changes? Can we say that any one of them, or a combination thereof is the *cause* of the rainstorms, snowstorms, hurricanes, and tornadoes? We cannot.

In fact, everyone knows the weather can *completely change* from day-to-day, however...

The geography did *not* change...
The supply of liquid water did *not* change...
The Earth's rotation rate did *not* change...
The sunlit area of the Earth did *not* change...
The nitrogen, oxygen and argon that make up the majority of the Earth's atmosphere did *not* change. These "five basic factors" of weather are constant on a day-to-day basis.

> FQ: What is causing the day-to-day changes in the weather?

This is the million-dollar meteorology question, and the answer to this question follows the Simple Truth Principle—in Nature the simple truth is—that truth is simple.

> The five factors of weather are not the **cause** of day-to-day changes in the weather, but they are **affected** by **other** weather factors.

These missing factors affect both day-to-day and long-term weather patterns.

Earthquake Weather

On a USGS website, one of the Frequently Asked Questions concerns Earthquake Weather. Know that as you read the mod-

CHAPTER 9 THE WEATHER MODEL

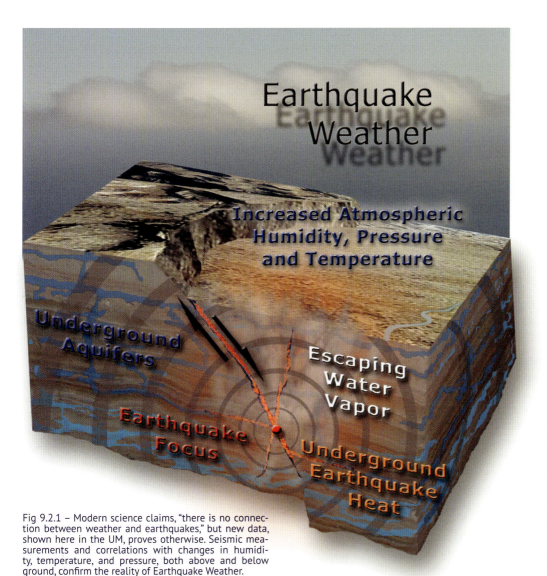

Fig 9.2.1 – Modern science claims, "there is no connection between weather and earthquakes," but new data, shown here in the UM, proves otherwise. Seismic measurements and correlations with changes in humidity, temperature, and pressure, both above and below ground, confirm the reality of Earthquake Weather.

ern USGS geologist's answer, this chapter will demonstrate that their answer is *incorrect*, and that the Aristotelian explanation the USGS dismisses as being outdated—is actually not too far from the truth:

"**Question**: Is there **earthquake weather**?

"**Answer**: In the 4th Century B.C., Aristotle proposed that earthquakes were caused by winds trapped in subterranean caves. Small tremors were thought to have been caused by air pushing up on the cavern ceilings, and large ones by the air breaking the surface. This theory lead to a belief in **earthquake weather**, that because a large amount of air was trapped underground, the **weather would be hot and calm before an earthquake**.

"Nowadays, **thanks to the advent of science, it has been shown there is no connection between weather and earthquakes. Earthquakes are the result of geologic processes within the earth** and can happen in any weather and **at any time during the year**." Note 9.2b

In subchapter 5.3, we discovered that although the USGS believes earthquakes are "the result of geological processes *within* the earth," they are, in reality, caused by grav-

"Nowadays, thanks to the advent of science, it has been shown there is no connection between weather and earthquakes."

itational processes *outside* the Earth. Instead of the idea that earthquakes do not follow cycles or have no predictability, we discovered 12 and 24-hour cycles, and an "earthquake season" related to the *Earthtide*, which is the daily vertical movement of the crust. This was a relatively new concept, even for many scientists.

How can modern geology state that, "there is **no** connection between **weather** and **earthquakes**" if geologists do not understand the origin of either weather **or** earthquakes?

Throughout this chapter, we will show that there is a direct connection between weather and earthquakes and more importantly, earthquake activity is *the cause of the Earth's weather patterns!*

The vast collection of data through constant observation coupled with first-hand weather experience makes the weather-earthquake connection easy to understand. In the *Handy Weather Answer Book,* we read that West Coast observers relate fair skies and high-pressure with major earthquakes, an observation not necessarily shared by meteorologists:

"Is there such a thing as **earthquake weather**? Some residents of the West Coast think that fair skies associated with a mild **high-pressure system that is stalled over the western United States are typical of conditions during major quakes** (as in the 1989 Loma Prieta quake). While some have theorized that the increased mass of the atmosphere over the region due to the higher atmospheric pressure might play some role in a quake, this is a question that **will require many years of research in order to be resolved**." Bib 66 p256

Perhaps the "years of research" will be shortened to minutes after reading the earthquake weather connection in this chapter!

The Origin of Weather

This subchapter will identify a new Weather Model to include the effects of tectonic forces within the Earth. A model where the origin of weather can be tested and confirmed using empirical evidence. Development and verification of the Weather Model was made possible because of processes identified elsewhere in this book. In fact, understanding the contributions of three important UM weather factors and their associated models was essential. They are:

1. Earthquake Heating – Lava-Friction Model
2. Abundant underground water – Hydroplanet Model
3. Earth's Energy Field – Geofield Model

The first two models were presented in earlier chapters, and the Geofield Model will be introduced later in this chapter. The Lava-Friction Model established the true origin of lava and explained where *all of the heat in the crust* originates—from earthquakes and earthtide. Knowing the origin of the heat in the crust and that it manifests itself in cycles is the first key to comprehending weather's origins and its cycles. The second key comes from the Hydroplanet Model, which documented the abundance of underground water. With these two keys, we can begin to unlock the mystery of weather by asking a simple FQ:

What happens to the vast expanses of water in the crust when it is subjected to cycles of heating?

When water is sufficiently heated, it becomes water vapor, and on a planet-sized scale, that creates **high-pressure areas** and **humidity**.

Fig 9.2.1 illustrates the process of underground water heated by earthquake friction. When liquid water is heated past its boiling point, it *expands **1,700 times** its liquid volume* at sea level. Beneath the surface of the Earth, pressure from overlying rock and other crustal material allows water to absorb an enormous amount of heat, without boiling, or becoming water vapor. However, as the water migrates toward the surface, the reduction in pressure allows the transition from liquid to gas (water vapor) to occur. If this happens rapidly, as it did during the Flood, and as it does during volcanic eruptions, a violent phreatic explosion is the result. Most of the time however, the event is benign. Heated water rises passively, vaporizing as it rises to the surface and into the atmosphere. As we will see in a moment, the addition of this new gas into the atmosphere can completely change weather conditions on the surface.

Water vapor rising from beneath the Earth's surface is invisible to the naked eye and has apparently been hidden from the eyes of researchers looking for answers about the weather and its origins. While it is true that meteorologists are aware of humidity and water vapor gains through evaporation, there is a significant difference between the formation of earthquake-water vapor and water vapor coming from ordinary surface evaporation. That difference is ***time***.

The evaporation of a good-sized puddle of water may take all day, or even longer, but toss the same amount of water onto rocks heated by a campfire and it would instantly be turned into steam. What this shows in terms of the Earth's surface is that in a very short period of time, large areas of the Earth can produce an outpouring of water vapor heated from earthquake friction from seemingly quiet seismic activity. This newly added gas can *create, or at the very least, alter high-pressure systems.* High-pressure systems are represented by a capital "H" on weather maps.

To our knowledge, this is the first time anyone has proposed this natural weather controlling mechanism, and with this new paradigm, many unexplained weather phenomena can be explained. For example, the cause of 'cold snaps' (cold bodies of air moving quickly from northern latitudes) has no clear explanation in meteorology, but with high-pressure creation from within the Earth, a large amount of new water vapor can be generated quickly, forcing blocks of cold air away, toward lower latitudes.

Because these high-pressure systems control weather patterns, an understanding of their origin is critical to atmospheric science. The current modern science explanation for the origin of high-pressure systems does not hold up under scrutiny, but we will get into that shortly. For now, we'll explore the new concept of Earthquake Heat—and find out just how real this missing weather factor is.

Earthquake Heat—The Missing Factor of Weather

The ***only*** heat source *atmospheric science* is concerned with today is the heat the Earth receives from the Sun. The big mistake is that it is *not* the *only* heat source on the planet—and every *geologist* intuitively knows this. Unfortunately, this is another excellent example of *scientific specialization* gone awry.

The key to understanding weather lies in the knowledge of Earth's **secondary** source of heat. In the modern science world of weather, meteorologists do not think that heat from under the crust has any effect on the weather. They suppose that all sub-crustal heat comes from magma deep in the Earth and that it therefore plays an insignificant role in weather formation. An apparent confirmation of this belief is that weather has distinct cycles, whereas magma has no ob-

Fig 9.2.2 –The fence in this photo warped because of surface movement during an earthquake. We are accustomed to ground movement and the destruction of surface structures during earthquakes, but few realize the impact earthquakes have on weather. All earthquakes produce heat in the crust. The release of that heat and the accompanying water vapor is a completely new phenomenon in modern science.

served cycles, which causes atmospheric scientists to dismiss the *heat* in the crust from their weather models.

Yet there is a surprising amount of heat in the Earth's crust. One wonders why atmospheric scientists have not considered the "inexhaustible supply of heat" from the Earth's interior. In the *Letters* section of the January 2007 edition of *Scientific American*, one reader expressed a very good question relating to the previous month's article, *Energy's Future: Beyond Carbon*:

"I was very disappointed that you did not mention **geothermal energy**. The earth has an **inexhaustible supply of heat**; one merely has to dig a well deep enough and pump the water to a heat exchanger or reticulate water from the surface. Why omit this perfectly clean source of energy that is abundant everywhere?" Note 9.2c

Although the reader probably assumed the inexhaustible supply of heat was magma from the Earth's interior, he was not too far afield of the truth! The Earth does have a geotherm where frictional heat is produced in certain areas at certain times throughout the astronomical cycle of the solar system. This concept was detailed somewhat in the Magma Pseudotheory chapter. The cyclical nature of earthquakes, moonquakes, rockbursts, and geysers are all a testament to a cyclical heat source that must be understood before weather patterns can be predicted.

Right now, atmospheric science is unable to predict weather much beyond what images the satellites can provide because heat in the Earth's atmosphere, the driving force behind the weather, is thought to come only from the Sun. This is rather ironic when one considers the heat that magma *should* be producing, but is not. Without the knowledge of the true source of underground heat—gravitational-friction—advances in atmospheric science will be stifled.

Earthquake Heating Detected by Satellites

This new weather factor—earthquake heat—was just waiting to be discovered with advances in technology. For several decades now, weather satellites have been gathering and transmitting data on a host of weather factors, including temperatures of the Earth's surface. Only recently have scientific investigators noticed a *correlation between earthquakes and an increase in surface temperatures of the area surrounding the earthquake*. Most of this research comes from several Asian countries including China, Japan, India, and in particular—Russia. Overall, the United States, who is by far the world's leader in scientific research and weather satellite technology, has failed to recognize the importance of earthquake heat. Only one small private firm, dedicated to finding a way to predict big earthquakes, seems to see any value in exploring this connection. The firm, Quakefinder, has been primarily involved in identifying the correlation between earthquakes and the Earth's energy field, but they also noted a relationship between infrared radiation (heat) and earthquake activity:

"**Infrared radiation detected by satellites** may also **prove to be a warning sign of earthquakes to come**. Researchers in China reported several instances during the past two decades of satellite-based instruments registering an infrared signature consistent with a **jump of 4 to 5°C before some earthquakes**. Sensors in NASA's Terra Earth Observing System satellite registered what NASA called a '**thermal anomaly**' on 21 January 2001 in Gujarat, India, just **five days before a 7.7-magnitude quake there; the anomaly was gone a few days after the quake.**" Note 9.2d

The American scientists began to realize that earthquakes were significantly affecting the heating of the Earth's crust, seeing the heat as an "anomaly" associated directly with seismic activity. We will soon demonstrate that such thermal anomalies are a fact of earthtide, and are far more meaningful than being just a "warning sign" of large earthquakes—the anomalies signify *the missing factor* of weather.

Atmospheric weather begins with
earthquake heating in the crust.

This is an *extraordinary claim*, and there is *extraordinary evidence* to back up the UM theory that earthquakes are significantly heating *the rock and water in the crust to a level that will increase the temperature and humidity of the surrounding area.*

1996 Russian Research Evidence

Fortunately for our research on earthquake heating, A.A. Tronin of the Russian Academy of Sciences, began to observe earthquakes from both space (via satellites) and the ground, back in the 1980s. Today, Tronin is not recognized as a leader among global atmospheric scientists and his work on the effects earthquake heat has on the weather remains vastly underappreciated. The odds are—this will soon change.

It was difficult at first for Tronin to understand how mechanical energy could be driving the heat engine, and after years of observation, he wrote in a 1996 article:

"**It is hard to assume that the observed increase of surface temperature is a result of the direct conversion of mechanical energy into heat.**" Note 9.2e

It was "hard" because his observations seemed *so contrary* to the prevailing view. Tronin was trying to understand where the heat he was observing was coming from, and the magmaplanet paradigm provided for no such heat source. Tronin, a great example of real investigative science, focused on observations of natural phenomena, and over a 10-year period, he analyzed over 10,000 images from the American NOAA weather satellites. He was able to document a "statistically significant correlation" between temperature increases and seismic activity in the region he studied:

"The National Oceanic and Atmospheric Administration (NOAA) series satellite thermal images (STI) study showed the presence of **positive anomalies** of the outgoing Earth radiation flux recorded at night time and **associated with largest linear structures and fault systems of the crust**. The analysis of a continuous series (100-250 days) of nightly STI data for a period of **10 years** allowed identification of a set of IR radiation anomalies in the Central Asian seismoactive region, Iran, Egypt, etc. About **10,000 NOAA images were analyzed**. It was **actually discovered that there was a statistically significant correlation between the activity of IR** [near-infrared temperature] **anomalies** (mean value of area per year or month) **and the seismic activation of the Central Asian seismoactive region**. At present the nature of stable and non-stable IR anomalies **is not clear**." Note 9.2e p1439

At the end of this statement, we see that Tronin was "not clear" about the nature of surface temperature changes, and was not sure how those changes were related to regional faulting.

Fig 9.2.3 – The image above shows the size of the area heated by the 8 June 1993 Kamchatsky earthquake. A Russian research team was among the first to establish a direct link between earthquakes, crustal heating, and atmospheric temperature increase. They observed a rise in temperature of **10° C (18° F)** that changed weather over the earthquake-heated area. Historically, the scientific connection between earthquakes and a change in weather parameters noted by A.A. Tronin is of great significance, although he was not yet aware of the effect earthquakes and earthtide have on global weather cycles.
Image adapted from the journal *Physics and Chemistry of the Earth*, 29, (2004) p502.

Even after the culmination of ten years of research, without the wisdom of the Frictional Heat law, the observations and data proved difficult to comprehend. One of the elements Tronin was puzzled about was the *time* involved. In the magmaplanet paradigm, heat changes via migration from the seismic focus should take 'millions' of years—yet observed heat changes took *only* "*several days*":

"**The fast development of the anomaly in the order of several days excludes the consideration of thermoconductivity or convective transfer as possible mechanisms of surface temperature change prior to an earthquake.** First of all, surface temperature alteration **cannot be explained by direct thermoconductivity because of the very large time period of thermal propagation**, from seismic source to the surface, which is several orders of magnitude more than characteristic periodicity of seismic events of 13-14K energy class in the Central Asian seismically active region. Characteristic propagation time of a thermal impulse from a depth of about 10km, may be estimated to be in the order of magnitude of **10^7-10^8 years**." Note 9.2e p1449

As he noted, increased temperature changes during the *short time frame* of several days "excludes the consideration of thermoconductivity or convective transfer"—in other words, the heat could *not* have come from magma deep inside the Earth. It simply happened too fast. Instead of 100,000,000 years, Tronin observed evidence of the transfer of several degrees of thermal radiation heat on the peninsula just northeast of Japan (see Fig 9.2.3) in only a few short days.

Where did the heat come from? The "shape" of the thousands-square-kilometers heated area was "linear," coinciding with the faults in the area:

"The positive anomaly of **several degrees Celsius** at the foot of Kopetdag has a **linear shape of 25-30km in width and about 500km in length**. This anomaly is related to the **Kopetdag Fault**—the boundary structure of the first order, separating Alpine Kopetdag formations from the Turan Plate. Besides, the anomaly **coincides with the 'thermal line' of the Kopetdag hot water basin**—a unique hydrogeological structure, described in detail by Nikshich (1925).

"The second anomaly about **50km in width and 300km in length** occurs at the foot of the Karatau Range. Spatially it coincides with the **Karatau Fault**—the first order structure separating the Turan Plate from the Central Asian folded zone. The Karatau Fault proves to be the extension of the deep Talas-Fergana Fault, which controls in many respects the geodynamics of the region. The **permanent relation of these anomalies to the large tectonic structures defines them as stable IR anomalies**." Note 9.2e p1441

A heated area that "coincides with the Karatau *Fault*" was clearly *not* from a global heat source as broad as the Sun. Despite this, Tronin did not propose how mechanical friction energy in the crust could produce the heat, nor did he suggest any other heat source capable of raising the surface temperatures an *incredible* **10° F (18° F**, from 32° to 50° F) over a very short period. Thus, Tronin and colleagues were left to conclude, "solar heating" was "the most probable source" responsible for the massive increases in temperature:

"Therefore, **the most probable source**, which provides sufficient energy for **surface temperature alteration is non-uniform solar heating** or heat loss." Note 9.2e p1450

Eventually, the solar heat source *theory* died. The area under study was heated by earthquake friction, but heating was not the only change observed; there was also an increase in *soil moisture and water vapor*:

"It has been reported that **pre-seismic activity alters the characteristics of soils**, including **soil moisture** (Sugisaki et al. 1980), **gas content and composition** (Rikitake 1976, Sugusaki et al. 1980). There are also **numerous** observations of surface and near surface temperature changes prior to the Earth's crust earthquakes. For example, **soil temperature anomalies of 2.5° C were measured** in the zone of preparation of the **Tangshan earthquake** (China, 1976—magnitude more than 7.0)."
Note 9.2e p1439

We will discuss the variety of gases released during earthquake heating activities shortly, but for now, consider how the introduction of a large quantity of water vapor into the air can *completely* change the weather. Moreover, such changes can produce several types of new weather, depending on the initial humidity, temperature, and barometric pressure.

CHAPTER 9 THE WEATHER MODEL

During Tronin's 1996 research, it is doubtful that he or any other earthquake researchers knew of earthtide; the GPS technology that established the diurnal crustal flexure was still in its infancy. Thus, the ramifications of frictional heat from *earthtide* was a thought far removed from investigators' minds. However, it was common knowledge that humidity and increased moisture content in the soil were "very important factors" affecting weather:

"**Moisture content in soil and humidity in air remain very important factors controlling surface temperature**. These parameters affect the run of such processes as evaporation and condensation of moisture—q_{ev}. Evaporation is most intensive in the daytime, when solar heating takes place, and it leads to a decrease of surface temperature. Moisture content in soil also alters its thermophysical properties and **affects the process of dew-fall**, which is **known to be associated with the release of heat**." Note 9.2e p1449

Tronin's research included the observation that CO_2 gas was released along with the moisture from the earthquakes. Surprisingly, we found little research addressing this particular source of CO_2 gas. Previously, the Universal Flood Model documented how heating of underground water enabled massive microbial growth, which created enormous calcium *carbonate* deposits. Decaying microbes also release carbon dioxide gas as a byproduct of decomposition, a process that is still happening today, when the crust is heated by earthquake friction. Regardless of its origin, astute individuals will realize that this one new observation has profound implications on the concept of increased 'global warming' due to carbon dioxide increases. Instead of being produced by humans, this source of natural carbon dioxide, essentially unknown to modern science, changes completely the global warming debate:

"Field measurements of the concentration of CO_2 in the near-surface atmosphere and soil across the Kopetdag fault are shown in figure 16 [not shown]. There is an **increase of CO_2 concentration by up to 0.3 per cent in soil and up to 0.1 per cent in the near-surface atmosphere with background values of the latter of about 0.03 per cent**. There was a correspondence between the zone of high CO_2 concentration, that of radon emanation and with the subsurface temperature at a depth of 1.5m. The region of **high surface temperature taken at 06:00am follows a zone of high subsurface temperature and gas concentration**." Note 9.2e p1451

The change of .03 to .1 percent is a noteworthy increase in atmospheric carbon dioxide across large regions; an increase from earthquakes and microbes below the surface—*not* from human activity, grazing animals, or fires! Tronin reports that both the increase in soil moisture and gas (CO_2) are coming from the same source and concludes "ground water motion" may be the source:

"In our opinion, mechanisms leading to the **increase of soil moisture and gas concentration are pretty much similar. Ground water motion** may be the single reason for **all effects**." Note 9.2e p1453

So far as is known, all natural terrestrial waters, both subterranean and on the surface, contain microbes; a fact of great significance left unnoticed for too long. In the Universal Model, microbes are one of the golden threads running through and influencing almost every field of science in ways never imagined.

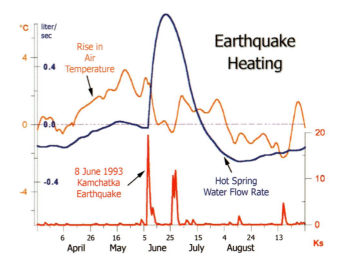

Fig 9.2.4 – The relationship between the 8 June 1993 Kamchatka (Russia) Earthquake (red), the hot-spring water flow rate (blue), and the rise and fall of air temperatures (orange) over the earthquake heated area shown in Fig 9.2.3 is depicted in the above graph. Only during the last decade have researchers had the luxury of simultaneously acquiring both satellite and ground based data, making it possible to establish the connection between earthquakes, changes in the soil temperature and moisture, changes in wells and hot-spring water, atmosphere temperature and humidity, and the discharge of CO_2 gas. Unfortunately, the research from this obscure group of physical science investigators failed to catch the eye of the atmospheric science community.
Graph and data adapted from the journal, *Physics and Chemistry of the Earth* 29 (2004) p503.

In a newer report from 2004, Tronin collaborated with four researchers, three Russian and one Italian, documenting further connections between Russian earthquakes and ground surface and atmosphere changes (see Fig 9.2.4).

2004 Russian Research Evidence

Tronin and his colleagues from Russia and Italy reported new findings that are important for several reasons. Ground and satellite observations were taken simultaneously, including data from wells, giving a much clearer picture of what was happening. The earthquake-heat seismic correlation became more significant and clearer after observing changes to pH levels in the well water, analogous to the changes observed at black and white smokers at the bottom of the ocean. Water wells emit acidic or alkaline waters according to the variety of microbial communities from which their subsurface water is drawn:

"**Air temperature, surface temperature**, retrieved from satellite data, **and well observations** on the Kamchatka peninsula, Far East, Russia were **jointly analyzed**. **Air temperature indicates correlation with seismic activity**. Satellite observations showed the presence of thermal anomalies on the earth surface in the basin of the Kamchatka River. Thermal anomalies' reactions on three strong earthquakes were recorded. **Water temperature, outflow and hydrogen ion exponent (pH) changes were observed as a response to seismic events**. Joint analysis indicates similarity [in] both satellite and ground observations related to earthquakes." Note 9.2f

As shown in Fig 9.2.4, many days before the 1993 Kamchatka earthquake, both air temperatures and water well temperatures began rising. Had the seismograph been more sensitive, we expect there would have been an increase in "silent earthquake" activity, discussed in subchapter 5.3. A surge in frictional heating on the Kamchatka peninsula increased the water temperature in the ground and the atmospheric temperature above the ground, but the greatest effect on the weather was the increased water vapor, which raised the humidity and barometric pressure of the area.

The researchers reported a "clear increase in air temperature" and "water flow" 40 days prior to the quake:

"**A clear increase of the air temperature was observed for these cases**. T_a [air temperature] variations at w1 and w2 show practically the same pattern, **demonstrating a large-scale temperature increase by +3–4 °C about 5–20 days before the earthquake**. Taking into account other cases we can extend the period—**up to 40 days**. In both cases we also see a **coseismic outburst of water in the hot springs and a insignificant increase of water flow about 20 days before the shock for 8 June 1993. A clear anomaly in the water flow was observed 10–40 days before the shock** in case of the 21 June 1996 earthquake." Note 9.2f p503

It was no small area affected; the earthquake-heated area measured thousands of square kilometers. Nor was the temperature change minuscule. The rise in air temperature from the earthquake swarm preceding the 'big slip' on June 8th was a whopping **10 °C (18 °F)**:

"The earthquake of 8 June 1993 had the **largest** magnitude of the shocks considered here. **The thermal anomaly recorded at 2 June 1993 covered large part of peninsula, had an unusual shape and intensity—up to 10 °C. Water temperatures also started to increase on this day and continued to grow up to the day of the earthquake**—8 June 1993. In this case we can also compare simultaneous satellite and ground observations in this case. **Both ground and satellite observations indicate an increase of air, water and surface temperature before the shock**." Note 9.2f p505

Another important factor Tronin and others observed was that the heat at the surface during the 8 June 1993 earthquake occurred because the magnitude 7.5 quake was only 70 km below the surface. An earlier (24 June 1983), 6.3-magnitude earthquake was *deeper*, at 180 km; it produced *no* thermal anomaly at the surface:

"We did **not find any thermal anomaly prior to and during the event of 24 June 1983 with a hypocenter depth 180 km**, regardless of the fact that this earthquake was located **closer** to Kamchatka river artesian basin then other shocks. We interpreted it by **big depth of epicentre**." Note 9.2f p505

The seismic tomographic evidence in the Magma Pseudotheory chapter showed the highest temperatures in a variety of deep-section scans were near the surface, near active faults, as opposed to being near the center where magma supposedly exists.

The investigators actually noted that rising fluid separated into water and gas. Besides water vapor, seven other gases were being released into the atmosphere from the earthquake heating activity. Additionally, pH levels in the wells were monitored and showed changes corresponding with the earthquake heating process:

"In any case **fluid rises to the earth surface**. Depending on geological and tectonic situations, **near the surface, at a depth of a few kilometres, the fluid is separated into water and gas**. The water causes change of debit, temperature and chemical composition in wells and springs. **Gas** (H_2, He, CH_4, CO_2, O_3, H_2S, Rn) **moves to the atmosphere** (Wakita, et al., 1978).

Depth and magnitude of the shock and geological conditions **determinate the mosaic character of these phenomena on the earth's surface**. This statement is **confirmed** by the **observations of water temperature, debit, pH in wells and thermal anomalies in Kamchatka**." Note 9.2f p505

The most important aspect of Tronin's research as it relates to this chapter is how earthquakes affect the atmosphere, which in turn, affects the weather. The release of a large quantity of hot water vapor from earthquake activity can *increase* the temperature of the air. The *greenhouse effect* of several of the gases released during that activity contributed to air temperature and humidity increases, as the researchers explain:

"We examined a few mechanisms of interaction. First—convection heat flux (**hot water and gas**) **changes the temperature** of the **earth surface and air**. Second—**change of the water levels** with usual temperature leads to **changes in soil moisture**, and consequently the physical properties of the soil. The difference in physical properties **means a different temperature on the surface**. Third—greenhouse **effect, when the optically active gases (CO_2, CH_4, water vapour) escape from the surface**. These gases absorb IR radiation, warm up and heat the surface. As a result of **gas and water appearing at the surface we expect to find changes in temperature, humidity and atmospheric pressure in surface air**." Note 9.2f p505-6

We will shortly examine how escaping gases (mostly water vapor) affect high and low pressure areas commonly denoted on meteorological weather maps, and how this affects weather patterns and their movement. Because the UM has established that the Earth's crust is in constant earthtide motion, and that large earthquakes release considerable heat and water vapor, we can begin to visualize the *invisible earthquake weather* in constant motion around us. As future observations record seismic occurrences, water pressure, temperature increases, and gas releases in greater detail, previously unknown weather patterns will begin to emerge.

India Studies Confirm Earthquake-Temperature Connection

Some critics of earthquake weather may suppose this phenomenon is limited only to Russia, but there was a case study of earthquake weather in India. In 2005, Indian scientists, utilizing US satellite data, 'confirmed' surface temperature spikes that preceded earthquakes:

"Indian scientists studying archived satellite data **have confirmed that earthquakes tend to be preceded by surface-temperature spikes in the immediate area**, suggesting that seismic events could one day be predicted from space.

"'Our study **was successful in detecting thermal anomalies prior to all these earthquakes**,' Arun K. Saraf and Swapnamita Choudhury of the Department of Earth Sciences at the Indian Institute of Technology in Roorkee, reported in the July issue of the *Journal of the Indian Geophysical Union*.

"Surface temperatures above the quake epicenters **increased between 4 and 10 degrees Celsius immediately before the events and returned to normal soon afterwards**, the scientists reported. The thermal record was compiled using data collected by U.S. environmental satellites." Note 9.2g

Documentation of surface temperature spikes of up to **10 °C (18 °F)** are direct confirmation of the Gravitational-Friction Law presented in subchapter 5.4 of the Magma Pseudotheory.

They establish the heating mechanism that drives the weather, according to the UM Weather Model. Without knowledge of the additional heat from earthtide, a Weather Model capable of predicting past and future weather events would be nearly impossible. Global subterranean heat created by friction is the key to understanding all major weather events on the Earth.

There are several new weather terms necessary to envision and understand new, upcoming weather concepts.

New Weather Terminology

With the new discovery that large earthquakes significantly heat the crust and atmosphere, we can expand that view to include smaller earthquake swarms and silent earthquakes. They too, heat the crust and impact the weather—just on a smaller scale, but over larger distances. Everyone knows that ocean waves never stop—but they do change, depending on the gravitational tidal forces of the Moon, Sun and other factors. Wind has a great effect on ocean waves, but the wind comes from high-pressure systems that are earthquake centered, which will be discussed shortly.

Tidal forces of Moon and Sun also cause **Earthtide Heating**, which is defined thus:

> **Earthtide Heating** - The constant frictional heating of the crust by gravitational tidal forces.

The great volume of subsurface water documented in the Hydroplanet Model is obviously impacted by earthtide heating in ways never before imagined—both physically and biologically. We've only recently learned about the environment that exists beneath the Earth's crust, an environment that will surely spawn a whole new field of yet unnamed scientific study. The heating of the Earth's subterranean water through seismic, or earthtide activity, is a new concept. By combining the three forces into one word, we have a term that will be used often throughout the remainder of this chapter—**Hyquatherm**.

> **Hyquatherm:** An earthquake-heated water system in the crust that generates pressure systems in the atmosphere that change the weather.

A Hyquatherm system takes place in the area of the upper crust that contains significant amounts of water and experiences a number of earthquakes (see Fig 9.2.5). The **Hyquathermal Process** produces gases that rise into the atmosphere and change the weather. In the *upper continental crust*, this earthquake-heated water vapor system produces high and low-pressure events that control weather cycles. In the *oceanic crust*, earthquake-heated ocean water affects global weather patterns and storms (like El Niño and La Niña).

A *hyquatherm* is very similar to the natural *hypretherm* described in the Hydroplanet Model; both are water environments within the crust that experience elevated temperatures and pressures. In the hyquatherm, gases, primarily water vapor in the continental crust, and warmed ocean water in the oceanic crust, escape and rise to the surface. These water systems are driven by astronomical cycles, and they in turn, drive many short and long-term weather cycles observed by mankind.

When water changes from a liquid state to a gas state, it is vaporized. The reason vaporized water is so important to atmospheric weather is that liquid water increases its volume *1,700 times* as it becomes gaseous at sea level. Understanding the origin of gaseous water vapor is the key to understanding the Earth's changing weather.

Evaporation is one type of vaporization, and in modern meteorology, there are two types of evaporation: evaporation from

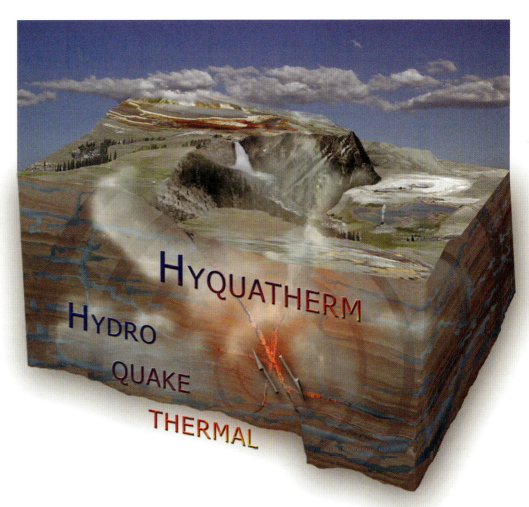

Fig 9.2.5 – The discovery of earthquake heating necessitates new terminology to describe the underground earthquake-heating water vaporization process. Drawing from the root words, hydro, quake, and thermal we derive the term **Hyquatherm** or **Hyquathermal Process**. In the upper continental crust, earthquake-heated water is vaporized and expelled, producing areas of high-pressure. In the oceanic crust, earthquake-heated ocean waters affect global weather patterns and storms.

standing water and transpiration from plants. **Transpiration** occurs when a plant's leaves, stems, flowers, or roots release water vapor into the atmosphere. **Evapotranspiration** is a term that includes both evaporation and transpiration, the two ways in which most water vapor is thought to enter the atmosphere. When water is in a solid form, such as ice or snow, it can also become water vapor through a process called **sublimation**. However, this is a very slow process and is not known to affect weather cycles significantly.

Evapotranspiration or sublimation does not account for the water vapor introduced into the atmosphere through the hyquathermal process just outlined. Therefore, a new term to describe the hyquathermal water vapor from beneath the crustal surface was devised—**endovaporization**. This newly discovered process is defined as:

> **Endovaporization** - The rapid vaporization and release of sub-crustal liquid water into the atmosphere.

Although evaporation, transpiration, and endovaporization directly affect the water cycle (see Fig 9.2.6), *only* endovaporization has the ability to change a significant quantity of liquid water into water vapor *quickly*, with an expansion rate of 1700 times at sea level. When a large volume of this gas is released through the hyqatherm process, we have the making of new weather patterns.

Endovaporization is a major component of meteorology, but atmospheric science needs a term to describe *all* of the natural water vaporization processes that occur on the Earth, a term that encompasses evaporation, transpiration, endovaporization, sublimation, and any other water vaporization process that affects the atmosphere, including the addition of water vapor from space. These can be included in the term: **omnivaporization**.

Water Vapor—the Key to Changing Weather

Having read that scientists documented earthquake weather and the formation and release of water vapor from hyqatherms, we can explore the role water vapor plays in weather systems. Meteorologists know just how important water vapor really is:

"**Water vapor** constitutes only a small fraction of the atmosphere, varying from as little as one-tenth of 1 percent up to about 4 percent by volume. But the **importance of water in the air is far greater than these small percentages would indicate. Indeed, scientists agree that** *water vapor* **is the most important gas in the atmosphere when it comes to understanding atmospheric processes.**" Bib 180 p103

Water vapor is by far "the most important gas in the atmosphere when it comes to understanding atmospheric processes" and how the weather works. This is *why* the new discovery of earthtide heating and the hyqatherm is so crucial when it comes to comprehending how weather is produced. Atmospheric scientists have said that:

"The **hydrologic cycle** is a gigantic system **powered by energy from the Sun** in which the atmosphere provides the vital link between the oceans and continents. Water from the oceans and, to a much lesser extent, from the continents, evaporates into the atmosphere. Winds transport this moisture-laden air, often over great distances." Bib 180 p98

Fig 9.2.6 illustrates two Water Cycle diagrams, one by the USGS, powered only by the Sun, the other by the UM, powered by both the Sun and the hyquathermal process. Heat from the Sun is very constant, and predictable, leaving meteorologists mystified and unable to identify cycles in the Sun's heat to account for all the weather cycles. Only through the hyquathermal process can we account for large, earthquake-heated areas that can change weather patterns.

As earlier cited earthquake researchers demonstrated, large quantities of water vapor are generated rapidly, (in minutes) by earthquakes across vast areas. The slow evaporation process cannot appreciably affect large storm systems or cause high and low pressure systems to appear suddenly—but the endovaporization process can.

Previously, the great volume of water vapor in the air from hyqatherms was unnoticed because background water vapor (or humidity) is invisible to the naked eye. There is actually *six times more water in the air* than is transported by all of the continent's rivers! From 2007 college textbook, *The Atmosphere*:

"Although the **amount of water vapor in the air** is just a tiny fraction of Earth's total water supply, the absolute quantities that are cycled through the atmosphere in a year are **immense**, some 380,000 cubic kilometers (91,000 cubic miles). This is enough to cover the Earth's surface uniformly to a depth of about 1 meter (3.3 feet). Estimates show that over North America almost **six times more water is carried within the moving currents of air than is transported by all the continent's rivers.**" Bib 180 p99

Although atmospheric scientists know that water vapor is the single most important gas in the atmosphere affecting the weather, they are unable to account for the water vapor generated by hyqatherms because they are ignorant of its existence. How then, can atmospheric science hope to comprehend global warming or the forces that drive it?

The truth is, they can't.

Further details about this problem will be dealt with in the upcoming Global Warming Pseudotheory subchapter.

> "Indeed, scientists agree that **water vapor** is the **most** important gas in the atmosphere when it comes to understanding atmospheric processes."
>
> Frederick K. Lutgens

The Weather Model

Having identified direct scientific evidence connecting astronomical cycles with earthquakes, frictional heating in the crust, and the presence of vast sub-crustal oceans, we have the groundwork to establish the first four principles of the **Weather Model**. There are four new weather principles and three new natural weather laws, each of which will be supported by empirical evidence presented throughout this chapter. These are the new principles and laws:

1. Hyquatherms change the Earth's weather systems; they are driven by Earthtide Heating, which is the constant frictional heating of the crust caused by gravitational tidal forces.

Creating a New Water Cycle

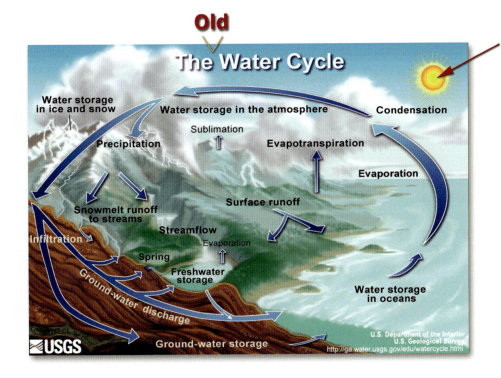

Only Heat Source

No engine for storms or new weather patterns.

Vaporization is **only** a slow process.

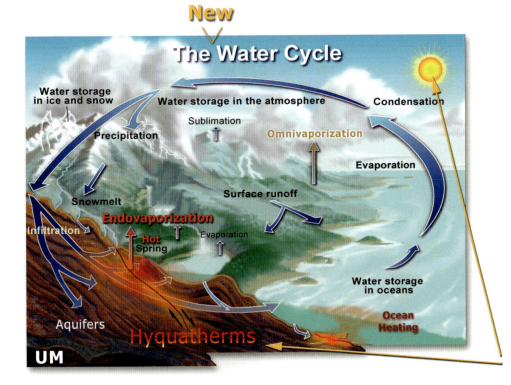

Hyquathermal processes create new storms and weather patterns.

Vaporization can be a slow **or** a rapid processes.

Two Heat Sources

Fig 9.2.6 - The modern science Water Cycle is illustrated in the top diagram. The bottom diagram is the *new* modified UM Water Cycle, which includes hyquatherms, the driving force behind weather because of its impact on pressure and humidity. Two new terms are included in the New Water Cycle: endo-vaporization and omnivaporization. **Endovaporization** describes a newly discovered water vaporization process going on beneath the crust. It is caused by earthquakes and is defined as the *rapid* conversion of liquid water in the crust to water vapor in the air. This occurs when the crust is heated by tidal friction, vaporizing the surrounding water, which is expelled into the atmosphere. **Omnivaporization** is a new term describing all the processes that introduce water vapor into the atmosphere. The New Water Cycle replaces the Old Water Cycle by including a new mechanism of storm and weather pattern creation—the hyquatherm. The "**inexact science**" of weather forecasting and the mysterious unsolved origins of weather are the result of not knowing about the *second* source of heat that drives weather; the hyquatherms.

SUBCHAPTER 9.2 THE ORIGIN OF WEATHER

2. Hyquathermal heating of the seas and underground water beneath the continents causes high pressure and temperature zones in the atmosphere, which changes the Earth's weather.

3. The Earth's weather follows patterns and earthtide cycles that originate from the astronomical positions of the Earth, Moon, and Sun.

4. The Earth's weather and the Earth's Geofield are interrelated, connected by Earthtide Heating and the piezoelectric field, which are both created by the constant gravitational tidal movement of the Earth's crust.

The fourth principle of the Weather Model includes the Earth's energy field, or **Geofield**, which will be discussed later in this chapter. This principle ties together important geophysical phenomena and weather phenomenon heretofore *not* a part of meteorological science. The Weather Model will show that the Earth has a piezoelectric field and that this field is connected to hyquathermal activity and earthtide.

The Three Laws of Weather

In the Weather Model, there are **Three Laws of Weather**, which are defined as follows:

The First Law of Weather
The Earth's weather is
changed by hyquatherms.

The Second Law of Weather
Hyquatherms are changed by
gravitational-astronomical cycles.

The Third Law of Weather
Earthtide-atmospheric pressure
and the Geofield are directly connected
through gravitational-astronomical cycles.

One exciting aspect of these new models and accompanying natural laws is that they can be tested repeatedly in situations never before considered. If the Weather Model and the Three Laws of Weather are in fact correct, we should be able to find confirmation of the laws by examining various aspects of everyday weather and in long-term weather patterns.

The reality that **Earthquake Weather** exists is one key aspect of the new Weather Model. The effects of earthtide, which are comprised of both silent earthquakes and large, identifiable earthquakes, and the central role the *hyquatherm* plays in developing weather patterns and pressure systems, are also important features of the new Weather Model.

The Atmospheric-Pressure Error

The "importance of atmospheric pressure" is expressed by the authors of the 2007 weather textbook, *The Atmosphere*:

"**The importance of atmospheric pressure to Earth's weather cannot be overemphasized**. As you shall see shortly, differences in air pressure **create global winds** that become organized into the **systems that 'bring us weather'**." Bib 180 p175

Atmospheric pressure cannot be overemphasized because this is where the Earth's weather comes from! In fact, if the following FQ can be answered correctly, it will reveal the true origin of weather:

What is the source of the Earth's
high and low-pressure systems?

Before we move on, remember the meteorological statement at the beginning of this chapter, that "forecasting is still an *inexact science*." *If* meteorologists *really* understood the source of barometric pressure systems, which they know "bring us weather," forecasting would *not* be such an inexact science. Therefore, we should question modern science's current atmospheric-pressure system origin:

"**Atmospheric pressure** at any point on the Earth is **caused by the weight of the column of air above that point**, as is measured with an instrument called a barometer." Bib 183 p30

According to this definition, the *cause* of atmospheric pressure is the result of *gravitational* force, which requires "sink-

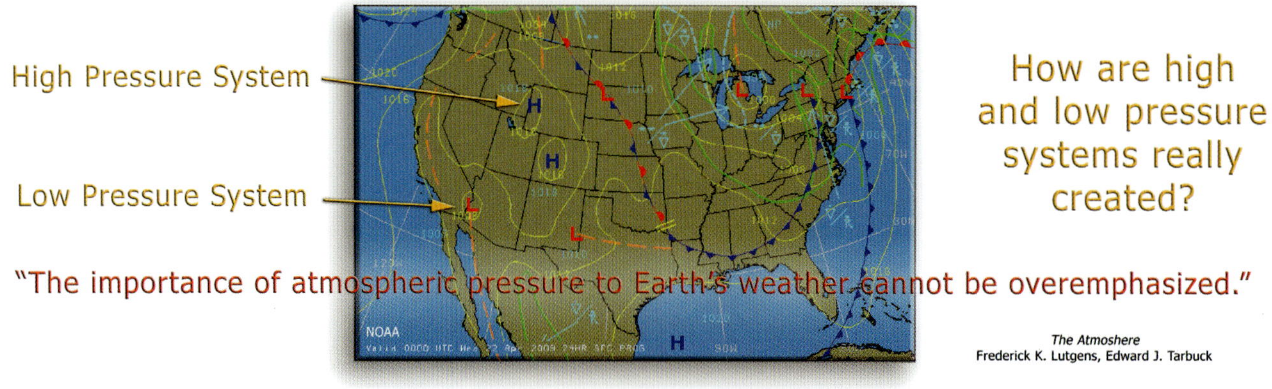

High Pressure System

Low Pressure System

How are high and low pressure systems really created?

"The importance of atmospheric pressure to Earth's weather cannot be overemphasized."

The Atmoshere
Frederick K. Lutgens, Edward J. Tarbuck

Fig 9.2.7 – On weather maps, H and L represent respectively, high and low atmospheric pressure systems. The barometric pressure difference is slight enough that humans rarely feel the change, however many animals can feel atmospheric pressure changes and react instinctively when low-pressure systems are forming, because storms are drawn toward low-pressure systems. High-pressure systems usually exhibit fair weather and are responsible for moving large bodies of air around the world, helping create the weather here on Earth. How are the high and low-pressure systems really created? Meteorology attempts to explain the mechanism behind the air movement by explaining that air is **sinking** over high-pressure systems, but cannot demonstrate the process. The UM Weather Model shows that air is *rising* over high-pressure systems, due to hyquatherms. The UM Weather Model and modern meteorological science share one important common point: both acknowledge, "The importance of atmospheric pressure to Earth's weather *cannot be overemphasized*." Atmospheric pressure changes create weather! Modern science missed the real origin of atmospheric pressure changes because it sees the Sun as the *only* heat source driving the weather systems, overlooking entirely the gravitational-friction heating of the crust and its importance.

ing air" over a high-pressure system and "rising air" over a low-pressure system. The National Oceanic and Atmospheric Administration's (NOAA) website includes an illustration and a good description of air *sinking* over a *"high"* pressure system:

> "What about the **diverging air near a high**? **As the air spreads away from the high, air from above must sink to replace it**." Note 9.2h

The NOAA site continues by describing air *rising* over *low*-pressure areas:

> "What happens to the converging winds near a **low**? A property called mass continuity states that mass cannot be created or destroyed in a given area. So **air cannot 'pile up' at a given spot**. **It has to go somewhere so it is forced to rise**." Note 9.2h

Such definitions of air movement between high and low-pressure systems are firmly entrenched throughout meteorology, but has this been observed?

> FQ: Have observations shown air "sinking" over highs and "rising" under low-pressure systems?

Our research revealed a surprising result—it seems no one had observed this. In fact, those familiar enough with weather systems know the idea of air *sinking* over area highs or rising under areas of low-pressure makes no sense, because:

> Air moves **away** from areas of **high** pressure **toward** areas of **low** pressure.

Moreover, air movement from high-pressure systems to low pressure systems happens in *all* directions. What mechanism would cause air to come from all directions and *move over* a high-pressure body of air? We have found none, nor have we ever seen an explanation for such atmospheric behavior. It is curious why meteorologists think air moves *away* from areas of high-pressure ("H" on their weather maps), yet think somehow, *at the same time*, air is supposedly moving *toward* the high-pressure where it can somehow "pile up." After all, they just said "air cannot pile up at a given spot." In other words, meteorologists cannot explain atmospheric pressure systems because the real origin of the pressure systems remains unknown (see Fig 9.2.8).

One way to visualize the interaction between high-pressure and low-pressure systems is shown in Fig 9.2.9. A container is filled with blue and red balloons. The larger red balloons represent areas of high-pressure; they are expanding and pressure is increasing because they are being heated. The smaller blue balloons are shrinking; pressure is decreasing because they are cooling. The heated balloons expand outward in all directions; the cooling balloons contract in the same manner. As heating and cooling occurs, the balloons expand or contract, interacting with each other in a predictable manner; the expanding balloons fill the space of the contracting balloons. The high and low-pressure systems in our atmosphere are doing the same thing. The process is based on the simple physics of heating and cooling gases.

The general gas law presented in the Air-Water Model (Chapter 23), states that PV ~ T (pressure times volume is proportional to the temperature of the gas) and is a general mathematical formula that describes the hyquathermal pressure changes in

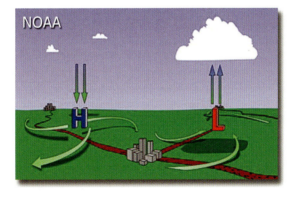

Although air is said to be sinking over high-pressure areas and rising under low-pressure areas, evidence for this claim is lacking.

Meteorology's Atmospheric Pressure Error

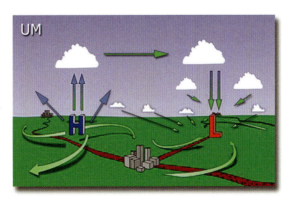

Atmospheric Pressure Reality

High-pressure causes air to expand in all directions, especially toward contracting, cool, low-pressure areas.

Fig 9.2.8 – The top diagram comes from NOAA. It illustrates the decades-old error still taught today in the classroom. This erroneous model is based on the faulty assumption that the Sun is the only heat source. No evidence has been shown that air is converging over high-pressure systems after spilling off the top of low-pressure systems. In fact, just the opposite is true, as seen in the lower UM diagram. High-pressure air *expands* in all directions—away from the heated high-pressure area. It moves toward cooler, *contracting* areas of low-pressure. This is illustrated with a simple experiment, seen in Fig 9.2.10. (NOAA diagram at http://www.srh.noaa.gov/jetstream//synoptic/wind.htm - Accessed 6.15.09)

our atmosphere. If we raise the temperature of a gas, pressure and volume increase. If the temperature is lowered, the pressure and volume of the gas is decreased. This will be demonstrated in the Air-Water Model chapter by heating a balloon with a small amount of water in it using a microwave oven. The balloon experienced an increase in volume and pressure (the same in all directions), based on the increased temperature.

Meteorology was forced to turn to illogical processes to describe the high and low-pressure systems simply because they saw only **one** source of atmospheric heat. Seeing the Sun as the only heat source meant ignoring or downplaying obvious problems. The Sun is not a discriminating source of heat; shining on half the Earth at all times, it **could not be** the primary source of energy behind the high and low-pressure systems.

Clouds provide the only possibility of reducing temperatures from the Sun, but cloud systems do not necessarily correlate with cold, low-pressure systems. In fact, as we will see shortly, some clouds form in high-pressure systems! Clouds and storms tend to flow toward areas of low-pressure because the air is contracting. Furthermore, low pressure facilitates condensation and cloud formation, but the changes in air pressure do ***not*** show a direct relationship to sunlight.

Creating a Simple Weather-Pressure System Experiment

One reason the comprehension of weather has been so fleeting is that we cannot *see* with our naked eye, the great abundance of water vapor in the air. In the early days of medical science, after the microscope allowed doctors to *see* microbes, the comprehension that illness was caused by unhealthy microorganisms was made possible. In the same way, *seeing* weather patterns in a simple experiment will help illustrate how natural weather patterns form.

Earthtide heating and cooling (the Missing Factor of Weather discussed at the beginning of this subchapter) is illustrated in Fig 9.2.10. A 10-gallon aquarium is divided into two sections and the top is covered with a sheet of clear plastic. Two two-inch squares are cut out of the top and bottom of the divider to allow for airflow. To reproduce the effect of an active hyquatherm, a beaker of heated water is placed on the left side, on an insulated pad. On the right, an ice-filled beaker reproduces a cooling hyquatherm of *contracting* air. An incense stick is lit and placed on the hot water side to show air movement. As the photos clearly demonstrate, expanding water vapor from the beaker of heated water (heated close to boiling) moves air into the right, cold-air side, which is an area of low-pressure.

In the lower photograph, the heated water was removed and an interesting air movement developed. The heated beaker had heated the surrounding glass and pad, apparently producing enough heat to cause the air to continue expanding, albeit at a much slower rate. This caused the air to clump, cloud-like, in the upper portion of the heated left side. On the right side, a clearly visible low-pressure area formed above the ice-filled beaker, with air streaming *towards* the container, just as it does in atmospheric low-pressure systems.

This easily repeatable experiment illustrates how air moves away from areas of high-pressure towards areas of low-pressure. Because of the expansion of hot air and the contraction of cold air, no mechanical circulation was required.

The experiment also demonstrates the First Law of Weather; the Earth's weather is changed by hyquatherms, by showing the movement of air systems according to the simple general gas law PV ~ T, where pressure changes from a change in temperature. The First Law of Weather is powerful because of its simplicity; knowing that hyquatherms create areas of high and low-pressure will make weather much easier to understand and comprehend.

High-Pressure Narrow-Ridge Evidences

If high-pressure areas are *formed by earthtide and hyquatherms*, we can expect high-pressure systems to be generally **narrow** in shape because earthquakes occur along predominantly linear fault lines. In fact, since many faults parallel mountain ridges, the high-pressure systems may appear to look like "high pressure ridges" themselves. A 2005 article titled, *Weather's Highs and Lows: Part 1 The High*, Keith C. Heidorn explains just such a phenomenon:

"Following formation, **most Highs are generally elliptical in shape**, and **often large and sprawling**. But as they interact with other air masses and topography, and are distorted by forces of the upper atmosphere, **high pressure cells often become long and narrow in shape**. When plotted on a surface weather map, **these elongated pressure patterns resemble mountain ridges on terrain maps**. Meteorologists therefore refer to them as **high pressure ridges** or simply **ridges**." Note 9.2i

This observation provides a significant piece of the high-pressure puzzle; high-pressure systems are "generally elliptical in shape" and are "long and narrow." This corresponds to the active, narrow earthquake zones that are expected to create the high-pressure systems. It is further evidence of the hyquathermal origin of high-pressure areas, and it shows that high-pres-

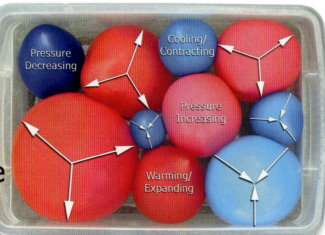

Fig 9.2.9 – Red and blue balloons illustrate areas of high and low-pressure that exist in our atmosphere. The larger red balloons represent an expanding high-pressure system because of warming. The smaller blue balloons illustrate the contracting nature of low-pressure systems. As high-pressure systems are created, they expand to fill the place of contracting low-pressure systems.

Expanding High and Contracting Low-Pressure Balloon Example

Fig 9.2.10– The Weather-Pressure System Experiment illustrates how air pressure is changed in a closed system by heating and cooling the air. A 10-gallon aquarium with a clear plastic top is divided into two compartments with a piece of cardboard. Two two-inch openings are cut to allow for airflow. A beaker of nearly boiling hot water is placed on the left side to illustrate the *expanding heated hyquatherm* from beneath the Earth's crust. A beaker of ice is placed on the right to illustrate a *contracting, cooling hyquatherm*. An incense stick is lit to show the movement of air in the mini-atmosphere. Airflow moves up and away from the high-pressure area created by the hot water, but moves downward, towards the low-pressure area. This is **opposite** the current weather theory. After the beaker of hot water was removed (bottom image), the air on the left clumps cloud-like in air above where the heated water was located, while a clearly visible flow develops above the cold, low-pressure beaker of ice. This experiment demonstrates the first Law of Weather: The Earth's weather is changed by hyquatherms.

Weather Pressure System Experiment

sure systems are *not* randomly formed weather systems with a 'column of heavier air.' As we see here, and in the next subchapter, high-pressure systems have a cyclical nature, often reoccurring in the same locations.

The Origin of Weather Summary

This subchapter introduced many important concepts, including the four main principles of the Weather Model and three new weather laws. It also revealed the missing factor of weather—earthquakes—without which, the origin of weather would remain a mystery.

Several previously introduced UM models, including the Lava-Friction Model and the Hydroplanet Model, made it possible to discern the source of heat—gravitational friction—that is pumping heat and water vapor into the atmosphere, changing the weather. The remaining subchapters will continue to establish this extraordinary claim.

Previously unacknowledged but important research from Russia and India is now comprehensible within the framework of the Weather Model, which sees Earthtide Heating and the Hyquatherm as the origin of weather. In the future, as researchers become aware of the Weather Model, new research will further refine the three new natural laws, shown to be involved in the development of high and low-pressure systems that control the weather process. Next, we will explore the Earthquake Cloud Evidence.

9.3 Earthquake Cloud Evidence

In the last subchapter, we discovered that changes in weather are a product of changes in atmospheric pressure. Barometric pressure changes are a result of hyquatherms, which are earthquake-heated water systems in the Earth's crust that are driven by earthtide. When the earthtide is accompanied with an above average swarm of earthquakes, a substantial increase in vaporization occurs.

Using earthquake clouds, some researchers have been able to predict the size, location, and timing of earthquake activity with some degree of accuracy. This subchapter identifies three new classifications of clouds and outlines a new mechanism of cumulus cloud formation. Connections between earthquakes, cumulus clouds, tornadoes, microbes, and lunar-weather cycles are also reviewed.

Earthquake Cloud Evidence from Thermal Research

The last subchapter introduced thermal satellite research from a project headed by Russian scientist A. A. Tronin who conclusively established the connection between earthquakes and the frictional heating of the crust. Observations were confirmed using thermal imaging from space and ground temperature measurements, including borehole temperatures and pressures. With the aid of the Hydroplanet Model, we have evidence of large quantities of water throughout the crust, including a vast quantity of sub-continental water never before envisioned by modern science.

Tronin and his fellow researchers were not specifically looking for a connection between earthquakes and area clouds, in fact, the clouds were considered a nuisance, getting in the way of thermal satellite measurements. But on 21 December 1996, a 5.7 magnitude earthquake struck at a depth of 44 km below Japan's Ibaraki prefecture, about 100 km north of Tokyo. Tronin recorded the following right "after the earthquake":

"**Immediately after the earthquake, [a] strange cloud appeared above the epicenter**. [The] next day (22 December) **a very large and clear thermal structure was present again**, from SW to NE." Note 9.3a

The cloud was "strange" because of its unexpected appearance. Though little is known about them, earthquake clouds are a very real and important element of the Weather Model. Because modern science does not understand earthquake clouds, the evidences they might have provided about future earthquakes has generally been rejected. Now, in the light of the UM Earth System, including the Weather Model, earthquake clouds make sense, and are important precursors to potentially dangerous earthquakes, and should be taken seriously.

Earthquake Cloud Predictor—Zhonghao Shou

Zhonghao Shou is a pioneer in the field of earthquake clouds. In 1990, Shou began by predicting a large earthquake in Iran that killed or injured 370,000 people. From the introduction section on his website, earthquakesignals.com, Shou tells how he got started:

"I have been predicting earthquakes since June 20, 1990 for the 7.7 Iran earthquake, which occurred 18 hours later and killed or injured 370,000 people. Because the earthquake was the only one bigger than 7 in the northwest direction of my home town Hangzhou (30.27 N, 120 E), China, for 333 days from May 31, 1990 till Apr. 28, 1991, **I believed that there must be a strong relationship between the cloud and the earthquake.**" Note 9.3b

Shou describes how he thinks earthquake clouds form, which is remarkably similar to the UM Weather Model explanation:

"According to Shou, **earthquake clouds are formed when underground water is converted into water vapor by the heat generated in the epicentric area of a fault rock, which is undergoing constant stress and friction**.

"**When this vapor escapes to the surface and rises through the atmosphere, it forms a cloud**. 'The shape of the gap and surface current may endow the cloud with a special configuration like a snake, a wave, a feather, or a lantern, which will be able to be distinguished from weather clouds,' says Shou." Note 9.3c

Shou, a retired chemist, devotes his time to the study of earthquake clouds for the sole purpose of helping people avoid disaster. Because Shou is not a geophysicist and does not work within the realm of 'established theories,' he has been ignored, even mocked, or insulted:

"Being the only one having predicted the Iran earthquake correctly, I felt my duty to develop the method, which is why I, a retired Chinese chemist, **have faced a strange field, ignored all of doubt, mock, and even insults**, and spent about $35,000

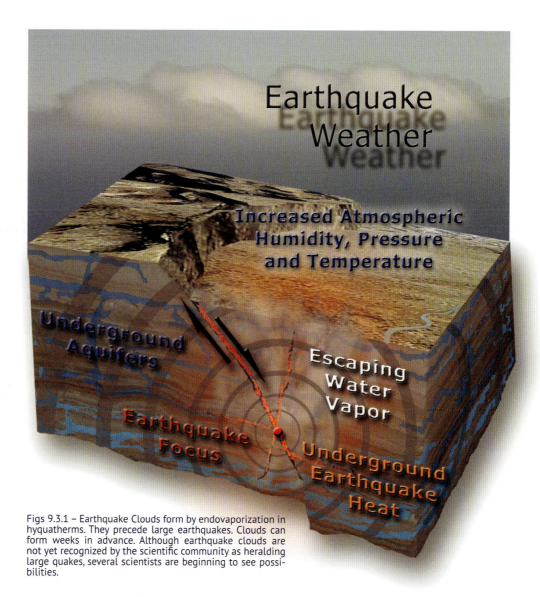

Figs 9.3.1 – Earthquake Clouds form by endovaporization in hyquatherms. They precede large earthquakes. Clouds can form weeks in advance. Although earthquake clouds are not yet recognized by the scientific community as heralding large quakes, several scientists are beginning to see possibilities.

Earthquake Clouds

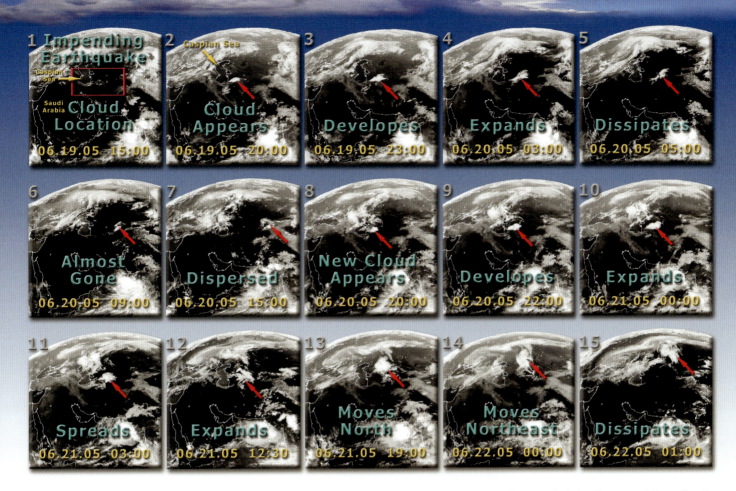

Figs 9.3.2 – Earthquake Clouds develop as water vapor created by earthquake heating is released. Zhonghao Shou, a retired chemist, successfully predicted dozens of large earthquakes, sending his predictions to the USGS before they occurred. Based on earthquake cloud formation that was seen to appear on two separate days, Shou predicted, on 24 June 2005, that an earthquake would strike just east of the Caspian Sea in Iran with a magnitude greater 4.0 within 103 days. Just 42 days later, on 31 July 2005, a 4.8 magnitude earthquake erupted, precisely where it had been predicted.

In photo #1 above, the area indicated is free of clouds. Five hours later, a cloud appears. The sequence shows the development, expansion, and dissipation of a cloud over a day's time. A day later, the entire cycle is repeated. During this period, swarms of small earthquakes heated underground water, vaporizing it in a hyquatherm, increasing humidity and forming the earthquake cloud. These clouds are notably different from typical evaporative clouds that take a long time to form. From the size, shape, and time it takes for the cloud to form, predictions can be made about earthquakes in the near future.
Images adapted from the NEODASS/Dundee University Satellite Receiving Station data.

living with my daughter Wenying, a Ph.D. student at Caltech, where I can observe more clouds and get earthquake data, satellite images, and literature." Note 9.3d

Even while Shou's website is blocked in his homeland of China, he continues to seek open scientific debate within the geophysical community. In 2004, Shou and an associate published a report on earthquake clouds under the Space Applications Program for the United Nations Office for Outer Space Affairs. The report, titled, *Bam Earthquake Prediction & Space Technology*, examined 50 earthquake predictions Shou made between 1994 and 2001:

"Based on observations of earthquake clouds and geoeruptions, both visible and infrared satellite images, **Shou submitted 50 earthquake predictions between 1994 and 2001 to be certified by the USGS.** Table 5 exhibits all of them, and their subsequent earthquakes, as reported in USGS databases. Assuming all earthquake data is without error, so-called 'Peer on', **34 predictions or 68% of them are correct in time, location, and magnitude**. They are called 'hits', while the others are called 'misses'." Note 9.3e

No other known earthquake prediction mechanism comes close to the record Shou accumulated. Even some of his "misses" were impressive; predictions of large-scale events didn't materialize, but there were often several small events. Other "misses" included events that fell outside the predicted narrow window of time, with an earthquake appearing sooner, or after, the predicted period. According to the report, the odds that random guessing could reproduce Shou's results were 16,000 to 1 for prediction results through the year 2001. "This clear statistical significance validates the prediction method."

Fig 9.3.2 is a set of sequential satellite images showing the formation, expansion, and dissipation of an earthquake cloud, twice over two consecutive days. From this, Shou predicted the location, magnitude, and timeframe of an earthquake that proved to be remarkably accurate. Without any formal funding or support from the scientific community, Shou accomplished what other scientists had been unable to do. The evidence that earthquake clouds are a precursor to some earthquakes is conclusive and may yet prove especially valuable as more data is accumulated. Of course, earthquake clouds may not always be visible. In hot climates or during summer months, when the air is very dry, hyquathermal water vapor can be absorbed into the air without condensation occurring, which means clouds will not form.

Despite earthquake cloud research being in its infancy, there is sufficient evidence that it could contribute to the preservation of human life by being able to warn of impending seismic activity, and the framework of the Weather Model will contribute significantly to this new field.

Earthquake Cloud Experience—Perry Dean Sessions

My oldest son, Perry Dean Sessions, was in Auckland, New Zealand, on 22 February 2011 when he reported the following:

"When the [Christchurch, New Zealand, 2011] earthquake happened on Tuesday the 22nd, I was with Elder Jeppsen. We were driving in the car at about 3pm and I looked at the massive clouds in the sky. They were HUGE. Tall, white on the top and dark gray on the bottom, they appeared to be spread randomly in big clumps across the sky. They were also very close to the ground... closer than normal for New Zealand clouds. After I studied the clouds, I turned to Elder Jeppsen and said, **'Looking at those clouds I wouldn't be surprised if we had an earthquake today.'** Surprised, he asked, 'why do you say that, and what do clouds have to do with it?' I explained how I had learned from my dad's scientific research that clouds could foretell earthquakes. Only moments later, a woman on the street asked, 'did you hear about the earthquake in Christchurch? My friend and I both looked at each other in awe." [Personal communication]

New Zealand's South Island had experienced a 6.3 magnitude earthquake on 22 February 2011, at 12:51 pm. Perry was 680 km northeast in the Hamilton area of the North Island. He had not felt the earthquake, but saw what he perceived to be earthquake clouds, which was later corroborated. A number of Perry's friends were eager to learn how Perry was able to know about an earthquake by simply looking at the clouds. Perry's earthquake cloud encounter shows that anyone can have an experience like this, once they become familiar with Nature's processes.

Once we realize the vastness of the water beneath the crust, and how it is heated, hyquatherms and their effects are easy to understand. Only through the research of hyquatherms can we comprehend earthquake phenomena more fully.

Fig 9.3.3 – The 6.3 magnitude Christchurch Earthquake occurred on 22 February 2011 on New Zealand's South Island in the Cantebury region. There were more than 172 deaths in the quake, the second-deadliest natural disaster in New Zealand's recorded history. Significant hydrosediment was expelled during earthquake liquefaction, and a number of hydrocraters were produced. On the day of the earthquake, earthquake clouds were reported on the North Island of New Zealand. As we learn more about hyquathermal cloud formation, earthquakes will become easier to predict in the future.

Reaction to Earthquake Cloud Research

Shou's research may hold great promise in the saving of lives and in the advancement of our understanding about the weather, but there are reasons why the modern science community has chosen to reject Shou's successful predictions. First and probably foremost is that earthquake clouds do not fit in the modern geology paradigm. Furthermore, one of modern geology's own did not discover the earthquake clouds, and though it might seem arcane, politics has been around a long time in modern science and has always played a role in determining what gets attention and what gets funding.

But not all scholars are sitting on the sidelines watching Shou predict earthquakes by observing earthquake clouds. S.N. Bhavsar, a Vedic scholar associated with the Physics Department of Pune University, discovered the writings of an ancient Indian astronomer, mathematician and philosopher, Varahamihira (505-587 AD) who associated earthquakes and 'unusual cloud formations' like Shou did, but who included a host of other phenomenon, which are included in the UM Models:

"The 32nd chapter of the manuscript is devoted to **signs of earthquakes and correlates earthquakes with cosmic and planetary influ-**

ences, underground water and undersea activities, unusual cloud formations, and the abnormal behavior of animals." Note 9.3f

Of course, many present-day researchers "find it rather odd" that Varahamihira would know something 1500 years ago that modern science missed, although it had been confirmed by independent researcher Zhonghao Shou:

"'**I find it rather odd** that the description of earthquake clouds in Brihat Samhita [Varahamihira's manuscript] **matches the observations made by Zhonghao Shou** at the Earthquake Prediction Centre in Pasadena, California,' said B D Kulkarni, head of the National Chemical Laboratory's Chemical Engineering Division." Note 9.3f

Bhavsar, the Vedic scholar, cites the ancient text that reveals many important characteristics of the Earth, characteristics that modern scientists cannot acknowledge under the current paradigm:

"'What needs to be acknowledged,' he said, 'is that 1500 years ago a celebrated astronomer-astrologer-mathematician sought to study earthquakes on the Indian subcontinent. He drew correlations between terrestrial earth, the atmosphere and planetary influences. **He described earth as a mass floating on water and spoke of unusual cloud formations and abnormal animal behaviour as precursors to earthquakes.**'" Note 9.3f

Many of the scientific truths presented in the Universal Model were known thousands of years ago by the 'ancients.' Some 1500 years ago, Varahamihira likely obtained scientific knowledge from texts written many hundreds of years before his time; ancient texts that are no longer available today. Historians search constantly to validate ancient texts, but most 'modern' historians, like most 'modern' scientists accept *very little* ancient history as factual, seeing instead myths and legends of less civilized people. This idea is refuted in the World History Model chapter, where it will be shown that many ancient scientific claims are valid and can be proven with new technology and new scientific discovery.

Japanese and Chinese Geophysicists Confirm Earthquake Clouds

The wonderful thing about the discovery of scientific truth is that once a new truth, like earthquake clouds, is found, evidence seems to appear everywhere. In 2007, Japanese geophysicists reported an earthquake cloud that had been captured by satellite cameras. Sequencing the images allowed observers to track its movement:

"**Thermal anomaly detected by satellite before earthquake has been widely reported**. Here we reported a **new anomaly** detected with geostationary satellite image. The geostationary satellite image series were combined together to make an animation, and it is found that the common cloud moved continuously, while **the earthquake cloud stayed closely to the epicenter. Here an earthquake cloud example in Japan was

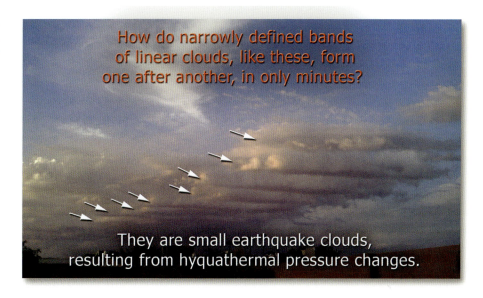

Fig 9.3.4 – A series of long, thin clouds formed in minutes, one after another before breaking apart and dissipating August 12, 2009 in Mesa, Arizona, USA. With no rain for months, and being in the middle of a desert with no water in the immediate vicinity to account for evaporation, how did they form? Clouds form because of changes in pressure to vapor-laden air, but these clouds formed one after another in a narrowly constricted band. Earthquake cloud formation processes from hyquathermal activity explains this common phenomenon. In instances such as this, cumulus clouds may form in linear bands corresponding with fault lines. This is manifested frequently along mountain ranges with underlying faults.

reported.**" Note 9.3g

The following year, in a 2008 *New Scientist* article, two Chinese geophysicists documented "distinctive cloud formations" that formed above an active fault just before two separate, large earthquakes struck:

"CAN unusual clouds signal the possibility of an impending earthquake? That's the question being asked following **the discovery of distinctive cloud formations above an active fault in Iran before each of two large earthquakes occurred**.

"Geophysicists Guangmeng Guo and Bin Wang of Nanyang Normal University in Henan, China, **noticed a gap in the clouds in satellite images from December 2004 that precisely matched the location of the main fault in southern Iran. It stretched for hundreds of kilometers, was visible for several hours and remained in the same place, although the clouds around it were moving**. At the same time, thermal images of the ground showed that the **temperature was higher along the fault. Sixty-nine days later**, on 22 February 2005, an **earthquake of magnitude 6.4 hit the area**, killing more than 600 people.

"In December 2005, a **similar formation again appeared** in the clouds for a few hours. **Sixty-four days later**, an **earthquake of magnitude 6 shook the region** (*International Journal of Remote Sensing*, vol 29, p 1921)." Note 9.3h

The geophysicists who authored this report are not connected with the private work Shou was doing; they are a good example of newly emergent, scientifically competent countries who appear more open minded than traditional Western science professionals, who have dominated modern science for hundreds of years. Even with *undeniable* physical evidence of earthquake clouds, U.S. geologists remain skeptical:

"The authors say that if recognizable cloud formations precede large quakes, they could be used for prediction, but **other seismologists are skeptical. 'There is no physical model that explains why something would suddenly occur two

Three Cloud Forms

Fig 9.3.5 – Clouds are divided into three main forms: Stratus, Cirrus and Cumulus. **Stratus** clouds are sheets of clouds that cover much of the sky. **Cirrus** clouds are thin and wispy. **Cumulus** clouds are much different. They are the common cauliflower-like, puffy, often flat-bottomed clouds, forming most often at lower altitudes. Using the Weather Model, these cloud forms and their origins can now be easily explained. Clouds are created by pressure changes to air saturated by one of *two* mechanisms—evaporation or endovaporization. Evaporation over large bodies of water produces *stratus* clouds, which take days or longer to develop. In contrast, *cumulus* clouds form rapidly over the continents as hyquatherms and endovaporization release water vapor into the atmosphere. Cumulus clouds may also form over the ocean as upwelling heated water rises from the ocean floor. After cumulus clouds form, they may eventually rise or be scattered by the wind to become cirrus clouds.

and endovaporization (the rapid conversion of liquid water in the crust to water vapor) must be included with the traditional evaporation model. Doing so leads to a new way of classifying clouds.

Clouds Defined

A popular meteorology textbook, *The Atmosphere*, explains that clouds can be defined by two criteria: form and height. There are three basic cloud forms:

"1. **Cirrus** clouds are **high, white, and thin**. They are separated or detached and form delicate veil-like patches or extended wispy fibers and often have a feathery appearance. (*Cirrus* is a Latin word meaning 'curl' or 'filament.')

"2. **Cumulus** clouds consist of **globular individual cloud masses**. Normally they **exhibit a flat base and appear as rising domes or towers**. Such clouds are frequently described as having a **cauliflower-like structure**.

"3. **Stratus** clouds are best described as **sheets** or layers (strata) that **cover much or all of the sky**. Although there may be minor breaks, there are no distinct individual cloud units."
Bib 180 p133

All clouds are classified into one of these three basic cloud forms, or combinations thereof, based entirely on their physical appearance (see Fig 9.3.5). Moreover, the three cloud forms can be divided into *two* classes. *Cirrus* and *stratus* clouds are typically high or sheet-like, covering much of the sky. In contrast, *cumulus* clouds are of a distinct shape, usually having a cauliflower, or cotton-like appearance, with flat bases. In many cases they form in lines or strands.

FQ: Is there a reason **cumulus** clouds are markedly different from other cloud forms?

Cumulus clouds differ from all other clouds because of the way in which they form, and in the time they take to do so (see Fig 9.3.6). Instead of forming slowly by evaporation and being carried across wide expanses of the sky, like other clouds, cumulus clouds can form quickly. Their distinctive cauliflower-like shape and well-defined flat bases are telltale signs of their origin: endovaporization and hyquatherms. In many instances, continental cumulus clouds are the *precursor of other cloud types*. After the formation of the cumulus cloud, wind and air movement disperse the cumulus cloud into different types of clouds.

months before an earthquake, and then shut off and not occur again,' says Mike Blanpied of the US Geological Survey's Earthquake Hazards Program." Note 9.3h

Modern meteorologists face the same challenge modern geologists and other scientists face; which is to be open minded enough to seriously consider the concepts presented here in the UM. Just because researchers *do not* understand natural phenomenon like earthquake cloud formation, does *not* mean the clouds are *not* forming based on yet unknown natural laws!

During the 1930s, as noted in the Rock Cycle Pseudotheory chapter, geologists noticed that certain sizes of sediment were missing in the sediment column. Without any mechanism to account for sediment formation other than erosion, they simply *ignored* the missing sediment sizes. But this did not make the missing sediment puzzle go away; there was an important reason certain sizes of sediment were missing, or rare. That reason would prove to be the undoing of the modern geology paradigm. It signified one of the most important events in the Earth's history—the Universal Flood.

In the same way, modern meteorology does not clearly understand cloud formation because it sees only one source of heat, one source of vaporization—the Sun. Now, the hyquatherm

This observation requires that we classify clouds into three new *types* of clouds, based on **how they formed**:

Evaporative Clouds:
Cirrus and stratus clouds formed from evaporation.

Earthtide Clouds:
Cumulus clouds formed from **minor** endovaporization.

Earthquake Clouds:
Clouds formed from **major** endovaporization.

Evaporative Clouds form slowly as water evaporates from oceans or lakes; they may be at high elevations or move inland from over the ocean. Stratus clouds are generally evaporative clouds. Evaporative clouds make up most of the clouds in the sky; they form primarily over the oceans and often dissipate before they reach land. The other two cloud types are the result of hyquathermal activity and endovaporization. They contribute to a large percentage of the clouds that form over the continents.

Earthtide Clouds are *cumulus* clouds – the most common type of cloud that forms over the continents (see Fig 9.3.6). Because of earthtide, the crust is continuously rising and lowering by as much as two feet at the equator. This movement produces a great quantity of frictional heat energy that is stored and periodically released through endovaporization. The escaping vaporized moisture forms both small and large groups of cumulus clouds that are not dependent on the evaporation of surface water. These clouds may eventually become *cirrus* clouds as they rise or are blown about by the wind.

Earthquake Clouds are a type of large *cumulus* cloud. These form when a large amount of vapor is expelled after the release of built up seismic stress that accumulated over time in the crust. The release may come from earthquake *swarms*, too small for humans to feel at the surface, but much stronger than the earthtide itself. The frictional heat and endovaporization is sufficient to form clouds large enough to be seen from satellites. Earthquake swarms loosen the surrounding fault structure and contribute to a situation where a large slip can take place, a slip large enough to cause severe infrastructure damage and loss of life.

Earthquake Clouds are a dramatic and illustrative way of demonstrating how the Weather Model works, but the number of large earthquakes that occur on a regular basis is relatively small, contributing minimally to overall weather patterns. Earthtide Clouds, *which make up* **most** *of the cumulus cloud formations*, play a much larger role in influencing the weather, contributing daily to fair weather around the world.

The Earthquake-Cumulus Cloud Connection

How do cumulus clouds form in a hot desert environment, where there is no water to evaporate? From where did the water come that made cloud formation possible? The following description, taken from a meteorology textbook, explains that "fair-weather" clouds are typically "cumulus" clouds with "flat bases." Furthermore, as noted earlier in this chapter, fair-weather conditions occur when high-pressure systems are over an area. These systems are created by hyquatherms, which produce uprising convective currents, or "thermals":

"These '**fair-weather cumulus**' speak of the end of a fine day. Cumulus clouds typically have rounded tops and flat bases. They are frequently **produced by localized convective cells ('thermals') that break away from surfaces that have been heated by the Sun and rise until they reach the condensation level—marked by the flat bases**. If the heating is particularly strong, they may go on to build much deeper clouds, towering cumulus, or even cumulonimbus clouds." Bib 182 p15

Cumulus clouds *frequently* form after the Sun has set, therefore requiring an explanation for heating of the surface other than the Sun; they also demand an explanation for the origin of the water that condenses into clouds at higher elevations.

Forecasters have long known that the cumulus family of clouds is frequently the bearer of "violent weather":

"**Cumulonimbus clouds** may produce **heavy rain and hail**, and turn into **thunderstorms**, regardless whether they have the typi-

In a matter of a few minutes, this cloud developed into this cloud.

Cumulus Clouds

Cumulus clouds are hyquatherm-formed clouds.

Fig 9.3.6 – The puffy cotton ball, or cauliflower-like cumulus clouds are one of two main types of clouds. They form primarily over the land and often have flat bottoms because of their rapid development as vaporized water condenses once the air mass reaches an altitude where pressure is reduced. Cumulus clouds are not evaporative clouds; they are endovaporization clouds that formed from hyquathermal activity. This is why they form and change rapidly, often within minutes. Gravitational friction from earthtide and other seismic events produce cumulus clouds by vaporizing and releasing water from underground aquifers.

cal anvil top. **They are the 'showers' frequently described in weather forecasts**. The largest clusters of active cells may become multicell storms or even supercell storms, which may **persist for many hours and produce violent weather, including spawning highly destructive tornadoes**." Bib 182 p34

Notice that cumulus clouds can develop into "vertical domes or towers" and that they "form on clear days." How do such vertical structures of saturated water form on clear days with no apparent source of water? Notice the "solar heating" sci-bi mentioned in the weather textbook, *The Atmosphere*:

"*Cumulus* clouds are individual masses that develop into **vertical domes or towers**, the tops of which often resemble cauliflower. **Cumulus clouds most often form on clear days** when **unequal surface heating causes parcels of air to rise convectively** above the lifting condensation level. This level is often apparent to an observer because the flat cloud bottoms define it.

"On days when cumulus clouds are present, we usually notice an increase in cloudiness into the afternoon as **solar heating** intensifies. Furthermore, because **small cumulus clouds** (*cumulus humilis*) rarely produce appreciable precipitation, and because **they form on 'sunny' days, they are often called 'fair-weather clouds.'**" Bib 180 p137

Fair-weather cumulus clouds form on "sunny" days because a hyquatherm created a high-pressure system. That system pushed out other weather in the area, leaving only "small cumulus clouds" to condense from the hyquathermal water vapor. Solar heating cannot explain how heating the dry ground can produce clouds!

When air is heated, the humidity is *reduced*—not increased (see Fig 9.5.6 – Law of Weather Parameters). High-humidity systems are typically low-pressure systems, where storms are attracted because pressure is being reduced. This causes further condensation, cloud development, and precipitation.

In a high-pressure system, with no ground water of any significance nearby, and with winds moving *away* from the system, how can the Sun create cumulus clouds that "grow dramatically" (over 20 kilometers or 12 miles) and "produce heavy precipitation" by merely heating the ground?

"Although **cumulus clouds are associated with fair weather**, they may, under the proper circumstances, **grow dramatically in height**. Once upward movement is triggered, acceleration is powerful, and **clouds with great vertical extent are formed**. As the cumulus enlarges, its top leaves the low height range, and it is called a *cumulus congestus*. Finally, when the cloud becomes **even more towering and rain begins to fall, it becomes a cumulonimbus**.

"*Cumulonimbus* are dark, dense, billowy clouds of considerable vertical extent in the form of huge towers. In its later stages of development, the upper part of a cumulonimbus turns to ice and appears fibrous. Furthermore, the tops of these clouds frequently spread out in the shape of an anvil. Cumulonimbus towers extend from a few hundred meters above the surface upward to 12 kilometers (7 miles) or, on rare occasions, **20 kilometers (12 miles). These huge towers produce heavy precipitation** with accompanying lightning and thunder and occasionally hail." Bib 180 p137

Clearly, there is something lacking with the current cumulus cloud formation theory. In fact, the deeper we look, the greater the connection between *earthquakes and cumulus clouds*.

Researchers know typical fair-weather cumulus clouds take only minutes to form:

"**Fair weather cumulus** have the appearance of floating cotton and have a **lifetime of 5-40 minutes**. Known for their flat bases and distinct outlines, fair weather cumulus exhibit only slight vertical growth, with the cloud tops designating the limit of the **rising air**. Given suitable conditions, however, harmless fair weather cumulus can later develop into towering cumulonimbus clouds associated with powerful thunderstorms." Note 9.3i

Only one type of vaporization occurs during a short 5-40 minute period that accounts for fair-weather cumulus clouds—endovaporization from a hyquatherm.

The Tornado-Cumuliform Cloud Connection

Another evidence of the hyquatherm pressure systems are tornadoes. An extreme form of weather, the tornadoes' formative mechanism continues to elude researchers:

"**Scientists still do not know the exact mechanisms by which most tornadoes form**, and occasional tornadoes still strike without a tornado warning being issued, especially in under-developed countries." Note 9.3j

The key to the discovery of where tornadoes originate lies in the *clouds* in *which* they originate:

"The *Glossary of Meteorology* defines a **tornado** as 'a violently rotating column of air, in contact with the ground, either pendant from a **cumuliform cloud** or underneath a **cumuliform cloud**, and often (but not always) visible as a funnel

cloud...'" Note 9.3k

How do powerful updrafts suddenly form, unless the air was suddenly heated at the surface? Solar heating cannot adequately explain this, but hyquathermal heating of air through rising water vapor can. The earthquake-cumulus cloud connection, unknown in atmospheric science, keeps the tornado forming mechanism a mystery.

Cumuliform clouds with a classic anvil shape generate powerful updrafts that can lead to tornadoes as the wind currents begin rotating:

"**Tornadoes** are commonly associated with a particularly severe type of storm known as a **supercell thunderstorm**. Such storms are usually characterized by **extremely powerful updrafts that sometimes extend to the top of the cloud**, producing a bulge in the **classic anvil shape**, called an overshoot. As the wind speed increases rapidly with height, and the wind changes direction, the **updraft** near the storm's center rotates rapidly—a phenomenon called a wind shear, this rotational characteristic is one of the main forces behind the savage, spinning energy of the tornado." Bib 183 p126

Tornadoes are not the only weather phenomenon related to cumulus clouds, which are in turn, directly related to earthquakes. In the British journal *Nature*, scientists report data they term "*unequivocal*" in identifying a connection between *typhoons and slow quakes*:

"Seismologists installed movement sensors in boreholes at depths of 200-270 metres (650-870 feet) in eastern **Taiwan**, monitoring a spot where two mighty plates, the Philippine Sea Plate and the Eurasian plate, bump and jostle in an oblique, dipping fault. **Over five years, researchers saw a remarkable link between tropical storms and 'slow' earthquakes**, a seismic beast first identified three decades ago. Slow quakes entail a slippage in the fault that unfolds progressively over hours or days, rather than a sudden, violent release of the kind that destroys buildings and lives. **The sensors noted 20 such slow earthquakes, 11 of which coincided with typhoons, during the study period**. The 11 quakes were all stronger and characterized by more complex seismic waveforms than other 'slow' events. '**These data are unequivocal in identifying typhoons as triggers of these slow quakes**. The probability that they coincide by chance is vanishingly small,' said co-author Alan Linde of the Carnegie Institution for Science in the United States." Note 9.3l

The researchers did not prove that typhoons caused earthquakes, but they did provide important evidence showing the strong relationship between the two phenomena. In the next subchapter, evidence will be revealed that heated hyquathermal water that is rising to the surface is the cause of the typhoons.

The further we look, the more evidence we find connecting earthquakes and weather, also confirming the Weather Model and the new weather laws. There is still much research to do, but with our understanding of tornadoes and typhoons, and with

"Scientists still do not know the exact mechanisms by which most tornadoes form."

Why would both these tornados occur in May, during 1981 and 1999 in Oklahoma, USA?

Fig 9.3.7 – Tornado's origins have long been anomalous for modern meteorologists. Why do they seem to occur in the same locations at the same time of year? Why are they associated with cumulus clouds, and why do they have a strong vertical uplift? To answer these questions we need more than modern meteorology can offer—we need the concepts in the new UM Weather Model.

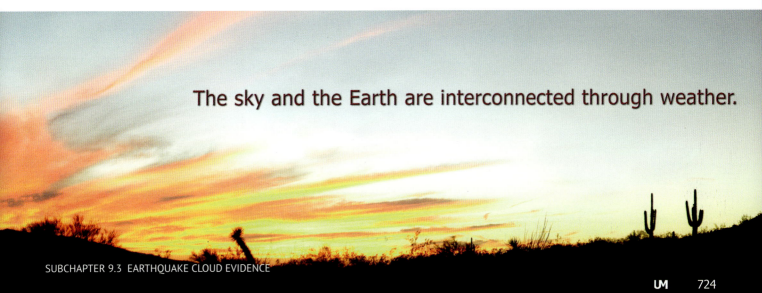

The sky and the Earth are interconnected through weather.

SUBCHAPTER 9.3 EARTHQUAKE CLOUD EVIDENCE

The beauty of weather is magnified

CHAPTER 9 THE WEATHER MODEL

when we understand the Weather Model.

SUBCHAPTER 9.3 EARTHQUAKE CLOUD EVIDENCE

UM 726

the Weather Model, we can look for answers in places never before considered—below ground is a good place to start! The water beneath the crust is like all other natural water found on Earth—teaming with *microbial life*! These microbes form an important relationship with the weather too, just as they did with the geology revealed in the Universal Flood Model.

The Biology-Climate Connection

There is definitely a biological component of weather! In a November 2008 journal, *Proceedings of the National Academy of Science*, researchers report finding a "link between biology and climate":

"The research gives scientists a **first glimpse into the link between biology and climate, and into how the tiny organisms globe-trot with the weather cycle**. The microbes—called ice nucleators—**are found in rain, snow, and hail throughout the world**, according to previous work by Brent Christner, a microbiologist at Louisiana State University. Christner had shown that, at a high enough concentration, these organisms may be efficient drivers for **forming ice in clouds, the first step in forming snow and most rain**. But he hadn't been able to pinpoint their source—until now. In the recent study, Christner and colleagues found that the critters hail from snow, soils, and young plant seedlings in such far-flung sources as Antarctica, Canada's Yukon Territory, and the French Alps. The bacteria may be part of **a constant feedback between these ecosystems and clouds**. 'This is sending ripples through the atmospheric science community,' Christner said. '**This idea would have been viewed as crazy 25 years ago, but these new findings have invigorated research ... in the role that biology may play in atmospheric processes**.'" Note 9.3m

The origin of the microbes that are helping to produce clouds is an important part of the Weather Model. Continuing, the researchers in the previous article said:

"In some places, the nucleators had come **mostly from soil and plant ecosystems**, the results showed." Note 9.3m

Microbes in the *soil* find their way into the atmosphere aboard the water vapor expelled in hyquatherms. In this new field of study, research suggests, "a whole host of other microbes" produce nucleators, and these too, will be found coming from hyquatherms. An important element of the Biology-Climate Connection is the temperature at which the microbes are active:

"Scientists still haven't identified most of the important ice nucleators in the atmosphere. For instance, **a whole host of other microbes**—as well as pollen grains, fungi, and other organisms—may be producing the ice nucleators detected, study author Christner added. **The vast majority of ice nucleators that are active at temperatures higher than 10 degrees Celsius (50 degrees Fahrenheit) have been found to be biological or bacterial**." Note 9.3m

The observation that ice nucleators active above 10 °C are biological is very important. Elevated soil temperatures are the result of hyquatherms, which stimulate microbial growth.

A connection between bacteria in clouds and hyquathermal activity is seen in the *cyclical* nature of *both* environments. Hyquatherms are affected by the astronomical cycles of the Sun and the Moon. These cycles affect cloud-borne bacteria, which influence cycles of rain and drought. Moreover, the bacteria from hyquathermal soil form a "feedback loop" with the bacteria in clouds, showing a direct connection to weather cycles:

"Scientists had already suspected that cloud bacteria may be linked to plants and soils in a 'feedback loop,' a system of exchange between ecosystems... The concept also ties into Sands's ongoing study of the idea that **drought cycles are connected to bacteria in clouds**. For instance, if people overgraze lands, 'these bacteria are without a home ... and can have serious consequences, possibly, for lack of rainfall,' Sands said. **Simply put, a lack of vegetation may lead to a lack of bacteria, which could limit clouds' ability to shed rain**." Note 9.3m

This new, previously unknown factor of soil microbes expelled through hyquatherms has an effect on the weather we are just beginning to understand.

Lunar-Weather Cycle Connection

Hyquatherms carry bacteria grown in heated interstitial water to the surface, releasing them into the atmosphere when the water is vaporized. Some microbes thrive and reproduce in extreme temperatures, which are generated by earthquake-frictional heat processes driven by the gravitational tug of the Sun and Moon. The celestial tug-of-war between the spheres is not the same day by day, but there are predictable cycles, which show up in weather patterns.

Everyone is familiar with *solar-weather* cycles and the seasons that are the result of Earth's tilted axis; a concept easily learned by grade-school children because of its simplicity. But what of the *lunar-weather cycles*, what is known about them? The Moon's size and gravitational pull is so much smaller than the Sun that it seems almost inconsequential, but its nearness to the Earth allows it to exert a gravitational influence nearly twice that of the Sun.

Louis M. Thompson, an Associate Dean and Professor of Agronomy at Iowa State University, compiled a large amount of research data on lunar, weather, and crop cycles, reporting his findings in the September 1988 journal of *Science and Food and Agriculture*:

"An analysis of corn and soybean yields in Illinois and Iowa indicates that there **have been five cycles in corn yields since 1891**. The normalized simulated yields for the five cycles have been superimposed in the accompanying figure. Records of soybean yields do not go back as far as 1891, but the simulated yields have been calculated back to that date. The clusters of years of low yields centered on 1899, 1917, 1936, 1954, and 1973. The years 1905, 1923, 1942, 1960 and 1979 were near the peak years of simulated yields. **The peak yields were 18 to 19 years apart, and the same was true for the minimum yields. The lengths of the weather cycles and the crop yield cycles correspond to the length of the lunar cycle**." Note 9.3n

The Moon does not exert its influence in just one area of the Earth's surface; its gravity affects the whole planet. One area of the crust experiences a *light* tidal pull based on the tilt and position of the Earth relative to the Moon and Sun, while elsewhere, the effects of *high* tidal pull are felt.

Thompson corroborated this in his report, citing observations from Southern Africa:

"**Significantly, the years of drought in the Corn Belt have corresponded to years of high rainfall in southern Africa and vice versa**. Tyson found that the rainfall in southern Africa, like that in the Corn Belt [central USA], **oscillates in cycles of about 18 years**. In a related statistical study, Tyson obtained **no evidence that the observed cycles could have arisen as a** result of chance." Note 9.3n

FQ: Why are lunar weather patterns neglected?

Thompson explains:

"The concept of a **relationship of weather to the 18.6-year lunar cycle** was proposed by Currie. The lunar cycle may be described as the change in angle between the moon's orbit around the earth and the earth's equator… **The problem in connecting lunar cycles with weather cycles and crop yields** is similar to that with solar cycles. **There are only hypotheses and no generally accepted theory to explain the empirical observations.**" Note 9.3o

Like the lunar weather cycle, many "empirical observations" have been swept under the scientific rug throughout the last century because, as Thompson explained:

The observations do **not** fit the theories.

Several climatologists were interviewed in 2006, including the 92-year-old Thompson, and S. Elwynn Taylor from Iowa State University. Although Taylor acknowledged that Thompson was the foremost expert on the 18-19 year crop cycles, he (Taylor) could see no *explanation* for the "empirical observations" and being ignorant of the UM Weather Model concepts, he dismissed the cyclical patterns as being merely "historical":

"It is well-known that the moon influences the tides of the ocean and the tides in the atmosphere, Taylor said, but there is scant scientific evidence about the direct influence of the moon on long-term patterns of rain or drought. There is a possible tie between the moon and the high winds in the atmosphere above the jet stream, Taylor said, but the correlation between the moon's orbit around the Earth and a wet and dry cycle of 18.6 years is **mostly historical**.

"**Tree ring records** going back 800 years in parts of the United States **show that the 18- to 19-year cycle has been amazingly consistent during that long time period**, Taylor said. Flood tides on the ocean also are on an 18.6-year cycle, Taylor said. 'That's well-known by coastal residents,' he said." Note 9.3p

In the upcoming Age Model, the dating technique exhibiting the greatest degree of accuracy is tree ring dating. If the 18-19 year cycle was "amazingly consistent" for 800 years, based on tree-ring growth records from across the United States, why were so few aware of it? Some of the interviewees had not even heard of the Lunar-Weather Cycle Connection:

"Charles Nelson, professor of physics and astronomy at Drake University, said he had never heard of Thompson's lunar weather cycle. However, said Nelson, **there is a cycle of lunar and solar eclipses that repeats itself every 18.6 years, just like the lunar crop cycle suggested by Thompson**. 'There are things we can observe that we cannot explain,' Nelson said." Note 9.3p

Sadly, many modern scientists have been doing just that for over a hundred years—observing many things that they cannot explain because they have been in the Dark Age of Modern Science.

Earthquake Clouds Confirm First and Second Law of Weather

The First Law of Weather states that the Earth's weather is *changed* by hyquatherms. Hyquatherms are the key to changing weather patterns because *other* weather factors remain essentially constant. Variations in solar radiation are minimal, being constant to within 1/1366 during its 11-year cycle, the Earth's rotation, and the amount of water in the Earth's water system remains virtually unchanged. However, changes based on the astronomical cycles of the Earth, Moon, and Sun cause hyquatherms. This is accounted for in the Second Law of Weather, which states that Hyquatherms are *changed* by astronomical cycles.

In this subchapter, earthquake clouds that preceded large earthquakes could be helpful at predicting where and when large earthquakes might occur in the future. Most meteorologists and geophysicists do not yet understand or accept the earthquake-cloud-precursor phenomenon, despite Japanese and Chinese scientist's confirmation of their existence and effectiveness.

With the discovery of hyquatherms and fair-weather cumulus clouds, all clouds can be classed as either evaporative clouds or hyquathermal clouds. Three new types of clouds were adopted, including **Evaporative** Clouds, **Earthtide** Clouds, and **Earthquake** Clouds. There were four different weather 'connections' examined: The Earthquake-Cumulus Cloud Connection explained that most of the clouds that form over continents are cumulus clouds, created in a hyquathermal environment of high-pressure and rising air (opposite modern meteorological theory). The Tornado-Cumuliform Cloud Connection showed that rising hot air from hyquatherms can cause tornadoes and typhoons. The Biology-Climate Connection discussed the effect bacteria from hyquatherms have on cloud formation. The Lunar-Weather Cycle Connection revealed the connection between lunar cycles, recurring weather patterns, and crop yields around the world.

In the next subchapter, we explore how large, global weather patterns contribute to the Weather Model.

9.4 The Global-Weather System Evidence

In the last subchapter, the events that produced earthquake clouds influenced continental weather patterns and pockets of local climate change. In this subchapter, we will explore heated ocean surface waters that affect Global-Weather Systems. One of the most well-known global weather systems is the Pacific Ocean El Niño and La Niña events that recur every several years. Another recurring global weather event is the North Atlantic Oscillation, which peaks every few years, producing large hurricanes in North America.

Changes in these global systems create worldwide flooding and widespread drought around the world. Just as climatologists and geophysicists have long thought the Sun was the sole heat source behind continental weather and high-pressure systems, they see the Sun as the source of heat driving the rise in ocean surface temperatures, but this subchapter will show that hyquatherms affect the oceanic crust in much the same way they do the continental crust. Upwelling mega-plumes of earthquake-friction heated water are a significant contributor to these Global-Weather Systems.

The El Niño-La Niña Weather Condition

El Niño and La Niña weather systems consist of cyclical fluctuations to the *temperatures and heights* of surface water that originate in the tropical Eastern Pacific Ocean. Fig 9.4.1

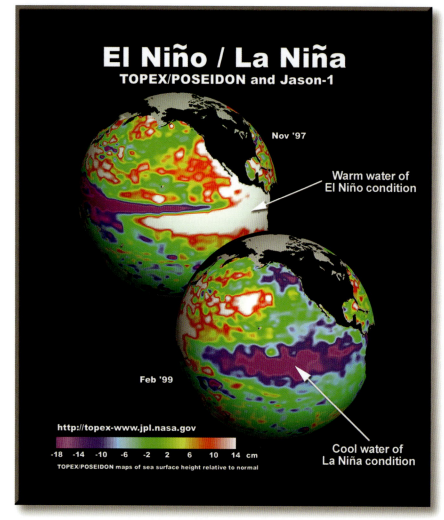

Fig 9.4.1 – Two satellite images showing sea surface height during El Niño and La Niña events as compared with normal elevations reveal hot and cold surface areas of the Pacific Ocean. The white area in the top image represents the 1997 El Niño condition showing high ocean water levels and increased temperature. The purple area in the lower image shows the 1999 La Niña condition, which represents both a depressed ocean surface level and lower than normal temperatures. Such weather conditions cause an imbalance in global weather patterns, often producing large storms or droughts. Modern meteorology struggles to identify the cause of these and other macro-scale global weather conditions.

"During last season's cycle, surface temperatures off the west coast of South America soared from a normal high of 74 degrees [Fahrenheit] to 86 degrees. This vast area of tepid water, **twice the size of the continental United States**, interacted with the atmosphere, creating storms and displacing high-altitude winds.

"El Niño brought rain that flooded the normally dry coastal areas of Ecuador, Chile and Peru. Meanwhile, devastating droughts struck Australia and Indonesia.

"Fires have destroyed some five million acres of Indonesian forest. The drought, along with an economic crisis, left about five million Indonesians in desperate need of food and water. These conditions helped set the stage for riots that led to the downfall of President Suharto.

"Closer to home, more than 30 inches of rain fell on Los Angeles from July 1, 1997, through June 30, 1998—twice the normal amount. San Francisco logged 47-plus inches of rain, almost 27 more than usual. Heavy precipitation devastated crops in many parts of California.

"El Niño also took the rap for severe drought and heat in Texas that raised temperatures in College Station to over 100 degrees for a record 30 days in a row last summer. The heat wave claimed some 125 lives statewide. While Florida experienced a wetter than normal winter, its lush vegetation turned to tinder during the unusually hot and dry summer, fueling wildfires that raged through the state." Note 9.4b

With such a disruption to human activity, understanding the causes and changes to these global-weather systems is critical!

While El Niño turns normal global-weather upside down, La Niña generally exaggerates normal conditions. But can meteorologists predict these global-weather systems?

El Niño-La Niña Weather Predictions

In 1997, during the largest El Niño in history, Michael Glantz worked for the National Center for Atmospheric Research. He noted how little was known at the time:

"'The discrepancy between what we **think we know** about El Niño and what **there is to know** may still be **quite large**.'" Note 9.4c

Four years later, the data evaluating computer model predictions of the 1997 El Niño event was in. Glantz reports the following in his 2001 book, *Currents of Change, Impacts of El Niño and La Niña on Climate and Society*:

"Interestingly, this was *the* most watched El Niño ever, with researchers and forecasters using all means available to determine the state of the Pacific Ocean some months in advance: satellites, buoys, ships of opportunity, and computer models. However, most observers held back on their forecasts until the sea surface temperatures began to warm noticeably. This was due in part to the fact **that the leading (some say flagship)**

contains two NASA satellite images of the Earth showing the Pacific Ocean during the 1997 El Niño and the 1999 La Niña events. The El Niño condition caused floods, droughts and other wide-ranging weather disturbances around the world.

In a *Time* magazine article titled, *Is it El Niño of the century?* the events of August 18, 1997 were chronicled as they happened:

"El Niño generally peaks around December, which is why Peruvian fishermen long ago gave the Christmastime weather visitor a name that in Spanish means 'Christ Child.' If the warming trend continues, scientists say, the incipient El Niño could pump so much heat into the ocean that **average sea-surface temperatures might rise 3.5°C, or 7°F**—and if this happens, the effects would be felt far into the new year. Among the disasters that would be likely to result are **landslides, flash floods, droughts and crop failures**." Note 9.4a

The Peruvian people are well aware how much El Niño can change their lives, but the rest of the world started paying attention only a few decades ago. The events predicted in 1997 were summarized in the November 1998 issue of Reader's Digest:

model for El Niño was projecting that a strong cold event (La Niña) would occur in 1997. It was soon discovered that this model **was in error** and that the ocean was really in a warming phase, heading quickly toward a major El Niño event.

"It seems that El Niño events continue to surprise researchers. Once an event passes, researchers analyze what happened and why, in order to determine why they had missed their forecasts. They then make appropriate adjustments and believe that they have pretty much solved the El Niño puzzle. **Some make excuses for their missed forecasts. Still others resort to spin doctoring: that is, presenting their erroneous projections as having been somehow correct**." Bib 188 Preface second edition

Glantz and another independent researcher conceded that *no model* predicted the strength of the El Niño event before it occurred:

"**No model predicted the strength of the 1997-98 El Niño** until it was already in the process of becoming very strong in the Northern Hemisphere **in the late spring of 1997**." Bib 188 p127

This admission is a telltale sign of the 20th century Scientific Dark Age. Despite having spent vast amounts of research funds, advances in technology, and thousands of investigative hours, oceanographers were unable to predict *when* or at what intensity these global-weather conditions will occur, simply because the wisdom of *why* they occur remains undiscovered. This however, did not stop some from misleading the public by claiming that their "models finally got it right," as Glantz continues:

"While progress continues to be made in observing, modeling, and forecasting various aspects of El Niño, our review showed that advancements in forecasting El Niño onset **have not been as good** as science reporter Kerr (1988) and news releases (NSF, 1998) had noted. **Questionable statements** of success such as, '**The Big Models finally got it right**' and '**El Niño and climate easier to predict than thought,' misled the public and policymakers about the state of the science** and has most likely heightened expectations for improved forecasts of the next El Niño. The sought-after high level of forecasting success **has yet to be achieved**. Statements that praise forecast success also tend to set up the public, media, decision makers, and even El Niño researchers, to be surprised by the next event, having been convinced by previous news headlines that the ability of the scientific community to forecast El Niño's onset has greatly improved." Bib 188 p128

Misleading the public on scientific matters is not a rare occurrence in modern science. One of the problems the public has is that there is nowhere to go to determine *the actual truth* on most major scientific topics. This is a primary goal of Millennial Science, to allow open access to scientific research and data for public review and debate. The Universal Model will usher in an entirely new era of science, where hundreds of new scientific discoveries will allow the public to become more involved in the sciences, and where the use of the Universal Scientific Method and full disclosure will be the norm. This will promote accountability for all publicly funded science programs.

The Ocean's Temperature is the Key

One thing meteorologists have correctly deduced respecting global-weather systems is that a variance from the norm in the ocean surface temperature determines when, where, and how large storms like El Niño are going to occur around the world:

"THE OCEAN'S TEMPERATURE IS THE KEY

"The best way to identify whether an El Niño or a La Niña is developing is to monitor the sea-surface temperature patterns across the tropical Pacific Ocean. During strong El Niños, sea-surface temperatures become unusually warm over the equatorial eastern Pacific Ocean, particularly around coastal Peru." Bib 183 p276

While the rising and falling of Pacific Ocean surface temperatures is a key element and indicator of developing El Niño or La Niña conditions, we are left to answer this FQ:

How were the ocean's surface waters heated?

For the modern meteorologist, the answer to this question *seems* simple enough—it was heated by the Sun. But when we look a little closer, this answer does not work. How did the Sun heat only the warm El Niño water, shown in white in Fig 9.4.2 (Top right of right page, Ocean-Surface Temperature Anomalies), while at the same time *not* heat the cooler La Niña surface waters shown in purple in the same image? The area along the equator receives heat from the Sun equally, and cloud cover did *not* correspond to heated or cooled areas.

The first clue as to how ocean-surface waters became heated comes from the satellite measurements used to create the image. The images in Fig 9.4.1 & 9.4.2 are *not* rendered from infrared or temperature readings, but are produced from data measuring sea surface *height* relative to normal!

Referring to the graph on the bottom left hand corner of Fig 9.4.1, we see that the sea surface experienced an elevation change of -18 to +14 cm of the mean (normal), for a total variation of 32 cm (about one foot). What does the ocean's surface elevation have to do with its temperature? As it turns out, colder water has a lower topographical elevation, whereas warmer than average water is elevated:

"This satellite [TOPEX/Poseidon] uses radar altimeters to bounce radar signals off the ocean surface to obtain precise measurements of the distance between the satellite and the sea surface. This information is combined with high-precision orbital data from the Global Positioning System (GPS) of satellites to produce maps of sea-surface height. Such maps show the topography of the sea surface. **Elevated topography ('hills') indicates warmer-than-average water, whereas areas of low topography ('valleys') indicate cooler-than-normal water.**" Bib 180 p228

Why would the height of the ocean water correspond to its temperature? The only option meteorologists have is wind, but piles of warm water from winds blowing towards the heat are not logical, and it is not supported by the data. In fact, just the opposite is true. As the Origin of Weather subchapter revealed, heated air expands causing winds to move away from the high-pressure system. Warm ocean waters do the same thing; high-pressure systems develop in the ocean, driving water upward, and eventually away. Researchers noted that wind is not necessarily the trigger:

"There are many more westerly wind bursts than El Niños, so it is **clear** these westerly **wind bursts do not always trigger an El Niño**." Note 9.4d

Ocean water expands only a small amount when it is warmed, but the major reason warm water exhibits an increase in elevation can be seen in Fig 9.4.2, bottom right hand corner. Ocean floor hyquatherms are the *cause of the heated water* and as up-

Ocean-Surface Temperature
(Based on Seawater Elevations)

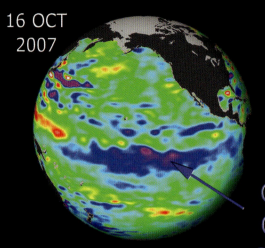

16 OCT 2007

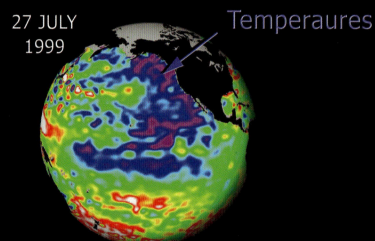

27 JULY 1999

Cooler Ocean Surface Temperaures

Why would the ocean surface temperatures become cool *just* on the equator in October (as seen on the left), and then colder in the summer along the whole Northern Hemisphere, as seen in the July image (lower left), if the Sun was responsible for the temperature change?

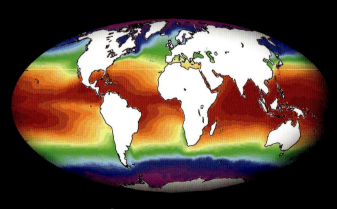

Sea-surface temperature °C

0   5   10   15   20   25   30

Clouds have *not* been shown to be the cause of the cooling. As the world sea-surface temperature map to the left illustrates, the Sun shines equally along the equator heating the ocean surface evenly. Therefore, there must be *another* source for the heating and cooling of these anomalous regions of the ocean.

Anomalies

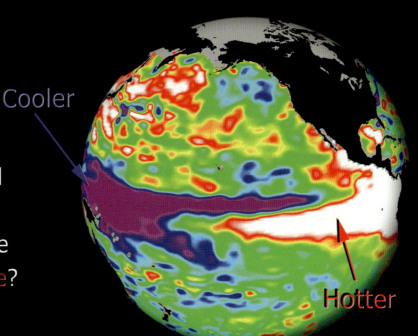

Cooler
Hotter

How could the Sun cool and heat these ocean-surface waters along the equator at the same time?

"El Niño events are not caused by global warming."

However, they are also not caused by the Sun.

Bruce Buckley
Edward J. Hopkins
Richard Whitaker
Bib 183 p277

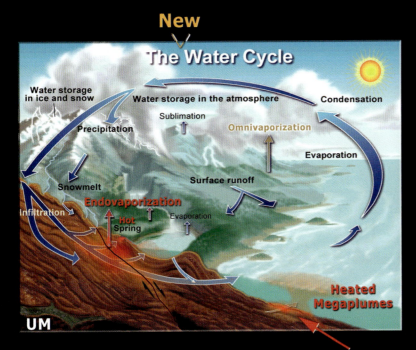

Hyquatherms under the ocean crust create plumes of hot water that rise to the surface and create global weather patterns.

Hyquatherm

Fig 9.4.2 - Ocean-Surface Temperature Anomalies occur because of hyquathermal activity beneath the ocean floor. Megaplumes of water that are heated by earthquake-friction rise toward the surface producing global weather patterns that are cyclical in nature. Images adapted from NASA and USGS.

SUBCHAPTER 9.7 THE AURORA EVIDENCE

welling water rises, it increases the height of the warmed water. Heated ocean water does not readily mix with colder ocean water, thus, as heated, deep water rises (it rises because it is less dense), it comes to the surface in plumes, 'floating' on top of the colder surrounding water, increasing both temperature and topographical surface elevation. As hyquathermal action ebbs and cooler water flows from hydrofountains on the ocean floor. This brings about a period of contraction, lowering surface elevation and temperature. These are conditions indicative of the La Niña cycle.

Notice that this atmospheric textbook describes what happens during the El Niño event, but has no explanation for why the sea level rises:

"Then when the Southern Oscillation [El Niño] occurs, the normal situation just described changes dramatically. **Barometric pressure rises in the Indonesian region**, causing the pressure gradient along the equator to weaken or **even reverse**. As a consequence, the once-steady trade winds diminish and may even change direction. This reversal creates a major change in the equatorial current system, with **warm water flowing eastward**. With time, **water temperatures in the central and eastern Pacific increase and sea level in the region rises**." Bib 180 p220-1

Meteorologists are still unable to explain *why* the sea level rises due to solar heating. Furthermore, researchers cannot explain why the El Niño and the La Niña events shown in Fig 9.4.3 happen based solely on heat from the Sun. These events make sense if a secondary heat source can be identified. The direct evidence for the secondary heat source—the hyquatherms—comes from deep-oceanic research that identified a connection between earthquakes and heated water.

Earthquake-Heated Water Connection

Warm oceanic surface water is the key to global weather change, but we cannot look to the Sun to explain the heating of *specific* areas of the ocean surface. The Sun is no respecter of any particular area of the ocean, heating the ocean's surface relatively evenly during the period of time in which the El Niño system forms. Application of the Self-Evident Principle suggests that the only way to heat a particular area of the ocean is from below.

Thanks to the Gravitational-Friction Law defined in the Magma Pseudotheory Chapter (5.3), we have a real scientific mechanism to explain ocean water heating from beneath the ocean floor. Frictional heating along fault lines correlated with the cycles of nearby celestial bodies, produce periodic temperature increases in the oceanic crust. Because hot water rises, the frictionally heated waters ascend to the surface of the ocean, destined to cause significant changes to weather patterns, such as El Niño.

If El Niño and other similar weather events have their origins in events described by the Gravitational-Friction Law, then measurable increases in ocean water temperature should show a relationship to earthquakes in the area. To see this, we need data showing increased temperatures along the deep-ocean fault lines, corresponding with seismic events and hot water emissions from vents in the vicinity. The problem—this was not possible until advances in technology over the last couple of decades made it so. With new research, the data is now available in support of the UM Weather Model claims.

One of the first locations to be included in new multi-field research was off the coast of California. A series of earthquakes began "migrating" along the Juan de Fuca Ridge under the ocean, causing excitement to percolate among marine scientists:

"Then as the cold war ended in the early 1990s, an invaluable tool suddenly became available to oceanographers: the top-secret listening devices operated by the U.S. Navy in the northern Pacific Ocean to detect foreign submarine traffic. On June 26, 1993, just four days after tuning in directly to the limited-access system, Christopher G. Fox and his colleagues, also at the Newport laboratory of NOAA, **detected a series of earthquakes migrating northward for a distance of nearly thirty miles in less than two days**. The **earthquake swarms** were taking place along a segment of the **Juan de Fuca Ridge**. Fox quickly alerted many of us in the community of **marine scientists**, who had expressed interest in chasing such eruptions in the dark." Note 9.4e

The researchers followed the earthquakes to a "large hot-water plume," a "fresh lava flow" oozing warm water already colonized with "bacterial mats":

"High above the northern end of the earthquake migration path, Thomson found large clouds of turbid water, **distinctly warmer than the usual temperature of the deep ocean**. Embley, also responding quickly, redirected a long planned research cruise to a site beneath the **large hot-water plume**. With the Canadian remotely operat-

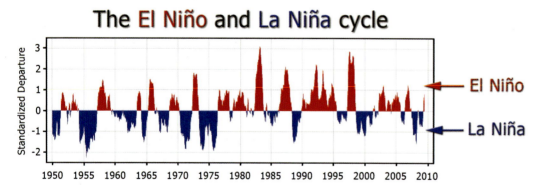

The El Niño and La Niña cycle

Obviously, these cycles do not follow any year-long solar cycle - but are the result of a longer cycle that includes lunar-earthtide heating.

Fig 9.4.3 – El Niño and La Niña cycles from 1950 to 2009 show on the above graph. If the Sun were the cause of El Niño heated surface water conditions, light cloud cover would relate directly to higher surface temperatures (red El Niño events) whereas dense cloud layers would correspond to colder surface temperatures (blue La Niña events). However, there appears to be no relationship. Instead, the cycles of heating and cooling over several *years* demonstrate that another factor is responsible for heating ocean surface waters. That factor is hyquathermal heating of seafloor water. Adapted from NOAA/ESRL/CIRES/CDC – Klaus Wolter (http://www.cdc.noaa.gov/people/klaus.wolter/MEI/).

ed vehicle known as ROPOS (for Remotely Operated Platform for Ocean Science), Embley and his coworkers discovered **a fresh lava flow, still oozing warm water and already colonized with bright-yellow bacterial mats**. Sixteen miles to the south of that flow, Embley and the other ROPOS users found a large, highly fractured zone along the path of the **migrating earthquakes. There, flocculated tufts of bacterial remnants were surging from below the seafloor**." Note 9.4e

This Earthquake-Heated-Water-Connection happened in the early 1990s sparking a completely new field of underwater research. By 1998, ten oceanographers and the submersible, Alvin, recorded the following measurements along the East Pacific Rise crest (East of Ecuador and South of Mexico) in 2,500 meters of water:

"Continuous temperature records and point measurements made by *Alvin*'s thermocouple probe show Bio9 vent fluids were stable for ~15 months at 365 ± 1°C, until March 26, 1995. **On March 26, an abrupt 7°C increase occurred over a period of eight days at this vent, and a maximum temperature of 372 ± 1°C persisted for 14 days**. The vent fluid cooled gradually over ~3.5 months to 366 ± 1°C..." Note 9.4f

The water temperature increase of 7°C (13°F) came from *exactly* the type of event and duration to explain weather patterns like El Niño. The 7°C increase in water temperature on the ocean floor need only cause a *one or two degrees rise* in ocean surface temperatures to affect global weather pattern changes. However, we need to establish an earthquake connection with the increased flow of vent fluids to demonstrate hyquatherm activity on the ocean floor. Researchers saw the changes as suggestive that "a crustal *event* occurred":

"The abrupt temperature increase at Bio9 vent, and coincident changes in faunal community structure, and geochemistry of vent fluids from this area **suggest that a crustal event occurred...**" Note 9.4f

The investigators likely supposed that a crustal "event" occurred just before the "abrupt temperature increase" of the ocean floor vent because it was *not* the first time such an event had been manifested. As seen previously in the Russian borehole and hot spring earthquake evidence (Fig 9.2.4), Pacific Ocean floor seismic instruments recorded the suspected "crustal event" just a few days before—an "earthquake swarm":

"...a small **earthquake swarm which occurred on March 22, 1995**..." Note 9.4f

Earthquake *swarms* produce excessive heat due to continuous friction over time. Heat can accumulate and may dissipate slowly after causing a rapid increase in heated water. The researchers reporting in *Planetary Science and Letters* observed:

"...pronounced temperature spike in March 1995 took **~3.5 months to decay**." Note 9.4f p420

Prolonged temperature spikes on the ocean floor can produce prolonged global weather system changes, like El Niño events. A prolonged *lack* of crustal earthquake events can also cause cooler vent waters, which contributes to cooler surface conditions driving La Niña event. There is no question among scientists that there are vents of heated water deep in the ocean that affect water temperatures, but this is the first time the mechanism behind the heating and cooling of global-weather systems has been attributed to deep-ocean hyquathermal events.

The researchers also found another vent, dubbed the "M vent" that showed a temperature increase from 320°C to 360°C (a 40°C (72°F) change) from 1992 to 1995 Ibid. p426. Changes in temperature of this magnitude across a large area of oceanic floor can cause dramatic shifts in the ocean's water temperature, its biosphere, and global atmospheric weather. Although these and other researchers have noted increases in water temperatures before, such as those at TAG Mound (presented in the Universal Flood Model Fig 8.9.12), the connection between them and the cyclical nature of gravitational-earthquake friction has been veiled by the old dogmagma of a molten Earth.

Seen here is one geophysicist's *theory* explaining the abrupt water temperature increase:

"Lister provided the first overall view of hydrothermal systems at MOR crests which included circulation of hot hydrothermal fluids through the crust **by convection into hot rock as a result of cooling and fracturing of the boundary layer between magma and circulating fluids**." Note 9.4f p427

Although the convection theory has no basis in fact, as the Magma Pseudotheory Chapter demonstrated, it continues to hold sway among many investigators in the geophysical community. Some geophysicists, able to look past the magma-convection theory, found evidence that the hyprethermal fluids did not come from "cooling of lava flows."

But first, let's put some of the El Niño puzzle pieces together that researchers have been unable to do:

"**Figuring out the science of El Niño is not unlike trying to piece together a jigsaw puzzle**. Some pieces are easy to identify—those with the flat edges that form the outside border of the puzzle. Others are identified by the pictures on the puzzle, and they are pieced together forming isolated clusters that by themselves provide a set of autonomous, unconnected pictures. The clusters are **not** yet attachable to each other or to the outside frame." Bib 188 p221

Research clearly shows that heated ocean surface waters evaporate much more rapidly than cooler areas. This creates high-pressure atmospheric systems that dramatically change continental weather conditions. Contrary to popular belief, the Sun is not the primary agent of change to heated surface waters where higher surface elevation demonstrates an increased water volume, but the heated water *came from **somewhere***. Looking *below* the ocean's surface instead of *above* it, we discover the direct connection between earthquakes and heated water on the seafloor. In fact, as noted in several of the previous blue quotes, it took 3.5 months for the heated ocean water to disperse after the 1995 earthquake at the East Pacific Rise. These long periods in which ocean water was heated can be correlated to similar long periods of El Niño events.

There is one more piece of the global-weather puzzle to install:

> FQ: Are large quantities of heated water rising from the ocean floor?

Before answering this question, know that few if any oceanographers have been able to look past the solar heating of El Niño waters and into possibilities that exist *below the surface* of the ocean. Michael McPhaden of NOAA came close, commenting that the "seeds" of El Niño's decay and the onset of La Niña's are to be found below the ocean's surface:

"McPhaden commented that a temperature drop of 8 degrees

Celsius at one buoy location in the central Pacific over a relatively brief 30-day span had not been observed before, making the decay of this El Niño as surprising as its relatively rapid onset. He noted that the **'seeds' for El Niño's decay, and potentially for the onset of a La Niña, were to be found below the ocean's surface"**. Bib 188 p123

The "seeds" he refers to are planted deep in the ocean floor—megaplumes of heated water.

The Hydrothermal Megaplume Evidence

During 1998, researchers in the East Pacific Rise were discovering the connection between earthquakes and heated ocean water, but previously, two researchers from the University of Bristol, UK, reported finding heated "***megaplumes***" rising from the ocean floor. The first observation of a megaplume occurred off the California coast above the Juan De Fuca ridge in 1986. 1,000 meters above the ocean floor, it had a thickness of 700 meters and a diameter of 20 kilometers (11 miles)! For the first time, great quantities of heated ocean water were observed rising from the ocean floor. The researchers reported in the journal, *Nature*:

"Plumes overlying vents of the 'black smoker' type typically rise 100-300 m above the sea floor and are emitted from sources that operate over periods of years; hence they are termed **'chronic plumes'. Recently, transient, detached, hydrothermal plumes have been observed up to 1,000 m above the sea floor.** These features are termed **megaplumes** as they are generated over a period of days **by a heat flux several orders of magnitude greater than that which generates chronic plumes**. Generations of megaplumes have been ascribed to **rapid emptying of a high-temperature (~350°C) hydrothermal reservoir in the crust, of the type underlying black smokers**, or the result of hydrothermal circulation within the oceanic crust initiated by dyke injection." Note 9.4g

Within the paradigm of the Weather Model, it is easy to see the pieces of the weather puzzle fitting snugly together; evidence of rising megaplumes of heated ocean water could certainly affect global-weather systems. But where were the megaplumes coming from? Using accumulated data, the researchers determined the heated water was *not* the result of ocean water's contact with cooling lava flows:

"The hypothesis **that megaplumes are derived from cooling of a lava flow has generally been discarded** on the grounds that; (1) plumes generated from a lava flow would **not** coalesce to form a single plume (that is, they would form a number of discrete plumes rising less far above the sea floor), (2) seawater interaction with a lava flow could **not** generate the **chemical signature observed in megaplumes**, and (3) pillow lavas show fresh surfaces that are **incompatible** with the extensive hydrothermal alteration required to produce megaplume chemical anomalies." Note 9.4g

If the water in the megaplume was not heated by lava flows, how was it heated? It was heated from beneath the surface in aquifers heated by earthquake-friction. The heated water contains the chemical properties of hyprethermal black smoker fluids including anhydrite, chalcopyrite, and silica minerals. Hydrofountains that spew mineral-rich heated water from the ocean floor are responsible, not only for much of the change in global-weather systems, but also for replenishing the nutrients needed by microorganisms in the ocean water that begin the oceanic food chain.

Megaplumes are actually only a small part of a much larger redistribution of the Earth's mass through water movement inside the Earth. When we examine global warming near the end of this chapter, we will discuss changes to the Earth's shape and mass distribution, which is brought about by the movement of water throughout the Earth. As we will see, lunar cycles come into play during the redistribution of underground water and the changing shape of the Earth.

The Sun or 'global warming' is not the only contribution to changing ocean surface temperatures—the *Moon* plays a significant role too. Heated megaplumes of seawater are part of long-term weather cycles that include the changing shape of the Earth.

North Atlantic Oscillation (NAO) Evidence

To understand weather changes we must comprehend how the water cycle works. A key component missing from modern science's version of the water cycle is the *secondary* heat source. Although the Sun is the primary source, hyquatherms on the

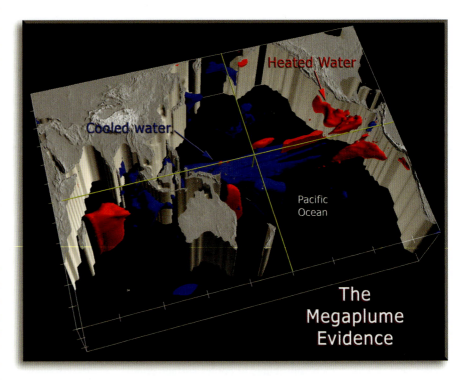

Fig 9.4.4 – This diagram illustrates hot and cold megaplumes of water that exist in the oceans. Only in recent years have plumes of hot water been identified rising from the ocean floor, where earthquake friction heated and released sub-crustal water. The first megaplume of heated water was discovered in 1986 off the coast of California, having a reported diameter of 20 kilometers (11 miles). The megaplume was 700 meters (2300 feet) thick and was 1,000 meters (3,300 feet) above the ocean floor. Megaplumes emanating from the ocean floor help explain why large areas of the ocean surface experience temperature change and how they impact global weather patterns. Megaplumes show that a secondary source of heat from hyquathermal heating resides beneath the ocean floor. Adapted from NOAA.

continents and under the ocean's crust are responsible for many weather changes.

In the Atlantic Ocean, only minor changes to surface temperatures are needed to cause significant atmospheric changes, including the spawning of hurricanes. On the side of the world opposite the Pacific Ocean, events somewhat like El Niño and La Niña systems form in the Atlantic Ocean. This series of high and low pressure systems are known as the North Atlantic Oscillation.

The North Atlantic Oscillation (NAO), or the **Bermuda or Azores High**, seen in Fig 9.4.5, includes high-pressure systems that form over Bermuda and the Azores Islands at different times. Here again, expanding high-pressure systems influence global-weather patterns. Just as we saw with the Pacific El Niño events, the high-pressure systems are brought on by heated ocean surfaces. The NAO system significantly influences weather and climate in Europe and in the Mediterranean area, and it drives the hurricanes that often batter North America. Researchers mistakenly think the Sun is responsible for heating the Atlantic waters:

Fig 9.4.5 – In the Northern Atlantic Ocean, high-pressure systems frequently develop around Bermuda and the Azores Islands. The highs form over heated surface waters that are tied to "deeper waters." When the surface waters warm to at least 28°C (82°F), hurricanes can form. They will move clockwise towards the Caribbean Islands and the Eastern United States.

"Normally, a large area of **high atmospheric pressure** sits firmly entrenched over the Atlantic Ocean. Called the **Bermuda or Azores High** because the center of the high pressure system **shifts back and forth** between the two islands, the system stirs air around the North Atlantic in a giant clockwise circle. **A high-pressure system pushes air down toward the surface, causing it to warm and dry out**. In the warmer, drier air, fewer clouds form to shield the ocean from the sun, allowing water temperatures to rise." Note 9.4h

Once again, high-pressure systems push air *out* in all directions, like a balloon when it is filled; thus, explanations with the Sun as the only heat source cannot be successful at explaining their origin. One "breakthrough" that allowed NASA scientists to devise a better model for the North Atlantic Ocean included data that showed changes come from "the *deeper* ocean temperatures":

"Several years ago **scientists made a breakthrough** when they confirmed through the use of computer models that part of this climatic memory **driving the NAO lies in the deeper ocean temperatures of the Atlantic and changes in these temperatures are largely responsible for variations in the NAO**." Note 9.4i

So, changes to the temperatures of surface waters are responsible for the large hurricanes of the Atlantic, and the "breakthrough" is that *warmer* water from the "*deeper ocean*" is largely responsible for the rise in surface water temperature. Meteorologists have simply been unable to see the link between **crustal heated** seawater and surface water. Some scientists suggest that deep ocean currents bring heated water (and a change in salinity) to the surface, but have yet to identify the source of the heated ocean water.

The Hurricane Evidence

Meteorologists have documented that destructive hurricanes form with only a slight increase in surface seawater temperature. They know that hurricanes are an "engine that runs on heat":

"'A **hurricane** is essentially an **engine that runs on heat**,' says Chris Landsea, a meteorologist at NOAA's Hurricane Research Division in Miami. 'The **warmer the sea-surface temperature** [it must be at least 80°F for a hurricane to start] and the more warm, moist air that's available, **the stronger a hurricane can become**.'" Note 9.4j

Fig 9.4.6 includes a graph of the Atlantic Multidecadal Oscillation (AMO) that is primarily the result of multi-decade changes to ocean surface temperatures. The solar cycle does not exhibit decades of significantly increased or decreased sunlight, but follows only the obvious annual cycle, which includes the Earth's four seasons. Researchers have been unable to identify the cause of the multi-decade variation:

"That debate will continue, but many scientists agree that the present hurricane surge is likely part of a **60-to-70-year cycle** that changes the strength of ocean currents distributing heat around the globe. **Researchers have used tree rings and ice cores to track this variability back hundreds of years**. We're now in a **fast-flowing mode of this up-and-down cycle**, named the Atlantic Multidecadal Oscillation (AMO), during which Atlantic sea-surface temperatures and wind conditions favor hurricane generation. Ten years from now, or perhaps thirty (the timetable is difficult to predict), the cycle should reverse, tending to suppress major hurricanes.

"**Why the variation? 'Frankly, no one can say with 100 percent certainty, but it appears to be a natural effect**,' says Thomas Delworth, a climate modeler at NOAA's Geophysical Fluid Dynamics Laboratory in Princeton, New Jersey." Note 9.4k

The so-called "natural effect" has caused many hurricane researchers to distance themselves from the notion that global warming is an anthropogenic (manmade) condition, but not all. Despite the highly politicized global warming movement, the 30-year *cooling* of Atlantic Ocean surface temperatures from 1963 to 1993 (Fig 9.4.6) cannot be dismissed, leaving investigators asking questions like the following:

"The summer of 2004 seemed like a major wake-up call: an

unprecedented four hurricanes hit Florida, and 10 typhoons made landfall in Japan—four more than the previous record in that region. Daunted, **scientists offered conflicting explanations for the increase in these tropical cyclones and were especially divided about the role of global warming in the upsurge**. Then Mother Nature unleashed a **record-breaking 2005 season** in the North Atlantic, capped by the devastating hurricanes Katrina and Rita. But in 2006, as insurance rates in the southeastern U.S. soared, the number of North Atlantic storms **dropped well below predictions. If global warming was playing a role, why was the season so quiet?**" Note 9.4l

The reason global warming is *not* playing a role in the warming of the Atlantic Ocean surface is the same reason it is not playing a role in the warming of the Pacific Ocean El Niño waters. Meteorologists already know anthropogenic global warming does not cause these events, as recorded in the book *Weather*:

"**El Niño events are not caused by global warming**; a variety of evidence (including archaeological) indicates that the phenomenon has existed for hundreds of years..." Bib 183 p277

The warming events are not caused "by global warming" *or solar heating*. They are caused by a rapid increase of effluent subterranean hot water from cyclical seismic activity.

Gravitationally heated, mineral rich water flows from countless deep water vents; a sudden increase in earthquake activity may stimulate elevated fluid emissions, creating giant megaplumes of rising warm water. Given enough energy, the rising megaplume can pierce oceanic thermoclines on its way to the surface where it can affect the weather.

This subchapter discussed two major Global-Weather Systems driven by increased surface temperatures in the Pacific and Atlantic Oceans. Solar heating cannot explain the irregular and erratic ocean surface heating, so deep-ocean sources were investigated. Although this is a relatively new field of research, several examples of earthquake-heated crustal waters, including rising megaplumes of heated water were documented. While there is still much to learn, the evidence supports the first three principles of the Weather Model. The fourth principle of the Weather Model deals with the Geofield; it is the subject of our next subchapter.

9.5 The Geofield Model

The Earth's weather and the Earth's Geofield are interrelated, connected by Earthtide Heating and the piezoelectric field, which are both created by the constant gravitational tidal movement of the Earth's crust.

The Earth's energy field is commonly referred to by modern science as Earth's "magnetic" field. This was first discussed in the Magma Pseudotheory. In this section, we propose a piezoelectric mechanism for the energy field that surrounds planetary bodies along with two new Planetary Piezofield Laws. Once the preliminary foundation is laid, we can then define the Geofield Model, along with one additional natural law, which is the Law of Weather Parameters. This law demonstrates the relationship between the Earth's energy field and other weather parameters.

Magnetic Field and Dynamo Theory

In subchapter 5.12, in the Magma Pseudotheory Chapter, we learned that modern scientists long thought that the Earth's magnetic field originated from the innards of the molten magmaplanet, in a theory called the dynamo theory. In that chapter, quotations from specialists working on the dynamo theory stated that all tests of the theory proved ambiguous. After reading the UM, the ambiguity is obvious, because there was never any direct evidence that magma itself even exists and, therefore, the dynamo theory is left without foundation.

However, for those not yet familiar with the UM, magma seems very real, and is still the center of all modern theory. One NASA web site explains the most widely accepted theory:

"The **magnetic field** near the Earth is **from a combination of three sources**:

- 97 - 99 % Main Field (From electric currents in the Outer Core)
- 1 - 2 % Crustal Field (From magnetized rock in the Crust)
- 1 - 2 % External Field (From ionized particles above the Earth)" Note 9.5a

One of the puzzling paradoxes for the modern geophysicist is to think that the Earth's magnetic field originates primarily from an extremely hot outer core, because heat *destroys* magnetism.

"Temperature sensitivity varies, but **when a magnet is heated to a temperature known as the Curie point, it loses all of its magnetism**, even after cooling below that temperature." Note 9.5b

The only way to understand this thinking is to realize that there has not been any *other* serious, comprehensive theory for the origin of the Earth's energy field in the history of science (which includes a mechanism for the energy field's production). This subchapter provides just that, including the necessary empirical evidences to support the newly identified mechanism of energy production. This extraordinary claim about the Earth's energy field is tied directly to numerous evidences documented in this and other chapters of this book. This puts one more piece of Nature's puzzle in place, improving the beautiful image of Nature, made possible because of a willingness to question old theories and open the mind to new ideas that are supported by empirical evidence.

Near the beginning of the Magma Pseudotheory chapter, we quoted a planetary scientist who had attended the National Science Foundation sponsored Study of the Earth's Deep Interior workshop. He noted that to understand the latest scientific observations, one must understand Earth's water cycle. Understanding the actual water cycle will help solve many magma problems, and knowing the significance of its role in hydroplanet Earth will help provide answers about the magnetic field:

"It seems likely that we will not understand the origin of **Earth's magnetic field** until we know how the mantle controls heat flow in the core. But we cannot understand the mantle side until we have a better understanding of plate tectonics. This may in turn **depend on understanding Earth's water cycle**." Note 9.5c

This is one primary reason the Earth's magnetic field is still not understood—the ***hydroplanet Earth*** is not yet understood. Having the physical evidence of water inside the Earth, we can begin to understand and explain how the energy field that surrounds the Earth is generated.

Earth's Energy-Field Questions and Name

In subchapter 5.12, Earth's Magnetic Field Pseudotheory, four questions must be answered if a model is to explain successfully the source of the Earth's energy field. These questions were:

1. What causes the field and why does it exist?
2. How does the field instantly fluctuate?
3. What causes the magnetic pole daily cycle?
4. Why do magnetic poles move annually?

Because there is no magma with which to create a 'dynamo,' there must be another mechanism for creating the energy field around planetary bodies, and this mechanism must be able to answer the above questions. A key factor affecting our comprehension of the 'magnetic field' is in its **name**. The word 'magnetic' implies a field generated by a *magnet*, which suggests a smooth, stable field. However, the field surrounding celestial bodies is neither 'smooth' nor 'stable,' which is one reason they are misunderstood.

Instead of a magnetic field, this subchapter will show that the Earth has an **electric field** surrounding it. This important distinction will allow us to learn *how* the *electric field* is generated.

The quest to uncover the origin of the Earth's energy field has not been a simple matter. In 1905, Einstein said of the problem surrounding the origin of the Earth's magnetic field:

"…one of the **most important unsolved problems** in physics." Bib 36 p17

From Magnetic Field to Piezofield

Fig 9.5.1 illustrates the Earth's field as it has long been described: a "magnetic" field similar to the field of a bar magnet, with a dipole configuration. The Earth's energy field has been associated with a magnet for decades, but there has never been a clear explanation as to where the field originates or why it fluctuates the way it does. Without magma, modern science has no other origin for this strange energy field.

By studying the Earth's energy field, two observations show up immediately that do *not* fit the 'magma-origin field':

1. The field is strongest at the surface.

2. Multi-lobed fields exist at the surface.

Why would the energy field, that supposedly comes from the core, be *stronger* at the surface? And why are there *multiple* polarity incidences at the Earth's surface? With no clear explanations, modern geology may be tempted to avoid the questions, but the answers are critical in identifying the source of the Earth's energy field. The mechanism must account for these two observations and must answer the four previous questions concerning the Earth's energy field.

Actually, the new energy-field mechanism was first discussed in subchapter 5.8; it involves the **piezoelectric effect**. Silica, or quartz rocks in the Earth's crust have piezoelectric properties. This means that electricity is being produced in the rocks of the crust by mechanical pressure. The quartz-based rocks in the Earth's crust could not have been melted nor could they have been heated above the Curie Point of 570° C or the piezoelectric property of the rocks would have been all but destroyed. We know that quartz-based rocks were never melted because they have strong piezoelectric properties, which means they were never a part of a magmatic-melt. One surprising property of the piezoelectric effect is its capacity to produce high voltage:

"For example, a 1 cm^3 cube of quartz with 2 kN (500 lbf) of correctly applied force can produce a **voltage of 12,500 V**." Note 9.5d

The *majority* of the Earth's crust consists of quartz-based rocks, which are capable of producing electricity when stressed, and it is important to investigate this energy-field mechanism.

The Piezofield Mechanism

We can now ask a very simple but profound FQ:

If the Earth's crust consists primarily of quartz-based piezoelectric rocks that are subjected to constant stresses induced by the tidal forces of the Sun and Moon, what would the stressed piezoelectric rocks produce?

The answer is a piezoelectric field!

The igniter button on a common household barbecue in Fig 9.5.2 is an example of a device that uses piezoelectricity. When the igniter button is depressed, it squeezes quartz material, creating a spark to light the grill. If you ever hold the end of the wire that comes from the piezoelectric igniter when it's pushed, you will never forget the shock! It generates quite a jolt, illustrating just how powerful a small piezoelectric device (less than an inch long) can be.

Fig 9.5.3 shows a thin piezoelectric film, which consists of a very thin layer of quartz between plastic with two leads connecting a light bulb. By simply flexing the thin film, piezoelectricity is generated, lighting the bulb. This is an easy demonstration to show the important piezoelectric property of rocks and minerals.

The piezoelectric effect is expanded in the experiment shown in Fig 9.5.4. The Piezorock Experiment involves applying and

The relationship between magma and the Earth's magnetic field, have never made sense because heat destroys magnetic fields.

CHAPTER 9 THE WEATHER MODEL

releasing pressure to various quartz-based materials with a C-clamp and measuring the current produced with a voltmeter.

The simple Piezorock Experiment shows that many of the Earth's crustal rocks are capable of producing an electrical current and a corresponding energy field when they are compressed and released. The experiment helps one to visualize how minerals produce electricity and a corresponding energy field after being subjected to constant stress by daily earthtide. This happens because the Earth's crust is being raised and lowered by about two feet (.6 meters) at the equator each day, as the Earth spins on its axis. Furthermore, the continual compress-and-release cycle in the crust is magnified because of lunar and solar influences.

Returning to the two important observations about the Earth's energy field, the reason the field is the strongest near the surface is *because the greatest field is being generated* **at fault lines in the crust** (which we will explore shortly). The multi-lobed electrical fields of the Earth can be explained by the geographic location of the continents. The multi-lobed poles of the Earth's energy field are generated by the *multiple continents*. The strength of each depends on the continent's size, mineral makeup, and distance from the equator. The closer the landmass is to the equator, the greater the earthtide movement, which translates to greater electrical current generation and stronger energy fields. Ultimately, the energy from multiple poles blends together to form a single dipole—*but* this is only when the electric field is viewed from a distance. When viewed or measured on the surface (where the field is the strongest), the Earth's piezoelectric field contains several distinctively separate poles. The multi-pole piezoelectric field of the Earth is called the **Geofield**. It is illustrated in Fig 9.5.5. The Geofield Map in Fig 9.5.5 shows the total measured intensity of the Earth's primary energy field as measured by NOAA in 2005. The various lines of strength indicate circular high and low fields of electrical strength, in much the *same way* that high and low atmospheric pressure maps are drawn.

FQ: Are the electric fields and the Earth's atmospheric pressure systems created by the **same mechanism**?

Fig 9.5.2 – The igniter button on this barbecue mechanically stresses quartz material, generating a piezoelectric spark, which lights the gas grill. Piezoelectric crustal rocks are also mechanically stressed through the constant daily motion of earthtide, which creates the Earth's electric field.

Fig 9.5.1 - This illustration is a typical rendition of Earth's 'magnetic' field as taught in modern astronomy and geology. However, it only portrays the outer field, which exhibits a dipole (two poles—one north and one south) field far from the Earth's surface. Closer to the surface of the Earth, the field exhibits multiple poles. The erratic non-dipole lines are typically left out of illustrations because they do not conform to the well-established 'magnetic' field. The bar magnet in the inset image does not have multiple poles as the Earth does.

The UM-created Geofield diagram on the right in Fig 9.5.5 illustrates the multiple paths of energy that the Geofield forms as it surrounds the Earth. Although the Geofield Map on the left has been in existence for many decades, no artistic rendering could be found. Instead, only the typical dipole diagram showing the simple north/south pole orientation is depicted, which misrepresents the true multi-lobed nature of the Earth's energy field. This shows how powerful dogmagma is at perpetuating the theorized notions of a magnetic 'iron' core with the magnetic field generation being from a supposed 'magma dynamo.'

The Two Planetary Piezofield Laws

The newly recognized Piezofield Mechanism of the Geofield led to the identification of two new **Planetary Piezofield Laws.** These two new natural laws explain what Einstein and modern science recognized as being "one of the most important unsolved problems in physics." In the next subchapter, we will have the opportunity to use the Universal Scientific Method to test these laws by observing the evidence directly and by evaluating data accumulated through years of observing the Earth's energy field and crustal movements.

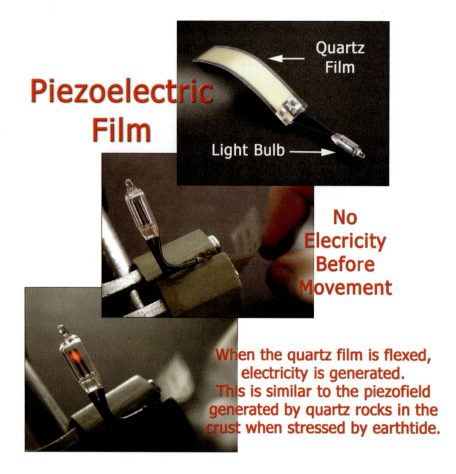

Piezoelectric Film

No Elecricity Before Movement

When the quartz film is flexed, electricity is generated. This is similar to the piezofield generated by quartz rocks in the crust when stressed by earthtide.

Fig 9.5.3 – Piezoelectric film is made with quartz. It is used in many electronic applications. The film produces electricity when it is mechanically stressed. A light bulb is connected to the piezoelectric film to show the flow of electricity. By flexing the film, electricity is generated, turning on the light. The piezoelectric mechanism is the basis of both Planetary Piezofield Laws, which will be presented shortly.

Magnetic Field: loosely-held atoms arranged in a crystalline lattice around an object according to **strong** lines of attraction.

Gravity Field: loosely-held atoms arranged in a crystalline lattice around all objects according to **weak** lines of attraction.

Electric or Energy Field: a continual transfer of energy through matter via the Domino Effect.

Piezofield: an energy field produced by the application of stress on piezoelectric minerals.

Geofield: the Earth's energy field or piezofield.

Energy Fields Defined

The Weather Model has new definitions for the various energy fields. They are presented here so they can be incorporated into the new weather paradigm. We will expand the new definitions as scientific evidence verifying the origin of the Geofield is introduced.

In the Energy-Matter Model (Chapter 22) we will define both energy and matter. Matter is all things in the Universe, whereas energy is accelerated matter. We also will discuss the dual nature of energy-matter: all matter has energy—from the atomic-scale where electrons are in motion around the nucleus to the Universe-scale, where matter is in motion around the center of the Universe.

The following *fields* of matter around an object are invisible to the naked eye. The Energy-Matter Model discusses the first three fields, but the other two are specific to this chapter.

The Geofield Model

The energy source of Earth's piezofield comes from the earthtide-induced movement and stress on piezoelectric minerals in the crust. Like the back-and-forth movement of the piezoelectric film in Fig 9.5.3, the constant gravitational stretching and compressing of the Earth's crust generates tremendous electricity and an accompanying electric field that surrounds our globe. We call this energy field the **Geofield**.

In the Geofield Model, the Piezofield Mechanism describes the origin of the Earth's energy field, its propagation, strength, and cycles. The following four principles based on the two previously stated Planetary Piezofield Laws define the **Geofield Model**:

1. The Earth's energy field, or 'Geofield,' is **generated** by piezoelectricity in the crust.

2. The Geofield is **propagated** and **controlled** by the gravitational tidal forces of the Sun and Moon.

3. The **strength** of the Geofield at any one location is determined by the makeup of the surrounding piezoelectric material and the magnitude of the tidal forces acting on that material.

4. Diurnal, annual, and millennial Geofield cycles are **controlled** by the astronomical cycles of the solar system and the Universe.

Fig 9.5.4 - (Next page) The Piezorock Experiment setup involves applying and releasing pressure to quartz-based rocks with a C-clamp (or a vise) and measuring the electrical current with a voltmeter. The Piezorock experiment shows that all quartz-based rocks, and also glass, produced some electricity through the application and release of pressure from the C-clamp. This simple experiment illustrates that most rocks in the Earth's crust are capable of producing electricity from the movement of earthtide, movement that is the result of the gravitational influence of the Moon and Sun. Anyone can perform this experiment. Copper sheets are cut slightly larger than the clamp ends. A piece of rubber acts as an insulator between the copper and the clamp. A rock is placed between the two pieces of copper and electrodes are connected to the copper. The voltage meter is set to the DC setting ≈ 2000 mV. Snug the clamp, then quickly tighten the clamp, watching the voltage meter for changes. Tighten and release the clamp quickly, about a ¼ turn 10 times, recording the highest voltage meter readings. Average the numbers for each rock sample. A control test can be made by pressing the copper sheets together to be sure no electricity is coming from them. Touch the electrodes between each measurement to be sure there is a zero reading before each test. Rock samples work best if they have flat surfaces against which the copper can press. Analyze the piezoelectric data and compare different types of minerals from the Earth's crust with man-made substances.

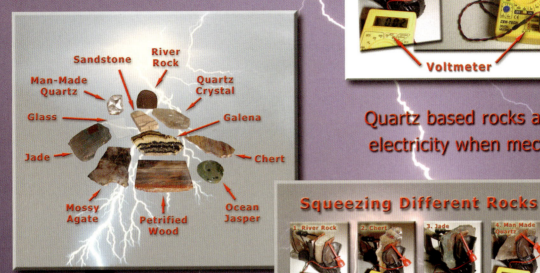

Piezorock Experiment

Experiment Setup
Quartz Crystal, C-Clamp, Electrodes, Copper Sheets, Voltmeter, Rubber Insulation

Quartz based rocks are shown to create electricity when mechanically stressed.

Rocks tested: Sandstone, River Rock, Man-Made Quartz, Quartz Crystal, Glass, Galena, Jade, Chert, Mossy Agate, Petrified Wood, Ocean Jasper

Squeezing Different Rocks
1. River Rock, 2. Chert, 3. Jade, 4. Man Made Quartz, 5. Mossy Agate, 6. Sandstone, 7. Ocean Jasper, 8. Petrified Wood, 9. Man-Made Quartz Cube, 10. Glass, 11. Galena, 12. Copper Control

Piezoelectricity from the earthtide creates the geofield.

Piezorock Experiment Results

Order	Rock Type	Average per 1/4 turn
1	River Rock	10.8 mV
2	Chert	9.7 mV
3	Jade	7.8 mV
4	Quartz	6.4 mV
5	Moss Agate	6.0 mV
6	Sandstone	5.6 mV
7	Ocean Jasper	5.5 mV
8	Petrified Wood	4.9 mV
9	Quartz Cube	4.3 mV
10	Glass	2.6 mV
11	Gallena	0.0 mV
12	Copper (control)	0.0 mV

Easy enough for gradeschool students to demonstrate.

SUBCHAPTER 9.5 THE GEOFIELD MODEL Fig 9.5.4

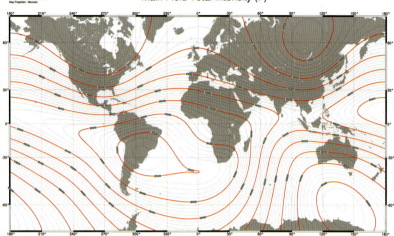
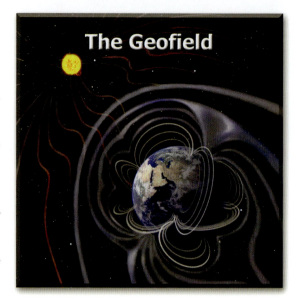

Fig 9.5.5 – The Geofield Map (left) shows measured energy field intensity lines around the globe. These were used to model the Geofield illustrated on the right. The Geofield illustration on the right is different from the traditional magnetic field diagram (seen in Fig 9.5.1 and in most textbooks). This illustration shows the north and south dipole far from the Earth's surface, and shows the multiple poles that appear on the continents as lunar and solar tides tug and release the Earth's crust, squeezing the minerals and creating a piezoelectric field. The concentric areas on the Geofield Map represent high and low areas of the energy field that are similar to the highs and lows of atmospheric pressure systems—which are also generated by earthtide. This is the first time the multiple fields of the Geofield have been explained with a clear mechanism for their origin. As we explore the evidences of the Geofield in the next subchapter, we will see how this piece of Nature's Puzzle fits with the other pieces of the Earth science puzzle previously revealed in the UM.

These four Geofield Model principles briefly answer the questions about the Earth's energy-field asked at the beginning of this subchapter. Detailed physical evidence will be required to firmly establish the veracity of this model, and the next subchapter, The Geofield Evidence outlines many such details. Following subchapters provide additional empirical evidences, each contributing to the new emerging picture of Nature's weather and the Geofield.

Meteorology and the Geofield

The Geofield and weather are directly related because of earthtide, but modern meteorology sees no connection "between magnetic field fluctuations and any weather parameter." We asked a scientist at the Atmospheric Research Section of the Argonne National Laboratory if there was any known relationship between the Earth's magnetic field and common weather variables such as temperature, pressure, humidity, oxygen, or carbon dioxide levels, etc. His response was simple and to the point:

"I know of **no connections between magnetic field fluctuations and any weather parameter**." Note 9.5e

Scientists at the National Oceanic and Atmospheric Administration (NOAA) were asked the same question; their reply was the same—no known relationships. One loosely connected phenomenon is the aurora, the shimmering bands of luminescence in the northern or southern regions of the Earth. Although meteorologists acknowledge that the aurora lights are connected to the Earth's energy field, the phenomenon is generally viewed as being astronomical in nature and not generally associated with weather. We'll investigate auroras following the Geofield Evidence; they are an interesting piece of nature's weather puzzle.

We know intuitively that weather changes are brought on by the seasons, which are essentially astronomical cycles. But the

The First Planetary Piezofield Law
The energy fields around planetary bodies are **created** by piezoelectricity in the body's crust.

The Second Planetary Piezofield Law
A planetary body's piezofield is **controlled** by the makeup of the piezoelectric materials in the crust and the astronomical tidal forces acting on the body.

connection between weather and Earth's energy field has continued to elude researchers—until now.

The Law of Weather Parameters (PG≈TH)

Chapter 23, the Air-Water Model will detail new gas and water laws that affect our understanding of the weather. Those laws will show that gas and water molecules are not merely moving about randomly, they are organized into an ordered crystalline matrix.

The order of the air-water crystal that will be described in the Air-Water Model provides the foundation of the Law of Weather Parameters. The discoveries in the Air-Water Model and the piezoelectric nature of the Geofield make it possible to describe the relationship between atmospheric pressure (P), the *Geofield* (G), temperature (T), and humidity (H). These relationships are defined by a new law called the **Law of Weather Parameters.**

The Law of Weather Parameters
Pressure and the Geofield are directly proportional to the temperature and humidity of air in a controlled sealed environment (PG≈TH).

With this new law and the previous weather laws, we have a completely new paradigm with which to view and evaluate weather conditions. The Earth's energy field is closely associated with weather in that Geofield measurements now have specific meaning.

Fig 9.5.6 shows the results of two weather experiments—a 24-hour weather jar and a 15-day outdoor weather experiment. Weather jar gas measurements will also be shown in the Air-Water Model but here include a new weather parameter—the Geofield. The four top graphs represent the outcome of a 24-hour test period, revealing that pressures and temperatures in the jar share a strong relationship. Also visually apparent is the proportionately linked Geofield and humidity. Moreover, the two upper graphs (pressure and temperature) are inversely proportional to the two lower graphs (Geofield and humidity), meaning they reflect movement in opposing directions.

The four bottom graphs show the results of a 15-day Outdoor Weather Experiment, which consisted of simply recording the ambient weather parameters of pressure, temperature, the Geofield, and humidity during a hot Arizona July. With the consistently low humidity and little weather activity associated with hot summer days in Arizona, the daily up-and-down cycle of temperature and pressure are remarkably consistent. The Geofield and humidity readings were also closely linked in their daily cycle. Cloud activity on the second and eighth days lowered the temperatures and affected the other weather parameters too. Most importantly, the same inverse relationship witnessed in the 24-Hour experiment is visible here; pressure and temperature are inversely proportional to the Geofield and humidity.

Surprisingly, we didn't see *simple* experiments of this type in the textbooks. The Law of Weather Parameters is a quantitative, easy-to-see way of understanding the relationship between weather parameters. Starting with relatively dry air in a Weather Jar, we can have complete control of the gases associated with weather so that we can study their behavior in controlled situations, allowing us to observe how they interact. This is analogous to one of the simplest examples of real weather—consistently hot days with little moisture in the atmosphere. When so little water is involved in the experiment, consistent results are easily obtained that show the relationship between the four weather parameters. However, we will be going into more detail in the Air-Water Model chapter, when water becomes a large part of the air-water crystal, the water significantly changes the behavior of air, which also changes the weather.

There are several factors involved with the weather parameters, but all are centered on *water*. First, a large quantity of water can be absorbed into the air-water crystal when the air temperature is increased. This is similar to water's ability to dissolve more salt or sugar as its temperature is increased. The invisible water vapor in the air is responsible for much of the storminess encountered as temperatures lower and pressure systems develop. One reason afternoon-evening showers are prevalent is that the setting Sun results in lower temperatures, which increases the humidity or water vapor that comes out of the air-water crystal. When enough of the water vapor escapes the air-water crystal, saturation occurs and clouds form, which may result in the precipitation of rain or snow. The large amount of water vapor in the air-water crystal expands, causing the volume of one liquid unit to become 1,700 gas units, which is the power behind the weather. The water vapor in the air contracts when the temperature decreases, becoming liquid again, which causes a sudden drop in atmospheric pressure.

Another factor capable of producing abrupt changes in weather parameters is the sudden *drop in pressure*. Areas of low-pressure draw in high-pressure systems, which can contribute moisture-laden air to a building storm system. When this occurs, it can have the effect of lowering pressure even further, resulting in excessively high winds at the high-pressure-low-pressure transition zone. High winds can be destructive and dangerous, so understanding where the winds originate is an important goal of meteorology.

High winds are a weather factor best understood if we know the original cause of the high and low-pressure systems that caused the high winds. In *global* weather patterns, subocean hyquatherms heat seawater that warms vast areas of the ocean's surface, which in turn creates zones of high-pressure that ultimately affect the continents. On the continents, small-scale hyquatherms produce bands of cumulus clouds that can collectively add to or change *local* weather.

One important aspect affecting weather parameters, but often overlooked, is the element of *time*. Weather is all about atmospheric water, but continental water evaporation is a slow process and cannot account for the fast-moving storms responsible for heavy flooding and wind damage. Rapidly developing storms are spawned by endovaporization processes occurring in crustal hyquatherms as underground water is vaporized.

In summary, this subchapter contrasted modern science's version of the Earth's magnetic field with the Geofield of the Universal Model. Two Planetary Piezofield Laws were introduced along with the Law of Weather Parameters. These laws broaden our understanding of the cause-and-effect processes that drive the Earth's weather systems. With that, we are poised for a quick segue into the scientific evidence supporting the Geofield Model.

9.6 The Geofield Evidence

The Weather Model, like all other UM models, relies upon the Universal Scientific Method to establish its credibility. This subchapter contains a number of Geofield evidences, supporting and substantiating the origin of weather already presented. The evidences confirm the First and Second Piezofield Laws and help explain the natural processes that affect weather. In so doing, we are better able to comprehend how weather works.

We will also discover why lightning is still such a mystery in modern science, and that fair-weather conditions originating from hyquatherms actually promote lightning activity. The Geofield-lightning connection, along with several little-known research projects from around the world show that earthquakes and the Geofield are closely connected. In fact, there are many phenomena sharing a direct connection to the Geofield, such as the ancient art of dowsing and previously unexplained Geofield cycles. We will explore why the Geofield has been so mysterious, and what impact the Geofield model will have on the scientific community.

Overloading the Geofield

There are at least three ways piezoelectricity produces or affects the Geofield. The first is the diurnal earthtide movement, which we have already discussed. The second is more subtle, involving a "thermally stimulated current." Piezoelectric rocks can be mechanically stressed by low temperature heating. A small rise in temperature causes the rock to *expand*, contributing a minor amount of electrical current:

"A **thermally stimulated current was observed for all the various rock samples** examined, and reached **a maximum around room temperature**." Note 9.6a

As previously mentioned, high temperatures dampen or destroy the piezoelectric current, but relatively minor temperature fluctuations caused by the thermal expansion and contraction of surface rocks contributes measurably to the Geofield. These two methods (earthtide and thermal expansion) of piezofield generation are *slow* and constant. But there is another way in which piezoelectricity impacts the Geofield, albeit much more rapidly. It happens with a *quick* charge or discharge of current. It is on this dramatically intense method of piezoelectric generation that we direct our attention on the following several pages.

Two scenarios will help illustrate the concept: Imagine running a large motor on a long, light-gauge extension cord. The cord will become hot because of *overloading*, meaning that there is too much electrical amperage being drawn through it. Additionally, hold two wires charged with electrical current close together and the electricity will discharge, arcing between the wires, jumping through the air with an accompanying bright flash.

As regions of the Earth's energy field are charged up by the constant movement of the daily earthtide and thermal expansion of the Earth's piezoelectric crust, it is but a matter of time before the electrical buildup reaches capacity. When that happens, there is an electrical overload and an electrical discharge. This discharge produces the hottest and brightest known natural flash of light that humans observe on Earth, and it occurs over 4 million times a day! It is known as **lightning**.

One primary reason the true origin of the Earth's Geofield remains mysterious is that the true *origin of lightning* remains a mystery. They are both a part of the geotidal electrical generator that keeps our planet bathed in electrical energy.

> "The true origin of lightning remains a mystery."
>
> Astrophysicist strikes blow to lightning theory,
> Nature, Science Update, 17 Nov 2003

"Lightning Remains a Mystery"

The 17 November 2003 edition of *Nature Science Update* outlines the most common lightning sci-bi:

"Most scientists **believe lightning is generated** when a giant electric field builds up **in the atmosphere**. **Although no one has actually seen such a field**, researchers **assumed** that this was simply because they hadn't looked hard enough." Note 9.6b

Because of assumption, the Question Principle is dismissed, and the origin of lightning is not questioned with an open mind. The *Nature* article continued explaining how new research showed the failings of old theories:

"**The conventional view of how lightning is produced is wrong**, according to a Florida-based physicist. **Electrical fields in the atmosphere simply cannot grow large enough to trigger lightning**, calculates Joseph Dwyer of the Florida Institute of Technology in Melbourne. '**This means back to the drawing board**,' he says." Note 9.6c

The biggest problem facing most researchers is that the drawing board they return to is not empty; it is littered with an excess of pseudotheories. Thinking outside the box is one of the most difficult things for anyone to do, especially after being steeped in tradition. But here is where the UM is different; it is able to ignore age-old pseudotheories because it is built on new natural laws and models that are backed by observable, empirical evidence confirmed within the scope of the project.

Joseph Dwyer authored a *Scientific American* article in 2005 titled, *A Bolt Out of the Blue*, that essentially explains why meteorology does not comprehend how lightning is created. Because lightning is seen so often with clouds, it is assumed that it is created in the clouds. But Dwyer's comment about lightning out of a *cloudless* blue sky is telling:

"**Lightning** is a particularly unsettling product of **bad weather**. It causes more deaths and injuries in the U.S. than either hurricanes or tornadoes do, and it strikes without warning, **sometimes with nothing but blue sky overhead**." Note 9.6d

What do we learn from this? The important point:

> Lightning is **not** generated **from** clouds.

It is understandable why researchers are confused. After all, lightning is usually *associated* with clouds, but lightning does *not* originate *from* clouds. In fact, satellites have documented millions of lightning flashes each day on Earth, and they have been seen on other planets too:

"Worldwide, lightning flashes about **four million times a day**, and bolts have even been observed **on other planets**. Yet despite its familiarity, **we still do not know what causes lightning**." Note 9.6d

Although, investigators admit not knowing the cause of lightning, Dwyer gives us another important clue:

> Lightning occurs on planets that do not have an atmosphere like Earth's; therefore, the **cause** of lightning is not linked solely to the Earth's unique atmosphere.

Dwyer's bolt of lightning out of a cloudless blue sky along with other anomalies left them wondering what they missed:

"These difficulties have led many researchers in the field, including me, to **wonder if we have missed something important**." Note 9.6d p66

"Difficulties" arise in part because of the *specialization* most researchers follow. Dwyer is an *astrophysicist*, which means his focus is generally on the world from above—yet the answer to lightning does not lie there—it comes from below, from the *geophysical* world. There should be the research that reveals the cause of lightning and the origin of the Geofield.

To crack the lightning mystery, we must be founded upon natural laws and observations. Some of these were established in the Magma Pseudotheory chapter. The Gravitational-Fric-

Law of Weather Parameters

24-Hour Weather Jar Experiment

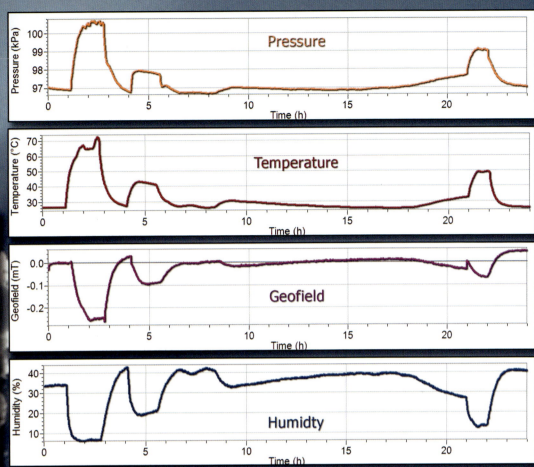

Fig 9.5.6

15-Day Outdoor Weather Experiment

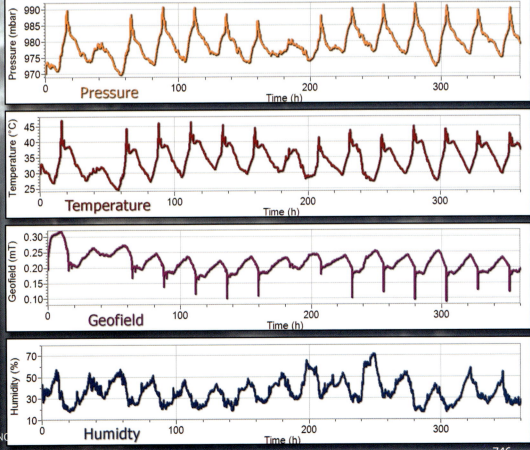

tion Law demonstrated that the gravitational influence exerted on the crust by other celestial bodies generated frictional heating. The process, called earthtide, is not limited to the production of frictional heating; it is the source of the piezoelectric energy surrounding the Earth, and it is the source of the electrical buildup in the atmosphere and in the crust, which is the primary cause of lightning. To explore this new lightning theory, we begin by examining the conditions present during electrification of the atmosphere.

The Fair-Weather Condition Evidence

Evidence of the Geofield's generation through electrical activity in the atmosphere was recorded as far back as the 1700s. But with James Hutton's magmaplanet pseudotheory coming into favor during the 1800s, the Earth's energy field became hopelessly entangled in the Magma Pseudotheory.

With the limited technology of the 1700s, researchers identified what became known as the "Fair Weather Condition." We first saw the fair weather concept in the Earthquake Cloud Evidence subchapter, where it was tied directly to high-pressure systems and cumulus earthquake clouds. The encyclopedic definition of **Fair weather condition** is:

"**Fair weather condition** concerns the **electric field** and the electric current **in the air** as well as the conductivity of the atmosphere."
Note 9.6e

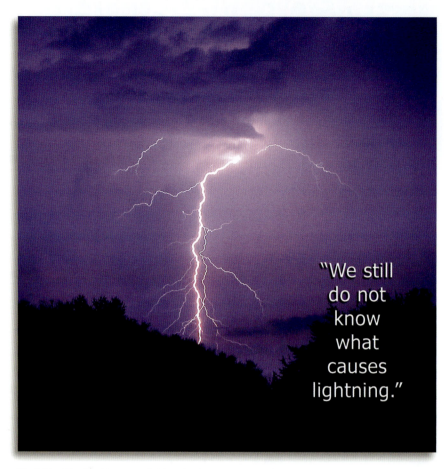

"We still do not know what causes lightning."

Fig 9.6.1 – The Earth's energy field and lightning are inseparably connected, but how? The Piezofield and Geofield Models remove the "mystery" surrounding lightning by demonstrating how lightning is generated and why it is usually, but not always associated with clouds and moisture.

The Earth's electric field was once investigated by Ben Franklin, famously in 1752 by flying a kite in a lightning storm. About the same time, L.G. Lemonnier noticed that there was a daily (diurnal) cycle in the energy field, along with periods of positive electric polarity during "fair weather conditions." So the question is—when do fair weather conditions exist?

> Fair weather exists during high-pressure development, which comes from hyquatherms.

So, the changes in weather (from developing high-pressure systems) and the Geofield (from piezoelectric earthtide forces) are *both* produced by the *same* gravitational forces. H.B. Saussur (1779) also noted that the Earth's electric field and the hyquathermal Fair-Weather Conditions occurred in *cycles*:

"Around June of 1752, Ben Franklin reportedly performed his famous kite experiment. L. G. Lemonnier (1752) reproduced Franklin's experiment with an aerial, but substituted the ground wire with some dust particles (testing attraction). He went on to document the *fair weather condition*, **the clear-day electrification of the atmosphere, and the diurnal variation of the atmosphere's electricity**. G. Beccaria (1775) confirmed Lemonnier's diurnal variation data and determined that the atmosphere's charge polarity was **positive in fair weather**. H. B. Saussure (1779) recorded data relating to a conductor's induced charge in the atmosphere. Saussure's instrument (which contained two small spheres suspended in parallel with two thin wires) was a precursor to the electrometer. **Saussure found that the fair weather condition had an annual variation**."
Note 9.6e

The "annual variation" comes, of course, from the Earth's yearly journey around the Sun (heating the Earth's electric field

Fig 9.6.3 – This lightning strike sequence over Toulouse France was recorded by Sebastien D'Arco. It shows the electrical charge from the Geofield discharging in a brilliant lightning strike. The sequence is spaced at .32 second intervals. Notice that the lightning begins as a dot in the air on the second photo.

with the Sun reduces the field's intensity). The daily cycle in both the weather *and* Earth's energy field are tied to the Earth's axial rotation. If we measure the Geofield over a 24-hour period (see Fig 9.5.6), we find, that in general, the Geofield strength decreases as the temperature increases. However, if large earthquake activity happens in localized areas, the general rule can change and the Geofield can be activated, even at high temperatures.

Later, the well-known scientist, Lord Kelvin recognized a connection between the Earth's energy field (atmospheric positive charges) and the Fair Weather Condition, but assumed it was the Earth's energy field causing the Fair Weather Condition. He did not recognize that the Geofield and the Fair Weather Condition **were both created** by the *gravitational effects of earthtide*, being directly connected in time and locale. Kelvin also noted there was not one, but several "electric fields" surrounding the Earth:

"Lord Kelvin (1860s) proposed that **atmospheric positive charges explained the** *fair weather condition* and, later, recognized the existence of **atmospheric electric fields**." Note 9.6e

Multiple, or multi-lobed, electric fields are not explained by a single, magma-Earth dynamo, but they are clearly evident in continental piezoelectric earthtide movements, and they play an important role in the fair weather condition. The *cumulus* clouds, already shown to be fair weather clouds (see Fig 9.3.4), are often the *only* clouds present during fair weather conditions because they are earthquake clouds.

Although researchers acknowledge the formative mechanism is "not fully understood," there is an obvious connection between cumulus clouds and lightning:

"The growth of **cumuliform clouds**, particularly cumulonimbus, may increase dramatically when heat (known as latent heat) is released when water vapor condenses into cloud droplets and water droplets freeze into ice crystals. The formation of ice crystals is an important mechanism in the creation of rain and—although the exact process is still **not fully understood**—appears to be **closely involved with the separation of electrical charges that leads to lightning**." Bib 182 p88

Instead of imagining that the electrical charges in cumulus clouds come from "ice crystals," the Geofield Model explains how the increase in atmospheric electrical charge and cumulus clouds are connected—they both originate from hyquatherms, which also explains why lightning can occur in "nothing but blue sky overhead." Clouds are not necessary to generate lightning—the electricity is being generated in the piezoelectric crust.

The electrical current created in the crust by earthtide movement also explains why there is an "electrification" of the atmosphere during high-pressure systems when hyquatherms are active. Built up stress in the crust from earthtide can eventually lead to earthquakes, revealing more dramatic evidence of the Geofield Model.

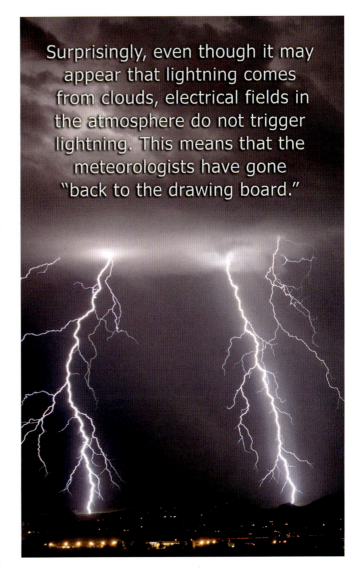

Fig 9.6.2 – We know that lightning does not come from clouds because lightning can occur **without** clouds. Other solar system bodies having atmospheres unlike that of the Earth also have lightning. Therefore, we should look to the one place capable of building such a charge—the crust.

The Earthquake Lightning Evidence

The Earth's geofield is produced by ongoing gravitational tidal forces of the Sun, Moon, and other celestial objects, but what happens to the energy field when there are large shifts— earthquakes—in the Earth's crust?

The answer is simple: large shifts generate large electrical disturbances. When there is an electrical charge overload, the electricity jumps between conductors, which is exactly what happens with *earthquake lightning*. Earthquake lightning may strike downward or *upward,* and *glowing lights* are commonly reported during nighttime earthquakes. These may be visible even if the observer is some distance from the earthquake zone.

Because most geologists still believe that earthquakes come

from magma, earthquake lights are just another "anomaly" left unexplained. For decades, reports of earthquake lights were dismissed as unreliable by most scientists, but a steady stream of observations accompanied with photographic evidence convinced researchers that they are real. A summary of this now recognized phenomenon is found in *The Handy Weather Answer Book*:

"We tend to think of earthquakes as only affecting the ground and things on it. **There are many confirmed reports, however, of the atmosphere glowing with a strange light before and during major earthquakes**. The skies over the Andes mountains appeared to be on fire the night of the great Chilean earthquake of 1906. In the Cerro Gordo quake in California in the 1860s, reliable witnesses reported sheets of 'flame' emerging from the rocky sides of the nearby Inyo Mountains. Sheets of what looked like lightning flashed across the sky just before the tragic Tangshan quake in China in 1976.

"**While there is little formal research available on the topic**, it appears that earthquakes generate substantial electrical disturbances. **Piezoelectric** effects result from compression and strain applied to many rocks and minerals. The suspected low frequency electromagnetic waves emanating from earthquake zones might also stimulate the atmosphere to produce optical emissions in ways **not yet understood**." Bib 66 p247

Without the Planetary Piezofield Laws, the importance of the lights and their relationship to the Earth's energy field would remain unknown. When the above book was written in 1997, the author reported that, "little formal research" was "available on the topic" of piezoelectric effects and earthquakes. Today however, there is much more information available about a variety of electromagnetic energy coming from earthquakes. For example, the 1989 Loma Prieta, California earthquake produced strong, "low-frequency radio emissions" and electrical disturbances that were associated with earthquakes:

"Quite by accident scientists at Stanford University **discovered that strong low-frequency radio emissions preceded the 1989 Loma Prieta earthquake** in California. There have been other reports suggesting ELF (Extremely Low Frequency waves) and even lower frequency **electrical disturbances are associated with earthquakes**." Bib 66 p246

In another incident, Hawaii's second largest earthquake in recorded history on 29 November 1975 produced a 7.2 tremor accompanied by "earthquake lights" as reported by the USGS:

"During and immediately after the main shock, '**earthquake lights**' of white to bluish flashes or glows lasting several seconds were reported by a number of observers. Earthquake lights are associated with major earthquakes and have been observed in Japan and California." Note 9.6f

Earthquakes are not the only event to produce such lights; USGS team members monitoring the famous 1980 eruption of Mt. St. Helens also reported blue lights in the volcano's crater:

"Our next chance for a clear look was on Sunday morning, March 30th, four days after the original eruption. During the night the United States Geological Survey (USGS) had **reported a glow visible in the crater with an arch flashing, like gas or lightning, of 'sharp blue color**.'" Note 9.6g

In 1943, a volcano was born in a Mexican cornfield. Named Paricutin, the volcano came out of nowhere after two bisecting fissures developed in a cornfield. There, the greatest amount of friction occurred after numerous small and large earthquakes shook the ground. Earthquakes occurred before and continued during the eruptions. Then on May 25, 1943, the following was observed and photographed:

"**Lightning flashes were frequent** in the dense, swiftly rising eruptive column." Note 9.6h

Geologists do not understand what caused flashes of lights viewed by thousands of Taiwanese residents during Taiwan's largest quake in over a century. *Geophysical Research Letters* reported that in 1999, a 7.6 quake killed over 2300 people and caused billions in damage was accompanied with earthquake lights:

"…consisted of four separate eruptions accompanied by **flashes of light as seen by the residents**." Note 9.6i

While American and European researchers seem slow to accept this ever-mounting evidence, open-minded Asian scientists appear more willing to accept the empirical evidences, some even coming close to describing the piezoelectric mechanism of the Geofield:

"… thought to be the product of **rocks squeezing and rubbing together**." Note 9.6j

Earthquake lightning evidence of the Geofield Model continues to grow and is becoming overwhelming. Scientists regularly watch for it, although they do not yet understand it. The Geofield accounts for and explains fair-weather conditions and earthquake lightning, leaving the next step of determining the direction of the electric charge that creates the lightning.

Measuring the Geofield-Lightning Connection

It was noted earlier that the atmosphere is incapable of holding the amount of charge that is released in lightning:

"Electrical fields in the **atmosphere simply cannot grow large enough** to trigger lightning, calculates Joseph Dwyer of the Florida Institute of Technology in Melbourne." Note 9.6k

We have also discussed the electrical charge buildup in the crust, but can the charge buildup be measured? Yes, it can, and it is fairly easy using current technology! In the December 2005 issue of *IEEE Spectrum*, the article, *Earthquake Alarm, impending earthquakes have been sending us warning signals—and people are starting to listen*, reported researchers' findings that "conductivity of the air over the quake zone caused by current welling up from the ground" can be easily measured:

"Besides detecting magnetic-field [Geofield] disturbances, ground-based sensors can record changes in the **conductivity of the air over the quake zone caused by current welling up from the ground**. These sensors can vary in form, but those we use are made from two 15-centimeters by 15-cm steel plates locked into position about **1 cm apart**. A 50-volt dc battery charges one plate; the other is grounded. A resistor and voltmeter between the battery and the first plate senses any flow of current.

"Normally, the air gap between the plates acts as an **insulator**, and no current flows. **If**, however, there are charged particles in the **air**, a current begins to flow, creating a voltage drop across the resistor that registers with the voltmeter. The currents created in this way are not large—on the order of millivolts—**but are detectable**." Note 9.6l

The first important point to note is that a change in the Geofield is caused by earthquake current "welling up from the ground." This is easily detected by measuring the amount of the charge buildup in a small area (roughly the amount of air between your hands held about one quarter inch apart). The electric current between the plates is not large because there is only 1 cm of charged air between the 15cm² (6-inch) plates. One million cubic meters of charged air (of which there may be many millions during a thunderstorm) generates a current *millions* of times that of the measured area—and that's enough to produce lightning.

How are millions of cubic meters of air charged?

By the mechanical strain of millions of cubic meters of piezoelectric rock in the crust.

This is why a clear understanding of the Piezofield Model is so important. It helps explain the origin of dangerous electrical storms and it explains the protective energy field surrounding the Earth. It also provides necessary navigation mechanisms for many animals and even affects the human body in ways we are just beginning to comprehend.

Turkey Earthquake Prediction Project

It is interesting to note that countries like Japan and Turkey, who experience large-scale earthquakes on a regular basis, are more interested in scientific research connected to the Piezofield Model than are their American or European counterparts. Researchers at Istanbul Technical University in Turkey are working a program called the *Earthquake Prediction System Based on Observing the Stress of Rocks by the Method of ULF Electric Field Measurement*. They have been taking measurements over the past several years and noticed that piezoelectric charges were being released during the rupture period following "a long term stress on the rocks at the fault zones":

"Depending upon their mineral structure, rocks consist of various quantities of piezoelectric characterized material such as quartz (SiO_2). It is thought that it will be possible to observe **piezoelectrically caused changes from the surface, that occur as a result of an increase in long term stress on the rocks at the fault zones and changes during the rupture period**."
Note 9.6m

Even with such findings, government funding remains sparse for predictive research like this, so researchers often work on their own dime, perhaps out of concern for their fellow citizens. Hopefully, the Geofield Model will raise the awareness of the true nature of earthquakes and their connection to the Geofield. Then better preventative measures can be developed well ahead of major quake events.

One private company, Quakefinder, leads the earthquake-prediction research pack by working with governmental agencies to identify ways of forecasting earthquakes.

Quakefinder Evidence

Some of the best evidence of the Geofield Model comes from earthquake researchers who are looking for ways to predict *large-scale* earthquakes. Quakefinder, a privately held firm in Palo Alto, California, conducts pioneering research in earthquake forecasting. The company works with NASA, UC Berkeley, Stanford, and The Weather Channel in a collaborative approach to monitor the relationship between large earthquakes and the Earth's energy field. Instead of trying to decode the theoretical earthquake-magma connection, Quakefinder engineers developed a technique to measure fluctuations in the Earth's energy field where large earthquakes occur. Quakefinder's researchers note that the occurrence of Extremely Low Frequency (ELF) fluctuations in the Geofield prior to large earthquakes has been known for some time:

"ELF signals have been studied for quite some time. In 1989, Stanford researcher, Dr. Tony Fraser Smith, detected ELF magnetic fields near Loma Prieta, CA. He found that ELF signals were **twenty times normal fourteen days prior to the 7.1 earthquake**. The same signals increased to **60 times normal 3**

Fig 9.6.4 – This photograph of the 1994 eruption of Mount Rinjani illustrates the connection between volcanic eruptions, earthquakes, and lightning. As shown in the Magma Pseudotheory subchapter 5.3, earthquakes precede lava flows because the melted rock is generated by frictional heat in seismically active fracture zones. Earthquake fracture zones also produce considerable piezoelectric energy that is discharged as lightning, as seen in this photo. Photograph by Oliver Spalt.

hours before the quake, and they persisted 2 months after the quake. In 1989, a Russian satellite (Cosmos 1809) collected earthquake-related signals associated with the Armenian quake. The French (M. Parrot) have also been pursuing this theory, and have conducted satellite statistical surveys since the 1990's."
Note 9.6n

Their observations are part of the growing evidence connecting earthquakes and the Geofield:

"**There is considerable and growing evidence that the crystalline rock compression produces electromagnetic energy**. And that energy may be detected, analyzed and verified.

"There is a **growing body of evidence that indicates strong connections between ULF [Ultra Low Frequency] waves and earthquakes** (US, Japan, Russian, French)." Note 9.6o

This private company has accumulated more data about the forecasting of earthquakes than *any* educational facility we found. Quakefinder's researchers, like Earthquake Cloud researcher Shou, are not traditional Earth sciences people. Instead, their background is in engineering, and they have a lot to offer. In a 2005, Quakefinder report:

"For decades, researchers have detected strange phenomena in the form of odd radio noise and eerie lights in the sky in the weeks, hours, and days preceding earthquakes. **But only recently** have experts started systematically monitoring those phenomena and correlating them to earthquakes.

"**A light or glow in the sky sometimes heralds a big earthquake**. On 17 January 1995, for example, there were 23 reported sightings in Kobe, Japan, of a white, blue, or orange light extending some 200 meters in the air and spreading 1 to 8 kilometers across the ground. **Hours later a 6.9-magnitude earthquake killed more than 5500 people**. Sky watchers and geologists have documented similar lights before earthquakes elsewhere in Japan since the 1960s and in Canada in 1988."
Note 9.6p

Scientists in other parts of the world are receptive to the Quakefinder research, some corroborating their work. Taiwanese researchers noted that the large Chi-Chi quake generated tremendous electrical energy:

"In Taiwan, sensors that continuously monitor Earth's normal **magnetic field registered unusually large disturbances in a normally quiet signal pattern shortly before the 21 September 1999 Chi-Chi, Taiwan, earthquake, which measured 7.7**. Using data from two sensors, one close to the epicenter, and one many kilometers away, researchers were able to screen out the background noise by subtracting one signal from the other, leaving only the magnetic field noise created by the imminent earthquake. Two teams, one in Taiwan and one in the United States, calculated that the **currents required to generate those magnetic-field disturbances were between 1 million and 100 million amperes**." Note 9.6p p25

This is a very large amount of energy. Existing GPS technology is another method being utilized to detect changes in the Geofield in hopes of predicting earthquakes:

"Even the existing global positioning system may serve as part of an earthquake warning system. Sometimes **the charged particles generated under the ground in the days and weeks before an earthquake change the total electron content of the ionosphere**—a region of the atmosphere above about 70 km, containing charged particles...

"Researchers in Taiwan monitored 144 earthquakes between 1997 and 1999, and they found that for those registering 6.0 and higher the electron content of **the ionosphere changed significantly one to six days before the earthquakes**." Note 9.6p p26

As we know, the test of truth is time, and the true connection between earthquakes and the Earth's energy field is "becoming increasingly solid":

"The **connection between large earthquakes and electromagnetic phenomena in the ground and in the ionosphere is becoming increasingly solid**. Researcher in many countries, including China, France, Greece, Italy, Japan and the United States, are now contributing to the data **by monitoring known earthquake zones**." Note 9.6p p26

In summary, Quakefinder's evidence demonstrates the direct connection between earthquakes and the geofield. New observations are being enabled by new *technology* and by open-minded scientists who are concerned with saving lives by understanding and predicting *large* earthquakes. Quakefinder's engineers and their associates continue working outside the 'black box of magma' as they seek to predict large earthquakes. Their technology enables them to measure the geofield in the Ultra Low Frequency (ULF) range.

The geofield and lava are connected
because earthquakes create both of them.

Despite the evidence, most American geologists are still resistant to the idea that earthquakes and the Earth's energy field are connected.

The KTB Borehole Piezoelectric Evidence

In addition to the Dynamo Theory, which sees magma as an electrical field source, some researchers think that the Earth's energy field comes from a so-called 'geobattery' (Sato and Moony 1960). An important aspect of their theory was that the source of the electrical field was considered to have "self-potential," meaning it came from within itself or the surrounding rock environment—***not** from the core of the Earth*. However, KTB (German super-deep borehole) researchers found that the surrounding electrolytic conductors (like acids in a battery) did ***not*** work. They could not satisfactorily explain the electrical fields around the borehole:

"Notwithstanding the general acceptance of the basic theory of *Sato and Moony* [1960], the large amplitudes of observed **self-potential anomalies** around the KTB site and elsewhere **could not be satisfactorily explained**." Note 9.6q

Without the "geobattery," German researchers were back to square one. "It was expected" that the one-third-billion-dollar borehole "would give important new insights into interpretation of such anomalies":

"The KTB superdeep borehole was sited on a strong magnetic **anomaly** and gravity high, **and it was expected** that borehole geophysics and laboratory measurements of rock samples obtained during drilling **would give important new insights** into interpretation of such **anomalies**." Note 9.6r

The word "anomaly," which means a deviation from the norm, traditional rule, or in this case scientific theory, is used throughout the journal articles. The theory is that the Earth's 'magnetic' field comes from magma, which means there should be a *gradual increase* in the intensity of the field the closer one gets to the magma source, but at the KTB site, this was not observed. And no correlation was found with magnetic-iron-rock types. The magnetic "anomaly" was recorded when large increases in the energy field were measured.

Researchers tried checking the rock layers (lithology) to see which layers might be restricting or impeding the electrical field of the Earth, but saw no correlation:

"The resistivity shows **no correlation with lithology**." Note 9.6s

They did not find what they were looking for because the German KTB geophysicists did the same thing the Russian researchers did at the KOLA borehole when analyzing the heat anomaly there—they "used the traditional **assumption**" that the direction of the electrical field gradient in the rock layers is *vertical* because they **assumed** it was coming from magma. *Tectonophysics*, 306, (1999) p357. It appears piezoelectric rock types were not studied at all.

The Piezofield Model predicts that the Earth's energy field is not unidirectional from a supposed magma-planet center, but is generated within the crystalline rock throughout the crust, by friction, as the Lava-Friction Model demonstrated in Fig 5.3.2. If the Earth's energy field is of a piezoelectric origin, then a *change* in the energy field should be evident at depth, depending on the piezoelectrical capacity of each rock layer. Increased pressure at depth should reflect an increase in the energy field. The German researchers noted:

"The KTB superdeep hole penetrated **many magnetic anomalies**, which have a vertical extent of some **meters up to about a hundred meters**. One of the most important of these anomalies, **with a strong decrease of the magnetic field intensity, occurs near the surface**. Below about 1200 m the total magnetic field intensity increases systematically with depth with a gradient of up to 200nT/km, which is much higher than the undisturbed Earth's magnetic field would produce (about 22nT/km). **This result requires a magnetic body at depth, the exact nature of which, however, remains unknown**." Note 9.6t

Their mysterious "magnetic body at depth" is reminiscent of Barringer's buried 'meteorite' in the Arizona Hydrocrater. He spent a fortune searching for something that never existed. Likewise, there is no "magnetic body at depth" beneath the KTB borehole. The reason researchers are confused about the energy field data from the borehole is that they are looking through magma-magnetic glasses, and do not recognize the importance of the piezoelectricity. The Piezofield Model is further confirmed by predicting the location of the strongest magnetic anomaly; it should occur in the proximity of faults. In the KTB borehole, the largest fault zones are 6.9 - 7.8 km below the surface:

"In the section of the **fault zones (6.9-7.8 km)**..." Note 9.6u

Just as is expected in the Piezofield Model, note the location of the "strongest magnetic anomaly" that was encountered along the borehole:

"Nevertheless, the deep magnetic data obtained with this tool **document that the strongest magnetic anomaly encountered in the borehole occurs between 7300m and 7900 m**." Note 9.6v

The "strongest magnetic anomaly" in the KTB borehole was *right at the fault zone*, but the geologists involved had no idea about the Piezofield Model. Neither could they know that the data from the KTB super-deep borehole would one day be strong evidence of the UM Geofield Model. It was no accident that researchers saw a correlation between the borehole's fault

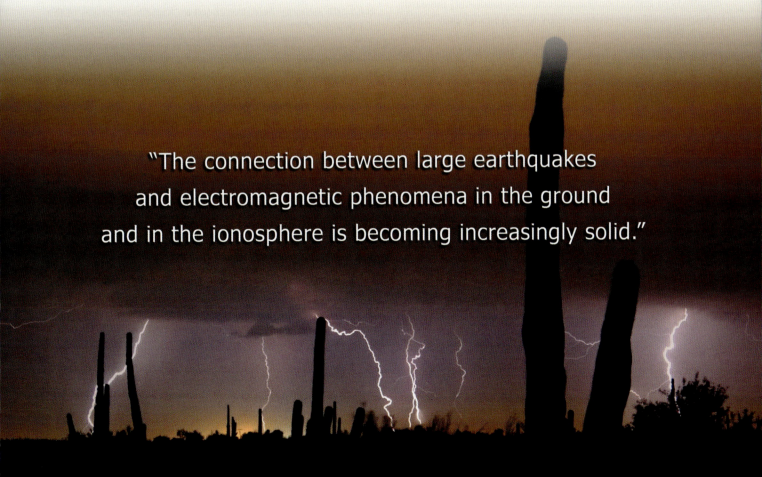

"The connection between large earthquakes and electromagnetic phenomena in the ground and in the ionosphere is becoming increasingly solid."

SUBCHAPTER 9.6 THE GEOFIELD EVIDENCE

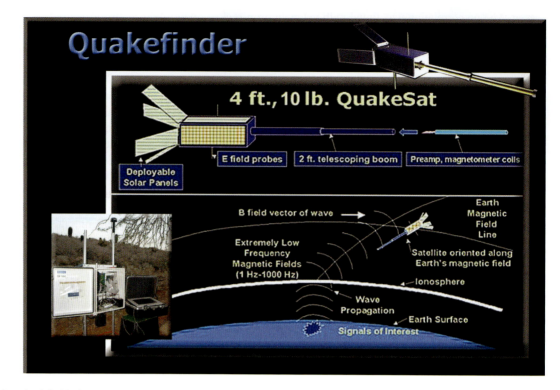

Fig 9.6.5 – Quakefinder, a private company, has had success identifying connections that exist between earthquakes and the Earth's energy field. Even though the mainstream geological community has yet to accept Quakefinder's data or their explanations behind the piezoelectric nature of the Earth's energy field, a "growing body of evidence" coincides directly with the Geofield Model outlined here in the UM. In this diagram, site-based electronic detectors measure changes in the Earth's energy field as well as seismic disturbances. One day, satellites that measure specific changes in the Earth's energy field will help correlate surface earthquake and Geofield data into one cohesive picture of the Earth's energy field that can help predict earthquakes.

zones and the strongest electrical fields because the Earth's energy field *originates* from the flexing of the Earth's crust via piezoelectricity generated at fault zones.

The Russian Geofield Evidence

Space constraints and brevity limit the inclusion of Geofield Evidence to a small fraction of that which is known. Among the hundreds of articles on the subject, one stands out; the earthquake-geofield connection seen by K.N. Abdullabekov, a Russian scientist who published a book in 1991 called, *Electromagnetic Phenomena in the Earth's Crust*. This little-known book contains some of the clearest and earliest recorded evidence about the true nature of the Earth's Geofield; evidence largely ignored by the scientific community.

For many decades, Abdullabekov and his colleagues investigated the relationship between earthquakes, faults, and the Geofield. Referring to the "field" of energy around the Earth as being "magnetic," the results of their studies are powerful:

"In many studies, a relationship between the field changes and the processes in the Earth's crust and the upper mantle was studied with the aid of repeated magnetic surveying… **All [energy field] anomalies have been associated with active deep-reaching faults**." Note 9.6w

The researchers had observed changes in the Geofield that were directly associated with active faults and earthquakes, confirming the First Planetary Piezofield Law, which is that the energy fields of planetary bodies are created by piezoelectricity in the body's crust. Abdullabekov continued with an important fact unconnected to any dynamo theory but directly connected to the Second Planetary Piezofield Law, which states that a planetary body's piezofield is controlled by the makeup of the piezoelectric materials in the crust and the *astronomical* tidal forces acting on the body. The astronomical tidal forces are related to celestial rotational cycles evident in the cycles seen in the Earth:

"The geological processes in the Earth are **cyclic**." Note 9.6w p10

On page 16 of Abdullabekov's book, he lists more than two dozen regions across the Earth where violent earthquakes occur in *cycles*. His list includes cycles that range from minutes, days, months, even hundreds of years. Dozens of scientific studies are referenced, yet most American geologists seem unaware or simply choose to ignore the research.

Some Russian scientists already recognize that variations in the Earth's energy field are related to "various processes in the Earth's *crust*":

"**Many cases of the variability of the Earth's magnetic field due to various processes in the Earth's crust are already well known**. The identified modes of variation are usually explained in terms of **excessive stresses prior to earthquakes** and thermal, chemical, **electric** or other factors." Note 9.6w p23

The **first example** from Abdullabekov shows that changes in the Geofield are from crustal movement by creating explosions in the crust. The explosions produced "irreversible changes" in the intensity of the Geofield where the explosions took place:

"The **irreversible changes**, with an intensity from tenths to first tens of nanoteslas were controlled by irreversible changes in rock magnetism in the disturbance zone. The field changes of the relaxation type have been linked to the relaxation of elastic stresses in the zone of explosion. The characteristic time varied from seconds to several hours and more, while the amplitude ranged from a few nT to 10 nT.

"Hence the experiments on the local variation of magnetic field due to explosions in rocks conducted in various regions were carried out in different geological-tectonic physico-mechanical and other natural and artificial conditions (weakly, medium and strongly magnetic rock at distance of 50 to 2000 m from the centre of explosion).

"The strength of the explosions varied from 100 to a few thousand kilograms of explosives. As a result, **a unique relationship has been obtained between the arising mechanical stresses and the changes in the magnetic field**. It has also been established that the intensity of the effect depends on the distance from the point of explosion and the rock magnetism." Note 9.6w p24

These results show that the energy field above the disturbed site changed when underlying rocks where cracked and relaxed during the explosion (supporting the First Planetary Piezofield Law). These findings were confirmed when American scientists detonated a *nuclear* device in 1972:

"As a result of the CANNIKIN **nuclear explosion**, the magnetic field several kilometers from the epicenter appears to have been permanently altered. Within 30 sec after detonation, a proton magnetometer 3 km away recorded a 9-gamma step increase in total magnetic field." Note 9.6x

The **second example** of crustal changes that led to changes in the Geofield involved the filling and discharge of water in the Charvak Reservoir, 70 km northeast of Tashkent, Uzbekistan. The dam forming the 12 km by 5 km reservoir is 168 meters high, and from May to July, the reservoir is filled by mountain stream runoff. From September through January, the water is discharged creating a cyclical stress on the surrounding rocks resulting in changes to the Geofield in the area. The top graph in Fig 9.6.6 is adapted from Abdullabekov's book; it shows a reduction in the intensity of the Geofield as the reservoir fills and an increase as the reservoir empties.

Abdullabekov recognized the "piezometric properties of rock" in the area:

"The **Charvak area** possesses the necessary natural conditions favoring experiments on the effect of varying pressure on the magnetic field of the Earth such as **variable cyclic conditions** of water level and volume variation, low background of natural and artificial noise, presence of rock with optimum magnetism, and **known piezometric properties of rock (from laboratory experiments)**." Note 9.6y

With scientific truth, the test of time can be applied to the observed changes in the Geofield and the water levels:

"**The relationship of the anomalous field variations and water level in the reservoir was clearly repeated from year to year**." Note 9.6y p26

The researchers go on to state:

"Hence we have **established experimentally a relationship between the change in magnetic field versus the water volume in the reservoir**." Note 9.6y p28

The **third example** is an underground gas storage facility where natural gas is pumped into and out of a natural underground chamber (the anticlinal aquifer):

"**Detailed magnetometric observations were conducted** in the region of an **underground gas storage**. Natural gas was pumped at 9-9.5 MPa into an anticlinal aquifer at a depth of 600-650 m. The anticlinal structure had a size of 6 x 1.5 km. Eleven points were deployed in that area, and repeated observations were conducted 3-5 times a year." Note 9.6y p30

The second set of graphs in Fig 9.6.6 shows the results after pumping gas into the crust. The investigators note that the rocks at the gas-pumping site have a "low magnetism" and the maximum excess pressure in the underground structure did not exceed 3 MPa, (about 435 psi).

They note that the "piezomagnetic effect" was significant and that the change to the Geofield was large:

"Hence **the piezomagnetic effect is not negligible as it is counted in tenths of nanotesla**. The temperature and chemical composition of the gas **remains practically constant** during the procedures of pumping in and out. Thus, the thermal chemical processes **cannot** bring about substantial anomalies in the magnetic field changes." Note 9.6y p31

The **fourth example** involved a change to the Geofield in the days before and after the 1978 magnitude 6.5 Alay earthquake in Tajikistan. The third graph in Fig 9.6.6 shows the change in the Geofield just prior to the earthquake. Researchers correctly forecasted the approximate place and time of the 1978 Alay earthquake in part because of data derived from the change in the Geofield:

"Studies on **the dynamics of the geomagnetic field** in the Fergana area **permitted the magnetologists to predict reliably the time and approximate place of the violent Alay earthquake** of 1st November 1978; this prediction was first in the USSR and embodied a number of other forecast methods." Note 9.6y p87

Stress in the crust can be accumulating or dispersing, causing a rise or fall in the intensity of the field. A **fifth example** is the "bay-type" drop in the intensity of the Geofield (bottom graph Fig 9.6.6) discovered after hundreds of analyses. Changes were accompanied by earthquakes only 70-80% of the time:

"The analysis has shown that in **70-80% of cases the anomalous changes were accompanied by earthquakes**, while in **20-30% of events anomalous changes in the geomagnetic field were not present**. The shape of the anomalous changes was primarily **bay-type**, with the negative sign." Note 9.6y p89

The most likely reason the energy field did not react to one fifth of the earthquakes is that earthquakes can occur in areas of non-crystalline geology (such as limestone), although they do so less frequently. Areas containing low-density quartz rock would produce less piezoelectric energy. Therefore, earthquakes in these areas would have less of an effect on the Geofield.

Even though the Russians had gathered all this data (some of which is shown in Fig 9.6.6) indicating the direct relationship between the movement of the crust and changes in the Earth's Geofield, two major factors kept the 'mystery' of the Earth's energy field alive. One was a matter of timing: the research was conducted and reported in 1991, *before* the advent of GPS and the subsequent discovery of earthtide. Because earthtide was unknown to the Russian research scientists, they did not have the luxury of knowing that the constant gravitational movement of the crust was driving the Earth's piezoelectric generator. The second factor keeping the Russian researchers in the dark was their assumptions about the factors that cause the Geofield:

"The secular **change in the geomagnetic field is caused by** various processes of **internal and external origin**, associated with the following features:

1. Processes in the **Earth's core**;
2. External processes in the **Earth's magnetosphere**;
3. Changes in the **Earth's crust** (independent or active, induced or passive).

"The global and large regional changes in the geomagnetic field are brought about by processes in the Earth's core and external sources, while **the local changes are caused by different processes in the Earth's crust**." Note 9.6y p17

Although "processes in the Earth's core" (Magma Pseudotheory) have never been ***shown*** to change the Geofield, the two other features listed *do in fact* change the Earth's Geofield. The magnetosphere is the energy field surrounding the Earth. The Sun's radiation pushes on this field, changing it when the Sun's radiation output changes. Furthermore, changes to the field at the surface are correctly attributed to changes in **the Earth's crust!** This huge leap in the geophysical world was made possible by the Russians because of the marvelous Geofield experiments they conducted. It unmistakably showed the connection between the Earth's energy field and piezoelectricity in the crust:

"Gokhberg et al. (1980) described the **sources of electromagnetic earthquake effects** as mechano-electrical (**piezoelectric effect**, Stepanov's effect, **processes of electric charging by friction**, failure or destruction of double electric layers) **which occur at earthquake focuses**, and electrokinetic phenomena, or electric eddy fields. **All mechano-electrical processes are sources of EMR [electromagnetic radiation-Earth's energy field] due to relaxation of the separated discharges**…In 1953 Volorovich & Parkhomenko (1954, 1955) recorded experimentally and explored theoretically the **piezoelectric effect in samples of granite, gneiss and vein quartz**." Note 9.6y p114

Endless Geofield-Earthquake Evidence

Once the Geofield Model is known, it is *easy* to see evidence of it in Earth science journals, and there are seemingly endless relevant examples throughout the scientific literature.

One example has several French researchers using the DORIS system, which consists of several ground-based radio beacons and several satellites monitoring the signals and changes in the ionosphere. They compared the data to earthquakes of magnitude 5.0 or larger. The results show there is an element of *predictability*:

"A statistical study was done with these estimated TEC [Total Electron Content] data **to check if the measurements can be influenced by the seismic activity**. The statistical results show that there are perturbations of the TEC **several days before the earthquakes and right after the time of earthquakes**.

"These two points are **well in agreement with previous works**. Ducic *et al.* (2003) presented observations of TEC variations just after the time of a powerful earthquake… It corresponds with the results of Liu *et al.* (2004) who observed variations between 18:00 and 22:00 LT within 5 days before earthquakes in Taiwan area." Note 9.6z

Another comes from an article in the 2002 *Journal of Geodynamics* titled, *Magnetic fields over active tectonic zones in ocean*.

Researchers noted that the strength of the Geofield varied by 0.1-1 nT (nanoteslas) on the ocean floor, but were surprised to discover that the overall strength of the Geofield was induced or changed by *1-40 nT* when they measured the field over a plume top. The aim of the investigator was to show that the Geofield (magnetic signals) could be induced "by hydrodynamic processes which were initially produced by tectonic phenomena":

"The aim of this new work is to show that **magnetic signals can be induced by hydrodynamic processes which were initially produced by tectonic phenomena**." Note 9.6aa

Because of the Hydroplanet Model, we can comprehend how gravitational-frictional heat can create hot plumes of water, and at the same time, generate the piezofield around the Earth. Although the researchers were unaware of these factors, they correctly deduced the fact that magnetic (Geofield) surveys of the ocean floor "must" consider the additional electromagnetic background signals forming the Geofield:

"These magnetic disturbances in the sea medium **create an additional natural electromagnetic background that must be considered when making detailed magnetic surveys**." Note 9.6aa

Moving from the ocean floor evidence of the Geofield, we turn our attention to the Sun-earthquake-Geofield connection. In the journal, *Physics and Chemistry of the Earth*, Austrian and Italian researchers documented the relationship with Mt. Vesuvius' seismicity cycles and the variations in the Geofield. Their article, *Seismicity Cycles in the Mt. Vesuvius Area and their Relation to Solar Flux and the Variations of the Earth's Magnetic Field*, clearly showed that variations in the solar *and* Geofield activity are "in a *close* relationship with *earthquake* activity":

"The present study indicates that the **variations of solar activity** and the **Earth's magnetic field** are in **close relation to earthquake activity**." Note 9.6ab

With the UM models, it becomes a simple matter to connect the relationships between the Sun, Moon, hydrothermal activities, earthquakes, and weather. Life here on Earth could not exist without these complex connections, and as we move into the upcoming chapters, many more connections will be uncovered, dispelling the notion that such are mere "random chance" as theorized by modern science. Amazingly, modern science *tries* to find the interconnectedness in Nature—but it denies the very order and design Nature so clearly exhibits.

The next subject brings such connectivity in view, although at first the relationship may not be apparent. Scorned by many, dowsing has mystified investigators for centuries, but the Universal Model paradigm and recent research removes the mystery of dowsing, making it possible to finally understand.

The Dowsing Evidence

Dowsing is an age-old way of finding objects under the ground, especially underground water or valuable ores. Records from sixteenth century Europe show dowsing was a well-known activity with roots founded in even earlier times. The art of dowsing involved the use of a divining rod, a Y-shaped branch, or one or two L-shaped metal wires or wooden twigs. Some dowsers used a pendulum, while some used only their body. Dowsing is still practiced today and, chances are, if you live in a rural area you know someone who is able to dowse the best location for a new water well.

I once encountered a dowser while living in the city. A utility worker came to our home to mark the location of underground utility lines. The individual got out of his truck, walked across the yard several times with his dowsing rod in hand, and within a few minutes, had marked the location of the lines on the ground. After he finished, I asked if he had always used the "rod." He said that he had an electronic device that did the same thing, but that it was usually quicker for him to use his rod. Not being acquainted with the technique, I found it to be a fascinating experience though, at the time, I had no idea that it would connect with the Weather Model.

Dowsing is often associated with mysticism, not scientific understanding and, as such, it is generally dismissed. Few people give dowsing serious attention, and an article in the *New Scientist* reveals the thoughts of most scientists:

Russian Geofield Evidence

Change in the Geofield

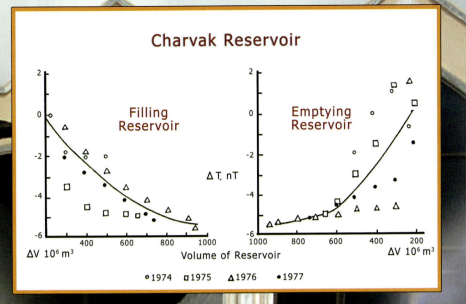

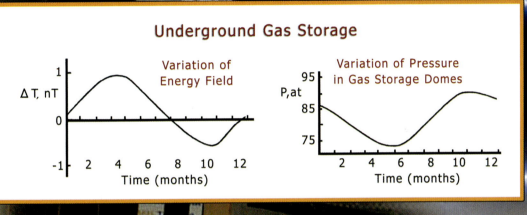

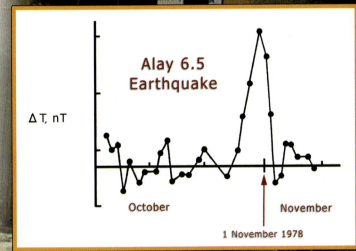

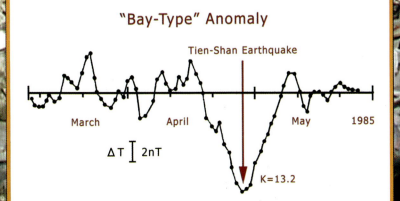

All four of these graphs show that a change in the movement of the crust, resulted in a change in the geofield.

Fig 9.6.6

"Most scientists accordingly view dowsing as **nothing more than self deception** resulting from autosuggestion and certainly not worth serious attention." Note 9.6ac

However, as it turns out, there are "magnetic" anomalies, actually 'electrical' in nature, that occur due to changes in the Earth's Geofield. The *New Scientist* continues:

"Gene Simmons, a geophysicist at the Massachusetts Institute of Technology, recently came up with a good example of how magnetic sensitivity could play a part in successful dowsing for water. Simmons conducted surveys of gravity and magnetic fields around two dowsed wells near Boston, Massachusetts. Unlike most other wells drilled in the crystalline rocks of the district, the holes yielded large quantities of water, at least 140,000 liters per hour. Simmons found that the dowsers had sited both holes within a narrow magnetic [electrical] anomaly only a few metres wide. **The anomaly resulted from a fracture zone that was channeling the flow of ground water, hence the exceptionally high yield of the wells**.

"The theory that people can respond to magnetic cues **could also explain how dowsers may find veins of metal ores**. These veins are usually associated with **faults or fracture zones**, which often produce magnetic [electrical] anomalies, and in some cases ore minerals are themselves magnetic. Support for the idea comes from the Soviet Union (*New Scientist*, 8 February 1979, p371). Since the early 1970s, geologists there have used dowsing on a large scale and have drilled thousands of tests bores on the basis of dowsing. The geologists report that dowsers show most frequent rod movements over sulphide or chromite ore bodies, diamond pipes, certain faults and fracture zones, steel pipelines, underground cavities and beneath electricity cables. **These are just the places where magnetic [electrical] changes occur**." Note 9.6ac p41

The Geofield Model reveals why "fractured zones" are so important to the dowser; the greatest amount of movement and piezoelectricity occurs in such zones, and it is where water is most likely to flow. Dowsing is an excellent example of how a natural technique can have success in an area where scientists' skepticism and false magnetic field-dynamo theory holds them back. The technologist will employ the use of natural phenomenon, if it works, whereas the scientist wants to try to explain the phenomenon before it can be accepted. Maybe that is why there are more technological advances each year than there are scientific advances. Civil engineers in England discovered value in learning to use this natural tool:

"If dowsers can detect small magnetic anomalies, they should also be able to help in surveying the ground beneath building sites, because pipes, cables, drains, culverts and the like all produce such changes. Many civil engineers and construction workers do use dowsing to trace these features—there is even a flourishing trade in the L-shaped metal rods that these workers prefer to the traditional water dowser's V-shaped tool. Ian Killip and his colleagues at the Department of Building and Civil Engineering at Liverpool Polytechnic have investigated the potential of dowsing in this application. Killip has been sufficiently impressed by the results to develop a systematic method of dowsing for the construction industry. He claims that **the technique offers a cheap and quick alternative to geophysical methods**." Note 9.6ac p41

This article supports the experience with the dowser at my home. Moreover, dowsing is a biological activity—that is, the human body is interacting with the Geofield in a way that can detect changes in the Geofield:

"Despite these intriguing results, scientists were at first reluctant to take the magnetic theory of dowsing seriously. In the early 1970s, the notion that any animals—let alone humans—could sense small magnetic changes was regarded with almost as much distaste as dowsing itself. **There has since been a revolution in scientific attitudes to this question. The concept of magnetic sensing has shed its aura of disreputability and become well established, having been demonstrated in an astonishing variety of organisms**. Bacteria living in the darkness of muds on the ocean floor use a magnetic sense to tell up from down. In the absence of visual cues, honeybees rely on their sense of magnetism to build their combs in a north-south direction. Scientists have trained tuna fish to swim through a frame in response to magnetic cues. Numerous species, including many birds—such as homing pigeons and robins—and apparently, some rodents, **use a magnetic sense to help them to navigate** when other cues, such as the position of the Sun or stars, are not available. Perhaps most intriguing of all, whales seem to use the magnetic stripes that were produced during the spreading of the ocean floors as cues in long-distance navigation." Note 9.6ac p41-2

The idea that the human body has a 'sixth' *physical* sense is a fascinating new concept for modern medical science, but has been in several ancient cultures for thousands of years. Some learned to use the body's natural energy aura in healing; others developed effective combat fighting techniques, such as karate. Cultivating and developing the body's extrasensory 'sixth sense' allowed the practitioner to detect the position of one's opponent and where the opponent was moving to, before he (or she) got there.

One aspect of dowsing that keeps science from being able to accept it is that the connection between underground water, the Earth's fracture zones, and the Geofield is *indirect*. Do dowsers find underground flowing water by detecting the water itself? The following study appears to make this clear:

"More recently, a [2004] study was undertaken in Kassel, Germany, under the direction of the Gesellschaft zur Wissenschaftlichen Untersuchung von Parawissenschaften (GWUP) [Society for the Scientific Investigation of the Parasciences]. The three-day test of some 30 dowsers involved plastic pipes through which a large flow of water could be controlled and directed. The pipes were buried 50 centimeters under a level field. On the surface, the position of each pipe was marked with a colored stripe, so all the dowsers had to do was tell whether there was water running through the pipe. All the dowsers signed a statement agreeing this was a fair test of their abilities and that they expected a 100 percent success rate. **However, the results were no better than what would have been expected by chance**." Note 9.6ad

This is the primary reason modern science dismisses dowsing as being unreliable—there seems to be no mechanism for the body to sense moving underground water directly. Researchers have difficulty accepting dowsing because they see the world through 'dogmagma' glasses; they believe the Earth's energy field is tied directly to magma, not the crystalline rocks beneath our feet. Therefore, with no logical connection between the dowser, water, and magma, the practice has no legitimacy. To the modern scientist, fault lines, piezoelectric current, and aquifers are just coincidental, but with the new UM evidence and the connection between these geological entities, the dowsing phenomena can now be understood.

Dowsing and Piezoelectric Effect in Bone

One of the best investigations into dowsing was conducted by No Nordell of the Department of Environmental Engineering at Luleå University of Technology in Sweden in 1988. The report: *The Dowsing Reaction Originates From Piezoelectric Effect in Bone* discussed the geological and biological aspects of dowsing:

"A Rumanian researcher, A Apostol, gave a lecture on dowsing at Luleå University of Technology. He suggested that **the dowsing reaction originates from mechanical stress concentrations underground. Stress concentrations occur in fractures and fissures and since such underground openings in most cases are water conducting the rod can indirectly indicate water**. In a field test, at Kallax moor where the bedrock is covered with 30 m of sand, he demonstrated his skill with the wooden twig by **detecting fracture zones in the bedrock. He could also tell the direction of the fracture zones. His results were found to be remarkably good compared to geophysical measurements carried out before his visit.**" Note 9.6ae

Apostol found the key as to why dowsers could not find water 100% of the time. Water *may* run along fault lines, but not always, and it is a *change in the Geofield* that the person using the dowsing rods feels—not necessarily a change in underground water flow. The rods act as conductors that, when coupled with the human body, form a 'conduit' into the ground allowing the slightest field changes to be detected.

As the dowser moves over an elevated energy field, the dowsing rods (L-shaped type) cross. The elevated energy field can include fault lines, even minor faults, or metal pipes, which are good conductors of energy.

One reason dowsing seems to work is that the piezoelectricity emitted by crustal rocks appears similar to that created in the bones. Nordell cited research on the piezoelectric properties of bone:

"Fukada and Yusuda (1957) discovered that **'dry bone' has piezoelectric qualities. Continued research showed that 'wet bone' and bone 'in vivo' are also piezoelectric**. This discovery resulted in treatment methods for electrically induced healing of bone. A considerable list of references is given in Herbst (1983)." Note 9.6ae p5

Nordell conducted laboratory tests of his own that revealed that stressed bones create the same type of energy field that is found in crustal rocks, and that the Geofield, when focused into the arms, could cause "involuntary muscle movements which cause the dowsing reaction":

"The piezoelectric properties of bone in the forearm could be the reason for **involuntary muscle movements which cause the dowsing reaction**. This could be explained both mechanically and electrically." Note 9.6ae p5

Another explanation is that the Geofield "influences the nerves which are operating these muscles":

"Another possible explanation is that the **generated piezoelectricity directly influences the nerves which are operating these muscles**. If the electromagnetic signals are alternating the reaction of the bone is also alternating at the same frequency. Consequently, this should entail quivering of the forearm muscles which is many times reported by dowsers." Note 9.6ae p6

Nordell also performed laboratory tests with pig forearm bones to verify the bone's piezoelectricity. The forearm bone ends were fixed in gypsum so that a load could be placed on the bone to measure stress. Contact pins and a voltmeter were used to measure the electricity produced by the bone as it was mechanically stressed. When the bone was struck, the voltage increased; strikes that were stronger resulted in even larger voltages, some being as much as five volts.

The most surprising aspect of the experiments was the ***sensitivity*** the bone exhibited to changes in the electrical field. An electric lamp cord was placed on the top of the test bone with the lamp switched off. When the lamp was turned on and off, a threefold increase in the voltage through the bone was ob-

Fig 9.6.7 – Details about dowsing are recorded at least as far back as the 16th century. This 18th century image illustrates a Frenchman using a dowsing stick, probably looking for water. The dowser was long-thought a mystic, achieving no legitimacy within the scientific community because it appeared there was no way to explain it, and because there were many fakes. Although new research has begun to change this perception, the Geofield Model can finally explain how dowsing works.

served:

"An electric cable, to a lamp, was placed on top of the test bone. The resolution of the oscilloscope was 1 mV/cm. Initially the cable was not conducting electricity and the oscilloscope showed the background voltage field of the room. When switching on the light the amplitude of the voltage curve **increased to three times its initial value**. When repeating this test by placing the cable under the test bone the resulting amplitude of the voltage curve was **decreased by 1/3 of its initial value**. This implies that the bone can detect the direction of the current." Note 9.6ae p8

The experiment demonstrated the sensitive nature of the piezoelectric properties in bone that would certainly allow the trained dowser to detect a change in the Geofield. The experiment also demonstrated that bone could detect other aspects of the field, such as current direction or strength.

Denying the Piezofield Evidences

In 1998, the United Nations published a 147-page *Manual on the Forecasting of Natural Disasters: Geomagnetic Methods*. The manual was written and submitted by five Chinese scientists, bringing new insights to a field some American scientists continue to view as "pseudoscience nonsense." One physicist rejected the findings outright, publishing a paper in the *EOS* journal titled, *Reply: U.N. Should Have Sought Expert Advice*. The paper began with the following sentence:

"The **legitimate** scientific community needs to be alerted to the expenditure of considerable public funds for **pseudoscientific projects** that build false hopes of protection from geophysical hazards." Note 9.6af

As expected, this scientist saw *himself* as being the "legitimate" scientist, and the five Chinese scientists as pseudo-scientists! One wonders what happened to "peer-review," or were all the editors at *Eos* (Transactions of the American Geophysical Journal) this biased towards a new understanding of earthquake prediction. The American went on to state what "recognized" researchers already know:

"Most **recognized** researchers are **skeptical of the premise of this research**, which is the **claim that earthquakes are preceded by electromagnetic precursors**." Note 9.6ag

The American concluded his paper, which was void of experiments or observations refuting the Chinese scientist's report, with his closing statement:

"...I believe that the publication **has no scientific value** as a document on the use of geomagnetic fields for the 'forecasting of natural disasters.' Instead, the manual provides clear and reliable evidence of the misdirection of public funds for **pseudoscientific nonsense**." Note 9.6ah

Thankfully, not all American scientists share his narcissistic view. Friedemann Freund, a NASA researcher, responded more open-mindedly in a paper to the *Eos* journal rebutting the skeptical American scientist's views:

"Lurking behind his angry words is an **enmity that many seismologists seem to harbor toward those who dare think about earthquake prediction**. For them, earthquakes are to be objectively and dispassionately dissected. They proudly point out how much seismology has contributed to our knowledge about the spasms of the Earth. Trying to predict such spasms is a dangerous game. Any ever-so-slightly optimistic assessment

Fig 9.6.8 – Although its popularity has waned, there are many present-day dowsers. Some are involved in mysticism, but others are strictly practical, using their dowsing skills to earn their living finding underground water, modern day utility lines, or mineral deposits. Recent research shows that human bone, such as those in the forearm, exhibits a piezoelectric reaction, responding to natural energy fields, like the Geofield. The sensitivity can apparently be developed so that the dowser can detect minute changes in the field by moving over the ground with the metal rods.

of mankind's chances to forecast such natural disasters may raise false hopes." Note 9.6ai

Freund went on to comment about the unfounded "pseudoscientific nonsense" moniker given to the research:

"Pseudoscientific nonsense? Wishing to understand the signals that come from beneath our feet **remains an important question of our time, inside and outside the United States.**" Note 9.6ai

In contrast to the American skeptic, who provided no scientific evidence himself to back his claim, Freund performed a series of "benchmark experiments" that in 2005 produced electric current from large blocks of granite and high-pressure presses:

"Earthquakes are the result of stresses that build up in the crust of the Earth, most often associated with plate motion. **Stresses on rock may produce other observable phenomena in advance of the actual rupture (earthquake)**. One of these observables may be caused by weak currents generated in the rock by increased pressure associated with the build up of stress. This has been the focus of research by **Friedemann Freund** and co-workers. They have approached this problem from both a theoretical and experimental basis, and **recently conducted a series of benchmark experiments which show that stresses on igneous rock (like granite) convert mechanical energy into electric current**." Note 9.6aj

Freund reported that "enormous" ground currents were produced from rocks under stress:

"Thus our model, simple as it may be, points to the possibility that the **often reported** pre-earthquake low frequency EM emissions arise from ground currents flowing deep in the Earth's crust. The ground currents may be very powerful. For instance, taking the currents flowing out of the squeezed end of the granite slab in our experiment, we may ask what would be the current flowing out of a cubic kilometer of granite or gabbro in the crust, all other conditions being the same. The answer is a surprisingly large value, somewhere **between 100,000 and 1,000,000 amperes. Since huge volumes of rocks – tens of thousands of cubic kilometers – come under increasing stress during the build-up of large earthquakes, the ground currents could indeed be enormous.**" Note 9.6ak

The Geofield Model could not be any clearer. Freund's lab experiments demonstrated that the kinds of current needed to produce the Earth's energy field could be created by gravitational-friction stresses on crustal rocks. All we need to do is to multiply the outcomes from the lab with the planet-sized stresses of earthtide.

It is only the recent discovery of the earthtide that made the Geofield Model possible. Before GPS satellites, scientists were unaware of how much movement the Earth's crust experienced every day. Such ignorance is the number one reason seismologists discount piezoelectricity as the origin of the geofield. Before the discovery of earthtide, researchers assumed crustal rocks were only capable of producing pulses of energy *when earthquakes occur*. And conventional thinking at the time was that earthquakes were relatively rare. However, because earthtide is a *continual,* daily vertical oscillation, it is *the* 'piezofield generator' of the Earth's Geofield.

Another interesting confirmation from Freund's research was the marked difference in piezoelectric current potential between *glass* and anorthosite (a common silicate crustal rock):

"Hence, the **currents** drawn through the **glass** tile under stress and through the **anorthosite** tile under the identical conditions **differ by a factor of 10**^{7}. " Note 9.6al

The Magma Pseudotheory chapter established that if the Earth was once melted, the crust would be relatively glassy, which is of course, not the case. The piezoelectricity in the crust is confirmation of this fact.

American scientists have been generally slow at seeing the connection between earthquakes, piezoelectricity, and the Geofield, but the Russians, Japanese, and other Europeans have known about the connection for years. They simply lacked a model to connect the pieces of the puzzle. A statement in the *European Seismological Congress 2002 Report* cites a *number* of Piezofield Model evidences:

"Survivors reported **lightning** spirits **before the Kocaeli earthquake** in 17th August 1999 and **the spark over the fault line above the sea and land during the earthquake**. Unusual behavior of some animals were recorded by the security cameras just before the earthquake. Some people determined that their **watches had stopped** without any technical reason a few days before the earthquake and those problems disappeared after the earthquake. These observations led us towards one of the major measurable precursor of the earthquakes that might be the change in electric field close to the surface since (a) the watches had **quartz crystals and piezoelectric property is reciprocal,** (b) some gases locally **lightens** because of electric discharge due to electric field strength inside the atmosphere (c) serotonin change is determined due to electric field change under the laboratory conditions where the serotonin is a behavior affecting hormone in **animals** (d) and it is known that long animals such as snakes tends to stay vertical to the electric fields in order to decrease the potential difference on its body." Note 9.6am

One of the greatest friends of science is time; it eventually reveals scientific truth. It may take decades or even centuries for old theories to pass away—but no one can change the truth. An entire book could easily (and surely will someday) be written to expound upon the manifold Geofield evidences, but there is one last evidence of the Earth's Geofield that we need to address before wrapping up this subchapter—the Geofield Cycle.

The Geofield Cycle Evidence

Fig 9.6.9 shows the Geofield Cycle diagram first seen in the Magma Pseudotheory chapter. Few people, including many scientists, are aware that the poles of the Earth's energy field (the 'magnetic' north and south poles) are not geographically stable. They move in a daily, elliptical course and drift in a hundreds-of-years cycle around the Earth.

When the Geofield Model was outlined in the last subchapter, we asked four FQ's:

1. What causes the field and why does it exist?
2. How does the field instantly fluctuate?
3. What causes the magnetic pole daily cycle?
4. Why do the magnetic poles move annually?

With the evidences of the Earth's Geofield established, we can answer these questions:

1. The 'magnetic' field is actually the Earth's Geofield. It exists because of the piezoelectric properties of the Earth's crust and effects of earthtide, which is the constant vertical oscillation of the Earth's crust, which is caused by the daily cycles of the Earth's rotation and the gravitational influences of the Sun and Moon.

2. The Geofield can fluctuate because of stress accumulation along faults distributed throughout the crust. When the built-up stress is released suddenly, the Geofield experiences an instant fluctuation. The accumulation of energy occurs because of the effects of earthtide and the sudden release causes earthquakes. Thus, earthquakes (both large and small) are a major cause of the Geofield fluctuations. Silent earthquakes and earthquake swarms that may last for days are also responsible for fluctuations in the Earth's energy field.

3. The Geofield poles experience daily cycles due to the diurnal rotation of the Earth. However, the Geofield does not move in the daily elliptical pattern simply because of solar wind as geoscientists have assumed. If this had been the case, the daily movements would be very similar from day to day—but they are *not*. They vary widely because of the Moon's gravitational contribution to the earthtide is twice that of the Sun. Thus, the Moon's daily progression of about thirteen degrees and the variation in its distance from the Earth affect the crust differently from day to day, which shows up in the daily cycle of Geofield. The variation amounts to about 53 miles (85 km – see Fig 9.6.9) each day.

4. The Geofield itself is moving across the surface of the globe in a long-term cycle tied to the astronomical cycles of the Earth, Moon, and Sun. The Geofield North Pole has travelled longitudinally just over ten degrees during the past one hundred years. If the path and rate continue, the pole will arrive in Russia after another century and circle the globe in about 3,000 years. But we have only the last century's measurements to go on. Continued observations and future astronomical tools will

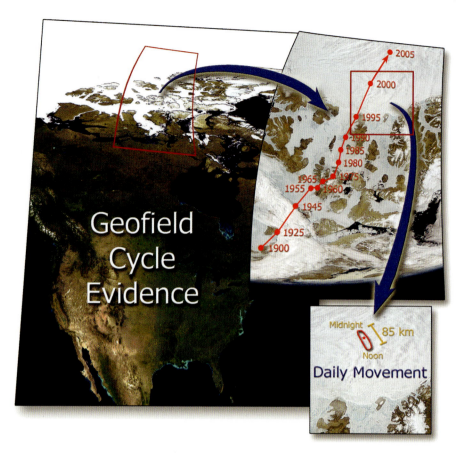

Fig 9.6.9 – This diagram illustrates the amazing Geofield Cycle, which consists of both daily and yearly movements of the Geofield poles (magnetic poles). Movements of the poles and Earth's entire energy field can be understood by applying the astronomical movements of the Earth, Sun, and Moon. Earthtide and the piezoelectric mechanism explain the cycles of the Earth's Geofield, whereas the magma-dynamo-theory cannot.

allow the predicting of the Earth, Moon, and Sun's position in the Geofield Model context, and a predicted movement of the Geofield North Pole.

Why Has the Origin of the Geofield Been Missed?

This question can be answered with one word—dogmagma. Old theories die hard, and though the magma dynamo theory was never proven, it is deeply ingrained in modern geology, despite being in opposition to any **known** method of energy field generation. And then again, no one has provided a better explanation of the planetary energy fields. The second reason the origin of the Geofield was missed is that technology only recently allowed GPS measurements to identify the earthtide. The degree of vertical movement of the crust was unknown prior to the advent of this important technology.

Now with the Piezofield Model, the crustal movement and stress buildup of earthtide *explain* the anomaly shown in the magma pseudotheory chapter. From an 1820 encyclopedia:

"Since that time [1640] it has been found that the magnetic needle not only varies after a considerable period, but that it is **continually fluctuating**, so that the variation of it may generally be observed within the period of an hour or two, and often in a much shorter time.

"The declination is not only subject to a continual variation in the same place, but it is different in different parts of the world. It also varies **differently** in each particular place; so much so, that notwithstanding the exertions of the greatest philosophers and mathematicians, **no theory nor rule has been discovered which might furnish the means of foretelling with accuracy the declination of the magnetic needle, for any future period, at any particular place**." Bib 123 under "Declination"

The Geofield is "continually fluctuating" because the astronomical locations of the Earth, Sun and Moon are constantly changing. As we begin to install sensors in the crust to measure the energy field and stress levels along with the astronomical positions of the celestial bodies that affect earthtide, earthquakes and changes in the Geofield will become easier to predict.

KTB borehole researchers realized that energy field "anomalies" were controlled by some "geological" process:

"Field studies have detected a large number of high-conductivity **anomalies** in the crust which **do indeed seem to be controlled by tectonic, magmatic, or other geological processes**. The **significance** and the **origin** of those electrical low-resistivity **anomalies** in the Earth's crust are **still a mater of debate**." Note 9.6an

Unfortunately, the researchers *were working from within the magma paradigm,* where everyone '*knew*' that the Earth's electric field was coming from magma. Therefore, the faults and **tectonic** areas that produced the strongest electric fields were simply overlooked, dismissed "anomalies" in the Earth's crust.

Fig 9.6.10 illustrates how the "ocean of iron" and direction of the "Magnetic North Pole" are illustrated incorrectly in modern science. There is no evidence that the energy field comes from magma or that the field is *rotating around the pole* as the NASA diagram illustrates.

It took investigators from outside the field of geology to move past the magma mindset into a more successful earthquake-technology mindset that measures the Geofield by monitoring known earthquake zones. Replacing magmatic crustal heating with gravitational-friction heating moves the friction-producing piezofield to center stage. Fortunately, Nature's show is beautifully simple and simply beautiful when the truth is known.

Geofield Evidence Summary

Knowledge of the Earth's energy field dates back at least to the invention of the compass. Several millennia ago, it was observed that certain iron rocks (magnetized by lightning strikes or formed in a strong piezofield during the Universal Flood) attracted other iron rocks. These were known as loadstone or 'magnetic' rocks. The magnetic rocks displayed the interesting characteristic of pointing in a certain direction when placed on a floating object in water. Today's compasses still use a magnetic needle freely suspended so that it can point towards 'magnetic' north.

This is a diagram of the Earth's magma dynamo and movement of the energy field's north pole as envisioned by NASA.

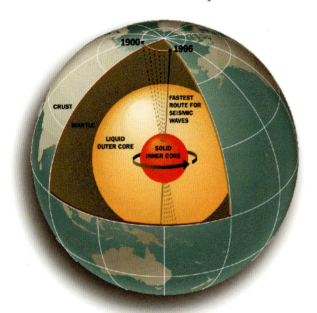

The problem is that the "ocean of iron" shown in orange and the direction of the pole movement are incorrect.

Fig 9.6.10 – This NASA diagram is from an article titled, "Earth's Inconstant Magnetic Field." The article states that, "Earth's magnetic field comes from this ocean of iron, which is an electrically conducting fluid in constant motion." However, if the heat is "about as hot as the surface of the Sun," it would destroy the energy field. Moreover, the circular motion of the field's north pole is also incorrect. The correct movement can be seen in the previous figure, Fig 9.6.9.
Diagram accessed 9.2.09 and adapted from:
http://science.nasa.gov/headlines/Y2003/29dec_magneticfield.htm

Magnets move other magnets, which is one reason the magnetic needles used in compasses contributed to the idea that the Earth had a 'magnetic' field. However, electric fields *also* move magnets, and the piezofield, which is an electric field created by the crust's piezorocks, now clearly explains the characteristics of the energy field of the Earth, whereas the existing modern science magnetic field theory cannot.

Scientists have long-known that discovering the origin of the Earth's 'magnetic' field was "important" and described the problem as being one of the most important unsolved problems in physics. Although the discovery of the piezofield in celestial bodies is significant, and though it changes the paradigm of many scientific fields that were dependent on the old magma-dynamo-theory, the Earth's piezofield, or Geofield discovery has special significance because we are just beginning to comprehend how important the Geofield is. Life on this planet depends on the protection the Geofield provides from the Sun's harmful radiation. Furthermore, today's technology, from the electric grid to satellites, depends on the Geofield acting in a predictable manner.

While "lightning remains a mystery" in meteorology, it too, is direct evidence of the Geofield, along with the fair-weather condition and the earthquake-energy field connection. Borehole Geofield evidence, the art of dowsing, and the Geofield cycles all validate the reality of the Geofield Model.

Even though the Geofield evidences are powerful, the Geofield's relationship with the Weather Model is equally significant. Hyquatherms *and* the piezofield both share the same origin—gravitational tidal friction. This interrelatedness is powerful testimony to the truthfulness of both models.

We will now present other Geofield Evidences that are *global* in Nature, evidences that will help us understand more fully the marvelous natural beauty and order of the Geofield surrounding us—the Earth's aurora.

9.7 The Aurora Evidence

This subchapter deals with one of Nature's most beautiful sights—the shimmering, magnificent aurora. The mysterious aurora first captivated humans long ago and they still do today. Believed to be caused by outbursts from the Sun, little is known about how they connect with the Earth's energy field. In this subchapter, we will introduce the piezofield evidence that helps explain their origin. Coastal auroras and their importance will be discussed as will the Geofield-Continent Connection. Both aurora evidences will take on new meaning when seen in the light of the Weather Model. The continent connection will become apparent after seeing its correlation with lightning strikes.

The Aurora Mystery

For most people not living near the north or south pole of the Earth, the Aurora is a phenomenon seen only second hand, on the pages of a magazine, or some other media, but these incredible displays, known also as 'northern or southern lights,' are a sight to behold and can only be appreciated in live splendor. Seen at night from the ground, Auroras form slow-moving sinuous sheets of red or green light reaching high into the atmosphere.

From space, the aurora, seen in Fig 9.7.1, encircles the north pole of the Earth. Auroras also encircle the poles of other planets. Saturn, also seen in Fig 9.7.1, displayed its auroras for the Hubble Space Telescope.

Being somewhat acquainted with these wonders of Nature, a simple, fundamental question comes to mind:

FQ: How is the aurora created?

On a small-scale, there is a measurable energy field around a magnet or a wire with electricity running through it. The energy field exhibits polar properties, but by itself, the field does not create lights at the poles, no matter how strong it is. The *interaction* with another energy field is required for this. Plan-

Aurora Evidence

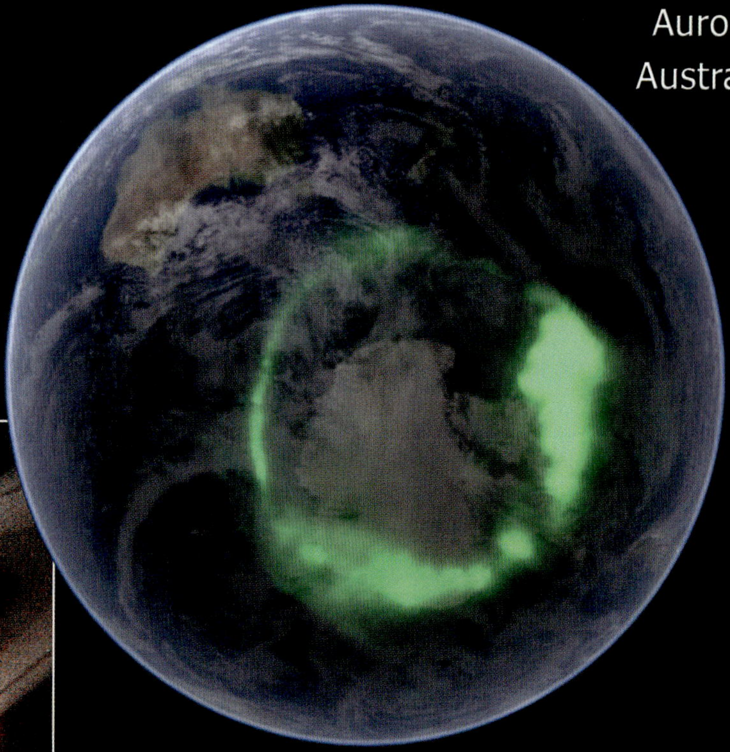

Aurora Australis

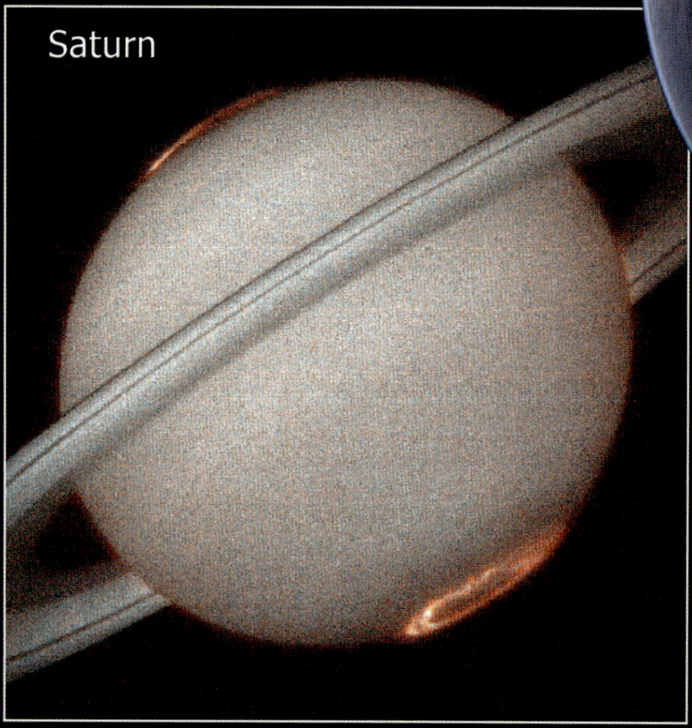

Saturn

Auroras are created by the interaction between the Sun's and Earth's energy fields.

However, we can't understand the aurora unless we comprehend how the energy fields are created.

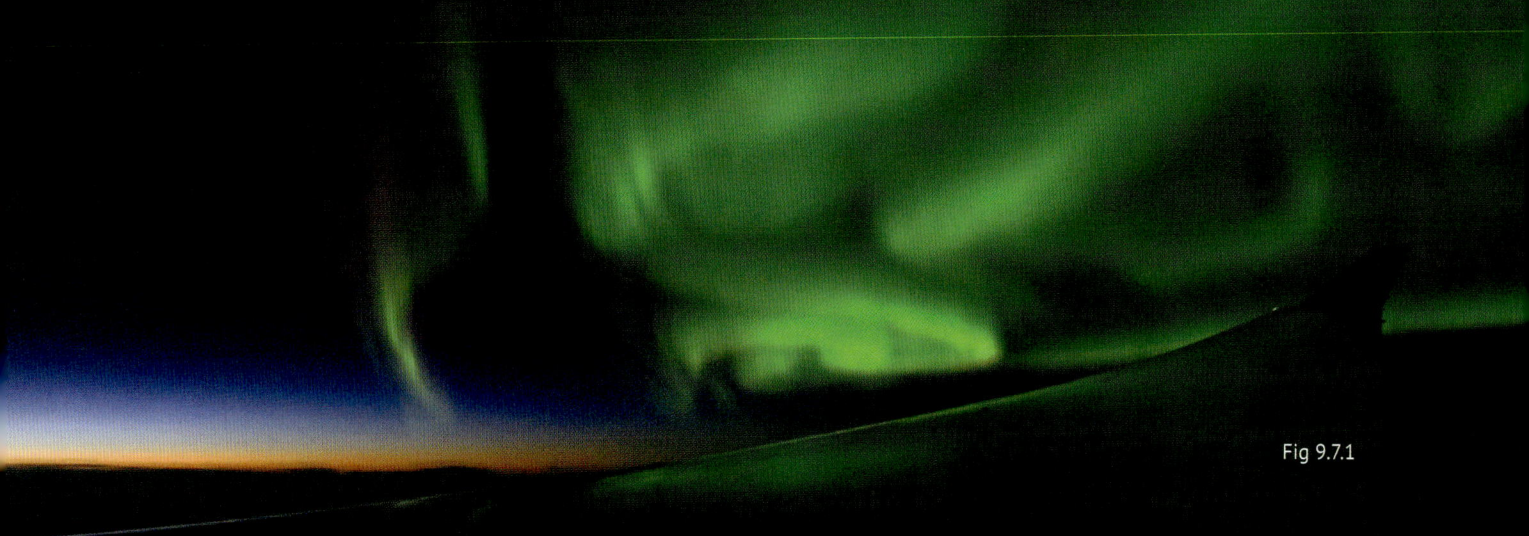

Fig 9.7.1

CHAPTER 9 THE WEATHER MODEL

The beauty of the aurora—

is only surpassed by understanding its true origin.

SUBCHAPTER 9.7 THE AURORA EVIDENCE

Coastline Auroras

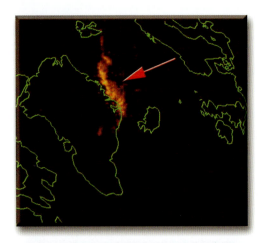

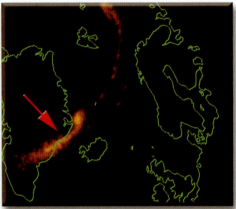
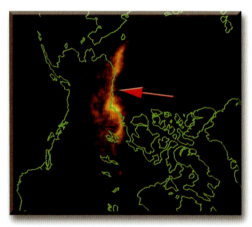

"No theory can presently account for the formation of such coastline auroras."

Fig 9.7.2 – Coastline Auroras, as their name implies, are aurora that are amplified along continental coastlines. Why would auroras be energized along coastlines? Researchers say no current theory can account for this phenomenon. But researchers are not aware of the piezofield that is generated along the coastlines where continental plates experience elevated movement due to earthtide. The frictional grinding of continental plates and fracture zones create the highest intensities of the Geofield, which drive auroral events.

lines of magnetic force into the Polar Regions. These particles are boosted in energy in Earth's upper atmosphere, and when they collide with oxygen and nitrogen atoms, they produce dazzling auroral light."
Note 9.7a

One problem with this explanation is that it does not deal with the true nature of the Earth's energy field. Furthermore, the Earth does ***not*** have a stable 'magnetic' field nor are auroras merely the result of coronal mass ejections. In reality, auroras, like all other weather phenomena, occur *in patterns associated with astronomical cycles*. Auroras occur most often each year during the spring and autumn when the Earth is near its equinoxes:

"Geomagnetic storms that ignite auroras **actually happen more often during the months around the equinoxes. It is not well understood why geomagnetic storms are tied to Earth's seasons** while polar activity is not. **But it is**

etary scientists have known for a long time that one energy field interacting with the Earth's energy field is the Sun's. However, the true origin of both energy fields continues to elude modern science. If we cannot correctly explain how these fields are generated, how can we expect to explain how the aurora is created?

We will use this subchapter to show that without a correct understanding of the Geofield, *the origin and nature of the aurora cannot be properly explained.* To begin, we read a summary of how auroras are formed, according to modern theory:

"**The origin of the aurora begins on the surface of the sun** when solar activity ejects a cloud of gas. Scientists call this a coronal mass ejection (CME). If one of these reaches earth, taking about 2 to 3 days, it collides with the Earth's magnetic field. This field is invisible, and if you could see its shape, it would make Earth look like a comet with a long magnetic 'tail' stretching a million miles behind Earth in the opposite direction of the sun.

"When a coronal mass ejection collides with the magnetic field, it causes complex changes to happen to the magnetic tail region. These changes generate currents of charged particles, which then flow along

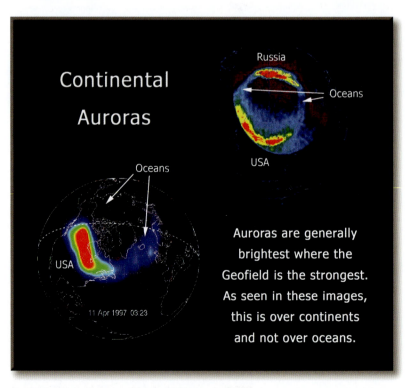

Auroras are generally brightest where the Geofield is the strongest. As seen in these images, this is over continents and not over oceans.

Fig 9.7.3 - These diagrams show the typical occurrence of northern auroras. The brightest incidences occur over continental areas (red and yellow) whereas the oceans are consistently weaker. This evidence does not support the dynamo theory, but it does support the Earth's Piezofield Model.

CHAPTER 9 THE WEATHER MODEL

known that during spring and autumn, the interplanetary magnetic field and that of Earth link up." Note 9.7b

What else happens around the world during these two times of the year—the monsoon season develops. The monsoon season is a rainy weather period often accompanied with large cumulus (hyquathermal) clouds, which develop during specific times of the year. But how are the aurora and monsoons connected?

The Aurora and monsoons share a common origin— gravitational tidal friction in the crust.

A number of evidences show a direct connection between the aurora lights and the Earth's piezofield.

A Piezofield Answer

The Crater Model (subchapter 7.9) demonstrated that a majority of the solar system's craters are hydrocraters rather than impact craters. However, impacts do occur, though much less frequently than impactologists have supposed. A large impact on a rocky, quartz-mineral surface would behave similar to a very large earthquake, disrupting the planet's piezofield. This prediction was verified on 16 July 1994, when the comet Shoemaker-Levy 9 sensationally impacted Jupiter. Gene Shoemaker, the comet's namesake and co-discoverer noted one "wholly unanticipated" observation:

"One **wholly unanticipated phenomenon**, observed 45 minutes **after the impact** of one of the brightest nuclei, was the **release of charged particles from the magnetosphere, which produced new auroral spots in both the northern and southern hemispheres.**" Note 9.7c

This event supports a piezofield origin of Jupiter's energy field, and is evidence of a rocky quartz surface. Obviously, no 'magma' inside Jupiter was affected by the impact, nor would a supposed magma dynamo explain new auroral spots created by the impact-stressed rocks on the surface. However, a Jovian piezofield easily explains why new spots would occur following a recent crustal impact.

Coastline Aurora Evidence

One fascinating piece of evidence supporting the true creation of auroras is the Coastline Aurora Evidence in Fig 9.7.2. Three investigators from the University of Iowa evaluated approximately 9,000 images taken by the Polar Visible Imaging System satellite in 1997. They found the following:

"Humans are not alone in showing a preference for coastlines. New Observations suggest that auroras—those brilliant curtains of light seen in the night sky of the polar regions—may also favor Earth's coastlines on occasion. A camera designed to image the Earth's aurora from NASA's Polar spacecraft has provided strong evidence that coastlines do sometimes influence the spatial distribution of these colorful, shimmering lights, **the photographs produced by Polar's state-of-the-art camera shows auroral arcs sometime following coastlines for hundreds of miles. At other times, the coastlines appear to deflect or dim the auroral arc. No theory can presently account for the formation of such coastline auroras.**" Note 9.7d

Why does no theory account for the formation of coastline auroras? The last paragraph of the article provides the answer:

"The present finding of auroral emissions which are affected by the presence of a coastline offers a further challenge to our understanding **because it has been always assumed that these lights were completely controlled by processes in and above the Earth's upper atmosphere, and not occasionally by the surface sea and land masses.**" Note 9.7d

Coastline auroras do not fit current aurora theory because researchers *assume* auroras are controlled mainly by the Sun's energy. In fact, auroras are directly connected to continental landmasses through the Piezofield and Geofield Models. Coastlines represent the generalized edge of continental plate boundaries where earthquakes of increased frequency and intensity occur, and where piezofield energy is produced. It is here, directly above these coastal regions where the atmosphere is excited. The images in 9.7.2 and 9.7.3 show that the Geofield, as well as the Sun's energy field, are contributory to the auroras. It is also possible to see why the Earth's position twice a year during the equinoxes causes auroral lights and monsoonal storms. The unique stresses present during these periods produce gravitationally induced earthtides that are higher than normal.

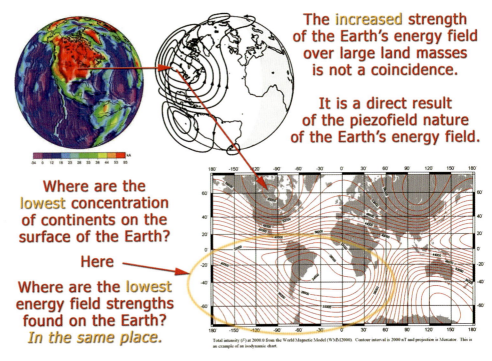

The Geofield-Continent Connection

The increased strength of the Earth's energy field over large land masses is not a coincidence.

It is a direct result of the piezofield nature of the Earth's energy field.

Where are the lowest concentration of continents on the surface of the Earth?
Here →

Where are the lowest energy field strengths found on the Earth?
In the same place.

Fig 9.7.4 – This diagram illustrates high and low areas of the geofield on a global basis. The images and map are from the USGS and NASA; they show field strengths during 2000. The areas of highest strength (red in the upper left globe) are over the landmasses of North America, Russia and an area south of Australia, whereas the area of low strength is circled in yellow. The area of lowest field strength corresponds to the area **lowest in continental landmass**. This is not a coincidence. Large landmasses produce the planet's piezofield because of elevated concentrations of piezoelectric rocks in the crust.

SUBCHAPTER 9.7 THE AURORA EVIDENCE

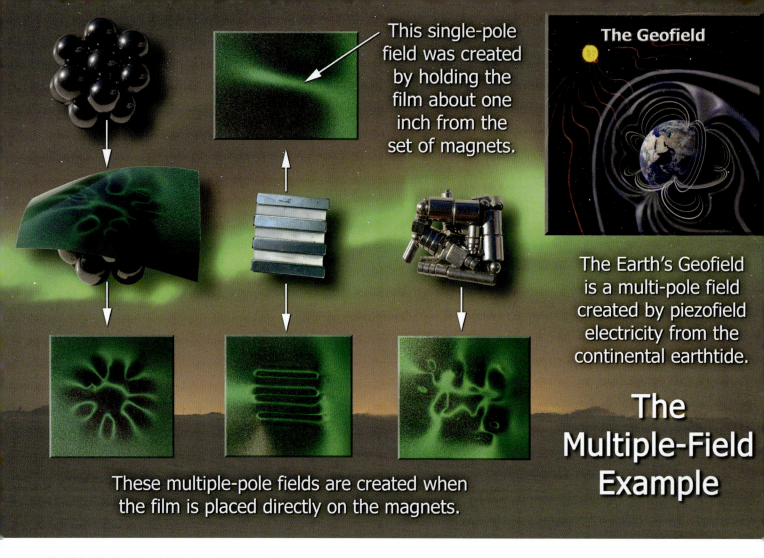

Fig 9.7.5 – The Earth's multi-lobed geofield is easily illustrated by using groups of magnets and a special green, magnetically reactive film. When the film is placed directly over the magnets, we observe multiple fields, which can be seen in the three bottom inset images. The Earth's multiple energy fields and their origin have been a mystery since the Earth's Geofield was first discovered. They are formed by the gravitational stressing of piezoelectric continental rocks caused by the effects of the Moon and Sun. The familiar single-pole field, shown in the top (center inset) image, is found in most textbooks, but does not represent the true multi-lobed nature of the Earth's energy field.

Fig 9.7.2 shows only three images, but the researchers identified hundreds of instances where auroras were "indeed truly aligned with the coastline":

"After eliminating coincidental alignments, they found **100-to-200 cases where the aurora was indeed truly aligned with the coastline**." Note 9.7d

Another example showing the connection between the aurora and the Earth's piezofield is seen in Fig 9.7.3. Two enhanced auroral images are superimposed over Earth backgrounds clearly showing that the brightest portions of the aurora are located over landmasses.

The Geofield-Continent Evidence

Knowing how important the continents are to auroras, we can look a little deeper into why this is so. If the geofield is a piezoelectric field generated in the Earth's crust, the field should be strongest where the greatest concentration of piezoelectric quartz rocks is located.

FQ: Where is the greatest concentration of piezoelectric quartz rocks?

Not only are the oceanic crusts far from the surface, they are generally made up of dark, basalt-like rocks containing only a modest amount of quartz. Because of this, the oceanic geofield should be generally weaker than the continental geofield.

Therefore, we can make a prediction: if the geofield (responsible for auroras) is actually generated primarily by continental piezoelectric rocks, then, the area of greatest intensity should lie generally over the continents. Fig 9.7.4 shows that this is *exactly* what is found.

The Geofield-Continent Connection shows up in Fig 9.7.4. Areas of lowest field strength correlate to the area with the lowest concentration of continental landmass: over the Pacific and Western Atlantic Ocean areas. The colored globe toward the top left of Fig 9.7.4 shows the high strength of the field (shown in red) over North America. It provides a stark contrast to the low-strength oceanic areas (blue and purple) around the world.

Of course, the geofield does move around, but the movement can be explained by the astronomical movements of the Moon and Earth with respect to the Sun. Even so, the continents still dictate where the majority of the high and low strength energy-field areas will be.

The Multiple-Field Example

To help visualize the multi-lobed Geofield-Continent Connection, we will look at the magnetic field created by a group

Magsat Evidence

Original Image

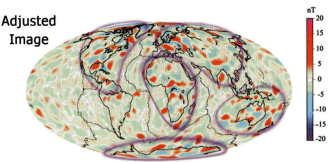

Adjusted Image

In this adjusted image, the mid-range colors have been muted, leaving only the strongest and weakest energy fields. This reveals the **source** of the energy fields, which is clearly the continents.

Fig 9.7.6 – Magsat was a 1980s satellite sent to measure the "magnetic" field of the Earth. The original image, at the top of this NASA diagram is difficult to interpret because the colors are seemingly evenly distributed. The areas of highest and lowest field strength (shown in the red and blue areas) are hard to observe because of the mid-strength fields shown in green and yellow. By muting the green and yellow areas, a **compelling new** image of the Earth's energy fields was revealed. The adjusted image exposed five areas (circled in purple) with the highest and lowest individual field strengths (red and blue areas). These were the strongest sources of the overall geofield. These five areas also correspond to the primary continents, the exception being South America, which is dominated by the two largest oceanic areas on either side of the continent.

Frequency of Lightning Strikes

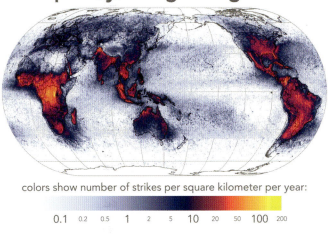

Fig 9.7.7 – Auroras and lightning strikes are the two largest and most common electrical phenomena observed in the atmosphere. Both are connected to the energy field of the Earth. What do they have in common? As is easily seen above, lightning strikes occur primarily **over continents** just like the auroras do! Once again, this refutes the dynamo theory but supports the Geofield Model. Image based on data from http://thunder.nsstc.nasa.gov/data/ and created by Citynoise @ Wikipedia.

of individual magnets, illustrated in Fig 9.7.5. The magnetic fields of different groups of rare-earth neodymium and hematite magnets are visible using a special magnetically reactive green film. In the diagram, the magnetic field viewed about one inch from the magnets displays a single-pole magnetic field. However, when the film is placed directly over the magnetic clusters, multiple magnetic fields appear in a sinuous, multi-lobed mass.

As was noted in Fig 9.5.5, the 's actual energy field is not a single-pole field; it is composed of countless fields for which modern science has no effective origin. However, multiple fields are generated as continuous lunar and solar gravitational forces deform the Earth's crust, each tug and flex generating a piezofield all across the continental crust.

In our illustration, each magnet is representative of continental land areas. On the continental crust, gravitationally induced flexing compresses, moves, and decompresses quartz-based rocks, creating a piezoelectric energy field, which is similar to the magnetic field surrounding the magnets in the illustration. In modern science, the Earth's 'electromagnetic field' is typically shown as a single-pole field, supposedly powered by magma and a spinning iron core. But this pseudotheory is based wholly on the assumption that there is magma in the deep Earth, a failed notion that should be abandoned.

Another factor that seems to have kept the Geofield hidden from the view of the researchers is the constantly moving nature of the energy field. But this is easily explained by the astronomical movements of the Earth, Moon, and Sun. Each has several cycles that affect the Earth's multi-lobed piezofield in a predictable manner, including the diurnal oscillation and the annual migration of the 'magnetic' North and South Poles. Technically speaking, the poles should be more accurately designated **piezoelectric poles**—not magnetic poles.

Because the Geofield energy lines are constantly moving, areas of low field strength, such as those circled in Fig 9.7.4, can sometimes move over continental landmasses that normally exhibit higher field intensities. Because the ocean surrounding South America is so large, the overall field strength appears to be lower, probably because the strength of the continental area is overshadowed by the vastness of the lower oceanic field.

The Multiple-Field Example helps show the multi-lobed energy fields that exist near the surface of the Earth. It also helps explain how the energy fields affect the beautiful auroral displays over the polar continents. Another observational source documenting the Geofield-Continent Connection is the Magsat Satellite Evidence.

The Magsat Satellite Evidence

The **Mag**netic Field **Sat**ellite, or **Magsat**, was launched in 1979 for a short mission that was completed in 1980. The original image (Fig 9.7.6) included colors representing high, low, and mid-field strengths, making the image difficult to interpret. After removing the mid-range colors, (shown in the lower adjusted image), another direct confirmation of continental energy field is revealed.

Five areas with elevated field strengths emerge in the image. These are highlighted in purple and they correspond to five of the six continents, the exception being South America, which is unique in that it is surrounded by vast oceanic crust. Perhaps, as previously mentioned, the oceanic area surrounding South America contributes to a lessening of the overall field strength.

The aurora-continent-geofield connections are amazing, and it is interesting that the correlations have gone unnoticed for so long. The Magsat project took place over three decades ago with the data being public record since then. The Weather Model and the Geofield Model reveal these important new connections, which help explain much of the Earth's weather.

Lightning-Strike Geography Evidence

Auroras are quite common, but are second in overall electrical disturbances recorded in our atmosphere. Lightning strikes are far more common, with millions of lightning storms happening each year. One reason lightning remains such a mystery is that lightning strikes are believed to be cloud generated. Now that the UM has shown that this is clearly not the case—atmospheric science has nowhere else to turn to identify the origin of the electrical charge. According to modern science theory, the Earth's energy field does not fit into the meteorological equation because it is connected to a magmatic dynamo, which obviously has no connection to lightning strikes. In fact, if lightning were connected to a spinning, mid-earth dynamo, it would be *distributed* relatively *evenly across* the surface of the Earth.

This is certainly not the case. Fig 9.7.7 shows just how **unevenly distributed** the lightning strikes really are. The diagram plainly manifests that the **continents** have everything to do with lightning strikes. Clouds are distributed over the continents *and* the oceans, but the vast majority of the lightning activity occurs over land and near the equator, where the piezofield is at its greatest strength.

Continuing our discussion of atmospheric phenomenon, we turn now to a puzzling northern latitude characteristic of the Earth's atmosphere—the so-called ozone hole.

9.8 The Ozone-CFC Pseudotheory

Ozone (O_3) is comprised of three oxygen atoms that form a molecule harmful to organisms in high concentrations in the lower atmosphere. However, in the upper atmosphere, ozone helps protect the Earth from the Sun's harmful radiation. During the latter part of the 20th century, debate raged over a reduction in the upper ozone layer and a newly discovered (in the 1970s) ozone hole over the Antarctic thought to have been caused by manmade chemicals.

This subchapter addresses some of the misconceptions about the ozone hole and the manmade CFC chemicals. Today, re-

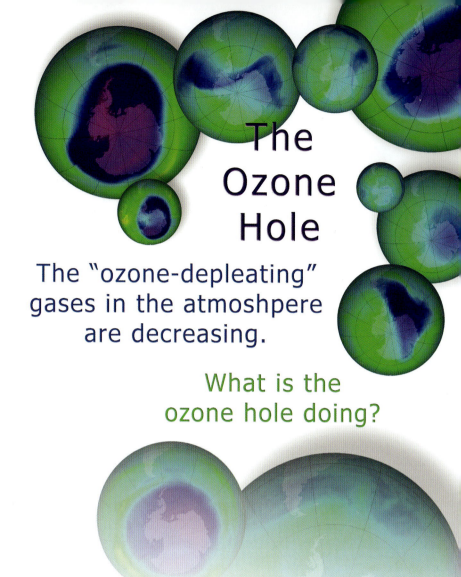

The Ozone Hole

The "ozone-depleating" gases in the atmoshpere are decreasing.

What is the ozone hole doing?

Fig 9.8.1 – The purple and blue areas represent the ozone hole in the upper atmosphere during different times of the year, and during different years. Harmful UV rays can pass through a weakened ozone layer causing illness to humans. Alarmed scientists and media hype led to a worldwide ban on manmade chemicals, called CFCs over two decades ago because it was thought that they were the cause of the ozone hole. However, the science behind the ozone hole is as young as the discovery of the hole itself just a few decades ago. Today there is no solid scientific evidence that the ban on CFCs has had any impact on the size and magnitude of the ozone hole. Instead, it appears that the hole is a naturally occurring, cyclical phenomenon.

searchers continue to be confused about whether the hole is increasing or decreasing in size. Ozone research is relatively new. Unproven claims coupled with political influence and media hype incite fear, spurring action that may have been unwarranted based on the science.

The Man-Made Ozone Hole Pseudotheory

During the 1980s, the "ozone hole" became a familiar talking point around the world. Researchers noticed an area of depleted ozone gas in the atmosphere over the Antarctic. The media jumped on the idea that the depleted ozone 'hole' was manmade from CFC chemicals. Three researchers wrote a book, *Weather: a Visual Guide* (2004) summarized the ozone controversy:

"**Ozone**—a gas that in the stratosphere absorbs ultraviolet (UV) rays from the Sun—plays a vital role in protecting life on Earth from the harmful effects of UV radiation. High levels of exposure to UV light cause skin cancers in humans and also harm many other forms of life. Since the late 1970s, scientists have noticed that each year during late spring a large area of severely depleted ozone, known as the 'ozone hole,' forms over the Antarctic continent; a smaller hole appeared for the first time over northern polar skies in the late 1990s. **The size of these holes is increasing each year**. Investigations have shown that man-made chemicals—chlorofluorocarbons (CFCs)—rise high into the stratosphere and **deplete the ozone layer**. An international treaty—the Montreal Protocol—was signed in 1987 (and later amended) to eliminate certain CFCs from industrial production." Bib 183 p272

An ongoing puzzle is that if CFC chemicals causing the ozone hole were banned several decades ago, why would the size of the holes (over both poles) be "increasing each year?" Before answering this question, two other researchers, in the textbook, *The Atmosphere* (2007), wrote about the "total abundance of ozone-depleting gases in the atmosphere":

"The Montreal Protocol represents a positive international response to a global environment problem. As a result of the action, **the total abundance of ozone-depleting gases in the atmosphere has started to decrease in recent years**. If the nations of the world continue to follow the provisions of the protocol, **the decreases are expected to continue** throughout the twenty-first century." Bib 180 p23

This statement is the basis of this important question:

> FQ: If ozone-depleting gases in the atmosphere have **decreased**, what should the ozone hole reflect?

It doesn't take much to figure this one out—if ozone-depleting gases in the atmosphere have **decreased**, the ozone hole should have also **decreased** in size. That apparently is not the case, as the previous researchers stated, "The size of the hole is *increasing* each year." Of course, the ozone hole cannot be both increasing *and* decreasing in size each year. The researchers are in conflict, but the truth is, both groups are incorrect; the statements from each research group are false.

Ozone-Hole Facts

Take the second statement; "ozone-depleting gases in the atmosphere [have] started to *decrease* in recent years." This was published in 2007, but a NASA article on the subject was written a year earlier, in 2006. The article, NASA *and NOAA Announce Ozone Hole is a Double Record Breaker* states:

"NASA and National Oceanic and Atmospheric Administration (NOAA) scientists report this year's **ozone hole in the polar region of the Southern Hemisphere has broken records for area and depth**." Note 9.8a

If atmospheric ozone-depleting gases, which are allegedly the cause of the ozone hole, "started to *decrease* in recent years" (from the 2007 publication), why did the ozone hole break the all-time "records for area and depth" in 2006?

Apparently, the authors of the textbook *The Atmosphere* had been mistaken because the ozone gases and the hole had *not* decreased in the year prior to 2007, an all-time record-breaking year for the size of the hole! What about the previous claim where researchers said, in the 2004 article that the size of the hole was *increasing each year*—were they correct?

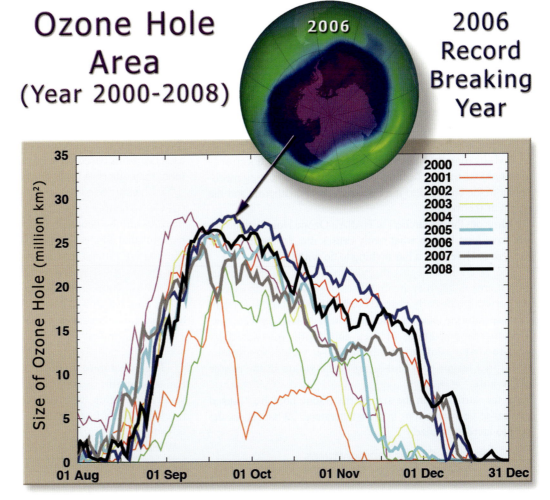

Fig 9.8.2 – Some researchers claim that the ozone hole is increasing in size each year; others say that the ozone-depleting gases in the upper atmosphere have decreased in recent years, but neither of these claims are true. In reality, investigators are confused because there is no indication that the ban on ozone-depleting substances has had any impact on the ozone hole. The graphs to the right show that the size of the ozone hole is both increasing *and* decreasing from year to year, except the record-breaking year of 2006. The real cause of the ozone hole is the change in temperatures over the Antarctic that vary from year to year.

Adapted from a KNMI graph based on data from instruments on the GOME and SCIAMACHY satellites.

SUBCHAPTER 9.8 THE OZONE-CFC PSEUDOTHEORY

We find answers to these questions on a graph from the World Meteorological Organization, shown in Fig 9.8.2. In that figure, the size of the ozone hole is plotted for the years 2000 through 2008. The graph shows data from August to December each year because that is when the hole is the largest.

The authors of the 2004 article who reported that the "size of the hole is *increasing each year*" were incorrect. A careful review of Fig 9.8.2 reveals that each year the size of the ozone hole is oscillating; it is growing **both** larger and smaller from year to year.

Oscillating patterns in weather are a natural phenomena—not man-made cycles.

The graph in Fig 9.8.2 is summarized as follows:
2001 (red) is *up* from 2000 (purple).
2002 (orange) is ***down*** from 2001 (red).
2003 (yellow) is *up* from 2002 (orange).
2004 (green) is ***down*** from 2003 (yellow).
2005 (turquoise) is *up* from 2004 (green).
2006 (blue) is *up* from 2005 (turquoise).
2007 (grey) is ***down*** from 2006 (blue).
2008 (black) is *up* from the 2007 (grey).

From this analysis, we can easily see why 2006 was a record-sized ozone-hole year. That year deviated from the up-down *cycle* and instead of being a down year, was up for the second year in a row—leading to a record-breaking year.

The ozone in our atmosphere is similar to the water vapor and other weather phenomena discussed previously—they exhibit *cycles*.

How are the cycles driven and why are they never quite the same?

Atmospheric and weather phenomena occur in cycles because they are governed by the *astronomical* cycles of the Earth, Moon, Sun, and possibly other celestial bodies. The weather patterns and cycles are never exactly the same because the astronomical cycles and positions of the celestial bodies affecting them vary so widely. It's that simple.

This important new Weather Model discovery came about after investigating the true cause of the heat in the Earth's crust—gravitational-friction heating. Knowledge of the heating mechanism in the crust led to the discovery of hyqatherms and the origin of the weather.

It turns out the researchers at NASA's Ozone Hole Watch web site have long known what really causes the change in size of the Ozone Hole; and it has nothing to do with CFC's:

"The ozone hole begins to grow in August and reaches its largest area in depth in the middle of September to early October period. In the early years (before 1984) the hole was small because chlorine and bromine levels over Antarctica were low. **Year-to-year variations in area and depth are caused by year-to-year variations in temperature. Colder conditions result in a larger area and lower ozone values in the center of the hole.**" Note 9.8b

So, the variations in the size of the Ozone hole are "caused by year-to year variations in the temperature." What causes the hole to increase in size?

Colder temperatures
increase the size of the ozone hole!

The fact that "colder conditions" have existed over Antarctica during the last several years will be important in the next sub-chapter—The Global Warming Pseudotheory.

The Real Ozone Connection—Politics

When a large number of scientists and politicians gather to make laws based on an unfounded and unproven scientific theory, what are we to expect? *Correct* theories allow for *correct* predictions, but bad theories often lead to failed predictions and unnecessary consequences. This is just what happened with estimated ozone losses:

"**Predictions of ozone levels remain difficult**. The World Meteorological Organization Global Ozone Research and Monitoring Project - Report No. 44 comes out strongly in favor for the Montreal Protocol, but notes that a UNEP 1994 Assessment **overestimated ozone loss for the 1994–1997 period**." Note 9.8c

The 1990s were "difficult" years in ozone prediction, but the following decade proved to be no better. Note that a 'no-ozone-hole' episode in 2002 was followed by a near record breaking 2003 hole:

"The appearance of the **near-record size of the 2003 ozone hole confirmed that the 'no-ozone-hole' episode** observed in the year 2002 does **not** denote a **recovery of the ozone layer**. **Despite** the current successful attempts to get a sufficient understanding for the genesis of both extraordinary events, **more observations and further modeling efforts are necessary** to more reliably assess the contribution of various dynamic mechanisms to the recently observed tropo-stratospheric surprises." Note 9.8d

Can "more observations and further modeling efforts" really give the climatologists sufficient understanding of the ozone hole if their ozone-hole origin theory is incorrect?

Unfortunately, the only way for modern-science-based-theory to become natural law is to completely change the theory, which means a complete change in paradigm because modern science has operated in a Dark Age for at least a century. Such a change will only come about after the public is made aware of the egregious errors, even deceptions, demanding change through their elected representatives. Until then politically biased decisions based on unproven science affect all of us, some more than others. A 2008 *Scientific American* article explains:

"A federal ban on ozone-depleting chlorofluorocarbons (CFCs), to conform with the Clean Air Act, is, ironically, affecting 22.9 million people in the U.S. who suffer from asthma. Generic inhaled albuterol, which is the most commonly prescribed short-acting asthma medication and requires CFCs to propel it into the lungs, will no longer be legally sold after December 31, 2008. **Physicians and patients are questioning the wisdom of the ban**, which will have an insignificant effect on ozone but a measureable impact on wallets: the reformulated brand-name alternatives can be three times as expensive, raising the cost to about $40 per inhaler...

"'**The decision to make the change was political, not medical or scientific**,' says pharmacist Leslie Hendeles of the University of Florida, who co-authored a 2007 paper in the *New England Journal of Medicine* explaining the withdrawal and transition." Note 9.8e

When we-the-people in democratic societies demand that true science, based on real observation instead of theory, be used in public decision matters, we-the-people will be best served.

There have been few investigators who understand how theoretical the data on ozone depletion really is, and they are asking for "new plausible mechanisms" that might explain the ozone holes as expressed in a 2008 article in the journal of *Environmental Science and Pollution Research*:

"In the light of these very important conclusions, **it is time that atmospheric scientists begin to think about new plausible mechanisms**, which could fill the aforementioned **gap between the observed and theoretical data of ozone depletion**, thus, improving our understanding of how ozone holes are formed." Note 9.8f

The "gap" between observation and theory remains far too wide in this and many other scientific fields and has for a long time.

Why the Ozone-CFC Theory is a Pseudotheory

During the decades researchers claimed CFCs were creating the ozone hole, they scared the public into thinking that everyone would be burned up by Sun's UV rays if something was not done. But the Ozone-CFC theory was *not* proven through experimentation duplicating what the scientists said was happening in the atmosphere. The experiments replicated only a portion of the theory, leaving the rest to be "modeled" by computers.

There is great danger in not being able to replicate the processes supposedly taking place high above our heads. Assumptions have to be made and this is where mistakes are often found. For example, in a 2007 *Nature* article, a group of scientists from the Jet Propulsion Laboratory (JPL) at the California Institute of Technology found a problem with previous ozone laboratory experiments. Newer ozone experiments produced results different than had been expected. This was reported in an article titled, *Chemists poke holes in ozone theory.* Here are some highlights of that report:

"'Our understanding of chloride chemistry has **really been blown apart**,' says John Crowley, an ozone researcher at the Max Planck Institute of Chemistry in Mainz, Germany.

"'**Until recently everything looked like it fitted nicely**,' agrees Neil Harris, an atmosphere scientist who heads the European Ozone Research Coordinating Unit at the University of Cambridge, UK. '**Now suddenly it's like a plank has been pulled out of a bridge**.'" Note 9.8g

Another researcher noted the seriousness of incorrect measurements:

"'If the measurements are correct we can basically **no longer say we understand how ozone holes come into being**.'" Note 9.8g

Only one experiment showed serious flaws in the CFC-ozone depletion theory, making the evidence that CFCs are the cause of ozone holes very, very weak. The JPL researcher's experiment indicated that ozone reactions with CFCs were much slower than previously stated. In 2009, researchers from Taiwan performed similar experiments claiming their results showed that the earlier readings were still correct, that the rate was faster than the experimenters at JPL had seen. The debate will certainly go on when the evidence is as weak as it is in this situation. However, it is hard to swallow comments like the following after new research indicates that the ozone-depleting CFC theory might actually be incorrect:

"'**We are starting to see the benefits of the protocol** [banning CFCs], but we need to keep the pressure on.'" Note 9.8g p383

What "benefits" were shown in Fig 9.8.2? The size of the ozone hole fluctuates each year with 2006 being the *largest* in the decades since the Montreal "Protocol" was signed in 1987! Where are the benefits the scientists are talking about—especially since many of us have to pay more for air-conditioning and asthma inhalers? It is not clear why we should continue keeping the "pressure on" to reduce *alleged* ozone-depleting substances.

Ozone abundance is cited in the 1993 book, *Investigating the Ozone Hole*:

"Compared to nitrogen and oxygen, **the amount of ozone in the atmosphere is very small—it makes up only 0.01 percent of the atmosphere.**" Note 9.8h

The problem with the statement is that it shows just how much modern science does not know. An entire book on one subject—the ozone hole—and the author completely missed the mark on how much ozone is even in the atmosphere. The "0.01 percent" figure stated is only *one out of ten thousand*, but a newer, widely accepted 2007 textbook cites a much lower amount:

"There is very little **ozone in the atmosphere**. Overall, it represents just **three out of every 10 million molecules**." Bib 180 p19

This figure is *several orders of magnitude smaller* than the figure from *Investigating the Ozone Hole*. The point is that *what we think we know* about the basics of many modern science subjects has turned out to be incorrect. The reason errors show up in such great numbers in modern science is that it is laboring under the Dark Age of science, and has failed to question everything with an open mind.

> FQ: If meteorology does not understand the origin of weather, how can meteorology understand the origin of ozone holes?

This question deserves some deep pondering.

Ozone Hole to Global Warming

The Weather Model's completely new model for understanding the weather and its origin sheds new insights about the fallacy that humans are influencing global weather. Reasonable restrictions on manmade chemicals harmful to the environment should always be considered—*but with data from real science.* The manufacture of CFCs, or any harmful chemical for that matter, must be subject to scrutiny and safety procedures before, during, and after manufacture *if they are shown scientifically* to be of significant harm to the Earth.

The ozone hole is a pebble in a large pond compared to the idea of global warming. Banning CFCs did affect everyone economically, to some degree, but probably minimally. Implementing a ban on carbon dioxide production because duped scientists and politicians *believe* (sound familiar?) that the world is warming because of manmade CO_2 emissions will severely harm the economic condition of every nation.

> If the science used to understand the ozone hole was incorrect, why would the **same science** be any different about global warming?

9.9 The Global-Warming Pseudotheory

What other science topic today has caused more confusion and debate than global warming? If the 'modern science' behind the global warming issue is so clear-cut, why are there so many people questioning it?

This subchapter will show that the scientific attitude towards the so-called manmade warming of the Earth has changed. As research continues to come forth, long-term global warming and cooling trends are becoming more apparent. Carbon dioxide and sea level changes will be discussed, along with unique insights that come only from a Universal Model paradigm. Finally, we should be able to answer whether or not the climb to the top of the global warming molehill was worthwhile.

Manmade-Global-Warming Pseudotheory Defined

The global warming theory, like so many other modern science theories, mingles truth with error. Separating the truth can be challenging, but with the Weather Model, we have new tools to help make it easier. Changes to global temperatures, whether warming or cooling, have taken place since the Earth was first formed. The current global-warming theory includes the *assumption* that humanity has somehow influenced the weather enough to *cause* global warming. Global warming alarmists claim that the warming trend is **anthropogenic**, or caused by humans, while other scientists disagree. The debate centers on the question of whether or not this is true, although some closed-minded individuals and politicians claim, "the debate is over":

> "If you look at the peer reviewed scientific literature, **the debate is over**." Al Gore

However, a 2007 report from the Intergovernmental Panel on Climate Change (IPCC) is not so sure:

> "'Most of the observed increase in global average temperatures since the mid-20th century is **very likely** due to the observed increase in **anthropogenic** greenhouse gas concentrations.'" Note 9.9a

Even the IPCC, an organization dominated by global warming advocates, could only say that global warming is "very likely" caused by human activity. The debate is far from over because the science has not been subject to the standardization of the Universal Scientific Method. Until reproducible results can show that manmade carbon dioxide is causing the warming of the Earth, the debate will continue. Of course, it would be logical for the scientists claiming to know how the world is warming, to first comprehend how weather really works.

Like all scientific truth, the "truth" in Al Gore's 2006 *Inconvenient Truth* documentary, must stand the test of time. Unfortunately, for those like Gore who have publicized their manmade global warming view, time has not bolstered their side of the debate. Each year more scientists are questioning the claim that humanity is the cause of the latest warming trend. The Internet is now overflowing with "inconvenient truths" about Gore's documentary. See http://ben-israel.rutgers.edu/711/monckton-response-to-gore-errors.pdf for *35 Inconvenient Truths* as an example.

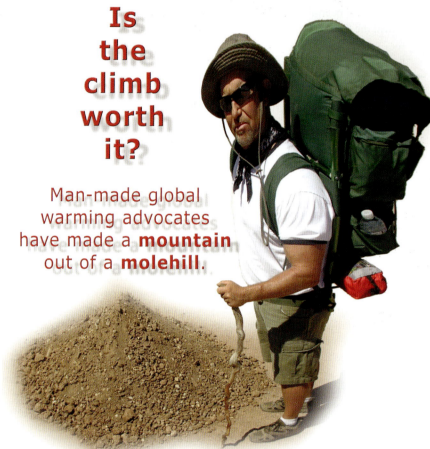

Is the climb worth it?

Man-made global warming advocates have made a **mountain** out of a **molehill**.

Most of us see Nature as a beautiful thing and agree that the world we live in is a magnificent place that should be protected and preserved for future generations. Many well-intentioned people find that being involved in a cause greater than oneself brings happiness and fulfillment and for some, environmental causes fulfill this need. There are environmental 'extremists' that succeed in their quests such as eliminating the building of new nuclear generators in the USA, but these folks are few compared to the majority who want a balanced environmental approach. There will always be many sides to what that balanced approach is, but the manmade, alarmist global warming juggernaut and the resulting 'cap and trade' solution are anything but balanced. Global warming has become one of the causes to which many with good intentions have clamored, and we should applaud those who seek to do good and to make our planet a better place to live.

However, it does not matter how good someone's intentions are, if they are wrong or if they cause harm. As the Universal Model continues to point out, modern science, including meteorology, is in a Dark Age. To escape this, climatologists will have to adjust their theories to include the newly discovered truths of the Weather Model. How big those adjustments will be depends to a degree on how "confident" they are that human activities are the cause of Earth's warming.

In a 2007 *Scientific American* article titled, *The Physical Science Behind CLIMATE CHANGE*:

> "Why are climatologists so **highly confident** that human activities are dangerously warming the earth?" Note 9.9b

We can answer this question with our own FQ:

Since global warming is a *weather* phenomenon, how can climatologists be "highly confident" that human activity is warming the Earth, when they do **not understand** the origin of weather itself?

The "highly confident" climatologists go on to acknowledge that some of the "*most critical components* of the climate processes are *less* well understood":

"Using many models helps to quantify the effects of uncertainties in various climate processes on the range of model simulations. Although some processes are **well understood** and well represented by physical equations (the flow of the atmosphere and ocean or the propagation of **sunlight and heat**, for example), **some of the most critical components of the climate system are less well understood**, such as **clouds**, ocean eddies and transpiration of vegetation." Note 9.9b p69

The heating process of the Earth, which is assumed to be only by the Sun, is supposedly "well understood." But therein lies one of the chief causes of the misunderstanding about global warming. This is typical of what happens: we *think we already know* and that blinds us from finding the truth. Without the knowledge of the First and Second Laws of Weather, which explain the Earth's secondary source of heat—hyquatherms—how can climatologists correctly model the Earth's weather. Or how can they know if "human activities are dangerously warming the earth?" The truth about anthropogenic global warming can *only* come out of a science that explains clearly and simply, how and why climate processes perform the way they do. Even those with good intentions who try to convince the public of manmade global warming will ultimately be largely unsuccessful.

A survey of *Top Priorities for 2009* (http://people-press.org/report/485/economy-top-policy-priority) listed global warming at the very *bottom* of a twenty-item list for the United States. Elsewhere, in 2007-2008 Gallup Poll, "...more than a third of the world's population has **never heard of global warming**." Note 9.9c

If the planet is experiencing significant warming over recent decades, why isn't everyone feeling it and why aren't they talking about it? Are people really unconcerned, or do they believe the global warming 'science' is a political pocket-liner promoted by those looking to make money from hidden 'taxes' on energy?

In one highly successful and widely distributed documentary, the notion of global climate change was laid out by Vice President Al Gore in his 2006 film, *An Inconvenient Truth*. Eventually, Gore won an Academy Award for the production, even though the movie was filled with unfounded and biased data, which in many cases was completely fabricated. Although well presented, the underlying inconvenient truth in Al Gore's documentary on global warming is that he lost the 2000 presidential election. When one presents more images of himself and his life than solid scientific evidence of anthropogenically triggered global warming, it is hard not to see that the *real* inconvenient truth was not about the planet, it was about Al Gore. Perhaps those who backed him were more interested in the next research-grant handout rather than the scientific truth.

"There is No Convincing Scientific Evidence"

Global warming, or as it is more politically correct to say, climate change, is one of the most discussed scientific subjects among leading modern science institutions. It has been foisted on the public through the media and in the schools during the first decade of the new millennium, but it has been given relatively little attention here in the Universal Model. The reason is that the global warming pseudotheory has been refuted by a growing number of scientists. This was not the case during the 1990s, but by the new millennium, thousands of scientists around the world, including many climatologists, began to see serious issues, discovering that the manmade-global warming science was completely wanting and without merit.

Since then a petition (http://www.petitionproject.org/) has been signed by **over 31,000 American scientists** who believe that "there is no "convincing scientific evidence" that the human release of carbon dioxide is causing catastrophic heating of the Earth:

"We urge the United States government to **reject the global warming agreement** that was written in Kyoto, Japan in December, 1997, **and any other similar proposals**. The proposed limits on greenhouse gases would harm the environment, hinder the advance of science and technology, and damage the health and welfare of mankind.

"**There is no convincing scientific evidence that human release of carbon dioxide, methane, or other greenhouse gases is causing or will, in the foreseeable future, cause cat-**

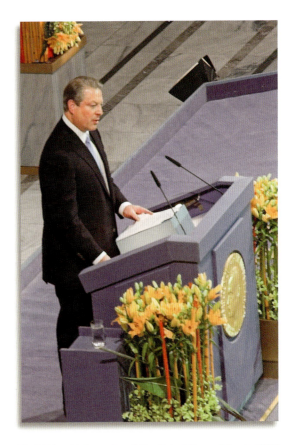

Fig 9.9.1 – Al Gore shared the Nobel Peace Prize in 2007 for his "efforts to build up and disseminate greater knowledge about man-made climate change." Unfortunately, the information disseminated in Gore's *Inconvenient Truth* documentary film and book led to widespread misunderstanding of the facts involved with global warming. *An Inconvenient Truth* was one of the most unscientific documentaries ever produced. It has been highly criticized by many leading scientists for the errors contained within it.

> **If science said global cooling was manmade a couple of decades ago, why should we believe that same science that now says global warming is manmade?**

astrophic heating of the Earth's atmosphere and disruption of the Earth's climate. Moreover, **there is substantial scientific evidence** that increases in atmospheric carbon dioxide produce **many beneficial effects** upon the natural plant and animal environments of the Earth." Note 9.9d

Unfortunately, the Global Warming debate has become *less scientific* and a whole lot *more political*. Interestingly, in the United States, the political left, which includes the mainstream media, has taken the side of manmade global warming, whereas the political right tends to see the warming trend as being driven by natural causes. For the media and well-known figureheads like Al Gore, the debate is presumed to be over and the 31,000 scientists are completely ignored while many global warming activists routinely claim that no legitimate climatologists think global warming is naturally caused.

Among modern scientists, many claim 'consensus' of their colleagues as proof that their theories are valid. But where is the list of thousands of scientists putting their names on a petition supporting man-made global warming? In 1989, the Union of Concerned Scientists assembled a list of 700, but only a couple of climatologists were included. An updated review of their website (ucsusa.org) reveals no list of scientists, although they claim the list has grown to 1,700, including some *economists*. A search for any other global warming advocacy group with a large list of adherents proved fruitless.

Meanwhile, each year produces more scientists, like William Gray, from the Department of Atmospheric Sciences at Colorado State University, who disagree with the global warming camp. Studying hurricanes, Gray says that President Obama and other politicians are incorrect when they try to imply that global warming is causing an increase in the storms. In 2008, Gray stated:

"President-Elect Barack Obama said last week that '**storms are growing stronger with each passing hurricane season**' (implying that this is due to CO_2 increases). He is repeating what Al Gore has been saying for years and what was **implied by thousands of media reports after the damaging Atlantic seasons of 2004-2005**. Polls have shown that a relatively high percentage of US citizens think that human-induced global warming has increased hurricane activity.

"Yes, the Atlantic has seen a very large increase in major hurricanes during the 14-year period of 1995-2008 (average 3.9 per year) in comparison to the prior 25-year period of 1970-1994 (average 1.5 per year). But, have rises in CO_2 been, in any way, been responsible for the recent large upswing in Atlantic basin major hurricanes since 1995?

"**I and a number of my colleagues** believe that this large increase in Atlantic major hurricanes is primarily due to the multi-decadal increase in the Atlantic Ocean Thermohaline Circulation (THC) that is driven by Atlantic salinity variations. These Atlantic multi-decadal changes have also been termed the Atlantic Multidecadal Oscillation (AMO). **These increases are not a result of global surface temperatures or CO_2 increases**.

"Although global surface temperatures have increased over the last century and over the last 30 years, there are now **many observational studies which indicate there has not been any significant long term increase in hurricane frequency or intensity in any of the globe's tropical cyclone basins**." Note 9.9e

Amazingly, even though tens of thousands of scientists continue to state in the most straightforward terms that government should not *try* to control global warming science, world governments continue to implement global warming taxes by issuing stiff regulations on utility and other related companies. These companies simply pass on this hidden tax in the form of higher costs to consumers.

There are literally thousands of websites discussing the fact that global warming is a natural phenomenon. This subchapter will only deal with some of the issues and the unique insights that are connected to the Weather Model. For those who would like to look into objective global warming information, a few of the more scientific global warming websites for review are listed in Note 9.9f.

Global Cooling Was Not Manmade

The debate over global warming is incomplete without knowing the history of global *cooling*. Global cooling is not a topic global warming alarmists care to discuss today, but like the ozone hole incident, there is a lesson to be learned, one we are destined to repeat if we do not learn from it.

From the 1940s to the 1970s, the Earth experienced a period of global cooling, which many investigators thought at the time was an indicator that the Earth was headed toward the next 'ice age.' The cooling alarmists suffered from the same malady today's warming enthusiasts do. After all, modern climatology has proven that it does not have the scientific wherewithal to *predict* effectively short-term weather patterns, let alone long ones, so how can it predict the origin of warming or cooling trends?

Modern investigators are well aware of this, as expressed in this perspective from an article on global climate change in *Scientific American* from 2005:

"Part of the reason that policymakers had trouble embracing the initial predictions of global warming in the 1980s was that **a number of scientists had spent the previous decade telling everyone almost exactly the opposite—that an ice age was on its way**." Note 9.9g

Global *cooling* was thought to be manmade just a couple decades ago, now we are expected to rely on the same science to prove global *warming* is also manmade—where is the logic? The real issue isn't global warming or global cooling, it is one of long term weather cycles and temperature variation. These changes have been ongoing throughout the Earth's existence. The question is whether or not mankind has had a significant impact on the current warming trend by the burning of fossil fuels or by other 'carbon contributing' activity. This subchapter explores several lines of evidence that demonstrate global warming and global cooling are cyclical natural events, and that human activity does not have a significant effect.

Hopefully, new evidence in the Weather Model regarding the origin of weather, El Niño, La Niña, hurricanes, Earthquake Clouds, and a host of other weather phenomena will bring about a shift in the Global Warming and Cooling debate. The true origins of these weather conditions would have remained hidden without recognizing the falseness of the magma pseudotheory, and without the discovery of the Hydroplanet Model upon which the Geofield Model and the Weather Model are based.

Many scientists know that Earth's long-term weather involves cycles of warming and cooling on a global scale, yet no mechanism to account for such changes has been identified by them. The Weather Model mechanism recognizes that a heat source secondary to that of the Sun drives crustal *hyquatherms*, and its effects play an important role in global temperature change.

Long-Term Global Warming and Cooling

In addition to the decadal cycles of global cooling and heating, much longer temperature-change trends have been scientifically recorded. Before looking at these trends, consider the important, not-to-be-overlooked factor of our planet's climate—the constancy of the Sun's heat. The Sun's average yearly radiation is extraordinarily constant, showing no appreciable change since humans first began monitoring it. In fact, the Sun's irradiance varies by less than 1/1,366 during its 11-year sunspot cycle, with occasional, very short term blips. The consistency of radiation finds analog with the Laws of Constant Time that will be discussed in the Time Model, Chapter 20.

Because the Sun's radiation remains so consistent and because it is seen as the *only* natural source of energy affecting the Earth's atmosphere, global warming advocates refer to the greenhouse effect to explain global warming.

In a 2005 edition of the journal, *Nature*, five investigators produced one of the most comprehensive studies on global temperature changes to date. The researchers investigated a large number of physical temperature identifiers spanning the last two millennia. These identifiers included tree rings, ice layers, boreholes, pollen samples, shells, diatoms, stalagmites, and others. Fig 9.9.2 displays the mean temperature trend derived from the research. The graph clearly shows two well-known periods of marked temperature change: the Medieval Warm Period and the Little Ice Age. Both periods are well documented in European history and their effects caused tremendous change to the European economy and to the lifestyle of the folks who lived there.

The researchers acknowledged that such changes in temperatures had a "large natural variability…that is likely to continue":

"According to our reconstruction, **high temperatures**—similar to those observed in the twentieth century before 1990—**occurred around AD 1000 to 1100**, and **minimum temperatures** that are about 0.7K below the average of 1961–90 **occurred around AD 1600. This large natural variability in the past suggests an important role of natural multicentennial variability that is likely to continue**." Note 9.9h

Toward the end of the graph in Fig 9.9.2, there is a section colored green, which is labeled "Instrumental Observation Error?" It indicates an exponential rise in temperature during a remarkably short period, but is the temperature increase properly recorded as it relates to previous periods—periods where temperatures were indicated by means *other* than modern technological instruments? Since the 1900s, modern technology has

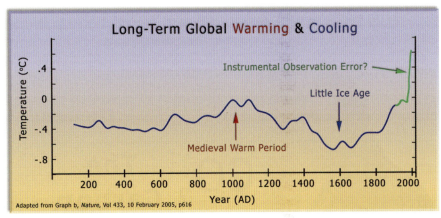

Fig 9.9.2 – If we are to comprehend global warming and cooling we must take into account the *long-term trends* of global temperature change. This graph clearly shows two periods where warming and cooling took place, and these events are documented in history. The green area on the right side of the graph appears to show an exponential increase in the temperature, which has been attributed to man-made CO_2 by some researchers. However, man-made CO_2 did *not* actually rise *exponentially* during this time-period, therefore, either the observations are incorrect (we see more of an increase because we are looking for it) or the temperature increase in recent years is real but is the result of natural processes, as explained in this subchapter.

allowed for the collection of highly accurate data from more places on the Earth than at any other time in history. We have the ability to view the world from orbiting satellites and with highly sophisticated equipment, but that means we are accumulating data of a completely different nature than was collected for the earlier periods on the graph. Thus, there may be an unintended Instrumental Observational Error when comparing modern technologically derived measurements with ancient indicators.

Another possibility is that we may be living through one of the warmest periods of the last two millennia. Even then, the warming averages less than one degree annually, barely perceptible by humans. Assuming that this is the case, there is a simple way to know that recent warming is *not* caused by the burning of fossil fuels.

If CO_2 was the cause of the sudden temperature spike (green line in Fig 9.9.2), the exponential rise must be accompanied by an equivalent rise in CO_2 gas over the same period—but this was *not* recorded.

Therefore, it is *highly unlikely* that a minute, steady increase of atmospheric CO_2 is causing the warming. It must be remembered that atmospheric CO_2 is a very small part of atmospheric gas (+/- 0.3%) as compared with water vapor, argon, oxygen and nitrogen. Let us look at some of the carbon dioxide evidence.

Carbon Dioxide Evidence

There are two main issues with respect to carbon dioxide and global warming. The first is the little-known fact that a rise in air temperature *increases* CO_2 readings. We discovered this over a decade ago while performing 'weather jar' experiments. Heating and cooling air in a *sealed* jar by placing the jar in or out of direct sunlight caused the CO_2 readings in the jar to rise or fall *according to the temperature*.

The amount of CO_2 gas in the sealed jar experiment *did not change*. Therefore, the jar did not experience a "greenhouse effect" because of increased CO_2, but rather, as the temperature rose, there was corresponding increase in CO_2 *readings*!

This may be another factor of the sudden spike in the graph in Fig 9.9.2. In addition to the errors related to comparisons of present-day temperatures with those of earlier periods because of the nature of the indicators used to derive them, the carbon dioxide readings from the same periods may have inherited similar errors. One astute researcher and climatologist, Vincent R. Gray, who held a seat on the Intergovernmental Panel on Climate Change (IPPC), recognized the problem of carbon dioxide and temperature variability, declaring that because of it, "It is back to the drawing board for carbon cycle models":

"**It is back to the drawing board for carbon cycle models**. Atmospheric carbon dioxide concentration varies in a manner **which has not been predicted successfully by existing models. There was significant variability before there could have been a human contribution**. This variability appears to **have followed temperature changes, rather than being responsible for them**. Although there has been an increase during the period of industrial development, the increase has not been uniform. Thus, the period between 1935-45 showed **no change**. The period since 1972, when the increase has been **linear** despite an increase of over 45% in emissions, suggests that there are new carbon sinks being established in the ocean and in the terrestrial biosphere to absorb the increases. This behavior plays havoc with previous predictions of global warming, but it is difficult to know how long the present apparently stable rate of increase will continue." Note 9.9i

Gray carried on a discussion with Richard Courtney regarding the change in carbon dioxide due to the change in temperature:

"**The fact that the Little Ice Age caused a fall in carbon dioxide means that at least part of the subsequent rise in carbon dioxide must have been caused by the rise in temperature**, as you have repeatedly pointed out by your reference to the paper by Kuo et al 1990. I am not sure that I can go along with the contention of Nigel Calder that the temperature increase is entirely responsible for the rise in carbon dioxide, **but some of it must be, and people are going to have a hard time explaining the rest**."

Note 9.9i, email discussion between Vincent Gray and Richard Courtney.

"It is back to the drawing board for carbon cycle models."
Vincent R. Gray, Climatologist

Researchers still have a hard time understanding or explaining the rise and fall of carbon because the carbon cycle has not been wholly understood, a problem that will be outlined in the Carbon Cycle Pseudotheory subchapter of the Chemical Model chapter. This brings us to the second of the Carbon Dioxide Evidences. The primary reason the carbon cycle remains misunderstood is ignorance of the Hydroplanet Model, which includes abundant water and *microbial* material in the crust. Once we comprehend the Hydroplanet Model and the biological load in the crust as evidenced in the Universal Flood Model, we come to a point where we can understand that a *significant* source of atmospheric CO_2 resides *underneath the crust*.

In subchapter 9.2, the research data from A.A. Tronin, the Russian scientist, revealed an increase in air temperature of 10° C (18° F) that was the result of earthquake frictional heating over a wide area. This one observation completely changes the heat-origin possibilities that drive global warming. Besides the continental heat increase from hyquatherms, oceanic crustal heating is a significant global heating contributor. More on this later, but for now we will focus on the carbon dioxide that is *supposedly* coming from human activity. There is a source of **natural** carbon dioxide relatively *unknown* to modern science, and it changes the entire global warming debate. Tronin reported:

"Field measurements of the concentration of CO_2 in the near-surface atmosphere and soil across the Kopetdag fault zone are shown in figure 16. **There is an increase of CO_2 concentration by up to 0.3 per cent in soil and up to 0.1 per cent in the near-surface atmosphere with background values of the latter of about 0.03 per cent**. There was a correspondence between the zone of high CO_2 concentration, that of radon emanation and with the subsurface temperature at a depth of 1.5m. The region of high surface temperature taken at 06:00am follows a zone of high subsurface temperature and gas concentration." Note 9.9j

An increase of 0.1 percent CO_2 in the air was a *threefold* increase in atmospheric carbon dioxide over the affected region spanning thousands of square kilometers. In other words, that one event *tripled the CO_2 in the region's atmosphere.*

The source of the CO_2 was **microbes** from **subsurface hyquatherms**—not automobiles, energy plants or farm animals! Although detectable, the microbes are easily overlooked, even if you know what you are looking for. Tiny bacterium have a very short lifespan, lasting perhaps hours or even minutes. They can lie dormant in relatively small numbers throughout the soil, until environmental temperatures and humidity increases to reproductive favorable levels brought on by small earthquake swarms. Once the temperature increases in a hyquatherm, it becomes a veritable microbial springtime. The subterranean microbial world comes to life, reproducing exponentially until the temperature drops and the microbes go dormant again. This process identifies a heretofore-unrecognized source of methane and carbon dioxide (CH_4 and CO_2) that must be accounted for in all carbon cycle models.

The Russians were not the only ones to notice "greenhouse gases" coming from hyquatherms. When *Chinese* researchers began recording surface levels of CH_4 and CO_2 simultaneously with satellite thermal infrared temperature readings, they found a startling correlation between seismically active areas, a rise in air temperature, and the release of natural CH_4 and CO_2 gas:

"Mechanism of satellitic thermo-infrared brightness temperature and **temperature increasing is studied**. Experiments are made with a **gas sample taken around the epicenter area**. **The gas sample is proved to contain green house gases such as CH_4 and CO_2 which have increased by tens of thousands of times**. In addition, lab research also proves that CH_4 and CO_2 can obtain energy under the action of transient electric field and release heat, **thus resulting in a temperature increase of 2-6°C**." Note 9.9k

The Chinese team noted an increase in CO_2 that was "tens of thousands of times" higher than background CO_2 levels. The importance of these exciting observations seemed to slip completely under modern science's radar. An entirely new field of research will open up once this new source of **natural** "greenhouse gases" is recognized by the scientific community. We will continue to learn about the importance of microbes to all living things in the Microbe Model, subchapter 13.3. The significance of the hyquathermal microbe community and its effect on the planet remains largely unknown, but once it is recognized, the increase in global atmospheric carbon dioxide can be evaluated *objectively* and its influence on global warming correctly studied.

The Sea-Level Rise Scare

Neither the rise in carbon dioxide levels nor the increase in temperature is significant enough to detect with our human senses. Because of this, the global warming advocates have had to turn to scare tactics to make the global warming idea real. Their tactic?—rising sea levels.

It turns out that this *should* be one of the easiest signs of global warming to evaluate. Al Gore's documentary: *An Inconvenient Truth* made it very clear that by the year 2100, the world's sea-levels will have risen over 20 feet (6 meters); this was of course, going to displace millions of people living near the coast. But what exactly has the sea level been doing over the past 50 years?—was Gore right? Will millions of people be displaced in just one generation if the world keeps warming? This is one question scientists would certainly like to find the answer to.

Five researchers collaborated on a report titled, *Satellite Measurements of Sea Level Change: Where Have We Been and Where Are We Going, 15 years of Progress in Radar Altimetry*. The researchers were quick to note that:

"Our understanding of sea level change has **improved** considerably over the last decade." Note 9.9l

What do we really *know* about sea level change, and how is the alleged *improvement* measured?

> FQ: How **accurately** have scientists been able to measure sea level change?

There are two methods for measuring the sea level, *tidal gauges* (which measure the average height of the water) and *satellite* telemetry (which measure the distance between the satellite and the ocean's surface). The *same* problem regarding long-term temperatures and CO_2 levels related to the graph in Fig 9.9.2 is inherent in the historical data of sea levels—tidal gauges data extend to over a century but satellite measurements are only available for about one decade. The accuracy of satellite measurements are why the researchers consider themselves as having "improved" in their understanding of sea level change. However, this method is too new to provide any perspective of real change over time.

Most tidal gauge records are newer than 50 years so the researchers have had to extrapolate the numbers to arrive at an average over the past century. Nevertheless, they acknowledge in a 2006 report on sea levels that:

"**No significant acceleration of sea level rise** has been detected in the tide gauge data…" Note 9.9l

Although they declare "no significant acceleration" has taken place, they provide a figure of 1.8 mm/year for the extrapolated tidal gauge number. But no error rate is given, so we are therefore left to guess how accurate the number really is.

The researchers, of course, intended to emphasize the new

satellite sea level data even though only a few years of measurement are available—and the data does not correlate with the tidal gauge records. They reported an average annual increase of 3.2 ±0.4mm/year based on satellite measurements. But there are several problems with the assumption that such measurements represent a change caused by anthropogenic global warming.

First, researchers acknowledge that presumably, some of the rise in sea levels is because of polar and glacial ice melt, but the rest is from the *thermal expansion* of ocean waters. No one is sure how much comes from where. One reason this is not known by the modern-science climate-change scientists is that ocean warming is *not caused solely by the Sun*, as we have already demonstrated.

Another problem with the supposed sea level rise is the margin of error. Scientists acknowledge that satellite instruments have an accuracy range of ±4-5 mm, which places the figure of 3.2 mm below the range of accuracy, greatly increasing the error rate. This fact is borne out in a 2012 report from research data obtained from the GRACE satellite. The data showed a rise of only "1.5 millimeters" per year from 2003 to 2010, well below the range of accuracy.

In the article titled, *Himalayan glaciers have lost no ice in the past 10 years, new study reveals*, the authors showed that previous forecasts of glacier melting had been misguided and new data "shocked scientists":

"The U.N. got it wrong on Himalayan glaciers–and the proof is finally here. The authors of the U.N.'s climate policy guide were red-faced two years ago when it was revealed that they had **inaccurately forecast that the Himalayan glaciers would melt completely in 25 years, vanishing by the year 2035**. Rajendra Pachauri, head of the U.N.'s Intergovernmental Panel on Climate Change (IPCC) and director general of the Energy and Resources Institute (TERI) in New Dehli, India, ultimately issued a statement offering regret for what turned out to be a poorly vetted statement. A new report published Thursday, Feb. 9, in the science journal Nature offers the first comprehensive study of the world's glaciers and ice caps, and one of its conclusions has **shocked scientists**. Using GRACE, a pair of orbiting satellites racing around the planet at an altitude of 300 miles, it comes to the eye-popping conclusion that **the Himalayas have barely melted at all in the past 10 years**." Note 9.9m

Bristol University glaciologist Jonathan Bamber stated that the results were "very unexpected":

"'The **very unexpected result** was the negligible mass loss from high mountain Asia, **which is not significantly different from zero**,' he told the Guardian." Note 9.9l

Some of the inconsistencies and irregularity may be negated with a data-set covering a longer period, but even then, there are bigger problems when it comes to measuring changes in the sea level.

Large-Scale Mass Redistribution of Water

Two of the previous five researchers also published a paper in Science in 2002 titled, *Redistributing Earth's Mass*. They discussed recent satellite measurements of sea levels that simply did not correlate with ice melt observations.

They refer to a factor called J_2, which defines the Earth's dynamic oblateness, or in other words, J_2 is a measurement of how much the Earth is flattened at the poles. Fig 9.9.3 helps illustrate the Earth's oblate shape, which is a little like a squashed ball—wider at the equator than the poles by about .3%. The illustration is exaggerated to help visualize this effect.

The researchers note that none of the current global warming theories "can explain the observations":

"Cox and Chao report satellite laser ranging data to numerous satellites from 1979 to 2001. For most of the past two decades, J_2 has been **steadily decreasing. But in early 1998 it suddenly started to increase substantially, indicating a large-scale mass redistribution from high latitudes to the equatorial regions**.

"Cox and Chao discuss several mechanisms that might explain these observations: melting of the polar ice caps, melting of Arctic sea ice. According to current knowledge, however, **none of these can explain the observations**. Ice cap melting should indeed lead to a rise in the global mean sea level, **but the observed sea level rise since 1992 is incompatible with the amount of ice melting required to explain the observed J_2 change** (even if the observed rise is attributed entirely to ice melting, which is not the case).

"What, then, is causing the change?" Note 9.9n

As mentioned previously in the magma pseudotheory chapter, the continents are "floating," but researchers have yet to figure out on what they are floating. One of the evidences that the continents are floating is glacial rebound. When glaciers or large areas of ice and snow melt, the reduction in weight causes a 'rebound' in continental elevation, or in other words, continents *rise* as ice melts.

According to observations, the Earth's J_2 was *decreasing* for

Earth's Oblateness

Spherical Earth

J_2 Increase

J_2 Decrease

Fig 9.9.3 – The Earth's Oblateness or flattening of the poles is a measurement described as nodal precession, or J_2. The Earth is slightly non-spherical, having an equatorial bulge of 43 km (27 miles) that has only recently been studied closely using new satellite and underwater pressure data. For 25 years, the Earth's J_2 had been decreasing, and geoscientists thought the decrease was the result of glacial rebound as glaciers melted because of global warming. However, in 1997 the J_2 began *increasing!* Was this a result of human-caused global cooling? Global temperatures did not change drastically, proving that the melting ice could not have caused a sudden decrease or increase in the Earth's J_2. In fact, the **100 km³** of pre-1997 glacier melt that supposedly caused "glacial rebound" would have had to increase suddenly, melting **700 km³ of ice** to account for the recent change in J_2. No observations support this idea, but a "Large–Scale Mass Redistribution" had apparently taken place and the geoscientists could not account for where the mass came from. They knew it did **not** come from melting polar ice or mountain glacial wasting. The redefined hydrological cycle in chapter 7.7 holds the answer to this fascinating event.

more than two decades, causing geoscientists to deduce that the decrease was the result of melting ice at the poles, which caused the landmasses to rise because of "postglacial rebound":

"Earth's dynamic oblateness (J_2) has been **decreasing** due to **postglacial rebound** (PGR)." Note 9.9o

However, a sudden increase in the Earth's oblateness (J_2) (see previous citation) meant that the equivalent of seven times more ice (100 km3 to 700 km3) would have had to melt to account for the sudden change. This amounted to an apparent large-scale mass redistribution, which was completely unsupported by ice sheet and glacial melt observations. Cox and Chao discuss this in their 2002 paper in the journal, *Science*:

"Recent studies indicated an acceleration of the mass wastage of mountain glaciers. The average loss rate for the subpolar glaciers had been **~100 km³ of water per year before 1997**, with accelerated rates in the past decade. **For the observed J_2 rate change to be explained by additional glacier mass loss, an additional water mass loss of ~700 km³ per year would be required**. The resultant GSL [Global Sea Level] increase of 2.0 mm/year over the pre-1998 rate has **not** been observed. It is **unlikely** that transport of terrestrial water mass to the oceans **can explain the J_2 changes**; however, more recent glacier and ice height data are needed to definitively rule this out." Note 9.9p

At first, such a rapid movement of mass from the poles toward the equator seems to support the notion of global warming, but the observations tell an entirely different story. Just as the exponential rise in temperature did not correspond to the linear rise in CO_2 in long-term temperature trends, the large-scale mass distribution does not correspond to observed ice loss at the poles. There simply was not enough ice to account for such rapid loss. The question is:

FQ: What caused the sudden reversal in the Earth's J_2 (oblateness) in 1998?

In the Hydroplanet Model Chapter, 7.7, the hydrological cycle was redefined to include the massive aquifers deep in the Earth. There is significant evidence that aquifers thousands of miles apart are connected and that great quantities of water can be moved through them. Furthermore, the sudden reversal of the Earth's oblateness is indicative of underlying cyclical events. Although we do not know exactly how much water can be redistributed through sub-crustal aquifers, the Hydroplanet Model is the only model than can account for the observations recorded by Cox and Chao.

The Sea-Level Rise Scare employed by Gore and other activists will forever be remembered as politically motivated science with no basis in fact. The prediction of a 20-foot rise in sea level over the next century is totally unfounded. Even *if* we assume the highest level of sea rise (3 mm/year) detected over the last decade by satellite measurement, it equates to only 300 mm, *less than one foot* in over a century. The cyclical nature of the Earth's J_2 is clearly associated with astronomical events, and so are the seemingly anomalous changes observed by Cox & Chao.

Global Warming Science-One Big Mistake

Unfortunately, Gore's 20-foot rise in sea level claim is not the only mistake when it comes to the science of 'Global Warming.' Another example comes from the director of NASA's Goddard Institute for Space Studies in an article in the March 2004 edition of *Scientific American*. The director asserts his opinion that the only "dominant issue" in global warming is—"sea-level change":

"**The dominant issue in global warming**, in my opinion, **is sea-level change** and the question of how fast ice sheets can disintegrate."
Note 9.9q

This pro-global warming researcher makes a big deal about the IPCC (International Panel on Climate Change) estimate of only "several tens of centimeters in 100 years" and he thinks we should all be called to action. However, in the last several sentences of his paper, he states:

"The peak rate of deglaciation following the last ice age was a sustained rate of melting of more than 14,000 cubic kilometers a year—about **one meter of sea-level rise every 20 years**, which was maintained for several centuries." Note 9.9q

The glaring point the researcher makes with respect to global warming is that sea levels were rising—*long before modern humans*— at a rate of one meter every 20 years. Apparently, the pre-human sea-level rise is *completely natural!* How can it be said that on one hand, humans are causing the sea-level to rise while on the other, it is a long-term, naturally occurring event that was once happening at an even greater rate than is happening today?

Moreover, modern climatologists admit that they *do not know what caused* the last 'ice age' nor can they identify the cause of the warming trend that eventually melted the ice.

There are far more errors in the global warming science arena than can possibly be included in this book, but those centering on the 'sea-level rise' hysteria are especially poignant. One such comes from researcher Robin Bell at Columbia University in a 2008 *Scientific American* article:

"Abundant liquid water newly discovered underneath the world's great ice sheets could intensify the destabilizing effects of global warming on the sheets. Then, even without melting, **the sheets may slide into the sea and raise sea level catastrophically**." Note 9.9r

On the surface, this claim seems simple—ice falls into water and water levels rise, but when we look closer, the error becomes evident. In Fig 9.9.4, Bell's illustrative example of three glasses is reproduced. She explained that the "global sea level rises the same way" the water does in the glasses:

"Water level in the glass at the left rises when ice is added (center). When the ice melts, the water level remains unchanged (right). **Global sea level rises the same way when ice slides off land and into the ocean**." Note 9.9r p61

But the researcher forgot one significant structure in the illustration—the *continent!*

Is the world solid matter or liquid water?

The way in which the illustration should have been presented is shown on the right side of Fig 9.9.4, showing a bowl of water with a piece of floating pumice to represent the landmass. Adding ice, or in a little trickier move, getting the ice to slide off the pumice and into the water (which represents ice sheets sliding into ocean), shows what would happen.

FQ: Does the water level relative to the floating pumice rise when ice is added to the water?

Illustrating Sea Level Rise

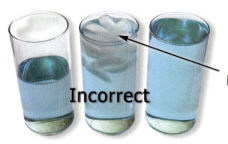

"Global sea level rises the same way when ice slides off land and into the ocean."

This piece of pumice represents a floating continent. Adding ice to the water causes the water level and pumice to rise **together**.

Fig 9.9.4 – The glasses on the left illustrate the concept researchers adhere to when discussing rising sea levels when ice sheets fall into the ocean. What makes this illustration incorrect is the absence of floating continents! The Earth's surface is more than just water. A more correct way of showing things as they are in nature includes a simulated continent—a pumice rock. Adding ice to the water does not cause the water level to rise on the rock; both rock and water levels rise together. Glacial rebound can also be simulated by loading the pumice rock with ice. The ice-loaded rock rides lower in the water, rising as the ice melts. This is an easy demonstration showing the reason global sea levels show no significant increases on the continents.

Adding water to the bathtub never causes the water level to rise up the side of a floating rubber ducky, and neither does adding melt water from glacial or ice sheet wasting. It doesn't take a scientist to answer this question, so why have they apparently missed the simple answer?

First, we must understand what the geologists do know—that the continents "float," and that they have the ability to rise or sink; this is called "isostasy":

"The idea that **continents are less dense than the mantel and float on it**, as a life jacket or an iceberg floats on the ocean, is the **principle of isostasy**." Bib 59 p490

Isostasy has been proven by direct observation, and we showed this concept earlier in the Hydroplanet Model, showing a metal paperclip "floating" on the water due to surface tension. There is real-time continental evidence, as reported in May 2009 about the fastest *rising* land area in the world:

"Global warming **conjures images of rising seas that threaten coastal areas. But** in Juneau [Alaska], as almost nowhere else in the world, climate change is **having the opposite effect**: As the glaciers here melt, **the land is rising**, causing the sea to retreat.

"Morgan DeBoer, a property owner, opened a nine-hole golf course at the mouth of Glacier Bay in 1998, on land that was underwater when his family first settled here 50 years ago.

"'The highest tides of the year would come into what is now my driving range area,' Mr. DeBoer said.

"Now, with the high-tide line receding even farther, he is contemplating adding another nine holes.

"'**It just keeps rising**,' he said.

"In Gustavus, where Mr. DeBoer's property is, the land is rising almost **three inches a year**, Dr. Molnia said, making it 'the fastest-rising place in North America.'" Note 9.9s

Here we see the continental landmass behaving like a ship: Add weight and the ship begins to sink into the water, remove weight and the ship rises, just like the land in Alaska is doing. What seems intuitive is not to the geologists. For them:

"The geology is **complex**..." Note 9.9s

The geology is "complex" to those who think the liquid upon which the continents are floating is magma, but the short period in which the continents rebound, often within days, does not fit within the viscous-magma theory.

Fig 9.9.5 illustrates the isostasy of a floating iceberg. Removing ice from the iceberg does not cause the sea to rise on its side; it simply rises like a ship relieved of some of its cargo. By recognizing the width of the continents in relation to their thickness, which can be compared to the skin on an apple, the wide but thin landmass would easily float. Like the paperclip, surface tension plays an important role, the dense but thin continent rides above the subcrustal water. After realizing that there is no magma, and that apple-skin-thin continents are *floating on water* instead of magma, the rising Alaskan landscape, and the **nonexistent** sea level rise makes perfect sense. Note 9.9t

Antarctic Warming Evidence

It was not surprising to find evidence substantiating the Weather

Fig 9.9.5 – Icebergs are good examples of the Isostasy Principle, which is that large bodies such as continents are floating in a state of equilibrium, balancing the downward force of gravity with the upward force of buoyancy. What would happen if we raised the water level surrounding this iceberg? Would the water level rise over the iceberg? No. In like manner, if all the ice and glaciers were to melt, there would be no substantial change to ocean levels. This is why there has never been any real evidence based on actual measurements showing significant rising or falling of the world's ocean levels.

Model in every corner of the globe—including the Antarctic. Fig 9.9.6 is a diagram showing the Antarctic warming trend from 1957 through 2006. Notice that the western side of Antarctica has experienced significantly more heat (area in red) than the eastern portion of the continent. The six NASA scientists, who compiled the data for this image, the most accurate to date, said this:

"This warming trend is **difficult to explain** without the radiative forcing associated with increasing greenhouse-gas concentrations." Note 9.9u

Why was the warming trend in Fig 9.9.6 *difficult* to explain without the "radiative forcing" associated greenhouse-gas concentrations? Because radiative forcing (i.e. solar irradiance as influenced by external factors) is based entirely on *heat from the Sun*, the scientists are left with no basis to explain the warming trend.

This poses a major dilemma for the researchers. CO_2 and other so-called greenhouse gases are generally distributed *evenly* throughout the atmosphere, including over the Antarctic. Therefore, the Sun's irradiance and the "radiative forcings" alone can *not* account for the dissimilar Antarctic heating trends:

"Radiative forcings alone are **inadequate to account for the observations**." Note 9.9u p461

Amazingly, this was the only sentence in four pages of technical reporting in the *Nature* journal article that indicated the real issue: why was the western portion of the Antarctic continent heating up more than the eastern portion? Because the heating cannot be accounted for by Sun-driven "radiative forcings," we look to the next most logical source of heat—beneath the surface. Look closely at Fig 9.9.6 and the evidence becomes apparent through the paradigm of the Weather Model.

The heated western third of the continent is bounded generally by coastline and mountain ridges indicative of plate boundaries, the location where gravitational-frictional heating is most likely to occur. The greater the plate movement, the more likely we are to see frictional heat buildup. This new data is evidence of a *division* between the Antarctic continental plates into a **Western Antarctic Plate** and an **Eastern Antarctic Plate**, whereas today, Antarctica is recognized as being only a single continental plate.

The plates appear to be divided along a mountainous boundary evident in the image. The Western Antarctic Plate may be made of rock that is crystalline in nature, more so than that of the Eastern Antarctic Plate. This would account for the increased heat accumulation from earthtide activity. Future excavations and drilling will eventually provide conclusive evidence of the geological difference between these two plates that lie deep beneath an icy covering. It is also of interest that the ice sheet covering the Antarctic extends far beyond the continental shelf, yet this ice sheet itself did not appear to reflect the increase in temperature that the continent did. If the Sun was responsible for the heating, the entire ice sheet would have certainly been more equally heated.

The Antarctic Warming Evidence shows up in another report from a 2006 conference on radar altimetry in Italy. The report noted that the East Antarctica saw an ice mass "*increase*" between the years 1992 and 2003, while the Western Antarctica ice sheet showed a "significant *decrease*":

"Perhaps the least is known about the contributions to sea level change from Greenland and Antarctica ice melt. Even during the satellite era, the sea level contributions from Greenland and Antarctica are uncertain, in particular because measurements do not provide complete coverage of the ice sheets. Recent radar altimetry measurements over Antarctica report ice mass **increase** between 1992-2003 **over the East sector**, equivalent to 0.12 ± 0.02 mm/yr sea level drop. In contrast, radar altimetry as well as interferometric SAR show significant **decrease** in the **Western part of the ice sheet**, of roughly a similar amount for the last decade." Note 9.9v

What did the researchers think when they observed such a significant difference in changes to the ice mass to both the Eastern and Western Antarctica Plates? If the Sun was the sole heat source, and ice was being added to the east side of Antarctica but melted on the west side *during the same period*, what

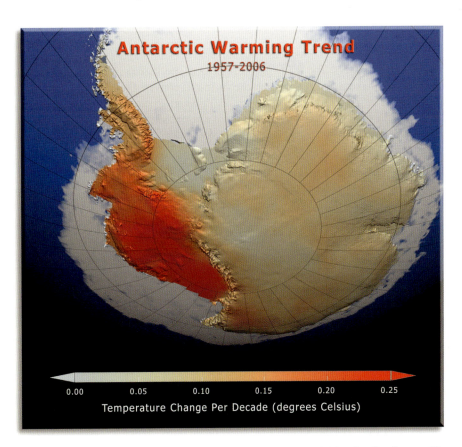

Fig 9.9.6 – In 2009, the most accurate measurements of the Antarctic Warming Trend over a 50-year time span were collected and analyzed. They are illustrated above. Western Antarctica (in red) experienced a considerable increase in temperature as compared with the eastern side of the continent. Researchers noted that the Sun's radiation "alone are inadequate to account for the observations." Why would just the western side of the continent receive more heat than the eastern side, if the Sun were doing all the heating? With the Weather Model and with an understanding of hyquatherms, we can explain why this is happening.

would you think?:

> "Thus, **additional research and observations are needed** to resolve these apparent discrepancies." Note 9.9v

Of course 'we need more observations,' but if the theory we built on was the *cause* of the "discrepancies," what then? Many scientists already recognize that the notion of anthropogenic global warming is without merit, and new insights in the Weather Model further support its falsity. In fact, the Antarctic Warming Trend does not prove that 'humans' are causing the current warming trend—it disproves it.

> Once we realize that global warming/cooling is a natural cycle driven by **earthtide heating** there will no longer be a reason for the global warming debate.

Global Warming—A "Secular Religion"

Ian Plimer has been called 'Australia's top earth scientist' and has received a number of scientific awards throughout his career. He sees himself as an 'old-fashioned scientist,' or one who questions the orthodoxy, including global warming:

> "'**Global warming has become the secular religion of today**,' he writes in the powerful conclusion to his book. Logic, questioning or contrary data are not permitted. To thumb your nose at the prevailing orthodoxy is to risk being branded a climate-change denier, a scientific knuckle-dragger, or worse. Plimer doesn't let it worry him. 'My job is to profess my discipline and, if people don't like that, bad luck,' he says." Note 9.9w

Although Plimer is a controversial figure and quite outspoken, one thing is sure, you do not have to be a top Earth scientist to recognize that global warming has become a secular religion. All one need do is learn how to ask fundamental questions and seek the real answers. Too many people; scientists, politicians, and activists have made it their purpose in life to convert the world to the religion of global warming, taking as an article of their faith the belief that humans are the chief cause of the planet's woes. We will visit this again briefly in the Human Model chapter.

Climbing a Molehill Instead of a Mountain

Global warming is one clear example of how many scientific investigators are in the dark. Michael Shermer, a self-proclaimed 'master skeptic' writes a Skeptic column each month for *Scientific American*. In his June 2006 column, Shermer discusses his "flip" from being a skeptic of global warming to becoming an activist:

> "Because of the complexity of the problem, environmental skepticism was once tenable. **No longer. It is time to flip from skepticism to activism**." Note 9.9x

What caused the consummate skeptic to "flip?" The answer is couched in a much bigger question as to why so many scientists and investigators have been deceived.

Shermer acknowledges that he once **believed** the 'experts' who had said in the 1990s that overpopulation would lead to "worldwide starvation and the exhaustion of key minerals, metals, and oil." But he admits that these predictions "failed utterly." Herein lies the core of the problem—too many people merely "believe" the so-called experts.

Instead of reading what the foremost *climatologists* had to say (or better yet, reading the data for himself), Shermer lists four books on global warming—books written by an *archeologist*, a *geographer*, a *journalist*, and *biologist*, but states the "finest summation of evidence for global warming" he had *ever* heard, the evidence that put him over the edge, was from none other than Al Gore—a *politician*. This is what caused Shermer's "flip from skepticism to activism!" Ironically, this is exactly what global warming advocates 'say' the anti-global warming crowd is doing—listening to everyone except the weather scientists.

Time will eventually clear the muddied waters. The worldwide starvation and mineral resource exhaustion alarmists of the 1980s and 90s, are now the man-made global warming alarmists. The cause has changed but their agenda remains the same. Shouldn't we demand more than shaky theories before over-taxing everyone for energy usage and their carbon impact?

The global warming molehill has been made into a mountain. It has had its temperature and humidity measured, it has been photographed, chemically analyzed, weighed, observed with satellites, microscopes, and just about anything else taxpayers' money can buy. After all the wasted resources and false alarms that accompany the man-made global warming hype, there is *one positive lesson* we can extract:

> If modern science accepts and teaches as fact a false theory based on such tenuous and obviously erroneous data, what stops it from teaching complicated false theories in **all** areas of science?

The ozone hole is a typical example; modern science has never known what formed upper atmosphere ozone in the first place. Governments acted hastily on poor science following the discovery of the ozone hole. Since then, science has failed to show that the ozone hole is closing because of governmental actions. Are we headed in the same direction with global warming and greenhouse gas theories?

Since both *water* and CO_2 are "greenhouse gases" and since there is much more water than CO_2 pumped into the atmosphere, especially when anything is burned, perhaps the next super-hyped panic will be based on 'artificial excess' water va-

How can the global warming debate ever be settled if we include only the Sun as our planet's heat source?

por in the atmosphere because of anthropogenic burning. Perhaps there should be a ban on anything that vaporizes water. Laughable? Maybe, but this is the type of thinking that leads to political turmoil, an over-taxed economy, and outright failure at addressing bonafide environmental concerns and opportunities. The multi-billion dollar climate-change juggernaut is a mountain made from a molehill that needs to be discharged. There are many promising opportunities with which to invest our valuable time and resources: resources currently being wasted on global warming.

Summing Up the Global Warming Pseudotheory

The global warming issue is flawed at its very base, in part because it is founded on flawed weather theory, which recognizes only a single source of heat—the Sun. The Weather Model chapter includes another source of heat that researchers in the global warming debate have been unable to recognize—the gravitational-friction heat in the crust. Two processes produce this heat:

1. Continental heating, one example being the Russian Earthquake Heating Evidence in Fig 9.2.3 & 4.
2. Oceanic heating, including heated plumes that rise from the bottom of the ocean, shown in Fig 9.4.4.

The two heating methods encompass the entire globe and are the direct result of hyquathermal heating. Only with this new endothermic heating source can we correctly explain weather changes, including global temperature trends and long-period cycles, which allow the correct study of the Earth's warming and cooling periods.

Even with constant media pressure and despite the fact that modern science is stumbling about in a theoretical dark age, there are *ten times more* scientists who are willing to sign a petition that global warming is caused *naturally*, than the number of scientists willing to sign a petition that the warming is *human*-caused.

In this subchapter, we showed global warming and global cooling are long-term weather patterns that climatology does not understand because of their ignorance about the origin of weather. The actual warming trend and the identification of additional CO_2 have introduced uncertainties in their data collection and interpretation, and previously unknown natural causes have been left out of the equation. The rising sea level is the only accurately measurable consequence of warming, and thus far, nothing significant has come of it, except hype and scare tactics. There is no conclusive evidence that sea levels have been rising over time, yet politicians and unknowing scientists have tried to scare the public with biased data and false claims. However, the continents *are floating* and the Alaska shoreline is *rising, 3-inches a year* thanks to isostatic adjustments made possible because they are floating. The reality is; some continents are **rising**—but **none** are being inundated by water.

The Antarctic Warming Trend suggests that the poles are melting naturally by a hyquathermal process, not by a solar process. Future investigation of the active faults along the Western side of the continent will likely show that the warming of the western side of the continent is coming from below, not from above. Finally, the religion of Global Warming must be dismissed for what it really is—a pseudotheory.

9.10 Weather and Geofield Prediction

To finish the chapter, we look toward the futurity of meteorology and the geofield. Predicting weather and changes to the Earth's geofield has always been a sticky subject, but why is this so? Until one comprehends how the weather and geofield are created, predicting even significant changes to these phenomena will remain nothing more than an elusive hope. The Weather Model gives new direction to help solve some very old problems. Throughout the Dark Age of Science, powerful computers and satellites have been of limited use because the data they produce is filtered through incorrect theories. Almost everyone knows that weather predictions are often unreliable. We will also look at several examples where magma-based theories have failed to predict the presence or strength of planetary energy fields, and how the Piezofield Model has had success. Combining geology and meteorology in this new way gives rise to a new field called **Geometeorology**, which combines new discoveries in both fields to aid in predicting the changes to the weather and geofield.

The Weather Model Summary

Having established direct scientific evidence that astronomical cycles cause earthquakes, that earthquakes generate frictional heating in the crust, and that the crust contains massive amounts of water, the following four Weather Model principles mentioned at the beginning of this chapter have greater meaning. They are here for review:

1. Hyquatherms change the Earth's weather systems; they are driven by Earthtide Heating, which is the constant frictional heating of the crust caused by gravitational tidal forces.

2. Hyquathermal heating of the seas and underground water beneath the continents causes high pressure and temperature zones in the atmosphere, which changes the Earth's weather.

3. The Earth's weather follows patterns and earthtide cycles that originate from the astronomical positions of the Earth, Moon, and Sun.

4. The Earth's weather and the Earth's Geofield are interrelated, connected by Earthtide Heating and the piezoelectric field, which are both created by the constant gravitational tidal movement of the Earth's crust.

Let us apply these four Weather Model principles by predicting changes to the weather and geofield.

Predicting with Weather Patterns

Previously, we discussed the aurora and its connection with the Earth's monsoon season. As we learn more about their origin—gravitational tidal friction in the crust, we will be able to predict their occurrences with more accuracy.

The connection between the aurora and the monsoons is just one example of how seemingly unrelated weather phenomena are connected, driven by *crustal astronomical tidal cycles*. This cycle is evident with auroras as they most often occur each year during the spring and autumn equinoxes:

"Geomagnetic storms that **ignite auroras actually happen more often during the months around the equinoxes… it is** known that during spring and autumn, the interplanetary magnetic field and that of Earth link up." Note 9.10a

This is also the monsoon season for many areas around the world, which is a rainy season usually accompanied by large cumulus (hyquathermal) clouds.

The Weather Model predicts *two* periods each year as the crust experiences a period of transition. These two periods occur near the **equinoxes** as the Earth's orbit transitions between apogee and perigee (the Earth's most distant and least distant position from the Sun). During this transition (at the equinoxes), the crust experiences a greater amount of flexure resulting in increased hyquathermal activity. These active hyquatherms in turn create the *two rainy seasons* that affect most areas around the world. These two periods also coincide with periods of geofield activity—thus increasing the incidence and intensity of auroras during the equinoxes.

We can also predict and explain why some rainy seasons and auroral periods are more dramatic than others—by overlaying lunar earthtide cyclical data. Each equinox occurs with an associated lunar positioning, which translates to a varying degree of gravitational influence each year. Thus, by using Kepler's astronomical laws, we can predict future years' auroral and monsoonal activity, based on the effect the Moon will have on the crust. This effect may be magnified or diminished depending on distance from the Earth and its location in relation to the Sun.

Other celestial bodies, such as the UF Comet can also dramatically affect the weather of our planet. No other time in recorded history did an astronomical event have such a significant influence on the Earth's weather as it did during the Universal Flood. These insights add a new perspective on future comets that will surely approach the Earth affecting its weather. With modern technological advances in astronomy, the ability to detect such a comet before it approaches Earth will allow for some warning and the prediction of possible weather changes and other consequences.

Future Weather Predicting Models

One thing meteorologists are proud of is their models—their *computer* models that is!

"At the beginning of the twentieth century scientists **believed it was possible to predict how the weather would behave using mathematical equations**. Unfortunately these equations were so complex that the experts were **unable to prove their theories**. The invention of the **computer solved this problem**, and computerized weather models of ever-increasing complexity and detail are now run by national weather services several times each day. The predictions from these computer models have become one of meteorologists most powerful forecasting tools." Bib 183 p180

The question remains:

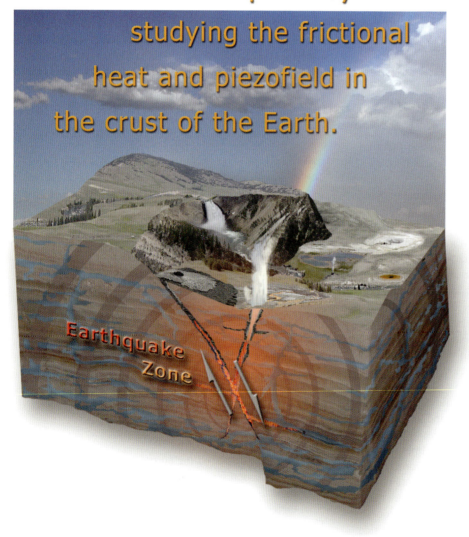

Predicting changes in the weather and geofield can take place by studying the frictional heat and piezofield in the crust of the Earth.

Fig 9.10.1 – Future predictability of the weather and the geofield will come from a study of hyquatherms. The Weather Model has identified a mechanism that causes weather changes and generates energy field changes around the Earth. With this new mechanism, researchers can place sensors in the crust designed to monitor the movement, gas emissions, and crustal energy levels. The data can then be analyzed with respect to the Weather Model and astronomical movements of the Earth, Moon, and Sun, shedding new light on earthquakes and their cycles and the geological connection to the weather and the Earth's energy field.

A century ago, scientists believed it possible to predict weather with mathematical equations. How well have they accomplished that goal today with their computers?

At the beginning of this chapter we discussed how new technology allowed satellites to look down on weather patterns as they develop, which dramatically improved forecasting simply by giving us a 'bird's eye view of storm systems from above.' Beyond the ability to see weather systems from far off, how far has meteorology progressed *without knowing the source* of weather change? The biggest and fastest computers are of no use if they continue to compute based on false models and equations that omit the second source of heat in the crust.

It is a matter of mirth throughout the world, where local folks sometimes say:

> "If you don't like the weather in (Florida, Melbourne or wherever you are) just wait five minutes."

Computers will always have difficulty trying to model local weather patterns in real time, but animals are aware of storms before they come and regularly flee to safety. Why shouldn't we be able to do the same? Human senses do not read the geofield well, with some exception, nor do we feel atmospheric pressure changes easily. Nevertheless, the wisdom behind why these changes occur can facilitate the placement of sensors that can monitor such changes on a continuous basis, transmitting the data directly for processing and interpretation.

Moreover, by comprehending the different types of clouds and the process by which cumulus clouds form (the hyquathermal process in the crust), we can visually evaluate the environment to determine whether a high or low-pressure system is building. 'Fair weather' is indicative of heat and high pressure coming from the crust. If a low-pressure system is developing, the crust is cooling and contracting, and if it is rapid enough, the lower pressure will draw moisture from high-pressure areas toward it. With lower temperatures, moisture can condense into rain or snow.

We can now explain, to even our children, how weather really works instead of merely shrugging our shoulders. Now, when someone asks, why it is so hot or why is a storm coming, the sky holds new answers. Hopefully we will never look at the sky the same again!

Magma Theory Fails in Predicting "Magnetic Fields" in the Solar System

The Weather Model enables predictions about the energy fields surrounding the other planets in our solar system and the Planetary Piezofield Laws make it possible to understand and explain the fields more clearly. But first, let's establish what scientists say about current theory and the 'magnetic fields' surrounding some of the planets.

Margaret G. Kivelson of UCLA gave an insightful lecture on planetary energy fields based on new discoveries from the Galileo spacecraft, which had visited Jupiter in the late 1990s. She admits that the way in which these energy fields are produced is "only partially understood":

"We know a great deal about magnetic fields and how they affect magnetized materials and moving electrically-charged particles. **But** the way in which **planetary and stellar magnetic fields are generated is, surprisingly, only partially understood**. As scientists are attracted by puzzles, there are quite a few people working on the question of how planets produce magnetic fields." Note 9.10b

Even in ignorance, it's human nature to think we are on the verge of figuring it all out; unfortunately, this has the unintended consequence of keeping one's mind from being open to what really is. With a perspective based in dogmagma, the investigators are compelled to ask why Venus, though similar in size to the Earth, lacks a magnetic field. Shouldn't the magma-dynamo be working inside of Venus?

"But not all of the planets have magnetic fields. The field at Venus was studied by Chris Russell of UCLA using the Pioneer Venus Orbiter. He discovered that Venus has no planetary magnetic field, or one so small (less than 0.01% of Earth's) that it doesn't matter. **Venus is not much smaller than Earth. Why doesn't it have a magnetic field**?" Note 9.10b

The next statement by Kivelson really expresses well the frustration researchers have had with Venus' lack of an energy field. In the scientists' minds, Venus "should not have cooled so much that its interior is solid."

"It is **hard** to understand why Venus doesn't have a magnetic field. **It should not have cooled so much that its interior is solid**. People have **proposed** explanations why Venus should have no field and have proposed explanations why Mercury, though very small, **should** have one. Still, **physicists have not successfully predicted in advance of the measurements which planets had magnetic fields**. That, you will agree, suggests that **our understanding of planetary magnetism has a way to go**." Note 9.10b

This declaration gets right to the heart of the matter: even though we read otherwise in the textbooks, or hear differently on the Discovery Channel, the modern science researchers working in the trenches know that understanding planetary magnetism has a long way to go!

Having looked unsuccessfully for a pattern in the predictability of planetary energy fields, only faulty theories based on magma-planet assumptions remain. Kivelson continues:

"With the fields of eight planets known (or almost known), **can we find a pattern**? We think that we can. In our description of this pattern, we have to **assume** that we understand the interior structure of the planets. We **think** that planets **melt** right through in the early stages of their development, allowing the heavy metallic elements to settle to the center under the pull of gravity. The planets then begin to cool. The outside layers cool most, forming solid surfaces in the case of the smaller planets. Within the deep interior there may be both solid and liquid heavy metals. These heavy metals of the deep interior are good conductors of electricity. The central core of the earth, for example, is solid but it is surrounded by a zone of **melted iron-rich metal**." Note 9.10b

The energy-field research confirms the pseudotheoretical nature of the modern science magma theory:

"**The generation of planetary magnetic fields is an unsolved problem that has been with us for some time**. There are claims that Einstein identified the origin of Earth's field as one of the five most important **unsolved problems in physics**. Why **must** I say that **we haven't solved the problem of how Earth's magnetic field is generated**? It is **because** the **details are hard** to work out **mathematically** and so it remains **slightly uncertain** that the analysis is heading in **precisely the right direction**." Note 9.10b

Modern science has followed this pattern repeatedly during the 20th century Dark Age of Science. The Earth's energy field origin is a ***completely*** *unsolved problem* in geophysics with investigators only "slightly uncertain" they are headed in the right direction because the mathematical details "are hard to work out." But this chapter demonstrates again that hard questions are easy—when you know the answer. It never was a 'mathematical problem'—the pseudotheories that the modern research is based on are just wrong: magma does not exist and energy fields are not being generated by melted iron.

We can evaluate a simple, logical explanation for the planetary energy fields with the Geofield Model and we can make predictions using the Universal Scientific Method to test the predictions with actual observations of the planets' and moons' energy fields.

Geofield Model Predicts Piezofields in Solar System

Venus is often referred to as Earth's sister or twin because of its size and mass similarities, but it has a small, almost nonexistent energy field—why? The answer is simple.

Venus has no moon.

> Venus has no moon and therefore has a very weak energy field!

Without another body to produce sufficient tidal forces like the Moon produces on Earth, Venus is unable to produce a continuous piezoelectric energy field. Next, look at **Mercury**. Does Mercury have a moon? Does it have an energy field like the Earth? Mercury is the planet closest to the Sun and is therefore affected by the tidal forces of the Sun, but it has no moon. Its close proximity does affect Mercury's crust, but only to the degree that it produces an energy field equal to about 1% of the Earth's geofield.

Looking past the Earth to **Mars**, we find another planet that shows little sign of a current energy field, but the fact that parts of the crust have been magnetized indicates that it had one in the past. Mars has two very small moons, but they are too small to create much of a pull on the crust, which is three times as thick as the Earth's. Therefore, this planet, with its lack of an energy field, also supports the Planetary Piezofield Laws.

Jupiter has an energy field 14 times stronger than the Earth's. It has 63 named satellites, including four large interior moons, three of which are *larger* than Earth's moon and the fourth almost as large. Jupiter and its moons create the largest energy field of any of the planets because of their size and proximity. The shorter orbital period of the moons are due to their closeness to the planet: Io is under 2 days, Europa is less than 4 days, Ganymede is just over 7 days and Callisto is about 17 days; all quite short when compared with the Earth's lunar orbit of 28 days. Thus, Jupiter also supports the Piezofield Model of energy field production, and there is even more support for the Piezofield Model when we look at Jupiter's moons and their energy fields, which will be discussed shortly.

Saturn's energy field is about the same strength as the Earth's; the tidal pull from its moons is consistent with the Piezofield Model. Although Saturn has at least 61 moons, Titan, the largest, makes up 90 percent of the mass of all Saturn's moons and yet is not even as big as the largest of Jupiter's moons. Moreover, it has an orbital period of approximately 16 days, which places it much further away than Jupiter's largest moons. Because of this, a much smaller energy field is predicted.

The Earth's geofield ranges from .3 to .6 gauss, whereas **Uranus** sports an energy field that varies between .1 to 1.1 gauss. Although the range is greater, the overall average field strength is somewhat similar. What then should we expect to discover about the tidal forces acting on the planet? Something similar to Earth, and in fact, they appear to be so. Uranus has 27 known moons with five satellites that are close to the planet in an orbital resonance similar to Jupiter's moons. The largest moon, Titania, is about half the diameter of Earth's moon.

Neptune has an energy field strength of about 14 microteslas (.14 gauss), which is about one fourth that of Earth's. The Neptunian and Saturnian energy fields do not come from the center

> "The generation of planetary magnetic fields is an unsolved problem that has been with us for a long time."
> Margaret G. Kivelson
> Astrophysicist

Fig 9.10.2 – 'Magnetic fields' in the solar system are called magnetospheres by modern science. However, because the energy fields are not coming from magnets or magma dynamos, researchers have failed to predict when or why they occur. With the Piezofield Model, planetary energy fields are predicted to occur where the tidal forces of planets and moons interact.

of the planets because they are tilted with respect to their axis of rotation. This is difficult to square with the dynamo theory, which is based essentially on a spherical model, but it aligns well with the Piezofield Model because landmasses on the surface are usually comprised of differing minerals that generate piezofields of varying strength based on structure and content. This is somewhat analogous to the Earth's energy field, which has an elevated energy level where there are continents as compared with the oceans. Neptune's energy field also displays several non-dipole characteristics, with lobes exhibiting fields much stronger than the primary field. There are 13 known moons orbiting Neptune. The largest, Triton, contains more than 99% of the total mass of Neptune's moons. It is about three-quarters the size of the Earth's moon and has an orbit just under six days. Based on what we know about Neptune and its tidal forces, this planet also supports the Piezofield Model.

The moons themselves also support the Piezofield Model. Looking first at our own **Moon** and its energy field, the Piezofield Model predicts that the Moon should have a very weak, but present energy field. The reasons are that the Moon has a very thick crust that does not move much at all even though the tidal pull of the nearby Earth is strong. We know that there is *some* movement because of the moonquakes measured during the Apollo missions. The nature of the moonquakes indicates that the Moon *should have* at least a small energy field, and it does:

"…one to a hundred nanotesla—**less than one hundredth of the Earth**…"
Note 9.10c

Once again, our own Moon supports the Piezofield Model of planetary energy field production.

Although the Piezofield Model has not found its way into the scientific world prior to the publication of the Universal Model, it is amazing how close to the truth many investigators came by studying other moons, like Jupiter's Ganymede and Io.

When the Galileo spacecraft flew by Ganymede, researchers did not expect to find an energy field on such a small celestial object (Ganymede is 1.5 times the diameter of the Earth's Moon) because the "interior was thought to have frozen solid" long ago. However, the scientists "were all wrong":

"Ganymede's orbit lies much outside of Io's. **Tidal forces** weaken with distance and tidal heating is not important at Ganymede. This explains why Ganymede has no volcanoes and no other significant sign of being hot in its interior. It was thought possible that the interior had not ever separated into different layers, and even if it had separated, its **interior was thought to have frozen solid over the lifetime of the solar system**. So there was **little reason** to expect that it would have a **magnetic field, but we were all wrong**. At Ganymede, there is **unambiguously an internal magnetic field** that opposes Jupiter's. Galileo has now completed four close passes of Ganymede, approaching it from different directions, and our measurements confirm unambiguously that the magnetic field is produced by currents flowing within Ganymede." Note 9.10d

This should have caused a complete rewrite of current planetary formation theory, but at least it left researchers knowing that "something needs to be modified":

"The discovery of Ganymede's magnetic field **shows that something needs to be modified in the accepted description of the evolution of the solar system**. There is a **new puzzle to solve. We still don't know just how Ganymede's field is produced**." Note 9.10d

"The discovery of Ganymede's magnetic field shows that something needs to be modified in the accepted description of the evolution of the solar system."

The Universal Model was been written to solve puzzles like these, and an even bigger puzzle; the energy generated by Io is not small:

"Io actually generates as much wattage as about **1,000 nuclear power plants**." Note 9.10e

This really confused the investigators. From a seemingly frozen little moon, all this energy was being produced, yet they still turn to a theory that has no basis and does not work—"Io's molten iron core"—even after acknowledging that Io is "being heated from the *outside*, *by tidal flexing* of the layers around it, *rather than being heated from the center*":

"'There's no intrinsic field,' Kivelson said. 'We can put that question to rest.' That means **Io's molten iron core** does not have the same type of convective overturning by which Earth's molten core generates Earth's magnetic field. **Lack of that overturning fits a model of Io's core being heated from the outside, by tidal flexing of the layers around it, rather than being heated from the center**.'" Note 9.10e

The heat on the *outside* of Io clearly demonstrates how lava is really generated—frictional heating. And the same mechanism—"tidal flexing"—is creating Io's piezofield. It has been right there all along, just waiting to be discovered; yet modern science's tenacious adherence to the Magma Pseudotheory has prevented it.

> A **model** of Io's core:
> "heated from the **outside, by tidal flexing** of the layers around it, **rather than** being heated from the **center**."
>
> Margaret Kivelson, Astrophysicist

Geometeorology

Weather origins and the geofield have taken on completely new meaning in the Weather Model chapter and for good reason. Six new natural laws about how and why weather and the geofield form and how they are interrelated made new discoveries possible. There are more new natural laws in the Weather Model than any other chapter of the Universal Model. Numerous empirical evidences of these new laws were presented and explained, yet this chapter is just the beginning of a completely new field of study called Geometeorology.

Geometeorology is the study of the geofield and weather with the Gravitational-Friction Law and the Planetary Piezofield Laws. With the six new geometeorological laws, researchers can now see the connections between geology, weather, and the geofield. Much evidence exists already—yet perhaps hundreds of times more is waiting to be mined by those who will enter this exciting new scientific field.

With geometeorology, we can see the causes of weather and changes to the geofield. The effects are obvious within the new Weather Model paradigm. For example, on 11 March 2011, the great magnitude 9.0 Tohoku earthquake (the most powerful to ever to hit Japan, that killed over 16,000 individuals, see the images in the following page), produced significant geofield changes, and hyquathermal activity several days before the quake struck:

"Today, Dimitar Ouzounov at the NASA Goddard Space Flight Centre in Maryland and a few buddies present the data from the Great Tohoku earthquake which devastated Japan on 11 March. Their results, although preliminary, are **eye-opening**. They say that before the M9 earthquake, the total **electron content of the ionosphere increased dramatically over the epicentre**, reaching a maximum **three days before the quake struck**. At the same time, satellite observations showed a **big increase in infrared emissions from above the epicentre**, which peaked in the **hours before the quake**. In other words, the atmosphere was heating up." Note 9.10f

The "electron content" of the ionosphere dramatically increased (geofield activity) *and* the "infrared emissions" (hyquatherm activity) peaked directly over the epicenter of the earthquake in the hours before the quake. How much better of an example for the Weather Model could we ask for?

The main purpose of Geometeorology should be to reach beyond the mere understanding of a celestial body's energy field and weather—it should move towards the *prediction* of those fields, which in turn can be used to predict life-threatening geophysical activity. If the geofield and hyquatherm activity were clearly understood before the Japanese earthquake had occurred, perhaps time could have been given to those who were affected the most and enough time to prepare and avoid such a tragic loss of life.

Certainly, the Weather Model has great potential in helping humankind comprehend how, where, and when weather changes will occur. This will be accomplished because the energy fields that drive weather patterns move in *cycles* and these cycles can now be comprehended and predicted, just as the movement of the planets in Kepler's day became predictable after we came to understand the natural laws by which they moved. The predictable celestial movements of our Earth, the Moon, and the Sun, along with new gas, temperature, and geofield sensors located in the crust, will be monitored with satellites that are equipped with their own sensors and other technology. All of this will surely change our whole understanding of Geometeorology and future weather and geofield forecasting.

The beauty of nature is only surpassed by comprehending how it works.

Acknowledgements

These acknowledgements, as most things in the Universal Model, come from a unique perspective. We would like to thank all the special people and acknowledge the situations that helped create the UM over the last four decades. In addition, the author wishes to thank individuals who never knew they were contributing to the UM.

The Universal Model began when there was no Internet, digital photography, or even cell phones. The advent of all of these dramatically accelerated the research and writing of the UM and no doubt changed the way we do science. From my Rose Lane Elementary School second grade teacher, Mrs. Papa, to my 7th & 8th grade shop teacher Mr. Holland, I was instilled with a desire to learn about the world around me. I would like to give credit to my Central High School physics teacher, Mr. Hart, who excelled in his teaching and love of science. It is through his example that I came to enjoy science, especially physics. My chemistry teacher, Mr. Benson, made Chemistry fun and also created a unique desire in me to learn more of the world around me.

It was not long after graduating from college and entering the business world that my love of science began to immerge again. In the early 1990s a study of archeology and geology took me to the offices of more than a dozen professors with questions they could not answer. I spent years visiting libraries and looking through card catalogs and indexes in search of answers to my questions and for evidence supporting the answers that I came to understand. It became much easier to find out what the world thought regarding scientific matters once Wikipedia came online. High school teachers and college professors at first derided this knowledge bank, but it has become one of the largest sources of knowledge in the world, for which I am grateful.

I acknowledge the many scientists who contributed significantly to our understanding of natural phenomena. We included a number of them with their accomplishments in the UM. We would like to acknowledge all sources of truth that assisted in the development of the UM.

The number of supporters and contributors, including both family and friends, who have helped with this project over the years has grown to such an extent that one page could not contain them all. However, I wish to give special attention to some of them. During the early research, longtime friends, Kurt Kleinman and his wife Robyn, provided instrumental support and encouragement. Jim and Judy Michaud, two other choice friends have contributed much time and effort towards this work. College roommate and longtime friend, Nick Margaritas worked with us during our early research efforts as well as Rod Meldrum, who along with his wife, Tonya, contributed many years to the UM effort. Their friendship has been a support to us as well.

Russ and Heidi Barlow are two extraordinary people, along with their beautiful family, who have given much of their time and talents towards the UM over the years. Russ fills the role of chief editor and consultant and has contributed in so many ways that they cannot all be listed. We thank them from the bottom of our hearts, for the polished words of the UM would not exist without their editing and proofreading gifts.

We offer special thanks to Courtney and De Ette Snell, who have been close friends and supporters of the Universal Model for many years. Courtney was responsible for the desktop publishing of both the digital and print versions of the book along with many other business activities and consulting.

Daren Ortiz, Ken Krogue, and Boyd Tuttle, also wonderful friends who have given of their time and talents helping with the marketing, business and consulting aspects of the project. Their support and insights have been invaluable.

A number of researchers, experimentalists, photo, audio and video assistance, and business management assistants have helped over the years; including Greg McIver, Douglas Harkey, Bruce Ross, Bryan Illguth, Jacob Householder, Sam Boyle, Lynn Howlett Photography, Shane Kester, and LeAnn and Pat Hord.

UM Science Teachers, educators, reviewers, proofreaders, publishers, printers and also include but are not limited to: Brian Mills, Bob Gardner and his son Michael Robert (Mr), Carter Brown, Cody and Rachel Harper, Daniel Burdett, Darrel Barger, Emily and Spencer Daines, Gary Wright, John and Sean A'lee Bevell, Lareme and Dawn Fessler, Larry and Brenda Farris, Marcus Frenzel, Steve and Nancy Wilkes, Pam Bolla, Rick Durfee, Sara Calton, Tom Hummel, Vic Crane, Wade and Michelle Kohlhase, Allan and Rose Wade, Rod Boone, Aaron and Monica Sessions, Ruthanne Christensen, David Barker, David, Cindy, and James Hampton.

Science, technology, and business consultants include but are not limited to: Dallas Gale, Darryl Barger, Clint Rogers, David Allan, Donny Harrop, Frank and Sally Johnson, Fred and Peggy Ash, Johnathan and Beverly Nevill, David and Darelyn Peterson, Steve and Linda Allen, Nathan Welch, Robert Byrnes, and Joshua Bishop. Jae Heiner, a special friend and colleague, has been such a strength to me in helping to review the UM and consult on numerous scientific topics. All of these individuals in their special way have contributed to the discoveries of the Universal Model.

The most amazing support we have received for all of our lives is from our family. Both of my wife's parents, James and Nita Caffrey, and my parents Vernal and Sally Sessions and all of their children have supported us in one way or another for many years. These brothers, sisters and parents have been instrumental in the progress of the UM. In particular, I would like to thank my brother Mark for his vision of the UM and its future. Our oldest daughter Kara and her husband Bryan have helped with photography and video. Our Brooke and her husband Daniel have been involved in teaching the UM and writing the UM Introduction book. Aubrey and her husband Jeremy, were some of the earliest full-time contributors to the UM as we prepared to go public and have been instrumental in preparing the UM document for publication. Our oldest son Perry and his wife Karen have helped in countless ways by providing the technical knowhow for our website and in helping manage our business. Jarom and his soon to be wife, Braquell, have also been instrumental in building our website, including the photo credits, notes, blogs and reviews.

Likewise, the joy of our family's science journey continued with our loving grandchildren: Miles, Benson, Owen, Sterling, Chloe, Barrett, Reggie, Navy, Macy, and Baby Hammer. Because of their enthusiasm for learning, they have helped keep the wonder of a child and an open mind alive and well in all of us by their continual questioning.

The last person I wish to thank is my sweet companion Danette. Without her there would be no UM. She has stood at my side for so many years, contributing in ways only she and I know. I deeply appreciate all of her support, wisdom, council, and love.

Ending with the beginning, my greatest acknowledgement is to The Creator, who is both omnipotent and omniscient. For without His guiding light, the Universal Model, a New Millennial Science, would not exist.

Dean W. Sessions

15 October 2016

Editor's Note

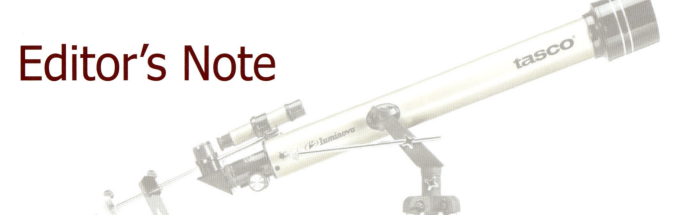

Answers come from questions. Although easy to understand, I find that simple sentence curiously liberating, and yet difficult all at once. How does one know what questions to ask, or in which order to ask them? The Universal Model did not come to Dean, its author, in a linear think-of-a-question then go find the answer ribbon of knowledge; rather, meditative pondering often brought the question—the right question—into focus. That was usually the answer; to ask the *right* question, but sometimes, that is not easy do.

As a young boy, I stared up at the night sky, marveling at the countless points of light I could see through my tiny Tasco telescope, and then peered tirelessly at the Sun, fascinated by the march of the sunspots as they traversed its face. I drew out the constellations for my own book, read every printed word about the Mars Viking missions, and worried—actually, really worried terribly—that I would not live to know how it all actually happened. I wondered how trees got to the tops of the mountains, why Indians chose to live in the desert, and how dinosaurs turned into oil.

I grew up learning about things in the Bible: the Creation, Noah's Flood, and Moses splitting the Red Sea. Then I learned about other things in school: magma erupting from volcanoes, exploding stars that made all the elements, billions-of-years old Earth, millions-of-years old dinosaurs, and Lucy, the oldest hominid, our ancestral ape-like relative from Africa. I blended all this and more into an intellectual, metaphysical smoothie that allowed me to reverence the Creator while paying homage to the fathers of astronomy, geology, the space program, and Star Trek. I admit I hoped most of all for Scottie to "beam me up" so I could "boldly go" where Captain Kirk and his crew were headed.

Then life, marriage, children, and job happened. Happy though they were, like almost everyone who ever lived, I packed away youthful dreams and wonder in favor of living indoors and eating three times a day with cute little people. The ebb and flow of business tides brought me to a point where one day I did not know if I could continue to do what I was doing and so, in a moment of temporary desperation, I headed to the local community college where I matriculated in business management.

As with all degrees, the required coursework included science with accompanying lab work, and so I chose geology. Then, as fate would have it, I enrolled concurrently in an archaeology class to satisfy the requisite humanities credit. The two complemented each other in extraordinary ways, catalyzing in my mind an astounding desire to learn more. So engaged in the subject was I that the lab professor organized extra field trips into the surrounding landscape, and it was during one of those excursions that I experienced something profound. Tromping through exquisite sandstone formations that graced Red Rock State Park in Nevada, we stopped to examine up close, banded red and white sandstone, presented in an undulating waveform of no particular order. As I stood there with my hand on the coarse wall, the professor asked, "What formed this undulating white and red sandstone?" She broke the silence with an unexpected answer, "No one knows!" Unknown to anyone else there, I experienced a moment of serendipity wherein I somehow knew that I would learn how sandstone formed—specifically how *this* sandstone actually formed.

Geology and archaeology had awakened an inner passion, stirring an unquenchable thirst for knowledge about all things science. I continued my quest by attending classes in other fields, including chemistry, astronomy, biology, meteorology, physics, environmental science, and so on, always engaging whenever possible in extracurricular activities associated with the college, the USGS, lectures at UNLV, and a host of other opportunities. In all this, I noticed a disturbing trend among all the sciences, a trend made all the more obvious when I shared my excitement with my sweetheart (whom I had convinced I needed to buy my own lab equipment), when she would ask simply, "How do they know that?" Too often, I could not answer her with empirical evidence because there was none; almost everything hung on one theory or another, many of them unsubstantiable.

Then one evening in the fall of 2004, I received a call from an old friend seeking to update his Christmas card list. As our conversation progressed from the perfunctory small talk of long absent acquaintanceship to a discussion about our current interests, the shock that we had both diverged from our common career paths into science—me merely an avocation but for him full-time research—led to a discussion that lasted into the wee hours of the next morning. The die was cast, he had tantalized me with UM principles, with a garage-sized laboratory, and with a project that included among other things, an opposing view about how sandstone actually forms, directly contradicting the story in all the textbooks.

The next year brought about remarkable learning opportunities; a family vacation deep into the heart of Yellowstone took us to Fossil Ridge; an excursion to Denali in Alaska and a survey of Ruth Glacier; and an expedition to Mount St Helens with my friend Rod, the researcher and my newly introduced friend, Dean, the UM's author. We each brought a son along as we surveyed up close and personal Mount St Helens on the 25-year anniversary of its eruption. On that trip I listened and learned, and read things that rocked my world. Here were answers, but here also were challenges to my modern science trained worldview. Still, I could not put it down and during the next few months, I read incredible things that answered so many of my questions. The ideas, answers, and concepts thrilled me, but I struggled with its written form, until one day I sat down to type out a page or two simply so that I could enjoy the read, even if only for a single page. That small act of editing portended monumental consequences as I suddenly knew my role in this extraordinary project.

I am not a professional editor, and my first foray into this new endeavor proved surprisingly difficult, taking nearly a year to process the first 65 pages of editing. A good part of that time involved reading and a complete immersion into all available material as I sought to grasp more fully the magnitude of this project. It was enormous; in fact, the most remarkable aspect of the Universal Model project is its scope. No science book, indeed no science project that I knew of ever covered more territory than the UM, in both breadth—across many fields—and depth, with its mutlitudinal concepts within each field. As for myself, reading, editing, proofing, and reviewing represented an act of love and a fulfillment of the incredible thirst for knowledge I had encountered during my days sitting in class.

Now I know the incredible processes and the unique environment that produced and shaped many of the Earth's rocks, minerals, and landforms. No longer speculative because I have seen the work, the experiments, and the empirical evidence that back these claims, I write, or rather edit, this work with confidence. This work is true, profoundly so, and I know it. The project is the foundation, the basis of a new Millennial Science movement destined to flood the world as it reaches out to aspiring minds unfettered by long standing dogma, and restoring science to its empirical and experimental roots from which we can build a clearer picture of Nature and the natural world.

Now I know dinosaurs did not turn into oil, but I know their demise and oil's origin are related. There is no magma driving the Earth's volcanoes, but I know where that heat originates. I know Lucy is not a human ancestor, but I also know the actual migration of human civilization from its earliest point in the First World. And, quite significantly, I know how undulating red and white sandstone formed, not from erosion, but through extraordinary processes involving matter ubiquitous throughout the universe, a precious substance upon which we all rely—water. These answers and many thousands of others fill the pages of this book, awaiting only the perusal of the truthseeker. What Dean has done, no person could do entirely on his own, and I know this because of my own firsthand experiences; there are great forces at work here.

What will you read and learn in the UM?

As for me, I learned the Truth.

Russell H Barlow
Chief Editor

Banded Red and White Sandstone in Red Rock State Park, Nevada

NOTES

These Notes comprise a list of references taken from the text of the UM. They supplement the bibliography (Bib) references, which follow, from which multiple citations are drawn throughout the book. There are approximately 3,000 Notes in this section and about the same number of additional Bibliography references for a total of approximately 6,000 entries in all three UM volumes.

The Notes numbering convention includes chapter, subchapter, and alpha character. For example, **Note 21.8a** references the first note (a) in Chapter 21, subchapter 8. References come from scientific journal articles, Internet sites, and other scientific references. When possible, the page number from whence the quote was taken is included in the reference. We sometimes include additional support material to address items more technical in nature.

We provide these references—the UM 'blue quotes'—to encourage independent study and individual verification. Some of the references are out of print or the internet site is no longer available; in these cases, we provide the last known reference. We include the 'Accessed' date following internet citations to account for frequent and dynamic content changes to internet websites.

One final note; the authors of the citations included in the UM may or may not agree with the usage, commentary, or message of the Universal Model, and no endorsement is intended. Universal Model author, reviewers, editor, and researchers have tried to keep within the context of the citation's subject matter. Because the Universal Model is the foundation of a new Millennial Science, we expect the application of this newfound knowledge to effect a shift in the paradigm within which the cited quotes were originally made.

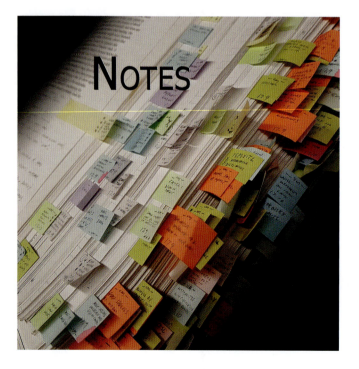

1 Introduction to The Universal Model

1.1 Answers to Questions
1.1a *Clouds of Venus, Sky & Telescope*, June 1999, p60

1.2 The Question Principle
1.2a https://www.youtube.com/results?search_query=ben+stein+expelled+-full+movie
1.2b http://www.ldolphin.org/scimyth.html - Accessed 10.4.16, Skeptical Inquirer, Theodore Schick Jr., 13 March 1997, No. 2, Vol. 21, p36

1.5 Truth Defined
1.5a George Washington, The Writings of George Washington, vol. XII (1790-1794) [1891]) http://oll.libertyfund.org/titles/washington-the-writings-of-george-washington-vol-xii-1790-1794
1.5b Truth Ignored and Rationalized

How can we teach truth in science if truth has never been defined in the scientific textbooks? The word 'truth' is avoided in most scientific circles today with its implications toward 'absolute truth'. For most of us, examples of truth establish its existence. Truth is self-evident for all to see through examples. Several examples of scientific truth have been given in the Universal Model (such as 2 x 3 = 6), however, if we really try hard we can leave reality by rationalizing everything.

The rationalizing attitude of some scientists was literally shocking the first time it was encountered. When asked if 2 x 3 did not always equal 6, a bright professor said, "No, it does not when you are not in base 10." The professor was correct, if a person does know about other base number systems, 2 x 3 could equal another number besides 6. (Some ancient cultures had different numbering systems. For example, 9 inches is not 9/10 of a foot because a foot has 12 inches). However, the denial of truth ends there. For if a person only knows and uses a base 10 numbering system (like the vast majority of the world today), 2 x 3 has always, does, and always will be 6 for that person. Truth is to "know"… If we do not know something, the truth of that something does not exist to us. The vast majority of the public only counts in a base 10 system, so the simple example of 2 x 3 was used. Other bases besides base 10 were not included in the discussion of the main body of this book. However, the truth example of 2 x 3 = 6 can be reworded to include the base 10 system I the example. (In a base 10 numbering system, 2 x 3 was, is, and will be equal to 6).

The point we are making here is that there is an exception to any 'truth statement', but the exception can usually be added into the statement, so that the new statement can become true, until we add another exception. Thus we really never reach absoluteness with only the five physical senses and logic. The most important principle in the UM is to question everything with an open mind. Thus, we are always watching for a situation in that a scientific truth statement will not hold. However, when such a situation is found, the truth statement can typically be adjusted to include the new information. If not, the statement is no longer true. As noted in Chapter 3.1, natural law is a statement of truth, and law is defined in Webster's Dictionary as:

"A statement of an order or relation of phenomena that, so far is known, is invariable under the given conditions." Bib 8 p478

Another example of rationalization could include someone noting that the statement, "A living human body contains water", is not true because it should read, "salt water" instead of just "water". However, the human body does not have just one 'salt' in its water, and it also has many more minerals. Nevertheless, the living human body still does contain water. From all of our known information, the human body always has, does, and will always contain water. The day a scientist finds a living human body that doesn't contain water—is the day he or she can say the statement is not true. Until then, all rational people, including real scientists, must agree that the statement is scientifically true.

It is important that when trying to make true statements, that we be specific in what we say because truth by its definition is absolute. If there can be found any situation in which the statement does not hold, we can say it is not true. For example, a hollow cube has six sides on the outside as well as the inside. The rainbow, if found somewhere else besides the Earth, say in a different planet's atmosphere, may have less colors or change the speed of light in a way that changes the order of the colors. However, until some new change in order of the color in the rainbow is observed, our understanding is that the order of the color is always the same. These 'truths' we have defined such as "a solid cube has six sides", or natural phenomenon that is self-evident such as "a living human body contains water", give science a foundation of truth to build on. Without this foundation of truth, science is building on sand. The higher you build, the bigger the fall will be, because sand just can not hold the forces that nature will put on any establishment that is not built on truth.

This is why there is a restoration of truth in the UM, because the foundation is no longer made out of natural laws that change very little with time. During the 17th, 18th and 19th centuries, natural laws were considered the foundation of scientific thinking. However, beginning with popularization of the evolution theory in the 20th century, theories became the "end points of science", and established the Scientific Dark Age of the last century.

What other element has a density of 19.320 /kg m-3 at 293K, a melting point of 1337.58K, a boiling pint of 3080K, and looks like gold? Who is going to say that the statement: "Only one element contains all the physical properties of gold", is not a true scientific statement? Only an irrational person. The UM is not written to such people, and science is not for such people. The voice of reason demands a new standard of absoluteness for the discovery of natural law. We should be searching for new laws that will not change if we want to really discover how nature is.

Experiences with people that do not believe in the existence of truth has eventually brought the conversation to a point where the individual denys their own existence. When a person does this, when they will not admit they exist while they are talking to you, it's time to leave. The appalling thing is that we have allowed such people (especially scientists) to teach such a false philosophy under the umbrella of science and 'higher education'. Absolute truth is taught properly under religion, not science, but scientific truth is the foundation of real science.

As Chapter 22 the Evolution Pseudotheory establishes, the scientific leadership has had an agenda over the last century, and taking truth out of science has been part of it. If we cannot know the fact that we exist is true, than we cannot know anything is true. This is exactly what some people want, so they can justify in their mind that they really don't know if they exist, thus it doesn't matter what they do. Some even call this 'intellectual freedom'. However, all false philosophies lead down the same road. Human behavior that goes against natural law, has the same consequence whether rationalized or not. Rationalization cannot change truth.

Where did science get the false philosophy that truth does not exist? Chapter 9 The Relativity Pseudotheory, details some of the history. Heisenberg and his uncertainty principle (which you don't hear much about these days due to technologists being able to isolate individual atoms and photons which proves the principle is wrong) and Einstein's Relativity Pseudotheory were the seeds for taking reality out of science. However, as Chapter 11 establishes, relativity was always a Pseudotheory, and with this knowledge we can begin to restore truth back into science.

1.5c *One Hundred Years Without Darwin Are Enough*, School Science and Mathematics 59, H. J. Muller, 1959, p304-5, http://www.talkorigins.org/faqs/evolution-fact.html – Accessed 10.4.16

2 Methods Models and Pseudotheories

2.2 Against Method
2.1a http://en.wikipedia.org/wiki/Paul_Feyerabend – Accessed 10.6.16

2.3 The Modern Science Method of Confusion
2.3a http://www-internal.sandi.net/standards/HTML/SciK12.html, Site is no longer available
2.3b http://depts.washington.edu/rural/design/scimethod.html, Site no longer available
2.3c http://www.ucmp.berkeley.edu/fosrec/Lipps.html – Accessed 10.6.16
2.3d Newton's "Method of Analysis". The following quote are some of Isaac Newton's last words that he ever wrote before he died. They come from his book, Opticks, which was published just before his death. The second to last paragraph of the book is as follows:

"As in Mathematicks, so in Natural Philosophy, the Investigation of difficult Things by the Method of Analysis, ought ever to precede the Method of Composition. This Analysis consists in making Experiments and Observation, and in drawing general Conclusions from them by Induction, and admitting of no Objections against the Conclusions, but such as are taken from Experiments, or other certain Truths. For Hypotheses are not to be regarded in experimental Philosophy. And although the arguing from Experiments and Observations by Induction be no Demonstration of general Conclusions; yet it is the best way of arguing which the Nature of Things admits of, and may be looked upon as so much the stronger, by how much the Induction is more general. And if no Exception occur from Phenomena, the Conclusion may be pronounced generally. But if at any time afterwards any Exception shall occur from Experiments, it may then begin to be pronounced with such Exceptions as occur. By this way of Analysis we may proceed from Compounds to Ingredients, and from Motions to the Forces producing them; and in general, from Effects to their Causes, and from particular Causes to more general ones, till the Argument end in the most general. This is the Method of Analysis: And the Synthesis consists in assuming the Causes discover'd and establish'd as Principles, and by them explaining the Phenomena proceeding from them, and proving the Explanations." Bib 2 p404

2.4 Controversial Science
2.4a *Darwin's Influence on Modern Thought*, *Scientific American*, July 2000, p81
2.4b Teaching About Evolution and the Nature of Science, Chapter 1, NAS 1998, http://search.nap.edu/html/evolution98/evoll.html – Accessed 10.6.16

2.5 The Universal Scientific Method (USM)
2.5a *Sky & Telescope*, October 1999, p86

3 The Dark Age of Science

3.1 The History of Science
3.1a http://en.Wikipedia.org/wiki/Science – Accessed 5.25.06, Site has since been modified
3.1b http://plato.stanford.edu/entries/kepler/#Emp – Accessed 10.7.16

3.2 The Dark Age of Science
3.2a *Collective Electrodynamics*, Carver A. Mead, 2000, p1, *American Spectator* Interview, Sep/Oct 2001, Vol. 34, Issue7 p68,
3.2b http://www2.uah.es/farmamol/The%20Pharmaceutical%20Century/Ch1.html – Accessed 10.7.16

3.3 Modern Science Today
3.3a http://history.nasa.gov/ap11-35ann/top10sci.html – Accessed 10.7.16
3.3b http://www.madsci.org/posts/archives/mar97/854234635.Ph.r.html - Accessed 10.7.16

3.4 Modern Science and Truth
3.4a *National Geographic*, April 2000, p115

3.5 The Big Picture of Modern Science
3.5a *National Geographic*, March 1998, p80
3.5b The Arizona Republic, (Cox News Service) Seeds of life are possible in space, A8, Jan 30, 2001
3.5c *Science*, Vol. 284, 25 June 1999, p2111
3.5d *Scientific American*, October 1994, p36
3.5e Michael J. Pidwirny, PH.D., Department of Geography, Okanagan University College, http://www.geog.ouc.bc.ca/physgeog/contents/5a.html
3.5f *What Life Means to Einstein*: An Interview by George Sylvester Viereck, *The Saturday Evening Post*, Vol. 202 (26 October 1929), p117, also in http://en.wikiquote.org/wiki/Albert_Einstein

3.6 How Popular is Modern Science?
3.6a THE DECLINE OF REASON? Jere H. Lipps, http://www.ucmp.berkeley.edu/fosrec/Lipps.html – Accessed 10.7.16

3.7 Science, Technology and Mathematics
3.7a http://www.scalewatcher.com/us/ – Accessed 12.7.09, Site has since been changed

4 Scientific Revolutions

4.2 What is a Scientific Revolution?
4.2a *Crisis in the Cosmos*, Sam Flamsteed, *Discover*, March 1995, p66, http://discovermagazine.com/1995/mar/crisisinthecosmo478/ – Accessed 10.8.16
4.2b *Slaying the Age Paradox*, *Scientific American*, Dec 1997

4.3 Why Will a Scientific Revolution Take Place?
4.3a *Essay Concerning Human Understanding*, Locke, 1690, Book IV

5 The Magma Pseudotheory

5.1 Magma Defined

5.1a *What Lies Beneath*, Brad Lemley, *Discover*, August 2002, p35
5.1b *Volcanoes*, Gordon A. Macdonald, University of Hawaii, Prentice-Hall, Inc., 1972, p399
5.1c *Asimov's Chronology of Science & Discovery*, Isaac Asimov, Harper & Row, 1989, p290
5.1d *Mantle Convection and Plate Tectonics: Toward an Integrated Physical and Chemical Theory*, Paul J. Tackley, *Science*, 16 June 2000, Vol. 288, p2002

5.2 The Magmaplanet Belief

5.2a http://www.scientificamerican.com/article/why-is-the-earths-core-so/ – Accessed 4.13.16
5.2b https://books.google.com/books?id=ckizOzrwEuIC&pg=PA183&lpg=PA183&dq=#v=onepage&q&f=false – Accessed 4.13.16

5.3 The Lava-Friction Model

5.3a This question and answer appeared prior to 2004 on the following USGS website but was later removed. http://www.usgs.gov/faq/?q=taxonomy/term/9830
5.3b http://vulcan.wr.usgs.gov/Glossary/Seismicity/what_causes_earthquakes.html – Accessed 5.29.13, Site no longer available
5.3c *Mount Etna's Ferocious Future*, Tom Pfeiffer, *Scientific American*, April 2003, p60
5.3d *Volcanoes in the Sea, the Geology of Hawaii*, Gordon A. Macdonald and Agatin T. Abbott, University of Hawaii Press, 1970, p57
5.3e http://www.avo.alaska.edu/dbases/swarmcat/ofr/swarmcat.html - Accessed around 2002
5.3f In addition, only 6% of the total swarms (Type 2c - 41% of 15% of the total) reported eruption before earthquakes. However, this and any other inconsistency in the data could be easily due to the poor recording techniques and lack of technology that existed in some of the data back to 1979. The researchers reported that the older data were not as reliable. Although the researchers of course mention 'magma' as the primary cause of the earthquakes, they do note "other factors" that are involved in sequences of earthquakes. These factors include "earth and ocean tidal stresses", and "seasonal ocean-loading or changes in barometric pressure".
5.3g *Frictional Melting During The Rupture Of The 1994 Bolivian Earthquake*, Kanmori, Anderson & Heaton, *Science*, Vol. 279, 6 February 1998, p839
5.3h *Crustal Heat Flow: a guide to measurement and modeling*, G. R. Beardsmore, J. P. Cull, Cambridge University Press, 2001, p41
5.3i *Mount Etna's Ferocious Future*, Tom Pfeiffer, *Scientific American*, April 2003, p63
5.3j *Frictional Melting During The Rupture Of The Bolivian Earthquake*, Kanmori, Anderson & Heaton, *Science*, Vol. 279. 6 February 1998, p839
5.3k Abstract, *Evidence for melt Lubrication during large earthquakes*, *Geophysical Research Letters*, VOL. 32, 5 April 2005
5.3l *The Threat of Silent Earthquakes*, Peter Cervelli, *Scientific American*, Peter Cervelli, March 2004, p87
5.3m http://spaceplace.nasa.gov/io-tides/en/ – Accessed 4.14.16
5.3n *Tidal Triggering of Earthquake Swarms at Kilauea Volcano, Hawaii*, Rydelek, Davis, Koyanagi, *Journal of Geophysical Research*, Vol. 93, NO. B5, p4401, May 10, 1988
5.3o *Mount Etna's Ferocious Future*, Tom Pfeiffer, *Scientific American*, April 2003, p63
5.3p *Volcanic Seismology*, P. Gasparini, R. Scarpa, K. Aki, 1992, p62
5.3q http://news.nationalgeographic.com/news/2002/02/0215_020215_volcanohunter.html – Accessed 4.14.16
5.3r *The Threat of Silent Earthquakes*, Peter Cervelli, *Scientific American*, Peter Cervelli, March 2004, p91
5.3s *Earthly Circuitry, Breathing, and Shakes*, Richard A. Kerr, *Science*, 26 January 2001, Vol. 291, p584
5.3t *Moonquakes, meteoroids, and the state of the lunar interior*, G. Latham, J. Dorman, F. Duennebier, M. Ewing, D. Lammlein, Y. Nakamura, Proceedings of the Fourth Lunar Science Conference, 1973, Vol. 3, p2521-22
5.3u http://vulcan.wr.usgs.gov/Glossary/Seismicity/what_causes_earthquakes.html – Accessed 5.29.13, Site no longer available
5.3v *Annual Modulation of Triggered Seismicity Following the 1992 Landers Earthquake in California*, Stephen S. Gao, Paul G. Silver, Alan T. Linde andI. Selwyn Sacks, *Nature*, Vol. 406, 3 August 2000, p500
5.3w *Melt may trigger quakes*, Tom Clarke, *Nature, Science Update*, 15 February 2003
5.3x *Scientific Information System for the world's deepest borehole*, Kola SDP-3, F. Mitrofanov, A.J. Kumpel, F. Gorbatsevich, UNESCO & International Union of Geological Sciences, IGCP480: *Rocks and Minerals at Great Depth and on the Surface*, p9
5.3y *Tidal Triggering of Microearthquakes on the Juan de Fuca Ridge*, William S. D. Wilcock, *Geophysical Research Letters*, October 15, 2001, Vol. 28, No.20, p3999
5.3z *Statistical Test of the Tidal Triggering of Earthquakes: Contribution of the Ocean Tide Loading Effect*, Hiroshi Tsuruoka, Masakazu Ohtake and Haruo Sato, *Geophysical Journal International*, 1995, Vol.122, p183
5.3aa *Proceedings of the Seventh International Symposium on Earth Tides*, edited by Gyula Szadeczky-Kardoos, 1976, p181
5.3ab *Water Thrown on Earthquake Prediction*, Tom Clark, *Nature Science Updates*, 23 August 2001
5.3ac *Injection-Induced Earthquake and Crustal Stress at 9 km Depth at the KTB Deep Drilling Site, Germany*, Mark D. Zoback and Hans-Peter Harjes, *Journal of Geophysical Research*, Vol. 102, No. B8, August 10, 1997, p18,477 - 18,491
5.3ad *Earthquakes and Geological Discovery*, Bruce A. Bolt, *Scientific American Library*, 1993, p114
5.3ae *Earth Science*, Richard Monastersky, *Science News*, September 5, 1998, Vol. 154, p155
5.3af *Earthquakes and Geological Discovery*, Bruce A. Bolt, *Scientific American Library*, 1993, p114
5.3ag *Huge Rock Eruption Caused By the 1999 Chi-Chi Earthquake in Taiwan*, Shih-Wei Huang and Jiin-Shuh Jean and Jyr-Ching Hu, *Geophysical Research Letters*, Vol. 30, No. 16, 1858, doi:10.1029/2003GL017234, 2003
5.3ah http://earthquake.usgs.gov/learning/faq.php?categoryID=6&faqID=17 – Accessed 9.1.08, Site no longer available
5.3ai http://science.nasa.gov/headlines/y2000/ast04may_1m.htm – Accessed 4.15.16
5.3aj http://spaceplace.jpl.nasa.gov/gll_io_fact.htm – Accessed 4.15.16
5.3ak http://www.spacedaily.com/news/galileo-00n.html – Accessed 4.15.16

5.4 Magma Theory Defies Heat Flow Physics

5.4a *Microbes Deep Inside the Earth*, James K. Fredickson, Tullis C. Onstott, *Scientific American*, October 1996, p13
5.4b *Geomechanical Causes of the KTB Borehole Instabilities*, *Journal of Geophysical Research*, Vol. 102, No. B8, August 10, 1997, p18,503
5.4c *Riding the Wilson Cycle, How the Theory of Plate Tectonics continues to Evolve*, Paul Silver, *Geotimes*, July 2007, p31, http://www.geotimes.org/july07/article.html?id=feature_wilson.html – Accessed 4.18.16
5.4d *Stirring the Oceanic Incubator*, Robert P. Dziak and H. Paul Johnson, *Science*, Vol. 296, 24 May 2002, p1406
5.4e *The World's Deepest Well*, Ye. A. Kozlovsky, *Scientific American*, 251, 1984, p104
5.4f *Inner Space: The Soviet Union and West Germany Delve into Earth's Deep Secrets as the United States Looks On*, Richard Monastersky, *Science News*, Oct 21, 1989, Vol.136, No. 17, p266
5.4g *Scientific Information System for the world's deepest borehole*, Kola SDP-3, UNESCO & International Union of Geological Sciences, IGCP480: *Rocks and Minerals at Great Depth and on the Surface*, F. Mitrofanov, A.J. Kumpel, F. Gorbatsevich, p6
5.4h *New Geothermal Data From the Superdeep Well SG-3*, Yuri A. Popov, *Tectonophysics*, Vol. 306, 1999, p357
5.4i *Scientific Information System for the world's deepest borehole*, Kola SDP-3, UNESCO & International Union of Geological Sciences, IGCP480: *Rocks and Minerals at Great Depth and on the Surface*, F. Mitrofanov, A.J. Kumpel, F. Gorbatsevich, p6
5.4j *New Geothermal Data From the Superdeep Well SG-3*, Yuri A. Popov, *Tectonophysics*, Vol. 306, 1999, p357

5.5 The Accretion Theory

5.5a *Forging the Planets*, John A. Wood, *Sky & Telescope*, Jan 1999, p36
5.5b *The Evolution of the Earth*, Claude J. Allegre and Stephen H. Schneider, *Scientific American*, October 1994 & Bib 37, p5
5.5c *The Small Planets*, Erik Asphaug, *Scientific American*, May 2000, p54
5.5d *Meteorites: The long trip to earth*, Clark R. Chapman, *Nature*, October 5, 2000, Vol. 407, p573-576

5.6 The Radioactive Myth

5.6a http://volcanoworld.wordpress.com/2008/08/26/how-is-lava-made/ – Accessed 4.20.16
5.6b *Earth's Core and the Geodynamo*, Bruce A. Buffet, *Science*, Vol. 288, 16 June 2000, p2009
5.6c J. Marvin Herndon, PhD private e-mail communication on May 27, 2003
5.6d *The World's Deepest Well*, Ye. A. Kozlovsky, *Scientific American*, 251, 1984, p104
5.6e *Introduction to Special Section: The KTB Deep Drill Hole*, Boler Haak and Alan G. Jones, *Journal of Geophysical Research*, Vol. 102, No. B8, 200, August 10, 1997, p18
5.6f *The Core-Mantle Boundary Region*, Anderson, *Geodynamics Series*, Vol. 28, American Geophysical Union, 1998, p258

5.7 Glass is Not Quartz

5.7a http://www.galleries.com/Quartz – Accessed 4.20.16
5.7b Thermal Conductivity reference is Bib 5 p12-177-8. Refraction of quartz is based on quartz being parallel to c axis. Density of both quartz and glass (fused silica) based at 20° C., http://www.olympusmicro.com/primer/lightandcolor/polarizedlightintro.html – Accessed 9.1.08, Site no longer available
5.7c *Experimental Study Into the Rheology of Synthetic Polycrystalline Coesite Aggregates*, J. Renner, B. Stöckhert, A. Zerbian, K. Röller, and F. Rummel, *Journal of Geophysical Research*, Vol. 106, NO. B9, September 10, 2001, p19,422
5.7d *Hydrothermal Synthesis of Crystals*, Robert A. Laudise, *C&EN*, September 28, 1987, p32

5.8 The Piezoelectric Evidence

5.8a *Hydrothermal Synthesis of Crystals*, Robert A. Laudise, *C&EN*, September 28, 1987, p32-4
5.8b *Understanding Piezoelectric Quartz Crystals*, Louis Bradshaw, *RF Time and Frequency*, August 2000, p50, http://images.rfdesign.com/files/4/0800Bradshaw50.pdf – Accessed 4.20.16

5.9 The Non-Iron Core Evidence

5.9a *Mineralogical Society of America, Reviews in Mineralogy Volume 37, Ultrahigh-Pressure Mineralogy: Physics and Chemistry of the Earth's Deep Interior*, 1998, Stixrude & Brown: *The Earth's Core*, p273. See also *Geophysical Research Letters*, Shen, Mao, and Hemley, Feb 1, 1998, Vol. 25, No. 3. p373
5.9b *Science*, Vol 267, 31 March 1995, p1972

5.10 Deep Earthquake Evidence

5.10a *Deep Earthquakes*, Cliff Frohlich, *Scientific American*, January 1989, p48-9
5.10b *Earthquakes and Geological Discovery*, Bruce A. Bolt, *Scientific American Library*, 1993, p85

5.11 The Drilling Evidence

5.11a *Looking-Deeply-Into the Earth's Crust in Europe*, Richard A. Kerr, *Science*, Vol. 261, 16 July 1993, p295, quote stated by Alfred Duba
5.11b *Long Valley Exploration Fact Sheet*, Sandia National Laboratories, http://www.sandia.gov/geothermal/Programs/lvf.htm – Accessed 8.11.16
5.11c *Magma Drilling Comes Up Short*, *Science*, 25 September 1998, 281:1951
5.11d *Phase 3 Drilling Operations at the Long Valley Exploratory Well LVF51-20 Sandia Report SAND99-1279*, *Sandia National Laboratories*, June 1999, p14, http://www.prod.sandia.gov/cgi-bin/techlib/acess-control.pl/1999/991279.pdf, Site no longer available
5.11e Private email communication with William L. Ellsworth, U. S. Geological Survey, March 25, 2002
5.11f Photos of Hot Springs and Thermal Features in the Long Valley Area, *California*, *USGS*, http://lvo.wr.uss.gov/gallery/Thermal_1.html, site no longer available
5.11g *Scientific Drilling in Long Valley, California– What Will We Learn?*, Susan S. Priest, John H. Sass, Bill Ellsworth, Christopher D. Farrar, Michael L. Sorey, David P. Hill, Roy Bailey, Ronald D. Jacobson, John T. Finger, Vicki Sue McConnell, and Mark Zoback, *USGS* Fact Sheet-077-98
5.11h http://science.sciencemag.org/content/285/5436/2119 – Accessed 4.21.16
5.11i *Volcanic Seismology*, P. Gasparini, R. Scarpa, K. Aki, 1992, p301
5.11j http://adsabs.harvard.edu/abs/2002EGSGA..27.3916K – Accessed 4.21.16
5.11k *Looking-Deeply-Into the Earth's Crust in Europe*, Richard A. Kerr, *Science*, Vol. 261, 16 July 1993, p295
5.11l Peter Kehrer, *Science*, Vol. 266, 28 October 1994, p545

5.12 Earth's Magnetic Field Pseudotheory

5.12a *Life's ups and downs*, Philip Cohen, *New Scientist*, 13 April 2002, p11
5.12b http://www.ngdc.noaa.gov/IAGA/vmod/igrfhw.html – Accessed 8.11.16
5.12c *Probing the Geodynamo*, Gary A. Glatzmaier and Peter Olson, *Scientific American*, April 2005, p55
5.12d *Power requirement of the geodynamo from ohmic losses in numerical and laboratory dynamos*, Ulrich R. Christensen and Andreas Tilgner, *Nature*, Vol. 429, 13 May 2004, p169
5.12e *The Riga Dynamo Experiment*, Agris Gailitis, Olgerts Lielausis, Ernests Platacis, Gunter Gerbeth, and Frank Stefani, *Surveys in Geophysics* 24: 247-267, 2003, p253
5.12f *Laboratory experiments on hydromagnetic Dynamos*, Agris Gailitis, Olgerts Lielausis, Ernests Platacis, Gunter Gerbeth, and Frank Stefani, *Reviews of Modern Physics*, Vol. 74, October 2002, p973

5.13 The Continental Uplift Pseudotheory

5.13a The email conversation with Frank Press, Author of Understanding Earth, took place on June 19, 2004
5.13b http://www.johnmartin.com/earthquakes/eqpapers/00000030.htm – Accessed 4.22.16, also stated to come from *California Geology*, March 1974, Vol. 27, No.3, Crustal Movement Map of USA
5.13c *An Experimental Study into the Rheology of Synthetic Polycrystalline Coesite aggregates*, J. Renner, B. Stockhert, A. Zerbian, K. Roller, and F. Rummel, *Journal of Geophysical Research*, Vol. 106, No. B9, September 10, 2001, p19,411
5.13d http://earthobservatory.nasa.gov/IOTD/view.php?id=5449 – Accessed 4.22.16
5.13e *Geology of the Grand Canyon*, William J. Breed and Evelyn C. Roat, *Museum of Northern Arizona/Flagstaff Grand Canyon Natural History Association*, 1974, pX
5.13f *Technology Lifts Everest to New Official Height*, Barbara Fallon, *National Geographic*, http://www.directionsmag.com/pressreleases/technology-lifts-everest-to-new-official-height/97035 – Accessed 4.22.16
5.13g *Enhancing Earthscope by Constraining Vertical Motions of the Continental Crust and Surface*, J. A. Spotila, *Virginia Tech*, http://www-scec.usc.edu/news/01news/es_abstracts/spotila.abstract.e-scope.pdf – Accessed 4.22.16
5.13h Bib 59, *Understanding Earth*, Frank Press and Raymond Siever, Fig 21.21, p556, 2nd Edition, and Fig 21.19, p500, Third Edition
5.13i http://www.johnmartin.com/earthquakes/eqpapers/00000030.htm – Accessed 4.22.16
5.13j *Enhancing Earthscope by Constraining Vertical Motions of the Continental Crust and Surface*, J. A. Spotila, *Virginia Tech*, http://www-scec.usc.edu/news/01news/es_abstracts/spotila.abstract.e-scope.pdf – Accessed 4.22.16
5.13k http://woodshole.er.usgs.gov/projects/project_get.php?proj=29210EN&style=html – Accessed 4.22.16
5.13l *Essentials in Geology*, Frederick K Lutgen & Edward J Tarbuck, 7th Ed, 2002, p17-2
5.13m *IGCP480: Rocks and Minerals at Great Depth and on the Surface*, F. Mitrofanov, A.J. Kumpel, D. Guberman, and F. Gorbatsevich, *UNESCO & International Union of Geological Sciences*, p5, http://acoustpol.narod.ru/intas/out/item_79.html – Accessed 6.23.16

5.14 Other Magma Pseudotheories

5.14a https://en.wikipedia.org/wiki/Bowen%27s_reaction_series – Accessed 4.22.16
5.14b Hall's actual words about his experiments that he presented to the Royal Society of London can be read on the web at http://tigger.uic.edu/~rdemar/geol107/hall.htm – Accessed 4.22.16
5.14c Hall described is sandstone experiments in greater detail. For those experiments, he used salt and salt water. The sandstone that consolidated from the salt fumes obviously would have fallen apart in water, although he suggests otherwise in some of his results. Instead of using a lower heat of 350-400° C and high pressure, it appears that he used a high, almost melting temperature of 1700+° C and standard atmospheric pressure. Hall's work identified some of the first steps in the making of real sandstone. Interestingly, when he was using common salt water (probably from the ocean), he may have had in place, most of the ingredients for making real sandstone as we

describe later in the Crystallization Process in the Hydroplanet Model chapter.

5.14d *Mantle Convection and Plate Tectonics: Toward an Integrated Physical and Chemical Theory*, Paul J. Tackley, *Science*, Vol. 288, 16 June 2000 p2002

5.14e *Volacnic Bomshell*, Nicola Jones, *New Scientist*, 8 March 2003, p32

5.14f *Fixed Hotspots Gone With the Wind*, Ulrich Christensen, *Nature*, Vol. 391, 19 February 1998 p740

5.14g *Flood basalts and large igneous provinces from deep mantle plumes: fact, fiction, and fallacy*, H.C. Sheth, *Tectonophysics*, Vol. 311, 1999, p1

5.14h *Volacnic Bomshell*, Nicola Jones, *New Scientist*, 8 March 2003, p37

5.14i www.lelandacademy.org/gillis/seismic/resources – Accessed 11.22.05, Site no longer available

5.14j *Looking-Deeply-Into the Earth's Crust in Europe*, Richard A. Kerr, *Science*, Vol. 261, 16 July 1993, p295

5.14k *Mantle Convection and Plate Tectonics: Toward an Integrated Physical and Chemical Theory*, Paul J. Tackley, *Science*, Vol. 288, 16 June 2000, p2002

5.15 The 'Smoking Gun' - Tomography

5.15a *Earthquakes and Geological Discovery*, Bruce A. Bolt, *Scientific American Library*, 1993, p151

5.15b *Radiogenic Heat Produced in the Upper Third of Continental Crust from KTB*, Dan F.C. Pribnow & Helmuth R. Winter, *Geophysical Research Letters*, Vol. 24, No. 3, February 1, 1997, p349

5.16 The Magma Freeze

5.16a http://www.geology.yale.edu/~jjpark/AGIvolcano.html – Accessed 9.4.08, Site no longer available

6 The Rock Cycle Pseudothery

6.1 The Most Abundant Mineral Mystery

6.1a http://oceanservice.noaa.gov/facts/whysalty.html – Accessed 6.28.16

6.1b *Why is the Ocean Salty?* Herbert Swenson, *US Geological Survey Publication,* http://ponce.sdsu.edu/usgs_why_is_the_ocean_salty/usgs_why_is_the_ocean_salty.html – Accessed 4.27.16

6.1c *The Igneous Rocks in the Light of High-Temperature Research*, Norman L. Bowen, *Carnegie Institution of Washington*, No. 14, March 26, 1935, p7

6.1d *Surface Mining*, Bruce A. Kennedy, 1990, p174

6.1e *Origin of Saline Giants: A Critical Review after the Discovery of the Mediterranean Evaporite*, K. J. Hsü, *Earth-Science Reviews*, 1972, p371-2

6.1f *Geochemistry of a Modern Marine Evaporite: Bocana De Virrila, Peru*, Susan L. Brantley, David A.Crerar, Nancy E. Moller, John H. Weare, *Journal of Sedimentary Petrology*, Vol. 54, No.2, June 1984, Introduction

6.1g *The Evaporation Path of Seawater and the Coprecipitation of Br and K+ with Halite*, M. A. McCaffrey, B. Lazar, H. D. Holland, *Journal of Sedimentary Petrology*, Vol. 57, No. 5, 5 September 1987, Abstract

6.1h *Why is the Ocean Salty?* Herbert Swenson, *US Geological Survey Publication,* http://ponce.sdsu.edu/usgs_why_is_the_ocean_salty/usgs_why_is_the_ocean_salty.html – Accessed 4.27.16

6.1i *Origin of Saline Giants: A Critical Review after the Discovery of the Mediterranean Evaporite*, K. J. Hsü, *Earth-Science Reviews*, 1972, p382

6.1j *Geochemistry of a Modern Marine Evaporite: Bocana De Virrila, Peru*, Susan L. Brantley, David A.Crerar, Nancy E. Moller, John H. Weare, *Journal of Sedimentary Petrology*, Vol. 54, No.2, June 1984, Introduction

6.1k http://supercow.uoregon.edu/~mstrick/RogueComCollege/RCC_Lectures/Sedimentary.html – Accessed 6.28.16

6.1l *Arizona Has Salt!*, Steven L. Rauzi, *Arizona Geological Survey*, January 2002, p19

6.1m *Origin of Saline Giants: A Critical Review after the Discovery of the Mediterranean Evaporite*, K. J. Hsü, *Earth-Science Reviews*, 1972, p378

6.1n *Salt Deposits, Their Origin and Composition*, O. Braitsch, in consultation with A. G. Herrmann and R. Evans, *Springer*, 1971, p254

6.1o *Gypsum and Anhydrite in the United States*, C. F. Withington, *Department of the Interior United States Geological Survey*, 1962, p1

6.1p *Geochemistry of a Modern Marine Evaporite: Bocana De Virrila, Peru*, Susan L. Brantley, David A.Crerar, Nancy E. Moller, John H. Weare, *Journal of Sedimentary Petrology*, Vol. 54, No.2, June 1984, Introduction

6.1q *Gypsum and Anhydrite in the United States*, C. F. Withington, *Department of the Interior United States Geological Survey*, 1962, p1

6.1r *Salt Deposits, Their Origin and Composition*, O. Braitsch, in consultation with A. G. Herrmann and R. Evans, *Springer*, 1971, p271

6.1s *Origin of Saline Giants: A Critical Review after the Discovery of the Mediterranean Evaporite*, K. J. Hsü, *Earth-Science Reviews*, 1972, p393

6.2 The Real History of Geology

6.2a http://www.gemsociety.org/info/igem17.htm – Accessed 4.28.16

6.2b *The Spirit of Exploration, NASA's Rover Fights the Curse of the Angry Red Planet*, George Musser, *Scientific American,* March 2004, p56

6.2d *The Open Question Selenology*, David W. Hughes, *Nature*, Vol. 327, 28 May 1987, p291

6.3 The Sand Mystery

6.3a *Geology Of Sand Dunes*, John Mangimeli, *Journal of Geology*, Vol. 71, No. 5, https://www.nps.gov/whsa/learn/nature/upload/Geology%20of%20Sand%20Dunes.pdf – Accessed 4.28.16

6.3b *Will the Dunes March Once Again?*, Daniel Jack Chasan, *Smithsonian*, December 1997, Vol. 28, No. 9, p75-77

6.3c *Sand*, *Scientific American,* Vol. 202, No 4, April 1960, p95

6.3d *The Importance of Eolian Abrasion in Supermature Quartz Sandstones and the Paradox of Weathering on Vegetation-Free Landscapes,* R.H. Dott Jr., *Journal of Geology*, 2003, Vol. 111, p387

6.3e *Earthquakes, Gases, and Earthquake Prediction*, Thomas Gold, 1994, p6, http://ellenwhite.org/content/file/earthquakes-gases-and-earthquake-prediction#document – Accessed 4.28.16

6.3f *Geology of Sand Dunes*, John Mangimeli, *Journal of Geology*, Vol. 71, No. 5, *Geology Of Sand Dunes* , https://www.nps.gov/whsa/learn/nature/upload/Geology%20of%20Sand%20Dunes.pdf – Accessed 4.28.16

6.3g http://en.wikipedia.org/wiki/Sandstone – Accessed 4.28.16

6.3h *International Association of Sedimentologists*, *Quartz Cementation in Sandstones*, Special Publication Number 29, 2000, p1

6.3i *Grand Canyon Born on East Coast*, Betsy Mason, *Nature News*, 16 September 2003

6.3j *Rock and Roll Geology*, Richard Allen, *Northwest Valley Lifestyles Magazine*, August 2003, http://www.gemland.com/geostories/rock_roll.pdf – Accessed 4.28.16

6.3k Natural Expressions Inc, Gilbert, AZ

6.3l *The Geology of Grand Canyon,* Edwin D. McKee, 1931, http://grandcanyontreks.org/geology4.htm – Accessed 4.28.16

6.4 The Quartz Mystery

6.4a *Origins of Igneous Rocks*, Paul C. Hess, *Harvard University Press*, 1989, p70

6.4b *How Do Agates Form? Interview with Peter Heaney at Penn State*, http://www.rps.psu.edu/0109/form.html – Accessed 4.28.16

6.4c *Geodes—Small Treasure Vaults in Illinois*, David L. Reinertsen, D. Scott Beaty, and Jonathan H. Goodwin, *Illinois State Geological Survey*, http://www.isgs.uiuc.edu/maps-data-pub/publications/geobits/geobit3.shtml – Accessed 4.28.16

6.4d *How Do Agates Form? Interview with Peter Heaney at Penn State*, http://www.rps.psu.edu/0109/form.html – Accessed 4.28.16

6.4e http:en.wikipedia.org/wiki/Shiprock – Accessed 10.3.16

6.4f Pegmatites, http://geology.csupomona.edu/drjessey/class/gscree/pegmatites.htm, p1 – Accessed 5.7.07, Site no longer available

6.4g *The Nature and Origin of Granite*, Wallace Spencer Pitcher, Chapman & Hall, 1997, p43

6.4h *Mind Over Magma, The Story of Igneous Petrology*, Davis A. Young, *Princeton University Press*, 2003, p82

6.5 The Basalt Mystery

6.5a http://en.wikipedia.org/wiki/Basalt – Accessed 8.14.06, Site has since changed

6.5b http://www.nps.gov/deto/naturescience/geologicformations.htm – Accessed 2.28.16

6.6 The Obsidian Mystery

6.6a *Volcanoes, A Planetary Perspective*, Peter Francis, *Clarendon Press*, 1993, p161

6.8 The Ore Mystery

6.8a *The Core-Mantle Boundary Region*, Anderson, *Geodynamics Series,* Vol. 28, American Geophysical Union, 1998, p258

6.8b *Genesis of the World's Largest Gold Deposits*, Hartwig E. Frimmel, *Science*, Vol. 297, 13 September 2002, p1815

6.9 The Carbonate Mystery

6.9a http://www.hardwater.org – Accessed 4.28.16

6.9b 2001 Annual Drinking Water Quality Report, City of Mesa, Arizona

6.9c *Encyclopedia of Caves and Karst Science*, John Gunn, 2004, p185

6.9d The term lime is used here to describe the calcium carbonate 'slurry' from which limestone is formed. Actual 'lime' is calcium oxide, (CaO) which is a manmade product and does not occur naturally.

6.9e *Fundamentals of Ecology*, M. C. Dash, 2001, p197

6.9f http://en.wikipedia.org/wiki/Diagenesis – Accessed 4.29.16

6.9g https://en.wikipedia.org/wiki/Flint – Accessed 4.29.16

6.9h http://www.marblejunction.com/m_intro.htm – Accessed 1.7.02, Site no longer available

6.9i http://en.wikipedia.org/wiki/Cave – Accessed 4.29.16

6.9j *Hydrology of a Large, High-Relief, Sub-Tropical Cave System: Sistema Purificación*, Hose, Louise D., Tamaulipas, Mexico, *Journal of Caves and Karst Studies*, 58(1):22-29

6.9k *Geochemistry of Carlsbad Cavern Pool Waters*, Jeffrey R. Forbes, Guadalupe Mountains, New Mexico, *Journal of Cave and Karst Studies*, 62(2): 127-134

6.9l *Cavern of Sleep Lake*, 14 readings between 6.4.1990 to 2.15.1992, Data from National Park Service email containing a chart of all pH readings

6.9m *Measurement of pH for field studies in karst areas*, Ira D. Sasowsky and Cory T. Dalton, *Journal of Cave and Karst Studies*, August 2005, Vol. 67, No. 2, p127

6.9n Selected Abstracts From The 2003 National Speleological Society Convention In Porterville, California, M.N. Spilde, P.J. Boston, D.E. Northup, Journal of Cave and Karst Studies, December 2003, p188

6.9o *Hydrochemical Interpretaion of Cave Patterns in the Guadalupe Mountains, New Mexico*, Arthur N. Palmer, Margaret V. Palmer, *Journal of Cave and Karst Studies*, August 2000 • 94, Caves of the Guadalupe Mountains • 44

6.9p http://www.nps.gov/wica/naturescience/wind-cave-geology.htm – Accessed 4.29.16

6.9q *Hydrochemical Interpretaion of Cave Patterns in the Guadalupe Mountains, New Mexico*, Arthur N. Palmer, Margaret V. Palmer, *Journal of Cave and Karst Studies*, August 2000 • 94, Caves of the Guadalupe Mountains • 44

6.9r Kartchner Caverns State Park brochure – 2005, http://www.azparks.gov/Parks/parkhtml/kartchner.html – Accessed 9.5.06, Site no longer available

6.9s *Oddities of the Mineral World*, William B. Sanborn, *Van Nostrand Reinhold Co.*, 1976, p110

6.9t *Fresh Natro Carbonatite Lava From Oldoinyo L'Engai*, Dr. C. G. Du Bois, J. Furst, Dr. N. J. Guest and D. J. Jennings, *Nature*, February 2, 1963, Vol. 197, p445

6.9u http://www.worldwide-safaris.com/mount-oldoinyo-lengai/ – Accessed 4.29.16

6.9v http://wikipedia.org/wiki/Caliche – Accessed 3.30.06, Site has since changed

6.9w http://www.galleries.com/dolomite – Accessed 4.29.16

6.10 The Loess Mystery

6.10a The following are map references for the Loess World Map:
United States Quaternary Eolian deposits, http://www.geog.umn.edu/courses/5441/LoessMap2.gif
A Complex Origin for Late Quaternary Loess in Central Alaska, Daniel R. Muhs, James R. Budahn, USGS, http://www.colorado.edu/INSTAAR/AW2004/get_abstr.html?id=28
Loess: Geomorphological Processes and Hazards, Setsuo Okuda, 1991, p120, (locations of loess in China only).
Physical Geology: Exploring the Earth, James S. Monroe, Reed Wicander, 1992, p559
Loess of the World Map, http://www.physicalgeography.net/fundamentals/images/loess_deposits.gif

6.10b *A Complex Origin for Late Quaternary Loess in Central Alaska, USA*, Daniel R. Muhs, James R. Budahn, 2003, http://www.colorado.edu/INSTAAR/AW2004/get_abstr.html?id=28 – Accessed 10.11.16

6.10c *Sand*, Philip H. Kuenen, *Scientific American*, Vol. 202, No. 4, April 1960, p101

6.10d The Great Wind-or-Water Debate, http://www.backyardnature.net/loess/debate.htm – Accessed 4.29.16

6.10e *Sand*, Philip H. Kuenen, *Scientific American*, Vol. 202, No. 4, April 1960, p104

6.10f Kuenen referred to 'chemical rounding' as another possible mechanism responsible for the rounding of sand grains. In this process, the rounding of grains was thought to come from quartz growth in an aqueous solution, which filled the cavities in the grains. Although a very small amount of quartz may precipitate out of water at room temperature, it would not be enough to make a measurable difference in sand or loess deposits. Kuenen further notes: "In areas where all quartz sand is newly produced from crystalline rocks and where only fluviatile action has taken place, the grains must be examined for increase in roundness by transportation. If this is detectable, it would indicate chemical rounding. The chief argument against rounding by solution is that the smaller grains in a sand are almost invariably the least rounded and particles below 0.050 mm. are sharp cornered." *Journal of Geology*, Philip H. Kuenen, 68 (4) 1960, p428

6.10g *Experimental Abraision 4: Eolin Action*, Philip H. Kuenen, *Journal of Geology*, 68 (4) 1960, p446

6.10h *Origin Of Desert Loess From Some Experimental Observations*, W. B. Whalley, J. R. Marshall, B. J. Smith, *Nature*, 300, 02 December 1982, p434

6.10i *Aeolian Abrasion And Modes Of Fine Particle Production From Natural Red Dune Sands: An Experimental Study*, J. E. Bullard, G. H. McTainsh, C. Pudmenzky, *Sedimentology*, 51, 2004, p1124

6.10j *Formation of Quartz Sand*, I. J. Smalley, *Nature*, July 30, 1966, Vol. 211, p478

6.10k *Size Distribution of Sedimentary Populations*, *Science*, Vol. 141, 1963, p802

6.10l *Journal of Sedimentary Petrology*, Vol. 28, No. 3, Sept. 1958, p374

6.10m *Sedimentary Rocks*, Francis John Pettijohn, 1957, p50

6.11 The Erosion Mystery

6.11a http://www.protrails.com/trail/821/arches-national-park-balanced-rock – Accessed 4.29.16

6.11b *Paleosols: Their Recognition and Interpretation*, Blackwell Scientific Publications, 1986, pxi

6.11c http://en.wikipedia.org/wiki/planation – Accessed 10.26.09, Site has since been updated

6.11d *Arizona Has Salt!*, Arizona Geological Survey, January 2002, Steven L. Rauzi, p21

6.12 The Earth Crust Mystery

6.12a http://en.wikipedia.org/wiki/Alfred_Wegener – Accessed 9.5.06, Significantly modified since this date

6.12b The Meteorologist Who Started a Revolution, by Patrick Hughes, http://pangaea.org/wegener.htm – Accessed 5.2.16

6.12c *Alfred Wegener. The Origins of Continents and Oceans, 4th Edition*, http://www.ucmp.berkeley.edu/history/wegener.html – Accessed 5.2.16

6.12d *Deep Secret*, James Bishop Jr., *Phoenix Magazine*, November 2002, p76

6.12e *Arizona Geology*, Arizona Geological Survey, Vol. 35, No. 4, Winter 2005, p1

6.12f *Deep Secret*, James Bishop Jr., *Phoenix Magazine*, November 2002, p76

6.12g Email from Robert Jackson, http://www.combustionresources.com

6.12h http://en.wikipedia.org/wiki/Peat –Accessed 10.23.06, Site has since been changed

6.12i http://www.fe.doe.gov/education/energylessons/coal/gen_howformed.html – Accessed 5.2.16

6.12j http://en.wikipedia.org/wiki/Coal – Accessed 10.19.06, Site has since been changed

6.12k *Sedimentation in Barred Basins and Source Rocks of Oil*, *Origin of Evaporites*, W. G. Woolnough, *American Assoc. of Petroleum Geologists*, 1971, p6

6.12l *Erratic Boulders in Sewell Coal of West Virginia*, Paul H. Price, *Journal of Geology*, 1932, Vol. 40, p67

6.12m *The Non-Organic Theory of the Genesis of Petroleum*, Samar Abbas, Utkal University, India, 15 Oct 1996, p2, http://arxiv.org/PS_cache/physics/pdf/9610/9610011v1.pdf – Accessed 5.2.16

6.13 Geotheoretical to Geological

6.13a *Sedimentation in Barred Basins, and Source Rocks of Oil*, W. G. Woolnough, *Bulletin of the American Association of Petroleum Geologists (AAPG)*, Vol. 21, No.9, September 1937, p 1102, Entire article also reprinted in *Origin of Evaporites*, AAPG Bulletin, 1971, p 3-59

6.13b It should be noted that over the past decade or two, with the advent of the theory that dinosaurs were exterminated by a meteorite impact, some scientists have toned down their absolute belief in the principle of uniformity. Note the 'softening' of the words (emphasis added) in the following quote from a university level geology textbook published 1998:

    "A fundamental principle of geology, advanced in the eighteenth century by the Scottish physician and geologist James Hutton, is that "the present is the key to the past." This principle of uniformitarianism, as it is now known, holds that the geologic processes we see in operation as they modify Earth's crust today have worked **in much the same way** over geological time." Bib 59 p4

Although there is a trend toward *softening* the words toward many of the geological processes that are not happening today, it is important to recognize that, *the most significant geological events in Earth's geological history are not happening today*. The Magma Pseudotheory and Rock Cycle Pseudotheory Chapters have laid out empirical evidence to show just how illogical modern geology is and how much theory exists without fact or observation. The Hydroplanet, Universal Flood, and other Model Chapters all demonstrate, using the Universal Scientific Method, a refreshingly new and clear geological history of the Earth.

Of further note, the establishment of the uniformity principle was not widely accepted in the beginning. With its acceptance came a secondary reason for adherence to its principles – its support of evolution:

"Unfortunately, Hutton's ideas were not widely disseminated or accepted. In 1830, however, Charles Lyell published a landmark book, Principles of Geology, in which he championed Hutton's concept of uniformitarianism. Instead of relying on catastrophic events to explain various features of the Earth, Lyell recognized that imperceptible changes brought about by present-day processes could, over long periods of time, have tremendous cumulative effects. Through his writings, Lyell firmly established uniformitarianism as the guiding philosophy of geology. **Furthermore, the recognition of virtually limitless time was also necessary for, and instrumental in, the acceptance of Darwin's 1859 theory of evolution.**" Bib 54 p219

6.13c *The Mid-Oceanic Ridges*, Claude J. Allegre, Forward, pix Bib 120

7 The Hydroplanet Model

7.0

17.0a *Strange Universe*, Bob Berman, 2003, p117
17.0b *Geology of the Sierra Nevada*, Mary Hill, 1975, p162

7.1 Magmaplanet to Hydroplanet

7.1a *Earth Science: Inside History in Depth*, David Stevenson, *Nature*, 01 April 2004, 428, p476
7.1b *A Cool Early Earth*, John W. Valley, William H. Peck, Elizabeth M. King, Simon A. Wilde, 2002, *Geology* 30, p353
7.1c *Plume IV: Beyond the Plume Hypothesis, Tests of the plume paradigm and alternatives*, Gillian R. Foulger, James H. Natland, and Don L. Anderson, *Scientific report: GSA Today*, Penrose Conference, 14, 2004, p26-28
7.1d http://www.geography4kids.com/files/earth_tectonics.html – Accessed 5.5.16
7.1e *Water for the Rock*, Ben Harder, *Science News*, 23 March 2002, Vol. 161, No. 12, p184, http://phschool.com/science/science_news/articles/water_for_the_rock.html, p2 – Accessed 5.5.16
7.1f http://spaceresearch.nasa.gov/sts-107/resourceguide/content/images/summary/images_gallery/dropphysics.html – Accessed 5.20.08, Site no longer available
7.1g *Subsurface Combustion in Mali: Refutation of the Active Volcanism Hypothesis in West Africa*, *Geology*, July 2003, Vol. 31, No.7, p581

7.2 Celestial Water

7.2a *The Ice of Life*, David F. Blake, Peter Jenniskens, *Scientific American*, August 2001, p46-7
7.2b *Boiling Hot? Maybe Not.*, Heidi Schultz, *National Geographic*, May 2004, p1
7.2c *The Ice of Life*, David F. Blake, Peter Jenniskens, *Scientific American*, August 2001, p50
7.2d http://en.wikipedia.org/wiki/Maser#Hydrogen_maser – Accessed 1.16.07, 10.11.16
7.2e *Water on the Sun*, L. Wallace, et al., *Science*, Vol. 268, 26 May 1995, p1155
7.2f *Water's Role in Making Stars*, Brunella Nisini, *Science*, Vol. 290, 24 November 2000, p1513
7.2g *Wet Stellar System Like Ours Found*, Richar A Kerr, *Science*, Vol. 293, 20 July 2001, p407-8
7.2h *Water's Role in Making Stars*, Brunella Nisini, *Science*, Vol. 290, 24 November 2000, p1513
7.2i *Water on the Sun*, L. Wallace, et al., *Science*, Vol. 268, 26 May 1995, p1155
7.2j *Water on the Sun: Molecules Everywhere,* Takeshi Oka, *Science*, Vol. 277, 18 July 1997, p328
7.2k http://www.psrd.hawaii.edu/Jan97/MercuryUnveiled.html – Accessed 5.5.16
7.2l http://www.nytimes.com/2012/11/30/science/space/mercury-home-to-ice-messenger-spacecraft-findings-suggest.html?_r=0 – Accessed 5.5.16
7.2m *Water, Water Everywhere*, *Sky & Telescope*, February 1999, p34
7.2n http://adsabs.harvard.edu/abs/1999DPS....31.5104R – Accessed 5.5.16
7.2o http://photojournal.jpl.nasa.gov/catalog/PIA05421 – Accessed 5.6.16
7.2p http://www.nasa.gov/worldbook/uranus_worldbook.html – Accessed 1.24.07, Site no longer available
7.2q http://www.nasa.gov/worldbook/neptune_worldbook.html – Accessed 1.24.07, Site no longer available
7.2r *Space*, Elon Musk, *Discover*, December 2005, p69
7.2s *Water at the Equator*, *National Geographic*, January 2004, p17
7.2t *Martian Lake View*, JR Minkel, *Scientific American*, May 2005, p34
7.2u http://www.bbc.com/news/science-environment-29343987 – Accessed 5.6.16
7.2v http://www.exeter.ac.uk/news/research/title_414371_en.html – Accessed 6.21.16
7.2w *The Volume Library*, Educators Association, Inc., 1946, p1086
7.2x *Cheapest Mission Finds Moon's Frozen Water*, Richar A. Kerr, *Science*, Vol. 279, 13 March 1998, p1628
7.2y *Neutron- star Extrema: Citius, Altius, Fortius*, *Sky & Telescope*, February 1999, p35
7.2z *Science*, Vol. 282, 4 December 1998, p1811
7.2aa *Surfing the Solar System, It's wetter than you think*, Michael Milstein, *Air & Space*, December 1997/January 1998, p56
7.2ab *Searching For the Molecules of Life in Space*, Steve Nadis, *Sky & Telescope*, January 2002, p32
7.2ac http://csep10.phys.utk.edu/astr162/lect/sun/composition.html – Accessed 5.6.16
7.2ad *Searching For the Molecules of Life in Space*, Steve Nadis, *Sky & Telescope*, January 2002, p36
7.2ae http://sci.esa.int/science-e/www/object/printfriendly.cfm?fobjectid=31364 – Accessed 5.9.16

7.3 Hydrospheres

7.3a http://photojournal.jpl.nasa.gov/catalog/?IDNumber=pia06140 – Accessed 5.9.16
7.3b http://photojournal.jpl.nasa.gov/catalog/?IDNumber=pia07744 – Accessed 5.9.16
7.3c http://photojournal.jpl.nasa.gov/catalog/?IDNumber=pia07746 – Accessed 5.9.16
7.3d http://photojournal.jpl.nasa.gov/catalog/?IDNumber= pia06162 – Accessed 4.7.07, Site no longer available
7.3e *Titan Reveals Methane Rain and Rocks of Water*, Achim Schneider, *Nature*, 21 January 2005
7.3f http://www.nasa.gov/mission_pages/cassini/multimedia/pia09184.html – Accessed 6.23.16
7.3g http://www.nasa.gov/mission_pages/cassini/multimedia/pia06440.html – Accessed 6.23.16
7.3h https://saturn.jpl.nasa.gov/resources/2562/ – Accessed 6.23.16
7.3i http://photojournal.jpl.nasa.gov/catalog/?IDNumber=pia07767 – Accessed 5.9.16
7.3j *Amalthea's Destiny is Less Than That of Water*, *Science*, Vol. 308, 27 May 2005, p1291
7.3k http://photojournal.jpl.nasa.gov/catalog/PIA01082 – Accessed 5.9.16
7.3l *Court of King Jupiter*, William R. Newcott, *National Geographic*, September 1999, p138
7.3m http://photojournal.jpl.nasa.gov/catalog/PIA01478 – Accessed 5.9.16
7.3n http://photojournal.jpl.nasa.gov/catalog/?IDNumber=pia01656 – Accessed 5.9.16
7.3o http://photojournal.jpl.nasa.gov/catalog/?IDNumber=pia03455 – Accessed 5.9.16
7.3p *Court of King Jupiter*, William R. Newcott, *National Geographic*, September 1999, p136
7.3q *Evidence for a Subsurface Ocean on Europa*, *Nature*, Vol. 391, 22 January 1998, p363
7.3r http://photojournal.jpl.nasa.gov/catalog/PIA03002 – Accessed 5.9.16
7.3s *Surfing the Solar System, It's wetter than you think*, Michael Milstein, *Air & Space*, December 1997/January 1998, p59
7.3t http://photojournal.jpl.nasa.gov/catalog/PIA00347 – Accessed 5.9.16
7.3u http://solarsytem.nasa.planets/profile.cfm?Object=Enceladus – Accessed 1.18.07, Site no longer available
7.3w http://photojournal.jpl.nasa.gov/catalog/PIA07759 – Accessed 5.9.16

7.3x *What is a Planet?*, Steven Soter, *Scientific American*, January 2007, p34

7.4 The Crystallization Process

7.4a http://en.wikipedia.org/wiki/Crystal – Accessed 4.24.07
7.4b http://en.wikipedia.org/wiki/Precipitate – Accessed 4.25.07
7.4c *Origin of Granite in the Light of Experimental Studies in the System*, Na AlSi3O8-KalSi3O8-SiO2--H2), O. F. Tuttle and N. L. Bowen, *The Geological Society of America*, 1958, p67-69
7.4d http://en.wikipedia.org/wiki/Evaporite – Accessed 5.9.16
7.4e http://www.ogj.com/articles/save_screen.cfm?ARTICLE_ID=282114, – Accessed 5.9.16
7.4f http://www.ogj.com/articles/save_screen.cfm?ARTICLE_ID=282114, – Accessed 5.9.16
7.4g http://www.ogj.com/articles/save_screen.cfm?ARTICLE_ID=282114, – Accessed 5.9.16
7.4h *Oddities of the Mineral World*, William B. Sanborn, *Van Nostrand Reinhold Co.*, 1976, p51
7.4i *Journal of Geology*, 1982, Vol. 90, p302
7.4j *What Heated the Astroides?*, Alan E. Rubin, *Scientific American*, May 2005, p83
7.4k *Mind Over Magma, The Story of Igneous Petrology*, Davis A. Young, *Princeton University Press*, 2003, p86
7.4l *No Water, No Granites- No Oceans, No Continents*, I. H. Campbell and S. R. Taylor, *Geophysical Research Letters*, Vol. 10, No. 11, November 1983, p1061
7.4m *Origins of Igneous Rocks*, Paul C. Hess, *Harvard University Press*, 1989, p69
7.4n *The Nature and Origin of Granite*, Wallace Spencer Pitcher, *Chapman & Hall*, 1997, p33
7.4o *Handbook of Hydrothermal Technology A Technology for Crystal Growth and Materials Processing*, K. Brappa, Masahiro Yoshimura, *Noyes Publications*, 2001, p1
7.4p *Hydrothermal Sysnthesis of Crystals*, Robert A. Laudise, *Chemical & Engineering News*, September 28, 1987, p38
7.4q http://geology.csupomona.edu/drjessey/class/gscree/pegmatites.htm, p1, Site no longer available
7.4r *Origins of Igneous Rocks*, Paul C. Hess, *Harvard University Press*, 1989, p244
7.4s *An experimental study of B-, P- and F-rich synthetic granite pegmatite at 0.1 and 0.2 GPa*, Ilya V. Veksler, Rainer Thomas, *Contributions to Mineralology and Petrology*, 2002, p675
7.4t *Novel TEM approaches to Imaging of Microstructures in Carbonates: Clues to Growth Mechanisms in Calcite and Dolomite*, J. Paquette, H. Vali, E. Mountjoy, *American Mineralogist*, 1999, Vol. 84, p1940
7.4u *Nanotopography of Synthetic and Natural Dolomite Crystals*, Lisa A. Kessels, Duncan F. Sibley and Stephan H. Nordeng, *Sedimentology*, 2000, 47, p173
7.4v *Calcite-Dolomite-Magnesite Stability Relations in Solutions at Elevated Temperatures*, P. E. Rosenberg, H. D. Holland, *Science*, 14 August 1964, p700-1
7.4w http://www.glossary.oilfield.slb.com/Terms/d/diagenesis.aspx?p=1 – Accessed 6.21.16
7.4x *Hydrothermal Synthesis and Optical Properties of Calcite Single Crystals*, I. V. Nefyodova, N. I. Leonyuk, I. A. Kamenskikh, *Journal of Optoelectronics and Advanced Materials* Vol. 5, No. 3, September 2003, p609-613, *Pinacoidal growth and optical properties of calcite crystals*, I. V. Nefyodova, V. I. Lyutin, V. L. Borodin, P. P. chvanski, N. I. Leonyuk, *Progress in Crystal Growth and Characterization of Materials*, 2000, p263-271
7.4y *Hydrothermal Alteration of Olivine in a Flow-Through Autoclave: Nucleation and Growth of Serpentine Phases*, Charles Normand, *American Mineralogist*, 2002, Vol. 87, p1699
7.4z *Water-Induced Fabric Transitions in Olivine*, Haemyeong Jung and Shun-ichiro Karato, *Science*, 24 August 2001, Vol. 293, p460
7.4aa *Hydrothermal Sysnthesis of Crystals*, Robert A. Laudise, *Chemical & Engineering News*, September 28, 1987, p40

7.5 A New Geology

7.5a *Relation of nucleation and Crystal-Growth Rate to the Development of Granitic Textures*, Samuel E. Swanson, *American Mineralogist*, 1977, Vol. 62, p977

7.6 The Hydroplanet Earth

7.6a http://www.ufn.ru/en/articles/2004/7/b/ – Accessed 5.9.16
7.6b *Ultrahigh-Pressure Mineralogy: Physics and Chemistry of the Earth's Deep Interior*, Russell J. Hemley, *Reviews in Mineralogy*, Mineralogical Society of America, 1998, Vol. 37, Preface
7.6c *Earth-like Planets Probably Water-Logged*, Philip Ball, *Nature*, 21 August 2003
7.6e *Deep Waters, New Scientist*, 30 Aug. 1997, p4
7.6f *Is the transition zone an empty water reservoir? Inferences from numerical model of mantle dynamics*, G. Richard, M. Monnereau, J. Ingrin, *Earth and Planetary Science Letters 205*, 2002, p37
7.6g *Scientists make special snowball*, L. Sanders, *Science News*, 10 Oct. 2009, p10
7.6h http://www.nbcnews.com/id/7432966/ns/technology_and_science-science/t/earths-oldest-known-object-display/#.VzEf54SDFBc –Accessed 5.9.16, In addition, see the journal *Geology, A cool early Earth*, 2002, 30: 351-354
7.6i *A Cool Early Earth*, John W. Valley, *Scientific American*, October 2005, p59
7.6j http://www.scientificamerican.com/article/what-do-we-know-about-the/ – Accessed 5.9.16
7.6k Private email communication with William L. Ellsworth, U. S. Geological Survey, March 25, 2002
7.6l *Phase 3 Drilling Operations at the Long Valley Exploratory Well LVF51-20 Sandia Report SAND99-1279*, Sandia National Laboratories, June 1999, p14, http://www.prod.sandia.gov/cgi-bin/techlib/acess-control.pl/1999/991279.pdf – Site is no longer available
7.6n *Orgin and Nature of Crustal Reflections: Results From Integrated Seismic Measurements at the KTB Super Deep Drilling Site*, K. Bram, *Journal of Geophysical Research*, Vol. 102, No. B8, p18, 267
7.6o *Hydraulic pathways in the crystalline rock of the KTB*, Zimmermann, Gunter, Korner, Alexander, Burkhardt, Hans, *Geophysical Journal International*, Vol. 142, Issue 1, 2000, p13
7.6p *Scientists Ponder Deep Slabs, Small Comets, Hidden Oceans*, Richard A. Kerr, *Science*, Vol. 280, 12 June 1998, p1694-5, http://science.sciencemag.org/content/280/5370/news-summaries – Accessed 5.10.16
7.6q http://en.wikipedia.org/wiki/Earth – Accessed 5.10.16
7.6r *Looking-Deeply-Into the Earth's Crust in Europe*, *Science*, Vol. 261, 16 July 1993, p296
7.6s *German Super-Deep Hole Hit Bottom*, Richard A. Kerr, *Science*, Vol. 266, 28 October 1994, p545
7.6t *The Earth's Mantle, Composition, Structure, and Evolution*, Edited by Ian Jackson, *Australian National University*, Cambridge University Press, 1998, p193
7.6u *Earth's Core and the Geodynamo*, Bruce A. Buffett, *Science*, Vol. 288, 16 June 2000, p2007

7.7 Earth's Hydrology Redefined

7.7a http://en.wikipedia.org/wiki/Hydrology – Accessed 5.10.16
7.7b *Mantle Convection and Plate Tectonics*, Paul J, Tackley, *Science*, June 16, 2000, p2002
7.7c Private Email, October 2003, Paul J. Tackley
7.7d *Scientific American*, January 1989, p54
7.7e *Localized Polymorphic Phase Transformation in High-Pressure Faults and Applications to the Physical Mechanism of Deep Earthquakes*, Stephen H. Kirby, *Journal of Geophysical Research*, Vol. 92, No. B13, 1987, p13,789-13,800
7.7f *Scientific American*, January 1989, p55
7.7g *Possible presence of high-pressure ice in cold subducting slabs*, Craig R. Bina & Alexandra Navrotsky, *Nature*, 14 December 2000, p844-845
7.7h *Stirring the Oceanic Incubator*, Robert P. Dziak and H Paul Johnson, *Science*, Vol. 296, 24 May 2002, p1406
7.7i *Triggered Swarms: How Big Quakes Can Cause Small Quakes Far Away*, Naomi Lubick, *Scientific American*, March 2003, p32, http://www.scientificamerican.com/article.cfm?id=triggered-swarms – Accessed 5.10.16
7.7j *Continental-scale Links Between The mantle and Groundwater Systems of the Western United States*, Dennis L. Newell, Laura J. Crossey, Karl E. Karlstrom, Tobias P. Fischer, David R. Hilton, *GSA Today*, December 2005, p9
7.7k *Periodically Triggered Seismicity at Mount Wrangell, Alaska, After the Sumatra Earthquake*, Michael West, John J. Sanchez, Stephen R. McNutt, *Science*, 20 May 2005, Vol 308, p1145
7.7l *Detection of Widespread Fluids in the Tibetan Crust by Magnetotelluric Studies*, *Science*, Vol. 292, 27 April 2001, p716
7.7m *INDEPTH Wide-Angle Reflection Observation of P-Wave-to-S-Wave Conversion from Crustal Bright Spots in Tibet*, *Science*, Vol. 274, 6 December 1996, p1690
7.7n *Detection of Widespread Fluids in the Tibetan Crust by Magnetotelluric*

Studies, Science, Vol. 292, 27 April 2001, p716

7.7o *Scientific American*, January 1989, p54

7.7p *Tomography of the Source Area of the 1995 Kobe Earthquake: Evidence for Fluids at the Hypocenter?*, Dapeng Zhao, Hiroo Kanamori, Hiroaki Negishi, Douglas Wiens, *Science,* Vol. 274, 13 December 1996, p1892

7.7q *Detection of Hydrothermal Precursors to Large Northern California Earthquakes*, Paul G. Silver, Nathalie J. Valette-Silver, *Science*, Vol. 257, 4 September 1992, p1363

7.7r Email personal communication with Bill Ellsworth, USGS Menlo Park, 25 March 2002

7.7s *The Permeability of Young Oceanic Crust east of Juan De Fuca Ridge Determined Using Borehole Thermal Measurements*, Fisher, Becker, Davis, *Geophysical Research Letters*, Vol. 24, No. 11. June 1, 1997, p1311

7.7t *Stirring the Oceanic Incubator*, Robert P. Dziak and H Paul Johnson, *Science*, Vol. 296, 24 May 2002, p1406-7

7.7u *Seismic reflectors, conductivity, water and stress in the continental crust*, D. Ian Gough, *Nature*, Vol. 323, 11 September 1986, p143

7.7v http://farshores.org/n05lake.htm – Accessed 7.23.07, Site is currently not available

7.7w *Resistivity and self-potential changes associated with volcanic activity: The July 8, 2000 Miyake-jima eruption (Japan)*, 2003, J. Zlotnicko, *Earth and Planetary Science Letters* 2005, p151

7.7x http://pubs.usgs.gov/fs/1997/fs113-97/ – Accessed 5.11.16

7.7y *Remote sensing of CO_2 and H_2O emission rates from Masaya volcano, Nicaragua*, Michael R. Burton, Clive Oppenheimer, Lisa A. Horrocks, Peter W. Francis, *Geology*, October 2000, Vol. 28, No. 10, p915

7.7z *Remote sensing of CO_2 and H_2O emission rates from Masaya volcano, Nicaragua*, Michael R. Burton, Clive Oppenheimer, Lisa A. Horrocks, Peter W. Francis, *Geology*, October 2000, V 28, No 10, p915

7.7aa *Volcanic Eruptions of 1980 at Mount St. Helens: The First 100 Days*, Foxworthy and Hill, 1982*, USGS Professional Paper 1249*, p125

7.7ab http://news.nationalgeographic.com/news/2004/10/1007_041007_mts-thelens_recap_2.html – Accessed 5.11.16

7.7ac http://en.wikipedia.org /wiki/Lahar – Accessed 7.27.07

7.7ad *Volcanoes in the Sea, the Geology of Hawaii*, Gordon A. Macdonald and Agatin T. Abbott, University of Hawaii Press, 1970, p81

7.7ae http://saturn.jpl.nasa.gov/news/press-release-details.cfm?newsID=619 – Accessed 12.13.05, Site no longer available

7.7af *New Madrid in Motion*, Martitia P. Tuttle, *Nature*, Vol. 435, 23 June 2005, p1037

7.7ag *Encyclopedia Britannica*, Vol. XXVIII, Hugh Chisholm, Cambridge University Press, 1911, p186

7.7ah *Earthquakes Fired Steam-powered Boulders*, Philip Ball, *Nature*, Science Update, 3 September 2003

7.7ai *Fluid Overpressure in Layered Intrusions: Formation of a Breccia Pipe in the Eastern Bushveld Complex, Republic of South Africa, Minerallium Deposita*, 2003, p38:363

7.8 Hydrocrater Model

7.8a *Field Geology and Petrology of the Minette Diatreme at Buell Park, Apache County, Arizona*, Michael Frank Roden, Thesis, *University of Texas at Austin*, 1977, p1

7.8b http://en.wikipedia.org/wiki/Xenolith – Accessed 5.11.16

7.8c *Desert Heart, Chronicles of the Sonoran Desert*, William K. Hartmann, *Fisher Books*, 1989, p152

7.8d http://www.nps.gov/crla/ – Accessed 5.11.16

7.8e *On the Terrestrial Heat Flow and Physical Limnology of Crater Lake, Oregon*, David L. Williams, Richard P. Von Herzen, *Journal of Geophysical Research*, February 10, 1983, Vol. 88, No. B2, p1103

7.8f *The Birth of Nilahue, A New Maar Type Volcano at Riñahue, Chile*, G. Müller and G. Veyl, *International Geological Congress,* 1957 Report 20, Sect. 1, p380

7.8g *Deep Sea Pockmark Environments in the Eastern Mediterranean*, Lyobomir Dimitrob, John Woodside, *Marine Geology*, 195, 2003, p263-4

7.8h *Can We Buy Global Warming?*, Robert H. Socolow, *Scientific American*, July 2005, p54

7.9 The Crater Debate

7.9a *The Cosmic Serpent*, Clube and Napier, 1982, p78

7.9b Ralph Baldwin as Quoted in *The Modern Moon*, Charles A Wood, 2003, p11

7.9c *The Open Question in Selenology*, David W. Hughes, *Nature*, Vol. 327, 28 May 1987, p291

7.9d *From Magma to Tephra*, Chapter 7, Pyroclastic Surges and Compressible Two-Phase Flow, Kenneth H. Wohletz, Los Alamos National Laboratory, Elsevier, 1998, p253-4, http://www.lanl.gov/orgs/ees/geodynamics/Wohletz/Pyroclastic%20Surges.pdf – Accessed 5.11.16

7.9e *A Cool Early Earth*, John W. Valley, William H. Peck, Elizabeth M. King, Simon A. Wilde, 2002, *Geology* 30, p353

7.9f *The Modern Moon*, Charles A Wood, 2003, p12

7.9g *Hubble's Top 10*, Mario Livil, *Scientific American*, July 2006, p44

7.9h *Oblique incidence hypervelocity impacts on rock*, M. J. Burchell, L. Whitehorn, *Mon. Not. R. Astron. Soc.* 341, 2003, p192

7.9i http://gsa.confex.com/gsa/2001AM/ finalprogram/abstract_28367.htm – Accessed 9.4.07, Site no longer available

7.9j http://nssdc.gsfc.nasa.gov/planetary/comet.html – Accessed 5.12.16

7.9k *Ultrahigh-Pressure Mineralogy, Physics and Chemistry of the Earth's Deep Interior*, Reviews in Mineralogy, Vol. 37, Mineralogical Society of America, 1998, p5

7.9l *Observation of expanding vapor cloud generated by hypervelocity impact*, Toshihiko Kadono, Akira Fujiwara, *Journal of Geophysical Research*, Vol. 101, No. E11, November 25, 1996, p26,106

7.9m *A Reevaluation of Impact Melt Production*, E. Pierazzo, A. M. Vickery, H. J. Melosh, ICARUS 127, June 1997, p408

7.9n *Reduction of $SiO2$ to Si and Metallurgical Transformations in Al by Hypervelocity Impact of Al-Projectiles into Quarts Sand, Lunar and Planetary Science XIV*, 03/1983, p347, http://www.hou.usra.edu/meetings/gap2015/pdf/1071.pdf – Accessed 6.23.16

7.9o *Morphology and chemistry of projectile residue in small experimental impact crater*, Proceedings of the Fourteenth Lunar and Planetary Science Conference, Par 1 Journal of Geophysical Research*, Vol. 88, Supplement, November 15, 1983, pB353, http://ntrs.nasa.gov/search.jsp?R=19840035697 – Accessed 5.12.16

7.9p *Crater Morphology at Impact Velocities Between 8 and 17 km/s*, H. Iglseder, E. Igenbergs, *International Journal of Impact Engineering*, Vol. 10, 1990, p271

7.9q *Observation of high-velocity, weakly shocked ejecta from experimental impacts*, Andrew J. Gratz, William J. Nellis, Neal A. Hinsey, *Nature*, Vol. 363, 10 June 1993, p522-4

7.9r *Ultra-High Velocity Impacts: Cratering Studies of Microscopic Impacts from 3 km/s to 30 km/s*, G. L. Stradling, G. C. Idzorck, B. P. Shafer, H. L. Curling, Jr., M. T. Collopy, A. A. Blossom, S. Fuerstenau, *International Journal of Impact Engineering*, Vol. 14, 1993, p722

7.9s *A Reevaluation of Impact Melt Production*, E. Pierazzo, A. M. Vickery, H. J. Melosh, ICARUS 127, June 1997, p408

7.9t *An analysis of differential impact melt-crater scaling and implications for the terrestrial impact record*, R. A. F. Grieve, M. J. Cintala, *Meteoritics* 27, 1992, p533

7.9u *Day of Trinity*, Lansing Lamont, *Atheneum*, New York, 1985, p246

7.9v *Earthquakes and Geological Discovery*, Bruce A. Bolt, *Scientific American Library*, 1993, p89

7.9w http://barringercrater.com/science/ – Accessed 9.12.07, Site is no longer available

7.9x *Are Cryptovolcanic Structures Due to Meteoric Impact?*, G.J. H. McCall, *Nature*, No. 4916, January 18, 1964, p252

7.9y *Secrets of the Wabar Craters*, Wynn J. C. and Shoemaker E. M., *Sky and Telescope*, 1997, p44-48

7.9z *Geology of the Wabar Meteorite Craters, Saudi Arabia*, E. M. Shoemaker, J. C. Wynn, *Lunar and Planetary Science XXVII*, 1995, USGS and Lowell Observatory, p1

7.9aa *Secrets of the Wabar Craters*, Wynn J. C. and Shoemaker E. M., *Sky and Telescope*, 1997, p48

7.9ab *Geology of the Wabar Meteorite Craters, Saudi Arabia*, E. M. Shoemaker, J. C. Wynn, *Lunar and Planetary Science XXVII*, 1995, USGS and Lowell Observatory, p1

7.9ac *Experimental Studies of Oblique Im*pact, Donald E. Gault, John A. Wedekind, *Proc. Lunar Planet. Sci. Conf.* 9th, 1978, p3860

7.9ad http://history.nasa.gov/ap11ann/top10sci.htm – Accessed 5.12.16

7.9ae *Origin of the Earth and Moon, Chronology and Isotopic Constraints on Lunar Evolution*, University of Arizona Press, G. A. Snyder, L. E. Borg, L. E. Nyquist, and L. A. Taylor, 2000, p381

7.9af http://www.clavius.org/envsoil.html – Accessed 9.28.07

7.9ag http://setas-www.larc.nasa.gov/LDEF/MET_DEB/md_chemistry.html – Accessed 5.12.16

7.9ah http://www.unb.ca/passc/ImpactDatabase/index.html – Accessed 5.12.16

7.9ai *The Structure of Scientific Revolutions,* Kuhn, T. S. (1970). Chicago, University of Chicago Press, Chapter 6

7.9aj *Impact, Cryptoexplosion, or Diapric Movements?*, G.C. Amstutz, *Transactions of the Kansas academy of science*, Vol. 67, 1964, p349

7.9ak *Are Cryptovolcanic Structures Due to Meteoric Impact?*, G. J. H. McCall, *Nature*, No. 4916, January 18, 1964, p252

7.9al *Silicified Cone-in-Cone Structures from Erfoud (Morocco): A Comparison with Impact-Generated Shatter Cones*, C. Koeberi, H. Henkel, Stefano Lugli, Wolf U. Reimold, Christian Koeberl, Editors, *Impact Tectonics, Impact Studies*, 2005, p81

7.9am *Argument supporting explosive igneous activity for the origin of "cryptoexplosion" structures in the midcontinent, United States*, John Luczaj, *Geology,* April 1998; Vol. 26; No. 4, p295

7.9an *Craters Unchanged*, H. J. Melosh, *Nature*, Vol. 394, 16 July 1998, p223

7.9ao *Argument supporting explosive igneous activity for the origin of "cryptoexplosion" structures in the midcontinent, United States*, John Luczaj, *Geology,* April 1998; Vol. 26; No. 4, p297

7.9ap *Cryptoexplosion Structures: A Discussion*, Robert S. Dietz, *American Journal of Science*, Vol. 262, Summer 1963, p650

7.9aq *Are Cryptovolcanic Structures Due to Meteoric Impact?*, G. J. H. McCall, *Nature*, No. 4916, January 18, 1964, p253

7.9ar *Hypervelocity Impact of Steel Into Coconino Sandstone*, E. M. Shoemaker, D. E. Gault, H. J. Moore, and R. V. Lugn, *American Journal of Science*, Vol. 261, Summer 1963, p673

7.9as *A Coesite-Sanidine Grospydite from the Roberts Victor Kimberlite*, Joseph R. Smyth, *Earth and Planetary Science Letters*, 34, 1977, p284

7.9at *An Alternative Origin for Coesite from the Richat Structure, Mauritania*, S. Master, J. Karfunkel, *Meteoritics & Planetary Science*, Vol. 36, No. 9, Supplement, 2001, pA125

7.9au *Coesite from Wabar Crater, near Al Hadida, Arabia*, E. C. T. Chao, J.J. Fahey, Janet Littler, *Science*, New Series, Vol. 133, No. 3456, March 24, 1961, p882-3

7.9av *Metallic Spherules in Impactite and Tektite Glasses*, Robin Brett, *The American Mineralogist*, Vol. 52, May-June, 1967, p723

7.9aw *Libyan Desert Glass* by Gerhard Muehle, February 1998, *Meteorite Magazine*, http://meteoritemag.uark.edu/608.htm – Accessed 11.2.07, Site no longer available

7.9ax *Non random Distribution of Lunar Craters*, W. E. Elston, M. J. Aldrich, E. I. Smith, R. C. Rhodes, *Journal of Geophysical Research*, August 10, 1971, Vol. 76, No. 23, p5675

7.9ay *Martian Doublet Craters*, Verne R. Oberbeck, Michio Aoyagi, *Journal of Geophysical Research*, May 10, 1972, Vol. 77, No. 14, p2419

7.9az http://photojournal.jpl.nasa.gov/catalog/?IDNumber=pia00334 – Accessed 5.16.16

7.9ba *Explosion Cratering and the Formation of Central Uplifts and Multi-rings*, David J. Roddy, USGS, G. H. S. Jones, Scientific Advisor, Emergency Planning Canada, Abstracts of Papers Presented to the Conference on Multi-ring Basins: Formation and Evolution, Lunar and Planetary Institute, November 10-12, 1980, p68

7.9bb *Marine Geology*, 195, 2003, p273

7.9bc http://photojournal.jpl.nasa.gov/catalog/PIA06254 – Accessed 5.16.16

7.9bd http://photojournal.jpl.nasa.gov/catalog/PIA06248 – Accessed 5.16.16

7.9be *Geological Evidence for solid-state convection in Europa's ice shell*, R.T. Pappalardo, *Nature*, Vol. 391, 22 January 1998, p365

7.9bf *Court of King Jupiter, Callisto*, William R. Newcott, *National Geographic*, September 1999, p139

7.9bg http://solarsystem.nasa.gov/planets/profile.cfm?Object=Hyperion, 2008 – Accessed 5.16.16, Quote has been updated

7.9bh http://photojournal.jpl.nasa.gov/catalog/PIA07761 – Accessed 5.16.16

7.9bi *Close-Up of Saturn's Moon Hyperion*, Jessica Ruvinsky, *Discover*, December 2005, p8, http://discovermagazine.com/2005/dec/saturn-moon-cassini – Accessed 5.16.16

7.9bj *Cryptoexplosion Structures caused From Without or From Within the Earth? ("Astoblemes" or "Geoblemes?")*, Walter H. Butcher, *American Journal of Science*, Vol. 261, Summer 1963, p597

7.9bk *Impact, Cryptoexplosion, or Diapric Movements?*, G.C. Amstutz, *Transactions of the Kansas Academy of Science*, Vol. 67, 1964, p351

7.10   The Meteorite Model

7.10a http://en.wikipedia.org/wiki/Meteoritics – Accessed 5.16.16

7.10b http://www.nhm.ac.uk/nature-online/space/meteorites-dust/collecting-identifying-meteorites/index.html – Accessed 11.26.07, Site is no longer available

7.10c *Fireballs Photographed in Central Europe*, Z. Ceplecha, Bull. Astron. Inst. Czech 20(1977), p3228-335

7.10d *Photographic observations of Neuschwanstein, a second meteorite from the orbit of the Pribram chondrite*, Pavel Spurny, Jurgen Oberst, Dieter Heiniein, *Nature*, Vol. 423, 8 May 2003, p151

7.10e *Criteria for Identification of Ablation Debris From Primitive Meteoric Bodies*, D. E. Brownlee, et al, *Journal of Geophysical Research*, Vol. 80, No. 35, December 10, 1975, p4917

7.10f http://nau.edu/cefns/labs/meteorite/about/meteorite-minerals/ – Accessed 5.17.16

7.10g *The composition of Stony Meteorites III. Some Inter-Element Relationships*, L. H. Ahrens, H. Von Michaelis, *Earth and Planetary Science Letters 5*, North-Holland Publishing Comp., Amsterdam, 1969. p395-400, http://adsabs.harvard.edu/abs/1968E%26PSL...5..395A – Accessed 5.17.16

7.10h *The Penetration of Iron Meteorites into the Ground*, Willard J. Fisher, Harvard College Observatory, *Astronomy*, Proc. N.A.S., Vol. 19, 1933, February 7, p290, http://www.pnas.org/content/19/3/286.full.pdf – Accessed 5.17.16

7.10i *The Cooling Rates of Iron Meteorites: A Revised Model to simulate the growth of the Widmanstätten pattern in Iron Meteorites*, W. D. Hopfe, J. I Goldstein, *Meteoritics & Planetary Science*, Vol. 34, No. 4, Supplement, 1999, pA56

7.10j *New Metallographic Constraints on Meteoritic Widmanstätten structure Formation*, P. Z. Budka, J. R. M. Viertl, *Meteoritics & Planetary Science*, Vol. 32, No. 4, Supplement, 1997, pA23, http://www.lpi.usra.edu/meetings/metsoc97/pdf/5082.pdf – Accessed 5.17.16

7.10k *The metallographic cooking rate method revised: Application to iron meteorites and mesosiderites*, W. D. Hopfe, J. I. Goldstein, *Meteoritics & Planetary Sciences*, 36, 2001, p135

7.10l *The Meteoritic Widmanstatten Structure: A Modern Metallurgical Reevaluation*, P. Z. Budka1 and J.R.M. Viertl, Lunar and Planetary Science XXVII, http://www.lpi.usra.edu/meetings/lpsc97/pdf/1126.PDF – Accessed 5.17.16

7.10m *Alteration of the Widmanstätten structure of Meteorites by Heating*, Reed Knox, Jr., *Meteoritics*, Vol. 1, No. 2, 1954, p204

7.10n http://epsc.wustl.edu/admin/resources/meteorites/meteorwrongs/metal.htm Accessed 12.12.06 – changed to http://meteorites.wustl.edu/id/metal.htm, Sites no longer available

7.10o *Manual of the Natural History, Geology, and Physics of Greenland, and the Neighbouring Regions*, T. Rupert Jones, p445

7.10p *Meteorites, A Petrologic, Chemical and Isotopic Synthesis*, Robert Hutchison, *Cambridge University Press*, 2004, p15

7.10q *The Geological Magazine*, A. E. Nordenskiöld, No. XCVII, July 1872, Account of an Expedition to Greenland in the Year 1870, p461

7.10r http://www.1911encyclopedia.org/Greenland – Accessed 1.29.08, Site is no longer available

7.10s *The mineralogy and origin of Josephinite*, Russell A. Morley, *Popular Astronomy*, Vol. 57, 1949, p93

7.10t *Josephinite: Crystal Structures and Phases Relations of the Metals*, Basset, Bird, Weathers, Kohlstedt, *Physics of the Earth and Planetary Interiors*, 23 (1980), p255

7.10u http://www.usgennet.org/alhnorus/ahorclak/MeteorTreasures.html – Accessed 5.17.16

7.10v *A Man Named Bear, and His Meteorite. The Fall and Rise of the Pallas Nugget*, Part 1, A. R. Gallant, *Meteorite*, 1998, Vol. 4, p8-11

7.10w *Solidification History of the Kitdlît Lens: Immiscible Metal and Sulphide Liquids from a Basaltic Dyke on Disko, Central West Greenland, Journal of Petrology*, FINN ULFF-MØLLER, Vol. 26, No. 1, Abstract, 1985, http://petrology.oxfordjournals.org/cgi/content/abstract/26/1/64 – Accessed 5.17.16

7.10x http://meteorites.wustl.edu/ id/fusioncrust.htm – Accessed 1.4.08, Site is no longer available

7.10y http://meteorites.wustl.edu/id/regmaglypts.htm – Accessed 5.17.16

7.10z *Popular Astronomy*, Russell A. Morley, 1948, Vol. 56, p558

7.10aa 10.05.07 meeting between Dean W. Sessions, Jae Heiner, and Dr. Carlton Moore of Arizona State University, Mesa, Arizona

7.10ab *Artificial Meteor Ablation Studies: Olivine*, Maxwell B. Blanchard, Gary G. Cunningham, *Journal of Geophysical Research*, Vol. 79, No. 26, September 10, 1974, p3979

7.10ac http://en.wikipedia.org/wiki/Pallasite – Accessed 5.18.16

7.10ad http://www.meteoritestudies.com/protected_SHIROKOV.HTM – Accessed 2.5.08, Site no longer available

7.10ae http://meteoriticalsociety.org/?page_id=107, – Accessed 5.18.16

7.10af www.meteoritestudies.com/protected_HUCKITTA.HTM – Accessed 6.23.16

7.10ag http://www.passc.net/EarthImpactDatabase/index.html – Accessed 5.18.16

7.10ah *Introduction to Planetary Geology*, Billy P. Glass, *Cambridge University Press*, 1982, p130

7.10ai *What Heated the Asteroids?*, Alan E. Rubin, *Scientific American*, May 2005, p83

7.10aj *Interstellar water in meteorites?* Etienne Deloule, Francois Robert, *Geochimica et Cosmochimica Acta*, Vol. 59, No. 22, p4704

7.10ak *Early equeous activity on primitive meteorite parent bodies*, Magnus Endress, Ernst Zinner, Adolf Bischoff, *Nature*, Vol. 379, 22 February 1996, p701

7.10al *Asteroidal Water Within Fluid Inclusion-Bearing Halite in an H5 Chondrite, Monahans (1998)*, Zolensky, etal, *Science*, Vol. 285, 27 August 1999, p1377

7.10am *Water soluble ions in the Nakhla Martian meteorite*, Douglas J. Sawyer, Michael D. McGehee, Julie Canepa and Carleton B. Moore, *Meteoritics & Planetary Science* 35, 2000, p745

7.11 The Arizona Hydrocrater

7.11a *Exploration at Meteor Crater*, Daniel Moreau Barringer, *Engineering and Mining Journal-Press*, Vol. 121, Issue 2, 9 January 1926 p59

7.11b *Cryptoexplosion Structures: A Discussion*, Robert S. Dietz, *American Journal of Science*, Vol. 262, Summer 1963, p650

7.11c http://www.barringercrater.com/about/history_6.php – Accessed 5.19.16

7.11d *Meteor Crater Formed by Low-velocity Impact*, H. J. Melosh, G. S. Collins, *Nature*, Vol. 434, 10 March 2005, p157

7.11e *Secrets of the Wabar Craters*, Wynn J. C. and Shoemaker E. M., *Sky and Telescope*, 1997, p44-48

7.11f *Hypervelocity Impact of Steel Into Coconino Sandstone*, E. M. Shoemaker, D. E. Gault, H. J. Moore, and R. V. Lugn, *American Journal of Science*, Vol. 261, Summer 1963, p681

7.11g *Shock Melting of the Canyon Diablo Impactor: Constraints from Nickel-59 Contents and Numerical Modeling*, C. Schnabel, et al, *Science*, Vol. 285, 2 July 1999, p87

7.11h *The Origin of Hypotheses, Illustrated by the Discussion of a Topographic Problem*, G. K. Gilbert, *Science*, Vol. 3 No. 53, 3 January 1896, p12

7.11i *Meteor Crater formed by low-velocity impact*, H. J. Melosh, G. S. Collins, *Nature*, Vol. 434, 10 March 2005

7.11j *Hypervelocity Impact of Steel Into Coconino Sandstone*, E. M. Shoemaker, D. E. Gault, H. J. Moore, and R. V. Lugn, *American Journal of Science*, Vol. 261, Summer 1963, p674

7.11k *Penetration Mechanics of High Velocity Meteorites, Illustrated by Meteor Crater, Arizona*, Eugene M. Shoemaker, *International Geological Congress* XXI Session 18, 1960, p418-20

7.11l *Meteorite Craters as Topographical Features on the Earth's Surface: Discussion*, L. J. Spencer, *The Geographical Journal*, Vol. 81, No.3, March 1933, p229

7.11m *Penetration Mechanics of High Velocity Meteorites, Illustrated By Meteor Crater, Arizona*, Eugene M. Shoemaker, *International Geological Congress* XXI Session 18, 1960, p424

7.11n *A Seismic Refraction Technique Used for Subsurface Investigations at Meteor Crater, Arizona*, H. D. Akermann and R. H. Godson, J. S. Watkins, *Journal of Geophysical Research*, February 10, 1975, p766

7.11o *Penetration Mechanics of High Velocity Meteorites, Illustrated By Meteor Crater, Arizona*, Eugene M. Shoemaker, *International Geological Congress* XXI Session 18, 1960, p424

7.11p *A Seismic Refraction Technique Used for Subsurface Investigations at Meteor Crater, Arizona*, H. D. Akermann and R. H. Godson, J. S. Watkins, *Journal of Geophysical Research*, February 10, 1975, p765

7.11q *Gravity and Magnetic Investigations of Meteor Crater, Arizona*, Robert D. Regan, William J. Hinze, *Journal of Geophysical Research*, Vol. 80 No. 5, February 10, 1975, p786

7.11r *The Origin of Hypotheses, Illustrated by the Discussion of a Topographic Problem*, G. K. Gilbert, *Science*, Vol. 3 No. 53, 3 January 1896, p10

7.11s Meteor Crater Brief History, current handout given at the Meteor Crater Museum for a number of years

7.11t *Penetration Mechanics of High Velocity Meteorites, Illustrated by Meteor Crater, Arizona*, Eugene M. Shoemaker, *International Geological Congress* XXI Session 18, 1960, p426

7.11u *Significance of the Norton, Kansas, Meteorite*, H. H. Nininger, reported by G. W. Stevens, *Transactions of the Kansa Academy of Science*, Vol. 52 No. 1, March 1949, p113, (The quote has been adapted from "Only nature is authority" to "Nature is the Authority")

7.11v *Origin of Diamonds in the Canyon Diablo and Novo Urei Meteorites*, Neville L. Carter, George C. Kennedy, *Journal of Geophysical Research*, Vol. 71, No.2, January 15, 1966, p2419

7.11w Critique of Paper by N.L. Carter and G.C. Kennedy, 'Origin of Diamonds in the Canyon Diablo and Novo Urei Meteorites', *Journal of Geophysical Research*, Edward Anders, Michael E. Lipschutz, Vol. 71, No.2, January 15, 1966, p643

7.11x *Sedimentation and Volcanism in the Hopi Buttes, Arizona*, John T. Hack, *Bulletin of the Geological Society of America*, Vol. 53, February 1, 1942, p364

7.11y *Impactite Slag at Barringer Crater*, H. H. Nininger, *American Journal of Science*, 252, May 1954, p288 and also in *The Published Papers of Harvey Harlow Nininger, Biology and Meteoritics*, Publication No. 9 by the Center for Meteorite Studies, ASU, June, 1971, p658

7.11z http://www.jpl.nasa.gov/news/news.cfm?release=2007-144 – Accessed 5.19.16

7.11aa *A Hazard of Lunar Exploration*, H. H. Nininger, *The Published Papers of Harvey Harlow Nininger, Biology and Meteoritics*, No. 9, Center for Meteorite Studies, ASU, June 1971, p665

7.12 The Impact to Hydrocrater Evidence

7.12a *Impact From the Deep*, Peter D. Ward, *Scientific American*, October 2006, p65

7.12b *Introduction to Planetary Geology*, Billy P. Glass, *Cambridge University Press*, 1982, p130

7.12c http://www.passc.net/EarthImpactDatabase/Scientists.html – Accessed 5.21.16

7.12d The web site of the Australian Wolfe Creek Crater National Park that this quote came from (2004) no longer exists, however some information can be found on westernaustralia.com about the crater.

7.12e *Gravity Investigation of Wolfe Creek Crater, Western Australia*, R. F. Fudali, *Journal of Geology*, 1979, Vol. 87, p58

7.12f *The Wolfe Creek, Western Australia, Meteorite Crater*, William A. Cassidy, *Meteoritics*, Vol. 1, 1954, p198

7.12g *Wolfe Creek Meteorite Crater, Western Australia*, D. J. Guppy, R. S. Matheson, *Journal of Geology*, Vol. 58, 1950, p36

7.12h *The Wolfe Creek, Western Australia, Meteorite Crater*, William A. Cassidy, *Meteoritics*, Vol. 1, 1954, p197

7.12i http://www.tec.army.mil/research/ products/desert_guide/lsmsheet/lslater.htm – Accessed 3.25.08, Site is no longer available, http://www.enggjournals.com/ijet/docs/IJET10-02-04-04.pdf – Accessed 6.23.16

7.12j *Meteoritic Material From the Wolfe Creek, Western Australia, Crater*, Lincoln Lapaz, *Meteoritics*, Vol. 1, 1954, p201

7.12k *Wolfe Creek Meteorite Crater, Western Australia*, D. J. Guppy, R. S. Matheson, *Journal of Geology*, Vol. 58, 1950, p34

7.12l http://webmineral.com/data/Bunsenite.shtml – Accessed 5.20.16

7.12m *An interpretation of some recent Geophysical Data of Wolfe creek Crater, south of Halls Creek, Kimberly region, Western Australia*, Louis A. G. Hissink, *Consulting Diamond Geologist*, December 2002, http://www.kronia.com/thoth/thoVII03.txt – Accessed 3.26.08, Site no longer available

7.12n http://en.wikipedia.org/wiki/Lamproite – Accessed 5.20.16

7.12o *An interpretation of some recent Geophysical Data of Wolfe creek Crater, south of Halls Creek, Kimberly region, Western Australia*, Louis A. G. Hissink, Consulting Diamond Geologist, December 2002, http://www.kronia.com/thoth/thoVII03.txt – Accessed 3.26.08, Site no longer available

7.12p *Historical Notes on the Odessa Meteorite Crater*, Brandon Barringer, *Meteoritics*, Vol. 3, No. 4, December 1967, p162

7.12q *Statement of Progress of Investigation at Odessa Meteor Craters*, E. H. Sellards, Glen Evans, *University of Texas, Bureau of Economic Geology*, September 1, 1941, p7

7.12r *Historical Notes on the Odessa Meteorite Crater*, Brandon Barringer, *Meteoritics*, Vol. 3, No. 4, December, 1967, p166

7.12s *Statement of Progress of Investigation at Odessa Meteor Craters*, E. H. Sellards, Glen Evans, University of Texas, Bureau of Economic Geology, September 1, 1941, p5

7.12t *Historical Notes on the Odessa Meteorite Crater*, Brandon Barringer, *Meteoritics*, Vol. 3, No. 4, December 1967, p166. The sign posted at the crater in 2007 stated that the "meteor fall was at least 50,000 years ago." This is a good example of a date given that has no basis in fact. Not only does the scientific literature state half this age, the unsuspecting public are purposely mislead by these dates, which have no meaning and no science to back the false claim. This crater was made in the surrounding sediment that was laid down during the end of the Universal Flood; therefore, the crater could be no older than the time of the Universal Flood. The worldwide flood is demonstrated in the following chapter to be a very recent event of only several thousand years ago and the World History Model verifies this by the 4,300-year-old flood date recorded by many global societies.

7.12u http://en.wikipedia.org/wiki/Septarian_concretion#Septarian_concretions – Accessed 5.20.16

7.12v *Jackpine Creek magnetic anomaly: Identification of a buried meteorite impact structure*, S. A. Goussev, R. A. Charters, J. W. Peirce, and W. E. Glenn, *GEDCO, The Leading Edge*, August 2003

7.12w *Upheaval Dome, Canyonlands, Utah: Strain Indicators that Reveal an Impact Origin*, Peter W. Huntoon, *Geology of Utah's Parks and Monuments 2000 Utah Geological Association Publication 28*, p1, http://www.utahgeology.org/pub28_pdf_files/UpheavalDome.pdf – Accessed 5.20.16

7.12x *Geological and Geophysical Studies of the Upheaval Dome Impact Structure*, Utah, K. E. Herkenhoff, R. Giegengack, B. J. Kriens, J. N. Louie, G. I. Omar, J. B. Plescia, E. M. Shoemaker, *Lunar and Planetary Science, 30th Conference*, March 15-29, 1999, abstract 1932

7.12y *Roberts Rift, Canyonlands, Utah, A Natural Hydraulic Fracture Caused by Comet or Asteroid Impact*, Peter W. Huntoon, Eugene M. Shoemaker, *Ground Water*, IGDW, Vol. 33, No.4, p563

17.12z *Geology of Upheaval Dome impact structure, southeast Utah*, Bryan J. Kriens, Eugene M. Shoemaker, Ken E. Herkenhoff, *Journal of Geophysical Research*, Vol. 104, No. E8, August 25, 1999, p18,879

7.12aa *A petrographical and geochemical study of quartzose nodules, country rocks, and dike rocks from the Upheaval Dome structure*, Utah, Christian Koeberl, J. B. Plescia, Chris L. Hayward, Wolf Uwe Reimold, *Meteoritics & Planetary Science* 34, 1999, p861, Abstract

7.12ab http://en.wikipedia.org/wiki/Upheaval_Dome – Accessed 5.20.16

7.12ac *Upheaval Dome, Canyonlands, Utah: Strain Indicators that Reveal an Impact Origin*, Peter W. Huntoon, *Geology of Utah's Parks and Monuments 2000 Utah Geological Association Publication 28*, p7, http://www.utahgeology.org/pub28_pdf_files/UpheavalDome.pdf – Accessed 5.20.16

7.12ad *Upheaval Dome Impact Structure, Utah*, E. M. Shoemaker, *Lunar and Planetary Science* 15, 1984, p779

7.12ae *Richat and Semsiyat Domes (Mauritania): Not Astroblemes*, Robert S. Dietz, Robert Fudali, William Cassidy, *Geological Society of America Bulletin*, Vol. 80, July 1969, p1370

7.12af *Friction melt distribution in a multi-ring impact basin*, John G. Spray, Lucy M. Thompson, *Nature*, Vol. 373, 12 January 1995, p130

7.12ag *Pachauri Defeats Watson in New Chapter for Global Panel*, Andrew Lawler, *Science*, 26 April 2002, Vol. 296, No. 5568, p647-8

7.12ah http://photojournal.jpl.nasa.gov/catalog/PIA01648 – Accessed 5.20.16

7.12ai *On the Origin of Some 'Recent' Craters on the Canadian Shield*, K. L. Currie, *Meteoritics*, Vol. 2, No. 2, February 1964, p93

7.12aj http://photojournal.jpl.nasa.gov/catalog/?IDNumber=pia01630 – Accessed 5.20.16

7.13 The Hydromoon Evidence

7.13a *Lunar Boulders Seen at Very High Resolution: Implications for 433 Eros*, B. Wilcox, M.S. Robinson, P.C. Thomas, *Lunar and Planetary Science 33*, 2002, p1637

7.13b *Moonquakes, Meteoroids, and the State of the Lunar Interior*, Latham, Dorman, Duennebier, Ewing, Lammlein, Nakamura, *Proceedings of the Fourth Lunar Science Conference*, 1973, Vol. 3, p2517-18

7.13c *Farside Deep Moonquakes and Deep Interior of the Moon*, Yosio Nakamura, *Journal of Geophysical Research*, Vol. 110, 2005, p2

7.13d *Further Constraints on the Deep Lunar Interior*, A. Khan & K. Mosegaard, *Geophysical Research Letters*, Vol. 32, l22203, doi: 10.1029.2005GL023985, 2005, p4

7.13e *The New Moon*, Paul D. Spudis, *Scientific American*, December 2003, p86

7.13f http://www.nature.com/news/2005/050317/full/news050314-15.html, – Accessed 5.23.16

7.13g *The New Moon*, Paul D. Spudis, *Scientific American*, December 2003, p92

7.13h *Impact, Cryptoexplosion, or Diapric Movements?*, G.C. Amstutz, *Transactions of the Kansas academy of science*, Vol. 67, 1964, p346

7.13i *Strange World of the Moon*, V. A. Firsoff, *Basic Books*, 1959, p62

7.13j http://gsa.confex.com/gsa/2001AM/finalprogram/abstract_28367.htm – Accessed 5.23.16

7.13k *The New Moon*, Paul D. Spudis, *Scientific American*, December 2003, p90

7.14 The Hydrocomet Evidence

7.14a http://photojournal.jpl.nasa.gov/catalog/?IDNumber=PIA02107 – Accessed 5.4.08

7.14b *Stardust Memories*, Dale P. Cruikshank, *Science*, Vol. 275, 28 March 1997, p1896

7.14c *On the Role of Dust Mass Loss in the Evolution of Comets and Dudty Disk Systems*, Carey Lisse, *Earth, Moon and Planets*, 90, 2002, p497

7.14d *Water Production of Comet C/199 S4 (LINEAR) Observed with the SWAN Instrument*, *Science*, 18 May 2001, Vol. 292, p1326

7.14e *Water for the Rock,* Ben Harder, *Science News*, 23 March 2002, Vol. 161, p184, http://phschool.com/science/science_news/articles/water_for_the_rock.html, p2 – Accessed 5.23.16

7.14f *Strange Comet Unlike Anything Known, Comet Wild 2's Jets and Craters Astound Astronomers*, Robert Roy Britt, http://www.msnbc.msn.com/id/5234058/ – Accessed 5.23.16

7.14g http://www.newscientist.com/article.ns?id=dn7961, – Accessed 5.23.16, *Deep Impact collision ejected the stuff of life*, Maggie McKee, *NewScientist Magazine,* 07 September 2005

17.14h *Deep Impact: Sifting Through the Debris*, Mark Peplow, *Nature*, Vol. 436, 14 July 2005, p159

17.14i http://www.newscientist.com/article.ns?id=dn7961, – Accessed 5.23.16, *Deep Impact Collision Ejected the Stuff of Life*, Maggie McKee, *NewScientist Magazine,* 07 September 2005

17.14j http://www.newscientist.com/article.ns?id=dn7971 – Accessed 8.11.16, *Comet's minerals hint at liquid water*, *NewScientist Magazine*, Maggie McKee, 08 September 2005

17.14k *The Great Comet Crash, The impact of Comet Shoemaker-Levy on Jupiter*, Edited by John R. Spencer and Jacqueline Mitton, Forward by Gene and Carolyn Shoemaker, 1995, *Cambridge University Press*

7.14l *New Class Of Comets May be the Source of Earth's Water*, Henry Hsieh and David Jewitt (interview), *University of Hawaii Institute for Astronomy*, March 22, 2006, http://www.ifa.hawaii.edu/~hsieh/mbc-release.html, Site is no longer available

7.14m *The Deep Space 1 Encounter with Comet 19P/Borrelly*, D. C. Boice, et al., *Earth, Moon and Planets*, 89, 2002, p303

7.15 The Hydroid Evidence

7.15a *The "New" Solar System*, *Scientific American*, Vol. 296, January 2007, p38

7.15b *Asteroidal Catastrophic Collisions Simulated by Hypervelocity Impact Experiments*, F. Capaccioni, et al., *ICARUS* 66, 1986, p495

7.15c *The Small Planets*, Erik Asphaug, *Scientific American*, Vol. 282, May 2000, p47

7.15d http://en.wikipedia.org/wiki/253_Mathilde – Accessed 5.23.16

7.15e *Meteorites: The long trip to earth*, Clark R. Chapman, *Nature*, 407, October 5, 2000, p573-576

7.15f *Detailed Images of Asteroid 25143 Itokawa from Hayabusa*, J. Saito, et al, *Science* Vol. 312, 2 June 2006, p1342

7.15g *The Rubble-Pile Asteroid Itokawa as Observed by Hayabusa*, A. Fujiwara, et al., *Science*, Vol. 312, 2 June 2006, p1343

7.15h http://www.astronomy.com/news-observing/news/2005/09/an%20icy%20interior%20for%20ceres – Accessed 5.23.16, Additional details from this web site:

"Measurements of the asteroid's shape also provide information about its interior. The equatorial region of any body that spins fast enough bulges outwards. Instead of looking like a perfect sphere, the star or planet (or asteroid, in the case of Ceres) looks "flattened." By measuring the distances from the body's center to its equator and poles, scientists can determine the amount of flattening, which constrains the interior's possible structure.

"Ceres has a mean density of about 2.077 grams per cubic centimeter (roughly twice that of water), and a uniform body of this density should have a polar radius about 23.8 miles (39.7 km) smaller than the equatorial radius. Ceres' polar radius is only 19.6 miles (32.6 km) smaller. This is strong evidence its interior consists of different layers. In fact, the smaller amount of flattening indicates the presence of a mantle and a core."

7.15i http://www.esa.int/Our_Activities/Space_Science/Herschel/Herschel_discovers_water_vapour_around_dwarf_planet_Ceres – Accessed 5.23.16

7.15j *Small bodies of the Solar System*, Don Yeomane, *Nature*, 404, 20 April 2000, p829

7.15k *New class of comets may be the source of Earth's water*, Henry Hsieh and David Jewitt (interview), University of Hawaii Institute for Astronomy, March 22, 2006, https://manoa.hawaii.edu/news/article.php?aId=1379 – Accessed 5.23.16

7.16 More Hydroplanet Evidence

7.16a *Magma to Tephra*, Chapter 7, *Pyroclastic Surges and Compressible Two-Phase Flow*, Kenneth H. Wohletz, *Los Alamos National Laboratory*, Elsevier, 1998, p254

7.16b *Discover*, Laurence A. Soderblom, Research Geophysicist, Astrogeology Program of the U. S. Geological Survey, December 2005, p70

7.16c *The Many Faces of Mars*, Philip R. Christensen, *Scientific American*, July 2005, p38

7.16d http://www.esa.int/Our_Activities/Space_Science/Mars_Express/Lava_tubes_on_Pavonis_Mons/(print) – Accessed 5.24.16

7.16e *The Many Faces of Mars*, Philip R. Christensen, *Scientific American*, July 2005, p38

7.16f http://en.wikipedia.org/wiki/Phobos_%28moon%29 – Accessed 5.24.16

7.16g http://en.wikipedia.org/wiki/Ganymede_%28moon%29 – Accessed 5.24.16

7.17 The Hydroplanet Frontier

7.17a *To the Center of the Earth*, Joel Achenbach, *National Geographic*, April 2004

8 The Universal Flood Model

8.1 The Universal Flood (UF) History

8.1a *Sedimentation in Barred Basins, and Source Rocks of Oil*, W. G. Woolnough, *Bulletin of the American Association of Petroleum Geologists* (AAPG), Vol. 21, No.9, September 1937, p1101, Entire article also reprinted in Origin of Evaporites, AAPG Bulletin, 1971, p3-59

8.1b *Sedimentation in Barred Basins, and Source Rocks of Oils*, W. G. Woolnough, Bulletin of the American Association of Petroleum Geologists, V. 21 No. 9, September 1937, p1102-3

8.1c *Sedimentation in Barred Basins, and Source Rocks of Oils*, W. G. Woolnough, Bulletin of the American Association of Petroleum Geologists, V. 21 No. 9, September 1937, p1103

8.2 The Acknowledged Flood

8.2a http://facstaff.gpc.edu/~pgore/Earth&Space/Salini-ty.html – Accessed 3.5.09, *Salinity*, Pamela J. W. Gore

8.2b *Geology of Grand Canyon National Park, North Rim*, Annabelle Foos, Geology Department, University of Akron, p4, http://www.nature.nps.gov/geology/education/foos/grand.pdf – Accessed 12.2.15

8.2c *Boiling seas linked to mass extinction*, Tom Clark, *Nature News*, 22 August 2003, http://www.nature.com/news/2003/030822/full/news030818-16.html – Accessed 12.3.15. Another related article is entitled: *How to kill (almost) all life: the end-Permian extinction event*, Jichael J. Benton, Richard J. Twitchett, TREE, Vol. 18, No.7, July 2003

8.2d *The Day the World Burned*, David A. Kring, Daniel D. Durda, *Scientific American*, December 2003, p104

8.2e *Digging Dinosaurs*, John R. Horner, James Gorman, Workman Publishing, 1988, p122

8.2f http://science.nasa.gov/headlines/y2002/28jan_extinction.htm – Accessed 12.8.15

8.2g http://science.nasa.gov/headlines/y2002/28jan_extinction.htm – Accessed 12.8.15

8.2h *Sedimentation in Barred Basins, and Source Rocks of Oils*, W. G. Woolnough, *Bulletin of the American Association of Petroleum Geologists*, V. 21, No. 9, September 1937, p1110-1111

8.2i *Arizona Geology*, Jon E. Spencer, Philip A. Pearthree, *Arizona Geological Survey*, Vol. 35, No. 4, Winter 2005, p4

8.2j *Arizona Geology*, *Arizona Geological Survey*, Vol. 35, No.4, Winter 2005, p1

8.2k *Surfing the Solar System, It's wetter than you think*, Michael Milstein, *Air & Space*, December 1997/January 1998, p56

8.2l *National Geographic*, January 2004, p19

8.2m *Blazing the Trail to the Red Planet*, Robert Naeye, *Astronomy*, October 1997, p51

8.2n *The Mars Pathfinder Mission*, Matthew P. Golombek, *Scientific American*, July 1998, p46

8.2o *A Mars Never Dreamed Of*, Kath Sawyer, *National Geographic*, February 2001, p38

8.2p *Making a Splash on Mars*, Charles W. Petit, *National Geographic*, July 2005, p71

8.2q *Making a Splash on Mars*, Charles W. Petit, *National Geographic*, July 2005, p76

8.2r *Blazing the Trail to the Red Planet*, Robert Naeye, *Astronomy*, October 1997, p51-2

8.2s *A Mars Never Dreamed Of*, Kath Sawyer, *National Geographic*, February 2001, p49

8.2t *Study: Megafloods Made Britain An Island*, S. Alfano, July 2007, http://www.cbsnews.com/stories/2007/07/19/tech/main3074619.shtml – Accessed 12.9.15

8.2u *Ancient Megafloods Created English Channel*, D. Mosher, July 2007 http://www.foxnews.com/story/0,2933,289812,00.html – Accessed 12.10.15

8.2v *Catastrophic flooding origin of shelf valley systems in the English Channel*, Sanjeev Gupta, Jenny S. Collier, Andy Palmer-Felgate, Graeme Potter, *Nature*, Vol. 448, 19 July 2007, p344

8.3 The Universal Flood Mechanisms

8.3a *Unvorgreifliche Kometen-Gedanke* (1744), S. Suschken, p8 – Quoted in Imanuel Velikovsky's book, *In the Begining*, http://www.varchive.org/itb/ecwhist.htm – Accessed 1.11.16

8.3b http://en.wikipedia.org/wiki/90377_Sedna – Accessed 1.11.16

8.3c http://en.wikipedia.org/wiki/Comet#Long_period – Accessed 1.11.16

8.3d *Earthquake-induced changes in a hydrothermal system on the Juan de Fuca mid-ocean ridge*, H. Paul Johnson, et al, *Nature*, Vol. 407, 14 September 2000, p174

8.3e *Earthquakes' Impact on Hydrothermal Systems May Be Far-reaching*, H. Paul Johnson, *EOS*, May 22, 2001, p234

8.3f http://www.rockhounds.com/grand_hikes/geology/cardenas_lava.shtml – Accessed 1.12.16

8.3g *Tsunami: Wave of Change*, Eric L. Geist and Vasily V. Titov and Costas E. Synolakis, *Scientific American*, January 2006, p58

8.3h http://www.rockhounds.com/grand_hikes/geology/cardenas_lava.shtml – Accessed 1.12.16

8.3i http://en.wikipedia.org/wiki/Waimea_Canyon – Accessed 1.12.16

8.4 The Hydrofountain Mark

8.4a *Experimental Evidence for Fluidization Processes in Breccia Pipe Formation*, M. E. McCallium, *Economic Geology*, Vol. 80, 1985, p1527

8.4b *Experimental Evidence for Fluidization Processes in Breccia Pipe Formation*, M. E. McCallium, *Economic Geology*, Vol. 80, 1985, p1528

8.4c *Isotopic evidence (He, B, C) for deep fluid and mud mobilization from mud volcanoes in the Caucasus continental collision zone*, Achim Kopf, et al., *Int J Earth Sci (Geol Rundsch)*, 2003, 92, p408

8.4d *Mud volcanoes in the Alboran Sea: evidence from micropaleontological and geophysical data*, A. Sautkin, et al., *Marine Geology* 195, 2003, p237. Another researcher (A. V. Milkov) states: "…a submarine mud volcano is considered as a topographically expressed seafloor edifice from which mud and fluid (water, brine, gas, oil) flow or erupt." *Marine Geology* 167, 2000, p29

8.4e *Marine Geology* 195, 2003, p238, http://citeseerx.ist.psu.edu/viewdoc/download?doi=10.1.1.378.7126&rep=rep1&type=pdf – Accessed 1.12.16

8.4f *Mud Volcanoes: Distribution Regularities and Genesis (Communication 2. Geological-Geochemical Peculiarities and Formation Model)*, V. N. Kholodov, *Lithology and Mineral Resources*, Vol. 37, No. 4, 2002, p297

8.4g *Worldwide distribution of submarine mud volcanoes and associated gas hydrates*, A. V. Milkov, *Marine Geology*, 167, 2000, p29

8.4h *Mud Volcanoes: Distribution Regularities and Genesis (Communication 2. Geological-Geochemical Peculiarities and Formation Model)*, V. N. Kholodov, *Lithology and Mineral Resources*, Vol. 37, No. 4, 2002, p298

8.4i *Worldwide distribution of submarine mud volcanoes and associated gas hydrates*, A. V. Milkov, *Marine Geology* 167, 2000, p29

8.4j *Mud Volcanoes: Distribution Regularities and Genesis (Communication 2. Geological-Geochemical Peculiarities and Formation Model)*, V. N. Kholodov, *Lithology and Mineral Resources*, Vol. 37, No. 4, 2002, p293

8.4k *Drowning in Mud*, Andrew Marshall, *National Geographic*, May 2008, p60

8.4l http://www.nps.gov/archive/badl/exp/home.htm – Accessed 8.1.08, Site is no longer available

8.4m *Ultramafic and Related Rocks*, edited by P. J. Wyllie, written by K. D. Watson, 1967, p263

8.4n *Ultramafic and Related Rocks*, edited by P. J. Wyllie, written by K. D. Watson, 1967, p267

8.4o *Supplement of the Nashville News*, Nashville, Ark. (No date, original published ca. 1912). This appeared in a report of Crater of Diamonds State Park, and seems to be a layperson's scientific appraisal of the origin of kimberlite diatremes, perhaps from a scientist not steeped in magma pseudotheory. It appears to have been written about a century ago, in 1912. Whether it was written around 1912 or not, it illustrates how easily answers to nature's questions are found if one is objective and not caught up in the false modern science paradigm., http://www.pcahs.org/pcaolr/places01/crater01.htm – Accessed 6.23.16

8.4p http://en.wikipedia.org/wiki/Kimberlite – Accessed 1.22.16

8.4q *Wood Found in Ekati Kimberlite*, NAPEGG Newsletter, Oct 1999,

Vol. 16, No. 13, p10, http://www.napegg.nt.ca/newsletters/oct99/oct99.pdf – Accessed 7.23.08

8.4r *Experimental Evidence for Fluidization Processes in Breccia Pipe Formation*, M. E. McCallium, *Economic Geology*, Vol. 80, 1985, p1528-9

8.4s *Experimental Evidence for Fluidization Processes in Breccia Pipe Formation*, M. E. McCallium, *Economic Geology*, Vol. 80, 1985, p1529

8.4t *Earthquake-induced clastic dikes detected by anisotropy of magnetic susceptibility*, Tsafrir Levi, Ram Weinberger, Tahar Aifa, Uehuda Eyal, Shmuel Marco, *Geology*, February 2006, Vol. 34, No. 2, p69

8.4u *Fluid overpressure in layered intrusions: formation of a breccia pip in the Eastern Bushveld Complex, Republic of South Africa*, Sonja L. Boorman, James B. McGuire, Alan E. Boudreau, F. Johann Kruger, *Mineralium Deposita* (2003), 38, p363

8.4v *Capitol Reef Black Boulders,* June 2006 Bulletin written by Richard Waitt of the USGS

8.4w http://en.wikipedia.org/wiki/Geysers_on_Mars – Accessed 1.25.16

8.4x *National Geographic*, March 2008, p79

8.5 The Sand Mark

8.5a *Sand*, PH. H. Kuenen, *Scientific American*, Vol. 202, No. 4, April 1960, p95

8.5b *Origin of Granite in the Light of Experimental Studies in the System, Na AlSi3O8-KalSi3O8-SiO2--H2*, O. F. Tuttle and N. L. Bowen, The Geological Society of America, 1958, p69

8.5c *Crystallization: How come you look so good?,* Roger J. Davey, *Nature*, 25 March 2004, Vol. 428, Issue 6981, p374

8.5d *Size Distribution of Sedimentary Population*, *Science*, Vol. 141, 1963, p802

8.5e *Sedimentary Rocks*, Francis John Pettijohn, 1957, p50

8.5f *Sedimentary Rocks,* Francis John Pettijohn, 1957, p652

8.5g *The occurrence and significance of biogenic opal in the regolith*, Jonathan Clarke, *Earth-Science Reviews*, Vol. 60, 2003, p175

8.5h *Diagenetic Origin of Quartz Silt in Mudstones and Implications for Silica Cycling*, Jürgen Schieber, Dave Krinsley, and Lee Riciputi, *Nature*, Vol. 406, 2000, p981-985

8.5i Personal email from Dr Jürgen Schieber on 2.4.03

8.5j *Diagenetic Origin of Quartz Silt in Mudstones and Implications for Silica Cycling*, J. Schieber, D. Krinsley, L. Riciputi, *Nature*, Vol. 406, 2000, p981-985

8.5k *Homegrown Quartz Muddies the Water*, Erik Stokstad, *Science*, Vol. 289, 1 September 2000, p1449, http://www.sciencemag.org/news/2000/08/homegrown-quartz-muddies-water – Accessed 1.28.16

8.5l http://en.wikipedia.org/wiki/Clay – Accessed 1.28.16

8.5m http://en.wikipedia.org/wiki/Places_of_interest_in_the_Death_Valley_area#Mesquite_Sand_Dunes – Accessed 1.28.16

8.5n http://en.wikipedia.org/wiki/Death_Valley – Accessed 1.28.16

8.5o *Sandstone Diagenesis Recent and Ancient*, Reprint Series of Vol. 4 of the International Association of Sedimentologists, 2003, p323

8.5p *Quartz Cementation in Sandstones*, International Association of Sedimentologists, Special Publication Number 29, 2000, pvii

8.5q *Sedimentary Rocks*, F. J. Pettijohn, 1957, p656-7

8.5r *Sandstone Diagenesis Recent and Ancient, a Reprint Series of Volume 4 of the International Association of Sedimentologists*, 2003, p36

8.5s Sedimentary Rocks, F. J. Pettijohn, 1957, p657

8.5t *Sandstone Diagenesis Recent and Ancient*, Reprint Series of Vol. 4 of the International Association of Sedimentologists, 2003, p323

8.5u *Quartz Cementation in Sandstones*, International Association of Sedimentologists, Special Publication Number 29, 2000, p2

8.5v *Sandstone Diagenesis Recent and Ancient*, Reprint Series of Vol. 4 of the International Association of Sedimentologists, 2003, p37

8.5w *Fluid Inclusions*, Edwin Roedder, *Reviews in Mineralogy*, Mineralogical Society of America, Vol. 12, 1984, p321

8.5x *Quartz Cementation in Sandstones*, International Association of Sedimentologists, Special Publication Number 29, 2000, p1

8.5y *Quartz Cementation in Sandstones*, International Association of Sedimentologists, Special Publication Number 29, 2000, p1

8.5z *Sedimentary Rocks*, by Francis John Pettijohn, 1957, p652

8.5aa *Sedimentary Rocks*, F. J. Pettijohn, 1957, p653

8.5ab *Sedimentary Rocks*, F. J. Pettijohn, 1957, p659

8.5ac *Sedimentary Rocks*, F. J. Pettijohn, 1957, p658

8.5ad *Quartz Cementation in Sandstones*, International Association of Sedimentologists, Special Publication Number 29, 2000, p33

8.5ae *Quartz Cementation in Sandstones*, International Association of Sedimentologists, Special Publication Number 29, 2000, p37

8.5af *Science*, Vol. 173, 6 August 1971, p534

8.5ag *Evaluation of pH and Granite/Hot water Interactions under Geothermal Conditions*, Y. Wang, et al., *High Pressure Research*, Vol. 20, 2001, p337-8,

8.6 The Erosion Mark

8.6a *Recent Debris Flows and Floods in Southern Arizona*, Philip A. Pearthree and Ann Youberg, *Arizona Geology,* Arizona Geological Survey, Vol. 36, No. 3, Fall 2006

8.6b *Lectures on Soil Erosion: Its Extent and Meaning and Necessary Measures of Control, Subchapter - Time Required to Build the Topsoil*, H. H. Bennett, November 4, 1932, http://www.nrcs.usda.gov/wps/portal/nrcs/detail/national/about/history/?cid=nrcs143_021396 – Accessed 2.17.16

8.6c http://en.wikipedia.org/wiki/Kaolinite – Accessed 2.17.16

8.6d *Experimental study of the hydrothermal formation of kaolinite*, F. Javier Huertas, Saverio Fiore, Francisco Huertas, Jose Linares, *Chemical Geology*, 156, 1999, p171

8.6e *Turbidity Currents as a Cause of Graded Bedding*, PH. H. Kuenen, C. I. Migliorini, *The Journal of Geology*, Vol. 58, No. 2, March 1950, p91

8.6f *Ten turbidite myths*, G. Shanmugam, *Earth-Science Reviews 58,* 2002, p311

8.6g http://earthobservatory.nasa.gov/Study/Earthquake/ – Accessed 2.17.16

8.6h *Bhuj, India Earthquake of January 26, 2001 Reconnaissance Report*, Sudhir K. Jain, et al., *Earthquake Spectra*, Publication 2002-01, p97

8.6i http://en.wikipedia.org/wiki/Soil_liquefaction – Accessed 2.17.16

8.6j http://en.wikipedia.org/wiki/Agathla – Accessed 2.17.16

8.6k http://en.wikipedia.org/wiki/Volcanic_plug – Accessed 2.17.16

8.6l http://www.hyden.asn.au/geo_explan.htm – Accessed 9.9.08, Site no longer available

8.6m *Recent Debris Flows and Floods in Southern Arizona*, Philip A. Pearthree and Ann Youberg, *Arizona Geology,* Arizona Geological Survey, Vol. 36, No. 3, Fall, 2006, http://www.azgs.az.gov/Fall%2006.pdf – Accessed 2.17.16

8.7 The Depth Mark

8.7a Origins of Igneous Rocks, Paul C. Hess, Harvard University Press, 1989, p69

8.7b *http://dictionary.reference.com/browse/obsidian* – Accessed 2.17.16, p2

8.7c *Encyclopedia of Glass, Ceramics, and Cement*, Editor: Martin Grayson, John Wiley & Sons, Encyclopedia Reprint Series, 1985, p463

8.7d *Summary of the 1953-57 Eruption of Tuluman Volcano, Papua New Guinea*, M. A. Reynolds and J. G. Best, *Volcanism in Australasia*, 1976, p287

8.7e *http://dictionary.reference.com/browse/basalt* – Accessed 2.23.16

8.7f *Vesicles: A Fundamental Characteristic of Planetary Surface Rocks*, L. S. Crumpler, K. Cashman, R. Schultz, *Lunar and Planetary Science XXX*, 2002

8.7g *Volcanoes A Planetary Perspective*, Peter Francis, *Clarendon Press*, 1993, p326

8.7h *Explosive Subaqueous Volcanism, Introduction: A Deductive Outline and Topical Overview of Subaqueous Explosive Volcanism*, J. White, J. Smellie, D. Clague, *American Geophysical Union, Geophysical Monograph 140*, 2003, p3

8.7i *Magma-water interaction in subaqueous and emergent basaltic volcanism*, Peter Kokelaar, *Volcanology*, 1986, 48, p280

8.7j *Hawaiian Volcanoes, Deep Underwater Perspectives, Volcanic Morphology of the Submarine Puna Ridge, Kilauea Volcano*, D. Smith, L. Kong, K. Johnson, J. Reynolds, *American Geophysical Union, Geophysical Monograph 128, 2002,* p125

8.7k http://www.cr.nps.gov/history/online_books/deto/sec1.htm p8 – Accessed 9.28.08, Site is no longer available.

8.7l http://www.nps.gov/deto/geology.htm – Accessed 9.23.08. as of 3.2.11 no longer available

8.7m *Experimental simulation of basalt columns*, Gerhard Muller, *Journal of Volcanology and Geothermal Research*, 86, 1998, p96

8.7n *No Water, No Granites – No Oceans, No continents*, I. H. Campbell & S. R. Taylor, *Geophysical Research Letters*, 1983, Vol. 10, No. 11. p1062

8.8 The Carbonate Mark

8.8a *A Neoproterozoic Snowball Earth*, Paul F. Hoffman, Alan J. Kaufman, Galen P. Halverson, Daniel P. Schrag, *Science*, 28 August 1998, Vol 281, p1342 (Abstract)

8.8b *Microbes Deep inside the Earth*, J. K. Fredrickson & T. C. Onstott, *Scientific American*, October 1996. Reprinted in Bib 148, p11

8.8c *To Hell and Back*, Kevin Krajick, *Discover*, July 1999, p78

8.8d *Extreme Environments on Earth: Analogs for Exobiology*, Faculty: Baross, Delaney, Deming, Frederickson, Gammon, Leigh, Rodrigo, Staley, p2, http://www.astro.washington.edu/astrobio/extreme.html – Accessed 10.1.08

8.8e *Journal of Geophysical Research*, Vol. 102, No. B8, August 10, 1997, p18,235

8.8f *Abiogenic formation of alkanes in the Earth's crust as a minor source of global hydrocarbon reservoirs*, Nature, Vol. 416, 4 April 2002, p522, http://www.nature.com/nature/journal/v416/n6880/abs/416522a.html – Accessed 2.25.16

8.8g http://www.infoplease.com/encyclopedia/science/chemosynthesis.html – Accessed 6.14.16

8.8h http://en.wikipedia.org/wiki/Calcium_carbonate – Accessed 2.16.11

8.8i *Calcification of cyanobacterial mats in Solar Lake, Sinai*, W. Berry Lyons, et al., *Geology*, Vol. 12, October 1984, p625

8.8j *Production of carbonate sediments by a unicellular green alga*, Kimberly K. Yates, Lisa L. Robbins, *American Mineralogist*, Vol. 83, 1998, p1508

8.8k http://en.wikipedia.org/wiki/Ocean_acidification – Accessed 10.21.08

8.8l *Is the contribution of bacteria to terrestrial carbon budget greatly underestimated?*, Olivier Braissant, Eric P. Verrecchia, Michel Aragno, *Naturwissenschaften*, Vol. 89, 2002, p368

8.8m *Induced Calcite Precipitation by Cyanobacteria Synechococccus*, Maria Dittrich, Beat Muller, Denis Mavrocordatos, Bernhard Wehrli, *Acta hydrochim. hydrobiol.*, Vol. 31, 2003, p168

8.8n *Floor Show*, John R. Delaney, *The Sciences*, New York Academy of Sciences, July/Aug 1998, p30, http://www.nyas.org/publications/sciences/pdf/ts_07_98.pdf – Accessed 10.7.08, Site no longer available

8.8o *Where Biosphere Meets Geosphere*, Naomi Lubick, *Scientific American*, January 28, 2002, http://www.sciam.com/article.cfm?id=where-biosphere-meets-geo – Accessed 3.0.16

8.8p *The great black spot*, Mark Schrope, *New Scientist*, 6 April 2002, p11

8.8q *The decline and fate of an iron-induced subarctic phytoplankton bloom*, Philip w. Boyd, et al., *Nature*, Vol. 428, 1 April 2004, p549

8.8r *Stirring the Oceanic Incubator*, Robert P. Dziak and H. Paul Johnson, *Science*, Vol. 296, 24 May 2002, p1406

8.8s http://people.ku.edu/~stalder/KS-limestone.html, – Accessed 3.1.16

8.8t http://www.galleries.com/minerals/carbonat/dolomite/dolomite.htm – Accessed 3.2.16

8.8u *Where Biosphere Meets Geosphere*, Naomi Lubick, *Scientific American*, January 28, 2002, http://www.sciam.com/article.cfm?id=where-biosphere-meets-geo – Accessed 3.1.16

8.8v *Microbial mediation as a possible mechanism for natural dolomite formation at low temperatures*, Crisogono Vasconcelos, Judith A. McKenzie, Stefano Bernasconi, Djordje Grujic, Albert J. Tien, *Nature*, Vol. 377, 21 September 1995, p221

8.8w http://en.wikipedia.org/wiki/Dolomite – Accessed 10.1.08

8.8x *Fluid Inclusions*, Edwin Roedder, *Reviews in Mineralogy*, Mineralogical Society of America, Vol.12, 1984, p320

8.8y *Nanotopography of synthetic and natural dolomite crystals*, Lisa A. Kessels, Duncan F. Sibley, Stephan H. Nordeng, *Sedimentology*, Vol. 47, February 2000, p173

8.8z *Sand*, Ph. H Kuenen, *Scientific American*, Vol. 202, No. 4, April 1960, p106

8.8aa *Loess, Geomprphological Hazards and Processes*, K. Rogner & W. Smykatz-Kloss, *Catena Supplement 20, The Deposition of Eolian Sediments in Lacustrine and Fluvial Environments of Central Sinai (Egypt)*, 1991, p84

8.8ab *Sources of non-glacial, loess-size quartz silt and the origins of "desert loess"*, B. J. Smith, J. S. Wright, W. B. Whalley, *Earth-Science Reviews*, 59, 2002, p6

8.8ac *Climatic periodicity during the late Pleistocene from a loess-paleosol sequence in northwest Argentina*, Alfred J. Zinck, Jose Manuel Sayago, Oral presentation in Symposium 16, p1. Also published in *Quaternary International*, 2001, http://natres.psu.ac.th/Link/SoilCongress/bdd/symp16/117-t.pdf – Accessed 3.3.16

8.8ad *Caves*, Stephen Kramer, *Carolrhoda Books, Inc.*, 1995,22

8.8ae *Morphoanalysis of Bacterially Precipitated Subaqueous Calcium Carbonate from Weebubbie Cave, Australia*, A. K. Contos, et al., *Geomicrobiology Journal*, 18, 2001, p331

8.8af *Microbial Activity in Caves—A Geological Perspective*, Brian Jones, *Geomicrobiology Journal*, 18, 2001, p348-9

8.8ag *Morphoanalysis of Bacterially Precipitated Subaqueous Calcium Carbonate from Weebubbie Cave, Australia*, A. K. Contos, et al., *Geomicrobiology Journal*, 18, 2001, p334

8.8ah http://www.dec.wa.gov.au/hotproperty/property/national-parks/nambung-national-park-pinnacles.html – Accessed 10.28.08, Site no longer available

18.8ai *Pleistocene Algal Pinnacles at Searles Lake*, California, David E. Scholl, *Journal of Sedimentary Petrology*, Vol. 30, No. 3 September 1960, p414

8.8aj *Algae, Contributors to the Formation of Calcareous Tufa, Mono Lake, California*, David W. Scholl, William H. Taft, *Journal of Sedimentary Petrology*, Vol. 30, No. 3 September 1960, p309

8.8ak *Calcareous Tufa Formations, Searles Lake and Mono Lake*, Ted Rieger, *California Geology*, July/August 1992, p99

8.9 The Salt Mark

8.9a *Sedimentation in Barred Basins, and Source Rocks of Oils*, W. G. Woolnough, *Bulletin of the American Association of Petroleum Geologists*, Vol. 21 No. 9, September 1937, p1104

8.9b *Sedimentary Rocks*, by Francis John Pettijohn, 1957, p8

8.9c http://cementamericas.com/mag/cement_advantages_synthetic_gypsum/ – Accessed 10.4.08, Site no longer available

8.9d http://en.wikipedia.org/wiki/Hydrogen_sulphide – Accessed 3.4.16

8.9e *Deep-sea mission finds life in the Lost City*, J. Ebert, 3 March 2005, http://www.nature.com/news/2005/050228/full/news050228-14.html – Accessed 3.4.16

8.9f *Experimental evidence of three coexisting immiscible fluids in synthetic granitic pegmatite*, Ilya V. Veksler, et al., *American Mineralogist*, Vol. 87, 2002, p775

8.9g *Evaporation of Seawater: Calculated Mineral Sequences*, Charles E. Harvie and John H Weare, *Science*, Vol. 208, 2 May 1980, p498

8.9h *Annual Review Earth Planet Science*, Lawrence A Hardie, *On Significance Of Evaporites*, 1991, 19:131-68, p132

8.9i *Salt orgin without evaporation proposed*, *Oil and Gas Journal*, Vol. 105, Issue 3, January 15, 2007 http://www.ogj.com/articles/save_screen.cfm?ARTICLE_ID=282114 – Accessed 3.4.16

8.9j *Salt formation association with sub-surface boiling and super critical water*, M. Hovland, H.G. Rueslatten, J.K. Johnson, B. Kvamme, T.Kuznetsova, *Marine and Petroleum Geology* 23, 2006, p866

8.9k *Origin of Ancient Potash Evaporites: Clues from the Modern Nonmarine Qaidam Basin of Western China*, Tim K. Lowenstein, Ronald J. Spencer, Zhang Pengxi, *Science*, 13 July 1989, Vol. 245, p1090

8.9l *Industrial Minerals and Rocks*, Sherilyn C. Williams-Stroud, James P. Searls, Robert J. Hite, *SME*, Chapter 69, *Potash Resources*, p21, – Accessed 9.20.01, Site no longer available

8.9m *Syndepositional Origin of Potash Evaporites: Petrographic and Fluid Inclusion Evidence*, Tim K. Lowenstein, Ronald J. Spencer, *American Journal of Science*, Vol. 290, January 1990, p1

8.9n *Salt formation association with sub-surface boiling and super critical water*, M. Hovland, H.G. Rueslatten, J.K. Johnson, B. Kvamme, T.Kuznetsova, *Marine and Petroleum Geology* 23, 2006, p860

8.9o *Formation of gigantic gypsum crystals*, J. Garcia-Guinea, S. Morales, A. Delgado, C. Recio, J. M. Calaforra, *Journal of the Geological Society*, 2002, Vol. 159, No. 4, Abstract, http://giantcrystals.strahlen.org/europe/pilar.htm – Accessed 3.4.16

8.9p *Salt formation association with sub-surface boiling and super critical water*, M. Hovland, H.G. Rueslatten, J.K. Johnson, B. Kvamme, T.Kuznetsova, *Marine and Petroleum Geology* 23, 2006, p860

8.9q *Salt formation association with sub-surface boiling and super critical water*, M. Hovland, H.G. Rueslatten, J.K. Johnson, B. Kvamme, T.Kuznetsova, *Marine and Petroleum Geology* 23, 2006, p863

8.9r http://earthobservatory.nasa.gov/IOTD/view.php?id=6465 – Accessed 3.4.16

8.9s *Deep sea pockmark environments in the eastern Mediterranean*, Lyobomir Dimitrov, John Woodside, *Marine Geology* 195, 2003, p268

8.9t *Fluid Mixing and Anhydrite Precipitation Within the TAG Mound*, Rachel A. Mills, Damon A. H. Teagle, Margaret K. Tivey, *Proceedings of the Ocean Drilling Program, Scientific Results*, Vol 158, 1998, p119

8.9u *Temperature and Salinity of Fluid Inclusions in Anhydrite as Indicators of Seawater Entrainment and Heating in the TAG Active Mound*, Margaret Kingston Tivey, Rachel A. Mills, Damon A. H. Teagle, *Proceedings of the Ocean Drilling Program, Scientific Results*, Vol. 158, 1998, p180

8.9v *Sedimentation in Barred Basins, and Source Rocks of Oils*, W. G. Woolnough, Bulletin of the American Association of Petroleum Geologists, Vol. 21, No. 9, September 1937, p1104

8.10 The Oil and Gas Mark

8.10a http://en.wikipedia.org/wiki/Petroleum – Accessed 11.17.08

8.10b *The Origin of Natural Gas and Petroleum, and the Prognosis for Future Supplies*, Thomas Gold, *Annual Review of Energy*, November 1985, Vol. 10, p55

8.10c *Inorganic Origin of Petroleum*, V. B. Porfir'ev, *The American Association of Petroleum Geologists Bulletin*, Vol. 58, No. 1, January 1974, p13, More information on Abiogenic Theory can be found at: http://oilismastery.blogspot.com/2008_06_01_archive.html – Accessed 11.21.08

8.10d *The Origin of Methane (and Oil) in the Crust of the Earth*, Thomas Gold, USGS Professional Paper 1570, The Future of Energy Gases, 1993, http://thinkorbebeaten.com/Energy/The%20Origin%20of%20Methane%20(and%20Oil)%20in%20the%20Crust%20of%20the%20Earth.pdf – Accessed 3.4.16

8.10e *The Microbial Origin of Fossil Fuels*, Guy Ourisson, Pierre Albrecht, Michel Rohmer, *Scientific American*, 1984, 251:2, p44

8.10f *Petroleum generation by laboratory-scale pyrolysis over six years simulating conditions in s subsiding basin*, J. D. Saxby, k. W. Riley, *Nature*, Vol. 308, 8 March 1984, p177

8.10g *Laboratory Simulation of Petroleum Formation*, *Hydrous Pyrolysis*, M. D. Lewan, *Organic Geochemistry Principles and Applications*, Edited by Michael H. Engel, Stephen A. Macko, Plenum Press, 1993, p419

8.11 The Coal Mark

8.11a *From lignin to coal in a year*, John Larsen, *Nature,* Vol. 314, 28 March 1985, p316

8.11b *From lignin to coal in a year*, John Larsen, *Nature*, Vol. 314, 28 March 1985, p316

8.11c *Chemical Structure and Properties of Coal XXVI—Studies on Artificial Coalification*, J. P. Schumacher, F. J. Juntjens, D. W. van Krevelen, Fuel, Vol. 39, 1960, p223

8.11d *From lignin to coal in a year*, John Larsen , *Nature*, Vol. 314, 28 March 1985, p316

8.11e *Artificial Coalification study: Preparation and characterization of synthetic Macerals*, R. Hayatsu, et al., Org. Geochem., 1984, Vol. 6, p467

8.11f *From lignin to coal in a year*, John Larsen, *Nature*, Vol. 314, 28 March 1985, p316

8.11g *Artificial Coalification study: Preparation and characterization of synthetic Macerals*, R. Hayatsu, et al., Org. Geochem. Vol. 6, 1984, p463

8.11h *Nuclear magnetic resonance studies of ancient buried wood—II. Observations on the origin of coal from lignite t bituminous coal*, P. G. Hatcher, I. A. Breger, N. Szeverenyi, G. E. Maciel, *Org. Geochem.* Vol. 4, 1982, p16

8.11i *From lignin to coal in a year*, John Larsen, *Nature*, Vol. 314, 28 March 1985, p316

8.11j http://www.ipcc.ie/cbformation.html – Formation of Cutover and Cutaway Bogs – Accessed 11.21.08, Site no longer available

8.11k *Ecological gradients within a Pennsylvanian mire forest*, William A. DiMichele, et al., *Geology*, Vol. 35, No. 5, May 2007, p416

8.11l *Argonne Scientists Make Artificial Coal*, Joseph Haggin, *C&EN*, November 21, 1983, p42

8.11m *Sedimentation in Barred Basins, and Source Rocks of Oil*, W. G. Woolnough, *Bulletin of the American Association of Petroleum Geologists (AAPG)*, Vol. 21, No.9, September 1937, p1101, Entire article also reprinted in *Origin of Evaporites*, AAPG Bulletin, 1971, p 3-59

8.11n Carbonate Map – *US Karst Map*, American Geological Institute, Veni, et al., 2001 , https://caves.org/pub/journal/PDF/V64/v64n1-Veni.pdf – Accessed 3.15.16

Coal Map – *Coal Fields of the Conterminous United States*, 1996, John Tully, USGS, Open File Report No. 96-92, http://pubs.usgs.gov/of/1996/of96-092/other_files/us_coal.pdf – Accessed 3.16.16

Salt Map – *Handbook of World Salt Resources*, Stanley J. Lefond, Plenum Press, New York, 1969, p2, http://www.springer.com/us/book/9781468407051 – Accessed 3.16.16

Loess and Sand Maps – *Quaternary Eolian Deposits*, Pleistocene Eolian Deposits of the United States, Thorp, et al., 1951
US Department of the Interior United States Aquifer map, USGS, http://gec.cr.usgs.gov/archive/eolian/pdfs/Bettis2003QSR.pdf – Accessed 3.16.16, http://geochange.er.usgs.gov/sw/impacts/geology/sand/swsand_fig1.gif – Accessed 3.16.16

8.12 The Pyrite Mark

8.12a *Magnetite from magnetotactic bacteria: Size distributions and twinning*, Bertrand Devouard, et al., *American Mineralogist*, Vol. 83, Abstract, 1998, p1387

8.12b *The Impact of Bacteria on the Deposition of Ironformations*, D. Ann Brown, Gordon A. Gross, *Canadian Society of Exploration Geophysicists 2000 Conference*, Abstract

8.12c *Thermophilic, anaerobic bacteria isolated from a deep borehole in granite in Sweden*, Ulrich Szwzyk, Regine Szewzyk, Thor-Axel Stenstrom, *Microbiology, Proc. Natl. Acad. Sci. USA*, Vol. 91, March 1994, p1812

8.12d The following is a more complete quote from *Soils of the Past*, G. J. Retallack, stating evidence that meteorites come from soils rather than gas, liquid or dust in space. "The most convincing evidence for the origin of carbonaceous chondrites [meteorites] as soils comes from their petrograhical and mineralogical features. If they represented frozen solar nebular material, then the grains and chondrules would be little altered and have sharp boundaries. In contrast, carbonaceous chondrites contain pseudomorphs of chondrules, clasts, and aggregates **replaced by calcite, septechlorite clays, and iron oxides, which imply extensive chemical alteration in aqueous solution** (Bunch & Chang 1980). There are also cross-cutting veins, cavity-lining concentric zones of colloidal material (colloform structure), interlocking crystal growth so septechlorite clays, clayey alteration along cleavage planes of mineral grains, and mesh-like bright clay microfabrics (lattisepic plasmic fabric). **These are all more like alteration of material by water in a soil than individual reactions between dispersed gas, liquid, and dust.**" Bib 118 p319

8.12e http://en.wikipedia.org/wiki/Sulfur – Accessed 3.18.16

8.12f *Pre-Eruption Vapor in Magma of the climactic Mount Pinatubo Eruption: Source of the Giant Stratospheric Sulfur Dioxide Cloud*, T. M. Gerlach, H. R. Westrich, R. B. Symonds, USGS
http://pubs.usgs.gov/pinatubo/gerlach/ – Accessed 3.18.16

8.12g *On the Origin and Significance of Pyrite Spheres in Devonian Black Shales of North America*, Jürgen Schieber and Gordon Baird, *Journal of Sedimentary Research*, Vol. 71, No. 1, January 2001, p158

8.12h *On the Origin and Significance of Pyrite Spheres in Devonian Black Shales of North America*, Jürgen Schieber and Gordon Baird, *Journal of Sedimentary Research*, Vol. 71, No. 1, January 2001, p165

8.12i *Impact From the Deep*, Peter D. Ward, *Scientific American*, October 2006, p65

8.13 The Ore Mark

8.13a *Volcanoes A Planetary Perspective*, Peter Francis, Clarendon Press, 1993, p333

8.13b *The Origin of Cu/Au Ratios in Porphyry-ype Ore Deposits*, W. E. Halter, T. Pettke, C. A. Heinrich, *Science*, 7 June 2002, p1844

8.13c *Origins of Hydrothermal Ores*, H.L. Barns, A. W. Rose, *Science*, Vol. 279, 27 March 1998, p2064

8.13d *Salt formation association with sub-surface boiling and super critical water*, M. Hovland, H.G. Rueslatten, J.K. Johnson, B. Kvamme, T.Kuznetsova, *Marine and Petroleum Geology* 23, 2006, p862

8.13e *Leg 158 Explores the Mid-Atlantic Ridge*, press release, *Hot Springs Create Mineral Deposits On The Ocean Floor*,
 http://www-odp.tamu.edu/public/pressrel_html/leg158pr.html – Accessed 3.18.16

8.13f *The Donoso copper-rich, tourmaline-bearing breccia pipe in central Chile*, M. Alexandra Skews, Carmen Holmgren, Charles R. Stern, *Mineralium Deposita 33,* 2003, Abstract

8.13g *Metal – Microbe Interactions*, Edited by Robert K. Poole, Geoffrey M. Gadd, *Special Publications of the Society for General Microbiology*, Volume 26, IRL Press, 1989, p104

8.13h *Alteration of microbially precipitated iron oxides and hydroxides*, D. Ann Brown, J. A. Sawicki, Barbara L. Sherriff, *American Mineralogist*, Vol. 83, 1998, p1420

8.13i *Metal – Microbe Interactions*, Edited by Robert K. Poole, Geoffrey M. Gadd, *Special Publications of the Society for General Microbiology*, Volume 26, IRL Press, 1989, p33

8.13j *Occurrence of Uranium in Diatremes on the Navajo and Hopi Reservations, Arizona, New Mexico, and Utah*, Eugene M. Shoemaker, *USGS Professional Papers No. 300*, 1956, p179

8.13k *Hydrothermal Uranium Deposits*, Robert A. Rich, Heinrich D. Holland, Ultrich Petersen, *Developments in Economic Geology*, 6, Elsevier Scientific Publishing Co., 1977, p11

8.13l *Uranium, Where It Is and How To Find It*, Paul Dean Proctor, Edmond P. Hyatt, Kenneth C. Bullock, Eagle Rock Publishers, 1954, p21

8.13m *Mineralogy at a Crossroads*, Russell J. Hemley, *Science*, 13 August 1999, Vol. 285, p1026

8.13n *Impact, The Threat of Comets and Asteroids*, Gerrit L. Verschuur, *Oxford University Press*, 1996, p9

8.13o *Cretaceous-Tertiary boundary event: Evidence for a short time scale*, Iain Gilmour, Edward Anders, *Geochimica et Cosmochimica Acta*, Vol. 53, Issue 2, Feb 1989, Abstract

8.13p *Microbes complicate the K-T mystery*, R. Monastersky, *Science News*, Nov 25, 1989, p341

8.13q http://en.wikipedia.org/wiki/Iridium – Accessed 7.22.08, Quote has since been changed

8.14 The Surface Mark

8.14a *A proposed mechanism for the growth of chalcedony*, Peter J. Heaney, *Contributions to Mineralogy and Petrology*, 1993, Vol. 115, p67

8.14b *A proposed mechanism for the growth of chalcedony*, Peter J. Heaney, *Contributions to Mineralogy and Petrology*, 1993, Vol. 115, p68

8.14c The examples of Surface Chalcedony given here are general in nature and apply only to areas of the desert not heavily eroded during the last several thousand years. Erosion is, of course, an ongoing process, and the surfaces of many areas that once had Surface Chalcedony might have been washed clean; the specimens carried away and buried in sediment layers where it rests today. In those places, Surface Chalcedony would obviously be found below the surface, probably in similar abundance. However, we have yet to find any areas where this took place in any significance. Only in the occasional dry streambeds was chalcedony found buried in the various layers, but this represents a **disturbed** area, not the unweathered desert areas described herein. For those that want to search for Surface Chalcedony specimens, knowing about the other Surface Marks, such as desert varnish will be a great help in identifying potential areas where Surface Chalcedony can be found.

8.14d *The Chalcedony Localities in Lopburi and Kanchanaburi Provinces*, Panjai Saraphanchotwitthaya, *International Gem & Jewelry Conference (GIT 2008)*, December 11-14, 2008, Bangkok and Kanchanaburi, Thailand

8.14e *The nature of crystalline silica from the TAG submarine hydrothermal mound, 26 N Mid Atlantic Ridge*, Laurence Hopkinson, Stephen Rovers, Richard Herrington, Jamie Wilkinson, *Contributions to Mineralogy and Petrology*, Vol. 137, No. 4, December 1999, Abstract

8.14f http://www.zianet.com/GEODEKID/index.html – Accessed 3.31.16

8.14g http://en.wikipedia.org/wiki/Caliche_(mineral) – Accessed 3.31.16

8.14h *Diversity of Microorganisms within Rock Varnish in the Whipple Mountains, California*, K. R. Kuhlman, et al., *Applied and Environmental Microbiology*, Vol. 72, No. 2, February 2006, p1708

8.14i http://en.wikipedia.org/wiki/Desert_varnish – Accessed 3.31.16

8.14j http://www.geotimes.org/sept06/NN_Mojave.html – Accessed 3.31.16

8.14k *A Review Of The History Of Dating Rock Varnishes*, Alan Watchman, *Earth-Science Reviews*, 49, 2000, p262

8.14l http://en.wikipedia.org/wiki/Desert_varnish – Accessed 2.3.09, Since then some words have been changed – Accessed 6.23.16

8.14m *Microbial Origin of Desert Varnish*, R. I. Dorn, T. M. Oberlander, *Science*, Vol. 213, 11 September 1981, p1245

8.14n http://www.geotimes.org/sept06/NN_Mojave.html – Accessed 3.31.16, for amino acids in varnish see: *Amino Acid Analyses of Desert Varnish from the Sonoran and Mojave Deserts*, Randal S. Perry, et al., *Geomicrobiology Journal*, Vol. 20, No. 5, Sep-Oct 2003, p427-38

8.14o *Diversity of Microorganisms within Rock Varnish in the Whipple Mountains, California*, K. R. Kuhlman, et al., *Applied and Environmental Microbiology*, Vol. 72, No. 2, February 2006, p1710

8.14p http://www.geotimes.org/sept06/NN_Mojave.html – Accessed 3.31.16

8.14q http://www.geotimes.org/sept06/NN_Mojave.html – Accessed 3.31.16

8.14r *Ambiguities in Direct Dating of Rock Surfaces Using Radiocarbon Measurements*, W. Beck, et al., *Science*, Vol. 280, No. 5372, 26 June 1998, p2132

8.14s *Ambiguities in Direct Dating of Rock Surfaces Using Radiocarbon Measurements*, W. Beck, et al., *Science*, Vol. 280 No. 5372, 26 June 1998, p2132

8.14t *Rock Dates Thrown Into Doubt, Researcher Under Fire*, David Malakoff, *Science*, Vol. 280, No. 5372, 26 June 1998, p2041

8.14u *Ambiguities in Direct Dating of Rock Surfaces Using Radiocarbon Measurements*, W. Beck, et al., *Science*, Vol. 280, No. 5372, 26 June 1998, p2136

8.14v *A Review Of The History Of Dating Rock Varnishes*, Alan Watchman, *Earth-Science Reviews*, 49, 2000, p271

8.14w *Dust settles on defamation case*, Rex Dalton, *Nature*, Vol. 411, 31 May 2001, p511

8.14x *Rock Dates Thrown Into Doubt, Researcher Under Fire*, David Malakoff, *Science*, Vol. 280 No. 5372, 26 June 1998, p2042

8.14y *Baking black opal in the desert sun: The importance of silica in desert varnish*, Randall S. Perry, et al., *Geology*, Vol. 34, No. 5, July 2006, p540

8.14z *Making silica rock coatings in the lab: synthetic desert varnish*, Randall S. Perry, et al., *Astrobiology and Planetary Missions*, Proc. SPIE Vol. 5906, September 2005, p273

8.14aa *Manganese mineral formation by bacterial spores of the marine Bacillus, strain SG-1: Evidence for the direct oxidation of Mn (II) to Mn (IV)*, Kevin W. Mandernack, Jeffrey Post, Bradley M. Tebo, *Geochimica et Cosmochimica Acta*, Vol. 59, No. 21, 1995, p4393

8.14ab *Manganese mineral formation by bacterial spores of the marine Bacillus, strain SG-1: Evidence for the direct oxidation of Mn (II) to Mn (IV)*, Kevin W. Mandernack, Jeffrey Post, Bradley M. Tebo, *Geochimica et Cosmochimica Acta*, Vol. 59, No. 21, 1995, p 4404, Conclusions

8.14ac *Hydrothermal activity at mid-ocean ridge axes*, J. M. Edmond, *Nature*, Vol. 290, 12 March 1981, p87

8.14ad *The Manganese Nodule Belt of the Pacific Ocean, Geological Environment, Nodule Formation, and Mining Aspects*, Edited by Ferdinand Enke Verlag Stuttgart, 1988, p48

8.14ae *Trace-element evidence for the origin of desert varnish by direct aqueous atmospheric deposition*, Nivedita Thiagarajan, Cin-Ty Aeolus Lee, *Earth and Planetary Science Letters* 224, 2004, p131

8.14af *Microbial Fossils Detected In Desert Varnish*, B.E. Flood, C. Allen, T. Longazo, *Lunar and Planetary Science* XXXIV, 2003, 1633.pdf

8.14ag *Mysteries of Sandstone Colors and Concretions in Colorado Plateau Canyon Country*, Chan, M.A. and W.T. Parry, *Utah Geological Survey Public Information Series.*, 2002, No. 77, p10, http://geology.utah.gov/online/pdf/pi-77.pdf – Accessed 3.31.16

8.14ah http://en.wikipedia.org/wiki/Moqui_marbles – Accessed 3.31.16

8.14ai http://ijolite.geology.uiuc.edu/02SprgClass/geo117/lectures/Lect12.html – Accessed 3.31.16

8.14aj http://cps-amu.org/sf/notes/lect10.htm – Accessed 2.19.09

8.14ak *The Manganese Nodule Belt of the Pacific Ocean, Geological Environment, Nodule Formation, and Mining Aspects*, Edited by Ferdinand Enke Verlag Stuttgart, 1988, p14

8.15 The Diamond Mark

8.15a *Microscopic diamonds crack geologic mold*, R. Monastersky, *Science News*, Vol. 148, July 8, 1995, p22

8.15b *Microdiamond in high-grade metamorphic rocks of the Western Gneiss region, Norway*, Larissa F. Dobrzhinetskaya, et al., *Geology*, July 1995, p600

8.15c *Microscopic diamonds crack geologic mold*, R. Monastersky, *Science News*, Vol. 148, July 8, 1995, p22

8.15.d *Fluid Inclusions in Minerals: Methods and Applications*, Edited by Benedetto De Vivo, Maria Luce Frezzotti, Virginia Tech, Siena Italy Workshop, 1-4 September 1994, *Fluid Inclusion Evidence of Mantle Fluids*, Edwin Roedder, Harvard University, p286

8.15e http://www.ens-newswire.com/ens/may2003/2003-05-01-01.asp – Accessed 3.31.16

8.15f *Finding Diamonds in your Backyard?*, Kathryn Hansen, *Geotimes*, July 2007, p19

8.16 The Inclusion Mark

8.16a *Hydrothermal Synthesis of Crystals*, Robert A. Laudise, AT&T Bell Laboratories, *Chemical & Engineering News*, September 28, 1987, p31

8.16b *Fluid Inclusions*, Edwin Roedder, *Reviews in Mineralogy*, Mineralogical Society of America, Vol.12, 1984, p37

8.16c *Fluid Inclusions in Minerals: Methods and Applications*, *Fluid Inclusion Evidence of Mantle Fluids*, Edwin Roedder, Virginia Tech, 1994, p288

8.16d *The Composition and Significance of Gas Released from Natural Diamonds from Africa and Brazil*, Charles E. Melton, A. A. Giardini, *American Mineralogist*, 1974, Vol. 59, p780, Methane is also reported in diamonds in the July 1-4, 1997 XIV ECROFI meeting in Nancy, France, *Fluid Components of the Diamond Crystallization Environment*, p1

8.16e *Petrography and uplift history of the Quaternary Takidani Granodiorite: could it have hosted a supercritical (HDR) geothermal reservoir?*, Masatoshi Bando, et al., *Journal of Volcanology and Geothermal Research*, 120, 2003, Abstract

8.16f *New Geothermal data from the Kola superdeep well SG-3*, Yuri A. Popov, et al., *Tectonophysics*, 306, 1999, p354, Fig 5

8.16g *Ion microporobe analysis of 18O/16O in authigenic and detrital quartz in the St. Peter Sandstone, Michigan Basin and Wisconsin Arch, USA: Contrasting digenetic histories*, Colin M. Graham, John W. Valley, Bryce L. Winter, *Geochimica et Cosmochimica Acta*, Vol. 60, No. 24, 1996, p5112

8.16h *Fluid Inclusions*, Edwin Roedder, *Reviews in Mineralogy*, Mineralogical Society of America, 1984, Vol. 12, p8

8.16i *Oddities of the Mineral World*, William B. Sanborn, Van Nostrand Reinhold Co., 1976 p52-54

8.16j *Fluid Inclusions*, Edwin Roedder, *Reviews in Mineralogy*, Vol. 12, Mineralogical Society of America, 1984, p322

8.16k *The Story of Anthraxolite*, Brian Bell, Brent Duse, Tim Venne, *Chelmsford Valley District Composite School, Sudbury Region*, http://www.earth.uwaterloo.ca/services/whaton/waton/s903.html – Accessed 3.2.09, Site no longer available

8.16l *Fluid Inclusions*, Edwin Roedder, *Reviews in Mineralogy*, Vol. 12, 1984, *Mineralogical Society of America*, p324

8.16m Although the researchers give excellent descriptions of a variety of Flood conditions, including salinity, temperature and the presence of organic substances involved in the mineralizing environment, the investigators generally fail to mentioned that a higher *pressure* is involved. Computing *pressures* along with temperatures and the dissolved solids associated with the inclusions are required to know the actual environment of the Hypretherm. It is also critical knowledge for identifying UF conditions involved in the formation of all the geological deposits on the surface. This is an area where Millennial Science will make a great contribution and may find that the lower temperatures that were reported in the forming of natural coals came about from the higher pressures that were present during the UF. Pressure cookers once used in the common kitchen for home-canning purposes illustrate conclusively that cooking under pressure lowers the temperature and shortens cooking times than would otherwise be necessary.

9 The Weather Model

9.1 The Mystery of Weather

9.1a *Weather Forecasting, National Geographic*, June 2005, p95
9.1b *The Decline of Reason?*, Jere H. Lipps, http://www.ucmp.berkeley.edu/fosrec/Lipps.html – Accessed 6.8.16

9.2 The Origin of Weather

9.2a *The Western Weather Guide*, Richard A. Keen, Fulcrum Inc., *Skywatch*, 1987, p3
9.2b http://www.usgs.gov/faq/list_faq_by_category/get_answer.asp?id=148 – Accessed 4.1.09, Site no longer available
9.2c *Letters, Scientific American*, January 2007, p12
9.2d *Earthquake Alarm, impending earthquakes have been sending us warning signals—and people are starting to listen*, Tom Bleier, Friedemann Freund, *IEEE Spectrum*, December 2005, p26 (http://spectrum.ieee.org/dec05/2367 – Accessed 9.28.15
9.2e *Satellite Thermal Survey—A New Tool for the Study of Seismoactive Regions*, A. A. Tronin, *Int. J. Remote Sensing*, 1996, vol. 17, no. 8, p1449
9.2f *Temperature variations related to earthquakes from simultaneous observation at the ground stations and by satellites in Kamchatka area*, A.A. Tronin, et al., *Physics and Chemistry of the Earth* 29, 2004, p501 Abstract
9.2g *Space News*, November 21, 2005, p15
9.2h *Orgin of Wind*, http://www.srh.noaa.gov/jetstream//synoptic/wind.htm – Accessed 9.29.15
9.2i *Weather's Highs and Lows: Part 1 The High*, Keith C. Heidorn, PhD, 2005, http://www.islandnet.com/~see/weather/elements/high.htm, – Accessed 9.29.15

9.3 Earthquake Cloud Evidence

9.3a *Thermal IR satellite data application for earthquake research in Japan and China*, Andrew A. Tronin, Masashi Hayakawa, Oleg A. Molchanov, *Journal of Geodynamics 33*, 2002, p529√
9.3b http://www.earthquakesignals.com/zhonghao296/introduction.html – Accessed 6.8.16
9.3c *The Times of India*, Delihi, Abhay Vaidya, 28 April 200, http://farshores.org/amskrit.htm – Accessed 6.25.09
9.3d http://www.earthquakesignals.com/zhonghao296/introduction.html
9.3e *Bam Earthquake Prediction & Space Technology*, Zhonghao Shou, Darrell Harrington, 2004, (found at earthquakesignals.com
9.3f *The Times of India*, Delihi, Abhay Vaidya, 28 April 2001, http://farshores.org/amskrit.htm – Accessed 6.25.09
9.3g *Earthquake cloud over Japan detected by satellite*, G. Gup, G. Xie, *International Journal of Remote Sensing*, Vol. 28, Issue 23, January 2007, Abstract
9.3h *New Scientist*, Lynn Dicks, 11 April 2008, Issue 2651, http://www.newscientist.com/article/mg19826514.600-curious-cloud-formations-linked-to-quakes.html – Accessed 9.29.15
9.3i Department of Atmospheric Sciences (DAS) at the University of Illinois at Urbana-Champaign, http://ww2010.atmos.uiuc.edu/(Gh)/wwhlpr/fair_cumulus.rxml – Accessed 9.29.15
9.3j http://en.wikipedia.org/wiki/Tornado – Accessed 5.11.09
9.3k http://en.wikipedia.org/wiki/Tornado – Accessed 5.11.09
9.3l *Typhoons trigger earthquakes on Taiwan*, Breitbart.com, 10 June 2009, http://www.breitbart.com/print.php?id=CNG.0e72e8e889357fc5b4c54b72f3332e06.1a1&show_article=1 – Accessed 7.9.2009, Site no longer available
9.3m *Rainmaking Bacteria Ride Clouds to "Colonize" Earth?*, Christine Dell'Amore, National Geographic News, 12, January 2009, http://news.nationalgeographic.com/news/pf/64122879.html – Accessed 7.27.09, Site no longer available
9.3n *Weather Cycles, Crops, and Prices*, Louis M. Thompson, *Science of Food and Agriculture*, September 1988
9.3o *Weather Cycles, Crops, and Prices*, Louis M. Thompson, *Science of Food and Agriculture*, September 1988 (The January 1994 issue was the last publication of this journal. R. G. Currie is referenced by Thompson in the 1981 *Journal of Geophysical Research*, *Evidence for 18.6 year MN signal in temperature and drought conditions in North America since AD 1800*)
9.3p http://reykr.livejournal.com/383099.html – Accessed 9.29.15

9.4 The Global-Weather System Evidence

9.4a *Is it El Niño of the Century?*, J. Madeleine Nash, *Time*, 18 August 1997, p56
9.4b *Here Comes More Weird Weather*, Per Ola, Emily dAulaire, *Reader's Digest*, November 1998, p139
9.4c *Is it El Niño of the Century?*, J. Madeleine Nash, *Time*, 18 August 1997, p57
9.4d *The Oryx Resource Guide to El Niño and La Niña*, Joseph S. D'Aleo, *Oryx Press*, 2002, p16
9.4e *Floor Show*, John R. Delaney, *The Sciences*, July/Aug 1998, p29, New York Academy of Sciences
9.4f *Time-series temperature measurements at high-temperature hydrothermal vents, East Pacific Rise 9°49'-51'N: evidence for monitoring a crustal cracking event*, D. J. Fornari, et. al., *Earth and Planetary Science Letters*, 160 (1998), p419
9.4g *Generation of Hydrothermal Megaplumes by Cooling of Pillow Basalts at Mid-ocean Ridges*, M. R. Palmer , G. G. J. Ernst, Nature, 393, 14 April 1998, p643-647
9.4h http://visibleearth.nasa.gov/view_rec.php?id=17268 – Accessed 5.7.09
9.4i http://earthobservatory.nasa.gov/Features/NAO/ – Accessed 10.2.15
9.4j *In Hot Water*, Chris Carroll, *National Geographic*, August 2005, p79
9.4k *In Hot Water*, Chris Carroll, *National Geographic*, August 2005, p78-9
9.4l *Warmer Oceans, Stronger Hurricanes*, Kevin E. Trenberth, *Scientific American*, July 2007, p45

9.5 The Geofield Model

9.5a http://denali.gsfc.nasa.gov/research/mag_field/conrad/explain.html – Accessed 6.8.16
9.5b http://en.wikipedia.org/wiki/Magnet#Temperature – Accessed 10.2.15
9.5c *Inside History in Depth*, David Stevenson, *Nature*, Vol 428, 1 April 2004, p477
9.5d http://en.wikipedia.org/wiki/Piezoelectricity – Accessed 10.2.15
9.5e Email from David R. Cook, Atmospheric Research Section, Environmental Research Division, Argonne National Laboratory, 9 April 2001. Cook also said "the magnetic field can have some affect on upper atmosphere phenomenon like the aurora."

9.6 The Geofield Evidence

9.6a *Thermally Stimulated Currents in Rocks. II.*, E. Dologlou, *Tectonophysics*, 224, 1993, p177
9.6b *Astrophysicist Strikes Blow to Lightning Theory*, Betsy Mason, *Nature Science Update*, 17 November 2003, *A fundamental limit on electric fields in air*, Dwyer, J. R., *Geophysical Research Letters*, 30, 2055, 2003
9.6c *Astrophysicist Strikes Blow to Lightning Theory*, Betsy Mason, *Nature Science Update*, 17 November 2003, *A fundamental limit on electric fields in air*, Dwyer, J. R., *Geophysical Research Letters*, 30, 2055, 2003
9.6d *A Bolt Out of the Blue*, Joseph R. Dwyer, *Scientific American*, May 2005, p65
9.6e http://en.wikipedia.org/wiki/Fair_weather_condition – Accessed 10.5.15
9.6f http://hvo.wr.usgs.gov/earthquakes/destruct/1975Nov29/ – Accessed 8.11.16
9.6g *Mt. St. Helens, The Volcano Explodes!*, Prof. Leonard Palmer, Norwest Illustrated, 1980, p24
9.6h *Paricutin, The Volcano Born in a Mexican Cornfield*, James F. Luhr and Tom Simkin, Editors, Geoscience Press, Inc., 1993, p73
9.6i *Geophysical Research Letters*, Vol. 30, No. 16, 1858, doi

10.1029/2003GL017234, 2003, p11

9.6j *Nature*, Science Update, 3 September 2003, http://wwwww.nature.com/nsu/nsu_pf/030901/030901-3.html

9.6k *A fundamental limit on electric fields in air*, Dwyer, J. R., *Geophysical Research Letters*, 30, 2003, p2055

9.6l *Earthquake Alarm, impending earthquakes have been sending us warning signals—and people are starting to listen*, Tom Bleier, Friedemann Freund, *IEEE Spectrum*, December 2005, p23, http://spectrum.ieee.org/computing/hardware/earthquake-alarm – Accessed 10.5.15

9.6m http://www.deprem.cs.itu.edu.tr/harici-e.html – Accessed 6.8.16

9.6n http://www.quakefinder.com/2004results/UsingGroundSensors.pdf – Accessed 6.28.06, Site no longer available

9.6o http://www.quakefinder.com/fppt/LowFreqAn.htm – Accessed 6.28.06

9.6p *Earthquake Alarm, impending earthquakes have been sending us warning signals—and people are starting to listen*, Tom Bleier, Friedemann Freund, *IEEE Spectrum*, December 2005, p23, http://spectrum.ieee.org/dec05/2367 – Accessed 6.8.16

9.6q *Journal of Geophysical Research*, Vol. 102, No. B8, p18, 300 August 10, 1997

9.6r *Journal of Geophysical Research*, Vol. 102, No. B8, p18,192, August 10, 1997

9.6s *Journal of Geophysical Research*, Vol. 102, No. B8, p18,296, August 10, 1997

9.6t *Journal of Geophysical Research*, Vol. 102, No. B8, p18, 192, August 10, 1997

9.6u *Journal of Geophysical Research*, Vol. 24, No. 3, p350, February 1, 1997

9.6v *Journal of Geophysical Research*, Vol. 102, No. B8, p18, 192 August 10, 1997.

9.6w *Electromagnetic Phenomena in the Earth's Crust*, K. N. Abdullabekov, *Geotechnika* 1, 1991, p5

9.6x *Quasi-Static Magnetic Field Changes Associated with the Cannikin Nuclear Explosion*, W. P. Hasbrouck, J. H. Allen, *Bulletin of the Seismological Society of America*, Vol. 62, No. 6, p1479

9.6y *Electromagnetic Phenomena in the Earth's Crust*, K. N. Abdullabekov, *Geotechnika* 1, 1991, p25

9.6z *Study of the TEC data obtained from the DORIS stations in relation to seismic activity*, Feng Li, Michel Parrot, *Annals of Geophysics*, Vol. 50, No. 1, February 2007, p49

9.6aa *Magnetic fields over active tectonic zones in ocean*, Yu. A. Kopytenko, et al., *Journal of Geodynamics*, 33, 2002, p489

9.6ab *Seismicity Cycles in the Mt. Vesuvius Area and their Relation to Solar Flux and the Variations of the Earth's Magnetic Field*, G. Duma, G. Vilardo, *Physics and Chemistry of the Earth*, Vol. 23, No. 9-10, p927

9.6ac *A sense of direction for dowsers?*, Tom Williamson, *New Scientist*, 19 March 1987, p40

9.6ad http://en.wikipedia.org/wiki /Dowsing – Accessed 8.25.09

9.6ae *The Dowsing Reaction Originates From Piezoelectric Effect in Bone*, No Nordell, Presented at the 6th International Symposium of Ecological Design, May 19-21, 1988, Svedala, Sweden, p2, http://www.ltu.se/polopoly_fs/1.5014!dowsing.pdf – Accessed 10.9.15

9.6af http://www.globalwatch.org/ungp/EOS_98.htm – Accessed 10.9.15

9.6ag http://www.globalwatch.org/ungp/campbell_and_geller.htm – Accessed 10.9.15

9.6ah http://www.globalwatch.org/ungp/EOS_98.htm – Accessed 6.8.16

19.6ai http://www.globalwatch.org /ungp/friedemann98.htm – Accessed 10.9.15

9.6aj http://denali.gsfc.nasa.gov/sci_hi/sci_hi_2005_01/2005_01b.html – Accessed 6.8.16

9.6ak *Cracking the Code of Pre-Earthquake Signals*, Friedemann Freund, *SETI Institute*, 20 September 2005, http://www.space.com/1695-cracking-code-pre-earthquake-signals.html – Accessed 6.8.16

9.6al *Electric Properties of Igneous Rock under Pre-Earthquake Conditions*, Joshua J. Mellon, et al., working under Friedemann T. Freund an unpublished thesis, p6

9.6am *Earthquake Prediction Using a New Monopolar Electric Field Probe*, B. Ustundag, S. Ozerdem, *European Seismological Congress* (ESC2002) IASPEI/IUGG, GENOA, September 2002

9.6an *Journal of Geophysical Research*, Vol. 102, No. B8, August 10, 1997, p18,289

9.7 The Aurora Evidence

9.7a http://www.loc.gov/rr/scitech/mysteries/northernlights.html – Accessed 10.10.15

9.7b http://en.wikipedia.org/wiki/Aurora_%28astronomy%29 – Accessed 9.3.09

9.7c *The Great Comet Crash, The impact of Comet Shoemaker-Levy on Jupiter*, Edited by John R. Spencer and Jacqueline Mitton, Forward by Gene and Carolyn Shoemaker, 1995, Cambridge University Press, pix

9.7d http:www-pi.physics.uiowa.edu/vis-data/coastlines/ – Accessed 10.10.15

9.8 The Ozone-CFC Pseudotheory

9.8a http://www.nasa.gov/vision/earth/lookingatearth/ozone_record.html – Accessed 10.10.15

9.8b http://ozonewatch.gsfc.nasa.gov/meteorology/index.html – Accessed 10.10.15

9.8c http://en.wikipedia.org/wiki/Ozone_hole – Accessed 9.9.09

9.8d *The extraordinary events of the major, sudden stratospheric warming, the diminutive antarctic ozone hole, and its split in 2002*, Varotsos C., *Environ Sci Pollut Res Int.*, 2004, 11(6), Abstract

9.8e *Change in the Air, Banning CFC-driven inhalers could levy a toll on asthma sufferers*, Emily Harrison, *Scientific American*, August 2008, p20

9.8f *The 20th anniversary of the Montreal Protocol and the unexplainable 60% of ozone loss*, Costas A. Varotsos, Environ Sci Pollut Res (2008) 15:448–449, http://www.springerlink.com/content/u48l035jg18th948/fulltext.pdf – Accessed 10.10.15

9.8g *Chemists poke holes in ozone theory*, Quirin Schiermeier, *Nature*, Vol. 449, 27 September 2007, p382

9.8h *Investigating the Ozone Hole*, Rebecca L. Johnson, Lerner Publications Company, 1993, p18

9.9 The Global-Warming Pseudotheory

9.9a http://en.wikipedia.org/wiki/IPCC_Fourth_Assessment_Report – Accessed 10.22.15

9.9b *Physical Science Behind CLIMATE CHANGE*, William Collins, et al., *Scientific American*, August 2007, p64

9.9c http://www.gallup.com/poll/117772/Awareness-Opinions-Global-Warming-Vary-Worldwide.aspx – Accessed 10.22.15

9.9d http://www.petitionproject.org/ – Accessed 10.22.15

9.9e http://icecap.us/images/uploads/Gray12-08.pdf – Accessed 10.22.15

9.9f Tropical Meteorology Project - http://tropical.atmos.colostate.edu/ – Accessed 10.22.15

Science & Public Policy Institute - http://scienceandpublicpolicy.org/ – Accessed 10.22.15

Cato Institute - http://www.cato.org/subtopic_display_new.php?topic_id=27&ra_id=4 – Accessed 10.22.15

CO2 Science - http://co2science.org/index.php – Accessed 10.22.15

9.9g *How Did Humans First Alter Global Climate?*, William F. Ruddiman, *Scientific American*, March 2005, p53

9.9h *Highly Variable Northern Hemisphere Temperatures Reconstructed From Low- and High-ResolutionProxy Data*, Anders, Moberg, *Nature*, Vol. 433, 10 February 2005, p614

9.9i http://www.john-daly.com/bull120.htm – Accessed 6.8.16

9.9j *Satellite Thermal Survey—A New Tool for the Study of Seismoactive Regions*, A. A. Tronin, *Int. J. Remote Sensing*, 1996, Vol. 17, No. 8, p1451

9.9k *Satellitic thermal infrared brightness temperature anomaly image—short-term and impending earthquake precursors*, Zuji Qiang, et al., *Science in China Series D: Earth Sciences*, Vol. 42, No. 3, June 1999, Abstract

9.9l *Satellite Measurements of Sea Level Change: Where Have We Been and Where Are We Going, 15 Years of Progress in Radar Altimetry*, Venice, Italy, March 13-18, 2006, R. S. Nerem, D. P. Chambers, E. W. Leuliette, G. T. Mitchum, and A. Cazenave,

9.9m http://www.foxnews.com/scitech/2012/02/09/himalayan-glaciers-have-lost-no-ice-in-past-10-years-new-study-reveals/?intcmp=features – Accessed 10.23.15

9.9n *Redistributing Earth's Mass*, Anny Cazenave, R. Steven Nerem, *Science*, Vol. 297, 2 August 2002, p783

9.9o *Recent Earth Oblateness Variation: Unraveling Climate and Postglacial Rebound Effects*, Jean O. Dickey, et al., *Science*, Vol. 298, 6 December 2002, p1975

9.9p *Detection of a Large-Scale Mass redistribution in the Terrestrial System Since 1998*, Christopher M. Cox, Benjamin F. Chao, *Science*, 2 August 2002, p831

9.9q *Defusing the Global Warming Time Bomb*, James Hansen, *Scientific American*, March 2004, p73

9.9r *The Unquiet Ice*, Robin E. Bell, *Scientific American*, February 2008, p60

9.9s http://www.nytimes.com/2009/05/18/science/earth/18juneau.html?_r=1 – Accessed 10.26.15

9.9t It should be noted that some areas of the crust with deep roots, like the Himalayan Mountain range, are thought *not* to be floating. These areas probably reach deep into the underlying parts of the Earth in significant ways and may even be connected with the mantle of the Earth such that excess ice is unable to sink or push the continent in the same way it does with other thinner areas of the crust that are floating on the mantle. Therefore, these deep-rooted areas appear unable to rebound as ice is melted away.

9.9u *Warming of the Antarctic ice-sheet surface since the 1957 International Geophysical Year*, Eric J. Steig, et al., *Nature*, Vol. 457, 22 January 2009, p462

9.9v *Satellite Measurements of Sea Level Change: Where Have We Been and Where Are We Going, 15 Years of Progress in Radar Altimetry*, Venice, Italy, March 13-18, 2006, R. S. Nerem, D. P. Chambers, E. W. Leuliette, G. T. Mitchum, and A. Cazenave, http://earth.esa.int/workshops/venice06/participants/1092/paper_venice06.pdf – Accessed 10.26.15

9.9w http://www.theaustralian.news.com.au/story/0,25197,25348271-11949,00.html – Accessed 8.11.16

9.9x *The Flipping Point*, Michael Shermer, *Scientific American*, June 2006, p28

9.10   Weather and Geofield Prediction

9.10a http://en.wikipedia.org/wiki/Aurora_%28astronomy%29 – Accessed 9.3.09

9.10b http://www.igpp.ucla.edu/mpg/lectures/mkivelson/faculty97/lecture.html – Accessed 9.22.09, Site no longer available

9.10c http://en.wikipedia.org/wiki/Moon#Magnetic_field – Accessed 9.29.09

9.10d http://www.igpp.ucla.edu/mpg/lectures/mkivelson/faculty97/lecture.html – Accessed 9.22.09, Site no longer available

9.10e *Jupiter's Io Generates Power and Noise, But No Magnetic Field*, Guy Webster, 10 December 2001 Press Release, http://www.jpl.nasa.gov/releases/2001/release_2001_240.html – Accessed 6.8.16

9.10f *Atmosphere Above Japan Heated Rapidly Before M9 Earthquake*, 05.18.11, Technology Review, http://www.technologyreview.com/blog/arxiv/26773/ – Accessed 6.8.16

Bibliography

Each reference below has a bibliography number (Bib #), which are located on the left of the page and correspond to the adjacent reference. This bibliography is a partner to the Notes Section of this book that contains thousands of journal articles and references found in the UM. This bibliography is also listed alphabetically by author in a seperate bibliography list below the Bib # list.

Where possible, every bibliography reference and quote in the UM has also been given a page number to help in finding the quote. Using the references will not only increase our understanding of the Universal Model and confirm the context in which the quote was taken, but from original research we earn the right to ask even more questions, which leads to more answers when we build upon the new scientific discoveries in the UM.

Bib 1   **CRC Handbook of Chemistry and Physics**: *Robert C. Weast,* 57th Edition, CRC Press, Cleveland, Ohio, 1976
Bib 2   **Opticks**: *Sir Isaac Newton,* Preface by I. Bernard Cohen, Dover Publications, New York, N.Y., 1979
Bib 3   **Principia**: *Sir Isaac Newton,* Prometheus Books, Translated by Andrew Motte – Great Mind Series, 1995
Bib 4   **On the Revolutions of Heavenly Spheres**: *Nicolaus Copernicus,* Prometheus Books, 1995
Bib 5   **CRC Handbook of Chemistry and Physics**: *David R. Lide,* 77th Edition, CRC Press, Boca Raton, Florida, 1996
Bib 6   **Dialogues Concerning Two New Sciences**: *Galileo Galilei,* Dover Publications, 1954, National Edition, 1638
Bib 7   **Webster's Universal College Dictionary**: Random House, Inc., 1997
Bib 8   **Webster's Seventh New Collegiate Dictionary**: G. & C. Merriam Co., 1963
Bib 9   **Webster's New International Dictionary, Second Edition Unabridged**: G. & C. Merriam Company, Publishers, 1957
Bib 10 **Physics Fourth Edition**: *Cutnell & Johnson,* John Wiley & Sons, Inc., 1998
Bib 11 **From Copernicus to Einstein**: *Hans Reichenbach,* Dover Publication, 1980, Copyright in 1942, originally published in 1927
Bib 12 **McGraw-Hill Concise Encyclopedia of Science and Technology - Fourth Edition**: *Sybil P. Parker, Editor,* Lakeside Press, 1998
Bib 13 **Life Science Library - Water**: *Rene Dubos, Henry Margenau, C. P. Snow,* Time-Life Books, New York, 1970
Bib 14 **Life Science Library - The Scientist**: *Rene Dubos, Henry Margenau, C. P. Snow,* Time-Life Books, New York, 1964
Bib 15 **Rocks, Minerals & Gemstones**: *Walter Schumann,* HarperCollins Publishers and Houghton Mifflin Company, 1993
Bib 16 **Meteors - Sky & Telescope Observer's Guides**: *Neil Bone,* Sky Publishing Corporation, 1993
Bib 17 **New Guide to the Moon**: *Patrick Moore,* 1976
Bib 18 **The Truth of Science**: *Roger G. Newton,* Harvard University Press, 1997
Bib 19 **Isaac Newton - The Last Sorcerer**: *Michael White,* Perseus Books, 1997
Bib 20 **The End of Science - Facing the Limits of Knowledge in the Twilight of the Scientific Age**: *John Horgan,* Little, Brown & Co., 1996
Bib 21 **The End of Physics - The Myth of a Unified Theory**: *David Lindley,* Basic Books, 1993
Bib 22 **Shattering the Myths of Darwinism**: *Richard Milton,* Park Street Press, ????
Bib 23 **Einstein's Greatest Blunder?**: *Donald Goldsmith,* Harvard University Press, 1995
Bib 24 **The Ultimate Einstein**: *Donald Goldsmith and Robert Libbon,* Pocket Books, 1997
Bib 25 **Voyage to the Great Attractor**: *Alan Dressler,* Vintage Books, 1994
Bib 26 **How the Laser Happened - Adventures of a Scientist**: *Charles H. Townes,* Oxford University Press, 1999
Bib 27 **Early Astronomy**: *Hugh Thurston,* Springer – Verlag, New York, Inc., 1994
Bib 28 **Secrets of the Night Sky**: *Bob Berman,* William Morrow and Company, Inc., 1995
Bib 29 **Rocks From Space**: *O. Richard Norton,* Mountain Press Publishing Company, 1994
Bib 30 **The Common Sense of Science**: *J. Bronowski,* Vintage Books, 1951
Bib 31 **Predictions - 30 Great Minds on the Future**: *Edited by Sian Griffiths,* Oxford University Press, 1999
Bib 32 **Elements of Philosophy An Introduction**: *Samuel E. Stumpf,* McGraw-Hill, Inc., 1979
Bib 33 **The Structure of Scientific Revolutions**: *Thomas S. Kuhn,* The University of Chicago Press, 1996
Bib 34 **The Facts On File Dictionary of Biology**: *Robert Hine,* Third Edition, Market House Books, 1999
Bib 35 **100 Scientists Who Shaped World History**: *John Hudson Tiner,* Bluewood Books, 2000
Bib 36 **The Magnetic Field of the Earth**: *Merril, McElhinny, McFadden,* Academic Press, 1996
Bib 37 **Revolutions in Science - The Evolution of the Earth**: *Claude J. Allegre and Stephen H. Schneider, Scientific American,* 1999
Bib 38 **Waves and Grains**: *Mark P. Silverman,* Princeton University Press, 1998
Bib 39 **Constitution of the United States with Index and The Declaration of Independence**: Commission on the Bicentennial of the United States Constitution Washington, D.C. (Forward by Warren E. Burger, Chief Justice of the United States, 1969-1986)
Bib 40 **Popper Selections Edited by David Miller**: *Sir Karl Popper,* Princeton University Press, 1985, (excerpts date from 1934-1977)
Bib 41 **The Origin of Species**: *Charles Darwin,* Bantam Classic Book, 1999, (original text year 1859)
Bib 42 **Chaos Making a New Science**: *James Gleick,* Penguin Books, 1987
Bib 43 **Farce of Physics**: *Bryan G. Wallace,* WindSpiel Company, 1993, (http://surf.de.uu.net/bookland/sci/farce)
Bib 44 **Crystals and Crystal Growing**: *Alan Holden & Phylis Morrison,* MIT Press edition 1982, 1997 eleventh printing
Bib 45 **Heath Chemistry**: *Herron, Frank, Sarquis, Sarquis, Schrader, Kukla,* D. C. Health and Company, 1993
Bib 46 **Modern Physics**: *Holt, Rinehart, Winston,* Publishers, 1984
Bib 47 **Petrified Forest Trails**: *Jay Ellis Ransom,* Mineralogist Publishing Company, 1955
Bib 48 **Chemistry**: *Steven S. Zumdahl,* D. C. Heath and Company, 1986
Bib 49 **Psychology in Action** - Fourth Edition: *Karen Huffman, Mark Vernoy, Judith Vernoy,* John Wiley & Sons, Inc., 1997
Bib 50 **Chemistry**: *James P. Birk,* Houghton Mifflin Company, 1994
Bib 51 **Earth Science**: *F. Martin Brown, Wayne Bailey,* Silver Burdett Company, 1978

Bib 52	**Investigating the Earth** - Fourth Edition: *Matthews, Roy, Stevenson, Harris Hesser, Dexter*, Houghton Mifflin Company, 1984	
Bib 53	**General Science**: *Watkins, Emiliani, Chiaverina, Harper, LaHart*, Harcourt Brace Jovanovich, Inc., 1989	
Bib 54	**Physical Geology**: *James S. Monroe, Reed Wicander*, West Publishing Company, 1992	
Bib 55	**Chemistry**: *James P. Birk*, Houghton Mifflin Company, 1994	
Bib 56	**The General Pattern of the Scientific Method SM-14**: *Norman W. Edmund*, 1994 (#52038 Edmund Scientific, No longer selling)	
Bib 57	**Contemporary College Physics**: *Edwin R. Jones, Richard L. Childers*, Addison-Wesley Publishing Company, Inc., 1993	
Bib 58	**Physics** -Third Edition: *John D Cutnell, Kenneth W. Johnson*, John Wiley & Sons, Inc., 1995	
Bib 59	**Understanding Earth** - Second Edition: *Frank Press, Raymond Siever*, W. H. Freeman and Company, 1998	
Bib 60	**Physics A General Introduction** - Second Edition: *Alan Van Heuvelen*, Little, Brown and Company, 1986	
Bib 61	**The Use and Care of a Balance**: *Peter J. Krayer*, The Chemical Publishing Company, Easton, PA., 1913	
Bib 62	**A Course of Instruction in QUANTITATIVE CHEMICAL ANALYSIS for Beginning Students**: *George McPhail Smith*, 1921 The MacMillan Co., NY, p 7-21. (http://humboldt.edu/scimus/LitIndex.html) Richard A. Paselk Scientific Instrument Museum	
Bib 63	**The Heart of the Earth**: *O. M. Phillips*, Freeman, Cooper & Company, 1968	
Bib 64	**E=mc²** - a biography of the world's most famous equation: *David Bodanis*, 2000	
Bib 65	**Seeing Red, Redshifts, Cosmology and Academic Science**: *Halton Arp*, Apeiron Publishing, 1998 (redshift.vif.com)	
Bib 66	**The Handy Weather Answer Book**: *Walter A. Lyons*, Accord Publishing Ltd., 1997	
Bib 67	**Ancient North America**: *Brian M. Fagan*, Thames and Hudson Ltd, London 1991	
Bib 68	**The Amber Book**: *Ake Dahlstrom, Leif Brost*, Geoscience Press, Inc., 1996	
Bib 69	**Science**: American Association for the Advancement of Science, Vol.284, No. 5423, Pages 2045-2220, 25 June 1999	
Bib 70	**The Ice Finders**: *Edmund Blair Bolles*, Counterpoint, 1999	
Bib 71	**Marvels and Mysteries of The World Around Us**: General Consultant: *Rhodes W. Fairbridge*, Professor Of Geology, Columbia University, The Reader's Digest Association, Inc., 1972	
Bib 72	**The Descent of Man**: *Charles Darwin*, Intro by *H. James Birx*, Prometheus Books, 1998, originally published in 1871	
Bib 73	**GX Series (GX-6100) Multi-Function Balance Instruction Manual**: A & D Company, Limited Ltd., 2000, p7-8	
Bib 74	**Life on the Mississippi**: *Mark Twain*, originally published by James R. Osgood & Co., 1883, (Http://docsouth.unc.edu/twainlife/twain.html)	
Bib 75	**The Quest for Life in Amber**: *George Poinar, Roberta Poinar*, 1994	
Bib 76	**The Golem** - What you Should Know about Science - 2nd Edition: *Harry Collins & Trevor Pinch*, Cambridge University Press, 1998	
Bib 77	*TIME*: (The Weekly Newsmagazine) Time Inc., December 31, 1999	
Bib 78	**Amber - Window to the Past**: *David A. Grimaldi*, Harry N. Abrams, Inc., with the American Museum of Natural History, 1996	
Bib 79	**Amber, Resinite, and Fossil Resins**: *Ken B. Anderson, John C. Crelling* (Editors of ACS Symposium Series 617.) American Chemical Society, 1995	
Bib 80	**Amber, The Golden Gem Of The Ages**: *Patty C. Rice*, Litton Educational Publishing, Inc., 1980	
Bib 81	**Life in Amber**: *George O. Poinar, Jr.*, Stanford University Press, 1992	
Bib 82	**Amber World - The secrets of Dominican Amber**: *George Caridad*, publisher,1998	
Bib 83	**The Facts on File Dictionary of Chemistry** - Third Edition: *John Daintith*, Checkmark Books, 1999	
Bib 84	**The Facts on File Dictionary of Physics** - Third Edition: *John Daintith & John O. E. Clark*, Checkmark Books, 1999	
Bib 85	**Darwin's Influence on Modern Thought**: *Ernst Mayr*, Scientific American, July 2000 p80-83	
Bib 86	**The Making of the Modern Mind**: *John Herman Randall, Jr.*, Columbia University Press, 1940	
Bib 87	**Magmas and Magmatic Rocks**: *Eric A J Middlemost*, Longman Group Limited, 1985	
Bib 88	**Darwin's Black Box** - The Biochemical Challenge to Evolution: *Michael J. Behe*, Touchstone by Simon & Schuster, 1996	
Bib 89	**The Autobiography of Charles Darwin**: Original omissions restored by grand-daughter: *Nora Barlow*, W. W. Norton & Company, 1969	
Bib 90	**Einstein The Life and Times**: *Ronald W. Clark*, World Publishing Company, 1971	
Bib 91	**Science, Money, and Politics, Political Triumph and Ethical Erosion**: *Daniel S. Greenberg*, The University of Chicago Press, 2001	
Bib 92	**Open Questions in Relativistic Physics**: Edited by *Franco Selleri*, Apeiron, 1998	
Bib 93	**Darwin on Trial**: *Phillip E. Johnson*, Inter Varsity Press, 1993	
Bib 94	**More Letters of Charles Darwin, Vol. I**: *Charles Darwin*, D. Appleton and Company, 1903	
Bib 95	**Scientists and Religion in America**: *Edward J. Larson, Larry Witham*, Scientific American, September 1999, p89-93 (Also see companion article in *Nature*, Vol. 394, July 23, 1998)	
Bib 96	**Ideas and Opinions**: *Albert Einstein*, Bonanza Books, New York, 1954	
Bib 97	**In Search of Deep Time**: *Henry Gee*, The Free Press, 1999	
Bib 98	**Webster's Unified Dictionary and Encyclopedia**: *Adams, Teall, Taylor, Bokkelen*, H. S. Stuttman Company, 1960	
Bib 99	**The Los Alamos Primer**: *Robert Serber*, University of California Press, 1992	
Bib 100	**The Evolutionary Tale of Moths and Men, The untold story of science and the peppered moth**: *Judith Hooper*, W. W. Norton & Company, 2002	
Bib 101	**The Blind Watchmaker, Why the evidence of evolution reveals a universe without design**: *Richard Dawkins*, W. W. Norton & Company, 1987	
Bib 102	**Fossils**: *Cyril Walker and David Ward*, A Dorling Kindersley book, 1992	
Bib 103	**Continental Drift**: *Don and Maureen Tarling*, Doubleday & Company, Inc., 1971	
Bib 104	**Gems Made by Man**: *Kurt Nassau*, Gemological Institute of America, 1980	
Bib 105	**Einstein A Life in Science**: *Michael White and John Gribbin*, Dutton, Published by the Penguin Group, 1993	
Bib 106	**Was Einstein Right?**: *Clifford M. Will*, BasicBooks, A member of the Perseus Group, 1993	
Bib 107	**Icons of Evolution, Science or Myth**?: *Jonathan Wells*, Regnery Publishing, Inc., 2000	
Bib 108	**A Dictionary of Astronomy**: *Ian Ridpath*, Oxford University Press, 1997	
Bib 109	**God in the Equation**: *Corey S. Powell*, Free Press, 2002	
Bib 110	**Darwin, the Life of a Tormented Evolutionist**: *Adrian Desmond & James Moore*, W. w. Norton & Company, 1991	
Bib 111	**Moon Atlas**: *V. A. Firsoff*, N.V. Cartografisch, 1961	
Bib 112	**The Miracle Planet**: *Bruce Brown and Lane Morgan*, Gallery Books, 1990	
Bib 113	**Earth, The Making, Shaping and Working of a Planet**: *Derek Elsom*, Macmillan Publishing Company, 1992	
Bib 114	**Planetary Geology**: *Nicholas M. Short*, Prentice-Hall, Inc., 1975	
Bib 115	**Comets, Creators and Destroyers**: *David H. Levy*, Touchstone, 1998	
Bib 116	**The Ocean's Invisible Forest**: *Paul G. Falkowski*, Scientific American, August 2002	
Bib 117	**Alternative Science, Challenging the Myths of the Scientific Establishment**: *Richard Milton*, Park Street Press, 1994 &1996	
Bib 118	**Soils of the Past, An Introduction to Paleopedology**: *G.J. Retallack*, Boston Unwin Hyman, 1990	
Bib 119	**Meteorite Craters**: *Kathleen Mark*, The University of Arizona Press, Tucson, 1987	
Bib 120	**The Mid-Oceanic Ridges, Mountains Below Sea Level**: *Adolphe Nicolas*, Springer-Verlag Berlin Heidelberg, 1995	
Bib 121	**Fossils, the Key to the Past**: *Richard Fortey*, Harvard University Press Cambridge, Massachusetts, 1991	

Bib 122	**Ore Deposits of the Western States, Rocky Mountain Fund Series**: *Edited by the Committee on the Lindgren Volume*, Published by the American Institute of Mining and Metallurgical Engineers, 1933	
Bib 123	**The Cyclopedia or Universal Dictionary of Arts, Sciences, and Literature**: *Abraham Rees*, Vol XI, 1st American Edition, 1820	
Bib 124	**Mathematics for the Nonmathematician**: *Morris Kline*, Dover Publications, 1967	
Bib 125	**Textbook of Geology Part 1**: *Longwell, Knopf Flint*, John Wiley and Sons, Second Edition, 1939	
Bib 126	**Arizona Meteor Crater, Past—Present--Future**: *H. H. Nininger*, American Meteorite Museum—Sedona, Arizona, World Press, 1956	
Bib 127	**The Restless Earth**: *Nigel Calder*, The Viking Press, 1972	
Bib 128	**Geology Underfoot**: *Robert P. Sharp & Allen F. Glazner*, Mountain Press Publishing Company, 1997	
Bib 129	**The Once and Future Moon**: *Paul D. Spudis*, Smithsonian Institution Press, 1996	
Bib 130	**The Handy Science Answer Book**: Compiled by the Science and Technology Department of the Carnegie Library of Pittsburgh, 1997	
Bib 131	**History Reborn**: *Vicky Jo Anderson*, Zichron Historical Research Institute, 1994	
Bib 132	**Gem Stones**: *Cally Hall*, Dorling Kindersley Limited, London, 1994	
Bib 133	**The Age of the Earth**: *G. Brent Dalrymple*, Stanford University Press, 1991	
Bib 134	**The Geysers of Yellowstone**, Third Edition, *T. Scott Bryan*, University Press of Colorado, 1995	
Bib 135	**Can You Tell Me Anything about Evolution?**: Transcript of *Colin Patterson's* November 5, 1981 presentation at the American Museum of Natural History, New York City, p6. Both the transcript and tape recording can be obtained at www.arn.org	
Bib 136	**Melting the Earth, The History of Ideas on Volcanic Eruptions**: *Haraldur Sigurdsson*, Oxford University Press, 1999	
Bib 137	**Neglected Geological Anomalies**, A Catalog of Geological Anomalies: *William R. Corliss*, The Sourcebook Project, 1990	
Bib 138	**Radiometric Dating, Geologic Time, And The Age Of The Earth: A Reply To "Scientific Creationism"**: *G. Brent Dalrymple*, USGS, Open-File Report 86-110, 2.17.82 (http://pubs.er.usgs.gov/usgspubs/ofr/ofr86110 – Accessed 2.24.10)	
Bib 139	**Personality Psychology, Domains of Knowledge About Human Nature**: *Randy J. Larsen & David M. Buss*, McGraw-Hill, 2005	
Bib 140	**Plate Tectonics**: *Arthur N. Strahler*, Geo-Books, 1998	
Bib 141	**The Origin of Mountains**: *Cliff Ollier and Colin Pain*, Routledge – imprint of the Taylor and Francis Group, 2000	
Bib 142	**Lessons in Physical Geography**: *Charles R. Dryer*, American Book Company, 1901	
Bib 143	**Handbook of World Salt Resources**: *Stanley J. Lefond,* Plenum Press, 1969	
Bib 144	**Salt Domes Gulf Region, United States & Mexico**: *Michel T. Halbouty*, Gulf Publishing Company, 1979	
Bib 145	**Science Desk Reference, New York Public Library**: *Patricia Barnes-Svarney*, Editorial Director, A Stonesong Press Book, 1995	
Bib 146	**The Farce of Physics**: *Bryan G. Wallace*, http://surf.de.uu.net/bookland/sci/farce/farce_toc.html	
Bib 147	**Experimental syntheses of framboids—a review**: *Hiroaki Ohfuji, David Rickard*, Earth-Science Reviews 71, (2005) p147-170	
Bib 148	**Earth From the Inside Out**: Various authors from March 1996 to December 1999, Scientific American, 2000	
Bib 149	**Mission to the Planets**: *Patrick Moore*, Cassell Publishers Limited, 1995	
Bib 150	**Dynamic Astronomy:** *Robert T. Dixon*, Prentice-Hall Inc.,1971	
Bib 151	**The Complete Guide To Rocks & Minerals**: *John Farndon*, Anness Publishing, 2006	
Bib 152	**Biological Transmutations**: *C. Louis Kervran,* Happiness Press, 1998	
Bib 153	**Charles Darwin, A Scientific Biography**: *Sir Gavin de Beer,* The Natural History Library & Anchor Books, 1963	
Bib 154	**The Man Who Found Time, James Hutton and the Discovery of Earth's Antiquity**: *Jack Repcheck*, Perseus Publishing, 2003	
Bib 155	**Fluid Inclusions**: *Edwin Roedder*, Reviews in Mineralogy, Volume12, Mineralogical Society of America, 1984	
Bib 156	**Handbook of Hydrothermal Technology, A Technology of Crystal Growth and Materials Processing**: *K. Byrappa, Masahiro Yoshimura*, Noyes Publications, 2001	
Bib 157	**Catastrophic Flooding, The Origin of the Channeled Scabland**: *Victor R. Baker*, Benchmark Papers in Geology/55, 1981	
Bib 158	**Gem Stones**, The visual guide to more than 130 gemstone varieties: *Cally Hall*, A DK Publishing Book, 1994	
Bib 159	**Sedimentary Rocks**: *F. J. Pettijohn,* Harper & Brothers, 1957	
Bib 160	**Impact and Explosion Cratering**, Planetary and Terrestrial Implications, Proceedings of the Symposium on Planetary Cratering Mechanics: Flagstaff, AZ, September 13-17 1976, Edited by *D. J. Roddy, R. O. Pepin, R.B. Merrill*, Lunar Science Institute, Pergamon Press, 1977	
Bib 161	**Impact Cratering, A Geological Process**: *H. J. Melosh*, Lunar and Planetary Laboratory, UofA, Oxford University Press, 1989	
Bib 162	**Planetary Science: A Lunar Perspective**: *Stuart Ross Taylor*, Lunar and Planetary Institute Houston, Texas, USA, 1982	
Bib 163	**Patrick Moore on the Moon**: *Patrick Moore*, Cassell & Co, 2001	
Bib 164	**Bones, Rocks and Stars, The Science of When Things Happened**: *Chris Turney*, Macmillan, 2006	
Bib 165	**Coon Mountain Controversies**: *William Graves Hoyt*, The University of Arizona Press, 1987	
Bib 166	**Meteorites and Their Parent Planets**: *Harry Y. McSween*, Jr., Cambridge University Press, 1999	
Bib 167	**Giant Meteorites**: *E. L. Krinov*, Pregamon press, 1966	
Bib 168	**Australia's Meteorite Craters**: *Alex Bevan and Ken McNamara*, Western Australian Museum, December 1993	
Bib 169	**The Odessa Meteor Craters and Their Geological Implications**: *Glen L. Evans, Charles E. Mear*, Occasional Papers of the Strecker Museum No. 5, Baylor University, 2000	
Bib 170	**Language in Thought and Action**: Fifth Edition, *S.I. Hayakawa, Alan R. Hayakawa*, Harcourt Inc., 1990	
Bib 171	**Science, The Moon Issue**: AAAS, Vol. 167, No. 3918, 30 January 1970	
Bib 172	**Essentials of Geology**: Eighth Edition, *Frederick K. Lutgens, Edwards J. Tarbuck*, Prentice Hall, 2003	
Bib 173	**Glossary of Geology**: Fifth Edition, *Klaus K. E. Neuendorf, James P. Mehl, Jr., Julia A. Jackson*, American Geological Institute, 2005	
Bib 174	**The Devil's Delusion**, Atheism and its Scientific Pretensions: *David Berlinski*, Crown Forum, 2008	
Bib 175	**A Thorough and Accurate History of Genuine Diamonds in Arkansas**: *Glenn W. Worthington*, Mid-America Prospecting, 2003	
Bib 176	**The Complete Works of Josephus**: *Flavius Josephus*, Kregel Publications, © 1981, printed 2000	
Bib 177	**Fabulous Science, Fact and Fiction in the History of Scientific Discovery**: *John Waller*, Oxford University Press, 2002	
Bib 178	**The Trouble with Physics, The Rise of String Theory, the Fall of Science, and What Comes Next**: *Lee Smolin*, Houghton Mifflin Co., 2006	
Bib 179	**Electromagnetic Phenomena in the Earth's Crust**: *K. N. Abdullabekov*, Translated from Russian/edited by R. B. Zeidler, Geotechnika 1, A. A. Balkema Publishers, 1991	
Bib 180	**The Atmosphere An Introduction to Meteorology**: *Frederick K. Lutgens, Edward J. Tarbuck*, Pearson Prentice Hall, Tenth Edition, 2007	
Bib 181	**The Origin of Continents and Oceans**: *Alfred Wegener*, Translated from the 4th German Ed by John Biram, Dover Publications, 1966	
Bib 182	**Weather, Spectacular Images of the World's Extraordinary Climate**: *Storm Dunlop*, Thunder Bay Press, 2006	
Bib 183	**Weather, A Visual Guide**: *Bruce Buckley, Edward J. Hopkins, Richard Whitaker*, Firefly Books Ltd., 2004	
Bib 184	**The Elements**: *John Emsley*, Third Edition, Oxford University Press, 1998	
Bib 185	**Against Method**: *Paul Feyerabend*, Third Edition, Verso, 1997	
Bib 186	**A Slice Through Time**, Dendrochronology and precision dating: *M. G. L. Baillie*, B.T. Batsford Ltd, London, 1995	
Bib 187	**Exodus to Arthur**, Catastrophic Encounters with Comets: *Mike Baillie*, B.T. Batsford Ltd, London, 2000	
Bib 188	**Currents of Change, Impacts of El Niño and La Niña on Climate and Society**: *Michael Glantz*, Cambridge University Press, 2001	

Bib 189 **The Big Splash**: *Louis A. Frank with Patrick Huyghe*, Birch Lane Press Book, 1990
Bib 190 **Interview With Clair C. Patterson**: *Shirley K. Cohen*, California Institute of Technology, Oral History Project, March 1995
(http://oralhistories.library.caltech.edu/32/00/OH_Patterson.pdf)
Bib 191 **Numbers, The Universal Language**: *Denis Guedj*, Discoveries, Harry N. Abrams, Inc., 1997
Bib 192 **Fossil Evidence**, The human evolution journey: *Frank E. Poirier*, Second Edition, C. V. Mosby Company, 1977
Bib 193 **Relativity, The Special and the General Theory**: *Albert Einstein*, Three Rivers Press, 1961
Bib 194 **Petrified Wood, The World of Fossilized Wood, Cones, Ferns, and Cycads**: *Frank J. Daniels*, Western Colorado Publishing Co., 1998
Bib 195 **After the Flood**, The early post-flood history of Europe traced back to Noah: *Bill Cooper*, New Wine Press, 1995
http://www.ldolphin.org/cooper/contents.html
Bib 196 **Ancient History**: *Philip Van Ness Myers*, Ginn and Company, 1904, p2
Bib 197 **Archaeology**, Original Readings in Method and Practice: Edited by *Peter N. Peregrine, Carol R. Ember, Melvin Ember*, Chapter 11, When and How Did Humans Populate the New World, *William J. Parry*, Prentice Hall, 2002
Bib 198 **Clovis Revisited, New Perspectives on Paleoindian Adaptations from Blackwater Draw, New Mexico**: *Anthony T. Boldurian, John L. Cotter*, The University Museum, University of Pennsylvania, Philadelphia, 1999
Bib 199 **American Genesis**: *Jeffrey Goodman*, Summit Books, 1981
Bib 200 **Clovis Settlement Patterns, Nottoway River Survey Part 1**: *Joseph M. McAvoy*, Dietz Press, 1992
Bib 201 **The Prehistoric Men of Kentucky**: *Colonel Bennett H. Young*, J. P. Morton & Co., Printers to the Filson Club, 1910
http://books.google.com/books/about/The_prehistoric_men_of_Kentucky.html?id=z6cAk5SiLt4C
Bib 202 **1491, New Revelations of the Americas Before Columbus**: *Charles C. Mann*, Vintage Books, 2006
Bib 203 **Iron Age America**: *William D. Conner*, Coachwhip Publications, 2009
Bib 204 **First Hunters, Ohio's Paleo-Indian Artifacts**: *Lar Hothem*, Hothem House Books, 1990
Bib 205 **The Great Journey, the Peopling of Ancient America**: *Brian M. Fagan*, Thames and Hudson, 1987
Bib 206 **Concepts of Mass in Classical and Modern Physics**: *Max Jammer*, Dover Publications, Inc., 1997
Bib 207 **Chemistry, The Central Science**: *Theodore L. Brown, H. Eugene Lemay, Jr., Bruce E. Bursten*, Prentice Hall, 2000
Bib 208 **The Sciences, An Integrated Approach, Fifth Edition**: *James Trefil, Robert M. Hazen*, John Wiley & Sons, Inc., 2007
Bib 209 **The Field, The Quest for the Secret Force of the Universe**: *Lynne McTaggart*, Quill, 2003
Bib 210 **The Vaccine Book, Making the Right Decisions for Your Child**: *Robert W. Sears*, Little, Brown and Company, 2011
Bib 211 **UNCERTAINTY, Einstein, Heisenberg, Bohr, and the Struggle for the Soul of Science**: *David Lindley*, Doubleday, 2007
Bib 212 **World in the Balance, The Historic Quest for an Absolute System of Measurement**: *Robert P. Crease*, W. W. Norton & Company, 2011
Bib 213 **The Light Beyond, New Explorations by the Author of Life After Life**: *Raymond A. Moody, Jr.*, Bantam Books, 1989
Bib 214 **Proof of Heaven**: *Eben ALexander*, Simon & Schuster, 2012
Bib 215 **Cavendish, The Experimental Life**: *Christa Jungnickel, Russell McCormmach*, Bucknell, 1999
Bib 216 **The Grand Design, New Answers to the Ultimate Questions of Life**: *Stephen Hawking, Leonard Mlodinow*, Bantam Books, 2010
Bib 217 **The Measure of all things, The Seven-Year Odyssey and Hidden Error That Transformed the World**: *Ken Alder*, The Free Press, 2002
Bib 218 **From Eternity to Here, The Quest for the Ultimate Theory of Time, Sean Carroll, Dutton**: The Penguin Group, 2010
Bib 219 **PHYSICS AND PHILOSOPHY, the Revolution in Modern Science**: *Werner Heisenberg*, Harper Collins Publishers, 1958
Bib 220 **Crystals and Light, An Introduction to Optical Crystallography**: *Elizabeth A. Wood*, Dover Publications, 1977
Bib 221 **13 Things That Don't Make Sense, The Most Baffling Scientific Mysteries of Our Time**: *Michael Brooks*, Vintage Books, 2008

Bibliography Listed Alphabetically By Author

Abdullabekov, K. N., **Electromagnetic Phenomena in the Earth's Crust**: Translated from Russian/edited by R. B. Zeidler, Geotechnika 1, A. A. Balkema Publishers, 1991, Bib 179
Adams, Teall, Taylor, Bokkelen, **Webster's Unified Dictionary and Encyclopedia**: H. S. Stuttman Company, 1960, Bib 98
Alder, Ken, **The Measure of all things, The Seven-Year Odyssey and Hidden Error That Transformed the World**: The Free Press, 2002, Bib 217
Alexander, Eben, **Proof of Heaven**: Simon & Schuster, 2012, Bib 214
Allegre, Claude J. & Schneider, Stephen H., **Revolutions in Science - The Evolution of the Earth**: Scientific American, 1999, Bib 37
American Association for the Advancement of Science, **Science, The Moon Issue**: Vol. 167, No. 3918, 30 January 1970, Bib 171
American Association for the Advancement of Science, **Science**: Vol. 284, No. 5423, Pages 2045-2220, 25 June 1999, Bib 69
American Institute of Mining and Metallurgical Engineers, 1933, Bib 122
Anderson, Ken B., **Amber, Resinite, and Fossil Resins**: John C. Crelling (Editors of ACS Symposium Series 617.) American Chemical Society, 1995, Bib 79
Anderson, Vicky Jo, **History Reborn**: Zichron Historical Research Institute, 1994, Bib 131
Arp, Halton, **Seeing Red, Redshifts, Cosmology and Academic Science**: Apeiron Publishing, 1998 (redshift.vif.com), Bib 65
Baillie Mike, B.T. & Batsford Ltd, **Exodus to Arthur**: Catastrophic Encounters with Comets, London, 2000, Bib 187
Baillie, M. G. L. & Batsford, B.T. Ltd, **A Slice Through Time**: Dendrochronology and precision dating, London, 1995, Bib 186
Baker, Victor R., **Catastrophic Flooding, The Origin of the Channeled Scabland**: Benchmark Papers in Geology/55, 1981, Bib 157
Barlow, Nora, **The Autobiography of Charles Darwin**: Original omissions restored by grand-daughter: W. W. Norton & Company, 1969, Bib 89
Barnes-Svarney, Patricia, **Science Desk Reference, New York Public Library**: Editorial Director, A Stonesong Press Book, 1995, Bib 145
Behe, Michael J., **Darwin's Black Box**: The Biochemical Challenge to Evolution: Touchstone by Simon & Schuster, 1996, Bib 88
Berlinski, David, **The Devil's Delusion**: Atheism and its Scientific Pretensions, Crown Forum, 2008, Bib 174
Berman, Bob, **Secrets of the Night Sky**: William Morrow and Company, Inc., 1995, Bib 28
Bevan, Alex & McNamara, Ken, **Australia's Meteorite Craters**: Western Australian Museum, December 1993, Bib 168
Birk, James P., **Chemistry**: Houghton Mifflin Company, 1994, Bib 50
Birk, James P., **Chemistry**: Houghton Mifflin Company, 1994, Bib 55
Bodanis, David, **E=mc²**: a biography of the world's most famous equation: 2000, Bib 64
Boldurian, Anthony T. & Cotter, John L., **Clovis Revisited, New Perspectives on Paleoindian Adaptations from Blackwater Draw, New Mexico**: The University Museum, University of Pennsylvania, Philadelphia, 1999, Bib 198
Bolles, Edmund Blair, **The Ice Finders**: Counterpoint, 1999, Bib 70
Bone, Neil, **Meteors - Sky & Telescope Observer's Guides**: Sky Publishing Corporation, 1993, Bib 16
Bronowski, J., **The Common Sense of Science**: Vintage Books, 1951, Bib 30
Brooks, Michael, **13 Things That Don't Make Sense, The Most Baffling Scientific Mysteries of Our Time**: Vintage Books, 2008, Bib 221
Brown, Bruce & Morgan, Lane, **The Miracle Planet**: Gallery Books, 1990, Bib 112
Brown, F. Martin & Bailey, Wayne, **Earth Science**: Silver Burdett Company, 1978, Bib 51
Brown, Theodore L., Lemay, H. Eugene, Jr.,& Bursten, Bruce E., **Chemistry, The Central Science**: Prentice Hall, 2000, Bib 207
Bryan, T. Scott, **The Geysers of Yellowstone**: Third Edition, University Press of Colorado, 1995, Bib 134
Buckley, Bruce, Hopkin,s Edward J., & Whitaker, Richard, **Weather, A Visual Guide**: Firefly Books Ltd., 2004, Bib 183
Byrappa, K. & Yoshimura, Masahiro, **Handbook of Hydrothermal Technology**: A Technology of Crystal Growth and Materials Processing, Noyes Publications, 2001, Bib 156
Calder, Nigel, **The Restless Earth**: The Viking Press, 1972, Bib 127
Caridad, George, **Amber World - The secrets of Dominican Amber**: 1998, Bib 82
Clark, Ronald W., **Einstein The Life and Times**: World Publishing Company, 1971, Bib 90
Cohen, Shirley K., **Interview With Clair C. Patterson**: California Institute of Technology, Oral History Project, March 1995, http://oralhistories.library.caltech.edu/32/00/OH_Patterson.pdf, Bib 190
Collins, Harry & Pinch, Trevor, **The Golem**: What you Should Know about Science - 2nd Edition: Cambridge University Press, 1998, Bib 76
Committee on the Lindgren Volume, Editor, **Ore Deposits of the Western States, Rocky Mountain Fund Series**: Published by the
Conner, William D., **Iron Age America**: Coachwhip Publications, 2009, Bib 203
Constitution of the United States with Index and The Declaration of Independence: Commission on the Bicentennial of the United States Constitution Washington, D.C. (Forward by Warren E. Burger, Chief Justice of the United States, 1969-1986), Bib 39
Cooper, Bill, **After the Flood, The early post-flood history of Europe traced back to Noah**: New Wine Press, 1995, http://www.ldolphin.org/cooper/contents.html, Bib 195
Copernicus, Nicolaus, **On the Revolutions of Heavenly Spheres**: Prometheus Books, 1995, Bib 4
Corliss, William R., **Neglected Geological Anomalies**: A Catalog of Geological Anomalies, The Sourcebook Project, 1990, Bib 137
Crease, Robert P., **World in the Balance, The Historic Quest for an Absolute System of Measurement**: W. W. Norton & Company, 2011, Bib 212
Cutnell & Johnson, **Physics Fourth Edition**: John Wiley & Sons, Inc., 1998, Bib 10
Cutnell, John D. & Johnson, Kenneth W., **Physics**: Third Edition: John Wiley & Sons, Inc., 1995, Bib 58
Dahlstrom, Ake & Brost, Leif, **The Amber Book**: Geoscience Press, Inc., 1996, Bib 68
Daintith, John & Clark, John O. E., **The Facts on File Dictionary of Physics**: Third Edition: Checkmark Books, 1999, Bib 84
Daintith, John, **The Facts on File Dictionary of Chemistry**: Third Edition: Checkmark Books, 1999, Bib 83
Dalrymple, G. Brent, **Radiometric Dating, Geologic Time, And The Age Of The Earth: A Reply To "Scientific Creationism"**: USGS, Open-File Report 86-110, 2.17.82 (http://pubs.er.usgs.gov/usgspubs/ofr/ofr86110 – Accessed 2.24.710), Bib 138
Dalrymple, G. Brent, **The Age of the Earth**: Stanford University Press, 1991, Bib 133
Daniels, Frank J., **Petrified Wood, The World of Fossilized Wood, Cones, Ferns, and Cycads**: Western Colorado Publishing Co., 1998, Bib 194
Darwin, Charles, Intro by *Birx, H. James*, **The Descent of Man**: Prometheus Books, 1998, originally published in 1871, Bib 72
Darwin, Charles, **More Letters of Charles Darwin, Vol. I**: D. Appleton and Company, 1903, Bib 94
Darwin, Charles, **The Origin of Species**: Bantam Classic Book, 1999, (original text year 1859), Bib 41
Dawkins, Richard, **The Blind Watchmaker, Why the evidence of evolution reveals a universe without design**: W. W. Norton & Company, 1987, Bib 101
de Beer, Sir Gavin, **Charles Darwin, A Scientific Biography**: The Natural History Library & Anchor Books, 1963, Bib 153
Desmond, Adrian & Moore, James, **Darwin, the Life of a Tormented Evolutionist**: W. w. Norton & Company, 1991, Bib 110
Dixon, Robert T., **Dynamic Astronomy**: Prentice-Hall Inc.,1971, Bib 150
Dressler, Alan, **Voyage to the Great Attractor**: Vintage Books, 1994, Bib 25
Dryer, Charles R., **Lessons in Physical Geography**: American Book Company, 1901, Bib 142
Dubos, Rene, Margenau, Henry, Snow, C. P., **Life Science Library - Water**: Time-Life Books, New York, 1970, Bib 13

Dubos, Rene, Margenau, Henry, Snow, C. P., **Life Science Library - The Scientist**: Time-Life Books, New York, 1964, Bib 14
Dunlop, Storm, **Weather, Spectacular Images of the World's Extraordinary Climate**: Thunder Bay Press, 2006, Bib 182
Edmund, Norman W., **The General Pattern of the Scientific Method SM-14**: 1994 (#52038 Edmund Scientific, No longer selling), Bib 56
Einstein, Albert, **Ideas and Opinions**: Bonanza Books, New York, 1954, Bib 96
Einstein, Albert, **Relativity, The Special and the General Theory**: Three Rivers Press, 1961, Bib 193
Elsom Derek, **Earth, The Making, Shaping and Working of a Planet**: Macmillan Publishing Company, 1992, Bib 113
Emsley, John, **The Elements**: Third Edition, Oxford University Press, 1998, Bib 184
Evans, Glen L. & Mear, Charles E., **The Odessa Meteor Craters and Their Geological Implications**: Occasional Papers of the Strecker Museum No.5, Baylor University, 2000, Bib 169
Fagan, Brian M., **Ancient North America**: Thames and Hudson Ltd, London 1991, Bib 67
Fagan, Brian M., **The Great Journey, the Peopling of Ancient America**: Thames and Hudson, 1987, Bib 205
Fairbridge, Rhodes W., **Marvels and Mysteries of The World Around Us**: General Consultant: Professor Of Geology, Columbia University, The Reader's Digest Association, Inc., 1972, Bib 71
Falkowski, Paul G., **The Ocean's Invisible Forest**: Scientific American, August 2002, Bib 116
Farndon, John, **The Complete Guide To Rocks & Minerals**: Anness Publishing, 2006, Bib 151
Feyerabend, Paul, **Against Method**: Third Edition, Verso, 1997, Bib 185
Firsoff, V. A., **Moon Atlas**: N.V. Cartografisch, 1961, Bib 111
Fortey, Richard, **Fossils, the Key to the Past**: Harvard University Press Cambridge, Massachusetts, 1991, Bib 121
Frank, Louis A. & Huyghe, Patrick, **The Big Splash**: Birch Lane Press Book, 1990, Bib 189
Galilei, Galileo, **Dialogues Concerning Two New Sciences**: Dover Publications, 1954, National Edition, 1638, Bib 6
Gee, Henry, **In Search of Deep Time**: The Free Press, 1999, Bib 97
Glantz, Michael, **Currents of Change, Impacts of El Niño and La Niña on Climate and Society**: Cambridge University Press, 2001, Bib 188
Gleick, James, **Chaos Making a New Science**: Penguin Books, 1987, Bib 42
Goldsmith, Donald & Libbon, Robert, **The Ultimate Einstein**: Pocket Books, 1997, Bib 24
Goldsmith, Donald, **Einstein's Greatest Blunder?**: Harvard University Press, 1995, Bib 23
Goodman, Jeffrey, **American Genesis**: Summit Books, 1981, Bib 199
Greenberg, Daniel S., **Science, Money, and Politics, Political Triumph and Ethical Erosion**: The University of Chicago Press, 2001, Bib 91
Griffiths, Sian, Editor, **Predictions - 30 Great Minds on the Future**: Oxford University Press, 1999, Bib 31
Grimaldi, David A., **Amber - Window to the Past**: Harry N. Abrams, Inc., with the American Museum of Natural History, 1996, Bib 78
Guedj, Denis, **Numbers, The Universal Language**: Discoveries, Harry N. Abrams, Inc., 1997, Bib 191
GX Series (GX-6100) Multi-Function Balance Instruction Manual: A & D Company, Limited Ltd., 2000, p7-8, Bib 73
Halbouty, Michel T., **Salt Domes Gulf Region, United States & Mexico**: Gulf Publishing Company, 1979, Bib 144
Hall, Cally, **Gem Stones**: The visual guide to more than 130 gemstone varieties, A DK Publishing Book, 1994, Bib 158
Hall, Cally, **Gem Stones**: Dorling Kindersley Limited, London, 1994, Bib 132
Hawking, Stephen & Mlodinow, Leonard, **The Grand Design, New Answers to the Ultimate Questions of Life**: Bantam Books, 2010, Bib 216
Hayakawa, S.I. & Hayakawa, Alan R., **Language in Thought and Action**: Fifth Edition, Harcourt Inc., 1990, Bib 170
Heisenberg, Werner, **PHYSICS AND PHILOSOPHY, the Revolution in Modern Science**: Harper Collins Publishers, 1958, Bib 219
Herron, Frank, Sarquis, Sarquis, & Schrader, Kukla, **Heath Chemistry**: D. C. Health and Company, 1993, Bib 45
Hine, Robert, **The Facts On File Dictionary of Biology**: Third Edition, Market House Books, 1999, Bib 34
Holden, Alan & Morrison, Phylis, **Crystals and Crystal Growing**: MIT Press edition 1982, 1997 eleventh printing, Bib 44
Holt, Rinehart, & Winston, **Modern Physics**: Publishers, 1984, Bib 46
Hooper, Judith, **The Evolutionary Tale of Moths and Men, The untold story of science and the peppered moth**: W. W. Norton & Company, 2002, Bib 100
Horgan, John, **The End of Science - Facing the Limits of Knowledge in the Twilight of the Scientific Age**: Little, Brown & Co., 1996, Bib 20
Hothem, Lar, **First Hunters, Ohio's Paleo-Indian Artifacts**: Hothem House Books, 1990, Bib 204
Hoyt, William Graves, **Coon Mountain Controversies**: The University of Arizona Press, 1987, Bib 165
Huffman, Karen, Vernoy, Mark, & Vernoy, Judith, **Psychology in Action**: Fourth Edition: John Wiley & Sons, Inc., 1997, Bib 49
Jammer, Max, **Concepts of Mass in Classical and Modern Physics**: Dover Publications, Inc., 1997, Bib 206
Johnson, Phillip E., **Darwin on Trial**: Inter Varsity Press, 1993, Bib 93
Jones, Edwin R. & Childers, Richard L., **Contemporary College Physics**: Addison-Wesley Publishing Company, Inc., 1993, Bib 57
Josephus, Flavius, **The Complete Works of Josephus**: Kregel Publications, © 1981, printed 2000, Bib 176
Jungnickel, Christa & McCormmach, Russell, **Cavendish, The Experimental Life**: Bucknell, 1999, Bib 215
Kervran, C. Louis, **Biological Transmutations**: Happiness Press, 1998, Bib 152
Kline, Morris, **Mathematics for the Nonmathematician**: Dover Publications, 1967, Bib 124
Krayer, Peter J., **The Use and Care of a Balance**: The Chemical Publishing Company, Easton, PA., 1913, Bib 61
Krinov, E. L., **Giant Meteorites**: Pregamon press, 1966, Bib 167
Kuhn, Thomas S., **The Structure of Scientific Revolutions**: The University of Chicago Press, 1996, Bib 33
Larsen, Randy J. & Buss, David M., **Personality Psychology, Domains of Knowledge About Human Nature**: McGraw-Hill, 2005, Bib 139
Larson, Edward J. & Witham Larry, **Scientists and Religion in America**: Scientific American, September 1999, p89-93 (Also see companion article in *Nature*, Vol. 394, July 23, 1998), Bib 95
Lefond, Stanley J., **Handbook of World Salt Resources**, Plenum Press, 1969, Bib 143
Levy, David H., **Comets, Creators and Destroyers**: Touchstone, 1998, Bib 115
Lide, David R., **CRC Handbook of Chemistry and Physics**: 77th Edition, CRC Press, Boca Raton, Florida, 1996, Bib 5
Lindley, David, **The End of Physics - The Myth of a Unified Theory**: Basic Books, 1993, Bib 21
Lindley, David, **UNCERTAINTY, Einstein, Heisenberg, Bohr, and the Struggle for the Soul of Science**: Doubleday, 2007, Bib 211
Longwell, Knopf Flint, **Textbook of Geology Part 1**: John Wiley and Sons, Second Edition, 1939, Bib 125
Lutgens, Frederick K. & Tarbuck, Edward J., **The Atmosphere An Introduction to Meteorology**: Pearson Prentice Hall, Tenth Edition, 2007, Bib 180
Lutgens, Frederick K. & Tarbuck, Edwards J., **Essentials of Geology**: Eighth Edition, Prentice Hall, 2003, Bib 172
Lyons, Walter A., **The Handy Weather Answer Book**: Accord Publishing Ltd., 1997, Bib 66
Mann, Charles C., **1491, New Revelations of the Americas Before Columbus**: Vintage Books, 2006, Bib 202
Mark, Kathleen, **Meteorite Craters**: The University of Arizona Press, Tucson, 1987, Bib 119
Matthews, Roy, Stevenson, Harris & Hesser, Dexter, **Investigating the Earth**: Fourth Edition: Houghton Mifflin Company, 1984, Bib 52
Mayr, Ernst, **Darwin's Influence on Modern Thought**: Scientific American, July 2000 p80-83, Bib 85
McAvoy, Joseph M., **Clovis Settlement Patterns, Nottoway River Survey Part 1**: Dietz Press, 1992, Bib 200
McSween, Harry Y., **Meteorites and Their Parent Planets**: Cambridge University Press, 1999, Bib 166

McTaggart, Lynne, **The Field, The Quest for the Secret Force of the Universe**: Quill, 2003, Bib 209
Melosh, H. J., **Impact Cratering, A Geological Process**: Lunar and Planetary Laboratory, UofA, Oxford University Press, 1989, Bib 161
Merril, McElhinny, & McFadden, **The Magnetic Field of the Earth**: Academic Press, 1996, Bib 36
Middlemost, Eric A J, **Magmas and Magmatic Rocks**: Longman Group Limited, 1985, Bib 87
Milton, Richard, **Alternative Science, Challenging the Myths of the Scientific Establishment**: Park Street Press, 1994 &1996, Bib 117
Milton, Richard, **Shattering the Myths of Darwinism**: Park Street Press, Bib 22
Monroe, James S. & Wicander, Reed, **Physical Geology**: West Publishing Company, 1992, Bib 54
Moody, Raymond A., Jr., **The Light Beyond, New Explorations by the Author of Life After Life**: Bantam Books, 1989, Bib 213
Moore, Patrick, **Mission to the Planets**: Cassell Publishers Limited, 1995, Bib 149
Moore, Patrick, **New Guide to the Moon**: 1976, Bib 17
Moore, Patrick, **Patrick Moore on the Moon**: Cassell & Co, 2001, Bib 163
Nassau, Kurt, **Gems Made by Man**: Gemological Institute of America, 1980, Bib 104
Neuendorf, Klaus K. E., Mehl, James P. Jr., & Jackson Julia A., **Glossary of Geology**: Fifth Edition, American Geological Institute, 2005, Bib 173
Newton, Roger G., **The Truth of Science**: Harvard University Press, 1997, Bib 18
Newton, Sir Isaac, **Opticks**: Preface by I. Bernard Cohen, Dover Publications, New York, N.Y., 1979, Bib 2
Newton, Sir Isaac, **Principia**: Prometheus Books, Translated by Andrew Motte – Great Mind Series, 1995, Bib 3
Nicolas, Adolphe, **The Mid-Oceanic Ridges, Mountains Below Sea Level**: Springer-Verlag Berlin Heidelberg, 1995, Bib 120
Nininger, H. H., **Arizona Meteor Crater, Past—Present--Future**: American Meteorite Museum—Sedona, Arizona, World Press, 1956, Bib 126
Norton, O. Richard, **Rocks From Space**: Mountain Press Publishing Company, 1994, Bib 29
Ohfuji, Hiroaki & Rickard, David, **Experimental syntheses of framboids—a review**: Earth-Science Reviews 71, (2005) p147-170, Bib 147
Ollier, Cliff & Pain, Colin, **The Origin of Mountains**: Routledge – imprint of the Taylor and Francis Group, 2000, Bib 141
Parker, Sybil P., Editor, **McGraw-Hill Concise Encyclopedia of Science and Technology - Fourth Edition**: Lakeside Press, 1998, Bib 12
Parry, William J., **Archaeology, Original Readings in Method and Practice**: Edited by Peter N. Peregrine, Carol R. Ember, Melvin Ember, Chapter 11, *When and How Did Humans Populate the New World*, Prentice Hall, 2002, Bib 197
Paselk, Richard A., **Scientific Instrument Museum**: The MacMillan Co., NY, p7-21. (http://humboldt.edu/scimus/LitIndex.html), Bib 62
Pettijohn, F. J., **Sedimentary Rocks**: Harper & Brothers, 1957, Bib 159
Phillips, O. M., **The Heart of the Earth**: Freeman, Cooper & Company, 1968, Bib 63
Poinar, George & Poinar, Roberta, **The Quest for Life in Amber**: 1994, Bib 75
Poinar, George O. Jr., **Life in Amber**: Stanford University Press, 1992, Bib 81
Poirier, Frank E., **Fossil Evidence, The human evolution journey**: Second Edition, C. V. Mosby Company, 1977, Bib 192
Popper, Sir Karl, **Popper Selections Edited by David Miller**: Princeton University Press, 1985, (excerpts date from 1934-1977), Bib 40
Powell, Corey S., **God in the Equation**: Free Press, 2002, Bib 109
Press, Frank & Siever, Raymond, **Understanding Earth**: Second Edition: W. H. Freeman and Company, 1998, Bib 59
Randall, John Herman, **The Making of the Modern Mind**: Jr., Columbia University Press, 1940, Bib 86
Ransom, Jay Ellis, **Petrified Forest Trails**: Mineralogist Publishing Company, 1955, Bib 47
Rees, Abraham, **The Cyclopedia or Universal Dictionary of Arts, Sciences, and Literature**: Vol. XI, 1st American Edition, 1820, Bib 123
Reichenbach, Hans, **From Copernicus to Einstein**: Dover Publication, 1980, Copyright in 1942, originally published in 1927, Bib 11
Repcheck, Jack, **The Man Who Found Time**: James Hutton and the Discovery of Earth's Antiquity, Perseus Publishing, 2003, Bib 154
Retallack, G.J., **Soils of the Past, An Introduction to Paleopedology**: Boston Unwin Hyman, 1990, Bib 118
Rice, Patty C., **Amber, The Golden Gem Of The Ages**: Litton Educational Publishing, Inc., 1980, Bib 80
Ridpath, Ian, **A Dictionary of Astronomy**: Oxford University Press, 1997, Bib 108
Roddy, D. J., Pepin, R. O., & Merrill, R.B., Editors, **Impact and Explosion Cratering**: Planetary and Terrestrial Implications, Proceedings of the Symposium on Planetary Cratering Mechanics, Flagstaff, AZ, September 13-17 1976, Lunar Science Institute, Pergamon Press, 1977, Bib 160
Roedder, Edwin, **Fluid Inclusions**: Reviews in Mineralogy, Vol. 12, Mineralogical Society of America, 1984, Bib 155
Schumann, Walter, **Rocks, Minerals & Gemstones**: HarperCollins Publishers and Houghton Mifflin Company, 1993, Bib 15
Sean Carroll, Dutton, **From Eternity to Here, The Quest for the Ultimate Theory of Time**: The Penguin Group, 2010, Bib 218
Sears, Robert W., **The Vaccine Book, Making the Right Decisions for Your Child**: Little, Brown and Company, 2011, Bib 210
Selleri, Franco, Editor, **Open Questions in Relativistic Physics**: Apeiron, 1998, Bib 92
Serber, Robert, **The Los Alamos Primer**: University of California Press, 1992, Bib 99
Sharp, Robert P. & Glazner, Allen F., **Geology Underfoot**: Mountain Press Publishing Company, 1997, Bib 128
Short, Nicholas M., **Planetary Geology**: Prentice-Hall, Inc., 1975, Bib 114
Sigurdsson, Haraldur, **Melting the Earth, The History of Ideas on Volcanic Eruptions**: Oxford University Press, 1999, Bib 136
Silverman, Mark P., **Waves and Grains**: Princeton University Press, 1998, Bib 38
Smith, George McPhail, **A Course of Instruction in QUANTITATIVE CHEMICAL ANALYSIS for Beginning Students**: 1921
Smolin, Lee, **The Trouble with Physics, The Rise of String Theory, the Fall of Science, and What Comes Next**: Houghton Mifflin Co., 2006, Bib 178
Spudis, Paul D., **The Once and Future Moon**: Smithsonian Institution Press, 1996, Bib 129
Strahler, Arthur N., **Plate Tectonics**: Geo-Books, 1998, Bib 140
Stumpf, Samuel E., **Elements of Philosophy An Introduction**: McGraw-Hill, Inc., 1979, Bib 32
Tarling, Don & Maureen, **Continental Drift**: Doubleday & Company, Inc., 1971, Bib 103
Taylor, Stuart Ross, **Planetary Science: A Lunar Perspective**: Lunar and Planetary Institute Houston, Texas, USA, 1982, Bib 162
The Handy Science Answer Book: Compiled by the Science and Technology Department of the Carnegie Library of Pittsburgh, 1997, Bib 130
Thurston, Hugh, Springer, Verlag, **Early Astronomy**: New York, Inc., 1994, Bib 27
TIME, (The Weekly Newsmagazine) Time Inc., December 31, 1999, Bib 77
Tiner, John Hudson, **100 Scientists Who Shaped World History**: Bluewood Books, 2000, Bib 35
Townes, Charles H., **How the Laser Happened - Adventures of a Scientist**: Oxford University Press, 1999, Bib 26
Transcript of *Colin Patterson's* November 5, 1981 presentation at the American Museum of Natural History, **Can You Tell Me Anything about Evolution?**: New York City, p6. Both the transcript and tape recording can be obtained at www.arn.org, Bib 135
Trefil, James & Hazen, Robert M., **The Sciences, An Integrated Approach, Fifth Edition**: John Wiley & Sons, Inc., 2007, Bib 208
Turney, Chris, **Bones, Rocks and Stars, The Science of When Things Happened**: Macmillan, 2006, Bib 164
Twain, Mark, **Life on the Mississippi**: originally published by James R. Osgood & Co., 1883, http://docsouth.unc.edu/twainlife/twain.html, Bib 74
Van Heuvelen, Alan, **Physics A General Introduction**: Second Edition: Little, Brown and Company, 1986, Bib 60
Van Ness Myers, Philip, **Ancient History**: Ginn and Company, 1904, p2, Bib 196
Various authors from March 1996 to December 1999, **Earth From the Inside Out**: Scientific American, 2000, Bib 148
Walker, Cyril & Ward, David, **Fossils**: A Dorling Kindersley Book, 1992, Bib 102

Wallace, Bryan G., **Farce of Physics**: WindSpiel Company, 1993, (http://surf.de.uu.net/bookland/sci/farce), Bib 43
Wallace, Bryan G., **The Farce of Physics**: http://surf.de.uu.net/bookland/sci/farce/farce_toc.html, Bib 146
Waller, John, **Fabulous Science**: Fact and Fiction in the History of Scientific Discovery, Oxford University Press, 2002, Bib 177
Watkins, Emiliani, Chiaverina, Harper, LaHart, **General Science**: Harcourt Brace Jovanovich, Inc., 1989, Bib 53
Weast, Robert C., **CRC Handbook of Chemistry and Physics**: 57th Edition, CRC Press, Cleveland, Ohio, 1976, Bib 1
Webster's New International Dictionary, Second Edition Unabridged: G. & C. Merriam Company, Publishers, 1957, Bib 9
Webster's Seventh New Collegiate Dictionary: G. & C. Merriam Co., 1963, Bib 8
Webster's Universal College Dictionary: Random House, Inc., 1997, Bib 7
Wegener, Alfred, **The Origin of Continents and Oceans**: Translated from the 4th German Ed by John Biram, Dover Publications, 1966, Bib 181
Wells, Jonathan, **Icons of Evolution, Science or Myth**?: Regnery Publishing, Inc., 2000, Bib107
White, Michael & Gribbin, John, **Einstein A Life in Science**: Dutton, Published by the Penguin Group, 1993, Bib 105
White, Michael, **Isaac Newton - The Last Sorcerer**: Perseus Books, 1997, Bib 19
Will, Clifford M., **Was Einstein Right?**: BasicBooks, A member of the Perseus Group, 1993, Bib 106
Wood, Elizabeth A., **Crystals and Light, An Introduction to Optical Crystallography**: Dover Publications, 1977, Bib 220
Worthington, Glenn W., **A Thorough and Accurate History of Genuine Diamonds in Arkansas**: Mid-America Prospecting, 2003, Bib 175
Young, Colonel Bennett H., **The Prehistoric Men of Kentucky**: J. P. Morton & Co., Printers to the Filson Club, 1910,
  http://books.google.com/books/about/The_prehistoric_men_of_Kentucky.html?id=z6cAk5SiLt4C, Bib 201
Zumdahl, Steven S., **Chemistry**: D. C. Heath and Company, 1986, Bib 48

Image Credits

Most of the photos and diagrams for the UM have been produced by the author and have no credits given under the illustration and are not listed below. Researcher Rod L. Meldrum, Lynn Howlett Photography, Flickr.com, Wikipedia.com, Foter.com, Unsplash.com and Shutterstock.com have also contributed towards some of the photography. Other illustrations and permissions have been kindly given or donated by the following individuals or entities listed below.

The attribution given to each author below does not suggest in any way that these authors endorse the author of the UM or use of their work in any way. Each image is categorized by the chapter, subchapter and image number as listed in the text. They may also have been adapted in their presentation for this book.

Extensive effort has been made to get permission for all copyrighted material utilized. Every attempt to follow fair use laws has been made and no known use of copyrighted material has been done without permission. When "CC" is used after the source below, it refers to the Creative Commons license agreement found at CreativeCommons.org. The Creative Commons Attribution license agreement is what most of the images in the UM using CC were designated. All of the Wikipedia and Flickr image credits use the Creative Commons license. Any attribution not included in this list has been included with the image description located next to the image.

1 Introduction to The Universal Model

1.1 Universe, Shutterstock, CC; Bryce Canyon, Unsplash, CC; Waterfall, Unsplash, CC
1.2 Question Marks, Shutterstock, CC
1.3.1 Background book image, Shutterstock
1.5 Truth defined, sanddunes, Shutterstock, CC
1.8.1 Green hills with clouds (bottom right), Flickr, Jay Huang; Purple clouds (top left), Unsplash, CC; Universe and car tail lights (top right), White cliffs and water (bottom left), T-rex, Chemist, Parthonon, Shutterstock, CC; Fossil Fish, Wiki, By Didier Descouens, CC; Ruler, wiki, Gowolves109, CC;

2 Methods Models and Pseudotheories

2.1 Scientist in a lab, courtesy of Lynn Howlett
2.3 2 scientists in front of screen, Sandia National Laboratories
2.4 3 scientists in the Houston Space Center, NASA
2.4 Hawaii landscape, Bryan Illguth
2.5 Scientist with microscope, courtesy of Lynn Howlett
2.6.2 Car Model, Shutterstock, CC
2.6.3 Model house, Shutterstock, CC
2.11 Pseudotheory, Shutterstock, CC
2.11 Scale solar system, Wiki, courtesy of Lsmpascal, CC

3 The Dark Age of Science

3.2.2 100 Scientists That Shaped World History, John Hudson Tiner
3.4 Man thinking, Shutterstock, CC
3.7 RHIC, courtesy if Brookhaven National Laboratory; Man with small electronic devices, Sandia National Laboratories, Sea Shore image Lynn Howlett Photography

4 Scientific Revolutions

4.1 Thomas S. Kuhn, courtesy of Jehane Kuhn
4.1 Revolution, Shutterstock, CC
4.1.1 Linear Science, Shutterstock, CC
4.4 Unique UM Claim image, Shutterstock, CC

5 The Magma Pseudotheory

5.1.1 Extrusive lava flow over cliff by ocean, USGS
5.1.2 Actual lava flow in Hawaii, USGS
5.1.3 James Hutton, Abner Lowe, Wiki, USGS, CC
5.1.4 Hutton's drawing of theoretical magma chamber, USG
5.2 Paricutin, Mexico, R.E. Wilcox, U.S. Geological Survey, PD-USGov-NOAA
5.3.3 Lava Flow Along a Fault in Hawaii, USGS
5.3.6 Frictional Welding, American Frictional Welding, Inc
5.3.8 Mt. Etna plume, NASA, Sicily
15.3.12 Aurum Geyser in Yellowstone, courtesy of Jim Peaco
5.3.14 Ring of Fire Map, USGS
5.6 Lava flow by ocean in Hawaii, USGS
5.11.1 Long Valley Exploratory Well, USGS
5.11.2 Long Valley Caldera Crossection, Wiki
5.11.4 KTB Borehole, Germany, KTB Information System

6 The Rock Cycle Pseudotheory

6.1.14 Bolivia Salt Piles, Shutterstock, CC
6.4.9 Hawaii Lava, USGS
6.4.10 Shiprock Mountain in New Mexico, by Southwest Desert Lover
6.4.11 The Orchard Quarry Deposit in Buckfield, Main, USA, Courtesy of Gary Freeman
6.5.1 Stop sign with lava flow, Kilauea flow of January 2004, Volcano, USGS;
6.8.3 Courtesy of the Kennecott Copper Mine
6.9.2 Water Hardness, USGS
6.9.3 Annual Average Precipitation, USGS
6.9.9 Reed Quarry, Courtesy of the Kentucky Geological Survey
6.9.10 Huge Hauler, by Kees DeJong
6.9.12 Bahama Banks, NASA
6.9.19 Dolomites, Italy, Shutterstock, CC
6.11.9 Grand Canyon, by Grand Canyon Helicopter
6.11.14 Pinnacles at sunset, Guido Claesse
6.11.15 Mountains in purple ocean, Shutterstock, CC

7 The Hydroplanet Model

7.1 Hawaii lava tube with person, USGS
7.1.3 Water forms into a sphere, NASA
7.2.1 Space scene with Sun and Earth, Shutterstock, CC
7.3.1 Planets and Moons, NASA
7.3.4 The Titan Hydrosphere, NASA
7.3.5 Icy Moons of Saturn, NASA
7.3.10 The Europa Hydroshpere, NASA
7.4.16 Synthetic Emeralds, Wiki, CC
7.7 Honeystick, Shutterstock, CC
7.7.1 Glacier ice channels, Shutterstock, CC
7.7 Fallen Building on fire, NOAA
7.7.8 Mount Pinatubo, USGS
7.7.11 Mt. Saint Helens Steam Eruptions, USGS
7.7.13 Saturn and its Moon, Hubble Space Telescope March 22, 2004
7.7.14 CA sand blow, USGS; Railroad tracks dangling, NOAA
7.7.15 Chi-Chi Earthquake, NOAA
7.8.5 Buell Crater, Elegante Crater, NASA; Panum Crater, USGS
7.8.9 Pinacate Craters, USGS
7.8.10 Crater Lake, NASA
7.8.12 Mt. Saint Helens craters, USGS; Phreatic craters, NASA
7.9.1 Comet Tempel 1, NASA
7.9.3 Jupiter Impact, NASA
7.9.4 Mars Impact Site, NASA
7.9.5 Genesis crash, Mars slow-speed impact, NASA
7.9.6 Mars Impacts 2006, NASA
7.9.7 Sedan Crater, DOE
7.6.8 Moon Craters, NASA
7.9.9 Moon Crater, NASA
7.9.13 Trinity Glass, USGS, Trinity Atomic Explosion, Courtesy of Jack Aeby and Los Alamos National Laboratory
7.9.15 Sedan nuclear explosion ejecta, DOE
7.9.20 Orange Glass Soil, NASA
7.9.21 Definitive Impact Criteria, NASA
7.9.24 Ed Chao and Gene Shoemaker, USGS
7.9.27 Crater Doublet Evidence, NASA
7.9.28 Water drop, Shutterstock, CC
7.9.29 Crater Lake, NASA; Crater Lake Sunset, Flickr, Scott Johnson, CC; Luner Crater, NASA

7.9.33 Crater Chains, NASA
7.9.37 Europa, NASA
7.9.39 Hyperion, NASA
7.10.3 Comet Tempel 1, NASA
7.10.14 Columbia Fuselage Structure, NASA
7.10.26 Astronaut in space, NASA
7.11.8 Meteoric material around the Arizona Meteor Crater, by Holsinger
7.11.15 David Roddy and George Herman with CBS, USGS
7.11.18 Bowl Shock Wave, Sandia National Laboratories
7.11 Gilbert horse and crater 1891, USGS
7.11.22 Hopi Buttes, Sunset Crater topo map, USGS; Meteor Crater, NASA
7.11 Flagstaff crater making by Shoemaker, USGS
7.12.3 Wolfe Creek, by Stephen B. Andy
7.12.18 Panum Crater, USGS; Upheaval Dome, Wiki by NASA, CC
7.12.23 Richat Structure, NASA, USGS
7.12.24 Bushveld Complex, NASA
7.12.26 Doh Crater, NASA
7.13.2 Apollo 11 footprint in lunar soil, Apollo 16 flag salute, NASA
7.13.3 Boulder Tracks, NASA
7.13.4 Eros surface, NASA
7.13.8 Kilauea flow of January 2004. Orange flow Hawaii Volcano, USGS; Moon surface, NASA
7.13.10 Full Moon, Lowell Crater of Mars, NASA
7.13.12 Copernicus Crater, NASA
7.13.13 Secondary Crater, USGS
7.13.14 Humboldt Hydrocrater, NASA
7.13 Lunar Hydrocrater out-channels, NASA
7.13.16 Lunar Gravity Diagrams, NASA
7.14.4 Halley Comet, Wild 2 Comet, NASA
7.15 Hydroid Eros backside, NASA
7.16.8 Juntion lava tube, Lava fissure, Hawaii lava tube, Pu'u'O'o volcano lava tube, USGS
7.16.14 Comet Shoemaker-Levy, NASA
7.17.1 Hidden ocean under sea ice, NOAA
7.17.2 Alvin submersible underwater, NOAA

8 The Universal Flood Model

8.2.3 Marmolada, Wiki, CC; Folded Layers, Fred Webb
8.2.4 Subducting Plate, NOAA
8.2.5 Mars surface panorama, NASA
8.2.6 English Channel, NASA
8.2.7 English Channel, NASA
8.2.8 Columbia Gorge. Courtesy of Lynn Howlett; Scabland, Scablands National Park Service
8.3.2 Surface Tension, Wiki by Armin Kuelbeck, CC
8.3.3 William Whiston, Wiki, CC
8.3.4 Kuiper Belt, Sedna Orbit, Sedna Comparis, NASA
8.3.7 Enceladus, Enceladus fountain, NASA
8.4.2 Castle Geyser, White Dome Geyser, J Schmidt
8.4.7 Mt. Saint Helens mudflow, USGS
8.4.8 Mudfountains, Courtesy of Harry Roberts
8.5.9 Lusi mudflow, Wiki by Fragrag, CC
8.4.13 American Diamond Rush, Courtesy of Glenn Worthington from his book titled, A Thorough and Accurate History of Genuine Diamonds in Arkansas
8.4.17 Israel Fountain, by Sergeev
8.4.22 Mars surface and geyser, NASA
8.5.6 Diatoms, NOAA
8.6.1 Boulder Debris on Sabino Canyon tram road, Arizona, Arizona Geology Survey
8.6.2 Devil's Marble split up, by Guido Claessen
8.6.3 Landscape Arch, Wiki by Lorax, CC; Landscape Arch from the air, by Louis J. Maher, Jr.; Wall arch after fall, Wall arch before fall, NPS by Tim Conners
8.6.6 Top of Uluru; Ayers Rock, by Guido Claessen
8.6.7 Pinnacles with rainbow, Guido Claessen
8.6.8 Death Valley Alluvial fan; Mt Saint Helens, USGS
8.6.9 Mini Alluvial Fan, Wiki by Wing-Chi Poon, CC
8.6.10 Kaolinite, Bularia, Wiki by Nikila Gruev
8.6.12 Coast of California, USGS
8.6.15 Japan building, NOAA
8.6.16 Chuetsu earthquake, Wiki
8.6 Erosion at Mt. Saint Helens, USGS

8.7.5 People swinging on chairs, Shutterstock, CC
8.7.11 Giants Causeway, Shutterstock. CC; Giants Causeway walkway, chucklohr.com
8.8.4 Bacteria; Algae, ARSUSDA.gov
8.8.6 The submersible Alvin, NOAA
8.8.7 Newfoundland Bloom, NASA
8.8.8 Florida Blackwater, NASA
8.8 Massive bloom in Alaska, NASA; Tzin Valley, Sergeev
8.8.13 Alabama cave, Wiki, CC
8.8 Big room domes; The Klansman, NPS
8.8.17 Lost City, courtesy of IFE,URI-IAO, UW, NOAA, and University of Washington; Pinnacles in Australia, Nambung National Park
8.9.2 Michigan salt mine, Michigan Museum
8.9.4 Black Smoker, NOAA
8.9.6 Nazi loot; Salt mine, Ntl Archives
8.9.10 Bill Hill Salt Dome in Texas, USA, Courtesy of Sandia National Labs, Albuquerque and Continum Resources International Corporation, Houston
8.9.11 Iran Salt Dome; Salt Glaciers in Zagros mountains of Iran, NASA
8.10.2 Petroleum, Wiki, CC
8.10.6 Hydrous experiment, USGS
8.11.1 Coal, Wiki, CC
8.12.7 Framboids, USGS
8.13.1 Copper mine, Shutterstock, CC
8.13.6 K-T boundary, Wiki by Glen Larson, CC
8.14.11 Thunderegg Formation Sequence, Courtesy of Robert Colburn (Geode Kid)
8.14.20 Mars limesurface, NASA
8.14.27 Petroglyph, Wiki, CC
8.14.37 Nodules, USGS; ROV submersible, Soest, Hawaii.edu
8.14 Geyserite Nodules, USGS
8.15 Diamond, Shutterstock, CC
8.16.3 Japanese Alps, Wiki, CC
8.16.4 St. Peter Sandstone, Wiki, CC

9 The Weather Model

9.1.1 Rocky Point water spout, by Sergio Antunes
9.1.1 GPS Satellite, NASA; Hurricane Floyd perspective, NASA
9.1 Hot Springs, NPS
9.2.2 Yellowstone 1959 Earthquake and Geyser Connection, NPS
9.2.7 High Low Map, NOAA
9.3.3 Christchurch-quake, Wiki by Tim, CC; Wiki by Martin Luff, CC
9.3.7 Dust Fixed NASA Tornado Oklahoma May 3 1999, by NSSL; Dust Fixed Tornado Oklahoma May 22 1981, by NSSL
9.4.5 Hurricane Heated Water World Map, NASA
9.5.1 Earth's Magnetic Field, NASA
9.5.4 Cactus with Lighting, Shutterstock, CC
9.5.5 Geofield Map, NOAA; Solar Wind, NOAA
9.6.1 Purple Lightning, Shutterstock, CC
9.6.2 Lightning City, Shutterstock, CC
9.6.3 Atmospheric Electricity, Wiki. CC
9.6. Lightning and Cactus, Shutterstock, CC
9.6.5 Quakefinder Diagram, adapted from quakefinder.com
9.6.6 Geofield evidence, adapted from K. N. Abdullabekov's book entitled, Electromagnetic Phenomena in the Earth's Crust
9.6.9 North Pole and North America satellite view, NASA
9.7.1 Aurora Australis 2005, Wiki by NASA; Aurora Borealis, Wiki, CC; Saturn Aurora, NASA
9.7.2 Aurora, Polar Pixie NASA; Aurora, NASA G. E. Parks and UVI Team
9.7.3 Continental Auroras, NASA
9.7.4 Earths Geofield, NASA and USGS
9.7.5 Auroras at Poker Flats, Alaska, NASA by James Spann
9.8.1 Ozone Holes, NASA
9.9.1 Al Gore Nobel Peace Prize 2007. Wiki by Kjetil, CC
9.9 Global Cooling, NASA
9.9 Lecturer #2, Shutterstock
9.9.6 Antarctic Warming Map, NASA
9.10.2 Ganymede, NASA by M. Carroll
9.10 Jupiter Moons, Wiki, CC
9.10 Aurora, Wiki by US Air Force Senior Airman Joshua Strang, CC

IMAGE CREDITS

UM Index

Please note that all entries with more than twenty pages listed may have additional pages where the entry may be found. Also, each page listed may have multiple ocurrances of the entry.

A

Abiogenic Oil Verses Biogenic Oil 610
ablation 371, 372, 373, 374, 375, 419
Absolute Truth 5, 19
Accretion 4, 69, 96, 362, 379
accretion 77, 97, 100, 103, 105, 232, 243, 244, 246, 252, 280, 321, 361, 376, 443, 453, 459, 461, 651, 681
Acid-Base Biosalt Origin 595, 596, 598, 599, 609
Acid-Base Neutralization Process 597
Acknowledged Flood 475, 478, 577
Agates 161, 162, 650
Agathla Peak Hydromountain Evidence 557
algal blooms 506, 546, 594
Allegre, Claude J. 227
All Mountains Covered by the Sea 482
Alluvial Fan Hydrofountain Evidence 551
Alluvial Fan Mystery 215, 216
Alluvial Fans 207, 215, 216, 217, 546
Amalthea Hydromoon Evidence 247
America's Diamond Rush 517, 518
amorphous mineral 261
Ancient Salt Deposits Are Prethermites Not Evaporites 599
Anhydrite 137, 140, 144, 602, 603, 606
Anhydrous Moon Myth 437
Answers to Questions 5, 6
Antarctic Warming Evidence 781, 782
anthracite 221, 222, 617, 621
anthropogenic 736, 737, 773, 774, 779, 783, 784
Apache Tears Mystery 176
Arches National Park 156, 207, 208, 209, 548
Arch Formation 207, 208, 546, 548
Arch Formation Evidence 548
Argentine Loess Evidence 588
Arizona Copper 182
Arizona Crater Brief History 383
Arizona Crater diamonds 407
Arizona Crater Diatreme 401
Arizona Crater Volcanic District 404
Arizona Hydrocrater 229, 339, 382, 383, 393, 396, 397, 401, 402, 404, 406, 407, 411, 413, 416, 417, 421, 424, 427, 436, 466, 513, 536, 752
 1. Water Source Evidence 397
 2. Bisecting Fault Evidence 397
 3. A Diatreme - the Smoking Gun 398
 4. Volcanic District Evidence 397, 403
 5. Shale Ball Evidence 405
 6. Diamond Evidence - Diatreme Formation 406
 7. Pure Silica - The Second 'Smoking Gun' 407
Arizona Meteor Crater 307, 332, 333, 340, 343, 367, 382, 387, 396, 411, 439, 686
Arizona meteorite 385, 390
Arkansas Diamond Diatreme Wood Evidence 518
Artists Pallet Evidence 521
Asteroid Impact Menaces 460
asteroids 95, 109, 234, 240, 243, 260, 320, 359, 365, 376, 382, 414, 454, 455, 458, 460, 462, 463, 633, 644
Atmospheric-Pressure Error 713
Aurelia Crater on Venus 463
Aurora Evidence 699, 762, 766
Aurora Mystery 762
authigenic 529, 530, 535, 536, 569, 578, 590, 626, 681, 689
Ayers Rock 213, 550, 551

B

bacteria 53–60, 65–68, 139–228, 480–698, 727–790
Bacterial blooms 580
Badlands Mud Evidence 513
Bahama Banks 189, 190
Baker, Victor R. 490
barometric pressure 707, 709, 713
Barringer 332, 333, 383, 392, 397, 399, 400, 401, 403, 405, 406, 407, 410, 412, 413, 419, 421, 423, 425, 752
Barringer Crater Company 332
Barringer, Daniel M. 383
Barringer, D. Moreau 332
Basalt 131, 170, 172, 174, 215, 233, 335, 337, 351, 437, 442, 445, 525, 560, 565, 568, 570, 573, 574, 654
Basalt and Obsidian Mysteries Solved 560
Basalt-Carbonate Connection 573, 574
Basalt Column Mystery 172
Basalt Defined 170
Basalt Mystery 131, 170, 174, 175, 335, 337, 351, 437, 442, 445, 525, 560, 566, 568, 570, 572
Basic Modern Geology 146
Big Picture of Modern Science 20, 39, 51, 52, 53, 54, 57, 60, 64, 128
Biogenic 275, 277
biogenic 226, 313, 375, 419, 428, 434, 499, 506, 530, 556, 573, 594, 602, 622, 631, 643, 656, 672, 681, 688
biogenic deposits 584
Biogenic Meteorite Evidence 628
Biogenic Minerals 277
Biogenic Opal Evidence 534
Biogenic Origin of the Earth's Oceanic Crust 572
Biological Origin of Ore 639
Biology-Climate Connection 727, 728
Biomarker Evidence 632
biomarkers 632, 633, 634
Biomineralization 573, 625, 635, 642
biosalt synthesis 598, 599, 600, 605
biosand 544, 546
biotrans 574, 575, 578
Bituminous 221
black iron magnetite 627
black smokers 500, 580, 598, 607, 608, 636, 677, 678, 681, 682, 735
Boiling seas 479
Borehole 94, 100, 112, 113, 292, 751, 762
Bore Hole 112
Boreholes 100
boulders in coal 224, 225
Boulder Track Evidence 439
Bowen's Reaction Series Pseudotheory 122
Boyle, Robert 42
breached dam theory 505
breccia 113, 313, 334, 337, 396, 401, 430, 433, 449, 509, 520, 521, 524, 557, 607, 608, 638, 640
Bretz, J. Harlen 490
Buell Hydrocrater 312, 414, 422, 431, 516, 517
Buell Hydrofountain Evidence 516
Buell Park 312, 313
bunsenite 419, 420
Bushveld Complex 303, 340, 341, 433, 434, 524
Bushveld Complex Pseudotheory 433
Butte Fault 502, 504

C

calcite 138, 141, 161, 184, 188, 190, 195, 257, 271, 277, 376, 534, 541, 552, 577, 585, 592, 600, 628, 652
Calcite Hypretherm Evidence 271
Calcite-Lava Mystery 196
calcium carbonate 137, 140, 154, 165, 184, 188, 196, 408, 573, 575, 578, 581, 589, 592, 598, 631, 656, 677, 682, 708
caldera 70, 88, 103, 110, 111, 294, 314, 315, 319, 348
Caliche Mystery 197, 654
Caliche Rocks 645
Caliche Surface Mark 654
Callisto 90, 248, 249, 355, 356, 358, 435, 437, 787
Callisto Hydrocraters Evidence 355, 358
Callisto Hydromoon Evidence 249
Campo del Cielo iron 371, 372
Canyon Diablo Crater 383
Canyon Diablo meteorites 367, 407
Capitol Reef Basalt Hydrofountain Evidence 525
Capitol Reef National Park 525
Cap Rock Mystery 137
cap rock 137, 138, 307, 522, 605, 606
Cap Rocks 137, 575, 576
Cap Rocks from "Snowball Earth" Theory 575
Carbonate Defined 183
Carbonate Mark 57, 271, 475, 529, 545, 569, 573, 574, 583, 594, 595, 597, 600, 612, 631, 635, 636, 654, 655, 687
Carbonate Mark Summary 594
Carbonate Mystery 57, 131, 183, 184, 226, 575, 587
Carbonate Pinnacles 594
Carbonates Produced Experimentally by Microbes 578
Carbonate Surface Marks 645, 654, 660, 662
Carbon Cycle Pseudotheory 187, 189, 574, 575, 595, 777
Carbon Dioxide Evidence 777
Carbonic Acid Cave Formation Pseudotheory 192, 194, 537, 590
Carbonic Acid Cave Pseudotheory Debunked 590
Carbonic Acid Pseudotheory 195, 584, 585, 590, 592, 654
Cardenas Lava 502, 504
Cassini-Huygens' spacecraft 238
Cassini 238, 244, 245, 246, 251, 252, 300, 354, 355, 448
catastrophic disappearance 479
catastrophic event 478, 479, 481, 482, 488, 501, 574, 620, 644, 645, 695
catastrophic flood 488, 490, 491
catastrophic mud flow 480
Catastrophic Nature of the UF 491
Catastrophic Principle 73, 118, 492, 504, 505, 645, 690
Cause and effect 34
Celestial Body Formation 243
Celestial Spheres 233
Celestial Water 229, 234, 235, 282
Celestial Water Universe 235
cementation 138, 146, 154, 155, 159, 162, 197, 510, 511, 528, 529, 537, 540, 541, 542, 543, 600, 689, 690
Center for Meteorite Studies 399, 813
Central Peak Hydrocraters 306, 307
central peaks 348, 350, 352, 416, 453, 465
Central Universal Energy 493, 499
central uplift formation 350

Centrifugal force 495
Ceres Hydroid Evidence 462
Channeled Scabland Megaflood Controversy 489
Chevelon Collapse Hydrocraters 308
Chevelon Craters 314
Chevelon Hydrocrater 405
Chevelon Hydrocraters 465, 466
Chicxulub 319, 414, 480
Chlorine - the Missing Element 134
Chondrites 375
Cirrus clouds 721
Clay Evidence 536, 537
climatology 700
Clouds Defined 721
coal balls 224
coal beds 221, 222, 223, 224, 619
Coal Mark 475, 615, 620, 635, 687
Coal Mystery 221, 616, 620
Coal Mystery Unveiled 616
Coal/Sphalerite Inclusion Evidence 619
Coastline Aurora Evidence 766
cobblestones 307, 406, 430, 431, 486, 487, 520, 522, 525, 538, 574, 658, 660
coesite 102, 103, 339, 342, 343, 344, 395, 396, 398, 401, 402, 407, 422, 423, 429, 433, 436, 685, 686
Coesite Impact Criterion Myth 342, 358
collapsed lava tubes 466, 467, 468, 469
Collapse Hydrocrater 307
Colorado Plateau 118, 155, 213, 219, 312, 428, 483, 492, 501, 508, 514, 523, 536, 546, 556, 559, 603, 642, 657, 677
Colorado Plateau Liquefaction Evidence 556
Colorado Plateau Sand Mystery 156
Colton Hydrocrater 308
Comets Defined 454
Comet Tempel 1 321, 322, 361, 456, 457
Comet Wild 2 455, 456, 457, 814
Commercial Quartz Growth 266
Continental Crust Water Evidence 293
Continental Uplift 69, 117, 118, 119, 189, 637
Copernican Revolution 9, 62
Copernicus lunar crater 448
Copernicus, Nicholas 41
Copernicus 9, 39, 41, 43, 47, 50, 62, 63, 69, 232, 351, 448, 449
Cracks in the surface 252, 503
crater 82, 139, 174, 197, 230, 249, 294, 302, 315, 339, 352, 377, 392, 420, 443, 465, 517, 568, 607, 640, 686, 749
Crater Chains and Channel Evidence 352, 358
Crater Chemical Composition Evidence 353, 358
Crater Debate 229, 230, 318, 319, 320, 327, 333, 357, 358, 359, 382, 412, 416
Crater Doublet Evidence 346, 348, 358
Crater Lake 307, 313, 314, 315, 327, 348, 354, 450
Crater Lake caldera 314
Crater Mountain 308
crater of green 331
Crater Peaks Evidence 348, 358
crater rim 313, 314, 325, 334, 347, 385, 403, 408, 420, 421, 424, 470
craton 479
Creation Hypretherm 688
Critical Depth Fundamental Answer 567
Cross-Bedded Sandstone Mystery 159
crustal formation 303, 501
Crust Material Mystery 217
Crusts Collapse, Continents Submerged 497
Crust Thickness Mystery 217
crystalline minerals 243, 254, 263, 272, 275, 442
crystalline rock 261
Crystallization - Making Rocks 253

Crystallization Process 123, 229, 253, 254, 282, 376, 377, 396, 445, 509, 530, 584, 599, 601, 608, 684
Crystallization 139, 254, 256, 269
Cumuliform clouds 724
cumulus cloud formation 716, 723
Cumulus clouds 721
Curie Point 106, 739

D

Dark Age of Geology 475, 477, 622, 699
Dark Age of Science 32, 33, 39, 43, 45, 49, 50, 57, 59, 62, 182, 227, 253, 268, 341, 438, 477, 576, 590, 628, 635, 637, 674, 692, 694, 699, 784, 786
Darwin, Charles 23, 24, 43, 49, 53, 54, 60, 131, 269, 662
Death Valley, California 213, 214, 215, 216, 310, 411, 521, 538, 551, 672, 673
Death Valley Sand Dune and Clay Evidence 537
Deep Basin Salt Mystery 141
Deep Earthquake Evidence 69, 109, 289
deep earthquakes 281, 287, 288, 289
Deep Impact 321, 322, 361, 456, 457
deep mantle materials 687
deep mantle minerals 688
Definition Principle 15, 23, 43
Delicate Arch 209, 548
Depth Mark 475, 560, 572, 573, 574, 593, 602
Devils Punchbowl 308
Devils Tower Mark 568
Devils Tower Mystery 173
diagenesis 271
Diagenesis 271
Diagenesis Pseudotheory 189
Diamond bearing Kimberlite pipes 685
Diamond Inclusion Mark 685
Diamond Mark 475, 516, 682, 684, 686, 687
Diamond Origin Pseudotheory 684
diamonds 276, 322, 343, 363, 395, 401, 406, 516, 518, 519, 533, 559, 596, 603, 682, 684, 685, 686, 688, 691
Diatoms 534
Diatreme 306, 397, 398, 400, 401, 402, 406, 427, 518, 519, 640, 642
diatreme pipes 343, 453, 516, 520, 640, 642
Dickinson Hydrocrater on Venus 463
Dietz, Robert S. 342
Dike Intrusions 161
dinosaur bones 162, 302, 653
dinosaurs 359, 414, 478, 480, 481, 610, 616, 644
Dione 244, 245, 246, 252
Dione Hydrosphere 244, 246
Discovering and Confirming the Origin of Weather 702
Dissolved Rocks 645, 654
Dissolved Surface Mark 658
Doctrine of Uniformity 226, 476, 478, 490
Dogmagma 230, 283, 293, 635, 645, 650
Dogmagma Ore to Hyprethermal Ore 635
Dolomite 197, 198, 270, 271, 484, 584, 585, 586
Dolomite Hypretherm Evidence 270
Dolomite Mark 586
Dolomite Problem 197, 198, 585
Dolomites of Italy 484
Dome Crater Pseudotheory 435
Dorn Controversy 668, 670, 675
double craters 350
Double Terminated Crystal Mystery 168, 531
double terminated crystals 168, 170, 532
Double Terminated quartz crystals 169, 170
Dover Strait 488, 489
Dowsing and Piezoelectric Effect in Bone 758

Dowsing Evidence 755
drainage channels 314, 588
Drilling Evidence 69, 109
dwarf planet 462, 496
Dynamo Theory 115, 737, 751

E

Earth Crust Mystery 131, 217, 499, 566, 567, 572
Earth Impact Database 339, 341, 342, 377, 415, 433, 435, 436
Earthquake Cloud Evidence 699, 716, 747
Earthquake Cloud Evidence from Thermal Research 716
Earthquake Cloud Predictor - Zhonghao Shou 717
earthquake clouds 716, 717, 718, 719, 720, 728, 747, 748
Earthquake Clouds 291, 717, 718, 720, 722, 728, 776
Earthquake-Cumulus Cloud Connection 722, 728
earthquake-heated ocean water 710
Earthquake-Heated Water Connection 733
Earthquake Heating 705, 706, 784
Earthquake Heating Detected by Satellites 706
earthquake-heat seismic correlation 708
Earthquake Heat - The Missing Factor of Weather 705
earthquake-induced thermal vents 501
earthquake lightning 748, 749
Earthquake Lightning Evidence 748
earthquake prediction mechanism 718
Earthquake Swarm 79
earthquake swarm friction 294
Earthquake Weather 703, 704, 713
Earth's Density 107
Earth's Energy Field 116, 117, 705
Earth's energy field 105, 114, 115, 402, 706, 713, 737, 741, 743, 745, 749, 751, 755, 761, 765, 767, 769, 785, 786, 788
Earth's Energy-Field Questions and Name 739
Earth's Hydrology Redefined 287
Earth's Hydroplumbing System 289, 290, 291
Earth's mass extinction 479
Earth's Megaflood Evidence 487
Earth's multi-lobed geofield 767
Earth's oceans 230, 231, 243, 250, 279, 282, 317, 328, 354, 358, 380, 441, 446, 447, 450, 455, 468, 472
Earth's original minerals 688
Earth's Rotation Rate Reduced 496
Earth's secondary source of heat 705, 774
Earth's species disappeared 479
Earth's weather and the Earth's Geofield are interrelated 713, 737, 784
Earth's weather follows earthtide cycles 713, 784
Earth's weather follows patterns 713, 784
Earth's weather systems 711, 744, 784
Earth System 5, 20, 69, 73, 717
Earthtide 81, 82, 83, 84, 85, 86, 88, 89, 93, 112, 129, 285, 286, 290, 295, 704, 710, 711, 713, 715, 716, 722, 728, 737, 761, 784
Earthtide-atmospheric pressure 713
Earthtide Clouds 722, 728
Earthtide Heating 710, 711, 713, 716, 737, 784
ejecta 244, 249, 295, 313, 320, 345, 387, 408, 410, 415, 422, 439, 440, 447, 460, 463, 464, 465, 470
ejecta blanket 306, 323, 325, 327, 328, 332, 333, 334, 415, 451
Ejecta Evidence 325, 358, 464
ejecta material 300, 313, 323, 327, 333, 389, 391, 465, 470
ejecta patterns 323, 325, 327, 464

ejectites 376, 377, 378, 379, 380, 381, 389, 390, 393, 395, 396, 404, 406, 419, 424, 425, 427, 467, 628, 644, 686
Electric or Energy Field 741
El Niño and La Niña cycles 733
El Niño-La Niña Weather Condition 728
El Niño-La Niña Weather Predictions 729
Empty Cavity Evidence 293
Enceladus 300-Mile High Hydrofountains 300
Enceladus Hydrofountain 252, 300
Enceladus' Large Hydrofountain 354, 358
Enceladus' Water Fountain Evidence 448
Endless Flood Varnish Evidence 674
Endobiosphere 577, 578, 579
endobiosphere 577, 579, 580, 581, 584, 594, 595, 626, 634
Endobiosphere Environment Discovered 579
Endobiosphere Evidence 577
endoerosion 277, 487, 521, 522, 523, 524, 529, 650
Endoprethermic 275, 276
Endoprethermic Minerals 276
Endovaporization 711, 712
Energy Fields Defined 741
English Channel Megaflood 488, 492
English Channel Megaflood Evidence 488
enhydro 257, 258, 259, 260, 263, 279, 386
Enhydro Evidence 258, 378
Enhydro Sci-bi 259
Enigma of Chondrules 375
epeirogeny 120
Equatorial Band Evidence 681
Equatorial Bulge Evidence 284
Erosionary Sediment 275, 277, 529
Erosion Defined 206
Erosion Mark 121, 475, 503, 546, 547, 590, 646
Erosion Mingle 206
Erosion Mysteries 274, 307
Erosion Mystery 121, 131, 205, 206, 207, 214, 217, 221, 226, 301, 546
Europa 248, 249, 250, 300, 355, 356, 358
Europa Hydrocrater Features Evidence 355, 358
Europa Hydrosphere 250
Evaporate Rock Pseudotheory 255
Evaporation 140, 143, 256, 257, 314, 598, 602, 708, 710, 721
evaporation 138, 144, 189, 255, 258, 278, 287, 303, 313, 450, 540, 596, 603, 604, 606, 608, 656, 705, 711, 720, 722, 744
Evaporative Clouds 722, 728
Evaporite Crisis 145
evaporites 132, 138, 139, 140, 141, 142, 145, 255, 256, 257, 380, 599, 600, 601, 602, 604, 628, 640
Evapotranspiration 711
Evidences Against the Arizona Crater Impact Theory 384
  1. Lack of impact glass 385
  2. Lack of melt-evident meteorites 386
  3. Lack of residual vaporized material 387
  4. Presence of Widmanstätten pattern in meteorites 389
  5. Lack of shrapnel fragments 390
  6. Lack of crater embedded non-vaporized meteorites 391
  7. Multiple iron sources require multiple impactors—and multiple craters 392
  8. No elliptical meteorite strewn field 393
  9. Limestone at the crater has not been heated 395
  10. Absence of shatter cones 395
  11. The amount of iron at the crater is insufficient to produce the crater itself 396
extraordinary claim 77, 104, 170, 292, 319, 329, 371, 478, 504, 532, 560, 596, 609, 626, 687, 706, 716, 737
extraordinary claims 1, 6, 21, 22, 45, 59, 62, 65, 69, 121, 357, 359, 379, 382, 436, 654, 657
Extraordinary Discovery 22
Extraordinary Evidence 22
Extreme Ultraviolet Imaging Telescope 240
Extrusive lava 70
extrusive lava 74, 77, 79, 89, 113, 170, 276, 485

F

Fair-Weather Condition Evidence 747
father of geology 72, 131, 145
Fault Line 306
Final Look at the Depth Mark 574
First Law of Weather 713, 715, 728
First Planetary Piezofield Law 743, 753, 754
Flat Crater Floor Evidence of Water 351
Flawed Impact Criteria 339, 358
Flint 164, 191
Floating Continents 231
Flood catastrophe 481
Fluidization 509, 524
Fluidization Evidence 509
Fluvial erosion 303
fluvial transport 166, 304
formed underwater 327, 336, 337
Fossil Creek Formation Mystery 186
Fossil Creek limestone deposits 186
Fossil Hydrofountain 303, 304, 307, 308, 508
Fossil Hydrofountains 302, 307
fossil hydrofountains 507, 508, 509, 527, 560, 592, 637
Fossilized seashells 483
Fossilized Silica Hydrofountain 310
fossil record 414, 482
fossils 48–60, 73–130, 143–228, 278–474, 478–698
Four-Stage Coal Formation Pseudotheory 221, 222, 223
fragmented meteorite shrapnel 390
Framboid Pyrite Evidence 629, 630
freshwater lakes 303
frictional heating 77, 90, 112, 129, 240, 257, 270, 289, 316, 331, 361, 387, 465, 499, 530, 604, 650, 685, 702, 717, 778, 789
Frictional-Heat Law 77, 81, 86, 88, 90, 111, 128, 299, 506, 689, 702
Fukang Pallasite 374
Fusion Crust 370, 371
Fusion Crust Enigma 371
Future Weather Predicting Models 785
Futurity 2, 20

G

Galilei, Galileo 41
Galileo 9, 16, 35, 50, 63, 69, 90, 218, 232, 238, 249, 318, 355, 435, 437, 444, 446, 473, 786, 788
Galileo spacecraft 238, 250, 355, 437
Ganymede 90, 247, 248, 249, 250, 300, 350, 351, 470, 471, 787, 788
Ganymede Hydromoon Evidence 470
Genesis impactor 323
Geode Crystals 269
Geode Mark 648
Geodes 161, 162, 259, 269, 645, 646, 648, 650, 652, 653, 675, 691
Geofield 4, 268, 699, 705, 713, 737, 741, 743, 745, 747, 749, 753, 757, 761, 765, 767, 769, 776, 784, 787

Geofield-Continent Evidence 767
Geofield Cycle Evidence 760, 761
Geofield-Earthquake Evidence 755
Geofield Evidence 699, 743, 744, 753, 761
Geofield-lightning connection 744
Geofield Map 743
Geofield Model 268, 699, 705, 737, 741, 743, 744, 748, 749, 750, 752, 753, 755, 757, 758, 760, 761, 762, 768, 769, 776, 787
Geofield Model Predicts Piezofields in Solar System 787
Geological Events Supporting UF Mechanisms 500
Geometeorology 784, 789
Geophysicists Confirm Earthquake Clouds 720
Geotheoretical 131, 226, 692
Geyser Hydrofountains 507
geyserite 264, 275, 411, 412, 507, 508, 537, 627, 647, 682
Geyser Nodules 682
Gilbert, Grove K. 383, 403
glass covered impact crater 331
Glass Evidence 327, 328, 331, 335, 344, 358
Global Evidence of Water on Mars 487
Global Flooding 479
Global Warming - A "Secular Religion" 783
Global Warming Pseudotheory 487, 699, 711, 771, 784
global warming 708
Global Warming Science-One Big Mistake 780
Global-Weather System Evidence 699, 728
Gold Mystery 183
Grand Canyon 118–130, 156–228, 327–474, 479–698
Grand Canyon Crustal Layers Mystery 219
Grand Canyon layers 221, 504
Grand Canyon's Earthquake Origin 500, 501, 502
Granite Boulder Evidence 546
Granite Mystery 167
Granites 161
Granite 161, 167, 207, 256, 261, 263, 546, 547, 548, 688, 689
gravitational-astronomical cycles 713
gravitational constant 106, 107
Gravitational-Friction Law 77, 84, 85, 86, 89, 90, 113, 250, 251, 252, 300, 397, 435, 442, 448, 498, 527, 709, 733, 745, 789
Gravitational Friction Law 92, 126, 455, 462
Gravity Field 741
Great Salt Lake Sand Evidence 660, 662
Green Knobs 312
Greenland Native Iron-Nickel Evidence 368
Gypsum/Anhydrite Evidence 603
Gypsum/Anhydrite Mystery 144

H

Halley's Comet 455, 495
Hard Question Principle 12, 321, 353, 411
Hard Water Mystery 185
Haughton Hydrocrater 435
Heat Flow Physics 69, 90, 91, 113
Helictite Mystery 195, 196
Herkimer Diamonds 533
Herkimer Quartz Inclusion Evidence 691
high-pressure areas 705, 715, 786
High-Pressure Narrow-Ridge Evidences 715
high velocity impact 328, 331
high-velocity impacts 329
History of Geology 131, 145, 274
History of Science 23, 29, 39, 43, 45, 46, 692, 693, 804
History of Science Table 43
Hoba meteorite 364

Holsinger Map 392
Homegrown Authigenic Quartz 534, 535
Homestead Meteorite 393
homogeneous sand 151, 310, 502, 540, 544
hot smokers 599
hotspots 99, 124
how Saturn got its rings 300
How to Make Sandstone 542
Huckitta pallasite 375, 377
Human Touch 5, 19, 20
Humboldt Hydrocrater 450
humidity 137, 196, 238, 295, 700, 704, 705, 706, 707, 708, 709, 711, 712, 718, 723, 743, 744, 778, 783
Hurricane Evidence 736
Hutton, James
James Hutton 71, 72, 120, 123, 131, 145, 146, 229, 233, 234, 268, 273, 484, 487, 636, 692, 747
Hydrocomet 229, 453, 455, 456
Hydrocomet Evidence 229, 453, 455, 456
Hydrocrater 229, 305, 315, 337, 352, 377, 382, 393, 407, 413, 420, 431, 444, 450, 462, 513, 527, 536, 632, 752
Hydrocrater Chains 465, 466
Hydrocrater Formation 306
Hydrocrater Model 229, 305, 352, 363, 377, 382, 385, 395, 398, 402, 406, 416, 419, 426, 436, 441, 446, 449, 459, 462
Hydrofountain History 507
Hydrofountain-Hydrocrater Connection 527
Hydrofountain Mark 475, 493, 506, 507, 551, 592
Hydrofountain Origin of Meteorites 377, 378
Hydrofountains 246, 250, 299, 303, 305, 355, 465, 487, 498, 506, 509, 516, 527, 551, 581, 595, 650, 674, 735
Hydrofountains at Plate Boundaries 498
Hydrofountains Defined 299
Hydrofountain Sediment 303, 304, 430
Hydroid 229, 458, 462
Hydroid Defined 458
Hydroid Evidence 229, 458, 462
hydrological cycle 303, 304, 779, 780
Hydrology 229, 287, 305
Hydromoon 84, 229, 247, 248, 249, 300, 328, 338, 437, 438, 441, 442, 443, 444, 446, 448, 449, 450, 453, 459, 469, 470, 672
Hydromoon Features 443, 444
Hydromoon Fundamental Questions 438
Hydroplanet 16, 49, 94, 144, 162, 183, 227, 243, 262, 290, 312, 336, 350, 396, 441, 482, 553, 614, 692, 776
Hydroplanet Earth 108, 229, 248, 262, 279, 280, 284, 305, 396, 431, 451, 492, 493, 694
Hydroplanet Earth Origin 493
Hydroplanet Model and the Weather Model 702
Hydroplanet Model of the Earth 279, 280
hydroplumbing 289, 291, 292, 294, 298, 304, 305, 529
Hydrorock Fountain Evidence 301
Hydrorock Fountains Evidence 521
Hydrorock Fountains 302, 304, 521
Hydrosand Fountain Evidence 301
Hydrosand Fountains 301, 408
Hydrosediment 275, 277, 529, 530
Hydrosphere Defined 243
Hydrospheres 229, 243, 244, 246, 252, 459
hydrothermal 84–130, 166–228, 256–474, 501–698, 735–790
Hydrothermal 263, 264, 268, 271, 273, 275, 277, 291, 295
hydrothermal circulation 293, 501, 735
Hydrothermal History 268

Hydrothermal Megaplume Evidence 735
Hydrothermal Minerals 275
Hydrothermal Precursors to Earthquakes 291
Hydrothermal Process 263
Hydrothermic 275
Hydrothermic Minerals 275
Hydrovalley channel structures 469
Hydrovolcanoes 294, 295, 296, 300, 304
Hyperion Hydrocrater Evidence 356
hypervelocity 96, 320, 322, 327, 329, 331, 333, 335, 338, 339, 364, 386, 394, 401, 422, 427, 429, 449, 457, 458
hypervelocity impact 321, 325, 327, 329, 330, 331, 332, 333, 334, 335, 339, 394, 401, 422, 427, 429, 458
Hypervelocity Laboratory Impact Studies 329
Hypretherm 264, 265, 268, 270, 271, 272
hyprethermal 264, 265, 266, 267, 268, 269, 287, 302, 310, 328, 341, 354, 366, 375, 390, 419, 428, 442, 445, 449
Hyprethermal 265, 275
Hyprethermal Conditions 498, 609
hyprethermal environment 265, 266, 268, 270, 277, 302, 328, 366, 367, 378, 419, 428, 450, 453
Hyprethermal Marble Evidence 593
Hyprethermal Minerals 275, 276
Hyprethermal Origin of Ore 636
Hyprethermal Potash Evidence 602
Hyprethermal Salt Plug Evidence 605
Hyprethermal Sand and Sediment Formation Mechanism 530, 531
Hyprethermal Sand Origin 531, 532
Hyprethermal Sedimentation Model 531, 532, 557
Hyprethermal Sulfur Evidence 628
Hyprethermic 275, 276
Hyprethermic Minerals 276
hyprethermic minerals 507
Hyquatherm 710, 711, 716
Hyquathermal heating 713, 784
Hyquathermal Process 710

I

Icequakes 288
Igneothermic 275, 276
igneothermic minerals 276, 507
Igneothermic Minerals 276
Igneous 269, 270, 273, 274
Igneous Rock 146
igneous rocks 123, 133, 134, 145, 171, 181, 262, 279, 336, 516, 557, 575, 577, 581, 595, 597, 639, 688
Ignored Historical Basalt Evidence 570
Ignored Modern Day Empirical Evidence 572
Ignoring the Coal Reality 620
Impact Boulder Mystery 441
Impact Crater 229, 319, 330, 341, 351, 361, 373, 383, 397, 411, 421, 432, 441, 450, 460, 471, 480, 501, 626, 632
impact explosion 328, 333, 340, 342, 387, 388, 389, 391, 392, 408, 410, 430, 464
Impact Fad 415
impact model 320
impactologists 312, 325, 328, 336, 342, 350, 351, 383, 384, 386, 390, 396, 404, 416, 417, 418, 424, 433, 449, 466
Impactologists 340, 344, 358, 399, 427, 428, 429, 460, 470, 685
impactology 327, 341, 350, 383, 384, 402, 415, 436
Impact Paradigm Shift 437
impact theory 249, 319, 320, 321, 337, 350, 354,

356, 357, 383, 391, 393, 394, 408, 412, 421, 430, 441, 456, 470
Imperial Dunes Mystery 151
Importance of Definitions 5, 15
Inclusion Mark 475, 636, 676, 685, 686, 687, 691, 692
Inclusions Defined 686
increase in soil moisture and water vapor 707
increase in vaporization 716
India Studies Confirm Earthquake-Temperature Connection 709
Intrusive lava 70, 112
Intuition 19–22, 54–60
Inyo Hydrocraters 307, 308, 565
Io 89, 90, 126, 245, 300, 787, 788, 789
Iron Core 69, 106, 108, 109, 178
iron core 114, 115, 116, 183, 281, 368, 375, 444, 768, 789
iron-ejecta 387, 391
Iron, Iron, Everywhere 622
Iron Mystery 131, 177, 226, 629
irons 360, 364, 365, 371, 387, 389, 390, 391, 392, 393, 394, 405, 406, 407, 413, 419, 422, 426, 427, 458
Iron Sediment Experiment 178, 180
Iron Soil Mystery 177
Isostatic Adjustment 120
Itokawa Evidence 461

J

Japanese Alps Granite Inclusion Evidence 688
Josephinite Native Iron-Nickel Evidence 369
Joshua Tree National Park 206, 207, 208
Juan de Fuca Ridge Events 500
Jupiter 41, 72, 89, 90, 126, 238, 280, 299, 321, 328, 335, 350, 367, 435, 458, 470, 471, 482, 766, 786
Jupiter Impact 322, 471
Jupiter's moons 238, 247, 249, 250, 300, 435, 787
Jupiter's Three Large Hydromoons 248

K

Kamchatka earthquake 708
Kaolinite Evidence 553
Kartchner Caverns Mystery 194
Kepler, Johannes 41
Kimberlite 312, 343, 516, 517, 518, 519, 685, 686
Kimberlite diatremes 343
Kimberlite Hydrofountains 516
Kimberlite pipes 343, 686
Kimberly Diamond Mine 686
Kirin fall 364
Kobe Earthquake Evidence 291
Kodachrome Basin State Park 303, 307
Kodachrome Fossil Hydrofountain Evidence 508
KREEP Evidence 441
KTB 87, 93, 95, 100, 110, 113, 114, 125, 283, 286, 293, 577, 751, 752, 761
KTB Borehole Piezoelectric Evidence 751
KTB Evidence for Water Boundary Layers 283
K-T Boundary Evidence 644, 645
Kuenen, Philip Henry 200
Kuenen 149, 200, 201, 202, 203, 204, 554, 556
Kuhn, Thomas S. 61
Kuiper belt 252, 253

L

lamproites 420, 519
Large Meteorites Missing Craters Evidence 364

laterite 419, 420
Lava-Friction Model 69, 77, 81, 85, 88, 89, 90, 96, 295, 705, 716, 752
Law of Hydrobiogenesis 243
Law of Hydroformation 243, 253, 263, 275, 328, 379, 442, 676, 682
Law of Hydrogenesis 243
Law of Paragenesis 273, 278
Law of Primordial Matter 243
Law of Weather Parameters 723, 737, 743, 744
Learning From Enhydros 259
Learning Process 5, 12, 13, 15, 31, 32, 487, 690
Libyan Desert glass 344
Libyan Desert Glass 344, 345, 358
lightning 103, 343, 385, 700, 723, 744, 745, 747, 748, 749, 750, 760, 761, 762, 768, 769
Lightning-Strike Geography Evidence 769
lignite 221, 222, 617, 621
Limestone 137, 152, 161, 164, 184, 186, 188, 191, 395, 413, 426, 483, 484, 544, 584, 585, 588, 592, 640, 675
Limestone Deposits 188, 190, 588
Limestone-Dolomite-Loess Connection 584
Limestone Origin Mystery 187, 188
Limestone Quartz Chert 161
Limestone-Quartz Chert Mystery 164, 675
limesurface rocks 656, 657, 658, 660, 674, 681
Limesurface Rocks 645
limonite 422, 426, 427
liquefaction 298, 301, 350, 408, 498, 556, 557, 559, 561, 719
Liquid Sphere 232
liquid water in space 235
lithification 155, 189, 190, 271, 537, 584
Little Colorado Canyon 503
Living System 5, 20
Loess Defined 198
Loess Map 199
Loess Mark 587
Loess Model 590
Loess Mystery 131, 198, 199, 204, 206, 226, 531, 537, 554, 587
Long-Term Global Warming and Cooling 776
Long Valley 110, 111, 112, 282, 283, 292
Long Valley California Water Evidence 282
low-speed impact 385, 386, 387, 391, 392, 393, 403, 413
Lunar Basalt Mystery 174
Lunar Core Evidence 442
Lunar Crater Origins Evidence 327
lunar craters 17, 318, 320, 325, 328, 331, 335, 347, 351, 353, 356, 359, 363, 381, 382, 403, 416, 444, 446, 449
Lunar Doublets 347, 348
Lunar Glass Evidence 335, 358
Lunar landings 327
Lunar Mare Basin Evidence 445
lunar rocks 105, 336, 337, 438, 439, 440, 441, 442, 447, 688
Lunar Salt Evidence 381
Lunar-Weather Cycle Connection 727, 728
Lusi, the Java Mudfountain 513
Lyell, Charles
Charles Lyell 43, 72, 122, 131, 226, 268, 484

M

Magma 69, 92, 130, 161, 232, 272, 292, 375, 401, 454, 473, 517, 528, 626, 632, 684, 692, 733, 745, 760
Magma-based Tectonic Plate Pseudotheory 122, 125
Magma Convection Pseudotheory 122, 123, 124
magma convection 124, 272

magma oceans 230
Magmaplanet 69, 73, 76, 82, 128, 129, 229, 230
magmaplanet geotherm 127, 128
Magma Pseudotheory 69-130, 131-228, 230-474, 479-698, 699-790
Magma Pseudotheory and the Weather Model 701
Magma Theory 90, 109, 786
magma theory 70, 83, 96, 108, 110, 115, 122, 130, 281, 289, 320, 337, 404, 457, 484, 636, 693, 781
Magnetic Diatreme Evidence 402
Magnetic Field 69, 114, 115, 116, 737, 739, 741, 755, 762, 768
Magnetic Field and Dynamo Theory 737
Magnetic Field Movement 116
Magnetic Field Pseudotheory 69, 114, 739
Magsat Satellite Evidence 768
Maiasaur dinosaur fossils 480
maiasaur 481
Manganese Iron Nodule Evidence 677
Manmade-Global-Warming Pseudotheory Defined 773
Man-Made Ozone Hole Pseudotheory 769
Mantle Plume Pseudotheory 122, 124
mantle plume theory 124
Mare Orientale 447
Maria Basalt Hydroevidence 444
maria basins 328, 446
Marine fossils 224
marine species disappeared 479
Mars 105, 126, 147, 238, 317, 339, 346, 362, 376, 411, 416, 435, 447, 458, 463, 486, 490, 527, 658, 675, 787
Mars Doublets 347, 348
Mars Geyser-Hydrofountain Evidence 527
Mars Hrad Vallis water channel 464, 465
Mars Hydrocrater Chain Evidence 466
Mars Hydroplanet 465
Mars Hydrovalley Evidence 467
Mars Impact Events 321, 322
Mars Impacts 323, 325, 327, 328
Mars Opportunity 323
Mars Pathfinder 486
Martian Deluge 478, 486
mass extinction 414, 458, 479, 633, 634
Mass grave 480
Mathilde 460, 461
Measuring the Geofield-Lightning Connection 749
mega-earthquake 502
Megaflood Evidence on Mars 486
megafloods 487, 489, 492, 503
melting glaciers 479
Melt processes 253
Mercury 105, 238, 241, 244, 245, 247, 321, 328, 346, 352, 376, 428, 452, 470, 786, 787
Messenger spacecraft 238
Metamorphic 273
Metamorphic Rock 146
metamorphism 118, 123, 130, 190, 191, 221, 261, 262, 271, 342, 407, 422, 430, 516, 593, 594
Meteorite 75, 97, 109, 229, 308, 329, 358, 361, 371, 381, 392, 405, 412, 419, 420, 436, 446, 458, 628, 644
meteorites 376
Meteorite Enhydro Evidence 378
Meteorite Mineral Enigma 362, 363
Meteorite Model 75, 97, 229, 329, 358, 359, 373, 382, 389, 390, 392, 416, 436, 628, 644
meteorite projectile 329
Meteorites Are Ejectites 376
Meteorites come from space? 361

Meteorites Defined 359
meteoritics 359, 361, 367, 378, 405, 429, 436
Meteorology 700
Meteorology and the Geofield 743
Methane Inclusion Evidence 687
Microbe Dogma 577
microbes 500, 530, 544, 567, 577, 590, 600, 611, 620, 631, 640, 656, 662, 668, 673, 681, 688, 708, 727
microbial blooms 583, 584, 585, 590
microbial carbonates 578
microbial waste 613
microcrystalline quartz 162, 164, 590
Micro-Fracture 121
microgravity 232, 367, 439
micro-impacts 327, 331
micrometeorites 338, 440, 441, 453
microorganisms 53, 138, 187, 192, 315, 428, 508, 534, 535, 546, 569, 580, 583, 584, 594, 595, 597, 607, 612, 625, 626, 644, 715, 735
Microscopic Sand Mystery 149
Mystery of Microscopic Sand 151
Millennial Science 5, 20, 21, 22, 31, 37, 43, 51, 59, 67, 221, 230, 315, 358, 403, 412, 436, 471, 476, 552, 580, 599, 654, 730, 791
mini alluvial fan 552, 553
Missing Continents 121
Missing Factor of Weather 703, 705, 715
missing factor of weather 706, 716
missing iron 390
Missing Pebble and Sand sizes 532
Missing Sand and Silt segments 203
Missing Sediment Mystery 203, 204, 206, 310, 587
Miyake-Jima Hydrovolcano Evidence 294
Modern Science 7, 20, 23, 25, 39, 43, 48, 49, 50, 51, 52, 53, 54, 57, 58, 59, 60, 64, 67, 73, 121, 128, 133, 478, 728
Mono Lake hydrocrater 308
Montezuma Hydrocrater 310, 405
Montezuma Well 310
Monument Valley 540, 558, 559, 560
Moon and its origin 147
Moon craters 320, 353
Moondust 439, 440
Moonquake 83, 89, 442
Moon's Gravitational "Anomalies" 451
Moon's surface 325, 330, 335, 336, 438, 439, 445, 449, 450, 451
Moqui Marble Nodule Evidence 676
Most Abundant Mineral 131, 132
most abundant substance in space 236
Most Important Discovery in the last 25 Years 239
Mountain Liquefaction 559, 560, 561
Mountain Liquefaction Evidence 559, 561
Mount Everest 119, 467, 484
Mt. Pinatubo Hydrovolcano Evidence 295
Mt. Saint Helens 206, 296, 297, 298, 304, 316, 342, 354, 464, 466, 467, 492, 512, 553
Mount Saint Helens 347
Mt Saint Helens 216, 226, 297, 300, 316, 467, 468, 551, 552
Mt. Saint Helens Hydrovolcano Evidence 296
Mt. Timpanogos 483
mudflows 295, 296, 297, 298, 328, 354, 464, 467
Mudfountains 506, 511, 512, 513, 516
Mudstone Mystery 158
Multilobed Ejecta Evidence 464
Multiple electric fields 748
Multiple-Field Example 767, 768
Museum of the Rockies 480
Mystery of Ore Revealed 643

Mystery of Paragenetic Ore 183
Mystery of Weather 699, 700

N

Nakhla meteorite 381
NASA 49, 81, 116, 232, 294, 300, 320, 335, 350, 411, 435, 448, 452, 468, 557, 574, 581, 607, 706, 729, 736, 750, 766, 780
Natural Hypretherm Growing Conditions 268
Natural Law 23, 24, 29, 31, 32, 33, 278
Nature's Puzzle 20, 285, 344, 437, 506, 530, 610, 691, 743
Nature's puzzle 7, 8, 20, 43, 213, 222, 260, 631, 662, 675, 684, 737
near-vertical fault 310
negative magnetic anomaly 403
Neglected One-Sided Limesurface Mark 656
Neptune 239, 240, 241, 252, 253, 787, 788
Neptunism 232, 233, 262
Neptunists 232, 233, 234, 571
New Geologic Time Scale 278
New Geology 229, 239, 273, 274, 275, 276, 277, 279, 306, 310, 336, 363, 438, 471, 492, 529, 597, 626, 693, 694, 699
New Mineral Classification 275
Newtonian Revolution 62
Newton, Isaac 42, 51
New Weather Terminology 710
Nilahue Hydrocrater 315, 397
Nine Classifications of Minerals 275
Nininger, Harvey H. 382, 383
nodules 161, 174, 191, 343, 368, 377, 405, 422, 430, 466, 584, 627, 645, 672, 676, 677, 678, 681, 682, 814
Nodule Surface Mark 646, 675, 682
non-impact craters 340, 342
Non-Iron Core Evidence 69, 106, 178
North Atlantic Oscillation (NAO) Evidence 735
North Rim of the Canyon 502
Norwegian Salt Formation Evidence 601
Nuclear Crater Evidence 332, 358
nuclear explosion 331, 333, 350, 754, 823

O

observation 10–22, 28–38, 39–60, 64–68, 76–130, 137–228, 191, 230–474, 477–698, 700–790
Obsidian Defined 175
Obsidian Mystery 131, 175, 176, 560, 565
Ocean Depth Mystery 142
Oceanic Crust Earthquakes Producing Microbes Today 583
Oceanic Crust Hydroplumbing System 292
Oceanic Hydrocraters 354
Ocean's Invisible Forest 581
Ocean's Temperature is the Key 730
Odessa Flat Floor Evidence 424
Odessa Hydrocrater Evidences 423
Odessa Impact Crater Myth 421
Odessa Iron Ejectite Evidence 426
Odessa Kaolinite/Mercury Evidence 428
Odessa Oil/Salt Diatreme Evidence 427
Odessa Rock Flour Evidence 427
Oil and Gas Mark 475, 601, 610, 615, 635
Oil and Gas Model 610, 612
Oil/Gas and Salt Relationship 612
Oil/Gas Debate 610, 612
Oil Mystery 225
Old Classification of Rocks 273
Old Geology 273, 275
Olivine Crystals Evidence 374, 375
Olivine Hypretherm Evidence 272

omnivaporization 711, 712
Open Fossil Hydrofountain 307, 308
orange lunar soil 338
Ore Defined 180
Ore deposits 181, 182
Ore-Fossil Evidence 642
Ore Mark 101, 475, 500, 522, 569, 573, 584, 593, 601, 625, 635, 643, 644, 645, 668, 672, 681
Ore Mark Summary 645
Ore Mystery 131, 180
Organic Eruptions Explained 580
organic sediment 210, 534
organic soil 209, 210, 220, 548, 550
Origin of Basalt 566
origin of Carbonates 185
Origin of Carbonates 574
Origin of Comets 453
Origin of Floodwaters 492
origin of granite 168, 270
Origin of Granite 261
origin of lightning 745
origin of limestone and dolomite 594
origin of loess 198, 199, 202, 203, 277, 574, 589, 594, 621
origin of meteorite-like rocks 359, 362
Origin of Obsidian 565
origin of salt 132, 134, 136, 138, 140, 142, 226, 255, 256, 596, 597, 605, 606, 609, 612, 628
Origin of Sand 149, 529, 533
Origin of Sandstone 538
origin of shatter cones 341
Origin of Weather 699, 702, 703, 704, 716, 730
Origin of Weather Summary 716
Orion water line 236
out-channel 414, 432
Overblown Meteorite Number 360, 361
Overloading the Geofield 745
Ozone-CFC Pseudotheory 699, 769
Ozone-Hole Facts 770

P

paleosols 209, 210, 588
pallasite 370, 374, 375, 377, 445
Pangaea 217, 218, 219, 499, 500, 567, 572, 573
Pangaea Mystery 217, 219, 567
Pangaea super-continent 499
Panum Hydrocrater 308, 429
Paragenesis Fundamental Answer 277
PDFs 339, 342, 429, 436
Peat 221, 222, 223, 619
Peat Mystery 222, 223, 619
Peat Mystery to UF Peat 619
Pedestal Evidence 551
Pedestal Formations 207, 546
Pedestal Mystery 215, 509, 552
Pedestals 214
Pegmatite Mystery 270
Pegmatites 161, 166, 270, 278
Pettijohn, F. J. 274
Phillips hole 214
pH level 192, 194, 546, 579
Phobos Hydromoon Evidence 470
Phreatic 315
phreatic 295, 299, 305, 312, 328, 333, 343, 347, 354, 357, 390, 398, 404, 410, 411, 416, 419, 424, 469, 705
Phytoplankton blooms 582
piezoelectric 268, 274, 403
Piezoelectric 69, 105, 740, 741, 745, 749, 751, 758
piezoelectric effect 105, 106, 107, 191, 268, 274,

403, 739, 755
Piezoelectric Evidence 69, 105, 751, 806
piezoelectric field 713, 737, 739, 740, 743, 767, 784
piezoelectricity 105, 106, 739, 741, 743, 745, 752, 753, 755, 757, 758, 760
piezoelectric poles 768
Piezofield 737, 739, 740, 741, 744, 747, 749, 750, 752, 753, 759, 760, 761, 765, 766, 784, 786, 787, 788, 789
Piezofield Mechanism 739, 740, 741
Piezorock Experiment 739, 740, 741, 742
pillars 10, 48, 67, 131, 134, 214, 215, 216, 229, 235, 506, 508, 510, 551, 552, 554, 559, 592, 593, 694
Pillow lava 483
Pillow Lava Mountain Formations 483
Pinacate Hydrocraters 313
pipes of rock 304
planar deformation features 339
Planation 207, 212, 213, 214, 215, 503, 546, 550
Planation Evidence 550
Planetary Craters 306
Planetary Piezofield Laws 737, 740, 741, 744, 749, 786, 787, 789
plankton blooms 581, 594
Plankton Blooms Decline as Fast as Created 582
plate tectonics 73, 78, 124, 125, 126, 218, 469, 484, 684, 693, 737
Plate Tectonic Theory 125
Plate Tectonic Uplift Deception 484
Plutonism 232
Plutonists 233, 234, 262
Pockmark Evidence 316
Potato Starch Column Experiment 570
Precipitate 254, 255, 256
Precipitate Salt Deposit Model 255, 256
Precipitation 254, 255
Precipitation Redefined 254
pre-flood Pangaea 499
prethermation 257, 258, 278, 530, 531, 598, 599, 608, 610, 636
Prethermation Process 256, 257
prethermite 257, 275, 277, 278
Pretoria Salt Pan 436
primeval ocean 232, 233
Principle of Resonance 493
projectile melt 329, 330
pseudometeorite 375
Pseudotheory 4, 23, 74, 106, 161, 209, 268, 282, 335, 388, 401, 473, 511, 531, 590, 632, 654, 701, 760, 784
Puna Ridge Subaqueous Basalt Columns 568
Pure Salt Evaporation Mystery 143
Pure Silica Evidence 410, 411
Pure silica sediment deposits 412
Purpose of Science 40
Pyrite Fossil Sphere Evidence 631
Pyrite Mark 475, 622, 631, 632, 635, 639
Pyrite Mystery 177
Pyrite Sun Mystery 177
Pyrite Sun UF Evidence 634

Q

Quakefinder Evidence 750
Quartz 4, 69–130, 131–228, 253–474, 507–698
quartz crystals 106, 168, 170, 171, 253, 258, 265, 272, 510, 529, 530, 542, 599, 604, 646, 653, 655, 684, 691, 760
Quartz Mystery 102, 131, 161, 195, 259, 532, 646
quartz sand 139, 152, 201, 256, 329, 335, 408, 410, 428, 529, 530, 531, 535, 536, 537,

557, 593, 602, 626
Quartz Sediments Formed in a Hot-Deep Ocean 690
quartz veins 269
Question Principle 5, 9, 69, 171, 194, 321, 353, 401, 402, 411, 554, 745

R

radioactive 233, 252, 274, 280, 362, 420, 441
Radioactive Myth 69, 97, 98, 108
Radioactivity 100, 101
rapid erosion 483, 492
ray patterns 327, 328, 335
rays 288, 300, 323, 325, 327, 331, 332, 333, 388, 443, 447, 453
Real Origin of Carbonate Caves 592
Real Origin of Coal 616
Real Origin of Diamonds 685
Real Origin of Dolomite Deposits 585
Real Origin of Meteorites 365
Regmaglypt Pseudotheory 371, 373
regmaglypts 370, 371, 372, 373, 374, 378, 379, 390, 391, 405, 427
Richat Hydrocrater Evidence 433
Ries Crater 344, 449
Ring of Fire 87, 88, 89, 628
Rock Cycle 335
rock cycle crisis 272
Rock Cycle diagram 131
rock cycle paradigm 261, 517, 568
Rock Cycle Pseudotheory 69–130, 131, 133–228, 230–474, 499–698, 721–790
Rock Cycle theory 131, 167, 168, 189, 217
rock cycle theory 103, 130, 146, 183, 273
rock flour 410, 422, 423, 424, 427, 428, 509
rock pillars 506, 508, 510
Rock Pipes 509, 511
Rock Tree Mystery 146
Rock Varnish Defined 662
Rock Varnish Research History 667
Rock Varnish Trace-Element Smoking Gun 672
Role of Organics 272
Rotating Core Evidence 280
Rotation Rate Increases, Flood Waters Recede 499
rounded boulders 207, 303, 304, 312, 527, 547, 548, 559, 560
rounded pebbles 310, 486, 487, 522
Russia Earthquake Heating Evidence 707
Russian Geofield Evidence 753
Russian Research Evidence 706, 708

S

salt-by-evaporation theory 139
salt crystals 10, 138, 139, 145, 255, 380, 600
Salt Dome 135, 605, 606, 613
Salt Dome Mystery 135
salt dome 134, 135, 136, 137, 138, 145, 278, 596, 602, 605, 606, 607
Salt Evaporation Mystery 138, 143
Salt Evaporation Pseudotheory 140
salt is a biogenic mineral 597
Salt Mark 256, 475, 541, 595, 596, 598, 600, 607, 609, 612, 617, 628, 631, 635, 688
Salt Mystery 132, 141, 144, 258, 604
salt plug 136, 605, 606
Salt's biogenic origin 612
Sand Abrasion by Water 200
Sand, Carbonate, Salt, Oil and Coal Connection 621
Sand Crystal Mystery 157, 604
Sand Crystal Mystery Answered 604

Sand Defined 149
Sand Mark 310, 439, 475, 521, 528, 544, 556, 557, 559, 569, 667, 689
Sand Mystery 131, 148, 149, 156, 160, 177, 195, 197, 206, 225, 226, 439, 528, 529, 530, 532, 625
Sandstone and Limestone Connection 544
sandstone-growing experiments 545
Sandstone Origin Mystery 154
Satellite Impact Evidence 338, 358
Saturn 238, 241, 244, 247, 248, 251, 252, 299, 300, 301, 305, 354, 355, 356, 357, 376, 448, 498, 762, 787
Saturnian ring system 251
Saturn's Hydrospheres Evidence 244
Saturn's moons 238, 245, 246, 300, 787
Saturn's rings 238, 251, 300, 301
Scabland 488, 489, 490, 491, 492, 503, 504
schreibersite 370, 371
sci-bi 74
"Science" Backs Meteor Crater 383
scientific agenda 58, 59, 493
Scientific Evaluation 33
scientific investigator 227, 412
Scientific Models 23, 34
Scientific observation 32, 506
Scientific Prediction 32
Scientific Revolution 61, 62, 63, 65
Scientific Testing 32
Scientific Theory 31
Scientific Truth 19, 24, 35, 46
Scientific truths 17
Sea-Level Rise Scare 778, 780
Sea Mystery 231
Sea Salts Mystery 142, 602
Sea shells 483
Seashells 188, 519
Sea Shells and Limestone on Mountain Tops 484
Seawater 133, 188, 189, 380
Seawater Evidence 380
Secondary Flow Channels 489
secondary impact craters 443, 449
Secondary Impact Evidence 448
Second Law of Weather 713, 728
Second Planetary Piezofield Law 743, 753
Sedan Crater 325, 327, 333
Sedan nuclear explosion 333, 823
Sediment Abrasion by Wind 201
Sedimentary 273, 274
Sedimentary Rock 146
sedimentary rock mystery 274
Sedimentology 200, 204, 271, 587, 590
Sediment Size Table 203, 204, 205, 206, 221, 439, 590
Sedna 495, 496
Seismic Evidence of a Diatreme beneath the Arizona Crater 400
seismic tomography 124, 127, 128
Seismic Waves 75
Self-Evident Principle 17, 263, 690, 733
septarian concretions 427
Seventh Sense 19
Shallow Cold Sea Pseudotheory Overturned 637
Shatter Cone and PDF Impact Criterion Myths 340
shatter cones 339, 340, 341, 342, 344, 385, 395, 396, 398, 413, 417, 422, 423, 429, 436
shear waves 75, 76, 290, 442, 443
Shirokovsky pallasite 374
shocked quartz 341, 342, 344, 415
Shocked rocks 342
shock-wave compression 329
Shoemaker 72, 321, 328, 340, 353, 367, 382, 396, 412, 425, 440, 458, 470, 482, 493, 517,

520, 640, 766
Shoemaker, Eugene M. 321, 383
Shoemaker-Levy comet 321, 367
Shorty Lunar Hydrocrater Evidence 337, 358
Shou, Zhonghao 717
Significance of Double Termination 532
Sikhote-Alin fall 364, 373, 390, 403, 427
Sikhote-Alin 365, 372, 391, 392, 422
Silica Hydrofountain 310
Silica Phase Diagram 102, 103, 104, 268, 401, 632
Simple Truth Principle 8, 53, 58, 417, 703
Simplicity Principle 31, 34, 48, 632
Sinking Subduction Myth 485
Six Senses 19
sixth sense 19, 41, 757
Skipperock Hydrofountain Evidence 523
Skipperock Mystery 211, 523, 524, 550
Skipperock Origin Diagram 524, 550
Skipperocks 207, 211, 212, 523, 524, 546, 550
Skipperocks Evidence 550
Slow-Speed impact 323
Small Hydrobodies 462, 463
Smoking Gun 69, 126, 128, 286, 398, 407, 410, 413, 647, 672
smoking gun 127, 248, 286, 328, 345, 396, 397, 398, 400, 402, 407, 413, 427, 429, 606, 647, 648, 672
Smoking-Gun 328
SOHO 240, 241, 439
Soil formation 207
Soil Formation Evidence 548, 590
Solar System 6, 35, 41, 90, 96, 97, 242, 243, 244, 247, 251, 252, 280, 315, 359, 379, 457, 460, 498, 786
source of the Earth's high and low-pressure systems 713
South Africa kimberlite diatreme 343
spherical shape 233, 243, 244, 246, 461, 462, 471, 472, 651, 676, 682
spheroids 375, 387, 388, 389
Spouting Processes of fountains 506
Spouting Process Evidence 519
star formation 237, 243
steam-eruption crater 398
steam-explosion hydrocrater 417
steam explosions 80, 295, 298, 313, 315, 316, 340, 376, 377, 397, 405, 411, 429, 456, 457, 528
Steam-Produced Pumice Evidence 410
stishovite 103, 339, 342, 401, 429
St. Peter Sandstone Inclusion Evidence 689
Strategic Petroleum Reserve 134, 135
Stratus clouds 721
strewn-field 393, 413
Subcrustal Hypretherm 530
subduction 118, 122, 191, 195, 213, 214, 272, 289, 293, 478, 483, 485, 486, 533, 550, 622, 688
sublimation 711
submerged continents 122, 189, 224
subocean 248, 249, 250, 285, 566, 744
subsidence 117, 118, 120, 122, 213, 224, 320, 483, 484, 485, 498, 513, 575, 594, 614, 615, 620, 643
Subsurface Erosion 529
Sudbury Impact Pseudotheory 434
sulfuric acid 192, 193, 194, 599
Sulfuric Acid Cave Formation Pseudotheory 192
Summarizing the Arizona Hydrocrater 413
superheated water 261, 262, 299, 301, 305, 312, 343, 498, 545, 565
Surface Chalcedony 161, 163, 164
Surface Chalcedony Mark 646

Surface Chalcedony Smoking Gun 647
surface erosion 210, 315, 319, 378, 425, 467, 517, 529, 535, 536, 540, 544, 551, 552, 553, 557, 660
Surface Erosion 529
surface explosions 339, 340
Surface Hypretherm 530
Surface Mark 307, 373, 475, 531, 536, 542, 566, 593, 645, 646, 652, 653, 654, 658, 670, 675, 676, 677, 678, 682, 684, 691
Surface Marks Defined 645
Surface Mark Summary 682
swarm of earthquakes 716

T

TAG Hyprethermal Mound Evidence 607
Tektite and Libyan Desert Glass Evidence 344, 358
Tektites 344, 345
test of time 16, 17, 24, 25, 28, 33, 35, 47, 48, 51, 273, 321, 339, 421, 613, 693, 754, 773
Tethys 244, 245, 484
Tethys Hydrosphere 245
The Great Dying 481
Third Law of Weather 713
Three Laws of Weather 713
Thunderegg Evidence 650
Tibetan Hydroplumbing System Evidence 290
Titan 245, 246, 247, 357, 787
Titan's largest lake 245
Tomography 126, 127, 286
Tornado-Cumuliform Cloud Connection 723, 728
Townes, Charles 235
Transpiration 711
Trinity Glass Evidence 331, 358
Trinity nuclear explosion 333
Tronin, A.A. 706
Truth Defined 5, 11, 15
Tufa Pinnacle Evidence 592
Turbidites 556
Turbidity defined 554
Turbidity Model 548, 554, 555, 556
Turkey Earthquake Prediction Project 750
Two Forces Hold Earth's Crust in Equilibrium 493
Two Forces Return to Equilibrium 499
Tyco Crater 328
Tyson Creek Geode Inclusion Evidence 690

U

Ubehebe Hydrocraters 310, 313
Ubehebe Silica-Dike Evidence 411
UF Comet Passes Close to the Earth 495
UF Connection 694
UF Creation 694
UF Environmental Conditions 506
UF Erosion Evidence 559
UF Geological Map Evidence 621
UF Hypretherm 530, 535, 541, 556, 574, 583, 590, 594, 599, 608, 622, 626, 630, 637, 642, 647, 650, 653, 682, 695
UF Sandstone Experiment 543
UF Summary 475, 692
Ukinrek Hydrocrater 315
Underground Slabs Evidence 283
Understanding Hypervelocity 328
Unequivocal Seismic Evidence of a Diatreme 400, 401
Unified Field Theory 37
uniformitarianism 72, 76, 131, 145, 146, 175, 215, 226, 227, 483, 485, 486, 490, 493, 529, 621

Uniformity 72, 156, 166, 170, 189, 223, 226, 227, 257, 345, 414, 431, 464, 468, 476, 490, 492, 511, 637, 690
Uniformity Myth 72, 173, 223, 226, 345, 347, 414, 464, 468, 492, 511
Uniformity Principle 72, 156
Universal Concept of Water 243, 253
Universal Energy Laws 493
Universal Flood 69, 144, 195, 202, 214, 271, 275, 344, 431, 480, 500, 543, 566, 582, 621, 642, 662, 695, 721, 761
Universal Flood Defined 476
Universal Flood Evidence 473
universal flooding 479, 497
Universal Flood Mechanism Diagram 493, 496
Universal Flood Mechanisms 475, 492, 497, 556
Universal Flood Salt Model 596, 609
Universal Laws of Water 243
Universal Model 5, 11, 31, 64, 121, 177, 194, 218, 227, 246, 382, 415, 489, 535, 542, 622, 643, 720, 773, 788
Universal Scientific Method 19, 23, 25, 28, 29, 31, 33, 34, 36, 64, 89, 95, 117, 126, 136, 203, 227, 278, 306, 357, 365, 375, 477, 514, 543, 699, 703, 730, 740, 744, 773, 787
Universe System 5, 20, 493
Unseen Water in Rocks 260
Untold Story of Sedimentology 200, 204, 587, 590
Upheaval Dome 308, 414, 428, 429, 430, 431, 432
Upheaval Dome Crater 428, 429, 431, 432
Upheaval Dome Impact Myth 428
Upheaval Dome Out-Channel Evidence 432
Upheaval Dome Quartz Nodule Evidence 430
Upheaval Dome's Hydrocrater Evidences 430
Uplift 69, 117, 118, 119, 120, 121, 165, 189, 194, 223, 479, 484, 485, 486, 513, 536, 637, 653, 686
Uplift Deception 484
Uplift Pseudotheory 4, 69, 117, 118, 119, 120, 121, 189, 194, 223, 479, 485, 513
Uranium Ore Diatreme Evidence 640
Uranus 229, 239, 241, 787
USGS 77, 84, 92, 110, 119, 140, 184, 199, 222, 283, 295, 314, 383, 400, 465, 501, 525, 611, 629, 642, 703, 718, 749, 766

V

vaporized 89, 244, 276, 300, 330, 334, 345, 367, 383, 388, 390, 412, 413, 422, 438, 449, 710, 722, 727, 744
vaporized meteorite 334, 345, 387, 388, 422
Varnish Surface Mark of the Flood 670
Venus 7, 41, 126, 238, 244, 245, 321, 376, 463, 464, 786, 787
Veology 472, 473
vertical pipe 167, 430
vertical pipes 304, 307, 310, 524, 606, 673
Vesalius, Andreas 46
vesicles 261, 276, 335, 345, 567
Vesicular Basalt Nodule Mystery 174
V-Impact 323
V-Impact Signature 325, 327, 358
volcanic ash 173, 295, 338, 388, 389, 513, 514, 570
Volcanic craters 389
volcanic explosion 327, 342, 401, 403
volcanic origin 172, 233, 302, 305, 318, 335, 341, 357, 389, 399, 404, 514, 571, 662
Volcanic Rocks Contain Water 261
Volcanic Spheroids 387

Volcanic Theory 319
volcano 70, 81, 110, 168, 196, 246, 291, 305, 313, 319, 336, 404, 438, 467, 511, 519, 565, 607, 629, 749
volume of salts 142
Voyager 2 244, 252
Vulcanists 233, 234

W

Wabar Impact Crater 333, 334, 358
Waimea Canyon Evidence 500, 505
water content of meteorites 380
Water Cycle 703, 711, 712
Water exists in space 241
Water Hardness 185
Water in Comets 240
water in rocks 258, 260, 261, 379, 437
Water in the Stars 236
water medium 254
Water on Exoplanets 240
Water on the Moon 240
Water on the Planets 238
Water On the Sun 237
Water processes 253
Water Vapor - the Key to Changing Weather 711
Water, Water Everywhere 241
Weather Defined 700
Weather-Pressure System Experiment 715, 716
Webster, Noah 42
Wegener, Alfred 217
Werner, Abraham Gottlob 233
What Happened to the Dinosaurs 480
Where Mineral Deposits Get Their Colors From 584
White Sands Gypsum Evidence 604
White Sands Gypsum Salt Mystery 144, 604
White Sands National Monument 143, 144, 145
Why the UF Rock Varnish Model Works 674
Widmanstätten Crystalline Pattern Evidence 366
Widmanstätten pattern 366, 367, 368, 369, 373, 381, 384, 389, 390, 413, 422
Widmanstätten structure 366, 367, 368
Wieliczka Salt Mine 597
Willamette meteorite 364, 365, 371
Wizard Island 314, 315
Wolfe Creek Crater 405, 416, 417, 418, 419, 420, 428
Wolfe Creek Crater National Park 416
Wolfe Creek Hydrocrater Evidences 417
Wolfe Creek Magnetic Survey Evidence 420
Wolfe Creek Nickel-Uranium Evidence 419
Wonderstone Mystery 158
Woolnough, W. G. 227, 476, 596
world's largest meteorites have no craters 365

X

xenoliths 312, 516

Y

Yellowstone National Park 85, 86, 88, 277, 289, 306, 389, 398, 507, 574, 585, 627, 682
Yucatan peninsula 319, 414
Yuty Hydrocrater on Mars 464

Z

Zebra Rock Mystery 158
Zuni Hydrocrater 308
Zuni Lake 139
Zuni Salt Lake crater 308, 310